# *ARABIDOPSIS*

Edited by

**Elliot M. Meyerowitz**
California Institute of Technology, Pasadena

**Chris R. Somerville**
Carnegie Institution, Stanford, California

**COLD SPRING HARBOR LABORATORY PRESS**
**1994**

***ARABIDOPSIS***

Monograph 27
© 1994 by Cold Spring Harbor Laboratory Press
All rights reserved
Printed in the United States of America
Book design by Emily Harste

ISBN 0-87969-428-9
ISSN 0270-1847
LC 94-79111





All Cold Spring Harbor Laboratory Press publications may be ordered directly from Cold Spring Harbor Laboratory Press, 10 Skyline Drive, Plainview, New York 11803. Phone: 1-800-843-4388 in Continental U.S. and Canada. All other locations: (516) 349-1930. FAX: (516) 349-1946.

# Contents

# Preface

During the past decade, plant biology has been revolutionized by application of the methods and concepts of molecular genetics. During this period, the small mustard *Arabidopsis thaliana* has emerged as the most widely used and facile experimental organism for studies of a broad range of problems in development, metabolism, genetics, environmental adaptation, pathogen interactions, and many other areas. Although it is a typical higher plant in most respects, and is therefore a good model for approximately 250,000 other species, *Arabidopsis* has an unusually small genome, can be readily grown in confined laboratory environments, has convenient genetic properties such as a short generation time and prolific seed production, and is easily transformed. For these and related reasons, biological and genetic information about *Arabidopsis* is in a phase of explosive growth, and *Arabidopsis* has been chosen as the major organism for plant genome projects.

This book represents a synthesis of currently available information about the biology of *Arabidopsis*, written by a consortium of authors and groups of authors representing the major fields of enquiry in plant biology. It provides up-to-date reviews, descriptions of recent and ongoing research in specific areas, and reference material. Plant molecular biology and, in particular, *Arabidopsis* molecular biology are rapidly changing fields, and the pace of change is increasing. Because of this, the authors of each chapter have been asked to include unpublished information as well as published work and to indicate the expected future directions of research in each area.

Since *Arabidopsis* is of interest as a model for all higher plants, the book was conceived not only as a reference for work on *Arabidopsis thaliana*, but as a discussion of plant molecular biology in general, with an emphasis on specific examples from the recent work on *Arabidopsis*. As such, the book should serve as a useful introduction to modern concepts in higher plant biology for students and researchers who work on other plants or on organisms in other kingdoms, as well as a compendium of specific information and reference material of interest to *Arabidopsis* researchers.

We thank those who have made this book possible. Most important are the authors of the different chapters. The *Arabidopsis* research community, although it has grown from a handful of laboratories to hundreds of laboratories in the past ten years, has remained a collegial, cooperative, and friendly group. This is reflected in the authorship of many of the chapters, which were written by groups of researchers who share common interests, but who work at widely separated institutions. We thank all of the authors for their contributions to research and to the writing of this book, and we particularly thank the authors who were willing to work collaboratively across long distances to produce excellent and up-to-date summaries of their research areas. We also thank Patricia Barker, Joan Ebert, and the staff at the Cold Spring Harbor Laboratory Press for teaching us about book editing and for doing much of the work. Finally, we thank our colleagues around the world for performing the research described in this book.

**E.M. Meyerowitz**
**C.R. Somerville**

# Introduction

**Chris R. Somerville**
Department of Plant Biology
Carnegie Institution
Stanford, California 94305

**Elliot M. Meyerowitz**
Division of Biology, 156-29
California Institute of Technology
Pasadena, California 91125

The contents and scope of the chapters in this volume attest to the fact that *Arabidopsis thaliana* is a useful and popular model organism for investigating a broad range of topics in plant biology. The rapidity and magnitude of growth in the use of *Arabidopsis* as an experimental organism during the past 10 years is undoubtedly due to a variety of factors. The many advantages of *Arabidopsis* for studies in classical genetics, and in large-scale mutant screens, have long been known and stimulated a burst of enthusiasm for its use as a model for plant genetics in the 1960s. However, perhaps because of the perceived linkage between plant breeding and genetics and the absence of any compelling reasons to work on *Arabidopsis* before the advent of gene cloning, crop species such as maize, tomato, pea, and barley remained the major organisms for genetic studies of plants. A cogent review of the virtues of *Arabidopsis* (Rédei 1975) and the first stirrings of plant molecular biology attracted a small coterie of students in the late 1970s (see, e.g., Meinke and Sussex 1979; Somerville and Ogren 1979; Koornneef et al. 1980). The first genetic map of the organism, which showed the power of large-scale mutagenesis and the ease of performing classical genetic manipulations, was published in 1983 (Koornneef et al. 1983).

The advantages of the organism for studies in molecular biology, which became apparent after the small size and simple structure of the nuclear genome was determined (Leutwiler et al. 1984; Pruitt and Meyerowitz 1986), were spelled out in several reviews (Meyerowitz and Pruitt 1985; Meyerowitz 1987). These papers, and the early and enthusiastic reports of applications of modern methods to *Arabidopsis* (see, e.g., North 1985), attracted large numbers of plant biologists who had only recently begun to incorporate the techniques of molecular biology into their research programs. They also attracted biologists who had been trained in yeast, *Drosophila*, bacteria, or vertebrates but were interested in moving to plants. Reports describing methods for *Arabidopsis* transformation (An et al. 1986; Lloyd et al. 1986; Feldmann and Marks 1987), and examples of the uses of mutants for physiological, biochemical, and developmental studies (Estelle and Somerville 1986), stimulated and consolidated interest at that time.

A second wave of interest in *Arabidopsis* developed several years later as the new discipline of "plant molecular biology" began to mature. As plant biologists became increasingly familiar with genes, a widespread appreciation for the methodological approaches and power of genetics developed in parallel. Although there were large numbers of interesting mutations and a rich heritage of classical genetics in plants such as maize, pea, *Antirrhinum*, and tomato, the technical advantages of *Arabidopsis* for many genetic methods were persuasive. In addition, largely due to the work of Maarten Koornneef and his colleagues at Wageningen, a collection of important mutants affecting floral morphology, phytochrome, phytohormone responses, and flowering time was freely available. These and other mutants from the collections of Albert Kranz and George Rédei, among others, served as a kind of "starter kit" that allowed many laboratories to begin genetic investigations of problems of broad importance without the delays associated with isolating and characterizing new mutants. Perhaps the clinching attraction was the development of a method for producing large numbers of T-DNA insertions (Feldmann and Marks 1987) and the dissemination of a large collection of T-DNA insertion mutants by Ken Feldmann and colleagues. The ease with which genes could be cloned from this collection proved irresistible to the large number of people fortunate enough to find their favorite mutation in the collection.

Whatever the precise reasons for the large number of converts, one effect of having large numbers of scientists using *Arabidopsis* was that it became worthwhile to develop organism-specific tools and infrastructure. The development of molecular marker maps, recombinant inbred lines, and YAC libraries could be justified by the large number of potential users. Once such resources were available, they became additional reasons to use *Arabidopsis* as an experimental organism. It was fortunate for plant biology that a number of granting agency administrators around the world also realized the potential of *Arabidopsis* at an early stage and provided both financial and administrative support not only for individual research projects, but also for workshops and community-wide resources such as stock centers, databases, and the EST projects. Some of the U.S. administrators who deserve special mention in this respect are

Machi Dilworth, Delill Nasser, and Mary Clutter of the National Science Foundation and Bob Rabson and Greg Dilworth of the Department of Energy. Étienne Magnien of the Commission of the European Communities, Directorate of Biology, played a similarly influential role in Europe.

As a result, we are now in a "golden age" of discovery in plant biology. Problems that have been intractable for decades are yielding to the application of modern methods in molecular and cellular biology. The formula for much of this success is conceptually simple: Isolate a mutation that affects the process or structure of interest, clone the gene, find out where and when it is expressed, where the gene product is located, what it does, and what it interacts with, directly or indirectly. The adaptation of this "molecular genetic paradigm" to problems in plant biology is not species-specific but has been greatly facilitated by the widespread adoption of *Arabidopsis* as a model species. Large numbers of interesting and informative mutations are now available, and the list of mutations is not only increasing rapidly but is increasing in the sophistication of the criteria used to identify the mutations. Now that the promise of map-based cloning has been realized by a number of laboratories, it seems that everything is within grasp. Although it is not necessarily easy, any gene that can be marked by a mutation can be cloned. This is a qualitatively different situation from anything that has ever before existed in plant biology and is, for the moment, unique to *Arabidopsis*.

Having come this far, it seems an opportune time to pose the question of where we are bound. Because the *Arabidopsis* community is now organized on an international scale with elected representatives and an electronic forum for rapid dissemination of information and discussion of issues (Dennis et al. 1993), it is possible to undertake large-scale projects that serve the community as a whole or that require broad participation. In this respect, the goal of sequencing the entire genome is a logical next step. Indeed, the first tentative steps toward large-scale genomic sequencing have been taken in Europe, and representatives of the U.S. *Arabidopsis* community have endorsed the initiation of a complementary U.S. project. Thus, we may already imagine that by 2004 it will be possible to sit down at a computer and call up the entire sequence of a plant genome. It is not too early to begin thinking about the consequences and possibilities inherent in this eventuality.

An attractive model for large-scale sequencing is the *Caenorhabditis elegans* sequencing project. The members of this project release the data to the community as it is produced (as opposed to the yeast sequencing projects, which release whole chromosomes). Thus, the community can exploit the information as it is produced. Although it will be several years before the large-scale *Arabidopsis* sequencing is producing comparable levels of useful information, in the meantime, it seems likely that sequence information will be available for the vast majority of expressed genes via the EST sequencing projects (Höfte et al. 1993; Newman et al. 1994). These projects have already produced more than 8000 partial cDNA clones that have, in turn, permitted the identification of more than 1000 genes of identifiable function by reference to the sequences in the public databases. It also seems likely that most of the ESTs will be placed on the genetic map and that a set of overlapping YAC clones will soon be available.

Having the sequence of the genome will not identify the function of all the genes, but it will certainly provide new tools for analyzing gene function. Presumably, we will find ourselves in a situation analogous to the current situation with the *Saccharomyces* genome project. Sequencing of the first few yeast chromosomes revealed that mutational analysis of yeast had identified only about 20% of the genes. Thus, much of the mutational analysis of yeast will now proceed by directed mutagenesis. The availability of the total genomic sequence is essential to this because it permits identification of other copies of structurally related genes in the genome that may also need to be inactivated in order to assess the gene function. Similarly, we may expect that directed gene inactivation of *Arabidopsis* will become an increasingly important aspect of research on *Arabidopsis*. One efficient way of accomplishing this may be the development of a comprehensive set of insertional mutants that can be screened by PCR methods to identify the desired insertion. Under optimal conditions, this approach could permit one to have seed of a desired insertional mutant in hand within a few days from the start of an experiment. There is no theoretical reason why directed gene knockouts cannot be done as well, although methods for doing these efficiently have yet to be found.

In parallel with the elucidation of genome structure, we envision the development of new generations of computer resources. The role of relatively simple databases in managing sequence and mapping information can be seen from contemporary models. However, there is no apparent reason why all other information about *Arabidopsis* cannot also be organized in a database. Imagine the advantages of having all the information *summarized* in this volume available in a relational database! In principle, the computing tools already exist so that all overlapping information can be linked.

As our knowledge of plant biology grows and is refined, it will increasingly be necessary to connect the artificially fragmented aspects of biological knowledge into a seamless and interactive whole that explains the overall properties of the organism. Our goal should be to have such a deep understanding of the biology that we can confidently predict, from first principles, how a change in one environmental parameter, or the activity of a particular gene, will affect the growth and development of the organism. Such methods could also be applied to learning how a change in a gene or in its activity might have had (or might yet have) a particular impact during the evolution of the organism. It is to be hoped that, as one of the forefront biological systems, *Arabidopsis* will continue to attract researchers with a primary interest in bioinformatics and knowledge systems. Eventually, it should be possible to sit down in front of a computer and ask complex "What if" questions and get back synthetic responses in which the computer synthesizes specific answers from predictive models of all aspects of the biology of the organism. The inability of a computer to answer such a query might become a criterion for pursuing an experiment.

The purpose of an organism as an experimental "model system" is to efficiently understand the organisms being modeled. *Arabidopsis* differs from several of the other important model systems in at least one key respect. Unlike yeast, which is taken as a model for all eukaryotes, or *C. elegans* or *Drosophila*, which are considered to be useful models for animal development, *Arabidopsis* is closely related to the species it models. All angiosperms share similar life-styles, environmental challenges, modes of reproduction, and body plans. The roughly 250,000 species of angiosperms are thought to have evolved from a common ancestor within the last 150 million years. Because of this relatively recent evolution, the average *Arabidopsis* gene can be confidently expected to functionally replace a homolog in many other angiosperms and can generally be used as a heterologous hybridization probe to isolate the corresponding gene. One implication of this high degree of similarity between the model and the modeled is that all aspects of *Arabidopsis* biology—developmental, metabolic, biochemical, environmental, and so on—are worthy of investigation because of the broad applicability of the information. A second implication is that there are many opportunities to exploit detailed knowledge about *Arabidopsis* to gain insights into similar processes in other plants. Thus, biologists with a primary interest in another plant species will find it increasingly useful to have a deep understanding of the resources and knowledge produced through research on *Arabidopsis*. This volume is a step in attempting to make this knowledge accessible to all plant biologists and, in fact, to make it available to all biologists interested in eukaryotic organisms, in the interactions of eukaryotes with prokaryotes, and in the profound differences between plants and animals.

## REFERENCES

An, G., B.D. Watson, and C.C. Chiang. 1986. Transformation of tobacco, tomato, potato, and *Arabidopsis thaliana* using a binary Ti vector system. *Plant Physiol.* **81:** 301–305.

Dennis, L., C. Dean, R. Flavell, H. Goodman, M. Koornneef, E. Meyerowitz, Y. Shimura, M. van Montagu, and C. Somerville. 1993. The multinational coordinated *Arabidopsis thaliana* genome research project progress report: Year three. *U.S. National Science Foundation Publ.* NSF 93-173.

Estelle, M.A. and C.R. Somerville. 1986. The mutants of *Arabidopsis. Trends Genet.* **2:** 89–93.

Feldmann, K.A. and M.D. Marks. 1987. *Agrobacterium*-mediated transformation of germinating seeds of *Arabidopsis thaliana*: A non-tissue culture approach. *Mol. Gen. Genet.* **208:** 1–9.

Höfte, H., T. Desprez, J. Amselem, H. Chiapello, M. Caboche, A. Moisan, M.F. Jourjon, J.L. Charpenteau, P. Berthomieu, D. Guerrier, J. Giraudat, F. Quigley, F. Thomas, D.Y. Yu, R. Mache, M. Raynal, R. Cooke, F. Grellet, M. Delseny, Y. Parmentier, G. Marcillac, C. Gigot, J. Fleck, G. Philipps, M. Axelos, C. Bardet, D. Tremousaygue, and B. Lescure. 1993. An inventory of 1152 expressed sequence tags obtained by partial sequencing of cDNAs from *Arabidopsis thaliana. Plant J.* **4:** 1051–1061.

Koornneef, M., E. Rolff, and C.J.P. Spruit. 1980. Genetic control of light-induced hypocotyl elongation in *Arabidopsis thaliana* (L.) Heynh. *Z. Pflanzenphysiol.* **106:** 147–160.

Koornneef, M., J. Van Eden, C.J. Hanhart, P. Stam, F.J. Braaksma, and W.J. Feenstra. 1983. Linkage map of *Arabidopsis thaliana. J. Hered.* **74:** 265–272.

Leutwiler, L.S., B.R. Hough-Evans, and E.M. Meyerowitz. 1984. The DNA of *Arabidopsis thaliana. Mol. Gen. Genet.* **194:** 15–23.

Lloyd, A.M., A.R. Barnason, S.G. Rogers, M.C. Byrne, R.T. Fraley, and R.B. Horsch. 1986. Transformation of *Arabidopsis thaliana* with *Agrobacterium tumefaciens. Science* **234:** 464–466.

Meinke, D.W. and I.M. Sussex. 1979. Embryo-lethal mutants of *Arabidopsis thaliana*: A model system for genetic analysis of plant embryo development. *Dev. Biol.* **72:** 50–61.

Meyerowitz, E.M. 1987. *Arabidopsis thaliana. Annu. Rev. Genet.* **21:** 93–112.

Meyerowitz, E.M. and R.E. Pruitt. 1985. *Arabidopsis thaliana* and plant molecular genetics. *Science* **229:** 1214–1218. (Reprinted [1986] in *Biotechnology: The renewable frontier* [ed. D.E. Koshland, Jr.], pp. 311–320. AAAS, Washington, D.C.)

Newman, T., F.J. de Bruijn, P. Green, K. Keegstra, H. Kende, L. McIntosh, J. Ohlrogge, N. Raikhel, S. Somerville, M. Thomashow, E. Retzel, and C.R. Somerville. 1994. Genes galore: A summary of methods for accessing results from large-scale partial sequencing of anonymous *Arabidopsis* cDNA clones. *Plant Physiol.* (in press).

North, G. 1985. A plant joins the pantheon at last? *Nature* **315:** 366–367.

Pruitt, R.E. and E.M. Meyerowitz. 1986. Characterization of the genome of *Arabidopsis thaliana. J. Mol. Biol.* **187:** 169–183.

Rédei, G.P. 1975. *Arabidopsis* as a genetic tool. *Annu. Rev. Genet.* **9:** 111–127.

Somerville, C.R. and W.L. Ogren. 1979. A phosphoglycolate phosphatase-deficient mutant of *Arabidopsis. Nature* **280:** 833–836.

# 1

# Systematic Relationships of *Arabidopsis:* A Molecular and Morphological Perspective

**Robert A. Price[1] and Jeffrey D. Palmer**
Department of Biology, Indiana University
Bloomington, Indiana 47405

**Ihsan A. Al-Shehbaz**
Missouri Botanical Garden
St. Louis, Missouri 63110

Our rapidly expanding knowledge of the molecular and developmental biology of *Arabidopsis thaliana* is of fundamental importance to a better understanding of biological systems in crops and other plants. This knowledge will be particularly valuable in future examinations of evolutionary processes at the generic, specific, and population levels. To effectively utilize the knowledge gained from *A. thaliana* in an evolutionary context, information concerning its phylogenetic relationships is essential. However, despite the great utility of *A. thaliana* for comparative studies, remarkably little basic information is available concerning the systematics and evolution of the genus *Arabidopsis* and its relationships within the mustard family. The boundaries of the genus *Arabidopsis* are very poorly known, and the majority of species now assigned to the genus are not yet accessible for scientific study because of their restricted geographic distributions in remote mountainous areas of Asia. In the following sections, we summarize available information concerning the systematics and phylogenetic position of *A. thaliana* both from the standpoint of traditional morphologically based classifications and from new data from molecular comparisons. We review the geographic distribution and chromosome numbers of the other species currently placed in the genus *Arabidopsis* and the problematic nature of its delimitation and tribal placement in the Brassicaceae. We also present new information on the phylogenetic relationships of *A. thaliana* within the family obtained

[1]Current address: Department of Botany, University of Georgia, Athens, Georgia 30602.

*Arabidopsis*
© 1994 Cold Spring Harbor Laboratory Press 0-87969-428-9/94 $5 + .00

from sequence comparisons of the chloroplast *rbc*L gene and from preliminary analyses of restriction site data from the chloroplast genome.

## SYSTEMATIC RELATIONSHIPS OF *ARABIDOPSIS*

### Familial and Tribal Classification of *Arabidopsis*

The genus *Arabidopsis* belongs to the mustard or crucifer family (Brassicaceae or Cruciferae), a widely distributed family of approximately 340 genera and 3350 species, with greatest abundance of species and genera in the temperate zone of the northern hemisphere (Al-Shehbaz 1984). Members of the family are particularly concentrated in southwestern and central Asia (~900 species, with 530 endemic) and in the countries of the Mediterranean region (~630 species, with 290 endemic), with secondary centers of diversity in the arctic, western North America, and the mountains of South America. Crucifers are primarily annual to perennial herbs and are characterized by their cross-shaped corolla, usually tetradynamous stamens (four long plus two short), and their specialized capsular fruit (silique or silicle), usually consisting of two valves separating at maturity from a central partition. Along with the related families in the order Capparales, the mustard family is noted for its possession of mustard oils (glucosinolates and their breakdown products), which provide the pungent flavors associated with the food and spice plants in the group.

The family is of significant economic importance as a source of vegetable crops, oil seeds, and spices, and, to a lesser extent, of ornamentals. Much of its agricultural importance derives from the genus *Brassica*, whose half dozen or so crop species provide an immense range of economic forms (various mustards, canola, cabbage, broccoli, cauliflower, kale, turnip, etc.) (Tsunoda et al. 1980; Williams and Hill 1986). Other vegetable and spice crops in the family include radish (*Raphanus*) and horseradish (*Armoracia*). Species cultivated for their ornamental flowers include dame's rocket (*Hesperis*), stock (*Matthiola*), wallflower (*Erysimum*), and sweet alyssum (*Lobularia*); cabbages and related forms of *Brassica* are sometimes cultivated for their ornamental foliage.

There is good evidence from phylogenetic analyses of both morphological (Rodman 1991a,b) and molecular data (Rodman et al. 1993) that virtually all of the mustard-oil-containing families of flowering plants form a single monophyletic group, which can be treated as an expanded order, Capparales. Sequence data from the *rbc*L gene (Rodman et al. 1993; R.A. Price et al., unpubl.) agree with overall morphology in indicating that the Brassicaceae is nested well within the clade of mustard-oil-containing families and that its closest relatives belong to the caper family (Capparaceae). Figure 1 presents the single minimum length tree produced by a maximum parsimony analysis of nucleotide sequences of the chloroplast *rbc*L gene (encoding the large subunit of ribulose bisphosphate carboxylase) from representatives of several tribes of the Brassicaceae and outgroup taxa representing the two main subfamilies of the Capparaceae. Bootstrap sampling (Felsenstein 1985) provides an assessment of the support for individual shared branches on a cladogram, as judged by the frequency of retention of particular groups in repeated analyses using random resampling of the data. The cladogram in Figure 1 indicates strong support for the Brassicaceae as a monophyletic group (a bootstrap value of 99 out of 100 replicates) and suggests that the Brassicaceae may be derived from within the Capparaceae. This would be consistent with the similarity in details of fruit morphology between the mustard family and the subfamily Cleomoideae of the Capparaceae (Al-Shehbaz 1973; Rodman 1991a).

The mustard family has been divided into a widely varying number of tribes, primarily on the basis of a limited number of characters  from fruit

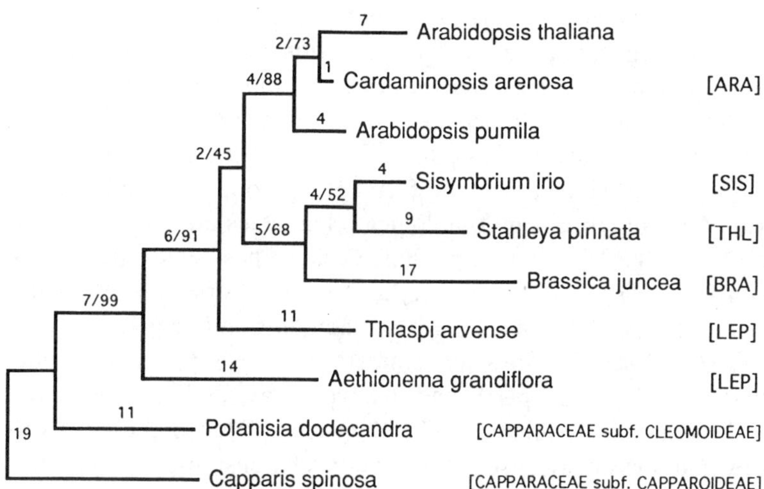


*Figure 1* Molecular phylogeny of some representative crucifers based on *rbc*L gene sequences. The tree shown is the single minimum length tree from a maximum parsimony analysis of a 1347-bp-long region of the *rbc*L gene of the 10 species shown. The number of inferred nucleotide changes is given for each branch, followed by the bootstrap value out of 100 replications for all branches subtending two or more taxa. Assignments of taxa to tribes of Brassicaceae in the classification of Al-Shehbaz (1984) are indicated as follows: Arabideae (ARA), Brassiceae (BRA), Lepidieae (LEP), Sisymbrieae (SIS), and Thelypodieae (THL).

and seed morphology. The most comprehensive worldwide treatment of the family (Schulz 1936) recognized 19 tribes, several of which have been merged in more recent treatments (see, e.g., Janchen 1942; Hedge 1976; Al-Shehbaz 1984). Eight major tribes, seven of which occur in the northern hemisphere, are recognized in the treatment of Al-Shehbaz (1984) on the basis of overall morphology, but it has long been suspected that few of these are actually monophyletic groups. Preliminary data from restriction site comparisons of chloroplast DNA (Fig. 2) (R.A. Price and J.D. Palmer, unpubl.) indicate that few of the tribes are monophyletic in their current delimitations, with apparent exceptions being the Brassiceae and Thelypodieae.

The genus *Arabidopsis* has often been placed in the tribe Sisymbrieae following the treatments of Schulz (1924, 1936), but it has been suggested on the basis of morphological comparisons that the tribe is a highly artificial group (Al-Shehbaz 1984, 1988). The tribes Anchonieae (including Hesperideae and Matthioleae), Arabideae, and Sisymbrieae each tend to have long, narrow fruits and have been separated primarily on the basis of the position of the embryonic root relative to the cotyledons in the seed, a character that is subject to substantial variation within certain species groups in these tribes (Al-Shehbaz 1984). The parsimony cladogram in Figure 2, based on restriction site mapping of the chloroplast genome, suggests that the tribe Sisymbrieae in the sense of Schulz (1936) is indeed a highly unnatural group. On the basis of this tree, it would appear that *Arabidopsis* and several other genera previously placed in tribe Sisymbrieae in fact belong to an expanded tribe, Arabideae. One significant morphological character that appears to correlate well with the DNA-based phylogenies, but has been little emphasized in tribal classification to date, is trichome structure. Branched trichomes occur widely in the group including the expanded tribe Arabideae, whereas trichomes are almost always simple in the tribes Brassiceae, Lepidieae (in part), and Thelypodieae, and also in *Sisymbrium* and its immediate relatives. The chloroplast DNA-based trees suggest a new and previously unsuspected set of relationships, with a portion of the tribe Lepidieae, including the genus *Aethionema*, occupying a basal position in the family and the tribes Brassiceae, Sisymbrieae, and Thelypodieae possibly comprising a previously unrecognized natural group.

The cladogram shown in Figure 2 is a preliminary analysis from a much more comprehensive study of chloroplast DNA restriction site variation in over 100 species of crucifers, whose main goals are to provide new assessments of tribal delimitation and generic relationships in the Brassicaceae that are independent of the earlier morphological classi-

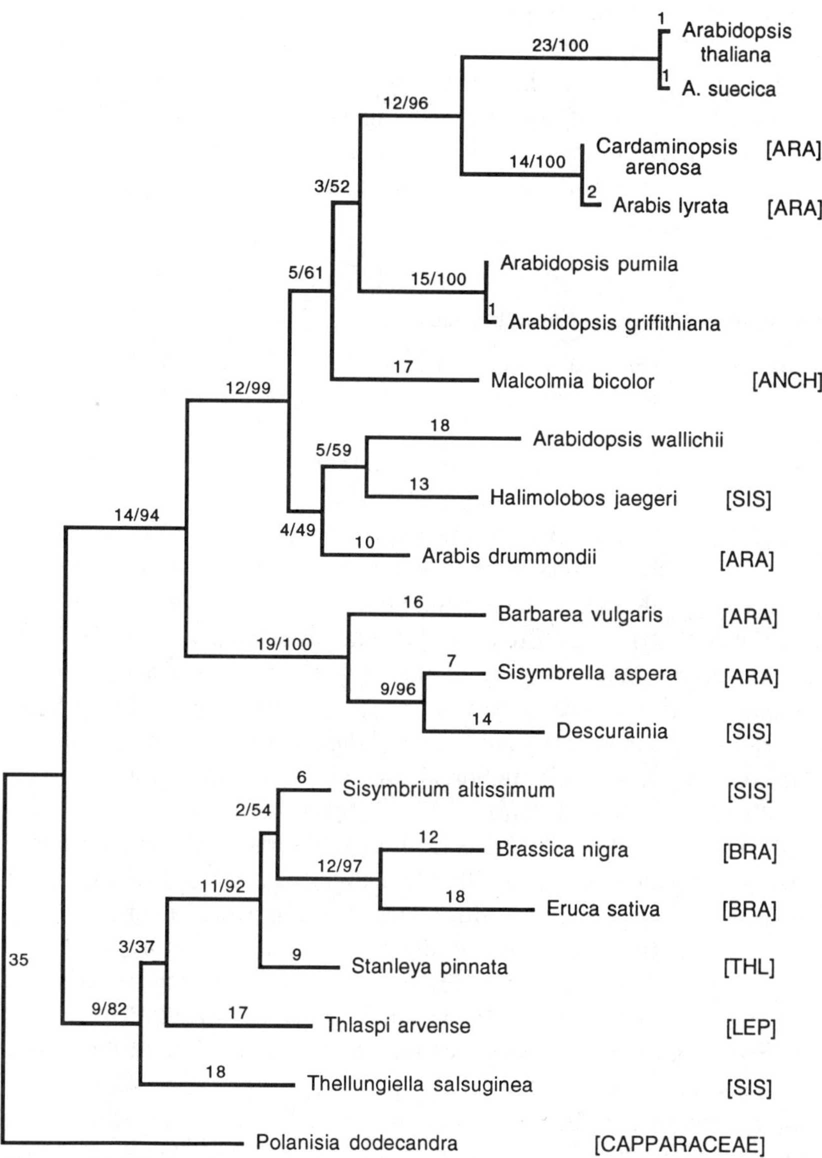



*Figure 2*    Molecular phylogeny of some representative crucifers based on chloroplast DNA restriction site variation. The tree shown is a bootstrap consensus tree from parsimony analysis of 345 variable restriction sites. The number of inferred restriction site changes is given for each branch, followed by the bootstrap value out of 100 replications for all branches subtending two or more taxa. Tribal assignments of taxa are as in Fig. 1, with the addition of Anchonieae (ANCH).

fications. Major phylogenetic studies of chloroplast DNA restriction site variation have already been published for the tribe Brassiceae (Palmer et al. 1983; Warwick and Black 1991, 1993). Most notably, these studies indicate that the genus *Brassica* is a highly unnatural group in its current delimitation. In addition, the laboratory of H. Hurka and K. Mummenhoff at the University of Osnabrück is engaged in an extensive restriction site survey of the tribe Lepidieae and possible relatives which promises to contribute very significantly to our knowledge of phylogenetic relationships in the family.

### Generic Relationships of *Arabidopsis*

Schulz (1924, 1936) assigned the genus *Arabidopsis* to the subtribe Arabidopsinae of the tribe Sisymbrieae, along with eight genera from Australia, four from the Americas, and two from Asia. The subtribe was separated from the remainder of the tribe largely on the basis of the possession of confluent nectar glands and mucilaginous seeds. These characters occur widely elsewhere in the family, however, and in some cases vary significantly within genera. On the basis of overall morphology, various other authors (see, e.g., Ball 1964; Hedge 1968, 1976; Jafri 1973; Al-Shehbaz 1988) have associated the genus *Arabidopsis* with the European *Cardaminopsis* (~8 spp.); the Asian *Cymatocarpus* (3 spp.), *Drabopsis* (1 sp.), *Microsisymbrium* (6 spp.), *Nasturtiopsis* (2 spp.), and *Neotorularia* (25 spp.); the American *Halimolobos* (~18 spp.) and *Pennellia* (10 spp.); the circumboreal *Braya* (~16 spp.); and the very widespread *Arabis* (~170 spp.).

Schulz separated *Halimolobos* from *Arabidopsis* solely on the basis of the latter being slender herbs with styles slightly narrower than the ovary, as opposed to coarser herbs with styles much narrower than the ovary. These alleged differences have not held up to further scrutiny, as several species now assigned to one or the other do not exhibit these combinations of characters. Hedge (1968) suggested that *Arabis* differs from *Arabidopsis* only in the position of the cotyledons relative to the radicle in the seed and that the Himalayan species *Arabidopsis wallichii* is essentially intermediate between the two. Jafri (1973) also called attention to the similarity between the two genera in the flora of Pakistan, and Jafri (1973) and An (1987) transferred several species from *Arabis* to *Arabidopsis*. The overall morphological similarity of *Cardaminopsis* to *Arabidopsis* has also been widely noted (see, e.g., Ball 1964, 1993), and it has been suggested that *Arabidopsis suecica* is a natural amphidiploid hybrid of *Arabidopsis thaliana* and *Cardaminopsis arenosa* (Hylander 1957; Löve 1961; Berger 1968). In addition, the widespread boreal species *Arabis lyrata* has often recently been assigned to the genus *Cardaminopsis*, which it closely resembles in morphology.

The chloroplast DNA data in Figures 1 and 2 suggest that *Cardaminopsis arenosa* and *Arabis lyrata* are indeed more closely related to *Arabidopsis thaliana* than are some other species now assigned to *Arabidopsis*. The molecular data are also consistent with relatively close relationships among the genera *Arabis,* *Halimolobos,* and *Arabidopsis*. The generic limits of *Arabidopsis* are not well defined on the basis of morphology (see discussion in Al-Shehbaz 1988), and chloroplast DNA data imply that the traditional delimitation of the genus is highly artificial. The overall habit and fruit shape of *Arabidopsis thaliana* and several putative relatives are shown in Figure 3. The paucity of clear-cut morphological characters distinguishing the genus *Arabidopsis* from other crucifers is illustrated by the circumboreal species *Thellungiella salsuginea*, which has sometimes been placed in *Arabidopsis* (see Al-Shehbaz 1988) and differs primarily in having clasping leaves and lacking trichomes. Restriction site comparisons (Fig. 2) indicate that *T. salsuginea* is only distantly related to *Arabidopsis* or to the tribe Arabideae as a whole.

## Distribution of the Genus *Arabidopsis*

The geographic distributions of 27 species that have been widely assigned to the genus *Arabidopsis* are given in Table 1, along with chromosome numbers that have been reported in the literature. One Himalayan taxon that has sometimes been treated as distinct and has been available for experimental work, *A. griffithiana*, is here included within the limits of *A. pumila* following the treatment of Jafri (1973). Of the species assigned to the genus, 22 occur within central Asia and adjacent areas of the Himalayas, which is evidently the center of greatest diversity of the genus in its current circumscription. This region is very rich in apparently indigenous populations of *A. thaliana*, now a cosmopolitan weed, and it also represents the area of greatest morphological variability within several other species of the genus. Because of their restricted occurrence in isolated mountain ranges in this area of Asia, many species referred to the genus *Arabidopsis* are very poorly collected and are currently unavailable for experimental work, although expeditionary work is being undertaken by I.A. Al-Shehbaz for several of the critical areas. Individual species referred to the genus *Arabidopsis* are endemic to Egypt (*A. kneuckeri*), Saudi Arabia (*A. erysimoides*), eastern Asia (*A. tschuktschorum*), northern Siberia (*A. bursifolia*), and northern

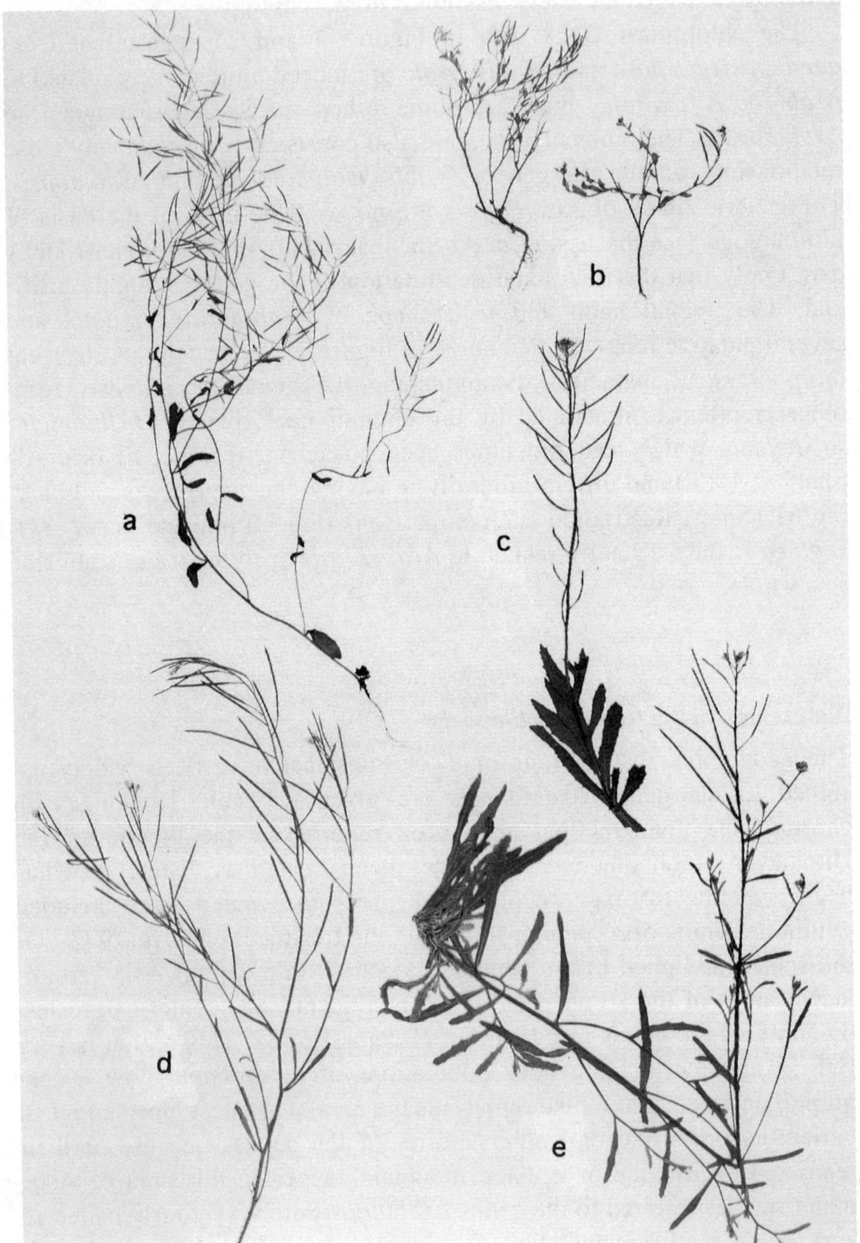

*Figure 3* Habit of *Arabidopsis thaliana* and possible relatives. (a) *Arabidopsis thaliana*; (b) *Thellungiella salsuginea*; (c) *Arabidopsis pumila* var. *griffithiana*; (d) *Arabis lyrata*; (e) *Arabidopsis wallichii*.

*Table 1*  Geographic distribution and chromosome numbers of *Arabidopsis* species

| Species | Distribution | Chromosome numbers ($2n$) |
| --- | --- | --- |
| *A. bactriana* | C. Asia | |
| *A. brevicaulis* | Pakistan | |
| *A. bursifolia* | N. Siberia | 16 |
| *A. drassiana* | Kashmir | |
| *A. erysimoides* | Saudi Arabia | |
| *A. eseptata* | Afghanistan | |
| *A. gamosepala* | Afghanistan | |
| *A. himalaica* | China, W. Himalaya, Pakistan | 16 |
| *A. kneuckeri* | Egypt | |
| *A. korshinskyi* | C. Asia | 48 |
| *A. lasiocarpa* | W. Himalaya, Pakistan | |
| *A. mollissima* | Afghanistan, C. Asia, China, W. Himalaya, Pakistan | 16 |
| *A. monachorum* | W. China | |
| *A. ovczinnikovii* | C. Asia | |
| *A. parvula* | S.E. Russia, C. Asia, Turkey | |
| *A. pumila* | Afghanistan, C. & S.W. Asia, China, Iran, Pakistan, E. Russia | 16, 32 |
| *A. russelliana* | Pakistan | |
| *A. sarbalica* | Kashmir | |
| *A. stricta* | China, Pakistan, W. Himalaya | |
| *A. suecica* | N. Europe | 26 |
| *A. taraxacifolia* | Pakistan | |
| *A. thaliana* | Eurasia, N. Africa; widely introduced elsewhere | 10 |
| *A. toxophylla* | Afghanistan, C. Asia, China, S.E. Russia, W. Siberia | 12 |
| *A. tschuktschorum* | Far East | 16 |
| *A. tuemurica* | W. China | |
| *A. wallichii* | Afghanistan, C. Asia, China, W. Himalaya, Iran, Pakistan | 14, 16, 18 |
| *A. yagodensis* | W. China | |

Information on geographic ranges has been compiled primarily from regional floras, and chromosome numbers have been compiled from the primary literature.

Europe (*A. suecica*). Inclusion of the genus *Cardaminopsis* within *Arabidopsis* would make Europe a secondary center of diversity for the genus and possibly a major center of diversity for taxa closely related to *A. thaliana* itself.

## Chromosome Numbers

Chromosome numbers have been reported in the literature for 10 species of *Arabidopsis* (Table 1, which is taken from an unpublished compilation for the family by I.A. Al-Shehbaz; see also Bolkhovskikh et al. 1969). Most of the reported chromosome numbers are based on single counts, and no counts are available for the other 17 species referred to the genus. In contrast, numerous chromosome counts have been made for *A. thaliana* (the great preponderance of them being $2n = 10$), and meiosis has been studied in detail in this species (see Klasterska and Ramel 1980). *Arabidopsis korshinskyi* is a hexaploid based on $x = 8$ (Ginter and Ivanov 1968), whereas *A. pumila* has both diploid and tetraploid populations with the same basic number (see, e.g., Ginter and Ivanov 1968; Polatschek 1971; Aryavand 1975, 1983). Except for the presumed allotetraploid hybrid *A. suecica*, the remaining species that have been counted are diploid, with basic numbers ranging from $x = 9$ down to $x = 5$ in *A. thaliana*. *Arabidopsis wallichii* has been variously reported to have basic numbers of 7, 8, and 9 (Ginter and Ivanov 1968; Polatschek 1968; Podlech and Bader 1974; Naqshi and Javeid 1976). Six species of *Arabidopsis* have the basic number $x = 8$, which is also the case in *Cardaminopsis* (7 spp. counted) and *Arabis lyrata*, and in most species of *Halimolobos* (5 out of 6 counted). Eight is also the modal basic number for the tribe Arabideae as a whole, suggesting as a working hypothesis that this is the primitive state for the genus *Arabidopsis*, from which reduced numbers such as that in *A. thaliana* have been derived. In contrast, the modal basic chromosome number for *Sisymbrium* is $x = 7$ (Bolkhovskikh et al. 1969), and this number is also seen in all species of *Braya*, *Neotorularia*, and *Thellungiella* that have been counted to date. Chromosome base number is subject to substantial variation within several large and seemingly natural genera in the family (e.g., in *Draba* and *Erysimum*) as well as in *Arabis* and *Arabidopsis*, so caution is warranted in its use as a phylogenetic character.

An allotetraploid origin of *Arabidopsis suecica* ($2n = 26$) from *A. thaliana* ($2n = 10$) and *Cardaminopsis arenosa* ($2n = 16$) was suggested by Hylander (1957) and Löve (1961) on the basis of morphological resemblance, geographic proximity, and experimental reconstitution of the hybrid, although some other workers have subsequently challenged the hypothesis (see Laibach 1958; Rédei 1974). The restriction fragment patterns of chloroplast DNA for nine six-base cutting enzymes in *A. suecica* and *A. thaliana* are almost identical (differing in two documented site changes) and are much more similar than they are to any other species compared to date, including *C. arenosa* (R.A. Price and J.D. Palmer, unpubl.). Because *A. suecica* is restricted to northern

Europe, where *A. thaliana* is the only other representative of the genus, the similarity in chloroplast DNA suggests that *A. thaliana* is the maternal parent of *A. suecica*, although biparentally inherited markers will need to be used to more rigorously establish the parentage of the latter species.

## CONCLUSIONS AND PERSPECTIVES

To obtain an evolutionary perspective on the unusual features of *Arabidopsis thaliana*, such as its small nuclear genome (and such correlates as its relative lack of repetitive DNA elements, reduced copy number of multigene families, and reduced size of introns and intergenic spacers), and to make best use of the information being obtained about its gene sequences, genomic structure, and the genetic control of its developmental and biochemical pathways, a phylogenetic framework is needed so that the species can be compared to its near relatives. Preliminary phylogenies based on comparisons of chloroplast DNA indicate that *Arabidopsis thaliana* has traditionally been placed in the wrong tribe. Moreover, other crucifers, such as the European genus *Cardaminopsis* and at least some species of the widespread north temperate genus *Arabis*, may be closer relatives of *A. thaliana* than are other species traditionally placed in *Arabidopsis*. Comparisons of nuclear genome size across a broad range of taxa in the mustard family are also in progress to investigate the evolution of this character (R.A. Price et al., unpubl.). Further sequence comparisons utilizing other, more rapidly evolving chloroplast genes and regions of nuclear DNA such as the internal transcribed spacer regions (ITS) of ribosomal DNA are in progress and should resolve more fully the relationships in the phylogenetic neighborhood of *A. thaliana*. To adequately address these phylogenetic questions, however, it is of critical importance that a wider sampling of species of the genus *Arabidopsis* and possible relatives be collected and made available for scientific use. We recommend that seed for these taxa be maintained in the seed bank facilities that have already been established for *A. thaliana* in Columbus, Ohio in the United States and in Nottingham, England.

This survey of our knowledge of purported relatives of *Arabidopsis thaliana* emphasizes how little we know of the basic biology of the group, and the great need for the collection and preservation of germ plasm of the wild relatives for use in comparative studies. The wild relatives should be rich sources of allelic variation for genes now under investigation in *A. thaliana* and should greatly expand the scope of experimental work that can be conducted using the *Arabidopsis* model system.

## ACKNOWLEDGMENTS

We are grateful to the National Science Foundation for support of our research through grants BSR-8996262 to J.D.P. and DEB-9208433 to I.A.S.; to Elizabeth Kellogg for seed samples of *Arabidopsis* and help in instigating the project; and to Jennifer Thomas and Jonathan Dowdy-Jafari for their assistance in *rbc*L sequencing.

## REFERENCES

Al-Shehbaz, I.A. 1973. The biosystematics of the genus *Thelypodium* (Cruciferae). *Contrib. Gray Herb.* **204:** 3–148.

————. 1984. The tribes of Cruciferae (Brassicaceae) in the southeastern United States. *J. Arnold Arbor.* **65:** 343–373.

————. 1988. The genera of Sisymbrieae (Cruciferae; Brassicaceae) in the southeastern United States. *J. Arnold Arbor.* **69:** 213–237.

An, Z.-X. 1987. *Arabidopsis. Fl. Reipubl. Popularis Sinica* **33:** 280–288.

Aryavand, A. 1975. Contribution à l'étude cytotaxonomique de quelques angiospermes de l'Iran. *Bot. Not.* **128:** 299–311.

————. 1983. Contribution à l'étude cytotaxonomique des Cruciferes de l'Iran. III. *Bull. Soc. Neuchatel. Sci. Nat.* **106:** 123–130.

Ball, P.W. 1964. *Arabidopsis. Fl. Europaea* **1:** 267–268.

————. 1993. *Arabidopsis. Fl. Europaea* (ed. 2) **1:** 322–323.

Berger, B. 1968. Entwicklungsgeschichtliche und chromosomale Ursachen der verschiedenen Kreuzungsvertraglichkeit zwischen Arten des Verwandtschaftskreises *Arabidopsis. Beitr. Biol. Pflanz.* **45:** 171–212.

Bolkhovskikh, Z., V. Grif, T. Natvejeva, and O. Zakhareva. 1969. *Chromosome numbers of flowering plants.* Academy of Sciences of the USSR, Leningrad.

Felsenstein, J. 1985. Confidence limits on phylogenies: An approach using the bootstrap. *Evolution* **39:** 783–791.

Ginter, T.N. and V.I. Ivanov. 1968. Chromosome numbers in *Arabidopsis* species. *Arabidopsis Inf. Serv.* **5:** 23.

Hedge, I.C. 1968. *Arabidopsis. Fl. Iranica* **57:** 328–334.

————. 1976. A systematic and geographical survey of the old world Cruciferae. In *The biology and chemistry of the Cruciferae* (ed. J.G. Vaughan et al.), pp. 1–45. Academic Press, London.

Hylander, N. 1957. *Cardaminopsis suecica* (Fr.) Hiit., a northern amphidiploid species. *Bull. Jard. Bot. Etat Brux.* **27:** 591–604.

Jafri, S.M.H. 1973. Brassicaceae. *Fl. Pakistan* **55:** 1–308.

Janchen, E. 1942. Das System der Cruciferen. *Oesterr. Bot. Z.* **91:** 1–28.

Klasterska, I. and C. Ramel. 1980. Meiosis in PMCs of *Arabidopsis thaliana. Arabidopsis Inf. Serv.* **17:** 1–10.

Laibach, F. 1958. Über den Artbastard *Arabidopsis suecica* (Fr.) Norrl. X *A. thaliana* (L.) Heynh. und die Beziehungen zwischen den Gattungen *Arabidopsis* Heynh. und *Cardaminopsis* (C.A. Meyer) Hay. *Planta* **51:** 148–166.

Löve, A. 1961. *Hylandra*—A new genus of Cruciferae. *Sven. Bot. Tidskr.* **55:** 211–217.

Naqshi, A.R. and G.N. Javeid. 1976. In IOPB chromosome number reports. LIV. *Taxon* **25:** 631–649.

Palmer, J.D., C.R. Shields, D.B. Cohen, and T.J. Orton. 1983. Chloroplast DNA evolution and the origin of amphidiploid *Brassica* species. *Theor. Appl. Genet.* **65:** 181–189.

Podlech, D. and O. Bader. 1974. Chromosomenstudien an afghanischen Pflanzen II. *Mitt. Bot. Munchen* **11:** 457–488.

Polatschek, A. 1968. Cytotaxonomische Beitrage zur Flora Iranica. I. *Ann. Naturhist. Mus. Wien* **72:** 581–586.

————. 1971. Cytotaxonomische Beitrage zur Flora Iranica. III. *Ann. Naturhist. Mus. Wien* **75:** 173–182.

Rédei, G.P. 1974. Is *Hylandra* an amphidiploid of *Arabidopsis* and *Cardaminopsis arenosa*? *Arabidopsis Inf. Serv.* **11:** 5.

Rodman, J.E. 1991a. A taxonomic analysis of glucosinolate-producing plants. 1. Phenetics. *Syst. Bot.* **16:** 598–618.

————. 1991b. A taxonomic analysis of glucosinolate-producing plants. 2. Cladistics. *Syst. Bot.* **16:** 619–629.

Rodman, J.E., R. Price, K. Karol, E. Conti, K. Sytsma, and J. Palmer. 1993. Nucleotide sequences of the *rbc*L gene indicate monophyly of mustard oil plants. *Ann. Mo. Bot. Gard.* **80:** 686–699.

Schulz, O.E. 1924. Cruciferae-Sisymbrieae. *Pflanzenreich IV.* **105**(Heft 86)**:** 1–388.

————. 1936. Cruciferae. *Nat. Pflanzenfam.* (ed. 2.) **17B:** 227–658.

Tsunoda, S., K. Hinata, and C. Gomez-Campo, eds. 1980. Brassica *crops and wild allies.* Japan Science Society Press, Tokyo.

Warwick, S.I. and L.D. Black. 1991. Molecular systematics of *Brassica* and allied genera (subtribe Brassicinae, Brassiceae)—Chloroplast genome and cytodeme congruence. *Theor. Appl. Genet.* **82:** 81–92.

————. 1993. Molecular relationships in subtribe Brassicinae (Cruciferae, tribe Brassiceae). *Can J. Bot.* **71:** 906–918.

Williams, P.H. and C.B. Hill. 1986. Rapid-cycling populations of *Brassica*. *Science* **232:** 1385–1389.

# 2

# Structure and Organization of the *Arabidopsis thaliana* Nuclear Genome

**Elliot M. Meyerowitz**

Division of Biology 156-29
California Institute of Technology
Pasadena, California 91125

The nuclear genome of *Arabidopsis thaliana* is unusually small for a flowering plant and has remarkably little dispersed repetitive DNA. These properties facilitate a series of different types of experiments in molecular genetics and have allowed facile cloning of many *Arabidopsis* genes by methods that would be difficult or impossible if the genome were larger or more typical in its content of repetitive sequences. Despite the unusual size and structure of the *Arabidopsis* nuclear genome, the structure of individual genes, the structure of chromosomes, the genetic properties, and the overall complement of genes in the genome are typical of those of other flowering plants.

## GENOME SIZE

That *A. thaliana* is among the higher plants with the smallest genomes has been known for several decades: Sparrow and Miksche (1961) showed that radiation sensitivity and DNA content are related in plants. They then found that *Arabidopsis* is highly resistant to ionizing radiation, showing half-inhibition of vegetative growth only when exposed to 4,000 r/day of ionizing radiation. This is many-fold higher than the dose required for similar inhibition of the other plants tested. Sparrow et al. (1972) later showed a correlation of nuclear volume and genome size in plants and found *A. thaliana* to have the smallest nuclear volume among the angiosperms examined.

Subsequent measurements are more readily converted to quantitative estimates of nuclear genome size. Microspectrophotometry of *Arabidopsis* nuclei specifically stained for DNA with the Feulgen reaction indicates a haploid nuclear genome size of around 0.2 pg (~200 Mb) (Bennett and Smith 1976, 1991). DNA reassociation analysis originally showed a size of around 70 Mb (Leutwiler et al. 1984). The initially reported value of 70 Mb was the midpoint of a range of values obtained from different types of measurements of reassociation; the range was from approximately 50 Mb to 90 Mb. Recent reassessments of the size of the *Escherichia coli* genome (it used to be considered 4,100 kb, it is now thought to be 4,700 kb), which was the kinetic standard used for the reassociation calculations, indicate that the range measured should now be considered 60 Mb to 100 Mb, with a midpoint of 80 Mb (Meyerowitz 1992). The reassociation studies of Leutwiler et al. (1984) included experiments in which equal amounts of *A. thaliana* and *Drosophila melanogaster* DNA were allowed to reassociate in the same tube. In these experiments, the low-copy *Arabidopsis* DNA clearly reassociated more rapidly than the fly DNA, indicating that *Arabidopsis* has a smaller genomic complexity than *Drosophila*. Quantitative gel blot hybridization gives a size of 50 Mb (Francis et al. 1990), and electron-microscopic measurement of chromosome volume indicates a haploid nuclear genome of about 100 Mb (Heslop-Harrison and Schwarzacher 1990). Flow cytometry measurements give an *A. thaliana* haploid genome size of 86 Mb when the yeast *Saccharomyces cerevisiae* is used as a size standard (Galbraith et al. 1991) and a size of 145 Mb when chicken red blood cells are used as a size standard (Arumuganathan and Earle 1991; Galbraith et al. 1991). This large variation in calculated genome size from measurements done by the same method is thought to result from a degree of base composition or base sequence specificity in the binding of fluorescent dyes to DNA (Galbraith et al. 1991), and since it is not known which of the two standards is more appropriate for an *Arabidopsis* genome size estimate, it is not clear which (if either) of these measurements reflects the actual size of the *Arabidopsis* nuclear genome. None of the many different types of measurement, thus, gives us an accurate absolute size for the nuclear genome of *A. thaliana*, but they do indicate quite clearly a range of sizes (around 50–150 Mb) within which one can expect the *Arabidopsis* genome to fall. For convenience, the size of the haploid nuclear genome is often considered to be 100 Mb.

This is one of the smallest genomes known among angiosperms (whose nuclear genomes are typically severalfold to 100 times larger than that of *Arabidopsis*; Bennett and Smith 1976, 1991; Arumuganathan and Earle 1991). One can use the 100 Mb figure to calculate a variety of useful values relating to molecular biological experiments with *Arabidopsis*. Since the genetic map (calculated either from visible genetic markers alone [Koornneef et al. 1983], from restriction fragment length polymorphism [RFLP] maps [Chang et al. 1988; Nam et al. 1989], or from the combined genetic-RFLP map [Hauge et al. 1993]) is on the order of 500 centimorgans, 1 cM of genetic map distance is on average 200 kilobase pairs of DNA. A genomic library made using a bacteriophage λ vector that can hold 20-kb inserts will require only 23,000 random clones to have a 99% probability of representing any individual sequence, and 15,000 clones to have a 95% chance of containing any genomic segment. A cosmid library made with a vector that can hold 40-kb inserts will require half as many clones in either case. A yeast artificial chromosome (YAC) library with average 150-kb inserts (such as those made by Ward and Jen [1990] and Grill and Somerville [1991]) will require 2,000 independent clones to achieve a 95% chance that any chosen sequence will be present in the library, and 3,100 clones for a 99% chance. These figures show the utility of a plant with such a small genome: Chromosome walks, which require repeated screening of genomic libraries, and library screens in general can be done with much less effort than with plants that have genomes of a more typical size (Meyerowitz and Pruitt 1985; Meyerowitz 1987).

## CHROMOSOMES

The nuclear DNA of *A. thaliana* is packaged into five types of chromosomes (this is the haploid number; the plant is typically diploid), as revealed both by microscopic observations of metaphase chromosomes (Laibach 1907; Lee-Chen and Steinitz-Sears 1967; Ambros and Schweizer 1976; Schweizer et al. 1988; Maluszynska and Heslop-Harrison 1991; Heslop-Harrison and Maluszynska, this volume) and by the five well-established linkage groups of the genetic map (Koornneef et al. 1983; Hauge et al. 1993). The relative sizes of the chromosomes have been measured in Giemsa C-banded material, with the individual chromosomes showing lengths of 25.4%, 22.2%, 19.6%, 19.4%, and 13.4% of the total chromosome length (Schweizer et al. 1988). Using 100 Mb as the haploid genome size, we can calculate that the chromosomes range between 13.4 Mb and 25.4 Mb, a bit too large for us to expect to be able to display them on current pulsed-field gels. A more complete description of *Arabidopsis* chromosomes and the methods used to study them is found in Heslop-Harrison and Maluszynska (this volume).

## GENE-CODING PROPERTIES

Estimates of the total number and average size of distinct transcripts in higher plants range from 15,000 transcripts averaging 1.2 kb (in parsley

[Flavell 1980]) through between 13,700 and 35,000 transcripts of average size 1.4 kb (in parsley root callus, depending on the method of measurement [Kiper et al. 1979]), to 60,000 different transcripts of average size 1.24 kb, not including the poly(A) tail (in the combination of tobacco leaf, root, stem, petal, anther, and ovary [Kamalay and Goldberg 1980]). There is no reason to think that *A. thaliana* has fewer or different genes than other, related plants; its morphology, anatomy, growth, development, and environmental responses are all typical of flowering plants, and it is known that closely related plants can have very different genome sizes (Bennett and Smith 1976, 1991). If *Arabidopsis* falls within the range of gene numbers estimated for parsley and tobacco, around one-quarter to three-quarters of the total *A. thaliana* nuclear genome is genic and transcribed, and if the DNA in which genes are found includes the entire genome, a gene should be found on average in every 2–7 kb of the nuclear DNA.

## REPEATED DNA SEQUENCES

There are portions of the *Arabidopsis* nuclear genome that are not genic; therefore the gene spacing is probably even closer than 2–7 kb. Reassociation kinetic studies have shown approximately 10% of the nuclear genome to consist of highly repeated or foldback sequences (Leutwiler et al. 1984), which are presumably not protein-coding, and other studies (Pruitt and Meyerowitz 1986) have shown a sizable fraction of the genome to be tandem ribosomal DNA repeats. There are also dispersed repetitive elements, some of which are dispersed gene families, and some transposable elements.

Three different families of short and high-copy-number repeats have been described. They include a 180-bp sequence repeated around 5,000 times in largely tandem arrays; a 500-bp repeat present in approximately 500 copies, also largely in tandem; and a 160-bp repeat present in around 1500 copies, again, largely in tandem arrays (Martinez-Zapater et al. 1986; Simoens et al. 1988; Richards et al. 1991). At least one class of these repeats (the 180-bp class) has been shown by in situ hybridization to be present in the centromeric heterochromatin of all five chromosomes (Maluszynska and Heslop-Harrison 1991), and RFLP mapping has shown some repeats of the 500-bp class to reside near the centromere of chromosome 1 (Richards et al. 1991). Since karyotype analysis has shown that approximately 12.5% of the genomic DNA is found in centromeric heterochromatin (Schweizer et al. 1988), it is not unreasonable to think that most or all of the highly repeated sequences are components of this heterochromatin and, conversely, that the centromeric heterochromatin consists largely of high-copy tandem repeats.

There are also other tandemly repeated sequences in the *Arabidopsis* nuclear genome. Ribosomal DNA repeats that vary in size, but that are generally around 10 kb in length, are repeated approximately 570 times per haploid genome (Pruitt and Meyerowitz 1986). They consequently comprise around 6% of the total nuclear DNA. The rDNA repeats have been shown by in situ hybridization to reside in clusters on chromosomes 2 and 4 (Maluszynska and Heslop-Harrison 1991). There are also repeated sequences at the telomeres of the *Arabidopsis* chromosomes (Richards and Ausubel 1988). These are found in blocks of tandem repeats that are made, largely, of the sequence 5′-CCCTAAA-3′(on the other strand, 5′-TTTAGGG-3′), with up to 350 of these 7-bp repeats present at each of the ten telomeres.

If we subtract the centromeric heterochromatin, the telomeric repeats, and the ribosomal DNA repeats from our estimate of the total genome size, we are left with around 80 Mb of nuclear DNA that might contain single-copy or low-copy-number genes. If the highest of the estimates of plant gene numbers is correct, that is, if there are 60,000 transcripts of average size 1.24 kb in *Arabidopsis*, then they occupy 93% of the non-heterochromatic DNA and would be separated from each other by less than 100 base pairs. Since this calculation assumes that there are no introns, whereas there are indeed many introns in *Arabidopsis* genes (see below), one must conclude that there is not enough room in the *Arabidopsis* genome for 60,000 genes, coding on average 1.24 kb processed transcripts. The lowest of the estimated figures for coding capacity, 15,000 transcripts of a processed size 1.2 kb, leaves enough room both for introns and for space between genes, although it leaves only a total of slightly more than 4 kb per gene that can be devoted to introns and distance from neighboring genes, including upstream control sequences. Even with the lower estimate of gene number, then, the euchromatic *Arabidopsis* genome must be tightly packed with coding sequences.

In the euchromatic regions of the genome, there are dispersed repeats as well as single-copy sequences. Some of the dispersed repeats have sequence similarity to retrotransposons (Voytas and Ausubel 1988; Konieczny et al. 1991), although each type of these is only present in one to several copies per haploid genome. The only known endogenous element in *Arabidopsis* that has been shown to transpose under experimental conditions, Tag1, is present in three copies in the Landsberg *erecta* ecotype and is thus a member of a small dispersed repeat family in that strain. Related elements are absent in at least two other strains, however, pointing out that specific sequences can be repetitive in some ecotypes and nonrepetitive or absent in others (Tsay et al. 1993). Additional dispersed, moderately repetitive elements have been described (Pruitt and Meyerowitz 1986), but an analysis of genomic λ clones representing almost 1% of the nuclear genome revealed only five of these, which seem to exist with an average spacing of one every 125 kb. That some such repeats are members of small gene families and not transposons or transposon relatives has been demonstrated many times, for example in the analysis of the α-tubulin genes, of which there are at least six, and the β-tubulin genes, of which there at least nine (Kopczak et al. 1992; Snustad et al. 1992), or in the analysis of the H3 and H4 histone gene families, each of which contains five to seven individual histone-coding genes, some of which are only a few kilobases apart from each other in the genome (Chaubet et al. 1987). A number of genes have been shown to be members of even smaller gene families, represented either as tandem or inverted repeats (e.g., three almost-identical LHCPII genes [Leutwiler et al. 1986]) or as dispersed repeats (e.g., the two nitrate reductase genes, which are on different chromosomes [Cheng et al. 1988]). McGrath et al. (1993) have recently estimated that 17% of randomly selected *Arbidopsis* cDNA clones hybridize to two or more distinct genomic regions. The gene-containing parts of the *Arabidopsis* chromosomes thus consist of a mixture of single-copy genes and small gene families, which can have dispersed or nearly tandem members.

## CLONING METHODS FACILITATED BY THE *ARABIDOPSIS* GENOME STRUCTURE

The predominant pattern of repeated DNA organization in *Arabidopsis* euchromatin, with long stretches of single- or low-copy DNA separated at infrequent intervals by very short (a few kb) regions of moderately repeated DNA is, so far, unique among the flowering plants. The typical case is for the predominant classes of DNA to be repetitive, with only very small regions of single-copy or low-copy DNA separated by large amounts of repetitive sequence (Meyerowitz and Pruitt 1985). Although the evolutionary advantage or disadvantage (and the evolutionary origin) of the *Arabidopsis* type of organization is unknown, the practical advantage is evident: It is possible to perform chromosome walking in *Arabidopsis*, that is, the isolation of overlapping cloned fragments, so as to proceed from a given starting point to any gene of interest (Meyerowitz and Pruitt 1985; Meyerowitz 1987). Repetitive DNA makes such experiments difficult, if not impossible, in other plant species. In fact, several genes have now been cloned by chromosome walking in *Arabidopsis* (see, e.g., Arondel et al. 1992; Giraudat et al. 1992; Chang et al. 1993; Leyser et al. 1993). The small genome size has also potentiated several other methods of gene cloning. One is subtractive hybridization, in which deletion mutations are obtained and the DNA deleted is identified by hybridizing an excess of mutant plant DNA to wild-type sequences, and then recovering the wild-type fraction not hybridized by the mutant genome. This depends to a considerable degree on having a small genome, since the rate of DNA hybridization is proportional to the square of genomic complexity (Wetmur and Davidson 1968). By this method, the GA1 gene (involved in gibberellin biosynthesis) has been cloned (Straus and Ausubel 1990; Sun et al. 1992). Such methods, presently feasible for *Arabidopsis*, are only now beginning to be extended, by use of intermediate polymerase chain reaction steps, to test cases involving large-genome organisms (Lisitsyn et al. 1993). Another cloning method that apparently benefits from the small *Arabidopsis* genome is T-DNA tagging, in which *Agrobacterium*-mediated transformation results in a collection of plants with different locations of integrated, foreign DNA. Presumably because the *Arabidopsis* genome is largely genic, a high proportion of such transformants, when made genetically homozygous for the inserted foreign DNA, show new mutations. These are often caused by the introduced DNA, allowing molecular cloning of the mutated gene by use of probes to the foreign DNA (Feldmann et al. 1989; Feldmann 1991, 1992; Koncz et al. 1992).

## PHYSICAL MAPPING

The small size of the *Arabidopsis* genome has made approachable the goals of obtaining very high physical density RFLP maps and of obtaining a complete physical map of the genome. The present *Arabidopsis* RFLP map has over 300 molecular markers (Hauge et al. 1993), which is on average one genetically mapped fragment every 300 to 350 kb. Starting points are thus available for chromosome walks to virtually every genomic region, and the walk to any gene is on average only one YAC clone. Work toward providing a complete physical map of the *A. thaliana* genome is in progress, with the genome presently available as 750 groups of contiguous cosmid clones that are thought to represent around 90–95% of the genomic DNA (Hauge and Goodman 1992 and pers. comm.), and also as a collection of 296 ordered YAC clones (obtained by finding YAC clones that cross-hybridize with 125 genetically mapped RFLP markers) that include 30% of the nuclear genome (Hwang et al. 1991). Ongoing work is expanding this number, and physical maps of many genomic regions are presently under assembly by chromosome walking with YAC clones (Putterill et al. 1993; Schmidt and Dean 1993) and by long-range restriction mapping using pulsed-field gel electrophoresis (Bancroft et al. 1992). It is the small genome and generally single- or low-copy nature of the DNA of *Arabidopsis* that has made rapid progress in physical mapping possible with this species.

One question raised by RFLPs, which are identified between different *A. thaliana* ecotypes, is the nature of the nucleotide differences between different strains. Chang et al. (1988) selected as RFLP probes those low-copy clones that showed clear polymorphisms (between the ecotypes Nd-0, Col-0, and the laboratory strain Landsberg *erecta*) in which both alleles could be detected in genomic DNA blots. The genetic variation could have been due to single nucleotide differences that altered the restriction sites under observation or due to insertions in one strain relative to another. The presence of a sequence in one strain and its absence in another also appeared to occur, although clones for which heterozygotes gave the same blot pattern as one of the parents (i.e., dominant/null polymorphisms) were not utilized in the construction of the RFLP map and were not analyzed sufficiently to be certain that they revealed presence-absence polymorphism. A survey of the RFLPs found in that study allowed the calculation that Nd-0 and L*er* differ in approximately 1.4% of nucleotides in low-copy DNA, and that Col-0 and L*er* differ in around 1.1% (Chang et al. 1988). This calculation involved several assumptions (that each RFLP probe detected a single substitution; that each surveyed four restriction sites for each enzyme; that all sites counted were nonoverlapping), so these numbers can only be considered approximations. A different calculation of the sequence divergence found between the L*er* and Col-0 lines was based on the frequency of polymorphic changes found in a set of PCR-amplified sequences (Konieczny and Ausubel 1993). These experiments gave a value of 0.4% divergence. An analysis of 28 ecotypes with 25 different RFLP probes has been reported and has revealed that there are subgroups among ecotypes, with strains in each subgroup more closely related to each other than to strains in other subgroups (King et al. 1993).

In addition to single-base substitutions, there are reports of deletion/substitution differences between ecotypes. The case of variable Tag1 element number in different strains, as described above, is an example; another example is the presence of two tandem copies of the 12S seed storage protein gene *CRA* in the Landsberg genetic background, whereas there is only one copy at the cognate genomic position in the Col-0 strain (Pang et al. 1988). In this case, the single gene in Col-0 can be seen from its restriction map to be composed of the 5′ half of the upstream gene of L*er* and the 3′ half of the downstream L*er* gene. The Col-0 condition is thus clearly derived from the ancestral L*er* condition by deletion (presumably mediated by unequal crossing-over or by intrachromosomal homologous recombination). A different type of deletion/substitution polymorphism, that found in repeated simple sequence (microsatellite) regions of the *Arabidopsis* genome, has been found to occur with high frequency in the L*er* and Columbia strains (Bell and Ecker 1994).

## CLONED GENES AND SEQUENCING PROJECTS

More traditional means of gene cloning have also been applied to *Arabidopsis*; in particular, cloning by homology with known genes, cloning from cDNA libraries, and cloning from protein sequences (Bauw et al. 1992; Meyerowitz 1992). These have resulted in a collection of 5,770 separate sequence entries in the GenBank DNA sequence database (release 82.0), which represent over 3 Mb of nuclear DNA sequence. In fact, *Arabidopsis* is 10th on the list of "most sequenced" organisms and is the top plant on that list. The next plants are rice and maize, which are in 11th and 16th places, respectively, in the overall list. Rice has approximately two-thirds as many nucleotides of nuclear DNA sequenced as does *Arabidopsis*, maize approximately one-third as many. In addition, single-pass sequencing of *Arabidopsis* cDNA clones has resulted in a collection of 5,047 expressed sequence tags available from the dbEST database (version 2.2). *Arabidopsis* has more sequence entries in this collection than any other organism except for *Homo sapiens*.

From the sequence database, tables of the biased codon usage of *Arabidopsis* genes, of the biased representation of noncoding base triplets, and of intron boundary consensus sequences have been made (Cherry and Cartinhour 1992; Meyerowitz 1992). None of the sequences has shown *Arabidopsis* genes to be different from the genes of other higher plants in their general coding and structural properties: *Arabidopsis* genes are not extraordinarily shorter than the homologous genes of large-genome plants (see, e.g., Chang and Meyerowitz 1986), and they are interrupted by introns just as are the genes of other plants. Indeed, one of the largest known plant introns is in an *Arabidopsis* gene (2,985 bases, in the *AGAMOUS* floral homeotic gene [Yanofsky et al. 1990]), and as many as 24 introns have been found in a single *Arabidopsis* gene (in the gene for the second largest subunit of RNA polymerase II [Larkin and Guilfoyle 1993]).

Given the small size of the *Arabidopsis* genome, it seems likely that it will be the first plant nuclear genome whose physical map and DNA sequence will be completely known. An international effort to achieve the goals of complete physical mapping and complete sequencing of the genome of *A. thaliana* has been organized (National Science Foundation 1990, 1991, 1992, 1993).

## ADDITIONAL PROPERTIES

Characteristics of the *A. thaliana* nuclear genome other than its size and sequence structure that have been reported include a G+C content estimated at 41.4% (Leutwiler et al. 1984), a typical figure for a member of the mustard family (Thomas and Sherratt 1956; Vanyushin and Belozerskii 1959). Cytosine methylation has also been reported, amounting to approximately 6% of nuclear cytosine on a molar basis (Leutwiler et al. 1984), a rather low figure for an angiosperm (Meyerowitz 1992).

DNA replication origins have been studied by observing tritiated thymidine labeling of DNA in cells of seedlings (Van't Hof et al. 1978). The average cell cycle in these cells was measured to be 8.5 hours, with the S phase (DNA synthesis) occupying only 2.8 hours. The average size of replicons observed was 24 μm, or around 70 kb, and the average single fork rate (the rate of DNA replication) was 5.8 μm per hour, or around 17 kb per hour. Since each replicon replicates its DNA bidirectionally, an average replicon exists for about 2 hours, leaving little time in a 2.8-hour S phase for separate initiation times of different classes of replicons. Van't Hof et al. (1978) interpreted their data as best fitting a model of replication in which there are two classes of replicons, with 36 minutes separating the initiation of replication in the two classes.

Finally, ploidy levels of cells in the plant have been shown to vary. Maluszynska and Heslop-Harrison (1991) reported variation in nuclear size and in chromosome fluorescence, stained both with general DNA dyes and during in situ hybridization to repetitive sequences. This indicates the possibility of endopolyploidization of chromosomes in some nuclei. This possibility was confirmed with the flow cytometric measurements of nuclear genome size described above (Arumuganathan and Earle 1991; Galbraith et al. 1991). In these measurements, it was observed that the nuclear genome sizes in different nuclei varied by multiples of two, with nuclei containing apparent 2C, 4C, 8C, 16C, and even 32C amounts of DNA. The proportion of polyploid nuclei varies in different tissues and different stages of the life cycle. Nuclei are largely 2C and 4C in newly germinated seedlings, although some 8C nuclei are found in these early stages. Older plants show higher proportions of multiploid nuclei, with the oldest parts of plants showing the highest ratio of multiploids: Older rosette leaves have the highest levels, with decreasing levels in younger rosette leaves, stems, and cauline leaves. Floral buds show only 2C and 4C cells, and thus no evidence of endopolyploidy. A more detailed anatomical view of the nature of the polyploid cells shows that leaf trichome cells have elevated ploidy levels from 4C to 64C, and that stem and leaf epidermal cells can also have elevated ploidy levels, but only as high as 16C in leaf epidermis and 8C in stem (Melaragno et al. 1993). Correlation of ploidy and developmental stage in the maturation of trichomes has been made by Hülskamp et al. (1994). Despite this apparent chromosomal DNA endoreduplication, the genetic properties of the plant show that its gametophytes are haploid and its zygotes diploid, except in cases where polyploidy has been deliberately induced and maintained (Koornneef, this volume).

## CONCLUDING REMARKS

The utility of *A. thaliana* for experiments in molecular biology comes in part from the unusual size and structure of its nuclear genome, which reduces the expense and effort required for many types of experiments in molecular genetics. In addition to the interest in the *Arabidopsis* genome that results from its contribution to the ease of use of this plant in the laboratory, the genome is interesting in its own right. One key question raised by closely related plants with very different genome sizes is, What is the DNA that is "extra" in the plants with large genomes? This question is unanswered in the case of *Arabidopsis* and its larger-genome relatives such as *Brassica oleracea*, with a haploid genome estimated at 800–900 Mb (Bennett and Smith 1976, 1991). Recent work with grasses (for review, see Bennetzen and Freeling 1993) shows that the genomes of sorghum and maize (with the maize nuclear genome 3.5 times the size of the sorghum genome) differ by their content of repetitive DNA. More than 95% of single-copy maize probes easily cross-hybridize with the sorghum genome, whereas most repetitive sequences cross-hybridize poorly, or not at all. The RFLP maps of the two species show that single-copy probes that are genetically linked in maize are also generally linked in sorghum, and thus that the single-copy DNA has not been extensively rearranged in the 20 million years since the common ancestor of the two species. A similar conservation has been shown between rice and wheat, with a 35-fold difference in nuclear DNA content and a divergence time of 60 to 70 million years. Since the repeated DNA is interspersed between single-copy (genic) sequences in these plants, it seems that plants can gain or lose large arrays of noncoding, intergenic repetitive sequences in the course of evolution. Indeed, it seems likely that each plant lineage has the ability both to gain and to lose repetitive arrays and that this ability accounts for the large variations in nuclear genome size between closely related plants. The ease with which the genes of *Arabidopsis* and those of other flowering plants (including monocots such as maize) can be cross-hybridized (see, e.g, Chang and Meyerowitz 1986) implies that over the 120 million years of flowering plant evolution, there has been little change in the basic gene complement: The differences are in noncoding and intergenic repeats. Given a choice, then, one should choose to work with a plant like *Arabidopsis*, whose complement of genes is very likely the same as that of other flowering plants, but whose nuclear genome (at this time in its evolution) is near the minimal size that can contain this gene complement. It is easier to clone genes from plants such as *Arabidopsis*, and if necessary, to use these cloned genes as hybridization probes to obtain homologous sequences from plants with larger genomes.

## ACKNOWLEDGMENTS

My laboratory's work on *Arabidopsis* has enjoyed long-term support from the U.S. National Science Foundation, the U.S. Department of Energy, and the U.S. National Institute of General Medical Sciences of the National Institutes of Health.

## REFERENCES

Ambros, P. and D. Schweizer. 1976. The Giemsa C-banded karyotype of *Arabidopsis thaliana* (L.) Heynh. *Arabidopsis Inf. Serv.* **13:** 167–171.

Arondel, V., B. Lemieux, I. Hwang, S. Gibson, H.M. Goodman, and C.R. Somerville. 1992. Map-based cloning of a gene controlling omega-3 fatty acid desaturation in *Arabidopsis. Science* **258:** 1353–1355.

Arumuganathan, K. and E.D. Earle. 1991. Nuclear DNA content of some important plant species. *Plant Mol. Biol. Rep.* **9:** 208–218.

Bancroft, I., L. Westphal, R. Schmidt, and C. Dean. 1992. PFGE-resolved RFLP analysis and long range restriction mapping of the DNA of *Arabidopsis thaliana* using whole YAC clones as probes. *Nucleic Acids Res.* **20:** 6201–6207.

Bauw, G., M. Van Montagu, and D. Inzé. 1992. Microsequence analysis of *Arabidopsis* proteins separated by two-dimensional polyacrylamide gel electrophoresis: A direct linkage of proteins and genes. In *Methods in* Arabidopsis *Research* (ed. C. Koncz et al.), pp. 357–377. World Scientific, Singapore.

Bell, C.J. and J.R. Ecker. 1994. Assignment of 30 micosatellite loci to the linkage map of *Arabidopsis. Genomics* **19:** 137–144.

Bennett, M.D. and J.B. Smith. 1976. Nuclear DNA amounts in angiosperms. *Philos. Trans. R. Soc. Lond. B Biol. Sci.* **274:** 227–274.

————. 1991. Nuclear DNA amounts in angiosperms. *Philos. Trans. R. Soc. Lond. B Biol. Sci.* **334:** 309–345.

Bennetzen, J.L. and M. Freeling. 1993. Grasses as a single genetic system: Genome composition, collinearity and compatibility. *Trends Genet.* **9:** 259–261.

Chang, C. and E.M. Meyerowitz. 1986. Molecular cloning and DNA sequence of the *Arabidopsis thaliana* alcohol dehydrogenase gene. *Proc. Natl. Acad. Sci.* **83:** 1408–1412.

Chang, C., S.F. Kwok, A.B. Bleecker, and E.M. Meyerowitz. 1993. *Arabidopsis* ethylene response gene *ETR1:* Similarity of product to two-component regulators. *Science* **262:** 539–544.

Chang, C., J.L. Bowman, A.W. DeJohn, E.S. Lander, and E.M. Meyerowitz. 1988. Restriction fragment length polymorphism linkage map for *Arabidopsis thaliana. Proc. Natl. Acad. Sci.* **85:** 6856–6860.

Chaubet, N., M.-E. Chaboute, G. Philipps, and C. Gigot. 1987. Histone genes in higher plants: Organization and expression. *Dev. Genet.* **8:** 461–473.

Cheng, C., J. Dewdney, H. Nam, B.G.W. den Boer, and H.M. Goodman. 1988. A new locus (*NIA1*) in *Arabidopsis thaliana* encoding nitrate reductase. *EMBO J.* **7:** 3309–3314.

Cherry, J.M. and S. Cartinhour. 1992. Codon usage and splice-site tables available from AAtDB. *Probe* **2:** 10–11.

Feldmann, K.A. 1991. T-DNA insertion mutagenesis in *Arabidopsis*: Mutational spectrum. *Plant J.* **1:** 71–82.

―――. 1992. T-DNA insertion mutagenesis in *Arabidopsis*: Seed infection/transformation. In *Methods in* Arabidopsis *research* (ed. C. Koncz et al.), pp. 274–289. World Scientific, Singapore.

Feldmann, K.A., M.D. Marks, M.L. Christianson, and R.S. Quatrano. 1989. A dwarf mutant of *Arabidopsis* generated by T-DNA insertion mutagenesis. *Science* **243:** 1351–1354.

Flavell, R. 1980. The molecular characterization and organization of plant chromosomal DNA sequences. *Annu. Rev. Plant Physiol.* **31:** 569–596.

Francis, D.M., S.H. Hulbert, and R.W. Michelmore. 1990. Genome size and complexity of the obligate fungal pathogen, *Bremia lactucae. Exp. Mycol.* **14:** 299–309.

Galbraith, D.W., K.R. Harkins, and S. Knapp. 1991. Systemic endopolyploidy in *Arabidopsis thaliana. Plant Physiol.* **96:** 985–989.

Giraudat, J., B.M. Hauge, C. Valon, J. Smalle, F. Parcy, and H.M. Goodman. 1992. Isolation of the *Arabidopsis ABI3* gene by positional cloning. *Plant Cell* **4:** 1251–1261.

Grill, E. and C. Somerville. 1991. Construction and characterization of a yeast artificial chromosome library of *Arabidopsis* which is suitable for chromosome walking. *Mol. Gen. Genet.* **226:** 484–490.

Hauge, B.M. and H.M. Goodman. 1992. Genome mapping in *Arabidopsis*. In *Methods in* Arabidopsis *research* (ed. C. Koncz et al.), pp. 191–223. World Scientific, Singapore.

Hauge, B.M., S.M. Hanley, S. Cartinhour, J.M. Cherry, H.M. Goodman, M. Koornneef, P. Stam, C. Chang, S. Kempin, L. Medrano, and E.M. Meyerowitz. 1993. An integrated genetic/RFLP map of the *Arabidopsis thaliana* genome. *Plant J.* **3:** 745–754.

Heslop-Harrison, J.S. and T. Schwarzacher. 1990. The ultrastructure of *Arabidopsis thaliana* chromosomes. In *Abstracts from the 4th International Conference on* Arabidopsis *Research*, Vienna, p. 3. University of Vienna, Austria.

Hülskamp, M., S. Miséra, and G. Jürgens. 1994. Genetic dissection of trichome cell development in *Arabidopsis. Cell* **76:** 555–566.

Hwang, I., T. Kohchi, B. Hauge, H.M. Goodman, R. Schmidt, G. Cnops, C. Dean, S. Gibson, K. Iba, B. Lemieux, V. Arondel, L. Danhoff, and C. Somerville. 1991. Identification and map position of YAC clones comprising one-third of the *Arabidopsis* genome. *Plant J.* **1:** 367–374.

Kamalay, J.C. and R.B. Goldberg. 1980. Regulation of structural gene expression in tobacco. *Cell* **19:** 935–946.

King, G., J. Nienhuis, and C.E. Hussey. 1993. Genetic similarity among ecotypes of *Arabidopsis thaliana* estimated by analysis of restriction fragment length polymorphisms. *Theor. Appl. Genet.* **86:** 1028–1032.

Kiper, M., D. Bartels, F. Herzfeld, and G. Richter. 1979. The expression of a plant genome in hnRNA and mRNA. *Nucleic Acids Res.* **6:** 1961–1978.

Koncz, C., J. Schell, and G.P. Rédei. 1992. T-DNA transformation and insertion mutagenesis. In *Methods in* Arabidopsis *research* (ed. C. Koncz et al.), pp. 224–273. World Scientific, Singapore.

Konieczny, A. and F.M. Ausubel. 1993. A procedure for mapping *Arabidopsis* mutations using co-dominant ecotype-specific PCR-based markers. *Plant J.* **4:** 403–410.

Konieczny, A., D.F. Voytas, M.P. Cummings, and F.M. Ausubel. 1991. A superfamily of *Arabidopsis thaliana* retrotransposons. *Genetics* **127:** 801–809.

Koornneef, M., J. Van Eden, C.J. Hanhart, P. Stam, F.J. Braaksma, and W.J. Feenstra. 1983. Linkage map of *Arabidopsis thaliana*. *J. Hered.* **74:** 265–272.

Kopczak, S.D., N.A. Haas, P.J. Hussey, C.D. Silflow, and D.P. Snustad. 1992. The small genome of *Arabidopsis* contains at least six expressed $\alpha$-tubulin genes. *Plant Cell* **4:** 539–547.

Laibach, F. 1907. Zur Frage nach der Individualität der Chromosomen im Pflanzenreich. *Beih. Bot. Cbl.* (1 abt.) **22:** 191–210.

Larkin, R. and T. Guilfoyle. 1993. The second largest subunit of RNA polymerase II from *Arabidopsis thaliana*. *Nucleic Acids. Res.* **21:** 1038.

Lee-Chen, S. and L.M. Steinitz-Sears. 1967. The location of linkage groups in *Arabidopsis thaliana*. *Can. J. Genet. Cytol.* **9:** 381–384.

Leutwiler, L.S., B.R. Hough-Evans, and E.M. Meyerowitz. 1984. The DNA of *Arabidopsis thaliana*. *Mol. Gen. Genet.* **194:** 15–23.

Leutwiler, L.S., E.M. Meyerowitz, and E.M. Tobin. 1986. Structure and expression of three light-harvesting chlorophyll a/b-binding protein genes in *Arabidopsis thaliana*. *Nucleic Acids Res.* **14:** 4051–4064.

Leyser, H.M.O., C.A. Lincoln, C. Timpte, D. Lammer, J. Turner, and M. Estelle. 1993. *Arabidopsis* auxin-resistance gene *AXR1* encodes a protein related to ubiquitin-activating enzyme E1. *Nature* **364:** 161–164.

Lisitsyn, N., N. Lisitsyn, and M. Wigler. 1993. Cloning the difference between two complex genomes. *Science* **259:** 946–951.

Maluszynska, J. and J.S. Heslop-Harrison. 1991. Localization of tandemly repeated DNA sequences in *Arabidopsis thaliana*. *Plant J.* **1:** 159–166.

Martinez-Zapater, J.M., M.A. Estelle, and C.R. Somerville. 1986. A highly repeated DNA sequence in *Arabidopsis thaliana*. *Mol. Gen. Genet.* **204:** 417–423.

McGrath, J.M., M.M. Jancso, and E. Pichersky. 1993. Duplicate sequences with a similarity to expressed genes in the genome of *Arabidopsis thaliana*. *Theor. Appl. Genet.* **86:** 880–888.

Melaragno, J.E., B. Mehrotra, and A.W. Coleman. 1993. Relationship between endopolyploidy and cell size in epidermal tissue of *Arabidopsis*. *Plant Cell* **5:** 1661–1668.

Meyerowitz, E.M. 1987. *Arabidopsis thaliana*. *Annu. Rev. Genet.* **21:** 93–111.

______. 1992. Introduction to the *Arabidopsis* genome. In *Methods in* Arabidopsis *research* (ed. C. Koncz et al.), pp. 100–118. World Scientific, Singapore.

Meyerowitz, E.M. and R.E. Pruitt. 1985. *Arabidopsis thaliana* and plant molecular genetics. *Science* **229:** 1214–1218. (Reprinted in *Biotechnology: The renewable frontier* [ed. D.E. Koshland, Jr.], pp. 311–320. 1986, AAAS]).

Nam, H.-G., J. Giraudat, B. den Boer, F. Moonan, W.D.B. Loos, B.M. Hauge, and H.M. Goodman. 1989. Restriction fragment length polymorphism linkage map of *Arabidopsis thaliana*. *Plant Cell* **1:** 699–705.

National Science Foundation. 1990. A long-range plan for the multinational coordinated *Arabidopsis thaliana* genome research project. *Natl. Sci. Found. NSF Publ.* **90-80**.

————. 1991. The multinational coordinated *Arabidopsis thaliana* genome research project progress report: Year one. *Natl. Sci. Found. NSF Publ.* **91-60.**

————. 1992. The multinational coordinated *Arabidopsis thaliana* genome research project progress report: Year two. *Natl. Sci. Found. NSF Publ.* **92-112.**

————. 1993. The multinational coordinated *Arabidopsis thaliana* genome research project progress report: Year three. *Natl. Sci. Found. NSF Publ.* **93-173**

Pang, P.P., R.E. Pruitt, and E.M. Meyerowitz. 1988. Molecular cloning, genomic organization, expression and evolution of 12S seed storage protein genes of *Arabidopsis thaliana. Plant Mol. Biol.* **11:** 805–820.

Pruitt, R.E. and E.M. Meyerowitz. 1986. Characterization of the genome of *Arabidopsis thaliana. J. Mol. Biol.* **187:** 169–183.

Putterill, J., F. Robson, K. Lee, and G. Coupland. 1993. Chromosome walking with YAC clones in *Arabidopsis*: Isolation of 1700 kb of contiguous DNA on chromosome 5, including a 300 kb region containing the flowering-time gene *CO. Mol. Gen. Genet.* **239:** 145–157.

Richards, E.J. and F.M. Ausubel. 1988. Isolation of a higher eukaryotic telomere from *Arabidopsis thaliana. Cell* **53:** 127–136.

Richards, E.J., H.M. Goodman, and F.M. Ausubel. 1991. The centromere region of *Arabidopsis thaliana* chromosome 1 contains telomere-similar repeats. *Nucleic Acids Res.* **19:** 3351–3357.

Schmidt, R. and C. Dean. 1993. Towards construction of an overlapping YAC library of the *Arabidopsis thaliana* genome. *BioEssays* **15:** 63–69.

Schweizer, D., P. Ambros, P. Gründler, and F. Varga. 1988. Attempts to relate cytological and molecular chromosome data of *Arabidopsis thaliana* to its genetic linkage map. *Arabidopsis Inf. Serv.* **25:** 27–34.

Simoens, C.R., J. Gielen, M. Van Montagu, and D. Inzé. 1988. Characterization of highly repetitive sequences of *Arabidopsis thaliana. Nucleic Acids Res.* **14:** 6753–6766.

Snustad, D.P., N.A. Haas, S.D. Kopczak, and C.D. Silflow. 1992. The small genome of *Arabidopsis* contains at least nine expressed β-tubulin genes. *Plant Cell* **4:** 549–556.

Sparrow, A.H. and J.P. Miksche. 1961. Correlation of nuclear volume and DNA content with higher plant tolerance to chronic radiation. *Science* **134:** 282–283.

Sparrow, A.H., H.J. Price, and A.G. Underbrink. 1972. A survey of DNA content per cell and per chromosome of prokaryotic and eukaryotic organisms: Some evolutionary considerations. *Brookhaven Symp. Biol.* **23:** 451–494.

Straus, D. and F.M. Ausubel. 1990. Genomic subtraction for cloning DNA corresponding to deletion mutations. *Proc. Natl. Acad. Sci.* **87:** 1889–1893.

Sun, T., H.M. Goodman, and F.M. Ausubel. 1992. Cloning the *Arabidopsis GA1* locus by genomic subtraction. *Plant Cell* **4:** 119–128.

Tsay, Y., M.J. Frank, T. Page, C. Dean, and N.M. Crawford. 1993. Trapping of an active endogenous transposon in *Arabidopsis thaliana. Science* **260:** 342–344.

Thomas, A.J. and H.S.A. Sherratt. 1956. The isolation of nucleic acid fractions from plant leaves and their purine and pyrimidine composition. *Biochem. J.* **62:** 1–4.

Van't Hof, J., A. Kuniyuki, and C.A. Bjerknes. 1978. The size and number of replicon families of chromosomal DNA of *Arabidopsis thaliana. Chromosoma* **68:** 269–285.

Vanyushin, B.F. and A.N. Belozerskii. 1959. Nucleotide composition of the desoxyribonucleotides of higher plants. *Dokl. Akad. Nauk. USSR* **129:** 944–946.

Voytas, D.F. and F.M. Ausubel. 1988. A copia-like transposable element family in *Arabidopsis thaliana. Nature* **336:** 242–244.

Ward, E.R. and G.C. Jen. 1990. Isolation of single-copy-sequence clones from a yeast artificial chromosome library of randomly-sheared *Arabidopsis thaliana* DNA. *Plant*

*Mol. Biol.* **14:** 561–568.

Wetmur, J.G. and N. Davidson. 1968. Kinetics of renaturation of DNA. *J. Mol. Biol.* **31:** 349–370.

Yanofsky, M.F., H. Ma, J.L. Bowman, G.N. Drews, K.A. Feldmann, and E.M. Meyerowitz. 1990. The protein encoded by the *Arabidopsis* homeotic gene *AGAMOUS* resembles transcription factors. *Nature* **346:** 35–39.

# 3

# Chloroplast and Mitochondrial DNAs of *Arabidopsis thaliana:* Conventional Genomes in an Unconventional Plant


**Jeffrey D. Palmer, Stephen R. Downie,[1]
and Jacqueline M. Nugent[2]**
Department of Biology, Indiana University
Bloomington, Indiana 47405

**Petra Brandt, Michael Unseld,
Mathieu Klein, Axel Brennicke,
and Wolfgang Schuster**
Institut für Genbiologische Forschung
D-1000 Berlin 33, Federal Republic of Germany

**Thomas Börner**
Fachbereich Biologie, Humboldt Universität
D-1040 Berlin, Federal Republic of Germany


Many of the unusual features that make *Arabidopsis thaliana* such an attractive model system for flowering plants relate, either directly or indirectly, to its unconventional nuclear genome. Of paramount importance, the notably small size of this genome, together with the relative lack of repeated sequences and reduced size of many introns and gene families, directly facilitates efforts to map, isolate, and characterize genes. Indirectly, the small size of the nuclear genome may be related to, perhaps in some measure responsible for, the reduced size of the plant and length of its life cycle, features that considerably enhance its genetic tractability.

The chloroplast and mitochondrial genomes of *Arabidopsis* have not shared the limelight with its nuclear genome. This is in part because these organellar genomes have not, until recently, been subject to much experimental study and also because they appear, in distinct contrast to the nuclear genome, to be utterly conventional in all important respects. The primary goal of this chapter is to show how the chloroplast DNA (cpDNA) and mitochondrial DNA (mtDNA) of *Arabidopsis* are typical of flowering plant organellar genomes in terms of size, organization,


Present addresses: [1]Department of Plant Biology, University of Illinois, Urbana, Illinois 61801; [2]John Innes Institute, Colney Lane, Norwich NR4 7UH, United Kingdom.

content of repeats and genes, gene structure, and, probably, modes of expression and regulation. In terms of direct molecular characterization, the *Arabidopsis* mitochondrial genome should soon assume primacy among flowering plant mtDNAs, thanks to a genome-sequencing project that is now more than half complete.

A secondary goal of this chapter is to describe what we view as the proper and rightful swing of the nuclear spotlight onto the organellar genomes of *Arabidopsis*, a swing that is just starting and which should soon accelerate. This swing is a natural reflection of the intricate and extensive interdependency of the nuclear and organellar genomes in controlling organellar biogenesis and function: The great majority of chloroplast and mitochondrial proteins are encoded by nuclear genes, and virtually all regulation of organellar gene expression is mediated by nuclear gene products. Hence, the power of *Arabidopsis* nuclear molecular genetics ought to be fully unleashed onto questions of organellar function and biogenesis.

The first two sections of this chapter summarize our current knowledge of *Arabidopsis* cpDNA and mtDNA with respect to their organization, gene content, and expression. This information is placed in the context of our overall knowledge of organelle genomes in land plants. Finally, we review published studies and describe promising future directions of integrated work on nuclear/organellar genetic interactions in *Arabidopsis*.

## THE CHLOROPLAST GENOME

### Tobacco as Paradigm

The best perspective for viewing the *Arabidopsis* chloroplast genome is as a tobacco chloroplast genome only slightly modified by the roughly 100 million years of evolution that separate the two plants. This statement follows from three observations: First, with some exceptions, land plant chloroplast genomes are highly conserved in all important aspects of size, organization, and sequence (Palmer 1991, 1992; Wolfe et al. 1991; Downie and Palmer 1992; Sugiura 1992). Second, the tobacco chloroplast genome is the type chloroplast genome for seed plants by virtue of being the only completely sequenced genome (Shinozaki et al. 1986; Shimada and Sugiura 1991) that has the ancestral gene order for this group (Palmer and Stein 1986; Raubeson and Jansen 1992). Third, a series of comprehensive hybridizations between the two genomes shows that *Arabidopsis* and tobacco cpDNAs are virtually identical in size, gene order, and gene content (Fig. 1).

The tobacco chloroplast genome is 156 kb in size and contains the characteristic cpDNA "inverted repeat," an inverted duplication of 25 kb that separates large and small single-copy regions of 87 kb and 18 kb, respectively (Fig. 1) (Shinozaki et al. 1986). The 113 known genes in the tobacco genome can be grouped into four classes according to function (Shimada and Sugiura 1991; Sugiura 1992). More than half of the genes encode components of the machinery for gene expression, including 4 subunits of RNA polymerase, a putative splicing factor, 4 ribosomal and 30 transfer RNAs, 21 ribosomal proteins, and a translation initiation factor (although the gene for the latter may be a pseudogene in tobacco; Wolfe et al. 1992b). Another 40 genes encode either known components of the photosynthetic apparatus (29 genes) or subunits of a chlororespiratory NADH dehydrogenase (11 genes) thought to be associated with photosynthetic metabolism (dePamphilis and Palmer 1990; Wolfe et al. 1992a). The functions of another 10 genes are completely unknown. Finally, only 2 genes, encoding subunits of the clpP protease and acetyl-CoA carboxylase of fatty acid synthesis, are known to specify proteins that function in some process other than gene expression or photosynthesis. In this respect, tobacco and other angiosperm cpDNAs are notably poor; even the bryophyte *Marchantia polymorpha* contains several additional chloroplast genes, for sulfate transport and light-independent chlorophyll biosynthesis (Wolfe et al. 1991), whereas the genomes of nongreen algae encode a remarkable number and diversity of proteins involved in both genetic and metabolic activities (see, e.g., Reith and Munholland 1993).

The chloroplast genes of tobacco and other land plants are organized and expressed in an intriguing melange of ways, some primitive and prokaryotic, others derived subsequent to the endosymbiotic origin of the chloroplast some 1 billion years ago (Gray 1989, 1992). On the one hand, most genes are tightly packed into eubacterial-like operons (Fig. 1). Many transcription units contain eubacterial-like "-35" and "-10" promoter elements, and the genome encodes all four core subunits of a eubacterial-like RNA polymerase (Igloi and Kössel 1992). Chloroplast ribosomes are highly similar to eubacterial 70S ribosomes in size, subunit composition, and antibiotic resistance properties (Subramanian et al. 1991). On the other hand, some genes, including tRNA genes with internal promoters, do not contain obviously bacterial promoters, and there is increasing evidence that a second, nuclear-encoded RNA polymerase of probable nonbacterial ancestry is also active in transcribing chloroplast genes (Morden et al. 1991; Igloi and Kössel 1992; Wolfe et al. 1992a; Hess et al. 1993; Lerbs-Mache 1993). RNA processing is much more complex in plastids than in bacteria. Tobacco and most angiosperms con-

**A**

Gene labels (top): trnH psbA trnK*(matK) rps16* trnQ trnS psbK psbI trnG* trnR atpA atpF* atpH atpI rps2 rpoC2 rpoC1* rpoB trnC ORF29 psbM trnD trnY trnE trnT psbD psbC ORF62 trnG psbN trnfM rps14 psaB psaA ORF168* rps4 trnT ndhJ ndhK ndhC trnV* atpE trnL* trnF trnS trnM

Scale (kb): 0 · 5 · 10 · 15 · 20 · 25 · 30 · 35 · 40 · 45 · 50 · 55

| | | |
|---|---|---|
| Tobacco | 1 2 3 4 5 6 7 8 9 10 11 12 13 14 15 16 17 18 19 20 21 22 23 24 25 26 27 28 29 30 31 32 33 34 35 36 37 38 39 40 | |
| Brassica | 2.2 5.8 5.5 2.1 9.0 9.8 1.15 4.0 11.5 1.05 | |
| BamHI | 4.9 .5 6.0 1.4 4.5 6.2 4.9 6.8 4.6 1.5 3.2 2.4 .5 4.0 4.7 | |
| HindIII | 4.5 3.5 2.0 .8 .2 .6 1.7 2.2 6.8 7.2 1.5 9.0 5.5 6.5 | |
| SalI | 9.3 9.9 28 18 | |
| BglI | 27 21 32 | |
| KpnI | 32 9.3 2.5 15 11 9.4 | |
| PstI | 15 16 15 1.1 19 | |

**B**

Gene labels: atpB rbcL accD psaI ORF229 petA psbJ psbL psbF psbE petG trnW trnP ORF31 psaJ rpl33 rps18 rpl20 rps12* clpP** psbB psbH petB* petD* ORF34 psbN rps11 rpl36 rps8 rpl14 infA rps3 rpl22 rpl2* rpl23 rpoA rpl16* trnI ORF2280 trnV 16S rDNA trnI* trnA* rDNA rDNA 23S 4.5S 5S trnL ndhB* rps7 3'rps12*

Scale (kb): 55 · 60 · 65 · 70 · 75 · 80 · 85 · 90 · 95 · 100 · 105 · 110

| | | |
|---|---|---|
| Tobacco | 41 42 43 44 45 46 47 48 49 50 51 52 53 54 55 56 57 58 59 60 61 62-78, see below 79 80 81 82 83 84 85 86 87 88 89 90 91 92 93 94 | |
| Brassica | 1.1 15 2.0 1.6 12.3 .3 12.3 5.8 3.5 | |
| BamHI | 4.5 9.3 3.8 9.3 1.8 3.0 1.6 2.9 .3 1.7 1.4 .2 2.0 .3 2.9 1.4 7.3 | |
| HindIII | 6.5 .4 4.9 7.5 4.6 2.5 .9 .5 2.5 4.5 1.2 .6 .9 1.2 1.6 4.8 3.7 .3 1.1 1.3 | |
| SalI | 18 11.3 11 12 4.0 22 46 | |
| BglI | 9.4 21 6.9 7.3 28 4.8 30 | |
| KpnI | 18 22 .7 | |
| PstI | 30 14 19 | |

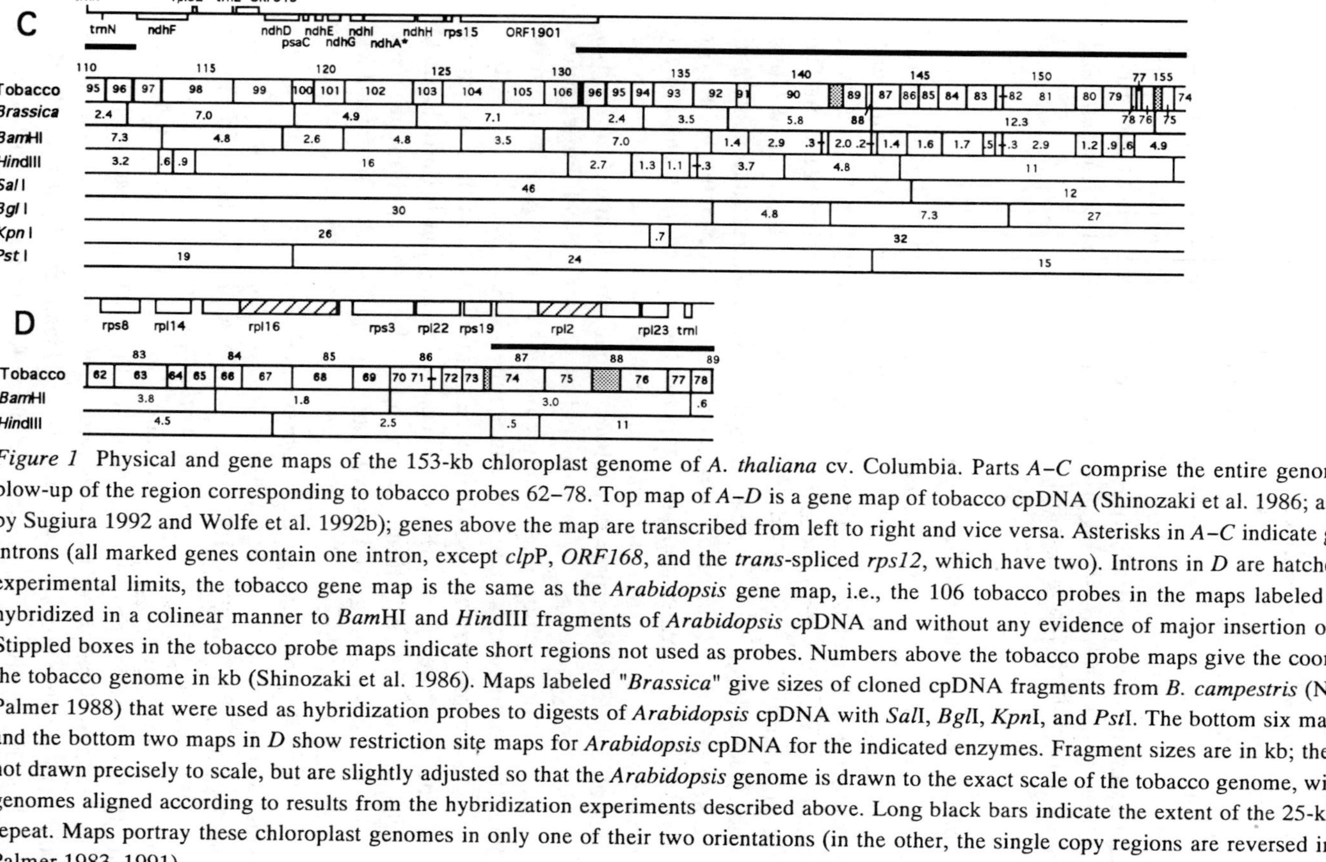

*Figure 1* Physical and gene maps of the 153-kb chloroplast genome of *A. thaliana* cv. Columbia. Parts *A–C* comprise the entire genome; *D* is a blow-up of the region corresponding to tobacco probes 62–78. Top map of *A–D* is a gene map of tobacco cpDNA (Shinozaki et al. 1986; as modified by Sugiura 1992 and Wolfe et al. 1992b); genes above the map are transcribed from left to right and vice versa. Asterisks in *A–C* indicate genes with introns (all marked genes contain one intron, except *clp*P, *ORF168*, and the *trans*-spliced *rps12*, which have two). Introns in *D* are hatched. Within experimental limits, the tobacco gene map is the same as the *Arabidopsis* gene map, i.e., the 106 tobacco probes in the maps labeled "tobacco" hybridized in a colinear manner to *Bam*HI and *Hind*III fragments of *Arabidopsis* cpDNA and without any evidence of major insertion or deletion. Stippled boxes in the tobacco probe maps indicate short regions not used as probes. Numbers above the tobacco probe maps give the coordinates of the tobacco genome in kb (Shinozaki et al. 1986). Maps labeled "*Brassica*" give sizes of cloned cpDNA fragments from *B. campestris* (Nugent and Palmer 1988) that were used as hybridization probes to digests of *Arabidopsis* cpDNA with *Sal*I, *Bgl*I, *Kpn*I, and *Pst*I. The bottom six maps in *A–C* and the bottom two maps in *D* show restriction site maps for *Arabidopsis* cpDNA for the indicated enzymes. Fragment sizes are in kb; the maps are not drawn precisely to scale, but are slightly adjusted so that the *Arabidopsis* genome is drawn to the exact scale of the tobacco genome, with the two genomes aligned according to results from the hybridization experiments described above. Long black bars indicate the extent of the 25-kb inverted repeat. Maps portray these chloroplast genomes in only one of their two orientations (in the other, the single copy regions are reversed in polarity; Palmer 1983, 1991).

tain 21 introns; 20 of these are group II introns of likely postendosymbiotic acquisition by the green alga ancestors of land plants (Palmer 1991), whereas only a single intron, a group I intron in a tRNA-Leu gene, appears to be of cyanobacterial ancestry (Kuhsel et al. 1990; Xu et al. 1990). In addition to splicing, chloroplast transcripts are subject to complex pathways of 5′ and 3′ trimming and, for polycistronic transcripts, internal cutting (Sugiura 1991, 1992; Herrmann et al. 1992). In contrast to bacteria, where most control of gene activity is exerted at the level of transcription, chloroplast gene activity is also thought to be regulated extensively at posttranscriptional levels of RNA processing, RNA stability, and translation (Link 1991; Rochaix 1992a,b; Gruissem and Tonkyn 1993).

### *Arabidopsis* and Tobacco cpDNAs Compared

The only published studies on *Arabidopsis* cpDNA report the unremarkable sequences of a few of its genes (Chen et al. 1988; Luschnig and Schweizer 1992) and observations on duplicated cpDNA sequences in *Arabidopsis* mtDNA (Nugent and Palmer 1988; discussed below in The Mitochondrial Genome). In addition, the chapter by Price et al. (this volume) describes phylogenetic studies that have used restriction site and *rbc*L sequence data from *Arabidopsis* cpDNA. In this chapter, we report the first complete restriction site map for the *Arabidopsis* chloroplast genome and a complete gene map as deduced from a comprehensive series of hybridizations between tobacco gene probes and *Arabidopsis* cpDNA. This gene map was constructed as part of a much larger study on the evolution of chloroplast genome organization and gene and intron content in angiosperms (Downie and Palmer 1992). We are using this chapter as the primary and original place of publication of these physical and gene maps; i.e., they will not be reported in any journal article, and, hence, this chapter constitutes the proper place for citation of these results. These mapping studies were conducted by S.R. Downie, J.M. Nugent, and J.D. Palmer.

A cleavage site map of *Arabidopsis* cpDNA for six restriction enzymes was constructed in two ways. Sites for *Bam*HI and *Hind*III were mapped by hybridizing the 106 tobacco cpDNA probes shown in Figure 1 to filter-blots containing digests of *Arabidopsis* cpDNA with these two enzymes. Sites for *Sal*I, *Bgl*I, *Kpn*I, and *Pst*I were mapped by hybridizing the 22 *Brassica* probes also shown in Figure 1 to digests with these four enzymes. The use of two unconnected sets of hybridizations means that there is some uncertainty in the exact placement of sites for *Bam*HI and *Hind*III relative to those for the other four enzymes. Furthermore, because all hybridizations were to single digests of *Arabidopsis* cpDNA, there is also uncertainty in the exact registry of the single enzyme maps within each of the two sets of maps. However, this uncertainty is minimal for *Bam*HI and *Hind*III compared to the other four enzymes, due to the small sizes of the tobacco fragments used as probes and of the *Arabidopsis* fragments being mapped.

The restriction site map (Fig. 1) shows that the *Arabidopsis* chloroplast genome is a circular chromosome 153 kb in length. Thus, the *Arabidopsis* genome is very similar in size to genomes in *Brassica* and related crucifers (Palmer et al. 1983; Warwick and Black 1991) and less than 3 kb smaller than the tobacco genome (Shinozaki et al. 1986). The four sectors of the *Arabidopsis* chloroplast genome—the two inverted repeats and the large and small single-copy regions—are also of a typical, tobacco-like size.

The 106 tobacco probes contain 98.1% of the 130.5-kb sequence complexity of the tobacco genome and average 1.2 kb in size. Fifty-five probes are internal to or contain part of but a single gene, 30 contain all or part of two genes, and 21 probes contain parts of more than two genes (Fig. 1). This level of coverage of the entirely sequenced tobacco genome allows a fine level of inference about gene content and location in *Arabidopsis* cpDNA (Fig. 1). Each of the 106 tobacco probes showed significant hybridization to *Arabidopsis* cpDNA as normalized relative to 15 other angiosperm cpDNAs present on the same set of replica filter blots. That is, under the same conditions, a number of probes failed to hybridize to one or more of the other cpDNAs, leading to inferences (in some cases proven by subsequent sequence analysis) of specific losses of individual genes or introns (see, e.g., Gantt et al. 1991; Downie et al. 1991; for review, see Downie and Palmer 1992). Moreover, the 106 tobacco probes hybridized in a colinear manner to *Arabidopsis* cpDNA (Fig. 1). Thus, gene order and content appear to be the same in tobacco and *Arabidopsis* cpDNAs.

The limits of these experiments must be emphasized. Part or all of a small gene that comprises only a short region of a particular tobacco probe fragment could in fact be absent from the *Arabidopsis* genome, and yet we would have no basis for suspecting such an absence. Even a gene that occupies the entire length of a strongly hybridizing tobacco fragment could be nonfunctional, if it contained a frameshift or other subtle mutation that would only be detectable by sequencing. However, because all tobacco probes hybridize so well, because the two genomes are so similar in size, and because their aligned maps show no evidence of gene deletion in *Arabidopsis*, our estimate is that few if any genes present in tobacco cpDNA have been lost from or are nonfunctional in

*Arabidopsis* cpDNA.

The 21 introns in tobacco cpDNA range in size from 503 bp to 1148 bp. Only 2 of these, from *rpl2* (also see Downie et al. 1991) and *rpl16*, were represented by intron-specific probes among the 106 probes shown in Figure 1. In other experiments, we have also tested specifically for the presence of the *cis*-spliced introns in *trnI* and *rps12*. All 4 of these introns are present in their respective genes in *Arabidopsis* cpDNA. These results, together with the above-stated conclusion that there is no evidence for any major deletions from the *Arabidopsis* chloroplast genome (relative to tobacco), lead us to tentatively conclude that intron content in *Arabidopsis* cpDNA is likely to be very similar, possibly identical, to that in tobacco.

In summary then, within the limits of detection of these mapping studies, the *Arabidopsis* chloroplast genome is identical to that of tobacco in gene order, gene content, and intron content. This conclusion should come as no surprise, given the large body of evidence indicating that land plant cpDNAs are highly conserved in their evolution (see beginning of preceding section).

## An *Arabidopsis* Chloroplast Genome Project?

It seems inevitable that the *Arabidopsis* chloroplast genome will eventually be entirely sequenced, so the relevant question is not whether, but when. Given the inherent conservatism of cpDNA evolution and the (unremarkable) similarity between *Arabidopsis* cpDNA and the fully sequenced genome of tobacco, we would argue that sequencing of the former genome should not be a high priority. Instead, it would seem appropriate to sequence this 153-kb genome only as part of a much larger project to sequence the nuclear genome of *Arabidopsis*; i.e., it would be a poor investment to sequence the chloroplast genome until sequence technologies and funding allow efficient sequencing of the approximately 1000 times larger nuclear genome. In the meantime, any sequencing of *Arabidopsis* cpDNA should continue to be carried out in a piecemeal fashion by those scientists who need to know more about a particular region of the genome.

## THE MITOCHONDRIAL GENOME

The situation with respect to the mitochondrial genome of *Arabidopsis* and other land plants stands in direct contrast to that for the corresponding chloroplast genomes. Plant mtDNAs are spectacularly variable in size and structure and are relatively poorly understood (only a single genome, from the bryophyte *Marchantia polymorpha*, is completely sequenced [Oda et al. 1992b], and little is known about mitochondrial gene expression), whereas *Arabidopsis* mtDNA is now one of the best-characterized genomes. These differences dictate a completely different organization of this section compared to the preceding one. Here, we present a straightforward review of what is known about *Arabidopsis* mtDNA—its structure, genes, and expression—and interweave this material with general comparisons to other plant mitochondrial genomes.

### The *Arabidopsis* Mitochondrial Genome Project: Overview of Genome Structure

The laboratory of A. Brennicke is now about 60% through a project to determine the entire sequence of the approximately 372-kb mitochondrial genome of *A. thaliana*. This is likely to be the first complete mtDNA sequence from any flowering or even vascular plant. The ultimate goal of this project is to provide a major part of the necessary background information for studies on plant mitochondrial gene expression and regulation. Indeed, as described below, a substantial start in this direction has already been made. An ancillary goal is to further our understanding of the processes that have led to the extraordinarily large sizes of plant mtDNAs, some of which are more than 100 times larger than animal mtDNAs.

Analysis of more than 300 individual cosmids has allowed construction of a single, circular linkage map that includes their entire sequence complexity (Fig. 2). This map contains two different pairs of large (several kb) repeated sequences that meet the definition (Stern and Palmer 1984) of "recombination repeats," i.e., sequences that recombine at a high enough frequency to generate substantial levels of genomic isomers that differ only with respect to the combinations of unique sequences that flank the repeats. Cosmids containing all four possible combinations of repeat-flanking sequences have been isolated for each set of repeats. The particular representation of the *Arabidopsis* mitochondrial genome shown in Figure 2 features one pair of direct repeats interspersed with one pair of inverted repeats. All four genomic isomers resulting from this arrangement of recombination repeats are formally diagramed in Figure 1C of Lonsdale (1984). For *Arabidopsis*, these would include three different master chromosomes of 372 kb, which interconvert with each other via recombination between inverted repeats, and one set of subgenomic circles (of ~234 kb and ~138 kb), which are generated from two of the master circles via recombination between direct repeats. In complexity of organization, *Arabidopsis* mtDNA ranks midway between the

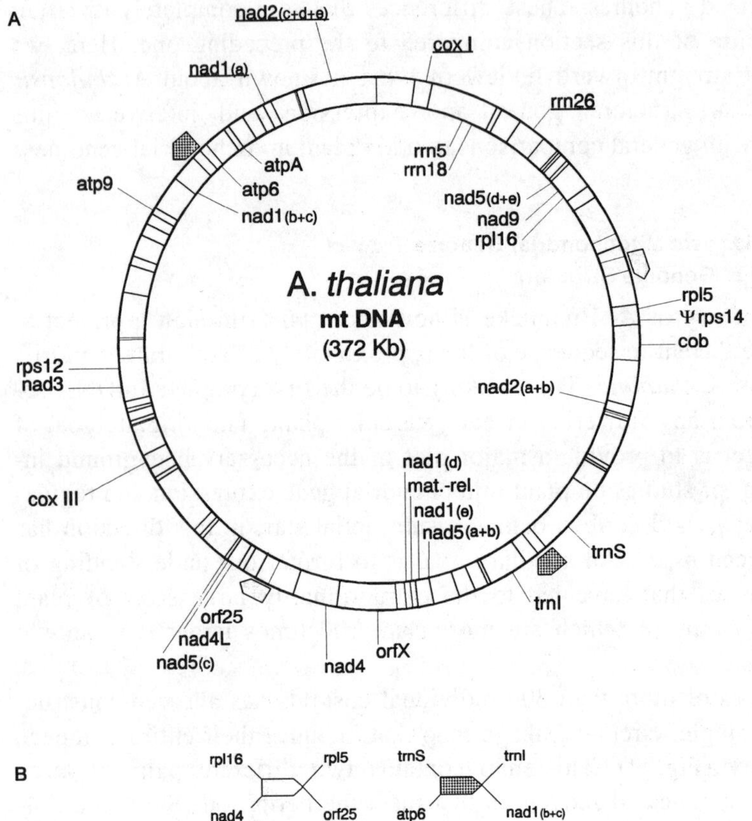


*Figure 2*   Physical and gene maps of the 372-kb mitochondrial genome of *A. thaliana* line C24. (*A*) Circular map containing the entire sequence content of the genome. Radial lines indicate *Bam*HI sites. Stippled and cross-hatched thickened arrows indicate the positions of the two pairs of large (several kb) repeated elements in the genome. Recombination between members of each pair, present in either direct or indirect orientation, leads to three other isomeric forms of the genome besides the one shown (see text and Fig. 1C of Lonsdale et al. 1984). Only those genes whose positions on the map are precisely known are shown (cf. Table 1). The positions of genes and of exons of three *trans*-spliced genes are marked outside the circle for sequences transcribed clockwise and vice versa. (*B*) Arrangements of the four unique sequences flanking each of the two different major mtDNA repeats. For example, the genome representation in *A* shows the *nad4-orf25* and *rpl16-rpl5* configurations of the stippled repeat, whereas the other two configurations of this repeat (flanked by *nad4-rpl5* and *rpl16-orf25*) are present in other isomeric forms of the genome that result from recombination between the two copies of this repeat.

fairly simple genomes of spinach, sunflower, and most other crucifers, which exist as one master chromosome and one pair of subgenomes, and the highly complex genomes of maize, wheat, sugar beet, and petunia (Palmer 1992).

For purposes of discussion, the *Arabidopsis* mitochondrial genome can be broken down into five overlapping categories of sequences: genes, introns, repeated elements, sequences of chloroplast origin, and everything else. Genes and introns are discussed in the following two sections. The two largest repeats, highly active in recombination, are discussed in the preceding paragraph. A large number of other, much shorter repeats (up to 160 bp in size) are dispersed throughout the genome. These include duplications of both noncoding regions and fragments of genes, such as portions of *nad3*, exon A of *nad5*, and the 18S and 26S rRNA genes (A. Brennicke et al., unpubl.). The potential function of these short repeats with respect to transcription and translation signals needs further investigation (Wissinger et al. 1991a).

Extrapolation from the sequenced portions of the *Arabidopsis* mitochondrial genome suggests that roughly 15% of the genome exists as repeat sequence DNA. Similar estimates have been made for the much larger genomes of cucurbits by reassociation kinetics (Ward et al. 1981) and for the smaller, 218-kb *Brassica campestris* genome by comprehensive Southern hybridization and partial sequencing studies (M. Shirzadegan and J. Palmer, unpubl.). In the latter case, an estimated 66 distinct families of repeats are present, with each kilobase of the genome containing, on average, one repeat element. Overall then, although sequence duplication contributes a significant fraction to the inordinately large sizes of angiosperm mitochondrial genomes, additional forces must also be responsible.

One well-known way in which plant mtDNAs grow larger is by the fairly frequent and stable uptake of cpDNA-derived sequences (Schuster and Brennicke 1988; Palmer 1992). A low-resolution, hybridization study showed that *Arabidopsis* mtDNA contains only about half as many cpDNA sequences (5–7 kb; ≤2% of the mitochondrial genome) as most other crucifer genomes of substantially smaller size (Nugent and Palmer 1988). These sequences include portions of the chloroplast *rbc*L and *psa*A genes. Several cpDNA-derived regions have been sequenced, including tRNA[Ser], tRNA[Met], and part of 23S rRNA (Wintz et al. 1988; Brandt et al. 1993; P. Brandt and M. Unseld, unpubl.). Unlike all other cpDNA-derived sequences, most chloroplast tRNA genes appear to be functional within angiosperm mitochondria (see, e.g., Joyce and Gray 1989).

Although exemplary of flowering and, perhaps, vascular (Palmer et al. 1992) plants in all of the above respects, *Arabidopsis* mtDNA differs substantially from the fully sequenced genome of the bryophyte *Marchantia polymorpha* (Oda et al. 1992b). At 187 kb, the *Marchantia* genome is almost exactly half the size of the *Arabidopsis* genome; although this size is large for non-plants, it is smaller than that of any flowering plant. Although it contains "quite a few" repetitive sequences (Oda et al. 1992b), *Marchantia* mtDNA lacks any large recombination repeats and exists as a simple, unicircular chromosome, in contrast to all but one characterized angiosperm mtDNA (Palmer and Herbon 1987; Palmer 1992). Finally, *Marchantia* mtDNA does not appear to contain any cpDNA-derived sequences (Oda et al. 1992a,b).

## Gene Content and Organization

About 50 genes have been identified thus far on the *Arabidopsis* mitochondrial genome. Most have been identified by complete sequencing of the gene locus, a few by partial sequencing, and several by heterologous hybridization only. These include a full set of three rRNA genes and about a dozen tRNA genes (Table 1). As in other angiosperms, translation in *Arabidopsis* mitochondria probably uses three sets of tRNAs of distinct genetic origins, a set of about 15 tRNAs of mitochondrial origin, several tRNAs encoded by mtDNA sequences of chloroplast origin (see above), and several imported tRNAs encoded by nuclear genes of eukaryotic origin (Joyce and Gray 1989; Dietrich et al. 1992).

Although only approximately 60% sequenced, *Arabidopsis* mtDNA is already known to contain substantially more protein genes (>30) than the fully sequenced genomes of animals (12–13), fungi (7–14), the green alga *Chlamydomonas reinhardtii* (7), the ciliate *Paramecium aurelia* (16), the trypanosome *Trypanosoma brucei* (12), and the apicomplexan *Plasmodium falciparum* (3) (Gray 1992). The only mitochondrial genome known to encode many more proteins (60–70) is that of *Marchantia* (Oda et al. 1992b). However, we expect that *Arabidopsis* will turn out to contain a similar number of mitochondrial protein genes, given that so much of the genome remains to be sequenced. The mystery of why plant mtDNAs encode so many proteins is really just part of the larger mysteries of why gene content is so extraordinarily variable among mitochondria as a whole (see above), and why so many different lineages (e.g., *Saccharomyces, Chlamydomonas, Plasmodium,* with only 3–7 encoded metabolic proteins) have independently experienced near-extinctions of their mitochondrial genomes. This unusual abundance of genes is clearly an additional factor responsible for the large baseline size of plant mitochondrial genomes.

*Table 1*  Gene and intron content of *Arabidopsis* mtDNA

| Class of gene product | Gene name |
|---|---|
| Ribosomal RNA | *rrn26, rrn18, rrn5* |
| Transfer RNA | *trnE, trnF, trnF, trnG, trnH, trnI, trnK, trnN, trnS, trnS, trnT, trnY* |
| Small subunit ribosomal protein | *rps3*(1)*, rps7, rps12, ψrps14, ψrps19* |
| Large subunit ribosomal protein | *rpl2, rpl5, rpl16* |
| ATP synthase subunit | *atpA, atp6, atp9, ψatp9* |
| Cytochrome *c* oxidase subunit | *coxI, coxII, coxIII* |
| Cytochrome *b* subunit | *cob* |
| NADH dehydrogenase subunit | *nad1*(4)*, nad2*(4)*, nad3, nad4*(3)*, nad4L, nad5*(4)*, nad6, nad7*(4)*, nad9* |
| Open reading frames | *orfX, orf25, ψorf86A, orf169, orf206, orf322*(1)*, orf509, mat-r* |

For all genes with introns, the number of introns is given in parentheses after the gene name. References for published gene sequences: Wintz et al. 1988; Chen et al. 1989; Binder et al. 1991; Knoop et al. 1991; Wissinger et al. 1991a; Aubert et al. 1992; Brandt et al. 1992, 1993; Martínez-Zapater et al. 1992; Schuster et al. 1993. All other sequences are unpublished data of A. Brennicke et al. Also see Grohmann et al. 1993; Wissinger et al. 1991b.

In terms of function, the *Arabidopsis* mitochondrial protein genes can be conveniently divided into three groups (Table 1). Sixteen genes encode subunits of four respiratory chain complexes; it is members of this class of genes that are most commonly found throughout the spectrum of mitochondrial genomes. Eight genes are homologous to eubacterial ribosomal protein genes. However, two of these (*rps14* and *rps19*) are pseudogenes, and no sequences at all for a third (*rps13*) were detectable by hybridization (Aubert et al. 1992; Brandt et al. 1993; Nugent and Palmer 1993; A. Brennicke, unpubl.). Functional copies of all three of these ribosomal protein genes may have been recently relocated to the nucleus in *Arabidopsis*, as has been demonstrated for *coxII* in legumes (Nugent and Palmer 1991; Covello and Gray 1992) and *rps12* in *Oenothera* (Grohmann et al. 1992). Finally, eight genes are open reading frames (most with conserved counterparts known in other mitochondrial genomes) of poorly known or unknown function (Table 1). One of these (*orf509*) is homologous to a bacterial gene whose product is involved in cytochrome *c* biogenesis (Schuster et al. 1993).

As in other angiosperm mtDNAs, genes in *Arabidopsis* mtDNA are scattered randomly in the genome and, for the most part, transcribed singly. Reflecting the extremely rapid rate of inversions and other rearrangements in plant mtDNA, the order of genes in *Arabidopsis* is quite different from that in other plants (Palmer et al. 1992). Compared to *Marchantia*, where 14 of 16 ribosomal protein genes are organized into two clusters (putative operons) of eubacterial organization (Takemura et al. 1992), most ribosomal protein genes are scattered in *Arabidopsis*, with the only organizational trace of their eubacterial ancestry being the cotranscribed *rpl5* and ψ*rps14* genes.

### Introns and *cis*- and *trans*-Splicing

Little is known about the transcription of *Arabidopsis* mitochondrial genes, although some progress has been made on this subject in other plant systems (for review, see Gray et al. 1992). However, as described in this and the following section, much is known about the unusual RNA processing of transcripts in mitochondria of *Arabidopsis* and other plants. A total of 21 introns have been identified in *Arabidopsis* mtDNA, most by gene sequencing (Knoop et al. 1991; Wissinger et al. 1991b; A. Brennicke, unpubl.) and a few by heterologous hybridization (Nugent and Palmer 1993 and unpubl.). All 21 introns are group II introns, compared to 25 group II introns and 7 group I introns in *Marchantia* mtDNA (Ohta et al. 1993). Only one of these introns, intron C in *nad2*, is homologous in position in *Arabidopsis* and *Marchantia* mtDNAs (Nozato et al. 1993), yet most of these introns are present broadly in all monocots and dicots examined (Nugent and Palmer 1993 and unpubl.). It thus seems likely that angiosperms acquired most of their introns after their divergence from a common ancestor with bryophytes.

The mitochondrial introns in *Arabidopsis* and other angiosperms are largely restricted to a specific subset of protein genes (Table 1). No introns are present in any tRNA or rRNA genes or in 27 of 34 completely sequenced protein genes, and 2 genes (*rps3* and *orf322*) each contain a single intron. On the other hand, the remaining 19 introns are clustered within five NADH dehydrogenase genes, four of which have four introns each (Table 1). *Marchantia* mtDNA has a more random distribution of introns in both respects (Oda et al. 1992b), and the significance, if any, of the angiosperm pattern is unclear.

The most striking aspect of angiosperm mitochondrial introns is the extent to which they have been fractured by rearrangement events that have created *trans*-spliced introns from formerly *cis*-spliced ones. Only two *trans*-spliced group II introns are known outside of angiosperm mitochondria (one in *rps12* of land plant chloroplasts [Fig. 1] and the other in *psaA* of *Chlamydomonas* chloroplasts [Kück et al. 1987]), whereas 5 of the 12 group II introns in the *Arabidopsis nad1*, *nad2*, and *nad5* genes are *trans*-spliced (Knoop et al. 1991; Wissinger et al. 1991b; A. Brennicke, unpubl.), and an additional *nad1* exon is *trans*-spliced in wheat and petunia (Chapdelaine and Bonen 1991; Conklin et al. 1991).

With the exception of this *nad1* intron, the *trans*-spliced arrangements of these three genes are highly conserved among monocots and dicots (Knoop et al. 1991; Wissinger et al. 1991b; Nugent and Palmer 1993 and unpubl.), indicating that these introns were fractured prior to the emergence of angiosperms. Particularly striking is the structure of the *nad5* gene, which requires two *trans*-splicing events to join the central exon of only 22 nucleotides to its two flanking exons (Knoop et al. 1991).

The mechanism of *trans*-splicing of group II introns is thought to be essentially the same as for *cis*-splicing, with the folded secondary structure of the uninterrupted *cis*-spliced intron (Michel et al. 1989) mimicked by the noncovalent association of base-paired halves of the interrupted group II intron (Wissinger et al. 1992). Interestingly, all of the *trans*-spliced fractures in *Arabidopsis* mitochondrial genes occur within domain IV of the group II intron, and the specificity of *trans*-splicing may be associated with a somewhat longer base-pairing stem of this interrupted domain. The RNA editing events that characterize most plant mitochondria mRNAs (see next section) also occur within introns, and some of these edits may improve the extent of intronic base-pairing and thereby facilitate *trans*-splicing (Wissinger et al. 1991b, 1992; Binder et al. 1992). The relative timing of splicing and editing events has been investigated for *nad1* in petunia (Sutton et al. 1991) and *coxII* in maize (Yang and Mulligan 1991). It appears that intron excision is not required for editing; editing can precede splicing at any editing site, yet unspliced RNAs are generally less completely edited than spliced RNAs. Thus splicing and editing, although independent, are temporally correlated.


## RNA Editing

Extensive mRNA editing has been observed in mitochondria of all flowering plants and at least one conifer. Seed plants again differ, and again in a derived manner, from *Marchantia*, in which mtRNA editing is thought to be absent or very infrequent (Oda et al. 1992b). Sequence comparisons of cDNAs and genomic DNA show that transcripts from each of eight *Arabidopsis* protein genes investigated (*nad3*, *nad4L*, *nad5*, *cob*, *rps12*, ψ*rps14*, *rpl5*, and *orf25*) are edited. All edits found to date in *Arabidopsis* mitochondria are C to U transitions; in other flowering plants a few U to C changes have also been observed (Bonnard et al. 1992; Wissinger et al. 1992; Gray and Covello 1993). Most editings occur at replacement positions, i.e., they change the encoded amino acid. For example, all 9 *nad4L* edits in *Arabidopsis* lead to amino acid changes, whereas only 1 of 10 edits for *rpl5* is silent (Brandt et al. 1992).

With the exception of cytochrome *b*, for which all examined cDNA clones show identical, apparently complete, editing patterns, each of the other seven *Arabidopsis* genes studied shows variably complete editing among sequenced cDNAs. Editing frequencies at conserved sites vary between *Arabidopsis* and other plant species. In *Arabidopsis*, approximately 90% of the transcripts containing the *nad3* and the *rps12* reading frames are edited at all identified positions, whereas some sites in the *nad3* gene are altered in only about 50% of cDNA clones from *Oenothera* (Schuster et al. 1990). The *Arabidopsis rps14* pseudogene, although interrupted by a stop codon and a genomic frameshift, is transcribed and edited at three positions (Aubert et al. 1992). A gene requiring unusually little RNA editing is *atp6*, for which all 20 non-silent editing positions described for *Oenothera* (Schuster and Brennicke 1991) are already encoded as Ts in *Arabidopsis* mtDNA. However, the *Arabidopsis atp6* mRNA is expected to be edited at one additional site that is edited in *Brassica* (Handa and Nakajima 1992) but not *Oenothera*.

The mechanism(s) of and sources of information and specificity for RNA editing in plant mitochondria are entirely unknown. Determination of the complete sequence of the *Arabidopsis* mitochondrial genome should allow searches for potential *trans*-acting factors (e.g., something akin to guide RNAs of trypanosomes; Cattaneo 1991; Stuart 1991) of the RNA editing process.

## ORGANELLE DNA MUTATIONS AND NUCLEAR MUTATOR GENES

It is well established that alterations in angiosperm mtDNA may lead—in combination with a certain nuclear genotype—to cytoplasmic male sterility (CMS). Although CMS mutations are known from a large number of plants (most prominently, maize and petunia [Hanson 1991; Braun et al. 1992; Levings and Siedow 1992]), including several cultivated crucifers (Makaroff and Palmer 1988; Makaroff et al. 1989; Singh and Brown 1991; Bonhomme et al. 1992), none has yet been reported in *Arabidopsis*. Until recently, the only other phenotypic class of plant mtDNA mutations were the nonchromosomal stripe (NCS) mutants of maize, which have variegated, striped leaves (Newton et al. 1989). In contrast, numerous cpDNA mutants have been analyzed. These may occur spontaneously, may be induced by mutagens, or may occur with usually high frequency in response to certain nuclear mutations (for review, see Börner and Sears 1986; Hagemann 1992).

In *Arabidopsis*, a large number of nuclear mutants have been selected that affect morphology, specific components, development, differentiation, and functions of plastids (Rédei 1992; Chory and Susek, this volume). This contrasts with a surprisingly low number of cpDNA mutants. No spontaneous mutation is known in *Arabidopsis* that is inherited in the uniparental, maternal manner typical for mutations in the plastid (and mitochondrial) genome. However, Röbbelen (1962) reported on the appearance of cpDNA mutants after irradiation of *Arabidopsis* with X-rays.

The way in which a nuclear gene induces mutations in organellar DNA is unknown. However, the available data indicate the existence of at least two mechanisms. One category of nuclear mutator genes always induces a single type of organellar mutation (deduced from identical phenotypes), such as *iojap* and *chloroplast mutator* of maize (Walbot and Coe 1982; Thompson et al. 1983; Han et al. 1992); *Okina-muki, albostrians*, and *striata 4* of barley (Imai 1928; von Wettstein 1961; Hagemann and Scholz 1962; Hess et al. 1993); and, probably, *albomaculans* of *Arabidopsis* (Röbbelen 1966). Another category of nuclear mutator genes induces various kinds of mutations (again deduced from the phenotype, which in this case is variable). To this group belong the *mp1* and *mp2* alleles of *Epilopium hirsutum* (Michaelis 1968a,b), the cpDNA mutator of *Oenothera hookeri* (Epp 1973; Chui et al. 1990), and the *chm* locus of *A. thaliana* (Rédei 1973). As described in the following section, mutations at this last locus have been extensively studied and have led to some recent surprises.

## The Choroplast Mutator of *Arabidopsis* Causes mtDNA Rearrangements

Rédei (1973) discovered a nuclear gene (*chm*) that acts as a chloroplast mutator in *A. thaliana*. He showed that the spontaneous rate of appearance of new plastid phenotypes in leaf cells was increased about $10^6$-fold in plants that were homozygous for certain mutant alleles at the *chm* locus. Once induced, the new phenotypes were inherited independently of the mutator gene in the uniparental, maternal manner characteristic of cytoplasmic mutations.

The mutator gene induces a variety of phenotypes: discolored leaf sectors (white, yellow, pale green), distorted leaf differentiation ("rough and ragged" leaves), and impaired fertility (Rédei and Plurad 1973). Electron microscopy revealed an almost continuous series of plastid aberrations, affecting both overall plastid morphology and internal membrane architecture, in cells of the discolored sectors (Rédei and Plurad 1973; Rédei 1974; Mourad and White 1992). The occurrence of two or more types of plastids (both normal chloroplasts and variously aberrant plastids) within a single, "mixed" cell is regarded as good evidence that the plastid genome is the site of the mutation (Rédei and Plurad 1973;

Mourad and White 1992). Nuclear (and mitochondrial) mutations that affect plastids should lead to the same aberration in all plastids of the affected cells (Hagemann, 1964; Rédei, 1967). In addition, the ultrastructure of mitochondria was normal in chm cells with aberrant plastids (Rédei and Plurad 1973).

As pointed out by Rédei and Plurad (1973), the observation of variable plastid phenotypes does not necessarily prove that different plastid mutations were responsible for each type of alteration. Therefore, the isolation of different, apparently homoplastidic mutants with color variations or rough-leaf phenotypes provides not only excellent material for further investigations into the molecular basis of the mutations, but also evidence for the induction of different organellar DNA mutations (Rédei 1973; Mourad and White 1992).

Martínez-Zapater et al. (1992) studied a new variegated mutant of *Arabidopsis* that results from mutation at the chm locus. As in Rédei's (1973) analyses, the new allele (designated *chm-3*) induces clearly cytoplasmic mutations in all homozygous plants. Southern hybridizations of total DNA from normal and variegated plants with several cpDNA probes showed identical band patterns. This is consistent with studies on cpDNA isolated from normal and homoplastidic mutant leaves (Mourad and White 1992; G.P. Rédei, pers. comm.). Surprisingly, however, two of four mtDNA probes hybridized to additional mtDNA fragments in variegated plants compared to normal plants (Martínez-Zapater et al. 1992). Furthermore, these new bands cosegregate with the mutant phenotype, and their abundance appears to correlate with the extent of variegation. Sequence analysis of one of the new mtDNA fragments showed that it is rearranged relative to wild-type mtDNA (Martínez-Zapater et al. 1992). A number of examples of specific nuclear gene effects on mitochondrial genome sorting-out and/or rearrangement have also been described in maize and bean (see Martínez-Zapater et al. 1992).

Mitochondria and chloroplasts are known to be  intimately related metabolically and sometimes even physically, and therefore mutations in the DNA of one organelle might be expected to affect the other organelle. The NCS mutants of maize (see above) are prime examples in which various mutations in mtDNA lead to altered ultrastructure and physiology of chloroplasts (see, e.g., Roussell et al. 1991). Conversely, a nuclear-gene-induced cpDNA mutant of barley that lacks plastid ribosomes has altered levels of mtDNA and mitochondrial transcripts (Börner and Hess 1993 and unpubl.).

The data of Martínez-Zapater et al. (1992) clearly support the idea that the observed changes in the *Arabidopsis* mitochondrial genome are due to the action of the *chm* gene. However, it is less clear whether these mtDNA changes are responsible for leaf variegation and the underlying plastid alterations. One possibility that cannot be ruled out is that the *chm* locus induces mutations in both mtDNA and cpDNA (e.g., point mutations or small insertions/deletions in cpDNA would have gone undetected in the studies cited above). As stated above, the existence of mixed cells with different types of plastids is more readily explained by a direct effect of the *chm* locus on the plastid genome than by an indirect effect on the mitochondrial genome. This intriguing set of nuclear/cytoplasmic mutants clearly deserves further study.

## CONCLUSIONS AND PERSPECTIVES

Precisely because they are so conventional in organization, gene content, and, apparently, expression, the organelle genomes of *Arabidopsis* should serve as exemplars for angiosperms in general and, in the case of the chloroplast, for land plants as a whole. Although we currently have relatively little direct knowledge about the molecular biology of the *Arabidopsis* chloroplast, the studies newly presented in this chapter suggest that much of the extensive information already available about the chloroplast genomes of tobacco and other land plants will be directly transferable to *Arabidopsis*. Conversely, as the progress report in this chapter documents, *Arabidopsis* mtDNA is rapidly becoming one of the best-characterized mitochondrial systems in angiosperms: The genome seems likely to become the first of the large angiosperm mitochondrial genomes to be completely sequenced and is also extensively characterized in terms of basic aspects of RNA editing and splicing.

Two developments must be realized, however, in order for the study of organelle genomes in *Arabidopsis* to truly blossom. First, the full power of the facile molecular genetics of the *Arabidopsis* nuclear genome must be brought to bear on the organelles. More than 80–90% of chloroplast and mitochondrial proteins are nuclear gene products. These include virtually all proteins of primary organelle metabolism except for a limited number of photosynthetic and respiratory polypeptides, and many of these proteins and metabolic systems have already been investigated in some detail in *Arabidopsis* (see various chapters in this volume). However, the parallel application of nuclear molecular genetics to the study of basic processes of organelle molecular biology—such as organelle DNA replication, recombination, transcription, splicing, RNA editing, translation, and the whole control of gene expression during development—is barely touched. The recent study by Martínez-Zapater et al. (1992), on a nuclear mutator gene that seems to affect both chloroplasts and mitochondria, is a promising start in this direction. An impressive number of nuclear gene products control and interact with mitochondrial and chloroplast genetic systems in yeast and *Chlamydomonas*, respectively (Gillham 1994), and one can expect a still greater complexity of nuclear/cytoplasmic genetic interactions in a more complex, multicellular eukaryote.

The second development that must be realized is a technical one: the establishment of efficient systems for the transformation of both organelle genomes of *Arabidopsis*. Organellar transformation is currently facile only for the two above-mentioned microbial systems, to the extent that many workers on land plant chloroplasts have expanded their efforts to include the *Chlamydomonas* system (Boynton et al. 1992). A promising start on chloroplast transformation has, however, been made in tobacco (Maliga 1993; Svab and Maliga 1993), and with luck, the *Arabidopsis* organelles will not be far behind.

## ACKNOWLEDGMENTS

Work at Indiana University was supported by the National Institutes of Health (GM-35087), and work at the Institut für Genbiologische Forschung was supported by the Bundesministerium für Forschung und Technologie.

## REFERENCES

Aubert, D., C. Bisanz-Seyer, and M. Herzog. 1992. Mitochondrial *rps14* is a transcribed and edited pseudogene in *Arabidopsis thaliana*. *Plant Mol. Biol.* **20:** 1169–1174.

Binder, S., V. Knoop, and A. Brennicke. 1991. Nucleotide sequences of the mitochondrial genes *trnS*(TGA) encoding tRNA-Ser-TGA in *Oenothera berteriana* and *Arabidopsis thaliana*. *Gene* **102:** 245–247.

Binder, S., A. Marchfelder, A. Brennicke, and B. Wissinger. 1992. RNA editing in intron sequences may be required for *trans*-splicing of *nad2* transcripts in *Oenothera* mitochondria. *J. Biol. Chem.* **267:** 7615–7623.

Bonhomme, S., F. Budar, D. Lancelin, I. Small, M.-C. Defrance, and G. Pelletier. 1992. Sequence and transcript analysis of the *Nco2.5* Ogura-specific fragment correlated with cytoplasmic male sterility in *Brassica* cybrids. *Mol. Gen. Genet.* **235:** 340–348.

Bonnard, G., J.M. Gualberto, L. Lamattina, and J.M. Grienenberger. 1992. RNA editing in plant mitochondria. *Crit. Rev. Plant Sci.* **10:** 503–524.

Börner, T. and W.R. Hess. 1993. Altered nuclear, mitochondrial and plastid gene expression in white barley cells containing ribosome-deficient plastids. In *Plant mitochondria* (ed. A. Brennicke and U. Kück), pp. 207–220. VCH Chemie, Weinheim.

Börner, T. and B.B. Sears. 1986. Plastome mutants. *Plant Mol. Biol. Rep.* **4:** 69–92.

Boynton, J.E., N.W. Gillham, S.M. Newman, and E.H. Harris. 1992. Organelle genetics and transformation of *Chlamydomonas*. In *Plant gene research. Cell organelles* (ed.

R.G. Herrmann), pp. 3–64. Springer Verlag, Wien.

Brandt, P., M. Unseld, U. Eckert-Ossenkopp, and A. Brennicke. 1993. An *rps14* pseudogene is transcribed and edited in *Arabidopsis* mitochondria. *Curr. Genet.* **24:** 330–336.

Brandt, P., S. Sünkel, M. Unseld, A. Brennicke, and V. Knoop. 1992. The *nad4L* gene is encoded between exon c of *nad5* and *orf25* in the *Arabidopsis* mitochondrial genome. *Mol. Gen. Genet.* **236:** 33–38.

Braun, C.J., G.G. Brown, and C.S. Levings III. 1992. Cytoplasmic male sterility. In *Plant gene expression. Cell organelles* (ed. R.G. Herrmann), pp. 219–245. Springer Verlag, Wien.

Cattaneo, R. 1991. Different types of messenger RNA editing. *Annu. Rev. Genet.* **25:** 71–88.

Chapdelaine, Y. and L. Bonen. 1991. The wheat mitochondrial gene for subunit I of the NADH dehydrogenase complex: A *trans*-splicing model for this gene-in-pieces. *Cell* **65:** 465–472.

Chen, H.-C., H. Wintz, J.-H. Weil, and D.T.N. Pillay. 1988. Nucleotide sequence of chloroplast CF1-ATPase ε-subunit and elongator tRNA$^{Met}$ genes from *Arabidopsis thaliana*. *Nucleic Acids Res.* **16:** 10372.

———. 1989. Three mitochondrial tRNA genes from *Arabidopsis thaliana*: Evidence for the conversion of a tRNA$^{Phe}$ gene into a tRNA$^{Tyr}$ gene. *Nucleic Acids Res.* **17:** 2613–2621.

Chui, W.-L., E.M. Johnson, S.A. Kaplan, K. Blasko, M.B. Sokalski, R. Wolfson, and B.B. Sears. 1990. *Oenothera* chloroplast DNA polymorphisms associated with plastome mutator activity. *Mol. Gen. Genet.* **221:** 59–64.

Conklin, P.L., R.K. Wilson, and M.R. Hanson. 1991. Multiple *trans*-splicing events are required to produce a mature *nad1* transcript in a plant mitochondrion. *Genes Dev.* **5:** 1407–1415.

Covello, P.S. and M.W. Gray. 1992. Silent mitochondrial and active nuclear genes for subunit 2 of cytochrome c oxidase (*cox2*) in soybean: Evidence for RNA-mediated gene transfer. *EMBO J.* **11:** 3815–3820.

dePamphilis, C.W. and J.D. Palmer. 1990. Loss of photosynthetic and chlororespiratory genes from the plastid genome of a parasitic flowering plant. *Nature* **348:** 337–339.

Dietrich, A., J.H. Weil, and L. Maréchal-Drouard. 1992. Nuclear-encoded transfer RNAs in plant mitochondria. *Annu. Rev. Cell Biol.* **8:** 115–131.

Downie, S.R. and J.D. Palmer. 1992. Use of chloroplast DNA rearrangements in reconstructing plant phylogeny. In *Molecular systematics of plants* (ed. P.S. Soltis et al.), pp. 36–49. Chapman and Hall, New York.

Downie, S.R, R.G. Olmstead, G. Zurawski, D.E. Soltis, P.S. Soltis, J.C. Watson, and J.D. Palmer. 1991. Six independent losses of the chloroplast *rpl2* intron in dicotyledons: Molecular and phylogenetic implications. *Evolution* **45:** 1245–1259.

Epp, M.D. 1973. Nuclear gene-induced plastome mutations in *Oenothera hookeri*. I. Genetic analysis. *Genetics* **75:** 465–483.

Gantt, J.S., S.L. Baldauf, P.J. Calie, N.F. Weeden, and J.D. Palmer. 1991. Transfer of *rpl22* to the nucleus greatly preceded its loss from the chloroplast and involved the gain of an intron. *EMBO J.* **10:** 3073–3078.

Gillham, N.W. 1994. *Organelle genes and genomes*. Oxford, New York. (In press.)

Gray, M.W. 1989. The evolutionary origin of organelles. *Trends Genet.* **5:** 294–299.

———. 1992. The endosymbiont hypothesis revisited. *Int. Rev. Cytol.* **141:** 233–357.

Gray M.W. and P.S. Covello. 1993. RNA editing in plant mitochondria and chloroplasts. *FASEB J.* **7:** 64–71.

Gray, M.W., P.J. Hanic-Joyce, and P.S. Covello. 1992. Transcription, processing and editing in plant mitochondria. *Annu. Rev. Plant Physiol. Plant Mol. Biol.* **43:** 145–175.

Grohmann, L., A. Brennicke, and W. Schuster. 1992. The gene for mitochondrial ribosomal protein S12 has been transferred to the nuclear genome in *Oenothera*. *Nucleic Acids Res.* **20:** 5641–5646.

———. 1993. Genes for mitochondrial ribosomal protein in plants. In *The translational apparatus* (ed. K. Nierhaus et al.), pp. 599–607. Plenum Press, New York.

Gruissem, W. and J.C. Tonkyn. 1993. Control mechanisms of plastid gene expression. *Crit. Rev. Plant Sci.* **12:** 19–55.

Hagemann, R. 1964. *Plasmatische Vererbung*. VEB Gustav Fischer Verlag, Jena.

———. 1992. Plastid genetics in higher plants. In *Plant gene research. Cell organelles* (ed. R.G. Herrmann), pp. 65–96. Springer Verlag, Wien.

Hagemann, R. and F. Scholz. 1962. Ein Fall geninduzierter Mutationen des Plasmotypus bei Gerste. *Züchter* **32:** 50–59.

Han, C.-D., E.H. Coe, Jr., and R.A. Martienssen. 1992. Molecular cloning and characterization of iojap (ij), a pattern striping gene of maize. *EMBO J.* **11:** 4037–4046.

Handa, H. and K. Nakajima. 1992. RNA editing of *atp6* transcripts from male-sterile and normal cytoplasms of rapeseed (*Brassica napus* L.). *FEBS Lett.* **310:** 111–114.

Hanson, M.R. 1991. Plant mitochondrial mutations and male sterility. *Annu. Rev. Genet.* **25:** 461–486.

Herrmann, R.G., P. Westhoff, and G. Link. 1992. Biogenesis of plastids in higher plants. In *Plant gene research. Cell organelles* (ed. R.G. Herrmann), pp. 275–349. Springer Verlag, Wien.

Hess, W.R., A. Prombona, B. Fieder, A.R. Subramanian, and T. Börner. 1993. Chloroplast *rps*15 and the *rpo*B/C1/C2 gene cluster are strongly transcribed in ribosome-deficient plastids: Evidence for a functioning non-chloroplast-encoded RNA polymerase. *EMBO J.* **12:** 563–571.

Igloi, G.L. and H. Kössel. 1992. The transcriptional apparatus of chloroplasts. *Crit. Rev. Plant Sci.* **10:** 525–558.

Imai, Y. 1928. A consideration of variegation. *Genetics* **13:** 544–563.

Joyce, P.B.M. and M.W. Gray. 1989. Chloroplast-like transfer RNA genes expressed in wheat mitochondria. *Nucleic Acids Res.* **17:** 5461–5476.

Knoop, V., W. Schuster, B. Wissinger, and A. Brennicke. 1991. *Trans*-splicing integrates an exon of 22 nucleotides into the *nad5* mRNA in higher plant mitochondria. *EMBO J.* **10:** 3483–3493.

Kück, U., Y. Choquet, M. Schneider, M. Dron, and P. Bennoun. 1987. Structural and transcription analysis of two homologous genes for the P700 chlorophyll a-apoproteins in *Chlamydomonas reinhardtii*: Evidence for *in vivo trans*-splicing. *EMBO J.* **6:** 2185–2195.

Kuhsel, M.G., R. Strickland, and J.D. Palmer. 1990. An ancient group I intron shared by eubacteria and chloroplasts. *Science* **250:** 1570–1573.

Lerbs-Mache, S. 1993. The 110-kDa polypeptide of spinach plastid DNA-dependent RNA polymerase: Single-subunit enzyme or catalytic core of multimeric enzyme complexes? *Proc. Natl. Acad. Sci.* **90:** 5509–5513.

Levings, C.S., III and J.N. Siedow. 1992. Molecular basis of disease susceptibility in the Texas cytoplasm of maize. *Plant Mol. Biol.* **19:** 135–147.

Link, G. 1991. Photoregulated development of chloroplasts. *Cell Cult. Somatic Cell Genet.* **7B:** 365–394.

Lonsdale, D.M. 1984. A review of the structure and organization of the mitochondrial genome of higher plants. *Plant Mol. Biol.* **3:** 201–206.

Luschnig, C. and D. Schweizer. 1992. Nucleotide sequence of *trn*I$^{(CAU)}$ and *rpl*23 from

*Arabidopsis thaliana* chloroplast genome. *Nucleic Acids Res.* **20:** 3511.

Makaroff, C.A. and J.D. Palmer. 1988. Mitochondrial DNA rearrangements and transcriptional alterations in the Ogura male-sterile cytoplasm of radish. *Mol. Cell. Biol.* **8:** 1474–1480.

Makaroff, C.A., I.J. Apel, and J.D. Palmer. 1989. The *atp6* coding region has been disrupted and a novel reading frame generated in the mitochondrial genome of cytoplasmic male-sterile radish. *J. Biol. Chem.* **264:** 11706–11713.

Maliga, P. 1993. Towards plastid transformation in flowering plants. *Trends Biotechnol.* **11:** 101–107.

Martínez-Zapater, J.M., P. Gil, J. Capel, and C.R. Somerville. 1992. Mutations at the *Arabidopsis* CHM locus promote rearrangements of the mitochondrial genome. *Plant Cell* **4:** 889–899.

Michaelis, P. 1968a. Beiträge zum Problem der Plastidenabänderungen. IV. Über das Plasma- und Plastidenabänderungen auslösende, Isotopen-32P-induzierte Kerngen mp1 von Epilobium. *Mol. Gen. Genet.* **101:** 257–306.

————. 1968b. Beiträge zum Problem der Plastidenabänderungen. V. Über eine weitere Isotopen-(35S)-induzierte Kernmutante, die Plastidenabänderungen hervorruft. *Theor. Appl. Genet.* **38:** 314–320.

Michel, F., K. Umesono, and H. Ozeki. 1989. Comparative and functional anatomy of group II catalytic introns—A review. *Gene* **82:** 5–30.

Morden, C.W., K.H. Wolfe, C.W. dePamphilis, and J.D. Palmer. 1991. Plastid translation and transcription genes in a nonphotosynthetic plant: Intact, missing, and pseudo genes. *EMBO J.* **10:** 3281–3288.

Mourad, G.S. and J.A. White. 1992. The isolation of apparently homoplastidic mutants induced by a nuclear recessive gene in *Arabidopsis thaliana. Theor. Appl. Genet.* **84:** 906–914.

Newton, K.J., E.H. Coe, Jr., S. Gabay-Laughnan, and J.R. Laughnan. 1989. Abnormal growth phenotypes and mitochondrial mutations in maize. *Maydica* **34:** 291–296.

Nozato, N., K. Oda, K. Yamato, E. Ohta, M. Takemura, K. Akashi, H. Fukuzawa, and K. Ohyama. 1993. Cotranscriptional expression of mitochondrial genes for subunits of NADH dehydrogenase, *nad5, nad4, nad2,* in *Marchantia polymorpha. Mol. Gen. Genet.* **237:** 343–350.

Nugent, J.M. and J.D. Palmer. 1988. Location, identity, amount and serial entry of chloroplast DNA sequences in crucifer mitochondrial DNAs. *Curr. Genet.* **14:** 501–509.

————. 1991. RNA-mediated transfer of the gene *coxII* from the mitochondrion to the nucleus during flowering plant evolution. *Cell* **66:** 473–481.

————. 1993. Evolution of gene content and gene organization in flowering plant mitochondrial DNA: A general survey and further studies on *COXII* gene transfer to the nucleus. In *Plant Mitochondria* (ed. A. Brennicke and U. Kück), pp. 163–170. VCH Chemie, Weinheim.

Oda, K., K. Yamato, E. Ohta, Y. Nakamura, M. Takemura, N. Nozato, K. Akashi, and K. Ohyama. 1992a. Transfer RNA genes in the mitochondrial genome from a liverwort, *Marchantia polymorpha*: The absence of chloroplast-like tRNAs. *Nucleic Acids Res.* **20:** 3773–3777.

Oda, K., K. Yamato, E. Ohta, Y. Nakamura, M. Takemura, N. Nozato, K. Akashi, T. Kanegae, Y. Ogura, T. Kohchi, and K. Ohyama. 1992b. Gene organization deduced from the complete sequence of liverwort *Marchantia polymorpha* mitochondrial DNA: A primitive form of plant mitochondrial genome. *J. Mol. Biol.* **223:** 1–7.

Ohta, E., K. Oda, K. Yamato, Y. Nakamura, M. Takemura, N. Nozato, K. Akashi, K.

Ohyama, and F. Michel. 1993. Group I introns in the liverwort mitochondrial genome: The gene coding for subunit 1 of cytochrome oxidase shares five intron positions with its fungal counterparts. *Nucleic Acids Res.* **21:** 1297–1305.

Palmer, J.D. 1983. Chloroplast DNA exists in two orientations. *Nature* **301:** 92–93.

———. 1991. Plastid chromosome: Structure and evolution. *Cell Cult. Somatic Cell Genet. Plants* **7A:** 5–53.

———. 1992. Chloroplast and mitochondrial genome evolution in land plants. In *Plant gene research. Cell organelles* (ed. R.G. Herrmann), pp. 99–133. Springer-Verlag, Wien.

Palmer, J.D. and L.A. Herbon. 1987. Unicircular structure of the *Brassica hirta* mitochondrial genome. *Curr. Genet.* **11:** 565–570.

Palmer, J.D. and D.B. Stein. 1986. Conservation of chloroplast genome structure among vascular plants. *Curr. Genet.* **10:** 835–841.

Palmer, J.D., D. Soltis, and P. Soltis. 1992. Large size and complex structure of mitochondrial DNA in two nonflowering land plants. *Curr. Genet.* **21:** 125–129.

Palmer, J.D., C.R. Shields, D.B. Cohen, and T.J. Orton. 1983. Chloroplast DNA evolution and the origin of amphidiploid *Brassica. Theor. Appl. Genet.* **65:** 181–189.

Raubeson, L.A. and R.K. Jansen. 1992. Chloroplast DNA evidence on the ancient evolutionary split in vascular land plants. *Science* **255:** 1697–1700.

Rédei, G.P. 1967. Biochemical aspects of a genetically determined variegation in *Arabidopsis. Genetics* **56:** 431–443.

———. 1973. Extra-chromosomal mutability determined by a nuclear gene locus in *Arabidopsis. Mutat. Res.* **18:** 149–162.

———. 1974. Genetic mechanisms in differentiation and development. In *Genetic manipulations with plant material* (ed. L. Ledoux), pp. 183–209. Plenum Press, New York.

———. 1992. Classical mutagenesis. In *Methods in* Arabidopsis *Research* (ed. C. Koncz et al.), pp. 16–82. World Scientific, New York.

Rédei, G.P. and S.B. Plurad. 1973. Hereditary structural alterations of plastids induced by a nuclear mutator gene in *Arabidopsis. Protoplasma* **77:** 361–380.

Reith, M. and J. Munholland. 1993. A high-resolution gene map of the chloroplast genome of the red alga *Porphyra purpurea. Plant Cell* **5:** 465–475.

Röbbelen, G. 1962. Plastommutationen nach Röntgenbestrahlung von *Arabidopsis thaliana* (L.) Heynh. *Z. Vererbungsl.* **93:** 25–34.

———. 1966. Chloroplastendifferenzierung nach geninduzierter Plastommutation bei *Arabidopsis thaliana* (L.) Heynh. *Z. Pflanzenphysiol.* **55:** 387–403.

Rochaix, J.-D. 1992a. Post-transcriptional steps in the expression of chloroplast genes. *Annu. Rev. Cell Biol.* **8:** 1–28.

———. 1992b. Control of plastid gene expression in *Chlamydomonas reinhardtii.* In *Plant gene research. Cell organelles* (ed. R.G. Herrmann), pp. 249–274. Springer Verlag, Wien.

Roussell, D.L., D.L. Thompson, S.G. Pallardy, D. Miles, and K.J. Newton. 1991. Chloroplast structure and function is altered in the NCS2 maize mitochondrial mutant. *Plant Physiol.* **96:** 232–238.

Schuster, W. and A. Brennicke. 1988. Interorganellar sequence transfer: Plant mitochondrial DNA is nuclear, is plastid, is mitochondrial. *Plant Sci.* **54:** 1–10.

———. 1991. RNA editing in ATPase subunit 6 mRNAs in *Oenothera* mitochondria: A new termination codon shortens the reading frame by 35 amino acids. *FEBS Lett.* **295:** 97–101.

Schuster, W., B. Combettes, K. Flieger, and A. Brennicke. 1993. A plant mitochondrial gene encodes a protein involved in cytochrome *c* biogenesis. *Mol. Gen. Genet.* **239:** 49–57.

Schuster, W., B. Wissinger, M. Unseld, and A. Brennicke. 1990. Transcripts of the NADH-dehydrogenase subunit 3 gene are differentially edited in *Oenothera* mitochondria. *EMBO J.* **8:** 263–269.

Shimada, H. and M. Sugiura. 1991. Fine structural features of the chloroplast genome: Comparison of the sequenced chloroplast genomes. *Nucleic Acids Res.* **19:** 983–995.

Shinozaki, K., M. Ohme, M. Tanaka, T. Wakasugi, N. Hayashida, T. Matsubayashi, N. Zaita, J. Chunwongse, J. Obokata, K. Yamaguchi-Shinozaki, C. Ohto, K. Torazawa, B. Y. Meng, M. Sugita, H. Deno, T. Kamogashira, K. Yamada, J. Kusuda, F. Takaiwa, A. Kato, N. Tohdoh, H. Shimada, and M. Sugiura. 1986. The complete nucleotide sequence of the tobacco chloroplast genome: Its gene organization and expression. *EMBO J.* **5:** 2043–2049.

Singh, M. and G.G. Brown. 1991. Suppression of cytoplasmic male sterility by nuclear genes alters expression of a novel mitochondrial gene region. *Plant Cell* **3:** 1349–1362.

Stuart, K. 1991. RNA editing in mitochondrial mRNA of trypanosomatids. *Trends Biochem. Sci.* **16:** 68–72.

Stern, D.B. and J.D. Palmer. 1984. Recombination sequences in plant mitochondrial genomes: Diversity and homologies to known mitochondrial genes. *Nucleic Acids Res.* **12:** 6141–6157.

Subramanian, A.R., D. Stahl, and A. Prombona. 1991. Ribosomal proteins, ribosomes, and translation in plastids. *Cell Cult. Somatic Cell Genet. Plants* **7A:** 191–215.

Sugiura, M. 1991. Transcript processing in plastids. Trimming, cutting, splicing. *Cell Cult. Somatic Cell Genet. Plants* **7A:** 125–137.

———. 1992. The chloroplast genome. *Plant Mol. Biol.* **19:** 149–168.

Sutton, C.A., P.L. Conklin, K.D. Pruitt, and M.R. Hanson. 1991. Editing of pre-mRNAs can occur before *cis-* and *trans*-splicing in *Petunia* mitochondria. *Mol. Cell. Biol.* **11:** 4274–4277.

Svab, Z. and P. Maliga. 1993. High-frequency plastid transformation in tobacco by selection for a chimeric *aadA* gene. *Proc. Natl. Acad. Sci.* **90:** 913–917.

Takemura, M., K. Oda, K. Yamato, E. Ohta, Y. Nakamura, N. Nozato, K. Akashi, and K. Ohyama. 1992. Gene clusters for ribosomal proteins in the mitochondrial genome of a liverwort *Marchantia polymorpha*. *Nucleic Acids Res.* **20:** 3199–3205.

Thompson, D., V. Walbot, and E.H. Coe, Jr. 1983. Plastid development in iojap- and chloroplast mutator-affected maize plants. *Am. J. Bot.* **70:** 940–950.

von Wettstein, D. 1961. Nuclear and cytoplasmic factors in the development of chloroplast structure and function. *Can. J. Bot.* **39:** 1537–1545.

Walbot, V. and E.H. Coe, Jr. 1982. Nuclear gene iojap conditions a programmed change to ribosome-less plastids in *Zea mays*. *Proc. Natl. Acad. Sci.* **76:** 2760–2764.

Ward, B.L., R.S. Anderson, and A.J. Bendich. 1981. The mitochondrial genome is large and variable in a family of plants (Cucurbitaceae). *Cell* **25:** 793–803.

Warwick, S.I. and L.D. Black. 1991. Molecular systematics of *Brassica* and allied genera (subtribe Brassicinae, Brassiceae)—Chloroplast genome and cytodeme congruence. *Theor. Appl. Genet.* **82:** 81–92.

Wintz, H., H.-C. Chen, and D.T.N. Pillay. 1988. Presence of a chloroplast-like elongator tRNA$^{Met}$ gene in the mitochondrial genomes of soybean and *Arabidopsis thaliana*. *Curr. Genet.* **13:** 255–260.

Wissinger, B., A. Brennicke, and W. Schuster. 1992. Regenerating good sense: RNA editing and *trans* splicing in plant mitochondria. *Trends Genet.* **8:** 322–328.

Wissinger, B., W. Schuster, and A. Brennicke. 1991a. *Trans*-splicing in *Oenothera* mitochondria: *nad1* mRNAs are edited in exon and *trans* splicing group II intron sequences. *Cell* **65:** 473–482.

Wissinger, B., R. Hiesel, W. Schobel, M. Unseld, A. Brennicke, and W. Schuster. 1991b. Duplicated sequence elements and their function in plant mitochondria. *Z. Naturforsch.* **46:** 709–716.

Wolfe, K.H., C.W. Morden, and J.D. Palmer. 1991. Ins and outs of plastid genome evolution. *Curr. Opin. Genet. Dev.* **1:** 523–529.

————. 1992a. Function and evolution of a minimal plastid genome from a nonphotosynthetic plant. *Proc. Natl. Acad. Sci.* **89:** 10648–10652.

Wolfe, K.H., C.W. Morden, S.C. Ems, and J.D. Palmer. 1992b. Rapid evolution of the plastid translational apparatus in a nonphotosynthetic plant: Loss or accelerated sequence evolution of tRNA and ribosomal protein genes. *J. Mol. Evol.* **35:** 304–317.

Yang, A.J. and R.M. Mulligan. 1991. RNA editing intermediates of *cox2* transcripts in maize mitochondria. *Mol. Cell. Biol.* **11:** 4278–4281.

Xu, M.-Q., S.D. Kathe, H. Goodrich-Blair, S.A. Nierzwicki-Bauer, and D.A. Shub. 1990. The prokaryotic origin of a chloroplast intron: A self-splicing group I intron in the gene for tRNA$^{Leu}$UAA of cyanobacteria. *Science* **250:** 1566–1570.

# 4

# Molecular Cytogenetics of *Arabidopsis*


**John S. Heslop-Harrison**
Karyobiology Group, Department of Cell Biology
John Innes Centre, Norwich NR4 7UH,
United Kingdom

**Jola Maluszynska**
Department of Plant Anatomy and Cytology
Silesian University
Jagiellonska 28, 40-032
Katowice, Poland


*Arabidopsis thaliana* (L.) Heynh. is widely used for studies of genetics and molecular genetics because of its many advantages: the small plant size, short life cycle, and small genome. The low genome size, measured as nuclear DNA content, is convenient for molecular analysis because some 60% of the genome consists of single-copy DNA sequences, most of which are genes or associated sequences (Leutwiler et al. 1984). However, the small genome makes the examination of the chromosomes, their activity, segregation, pairing, and aberrations in number or structure — that is, cytogenetics — difficult (Fig. 1). Nevertheless, using the highest powers of light microscopy (limited by the wavelength of light), the five pairs of chromosomes in the diploid plant ($2n=10$) have been investigated both at mitosis and meiosis for many years (Laibach 1907, named as *Stenophragma thalianum*; Steinitz-Sears 1963; Ambros and Schweizer 1976).

In other organisms, particularly humans and cereals, cytogenetic analysis has led molecular analysis and has been widely used to understand the structure, organization, and behavior of the genome (Heslop-Harrison 1991). In *Arabidopsis*, the chromosome size has made such analysis difficult; even for distinguishing between diploid and tetraploid strains, alternative methods such as weighing seeds or measuring pollen grains are widely used. The use of trisomic stocks with an extra chromosome ($2n=2x+1=11$; Sears and Lee-Chen 1970) has been of great value in gene mapping in *Arabidopsis* (Koornneef and van der Veen 1983; see Koornneef, this volume), as in other species. By examining the chromosomes at

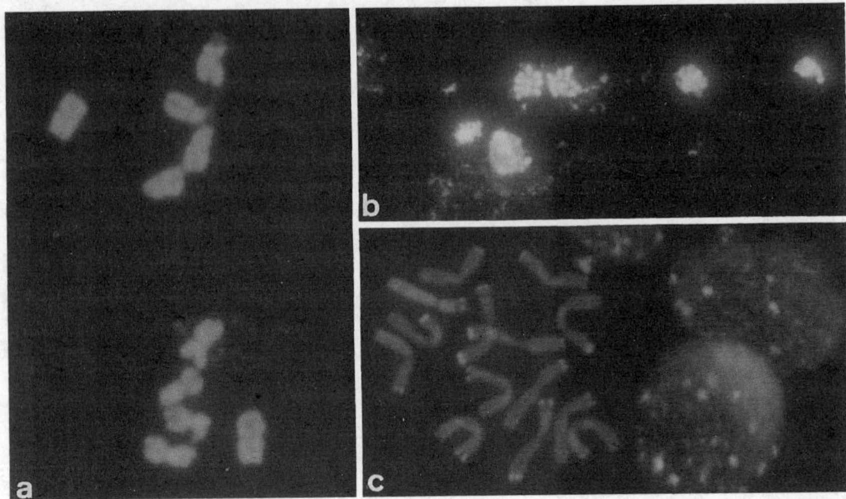

*Figure 1* Chromosomes of (*a, b*) *A. thaliana* fluorescing after staining with the fluorochrome DAPI. (*a*) Preparation from an unpretreated cell. (*b*) Preparation made using optimum techniques shows the ten chromosomes, their centromeres, and diffuse regions associated with the nucleolar organizing regions (NORs). For comparison, a rye metaphase is shown (*c*); the smaller arms of rye chromosomes contain about the same amount of DNA as the whole *Arabidopsis* nucleus. Magnification, 2250x.

mitosis, we can identify numerical aberrations, including polyploidy and trisomy. However, even with the excellent genetics, cytogenetic investigation of translocations, deletions, or other structural chromosome rearrangements has been virtually impossible because of the small size and similarity of all the five pairs of chromosomes.

During the last few years, several new techniques have been applied to the species and are making molecular cytogenetic investigations productive. First, the development of methods based on fluorescence microscopy (Fig. 1), including the use of DNA:DNA in situ hybridization to localize DNA sequences and the use of interphase cytogenetics, has permitted chromosome preparations to be analyzed easily at high resolution. Electron microscopy is allowing structural analysis of the nuclei, and flow cytometry is being applied to quantitative analysis of DNA contents. Finally, spreading of the proteinaceous cores of meiotic prophase axes, the synaptonemal complexes, has prospects for identifying chromosome rearrangements and analysis of extended chromosomes. Taken together, these techniques are now enabling us to investigate many aspects of the cytogenetics of *Arabidopsis*, although none of the methods is as yet fully exploited.

**THE CHROMOSOMES OF *ARABIDOPSIS***

Laibach (1907) first established the chromosome number of *Arabidopsis* as 2n=10 by counting chromocenters (darkly staining regions; see below) within interphase nuclei. The first karyotype was produced by Steinitz-Sears (1963) by analysis of paired chromosomes at late meiotic prophase (diplotene); after Giemsa staining, paired meiotic chromosomes were also investigated by Klasterska and Ramel (1980). An early idiogram based on somatic chromosomes was published by Měsiček (1967) after staining with aceto-carmin and aceto-orcein. The first Giemsa C-banding was carried out by Ambros and Schweizer (1976), and a banded karyotype was given by Schweizer et al. (1987).

The application of epifluorescent microscopy using fluorochromatic dyes which stain DNA specifically has considerably enhanced the value and ease of cytological analysis. DAPI (4′,6-diamidino-2-phenylindole), a DNA-specific dye with some A-T specificity, fluoresces blue when excited with ultraviolet light and has been of particular value for examination of metaphase chromosomes and interphase nuclei. Figure 1a shows a metaphase plate from a root-tip meristem after DAPI staining. Five pairs of homologous chromosomes, the positions of their centromeres, and differential staining of DNA associated with the nucleolar organizing regions (NORs) (see "Locations of Repetitive DNA" below) can be seen.

The chromosome designation used here, which we advocate, follows that used by Koornneef and van der Veen (1983) for genetic linkage maps. These differ from the numbers used by Schweizer et al. (1987); chromosome 1 (the longest) and 3 are the same in both systems, chromosome 4 Koornneef (carrying the most obvious NOR) corresponds to chromosome 5 Schweizer. The smallest chromosome, 2 Koornneef, corresponds to 4 Schweizer and also carries rDNA. Chromosome 5 Koornneef corresponds to chromosome 2 Schweizer.

Chromosomes of *Arabidopsis* have different morphologies, with chromosomes ranging from 1.5 μm to 2.8 μm in length, although preparation techniques and  tissue pretreatments can give considerable size variation. Often, three individual chromosome types can be distinguished in fluorochrome-stained metaphase preparations (Fig. 1); one pair, 4, has a conspicuous puffed region at the NOR, another is short with a smaller puffed region (2), and another chromosome is short and unequally armed (probably 3). C-banding of chromosomes shows that there are Giemsa-staining regions of heterochromatin (Ambros and Schweizer 1976), and DAPI-staining may reveal similar sites of heterochromatin in prophase or prometaphase chromosome preparations. The stained regions are located around the centromeres of all ten chromosomes, and extra sites may also be associated with the NORs.

Any tissues of the plant containing dividing cells are suitable for making chromosome preparations, although the small root and apical meristems contain few cell divisions relative to those in other species. In practice, somatic chromosomes are most often investigated from seedling root tips grown on agar mineral medium. Buds are obtainable in great numbers from single plants, which remain alive and have both somatic (mitotic) and meiotic divisions. Although the quality of individual chromosome preparations tends to be lower than those from roots, probably because of the metabolites accumulated in the cytoplasm, many mitoses can be observed. The first mitosis in pollen grains can give good quality metaphase spreads with 5 chromosomes, the haploid number (Fig. 2). "Hairy-root" cultures also provide suitable material for chromosome preparation.

The mitotic activity of the *Arabidopsis* root, measured as a number of cell divisions per meristem, increases from 12 hours after germination, when the incorporation of tritiated thymidine is first detectable in nuclei (Maluszynska 1982). A maximum number of cell divisions was found in roots about 1 mm long after 60 hours, but the number of divisions then declines to a steady state of about three per root (Fig. 3).

Although cytogenetics normally involves study of metaphase chromosomes, interphase is the most important stage of the cell cycle for gene expression (DNA transcription) and DNA replication. Interphase nuclei of *Arabidopsis* show differential staining, with (in DAPI-stained nuclei) dark areas overlying the nucleolus, and bright chromocenters consisting of condensed, heterochromatic DNA (see Fig. 6d,m,o,q). In stains used for transmitted light microscopy, such as Feulgen or aceto-carmine, these chromocenters stain darkly. The number of sites varies from 10 to 14, approximately the same as the number of C-bands or DAPI-positive bands seen at the centromeres and NORs in metaphase chromosome preparations.

## MEIOTIC ANALYSIS

Because of the relatively large size of the cells, analysis of pollen mother cells at metaphase I of meiosis can be straightforward (Steinitz-Sears 1963; Sears and Lee-Chen 1970). At metaphase I, the number of bivalents can be counted, and univalents arising from trisomic lines can be examined. Vieira et al. (1990) describe the techniques to find anthers with pollen mother cells at meiosis.

Analysis of pachytene chromosomes at prophase of meiosis is a powerful technique in many species to examine extended chromosomes with features along their length including centromeres, NORs and dark-

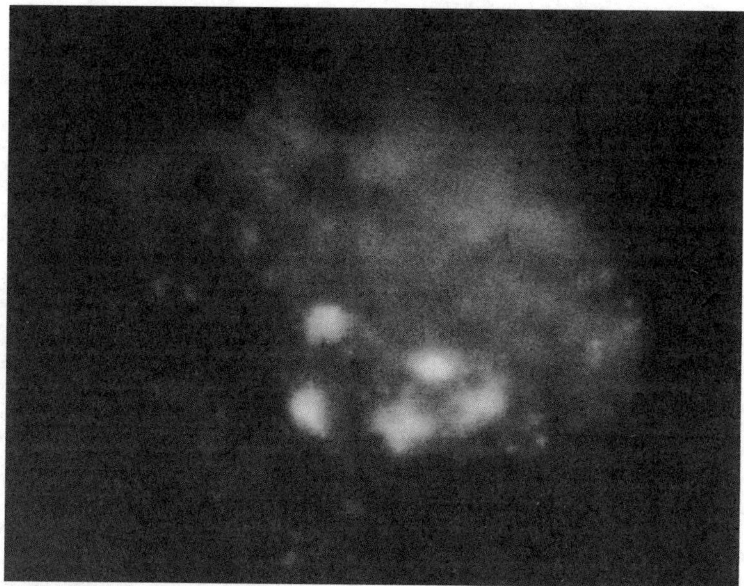

*Figure 2*  Developing pollen grain of *Arabidopsis* at first pollen mitosis. The five chromosomes in the haploid set are stained with DAPI. Magnification, 3000x.

staining, contracted regions (e.g., in tomato, Barton 1950; Ramanna and Prakken 1967). The short *Arabidopsis* chromosomes are hardly suitable for such analysis, but the method of surface spreading of synaptonemal complexes can be applied to pachytene nuclei (Albini 1994). During meiotic prophase, the proteinaceous axial cores of homologous pachytene chromosomes, to which the chromatin is attached in a series of loops, become aligned in parallel to form the synaptonemal complex (Wettstein et al. 1984). Since the surface spreading method was developed by Moses (Moses et al. 1977), it has been used to measure the sizes of pachytene bivalents and to examine translocations and recombination (Albini et al. 1984; Albini and Jones 1988; Jones et al. 1991; Albini and Schwarzacher 1992) because the length of the synaptonemal complexes is proportional to the DNA content (and hence length) of the chromosomes within a chromosome complement (Anderson and Chuanshan 1985). The technique has been used successfully for *Arabidopsis* by Dr. Susan Albini, whose preparations show the five bivalent axes in the electron microscope (Fig. 4) (Albini 1994).

Spread synaptonemal complex preparations have considerable potential because their extended length enables high-resolution measurement of morphological features, and surface spreading can be combined with

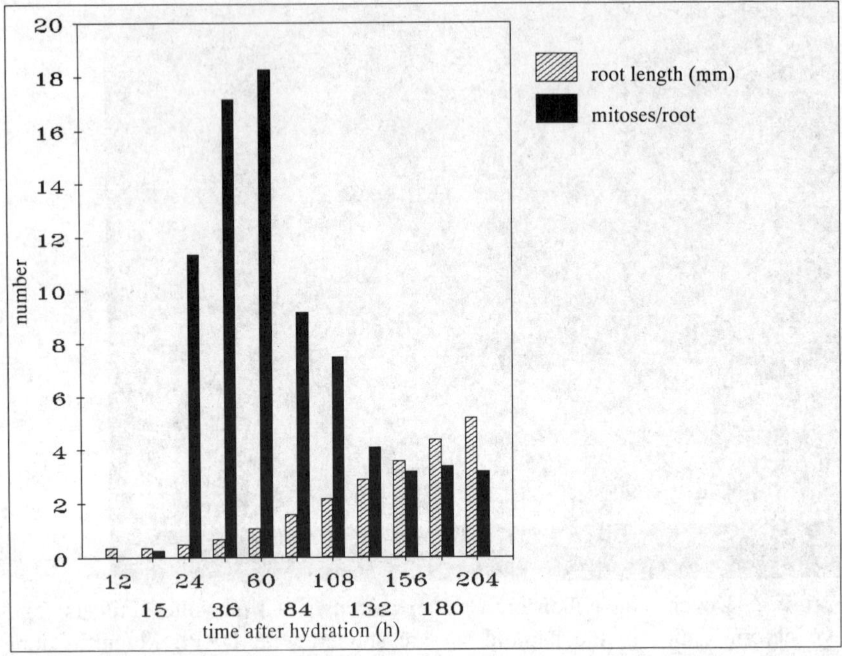


*Figure 3* Numbers of divisions in root tips of different lengths and at different times after germination. Seeds were presoaked for 24 hr in water and then germinated on an agar medium at 24°C (Maluszynska 1982) under steady light conditions.

in situ hybridization to show the physical locations of DNA sequences (see below) (Albini and Schwarzacher 1992). The method is also likely to be valuable in *Arabidopsis* for identifying chromosome rearrangements, including translocations or inversions.

## ELECTRON MICROSCOPY OF CHROMOSOMES

Chromosome morphology can be studied in the electron microscope following fixation in conventional, cross-linking fixatives. Chromosomes of *Arabidopsis* look similar to those of other species at metaphase and anaphase (Fig. 5). As with other species, serial sections can be made and complete nuclei can be reconstructed in three dimensions (Heslop-Harrison and Schwarzacher 1990; J.S. Heslop-Harrison and J. Maluszynska, in prep.). No high-resolution studies comparable to those of Wanner et al. (1991) in barley have been made, but they would enable the packing of single-copy and repeated DNA sequences into chromosomes to be investigated.

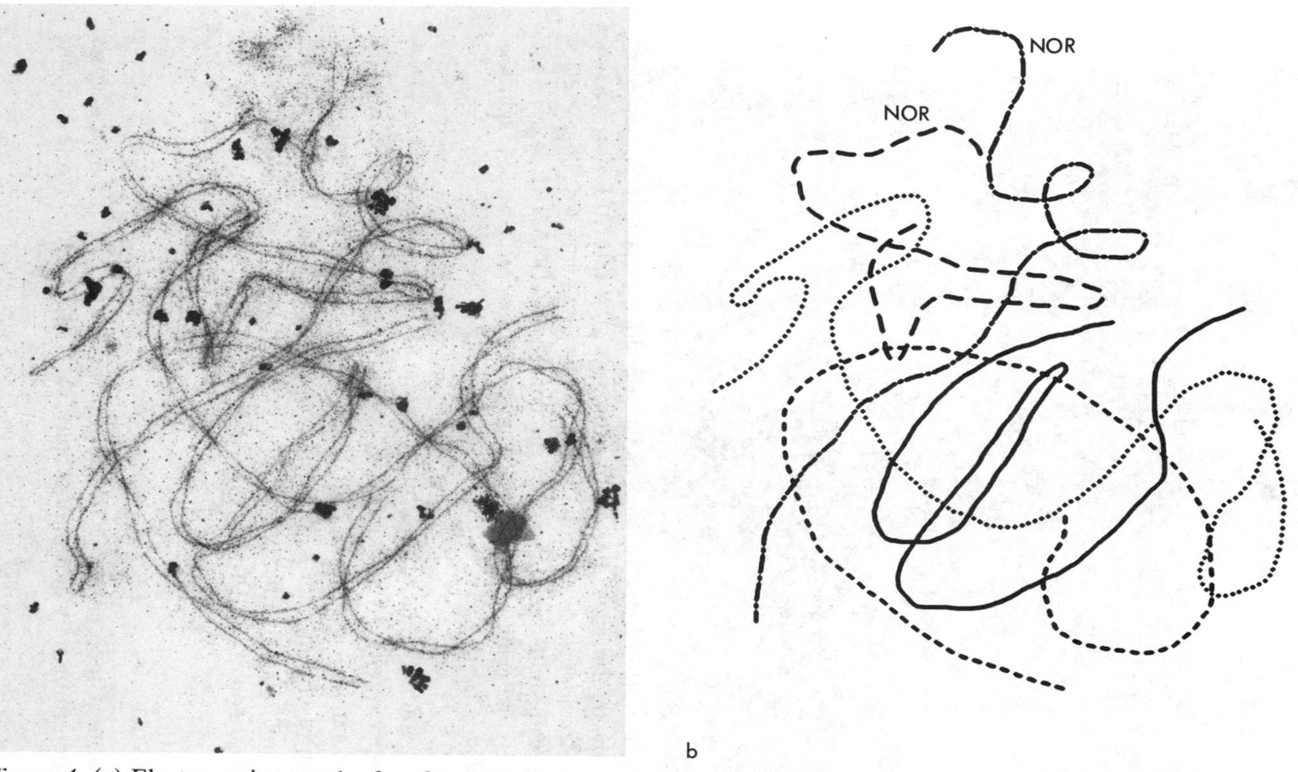


*Figure 4*  (*a*) Electron micrograph of surface spread synaptonemal complexes from *Arabidopsis* pollen mother cells at pachytene. (*b*) Interpretive diagram showing the five axes and sites of NORs. Prepared and kindly supplied by Dr. Susan Albini (for methods, see Albini and Schwarzacher 1992; Albini et al. 1984).

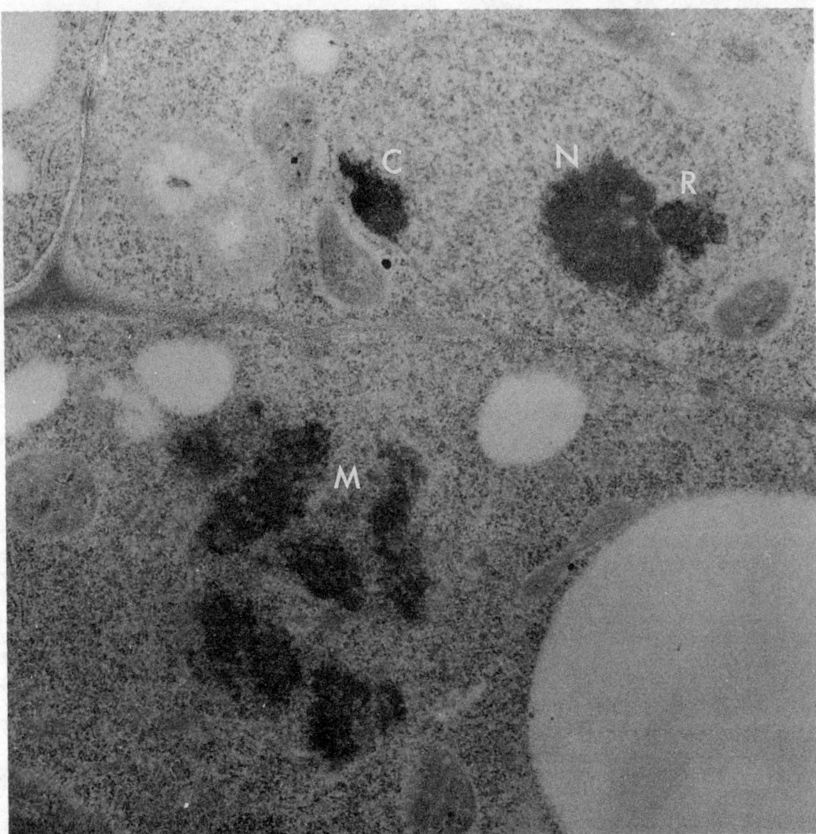

*Figure 5*  An interphase nucleus and metaphase chromosomes in two adjacent cells seen by transmission electron microscopy of a thin (0.1 μm) section through an *Arabidopsis* ovary wall. Metaphase chromosomes (M) appear electron dense and similar in appearance to condensed rDNA (R) next to the nucleolus (N) and centromeric DNA (C) in the interphase nucleus. Magnification, 19,500x.

Electron micrographs of interphase nuclei show that most of the DNA is decondensed, with a few condensed sites lying near the nucleolus and against the nuclear envelope (Fig. 5) (Nagl 1980). The decondensation pattern contrasts with that in plant species with a higher DNA content, where the nucleus contains a much higher proportion of condensed DNA (Heslop-Harrison et al. 1993).

### THE CELL CYCLE

The duration of the cell cycle in root-tip meristem cells of *Arabidopsis* has been determined by autoradiography. At 22°C, the cell cycle is about

8.5 hours, with $G_1$ lasting 1.7 hours (20%) and S phase lasting 2.8 hours (33%; Van't Hof et al. 1978). The relative duration of S is very similar to that of plants with higher DNA contents (*Crepis capillaris*, *Haplopapus gracillis*, *Zea mays*). The overall cycle time in *Arabidopsis* is slightly shorter than these other species, although the measurements have a high standard deviation because of the techniques used (conditions and intervals between samples) and the fact that different cells in the population sampled are cycling at different rates. Whether cell cycle mutants have effects on the relative and absolute times is not known, although the facility with which cell cycle timings can be measured by flow cytometry means this quantitative cytogenetic method should become important for mutant analysis, as it is already for analysis of the human cell proliferation in cancers and other syndromes (Landberg et al. 1990).

## PLOIDY LEVELS, ANEUPLOIDY, AND POLYPLOIDY

The haploid chromosome number of 5 is likely to be the basic, haploid chromosome number of *Arabidopsis*. Restriction fragment length polymorphism (RFLP) maps, show little duplication of loci (see Koornneef, this volume). RFLP data indicate that, in contrast to *Arabidopsis*, "diploid" *Brassica* species (*B. campestris*, $2n=20$; *B. nigra*, $2n=16$; *B. oleracea*, $2n=18$) include duplications of large parts of linkage groups in the genome, and the species are secondary polyploids derived from an ancestral genome(s) of fewer chromosomes (Kianian and Quiros 1992a) by duplication of chromosomes or large chromosome segments with subsequent structural rearrangements (Song et al. 1991). There is one report of an *Arabidopsis* plant with fewer than $2n=10$ chromosomes: Arnold and Cruse (1965) stated that they obtained a $2n=8$ plant following X-irradiation. They show a photograph of the plant which is bushy, but not dissimilar to other mutants, but no picture of the chromosomes is shown. Viable diploids with deficiencies are extremely rare (Finch 1984), and the paucity of examples in *Arabidopsis* provides corroborative evidence that the species is truly diploid. In pseudo-diploid species where there is considerable duplication of loci, lines with fewer than the euploid chromosome number are easily produced (e.g., maize).

The combination of mapping and the use of sequence probes isolated from a true diploid like *Arabidopsis* may enable the mechanisms of genome evolution to be examined and an ancestral genome of the crucifers, or perhaps all plants, to be reconstructed (Bennett 1984; Flavell et al. 1993). Knowledge of the types of changes that have happened during evolution provides a valuable indicator of the changes that could be introduced into a genome by chromosome or molecular engineering in the future (Heslop-Harrison and Schwarzacher 1993). In humans, sets of pooled chromosome-specific probes can be used for "painting" particular chromosomes uniformly along their entire lengths by in situ hybridization (Lichter et al. 1988) and can be used to identify translocations between primates (Wienberg et al. 1992). Perhaps similar probe sets could be derived from *A. thaliana* ($2n$=10) and used to paint the homologous segments of the "diploid" and "tetraploid" *Brassica* species, so that translocations and other rearrangements could be visualized and analyzed directly.

Haploid ($n$=$x$=5) plants of *Arabidopsis* were first produced by Gresshoff and Doy (1972) by regeneration of plants from haploid callus derived from anther microspores. The plants had the normal *Arabidopsis* stem and flower morphology but, as expected, were sterile and died about 15 days after flowering. Autopolyploid strains of *Arabidopsis* have been induced by colchicine treatment (Bouharmont 1969; Maluszynska et al. 1990). Hexaploid ($2n$=$6x$=30) and octoploid ($2n$=$8x$=40) plants were produced but were unstable and lost chromosomes when selfed (Bouharmont 1969). Tetraploid ($2n$=$4x$=20) plants are stable and can be maintained easily through seed; Figure 6f shows a metaphase from a plant with $2n$=$4x$=20 chromosomes. Some tetraploid "ecotypes," such as Stockholm, are used in laboratory studies. Triploid plants can be made by hybridization of diploid and tetraploid parents. There is a good correlation between ploidy level and both the size of pollen grains and the number of pollen colpae, the grooves or furrows in the exine of the pollen grain through one of which the pollen tube will emerge. Normally, haploid pollen derived from a diploid plant has three colpae, whereas diploid pollen has four colpae, and tetraploid pollen (from an octoploid plant) has six colpae. As an alternative to examination of chromosomes in root tips or buds, or at first pollen mitosis, microscopic inspection of pollen morphology is a quick and efficient method to screen plants for different ploidies.

In common with many other species, the size of nuclei of *Arabidopsis* can increase markedly during cell differentiation because of systemic polyploidy or multiploidy. Galbraith et al. (1991) systematically surveyed the relative numbers of nuclei at each ploidy level in different tissues of the plant by flow cytometry. Most somatic cells of the species appeared to be multiploid, with up to 16 times the unreplicated haploid ($1C$) DNA amount. The extensive DNA variation, probably caused by chromosome endoreduplication (Maluszynska and Heslop-Harrison 1991), seems to involve the complete genome and is not selective. Galbraith et al. (1991) supported the idea that differentiation may require a minimal mass of nuclear DNA to maintain specific regulatory and functional states; hence, there is a minimum physical size of the nucleus related to DNA content, and only indirectly related to gene numbers. However, multiploidy has not been observed in all species with small genomes (Arumuganathan and Earle 1991), so the phenomenon is not a universal feature of plants with a small genome size.

Variation in ploidy may be extremely significant for the success of various molecular procedures. For example, regeneration from diploid cells may be more reliable than from tetraploid, and scoring of somatic recombination events or DNA sequence excision may depend heavily on the ploidies found in particular tissues. Such factors have not been widely studied, and it is possible that careful correlation with tissue state and ploidy may explain some of the results obtained in a number of studies.

## LOCALIZATION AND ACTIVITY OF rDNA SITES

Epifluorescence microscopy and in situ hybridization allows localization of DNA sequences on particular chromosomes in *A. thaliana* (Maluszynska and Heslop-Harrison 1991) and within interphase nuclei (Bauwens et al. 1991; Maluszynska and Heslop-Harrison 1991). For many repetitive DNA sequences, in situ hybridization is the method of choice for sequence mapping, since the high and often variable copy number and location at several sites in the genome, combined with the presence of minor sequence variants within sites, makes mapping on the basis of segregation analysis of RFLPs difficult.

The DNA coding for the 18S-5.8S-26S rRNA multigene family (termed rDNA henceforth) occurs in about 570 copies in the *Arabidopsis* genome (Meyerowitz and Pruitt 1985; Pruitt and Meyerowitz 1986) and accounts for some 8% of the total genome. rDNA transcription occurs in a specialized nuclear structure at interphase, the nucleolus (Fig. 5). In situ hybridization to interphase nuclei and metaphase chromosomes of *Arabidopsis* shows that there are four sites of rDNA in a diploid nucleus (Fig. 6a-e) (Maluszynska and Heslop-Harrison 1991). In situ hybridization to trisomic lines enabled unequivocal assignment of rDNA sites to particular chromosomes (Maluszynska and Heslop-Harrison 1991). The sites are located on chromosomes 2 and 4, and the strength of hybridization indicates that copy number is similar at both pairs of sites. Meiotic analysis of trisomic lines also indicates that there are two NOR chromosome types in *A. thaliana* (Sears and Lee-Chen 1970), and the synaptonemal complex morphology confirms the result (Fig. 4) (S. Albini, pers. comm.).

Chromosomes with active rDNA often have a characteristic morphology: secondary constrictions or puffed regions at the sites of rDNA, the

NORs. Two such regions can be identified in the DAPI-stained preparation (Fig. 1), correlating with active NORs. Silver staining is a more accurate method to examine activity since it specifically stains the basic nucleolar proteins present within the nucleolus or attached to NOR chromosomes, and hence enables the activity of rDNA loci to be tested (Bloom and Goodpasture 1976; Miller et al. 1976). A maximum of four nucleoli can be detected within interphase nuclei, indicating that all four loci can be expressed (Fig. 6h).

In situ hybridization shows the organization of the rDNA sites and their decondensation during gene expression. In many cells, most rRNA genes or even loci are unexpressed, and expression has been correlated with dispersion at the locus during interphase (Wachtler et al. 1993), as shown in Figure 6b, where two loci are dispersed. However, many nuclei have no dispersed loci (Fig. 6e), although presumably some of the rRNA

*Figure 6*   Root-tip interphase nuclei and metaphase chromosomes of *Arabidopsis*. All blue images show chromosomes or nuclei stained with the DNA fluorochrome DAPI. DAPI-stained chromosomes are slightly deformed by the denaturation procedure so the morphology is not as clear as in Fig. 1a. Yellow-green images show the sites of in situ hybridization of probes detected by FITC fluorescence. Most probes were labeled with digoxigenin. Magnification, 1800x. (*a–g*) In situ hybridization of the rDNA probe pTa71 to (*a,b*) an interphase nucleus and metaphase chromosomes of *Arabidopsis* showing the four chromosomes carrying rDNA and dispersed hybridization signals at interphase. (*c*) Probe sites detected by an enzymatic label that oxidizes diaminobenzidine (DAB) to form a brown precipitate at hybridization sites. (*d,e*) Interphase nucleus showing the four rDNA sites with little decondensation. (*f,g*) Metaphase from a tetraploid (2n=4x=20) *Arabidopsis* plant. Eight chromosomes (*arrows*) show rDNA hybridization sites. (*h*) Interphase nuclei of *Arabidopsis* following silver staining of the nucleoli. A maximum of four nucleoli are seen in each nucleus, although most nuclei have lower numbers. Whole loci may be inactive, and hence produce no nucleolus, and nucleoli may fuse during the cell cycle. Thus, the maximum number is only detected in a few cells that have recently completed division. (*i,j*) Double target in situ hybridization showing labeling of a single chromosome with biotin-labeled rDNA detected with Texas red and (*i*) the centromeric probe pAL1 (see *k–r*) labeled with digoxigenin and detected with the yellow-green fluorochrome FITC. (*k–l*) Localization of the highly repetitive DNA sequence pAL1 along metaphase chromosomes and interphase nuclei of *Arabidopsis*. The probe hybridizes to the paracentromeric regions of all ten chromosomes. (*m–r*) Localization of the highly repetitive DNA sequence pAL1 in diploid (2n=2x=10; 10 hybridization sites, *m, n*), triploid (2n=3x=15; 15 hybridization sites; *o, p*), and tetraploid (2n=4x=20; 20 hybridization sites; *q, r*) nuclei of *Arabidopsis*.

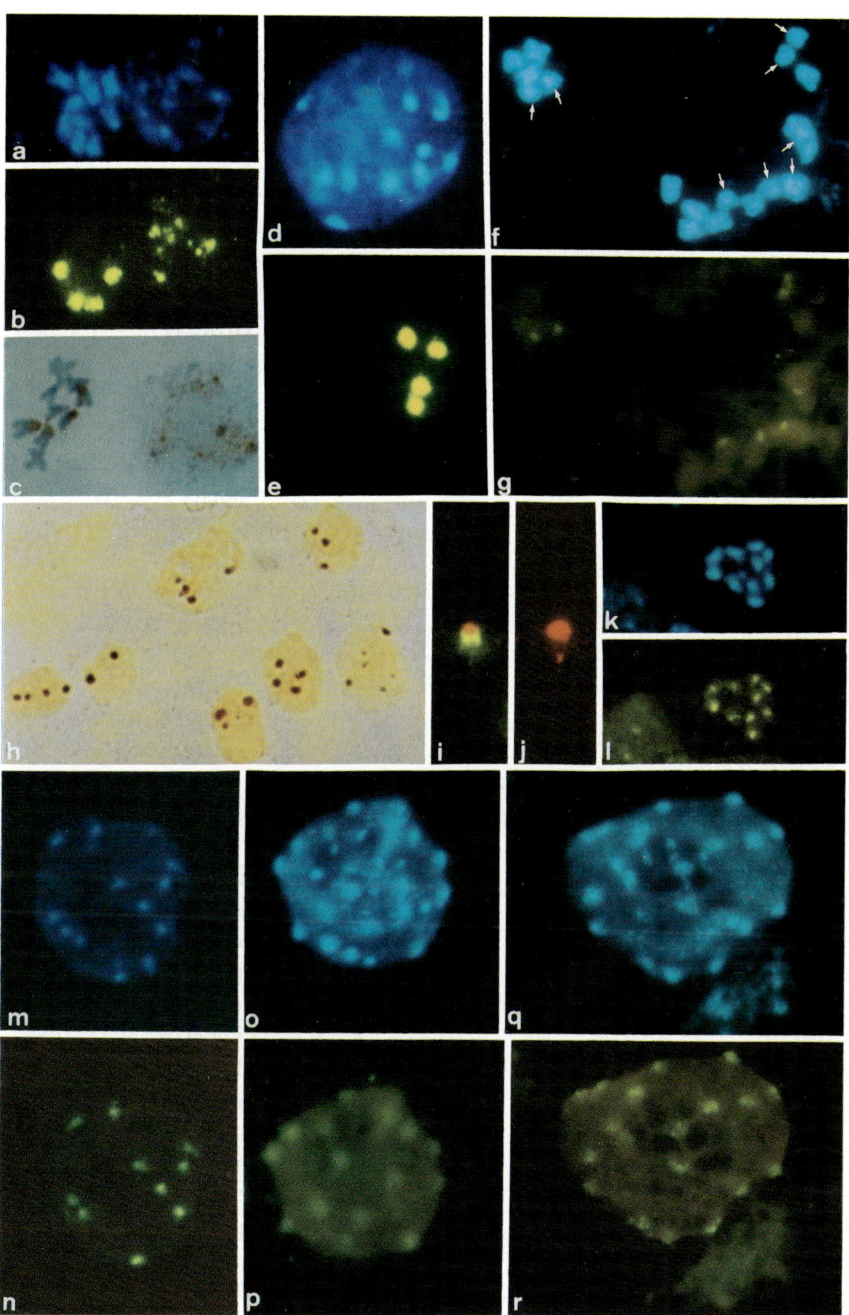

*Figure 6  (See facing page for legend.)*

genes are actively expressed. The appearance of the rDNA and its relationship to nucleoli vary considerably in *Arabidopsis*. Sometimes, condensed sites may be next to the nucleolus (Fig. 5), whereas in other cases, diffuse rDNA is evident throughout the nucleolus (see also Maluszynska and Heslop-Harrison 1991). Whether different states of decondensation relate to different activities of the rDNA, perhaps reflecting cell type, stage in the cell cycle, or endopolyploidy, is unknown. Although different decondensation patterns have been identified at the light microscope level, as yet, high-resolution electron microscope studies of expression and its modulation (Leitch et al. 1992) have not been completed in *Arabidopsis*. Detailed examination of the physical location and change in condensation and perhaps methylation of genes during expression in the *Arabidopsis* nucleolus may be able to link physical, developmental, and molecular studies of gene expression, and there are interesting possibilities to use similar methods to investigate the expression of other genes in the nucleus.

## LOCATIONS OF REPETITIVE DNA

Leutwiler et al. (1984) showed that the total DNA (nuclear, mitochondrial, and chloroplast) from *Arabidopsis* included 10–14% rapidly annealing sequences, 23–27% middle repetitive sequences, and 50–55% single copy; an unknown but high proportion of the middle repetitive sequences are from the chloroplast genome.

Representatives of a major class of repetitive DNA were cloned by Martínez-Zapater et al. (1986). The basic unit of the sequence, pAS1, is about 180 bp long, which is typical of many plant repetitive sequences, as it corresponds to the number of bases associated with a single nucleosome. Members of the family have a series of small modifications, including some at restriction sites which give repeats of two basic units (pAL1). The sequence has been shown to be present in tandem arrays (Martínez-Zapater et al. 1986), with a total copy number of about 5000. It then constitutes 1–1.5% of the genome.

The probe pAL1 hybridizes in situ to the paracentromeric region of all five chromosome pairs (Fig. 6k, l) and colocalizes with the centromeric blocks of heterochromatin, which stain brightly with DAPI (Fig. 6m–r) (see also Maluszynska and Heslop-Harrison 1991). Martínez-Zapater et al. (1986) suggested that pAL1 and related sequences lie within the heterochromatic blocks seen in *Arabidopsis* nuclei. The in situ evidence demonstrates that the sequence is present at all the centromeres and that all five pairs of chromosomes have a similar sequence in the centromeric heterochromatin. Hybridization signal strength is similar on all centromeric regions and approximately symmetrical on both sides of the centromere. In diploid interphase nuclei, 10 sites of pAL1 hybridization are detected (Fig. 6m,n), whereas 15 are seen in triploid nuclei (Fig. 6o,p) and 20 in tetraploid nuclei (Fig. 6q,r). Again, the sites of hybridization colocalize with the centromeric heterochromatin or chromocenters. The condensed chromocenters visible in the electron microscope away from the nucleolus at interphase also correspond to these centromeric chromosome regions.

It is clear that pAL1 is a member of a family of sequences, and it may not be the only family of sequences located at the centromeres; alternatively, there may be sequence variants at different centromeres. Although other families of middle-repetitive DNA have been isolated (Simoens et al. 1988), they have not been as extensively characterized, and it is possible that some will be dispersed in the genome. Further knowledge of their abundance and genomic locations are likely to be of great interest for understanding long-range genome organization. Analysis of dispersion patterns and sites with respect to methylation-free islands or genes (Moore et al. 1993) may be of significance in developing strategies for gene isolation by "walking" or "jumping."

## PHYSICAL MAPPING OF GENES AND SINGLE-COPY DNA

There are three different measurements of the length of a genome and the relative distance between linked markers: centiMorgans (cM) from genetic linkage measurements, kb from mapping of large clones, and μm from measurements of chromosomes. It is necessary to associate and understand the relationships of the three maps to enable description of the organization and behavior of the genome. Comparison of the physical and genetic maps is an important tool with which to study recombination and the dynamics of chromosome evolution in yeast (Oliver et al. 1993) and other species. In humans (Trask et al. 1993) and cereals (see, e.g., Leitch and Heslop-Harrison 1993), the discrepancies may be substantial; using a transposon tagging strategy, Dooner (1986) has shown that 1 cM in *Zea mays* is covered by only 14 kb of DNA in the *Bronze* locus, compared to an average of 2300 kb in the genome as a whole. In *Arabidopsis*, the number of recombinants known within a yeast artificial chromosome (YAC) clone is low, so standard deviations of measurements are very high; however, some YACs which have been analyzed would be expected to cover 1 cM from their length, but actually include markers almost 3 cM apart (Grill and Somerville 1991; Schmidt and Dean 1992).

DNA:DNA in situ hybridization using plasmid, cosmid, and YAC clones is able to localize single-copy and low-copy DNA sequences both along chromosomes and within interphase nuclei (see next section) in *Arabidopsis* (J. Maluszynska and J.S. Heslop-Harrison, unpubl.) and many other organisms. Mapping by in situ hybridization complements physical mapping by joining YAC clones, because distances between markers on one chromosome can be determined even where contiguous clones (contigs) have not been identified or where markers are quite widely separated. The method is unaffected by the presence of repetitive DNA near the single-copy sequences and can show the direction of extension of contigs. In particular, the presence of dispersed repetitive sequences may frustrate linking of contigs, and physical mapping in situ may be a key to joining them.

Technical problems still need to be overcome before single-copy in situ hybridization in *Arabidopsis* is routine. Although statistical methods can be applied, we believe accurate identification of the chromosome arms where hybridization sites are detected, and preferably the mapping of two or more sites relative to each other, are essential for accurate data analysis. Chromosome morphology can identify chromosomes, but after denaturation, this is usually a poor criterion because of distortion (cf. Figs. 1a and 6a, k). Hence, we think that multiple target in situ hybridization is a promising technique for chromosome identification. Detection of two different hybridization sites simultaneously with different labels and reporter molecules (Fig. 6i, j), or sequential labeling of the same metaphase (as in Heslop-Harrison et al. 1992), enables the relative locations of two sequences to be determined, and one sequence can be used for chromosome identification. Sadly, misinterpretation of sites of background and nonspecific hybridization now confuse the plant literature (see Lehfer et al. 1991).

Hybridization to extended interphase or prophase chromosomes enables clones to be mapped at a resolution of a few tens of kilobases in humans (Brandriff et al. 1991; Trask et al. 1992, 1993). We know that cereal chromosomes decondense in linear domains at interphase (Heslop-Harrison et al. 1990), but gene-rich regions seem to decondense more (Heslop-Harrison and Schwarzacher 1993). Our current work is examining interphase chromosome locations in *Arabidopsis* both for physical mapping and for the temporal and spatial localization of gene expression domains.

## THE GENUS *ARABIDOPSIS*: CYTOLOGY AND HYBRIDIZATION

About half of all flowering plants are polyploids, and the genus *Arabidopsis* is no exception: Cytological investigations show that the chromosome numbers vary from 10 (*A. thaliana*) up to at least 32 (*A. pumila*; Ginter and Ivanov 1968; Maluszynska and Heslop-Harrison 1993a).

In many genera, direct study of DNA enables relationships between different species to be examined. As examples, extensive sequence and restriction site comparisons of single nuclear gene families are being made in *Brassica* (e.g., the oleosins [Keddie et al. 1992]; rDNA sequences [Delseny et al. 1990]). Extensive comparisons of plastid gene sequences are also showing genetic relationships between all angiosperms (Chase et al. 1992). Investigations of the distribution and sequence of repetitive DNA sequences are also proving valuable to examine relationships and evolutionary patterns in both DNA sequence and species within the Triticeae (see, e.g., McIntyre et al. 1990; Moore et al. 1991; Mukai et al. 1991; Anamthawat-Jónsson and Heslop-Harrison 1993) and *Brassica* (Iwabuchi et al. 1991; Kianian and Quiros 1992b; Maluszynska and Heslop-Harrison 1993b). Such investigations are at an early stage in *Arabidopsis*, but in situ hybridization has shown that the number of major rDNA sites varies from one pair (*A. wallichii*, 2n=16) to three pairs (*A. pumila* or *A. griffithiana*, 2n=32) (Maluszynska and Heslop-Harrison 1993a). Although these species all have extensive centromeric heterochromatin, the DNA sequences do not seem to be related to pAL1 found in *A. thaliana*. More detailed investigation of the sequences involved in the heterochromatin in the polyploid species and analytical, gel-based, or dot-blot analysis of sequence homologies will further characterize the similarities and divergence between the species. Comparative analysis both of genes and of repeated sequences has considerable potential to help find common elements in genome structure, genetic synteny, and species or sequence relationships and may assist with many programs in genetic analysis by increasing the number of distinguishable alleles at various loci.

An analysis of meiotic chromosome pairing in *A. pumila* and a hybrid with *A. thaliana* has led to the conclusion that *A. pumila* is amphidiploid and contains a complete genome of *A. thaliana* (Berger 1968). The presence of three pairs of chromosomes with rDNA, and extra minor sites, observed in *A. pumila* and *A. griffithiana*, is consistent with their being polyploid species. However, the lack of in situ hybridization of the pAL1 sequence to the polyploid does not support the conclusion that the genome of *A. thaliana* is present in the polyploid species. Although evolution of centromeric sequences can be rapid, it would be surprising if the entire sequence was lost from an *A. thaliana* genome in the polyploid.

Hybrids between *A. thaliana* and *Brassica campestris* ("*Arabido-*

*brassica*") have been made by regeneration of plants from fused, isolated protoplasts by Gleba and Hoffmann (1980). They also obtained asymmetric hybrids where one or both parental genomes had lost some chromosomes. Analysis indicated that there was extensive recombination or interchange between chromosomes (Hoffmann and Adachi 1981). The plants are some of the relatively few intertribal hybrids, and hybridization has exciting possibilities by enabling transfer of genes between the economically important *Brassicas* and the extensively studied *Arabidopsis*.

## NUCLEAR ARCHITECTURE

Where do chromosomes lie at interphase? Where are genes within the interphase nucleus? Which parts of chromosomes condense and decondense, and how and when do they do so? Where is DNA transcribed and where are genes expressed? Where are transcripts processed? What structures are associated with DNA replication? The organization of the interphase nucleus and the importance of relative chromosome positions are subjects of considerable interest (see Heslop-Harrison and Bennett 1990; Schwarzacher et al. 1991; Heslop-Harrison 1992; Carter et al. 1993), and research is beginning to answer some of these questions. Some aspects of nuclear architecture are likely to be universal, but those species such as *Arabidopsis*, with low DNA amounts and a high proportion of single-copy sequences, must have a somewhat different nuclear organization from species with larger genomes. Learning about the physical organization and packing of DNA sequences in large and small genomes will enable general features of genome organization and nuclear architecture to be determined by comparative studies.

As yet, there are no complete three-dimensional reconstructions of *Arabidopsis* nuclei, even showing the locations of rDNA sequences or the heterochromatin visible in the DAPI-stained preparations. However, it is clear that the sites of centromeric heterochromatin are normally discrete and do not cluster in zones within the interphase nuclei from root tips (Fig. 6m–r), in contrast to cereals, where centromeres and associated sequences often tend to cluster (Anamthawat-Jónsson and Heslop-Harrison 1990).

Molecular cytogenetic analysis using in situ hybridization shows the actual locations of sequences at interphase. In humans, modulation of centromere position both during the cell cycle (Bartholdi 1991) and in disease syndromes (Borden and Manuelidis 1988) has been demonstrated, and work by Lawrence and colleagues has localized genes during expression and the sites where RNA transcript processing occurs

(Lawrence and Singer 1985; Lawrence et al. 1988; Carter et al. 1993; Xing et al. 1993). Understanding high-level genome structure of both interphase nuclei and metaphase chromosomes within individual nuclei is important for analysis of gene behavior and will potentially assist with the direct manipulation of transformed genes. Such an architectural view of genome organization may answer some problems in transgene expression and cosuppression (Matzke and Matzke 1990).

## CONCLUSIONS AND PERSPECTIVES

Techniques for cytogenetic investigation using epifluorescent microscopy and molecular cytogenetics have assisted greatly the investigation of species with small chromosomes. Metaphase analysis continues to be valuable, but application of in situ hybridization techniques for the study of interphase nuclei is also now an important cytogenetic technique. DNA probes previously localized on metaphase chromosomes can be used as markers in interphase nuclei, both for understanding nuclear architecture and its modulation and for routine determination of chromosome numbers (including trisomies or polyploidy) or ordering of probes on chromosome arms at interphase. Interphase cytogenetics can let us look for changes in differentiated, nondividing tissues, and the presence of micronuclei or abnormal division figures can indicate nuclear instability through chromosome loss. Developments in molecular, labeling, imaging, and analytical techniques, including quantitative analysis and flow cytometry, are driving research forward in these areas of molecular cytogenetics.

The impact of the cytogenetic work now in progress has been discussed in many sections above. Two new areas are also likely to be important in the future: the analysis of meiosis and the manufacture of artificial chromosomes. The meiotic spreading techniques described above will enable high-resolution mapping of sequences, and identification of chromosomal rearrangements by unequal pairing, which may be particularly valuable in transformed or mutagenized plants. Beyond this, in situ hybridization is enabling us to answer many of the long-standing questions about where and when homologous chromosomes pair during meiosis, what sequences are associated with recognition, and when they recombine (T. Schwarzacher, pers. comm.). YAC technology is important for genome analysis in *Arabidopsis* and other species (Flavell et al. 1993), but it is probable that artificial chromosomes can be constructed in other species, including *Arabidopsis*. The structural sequences at the telomeres have been identified (Richards and Ausubel 1988) and localized by in situ hybridization (Schwarzacher and Heslop-Harrison 1991), and now centromere- and telomere-associated sequences are being analyzed in *Arabidopsis* (Richards et al. 1991, 1993), while work on eukaryotic origins of replication is advancing rapidly (DePamphilis 1993). Soon, we will be able to assemble whole chromosomes and analyze their stability and potential to carry foreign DNA in plants.

Molecular cytogenetics uniquely facilitates comparative studies of large-scale genomic structures and rearrangements, since coding sequences are conserved between most species, whereas repetitive, noncoding DNA differs in copy number, locations, sequence, and distribution. The possibility to compare and contrast genome organization in *Arabidopsis* with that of other closely and more distantly related species is exciting, and it is clear that information can flow in both directions. Knowledge of nuclear architecture in *Arabidopsis*, which will largely come from direct observation of probed nuclei, will allow testing of hypotheses about genome and species evolution in both long (evolutionary) and short (e.g., plant breeding) time scales, giving insight into how genomes can be manipulated, hybridized, and selected.

## ACKNOWLEDGMENTS

We thank the Biotechnology and Biology Research Council for support of this work through a plant molecular biology grant. We are grateful to Dr. Trude Schwarzacher and Gill Harrison for helping with many aspects of the program and to Professor M.D. Bennett and Drs. D. Flanders and M. Roose for comments on the manuscript. We particularly thank Dr. Sue Albini for allowing us to use the figure showing a synaptonemal complex spread.

## REFERENCES

Albini, S.M. 1994. A karyotype of the *Arabidopsis thaliana* genome derived from synaptonemal complex analysis at prophase I of meiosis. *Plant J.* **5:** 665–672.

Albini, S.M. and G.H. Jones. 1988. Synaptonemal complex spreading in *Allium cepa* and *Allium fistulosum*. II. Pachytene observations: The SC karyotype and the correspondence of late recombination nodules and chiasmata. *Genome* **30:** 399–410.

Albini, S.M. and T. Schwarzacher. 1992. *In situ* localization of two repetitive DNA sequences to surface-spread pachytene chromosomes of rye. *Genome* **35:** 551–559.

Albini, S.M., G.H. Jones, and B.M.N. Wallace. 1984. A method for preparing two-dimensional surface-spreads of synaptonemal complexes from plant meiocytes for light and electron microscopy. *Exp. Cell Res.* **152:** 280–285.

Ambros, P. and D. Schweizer. 1976. The Giemsa C-banded karyotype of *Arabidopsis thaliana*. *Arabidopsis Inf. Serv.* **13:** 167–171.

Anamthawat-Jónsson, K. and J.S. Heslop-Harrison. 1990. Centromeres, telomeres and chromatin in the interphase nucleus of cereals. *Caryologia* **43:** 205–213.

————. 1993. Isolation and characterization of genome-specific DNA sequences in Triticeae species. *Mol. Gen. Genet.* **240:** 151–158.

Anderson, L.K. and Z. Chuanshan. 1985. The relationship between genome size and synaptonemal complex length in higher plants. *Exp. Cell Res.* **156:** 367–378.

Arnold, C.G. and D. Cruse. 1965. Eine 8 chromosomige Mutante von *Arabidopsis thaliana*. *Flora* **155:** 474–476.

Arumuganathan, K. and E.D. Earle. 1991. Nuclear DNA content of some important plant species. *Plant Mol. Biol. Rep.* **9:** 208–218.

Bartholdi, M.F. 1991. Nuclear distribution of centromeres during the cell cycle of human fibroblasts. *J. Cell Sci.* **99:** 255–263.

Barton, D.W. 1950. Pachytene morphology of the tomato chromosome complement. *Am. J. Bot.* **37:** 639–643.

Bauwens, S., P. Van Oostveldt, G. Engler, and M. Van Montague. 1991. Distribution of the rDNA and three classes of highly repetitive DNA in the chromatin of interphase nuclei of *Arabidopsis thaliana*. *Chromosoma* **101:** 41–48.

Bennett, M.D. 1984. The genome, the natural karyotype, and biosystematics. In *Plant biosystematics* (ed. W.F. Grant), pp. 41–66. Academic Press, Ottawa.

Berger, B. 1968. Entwicklungsgeschichtliche und chromosomale Ursachender verschiedenen Kreuzungsverträglichkeit zwischen Arten des Verwandtschaftskreises *Arabidopsis*. *Beitr. Biol. Pflanz.* **45:** 171–212.

Bloom, S.E. and C. Goodpasture. 1976. An improved technique for selective silver staining of nucleolar organizer regions in human chromosomes. *Hum. Genet.* **34:** 199–206.

Borden, J. and L. Manuelidis. 1988. Movement of the X chromosome in epilepsy. *Science* **242:** 1687–1691.

Bouharmont, J. 1969. Evolution of chromosome numbers in *Arabidopsis* polyploids. *Chromosomes Today* **2:** 197–201.

Brandriff, B., L. Gordon, and B. Trask. 1991. A new system for high-resolution DNA sequence mapping in interphase pronuclei. *Genomics* **10:** 75–82.

Carter, K.C., D. Bowman, W. Carrington, K. Fogarty, J.A. McNeil, F.S. Fay, and J.B. Lawrence. 1993. A three-dimensional view of precursor messenger RNA metabolism within the mammalian nucleus. *Science* **259:** 1330–1335.

Chase, M.W., D.E. Soltis, R.G. Olmstead, D. Morgan, D.H. Les, and 42 others. 1992. Phylogenetics of seed plants: An analysis of nucleotide sequences from the plastid gene rbcL1. *Ann. Mo. Bot. Gard.* **8:** 528–580.

Delseny, M., J.M. McGarth, P. This, A.M. Chevre, and C.F. Quiros. 1990. Ribosomal RNA genes in diploid and amphiploid *Brassica* and related species: Organization, polymorphism, and evolution. *Genome* **33:** 733–744.

DePamphilis, M.L. 1993. Eukaryotic origins of DNA replication. In *The chromosome* (ed. J.S. Heslop-Harrison and R.B. Flavell), pp. 75–102. Bios, Oxford, United Kingdom.

Dooner, H.K. 1986. Genetic fine structure of the *Bronze* locus in maize. *Genetics* **113:** 1021–1036.

Finch, R.A. 1984. Deficiency of 6L in diploid barley. *Barley Genet. Newsl.* **13:** 2–3.

Flavell, R.B., G. Moore, and C.M. Dean. 1993. Genome mapping and sequencing. In *The chromosome* (ed. J.S. Heslop-Harrison and R.B. Flavell), pp. 249–264. Bios, Oxford, United Kingdom.

Galbraith, D.W., K.R. Harkins, and S. Knapp. 1991. Systemic endopolyploidy in *Arabidopsis thaliana*. *Plant Physiol.* **96:** 985–989.

Ginter, T.N. and V.I. Ivanov. 1968. Chromosome numbers in *Arabidopsis* species. *Arabidopsis Inf. Serv.* **5:** 23.

Gleba, Y.Y. and F. Hoffmann. 1980. "*Arabidobrassica*": A novel plant obtained by protoplast fusion. *Planta* **149:** 112–117.

Gresshoff, P.M. and C.H. Doy. 1972. Haploid *Arabidopsis thaliana* callus and plants from anther culture. *Aust. J. Biol. Sci.* **25:** 259–264.

Grill, E. and C. Somerville. 1991. Construction and characterization of a yeast artificial chromosome library of *Arabidopsis* which is suitable for chromosome walking. *Mol. Gen. Genet.* **226:** 484–490.

Heslop-Harrison, J.S. 1991. The molecular cytogenetics of plants. *J. Cell Sci.* **100:** 15–21.

————. 1992. Nuclear architecture in plants. *Curr. Opin. Genet. Dev.* **2:** 913–917.

Heslop-Harrison, J.S. and M.D. Bennett. 1990. Nuclear architecture in plants. *Trends Genet.* **6:** 401–405.

Heslop-Harrison, J.S. and T. Schwarzacher. 1990. The ultrastructure of *Arabidopsis thaliana* chromosomes. In *4th International Conference on* Arabidopsis *Research*, Vienna (ed. D. Schweizer et al.), p. 3. University of Vienna, Austria.

————. 1993. Molecular cytogenetics—Biology and applications in plant breeding. *Chromosomes Today* **11:** 191–198.

Heslop-Harrison, J.S., G.E. Harrison, and I.J. Leitch. 1992. Reprobing of DNA:DNA *in situ* hybridization preparations. *Trends Genet.* **8:** 372–373.

Heslop-Harrison, J.S., A.R. Leitch, and T. Schwarzacher. 1993. The physical organization of interphase nuclei. In *The chromosome* (ed. J.S. Heslop-Harrison and R.B. Flavell), pp. 221–232. BIOS, Oxford, United Kingdom.

Heslop-Harrison, J.S., A.R. Leitch, T. Schwarzacher, and K. Anamthawat-Jónsson. 1990. Detection and characterization of 1B/1R translocations in hexaploid wheat. *Heredity* **65:** 385–392.

Hoffmann, F. and T. Adachi. 1981. "*Arabidobrassica*": Chromosomal recombination and morphogenesis in asymmetric intergeneric hybrid cells. *Planta* **153:** 586–593.

Iwabuchi, M., K. Itoh, and K. Shimamoto. 1991. Molecular and cytological characterization of repetitive DNA sequences in *Brassica*. *Theor. Appl. Genet.* **81:** 349–355.

Jones, G.H., S.M. Albini, and J.A.F. Whitehorn. 1991. Ultrastructure of meiotic pairing in B chromosomes of *Crepis capillaris*. II. 4B pollen mother cells. *Chromosoma* **100:** 193–202.

Keddie, J.S., E.-W. Edwards, T. Gibbons, C.H. Shaw, and D.J. Murphy. 1992. Sequences of an oleosin cDNA from *Brassica napus*. *Plant Mol. Biol. Rep.* **19:** 1079–1083.

Kianian, S.F. and C.F. Quiros. 1992a. Generation of a *Brassica oleracea* composite RFLP map: Linkage arrangements among various populations and evolutionary implications. *Theor. Appl. Genet.* **84:** 544–554.

————. 1992b. Genetic analysis of major multigene families in *Brassica oleracea* and related species. *Genome* **35:** 516–527.

Klasterska, I. and C. Ramel. 1980. Meiosis in PMCs of *Arabidopsis thaliana*. *Arabidopsis Inf. Serv.* **17:** 1–10.

Koornneef, M. and J.H. van der Veen. 1983. Trisomics in *Arabidopsis thaliana* and the location of linkage groups. *Genetica* **61:** 41–46.

Laibach, F. 1907. Zur Frage nach der Individualität der Chromosomen in Pflanzenreich. *Beih. Bot. Cbl. 1 Abt.* **22:** 191–210.

Landberg, G., E.M. Tan, and G. Ross. 1990. Flow cytometric multiparameter analyses of proliferating cell nuclear antigen/Cyeln and Ki-67 antigen: A new view of the cell cycle. *Exp. Cell Res.* **187:** 111–118.

Lawrence, J.B. and R.H. Singer. 1985. Quantitative analysis of *in situ* hybridization methods for the detection of actin gene expression. *Nucleic Acids Res.* **13:** 1777–1799.

Lawrence, J.B., C.A. Villnave, and R.H. Singer. 1988. Sensitive, high-resolution chromatin and chromosome mapping in situ: Presence and orientation of two closely integrated copies of EBV in a lymphoma line. *Cell* **52:** 51–61.

Lehfer, H., G. Wanner, and R.G. Herrmann. 1991. Physical mapping of DNA sequences on plant chromosomes by light microscopy and high resolution scanning electron microscopy. *Plant Mol. Biol.* **2:** 277–284.

Leitch, A.R., W. Mosgöller, M. Shi, and J.S. Heslop-Harrison. 1992. Different patterns of rDNA organization at interphase in nuclei of wheat and rye. *J. Cell Sci.* **101:** 751–757.

Leitch, I.J. and J.S. Heslop-Harrison. 1993. Physical mapping of four sites of 5S ribosomal DNA sequences and one site of the alpha-amylase 2 gene in barley (*Hordeum vulgare*). *Genome* **36:** 517–523.

Leutwiler, L.S., B.R. Hough-Evans, and E.M. Meyerowitz. 1984. The DNA of *Arabidopsis thaliana*. *Mol. Gen. Genet.* **194:** 15–23.

Lichter, P., T. Cremer, J. Borden, L. Manuelidis, and D.C. Ward. 1988. Delineation of individual human chromosomes in metaphase and interphase cells by in situ suppression hybridization using recombinant DNA libraries. *Hum. Genet.* **80:** 224–234.

Maluszynska, J. 1982. DNA synthesis and cell division in germinating seeds of *Arabidopsis thaliana*. *Arabidopsis Inf. Serv.* **19:** 42–46.

Maluszynska, J. and J.S. Heslop-Harrison. 1991. Localization of tandemly repeated DNA sequences in *Arabidopsis thaliana*. *Plant J.* **1:** 159–166.

———. 1993a. Molecular cytogenetics of the genus *Arabidopsis*: *In situ* localization of rDNA sites, chromosome numbers and diversity in centromeric heterochromatin. *Ann. Bot.* **71:** 479–484.

———. 1993b. Numbers and sites of rDNA in *Brassica* species. *Genome* **36:** 774–781.

Maluszynska, J., M. Maluszynski, G. Rebes, and E. Wietrzyk. 1990. Induced polyploids of *Arabidopsis thaliana*. In *4th International Conference on* Arabidopsis *Research*, Vienna (ed. D. Schweizer et al.), p.12. University of Vienna, Austria.

Martínez-Zapater, J.M., M.A. Estelle, and C.R. Somerville. 1986. A highly repeated DNA sequence in *Arabidopsis thaliana*. *Mol. Gen. Genet.* **204:** 417–423.

Matzke, M.A. and A.J.M. Matzke. 1990. Gene interactions and epigenetic variation in transgenic plants. *Dev. Genet.* **11:** 214–223.

McIntyre, C.L., S. Pereira, L.B. Moran, and R. Appels. 1990. New *Secale cereale* (rye) DNA derivatives for the detection of rye chromosome segments in wheat. *Genome* **33:** 635–640.

Měsiček, J. 1967. The chromosome morphology of *Arabidopsis thaliana* (L.) Heynh. and some remarks on the problem of *Hylandra suecica* (Fr.) Love. *Folia Geobot. Phytotaxon.* **2:** 433–436.

Meyerowitz, E.M. and R.E. Pruitt. 1985. *Arabidopsis thaliana* and plant molecular genetics. *Science* **229:** 1214–1218.

Miller, D.A., V.G. Dev, R. Tantravahi, and O.J. Miller. 1976. Suppression of human nucleolus organizer activity in mouse-human somatic hybrid cells. *Exp. Cell Res.* **101:** 235–243.

Moore, G., W. Cheung, T. Schwarzacher, and R. Flavell. 1991. BIS 1, a major component of the cereal genome and a tool for studying genomic organization. *Genomics* **10:** 469–476.

Moore, G., W. Cheung, T. Foote, M.D. Gale, R.M.D. Koebner, A.R. Leitch, I.J. Leitch, T. Money, P. Stancombe, M. Yano, and R. Flavell. 1993. Some key features of cereal genome organisation. *Genomics* **15:** 472–482.

Moses, M.J., G.H. Slatton, T.M. Gambling, and C.F. Starmer. 1977. Synaptonemal complex karyotyping in spermatocytes of the Chinese hamster (*Cricetulus griseus*). III.

Quantitative evaluation. *Chromosoma* **60:** 345–375.

Mukai, Y., T.R. Endo, and B.S. Gill. 1991. Physical mapping of the 18S.26S rRNA multigene family in common wheat: Identification of a new locus. *Chromosoma* **100:** 71–78.

Nagl, W. 1980. "*Arabidobrassica*": Evidence for intergeneric somatic hybrid nature from electron microscopic morphometry of chromatin. *Eur. J. Cell Biol.* **21:** 227–228.

Oliver, S.G., C.M. James, M.E. Gent, and K.J. Indge. 1993, The Great Wall: Sequencing the yeast genome. In *The chromosome* (ed. J.S. Heslop-Harrison and R.B. Flavell), pp.233–248. Bios, Oxford, United Kingdom.

Pruitt, R.E. and E.M. Meyerowitz. 1986. Characterization of the genome of *Arabidopsis thaliana*. *J. Mol. Biol.* **187:** 169–183.

Ramanna, M.S. and R. Prakken. 1967. Structure of and homology between pachytene and somatic metaphase chromosomes of the tomato. *Genetica* **38:** 115–133.

Richards, E.J. and F.M. Ausubel. 1988. Isolation of a higher eukaryotic telomere from *Arabidopsis thaliana*. *Cell* **53:** 127–136.

Richards, E.J., J. Giraudat, H.M. Goodman, and F.M. Ausubel. 1991. The centromere region of *Arabidopsis thaliana* chromosome 1 contains telomere-similar sequences. *Nucleic Acids Res.* **19:** 3351–3357.

Richards, E.J., A. Vongs, M. Walsh, J. Yang, and S. Chao. 1993. Substructure of telomere repeat arrays. In *The chromosome* (ed. J.S. Heslop-Harrison and R.B. Flavell), pp. 103–114. BIOS, Oxford, United Kingdom.

Schmidt, R. and C. Dean. 1992. Physical mapping of the *Arabidopsis thaliana* genome. In *Genome analysis: Strategies for physical mapping* (ed. K.R. Davies and S.M. Tilghman), vol.4, pp. 71–98. Cold Spring Harbor Laboratory Press, Cold Spring Harbor, New York.

Schwarzacher, T. and J.S. Heslop-Harrison. 1991. *In situ* hybridization to plant telomeres using synthetic oligomers. *Genome* **34:** 317–323.

Schwarzacher, T., A.R. Leitch, and J.S. Heslop-Harrison. 1991. *In situ* hybridization and the architecture of the nucleus. *Trans. R. Microsc. Soc.* **1:** 669–674.

Schweizer, D., P. Ambros, P. Gründler, and F. Varga. 1987. Attempts to relate cytological and molecular chromosome data of *Arabidopsis thaliana* to its genetic linkage map. *Arabidopsis Inf. Serv.* **25:** 27–35.

Sears, L.M.S and S. Lee-Chen. 1970. Cytogenetic studies in *Arabidopsis thaliana*. *Can. J. Genet. Cytol.* **12:** 217–223.

Simoens, C.R., J. Gielen, M. Van Montagu, and D. Inze. 1988. Characterization of highly repetitive sequences of *Arabidopsis thaliana*. *Nucleic Acids Res.* **16:** 6753–6766.

Song, K.M., J.Y. Suzuki, M.K. Slocum, P.H. Williams, and T.C. Osborn. 1991. A linkage map of *Brassica rapa* (syn. *campestris*) based on restriction fragment length polymorphism. *Theor. Appl. Genet.* **82:** 296–304.

Steinitz-Sears, L.M. 1963. Chromosome studies in *Arabidopsis thaliana*. *Genetics* **48:** 483–490.

Trask, B., H.F. Massa, and M. Burmeister. 1992. Fluorescence *in situ* hybridization establishes the order cen-DXS28(C7)-DXS67(B24)-DXS68(L1)-tel in human chromosome Xp21.3. *Genomics* **13:** 455–457.

Trask, B., A. Feritta, M. Christensen, J. Youngblom, A. Bergmann, A. Copeland, P. de Jong, H. Mohrenweiser, A. Olsen, A. Carrarno, and K. Tynan. 1993. Fluorescence *in situ* hybridization mapping of human chromosome 19: Cytogenetic band location of 540 cosmids and 70 genes or DNA markers. *Genomics* **15:** 133–145.

Van't Hof, J., A. Kuniyuki, and C.A. Bjerknes. 1978. The size and number of replicon families of chromosomal DNA of *A. thaliana*. *Chromosoma* **68:** 269.

Vieira, M.L.C., L.G. Briarty, and B.J. Mulligan. 1990. A method for analysis of meiosis in anthers of *Arabidopsis thaliana*. *Ann. Bot.* **66:** 717–719.

Wachtler, F., W. Mosgöller, C. Schöfer, J. Sylvester, P. Hozak, M. Derenzini, and A. Stahl. 1993. Ribosomal genes and nucleolar morphology. *Chromosomes Today* **11:** 63–77.

Wanner, G., H. Formanek, R. Martin, and R.G. Herrmann. 1991. High resolution scanning electron microscopy of plant chromosomes. *Chromosoma* **100:** 103–109.

Wettstein, D., P.B. Holm, and S.W. Rasmussen. 1984. The synaptonemal complex in genetic segregation. *Annu. Rev. Genet.* **18:** 331–413.

Wienberg, J., R. Stanyon, A. Jauch, and T. Cremer. 1992. Homologies in human and *Macaca fuscata* chromosomes revealed by in situ suppression hybridization with human chromosome specific DNA libraries. *Chromosoma* **101:** 265–270.

Xing, Y., C.V. Johnson, P.R. Dobner, and J.B. Lawrence. 1993. Higher level organization of individual gene transcription and RNA splicing. *Science* **259:** 1326–1330.

# 5

# *Arabidopsis* Genetics

**Maarten Koornneef**

Department of Genetics, Wageningen Agricultural University
6703 HA Wageningen, The Netherlands

*Arabidopsis thaliana* (L.) Heynh. is a true diploid plant and does not differ in its basic genetics from other diploid plants. Specific features of its reproductive biology, however, such as its being a self-fertilizing species with bisexual flowers, indicate that certain types of genetic analysis should be performed differently than with, e.g., maize. Genetic analysis requires genetic variation for specific traits. This variation can be induced by various types of mutagenic treatment or can be found as "natural" variation within and between wild populations.

The analysis of genetic variation should answer the following questions:

1. How many genes control the observed variation?
2. What are the dominance relations between the various alleles?
3. Are the genes allelic to previously described genes?
4. What are the epistatic relations with other genes?
5. Does the respective gene act in a cell-autonomous fashion or in a non-cell-autonomous way?
6. Where is the gene located on the linkage map?

The main features of *Arabidopsis* that enable an efficient (classical) genetic analysis are its short generation time and its small size, which provides a limited space requirement. *Arabidopsis* is almost 100% self-pollinating in greenhouse and climate chamber conditions. In natural populations, outcrossing is a rare phenomenon, although heterozygote individuals are sometimes found (Abbott and Gomes 1989). Studies with artificial populations constructed from mutant lines led to estimates of 1–2% outcrossing. However, under conditions of high pollinator activity and in genotypes with a certain degree of male sterility, higher outcrossing rates have been reported (Abbott and Gomes 1989 and references therein). The consequence of this is that the large number of small dry seeds almost exclusively originates by self-fertilization, especially when the plants are grown in a greenhouse or climate chamber. This, together with the ease of obtaining tens up to a few hundred seeds by manual crossing, makes handling the plant for genetic purposes not very labor intensive. Another genetic advantage of *Arabidopsis* is the low chromosome number, which allows a more efficient linkage analysis than in plants with more chromosomes. The high degree of polymorphism at the DNA level between ecotypes (Chang et al. 1988; King et al. 1993; Williams et al. 1993) allows the use of molecular markers in almost any population derived from ecotype crosses. The small size of its chromosomes, however, makes cytogenetics difficult enough that it cannot easily be used as a routine technique, at least at present (Heslop-Harrison and Maluszynska, this volume).

## SEGREGATION ANALYSIS

Segregation ratios obtained in the appropriate segregating populations provide information about the monogenic (or digenic, etc.) nature of the trait. In case a mutant has a complicated phenotype, this segregation analysis indicates pleiotropism (a single gene being responsible for distinct and seemingly unrelated traits) as the cause of the association of various characters found together in a mutant, or it indicates if this association is due to mutations in different genes. In the case of the *ttg* mutants, for example, two apparently unrelated traits (the absence of trichomes and yellow seeds) always occurred together in the progeny of crosses, indicating true pleiotropism. The finding of a number of independent allelic mutants with a similar phenotype supports this conclusion (Koornneef 1981).

Segregation ratios can deviate from the expected Mendelian ratios in a number of ways. Most frequently, a reduced transmission of the mutant phenotype is found. Extensive analysis of a large number of viable mutants (Dellaert 1980) indicated that reduced transmission through the male gametophyte due to competition in growth rate between pollen tubes of different genotypes (called certation) is the main cause of this phenomenon in *Arabidopsis*. Certation was demonstrated also for some embryo-lethal mutations, which in heterozygous plants were located more in the top half (near the stigma region) than in the lower half of the silique (Meinke 1982). However, reduced germinability of seeds and reduced viability, which in its extreme case is embryo lethality, are other causes. Examples of reduced germination are leaky gibberellin-requiring (*ga*) mutants (Koornneef and van der Veen 1980). Partial embryo lethality was observed in the selfed progeny of plants heterozygous for the *emb20-3* mutant, which segregated yellow-colored plants in a low frequency. These plants flowered normally and set seeds. However, most but not all seeds on these plants were shriveled, similar to many of the seeds harvested on the heterozygous genotype (M. Koornneef, unpubl.). Apparently most seeds did not develop to maturity. Röbbelen (1966) noticed that after prolonged storage the mutant frequency in the progeny of heterozygous plants decreased, because longevity of the mutants is less than that of the wild-type siblings.

Another reason for aberrant segregation ratios is linkage to alleles of genes that lead to reduced transmission through either the male or female gametes. This depends on whether the gametophyte factor occurs in the repulsion or in the coupling phase with the mutant; whether the mutant frequency is in excess or if it shows a deficit. Rédei (1965) described the *gf* gene, which prevents transmission through the female and which shows reduced transmission through the male gametophyte. A gene affecting transmission through the male gametophyte was described by Li (1967). This type of mutant may affect gametophyte development itself, which results in 50% sterile ovules or 50% lethal pollen, if the effects are specific for either the female or male gametophyte. Complications in the recognition of mutant phenotypes can be due to the segregation of modifiers in the same population and/or to reduced penetrance.

Modifier genes that enhance or suppress the mutant phenotype may play a role in ecotype crosses and are especially disturbing when the trait affected by the mutation also differs between the two ecotypes. For example, this is the case for flowering time, plant height, seed dormancy, and also for characters such as leaf shape. Even in the case of "qualitative" markers, this complication exists. In the W100 (multiple marker line in L*er* background) x Ws, the phenotypic expression of almost all markers was affected more or less (P.A. Scolnik, pers. comm.). The identification of the cauliflower (*cal*) allele in Ws as an extreme modifier of *apl* in the $F_2$ progeny of *apl* (L*er*) x Ws crosses (Bowman et al. 1993) is an extreme example of this.

Penetrance problems arise when the trait is modified strongly by environmental factors; e.g., light intensity or light quality. In *Arabidopsis* this is the case with genes affecting flowering time (Martínez-Zapater and Somerville 1990), seed dormancy (Hayes and Klein 1974), photomorphogenesis (Robson et al. 1993), and anthocyanin formation. Choosing a discriminative environment is the solution for penetrance problems.

Differences in the outcome of crosses that depend on the direction in which the cross was made (reciprocal differences) can be due to cytoplasmic inheritance or maternal inheritance. The only case of cytoplasmic mutants described in *Arabidopsis* are the white plastid mutations induced by the nuclear *chm* mutation (Rédei 1973), which leads to specific rearrangements of the mitochondrial genome (Martínez-Zapater et al. 1992). Maternal inheritance is found for genes affecting the seed coat such as the *tt* genes and *ttg* (Koornneef 1990b), which is simply explained by the fact that the testa, in contrast to the embryo, is derived from the maternal ovule integuments.

## GENETIC CHIMERAS AND SOMATIC RECOMBINATION

*Arabidopsis* meristems, like those of other dicotyledonous plants, have a three-layered meristem. The outermost $L_1$ layer gives rise to the epidermal structures, whereas the subepidermal $L_2$ layer gives rise to the subepidermal mesophyll in the center of the leaf and all of the mesophyll at the leaf margin. The germ-line cells originate exclusively from the $L_2$ layer. The inner $L_3$ layer gives rise to the center of the stem and a rather variable and irregular core in the leaves (Furner and Pumfrey 1992). The development of these meristems has been studied in genetic chimeras, which are plants composed of genetically different tissues. The two main mechanisms leading to chimeras are the occurrence of somatic mutations and somatic recombination. Since a seed is a multicellular structure, mutagen treatment of seeds results in chimeric $M_1$ plants, which has its implications for mutant isolation strategies (see Feldmann et al., this volume). In *Arabidopsis*, two cells of the embryo generally give rise to the germ line (for review, see Rédei and Koncz 1992), implying that on average the germ cell portion of the inflorescence is composed of two sectors after a seed mutagenesis. An important use of chimeras is in the analysis of the developmental fate of particular cells. Irish and Sussex (1992) and Furner and Pumfrey (1992) performed a clonal analysis of the dry seed meristem by using irradiation to induce dominant color mutations (Irish and Sussex 1992) or deletions in a plant heterozygous for the *alb1* mutation (Furner and Pumfrey 1992). The deletions uncovered the mutation and led to patches of white tissue of variable size derived from a cell in which the mutation occurred. These analyses of the dry seed meristem suggested that the $L_2$ layer in this meristem contains 36–38 (Irish and Sussex 1992) or 20 (Furner and Pumfrey 1992) cells. Developmental fate of these cells could only be predicted in a general and probabilistic way, because cell fate occurs continuously as the plant grows, in a position-dependent, lineage-independent fashion (Furner and Pumfrey 1992).

Another application of chimeras produced in a similar way to those described above is the analysis of whether a particular mutant phenotype is cell autonomous. For this, the gene to be studied should be distally located to a known cell-autonomous mutation, whereafter the phenotype of the sectors is analyzed for the trait under study (see, e.g., Hake 1992). Although several mapped cell-autonomous mutants (mainly affecting leaf pigmentation) are known in *Arabidopsis*, such studies have not yet been published for this species.

Mitotic recombination, the frequency of which can be increased by various treatments such as irradiation, has been studied by Hirono and Redei (1965). These authors subjected plants that were heterozygous for a number of chromosome-1 markers (*gi*, *ch1*, and *le* [=*pa*]) to irradiation of presoaked seeds with 13 krad of X-rays. This resulted in somatic sectors homozygous for *ch1* or *le* in 5–10% of the plants grown from the treated seeds. The presence of occasional "twin spots" and the analysis of the sexual progeny of a *ch1/ch1* sector indicated that a mitotic recombination event between *ch1* and *gi* had occurred. Somatic pairing, which seems a prerequisite for mitotic recombination, has been observed (Steinitz-Sears 1963). The occurrence of mitotic recombination due to tissue culture was described by Gaj and Maluszynski (1986).


## GENETICS OF CHROMOSOME NUMBER VARIANTS

### Polyploidy

Tetraploid ($2n=4x=20$) and hexaploid ($2n=3x=30$) *Arabidopsis* plants were found after treatment with colchicine (Rédei 1964; Bouharmont 1965) and after tissue culture (Negrutiu et al. 1978). A long-term callus suspension culture used by Gleba and Hoffmann (1978) was predominantly octaploid. The occurrence of polyploidy in tissue culture is no surprise, since extensive endoreduplication has been observed in somatic tissues of *Arabidopsis* plants. Polyploid cells were found in all tissues except those of the inflorescences (Galbraith et al. 1991). The morphological difference between diploid and tetraploid *Arabidopsis* plants is relatively small (Bouharmont and Macé 1972). The leaves of tetraploids are more round and have thicker stems, and the seeds and pollen grains are larger. Fertility is significantly reduced, although it still is fairly good (Bouharmont 1965). The reduced seedset per silique (40–60% of diploid L*er*) is partly due to fewer ovules per silique, and partly to fewer ovules fertilized. The size of the pollen grains can be conveniently used as a diagnostic characteristic for tetraploids (Bronckers 1963; Altmann et al. 1992).

Tetrasomic inheritance was demonstrated by Bouharmont (1969). van der Veen and Blankestijn-de Vries (1973) studied the phenomenon of double reduction (the occurrence of two sister-chromatid segments of a chromosome in one gamete) in the selfed progeny of heterozygotes of *alb1*. Since simplex heterozygotes (*ALB1/alb1/alb1/alb1*) for this gene have a distinct pale green phenotype, the coefficient of double reduction, which is the frequency that gametes contain the alleles of two sister chromatids, could be accurately estimated for this gene as 0.1.

Tetraploid *Arabidopsis* plants can be easily crossed with diploids, which results in vigorous triploid plants. After selfing or backcrossing with a diploid, triploid plants yield, apart from diploids and triploids, many aneuploid progeny including trisomics. Trisomics give almost exclusively diploid and trisomic progeny. The observation that crosses between diploids and tetraploids are relatively easy allows (e.g., in the case of tetraploid transformants) the introduction of a desired gene from a tetraploid into a diploid background. Hexaploid x tetraploid crosses resulted in progeny, in contrast to hexaploid x diploid crosses. It seems that gene transfer from hexaploids to diploids can be mediated through tetraploids (Rédei 1964).

## Aneuploidy

*Arabidopsis* plants with a chromosome number which is not a multiple of 5 are aneuploids. Among them, trisomics, with one extra chromosome, have been described by several groups (Steinitz-Sears 1963; Röbbelen and Kribben 1966; Koornneef and van der Veen 1983). The phenotype of a trisomic depends on which chromosome is involved (Table 1). In addition to these primary trisomics, telotrisomics with one extra chromosome arm have been isolated mainly as rare off-types in the progeny of primary trisomics (Sears and Lee-Chen 1970; Koornneef 1983; Koornneef and van der Veen 1983).

Especially in a uniform genetic background, the chromosome-specific trisomic phenotype allows an almost unambiguous scoring in populations that segregate for trisomics, without the need for a laborious and, in *Arabidopsis*, difficult cytogenetic analysis. Trisomics are generally derived from the progeny of triploids and can be maintained by selfing and/or by crossing the trisomic (as female parent) with the diploid. Because the male transmission of gametes with an extra chromosome is low (except for telotrisomics and Tr4; see Table 1), the selfed progeny mainly yield trisomics and diploids. Putative tetrasomics ($2n=2x+2$) can be observed at low frequency and are characterized by a phenotype similar to, but more extreme than, that of the corresponding primary trisomic; they are much more often sterile.

Markers segregating in the progeny of trisomics can be found to be associated with the extra chromosome by observation of specific trisomic segregation ratios. In the selfed progeny ($F_2$) of a duplex ($AAa$) plant, this is 1:0 among the trisomic progeny and 8:1 among the diploid prog-

*Table 1*  The phenotype and transmission of trisomics in *Arabidopsis* Landsberg *erecta*

| Trisomic | Transmission % female[a] | male[b] | Phenotype |
|---|---|---|---|
| Tr1 (F) | 22 | 0 | small rosette, which soon becomes necrotic, fragile plant with irregular flowers, highly sterile |
| Tr1A | 28 | 8 | dark green leaves, slightly lozenge, semi-dwarf, irregular petals |
| Tr1B | 27 | 7 | small rosette leaves, which soon become necrotic, slender plant type |
| Tr2 (R) | 16 | 4 | rosette leaves more round and slightly bent, flowers a few days later than wild type |
| Tr3 (Y) | 21 | 2 | narrow and grayish-green rosette leaves, irregular petals, rather sterile |
| Tr3A | 26 | 32 | like T3 but less extreme and much more fertile |
| Tr4 (C) | 29 | 16 | slightly smaller and more flat rosette, flower buds thickened |
| Tr5 (N) | 30 | 0 | narrow leaves, semi-dwarf, stigmata remain pronounced on siliques after flowering |
| Tr5A | 18 | 28 | like Tr5, but less extreme, leaf margins more serrated; better fertility than Tr5 |

Data from Koornneef and van der Veen (1983). Between parentheses is the code for the corresponding trisomic in Columbia background (Steinitz-Sears 1963).

[a]Percentage of trisomics in the progeny of the $(2n+1) \times 2n$ cross.

[b]Percentage of trisomics in the progeny of the $2n \times (2n+1)$ cross.

eny. On the basis of these ratios, genes can be located on particular chromosomes.

Telotrisomics have been used to locate centromeres in *Arabidopsis*. When one of two closely linked markers is associated with the telotrisomic and the other is not, the centromere lies between the two. By using four out of the ten possible telotrisomics, the centromeres of chromosomes 1, 3, and 5 were located (Koornneef and van der Veen 1983; Koornneef et al. 1983b). The segregation ratios for telotrisomics are slightly different from those of primary trisomics because the two complete chromosomes predominantly segregate to opposite poles at anaphase I. In case of a recessive marker located on the additional chromosome arm, this results upon selfing $A^t Aa$ in a 1:0 segregation among the trisomic progeny and 3:1 in the diploid progeny. The effect on the expected segregation ratios of crossing-over between a locus and the centromere, and the use of these data for the location of centromeres, has been discussed previously by Koornneef (1983).

Recombination between markers and a centromere can result in a low frequency of *aaa* plants in the progeny of duplex (*AAa*) primary trisomics. On the basis of this phenomenon (called double reduction), Sears and Lee-Chen (1970) localized the centromere of chromosome 2 above *hy1*.

## Translocations

Due to the difficulty of chromosome studies in *Arabidopsis*, chromosomal aberrations have not yet been studied in detail. One exception is translocations, which can be recognized on the basis of the semi-sterility of translocation heterozygotes. This semi-sterility can be assayed by determining pollen sterility using a pollen-staining procedure, or by the determination of ovule sterility (Sree-Ramulu and Sybenga 1979, 1985) with Müller's embryo test (Müller 1963). The presence of quadrivalents in meiosis and changed linkage relations of marker genes (Koornneef et al. 1982) can be used to confirm that semi-sterility in a certain genotype is due to a translocation. Sree-Ramulu and Sybenga (1979, 1985) showed that translocations occur relatively frequently after irradiation with X-rays and fission neutrons. About half of the translocations show a reduced frequency or even absence of translocation homozygotes in the selfed progeny of heterozygotes, presumably due to damage at the breakpoint (Sree-Ramulu and Sybenga 1985).

A set of characterized chromosome translocations between the different *Arabidopsis* chromosomes (Koornneef et al. 1982) can be used to locate genes on specific chromosomes. A marker gene will show linkage with the semi-sterility character for the different translocations involving genetically linked regions of its chromosome. Unfortunately, these translocations are not available anymore.

## LINKAGE ANALYSIS AND MAPPING

### General Outline of Linkage Analysis

The mapping of genes is based on the estimate of the average number of crossovers between two genes, which results in recombination between those genes. Although recombination can also occur between different mutations within a gene, as was demonstrated for the *ga1* locus (Koornneef et al. 1983a), linkage analysis almost exclusively deals with recombination between different loci. From a set of estimates of recombination values, the map position of a certain locus is determined. The various procedures used to perform linkage analysis and mapping in *Arabidopsis* have been described by Koornneef and Stam (1992) and Reiter et al. (1992a).

In many organisms, testcross populations are preferred for linkage analysis because of their efficiency due to the fact that each individual can be scored as either recombinant or parental. This results in an easy interpretation of the resulting data. However, in $F_2$ generations, where each individual is composed of two gametes and where, in the case of dominance, recessive alleles are not "visible" in heterozygotes, this analysis is more complex. A drawback of testcrosses is that they require double recessive genotypes and accurate crossings, which are not easy in some genotypes of *Arabidopsis*. For these practical reasons and because selfed seeds can be easily obtained in large quantities, mapping in *Arabidopsis* has been performed mainly with $F_2$ populations and inbred populations derived from these $F_2$ lines ($F_3$ lines and recombinant inbred lines, see below). Using $F_2$ populations implies that estimates of the average recombination percentage of the female and male meiosis are obtained. Recombination can differ strongly between sexes (Vizir and Korol 1990).

To map a new locus the (usually recessive) mutant is crossed with genotypes possessing previously localized marker genes. Linkage with a tester gene can be detected by a deviation from independent segregation in the relevant population (e.g., an $F_2$ population obtained after selfing the hybrid). The minimal population size required to detect linkage for different types of segregating populations has been described previously by Koornneef et al. (1987). To refine the crude map position obtained by an estimate of recombination percentages with one or two markers, subsequent crosses are made with genes located in the vicinity of the preliminary position, and the resulting segregating generations are analyzed.

Instead of using morphological or "classical" markers, linkage analysis can be performed by using molecular markers. In this case, the mutant has to be crossed with an ecotype different from the one in which the mutant was isolated.

From the observed segregation frequencies, the recombination fraction, $p$, or the recombination percentage, $r=100 \times p$, can be estimated using various procedures (Allard 1956; Koornneef and Stam 1992) for which computer programs are available. Reduced transmission of certain gametes (certation) may lead to slightly wrong estimates, but this parameter can be taken into account in maximum likelihood procedures, as was done in the RECF2 program developed by P. Stam (Koornneef and Stam 1987, 1992).

A problem with $F_2$ repulsion data, when the mutant and marker are in *trans*, especially in the case of close linkage, is the inaccuracy of the estimate of $r$. The analysis of $F_3$ lines derived from specific $F_2$ plants allows an accurate estimate of $r$, e.g., in cases where no double recessive genotypes are found in an $F_2$ of a few hundred plants (Koornneef and Stam 1992).

Lack of accuracy is less relevant when the heterozygote can be recognized. This is the case for codominant mutants, many isozymes, restriction length fragment polymorphisms (RFLPs), some of the polymerase chain reaction (PCR)-based markers, and embryo lethals or albinos where the heterozygosity is determined with Müller's embryo test. The large number of mapped embryonic lethals (Patton et al. 1991; D. Meinke, pers. comm.) are for this reason very suitable as linkage testers.

## Markers

Morphological or classical markers, which are mostly mutations, are simple to use and require no use of molecular biology but can suffer from ambiguous scoring and interference between the marker phenotype and the phenotype to be mapped. In addition, only a limited number of markers can be reliably scored in the progeny of a single cross. Lines have been developed that combine different markers located either on the same or on different chromosomes (Koornneef and Stam 1992). The W100 genotype combines nine markers located on all five *Arabidopsis* chromosomes (Koornneef et al. 1987). Since classical markers in general are recessive, the accuracy of the recombination fraction estimates is limited when the phenotype to be mapped is in repulsion (*trans*) with the marker.

Molecular markers can be isozyme variants, which are available in *Arabidopsis* and which have been used in population surveys (Grover 1975; Abbott and Gomes 1989) but not yet in mapping experiments. Markers detecting DNA sequence variation, on the other hand, have been used extensively and provide a much larger source of suitable markers. The advantages of molecular markers are (1) that they do not interfere with the mutant phenotype, (2) that many markers all over the genome are available (especially DNA markers), and (3) that in general they are codominant, resulting in accurate mapping.

RFLPs using random $\lambda$ (Chang et al. 1988), cosmid clones (Nam et al. 1989), and cloned genes (Hauge et al. 1993) have been used as probes in Southern blot hybridizations. The large inserts in these clones limit the possibilities for combining several of these RFLP markers. By subcloning only those fragments detecting the polymorphism, Fabri and Schäffner (1994) developed an *Arabidopsis* RFLP mapping set (ARMS) of 13 markers uniformly covering the *Arabidopsis* genome and scoreable in two successive Southern experiments by combining 8 and 5 probes, respectively, in the two hybridizations. PCR-based markers are attractive because they require only small amounts of DNA (10–100 ng) and the assay is nonradioactive and much less time-consuming than the analysis of traditional RFLP markers. Rapid amplified polymorphic DNAs (RAPDs) have been used for the construction of a linkage map in *Arabidopsis* by Reiter et al. (1992b). The RAPD amplification reaction is performed on a genomic DNA template and primed by an arbitrary oligonucleotide primer, resulting in the amplification of several discrete DNA products, and visualized by ethidium bromide staining (Williams et al. 1990; Reiter et al. 1992a). The RAPD markers have as disadvantages that they are usually dominant and their amplification is relatively sensitive to PCR conditions.

A more reproducible PCR marker can be developed using a larger and specific sequence as primer (for the various types of PCR-based markers, see Rafalski and Tingey 1993). By PCR amplification, using oligonucleotide primer pairs (19–27 bases per primer) derived from previously mapped sequenced genes in combination with the digestion of the amplified product with restriction enzymes, Konieczny and Ausubel (1993) developed codominant cleaved amplified polymorphic sequences (CAPS) as mapping markers. These authors designed 18 sets of these PCR primers, each of which in combination with a specific restriction enzyme gives a polymorphism between the Ler and Col ecotypes for loci with a previously described map position. Simple sequence length polymorphism (SSLP) or microsatellite sequences have been shown to be abundant and highly polymorphic in eukaryotic genomes, including plants (Rafalski and Tingey 1993). SSLP markers are analyzed by PCR amplification of a short genomic region containing the repeated sequence (mostly dinucleotide repeats) and using DNA sequences that flank the repeat as primers. Bell and Ecker (1994) identified such microsatellite markers in *Arabidopsis* and showed that especially $(GA)_n$ and $(AT)_n$ repeats could be amplified easily and were highly polymorphic in this species. Thirty of these microsatellite loci were mapped using the recombinant inbred line population developed by Lister and Dean (1993). Molecular and morphological markers have been combined in situations where morphological markers were used to find the approximate map position, after which fine mapping was performed with specific molecular markers located in that region. Morphological markers allow the isolation of many recombinants adjacent to the gene of interest from large populations. These recombinants are subsequently analyzed for their molecular marker phenotype. This procedure saves labor and can result in a high-resolution map around that gene, which is required for map-based cloning strategies (Giraudat et al. 1992; Putterill et al. 1993). The efficiency of mapping morphological mutants with molecular markers can be improved by procedures such as bulk segregrant analysis (Mitchelmore et al. 1991) and other pooling strategies (Patton et al. 1991; Reiter et al. 1992a; Churchill et al. 1993; Williams et al. 1993).

## Mapping Populations

In contrast to morphological markers, the segregation of molecular markers cannot be studied in a uniform genetic background and requires segregating populations derived from ecotype crosses. In *Arabidopsis*, crosses between L*er*, Col, Nz, and Ws have been used thus far. However, since *Arabidopsis* ecotypes differ in the degree of polymorphism (King et al. 1993), particular crosses will be more informative than others. A survey of 28 ecotypes by King et al. (1993) revealed that Nz was the most distantly related ecotype (related to the other ecotypes tested), which means that this is the most suitable parent to cross with a mutant (e.g., in L*er* or Col background) for the analysis of molecular markers. Williams et al. (1993) listed the presence of specific products amplified by their standard RAPD mapping set (selected for allelic variation between Ws and W100) for eight frequently used strains including the W100 multiple marker line. They found that W100 is not 100% L*er*, which can be explained by the use of "old" mutants isolated by Rédei to construct this marker line. Among a survey of six ecotypes, the microsatellite markers revealed on the average four different alleles per marker. Since L*er* x Col crosses are frequently used and markers specific for this cross have been developed (Konieczny and Ausubel 1993; Fabri and Schäffner 1994), and because a frequently used recombinant inbred population derives from this cross, it is important to know the amount of polymorphism between these ecotypes. Konieczny and Ausubel (1993) estimated it to be one in every 261 bp based on screening 5227 nucleotides that comprised the recognition sites surveyed for polymorphisms, when they developed their CAPS markers.

A potential complication of the use of different ecotype crosses could be that recombination is affected by the genetic background. However, a comparison between recombination data obtained from either L*er*/L*er* or L*er*/Nz under identical conditions did not show significant differences (Koornneef and Hanhart 1988).

The mapping populations used for molecular markers yield a large amount of linkage data, since many markers are assayed in the same population. Since a single plant does not yield enough material to extract DNA for many hybridizations, $F_3$ lines are preferred material for RFLP analysis. PCR-based procedures require less material; one leaf usually provides sufficient DNA. The use of $F_3$ lines has the additional advantage that each $F_2$ plant can be classified for homo/heterozygosity of dominant/recessive markers segregating in the same population. However, even when $F_3$ lines are used, the amount of plant material is limited and $F_3$ seeds still segregate. Recombinant inbred lines (RILs) produced by continually selfing the progeny of individual $F_2$ plants, using a single seed descent procedure, until homozygosity is achieved, do not suffer from these problems and provide a permanent mapping population (Burr and Burr 1991; Reiter et al. 1992a). The permanence of the population implies that any new information is additive and becomes automatically integrated with the data already obtained for that population. RILs will probably be used exclusively in the near future to map DNA sequences. Two sets of RILs have been developed for *Arabidopsis* at present. One is derived from a W100 (L*er*) x Ws cross (Reiter et al. 1992b) and another set from a L*er* x Col cross (Lister and Dean 1993).

In addition to mapping new DNA markers, RILs can be used to map any phenotypic trait that is polymorphic between the parents. This was demonstrated by the mapping of the *RPP5* locus by Parker et al. (1993) and the mapping of flowering time differences (C. Lister and C. Dean, pers. comm.).

### Construction of Linkage Maps

When estimates of the recombination percentage between at least three genes are available, these genes can be ordered in a linear linkage group. The construction of a genetic map requires a mapping function which relates recombination frequency with the average number of crossovers during meiosis between two loci (= map distance in Morgans) and corrects this recombination frequency for double crossovers. Different mapping functions correspond to different levels of crossover interference. For many organisms, the degree of interference is satisfactorily described by Kosambi's (1944) mapping function; this function has also been used for *Arabidopsis* (Koornneef et al. 1983b; Hauge et al. 1993).

A method for the simultaneous estimation of map distances from all available data has been developed by Jensen and Jörgensen (1975) and was used to construct the classical maps (Koornneef et al. 1983b; Koornneef 1990a). Chang et al. (1988) and Nam et al. (1989) used the MAPMAKER computer program developed by Lander (Lander et al. 1987) to construct two *Arabidopsis* RFLP maps. Although MAPMAKER can be used to combine data from several independent $F_2$s, it cannot combine estimates from distinct types of experiments (e.g., $F_2$s, backcrosses, recombinant inbred lines, recombination estimates with standard deviations). The integration of the data used to construct the two RFLP maps and the classical map together with additional data in one map (Hauge et al. 1993) has been performed with the program JOINMAP developed by P. Stam (1993). Although the statistical integration gives the best "average" for the genetic distances, the mutual order of closely linked markers can only be established unambiguously when these markers are studied in the same population. This is often not the case with classical markers or when the data of various populations used for the mapping of molecular markers are integrated.

This explains the differences between the statistically integrated map and data obtained after refining a map in a certain region or the use of "all" markers in one mapping population. The more common markers are used, the better the integration is. Since Reiter et al. (1992b) and Lister and Dean (1993) used many markers from both RFLP maps, the marker order in this map is the most likely one in situations where the order of the integrated map differs with these RI-based maps. The next generation of genetic maps will relate the location of contigs of cloned DNA to the genetic map. A first example of a physical genetic map in *Arabidopsis* was the publication by Putterill et al. (1993) of a 1700-kb contig spanning at least 5 cM on the top of chromosome 5.

## THE *ARABIDOPSIS* GENETIC MAP

Trisomics have been used to assign linkage groups to the chromosomes. Initially, six linkage groups were described by Rédei and Hirono (1964), and Lee-Chen and Steinitz-Sears (1967) assigned each linkage group to one of the five chromosomes. However, due to ambiguous data for group 4 (represented by the marker *chl*), it was assigned to the wrong chromosome. Further trisomic analysis (Koornneef and van der Veen 1983) and linkage studies by Feenstra (1978) and Koornneef et al. (1983b) clarified the situation. McKelvie (1965) also constructed linkage groups based on linkage studies with the mutants he isolated. The relation of these groups with chromosomes is discussed by Koornneef and van der Veen (1983).

The first comprehensive and internally consistent linkage map of *Arabidopsis* using exclusively morphological markers ("classical map") was published in 1983 (Koornneef et al. 1983b). Updated versions of this map were published by Koornneef (1990a) and Patton et al. (1991). The chromosome numbering was based on the linkage groups established by Rédei (Rédei and Hirono 1964) and has been generally accepted. The chromosome numbers do not correlate with the genetic or cytogenetic size order, nor is the orientation of the map related to short arms versus long arms (see Heslop-Harrison and Maluszynska, this volume).

An updated classical map, based on all pairwise recombination data obtained with classical markers by Wageningen and Stillwater (D. Meinke, pers. comm.), all data published in the literature, and data communicated by various scientists, is shown in Figure 1. This map contains 284 loci. Updated versions of the RFLP maps of Chang et al. (1988) and Nam et al. (1989) are shown in Figures 2 and 3. These maps are based on the data used to construct the integrated map (Fig. 4) described by Hauge et al. (1993). The mapping populations of these two RFLP maps consist of several subpopulations, which were not analyzed completely for all markers (especially in case of the classical markers on these maps). The maps shown in Figures 1–4 were constructed using the JOINMAP program (Stam 1993). The map based on the RILs constructed by Lister and Dean is shown in Figure 5. More than 500 new loci are currently being mapped using these RI lines (Lister and Dean 1993) including the microsatellite markers (Bell and Ecker 1994). The map published by Reiter et al. (1992b) and shown in Figure 6 is mainly based on RAPD markers and also contains 9 classical markers and 60 RFLP markers that were also used for the construction of the RFLP maps. The genetic length of the *Arabidopsis* genome based on the classical map (Fig. 1) totals 485 cM, ranging from 83 cM for chromosome 2 to 122 cM for chromosome 1. On the basis of individual chromosomes, the genetic length corresponds well with that of other plant chromosomes, although the chromosomes of *Arabidopsis* are much smaller. Based on the estimated genome size of 100,000 kb, this implies that the average centimorgan represents slightly more than 200 kb. It is very likely that this figure will differ from chromosome region to chromosome region. Maybe the regions around the centromere that show up cytologically as C-banded heterochromatin have a very different behavior in this respect.

## CONCLUSIONS AND PERSPECTIVES

Although it took a long time before the biological advantages, which were recognized as early as 1943, were fully exploited, *Arabidopsis* genetics has developed in such a way that one of the most complete and accurate genetic maps in higher plants has been constructed for this species. This map includes a large number of genes with functions known at the molecular level. This number will increase greatly in the coming years. The small size of the chromosomes has limited the use of cytogenetics, and this is expected to remain the case. Even the development of in situ hybridization and the analysis of synaptonemal complexes (see Heslop-Harrison and Maluszynska, this volume) will not result in as great a genetic accuracy as can be obtained with linkage analysis. How-

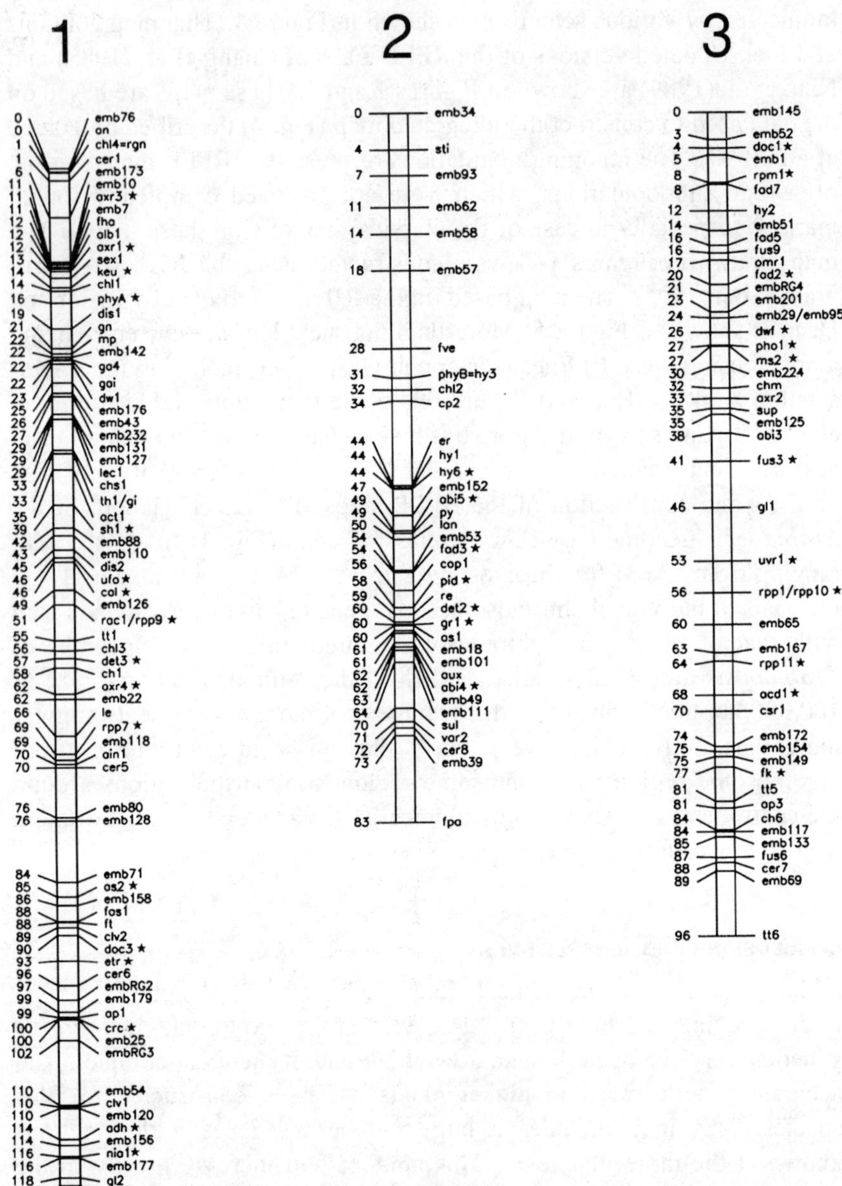

*Figure 1*  (*See facing page for legend.*)

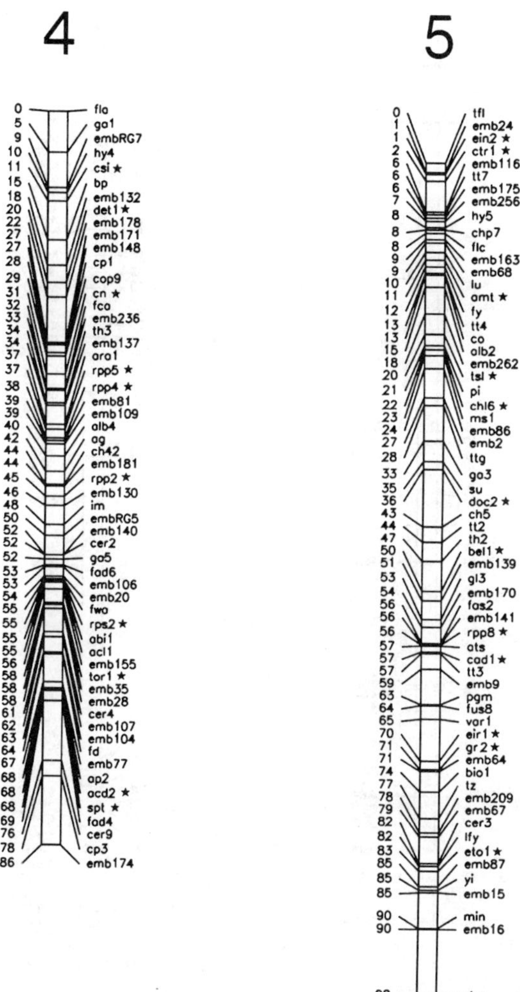


*Figure 1* Linkage map of *A. thaliana* based on linkage data obtained with classical markers. The map is based on data cited by Hauge et al. (1993). Genetic data published since January 1993, additional linkage data obtained in Wageningen (fla, flc, and fy) and data for emb mutants were provided by David Meinke (D. Meinke et al., in prep.). The map was constructed with the JOIN-MAP (Stam 1993) computer program. Gene symbols marked with an asterisk are placed on the map by hand and their position is based on a map distance with mostly one marker and the position of the gene in relation to this marker (D. Meinke et al., pers. comm.). When markers were only mapped in relation to molecular markers, the relative position of these molecular markers in relation to the classical markers was based on Hauge et al. (1993).

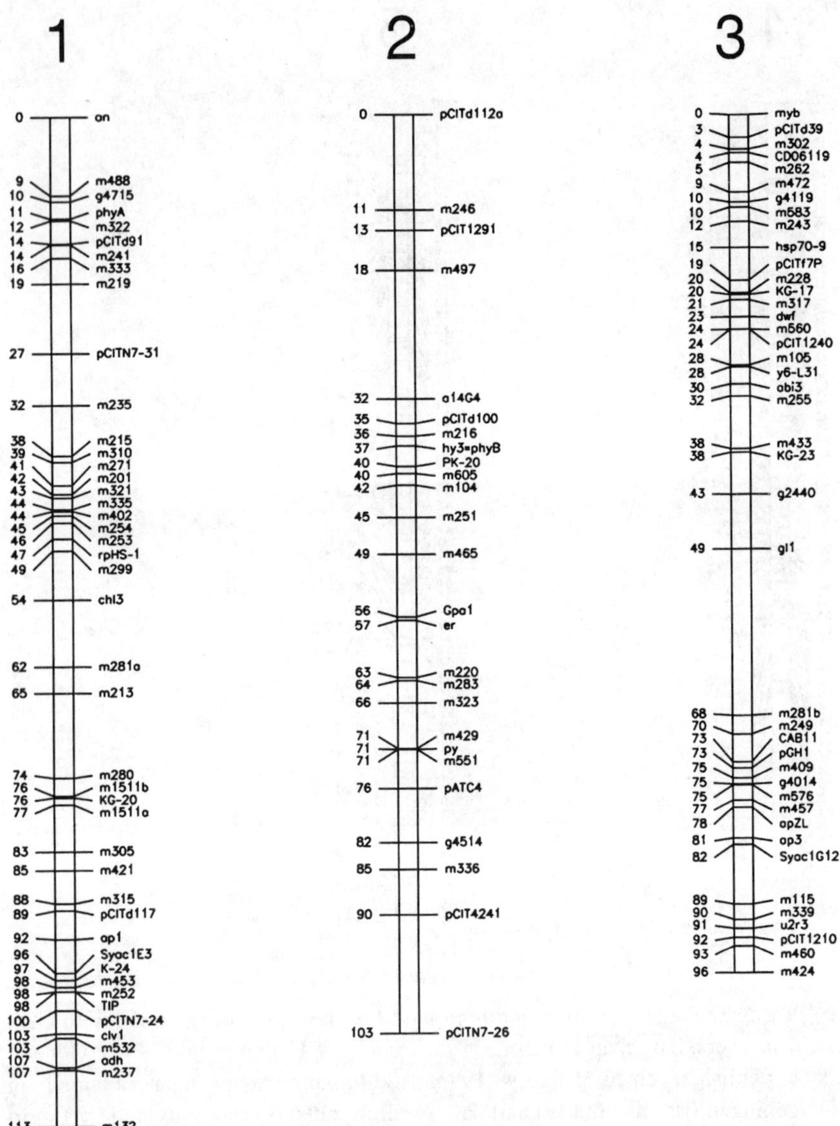

*Figure 2* Linkage map of *A. thaliana* based on linkage data obtained with RFLP markers obtained in the Meyerowitz lab. The map is based on the populations described by Chang et al. (1988) and Hauge et al. (1993). The same data were used for the integrated map published by Hauge et al. (1993) (see Fig. 4).

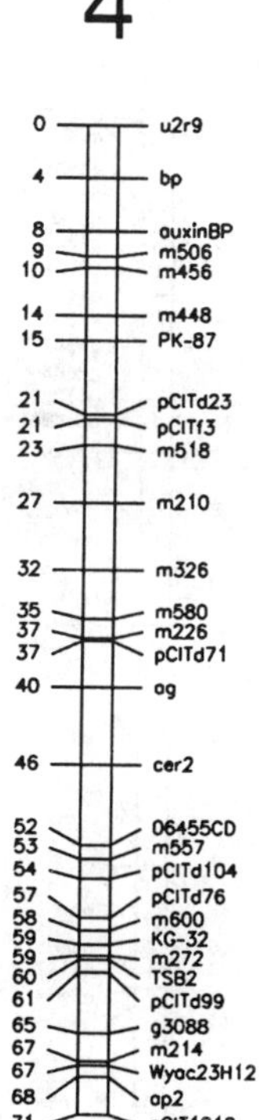

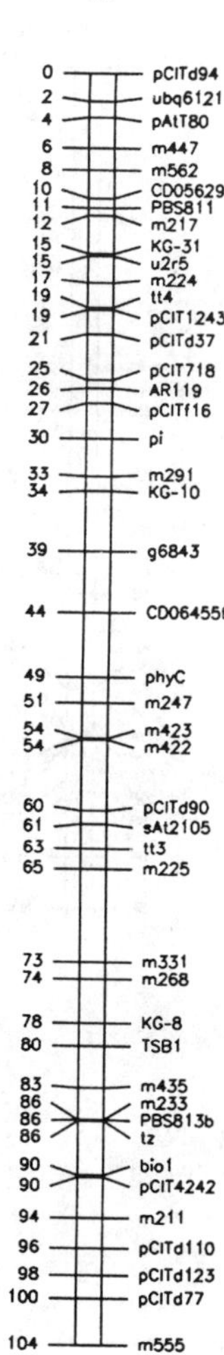

*Figure 2* (*continued*)

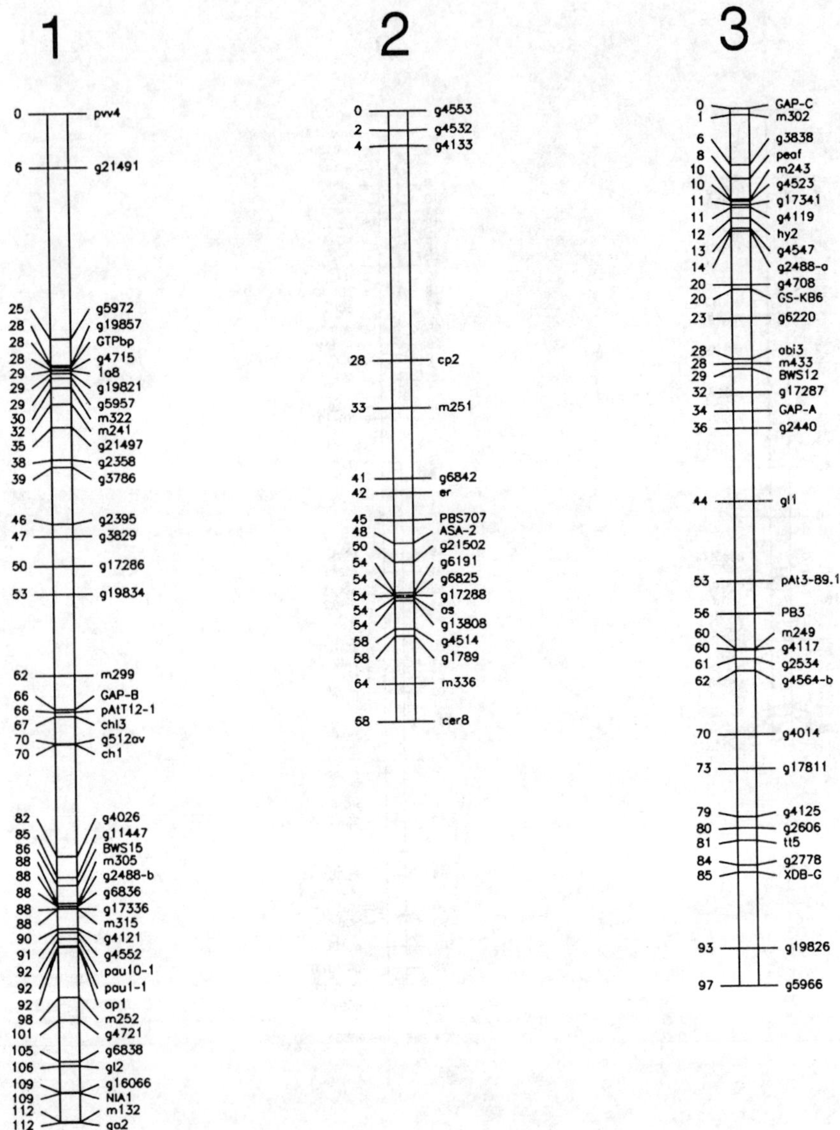

*Figure 3* Linkage map of *A. thaliana* based on linkage data obtained with RFLP markers obtained in the Goodman lab. The map is based on the populations described by Nam et al. (1989). The same data were used for the integrated map published by Hauge et al. (1993) (see Fig. 4).

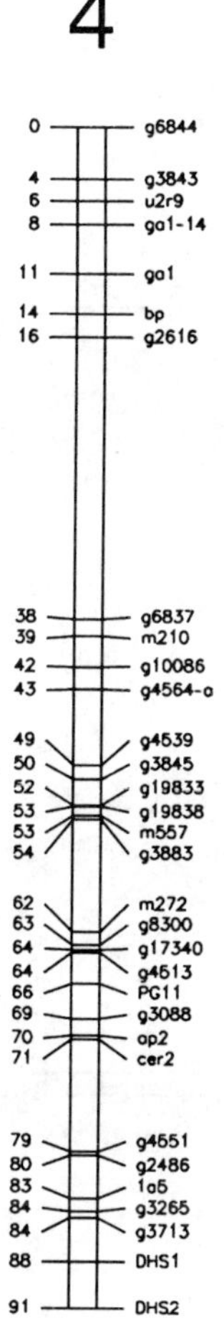


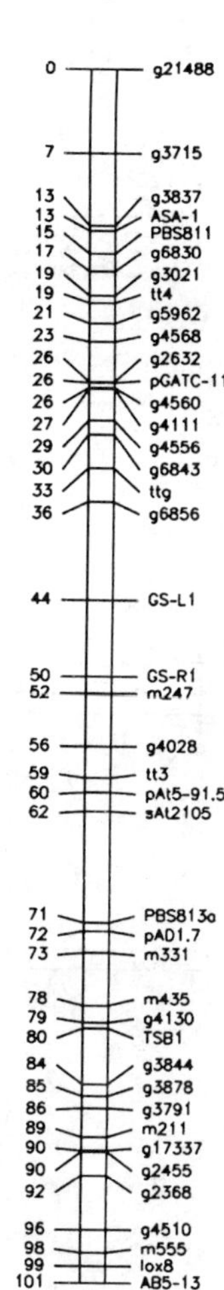


*Figure 3* (*continued*)

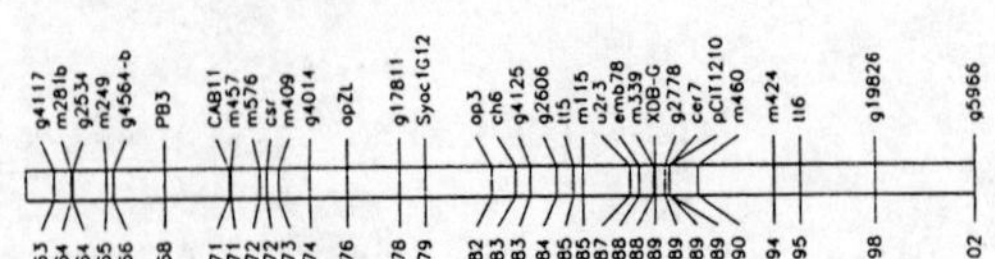
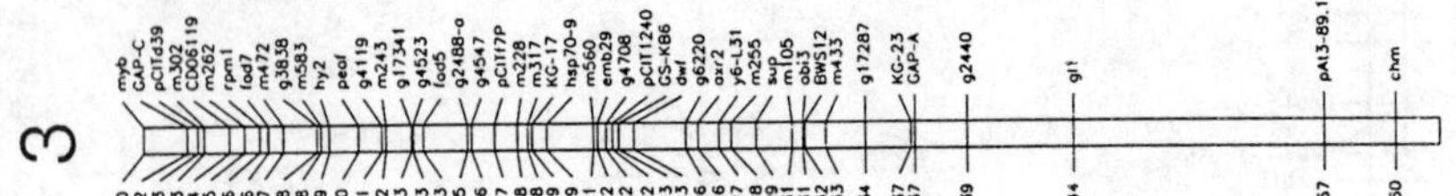
3
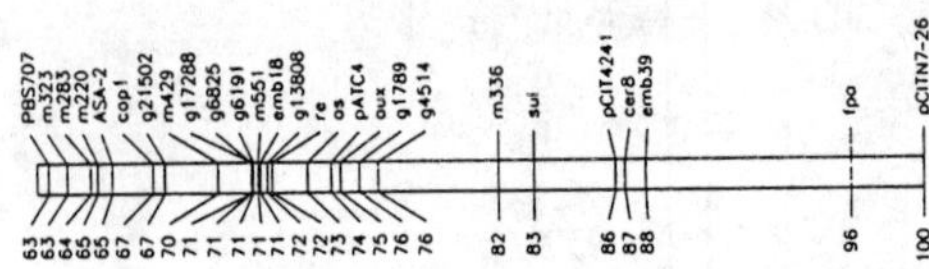
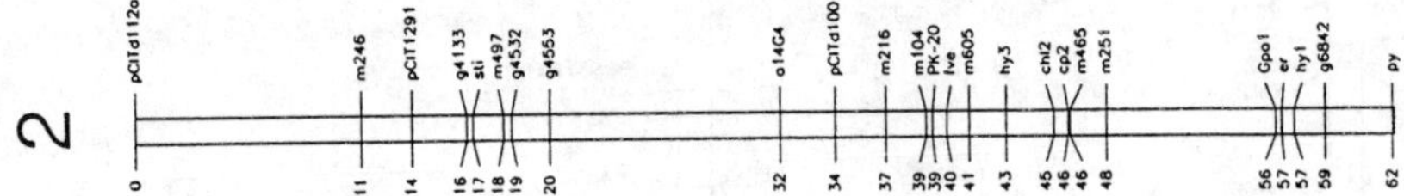
2
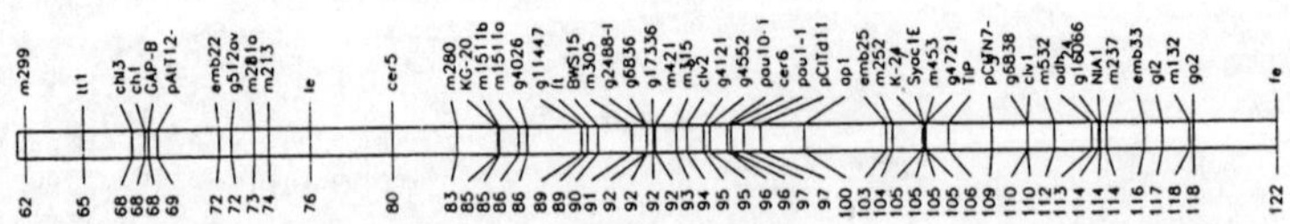
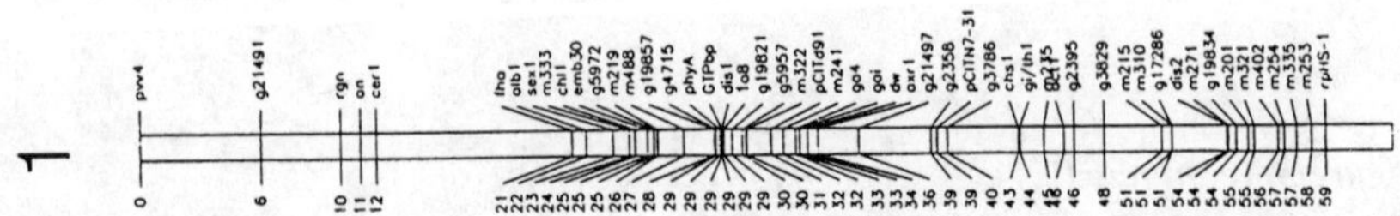
1

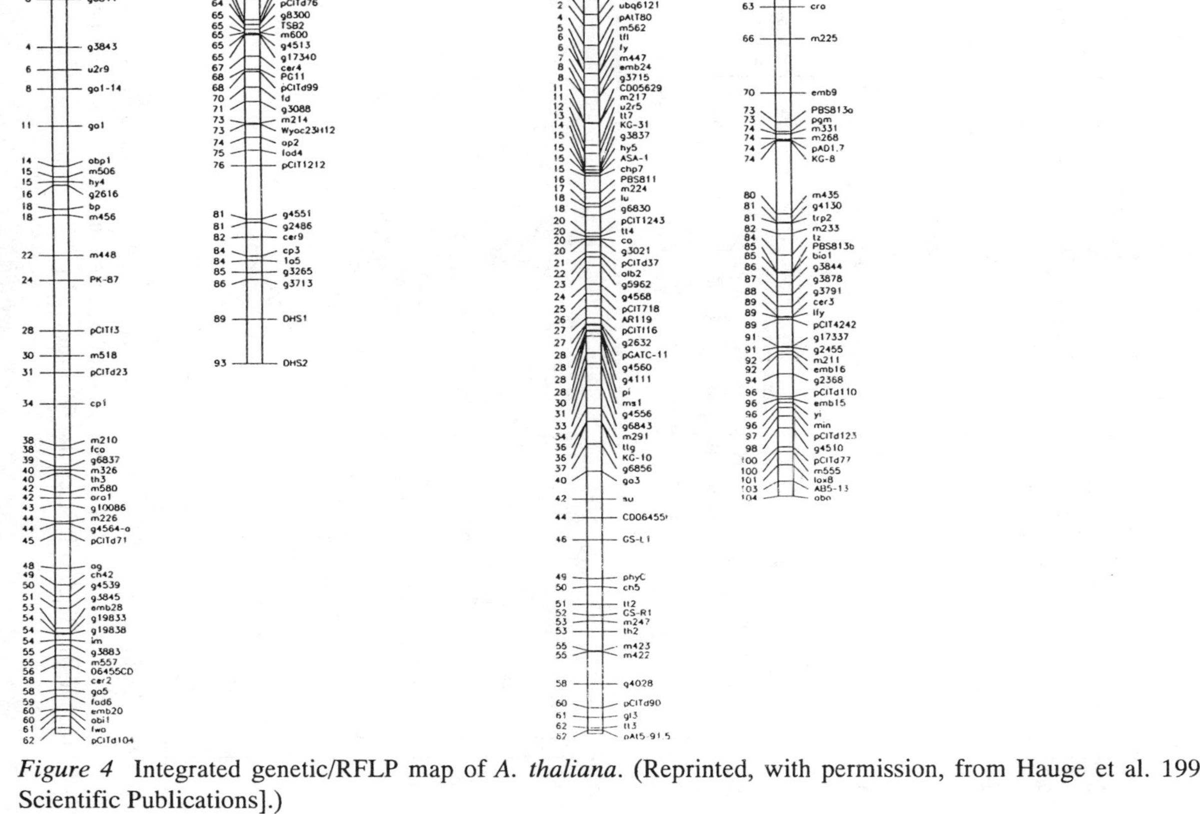

*Figure 4*  Integrated genetic/RFLP map of *A. thaliana*. (Reprinted, with permission, from Hauge et al. 1993 [copyright Blackwell Scientific Publications].)

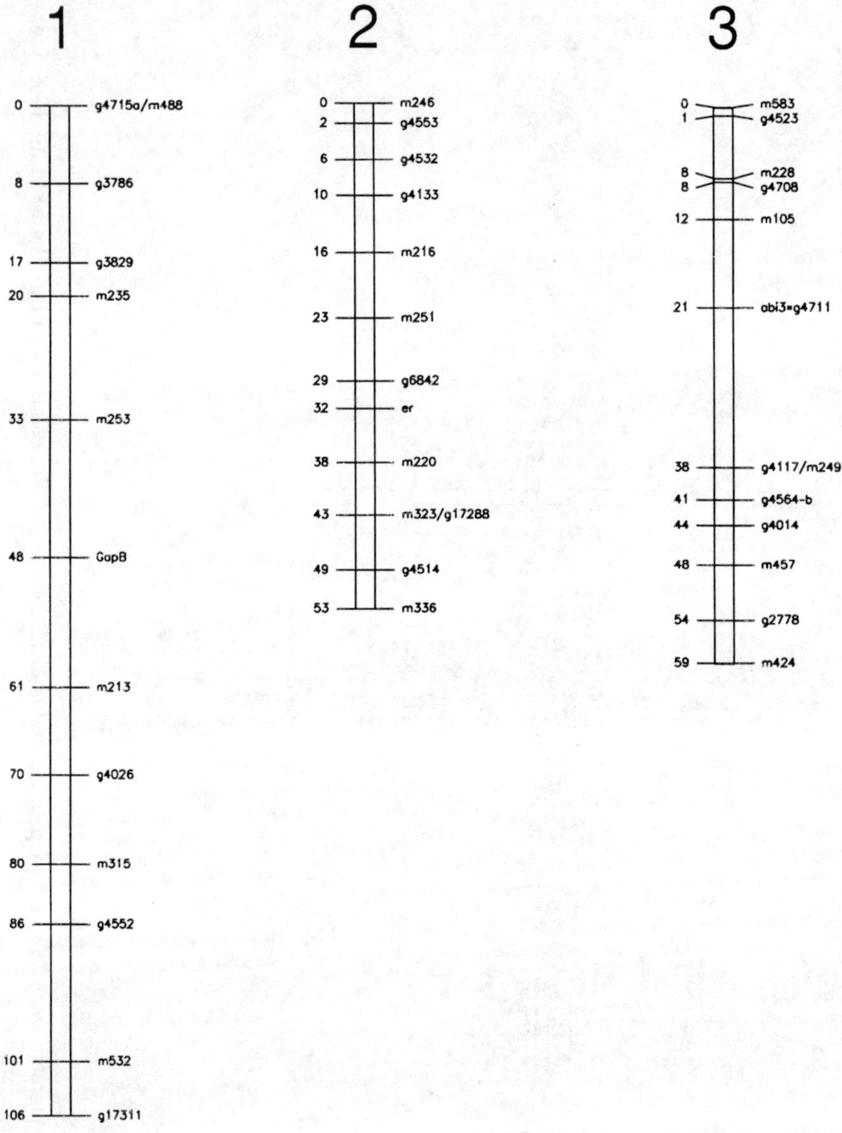



*Figure 5* Linkage map of *A. thaliana* based on recombinant inbred lines derived from the cross L*er* × Col. (Reprinted, with permission, from Lister and Dean 1993 [copyright Blackwell Scientific Publications].)

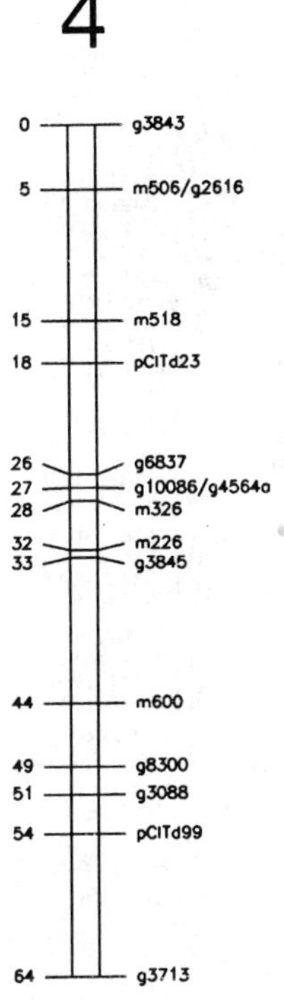



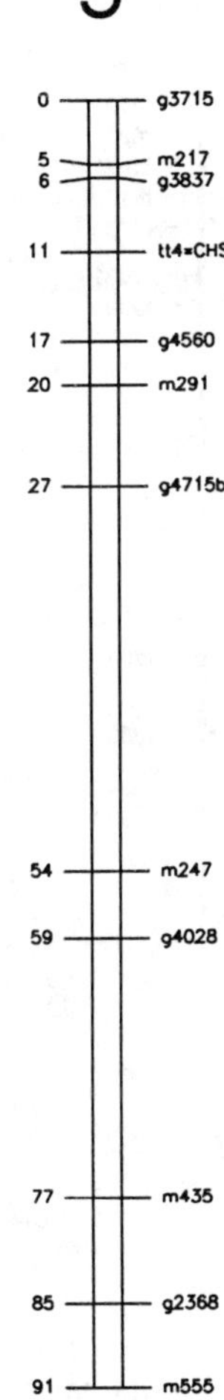



*Figure 5* (*continued*)

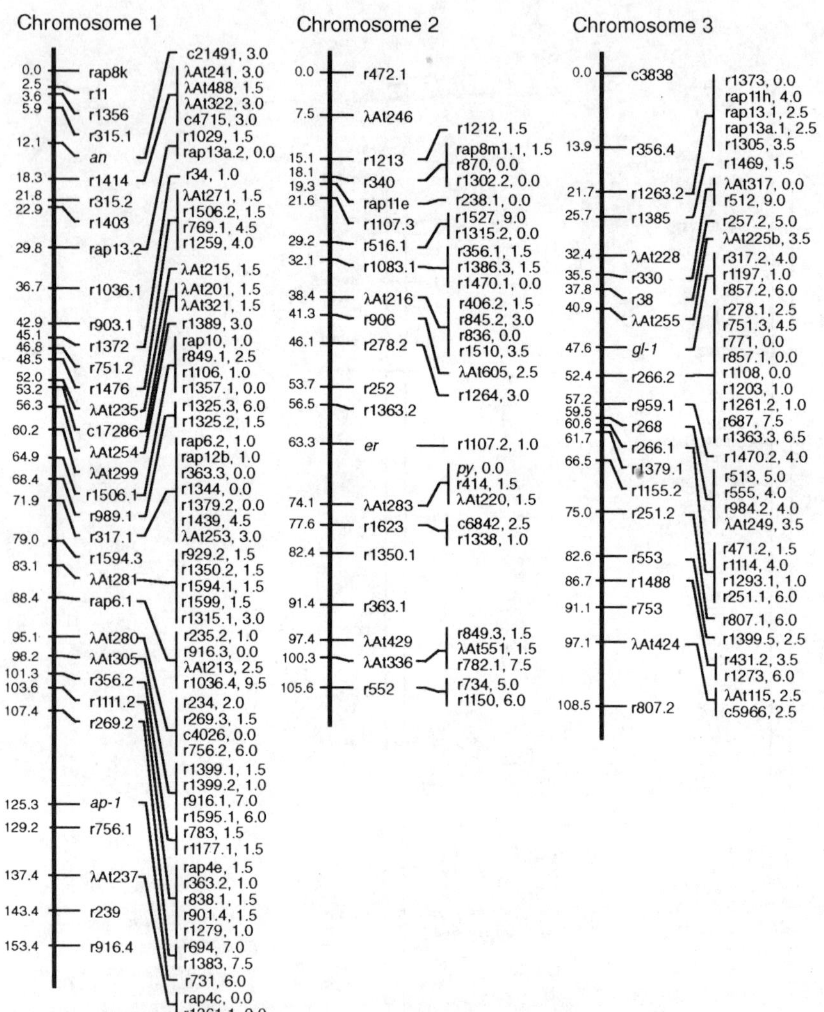



*Figure 6* Linkage map of *A. thaliana* based on recombinant inbred lines derived from the cross W100 × Ws. The 126 markers along the five vertical bars were ordered with LOD score differences >3.0. Marker loci listed to the right of each chromosome bar could not be ordered with equal confidence (LOD score differences <3.0). Markers to the right along with the approximate distance in cM from the markers placed on the LOD 3.0 map. Phenotypic markers are italicized; RFLP markers are designated as λ At ... (= m... on maps 2, 3, 4, and 5) or c ... (= g... on maps 2, 3, 4, and 5). The remaining markers are RAPDs as described by Reiter et al. (1992b). (Reprinted, with permission, from Reiter et al. 1992b.)

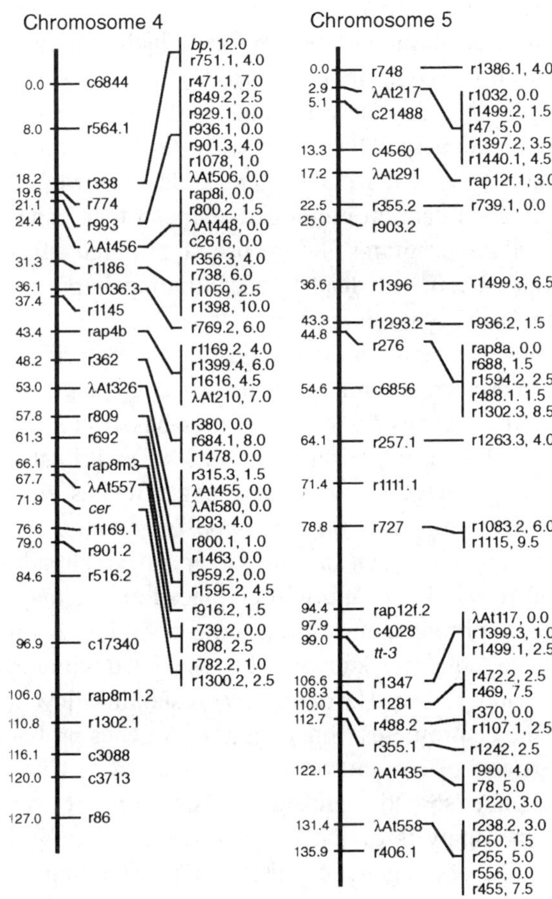


*Figure 6  (continued)*

ever, these techniques will provide the information about the relation between the genetic maps and chromosome arms, centromeres, etc.

The present genetic map still contains incorrect gene orders and inaccurate estimates of genetic distances. These are because of the use of different mapping populations and statistical inaccuracies due to limited population sizes. The continuous addition of new markers, including randomly isolated and sequenced cDNAs, cloned genes, SSLPs, and other PCR-based markers, along with many new mutants, will increase the number of mapped loci tremendously. This increase in marker number will also probably bring about more inaccurate gene orders. However, the accumulation of new detailed data on previously mapped loci, especially from chromosome-walking programs and the use of a limited number of mapping populations (RILs), will lead to a continuing improvement of the accuracy of the *Arabidopsis* linkage map.

The integration of the genetic map with the physical maps comprising a set of overlapping YAC and cosmid contigs has already been achieved to a large extent, and the first successful cases of map-based cloning have been reported (Arondel et al. 1992; Giraudat et al. 1992). Ultimately this will result in a clear picture of how genetic length relates with physical length in the various regions of the *Arabidopsis* genome. Sequences near the ends (telomeres) and near the centromeres have already been cloned (Richards et al. 1991, 1992) and will be placed on the map. The telotrisomic available for chromosomes 1, 3, and 5 combined with molecular markers, e.g., those sequences known to be associated with the centromeres (Maluszynska and Heslop-Harrison 1991), should allow a more accurate location of the centromeres. Since the rDNA genes probably represent a whole short arm of chromosome 4 (Sears and Lee-Chen 1970), mapping this sequence should indicate the structure of this chromosome. Ultimately, the analysis of the *Arabidopsis* genome will provide a very complete picture of the genetic makeup of a higher plant.

**ACKNOWLEDGMENTS**

I thank Dr. David Meinke for his assistance in the preparation of the classical linkage map and for making available many unpublished mapping data for this. I thank Dr. Piet Stam and Dr. Ton Peeters for their help in constructing the maps and Dr. Johan van Ooijen for making available the DRAWMAP program to prepare the graphic maps.

**REFERENCES**

Abbott, R.J. and N.F. Gomes. 1989. Population genetic structure and outcrossing rate of *Arabidopsis thaliana* (L.) Heynh. *Heredity* **62:** 411–418.

Allard, R.W. 1956. Formulas and tables to facilitate the calculation of recombination values in heredity. *Hilgardia* **24**: 235–278.

Altmann, T., B. Damm. U. Halfter, L. Willmitzer, and P.M. Morris. 1992. Protoplast transformation and methods to create specific mutants in *Arabidopsis thaliana*. In *Methods in* Arabidopsis *research* (eds. C. Koncz et al.), pp. 310–330. World Scientific, Singapore.

Arondel, V., B. Lemieux, I. Hwang, S. Gibson, H.M. Goodman, and C.R. Somerville. 1992. Map-based cloning of a gene controlling omega-3 fatty acid desaturation in *Arabidopsis*. *Science* **258**: 1353–1355.

Bell, C.J. and J.R. Ecker. 1994. Assignment of thirty microsatellite loci to the linkage map of *Arabidopsis*. *Genomics* **19**: 137–144.

Bouharmont, J. 1965. Fertility studies in polyploid *Arabidopsis thaliana*. In Arabidopsis *research* (ed. G. Röbbelen), pp. 31–36. Wasmund, Gelsenkirchen.

———. 1969. Etude de l'heredite tetrasomique des mutations induites chez *Arabidopsis thaliana*. In *Induced mutations in plants*, pp. 603–610. IAEA, Vienna.

Bouharmont, J. and F. Macé. 1972. Valeur competative des plantes autotetraploides d'*Arabidopsis thaliana*. *Can. J. Genet. Cytol.* **14**: 257–263.

Bowman, J.L., J. Alvarez, D. Weigel, E.M. Meyerowitz, and D.R. Smyth. 1993. Control of flower development in *Arabidopsis thaliana* by APETALA1 and interacting genes. *Development* **119**: 721–743.

Bronckers, F. 1963. Variations polliniques dans une série d'autopolyploides artificiels d'*Arabidopsis thaliana* (L.) Heynh. *Pollen Spores* **5**: 233–238.

Burr, B. and F.A. Burr. 1991. Recombinant inbreds for molecular mapping in maize: Theoretical and practical considerations. *Trends Genet.* **7**: 55–60.

Chang, C., J.L. Bowman, A.W. DeJohn, E.S. Lander, and E.M. Meyerowitz. 1988. Restriction fragment length polymorphism linkage map for *Arabidopsis thaliana*. *Proc. Natl. Acad. Sci.* **85**: 6856–6860.

Churchill, G.A., J.J. Giovannoni, and S.D. Tanksley. 1993. Pooled-sampling makes high-resolution mapping practical with DNA markers. *Proc. Natl. Acad. Sci.* **90**: 16–20.

Dellaert, L.M.W. 1980. Segregation frequencies of radiation-induced viable mutants in *Arabidopsis thaliana* (L.) Heynh. *Theor. Appl. Genet.* **57**: 137–143.

Fabri, C.O. and A.R. Schäffner. 1994. An *Arabidopsis thaliana* RFLP mapping set to localize mutations to chromosomal regions. *Plant J.* **5**: 149–156.

Feenstra, W.J. 1978. Contiguity of linkage groups 1 and 4 as revealed by linkage relationship of two newly isolated markers dis-1 and dis-2. *Arabidopsis Inf. Serv.* **15**: 35–38.

Furner, I.J. and J.E. Pumfrey. 1992. Cell fate in the shoot apical meristem of *Arabidopsis thaliana*. *Development* **115**: 755–764.

Gaj, M.D. and M. Maluszynski. 1986. Mitotic recombination in callus of *Arabidopsis thaliana* (L.) Heynh. after fast-neutron treatment. In *Nuclear techniques and in vitro culture for plant improvement*, pp. 147–153. IAEA, Vienna.

Galbraith, D.W., K.R. Harkins, and S. Knapp. 1991. Systemic endopolyploidy in *Arabidopsis thaliana*. *Plant Physiol.* **96**: 985–989.

Giraudat, J., B.M. Hauge, C. Valon, J. Smalle, F. Parcy, and H.M. Goodman. 1992. Isolation of the *Arabidopsis ABI3* gene by positional cloning. *Plant Cell* **4**: 1251–1261.

Gleba, Y.G. and F. Hoffmann. 1978. Hybrid cell lines *Arabidopsis thaliana* + *Brassica campestris*: No evidence for specific chromosome elimination. *Mol. Gen. Genet.* **165**: 257–264.

Grover, N.S. 1975. Characterization of *Arabidopsis thaliana* ecotypes on the basis of genetic variation at ten isozyme loci. *Arabidopsis Inf. Serv.* **12**: 19–21.

Hake, S. 1992. Unraveling the knots in plant development. *Trends Genet.* **8:** 109–114.

Hauge, B.M., S.M. Hanley, S. Cartinhour, J.M. Cherry, H.W. Goodman, M. Koornneef, P. Stam, C. Chang, S. Kempin, L. Medrano, and E.M. Meyerowitz. 1993. An integrated genetic/RFLP map of the *Arabidopsis thaliana* genome. *Plant J.* **3:** 745–754.

Hayes, G.R. and W.H. Klein. 1974. Spectral quality influence of light during development of *Arabidopsis thaliana* plants in regulating seed germination. *Plant Cell Physiol.* **15:** 643–653.

Hirono, Y. and G.P. Rédei. 1965. Induced premeiotic exchange of linked markers in the angiosperm *Arabidopsis*. *Genetics* **51:** 519–526.

Irish, V.F. and I.M. Sussex. 1992. A fate map of the *Arabidopsis* embryonic shoot apical meristem. *Development* **115:** 745–753.

Jensen, J. and J.H. Jörgensen. 1975. The barley chromosome 5 linkage map. *Hereditas* **80:** 5–16.

King, G., J. Nienhuis, and C. Hussey. 1993. Genetic similarity among ecotypes of *Arabidopsis thaliana* estimated by analysis of restriction fragment length polymorphisms. *Theor. Appl. Genet.* **86:** 1028–1032.

Konieczny, A. and F.M. Ausubel. 1993. A procedure for mapping *Arabidopsis* mutations using co-dominant ecotype-specific PCR-based markers. *Plant J.* **4:** 403–410.

Koornneef, M. 1981. The complex syndrome of *ttg* mutants. *Arabidopsis Inf. Serv.* **18:** 45–51.

―――. 1983. The use of telotrisomics for centromere mapping in *Arabidopsis thaliana* (L.) Heynh. *Genetica* **62:** 33–40.

―――. 1990a. Linkage map of *Arabidopsis thaliana* (2n=10). In *Genetic maps V* (ed. S.J. O'Brien), pp. 6.95–6.97. Cold Spring Harbor Laboratory, Cold Spring Harbor, New York.

―――. 1990b. Mutations affecting the testa colour in *Arabidopsis*. *Arabidopsis Inf. Serv.* **27:** 1–4.

Koornneef, M. and C.J. Hanhart. 1988. The effect of genetic background on recombination in *Arabidopsis*. *Arabidopsis Inf. Serv.* **26:** 73–78.

Koornneef, M. and P. Stam. 1987. Procedures for mapping by using $F_2$ and $F_3$ populations. *Arabidopsis Inf. Serv.* **25:** 35–40.

―――. 1992. Genetic analysis. In *Methods in* Arabidopsis *research* (ed. C. Koncz et al.), pp. 83–99. World Scientific, Singapore.

Koornneef, M. and J.H. van der Veen. 1980. Induction and analysis of gibberellin-sensitive mutants in *Arabidopsis thaliana* (L.) Heynh. *Theor. Appl. Genet.* **58:** 257–263.

―――. 1983. The trisomics of *Arabidopsis thaliana* (L.) Heynh. and the location of linkage groups. *Genetica* **61:** 41–46.

Koornneef, M., H.C. Dresselhuys, and K. Sree-Ramulu. 1982. The genetic identification of translocations in *Arabidopsis*. *Arabidopsis Inf. Serv.* **19:** 93–99.

Koornneef, M., C.J. Hanhart, E.P. van Loenen-Martinet, and J.H. van der Veen. 1987. A marker line that allows the detection of linkage on all *Arabidopsis* chromosomes. *Arabidopsis Inf. Serv.* **23:** 46–50.

Koornneef, M., J. van Eden, C.J. Hanhart, and J.A.M. de Jongh. 1983a. Genetic fine structure of the ga-1 locus in the higher plant *Arabidopsis thaliana*. *Genet. Res.* **41:** 57–68.

Koornneef, M., J. van Eden, C.J. Hanhart, P. Stam, F.J. Braaksma, and W.J. Feenstra. 1983b. Linkage map of *Arabidopsis thaliana*. *J. Hered.* **74:** 265–272.

Kosambi, D.D. 1944. The estimation of map distances from recombination values. *Ann. Eugenet.* **12:** 172–175.

Lander, E.S., P. Green, J. Abrahamson, A. Barlow, M.J. Daly, S.E. Lincoln, and L. Newberg. 1987. Mapmaker: An interactive computer package for constructing primary genetic linkage maps of experimental and natural populations. *Genomics* **1:** 174–181.

Lee-Chen, S. and L.M. Steinitz-Sears. 1967. The location of linkage groups in *Arabidopsis thaliana. Can. J. Genet. Cytol.* **9:** 381–384.

Li, S.L. 1967. A new segregation distorter factor in *Arabidopsis. Arabidopsis Inf. Serv.* **4:** 5–6.

Lister, C. and C. Dean. 1993. Recombinant inbred lines for mapping RFLP and phenotypic markers in *Arabidopsis thaliana. Plant J.* **4:** 745–750.

Maluszynska, J. and J.S. Heslop-Harrison. 1991. Localization of tandemly repeated DNA sequences in *Arabidopsis thaliana. Plant J.* **1:** 159–166.

Martínez-Zapater, J.M. and C.R. Somerville. 1990. Effect of light quality and vernalization on late-flowering mutants of *Arabidopsis thaliana. Plant Physiol.* **92:** 770–776.

Martínez-Zapater, J.M., P. Gill, J. Capel, and C.R. Somerville. 1992. Mutations at the *Arabidopsis CHM* locus promote rearrangements of the mitochondrial genome. *Plant Cell* **4:** 889–899.

McKelvie, A.D. 1965. Preliminary data on linkage groups in *Arabidopsis.* In Arabidopsis *research* (ed. G. Röbbelen), pp. 31–36. Wasmund, Gelsenkirchen.

Meinke, D.W. 1982. Embryo-lethal mutants of *Arabidopsis thaliana*: Evidence for gametophytic expression of the mutant genes. *Theor. Appl. Genet.* **63:** 381–386.

Mitchelmore, R.W., I. Paran, and R.V. Kesseli. 1991. Identification of markers linked to disease-resistance genes by bulked segregant analysis: A rapid method to detect markers in specific genomic regions by using segregating populations. *Proc. Natl. Acad. Sci.* **88:** 9828–9832.

Müller, A.J. 1963. Embryonentest zum Nachweis rezessiver Lethalfactoren bei *Arabidopsis thaliana. Biol. Zentralbl.* **82:** 133–163.

Nam, H.G., J. Giraudat, B. den Boer, F. Moonan, W.D.B. Loos, B.M. Hauge, and H.M. Goodman. 1989. Restriction fragment length polymorphism linkage map of *Arabidopsis thaliana. Plant Cell* **1:** 699–705.

Negrutiu, I., M. Jacobs, and D. Cachita. 1978. Some factors controlling in vitro morphogenesis of *Arabidopsis thaliana. Z. Pflanzenphysiol.* **86:** 113–124.

Parker, J.E., V. Szabò, B.J. Staskawicz, C. Lister, C. Dean, M.J. Daniels, and J.D.G. Jones. 1993. Phenotypic characterization and molecular mapping of the *Arabidopsis thaliana* locus *RPP5*, determining disease resistance to *Peronospora parasitica. Plant J.* **4:** 821–831.

Patton, D.A., L.H. Franzmann, and D.W. Meinke. 1991. Mapping genes essential for embryo development in *Arabidopsis thaliana. Mol. Gen. Genet.* **227:** 337–347.

Putterill, J., F. Robson, K. Lee, and G. Coupland. 1993. Chromosome walking with YAC clones in *Arabidopsis*: Isolation of 1700 kb of contiguous DNA on chromosome 5, including a 300 kb region containing the flowering-time gene CO. *Mol. Gen. Genet.* **239:** 145–157.

Rafalski, J.A. and S.V. Tingey. 1993. Genetic diagnostics in plant breeding: RAPDs, microsatellites and machines. *Trends Genet.* **9:** 275–280.

Rédei, G.P. 1964. Crossing experiences with polyploids. *Arabidopsis Inf. Serv.* **1:** 13.

———. 1965. Non-mendelian megagametogensis in *Arabidopsis. Genetics* **51:** 857–872.

———. 1973. Extra-chromosomal mutability determined by a nuclear gene locus in *Arabidopsis. Mutat. Res.* **18:** 149–162.

Rédei, G.P. and Y. Hirono. 1964. Linkage studies. *Arabidopsis Inf. Serv.* **1:** 9–10.

Rédei, G.P. and C. Koncz. 1992. Classical mutagenesis. In *Methods in* Arabidopsis *research* (ed. C. Koncz et al.), pp. 16–82. World Scientific, Singapore.

Reiter, R.S., R.M. Young, and P.A. Scolnik. 1992a. Genetic linkage of the *Arabidopsis* genome: Methods for mapping with recombinant inbreds and random amplified polymorphic DNAs (RAPDs). In *Methods in* Arabidopsis *research* (ed. C. Koncz et al.), pp. 170–190. World Scientific, Singapore.

Reiter, R.S., J.G.K. Williams, K.A. Feldmann, J.A. Rafalski, S.V. Tingey, and P.A. Scolnik. 1992b. Global and local genome mapping in *Arabidopsis thaliana* by using recombinant inbred lines and random amplified polymorphic DNAs. *Proc. Natl. Acad. Sci.* **89:** 1477–1481.

Richards, E.J., H.M. Goodman, and F.M. Ausubel. 1991. The centromere region of *Arabidopsis thaliana* chromosome 1 contains telomere-similar sequences. *Nucleic Acids Res.* **19:** 3351–3358.

Richards, E.J., S. Chao, A. Vongs, and J. Yang. 1992. Characterization of *Arabidopsis thaliana* telomeres isolated in yeast. *Nucleic Acids Res.* **20:** 4039–4046.

Röbbelen, G. 1966. Seed aging causes mutant deficit. *Arabidopsis Inf. Serv.* **3:** 7.

Röbbelen, G. and F.J. Kribben. 1966. Erfahrungen bei der Auslese von Trisomen. *Arabidopsis Inf. Serv.* **3:** 16–17.

Robson, P.R.H., G.C. Whitelam, and H. Smith. 1993. Selected components of the shade-avoidance syndrome are displayed in a normal manner in mutants of *Arabidopsis thaliana* and *Brassica rapa* deficient in phytochrome B. *Plant Physiol.* **102:** 1179–1184.

Sears, L.M.S. and S. Lee-Chen. 1970. Cytogenetic studies in *Arabidopsis*. *Can. J. Genet. Cytol.* **12:** 217–223.

Sree-Ramulu, K. and J. Sybenga. 1979. Comparison of fast neutrons and X-rays in respect to genetic effects accompanying induced chromosome aberrations and analysis of translocations in *Arabidopsis thaliana*. *Arabidopsis Inf. Serv.* **16:** 27–34.

———. 1985. Genetic background damage accompanying reciprocal translocations induced by X-rays and fission neutrons in *Arabidopsis* and Secale. *Mutat. Res.* **149:** 421–430.

Stam, P. 1993. Construction of integrated linkage maps by means of a new computer package: JOINMAP. *Plant J.* **3:** 739–744.

Steinitz-Sears, L.M. 1963. Chromosome studies in *Arabidopsis thaliana*. *Genetics* **48:** 483–490.

van der Veen, J.H. and H. Blankestijn-de Vries. 1973. Double reduction in tetraploid *Arabidopsis thaliana*, studied by means of a chlorophyll mutant with a distinct simplex phenotype. *Arabidopsis Inf. Serv.* **10:** 11–12.

Vizir, I.Y. and A.B. Korol. 1990. Sex differences in recombination frequency in *Arabidopsis*. *Heredity* **65:** 379–383.

Williams, J.G.K., R.S. Reiter, R.M. Young, and P.A. Scolnik. 1993. Genetic mapping of mutations using phenotypic pools and mapped RAPD markers. *Nucleic Acids Res.* **21:** 2697–2702.

Williams, J.G.K., A.R. Kubelik, K.J. Livak, J.A. Rafalski, and S.V. Tingey. 1990. DNA polymorphisms amplified by arbitrary primers are useful as genetic markers. *Nucleic Acids Res.* **18:** 6531–6535.

# 6

# Quantitative Genetics

**Randy Scholl**
Arabidopsis Biological Resource Center
Ohio State University
Columbus, Ohio 43210

**Kenneth A. Feldmann**
Department of Plant Sciences
University of Arizona
Tucson, Arizona 85721

**Andrew H. Paterson**
Department of Soil and Crop Science
Texas A & M University
College Station, Texas 77843-2474

Quantitative genetics, by the broadest interpretation, is the study of any trait that can be measured on some quantitative scale. However, quantitative traits traditionally have been defined by geneticists as those traits that can only be analyzed by the application of statistical approaches. These traits have a common set of tendencies: (1) The number of genes affecting them is relatively large and usually undetermined; (2) the effects of individual genes are small; (3) the effect of environment on phenotype is large—at least as large as the overall genetic influence; and (4) as a consequence, the observed distribution of phenotypes in segregating populations is continuous. Many traits important to plant breeding are quantitative, including yield of most crop commodities and traits defining crop quality. Hence, the study of quantitative traits represents one of the most important branches of genetics.

Much of the focus of plant quantitative genetics has been on the improvement of plant breeding practices. Basic studies, aimed at increasing the understanding of the inheritance mechanisms of quantitative traits, naturally complemented the applied studies. In all cases, experimental quantitative genetic studies involve careful statistical analysis of large populations. A major effort has been invested in exploring the most efficient methods for applying artificial selection to populations toward improving quantitative traits. Much of the basic quantitative genetics re-

*Arabidopsis*
© 1994 Cold Spring Harbor Laboratory Press  0-87969-428-9/94 $5 + .00

search has been conducted with crop plants, such as corn, wheat, soybeans, and cotton, and often by plant breeders. Nevertheless, the need has always existed for the study of model organisms, and *Arabidopsis* has, in some cases, been utilized to address important but complex basic problems.

Significant research in almost all major areas of quantitative genetics has been conducted with *Arabidopsis*, although it has remained less prominent than other areas of *Arabidopsis* research. Studies of natural variation, genotype x environment interaction, heterosis, and inter-genotypic competition have all been conducted. We have attempted to summarize the research utilizing *Arabidopsis* in all of these areas, emphasizing the aspects of the *Arabidopsis* system that are useful for quantitative genetics. An analysis of the potential for future *Arabidopsis* quantitative genetic study, specifically, contemporary quantitative trait locus (QTL) research, is included. The latter studies are especially interesting because they represent attempts to identify actual genes affecting quantitative traits (QTLs) through linkage to restriction fragment length polymorphisms (RFLPs) and their subsequent physical study, including cloning.

Several reviews of *Arabidopsis* quantitative genetics have been conducted. Rédei (1970, 1975) has comprehensively reviewed early literature, including quantitative genetics. Recently, Griffing and Scholl (1991) published a perspective on the subject.

## TRAITS OF *ARABIDOPSIS*

Several *Arabidopsis* traits have been utilized for the study of quantitative genetics, and their heritability has been estimated. Heritability is an important quantitative genetic parameter, since it is "the ratio of the genetically caused variability to the total variability of a character in a population" (Rieger et al. 1991). The heritability of a trait is often abbreviated as "$h^2$". Dobrovolná (1967) studied the number of days to appearance of first flower primordia, number of rosette leaves, and "number of rosette leaves per day." All of these were found to have relatively high heritability—mostly near 0.5 (on a scale of 0 to 1). Some similar measures of earliness of flowering have been studied by a number of authors (Härer 1947, 1951; Reinholz 1947; Napp-Zinn 1957; Rédei 1962a,b; Van der Veen 1965). This trait clearly must be treated as quantitative, although estimates of 5–6 genes affecting it have been reported. However, these types of estimates represent minimum numbers, and in no case preclude the existence of larger numbers of genes with small effects working together with a few major genes.

Another trait that has been used successfully in quantitative genetics of *Arabidopsis* is plant growth rate (Langridge and Griffing 1959; Griffing and Langridge 1963; Griffing 1989). The fresh weight (or dry weight) of single plants can be accurately determined. Numerous ecotypes can be shown to grow exponentially between germination and time of flowering. Hence, fresh weight measurement can be made once, at 3 weeks after germination, and utilized as an indicator of relative growth rate. Either plant weight or time to flowering, both being easily measured complex traits which are potentially influenced by many genes acting at different stages of plant development, should be well suited for quantitative genetic study, including QTL mapping research.

### SURVEYS OF NATURAL PHENOTYPIC VARIATION

Numerous ecotypes of *Arabidopsis* have been collected worldwide (Laibach 1943; Röbbelen 1965; Rédei 1970). These have been characterized for many traits, and they have been maintained (Kranz and Kirchheim 1987; Kranz 1990) and widely utilized in experiments. Hybrids and hybrid populations among these have been employed to excellent advantage for quantitative genetic study (see, e.g., Langridge and Griffing 1959).

Characterization of ecotypes has been conducted by many authors. One of the most interesting types of natural variation is flowering time and its response to cold temperatures (see, e.g., Burn et al. 1993). This trait has obvious fitness implications and is currently the subject of intense molecular study. Effects of temperature and vitamin supplementation on growth rates of ecotypes were studied by Langridge and Griffing (1959). They found substantial differences in growth rate (as distinguished from simple differences in plant size due to variations in flowering time). Clear interactions between temperature and vitamin supplement were identified; vitamin supplement was effective in counteracting the negative effects of high temperature on some ecotypes, but not others. Scholl and Dennison (1978) found differences in sulfanilamide resistance among callus cultures of ecotypes. This resistance was shown to be competitively antagonized by para-aminobenzoic acid, the substrate of the enzyme inhibited by sulfanilamide.

Clearly, variation occurs among *Arabidopsis* ecotypes for many traits. As demonstrated by recent research with pathogen resistance and flowering time, some such variations can be isolated and utilized for further genetic analysis. Intra-race variation, although less dramatic, also exists (see, e.g., Dobrovolná 1967; Effmertová and Cetl 1968 and others).

## HETEROSIS AND GENOTYPE X ENVIRONMENT INTERACTION

The extent of both heterosis and genotype x environment interaction are very important considerations to the formulation of plant breeding programs. Heterosis for a trait occurs when a hybrid is superior to its parental lines. The hybrid can be compared to the best parent, but the mid-parent value (mean of the two parents) is a more proper point of reference. When heterosis is great, the use of $F_1$ hybrids by growers is advantageous. Genotype x environment (GxE) interaction, on the other hand, can be defined as the differential performance of genotypes among environments. Stress tolerance and sensitivity are classical examples of this phenomenon. GxE interaction is a simple example of statistical interaction and should be analyzed as such. Many examples of incorrectly claimed interactions exist in the biological literature because proper analysis was not applied.

These two phenomena, often considered separately in quantitative studies, have been examined jointly by several *Arabidopsis* investigators, with the aim of elucidating the biological bases of heterosis. Specifically, the hypothesis that increased resistance to stress or "phenotypic stability" (the tendency for the expression of a trait to be less variable across environments, i.e., to be buffered against fluctuation of performance due to environmental variations) of hybrids may be responsible for actual observed heterosis under field conditions was examined. Specifically, under this hypothesis, the hybrid would be less subject to reductions in performance in adverse environments.

Griffing and Langridge (1963) investigated heterosis and phenotypic stability utilizing hybrids between *Arabidopsis* ecotypes in a 5-parent diallel cross. A "diallel" analysis represents the examination of the progenies of all possible hybrids arising from a set of parents. The number of crosses to be conducted, neglecting reciprocals, is $n(n-1)/2$ where $n$ = the number of parents (or 10 for the experiment in question.) These authors clearly demonstrated the existence of heterosis in *Arabidopsis*, analyzing the parents and $F_2$ progenies. Mean heterosis of growth rate (as defined above for fresh weight of whole plants, expressed on an exponential scale) was relatively small at optimum growth temperature (25°C) but increased dramatically for higher temperatures. Furthermore, the variability of growth rate was shown to be lower for the hybrids than for the parents. Similar results were obtained by Pederson (1968). Pederson examined 15 hybrids generated from a set of 10 parental *Arabidopsis* ecotypes. The effects of water, temperature, and light stress on heterosis were examined. For all three stresses, the heterosis increased with the degree of stress.

The effects of identifiable *Arabidopsis* genes, with known effects, also have been examined for their contribution to heterosis. Rédei (1962b) showed that heterozygotes for both the *angustifolia* and *erecta* loci grow more rapidly than either the corresponding wild type or the homozygous mutants. In both cases, the mutants were substantially reduced in fresh weight, stem length, and number of fruits. The wild-type and mutant lines utilized were isogenic, so the observed heterotic effects should have been due exclusively to the loci studied. Wricke (1955) also demonstrated overdominance effects for a chlorophyll-deficiency locus using tetraploid *Arabidopsis*. This is a significant finding because actual gene actions leading to heterosis are still poorly understood, and, specifically, the relative contributions of independent single-gene effects to heterosis as opposed to epistatic interaction effects among genes are difficult to assess. The existence of specific cases where the heterozygote for an identifiable gene is superior to both homozygotes (overdominance) represents clear evidence for this mode of gene action.

Single-locus heterosis has also been observed for the thiamine auxotrophs utilizing temperature-sensitive, leaky alleles of the *py* locus (Li and Rédei 1969a,b; Rédei and Li 1969). Heterozygous combinations of several of these alleles had greater fresh weight and number of flowers than the corresponding homozygous genotypes. High temperatures tended to increase the heterotic effects between the pairs of alleles.

Heterosis remains, like quantitative inheritance in general, one of the more intriguing unsolved biological phenomena. The study of *Arabidopsis* has contributed substantially to the present understanding. With the tools currently available for genetic analysis of *Arabidopsis*, fruitful study of heterosis should be possible.


**RADIATION GENETICS**

The use of radiation in plant breeding has been summarized by Gottshalk and Wolf (1983). The effect of radiation on the expression of quantitative variation has been studied in a number of plant species. The ultimate goal of these studies is the incorporation of mutagenic methods into plant breeding programs, but this has proven difficult. The best known success in this regard is the work of Gregory and colleagues (see, e.g., Gregory 1955). They incorporated X-ray mutagenesis into the plant breeding process to develop commercial peanut varieties. Radiation-treated and control populations were analyzed for variability of peanut yield, and selection of the best yielding lines was made, resulting in lines superior to those that could have been developed from the nonirradiated populations. It should be noted that the peanut is a true autotetraploid, for which genetic manipulations are substantially more difficult than for diploid species. In *Arabidopsis*, several basic studies of the effects of ionizing radiation on quantitative traits have contributed to the advancement of this field.

One of the most extensive radiation studies was conducted by Brock and co-workers (Brock 1967, 1968, 1970). A homozygous population was subjected to thermal neutron and $\gamma$ irradiation. Dry seeds were subjected to doses of $\gamma$ irradiation ($^{60}$Co) up to a maximum of 100 krad (rate = 336 rad/min) or thermal neutron doses up to $1.61 \times 10^{14}$ $N^{th}/cm^2$. In addition to many qualitative variants of different phenotypes, variation of a quantitative nature was induced for time to flowering. The mean flowering time was delayed approximately 0.6 days by $\gamma$ irradiation (mean time to flowering of the control, Estland, was 17.57 days). The variance of flowering time induced by thermal neutron irradiation was 0.55 (i.e., standard deviation of 0.74 days), with all visible mutants removed from the population. With visible mutants removed, the mean fresh weight was not significantly affected by irradiation. However, the population variance due to mutagenic effects was fivefold greater than the control variance. A negative correlation of induced variation between time to flowering and fresh weight was observed. The results indicated that the theoretical genetic predictions pertaining to the effects of radiation were generally upheld: The means were decreased (reduced growth rate = later flowering time) in irradiated populations and variance increased, with some segregants of the irradiated groups having superior performance to any lines of the control. Daly (1973) also examined the effect of $\gamma$ irradiation (maximum dose = 150 kR, rate = 410 kR/hr) of dry seeds on flowering time. Flowering time was increased a maximum of approximately one day by both fast neutron and $\gamma$ irradiation. The variance of flowering time was increased approximately twofold ($h^2 \sim 0.6$) for the highest doses. The increase of variance was proportional to dose applied for this trait. Lawrence (1968) conducted a selection experiment on flowering time of *Arabidopsis* in conjunction with the application of $\gamma$ irradiation. The rate of increase of flowering time per generation in response to the selection was increased in proportion to the dosage applied. In both early and late flowering selections, the numbers of seeds per silique was decreased by irradiation. Hence, the findings of all of these studies indicate a measurable increase in quantitative genetic variation resulting from ionizing radiation.

## INTERGENOTYPIC COMPETITION

Competitive effects in plant communities, including crop fields, can be significant. Interesting competitive effects can occur in bulk-propagated populations, including positive effects on overall seed yield (Suneson and Weibe 1942; Jensen 1952). Likewise, the spread of pathogens can be impeded by genetic diversity of crops. As shown theoretically by Griffing (1967, 1968, 1981), the overall competitive effects associated with genetic mixtures can be complex, and results totally unexpected from studies of pure stands, including strong negative effects, can occur. Hence, this represents an important although relatively undeveloped area of genetics. Furthermore, the potential complexity of the effects makes the utilization of models such as *Arabidopsis* necessary if understanding is to be achieved.

Griffing and Zsiros (1971) and Griffing (1989) analyzed the competitive effects in the context of varying environments and competition between hybrid genotypes and their parents. Plants were grown singly or in pairs in tubes to assess different planting densities and competitive effects. In one case, the ecotypes Ws, Wil, and their $F_2$ hybrid were grown at different temperature and nutrient levels. Fresh weight was measured at four harvest times ranging from 7 days to 16 days after planting. For each of the factorial combinations of the above treatments, 20 to 30 individual plants were observed, generating several thousand observations.

This experimental system allows analysis of simple heterosis, the additional heterotic effects resulting from environmental disturbance and the influence of competing genotypes on heterosis. The influence of the external factors on heterosis was as follows. Heterosis increased as temperature increased. From the experimental design, two types of group-dependent heterosis can be defined: (1) "direct" heterosis, which is the difference between $F_2$s and mid-parents $(F_2 - \{P_1 + P_2\}/2)$ in competition with a single genotype, and (2) "associate" heterosis, which is the difference between $F_2$s and the mid-parent competitors as they affect a single genotype being measured. Both types of heterosis were observed. Interestingly, both direct and associate heterosis were lower with the $F_2$ as a competing genotype as opposed to either parent at standard temperature but were much greater at high temperature. Hence, complex competitive effects, involving interaction between environmental factors and the competing genotype, were observed. The implications of these effects are significant for the design of quantitative experiments, especially when genotypes are being evaluated in close quarters, where competitive effects are possible. From such experiments, it also may be possible to predict conditions where heterosis is most likely to be greatest. For example, moderate temperatures and low nutritional status would in all probability not be conducive to heterotic effects. Similarly, hybrids are likely to perform best under widely varying, as well as high, temperatures.

## SELECTION EXPERIMENTS

One of the primary uses of model systems in quantitative genetics is the testing of genetic hypotheses by selection experiments. Typically, some segregating group is chosen as the base population, individuals of this population are assessed for the trait(s) of interest, and a fixed percentage consisting of those individuals scoring best for the trait(s) under selection are saved to be used as parents for the next generation. After intermating of these individuals, the progenies are evaluated as for the first generation, and a second cycle of selection is thus initiated. Selection can be practiced for any number of generations, and improvement for the selected traits should continue as long as the genetic variation for the trait has not been exhausted. The relative success of the selection represents the extent to which the selected trait is actually under genetic control.

One of the main difficulties in studying quantitative genetics with *Arabidopsis* is the effort required for the large numbers of cross-fertilizations required for standard selection experiments. Nevertheless, some selection experiments have been conducted.

Lawrence (1968) analyzed selection for flowering time based on genetic variation generated by repeated radiation treatments over eight generations of self-fertilization. Response to selection was not large. Song (1974) studied isozyme frequencies in *Arabidopsis* populations which had been subjected to either individual or group selection for fresh weight. The base population was an $F_2$ of a cross between two ecotypes. The population was self-fertilized through the $F_8$ generation under three regimes: no artificial selection (control), individual selection for increased weight, and selection among groups for fresh weight. In all populations, changes in the frequency of allozyme alleles occurred. The effects of natural selection and artificial selection could be resolved; in some cases the two acted in concert and in others they acted in opposition. The association of the response of the above-cited biochemical markers to the selected quantitative trait would seem to bode well for the application of QTL approaches utilizing RFLPs, rapid amplified polymorphic DNAs (RAPDs), or allozymes to *Arabidopsis*.

## QTLs AND MOLECULAR POLYMORPHISM

The recent development of several genetic maps of *Arabidopsis* based on DNA markers (Chang et al. 1988; Nam et al. 1989; Reiter et al. 1992) provides the essential tools for undertaking molecular analysis of quantitative inheritance. In light of the classical quantitative genetics research that has been done in *Arabidopsis*, as described in this chapter, it is somewhat surprising that only a limited amount of "molecular quantitative genetics" has been undertaken in this system. No doubt, the future holds an important role for *Arabidopsis* in elucidating basic principles underlying quantitative inheritance. Both the theoretical basis for this approach and the specific potential for its application to *Arabidopsis* are addressed here.

## Context of Molecular Analysis of Quantitative Traits

For a long time following the rediscovery in 1900 of Mendel's work, it was not clear how such discrete genetic factors could give rise to phenotypes that appeared to vary continuously (e.g., height, weight, yield, or other traits) (Provine 1971). In fact, Bateson, one of the leaders of the ensuing avalanche of genetic research, maintained that quantitative traits were not under genetic control. Such variation was ultimately explained by the independent actions of many discrete genetic factors, each having only a small effect on the overall phenotype. These "polygenes" (Thoday 1961) or QTLs (Geldermann 1975) were hypothesized to differ little from genes affecting simply inherited traits, i.e., they were expected to exhibit the fundamental Mendelian properties of segregation and recombination. Consequently, it seemed likely that one might identify linkages between genetic markers and QTLs and thus determine the locations of individual QTLs.

Linkage between a genetic marker and a QTL in plants was first reported by Sax (1923), who found that *Phaseolus* genotypes with different seed coat colors also differed in average seed size. Associations of many different traits with visible markers have been found in many other species, using morphological markers and isozymes (for a more detailed list, see Paterson et al. 1992).

Many of these early researchers had rather few genetic markers with which to work; this posed a serious limitation, as it could be shown that QTLs were in the vicinity of a particular marker, but there were seldom enough additional markers nearby to allow the location of the QTL to be pinpointed. As genetic maps came to include more markers, it became possible to more precisely estimate the location of a QTL, by studying several markers along a chromosome (Thoday 1961; Tanksley et al. 1982); Thoday (1961) emphasized that "the main practical limitation ... (to localizing QTLs) ... seems to be the availability of suitable markers." This limitation was remedied by the construction of complete RFLP linkage maps, permitting systematic searches of an entire genome for QTLs influencing a trait (Paterson et al. 1988). Furthermore, new algorithms for QTL mapping minimized the number of individuals as well as the number of genetic markers needed to map QTLs (Weller 1986;

Lander and Botstein 1989; Knapp et al. 1990).

Using DNA markers, QTLs can be described by their chromosomal location, gene dosage effect, phenotypic effect(s), and sensitivity to environment (Paterson et al. 1991). Such a description has long been possible for genes affecting simply inherited traits; however, only with the availability of high-density genetic maps has it become possible to obtain this information for individual QTLs. With a "complete" map of genetic markers, one can employ powerful statistical methods (Lander and Botstein 1989) to determine "likelihood intervals" for the locations of QTLs (Paterson et al. 1988). Having mapped a QTL to a likelihood interval, one might narrow the location of the QTL to perhaps 1/50–1/100 of a chromosome (about 0.1% of the genome in tomato) by comparing individuals which carry different portions of the likelihood interval (Paterson et al. 1990).

QTL mapping permits detailed investigation of a number of genetic phenomena that previously could only be studied at the whole-plant level. Transgression, the expression of more extreme phenotypes in the progeny than were observed in the parents, can be accounted for by alleles with opposing effects on a trait occurring in the same parent (Paterson et al. 1988). Heterosis, the superiority of $F_1$ hybrid individuals relative to either of their parents, can be accounted for by particular genomic regions (Stuber et al. 1992), although sufficient genetic map resolution has not yet been achieved to determine whether heterosis is due to single genes with true "heterozygote advantage" (East 1908; Shull 1908, 1911) or to groups of closely linked genes having simple dominant alleles in *trans* configuration (Bruce 1910; Keeble and Pellew 1910). Correlation between traits, a persistent problem in classical plant breeding, especially in introgression of traits from wild relatives, can be shown in some cases to be due to linkage of genes with desirable effects to genes with undesirable effects (Miller and Rawlings 1967). Epistasis, or nonlinear interaction between different genes, has long been believed to have a pronounced influence on phenotype (Wright 1968; Allard 1988); however, genetic mapping experiments have documented only a few cases of epistasis among many candidates that have been studied (Spickett and Thoday 1966; Edwards et al. 1987; Allard 1988; Paterson et al. 1988, 1990, 1991; Tanksley and Hewitt 1988; Weller et al. 1988). This indicates either that epistasis is less prominent than previously thought, or that genetic mapping experiments have insufficient statistical power to detect it.

To achieve a mechanistic understanding of quantitative inheritance, it will ultimately be necessary to clone individual determinants of quantitative variation. In a few fortuitous cases, this may be accomplished as a "fringe benefit" of cloning mutations with discrete effects and identifying additional alleles at these genetic loci with lesser effects. However, reverse genetics (Orkin 1986) offers a more certain path to cloning genetic determinants of traits that are poorly understood. The genetic maps, yeast artificial chromosome (YAC) libraries, and collections of phenotypic variants available in *Arabidopsis* provide ample opportunity to pursue map-based cloning of complex traits, thus extending the reach of molecular cloning into the class of traits most fundamental to agricultural productivity.

### QTL Analysis in *Arabidopsis*

The path to cloning of a QTL in *Arabidopsis* is clear, albeit tortuous. Assuming that no information is available concerning biochemical pathways or enzyme-mediated functions likely to be associated with the action of a QTL, a reverse genetics approach is warranted. The starting point for such an approach would in all probability include the use of a linkage map of DNA markers to determine an approximate genetic map position (likelihood interval) of the QTL(s) of interest (as per Paterson et al. 1988). Such a likelihood interval might span 20 cM, or perhaps 3 Mb in *Arabidopsis*.

Clearly, such an interval is too large to clone, so fine resolution of the map position of the gene would be warranted. Methods for high-resolution mapping of QTLs have been described previously (Paterson et al. 1990) but require a large number of DNA markers in the target region and prior knowledge of the linear order of these closely linked markers. Consequently, it would in all probability be necessary to enrich the target region for DNA markers, using recently developed strategies based on screening synthetic pools of genomic DNA from segregating progenies (Giovannoni et al. 1991; Michelmore et al. 1991) with polymerase chain reaction (PCR)-based candidate markers such as RAPDs (Williams et al. 1990) or arbitrarily primed PCR (Welsh and McClelland 1990). Since these markers would all be closely linked, it would probably be necessary to utilize new techniques for high-resolution linkage analysis (Churchill et al. 1993; S.P. Kowalski et al., in prep.) to determine their linear order.

After the region of the QTL was enriched for DNA markers and the QTL was fine-mapped to a smaller region of the genome, e.g., 1–3 cM, long-range restriction mapping of the relevant genomic region might then be employed to determine the feasibility of a "chromosome walk" across the gene of interest. Should the distance prove tractable (less than 500 kb), one could initiate a walk from YAC islands identified by flanking RFLP markers (Hwang et al. 1991), using high-resolution linkage analysis (Churchill et al. 1993; S.P. Kowalski et al., in prep.) of YAC ends to establish direction of the walk.

Having identified YACs spanning the target region, one faces the sticky problem of which gene, among the many likely to be present in even a single *Arabidopsis* YAC, is responsible for the phenotype. The problem is further complicated by the fact that a QTL, by definition, does not condition a discrete phenotype, thus the transformant carrying the target gene is not immediately obvious, unlike the classical mutant complementation scenario. We anticipate that one solution to this problem is a progeny test; that is, the phenotype of a transformant would be determined by the average phenotype of a group of its self-pollinated progeny. The number of individuals required can be calculated from the (previously determined) phenotypic effect of the target QTL and a predetermined estimate of the effect of nongenetic factors. Important among nongenetic factors might be somaclonal variation as a result of conducting transformations in tissue culture; thus, it would be preferable to use seed transformation (Feldmann et al. 1989), which seems to minimize somaclonal variation.

Although technological advances such as more detailed genetic maps, more extensive YAC islands and cosmid contigs, and whole-YAC transformation will expedite various steps of the above-described scenario (Paterson and Wing 1993), it is currently feasible, in principle, to consider cloning of QTLs in *Arabidopsis*. In recent experiments, a suitable target QTL has been mapped, a high-resolution mapping population has been constructed, and enrichment of the target region for DNA markers has begun (S.P. Kowalski et al., unpubl.).

The importance of quantitative genetics in agriculture, medicine, and evolution places considerable priority on developing an understanding of its mechanism. The next logical step toward this objective is to clone a QTL, and among plant species, there can be little doubt that *Arabidopsis* is the system of choice. We expressly do not suggest that *Arabidopsis* will directly permit identification of genes associated with processes explaining quantitative variation in major crops (e.g., maize or soybean yield, or cotton fiber quality). In fact, this would be a very speculative hypothesis. However, we do propose that *Arabidopsis* can be exploited as a facile model for "molecular" quantitative genetics, in addition to its widely recognized role in qualitative genetics and development. The necessary base of the classical quantitative genetic research and the molecular tools to address the problems are both in place. It can be anticipated that basic analysis of *Arabidopsis* quantitative genetics and QTLs will proceed with success.

## REFERENCES

Allard, R.W. 1988. Genetic changes associated with the evolution of adaptedness in cultivated plants and their wild progenitors. *J. Hered.* **79:** 225–238.

Brock, R.D. 1967. Quantitative variation in *Arabidopsis thaliana* induced by ionizing radiations. *Radiat. Bot.* **7:** 193–203.

————. 1968. Induced quantitatively inherited variation in *Arabidopsis thaliana*. Mutation in plant breeding. *Int. Atomic Energy Agency Panel Proc. Ser.* **2:** 57–58.

————. 1970. Mutations in quantitatively inherited traits induced by neutron irradiation. *Radiat. Bot.* **10:** 209–223.

Bruce, A.B. 1910. The Mendelian theory of heredity and the augmentation of vigor. *Science* **32:** 627–628.

Burn, J.E., D.J. Bagnall, J.D. Metzger, E.S. Dennis, and W.J. Peacock. 1993. DNA methylation and the initiation of flowering. *Proc. Natl. Acad. Sci.* **90:** 287–291.

Chang, C., J.L. Bowman, A.W. DeJohn, E.S. Lander, and E.M. Meyerowitz. 1988. Restriction fragment length polymorphism linkage map for *Arabidopsis thaliana*. *Proc. Natl. Acad. Sci.* **85:** 6856–6860.

Churchill, G.A., J.J. Giovannoni, and S.D. Tanksley. 1993. Pooled-sampling mapping makes high-resolution mapping practical with DNA markers. *Proc. Natl. Acad. Sci.* **90:** 16–20.

Daly, K. 1973. Quantitative variation induced by gamma rays and fast neutrons in *Arabidopsis thaliana*. *Radiat. Bot.* **13:** 149–154.

Dobrovolná, J. 1967. The variability of developmental characters in natural populations of *Arabidopsis thaliana* (L.) Heynh. *Arabidopsis Inf. Serv.* **4:** 6–7.

East, E.M. 1908. Inbreeding in corn. *Rep. Conn. Agric. Exp. Stn.*, pp. 419–428.

Edwards, M.D., C.W. Stuber, and J.F. Wendel. 1987. Molecular-marker-facilitated investigations of quantitative trait loci in maize. I. Numbers, genomic distribution, and types of gene action. *Genetics* **116:** 113–125.

Effmertová, E. and I. Cetl. 1968. The behaviour of progenies derived from a "winter annual" population of *Arabidopsis thaliana* (L.) Heynh. *Arabidopsis Inf. Serv.* **5:** 16.

Feldmann, K.A., M.D. Marks, M.L. Christianson, and R.S. Quatrano. 1989. A dwarf mutant of *Arabidopsis* generated by T-DNA insertion mutagenesis. *Science* **243:** 1351–1354.

Geldermann, H. 1975. Investigations on inheritance of quantitative characters in animals by gene markers. I. Methods. *Theor. Appl. Genet.* **46:** 319–330.

Giovannoni, J.J., R.A. Wing, M.W. Ganal, and S.D. Tanksley. 1991. Isolation of molecular markers from specific chromosomal intervals using DNA pools from existing mapping populations. *Nucleic Acids Res.* **19:** 6553–6558.

Gottschalk, W. and G. Wolf. 1983. Induced mutations in plant breeding. *Monogr. Theor. Appl. Genet.* **7:** 238.

Gregory, W.C. 1955. X-ray breeding of peanut (*Arachis hypogea* L.). *Agron. J.* **47:** 396–399.

Griffing, B. 1967. Selection in reference to biological groups. I. Individual and group selection applied to populations of unordered groups. *Aust. J. Biol. Sci.* **20:** 127–139.

————. 1968. Selection in reference to biological groups. II. Consequences of selection in groups of one size when valuated in groups of a different size. *Aust. J. Biol. Sci.* **21:** 1163–1170.

————. 1981. A theory of natural selection incorporating interactions among individuals. I.-V. *J. Theor. Biol.* **89:** 636–710.

————. 1989. Genetic analysis of plant mixtures. *Genetics* **122:** 943–956.

Griffing, B. and J. Langridge. 1963. Phenotypic stability of growth in the self-fertilized species, *Arabidopsis thaliana*. *Natl. Acad. Sci., Natl. Res. Council. Publ.* **982**: 368–394.

Griffing, B. and R.L. Scholl. 1991. Qualitative and quantitative studies of *Arabidopsis thaliana*. *Genetics* **129**: 605–609.

Griffing, B. and E. Zsiros. 1971. Heterosis associated with genotype environment interactions. *Genetics* **68**: 443–455.

Härer, L. 1947. Die Vererbung des Bluhalters fruher und spater sommereinjahriger Rassen von *Arabidopsis thaliana* (L.) Heynh. *Fiat Rep.* **1090**: 5–27.

————. 1951. Die Verebung des Bluhalters fruher und spater sommereinjahriger Rassen von *Arabidopsis thaliana* (L.) Heynh. *Deitr. Biol. Pflanzen.* **28**: 1–35.

Hwang, I., S. Hanley, B.M. Hauge, and H.M. Goodman. 1991. Identification of the map position of YAC clones comprising one-third of the *Arabidopsis* genome. *Plant J.* **1**: 367–374.

Jensen, N.F. 1952. Intravarietal diversification in oat breeding. *Agron. J.* **44**: 30–34.

Keeble, F. and C. Pellew. 1910. The mode of inheritance of stature and flowering time in peas (*Pisum sativum*). *J. Genet.* **1**: 47–56.

Knapp, S.J., W.C. Bridges, Jr., and D. Birkes. 1990. Mapping quantitative trait loci using molecular marker linkage maps. *Theor. Appl. Genet.* **79**: 583–592.

Kranz, A.R. 1990. Additions and corrections to the seed bank listing. *Arabidopsis Inf. Serv.* **27**: 89.

Kranz, A.R. and B. Kirchheim. 1987. Genetic resources in *Arabidopsis*. *Arabidopsis Inf. Serv.* **24**: 1–167.

Laibach, F. 1943. *Arabidopsis thaliana* (L.) Heynh. als objekt für genetische und entwicklungsphysioligische untersuchungen. *Bot. Archiv.* **44**: 265–455.

Lander, E.S. and D. Botstein. 1989. Mapping Mendelian factors underlying quantitative traits using RFLP linkage maps. *Genetics* **121**: 185–199.

Langridge, J. and B. Griffing. 1959. A study of high temperature lesions in *Arabidopsis thaliana*. *Aust. J. Biol. Sci.* **12**: 117–135.

Lawrence, C.W. 1968. Radiation-induced polygenic mutation in *Arabidopsis thaliana* (L.) Heynh: Selection for flowering time. *Heredity* **23**: 321–337.

Li, S.L. and G.P. Rédei. 1969a. Allelic complementation at the pyrimidine (*py*) locus of the crucifer *Arabidopsis*. *Genetics* **62**: 281–288.

————. 1969. Direct evidence for models of heterosis provided by mutants of *Arabidopsis* blocked in the thiamine pathway. *Theor. Appl. Genet.* **39**: 68–72.

Michelmore, R.W., I. Paran, and R.V. Kesseli. 1991. Identification of markers linked to disease resistance genes by bulked segregant analysis: A rapid method to detect markers in specific genomic regions using segregating populations. *Proc. Natl. Acad. Sci.* **88**: 9828–9832.

Miller, P.A. and J.O. Rawlings. 1967. Breakup of initial linkage blocks through intermating in a cotton breeding population. *Crop Sci.* **7**: 199–204.

Nam, H.-G., J. Giraudat, B.D. Boer, F. Moonan, W.D.B. Loos, B.M. Hauge, and H.M. Goodman. 1989. Restriction fragment length polymorphism linkage map of *Arabidopsis thaliana*. *Plant Cell.* **1**: 699–705.

Napp-Zinn, K. 1957. Weitere Untersuchungen uber die Beziehungen zwischen Atmungsintensitat und Bluhalter. *Planta* **48**: 683–695.

Orkin, S.H. 1986. Reverse genetics and human disease. *Cell* **47**: 845–850.

Paterson, A.H. and R.A. Wing. 1993. Genome mapping in plants. *Curr. Opin. Biotechnol.* **4**: 142–147.

Paterson, A.H., S.D. Tanksley, and M.E. Sorrells. 1992. DNA markers in crop improvement. *Adv. Agron.* **46**: 39–90.

Paterson, A.H., J.W. Deverna, B. Lanini, and S.D. Tanksley. 1990. Fine mapping of quantitative trait loci using selected overlapping recombinant chromosomes, in an interspecies cross of tomato. *Genetics* **124:** 735–742.

Paterson, A.H., E.S. Lander, J.D. Hewitt, S. Peterson, S.E. Lincoln, and S.D. Tanksley. 1988. Resolution of quantitative traits into Mendelian factors, using a complete linkage map of restriction fragment length polymorphisms. *Nature* **335:** 721–726.

Paterson, A.H., S. Damon, J.D. Hewitt, D. Zamir, H.D. Rabinowitch, S.E. Lincoln, E.S. Lander, and S.D. Tanksley. 1991. Mendelian factors underlying quantitative traits in tomato: Comparison across species, generations, and environments. *Genetics* **127:** 181–197.

Pederson, D.G. 1968. Environmental stress, heterozygote advantage and genotype-environment interaction in *Arabidopsis*. *Heredity* **23:** 127–138.

Provine, W.B. 1971. *The origins of theoretical population genetics*. University of Chicago Press, Illinois.

Rédei, G.P. 1962a. Genetics block of "vitamin thiazole" synthesis in *Arabidopsis*. *Genetics* **47:** 979.

————. 1962b. Single locus heterosis. *Z. Vererbungsl.* **93:** 164–170.

————. 1970. *Arabidopsis thaliana* (L.) Heynh. A review of the genetics and biology. *Bibliogr. Genet.* **20:** 1–151.

————. 1975. *Arabidopsis thaliana*. In *Handbook of genetics* (ed. R.C. King), pp. 151–180. Plenum Press, New York.

Rédei, G.P. and S.L. Li. 1969. Physiological resolution of the *py* locus of *Arabidopsis* by means of allelic complementation. *Proc. Int. Bot. Congr. (Abstr.)* **11:** 178.

Reinholz, E. 1947. Auslosung von Rontgenmutationen bei *Arabidopsis thaliana* (L.) Henyh. und ihre Bedeutung fur die Pflanzenzuchtung und Evolutionstheorie. *Fiat Rep.* **1006:** 1–70.

Reiter, R.S., J.G.K. Williams, K.A. Feldmann, J.A. Rafalski, S.V. Tingey, and P.A. Scolnik. 1992. Global and local genome mapping in *Arabidopsis thaliana* by using recombinant inbred lines and random amplified polymorphic DNAs. *Proc. Nat. Acad. Sci.* **89:** 1477–1481.

Rieger, R., A. Michaelis, and M.M. Green. 1991. *Glossary of genetics*, 5th edition. Springer-Verlag, Berlin.

Röbbelen, G. 1965. The Laibach standard collection of natural races. *Arabidopsis Inf. Serv.* **2:** 36–47.

Sax, K. 1923. The association of size differences with seed-coat pattern and pigmentation in *Phaseolus vulgaris*. *Genetics* **8:** 552–560.

Scholl, R.L. and K.M. Dennison. 1978. Sensitivity of cultured tissue of *Arabidopsis thaliana* races to sulfanilamide. *Physiol. Plant.* **43:** 321–325.

Shull, G.H. 1908. The composition of a field of maize. *Rep. Am. Breed. Assoc.* **4:** 296–301.

————. 1911. The genotypes of maize. *Am. Nat.* **45:** 234–252.

Song, C.-M. 1974. "Isozyme variation in *Arabidopsis thaliana*." Ph.D. thesis, Ohio State University, Columbus.

Spickett, S.G. and J.M. Thoday. 1966. Regular responses to selection 3: Interaction between located polygenes. *Genet. Res.* **7:** 96–121.

Stuber, C.W., S.E. Lincoln, D.W. Wolff, T. Helentjaris, and E.S. Lander. 1992. Identification of genetic factors contributing to heterosis in a hybrid from two elite maize inbred lines using molecular markers. *Genetics* **132:** 823–839.

Suneson, C.A. and G.A. Weibe. 1942. Survival of barley and wheat varieties in mixtures. *Agron. J.* **34:** 1052–1056.

Tanksley, S.D. and J.D. Hewitt. 1988. Use of molecular markers in breeding for soluble solids in tomato—A re-examination. *Theor. Appl. Genet.* **75:** 811–823.

Tanksley, S.D., H. Medina-Filho, and C.M. Rick. 1982. Use of naturally-occurring enzyme variation to detect and map genes controlling quantitative traits in an interspecific backcross of tomato. *Heredity* **49:** 11–25.

Thoday, J.M. 1961. Location of polygenes. *Nature* **191:** 368–370.

Van der Veen, J.H. 1965. Genes for late flowering. *Arabidopsis Inf. Serv.* **2:** 5–6.

Weller, J.I. 1986. Maximum likelihood techniques for the mapping and analysis of quantitative trait loci with the aid of genetic markers. *Biometrics* **42:** 627–640.

Weller, J.I., M. Soller, and T. Brody. 1988. Linkage analysis of quantitative traits in an interspecific cross of tomato (*L. esculentum* x *L. pimpinellifolium*) by means of genetic markers. *Genetics* **118:** 329–339.

Welsh, J. and M. McClelland. 1990. Fingerprinting genomes using PCR with arbitrary primers. *Nucleic Acids Res.* **18:** 7213–7218.

Williams, J.G.K., A.R. Kubelik, K.J. Livak, J.A. Rafalski, and S.V. Tingey. 1990. Oligonucleotide primers of arbitrary sequence amplify DNA polymorphisms which are useful as genetic markers. *Nucleic Acids Res.* **18:** 6531–6535.

Wricke, G. 1955. Ein Fall von Superdominanz bei einer Experimente 11 hergestellten autotetraploiden von *Arabidopsis thaliana*. *Z. Induk. Abstammungs. Vererbungsl.* **87:** 47–64.

Wright, S. 1968. *Evolution and the genetics of populations.* University of Chicago Press, Illinois.

# 7

# Mutagenesis in *Arabidopsis*


**Kenneth A. Feldmann**
Department of Plant Sciences, University of Arizona
Tucson, Arizona 85721

**Russell L. Malmberg**
Botany Department, University of Georgia
Athens, Georgia 30602

**Caroline Dean**
Department of Molecular Genetics, B.B.S.R.C., I.P.S.R.
John Innes Centre, Norwich, NR4 7UJ, United Kingdom


Mutagenesis can be described as the process of inducing any heritable change in the genetic material which is subsequently transmitted to daughter cells where it gives rise to a mutant cell or individual (Rieger et al. 1976). The use of classical mutagenic agents, e.g., ethylmethane sulfonate, nitrosoguanidine, nitrosourea, and X-ray, has resulted in thousands of mutants in *Arabidopsis* (see McKelvie 1962; Rédei 1970). Considerable insight has been gained about plant developmental and physiological processes from the characterization of some of these mutants. Recently developed technologies make it possible to clone genes from interesting mutants. The use of restriction fragment length polymorphism (RFLP) maps (Chang et al. 1988; Nam et al. 1989; Reiter et al. 1992), for example, makes it possible, although laborious, to isolate genes mutated with these agents (Arondel et al. 1992; Giraudat et al. 1992; Chang et al. 1993). Positional cloning will become increasingly expedient as more molecular markers are positioned on the map and as the physical map nears completion. Another recent technique, genomic subtraction, can be utilized to isolate the affected gene when deletion mutagens are employed, e.g., diepoxybutane and γ irradiation (Sun et al. 1992). However, current technical limitations may make it difficult to isolate large numbers of genes with this technique. Chemical and physical agents and their mutagenic activities are discussed because of their historical and continued usefulness.

Mutagens that have found wide popularity among molecular biologists are insertion mutagens, T-DNAs or transposons (van Sluys et al. 1987; Feldmann 1991). These mutagens provide a relatively facile means of isolating the affected gene (see, e.g., Yanofsky et al. 1990; Deng et al. 1992; Kieber et al. 1993; Tsay et al. 1993). Much of this review is directed toward these two mutagenic agents. In addition, we describe a procedure that will make it possible to identify an insertion mutant, based on insertion polymorphisms, for a specific DNA sequence of interest.

## CLASSICAL MUTAGENESIS

Most of the mutations available in *A. thaliana* have been induced by chemical mutagens or by some form of ionizing radiation. Furthermore, since the frequency of insertion-induced mutations is generally lower than with chemical mutagenesis or radiation, it may be advantageous to first define a locus by these classical methods. Once one knows that a particular kind of mutation is possible, it is then practical to plan a strategy for isolation of the corresponding gene. Chemical mutagenesis also offers some more subtle advantages, including the potential ability to identify temperature-sensitive mutants and to find a full range of mutant alleles with degrees of phenotypic severity. Chemical mutagenesis allows the investigator full flexibility in choice of starting strain and initial genotype and is easy and relatively inexpensive to perform. Finally, the cloning of genes defined only by point mutations will no longer be enormously slower than insertion-induced mutants, as the map-based cloning strategies become more powerful. One can genetically map a new phenotype after several sets of polymerase chain reactions (PCR) (Konieczny and Ausubel 1993; Williams et al. 1993), then frequently look up the yeast artificial chromosome (YAC) or contig clones that exist for that region.

### Choice of Mutagen and Dose

Good references for mutagens and appropriate doses of each are Rédei (1970) and Rédei and Koncz (1992). Ethyl methanesulfonate (EMS) topped the list of chemical mutagens for its high efficiency in inducing mutations; it has a relatively high induced mutation rate with less of the side effect of lethality/sterility, as compared to other mutagens. Some other methane sulfonates, hydroxyethyl methane sulfonate and methoxy methanesulfonate, worked as well as EMS, although not all tested methane sulfonates were as efficient. Nitrosomethylurea (NMU) also worked well, as did methyl nitrosoguanidine. Base analogs (bromo-deoxyuridine) did not work nearly as well as the alkylating agents EMS and NMU. Rédei (1970) also summarized information that fast-neutron bombardment was a more effective irradiation mutagen than X-rays.

Most mutagens have a variety of physiological and developmental effects on the just-treated plants, in addition to creating new mutants; these can include lethality and sterility. These effects need to be considered in both choosing the mutagen and selecting the appropriate dose of the mutagen. A simple theoretical model (Malmberg 1993) predicts that the optimum yield of mutations for a given mutagen should be the dose when 37% of the $M_1$ seeds survive to reproduction and contribute to the $M_2$. This may be a useful starting point, but the ultimate measure of success is the production of mutants. Two traits for calibrating the effectiveness of a mutagen are the frequencies of embryonic lethals and the production of albino seedlings; both traits are readily scored and should be abundant, as they represent multilocus targets. Embryonic lethals can be scored directly in the siliques of the $M_1$ plants, appearing as a 3:1 segregation of pale white embryos instead of the normal green. After EMS mutagenesis, the percentage of $M_1$ plants segregating embryonic lethals can be as high as 5–10% (D. Meinke, pers. comm.). The frequency of albinos in the $M_2$ may be as high as 1 in 250 after a successful EMS mutagenesis (R. Wilson, pers. comm.). Haughn and Somerville (1987) have estimated that an $M_1$ population size of roughly 125,000 plants, after EMS mutagenesis, will effectively saturate the genome for all possible EMS-inducible base changes; Rédei (1975) describes similar calculations to optimize mutant isolation strategies.

EMS has been the mutagen of choice for most investigators because it works well and it induces a wide spectrum of allele severities, running the gamut from amorph to hypermorph. Mutagens that induce deletions have the advantage that they are, almost by definition, amorphs. In addition, subtractive cloning methods have advanced to the point that it is possible to think about cloning a deletion mutation by direct comparison of the DNA with wild type. The chemicals diepoxyoctane and diepoxybutane, as well as ionizing irradiation, have been reported to cause deletion mutations in various eukaryotes (Rédei 1970; Brockman et al. 1984). At least one gene in *Arabidopsis* has already been cloned by subtractive hybridization, using a deletion mutant generated by fast-neutron bombardment mutagenesis (Sun et al. 1992). Since different mutagens may induce different patterns of base changes, the use of several different mutagens on separate seed populations may allow identification of a wider spectrum of alleles.

Lehle Seeds (6531 North Camino Katrina, Tucson, Arizona 85718) is a commercial supplier of *Arabidopsis* seeds, including some mutagenized with EMS and fast-neutron bombardment. This source has been useful to a number of investigators. A drawback of a commercial source is that one does not have as much control of the initial genotype used and the manner in which the $M_2$ seeds are pooled and collected for analysis, as discussed below.

## Plants from Treated Seeds Are Chimeras

Mutagenesis of *A. thaliana* is usually performed by treating the seed with the mutagen, letting the surviving seeds germinate, and then recovering the progeny for analysis. The generation that grows from the mutagenized seed is called the $M_1$. Progeny collected from the $M_1$ plants are the $M_2$ generation. Each cell within the original treated seed is mutagenized in its own unique way. The $M_1$ plant that sprouts from the seed is thus a chimera of genetically different cell lineages, each of which we would expect to be heterozygous for whatever new mutations have been induced. The population of $M_2$ seeds from a given $M_1$ plant contains both heterozygotes and homozygotes for any mutation that was induced in a cell whose descendants gave rise to gametes.

Fortunately, a small number of cells in the mature seed give rise to the body of the plant and subsequently contribute to the next generation. Based on patterns of genetic segregation in the $M_2$, Li and Rédei (1969) have estimated that there are two precursor cells to the gametes in the *Arabidopsis* seed, a number referred to as the genetically effective cell number (GECN). Several fate maps of the shoot apical meristem in the *Arabidopsis* seed have recently been completed (Furner and Pumfrey 1992; Irish and Sussex 1992). Although the studies show that there was significant variability and plasticity in cell fates, they also permit an estimate that from 1 to 4 cells in the seed give rise to the inflorescence. This observed variability in cell number is due to underlying biological variability, to the extent that the authors of these papers prefer to call their results probability maps rather than fate maps. Thus, the gametes of any given *Arabidopsis* plant may be derived from 1, 2, or higher numbers of progenitor cells. Immediate segregation of mutants from these $M_1$ plants may or may not conform to simple 3:1 or 7:1 ratios. Seeds collected from siliques on one portion of an $M_1$ shoot may be genetically different from those collected at another position on the same shoot.

Since the $M_1$ plant is expected to be a chimera, most individuals screen the $M_2$ generation, when homozygous recessive mutations should be segregating. A recent valuable example exists of screening the $M_1$ generation directly for expected semidominant or dominant mutant alleles. Peng and Harberd (1993) screened for revertants of the *gai* locus (gibberellin insensitivity) by irradiating a *gai/gai* homozygote. Since the *gai* phenotype includes dwarfing, it was possible to screen for semidominant revertants by looking for less-dwarfed $M_1$ plants. From 60,000 heavily irradiated seeds, they obtained 13 phenotypic revertants. For 9 of the 13, no inheritance was seen in the next generation; presumably the cell lineages for these 9 did not contribute to the germ line, although they did alter the phenotype of the $M_1$. The strategy of screening the $M_1$ for dominant mutations might work particularly well if the desired mutant was non-cell-autonomous, in which case the whole plant might be phenotypically mutant, although the mutant would segregate only if its lineage included the germ line.

## To Pool or Not to Pool?

In large-seeded plants such as maize, one collects seeds from a given plant and keeps track of which plants were the parents. If an interesting mutant is recovered as a recessive homozygote, then it is always possible to readily recover a heterozygote for the same allele by just planting more seeds from the same parental plant. In *Arabidopsis*, the plants and seeds are tiny, and multiple mutagenized $M_1$ seeds may be planted in a given flat. The biology of *Arabidopsis* thus tempts one to pool seeds from multiple $M_1$ plants. The obvious drawback of pooling $M_1$ seeds is that it may be quite difficult to recover a heterozygote for a mutant if, for example, the homozygote is sterile or lethal. A second drawback is that one may recover two noncomplementing mutant alleles and not know if they are independent of each other or identical. The drawback to collecting seeds on a plant-by-plant basis is that the total size of the $M_1$ population screened is likely to be much smaller. A compromise is possible in which one collects seeds from small pools of $M_1$ plants (on the order of 10–100 plants). It should then still be possible to identify a heterozygote sib for a homozygote mutant by doing progeny analysis on a number of the seeds collected in the appropriate $M_1$ seed pool. This two-step screening, first finding a mutant in seeds from an $M_1$ pool and then screening progeny from plants in the pool to identify the heterozygotes, can be optimized by using a pool size that is the square root of the planned total $M_1$ population.

## Genetic Analysis of Newly Isolated Mutants

Every $M_2$ plant that has been identified as possessing an interesting mutant phenotype will have additional multiple mutations elsewhere in the genome. The per-locus mutant frequency in the $M_2$, after successful EMS treatment, will be in the range of 1 in 1000 to 1 in 5000. Depending on assumptions made, we can estimate that each mutant plant identified contains 2–75 additional mutations elsewhere. The consequences of this are simple. Either every new mutant must be backcrossed to a wild type repetitively until the genetic background is cleaned up, or additional independently isolated alleles must be found at the locus. Until one of these procedures is done, one cannot be sure whether a given phenotype is due to a single locus or to multiple interacting loci. At least five cycles of backcrossing to wild type are needed to achieve a 32-fold reduction in the background of undesirable mutants. Backcrossing, however, only slowly breaks up linked mutations. If a mutation appears to be pleiotropic, it is also important to examine whether all the phenotypes cosegregate in a large population of progeny from a cross.

New mutants should be put on the genetic map. As mentioned above, besides standard genetic mapping with visible mutants, there are also newer methods based on molecular markers. Williams et al. (1993) have developed a strategy based on rapid amplified polymorphic DNA (RAPD)-PCR. A mutant in one genetic background is crossed to another genetic background which differs in having multiple RAPD-PCR markers that the mutant strain does not have. In the $F_2$ from this cross, the plants that are phenotypically mutant are pooled and the DNA is extracted and analyzed. RAPD bands that are absent in the pool of mutant DNAs are ones that are tightly linked to the mutant allele. The greater the number of RAPD markers employed and the larger the number of mutants pooled for DNA extraction, the greater will be the resolution of the mapping. Once this method is routinely working in a laboratory, mapping a new phenotype to a first approximation can be accomplished with a couple of sets of PCR reactions.

Konieczny and Ausubel (1993) have developed a PCR mapping strategy based on codominant cleaved amplified polymorphic sequences (CAPS). Sequences are amplified from characterized pairs of primers that give already-mapped amplification products. Digests with appropriate restriction endonucleases then identify ecotype-specific polymorphisms. Bell and Ecker (1994) have further simplified this approach by taking advantage of the simple sequence length polymorphisms (SSLPs) in microsatellite repeat sequences in *Arabidopsis*. They have assigned 30 microsatellites to the linkage map and provided polymorphism data for these 30 repeats in six ecotypes. The advantage of SSLPs over CAPS is that restriction endonuclease digestion is not required. (SSLP and CAP primers are available for nominal charge from Research Genetics, 2130 Memorial Parkway SW, Huntsville, AL 35801. Telephone in USA and Canada 800-533-4363; in UK 0-800-89-1393. FAX 205-536-9016.) Analysis of CAPS or SSLPs in the progeny of a cross of a new mutation with another ecotype thus allows one to map the new allele. These methods have the advantage of generating codominant restriction digest patterns, whereas RAPD amplification detects dominant alleles.

### Rates and Types

As the *A. thaliana* genetic map becomes richer, and sequences of mutants become more available, we will inevitably learn in detail the kinds of mutational spectra that are induced by the various mutagens in our favorite organism. In this section, we discuss a few references from diverse experimental systems that provide some information about mutation rates and the basis of the mutations induced.

A key base reference for *Arabidopsis* is Li and Rédei (1969), who estimated the mutation rates to thiamine auxotrophy and pale seedlings for both X-rays and EMS. At doses that gave 95% seed survival, the EMS-induced mutant frequency was about tenfold higher than the X-ray-induced frequency. This supports the idea that EMS induces mutations with less collateral physiological and genomic damage than other mutagens. The frequencies reported in this article for EMS induction, on the order of $10^{-5}$ to $10^{-7}$, are smaller than those found by most investigators who are looking for specific phenotypes, as opposed to trying to study the mutation rate. The strategy of having a heavily mutagenized genome with lower $M_1$ seed viability makes sense when the goal is to find a mutant that will illuminate a particular physiological or developmental process.

Jürgens et al. (1991) isolated mutants of *Arabidopsis* that altered pattern formation during embryogenesis. As part of this study, they classified all the types of phenotypes they observed in just-germinated seedlings. From 44,000 lines, they found 25,000 embryonic lethals and 5000 abnormal seedlings. Among the 5000 abnormal seedlings, 2500 had some form of pigmentation defect, including 800 albinos and 80 with abnormal early anthocyanin production (*fusca*). The other 2500 lines were classified as abnormal morphology mutants, of which 250 were initially scored as the desired pattern-formation mutations. These results suggest that the dose of EMS they used, the standard 0.3% for 8 hours that many investigators use, is highly effective. It also further reinforces the likelihood that this dose of EMS is inducing multiple mutations in each line.

In barley, Kahler et al. (1984) attempted to estimate the spontaneous mutation rates to morphological loci and new isozymes. They examined five enzyme loci in 84,126 seedlings and found no variants, permitting them to estimate an upper bound to the mutation rate for isozymes at $3.56 \times 10^{-6}$ per locus per gamete per generation. Similarly, they found no morphological variants with a corresponding upper bound for the mutation rate at $8.85 \times 10^{-7}$ per locus per gamete per generation. Neel et al. (1986) estimated in humans that the spontaneous mutation rate to altered isozyme mobility was $0.6 \times 10^{-5}$ per locus per generation, roughly 10-fold higher than the plant data.

To reference the barley data to mutagenized *Arabidopsis*, the frequency of new mutants for morphological or absence of enzyme activity is on the order of 1 in 1000 $M_2$ plants. Although no one has specifically measured the induced mutation rate to isozyme variants, the data of Kahler et al. (1984) suggest they would also probably be found in 1 in 1000 $M_2$ plants. The comparison of the induced mutation frequency in *Arabidopsis* to the upper bound of spontaneous mutations in barley also suggests that EMS is typically increasing the mutation rate about 1000-fold. This is a very high degree and may be the explanation for the success of EMS as a mutagen.

Freeling and Cheng (1978) characterized 69 radiation-induced mutants at the *Adh* locus in maize following allyl alcohol selection of pollen. By examining isozyme profiles, they were able to observe changes in amounts of ADH produced in each mutant and then to classify the mutants. Of the mutants, 48 (68%) showed a complete dysfunction or lack of stainable enzyme activity; one of these was shown to be a tissue-specific turn-off of gene activity. The remaining 21 mutants showed quantitative increases or decreases in amount of enzyme activity observed in one or another of the several isozyme bands. All of these mutants were shown to involve chromosomal aberrations that initiated outside the *Adh* gene, as opposed to intragenic deletions. This was in agreement with earlier work of Stadler and Roman (1948), who also failed to find evidence for X-ray-induced mutations within a gene, as opposed to chromosomal aberrations that encompassed the gene. Freeling (1977) earlier had estimated the spontaneous forward (to *adh*−) and reversion (to *Adh*+) mutation rates and found them to be less than $2 \times 10^{-7}$ and $5.7 \times 10^{-6}$, respectively; the radiation-induced rate was roughly 50-fold higher.

The nucleotide changes that have been caused by a particular mutagen have been determined for a number of mutant alleles in *Arabidopsis*. As examples: Weigel et al. (1992) determined the basis for seven EMS-induced alleles at the *leafy* locus, showing that they were point mutations leading to nonsense or missense changes in the sequence; two examples of T-DNA insertion mutants were also characterized. Wilkinson and Crawford (1993) have characterized γ-ray-induced deletions to nitrate reductase deficiencies (chlorate resistance) in *Arabidopsis* and found that the deletions were 3–30 kb. Information on both irradiation- and EMS-induced mutants in *Arabidopsis* is available from the work of Sun et al. (1992). Multiple alleles exist at the *GA1* locus, including some that are fast-neutron-bombardment-induced (fnbi) as well as EMS-induced. The fnbi *ga1-3* allele was shown to be due to a 5-kb deletion, and this was used as the basis for molecular cloning by genomic subtraction. Among the other two fnbi alleles, one was associated with a 3.4-kb insertion, and the other had no apparent gross changes. Three EMS-induced alleles were all shown to be single nucleotide changes. This study thus provides evidence that fnbi mutants have chromosomal rearrangements, some of which may be deletions, and that EMS is functioning in *Arabidopsis* in a way similar to other systems. The mechanism of EMS mutagenesis is that it is an alkylating agent; in human cell lines it primarily causes base substitutions of G:C to A:T (DuBridge and Calos 1987).

Temperature-sensitive alleles are of exceptional utility, so it would be helpful to know the frequency with which they are obtained. Most investigators screen for temperature sensitivity among already isolated alleles, as opposed to deliberately constructing mutant screens for temperature-sensitive alleles. We are not aware of specific numbers for the frequencies of temperature-sensitives among all alleles, but most investigators seem to find them at as high as 25–50% in their collections (for two recent examples, see Baskin et al. 1992; Tsukaya et al. 1993). Since EMS induces base-pair transitions, it is probably not too surprising that a high percentage of these are capable of altering the thermal denaturation profile of the afflicted protein.

There is recent evidence that EMS mutagenesis frequently leads to mutations which cause loss or alteration of the properties of the corresponding transcript. Complete loss of the *Axr1* transcript was observed by Leyser et al. (1993). Similarly, loss of the *Fad7* transcript was seen in two EMS-induced mutants by Iba et al. (1993). In some cases, alterations in mRNA processing have also been observed. Orozco et al. (1993) found defective intron processing in an EMS-induced mutation at the *RCA* locus, and a reduction in the size of the mRNA in an *act1* mutant was also observed (J.C. Schneider and C.R. Somerville, unpubl.).

To summarize, there is plenty of evidence that EMS is an exceptionally useful mutagen, capable of inducing point mutants with high frequency. Irradiation clearly can cause chromosomal rearrangements, including useful deletions, but irradiation is not as efficient as EMS, and not all radiation-induced alleles are of the useful sort. Discussions on the *Arabidopsis* electronic newsgroup make it clear that we can expect much more information on the molecular basis of mutations in the near future.

## INSERTIONAL MUTAGENESIS

Insertional mutagenesis is based on the inactivation of a gene via insertion of a known DNA fragment. Because it involves the insertion of some type of DNA fragment, the mutations generated are generally loss-of-function rather than gain-of-function mutations. However, there are several examples of insertions generating gain-of-function mutations (see, e.g., Koncz et al. 1990; Oppenheimer et al. 1991). Insertion mutagenesis has been very successful in bacteria (Kleckner 1981) and *Drosophila* (Cooley et al. 1988) and recently has become a powerful tool in several plant species (corn: e.g., Schmidt et al. 1987; *Arabidopsis*: e.g., Herman and Marks 1989; Koncz et al. 1990; *Antirrhinum*: e.g., Sommer et al. 1990). In plants there are two elements that have been used for mutagenesis: T-DNAs and transposons.

### T-DNA Insertional Mutagenesis

The T-DNA, or transfer DNA, of *Agrobacterium tumefaciens* is a defined segment of the tumor-inducing plasmid which, upon infection, is transferred to a susceptible plant cell where it can integrate into the plant genome (see Koncz et al. 1992a). Upon integration, the genes contained within the T-DNA are expressed. Depending on where integration occurs, a mutation may result such that the mutation and the marker in the T-DNA will cosegregate.

Several methods have been developed for introducing T-DNA into *Arabidopsis*. These include various tissue-culture and whole-plant techniques. These are discussed in the context of their utility for insertional mutagenesis.

### *Tissue Culture*

There have been a large number of papers published describing various tissue-culture-based *Agrobacterium*-mediated transformation protocols for *Arabidopsis* (for review, see Koncz et al. 1992a; Morris and Altmann, this volume). Most of these were not directed toward insertion mutagenesis but rather toward the introduction and expression of an engineered gene. However, several protocols have been established in an attempt to make insertion mutagenesis a feasible alternative to map-based gene cloning (Lloyd et al. 1986; Valvekens et al. 1988; Koncz et al. 1989).

Lloyd et al. (1986) utilized a modified leaf disk *Agrobacterium*-mediated transformation procedure (Horsch et al. 1985) to introduce a hygromycin-resistance marker into *Arabidopsis*. The generation of approximately 100 transformants and their subsequent phenotypic analyses resulted in numerous mutants, none of which cosegregated with the hygromycin-resistance marker (H. Klee and R. Horsch, pers. comm.). This result was assumed to be due to somaclonal variation presumably induced during long periods of exposure to exogenous hormones through the transformation and regeneration process (Larkin and Scowcroft 1981).

In another attempt to develop an insertion mutagenesis system and avoid long-term exposure to exogenous hormones, Feldmann and Marks (1986) developed a regeneration procedure for leaves and stems of *Arabidopsis* which involved a short exposure to a callus-inducing medium (5–7 days) followed by transfer to a shoot-inducing medium. This protocol was based on regeneration experiments in the field bindweed, *Convolvulus arvenesis* (Christianson and Warnick 1985). With this procedure, regenerated intact shoots could be removed from exogenous hormones in as little as 3 weeks. However, when an *Agrobacterium*-mediated transformation system was coupled with this procedure, it necessitated a longer exposure to callus-inducing medium (K.A. Feldmann and M.D. Marks, unpubl.) and thus the possibility of increased somaclonal variation.

Valvekens et al. (1988) followed up on this short preculture concept using root explants and observed high rates of transformation and plant regeneration. Again, the transformation and selection process meant longer exposure to exogenous hormones before regenerants were large enough to be transplanted. As a result, preliminary data from progeny of 60 root- or leaf-derived transformants showed that none of the resulting mutations were linked to the T-DNA (Valvekens et al. 1988). In a follow-up report (Valvekens and Van Montagu 1990), progeny from 101 root-derived regenerants (nontransformed) were found to segregate for seven recessive mutations whereas screening of progeny from 84 transformants resulted in an additional seven lines segregating for a recessive mutation; none of these latter lines appeared to be linked to the marker in the T-DNA. These limited data strongly suggest that the seven mutants in the transformed population are due to somaclonal variation rather than a mutagenic event induced by the transformation process.

Van Lijsebettens et al. (1991a) screened 110 leaf-derived transformants and found that only 5% (6/110) of the transformants displayed an alteration in phenotype. Only in one of these mutants, a leaf morphology mutant, did the T-DNA appear to cosegregate with a mutation. They have subsequently cloned this gene and shown, via molecular complementation, that it was tagged (Van Lijsebettens et al. 1991a,b). For four other mutants, the T-DNA and mutation were unlinked; one additional mutant needs further characterization. These 110 transformants contained about 150 inserts such that the "T-DNA mutation induction frequency" was between 0.5% and 1%.

In a much larger set of tissue-culture-based transformation experiments, Koncz et al. (1989) used stem, leaf, and root explants with either the hygromycin or kanamycin markers to generate approximately 3,000 transformants. Preliminary screening of 450 of these segregating families showed that only 0.2–1% "of the T-DNA induced mutations" caused a visible alteration in phenotype. One of these (*ch-42*, a yellow-green phenotype) was cloned and via molecular complementation shown to be the gene responsible for the altered phenotype (Koncz et al. 1990).

## *Seed Infection/Transformation*

To minimize the background mutation frequency due to somaclonal variation, Feldmann and Marks (1987) developed a protocol for plant transformation that avoided the tissue culture step. The protocol for this procedure has been recently reviewed (Feldmann 1991, 1992; Forsthoefel et al. 1992). Briefly, wild-type seeds ($T_1$) were incubated with *Agrobacterium* (containing a marker for $Kan^R$ in the T-DNA), and the plants were grown to maturity. The resulting progeny ($T_2$) were screened on medium containing kanamycin. Rare $Kan^R$ $T_2$ seedlings were isolated and grown to maturity; progeny ($T_3$) were collected from single $T_2$ plants and numbered chronologically. Approximately 14,000 independent transformants have been harvested in this manner (Feldmann and Marks 1987; Feldmann 1991; Forsthoefel et al. 1992 and unpubl.). A considerable amount of genetic and molecular data indicate a high probability that these 14,000 transformants are unique. To generate these transformants, less than $1.1 \times 10^6$ infected plants ($T_1$) have been grown to maturity, and less than $10^8$ seeds ($T_2$) from these plants have been tested on medium containing kanamycin.

To ascertain the average number of functional T-DNA inserts, $T_3$ seeds from about 1,000 $T_2$ plants have been tested on kanamycin-containing medium. Of the lines tested, 57% segregate for one insert (3 $Kan^R$:1 $Kan^S$), 25% for two linked (3:1 < x < 15:1) or unlinked inserts (15:1), 10% for 3 or more linked or unlinked inserts (x ≥ 63:1), and 9% segregated less than 3:1 ($Kan^R$:$Kan^S$; exceptional lines). These exceptional lines, in subsequent generations, were found to be segregating for 1–3 independent inserts (K.A. Feldmann and M.L. Christianson, unpubl.). Toward mutagenesis, these results show that the average number of inserts per transformant is 1.5. Thus, in the population of 14,000 transformants there are 21,000 functional (as assayed on kanamycin) inserts.

The original transformants, $T_2s$, are always scored for any visible alteration in phenotype. There are generally two phenotypes observed. The most common alteration in phenotype is reduced fertility. Depending on the environmental conditions during the $T_2$ selection process, several percent to 10 percent of the $T_2$ plants exhibit this phenotype to a degree. However, this trait has never been carried to the next generation in any one of these lines. Thus, this phenotype seems to be epigenetic and a result of the selection process. The second common phenotype observed in the $T_2s$ is late flowering. We scored 53 lines (0.38%) as late-flowering $T_2$ plants (Forsthoefel et al. 1992). All of those that have been tested gave rise to only true-breeding late-flowering plants (K.A. Feldmann and R. Amasino, unpubl.). In addition, these lines are segregating for the kanamycin-resistance marker, indicating that the late-flowering phenotype is not due to a T-DNA disruption. The origin of these late-flowering lines is not clear. Finally, we have observed a few $T_2$ plants that were "weird." This variable phenotype fails to segregate in the next generation and is likely a result of axenic culture during the early development of the $T_2$ seedling. To date, we have found no evidence for dominant mutations among the 14,000 transformed lines.

Segregating $T_3$ seed families have been screened for visible alterations in phenotype under two environmental regimes: on soil in the greenhouse and in vertically oriented agar plates (described in Feldmann 1991). The agar screen was primarily used to identify mutants in root and early seedling development and pigmentation defects. The greenhouse screen was employed to find mutants in seedling, mature plant, and embryo development. The types of mutants and their frequencies are described below (Table 1).

There are an increasing number of T-DNA insertion mutants being characterized from this population; this is primarily due to the availability of 6,400 transformants (available from the Arabidopsis Stock Centers). A number of mutants, in various pathways, have been characterized to date (dwarf: Feldmann et al. 1989; trichome: Marks and Feldmann 1989; flower: Yanofsky et al. 1990; Roe et al. 1993; embryo: Errampalli et al. 1991; Meinke 1992; chlorate-resistant: LaBrie et al. 1992; Tsay et al. 1993; vegetative meristem: Medford et al. 1992; light regulatory: Deng et al. 1992; ethylene-affected: Kieber et al. 1993; and epicuticular wax: McNevin et al. 1993). For several of these, the genes responsible for the phenotypes have been cloned and characterized (see Table 2).

Several techniques have been used to isolate the disrupted plant DNA including (1) the generation of genomic libraries from the mutants and screening with sequences homologous to the right or left border region

*Table 1* Frequencies of various classes of visible mutant phenotypes observed from the screen of 5000 transformed lines from seed infection/transformation (Arizona population) and 1340 transformed lines derived via tissue culture procedures

| Phenotype | Seed transformation (%) | Tissue culture (%) |
|---|---|---|
| Seedling-lethal | 1.1 | 0.59 |
| Size variant | 4.4 | 5.52 |
| Embryo-defective | 3.6 | 5.52 |
| Reduced-fertility | 1.1 | 0.36 |
| Pigment mutants | | |
|   yellow-green | 3.0 | 2.83 |
|   albino | 0.6 | 1.12 |
|   dark green | 0.26 | 0.52 |
| Dramatic mutants | | |
|   flower | 1.3 | 0.22 |
|   root mutants | 1.1 | 2.76 |
|   trichomes | 0.22 | 0.82 |
| Physiological mutants | | |
|   flowering time | 0.48 | 0.44 |
| Other | 1.8 | 4.31 |
| Total | 19 | 25 |

Data from Forsthoefel et al. (1992) and Koncz et al. (1992b).

(Oppenheimer et al. 1991), (2) plasmid rescue, utilizing the selectable markers in the T-DNA to isolate T-DNA-plant junctions in *Escherichia coli* (Yanofsky et al. 1990), and (3) inverse polymerase chain reaction (IPCR), utilizing primers made from the left or right border sequences (Deng et al. 1992). Proof that the correct gene has been tagged and isolated has been accomplished by molecular complementation, introducing the wild-type sequence into the mutants to show that the mutant is reverted to wild type (Herman and Marks 1989; Tsay et al., 1993), or sequencing a number of mutant alleles to correlate a change in the mRNA sequence with the severity of the mutant phenotype (Kieber et al. 1993).

Along with these tagged mutations there have been an increasing number of mutations that have been found that are not cosegregating with a kanamycin-resistance marker. This includes a large number of embryo-defective mutants (Errampalli et al. 1991; Meinke 1992; Castle et al. 1993), flower mutants (Weigel et al. 1992), epicuticular wax mutants (McNevin et al. 1993), and several dwarfs (K.A. Feldmann and Y. Wu, unpubl.). In fact, for embryo-defective, epicuticular wax, and

*Table 2* Genes cloned from a transformed population generated via seed infection/transformation

| Gene symbol | Function | Reference |
| --- | --- | --- |
| AG | transcription factor - MADS | Yanofsky et al. (1990) |
| CHL1 | nitrate transporter | Tsay et al. (1993) |
| COP1 | transcriptional regulator | Deng et al. (1992) |
| CTR1 | serine-threonine kinase | Kieber et al. 1993) |
| DWF1 | none | K. Feldmann (unpubl.) |
| FAD2 | fatty acid desaturase | Okuley et al. (1994) |
| FAD3 | fatty acid desaturase | Yadav et al. (1993) |
| FUS6 | none | Castle and Meinke (1994) |
| GL1 | transcription factor - myb | Oppenheimer et al. (1991) |
| HY3 | phytochrome B | Reed et al. (1993) |
| HY4 | microbial DNA photolyase | Ahmad and Cashmore (1993) |
| LD | bipartate nuclear local. sig. | Lee et al. (1994) |
| TSL1 | serine-threonine kinase | Roe et al. (1993) |

dwarf phenotypes, only about 40% of the mutants found in this population appear to be due to a disruption by a functional T-DNA; 6 of 13 (46%) for wax (McNevin et al. 1993), at least 6 of 16 for dwarf (38%), and 41 of 115 (36%) for embryo-defective mutants (Castle et al. 1993).

Our work (K.A.F.) on dwarfs and other mutants has shown that if a $T_3$ line ($F_2$ equivalent) is segregating 3:1 (wild type:mutant) and segregating in a Mendelian manner for a single insert, there is a much higher probability that the mutation will be due to a T-DNA disruption. Even if the line is segregating in a Mendelian manner for two inserts, there is a higher probability that it is tagged than if the segregation ratios for either of the markers are extremely distorted. This distortion, generally exhibited as a deficiency in mutant phenotypes in a segregating family, is due presumably to translocations or rearrangements of the inserts. We, in fact, have observed several dwarf lines in this category. For one of these, it has taken several generations to establish a line that has a single stable insert, but linkage has now been firmly established. Alternatively, somatic mutations may be induced in the $T_2$ plant during some stage of development so that the progeny do not segregate in a 3:1 manner. In fact, Weigel et al. (1992) sequenced a *LFY* allele from this population (*lfy7*) that segregated for a deficiency of mutants in the $T_3$ generation and found that it contained a simple base change. This would suggest that the mutation is not the result of an aborted integration event. Whenever a line with abnormal segregation ratios is observed, a more detailed genetic examination should be completed to ensure that the mutation is definitely not due to a T-DNA.

The greater frequency of mutants induced in the whole-plant transformation procedure than in transformants arising from tissue culture is surprising. One explanation for this low frequency in the tissue culture collections may be the way in which the lines were screened for visible alterations in phenotype. In the Feldmann collections, numerous researchers, mostly expert for the phenotype for which they were screening, participated in large screens where the plants were examined multiple times under at least two growth conditions. In support of this idea, Koncz et al. (1992b) have recently rescreened their original 450 transformants in the context of a larger population and with a large number of collaborators, and they have observed a much higher frequency of mutants; in fact, a frequency comparable to that previously reported for seed infection/transformation (Table 1) (Feldmann 1991). The similarities and differences in these populations are discussed in the following section.

## Mutation Spectrum of Whole-plant- and Tissue-culture-derived Populations

Feldmann et al. (1990, 1991) and Feldmann and Meyerowitz (1991) placed the observed mutants into six classes: seedling-lethal, size variant, embryo-lethal, reduced fertility, pigment, and dramatic. They estimated that the population segregated for 10%, 7%, 3–5%, 2.5–3%, 1%, and 1% for each of these mutant types, respectively. These were the results from screening 1,300 transformants, where most of the screening was accomplished by K.A.F. These were screened on soil under greenhouse conditions. The screening of 7,100 lines, including 200 of the 1,300 previously screened lines, on soil and in vertically oriented petri plates containing agar-solidified medium, with the help of 60 participants (acknowledged in Feldmann 1991), showed that the previously reported estimates for some classes were too high and for others too low (Feldmann 1991; Coomber and Feldmann 1993). The frequencies for the larger population were 3–5%, 3–5%, 2.5–3.5%, 1–2%, 2–3%, and 2.5–3% for seedling-lethal, size variant, embryo-defective, reduced fertile, pigment, and dramatic mutants, respectively. Suboptimal growth conditions may explain the high frequency of seedling- and embryo-lethal, size variant, and reduced fertility mutants in the screen of 1,300 lines. The increase in the frequency of dramatic mutants (from 1% to 2.5–3%) is due to the application of the agar screen as well as the greater number of researchers screening the lines. In addition, a seventh class of mutants was added, physiological mutants. These included the epicuticular wax, late-flowering, wilty, and high-fluorescence mutants (Feldmann 1991; Coomber and Feldmann 1993). The total number of lines noted that had some visible alteration in phenotype was between 15% and 26%.

Using what had been learned from the previous two populations, the screening of the next 5,000 lines was refined (Forsthoefel et al. 1992). The frequencies of the various classes of mutants are listed in Table 1. These numbers probably reflect the most accurate frequencies, as the screen was conducted in Arizona under winter greenhouse conditions in which, to a large extent, light, temperature, and humidity were well controlled. In addition, researchers who participated in this screen had participated in previous screens. It is noteworthy that the frequency of seedling-lethals has dropped further and the frequency of pigment mutants and dramatic mutants has increased (Table 1) (Forsthoefel et al. 1992).

A comparison of the mutational spectra and frequency for the 1,340 tissue-culture-derived lines that Koncz et al. (1992b) generated and the 5,000 whole-plant-derived transformants generated by Forsthoefel et al. (1992) indicates that for many of the phenotypes for which we could do a direct comparison, the frequencies are surprisingly similar, e.g., size variants, embryo-defectives, and pigment mutants. The slight variability in some of these classes could be due to the different environmental conditions used by each group to screen the transformants or the specific criteria we each used to classify a mutant. On the other hand, these influences may actually be minimizing the true differences in the two populations. The overall frequency of mutants in the tissue-culture-derived population (25%) is within the range that we had reported previously for seed infection/transformation (15–26%; Feldmann 1991), but higher than what was reported in our last screen (19%; Forsthoefel et al. 1992).

In summary, it does appear that the two different transformation regimes result in populations with similar mutants and mutational spectra. However, preliminary data indicate that a larger percentage of the mutants in the seed-derived population are tagged (~40%) in comparison to the tissue-culture-derived populations (~10%; Koncz et al. 1990; Van Lijsebettens et al. 1991a; C. Koncz, unpubl.). It seems somewhat improbable that the T-DNAs would insert differentially, i.e., inserting at a different frequency in transcription regions, in each of these two transformation systems. This is especially true since Koncz et al. (1989) have demonstrated that in the tissue-culture-derived populations, 30% of the T-DNAs insert into regions which are potentially transcriptionally active. It also does not seem plausible to suspect that the T-DNAs introduced by seed infection/transformation preferentially insert into genes that produce a visible alteration in phenotype, whereas with tissue-culture-derived transformation systems they insert into nonessential genes. Additional data from each of these two populations may show that the frequency of tagged mutants is actually closer than it now appears. The basis for the untagged mutants in either population remains unclear.

## Other Whole-plant Transformation Strategies

Another whole-plant transformation technique designed both for the introduction of new constructs and for insertion mutagenesis is the in planta transformation protocol of Hong-Gil Nam (Chang et al. 1990, 1994). In this system, young inflorescences are cut off near the base of soil-grown plants, and the wounded surfaces of the plant are inoculated with *Agrobacterium*. After the secondary inflorescences arise, they are cut off near the base and *Agrobacterium* is reapplied. The seeds are collected at maturity from the newly arising inflorescences and tested on the appropriate selective medium. With this technique, transformants have been generated utilizing both C58 and LBA4404 strains of *Agrobacterium* (see Coomber and Feldmann 1993). This protocol is being used successfully by a number of laboratories. R. Whittier has isolated 400 independent lines by means of this technique and has identified several visible alterations in phenotype (R. Whittier, pers. comm.). However, this technique is probably not going to be sufficiently useful for insertion mutagenesis. There are at least two reasons for this skepticism. First, because it is possible to generate large clones of transformed tissue in a single plant, due to the possibility of the incorporation of a transformed cell in a newly forming meristematic region, the seeds from each treated plant will have to be collected separately to prevent contamination of the putative insertion population with siblings. Second, this technique is very labor intensive. Despite this note of caution for insertion mutagenesis, this in planta technique does appear to be a very useful transformation system for introducing T-DNAs in an increasing number of laboratories.

Sangwan et al. (1993) used a zygotic embryo transformation procedure with C24 to generate 5,000 transformants. Preliminary genetic and molecular characterization of the first 1,000 transformants showed that they contained between one and five inserts, possessed a mutational spectrum similar to that described for seed infection/transformation, and contained a low frequency (<5%) of somaclonally induced mutations (Sangwan et al. 1993).

A whole-plant transformation procedure that will be very powerful for saturating the *Arabidopsis* genome was published by Bechtold et al.

(1993). For this protocol, plants are submersed in an *Agrobacterium* culture and put under vacuum. The plants are allowed to recover in the greenhouse, and after several weeks progeny are collected. The progeny containing T-DNA inserts are selected and grown to maturity. The authors estimate that they have generated >50,000 transformants utilizing this procedure and are in the process of characterizing the first 4,000. This population is very similar to the transformants generated by seed infection/transformation, i.e., there are approximately 1.5 inserts/transformant, the inserts are concatamers of T-DNAs, and there is a significant percentage of untagged mutants (M. Caboche, pers. comm.). This technique may prove not only to have solved the problem for saturation mutagenesis, but also to be useful for routine transformation.

### Mechanism of Transformation in the Whole-plant Transformation Procedures

The data collected on the transformed populations generated via the whole-plant transformation procedures described above indicate that the timing and location of the transformation event, i.e., the cell type that is transformed, are the same; it happens very late in the development of the flower such that each resulting transformant is unique. These data include the following: (1) There are many instances of treatments that generate one or a few transformants, indicating that an average plant is not chimeric for an insert (K. Feldmann, unpubl.); (2) in a single treatment, the T-DNA-generated, visible mutants are single and unique; (3) T-DNA-induced mutants are not observed in the progeny of an infected plant (they are not observed until the subsequent generation); and finally, and most compelling, (4) the transformed progeny from an infected plant are heterozygous for the selectable marker (Feldmann and Marks 1987; K. Feldmann unpubl.; M. Caboche; H.-G. Nam; both pers. comm.). These observations are consistent with the hypothesis that after *Agrobacterium* is introduced into the plant (interstitial spaces of the shoot apex) early in development, it is harbored in the developing tissue until sometime late in flower development, when it is induced to transfer the T-DNA to a cell that will give rise to the zygote, or to the zygote itself. If the transformational event occurred prior to the formation of the flower or too early in flower development, both gametes would contain the insert(s), and plants homozygous for the insert(s) would be generated. If the transformational event occurred after the first mitotic division of the zygote, a seedling that was resistant to the selective agent would probably not be generated. This is because the basal cell of the two-celled embryo gives rise to the root cap columella cells and the quiescent center, whereas the apical cell gives rise to the root initials, as well as the rest of the plant (Dolan et al. 1993). If the quiescent center and the root cap consist of untransformed cells, the seedling would not likely develop in a normal manner (P. Benfey, pers. comm.), which is not the case for the transformants generated with the whole-plant procedures. These various whole-plant protocols simply represent different ways of introducing *Agrobacterium* into the plants.

### Saturation Mutagenesis

For T-DNA insertion mutagenesis to be useful for isolating any gene in the *Arabidopsis* genome, it will be necessary to saturate the genome with T-DNAs. Given that (1) the *Arabidopsis* nuclear genome is approximately 100,000 kb in length (Meyerowitz, this volume), (2) insertion is random, (3) the average length of an *Arabidopsis* gene is 4 kb, and (4) there are 1.5 inserts per transformant, a population of 50,000 transformants will need to be generated to give a 95% probability of having an insert in an average gene.

It will take considerable time and data to definitively prove that T-DNA insertion is random, but several pieces of data give strong evidence that this is, in fact, the case. Utilizing a T-DNA that carried a promoterless fusion at one end, Koncz et al. (1989) have shown that 30% of the T-DNAs in *Arabidopsis* and *Nicotiana* insert into regions that are potentially transcriptionally active. For *Nicotiana* this suggests that insertion may not be random and that there may be some preference for transcriptionally active regions. However, for *Arabidopsis* with its much smaller genome and limited repetitive DNA, 30% is close to what might be predicted for random insertion. Within the gene, the insertions also appear to be random. For 13 T-DNA-tagged mutants that have been characterized, inserts were found in introns and exons as well as in the 5' and 3' ends of the gene (see references in Table 2). The average number of introns was 6–7 and the average total intron plus exon length for these 13 genes was 4.0 kb. At the genome level, mapping of more than 40 T-DNAs or T-DNA-tagged mutants has shown that these are randomly distributed over the five chromosomes of *Arabidopsis* (references in Table 2; K.A. Feldmann, unpubl.; D. Meinke, pers. comm.). Finally, when the frequency of various mutants isolated from these large T-DNA-generated populations is examined, it is clear that the mutants which are observed do not represent hot spots for the T-DNA (K.A. Feldmann, unpubl.).

The average length of a gene calculated above, 4.0 kb, does not include the 5' and 3' regions of the gene, which if disrupted can also affect gene function and result in an altered phenotype. So even if this sample is small and potentially biased toward longer genes, it is a close approximation. Finally, it has been shown genetically that the average number of inserts in each transformant equals 1.5. Thus, the present data appear to support these four assumptions.

Through seed infection/transformation, 14,000 transformants have been generated (Feldmann and Marks 1987; Feldmann 1991 and unpubl.; Forsthoefel et al. 1992), and with tissue-culture-based procedures there are 4,000 transformants (Koncz et al. 1990; L. Willmitzer, pers. comm.). There are certainly thousands of additional *Arabidopsis* transformants in other laboratories around the world. This still represents less than half the population that is required. It does, however, constitute a population that, given the conditions listed above, presents a high probability (>70%) of having an insert in any average gene. To generate the additional 30,000 transformants to achieve saturation will take a substantial amount of work, but Bechtold et al. (1993) and Bouchez et al. (1993) are well on the way to this target. However, for certain mutant phenotypes, this population may still take a considerable effort to screen. A technique that may complement this approach is transposon mutagenesis.

## TRANSPOSON MUTAGENESIS

The preceding sections in this chapter clearly demonstrate that insertional mutagenesis using T-DNA has provided many useful mutations. However, most laboratories have not yet been able to generate large populations of transformants using the seed transformation or in planta transformation procedures, and the high ratio of non-tagged to tagged mutants in transformants generated from tissue culture does not make this approach an attractive option. These factors have provided the incentive for the development of transposon tagging systems in *Arabidopsis*. The availability of such a system would bring a number of advantages. First, all interested laboratories could easily generate new transposition events. Second, modified transposons could be used to increase tagging efficiency and extend the range of genes that could be isolated. Third, the criteria for identifying a tagged mutation could be reversion to a wild-type phenotype rather than complementation experiments. Fourth, the various properties of the transposons could be taken advantage of; that is, transposition to linked sites could be used to target certain loci, and the presence of a "footprint" after the element has excised could be used to generate an allelic series at the locus of interest.

### Endogenous Transposable Elements

The search for endogenous elements in *Arabidopsis* has revealed a number of families. A superfamily of copia-like retrotransposons *Ta1-Ta10* has been identified (Konieczny et al. 1991) and differences in their positions in the different ecotypes suggest that transposition has occurred since the ecotypes diverged. Whether they are still active has not been demonstrated. An insertion identified in an *S*-adenosylmethionine gene showed characteristics of a transposable element (Peleman et al. 1991). This element, termed *Tat1*, is only 431 bp and contains no open reading frames, so it is unlikely to encode the proteins necessary for its own transposition. Again, different hybridization patterns of the four genomic copies of *Tat1* were observed in different ecotypes, but no direct evidence has been obtained that these elements are still active.

Recently, a very active endogenous transposable element has been identified in *Arabidopsis*. A chlorate-resistance screen was used to trap the *Tag1* element in the *CHL1* gene (Tsay et al. 1993). This element, of which there are three cross-hybridizing fragments in the Landsberg *erecta* genome and none in the Columbia or WS genomes, is 3.3 kb, has 24-bp inverted repeats, and bears only little sequence homology with other known transposons. It is very active, as demonstrated by the high reversion of the mutation to chlorate sensitivity. *Tag1* was discovered in lines containing introduced active copies of the maize transposable element *Ac*. It appears that activity of *Tag1* may be increased by the presence of *Ac* transposase (A. Bhatt and C. Dean, unpubl.). All three *Tag1* hybridizing fragments cosegregate and map to the lower half of chromosome 1 (A. Bhatt et al., unpubl.). The characteristics of transposition of this element are currently being analyzed. It will be important to understand these before the full potential of this element to tag other genes in *Arabidopsis* can be assessed.

### Use of Heterologous Transposons

The lack of well-characterized endogenous transposons in *Arabidopsis* stimulated a number of groups to introduce elements from other species (for review, see Balcells et al. 1991; Bhatt and Dean 1992; Coupland 1992). These included the maize *Ac*, *Ds*, *Spm*, and *Mu1* elements and the *Antirrhinum Tam3* element. The activity of all but *Mu1* has been demonstrated in transgenic plants. Most work and success to date has been with the maize *Ac/Ds* elements. The *Ac* element is an autonomous element producing one protein termed transposase (Kunze et al. 1987). The nonautonomous *Ds* carries all the sequences necessary for transposition in *cis*, but does not produce functional transposase. Mutagenesis strategies have used either the single autonomous *Ac* element or a two-element system with *Ds trans*-activated by a transposase source which is itself not capable of transposition.

## *Activity of* Ac/Ds *Elements in* Arabidopsis

*Ac* was initially shown to transpose in *Arabidopsis* callus by van Sluys et al. (1987). Schmidt and Willmitzer (1989) extended this analysis and demonstrated *Ac* transposition in regenerated *Arabidopsis* plants and their progeny. They estimated the minimal germinal excision frequency to be 0.2–0.5%. Dean et al. (1992) and Keller et al. (1992) then analyzed somatic and germinal excision frequencies, reintegration frequencies, and *Ac* dosage and heritability using an excision assay based on resistance to streptomycin (Jones et al. 1989).

One feature of *Ac* activity in *Arabidopsis* is the relatively low frequency of transposition as compared to tobacco (Jones et al. 1989) and tomato (Yoder 1988). Different approaches have been taken to try to increase *Ac* activity in *Arabidopsis* to a level more useful for gene tagging. The excision frequency was increased approximately fivefold by deletion of 537 bp within the 5′ untranslated leader of the transposase transcript (Lawson et al. 1994). This also increased the *trans*-activation of a *Ds* element (Bancroft et al. 1992). Increased *trans*-activation of *Ds* in *Arabidopsis* has been achieved by utilizing fusions of the transposase gene with a strong plant promoter—the 35S promoter from cauliflower mosaic virus (CaMV) (Grevelding et al. 1992; Swinburne et al. 1992; Honma et al. 1993). Other promoter fusions designed to give transposase expression at specific developmental stages, e.g., pollen-specific or heat shock-induced, are also being developed (Balcells et al. 1994; R. Scott and J. Draper, pers. comm.). The ultimate aim of these experiments is to find transposase fusions that give high levels of independent *Ds* transpositions. Reintegration of the *Ds* elements into the genome can be selected for by including a dominant selectable marker within the *Ds* element. Fusions conferring resistance to methotrexate, hygromycin, chlorsulfuron, and kanamycin have proved effective for this purpose (Masterson et al. 1989; Bancroft et al. 1992; Honma et al. 1993; B. Sundaresan and R. Martienssen, pers. comm.).

## Arabidopsis *Loci Affecting* Ac *Activity*

A mutagenesis experiment has suggested that *Ac* activity is low in *Arabidopsis* due to suppression of *Ac* transposition in the *Arabidopsis* genome. An *Arabidopsis* transformant carrying a single *Ac* element showing a typically low level of activity (Dean et al. 1992) was mutagenized with γ irradiation (F. Belzile et al., unpubl.). The $M_2$ seeds were analyzed for somatic *Ac* activity using the streptomycin-resistance assay. Of 200,000 individuals, 25 showed a 50-fold higher level of *Ac* activity. These were allelic, recessive, and unlinked to the T-DNA carrying the original low activity *Ac* element. This mutation also results in improved penetrance and expressivity of the *Ac* locus. Analysis of these mutants and, ultimately, the cloning of the locus should reveal how the host factor, encoded by this locus, *IAE* (for *increased Ac excision*), influences *Ac* activity.

### *Pattern of Transposition in the* Arabidopsis *Genome*

Recent data from Keller et al. (1993) and Bancroft and Dean (1993) show that the pattern of *Ac* and *Ds* transposition in *Arabidopsis* is similar to that in maize and tobacco; that is, preferential to linked sites (Greenblatt 1984; Dooner and Belachew 1989; Dooner et al. 1991). The distribution of transposed *Ds* (*tDs*) elements from four T-DNA insertions in the *Arabidopsis* genome is shown in Figure 1. Since a large number of transpositions are to linked sites, the targeting of a specific locus will be most efficient from a linked T-DNA. It will also be necessary to start from T-DNAs integrated at many sites distributed throughout the genome to mutagenize efficiently the whole genome in non-targeted tagging experiments. For this reason, several groups are mapping T-DNA integrations carrying *Ds* elements onto the *Arabidopsis* RFLP map (see, e.g., Bancroft and Dean 1993; C. Lister and C. Dean; B. Osborne and B. Baker; H. Dooner; R. Masterson; J. Goodrich and G. Coupland; all pers. comm.).

### *Transposon-induced Mutations*

Screening for mutations in plants carrying transposed elements is now a major activity in several laboratories. The first *Ds*-induced mutation in *Arabidopsis* was obtained in an experiment in which five mutations were identified from 430 families, representing at least 240 independent *Ds* reinsertion events (Bancroft et al. 1993). Only three of the five mutants (shown in Fig. 2) contained a transposed *Ds*, so two mutants could not have been caused by insertion of the element. All three mutations showed linkage to the *tDs*. Revertants were isolated from two of them, and Southern blot and sequence analysis of one proved that the mutation was caused by insertion of the *Ds* element. This mutant, called *drl-1* (*deformed roots and leaves*), has long and narrow leaves, abnormal phyllotaxy, stunted roots, and the stable mutant never flowers. Several more *Ds* or *Ac*-induced mutations have now been identified (Long et al. 1993; A. Bhatt et al., unpubl.) and in addition, the maize *En/spm* element system has been used to tag a gene involved in pollen development (Aarts et al. 1993).

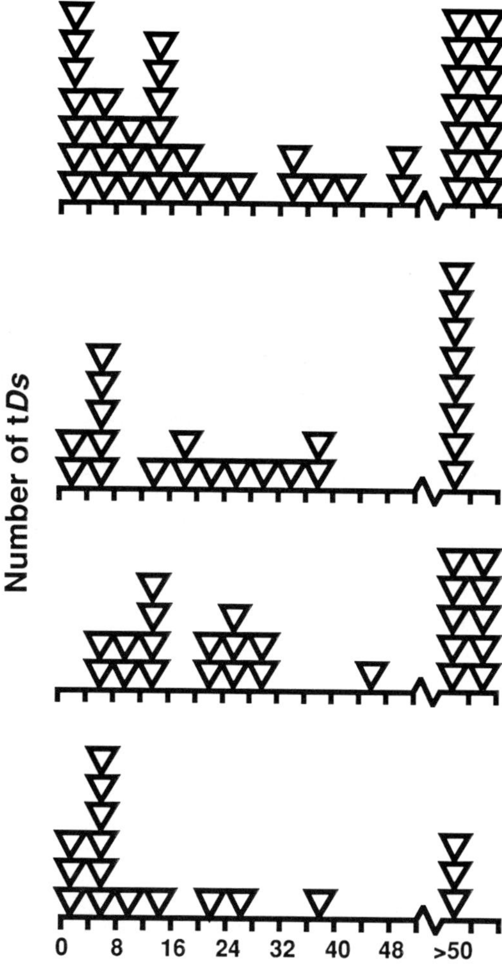



## Map distance (cM) from SPT locus

*Figure 1* Transposition pattern of *Ds* elements at four different locations in the *Arabidopsis* genome. Each triangle represents an independent transposition event. The transposition distance of the *Ds* away from the T-DNA (at 0) was estimated by measuring the recombination frequency between antibiotic resistance markers on the T-DNA and the *Ds* element (Bancroft and Dean 1993).

The apparent success of these experiments contrasts with previous ones where mutants had been obtained, but after analysis it was clear that the mutation had not been caused by a *tAc* or *tDs*. In an early screen, approximately 200 families carrying transposed *Ac* elements were screened for segregating mutations caused by *Ac* insertion (A. Bhatt et al., un-

*Figure 2* Mutations identified in families carrying *tDS* elements. The three mutants *drl1*, *slc1*, and *pcm1* are shown above their wild-type siblings (Bancroft et al. 1993).

publ.). Eight mutants were found with phenotypes that included pale green leaves and cotyledons, dwarf plants, and albinos. Segregation analyses of these with the transposed *Ac* element showed that none of them had been caused by insertion of the *Ac* element. These mutations were also not caused by insertion of the endogenous *Tag1* transposon. What the basis for these mutant phenotypes is has not been established, but it may be a mechanism similar to that which caused the background mutations in the T-DNA tagging experiments (both tissue culture and non-tissue culture). Interestingly, the early population contained plants with multiple T-DNA insertions whereas the latter ones contained plants with only one T-DNA insertion. Thus, rearrangement or interactions of multiple T-DNA insertions in the genome may be a cause of the relatively high frequency of non-tagged mutations.

## Transposon Derivatives to Extend the Range of Mutations

As with the T-DNA elements (Topping et al. 1991), derivatives of transposable elements have been made in order to extend the range of mutations that can be obtained. The wealth of knowledge about the structure and function of the *Ac/Ds* element system made these elements the system of choice for this approach. Following the successful examples in *Drosophila* and mouse (Bier et al. 1989; Wilson et al. 1989; Skarnes 1990), enhancer (Fedoroff and Smith 1993; V. Sundaresan et al.; J. Jones.; G. Coupland; all pers. comm.) and gene trap (V. Sundaresan et al.; G. Coupland; both pers. comm.) elements have been constructed by positioning a β-glucuronidase (*GUS*) fusion at one end of a *Ds* element. Enhancer-trap elements have a basal promoter requiring an enhancer element nearby to increase expression to detectable levels. A gene-trap element has no promoter and includes splice acceptor sites at the beginning of the *GUS* coding region (often in all three reading frames to enhance the success rate of an in-frame fusion). Transposition of the gene-trap element into an intron, and subsequent splicing using the gene's splice donor site and the element's splice acceptor sites, generates a *GUS* fusion under the control of a new promoter. Lines carrying a transposed enhancer or gene-trap *Ds* element are screened for novel patterns of *GUS* expression in addition to mutant phenotypes. These elements provide a number of advantages over unmodified *Ds* elements. First, they allow genes to be detected which are duplicated, and thus would not give a mutant phenotype upon insertion of an unmodified *Ds* element. Second, they allow the pattern of gene expression to be correlated with a mutant phenotype, when a mutation is caused by the insertion. Third, functions of genes, which lead to embryo lethality in the mutant, can be determined in the heterozygote.

Another *Ds* derivative that is currently being tested carries the strong 35S CaMV promoter, reading out of one end of the element (R. Masterson; G. Coupland; both pers. comm). This is designed to make dominant mutations by overexpression or ectopic expression of genes, as recently demonstrated with a T-DNA version of this construct (Hayashi et al. 1992).

## IDENTIFICATION OF AN INSERTION IN A GENE OF CHOICE

It is now possible to isolate a mutant, using one of several techniques described in this review, and to subsequently isolate the corresponding gene using one of several approaches. However, the reverse approach, starting with a sequence and attempting to elucidate its function, remains laborious and unpredictable. Techniques currently being used to determine the functions of cloned genes are antisense or cosuppression inhibition (see, e.g., Smith et al. 1988; Napoli et al. 1990). However, these techniques often do not give complete null phenotypes. These problems can be circumvented by selecting transformants carrying an insertion (either T-DNA or transposon) in the gene of interest. This reverse genetics procedure has been used successfully in *Drosophila* with *P* elements (see, e.g., Ballinger and Benzer 1989; Kaiser and Goodwin 1990; Hamilton et al. 1991) and in *Caenorhabditis elegans* with the *Tc1* element (Plasterck 1993; Rushforth et al. 1993). PCR primers were made from the transposon and the gene of interest to generate a PCR fragment. Sib selection was used to isolate the disruption mutant. N. Ali and K. Feldmann (unpubl.) are currently taking advantage of the thousands of independent T-DNA insertion lines in *Arabidopsis* to identify insertional mutations for sequenced genes or cDNAs. DNA from small pools of transformants is being harvested and used in PCR reactions or transferred to hybridization filters. The filters will be probed with single and pooled DNA sequences. Detection of an RFLP in one of the pools indicates that one of the transformants contains a disrupted locus for that specific probe. This polymorphism is verified on blots in which the DNA has been restricted with a second enzyme. An alternative approach is to use PCR screening and sib selection to identify the individual plant containing the insertional mutation. For this procedure DNA from 1,000 lines is being pooled in a 10 x 10 matrix such that 20 PCR reactions, each containing DNA from 10 pools of 10 transformants, will indicate which pool of 10 transformants among the 1,000 lines contains the disruption mutant. Because of the concatameric nature of the inserts, and the need to use both primers derived from left and right borders simultaneously with several primers derived from the gene of interest, it is anticipated that numerous PCR products will be generated from each pool of DNA from 100 transformed lines. As such, the reaction products will be transferred to a hybridization filter and probed with the gene from which the primers were derived to identify PCR products that contain part of the gene of interest. Since each transformant is represented twice in the 20 PCR reactions, the same band should be visible in two lanes. The intersecting pool of 10 transformants will contain the insertion mutant. A second round of PCR on these 10 lines will indicate which of the lines is the insertion mutant.

The most important advantage of this technique is that it will be possible to identify a phenotype for a specific gene sequence. This should be key to understanding how that gene functions in the plant. These mutants will generally be loss-of-function mutants, so a true mutational analysis will be possible. Second, it will be possible to knock out individual members of gene families, for example, actins, tubulins, heat shock genes, or ATPases, and then cross the lines to generate mutants that have several genes in the family knocked out. Third, this approach does not require that the T-DNA insert be functional; i.e., a suppressed or truncated insert will still produce a detectable polymorphism and an identifiable mutant.

Finally, this technique will not be limited to viable mutants but will make it possible to identify lethals in the population due to the dominance nature of the polymorphisms.

With the currently available T-DNA transformants (~18,000 lines representing 27,000 inserts), and using the previous assumptions, i.e., 100,000 kb/haploid genome, an average of 1.5 inserts/transformant, an average gene length of 4 kb, and random insertion, approximately 70% of all the DNA probes would be predicted to detect a corresponding T-DNA mutant. This should prove to be a very powerful approach for identifying a function in the plant for a specific gene sequence.

## CONCLUSIONS AND PERSPECTIVES

There are many hundreds of laboratories undertaking mutagenesis experiments in *Arabidopsis* to try to identify mutations in an ever increasing range of processes. There is an increasing list of mutagens to choose from in conjunction with different methods to isolate the mutated gene. Different strategies can be adopted for different mutations. For rare mutations or mutations for which it is difficult to screen, it is probably advisable to use a heavily mutagenized population (with one of the classical mutagens, EMS, $\gamma$ rays, or X-rays), map the mutation, and then isolate the gene using chromosome walking, or reisolate the mutation using a directed transposon tagging strategy. This would involve starting from a line containing a transposon in a linked T-DNA. Alternatively, populations carrying random T-DNA or transposon insertions can be screened for desired mutations. This process has been made universally available since the Arabidopsis Stock Centers have been distributing these lines. Strategies aimed at generating deletion or rearrangement alleles, followed by cloning using PCR-based techniques, are likely to become more popular in the future, given the large effort currently in progress developing techniques to isolate differences between complex genomes (see, e.g., Strauss and Ausubel 1990; Lisitsyn et al. 1993).

Because mutants help provide the essential link between genotype and phenotype, the *Arabidopsis* community must aim for saturation mutagenesis of the genome. This will allow researchers to fully interpret the genomic sequence as it becomes available and facilitate an understanding of the working of this model flowering plant.

## ACKNOWLEDGMENTS

K.A.F. acknowledges the support of a grant from the National Science Foundation (DIR-9108442). Research in the laboratory of R.L.M. is supported by a grant from Department of Energy BioSciences (DEFG-0991ER-20034). C.D. acknowledges support from the Agricultural and Food Research Council Plant Molecular Biology Program. We thank Chris Somerville for useful discussions about the mutation data. The authors also thank their many colleagues who shared unpublished results for this review.

## REFERENCES


Aarts, M.G.M., W.G. Dirkse, W.J. Stiekema, and A. Pereira. 1993. Transposon tagging of a male sterility gene in *Arabidopsis*. *Nature* **363:** 715–717.

Ahmad, M. and A.R. Cashmore. 1993. *HY4* gene of *A. thaliana* encodes a protein with characteristics of a blue-light photoreceptor. *Nature* **366:** 162–164.

Arondel, V., B. Lemieux, I. Hwang, S. Gibson, H.M. Goodman, and C.R. Somerville. 1992. Map-based cloning of a gene controlling *omega-3* fatty acid desaturation in *Arabidopsis*. *Science* **258:** 1353–1355.

Balcells, L., E. Sundberg, and G. Coupland. 1994. A heat-shock promoter fusion to the *Ac* transposase gene drives inducible transposition of a *Ds* element during *Arabidopsis* embryo development. *Plant J.* (in press).

Balcells, L., J. Swinburne, and G. Coupland. 1991. Transposons as tools for the isolation of plant genes. *Trends Biotechnol.* **9:** 31–37.

Ballinger, D.G. and S. Benzer. 1989. Targeted gene mutations in *Drosophila*. *Proc. Natl. Acad. Sci.* **86:** 9402–9406.

Bancroft, I. and C. Dean. 1993. Transposition pattern of the maize element *Ds* in *Arabidopsis thaliana*. *Genetics* **134:** 1221–1229.

Bancroft, I., J. Jones, and C. Dean. 1993. Heterologous transposon tagging of the DRL1 locus in *Arabidopsis*. *Plant Cell* **5:** 631–638.

Bancroft, I., A.M. Bhatt, C. Sjodin, S. Scofield, J.D.G. Jones, and C. Dean. 1992. Development of an efficient two-element transposon tagging system in *Arabidopsis thaliana*. *Mol. Gen. Genet.* **233:** 449–461.

Baskin, T.I., A.S. Betzner, R. Hoggart, A. Cork, and R.E. Williamson. 1992. Root morphology mutants in *Arabidopsis thaliana*. *Aust. J. Plant Physiol.* **19:** 427–437.

Bechtold, N., J. Ellis, and G. Pelletier. 1993. *In planta Agrobacterium* mediated gene transfer by infiltration of adult *Arabidopsis thaliana* plants. *C.R. Acad. Sci.* **316:** 1194–1199.

Bell, C.J. and J.R. Ecker. 1994. Assignment of 30 microsatellite loci to the linkage map of *Arabidopsis*. *Genomics* **19:** 137–144.

Bhatt, A.M. and C. Dean. 1992. Development of tagging systems in plants using heterologous transposons. *Curr. Opin. Biotechnol.* **3:** 152–158.

Bier, E., H. Vaessin, S. Shepherd, K. Lee, K. McCall, S. Barbel, L. Ackerman, R. Carretto, T. Uemera, F. Grell, I.Y. Jan, and Y.N. Jan. 1989. Searching for pattern and mutation in the *Drosophila* genome with a P-*lacz* vector. *Genes Dev.* **3:** 1273–1287.

Bouchez, D., C. Camilleri, and M. Caboche. 1993. A binary vector based on Basta resistance for *in planta* transformation of *Arabidopsis thaliana*. *C.R. Acad. Sci.* **316:** 1188–1193.

Brockman, H.E , F.J. deSerres, T. Ong, D.M. DeMarini, A.J. Katz, A.J.F. Griggiths, and R.S. Stafford. 1984. Mutation tests in *Neurospora crassa*. *Mutat. Res.* **133:** 87–134.

Castle, L. and D.W. Meinke. 1994. A *FUSCA* gene of *Arabidopsis* encodes a novel protein essential for plant development. *Plant Cell* **6:** 25–41.

Castle, L., D. Errampalli, T.L. Atherton, L.H. Franzmann, E.S. Yoon, and D.W. Meinke. 1993. Genetic and molecular characterization of embryonic mutants identified following seed transformation of *Arabidopsis*. *Mol. Gen. Genet.* **241:** 504–541.

Chang, C., S.F. Kwok, A.B. Bleecker, and E.M. Meyerowitz. 1993. *Arabidopsis* ethylene-response gene *ETR1*: Similarity of product to two-component regulators. *Science* **262:** 539–544.

Chang, C., J.L. Bowman, A.W. DeJohn, E.S. Lander, and E.M. Meyerowitz. 1988. Restriction fragment length polymorphism linkage map for *Arabidopsis thaliana*. *Proc. Natl. Acad. Sci.* **85:** 6856–6860.

Chang, S.-S, S.-K Park, and H.-G. Nam. 1990. Transformation of *Arabidopsis* by *Agrobacterium* inoculation on wounds. In *Abstracts from the 4th International Conference on* Arabidopsis *Research*, Vienna, Austria, p. 28. University of Vienna, Austria.

Chang, S.S., S.K. Park, B.C. Kim, B.J. Kang, D.U. Kim, and H.G. Nam. 1994. Stable genetic transformation of *Arabidopsis thaliana* by *Agrobacterium* inoculation in planta. *Plant J.* **5:** 551–558.

Christianson, M.L. and D.A. Warnick. 1985. Temporal requirement for phytohormone balance in the control of organogenesis *in vitro. Dev. Biol.* **112:** 494–497.

Cooley, L., R. Kelley, and A. Spradling. 1988. Insertional mutagenesis of the *Drosophila* genome with single P elements. *Science* **239:** 1121–1128.

Coomber, S.A. and K.A. Feldmann. 1993. Gene tagging in transgenic plants. In *Transgenic plants: Engineering and utilization* (ed. S.-d. Kung and R. Wu), vol. 1, pp. 225–240. Academic Press, New York.

Coupland, G. 1992. Transposon tagging in Arabidopsis. In *Methods in* Arabidopsis *research* (ed. C. Koncz et al.), pp. 291–309. World Scientific, London.

Dean, C., C. Sjodin, T. Page, J. Jones, and C. Lister. 1992. Behaviour of the maize transposable element *Ac* in *Arabidopsis thaliana*. *Plant J.* **2:** 69–81.

Deng, X.-W., M. Matsui, N. Wei, D. Wagner, A.M. Chu, K.A. Feldmann, and P.H. Quail. 1992. *COP1*, an *Arabidopsis* photomorphogenic regulatory gene, encodes a novel protein with both a Zn-binding motif and a domain homologous to the β-subunit of trimeric G-proteins. *Cell* **71:** 791–801.

Dolan, L., K. Janmaat, V. Willemsin, P. Linstead, S. Poethig, K. Roberts, and B. Scheres. 1993. Cellular organisation of the *Arabidopsis thaliana* root. *Development* **119:** 71–84.

Dooner, H.K. and A. Belachew. 1989. Transposition pattern of maize element *Ac* from the *bz-m2(Ac)* allele. *Genetics* **122:** 447–457.

Dooner, H.K., J. Keller, E. Harper, and E. Ralston. 1991. Variable patterns of transposition of the maize element *Activator* in tobacco. *Plant Cell* **3:** 473–482.

DuBridge, R.B. and M.P. Calos. 1987. Molecular approaches to the study of gene mutation in human cells. *Trends Genet.* **3:** 293–297.

Errampalli, D., D. Patton, L. Castle, L. Mickelson, K. Hansen, J. Schnall, K. Feldmann, and D. Meinke. 1991. Embryonic lethals and T-DNA insertional mutagenesis in *Arabidopsis*. *Plant Cell* **3:** 149–157.

Fedoroff, N.V. and D.L. Smith. 1993. A versatile system for detecting transposition in *Arabidopsis*. *Plant J.* **3:** 273–290.

Feldmann, K.A. 1991. T-DNA insertion mutagenesis in *Arabidopsis*: Mutational spectrum. *Plant J.* **1:** 71–82.

———. 1992. T-DNA insertion mutagenesis in *Arabidopsis:* Seed infection/transformation. In *Methods in* Arabidopsis *research* (ed. C. Koncz et al.), pp. 274–289. World Scientific, London.

Feldmann, K.A. and M.D. Marks. 1986. Rapid and efficient regeneration of plants from explants of *Arabidopsis thaliana*. *Plant Sci.* **47**: 63–69.

————. 1987. Agrobacterium-mediated transformation of germinating seeds of *Arabidopsis thaliana*: A non-tissue culture approach. *Mol. Gen. Genet.* **208**: 1–9.

Feldmann, K.A. and E.M. Meyerowitz. 1991. Tagging floral structure genes. In *Genetics and breeding of ornamental species* (ed. J. Harding et al.), pp. 271–283. Kluwer Academic Publishers, Dordrecht, The Netherlands.

Feldmann, K.A., M.D. Marks, M.L. Christianson, and R.S. Quatrano. 1989. A dwarf mutant of *Arabidopsis* generated by T-DNA insertion mutagenesis. *Science* **243**: 1351–1354.

Feldmann, K.A., A.M. Wierzbicki, R.S. Reiter, and S.A. Coomber. 1991. T-DNA insertion mutagenesis in *Arabidopsis*. In *Plant molecular biology* (ed. R. Hermann and B. Larkins), vol. 2, pp. 563–574. Plenum Press, New York.

Feldmann, K.A., T.J. Carlson, S.A. Coomber, C.E. Farrance, M.A. Mandel, and A.M. Wierzbicki. 1990. T-DNA insertional mutagenesis in *Arabidopsis thaliana*. In *Horticultural biotechnology* (ed. A.B. Bennett and S.D. O'Neill), pp. 109–120. Wiley-Liss, New York.

Forsthoefel, N.R., Y.Wu, B. Schulz, M.J. Bennett, and K.A. Feldmann. 1992. T-DNA insertion mutagenesis in *Arabidopsis:* Prospects and perspectives. *Aust. J. Plant Physiol.* **19**: 353–366.

Freeling, M. 1977. Spontaneous forward mutation vs. reversion frequencies for maize *Adh1* in pollen. *Nature* **267**: 154–156.

Freeling, M. and D. Cheng. 1978. Radiation induced alcohol dehydrogenase mutants in maize following allylalcohol selection of pollen. *Genet. Res.* **31**: 107–129.

Furner, I.J and J.E. Pumfrey. 1992. Cell fate in the shoot apical meristem of *Arabidopsis thaliana*. *Development* **115**: 755–764.

Giraudat, J., B.M. Hauge, C. Valon, J. Smalle, F. Parcy, and H.M. Goodman. 1992. Isolation of the *Arabidopsis ABI3* gene by positional cloning. *Plant Cell* **4**: 1251–1261.

Greenblatt, I.M. 1984. A chromosome replication pattern deduced from pericarp phenotypes resulting from movements of the transposable element modulator in maize. *Genetics* **108**: 471–485.

Grevelding, C., D. Becker, R. Kunze, A. von Menges, V. Fantes, J. Schell, and R. Masterson. 1992. High rates of *Ac/Ds* germinal transposition in *Arabidopsis* suitable for gene isolation by insertional mutagenesis. *Proc. Natl. Acad. Sci.* **89**: 6085–6089.

Hamilton, B.A., M.J. Palazzolo, J.H. Chang, K. VijayRaghavan, C.A. Mayeda, M.A. Whitney, and E.M. Meyerowitz. 1991. Large scale screen for transposon insertions into cloned genes. *Proc. Natl. Acad. Sci.* **88**: 2731–2735.

Haughn, G.W. and C.R. Somerville. 1987. Selection for herbicide resistance at the whole plant level. In *Biotechnology in agricultural chemistry* (ed. H.M. LeBaron et al.), pp. 98–108. American Chemical Society, Washington, D.C.

Hayashi, H., I. Czaja, H. Lubenow, J. Schell, and R. Walden. 1992. Activation of a plant gene by T-DNA tagging: Auxin-independent growth in vitro. *Science* **258**: 1350–1353.

Herman, P.L. and M.D. Marks. 1989. Trichome development in *Arabidopsis thaliana*. II. Isolation of complementation of the *GLABROUS1* gene. *Plant Cell* **1**: 1051–1055.

Honma, M., B.J. Baker, and C.S. Waddell. 1993. High frequency germinal transposition of $Ds^{ALS}$ in *Arabidopsis*. *Proc. Natl. Acad. Sci.* **90**: 6242–6246.

Horsch, R.B., J.E. Fry, N.L. Hoffmann, D. Eichholtz, S.G. Rogers, and R.T. Fraley. 1985. A simple and general method for transferring genes into plants. *Science* **227**: 1229–1231.

Iba, K., S. Gibson, T. Nishiuchi, T. Fuse, M. Nishimura, V. Arondel, S. Hugly, and C.R. Somerville. 1993. A gene encoding a chloroplast *omega-3* fatty acid desaturase complements the fatty acid composition and chloroplast copy number defects for the *fad7* mutant of *Arabidopsis thaliana*. *J. Biol. Chem.* **268:** 24099–24105.

Irish, V.F. and I.M. Sussex. 1992. A fate map of the *Arabidopsis* embryonic shoot apical meristem. *Development* **115:** 745–753.

Jones, J.D.G., F.M. Carland, P. Maliga, and H.K. Dooner. 1989. Visual detection of transposition of the maize element *Activator* (*Ac*) in tobacco seedlings. *Science* **244:** 204–207.

Jürgens, G., U. Mayer, R.A. Torres-Ruiz, T. Berleth, and S. Misera. 1991. Genetic analysis of pattern formation in the *Arabidopsis* embryo. *Dev. Suppl.* **1:** 27–38.

Kahler, A.L., R.W. Allard, and R.D. Miller. 1984. Mutation rates for enzyme and morphological loci in barley. *Genetics* **106:** 729–734.

Kaiser, K. and S.F. Goodwin. 1990. "Site-selected" transposon mutagenesis in *Drosophila*. *Proc. Natl. Acad. Sci.* **87:** 1686–1690.

Keller, J., E. Lim, and H.K. Dooner. 1993. Preferential transposition of *Ac* to linked sites in *Arabidopsis*. *Theor. Appl. Genet.* **88:** 585–588.

Keller, J., E. Lim, D.W. James, Jr., and H.K. Dooner. 1992. Germinal and somatic activity of the maize element *Activator* (*Ac*) in *Arabidopsis*. Genetics **131:** 449–459.

Kieber, J.J., M. Rothenberg, G. Roman, K.A. Feldmann, and J.R. Ecker. 1993. The ethylene response pathway in *Arabidopsis thaliana* is negatively regulated by *CTR1*, a predicted member of the *Raf* family of protein kinases. *Cell* **72:** 427–441.

Kleckner, N. 1981. Transposable elements in prokaryotes. *Annu. Rev. Genet.* **15:** 341–404.

Koncz, C., J. Schell, and G.P. Rédei. 1992a. T-DNA transformation and insertion mutagenesis. In *Methods in* Arabidopsis *research* (ed. C. Koncz et al.), pp. 224–273. World Scientific, London.

Koncz, C., K. Nemeth, G.P. Rédei, and J. Schell. 1992b. T-DNA insertional mutagenesis in *Arabidopsis*. *Plant Mol. Biol.* **20:** 963–976.

Koncz, C., N. Martini, R. Mayerhofer, Z. Koncz-Kalman, H. Korber, G.P. Rédei, and J. Schell. 1989. High-frequency T-DNA-mediated gene tagging in plants. *Proc. Natl. Acad. Sci.* **86:** 8467–8471.

Koncz, C., R. Mayerhofer, Z. Koncz-Kalman, C. Nawrath, B. Reiss, G.P. Rédei, and J. Schell. 1990. Isolation of a gene encoding a novel chloroplast protein by T-DNA tagging in *Arabidopsis thaliana*. *EMBO J.* **9:** 1337–1346.

Konieczny, A. and F.M. Ausubel. 1993. A procedure for mapping *Arabidopsis* mutations using codominant ecotype-specific PCR-based markers. *Plant J.* **4:** 403–410.

Konieczny, A., D.F. Voytas, M.P. Commings, and F.M. Ausubel. 1991. A superfamily of *Arabidopsis thaliana* retrotransposons. *Genetics* **127:** 801–809.

Kunze, R., U. Stochaj, J. Laugs, and P. Starlinger. 1987. Transcription of transposable element *Activator* (*Ac*) of *Zea mays* L. *EMBO J.* **6:** 1555–1563.

LaBrie, S.R., J.Q. Wilkinson, Y.-F. Tsay, K.A. Feldmann, and N.M. Crawford. 1992. Identification of two tunstate-sensitive molybdenum cofactor mutants, *chl2* and *chl7*, of *Arabidopsis thaliana*. *Mol. Gen. Genet.* **233:** 169–176.

Larkin, P.J. and W.R. Scowcroft. 1981. Somaclonal variation—A novel source of variability from cell cultures for plant improvement. *Theor. Appl. Genet.* **60:** 197–214.

Lawson, E., S. Scofield, C. Sjodin, J.D.G. Jones, and C. Dean. 1994. Modification of the 5' untranslated leader region of the maize *Activator* element leads to increased activity in *Arabidopsis*. *Mol. Gen. Genet.* (in press).

Lee, I., M.J. Aukerman, S.L. Gore, K.N. Lohman, S.D. Michaels, L.M. Weaer, M.C. John, K.A. Feldmann, and R.M. Amasino. 1994. Isolation of *Luminidependens*: A gene involved in the control of flowering time in *Arabidopsis*. *Plant Cell* **6:** 75–83.

Leyser, H.M.O., C.A. Lincoln, C. Timpte, D. Lammer, J. Turner, and M. Estelle. 1993. *Arabidopsis* auxin-resistance gene *AXR1* encodes a protein related to ubiquitin-activating enzyme E1. *Nature* **364:** 161–164.

Li, S. and G.P. Rédei. 1969. Estimation of mutation rates in autogamous diploids. *Radiat. Bot.* **9:** 125–131.

Lisitsyn, N., N. Lisitsyn, and M. Wigler. 1993. Cloning the differences between two complex genomes. *Science* **259:** 946–951.

Lloyd, A.M., A.R. Barnason, S.G. Rogers, M.C. Byrne, R.T. Fraley, and R.B. Horsch. 1986. Transformation of *Arabidopsis thaliana* with *Agrobacterium tumefaciens*. *Science* **234:** 464–466.

Long, D., M. Martin, E. Sundberg, J. Swinburne, P. Puangsomlee, and G. Coupland. 1993. The maize transposable element system *Ac/Ds* as a mutagen in *Arabidopsis*: Identification of an albino mutation induced by *Ds* insertion. *Proc. Natl. Acad. Sci.* **90:** 10370–10374.

Malmberg, R.L. 1993. Production and analysis of plant mutants, emphasizing *Arabidopsis thaliana*. In *Methods in plant molecular biology and biotechnology* (ed. B.R. Glick and J.E. Thompson), pp. 11–28. CRC Press, Boca Raton, Florida.

Marks, M.D. and K.A. Feldmann. 1989. Trichome development in *Arabidopsis thaliana*. I. T-DNA tagging of the *GLABROUS1* gene. *Plant Cell* **1:** 1043–1050.

Masterson, R.V., D.B Furtek, C. Grevelding, and J. Schell. 1989. A maize *Ds* transposable element containing a dihydrofolate reductase gene transposes in *Nicotiana tabacum* and *Arabidopsis thaliana*. *Mol. Gen. Genet.* **219:** 461–466.

McKelvie, A.D. 1962. A list of mutant genes in *Arabidopsis thaliana* (L). *Heynh. Radiat. Bot.* **1:** 233–241.

McNevin, J.P., W. Woodward, A. Hannoufa, K.A. Feldmann, and B. Lemieux. 1993. Isolation and characterization of *Eceriferum* (*cer*) mutants induced by T-DNA insertions in *Arabidopsis thaliana*. *Genome* **36:** 610–618.

Medford, J.I., F. J. Behringer, J.D. Callos, and K.A. Feldmann. 1992. Normal and abnormal development in the *Arabidopsis* vegetative shoot apex. *Plant Cell* **4:** 631–643.

Meinke, D.M. 1992. A homeotic mutant of *Arabidopsis thaliana* with leafy cotyledons. *Science* **258:** 1647–1650.

Nam, H.-G., J. Giraudat, B. den Boer, F. Moonan, W.D.B. Loos, W.B.M. Hauge, and H.M. Goodman. 1989. Restriction fragment length polymorphism linkage map of *Arabidopsis thaliana*. *Plant Cell* **1:** 699–705.

Napoli, C., C. Lemieux, and R. Jorgenson. 1990. Introduction of a chimeric chalcone synthase gene into *Petunia* results in reversible co-suppression of homologous genes in trans. *Plant Cell* **2:** 279–289.

Neel, J.V., C. Satoh, K. Goriki, M. Fujita, N. Takahashi, J. Asakawa, and R. Hazama. 1986. The rate with which spontaneous mutation alters the electrophoretic mobility of polypeptides. *Proc. Natl. Acad. Sci.* **83:** 389–393.

Okuley, J., J. Lightner, K. Feldmann, N. Yadav, and J. Browse. 1994. The *Arabidopsis FAD2* gene encodes the enzyme that is essential for polyunsaturated lipid synthesis. *Plant Cell* **6:** 147–158.

Oppenheimer, D.G., P.L. Herman, S. Sivakumaran, J. Esch, and M.D. Marks. 1991. A *myb* gene required for leaf trichome differentiation in *Arabidopsis* is expressed in stipules. *Cell* **67:** 483–493.

Orozco, B.M., C.R. McClung, J.M. Werneke, and W.L. Ogren. 1993. Molecular basis of the ribulose-1,5-bisphosphate carboxylase/oxygenase activase mutation in *Arabidopsis thaliana* is a guanine to adenine transition at the 5′-splice junction of intron 3. *Plant Physiol.* **102:** 227–232.

Peleman, J., B. Cottyn, W. van Camp, M. Van Montagu, and D. Inze. 1991. Transient oc-currence of extrachromosomal DNA of an *Arabidopsis thaliana* transposon-like ele-ment, *Tat1*. *Proc. Natl. Acad. Sci.* **88:** 3618–3622.

Peng, J. and N.P. Harberd. 1993. Derivative alleles of the *Arabidopsis* gibberellin insensi-tive (*gai*) mutation confer a wild type phenotype. *Plant Cell* **5:** 351–360.

Plasterck, R.H.A. 1993. Reverse genetics of *Caenorhabditis elegans*. *BioEssays* **14:** 629–633.

Rédei, G.P. 1970. *Arabidopsis thaliana*: A review of the genetics and biology. *Bibliogr. Genet.* **20:** 1–151.

————. 1975. *Arabidopsis* as a genetic tool. *Annu. Rev. Genet.* **9:** 111–151.

Rédei, G.P. and C. Koncz. 1992. Classical mutagenesis. In *Methods in* Arabidopsis *re-search* (ed. C. Koncz et al.), pp. 16–82. World Scientific, London.

Reed, J., P. Nagpal, D. Poole, M. Furuya, and J. Chory. 1993. Mutations in the gene for the red far-red light receptor phytochrome-B alter cell elongation and physiological responses throughout *Arabidopsis* development. *Plant Cell* **5:** 147–157.

Reiter, R.S., J.G. Williams, K.A. Feldmann, J.A.Rafalski, S.V. Tingey, and P.A. Scolnik. 1992. Global and local genome mapping in *Arabidopsis thaliana* by using recombinant inbred lines and random amplified polymorphic DNAs. *Proc. Natl. Acad. Sci.* **89:** 1477–1481.

Rieger, R., A. Michaelis, and M.M. Green. 1976. *Glossary of genetics and cytogenetics*. Springer-Verlag, New York.

Roe, J.L., C.J. Rivin, R.A. Sessions, K.A. Feldmann, and P.C. Zambryski. 1993. The *TOUSLED* gene in *Arabidopsis* encodes a protein kinase homologue and is required for leaf and flower development. *Cell* **75:** 939–950.

Rushforth, A.M., B. Saari, and P. Anderson. 1993. Site-selected insertion of the transposon *Tc1* into a *Caenorhabditis elegans* myosin light chain gene. *Mol. Cell. Biol.* **13:** 902–910.

Sangwan, R.S., I. Velu, P. Cobanov, B. Vilcot, Y. Bourgeois, and B.S. Sangwan-Noreel. 1993. Spectrum of insertional mutagenesis in *Arabidopsis* using *Agrobacterium*-mediated zygotic embryo transformation methodology. In *Abstracts from the 5th Inter-national Conference on* Arabidopsis *Research*, Columbus, Ohio, p. 39. University of Ohio, Columbus.

Schmidt, R. and L. Willmitzer. 1989. The maize element *Activator* (*Ac*) shows a minimal germinal excision frequency of 0.2–0.5% in transgenic *Arabidopsis* plants. *Mol. Gen. Genet.* **220:** 11–24.

Schmidt, R.J., F.A. Burr, and B. Burr. 1987. Transposon tagging and molecular analysis of the maize regulatory locus *opaque-2*. *Science* **238:** 960–963.

Skarnes, W.C. 1990. Entrapment vectors: A new tool for mammalian genetics. *BioTech-nology* **8:** 827–831.

Smith, C.J.S., C.F. Watson, J. Ray, C.R. Bird, P.C. Morris, W. Schuch, and D. Grierson. 1988. Antisense RNA inhibition of polygalacturonase genes in transgenic tomatoes. *Nature* **334:** 724–726.

Sommer, H., Z.J.-P. Beltran, P. Huijser, H. Pape, W.-E. Lonnig, H. Saedler, and Z. Schwarz-Sommer. 1990. *Deficiens*, a homeotic gene involved in the control of flower morphogenesis in *Antirrhinum majus*: The protein shows homology to transcription factors. *EMBO J.* **9:** 605–613.

Stadler, L.J. and H. Roman. 1948. The effect of X-rays upon mutation of the gene *A* in maize. *Genetics* **33:** 273–303.

Strauss, D. and F.M. Ausubel. 1990. Genomic subtraction for cloning DNA correspond-ing to deletion mutations. *Proc. Natl. Acad. Sci.* **87:** 1889–1893.

Sun, T.-P., H.M. Goodman, and F.M. Ausubel. 1992. Cloning the *Arabidopsis GA1* locus by genomic subtraction. *Plant Cell* **4:** 119–128.

Swinburne, J., L. Balcells, S.R. Scofield, J.D.G. Jones, and G. Coupland. 1992. Elevated levels of.*Ac* transposase mRNA are associated with high frequencies of *Ds* excision in *Arabidopsis*. *Plant Cell* **4:** 583–595.

Topping, J.F., W. Wei, and K. Lindsey. 1991. Functional tagging of regulatory elements in the plant genome. *Development* **112:** 1009–1019.

Tsay, Y.-F., J. Schroeder, K.A. Feldmann, and N. Crawford. 1993. The herbicide sensitivity gene, *chl1*, of *Arabidopsis* encodes a nitrate-inducible nitrate transporter. *Cell* **72:** 705–713.

Tsukaya, H., S. Naito, G.P. Rédei, and Y. Komeda. 1993. A new class of mutations in *Arabidopsis thaliana*, *acaulis 1*, affecting the development of both inflorescences and leaves. *Development* **118:** 751–764.

Valvekens, D. and M. Van Montagu. 1990. Spontaneous mutagenesis associated with *Arabidopsis* root regeneration. In *Abstracts from the 4th International Conference on* Arabidopsis *Research*, Vienna, Austria, p. 46. University of Vienna, Austria.

Valvekens, D., M. Van Montagu, and M. Lijsebettens. 1988. *Agrobacterium tumefaciens*-mediated transformation of *Arabidopsis* root explants using kanamycin selection. *Proc. Natl. Acad. Sci.* **85:** 5536–5540.

Van Lijsebettens, M., R. Vanderhaeghen, and M. Van Montagu. 1991a. Insertional mutagenesis in *Arabidopsis thaliana*: Isolation of a T-DNA-linked mutation that alters leaf morphology. *Theor. Appl. Genet.* **81:** 277–284.

Van Lijsebettens, M., B. den Boer, J.-P. Hernalsteens, and M. Van Montagu. 1991b. Insertion mutagenesis in *Arabidopsis thaliana*. *Plant Sci.* **80:** 27–37.

van Sluys, M.A., J. Tempe, and N. Fedoroff. 1987. Studies on the introduction and mobility of the maize *Activator* element in *Arabidopsis thaliana* and *Daucus carota*. *EMBO J.* **13:** 3881–3889.

Weigel, D., J. Alvarez, D.R. Smyth, M.F. Yanofsky, and E.M. Meyerowitz. 1992. LEAFY controls floral meristem identity in *Arabidopsis*. *Cell* **69:** 843–859.

Wilkinson, J.Q. and N.M. Crawford. 1993. Identification and characterization of a chlorate resistant mutant of *Arabidopsis thaliana* with mutations in both nitrate reductase structural genes *Nia1* and *Nia2*. *Mol. Gen. Genet.* **239:** 289–297.

Williams, J.G.K., R.S. Reiter, R.M. Young, and P.A. Scolnik. 1993. Genetic-mapping of mutations using phenotypic pools and mapped RAPD markers. *Nucleic Acids Res.* **21:** 2697–2702.

Wilson, C., R.K. Pearson, H.J. Belien, C.J. O'Kane, U. Grossniklaus, and W.J. Gehring. 1993. P-element mediated enhancer detection: An efficient method for isolating and characterizing developmentally regulated genes in *Drosophila*. *Genes Dev.* **3:**1301–1313.

Yadav, N., A. Wierzbicki, M. Agatier, C. Caster, L. Perez-Grau, A.J. Kinney, W.D. Hitz, J.R. Booth, B. Schweiger, K.L. Stecca, S.M. Allen, M. Blackwell, R.S. Reiter, T.J. Carlson, S. Russell, K.A. Feldmann, J. Pierce, and J. Browse. 1993. Cloning of higher plant *omega-3* fatty acid desaturases. *Plant Physiol.* **103:** 467–476.

Yanofsky, M.F., H. Ma, J.L. Bowman, G.N. Drews, K.A. Feldmann, and E.M. Meyerowitz. 1990. The protein encoded by the *Arabidopsis* homeotic gene *agamous* resembles transcription factors. *Nature* **346:** 35–39.

Yoder, J.I., J. Palys, K. Alpert, and M. Lassner. 1988. *Ac* transposition in transgenic tomato plants. *Mol. Gen. Genet.* **213:** 291–296.

# 8

# Tissue Culture and Transformation

**Peter C. Morris[1] and Thomas Altmann**
Institut für Genbiologische Forschung Berlin GmbH
14195 Berlin, Germany

## INTRODUCTION

### Plant Tissue Culture

The basic property of totipotency that enables plant cells derived from various differentiated organs like leaves, roots, cotyledons, or hypocotyls to regenerate whole plants allows the development of a variety of tissue culture/regeneration procedures, as well as methods for the efficient generation of genetically transformed plants. Crucial in these processes is the use of certain phytohormone combinations, basically auxins and cytokinins, in synthetic growth media that induce the initial dedifferentiation of cells and the subsequent redifferentiation into organized structures. Typically, explants like leaf disks or stem, root, or hypocotyl segments, or protoplasts isolated from leaves, are first induced to form callus that later, via organogenesis, regenerates into shoot and root structures or may form embryos via somatic embryogenesis. After the formation of the general structure of the plant body, consisting of an organized shoot with a developed root system, the regenerants may be transferred from the axenic culture to soil. The regenerated plants usually complete their normal life cycle with the formation of seeds or other propagative structures like tubers (for comprehensive presentations, see Vasil 1984, 1985, 1986; Dixon 1985; Lindsey 1991, 1992). The broad spectrum of in vitro culture techniques is reflected in the variety of applications, a few of which we briefly mention here.

A large number of plant species have been introduced into cell suspension culture. Although this technology is expected to allow the industrial production of plant (secondary) metabolites (Constabel and Vasil 1988), it is to date almost exclusively applied on a laboratory scale to serve as a constant supply of cell material for various purposes, including physiological studies. The use of embryogenic suspension culture has allowed the regeneration and genetic transformation of a number of impor-

---

[1]Present address: CNRS Institut des Sciences Végétales, Avenue de la Terrasse, 91198 Gif sur Yvette Cedex, France.

*Arabidopsis*
© 1994 Cold Spring Harbor Laboratory Press 0-87969-428-9/94 $5 + .00

tant crop plants (Rhodes et al. 1988; Zhang and Wu 1988; Vasil et al. 1990; Jähne et al. 1991). The regeneration of numerous plants from a few or even a single progenitor plant allows the rapid generation of large numbers of genetically identical siblings. This in vitro propagation is applied in the mass production of ornamental plants (Levin et al. 1988). It is also used to generate genetic variability due to increased mutation rates during the process of de-/redifferentiation in the course of the in vitro regeneration of plants, a phenomenon called somaclonal variation (Larkin et al. 1989). In contrast, the organized apical meristems of plants are used as regeneration explants (meristem culture) to cure virus-infected plants (Kartha 1986). An even more highly organized structure, the plant embryo, is cultured in vitro or used as regeneration explant to overcome blocks in the development of embryos derived from wide crosses (Monnier 1990). This technique thus allows the incorporation of less-related genetic material into breeding programs aimed at the improvement of crop species. Wider steps in the combination of alien genomes can be achieved by the generation of somatic hybrids through protoplast fusions and regeneration of the fusion products (Schieder and Kohn 1986). Of particular interest are asymmetric fusions where one of the fusion partners (the donor) is X- or γ-irradiated prior to fusion, resulting in a partial genome transfer (Zelcer et al. 1978; Gupta et al. 1984; Somers et al. 1986; Dudits et al. 1987). Novel combinations of the nuclear genome of a certain variety with the plastids or the mitochondria of a different variety can also be achieved by fusion of pretreated donor (e.g., after enucleation or irradiation) and recipient protoplasts, by transferring isolated nuclei, by subprotoplast/protoplast microfusion, or even through organelle transfer using microinjection (Aviv and Galun 1980; Sidorov et al. 1981; Menczel et al. 1982; Saxena et al. 1986; Ichikawa et al. 1987; Smith et al. 1989; Eigel and Koop 1992; Verhoeven and Blaas 1992), resulting in so-called cybrids. This technique is of particular interest if cytoplasmic male sterility (cms), which is highly useful for hybrid seed production, can be transferred to a particular cultivar (Kyozuka et al. 1989). Regeneration of plants from micro- or macrospores (pollen, ovule, respectively) can lead to haploid plants which in turn are a source of doubled haploids. These diploid plants with homozygous genomes are very useful in breeding programs (Jensen 1986; Snape et al. 1986; Morrison and Evans 1988).

### Transgenic Plants

The combination of regeneration procedures with those for transfer and integration of foreign DNA into plant cells allows the generation of genetically transformed plants. Among other systems like microinjection, liposome fusion, or the use of a particle gun, the most widely applied procedures are *Agrobacterium tumefaciens*-mediated gene transfer and direct gene transfer to protoplasts (Göbel and Lörz 1988; Walden and Schell 1990; Potrykus 1991; van Wordragen and Dons 1992). *Agrobacterium*-mediated gene transfer (Weising et al. 1988; Klee and Rogers 1989; Hooykaas and Schilperoort 1992; Zambryski 1992) makes use of a naturally evolved system that transfers DNA fragments through the action of bacterial factors from the bacterium to the plant cell nucleus where they are integrated into the chromosomal DNA, probably by a mode of illegitimate recombination (Gheysen et al. 1991; Mayerhofer et al. 1991). In contrast, the integration of DNA fragments taken up during polyethylene glycol (PEG)-mediated direct gene transfer to protoplasts does not involve factors others than those present in the plant cell (Davey et al. 1989; Paszkowski et al. 1989). The possibility of generating transgenic plants in a variety of plant species has been of importance in allowing the start of several major areas of research:

1. The analysis of gene regulation by the use of reporter gene constructs that allow easy assessment of the transcriptional activity conferred by *cis*-acting elements present on particular fragments of the plant promoter under analysis (Schell 1987; Willmitzer 1988; Benfey and Chua 1989)
2. The functional analysis of physiological or developmental processes by overexpression or ectopic expression of a particular gene product by a strong plant promoter (Schmülling et al. 1988; Sonnewald et al. 1991; Mandel et al. 1992) or by reduction of the expression of endogenous genes by antisense inhibition (Smith et al. 1988; van der Krol et al. 1988; Müller-Röber et al. 1992)
3. The establishment of gene tagging systems to isolate genes only characterized by a mutant phenotype or by a characteristic expression pattern (André et al. 1986; Teeri et al. 1986; Feldmann et al. 1989; Koncz et al. 1989; Feldmann 1991; Forbert et al. 1991; Topping et al. 1991; Van Lijsebettens et al. 1991; Walden et al. 1991; Bhatt and Dean 1992)
4. The generation of new traits (Gasser and Fraley 1989) like herbicide resistance (Botterman and Leemans 1988; Padgette et al. 1989), virus resistance (Beachy et al. 1990), resistance to insects (Vaeck et al. 1989), male sterility (Mariani et al. 1990), resistance to phytopathogenic bacteria (de la Fuente-Martínez et al. 1992), improved protein quality, antibody production, enkephalin production (Willmitzer and Töpfer 1992), altered flavor (Peñarrubia et al. 1992), human serum albumin production (Sijmons et al. 1990), polyhydroxy-

butyrate production (Poirier et al. 1992), altered starch or fatty acid composition (Willmitzer and Töpfer 1992)

In contrast to the well-established integration of foreign DNA into the plant nuclear DNA, genetic transformation of organelles (plastids, mitochondria) has rarely been reported in higher plants (Howe 1985; Venkateswarlu and Nazar 1991). Since the development of the micro-projectile bombardment procedure, however, transformation of several eukaryotic organelles (Butow and Fox 1990), as well as higher plant plastids (Svab et al. 1990; Staub and Maliga 1992), has been achieved.

## TISSUE CULTURE AND TRANSFORMATION IN *A. THALIANA*

*Arabidopsis* is also amenable to the techniques and methods used for other species in tissue culture and transformation. The development and current status of these methodologies with respect to *Arabidopsis* are detailed below.

### Tissue Culture

### Axenic Culture

Important initial steps toward tissue culture and transformation include protocols for seed sterilization, axenic culture of whole plants on artificial media, and phytotron growth conditions. Seed sterilization methods for *Arabidopsis* do not differ substantially from methods used with other species and include the use of hydrogen peroxide and ethanol (Langridge 1955, 1957), calcium hypochlorite (Rédei 1962), and sodium hypochlorite (Feldmann and Marks 1986). Axenic culture on artificial media was first described by Laibach (1943), who exploited the small size of *Arabidopsis* by growing the plant throughout its life cycle in glass test tubes. Langridge (1957) and Veleminsky and Gichner (1964) described these methods in greater detail. An important point for the culture of *Arabidopsis* in tubes is that there be sufficient gas exchange by means of, e.g., loose cotton wool bungs so that the humidity in the tube does not inhibit seed set (Langridge 1957). Many refinements on aseptic culture techniques, including growth on perlite (Feenstra 1965) or sand (Bounias 1973) and hydroponic culture (Ernst et al. 1981), are reported in the literature. Additionally, several workers have reported on methodologies for simple vegetative propagation of *Arabidopsis* (see, e.g. Reinholz 1972).

Phytotron growth conditions were investigated by Langridge (1957), who found that 25°C, continuous illumination or a 16-hour day at about 10,000 lux, and relative humidity above 60% were optimal for aseptic culture on artificial media. Essentially similar conditions are also used for growth in soil (Pang and Meyerowitz 1987). *Arabidopsis* will in general tolerate a wide range of conditions, but growth rate, habit, and flowering time can be highly influenced by the environment and are of course also distinct properties of the ecotype. Current practice is to employ phytotron conditions essentially as described by Langridge (1957) for both soil-grown plants and tissue culture, and as the basis for axenic culture either Gamborg's B5 medium (Gamborg et al. 1968) or Murashige and Skoogs (MS) (1962) revised medium at half or full concentration with 1–3% sucrose or glucose as the carbon source.

## Callus Formation

*Arabidopsis* may also be propagated in tissue culture in the form of undifferentiated callus, as reported by Loewenberg (1965), who grew seedlings on a medium containing kinetin and parachlorphenoxyacetic acid with subculture of the resulting callus at monthly intervals. Ziebur (1965) cultured hypocotyl segments on media containing 2,4-dichlorphenoxyacetic acid (2,4-D) and coconut milk, with successful results using a number of nutrient media, including the revised medium of Murashige and Skoog (1962). Similarly, Yokoyama and Jones (1965) obtained callus using MS medium as a basis. Shen and Sharp (1966) reported the production of callus from germinating seed material, and Anand (1966) analyzed the optimum levels of auxin and coconut milk (0.22 mg $l^{-1}$ 2,4-D and 15% coconut milk) for seedling-derived callus production. Negrutiu et al. (1975a,b) studied the ability of different *Arabidopsis* tissues under varying conditions to induce callus in vitro. Gamborg's B5 medium was used, and hormone composition and ratio were varied. Stems were found to give a better callus induction than either leaves or seeds and responded within a range of 2,4-D concentration of between 0.05 mg $l^{-1}$ and 2 mg $l^{-1}$, although the callus did not appear fully undifferentiated until 1 mg $l^{-1}$ 2,4-D was used. Callus morphology was influenced by the addition of kinetin, with 0.05 mg $l^{-1}$ giving a friable callus and high amounts (1.5 mg $l^{-1}$) giving a yellowish nonfriable callus.

The callus-forming capacities of different tissues from ecotype C24 were also analyzed in the authors' laboratory; root, hypocotyl, cotyledon, petiole, and leaf explants were cultured on half concentrated MS medium with 1 mg $l^{-1}$ benzyl aminopurine (BAP) and varying amounts of 2,4-D. Different tissues had different auxin optima for callus formation, with roots being most responsive to low levels and cotyledons and petioles requiring relatively high levels (Fig. 1). Leaves and hypocotyls were intermediate in their response (B. Regierer et al., unpubl.).

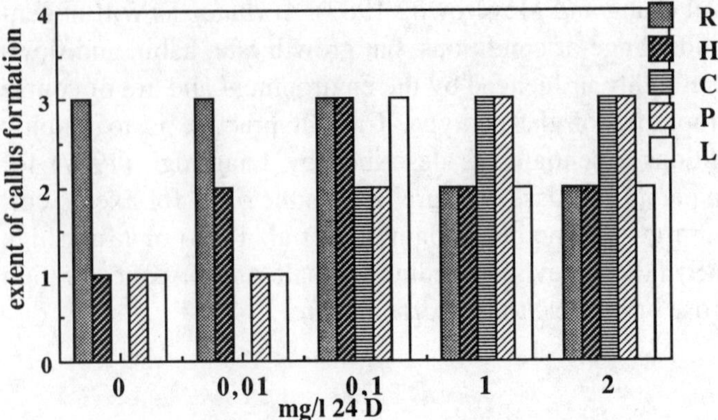

*Figure 1* The callusing response of various tissues from ecotype C24 when cultured on 1 mg l⁻¹ BAP and different amounts of 2,4-D. The extent of callus formation was graded from 0 (none) to 4 (high). (R) Roots, (H) hypocotyl, (C) cotyledon, (P) petiole, (L) leaf.

## Suspension Culture

Callus produced in this way can be further grown in liquid culture to produce a suspension culture; this was first described by Anand (1966), who used similar conditions to those described for his solid growth media. Corcos and Lewis (1971) reported that roots cultured in liquid vegetable juice medium or modified White's medium containing 2,4-D and coconut milk gave callus and slow-growing balls of tissue some 1 mm in diameter. Using whole plants as inoculum, fast-growing balls of callus some 1 cm in diameter were produced which contained numerous roots, stems, and leaves. Negrutiu et al. (1975b) studied growth kinetics and parameters influencing growth rates in cell suspension cultures. More recently, Ford (1990) has described the induction of suspension culture from roots of ecotype Wassilewskija by placing finely subdivided roots in liquid media containing 2,4-D and kinetin. Subculturing was performed every 8–12 days, and the culture was reported to have survived over 2 years.

## Root Culture

Roots have been used not only as a source for callus, but also to culture in their own right in the absence of exogenous plant growth regulators. This was first described for *Arabidopsis* by Neales (1968a,b). Roots excised from seedlings of ecotypes Estland and Pitztal were cultured in modified liquid White's nutrient medium with sucrose as carbon source, and the rapidly growing root mass was subcultured every 14 days.

Weiland and Müller (1972a,b) described the conditions needed for liquid culture of roots from the ecotype Dijon and of normally lethal *fusca* mutants. Critical parameters for root growth appeared to be temperature (27°C) and inclusion of thiamine in the medium, although roots of other genotypes were less demanding in their requirements.

## Regeneration

An important prerequisite for most plant transformation strategies is the capability to regenerate fertile plants from explants or callus. Regeneration from cotyledon, leaf, root, stem, and callus tissues was reported by some of the earliest investigators of *Arabidopsis* tissue culture; Yokoyama and Jones (1965) could induce regeneration from callus by inclusion of coconut milk and kinetin in the medium. Nitsch (1967) showed that hypocotyl sections could regenerate in the presence of kinetin and adenine to produce adventitious buds and flowers. Corcos et al. (1973) also demonstrated the regeneration of roots, leaves, stems, and fertile flowers from hypocotyl callus tissue.

Since then, many different workers have described the regeneration of fertile *Arabidopsis* plants from diverse tissues and explants of many different ecotypes. Frequently, such innovations in regeneration protocols are described in connection with genetic transformation methodologies. The *Arabidopsis* ecotypes, tissues, or explants, and media and hormones employed in a number of these reports are summarized in Table 1. It is very apparent from these different reports that although *Arabidopsis* is in general fairly easily regenerable by many approaches, the efficiency and nature of regeneration are highly influenced by the ecotype, the source of the explant, and the medium and hormones employed, thus making generalizations rather difficult.

It is clear that certain ecotypes, such as C24 (Valvekens et al. 1988), Nossen (Chaudhury and Signer 1989a), RLD (Rschew), and Wassilewskija (Feldmann and Marks 1986) do seem to have better regeneration properties than others. Regeneration can proceed directly from the explant (see, e.g., Lloyd et al. 1986) or via the induction of callus prior to regeneration, which seems to be more efficient (see, e.g., Feldmann and Marks 1986; Schmidt and Willmitzer 1988). In general, callus is induced by culture on auxin-containing media, sometimes using a tobacco cell suspension feeder layer (Lloyd et al. 1986; Schmidt and Willmitzer 1988), and regeneration is induced by subsequent culture on media containing reduced auxin levels and enhanced cytokinin levels. Root induction is generally carried out after shoot induction, by means of culture on intermediate levels of auxin. If, however, the regenerated shoots are re-

*Table 1* Regeneration strategies for *Arabidopsis*

| Ecotype | Tissue | Medium | Hormones (mg l$^{-1}$) | Regenerant | References |
|---|---|---|---|---|---|
| | seedling callus | MS | IAA, Kin (undefined amounts) | roots, shoots | Yokoyama and Jones (1965) |
| Wassilewskija Estland Martuba | anther | supplemented Gamborg salts | *GM2* 1 NAA 0.03 Kin *then* | callus | Gresshoff and Doy (1972) |
| | | supplemented Blaydes salts | *GM3* 0.5 NAA 10 Kin | shoots | |
| | hypocotyl callus | V+G | 0.05–1.5 Kin 0.003–0.3 IAA 1–1.5 Kin 0.03–0.3 IAA | roots shoots and flowers | Corcos et al. (1973) |
| Estland Columbia Coïmbra Wilna | callus derived from stems, anther, seeds, leaves | PG3 | 0.03 IAA 1 Kin | leaves, shoots, and flowers | Negrutiu et al. (1975a,b, 1978a,b) |
| Columbia | suspension-culture protoplasts | B5 | 1 2,4-D 0.15 BAP *then* 1 IAA 0.1 BAP | callus shoots | Xuan and Menczel (1980) |
| Estland | leaf and stem explants | B5 | 1 NAA 0.3 BAP | shoots, roots, and callus | Gotô (1980) |

| | | | | | |
|---|---|---|---|---|---|
| Wassilewskija × Han-noverisch-Münden $F_1$ | anthers | solid or liquid PG1<br><br>PG3 | 100 µM picloram<br>0.2 kinetin<br>*then*<br>0.005 IAA<br>0.2 Kin | callus<br><br>shoots | Keathley and Scholl (1982) |
| Columbia Langridge | callus | MS or B5 | 0.5 µM BAP<br>2 µM NAA | shoots | Acedo (1986) |
| Columbia | leaf explants | MS<br><br>1/2 MS | 1 BAP<br>0.1 NAA<br>*then* 1 IAA | shoots<br><br>roots | Lloyd et al. (1986) |
| Columbia | leaf callus | MS | 5 NAA<br>0.5 BAP<br>*then* 1 IPA<br>0.1 IAA | shoot induction<br>shoot formation | Zhang and Somerville (1987) |
| Wassilewskija | leaf explants | MS | *CIM*<br>5 IAA<br>0.5 2,4-D<br>0.5 zeatin<br>*then SIM*<br>0.05 IAA<br>7 2iP<br>*then RIM*<br>4 IAA | callus<br><br><br><br>shoots<br><br>roots | Sheikholeslam and Weeks (1987) |

*Table 1  (continued)*

| Ecotype | Tissue | Medium | Hormones (mg l$^{-1}$) | Regenerant | References |
|---|---|---|---|---|---|
| Columbia | immature cotyledons | MS salts, B5 vitamins | 1 BAP 0.1 NAA | shoots | Patton and Meinke (1988) |
| C24 | root explants | B5 | 0.5 2,4-D 0.05 Kin *then* | callus | Valvekens et al. (1988) |
| | | | 0.15 IAA 5 2iP | shoots | |
| C24 Wassilewskija Landsberg *erecta* Estland | leaf protoplasts | alginate embedded plus liquid B5 | 1 2,4-D 0.15 BAP *then* | callus induction | Damm and Willmitzer (1988) |
| | | MII | 0.2 2,4-D 1 Kin *then* | callus growth | |
| | | solid MS | 7 2iP 0.05 IAA *then* | shoot induction | |
| | | MS | 0.1 NAA 1 BAP 0.1 GA$_3$ *then* | shoot elongation | |
| | | MS | 1 IBA | root induction | |

| C24 Landsberg *erecta* Wassilewskija | leaf explants, cotyledons | MS | CIM 1 2,4-D 0.2 Kin | callus induction | Schmidt and Willmitzer (1988) |
| | | | *then* *SIM I* 1 BAP 0.4 NAA | shoot induction | |
| | | | *then* *SIMII* 7 2iP 0.05 IAA | shoot induction | |
| | | | *then* *SEM* 1 BAP 0.4 NAA 0.1 $GA_3$ | shoot elongation | |
| | | | *RIM* 1 IBA | root induction | |
| Columbia | root, leaf, and stem explants | R4 | 0.025 2,4-D 1.5 2iPR | callus induction | Rédei et al. (1988) |
| | | | *then* 2 2iPR 0.1 NAA | shoot induction | |
| | | | *then* 1.5 2iPR 0.1 NAA | shoot elongation | |

*Table 1  (continued)*

| Ecotype | Tissue | Medium | Hormones (mg l$^{-1}$) | Regenerant | References |
|---|---|---|---|---|---|
| Columbia | cell suspension | MS or B5 with gelrite | 2 thidiazuron | shoot induction | Gleddie (1989) |
|  |  |  | *then* 0.1 NAA | root induction |  |
| Columbia Landsberg *erecta* Bensheim Wassilewskija Nossen | leaves, cotyledons, stems, epidermal peels | MS | 1 BAP 0.1 NAA | shoots | Chaudhury and Signer (1989a) |
| Columbia | stem explants | Modified MS | *MSAR1* 2 IAA 0.5 2iPR 0.2 BAP 0.2 2,4-D | callus induction | Koncz et al. (1990) |
|  |  |  | *then MSAR2* 2 2iPR 0.05 NAA | shoot induction |  |
|  |  |  | *then MSAR3* 1.5 2iP 0.05 NAA | shoot elongation |  |
|  |  |  | *then MSAR4* 1 IBA 0.05 NAA | root induction |  |

| | | | | | |
|---|---|---|---|---|---|
| Wassilewskija | root-derived suspension culture protoplasts | Kao salts, MS micronutrients B5 vitamins | 0.05 Kin<br>0.5 2,4-D<br>*changing to*<br>0.2 Kin<br>0.2 2,4-D<br>over 7 weeks<br>*then*<br>0.05 NAA<br>1 Kin<br>*then*<br>0.05 IPA<br>0.1 $GA_3$ | callus<br><br><br><br><br><br><br><br><br>shoot induction<br><br>shoot elongation | Ford (1990) |
| Rschew Columbia | root explants | MS | *ARM I*<br>3 IAA<br>0.15 2,4-D<br>0.6 BAP<br>0.3 2iP<br>*then*<br>*ARM II*<br>0.1 NAA<br>2 2iP<br>1 2iPR<br>25 Ag $NO_3$<br>*then*<br>*ARM III*<br>0.5 IBA<br>0.2 zeatin<br>1 IAA<br>1 $GA_3$ | callus induction<br><br><br><br><br><br>shoots<br><br><br><br><br><br>roots | Márton and Browse (1991) |

*Table 1* (continued)

| Ecotype | Tissue | Medium | Hormones (mg l$^{-1}$) | Regenerant | References |
|---|---|---|---|---|---|
| Columbia<br>C24<br>Landsberg *erecta* | immature<br>embryos | MS salts<br>N+N micros | BM2<br>1 BAP<br>1 NAA<br>then | callus induction | Sangwan et al. (1991) |
| | | | *BM4*<br>1 BAP<br>0.1 IAA<br>*then* | callus and shoot | |
| | | | *BM6*<br>1 BAP<br>0.4 NAA<br>0.1 GA$_3$ | shoot elongation | |
| Zürich | leaf protoplasts | alginate<br>embedded plus<br>liquid MS | 1 2,4-D<br>0.15 Kin<br>*then* | callus initiation | Karesch et al. (1991a) |
| | | MS | 0.05 2,4-D<br>2 NAA<br>0.15 Kin<br>*then* | callus growth | |
| | | solid 1/2 MS | 0.15 IAA<br>5 2iP<br>*then* | shoot induction | |
| | | 1/2 MS | 1 NAA | root induction | |

| | | | | | |
|---|---|---|---|---|---|
| Columbia Nossen | liquid grown roots | MS | *ARA II*<br>1 BAP<br>0.1 NAA<br>0.5 2,4-D<br>*then*<br>*ARA I*<br>1 BAP<br>0.1 NAA<br>*then*<br>*ARA III*<br>5 2iP<br>0.15 IAA | callus induction<br><br><br><br><br>shoot induction<br><br><br><br>shoot elongation | Kemper et al. (1992) |
| Zürich C24 | immature embryos | B5 with 400 mg l⁻¹ glutamine | 0.6 BAP<br>0.05 Kin<br>*then*<br>0.1BAP<br>0.05 Kin | growth of explant<br><br><br>shoot induction | Kost et al. (1992) |
| Landsberg *erecta* | immature embryos | B5<br>*then* | 1 2,4-D<br><br>0.5 BAP<br>2 NAA<br>0.5 2,4-D<br>*then*<br>0.1 BAP<br>0.5 NAA<br>0.1 2,4-D | globular callus<br><br><br><br><br><br>somatic embryos | Wu et al. (1992) |

*Table 1  (continued)*

| Ecotype | Tissue | Medium | Hormones (mg l$^{-1}$) | Regenerant | References |
|---|---|---|---|---|---|
| Bensheim Columbia | hypocotyl | B5 | CIM | callus | Akama et al. (1992) |
| Landsberg *erecta* | | | 0.05 Kin | | |
| Wassilewskija | | | 0.5 2,4-D | | |
| | | | *then* | | |
| | | | SIM | shoots | |
| | | | 5 2iP | | |
| | | | 0.15 IAA | | |
| | | | *then* | | |
| | | MS | RIM | | |
| | | | 0.02 IBA | roots | |

Under column "Medium," the following references apply: Blaydes salts, Blaydes (1966); Gamborg salts, Gamborg and Eveleigh (1968); Kao salts, Kao et al. (1974); MS, Murashige and Skoog (1962); V+G, Velemínsky and Gichner (1964); PG1, PG3, Negrutiu et al. (1975a); B5, Gamborg et al. (1968); N+N, Nitsch and Nitsch (1969); MII, Li and Kohlenbach (1982); R4, Rédei et al. (1988). Under "Hormones," the following abbreviations apply: 2,4-D, 2,4-dichlorphenoxyacetic acid; 2iP, 6-(γ,γ-dimethylallylamino) purine; 2iPR, 6-(γ,γ-dimethylallylamino) purine riboside; BAP, benzylaminopurine; CM, coconut milk; GA$_3$, gibberellic acid; IAA, indole-3-acetic acid; IBA, indole-3-butyric acid; IPA, indole-3-propionic acid; Kin, kinetin; NAA, naphthaleneacetic acid.

quired for seed production in axenic culture, then root initiation is not generally a necessity.

An extensive analysis of the factors influencing regeneration of *Arabidopsis* has been carried out by Negrutiu and co-workers. Thus, Negrutiu et al. (1975b) demonstrated regeneration of solid- or liquid-grown callus by transfer to Gresshoff and Doy's (1972) growth medium 2 supplemented with IAA and kinetin, followed by a transfer to a basal medium lacking growth regulators as soon as differentiation had commenced. Differences were apparent with the ecotype employed, with Estland giving high shoot regeneration and Columbia favoring root formation. Origin of the explant also influenced the outcome, with primary callus from anther, seeds, and leaves giving the highest frequencies of shoot formation. A 10:1 kinetin to IAA ratio was optimal for shoot induction with an extremely wide range of hormone combinations favoring root formation. Low-kinetin callus-induction medium allowed a better subsequent regeneration than high-kinetin callus-induction medium, and low light conditions (<2000 lux) also improved efficiency of regeneration. Cold treatment at 4°C for 24 hours appeared to enhance the regenerative capacity of the calli, and the age of the callus was also a factor; the older the callus culture, the less morphogenic its response, but this could be restored by increasing the glucose content of the medium from 2% to 6% (for review, see Negrutiu 1976).

The same workers compared regeneration from callus derived from seed, anther, stem, and leaf segments (Negrutiu et al. 1978a,b; Negrutiu and Jacobs 1978). Again, leaf- and anther-derived callus showed the highest morphogenic capacity, and the hormone regime used to initiate the callus was also influential. Calli initiated under different hormonal conditions had different phenotypic appearances and also different morphogenic properties. Although regeneration could be achieved with a relatively wide range of hormone ratios and concentrations, it was found that each type of explant required differing amounts of exogenous growth regulators for optimum regeneration, with low levels of the cytokinin zeatin being more effective than kinetin or BA.

Gotô (1980) studied the regenerative capabilities of leaf and stem tissue from ecotype Estland and three mutant lines under different hormone regimes. The dwarf mutants F and le, and an undefined *apetala* mutant, had a reduced regenerative capacity, but it is not clear if these were also in the Estland background. No significant differences in the regenerative capacity of leaves of different ages were found. However, older stems were not as responsive as young stems. Regeneration occurred under a range of hormone concentrations: Leaf disks were found to produce both fertile shoots and roots or callus with an optimum hormone concentration of 1 mg l$^{-1}$ NAA and 0.3 mg l$^{-1}$ BA, whereas stem cuttings produced only shoots or callus within a wide range of NAA and BA concentrations. The auxin NAA was more effective in root initiation in leaf disks than IAA. Regeneration from leaf-derived callus tissue was also demonstrated, and here shoots and roots were produced using a combination of 0.3 mg l$^{-1}$ NAA and BA.

Plants can also be regenerated from root explants, as described by Valvekens et al. (1988). This method has the advantages of simplicity and speed and is frequently employed for *Agrobacterium*-mediated transformation. A similar approach was employed by Márton and Browse (1991) with the morphogenic response enhanced by a cool period for the root explants or donor seedlings and inclusion of silver nitrate at 25 mg l$^{-1}$ in the shoot induction medium to inhibit the adverse effect of ethylene on the regenerating shoots.

Akama et al. (1992) compared the regeneration of various tissues (cotyledon, hypocotyl, and root) from ecotypes Bensheim, Columbia, Landsberg *erecta*, and Wassilewskija, and found that hypocotyls had the highest regeneration efficiency for all four ecotypes. To summarize: Regeneration from callus or explants in *Arabidopsis* appears to be an efficient and simple procedure, but the method to be used (i.e., which combination of plant growth regulators is to be utilized) is highly dependent on the ecotype employed and the nature of the explant or callus which is to be regenerated.

## Protoplasts

Negrutiu et al. (1975b) could obtain protoplasts from leaf, callus, and suspension culture material. Young actively dividing cells in callus and suspension cultures gave the highest protoplast yields; however, very few cell divisions could be induced in the cultured protoplasts. Bhalla-Sarin et al. (1976) described the isolation of protoplasts from leaf, shoot, root, flower, bud, and callus tissues of *Arabidopsis*; roots proved to be a poor source of protoplasts. Protoplast fusion experiments carried out using PEG gave multinucleate protoplasts. Gresshoff (1976) described protoplast isolation from a regenerable callus line. Culture in hanging drops at a density of 1000 protoplasts per 50 μl together with 1.3 mg l$^{-1}$ 2,4-D and 0.01 mg l$^{-1}$ kinetin gave a regeneration frequency of 0.1%. Microcalli transferred to solid regeneration media could give rise to plantlets. Gleba and Hoffman (1978) prepared protoplasts from callus derived from an *albino* mutant in ecotype Enkheim and carried out fusion experiments together with protoplasts of *Brassica campestris*. The hybrid protoplasts proved to be highly proliferating, whereas both parents gave little division. Cytological examination showed that all the chromosomes of both species were retained in some of the hybrid callus. Protoplast fusion experiments were also carried out by Bauer-Weston et al. (1993) who produced and regenerated somatic hybrids between *Arabidopsis* and *Brassica napus*. Xuan and Menczel (1980) also prepared protoplasts from suspension cultures from ecotype Columbia. The protoplasts were cultured at a density of $3 \times 10^5$ per ml in liquid medium with 1 mg l$^{-1}$ 2,4-D and 0.15 mg l$^{-1}$ BA, and 0.4 M glucose as osmoticum, for 2 days followed by addition of medium containing agar at a final concentration of 0.6%. The highest plating efficiency obtained was 63%. Transfer of the calli to solid regeneration medium with 1 mg l$^{-1}$ IAA and 0.1 mg l$^{-1}$ BA gave shoot induction in 40% of the calli.

Damm and Willmitzer (1988) described protocols for the regeneration of fertile plants from leaf protoplasts of ecotypes C24 (erroneously described as Columbia in the original publication), Wassilewskija, Landsberg *erecta*, and Estland. The critical feature of this protocol was the embedding of the protoplasts in an alginate matrix; this allowed sustained division of about 0.5% of the protoplasts, of which over 80% could be induced to make shoots when removed from the alginate and cultured further on shoot induction medium. Similarly, Karesch et al. (1991a) described a protocol for regeneration of protoplasts from ecotype Zürich that also involves embedding of protoplasts in alginate or pectinate gels. Ford (1990) could regenerate protoplasts from cell suspension cultures of ecotype Wassilewskija without the need for embedding. The protoplasts were incubated initially at $2 \times 10^5$ per ml in media containing 0.05 mg l$^{-1}$ kinetin and 0.5 mg l$^{-1}$ 2,4-D, and over a period of 48 days, the cell density was reduced to 6650 per ml, glucose concentration was reduced from 0.4 to 0.1 M, the kinetin concentration increased to 0.2 mg l$^{-1}$, and the auxin concentration was reduced to 0.2 mg l$^{-1}$. This resulted in microscopic calli which could be plated on solid medium for plant regeneration. These methods did not work for protoplasts derived from leaves or roots of axenically grown plants.

Masson and Paszkowski (1992) have optimized a number of parameters for the high-frequency regeneration of protoplasts. It was found that the growth conditions of the donor plants were especially important, and a 10-hour photoperiod in a light spectrum biased to the blue end was optimal. The medium used to culture the embedded protoplasts was also investigated, and the use of nutrient media in which ammonium succinate replaced ammonium sulfate led to much improved regeneration frequencies. J. Siemens et al. (1993) found that leaf-derived protoplasts from different *Arabidopsis* ecotypes showed differences in their regeneration capacity, depending on the nutrient medium employed for both callus induction and regeneration. Investigations in the authors' laboratory has shown that mesophyll protoplasts derived from ecotypes C24, Columbia, Landsberg *erecta*, and the mutant *angudistri* in a Landsberg *erecta* background (obtained from Arabidopsis Information Service, Frankfurt, Germany) will regenerate to form callus under a wide range of hormonal concentrations, but also with differences in the callus-forming ability of the different genotypes (Table 2) (B. Regierer et al., unpubl.).

## Somatic Embryogenesis

There are few accounts in the literature of somatic embryogenesis in *Arabidopsis*. Huang and Yeoman (1983) reported on the induction of nodular callus from seeds, which could, under the appropriate circumstances, differentiate into what the authors termed cotyledon-stage embryos. However, as the authors state, the structures formed do not look like normal embryos, with predominantly fascicular leafy structures formed. Koncz et al. (1990) and Márton and Browse (1991) reported on regeneration methods from stem and root explants in which the plants were thought to derive from somatic embryos. Some caution is needed in this interpretation, since apart from a superficial resemblance to zygotic embryos, no proof was offered that these structures were anything other than small shoots. A visually more convincing example of somatic embryogenesis is that of Wu et al. (1992), in which culture of torpedo stage zygotic *Arabidopsis* embryos in liquid B5 media containing 1 mg l$^{-1}$ 2,4-D with subsequent transfer after callus induction into media containing 0.5 mg l$^{-1}$ 2,4-D, 2 mg l$^{-1}$ NAA, and 0.5 mg l$^{-1}$ BAP resulted in the growth of secondary embryos.

*Table 2* Influence of ecotype and auxin:cytokinin ratio of regeneration medium on callus production of alginate embedded protoplasts.

| Ratio 2,4-D:cytokinin (mg l$^{-1}$) | C24 | Columbia | Landsberg *erecta* | *angudistri* (Landsberg *erecta* background) |
|---|---|---|---|---|
| 1:0.15 ZR | +++ | ++ | ++ | 0 |
| 1:1 ZR | +++ | ++ | ++ | + |
| 1:3 ZR | +++ | 0 | 0 | 0 |
| 1:0.15 BAP | +++ | +++ | ++ | 0 |
| 0.1:0.3 BAP | +++ | 0 | +++ | + |
| 0.5:3 BAP | ++ | n.d. | n.d. | + |
| 0.05:0.15 BAP | +++ | n.d. | n.d. | + |
| 3:0.5 BAP | ++ | n.d. | n.d. | + |

(ZR) Zeatin riboside. (BAP) Benzyl aminopurine. Callus production was graded from 0 for no response to +++ for large numbers of individual calli. n.d. indicates not determined.

## Ploidy

Of importance to many transgenic plant experiments is that the regenerated plants are genetically normal diploids. It has long been recognized that tissue culture induces a certain degree of chromosomal instability and that regenerants can be diploid, aneuploid, or polyploid. This has also been recognized in *Arabidopsis*, although the small size of the chromosomes has made karyotypic analysis rather difficult. Negrutiu et al. (1975b) observed a wide range of heteroploidy in tissue culture of *Arabidopsis* with the number of chromosomes per cell ranging from 10 to 60. Leaf and stem callus were found to be more prone to polyploidy than seed-derived callus, although this also increased in ploidy with age of the culture. The concentration of 2,4-D in the culture media did not seem to have an effect on ploidy, but high kinetin (above 1.5 mg l$^{-1}$) led to high ploidy levels. Galbraith et al. (1991) have reported that somatic tissues of normal *Arabidopsis* plants undergo extensive endoreplication, resulting in tissues comprising mixtures of polyploid cells, even at the seedling stage. The only tissues not influenced by polyploidy were the inflorescences. In this context, it is of interest that Sangwan et al. (1991) report that transgenic plants derived from culture of immature embryos are diploid.

Irrespective of whether polyploidy in cultured material is due to the ploidy of the explant tissue or to the tissue culture conditions, the fact is that a high frequency (up to 40%) of transgenic *Arabidopsis* plants are tetraploid in character, with lowest levels of tetraploids reported from root-derived transgenic plants. A simple test for the ploidy of the plant is to measure the pollen; there is a good correlation between ploidy level of the plant and length or breadth of the mature pollen grains (T. Altmann et al., in prep.).

## Haploidy

A powerful addition to the *Arabidopsis* repertoire is the ability to produce haploid plants and cell lines, since selection for recessive mutations in the $M_1$ generation is in principle possible, with potentially great savings of space and time. Immature microspores represent a reservoir of haploid cells, which if persuaded to regenerate, can result in haploid plants. Gresshoff and Doy (1972) reported the production of haploid callus and plants of ecotypes Wassilewskija, Estland, and Martuba by means of anther culture. The calli were reported to still be haploid after one year of culture. Attempts with numerous other ecotypes did not succeed, however. Amos and Scholl (1978) and Amos (1978) reported success with anther culture of eight different ecotypes, including some with which Gresshoff and Doy had no success, as well as three species related to *Arabidopsis* (*A. griffithiana*, *A. korshinskyi*, and *A. pumila*) and found that medium containing 6 mg l⁻¹ IAA, 2 mg l⁻¹ 2,4-D, and 1.5 mg l⁻¹ kinetin gave calli in up to 87% of the anthers. Only 5% of the calli were found to be diploid; the rest were either haploid or aneuploid. Using an allozyme marker locus (acid phosphatase) and heterozygous donor plants, Scholl and Amos (1980) were able to demonstrate that the anther callus was haploid in nature; of the regenerated shoots, however, some 50% of diploid regenerants were doubled haploids whereas the rest were maternal clones. Keathley and Scholl (1982) used an $F_1$ hybrid of ecotypes Wassilewskija and Hannovrisch-Münden to carry out anther culture in solid or liquid media. Callus induction on solid media was faster, but liquid media gave in the long term (4–5 weeks) a significantly greater number of haploid (30%) calli. An analysis of the influence of exogenous plant growth regulators showed maximal callus induction (90% of anthers) occurring with $10^{-4}$ M picloram and 0.2 mg l⁻¹ M kinetin. The regenerative capacity of the calli fell off rapidly with age. Scholten and Feenstra (1984) also carried out anther culture on 41 ecotypes according to the method of Greshoff and Doy (1972), but all of the calli turned out to be diploid. Attempts to initiate direct haploid embryogenesis from the anthers gave rise to a limited number of "embryoid-like structures" from 4 ecotypes, of which some were found to be haploid.

In general, it seems that successful production of haploids from anther culture is more dependent on the stage of development of the anthers (buds 0.5 mm in length) than on any specific medium or growth hormone composition. Complete removal of the anther stalk is recommended, since this quickly gives rise to diploid callus. A wide range of auxin and cytokinin concentrations seem capable of inducing callus from anthers, e.g., 2.2 mg l⁻¹ 2,4-D and 0.05 mg l⁻¹ kinetin. Furthermore, unlike the situation for *Brassica* species, it appears that strict control of environmental parameters during growth of the donor plants is not a prerequisite for *Arabidopsis* (for review, see Scholl and Feldmann 1990). We have tried to induce haploid embryo formation from cultured microspores as described for *Brassica* species (see, e.g., Lichter 1982; Takahata and Keller 1991), with limited success. On rare and unpredictable occasions, what appeared to be somatic embryos could be induced, but these were difficult to culture further, and when this was successful, the regenerants turned out to be diploid.

### Transformation

Early experiments on *Arabidopsis* transformation described the ability of imbibing seeds to take up exogenously supplied DNA and the fate of the

DNA in the plant (for review, see Kleinhofs and Behki 1977). Thus, Ledoux and Jacobs (1969) and Ledoux et al. (1971) reported that tritiated *Escherichia coli* DNA was taken up by imbibing *Arabidopsis* seeds, and radioactively labeled DNA of a density corresponding to that of *E. coli* DNA integrated into the plant genome could be recovered from the plants once the seeds had germinated, and even in seeds and plants of the subsequent generation. These conclusions were based on the analysis of isopycnic density gradient centrifugation with CsCl, and as such have been criticized on the basis that the results were very likely due to bacterial contamination of the plant material, plant viruses, or other artifacts of CsCl gradient analysis (see Kleinhofs and Behki 1977). Lurquin and Hotta (1975) investigated the fate of bacterial DNA added to callus cells of *Arabidopsis*; they found no evidence for integration.

Ledoux et al. (1974) reported that *Arabidopsis* thiamine *py* mutants could be complemented by treatment of seeds with heterologous DNA from various sources, including calf thymus DNA. DNA from an *E. coli* thiamine-negative mutant would not complement the plant mutation. The corrected mutant phenotype was reported to be stably inherited, but in a non-Mendelian fashion. However, Feenstra et al. (1973) could not duplicate this result. Furthermore, Rédei et al. (1977), working with the corrected lines, could not find any evidence for DNA-mediated correction. Maluszynski et al. (1978) repeated their findings with the same *py* mutant as used by Feenstra, but here reported only a partial restoration of the *py* deficiency. No data were presented on inheritance of the corrected phenotype. Charles and Remy (1980) presented data in which DNA derived from progeny of corrected *py* plants was shown by Southern blotting to contain bacterial sequences, but again, this result might be caused by contaminated plant material.

In further work, Ledoux et al. (1985) allowed *Arabidopsis* seeds to germinate in the presence of circular plasmid DNA encoding antibiotic resistance genes (pBR325). Some of the treated seed produced plants with teratomas and other abnormalities, and β-lactamase enzyme activity could be detected in these plants. When progeny of these plants were analyzed, they also showed β-lactamase activity. Southern blot analysis of DNA from $F_2$ progeny suggested the presence of plasmid-derived DNA in the plant genome. However, attempts to produce kanamycin-resistant calli by this method did not succeed.

In other experiments, bacteriophage λ uptake into *Arabidopsis* protoplasts was induced by means of poly-L-ornithine or PEG treatment. Phage containing the *lacZ* gene enabled *Arabidopsis* cells to grow for three or four subcultures on lactose-containing media, whereas the uninoculated controls died. Similar results were reported using the *galZ* gene and galactose. Viable plaque-forming units could be recovered for a period of up to 6 days after transformation; however, integration of phage DNA into the plant genome was not demonstrated (Doy et al. 1973; Gresshoff and Rolfe 1976).

More recently, DNA-mediated gene transfer into protoplasts has been used as a powerful technique for transient assay or for the production of transgenic plants (for review, see Altmann et al. 1992). The methodologies described by Damm et al. (1989), Damm and Willmitzer (1991), and Karesch et al. (1991b) are based on the PEG-mediated direct gene transfer method of Negrutiu et al. (1987), initially established for tobacco. Essentially, protoplasts are incubated with magnesium ions before the addition of plasmid DNA, carrier DNA, and PEG. Subsequent to transformation, the protoplasts can be incubated for a few days in liquid culture to monitor for transient expression (Axelos et al. 1992; Doelling and Pikaard 1993; Hoffmann et al. 1994) or embedded in alginate gels as described previously to regenerate callus and, subsequently, plants (Damm and Willmitzer 1988). An absolute transformation frequency (transgenic plants per number of protoplasts) of about $1 \times 10^{-4}$ can be routinely achieved by this means, but as yet this has only been described for ecotypes Zürich and C24.

Transient expression in *Arabidopsis* has also been achieved by means of the particle gun; Seki et al. (1991a) used the *GUS* gene to monitor gene expression in both leaves and roots. Using the same approach, stable transformation of *Arabidopsis* has been reported (Seki et al. 1991b).

## T-DNA

Probably the most widely used approach to generate transgenic plants is to harness the capability of the plant pathogenic bacterium *Agrobacterium tumefaciens* to integrate sequences of its tumor-inducing (Ti) plasmid DNA into the genomic DNA of the plant host. Preliminary work showed that *Arabidopsis* is a good host for the bacterium. Thus, Corcos and Krupka (1979) showed a systemic infection of seedlings inoculated by *A. tumefaciens*, and Aerts (Aerts 1978; Aerts et al. 1979) showed that tumors could be initiated in *Arabidopsis* by inoculation with *Agrobacterium* strains T37, C58, and ACH5. The tumors could be excised and cultured in vitro without the addition of exogenous plant growth regulators, whereupon the tumors developed into teratomas and plants. Aerts (1980) isolated protoplasts from *A. tumefaciens*-induced tumors and demonstrated that exogenous plant growth regulators were not required for cell division and callus formation. Pavingerová et al. (1984) demonstrated that regenerants from *Arabidopsis* hairy root tumors caused by *A. rhizogenes* and *A. tumefaciens* containing Ri plasmids transmitted the capacity for mannopine and agropine synthesis into the next generation, indicating that the DNA was stably integrated into cells leading to the germ line.

The first accounts of *Arabidopsis* transformation using avirulent Ti plasmids to transfer antibiotic resistance to the plant were given by Lloyd et al. (1986), who inoculated leaf disks with *A. tumefaciens* containing the disarmed Ti plasmid pMON404. By means of a simplified callus induction-regeneration regime that employed only one medium, they could regenerate shoots whose progeny were hygromycin resistant, could synthesize nopaline, and contained the T-DNA in the plant genome as shown by Southern blotting. Tobacco feeder layer culture plates were found to reduce damage to the leaf disks by the bacteria. The authors suggested, however, that kanamycin-resistance selection was not very satisfactory in *Arabidopsis*. In 1986, An et al., using the binary Ti plasmid vector pGA472 in *Agrobacterium* lines containing the non-disabled helper Ti plasmid pTiBo542, could transform stem cuttings of *Arabidopsis* ecotype Columbia and demonstrated neomycin phosphotransferase activity in the kanamycin-resistant calli. However, transformation did not prove possible when a non-oncogenic helper Ti plasmid was used (pAL4404).

Zhang and Somerville (1987) reported the transformation of *Arabidopsis* callus tissue by cultivation with *A. tumefaciens* containing the Ti plasmid pGV3850 carrying the NPTII gene. Kanamycin-resistant callus was obtained and shoots regenerated, but only at a very low frequency, which was attributed to the age of the callus culture. Sheikholeslam and Weeks (1987) published a transformation protocol in which transformation rates of 63% were achieved with leaf explants using the neomycin phosphotransferase gene as a selectable marker. Critical to high transformation frequencies in their protocol was the addition of acetosyringone to the *Agrobacterium* cultures prior to infection of the leaf explants. Schmidt and Willmitzer (1988) developed an efficient cotyledon or leaf disk transformation method in which acetosyringone did not prove necessary. The significant point in the method was a precultivation of the explants for 2 days on callus induction medium prior to infection with *Agrobacteria*. Using a two-phase shoot regeneration protocol, 70–100% of explants were found to give rise to antibiotic resistant calli, of which more than 80% gave rise to shoots. Seeds could be harvested as early as 2 months after the start of transformation. The protocol was successfully used with ecotypes C24, Landsberg *erecta*, and Wassilewskija, and selection could be carried out using hygromycin as well as kanamycin or G418.

Feldmann and Marks (1987) described a transformation protocol in which seeds (ecotype Wassilewskija) were allowed to imbibe in the presence of *A. tumefaciens*. This led in the $T_2$ to plants stably transformed with the T-DNA at rates of up to 1 in 300. This topic is dealt with in more detail by Feldmann et al. (this volume).

An efficient transformation protocol employing root explants was developed by Valvekens et al. (1988). Using the ecotypes C24, Columbia, and Landsberg *erecta*, and a variety of *Agrobacterium* strains containing Ti plasmids conferring resistance to kanamycin and also the herbicide Basta, transformation efficiencies of between 20% and 80% (root explants producing seed-bearing transformants per total number of infected root explants) were achieved. Root explants were initially subjected to a short callus induction treatment, followed by infection with *Agrobacteria* and further culture on shoot induction medium. A number of variations and improvements of the root explant method have been published, for example, by Márton and Browse (1991), in which more complex callus and shoot induction media, together with a cold pretreatment of the explant donor plants, are described. These gave rise to faster and higher efficiencies of regeneration, or nondestructive transformation methods in which root explants were excised while the donor plant was still retained (Chaudhury and Signer 1989b). Kemper et al. (1992) also published a root explant transformation protocol in which a modified hormone cocktail is described for callus and shoot induction media: The use of liquid-grown plants as a source for root explants is also described, as is the selection of transformants with methotrexate by means of the dihydrofolate reductase gene. Clarke et al. (1992) described the use of silver thiosulfate in regeneration media to enhance regeneration frequencies.

Bechtold et al. (1993) have recently reported a transformation methodology based on vacuum infiltration of *Agrobacterium* into mature *Arabidopsis* plants shortly before flowering. In the progeny of these plants, a frequency of up to 5 independent transformants per inoculated plant was found, and the first 4000 transformants have been generated for analysis of insertion mutants. An interesting transformation approach was reported by Chang et al. (1990) in which *Agrobacteria* harboring the pBI121 plasmid were inoculated onto a wound site produced by removing the primary inflorescence of *Arabidopsis* plants. Seeds were harvested from subsequently regenerated shoots. When *GUS* was used as a marker gene, 30–60% of $T_2$ plants were found to be *GUS*-positive, and Southern data showed insertion of the T-DNA into the plant DNA (H.G. Nam, pers. comm.).

A transformation procedure employing zygotic embryos of *Arabidopsis* has been published in which it is claimed that a very low incidence of somaclonal variation occurs (Sangwan et al. 1991). Sangwan et al. (1992), using an intron-containing β-glucuronidase gene as a marker, have analyzed some of the early events in the *Arabidopsis–Agrobacterium* interaction. They found that one of the most critical stages was the preculture on callus induction medium prior to the infection by *Agrobacterium*. The cells that were susceptible to transformation in cotyledon explants were de-differentiating mesophyll cells. In root explants, the transformed cells were found to be de-differentiating pericycle cells.

Akama et al. (1992) studied the interaction of various *Agrobacterium* strains (EHA101, C58C1/pTiR225, and LBA4404) and different *Arabidopsis* ecotypes (Bensheim, Columbia, Landsberg *erecta*, and Wassilewskija) on transformation efficiency. They found that time of preincubation on callus-inducing medium had a strong influence on subsequent transformation frequency, with 7–8 days preincubation being optimal. The most efficient combination of plant and bacterium was ecotype Wassilewskija with strain EHA101; however, for Landsberg *erecta*, the strain C58C1 was found to be better. Thus, there may be some specific interaction between plant ecotype and bacterial strain.

An important aspect of any *Agrobacterium*-mediated transformation protocol is the use of antibiotic agents to select for transformed plant tissue and to remove the *Agrobacteria* after transformation to allow for successful regeneration. Lloyd et al. (1986) used hygromycin as a selectable marker, because they experienced problems when attempting to select using kanamycin. Sheikholeslam and Weeks (1987) also had problems using kanamycin sulfate on leaf explants; at levels high enough to inhibit the growth of untransformed tissue, the transformed tissues were also inhibited. However, using G418 (geneticin) instead of kanamycin allowed the use of the NPTII gene for selective purposes. In contrast, they found that when selecting for transformation of the $T_2$, kanamycin gave a clearer phenotype to the germinating seedlings than did G418. Patton and Meinke (1988) investigated the sensitivity of nontransformed cotyledon and leaf explants to a number of antibiotics commonly used for selection and found effective inhibition of growth and regeneration with 50–200 mg l$^{-1}$ kanamycin sulfate or G418, although the initial growth of leaf explants in culture was not as well controlled with kanamycin. Callus growth was also inhibited by 15-30 mg l$^{-1}$ hygromycin. Carbenicillin and claforan (cefotaxime), used to inhibit growth of *Agrobacteria*, did not inhibit the growth of leaf explants or mature cotyledons, but immature cotyledon growth was severely inhibited by these compounds. Vancomycin, however, at up to 500 mg l$^{-1}$, allowed growth and regeneration of the immature cotyledons. Valvekens et al. (1988) examined the influence of antibiotics on root regeneration; they found 50 mg $l^{-1}$ kanamycin, 10 mg $l^{-1}$ G418, or 25 mg $l^{-1}$ hygromycin could be used for selective purposes. They found that claforan strongly and carbenicillin slightly inhibited regeneration of roots, but the antibiotics triacillin and vancomycin were acceptable. Some authors have also noted an undesirable interaction between claforan and kanamycin, leading to damage or death of the *Arabidopsis* tissues (Rédei et al. 1988). Kemper et al. (1992) found that roots could be selected for transformation using methotrexate by employing the dihydrofolate reductase gene. Similar transformation frequencies were obtained to those using the NPTII gene and kanamycin selection or the *HPT* gene and hygromycin.

## APPLICATION OF TISSUE CULTURE AND TRANSFORMATION IN *ARABIDOPSIS*

Although the usefulness of *Arabidopsis* in genetic studies was readily appreciated and the simple constitution of its nuclear genome led to the isolation of numerous genes (Meyerowitz 1992), the application of tissue culture and transformation in *Arabidopsis* lagged considerably behind that of other plant species like tobacco. However, as discussed above, the view of *Arabidopsis* as a recalcitrant plant in terms of regeneration or transformation has changed completely since several highly efficient protocols for regeneration and transformation of cotyledons, leaves, hypocotyls, roots, and protoplasts were developed over the past years. Accordingly, the number of studies applying these techniques steadily increases. To outline the usefulness of tissue culture and transformation techniques in *Arabidopsis*, a number of successful applications are reported below.

### Propagation and Analysis of Mutants in Tissue Culture

A variety of mutants with severe defects may either abort early in their development, as exemplified by the large group of embryo-lethal mutants (Müller 1963), or they may be infertile. To maintain recessive mutants of this type, usually it is sufficient to propagate heterozygous plants; dominant mutants, however, will rapidly be lost. Furthermore, early abortion of mutants can cause severe problems in obtaining sufficient plant material for biochemical or molecular genetic analysis. In many cases, it may therefore be advantageous if the mutant tissue can be propagated and/or regenerated in vitro. Tissue culture may even allow one to test the biochemical basis for the mutant phenotype by either influencing or complementing it to wild type by the addition of certain compounds to the culture medium. This technique was applied to culture and regenerate a variety of embryo-lethal mutants arrested in different stages of development (Baus et al. 1986; Franzmann et al. 1989). In vitro propagation and regeneration were achieved by culture of arrested embryos on mineral nutrient medium supplemented with vitamins and different combinations of phytohormones. Mutants arrested at a globular stage produced callus but failed to turn green or to regenerate. Others, arrested at the heart-to-cotyledon stages of development, could be regenerated to plants but often were pale with abnormal leaves, rosettes, or inflorescences. Phenotypically normal plants produced siliques containing 100% aborted seeds, as expected for a homozygous embryo-lethal mutant after self-pollination. The cause for the developmental arrest of a further embryo-lethal mutant could be elucidated by the rescue of mutant embryos cultured in vitro through the addition of biotin to the culture medium (Schneider et al. 1989). Furthermore, mutant plants rescued in culture and supplemented with biotin during growth in soil produced normal seeds and thus allowed the conclusion that the analyzed embryo-lethal mutant was a biotin auxotroph.


**Mutant Isolation in Tissue Culture**

Cell culture can be used as the basis for a mutagenesis and selection program, either by applying a mutagen or by exploiting the normally occurring somaclonal variation that occurs in tissue culture, but this approach does not seem to have been much applied with *Arabidopsis* as yet. Negrutiu and Jacobs (1976) describe the isolation of mutant lines resistant to the toxic amino acid analogs SAEC and 5-MT from *Arabidopsis* cell suspension culture treated with ethyl methane sulfonate (EMS). Gaj and Maluszynski (1987) regenerated a large number of plants from callus and found that up to 12% of embryos derived from the regenerated plants exhibited lethality and 0.3% had chlorophyll deficiencies. Siliques of control plants contained only 1.2% lethal embryos. Increasing mutation frequency was observed with age of the cultures, as was an increase in ploidy.


**Cell Culture**

Only a few studies have so far taken the opportunity to use cell material from cell (suspension) cultures of *A. thaliana*, which could easily be obtained in large quantities (see above). Scholten et al. (1982) analyzed a chlorate-resistant mutant unable to grow in soil by means of the establishment of a cell line derived from mutant plants. By assaying for nitrate reductase, cytochrome-*c*-reductase, and xanthine dehydrogenase activities in extracts of the cell line, they concluded that the mutation causes a nitrate reductase deficiency due to the lack of a Mo-containing cofactor shared by the nitrate reductase and the xanthine dehydrogenase. Ferl and Laughner (1989) made use of the uniform accessibility of the suspension culture cells to the methylating agent dimethyl sulfate (DMS) to perform in vivo footprinting experiments on the constitutively expressed *Adh* gene. Braam (1992) used suspension culture cells to analyze the induction of touch (TCH) genes by stimuli like increased exogenous calcium or heat shock. Likewise, Curie et al. (1991) and McKendree and Ferl (1992) performed promoter deletion analysis of the *Adh* promoter and the promoter of the elongation factor EF-1α A1 gene, respectively, by testing for the transient expression of the β-glucuronidase (*GUS*) reporter gene in protoplasts derived from suspension culture cells. Transient gene expression in *Arabidopsis* protoplasts was also used by von Schaewen et al. (1990) to show the efficient secretion of fusion proteins between the potato proteinase inhibitor II and the yeast invertase. These findings were further proved by the analysis of protoplasts isolated from stably transformed *Arabidopsis* plants constitutively expressing the fusion proteins. After 68 hours of culture, 96% of the invertase activity detected was present in the culture medium, whereas 80–90% of α-mannosidase activity, a vacuolar marker enzyme, was found to be intracellularly localized, excluding that excessive breakage of the cells had occurred.

### Analysis of Gene Expression and Regulation
### in Transgenic *Arabidopsis*

Further studies on gene expression and regulation were performed with stably transformed *Arabidopsis* plants: Promoter deletion analysis has been performed on the *Arabidopsis* genes coding for the phenylalanine ammonia-lyase gene (PAL), the chalcone synthase (CHS) and the H4A748 histone (Ohl et al. 1990; Feinbaum et al. 1991; Atanassova et al. 1992). Tsukaya et al. (1991) showed that expression driven by the CHS-A promoter was enhanced by sucrose, glucose, and fructose. Using a fusion of the *cab1* promoter to the *tms2* reporter gene, Brusslan and Tobin (1992) showed that the light-independent developmental regulation of *cab* gene expression is due to transcriptional regulation. A series of further studies analyzed the tissue-specific expression of reporter genes directed by promoters isolated either from *Arabidopsis* or from other plant species. A promoter of an *Arabidopsis* gene corresponding to a cDNA clone isolated from cauliflower and characterized as being highly expressed in meristematic tissue was found to direct expression in the meristematic dome and in branching points in the shoot and the root of *Arabidopsis* plants (Medford et al. 1991). By fusion of the promoter of a *Brassica oleracea* S locus glycoprotein gene to the *GUS* reporter gene, Toriyama et al. (1991) showed expression of this promoter in stigmas and anthers of transgenic *Arabidopsis* plants. Using the same experimental approach, a promoter fragment of a *myb* homologous gene of *A. thaliana*, *gl1*, was shown to drive expression in stipules (Oppenheimer et al. 1991), and the promoter of a seed protein gene (USP) of *Vica faba* exhibited expression in the embryos of transgenic *Arabidopsis* plants (Bäumlein et al. 1991). Czako et al. (1992) created a phenotypic assay for the spatial and temporal expression pattern of a pea vicilin promoter by its fusion to the coding region of a diphtheria toxin A chain.

A phenomenon of reversible inactivation of a stably integrated hygromycin-resistance gene introduced into the nuclear genome by PEG-mediated direct gene transfer to protoplasts was observed in 50% of the transformed lines (Mittelsten Scheid et al. 1991). The sensitivity of the progeny plants was attributed to a reduced transcript level rather than to loss of the transgene. Occasional reactivation of the resistance gene appeared after outcrossing with wild-type plants or with progenies of other sensitive transformants. Progressive inactivation of a kanamycin-resistance gene introduced into *Arabidopsis* by *Agrobacterium*-mediated gene transfer was also reported (Kilby et al. 1992). Of seven families, each homozygous for a single insertion event, two showed inactivation of the resistance gene that correlated with methylation of its promoter region. Treatment of excised roots of plants from inactivated lines with the demethylating agent 5-azacytidine restored their ability to form roots on kanamycin-containing medium, supporting the suggestion that transgene inactivation is influenced by the methylation state of the DNA.

## Complementation of Mutants with Isolated Genes

The final proof that a gene cloned by any means indeed corresponds to a mutant locus identified by a certain altered phenotype can be obtained by complementation of the mutation through transformation of the mutant. In this way, the *glabrous1* mutation, which is characterized by the absence of leaf and stem trichomes, was complemented by *Agrobacterium*-mediated transformation of a DNA fragment isolated from a wild-type *Arabidopsis* genomic library (Herman and Marks 1989) using DNA sequences flanking a T-DNA insertion that caused a mutation in this gene (Marks and Feldmann 1989). Similarly, the *chlorata* (*ch-42*) mutation was complemented with a cDNA inserted into a plant expression vector used to transform mutant plants (Koncz et al. 1990). The corresponding cDNA was isolated using flanking sequences of a T-DNA insert that caused the *cs* (pale) mutation, shown to be allelic to *ch-42*. Furthermore, the wild-type *AGAMOUS* gene was isolated by means of T-DNA tagging and used to complement mutant plants homozygous for another agamous allele, *ag-2*, by *Agrobacterium*-mediated transformation (Yanofsky et al. 1990).

Recently, the *Arabidopsis ABI3* gene was cloned after mapping closely linked restriction fragment length polymorphism (RFLP) markers and chromosome walking with overlapping cosmid clones (Giraudat et al. 1992). Three overlapping fragments of an isolated cosmid very likely containing the *ABI3* gene were inserted into a T-DNA vector carrying a kanamycin-resistance marker to transform wild-type plants by *Agrobacterium*-mediated DNA transfer. Analysis of the segregation of abscisic acid insensitivity and kanamycin resistance in $F_2$ progenies of crosses between the *abi3-1* mutant and different transformants proved functional complementation of the mutation by one of the three fragments tested. Similarly, the *fad3* gene encoding an ω-3 fatty acid desaturase was cloned on the basis of the map position of a mutation affecting membrane and storage lipid fatty acid composition (Arondel et al. 1992). By using a yeast artificial chromosome (YAC) mapped to the chromosomal region containing the *fad3* gene, a *Brassica napus* cDNA library was screened, and from the positive plaques, a 1.4-kb cDNA clone representing an abundant transcript was selected for a genetic complementation test. This was carried out by the use of an *A. tumefaciens* strain carrying a Ri plasmid from *Agrobacterium rhizogenes*. The cDNA placed under the control of the cauliflower mosaic virus (CaMV) 35S promoter on a binary vector was transferred to cells of stem explants of the mutant that developed into kanamycin-resistant rooty tumors after transformation. Functional complementation of the *fad3* mutation by the introduced *B. napus* cDNA was shown by analysis of the fatty acid composition of total lipid extracts of the transformed roots. By transformation, Leyser et al. (1993) identified a DNA fragment coding for the *Arabidopsis* auxin-resistance gene *AXR1*. After identification of a YAC covering the genomic region around the *AXR1* gene, seven contiguous cosmid clones spanning this region were used to transform *axr1* mutants. Two cosmids were found to complement the *axr1* mutation, one of which was used to identify two corresponding transcripts of 0.8 kb and 1.8 kb, respectively. By analysis of mRNA levels in mutants representing five different *axr1* alleles, and by sequence determination of two of these alleles, the 1.8-kb transcript was confirmed to be the product of the *AXR1* gene.

Applying a new technology called genomic subtraction, the *GA1* gene has been cloned (Sun et al. 1992). Mutants of the *GA1* locus show an inhibition of germination in the absence of exogenously applied gibberellic acid, they are male sterile, and they are strongly reduced in size (Koornneef and van der Veen 1980). This phenotype can be converted to the wild type by repeated application of gibberellic acid. Using a cosmid clone carrying a 20-kb insert of wild-type DNA located between the left and right border sequences of the *A. tumefaciens* Ti plasmid, Sun et al. (1992) transformed mutant plants carrying the *ga1-3* allele and complemented the dwarf phenotype to wild type (GA1$^+$). Furthermore, they showed close linkage of the kanamycin-resistance marker located on the T-DNA to the GA1$^+$ phenotype.

Using an *E. coli* mutant (*trpD$^-$*) defective in phosphoribosylanthranilate transferase (PAT), a corresponding cDNA from *A. thaliana* was isolated through its capability to suppress the *trpD$^-$* mutation (Elledge et al. 1991). This cDNA was used to isolate a genomic clone from an *A. thaliana* λ library in order to clone the *PAT1* gene, mutations in which, according to biochemical evidence, are represented by the *trp1-1* and *trp1-100* mutants (Rose et al. 1992). Three overlapping subfragments of isolated genomic DNA were introduced into the *trp1-1* and *trp1-100* mutants via *Agrobacterium*-mediated transformation and were shown to complement the phenotype of both mutants to wild type, giving the ultimate proof that the *trp1* locus encodes the *A. thaliana PAT1* gene.

### Analysis of Developmental or Physiological Processes Using Transgenic *Arabidopsis* Plants

Transgenes either may serve as an assay system to report the function of endogenous regulatory factors or may themselves influence endogenous processes which thus become accessible to analysis. Examples for the first approach are the use of chimeric *GUS* reporter genes fused to the promoters of light-regulated genes that were introduced into mutants altered in their photomorphogenic response (Chory and Peto 1990; Chory et al. 1991; Deng et al. 1991). The mutations in the loci *det1*, *det2*, and *cop1*, respectively, conferred a morphological appearance of dark-grown mutant plants similar to wild-type plants grown in the light. The analysis performed with the reporter constructs allowed a test of whether mutations in genes involved in light-controlled plant development also affect the light dependence and the spatial pattern of expression of photoregulated genes on the transcriptional level. In this way, it was shown that the *det1*, *det2*, and *cop1* mutations caused derepression of light-regulated gene expression in darkness. The *det1* mutation, furthermore, caused ectopic expression of the chalcone synthase gene (*chs*) and the chlorophyll *a/b* binding protein genes (*cab1*, *cab2*, *cab3*) in roots, as well as an altered cell-type-specific *chs* expression in leaves.

The second approach was used in studies that analyzed the effect of overexpression or ectopic expression of genes coding for regulatory factors. The overexpression of either phytochrome A or phytochrome B in transgenic *Arabidopsis* plants resulted in a short-hypocotyl phenotype of light-grown seedlings that permitted the study of the biological activity of mutagenized phytochrome A polypeptides and provided direct evidence that phytochrome B is a biologically functional photoreceptor (Boylan and Quail 1991; Wagner et al. 1991). More dramatic morphological alterations were induced by the ectopic expression of the *AGAMOUS* gene (*AG*), which is required for the development of the reproductive floral organs (Mizukami and Ma 1992). From genetic studies, antagonistic functions of *AG* and another floral homeotic gene, *APETALA2* (*AP2*), have been proposed. The expression of *AG* (normally restricted to stamens and carpels) in all floral organs resulted in an aberrant flower development very similar to that of mutants defective in the *AP2* function, thus proving the proposed antagonism between the *AG* and *AP2* functions and giving further evidence for the inhibition of the *AP2* function by *AG*.

## Gene Targeting

Protoplast transformation allows one to handle very many individual transformed cell clones and thus is an ideal system to isolate transformants that arise from very rare events, provided the experimental approach allows an early identification or a selection for this event. In a model system, Halfter et al. (1992) tested the frequency of gene targeting events in *Arabidopsis* that would allow inactivation or substitution for modified copies of endogenous genes in their natural chromosomal position. Using a defective hygromycin phosphotransferase gene (*hpt*) rendered nonfunctional by an internal deletion as a stably integrated chromosomal target, restoration of hygromycin resistance through homologous recombination was achieved by transformation of protoplasts with the intact *hpt* coding region. In these experiments, the maximal ratio of homologous to nonhomologous transformation was reported to be $1 \times 10^{-4}$.

## Strategies for the Identification of Novel Mutants

Mutagenesis of *Arabidopsis* results in a wide variety of mutants, often showing very clear phenotypes that are easily identified. In other cases, however, the mutation may result only in a very minor alteration of the phenotype or may not cause any phenotypic alteration under the particular environmental conditions used. Furthermore, the apparent mutant phenotype might be completely different from what would be expected assuming a particular gene defect or modification. Therefore, new strategies have been developed to select or identify specific mutants by a scorable phenotype resulting from an altered expression pattern of an introduced transgene. Karlin-Neumann et al. (1991) created a new conditional lethal phenotype by fusion of the *tms2* gene from the *A. tumefaciens* T-DNA to a phytochrome-regulated promoter from an *Arabidopsis cab* gene (*cab140*). When introduced into transgenic *Arabidopsis* plants, this construct causes a strong inhibition of seedling growth in the presence of auxin amides. As the expression of the *tms2* gene is modulated by phytochrome, it can serve as a marker to select for mutants that lost the phytochrome-mediated light-responsive induction of *cab* gene expression. This would allow the isolation of mutants impaired in the phytochrome signal transduction pathway. Using the *GUS* and the firefly luciferase (*LUC*) reporter genes, the activities of which can easily be monitored, further systems have been developed to identify novel mutants: Using fusions of the promoter of the heat shock gene *HSP18.2* to the *GUS* reporter gene allowed Takahashi et al. (1992) to analyze the transcriptional activation at different temperatures and the tissue-specific expression of this heat shock gene and to isolate two mutants that showed a strong reduction of *GUS* activity upon heat shock compared to the parental transgenic line. Complementation analysis revealed that the mutations were *trans*-recessive and nonallelic. Martin et al. (1992) established a nondestructive assay system for detection of *GUS* activity in transgenic plants and created transgenic *Arabidopsis* plants carrying a fusion of a patatin class I promoter (B33G) isolated from potato to *GUS*. As this promoter in transgenic potato plants drives gene transcription in tubers, is inducible by primary metabolites (Rocha-Sosa et al. 1989), and is expressed in roots of transgenic *Arabidopsis* plants in dependence on the sucrose concentration in the medium, it provides a tool to identify components of a sink-related signal transduction pathway by mutagenesis (Frommer et al. 1991). Similarly, Millar et al. (1992) proposed the use of a fusion of the *cab2* promoter, which exhibits regulation by phytochrome and which shows a strong circadian cycling of its transcriptional activity (Millar and Kay 1991) to the *LUC* reporter gene to identify mutants in the circadian clock.

### *Arabidopsis* as a Model for Applied Aspects

*Arabidopsis* may not only serve as a model system to understand developmental or physiological processes in higher plants, but may also allow one to study the possibility of using higher plants for biotechnological applications that may be further applied to related crop species like *B. napus* or *B. oleracea*. To test the possibility of increasing the content of essential amino acids in the seed storage proteins, modified *Arabidopsis* 2S albumin (*AT2S1*) genes with 5–12 additional methionine codons and a Brazil nut 2S albumin gene fused to the *AT2S1* promoter were introduced into *Arabidopsis* (De Clercq et al. 1990). Only a low expression of the Brazil nut 2S albumin was found, but the expression levels of the introduced modified 2S albumin genes were high enough to yield 1–2% of the total protein in a high-salt seed extract. This level of expression of the modified *AT2S1* genes, however, would only increase the methionine level by 5% and needs further improvement. Modified 2S albumin genes were also used to test whether the neuropeptide Leu-enkephalin can be produced in transgenic plants. Per gram of seed, up to 200 nmole of enkephalin peptide were recovered from transgenic *Arabidopsis* plants, showing the possibility of producing peptides of interest in transgenic plants.

With the aim of shifting the seed fatty acid composition from long-chain to medium-chain fatty acids, the gene coding for a medium-chain acyl-carrier protein thioesterase (*BTE*), which was isolated from *Umbellularia californica*, was introduced into *Arabidopsis* under the control of a seed storage protein (napin) promoter from *B. rapa* (Voelker et al. 1992). In the transgenic plants, the medium-chain fatty acid laurate (12:0) that is absent from untransformed plants was present as the most abundant fatty acid species, reaching up to 23.5% of the total fatty acid content in seeds. The medium-chain fatty acids were found to be accumulated at the expense of long-chain fatty acids. The approach of expressing foreign genes in a higher plant to produce a new, valuable compound was also applied by Poirier et al. (1992). Two genes from the bacterium *Alcaligenes eutrophus*, an acetoacetyl-CoA reductase and a polyhydroxy-butyrate synthase, were placed under the control of the CaMV 35S promoter and were separately introduced into *Arabidopsis* through *A. tumefaciens*-mediated transformation. When both genes were present in combination in plants derived from crossings of transgenic lines carrying either of the two genes, polyhydroxybutyrate (PHB), a high-molecular-weight polyester, was synthesized. In leaves, up to 100 μg PHB per gram fresh weight was accumulated, demonstrating the possibility of producing novel biopolymers in plants through genetic engineering.

The strategy of introducing single genes into a plant species of interest can only be applied if the genetic basis for a particular trait has been uncovered at the molecular level. In contrast, transfer of large portions of a donor genome into a recipient might be desired if the trait of interest is due to the combinatorial action of many genes (polygenic), or if the responsible gene(s) has not been isolated. The successful combination of genetic material of plant species of taxonomically different tribes, *A. thaliana* and *B. campestris*, was achieved through fusion of protoplasts isolated from the two parents (Gleba and Hoffmann 1979). The resulting hybrid cell lines were shown to contain chromosomes of both parent species, although some chromosome rearrangements were observed and unspecific loss of chromosomes happened during prolonged culture. Different cell lines showed different morphogenetic potential, and from two of six hybrid lines, shoots could be regenerated, some of which formed (sterile) flowers. The regenerants were more or less heterogeneous in morphology, showing characteristics of both parent species, depending on the hybrid parent cell line from which they were regenerated. Asymmetric fusion between X-irradiated chlorsulforon-resistant *Arabidopsis* protoplasts and *B. napus* protoplasts and regeneration performed by Bauer-Weston et al. (1993) yielded male sterile somatic hybrids with vegetative morphologies intermediate between *Arabidopsis* and *Brassica*. They showed the presence of phosphoglucomutase, esterase, and peroxidase isozymes and the ribosomal genes of both parents; their chloroplast genome contents, however, were solely derived from the *B. napus* parent.

## FURTHER DEVELOPMENTS AND APPLICATIONS

As outlined in the previous sections, a number of methods like those for the setup of cell cultures, for plant regeneration from complex explants as well as from protoplasts, and for the generation of transgenic plants can be applied routinely and are readily used for various purposes. However, several procedures efficiently used in other (plant) systems have not yet been established satisfactorily for *Arabidopsis*. A few examples are given below.

The induction of somatic embryogenesis is a powerful tool to study the basic processes during the formation and maturation of the plant embryo. Especially, the combination of this technique applied to *Arabidopsis* along with the analysis of the numerous mutants isolated that are defective in their embryo development will be highly useful. Although reported to occur in a few studies dealing with plant regeneration or transformation (see above), as yet somatic embryogenesis has not clearly been proven to occur. In addition, culture conditions that reproducibly allow the induction of embryogenesis are not known. Gene replacement has been developed in animal systems and has found broad application in the modification or inactivation of endogenous genes (Bollag et al. 1989). This procedure allows one to analyze the function of gene products as well as to elucidate their structure/function relations in vivo. Only a very few examples of successful gene replacement through homologous recombination in plants, including *Arabidopsis*, have been reported to date (S. Ohl, in prep.). The further development of this approach will focus on the establishment of procedures suitable to reduce the number of nonhomologous integration events versus those resulting from homologous recombination. The establishment of conditions allowing selection against random integration (negative selection) and the creation of donor molecules more prone to homologous recombination, such as constructs with increased regions of homology to the chromosomal target, are the main lines to be followed.

The transfer and the integration of large DNA molecules into the genome of *Arabidopsis* by transformation will not only be useful in terms of increasing the rate of homologous recombination, but will also facilitate experiments aimed at the identification of genes on fragments of cloned DNA. In this respect, genomic cosmid libraries cloned into transformation-competent *Agrobacterium* vectors (Olszewski et al. 1988; Lazo et al. 1991) will surely be of great use. However, the identification of genes mapped to a certain chromosomal region may be simplified further by the complementation of the corresponding mutant by transformation using a set of mapped large, overlapping genomic clones (e.g., YACs) followed by sets of smaller subfragments. Such a procedure would be complementary to a fine-mapping program, probably speeding up the process of isolating phenotypically characterized genes.

The previously described advantages of *Arabidopsis* for molecular genetic studies also lend this plant an advantage in the analysis of nucleus/organelle interactions. However, to this end, it is important that the methods for organelle transformation, as demonstrated in tobacco (see above), be established in this species.

## CONCLUSIONS AND PERSPECTIVES

This review shows that tissue culture and transformation technology in *Arabidopsis* have developed tremendously over the past years. These advances have made possible many new insights into how plants function at the molecular level. In particular, our understanding has been advanced by the ability to dissect mutants at the level of DNA by means of gene cloning, complementation through transformation, and the subsequent analysis of the gene in transgenic *Arabidopsis* plants by means of promoter deletion studies, ectopic expression, and antisense repression. It is clear that the combination of tissue culture and molecular genetic techniques, together with the unique advantages of *Arabidopsis*, will open up many new and exciting possibilities for the study of higher plants in the future.

## REFERENCES

Acedo, G.N. 1986. Regeneration of *Arabidopsis* callus in vitro. *Plant Cell Tissue Organ Cult.* **6:** 109–114.

Aerts, M. 1978. *In vitro* culture and regeneration of *Arabidopsis thaliana* crown gall tumors. *Arabidopsis Inf. Serv.* **15:** 29–30.

———. 1980. Protoplast culture as a cloning technique for crown gall tumor cells. *Arabidopsis Inf. Serv.* **17:** 68–69.

Aerts, M., M. Jacobs, J.-P. Hernalsteens, M. Van Montagu, and J. Schell. 1979. Induction and *in vitro* culture of *Arabidopsis thaliana* crown gall tumors. *Plant Sci. Lett.* **17:** 43–50.

Akama, K., H. Shiraishi, S. Ohta, K. Nakamura, K. Okada, and Y. Shimura. 1992. Efficient transformation of *Arabidopsis thaliana*: Comparison of the efficiencies with various organs, plant ecotypes and *Agrobacterium* strains. *Plant Cell Rep.* **12:** 7–11.

Altmann, T., B. Damm, U. Halfter, L. Willmitzer, and P.C. Morris. 1992. Protoplast transformation and methods to create specific mutants in *Arabidopsis thaliana*. In *Methods in* Arabidopsis *research* (ed. C. Koncz et al.), pp. 310–330. World Scientific, Singapore.

Amos, J.A. 1978. Anther culture of four *Arabidopsis* species. *Arabidopsis Inf. Serv.* **15:** 20–23.

Amos, J.A. and R.L. Scholl. 1978. Induction of haploid callus from anthers of four species of *Arabidopsis*. *Z. Pflanzenphysiol.* **90:** 33–43.

An, G.B., D. Watson, and C.C. Chiang. 1986. Transformation of tobacco, tomato, potato, and *Arabidopsis thaliana* using a binary Ti vector system. *Plant Physiol.* **81:** 301–305.

Anand, R. 1966. Preliminary studies on callus culture of *Arabidopsis thaliana*. *Arabidopsis Inf. Serv.* **3:** 15.

André, D., D. Colau, J. Schell, M. Van Montagu, and J.-P. Hernalsteens. 1986. Gene tagging in plants by a T-DNA insertion mutagen that generates APH(3′)II-plant gene fusions. *Mol. Gen. Genet.* **204:** 512–518.

Arondel, V., B. Lemieux, I. Hwang, S. Gibson, H.M. Goodman, and C.R. Somerville. 1992. Map-based cloning of a gene controlling omega-3 fatty acid desaturation in *Arabidopsis*. *Science* **258:** 1353–1355.

Atanassova, R., N. Chaubet, and C. Gigot. 1992. A 126 bp fragment of a plant histone gene promoter confers preferential expression in meristems of transgenic *Arabidopsis*. *Plant J.* **2:** 291–300.

Aviv, D. and E. Galun. 1980. Restoration of fertility in cytoplasmic male sterile (CMS) *Nicotiana sylvestris* by fusion with X-irradiated *N. tabacum* protoplasts. *Theor. Appl. Genet.* **58:** 121–127.

Axelos, M., C. Curie, L. Mazzolini, C. Bardet, and B. Lescure. 1992. A protocol for transient gene expression in *Arabidopsis thaliana* protoplasts isolated from cell suspension cultures. *Plant Physiol. Biochem.* **30:** 123–128.

Bauer-Weston, B., W. Keller, J. Webb, and S. Gleddie. 1993. Production and characterisation of asymmetric somatic hybrids between *Arabidopsis thaliana* and *Brassica napus. Theor. Appl. Genet.* **86:** 150–158.

Bäumlein, H., W. Boerjan, I. Nagy, R. Bassüner, M. Van Montagu, D. Inzé, and U. Wobus. 1991. A novel seed protein gene from *Vicia faba* is developmentally regulated in transgenic tobacco and *Arabidopsis* plants. *Mol. Gen. Genet.* **225:** 459–467.

Baus, A.D., L. Franzmann, and D. Meinke. 1986. Growth *in vitro* of arrested embryos from lethal mutants of *Arabidopsis thaliana. Theor. Appl. Genet.* **72:** 577–586.

Beachy, R.N., S. Loesch-Fries, and N.E. Tumer. 1990. Coat protein-mediated resistance against virus infection. *Annu. Rev. Phytopathol.* **28:** 451–474.

Bechtold, N., J. Ellis, and G. Pelletier. 1993. *In planta Agrobacterium* mediated gene transfer by infiltration of adult *Arabidopsis thaliana* plants. *C.R. Acad. Sci.* **316:** 1194–1199.

Benfey, P.N. and N.-H. Chua. 1989. Regulated genes in transgenic plants. *Science* **244:** 174–181.

Bhalla-Sarin, N., S.K. Sopory, and S. Guha-Mukherjee. 1976. Studies on the isolation and fusion of protoplasts of *Arabidopsis. Arabidopsis Inf. Serv.* **13:** 200–204.

Bhatt, A.M. and C. Dean. 1992. Development of tagging systems in plants using heterologous transposons. *Curr. Opin. Biotechnol.* **3:** 152–158.

Blaydes, D.F. 1966. Interaction of kinetin and various inhibitors in the growth of soybean callus. *Physiol. Plant.* **19:** 748–735.

Bollag, R.J., A.S. Waldman, and R.M. Liskay. 1989. Homologous recombination in mammalian cells. *Annu. Rev. Genet.* **23:** 199–225.

Botterman, J. and J. Leemans. 1988. Engineering herbicide resistance in plants. *Trends Genet.* **4:** 219–222.

Bounias, M. 1973. Culture aseptique d'Arabidopsis sur sable et liquide minéral "minimum." *Arabidopsis Inf. Serv.* **10:** 38–40.

Boylan, M.T. and P.H. Quail. 1991. Phytochrome A overexpression inhibits hypocotyl elongation in transgenic *Arabidopsis. Proc. Natl. Acad. Sci.* **88:** 10806–10810.

Braam, J. 1992. Regulated expression of the calmodulin-related *TCH* genes in cultured *Arabidopsis* cells: Induction by calcium and heat shock. *Proc. Natl. Acad. Sci.* **89:** 3213–3216.

Brusslan, J.A. and A. Tobin. 1992. Light-independent developmental regulation of *cab* gene expression in *Arabidopis thaliana* seedlings. *Proc. Natl. Acad. Sci.* **89:** 7791–7795.

Butow, R.A. and T.D. Fox. 1990. Organelle transformation: Shoot first, ask questions later. *Trends Biol. Sci.* **15:** 465–468.

Chang, S.-S., S.-K. Park, and H.-G. Nam. 1990. Transformation of *Arabidopsis* by *Agrobacterium* inoculation on wounds. In *4th International Conference on* Arabidopsis *Research* (ed. D. Schweizer et al.), Abstr. S3/O. University of Vienna, Austria.

Charles, P. and J. Remy. 1980. Presence of donor DNA sequences in the progenies of corrected *Arabidopsis* mutants. *Arabidopsis Inf. Serv.* **17:** 79–83.

Chaudhury, A.M. and E.R. Signer. 1989a. Relative regeneration proficiency of *Arabidopsis thaliana* ecotypes. *Plant Cell Rep.* **8:** 368–369.

―――. 1989b. Non-destructive transformation of *Arabidopsis thaliana. Plant Mol. Biol. Rep.* **7:** 258–265.

Chory, J. and C.A. Peto. 1990. Mutations in the *DET1* gene affect cell-type-specific expression of light-regulated genes and chloroplast development in *Arabidopsis. Proc. Natl. Acad. Sci.* **87:** 8776–8780.

Chory, J., P. Nagpal, and C.A. Peto. 1991. Phenotypic and genetic analysis of *det2*, a new mutant that affects light-regulated seedling development in *Arabidopsis. Plant Cell* **3:** 445–459.

Clarke, M.C., W. Wei, and K. Lindsey. 1992. High frequency transformation of *Arabidopsis thaliana* by *Agrobacterium tumefaciens. Plant Mol. Biol. Rep.* **10:** 178–189.

Constabel, F. and I.K. Vasil, ed.. 1988. *Cell culture and somatic cell genetics of plants,* vol. 5. Academic Press, San Diego.

Corcos, A. and L. Krupka. 1979. Crown gall and *Arabidopsis thaliana*: A preliminary result. *Arabidopsis Inf. Serv.* **15:** 28.

Corcos, A. and R. Lewis. 1971. Preliminary report on callus formation and growth in vegetable juice and White's liquid medium. *Arabidopsis Inf. Serv.* **8:** 35–36.

Corcos, A., B. Piper, and R. Lewis. 1973. Redifferentiation of normal *Arabidopsis* plants from callus culture. *Arabidopsis Inf. Serv.* **10:** 10.

Curie, C., T. Liboz, C. Bardet, E. Gander, C. Médale, M. Axelos, and B. Lescure. 1991. *Cis* and *trans*-acting elements involved in the activation of *Arabidopsis thaliana* A1 gene encoding the translation elongation factor EF-1a. *Nucleic Acids Res.* **19:** 1305–1310.

Czako, M., J.C. Jang, J.M. Herr, Jr., and L. Márton. 1992. Differential manifestation of seed mortality induced by seed-specific expression of the gene for diphtheria toxin A chain in *Arabidopsis* and tobacco. *Mol. Gen. Genet.* **235:** 33–40.

Damm, B. and L. Willmitzer. 1988. Regeneration of fertile plants from protoplasts of different *Arabidopsis thaliana* genotypes. *Mol. Gen. Genet.* **213:** 15–20.

–––––––. 1991. *Arabidopsis* protoplast transformation and regeneration. In *Plant tissue culture manual* (ed. K. Lindsey), pp. A7/1–A7/20. Kluwer Academic Publishers, Dordrecht, The Netherlands.

Damm, B., R. Schmidt, and L. Willmitzer. 1989. Efficient transformation of *Arabidopsis thaliana* using direct gene transfer to protoplasts. *Mol. Gen. Genet.* **217:** 6–12.

Davey, M.R., E.L. Rech, and B.J. Mulligan. 1989. Direct DNA transfer to plant cells. *Plant Mol. Biol.* **13:** 273–285.

De Clercq, A., M. Vandewiele, J. Van Damme, P. Guerche, M. Van Montagu, J. Vandekerckhove, and E. Krebbers. 1990. Stable accumulation of modified 2S albumin seed storage proteins with higher methionine contents in transgenic plants. *Plant Physiol.* **94:** 970–979.

de la Fuente-Martínez, J.M., G. Mosqueda-Cano, A. Alvarez-Morales, and L. Herrera-Estrella. 1992. Expression of a bacterial phaseolotoxin-resistant ornithyl transcarbamylase in transgenic tobacco confers resistance to *Pseudomonas syringae* pv. *phaseolicola. Bio/Technology* **10:** 905–909.

Deng, X.-W., T. Caspar, and P.H. Quail. 1991. *Cop1*: A regulatory locus involved in light-controlled development and gene expression in *Arabidopsis. Genes Dev.* **5:** 1172–1182.

Dixon, R.A., ed. 1985. *Plant cell culture: A practical approach.* IRL Press, Oxford.

Doelling J.H. and C.S. Pikaard. 1993. Transient expression in *Arabidopsis thaliana* protoplasts derived from rapidly established cell suspension cultures. *Plant Cell Rep.* **12:** 241–244.

Doy, C.H., P.M. Gresshoff, and B.G. Rolfe. 1973. Biological and molecular evidence for the transgenesis of genes from bacteria to plant cells. *Proc. Natl. Acad. Sci.* **70:** 723–726.

Dudits, D., E. Maroy, T. Praznovszky, Z. Olah, J. Gyorgyey, and R. Cella. 1987. Transfer of resistance traits from carrot into tobacco by asymmetric somatic hybridization: Regeneration of fertile plants. *Proc. Natl. Acad. Sci.* **84:** 8434–8438.

Eigel, L. and H.-U. Koop. 1992. Transfer of defined numbers of chloroplasts into albino protoplasts by subprotoplast/protoplast microfusion: Chloroplasts can be "cloned", by using suitable plastome combinations or selective pressure. *Mol. Gen. Genet.* **233:** 479–482.

Elledge, S.J., J.T. Mulligan, S.W. Ramer, M. Spottswood, and R.W. Davis. 1991. λYES: A multifunctional cDNA expression vector for the isolation of genes by complementation·of yeast and *Escherichia coli* mutants. *Proc. Natl. Acad. Sci.* **88:** 1731–1735.

Ernst, P.J., K.D. Rodecap, and D.T. Tingey. 1981. A hydroponic method for culturing populations of *Arabidopsis. Arabidopsis Inf. Serv.* **18:** 1–14.

Feenstra, W.J. 1965. Remarks on sterile culturing. *Arabidopsis Inf. Serv.* **2:** 33.

Feenstra, W.J., D.L. De Heer, and F.J. Oostindier-Braaksma. 1973. Negative results of treatment with bacterial DNA on "repair" of mutants of *Arabidopsis. Arabidopsis Inf. Serv.* **10:** 33–34.

Feinbaum, R.L., G. Storz, and F.M. Ausubel. 1991. High intensity and blue light regulated expression of chimeric chalcone synthase genes in transgenic *Arabidopsis thaliana* plants. *Mol. Gen. Genet.* **226:** 449–456.

Feldmann, K.A. 1991. T-DNA insertion mutagenesis in *Arabidopsis*: Mutational spectrum. *Plant J.* **1:** 71–82.

Feldmann, K.A. and M.D. Marks. 1986. Rapid and efficient regeneration of plants from explants of *Arabidopsis thaliana. Plant Sci.* **47:** 63–69.

———. 1987. *Agrobacterium*-mediated transformation of germinating seeds of *Arabidopsis thaliana*: A non-tissue culture approach. *Mol. Gen. Genet.* **208:** 1–9.

Feldmann, K.A., M.D. Marks, M.L. Christianson, and R.S. Quatrano. 1989. A dwarf mutant of *Arabidopsis* generated by T-DNA insertion mutagenesis. *Science* **243:** 1351–1354.

Ferl, R.J. and B.H. Laughner. 1989. *In vivo* detection of regulatory factor binding sites of *Arabidopsis thaliana Adh. Plant Mol. Biol.* **12:** 357–366.

Forbert, P.R., B.L. Miki, and V.N. Iyer. 1991. Detection of gene regulatory signals in plants revealed by T-DNA-mediated fusions. *Plant Mol. Biol.* **17:** 837–851.

Ford, K.G. 1990. Plant regeneration from *Arabidopsis thaliana* protoplasts. *Plant Cell Rep.* **8:** 534–537.

Franzmann, L., D.A. Patton, and D. Meinke. 1989. In vitro morphogenesis of arrested embryos from lethal mutants of *Arabidopsis thaliana. Theor. Appl. Genet.* **77:** 609–616.

Frommer, W.B., T. Martin, R. Schmidt, S. Hummel, and L. Willmitzer. 1991. Patatin promoters as a tool to identify sink-related signal transduction pathways. In *Recent advances in phloem transport and assimilate compartmentation* (ed. J.L. Bonnemain et al.), pp. 254–257. Quest Editions, Nantes.

Gaj, M.D. and M. Maluszynski. 1987. Genetic variation in callus culture of *Arabidopsis thaliana* (L.) Heynh. *Arabidopsis Inf. Serv.* **23:** 1–8.

Galbraith, D.W., K.R. Harkins, and S. Knapp. 1991. Systemic endopolyploidy in *Arabidopsis thaliana. Plant Physiol.* **96:** 985–989.

Gamborg, O.L. and D.E. Eveleigh. 1968. Culture methods and detection of glycanases in suspension culture of wheat and barley. *Can. J. Biochem.* **46:** 417–421.

Gamborg, O.L., R.A. Miller, and K. Ojima. 1968. Nutrient requirements of suspension culture of soybean root cells. *Exp. Cell Res.* **50:** 151–158.

Gasser, C.S. and R.T. Fraley. 1989. Genetically engineering plants for crop improvement. *Science* **244:** 1293–1299.

Gheysen, G., R. Villarroel, and M. Van Montagu. 1991. Illegitimate recombination in plants: A model for T-DNA integration. *Genes Dev.* **5:** 287–297.

Giraudat, J., B.M. Hauge, C. Valon, J. Smalle, F. Parcy, and H. Goodman. 1992. Isolation of the *Arabidopsis ABI3* gene by positional cloning. *Plant Cell* **4:** 1251–1261.

Gleba, Y.Y. and F. Hoffman. 1978. Hybrid cell lines *Arabidopsis thaliana* + *Brassica campestris*: No evidence for specific chromosome elimination. *Mol. Gen. Genet.* **165:** 257–264.

———. 1979. "*Arabidobrassica*": Plant-genome engineering by protoplast fusion. *Naturwissenschaften* **66:** 547–554.

Gleddie, S. 1989. Plant regeneration from cell suspension cultures of *Arabidopsis thaliana* Heynh. *Plant Cell Rep.* **8:** 1–5.

Göbel, E. and H. Lörz. 1988. Genetic manipulation of cereals. *Oxf. Surv. Plant Mol. Cell Biol.* **5:** 1–22.

Gotô, N. 1980. Organogenetic capacity and its heritability of *Arabidopsis thaliana*. *Arabidopsis Inf. Serv.* **17:** 49–57.

Gresshoff, P.M. 1976. Protoplast and callus regeneration of *Arabidopsis thaliana*. *Arabidopsis Inf. Serv.* **13:** 211–214.

Gresshoff, P.M. and C.H. Doy. 1972. Haploid *Arabidopsis thaliana* callus and plants from anther culture. *Aust. J. Biol. Sci.* **25:** 259–264.

Gresshoff, P.M. and B.G. Rolfe. 1976. Bacteriophage uptake in *Arabidopsis thaliana* protoplasts. *Arabidopsis Inf. Serv.* **13:** 215–223.

Gupta, P.P., O. Schieder, and M. Gupta. 1984. Intergeneric nuclear gene transfer between somatically and sexually incompatible plants through asymmetric protoplast fusion. *Mol. Gen. Genet.* **197:** 30–35.

Halfter, U., P.-C. Morris, and L. Willmitzer. 1992. Gene targeting in *Arabidopsis thaliana*. *Mol. Gen. Genet.* **231:** 186–193.

Herman, P.L. and M.D. Marks. 1989. Trichome development in *Arabidopsis thaliana*. II. Isolation and complementation of the *GLABROUS1* gene. *Plant Cell* **1:** 1051–1055.

Hoffmann, A., U. Halfter, and P.-C. Morris. 1994. Transient expression in leaf mesophyll protoplasts of *Arabidopsis thaliana*. *Plant Cell Tiss. Org. Cult.* **36:** 53–58.

Hooykaas, P.J.J. and R.A. Schilperoort. 1992. *Agrobacterium* and plant genetic engineering. *Plant Mol. Biol.* **19:** 15–38.

Howe, C. 1985. Chloroplast transformation by *Agrobacterium tumefaciens*. *Trends Genet.* **1:** 217–218.

Huang, B.C. and M.M. Yeoman. 1983. Formation of somatic embryos in tissue cultures of *Arabidopsis thaliana*. *Arabidopsis Inf. Serv.* **20:** 73–77.

Ichikawa, H., L. Tanno-Suenaga, and J. Imamura. 1987. Selection of *Daucus carota* cybrids based on metabolic complementation between X-irradiated *D. capillifolius* and iodoacetamide-treated *D. carota* by somatic cell fusion. *Theor. Appl. Genet.* **74:** 746–752.

Jähne, A., P.A. Lazzeri, and H. Lörz. 1991. Regeneration of fertile plants from protoplasts derived from embryogenic cell suspensions of barley (*Hordeum vulgare* L.). *Plant Cell Rep.* **10:** 1–6.

Jensen, C.J. 1986. Haploid induction and production in crop plants. In *Genetic manipulation in plant breeding* (ed. W. Horn et al.), pp. 231–256. de Gruyter, Berlin.

Kao, K.N., F. Constabel, M.R. Michayluk, and O. L. Gamborg. 1974. Protoplast fusion and growth of intergenic hybrid cells. *Planta* **120:** 215–227.

Karesch, H., R. Bilang, and I. Potrykus. 1991a. *Arabidopsis thaliana*: Protocol for plant regeneration from protoplasts. *Plant Cell Rep.* **9:** 575–578.

Karesch, H., R. Bilang, O. Mittelsten Scheid, and I. Potrykus. 1991b. Direct gene transfer to protoplasts of *Arabidopsis thaliana*. *Plant Cell Rep.* **9:** 571–574.

Karlin-Neumann, G.A., J.A. Brusslan, and E. Tobin. 1991. Phytochrome control of the *tms2* gene in transgenic *Arabidopsis*: A strategy for selecting mutants in the signal transduction pathway. *Plant Cell* **3:** 573–582.

Kartha, K.K. 1986. Production and indexing of disease-free plants. In *Plant tissue culture and its agricultural applications* (ed. L.A. Withers and P.G. Alderson), pp. 219–238. Butterworths, London.

Keathley, D.E. and R.L. Scholl. 1982. Culture of *Arabidopsis thaliana* anthers on liquid medium. *Z. Pflanzenphysiol.* **106:** 199–212.

Kemper, E., C. Grevelding, J. Schell, and R. Masterson. 1992. Improved method for the transformation of *Arabidopsis thaliana* with chimeric dihydrofolate reductase constructs which confer methotrexate resistance. *Plant Cell Rep.* **11:** 118–121.

Kilby, N.J., H.M.O. Leyser, and I.J. Furner. 1992. Promoter methylation and progressive transgene inactivation in *Arabidopsis*. *Plant Mol. Biol.* **20:** 103–112.

Klee, H.J. and S.G. Rogers. 1989. Plant gene vectors and genetic transformation: Plant transformation systems based on the use of *Agrobacterium tumefaciens*. *Cell Cult. Somatic Cell Genet. Plants* **6:** 1–23.

Kleinhofs, A. and R. Behki. 1977. Prospects for plant genome modification by non-conventional means. *Annu. Rev. Genet.* **11:** 79–101.

Koncz, C., N. Martini, R. Mayerhofer, Z. Koncz-Kalman, H. Körber, G.P. Redei, and J. Schell. 1989. High-frequency T-DNA-mediated gene tagging in plants. *Proc. Natl. Acad. Sci.* **86:** 8467–8471.

Koncz, C., R. Mayerhofer, Z. Koncz-Kalman, C. Nawrath, B. Reiss, G.P. Redei, and J. Schell. 1990. Isolation of a gene encoding a novel chloroplast protein by T-DNA tagging in *Arabidopsis thaliana*. *EMBO J.* **9:** 1337–1346.

Koornneef, M. and J.H. van der Veen. 1980. Induction and analysis of gibberellin sensitive mutants in *Arabidopsis thaliana* (L.) Heynh. *Theor. Appl. Genet.* **58:** 257–263.

Kost, B., I. Potrykus, and G. Neuhaus. 1992. Regeneration of fertile plants from excised immature zygotic embryos of *Arabidopsis thaliana*. *Plant Cell Rep.* **12:** 50–54.

Kyozuka, J., T. Kaneda, and K. Shimamoto. 1989. Production of cytoplasmic male sterile rice (*Oryza sativa* L.) by cell fusion. *Bio/Technology* **7:** 1171–1174.

Laibach, F. 1943. *Arabidopsis thaliana* (L.) als Object für genetische und entwicklungsphysiologische Untersuchungen. *Bot. Arch.* **44:** 439–455.

Langridge, J. 1955. Biochemical mutants in the crucifer *Arabidopsis thaliana* (L.) Heynh. *Nature* **176:** 260–261.

————. 1957. The aseptic culture of *Arabidopsis thaliana* (L.) Heynh. *Aust. J. Biol. Sci.* **10:** 243–252.

Larkin, P.J., P.M. Banks, R. Phasic, RS. Brettell, P.A. Davies, S.A. Ryan, W.R. Scowcroft, L.H. Spindler, and G.J. Tanner. 1989. From somatic variation to variant plants: Mechanisms and applications. *Genome* **31:** 705–711.

Lazo, G.R., P.A. Stein, and R.A. Ludwig. 1991. A DNA transformation-competent *Arabidopsis* genomic library in *Agrobacterium*. *Bio/Technology* **9:** 963–967.

Ledoux, L. and M. Jacobs. 1969. Experimental conditions for the study of the uptake of foreign DNA by *Arabidopsis thaliana*. *Arabidopsis Inf. Serv.* **6:** 5–11.

Ledoux, L., R. Huart, and M. Jacobs. 1971. Fate of exogenous DNA in *Arabidopsis thaliana*. *Eur. J. Biochem.* **23:** 96–108.

————. 1974. DNA-mediated genetic correction of thiamineless *Arabidopsis*. *Nature* **249:** 17–21.

Ledoux, L., L. Diels, M. E. Thiry, R. Hooghe, Y. Maluszynska, C. Merckaert, J.M. Piron, A.M. Ryngaert, and J. Remy. 1985. Transfer of bacterial and human genes to germinating *Arabidopsis thaliana*. *Arabidopsis Inf. Serv.* **22:** 1–11.

Levin, R., V. Gaba, B. Tal, S. Hirsch, and D. DeNola. 1988. Automated plant tissue culture for mass propagation. *Bio/Technology* **6:** 1035–1040.

Leyser, H.M.O., C.A. Lincoln, C. Timpte, D. Lammer, J. Turner, and M. Estelle. 1993.

*Arabidopsis* auxin-resistance gene *AXR1* encodes a protein related to ubiquitin-activating enzyme E1. *Nature* **364**: 161–164.

Li, L.-C. and H.W. Kohlenbach. 1982. Somatic embryogenesis in quite a direct way in cultures of mesophyll protoplasts of *Brassica napus* (L.). *Plant Cell Rep.* **1**: 209–221.

Lichter, R. 1982. Induction of haploid plants from isolated pollen of *Brassica napus*. *Z. Pflanzenphysiol.* **105**: 427–434.

Lindsey, K., ed. 1991. *Plant tissue culture manual. Manual* (suppl. 1). Kluwer Academic Publishers, Dordrecht, The Netherlands.

————. 1992. *Plant tissue culture manual* (suppl. 2). Kluwer Academic Publishers, Dordrecht, The Netherlands.

Lloyd, A.M., A. Barnason, S.G. Rogers, M.C. Byrne, R.T. Fraley, and R.B. Horsch. 1986. Transformation of *Arabidopsis thaliana* with *Agrobacterium tumefaciens*. *Science* **234**: 464–466.

Loewenberg, J.R. 1965. Callus culture of *Arabidopsis*. *Arabidopsis Inf. Serv.* **2**: 34.

Lurquin, P.F. and Y. Hotta. 1975. Reutilization of bacterial DNA by *Arabidopsis thaliana* cells in tissue culture. *Plant Sci. Lett.* **5**: 103–112.

Maluszynski, M., J. Maluszynska, and L. Ledoux. 1978. Progeny of *py, er, gl* plants corrected with bacterial DNA. *Arabidopsis Inf. Serv.* **15**: 1–3.

Mandel, M.A., J.L. Bowman, S.A. Kempin, H. Ma, E.M. Meyerowitz, and M.F. Yanofsky. 1992. Manipulation of flower structure in transgenic tobacco. *Cell* **71**: 133–143.

Mariani, C., M. De Beuckeleer, J. Truettner, J. Leemans, and R.B. Goldberg. 1990. Induction of male sterility in plants by a chimaeric ribonuclease gene. *Nature* **347**: 737–741.

Marks, M.D. and K.A. Feldmann. 1989. Trichome development in *Arabidopsis thaliana*. I. T-DNA tagging of the *GLABROUS1* gene. *Plant Cell* **1**: 1043–1050.

Martin, T., R. Schmidt, T. Altmann, and W.B. Frommer. 1992. Non-destructive assay system for detection of β-glucuronidase activity in higher plants. *Plant Mol. Biol. Rep.* **10**: 37–46.

Márton, L. and J. Browse. 1991. Facile transformation of *Arabidopsis*. *Plant Cell Rep.* **10**: 235–239.

Masson, J. and J. Paszkowski. 1992. The culture response of *Arabidopsis thaliana* protoplasts is determined by the growth conditions of donor plants. *Plant. J.* **2**: 829–833.

Mayerhofer, R., Z. Koncz-Kalman, C. Nawrath, G. Bakkeren, A. Crameri, K. Angelis, G.P. Rédei, J. Schell, B. Hohn, and C. Koncz. 1991. T-DNA integration: A mode of illegitimate recombination in plants. *EMBO J.* **10**: 697–704.

McKendree, W.L. and R.J. Ferl. 1992. Functional elements of the *Arabidopsis Adh* promoter include the G-box. *Plant Mol. Biol.* **19**: 859–862.

Medford, J.I., J.S. Elmer, and H. Klee. 1991. Molecular cloning and characterization of genes expressed in shoot apical meristems. *Plant Cell* **3**: 359–370.

Menczel, L., G. Galiba, F. Nagy, and P. Maliga. 1982. Effect of radiation dosage on efficiency of chloroplast transfer by protoplast fusion in *Nicotiana*. *Genetics* **100**: 487–495.

Meyerowitz, E.M. 1992. Introduction to the *Arabidopsis* genome. In *Methods in Arabidopsis research* (ed. C. Koncz et al.), pp. 100–118. World Scientific, Singapore.

Millar, A.J. and S.A. Kay. 1991. Circadian control of *cab* gene transcription and mRNA accumulation in *Arabidopsis*. *Plant Cell* **3**: 541–550.

Millar, A.J., S.R. Short, N.-H. Chua, and S.A. Kay. 1992. A novel circadian phenotype based on firefly luciferase expression in transgenic plants. *Plant Cell* **4**: 1075–1087.

Mittelsten Scheid, O., J. Paszkowski, and I. Potrykus. 1991. Reversible inactivation of a transgene in *Arabidopsis thaliana*. *Mol. Gen. Genet.* **228:** 104–112.

Mizukami, Y. and H. Ma. 1992. Ectopic expression of the floral homeotic gene *AGAMOUS* in transgenic *Arabidopsis* plants alters floral organ identity. *Cell* **71:** 119–131.

Monnier, M. 1990. Culture of zygotic embryos of higher plants. In *Methods in molecular biology: Plant cell and tissue culture* (ed. J.W. Pollard and J.M. Walker), vol. 6, pp. 129–139. Humana Press, Clifton, New Jersey.

Morrison, R.A. and D.A. Evans. 1988. Haploid plants from tissue culture: New plant varieties in a shortened time frame. *Bio/Technology* **6:** 684–690.

Müller, A.J. 1963. Embryonentest zum Nachweis rezessiver Letalfaktoren bei *Arabidopsis thaliana*. *Biol. Zentralbl.* **82:** 133–163.

Müller-Röber, B., U. Sonnewald, and L. Willmitzer. 1992. Inhibition of the ADP-glucose pyrophosphorylase in transgenic potatoes leads to sugar-storing tubers and influences tuber formation and expression of tuber storage protein genes. *EMBO J.* **11:** 1229–1238.

Murashige, T. and F. Skoog. 1962. A revised medium for rapid growth and bioassays with tobacco tissue cultures. *Physiol. Plant.* **15:** 473–497.

Neales, T.F. 1968a. The nutritional requirements of excised roots of three genotypes of *Arabidopsis thaliana* (L.) Heynh. *New Phytol.* **67:** 159–165.

————. 1968b. Effects of high temperature and genotype on the growth of excised roots of *Arabidopsis thaliana*. *Aust. J. Biol. Sci.* **21:** 217–223.

Negrutiu, I. 1976. In vitro morphogenesis in *Arabidopsis thaliana*. *Arabidopsis Inf. Serv.* **13:** 180–187.

Negrutiu, I. and M. Jacobs. 1976. Induction of mutations in *Arabidopsis thaliana* for resistance to amino acid analogues. *Arabidopsis Inf. Serv.* **13:** 125–132.

————. 1978. Restoration of the morphogenic capacity in long-term callus cultures of *Arabidopsis thaliana*. *Z. Pflanzenphysiol.* **90:** 431–441.

Negrutiu, I., B. Beeftink, and M. Jacobs. 1975a. *Arabidopsis* as a model system in somatic cell genetics. I. Cell and tissue culture. *Plant Sci. Lett.* **5:** 293–304.

————. 1975b. "In vitro" culture of *Arabidopsis thaliana*. *Arabidopsis Inf. Serv.* **12:** 27–29.

Negrutiu, I., M. Jacobs, and D. Cachita. 1978a. Some factors controlling *in vitro* morphogenesis in *Arabidopsis thaliana*. *Z. Pflanzenphysiol.* **86:** 113–124.

Negrutiu, I., M. Jacobs, and W. de Greef. 1978b. *In vitro* morphogenesis of *Arabidopsis thaliana*: The origin of the explant. *Z. Pflanzenphysiol.* **90:** 363–372.

Negrutiu, I., R. Shillito, I. Potrykus, G. Biasini, and F. Sala. 1987. Hybrid genes in the analysis of transformation conditions. I. Setting up a simple method for direct gene transfer in plant protoplasts. *Plant Mol. Biol.* **8:** 363–373.

Nitsch, J.P. 1967. Toward a biochemistry of flowering and fruiting: Contributions of the "in vitro" technique. *Proc. Int. Hortic. Congr.* **3:** 291–308.

Nitsch, J.P. and C. Nitsch. 1969. Haploid plants from pollen grains. *Science* **163:** 85–87.

Ohl, S., S.A. Hedrick, J. Chory, and C.J. Lamb. 1990. Functional properties of a phenylalanine ammonia-lyase promoter from *Arabidopsis*. *Plant Cell* **2:** 837–848.

Olszewski, N.E., F.B. Martin, and F.M. Ausubel. 1988. Specialized binary vector for plant transformation: Expression of the *Arabidopsis thaliana AHAS* gene in *Nicotiana tabacum*. *Nucleic Acids Res.* **16:** 10765–10782.

Oppenheimer, D.G., P.L. Herman, S. Sivakumaran, J. Esch, and M.D. Marks. 1991. A *myb* gene required for leaf trichome differentiation in *Arabidopsis* is expressed in stipules. *Cell* **67:** 483–493.

Padgette, S.R., G. della-Cioppa, D.M. Shah, R.T. Fraley, and G.M. Kishore. 1989. Selective herbicide tolerance through protein engineering. *Cell Cult. Somatic Cell Genet. Plants* **6:** 441–476.

Pang, P.P. and E.M. Meyerowitz. 1987. *Arabidopsis thaliana*: A model system for plant molecular biology. *Bio/Technology* **5:** 1177–1181.

Paszkowski, J., M.W. Saul, and I. Potrykus. 1989. Plant gene vectors and genetic transformation: DNA-mediated gene transfer to plants. *Cell Cult. Somatic Cell Genet. Plants* **6:** 51–68.

Patton, D.A. and D.W. Meinke. 1988. High frequency plant regeneration from cultured cotyledons of *Arabidopsis thaliana*. *Plant Cell Rep.* **7:** 233–237.

Pavingerová, D., R. Bískova, and M. Ondrej. 1984. Gametic transmission of mannopine and agropine synthesis in *Arabidopsis thaliana* hairy root tumor regeneration. *Arabidopsis Inf. Serv.* **21:** 1–4.

Peñarrubia, L., R. Kim, J. Giovannoni, S.-H. Kim, and R.L. Fischer. 1992. Production of the sweet protein monellin in transgenic plants. *Bio/Technology* **10:** 561–564.

Poirier, Y., D.E. Dennis, K. Klomparens, and C. Somerville. 1992. Polyhydroxybutyrate, a biodegradable thermoplastic, produced in transgenic plants. *Science* **256:** 520–523.

Potrykus, I. 1991. Gene transfer to plants: Assessment of published approaches and results. *Annu. Rev. Plant Physiol. Plant Mol. Biol.* **42:** 205–225.

Rédei, G.P. 1962. Genetic block of "vitamin thiazole" synthesis in *Arabidopsis. Genetics* **47:** 979.

Rédei, G.P., C. Koncz, and J. Schell. 1988. Transgenic *Arabidopsis*. In *Chromosome structure and function* (ed. J.P. Gustafson and R. Appels), pp. 175–200. Plenum Press, New York.

Rédei, G.P., G. Acedo, H. Weingarten, and L.D. Kier. 1977. Has DNA corrected genetically thiamineless mutants of *Arabidopsis*? In *Cell genetics in higher plants* (ed. D. Dudits et al.), pp. 91–94. Akademiai Kiadó, Budapest.

Reinholz, E. 1972. Vegetative reproduction. *Arabidopsis Inf. Serv.* **9:** 37.

Rhodes, C.A., D.A. Pierce, I.J. Mettler, D. Mascarenhas, and J.J. Detmer. 1988. Genetically transformed maize plants from protoplasts. *Science* **240:** 204–207.

Rocha-Sosa, M., U. Sonnewald, W. Frommer, M. Stratmann, J. Schell, and L. Willmitzer. 1989. Both developmental and metabolic signals activate the promoter of a class I patatin gene. *EMBO J.* **8:** 23–29.

Rose, A.B., A.L. Casselman, and R. Last. 1992. A phosphoribosylanthranilate transferase gene is defective in blue fluorescent *Arabidopsis thaliana* tryptophan mutants. *Plant Physiol.* **100:** 582–592.

Sangwan, R.S., Y. Bourgeois, and B.S. Sangwan-Norreel. 1991. Genetic transformation of *Arabidopsis thaliana* zygotic embryos and identification of critical parameters influencing transformation efficiency. *Mol. Gen. Genet.* **230:** 475–485.

Sangwan, R. S., Y. Bourgeois, S. Brown, G. Vasseur, and B. Sangwan-Norreel. 1992. Characterization of competent cells and early events of *Agrobacterium*-mediated genetic transformation in *Arabidopsis thaliana*. *Planta* **188:** 439–456.

Saxena, P.K., M. Mii, W.L. Crosby, L.C. Fowke, and J. King. 1986. Transplantation of isolated nuclei into plant protoplasts. *Planta* **168:** 29–35.

Schell, J.S. 1987. Transgenic plants as tools to study the molecular organization of plant genes. *Science* **237:** 1176–1183.

Schieder, O. and H. Kohn. 1986. Protoplast fusion and generation of somatic hybrids. *Cell Cult. Somatic Cell Genet. Plants* **3:** 569–588.

Schmidt, R. and L. Willmitzer. 1988. High efficiency *Agrobacterium tumefaciens*-mediated transformation of *Arabidopsis thaliana* leaf and cotyledon explants. *Plant*

*Cell Rep.* **7:** 583–586.

Schmülling, T., J. Schell, and A. Spena. 1988. Single genes from *Agrobacterium rhizogenes* influence plant development. *EMBO J.* **7:** 2621–2629.

Schneider, T., R. Dinkins, K. Robinson, J. Shellhammer, and D.W. Meinke. 1989. An embryo-lethal mutant of *Arabidopsis thaliana* is a biotin auxotroph. *Dev. Biol.* **131:** 161–167.

Scholl, R.L. and J.A. Amos. 1980. Isolation of doubled-haploid plants through anther culture in *Arabidopsis thaliana. Z. Pflanzenphysiol.* **96:** 407–414.

Scholl, R.A. and K.A. Feldmann. 1990. *Arabidopsis thaliana* (L.): In vitro production of haploids. *Biotechnol. Agric. For.* **12:** 309–321.

Scholten, H.J. and W.J. Feenstra. 1984. Embryonid-like structures in anther culture of *Arabidopsis thaliana. Arabidopsis Inf. Serv.* **21:** 65–68.

Scholten, H.J., W.J. Feenstra, H. Nijdam, and G. Datema. 1982. The use of cell culture of *Arabidopsis thaliana* for the study of nitrate reductase-less mutants which are lethal as intact plants. *Arabidopsis Inf. Serv.* **19:** 108–110.

Seki, M., Y. Komeda, A. Iida, Y. Yamada, and H. Morikawa. 1991a. Transient expression of β-glucuronidase in *Arabidopsis thaliana* leaves and roots and *Brassica napus* stems using a pneumatic particle gun. *Plant Mol. Biol.* **17:** 259–263.

Seki, M., N. Shigemoto, Y. Komeda, J. Imamura, and H. Morikawa. 1991b. Transgenic *Arabidopsis thaliana* plants obtained by particle bombardment mediated transformation. *Appl. Microbiol. Biotechnol.* **28:** 228–230.

Sheikholeslam, S.N. and D.P. Weeks. 1987. Acetosyringone promotes high efficiency transformation of *Arabidopsis thaliana* explants by *Agrobacterium tumefaciens. Plant Mol. Biol.* **8:** 291–298.

Shen, M.J. and W.R. Sharp. 1966. An improved medium for rapid initiation of *Arabidopsis* tissue culture from seed. *Bull. Torrey Bot. Club* **93:** 68–69.

Sidorov, V.A., L. Menczel, F. Nagy, and P. Maliga. 1981. Chloroplast transfer in *Nicotiana* based on metabolic complementation between irradiated and iodoacetate treated protoplasts. *Planta* **152:** 341–345.

Siemens, J., M. Torres, M. Morgner, and M.D. Sacristán. 1993. Plant regeneration from mesophyll-protoplasts of four different ecotypes and two marker lines from *Arabidopsis thaliana* using a unique protocol. *Plant Cell Rep.* **12:** 569–572.

Sijmons, P.C., B.M.M. Dekker, B. Schrammeijer, T.C. Verwoerd, P.J.M. van den Elzen, and A. Hoekema. 1990. Production of correctly processed human serum albumin in transgenic plants. *Bio/Technology* **8:** 217–221.

Smith, C.J.S., C.F. Watson, J. Ray, C.R. Bird, P.C. Morris, W. Schuch, and D. Grierson. 1988. Antisense RNA inhibition of polygalacturonase gene expression in transgenic tomatoes. *Nature* **334:** 724–726.

Smith, M.A., A. Pay, and D. Dudits. 1989. Analysis of chloroplast and mitochondrial DNAs in asymmetric somatic hybrids between tobacco and carrot. *Theor. Appl. Genet.* **77:** 641–644.

Snape, J.W., E. Simpson, B.B. Parker, W. Fried, and B. Foroughi-Wehr. 1986. Criteria for the selection and use of doubled haploid systems in cereal breeding programmes. In *Genetic manipulation in plant breeding* (ed. W. Horn et al.), pp. 217–229. de Gruyter, Berlin.

Somers, D.A., K.R. Narayanan, A. Kleinhofs, S. Cooper-Bland, and E.C. Cocking. 1986. Immunological evidence for transfer of the barley nitrate reductase structural gene to *Nicotiana tabacum* by protoplast fusion. *Mol. Gen. Genet.* **204:** 296–301.

Sonnewald, U., M. Brauer, A. von Schaewen, M. Stitt, and L. Willmitzer. 1991. Transgenic tobacco plants expressing yeast-derived invertase in either the cytosole, vacuole or apoplast: A powerful tool for studying sucrose metabolism and sink/source interactions. *Plant J.* **1:** 95–106.

Staub, J.M. and P. Maliga. 1992. Long regions of homologous DNA are incorporated into the tobacco plastid genome by transformation. *Plant Cell* **4:** 39–45.

Sun, T., H.M. Goodman, and F.M. Ausubel. 1992. Cloning the *Arabidopsis GA1* locus by genomic subtraction. *Plant Cell* **4:** 119–128.

Svab, Z., P. Hajdukiewicz, and P. Maliga. 1990. Stable transformation of plastids in higher plants. *Proc. Natl. Acad. Sci.* **87:** 8526–8530.

Takahashi, T., S. Naito, and Y. Komeda. 1992. The *Arabidopsis HSP18.2* promoter/*GUS* gene fusion in transgenic *Arabidopsis* plants: A powerful tool for the isolation of regulatory mutants of the heat-shock response. *Plant J.* **2:** 751–761.

Takahata, Y. and W.A. Keller. 1991. High frequency embryogenesis and plant regeneration in isolated microspore culture of *Brassica oleracea* L. *Plant Sci.* **74:** 235–242.

Teeri, T.H., L. Herrera-Estrella, A. Depicker, M. Van Montagu, and E.T. Palva. 1986. Identification of plant promoters *in situ* by T-DNA-mediated transcriptional fusions to the *npt*-II gene. *EMBO J.* **5:** 1755–1760.

Topping, J.F., W. Wei, and K. Lindsey. 1991. Functional tagging of regulatory elements in the plant genome. *Development* **112:** 1009–1019.

Toriyama, K., M.K. Thorsness, J.B. Nasrallah, and M.E. Nasrallah. 1991. A *Brassica* S locus gene promoter directs sporophytic expression in the anther tapetum of transgenic *Arabidopsis. Dev. Biol.* **143:** 427–431.

Tsukaya, H., T. Ohshima, S. Naito, M. Chino, and Y. Komeda. 1991. Sugar-dependent expression of the *CHS-A* gene for chalcone synthase from petunia in transgenic *Arabidopsis. Plant Physiol.* **97:** 1414–1421.

Vaeck, M., A. Reynaerts, and H. Höfte. 1989. Protein engineering in plants: Expression of *Bacillus thuringiensis* insecticidal protein genes. *Cell Cult. Somatic Cell Genet. Plants* **6:** 425–439.

Valvekens, D., M. Van Montagu, and M. Van Lijsebettens. 1988. *Agrobacterium tumefaciens*-mediated transformation of *Arabidopsis thaliana* root explants by using kanamycin selection. *Proc. Natl. Acad. Sci.* **85:** 5536–5540.

Van Lijsebettens, M., R. Vanderhaegen, and M. Van Montagu. 1991. Insertional mutagenesis in *Arabidopsis thaliana*: Isolation of a T-DNA-linked mutation that alters leaf morphology. *Theor. Appl. Genet.* **81:** 277–284.

van der Krol, A.R., P.E. Lenting, J. Veenstra, I.M. van der Meer, R.E. Koes, A.G. M. Gerats, J.N.M. Mol, and A.R. Stuitje. 1988. An anti-sense chalcone synthase gene in transgenic plants inhibits flower pigmentation. *Nature* **333:** 866–869.

van Wordragen, M.F. and H.J.M. Dons. 1992. *Agrobacterium tumefaciens*-mediated transformation of recalcitrant crops. *Plant Mol. Biol. Rep.* **10:** 12–36.

Vasil, I.K., ed. 1984. *Cell culture and somatic cell genetics of plants*, vol. 1. Academic Press, Orlando.

———. 1985. *Cell culture and somatic cell genetics of plants*, vol. 2. Academic Press, Orlando.

———. 1986. *Cell culture and somatic cell genetics of plants*, vol. 3. Academic Press, Orlando.

Vasil, V., F. Redway, and I.K. Vasil. 1990. Regeneration of plants from embryogenic suspension culture protoplasts of wheat (*Triticum aestivum* L.). *Bio/Technology* **8:** 429–434.

Veremínsky, J. and T. Gichner. 1964. Sterile culture of *Arabidopsis* on agar medium. *Arabidopsis Inf. Serv.* **1:** 34–35.

Venkateswarlu, K. and R.N. Nazar. 1991. Evidence for T-DNA mediated gene targeting to tobacco chloroplasts. *Bio/Technology* **9:** 1103–1105.

Verhoeven, H.A. and J. Blaas. 1992. Direct cell to cell transfer of organelles by micro-injection. *Plant Cell Rep.* **10:** 613–616.

Voelker, T.A., A.C. Worrell, L. Anderson, J. Bleibaum, C. Fan, D.J. Hawkins, S.E. Radke, and H.M. Davies. 1992. Fatty acid biosynthesis redirected to medium chains in transgenic oilseed plants. *Science* **257:** 72–74.

von Schaewen, A., M. Stitt, R. Schmidt, U. Sonnewald, and L. Willmitzer. 1990. Expression of a yeast-derived invertase in the cell wall of tobacco and *Arabidopsis* plants leads to accumulation of carbohydrate and inhibition of photosynthesis and strongly influences growth and phenotype of transgenic tobacco plants. *EMBO J.* **9:** 3033–3044.

Wagner, D., J.M. Tepperman, and P.H. Quail. 1991. Overexpression of phytochrome B induces a short hypocotyl phenotype in transgenic Arabidopsis. *Plant Cell* **3:** 1275–1288.

Walden, R. and J. Schell. 1990. Techniques in plant molecular biology—Progress and problems. *Eur. J. Biochem.* **192:** 563–576.

Walden, R., H. Hayashi, and J. Schell. 1991. T-DNA as a gene tag. *Plant J.* **1:** 281–288.

Weiland, U. and A.J. Müller. 1972a. Nutritional requirements of excised and non-excised roots of strain "Dijon G". *Arabidopsis Inf. Serv.* **9:** 7–9.

———. 1972b. *In-vitro*-Kultur der Wurzeln von letalen *fusca*-Mutanten von *Arabidopsis thaliana*. *Kulturpflanze* **20:** 151–161.

Weising, K., J. Schell, and G. Kahl. 1988. Foreign genes in plants: Transfer, structure, expression, and applications. *Annu. Rev. Genet.* **22:** 421–477.

Willmitzer, L. 1988. The use of transgenic plants to study plant gene expression. *Trends Genet.* **4:** 13–18.

Willmitzer, L. and R. Töpfer. 1992. Manipulation of oil, starch and protein composition. *Curr. Opin. Biotechnol.* **3:** 176–180.

Wu, Y., G. Haberland, C. Zhou, and H.-U. Koop. 1992. Somatic embryogenesis, formation of morphogenetic callus and normal development in zygotic embryos of *Arabidopsis thaliana* in vitro. *Protoplasma* **169:** 89–96.

Xuan, L.T. and L. Menczel. 1980. Improved protoplast culture and plant regeneration from protoplast derived callus in *Arabidopsis thaliana*. *Z. Pflanzenphysiol.* **96:** 77–80.

Yanofsky, M.F., H. Ma, J.L. Bowman, G.N. Drews, K.A. Feldmann, and E.M. Meyerowitz. 1990. The protein encoded by the *Arabidopsis* homeotic gene *agamous* resembles transcription factors. *Nature* **346:** 35–39.

Yokoyama, K. and W.H. Jones. 1965. Tissue cultures of *Arabidopsis thaliana*. *Plant Physiol.* (suppl.) **40:** 77.

Zambryski, P.C. 1992. Chronicles from the *Agrobacterium*-plant cell DNA transfer story. *Annu. Rev. Plant Physiol. Plant Mol. Biol.* **43:** 465–490.

Zelcer, A., D. Aviv, and E. Galun. 1978. Interspecific transfer of cytoplasmic male sterility by fusion between protoplasts of normal *Nicotiana sylvestris* and X-ray irradiated protoplasts of male sterile *N. tabacum. Z. Pflanzenphysiol.* **90:** 397–407.

Zhang, H. and C.R. Somerville. 1987. Transfer of the maize transposable element *Mul* into *Arabidopsis thaliana*. *Plant Sci.* **48:** 165–173.

Zhang, W. and R. Wu. 1988. Efficient regeneration of transgenic plants from rice protoplasts and correctly regulated expression of the foreign gene in the plants. *Theor. Appl. Genet.* **76:** 835–840.

Ziebur, N.K. 1965. Tissue culture. *Arabidopsis Inf. Serv.* **2:** 34–35.

# 9
# Genetic Studies with *Arabidopsis:* A Historical View

**Csaba Koncz**

Max-Planck-Institut für Züchtungsforschung,
Köln 30, D-50829, Germany

**George P. Rédei**

Columbia, Missouri 65203-0906

In recent years, *Arabidopsis* has become a most popular tool for plant biological studies. Although it is well suited for a wide range of research areas, we focus primarily on its use for genetic analyses. It is often heard: *Arabidopsis genetics*, but we should rather say *genetics with Arabidopsis*. The interest is not in the specificities of this plant but, rather, in what it can reveal about basic biology. The primary advantages of *Arabidopsis* for studying basic biological phenomena (Rédei 1970, 1975a, 1992; Meyerowitz 1987, 1989; Somerville 1986, 1989; Rédei and Koncz 1992) were first summarized by Laibach (1943).

In comparison, Mendel's peas were very advantageous in the 19th century because spontaneous recessive variants were available in this autogamous plant, and the second generation could be classified within the pods of the $F_1$ plants. *Drosophila* has a 2-week life cycle and can be raised in large numbers in small milk bottles, and the polytene chromosomes of the salivary glands display 5000 landmarks. Genetic segregation in maize can be followed by the large number of individual kernels immobile on a single cob. In addition, at the pachytene stage, the extended chromosomes display discrete chromomeres. Crossing is easy because about 50 million pollen grains may be released by a single monoecious plant. Some principal advantages of *Neurospora* are the linear arrangement of the meiotic products in the asci and the less than 2-week life cycle, with vegetative cultures requiring only 4–5 days. Yeast (*Saccharomyces cerevisiae*) can be manipulated similarly to prokaryotes because the size of individual cells is only a little larger than that of bacteria; both haploid and diploid phases are available, and its life cycle is barely longer than 1 hour under most favorable conditions. The approximate genome size of these organisms varies greatly: maize $7 \times 10^9$ bp, *Drosophila* $1.6 \times 10^8$ bp, *Arabidopsis* $9 \times 10^7$ bp, *Neurospora* $4.2 \times 10^7$

bp, yeast $1.4 \times 10^7$ bp, *Escherichia coli* $4.5 \times 10^6$ bp.

Sturtevant (1971) was wise in pointing out that the right choice of a genetic organism depends on the nature of the problem, the historical time, and the experimenter's skill. We believe that *Arabidopsis* may be the best choice for many at the end of our century and beyond.

## *ARABIDOPSIS*: A LATE-COMER TO FORMAL GENETICS

Botanical description of *Arabidopsis* began with the Saxonian physician Johannes Thal, who identified this species before Linnaeus (1753) listed it as *Pilosella siliquosa minor*. During the centuries that followed, several other names were given to this monotypic genus (Rédei 1970). The taxonomic status of *Arabidopsis* is somewhat controversial (Berger 1968). *Arabidopsis suecica* had been renamed to *Hylandra suecica* by Löve (1961) because Hylander (1957) suggested that this species is an amphiploid of *Cardaminopsis arenosa* and *Arabidopsis thaliana*. All early efforts to produce an actual hybrid failed because *C. arenosa* ($2n=32$) is probably an autotetraploid and the *Arabidopsis* ($2n=10$) parent used was diploid. Měsiček (1967) successfully crossed *C. petraea* ($2n=16$) with *Arabidopsis* and obtained a sterile hybrid. Rédei (1974a) obtained fertile ($2n=26$) hybrids of *C. arenosa* ($2n=32$) and tetraploid *Arabidopsis*. Thus, a synthetic *H. suecica* was obtained. However, the problem remained, because the fertile hybrid did not resemble *H. suecica*, as expected. Rédei also obtained trigeneric (and other complex) hybrids with variable fertility. Yet, the evolutionary relationships remained unresolved; perhaps future molecular studies will shed more light on the origin of these species.

The first Mendelian segregation studies with *Arabidopsis* were conducted during World War II by Reinholz (1945). Mapping of genes started belatedly in *Arabidopsis*. The first linkage information, involving 21 loci, was obtained by Rédei and Hirono (1964). As expected, the number of linkage groups exceeded the chromosome number because syntenic gene clusters, far apart, recombined freely. To overcome this problem, trisomics were developed (Lee-Chen and Bürger 1967; Lee-Chen and Steinitz-Sears 1967). These permitted in *Arabidopsis* a correlation between genetic maps and chromosomes and facilitated the chromosomal assignment of genes (see Rédei et al. 1988). On the basis of double reduction frequencies, the relation of some genes and centromeres could be deduced (Lee-Chen and Steinitz-Sears 1967; Sears and Lee-Chen 1970). McKelvie (1965) reported a linkage group of about 150 map units, and later Koornneef et al. (1983) developed mapping information for all five chromosomes with a total length of about 500 map units. Hirono (1964) developed mathematical procedures based on the relationship between marker transmission, distorted segregation ratios, and map distances, thereby localizing deletions within the genetic map. Li (1968) simplified the procedure by considering the reduction of transmission through only one of the sexes and still obtained good estimates on the relative position of deficiencies. For several years, *Arabidopsis* geneticists relied on the product ratio method for calculating linkage intensities (Fisher and Balmakund 1928; Stevens 1939). The product method is fully efficient and gives estimates similar to those obtained by the maximum likelihood method even if the transmission of both recessive markers is reduced, but not when only one is afflicted (Bailey 1961). Both methods were developed by Fisher, but the so-called maximum likelihood procedure is considered superior. Linkage intensities were then converted to map units by either Haldane's (1919) or Kosambi's (1944) mapping functions; the former was more aesthetic, the latter perhaps more practical. In higher plants, mapping functions were first generally used in *Arabidopsis*. To estimate linkage, including restriction fragment length polymorphisms (RFLPs), the MAPMAKER and the JoinMap computer programs are used today (Stam 1993). These computer methods are based on the LOD scores and estimates by maximum likelihood, as originally developed for human genetics by Morton (1962).

**GENETIC NOMENCLATURE**

Gene symbolization began with the development of *Drosophila* and maize genetics. When new organisms, such as *Neurospora*, *E. coli*, and *S. cerevisiae* were introduced, the nomenclature was modified to meet special needs (Boyes et al. 1973). The *Arabidopsis* research beginning in the 1950s did not opt for a special symbolism until the 3rd International Meeting on *Arabidopsis* at Michigan State University. Unfortunately, the term "ecotype" is still incorrectly used in the literature. Ecotype (Turesson 1922) is a genetically ambiguous term, because, despite the homogeneity of the genetic basis of some of the adaptive traits, the genetic background may vary, as was obvious with the original Landsberg "race" (Rédei 1992). Columbia wild type and the so-called Landsberg *erecta* were both isolated from Laibach's Landsberg ecotype, and except for the presence of the X-ray-induced *er* mutation, their RFLP maps are obviously not identical. In addition, *er* may have several "non-wild-type" alleles and rearranged nucleotide sequences because of the exposure to ionizing radiation in 1957.

## CHROMOSOMAL THEORY OF INHERITANCE
## AND *ARABIDOPSIS*

Hybridization studies with plants originated before the nature and function of the gametes were fully understood (Stubbe 1965). The role of the nucleus in heredity and chromosomal mechanics (Coleman 1965) developed steadily during the period between Mendel's discovery (1866) and his rediscovery in 1900 (see Rédei 1974c). To the satisfaction of botanists, the terms pro-, meta-, ana-, and telophase were coined by Strasburger (1884). Strasburger was the major professor of Friedrich Laibach at the University of Bonn. Laibach became the forefather of *Arabidopsis* research by following his mentor's footsteps to study experimentally the continuity of chromosomes. Laibach (1907) used *Arabidopsis* to gain more insight into the problem. Most crucifers display heterochromatic bodies in their interphase nuclei. Laibach found that in the somatic cells of *Arabidopsis* there are 10 prochromosomes, corresponding to the $n = 5$ number of chromosomes observed cytologically during meiosis. He concluded that Mendelian inheritance can be explained on the basis of behavior of chromosomes. In contrast to Laibach, Morgan was very critical of Mendelism, until he discovered the white-eye gene in *Drosophila* in 1910. Thus, at the dawn of genetics, the work with *Arabidopsis* was ahead of that in *Drosophila* regarding the mechanism of heredity.

Laibach also made another important conclusion in 1907. The size of the *Arabidopsis* nuclei appeared to be only about one third of that in *Brassica*. Yet, the former species does not have fewer functions, thus indicating to Laibach that the total amount of chromatin may not be necessary for normal performance. The chromosomes of *Arabidopsis* are smaller than those of other organisms. This might have caused the delay in acceptance of *Arabidopsis* in an era when cytogenetics was prevalent. With the old cytological staining techniques, little was detectable about the structure of the *Arabidopsis* chromosomes. Today, Giemsa C-banding, fluorescent staining, and in situ hybridization permit high-resolution studies in *Arabidopsis* (Ambros and Schweizer 1976; Maluszynska and Heslop-Harrison 1991).

Laibach in *Sisymbrium strictissimum* observed somatic association of the chromosomes. This is apparently the second such observation in plants after Strasburger (1904). Somatic association of chromosomes of *Arabidopsis* was confirmed later (Steinitz-Sears 1962). This may have facilitated the premeiotic exchanges of linked markers in *Arabidopsis* (Hirono and Rédei 1965). In comparison, somatic recombination was first demonstrated in *Drosophila* by Stern in 1936. *Arabidopsis* body sectors occasionally contribute to the inflorescence, and, thus, somatic cross-over products can be recovered in seed progeny. The experiments showed that the mechanism of exchange in somatic cells is, however, not identical to that in meiosis, because the exchanged strands generally displayed reduced transmission.

Chromosomal aberrations in *Arabidopsis* were first demonstrated cytologically in the early 1960s. Tetraploidy and hexaploidy were first observed in *Arabidopsis* by Rédei (1964), and this material facilitated the isolation of the first set of trisomics (Steinitz-Sears 1963). Trisomics were also isolated by Röbbelen and Kribben (1966) and used for centromere mapping (Lee-Chen and Steinitz-Sears 1967; Sears and Lee Chen 1970; Koornneef 1983). Bouharmont (1965) described tetra-, hexa-, and octaploid *Arabidopsis* obtained by radiation and colchicine. Systemic endopolyploidy was observed by Galbraith et al. (1991). Translocations were induced in *Arabidopsis* (Sree Ramulu and Sybenga 1985) and used for mapping of breakage points (Koornneef et al. 1982b), as was done in maize (Brink and Cooper 1931).

## NUCLEAR GENOME SIZE AND STRUCTURE

Sparrow et al. (1972) examined Feulgen-stained nuclei of *Arabidopsis* by cytophotometric techniques and concluded that the genome size is $1 \times 10^9$ bp. Now, this appears too high (Schmidt and Dean 1992), yet these data proved that *Arabidopsis* has less DNA in its nucleus than other angiosperms. Leutwiler et al. (1984), using reassociation kinetics, arrived at an estimate of $7 \times 10^7$ bp that was recently revised to about $8 \times 10^7$ to $10 \times 10^7$ bp with a G+C content of 41.8% (Meyerowitz 1992). *Arabidopsis* has only about 2–5 times as much DNA as the genetically most-used ascomycetes, and only about 20 times more than *E. coli*. Pruitt and Meyerowitz (1986) have demonstrated very low redundancy in the *Arabidopsis* genome. The repetitive component is about 30%, leaving 65–80 Mbp to low-copy-number components. By mutation studies, the minimum number of genes of *Arabidopsis* was estimated at about 28,000 (for review, see Rédei and Koncz 1992). Thus, approximately 2,300 bp may be allocated to an "average" individual gene, including introns and regulatory sequences. Although this may be an underestimate, so far, the putative size of the genome and the number of genes are in reasonably good agreement. So far, about 0.2–0.3% of the transcribed DNA has been sequenced. Thus, a larger fraction of the genome is known in *Arabidopsis* than in any other higher plant. *Arabidopsis* has similar codon usage as other dicots. At the third position, G+C is more frequent when at the second place there is an A or a T. *Arabidopsis* also appears to use the XCG codons more frequently than other dicots (Gasch et al. 1992). These differences may not be significant because of the relatively small number of genes sequenced, or they may indicate the frequency of methylation, and/or differences in the reliance on particular functions.

## MUTAGENESIS OF *ARABIDOPSIS*

### Mutation Rates

The availability of genetic variation is the most important requisite for inheritance studies despite Linneaus's famous admonition to his students: "*varietates levissimas non curat botanicus.*" The frequency of overall spontaneous mutations in *Arabidopsis* is low and varies from 0.01% to 3% as measured by the various investigators. These figures are affected by the genes studied, the method of estimation, and the rigor of the classification (Rédei and Koncz 1992). Spontaneous mutation frequencies at specific loci have been recorded (Rédei 1982a), and induced mutation rates for 15 loci of *Arabidopsis* have also been calculated (Koornneef et al. 1982a). Spontaneous mutation rates are greatly influenced by the power of the genetic resolution. If we accept the gene number of *Arabidopsis* as about 30,000, the locus-specific average mutation rate appears to be in the $10^{-7}$ range. The gene number estimate used is within the range of that of other plants, inferred from single copy sequences (Meyerowitz 1992). For comparison, the latest estimate of human gene number is about 75,000 (Macilvain 1993). The low spontaneous mutability of *Arabidopsis* explains why mutations at "good genes" are rare in natural populations, unlike *Drosophila* and maize, where indigenous active transposable elements may rapidly reorganize the genome. The low mutability might have been the cause of the delays in using *Arabidopsis* for classical genetic studies.

### Mutagenesis Techniques

Despite the low spontaneous mutability, *Arabidopsis* responds well to mutagens. Reinholz (1945) found that even low doses of X-rays yielded 36 characterized mutations in a population of 1,600 families. This frequency, on a genome basis, corresponds to an overall mutation rate of about $6 \times 10^{-3}$. This was thus an increase by three to four orders of magnitude over the spontaneous rate. Reinholz's experiments also revealed that X-ray doses up to 2,000 kR applied to dry seeds failed to prevent germination. Similar experiments carried out by γ-radiation, combined with measurements of the nuclear volume, also suggested that *Arabidopsis* has the lowest DNA content among angiosperms (Sparrow et al. 1972). Practically all *Arabidopsis* mutants obtained before 1962 were induced by ionizing radiation. The majority of these were considered to be point mutations, because they were not associated with detectable chromosomal alterations and displayed quasi-normal recombination rates and meiotic transmission. The effect of X-rays on the genome was controversial for decades. Some of the *Drosophila* geneticists, including H.J. Muller, considered many of the X-ray mutations to be indistinguishable from the spontaneous ones, whereas the maize geneticists, including L.J. Stadler, concluded that apparently none of the X-ray mutations represented minute intragenic alterations and therefore were of minimal significance for the study of the gene. Today, X-ray mutations have become valuable tools following the development of the genomic subtraction method for gene isolation (Shirley et al. 1992; Sun et al. 1992). This type of gene isolation technique is particularly adaptable to *Arabidopsis* because of its small genome size, and it was used for the first time among higher eukaryotes. These recent studies lend support to Stadler's (1944) conclusion that most X-ray mutants indicated a loss of genetic material. Some other mutants of *Arabidopsis* obtained by ionizing radiation may be true base substitutions, just as appeared to be the case in *Neurospora* (Malling and de Serres 1973).

It took almost 20 years after the discovery of the mutagenic effect of X-rays to induce mutations in *Arabidopsis*. The use of chemical mutagens was also somewhat delayed (for review, see Rédei and Koncz 1992). Ethylmethane sulfonate (EMS), an alkylating agent, was first applied to *Arabidopsis* by Röbbelen (1962a), and McKelvie (1963) used EMS and ethyleneimine. EMS turned out to be a real "supermutagen," a term coined by Rapoport et al. (1946) for chemicals that on a molar basis have low toxicity yet are highly mutagenic without causing much chromosome breakage. Müller, Gichner, and Veleminsky tested hundreds of compounds for mutagenicity in *Arabidopsis*, and they found several mutagens of about the same efficiency as EMS (for review, see Rédei 1970). An assay for mutations was developed by Müller (1963), based on exactly the same principle as Mendel's use of the yellow versus green, wrinkled versus smooth, embryo characters to identify segregation in the pods of heterozygotes. This procedure is valuable for the rapid identification of mutagens, and indirectly carcinogens, with about 88% effectiveness (Rédei et al. 1980). Reducing the costs of mutation experiments by effective planning, as well as the statistical bases of economical mutant screening in *Arabidopsis* and other autogamous plants, were worked out (for review, see Rédei and Koncz 1992). *Arabidopsis* was the first plant in which mutation rate was expressed on a genome basis in a generally applicable form, and, thus, mutation rates of homoeologous genes could be compared across phylogenetic boundaries (Li and Rédei 1969a).

## Mutant Selection Techniques

A most effective selective mutant isolation technique involves allyl alcohol (AA), which is converted to highly toxic acrylaldehyde by alcohol dehydrogenase (ADH). Mutants lacking ADH thus survive, whereas the wild type dies on AA. This type of mutant isolation was first applied to yeast, then adapted to maize pollen, and proved to be successful also in *Arabidopsis* (Megnet 1967; Schwartz and Osterman 1976; Jacobs et al. 1988). *ADH* became the first cloned and sequenced *Arabidopsis* locus (Chang and Meyerowitz 1986), and base substitutions were first identified in *adh* alleles (Dolferus et al. 1990).

Somerville and Ogren (1982) developed very successful selection schemes in a controlled carbon dioxide atmosphere for screening of photorespiratory mutants. By 1986, *Arabidopsis* emerged as a premier organism in the field (Somerville 1986), attesting to the joint power of genetic selection, biochemistry, and molecular biology. The efficiency of selection was shown by the isolation of sulfonylurea herbicide-resistant mutants by plating up to 10,000 seeds per petri dish (Haughn and Somerville 1986). Although this population is smaller than in bacterial mutagenicity tests ($10^8$ cells), it is much larger than the approximately 500 kernels per ear of maize. In addition, the *Arabidopsis* screening may be repeated several times per month (for reviews, see Rédei 1970, 1974b; Meyerowitz 1987; Rédei and Koncz 1992 and references therein).

## GENETIC TRANSFORMATION OF *ARABIDOPSIS*

Genetic transformation of higher plants has a controversial history. Ledoux and Huart (1961) considered the introduction of various macromolecules into growing barley embryos. Subsequently, the genetic correction of thiamine auxotrophs of *Arabidopsis* was reported after treating seeds with bacterial DNA containing the thiamine gene cluster (Ledoux and Jacobs 1974). The transfer and expression of the β-galactosidase gene of *E. coli* has been reported under the term *transgenosis* by Doy et al. (1973). The correction of the thiamine deficiency carried out in the Ledoux laboratory could not be confirmed by either genetic or molecular analyses (Lurquin 1976; Rédei et al. 1976). "If the work is critically evaluated, one is tempted to find more explanations than corrections by exogenous DNA" (Hess 1977).

Direct DNA transfer to plant protoplasts succeeded by employing Ti plasmid DNA (Davey et al. 1980; Draper et al. 1982; Krens et al. 1982). *Arabidopsis* protoplasts were first transformed by plasmid DNA in 1989 using polyethylene glycol treatment (Damm et al. 1989). Delivering functional DNA segments into plant cells by microprojectiles was discovered by Klein et al. (1987), and the method was successfully applied to *Arabidopsis* by Seki et al. (1991). Although eventually protoplast transformation became the most important means of gene transfer during the 1980s, today gene transfer vectors based on the T-DNA of *Agrobacterium* Ti or Ri plasmids are used preferentially. The plant pathogen *Agrobacterium* induces crown gall by a plasmid, called TIP (tumor inducing principle) by Braun (1947), before anything was known about its substance. In 1974, the laboratory of Schell demonstrated that the tumor induction by *Agrobacterium* is due to its large Ti plasmid (Van Larebeke et al. 1974). Chilton et al. (1977) have shown that crown-gall formation is the direct consequence of T-DNA incorporation into the plant genome. An *Arabidopsis* tumor line (Aerts et al. 1979) provided important basic information on the T-DNA structure (DeBeuckeleer et al. 1981). In a systematic approach, the pGV3850 vector was constructed (Joos et al. 1983; Zambryski et al. 1983) by deletion of oncogenes from the natural nopaline Ti plasmid pTiC58. The new plasmid carried pBR322 sequences, allowing homologous recombination between the Ti plasmid and common *E. coli* cloning vectors. This Ti plasmid vector was suitable for introduction of any gene into plant cells. In the same laboratory, within the same year, antibiotic resistance genes (kanamycin, methotrexate, chloramphenicol) of bacterial transposons were linked to the nopaline synthase promoter and polyadenylation signals to facilitate large-scale selection of transgenic plant cells (Herrera-Estrella et al. 1983a,b). Simultaneously, Schilperoort's laboratory constructed a non-oncogenic Ti plasmid vector (Hille et al. 1983).

Transformation of *Arabidopsis* with disarmed Ti plasmid vectors (Hooykaas 1989; Walden et al. 1990) started relatively late. The first successful transformation of *Arabidopsis* was reported by the Monsanto group (Lloyd et al. 1986). The same year, transgenic *Arabidopsis* plants, produced by infection of leaves, stems, and roots with different binary vectors, were obtained (An et al. 1986; Rédei et al. 1988). A root transformation technique was developed by Valvekens et al. (1988). Feldmann and Marks (1987) infected seeds of *Arabidopsis* (Wassilewskija) with a C58 *Agrobacterium* strain carrying the cointegrate vector pGV3850::1003 (Velten et al. 1984) and obtained large numbers of transformants. In the laboratory of Pelletier and Caboche, an extremely efficient procedure has been developed for the transformation of soil-grown *Arabidopsis* by infiltration with agrobacteria (Bechtold et al. 1993).

The advantages of transformation are manifold: It can be used to tag genes, as well as to identify cloned genes by reintroducing them into mutant hosts. From the viewpoint of basic genetics, insertional mutagenesis was probably the most important gain from transformation. Insertional mutations are quite common in plants, and many of the recessive genes contain inserts. One of the most famous plant genes, the *r* allele of peas, responsible for the wrinkled-seed character studied by Mendel (1866), contains an insert of about 0.8 kb showing homology with the *Ac* (maize), *Tam3* (snapdragon), and *Tpc1* (parsley) transposable elements (Bhattacharyya et al. 1990). In *Arabidopsis*, silent retrotransposons were observed in several ecotypes (Voytas and Ausubel 1988; Voytas et al. 1990; Peleman et al. 1991). The *Tag1* element, recently discovered in *er* background, was mobilized in *Arabidopsis* after transformation with the *Ac* transposable element of maize. The authors concluded: "We think it unlikely that the *Ac* transposase directly mobilizes *Tag1*, as no *Ac* transposase binding site (AAACGG) is found adjacent to the inverted repeats of *Tag1* as it is in *Ac*" (Tsay et al. 1993).

## T-DNA TAGGING

For gene tagging by insertional mutagenesis, two alternatives were used until 1993: T-DNA or alien transposable elements. Teeri et al. (1986) and André et al. (1986) tagged *Nicotiana* genes by promoterless *aph(3')II* genes and generated in vivo fusions of reporter genes. In *Arabidopsis*, transcriptional and translational in vivo gene fusion vectors provided highly efficient gene tagging: About one third of the hundreds of transformants expressed the reporter gene (Koncz et al. 1989). Feldmann et al. (1989), using seed transformation, reported 36 different mutations that cosegregated with T-DNA. From thousands of morphological mutations obtained through seed infection, several have already been analyzed at various levels of depth (Herman and Marks 1989; Marks and Feldmann 1989; Yanofsky et al. 1990; Feldmann 1991; Oppenheimer et al. 1991). Using tissue culture transformation with a binary gene fusion vector, Koncz et al. (1990) have shown with a comprehensive analysis that T-DNA can indeed induce mutations by an insertion at the 3' end of the *CH-42* locus. This analysis included genetic recombination with a resolving power of less than 10 kb and in vivo complementation of the T-DNA-induced *cs* mutation through retransformation by the wild-type gene.

## GENE TAGGING BY ALIEN TRANSPOSABLE ELEMENTS

Zhang and Somerville (1987) introduced the *Mu1* mutator element of maize (Robertson 1978; Barker et al. 1984) into *Arabidopsis*. Although the transformation was successful, *Mu1* failed to increase the mutation rate in *Arabidopsis*. Van Sluys et al. (1987) introduced into *Arabidopsis* the *Ac* transposable element of maize using an Ri plasmid T-DNA vector. The *Ac* element (identical to *Mp*, described by Brink and Nilan 1952) was discovered by Barbara McClintock (1952) and isolated by Fedoroff et al. (1984). It was used first for transformation of alien (tobacco) cells by Baker et al. (1986). *Ac* induced no visible mutation in the first experiments with *Arabidopsis*, but the DNA analysis indicated excision and transposition of *Ac*. Schmidt and Willmitzer (1989) were first to genetically detect the movement of *Ac* in *Arabidopsis*. The vector, developed by Baker et al. (1986) and Coupland (1992), contained an *aph(3')II* gene, inactivated by an *Ac* insertion in the 5' untranslated sequences. Unless *Ac* left the leader sequence of this reporter gene, the transgenic plants stayed kanamycin-sensitive. In about 0.2–0.5% of the population, resistance was observed, indicating the excision of *Ac*. A similar procedure, using a streptomycin resistance gene (Dean et al. 1992), and *gus* (*uidA*) with the *Ac-Ds* (Bancroft et al. 1992), also indicated high *Ac* transposition in *Arabidopsis*. The transposed *Ac-Ds* elements tend to remain in the vicinity of their original insertion site in *Arabidopsis* (Bancroft and Dean 1993), as it was first observed in maize (Van Schaik and Brink 1959). The state of methylation of *Ac* apparently did not affect much the transposition frequency (Keller et al. 1992). The *En-I* system was also used with success to tag a male sterility locus (Aarts et al. 1993), and by application of the *Cre-Lox* site-specific recombination system of *E. coli*, high frequencies of chromosomal aberrations were obtained in *Arabidopsis* (Osborne et al. 1993).


## GENETICS OF SOMATIC CELLS

Genetic study of somatic cells is of particular interest in low-fecundity multicellular organisms and/or without means of controlled matings. *Arabidopsis* has neither of these problems, and body sectors may develop into inflorescence tissue, thus allowing somatic chromosome exchanges to be analyzed in the generative offspring (Hirono and Rédei 1965). When plants heterozygous for eight "visible" markers were X-irradiated, all displayed somatic sectoring, indicating that the "visible" mutations are generally cell-autonomous and involve nondiffusible gene products (Rédei 1967).

Barski et al. (1960) used another approach to cell genetics: They fused different cultured mammalian cells. Later, the laboratory of Cocking (Power et al. 1970) fused plant protoplasts, and Carlson et al. (1972) produced interspecific tobacco hybrids by protoplast fusion. An *Arabidopsis*/turnip somatic hybrid was obtained by Gleba and Hoffmann

(1978). In animal cells, somatic hybridization by cell fusion (Ephrussi and Weiss 1965) became one of the most important tools for the analysis of synteny of genes. In plants, the lack of efficient means of chromosome elimination prevented the use of this approach.

Mutant isolation in haploid *Antirrhinum* cell suspension cultures was reported quite early (Melchers and Bergmann 1959), and by the 1980s a wide variety of mutants were claimed in plant cells (for review, see Maliga 1984). Resistance to metabolite analogs (Negrutiu et al. 1978) and nitrate reductase deficiency (Scholten and Feenstra 1986) were selected in suspension cultures of *Arabidopsis*. Through radiation induction, exogenous hormone-independence has been observed (Campell and Town 1991; Persinger and Town 1991). Protoplasts permit the large-scale transformation and regeneration of *Arabidopsis* cells (Damm et al. 1989) and gene replacement (Halfter et al. 1991; Altmann et al. 1992).

## GENETICS OF CHLOROPLASTS AND MITOCHONDRIA

Röbbelen found that nearly 1 in 100,000 cells of *Arabidopsis* had two or more types of plastids, and after X-irradiation their frequency increased by two orders of magnitude (Röbbelen 1962b). The first plastid mutator gene was discovered in barley (Sô 1921), then the *iojap* mutation of maize was characterized (Rhoades 1943). Röbbelen (1964) discovered an X-ray mutation (*am*) that acted as a plastid mutator. The plastid mutations were maternally inherited until they were thinned out by selection. A similar mutation (*chm*) was obtained by Li and Rédei, and subsequently two additional alleles of the locus were discovered in chromosome 3 (Rédei 1973). Genes *am* and *chm* were never tested for allelism, but they do not appear identical (Rédei and Plurad 1973). By the removal of the mutator gene, apparently homoplastidic mutants could be fixed (Rédei 1975b). It is intriguing that genetic defects induced by *chm* were attributed to rearrangements in the mitochondrial DNA (Martínez-Zapater et al. 1992). Plastid gene recombination has not been observed yet in *Arabidopsis*, but in the unicellular green plant, *Chlamydomonas*, map construction based on genetic recombination of chloroplast genes has long been feasible (Sager and Ramanis 1970). Recently, image analysis permitted the screening for chloroplast mutants involved in control of plastid division and DNA replication (Pyke and Leech 1992).

## GENETICS IN METABOLIC AND DEVELOPMENTAL PATHWAYS

The first auxotroph of any higher plant was discovered in *Arabidopsis* by Langridge (1955), and subsequently, Rédei's laboratory identified more than 200 mutations in several steps of the thiamine pathway. Fink's laboratory isolated the first amino acid auxotrophs of plants in *Arabidop-*

*sis* (for review, see Rédei and Koncz 1992). Space limitations allow the inclusion of only a few of the recent milestones.

Studies of cell cycle and signaling emerged by the late 1980s. Cell cycle regulatory genes, such as protein kinases, phosphatases, and cyclins, were identified first in yeast, *Xenopus*, and mammals, and are now found and studied also in *Arabidopsis* (see, e.g., Ferreira et al. 1991; Hemerly et al. 1992; Nitschke et al. 1992). An analysis of tubulin genes revealed that the small genome of *Arabidopsis* has at least six expressed α-tubulin genes and nine β-tubulin genes, more than the much larger mammalian genomes (Snustad et al. 1992). The studies of the central elements of signaling cascades revealed the existence of small GTP-binding proteins (Anai et al. 1991), the α-subunit of trimeric G proteins (Ma et al. 1990), several tyrosine-, serine/threonine-specific protein kinases (see, e.g., De Guen et al. 1992; Kohorn et al. 1992), and a rapidly increasing number of transcription factors acting on hundreds of different genes in *Arabidopsis*. Among these, the TATA-box-binding general transcription factor, TFIID, isolated from *Arabidopsis* became the first plant transcription factor with the crystalline structure determined (Nicolov et al. 1992).

As a part of impressive progress in studies of metabolic pathways, different techniques for isolation of lipid mutants were developed, and the function and regulation of many such mutants were characterized (Browse and Somerville 1991; Somerville and Browse 1991). Mutants in the phytohormone pathways became essential tools for studies of metabolism and signaling. Following the pioneering efforts of Maher's laboratory (Maher and Martindale 1980), a large number of mutants displaying auxin resistance, anomalous gravitropic response, and/or pleiotropic cross-resistance to auxin, abscisic acid, and ethylene were isolated and characterized (see, e.g., Pickett et al. 1990; Okada et al. 1991). In addition to several genes encoding putative auxin receptors (Palme 1992), the auxin resistance mutation, *axr1*, was cloned (Leyser et al. 1993). Because of the pioneering work of Koornneef et al. (1982c, 1984, 1985), the gibberellin biosynthetic pathway of *Arabidopsis* is well mapped by mutations, and major functions in the abscisic acid response are defined (Finkelstein and Somerville 1990; Talón et al. 1990). Genes *GA1* and *ABI3* are the first from these pathways that were characterized molecularly (Giraudat et al. 1992; Sun et al. 1992). The analysis of the ethylene biosynthesis and signaling pathways provides an example of how a complex signaling cascade can be effectively rationalized by the use of combined biochemical and genetic methods (for review, see Chang et al. 1993; Kieber et al. 1993; Kieber and Ecker 1993). Although a number of interesting mutations affecting cytokinin biosynthesis and/or regulatory action were isolated (see, e.g., Moffatt et al. 1991; Su and Howell 1992), a closer insight into a signaling pathway(s) in *Arabidopsis* is still awaited. Many mutations influencing cytokinin action have a pleiotropic effect on the mechanisms involved in light response.

Various effects of light, other than photosynthesis, have been studied since the middle of the 19th century. Photoperiodism was discovered in 1918, phytochromes were identified in 1952 (Borthwick and Hendricks 1960). The first mutations affecting photomorphogenesis and photosynthesis in *Arabidopsis* were isolated in the 1960s. Mutants displaying long hypocotyl (*hy*) and etiolated or green color in light were isolated by Rédei (1965). Koornneef et al. (1980b) found five new loci including a series of *hy1* to *hy8*. Genetic dissection of photomorphogenic and skotomorphogenic responses advanced with great speed. New classes of regulatory mutants were isolated, such as the *det, cop,* and *blu* series (for review, see Liscum and Hangarter 1991; Quail 1991; Deng and Quail 1992; Deng et al. 1992; Chory 1993). Epistatic relations between these mutants facilitated the formulation of the first models of the pathways of light signaling (Chory 1992).

The *im* mutant causing red-light-dependent formation of white leaf sectors (Rédei 1963; Röbbelin 1968) has remained so far resistant to molecular approaches. Interestingly, the white leaf sectors of *im* can be suppressed by X-irradiation and azauracil without reverting the mutation. The white sectors appear to overproduce a ribonuclease and can be normalized by 6-azauracil, although the ultrastructure of plastids remains modified. Azauracil suppresses orotidylic acid pyrophosphorylase and inhibits orotydilic acid decarboxylase (Chung and Rédei 1974). Perhaps *im* will soon be better understood through progress in the characterization of ribonucleases of *Arabidopsis* (Yen and Green 1991).

Müller (1963) provided the first detailed study of the embryogenesis of *Arabidopsis*. Although the embryos of *Arabidopsis* are very small, the expression of approximately 80% of all "visible" mutations can be detected before germination of the seeds (Rédei 1981). Meinke and Sussex (1979) began to use lethal embryo mutations for the analysis of early development of *Arabidopsis*. By the early 1990s, more than 300 different mutations influencing morphogenesis, nutritional requirements, and homeotic changes were isolated and mapped (Jürgens et al. 1991; Mayer et al. 1991; Meinke 1991, 1992; Patton et al. 1991). From the dozen seed coat color mutations listed by Bürger (1971) and by Koornneef (1990), the *tt3* (dihydroflavonol-4-reductase), *tt4* (chalcone synthase), and *tt5* (chalcone flavonone isomerase) loci were molecularly characterized (Feinbaum and Ausubel 1988; Shirley et al. 1992). Complementation of the hairless phenotype of the *ttg* (*transparent testa* ) mutation by the *R* gene of maize has raised an intriguing question about complex regulatory interactions between anthocyanin biosynthesis, trichome development, light, and metabolic regulation (Lloyd et al. 1992). A large number of mutants affecting root and root hair development were collected (Schiefelbein and Somerville 1990), and methods were developed for the isolation of mutants with defects in geotropic and touch responses of root (Okada and Shimura 1990). Studies of cell lineages of the shoot apex (for review, see Rédei and Li 1969; Rédei 1970, 1992) were continued (Irish and Sussex 1992; Medford et al. 1992).

Laibach (1940, 1951) recognized that *Arabidopsis* is a long-day plant without a critical day length. Both early- and late-flowering mutants were observed by Reinholz (1945). At three loci, five different flowering time mutants were isolated by Rédei (1962a). These late-flowering mutants were demonstrated to flower very early in darkness in sugar-containing liquid media, as well as in the presence of halogenated pyrimidine nucleoside analogs. On the basis of this information, the conclusion was made that the onset of flowering in *Arabidopsis* is constitutively regulated in darkness, but under short-day conditions, a repressor is made that decays under continuous illumination; i.e., the process is under negative control (Rédei et al. 1974). These data, together with the characterization of a series of late-flowering mutants isolated by Koornneef et al. (1991), may provide the initial material for current approaches aiming at the isolation of genes regulating the process of photoperiodism and vernalization. Klaus Napp-Zinn, John Langridge, John Brown, and others contributed valuable information to this area in the early years (for review, see Rédei 1970).

Attention to flower differentiation dates back to Braun, who described the first *agamous* mutation in 1873 (cited by Yanofsky et al. 1990). Homeotic mutations *ap1* (McKelvie 1962), *ap2* (Koornneef et al. 1980a), *ap3* (Bowman et al. 1989), *pi* (Koornneef et al. 1983), *pin* (Gôto et al. 1987), *lfy* (Haughn and Somerville 1988), *sup/flo10* (Schultz et al. 1991; Bowman et al. 1992), and *tfl* (Shannon and Meeks-Wagner 1991) became central targets of developmental studies. Based on a simple genetic model (Meyerowitz et al. 1991), recent advances in characterization of the function of these genes provide insight into the complex regulatory interplay of functions determining cell fate and organ identity during flower development (Weigel and Meyerowitz 1993) and may be a "trendsetter," as Laibach predicted.

The process of fertilization and elements involved are also subjects of molecular genetic studies. The role and structure of the stigmatic surface is anatomically described (Elleman et al. 1992), and the early characterization of ovule and embryo sac (Vandendries 1909) is now well extended (Mansfield and Briarty 1991a,b). Sterile ovule mutants were isolated, and both ovule and megagametophyte mutations were observed (Robinson-Beers et al. 1992). These studies follow the earlier analysis of an unusual female gametophyte mutation characterized by lack of female transmission, significantly higher than 50% seed set, and the histological observation of twin megaspore tetrads (Rédei 1965). This feature permits megaspore selection when the tetrads are in opposite orientation and therefore produce 64% seed set in the complete absence of female transmission. The time course of microsporogenesis and cytochemical analysis of pollen development were also described (Polyakova 1964; Regan and Moffatt 1990). Androgenesis was induced by pollinating *Cardaminopsis arenosa* ($2n=32$) by diploid *Arabidopsis* (Měsiček 1971), and for safe identification of androgenic seedlings, a genetic construct was developed (Barabás and Rédei 1971). The frequency of androgenesis seems to be somewhat lower in *Arabidopsis* than in maize. In the latter, the frequency is increased by orders of magnitude through a single mutation (Kermicle 1969). Haploids can also be obtained in *Arabidopsis* from anther culture, although their use is so far very limited (Amos and Scholl 1978). Male gametophytic factors occur commonly after treatments with mutagens that cause chromosomal rearrangements and deletions (for review, see Rédei and Koncz 1992). These defects cause deficiency or excess of the recessive class, depending on the linkage phase of the breakage points. Male sterility caused by point mutations or insertions is characterized by high penetrance and expressivity in the male, but good fertility in the female organs (Van der Veen and Wirtz 1968; Aarts et al. 1993).

## GENETICS OF POPULATIONS

Theoretical population genetics got off the ground with the recognition of the Hardy (1908) and Weinberg (1908) equilibrium. The law is based on panmixis, and almost all the theoretical developments since deal with panmictic populations. *Arabidopsis*, an autogamous species, did not fit well into the framework based on allogamy. Many of the theoretical solutions to autogamous populations deserve future studies. The topic of this section was reviewed by Rédei (1975a) and Griffing and Scholl (1991).

### QUANTITATIVE INHERITANCE

Genetics of quantitative characters does not discriminate against the breeding system of *Arabidopsis* (Rédei 1975a; Griffing and Scholl 1991). Griffing and Langridge (1963) selected thus for phenotypic stability of 38 *Arabidopsis* ecotypes at six different environments. Later, an even more precise analysis of heterozygote advantage (Pederson 1968) followed. Competition studies were also conducted with a number of controlled variables (Griffing and Zsiros 1971). Langridge (1961) interpreted heterosis on the basis of interaction of temperature-sensitive alleles. Overdominance in isogenic background was demonstrated (Rédei 1962b), and biochemical evidence was provided on the basis of allelic complementation of temperature-sensitive pyrimidine genes (Li and Rédei 1969b). It was calculated that overdominance at larger numbers of loci cannot be maintained without danger of extinction, although at a few loci, it may be responsible for heterosis (Rédei 1982b). For a better understanding, the cloning of genes encoding quantitative characters will be required. Alternatively, a better insight into signal transduction mechanisms may help resolve the relations of epistasis, pleiotropy, and polygenic systems.

## THE FUTURE OF *ARABIDOPSIS* AS A BIOLOGICAL TOOL

Predictions regarding the future of science are risky because basic science is concerned with the unknown. Friedrich Laibach (1965), the founder of research with *Arabidopsis,* was right by pointing out, "I am certain, the small *Arabidopsis...* will be a trendsetter." Laibach's prediction appears to have been fulfilled. The road to success was not easy, but the goal remained clear and the progress has been impressive.

The major contributions of *Arabidopsis* are in biochemical genetics and in the control of differentiation and development. In the future, the most significant progress may come on a broad front of basic biology. We can expect *Arabidopsis* to be the first higher plant to have its total genome sequenced. The functions of many more genes will be understood in *Arabidopsis* than in other higher plants. It is predictable that a new level of understanding will be gained on the coordinated system of genes involved in receiving and transmitting signals and executing instructions of the genetic blueprint. Although in the past, genetics was characterized by biochemical, morphological, and statistical type dissections, the future will permit an understanding of the functional integration of the genetic elements. In short, the major developments will come in developmental genetics. Yet, today, 30 years later, the question of Chargaff (1963) is still unanswerable: "Will man discover the 'molecular structure of God'?"

## ACKNOWLEDGMENTS

We bequeath this review to the critics (as Goethe did with Faust) so they may have fun in finding the mistakes incurred by omission or judgment, but never by intention. We are particularly indebted to Dr. R.L. Scholl for bibliographical assistance. This work was supported by NATO grant 910856.

## REFERENCES

Aarts, M.G.M., W.G. Dirkse, W.J. Stiekema, and A. Pereira. 1993. Transposon tagging of a male sterility gene in *Arabidopsis*. *Nature* **363:** 715–717.

Aerts, M., M. Jacobs, J.-P. Hernalsteens, M. Van Montagu, and J. Schell. 1979. Induction and *in vitro* culture of *Arabidopsis thaliana* crown gall tumours. *Plant Sci. Lett.* **17:** 43–53.

Altmann, T., B. Damm, U. Halfter, L. Willmitzer, and P.-C. Morris. 1992. Protoplast transformation and methods to create specific mutants in *Arabidopsis thaliana*. In *Methods in* Arabidopsis *research*. (ed. C. Koncz et al.), pp. 310–330. World Scientific, Singapore.

Ambros, P. and D. Schweizer. 1976. The Giemsa C-banded karyotype of *Arabidopsis thaliana* (L.) Heynh. *Arabidopsis Inf. Serv.* **13:** 167–171.

Amos, J.A. and R.L. Scholl. 1978. Induction of haploid callus from anthers of 4 species of *Arabidopsis*. *Z. Pflanzenphysiol.* **90:** 33–40.

An, G., B.D. Watson, and C.C. Chang. 1986. Transformation of tobacco, potato and *Arabidopsis thaliana* using binary Ti plasmid vector system. *Plant Physiol.* **81:** 301–305.

Anai, T., K. Hasegawa, Y. Watanabe, H. Uchimiya, R. Ishizali, and M. Matsui. 1991. Isolation and analysis of cDNAs encoding small GTP-binding proteins of *Arabidopsis thaliana*. *Gene* **108:** 259–264.

André, D., D. Colau, J. Schell, M. Van Montagu, and J.-P. Hernalsteens. 1986. Gene tagging in plants by T-DNA insertion mutagen that generates *aph(3')II* plant gene fusions. *Mol. Gen. Genet.* **204:** 512–518.

Bailey, N.T.J. 1961. *Introduction to the mathematical theory of genetic linkage*. Clarendon Press, Oxford, United Kingdom.

Baker, B., J. Schell, H. Lörz, and N.V. Fedoroff. 1986. Transposition of the maize controlling element *Activator* in tobacco. *Proc. Natl. Acad. Sci.* **83:** 4844–4848.

Bancroft, I., and C. Dean. 1993. Transposition pattern of the maize *Ds* in *Arabidopsis thaliana*. *Genetics* **134:** 1221–1229.

Bancroft, I., A.M. Bhatt, C. Sjodin, S. Scofield, J.D.G. Jones, and C. Dean. 1992. Development of an efficient two-element transposon tagging system in *Arabidopsis thaliana*. *Mol. Gen. Genet.* **233:** 449–461.

Barabás, Z. and G.P. Rédei. 1971. Frequency of androgenesis. *Arabidopsis Inf. Serv.* **8:** 9–10.

Barker, R.F., D.V. Thompson, D.R. Talbot, J. Swanson, and J.L. Benetzen. 1984. Nucleotide sequence of the maize transposable element *Mu1*. *Nucleic Acids Res.* **12:** 5955–5967.

Barski, G., S. Sorieul, and F. Cornefert. 1960. Production dans des cultures *in vitro* de deux souches cellulaires en association de cellules de caractère "hybride". *C. R. Acad. Sci.* **251:** 1825–1827.

Bechtold, N., J. Ellis, and G. Pelletier. 1993. In planta *Agrobacterium* gene transfer by infiltration of adult *Arabidopsis* plants. *C.R. Acad. Sci.* **316:** 1194–1199.

Berger, B. 1968. Entwicklungsgeschichtliche und chromosomale Ursachen der verschiedenen Kreutzungsverträglichkeit zwischen Arten der Vervandtschaftskreise *Arabidopsis. Beitr. Biol. Pflanz.* **45:** 171–212.

Bhattacharyya, M.K., A.M. Smith, T.H.N. Ellis, C. Hedley, and C. Martin. 1990. The wrinkled-seed character of pea described by Mendel is caused by a transposon-like insertion in a gene encoding starch-branching enzyme. *Cell* **60:** 115–122.

Borthwick, H.A. and S.B. Hendricks. 1960. Photoperiodism in plants. *Science* **132:** 1223–1228.

Bouharmont, J. 1965. Fertility studies in polyploid *Arabidopsis thaliana*. In *Reports from the 1st International Conference on* Arabidopsis *Research*, (ed. G. Röbbelen), pp. 31–36. Univerity of Göttingen, Germany.

Bowman, J.L., D.R. Smyth, and E.M. Meyerowitz. 1989. Genes directing flower development in *Arabidopsis. Plant Cell* **1:** 37–52.

Bowman, J.L., H. Sakai, T. Jack, D. Weigel, U. Mayer, and E.M. Meyerowitz. 1992. *SUPERMAN*, a regulator of flower homeotic genes in *Arabidopsis. Development* **114:** 599–615.

Boyes, J.W., Y. Nishimura, and B.C. Boyes. 1973. *References to nomenclature and publications in genetics and cytology*. International Genetics Federation, Montreal, Canada.

Braun, A.C. 1947. Thermal studies on the factors responsible for tumor initiation on crown-gall. *Am. J. Bot.* **34:** 234–240.

Brink, R.A. and D.C. Cooper. 1931. The association of *semisterile-1* in maize with two linkage groups. *Genetics* **16:** 595–628.

Brink, R.A. and R.A. Nilan. 1952. The relation between light variegated and medium variegated pericarp in maize. *Genetics* **37:** 519–544.

Browse, J. and C.R. Somerville. 1991. Glycerolipid synthesis: Biochemistry and regulation. *Annu. Rev. Plant Physiol. Plant Mol. Biol.* **42:** 467–506.

Bürger, D. 1971. Die morphologische Mutanten des Göttinger *Arabidopsis*-Sortiments, einschlieslich der Mutanten mit abweichender Samenfarbe. *Arabidopsis Inf. Serv.* **8:** 36–42.

Campell, B.R. and C.D. Town. 1991. Physiology of hormone autonomous tissue lines derived from radiation-induced tumors of *Arabidopsis thaliana. Plant Physiol.* **97:** 1166–1173.

Carlson, P.S., H.H. Smith, and R.D. Dearing. 1972. Parasexual interspecific plant hybridization. *Proc. Natl. Acad. Sci.* **69:** 2292–2294.

Chang, C. and E.M. Meyerowitz. 1986. Molecular cloning and DNA sequence of the *Arabidopsis* alcohol dehydrogenase gene. *Proc. Natl. Acad. Sci.* **83:** 1408–1412.

Chang, C., F. S. Kwok, A.B. Bleecker, and E.M. Meyerowitz. 1993. *Arabidopsis* ethylene-response gene *ETR1*: Similarity of product to two-component regulators. *Science* **262:** 539–544.

Chargaff, E. 1963. *Essays on nucleic acids.* Elsevier, New York.

Chilton, M.-D., M.J. Drummong, D.J. Merlo, D. Sciaky, A.L. Montoya, M.P. Gordon, and E.W. Nester. 1977. Stable incorporation of plasmid DNA into higher plant cells: The molecular basis of crown gall tumorigenesis. *Cell* **11:** 263–271.

Chory, J. 1992. A genetic model for light-regulated seedling development in *Arabidopsis. Development* **115:** 337–354.

———. 1993. Out of darkness: Mutants reveal pathways controlling light-regulated development in plants. *Trends Genet.* **9:** 167–172.

Chung, S.C. and G.P. Rédei. 1974. An anomaly of the genetic regulation of the *de novo* pyrimidine pathway. *Biochem. Genet.* **11:** 441–453.

Coleman, W. 1965. Cell nucleus and inheritance: A historical study. *Proc. Am. Philos. Soc.* **109:** 124–158.

Coupland, G. 1992. Transposon tagging in *Arabidopsis*. In *Methods in* Arabidopsis *research* (ed. C. Koncz et al.), pp. 290–309. World Scientific, Singapore.

Damm, B., R. Schmidt, and L. Willmitzer. 1989. Efficient transformation of *Arabidopsis thaliana* using direct gene transfer to protoplasts. *Mol. Gen. Genet.* **213:** 15–20.

Davey, M.R., E.C. Cocking, J. Freeman, N. Pearce, and I. Tudor. 1980. Transformation of petunia protoplasts by isolated *Agrobacterium* plasmids. *Plant. Sci. Lett.* **18:** 307–313.

Dean, C., C. Sjordin, T. Page, J. Jones, and C. Lister. 1992. Behaviour of the maize transposable element *Ac* in *Arabidopsis thaliana*. *Plant J.* **2:** 69–81.

DeBeuckeleer, M., M. Lemmers, G. DeVos, L. Willmitzer, M. Van Montagu, and J. Schell. 1981. Further insight in the transferred-DNA of octopine crown-gall. *Mol. Gen. Genet.* **183:** 283–288.

De Guen, L., M. Thomas, M. Bianchi, N.G. Halford, and M. Kreis. 1992. Structure and expression of a gene from *Arabidopsis thaliana* encoding a protein related to *SNF1* protein kinase. *Gene* **120:** 249–254.

Deng, X.-W. and P.H. Quail. 1992. Genetic and phenotypic characterization of *cop1* mutants of *Arabidopsis thaliana*. *Plant J.* **2:** 83–95.

Deng, X.-W., M. Matsui, N. Wei, D. Wagner, A.M. Chu, K.A. Feldmann, and P.H. Quail. 1992. *COP1*, an *Arabidopsis* photomorphogenic regulatory gene, encodes a novel protein with both a Zn-binding motif and a domain homologous to the β-subunit of trimeric G-proteins. *Cell* **71:** 791–801.

Dolferus, R., D. von den Bossche, and M. Jacobs. 1990. Sequence analysis of two null mutant alleles of the single *Arabidopsis adh* locus. *Mol. Gen. Genet.* **224:** 297–302.

Doy, C.H., P.M. Gresshoff, and B.G. Rolfe. 1973. Biological and molecular evidence for the transgenosis of genes from bacteria to plant cells. *Proc. Natl. Acad. Sci.* **70:** 723–726.

Draper, J., M.A. Davey, J.P. Freeman, E.C. Cocking, and B.G. Cox. 1982. Ti plasmid homologous sequences present in the tissue from *Agrobacterium* plasmid transformed petunia protoplasts. *Plant Cell Physiol.* **23:** 451–458.

Elleman, C.J., V. Franklin-Tong, and H.G. Dickinson. 1992. Pollination in species with dry stigmas: The nature of the early stigmatic response and the pathway taken by pollen tubes. *New Physiol.* **121:** 413–424.

Ephrussi, B. and M.C. Weiss. 1965. Interspecific hybridization of somatic cells. *Proc. Natl. Acad. Sci.* **53:** 1040–1042.

Fedoroff, N., D. Furtek, and O. Nelson. 1984. Cloning of the *Bronze* locus in maize by a simple and generalizable procedure using transposable controlling element *Ac*. *Proc. Natl. Acad. Sci.* **81:** 3825–3828.

Feinbaum, R.L. and F.M. Ausubel. 1988. Transcriptional regulation of the *Arabidopsis thaliana* chalcone synthase gene. *Mol. Cell. Biol.* **8:** 1985–1992.

Feldmann, K.A. 1991. T-DNA insertional mutagenesis in *Arabidopsis*: Mutational spectrum. *Plant J.* **1:** 71–82.

Feldmann, K.A. and M.D. Marks. 1987. *Agrobacterium*-mediated transformation of germinating seeds of *Arabidopsis thaliana*: A non-tissue culture approach. *Mol. Gen. Genet.* **208:** 1–9.

Feldmann, K.A., M.D. Marks, M.L. Christianson, and R.S. Quatrano. 1989. A dwarf mutant of *Arabidopsis* generated by T-DNA insertion mutagenesis. *Science* **243:** 1351–1354.

Ferreira, P.G.G., A.S. Hemerly, R. Villaroel, M. Van Montagu, and D. Inzé. 1991. The *Arabidopsis* functional homolog of the $p34^{cdc2}$ protein kinase. *Plant Cell* **3:** 351–360.

Finkelstein, R.R. and C.R. Somerville. 1990. Three classes of abscisic acid (ABA)-

insensitive mutations in *Arabidopsis* define genes that control overlapping subsets of ABA responses. *Plant Physiol.* **94:** 1172–1179.

Fisher, R.A. and B. Balmakund. 1928. The estimation of linkage from the offspring of selfed heterozygotes. *J. Genet.* **20:** 79–92.

Galbraith, D.W., K.P. Harkins, and S. Knapp. 1991. Systemic endoploidy in *Arabidopsis thaliana*. *Plant Physiol.* **96:** 985–989.

Gasch, A., T. Aoyama, R. Foster, and N.-H. Chua. 1992. Gene isolation with polymerase chain reaction. In *Methods in* Arabidopsis *research* (ed. C. Koncz et al.), pp. 342–356. World Scientific, Singapore.

Gleba, Y.Y. and F. Hoffmann. 1978. Hybrid cell lines *Arabidopsis thaliana* + *Brassica campestris*. *Mol. Gen. Genet.* **165:** 257–264.

Giraudat, J., C. Valon, J. Smalle, F. Parcy, and H.M. Goodman. 1992. Isolation of the *Arabidopsis ABI3* gene by positional cloning. *Plant Cell* **4:** 1251–1261.

Gôto, N., M. Starke, and A.R. Kranz. 1987. Effects of gibberellins on flower development of the *pin-formed* mutant of *Arabidopsis thaliana*. *Arabidopsis Inf. Serv.* **23:** 66–71.

Griffing, B. and J. Langridge. 1963. Phenotypic stability of growth in the self-fertile species *Arabidopsis thaliana*. In *Statistical genetics and plant breeding* (ed. W.D. Hanson and H.F. Robinson), publ. 982, pp. 368-394. National Research Council, Washington, D.C.

Griffing, B. and R.L. Scholl. 1991. Qualitative and quantitative genetic studies of *Arabidopsis thaliana*. *Genetics* **129:** 605–609.

Griffing, B. and E. Zsiros. 1971. Heterosis associated with genotype-environmental interactions. *Genetics* **68:** 443–455.

Haldane, J.B.S. 1919. The combination of linkage values and calculation of distances between loci and linked factors. *J. Genet.* **8:** 299–309.

Halfter, U., P.-C. Morris, and L. Willmitzer. 1991. Gene targeting in *Arabidopsis thaliana*. *Mol. Gen. Genet.* **231:** 186–193.

Hardy, G.H. 1908. Mendelian proportions in a mixed population. *Science* **67:** 562–563.

Haughn, G.W. and C.R. Somerville. 1986. Sulfonylurea-resistant mutants of *Arabidopsis thaliana*. *Mol. Gen. Genet.* **204:** 430–434.

———. 1988. Genetic control of morphogenesis in *Arabidopsis*. *Dev. Genet.* **9:** 1081–1085.

Hemerly, A., C. Bergounioux, M. Van Montagu, D. Inzé, and P. Ferreira. 1992. Genes regulating the plant cell cycle: Isolation of a mitotic-like cyclin from *Arabidopsis thaliana*. *Proc. Natl. Acad. Sci.* **89:** 3295–3299.

Herman, P.L. and M.D. Marks. 1989. Trichome development in *Arabidopsis thaliana*. II. Isolation and complementation of the *GLABROUS* gene. *Plant Cell* **1:** 1051–1055.

Herrera-Estrella, L., A. Depicker, M. Van Montagu, and J. Schell. 1983a. Expression of chimeric genes transferred into plant cells using Ti-plasmid-derived vector. *Nature* **303:** 209–213.

Herrera-Estrella, L., M. De Block, E. Messens, J.-P. Hernalsteens, M. Van Montagu, and J. Schell. 1983b. Chimeric genes as dominant selectable markers in plant cells. *EMBO J.* **2:** 987–989.

Hess, D. 1977. Cell modification by DNA uptake. In *Applied and fundamental aspects of plant cell, tissue and organ culture* (ed. J. Reinert and Y.P.S. Bajaj), pp. 506–577. Springer Verlag, Berlin.

Hille, J., G. Wullems, and R.A. Schilperoort. 1983. Non-oncogenic T-region mutants of *Agrobacterium tumefaciens* do transfer T-DNA into plant cells. *Plant Mol. Biol.* **2:** 155–163.

Hirono, Y. 1964. A genetic method for localization of chromosome defects. *Arabidopsis Inf. Serv.* **1:** 13–14.

Hirono, Y. and G.P. Rédei. 1965. Induced premeiotic exchange of linked markers in the angiosperm *Arabidopsis. Genetics* **51:** 519–526.

Hooykaas, P.J.J. 1989. Transformation of plant cells via *Agrobacterium. Plant Mol Biol.* **13:** 327–336.

Hylander, N. 1957. *Cardaminopsis suecica* (Fr.) Hiit., a northern amphidiploid species. *Bull. Jard. Bot. Etat Brux.* **27:** 591–597.

Irish, V.F. and I.M. Sussex. 1992. A fate map of the *Arabidopsis* embryonic shoot apical meristem. *Development* **115:** 745–753.

Jacobs, M., R. Dolferus, and D. van den Bossche. 1988. Isolation and biochemical analysis of ethyl methanesulfonate-induced alcohol dehydrogenase null mutants of *Arabidopsis thaliana* (L.) Heynh. *Biochem. Genet.* **26:** 105–122.

Joos, H., D. Inzé, A. Caplan, M. Sormann, M. Van Montagu, and J. Schell. 1983. Genetic analysis of T-DNA transcripts in nopaline crown-galls. *Cell* **32:** 1057–1067.

Jürgens, G., U. Mayer, R.A. Torres Ruiz, T. Berleth, and S. Misera. 1991. Genetic analysis of pattern formation in the *Arabidopsis* embryo. *Dev. Suppl.* **91.1:** 27–38.

Keller, J., E. Lim, D.W. James, and H.K. Dooner. 1992. Germinal and somatic activity of maize element activator (*Ac*) in *Arabidopsis. Genetics* **131:** 449–459.

Kermicle, J.L. 1969. Androgenesis conditioned by a mutation in maize. *Science* **166:** 1422–1424.

Kieber, J.J. and J.R. Ecker. 1993. Ethylene gas: It's not just for ripening any more! *Trends Genet.* **9:** 356–361.

Kieber, J.J., M. Rothenberg, G. Roman, K.A. Feldmann, and J.R. Ecker. 1993. *CTR1*, a negative regulator of the ethylene response pathway in *Arabidopsis*, encodes a member of the RAF family of protein kinases. *Cell* **72:** 427–441.

Klein, T.M., E.D. Wold, R. Wu, and J.C. Sanford. 1987. High velocity microprojectiles for delivering nucleic acids into living cells. *Nature* **327:** 70–73.

Kohorn, B.D., S. Lane, and T.A. Smith. 1992. An *Arabidopsis* serine/threonine kinase homologue with an epidermal growth factor repeat selected in yeast for its specificity for a thylakoid membrane protein. *Proc. Natl. Acad. Sci.* **89:** 10989–10992.

Koncz, C., R. Mayerhofer, Z. Koncz-Kalman, H. Körber, G.P. Rédei, and J. Schell. 1989. High frequency T-DNA-mediated gene tagging in plants. *Proc. Natl. Acad. Sci.* **86:** 8467–8471.

Koncz, C., R. Meyerhofer, Z. Koncz-Kalman, C. Nawrath, B. Reiss, G.P. Rédei, and J. Schell. 1990. Isolation of a gene encoding a novel chloroplast protein by T-DNA tagging in *Arabidopsis thaliana. EMBO J.* **9:** 1337–1346.

Koornneef, M. 1983. The use of telotrisomics for centromere mapping in *Arabidopsis thaliana* (L.) Heynh. *Genetica* **6:** 33–40.

———. 1990. Mutations affecting the testa colour in *Arabidopsis. Arabidopsis Inf. Serv.* **27:** 1–4.

Koornneef, M. and J.A.D. Zeevaart. 1985. A gibberellin insensitive mutant of *Arabidopsis thaliana. Plant Physiol.* **65:** 33–39.

Koornneef, M., J.H. DeBruine, and P. Goetsch. 1980a. A provisional map of chromosome 4 of *Arabidopsis. Arabidopsis Inf. Serv.* **17:** 11–18.

Koornneef, M., L.W.M. Dellaert, and J.H. Van der Veen. 1982a. EMS- and radiation-induced mutation frequencies at individual loci in *Arabidopsis thaliana* (L.) Heynh. *Mutat. Res.* **93:** 109–123.

Koornneef, M., H.C. Dressehuys, and K. Sree Ramulu. 1982b. Genetic identification of translocations in *Arabidopsis. Arabidopsis Inf. Serv.* **19:** 93–99.

Koornneef, M., C.J. Hanhart, and J.H. Van der Veen. 1991. A genetic and physiological analysis of late flowering mutants in *Arabidopsis thaliana. Mol. Gen. Genet.* **229:** 57–66.

Koornneef, M., G. Reuling, and C.M. Karssen. 1984. The isolation and characterization of abscisic acid-insensitive mutants of *Arabidopsis thaliana. Physiol. Plant.* **61:** 377–383.

Koornneef, M., E. Rolff, and C.J.P. Spruit. 1980b. Genetic control of light-inhibited hypocotyl elongation in *Arabidopsis thaliana* (L.) Heynh. *Z. Pflanzenphysiol.* **100:** 147–160.

Koornneef, M., M.L. Jorna, D.L.C. Brinkhorst-Van der Swan, and C.M. Karssen. 1982c. The isolation of abscisic acid (ABA) deficient mutants by selection of induced revertants in non-germinating gibberellin-sensitive lines of *Arabidopsis thaliana* (L.) Heynh. *Theor. Appl. Genet.* **61:** 385–393.

Koornneef, M., A. Elgersma, C.J. Hanhart, E.P. Van Loenen-Martinet, L. Van Rijn, and J.A.D. Zeevaart. 1985. A giberellin insensitive mutant of *Arabidopsis. Plant Physiol.* **65:** 33–39.

Koornneef, M., J. Van Eden, C.J. Hanhart, P. Stam, F.J. Braaksma, and W.J. Feenstra. 1983. Linkage map of *Arabidopsis thaliana. J. Heredity* **74:** 265–272.

Kosambi, D.D. 1944. The estimation of map distances from recombination values. *Ann. Eugen.* **12:** 172–175.

Krens, F.A., L. Molendijk, G.J. Wullems, and R.A. Schilperoort. 1982. *In vitro* transformation of plant protoplasts with Ti plasmid DNA. *Nature* **296:** 72–74.

Laibach, F. 1907. Zur Frage nach der Individualität der Chromosomen in Pflanzenreich. *Beih. Bot. Cbl.* (1 Abt.) **22:** 197–210.

———. 1940. Die Ursachen der Blütenbildung und das Blühhormon. *Natur. Volk.* **70:** 55–65.

———. 1943. *Arabidopsis thaliana* (L.) Heynh. als Objekt für genetische und entwicklungs-physiologische Untersuchungen. *Bot. Arch.* **44:** 439–455.

———. 1951. Über sommer- und winter-annualle Rassen von *Arabidopsis thaliana* (L.) Heynh. Ein Beitrag zur Ätiologie der Blütenbildung. *Beitr. Biol. Pflanz.* **28:** 173–210.

———. 1965. 60 Jahre *Arabidopsis* Forschung 1905–1965. In *Reports from the 1st International Conference on* Arabidopsis *Research* (ed. G. Röbbelen), pp. 14–18. University of Göttingen, Germany.

Langridge, J. 1955. Biochemical mutations in the crucifer *Arabidopsis thaliana* (L.) Heynh. *Nature* **76:** 260–261.

———. 1961. A genetic and molecular basis for heterosis in *Arabidopsis* and *Drosophila. Am. Nat.* **96:** 5–27.

Lee-Chen, S. and D. Bürger. 1967. The location of linkage groups on the chromosomes of *Arabidopsis* by the trisomic method. *Arabidopsis Inf. Serv.* **4:** 4–5.

Lee-Chen, S. and L.M. Steinitz-Sears. 1967. The location of linkage groups in *Arabidopsis thaliana. Can. J. Genet. Cytol.* **9:** 381–384.

Ledoux, L. and R. Huart. 1961. Sur la possibilité d'un transfert d'acides ribo- et desoxyribonucléique et de proteins dans les embryons d'orge en croissance. *Arch. Int. Physiol. Biochem.* **69:** 598.

Ledoux, L. and M. Jacobs. 1974. DNA-mediated correction of thiaminless *Arabidopsis thaliana. Nature* **249:** 17–21.

Leutweiler, L.S., B.R. Hough-Evans, and E.M. Meyerowitz. 1984. The DNA of *Arabidopsis thaliana. Mol. Gen. Genet.* **194:** 15–23.

Leyser, H.M., C.A. Lincoln, C. Timpte, D. Lammer, J. Turner, and M. Estelle. 1993. *Arabidopsis* auxin-resistance gene *AXR1* encodes a protein related to ubiquitin-activating enzyme E1. *Nature* **364:** 161–164.

Li, S.L. 1968. "Genetics of thiamine metabolism in *Arabidopsis.*" Ph.D. thesis, University of Missouri, Columbia.

Li, S.L. and G.P. Rédei. 1969a. Estimation of mutation rate in autogamous diploids. *Radiation Bot.* **9:** 125–131.

———. 1969b. Direct evidence for models of heterosis provided by mutants of *Arabidopsis* blocked in the thiamine pathway. *Theor. Appl. Genet.* **39:** 68–72.

Linnaeus, C. 1753. *Species plantarum.* Lawrentius Salvius, Holmiae.

Liscum, E. and R.P. Hangarter. 1991. *Arabidopsis* mutants lacking blue light-dependent inhibition of hypocotyl elongation. *Plant Cell* **3:** 685–694.

Lloyd, A.M., A.R. Barnason, S.G. Rogers, M.C. Byrne, R.T. Fraley, and R.B. Horsch. 1986. Transformation of *Arabidopsis thaliana* with *Agrobacterium tumefaciens. Science* **234:** 464–466.

Lloyd, A.M., V. Walbot, and R.W. Davis. 1992. *Arabidopsis* and *Nicotiana* anthocyanin production activated by maize regulators *R* and *C1. Science* **258:** 1773–1775.

Löve, Å. 1961. *Hylandra*—A new genus of the cruciferae. *Sven. Bot. Tidskr.* **55:** 211–217.

Lurquin, F.P. 1976. Integration of exogeneous DNA in plants: A hypothesis awaiting clear-cut demonstration. In *Cell genetics of higher plants* (ed. D. Dudits et al.), pp. 77–90. Akadémiai Kiadó, Budapest, Hungary.

Ma, H., M.F. Yanofsky, and E.M. Meyerowitz. 1990. Molecular cloning and characterization of *GPA1*, a G-protein alpha subunit gene from *Arabidopsis thaliana. Proc. Natl. Acad. Sci.* **87:** 3821–3825.

Macilvain, C. 1993. Genome project 'to be done by 1994'. *Nature* **362:** 488.

Maher, E.P. and S.J.B. Martindale. 1980. Mutants of *Arabidopsis thaliana* with altered responses to auxins and gravity. *Biochem. Genet.* **18:** 1041–1053.

Maliga, P. 1984. Isolation and characterization of mutants in plant cell culture. *Annu. Rev. Plant Physiol.* **35:** 519–542.

Malling, H.V. and F.J. de Serres. 1973. Genetic alterations at the molecular level in X-ray induced *ad-3B* mutants of *Neurospora crassa. Radiat. Res.* **53:** 77–87.

Maluszynska, J. and J.S. Heslop-Harrison. 1991. Localization of tandemly repeated DNA sequence in *Arabidopsis thaliana. Plant J.* **1:** 159–166.

Mansfield, S.G. and L.G. Briarty. 1991a. Early embryogenesis in *Arabidopsis.* I. The mature embryo sac. *Can. J. Bot.* **69:** 447–460.

———. 1991b. The early embryogenesis in *Arabidopsis thaliana.* II. The developing embryo. *Can. J. Bot.* **69:** 461–476.

Marks, M.D. and K.A. Feldmann. 1989. Trichome development in *Arabidopsis thaliana.* I. T-DNA tagging of the *glabrous 1* gene. *Plant Cell* **1:** 1043–1050.

Martínez-Zapater, J.M., P. Gil-Viñuelas, J. Carpel, and C.R. Somerville. 1992. Mutations at the *Arabidopsis chm* locus promote rearrangements of the mitochondrial genome. *Plant Cell* **4:** 889–899.

Mayer, U., R.A. Thorres Ruiz, S. Misera, and G. Jürgens. 1991. Mutations affecting body organization in *Arabidopsis* embryo. *Nature* **353:** 402–407.

McClintock, B. 1952. Chromosome organization and genic expression. *Cold Spring Harbor Symp. Quant. Biol.* **16:** 13–47.

McKelvie, A.D. 1962. A list of mutant genes in *Arabidopsis thaliana* (L.) Heynh. *Radiat. Bot.* **1:** 233–241.

———. 1963. Studies on the induction of mutations in *Arabidopsis thaliana* (L.) Heynh. *Radiat. Bot.* **3:** 105–123.

———. 1965. Preliminary data on linkage groups in *Arabidopsis.* In *Reports from the*

*1st International Conference on* Arabidopsis *Research* (ed. G. Röbbelen), pp. 79–84. University of Göttingen, Germany.

Medford, J.I., F.J. Behringer, J. Cellos, and K.A. Feldmann. 1992. Normal and abnormal development in the *Arabidopsis* vegetative shoot apex. *Plant Cell* **4:** 631–645.

Megnet, R. 1967. Mutants potentially deficient in alcohol dehydrogenase in *Schizosaccharomyces pombe. Arch. Biochem. Biophys.* **121:** 194–201.

Meinke, D.W. 1991. Embryonic mutants of *Arabidopsis thaliana. Dev. Genet.* **12:** 382–392.

―――――. 1992. A homeotic mutant of *Arabidopsis thaliana* with leafy cotyledons. *Science* **258:** 1647–1650.

Meinke, D.W. and I.M. Sussex. 1979. Embryo lethal mutants of *Arabidopsis thaliana.* A model system for genetic analysis of plant embryo development. *Dev. Biol.* **12:** 50–61.

Melchers, G. and L. Bergmann. 1959. Untersuchungen an Kulturen von Haploiden Geweben von *Anthirrhinum majus. Ber. Dtsch. Bot. Ges.* **78:** 21–29.

Mendel, G. 1866. Versuche über Pflanzenhybriden. *Verh. Naturforsch. Ver. Brünn* **4:** 3–47.

Mĕsicĕk, J. 1967. The chromosome morphology of *Arabidopsis thaliana* (L.) Heynh. and some remarks on the problem of *Hylandra suecica* (Fr.) Löve. *Folia Geobot. Phytotax.* **2:** 433–436.

―――――. 1971. Androgenesis in *Arabidopsis thaliana* (L.) Heynh. *Arabidopsis Inf. Serv.* **8:** 8–9.

Meyerowitz, E.M. 1987. *Arabidopsis thaliana. Annu. Rev. Genet.* **21:** 93–111.

―――――. 1989. *Arabidopsis*, a useful weed. *Cell* **56:** 263–269.

―――――. 1992. Introduction to the *Arabidopsis* genome. In *Methods in* Arabidopsis *research* (ed. C. Koncz et al.), pp. 100–118. World Scientific, Singapore.

Meyerowitz, E.M., J.L. Bowman, L.L. Beckman, G.N. Drews, T. Lack, L.E. Sieburth, and D. Weigel. 1991. A genetic and molecular model for the flower development in *Arabidopsis thaliana. Dev. Suppl.* **1:** 157–167.

Moffatt, B., C. Perthe, and M. Lalou. 1991. Metabolism of benzyladenine is impaired in a mutant of *Arabidopsis thaliana* lacking adenine phosphoribosyl transferase activity. *Plant Physiol.* **95:** 900–908.

Morton, N.E. 1962. Segregation and linkage. In *Methodology of human genetics* (ed. W.J. Burdette), pp. 17–52. Holden-Day, New York.

Müller, A. 1963. Embryonentest zum Nachweis rezessiver Letalfaktoren bei *Arabidopsis thaliana. Biol. Zentralbl.* **83:** 133–163.

Negrutiu, I., A. Cattoir-Reynaerts, and M. Jacobs. 1978. Selection and characterization of cell lines of *Arabidopsis thaliana* resistant to amino acid analogues. *Arch. Int. Physiol. Biochim.* **86:** 442–443.

Nicolov, D.B., S.-H. Hu, J. Lin, A. Gasch, A. Hoffmann, M. Horikoshi, N.-H. Chua, R.G. Roeder, and S.K. Burley. 1992. Crystal structure of TFIID TATA-box binding protein. *Nature* **360:** 40–46.

Nitschke, K., U. Fleig, J. Schell, and K. Palme. 1992. Complementation of the *cs disc3-11* cell cycle mutant of *Schizosaccharomyces pombe* by a protein phosphatase from *Arabidopsis thaliana. EMBO J.* **11:** 1327–1333.

Okada, K. and Y. Shimura. 1990. Reversible root tip rotation in *Arabidopsis* seedlings induced by obstacle-touching stimulus. *Science* **250:** 274–276.

Okada, K., J. Ueda, M.K. Komaki, C.J. Bell, and Y. Shimura. 1991. Requirement of the auxin polar transport system in early stages of *Arabidopsis* floral bud formation. *Plant Cell* **3:** 677–684.

Oppenheimer, D.G., P.L. Herman, S. Sivakumaran, J. Esch, and M.D. Marks. 1991. A *myb* gene required for leaf trichome differentiation in *Arabidopsis* is expressed in stipules. *Cell* **67:** 483–493.

Osborne, B.I., U. Wirtz, and B. Baker. 1993. A method for mutation, genomic rearrangement and molecular cloning based on *Ds* transposon and the *Cre-Lox* recombination system. In *Abstracts from the 5th International Conference on* Arabidopsis *Research*, Columbus, Ohio, p. 188. Ohio State University, Columbus

Palme, K. 1992. Molecular analysis of plant signalling elements: Relevance of eukaryotic signal transduction models. *Int. Rev. Cytol.* **132:** 223–283.

Patton, D.A., L.H. Franzmann, and D.W. Meinke. 1991. Mapping genes essential for embryo development in *Arabidopsis thaliana. Mol. Gen. Genet.* **227:** 337–340.

Pederson, D.G. 1968. Environmental stress, heterozygote advantage and genotype-environment interaction. *Heredity* **23:** 127–138.

Peleman, J., B. Cottyn, W. Van Camp, M. Van Montagu, and D. Inzé. 1991. Transient occurence of extrachromosomal DNA of an *Arabidopsis* transposon-like element. *Proc. Natl. Acad. Sci.* **88:** 3618–3622.

Persinger, S.M. and C.D. Town. 1991. Isolation and characterization of hormone autonomous tumours of *Arabidopsis thaliana. J. Exp. Bot.* **42:** 1363–1370.

Pickett, F.B., A.K. Wilson, and M. Estelle. 1990. The *aux1* mutation confers both auxin and ethylene resistance. *Plant Physiol.* **94:** 1462–1466.

Polyakova, T.F. 1964. The development of the male and female gametophytes of *Arabidopsis thaliana* (L.) Heynh. *Issled. Genet.* **2:** 125–133.

Power, J.B., S.E. Cummings, and E.C. Cocking. 1970. Fusion of isolated plant protoplasts. *Nature* **225:** 1016–1018.

Pruitt, R.E. and E.M. Meyerowitz. 1986. Characterization of the genome of *Arabidopsis thaliana. J. Mol. Biol.* **187:** 169–183.

Pyke, K.A. and M.R. Leech. 1992. Chloroplast division and expansion is radically altered by nuclear mutations in *Arabidopsis thaliana. Plant Physiol.* **99:** 1005–1008.

Quail, P.H. 1991. Phytochrome: A light-activated molecular switch that regulates plant gene expression. *Annu. Rev. Genet.* **25:** 389–409.

Rapoport, J.A. 1946 Carbonyl compounds and the chemical mechanism of mutation. *Dokl. Akad. Nauk.* **54:** 65–67.

Rédei, G.P. 1962a. Supervital mutants of *Arabidopsis. Genetics* **47:** 443–460.

————. 1962b. Single locus heterosis. *Z. Vererbungsl.* **93:** 164–170.

————. 1963. Somatic instability caused by a cystein-sensitive gene in *Arabidopsis. Science* **139:** 767–769.

————. 1964. Crossing experiments with polyploids. *Arabidopsis Inf. Serv.* **1:** 13.

————. 1965. Non-Mendelian megagametogenesis in *Arabidopsis. Genetics* **51:** 857–872.

————. 1967. Genetic estimate of cellular autarky. *Experientia* **23:** 584.

————. 1970. *Arabidopsis thaliana* (L.) Heynh. A review of genetics and biology. *Bibliogr. Genet.* **20:** 1–150.

————. 1973. Extra-chromosomal mutability determined by a nuclear gene locus in *Arabidopsis. Mutat. Res.* **18:** 149–162.

————. 1974a. The origin of *Hylandra suecica* (Fr.) Löve. In *Abstracts from the International Symposium on* Cruciferae, London, p.20.

————. 1974b. *Arabidopsis* as a genetic tool. *Annu. Rev. Genet* **9:** 111–127.

————. 1974c. Steps in evolution of genetic concepts. *Biol. Zentralbl.* **93:** 385–424.

————. 1975a. *Arabidopsis thaliana.* In *Handbook of genetics* (ed. R.C. King), vol. 2, pp. 151–180. Plenum Press, New York.

————. 1975b. Genetic mechanisms in differentiation and development. In *Genetic manipulations with plant material* (ed. L. Ledoux), pp. 183–209. Plenum Press, New York.

————. 1981. *Arabidopsis* assay for environmental mutagens. In *Short-term bioassays in the analysis of complex environmental mixtures* (ed. M. Waters et al.), vol. 2, pp. 211–231. Plenum Press, New York.

————. 1982a. *Genetics.* Macmillan, New York.

————. 1982b. Dominance versus overdominance and the system of breeding. *Cereal Res. Commun.* **10:** 5–9.

————. 1992 A heuristic glance at the past of *Arabidopsis* genetics. In *Methods in* Arabidopsis *research* (ed. C. Koncz et al.), pp. 1–15. World Scientific, Singapore.

Rédei, G.P. and Y. Hirono. 1964. Linkage studies. *Arabidopsis Inf. Serv.* **1:** 9–10.

Rédei, G.P. and C. Koncz. 1992. Classical mutagenesis. In *Methods in* Arabidopsis *research* (ed. C. Koncz et al.), pp. 16–82. World Scientific, Singapore.

Rédei, G.P. and S.L. Li. 1969. Effects of X-rays and ethylmethane-sulfonate on the chlorophyll b locus in the soma and the thiamine loci in the germline of *Arabidopsis. Genetics* **61:** 453–459.

Rédei, G.P. and S.B. Plurad. 1973. Hereditary structural alterations of plastids induced by a nuclear mutator gene in *Arabidopsis. Protoplasma* **77:** 361–380.

Rédei, G.P., G. Acedo, and G. Gavazzi. 1974. Flower differentiation in *Arabidopsis. Stadler Genet. Symp.* **6:** 135–168.

Rédei, G.P., C. Koncz, and J. Schell. 1988. Transgenic *Arabidopsis. Stadler Genet. Symp.* **18:** 175–200.

Rédei, G.P., H. Weingarten, and L.D. Kiehr. 1976. Has DNA corrected genetically thiaminless mutants of *Arabidopsis?* In *Cell genetics of higher plants* (ed. D. Dudits et al.), pp. 91–94. Akadémiai Kiadó, Budapest, Hungary.

Rédei, G.P., M.M. Rédei, W.R. Lower, and S.S. Sandhu. 1980. Identification of carcinogens by mutagenicity for *Arabidopsis. Mutat. Res.* **74:** 469–475.

Regan, S.M. and B.A. Moffatt. 1990. Cytochemical analysis of pollen development in wild-type *Arabidopsis* and male sterile mutant. *Plant Cell* **2:** 877–879.

Reinholz, E. 1945. "Aulösung von Röntgenmutationen bei *Arabidopsis thaliana* (L.) Heynh., und ihre Bedeutung für die Pflanzenzüchtung und Evolutionstheorie." Ph.D. thesis, University of Frankfurt, Germany.

Rhoades, M.M. 1943. Genic induction of an inherited cytoplasmic difference. *Proc. Natl. Acad. Sci.* **29:** 327–329.

Robertson, D.G. 1978. Characterization of a mutator system in maize. *Mutat. Res.* **51:** 21–28.

Robinson-Beers, K., R.S. Pruitt, and C.S. Gasser. 1992. Ovule development in wild type *Arabidopsis* and two female-sterile mutants. *Plant Cell* **4:** 1237–1249.

Röbbelen, G. 1962a. Wirkungsvergleich zwischen Äthylmethansulfonat und Röntgen-Strahlen in Mutationsversuch mit *Arabidopsis thaliana. Naturwissenschaften* **49:** 65.

————. 1962b. Plastommutationen nach Röntgenbestrahlung von *Arabidopsis thaliana* (L.) Heynh. *Z. Vererbungsl.* **93:** 25–34.

————. 1964. A gene induced plastome mutant. *Arabidopsis Inf. Serv.* **1:** 12–14.

————. 1968. Genbedingte Rotlicht-Empfindlichkeit der Chloroplast Differenzierung bei *Arabidopsis. Planta* **80:** 237–254.

Röbbelen, G. and F.J. Kribben. 1966. Einfahrungen bei der Auslese von Trisomen. *Arabidopsis Inf. Serv.* **3:** 16–17.

Sager, R. and Z. Ramanis. 1970. A genetic map of non-Mendelian genes in *Chlamydomonas. Proc. Natl. Acad. Sci.* **65:** 593–600.

Schiefelbein, J.W. and C.R. Somerville. 1990. Genetic control of root hair development in *Arabidopsis thaliana. Plant Cell* **2:** 235–243.

Schmidt, R. and C. Dean. 1992. Physical mapping of *Arabidopsis thaliana* genome. In *Genome analysis: Strategies for physical mapping* (ed. K.E. Davis and S.M. Tilghman), vol. 4, pp. 71–97, Cold Spring Harbor Laboratory Press, Cold Spring Harbor, New York.

Schmidt, R. and L. Willmitzer. 1989. The maize autonomous element *Activator* (*Ac*) shows a minimal germinal excision frequency of 0.2-0.5% in transgenic *Arabidopsis thaliana* plants. *Mol. Gen. Genet.* **220:** 17–24.

Scholten, H.J. and W.J. Feenstra. 1986. Expression of mutant character of chlorate-resistant mutants of *Arabidopsis thaliana* in cell culture. *J. Plant Physiol.* **123:** 45–54.

Schultz, E.A., F.B. Pickett, and G.W. Haughn. 1991. The *FLO10* gene product regulates the expression domain of homeotic genes *AP3* and *PI* in *Arabidopsis* flowers. *Plant Cell* **3:** 1221–1237.

Schwartz, D. and J. Osterman. 1976. A pollen selection system for alcohol dehydrogenase negative mutants in plants. *Genetics* **83:** 63–65.

Sears, L.M.S. and S. Lee-Chen. 1970. Cytogenetic studies in *Arabidopsis thaliana. Can. J. Genet. Cytol.* **12:** 217–223.

Seki, M., N. Shigemoto, Y. Komeda, J. Imamura, and H. Morikawa. 1991. Transgenic *Arabidopsis thaliana* plants obtained by particle bombardement mediated transformation. *Appl. Microbiol. Biotechnol.* **26:** 228–230.

Shannon, S. and D.R. Meeks-Wagner. 1991. A mutation in the *Arabidopsis TFL1* gene affects inflorescence meristem. *Plant Cell* **3:** 877–892.

Shirley, B.W., S. Hanley, and H.M. Goodman. 1992. Effects of ionizing radiation on a plant genome: Analysis of two *Arabidopsis transparent testa* mutations. *Plant Cell* **4:** 333–347.

Snustad, D.P., N.A. Haas, S.D. Kopczak, and C.D. Silflow. 1992. The small genome of *Arabidopsis* contains at least nine expressed β-tubulin genes. *Plant Cell* **4:** 549–556.

Sô, M. 1921. On the inheritance of variegation in barley. *Jpn. J. Genet.* **1:** 21–36.

Somerville, C.R. 1986. Analysis of photosynthesis with mutants of higher plants and algae. *Annu. Rev. Plant Physiol.* **37:** 467–507.

———. 1989. *Arabidopsis* blooms. *Plant Cell* **1:** 1131–1135.

Somerville, C.R. and J. Browse. 1991. Plant lipids: Mutants, metabolism and membranes. *Science* **252:** 80–87.

Somerville, C.R. and W.L. Ogren. 1982. Isolation of photorespiratory mutants of *Arabidopsis*. In *Methods in chloroplast molecular biology* (ed. M. Edelman et al.), pp. 129–139. Elsevier, Amsterdam, The Netherlands.

Sparrow, A.H., H.J. Price, and A. Underbrink. 1972. A survey of DNA content per cell and per chromosome of prokaryotic and eukaryotic organisms: Some evolutionary considerations. *Brookhaven Symp. Biol.* **23:** 451–494.

Sree Ramulu, K. and J. Sybenga. 1985. Genetic background damage accompanying reciprocal translocations induced by X-rays and fusion neutrons in *Arabidopsis* and *Secale. Mutat. Res.* **149:** 421–430.

Stadler, L.J. 1944. The effect of X-rays upon dominant mutation in maize. *Proc. Natl. Acad. Sci.* **30:** 123–128.

Stam, P. 1993. Construction of integrated genetic linkage maps by means of a new computer package: JoinMap. *Plant J.* **3:** 739–744.

Steinitz-Sears, L.M. 1962. Chromosome studies in *Arabidopsis. Am. J. Bot.* **49:** 633–634.

———. 1963. Chromosome studies with *Arabidopsis. Genetics* **48:** 483–490.

Stern, C. 1936. Somatic crossing over and segregation in *Drosophila melanogaster. Genetics* **21:** 625–730.

Stevens, W.L. 1939. Tables of recombination fraction estimated from the product ratio. *J. Genet.* **39:** 171–180.

Strasburger, E. 1884. *Neue Untersuchungen über den Befruchtungsvorgang bei den Phanerogamen als Grundlage für eine Theorie der Zeugung.* Jena, Germany.

———. 1904. Über Reduktionsteilung. *Sitzungsber. König. Preuss Akad. Wiss.* **18:** 1–28.

Stubbe, H. 1965. *History of genetics from the pre-historic times to the rediscovery of Mendel's laws.* Waters, Cambridge, Massachusetts.

Sturtevant, A.H. 1971. On the choice of material for genetical studies. *Stadler Genet. Symp.* **1:** 51–57.

Su, W. and S.H. Howell. 1992. A similar genetic locus, *ckr1* defines *Arabidopsis* mutants in which root growth is resistant to low concentrations of cytokinin. *Plant Physiol.* **99:** 1569–1574.

Sun, T.-P., H.M. Goodman, and F.M. Ausubel. 1992. Cloning the *Arabidopsis GA1* locus by genomic subtraction. *Plant Cell* **4:** 119–128.

Talón, M., M. Koornneef, and J.A.D. Zeevaart. 1990. Endogeneous gibberellins in *Arabidopsis thaliana* and possible steps blocked in the biosynthetic pathways of the semidwarf *ga4* and *ga5* mutants. *Proc. Natl. Acad. Sci.* **87:** 7983–7987.

Teeri, T.H., L. Herrera-Estrella, A. Depicker, M. Van Montagu, and E.T. Palva. 1986. Identification of plant promoters *in situ* by T-DNA mediated transcriptional fusions to the *nptII* gene. *EMBO J.* **5:** 1755–1760.

Tsay, Y.-F., M.J. Frank, T. Page, C. Dean, and N.H. Crawford. 1993. Identification of a mobile endogeneous transposon in *Arabidopsis thaliana. Science* **260:** 342–344.

Turesson, G. 1922. The genotypical response of plant species to the habitat. *Hereditas* **3:** 211–250.

Valvekens, D., M. Van Montagu, and M. Van Lijsebettens. 1988. *Agrobacterium tumefaciens*-mediated transformation of *Arabidopsis* root explants by using kanamycin selection. *Proc. Natl. Acad. Sci.* **85:** 5536–5540.

Vandendries, R. 1909. Contribution à l'histoire du dévelopment des crucifères. *La Cellule* **25:** 415–458.

Van der Veen, J.H. and P. Wirtz. 1968. EMS induced male sterility in *Arabidopsis thaliana* (L.) Heynh.: A model selection experiment. *Euphytica* **17:** 371–377.

Van Larebeke, N., G. Engler, M. Holsters, S. Van den Elsacker, I. Zaenen, R.A. Schilperoort, and J. Schell. 1974. Large plasmid in *Agrobacterium tumefaciens* essential for crown-gall inducing ability. *Nature* **252:** 169–170.

Van Schaik, N.W. and R.A. Brink. 1959. Transposition of modulator, component of the variegated pericarp allele of maize. *Genetics* **44:** 725–738.

Van Sluys, M.A., J. Tempé, and N. Fedoroff. 1987. Studies on the introduction and mobility of the maize activator element in *Arabidopsis thaliana* and *Daucus carota. EMBO J.* **6:** 3881–3889.

Velten, J., L. Velten, R. Hain, and J. Schell. 1984. Isolation of a dual promoter fragment from the Ti plasmid of *Agrobacterium tumefaciens. EMBO J.* **12:** 2723–2730.

Voytas, D.M. and F.M. Ausubel. 1988. A *copia*-like transposable element family in *Arabidopsis thaliana. Nature* **336:** 242–244.

Voytas, D.M., A. Konieczny, M.P. Cummings, and F.M. Ausubel. 1990. The structure, distribution and evolution of the *Ta1* retrotransposable element family of *Arabidopsis thaliana. Genetics* **126:** 713–721.

Walden, R., C. Koncz, and J. Schell. 1990. The use of gene vectors in plant molecular biology. *Methods Mol. Cell. Biol.* **1:** 175–194.

Weigel, D. and E.M. Meyerowitz. 1993. Activation of floral homeotic genes in

*Arabidopsis. Science* **261:** 1723–1726.

Weinberg, W. 1908. Über den Nachweis der Vererbung bei Menschen. *Jahrb. Ver. Vaterl. Naturk.* **64:** 369–382.

Yanofsky, M.F., H. Ma, J.L. Bowman, G.N. Drews, K.A. Feldmann, and E.M. Meyerowitz. 1990. The protein encoded by the *Arabidopsis* homeotic gene *agamous* resembles transcription factors. *Nature* **346:** 35–39.

Yen, Y. and P.J. Green. 1991. Isolation and properties of the major ribonucleases of *Arabidopsis thaliana. Plant Physiol.* **97:** 1487–1493.

Zambryski, P., H. Joos, C. Genetello, J. Leemans, M. Van Montagu, and J. Schell. 1983. Ti plasmid vector for the introduction of DNA into plant cells without alteration of their normal regeneration capacity. *EMBO J.* **2:** 2143–2150.

Zhang, H. and C.R. Somerville. 1987. Transfer of the maize transposable element *Mu1* into *Arabidopsis thaliana. Plant Sci.* **48:** 165–173.

# 10

# Seed Development in *Arabidopsis thaliana*

**David W. Meinke**
Department of Botany
Oklahoma State University
Stillwater, Oklahoma 74078

Seed production is perhaps the most impressive feature of the life cycle in *Arabidopsis thaliana*. Anyone who has ventured into an *Arabidopsis* growth room or transported mature plants from one place to another has observed the large number of seeds scattered everywhere. A single plant under optimal conditions can produce over 20,000 seeds within a few weeks. This remarkable efficiency in seed production was one of the features that first attracted scientists to *Arabidopsis* as a model system for genetic analysis (Rédei 1970). However, the view that *Arabidopsis* might serve as a model system for genetic and molecular analysis of seed development (Meinke and Sussex 1979a) was initially greeted with skepticism. With so many crop plants and large seeds to choose from, it seemed unnecessary for plant biologists to divert attention to a laboratory organism with small seeds. The amount of effort required to isolate sufficient material for biochemical assays seemed to discourage even *Arabidopsis* enthusiasts.

Fortunately, there were also several attractive features of seed development in *Arabidopsis* that eventually became apparent to plant biologists. Classical botanists had long recognized that in crucifers, the arrangement of fruits (siliques) in a developmental progression along the length of the stem simplified the identification of seeds at desired stages of development. They also valued small seeds because relatively few sections were required to visualize internal structures. The translucent nature of the seed coat prior to desiccation allowed direct observation of many basic features of embryo development, and the ability to distinguish unfertilized ovules, aborted seeds, and normal seeds at different stages of development, as first described by Müller (1963), made it feasible to isolate and characterize a wide range of mutants defective in seed development. The establishment of sensitive biochemical assays and molecular techniques applicable to small samples also alleviated concerns about the feasibility of analyzing gene expression during embryo development in *Arabidopsis*. The result has been a gradual realization that *Arabidopsis* can serve as a model system not only for plant molecular genetics, but also for the analysis of seed development.

The purpose of this review is to provide a practical guidebook and detailed reference to seed development in both mutant and wild-type plants. The approach will be to (1) describe basic features of normal seed development; (2) summarize descriptive and experimental studies of this critical developmental pathway; and (3) review the spectrum of mutants identified with defects in seed development. The genetic analysis of embryo development in higher plants including *Arabidopsis* has recently been reviewed elsewhere (Meinke 1986, 1991a,b, 1995; Castle and Meinke 1993). Additional information on seed development can be obtained from reviews on experimental plant embryogenesis (Maheshwari 1950; Raghavan 1976, 1986; Johri 1984; Williams and Maheswaran 1986; Goldberg et al. 1989; West and Harada 1993) and plant development (Steeves and Sussex 1989; Lyndon 1990). The generation of body pattern during embryo development in *Arabidopsis* is discussed by Gerd Jürgens (this volume). The reader is also referred to a special issue of *The Plant Cell* (October, 1993) devoted to plant reproduction.

## OVERVIEW OF SEED DEVELOPMENT IN *ARABIDOPSIS*

### General Features of Siliques

The seeds of *Arabidopsis* and many other crucifers are produced in fruits known as siliques. Any discussion of seed development must therefore include a consideration of silique development. Each silique contains two carpels and a central septum (known as a replum in the Cruciferae) that separates two long rows of seeds (Fig. 1). Seeds are attached to the central replum through a funiculus that guides pollen-tube entry during fertilization and provides a pipeline for nutrients from the maternal plant throughout embryo development. The funiculus attaches the chalazal end of the seed to either the top or bottom rail of the replum. Seeds are usually oriented with the micropylar end facing the tip (stigma surface) of the silique. A typical silique produced by plants grown under optimal conditions contains 40–60 seeds. The number of viable seeds present at maturity is determined by the number of ovules produced, the percentage of ovules fertilized, and the frequency of spontaneous seed abortion. The number of ovules produced by a single ovary (silique before fertilization) is usually determined by plant vigor. Plants with large rosettes typically produce siliques with approximately 60 ovules. Most siliques contain at least a few unfertilized ovules following self-pollination. Incomplete

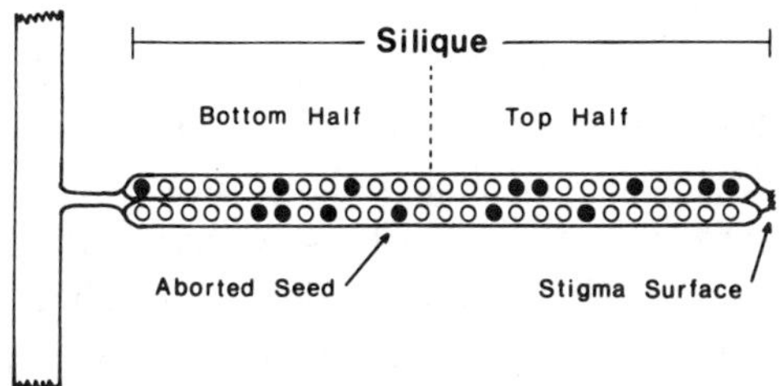



*Figure 1* Silique of *A. thaliana* produced by a plant heterozygous for a recessive embryo-defective mutation. Approximately 25% of the seeds produced following self-pollination of this plant are homozygous for the mutation. Mutant seeds shown here are distributed randomly along the length of the silique. Some mutations disrupt both seed development and pollen-tube growth, resulting in a non-random distribution of aborted seeds within the silique (Meinke 1982, 1985). (Reprinted, with permission, from Meinke 1991b.)

transfer of pollen to the stigma surface is usually responsible for the presence of these unfertilized ovules. Although the distribution of unfertilized ovules within a silique often appears random, clusters of unfertilized ovules are more often found at the tip. Failure of stamens to elongate sufficiently during flowering is most often responsible for the complete absence of fertilized ovules in some wild-type siliques. The frequency of spontaneous seed abortion is influenced by growth conditions, but it can be as low as 0.5% of fertilized ovules scored at maturity. Seed abortion is particularly amenable to analysis in *Arabidopsis* because unfertilized ovules remain white as the silique matures, whereas seeds arrested at early stages of development turn brown.

Reproductive development from fertilization to seed desiccation is usually completed in 2 weeks when plants are grown at 23°C in 16 hour/8 hour light/dark cycles. The developmental stage of an immature silique can be estimated by noting its position along the length of the stem. This is a particularly attractive feature for developmental studies because it allows the construction of a complete developmental progression from a single stem. Young siliques are located at the tip of each stem, whereas old siliques are located at the base. The precise location of siliques at desired stages of development is determined by the number of new flowers produced each day. Wild-type plants grown in my laboratory typically produce 2–3 new flowers per day. Thus, a single stem often contains 40 siliques at slightly different stages of development. Seventy or more siliques may be produced per stem before the inflorescence terminates. Axillary branches typically produce fewer siliques than the main stem. Plants grown under optimal conditions can nevertheless produce as many as 500–600 siliques containing a total of 20,000 or more seeds.

The length of individual siliques varies with both developmental age and number of seeds (Meinke and Sussex 1979a). Siliques begin to elongate from the center of a flower within 24 hours after fertilization and reach a mature length 3–4 days later, at the globular to heart stages of embryo development. Floral organs (petals, sepals, stamens) usually become detached during this period. Siliques with only unfertilized ovules remain attached to the stem but fail to extend beyond 3.5 mm in length. Male sterility and unsuccessful genetic crosses that fail to produce any seeds can therefore be identified by the presence of extremely short siliques. The relationship between seed set (number of seeds per silique) and silique length at maturity has been described by Meinke and Sussex (1979a). In the Columbia ecotype grown extensively in my laboratory, siliques with 60–70 seeds reach a maximal length of 14–16 mm at maturity. Siliques with fewer seeds are considerably shorter. Siliques remain green until late in development, when they turn yellow and then brown. Mature siliques exhibit varying degrees of dehiscence. The valves may either remain closed following desiccation, separate slightly along their junction, or become detached and release the seeds. Mature siliques can usually be removed from a plant without seed loss. Physical removal of valves from a mature silique usually results in seed dispersal. Mature seeds can be removed in order of their location within the silique by first soaking the silique on moistened filter paper.

Several well-known recessive mutations alter the shape of developing siliques. The *clavata* (*clv*) mutants produce broad siliques with additional carpels (Clark et al. 1993; Crone and Lord 1993). Seed distribution and development of the central replum in part of the silique are also disrupted in these mutants. The additional carpels and valves typically do not extend for the entire length of the silique. Similar defects are occasionally observed in wild-type siliques. The *brevipedicellus* (*bp*) mutants produce siliques that point downward toward the soil (Chu et al. 1993). This defect is especially pronounced when the mutation is present in a Landsberg *erecta* (*er*) genetic background. The *erecta* mutation itself also causes the tip of the silique to become more blunt than in a wild-type background. A number of other mutations cause semisterility and consequently reduce the average length of siliques at maturity. Some mutations affecting floral development also disrupt formation of the ovary and thus lead to deformed siliques. Development of wild-type siliques is not usually disrupted by environmental factors, excluding insects and plant pathogens. Anthocyanin begins to accumulate on the surface of wild-type siliques when plants are grown at low temperatures and high illumination. Other changes in pigmentation are rare except during the final stages of senescence.

## General Features of Seeds

The most striking feature of *Arabidopsis* seeds is their small size. Wild-type seeds are only 0.5 mm long at maturity and weigh about 20–30 µg when dried. Although seed size can vary slightly with plant growth conditions and genetic backgrounds, seeds within a single silique are usually the same size. Seed expansion occurs during the first 3–4 days after pollination (Müller 1963; Meinke and Sussex 1979a). Seeds at the heart stage are thus the same size as mature seeds prior to desiccation. Seed size decreases slightly during the final stages of maturation. Identical patterns of seed development are found among the different *Arabidopsis* ecotypes.

Characteristic stages of seed development in *Arabidopsis* are illustrated in Figure 2. Fertilized ovules are white, nearly transparent, and difficult to manipulate because of their small size. Seeds remain white until the late globular stage, when both the seed coat and embryo begin to turn pale green. Embryos beyond a torpedo stage of development are completely green. Seeds at these later stages appear green because the embryo is visible through the translucent seed coat. The developmental age of immature seeds prior to the globular stage can be judged by seed size, silique length, and silique location relative to the inflorescence. The age of seeds between the globular and cotyledon stages can be estimated more accurately by seed color and embryo dissection. Development of seeds within a silique is generally synchronous, although minor differences can be observed around the torpedo stage, when rapid changes occur in embryo size and morphology.

Seeds are complex structures composed of cells representing different stages of the life cycle. Immature seeds have a small amount of haploid tissue derived from the female gametophyte. Mutations that block the functions of these haploid cells during early stages of embryogenesis may prevent normal development of the seed and should be inherited as recessive female gametophytic factors. The most likely targets for these mutations are the antipodal and synergid cells within the megagametophyte. In contrast, the seed coat is maternal tissue derived from the integument of the ovule. Mutations that block pigment accumulation in the seed coat are therefore maternal in effect because it is the genotype of the maternal plant and not that of the embryo or endosperm that determines the phenotype. The maternal nature of the seed coat must also be considered when analyzing many embryo-defective mutants, because the homozygous mutant embryo is surrounded by a heterozygous seed coat. A mutation that blocks the production of a diffusible substance present in both the embryo and seed coat may therefore escape detection if the mutant embryo is rescued by the heterozygous seed coat. The shape of

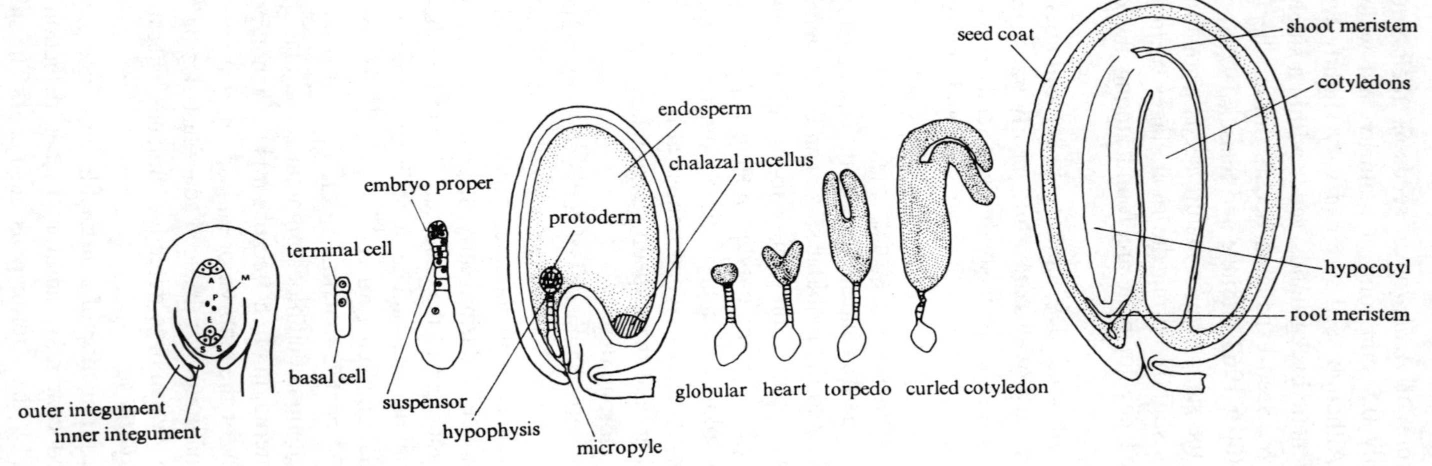



*Figure 2* Overview of embryogenesis and seed development in *Arabidopsis*. The ovule contains a megagametophyte (M) composed of three antipodals (A), two polar nuclei (P), two synergids (S), and the egg (E). The funiculus attaches the ovule to the central replum of the silique. The micropyle is located between the inner integuments, close to the synergids. Double fertilization leads to formation of a diploid zygote and triploid endosperm tissue. The suspensor connects the embryo proper (stippled) to maternal tissues (seed coat; integuments). (Reprinted, with permission, from Castle and Meinke 1993a.)

*Arabidopsis* seeds at maturity is also determined in part by the maternal genotype, as demonstrated by the recent identification of a mutant defective in development of the integument (Leon-Kloosterziel et al. 1994).

The seed coat (testa) of wild-type *Arabidopsis* seeds contains a brown pigment derived from the flavonoid biosynthetic pathway. Mutations in eleven different genes disrupt accumulation of this pigment (Koornneef 1990). Plants homozygous for these recessive *transparent testa* (*tt*) mutations produce mature seeds that lack pigmentation in the seed coat and are consequently pale yellow rather than brown. The molecular basis of this defect in pigmentation has recently been established for several *transparent testa* mutants. The *TT3* locus codes for chalcone isomerase, *TT4* corresponds to chalcone synthase, and *TT5* produces dihydro-flavonol 4-reductase (Feinbaum and Ausubel 1988; Shirley et al. 1992). Another locus known as *transparent testa glabra* (*ttg*) is particularly intriguing because it controls trichome formation in addition to seed pigmentation. It now appears that this locus may encode a regulatory factor related to the R gene of maize because both the trichome and pigmentation defects in *ttg/ttg* plants are corrected by constitutive expression of a wild-type R allele introduced through transformation (Lloyd et al. 1992).

The zygote in *Arabidopsis* divides to form an embryo composed of two parts, the embryo proper and the suspensor. Characteristic stages of embryogenesis are illustrated in Figure 2. The suspensor in angiosperms functions during early stages of development as a pipeline for nutrient transport from maternal tissues to the developing embryo proper and as a source of growth factors required for continued growth of the embryo proper (Yeung and Meinke 1993). The suspensor in *Arabidopsis* is a filamentous structure composed of an enlarged basal cell and a single file of 6–8 additional cells. Formation of the suspensor is complete within the first 3 days of seed development. The suspensor begins to degenerate around the heart stage of development, is crushed by the embryo proper during later stages, and is not present in the mature seed. The embryo proper first becomes visible beneath the dissecting microscope at the globular to heart stages of development. Several critical events occur at this stage. The embryo proper begins to turn green and becomes autotrophic in that it can survive growth in culture on a simple defined medium; symmetry in the embryo proper is altered with the initiation of cotyledons; the formation of specialized cell types in the embryo proper becomes more apparent with the establishment of provascular tissue; and the triploid endosperm tissue starts to become cellular. Morphogenesis in the embryo proper is usually completed at the end of the first week of seed development. A second week is then required for seed maturation and desiccation. Seed development in *Arabidopsis* can therefore be divided into two major phases: (1) an initial phase characterized by active cell division and morphogenesis and (2) a subsequent phase involving the accumulation of storage materials and preparation for dormancy and germination. Additional features of embryo development are discussed by Gerd Jürgens (this volume).

The triploid endosperm tissue is the other product of double fertilization in angiosperms and is another important factor to consider in seed development. The primary endosperm nucleus in *Arabidopsis* divides at least once before division of the zygote and then forms a syncytium of free nuclei distributed along the inner surface of the integument. The endosperm starts to become cellular at the early heart stage, then degenerates during the subsequent cotyledon stages of development and is not present in the mature seed. Storage proteins and lipids essential for early stages of seedling development are stored in the embryonic cotyledons. In contrast to the situation in maize, where gene expression in the developing endosperm has been studied in detail, very little information is available on genes expressed in the endosperm of *Arabidopsis*. The number of endosperm mutants included among embryonic lethals and defectives is also difficult to estimate. Any regulatory functions that the endosperm might play in *Arabidopsis* are likely to occur at early stages of development, because isolated embryos at later stages can be cultured to maturity on a relatively simple nutrient medium.

**Methods of Observation**

The first requirement for studying seed development in *Arabidopsis* is a quality dissecting microscope. There is no need to invest in a microscope equipped with transillumination; reflected light is adequate and often preferable. A reversible stage plate with black and white surfaces is critical for different types of observations. Dissections should be performed on a glass slide or coverslip to protect the surface of the plate. The second requirement for studying seed development in *Arabidopsis* is fine-tipped forceps. These may range in price from $10 to $20 per pair, but the investment is essential. My laboratory routinely uses Dumont #4 and #5 forceps for the most delicate work, and #3C forceps for removing siliques from the plant and grasping the base of the silique (pedicel) during dissections. Forceps of this type do not usually survive any type of significant fall. Care should therefore be taken to protect their fragile tips. A number of stones and sharpening devices are available, but we have found that bent forceps are almost impossible to repair.

Siliques at all stages of development can be removed from the stem by grasping and severing the pedicel with two pairs of forceps. The base of the pedicel is left attached to the stem to show where siliques have been removed. Fine forceps are then used to split the silique longitudinally along lines of natural dehiscence. One pair of forceps holds the pedicel in place while the other is used to separate the valves along both sides of the central replum, moving from the base of the silique to the tip. With experience, this can be performed without damage to seeds or silique tissue, provided the silique has reached at least a globular stage of embryo development. Siliques at earlier stages require more effort and may be damaged slightly during the dissection process. Splitting of siliques is most readily performed under a dissecting microscope at intermediate magnifications (10x to 20x) where most of the silique can be viewed at once. Higher magnifications are required to dissect seeds and recover embryos at early stages of development.

Wild-type embryos at the torpedo to curled cotyledon stages of development can be readily removed from seeds by puncturing the seed coat with forceps and recovering the extruded embryo. It often helps to nick the seed coat at one end and force out the embryo by applying pressure to the other end. Seeds nearing desiccation are particularly difficult to handle because the seed coat at this stage becomes sticky and closely bound to the embryo. Wild-type globular embryos are extremely difficult to recover because of their small size. Isolated embryos can be preserved for detailed observation by performing seed dissections at the edge of a coverslip placed on the surface of an agar plate. Embryos can then be moved onto the agar surface to prevent desiccation.

Seeds can be prepared in a variety of ways for light microscopy. One simple approach is to squash a large number of immature seeds in a drop of stain beneath a coverslip on a microscope slide and examine the contents for any embryos that might have been expelled. This technique is particularly useful for examining seeds at preglobular stages of development. Although a variety of stains can be employed, we have found a mixture containing safranin, basic fuchsin, and crystal violet (0.25%/0.25%/0.10%) in 50% ethanol to be particularly useful. Nomarski optics is another rapid method used to reconstruct the three-dimensional structure of developing seeds. This technique requires specialized optical attachments to a standard compound microscope, but the results are impressive and informative. Examples of Nomarski pictures of wild-type seeds at four different stages of development are shown in Figure 3. These seeds were viewed with an Olympus BHS microscope equipped with a relatively inexpensive Nomarski attachment. This method is therefore within the budget of most research laboratories with an interest in light microscopy. Seeds vary considerably in the amount of time

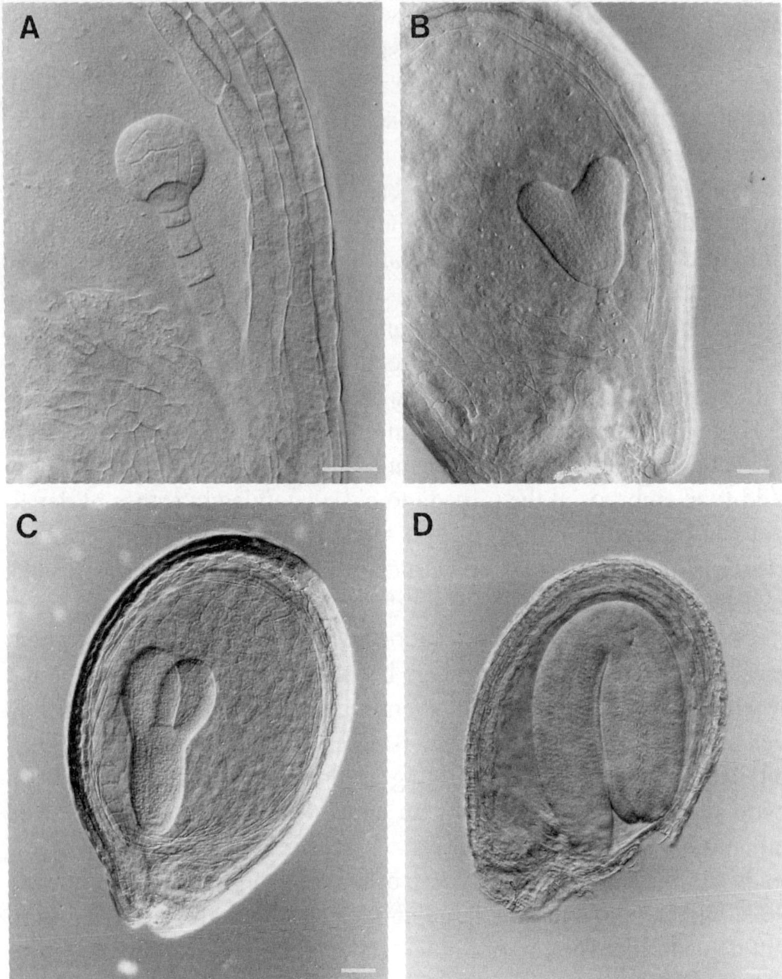

*Figure 3* Light microscopy of intact wild-type seeds viewed with Nomarski optics. Seeds were fixed for 15–30 min in Histochoice tissue fixative (Amresco Company, Solon, Ohio) and cleared for several hours in a modified Hoyer's solution containing 7.5 g of gum arabic, 100 g of chloral hydrate, 5 ml of glycerin, and 30 ml of water. Developmental stages: Early globular (*A*), heart (*B*), torpedo (*C*), and mature cotyledon (*D*). Photographs taken by Brian Schwartz. Bars, 20 μm (*A,B*); 40 μm (*C,D*).

required for clearing, the optimal time for observations, and the quality of the image produced, but with a little experience this method can provide a wealth of information on the internal organization of both mutant and wild-type seeds. One of the major advantages of Nomarski optics is that it allows one to focus on several different optical planes within a seed. It is therefore possible to focus first on integument layers on the top surface, then the endosperm tissue, then different layers of cells in the embryo proper, and finally the integument on the lower surface of the seed. Nomarski optics is most appropriate for immature seeds before the torpedo stage of development. Increased pigmentation in the seed coat and larger numbers of cells in the embryo make observations at later stages more difficult.

Detailed examination of changes in cell structure during embryogenesis requires a more traditional protocol of fixation, dehydration, infiltration, embedding, sectioning, and staining for light or electron microscopy. We have found that infiltration is greatly facilitated by nicking or puncturing the seed coat gently with the broken edge of a razor blade. This allows embedding materials to move through the inner integument layer that begins to accumulate an impermeable material at the globular stage of development. We place seeds to be fixed on a piece of filter paper saturated with fixative and then use forceps to transfer clumps of seeds to vials for further processing. Seeds floating on the surface of the fixative can be submerged by dropping fixative from above. The challenge throughout tissue preparation is to avoid losing track of the small seeds. It is therefore helpful to start with an excess of seeds and use caution to avoid decanting them during subsequent steps.

## Analysis of Seed Development

A variety of descriptive, experimental, and biochemical approaches have been used to study seed development in *Arabidopsis* and related crucifers. A detailed discussion of these studies is beyond the scope of this review. Instead, it is hoped that references cited below will direct the reader to appropriate original publications and reviews. Numerous papers have been published on the morphology of embryogenesis in *A. thaliana* (Misra 1962; Müller 1963; Yakolev and Alimova 1976; Meinke and Sussex 1979a; Marsden and Meinke 1985; Patton and Meinke 1990; Mansfield and Briarty 1990a,b, 1991; Jürgens et al. 1991; Mayer et al. 1991; Webb and Gunning 1991; Dolan et al. 1993) and its relative *Capsella bursa-pastoris* (Souèges 1914, 1919; Schulz and Jensen 1968a,b,c, 1969, 1971, 1973, 1974). The similarity of seed development in these two crucifers is significant because *Capsella* has long been used as a model system for descriptive studies of plant embryogenesis. Detailed anatomical studies have also been published on seed development in *Brassica napus* (Tykarska 1976, 1979, 1987a,b; Höglund et al. 1991) and *Sinapis alba* (Rest and Vaughan 1972; Werker and Vaughan 1974; Bergfeld et al. 1980; Bergfeld and Schopfer 1984, 1986). Development of the ovule prior to fertilization has been examined in several crucifers including

*Arabidopsis* (Webb and Gunning 1990; Mansfield et al. 1991; Robinson-Beers et al. 1992; Haughn et al. 1993), *Capsella* (Schulz and Jensen 1981, 1987), and *Brassica* (Sumner and Van Caeseele 1988). These descriptive studies have provided a solid foundation for genetic and molecular studies of seed development in *Arabidopsis*.

Several experimental approaches have been pursued in the analysis of seed development in *Arabidopsis*. Detailed studies on the growth of young *Capsella* embryos in culture (Monnier 1984) have led to similar studies with *Arabidopsis* (Wu et al. 1992). Plant regeneration from cotyledon explants (Patton and Meinke 1988), transformation of cultured embryos (Sangwan et al. 1991), growth in vitro of immature seeds (Meinke 1979), and occasional somatic embryogenesis (Wu et al. 1992) have also been described. Unfortunately, an efficient method of producing large numbers of somatic embryos and microspore-derived haploid embryos in *Arabidopsis* remains to be established. However, dramatic advances have recently been reported in culturing and manipulating *Brassica* zygotic proembryos at early stages of development (Liu et al. 1993). Developmental abnormalities observed following treatment of immature *Brassica* embryos with inhibitors of auxin transport (Liu et al. 1993) are similar to those seen in both the *pinformed* mutant of *Arabidopsis*, which is defective in polar auxin transport (Okada et al. 1991), and an embryo-defective mutant of *Arabidopsis* (*emb30/gnom*) with fused cotyledons and defects in apical-basal pattern formation early in development (Meinke 1985; Baus et al. 1986; Mayer et al. 1993; Shevell et al. 1994). Experimental manipulation of *Brassica* proembryos in culture may therefore complement ongoing genetic studies of embryo-defective mutants of *Arabidopsis*.

Many abnormalities in seed development have been observed following irradiation of immature siliques (Gerlach-Cruse 1969; Akhundova et al. 1978, 1979). Degeneration of the embryo proper was in some cases associated with abnormal growth of the suspensor, consistent with the view that growth of the suspensor during normal development is inhibited by the embryo proper (Yeung and Meinke 1993). Seed development has also been disrupted by expression of a foreign cytotoxin gene in developing embryos of transgenic *Arabidopsis* plants carrying a diphtheria toxin gene fused to a pea vicilin promoter (Czakó et al. 1992). A similar approach has been used to disrupt seed development in transgenic *Brassica* plants carrying a *Pseudomonas aeruginosa* exotoxin sequence fused to a napin promoter (Koning et al. 1992). Ablation studies of this type will become particularly informative when the cytotoxin genes are fused to promoters that direct expression in small groups of cells. Principles of clonal analysis have also been used to study the normal fates of cells in the shoot apical meristem of mature embryos (Irish and Sussex 1992).

The accumulation of storage materials in developing *Arabidopsis* seeds has recently been examined in some detail. The most extensive studies have dealt with the synthesis of major seed storage proteins (Heath et al. 1986; Krebbers et al. 1988; Pang et al. 1988; Guerche et al. 1990; Van der Klei et al. 1993). These proteins accumulate in mem- brane-bound protein bodies present in the hypocotyl and cotyledons of developing embryos (Patton and Meinke 1990). Several types of muta- tions have been shown to disrupt the accumulation of seed storage proteins, presumably by interfering with embryonic maturation (Heath et al. 1986; Koornneef et al. 1989; Meinke 1992; Nambara et al. 1992; Bäumlein et al. 1994; Keith et al. 1994; Meinke et al. 1994). Transgenic *Arabidopsis* plants have been constructed that produce either modified 2S albumins with higher methionine content (de Clercq et al. 1990) or a chimeric protein in which an internal region of the 2S albumin has been replaced by a neuropeptide (leu-enkephalin) flanked by tryptic cleavage sites (Vandekerckhove et al. 1989). This study was designed to show that plant seeds could be genetically modified to produce bioactive peptides with important applications to the pharmaceutical industry. Transgenic plants have also been used to study the expression of legume storage- protein genes in heterologous systems (Bäumlein et al. 1991a,b). The use of promoter trapping to identify genes expressed during seed develop- ment in *Arabidopsis* has been described previously (Devic et al. 1992; Lindsey and Topping 1993) and offers a promising alternative to muta- tional analysis. Cellular details of storage-protein synthesis in *Brassica* have been examined with immunolocalization (Höglund et al. 1992) and in situ hybridization (Fernandez et al. 1991). Other seed proteins ana- lyzed in some detail include lipoxygenase (Melan et al. 1993), myrosinase (Thangstad et al. 1990; Höglund et al. 1991; Xue et al. 1992), acyl carrier protein (Hlousek-Radojcic et al. 1992), oleosins (Batchelder et al. 1991; Van Rooijen et al. 1992), and late embryogenesis abundant (lea) proteins (Gilmour et al. 1992). Many studies have been published on the accumulation of seed lipids (Browse and Somerville 1991; Post- Beittenmiller et al. 1992). Several mutants of *Arabidopsis* with altered seed lipid composition have recently been identified (James and Dooner 1990, 1991; Lemieux et al. 1990; Arondel et al. 1992).

## MUTANTS DEFECTIVE IN SEED DEVELOPMENT

### General Classes of Mutants

Several reviews on embryo-defective mutants of *Arabidopsis* have recently been published (Meinke 1991a,b,c, 1994, 1995; Castle and

Meinke 1993; Lindsey and Topping 1993; Weigel 1993; West and Harada 1993; Yeung and Meinke 1993). This chapter does not attempt to duplicate those reviews. Instead, it briefly summarizes what is known about the full spectrum of mutants identified in *Arabidopsis* with defects in seed development. One issue that must be addressed concerns nomenclature of mutants. Many different names and classification systems have been used to describe the wide variety of mutants available. These include lethals, defectives, pattern mutants, pigment mutants, endosperm mutants, regulatory mutants, housekeeping mutants, homoeotic mutants, female-sterile mutants, auxotrophic mutants, and mutants defective in various aspects of seed maturation, shoot development, root development, and hormone response. This diversity of mutant types is not surprising in light of the large number of genes expressed at this stage of the life cycle, the number of different laboratories working on seed development, and the wide range of developmental and metabolic defects that are first detected during embryogenesis. For example, mutations that alter plant pigmentation are frequently identified by screening at the seedling stage and are usually thought to alter only vegetative development. However, these mutations can also be viewed as embryonic defectives because they disrupt the accumulation of pigments during embryogenesis and can be identified by screening siliques for the presence of white or pale green seeds. Similarly, mutants chosen for study in one laboratory because they exhibit defects in root or shoot development may be classified as embryonic defectives in another laboratory because development of mutant seeds is also disrupted.

The broadest definition of mutants defective in seed development might therefore include the following categories: (1) mutants with defects limited primarily to pigmentation; (2) mutants with defects limited primarily to the accumulation of storage materials (lipids, proteins, and carbohydrates); (3) mutants defective primarily in seed maturation and the transition from embryo development to germination; (4) mutants with defects in plant metabolism and cell function that indirectly disrupt or terminate morphogenesis; and (5) mutants with fundamental defects in plant morphogenesis. Even this classification system has two major limitations. Some mutants may qualify for more than one category and other mutants may not be readily classified on the basis of initial observations. All of the embryonic mutants identified in my laboratory were initially referred to as lethals, following the tradition of Müller (1963), despite the fact that mutant seeds often germinated in culture to produce defective seedlings. This led to the false impression in the literature that all of these mutants were simply defective in basic housekeeping functions unrelated to the regulation of embryo development. Our initial collection of lethals (Meinke and Sussex 1979b; Meinke 1985) actually included several albinos, a number of mutants that produced seeds arrested at early stages of development, several embryonic defectives that produced embryos capable of continued growth in culture (Franzmann et al. 1989), and a particularly interesting mutant with fused cotyledons (112A-2A; *emb30*) that formed rootless plants in culture (Baus et al. 1986) and was found to be allelic to the *gnom* pattern mutant analyzed extensively in the Jürgens laboratory (Mayer et al. 1991, 1993). We have now isolated a much wider range of mutants, revised our system of nomenclature, and called all of the mutants in our collection embryo defectives (*emb* mutants) in order to minimize the problem of placing mutants in specific classes before anything is known about the molecular and developmental factors responsible for their phenotypes.

### Identification of Mutants

Embryo-defective mutants can be identified either by screening germinated seedlings for defects in morphology or by screening immature siliques for the presence of defective seeds following self-pollination. Lethal mutations that result in desiccation intolerance and thus interfere with germination are not recovered with the first method. Protocols for both screening methods have been published (Müller 1963; Meinke and Sussex 1979a; Meinke 1985, 1991b; Jürgens et al. 1991). When large numbers of mutagenized families are screened at the seedling stage, it is often impractical to surface sterilize many different samples of seeds in preparation for germination in culture. Instead, seeds are usually germinated under nonsterile conditions on filter paper or agar without supplemental nutrients. Observations are then made within the first several days of germination.

My laboratory has always identified embryo-defective mutants by screening immature siliques for 25% defective seeds following self-pollination. We usually examine siliques that contain phenotypically normal, green seeds at a cotyledon stage of development. Mutant seeds at this stage can be recognized by their distinctive size, color, shape, or embryo morphology. Mutant and wild-type seeds are often indistinguishable at earlier stages of development. Examples of green siliques with normal and aborted seeds are shown in Figure 4A. Mutant and wild-type seeds can also in many cases be distinguished at maturity. We prefer to screen immature siliques because more information can be obtained on seed color and embryo morphology and because seeds can be easily counted without dispersal. Examples of mature seeds produced by a plant segregating for a *fusca* mutation that allows completion of morphogenesis but results in the accumulation of anthocyanin in cotyledons are shown in Figure 4B. Mutant seeds arrested early in development become deflated at maturity and can be identified by their small size relative to wild-type seeds.

Mutations affecting seed development are frequently observed following seed mutagenesis in *Arabidopsis*, presumably because a large number of genes perform essential functions at this stage of the life cycle. These mutations can be identified by screening chimeric $M_1$ plants produced directly from mutagenized seeds, $M_2$ plants descended from these $M_1$ plants, or heterozygous plants obtained in subsequent generations. Recessive embryo-defective mutants are generally maintained as heterozygotes. Mutant lines are propagated by planting the phenotypically normal seeds produced following self-pollination and finding desired heterozygotes once the first siliques have been produced. Wild-type plants identified through this process are usually discarded.

When analyzing embryo-defective mutants, care must be taken to distinguish between mutant seeds and spontaneous abortants resulting from physiological or environmental stress. These spontaneous abortants may be common in siliques produced by plants grown under certain conditions. Some mutant lines produce defective embryos with distinctive phenotypes that are distinguishable from spontaneous abortants. Aborted seeds that do not match the expected mutant phenotype can then be ignored. In other cases, a few defective seeds may need to be classified as unresolved because they cannot be clearly identified as either mutant seeds or spontaneous abortants. The best solution is to work exclusively with plants that exhibit a low frequency of spontaneous seed abortion. Wild-type plants growing in the same pot can be used to estimate abortion rates in the absence of an embryonic mutation.

### Diversity of Mutants Available

Several large collections of mutants defective in seed development are currently being analyzed. Table 1 presents an overview of the 250 mutants maintained in my laboratory at Oklahoma State University. Detailed information on each mutant line is presented in Appendix 1. Seeds for many of these lines are available through the *Arabidopsis* Biological Resource Center (ABRC) at Ohio State University. Other mutant collections have been established by Gerd Jürgens at the University of Munich (Jürgens et al. 1991; Mayer et al. 1991), Robert Goldberg and colleagues at the University of California (West et al. 1993), Ben Scheres at the University of Utrecht (Scheres et al. 1994), and Michel Delseny at the University of Perpignan in collaboration with Michel

*Figure 4* Embryo-defective mutants of *Arabidopsis*: Examples of mutant seed phenotypes. (*A*) Heterozygous siliques with normal (green) and mutant (pale) seeds prior to desiccation. The mutant seeds in this line contain a globular embryo. (*B*) Mature seeds produced by a plant heterozygous for a *fusca* mutation. The dark pigmentation of mutant seeds results from anthocyanin in the cotyledons. Seeds are ~0.5 mm in length.

*Figure 5* Examples of wild-type (*A,B*) and mutant (*C–Y*) embryos produced by plants segregating for mutations affecting seed development. Embryos were removed from seeds before desiccation, placed on agar plates, and photographed at the same magnification under a dissecting microscope. Bar, 200 μm. Included are normal embryos at the torpedo (*A*) and cotyledon (*B*) stages, and examples of mutant embryos with abnormal suspensors (*D–E*), a reduced hypocotyl (*F–G*), and reduced, single, duplicated, or distorted cotyledons (*H–W*). Examples of *leafy cotyledon* (*X*) and *fusca* (*Y*) mutant embryos are also included.

*Table 1* Overview of embryo-defective mutants analyzed in Meinke laboratory

| | Number of embryonic mutants identified | | | |
| | method of seed mutagenesis | | | |
| Mutant class[a] | EMS | X-ray | T-DNA[b] | Total |
|---|---|---|---|---|
| Preglobular | 8 | 3 | 49 | 60 |
| Globular | 11 | 5 | 63 | 79 |
| Transition | 9 | 3 | 26 | 38 |
| Cotyledon | 23 | 9 | 24 | 56 |
| Fusca | 0 | 0 | 6 | 6 |
| Other | 1 | 0 | 10 | 11 |
| Total | 52 | 20 | 178 | 250 |

[a]Based on size and shape of mutant embryos at maturity.
[b]The tagging status of these mutants is shown in Table 2.

Caboche at INRA in Versailles. Other laboratories are likely to begin screening for additional mutants in the future.

Although most of the mutants analyzed in my laboratory appear to define different genes, a number of duplicate alleles have been identified. The most efficient approach to finding alleles when large numbers of target genes are involved is to first map the chromosomal locations of mutant genes and then perform complementation tests only when two genes appear to be closely linked. The alternative approach of crossing mutants with similar phenotypes before any mapping data are obtained requires considerable effort and assumes that different alleles will have the same effect on seed development, which may not be true in many cases. My laboratory began to map embryo-defective mutations several years ago (Patton et al. 1991) and has now obtained linkage data for over 160 mutants with various phenotypes (Franzmann et al. 1994). One hundred of these mutant genes have been placed on the genetic map. Embryo defectives are therefore by far the most common class of visible marker on the genetic map of *Arabidopsis*. Seventeen examples of duplicate alleles have now been uncovered through our mapping studies. The probability of identifying new alleles at other loci should increase dramatically as the map becomes more saturated. Although the precise number of target genes essential for embryogenesis remains to be determined, we estimate, based on the frequency of embryo defectives among transgenic lines and the number of duplicate alleles identified to date, that approximately 500 genes can readily mutate in *Arabidopsis* to give a recessive embryo-defective phenotype, excluding albino and pale green mutants. This contrasts with estimates based on molecular hybridization studies that many thousands of genes are expressed during plant embryogenesis

(Goldberg et al. 1989). We therefore believe that many genes expressed during seed development in *Arabidopsis* have redundant functions that can be eliminated without resulting in a mutant phenotype.

## T-DNA Insertional Mutants

Chromosome walking is not presently a feasible approach to gene isolation when large numbers of genes are involved. T-DNA insertional mutagenesis offers a promising alternative to gene isolation and molecular analysis of gene expression during seed development in *Arabidopsis* (Errampalli et al. 1991). My laboratory has recently completed a large screen of 5000 transgenic families produced by Ken Feldmann following *Agrobacterium*-mediated seed transformation (Castle et al. 1993; Meinke 1994). The tagging status of 178 embryo-defective mutants identified from this population is summarized in Table 2. Approximately 35% of the 115 mutants resolved with respect to tagging appear to have a T-DNA insert associated with the mutant locus. The presence of a large number of untagged mutants suggests that the infection and transformation methods employed are also mutagenic. Many of the tagged mutants obtained from this collection contain complex T-DNA inserts with truncations, duplications, and rearrangements of T-DNA segments. There is also evidence from mapping studies of frequent chromosomal rearrangements associated with the mutant locus (Castle et al. 1993). Plasmid rescue is currently being used to recover plant sequences flanking T-DNA inserts and to isolate corresponding genomic sequences from wild-type plants. A similar approach is being taken by Robert Goldberg and colleagues in the analysis of a complementary group of tagged mutants obtained from other families in the Feldmann collection. It is likely that a significant number of genes with essential functions during embryogenesis in *Arabidopsis* will be isolated and characterized over the next several years. Other genes not represented among existing collections of T-DNA insertional mutants may eventually be tagged with transposable elements. Molecular analysis of these genes should help to establish the relationship between gene function and seed development.

## Diversity of Phenotypes

Abnormalities in seed development are often more difficult to characterize than defects in vegetative development because the structures involved are extremely small and thus require detailed examination with a compound light microscope. All of the mutants isolated in my laboratory have therefore been placed in general categories based on the size and

*Table 2*  Tagging status of embryo-defective mutants identified in transgenic families

| Mutant class[a] | Number of embryonic mutants identified | | | |
|---|---|---|---|---|
| | tagged[b] | not tagged[c] | unresolved[d] | total |
| Preglobular | 9 | 24 | 16 | 49 |
| Globular | 14 | 19 | 30 | 63 |
| Transition | 7 | 12 | 7 | 26 |
| Cotyledon | 6 | 11 | 7 | 24 |
| Fusca | 4 | 1 | 1 | 6 |
| Other | 1 | 7 | 2 | 10 |
| Total | 41 | 74 | 63 | 178 |

[a]Based on size and shape of mutant embryos at maturity.
[b]Kanamycin and nopaline results consistent with tagging.
[c]Kanamycin and nopaline results inconsistent with tagging.
[d]Many of these mutants contain additional inserts.

shape of mutant embryos at maturity as viewed under a dissecting microscope. This system of classification is useful but potentially misleading. For example, defective embryos from cotyledon mutants by definition reach a cotyledon stage, but the initial defect may occur much earlier in development. Arrested embryos from globular mutants do not generally resemble wild-type embryos; they may have a large suspensor, irregular protoderm, altered hypophysis, unusually large number of cells, or inappropriate cell types. A biotin auxotroph (*bio1*) with a defect in a general housekeeping function required for embryogenesis (Schneider et al. 1989), an allele of the *emb30/gnom* pattern mutant described by Mayer et al. (1993), and a homoeotic mutant with leafy cotyledons (Meinke 1992) are three examples of cotyledon mutants with particularly interesting phenotypes not reflected in Table 1. Preglobular mutants are likely to differ widely in patterns of development, but examination of fixed material will be required before consistent abnormalities are revealed. One goal of future research with embryo-defective mutants will be to describe in more detail the nature and origin of developmental abnormalities in large numbers of mutants.

Many different examples of abnormal development have nevertheless already been identified. Several particularly intriguing mutants are described in the chapter by Gerd Jürgens on embryonic pattern formation. Another mutation that interferes with development of the shoot apical meristem during embryogenesis has been described by Barton and Poethig (1993). Examples of mutant embryos with abnormal suspensors, an altered hypocotyl, distorted cotyledons, and modified pigmentation are shown in Figure 5. We have identified mutants with defects in cell wall formation during seed development (*emb101* and *emb173*), mutants with a defective embryo that protrudes through the seed coat late in maturation (*emb71*), and mutants that produce defective embryos with an enlarged shoot apical meristem (*emb152*), highly irregular protoderm (*emb145*), giant suspensor (*emb177/sus2*), altered number of cotyledons (*emb209*; *emb232*), fused cotyledons (*emb30*: *gnom*), extremely reduced cotyledons (*emb163*), leafy cotyledons (*lec1*), cotyledons that accumulate anthocyanin during embryonic maturation (*emb78/fus6*), disoriented structures that resemble a green blimp (*emb22*), roots that protrude through the seed coat (*lec1*), and interesting defects in morphology following germination in culture. One mutant with a unique pattern of inheritance (*emb173*) produces siliques with 50% defective seeds regardless of pollen genotype. Mutant seeds are larger than normal, often contain an embryo arrested at the heart stage, and occasionally germinate to form a plant that appears normal except for the presence of siliques with 100% defective seeds. The mutant endosperm tissue fails to form cell walls and remains instead as a syncytium of free nuclei. We have proposed two alternative models to explain the unusual phenotype and inheritance pattern in this mutant: (1) the gene is expressed in the haploid megagametophyte prior to fertilization, and the mutation is therefore a female gametophytic factor with a delayed effect on seed development or (2) the gene is expressed in the endosperm after fertilization, but a single wild-type allele introduced through the pollen is insufficient to rescue two mutant copies contributed by the polar nuclei in the megagametophyte. It may be possible to distinguish between these models by introducing pollen from a telotrisomic line with two copies of the wild-type allele.

## Fusca *Mutants*

The fusca phenotype as first described by Müller (1963) refers to inappropriate accumulation of anthocyanin in developing cotyledons of mutant embryos. Twelve complementation groups of *fusca* mutants have been identified in *Arabidopsis* (Castle and Meinke 1994; Miséra et al. 1994). Similar mutants have not been described in other plants. Mutant embryos complete morphogenesis but fail to develop into viable plants; most *fusca* alleles result in seedling lethality. Anthocyanin accumulation is not the direct cause of seedling lethality because double mutant seeds lacking anthocyanin (*fus/fus*; *tt/tt*) still exhibit the same defects in seedling development. Isolated roots from some *fusca* seedlings are capable of extended growth in culture (Weiland and Müller 1972). Several *fus* mutants identified following seed transformation have recently been examined in detail (Table 2; Castle and Meinke 1994). One mutant (*emb168*) representing the *fus1* complementation group is identical to the *cop1* photomorphogenic mutant of Deng and Quail (1992). This gene has recently been shown to encode a protein with a novel combination of regulatory domains (Deng et al. 1992). Another *fusca* (*fus2*) is allelic to a de-etiolated mutant (*det1*) analyzed in detail by Chory (Chory 1992; Pepper et al. 1994). A third *fusca* (*emb143*; *fus7*) corresponds to another photomorphogenic mutant (*cop9*) described by Wei and Deng (1992). Several *fusca* mutants have therefore been identified previously as mutants with altered light response pathways without reference to the fusca phenotype. Mosaic *fus1* plants have also been constructed to study questions of cell autonomy and cell function during later stages of development (Miséra 1993). My laboratory has recently cloned another *fusca* gene (*FUS6*) utilizing two tagged alleles (*emb78-1* and *emb78-2*) identified following seed transformation (Castle et al. 1993; Castle and Meinke 1994). Analysis of this mutant supports a model that *fuscas* are not simply defective in photomorphogenesis, but rather exhibit a wide range of defects in signal transduction pathways associated with various environmental and developmental factors. Another *fusca* mutant (*fus3*) exhibits a defective seed phenotype that closely resembles *leafy cotyledon* (Meinke 1992; Bäumlein et al. 1994; Keith et al. 1994; Meinke et al. 1994). *Fusca* mutants are therefore an intriguing class of embryonic defective that seem to be disrupted in important regulatory processes during embryonic maturation.

## Suspensor Mutants

Abnormal growth of the suspensor is a frequent phenomenon in embryo-defective mutants of *Arabidopsis* (Yeung and Meinke 1993). Apparently, the full developmental potential of the suspensor is revealed when inhibitory effects of the embryo proper are indirectly removed through mutation. Several tagged suspensor mutants have recently been identified in *Arabidopsis* following *Agrobacterium*-mediated seed transformation (Castle et al. 1993; Chasan 1993; West et al. 1993; Schwartz et al. 1994). Abnormal suspensors in aborted seeds from some of these *sus* mutants are composed of several hundred cells. Occasionally, cells of the mutant suspensor appear to assume characteristics of the embryo proper (e.g., formation of protein bodies and lipid bodies). The most dramatic example of abnormal suspensors has recently been identified in my laboratory (Vernon and Meinke 1994). This *twin* mutant was originally identified as an embryo defective based on the phenotype of mutant seeds, but it was subsequently found to produce a moderate frequency of twin seedlings following germination. Approximately 9% of the mutant seeds produced by homozygous mutant plants germinated to form twin seedlings in culture. Nomarski optics of developing seeds revealed a variety of abnormalities in the embryo proper, but also clearly demonstrated that the twin was produced from the suspensor of the first embryo. The *twin* mutation therefore results in a surprisingly high frequency of suspensor polyembryony. A number of triplet embryos produced through abnormal growth of the suspensor have also been found in mutant seeds. The secondary embryo appears in some cases to duplicate the pattern of cell orientation normally found in the early embryo proper. These results demonstrate that the suspensor in *Arabidopsis* has the potential not only to resume growth following developmental arrest of the embryo proper, but also to repeat the pattern of early development characteristic of the embryo proper. Furthermore, the orientation of the second embryo relative to the first appears to be variable such that in some cases the twin is oriented in the same direction, whereas in other cases the polarity is reversed, with the hypophysis of the twin facing the hypophysis of the original embryo proper. The *TWIN* gene product may therefore play an important function in maintaining cell identity and repressing embryonic potential in the suspensor.

## Maturation Mutants

Several mutations have been found that disrupt the final stages of embryonic maturation and seed development in *Arabidopsis*. A primary defect in some mutants is ABA response (Koornneef et al. 1989; Finkelstein and Somerville 1990; Meurs et al. 1992; Nambara et al. 1992). Mutations in three different genes (*abi1*, *abi2*, *abi3*) allow seeds to germinate in the presence of abscisic acid. The effects of *abi3* are limited primarily to seed development. The *ABI3* gene of *Arabidopsis* has recently been isolated by positional cloning (Giraudat et al. 1992) and shown to have distinct regions of high sequence homology with the *VP1* locus of maize (McCarty et al. 1991; Hattori et al. 1992). A putative null allele results in the formation of mutant seeds that remain green late in development, lack dormancy and desiccation tolerance, fail to accumulate storage proteins, and germinate in the presence of a GA biosynthetic inhibitor (Nambara et al. 1992). Additional screens are currently in progress to identify other mutations that cause seeds to remain green late in development and thus interfere with seed maturation.

The most severe disruption of seed maturation in higher plants has been observed with the *leafy cotyledon* mutant (*lec1*) of *Arabidopsis* (Meinke 1992; Meinke et al. 1994). Mutant seeds are desiccation intolerant and remain green late in development but also exhibit a variety of developmental abnormalities not found in ABA-insensitive mutants. The cotyledons prior to desiccation are unusually rounded and often accumulate anthocyanin, the hypocotyl is reduced and almost transparent, and mutant seeds are occasionally viviparous with a protruding root visible within the silique. The most striking feature of abnormal development is revealed as mutant embryos are allowed to germinate precociously in culture. The resulting seedlings develop into viable plants that are perfectly normal except for their cotyledons, which have trichomes on their adaxial surface, a characteristic of wild-type leaves. An example is shown in Figure 6. The vascular pattern and internal anatomy of mutant cotyledons is also rather leaflike. This homoeotic mutation therefore transforms cotyledons into structures that resemble leaves. Our current model is that the wild-type gene (*LEC1*) functions to activate a wide range of embryo-specific functions during seed maturation. In the absence of normal gene function, mutant cotyledons revert to a more primitive (leaflike) developmental state. We have recently demonstrated that *lec1* embryos exhibit normal sensitivity to abscisic acid, indicating that ABA is necessary but not sufficient for seed maturation (Meinke et al. 1994). We have also found several other mutants with related phenotypes, one of which appears to be a tagged allele of *lec1*, studied the developmental anatomy mutant embryos in more detail, and analyzed the phenotypes of double mutant constructs (Meinke et al. 1994). Further

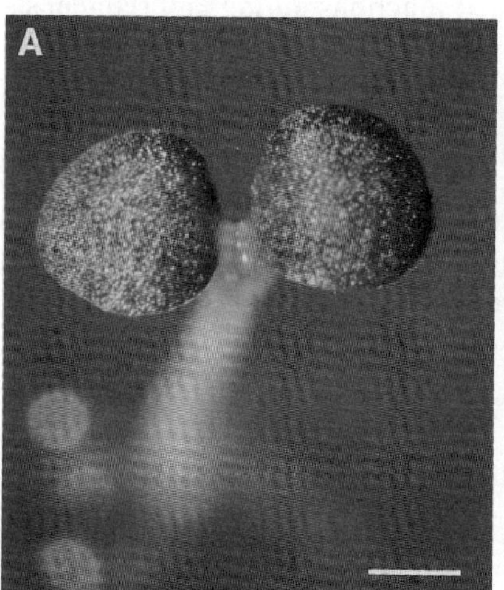

*Figure 6* Cotyledons of a wild-type seedling shortly after germination (*A*) and a *leafy cotyledon* (*lec*) mutant seedling produced following precocious germination of an immature mutant seed (*B*). Numerous trichomes are present on the surface of the mutant cotyledons. Bar, 1 mm.

analysis of these *leafy cotyledon* mutants should help to elucidate important regulatory networks that function during late embryogenesis in *Arabidopsis*.

## Housekeeping Versus Regulatory Mutants

Many embryo-defective mutants are likely to be defective in cellular functions that first become essential during seed development. The best example of a housekeeping defect that results in embryo lethality is a biotin auxotroph (*bio1*) of *Arabidopsis* (Schneider et al. 1989). Arrested embryos from this mutant range from the globular to cotyledon stages, contain reduced levels of biotin, and appear to be defective in the conversion of 7-keto-8-aminopelargonic acid to 7,8-diaminopelargonic acid (Shellhammer and Meinke 1990; Shellhammer 1991). Mutant embryos develop into normal plants in the presence of biotin, desthiobiotin, or 7,8-diaminopelargonic acid. Mutant embryos fail to accumulate lipid bodies in the absence of supplemental biotin, presumably from reduced activity of a biotin-dependent enzyme (acetyl CoA carboxylase) required for an initial step in fatty acid biosynthesis. Lipid bodies can be restored when mutant plants are rescued with biotin. The variable phenotypes of mutant embryos at maturity may result from differences in the amount of biotin provided by the seed coat or from different levels of functional gene product within the embryo itself. Auxotrophic mutations will result in embryo lethality only if the mutant pollen is viable and contributes to fertilization, duplicate genes with related functions during embryogenesis are not expressed, and the maternal plant cannot rescue the defective embryo. The thiamine and tryptophan auxotrophs of *Arabidopsis*, as described elsewhere in this book, were identified at the seedling stage and do not exhibit embryo lethality.

Many other examples of defects in housekeeping functions are likely to be found among existing collections of embryo-defective mutants. Genes with key regulatory functions may in some cases be identified by a distinctive phenotype such as that encountered with the *leafy cotyledon* and *twin* mutants. Mutations in other genes with important developmental functions may not result in such a distinctive phenotype. For example, some globular mutants may be defective in genes that help to regulate the initiation of cotyledons during embryogenesis, whereas others are disrupted in more general housekeeping functions not directly related to cotyledon formation. The *fusca* mutants illustrate that interesting regulatory mutants do not always exhibit dramatic morphological abnormalities. We need to learn more about the normal functions of a large number of essential genes before any conclusions can be drawn about which genes perform important developmental functions. The chances of discovering the normal function of a mutant gene through map-based correlation should increase significantly as more cloned genes with known housekeeping functions are mapped to the same region as *EMB* genes. We may then be able to predict with some certainty what embryo phenotypes should result from specific types of metabolic and regulatory defects.

## CONCLUSIONS AND PERSPECTIVES

The analysis of seed development in *Arabidopsis* has started to attract increasing attention from research laboratories world wide. Large numbers of mutants defective in various aspects of seed development have been generated following chemical and T-DNA insertional mutagenesis. The existence of tagged mutants should greatly facilitate gene isolation and molecular studies of gene expression during plant embryogenesis. Several critical questions need to be addressed in the future: (1) How many genes perform essential functions during plant embryogenesis and what specific functions do these genes perform? (2) How do different parts of the seed (megagametophyte, nucellus, zygote, embryo proper, suspensor, endosperm, seed coat) interact throughout development? (3) How does the chemical and physical environment of the seed change throughout development and how are these changes regulated? and (4) Which features of seed development are unique to *Arabidopsis* or crucifers in general and which can serve as models for other angiosperms? More attention will need to be directed to the structure and function of the endosperm, nucellus, and integument tissues throughout seed development, and to signal transduction pathways that allow seeds to respond appropriately to developmental and environmental signals. The fusion of cell-specific promoters to cytotoxin genes should lead to a wide range of genetic ablation studies designed to study cell interactions during seed development. The establishment of more sensitive analytical techniques should allow detailed chemical descriptions of the seed environment to be completed. Computer-aided reconstructions of sections through mutant and wild-type seeds should provide insights into questions of cell structure and morphology during embryogenesis. *Arabidopsis* should thus play an increasingly central role in experimental and molecular studies of seed development.

## ACKNOWLEDGMENTS

Research on embryo-defective mutants of *Arabidopsis* at Oklahoma State University has been supported by grants from the National Science Foundation, U.S. Department of Agriculture, and S.R. Noble Foundation. A large number of people in my laboratory have contributed to this research during the past 5 years: Deena Errampalli, Linda Castle, Brian Schwartz, Dan Vernon (insertional mutants); David Patton, Linda Franzmann, Elizabeth Yoon (mapping project); Joe Shellhammer (biotin auxotroph); and many qualified undergraduate students. Primary research collaborators have been Ken Feldmann (University of Arizona, Tucson) and Ed Yeung (University of Calgary).

# APPENDIX

## Embryo-defective Mutants Identified by Meinke Laboratory

| Mutant[a] | Mutagen | Tag[b] | Class[c] | Pigmentation[d] Seed | Embryo | % Mutant[e] | Top Half[f] | Linkage[g] | Phenotype of Mutant Embryos |
|---|---|---|---|---|---|---|---|---|---|
| emb1 | EMS | NAP | P | 1 | – | 25.5 | 51.8 | 3 | Very early preglobular |
| emb2 | EMS | NAP | P | 1 | – | 23.9 | 61.5* | 5 | Early preglobular |
| emb5 | EMS | NAP | P | 1 | – | 23.4 | 58.9* | x | Variable preglobular |
| emb6 | EMS | NAP | P | 1 | – | 28.0 | 52.1 | x | Variable preglobular |
| emb7 | EMS | NAP | P | 1 | – | 26.1 | 49.4 | – | Variable preglobular |
| emb8 | EMS | NAP | P | 1 | – | 21.0 | 64.5* | x | Variable preglobular |
| emb9 | EMS | NAP | P | 1 | – | 23.9 | 59.9* | 5 | Late preglobular |
| emb10 | EMS | NAP | G | 1-2 | 1 | 25.8 | 53.3 | 1 | Early globular |
| emb11 | EMS | NAP | G | 1-2 | 1 | 24.1 | 52.9 | x | Early globular |
| emb12 | EMS | NAP | G | 1-2 | 1 | 27.6 | 52.1 | x | Early globular |
| emb13 | EMS | NAP | G | 1-2 | 1 | 23.1 | 52.9 | x | Early globular |
| emb15 | EMS | NAP | G | 1 | 1 | 26.3 | 47.9 | 5 | Globular |
| emb16 | EMS | NAP | G | 1 | 1 | 24.5 | 47.8 | 5 | Globular |
| emb17 | EMS | NAP | G | 1 | 1 | 25.3 | 51.0 | 1 | Globular |
| emb18 | EMS | NAP | G | 1-2 | 1 | 20.3 | 62.8* | 2 | Globular, suspensor |
| emb19 | EMS | NAP | T | 1-2 | 1-2 | 21.5 | 57.7* | 1 | Globular-heart, suspensor |
| emb20-1 | EMS | NAP | T | 1 | 1 | 26.7 | 50.7 | 4 | Globular-heart |
| emb20-2 | EMS | NAP | T | 1 | 1 | 24.3 | 50.7 | [4] | Globular-heart |
| emb20-3 | EMS | NAP | C | 1-2 | 1-2 | 25.0 | 52.0 | 4 | Globular-cotyledon |
| emb21 | EMS | NAP | T | 1 | 1 | 29.8 | 48.1 | x | Globular-heart |
| emb22 | EMS | NAP | O | 3-4 | 3-4 | 18.7 | 62.8* | 1 | Green blimp |
| emb23 | EMS | NAP | C | 1-3 | 1-4 | 25.2 | 65.9* | x | Variable linear |
| emb24 | EMS | NAP | C | 2-3 | 2-3 | 25.8 | 53.0 | 5 | Globular-cotyledon |
| emb25 | EMS | NAP | C | 1 | 1 | 25.6 | 50.2 | 1 | Globular-cotyledon |
| emb27 | EMS | NAP | C | 1 | 1 | 28.7 | 47.8 | 1 | Globular-cotyledon |
| emb28-1 | EMS | NAP | C | 1 | 1 | 25.8 | 49.3 | 4 | Globular-cotyledon |
| emb29 | EMS | NAP | C | 1 | 1 | 26.7 | 51.3 | 3 | Globular-linear cotyledon |
| emb30-1 | EMS | NAP | C | 3-4 | 4 | 26.1 | 47.9 | 1 | Fused cotyledon, rootless |
| emb30-2 | EMS | NAP | C | 3-4 | 4 | 27.1 | 47.6 | 1 | Fused cotyledon (R.Dinkins) |
| emb30-3 | EMS | NAP | C | 3-4 | 4 | 24.9 | 46.1 | [1] | Fused cotyledon, rootless |
| emb31 | EMS | NAP | C | 3-4 | 3-4 | 23.3 | 58.1* | (2) | Reduced/delayed cotyledon |
| emb32 | EMS | NAP | P | 1 | – | 25.0 | | x | Preglobular |

*(Continued on following pages.)*

| Mutant | Mutagen | Tag | Class | Pigmentation Seed | Pigmentation Embryo | % Mutant | Top Half | Linkage | Phenotype of Mutant Embryos |
|---|---|---|---|---|---|---|---|---|---|
| emb34-1 | EMS | NAP | T | 1 | 1 | 26.4 | - | 2 | Globular-heart |
| emb34-2 | XRAY | NAP | G | 1 | 1 | 24.1 | 47.5 | 2 | Globular (emb55) |
| emb35 | EMS | NAP | C | 1 | 1 | 25.2 | - | 4 | Globular-linear cotyledon |
| emb37 | EMS | NAP | (T) | 1-2 | 1-2 | 28.4 | - | x | Variable elongate, suspensor |
| emb38 | EMS | NAP | C | 1 | 1 | 25.9 | - | 2 | Globular-distorted cotyledon |
| emb39 | EMS | NAP | C | 1 | 1-2 | 26.0 | - | 2 | Distorted cotyledon |
| emb40 | EMS | NAP | C | 4 | 4 | 19.9 | - | 5 | Linear cotyledon |
| emb41 | EMS | NAP | C | 1-3 | 1-3 | 27.4 | - | 1 | Variable cotyledon |
| emb42 | EMS | NAP | C | 1-4 | 1-4 | 25.5 | - | 4 | Globular-distorted cotyledon |
| emb43 | EMS | NAP | C | 2-3 | 1-4 | 14.1 | - | 1 | Fat hypocotyl |
| emb44 | EMS | NAP | C | 3 | 3-4 | 24.2 | - | x | Slightly distorted cotyledon |
| emb45 | EMS | NAP | C | 3-4 | 3-4 | 26.9 | - | x | Delayed cotyledon |
| emb47 | EMS | NAP | C | 2-4 | 2-4 | 14.3 | 66.2* | x | Variable fused cotyledon |
| emb48 | ? | NAP | C | 3-4 | 3-4 | Double mutant | | x | Abnormal linear; pale mature |
| emb49 | ? | NAP | T | 2 | 1-2 | 26.0 | 48.0 | 2 | Globular, heart, early linear |
| emb50 | ? | NAP | T | 2 | 1-2 | - | - | x | Globular-elongate-heart |
| emb51 | XRAY | NAP | C | 2 | 2 | 26.0 | 51.4 | 3 | Heart, torpedo |
| emb52 | XRAY | NAP | T | 1 | 1 | 26.2 | 48.7 | 3 | Elongate, blimp, torpedo |
| emb53 | XRAY | NAP | C | 2-4 | 3-4 | 26.7 | 51.2 | 2 | Globular, distorted cotyledon |
| emb54 | XRAY | NAP | P | 1 | - | (24.8) | (48.0) | 1 | Variable preglobular |
| emb56 | XRAY | NAP | G | 1 | 1 | 21.1 | 53.9 | 4 | Globular |
| emb57 | XRAY | NAP | C | 1 | 1 | 17.9 | 54.6 | 2 | Variable preglobular-linear |
| emb58 | XRAY | NAP | P | 1 | - | 23.1 | 51.5 | 2 | Variable preglobular |
| emb59 | XRAY | NAP | C | 2-4 | 2-4 | 24.0 | 56.0 | 5 | Distorted, fused cotyledons |
| emb62 | XRAY | NAP | G | 1 | 1 | 27.6 | 50.8 | 2 | Early globular |
| emb63 | XRAY | NAP | C | 1 | 1 | 26.7 | 45.0 | x | Linear-mature cotyledon |
| emb64 | XRAY | NAP | C | 1 | 1 | 23.9 | 50.3 | 5 | Globular-heart, linear |
| emb65 | XRAY | NAP | C | 1-2 | 1-3 | 14.0 | 84.5* | 3 | Varible globular-cotyledon |
| emb66 | XRAY | NAP | P | 1 | - | 24.9 | 56.2 | 2 | Preglobular |
| emb67 | XRAY | NAP | T | 1 | 1 | 22.7 | 61.0* | 5 | Globuar-heart |
| emb68 | XRAY | NAP | G | 1 | 1 | 19.8 | 53.6 | 5 | Early globular |
| emb69 | XRAY | NAP | T | 1 | 1 | 21.6 | 53.0 | 3 | Globular-heart |
| emb70 | XRAY | NAP | C | 1-2 | 1-3 | 24.4 | 51.5 | 4 | Globular-curled cotyledon |

| Mutant | Mutagen | Tag | Class | Pigmentation Seed | Embryo | % Mutant | Top Half | Linkage | Phenotype of Mutant Embryos |
|---|---|---|---|---|---|---|---|---|---|
| emb71 | EMS | NAP | C | 1-4 | 1-4 | 24.6 | - | 1 | External embryo; distorted |
| emb72 | T-DNA | U | P | 1 | 1 | 25.7 | - | x | Preglobular |
| emb73 | T-DNA | U | P | 1 | (-) | 30.4 | 52.9 | x | Late preglobular |
| emb75 | T-DNA | U | G | 1 | 1 | 21.9 | - | x | Globular |
| emb76-1 | T-DNA | Y | G | 1-2 | 1-2 | 24.4 | 50.5 | 1 | Globular, suspensor (sus1-1) |
| emb76-2 | T-DNA | N | G | 2 | 1 | 25.8 | 49.3 | 1 | Globular, suspensor (sus1-2) |
| emb76-3 | XRAY | NAP | G | 1-3 | 2 | 23.5 | 56.9 | 1 | Globular (emb60; sus1-3) |
| emb77 | T-DNA | N | P | 1 | - | 24.1 | 52.1 | 4 | Late preglobular |
| emb78-1 | T-DNA | Y | F | 4-5 | 4-5 | 23.0 | 50.2 | 3 | Fusca (fus6) |
| emb78-2 | T-DNA | (Y) | F | 4-5 | 4-5 | 24.4 | 53.1 | 3 | Fusca (fus6) |
| emb79 | T-DNA | N | P | 1 | - | 18.6 | 53.2 | 1 | Late preglobular |
| emb80 | T-DNA | N | P | 2 | - | 23.4 | 50.4 | 1 | Late preglobular |
| emb81 | T-DNA | N | T | 1 | 1 | 25.3 | 43.8 | 4 | Early heart |
| emb82 | T-DNA | Y | - | 1 | 1 | 23.1 | 50.9 | x | Albino mature |
| emb83 | T-DNA | (N) | T | 1-2 | 1-2 | 23.6 | 49.7 | 1+4 | Globular-heart |
| emb84 | T-DNA | Y | G | 2 | 1 | 25.6 | 42.1 | 3+4 | Globular, suspensor |
| emb86 | T-DNA | Y | T | 1 | 1 | 25.8 | 50.4 | 5 | Globular-heart |
| emb87 | T-DNA | Y | G | 1 | 1 | 25.1 | 52.5 | 5 | Globular |
| emb88 | T-DNA | Y | G | 1 | 1 | 27.7 | 48.4 | 1 | Globular |
| emb89 | T-DNA | N | C | 1 | 1 | 25.7 | 47.4 | 1 | Globular-mature cotyledon |
| emb90 | T-DNA | N | T | 1 | 1 | 26.5 | 49.1 | 3 | Globular-heart |
| emb91 | T-DNA | N | G | 1 | 1 | 25.4 | 50.3 | (2) | Early globular |
| emb93 | T-DNA | Y | T | 1 | 1 | 23.4 | 49.1 | 2 | Globular-cotyledon |
| emb94 | T-DNA | N | T | 1 | 1 | 24.6 | 47.0 | 2 | Globular-heart |
| emb95 | T-DNA | U | G | 1 | 1 | 23.0 | 56.9 | 3 | Early globular |
| emb98 | T-DNA | U | P | 1 | - | 24.4 | 47.5 | x | Preglobular |
| emb99 | T-DNA | U | C | 1/3 | 3-4 | 24.8 | 54.8 | x | Linear, transparent seed |
| emb101-1 | T-DNA | N | O | 2 | 2-3 | 24.3 | 53.0 | 2 | Large fleshy cotyledons |
| emb101-2 | T-DNA | N | O | 2 | 1-2 | 24.8 | 47.2 | 2 | Large bloated heart |
| emb102 | T-DNA | U | G | 1-2 | 1-2 | 29.0 | 42.3 | x | Variable early globular |
| emb103 | T-DNA | U | P | 1 | - | 24.2 | 52.4 | x | Late preglobular |
| emb104 | T-DNA | Y | C | 1-2 | 1-2 | 23.0 | 54.3 | 4 | Variable cotyledon, monocot |
| emb105 | T-DNA | N | G | 1 | 1 | 24.8 | 49.2 | 1 | Early globular |

| Mutant | Mutagen | Tag | Class | Pigmentation | | % Mutant | Top Half | Linkage | Phenotype of Mutant Embryos |
|---|---|---|---|---|---|---|---|---|---|
| | | | | Seed | Embryo | | | | |
| emb106-1 | T-DNA | N | E | 2 | 1-2 | 25.6 | 50.8 | 4 | Globular, elongate, triangle |
| emb106-2 | T-DNA | N | T | 1-2 | 1 | 24.6 | 51.6 | 4 | Elongate, heart (emb155) |
| emb107 | T-DNA | Y | P | 1 | - | 18.9 | 53.3 | 4 | Preglobular, watery seeds |
| emb108 | T-DNA | N | P | 1 | - | 19.5 | 56.8 | (5) | Preglobular |
| emb109 | T-DNA | N | P | 1 | - | 22.5 | 51.3 | 4 | Late preglobular |
| emb110 | T-DNA | U | G | 1 | 1 | 15.9 | 55.1 | 1 | Globular |
| emb111 | T-DNA | Y | T | 2 | 1 | 24.0 | 50.2 | 2 | Globular, elongate |
| emb113 | T-DNA | U | O | 1-2 | 1-2 | - | - | x | Globular-elongate, suspensor |
| emb115 | T-DNA | Y | G | 2 | 1 | 25.6 | 46.7 | 4 | Large globular |
| emb117 | T-DNA | (N) | G | 1 | 1 | 23.4 | 46.7 | 3 | Globular, suspensor |
| emb118 | T-DNA | N | C | 1 | 1 | 20.5 | 60.5* | 1 | Small heart-torpedo |
| emb119 | T-DNA | N | T | 1 | 1 | 24.2 | 50.2 | x | Globular-heart, triangle |
| emb120 | T-DNA | N | C | 2-3 | 2-3 | 20.5 | 56.3 | 1 | Globular, fat linear |
| emb122 | T-DNA | N | C | 1 | 3-4 | 18.8 | 48.3 | 1+3 | Variable linear, watery |
| emb123 | T-DNA | Y | P | 1 | - | 16.1 | 50.9 | 1+2 | Late preglobular |
| emb124 | T-DNA | U | P | 2-3 | - | 13.9 | 61.8* | x | Variable late preglobular |
| emb125 | T-DNA | N | G | 1 | 1 | 28.0 | 52.9 | 3 | Globular |
| emb126 | T-DNA | U | G | 1 | 1 | 20.1 | 45.3 | 1 | Globular |
| emb127 | T-DNA | N | P | 1 | - | 28.1 | 48.7 | 1 | Late preglobular, watery |
| emb128 | T-DNA | N | P | 1 | - | 25.2 | 57.1* | 1 | Preglobular |
| emb129 | T-DNA | Y | P | 1 | - | 14.8 | 54.0 | 3+5 | Late preglobular |
| emb130 | T-DNA | Y | P | 1 | - | 25.1 | 52.4 | 4 | Early preglobular |
| emb131 | T-DNA | N | T | 2 | 1-2 | 21.3 | 70.0* | 1 | Globular, elongate |
| emb132 | T-DNA | (N) | G | 1 | 1 | 18.6 | 55.1 | 4 | Variable early globular |
| emb133 | T-DNA | N | O | 1/3 | ? | 23.4 | 47.8 | 3 | Preglobular, watery seeds |
| emb134 | T-DNA | U | F | 4-5 | 4-5 | 22.2 | 53.3 | 5 | Fusca (fus8) |
| emb135 | T-DNA | U | ? | 1 | 1 | 25.2 | 45.6 | 2 | Globular-albino mature |
| emb136 | T-DNA | N | P | 1 | - | 21.6 | 57.1* | 1 | Late preglobular |
| emb137 | T-DNA | U | P | 1 | - | 20.5 | 52.4 | 4 | Early preglobular |
| emb139 | T-DNA | N | G | 1 | 1 | 24.5 | 60.7* | 5 | Globular |
| emb140 | T-DNA | Y | G | 1-2 | 1 | 24.5 | 51.2 | 4 | Early globular |
| emb141 | T-DNA | N | P | 1 | - | 24.6 | 49.1 | 5 | Late preglobular |
| emb142 | T-DNA | Y | G | 2 | 1-2 | 27.3 | 47.3 | 1 | Early globular; protoderm |

| Mutant | Mutagen | Tag | Class | Pigmentation | | % Mutant | Top Half | Linkage | Phenotype of Mutant Embryos |
| | | | | Seed | Embryo | | | | |
| --- | --- | --- | --- | --- | --- | --- | --- | --- | --- |
| emb143 | T-DNA | Y | F | 4-5 | 4-5 | 23.3 | 56.2 | 4 | Pale fusca (fus7, cop9) |
| emb144 | T-DNA | N | F | 4-5 | 4-5 | 26.4 | 49.5 | 3 | Fusca (fus9) |
| emb145 | T-DNA | Y | G | 2 | 1-2 | 26.3 | 47.5 | 3 | Globular, irregular protoderm |
| emb146 | T-DNA | Y | P | 1 | - | 23.8 | 60.3* | 2 | Preglobular |
| emb148 | T-DNA | N | P | 1 | - | 24.3 | 58.0* | 4 | Preglobular |
| emb149 | T-DNA | N | T | 2 | 2 | 23.0 | 55.0 | 3 | Globular-heart |
| emb150 | T-DNA | N | (T) | 2-3 | 2-3 | 24.1 | 50.9 | 4 | Distorted torpedo-linear |
| emb151 | T-DNA | Y | T | 2 | 1-2 | 22.7 | 58.3* | 3+5 | Globular, heart |
| emb152 | T-DNA | N | C | 1 | 1 | 25.5 | 52.5 | 2 | Linear-curled, large SAM |
| emb153 | T-DNA | N | G | 1 | 1 | 26.9 | 51.4 | 3 | Globular |
| emb154 | T-DNA | N | T | 1 | 1 | 26.3 | 55.2 | 3 | Large globular-heart |
| emb156-1 | T-DNA | N | G | 1 | 1 | 23.4 | 53.2 | 1 | Globular |
| emb156-2 | EMS | NAP | T | 1 | 1 | 24.3 | - | 1 | Heart; no hypocotyl (emb36) |
| emb158 | T-DNA | N | O | 1 | 1 | 21.8 | 53.6 | 1 | Elongate, suspensor (sus3) |
| emb160 | T-DNA | N | G | 1 | 1 | 22.1 | 42.3 | 4 | Early globular |
| emb161 | T-DNA | N | P | 1 | - | 21.2 | 51.2 | 5 | Preglobular, watery seeds |
| emb162 | T-DNA | N | - | 1 | - | 18.4 | 56.6 | x | Variable preglobular |
| emb163 | T-DNA | N | C | 2-3 | 2-3 | 26.2 | 51.1 | 5 | Linear; reduced cotyledons |
| emb164 | T-DNA | U | P | 1-2 | - | 36.1 | 52.2 | x | Variable preglobular |
| emb165 | T-DNA | U | G | 1-2 | 1 | 25.5 | 47.3 | x | Early globular |
| emb166 | T-DNA | U | G | 1-2 | 1 | 23.5 | 47.4 | 5 | Diffuse globular |
| emb167 | T-DNA | Y | P | 1 | - | 22.1 | 59.2* | 3 | Preglobular |
| emb168 | T-DNA | Y | F | 4-5 | 4-5 | 23.7 | 50.2 | 2 | Dark fusca (fus1, cop1) |
| emb170-1 | T-DNA | N | P | 1 | - | 24.2 | 55.2 | 5 | Preglobular |
| emb171 | T-DNA | N | G | 2 | 1 | 25.9 | 51.1 | 4 | Globular |
| emb172 | T-DNA | (N) | P | 1 | - | 22.7 | 49.8 | 3 | Preglobular |
| emb173 | T-DNA | N | O | 1/3 | 2-3 | 50.6 | 47.0 | 1 | Defective endosperm |
| emb174 | T-DNA | N | G | 1 | 1 | 23.1 | 50.3 | 4 | Variable globular |
| emb175 | T-DNA | Y | G | 1 | 1 | 27.5 | 46.9 | 5 | Globular |
| emb176 | T-DNA | N | P | 1 | - | 21.1 | 64.6* | 1 | Very early preglobular |
| emb177-1 | T-DNA | Y | T | 2 | 1-2 | 25.9 | 49.8 | 1 | Transition, suspensor (sus2-1) |
| emb177-2 | EMS | NAP | G | 1-2 | 1-2 | 23.1 | 40.6 | 1 | Globular (emb14; sus2-2) |
| emb177-3 | EMS | NAP | G | 2 | 1-2 | 24.8 | - | 1 | Globular, (emb33; sus2-3) |

| Mutant | Mutagen | Tag | Class | Pigmentation | | % Mutant | Top Half | Linkage | Phenotype of Mutant Embryos |
|---|---|---|---|---|---|---|---|---|---|
| | | | | Seed | Embryo | | | | |
| emb178 | T-DNA | U | G | 1 | 1 | 23.1 | 51.4 | 4 | Globular |
| emb179 | T-DNA | Y | C | 2 | 2-3 | 19.3 | 64.1* | 1 | Variable globular-linear |
| emb180 | T-DNA | (N) | P | 1 | - | 25.2 | 53.6 | 5 | Preglobular |
| emb181 | T-DNA | - | P | 1 | - | - | - | 4 | Early preglobular |
| emb201 | T-DNA | Y | C | 2 | 1-2 | 22.0 | 51.0 | 3 | Linear, reduced cotyledons |
| emb202 | T-DNA | N | G | 1-2 | 1 | 23.7 | - | - | Small globular |
| emb203 | T-DNA | U | P | 1 | - | 23.6 | - | x | Very early preglobular |
| emb204 | T-DNA | U | G | 1-2 | - | 28.2 | - | x | Early globular |
| emb205 | T-DNA | U | G | 1 | 1 | 28.0 | - | x | Globular |
| emb206 | T-DNA | U | G | 1 | 1 | 19.9 | - | x | Early globular |
| emb207 | T-DNA | U | P | 1 | - | 26.3 | - | x | Preglobular |
| emb208 | T-DNA | N | P | 1 | - | 20.9 | - | - | Early preglobular |
| emb209 | T-DNA | N | C | 2 | 2-3 | 23.9 | 49.3 | 5 | Globular-linear-monocot-tricot |
| emb210 | T-DNA | Y | C | 2 | 1-3 | 23.8 | 52.9 | 4 | Globular, linear, monocot |
| emb211 | T-DNA | U | P | 1 | - | 23.3 | 54.1 | x | Very early preglobular |
| emb213 | T-DNA | N | T | 1 | 1 | 22.2 | - | - | Globular, small elongate |
| emb215 | T-DNA | Y | G | 1 | 1 | 33.0 | - | - | Variable preglobular, globular |
| emb217 | T-DNA | N | G | 1 | 1 | 24.8 | - | - | Late preglobular, globular |
| emb218 | T-DNA | N | P | 1 | (1) | 25.0 | - | - | Late preglobular |
| emb219 | T-DNA | U | C | 2 | 2-3 | 23.5 | - | x | Globular, distorted linear |
| emb220 | T-DNA | U | G | 1-2 | 1 | 22.3 | - | x | Early globular |
| emb221 | T-DNA | U | P | 1 | - | 23.2 | - | x | Late preglobular |
| emb222 | T-DNA | N | T | 1 | 1 | 24.2 | 45.3 | - | Globular, heart, elongate |
| emb223 | T-DNA | U | G | 2 | 1-2 | 26.3 | - | x | Globular |
| emb224 | T-DNA | Y | T | 1 | 1 | 24.7 | 54.8 | 3 | Large heart |
| emb225 | T-DNA | N | G | 2-3 | 1-2 | 25.3 | 53.6 | 2 | Small globular, suspensor |
| emb226 | T-DNA | U | G | 1 | 1 | 19.8 | - | x | Globular |
| emb227 | T-DNA | U | G | 1-2 | 1 | 18.3 | - | x | Small globular |
| emb228 | T-DNA | U | T | 1 | 1-2 | 23.6 | 53.3 | x | Late globular-heart |
| emb229 | T-DNA | Y | G | 1 | 1 | 22.8 | - | - | Globular |
| emb230 | T-DNA | N | G | 1 | 1 | -- | - | - | Early globular |
| emb231 | T-DNA | U | G | 2 | - | 25.0 | - | x | Early globular |
| emb232 | T-DNA | N | C | 1-2 | 1 | 25.5 | 53.8 | 1 | Distorted heart-linear, tricot |

| Mutant | Mutagen | Tag | Class | Pigmentation | | % Mutant | Top Half | Linkage | Phenotype of Mutant Embryos |
|---|---|---|---|---|---|---|---|---|---|
| | | | | Seed | Embryo | | | | |
| emb233 | T-DNA | Y | G | 1 | 1 | 26.6 | - | - | Globular |
| emb234 | T-DNA | U | T | 1 | 1 | 27.4 | 50.8 | x | Globular, heart, triangle |
| emb236 | T-DNA | Y | T | 2 | 2 | 24.5 | - | 4 | Globular, rare hearts |
| emb237 | T-DNA | N | P | 1 | - | 24.6 | - | - | Preglobular |
| emb238 | T-DNA | U | P | 1 | - | 20.2 | - | x | Preglobular |
| emb239 | T-DNA | N | G | 1 | 1 | 20.7 | - | - | Globular |
| emb240 | T-DNA | U | G | 1 | 1 | 24.9 | - | x | Small globular |
| emb242 | T-DNA | U | G | 1 | 1 | 25.2 | - | x | Globular |
| emb243 | T-DNA | U | G | 1 | 2 | 19.3 | 59.5* | x | Globular, suspensor |
| emb244 | T-DNA | Y | G | 1 | 1 | 26.2 | - | - | Small globular, suspensor |
| emb245 | T-DNA | U | T | 2 | 1-2 | 23.9 | - | x | Globular, elongate, heart |
| emb246 | T-DNA | N | T | 1 | 1 | 23.9 | - | - | Globular, heart |
| emb247 | T-DNA | U | P | 1 | - | 23.4 | - | x | Early preglobular |
| emb250 | T-DNA | U | C | 3 | 2 | 20.3 | - | x | Linear-curled, not distorted |
| emb251 | T-DNA | Y | P | 1 | - | 24.5 | - | 5 | Late preglobular |
| emb252 | T-DNA | U | P | 1 | - | 15.6 | - | x | Variable preglobular |
| emb253 | T-DNA | U | T | 1 | 1 | 27.6 | 44.2 | x | Heart |
| emb254 | T-DNA | U | T | 1 | 1 | 23.7 | - | x | Small globular-heart |
| emb256-1 | T-DNA | N | C | 1 | 1-2 | 23.8 | 54.2 | 5 | Large heart, no hypocotyl |
| emb256-2 | T-DNA | Y | P | 1 | - | 14.3 | 52.9 | 5 | Late preglobular (emb116) |
| emb257 | T-DNA | N | G | 1 | 1 | 31.4 | - | - | Early globular |
| emb258 | T-DNA | U | P | 1 | - | 25.2 | - | x | Preglobular |
| emb259 | T-DNA | U | G | 1 | (-) | 25.6 | - | x | Early globular |
| emb260 | T-DNA | N | P | 1 | - | 24.6 | - | - | Late preglobular |
| emb261 | T-DNA | U | P | 1 | - | 22.7 | - | x | Preglobular |
| emb262 | T-DNA | Y | C | 3 | 1-2 | 26.7 | 54.6 | 5 | Linear-cotyledon |
| emb263 | T-DNA | U | C | 1 | 1-2 | 25.5 | 45.8 | 5 | Globular, distorted linear |
| emb264 | T-DNA | U | G | 1 | 1 | 23.5 | 60.6* | x | Globular |
| emb265 | T-DNA | U | G | 1-2 | 1-2 | 18.6 | - | x | Globular |
| emb266 | T-DNA | U | C | 1 | 3-4 | 22.5 | 47.4 | 5 | Heart, distorted cotyledon |
| emb267 | T-DNA | N | C | 1-4 | 1-4 | 20.6 | - | x | Variable |
| emb268 | T-DNA | U | G | 1 | 1 | 24.8 | - | x | Globular |
| emb269 | T-DNA | N | G | 1 | 1 | 25.0 | - | - | Globular |

| Mutant | Mutagen | Tag | Class | Pigmentation | | % Mutant | Top Half | Linkage | Phenotype of Mutant Embryos |
|---|---|---|---|---|---|---|---|---|---|
| | | | | Seed | Embryo | | | | |
| emb270 | T-DNA | Y | C | 1-3 | 1-3 | 25.9 | 53.0 | 1+4 | Globular, distorted, tricot |
| emb271 | T-DNA | U | G | 1 | 1 | 26.8 | - | x | Globular, suspensor |
| emb274 | T-DNA | N | P | 1 | - | 23.9 | - | - | Variable preglobular |
| emb275 | T-DNA | U | G | 1 | 1 | 23.2 | - | x | Globular |
| emb276 | T-DNA | U | P | 1 | - | 21.2 | - | x | Very early preglobular |
| emb277 | T-DNA | N | P | 1 | - | 21.2 | - | - | Late preglobular |
| emb278 | T-DNA | U | G | 1 | (-) | 25.5 | - | x | Early globular |
| emb279 | T-DNA | U | T | 1 | 1 | 23.1 | 51.5 | x | Large globular-heart |
| emb280 | T-DNA | U | G | 2 | 1 | 25.1 | - | x | Small globular |
| emb282 | T-DNA | U | P | 1 | - | 26.3 | - | x | Early preglobular |
| emb283 | T-DNA | U | G | 1 | 1 | 27.5 | - | x | Globular |
| emb285 | T-DNA | U | G | 1-2 | (-) | 24.7 | - | x | Globular |
| emb286 | T-DNA | Y | G | 1 | 1 | 26.6 | - | - | Early globular |
| emb287 | T-DNA | Y | P | 1 | - | 23.2 | - | - | Late preglobular |
| emb288 | T-DNA | U | G | 1-2 | 1 | 23.7 | - | x | Globular, elongate |
| emb289 | T-DNA | N | P | 2 | - | 15.4 | - | x | Variable preglobular |
| emb290 | T-DNA | N | P | 1 | - | 16.4 | - | - | Preglobular |
| emb291 | T-DNA | U | C | 4 | 4 | 21.1 | - | x | Variable |
| emb292 | T-DNA | U | G | 1 | 1 | 26.0 | - | x | Globular |
| emb293 | T-DNA | N | P | 1 | - | 22.2 | - | - | Late preglobular |
| emb294 | T-DNA | U | P | 1 | - | 21.3 | - | x | Very early preglobular |
| emb295 | T-DNA | U | C | 2-3 | 2-4 | - | - | x | Variable |
| bio1-1 | EMS | NAP | C | 2-3 | 1-2 | 26.5 | 50.2 | 5 | Globular-cotyledon; biotin |
| lec1-1 | T-DNA | N | C | 3-4 | 3-4 | 26.2 | 48.5 | 1 | Homeotic leafy cotyledon |
| lec1-2 | T-DNA | (Y) | C | 3-4 | 3-4 | - | - | [1] | Homeotic leafy cotyledon |
| lec2-1 | T-DNA | U | C | 4-5 | 4-5 | - | - | (1) | Homeotic leafy cotyledon |
| twn | T-DNA | (N) | C | 3 | 4 | 13.8 | 50.5 | 5 | Twin; suspensor polyembryony |

[a]All mutants currently being maintained in the Meinke laboratory at Oklahoma State University. Seeds for many of these lines are being transferred to the *Arabidopsis* Biological Resource Center at Ohio State University.

[b]Genetic studies indicate that mutants are either tagged with T-DNA (Y), not tagged (N), or unresolved (U) with respect to tagging (Castle et al. 1993). Tagging status not applicable (NAP) to EMS- and X-ray-induced mutations.

[c]Initial classification based on general morphology of mutant embryos at maturity: preglobular (P); globular (G); transition (globular-heart-elongate) (T); cotyledon (C); fusca (F); other (O).

[d]Color of mutant seeds and embryos prior to desiccation: white (1); pale yellow-green (2); pale green (3); green (4); mixture of green and dark red (5).

[e]% Mutant seeds in heterozygous siliques following self-pollination.

[f]% Total aborted seeds located in top half of silique. A nonrandom (*) distribution of aborted seeds (>> 50%) suggests that the mutant allele disrupts pollen development or pollen-tube growth.

[g]Linkage group assignment based on mapping with visible markers. Symbols: ((1)) tentatively assigned to linkage group #1; (x) not included in mapping project; (–) not yet assigned to linkage group; ([4]) linkage data based on mapping of other alleles.

## REFERENCES

Akhundova, G.G., L.I. Grinikh, and V.V. Shevchenko. 1978. Development of *Arabidopsis thaliana* embryos after gamma irradiation of plants in the generative phase. *Ontogenez* **9**: 514–519.

Akhundova, G.G., V.V. Shevchenko, and L.I. Grinikh. 1979. Twin plants of *Arabidopsis thaliana* induced by gamma irradiation. *Soviet Genet.* **15**: 428–443.

Arondel, V., B. Lemieux, I. Hwang, S. Gibson, H.M. Goodman, and C.R. Somerville. 1992. Map-based cloning of a gene controlling omega-3 fatty acid desaturation in *Arabidopsis*. *Science* **258**: 1353–1355.

Barton, M.K. and R.S. Poethig. 1993. Formation of the shoot apical meristem in *Arabidopsis thaliana*: An analysis of development in the wild type and in the shoot meristemless mutant. *Development* **119**: 823–831.

Batchelder, C., E.W. Edwards, and D.J. Murphy. 1991. Molecular genetics of oil body membrane proteins in *Brassica napus* and *Arabidopsis. J. Exp. Bot.* **42S**: 47.

Bäumlein, H., W. Boerjan, I. Nagy, R. Panitz, D. Inze, and U. Wobus. 1991a. Upstream sequences regulating legumin gene expression in heterologous transgenic plants. *Mol. Gen. Genet.* **225**: 121–128.

Bäumlein, H., W. Boerjan, I. Nagy, R. Bassüner, M. Van Montagu, D. Inze, and U. Wobus. 1991b. A novel seed protein gene from *Vicia faba* is developmentally regulated in transgenic tobacco and *Arabidopsis* plants. *Mol. Gen. Genet.* **225**: 459–467.

Bäumlein, H., S. Miséra, H. Luersen, K. Kölle, C. Horstmann, U. Wobus, and A.J. Müller. 1994. The *FUS3* gene of *Arabidopsis thaliana* is a regulator of gene expression during late embryogenesis. *Plant J.* (in press).

Baus, A.D., L. Franzmann, and D.W. Meinke. 1986. Growth *in vitro* of arrested embryos from lethal mutants of *Arabidopsis thaliana. Theor. Appl. Genet.* **72**: 577–586.

Bergfeld, R. and P. Schopfer. 1984. Formation of endoplasmic reticulum from osmiophilic globules during early embryogenesis in mustard (*Sinapis alba* L.). *Eur. J. Cell Biol.* **35**: 8–11.

———. 1986. Differentiation of a functional aleurone layer within the seed coat of *Sinapis alba* L. *Ann. Bot.* **57**: 25–33.

Bergfeld, R., T. Kühnl, and P. Schopfer. 1980. Formation of protein storage bodies during embryogenesis in cotyledons of *Sinapis alba* L. *Planta* **148**: 146–156.

Browse, J. and C. Somerville. 1991. Glycerolipid synthesis biochemistry and regulation. *Annu. Rev. Plant Physiol. Plant Mol. Biol.* **42**: 467–506.

Castle, L.A. and D.W. Meinke. 1993. Embryo-defective mutants as tools to study essential functions and regulatory processes in plant embryo development. *Semin. Dev. Biol.* **4**: 31–39.

———. 1994. A *FUSCA* gene of *Arabidopsis* encodes a novel protein essential for seedling development. *Plant Cell* **6**: 25–41.

Castle, L.A., D. Errampalli, T.L. Atherton, L.H. Franzmann, E.S. Yoon, and D.W. Meinke. 1993. Genetic and molecular characterization of embryonic mutants identified following seed transformation in *Arabidopsis. Mol. Gen. Genet.* **241**: 504–514.

Chasan, R. 1993. Evolving developments. *Plant Cell* **5**: 363–369.

Chory, J. 1992. A genetic model for light-regulated seedling development in *Arabidopsis. Development* **115**: 337–354.

Chu, N.M., C. Yoon, A. Chew, and M. Bournias. 1993. *Brevipedicellus* (*bp*), locus responsible for shape and size of epidermal cells in the pedicel of *Arabidopsis thaliana*. In *Abstracts from the 5th International Conference on* Arabidopsis *Research*, Ohio State University, Columbus, p. 152.

Clark, S.E., M.P. Running, and E.M. Meyerowitz. 1993. *Clavata1*, a regulator of meristem and flower development in *Arabidopsis*. *Development* **119**: 397–418.

Crone, W. and E.M. Lord. 1993. Flower development in the organ number mutant *clavata1-1* of *Arabidopsis thaliana* (Brassicaceae). *Am. J. Bot.* **80**: 1419–1426.

Czakó, M., J.C. Jang, J.M. Herr, Jr., and L. Márton. 1992. Differential manifestation of seed mortality induced by seed-specific expression of the gene for diphtheria toxin A chain in *Arabidopsis* and tobacco. *Mol. Gen. Genet.* **235**: 33–40.

De Clercq, A., M. Vandewiele, J. Van Damme, P. Guerche, M. Van Montagu, J. Vandekerckhove, and E. Krebbers. 1990. Stable accumulation of modified 2S albumin seed storage proteins with higher methionine contents in transgenic plants. *Plant Physiol.* **94**: 970–979.

Deng, X.W. and P.H. Quail. 1992. Genetic and phenotypic characterization of *cop1* mutants of *Arabidopsis thaliana*. *Plant J.* **2**: 83-95.

Deng, X.W., M. Matsui, N. Wei, D. Wagner, A.M. Chu, K.A. Feldmann, and P.H. Quail. 1992. *Cop1*, an *Arabidopsis* regulatory gene, encodes a protein with both a zinc-binding motif and a $G_B$ homologous domain. *Cell* **71**: 791–801.

Devic, M., M. Delseny, and P. Gallois. 1992. Promoter trapping in *Arabidopsis thaliana*: Searching for embryo-specific genes. *C.R. Seances Soc. Biol. Fil.* **186**: 541–549.

Dolan, L., K. Janmaat, V. Willemsen, P. Linstead, S. Poethig, K. Roberts, and B. Scheres. 1993. Cellular organisation of the *Arabidopsis thaliana* root. *Development* **119**: 71–84.

Errampalli, D., D. Patton, L. Castle, L. Mickelson, K. Hansen, J. Schnall, K. Feldmann, and D. Meinke. 1991. Embryonic lethals and T-DNA insertional mutagenesis in *Arabidopsis*. *Plant Cell* **3**: 149–157.

Feinbaum, R.L. and F.M. Ausubel. 1988. Transcriptional regulation of the *Arabidopsis thaliana* chalcone synthase gene. *Mol. Cell. Biol.* **8**: 1985–1992.

Fernandez, D.E., F.R. Turner, and M.L. Crouch. 1991. *In situ* localization of storage protein mRNAs in developing meristems of *Brassica napus* embryos. *Development* **111**: 299–313.

Finkelstein, R.R. and C.R. Somerville. 1990. Three classes of abscisic acid (ABA)-insensitive mutations of *Arabidopsis* define genes that control overlapping subsets of ABA responses. *Plant Physiol.* **94**: 1172–1179.

Franzmann, L., D.A. Patton, and D.W. Meinke. 1989. *In vitro* morphogenesis of arrested embryos from lethal mutants of *Arabidopsis thaliana*. *Theor. Appl. Genet.* **77**: 609–616.

Franzmann, L.H., E.S. Yoon, and D.W. Meinke. 1994. Saturating the genetic map of *Arabidopsis thaliana* with embryonic mutations. *Plant J.* (in press).

Gerlach-Cruse, D. 1969. Embryo- und Endospermentwicklung nach einer Röntgen-bestrahlung der Fruchtknoten von *Arabidopsis thaliana* (L.) Heynh. *Radiat. Bot.* **9**: 433–442.

Gilmour, S.J., N.N. Artus, and M.F. Thomashow. 1992. cDNA sequence analysis and expression of two cold-regulated genes of *Arabidopsis thaliana*. *Plant Mol. Biol.* **18**: 13–21.

Giraudat, J., B.M. Hauge, C. Valon, J. Smalle, F. Parcy, and H.M. Goodman. 1992. Isolation of the *Arabidopsis ABI3* gene by positional cloning. *Plant Cell* **4**: 1251–1261.

Goldberg, R.B., S.J. Barker, and L. Perez-Grau. 1989. Regulation of gene expression during plant embryogenesis. *Cell* **56**: 149–160.

Guerche, P., C. Tire, F. Grossi de Sa, A. De Clercq, M. Van Montagu, and E. Krebbers. 1990. Differential expression of the *Arabidopsis* 2S albumin genes and the effect of increasing gene family size. *Plant Cell* **2**: 469–478.

Hattori, T., V. Vasil, L. Rosenkrans, L.C. Hannah, D.R. McCarty, and I.K. Vasil. 1992. The *viviparous-1* gene and abscisic acid activate the C1 regulatory gene for anthocyanin biosynthesis during seed maturation in maize. *Genes Dev.* **6:** 609–618.

Haughn, G.W., Z. Modrusan, L. Reiser, R. Fischer, and K. Feldmann. 1993. The *fruitless* gene regulates ovule morphogenesis in *Arabidopsis thaliana. J. Cell. Biochem.* **17B:** 14.

Heath, J.D., R. Weldon, C. Monnot, and D.W. Meinke. 1986. Analysis of storage proteins in normal and aborted seeds from embryo-lethal mutants of *Arabidopsis thaliana. Planta* **169:** 304–312.

Hlousek-Radojcic, A., M.A. Post-Beittenmiller, and J.B. Ohlrogge. 1992. Expression of constitutive and tissue-specific acyl carrier protein isoforms in *Arabidopsis. Plant Physiol.* **98:** 206–214.

Höglund, A.S., M. Lenman, A. Falk, and L. Rask. 1991. Distribution of myrosinase in rapeseed tissues. *Plant Physiol.* **95:** 213–221.

Höglund, A.S., J. Rödin, E. Larsson, and L. Rask. 1992. Distribution of napin and cruciferin in developing rape seed embryos. *Plant Physiol.* **98:** 509–515.

Irish, V.F. and I.M. Sussex. 1992. A fate map of the *Arabidopsis* embryonic shoot apical meristem. *Development* **115:** 745–753.

James, D.W. and H.K. Dooner. 1990. Isolation of EMS-induced mutants in *Arabidopsis* altered in seed fatty acid composition. *Theor. Appl. Genet.* **80:** 241–245.

––––––. 1991. Novel seed lipid phenotypes in combinations of mutants altered in fatty acid biosynthesis in *Arabidopsis. Theor. Appl. Genet.* **82:** 409–412.

Johri, B.M., ed. 1984. *Embryology of angiosperms.* Springer-Verlag, Berlin.

Jürgens, G., U. Mayer, R.A. Torres Ruiz, T. Berleth, and S. Miséra. 1991. Genetic analysis of pattern formation in the *Arabidopsis* embryo. In *Molecular and cellular basis of pattern formation* (ed. K. Roberts et al.), pp. 27–38. Company of Biologists, Cambridge, United Kingdom.

Keith, K., M. Kraml, N.G. Dengler, and P. McCourt. 1994. *fusca3:* A heterochronic mutation affecting late embryo development in *Arabidopsis. Plant Cell* **6:** 589–600.

Koning, A., A. Jones, J.J. Fillatti, L. Comai, and M.W. Lassner. 1992. Arrest of embryo development in *Brassica napus* mediated by modified *Pseudomonas aeruginosa* exotoxin A. *Plant Mol. Biol.* **18:** 247–258.

Koornneef, M. 1990. Mutations affecting the testa color in *Arabidopsis. Arabidopsis Inf. Serv.* **27:** 1–4.

Koornneef, M., C.J. Hanhart, H.W.M. Hilhorst, and C.M. Karssen. 1989. *In vivo* inhibition of seed development and reserve protein accumulation in recombinants of abscisic acid biosynthesis and responsiveness mutants in *Arabidopsis thaliana. Plant Physiol.* **90:** 463–469.

Krebbers E., L. Herdies, A. De Clercq, J. Seurinck, J. Leemans, J. Van Damme, M. Segura, G. Gheysen, M. Van Montagu, and J. Vandekerckhove. 1988. Determination of the processing sites of an *Arabidopsis* 2S albumin and characterization of the complete gene family. *Plant Physiol.* **87:** 859–866.

Lemieux, B., M. Miquel, C. Somerville, and J. Browse. 1990. Mutants of *Arabidopsis* with alterations in seed lipid fatty acid composition. *Theor. Appl. Genet.* **80:** 234–240.

Leon-Kloosterziel, K.M., C.J. Keijzer, and M. Koornneef. 1994. A seed shape mutant of *Arabidopsis* that is affected in integument development. *Plant Cell* **6:** 385–392.

Lindsey, K. and J.F. Topping. 1993. Embryogenesis: A question of pattern. *J. Exp. Bot.* **44:** 359–374.

Liu, C., Z. Xu, and N.H. Chua. 1993. Auxin polar transport is essential for the establishment of bilateral symmetry during early plant embryogenesis. *Plant Cell* **5:** 621–630.

Lloyd, A.M., V. Walbot, and R.W. Davis. 1992. *Arabidopsis* and *Nicotiana* anthocyanin production activated by maize regulators R and C1. *Science* **258:** 1773–1775.

Lyndon, R.F. 1990. *Plant development.* Unwin Hyman, London.

Maheshwari, P. 1950. *An introduction to the embryology of angiosperms.* McGraw-Hill, New York.

Mansfield, S.G. and L.G. Briarty. 1990a. Development of the free-nuclear endosperm in *Arabidopsis thaliana* L. *Arabidopsis Inf. Serv.* **27:** 53–64.

———. 1990b. Endosperm cellularization in *Arabidopsis thaliana* L. *Arabidopsis Inf. Serv.* **27:** 65–72.

———. 1991. Early embryogenesis in *Arabidopsis thaliana.* II. The developing embryo. *Can. J. Bot.* **69:** 461–476.

Mansfield, S.G., L.G. Briarty, and S. Erni. 1991. Early embryogenesis in *Arabidopsis thaliana.* I. The mature embryo sac. *Can. J. Bot.* **69:** 447–460.

Marsden, M.P.F. and D.W. Meinke. 1985. Abnormal development of the suspensor in an embryo-lethal mutant of *Arabidopsis thaliana. Am. J. Bot.* **72:** 1801–1812.

Mayer, U., G. Büttner, and G. Jürgens. 1993. Apical-basal pattern formation in the *Arabidopsis* embryo: Studies on the role of the *gnom* gene. *Development* **117:** 149–162.

Mayer, U., R.A. Torres Ruiz, T. Berleth, S. Miséra, and G. Jürgens. 1991. Mutations affecting body organization in the *Arabidopsis* embryo. *Nature* **353:** 402–407.

McCarty, D.R., T. Hattori, C.B. Carson, V. Vasil, and I.K. Vasil. 1991. The *viviparous-1* developmental gene of maize encodes a novel transcriptional activator. *Cell* **66:** 895–905.

Meinke, D.W. 1979. "Isolation and characterization of embryo-lethal mutants of *Arabidopsis thaliana.*" Ph.D. thesis, Yale University, New Haven, Connecticut.

———. 1982. Embryo-lethal mutants of *Arabidopsis thaliana*: Evidence for gametophytic expression of the mutant genes. *Theor. Appl. Genet.* **63:** 381–386.

———. 1985. Embryo-lethal mutants of *Arabidopsis thaliana*: Analysis of mutants with a wide range of lethal phases. *Theor. Appl. Genet.* **69:** 543–552.

———. 1986. Embryo-lethal mutants and the study of plant embryo development. *Oxf. Surv. Plant Mol. Cell. Biol.* **3:** 122–165.

———. 1991a. Perspectives on genetic analysis of plant embryogenesis. *Plant Cell* **3:** 857–866.

———. 1991b. Embryonic mutants of *Arabidopsis thaliana. Dev. Genet.* **12:** 382–392.

———. 1991c. Genetic analysis of plant development. In *Plant physiology: Growth and development* (ed. F.C. Steward and R.G.S. Bidwell), vol. 10, pp. 437–490. Academic Press, New York.

———. 1992. A homoeotic mutant of *Arabidopsis thaliana* with leafy cotyledons. *Science* **258:** 1647–1650.

———. 1994. Diversity of embryonic mutants identified following *Agrobacterium*-mediated seed transformation in *Arabidopsis thaliana. NATO ASI Ser. Ser. H Cell Biol.* **81:** 105–115.

———. 1995. Molecular genetics of plant embryogenesis. *Annu. Rev. Plant Physiol. Plant Mol. Biol.* **46:** (in press).

Meinke, D.W. and I.M. Sussex. 1979a. Embryo-lethal mutants of *Arabidopsis thaliana*: A model system for genetic analysis of plant embryo development. *Dev. Biol.* **72:** 50–61.

———. 1979b. Isolation and characterization of six embryo-lethal mutants of *Arabidopsis thaliana. Dev. Biol.* **72:** 62–72.

Meinke, D.W., L.H. Franzmann, T.C. Nickle, and E.C. Yeung. 1994. *Leafy cotyledon* mutants of *Arabidopsis. Plant Cell* **6:** (in press).

Melan, M.A., X. Dong, M.E. Endara, K.R. Davis, F.M. Ausubel, and T.K. Peterman. 1993. An *Arabidopsis thaliana* lipoxygenase gene can be induced by pathogens, abscisic acid, and methyl jasmonate. *Plant Physiol.* **101:** 441–450.

Meurs, C., A.S. Basra, C.M. Karssen, and L.C. van Loon. 1992. Role of abscisic acid in the induction of desiccation tolerance in developing seeds of *Arabidopsis thaliana. Plant Physiol.* **98:** 1484–1493.

Miséra, S. 1993. "Genetische und entwicklungsbiologische Untersuchungen an *fusca* Genen von *Arabidopsis thaliana.*" Ph.D. thesis, University of Tübingen, Germany.

Miséra, S., A.J. Müller, U. Weiland-Heidecker, and G. Jürgens. 1994. The *FUSCA* genes of *Arabidopsis:* Negative regulators of light response. *Mol. Gen. Genet.* (in press).

Misra, R.C. 1962. Contribution to the embryology of *Arabidopsis thaliana* (Gay and Monn.). *Agra. Univ. J. Res. Sci.* **11:** 191–199.

Monnier, M. 1984. Survival of young immature *Capsella* embryos cultured *in vitro. J. Plant Physiol.* **115:** 105–113.

Müller, A.J. 1963. Embryonentest zum Nachweis rezessiver Letalfaktoren bei *Arabidopsis thaliana. Biol. Zentralbl.* **82:** 133–163.

Nambara, E., S. Naito, and P. McCourt. 1992. A mutant of *Arabidopsis* which is defective in seed development and storage protein accumulation is a new *abi3* allele. *Plant J.* **2:** 435–441.

Okada, K., J. Ueda, M.K. Komaki, C.J. Bell, and Y. Shimura. 1991. Requirement of the auxin polar transport system in early stages of *Arabidopsis* floral bud formation. *Plant Cell* **3:** 677–684.

Pang, P., R.E. Pruitt, and E.M. Meyerowitz. 1988. Molecular cloning, genomic organization, expression and evolution of 12S seed storage protein genes of *Arabidopsis thaliana. Plant Mol. Biol.* **11:** 805–812.

Patton, D.A. and D.W. Meinke. 1988. High-frequency plant regeneration from cultured cotyledons of *Arabidopsis thaliana. Plant Cell Rep.* **7:** 233–237.

———. 1990. Ultrastructure of arrested embryos from lethal mutants of *Arabidopsis thaliana. Am. J. Bot.* **77:** 653–661.

Patton, D.A., L.H. Franzmann, and D.W. Meinke. 1991. Mapping genes essential for embryo development in *Arabidopsis thaliana. Mol. Gen. Genet.* **227:** 337–347.

Pepper, A., T. Delaney, T. Washburn, D. Poole, and J. Chory. 1994. *DET1*, a negative regulator of light-mediated development and gene expression in *Arabidopsis*, encodes a novel nuclear-localized protein. *Cell* **78:** 109–116.

Post-Beittenmiller, D., J.B. Ohlrogge, and C.R. Somerville. 1992. Regulation of plant lipid biosynthesis: An example of developmental regulation superimposed on a ubiquitous pathway. In *Control of plant gene expression* (ed. D.P.S. Verma), pp. 157–174. CRC Press, Boca Raton, Florida.

Raghavan, V. 1976. *Experimental embryogenesis in vascular plants.* Academic Press, New York.

———. 1986. *Embryogenesis in angiosperms.* Cambridge University Press, Cambridge, United Kingdom.

Rédei, G.P. 1970. *Arabidopsis thaliana* (L.) Heynh. A review of the genetics and biology. *Bibliogr. Genet.* **20:** 1–151.

Rest, J.A. and J.G. Vaughan. 1972. Development of protein and oil bodies in the seed of *Sinapis alba* L. *Planta* **105:** 245–262.

Robinson-Beers, K., R.E. Pruitt, and C.S. Gasser. 1992. Ovule development in wild-type *Arabidopsis* and two female-sterile mutants. *Plant Cell* **4:** 1237–1249.

Sangwan, R.S., Y. Bourgis, and B. Sangwan-Norreel. 1991. Genetic transformation of

*Arabidopsis thaliana* zygotic embryos and identification of critical parameters influencing transformation efficiency. *Mol. Gen. Genet.* **230:** 475–485.

Scheres, B., V. Willemsen, K. Janmaat, H. Wolkenfelt, L. Dolan, and P. Weisbeek. 1994. Analysis of root development in *Arabidopsis thaliana*. *NATO ASI Ser. Ser. H Cell Biol.* **81:** 41–50.

Schneider, T., R. Dinkins, K. Robinson, J. Shellhammer, and D.W. Meinke. 1989. An embryo-lethal mutant of *Arabidopsis thaliana* is a biotin auxotroph. *Dev. Biol.* **131:** 161–167.

Schulz, P. and W.A. Jensen. 1968a. *Capsella* embryogenesis: The early embryo. *J. Ultrastruct. Res.* **22:** 376–392.

———. 1968b. *Capsella* embryogenesis: The synergids before and after fertilization. *Am. J. Bot.* **55:** 541–552.

———. 1968c. *Capsella* embryogenesis: The egg, zygote, and young embryo. *Am. J. Bot.* **55:** 807–819.

———. 1969. *Capsella* embryogenesis: The suspensor and basal cell. *Protoplasma* **67:** 139–163.

———. 1971. *Capsella* embryogenesis: The chalazal proliferating tissue. *J. Cell Sci.* **8:** 201–227.

———. 1973. *Capsella* embryogenesis: The central cell. *J. Cell Sci.* **12:** 741–763.

———. 1974. *Capsella* embryogenesis: The development of the free nuclear endosperm. *Protoplasma* **80:** 183–205.

———. 1981. Pre-fertilization ovule development in *Capsella*: Ultrastructure and ultracytochemical localization of acid phosphatase in the meiocyte. *Protoplasma* **107:** 27–45.

———. 1987. Prefertilization ovule development in *Capsella*: The dyad, tetrad, developing megaspore, and two-nucleate gametophyte. *Can. J. Bot.* **64:** 875–884.

Schwartz, B.W., E.C. Yeung, and D.W. Meinke. 1994. Disruption of morphogenesis and transformation of the suspensor in abnormal *suspensor* mutants of *Arabidopsis*. *Development* (in press).

Shellhammer, A.J. 1991. "Analysis of a biotin auxotroph of *Arabidopsis thaliana*." Ph.D. thesis, Oklahoma State University, Stillwater.

Shellhammer, J. and D.W. Meinke. 1990. Arrested embryos from the *bio1* auxotroph of *Arabidopsis* contain reduced levels of biotin. *Plant Physiol.* **93:** 1162–1167.

Shevell, D.E., W.-M. Leu, C.S. Gillmor, G. Xia, K.A. Feldmann, and N.-H. Chua. 1994. *EMB30* is essential for normal cell division, cell expansion, and cell adhesion in *Arabidopsis* and encodes a protein that has similarity to Sec7. *Cell* (in press).

Shirley, B.W., S. Hanley, and H.M. Goodman. 1992. Effects of ionizing radiation on a plant genome: Analysis of two *Arabidopsis transparent testa* mutations. *Plant Cell* **4:** 333–347.

Souèges, R. 1914. Nouvelles recherches sur le développement de l'embryon chez les Crucifères. *Ann. Sci. Nat. Bot. (Ser. 9)* **19:** 311–339.

———. 1919. Les premières divisions de l'oeuf et les différenciations du suspenseur chez le *Capsella bursa-pastoris*. *Ann. Sci. Nat. Bot. (Ser. 10)* **1:** 1–28.

Steeves, T.A. and I.M. Sussex. 1989. *Patterns in plant development*, 2nd edition. Cambridge University Press, Cambridge, United Kingdom.

Sumner, M.J. and L. Van Caeseele. 1988. Ovule development in *Brassica campestris*: A light microscope study. *Can. J. Bot.* **66:** 2459–2469.

Thangstad, O.P., T.H. Iversen, G. Slupphaug, and O. Bones. 1990. Immunocytochemical localization of myrosinase in *Brassica napus* L. *Planta* **180:** 245–248.

Tykarska, T. 1976. Rape embryogenesis. I. The proembryo development. *Acta Soc. Bot.*

*Pol.* **45**: 3–15.

———. 1979. Rape embryogenesis. II. Development of the embryo proper. *Acta Soc. Bot. Pol.* **48**: 391–421.

———. 1987a. Rape embryogenesis. V. Accumulation of lipid bodies. *Acta Soc. Bot. Pol.* **56**: 573–584.

———. 1987b. Rape embryogenesis. VI. Accumulation of protein bodies. *Acta Soc. Bot. Pol.* **56**: 585–598.

Van der Klei, H., J. Van Damme, P. Casteels, and E. Krebbers. 1993. A fifth 2S albumin isoform is present in *Arabidopsis thaliana*. *Plant Physiol.* **101**: 1415–1416.

Van Rooijen, G.L., L.I. Terning, and M.M. Moloney. 1992. Nucleotide sequence of an *Arabidopsis thaliana* oleosin gene. *Plant Mol. Biol.* **18**: 1177–1179.

Vandekerckhove, J., J. Van Damme, M. Van Lijsebettens, J. Botterman, M. De Block, M. Vandewiele, A. De Clercq, J. Leemans, M. Van Montagu, and E. Krebbers. 1989. Enkephalins produced in transgenic plants using modified 2S seed storage proteins. *Bio/Technology* **7**: 929–932.

Vernon, D.M. and D.W. Meinke. 1994. Embryogenic transformation of the suspensor in *twin*, a polyembryonic mutant of *Arabidopsis*. *Dev. Biol.* **165**: (in press).

Webb, M.C. and B.E.S. Gunning. 1990. Embryo sac development in *Arabidopsis thaliana*. I. Megasporogenesis, including the microtubular cytoskeleton. *Sex. Plant Reprod.* **3**: 244–256.

———. 1991. The microtubular cytoskeleton during development of the zygote, proembryo, and free-nuclear endosperm in *Arabidopsis thaliana* (L.) Heynh. *Planta* **184**: 187–195.

Wei, N. and X.W. Deng. 1992. *Cop9*: A new genetic locus involved in light-regulated development and gene expression in *Arabidopsis*. *Plant Cell* **4**: 1507–1518.

Weigel, D. 1993. Patterning the *Arabidopsis* embryo. *Curr. Biol.* **3**: 443–445.

Weiland, U. and A.J. Müller. 1972. In-vitro-Kultur der Wurzeln von letalen *fusca*-Mutanten von *Arabidopsis thaliana*. *Kulturpflanze* **20**: 151–164.

Werker, E. and J.G. Vaughan. 1974. Anatomical and ultrastructural changes in aleurone and myrosin cells of *Sinapis alba* during germination. *Planta* **116**: 243–255.

West, M.A.L. and J.J. Harada. 1993. Embryogenesis in higher plants: An overview. *Plant Cell* **5**: 1361–1369.

West, M.A.L., K.L. Matsudaira, R.B. Goldberg, R.L. Fischer, and J.J. Harada. 1993. Genes essential for embryogenesis in *Arabidopsis thaliana*. *J. Cell. Biochem.* **17B**: 37.

Williams, E.G. and G. Maheswaran. 1986. Somatic embryogenesis: Factors influencing coordinated behaviour of cells as an embryogenic group. *Ann. Bot.* **57**: 443–462.

Wu, Y., G. Haberland, C. Zhou, and H.-U. Koop. 1992. Somatic embryogenesis, formation of morphogenetic callus and normal development in zygotic embryos of *Arabidopsis thaliana in vitro*. *Protoplasma* **169**: 89–96.

Xue, J., M. Lenman, A. Falk, and L. Rask. 1992. The glucosinolate-degrading enzyme myrosinase in Brassicaceae is encoded by a gene family. *Plant Mol. Biol.* **18**: 387–398.

Yakolev, M.S. and G.K. Alimova. 1976. Embryogenesis in *Arabidopsis thaliana* (L.) Heynh. (Cruciferae). *Bot. Zh.* **6**: 12–24.

Yeung, E.C. and D.W. Meinke. 1993. Embryogenesis in angiosperms: Development of the suspensor. *Plant Cell* **5**: 1371–1381.

# 11

# Pattern Formation in the Embryo

**Gerd Jürgens**
Institut für Genetik und Mikrobiologie
Lehrstuhl für Genetik, Universität München
D-80638 München, Federal Republic of Germany

Constructing a multicellular organism takes more than cell division and cell differentiation. A characteristic body organization has to be established such that tissues and organs are arranged in a structurally and functionally meaningful context, or pattern. To generate the body pattern, the cells of the developing organism have to exchange, and respond to, information about their relative positions. Although in animals pattern formation is largely confined to embryogenesis, flowering plants establish their body pattern during two distinct phases of the life cycle. Starting from a single cell, the zygote, embryogenesis lays down a primary body organization that is expressed in the seedling. Although dicot and monocot types can be distinguished, the body organization of the seedling is remarkably uniform across flowering plant species. Postembryonically, localized growth centers, or meristems, which occupy the opposite ends of the seedling axis, add new structures to the seedling body, producing species-specific adult forms (for review, see Steeves and Sussex 1989; Sussex 1989; Medford 1992). Since the meristems themselves originate in the embryo, the structurally simple seedling is a good choice for analyzing how essential features of the plant body organization are generated de novo.

The body organization of the seedling can formally be described as the superimposition of two patterns, an apical-basal pattern along the main axis of polarity and a radial pattern perpendicular to the axis (Fig. 1). The apical-basal pattern consists of a series of distinct elements which are, from top to bottom, shoot meristem, cotyledons, hypocotyl, embryonic root (radicle), and root meristem. The radial pattern, which is most clearly seen in the hypocotyl, is made up of concentric layers of the main tissue types: the outer epidermis, the ground tissue, and the vascular tissue located in the center.

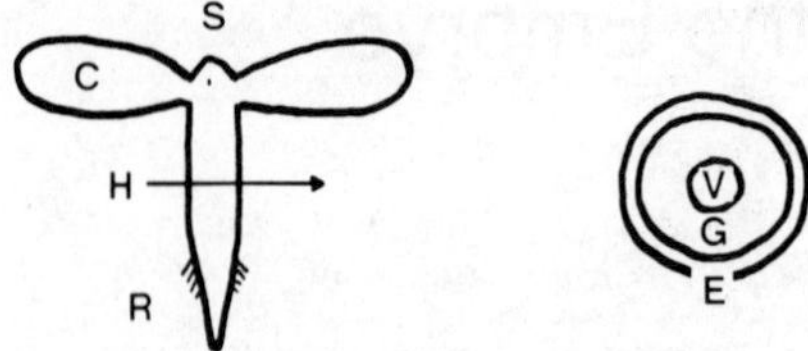


*Figure 1* Schematic representation of the body organization of the seedling. The apical-basal pattern is shown on the left, the radial pattern as seen in the hypocotyl on the right. (C) Cotyledons; (E) epidermis; (G) ground tissue; (H) hypocotyl; (R) root and root meristem; (S) shoot meristem; (V) vascular tissue. (Reprinted, with permission, from Mayer et al. 1993b.)

In this chapter, I attempt to tie up what is known about pattern formation in the *Arabidopsis* embryo, with special emphasis on genetic studies. How the seedling body organization develops in the embryo is briefly described to provide a reference for the discussion of mutant phenotypes and their implications. Then, an overview is given on the genetic approach, including the isolation and characterization of pattern mutants. Finally, genetic aspects of apical-basal pattern formation are discussed in some detail.

## EMBRYOGENESIS: HOW THE BODY ORGANIZATION DEVELOPS

*Arabidopsis* embryogenesis conforms to the crucifer type described earlier for *Capsella* and *Brassica* (Vandendries 1909; Müller 1963; Schulz and Jensen 1968; Tykarska 1976, 1979; Meinke and Sussex 1979; Mansfield and Briarty 1991; Jürgens and Mayer 1994). Within 14 days of fertilization, the zygote gives rise to a mature embryo which consists of 15,000–20,000 cells. The developing embryo passes through three distinct phases: (1) An early phase in which the basic body organization is laid down. This phase ends after about 30% of embryogenesis when the primordia of the principal seedling structures become morphologically recognizable. (2) A later phase in which the primordia grow and differentiate. (3) A final desiccation phase in which the mature embryo prepares for seed dormancy. The growing embryo undergoes changes in shape that are brought about by regional differences in mitotic rate, oriented cell divisions, and changes in cell shape (Lyndon 1990). Dicot plant embryos have been staged by their cell numbers and shapes, and a few stages are commonly distinguished: octant, globular, heart, torpedo, and bent-cotyledon (Natesh and Rau 1984). More recently, 20 stages of embryogenesis have been defined in *Arabidopsis* by using additional morphological criteria (Jürgens and Mayer 1994).

## Origin of Seedling Structures

The purpose of this section is to describe the embryonic origin of seedling structures. It has not been possible to follow directly the fate of individual marked cells from the early embryo. However, due to the lack of cell migration and the nearly invariant pattern of cell division, the seedling structures can be traced back to groups of cells in the early embryo. This "cell-lineage analysis" is similar to fate mapping in animal embryos: Cells at specific positions in the early embryo regularly give rise to specific body structures, but the predictability of developmental fate does not imply that the cells are firmly committed.

The early embryo consists of only a few cells, so different seedling structures originate from the same group of cells. As embryogenesis progresses, developmental fates successively segregate among the increasing cell population, and these events are reflected in a sequence of stage-specific morphological changes (Figs. 2, 3). For the sake of clarity, the development of the apical-basal pattern and that of the radial pattern are described separately, although they overlap in time.

## Development of the Apical-basal Pattern

During very early development, the zygote produces an incipient embryo and an extraembryonic suspensor that anchors the embryo to the micropylar pole of the ovule. Following fertilization, the zygote expands about threefold in the future apical-basal axis before dividing asymmetrically to give a small apical and a large basal cell (Fig. 2A). The apical cell gives rise to most of the mature embryo, whereas the basal cell only contributes the very basal end (Fig. 3A). The apical cell is partitioned by three rounds of cleavage divisions, two vertical and one horizontal, resulting in two tiers each of four cells. These upper and lower tiers make up the embryo proper of the octant stage, which is only slightly larger than the apical cell (Mansfield and Briarty 1991). The basal cell produces, by a series of transverse divisions, a file of 7–9 cells of which all but the uppermost one form the extraembryonic suspensor. The uppermost derivative of the basal cell becomes the hypophysis, which joins the embryo proper to complete the incipient embryo.

At the octant stage, three regions can be distinguished along the apical-basal axis: the upper and the lower tiers of the embryo proper, and the hypophysis (Fig. 2A). The upper tier gives rise to the shoot meristem and the cotyledons; the lower tier to the hypocotyl, the embryonic root (radicle), and the initials of the root meristem; and the hypophysis produces the center of the root meristem and the initials and layers of the central root cap (Fig. 3A). The different developmental fates of the two

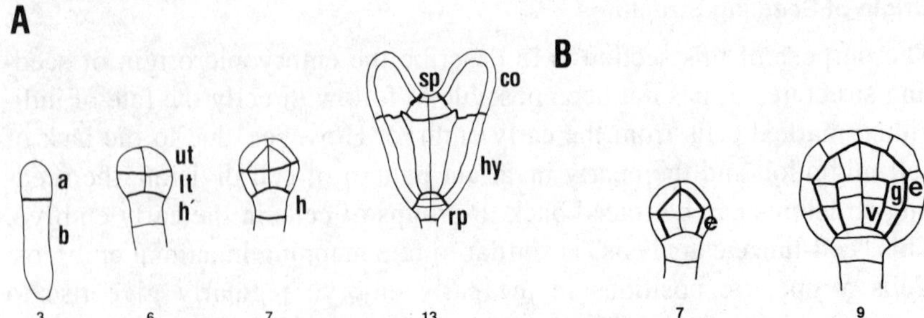


*Figure 2* Development of (*A*) apical-basal pattern, and (*B*) radial pattern during embryogenesis. (*A*) The zygote divides asymmetrically to give a small apical (a) and a large basal (b) cell. The apical cell produces, by cleavage divisions, the upper tier (ut) and the lower tier (lt) of the octant-stage embryo, whereas the precursor of the hypophysis (h′) is derived from the basal cell. The hypophysis (h) gives rise to the incipient root primordium (rp) of the mid-heart stage. The shoot primordium (sp) and the cotyledonary primordia (co) originate from the upper tier, the primordia of the hypocotyl and the root initials (hy′) from the lower tier. (See also Fig. 3A.) (*B*) The outer layer of the epidermis primordium (e; protoderm) is marked off at the dermatogen stage. At the mid-globular stage, the inner cell mass splits into the centrally located vascular primordium (v; procambium) and the ground tissue (g). Numbers refer to stages of embryogenesis according to Jürgens and Mayer (1994). (Adapted, with permission, from Mayer et al. 1993b.)

tiers correlate with different patterns of cell division which become increasingly clear during the globular stages: The derivatives of the upper tier divide nearly at random before the initiation of the cotyledonary primordia, whereas the derivatives of the lower tier form files of cells, thus elongating the body axis. The basally adjacent hypophysis generates the incipient root primordium of the heart-stage embryo and, again, a regular pattern of cell divisions presages developmental fate from the globular stages: The hypophysis divides asymmetrically to give an upper lens-shaped cell and a larger lower cell, both of which undergo two vertical divisions. The lens-shaped cell produces the four cells of the quiescent center and the lower cell gives rise to the initials of the central root cap. The formation of the embryonic root is delayed: It is not before the mid-heart stage that the derivatives of the lower tier that abut the quiescent center become the initials of the root meristem, which then also generate the embryonic root (Figs. 2A, 3A; see below).

## Development of the Radial Pattern

The main elements of the radial pattern, such as epidermis, ground tissue, and vascular tissue, are established during a brief period (Fig. 2B). Tan-

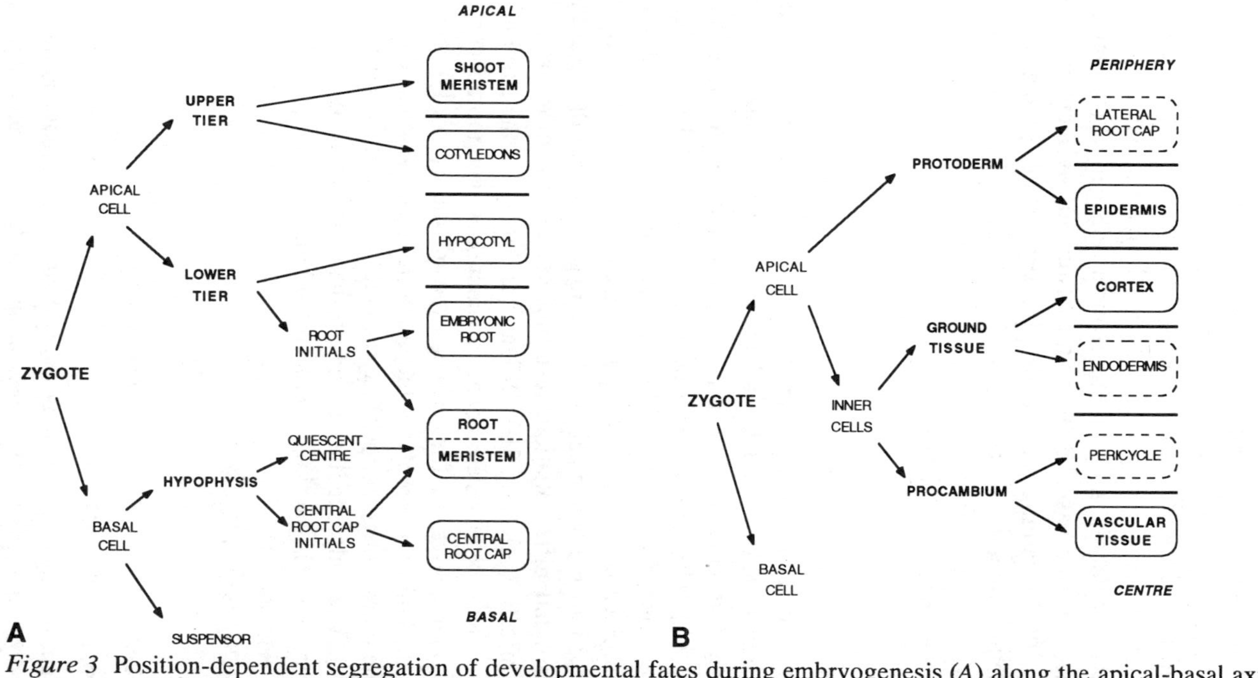



*Figure 3* Position-dependent segregation of developmental fates during embryogenesis (*A*) along the apical-basal axis of polarity and (*B*) perpendicular to the axis. Primary choices are printed in bold face. The pattern elements of the seedling are shown on the right; the meristems of the shoot and the root (*A*, in bold face) will produce adult structures during postembryonic development. Note that the root meristem is composite, with the upper layer of initials derived from the lower tier and the remainder from the hypophysis. The main tissue types (*B*, in bold face) are easy to score in whole-mount preparations of seedlings.

gential cell divisions transform the octant-stage into the dermatogen-stage embryo such that an outer layer of epidermal precursor cells (protoderm) is separated from the inner cell mass. Subsequently, cell divisions are anticlinally oriented (i.e., newly formed cell walls are perpendicular to the surface) in the surface layer, maintaining its integrity. The growing protoderm keeps up with the increase in size of the developing embryo and later forms a contiguous surface which includes the $L_1$ layer of the shoot meristem as well as the epidermis of cotyledons, hypocotyl, and root. Thus, the epidermal and subepidermal structures of the seedling derive from separate cell groups in the dermatogen-stage embryo. At the midglobular stage, the innermost cells can be recognized as the precursors of the vascular tissue (procambium) by their unequal cell divisions; the remaining inner cells give rise to the ground tissue (Fig. 2B). The procambial cells progressively elongate in the apical-basal axis and, by further cell divisions, produce a solid vascular primordium in the axis of the heart-shaped embryo.

### Refinements during Later Stages

The early phase of embryogenesis ends when the embryo turns heart-shaped. This transition from radial to bilateral symmetry is caused by the initiation of the two cotyledonary primordia near the apical end of the embryo (Fig. 2A). The embryo now consists of about 250 cells. Subsequently, the growing primordia of seedling structures show increasing differentiation. Each cotyledon produces a central vascular strand which then reticulates into the leaf blade. Both the hypocotyl and the root undergo further radial differentiation, resulting in the formation of additional layers of cell types (Fig. 3B) (Dolan et al. 1993). At the basal end of the embryo, the root meristem becomes functional: The initials above the quiescent center give off files of cells toward the hypocotyl which contribute to the embryonic root and the lateral root cap, whereas the initials below the quiescent center produce the cell layers of the central root cap (Fig. 3A). The only seedling structure that remains inconspicuous during embryogenesis is the shoot meristem, although its layered primordium can be seen as a small group of cells between the bases of the two cotyledonary primordia (Fig. 2A).

## MUTATIONAL DISSECTION OF PATTERN FORMATION

### The Genetic Approach: Rationale

Descriptive embryology tells us how the primary body organization develops but does not give any clues about underlying mechanisms.

Transplantation experiments which facilitated the analysis of pattern formation in animal embryos cannot easily be done because the plant embryo develops inside the ovule, which in turn is enclosed in the fruit, and because the embryo is very small during the critical early stages. An alternative approach, which is not constrained in the same way, involves the isolation and characterization of pattern mutants. The underlying assumption is that pattern formation is brought about by molecules that are themselves genetically specified or result from the activities of gene products. In either case, mutational inactivation of relevant genes should affect pattern formation in specific ways. Patterning genes that have been identified by mutant phenotype can in principle be cloned by RFLP mapping and chromosome walking, which provides a starting point for the molecular analysis of pattern formation. The feasibility of positional cloning has been demonstrated for other genes (Arondel et al. 1992; Giraudat et al. 1992; Leyser et al. 1993).

The genetic analysis relies on phenotypic criteria by which pattern mutants can be recognized. In flowering plants, many genes are required for the completion of embryogenesis (for review, see Meinke 1991). Inactivation of any one of these genes results in embryonic lethality, and the lethal phenotype does not enable distinction between genes specifically involved in pattern formation and those required for other processes, e.g., cell division, cell metabolism, or development of the suspensor or the endosperm. In contrast, mutations that do not interfere with the completion of embryogenesis but cause abnormal-seedling phenotypes are more likely to affect patterning genes, and distinctive phenotypes can be used to select putative pattern mutants from among the seedling mutants. Careful studies of developing mutant embryos would still be necessary to determine whether the pattern abnormalities of the seedling do indeed reflect defects in pattern formation.

## A Comprehensive Screen for Pattern Mutants

Attempts were made to isolate mutant alleles of all genes specifically involved in embryonic pattern formation (Jürgens et al. 1991). Mutations were induced by exposing seeds to ethylmethane sulfonate (EMS), which is known to cause predominantly point mutations, and mutant phenotypes were thus most likely due to single-gene changes. Plants that had developed from mutagenized seeds were individually screened for mutant seedling progeny to ensure that mutants with similar phenotypes had arisen independently. The scale of the screen was chosen such that each gene should be represented by five mutant alleles on average, which would correspond to statistical saturation of the genome. A total of

44,000 mutagenized lines were analyzed, yielding about 25,000 embryonic-lethal mutations but only 250 putative pattern mutants that were recognizable at the seedling stage (Jürgens et al. 1991). To assess whether saturation was achieved, mutants from two phenotypic classes were analyzed in complementation tests. One class comprised the *fusca* mutants, which accumulate high levels of anthocyanin in the seed (Müller 1963) and should therefore represent average non-pattern mutants; 52 of 84 *fusca* mutants have been assigned to nine complementation groups (Miséra et al. 1994). The other class was a selection of putative pattern mutants showing conspicuous phenotypes, and 73 mutants were initially assigned to nine complementation groups (Mayer et al. 1991). Thus, each gene from both classes was, on average, represented by more than 5 mutant alleles, indicating that the aim of statistical saturation was achieved.

## ANALYSIS OF PATTERN MUTANTS

Mutants are useful tools for dissecting pattern formation, and their phenotypes are studied from two different perspectives, which may be named process-oriented and gene-oriented. If the study is process-oriented, the mutant phenotype is taken to represent the outcome of an experiment in which pattern formation has been perturbed from within the embryo. Moreover, the system responds to the genetic removal of a single component, which is better defined than experimental manipulation of the developing embryo, and the resulting phenotype reveals which aspects of the process are affected and which are not. If the study of pattern mutants is gene-oriented, the aim is to define genes specifically involved in pattern formation, which often forms the basis for molecular analysis. In this case, the mutant phenotype suggests what role a gene may play in pattern formation, but this inference is only sound if the phenotypes of several mutant alleles can be compared and if the development of mutant embryos has been studied in some detail. Finally, the phenotypes of double mutants can give some idea about the logic of the process: If two genes act independently, their phenotypes add up to give the phenotype of the double mutant; alternatively, if two genes interact, a novel phenotype may result or only the two parental phenotypes are observed. In this section, I present general conclusions that can be drawn from mutant phenotypes. The role of specific genes is discussed later for apical-basal pattern formation.

Mutations in the nine genes identified by complementation analysis (see above) affect three different aspects of the seedling body organization: apical-basal pattern, radial pattern, and shape (Fig. 4). It has been

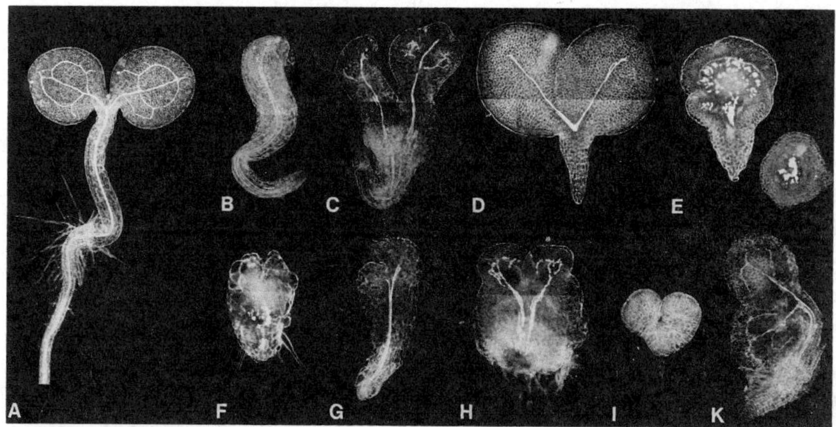

*Figure 4* Seedling phenotypes of mutations affecting different aspects of body organization. (*A*) Wild-type; apical-basal pattern deletions: (*B*) *gurke*, (*C*) *fackel*, (*D*) *monopteros*, (*E*) *gnom*; radial pattern defects: (*F*) *knolle*, (*G*) *keule*; shape changes: (*H*) *fass*, (*I*) *knopf*, (*K*) *mickey*. (Dark-field photographs of whole-mount preparations reprinted, with permission, from Mayer et al. 1991 [copyright Macmillan Magazines Ltd.].)

shown that the mutant seedling phenotypes result from changes in the early embryo rather than from defects during the later stages of growth (Mayer et al. 1991). Mutant embryos for any one of the genes are abnormal at the heart stage, indicating that the genes are required during the early phase of embryogenesis when the basic body organization is established. Furthermore, the early embryonic defects closely correspond, by position, with the structures that are missing or defective in the seedling.

Mutations in four genes (*gnom*, *monopteros*, *fackel*, and *gurke*) delete different regions of the apical-basal pattern by changing the developmental fate of cell groups (Fig. 4B–E; see below). In contrast, the tissue types of the radial pattern are present in the unaffected regions. Thus, the two patterns that make up the seedling body develop independently of each other. This is most clearly seen in *gn* mutant seedlings that show no morphological signs of apical-basal polarity and yet have a complete radial pattern (Fig. 4E).

Mutations in two genes (*knolle*, *keule*) change the radial pattern. The epidermis appears to be missing in *knolle* seedlings, whereas the epidermal cells look abnormal in *keule* seedlings (Fig. 4F,G). In both cases, seedling shape and apical-basal structures are also variably altered such that the radial pattern defect is not obvious. When traced back to the early-globular stage, the mutant phenotypes appear more specific: The primordium of the epidermis (protoderm) seems not to form in *knolle* embryos, whereas the epidermal precursor cells appear bloated in *keule* embryos (Mayer et al. 1991).

Mutations in three genes (*fass, knopf, mickey*) alter seedling shape but do not affect pattern formation (Fig. 4H–J). Nonetheless, the mutant phenotype of the *fass* (*fs*) gene reveals an interesting aspect of pattern formation in the embryo. Cell shape is abnormal from the very early stages such that no specific primordia, e.g., the epidermal layer, the vascular primordium, or the incipient root primordium, can be distinguished by morphological criteria (Mayer et al. 1991; R.A. Torres Ruiz and G. Jürgens, unpubl.). However, mutant *fs* seedlings differentiate the full complement of apical-basal and radial pattern elements and can even go on to produce adult plants that are morphologically abnormal (Mayer et al. 1993b; R.A. Torres Ruiz and G. Jürgens, unpubl.). Thus, cell shape seems to be irrelevant for pattern formation. It is rather more likely that cells normally take on characteristic shapes in a position-dependent manner in response to pattern formation.

## GENES FOR APICAL-BASAL PATTERN FORMATION

Pattern formation in the apical-basal axis of polarity generates an array of distinct elements: shoot meristem, cotyledons, hypocotyl, embryonic root (radicle), and root meristem. This pattern is altered by mutations in five genes: *gnom* (*gn*), *gurke* (*gk*), *fackel* (*fk*), and *monopteros* (*mp*) from the large-scale screen mentioned above (Mayer et al. 1991), and the *zwille* (*zll*) gene from a different mutagenesis experiment (Jürgens et al. 1994; T. Laux and G. Jürgens, unpubl.). Since each of these genes is represented by several mutant alleles, the gene-specific mutant phenotype probably results from reduced activity or complete inactivation of the gene (see below). The analysis of mutant phenotypes should help to assign each gene a specific role in apical-basal pattern formation.

### Gene-specific Phenotypes

Mutations in the *gn* gene affect the organization of the entire apical-basal axis, resulting in grossly abnormal seedlings (Mayer et al. 1993a). All mutant seedlings lack the root end, and the shoot end is variably reduced. In the extreme case, the seedlings show no morphological signs of apical-basal polarity (Fig. 4E). All 24 mutant alleles share essentially the same variable phenotype. Extensive complementation analysis revealed three different groups of mutant alleles: Group A and B alleles partially complemented each other, producing a short root, whereas group C alleles did not complement either of the other two groups. Thus, group C alleles behaved like null alleles in this assay, suggesting that the variability of the seedling phenotype does not reflect residual *gn* gene activity (Mayer et al. 1993a). The *gn* phenotype has been traced back to the zygote, which divides nearly symmetrically in the mutant such that the apical daughter cell is enlarged at the expense of the basal daughter cell (Fig. 5F). The enlarged apical cell divides obliquely or perpendicular, rather than in parallel, to the apical-basal axis and, due to this altered pattern of cell division, the *gn* octant-stage embryo has about twice the normal number of cells. In addition, the cell in place of the hypophysis does not undergo the regular divisions to produce the incipient root primordium, which accounts for the lack of the root in the mutant seedling.

The other genes seem to have more restricted domains of action. Mutations in the *mp* gene delete the basal structures of the seedling, such as hypocotyl, embryonic root, root meristem, and root cap, whereas the apical end of the seedling can be normal (Fig. 4D) (Mayer et al. 1991; Berleth and Jürgens 1993). The basal defect has been traced back to the early embryo: The cells in place of the lower tier and the uppermost derivative of the basal cell divide abnormally, suggesting that their patterns of cell division are directed by the *mp* gene product (Fig. 5G) (Berleth and Jürgens 1993). In mutant *fk* seedlings, the cotyledons appear to be directly attached to the root without any indication of an intervening hypocotyl (Fig. 4C) (Mayer et al. 1991), and this defect corresponds, by position, to abnormal cell divisions in the center of the globular embryo (Fig. 5H) (Jürgens et al. 1994). Mutant *gk* seedlings lack cotyledons and shoot meristem (Fig. 4B), and this defect becomes apparent when the cotyledonary primordia would normally emerge from the apical region of the early-heart-stage embryo (Fig. 5I) (Mayer et al. 1991; R.A. Ruiz and G. Jürgens, unpubl.). Mutations in the *zll* gene have the most localized effect on the seedling pattern, eliminating only the shoot meristem, and due to the small size of the shoot primordium in the embryo, this defect has only been observed after germination (Fig. 5K) (Jürgens et al. 1994; T. Laux and G. Jürgens, unpubl.).

A mutant seedling may lack a pattern element for either of two reasons: The gene may specifically be required during pattern formation in the embryo or, alternatively, the gene may also be necessary for the non-embryogenic formation of the same structure. To distinguish between these possibilities for the *gn* and *mp* genes, mutant seedlings, which lack root and root meristem, were assayed for their ability to form roots from a wound edge under root-inducing culture conditions. Wounded *mp* seedlings responded like bisected wild-type seedlings, forming a root from

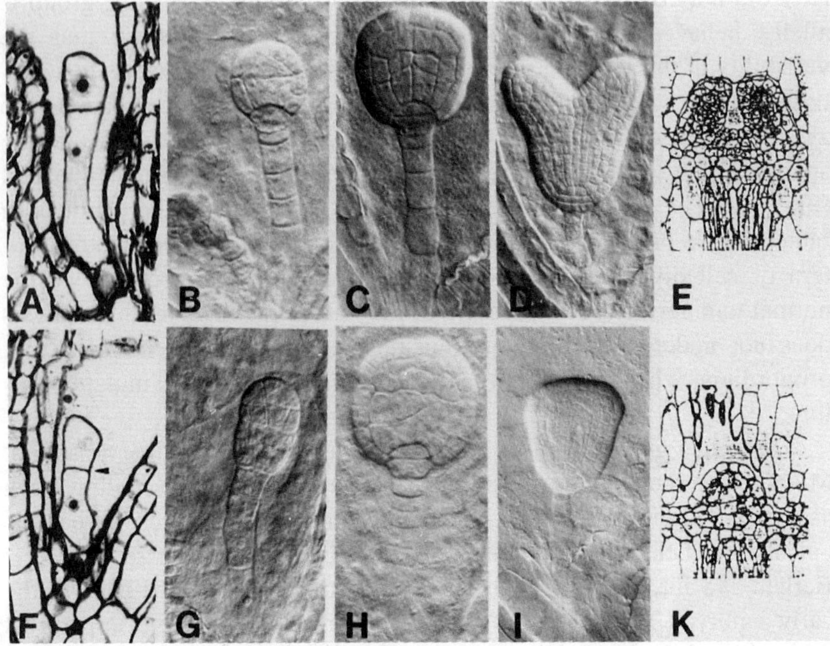

*Figure 5* Early phenotypes of apical-basal pattern mutants. (*A–E*) Wild-type, (*F*) *gnom*, (*G*) *monopteros*, (*H*) *fackel*, (*I*) *gurke*, (*K*) *zwille*. Corresponding developmental stages of wild-type and mutants are shown: (*A,F*) one-cell, (*B,G*) dermatogen, (*C,H*) mid-globular, (*D,I*) mid-heart stage embryos, (*E,K*) 2-day old seedlings. (*A,E,F,K*) Histological sections, (*B–D, G–I*) whole-mount preparations. (Reprinted, with permission, from Jürgens et al. 1994 [copyright Springer-Verlag].)

their basal end (Berleth and Jürgens 1993). In contrast, bisected *gn* seedlings were unable to form a root but produced callus instead (Mayer et al. 1993a). Thus, *mp* and *gn* represent two different classes of genes: *mp* is only necessary for root formation in the embryo, whereas the *gn* gene plays a more general role in root formation.

## Double Mutants Provide Evidence for Gene Interaction

The *gn* gene seems to act differently from other genes for apical-basal pattern formation. To determine directly how the *gn* gene relates to the other genes, relevant double mutants were analyzed phenotypically. In all combinations tested (*gn mp, gn gk, gn fk*), only parental phenotypes were observed, suggesting that one of the two genes is epistatic to the other. Since genotypes cannot be recognized independently of the mutant phenotype, the epistatic gene has not been identified except for one case.

The *gn* and *mp* genes are closely linked such that the proportion of double mutants among the seedling progeny depends on whether the mutant alleles are in *cis* or *trans* configuration in the parental doubly heterozygous plant. With the mutant alleles in *cis*, *gn* seedlings outnumbered *mp* seedlings, indicating that *gn* is epistatic (Mayer et al. 1993a). Thus, the *gn* gene has to be active for the *mp* gene to exert its function. Considering the developmental time at which the mutant embryos for the other genes deviate from normal, it is very likely that the action of those genes also depends on prior *gn* gene activity.

## A Model of Gene Action in Pattern Formation

Two aspects of pattern formation can conceptually be distinguished: partitioning and region-specific development (Fig. 6) (Jürgens et al. 1994). Early partitioning of the apical-basal axis into three major regions, apical, central, and basal, seems to result from the asymmetric division of the zygote, which requires *gn* gene activity. The regions show distinctive features such as characteristic patterns of cell division from their inception, and this region-specific development depends on the activities of other genes. For example, the *mp* gene directs the development of both the central and the basal regions, whereas the *gk* gene is required for normal development of the apical region. The apical region is later subdivided into the primordia of cotyledons and shoot meristem, with only the latter requiring the activity of the *zll* gene. The central region, which will give rise to the hypocotyl and the embryonic root, seems to need *fk* gene activity from the globular stage. However, *fk* mutations do not interfere with the development of the embryonic root, which suggests that the primordia of hypocotyl and root are genetically distinct long before they can be recognized morphologically. The basal region, which corresponds to the hypophysis, generates the incipient root primordium, including the center but not the initials of the root meristem. Both this region and the root initials derived from the central region require *mp* but not *fk* gene activity, and none of the other genes distinguishes between the two. It is conceivable that the root initials and thus the embryonic root are generated by induction.

The genes described here, with the exception of *gnom*, direct region-specific development but do not address the question of how the regions are established. Obviously, the model does not reflect the genetic complexity of the process. It is possible that only genes with pattern-specific functions have been identified, whereas genes involved in progressive partitioning may also be required for other processes, escaping phenotypic detection as pattern mutants.

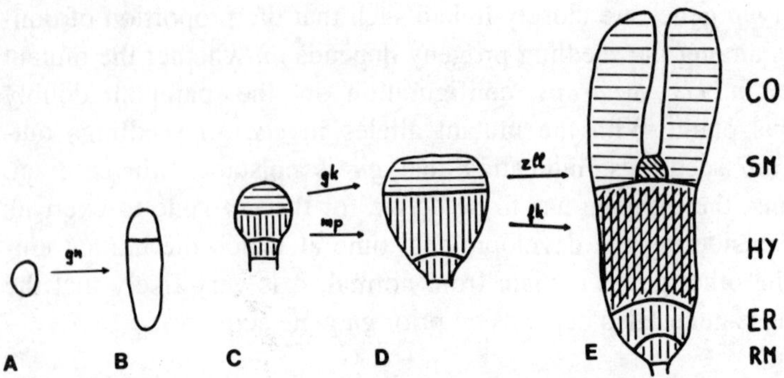



*Figure 6*  A model of gene action in apical-basal pattern formation. (*A–C*) Partitioning of the axis: *gn* gene action results, via asymmetric division of the zygote (*A,B*), in major subdivisions of the octant-stage embryo (*C*). Subsequent region-specific development (*C–E*) establishes, via the triangular stage (*D*), different primordia at the mid-torpedo stage (*E*): shoot meristem (SM), cotyledons (CO), hypocotyl (HY), embryonic root (ER), and root meristem (RM). Differently hatched regions correspond to putative domains of action of genes (*fk, gk, mp, zll*). For time of gene action, see text. (Reprinted, with permission, from Jürgens et al. 1994 [copyright Springer-Verlag].)

## CONCLUSIONS AND PERSPECTIVES

Until recently, pattern formation in the embryo was a neglected aspect of plant development. Genetic analysis has changed the situation by demonstrating that pattern formation can be dissected and that individual genes play specific roles in the process. This is especially clear for pattern formation along the apical-basal axis to which a few genes make major contributions. The next step will be to determine how these genes act at the molecular level. Since the genes are being cloned, we will soon learn whether their products are unique to plant pattern formation or whether they are related to familiar elements of animal pattern formation, e.g., transcription factors with known motifs or elements of one of the signal transduction cascades. It will also be interesting to compare the expression patterns of the region-specific genes with their domains of action, and if the two closely match, to determine how the expression is regulated. Furthermore, the molecular analysis of the genes identified may provide an explanation for the seemingly low genetic complexity of the process. It is unlikely that only five genes are sufficient to generate the apical-basal pattern of the seedling. However, it also seems unlikely that many more genes mutating to similar phenotypes exist but were not identified in the large-scale screen for pattern mutants. Could additional genes involved in pattern formation also be required for other developmental processes such that mutations cause embryonic or gametophytic lethality rather than pattern defects? If this is the case, molecular cloning of the genes identified should facilitate searches for "missing links" in embryonic pattern formation.

## ACKNOWLEDGMENTS

I thank my colleagues Martin Hülskamp, Ulrike Mayer, and Sara Ploense for critically reading the manuscript. The original research on which this review is based was supported by grants Ju-179/2-1 and Ju-179/2-2 from the Deutsche Forschungsgemeinschaft.

## REFERENCES

Arondel, V., B. Lemieux, I. Hwang, S. Gibson, H.M. Goodman, and C.R. Somerville. 1992. Map-based cloning of a gene controlling omega-3 fatty acid desaturation in *Arabidopsis. Science* **258**: 1353–1355.

Berleth, T. and G. Jürgens. 1993. The role of the *monopteros* gene in organising the basal body region of the *Arabidopsis* embryo. *Development* **118**: 575–587.

Dolan, L., K. Janmaat, V. Willemsen, P. Linstead, S. Poethig, K. Roberts, and B. Scheres. 1993. Cellular organisation of the *Arabidopsis thaliana* root. *Development* **119**: 71–84

Giraudat, J., B.M. Hauge, C. Valon, J. Smalle, F. Parcy, and H.M. Goodman. 1992. Isolation of the *Arabidopsis ABI3* gene by positional cloning. *Plant Cell* **4**: 1251–1261.

Jürgens, G. and U. Mayer. 1994. *Arabidopsis.* In *Embryos: A colour atlas of development* (ed. J.B.L. Bard), pp. 7–21. Wolfe Publishing, London.

Jürgens, G., U. Mayer, R.A. Torres Ruiz, T. Berleth, and S. Miséra. 1991. Genetic analysis of pattern formation in the *Arabidopsis* embryo. *Dev. Suppl.* **1**: 27–38.

Jürgens, G., R.A. Torres Ruiz, T. Laux, U. Mayer, and T. Berleth. 1994. Early events in apical-basal pattern formation in *Arabidopsis.* In *Molecular-genetic analysis of plant metabolism and development* (ed. G. Coruzzi and P. Puigdomènech), pp. 95–103. Springer-Verlag, Berlin.

Leyser, H.M.O., C.A. Lincoln, C. Timpte, D. Lammer, J. Turner, and M. Estelle. 1993. *Arabidopsis* auxin resistance gene *AXR1* encodes a protein related to ubiquitin-activating enzyme E1. *Nature* **364**: 161–164.

Lyndon, R.F. 1990. *Plant development. The cellular basis.* Unwin Hyman, London.

Mansfield, S.G. and L.G. Briarty. 1991. Early embryogenesis in *Arabidopsis thaliana.* II. The developing embryo. *Can. J. Bot.* **69**: 461–476.

Mayer, U., G. Büttner, and G. Jürgens. 1993a. Apical-basal pattern formation in the *Arabidopsis* embryo: Studies on the role of the *gnom* gene. *Development* **117**: 149–162.

Mayer, U., T. Berleth, R.A. Torres Ruiz, S. Miséra, and G. Jürgens. 1993b. Pattern formation during *Arabidopsis* embryo development. In *Cellular communication in plants* (ed. R.M. Amasino), pp. 93–98. Plenum Press, New York.

Mayer, U., R.A. Torres Ruiz, T. Berleth, S. Miséra, and G. Jürgens. 1991. Mutations affecting body organization in the *Arabidopsis* embryo. *Nature* **353**: 402–407.

Medford, J.I. 1992. Vegetative apical meristems. *Plant Cell* **4**: 1029–1039.

Meinke, D.W. 1991. Perspectives on genetic analysis of plant embryogenesis. *Plant Cell* **3:** 857–866.

Meinke, D.W. and I.M. Sussex. 1979. Embryo-lethal mutants of *Arabidopsis thaliana*: A model system for genetic analysis of plant embryo development. *Dev. Biol.* **72:** 50–61.

Miséra, S., A.J. Müller, U. Weiland-Heidecker, and G. Jürgens. 1994. The *FUSCA* genes of *Arabidopsis:* Negative regulators of light responses. *Mol. Gen. Genet.* **244:** (in press).

Müller, A.J. 1963. Embryonentest zum Nachweis rezessiver Letalfaktoren bei *Arabidopsis thaliana. Biol. Zentralbl.* **82:** 133–163.

Natesh, S. and M.A. Rau. 1984. The embryo. In *Embryology of angiosperms* (ed. B.M. Johri), pp. 377–443. Springer-Verlag, Berlin.

Schulz, R. and W.A. Jensen. 1968. *Capsella* embryogenesis: The egg, zygote, and young embryo. *Am. J. Bot.* **55:** 807–819.

Steeves, T.A. and I.M. Sussex. 1989. *Patterns in plant development.* Cambridge University Press, Cambridge, United Kingdom.

Sussex, I.M. 1989. Developmental programming of the shoot meristem. *Cell* **56:** 225–229.

Tykarska, T. 1976. Rape embryogenesis. I. The proembryo development. *Acta Soc. Bot. Pol.* **45:** 3–15.

―――. 1979. Rape embryogenesis. II. Development of the embryo proper. *Acta Soc. Bot. Pol.* **48:** 391–421.

Vandendries, R. 1909. Contribution à l'histoire du développement des crucifères. *Cellule* **25:** 412–459.

# 12

# Seed Dormancy and Germination

**Maarten Koornneef[1] and Cees M. Karssen[2]**
[1]Department of Genetics, NL-6703 HA
[2]Department of Plant Physiology, NL-6703 BD
Wageningen Agricultural University
Wageningen, The Netherlands

Seeds are designed to permit the mature embryo, surrounded by its seed coat, to survive the period between detachment from the parental plant and its establishment as a new seedling. Seeds have a number of characteristics related to their survival strategy. They are resistant to desiccation; they possess food reserves, assumed to be required for proper growth as a seedling; and they exhibit seed dormancy, a mechanism that prevents germination in periods unfavorable for seedling growth but allows germination at the proper time in the right environment. A well-adapted species only germinates at the beginning of the season that benefits seedling growth, which allows the completion of the life cycle. Thus, proper timing of germination is essential for survival of the species.

*Arabidopsis* is an annual distributed over a large part of the northern, mainly temperate, parts of the world. Since in many sites it flowers before the onset of a warm and dry summer, it can survive that summer period only in the seed stage. In general, conditions become favorable for germination during the fall when seeds germinate, and then the plants survive winter as seedlings or young nonflowering plants (Baskin and Baskin 1972). This is the typical behavior of a winter annual. In some cases, seed germination is postponed until the following spring, whereafter the plant completes its life cycle in the same year (Ratcliffe 1976). *Arabidopsis* is well suited for the study of the physiological mechanisms of dormancy and germination, because as a "wild" plant, it still contains all the regulatory circuits for dormancy control, in contrast to many crop plants. Combined physiological, genetic, and molecular analysis permits the identification of the biochemical and molecular mechanisms of these regulatory processes. In this chapter, seed development is described with special emphasis on the later stages that follow upon the completion of morphogenesis and cell division. This maturation phase is characterized

*Arabidopsis*
© 1994 Cold Spring Harbor Laboratory Press 0-87969-428-9/94 $5 + .00

by (1) the accumulation of nutritive reserves; (2) the arrest of tissue growth and development; (3) the induction of primary or innate dormancy, and (4) desiccation and the development of desiccation tolerance. The regulation of these various processes occurs either in an interrelated way, as a sequence of events, one resulting from another, or as independent parallel events.

## SEED DEVELOPMENT

The completion of seed development in *Arabidopsis* takes 18–21 days when plants grow in standard greenhouse conditions (temperature >20°C, sufficient light). In growth chambers with 24 hours of light at 22°C–24°C, this period is often reduced to 15–17 days. A mature seed is characterized by a brown testa and a yellowish embryo. The development of testa pigments and chlorophyll breakdown in the embryo are the last events of seed development. The yellow color of the embryo is clearly seen in transparent testa (*tt*) mutants that lack the brown testa pigments (Koornneef 1990).

The testa originates from the maternal ovule tissue, whereas the embryo and the thin layer of endosperm around it develop from the zygote. Therefore, the seed parts may differ genetically from each other.

### Embryogenesis

Embryogenesis proper takes less than 1 week in *Arabidopsis*. Therefore, the size and shape of seed and embryo hardly change in the period that follows, until complete maturity (Müller 1963; Mansfield and Briarty 1991). The embryo represents most of the volume of the mature seed. Many mutations affecting embryogenesis have been described previously (for review, see Meinke 1991) and are discussed in other chapters in this volume.

### Developmental and Biochemical Changes in the Embryo during Seed Maturation

The patterns of change in the content of a number of compounds during maturation have been described in *Arabidopsis*. In particular, the composition of compounds related to the "function" of maturation, such as storage and other proteins, lipids, carbohydrates, and hormones, has been investigated. The developmental patterns for all compounds analyzed thus far are very similar to those of related species such as *Brassica napus* (Norton and Harris 1975).

*Lipids*

As in most cruciferous plants, lipids stored in lipid bodies (Patton and Meinke 1990) are the major seed storage reserves. Most of the lipids are triacyl-glycerols, mainly containing 18:1 (oleic acid), 18:2 (linoleic acid), 18:3 (linolenic acid), and 20:1 (eicosenoic acid) fatty acids (Finkelstein and Somerville 1990; Lemieux et al. 1990). Mutants at a number of different loci (*fab1, fab2, fad2, fad3, ela1, fae1, fae2, rod1*) that variously affect the seed fatty acid composition (James and Dooner 1990; Lemieux et al. 1990) have been isolated. No effects of these mutations on seed morphology, seed size, or physiological characteristics of the seeds have been described. Eicosenoic acid levels are preferentially reduced in the *abi3-1* mutant (Finkelstein and Somerville 1990) and even more in the *aba abi3-1* double mutant (de Bruijn et al. 1993). The seed phospholipids showed a similar reduction of eicosenoic acid (Karssen and van Loon 1992; Ooms et al. 1993a) in this double mutant. This probably reflects the fact that eicosenoic acid synthesis is catalyzed by a seed-specific pathway, whereas the other fatty acids are catalyzed by a ubiquitous pathway that is not under *ABI3* control.


*Proteins*

The *Arabidopsis* storage proteins were first studied by Heath et al. (1986) and were shown to resemble the storage protein pattern of other Cruciferae. A set of 12S A and B subunits resembling the globulin or cruciferin proteins represent 50–55% of the proteins in mature seeds (Finkelstein and Somerville 1990). The other main storage proteins are the 2S proteins, resembling albumin or napin. The 2S albumins are found in mature seeds as heterodimers, consisting of two 3-kD and 9-kD subunits linked to each other by two disulfide bridges. These subunits are synthesized as 18-kD precursors that undergo posttranslational proteolytic processing (Krebbers et al. 1988; D'Hondt et al. 1993). The 12S genes have been cloned by Pang et al. (1988) and were shown to represent at least three subfamilies (*CRA, CRB,* and *CRC*), of which *CRA* has two members in Landsberg *erecta* (L*er*) (*CRA1* and *CRA2*) and only one in Columbia (Col). The 2S proteins are encoded by at least four genes (*AT2S1–AT2S4*), which are tightly linked in a tandem array (Krebbers et al. 1988). These four genes are expressed in both the hypocotyl and cotyledons, except *AT2S1*, which is not expressed in cotyledons (Guerche et al. 1990). A fifth isoform of the 2S proteins, probably encoded by a fifth *AT2S* gene, was recently described by van der Klei et al. (1993). The expression of both the 2S and 12S genes is restricted to the latter half of seed development. The storage proteins are stored in protein bodies of which the ultrastructure was described by Patton and Meinke (1990), and were found both in the hypocotyl and in the cotyledons. In situ hybridization analysis of the expression of the storage protein genes (Pang et al. 1988; Guerche et al. 1990) showed a similar localization.

During the later stages of seed maturation, when embryo water content is decreasing, specific genes are expressed which are called *late embryogenesis abundant* (*lea*) genes. The proteins encoded by these genes have a very hydrophilic amino acid composition and are supposed to function as desiccation protectants (Skriver and Mundy 1990). Two such genes (*ATEM1* and *ATEM6*), which are unlinked to each other and which are homologous to the wheat *Em* and cotton *D19* genes, have been cloned in *Arabidopsis* (Finkelstein 1993; Gaubier et al. 1993). In situ hybridization with the *ATEM1* probe demonstrated that the expression of this gene is located in the provascular tissues of the cotyledons and axis of the dry seed as well as in the epiderm and outer layers of the embryo cortex (Gaubier et al. 1993). A detailed electrophoretic analysis of storage protein accumulation during seed development showed that the seed maturation program is not completed in the *aba abi3-1* double mutant. On the contrary, germination patterns were initiated prematurely, as shown by the resemblance between the protein pattern in "mature" double mutants and that of germinating wild-type seeds (Meurs et al. 1992).

### Carbohydrates

*Arabidopsis* seeds accumulate starch when they are still green. Relatively late in development the starch level declines (Caspar et al. 1991; de Bruijn 1993; de Bruijn et al. 1993). Sucrose is the main soluble carbohydrate in developing *Arabidopsis* seeds, although in the second week after pollination equal amounts of glucose, fructose, and sucrose are present (de Bruijn et al. 1993).

### Plant Hormones

A partial time course of hormone levels during seed development has been published for abscisic acid (ABA) and gibberellins (GAs). ABA levels show a peak half-way through development, which is mainly of maternal origin (Fig. 1) (Karssen et al. 1983). ABA accumulates transiently in the embryo with a peak at 14–16 days after flowering. The relevance of ABA levels for the induction of dormancy is discussed below. Gibberellin biosynthesis, as measured by the *ent*-kaurene synthesizing capacity, is at its maximum during the first week of seed develop-

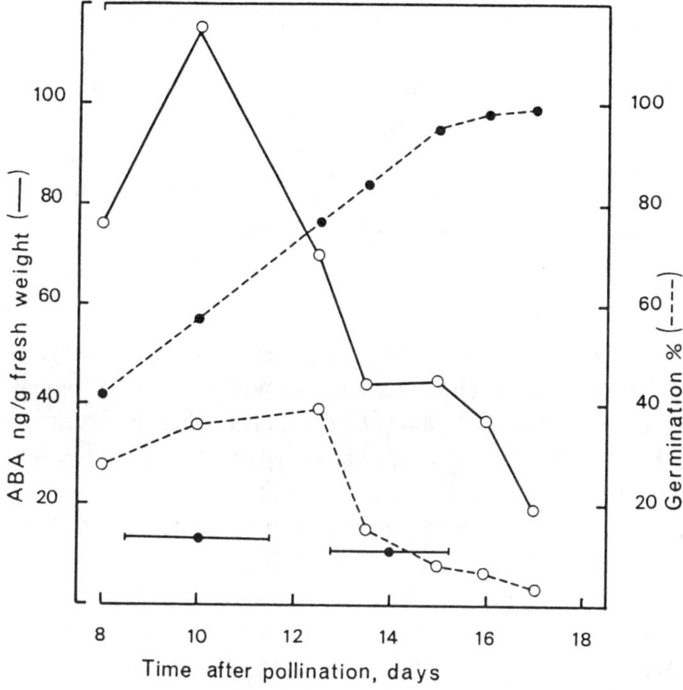

*Figure 1* Changes with time after pollination in ABA content (*solid line*) and in germination of isolated seeds of the *Arabidopsis* wild type (*open symbols*) and the *aba-3* mutant (*closed symbols*). (Modified, with permission, from Karssen et al. 1983.)

ment when cell divisions occur in the embryo together with rapid growth of the siliques (Barendse et al. 1986). Thereafter, this activity decreases.

## Development of Desiccation Tolerance

Excised immature seeds are able to germinate but, in contrast to mature seeds, cannot survive dry storage. Desiccation tolerance develops before the seeds are completely mature and have turned brown (Koornneef et al. 1989). Sensitivity to desiccation is one of the most obvious characteristics of the *aba abi3-1* double mutant (Koornneef et al. 1989) and the extreme *abi3* alleles (Nambara et al. 1992; Ooms et al. 1993b). In the double mutant, desiccation tolerance can be induced by weekly applications of ABA to the roots of the mother plant or by imbibition of the seeds in a mixture of ABA and 3% sucrose (Meurs et al. 1992). Damage by desiccation is essentially caused by damage to membranes, leading to leakage of cell constituents. Therefore, membrane components such as phospholipids and sugars, which are involved in the protection of membranes, were studied in desiccation-sensitive genotypes. However, both groups of compounds showed relatively high levels in the *aba abi3-1* double mutant instead of the low levels that were expected (de Bruijn et al. 1993; Ooms et al. 1993a). Another option is that membranes are protected by proteins. Late-embryogenesis-abundant proteins are indeed strongly reduced in *abi3-1* (Finkelstein 1993) and have been suggested as desiccation protectants previously (Skriver and Mundy 1990).

## Genetic Control of Seed Maturation

It is thought that there is a causal relationship between the appearance and disappearance of various chemical compounds and physiological changes during seed maturation. Mutants can be particularly useful in investigating the relationship between the various patterns of change because aberrations that occur simultaneously and that are due to a single gene mutation indicate a common regulation of the processes involved. Examples of such mutants are the *lec* mutant (Meinke et al. 1992), extreme *abi3* mutants (Nambara et al. 1992; Ooms et al. 1993b), and the *fusca C* (now *fus3*) mutants (Müller and Heidecker 1968), in which almost every aspect of seed maturation, including the accumulation of storage proteins, desiccation tolerance, and dormancy, is affected. The cotyledons of the leafy cotyledon (*lec*) mutant produce trichomes characteristic of leaves and exhibit a vascular pattern intermediate between that of leaves and cotyledons, together with an aberrant seed maturation (Meinke 1992). The *LEC* gene, located on chromosome 1, probably activates a wide range of embryo-maturation-specific pathways. The *fus3* mutants (Müller and Heidecker 1968), including the 10-2 or PG1 mutants described by Kraml et al. (1993), which, based on phenotype and map position (between *abi3* and *gl1* on chromosome 3) probably are allelic with *fus3* (S. Misera and P. McCourt, pers. comm.), are nondormant, lack storage proteins and lipids, accumulate anthocyanin during seed development, and have shriveled seeds when mature. The anthocyanin accumulation classifies them as *fusca* mutants. The *abi3* mutants are nondormant, have a reduced level of particular lipids and storage proteins, and do not lose chlorophyll when mature (green seeds). In addition, they are extremely resistant to the inhibitory effects of ABA on seed germination. As is the case with *lec* and *fus3* mutants, extreme *abi3* alleles such as *abi3-3* (Nambara et al. 1992), *abi3-4*, and *abi3-5* (Ooms et al. 1993b) are sensitive to desiccation. Mutants at all three loci give rise to completely normal looking plants when immature seeds are sown before desiccation has started, indicating that the expression of these genes plays no significant role during later stages of the plant's life cycle. It has been suggested that ABA is a major regulator of seed development. The phenotype of the ABA-insensitive (especially *abi-3* mutants) mutants, which are affected in many aspects of seed maturation, seem to confirm this. The *ABI3* gene has recently been cloned. It is one of the first examples of positional cloning in *Arabidopsis* (Giraudat et al. 1992) and shows a resemblance to the maize *VP1* gene, whose function seems to be the seed-specific enhancement of the effect of ABA (McCarty et al. 1991). The rather normal seed phenotype (except for the lack of dormancy) in ABA-deficient *aba* mutants (Koornneef et al. 1982) casts doubt on the general regulatory role of ABA in seed development. However, the dramatic phenotype of the *aba abi3-1* double mutant (Koornneef et al. 1989) indicates that ABA action is required for normal seed development in this genotype. A few observations argue against the exclusive role of ABA (mediated by the *ABA* and *ABI* genes) in the regulation of seed development. First, ABA deficiency, even when combined with ABA insensitivity conferred by the *abi1* mutant, does not result in a "green seed" phenotype (Koornneef et al. 1989). Second, even the leaky *abi3-1* mutant produces a slight reduction of storage protein and lipid accumulation (Finkelstein and Somerville 1990) and reduces Em mRNA levels (Finkelstein 1993), whereas these effects are not observed in *aba* and *abi1* mutants. Third, the double mutant *abi1 abi3-1* is 100-fold more resistant to ABA than the monogenic mutants, suggesting that the genes control parallel pathways (Finkelstein and Somerville 1990). The *ABI3* product apparently is required for an effective ABA response in seed development, maybe by transcriptional activation of ABA-induced genes as described for VP1 in maize (McCarty et al. 1991). The action of ABA is required for the induction of dormancy and the inhibition of seed germination by ABA, since *aba* and *abi* mutants are modified with respect to the latter process (Koornneef et al. 1984). However, in addition, *ABI3* may activate the maturation pathway as such. This pathway also requires *LEC* and *FUS3* without a direct relation to ABA. High levels of ABA may also affect these processes, but may not be required for them (Finkelstein 1993).

With the exception of those related to seed lipids and ABA, few other mutants affecting seed maturation or the chemical composition of *Arabidopsis* seeds have been described. However, some shriveled-seed mutants accumulate much less storage lipid and/or protein (C. Benning, pers. comm.). The *sex1* mutants, which affect starch degradation, accumulate starch in seeds as well as in other tissues (Caspar et al. 1991).

No role for ABA in the regulation of transport of assimilates to seeds could be shown using *aba* mutants together with reduced light conditions and starchless *pgm* mutants to reduce the availability of source material (de Bruijn and Vreugdenhil 1993). The physiological and biochemical background of genetic differences in seed weight, which shows a variation from 10 µg to 35 µg (Krannitz et al. 1991) between ecotypes, is not known. Mutants with larger seeds (K.M. Léon-Kloosterziel, pers. comm.) and mutants with small shriveled seeds (C. Benning, pers. comm.) have been isolated in *Arabidopsis*.

## Testa

The testa is of maternal origin, since it develops from ovule structures called integuments. The *Arabidopsis* testa resembles that of *Capsella bursa-pastoris* (Roth 1957; Bouman 1975; D. Valvekens, pers. comm.). At the apical end of the seed, it is composed of five cell layers. The testa determines to a large extent the shape of the seeds, since the aberrant testa shape (*ats*) mutant, which has only three cell layers and probably lacks one integument, is heart-shaped (K.M. Léon-Kloosterziel et al., in prep.). In the mature seeds, cell layers of the inner integument are compressed. The outer layer of the testa contains a mucilage composed of a pectic substance with rhamnose as its main neutral sugar (Goto 1985). It is released upon imbibition when the outer cell wall breaks. The *ttg gl2* (Koornneef 1981), *ats* (K.M. Léon-Kloosterziel et al., in prep.), and the *ap2* mutants (B. den Boer, pers. comm.) have an aberrant outer seed coat layer and produce no mucilage. The production of mucilage is apparently under the control of ABA, since *aba* mutants produce less mucilage than the wild type (Karssen et al. 1983).

The innermost cell layer of the inner integument contains brown pigments (Bouman 1975) derived from flavonolic compounds. This layer is colorless in anthocyanin-free mutants (Koornneef 1981, 1990).

The role of the testa in seed germination may be to prevent the protrusion of the embryo radicle (coat-imposed dormancy). Pricking seeds or removing the testa improves germination of *Arabidopsis* seeds (Kugler 1951; Dobrovolna and Cetl 1966; Goto 1982), although the effect is limited in some genotypes (Goto 1982). Most testa mutants indeed appear to have a reduced seed dormancy (K.M. Léon-Kloosterziel et al., in prep.).

## Development of Dormancy

### Terminology

Dormancy as used in this chapter is defined as "the temporary failure of a viable seed to germinate after a specified length of time in a particular set of environmental conditions that later evoke germination when the restrictive state has been terminated by either natural or artificial means" (Simpson 1990). Primary dormancy develops during seed maturation on the mother plant and can be relieved by dry storage (after ripening) or by imbibition at certain temperatures over a certain time period (Derkx and Karssen 1993).

When the environmental conditions are not suitable for germination (lack of water or oxygen, inappropriate temperature), seeds are in a state of so-called pseudo-dormancy, since germination occurs as soon as the environmental restrictions are terminated (Hilhorst and Karssen 1992). Prolongation of the environmental restrictions can result in a re-induction of dormancy, which is called secondary dormancy. Thus, dormancy is expressed in a certain set of conditions, and degrees of dormancy are often distinguished on the "strength" of the conditions necessary to alleviate or "break" dormancy. The consequence of this indirect assay for dormancy is that it is affected by the capacity to break dormancy.

*Genetic Analysis of Dormancy and Germination Characteristics*

Germination of an individual seed is an all-or-nothing event. Germination of a population of seeds is quantified by the proportion which germinates under a given set of conditions. The need for this quantification complicates genetic analysis, since one has to work with seed progenies of individual plants instead of individual seeds. The consequence of the strong environmental effects on seed-germination characteristics is that in order to observe genetic differences in degree of dormancy, the environmental factors acting during the development of dormancy on the mother plant and the storage conditions of the seeds and of the germination tests have to be kept identical. Under greenhouse conditions in Wageningen, freshly harvested seeds of wild type (L*er*) do not germinate in white light, whereas nondormant mutants germinate to a large extent in these conditions. However, the application of a cold treatment (4–7 days at 2–7°C) and dry storage for at least one month results in 100% germination of the wild type. Since germinability of nondormant mutants remains 100%, the genotypes cannot be discriminated any more following such treatments (Karssen and Lacka 1986). A high percentage of germination in darkness is often observed in less dormant seeds (Koornneef et al. 1984). However, even some seed batches of the wild-type L*er* show some dark germination (Cone and Spruit 1983). The various ways to "measure" dormancy have led to the use of different parameters to describe differences in dormancy such as light-requiring versus non-light-requiring genotypes (Kugler 1951), after-ripening-requiring and non-after-ripening-requiring genotypes (Laibach 1951; Evans and Ratcliffe 1972).

Genetic differences for seed germination characteristics between ecotypes have been reviewed by Rédei (1970). Kugler (1951) showed that the light dependency of germination of the ecotype Hannovrisch Münden (Hm) was recessive in crosses with the dark-germinating ecotypes Stockholm (St) and Haarlem (Haa). The comparison of reciprocal crosses indicated that this property was controlled by the genotype of the embryo. Napp-Zinn (1975), who analyzed $F_3$ lines derived from the Hm x St cross, suggested that three genes controlled the difference in light requirement between both parents and that this trait might be linked to flowering-time genes segregating in the same population. The polygenic inheritance of seed-germination characteristics in populations derived from crosses between ecotypes complicates the use of germination parameters in such populations, and such populations are necessary for RFLP mapping.

Mutants affecting dormancy and seed germination have been identified on the basis of their inability to germinate under normal conditions. Exogenously applied gibberellins induced germination and permitted their propagation (Koornneef and van der Veen 1980). These mutants are all GA-deficient (Zeevaart and Talon 1992). The same type of screen also yielded phytochrome-deficient mutants (Brusslan and Tobin 1991). This result was expected because *hy* mutants have a reduced capacity to respond to light, and this light requirement can be compensated by exogenous GAs (Koornneef et al. 1981).

By selecting for germinating revertants in the progeny of nongerminating GA-responsive *ga1* mutants, it was possible to isolate mutants that alleviated the GA requirement by a second-site suppressor mutation, which was in a gene (*aba*) that appears to control a step in ABA synthesis (Koornneef et al. 1982; Rock and Zeevaart 1991). This screen yielded several other mutants, one of which was identified as an *abi3* allele (M. Koornneef, unpubl.). Nambara et al. (1992) isolated the extreme *abi3-3* allele by a somewhat similar type of screen. They selected for mutants that germinated in the presence of the GA biosynthesis inhibitor uniconazol. In addition to ABA-related (wilty) mutants, Jacobsen and Olszewski (1993) isolated mutants at the *spindly* (*spy*) locus, which are characterized by a "GA overdose" or slender phenotype. The *spy* mutant is partially epistatic to all aspects of the *ga1-2* mutant (Jacobsen and Olszewski 1993). Mutants isolated by their ability to germinate at ABA concentrations that inhibit wild-type seeds were also identified as nondormant mutants. In these lines, mutated at the *abi1*, *abi2*, and *abi3* loci, the sensitivity to ABA is reduced (Koornneef et al. 1984). Velemínsky and Gichner (1967) reported an albino mutant that lacks dormancy. This mutant contained no detectable chlorophyll and carotenoids. The absence of the latter compounds may result in a reduced ABA level, since carotenoids are the precursors of ABA (Rock and Zeevaart 1991). The data mentioned above suggest that only ABA-related mutants may have a reduced dormancy. However, several mutants with this property, but with neither an altered sensitivity to ABA nor symptoms of excessive water loss, which are characteristic for ABA-deficient mutants, have been identified in *Arabidopsis* (K.M. Léon-Kloosterziel and M. Koornneef, unpubl.; P. McCourt, pers. comm.). Mutants capable of germination under saline and high-mannitol conditions (Saleki et al. 1993) may be affected in seed dormancy, although this was not tested.

## Development of Seed Dormancy

Primary seed dormancy develops on the mother plant during the latter stages of seed development. This process can be monitored by testing the germination (in water and light) of seeds excised at specific times after pollination (Fig. 1). Mature seeds of L*er* do not germinate under these conditions. Half-developed seeds do germinate, however. Seeds originating from genotypes that do not develop dormancy, such as ABA-deficient and ABA-insensitive mutants, gradually develop full germination capacity from this stage onward (Fig. 1) (Karssen et al. 1983; Karssen and Lacka 1986). Endogenous ABA levels peak halfway through development and, to a minor extent, once more later on in development (Fig. 1). Reciprocal crosses between wild type and the *aba* mutant showed that the first ABA peak only occurred when the wild type was used as the mother plant. This peak did not correlate with the induction of dormancy. The ABA present 14–17 *days after pollination* (dap) was mainly of embryonic origin and clearly correlated with dormancy induction (Karssen et al. 1983).

Developing seeds of *Arabidopsis* are rich sources of GAs, as in many other species. However, in the absence of GA synthesis, *ga1* seeds develop normally and show normal induction of primary dormancy (Karssen and Lacka 1986). The studies with *ga1* mutants, however, also showed that the GAs that are synthesized in the embryo have a stimulating effect on the growth of the surrounding silique (Barendse et al. 1986). In addition, as described above, GA synthesis or addition seems to be required for the induction of germination in wild-type seeds.

The large variation in dormancy between different seed lots of the same genotype must be attributed to the environment during seed development (Derkx and Karssen 1993). The effect of such environmental factors was demonstrated by McCullough and Shropshire (1970) and Hayes and Klein (1974), who showed that the spectral composition of the light source during growth of *Arabidopsis* plants influenced both the sensitivity of the seeds to red light and the level of dark germination. Seeds obtained from plants grown in fluorescent light (relatively high red/far-red [R/FR] ratio) had high dark germination but a low sensitivity to red-light-induced germination as compared to plants grown in conditions with additional incandescent bulbs (lower R/FR ratio). High dark germination related to the fraction of the far-red-light-absorbing form of phytochrome of the total phytochrome (Pfr/Ptot ratio) set during seed development, whereas the sensitivity for light for the promotion of germination relates to the degree of dormancy.

The phytochrome-deficient *hy1*, *hy2*, and *hy3* mutants (Koornneef et al. 1980) show a reduced responsiveness to light and have a reduced phytochrome content in their seeds (Spruit et al. 1980). The fluence response curves for the induction of germination by red light are more shallow in the mutants, but dark germination is relatively high (Cone and Kendrick 1985). These fluence response curves are compatible with a reduced phytochrome content combined with a so-called germination-promoting, light-independent "overriding factor" (Cone and Kendrick 1985). This can be interpreted as a reduced light requirement (reduced dormancy) due to phytochrome deficiency during seed development. This is then in agreement with the data obtained by Hayes and Klein (1974). The observation that phytochrome chromophore mutants *hy1* and *hy2*, which predominantly lack the phytochrome type PhyA, behave similarly to the PhyB-deficient *hy3* mutant suggests that total phytochrome and not a specific phytochrome species is responsible for the phytochrome effects on germination.

Karssen et al. (1990) and Derkx and Karssen (1993) reported dependency of the degree of dormancy on the growing season. Although not studied in *Arabidopsis*, factors such as temperature, day length, humidity, and nitrogen supply during seed formation, which affect dormancy in other species, probably also determine the degree of dormancy of *Arabidopsis* seeds (Derkx 1993; Derkx and Karssen 1993).

*Relief of Primary Dormancy*

The degree of dormancy induced during seed development can be relieved before germination starts. This relief of dormancy can occur during dry storage and in the imbibed state. Relief of dormancy is characterized by changes in a number of physiological parameters that affect the subsequent germination response. These are an increased sensitivity to germination-stimulating factors such as GAs, nitrate, cold and light treatments, and an increased temperature in which germination can occur. Virtually nothing is known about the biochemical mechanisms that underlie these phenomena (Hilhorst and Karssen 1992). The relief of dormancy by a period of dry storage, called after-ripening, is a very general phenomenon in many plant species. In *Arabidopsis*, after-ripening has been mentioned often in relation to the large genetic differences that exist between ecotypes (Laibach 1951) and even within populations (Evans and Ratcliffe 1972; Ratcliffe 1976).

Storage conditions, especially humidity, temperature, and light, determine the effectiveness of the after-ripening in breaking dormancy (Laibach 1956; Röbbelen and Kerstein 1965; Micke and Röbbelen 1966; Rehwaldt 1966). During dry storage of *Arabidopsis*, it was found that the sensitivity to GAs gradually increased both in wild-type and GA-deficient seeds (Karssen and Lacka 1986; Karssen et al. 1989; Derkx 1993). This illustrates that sensitivity to GA is independent of the capacity to synthesize GAs. The same conclusion was drawn from similar experiments in which chilling was used to break dormancy (Karssen and Lacka 1986; Karssen et al. 1989). In an experiment with *Arabidopsis* seeds that were imbibed and stored outdoors, it was found that the sensitivity of seeds to GA of both the wild type (L*er*) and the *ga1* mutant increased by a factor of 50–100 during breaking of primary dormancy (Derkx 1993).

*Secondary Dormancy*

Nondormant seeds that are exposed for some time to unfavorable germination conditions may enter a state of dormancy again. Seeds buried in the field annually pass through a cycle of dormancy-breaking and (re)induction (Karssen 1982). Such a seasonal dormancy has also been shown for *Arabidopsis* (Baskin and Baskin 1983). In temperate regions, temperature is the main regulatory factor in the control of these dormancy cycles of buried seeds. The effects of various temperatures on both the process of dormancy-breaking and induction of secondary dormancy are shown in Figure 2. The ecological significance of these processes is to reduce the chance of germination under unfavorable conditions, e.g., during a hot summer. The requirement for after-ripening is the second mechanism to achieve this (Ratcliffe 1976). Changes in GA sensitivity do not correlate with the annual dormancy pattern (Derkx 1993). However, fluctuations in the light requirement of the seeds show a strict correlation. The observation that in ABA-deficient *aba Arabidopsis* mu-

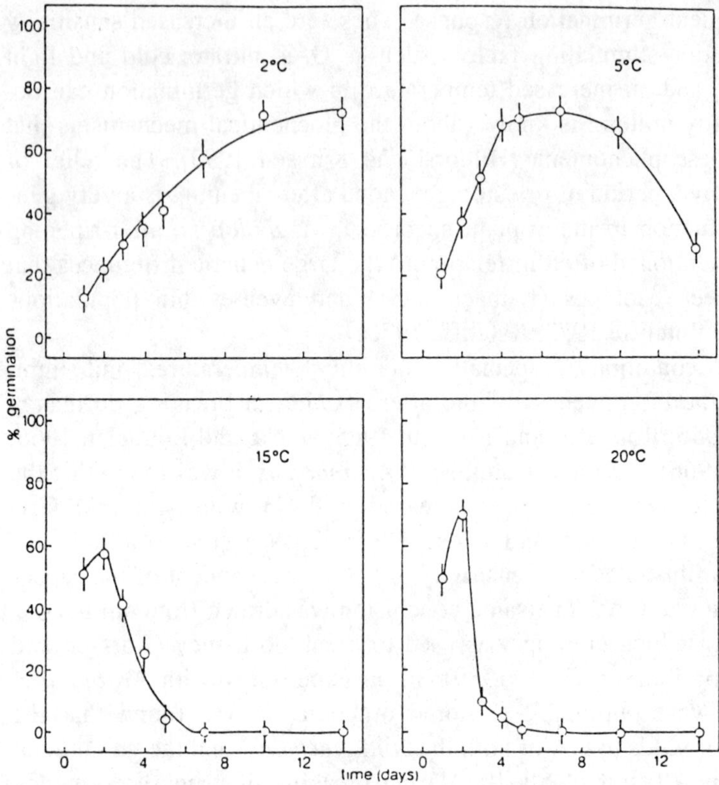



*Figure 2* Relationship between the time of imbibition and the percentage of germination at different imbibition temperatures for seeds of the L*er* ecotype. The seeds were imbibed in the dark and, after irradiation with red light (870 μmole m$^{-2}$), incubated for 4 days in darkness at 20°C. Nongerminating seeds are viable since strong dormancy-breaking treatments can induce germination. (Reprinted, with permission, from Cone and Spruit 1983.)

tants, secondary dormancy could be induced by a supraoptimal temperature of 35°C suggests that ABA is not involved in this process (Hilhorst and Karssen 1992).

## GERMINATION

A seed has germinated when the embryo (in *Arabidopsis*, the radicle) has protruded from the seed coat. Therefore, germination can be said to occur when embryo growth overcomes the mechanical restraint imposed by the testa. The stimulation of embryo growth is the consequence of a signal transduction chain that is induced after perception of the required environmental signals by nondormant seeds.

In light-requiring seeds such as *Arabidopsis*, the active form of phytochrome (Pfr) is assumed to be the initial trigger of this pathway, because typical phytochrome action spectra have been obtained in studies of *Arabidopsis* seed germination (Shropshire et al. 1961; Cone and Kendrick 1985). The period during which Pfr is required can be measured as the time that seeds need to escape from the red-light-reverting effect of far-red irradiation. This escape time varied from 1.5 hours to 9 hours, depending on the seed lot (Cone 1982), and correlated with the degree of seed dormancy. Certain pretreatments (e.g., 24 hours at 35°C) may induce an extreme responsiveness to Pfr called the *very low fluence response* (VLFR) (Cone et al. 1985). Some seed lots apparently do not require light and germinate in darkness. An explanation for this is that in these seeds the light (and GA, see later) requirement is so low that preexisting Pfr carried over from the mother plant is sufficient to satisfy this low light requirement.

Arguments have been provided (Karssen and Lacka 1986; Hilhorst and Karssen 1988; Karssen et al. 1989) that Pfr action is followed by GA synthesis. The absolute requirement for GA in germination is shown by the absence of germination in GA-deficient mutants (Koornneef and van der Veen 1980). Nambara et al. (1991) showed that the GA biosynthesis inhibitor uniconazol could not inhibit germination 18 hours after the start of imbibition, indicating that germination eventually becomes GA-independent. The observation that mutants without dormancy, such as the *aba* and extreme *abi3-3* mutant, germinate in a GA-deficient genetic background (Koornneef et al. 1982; Nambara et al. 1992) showed that GA is (only) necessary to alleviate the ABA-induced dormancy. The GA requirement for germination is lower than the GA requirement for elongation growth, since leaky alleles at the *ga1* locus do germinate without GAs but still are reduced in height (Koornneef et al. 1985). The roles of GA in seed germination are discussed by Hilhorst and Karssen (1992).

Other plant regulatory compounds also stimulate germination in *Arabidopsis*. An excess of ethylene bypasses the GA requirement of GA-deficient mutants (Goto 1983; Karssen et al. 1989), and high GA concentrations restored germination up to the wild-type level in the ethylene-insensitive *etr* mutant (Bleecker et al. 1988). However, the ethylene-induced germination in *ga1* mutants results in seedlings exhibiting aberrant seedling development (triple response), which indicates that the ethylene concentrations used do not occur normally (Karssen et al. 1989).

Cytokinins (Sankhla and Sankhla 1968) and certainly auxins hardly affect seed germination at physiological concentration and probably play no important role. Minor effects of polyamines on germination have been reported (Mirza and Bagni 1991). Nitrate, which is necessary for light-induced germination of the crucifer *Sisymbrium officinale* (Hilhorst and Karssen 1988), is not required in *Arabidopsis*, although it has a slight promoting effect on germination (Goto 1983; Derkx and Karssen 1993). Sometimes rather large differences in germination were obtained when different lots of filter paper were used during imbibition (Rehwaldt 1968; Spruit et al. 1981). Indications were obtained that polyethyleneimine present in some, but not all, batches of filter paper might be responsible for this germination-promoting effect (Spruit et al. 1981).

## SEED STORAGE AND SEED VIABILITY

Almost no information has been published about the length of time that stored *Arabidopsis* seeds remain viable. It is known that seeds survive storage at normal room temperature for a number of years, although in this respect, seed batches show considerable variation. Storage in dry, cold conditions increases the time during which crucifer seeds remain viable (Gomez-Campo 1976). It has been observed that the frequency of certain albino mutants decreased when seeds of the heterozygous plants were stored for a number of years (Röbbelen 1966), indicating that genetic differences exist for seed viability.

## CONCLUSIONS AND PERSPECTIVES

During the later stages of seed maturation, morphological changes are relatively limited, but physiological and biochemical changes are impressive and crucial for the survival of plants in the seed stage. Models concerning the role of hormones in seed development and germination have benefited strongly from the use of *Arabidopsis* mutants, especially those affecting GA and ABA biosynthesis and action. Such experiments have caused a revision of the hormone balance theory of seed dormancy (Karssen and Lacka 1986), where ABA and GA, instead of acting simultaneously in the control of seed dormancy and germination, are now thought to act in different periods of seed development. ABA during seed development causes dormancy, which determines the subsequent GA requirement. Genetics also demonstrated the importance of the site of synthesis of ABA in the dormancy induction process, since embryonic, but not maternal, ABA is effective (Karssen et al. 1983).

Probably the *LEC*, *FUS3*, and *ABI3* genes are crucial regulators of seed development. The *ABI3* gene might stimulate an ABA-induced signal transduction chain that is specific for this developmental phase. It is expected that regulation of various processes that are induced by ABA in combination with the *ABI3* gene product will be detected by the isolation of mutants and, subsequently, by cloning the respective genes. A mutant strategy that selects for variants in dormancy in combination with seed morphology and seed size seems most promising. It was successful in other species, such as cereals (King 1991) and pea (Wang and Hedley 1991). An additional fruitful approach may be the cloning of genes expressed in seeds in combination with subsequent experiments to find the regulatory proteins that activate the cloned genes. The analysis of genes cannot be performed without fully exploiting physiological analysis, using physiological, biochemical, and biophysical techniques. We expect that the combination of these different tools, along with the power of the molecular genetics of *Arabidopsis*, will impressively increase our knowledge of the chain of physiological and biochemical events that proceed during maturation and germination of seeds.

## ACKNOWLEDGMENTS

We thank Drs. Peter McCourt and Kallie Keith for their comments on the manuscript.

## REFERENCES

Barendse, G.W.M., J. Kepczynski, C.M. Karssen, and M. Koornneef. 1986. The role of endogenous gibberellins during fruit and seed development. Studies on gibberellin-deficient genotypes of *Arabidopsis thaliana*. *Physiol. Plant.* **67:** 315–319.

Baskin, J.M. and C.C. Baskin. 1972. Ecological life cycle and physiological ecology of seed germination of *Arabidopsis thaliana*. *Can. J. Bot.* **50:** 353–360.

———. 1983. Seasonal changes in germination responses of buried *Arabidopsis thaliana* and ecological interpretation. *Bot. Gaz.* **144:** 540–543.

Bleecker, A.B., M.A. Estelle, C. Somerville, and H. Kende. 1988. Insensitivity to ethylene conferred by a dominant mutation in *Arabidopsis thaliana*. *Science* **241:** 1086–1089.

Bouman, F. 1975. Integument initiation and testa development in some *Cruciferae*. *Bot. J. Linn. Soc.* **70:** 213–229.

Brusslan, J. and E. Tobin. 1991. The isolation of *Arabidopsis thaliana* mutants defective in phytochrome signal transduction. *Congress on Molecular Biology of Plant Development* (Abstr. 1479). A.R. Liss, New York.

Caspar, T., T.-P. Lin, G. Kakefuda, L. Benbow, J. Preiss, and C. Somerville. 1991. Mutants of *Arabidopsis* with altered regulation of starch degradation. *Plant Physiol.* **95:** 1181–1188.

Cone, J.W. 1982. The escape from photocontrol of seeds germination of *Arabidopsis thaliana*. *Arabidopsis Inf. Serv.* **19:** 35–38.

Cone, J.W. and R.E. Kendrick. 1985. Fluence-response curves and action spectra for promotion and inhibition of seed germination in wildtype and long-hypocotyl mutants of *Arabidopsis thaliana* L. *Planta* **163:** 43–54.

Cone, J.W. and C.J.P. Spruit. 1983. Imbibition conditions and seed dormancy of *Arabidopsis thaliana. Physiol. Plant* **59:** 416–420.

Cone, J.W., P.A.P.M. Jaspers, and R.E. Kendrick. 1985. Biphasic fluence-response curves for light induced germination of *Arabidopsis thaliana* seeds. *Plant Cell Environ.* **8:** 605–612.

de Bruijn, S.M. 1993. "Abscisic acid and assimilate partitioning during seed development." Ph.D. thesis, Wageningen Agricultural University, The Netherlands.

de Bruijn, S.M. and D. Vreugdenhil. 1993. Abscisic acid and assimilate partitioning to developing seeds. II. Does abscisic acid influence the sink strength of *Arabidopsis* seeds. *Physiol. Plant.* **88:** 583–589.

de Bruijn, S.M., J.J.J. Ooms, A.S. Basra, A.A.M. van Lammeren, and D. Vreugdenhil. 1993. Influence of abscisic acid on storage of lipids and carbohydrates in developing *Arabidopsis* seeds. In *4th International Workshop on Seeds—Basic and Applied Aspects of Seed Biology*, Angers, France (ed. D. Come and F. Corbineau), vol. 1, pp. 103–108. ASFIS, Paris.

Derkx, M.P.M. 1993. "Regulation of seasonal patterns in seed dormancy." Ph.D. thesis, Wageningen Agricultural University, The Netherlands.

Derkx, M.P.M. and C.M. Karssen. 1993. Variability in light-, gibberellin- and nitrate requirement of *Arabidopsis thaliana* seeds due to harvest time and conditions of dry storage. *J. Plant Physiol.* **141:** 574–582.

D'Hondt, K., J. van Damme, C. Van Den Bossche, S. Leejeerajumnean, R. De Rycke, J. Derksen, J. Vandekerckhove, and E. Krebbers. 1993. Studies of the role of the propeptides of the *Arabidopsis thaliana* 2S albumin. *Planta* **102:** 425–433.

Dobrovolna, J. and I. Cetl. 1966. An increase of germination of dormant seeds by pricking. *Arabidopsis Inf. Serv.* **3:** 33.

Evans, J. and D. Ratcliffe. 1972. Variation in "after-ripening" of seeds of *Arabidopsis thaliana* and its ecological significance. *Arabidopsis Inf. Serv.* **9:** 3–5.

Finkelstein, R.R. 1993. Abscisic acid-insensitive mutations provide evidence for stage-specific signal pathways regulating expression of an *Arabidopsis* late embryogenesis-abundant (*lea*) gene. *Mol. Gen. Genet.* **238:** 401–408.

Finkelstein, R.R. and C.R. Somerville. 1990. Three classes of abscisic acid (ABA)-insensitive mutations of *Arabidopsis* define genes that control overlapping subsets of ABA responses. *Plant Physiol.* **94:** 1172–1179.

Gaubier, P., M. Raynal, G. Hull, G.M. Huestis, F. Grellet, C. Arenas, M. Pages, and M. Delseny. 1993. Two different *Em*-like genes are expressed in *Arabidopsis thaliana* seeds during maturation. *Mol. Gen. Genet.* **238:** 409–418.

Giraudat, J., B.M. Hauge, C. Valon, J. Smalle, F. Parcy, and H.M. Goodman. 1992. Isolation of the *Arabidopsis ABI3* gene by positional cloning. *Plant Cell* **4:** 1251–1261.

Gómez-Campo, C. 1976. Conservation techniques of crucifer seed banks. *Arabidopsis Inf. Serv.* **13:** 18–21.

Goto, N. 1982. The relationship between characteristics of seed coat and dark-germination by gibberellins. *Arabidopsis Inf. Serv.* **19:** 29–38.

———. 1983. Germination responses to some promoting factors in different genotypes of *Arabidopsis thaliana. Arabidopsis Inf. Serv.* **20:** 115–118.

———. 1985. A mucilage polysaccharide secreted from testa of *Arabidopsis thaliana. Arabidopsis Inf. Serv.* **22:** 143–145.

Guerche, P., C. Tire, F. Grossi de Sa, A. de Clercq, M. van Montagu, and E. Krebbers. 1990. Differential expression of the *Arabidopsis* 2S albumin genes and the effect of increasing gene family size. *Plant Cell* **2:** 469–478.

Hayes, G.R. and W.H. Klein. 1974. Spectral quality influence of light during development of *Arabidopsis thaliana* plants in regulating seed germination. *Plant Cell Physiol.* **15:** 643–653.

Heath, J.D., R. Weldon, C. Monnot, and D.W. Meinke. 1986. Analysis of storage proteins in normal and aborted seeds from embryo-lethal mutants of *Arabidopsis thaliana*. *Planta* **169:** 304–312.

Hilhorst, H.W.M. and C.M. Karssen. 1988. Dual effect of light on the gibberellin- and nitrate stimulated seed germination of *Sisymbrium officinale* and *Arabidopsis thaliana*. *Plant Physiol.* **86:** 591–597.

————. 1992. Seed dormancy and germination: The role of abscisic acid and gibberellins and the importance of hormone mutants. *Plant Growth Regul.* **11:** 225–238.

Jacobsen, S.E. and N.E. Olszewski. 1993. Mutations at the spindly locus of *Arabidopsis* alter gibberellin signal transduction. *Plant Cell* **5:** 887–896.

James, D.W. and H.K. Dooner. 1990. Isolation of EMS-induced mutants in *Arabidopsis* altered in seed fatty acid composition. *Theor. Appl. Genet.* **80:** 241–245.

Karssen, C.M. 1982. Seasonal patterns in dormancy in weed seeds. In *The physiology and biochemistry of seed development, dormancy and germination* (ed. A.A. Khan), pp. 243–270. Elsevier Biomedical, Amsterdam.

Karssen, C.M. and E. Lacka. 1986. A revision of the hormone balance theory of seed dormancy: Studies on gibberellin and/or abscisic acid deficient mutants in *Arabidopsis thaliana*. In *Plant growth substances* (ed. M. Bopp), pp. 315–323. Springer Verlag, Heidelberg.

Karssen, C.M. and L.C. van Loon. 1992. Probing hormone action in developing seeds by ABA-deficient and -insensitive mutants. In *Progress in plant growth regulation* (ed. C.M. Karssen et al.), pp. 43–53. Kluwer Academic, Dordrecht, The Netherlands.

Karssen, C.M., H.W.M. Hilhorst, and M. Koornneef. 1990. The benefit of biosynthesis and response mutants to the study of the role of abscisic acid in plants. In *Plant growth substances 1988* (ed. R.P. Pharis and S.B. Rood), pp. 23–31. Springer Verlag. Berlin.

Karssen, C.M., S. Zagorski, J. Kepczynski, and S.P.C. Groot. 1989. Key role for endogenous gibberellins in the control of seed germination. *Ann. Bot.* **63:** 71–80.

Karssen, C.M., D.L.C. Brinkhorst-van der Swan, A.E. Breekland, and M. Koornneef. 1983. Induction of dormancy during seed development by endogeneous abscisic acid: Studies on abscisic acid deficient genotypes of *Arabidopsis thaliana* (L.) Heynh. *Planta* **157:** 158–165.

King, J. 1991. *The genetic basis of plant physiological processes.* Oxford University Press, New York.

Koornneef, M. 1981. The complex syndrome of *ttg* mutants. *Arabidopsis Inf. Serv.* **18:** 45–51.

————. 1990. Mutations affecting the testa colour in *Arabidopsis*. *Arabidopsis Inf. Serv.* **27:** 1–4.

Koornneef, M. and J.H. van der Veen. 1980. Induction and analysis of gibberellin sensitive mutants in *Arabidopsis thaliana* (L.) Heynh. *Theor. Appl. Genet.* **58:** 257–263.

Koornneef, M., G. Reuling, and C.M. Karssen. 1984. The isolation and characterization of abscisic acid-insensitive mutants of *Arabidopsis thaliana*. *Physiol. Plant.* **61:** 377–383.

Koornneef, M., E. Rolff, and C.J.P. Spruit. 1980. Genetic control of light-inhibited hypocotyl elongation in *Arabidopsis thaliana* (L.) Heynh. *Z. Pflanzenphysiol.* **100:** 147–160.

Koornneef, M., C.J. Hanhart, H.W.M. Hilhorst, and C.M. Karssen. 1989. In vivo inhibition of seed development and reserve protein accumulation in recombinants of abscisic acid biosynthesis and responsiveness mutants in *Arabidopsis thaliana. Plant Physiol.* **90:** 463–469.

Koornneef, M., M.L. Jorna, D.L.C. Brinkhorst-van der Swan, and C.M. Karssen. 1982. The isolation of abscisic acid (ABA) deficient mutants by selection of induced revertants in non-germinating gibberellin sensitive lines of *Arabidopsis thaliana* (L.) Heynh. *Theor. Appl. Genet.* **61:** 385–393.

Koornneef, M., J.H van der Veen, C.J.P. Spruit, and C.M. Karssen. 1981. Isolation and use of mutants with an altered germination behaviour in *Arabidopsis thaliana* and tomato. In *Induced mutations—A tool in plant breeding* (Symposium IAEA, Vienna), pp. 227–232.

Koornneef, M., J.W. Cone, C.M. Karssen, R.E. Kendrick, J.H. van der Veen, and J.A.D. Zeevaart. 1985. Planthormone and photoreceptor mutants of *Arabidopsis thaliana* and tomato. In *Plant genetics* (ed. M. Freeling), pp. 103–114. A.R. Liss, New York.

Kraml, M., K. Keith, and P. McCourt. 1993. A non-dormant *Arabidopsis* mutant which is sensitive to ABA. In *Abstracts from the 5th International Conference on* Arabidopsis *Research*, Ohio State University, Columbus, p. 49.

Krannitz, P.G., L.W. Aarssen, and J.M. Dow. 1991. The effect of genetically based differences in seed size on seedling survival in *Arabidopsis thaliana* (Brassicaceae). *Am. J. Bot.* **78:** 446–450.

Krebbers, E., L. Herdies, A. de Clercq, J. Seurinck, J. Leemans, J. van Damme, M. Segura, G. Gheysen, M. van Montagu, and J. Vandekerckhove. 1988. Determination of the processing sites of an *Arabidopsis* 2S albumin and characterization of the complete gene family. *Plant Physiol.* **87:** 859–866.

Kugler, I. 1951. Untersuchungen über das Keimverhalten einiger Rassen von *Arabidopsis thaliana* (L.) Heynh. Ein Beitrag zum Problem der Lichtkeimung. *Beitr. Biol. Pflanz.* **28:** 211–243.

Laibach, F. 1951. Uber sommer-und winterannuelle Rassen von *Arabidopsis thaliana* (L.) Heynh. Ein beitrag zur Atiologie der Blütenbildung. *Beitr. Biol. Pflanz.* **28:** 173–210.

———. 1956. Uber die Brechung der Samenruhe bei *Arabidopsis thaliana* (L.) Heynh. *Naturwissenschaften* **43:** 164.

Lemieux, B., M. Miquel, C. Somerville, and J.Browse. 1990. Mutants of *Arabidopsis* with alterations in seed lipid fatty acid composition. *Theor. Appl. Genet.* **80:** 234–240.

Mansfield, S.G. and L.G. Briarty. 1991. Early embryogenesis in *Arabidopsis thaliana.* II. The developing embryo. *Can. J. Bot.* **69:** 461–476.

McCarty, D.R., T. Hattori, C.B. Carson, V. Vasil, M. Lazar, and I.K. Vasil. 1991. The viviparous-1 developmental gene of maize encodes a novel transcriptional activator. *Cell* **66:** 895–905.

McCullough, J.M. and W. Shropshire. 1970. Physiological predetermination of germination responses in *Arabidopsis thaliana.* (L) Heynh. *Plant Cell Physiol.* **11:** 139–148.

Meinke, D.W. 1991. Embryonic mutants of *Arabidopsis thaliana. Dev. Genet.* **12:** 382–392.

———. 1992. A homeootic mutant of *Arabidopsis thaliana* with leafy cotyledons. *Science* **258:** 1647–1650.

Meurs, C., A.S. Basra, C.M. Karssen, and L.C. van Loon. 1992. Role of abscisic acid in induction of desiccation tolerance in developing seeds of *Arabidopsis thaliana. Plant Physiol.* **98:** 1484–1493.

Micke, A. and G. Röbbelen. 1966. The influence of seed storage in different atmospheric conditions on the rapidity of germination. *Arabidopsis Inf. Serv.* **3:** 5.

Mirza, J.I. and N. Bagni. 1991. Effects of exogenous polyamines and difluoromethylornithine on seed germination and root growth of *Arabidopsis thaliana*. *Plant Growth Regul.* **10**: 163–169.

Müller, A.J. 1963. Embryonentest zum Nachweis rezessiver Lethalfactoren bei *Arabidopsis thaliana*. *Biol. Zentralbl.* **82**: 133–163.

Müller, A.J. and U. Heidecker, 1968. Lebensfähige und lethale fusca-mutanten bei *Arabidopsis thaliana*. *Arabidopsis Inf. Serv.* **5**: 54–55.

Nambara, E., T. Akazawa, and P. McCourt. 1991. Effects of gibberellin biosynthetic inhibitor unicanozol on mutants of *Arabidopsis*. *Plant Physiol.* **97**: 736–738.

Nambara, E., S. Naito, and P. McCourt. 1992. A mutant of *Arabidopsis* which is defective in seed development and storage protein accumulation is a new *abi3* allele. *Plant J.* **2**: 435–441.

Napp-Zinn, K. 1975. On the genetical basis of light requirement in seed germination of *Arabidopsis*. *Arabidopsis Inf. Serv.* **12**: 10.

Norton, G. and J.F. Harris. 1975. Compositional changes in developing rape seed (*Brassica napus* L.). *Planta* **123**: 163–174.

Ooms, J.J.J., S.M. de Bruijn, and C.M. Karssen. 1993a. Acquisition of desiccation tolerance in seeds of *Arabidopsis thaliana*. In *4th International Workshop on Seeds—Basic and Applied Aspects of Seed Biology,* Angers, France (ed. D. Come and F. Corbineau), vol. 1, pp. 47–53. ASFIS, Pasis.

Ooms, J.J.J., K.M. Léon-Kloosterziel, D. Bartels, M. Koornneef, and C.M. Karssen. 1993b. Acquisition of desiccation tolerance and longevity in seeds of *Arabidopsis thaliana*: A comparative study using ABA-insensitive *abi3* mutants. *Plant Physiol.* **102**: 1185–1191.

Pang, P.P., R.E. Pruitt, and E.M. Meyerowitz. 1988. Molecular cloning, genomic organization, expression and evolution of 12S seed storage protein genes of *Arabidopsis thaliana*. *Plant Mol. Biol.* **11**: 805–820.

Patton, D.A. and D.W. Meinke. 1990. Ultrastructure of arrested embryos from lethal mutants of *Arabidopsis thaliana*. *Am. J. Bot.* **77**: 653–661.

Ratcliffe, D. 1976. Germination characteristics and their inter- and intra-population variability in *Arabidopsis*. *Arabidopsis Inf. Serv.* **13**: 34–45.

Rédei, G.P. 1970. *Arabidopsis thaliana* (L.) Heynh.: A review of the genetics and biology. *Biobliogr. Genet.* **20**: 1–151.

Rehwaldt, C.A. 1966. After-ripening in *Arabidopsis thaliana*. *Arabidopsis Inf. Serv.* **3**: 3.

———. 1968. Filter paper effect on seed germination of *Arabidopsis thaliana*. *Plant Cell Physiol.* **9**: 609–611.

Röbbelen, G. 1966. Seed aging causes mutant deficit. *Arabidopsis Inf. Serv.* **3**: 7.

Röbbelen, G. and H. Kerstein. 1965. Storage conditions, dormancy and germination of seeds. *Arabidopsis Inf. Serv.* **2**: 29–30.

Rock, C.D. and J.A.D. Zeevaart. 1991. The *aba* mutant of *Arabidopsis thaliana* is impaired in epoxy-carotenoid biosynthesis. *Proc. Natl. Acad. Sci.* **88**: 7496–7499.

Roth, I. 1957. Die Histogenese der integumente von Capsella bursa-pastoris und ihre morphologische Deutung. *Flora* **145**: 212–235.

Saleki, R., P.G. Young, and D.D. Lefebvre. 1993. Mutants of *Arabidopsis thaliana* capable of germination under saline conditions. *Plant Physiol.* **101**: 839–845.

Sankhla, N. and D. Sankhla. 1968. Interaction between growth regulators and (±)-Abscisin II in seed germination. *Z. Pflanzenphysiol.* **58**: 402–409.

Shropshire, W., W.H. Klein, and V.B. Elstad. 1961. Action spectra of photomorphogenetic induction and photoinactivation of germination in *Arabidopsis thaliana*. *Plant Cell Physiol.* **2**: 63–69.

Simpson, G.M. 1990. *Seed dormancy in grasses.* Cambridge University Press, Cambridge, United Kingdom.

Skriver, K. and J. Mundy. 1990. Gene expression in response to abscisic acid and osmotic stress. *Plant Cell* **2**: 503–512.

Spruit, C.J.P., G. Heeringa, and A. van den Boom. 1981. Effect of filter paper on light-induced germination of seeds of *Arabidopsis*. *Arabidopsis Inf. Serv.* **18**: 165–169.

Spruit, C.J.P., A. van der Boom, and M. Koornneef. 1980. Light induced germination and phytochrome content of seeds of some mutants of *Arabidopsis*. *Arabidopsis Inf. Serv.* **17**: 137–141.

van der Klei, H., J. Van Damme, P. Casteels, and E. Krebbers. 1993. A fifth albumin isoform is present in *Arabidopsis thaliana*. *Plant Physiol.* **101**: 1415–1416.

Velemínsky, J. and T. Gichner. 1967. An albina mutant without seed dormancy. *Arabidopsis Inf. Serv.* **4**: 14–15.

Wang, T.L. and C.L. Hedley. 1991. Seed development in peas: Knowing your three 'r's' (or four, or five). *Seed Sci. Res.* **1**: 3–14.

Zeevaart, J.A.D. and M. Talon. 1992. Gibberellin mutants in *Arabidopsis thaliana*. In *Progress in plant growth regulation* (ed. C.M. Karssen et al.), pp. 34–42. Kluwer Academic, Dordrecht, The Netherlands.

# 13

# Root Development in *Arabidopsis*

**John W. Schiefelbein**
Department of Biology
University of Michigan
Ann Arbor, Michigan 48109

**Philip N. Benfey**
Department of Biology
New York University
New York, New York 10003

The *Arabidopsis* plant, like most angiosperms, produces an extensive root system designed to function in the anchorage of the plant and absorption of water and mineral ions. The development of the plant root system is relatively easy to study, despite the fact that roots are generally subterranean organs. Several properties of roots make them amenable to developmental studies: (1) The root apical meristem is accessible and not enclosed by developing organs or primordia; (2) the root is free of pigments and therefore essentially transparent; (3) there are relatively few differentiated cell types in roots; (4) root morphogenesis in many plants occurs in a continuous and relatively uniform fashion without significant developmental transitions; and (5) cell files are easy to observe in longitudinal sections and their origin can be traced back to the apical meristem.

Our understanding of root morphology and development in *Arabidopsis* has come largely from studies of the seedling root system. The small size of the *Arabidopsis* seedling and its ability to be grown aseptically under defined conditions facilitates morphological, physiological, and genetic analyses. The genetic studies of root development in *Arabidopsis* have been particularly illuminating and have led to the isolation of many root developmental mutants. In this chapter, we review our current understanding of root development in *Arabidopsis* by describing research on roots of *Arabidopsis* as well as roots of other plants, where appropriate. For readers interested in a detailed discussion of physiological or functional aspects of roots, reviews can be found in Torrey and Clarkson (1975) and Gregory et al. (1987).

## OVERVIEW OF *ARABIDOPSIS* ROOT DEVELOPMENT

Root development in *Arabidopsis* is similar to that in other angiosperms and can be conveniently monitored by growing plants on nutrient agar in vertically oriented petri dishes (Fig. 1). Seed imbibition is followed by emergence of the radicle from the seed coat and activation of the root apical meristem, which is formed during embryogenesis. The structure of the early root is dominated by the radicle (or primary root), but the root system of *Arabidopsis* does not form a taproot system like some other dicots (e.g., alfalfa, dandelion, carrot). Instead, lateral roots begin to develop within 8–10 days (Fig. 1) and ultimately the primary root is difficult to distinguish from the other roots in the *Arabidopsis* root system.

The fundamental processes of cell division, cell expansion, and cell differentiation are largely confined to segments, or "zones," along the apical-basal axis of the root (Fig. 2). The root meristem, located at the root apex, includes a set of continuously dividing cells that produce the basic cell types and define their organization. The elongation zone (~1 mm long in *Arabidopsis*) includes small, densely cytoplasmic cells that are dividing and expanding in size. Controlled cell expansion in this region leads to the characteristic elongated shape of root cells, and it provides the force that drives the root apex forward. The specialization zone is located behind the elongation zone and contains cells that are differentiating into their final form and function. The most conspicuous cell type in this region is the root-hair cell (Fig. 2). It should be noted that the differentiation of some root cells (e.g., root-cap cells) does not occur in this region. In addition, the developmental zones of roots are not strictly separated; there is overlap between the cellular processes occurring in various zones.

In addition to the apical-basal polarity, roots possess radially organized layers of cells. In *Arabidopsis*, this organization is particularly simple, because each of the four outer cell layers (the epidermis, cortex, endodermis, and pericycle) is only one cell thick (Fig. 3). There is also a relatively constant number of cells within each layer of the primary root of *Arabidopsis* (discussed in a later section).

## ROOT DEVELOPMENT IN THE EMBRYO

### Cellular Organization

The remarkably simple organization of the *Arabidopsis* primary root has its origin in the embryonic root primordia. The zygote divides asymmetrically to generate an apical and a basal cell. The basal cell undergoes several transverse divisions to form the suspensor, which serves as a

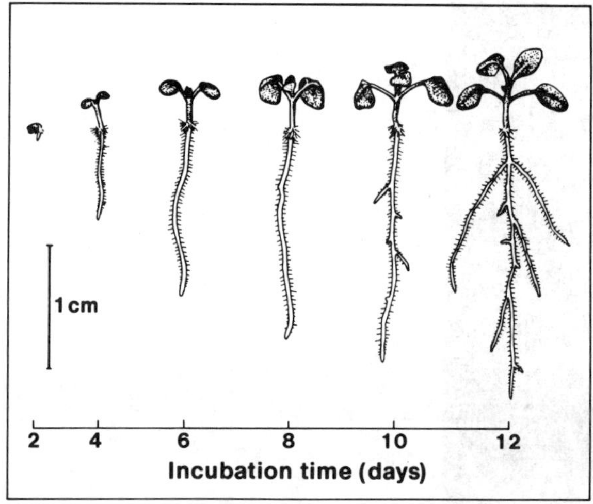

*Figure 1* Schematic drawing of a developmental profile derived from *Arabidopsis* seedlings axenically grown on agarose in sealed petri plates for various lengths of time. (Reprinted, with permission, from Schiefelbein and Somerville 1990 [copyright American Society of Plant Physiologists].)

nutrient conduit from the endosperm. The basal cell also forms the hypophysis, which is located at the top of the suspensor (Fig. 4). The apical cell gives rise to the rest of the embryo. During embryogenesis, the hypophysis divides to form a lens-shaped cell, which is the progenitor of a portion of the embryonic root meristem (Fig. 4) (Dolan et al. 1993). This lens-shaped cell gives rise to the four "central cells" (so named because of their position at the center of the embryonic root meristem) and to the initials for the portion of the root cap directly below the meristem. The other cells of the hypophysis divide to form the central portion of the root cap, which is known as the "columella" root cap. All the other cells that make up the embryonic root meristem appear to be derived from the apical cell (Dolan et al. 1993). The formation of the root meristem from cells of different clonal origin indicates that cell-cell interactions may serve to coordinate this process.

The cellular organization of the embryonic root meristem was revealed by anatomical studies of the root in *Arabidopsis* seeds (Dolan et al. 1993). A schematic figure of the meristem is provided in Figure 5. There appear to be 4 distinct types of cell initials organized around the 4 central cells. The 12 columella root-cap initials are located directly below the central cells (Fig. 5). Surrounding the central cells are 8 cells that appear to be the shared initials of the cortical and endodermal cell files. The epidermis and lateral root cap also appear to share a common precur-

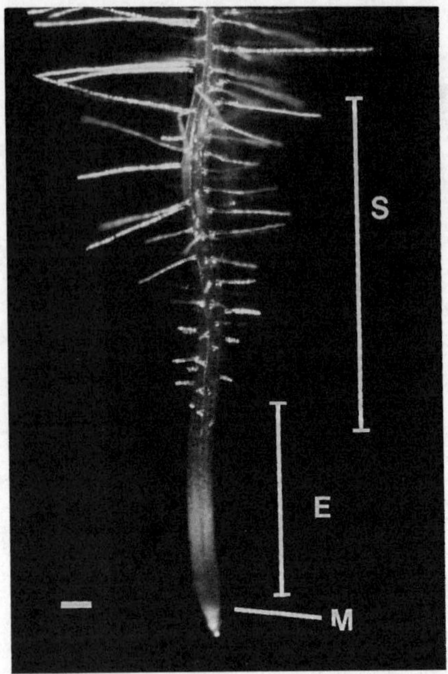


*Figure 2*  *A. thaliana* (ecotype Columbia) root with zones of development indicated. Seedling was grown on a vertically oriented petri plate under continuous light for 4 days. (M) Root meristem; (E) elongation zone; (S) specialization zone. Bar, 100 μm.

sor; these 16 initials surround the cortex/endodermal initials (Fig. 5). The progenitors of the stele cells (including pericycle and vascular tissue) are located above the central cells (Fig. 5).

To understand the functional properties of the cells in the root meristem, labeling with radioactive precursors of DNA synthesis has been carried out (Dolan et al. 1993). The central cells were found to have a very low incidence of labeling, suggesting that they may function as the "quiescent center" in *Arabidopsis*. The concept of the quiescent center was proposed for other plant roots to account for the observed population of cells within the meristem that divide very infrequently (Barlow 1976; Feldman 1984). The role of these cells is still unresolved. It has been suggested that they may be a source of replacement cells for the more rapidly cycling initials (Barlow 1976). Labeling of the cells surrounding the central cells in *Arabidopsis* was consistent with their inferred role as initials (Dolan et al. 1993).

Several major questions remain to complete our understanding of the formation of the embryonic root. Among these are, What is the sequence

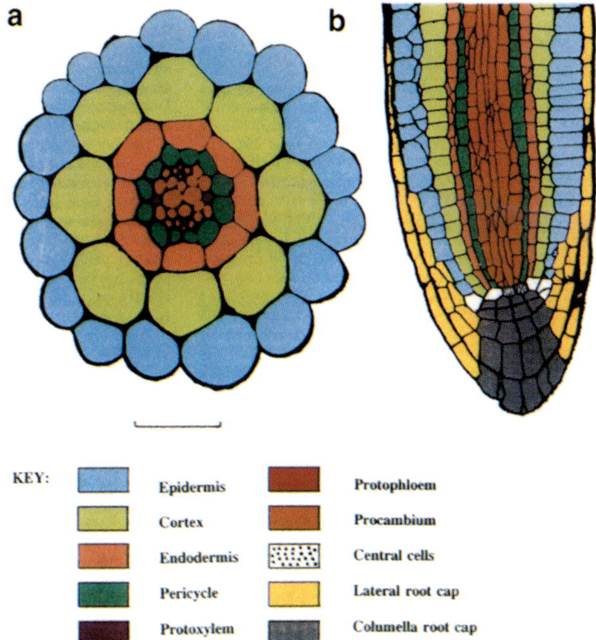



*Figure 3*  Cellular organization of the seedling root of *Arabidopsis*. (*a*) Organization of tissues in transverse section from the specialization zone. (*b*) Organization of tissues in longitudinal section. Initials for endodermis and cortex, and for epidermis and lateral root cap, are in white. Bar, 25 μm. (Modified, with permission, from Dolan et al. 1993 [copyright Company of Biologists].)

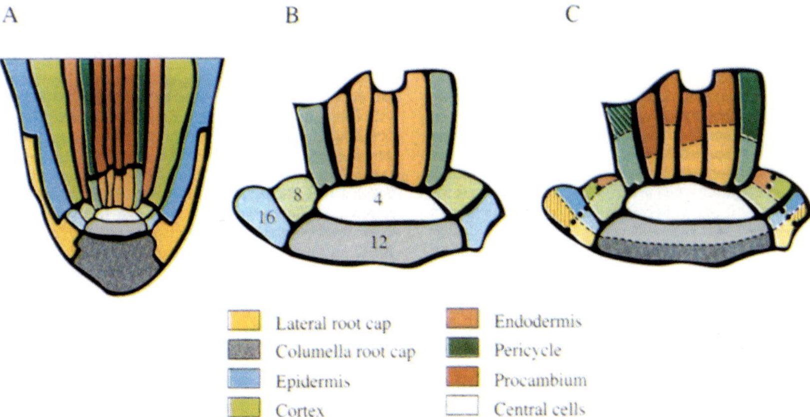



*Figure 5*  Organization of the embryonic root meristem of *Arabidopsis*. (*A*) Schematic representation of the initials and cell files in the embryonic root. (*B*) Magnified view of the region of the initials with the total numbers of each type of initial indicated. (*C*) Proposed division patterns of the initials that generate the cell files. The order of division is indicated as: first division, dashed lines; second division, large dots. (Reprinted, with permission, from Dolan et al. 1993 [copyright Company of Biologists].)

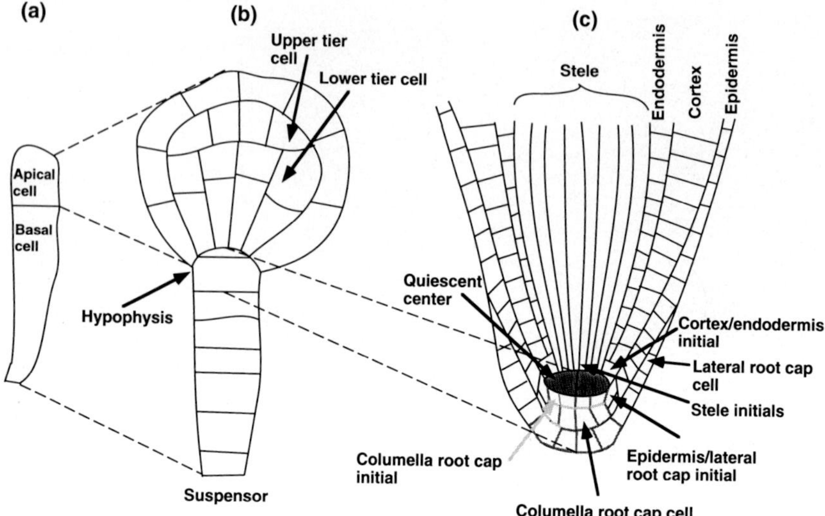


*Figure 4* Schematic diagram of embryonic root development in *Arabidopsis*. (*a*) The zygote undergoes an asymmetric division that results in a smaller apical cell and a larger basal cell. The basal cell forms the suspensor and hypophysis, and the apical cell forms most of the embryo proper. (*b*) The mid-globular stage in which lower-tier cells begin to elongate and no longer resemble upper-tier cells. The uppermost derivative of the basal cell, the hypophysis, divides to form the quiescent center and columellar root cap. (*c*) The embryonic root. To simplify the diagram, the individual cells of the stele cell files have not been depicted. The shading of the quiescent center and the cell walls of the columella root cap are to indicate that these cells are derived from the hypophysis. (Reprinted, with permission, from Benfey and Schiefelbein 1994 [copyright Elsevier Trends Journals].)

of cell divisions that occurs in the apical cell to form the stereotypical organization of the embryonic root meristem? How is the root meristem correctly positioned within the embryo? and How do cells of very different embryonic origins (e.g., apical cell and hypophysis) cooperate to form a functional meristem? The first question may be addressed through additional anatomical studies and the latter questions through genetic and molecular analyses.

## Genetic Analysis of the Embryonic Root

An extensive screen for mutations that affect embryonic development in *Arabidopsis* led to the identification of two mutants with altered root development (Mayer et al. 1991). The *gnom* mutant was initially described as lacking both a root and shoot apical meristem. Further characterization revealed that each of the *gnom* mutants displays a similar set of embryo phenotypes, ranging from a ball of cells to a seedling with a relatively normal hypocotyl but without a shoot or root meristem (Mayer et al. 1993). An explanation for this range of phenotypes was provided by the discovery of the earliest sign of abnormal embryogenesis in the mutant. Instead of an asymmetric division of the zygote, *gnom* embryos form an apical and a basal cell of approximately equal size (Mayer et al. 1993). Subsequent cell divisions are highly variable in the mutant, whereas in the normal embryo they follow a very predictable pattern. It was hypothesized that the variability in these later cell divisions leads to the range of seedling phenotypes (Mayer et al. 1993). This implies that correct placement of the first cell division plane is essential for proper embryonic root development as well as for correct development of other embryonic structures.

A second locus that alters embryonic root development, *monopteros*, appears to be somewhat more specific for the root. Weak alleles of *monopteros* have relatively normal aerial structures while completely lacking a root (Berleth and Jürgens 1993). Stronger alleles also are devoid of the hypocotyl. The earliest stage of embryogenesis in which abnormal development could be detected for this mutation was at the octant stage, when the apical cell has divided to form eight cells. Normally, the lower tier of four cells begins to divide and elongate to form the hypocotyl while the upper-tier cells remain relatively cuboidal in shape. In the mutant, the elongation of the lower-tier cells fails to occur. In addition, at this stage the normal hypophysis goes through a series of stereotyped longitudinal and transverse divisions to form the central cells and the columella root cap. In the mutant, the transverse divisions occur, but there are no longitudinal divisions of the hypophysis (Berleth and Jürgens 1993). Together, these observations suggest that the *MONOPTEROS* gene coordinates the activities of the lower tier and the hypophysis to form the embryonic root.

Identification of mutations that affect the later stages of embryonic root development should shed light on the process of organizing the root meristem. Molecular analysis of the affected genes and their products could elucidate the relative importance of cell-cell interactions and cell-autonomous genetic programing in the formation of the embryonic meristem. In this respect, a potentially informative mutant is *hobbit*, which appears to lack a root meristem even though the surrounding embryonic tissue appears normal (Scheres et al. 1994).

## MORPHOGENESIS OF THE PRIMARY ROOT

Upon germination, the cells in the root meristem must initiate a program of regulated cell division and expansion. Since there are no morphogenetic cell movements in plants, the final form of the root is primarily controlled by three parameters: the timing of cell division, the orientation of the plane of cell division, and the degree and direction of cell expansion. The detailed analysis of root morphogenesis is likely to lead to a better understanding of these fundamental cellular processes.

### Organization of Cell Layers

As described in a previous section, the files of cells in each of the *Arabidopsis* root cell layers can be traced to a particular set of initials in the root meristem. It appears that division is highly regulated within these initials to generate the proper number and organization of cells in each layer. Analysis of transverse sections of the primary root revealed a radial pattern in which there is a single layer of each of the outer cell types—epidermis, cortex, endodermis, and pericycle (Fig. 3). There is an invariant number of eight cortical and eight endodermal cells (Fig. 3), which may reflect the eightfold symmetry of the embryonic initials (Dolan et al. 1993). The number of cells in the epidermal layer and pericycle layer varies. The vascular cylinder also possesses a variable number of cells, but the normal root is always diarch (i.e., there are two xylem poles; Dolan et al. 1993).

The root cap of the seedling root encloses the root apex and consists of columella root-cap cells and lateral root-cap cells (Fig. 3). Longitudinal sections generally show three tiers of amyloplast-containing columella cells, which arise from initials situated just below the central cells. In transverse sections, 12 columella cells are observed in a pattern of 4 surrounded by 8 (Dolan et al. 1993). The columella cells abut three layers of lateral root-cap cells which extend along the side of the root for varying distances (Fig. 3).

### Genetics of Root Morphogenesis

Genetic screens for roots with abnormal shape or size have been performed in *Arabidopsis* to identify genes that regulate root morphogenesis. One of these screens was directed toward the isolation of temperature-sensitive root mutants, to facilitate the propagation of the mutant lines (Baskin et al. 1992). From this screen, three nonallelic mutants (*rsw1*, *rsw2*, and *rsw3*) have been described that exhibit radial swelling of the root apex at the restrictive temperature (Baskin et al. 1992). One of these mutants, *rsw3*, has expanded epidermal cells.

In an independent screen, several root-expansion mutants have been isolated that display preferential expansion in particular cell layers (Benfey et al. 1993; Hauser and Benfey 1994). The *cobra* mutant has the greatest expansion in epidermal cells (Fig. 6h), the *lion's tail* mutant has greatly expanded stele cells, and the *pom-pom* mutant displays grossly enlarged cortical and epidermal cells. Although the specific cell layer affected by these mutations differs, each has a common feature. The cell-expansion phenotype in each mutant is expressed only when the rate of root growth is maximal. When growth rate is reduced (e.g., by lowering the sucrose concentration in the medium or reducing the air temperature), the mutant roots display a nearly normal appearance (Benfey et al. 1993; Hauser and Benfey 1994). This suggests that there is a degree of functional redundancy in the regulation of cell expansion, and, when growth rate is maximal in the mutants, there are insufficient amounts of these gene products to enable normal expansion to occur. Consistent with the hypothesis that these gene products are limiting under some circumstances, the *cobra* mutation is semi-dominant although *lion's tail* and *pom-pom* display a simple recessive relationship to the wild-type alleles (Benfey et al. 1993; Hauser and Benfey 1994). Another root-expansion mutant, *sabre*, has abnormal expansion that is greatest in the cortex layer, but the phenotype is not conditional on the growth rate (Fig. 6g) (Benfey et al. 1993). The *SABRE* gene has recently been isolated, and a molecular analysis is in progress (R.A. Aeschbacher and P.N. Benfey, unpubl.).

The ability of a root to grow in a continuous fashion is dependent on the regulation of cell division and expansion as well as maintenance of a stem-cell population within the meristem. During the normal development of *Arabidopsis*, no predetermined cessation of root growth has been observed. However, several mutants have been identified that have roots that cease growing (Benfey et al. 1993). In the *short-root* mutant, roots cease growing (note the absence of the meristematic and elongation zones in Fig. 6b), and they lack the endodermal cell layer in the root and hypocotyl (Fig. 6, compare c and e) (Benfey et al. 1993). These results were confirmed using a monoclonal antibody that decorates endodermal cells (Fig. 6, compare d and f). At this time, it is not known whether the two defects in *short-root* are causally related or of independent origin.

## ROOT CELL DIFFERENTIATION

Cell differentiation in *Arabidopsis* roots occurs in a highly orchestrated manner. As new cells are generated by divisions in the root meristem, they become organized into cell files. The cells in each file differentiate

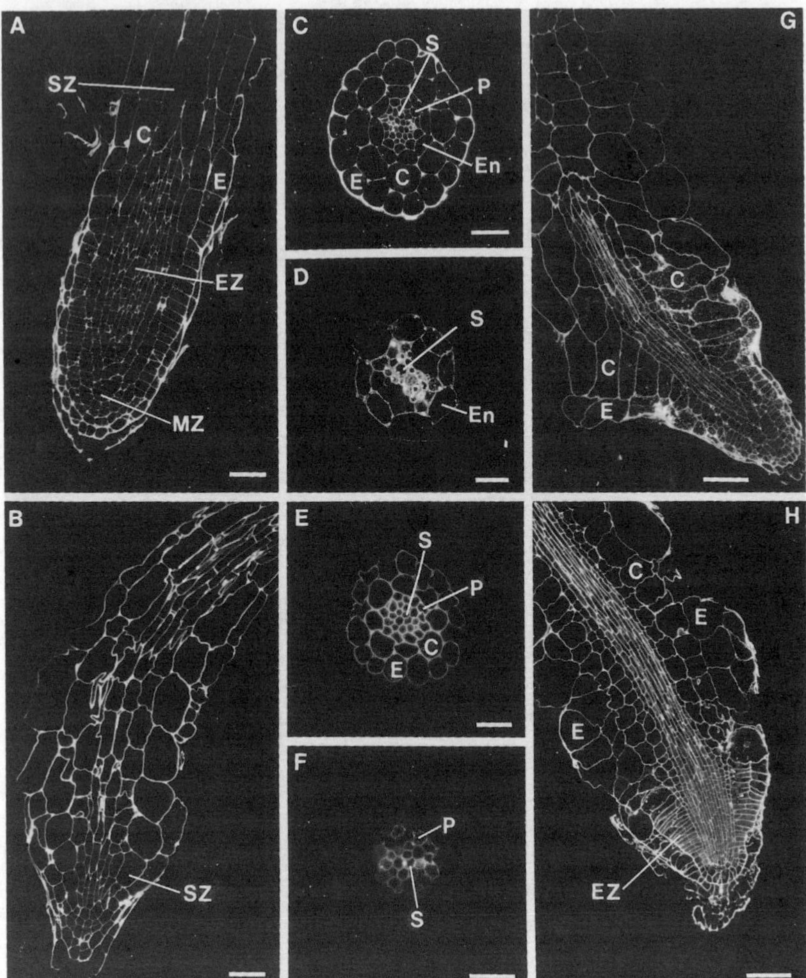



*Figure 6*    Antibody-stained sections of roots from wild type and root-morphogenesis mutants of *Arabidopsis*. (*A*) Longitudinal section of wild-type root stained with JIM 7, an antibody to pectin (a cell wall component). Bar, 25 μm. (*B*) Longitudinal section of *short-root* root tip stained with JIM 7. Bar, 25 μm. (*C*) Transverse section through the specialization zone of a wild-type root stained with JIM 7. Bar, 25 μm. (*D*) Transverse section through the specialization zone of a wild-type root stained with JIM 13, an antibody that stains endodermis and some stele cells. Bar, 10 μm. (*E*) Transverse section through the specialization zone of a *short-root* mutant root stained with JIM 7. Bar, 25 μm. (*F*) Transverse section through the specialization zone of a *short-root* mutant root stained with JIM 13. Bar, 25 μm. (*G*) Longitudinal section through *sabre* root tip stained with JIM 7. Note radial expansion of cortical cells. Bar, 50 μm. (*H*) Longitudinal section of *cobra* root stained with JIM 7. Bar, 50 μm. (MZ) Meristematic zone; (EZ) elongation zone; (SZ) specialization zone; (E) epidermis; (C) cortex; (En) endodermis; (P) pericycle; (S) stele. (Reprinted, with permission, from Benfey et al. 1993 [copyright Company of Biologists].)

in form and function from cells in other files. Although the fundamental mechanisms that control root-cell differentiation have not yet been defined, recent morphological and genetic studies have led to insights into the differentiation of specific root cells in *Arabidopsis*.

### Root Epidermal Cells

The *Arabidopsis* root epidermis consists of two distinct cell types: root-hair-bearing cells and hairless cells. Root hairs are long, tubular-shaped extensions of epidermal cells that serve to increase the surface area of the root for efficient water and mineral ion absorption (Fig. 2) (Cormack 1962). In wild-type *Arabidopsis*, root hairs begin to emerge from epidermal cells located approximately 1 mm behind the root meristem (Schiefelbein and Somerville 1990). The growth rate and mature length of root hairs depends largely on the environmental conditions, and the external $Ca^{++}$ concentration is an important factor (Schiefelbein et al. 1992). In *Arabidopsis* seedlings grown on the surface of agarose-solidified medium with 1 mM $Ca^{++}$, root hairs elongate at approximately 100 μm per hour and may reach a mature length of 1.5 mm (Schiefelbein et al. 1992).

Genetic analysis of root-hair cell differentiation in *Arabidopsis* is facilitated by the fact that root hairs are present on the surface of the root and are easy to observe with the aid of a low-power microscope. Another important feature is that root hairs are not required for the growth of *Arabidopsis* plants (Schiefelbein and Somerville 1990). This means that root-hair mutants of all types, even ones that lack root hairs entirely, may be grown and studied. The mutants that have been identified so far enable the process of root-hair differentiation to be divided into three stages: epidermal cell fate specification, root-hair initiation, and root-hair elongation.

### Epidermal Cell Fate Specification

A key factor controlling the fate of immature epidermal cells is the position of the cell relative to cells in the underlying cortical layer. In *Arabidopsis* and other crucifers, root-hair cells are present over the radial walls separating adjacent cortical cells, and hairless cells are located directly over the cortical cells (Cormack 1949; Bunning 1951; Barlow 1984; Dolan et al. 1993). Since there are eight files of cortical cells in *Arabidopsis* roots, there are eight files of root-hair-bearing epidermal cells (Fig. 3). The nature of this positional control over epidermal cell fate is not clear. Recently, mutants have been identified that alter the normal pattern of root-hair-bearing and hairless epidermal cells (M. Galway et al., unpubl.). The genes defined by these mutations are likely to provide, or respond to, positional signals to cause immature epidermal cells to differentiate into appropriate cell types.

Differentiating root-hair cells can be distinguished from differentiating hairless cells prior to the formation of root hairs. Near the onset of cell elongation, epidermal cells that are destined to produce root hairs display a delay in vacuolation relative to the differentiating hairless cells. This demonstrates that cell fate specification in the root epidermis occurs prior to cell elongation, and it may be influenced at the time of cell formation (by the epidermal initials) in the root meristem.

### Root-hair Initiation

The first outward sign of root-hair formation is the localized swelling of the epidermal cells at the site of root-hair emergence. In *Arabidopsis*, root hairs emerge at the apical end of the epidermal cell (the end nearest the root apex; Schiefelbein and Somerville 1990), which indicates that cell polarity influences hair initiation. One of the genes involved in hair initiation is *RHD1* (Schiefelbein and Somerville 1990). The epidermal cells of *rhd1* mutants swell excessively during root-hair initiation, indicating deregulation of epidermal cell expansion (Fig. 7). Mutations in the *REB1* gene of *Arabidopsis* also cause excessive swelling of root-hair cells (Baskin et al. 1992). However, it is not known whether *REB1* represents a second gene controlling root-hair initiation, since complementation tests have not been carried out between *rhd1* and *reb1* mutants.

### Root-hair Elongation

The elongation phase of root-hair development follows the initiation phase, and it normally proceeds by a process known as tip growth. Tip growth is a form of polarized cell expansion in which new cell growth is limited to one region (the tip) and generally leads to long, tubular-shaped cells (Sievers and Schnepf 1981; Schnepf 1986). In this mode of growth, secretory vesicles bearing new cell wall components and enzymes involved in cell expansion are directed to the tip of the growing cell. In addition to its role in root-hair development, tip growth is also responsible for the elongation of pollen tubes of plants and the hyphae of fungal cells.

At least four loci in *Arabidopsis* (*RHD2, RHD3, RHD4*, and *TIP1*) are required for the tip growth of root hairs (Schiefelbein and Somerville 1990). The *rhd2* mutants possess "stubby" hairs that are unable to elon-

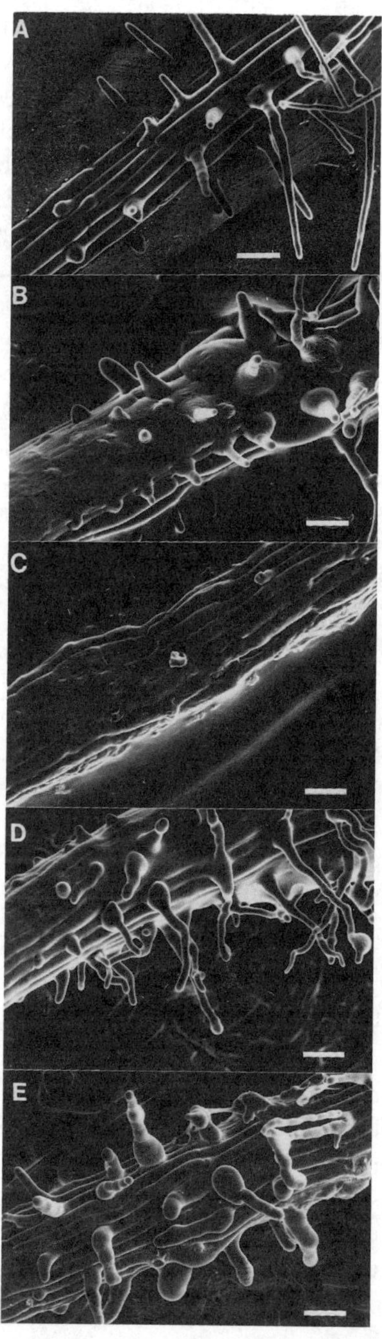

*Figure 7* Scanning electron micrographs of wild-type and mutant root hairs in *Arabidopsis*. Each panel displays a section of root that includes the root-hair development region. (*A*) Wild type. (*B*) *rhd1* mutant. (*C*) *rhd2* mutant. (*D*) *rhd3* mutant. (*E*) *rhd4* mutant. Bar, 50 μm. (Modified, with permission, from Schiefelbein and Somerville 1990 [copyright American Society of Plant Physiologists].)

gate (Fig. 7). The *RHD2* gene is probably involved in an early step of tip growth that allows the root hair to begin to expand beyond the initiation stage. The *RHD3* and *RHD4* genes appear to be required throughout the tip growth process. The root hairs of *rhd3* mutants are "wavy," which suggests that this gene controls the direction of cell expansion during tip growth (Fig. 7). It is possible that the *RHD3* gene product is required for the directed transport of secretory vesicles to the root-hair tip or for the proper arrangement of cytoskeletal components that orient cell expansion. The *rhd4* mutants possess "bulging" root hairs, which indicates that this gene normally regulates the degree of cell expansion (Fig. 7). For example, the *RHD4* gene product may control the production/delivery of secretory vesicles or may influence the stability of the growing tip.

The *tip1* mutant is particularly interesting, since it displays abnormalities in root-hair elongation and pollen-tube growth (Schiefelbein et al. 1993). The mutant root hairs are shorter and branched at their base, and the pollen tubes elongate more slowly than normal. Thus, the *TIP1* locus appears to encode a component that is essential for normal tip growth in both root-hair and pollen-tube cells. This indicates that there is some overlap in the genetic control of tip growth in different plant cells. However, it is difficult to estimate the degree of overlap between root-hair and pollen-tube growth from these experiments, since the desired mutant alleles are not transmitted efficiently through the pollen.

Other *Arabidopsis* loci appear to be necessary for normal root-hair development. The auxin-resistant mutants *dwf* and *axr2* produce altered root hairs that indicate a role for auxin in hair formation (Maher and Martindale 1980; Wilson et al. 1990). A cytokinin-resistant mutant (*ckr1*) produces shorter root hairs than the wild type (Su and Howell 1992). Interestingly, the *hy3* (long hypocotyl) mutant has been found to produce longer root hairs than the wild type when seedlings are grown under illumination (Reed et al. 1993).


## LATERAL ROOT DEVELOPMENT

In *Arabidopsis*, as in most angiosperms, lateral roots are initiated from cells in the pericycle layer. The process of lateral root formation is believed to involve the perception of a signal, redifferentiation of pericycle cells, organization of a root meristem, and growth of the nascent root through the endodermal, cortex, and epidermal layers of the existing root. At present, the regulation of this entire process is poorly understood.

Anatomical studies of lateral roots in *Arabidopsis* have revealed that, compared to the primary root, the lateral roots display more variability in cell number. For example, the number of cortical cells varies from seven to eleven rather than the invariant eight found in the primary root (Dolan et al. 1993). It is possible that the greater variability in the cellular organization of lateral roots is due to the absence of cell-cell interactions that may occur during embryogenesis.

The multistep process of lateral root formation raises the possibility that a mutational analysis could identify the genes that regulate each step. In addition, it would appear likely that loss of lateral root function should not greatly affect viability, since the primary root would still exist. However, to date, attempts to find mutants that are impaired in lateral root formation have been disappointing (I.M. Sussex; P.N. Benfey; both unpubl.). This may indicate that lateral root formation is regulated by the same genes as those that control embryonic root development or that are required for the formation of other organs.

## ENVIRONMENTAL EFFECTS ON ROOT DEVELOPMENT

The ultimate form of the plant root system depends on environmental conditions as well as genetic factors. Root growth can be profoundly affected by a variety of external stimuli, including gravity, light, temperature, moisture, aeration, and physical obstacles. These stimuli can alter the direction or degree of cell expansion, the amount of root branching, or the structure of root cells (for review, see Torrey 1976; Feldman 1984). Studies on the effects of environmental factors on *Arabidopsis* root development are described by Okada and Shimura (this volume).

## ROOT-SPECIFIC GENE EXPRESSION

There are a relatively small number of examples of root-specific gene expression in plants. Using differential screening methods, investigators have identified cDNAs from genes that are preferentially expressed in roots of tobacco (Conkling et al. 1990), pea (Evans et al. 1988), and Rhizobium-infected legumes (Gloudemans and Bisseling 1989).

In *Arabidopsis*, a few genes are known to display root expression. A β-tubulin isoform (Oppenheimer et al. 1988) and a plasma membrane $H^+$-ATPase isoform (Harper et al. 1990) are preferentially expressed in *Arabidopsis* roots. A tonoplast intrinsic protein (γ-TIP) gene of *Arabidopsis* appears to be expressed in the elongating cells of roots and may affect vacuole formation (Ludevid et al. 1992).

Promoter elements have been identified that appear to control root-specific expression. The cauliflower mosaic virus 35S promoter possesses a subdomain that directs root-specific expression and binds a factor (ASF1) that preferentially accumulates in root tissue (Katagiri et al. 1989; Benfey et al. 1990a). A subdomain of the petunia EPSP synthase also appears to confer root-specific gene expression (Benfey et al. 1990b). Flanking sequences at the 3' end of an oilseed rape gene are involved in gene expression in the developing cortical cells in roots and embryos (Dietrich et al. 1992). The promoter for the tobacco *TobRB7* gene directs *GUS* reporter gene expression in the root meristem and central cylinder region (Yamamoto et al. 1990, 1991).

## CONCLUSIONS AND PERSPECTIVES

As outlined in this chapter, the study of *Arabidopsis* roots has led to a better understanding of plant root development as well as new insights into fundamental morphogenetic processes. Of particular interest are the many loci identified that control various aspects of root development. These include genes affecting several levels of pattern formation: early apical-basal patterning in the embryo, radial patterns of specific cell types in the root, the activity of the root meristem, and the pattern of cells within specific root-cell layers. In addition, genes have been identified that influence root-cell differentiation or the response of roots to external stimuli.

In the future, the advantageous properties of *Arabidopsis* root development (simple morphology, small size, transparent organ) will undoubtedly lead to the continued use of this system for developmental studies. However, major advances in our understanding of root development will probably depend on the application of new experimental techniques. For example, there is a clear need to understand the nature of the products of the genes that have been defined by genetic analyses. Several approaches are currently being employed to try to clone these genes, including genetic-based strategies like insertional mutagenesis or chromosome walking. There are also potential approaches for identifying root-expressed genes, including differential hybridization and enhancer trapping experiments. Other techniques for studying root development are focused on the use of cell-specific markers to define the origin and fate of specific cell files (clonal analysis). Two types of markers are currently being developed for this purpose: antibodies that recognize cell-surface molecules (Knox et al. 1991) and transposable element-reporter gene constructs that rely on transposable element excisions to mark specific cell lineages.

## ACKNOWLEDGMENTS

We thank Moira Galway, Susan Ford, Yonca Ilkbahar, James Masucci, and Roger Aeschbacher for helpful comments and for communicating unpublished results. Research in our laboratories is supported by grants from the National Science Foundation (DCB-9004568 and IBN-9316409 to J.W.S.) and the National Institutes of Health (R01 G-743778 to P.N.B.).

## REFERENCES

Barlow, P.W. 1976. Towards an understanding of the behavior of root meristems. *J. Theor. Biol.* **57:** 433–451.

———. 1984. Positional controls in root development. In *Positional controls in plant development* (ed. P.W. Barlow and D.J. Carr), pp. 281–318. Cambridge University Press, Cambridge, United Kingdom.

Baskin, T.I., A.S. Betzner, R. Hoggart, A. Cork, and R.E. Williamson. 1992. Root morphology mutants in *Arabidopsis thaliana. Aust. J. Plant Physiol.* **19:** 427–437.

Benfey, P.N. and J.W. Schiefelbein. 1994. Getting to the root of plant development: The genetics of *Arabidopsis* root formation. *Trends Genet.* **10:** 84–88.

Benfey, P.N., L. Ren, and N.-H. Chua. 1990a. Tissue-specific expression from CaMV 35S enhancer subdomains in early stages of plant development. *EMBO J.* **9:** 1677–1684.

Benfey P.N., P.J. Linstead, K. Roberts, J.W. Schiefelbein, and R.A. Aeschbacher. 1993. Root development in *Arabidopsis*: Four mutants with dramatically altered root morphogenesis. *Development* **119:** 57–70.

Benfey, P.N., H. Takatsuji, L. Ren, D.M. Shah, and N.-H. Chua. 1990b. Sequence requirements of the 5-enolpyruvylshikimate 3-phosphate synthase promoter for tissue-specific expression in petals and seedlings. *Plant Cell* **2:** 849–856.

Berleth, T. and G. Jürgens. 1993. The role of the monopteros gene in organising the basal body region of the *Arabidopsis* embryo. *Development* **118:** 575–587.

Bunning, E. 1951. Ober die differenzierungsvorgange in der cruciferenwurzel. *Planta* **39:** 126–153.

Conkling, M.A., C.-L. Cheng, Y.T. Yamamoto, and H.M. Goodman. 1990. Isolation of transcriptionally regulated root-specific genes from tobacco. *Plant Physiol.* **93:** 1203–1211.

Cormack, R.G.H. 1949. The development of root hairs in angiosperms. *Bot. Rev.* **15:** 583–609.

———. 1962. Development of root hairs in angiosperms. II. *Bot. Rev.* **28:** 446–464.

Dietrich, R.A., S.E. Radke, and J.J. Harada. 1992. Downstream DNA sequences are required to activate a gene expressed in the root cortex of embryos and seedlings. *Plant Cell* **4:** 1371–1382.

Dolan, L., K. Janmaat, V. Willemsen, P. Linstead, S. Poethig, K. Roberts, and B. Scheres. 1993. Cellular organisation of the *Arabidopsis thaliana* root. *Development* **119:** 71–84.

Evans, I.M., R. Swinhoe, L.N. Gatehouse, J.A. Gatehouse, and D. Boulter. 1988. Distribution of root messenger RNA species in other vegetative organs of pea *Pisum sativum*

L. *Mol. Gen. Genet.* **214:** 153–157.

Feldman, L.J. 1984. Regulation of root development. *Annu. Rev. Plant Physiol.* **35:** 223–242.

Gloudemans, T. and T. Bisseling. 1989. Plant gene expression in early stages of Rhizobium-legume symbiosis. *Plant Sci.* **65:** 1–14.

Gregory, P.J., J.V. Lake, and D.A. Rose, eds. 1987. *Root development and function.* Cambridge University Press, Cambridge, United Kingdom.

Harper, J.F., L. Manney, N.D. DeWitt, M.H. Yoo, and M.R. Sussman. 1990. The *Arabidopsis thaliana* plasma membrane $H^+$-ATPase multigene family. *J. Biol. Chem.* **265:** 13601–13608.

Hauser, M.-T. and P.N. Benfey. 1994. Genetic regulation of root expansion in *Arabidopsis thaliana. NATO-ASI Plant Mol. Biol. Ser.* (in press).

Katagiri, F., E. Lam, and N.-H. Chua. 1989. Two tobacco DNA-binding proteins with homology to the nuclear factor CREB. *Nature* **340:** 727–730.

Knox, J.P., P.J. Linstead, J. Peart, C. Cooper, and K. Roberts. 1991. Developmentally regulated epitopes of cell surface arabinogalactan proteins and their relation to root tissue pattern formation. *Plant J.* **1:** 317–326.

Ludevid, D., H. Hofte, E. Himelblau, and M.J. Chrispeels. 1992. The expression pattern of the tonoplast intrinsic protein gamma-TIP in *Arabidopsis thaliana* is correlated with cell enlargement. *Plant Physiol.* **100:** 1633–1639.

Maher, E.P. and S.J.B. Martindale. 1980. Mutants of *Arabidopsis thaliana* with altered responses to auxins and gravity. *Biochem. Genet.* **18:** 1041–1053.

Mayer U., G. Buttner, and G. Jürgens. 1993. Apical-basal pattern formation in the *Arabidopsis* embryo: Studies on the role of the *gnom* gene. *Development* **117:** 149–162.

Mayer U., R.A. Torres Ruiz, T. Berleth, S. Misera, and G. Jürgens. 1991. Mutations affecting body organization in the *Arabidopsis* embryo. *Nature* **353:** 402–407.

Oppenheimer, D.G., N. Haas, C.D. Silflow, and D.P. Snustad. 1988. The beta-tubulin gene family of *Arabidopsis thaliana*: Preferential accumulation of the beta-1 transcript in roots. *Gene* **63:** 87–102.

Reed J.W., P. Nagpal, D.S. Poole, M. Furuya, and J. Chory. 1993. Mutations in the gene for the red/far-red light receptor phytochrome B alter cell elongation and physiological responses throughout *Arabidopsis* development. *Plant Cell* **5:** 147–157.

Scheres, B., V. Willemsen, K. Janmaat, H. Wolkenfelt, L. Dolan, and P. Weisbeek. 1994. Analysis of root development in *Arabidopsis thaliana. NATO-ASI Plant Mol. Biol. Ser.* (in press).

Schiefelbein, J.W. and C. Somerville. 1990. Genetic control of root hair development in *Arabidopsis. Plant Cell* **2:** 235–243.

Schiefelbein, J.W., A. Shipley, and P. Rowse. 1992. Calcium influx at the tip of growing root-hair cells of *Arabidopsis thaliana. Planta* **187:** 455–459.

Schiefelbein, J., M. Galway, J. Masucci, and S. Ford. 1993. Pollen tube and root-hair tip growth is disrupted in a mutant of *Arabidopsis thaliana. Plant Physiol.* **103:** 979–985.

Schnepf, E. 1986. Cellular polarity. *Annu. Rev. Plant Physiol.* **37:** 23–47.

Sievers, A. and E. Schnepf. 1981. Morphogenesis and polarity of tubular cells with tip growth. *Cell Biol. Monogr.* **8:** 265–299.

Su, W. and S.H. Howell. 1992. A single genetic locus, *Ckr1*, defines *Arabidopsis* mutants in which root growth is resistant to low concentrations of cytokinin. *Plant Physiol.* **99:** 1569–1574.

Torrey, J.G. 1976. Root hormones and plant growth. *Annu. Rev. Plant Physiol.* **27:** 435–459.

Torrey, J.G. and D.T. Clarkson, eds. 1975. *The development and function of roots*. Academic Press, New York.

Wilson, A.K., F.B. Pickett, J.C. Turner, and M. Estelle. 1990. A dominant mutation in *Arabidopsis* confers resistance to auxin, ethylene, and abscisic acid. *Mol. Gen. Genet.* **222:** 377–383.

Yamamoto, Y.T., C.-L. Cheng, and M.A. Conkling. 1990. Root-specific genes from tobacco and *Arabidopsis* homologous to an evolutionarily conserved gene family of membrane channel proteins. *Nucleic Acids Res.* **18:** 7449.

Yamamoto, Y.T., C.G. Taylor, G.N. Acedo, C.-L. Cheng, and M.A. Conkling. 1991. Characterization of *cis*-acting sequences regulating root-specific gene expression in tobacco. *Plant Cell* **3:** 371–382.

# 14

# Development of the Vegetative Shoot Apical Meristem

**June I. Medford, Joseph D. Callos,**
**Friedrich J. Behringer, and Bruce M. Link**
Department of Biology
The Pennsylvania State University
University Park, Pennsylvania 16802

The *Arabidopsis* shoot apical meristem starts as a tiny group of fewer than one hundred cells, yet this small group of cells is the source of the aboveground portion of the plant. This simple fact means that a remarkable amount of control (or perception of control signals) in the development of the plant is found in the apical meristem.

One reason so little is known about the vegetative apical meristem in *Arabidopsis* is that it is one of the smallest shoot meristems in the angiosperms (Vaughan 1952). In *Arabidopsis*, the extremely small size of the apical meristem may mean that some of the processes are regulated differently. For example, a hypothetical morphological gradient that extends over several cells in a large angiosperm meristem could be limited to a gradient in a smaller group of cells or perhaps one cell (i.e., intracellular gradient) in *Arabidopsis*. However, the size restriction of the *Arabidopsis* meristem is no longer limiting due to two advances. First, genetic studies using meristem mutants allow the genes and signals involved in various processes to be defined. Second, biochemical studies can be done using an interchangeable system, *Brassica oleracea*. Cauliflower or *B. oleracea* var. *botrytis* is a mutant that contains an immense amplification of shoot meristems (Sadik 1962; Medford et al. 1991). Cauliflower genes are highly homologous to *Arabidopsis* genes (typically 95% identity), allowing the systems to be interchanged (Medford et al. 1991). Hence, cauliflower (and other related *Brassicas*) can be used to collect biochemical quantities of materials, and these materials can then be analyzed in *Arabidopsis*.

The ability to circumvent the small size of the *Arabidopsis* meristem will be the key for defining the molecular and biochemical basis for apical meristem functions. The vegetative shoot apical meristem has four functions: organ initiation, tissue initiation, sending and receiving signals, and maintaining itself in a stem-cell-like state (Steeves and Sussex 1989; Medford 1992). It is important to note that these functions are found in the apical meristem, which is only a part of the shoot apex (Cutter 1965). The shoot apex consists of the shoot apical meristem, surrounding leaf primordia, and cells in the associated stem; the shoot apical meristem is the group of cells located distal to the youngest leaf primordia (Fig. 1).

Cytological differences within the apical meristem allowed early workers to define a series of zones in the apical meristem: the central zone, the peripheral zone, and the rib zone (Fig. 1B) (Steeves and Sussex 1989). The various meristem functions (above) are often preferentially found in different zones. For example, organ initiation takes place from the peripheral zone. Perhaps in anticipation of this event, cell division in the meristem is highest in the peripheral zone (Lyndon 1970). Frequent cell divisions can also be found in the base of the peripheral zone and the rib zone as the plant beings to bolt (Brown et al. 1964); these areas are responsible for tissue formation. The function of maintaining the meristem in a stem-cell-like state is localized to a small group of cells found in the central zone. However, early dissection studies suggested that when the meristem is disrupted, cells in all parts (including the peripheral zone by itself) can re-establish a new apical meristem. One possibility is that the stem-cell-like function is not exclusively localized in the central zone cells, but instead, cells located in this position perform this function by integrating signals. Furthermore, because the shape of the *Arabidopsis* meristem changes from a flattened rectangle to a rounded dome (see below), the relationship between cells in the various zones may change during development.

The fourth function of the apical meristem, signal communication, is not localized to any specific zone. One suggestion for signal communication in the apical meristem and shoot apex comes from the work of Wardlaw (1949). Wardlaw hypothesized that inhibitory signals originate from young leaf primordia and the center of the apical meristem, although he was never able to provide any proof for such a signal(s). A second suggestion of signal communication is the well-known observation that auxin, either from the apical meristem, young leaf primordia, or both, plays a large role in apical dominance in the shoot. Two more specific examples suggest that signal communication may play a large role in intra-meristem functions. For example, shear stress is the highest in the apical meristem at positions that form organs (leaf primordia) (Selker et al. 1992). In mammalian systems, shear stress has been shown to activate the phosphoinositol signaling pathway either directly or through a G-protein-mediated response (Hsieh et al. 1992, 1993). Hence,

**A.**

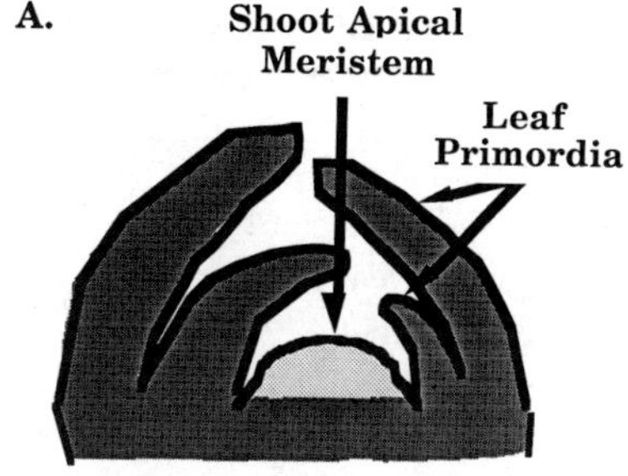


**B.**

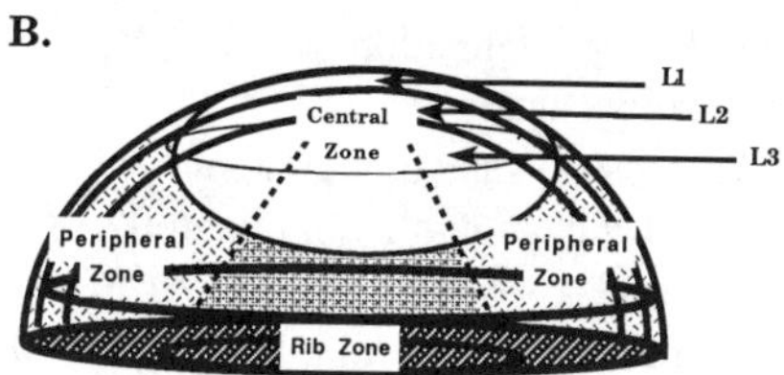


*Figure 1* Diagrams of the shoot apex (*A*) and shoot apical meristem (*B*). The entire diagram in *A* represents the shoot apex; the shoot apical meristem is the more lightly shaded region. *B* shows a series of zones and layers within the shoot apical meristem. In vivo the boundaries for the three zones (central, peripheral, and rib) are dynamic and not as distinct. L1, L2, and L3 correspond to the genetically defined three cell layers from the exterior of the meristem to the interior, respectively. Note that the apical meristem is flat in early development and becomes dome-shaped later in development. (Reprinted, with permission, from Medford et al. 1992.)

there may be activation of specific signaling pathways in the meristem between cells that will form organs and cells that will remain in the meristem. This possibility is further supported by recent data suggesting that a calcium influx (known to be released in the phosphoinositol signaling pathway) can be localized to the region of the *Graptopetalum* meristem that forms organs (leaf primordia) (Hush et al. 1991).

## DEVELOPMENTAL STAGES OF THE VEGETATIVE SHOOT APICAL MERISTEM

The vegetative shoot apical meristem forms the basal rosette through processes which at times follow a defined sequence and at other times simply repeat a single action. One way the processes can be understood is as proceeding through a series of stages, which, like the stages defined for floral development (Smyth et al. 1990), are not  meant to  be absolute

*Table 1* Developmental stages for the *Arabidopsis* vegetative shoot apex

| Stage | Defining process | Characteristics | Days[a] |
|---|---|---|---|
| 0 | Meristem forms in embryo | 3-layered region that is 7–8 cells in diameter | — |
| 1 | Mature seed | juvenile: bulges for leaf primordia 1 and 2 often apparent; cytological zonation indistinct | 0 |
| 2 | Leaf primordia 1 and 2 appear as distinct rounded knobs | juvenile: meristem is flat and rectangular in shape; cytological zonation apparent | 2 |
| 3 | Differentiation of leaf primordia 1 and 2 | juvenile: meristem still rectangular but now resembles loaf of bread | 4 |
| 4 | Initiation of leaf primordia 3 and 4 | juvenile: meristem adopts trapezoidal shape | 6 |
| 5 | Initiation of leaf primordia 5 and 6 | adult: transition stage; meristem symmetry changing from bilateral to radial | 9 |
| 6 | Initiation of leaf primordia in spiral phyllotactic pattern | adult: meristem has a rounded dome shape and radial symmetry | 10 |
| 7 | Transition to inflorescence meristem | adult: rapid extension of dome from cells at the base of the rib zone | 14 |

All stages are meant to define points of reference rather than absolute reference periods. After stage 5, processes forming leaves can occur repetitively until a stimulus to flower is perceived.

[a]Days refers to point in development when plants are grown under 18 hr of light at 22°C with low light (80–100 µE).

but to define points of reference for changes in the meristem. Stages for processes in the vegetative shoot apex are described in Table 1. When considering the details about the various stages (below), it is important to realize that the events represent the action of numerous genes whose expression may not necessarily be uniform throughout the meristem or the specific stage.

### Stage 0

Shoot apical meristems are formed during embryogenesis at the chalazal end of the embryo. Barton and Poethig (1993) have described formation of the shoot apical meristem during embryogenesis. Briefly, during late globular stage, the uppermost portion of the embryo forms three layers. The shoot apical meristem develops from these three layers in the torpedo-stage embryo. Further cell division in the nascent meristem results in a shoot apical meristem that is three cell layers deep and 7–8 cells across in the fully formed embryo (Barton and Poethig 1993).

**Stage 1**

Stage one is represented by the inactive meristem in the mature seed. Slight bulges for the first pair of leaf primordia are at times visible on either side of the meristem. At germination, the wild-type meristem has a square shape and contains approximately 70 cells in the WS (Wassilewskija) ecotype (Medford et al. 1992) or 110 cells in the Landsberg *erecta* ecotype (Irish and Sussex 1992). Figure 2 shows a section through a mature seed imbibed for 12 hours. The cells of the apical meristem have prominent nuclei and nucleoli and a conspicuous scarcity of protein storage bodies typical for cells of mature hypocotyls and cotyledons. The prominent nuclei are centrally located, and this central location is true of cells throughout the apical meristem. These characteristics seem to be typical of apical meristems that are not actively forming tissues and organs.

In the mature embryo, slight bulges or buttresses are often detected on either side of the apical meristem. These slight bulges correspond to the formation of the first two leaf primordia. Although it has not been rigorously established in *Arabidopsis*, the extent of development of the first two leaf primordia in the embryo may vary. In many species, differences can be seen when the meristems of numerous mature embryos are examined for the degree of development of the first leaf primordia (Steeves and Sussex 1989). In some seeds, the first leaf primordial bulges are quite prominent, whereas in other seeds, the bulges are not detectable. The fact that some mature embryos have bulges indicates that the meristem is active in embryogenesis. The two slight bulges in the shoot apex of a mature embryo suggest that the first two primordia are initiated nearly simultaneously. However, development of the first primordial pair is not completely uniform (see Stage 3, below). The difference in the development of the first primordial pair suggests that despite indistinguishable differences in the embryonic shoot apex, an asymmetry must exist in the mature seed. The factor(s) that produces the asymmetry is not known.

Embryogenesis is a continuous process interrupted by developmental arrest (Quatrano 1987), which means that the difference in the extent of development of the first two primordia is most likely a reflection of the point where the seed became arrested. Because morphological differentiation is always preceded by differential gene expression (e.g., in *Drosophila* embryogenesis) even though the bulges of leaf primordia one and two are sometimes not apparent, the molecular signals for the formation of the first pair of primordia must have already been provided prior to germination. Hence, development of the first leaf primordial pair may not require further stimulus after germination. Details like these may be

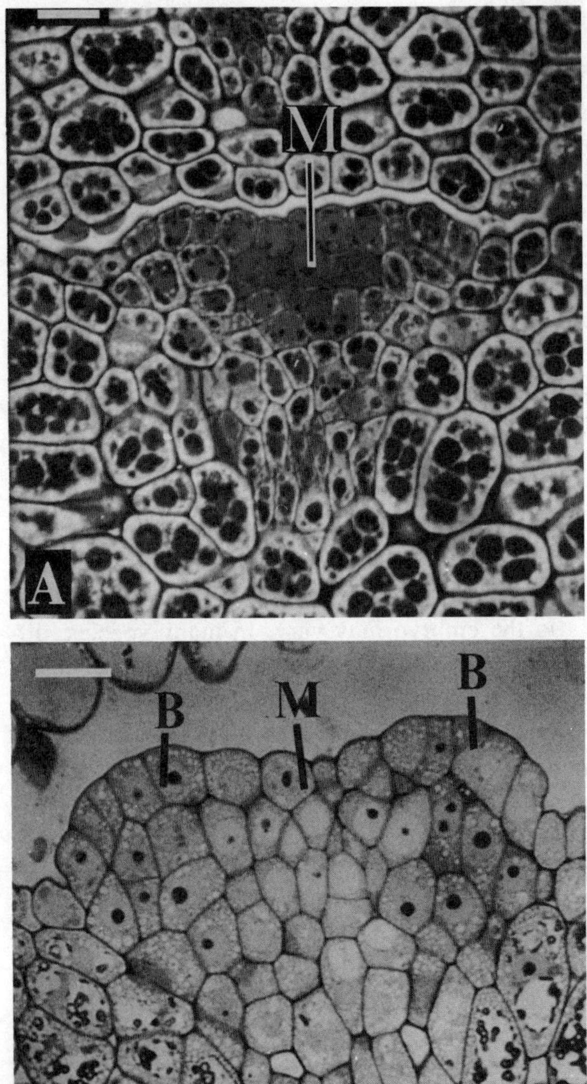

*Figure 2* (*A*) Stage 1: shoot apical meristem in the mature seed. Section through apical meristem imbibed for 12 hr. Nuclei of the meristem (M) are large and centrally located; zonation is not apparent. (*B*) Stage 2: shoot apical meristem at 2 days. Position of nucleus is no longer uniform in cells through the meristem; bulges (B) representing developing leaf primordia are on either side of the meristem. Bars, 10 μm. (Reprinted, with permission, from Medford et al. 1992.)

of importance in studies on seedling lethals or light-regulatory mutants (see, e.g., Deng and Quail 1992).

## Stage 2

Stage 2 begins with the development of the first pair of leaf primordia into radially symmetrical knob-like structures. Clonal analysis suggests that a leaf primordium in *Arabidopsis* is initiated from one rank of cells within the subepidermal layer (Irish and Sussex 1992; Furner and Pumfrey 1992). The number of cells within that rank can be somewhat variable (Irish and Sussex 1992). The primordia emerge as rounded knobs opposite to one another and have irregular-shaped cells on the surface. In the underlying cell layers (the corpus), both the nucleus size and the intracellular positioning have changed to a more variable pattern. Furthermore, the meristem now has 110–130 cells (WS ecotype), with the increase in cell number most apparent in a plane parallel to the cotyledons (Fig. 3A). This addition of cells results in a rectangular-shaped meristem.

## Stage 3

Stage 3 begins with external signs of differentiation in the first pair of leaf primordia. Development of the first two primordia is prominent on either long side of the rectangular meristem (Fig. 3B). Although the meristem is still rectangular in shape (in a plane parallel to the cotyledons), the cells of the meristem surface are now elongated. These cells are distinct from those seen in the meristem in Stage 2, where the cells are more or less isodiametric. The leaf primordia (now emerged) begin to flatten, and the first external signs of differentiation are apparent as the distal-most trichome and basal stipules emerge (Fig. 4A). *Arabidopsis* is somewhat distinct from other plants in the emergence of the first four-leaf primordia. In *Arabidopsis* the primordia emerge as rounded knobs with near-radial symmetry and only later flatten out, whereas in species such as tobacco, the early primordia have their adaxial surface (toward the meristem) already flattened (see, e.g., Poethig and Sussex 1985). In *Arabidopsis* the flattened adaxial surface correlates with the appearance of differentiated structures, suggesting that the initiation of primordia and differentiation are more temporally separated than in species such as tobacco. The distal-most trichome is one of the first external markers of differentiation, and its development is always apparent on one leaf primordium prior to the other. The meristem, although still rectangular in shape, begins to bulge up in a shape reminiscent of a loaf of bread (Fig. 4B). The leaf primordia con-

**A.**

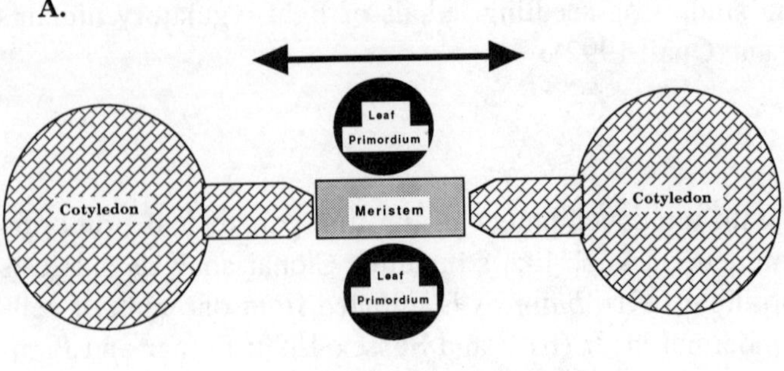

**B.**

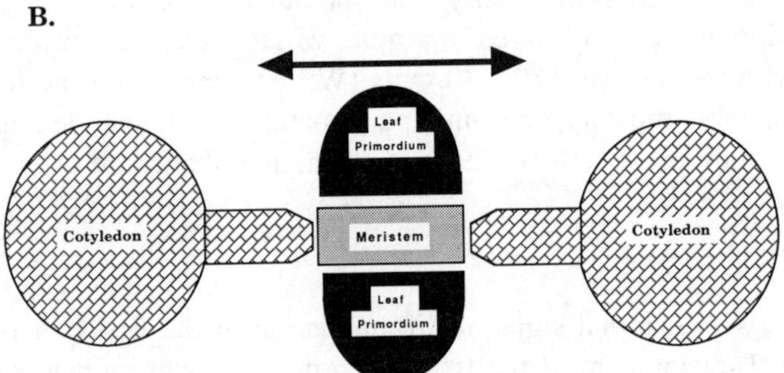

*Figure 3* Diagrammatic representation of changes in the shape of the apical meristem (*A*) and first two leaf primordia (*B*). (*A*) The number of cells in the meristem increases from ~70 to ~110 in a plane that is parallel to the cotyledons. This increase in cell number results in a rectangular-shaped meristem. (*B*) Leaf primordia 1 and 2 initially emerge as rounded knob-like structures and later flatten on the adaxial side (facing the meristem).

tinue to develop, and a polarity can now be detected in the surface cells of Stage 3. Leaf surface cells, particularly those at the margin, are rectangular and elongated in a plane parallel to the long axis of the developing leaf.

**Stage 4**

Formation of the third leaf primordium marks the beginning of Stage 4. The meristem changes from a rectangular to a trapezoidal shape (Fig. 5A) and is tightly surrounded by the first two leaf primordia and distinct club-shaped stipules. Cell number in the trapezoidal meristem has increased to approximately 170 cells (WS ecotype). The stipules in *Arabidopsis* contain about 20–30 cells and are formed at the base of each

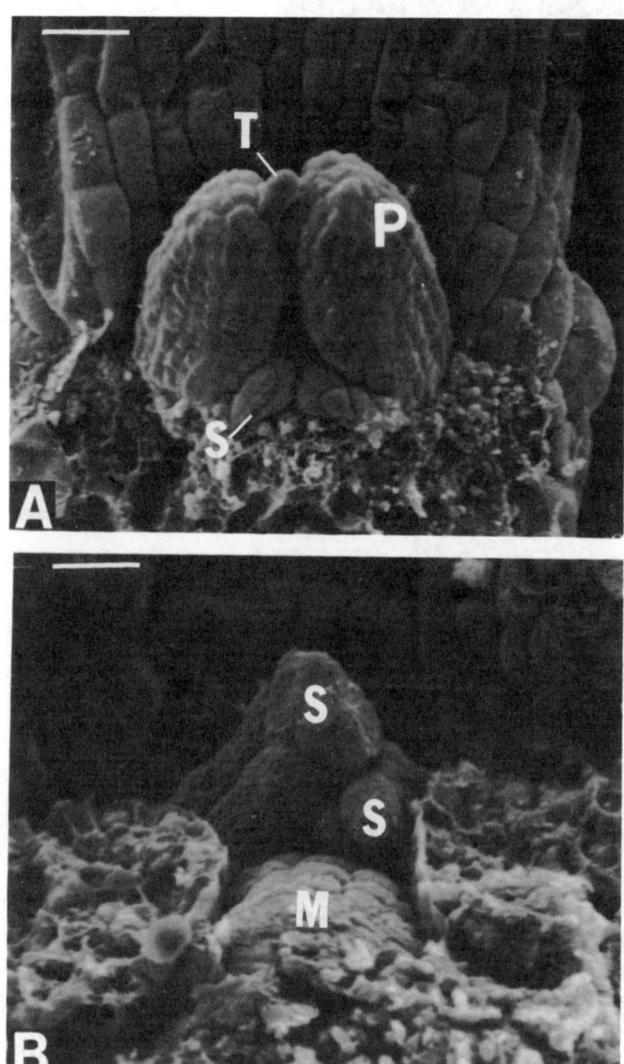


*Figure 4* Stage 3: differentiation of first pair of leaf primordia. (*A*) Scanning electron microscopy (SEM) showing the development of the distal trichomes concurrent with development of flattened (dorsiventral) surface on the leaf primordia. (*B*) SEM showing the rectangular-shaped apical meristem with a slight curvature; "loaf of bread." Leaf primordia 1 and 2 were located to either side of the meristem and removed for this view. Prominent stipules are seen at the back of the meristem. Bars, 10 μm. (M) Meristem; (S) stipule; (T) trichome; (P) primordium.

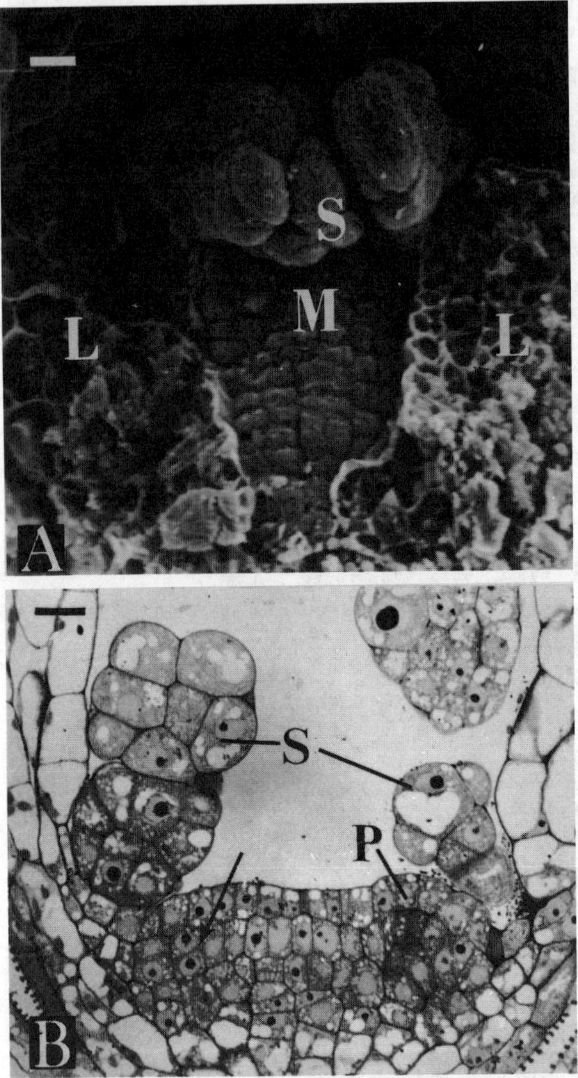

*Figure 5* Stage 4: trapezoidal meristem and initiation of leaf primordia 3 and 4. (*A*) SEM of meristem showing trapezoidal shape (cf. Fig. 4B). Leaf primordia 1 and 2 are at either side. (*B*) Section throughout the apical meristem. Centrally located cells with large prominent nuclei define the meristem's stem cells. Development of leaf primordia can be seen on either side of the meristem. Primordium 4 is just starting to form as evident by the periclinal divisions in $L_2$ (*arrow*). Periclinal cell divisions are also apparent in the rib zone, indicating that this region is active before the plant bolts. Bars, 10 μm. (M) Meristem; (S) stipule; (P) primordium; (L) leaf. (Reprinted, with permission, from Medford et al. 1992.)

leaf primordium. Data from other plants (see, e.g., Meicenheimer et al. 1983) suggest that the cells that form stipules are part of the original cells sequestered from the meristem to form the leaf primordia. The early stipules can vary greatly from a club shape to a flattened fan shape. In the meristem, the change to a trapezoidal shape may be in anticipation of the formation of the third primordium that arises from the wide end of the trapezoid. Initiation of the third and fourth primordia, like that of primordia one and two, is nearly simultaneous, although the third primordium is clearly initiated ahead of the fourth (Fig. 5B) (Medford et al. 1992). The initiation of *Arabidopsis* leaf primordia can be detected very early as a periclinal division in the second cell layer ($L_2$) of the meristem's peripheral zone. The third and fourth leaf primordia, like the first two leaf primordia, are initiated on opposite sides of the meristem. However, the relationship between primordia three and four and primordia one and two varies and is often not at a 90° angle. Importantly, the angle where leaf primordia are initiated and the angle where the primordia emerge can vary, presumably as a result of expanding primordia in the shoot apex. This point is important when considering genes responsible for organ position. For example, one possibility is that there is a gene or genes functioning in the meristem that control the point where leaf primordia are initiated and another gene(s) that controls growth parameters which determine where the leaf primordia emerge (macroscopically) in the shoot apex.

### Stage 5

Initiation of leaf primordium five marks a transition stage in the vegetative meristem. Before Stage 5, leaves are generally initiated on opposite sides of the meristem. However, subsequent leaves are initiated in a spiral pattern. The apical meristem continues to expand and contains approximately 450 cells. At the point when leaf five is initiated, the meristem's shape is intermediate between a rectangle-trapezoid and a rounded dome (Fig. 6). Although the plant has not yet bolted, extension of the apical meristem from the basal rosette is more apparent in Stage 5 than in previous stages. Moreover, because the change in the shape of the apical meristem also corresponds to a change in the shape of the leaf (e.g., from a more rounded shape to a serrated shape depending on the ecotype), the switch from a rectangle to rounded meristem most likely corresponds to a switch from a juvenile to adult stage. In *Arabidopsis* this is correlated with the appearance of trichomes on the lower surface of the petiole of adult leaves (Conway and Poethig 1993).

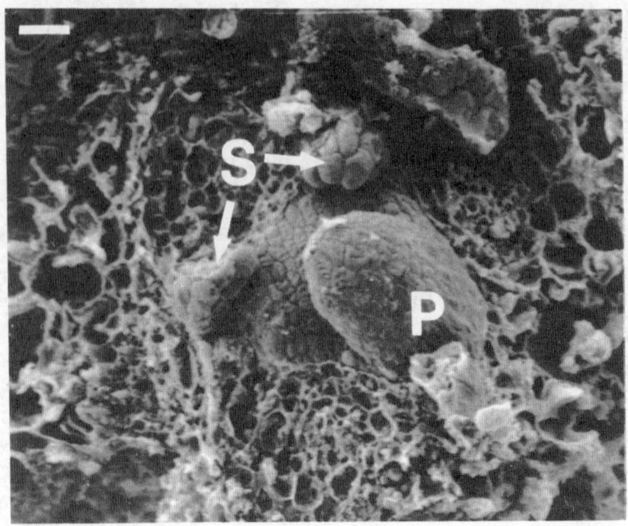

*Figure 6*   Stage 5: SEM showing the apical meristem of a late Stage 5 plant. Leaf primordium 5 is visible extending over the apical meristem. The meristem no longer has a rectangular shape, yet it does not have the domed triangular shape characteristic of Stage 6 meristems. Bar, 15 μm. (S) Stipule; (P) primordium.

## Stage 6

In Stage 6, the meristem has become a rounded dome, and leaf primordia arise in a spiral pattern (Fig. 7A,B). However, when viewed from above, the apex appears almost triangular in shape (Fig. 7A). This appearance is due to the presence of developing leaf primordia that are positioned nearly 137.5° apart around the apical meristem. This pattern of forming leaves continues throughout the remainder of vegetative development.

Under long days, the *Arabidopsis* meristem can convert from a vegetative to inflorescence state after forming only 5 leaves. The switch to the inflorescence meristem corresponds with an extensive activity of the rib zone, which until this stage had been mostly quiescent. The rib zone in *Arabidopsis* typically defines the cells at the base of the meristem and forms a transition between the meristem and the rest of the plant (Vaughan and Jones 1953). Its activation leads to bolting of the plant. Although the transition from the vegetative to inflorescence meristem can occur after only 5 leaves, in long days (16–18 hours of light), the transition typically occurs after 8 leaves, whereas in short days (8 or less hours of light), the *Arabidopsis* rosette contains approximately 40 leaves, although this number varies greatly depending on the ecotype and conditions of growth (temperature, light intensity, light quality, etc.).

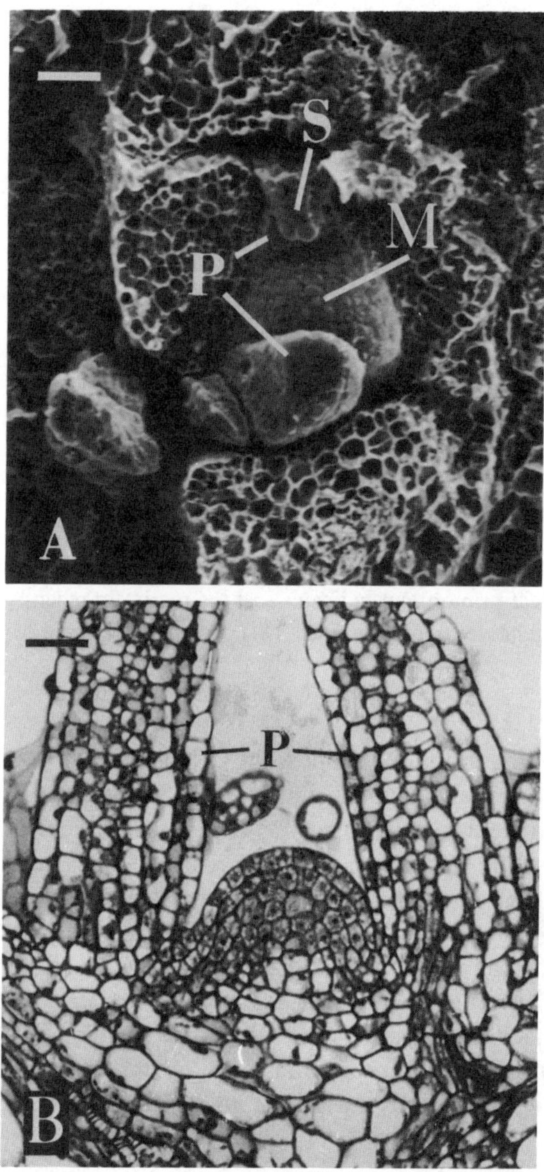

*Figure 7* Stage 6: (A) SEM showing a triangular-shaped shoot apex consisting of the apical meristem and overlapping primordia. (*B*) Section showing the meristem as a rounded dome. (M) Meristem; (S) stipule; (P) primordium.

## Stage 7

The transition from the adult vegetative to the inflorescence meristem marks the end of vegetative development. The apical meristem ceases producing rosette leaves and switches to the production of cauline leaves

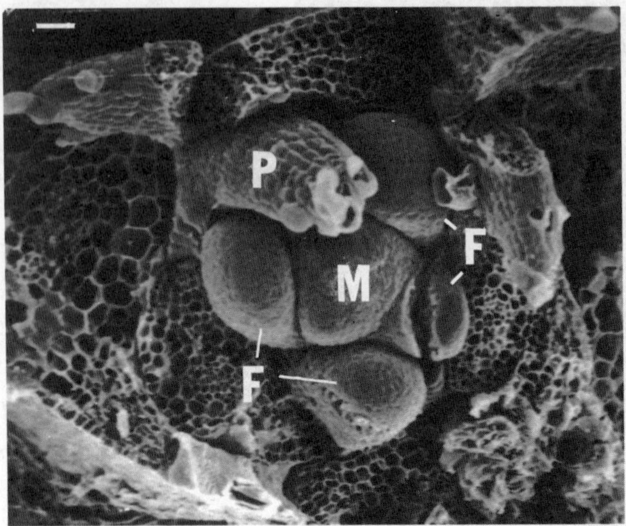



*Figure 8* Stage 7: SEM showing the apical meristem just past Stage 7 plant. The close proximity of a young leaf primordium indicates that the meristem recently made the transition from vegetative to inflorescence. (M) Meristem; (P) primordium; (F) floral meristem.

and floral meristems (Fig. 8). In the shoot apex, the distinction between cauline leaves and rosette leaves is subtle, with cauline leaf primordia having a slightly more triangular shape than the earlier rosette leaves. One possibility for the subtle distinction between rosette and cauline leaves is that the change from vegetative to an inflorescence meristem is gradual and may involve some overlapping programs. Perhaps the most noticeable change in the apical meristem is in the rib zone (Fig. 9). Cells in the rib zone no longer have the neat appearance that in vegetative development resembles a stack of bricks. During vegetative development this zone is largely inactive. With the transition to inflorescence meristem, the rib zone is very active, as the *Arabidopsis* rosette bolts and forms the floral stem.

## PATTERNS IN THE VEGETATIVE SHOOT APEX

In the vegetative shoot apex, the leaves are positioned with a remarkable precision that results in a spiral pattern. In plants with spiral phyllotaxy such as *Arabidopsis*, the angle between subsequent leaves (after leaf 5) approaches 137.51º, the "golden angle." Our calculations show that after the emergence of leaf 5 (leaves are numbered in order of their formation), the angle between *Arabidopsis* leaves is 136.4º ± 1.6 s.e. (*N* = 152),

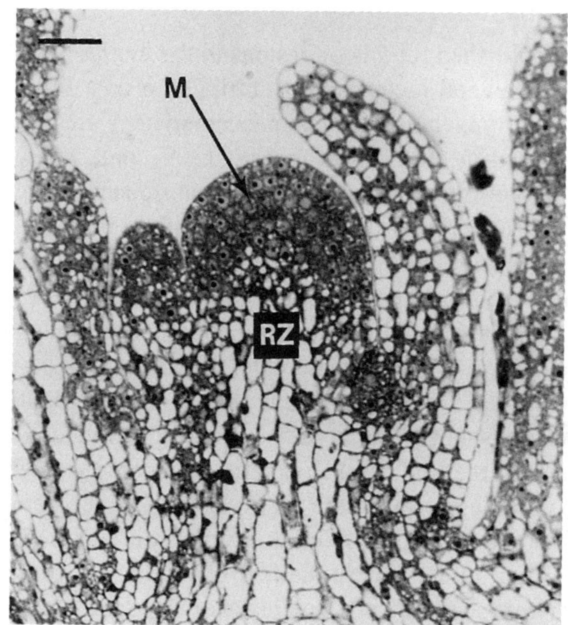

*Figure 9*  Stage 7 meristem. This apical meristem is undergoing a transition from vegetative to inflorescence development. Cells in the apical meristem are more uniform in size and appearance. The rib zone shows the greatest change. rib zone cells are now elongated in the plane of the stem. Bar, 25 μm. (M) Meristem; (RZ) rib zone.

in close agreement with the golden angle (Callos and Medford 1994). Further information about patterns in the shoot apex can be determined from data on the position of leaves. For example, the rate at which the shoot apex is expanding, referred to as the plastochron ratio, can be calculated from these data (Richards 1948). Initially, the plastochron ratio in *Arabidopsis* is somewhat variable, during the formation of the first four leaf primordia (Callos and Medford 1994). This means that the juvenile shoot apex is not always growing equally in all directions. However, as the plant enters adult phase, the meristem becomes radially symmetrical, leaves are produced in a spiral pattern, and growth of the shoot apex becomes stable. For example, after initiation of leaf primordium 5, the apex has a more or less constant plastochron ratio of $1.20 \pm 0.02$ ($N = 152$), whereas prior to this point, the plastochron ratio varies from 1.1 to 1.4. These values correspond to a pattern that can be described as a (3+5) phyllotaxy. This means that for any given leaf in the spiral pattern, the leaves contacting this leaf will always differ in number by three on one side and five on the other side.

## LESIONS IN MERISTEM FUNCTIONS

There have been very few published reports on lesions in the *Arabidopsis* vegetative apical meristem. Several factors may contribute to this. First, many mutants disrupting the vegetative shoot apical meristem are not fertile and must be maintained in the heterozygous state. Second, many vegetative mutants have an enormous amount of temperature sensitivity (most likely reflecting an underlying temperature-sensitive process), meaning that the phenotypes may be difficult to follow. Third, mutant recognition at times requires careful examination to distinguish between a slow-growing seedling and a mutant disrupting the shoot meristem.

In general, several macroscopic features are clues for mutations disrupting the vegetative shoot meristem. For example, in a 7-day-old plant, aberrant development in the shoot apex at times leads to cotyledons being abnormally separated so that the cotyledon petioles become almost parallel to the surface. Another outward clue of a vegetative meristem mutant is abnormal position of the leaf primordia or primordia that have aberrant leaf blades. However, alterations in the leaf blade that are from an apical meristem mutant are usually very severe. An important fact to consider when screening for shoot apical meristem mutants is that features visible to the eye originated at least several days earlier. Whatever the outward clue, all putative disruptions in the vegetative apical meristem should be confirmed with scanning electron micrographs and/or histological sections. This is particularly important because mutants disrupting leaf primordia do not necessarily disrupt the shoot apical meristem (Medford et al. 1992).

A mutant that does affect the apical meristem is *Forever young* (*fey*). Macroscopically, *fey* plants appear as small stunted plants with a shoot apex where the organs are abnormal. Furthermore, *fey* plants have an aberrant leaf shape and abnormalities in the apical meristem (Medford et al. 1992). Work is in progress to delineate the primary site of the *fey* lesion, yet *fey* leaf primordia are often formed with what seems to be an abnormally large number of cells, and previous experiments suggest that this could result from an alteration in the apical meristem. For example, in a series of experiments with potato, the apical meristem was severed from an incipient leaf primordium. When an aberrantly large number of cells were included in the incipient leaf primordium, bilobed or heart-shaped leaves resulted (Sussex 1964). Such bilobed leaves have been observed in *fey* mutants numerous times (Fig. 10A). Figure 10B shows that the position of *fey* leaves is also disrupted, in addition to the disruption of the initiation of leaf primordia. Analysis of the disruption in the formation and position of organs in *fey* plants may lead to a greater understanding of these processes.

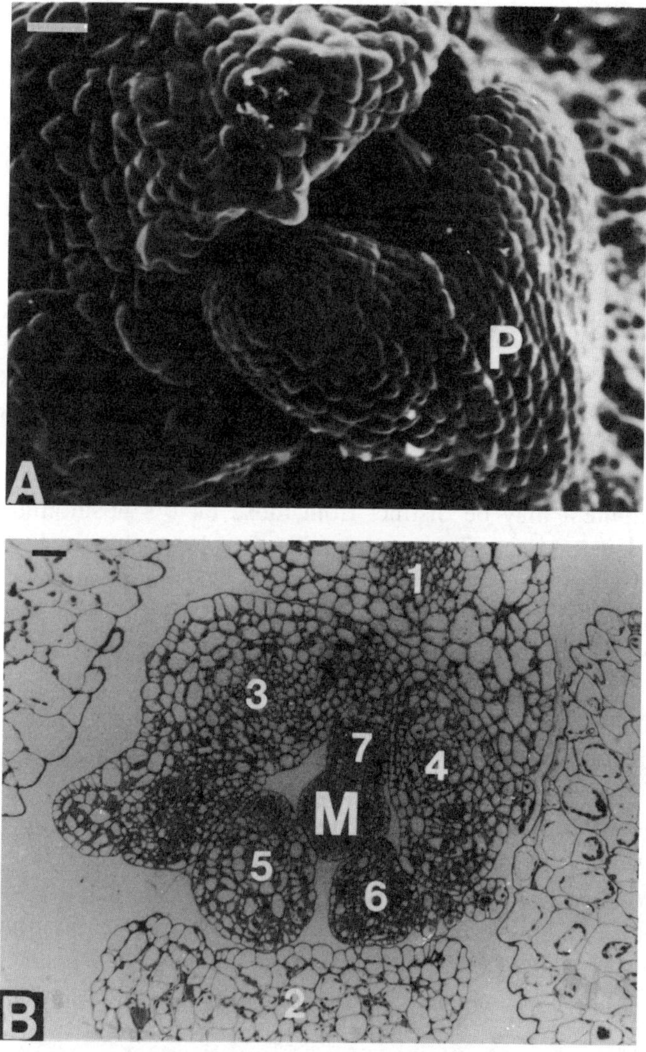



*Figure 10*  (*A*) SEM showing heart-shaped or bilobed leaf in a Fey plant. (*B*) Cross section through 21-day-old Fey plant (grown at 25°C) showing altered positions of leaves in the shoot apex. Bar, 10 μm.

A second mutant that has been characterized as disrupting functions of the vegetative meristem is *clavata1* (*clv1*) (Clark et al. 1993). The *clv1* mutant causes abnormalities in the size and shape of the vegetative apical meristem and produces club-like siliques. In addition, the *clv1* mutant, like the various *fasciated* (Leyser and Furner 1992) mutants, may cause a disruption in organ position. These results await confirmation, since the angle between leaf primordia at initiation and the angle between primordia at emergence can vary (Medford et al. 1992). If the altered positions can be verified, it will be interesting to see if the shape of the meristem causes leaves to form in new positions or if leaves forming in new positions cause the shape of the meristem to be altered. Although the role of the meristem shape in positioning organs is not clear, the fact that the *clv1* mutant has an enlarged meristem and normally shaped leaves suggests that the shape of the meristem is independent of the shape of the leaf. Clark et al. (1993) suggested two principal mechanisms by which CLV1 may act: through a role either in cell division patterns or in signaling between plant cells. Because the cell division patterns are likely to have a function both in apical meristem and the leaf primordia, and the *clv1* primordia are normal, it seems likely that CLV1 may play a role in cell signaling in the shoot apex.

Importantly, the factors regulating positioning of organs during vegetative development may be distinct from those factors positioning organs in floral development. For example, the *fey* plants show a less severe phenotype when grown at low temperatures, yet the position of the leaves is still abnormal (Medford et al. 1992). At 16°C, *fey* plants make weak inflorescence meristems and flowers. These flowers arrest at a point where all the organs are formed but still immature (approximately Stage 5 of Smyth et al. 1990). However, our preliminary observations indicate that the position of the floral organs is normal (J.I. Medford, unpubl.). Work is in progress to document these observations.

In the *fey* mutant, the abnormalities seen in the leaf primordia and the apical meristem often lead to meristems that are progressively smaller in size and to the eventual death of the shoot. Thus, it is possible that the *fey* mutant disrupts two functions of the apical meristem: organ formation and the ability of the apical meristem to remain in a stem-cell-like state. Alternatively, it is possible that at least these two meristem functions are linked (see below).

An important point to consider is that the vegetative apical meristem forms organs and also units known as phytomers, which consist of an axillary meristem, subtending leaf, and an internode (Irish and Sussex 1992; Furner and Pumfrey 1992). Axillary meristems are typically formed two to three plastochrons after the leaf is initiated. In the *Arabidopsis* rosette the axillaries are usually poorly developed. Little is known about how the axillaries function. The mutant analysis shows clearly that there is a difference between the axillary meristem and the main shoot apical meristem. For example, both the *schizoid* (*shz-1*) and *terminal flower* mutants (Shannon and Meeks-Wagner 1991; Medford et al. 1992) affect the main shoot apical meristem but have much less of an effect on the axillary meristem. The basis of this difference is unknown.

One possibility is that the axillary meristem bypasses, or has a shortened period in, the juvenile phase of development.

Clearly, the future is set for isolation of more mutants disrupting the vegetative shoot apex. Particularly useful would be mutants disrupting tissue formation and mutants altering signal communication. Lesions that alter fundamental and specific types of regulation may answer key questions about development of the vegetative portion of the plant.

## MODEL FOR VEGETATIVE DEVELOPMENT

The developmental stages in the vegetative shoot apical meristem represent several processes. For example, the formation of leaves 1–5 takes place regardless of the environmental conditions. Does this mean that this sequence is necessary for further development of the plant, or can it be bypassed? The Embryonic flower (*emf*) mutant (Sung et al. 1992) addresses this point. In the mature seed, the *emf* plants have a meristem with a shape that resembles that of the bread loaf in the Stage 3 meristem. Upon germination, weak *emf* mutants produce a series of leaf-like lateral appendages, then flower, whereas more severe *emf* mutants simply produce carpels. Sung et al. (1992) interpret the dominant wild-type EMF function to maintain the vegetative state of the meristem and without this product in *emf* mutants, the reproductive state results. These results suggest that at least some aspects of vegetative development are not necessary for further development of the plant. However, the *emf* meristem in the mature seed is distinct from the normal meristem (Barton and Poethig 1993) and resembles a normal Stage 3 meristem. Because the *emf* meristem differs from the normal meristem in the mature seed, it is possible that the normal sequence of changes seen in the meristem takes place during embryogenesis. Upon germination, a Stage 3 meristem could rapidly develop into a Stage 4 or Stage 5 meristem producing appendages (weak *emf* mutants) or carpels (strong *emf* mutants), respectively. Hence, a question still remains whether vegetative development is necessary for the further elaboration of the plant.

If at least some development from the vegetative meristem is required, it can be thought of as describing a developmental sequence. In the vegetative meristem the sequence of events can be seen as the progression in the shape of the apical meristem. The apical meristem progresses from a small square shape to a rectangle, to a trapezoidal shape, to a rounded dome. However, this progression can occur regardless of the complete development of any leaf. For example, in weak *emf* mutants, the lateral appendages are at times not fully formed, and in a to-

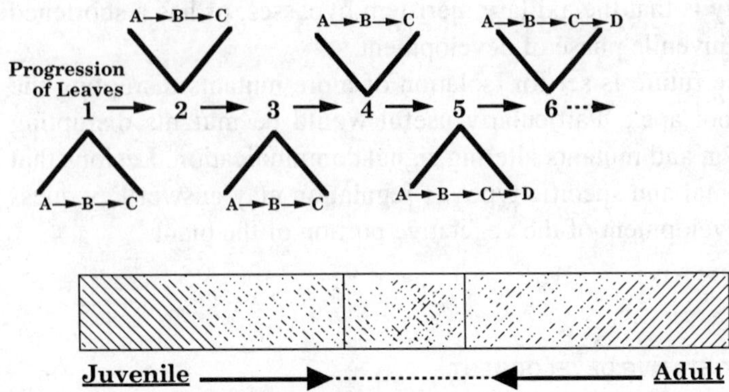



*Figure 11* Model for vegetative development. The sequence of leaves is indicated with the numbers 1–6. Formation of each leaf primordium in itself involves a set of genes indicated with letters (A→C, etc.). These functions repeat each time a leaf primordium is formed. Underlying the sequence with repetitive features is an age factor from the juvenile and adult programs. Transition from the juvenile to the adult program is gradual, with some overlap.

bacco mutant (McHale 1992), leaves develop without a differentiated blade. This suggests that there are multiple events in the process of leaf development which are independent of the progression of leaves. Figure 11 shows a diagrammatic representation of these processes. The progression of leaves is represented by the numerical sequence 1–6. Each leaf primordium must also have a series of processes independent of this and is diagramed A→C. (The number of processes here has not been defined.) The processes forming each leaf primordium repeat with each leaf. Furthermore, leaf morphology can become more elaborate by the addition of a single gene to the repetitive process (e.g., D is added for leaf primordium 5). The bladeless tobacco mutant (*lam*) (McHale 1992) would represent a disruption in a repetitive (A→C) function found in all leaf primordia. In contrast, the *fey* mutant (Medford et al. 1992), which disrupts the apical meristem as well as leaf formation and positioning, would represent a function acting either before the progression in the meristem is established or in a process common to the sequence of leaves.

If the formation of the *Arabidopsis* basal rosette is simply a sequence with repetitive features, what accounts for the differences in the shapes of leaf 1 and leaf 3? Another factor, perhaps an age factor represented by the juvenile and adult programs, could be superimposed on the repetition in leaf formation (Fig. 11). The transition between the juvenile and adult phases of development is not abrupt but gradual (Conway and Poethig

1993). This gradual transition could account for the increase in modifications seen between leaf 1 and the later leaves. As diagramed, the juvenile/adult age modification could have effects on either the repetitive process forming leaf primordia and/or the sequential processes. Because competence to flower is considered to be an adult feature, and flowering is not observed before formation of the fifth leaf in normal plants, *Arabidopsis* should not be completely adult until after the formation of leaf 5.

## CONCLUSIONS AND PROSPECTS

The analysis of the *Arabidopsis* vegetative shoot apical meristem holds much promise for understanding plant development. For example, the shoot apical meristem is formed during embryogenesis and is inactive at the completion of embryogenesis. What is the signal(s) that inactivates the meristem during embryogenesis? Is this the same signal(s) related to the onset of developmental arrest (e.g., ABA, dehydration)? Furthermore, what are the signals that lead to the reactivation of the shoot apical meristem after imbibition?

Two types of mutants show that at least part of the activation signal must involve perception of light. When germinated in darkness, *Arabidopsis* cotyledons do not separate, the hypocotyl remains downwardly curved, and there is no apparent development from the apical meristem. When either the *det1* or *cop1* mutant is germinated in darkness, there is a considerable amount of development from the apical meristem (Chory et al. 1991; Deng and Quail 1992). The *cop1-1* mutant even produces a pattern of leaves resembling the rosette produced in light-grown plants (Deng and Quail 1992). These two types of mutants imply that the apical meristem, although usually not directly receptive of light itself, must either directly or indirectly perceive light signals. Given the available details of morphological changes in the shoot apex (above), it would be interesting to see if any of the *det* or *cop* mutants allow development to proceed only to a certain point. For example, the *det1* mutant (Chory et al. 1991) may allow the development of the juvenile and not the adult apical meristem.

Another question about plant development is whether or not the various functions of the apical meristem are separate, coordinated, or linked. For example, data from numerous studies indicate that the formation of leaf primordia is a multicellular process (for review, see Steeves and Sussex 1989). Because leaf primordium formation is a multicellular process, it implies that several functions of the apical meristem are coordinated or linked. For example, as cells in the peripheral zone are sequestered into primordia, cells from the central zone must divide to replace them. The portion of the meristem that must be replaced after a leaf primordium is formed is considerable in plants such as *Arabidopsis* with low phyllotactic patterns (Richards 1948). Furthermore, the observation that the first two pairs of leaves are initiated nearly simultaneously implies that there is a communication between the most distal part of the meristem and the leaf primordia. The *fey* mutant seemingly disrupts two of the four meristem functions: initiation of organs and maintenance of itself as a formative region. Because *fey* segregates as a single recessive mutation, it suggests that these two functions are linked. If these two functions are linked, then are the other meristem functions also linked? The answer is unknown, but if the four major functions are in some way linked, it could provide a mechanism whereby the apical meristem could function as an organizing center for the entire shoot.

The vegetative meristem must have a series of factors that determine how, when, and to what extent the *Arabidopsis* basal rosette develops. Many important principles need to be tested. For example, a complete apical meristem is required for the production of organs, suggesting that cell communication is essential to the meristem. What are the types of cell-cell communication or factors responsible for this? Furthermore, what are the factors that control the remarkably precise position where plant leaves are positioned? The plant hormones (auxin and cytokinin) have been implicated in these processes (cf. Lyndon 1990; Sachs 1991), but due to the small size of the shoot apex, definitive data are still lacking. One idea suggested from previous studies is that the first steps of leaf primordium initiation can occur without involving cell division (Foard 1971). This suggests that genes involved in cell division (e.g., *CDC2* [Colasanti et al. 1991; Martinez et al. 1992]) may not play a regulatory role in organ formation. Furthermore, the apical meristem's influence does not end with organ formation. Even after an organ (leaf) is formed, other experiments suggest that the meristem is important for differentiation signals. For example, Sussex (1951) reported that potato leaves without dorsiventrality were formed when the incipient primordia were surgically severed from the apical meristem. This could result from a loss of a constant communication between the developing primordia and the apical meristem. Alternatively, it is possible that the determination process in leaf primordia is gradual and that there is constant communication with the meristem and the rest of the plant including older leaves, the stem, and root. The questions that remain about vegetative development are numerous. The answers should give us a greater understanding of fundamental principles in the process known as development.

**ACKNOWLEDGMENTS**

This work was supported by grants from the National Science Foundation and the U.S. Department of Agriculture. We thank Scott Poethig for a preprint of the work on juvenile and adult traits.

**REFERENCES**

Barton, M.K. and R.S. Poethig. 1993. Formation of the shoot apical meristem in *Arabidopsis thaliana*: An analysis of development in the wild type and in the *shoot meristemless* mutant. *Development* **119**: 823.

Brown, J.A.M., J.P. Miksche, and H.H. Smith. 1964. An analysis of $^3$H thymidine distribution throughout the vegetative meristem of *Arabidopsis thaliana* (L.) Heynh. *Radiat. Bot.* **4**: 107–113.

Callos, J.D. and J.I. Medford. 1994. Organ positions and pattern formation in the shoot apex. *Plant J.* (in press).

Chory, J., P. Nagpal, and C. Peto. 1991. Phenotypic and genetic analysis of *det2*, a new mutant the affect light-regulated seedling development in *Arabidopsis*. *Plant Cell* **3**: 445–459.

Clark, S.E., M.P. Running, and E.M. Meyerowitz. 1993. CLAVATA1, a regulator of meristem and flower development in *Arabidopsis*. *Development* **119**: 397–418.

Colasanti, J., M. Tyers, and V. Sundaresan. 1991. Isolation and characterization of cDNA clones encoding a functional p34$^{cdc2}$ homologue from *Zea mays*. *Proc. Natl. Acad. Sci.* **88**: 3377–3381.

Conway, L.J. and R.S. Poethig. 1993. Heterochrony in plant development. *Semin. Plant Dev.* **4**: 65–72.

Cutter, E.G. 1965. Recent experimental studies of the shoot apex and shoot morphogenesis. *Bot. Rev.* **31**: 7–113.

Deng, X.-W. and P.H. Quail. 1992. Genetic and phenotypic characterization of *cop1* mutants of *Arabidopsis thaliana*. *Plant J.* **2**: 83–95.

Foard, D.E. 1971. The initial protrusion of a leaf primordium can form without concurrent periclinal cell divisions. *Can. J. Bot.* **49**: 1601–1603.

Furner, I.J. and J.E. Pumfrey. 1992. Cell fate in the shoot apical meristem of *Arabidopsis thaliana*. *Development* **115**: 755–764.

Hsieh, H.-J., N.Q. Li, and J.A. Frangos. 1992. Shear-induced platelet-derived growth factor gene expression in human endothelial cells is mediated by protein kinase C. *J. Cell. Physiol.* **150**: 552–558.

———. 1993. Pulsatile and steady flow induces a c-*fos* expression in human endothelial cells. *J. Cell. Physiol.* **154**: 143–151.

Hush, J.M., R.L. Overall, and I.A. Newman. 1991. A calcium influx precedes organogenesis in *Graptopetalum*. *Plant Cell Environ.* **14**: 657–665.

Irish, V.F. and I.M. Sussex. 1992. A fate map of the *Arabidopsis* embryonic shoot apical meristem. *Development* **115**: 745–753.

Leyser, H.M.O. and I.J. Furner. 1992. Characterization of three shoot apical meristem mutants of *Arabidopsis thaliana*. *Development* **116**: 397–403.

Lyndon, R.F. 1970. Rates of cell division in the shoot apical meristem of *Pisum*. *Ann. Bot.* **34**: 1–17.

———. 1990. *Plant development. The cellular basis*. Unwin Hyman, Winchester, Massachusetts.

Martinez, M.C., J.-E. Jorgensen, M.A. Lawton, C.J. Lamb, and P.W. Doerner. 1992. Spatial pattern of *cdc2* expression in relation to meristem activity and cell proliferation during plant development. *Proc. Natl. Acad. Sci.* **89:** 7360–7364.

McHale, N.A. 1992. A nuclear mutation blocking initiation of the lamina in leaves of *Nicotiana sylvestris*. *Planta* **186:** 355–360.

Medford, J.I. 1992. Vegetative apical meristems. *Plant Cell* **4:** 1029–1039.

Medford, J.I., J.S. Elmer, and H.J. Klee. 1991. Molecular cloning and characterization of genes expressed in shoot apical meristems. *Plant Cell* **3:** 359–370.

Medford, J.I., F.J. Behringer, J.D. Callos, and K.A. Feldmann. 1992. Normal and abnormal development in the *Arabidopsis* vegetative shoot apex. *Plant Cell* **4:** 631–643.

Meicenheimer, R.D., F.J. Muehlbauer, J.L. Hindman, and E.T. Gritton. 1983. Meristem characteristics of genetically modified pea (*Pisum sativum*) leaf primordia. *Can. J. Bot.* **61:** 3430–3437.

Poethig, R.S. and I.M. Sussex. 1985. The cellular parameters of leaf development in tobacco: A clonal analysis. *Planta* **165:** 170–184.

Quatrano, R.S. 1987. The role of hormones during seed development. In *Plant hormones and their role in plant growth and development* (ed. P.J. Davies), pp. 494–514. Martinus Nijhoff, Boston.

Richards, F.J. 1948. The geometry of phyllotaxis and its origin. *Symp. Soc. Exp. Biol.* **2:** 217–245.

Sachs, T. 1991. *Pattern formation in plant tissues*. Cambridge University Press, New York.

Sadik, S. 1962. Morphology of the curd of cauliflower. *Am. J. Bot.* **49:** 290–297.

Selker, J.M.L., G.L. Steucek, and P. Green. 1992. Biophysical mechanisms for morphogenetic progressions at the shoot apex. *Dev. Biol.* **153:** 29–43.

Shannon, S. and D.R. Meeks-Wagner. 1991. A mutation in the *Arabidospsis TFL1* gene affects inflorescence meristem development. *Plant Cell* **3:** 877–892.

Smyth, D.R., J.L. Bowman, and E.M. Meyerowitz. 1990. Early flower development in *Arabidopsis*. *Plant Cell* **2:** 755–767.

Steeves, T.A. and I.M. Sussex. 1989. *Patterns in plant development*. Cambridge University Press, New York.

Sung, Z.R., A. Belachew, B. Shunong, and R. Bertrand-Garcia. 1992. *EMF*, an *Arabidopsis* gene required for vegetative shoot development. *Science* **258:** 1645–1647.

Sussex, I.M. 1951. Experiments on the cause of dorsiventrality in leaves. *Nature* **167:** 651–652.

———. 1964. The permanence of meristems: Developmental organizers or reactors to exogenous stimuli? In *Meristems and differentiation* (ed. J.P. Miksche et al.), pp. 1–12. Office of Technical Services (Department of Commerce), Washington, D.C.

Vaughan, J.G. 1952. Structure of the angiosperm apex. *Nature* **169:** 458–459.

Vaughan, J.G. and F.R. Jones. 1953. Structure of the angiosperm inflorescence apex. *Nature* **171:** 751–752.

Wardlaw, C.W. 1949. Experiments on organogenesis in ferns. *Growth* (suppl.) **13:** 93–131.

# 15

# Leaf Development in *Arabidopsis*

**Abby Telfer and R. Scott Poethig**
Plant Science Institute, Department of Biology
University of Pennsylvania
Philadelphia, Pennsylvania 19104-6018

Leaves are determinate structures whose primary function is to produce carbohydrates required for plant growth and development. From a developmental standpoint, leaves are interesting because they display a wide range of genetically and developmentally regulated variation in shape, and because their morphological simplicity and accessibility make them an ideal system for the analysis of basic problems in plant development. Growth control, pattern formation, and cell and tissue differentiation can all be conveniently studied in the context of leaf development. Although many of these problems have yet to be addressed in *Arabidopsis*, this situation is changing rapidly. This is due in part to the advantages of *Arabidopsis* for genetic and molecular analysis, but the relatively small size of the *Arabidopsis* leaf is an equally important advantage of this species because it permits microscopic analysis of an entire leaf at all stages of development.

Like most plants, *Arabidopsis* exhibits subtle heteroblasty (first described by Röbbelen 1957), in that leaves produced at different stages of shoot development vary in their morphology (Fig. 1). At a gross level, there are two distinct leaf types: rosette leaves and bracts. Rosette leaves are produced early in development and have long petioles and broad blades. Bracts (sometimes referred to as cauline leaves) form at the basal nodes of each branch in the inflorescence and differ from rosette leaves in that they lack petioles and have narrower, more oblong blades.

There are additional morphological differences within each of these leaf types. Early rosette leaves are small and have rounded blades with smooth margins. Rosette leaves produced later in development are larger, more elongated, and have slightly serrated edges. The position of the broadest portion of the leaf blade is more distal in later leaves, and narrow, winglike extensions of the blade form on the petiole, making the boundary between the blade and petiole less distinct. Bracts, too, exhibit a range of sizes and shapes, with successive bracts becoming smaller and narrower.

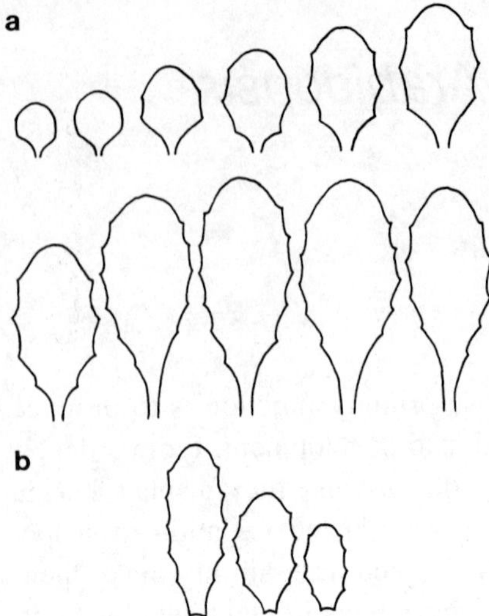



*Figure 1* Heteroblastic variation in leaf size and shape in the Antwerpen ecotype. Successive leaves and bracts are arranged left to right, top to bottom. (*a*) Vegetative leaves; (*b*) bracts. (Modified from Röbbelen 1957.)

In addition to their differences in overall shape, *Arabidopsis* leaves vary in the distribution of epidermal hairs, called trichomes, that form on the surfaces of the blade (A. Telfer and R.S. Poethig, unpubl.; J.M. Martínez-Zapater, pers. comm.). Trichomes develop on the adaxial, or upper, surfaces of all rosette leaves but typically do not occur on the abaxial, or lower, surfaces of the first few leaves. Trichomes are numerous on both surfaces of late rosette leaves and the first bracts, but they develop in reduced numbers on the adaxial surfaces of later bracts. In many ecotypes the last bracts to develop are completely devoid of adaxial trichomes. The number of trichomes that develop on the abaxial surfaces of later bracts is also reduced, but trichomes are not usually eliminated completely from these positions.

The first two leaf primordia form on opposite sides of the meristem and may be initiated simultaneously. Subsequent leaves, bracts, and flowers are initiated individually in a spiral phyllotaxis with successive organs being offset by approximately 140° (Furner and Pumfrey 1992). The number, size, and shape of rosette leaves and bracts produced on a plant vary among the ecotypes (Röbbelen 1965). In addition, these traits are sensitive to a variety of environmental factors (for review, see Rédei 1969; Martínez-Zapater and Somerville 1990; Koornneef et al. 1991) and to mutations that accelerate or delay flowering (Martínez-Zapater and Somerville 1990; Koornneef et al. 1991; Shannon and Meeks-Wagner 1991).

Leaf development in *Arabidopsis* has not been described in detail. To provide a more complete description, we examine what is known about this process in the context of the general understanding of leaf development in other dicotyledons. Our discussion focuses on the first two leaves, since development of later leaves has not been examined in *Arabidopsis*. It should be noted that because *Arabidopsis* undergoes heteroblastic development, the development of later leaves and bracts may differ substantially from the development of these early leaves. It is also likely that the details of leaf development vary between ecotypes. Finally, it should be noted that our description of leaf development is pieced together from observations of plants of different ecotypes raised under different environmental conditions, and so the reported timing of individual developmental events should be considered approximate.

## ORGANIZATION AND MERISTEMATIC ORIGIN OF LEAF TISSUES

The leaf blade, or lamina, is composed of the upper and lower epidermis and roughly five layers of internal tissue (Fig. 2). In the Columbia and Landsberg ecotypes, the average cell size in the upper epidermis is larger than that in the lower epidermis (Fig. 3), but this may not be the case for all ecotypes and genetic backgrounds (Rüffer-Turner and Napp-Zinn 1979). The palisade tissue, located immediately below the adaxial epidermis, consists of a single layer of regularly shaped cells that are elongated in the dorsiventral axis of the leaf and interspersed with small air spaces. The spongy mesophyll, located between the palisade layer and the lower epidermis, is made up of approximately four layers of relatively small, irregularly shaped cells and an extensive network of air spaces. The vascular network is embedded within the spongy mesophyll.

The meristematic origin of each of the tissue layers in the mature leaf has been determined in other dicotyledons by clonal analysis (Stewart 1978; Tilney-Bassett 1986). The upper and lower epidermis are derived from the $L_1$ layer of the shoot apical meristem, whereas the mesophyll and vascular tissues are derived from the $L_2$ and $L_3$ layers. The $L_2$ gives rise to the palisade and lower spongy mesophyll layers throughout the blade and to the inner spongy mesophyll layers near the leaf margins. Cells in the $L_3$ are the progenitors of the inner spongy mesophyll layers in the central region of the leaf.

Clonal analysis in other species also indicates that most of the cell divisions that occur during leaf expansion are oriented anticlinally (i.e., with the new cell wall perpendicular to the plane of the leaf). This restriction in the orientation of cell divisions has the effect of producing lineally related cells with similar cell fates (e.g., epidermal cells derived

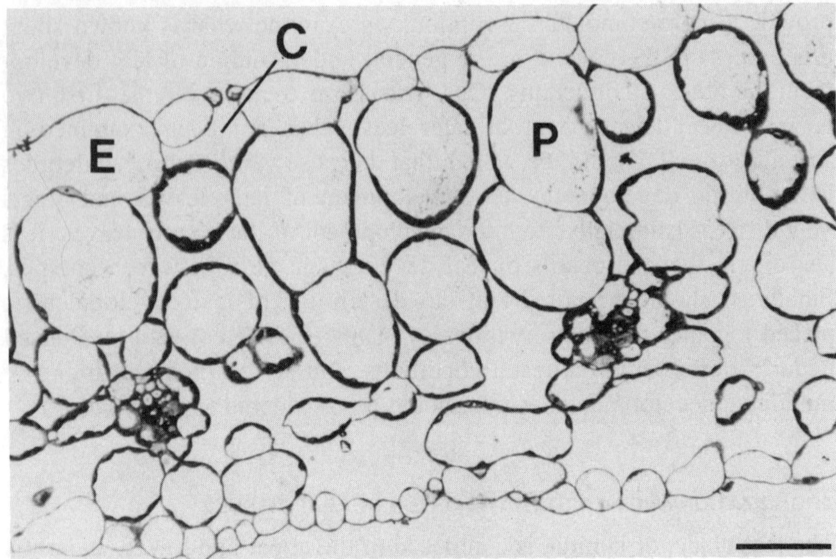

*Figure 2* Transverse section of a portion of a fully expanded first leaf. Notice the layered cell organization. The cells between the palisade layer and the lower epidermis form the spongy mesophyll. Two veins, embedded in the spongy mesophyll, are visible in this section. (E) Epidermal cell; (P) palisade cell; (C) substomatal chamber. Bar, 50 μm. (Courtesy of J.L. Marrison and R.M. Leech.)

from the $L_1$ meristem layer). However, it is important to note that cell lineage does not determine cell fate. If a cell from one tissue layer invades another layer as the result of a rare periclinal cell division (i.e., with the new cell wall forming parallel to the blade surface), its progeny cells assume fates appropriate to their new positions (Stewart and Dermen 1975). Although clonal analysis has not been used to study leaf morphogenesis in *Arabidopsis*, data from clonal analyses of *Arabidopsis* shoot growth (Furner and Pumfrey 1992; Irish and Sussex 1992) support the conclusions drawn from other species.

## DEVELOPMENT OF LEAF PRIMORDIA

In *Arabidopsis* the first two leaves are initiated during embryogenesis. The primordia of these leaves can be detected as slight bulges on opposite sides of the apical meristem in sectioned mature embryos (Medford et al. 1992; L. Conway, pers. comm.). Although the primordia initially appear to be the same size, one becomes slightly larger than the other several days after germination, suggesting that their initiation and/or development is not completely synchronous (Irish and Sussex 1992; Medford et al. 1992).

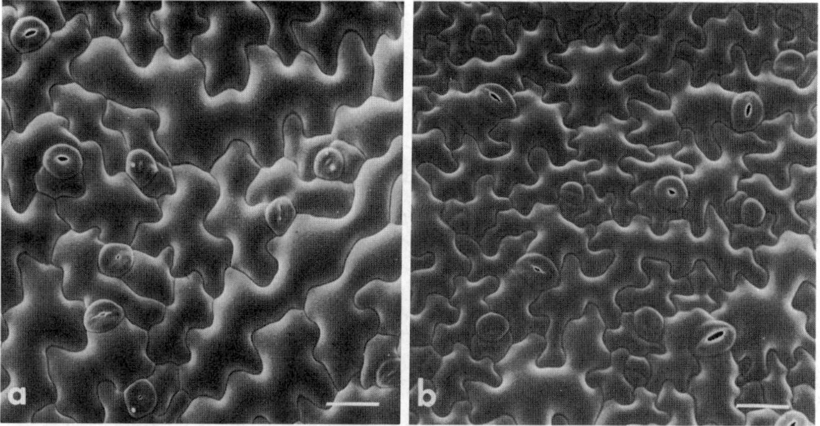

*Figure 3* Scanning electron micrographs of the adaxial and abaxial epidermis. From the blade of a fully expanded ninth rosette leaf, Columbia ecotype. (*a*) Adaxial epidermis; (*b*) abaxial epidermis. Bar, 20 μm. (Courtesy of Paul Linstead.)

The total number of cells recruited from the meristem into each leaf primordium has not been determined. However, Irish and Sussex (1992) concluded from clonal analysis experiments that at the dry seed stage of embryonic development, eight or nine cells in the $L_2$ layer of the shoot apical meristem are destined to contribute to each of the first two leaf primordia. We infer from the work of Furner and Pumfrey (1992) that cells in the shoot apical meristem continue to be recruited into the developing primordium after imbibition. They found that irradiated dry seeds occasionally gave rise to plants with mutant sectors that extended from one of the first leaves into other organs. Cells that were committed to form part of the first or second leaf primordium at the time of irradiation could not have given rise to such sectors because their progeny cells would all be contained in the same leaf. On the other hand, cells that were uncommitted at the time of irradiation could have given rise to daughter cells that were subsequently recruited into different primordia. Assuming that this result does not reflect the reorganization of leaf primordia as a consequence of radiation damage, the recruitment of cells into the first two leaf primordia must continue beyond the end of embryogenesis.

The feature that first distinguishes a developing leaf from a shoot is its dorsiventrality, which is reflected both in its external morphology and in its anatomy. Very young *Arabidopsis* leaf primordia appear triangular in transverse section (Fig. 4a) because the side of each primordium that faces the meristem is somewhat flattened. The flattened face forms the

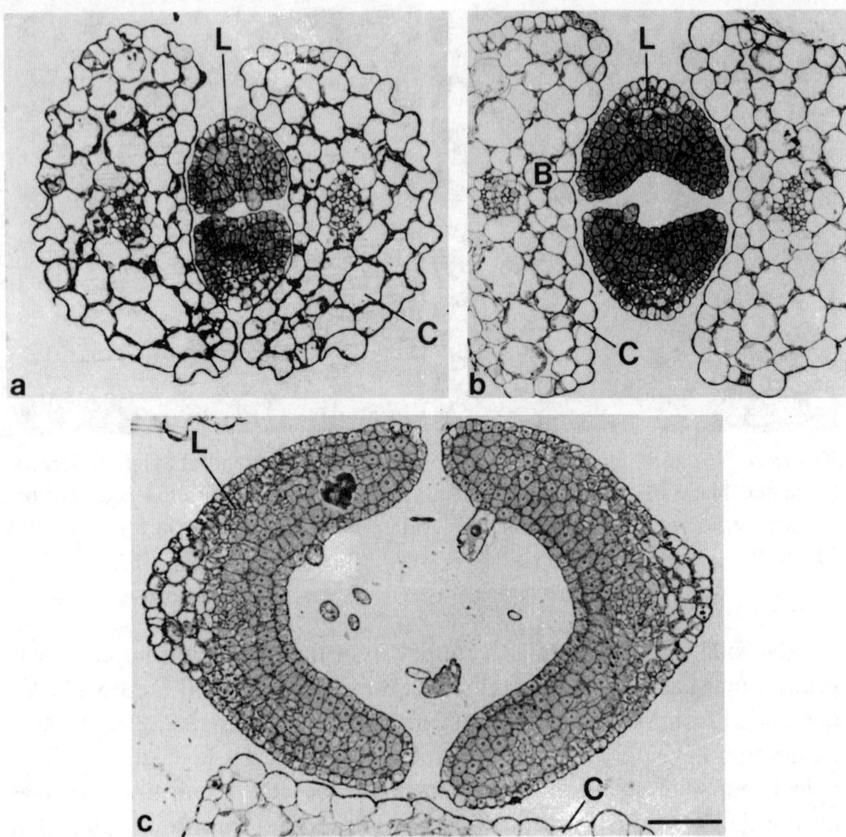

*Figure 4* Transverse sections of the first two leaves at three developmental stages. (*a*) Leaf primordia prior to blade initiation, harvested 5 days after sowing. (*b*) Primordia undergoing initial blade outgrowth, harvested 6 days after sowing. (*c*) Expanding leaves from plant harvested 8 days after sowing. The enlarged cells protruding from the adaxial surfaces of leaves in each panel are developing trichomes. (L) Leaf; (B) developing blade; (*C*) cotyledon. Bar, 50 μm. (Reprinted, with permission of Oxford University Press, from Pyke et al. 1991.)

presumptive adaxial surface of the leaf, and the leaf blade develops along its margin (see below). In the first few leaf primordia, the presumptive adaxial surface can also be identified by the presence of differentiating trichomes (Fig. 4a), which do not develop on the abaxial surfaces of the first leaves. Young leaf primordia also exhibit internal asymmetries; regions of the abaxial epidermis and adjacent mesophyll contain enlarged, highly vacuolated cells. During the subsequent development of the lamina (Fig. 4b,c), elongated palisade cells form below the adaxial epidermis, and the spongy mesophyll differentiates between the palisade layer and the abaxial epidermis.

An additional asymmetry can be detected upon the differentiation of the vasculature. Xylem elements are found on the adaxial side of the vascular bundles, and phloem differentiates on the abaxial side (A. Telfer and R.S. Poethig, unpubl.). It is not clear, however, whether asymmetry in the vasculature is specified by the same factors that specify the overall dorsiventrality of the leaf, because it has been shown in other plants that asymmetry can develop in the vasculature of abnormal leaves that are radially symmetric in all other respects. For example, tobacco plants infected with tobacco mosaic virus may produce "shoestring" leaves that are circular in transverse section and yet exhibit normal dorsiventrality in the vascular tissue (Tepfer and Chessin 1959). Vascular asymmetry has also been observed in "centric" leaves induced by surgical manipulation of the potato shoot apex (Sussex 1955). The absence of asymmetry in the other tissues of these leaves suggests that the origin of dorsiventrality in the vasculature may arise by a unique mechanism, perhaps as a topological extension of the vascular organization in the stem.

## DEVELOPMENT OF THE LEAF BLADE

### Blade Initiation

During leaf blade initiation, epidermal and mesophyll cells located along the margin of the leaf primordium must be recruited into the blade. Transverse leaf sections taken shortly after lamina initiation in *Arabidopsis* reveal that the number of cell layers in the young blade is nearly the same as the number of cell layers in the dorsiventral dimension of the leaf primordium prior to blade initiation (cf. Fig. 4a and 4b). This suggests that the blade is derived from a large fraction of the cells that lie at the edge of the *Arabidopsis* leaf primordium and is in contrast to tobacco, where only a small subset of the cells along the margin of the primordium give rise to the blade (Avery 1933; Poethig and Sussex 1985). Although it is possible that blade initiation occurs by the continuation of the growth patterns that create the flattened adaxial surface and establish dorsiventrality in the primordium, the identification of a mutation in *Nicotiana sylvestris* that blocks leaf blade initiation after the establishment of dorsiventrality (McHale 1992) suggests that blade initiation represents a distinct event in leaf development.

### Expansion of the Lamina

The kinetics of lamina expansion in the first two leaves of *Arabidopsis* has been examined (Pyke et al. 1991). The blade expands rapidly between the 7th and 15th day after sowing and then enlarges more slowly until the leaf reaches a maximum size of 30 mm$^2$ around day 18. Lamina maturation is characterized by the cessation of cell division, the formation of intercellular spaces, and cell enlargement. Paradermal sections of developing leaves (Fig. 5) reveal that blade maturation occurs basipetally (i.e., from the tip to the base of the leaf). In the first leaves of 9-day-old seedlings, mitotically active mesophyll cells are present at the leaf base and there is an acropetal (from the base to the tip) gradient of increasing cell size and enlarging air spaces. In contrast to other species (von Papen 1935), fully expanded first leaves of *Arabidopsis* maintain an acropetal gradient of increasing mesophyll cell size (Pyke et al. 1991). This suggests that the processes of cell division and cell expansion make different relative contributions to leaf blade growth in apical and basal regions, but this has not been confirmed.

Scanning electron micrographs of young leaf primordia shown by Irish and Sussex (1992) and Medford et al. (1992) indicate that the epidermal cell layers also follow a basipetal pattern of maturation. Developing trichomes and enlarging epidermal cells are present at the tips of young leaves at stages when the epidermal cells toward the leaf base remain small and show no external signs of differentiation. Unlike the mesophyll, however, the epidermis exhibits no obvious gradient in cell size in fully expanded leaves (our unpublished observation).

It was noted earlier that each tissue layer in the mature leaf has a distinct cellular organization. The differences in cell size and in the size of air spaces are thought to reflect differences in the duration of periods of cell division and cell enlargement in each layer (Dale 1988). In species where this has been carefully studied, cells in the epidermis are the first to cease dividing. Cell division stops later in the spongy mesophyll, and last in the palisade layer. The only data on the relative duration of cell division and expansion in the tissue layers of *Arabidopsis* leaves are from Pyke et al. (1991), who examined transverse sections taken near the leaf base at 9 days after sowing. At this stage, enlarged, vacuolated cells were detected in all cell layers except the palisade tissue. This is consistent with mitosis persisting in the palisade layer after it has ceased in the other layers.

To explain the ontogeny of air spaces in the mesophyll, Avery (1933) proposed that cell enlargement continues in the epidermal layers after cell division and expansion have ceased in the mesophyll, thus generating forces which pull on the mesophyll cells and separate them. More recently, Jeffree et al. (1986) suggested that although such forces might contribute to the expansion of existing air spaces, the first step in their formation is the enzymatic digestion of cell wall material at the junction between three or more mesophyll cells. Cellulase has been proposed to

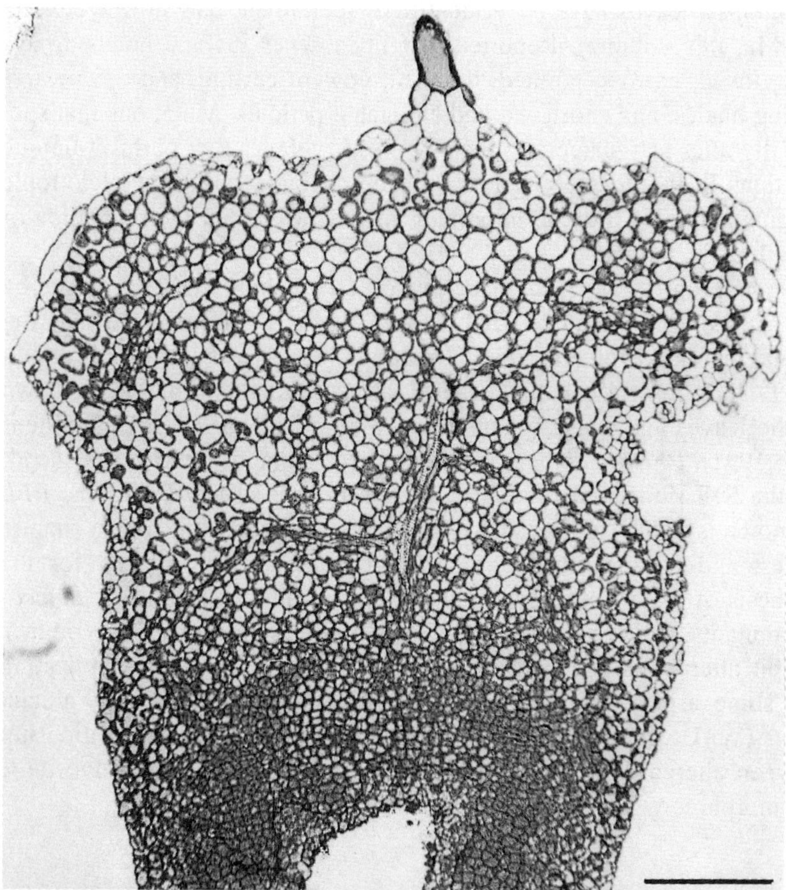

*Figure 5* Paradermal section from a first leaf harvested 9 days after sowing. The section, taken below the adaxial epidermis, covers about 75% of the leaf area and includes palisade tissue near the leaf tip. Note the acropetal gradient of increasing cell size and enlarging air spaces. The point at the tip of the section is a portion of a trichome cell. (Reprinted, with permission of Oxford University Press, from Pyke et al. 1991.)

play a role in this process, since increased cellulase activity has been detected during the formation and enlargement of air spaces in water-logged plants (Kawase 1979). Air spaces that form below the stomata may arise by a different mechanism (see below).

### Leaf Shape Mutations

The genetic regulation of leaf development may be investigated by the study of leaf shape mutations. Numerous mutations that alter the shape of

*Arabidopsis* leaves have been identified (see mutant lists in Meyerowitz and Ma, this volume). Reported leaf phenotypes include notched, serrated, lobed, narrow, pointed, rounded, upward-curling, and downward-curling blades, and shortened and elongated petioles. Mutations that specifically alter leaf shape are likely to identify genes that perform unique functions in the developing leaf, but the characterization of pleiotropic mutants may also provide important information about the regulation of growth in the lamina.

Some progress has been made toward establishing the roles played by two genes, *PFL* (pointed first leaves) and *LAN* (lanceolate), during normal leaf development. The *pfl* mutation, which was identified following T-DNA insertional mutagenesis, affects the shape of the first two rosette leaves and causes plants to grow more slowly (Van Lijsebettens et al. 1991). *PFL* was cloned and identified as one of three genes encoding the S18 ribosomal protein (Van Lijsebettens et al. 1994). The *PFL* promoter is active in meristems and leaf primordia, but not in mature leaves, and it is postulated that *PFL* expression is required for the synthesis of extra ribosomes in tissues undergoing cell division. *lanceolate* mutants, which have curled leaf edges (M. Koornneef, pers. comm.), exhibit alterations in cell wall monosaccharide composition, indicating that some aspect of polysaccharide biosynthesis is altered in mutant plants (W.-D. Reiter et al., pers. comm.). However, a causal relationship between altered cell wall composition and the characteristic *lanceolate* leaf morphology has yet to be demonstrated.

## CELLULAR DIFFERENTIATION

Several distinct tissues and numerous cell types differentiate during primordial growth and leaf expansion. The developing leaf provides a unique opportunity to study the differentiation of these tissues because numerous stages of development may be represented in a single immature leaf. In this section, we consider the development of the vasculature and of several epidermal cell types.

### Vascular Development

Fully expanded first leaves contain a primary midvein and several secondary veins that branch off the midvein and curve outward toward the margin and up toward the tip. Smaller, higher-order veins connect the secondary veins with each other and occasionally form junctions with the midvein (Fig. 6f). Our unpublished observations of leaves cleared at different developmental stages indicate the sequence in which this pattern

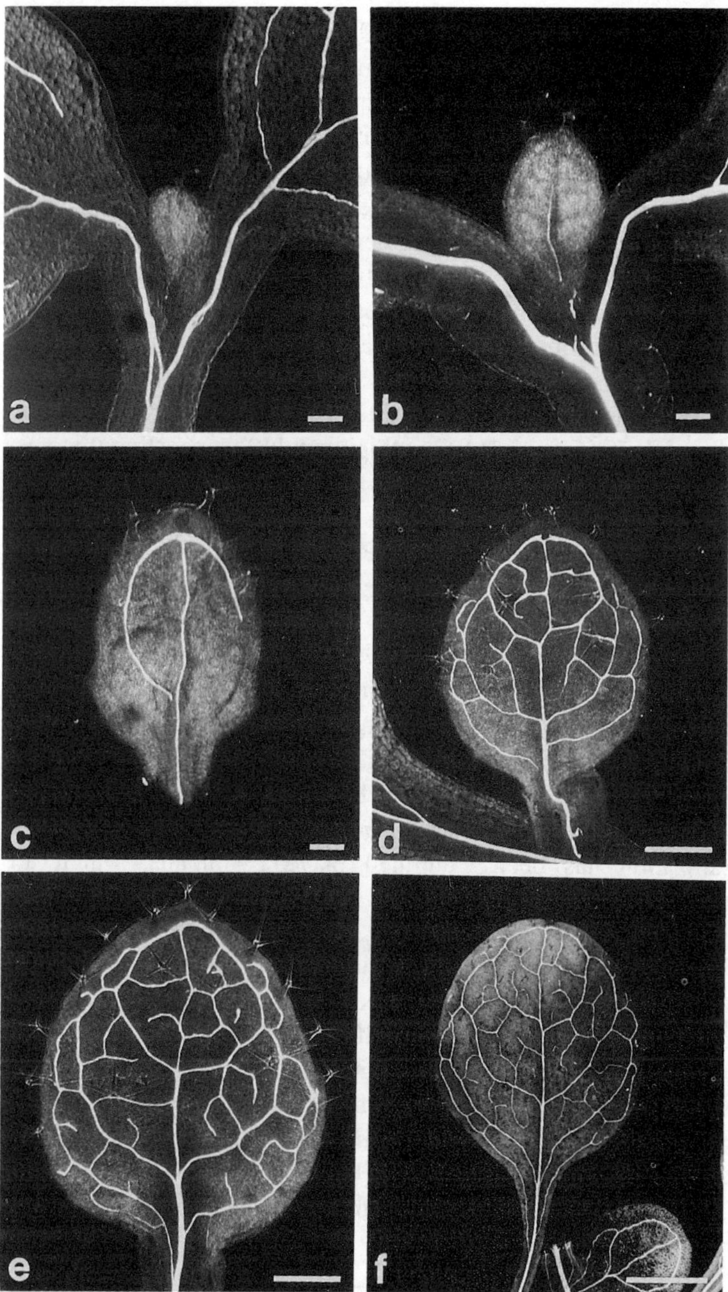



*Figure 6* Vascular development in expanding first leaves in the Columbia ecotype harvested (*a*) 4 days after sowing, (*b–e*) at several more advanced stages of development, and (*f*) when fully expanded, were cleared by incubating 30 min in 95% ethanol at 70°C followed by 30 min in lactic acid:phenol:glycerol:water (1:1:1:1) at 90°C, and photographed under dark-field illumination. Bars, *a–c*, 100 μm; *d–e*, 400 μm; *f*, 2 mm. (A. Telfer and R.S. Poethig, unpubl.)

develops. Primordia that have attained 250 µm in length contain differentiating provascular strands visible by Nomarski microscopy but no spiral lignification indicative of mature tracheary elements (Fig. 6a). The lignification of tracheary elements in the developing midvein begins before the primordia attain a length of 600 µm (Fig. 6b). In more mature primordia, the midvein bifurcates at the tip to form secondary veins which extend along the margins of the developing blade (Fig. 6c). Additional secondary and higher-order veins differentiate as the blade expands (Fig. 6d–f). Consistent with the observation that cells in the basal regions of the blade differentiate later than more distal cells (Pyke et al. 1991), secondary vascular elements are added basally as the blade expands. The ratio of total leaf area to total vein length increases during leaf expansion (A. Telfer and R.S. Poethig, unpubl.), suggesting that vascular development does not keep pace with the growth of other tissues in the leaf blade. This observation supports that of Pyke et al. (1991), who calculated from transverse sections that the vascular elements in the first leaves comprise a decreasing percentage of leaf volume as development progresses.

It is not known what signals control vascular differentiation in developing leaves or how the pattern of venation is determined. Work in other species indicates that the processes of phloem and xylem differentiation are sensitive to the concentration of various hormones, including auxin, and sugars (for review, see Aloni 1987), but the role of these compounds in vascular differentiation in *Arabidopsis* leaves remains unclear. Vascular organization has been examined in *axr1* mutants, which are resistant to exogenous auxin. Although vascular bundle differentiation is aberrant in the stems of mutant plants, vascular differentiation in the leaves does not appear to be affected (Lincoln et al. 1990).

Vascular development has not yet been the focus of genetic analysis in *Arabidopsis*, but mutants with specific vascular defects have been identified in other species (Postlethwait and Nelson 1957; Alldridge 1964), indicating that such an approach could provide substantial information about this process. One *Arabidopsis* mutation that affects vascular development has been reported (Rédei and Hirono 1964). This mutation, *reticulata*, causes a darkening of the vascular tissue but has not been thoroughly characterized.

### Epidermal Development

Three morphologically distinct classes of epidermal cells are readily detectable on the leaf blade. Most of the leaf surface is covered with pavement cells, and these are interspersed with stomatal guard cells and trichomes.

## Pavement Cells

The majority of the pavement cells on both the upper and lower surfaces of the lamina have highly convoluted anticlinal cell walls that create a jigsaw-puzzle-like pattern on the blade (Fig. 3). Pavement cells with straight anticlinal walls are restricted to the petiole, the leaf margin, the cells that ring the base of each trichome, and the base of the blade in the region over the midvein. Undulations develop in pavement cell walls relatively late in the expansion of the first leaves, arising first in cells near the leaf tip (A. Telfer and R.S. Poethig, unpubl.). They are detectable throughout the length of the blade when it has attained a length of 4 mm.

Sinuous pavement cell walls are common in plants, but the reason for their formation is not understood. Avery (1933), in an extension of his argument that epidermal cells pull the mesophyll apart during blade expansion, proposed that reciprocal forces exerted by the mesophyll pull on the epidermal cells, distorting their walls. This hypothesis could be tested in *Arabidopsis* by comparing the mesophyll organization in regions of the blade that contain convoluted pavement cells and regions that contain straight-walled cells.

Pavement cells vary considerably in size (Fig. 3). Large cells can be found interspersed among small ones throughout the blade, which suggests that they are not coordinated in their switch from mitosis to cell expansion. Pavement cells are also variably polyploid (Melaragno et al. 1993). Although approximately 36% of the pavement cells in the expanded leaf remain 2C, the nuclear DNA content in the remainder ranges up to 16C. The increased DNA content is correlated with increased cell volume. Neither the reason for the development of enlarged, polyploid cells, nor the lack of uniformity of this process is understood. It may be that epidermal growth can occur more efficiently and rapidly by cell expansion than by cell division, and endoreduplication is likely to facilitate rapid expansion. At the same time, it may be important for leaves to retain the capacity to differentiate new stomata or other cell types (including new pavement cells) within the expanding epidermis. Thus, the 2C cells, which are distributed throughout the epidermal pavement, may, in fact, be undifferentiated "stem cells" that serve as a source of new cells within the blade.


## Stomata

Stomatal development has not been described in *Arabidopsis*. The organization of stomatal complexes is somewhat variable (see Fig. 3), but the majority are anisocytic (i.e., each stoma is surrounded by three subsidiary cells of varying sizes). This arrangement is common in the crucifers (Metcalfe and Chalk 1950) and is thought to arise when a cell determined to be a stomatal progenitor undergoes a series of unequal cell divisions that ultimately give rise to the guard cell mother cell (Sachs 1984).

The factors controlling stomatal spacing are not understood. The stomatal index, which is calculated from the ratio of the number of stomata to the total number of epidermal cells in a representative area, is commonly used as a measure of stomatal spacing and is thought to reflect the proximity of stomatal progenitor cells at the time at which they become determined. The stomatal index is reported to be relatively constant within many species (Cutter 1978), but differences in stomatal spacing have been observed in *Arabidopsis* between plants of different ecotypes and genetic backgrounds, between the different leaves on individual plants, and between the adaxial and abaxial surfaces of individual leaves (Rüffer-Turner and Napp-Zinn 1979; L. Conway, pers. comm.).

Like other cell types in the leaf, stomata differentiate first at the leaf apex and later in basal regions. We detected guard cells near the tips of *Arabidopsis* first leaf primordia that had attained 600 µm in length, but none were visible near the base at that stage (A. Telfer and R.S. Poethig, unpubl.). When guard cells first differentiate, the surrounding epidermal cells have not completed their expansion. Thus, stomatal spacing in fully expanded leaves must be influenced by the amount of cell expansion that takes place within the epidermis. Indeed, stomatal spacing in *Arabidopsis* leaves is altered when cell elongation is inhibited by the application of exogenous ethylene or by mutations that cause a constitutive ethylene response (Kieber et al. 1993). Mutations that alter stomatal spacing by changing the number of stomata relative to the number of other epidermal cells have been identified (Yang and Sack 1993), and the genetic analysis of stomatal development is likely to add considerably to our understanding of this process.

The stomata overlie intercellular spaces within the mesophyll called substomatal chambers (see Fig. 2). The relative timing of formation of the stomata and substomatal chambers has not been determined in *Arabidopsis*, but in other species stomata differentiate at the same time as, or slightly after, substomatal chambers (Esau 1977). Chamber formation has been proposed to occur by the autophagy and lysis of the mesophyll cells that make contact with developing stomata (de Chalain and Berjak 1979). It will be interesting to discover whether the temporal and spatial coordination of stomatal differentiation and chamber development is achieved through mechanical interactions or molecular signals.

## Trichomes

Trichomes are unicellular hairs that develop on the leaves, sepals, and stems of *Arabidopsis* plants (Fig. 7d). Each trichome projects outward from the epidermal surface, attaining a height of 200–500 μm (Oppenheimer et al. 1993). Trichomes on the leaves form up to five branches, whereas those on the stems are predominantly unbranched. The cells have a distinctive, papillate surface and thickened cell walls (Haughn and Somerville 1988; Marks et al. 1991), and their nuclei are enlarged and highly polyploid (Melaragno et al. 1993; Hülskamp et al. 1994). Roughly a dozen rectangular support cells ring the base of each trichome. These cells also project outward from the plane of the surrounding epidermis.

As discussed previously, trichomes begin to differentiate very early in leaf development. They develop first at the leaf tip and only later in more basal regions of the blade. Additional trichomes can differentiate after this first wave of trichome development. These form on expanding leaves in the spaces between older trichomes and can be identified by virtue of their relatively early developmental stages (see below). Trichome spacing is regular, in that two trichomes almost never form in close proximity. In mature leaves, trichomes are separated by an average of approximately 30 other epidermal cells (Hülskamp et al. 1994).

Trichome development has been described previously in detail (Marks et al. 1991; Marks and Esch 1992; Hülskamp et al. 1994). During initiation, the nucleus of a protodermal cell expands and the cell begins to grow outward from the plane of the developing epidermis to form the trichome stalk (Fig. 7A). Branch development is initiated at several sites on the developing stalk (Fig. 7B), and both branches and stalk then undergo a period of expansion during which their length and diameter increase (Fig. 7C). Finally, the ring of support cells forms, the trichome cell wall thickens, and the surface becomes papillate.

Numerous mutations affecting trichome formation have been identified (Feenstra 1978; Koornneef 1981; Koornneef et al. 1982; Haughn and Somerville 1988; Marks et al. 1991; Marks and Esch 1992; Hülskamp et al. 1994). These mutations fall into several phenotypic classes, suggesting that a stepwise dissection of the genetic control of trichome development may be possible.

Mutations at two loci, *GL1* (glabrous) and *TTG* (transparent testa, glabrous), result in the absence of trichomes from the stem and from the surfaces of leaves and sepals (Koornneef et al. 1982). Mutations in *TTG* also block the production of anthocyanin and seed coat mucilage (Koornneef 1981). Nuclear enlargement, the first detectable event in trichome development, does not occur in *gl1* and *ttg* plants (Hülskamp et al. 1994), suggesting that the processes of trichome specification or initiation may

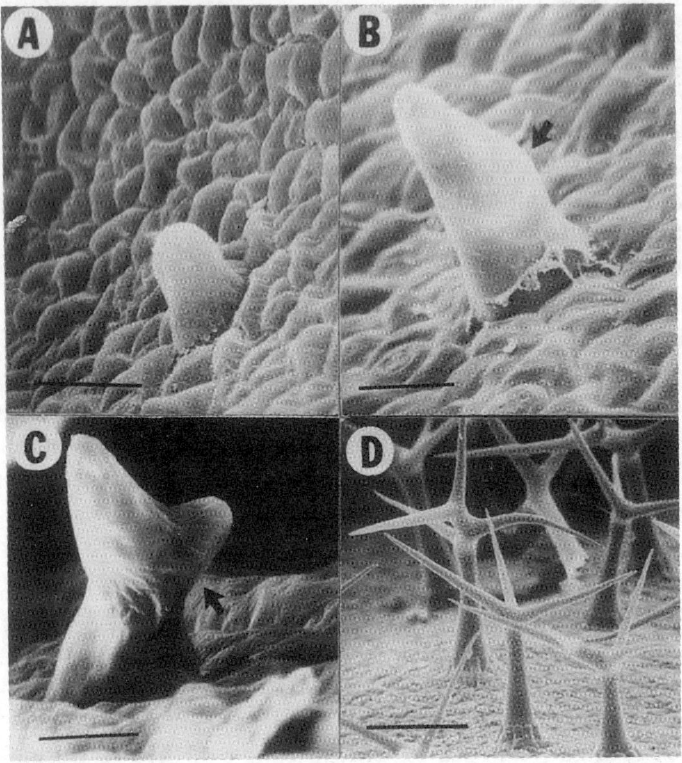

*Figure 7*   Scanning electron micrographs of developing and mature leaf trichomes. (*a*) Initiating trichome on immature protodermal tissue. (*b*) Elongating stalk with initiating branch (*arrow*). (*c*) Later stage with expanding stalk and branch (*arrow*). (*d*) Field of mature trichomes. Bars, 100 μm. (Reprinted, with permission of The Company of Biologists, from Marks et al. 1991.)

be affected. Mosaic analysis suggests that *GL1* acts locally because tissue expressing the *gl1* phenotype can develop even when surrounded by phenotypically wild-type cells (Hülskamp et al. 1994). Despite their profound effect on trichome development on the leaf blade, none of the identified mutations in *GL1* and *TTG* interfere with trichome initiation on the leaf margins or petioles. This indicates that the spatial regulation of trichome development is complex and may involve additional loci.

The regularity of trichome spacing is altered by the *Triptychon* mutation, which causes as many as four trichomes to form in a group, surrounded by a single ring of support cells (Hülskamp et al. 1994). Mutations affecting trichome outgrowth and shape define numerous loci that are required for trichome morphogenesis. Mutant phenotypes include small, aborted trichomes; reduced or increased numbers of branches;

shortened stalks; narrow or overexpanded stalks and branches; and wavy trichomes. As yet, no mutations have been found that specifically affect the development of trichome support cells.

GL1 was cloned by T-DNA tagging (Herman and Marks 1989) and found to encode a member of the *myb* family of transcriptional activators (Oppenheimer et al. 1991). Expression studies indicate that *GL1* is expressed in developing trichomes, under the control of 3′ regulatory sequences (Larkin et al. 1993). Although *TTG* has not been cloned, recent work provides indirect evidence of its role in *Arabidopsis* development. Transgenic *ttg* plants expressing the maize *R* gene produce trichomes and anthocyanin, indicating that *TTG* may be, or function through, a functional homolog of *R* (Lloyd et al. 1992). The *R* gene product regulates anthocyanin production in maize and contains sequence homology with mammalian transcriptional activators (Ludwig et al. 1989). It is thought to interact with the product of *C1* (Goff et al. 1992) which, like *GL1*, is a *myb* homolog (Paz-Ares et al. 1987). This raises the intriguing possibility that *TTG* and *GL1*, and *R* and *C1*, interact by similar mechanisms to regulate downstream genes.

## DETERMINATION AND THE ESTABLISHMENT OF ORGAN IDENTITY

A fundamental question concerning leaf development is when, and by what mechanism, a primordium becomes determined to differentiate as a leaf rather than an alternative organ type. Experiments in *Impatiens balsamina* indicate that determination is a gradual process that occurs cell by cell within developing leaf primordia (Battey and Lyndon 1988). Plants that have been induced to flower revert to vegetative growth upon transfer to noninductive conditions. If reverted plants are transferred back to growth under inductive conditions, leaf primordia smaller than 750 μm in length at the time of transfer develop as mosaic structures, with some cells expressing leaf traits and others expressing petal traits. Cells expressing petal traits are more likely to form toward the base of a mosaic organ, consistent with the normal basipetal pattern of cellular maturation in developing leaf primordia. These experiments provide some insight into how determination occurs within individual cells in the primordium. Intermediate cell types expressing both leaf and petal cell traits (such as epidermal cells with shapes characteristic of leaf cells but expressing petal pigments) may form, indicating that the cellular processes required to produce a particular cell type in a developing organ are not initiated in a coordinated fashion. They are, instead, fixed independently and may be influenced by different regulatory signals.

On the basis of the genetic analysis of flower development, Haughn and Somerville (1988) and Coen and Meyerowitz (1991) proposed that the leaf is the developmental ground state of floral organs and that the conversion of "leaf" primordia into flower parts requires the implementation of separate developmental programs under the control of the floral homeotic genes. A similar model proposed for the development of cotyledons is supported by the phenotype of the *leafy cotyledon* mutant of *Arabidopsis*, in which leaf traits are expressed in the cotyledons (Meinke 1992).

The identities of the first two *Arabidopsis* leaf primordia are altered by mutations in two genes, *XTC1* (extra cotyledon) and *XTC2*, which convert them into cotyledon-like structures (L. Conway and R.S. Poethig, unpubl.). Leaf development per se is not affected by these mutations because the third and later leaves develop normally. The phenotypic alteration of the first leaves appears to be a consequence, instead, of a change in developmental timing. The first leaf primordia normally undergo very limited development in the embryo, but in *xtc* embryos they develop to an abnormally advanced stage before the seed dries down. Thus, they may become partially determined to differentiate as cotyledons during their more extensive development in the embryonic environment.

## HETEROBLASTIC DEVELOPMENT AND LEAF IDENTITY

The formation of varied leaf forms during heteroblastic development raises additional questions about the establishment of leaf identity. Historically, vegetative development in heteroblastic plants has been thought to consist of a juvenile phase, in which developing leaves express one set of traits, followed by an adult phase characterized by the production of leaves expressing a different set of traits. However, this model does not explain the gradual changes in leaf morphology found in most heteroblastic plants, including *Arabidopsis*. To account for the full range of leaf forms, it has been proposed that juvenile and adult vegetative phases of development overlap, resulting in the formation of transition leaves that have mixed identities and exhibit traits characteristic of both phases (Poethig 1990).

A great deal of work has been done to identify factors that regulate the transition between the juvenile and adult phases (for review, see Cutter 1965), but few conclusions can be drawn. Röbbelen (1957), and later Medford et al. (1992), noted that the production of adult vegetative leaves in *Arabidopsis* coincides with an increase in the size of the shoot apical meristem and suggested that leaf shape is determined by the size of the meristem at the time of leaf initiation. Similar observations have been made in other species (Abbe et al. 1941; Crotty 1955), but a causal relationship between increased meristem size and greater leaf shape complexity has not been demonstrated.

Experimental investigations into the control of heteroblasty have focused on the influences of nutritional factors and various hormones (Allsopp 1967). Röbbelen (1957) observed a delay in the normal course of leaf shape change in several yellow-green mutants of *Arabidopsis*. Although he attributed the delay to the poor nutritional state of these mutants, the nature of any relationship between changes in nutrient levels and heteroblastic development in *Arabidopsis* has not been investigated. Exogenous applications of gibberellin (GA) have been shown to influence leaf form in several species, but the responses are not uniform. $GA_3$ inhibits the production of leaves with adult traits in *Centaurea* (Feldman and Cutter 1970), for example, but promotes the production of leaves with adult traits in *Eucalyptus* (Scurfield and Moore 1958). The effects of GA treatment on leaf shape and on the expression of other phase-specific traits in *Arabidopsis* have not been reported, nor has it been established whether changes in endogenous GA levels influence leaf morphology.

Recently, a genetic approach to the analysis of heteroblasty has been undertaken. Mutations in several maize genes prolong vegetative development and increase the number of leaves bearing juvenile traits, indicating that these genes may normally regulate the juvenile phase (Poethig 1988). A mutation that prolongs vegetative development and the production of leaves bearing juvenile traits has been identified in *Arabidopsis* (J.M. Martínez-Zapater, pers. comm.), and we have identified several *Arabidopsis* mutations that accelerate, rather than delay, normal sequences of morphological change observed in rosette leaves and bracts (A. Telfer and R.S. Poethig, unpubl.).

## CONCLUSIONS AND PERSPECTIVES

The complex problems of leaf development remain poorly understood. Many developmental programs must be executed coordinately during leaf growth, but few of these have been explored in detail, either at the phenomenological or experimental level. Important questions that need to be considered include: What controls leaf initiation and early outgrowth? How is dorsiventrality imposed on the primordium? What triggers leaf blade initiation? How are the rate, direction, and duration of cell division and expansion controlled during the growth of the blade? What signals direct the spatial and temporal regulation of cell differentiation in each of the tissue layers? How is pattern, as manifested in the spacing of stomata, trichomes, and veins, established? How is heteroblastic development regulated and executed?

The increasing utility of *Arabidopsis* for genetic and molecular analysis should provide the impetus for renewed investigation into the process of leaf development. Genetic analysis will be complicated because mutations in many genes are likely to affect leaf development. However, the coupling of genetic approaches with an intensive investigation into the basic biology of leaf development will allow the genes that control the key steps in leaf development to be identified and studied.

## ACKNOWLEDGMENTS

We gratefully acknowledge Laura Conway, Clint Chapple, Wolf-Dieter Reiter, Chris Somerville, José M. Martínez-Zapater, and Mieke Van Lijsebettens for communicating results prior to publication. We also thank M. David Marks, Paul Linstead, Liam Dolan, and R.M. Leech for providing photographs and figures; and Emily Lawson, Hilli Passas, and W.H. Telfer for their critical comments on the manuscript.

## REFERENCES

Abbe, E.C., L.F. Randolph, and J. Einset. 1941. The developmental relationship between the shoot apex and growth pattern of leaf blade in diploid maize. *Am. J. Bot.* **28:** 778–784.

Alldridge, N.A. 1964. Anomalous vessel elements in wilty-dwarf tomato. *Bot. Gaz.* **125:** 138–142.

Allsopp, A. 1967. Heteroblastic development in vascular plants. *Adv. Morphog.* **6:** 127–171.

Aloni, R. 1987. Differentiation of vascular tissues. *Annu. Rev. Plant Physiol.* **38:** 179–204.

Avery, G.S., Jr. 1933. Structure and development of the tobacco leaf. *Am. J. Bot.* **20:** 565–592.

Battey, N.H. and R.F. Lyndon. 1988. Determination and differentiation of leaf and petal primordia in *Impatiens balsamina. Ann. Bot.* **61:** 9–16.

Coen, E.S. and E.M. Meyerowitz. 1991. The war of the whorls: Genetic interactions controlling flower development. *Nature* **353:** 31–37.

Crotty, W.J. 1955. Trends in the pattern of primordial development with age in the fern *Acrostichum daneaefolium. Am. J. Bot.* **42:** 627–636.

Cutter, E.G. 1965. Recent experimental studies of the shoot apex and shoot morphogenesis. *Bot. Rev.* **31:** 3–113.

––––––––. 1978. *Plant anatomy*, 2nd edition. Addison-Wesley, Reading, Massachusetts.

Dale, J.E. 1988. The control of leaf expansion. *Annu. Rev. Plant Physiol. Plant Mol. Biol.* **39:** 267–295.

de Chalain, T.M.B. and P. Berjak. 1979. Cell death as a functional event in the development of the leaf intercellular spaces in *Avicennia marina* (Forsskål) Vierh. *New Phytol.* **83:** 147–154.

Esau, K. 1977. *Anatomy of seed plants*, 2nd edition. Wiley, New York.

Feenstra, W.J. 1978. Contiguity of linkage groups 1 and 4, as revealed by linkage relationships of two newly isolated markers *dis-1* and *dis-2*. *Arabidopsis Inf. Serv.* **15:** 35–38.

Feldman, L.J. and E.G. Cutter. 1970. Regulation of leaf form in *Centaurea solstitialis* L. I. Leaf development on whole plants in sterile culture. *Bot. Gaz.* **131:** 31–39.

Furner, I.J. and J.E. Pumfrey. 1992. Cell fate in the shoot apical meristem of *Arabidopsis thaliana*. *Development* **115:** 755–764.

Goff, S.A., K.C. Cone, and V.L. Chandler. 1992. Functional analysis of the transcriptional activator encoded by the maize *B* gene: Evidence for a direct functional interaction between two classes of regulatory proteins. *Genes Dev.* **6:** 864–875.

Haughn, G.W. and C.R. Somerville. 1988. Genetic control of morphogenesis in *Arabidopsis*. *Dev. Genet.* **9:** 73–89.

Herman, P.L. and M.D. Marks. 1989. Trichome development in *Arabidopsis thaliana*. II. Isolation and complementation of the *GLABROUS I* gene. *Plant Cell* **1:** 1051–1055.

Hülskamp, M., S. Miséra, and G. Jürgens. 1994. Genetic dissection of trichome cell development in *Arabidopsis*. *Cell* **76:** 555–566.

Irish, V.F. and I.M. Sussex. 1992. A fate map of the *Arabidopsis* embryonic shoot apical meristem. *Development* **115:** 745–753.

Jeffree, C.E., J.E. Dale, and S.C. Fry. 1986. The genesis of intercellular spaces in the developing leaves of *Phaseolus vulgaris* (L.). *Protoplasma* **132:** 90–98.

Kawase, M. 1979. Role of cellulase in aerenchyma development in sunflower. *Am. J. Bot.* **66:** 183–190.

Kieber, J.J., M. Rothenberg, G. Roman, K.A. Feldman, and J.R. Ecker. 1993. *CTR1*, a negative regulator of the ethylene response pathway in *Arabidopsis thaliana*, encodes a member of the *Raf* family of protein kinases. *Cell* **72:** 427–441.

Koornneef, M. 1981. The complex syndrome of *ttg* mutants. *Arabidopsis Inf. Serv.* **18:** 45–51.

Koornneef, M., L.W.M. Dellaert, and J.H. Van der Veen. 1982. EMS- and radiation-induced mutation frequencies at individual loci in *Arabidopsis thaliana*. *Mutat. Res.* **93:** 109–123.

Koornneef, M., C.J. Hanhart, and J.H. van der Veen. 1991. A genetic and physiological analysis of late flowering mutants in *Arabidopsis thaliana*. *Mol. Gen. Genet.* **229:** 57–66.

Larkin, J.C., D.G. Oppenheimer, S. Pollock, and M.D. Marks. 1993. *Arabidopsis GL1* gene requires downstream sequences for function. *Plant Cell* **5:** 1739–1748.

Lincoln, C., J.H. Britton, and M. Estelle. 1990. Growth and development of the *axr1* mutants of *Arabidopsis*. *Plant Cell* **2:** 1071–1080.

Lloyd, A.M., V. Walbot, and R.W. Davis. 1992. *Arabidopsis* and *Nicotiana* anthocyanin production activated by maize regulators *R* and *C1*. *Science* **258:** 173–175.

Ludwig, S.R., L.F. Habera, S.L. Dellaporta, and S.R. Wessler. 1989. *Lc*, a member of the maize *R* gene family responsible for tissue-specific anthocyanin production, encodes a protein similar to transcriptional activators and contains the *myc*-homology region. *Proc. Natl. Acad. Sci.* **86:** 7092–7096.

Marks, M.D. and J.J. Esch. 1992. Trichome formation in *Arabidopsis* as a genetic model system for studying cell expansion. In *Current topics in plant biochemistry and physiology* (ed. D.D. Randall), vol. 11, pp. 131–142. Interdisciplinary Plant Biochemistry and Physiology Program, University of Missouri. Columbia, Missouri.

Marks, M.D., J. Esch, P. Herman, S. Sivakumaran, and D. Oppenheimer. 1991. A model for cell-type determination and differentiation in plants. In *Molecular biology of plant development* (ed. G. Jenkins and W. Schuch), pp. 77–87. The Company of Biologists, Cambridge, United Kingdom.

Martínez-Zapater, J.M. and C.R. Somerville. 1990. Effect of light quality and vernalization on late-flowering mutants of *Arabidopsis thaliana*. *Plant Physiol.* **92:** 770–776.

McHale, N.A. 1992. A nuclear mutation blocking initiation of the lamina in leaves of *Nicotiana sylvestris*. *Planta* **186:** 355–360.

Medford, J.I., F.J. Behringer, J.D. Callos, and K.A. Feldman. 1992. Normal and abnormal development in the *Arabidopsis* vegetative shoot apex. *Plant Cell* **4:** 631–643.

Meinke, D.W. 1992. A homeotic mutant of *Arabidopsis thaliana* with leafy cotyledons. *Science* **258:** 1647–1649.

Melaragno, J., B. Mehrotra, and A.W. Coleman. 1993. Relationship between endopolyploidy and cell size in epidermal tissue of *Arabidopsis*. *Plant Cell* **11:** 1661–1668.

Metcalfe, C.R. and L. Chalk. 1950. *Anatomy of dicotyledons*. Clarendon Press, Oxford, United Kingdom.

Oppenheimer, D.G., J. Esch, and M.D. Marks. 1993. Molecular genetics of *Arabidopsis* trichome development. In *Control of plant gene expression* (ed. D.P.S. Verma), pp. 275–286. CRC Press, Boca Raton, Florida.

Oppenheimer, D.G., P.L. Herman, S. Sivakumaran, J. Esch, and M.D. Marks. 1991. A *myb* gene required for leaf trichome differentiation in *Arabidopsis* is expressed in stipules. *Cell* **67:** 483–493.

Paz-Ares, J., D. Ghosal, V. Wienand, A. Peterson, and H. Saedler. 1987. The regulatory *C1* locus of *Zea mays* encodes a protein with homology to *myb* proto-oncogene products and with structural similarities to transcriptional activators. *EMBO J.* **6:** 3553–3558.

Poethig. R.S. 1988. Heterochronic mutations affecting shoot development in maize. *Genetics* **119:** 959–973.

————. 1990. Phase change and the regulation of shoot morphogenesis in plants. *Science* **250:** 923–930.

Poethig, R.S. and I.M. Sussex. 1985. The developmental morphology and growth dynamics of the tobacco leaf. *Planta* **165:** 158–169.

Postlethwait, S.N. and O.E. Nelson, Jr. 1957. A chronically wilted mutant of maize. *Am. J. Bot.* **44:** 628–633.

Pyke, K.A., J.L. Marrison, and R.M. Leech. 1991. Temporal and spatial development of the cells of the expanding first leaf of *Arabidopsis thaliana* (L.) Heynh. *J. Exp. Bot.* **42:** 1407–1416.

Rédei, G.P. 1969. *Arabidopsis thaliana* (L.) Heynh. A review of the genetics and biology. *Bibliogr. Genet.* **21:** 1–151.

Rédei, G.P. and Y. Hirono. 1964. Linkage studies. *Arabidopsis Inf. Serv.* **1:** 9–10.

Röbbelen, G. 1957. Über Heterophyllie bei *Arabidopsis thaliana* (L.) Heynh. *Ber. Dtsch. Bot. Ges.* **70:** 39–44.

————. 1965. The Laibach standard collection of natural races. *Arabidopsis Inf. Serv.* **2:** 36–47.

Rüffer-Turner, M. and K. Napp-Zinn. 1979. Investigations of leaf structure in several genotypes of *Arabidopsis thaliana* (L.) Heynh. *Arabidopsis Inf. Serv.* **16:** 94–98.

Sachs, T. 1984. Controls of cell patterns in plants. In *Pattern formation* (ed. G.M. Malacinski and S.V. Bryant), pp. 367–391. Macmillan, New York.

Scurfield, G. and C.W.E. Moore. 1958. Effects of gibberellic acid on species of

*Eucalyptus. Nature* **181:** 1276–1277.

Shannon, S. and D.R. Meeks-Wagner. 1991. A mutation in the *Arabidopsis TFL1* gene affects inflorescence meristem development. *Plant Cell* **3:** 877–892.

Stewart, R.N. 1978. Ontogeny of the primary body in chimeral forms of higher plants. In *The clonal basis of development* (ed. I.M. Sussex and S. Subtelny), pp. 131–160. Academic Press, New York.

Stewart, R.N. and H. Derman. 1975. Flexibility in ontogeny as shown by the contribution of the shoot apical layers to leaves of periclinal chimeras. *Am. J. Bot.* **62:** 935–947.

Sussex, I.M. 1955. Morphogenesis in *Solanum tuberosum* L.: Experimental investigation of leaf dorsiventrality and orientation in the juvenile shoot. *Phytomorphology* **5:** 286–300.

Tepfer, S.S. and M. Chessin. 1959. Effects of tobacco mosaic virus on early leaf development in tobacco. *Am. J. Bot.* **46:** 496–509.

Tilney-Bassett, R.A.E. 1986. *Plant chimeras*. Edward Arnold, London.

Van Lijsebettens, M., R. Vanderhaeghen, and M. Van Montagu. 1991. Insertional mutagenesis in *Arabidopsis thaliana*: Isolation of a T-DNA-linked mutation that alters leaf morphology. *Theor. Appl. Genet.* **81:** 277–284.

Van Lijsebettens, M., R. Vanderhaeghen, M. De Block, G. Bauw, R. Villarroel, and M. Van Montagu. 1994. An S18 ribosomal protein gene copy, encoded at the *Arabidopsis PFL* locus, affects plant development by its specific expression in meristems. *EMBO J.* (in press).

von Papen, R. 1935. Beiträge zur Kenntnis des Wachstums der Blattspreite. *Bot. Arch. Z. Gesamte Bot.* **37:** 159–206.

Yang, M. and F.D. Sack. 1993. An *Arabidopsis* mutant with clustered stomata. *Int. Bot. Congr.* **15:** 433.

# 16

# The Transition to Flowering in *Arabidopsis*

**José M. Martínez-Zapater**
Departamento de Protección Vegetal
Centro de Investigación y Tecnología
Instituto Nacional de Investigación y Tecnología Agraria y Alimentaria
28040 Madrid, Spain

**George Coupland and Caroline Dean**
Department of Molecular Genetics
AFRC, IPSR, Cambridge Laboratory
John Innes Centre, Colney
Norwich, United Kingdom

**Maarten Koornneef**
Department of Genetics
Wageningen Agricultural University
6703 HA Wageningen, The Netherlands

Given the sessile life-style of plants, reproductive success depends on the correct timing of the transition from vegetative to reproductive development. Plants need to be able to recognize the most favorable environmental conditions in order to successfully complete their reproductive development (Murfet 1977). Light intensity, photoperiod, and temperature are environmental variables that change in a largely predictable pattern throughout the year, and different light conditions and temperatures induce flowering in a wide range of plant species (Evans 1969; Bernier et al. 1981; Halevy 1985). Consequently, the flowering process involves many steps, starting from perception of environmental conditions and finishing with the differentiation of three-dimensional structures, the flower primordia at the meristem (Zeevaart 1976).

Depending on their requirement for specific environmental conditions to flower, plant species can be considered as (1) autonomous, when apical meristems acquire competence and proceed from the vegetative to the reproductive  program independently of environmental conditions; (2) obligate, when the transition between the vegetative and reproductive developmental programs requires specific environmental conditions to proceed; or (3) facultative, when the transition between developmental programs is hastened or delayed by different environmental factors, but

*Arabidopsis*
© 1994 Cold Spring Harbor Laboratory Press 0-87969-428-9/94 $5 + .00

there is no obligate requirement (Bernier 1988), as is the case for *Arabidopsis* (Napp-Zinn 1985).

*Arabidopsis* has been the subject of physiological and genetic research on floral induction for several decades (Napp-Zinn 1969, 1985; Rédei 1970; Finkelstein et al. 1988; G.W. Haughn et al., in prep.). Because of the lack of specific requirements for floral transition, *Arabidopsis* has not been a preferred model for the physiological analysis of the flowering process. However, the high level of natural variation in flowering time existing in this species and the ability to produce and isolate mutants have made it, along with pea (Murfet 1989) and wheat (Law 1987), a preferred model for the genetic analysis of the flowering process. Understanding the genetic control of the floral transition process will help in understanding its physiology (Murfet 1977). Moreover, in species like *Arabidopsis*, the genetic identification of the loci involved in the floral transition will allow the molecular cloning of the corresponding DNA sequences.

A description of all the information available on the flowering process in *Arabidopsis* is outside the scope of this chapter; most of the information on the requirements of natural ecotypes has been summarized in several reviews by Napp-Zinn (1969, 1985). However, the number of mutations that affect floral transition is increasing rapidly and, so far, there is no comprehensive review of all the available information. Moreover, the availability of mutants in many other developmental responses and pathways provides the tools to analyze the effect of these on the flowering process, giving rise to valuable information on their genetic interactions.

The initial sections of this chapter provide an overview of the flowering process in *Arabidopsis* plants. This information forms the basis for the description, in later sections, of the phenotypes of mutants altered in the transition to flowering. The available results derived from the morphological, physiological, and genetic analysis of these mutants provide the frame for a speculative model for the genetic control of the transition to flowering. Finally, in the last section, we try to foresee how our understanding of the flowering process in this species will develop in the near future and to summarize the approaches currently being followed to investigate the molecular basis of this classic phenomenon in plant development.

## FLORAL INDUCTION IN WILD-TYPE PLANTS

*Arabidopsis* is an annual species in which the vegetative and reproductive developmental phases are temporally separated. *Arabidopsis* belongs to a characteristic group of plants that grow as rosettes during their vegetative growth, because of the reiterative production of leaves without internode elongation. Transition to the reproductive phase, at the apical meristem, gives rise to an open inflorescence typical of the *Brassicaceae* (Müller 1961). This transition is the result of two basic phenomena: the acquisition by the apical meristem of reproductive competence and the production of floral stimuli (probably in the leaves) as a consequence of appropriate environmental conditions. When both requirements are fulfilled, the vegetative meristem becomes an inflorescence meristem. The transition from the vegetative to the inflorescence meristem and the development of the first flower buds is immediately followed by the elongation of the internodes. As a result of this process, the *Arabidopsis* inflorescence is composed of two node types. The first basal nodes bear lateral indeterminate meristems, which will develop as coflorescences in the axils of cauline leaves, whereas later nodes bear flowers that are not associated with leaves (Schultz and Haughn 1991; G.W. Haughn et al., in prep.).

Before providing a description of the flowering process in wild-type plants, it is important to describe how flowering behavior can be measured. Then we summarize what is known about the environmental effects on the transition to flowering in *Arabidopsis*, followed by a description of the morphological changes that take place in the plant and in its apical meristem during the transition from vegetative to reproductive development.

## Measurement of Flowering

Different criteria have been used to measure flowering behavior in *Arabidopsis*, each one presenting specific advantages and disadvantages. In principle, the most commonly used criteria are flowering time (FT) and leaf number (LN). FT, considered as the time of appearance of the first visible flower bud, was initially used by Laibach (1951). However, other authors have preferred to consider FT as the time of opening of the first flower or the time of appearance of the first stigma, because these are easier to score (Napp-Zinn 1969). These approaches to measuring FT share a common problem related to the fact that the time from flower bud formation to flower opening can be very variable for different genotypes and environments (Bernier et al. 1981). For this reason, total LN, considered as the total number of rosette plus inflorescence leaves on the primary axis, is a more useful index (Koornneef et al. 1991). In very late genotypes, the quantification of LN can be extremely laborious and FT is used as an alternative. Rosette leaf number is frequently used in place of total leaf number, although these figures could not be strictly equivalent since they represent different developmental transitions: bolting (stem elongation) and production of lateral determinate floral meristems. Koornneef et al. (1991) have shown that, in the ecotype Landsberg *erecta*, (L*er*), both rosette LN and total LN are correlated with FT.

## Environmental Control

*Arabidopsis* can be considered as a facultative species and, generally, does not show an obligate requirement for any specific environmental condition. However, many different factors hasten or delay the transition to flowering in this species. Both photoperiod and cold temperatures can greatly accelerate flowering, and light quality (wavelength) and quantity (photon flux density) are also important factors in controlling the process. Other environmental conditions such as nutrient availability and growth temperature also seem to have some effect on the floral transition but have not been extensively analyzed. Finally, exogenous application of phytohormones and other compounds can mimic the effect of some environmental conditions. These are described at the end of this section.

### *Photoperiod*

*Arabidopsis* is generally found in temperate regions of the Northern Hemisphere, and most *Arabidopsis* ecotypes behave as facultative long-day (LD) plants whose flowering time is reduced when grown under LD photoperiods (Laibach 1951; for reviews in English, see Napp-Zinn 1969, 1985). Under short days (SD), flowering is delayed and plants produce a higher number of leaves, in both the rosette and the inflorescence (Koornneef et al. 1991). However, plants will ultimately flower even under very unfavorable conditions, with 4–5 hours as a minimum daylength (Laibach 1951). All ecotypes tested behave as LD plants (Laibach 1951; Karlsson et al. 1993), although there are differences in their degree of LD requirement (Gregory and Hussey 1953). For example, natural populations derived from subtropical regions, such as the ecotype Cvi (Cape Verde Islands), are less sensitive to SD (G. Coupland et al., unpubl.).

Columbia (Col) plants flower earlier under complete darkness than under long photoperiods. This was demonstrated when seedlings germinated and grown in liquid nutrient medium under continuous darkness flowered after producing only two pairs of leaves (Rédei et al. 1974). Similar results have been obtained by Araki and Komeda (1993b) using dark-liquid-shaken cultures. The positive effect of darkness on the transition to flowering has also been observed in other plant species (for references, see Rédei et al. 1974) and suggests that exposure to light delays a genetically programed default initiation of flowering (Rédei et al. 1974; Araki and Komeda 1993b).

## Light Quality

The spectral quality of light has an important effect on the flowering response of *Arabidopsis*. Meijer (1959) showed that a combination of blue (400–530 nm) and far-red (>700 nm) light accelerated floral initiation in all the ecotypes tested. Brown and Klein (1971), working with the early ecotype Estland (Est), and Eskins (1992) with the early ecotype Col, obtained similar conclusions when using different continuous monochromatic illumination of different wavelengths. They also showed that red light has a strong inhibitory effect on floral transition. By providing night-breaks of different wavelengths to plants grown under SD, Goto et al. (1991) obtained similar results. Blue and far-red light had a strong effect on floral transition when applied as a 1-hour night-break, and red light was the least effective. The effect of different red/far-red ratios on FT and LN of several *Arabidopsis* genotypes has also been analyzed. The results show that LN is increasingly reduced with decreasing red/far-red ratios (Whitelam and Smith 1991; Bagnall 1992, 1993). Taken together, these results show that the responses of *Arabidopsis* to different wavelengths of light are similar to the ones reported for other LD plants (Lane et al. 1965; Schneider et al. 1967; Imhoff et al. 1979) and suggest that the Pfr form of phytochrome represses floral transition (Goto et al. 1991).

## Photon Flux Density

In early *Arabidopsis* ecotypes, photon flux density (PFD) does not seem to have a very important effect and, in fact, Laibach (1951) showed that *Arabidopsis* is more sensitive to photoperiod length than to total light quantity. Recent experiments confirmed this observation for the early L*er* ecotype (Bagnall 1992). However, FT and LN in some of the late-flowering mutants, like *fca* (see below), were significantly reduced under higher PFD (Bagnall 1992).

## Vernalization

When different *Arabidopsis* ecotypes are grown under photoperiodically inductive conditions, they show large variations in flowering time within and between populations (Napp-Zinn 1969, 1985, 1987), and these differences are genetically determined (Laibach 1951; Napp-Zinn 1985).

When germinating seeds or vegetatively growing plants of late-flowering ecotypes such as Lund are exposed to cold temperatures (below 10ºC) for long periods of time, their flowering time and leaf number are drastically reduced, as shown in Figure 1. The effect of the cold treatment (vernalization) on the reduction of FT and LN depends on the genotype and the photoperiodic conditions of growth (Karlsson et al. 1993). Furthermore, the developmental stage of the plant and the temperature and length of the vernalization treatment are also very important. Napp-Zinn (1957) demonstrated that, for the ecotype Stockholm (St), imbibed seeds prior to germination and the first stages of vegetative development were the most responsive to vernalization. Moreover, longer exposures to cold (more than 4 weeks) and lower temperatures (between 0º and 4ºC) were the most effective (Napp-Zinn 1957).

The effects of light and vernalization on the floral transition are additive up to a point of saturation, from which stronger treatments do not produce a higher reduction in FT or LN (Martínez-Zapater and Somerville 1990). Long vernalization treatments eliminate or diminish both daylength (Chintraruck and Ketellapper 1969) and light quality responsiveness in *Arabidopsis* (Bagnall 1993). On the other hand, early ecotypes that demonstrate little or no cold response under LD show an acceleration of floral induction under SD if they are vernalized (for the L*er* response, see Wilson et al. 1992). An interesting feature of the vernalization response is that the vernalized condition can be lost if plants are exposed to heat (>30ºC) for 5 days (Napp-Zinn 1957). When

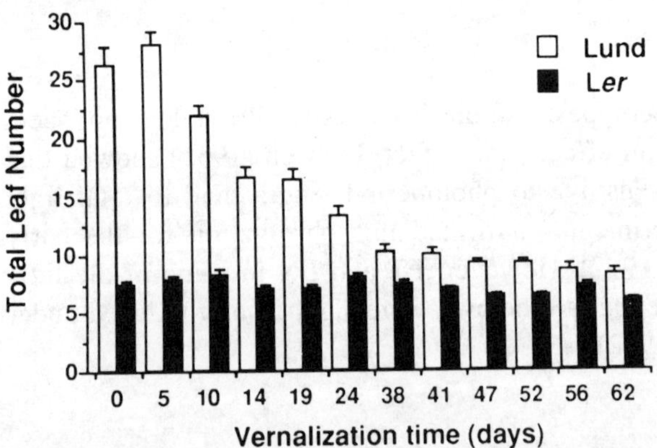


*Figure 1* Effect of increasing vernalization periods on the total LN of plants from the late ecotype Lund and the early ecotype L*er*. Imbibed seeds were vernalized at 4ºC in the dark for different periods of time before being grown at 20ºC under continuous fluorescent plus incandescent lights (as described in Martinez-Zapater and Somerville 1990).

plants are maintained at intermediate temperatures for more than 5 days after vernalization, the vernalization condition is stabilized and a consecutive heat treatment does not have any effect (Napp-Zinn 1957).

## Growing Temperature

Aside from the effect of vernalization, the growing temperature (over 10°C) also seems to have an effect on several aspects of the flowering process. Westerman and Lawrence (1971) grew seven different ecotypes of *Arabidopsis*, including Landsberg-1, at three different temperatures (15°C, 20°C, and 25°C) and found a significant effect on FT, LN, silique number, and height. In all the analyzed ecotypes, higher growing temperatures promoted a significant reduction in the first three variables and an increase in height. Araki and Komeda (1993a) have also shown similar effects of growing temperatures between 18°C and 28°C on the LN of Col and L*er* plants. High temperatures (28°C) produce a decrease in LN that is more pronounced in Col than in L*er* plants.

## Mineral Nutrition

The effect of nutrition on the FT of *Arabidopsis* has not been analyzed in detail. Plants grown at high density tend to produce small rosettes and to flower early, suggesting that poor nutrition might accelerate flowering. This is in agreement with the observation that very low nutrient levels lead to flowering with reduced LN (P. van Tienderen, pers. comm.). However, Myerscough and Marshall (1973) studied the characteristics of plants growing at different densities and provided with varying concentrations of a defined mineral solution. They reported that plants receiving the undiluted solution were larger and reached maturity earlier than those receiving dilutions of the solution. Unfortunately, they did not present a detailed analysis of flowering time, and more work of this type would be required to determine the effect of mineral nutrition on the floral transition.

## Chemical Treatments

Many substances suspected of having an effect on the floral transition have been tried on *Arabidopsis* (Rédei 1970). The rationale is that substances that promote floral transition, when applied exogenously, could be involved in the flowering process. Unfortunately, a major difficulty in interpreting the results of these experiments is that the final endogenous concentrations of the applied compounds are unknown.

The application of biologically active gibberellins (GAs) and gibberellin inhibitors indicated that these phytohormones can accelerate the floral transition. GA treatment of Est plants grown under SD significantly shortened their flowering time. However, almost no effect was observed when plants of this early ecotype were grown under continuous light (Langridge 1957). Later work showed that GA treatments could also accelerate flowering in nonvernalized cold-requiring ecotypes (Laibach 1958; Sarkar 1958; Napp-Zinn 1963). Moreover, treatments of imbibed seeds with increasing concentrations of the gibberellin biosynthesis inhibitor 2-chloroethyl-trimethylammoniumchloride (CCC) delayed flowering and reduced the effect of vernalization (Napp-Zinn 1969). Besnard-Wibaut (1981) also showed that gibberellin treatments were not effective on young apical meristems, suggesting that the apical meristem must be competent for the treatment to be effective.

The effects of cytokinins and auxins are not as strong as those of gibberellins. Two different cytokinins, kinetin and 6-benzyladenine, have been tried on the apical meristem of *Arabidopsis* rosettes grown under SD (Michniewicz and Kamienska 1965; Besnard-Wibaut 1981). They both produce a limited acceleration of the floral transition when applied on adult (competent) vegetative meristems (Besnard-Wibaut 1981). The auxin IAA produces a relatively limited acceleration of the floral transition when applied on nonvernalized cold-requiring plants of the St ecotype (Sarkar 1958).

Several laboratories have analyzed the effect of different base analogs on floral transition (for review, see Rédei 1970). Both 5-bromodeoxyuridine and 5-bromo-deoxycytidine produced a dramatic acceleration of flowering in wild-type Col plants and in a strongly delayed, nonvernalization-responsive, late-flowering mutant (*gi-2*), both under continuous light and SD. The flowering time difference between mutant and wild type was almost eliminated by the treatment. However, these base analogs did not have any effect on the FT of *ld*, a late-flowering vernalization-responsive mutant (Hirono and Rédei 1966). Another base analog (8-azaadenine) also accelerated flowering, but application of azauridine or azathymine had no effect (Hirono and Rédei 1965). The mechanism of action of these analogs on flowering is not well understood (Rédei 1970). Hirono and Rédei (1965, 1966) suggested that these compounds could act by inhibiting the synthesis of flowering inhibitors. Later, Brown (1968) proposed that the analogs could accelerate flowering by differentially affecting DNA synthesis in the cells of the apical meristem. Alternatively, base analogs could induce a stress response in the plant, resulting in the acceleration of the transition to flowering, as has been observed for other stresses (Laibach 1951).

Recently, Burn et al. (1993b) reported the substitution of the vernalization requirement of some late-flowering genotypes by treatments with 5-azacytidine, a base analog that inhibits cytosine methylation. Furthermore, 5-azacytidine did not have any effect in nonvernalization-responsive late mutants. On the basis of these results, they suggested that the cold treatment could activate the expression of genes involved in the synthesis of gibberellins, which may act as floral promoters, by preventing their methylation after DNA replication. The effect of 5-azacytidine on vernalization-responsive and nonresponsive genotypes seems to be opposite to the effect of 5-bromo-deoxyuridine and 5-bromo-deoxycytidine previously shown by Rédei (1970) and mentioned above.

## Morphological Changes Associated with the Transition to Flowering

Changes in leaf morphology during plant development are known as heteroblasty and are common in many plant species, where they have been correlated with different physiological states of the plant. Changes in leaf shape and phyllotaxy have been correlated with the transition between juvenile and adult vegetative meristematic phases (Poethig 1990). This transition has generally been associated with the acquisition of meristematic competence to respond to a floral stimulus (McDaniel et al. 1992), although both processes could be regulated through independent developmental programs (Bassiri et al. 1992).

In the rosette of the *Arabidopsis* Wassilewskija (Ws) ecotype, the first leaves are small, round, and entire and positioned in an opposite phyllotaxy, whereas later leaves are spatulate and serrate and positioned in a spiral phyllotaxy (Medford et al. 1992). Similar changes in leaf shape and phyllotaxy have also been observed in the ecotype L*er*, where an increase in trichome number on the adaxial surface of the leaves and the appearance of trichomes on the abaxial surface have also been correlated with the morphological changes (J.M. Martínez-Zapater et al., in prep.; A. Telfer and S. Poethig, pers. comm.). These changes in leaf shape, leaf trichome density, and phyllotaxy could mark, in *Arabidopsis*, the transition between the juvenile and adult vegetative meristematic phases (Fig. 2). In fact, wild-type Ws plants are not capable of reproductive development until formation of the fifth rosette leaf, supporting the above-mentioned association (Medford et al. 1992). However, Laibach (1951) reported flowering with one rosette leaf in Warschau plants grown under nutritional stress, indicating that the time of competence acquisition in *Arabidopsis* is under both genetic and environmental control.

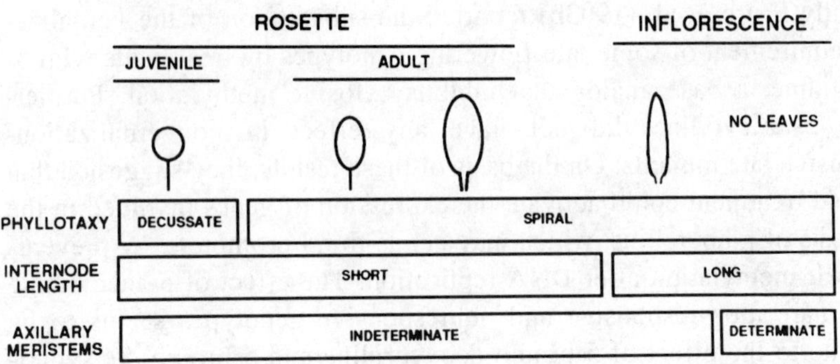



*Figure 2*   Morphological changes taking place during the development of *Arabidopsis*. The schematic representation illustrates the changes in leaf shape, phyllotaxy, internode length, and determination of the axillary meristems.

Transition from the adult vegetative to the inflorescence meristem is reflected at the plant level by elongation of internodes and the production of the inflorescence. Cauline leaves in the inflorescence are increasingly lanceolate, show a sharp decrease in the trichome density of their adaxial surface (J.M. Martínez-Zapater et al., in prep.), and bear axillary indeterminate coflorescence meristems (Schultz and Haughn 1991). Later, the inflorescence meristem gives rise to lateral determinate floral meristems that differentiate into flowers without subtending leaves (Schultz and Haughn 1991). Differentiation of cauline leaves and flower buds at the inflorescence meristem takes place prior to bolting. Bolting is often considered as synonymous with floral transition in rosette-like species; however, in many cases, bolting and floral transition can be experimentally separated by treatments with GAs or inhibitors of gibberellin biosynthesis (Zeevaart 1976). In *Arabidopsis* this separation is not clear. Most GA-deficient mutants, which are affected in the elongation of inflorescence internodes, show a close to normal flowering behavior under floral inductive conditions (Wilson et al. 1992; J.A. Jarillo et al., unpubl.). However, application of specific GAs, under noninductive conditions, produces both internode elongation and floral transition (Langridge 1957; Napp-Zinn 1963; Besnard-Wibaut 1981).

Morphological changes at the apical meristem during the transition from vegetative to reproductive development of *Arabidopsis* have been characterized in several ecotypes and under different environmental conditions (Vaughan 1955; Miksche and Brown 1965; Besnard-Wibaut 1977; Medford et al. 1992). Plants grown under noninductive conditions show a flat meristem that rapidly evolves into a dome-shaped meristem when they are shifted to inductive conditions (Vaughan 1955; Besnard-Wibaut 1977). This shape change takes place very fast when early ecotypes are grown under inductive conditions from germination (Miksche and Brown 1965; Medford et al. 1992). The transition from a flat to a dome meristem seems to be the result of changes in the mitotic activity of the cells in different meristematic regions (Besnard-Wibaut 1977) and has been associated with the transition from the vegetative to the inflorescence meristem (Vaughan 1955; Miksche and Brown 1965; Besnard-Wibaut 1977). In a recent report, Medford et al. (1992) interpret the meristematic shape transition that they observed in the early ecotype Ws under LD in a different way. They observed that the meristematic shape change was associated with a change in symmetry and phyllotaxy and proposed that all these changes represent the transition between juvenile and adult vegetative meristems. This apparent contradiction with previously reported results (Vaughan 1955; Miksche and Brown 1965; Besnard-Wibaut 1977) could be due to the differences in experimental conditions. Because early Ws plants were only grown under LD, the transition from the vegetative to the inflorescence meristem would take place as soon as the juvenile meristem has transformed into an adult meristem. Therefore, changes in phyllotaxy, which have typically been associated with a juvenility change (Poethig 1990), would take place almost at the same time as changes in meristem shape that have been associated with the vegetative to inflorescence transition (Vaughan 1955; Miksche and Brown 1965; Besnard-Wibaut 1977).

In L*er*, the duration of every meristematic phase (juvenile, adult, inflorescence) and the lengths of the inflorescence internodes are affected by environmental factors like photoperiod and temperature. Conditions that accelerate floral transition, like LD, also shorten each one of these phases and increase the length of the inflorescence internodes. Conditions that delay floral transition, like SD, extend all these phases and reduce the length of the inflorescence internodes (J.M. Martínez Zapater et al., in prep.). This effect of photoperiod and vernalization on elongation of the inflorescence internodes has been reported in other rosette-like species and seems to be the result of changes in GA biosynthesis and/or sensitivity (Zeevaart 1983; Pharis and King 1985).

## Reversion of Flowering

Reversion from reproductive to vegetative growth is under environmental control in many plant species. It is observed as flower or, more frequently, inflorescence reversion (Battey and Lyndon 1990). In *Arabidopsis*, reversion was described by Laibach (1951) in late ecotypes shifted to SD after they had flowered under LD. The reversion of flowering in these plants was expressed as the development of rosettes from the axillary meristems of the cauline leaves (inflorescence reversion) and, in some cases, even as the production of leaf-like floral organs in the flower perianth (Fig. 3). This suggests that floral inductive conditions are required throughout inflorescence and flower morphogenesis. The outgrowth of axillary rosettes after the daylength shift has been used as a tool to obtain rosettes useful for the vegetative propagation of *Arabidopsis* plants (Reinholz 1972). Reversion phenomena are sometimes observed in very delayed late-flowering mutants such as *F* (M. Koornneef et al. unpubl.; see later sections).

## GENETIC ANALYSIS OF FLORAL TRANSITION

The analysis of the genetic control of the floral transition in *Arabidopsis* is based on both the genetic variation present in natural populations and variation induced by mutagenesis. The analysis of the effect of mutations at different loci in the same genetic background enabled mutant analysis to progress faster and to be more informative. However, investigation of natural variation can lead to the identification of loci which are not revealed by mutagenesis of the commonly used laboratory ecotypes L*er* and Col. In this section, we first summarize the progress made on the use of natural variation, then describe the phenotype of mutants affected in their flowering behavior.

### Use of Natural Variation

Natural variation has been used to analyze the genetic basis of flowering time. The analysis has generally been performed in LD without vernalization and has involved ecotypes that are responsive to vernalization. By crossing late ecotypes with strong vernalization response with early ecotypes and studying the segregation of the parental types in $F_2$ progenies, it is possible to identify the number of genes responsible for the vernalization response in the parental lines. The number of genes involved varies depending on the ecotypes used, and little work has been done to identify different alleles of these genes in different ecotypes. Napp-Zinn (1962) identified at least four genes responsible for the vernalization response of the late ecotype St in a cross with the early ecotype Limburg-5 (Li-5). For one of them, *FRI*, the vernalization requirement was caused by a dominant allele, whereas for the three others, the requirement was elicited by recessive alleles with additive effects. Napp-Zinn (1961) also concluded that genes causing stronger cold requirements were epistatic to genes causing weaker cold requirements and that epistatic relationships could vary depending on light intensity and

*Figure 3* Reversion of flowering. Inflorescence and flower reversions taking place in a flowering St plant after it has been shifted to noninductive conditions. (Reprinted, with permission, from Laibach 1951.)

other environmental conditions. Several other late ecotypes also contained a dominant allele at the *FRI* locus responsible for their vernalization requirements (Napp-Zinn 1987). This locus has been mapped onto the top of chromosome 4 (Clarke and Dean 1993) and is not allelic to any of the late-flowering loci identified by mutagenesis in L*er* (Koornneef et al. 1991). It does, however, map very close to the *FLA* locus identified in a cross between San Feliu-2 (Sf-2) and Col (Lee et al. 1993). The dominant allele in Sf-2 confers late flowering and vernalization sensitivity, and this single gene accounts for most of the difference in FT between the two ecotypes. A dominant *FLA* allele also segregated in crosses between ecotypes Leiden-0 (Le-0) and Col (Lee et al. 1993). In addition, lateness in vernalization-responsive ecotypes Pitztal and Innsbruck was also controlled by dominant alleles at a locus mapping at the top of chromosome 4 and possibly allelic to the *FLA/FRI* locus (Burn et al. 1993a), hereafter named after the earliest description as the *FRI* locus. In most of these crosses, lateness would also depend on the segregation of specific alleles at locus *FLC*, not present in L*er*, but which have also been found in Col and Est (M. Koornneef et al., unpubl.). The *FLC* locus is located on top of chromosome 5 (S. Michaels et al.; J. Clarke and C. Dean; both unpubl.). In the relatively early ecotypes Dijon (Di) and Limburg-2 (Li-2), van der Veen (1965) identified two dominant genes, one in each ecotype, responsible for their vernalization response. However, in the ecotype Kiruna-2, lateness was due to a recessive allele at a previously unidentified locus (Burn et al. 1993a). Karlovska (1974) also found a few gene differences between early, medium, and late homozygote lines derived from natural ecotypes.

Some of the genes identified in $F_2$ progenies derived from crosses between different ecotypes are allelic to loci identified by mutagenesis in L*er* (Koornneef et al. 1991). In this way, differences in FT between ecotypes Li-2 and L*er* were found to be due to the effect of two genes, one of them allelic to the late-flowering locus *FVE* (Hussein 1968). J. Clarke and C. Dean (in prep.) have also detected five quantitative trait loci responsible for some of the flowering time differences between H51, a late line derived from ecotype St, and L*er*. Four of these loci are located near loci previously identified by mutagenesis of L*er*.

## Mutants in Early Ecotypes

Mutant analysis of floral transition has been performed on a few early ecotypes (Finkelstein et al. 1988). This approach suffers from only making use of a relatively small number of genes in these ecotypes that give a recognizable phenotype when mutated. The selection and characterization of mutants with delayed or accelerated flowering time have so far identified more than 20 independent loci that are involved in floral transition (Table 1). In most cases, these have been located on the *Arabidopsis* genetic map. Some mutations affecting flowering time also cause morphological alterations of the inflorescence and, less commonly, of the flowers. In this section, we classify the floral transition mutants into late and early classes according to their FT phenotype and describe their physiological and morphological features.

### Late Mutants

Mutants that are delayed in flowering time are frequently found when mutagenized plants are grown under LD. They have been isolated in the early ecotypes L*er* (Hussein 1968), Col (Rédei 1962), Est (McKelvie 1962), and Di (Vetrilova 1973). Many different mutations that reduce growth rate can also cause a delay in FT and, for this reason, only late-flowering mutants that show a correlative increase in LN have been con-

*Table 1*   Loci that have been involved in the transition to flowering in *Arabidopsis* and the phenotypic effect of their mutations

| Locus | Dominance | Phenotypic effect[a] | Daylength response[b] | Vernalization response[c] |
|---|---|---|---|---|
| *ABA* | R | early flowering under SD | [d] | n.d. |
| *ABI1* | D | early flowering under SD | [d] | n.d. |
| *CO* | S | late flowering | – | – |
| *DET2* | R | de-etiolated in darkness, late flowering | n.d. | n.d. |
| *ELF1* | R | early flowering | + | n.d. |
| *ELF2* | R | early flowering | + | n.d. |
| *ELF3* | R | early flowering | – | n.d. |
| *EMF* | R | early flowering, flowers upon germination | – | – |
| *EIN2* | R | ethylene insensitive, late flowering | n.d. | n.d. |
| *ETR1* | D | ethylene insensitive, late flowering | n.d. | n.d. |
| *FCA* | R | late flowering | + | + |
| *FD* | R | late flowering | +/– | – |
| *FE* | R | late flowering | +/– | +/– |
| *FHA* | R | late flowering | +/– | +/– |
| *FLC* | S | late flowering | n.d. | + |
| *FPA* | R | late flowering | + | + |
| *FRI*[e] | D | late flowering | + | + |
| *FT* | R | late flowering | +/– | +/– |
| *FVE* | R | late flowering | + | + |
| *FWA* | D | late flowering | +/– | – |
| *FY* | R | late flowering | + | + |
| *GA1* | R | GA deficient, late flowering under SD | + | – |
| *GAI* | S | GA insensitive, late flowering under SD | + | +/– |
| *GI* | R | late flowering | – | – |
| *HY1* | R | early flowering, long hypocotyl | – | n.d. |
| *HY2* | R | early flowering, long hypocotyl | – | n.d. |
| *HY3* | R | early flowering, long hypocotyl | – | n.d. |
| *LD* | R | late flowering | + | + |
| *SPY* | R | slender, early flowering | n.d. | n.d. |
| *TFL* | S | early flowering, determinate inflorescence | + | n.d. |

n.d. indicates not determined.

[a]Unless specified, phenotypic effects are considered under LD.

[b]Mutants are considered as daylength responsive if they are delayed by SD.

[c]Mutants are considered to be responsive to vernalization when the exposure of germinating seeds at 4°C during 2–4-week periods significantly reduces their FT.

[d]*aba* and *abi1* mutants are earlier than the wild type under SD and flower at similar time as wild type under LD.

[e]*FRI* and *FLA* are the same locus, and the *F* mutation is allelic to it.

sidered. Complementation analysis between 39 late-flowering mutants isolated in ecotypes L*er* and Col identified 11 loci (*CO, FCA, FD, FE, FHA, FPA, FT, FVE, FWA, FY*, and *GI*) that have been located on the *Arabidopsis* genetic map (Koornneef et al. 1991). The *ld* mutant isolated in Col (Rédei 1962; Lee et al. 1994) and the *F* mutant isolated in Est (McKelvie 1962) identify two additional loci. The *F* mutation could be allelic to the *FRI* locus identified in late ecotypes (M. Koornneef, unpubl.). Moreover, the L4, L5, and L6 mutants isolated in ecotype Di (Vetrilova 1973) are also the result of mutations at the *FRI* locus (M. Koornneef, unpubl.), whereas McKelvie's *f2* (McKelvie 1962) seems to be an allele of the *FHA* locus (M. Koornneef, unpubl.). Other late mutants isolated in L*er* (M. Koornneef, unpubl.) and Col (Y. Komeda, pers. comm.) still remain to be assigned to complementation groups. On the basis of the effect that photoperiod and vernalization have on the late-flowering phenotype of these mutants (Martínez-Zapater and Somerville 1990; Koornneef et al. 1991; Bagnall 1993), they can be provisionally classified into at least two different phenotypic groups (Koornneef et al. 1991; J.M. Martínez-Zapater et al., in prep.): (1) mutants that are delayed in FT under LD and SD and are strongly responsive to vernalization and (2) mutants that are more delayed in FT under LD and show a reduced vernalization response.

The first phenotypic group includes mutants that result from recessive mutations at loci *FCA, FPA, FVE, FY*, and *LD* and dominant mutations at *FRI*. Semidominant allelic variants identified at locus *FLC* could also be included in this phenotypic group. These mutants show a delay in flowering time under LD that is strongly increased under SD photoperiods (Koornneef et al. 1991; M. Koornneef, unpubl.). The late-flowering phenotype is also associated, in all the tested mutants (*fca, fpa, fve, fy*), with a significant reduction in the internode elongation of the inflorescence (J.M. Martínez-Zapater et al., in prep.). Vernalization treatment of germinating seeds produces a drastic decrease in FT and LN of these mutants (Martínez-Zapater and Somerville 1990; Koornneef et al. 1991 and unpubl.) and, when prolonged enough, can completely rescue the wild-type phenotype (Martínez-Zapater and Somerville 1990). Mutants in this phenotypic group (*fca, fpa, fve, fy*) are also more responsive to a low red/far-red ratio than other late-flowering mutants in the other phenotypic group (Bagnall 1993). All these common features have prompted their classification into a single physiological group. This is in agreement with the observation that double mutants obtained at Wageningen between *fca, fpa, fve*, and *fy* do not flower much later than the most extreme parent, which suggests that they are affected in the same developmental pathway (Koornneef et al. 1991; M. Koornneef, unpubl.).

The second phenotypic group would include those mutants that result from recessive mutations at loci *FD*, *FE*, *FHA*, *FT*, and *GI*, semi-dominant mutations at locus *CO*, and dominant mutations at locus *FWA* (Koornneef et al. 1991). In these mutants, flowering time is delayed under LD; however, they are not much more delayed by SD than wild-type plants, although differences in responsiveness are present (Koornneef et al. 1991; J.M. Martínez-Zapater et al., in prep.). Vernalization treatment has a limited effect on the reduction of flowering time and total LN in mutants *fe*, *fha*, and *ft* whereas *co*, *fd*, *fwa*, and *gi* are not responsive to short vernalization periods given to the imbibed seeds. Mutations at loci *CO*, *FD*, *FT*, *FWA*, and *GI* do not affect the elongation of the inflorescence internodes (J.M. Martínez-Zapater et al., in prep.), although *ft* and *fwa* mutations seem to produce a significant increase in the number of inflorescence internodes (Koornneef et al. 1991). Double mutants between any one of *co*, *fd*, *ft*, *fwa*, and *gi* do not flower later than the most extreme mutant parent, indicating that they are affected in the same developmental pathway. Furthermore, double mutants bearing mutations belonging to the two different phenotypic groups flower later than any one of the single mutants, supporting the hypothesis on the existence of at least two different pathways (Koornneef et al. 1991 and unpubl.).

Other mutations initially selected because they affect hormonal or photomorphogenic responses can also produce a late-flowering phenotype. Mutants at the *DET2* locus (Chory et al. 1991) have a light-independent morphogenetic response and flower around 10 days later than wild type with twice as many leaves. This mutant shows several morphogenetic responses in the dark that are triggered by light in wild-type plants. However, *cop1* mutants (*constitutively photomorphogenic*) that show a similar phenotype (Deng et al. 1991) are not delayed under LD. Mutations at five different loci (*GA1–GA5*) produce a GA-responsive dwarf phenotype as a consequence of defects in GA biosynthesis (Koornneef and van der Veen 1980; Talón et al. 1990a; Zeevaart and Talón 1992). Moreover, the *gai* mutation produces a dwarf phenotype that is insensitive to exogenously applied gibberellins (Koornneef et al. 1985; Talón et al. 1990b). Only extreme alleles of *GA1*, a locus involved in a very early step of GA biosynthesis (Koornneef and Van der Veen 1980; Zeevaart and Talón 1992), and mutations at locus *GAI* produce a small but significant delay in flowering time when plants are grown under LD (Wilson et al. 1992; J.A. Jarillo et al., unpubl.). Under short photoperiods, this delay is strongly enhanced, and plants homozygous for a deletion allele of *GA1* (*ga1-3*) are unable to flower under these conditions (Wilson et al. 1992). Moreover, these plants do not show a reduction in their flowering time under SD when exposed to low temperatures, suggesting that GA biosynthesis is required to respond to this treatment (Wilson et al. 1992). Furthermore, dominant *etr1* mutants and recessive *ein2* mutants that are insensitive to ethylene show a late-flowering phenotype when grown under LD (Bleecker et al. 1988; Guzmán and Ecker 1990).

### Early Mutants

Starting from early-flowering ecotypes, like L*er* or Col, it is also possible to select early-flowering mutants that flower earlier and with a lower LN than wild-type plants. All these mutations are recessive or semidominant, and on the basis of the mutant phenotype under different photoperiodic conditions they can be tentatively classified in two provisional phenotypic groups: (1) mutants that are earlier than the wild type both under LD and SD, but are sensitive to SD and show a delay in the transition to flowering under this photoperiod; (2) mutants that are earlier than wild type under LD and SD and behave as SD insensitive.

The first phenotypic group includes mutations at loci *TFL* (*TERMINAL FLOWER*) (Shannon and Meeks-Wagner 1991; Alvarez et al. 1992), *ELF1* (*EARLY FLOWERING*), and *ELF2* (Zagotta et al. 1992). Mutations at the *TFL* locus produce both a semidominant earliness and a recessive determinate inflorescence ending prematurely in a terminal flower (Shannon and Meeks-Wagner 1991). All *tfl* mutants isolated so far (more than 20 alleles, both in L*er* and Col) are sensitive to photoperiod, producing a higher number of vegetative and inflorescence leaves when grown under SD (Shannon and Meeks-Wagner 1991). Moreover, earliness in *tfl* mutants is strongly sensitive to growing temperature, being more extreme at 30ºC than at 15ºC (Alvarez et al. 1992). Recessive mutations at loci *ELF1* and *ELF2* have been isolated in Col. These mutations produce earliness without affecting the determinacy of the inflorescence. The transition to flowering is delayed under SD in *elf1* and *elf2* mutants, as it is in *tfl* mutants.

Two mutants, *elf3* and *emf1* (*EMBRYONIC FLOWERING1*), isolated in the Col background (Sung et al. 1992; Zagotta et al. 1992), can be included in the second phenotypic group containing those which do not respond to photoperiod. Homozygous *elf3* plants flower after the initiation of five rosette leaves whether grown in SD or LD conditions (Zagotta et al. 1992). In addition, *elf3* mutant plants show elongated hypocotyl and petioles and have a pale green color, due to reduced chlorophyll levels, when grown under SD (Zagotta et al. 1992). Complementation tests have indicated that the *elf3* mutation is not allelic to any of the *hy* loci tested (Zagotta et al. 1992). Seedlings homozygous for the *emf1* mutation flower upon germination. Mutant *emf* plants produce a short inflorescence without developing a vegetative rosette. The transition to flowering in these mutants is insensitive to photoperiod or vernalization given during the germination process. These and other recessive mutations that produce earliness could be allelic to dominant genes responsible for lateness, such as the loci which cause lateness in natural ecotypes (see above).

Four mutants (*esd1–4*) have been identified in the L*er* background in a screen for individuals which flower early under SD (S. Dash and G. Coupland, unpubl.). Unlike some of the long hypocotyl mutants (*hy*) that are early under SD (see below), none of these mutants shows an elongated hypocotyl. One of them, *esd1*, is daylength sensitive, flowering earlier under LD than SD, and is significantly earlier than wild-type plants under both daylengths. The other three mutants could belong to the SD-insensitive group. *esd2* and *esd3* flower at approximately the same time as wild type under LD but significantly earlier than wild type under SD. *esd4* is a very extreme early mutant, flowering under SD at around the same time as wild type under LD (S. Dash and G. Coupland, unpubl.).

Additional mutations that were initially isolated because of their effects on hormonal or photomorphogenic responses also accelerate the transition to flowering. Three long-hypocotyl mutants *hy1*, *hy2*, and *hy3* flower with a significantly lower LN than the wild type under SD and LD, and show a strong reduction in their sensitivity to SD as compared to the wild type (Goto et al. 1991). For this reason, they could be included in the same group as the *elf3* mutant. Mutants at loci *HY1* and *HY2* are deficient in phytochrome chromophore biosynthesis (Parks and Quail 1991) and are likely to contain a reduced level of all functional phytochromes, and *hy3* mutants are deficient in phytochrome B (Nagatani et al. 1991; Somers et al. 1991). Alternatively, mutants defective in phytochrome A are not earlier than wild type under LD (Nagatani et al. 1993; Parks and Quail, 1993; Whitelam et al. 1993).

Several early reports implicated abscisic acid (ABA) either as an activator or as a repressor of floral induction in different plant species (for review, see Zeevaart 1976). The ABA-deficient (*aba*) (Koornneef et al. 1982) or ABA-insensitive (*abi1*, *abi2*, *abi3*) (Koornneef et al. 1984) mutants of *Arabidopsis* do not show significant alterations in their FT and LN when grown under LD. However, under SD, the *aba* and the *abi1* mutants, lacking most of the ABA-mediated responses in vegetative tissues (Finkelstein and Somerville 1990), flower much earlier than wild type (J.A. Jarillo et al., unpubl.). These results are in agreement with previous reports suggesting that ABA could be a repressor of floral induction in LD plants when grown under SD (Addicott and Lyon 1969).

Consistent with the late-flowering phenotype of the GA-deficient and GA-insensitive mutants *ga1* and *gai*, slender mutants of *Arabidopsis*, carrying recessive mutations at the *SPINDLY* locus, show an early-flowering phenotype (Jacobsen and Olszewski 1993). The phenotype of these mutants is similar to that of wild-type plants repeatedly treated with $GA_3$, suggesting that the SPY gene product could regulate a portion of the GA signal transduction pathway (Jacobsen and Olszewski 1993).

## Second-site Mutations

Some of the late-flowering mutants in L*er* have been subjected to a second round of mutagenesis, and double mutants whose flowering was significantly earlier than the original late-flowering mutant have been isolated. These represent suppressor mutations of the original late-flowering phenotype. These mutations may be second-site mutations in the same gene or in other proteins which compensate for the original mutation. They should be very informative in the elucidation of the function of the late-flowering loci. Hussein (1968) isolated two semidominant mutations which flowered earlier than *fca*, but these mutants are no longer available. At least 10 independent mutations (from 1200 $M_2$ families) causing earlier flowering of *fca*, have recently been isolated (J. Chandler and C. Dean, unpubl.), as well as 1 mutation (from 650 $M_2$ families) in the *co* mutant (S. Dash and G. Coupland, unpubl.).

In addition to screening for earlier-flowering mutants in the *fca* line, J. Chandler and C. Dean (unpubl.) have taken advantage of the complete reversion of the late-flowering phenotype of the *fca* mutant by vernalization. They have isolated a group of mutants that show an altered response to the cold treatment. They have isolated 21 independent mutations (from 3000 $M_2$ families) which have been termed *vrn*. Two of these, representing independent loci, are recessive and result in an almost complete loss of response to the vernalization treatment.

## A MODEL FOR THE TRANSITION TO FLOWERING

The difficulty in performing grafting experiments in *Arabidopsis*, due to its growth habit, has precluded the analysis of the site of action of the genes identified by mutations. All the information we have to understand the mutant phenotype, and to draw conclusions about the function of the wild-type gene products, comes from the physiological characterization of some of the mutants and the phenotypic analysis of the double mutants. For these reasons, only preliminary conclusions can be drawn that require confirmation from future studies. Despite this, the existing information can be assembled into a general model attempting to describe the role of the *Arabidopsis* genes involved in the transition to flowering (Fig. 4). This model is based on two simple assumptions: (1) mutations that delay the transition to flowering are affecting genes involved in floral promotion and (2) mutations that accelerate floral transition are affecting genes involved in floral repression.

### The Transition to Flowering Can Be a Default Developmental Pathway

Most of the mutations that produce early flowering have only recently been isolated and, therefore, little information is available on the morphology and physiology of the corresponding mutants. Moreover, no genetic information on complementation analysis and double mutants has so far been reported. On the basis of the effect of these mutations on the transition to flowering, the *EMF1* locus stands apart from all the other "early" loci. The isolation of a recessive *emf1* mutation that allows flower development to proceed immediately after germination suggests that, in a wild-type plant, vegetative development takes place as the result of the repression of reproductive development. Therefore, reproductive development could be the default developmental pathway of the apical meristem (Fig. 4) (Sung et al. 1992). This repression is independent of the environment, since *emf* mutants flower after germination, irrespective of environmental conditions (Sung et al. 1992). Thus the EMF1 gene product could be directly or indirectly involved in the initiation of the juvenile vegetative phase at the apical meristem. Without active EMF1, genes required for the initiation of floral morphogenesis such as *LFY, CAL, AP1,* and *AP2* (Weigel and Meyerowitz 1993) would be activated in the apical meristem, giving rise to the differentiation of flowers.

The observation that Col seedlings grown in liquid sucrose medium under complete darkness flower with four leaves in the rosette and, therefore, earlier than seedlings growing in the light (Rédei et al. 1974; Araki and Komeda 1993b) suggests that (1) there is a repression pathway independent of the environmental conditions, which would be responsible for the production of the first leaves and could be mediated by EMF1, as suggested above; (2) the activity of this repression pathway would decay with time and/or successive cell divisions, allowing reproductive development to occur (after the fourth rosette leaf in Col); (3) extension of vegetative development after the fourth leaf stage would require the activity of other genes. In their absence, mutant plants would only develop a few leaves before floral transition, as is observed in most of the early-flowering mutants under LD.

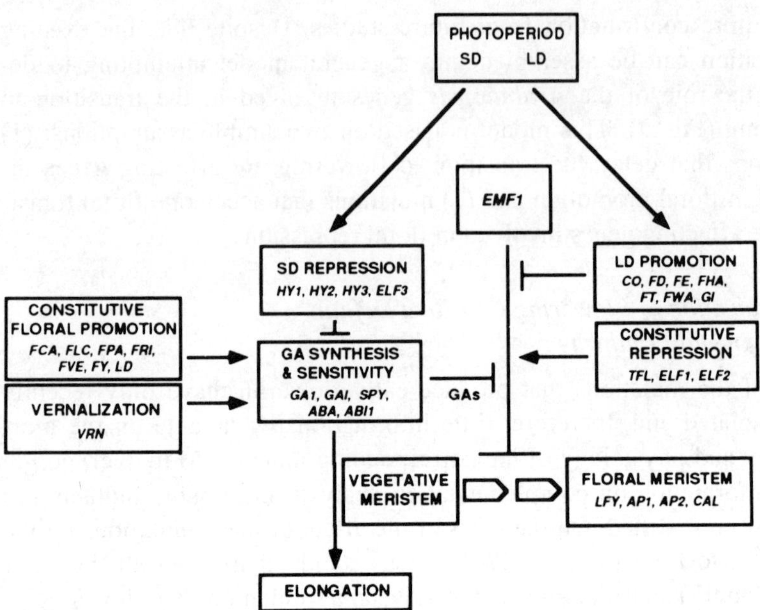

*Figure 4* Hypothetical model for the genetic and environmental control of the transition to flowering in *Arabidopsis*. Gene symbols indicate the action of the wild-type alleles. (→) Promotion effect; (⊣) repression effect. The model is similar to the one presented in G.W. Haughn et al. (in prep.), with slight modifications.

## Floral Repression: The Role of the "Early" Genes

The differences in the phenotypes described for early-flowering mutants suggest at least two floral repression pathways: a constitutive repression and a SD-dependent pathway (Fig. 4). Loci *TFL*, *ELF1*, and *ELF2* seem to be required to repress the transition to flowering independently from photoperiod. These early genes could delay the transition to flowering either by directly repressing the floral morphogenesis pathway or by promoting the activity of the EMF1 repressor (as shown in Fig. 4). In the absence of these genes, the transition to flowering would be accelerated because of the lower repression activity under both SD and LD. The activity of SD-dependent repression genes would delay flowering under SD, and mutations in these genes would render mutant plants almost insensitive to SD. Loci like *ELF3* and some of the genes involved in phytochrome biosynthesis (*HY1*, *HY2*, and *HY3*) could be considered within this pathway. The available phenotypic data of the *hy* mutants (Goto et al. 1991) and the effect of far-red on the transition to flowering (Brown and Klein 1971; Goto et al. 1991; Bagnall 1993) indicate that Pfr could be the phytochrome form responsible for SD-dependent repression of floral transition. The fact that *hy* mutants are also earlier under LD would indicate that Pfr also has a limited repression activity under LD. In many LD rosette-like species, photoperiod affects floral transition and internode elongation by controlling GA biosynthesis and/or sensitivity (Zeevaart 1983; Pharis and King 1985; Rood et al. 1989; Evans et al. 1990). On the basis of those results, it is attractive to think that genes in the SD-dependent repression pathway repress the floral transition by affecting synthesis and/or sensitivity to these hormones.

**Floral Promotion: The Role of the "Late" Genes**

Double mutant analysis and morphological and physiological characterization of late-flowering mutants distinguish at least two groups of late genes, suggesting the existence of at least two floral promotion pathways: a constitutive and a LD-dependent promotion pathway (Fig. 4). Both pathways would be required in an additive manner to produce early flowering under LD, since mutations in both of them cause a late-flowering phenotype.

The constitutive promotion pathway would include genes like *FCA*, *FLC*, *FPA*, *FRI*, *FVE*, *FY*, and *LD*, in which mutations produce a delay in FT both under SD and LD. The strong effect of vernalization on the phenotype of these mutants suggests that the outcome of this floral promotion pathway is similar to the outcome of the vernalization treatment itself. In *Thlaspi arvense*, another rosette-like LD plant in the mustard family, it has been suggested that the low-temperature treatment unblocks specific steps in the biosynthesis of C-13 desoxy gibberellins, responsible for low-temperature-induced stem elongation and floral transition (Hazebroek and Metzger 1990). In this way, the constitutive promotion pathway could be involved in the constitutive regulation of GA biosynthesis and/or sensitivity. In agreement with this suggested role, mutants at loci *FCA*, *FPA*, *FVE*, and *FY* show a reduction in internode elongation as compared to wild type. The activity of these genes could somehow interact with the effect of the SD-dependent repression pathway. GA synthesis seems to be the only way to overcome SD repression. In fact, completely deficient GA mutants like *ga1-3* (Koornneef et al. 1983) are unable to flower under SD, whereas the GA-insensitive mutants (*gai*) (Koornneef et al. 1985) are strongly delayed under this photoperiod (Wilson et al. 1992). These results indicate that, under SD, ent-kaurene synthesis (the step probably blocked in ga1-3; Zeevaart and Talón 1992) is required both for the activity of the constitutive promotion pathway and for the vernalization treatment. Under LD, other enzymatic steps could be derepressed which would make the plants less dependent on the activity of the *GA1* and *GAI* loci. The strong requirement of GA biosynthesis for floral transition under SD uncovers the effect that other mutations like *aba* or *abi1* could have on the GA-induced response, leading to a reduction in flowering time under these conditions (J.A. Jarillo et al., unpubl.). The important role played by gibberellins in the acceleration of floral transition is also supported by the early-flowering phenotype of the slender *spy* mutants (Jacobsen and Olszewski 1993).

The LD-dependent floral promotion pathway would function to reduce the activity of the floral repressor under LD, either by inhibition of the floral repressor (as shown in Fig. 4) or by directly promoting the default pathway. Loci *CO, FD, FE, FHA, FT, FWA,* and *GI* could play such a role. This promotion pathway would be almost inactive under SD, since mutations in the above genes do not produce strong delays under SD. Mutants in the LD-dependent promotion pathway would have an active constitutive promotion pathway and, therefore, vernalization would have very limited effect on the FT of these mutants. This interpretation would predict that *gai co* double mutants would be more delayed under LD than *gai fve* or *gai fca* double mutants, since in the first case the mutations affect two different floral promotion pathways, whereas in the second case they both affect the same pathway. The phenotypes of the double mutants agree with these predictions (G. Coupland et al.; J.M. Martínez-Zapater et al.; J. Chandler and C. Dean; all unpubl.).

In summary, floral transition at the apical meristem could be the result of the interaction of a few floral repression and floral promotion pathways under different environmental conditions. The fact that mutations at loci *CO, GI,* and *LD* are proposed to affect genes in both of the hypothesized promotion pathways (Fig. 4) and all of them flower at a similar time as the wild type under complete darkness (Rédei et al. 1974; Araki and Komeda 1993b) suggests that these promotion pathways are only required to counteract the effects of the repression pathways. According to this hypothesis, a reduction in the activity of the repression pathways would reduce the effect of the late-flowering mutants, and this is, in fact, the phenotype observed in *hy* late-flowering double mutants (Halliday et al. 1994).

### Floral Induction Interacts with Inflorescence and Flower Development

The inflorescence and floral reversions produced by shifting flowering plants from inductive to noninductive conditions (Fig. 3) (Laibach 1951) suggest that, in *Arabidopsis*, floral induction is not only required to trigger the inflorescence and floral developmental programs, but is also required throughout most of the stages of those developmental processes. Moreover, it also shows that floral promotion requires the constant presence of inductive environmental conditions.

Two environmental factors that induce flowering in wild-type plants, photoperiod and growing temperature (see previous sections), have an important effect on the phenotype of mutants at loci *LFY*, *AP1*, and *AP2*. Noninductive photoperiods (SD) produce a stronger mutant phenotype in *lfy*, *ap1* (Huala and Sussex 1992; Schultz and Haughn 1993), and *ap2* mutants (Komaki et al. 1988; Schultz and Haughn 1993), shifting the development of floral meristems toward inflorescence-indeterminate meristems. Growing temperature, another factor that affects floral transition, also has an effect on the phenotype of mutants *ap1* and *ap2* (Bowman et al. 1989). At low temperatures (16°C), the sepal primordia of the *ap2-1* mutant differentiate as leaves, whereas at high temperatures they differentiate as carpels. The similarity of the effect of SD photoperiods and low growth temperatures (16°C) on each of these mutants suggests that a factor involved in the transition to flowering regulates the expression or the activity of these genes. The outcome of the floral promotion/repression pathways interacts with the early floral homeotic genes. The flower position within the inflorescence also has a similar effect on the expression of the mutant phenotype. In the *ap2-1* mutant described above, later flowers in the inflorescence have more carpelloid structures than earlier flowers (Bowman et al. 1989; Kunst et al. 1989). Moreover, later flowers in the inflorescence of *ap1* or *lfy* mutants look more normal than the early flowers (Irish and Sussex 1990; Schultz and Haughn 1991). This suggests that the floral promotion activity at the apical meristem continues to increase along the development of the inflorescence after the transition to flowering has taken place.


## CONCLUSIONS AND PERSPECTIVES

The transition to flowering in *Arabidopsis* is a complex process with many levels of regulation, allowing large variations in flowering time in nature and a fine tuning to environmental factors. Because of the limited number of genotypes used as starting material for mutant screens and the emphasis on late-flowering mutants in earlier studies, it is probable that many other genes involved in floral transition still remain to be identified. Moreover, it is conceivable that some aspects of the regulation of the transition to flowering would not be amenable to genetic analysis due to genetic and/or biochemical redundancy. This could be the case for the complex network of GA biosynthetic reactions and their environmental regulation.

With the exception of several factors such as phytochrome, GA, and ABA, no other specific molecules predicted in the model have been identified. Fortunately, the available tools already allow the cloning of DNA sequences identified genetically. In this way, several late- and early-flowering genes are currently in the process of being cloned. A T-DNA tag has been used to clone the *LD* locus (Lee et al. 1993), and chromosome walking is being used to clone loci *CO* (Putterill et al. 1993), *FCA* (C. Dean), *FVE* (J.M. Martínez-Zapater), *FWA* (M. Koornneef), and *TFL* (R. Meeks-Wagner). Having access to the gene sequences will certainly help in obtaining additional information on the function of these genes and their interactions. However, only with a multidisciplinary approach involving physiology, genetics, and molecular biology will it be possible to dissect and understand this crucial and complex plant process.

### ACKNOWLEDGMENTS

We are indebted to Richard Amasino, David Bagnall, Elizabeth Dennis, George Haughn, Yoshibumi Komeda, Ry Meeks-Wagner, Scott Poethig, Manuel Talón, and Peter van Tienderen for sharing unpublished results with us. We also thank George Haughn, Elizabeth Schultz, and Julio Salinas for helpful discussions on the model in Figure 4, and José A. Jarillo for assistance with the figures. The research activity of the authors is currently supported by contract BIOT CT90-02-07 from the EEC and by grants CICYT PB91-0908 to J.M.M.Z., AFRC PG208/519 to G.C., and AFRC PG208/516 to C.D.

### REFERENCES

Addicott, F.T. and J.L. Lyon. 1969. Physiology of abscisic acid and related substances. *Annu. Rev. Plant Physiol.* **20:** 139–164.

Alvarez, J., C.L. Guli, X-H. Lu, and D.R. Smyth. 1992. *terminal flower*: A gene affecting inflorescence development in *Arabidopsis thaliana*. *Plant J.* **2:** 103–116.

Araki, T. and Y. Komeda. 1993a. Analysis of the role of the late-flowering locus, *GI* in the flowering of *Arabidopsis thaliana*. *Plant J.* **3:** 231–239.

———. 1993b. Flowering in darkness in *Arabidopsis thaliana*. *Plant J.* **4:** 801–811.

Bagnall, D.J. 1992. The control of flowering in *Arabidopsis thaliana* by light, vernalization and gibberellins. *Aust. J. Plant Physiol.* **19:** 401–409.

———. 1993. Light quality and vernalization interact in controlling late flowering in *Arabidopsis thaliana* ecotypes and mutants. *Ann. Bot.* **71:** 75–83.

Bassiri, A., E.E. Irish, and R.S. Poethig. 1992. Heterochronic effects of *Teopod 2* on the growth and photosensitivity of the maize shoot. *Plant Cell* **4:** 497–504.

Battey, N.H. and R.F. Lyndon. 1990. Reversion of flowering. *Bot. Rev.* **56:** 162–189.

Bernier, G. 1988. The control of floral evocation and morphogenesis. *Annu. Rev. Plant*

*Physiol.* **39:** 175–219.

Bernier, G., J.M. Kinet, and R.M. Sachs. 1981. *The physiology of flowering*, vol. 1. CRC Press, Boca Raton, Florida.

Besnard-Wibaut, C. 1977. Histoautoradiographic analysis of the thermoinductive processes in the shoot apex of *Arabidopsis thaliana* L. Heyhn, vernalized at different stages of development. *Plant Cell Physiol.* **18:** 949–962.

––––––––. 1981. Effectiveness of gibberellins and 6-benzyladenine on flowering in *Arabidopsis thaliana. Physiol. Plant.* **53:** 205–212.

Bleecker, A.B., M.A. Estelle, C. Somerville, and H. Kende. 1988. Insensitivity to ethylene conferred by a dominant mutation in *Arabidopsis thaliana. Science* **241:** 1086–1089.

Bowman J.L., D.R. Smyth, and E.M. Meyerowitz. 1989. Genes directing flower development in *Arabidopsis. Plant Cell* **1:** 37–52.

Brown, J.A.M. 1968. The role of competitive halogen analogs of thymidine in the induction of floral morphogenesis. In *Cellular and molecular aspects of floral induction* (ed. G. Bernier), pp. 22–25. Longman's Green, London.

Brown, J.A.M. and W.H. Klein. 1971. Photomorphogenesis in *Arabidopsis thaliana* (L.) Heyhn. *Plant Physiol.* **47:** 393–399.

Burn, J.E., D.R. Smyth, W.J. Peacock, and E.S. Dennis. 1993a. Genes conferring late flowering in *Arabidopsis thaliana. Genetica* **90:** 147–155.

Burn, J.E., D.J. Bagnall, J.D. Metzger, E.S. Dennis, and W.J. Peacock. 1993b. DNA methylation, vernalization and the initiation of flowering. *Proc. Natl. Acad. Sci.* **90:** 287–291.

Chintraruck, B. and H.J. Ketellapper. 1969. Interaction of vernalization, photoperiod and high temperature in flowering of *Arabidopsis thaliana* (L.) Heyhn. *Plant Cell Physiol.* **10:** 271–276.

Chory, J., P. Nagpal, and C.A. Peto. 1991. Phenotypic and genetic analysis of *det2*, a new mutant that affects light-regulated seedling development in *Arabidopsis. Plant Cell* **3:** 445–459.

Clarke, J. and C. Dean. 1993. Mapping *FRI*, a locus controlling flowering time and vernalization response in *Arabidopsis thaliana. Mol. Gen. Genet.* **242:** 81–89.

Deng, X.-W., T. Caspar, and P.H. Quail. 1991. *cop1*: A regulatory locus involved in light controlled development and gene expression in *Arabidopsis. Genes Dev.* **5:** 1172–1182.

Eskins, K. 1992. Light-quality effects on *Arabidopsis* development. Red, blue and far-red regulation of flowering and morphology. *Physiol. Plant.* **86:** 439–444.

Evans, L.T. 1969. *The induction of flowering.* Macmillan, Melbourne.

Evans, L.T., R.W. King, A. Chu, L.N. Mander, and R.P. Pharis. 1990. Gibberellin structure and florigenic activity in *Lolium temulentum*, a long day plant. *Planta* **182:** 97–106.

Finkelstein, R.R. and C.R. Somerville. 1990. Three classes of abscisic acid (ABA)-insensitive mutations of *Arabidopsis* define genes that control overlapping subsets of ABA responses. *Plant Physiol.* **94:** 1172–1179.

Finkelstein, R.R., M.A. Estelle, J.M. Martínez-Zapater, and C.R. Somerville. 1988. *Arabidopsis* as a tool for the identification of genes involved in plant development. In *Plant gene research: Temporal and spatial regulation of plant genes* (ed. D.P.S.Verma and R.B. Goldberg), vol. 5, pp. 1–25. Springer-Verlag, New York.

Goto, N., T. Kumagai, and M. Koornneef. 1991. Flowering responses to light-breaks in photomorphogenic mutants of *Arabidopsis thaliana*, a long day plant. *Physiol. Plant.* **83:** 209–215.

Gregory, F.G. and G.G. Hussey. 1953. Photoperiodic responses of *Arabidopsis thaliana.*

*Proc. Linn. Soc. Lond.* **164:** 137–147.

Guzmán, P. and J.R. Ecker. 1990. Exploiting the triple response of *Arabidopsis* to identify ethylene-related mutants. *Plant Cell* **2:** 513–523.

Halevy, A.H. 1985. *Handbook of flowering*, vols. I, II, III, and IV. CRC Press, Boca Raton, Florida.

Halliday, K.J., M. Koornneef, and G.C. Whitelam. 1994. Phytochrome B and at least one other phytochrome mediate the accelerated flowering response of *Arabidopsis thaliana* L. to low red:far-red ratio. *Plant Physiol.* (in press).

Hazebroek, J.P. and J.D. Metzger. 1990. Thermoinductive regulation of gibberellin metabolism in *Thlaspi arvense* L. *Plant Physiol.* **94:** 157–165.

Hirono, Y. and G.P. Rédei. 1965. Acceleration of flowering on the long-day plant *Arabidopsis* by 8-azaadenine. *Planta* **68:** 88–93.

————. 1966. Early flowering in *Arabidopsis* induced by DNA base analogs. *Planta* **71:** 107–112.

Huala, E. and I.M. Sussex. 1992. *LEAFY* interacts with floral homeotic genes to regulate *Arabidopsis* floral development. *Plant Cell* **4:** 901–913.

Hussein, H.A.S. 1968. Genetic analysis of mutagen-induced flowering time variation in *Arabidopsis thaliana* (L.) Heyhn. *Landbouwhogesch. Meded.* **68:** 1–88.

Imhoff, C., A. Lecharny, R. Jacques, and J. Brulfert. 1979. Two phytochrome dependent processes in *Anagallis arvensis* L.: Flowering and stem elongation. *Plant Cell Environ.* **2:** 67–72.

Irish, V.F. and I.M. Sussex. 1990. Function of the APETALA-1 gene during *Arabidopsis* floral development. *Plant Cell* **2:** 741–753.

Jacobsen, S.E. and N.E. Olszewski. 1993. Mutations at the *SPINDLY* locus of *Arabidopsis* alter gibberellin signal transduction. *Plant Cell* **5:** 887–896.

Karlovska, V. 1974. Genotypic control of the speed of development in *Arabidopsis thaliana* (L.) Heyhn. lines obtained from natural populations. *Biol. Plant.* **16:** 107–117.

Karlsson, B.H., G.R. Sills, and J. Nienhuis. 1993. Effects of photoperiod and vernalization on the number of leaves at flowering in 32 *Arabidopsis thaliana* (*Brassicaceae*) ecotypes. *Am. J. Bot.* **80:** 646–648.

Komaki, M.K., K. Okada, E. Nishino, and Y. Shimura. 1988. Isolation and characterization of novel mutants of *Arabidopsis thaliana* defective in flower development. *Development* **104:** 195–203.

Koornneef, M. and J.H. van der Veen. 1980. Induction and analysis of gibberellin sensitive mutants in *Arabidopsis thaliana* (L.) Heynh. *Theor. Appl. Genet.* **58:** 257–263.

Koornneef, M., M.L. Jorna, D.L.C. Brinkhorst-Van der Swan, and C.M. Karssen. 1982. The isolation of abscisic acid (ABA) deficient mutants by selection of induced revertants in non-germinating gibberellin sensitive lines of *Arabidopsis thaliana* (L.) Heynh. *Theor. Appl. Genet.* **61:** 382–393.

Koornneef, M., C.J. Hanhart, and J.H. van der Veen. 1991. A genetic and physiological analysis of late flowering mutants in *Arabidopsis thaliana*. *Mol. Gen. Genet.* **229:** 57–66.

Koornneef, M., G. Reuling, and C.M. Karssen. 1984. The isolation and characterization of abscisic-acid-insensitive mutants of *Arabidopsis thaliana*. *Physiol. Plant.* **61:** 377–383.

Koornneef, M., J. van Eden, C.J. Hanhart, and A.M.M. de Jongh. 1983. Genetic fine-structure of the *ga-1* locus in the higher plant *Arabidopsis thaliana* (L.) Heynh. *Genet. Res.* **41:** 57–68.

Koornneef, M., A. Elgersma, C.J. Hanhart, E.P. van Loenen-Martinet, L. van Rijn, and J.A.D. Zeevaart. 1985. A gibberellin insensitive mutant of *Arabidopsis thaliana*.

*Physiol. Plant.* **65:** 33–39.

Kunst, L., J.E. Klenz, J.M. Martínez-Zapater, and G.W. Haughn. 1989. *AP2* gene determines the identity of perianth organs in flowers of *Arabidopsis thaliana*. *Plant Cell* **1:** 1195–1208.

Laibach, F. 1951. Uber sommer und winterannuelle Rasse von *Arabidopsis thaliana* (L.) Heynh. Ein Beitrag zur Atiologie der Blutenbildung. *Beitr. Biol. Pflanz.* **28:** 173–210.

————. 1958. Uber den Artbastard *Arabidopsis suecica* (Fr.) Norrl. X A. *thaliana* (L.) Heyhn. und die Beziehungen zwischen den Gattungen *Arabidopsis* Heyhn and *Cardaminopsis* (C.A. Meyer) Hay. *Planta* **51:** 148–166.

Lane, H.C., H.M. Cathey, and L.T. Evans. 1965. The dependence of flowering in several long-day plants on the spectral composition of light extending the photoperiod. *Am. J. Bot.* **52:** 1006–1014.

Langridge, J. 1957. Effect of day-length and gibberellic acid on the flowering of *Arabidopsis*. *Nature* **180:** 36–37.

Law, C. 1987. The genetic control of day-length response in wheat. In *Manipulation of flowering* (ed. J.G. Atherton), pp. 225–240. Butterworths, London.

Lee, J., A. Bleecker, and R. Amasino. 1993. Analysis of naturally occurring late flowering in *Arabidopsis thaliana*. *Mol. Gen. Genet.* **237:** 171–176.

Lee, J., M.J. Aukerman, S.L. Gore, K.N. Lohman, S.D. Michaels, L.M. Weaver, M.C. John, K.A. Feldman, and R. Amasino. 1994. Isolation of *Luminidependens*: A gene involved in the control of flowering time in *Arabidopsis thaliana*. *Plant Cell* **6:** 75–83.

McDaniel, C.N., S.R. Singer, and S.M.E. Smith. 1992. Developmental states associated with the floral transition. *Dev. Biol.* **153:** 59–69.

McKelvie, A.D. 1962. A list of mutant genes in *Arabidopsis thaliana* (L.) Heyhn. *Radiat. Bot.* **1:** 233–241.

Martínez-Zapater, J.M. and C.R. Somerville. 1990. Effect of light quality and vernalization on late-flowering mutants of *Arabidopsis thaliana*. *Plant Physiol.* **92:** 770–776.

Medford, J.I., F.J. Behringer, J.D. Callos, and K.A. Feldmann. 1992. Normal and abnormal development in the *Arabidopsis* vegetative shoot apex. *Plant Cell* **4:** 631–643.

Meijer, G. 1959. The spectral dependence of flowering and elongation. *Acta Bot. Neerl.* **8:** 189–246.

Michniewicz, M. and A. Kamienska. 1965. Flower formation induced by kinetin and vitamin E treatment in long-day plant (*Arabidopsis thaliana*) grown in short day. *Naturwissenschaften* **52:** 623.

Miksche, J.P. and J.A. Brown. 1965. Development of vegetative and floral meristems of *Arabidopsis thaliana*. *Am. J. Bot.* **52:** 533–537.

Müller, A. 1961. Zur Charackterisierung der Bluten und Inflorreszenzen von *Arabidopsis thaliana* (L.) Heyhn. *Kulturpflanze* **9:** 364–393.

Murfet, I.C. 1977. Environmental interaction and the genetics of flowering. *Annu. Rev. Plant. Physiol.* **28:** 253–278.

————. 1989. Flowering genes in *Pisum*. In *Plant reproduction: From floral induction to pollination* (ed. E. Lord and G. Bernier), pp. 10–18. American Society of Plant Physiologists, Rockville, Maryland.

Myerscough, P.J. and J.K. Marshall. 1973. Population dynamics of *Arabidopsis thaliana* (L.) Heyhn. strain "Estland" at different densities and nutrient levels. *New Phytol.* **72:** 595–601.

Nagatani, A., J. Chory, and M. Furuya. 1991. Phytochrome B is not detectable in the *hy3* mutant of *Arabidopsis*, which is deficient in responding to end-of-day far-red light treatments. *Plant Cell. Physiol.* **32:** 1119–1122.

Nagatani, A., J.W. Reed, and J. Chory. 1993. Isolation and initial characterization of

432  J.M. Martínez-Zapater et al.

*Arabidopsis* mutants that are deficient in phytochrome A. *Plant Physiol.* **102:** 269–277.

Napp-Zinn, K. 1957. Die Abhängigkeit des Vernalisationseffektes bei *Arabidopsis thaliana* von der Dauer der Vorquellung der samen sowie von Alter der Pflanzen bei Beginn der Vernalisation. *Z. Bot.* **45:** 379–394.

————. 1961. Über die Bedeutung genetischer Untersuchungen an kältebedürftigen Pflanzen für die Aufklärung von Vernalisationserscheinungen. *Züchter* **31:** 128–139.

————. 1962. Über die genetischen Grundlagen der Vernalisationsbedürfnisses bei *Arabidopsis thaliana*. I. Die Zahl der beteiligten Faktoren. *Z. Vererbungsl.* **93:** 154–163.

————. 1963. Über den Einfluss von Genen und Gibberellinen auf die Blütenbildung von *Arabidopsis thaliana*. *Ber. Dtsch. Bot. Ges.* **76:** 77–89.

————. 1969. *Arabidopsis thaliana* (L.) Heyhn. In *The induction of flowering: Some case histories* (ed. L.T. Evans), pp. 291–304. Macmillan, Melbourne.

————. 1985. *Arabidopsis thaliana*. In *Handbook of flowering* (ed. H.A. Halevy), vol. 1, pp. 492–503. CRC Press, Boca Raton, Florida.

————. 1987. Vernalization. Environmental and genetic regulation. In *Manipulation of flowering* (ed. J.G. Atherton), pp. 123–132. Butterworths, London.

Parks, B.M. and P. H. Quail. 1991. Phytochrome-deficient *hy1* and *hy2* long hypocotyl mutant of *Arabidopsis* are defective in phytochrome chromophore biosynthesis. *Plant Cell* **3:** 1177–1186.

————. 1993. *hy8*, a new class of *Arabidopsis* long hypocotyl mutants deficient in functional phytochrome A. *Plant Cell* **5:** 39–48.

Pharis, R.P. and R.W. King. 1985. Gibberellins and reproductive development in seed plants. *Annu. Rev. Plant Physiol.* **36:** 517–568.

Poethig, R.S. 1990. Phase change and the regulation of shoot morphogenesis in plants. *Science* **250:** 923–930.

Putterill, J., F. Robson, K. Lee, and G. Coupland. 1993. Chromosome walking with YAC clones in *Arabidopsis*: Isolation of 1700 kb of contiguous DNA on chromosome 5, including a 300 kb region containing the flowering-time gene *CO*. *Mol. Gen. Genet.* **239:** 145–157.

Rédei, G.P. 1962. Supervital mutants of *Arabidopsis*. *Genetics* **47:** 443–460.

————. 1970. *Arabidopsis thaliana* (L.) Heyhn. A review of the genetics and biology. *Bibliogr. Genet.* **20:** 1–151.

Rédei, G.P, G. Acedo, and G. Gavazzi. 1974. Flower differentiation in *Arabidopsis*. *Stadler Genet. Symp.* **6:** 135–168.

Reinholz, E. 1972. Vegetative reproduction. *Arabidopsis Inf. Serv.* **9:** 37.

Rood, S.B., R. Mandel, and R.P. Pharis. 1989. Endogenous gibberellins and shoot growth and development in *Brassica napus*. *Plant Physiol.* **89:** 269–273.

Sarkar, S. 1958. Versuche zur Physiologie der Vernalisation. *Biol. Zentralbl.* **77:** 1–49.

Schneider, M.J., H.A. Borthwick, and S.B. Hendricks. 1967. Effects of radiation on flowering of *Hyoscyamus niger*. *Am. J. Bot.* **54:** 1241–1249.

Schultz, E.A. and G.W. Haughn. 1991. *LEAFY*, a homeotic gene that regulates inflorescence development in *Arabidopsis*. *Plant Cell* **3:** 771–781.

————. 1993. Genetic analysis of the floral initiation process (FLIP) in *Arabidopsis*. *Development* **119:** 745–765.

Shannon, S. and D.R. Meeks-Wagner. 1991. A mutation in the *Arabidopsis TFL1* gene affects inflorescence meristem development. *Plant Cell* **3:** 877–892.

Somers, D.E., R.A. Sharrock, J.M. Tepperman, and P.H. Quail. 1991. The *hy3* long hypocotyl mutant of *Arabidopsis* is deficient in phytochrome B. *Plant Cell* **3:** 1263–1274.

Sung, Z.R., A. Belachew, B. Shunong, and R. Bertrand-García. 1992. *EMF*, an *Arabidopsis* gene required for vegetative shoot development. *Science* **258**: 1645–1647.

Talón, M., M. Koornneef, and J.A.D. Zeevaart. 1990a. Endogenous gibberellins in *Arabidopsis thaliana* and the possible steps blocked in the biosynthetic pathway of the semi-dwarf *ga4* and *ga5* mutants. *Proc. Natl. Acad. Sci.* **87**: 7983–7987.

————. 1990b. Accumulation of $C_{19}$-gibberellins in the gibberellin-insensitive dwarf mutant *gai* of *Arabidopsis thaliana* (L.) Heyhn. *Planta* **182**: 501–505.

Van der Veen, J.H. 1965. Genes for late flowering in *Arabidopsis*. In Arabidopsis *Research* (Proceedings of the Göttingen Symposium) (ed. G. Röbbelen), pp. 162–171. Wasmund, Gelsenkirchen, Germany.

Vaughan, J.G. 1955. The morphology and growth of the vegetative and reproductive apices of *Arabidopsis thaliana* (L.) Heyhn., *Capsella bursa-pastoris* (L.) Medic. and *Anagallis arvensis* L. *J. Linn. Soc. Bot.* **55**: 279–300.

Vetrilova, M. 1973. Genetic and physiological analysis of induced late mutants of *Arabidopsis thaliana* (L.) Heyhn. *Biol. Plant.* **15**: 391–397.

Weigel, D. and E.M. Meyerowitz. 1993. Genetic hierarchy controlling flower development. In *Molecular basis of morphogenesis* (ed. M. Bernfield), pp. 93–107. Wiley-Liss, New York.

Westerman, J.M. and M.J. Lawrence. 1971. Genotype-environment interaction and developmental regulation in *Arabidopsis thaliana*. *Heredity* **25**: 609–627.

Whitelam, G.C. and H. Smith. 1991. Retention of phytochrome-mediated shade avoidance responses in phytochrome-deficient mutants of *Arabidopsis*, cucumber and tomato. *J. Plant Physiol.* **139**: 119–125.

Whitelam, G.C., E. Johnson, J. Peng, P. Carol, M.L. Anderson, J.S. Cowl, and N.P. Harberd. 1993. Phytochrome A *null* mutants of *Arabidopsis* display a wild-type phenotype in white light. *Plant Cell* **5**: 757–768.

Wilson, R.N., J.W. Heckman, and C.R. Somerville. 1992. Gibberellin is required for flowering but not for senescence in *Arabidopsis thaliana* under short days. *Plant Physiol.* **100**: 403–408.

Zagotta, M.T., S. Shannon, C. Jacobs, and D.R. Meeks-Wagner. 1992. Early-flowering mutants of *Arabidopsis thaliana*. *Aust. J. Plant Physiol.* **19**: 411–418.

Zeevaart, J.A.D. 1976. Physiology of flower formation. *Annu. Rev. Plant Physiol.* **27**: 321–348.

————. 1983. Gibberellins and flowering. In *The biochemistry and physiology of gibberellins* (ed. A. Crozier), vol. 2, pp. 333–374. Praeger Scientific, New York.

Zeevaart, J.A.D. and M. Talón. 1992. Gibberellin mutants in *Arabidopsis thaliana*. In *Current plant sciences and biotechnology in agriculture: Progress in plant growth regulation* (eds. C.M. Karssen et al.), vol. 13, pp. 34–42. Kluwer Academic, Amsterdam.

# 17

# *Arabidopsis* Flower Development

**Steven E. Clark and Elliot M. Meyerowitz**
Division of Biology
California Institute of Technology
Pasadena, California 91125

The *Arabidopsis* flower develops sixteen organs in four concentric rings, or whorls (Fig. 1). In the outermost whorl, whorl 1, four green sepals develop. Interior and alternate to the sepals are four white petals which occupy whorl 2. Whorl 3 contains six stamens: two pairs of medial stamens and two lateral stamens, which are shorter than the medial stamens. The center of the flower, whorl 4, consists of two fused carpels that comprise the gynoecium.

The development of wild-type flowers has been well characterized by Smyth et al. (1990), and readers are referred to that article for an in-depth analysis. All stages of development as described in this chapter are according to those outlined by Smyth et al. (1990). Flowers initially arise as undifferentiated bulges on the flank of the apical meristem (stage 1) (Fig. 1). Soon afterward, four sepal primordia develop along the edges of the floral meristem, establishing the first whorl (stage 3). The sepals grow to overlie the floral primordium (stage 4). Four petal and six stamen primordia initiate in whorls 2 and 3, respectively (stage 5), followed by carpel primordium development in the center of the floral meristem (stage 6).

Each floral organ has distinctive features that can be used to distinguish it from other organs. These characteristics are often used to identify mosaic or intermediate organ types seen in mutant flowers. Sepals are leaf-like, but display distinct differences from leaves (Fig. 2A, C, E). Sepals lack flanking stipules at their bases and develop simple, as opposed to stellate, trichomes. The outer face of a sepal has a distinctive cellular morphology with several cells more than 100 μm in length, and with characteristically ridged surfaces (Fig. 2C). Cells on the inner face of a sepal are longer than 100 μm and thin but lack significant ridge patterns on their surface (Fig. 2E). In contrast, petal epidermal cells are small (5–10 μm in diameter) and round in appearance and have a charac-

*Arabidopsis*
© 1994 Cold Spring Harbor Laboratory Press  0-87969-428-9/94 $5 + .00

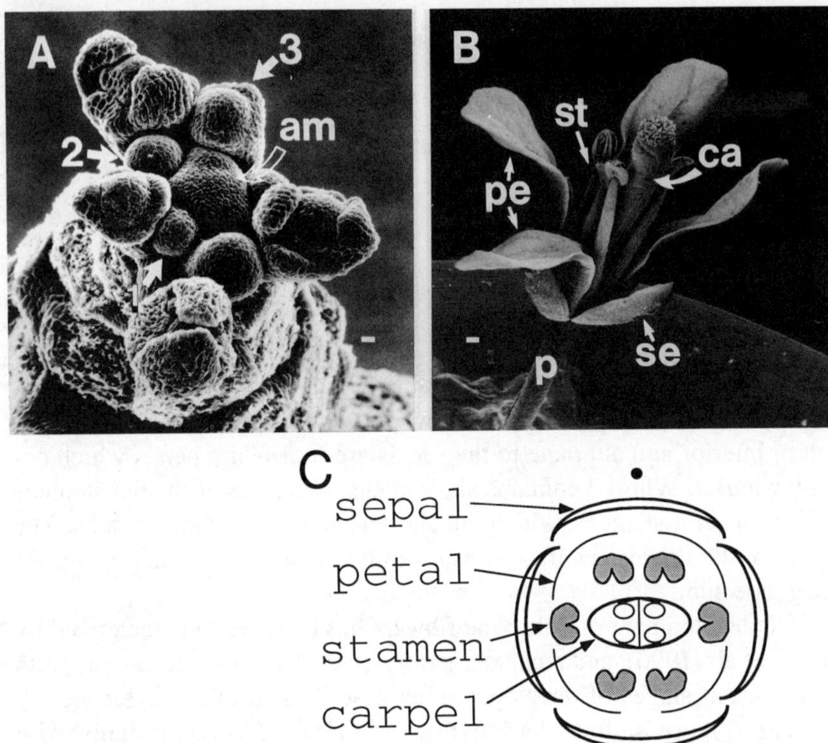



*Figure 1*   Wild-type flower development. The *Arabidopsis thaliana* apical meristem (am) continually develops floral primordia, leading to an inflorescence with flowers in a range of different stages (*A*). A wild-type ecotype Landsberg flower (*B*) develops 16 organs of four types characteristic of *A. thaliana* flowers (*C*). Stage numbers according to Smyth et al. (1990) are indicated near several flowers (*A*). Bars: 10 μm (*A*) and 100 μm (*B*).

teristically ridged and mounded cell wall. Petal cells facing the interior of the flower are more regular in both shape and ridge pattern compared to those facing out (Fig. 2D, E). Stamens consist of two portions, the filament and the anther. The filament is a column of cells on which the anther sits and which does not elongate until late in flower development. Filament cells are large with ridges running parallel and perpendicular to the axis of the filament (Fig. 3F). The anther is two-lobed, contains four locules in which pollen develops, and dehisces to release pollen (Fig. 3A, B, E). The surface cells of the anther are small and irregular in shape (Fig. 3C, D). The gynoecium is a complex organ composed of two fused carpels (Fig. 4A–C). The septum is a wall that forms between the two carpels on the interior of the gynoecium. Ovules develop at the edge of each carpel at the intersection with the septum.  The gynoecium is topped

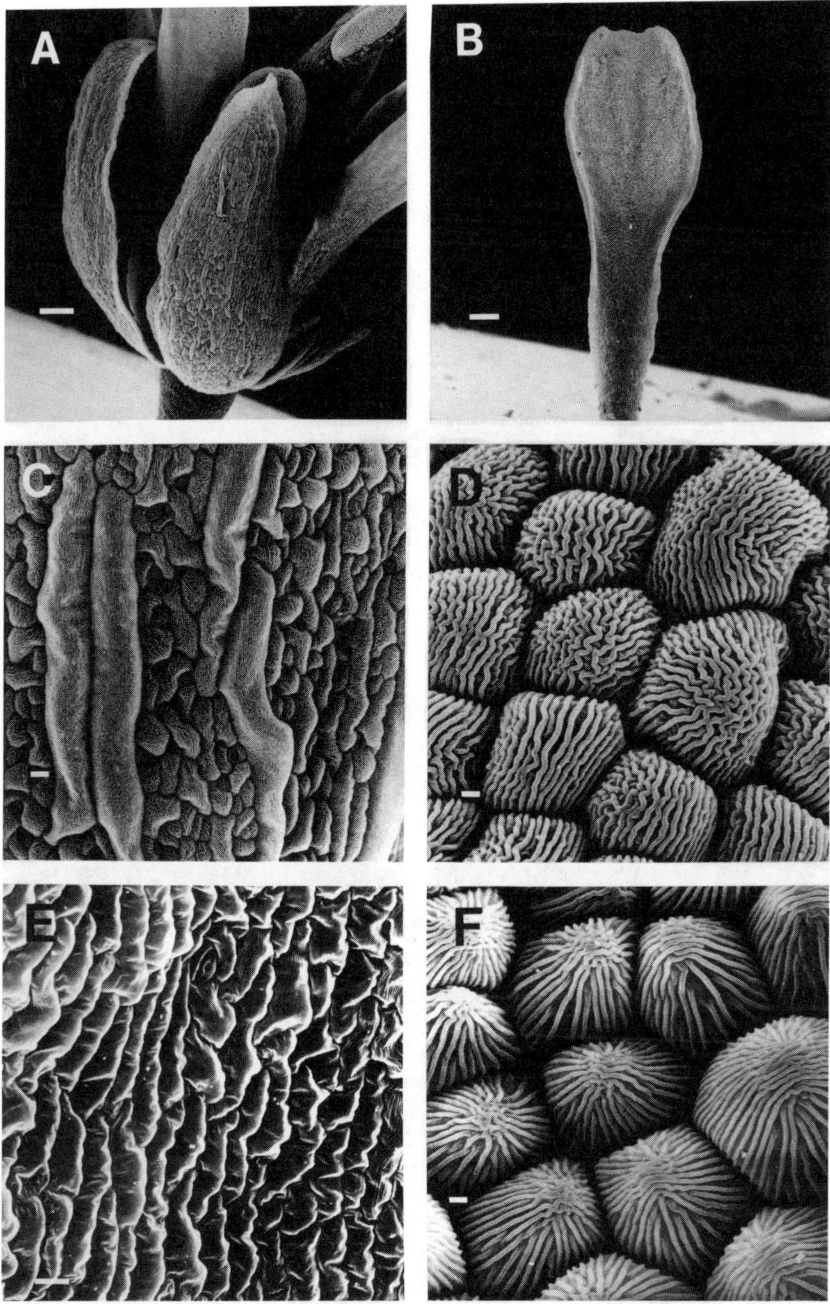

*Figure 2* Sepal and petal epidermal cell types. (*A*) An ecotype Landsberg flower is shown with facing and side views of the sepals visible. Cell morphology of outer (*C*) and inner (*E*) sepal cells are shown. A mature petal is shown in *B*, along with cell morphology of outer (*D*) and inner (*F*) petal cells. Bars: 1 μm (*D,F*), 10 μm (*C,E*), and 100 μm (*A,B*).

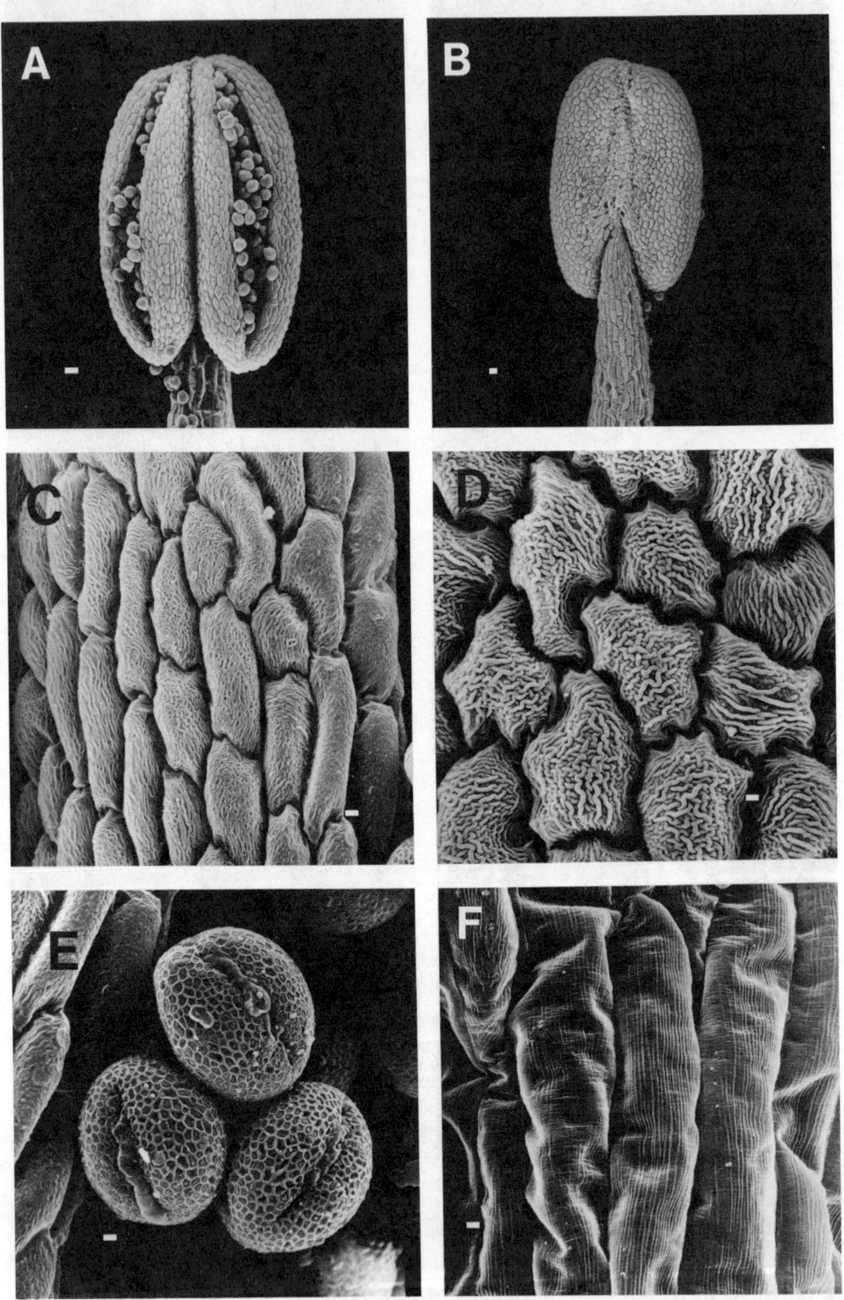

*Figure 3* Stamen surface cell types. Inner (*A*) and outer (*B*) views of anthers are shown, with cell morphology of each in *C* and *D*, respectively. Pollen grains are shown in *E*, and filament cells in *F*. Bars: 1 μm (*C–F*) and 10 μm (*A,B*).

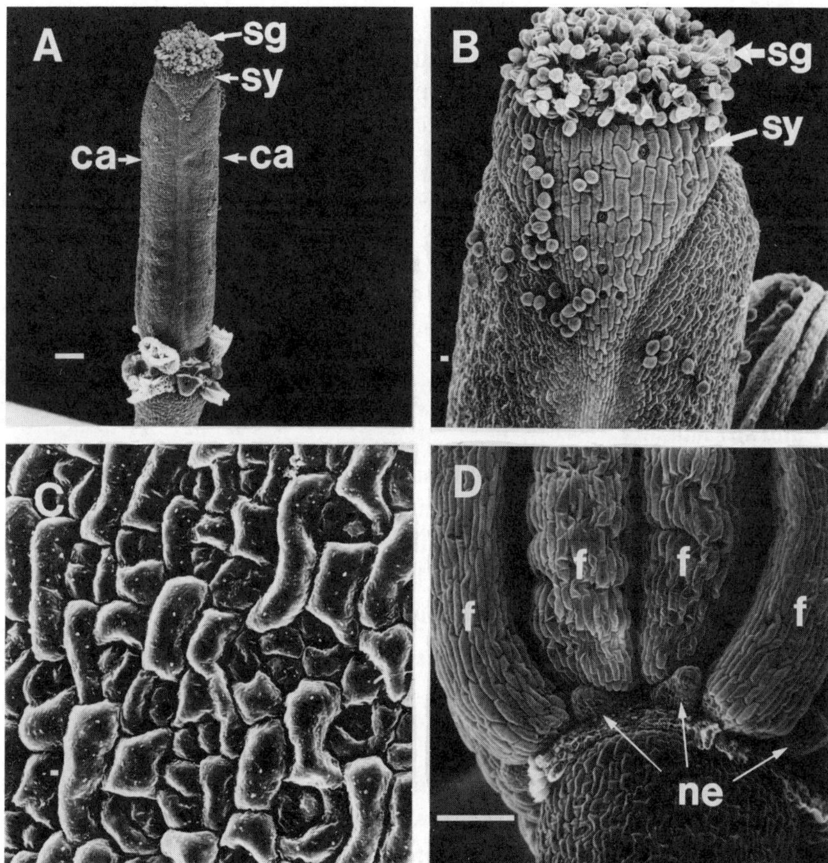

*Figure 4*  The gynoecium is a complex organ. The gynoecium (*A*) is composed of two carpels (ca) separated by a septum. (*B*) The top of the gynoecium develops stigmatic tissue (sg) above the style (sy). Carpel surface cells are shown in *C*. Nectaries (ne) develop near the base of the stamen filaments (f) (*D*; photo by J.L. Bowman). Bars: 1 μm (*C*), 10 μm (*B*), and 100 μm (*A,D*).

by stigmatic tissue beneath which is the style (Fig. 4B) (see also Pruitt and Hülskamp, this volume). All of these organs and cell types, and additional internal cell types, develop from a small collection of undifferentiated cells over the span of about 2 weeks at 25°C (Smyth et al. 1990).

Among the questions raised about flower development are, How are the fates of those first cells specified, so that the primordium develops as a flower, and not as a leaf or axillary meristem? What factors are involved in specifying the identity of the floral organs? How is the structure of the floral meristem controlled, such that the floral organs form in the correct numbers and positions?

## FLORAL IDENTITY

When a floral primordium is initiated by the shoot apical meristem, it must be given developmental cues to distinguish it from a rosette leaf or a cauline leaf, which are also initiated by the shoot apical meristem in a spiral phyllotactic pattern (Fig. 5). Two genes critical to imparting a floral identity to the primordium in *Arabidopsis* are *LEAFY* (*LFY*) (Schultz and Haughn 1991; Huala and Sussex 1992; Weigel et al. 1992; Weigel and Meyerowitz 1993) and *APETALA1* (*AP1*) (Irish and Sussex 1990; Mandel et al. 1992a; Bowman et al. 1993). Clues about their roles in this process have come from mutant analysis and RNA expression patterns.

### *lfy* and *ap1* Mutants

Plants mutant for *lfy* exhibit normal vegetative growth and a normal transition to inflorescence development. However, these plants generate more cauline leaves with axillary inflorescence shoots, compared to wild type, before developing flowers. These additional stems can be interpreted in one of two ways. They could be the result of a simple delay in the transition to developing flowers, or they could represent a complete transformation of the early flowers into shoot structures. Regardless of which interpretation is correct, *LFY* is necessary for the development of the early flowers, either through the correct timing of the shoot to flower transition, or through imparting floral meristem identity. *lfy* plants later develop flowers, although these flowers do exhibit a distinct, albeit partial, transformation to shoot structures (Weigel et al. 1992). First, these flowers are subtended by bracts, much the same way stems are subtended by cauline leaves (which are inflorescence bracts) (Fig. 6A, B). Second, *lfy* flowers often display a slight departure from the normal whorled pattern of floral organ initiation (Fig. 6A). Because shoot structures in *Arabidopsis* are characterized by a spiral pattern of organ initiation, the imperfect whorl patterning has been interpreted as a partial transformation to spiral patterning. Last, *lfy* flowers lack petals and stamens and are instead composed of sepals and poorly fused carpels.

Plants mutant for *ap1* also exhibit wild-type vegetative growth and an unaffected transition to inflorescence development (Irish and Sussex 1990). *ap1* plants generate a normal number of cauline leaves subtending axillary inflorescence shoots, and the next several primordia generated usually develop as flowers, although they can develop as axillary inflorescences (without a subtending bract) (Bowman et al. 1993). In addition, *ap1* flowers exhibit defects of two sorts (Fig. 6C). The first is a floral organ defect in which the sepals are converted into leaf-like structures or filamentous organs and the petals are absent or partly converted to

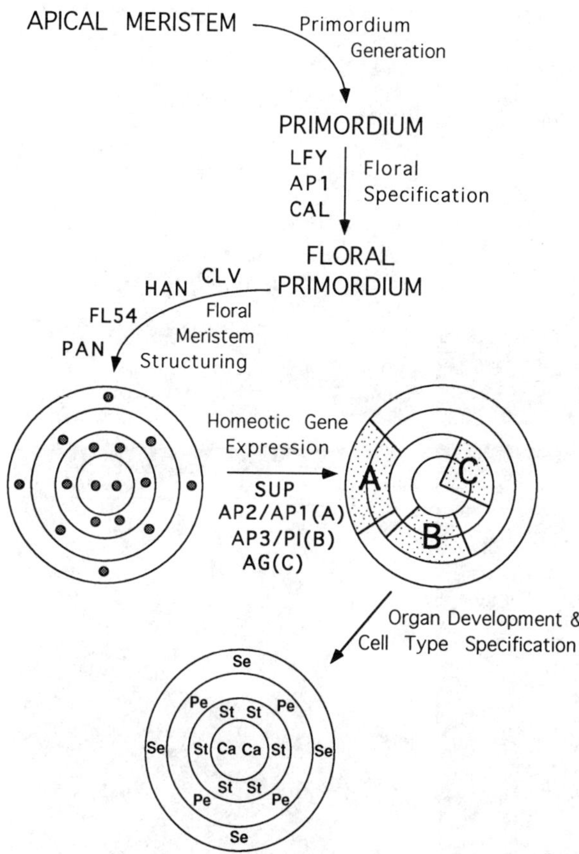



*Figure 5*  General pathway of flower development. As primordia are generated by the apical meristem, genes such as *LFY* and *AP1* act to impart floral identity on these primordia. Structure, including whorl and organ number, is then imparted on the floral meristem, potentially by genes such as *CLV*, *PAN*, *FL54*, and *HAN*. Homeotic genes are activated in specific whorls, presumably leading to the activation of hundreds of downstream genes responsible for cell type specification.

stamens. The second is the development of flowers in the axils of the first whorl organs. The appearance of leaf-like whorl-1 organs with axillary flowers is interpreted as an indication of the partial shoot transformation of *ap1* flowers.

## *LFY* and *AP1* Expression Patterns

Both *LFY* and *AP1* have been cloned—*LFY* by homology with the *Antirrhinum* ortholog *FLORICAULA* (Coen et al. 1990; Weigel et al.

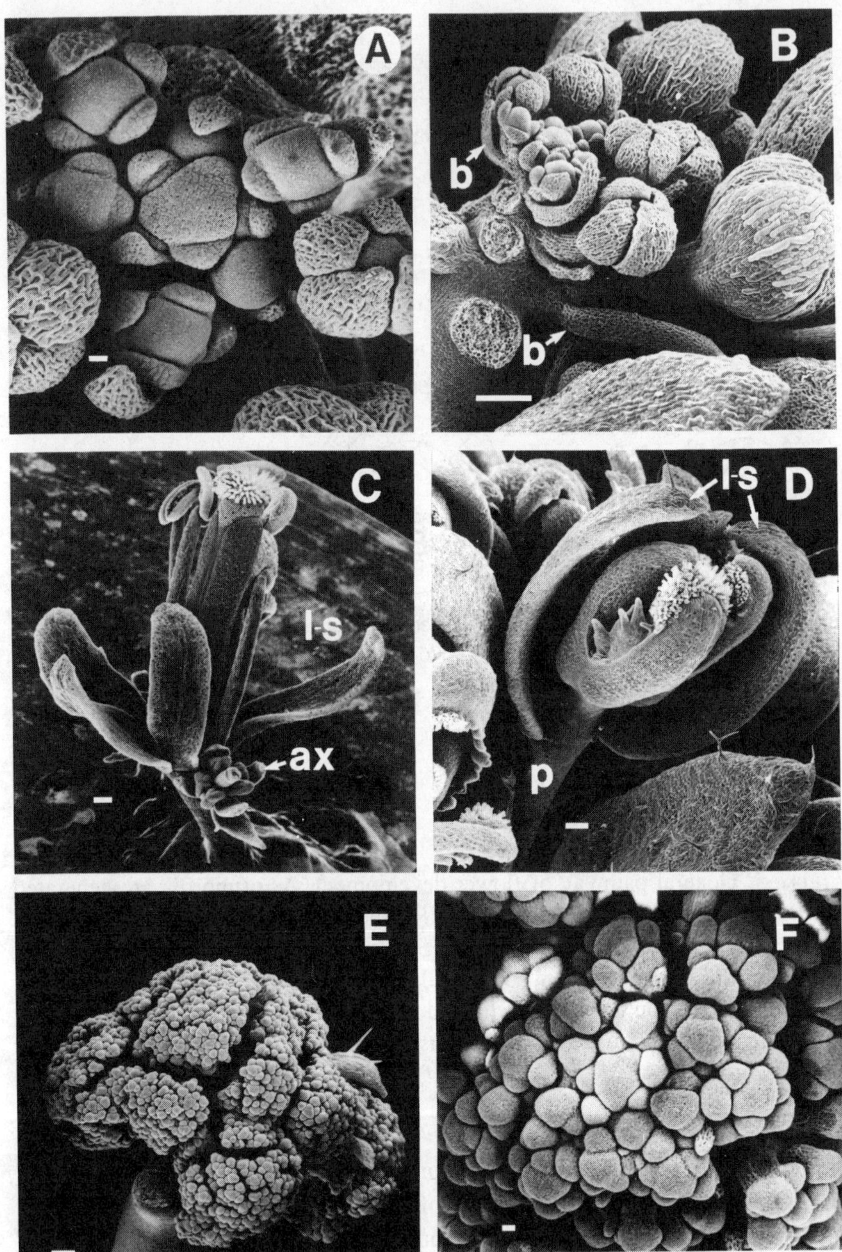

*Figure 6*  Loci imparting floral meristem identity. *lfy* inflorescences (*A,B*) develop flowers that are subtended by bracts (b) and lack proper whorl patterning. *ap1* flowers (*C*) have leaf-like first whorl organs (l-s) that occasionally subtend axillary flowers (ax). An *ap1 lfy* double mutant flower is shown in *D. ap1 cal* inflorescences (*E,F*) develop meristems in a highly reiterative process. (p) Flower pedicel. Bars: 10 μm (*A,F*) and 100 μm (*B–E*).

1992) and *AP1* by homology with the paralogous floral organ identity gene *AGAMOUS* (see below) (Mandel et al. 1992a). The *LFY* gene codes for a predicted protein of 412–424 amino acids (depending on the splice variant). The protein displays no sequence similarity to known proteins other than *FLORICAULA*, although it does contain a proline-rich and acidic domain found in transcriptional activators. Recent evidence that the LFY protein is localized to the nucleus implies that it might act as a transcription factor (Weigel and Meyerowitz 1993). *LFY* RNA is expressed in the floral anlage before the floral primordium protrudes as a bulge distinct from the apical meristem (Weigel et al. 1992). *LFY* expression is the earliest of any known floral-specific gene and suggests that floral specification begins at the earliest stages of primordium initiation, before the primordium becomes morphologically distinct from the apical meristem. Because *lfy* plants develop flowers subtended by bracts, this early *LFY* expression presumably acts to repress bract formation and provide developmental cues on whorl patterning.

*AP1* RNA is also expressed in early-developing flowers, but its expression is later than that of *LFY* and is not detectable until the floral primordium becomes morphologically distinct from the apical meristem (Mandel et al. 1992a). Initially, *AP1* is expressed throughout the floral primordium, and this expression is presumably related to its control of floral meristem identity. Later, its expression is restricted to the first and second whorls and is presumably linked to its organ identity function. This is discussed in more detail below. *AP1* is a member of the MADS box family, containing transcription factors SRF from humans (Norman et al. 1988) and *MCM1* (Passmore et al. 1988) and *ARG80* (Dubois et al. 1987) from yeast.

### *LFY-AP1* Interactions

Analysis of *LFY* and *AP1* expression patterns has also indicated that *AP1* does not control *LFY* expression (i.e., *LFY* is expressed normally in *ap1* plants; Weigel and Meyerowitz 1993). However, plants mutant for both *ap1* and *lfy* display synergistic transformations of flowers to shoot structures (Weigel et al. 1992). *ap1 lfy* plants develop many axillary inflorescence shoots subtended by bracts after the transition to flowering. Eventually, flowers develop but display extensive transformations to shoot structures (Fig. 6D). First, the organs initiate in a spiral pattern. The outer organs are leaf-like and often flanked by stipules. Inner organs are leaf-like, yet often develop ovules at their margins and stigmatic tissue at their tips. Additionally, the flowers are indeterminate. Many of these characteristics are not present in either the *ap1* or *lfy* single mutant plants, suggesting that *AP1* compensates for the loss of *LFY* activity in *lfy* plants to establish floral meristem characteristics, and vice versa. As described below, *LFY* and *AP1* act in concert to activate several of the homeotic floral organ identity genes (Weigel and Meyerowitz 1993).

## Other Floral Meristem Identity Genes

Several other genes seem to play a role in the floral meristem identity process. The *CAULIFLOWER* (*CAL*) gene appears to function in a redundant fashion with *AP1* (Bowman et al. 1993). First, mutations in *cal* only cause a mutant phenotype in *ap1* mutant plants; *cal* single mutant plants have been so far indistinguishable from wild type. Second, the predicted *CAL* coding region has the highest degree of similarity to the *AP1* coding region of any known *Arabidopsis* genes (S. Kempin and M. Yanofsky, pers. comm.). Plants carrying both *ap1* and *cal* mutations look quite similar to the vegetable cauliflower (*Brassica oleracea*, var. *botrytis*). After a normal transition to flowering, the primordia generated by *ap1 cal* inflorescences do not develop into flowers; instead, they generate new primordia on their flanks in a spiral pattern, and this process is reiterated many times (Fig. 6E, F). At 25°C, but not at 16°C, *ap1 cal* plants eventually develop flowers that are *ap1* in appearance. This difference is correlated with *LFY* expression: *ap1 cal* primordia express *LFY* at 25°C, but not at 16°C (Bowman et al. 1993).

The organ identity gene *AP2* (see Floral Organ Identity below) also seems to play a role in establishing floral meristem identity. Evidence for this comes from the ability of *ap2* mutants to enhance the phenotypes of both *lfy* and *ap1* mutants. Floral meristems of plants doubly mutant for *ap1* and a weak *ap2* allele develop an increased number of axillary flowers compared to *ap1* single mutants and are often transformed into inflorescence meristems (Irish and Sussex 1990; Bowman et al. 1993; Shannon and Meeks-Wagner 1993). For plants doubly mutant for *ap1* and a strong *ap2* allele, the number of axillary flowers is reduced compared to *ap1* plants, but, again, floral meristems are more often transformed into inflorescence meristems (Bowman et al. 1993). Reports of double mutant combinations between *lfy* and *ap2* are conflicting. The double mutant plants have been reported to be identical to *lfy* plants (Schultz and Haughn 1991), additive in phenotype (Weigel et al. 1992), and similar to *lfy ap1* plants (Huala and Sussex 1992). Much of the difference may be due to the different *ap2* alleles used in each study.

Mutations in another gene have curious interactions with *lfy* mutations. Plants with mutations at the *HANABA TARANU* (*HAN*) locus develop flowers with reduced numbers of floral organs (especially petals and stamens) and display disrupted ovule development leading to greatly reduced fertility (Fig. 7A) (H. Sakai and Meyerowitz, unpubl.). However, doubly mutant *lfy han* plants fail to develop any flowers at all; instead, the inflorescence meristem generates filamentous organs. It is unclear whether the filaments represent transformed flowers or some default state as the result of insufficient developmental cues.

Plants homozygous for strong mutant alleles of the *CLAVATA* loci (*CLV1*, *CLV2*, and *CLV3*) exhibit dramatic increases in apical and floral meristem size (see Floral Meristem Structure below) (Fig. 7C) (Leyser and Furner 1992; Clark et al. 1993; S.E. Clark et al., unpubl.). Flowers of *ap1 clv1* doubly mutant plants occasionally develop entire new inflorescences in the center of the flowers (Clark et al. 1993). In addition, plants doubly mutant for *clv1* and *lfy* exhibit a dramatic phenotype (Fig. 7E) (Clark et al. 1993). First, the inflorescences generate mainly filamentous structures and leaves. Second, the flowers that are generated display almost complete flower to shoot transformation: Organs are initiated spirally; outer organs are leaf-like, flanked by stipules, and bear stellate trichomes; inner organs are leaf-like with patches of stigmatic tissue; there is internode elongation between organs; and the flowers are indeterminate. Curiously, the *clv1 lfy* doubly mutant plants show greater enlargement of the apical meristem than do plants mutant for *clv1* alone.

*clv1 han* double mutant inflorescences also produce primarily filamentous organs and leaves (S.E. Clark and E.M. Meyerowitz, unpubl.). Plants singly mutant for *fl54* develop flowers with extra sepals and reduced and altered whorl-2 and -3 organs (Fig. 7B) (Komaki et al. 1988). Whereas inflorescences of *fl54* plants occasionally bear filamentous structures or leaves, those of *clv1 fl54* double mutant plants develop primarily filamentous organs and leaves (Fig. 7F).

Both *lfy ap1* and *lfy clv1* double mutant plants exhibit almost complete transformations of floral meristems into shoot meristems; however, the opposite phenotype is observed in plants with mutations in the *TERMINAL FLOWER* (*TFL*) gene (Shannon and Meeks-Wagner 1991, 1993; Alvarez et al. 1992). *tfl* mutant plants undergo the transition to flowering earlier than wild type, and the inflorescence develops only a few floral meristems before terminating in a flower. In this case, both the primary and secondary shoot meristems are transformed into flowers. *TFL* has been postulated to play a role in the putative inhibitory field purported to repress floral formation in the center of the apical meristem (Alvarez et al. 1992). Others have postulated that *TFL* plays an active role in suppressing *LFY* and *AP1* expression (Shannon and Meeks-Wagner 1993). These two ideas are not mutually exclusive. One feature of the *tfl* phenotype that complicates an interpretation is that the inflorescence

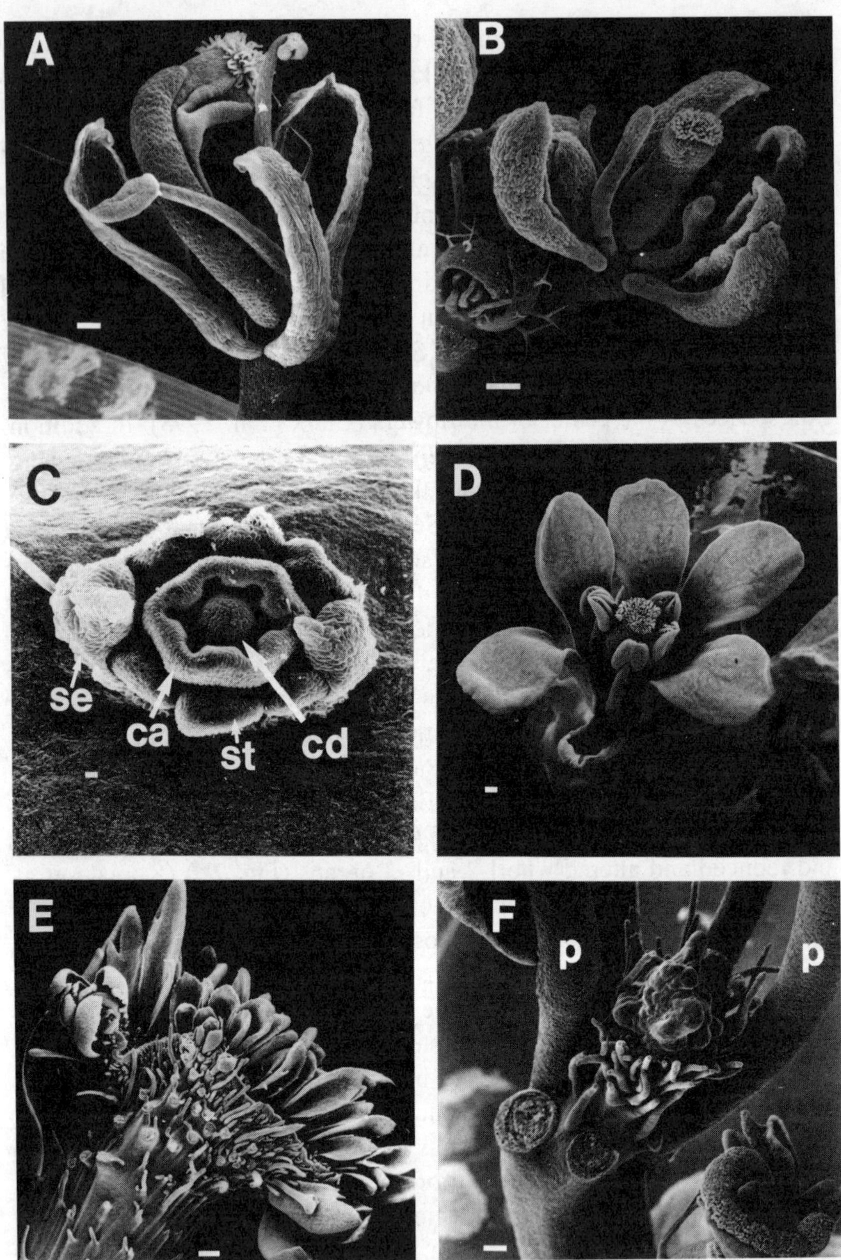



*Figure 7* Loci affecting organ number. *han* (*A*), *fl54* (*B*), *clv1-4* (*C*), and *pan* (*D*) mutants all develop flowers with altered numbers of organs (see text). In *C*, several sepal primordia (se) have been removed to reveal the additional stamens (st), carpels (ca), and still-proliferating central dome (cd). Inflorescences of *clv1-4 lfy-1* (*E*) and *clv1-4 fl54* (*F*) double mutants develop primarily filamentous organs and leaves. (p) Flower pedicels. Bars: 10 μm (*C*), 100 μm for all others.

termination is almost completely suppressed by growth at 16°C or under short-day conditions (Shannon and Meeks-Wagner 1991; Alvarez et al. 1992). Finally, triple mutant combinations have shown that *tfl* inflorescences do not terminate if floral specification is sufficiently disrupted either by *lfy* and *ap1* mutations (Shannon and Meeks-Wagner 1993) or by *lfy* and *clv1* mutations (S.E. Clark and E.M. Meyerowitz, unpubl.).

## Conclusions

Although the number of genes discovered to be necessary for floral meristem specification is growing, it is not clear that this is their primary role. It may be, instead, that normal processes of cell division or cellular communication (perhaps controlled by genes like *CLV1*) are required for floral specification and that the requirement for wild-type operation of these processes is more stringent when floral identity genes like *LFY* are mutant. This might explain why the effect on floral meristem identity of these genes is observed in double mutant combinations with *lfy*.

Another unknown involves the activation of the floral meristem identity genes. *LFY* is first expressed in a small group of cells on the flanks of the apical meristem and presumably initiates the floral development program in those cells. What activates *LFY*, so that it is expressed only in those cells, and only later in development after the leaves have been specified? Clearly, there must be some external floral signal. What is its source, and what genes are involved in the signaling? Although one imagines the late-flowering genes to be involved (see Martínez-Zapater et al., this volume), direct connections between flowering time genes and floral identity genes are not yet established.

## FLORAL ORGAN IDENTITY

Genes directing floral organ identity are discussed in the framework of a model developed to explain their interactions (Bowman et al. 1991b; Coen and Meyerowitz 1991; Meyerowitz et al. 1991). This model was developed on the basis of the phenotypes of *apetala2* (*ap2*), *apetala3* (*ap3*), *pistillata* (*pi*), and *agamous* (*ag*) mutations (Bowman et al. 1989, 1991b). The model is also supported by the expression patterns of three of these genes and by the identification of several new genes involved in organ identity.

## The Model

The model proposes that there are three main activities directing floral organ identity: **A**, **B**, and **C** in Figure 8. Each activity acts in two adjacent whorls and is partly or wholly responsible for the identity of organs in those two whorls. **A** activity functions in whorls 1 and 2, **B** activity functions in whorls 2 and 3, and **C** activity functions in whorls 3 and 4. The model proposes that the identity of organs that develop in a particular whorl depends on the combination of activities present in that whorl. Thus **A** activity alone yields sepals; **A** and **B** activities yield petals; **B** and **C** activities yield stamens; and **C** activity alone yields carpels.

A second feature of the model involves the regulation of **A** and **C**. Whereas the domains of **A** and **C** are complementary, it is further proposed that each acts to repress the other's activity. Thus, if **A** activity were to be removed, then **C** activity would be present in whorls 1, 2, 3, and 4, and if **C** activity were to be removed, then **A** activity would be present in all four whorls.

## The Mutants

The floral organ identity genes, mutant alleles of which were the basis for the model, are classified as **A**, **B**, or **C** function genes. *APETALA2*

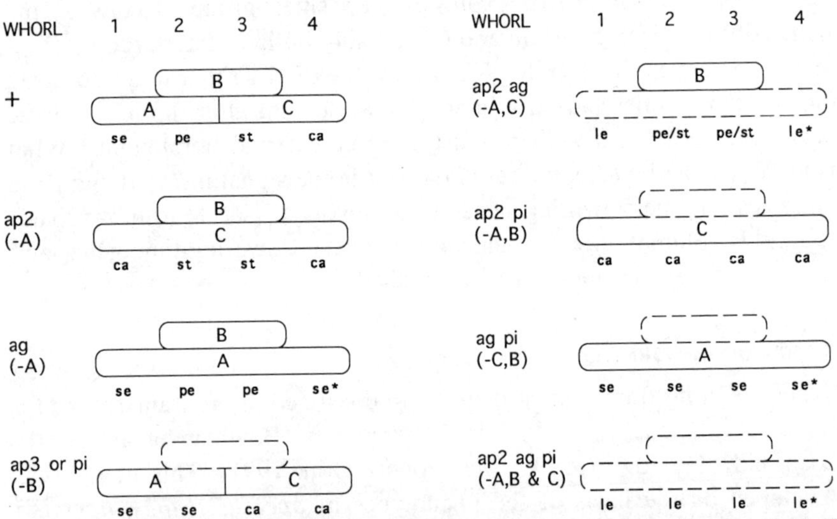


*Figure 8*  A model for the specification of floral organ identity. To the right of each genotype (+, wild type) the predicted domains in the four whorls of the **A**, **B**, and **C** function genes are shown. Below this is the resulting identity of the organs in each whorl. (se) Sepal, (pe) petal, (st) stamen, (ca) carpel, (pe/st) petal/stamen mosaic, (le) leaf. Various single, double, and triple mutants affecting **A**, **B**, and/or **C** function are presented, with the resulting changes in activity and organ identity shown to the right. *ag* mutants result in a new flower in the place of whorl 4, and this is indicated with an asterisk. (Modified from Bowman et al. 1991b.)

(*AP2*) is an **A** function gene. *APETALA3* (*AP3*) and *PISTILLATA* (*PI*) are **B** function genes. *AGAMOUS* (*AG*) is a **C** function gene. *APETALA1* (*AP1*; see above) has been added to this list as an **A** function gene (independently of its earlier action as a meristem identity gene).

A function genes are characterized by mutant alleles with changes in sepal and petal organ identity. Plants carrying a strong *AP2* mutant allele, *ap2-2*, develop flowers with carpels in whorl 1, no organs in whorl 2, and stamens and carpels in whorls 3 and 4, respectively (Fig. 9C) (Bowman et al. 1991b). Plants homozygous for a weak *AP2* mutant allele, *ap2-1*, develop leaves with occasional carpelloid features in whorl 1, staminoid petals or stamens in whorl 2, and stamens and carpels in whorls 3 and 4, respectively (Fig. 9A, B) (Bowman et al. 1989). Other alleles give various degrees of intermediate phenotypes, but the general effect of *ap2* mutants is to convert sepals to carpels and petals to stamens, along with reducing whorl-1, -2, and -3 organ number. *AP1* mutants are in some ways similar in phenotype (Irish and Sussex 1990; Bowman et al. 1993). *ap1* plants develop flowers with leaf-like sepals in whorl 1 (Fig. 6C). Whorl-2 organs in *ap1* flowers are absent or are staminoid petals or stamens. Except for the organ number defect, the homeotic conversions observed in *ap2* and *ap1* mutants fit with the model. For strong *ap2* alleles (e.g., *ap2-2*), loss of **A** function would allow the extension of **C** function to whorls 1 and 2, leading to carpels and stamens, respectively. For weak *ap2* alleles (*ap2-1*) and *ap1* mutants, partial loss of **A** function would allow the extension of **C** function only to whorl 2, leading to stamens, whereas the absence of **A**, **B**, and **C** function in whorl 1 might lead to a default pathway for organ development (i.e., leaves; see below).

Two **B** function genes have been identified, and mutant alleles of each have nearly identical phenotypes. Plants homozygous for strong mutant alleles of *AP3* (*ap3-3*) and *PI* (*pi-1*) develop flowers that exhibit petal-to-sepal conversion in whorl 2 and stamen-to-carpel conversion in whorl 3, whereas whorls 1 and 4 develop sepals and carpels, respectively Fig. 9D–F) (Bowman et al. 1989, 1991b; Jack et al. 1992). This agrees with the model, because, in *ap3* or *pi* mutant flowers, whorls 1 and 2 would both have **A** activity only, leading to sepals, and whorls 3 and 4 would have **C** activity only, leading to carpels (Fig. 8).

*AG* has two activities: one as a **C** function gene, the second to provide floral determinacy. Thus, plants mutant for *AG* develop flowers with normal whorls 1 and 2, but exhibit stamen-to-petal transformation in whorl 3. Whorl 4 is converted into an entirely new flower, transforming the normal (sepal, petal, stamen, carpel) organ arrangement to an indeterminate (sepal, petal, petal, sepal, petal, petal...) or (sepal, petal, pet-al)$_n$ phenotype (Fig. 10A, B) (Bowman et al. 1989). The model would

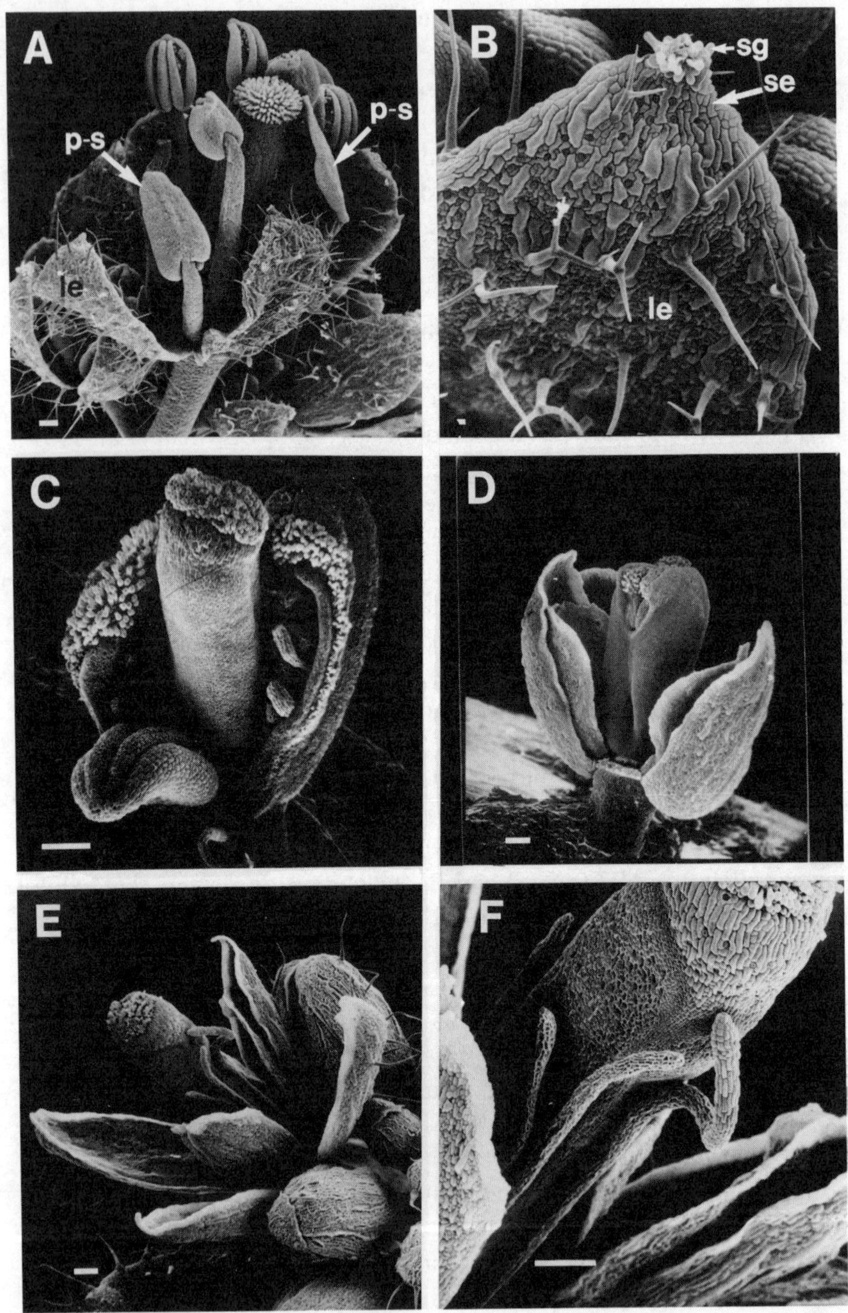



*Figure 9  (See facing page for legend.)*

predict that the lack of **C** activity from the *ag* mutation would result in **A** activity in all four whorls, leading to petals in whorl 3 and sepals in whorl 4.

### Double and Triple Mutants

The model makes specific predictions about double mutant combinations. Removing two activities through mutation would result in a single activity directing floral organ identity. For example, the removal of **A** and **C** activities (e.g., *ap2 ag* double mutant) would leave only **B** activity in whorls 2 and 3 (Fig. 8). The resulting flowers would develop with ground state organs (leaves) in whorls 1 and 4, and petal/stamen hybrids in whorls 2 and 3. This is indeed the phenotype of *ap2 ag* double mutants, with added indeterminacy due to *ag*: (leaves, petal/stamens, petal/stamens)$_n$ (Bowman et al. 1991b). **C** activity alone (e.g., *ap2 pi* double mutant) leads to flowers composed almost entirely of carpels (occasionally whorl-1 lateral organs develop as leaves) (Figs. 8, 10E) (Bowman et al. 1991b). **A** activity alone (e.g., *ag pi* double mutant) leads to an indeterminate flower consisting entirely of sepals (Fig. 8) (Bowman et al. 1991b). Flowers that develop on triple mutant plants in which all three activities have been removed would be predicted to lack the key developmental cues directing flower-specific organ identity. Examinations of *ap2 pi ag* or *ap2 ap3 ag* triple mutant flowers reveal that these flowers develop leaves in a whorled pattern, with occasional patches of stigmatic tissue present (Fig. 10F) (Bowman et al. 1991b). That the leaves are in a whorled pattern shows that organ identity, but not floral identity, is regulated by the **A**, **B**, and **C** activity genes.

### Unaddressed Phenotypes

One feature of the mutant phenotypes unaddressed by the model concerns the changes in organ number that result from the removal of **A**, **B**,

*Figure 9*  Homeotic mutants *ap2*, *pi*, and *ap3*. *ap2-1* flowers (*A,B*) exhibit altered identity of whorl-1 and -2 organs. *B* shows whorl-1 leaf-like organs with regions of leaf cells (le), sepal cells (se), and stigmatic tissue (sg). Stronger *ap2-2* mutants (*C*; photo by J.L. Bowman) develop carpels in whorl 1 and 4, and an occasional stamen in between. In *D* (photo by T. Jack), a whorl-1 and a whorl-2 sepal have been removed from this *ap3-3* flower to reveal sepals in whorls 1 and 2 and carpels in whorls 3 and 4. *pi-1* flowers (*E*) develop with sepals in whorl 2 and filamentous organs or carpels in whorl 3. A closer view of the filamentous organs is shown in *F*. (p-s) Petal-stamen hybrid organs. Bars: 10 μm (*B*), 100 μm for all others.

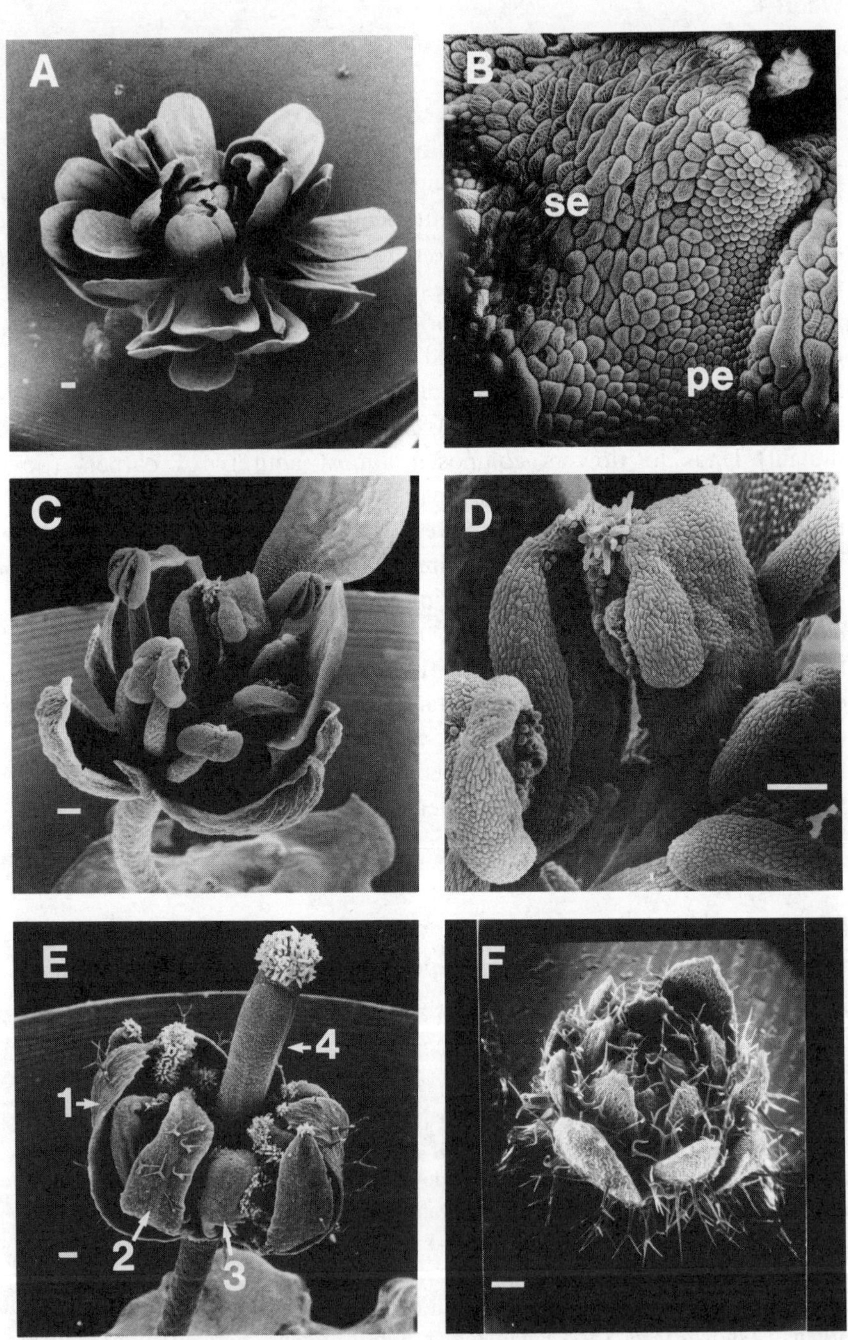



*Figure 10  (See facing page for legend.)*

or **C** activities. Loss of **A** function accompanies the most dramatic reductions in organ number. Strong *ap2* mutants, for example, develop flowers with reduced numbers of whorl-1 organs, and almost entirely absent whorl-2 and -3 organs, and weak *ap2* mutants display reduced numbers of whorl-2 and -3 organs. *ap1* mutants also exhibit reduced numbers of whorl-1 and -2 organs. The reduction of whorl-1 and -2 organ numbers in these mutants might be postulated, for two reasons, to be the result of ectopic expression of *AG* in these two whorls. First, the normal function of *AG* in whorl 4 appears to be to repress further growth of the floral meristem, so that ectopic expression of *AG* in whorls 1 and 2 may prematurely repress organ growth in those two whorls. Second, some of the reduction of organ number in *ap2* mutants can be reversed by eliminating AG activity (i.e., by crossing in a mutant *ag* allele), suggesting that the reduction in organ number in *ap2* mutants requires *AG* (Bowman et al. 1991b). One fact unexplained by this postulate is why *AG* would reduce whorl-3 organ number in the *ap2* mutant, since *AG* is normally expressed in whorl 3 in wild-type flowers without causing any reductions in organ number.

Another phenotype not explained by the model is the indeterminacy of *ag* flowers. One might postulate that *AG* is simply required to terminate the floral meristem, but, clearly, the phenotype is more complex than that. *ag* flowers in an *ERECTA* background, for example, display pedicel elongation after the development of each three whorls of organs, demonstrating that the fourth whorl is transformed into a new flower. The model does not explain why *ag* mutations cause a replacement of whorl 4 with flowers.

## The Genes

Most of the genes involved in the model have been cloned and analyzed. *AG* was the first, and it was cloned by T-DNA insertion mutagenesis

*Figure 10   ag, sup,* and double and triple mutants. *ag* flowers (*A,B*) exhibit stamen-to-petal transformation and replace the fourth whorl with a new flower, thus leading to an indeterminate flower. A mosaic organ from an *ag* flower containing regions of petal (pe) and sepal (se) tissue is shown in *B*. *sup* flowers (*C,D*) develop additional stamens at the expense of carpels. Mosaic stamen/carpel organs from a *sup* flower are shown in *D*. (*E*) An *ap2-1 ap3-1* double mutant flower, with carpelloid leaves developing whorls 1 (1), and 2 (2), staminoid carpels developing in whorl 3 (3), and carpels developing in whorl 4 (4). An *ap2-1 ap3-1 ag-1* triple mutant flower is shown in *F* (photo by J.L. Bowman). Note that all organs are leaves. Bars: 10 μm (*B*), 100 μm for all others.

(Yanofsky et al. 1990). The predicted coding region contains two identifiable domains. The amino-terminal domain shows homology with the yeast transcription factor MCM1 (Passmore et al. 1988), human transcription factor SRF (Norman et al. 1988), and the *Antirrhinum* floral organ identity gene DEF A (see below) (Schwarz-Sommer et al. 1992). Because of its homology, the region that codes for this domain was termed the MADS box (*M*CM1, *A*G, *D*EF A, *S*RF) (Schwarz-Sommer et al. 1990). This domain, in SRF and MCM1, has been shown to have DNA-binding and protein dimerization activities (Jarvis et al. 1989; Mueller and Nordheim 1991; Pollock and Treisman 1991). The MADS domain lacks homology with traditional DNA-binding domains, such as the helix-loop-helix, zinc-finger, or homeodomain.

The second domain has limited homology with the coiled-coil region of human keratin and is encoded by the K box (Ma et al. 1991). SRF, MCM1, and other MADS domain-containing proteins from yeast and animals do not have a K domain, but all plant MADS box proteins also contain this domain at a conserved distance carboxy-terminal to the MADS box. The homology of various plant K domains with both human keratin and each other is quite limited, but all retain the predicted ability to form amphipathic $\alpha$ helices, which are potentially involved in protein-protein interactions, such as seen in leucine-zipper proteins.

A site for SRF DNA binding, the serum response element (SRE), has been identified (Pollock and Treisman 1991), and a site for MCM1 DNA binding, the P-box, has also been identified (Jarvis et al. 1989). Significantly, proteins for the *Antirrhinum* orthologs for AP3 and PI, DEF A and GLO, are capable of binding as heterodimers to the yeast STE6 promoter, which contains a P-box sequence (Tröbner et al. 1992). This suggests that AP3 and PI may act as heterodimers to activate downstream genes as well as their own expression (see below). DNA-binding studies have identified consensus sequences to which the MADS domain portion of AG will bind (Shiraishi et al. 1993). This consensus sequence, 5′-TT(A/T/G)CC(A/T)$_6$GG(A/T/C)AA-3′, is quite similar to MCM1-binding sites in yeast. Furthermore, the AG MADS domain binds in vitro to the MCM1-binding site on the yeast STE6 promoter (Shiraishi et al. 1993). A separate analysis of AG binding using a nearly full-length protein identified a similar consensus sequence: 5′-TT(A/T)CC(A/T)(A/t)$_2$(T/A)NNGG(A/T/C)(A/t)$_2$-3′ (Huang et al. 1993).

The *Arabidopsis* family of MADS box genes is large and includes many of the critical loci involved in plant development. As mentioned above, *AG* was the first *Arabidopsis* MADS box gene cloned. Low-stringency hybridization of the *AG* MADS box to *Arabidopsis* genomic DNA revealed that it was part of a large gene family (Ma et al. 1991).

Both *AP1* and *CAL* have been cloned by their homology with *AG* (Mandel et al. 1992a; S. Kempin and M. Yanofsky, pers. comm.). Six other *AGL*s have been reported, and at least an equal number remain (Ma et al. 1991; G.N. Drews et al., unpubl.). *AP3* and *PI* are also MADS box proteins and were cloned by their homology with their *Antirrhinum* orthologs *DEF A* and *GLO* (Sommer et al. 1990; Jack et al. 1992; Tröbner et al. 1992; K. Goto and E.M. Meyerowitz, unpubl.).

Recently, the cloning of *AP2* has been reported (Okamuro et al. 1993). Although the sequence is not yet available, it is reported to code for a novel protein with no homology with known regulatory genes, but with a nuclear-localization signal. Its expression throughout the apical and floral meristem (Okamuro et al. 1993) suggests that it is subjected to some sort of posttranscriptional control, a feature already demonstrated for *AP3* regulation (see below).

### Expression Patterns

Having the genes responsible for floral identity in hand has allowed investigators to determine if the expression patterns correspond to the **A**, **B**, and **C** activities outlined in the model. There is no reason to assume, a priori, that RNA expression patterns mirror domains of activity, but the results show that some of the genes are primarily controlled at the transcription level. An additional problem with RNA expression analysis is that the homeotic genes analyzed to date begin their expression in stage-3 flowers. Stage-3 flowers are characterized by the initiation of sepal primordia, and at this stage, the entire floral meristem is only 50–60 μm across. This small size makes it difficult to accurately identify the precise whorl boundaries. Nonetheless, examinations of RNA expression patterns have provided useful information.

The domain of function for *AG* comprises whorls 3 and 4, based on the mutant phenotype. In situ hybridization experiments revealed that *AG* RNA is expressed exclusively in the anlage of whorls 3 and 4 early in floral development (Drews et al. 1991). Beginning at stage 3, with the initiation of the sepal primordia, *AG* RNA is detected in the center of the floral meristem, in the cells that will give rise to whorls 3 and 4. *AG* continues to be expressed in young stamens and carpels. In later flower development, its expression becomes confined to the stigmatic papillae and the inner cell layer of the inner integument around the embryo sac in the developing ovules (Bowman et al. 1991a).

*AP3* and *PI*, being **B** function genes, would be predicted to have overlapping expression in whorls 2 and 3 only. *AP3* is expressed starting at stage 3 in the anlage for whorls 2 and 3 (Jack et al. 1992). *PI* is expressed beginning at stage 3 in whorls 2, 3, and, to a lesser extent, 4 (K. Goto and E.M. Meyerowitz, unpubl.). Later, in wild-type flowers, *AP3* and *PI* expression becomes restricted to whorls 2 and 3.

The late pattern of *AP1* expression complements that of *AG*. Early, during stages 1 and 2, *AP1* is expressed in the entire floral primordium (Mandel et al. 1992a). This expression is probably related to its early role in establishing floral meristem identity (see above). Beginning at stage 3, *AP1* is expressed only in whorls 1 and 2, whereas *AG*, as noted above, is expressed in whorls 3 and 4. This later *AP1* expression is presumably related to its role in sepal and petal organ identity.

**Factors Controlling Expression**

According to the model, *AP2* represses *AG* activity in whorls 1 and 2. Thus, the model predicts that *AG* expression would expand into whorls 1 and 2 in *ap2* mutant flowers, and that is indeed what is observed (Drews et al. 1991). *AG* expression is also altered in *clv1* flowers (Clark et al. 1993).

*AP3* and *PI* are initially expressed independently of each other. In other words, *AP3* expression is initiated normally in *pi* flowers, and *PI* expression is initiated normally in *ap3* flowers (Jack et al. 1992; K. Goto and E.M. Meyerowitz, unpubl.). However, the expression of *AP3* and *PI* is not maintained unless both *AP3* and *PI* are present. Thus, in later *ap3* flowers, neither *AP3* nor *PI* is expressed (Jack et al. 1994; K. Goto and E.M. Meyerowitz, unpubl.). In later *pi* flowers, *PI* is not expressed, and *AP3* is only expressed in the second whorl (Jack et al. 1992; Goto and Meyerowitz 1994).

Another gene controlling *AP3* expression is *SUPERMAN* (*SUP*) (Schultz et al. 1991; Bowman et al. 1992). *sup* flowers are characterized by an increase in stamen number and a concomitant reduction in carpel tissue. This suggests that the domain of *AP3* expression is expanded to a portion of the fourth whorl in *sup* flowers. In situ hybridizations bear this out: In stage-3 *sup* flowers, *AP3* is expressed nearly to the center of the floral meristem.

*lfy* flowers are composed of sepals and carpels. Because these floral organs are characteristic of *ap3* flowers, it might be postulated that *LFY* is somehow required for *AP3* expression. Indeed, *AP3* and *PI* expression is reduced in *lfy* plants and almost entirely absent from *lfy ap1* double mutant plants (Weigel and Meyerowitz 1993). *lfy* and *ap1* mutations also influence *AG* expression. *AG* is expressed in the center of *lfy* and *ap1* flowers but is not expressed in *lfy ap1* double mutant flowers (Weigel and Meyerowitz 1993). This again indicates a role for *LFY* and *AP1* in the initial activation of the floral homeotic genes.

**Ectopic Expression and Posttranscriptional Control**

Ectopic expression of **A**, **B**, and/or **C** activities allows further tests of the model and the potential to custom-design flowers with nearly every possible combination of floral organs. The first examples of ectopic expression of floral organ identity genes were those of *AG* overexpression. In one study, *Arabidopsis AG* was expressed in *Arabidopsis* under control of a constitutive promoter (Mizukami and Ma 1992). This ectopic expression caused the flowers to display phenotypes similar to *ap2* mutants. This is consistent with the model, which predicts that *AG* expression in whorls 1 and 2 will repress *AP2* activity in those two whorls, leading to a phenotype similar to *ap2* mutants. Parallel studies in tobacco have also proved informative. In this case, the *AG* ortholog from *Brassica napus*, a species closely related to *Arabidopsis*, was expressed under a constitutive promoter in tobacco (Mandel et al. 1992b). This led to homeotic transformations of sepals to carpels and petals to stamens, identical phenotypes to *ap2* mutants in *Arabidopsis*. The ability of a floral organ identity gene to perform identical functions in a distantly related plant species implies that many features of flower development are likely to be widely conserved.

*AP3* has also been ectopically expressed in *Arabidopsis* (Jack et al. 1994). Plants with *AP3* under the control of a constitutive promoter exhibit *AP3* RNA expression in all four whorls. These flowers display carpel-to-stamen transformations, but have sepals in whorl 1. Because *PI* is transiently expressed in whorl 4, the addition of *AP3* expression in whorl 4 would establish the putative AP3-PI autoregulatory pathway, leading to stamens in whorl 4. Because there is no *PI* expression in whorl 1, sepals would be expected to develop there because both *AP3* and *PI* are required for B function. Surprisingly, in situ anti-AP3 antibody analysis of the flowers expressing *AP3* RNA in all four whorls (plus pedicels and stems) demonstrates that AP3 protein is only present at detectable levels in whorls 2, 3, and 4. Thus, there is likely some mechanism for posttranscriptional regulation of *AP3*. This adds a whole new dimension to the regulation of these genes and makes in situ RNA analysis a less-than-perfect predictor of actual protein activity.

**Other Genes**

Another potential **A** function gene is *LEUNIG* (Z. Liu et al., unpubl.). This locus was identified by a mutant allele that enhances the weak *ap2-1* phenotype, with the *leunig ap2-1* double mutant appearing similar to the strong *ap2-2* in phenotype. Plants mutant for *leunig* alone develop flowers with weak sepal-to-carpel and weak sepal-to-stamen transformations in whorl 1 and all other floral organs displaying an elongated phenotype. These transformations appear to be correlated with ectopic expression of *AG*.

Two newly identified loci are suspected of playing roles in carpel development. Mutant alleles of the two loci, *CRABS CLAW* and *SPATULA*, develop flowers with normal whorl-1, -2, and -3 organs but with slightly disrupted gynoecia (J. Alvarez and D. R. Smyth, pers. comm.). The *crabs claw spatula* double mutant, however, develops unfused leaf-like organs instead of carpels, indicating redundant roles for *CRABS CLAW* and *SPATULA* in carpel development. Because **C** function establishes the identity of stamens and carpels, these two new loci carry out a subset of **C** function activities. It is expected that many additional genes are involved in floral organ identity: as new genes of the **A**, **B**, and **C** classes; as regulators of these genes (such as *LFY* and *SUP*); and as downstream genes involved in the differentiation of specific organs and cell types.

### Cell Autonomous or Non-autonomous?

An important question that remains to be addressed concerns whether the genes or activities responsible for floral organ identity are cell autonomous or cell non-autonomous. Shoot apices of many plant species consist of three cell layers that, in general, remain clonally distinct throughout development. These same three cell layers are also often preserved in leaves and flowers (Satina et al. 1940). One question is whether expression of an organ identity gene in one layer, but not others, could induce correct development in all three layers.

There are several examples from "double" flower mutants (similar to *ag* in that there is an increase in petals often at the expense of whorl-3 and -4 organs) which indicate that the $L_1$ can have an inductive influence on the development of the $L_2$ and $L_3$ (i.e., the flower phenotype can be predicted by the $L_1$ genotype). In two cases, adventitious root cuttings from plants with double flowers gave rise to plants with single flowers (Bateson 1916; Zimmermann and Hitchcock 1951). Because adventitious roots usually derive from the $L_2$ or $L_3$, this suggests that the original plants with double flowers were chimeric, with the $L_1$ layer in each able to induce all cell layers to form a double flower. In another example, a chimera was generated between a wild-type *Camellia sasanqua* and a double *Camellia japonica*, with *C. sasanqua* $L_1$ cells and *C. japonica* $L_2$ and $L_3$ cells (Stewart et al. 1972). In this case, the wild-type $L_1$ cells were able to direct development of stamens and carpels, again indicating an inductive influence from the $L_1$.

**FLORAL MERISTEM STRUCTURE**

Although rapid progress has been made toward understanding floral organ specification, little is known about the process of organ initiation or how the structure of the floral meristem is established. Excellent work has been performed to describe morphologically the stages of wild-type development, but we know little of the genes involved in this process.

Mutations in the floral organ identity genes can affect floral organ number, and their proper expression is required for normal organ initiation. However, *ap2-1 pi-1 ag-1* triple mutant flowers develop with largely normal numbers of organs in the first three whorls, with, of course, a new flower in whorl 4 due to the mutation in *AG* (Bowman et al. 1991b). This suggests that an entirely separate set of loci is necessary for establishing floral meristem structure (Fig. 5).

How can we identify such loci? The most obvious way to identify genes that play a specific role in floral meristem structure is to look for mutants where the numbers of floral organs are altered but the identity of floral organs is unaffected. This approach rules out genes that play dual roles in floral meristem structure and floral organ or meristem identity, but it is a useful starting point.

The most extensively characterized loci in this category are the *CLAVATA* (*CLV*) genes: *CLV1*, *CLV2*, and *CLV3* (Koornneef et al. 1983; Leyser and Furner 1992; Clark et al. 1993; S.E. Clark et al, unpubl.). Mutations in any of the *CLV* loci result in two apparent defects in floral meristem structure (Fig. 7C). First, *clv* flowers display an increased number of organs in each of the four whorls, with the inner two whorls of stamens and carpels consistently exhibiting the greatest increases in organ number. Second, *clv* flowers develop one to several additional whorls of carpels interior to the gynoecium formed by the fourth whorl carpels. Analysis of very young *clv* flowers has revealed that, at the earliest stages of organ initiation (stage 3), *clv* flowers are much taller and slightly wider than wild-type flowers at the same stage. This increase in size is due to an increase in cell number. This suggests a number of things about organ initiation. First, a larger floral meristem can give rise to more floral organs, not larger floral organs. Despite the fact that a variable number of additional organs are formed in each whorl of *clv* flowers, very few mosaic or hybrid organs are observed. This suggests that whorl boundaries in *clv* flowers are rigidly maintained and that the additional organs are rarely formed straddling two whorls, or, if they are formed across two whorls, they are quickly assigned to a single whorl. Another interesting feature of young *clv* flowers is that although *AG* is expressed in the anlage for whorls 3 and 4, it is not expressed in the cells in the center of the floral meristem that give rise to the additional whorls

(Clark et al. 1993). This lack of *AG* expression may be correlated with the additional proliferation of these cells, because *AG* normally acts to make the floral meristem determinate.

As mentioned above in the discussion of floral meristem identity, double mutant combinations of *clv* with *lfy*, *han*, and *fl54* all result in inflorescence meristems that produce primarily filamentous organs and leaves (Clark et al. 1993; S.E. Clark and E.M. Meyerowitz, unpubl.). Many double mutant combinations between these loci (*CLV*, *LFY*, *HAN*, and *FL54*) result primarily in filamentous organs and leaves developing on the inflorescence. One interpretation of these phenotypes is that the other loci, namely *LFY*, *FL54*, and *HAN*, are also involved in establishing floral meristem structure, and that the additive disruption from two mutant loci prevents the floral primordium from developing (Fig. 5). Indeed, *lfy* and *han* have clear defects in floral meristem structuring. *LFY* plays a role in establishing the correct whorl patterning (Huala and Sussex 1992; Weigel et al. 1992), and *HAN* is necessary for proper organ numbers (H. Sakai and E.M. Meyerowitz, unpubl.). *fl54* flowers also display changes in organ number, with extra sepals and carpels, and reduced and altered petals and stamens (Fig. 7B) (Komaki et al. 1988). An alternative explanation is that these loci play roles in floral meristem identity, and that their defects in floral meristem structure are an indirect consequence of changes in floral meristem identity.

In addition, mutants for *CLV*, *HAN*, and *FL54* have defects in their apical meristem structure. The apical meristem is enlarged in *clv* mutants (Clark et al. 1993) and reduced in *han* mutants (H. Sakai et al., unpubl.), and flower phyllotaxis is disrupted in *fl54* mutants (Komaki et al. 1988). This suggests at least two possibilities. One is that the apical meristem utilizes some of the same loci that the floral meristem uses for establishing meristem structure. Alternatively, these loci control some underlying cellular process, such as cell division, that has indirect, albeit profound, effects on meristem structure and plant development in general.

A gene that codes for a protein kinase, *TOUSLED*, has been identified which also affects floral organ number without affecting organ identity (Roe et al. 1993). *TOUSLED* mutants (*tsl*) exhibit nearly random reductions in the number of organs in all four whorls. In addition, the normal bilateral symmetry of wild-type organ initiation is lost. *TSL* is unlikely to play a specific role in floral meristem structure, however, because it is expressed in most plant tissues and because the mutation alters leaf shape, timing of flowering, and size and shape of floral organs.

One feature of the Brassicaceae family, of which *Arabidopsis* is a member, is that the flowers are tetrameric, meaning that they have four sepals and four petals. The six stamens are viewed either as two whorls of four, with two not developing in the outer whorl, or one whorl of four, with two dividing to give a total of six. Either way, this floral structure is typical of the entire family. Pentamerous flowers are quite common in other families. Pentamerous flowers consist of five sepals, five petals, and one to several whorls of five stamens. Interestingly, mutations in a single locus of *Arabidopsis*, *PERIANTHIA* (*PAN*), transform *Arabidopsis* flowers from tetrameric to largely pentamerous (Fig. 7D) (M.P. Running and E.M. Meyerowitz, unpubl.). Analysis of this locus may provide insights into the way in which floral organ patterning is controlled in plants.

## CONCLUSIONS AND PERSPECTIVES

### Summary

The efforts of researchers during the past several years have brought us much closer to understanding the process of flower development in *Arabidopsis*. *LFY* and *AP1* are two critical genes involved in the process of floral meristem specification, and their activities set in motion the activation of several of the homeotic genes. *LFY* RNA expression patterns have revealed that the early floral primordium is composed of a few cells in the apical meristem. *AP1* is expressed later when the floral primordium is a differentiated bulge on the flank of the apical meristem, but *AP1* expression has a significant period of overlap with *LFY* expression. The simultaneous expression of *LFY* and *AP1*, as well as the phenotypes of *lfy ap1* double mutants, indicates that the respective proteins may work together to impart floral meristem identity.

The link between these floral identity genes and the organ identity genes remains unclear. The link may be direct, which would suggest that *LFY* and *AP1* act directly to activate the expression of *AG*, *AP3*, and *PI*, among others. However, *LFY* and *AP1* play their primary roles in floral meristem specification, leaving the possibility that this activation may be indirect. Thus, *lfy ap1* flowers may lack *AG* and *AP3* expression because the meristems have lost their floral character.

The roles of the homeotic genes are better defined. The proteins encoded by *AG*, *AP3*, and *PI* contain MADS boxes, with DNA-binding activity, and K boxes, with potential protein-interaction activity. This suggests these genes may act as homo- or heteromultimers to activate and/or repress the activities of other genes. Interestingly, *AP3* and *PI* are required to maintain their own expression, and their *Antirrhinum* orthologs DEF and GLO have been shown to bind DNA as heteromultimers. The roles of the homeotic genes fit a model hypothesizing the combinatorial action of homeotic genes to specify organ identity. This simple model has predicted accurately the phenotypes of double and triple mutants and has also predicted the domains of expression of the homeotic genes.

The investigation of the control of floral meristem structure that is responsible for the correct number and position of organs has just begun. Analyses of *clv1* mutants indicate that the *clv1* floral meristems are larger and, as a result, give rise to additional organs in each whorl, and to additional whorls of organs. The control of floral meristem structure by *CLV1* seems to be independent of the homeotic genes, as the double mutant phenotypes of *clv1* with *ap2*, *ap3*, *pi*, and *ag* mutations are all additive. Interestingly, mutations in all three *CLV* loci (*CLV1*, *CLV2*, and *CLV3*) result in larger floral meristems and larger apical meristems, suggesting that both meristems may use some of the same genes for their patterning.

## What Is Next?

Many questions remain about the activation and action of the known genes, and of the role that the many new floral genes may play: What activates the initial *LFY* expression in the floral primordium? What restricts *LFY* to small groups of cells in the apical meristem? Is the floral primordium already defined and *LFY* expression limited to the floral primordium, or is the floral primordium defined as those cells that express *LFY*? How does *LFY* act to impart floral specificity? Is it a transcription factor? What proteins act with *LFY* in the floral specification process? Eventually, many of these questions on the early steps of flower development are likely to require looking at the structure and development of the apical meristem, because many of these early steps take place in what was traditionally thought of as the apical meristem. The processes of apical meristem development and floral meristem development are certain to be linked.

We already know a great deal about the expression patterns of the homeotic genes, and in some instances, we know how the domains are established. However, certain things remain unanswered. For example, what prevents *AP3* and *PI* expression in whorl 1? How does *SUP* act to repress *AP3* in whorl 4? How is the posttranscriptional regulation of *AP3* controlled, and are other homeotic genes subjected to posttranscriptional regulation? Most importantly, how do the homeotic genes act to fulfill their postulated roles as **A**, **B**, or **C** activities? Do they all act as homo- or heteromultimers? These possible interactions are complex. For example, the potential dimers in whorl 3 include PI-PI, AP3-AP3, PI-AP3, AG-PI, AG-AP3, and AG-AG. Which, if any, of these function in vivo to specify stamen development? In addition, we know nothing of the downstream genes that are presumed targets of the homeotic (organ identity) genes. How many additional layers of regulatory gene activation separate the organ identity genes from the structural genes whose products distinguish different floral cell types (Fig. 5)?

## Other Systems

Some of the genes mentioned above have been studied both in *Antirrhinum* and *Arabidopsis* (Carpenter and Coen 1990; Coen et al. 1990; Sommer et al. 1990; Schwarz-Sommer et al. 1992; Tröbner et al. 1992; Bradley et al. 1993). These simultaneous approaches have not only benefited the field, but have also been informative from an evolutionary standpoint. Many of the *Antirrhinum* genes studied appear to have roles similar, if not identical, to those of their *Arabidopsis* counterparts. This implies that, despite the considerable evolutionary distance between the two species (a minimum of 70 Myr), mechanisms for flower development are very conserved. Similar genes are being identified in both tomato (Pnueli et al. 1991) and petunia (Angenent et al. 1992; Tsuchimoto et al. 1993; van der Krol et al. 1993). In fact, recent identification of a maize homolog for *AGAMOUS* suggests that it plays a similar role in the even further-diverged monocots (Schmidt et al. 1993). This wide conservation of floral regulators across divergent plant species makes investigations of *Arabidopsis* more significant, as they may provide general models for flower development in most species.

## ACKNOWLEDGMENTS

We dedicate this chapter to the memory of Christopher Bowersox (1985–1994). We thank our colleagues in the Meyerowitz laboratory for critical review of the manuscript. We also thank Hajime Sakai, Mark Running, Zhongchi Liu, David Smyth, and Marty Yanofsky for permission to reference unpublished data. Our work on flower development has been supported by grants from the National Science Foundation, the National Institutes of Health, the Department of Energy, and the Human Frontier Science Program. This material is based in part on work supported by the National Science Foundation under a fellowship awarded to S.E.C. in 1991.

## REFERENCES

Alvarez, J., C.L. Guli, and D.R. Smyth. 1992. *terminal flower*: A gene affecting inflorescence development in *Arabidopsis thaliana*. *Plant J.* **2:** 103–116.

Angenent, G.C., M. Busscher, J. Franken, J.N.M. Mol, and A.J. van Tunen. 1992. Differential expression of two MADS box genes in wild-type and mutant petunia flowers. *Plant Cell* **4:** 983–993.

Bateson, W. 1916. Root cuttings, chimeras and "sports." *J. Genet.* **6** (reprinted in *Scientific papers of William Bateson*, vol. II, 1928. Cambridge University Press, Cambridge, United Kingdom).

Bowman, J.L., G.N. Drews, and E.M. Meyerowitz. 1991a. Expression of the *Arabidopsis* floral homeotic gene *AGAMOUS* is restricted to specific cell types late in flower development. *Plant Cell* **3:** 749–758.

Bowman, J.L., D.R. Smyth, and E.M. Meyerowitz. 1989. Genes directing flower development in *Arabidopsis. Plant Cell* **1:** 37–52.

———. 1991b. Genetic interactions among floral homeotic genes of *Arabidopsis. Development* **112:** 1–20.

Bowman, H.L., J. Alvarez, D. Weigel, E.M. Meyerowitz, and D.R. Smyth. 1993. Control of flower development in *Arabidopsis thaliana* by *APETALA1* and interacting genes. *Development* **119:** 721–743.

Bowman, J.L., H. Sakai, T. Jack, D. Weigel, U. Mayer, and E.M. Meyerowitz. 1992. *SUPERMAN*, a regulator of floral homeotic genes in *Arabidopsis. Development* **114:** 599–615.

Bradley, D., R. Carpenter, H. Sommer, N. Hartley, and E. Coen. 1993. Complementary floral homeotic phenotypes result from opposite orientations of a transposon at the *plena* locus of *Antirrhinum. Cell* **72:** 85–95.

Carpenter, R. and E.C. Coen. 1990. Floral homeotic mutations produced by transposon-mutagenesis in *Antirrhinum majus. Genes. Dev.* **4:** 1483–1493.

Clark, S.E., M.P. Running, and E.M. Meyerowitz. 1993. *CLAVATA1*, a regulator of meristem and flower development in *Arabidopsis. Development* **119:** 397–418.

Coen, E.S. and E.M. Meyerowitz. 1991. The war of the whorls: Genetic interactions controlling flower development. *Nature* **353:** 31–37.

Coen, E.S., J.M. Romero, S. Doyle, R. Elliot, G. Murphy, and R. Carpenter. 1990. *floricaula*: A homeotic gene required for flower development in *Antirrhinum majus. Cell* **63:** 1311–1322.

Drews, G.N., J.L. Bowman, and E.M. Meyerowitz. 1991. Negative regulation of the *Arabidopsis* homeotic gene *AGAMOUS* by the *APETALA2* product. *Cell* **65:** 991–1002.

Dubois, E., J. Bercy, and F. Messenguy. 1987. Characterization of 2 genes, *ARGRI* and *ARGRIII* required for specific regulation of arginine metabolism in yeast. *Mol. Gen. Genet.* **207:** 142–148.

Goto, K. and E.M. Meyerowitz. 1994. Function and regulation of the *Arabidopsis* floral homeotic gene *PISTILLATA. Genes Dev.* **8:** 1548–1560.

Huala, E. and I.M. Sussex. 1992. *LEAFY* interacts with floral homeotic genes to regulate *Arabidopsis* floral development. *Plant Cell* **4:** 901–913.

Huang, H., Y. Mizukami, Y. Hu, and H. Ma. 1993. Isolation and characterization of the binding sequences for the product of the *Arabidopsis* floral homeotic gene *AGAMOUS. Nucleic Acids Res.* **21:** 4769–4776.

Irish, V.F. and I.M. Sussex. 1990. Function of the *apetala-1* gene during *Arabidopsis* floral development. *Plant Cell* **2:** 741–751.

Jack, T., L.L. Brockman, and E.M. Meyerowitz. 1992. The homeotic gene *APETALA3* of *Arabidopsis thaliana* encodes a MADS box and is expressed in petals and stamen. *Cell* **68:** 683–697.

Jack, T., G.L. Fox, and E.M. Meyerowitz. 1994. *Arabidopsis* homeotic gene *APETALA3* ectopic expression: Transcriptional and posttranscriptional regulation determine floral organ identity. *Cell* **76:** 703–716.

Jarvis, E.E., K.L. Clark, and G.F. Sprague. 1989. The yeast transcription activator PRTF, a homolog of the mammalian serum response factor, is encoded by the *MCM1* gene. *Genes Dev.* **3:** 936–945.

Komaki, M.K., K. Okada, E. Nishino, and Y. Shimura. 1988. Isolation and characterization of novel mutants *Arabidopsis thaliana* defective in flower development. *Development* **104:** 195–203.

Koornneef, M., J. van Eden, C.J. Hanhart, P. Stam, F.J. Braaksma, and W.J. Feenstra. 1983. Linkage map of *Arabidopsis thaliana*. *J. Hered.* **74:** 265–272.

Leyser, H.M.O. and I.J. Furner. 1992. Characterization of three shoot apical meristem mutants of *Arabidopsis thaliana*. *Development* **116:** 397–403.

Ma, H., M.F. Yanofsky, and E.M. Meyerowitz. 1991. *AGL1-AGL6*, an *Arabidopsis* gene family with similarity to floral homeotic and transcription factor genes. *Genes Dev.* **5:** 484–495.

Mandel, M.A., C. Gustafson-Brown, B. Savidge, and M.F. Yanofsky. 1992a. Molecular characterization of the *Arabidopsis* floral homeotic gene *APETALA1*. *Nature* **360:** 273–277.

Mandel, M.A., J.L. Bowman, S.A. Kempin, H. Ma, E.M. Meyerowitz, and M.F. Yanofsky. 1992b. Manipulation of flower structure in transgeneic tobacco. *Cell* **71:** 133–143.

Meyerowitz, E.M., J.L. Bowman, L.L. Brockman, G.N. Drews, T. Jack, L.E. Lieburth, and D. Weigel. 1991. A genetic and molecular model for flower development in *Arabidopsis thaliana*. *Dev. Suppl.* **1:** 157–167.

Mizukami, Y. and H. Ma. 1992. Ectopic expression of the floral homeotic gene *AGAMOUS* transgenic *Arabidopsis* plants alters floral organ identity. *Cell* **71:** 119–131.

Mueller, C.G.F. and A. Nordheim. 1991. A protein domain conserved between yeast MCM1 and human SRF directs ternary complex formation. *EMBO J.* **10:** 4219–4229.

Norman, C., M. Runswick, R. Pollock, and R. Treisman. 1988. Isolation and properties of cDNA clones encoding SRF, a transcription factor that binds to the c-*fos* serum response element. *Cell* **55:** 989–1003.

Okamuro, J.T., B.G.W. den Boer, and K.D. Jofuku. 1993. Regulation of *Arabidopsis* flower development. *Plant Cell* **5:** 1183–1193.

Passmore, S., G.T. Maine, R. Elble, C. Christ, and B. Tye. 1988. *Saccharomyces cerevisiae* protein involved in plasmid maintenance is necessary for mating of *MAT*α cells. *J. Mol. Biol.* **204:** 593–606.

Pnueli, L., M. Abu-Abeid, D. Zamir, W. Nacken, Z. Schwarz-Sommer, and E. Lifschitz. 1991. The MADS box gene family in tomato: Temporal expression during floral development, conserved secondary structure and homology with homeotic genes from *Antirrhinum* and *Arabidopsis*. *Plant J.* **1:** 255–266.

Pollock, R. and R. Treisman. 1991. Human SRF-related proteins—DNA-binding and potential regulatory targets. *Genes Dev.* **5:** 2327–2341.

Roe, J.L., C.J. Rivan, R.A. Sessions, K.A. Feldmann, and P.C. Zambryski. 1993. The *TOUSLED* gene in *A. thaliana* encodes a protein kinase homolog that is required for leaf and flower development. *Cell* **75:** 939–950.

Satina, S., A.F. Blakeslee, and A.G. Avery. 1940. Demonstration of the three germ layers in the shoot apex of *Datura* by means of induced polyploidy in periclinal chimeras. *Am. J. Bot.* **27:** 895–905.

Schmidt, R.J., B. Veit, M.A. Mandel, M. Mena, S. Hake, and M.F. Yanofsky. 1993. Identification and molecular characterization of *ZAG1*, the maize homolog of the

*Arabidopsis* floral homeotic gene *AGAMOUS. Plant Cell* **5:** 729–737.

Schultz, E.A. and G.W. Haughn. 1991. *LEAFY,* a homeotic gene that regulates inflorescence development in *Arabidopsis. Plant Cell* **3:** 771–781.

Schultz, E.A., F.B. Pickett, and G.W. Haughn. 1991. The *FLO10* gene product regulates the expression domain of homeotic genes *AP3* and *PI in Arabidopsis* flowers. *Plant Cell* **3:** 1221–1237.

Schwarz-Sommer, Z., P. Huijser, W. Nacken, H. Saedler, and H. Sommer. 1990. Genetic control of flower development: Homeotic genes in *Antirrhinum majus. Science* **250:** 931–936.

Schwarz-Sommer, Z., I. Hue, P. Huijser, P.J. Flor, R. Hansen, F. Tetens, W.-E. Lönnig, H. Saedler, and H. Sommer. 1992. Characterization of the *Antirrhinum* floral homeotic MADS-box gene *deficiens*: Evidence for DNA binding and autoregulation of its persistent expression throughout flower development. *EMBO J.* **11:** 251–263.

Shannon, S. and D.R. Meeks-Wagner. 1991. A mutation in the *Arabidopsis TFL1* gene affects inflorescence meristem development. *Plant Cell* **3:** 877–892.

———. 1993. Genetic interactions that regulate florescence development in *Arabidopsis. Plant Cell* **5:** 639–655.

Shiraishi, H., K. Okada, and Y. Shimura. 1993. Nucleotide sequences recognized by the *AGAMOUS* MADS domain of *Arabidopsis thaliana* in vitro. *Plant J.* **4:** 385–398.

Smyth, D.R., J.L. Bowman, and E.M. Meyerowitz. 1990. Early flower development in *Arabidopsis. Plant Cell* **2:** 755–767.

Sommer, H., J.-P. Beltrán, P. Huijser, H. Pape, W.-E. Lönnig, H. Saedler, and Z. Schwarz-Sommer. 1990. *Deficiens,* a homeotic gene involved in the control of flower morphogenesis in *Antirrhinum majus*: The protein shows homology to transcription factors. *EMBO J.* **9:** 605–613.

Stewart, R.N., F.G. Meyer, and H. Derman. 1972. *Camellia* + "Daisy Eagleson," a graft chimera of *Camellia sasanqua* and *C. japonica. Am. J. Bot.* **59:** 515–524.

Tröbner, W., L. Ramirez, P. Motte, I. Hue, P. Huijser, W.-E. Lönnig, H. Saedler, H. Sommer, and Z. Schwarz-Sommer. 1992. *GLOBOSA*—A homeotic gene which interacts with *DEFICIENS* in the control of *Antirrhinum* floral organogenesis. *EMBO J.* **11:** 4693–4704.

Tsuchimoto, S., A.R. van der Krol, and N.-H. Chua. 1993. Ectopic expression of *pMADS3* in transgenic petunia phenocopies in petunia *blind* mutant. *Plant Cell* **5:** 843–853.

van der Krol, A.R., A. Brunelle, S. Tsuchimoto, and N.-H. Chua. 1993. Functional analysis of petunia floral homeotic MADS-box gene *pMADS1. Genes Dev.* **7:** 1214–1228.

Weigel, D. and E.M. Meyerowitz. 1993. Activation of floral homeotic genes in *Arabidopsis. Science* **261:** 1723–1726.

Weigel, D., J. Alvarez, D.R. Smyth, M.F. Yanofsky, and E.M. Meyerowitz. 1992. *LEAFY* controls floral meristem identity in *Arabidopsis. Cell* **69:** 843–859.

Yanofsky, M.F., H. Ma, J.B. Bowman, G.N. Drews, K.A. Feldmann, and E.M. Meyerowitz. 1990. The protein encoded by the *Arabidopsis* homeotic gene *AGAMOUS* resembles transcription factors. *Nature* **346:** 35–49.

Zimmermann, P.W. and A.E. Hitchcock. 1951. Rose "sports" from adventitious buds. *Contrib. Boyce Thompson Inst.* **16:** 221–224.

# 18

# From Pollination to Fertilization in *Arabidopsis*


**Robert E. Pruitt and Martin Hülskamp**
Department of Cellular and Developmental Biology
Harvard University
Cambridge, Massachusetts 02138


The life cycle of plants comprises two alternating forms of the organism, one which is diploid (the sporophyte) and one which is haploid (the gametophyte). In higher plants, the gametophytic form of the organism has been greatly reduced. On the male side it includes the pollen grain, the pollen tube grown from it, and the two sperm cells that migrate down the pollen tube to participate in fertilization. The female gametophyte is usually a seven-celled structure, the embryo sac, embedded within the sporophytic tissue of the ovule. Although the size and developmental complexity of these gametophytes have been greatly reduced relative to the lower vascular plants, the function remains the same: to bring about the union of haploid gametes to form a sporophytic zygote and thus reinitiate the cycle. The term fertilization is usually defined as the final events of this process, that is, gamete and nuclear fusion. We define the term "fertilization process" to include all interactions between the male gametophyte and the sporophytic or gametophytic female tissues necessary to achieve the successful production of a zygote.

During the fertilization process, the male gametophyte must grow and develop in a precise manner to deliver the sperm cells to the embryo sac. The process begins with the arrival of the pollen grain on the surface of the stigma (pollination), where it undergoes hydration and germinates to grow a pollen tube. This tube penetrates the surface of the stigmatic cell and grows basally toward, and ultimately into, the ovary. During the first portion of this growth, the tube grows within a specialized tissue, the transmitting tract, before emerging on the surface of the transmitting tract and growing over this surface, up the surface of a funiculus to the micropyle of an individual ovule. The sperm cells are then delivered to the embryo sac, and the actual events of double fertilization take place. A great deal of descriptive and experimental work has been done on these processes in other plant species and has been comprehensively reviewed elsewhere (Heslop-Harrison 1987).


*Arabidopsis*
© 1994 Cold Spring Harbor Laboratory Press 0-87969-428-9/94 $5 + .00

The fertilization process represents a developmental sequence that is clearly regulated by a series of cellular interactions between the male and female reproductive systems. In attempting to define and analyze these interactions, it is valuable to identify discrete steps in the process that might be regulated by cellular communication (Fig. 1). One method of analysis is to examine species that exhibit self-incompatibility (SI) (see, e.g., Heslop-Harrison et al. 1975). These species can discriminate between developing pollen from genetically identical plants (self) and pollen from other members of the same species. The function of the SI systems in these plants is to effect the developmental arrest of grains that would lead to self-fertilization and, thus, to promote outcrossing. This type of analysis assumes that SI systems block the fertilization process at developmental steps that are a part of the normal pathway necessary to proceed to fertilization (Fig. 1). The first step depicted in Figure 1 in which water is transferred from the stigma to the pollen grain is the primary site of regulation in many sporophytic SI systems. Recognition of self-pollen leads to a block in the transfer of water from the stigmatic papillar cell to the pollen grain. In some sporophytic systems (e.g., *Brassica*), self-pollen that escapes from this primary control point can be arrested by a second regulatory system that blocks penetration of the cuticle or arrests the growth of the tube shortly after penetration (Ockendon 1972). Gametophytic SI systems are useful in defining a later control point, since they lead to the arrest of self-pollen tubes soon after they begin their growth within the transmitting tract. SI systems are of limited utility in trying to understand all the interactions that take place in the fertilization process, however, because they probably define a special set of inhibitory interactions that are present in addition to the interactions present in both self-compatible and self-incompatible plant species. Other steps in the pathway can be defined by the examination of points of inhibition in interspecies crosses. Experiments conducted with interspecific crosses in the genus *Oenothera* demonstrate the failure of several required interactions that correspond to the steps diagramed in Figure 1 (Glenk 1964). Depending on the species used in the cross, development of the male gametophyte can be arrested or altered in the following ways: (1) failure of the pollen tubes to penetrate the stigma; (2) arrest of pollen tube growth within the transmitting tract; (3) failure of pollen tubes to leave the transmitting tract; and (4) failure of pollen tubes that do leave the transmitting tract to be guided with any efficiency to the ovules. Interspecific crosses between other crucifers and *Arabidopsis* demonstrate that most crucifer pollen hydrates on the *Arabidopsis* stigma, but some species fail to germinate whereas others fail at later points in the fertilization process (R.E. Pruitt, unpubl.). Analysis using

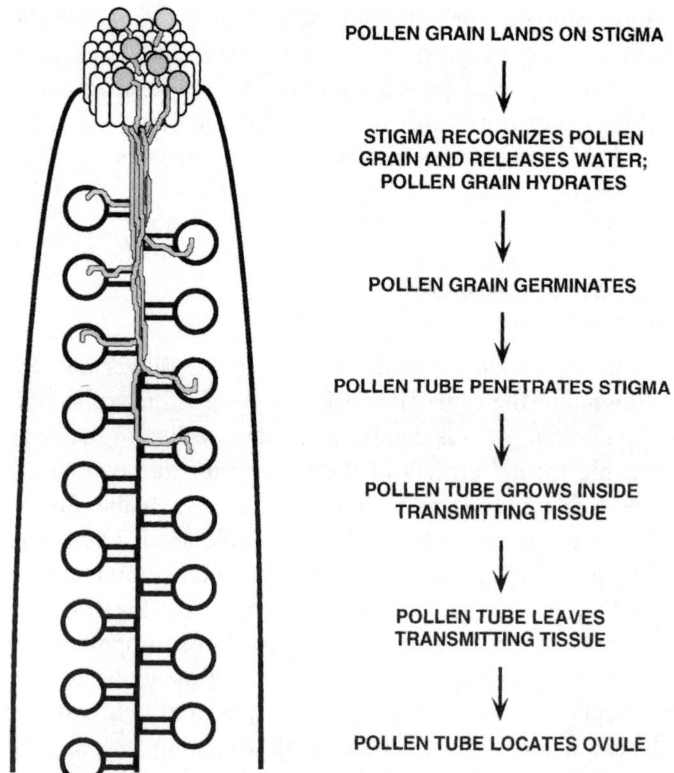



*Figure 1*  Schematic representation of the fertilization process in *Arabidopsis thaliana*. The drawing represents the reproductive system of *Arabidopsis* showing the approximate path taken by the growing pollen tubes (*light gray*). The flow chart depicts the various steps in the process which are likely to be regulated by cellular interactions as described in the text.

interspecific crosses also has its limitations because it can only detect interactions that have diverged sufficiently between species for the interspecific cross to fail. Many interspecific crosses within the crucifer family fail either at fertilization per se or postzygotically, indicating that many of the systems required for guidance of the pollen tubes to the ovules are reasonably conserved between these species (R.E. Pruitt, unpubl.).

To define as many of the cellular interactions required for fertilization as possible, we have undertaken a genetic analysis of the process in *Arabidopsis*. Even though *Arabidopsis* is self-compatible, the growth and development of the male gametophyte still depend on specific interactions with the female reproductive system. Some of these interactions serve as barriers to interspecific pollinations, and others are essential simply for the function of the reproductive system itself. In this chapter, we describe the events that occur between pollination and fertilization in wild-type *Arabidopsis* and use these to define specific steps that may depend on interactions between the male and female reproductive systems.

## STRUCTURE OF THE MALE AND FEMALE REPRODUCTIVE SYSTEMS

Mature pollen grains are produced in the anthers of the flower and are deposited on the stigma of the same flower in the process known as pollination. The mature pollen grain is composed of three cells: a vegetative cell that is responsible for the growth of the pollen tube and two sperm cells that travel down the growing pollen tube and are responsible for the double fertilization event required to produce a viable seed. The entire grain is enclosed by a wall of complex structure, the outer components of which contact and interact with the stigmatic cell (Heslop-Harrison 1975). The outer part of this wall (exine) is composed of sporopollenin, a substance that consists primarily of polymerized carotenoids, and is sculpted into a reticulate network of ridges with irregularly shaped pits between them. These depressions are filled with a coating (tryphine or pollenkitt) composed of lipids and proteins. Both the exine and the tryphine are produced by sporophytic cells during the development of the pollen and therefore reflect the diploid genotype of the parent plant rather than the genotype of the individual pollen grain.

The structure of the female reproductive system is much more complex. The gynoecium is an ovary composed of two carpels separated by a septum. The ovary is capped by a stigma composed of elongated papillar cells which are of the "dry" type, similar to those found in other members of the crucifer family (Heslop-Harrison and Shivanna 1977; Elleman et al. 1992). The transmitting tract is located in the septum that divides the two carpels, and the ovules are attached to the septum by funiculi at its lateral edges. Each carpel contains two rows of interleaved ovules, one coming from each edge of the septum. The path the pollen tubes normally follow is thus confined to the central parts of the ovary, and the outer covering does not normally come into contact with the growing tubes.

The ovules themselves are also complex structures, composed of a nucellus and two integuments that develop from the base of the nucellus and eventually surround it (Robinson-Beers et al. 1992). Within the nucellus, a single megasporocyte arises from a hypodermal cell and undergoes meiosis to form a multiplanar tetrad of megaspores (Webb and

Gunning 1990). This meiotic division is somewhat atypical in that both nuclear divisions precede cytokinesis, which then separates all four megaspores (Webb and Gunning 1990). Following wall formation, the megaspore nearest to the chalazal end develops into a functional megaspore, and the other three meiotic products degenerate. The megaspore then undergoes a series of three mitotic nuclear divisions followed by cellularization to form the seven-celled embryo sac. The structure of the female gametophyte has been extensively described previously (Mansfield et al. 1990; Murgia et al. 1993). As is the case in many angiosperms, the mature embryo sac is made up of two synergid cells and the egg cell at the micropylar end, a large central cell which at maturity usually contains a single diploid nucleus derived from two of the haploid mitotic products, and three antipodal cells located at the chalazal end. Some variability in this structure has been observed. It appears that in some lines the antipodal cells degenerate prior to pollination (Murgia et al. 1993).

The structures of the reproductive tissues and their derivation are important, because the cellular interactions that take place involve not only the haploid cells found in the gametophytes, but also diploid sporophytic tissues of the parent plant. The pollen grain itself is haploid, but the outer coating it bears is synthesized in the diploid tapetal cells. It is this coating that determines the nature of interactions at the stigma surface. Later interactions between the growing pollen tube and female reproductive tissues presumably depend on the haploid genotype of the pollen grain. On the female side, all of the tissues involved (stigma, transmitting tissue, funiculus, ovule) are diploid with the exception of the seven-celled embryo sac located within the ovule. Thus, most of the interactions required for successful fertilization are probably sporophytic on the female side, although the embryo sac itself may be involved in the final stages of the guidance of the pollen tube and is obviously involved in the recognition events that lead to fertilization itself.

## INTERACTIONS AT THE STIGMA SURFACE

Mature *Arabidopsis* pollen has undergone desiccation and contains relatively little water. In order for its development to proceed, it first must hydrate; in vivo, that water must come from the stigmatic papillar cell on which the pollen lands. To demonstrate that this process is regulated in *Arabidopsis*, pollen grains from unrelated plant species are placed onto the stigma, where they fail to hydrate. This stigmatic barrier to the release of water can also be demonstrated by placing dehydrated polyacrylamide beads onto the stigma; these also fail to extract water from the stigmatic cell. This water transfer process therefore does not reflect a purely osmotic removal of water from the stigmatic cell by the pollen grain, but instead requires some type of recognition between the two cells that leads to a localized change in water permeability of the stigmatic cell surface. The nature of the recognition system used is unknown, although clearly it can be disrupted by mutations acting sporophytically on the male side (Pruitt et al. 1991 and unpubl.; Preuss et al. 1993), which implies that the recognition substances are carried on the outside of the pollen grain and are produced in the tapetum. This is reminiscent of the situation in sporophytic self-incompatibility, and it will be interesting to see if the two systems bear more than a superficial resemblance.

Once the pollen grain has been recognized by the stigma, hydration commences. The hydration process can be readily observed by the change in shape of the pollen grain. The mature grain is elongated and contains three furrows, whereas the hydrated grain is nearly spherical. In *Arabidopsis* this process is quite rapid, requiring less than 10 minutes to reach completion (Fig. 2). As shown in Figure 3, the process is approximately linear and proceeds without a noticeable lag phase. Thus, the recognition process takes place on a time scale which is much shorter

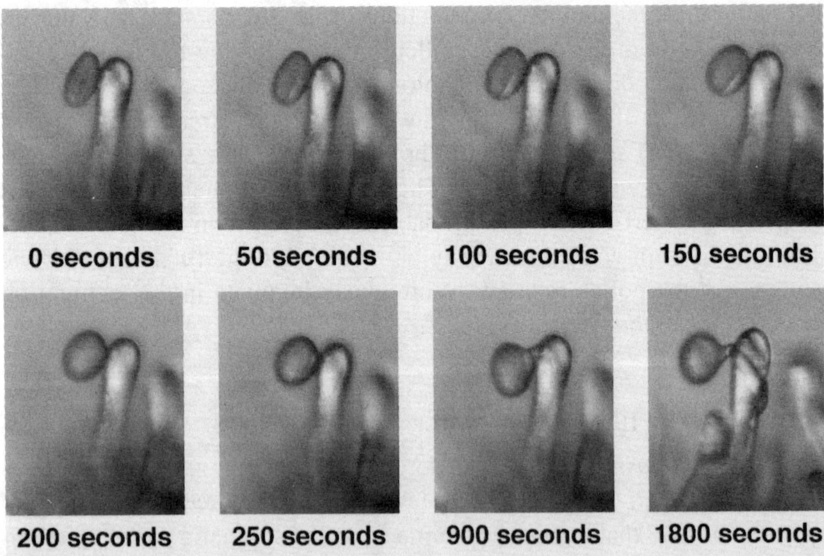


*Figure 2* Series of light micrographs showing the hydration and germination of a single pollen grain placed on a stigmatic papillar cell. Hydration can be followed both by the change of shape seen in the pollen grain and by the loss of the furrow running longitudinally down the grain. Germinated pollen tube is clearly visible after 900 sec and can be seen growing down the papillar cell by 1800 sec.

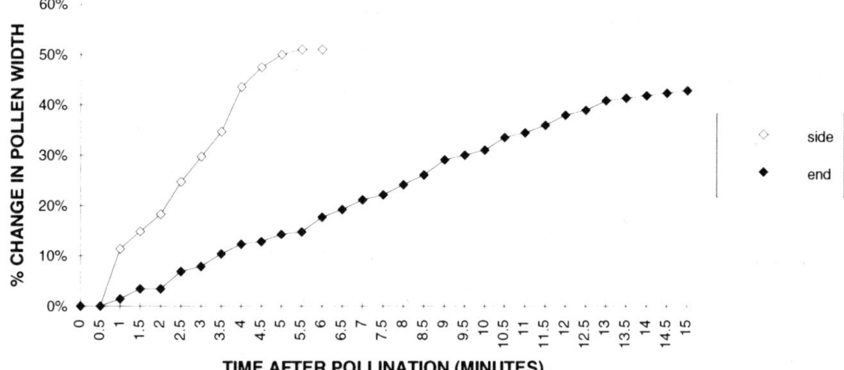


*Figure 3*  Graph showing the time course of hydration for pollen grains placed with either their sides or ends in contact with the stigma. The time course is approximately linear with very little lag evident prior to the beginning of hydration. Hydration is clearly slower when only the end is in contact with the stigma, probably reflecting the smaller area of contact between the two cells.

than the process of hydration itself. The process of hydration seems to depend on the positional relationship of the pollen grain to the stigma. Pollen grains placed with their sides in contact with the stigma hydrate with a rapid and reproducible time course. Pollen grains placed with their ends in contact with the stigma hydrate more slowly and show a good deal more variation. This may reflect that the rate of hydration is dependent on the surface area of the pollen grain that is in contact with the stigmatic cell, in turn implying that the area of the stigmatic cell that can become permeable to water is also variable and is controlled by contact with the pollen grain. Although this process is typically referred to as hydration and does clearly involve the transfer of water, it may also involve the transfer of other unidentified substances that are involved in further signaling between the male and female reproductive systems.

Since both self-incompatibility in the crucifer family and interspecific incompatibility appear to involve the control of the hydration process, it is natural to ask how the two systems are related. Classically, this has been addressed by Lewis and Crowe (1958), who have shown that species which are self-incompatible have far more precise interspecific incompatibility systems. The characterization of self-incompatibility in *Brassica oleracea* has led to the identification of a number of gene products believed to be involved in the recognition between pollen grain and stigmatic cell (for review, see Nasrallah et al. 1991; Dzelzkalns et al. 1992). These gene products appear to be encoded by the genetically defined *S* locus that specifies the identity of individual plants. The genes encoding these products are members of a large multigene family that encodes both small extracellular glycoproteins and membrane-bound receptor protein kinases. The receptor portion of the protein kinases is extremely similar in structure to the secreted glycoproteins. In the case of self-incompatibility, two genes have been singled out as having possible involvement in recognition: *SLG*, which encodes an extracellular glycoprotein found on the outside of the stigmatic papillar cells (Nasrallah et al. 1985; Kandasamy et al. 1989), and *SRK*, which encodes a receptor protein kinase also located on stigmatic cells (Stein et al. 1991). Both *SLG* and *SRK* also appear to be expressed in developing anthers, although at much lower levels than in stigmatic papillar cells. The precise roles these proteins play in recognition are not defined, although it is clear that they are present in a milieu of proteins encoded by other members of the gene family (Lalonde et al. 1989; Trick and Flavell 1989). Similar genes have been identified in *Arabidopsis* (Dwyer et al. 1992; Tobias et al. 1992; Walker 1993; S.E. Ploense and R.E. Pruitt, unpubl.), although only a single member of the family shows a similar expression pattern to those genes involved in SI. This particular gene (*AtS1* or $\Sigma 5$) shows an expression pattern extremely similar to that of *SLG*, but its sequence is much more like that of another member of the gene family, *SLR1*, which has a similar pattern of expression in *Brassica* but is not encoded at the *S* locus. No protein-kinase-encoding genes preferentially expressed in reproductive tissues have been identified in *Arabidopsis* (S.E. Ploense and R.E. Pruitt, unpubl.). In fact, most members of this gene family in *Arabidopsis* are expressed in the vegetative tissues of the plant, where they must play a role far removed from the fertilization process. Although it is clear that *Arabidopsis* does contain at least one gene that has both sequence similarity and a similar expression pattern to those which are thought to control recognition in SI, none of the available data shed any light on the role that gene might play. The fact that the *Arabidopsis* gene is more similar to *Brassica* genes which are not involved in SI may reflect the common involvement of this group of genes in interspecific incompatibility, or they may play some other role altogether.

### POLLEN GERMINATION

After the hydration process is complete, the pollen grain germinates and produces a pollen tube. This process is also quite rapid in *Arabidopsis*, and the tube is frequently visible within 15 minutes following pollination (see Fig. 2). When the pollen tube emerges, it does so from one of three apertures located in the regions that are furrowed in the mature pollen grain. The tube emerges from the aperture located on the side of the grain facing the stigmatic papillar cell. In the short interval between pollination and germination, the initially threefold symmetrical pollen grain acquires asymmetry with respect to the cellular machinery necessary to initiate the production of a pollen tube. The positional reference needed to establish this asymmetry is clearly provided by the stigmatic cell, although the identity of the signal and the mechanism of signaling are unknown. Experiments have been performed in other species to examine the rearrangement of the cytoskeleton prior to pollen tube emergence (for review, see Pierson and Cresti 1992). These experiments demonstrate a change in the actin microfilament network from an initially symmetric form to an asymmetric form with a preponderance of actin at the site where the pollen tube will emerge from the pollen grain. This actin accumulation is then maintained in the growing pollen tube tip, where it presumably plays a role in the elongation of the pollen tube. Unfortunately, many of these experiments were performed on pollen grains that were germinated in vitro, and hence it is impossible to examine the role the stigma might have in determining the polarity of the actin microfilament network.

The emergent pollen tube immediately contacts the surface of the papillar cell and penetrates the cuticle and, usually, one of two layers found in the primary cell wall (Elleman et al. 1992). The tube then grows basally in between the layers of the cell wall, eventually penetrating the surface of the stigma proper and entering the transmitting tissue. The growth of the pollen tube directionally with respect to the apical-basal axis of the papillar cell also presumably depends on signals coming from the female reproductive system, but no definitive information is available on what types of molecules are involved.

## POLLEN TUBE GROWTH WITHIN THE OVARY

Once the pollen tube has entered the transmitting tract, it grows toward the base of the ovary, passing through the intercellular spaces in this tissue. Experiments by Sanders and Lord (1989) indicate that some (or all) of the force needed for this growth may be provided by the female reproductive system, since latex beads of similar diameter to a pollen tube are translocated through the transmitting tissue. Although this has not been directly demonstrated for *Arabidopsis*, it has been shown to be the case in the crucifer *Raphanus raphanistrum*. Pollen tubes grow for some distance within the transmitting tract and then enter one of the two carpels of the *Arabidopsis* ovary by coming to the surface of the tract. The tubes then grow on the surface of the tract, either continuing their basipetal growth or growing laterally to a funiculus. Tubes that reach the funiculus proceed up it to the micropyle. The growth patterns of tubes can be seen clearly in Figure 4, which shows tubes growing in two planes of focus. The first focal plane (Fig. 4a) is within the transmitting tissue itself, where the tubes grow in remarkably straight lines, presumably constrained by the surrounding files of cells. Interestingly, the tubes also appear to be of smaller diameter in this plane of focus than at the surface of the transmitting tract. This might be expected if the tubes are being assisted by a pulling force from the female reproductive system. The second plane (Fig. 4b) shows tubes growing on the surface of the transmitting tract. Here the tubes growing on the surface are characterized by a growth pattern that is no longer completely straight but wanders back and forth on the surface of the cells. This growth pattern is more like that seen when pollen tubes are grown in vitro and may reflect the growth of the pollen tube under its own power as opposed to growth with the assistance of the female tissues. Note that the growth of pollen tubes following their emergence from the transmitting tract in wild-type ovaries is limited to the surface of the transmitting tract and the funiculus. Tubes are not seen growing on the interior surfaces of the carpel covers or on the surface of the ovules. Thus, the process of pollen tube guidance within the ovary is quite precise, with each tube taking a relatively direct course to an ovule. Figure 4c shows a surface view of the transmitting tract similar to that seen in Figure 4b, but with only a few pollen tubes present in the ovary. Under these circumstances, it is possible to follow the course of individual pollen tubes from their point of emergence to the ovule that they fertilize. Analysis of many such pollinations has allowed us to examine the relationships between the points of pollen tube emergence and their ultimate destinations. This analysis has revealed that the points at which pollen tubes emerge from the transmitting tract do not

*Figure 4* Whole *Arabidopsis* ovary with the outer covering removed; stained with aniline blue to reveal pollen tubes. (*a*) Plane of focus is inside the transmitting tract revealing very smooth, narrow pollen tubes. (*b*) Plane of focus is at the surface of the transmitting tract showing tubes of wider diameter than in *a* and "wandering" pattern of growth. (*c*) Plane of focus as in *b* but with fewer tubes present, allowing individual tubes to be followed. Single tube can be easily followed from its point of emergence to its termination in the ovule. (*d*) Composite photograph showing one half of an *Arabidopsis* ovary dissected open and with the ovules folded back off the transmitting tract. Numerous pollen tubes can be seen growing from the stigma down toward the base of the ovary. Single pollen tubes can be seen growing to each ovule, although some of them have been slightly dislodged from the funiculus in the process of folding the ovules.

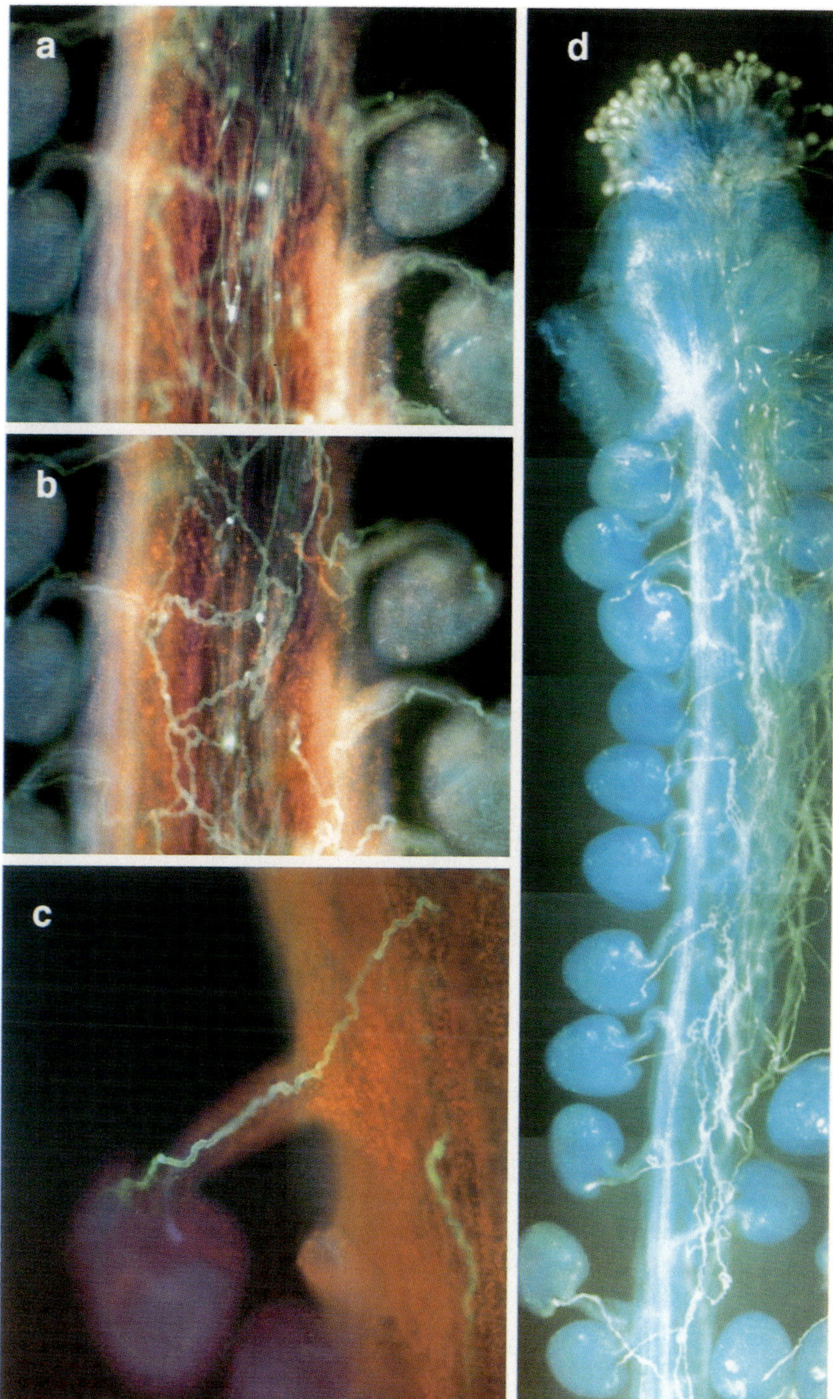

*Figure 4  (See facing page for legend.)*

appear to be predetermined, nor does the spacing of emergence points appear to be controlled in any obvious manner. Pollen tubes emerging from the transmitting tract may proceed directly to the nearest funiculus and grow to that ovule, or they may proceed basally over the surface of the transmitting tract, bypassing one or more ovules before growing laterally to a funiculus. The behavior of the tubes on the surface of the transmitting tract does not seem to be governed by the availability of unfertilized ovules, since many of the ovules that are bypassed do not have pollen tubes associated with them.

A broader perspective on the paths followed by pollen tubes generally can be seen in Figure 4d. This composite photograph shows an ovary in which the carpel covering has been dissected off and the ovules have been folded back away from the flat surface of the transmitting tract. The tissue has been stained with aniline blue, which reveals the callose-containing pollen tubes. Only one half of one carpel is shown in the figure; the other half would be a mirror image of this one. Although it is difficult to see in this type of image, many tubes are growing both within the transmitting tract proper and along its surface. These tubes can be distinguished by their relative appearances as was described for Figure 4. Another interesting phenomenon associated with pollen tube guidance can also be seen in Figure 4d. Despite the presence of an excess of pollen tubes with respect to the number of ovules, only a single pollen tube is associated with each ovule. This is true not only with respect to pollen tubes reaching the micropyle, but also with respect to pollen tubes associated with each funiculus. Thus, there appears to be an exclusion mechanism that limits each ovule to attracting a single pollen tube, and the exclusion mechanism acts early enough to prevent multiple pollen tubes from approaching a single funiculus.

As discussed in the next section, there is little definitive evidence about the signals that are involved in the guidance of the growing pollen tube or about the mechanics of pollen tube guidance. A great deal of work has been done characterizing the growth of the pollen tube in vitro, with particular emphasis on the involvement of the cytoskeleton. This work has shown that the pollen tube grows by the elongation of the growing tip and that there is a localization of actin in the growing region. The function of actin appears to be essential for the elongation of the pollen tube, since drugs that interfere with actin function block the elongation of the pollen tube in vitro (Franke et al. 1972; Mascarenhas and Lafountain 1972). A calcium gradient has also been shown to be present in the growing region of the tip (Jaffe et al. 1975) and is probably also essential for elongation, since calcium is an obligate part of all in vitro pollen germination media. Both calcium and actin seem to be common requirements for tip growing cells in plants, as both are also found in the tips of growing root hairs. The pollen tube also contains a microtubular cytoskeleton, although the function of this structure is unknown since the microtubule disrupter colchicine seems to have no effect on the growth of pollen tubes in vitro (Franke et al. 1972). This does not, of course, rule out a role for microtubules in pollen tube guidance. Because of the difficulty of working with pollen tubes grown in vivo and the lack of an in vitro pollen tube guidance system, the internal systems that control the directional growth of the pollen tube remain a mystery.

## MODELS FOR POLLEN TUBE GUIDANCE

Examination of the path that the pollen tube follows shows its growth to be divided into discrete steps. The pollen tube first grows in contact with the stigmatic papillar cell, growing within the cell wall to the base of the cell. It then grows intercellularly in the transmitting tract in what appears to be a straight line. Following this, the tube grows to the surface of the tract and then grows across this surface. During growth on the surface, the tube may immediately approach the nearest funiculus, or it may grow toward the base of the ovary before approaching a funiculus. In either case, upon reaching the base of the funiculus, the tube proceeds up it to the micropyle. Although these different phases of pollen tube growth seem to be quite dissimilar from one another, are the mechanisms used to guide the pollen tube really different from one phase to another? To ask the same question in different terms, What sort of guidance mechanism is used by the pollen tube to reach each of its intermediate goals?

Three types of models have been proposed for the directed growth of pollen tubes (for review, see Heslop-Harrison 1987). The first is that guidance is provided entirely by the female reproductive system literally pulling the tube to the desired destination. Although there does appear to be a contribution from the female side inside the transmitting tract, the fact that only small numbers of the latex beads used in the experiments of Sanders and Lord (1989) were properly guided outside the transmitting tract makes it unlikely that this system is solely responsible for the directed growth of the pollen tube. In this regard, the difference in morphology between tubes growing inside the transmitting tract and those growing along its surface (Fig. 4) is particularly interesting. The tubes inside the tract appear as virtual narrow straight lines, whereas those outside the tract exhibit a growth pattern that appears to wander back and forth, perhaps reflecting an alternate guidance mechanism used from this point onward. The second type of model that could be envisioned is a model where individual pollen tubes follow defined "tracks" of some surface molecule to their destinations. This model could easily account for the limitation of each ovule to a single pollen tube. In this type of model, however, one would expect the pollen tubes to follow invariant paths to the ovules, whereas, in fact, the pollen tube paths are anything but invariant. The points of emergence of the pollen tubes from the transmitting tract do not appear to be predefined nor do they have any regular spacing. Pollen tubes emerging from the tract do not always grow to the nearest ovule but seem to have a reasonable degree of freedom in choosing their subsequent path. In fact, the path that the pollen tube follows up the funiculus is also variable, with tubes growing up different sides of the funiculus relative to the micropyle. All these facts taken together make it unlikely that the growth of the pollen tube after it leaves the transmitting tract is guided by a "track" of some surface constituent. The last type of model that can be imagined is a chemotactic one. In this case, the growth of the pollen tube after emergence from the transmitting tract would be guided by a gradient of one or more signal molecules, which would serve to guide the tube to the micropyle. This type of model would be consistent with the variable behavior of the tubes following their emergence on the surface of the transmitting tissue and would also be consistent with the apparent "wandering" growth of the tubes, which is similar to the paths followed by other eukaryotic cells undergoing chemotaxis. A simple chemotactic model does not readily account for the exclusion of other pollen tubes from an already occupied funiculus, especially since it would require several minutes for the pollen tube to grow up the funiculus and reach the source of the presumed signal. In a heavily pollinated ovary, this would allow plenty of time for a second pollen tube to begin to grow up the same funiculus. Many possible additions to the simplest chemotactic models could account for the exclusion of other pollen tubes: the production of a second negative chemotactic substance by the pollen tube after it acquires the positive signal, for example.

Although the purely descriptive information detailed above leads us to favor a chemotactic model for the last stages of pollen tube guidance, it does little to enlighten us about the earlier stages of the growth process. The straight growth of the pollen tubes within the transmitting tract is consistent with a female force that pulls the tubes along, but what causes the tubes to ultimately reach the surface of the transmitting tract? The sites of tube emergence do not appear to be predetermined; what guides the tubes out of the transmitting tract? If the same signals that guide the tubes to the ovules were involved, one would expect to see a definite correlation between the points of pollen tube emergence and the ovules that were ultimately fertilized. Experiments indicate that this is not the case. Pollinations with small numbers of pollen grains indicate that all these tubes emerge from the transmitting tract near the top of the ovary. When more tubes are involved, some tubes proceed further inside the transmitting tract. What process determines where the tubes emerge, and how is it modified by the presence of other pollen tubes? Even earlier than this there is the question of how the tube begins its directional growth after contact with the stigmatic cell. The signal(s) responsible for this ability to determine apical-basal polarity is unknown.

Perhaps one of the most intriguing aspects of the problem of the guidance of pollen tubes is that the rules which govern the behavior of the tubes do not seem to be very hard and fast, and yet the end result is an efficient system for providing each ovule with a single pollen tube. The data described above make it clear that the points at which the pollen tubes emerge from the transmitting tract are not fixed with respect to other landmarks (ovule attachment points, for example), nor do they appear to bear any obvious relationship to one another. The same types of observations can be made with respect to many different aspects of the directed growth of the pollen tube. Individually, the pollen tubes seem to be constrained only by a very broad set of rules, and yet this leads to a very precise result.

## CONCLUSIONS AND FUTURE PERSPECTIVES

The bulk of this chapter has been devoted simply to a description of the reproductive process in *Arabidopsis*. Within that context we have discussed the problems of how interactions between the male and female reproductive systems might be involved in the regulation of the fertilization process. As is obvious from the above discussion, this system raises a great many questions, which have only been addressed in a descriptive way. We are presently involved in an attempt to dissect some of these processes genetically to better understand at both the cellular and molecular levels the functioning of some of these systems. To this end we and other workers (Pruitt et al. 1991 and unpubl.; Robinson-Beers et al. 1992; Preuss et al. 1993) have isolated mutations that disrupt the fertilization process and are characterizing those mutations to see what roles the genes may play in the wild-type process. Characterization of a single mutation by Preuss et al. (1993) has shown that defects in the biosynthesis of long-chain lipids can lead to defects in pollen/stigma recognition. Unfortunately, this mutation leads to the complete absence of tryphine on the outside of the pollen grain and thus does not indicate whether these lipids play a direct role in recognition or are required merely as a binding agent to attach the tryphine layer to the exine. Examination of a broader spectrum of mutations may help to address this question.

So far these genetic efforts have been concentrated on isolating mutations that are active in the sporophytic tissues and hence have been limited to identifying male genes acting at the level of the pollen/stigma interface and female genes which are required in tissues other than the embryo sac. We hope soon to extend these screens to gametophytic mutations in order to obtain a more complete perspective of the classes of mutant phenotypes that can be obtained. By characterizing both the cellular and molecular defects in these mutant plants, we hope to arrive at an understanding of the interactions that regulate the germination, growth, and guidance of the pollen tube.

## ACKNOWLEDGMENTS

This work was supported by National Science Foundation grants DMB-8718570 and DCB-9018889 to R.E.P. M.H. was supported by an EMBO long-term postdoctoral fellowship.

## REFERENCES

Dwyer, K.G., B.A. Lalonde, J.B. Nasrallah, and M.E. Nasrallah. 1992. Structure and expression of *AtS1*, an *Arabidopsis thaliana* gene homologous to the *S*-locus related genes of *Brassica*. *Mol. Gen. Genet.* **231:** 442–448.

Dzelzkalns, V.A., J.B. Nasrallah, and M.E. Nasrallah. 1992. Cell-cell communication in plants: Self-incompatibility in flower development. *Dev. Biol.* **153:** 70–82.

Elleman, C.J., V. Franklin-Tong, and H.G. Dickinson. 1992. Pollination in species with dry stigmas: The nature of the early stigmatic response and the pathway taken by pollen tubes. *New Phytol.* **121:** 413–424.

Franke, W.W., W. Herth, W.J. VanDerWoude, and D.J. Morré. 1972. Tubular and filamentous structures in pollen tubes: Possible involvement as guide elements in protoplasmic streaming and vectorial migration of secretory vesicles. *Planta* **105:** 317–341.

Glenk, H.O. 1964. Untersuchungen über die sexuelle affinität bei Oenotheren. In *Pollen physiology and fertilization* (ed H.F. Linskens), pp. 170–181. North-Holland, Amsterdam.

Heslop-Harrison, J. 1975. The physiology of the pollen grain surface. *Proc. R. Soc. Lond. B Biol. Sci.* **190:** 275–299.

―――. 1987. Pollen germination and pollen-tube growth. *Int. Rev. Cytol.* **107:** 1–78.

Heslop-Harrison, J., Y. Heslop-Harrison, and J. Barber. 1975. The stigma surface in incompatibility responses. *Proc. R. Soc. Lond. B Biol. Sci.* **188:** 287–297.

Heslop-Harrison, Y. and K.R. Shivanna. 1977. The receptive surface of the angiosperm stigma. *Ann. Bot.* **41:** 1233–1258.

Jaffe, L.A., M.H. Weisenseel, and L.F. Jaffe. 1975. Calcium accumulations within the growing tips of pollen tubes. *J. Cell Biol.* **67:** 488–492.

Kandasamy, M.K., D.J. Paolillo, C.D. Faraday, J.B. Nasrallah, and M.E. Nasrallah. 1989. The *S*-locus specific proteins of *Brassica* accumulate in the cell wall of developing stigma papillae. *Dev. Biol.* **134:** 462–472.

Lalonde, B.A., M.E. Nasrallah, K.G. Dwyer, C.-H. Chen, B. Barlow, and J.B. Nasrallah. 1989. A highly conserved *Brassica* gene with homology to the *S*-locus-specific glycoprotein structural gene. *Plant Cell* **1:** 249–258.

Lewis D. and L.K. Crowe. 1958. Unilateral interspecific incompatibility in flowering plants. *Heredity* **12:** 233–256.

Mansfield, S.G., L.G. Briarty, and S. Erni. 1990. Early embryogenesis in *Arabidopsis thaliana*. I. The mature embryo sac. *Can. J. Bot.* **69:** 447–460.

Mascarenhas, J.P. and J. Lafountain. 1972. Protoplasmic streaming, cytochalasin B, and the growth of the pollen tube. *Tissue Cell* **4:** 11–14.

Murgia, M., B.-Q. Huang, S.C. Tucker, and M.E. Musgrave. 1993. Embryo sac lacking antipodal cells in *Arabidopsis thaliana* (*Brassicaceae*). *Am. J. Bot.* **80:** 824–838.

Nasrallah, J.B., T. Nishio, and M.E. Nasrallah. 1991. The self-incompatibility genes of *Brassica*: Expression and use in genetic ablation of floral tissues. *Annu. Rev. Plant Physiol. Plant Mol. Biol.* **42:** 393–422.

Nasrallah, J.B., T.-H. Kao, C.-H. Chen, M.L. Goldberg, and M.E. Nasrallah. 1985. A cDNA clone encoding an *S*-locus-specific glycoprotein from *Brassica oleracea*. *Nature* **318:** 263–267.

Ockendon, D.J. 1972. Pollen tube growth and the site of the incompatibility reaction in *Brassica oleracea*. *New Phytol.* **71:** 519–522.

Pierson, E.S. and M. Cresti. 1992. Cytoskeleton and cytoplasmic organization of pollen and pollen tubes. *Int. Rev. Cytol.* **140:** 73–125.

Preuss, D., B. Lemieux, G. Yen, and R.W. Davis. 1993. A conditional sterile mutation eliminates surface components from *Arabidopsis* pollen and disrupts cell signalling during fertilization. *Genes Dev.* **7:** 974–985.

Pruitt, R.E., T.F. Horejsi, B.K. Pierskalla, and S.E. Ploense. 1991. Genetic analysis of cellular interactions during fertilization of *Arabidopsis thaliana*. *J. Cell. Biochem.* **15A:** 137.

Robinson-Beers, K., R.E. Pruitt, and C.S. Gasser. 1992. Ovule development in wild-type *Arabidopsis* and two female-sterile mutants. *Plant Cell* **4:** 1237–1249.

Sanders, L.C. and E.M. Lord. 1989. Directed movement of latex particles in the gynoecia of three species of flowering plants. *Science* **243:** 1606–1608.

Stein, J.C., B. Howlett, D.C. Boyes, M.E. Nasrallah, and J.B. Nasrallah. 1991. Molecular cloning of a putative receptor protein kinase gene encoded at the self-incompatibility locus of *Brassica oleracea*. *Proc. Natl. Acad. Sci.* **88:** 8816–8820.

Tobias, C.M., B. Howlett, and J.B. Nasrallah. 1992. An *Arabidopsis thaliana* gene with sequence similarity to the *S*-locus receptor kinase of *Brassica oleracea*: Sequence and expression. *Plant Physiol.* **99:** 284–290.

Trick, M. and R.B. Flavell. 1989. A homozygous *S* genotype of *Brassica oleracea* expresses two *S*-like genes. *Mol. Gen. Genet.* **218:** 112–117.

Walker J.C. 1993. Receptor-like protein kinase genes of *Arabidopsis thaliana*. *Plant J.* **3:** 451–456.

Webb, M.C. and B.E.S. Gunning. 1990. Embryo sac development in *Arabidopsis thaliana*. I. Megasporogenesis, including the microtubular cytoskeleton. *Sex. Plant Reprod.* **3:** 244–256.

# 19

# Ethylene: A Unique Plant Signaling Molecule


**Joseph R. Ecker**
Department of Biology
University of Pennsylvania
Philadelphia, Pennsylvania 19104

**Athanasios Theologis**
Plant Gene Expression Center
Albany, California 94710


Mammalian neurobiologists and neurochemists have been fascinated to discover that their experimental systems use simple gases such as nitric oxide (NO) and carbon monoxide (CO) as neuronal messengers (Snyder 1992; Verma et al. 1993). Recent advances in neurobiology have revealed that humans use NO in a variety of physiological responses. NO plays a role in blood pressure maintenance as a vasodilator, helps kill foreign invaders in the immune response, is a major biochemical mediator of penile erection, and is probably a major biochemical component of long-term memory (Snyder 1992).

Another simple gas, ethylene ($C_2H_4$, the simplest olefin), has fascinated plant biologists for more than a century by its spectacular effects on plant growth and development (Abeles et al. 1992; Theologis 1993). Plants do not have brains or sophisticated reproductive organs like humans, but the gas ethylene plays an important part in sex determination in some monoecious species (Abeles et al. 1992). This hydrocarbon gas, well known as the fruit-ripening hormone, is biologically active in trace amounts (as little as 10 nanoliters per liter of air). It promotes leaf and flower senescence and abscission, and causes the loss of geotropic sensitivity, root initiation, the onset of epinastic curvatures, the acceleration of respiration, and the modification of leaf and fruit pigments (Burg 1962; Abeles et al. 1992). Ethylene is thought to be a key regulator of the ability of rice to grow in the deepwater regions of southeast Asia and of most hypoxia-induced plant adaptations (Jackson 1985; Kende 1987). It also controls the plumular expansion and maintains the plumular hook structures that facilitate the emergence of germinating seedlings through the soil, a process vital to successful germination. In pea, exogenous ethylene exaggerates the curvature of the apical hook, inhibits stem elongation, and prevents a normal geotropic response, effects known as "the triple response" (Neljubov 1901; Abeles et al. 1992). Ethylene is also induced by a variety of external factors such as wounding, anaerobiosis, viral infection, auxin treatment, chilling, injury, drought, and $Cd^{++}$ and $Li^+$ (Yang and Hoffman 1984; Abeles et al. 1992).

Because of its commercial importance and its profound effects on plant growth, the biosynthesis of ethylene and mechanisms of its action have been intensely investigated. In this chapter, we summarize the recent advances in knowledge of the mechanisms by which ethylene exerts its biological effects on plants.

## ETHYLENE BIOSYNTHESIS

Elucidation of the pathway for ethylene synthesis by Yang and his associates (Yang and Hoffman 1984) during a 20-year-long experimental endeavor is one of the most significant contributions to plant biology. Although the simplicity of its chemical structure raised the possibility that many compounds could be potential precursors of ethylene, it is now clear that only methionine serves this role in higher plants and that carbons C-3,4 of methionine give rise to ethylene (Fig. 1). The rate-limiting step in the pathway is the formation of the amino acid 1-aminocyclopropane-1-carboxylic acid (ACC) from $S$-adenosyl-L-methionine (AdoMet), catalyzed by ACC synthase (reaction 2 in Fig. 1). The final step in the pathway is the conversion of ACC to $C_2H_4$, HCN, and $CO_2$ catalyzed by ACC oxidase (reaction 3 in Fig. 1). An alternative fate of ACC is its conjugation into malonyl ACC (MACC) (Fig. 1), which serves as a control mechanism for limiting ACC availability for $C_2H_4$ formation. The biochemical design of the pathway allows high rates of ethylene production without high intracellular concentration of methionine, a less abundant amino acid. This is achieved by recycling 5′-methylthioadenosine (MTA) to methionine (Fig. 1). The overall result is that the ribose moiety of ATP gives rise to the 4-carbon skeleton of methionine from which ethylene is derived. The $CH_3S$ group, however, is conserved for continued regeneration of methionine. Thus, with a constant pool of the $CH_3S$ group and available ATP, high rates of ethylene production can be achieved (Yang and Hoffman 1984).

The enzymes that catalyze ethylene biosynthesis have been difficult to purify to homogeneity because of their low abundance (Kende 1989, 1993). Recently, the purification of ACC oxidase (reaction 3 in Fig. 1) to near homogeneity has been reported (Dong et al. 1992a). Molecular cloning approaches and expression in heterologous systems allowed the iso-

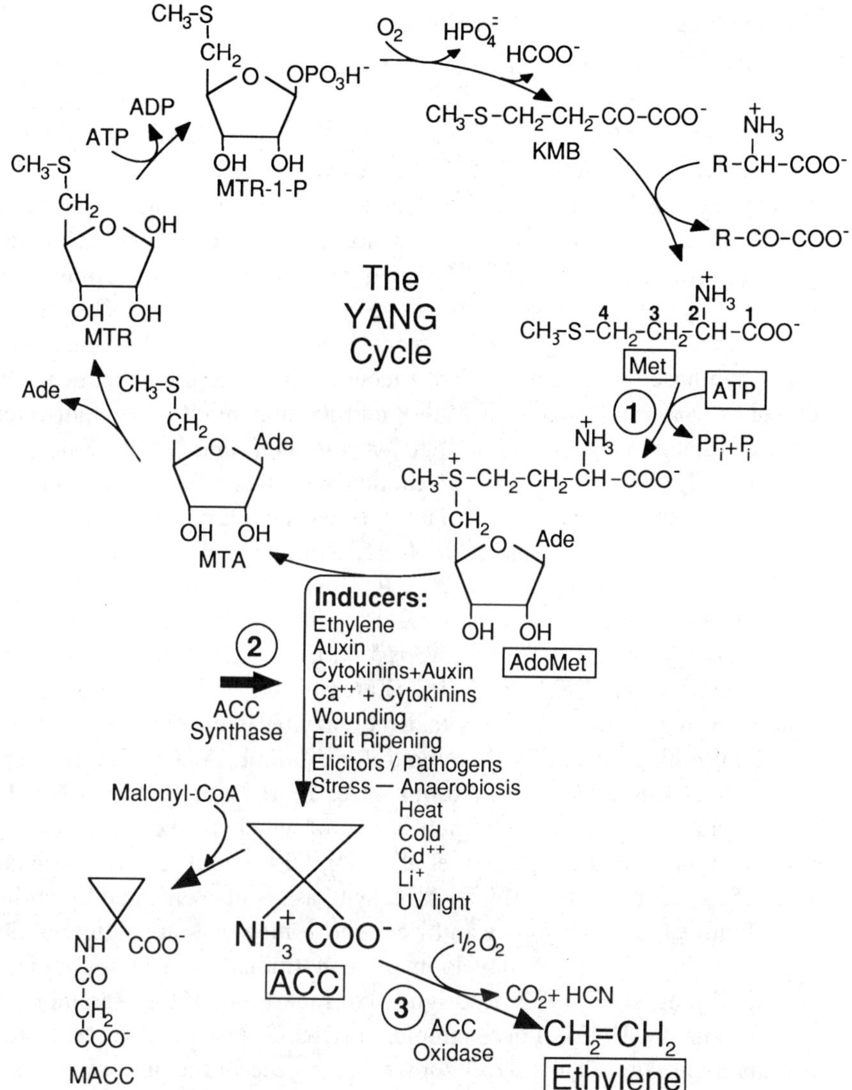



*Figure 1*  The ethylene biosynthetic pathway of higher plants (Yang and Hoffman 1984). Abbreviations: (AdoMet) *S*-adenosyl-L-methionine; (ACC) 1-aminocyclopropane-1-carboxylic acid; (KMB) 2-keto-4-methylthiobutyrate; (MACC) malonyl-ACC; (MTA) 5′-methylthio-adenosine; (MTR) 5′-methyl-thioribose; (MTR-1-P) 5′-methylthioribose-1-phosphate; (PG) polygalacturonase. (Reprinted, with permission, from Theologis 1992.)

lation of the genes encoding three of the enzymes, AdoMet synthase (Peleman et al. 1989), ACC synthase (Sato and Theologis 1989), and ACC oxidase (Slater et al. 1985; Hamilton et al. 1991; Spanu et al. 1991).

## ACC Synthase

After the discovery that ACC is the immediate precursor of ethylene, it became obvious that the enzyme whose activity limits ethylene biosynthesis is ACC synthase (Kende 1989). The induction of ethylene production by a variety of inducers and environmental stimuli is due to de novo synthesis of this enzyme (Kende 1989). A fundamental question arises concerning its regulation at the molecular level: Are there as many genes as inducers, or is there only one gene and somehow its promoter is activated by all the inducers? Consequently, cloning the ACC synthase gene(s) became a major effort in many laboratories. A cDNA encoding ACC synthase was cloned from zucchini using immunochemical approaches (Sato and Theologis 1989), and its authenticity was confirmed by expression experiments in *Escherichia coli* and yeast (Sato and Theologis 1989; Sato et al. 1991). Immediately thereafter, ACC synthase was cloned from a variety of plant species including tomato (Van der Straeten et al. 1990; Olson et al. 1991; Rottmann et al. 1991), squash (Nakajima et al. 1990; Nakagawa et al. 1991), mung bean (Botella et al. 1992, 1993), apple (Dong et al. 1991), carnation (Park et al. 1992), tobacco (Bailey et al. 1992), *Arabidopsis thaliana* (Liang et al. 1992; Van der Straeten et al. 1992), and rice (Zarembinski and Theologis 1993). The emerging picture indicates that the enzyme is encoded by a highly divergent multigene family. In tomato, for example, ACC synthase is encoded by at least nine genes (Rottmann et al. 1991; Yip et al. 1992; K. Kawakita and A. Theologis, unpubl.), two of which are expressed during fruit ripening (Van der Straeten et al. 1990; Olson et al. 1991; Rottmann et al. 1991; Lincoln et al. 1993). ACC synthase is also encoded by multigene families in *Arabidopsis* and rice whose members are differentially expressed in response to developmental, hormonal, and environmental stimuli (Liang et al. 1992; Rodrigues-Pousada et al. 1993; Zarembinski and Theologis 1993). For example, an *ACS2* promoter *GUS* fusion reveals expression of the *Arabidopsis ACS2* gene in the abscission zone of the petals and sepals (base of the silique, see Fig. 2). The plant hormone auxin, a known inducer of ethylene production (Yang and Hoffman 1984), regulates specific members of each multigene family in a tissue-specific manner (Huang et al. 1991; Nakagawa et al. 1991; Kim et al. 1992; Yip et al. 1992; K. Kawakita and A. Theologis, unpubl.).

Phylogenetic analysis using the PHYLIP programs of the nucleic acid sequences for 20 ACC synthase genes yields an unrooted phylogenetic tree shown in Figure 3. The tree indicates early trifurcation of the ACS genes into three main branches with subsequent divergence within each class accompanying the evolution of monocots and dicots. One major lineage contains the tomato *LE-ACS3* gene along with 5 other genes from

*Figure 2*  Tissue-specific expression of the *Arabidopsis* ACC synthase gene *ACS2*. *GUS* expression driven by the ACS2 promoter in the abscission zone of petals and sepals (X.W. Liang and A. Theologis, unpubl.).


apple (Dong et al. 1991), squash (Nakagawa et al. 1991), *Arabidopsis* (Liang et al. 1992), and rice (Zarembinski and Theologis 1993). On the basis of these results, it appears that the polymorphism of ACC synthase arose prior to the divergence of monocots and dicots. Furthermore, most of the genes in this sublineage are auxin-regulated in vegetative tissue, indicating a striking correlation between their phylogenetic relationship and their pattern of expression.

ACC synthase is a pyridoxal phosphate-requiring enzyme, and most such enzymes have a lysine residue in their active site (Yip et al. 1990). Lys-278 of a tomato isoenzyme, which is conserved in all ACC synthases so far cloned, has been shown by Yip et al. (1990) to be the site of pyridoxal phosphate attachment. Interestingly, the pyridoxal phosphate-

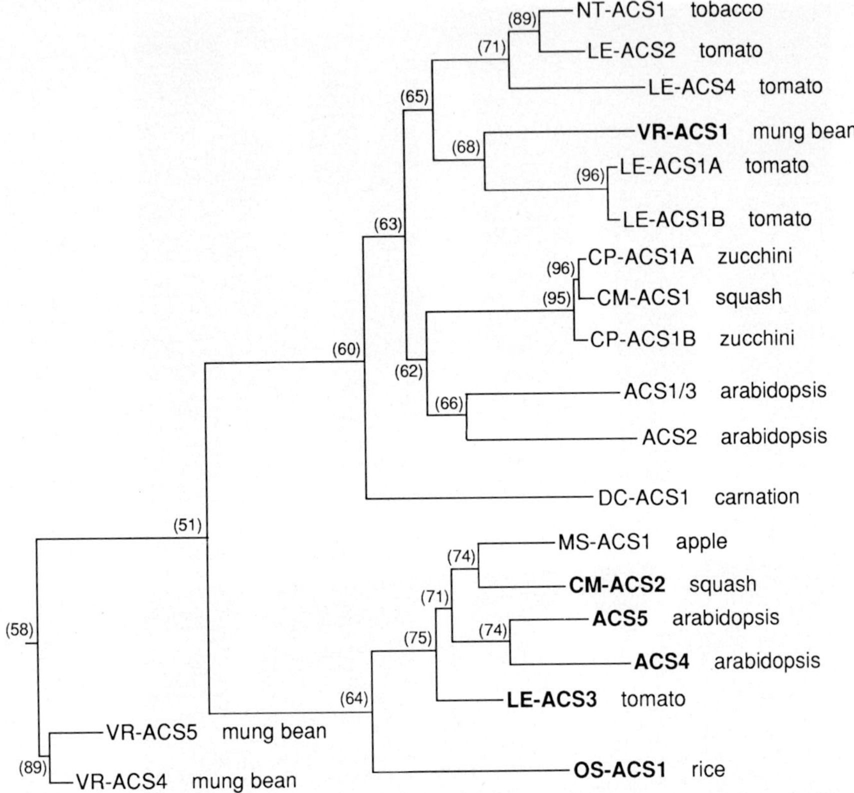



*Figure 3* Phylogenetic tree of ACC synthase constructed on the basis of the entire nucleic acid sequence identity of 20 ACC synthase isoenzymes from various plant species. (Reprinted, with permission, from Lincoln et al. 1993.)

binding site of several aminotransferases contains some of the residues surrounding Lys-278. Furthermore, among various aminotransferases, only 12 amino acid residues are completely conserved, and all but one of these residues are present in the appropriate places in all ACC synthases so far identified. This suggests that aminotransferases and ACC synthase may be evolutionarily related (Rottmann et al. 1991).

### ACC Oxidase

The enzyme is constitutively expressed in most vegetative tissues (Yang and Hoffman 1984) and is induced during fruit ripening (Gray et al. 1992), during carnation flower senescence (Woodson et al. 1992), and by a fungal elicitor (Spanu et al. 1991). A cDNA encoding ACC oxidase was first cloned from tomato fruit by differential screening (Slater et al.

1985), and its authenticity was confirmed by antisense experiments in transgenic plants (Hamilton et al. 1990) and by expression experiments in yeast (Hamilton et al. 1991) and *Xenopus* oocytes (Spanu et al. 1991). ACC oxidase was subsequently cloned from a variety of plant species including avocado (McGarvey et al. 1990, 1992), carnation (Wang and Woodson 1991), apple (Dong et al. 1992b; Ross et al. 1992), peach (Callahan et al. 1992), and orchid (Nadeau et al. 1993). The enzyme is encoded by a multigene family in tomato that is differentially expressed during plant development (Gray et al. 1992). Certain members of the family are constitutively expressed or induced by wounding and during fruit and flower senescence. Recent experimental evidence indicates that ACC oxidase is a dioxygenase that belongs to the superfamily of $Fe^{++}$/ascorbate oxidases and requires $CO_2$ for its activity (Dong et al. 1992a; McGarvey et al. 1992).

## ETHYLENE AND FRUIT SENESCENCE: INTRODUCTION AND HISTORICAL PERSPECTIVE

Folklore says that sealing fruits in a bag encourages them to ripen. Folklore is correct in this case, because the bag traps the ethylene released by the fruit, and ethylene enhances ripening. The earliest record of human manipulation causing fruit to ripen is found in the Old Testament, where the prophet Amos (8th century B.C.) described himself as a "piercer" of sycamore fig fruit. The Greek philosopher Theophrastus (3rd century B.C.) recognized that sycamore figs did not ripen unless they were scraped with an iron claw (Blanpied 1985). We now know, 23 centuries later, that wounding induces ethylene production, resulting in ripening of the fruit (Abeles et al. 1992). Denny (1924) identified ethylene as the active component in the combustion fumes of the kerosene stoves that caused lemon degreening in California and described its use as a ripening agent.

Senescence is a universal phenomenon in living organisms and is viewed as the final phase of development and differentiation (Varner 1961). In plants, ripening of a fruit is the prelude to senescence and involves coordinated changes in various biochemical pathways (Biale 1960, 1964; Burg and Burg 1962, 1965). Fruits are divided into two groups on the basis of their respiratory behavior during ripening: Climacteric fruits such as tomato, avocado, pear, apple, and banana undergo a burst of respiration (climacteric rise) accompanied by marked changes in composition and texture; non-climacteric fruits such as orange, lemon, and strawberry show no changes in fruit composition (Biale and Young 1981). In climacteric fruits, a sharp increase in ethylene production is observed concomitantly with the respiratory upsurge. Fruits of this group are also induced to ripen by treatment with exogenous ethylene at concentrations above 0.1 µl per liter of air. In both cases, once ripening is initiated, the endogenous ethylene production rises autocatalytically (Yang and Hoffman 1984).

Throughout the years, two major theories have emerged to explain the climacteric rise in respiration during the ripening of fruits. On the one hand, the climacteric has been attributed to a surge of protein synthesis (Hulme 1954). On the other hand, the climacteric has been viewed as a consequence of the breakdown in "organization resistance" (Blackman and Parija 1928; Solomos and Laties 1973). Ethylene is considered to be the causative endogenous ripening agent (Kidd and West 1930; Burg and Burg 1962), or according to a more recent interpretation by Biale (1960, 1964), ethylene is a by-product of the ripening process.

The advent of recombinant DNA technology allowed the cloning of genes induced during ripening (Gray et al. 1992 and references therein) and strengthened the view that de novo protein synthesis is a prerequisite for fruit senescence (Hulme 1954). It also led to the proposition that ethylene regulates fruit ripening by coordinating the expression of genes responsible for enhancing the respiratory rise, autocatalytic ethylene production, chlorophyll degradation, carotenoid lycopene synthesis, conversion of starch to sugars, and increased activity of cell-wall-degrading enzymes (Gray et al. 1992 and references therein).

## FRUIT RIPENING AND ITS INHIBITION BY REVERSE GENETIC APPROACHES

Because of the effects of ethylene on plant senescence, large losses of fruits and vegetables occur annually worldwide. Consequently, its role in fruit ripening has been extensively studied, and it has long been a goal of plant biologists to prevent or delay fruit ripening in a reversible manner. To fulfill this goal, various methods have been used, such as ventilation with air under hypobaric pressures (Burg and Burg 1966) or use of inhibitors of ethylene action (Yang and Hoffman 1984). Controlled atmosphere storage has been generally successful for some fruits such as apple, but it is expensive. Inhibitors of ethylene action do not prevent senescence satisfactorily. A more desirable approach would be the construction of mutant plants whose fruits do not ripen until treated with ethylene. Ripening mutants such as *rin*, *nor*, and *Nor* exist in tomato, but their phenotype is not reverted by ethylene (Tigchelaar et al. 1978). Furthermore, ripening mutants have the potential to offer answers about the mechanisms of $C_2H_4$ action and its role in the ripening process.

The cloning of genes induced during fruit ripening (Gray et al. 1992) and of genes involved in ethylene biosynthesis (Theologis 1992; Kende 1993) allowed the construction of ripening mutants in tomato using reverse genetics. Since gene replacement technology is in its infancy in plants, antisense RNA technology became the tool of choice. Initially, attempts were made to inhibit tomato fruit softening by antisense polygalacturonase (*PG*) RNA, a gene that has been widely considered to be responsible for cell wall loosening during ripening (Sheehy et al. 1988; Smith et al. 1988). Unfortunately, *PG* antisense failed to give a strong effect. *PG* mRNA accumulation and enzyme activity were severely inhibited in antisense fruits, but the fruits still soften, suggesting that PG is not the only determinant of cell wall loosening (see below).

Attempts either to metabolize ACC by overexpressing the *Pseudomonas* ACC deaminase gene (Klee et al. 1991) or to inhibit ACC oxidase with antisense RNA (Hamilton et al. 1990) were partially successful in inhibiting fruit senescence. Both mutants produce basal levels of ethylene (>0.5 nl/g fresh weight) that are sufficient to initiate fruit ripening. However, tomato mutants constructed by expressing antisense RNA to the ACC synthase gene (LE-ACS2), which is expressed during ripening, were found to be less leaky and severely inhibited fruit senescence. Two different promoters were used to express the LE-ACS2 antisense RNA, the constitutive CaMV *35S* and the fruit-specific *E8* (Theologis et al. 1993). Severe inhibition of ethylene production (<0.1 nl/g per hr) was observed in some transgenic lines expressing antisense ACC synthase with the CaMV *35S* promoter (Oeller et al. 1991; Theologis et al. 1993). When ethylene production is above 0.1 nl/g per hour, the ripening process is not inhibited. Antisense fruits accumulate large amounts of antisense *LE-ACS2* RNA, resulting in complete inhibition of the two fruit-ripening-inducible ACC synthase mRNAs, *LE-ACS2* and *LE-ACS4* (Oeller et al. 1991). The inhibition of *LE-ACS4* mRNA accumulation by the *LE-ACS2* antisense RNA is attributed to a stretch of 180 bp of near complete sequence identity between the two mRNAs near the amino terminus (Rottmann et al. 1991).

Detached, mature LE-ACS2 antisense fruits kept at 20°C never ripen. They develop a yellow/orange color as time progresses but never turn red and soft or develop an aroma. Antisense fruits do not show the climacteric rise of respiration even when they are 95 days old (Oeller et al. 1991). The antisense LE-ACS2 phenotype is reversed by treatment with exogenous $C_2H_4$ or $C_3H_6$, an ethylene analog, with a concomitant rise in respiration (Oeller et al. 1991). Treated fruits are indistinguishable from naturally ripened fruits with respect to texture, color, aroma, and compressibility (Oeller et al. 1991). The reversibility of the antisense phenotype strongly suggests that the inhibition of fruit ripening is due to the specific inactivation of the target gene (ACC synthase) and not due to nonspecific inactivation of regulatory genes required for fruit ripening (Woolf et al. 1992). The inability of exogenous $C_2H_4$ to fully reverse the phenotype of fruits expressing antisense ACC oxidase RNA (Picton et al. 1993) may be due to nonspecific effects of this antisense RNA. Our inability to detect sense RNA in antisense fruits suggests that the mechanism of gene inactivation by antisense RNA involves increased mRNA degradation rather than inhibition of the translatability of the target mRNA (Inouye 1988).

The reversibility of the antisense LE-ACS2 phenotype by exogenous $C_2H_4$ or $C_3H_6$ is complete after 6 days of treatment (Oeller et al. 1991). Shorter treatments are insufficient for full reversal. Three major conclusions can be deduced from these results. First, ethylene-mediated ripening requires continuous transcription of short-lived mRNAs or proteins encoded by them. Second, ethylene is indeed autocatalytically regulated. Finally, the hormone acts as a rheostat rather than as a switch for controlling the ripening process. Furthermore, ethylene is the key regulatory molecule for fruit ripening and senescence, not the by-product of ripening (Biale 1960, 1964; Burg and Burg 1965). The increase in respiration during ripening is viewed as a consequence of the ripening process that sustains the energy demand due to enhanced transcription and protein synthesis required for the ethylene-mediated ripening (Theologis 1992).

### ETHYLENE AND GENE EXPRESSION DURING FRUIT RIPENING

Fruit ripening is associated with dramatic changes in gene expression (Gray et al. 1992 and references therein). ACC synthase antisense fruits producing low levels of ethylene have been useful in assessing which ripening-induced genes are indeed ethylene inducible. The expression of the ACC oxidase gene (*pTOM13*) is found to be ethylene independent and precedes ACC synthase gene expression (Oeller et al. 1991). ACC oxidase activity is known to increase earlier and in greater magnitude than that of ACC synthase (Yang 1987).

Similarly, *PG* gene expression is ethylene independent (Oeller et al. 1991; Theologis et al. 1993). Although antisense fruits express large amounts of *PG* mRNA, they fail to accumulate the *PG* polypeptide, indicating that ethylene controls the translatability of *PG* mRNA or the stability of the *PG* polypeptide. Ethylene is known to enhance polyribose formation in carrot roots (Christoffersen and Laties 1982) and to post-translationally regulate polygalacturonase and cellulase enzyme activities in avocado fruits (Buse and Laties 1993). Interestingly, transgenic *rin* tomatoes (low ethylene producers) express large amounts of *PG* mRNA from an *E8-PG* chimeric construct without an increase in *PG* protein during air treatment. Propylene treatment, however, results in *PG* protein accumulation (Giovannoni et al. 1990), suggesting also a translational or posttranslational role of ethylene in *PG* expression. The accumulated experimental evidence indicates that *PG* gene expression is developmentally regulated and its enzyme activity is regulated by ethylene at the translational or posttranslational level. Furthermore, the results with the antisense fruits and the *rin* transgenes indisputably demonstrate that *PG* is not the sole determinant of fruit softening (Giovanonni et al. 1989, 1990; Theologis 1992).

Additional studies also show that among the five anonymous ripening-associated genes, *E4, E8, J49, E17,* and *D2,* only *E4* is ethylene regulated. The others are developmentally regulated and their expression is quite complex (Theologis et al. 1993). Surprisingly, *E4* gene expression is not restored in LE-ACS-2 antisense fruits by $C_3H_6$ treatment, suggesting that a transient developmental factor(s) may be involved in its ethylene regulation. However, antisense fruits treated with $C_3H_6$ ripen normally in the absence of *E4* gene expression, suggesting that *E4* protein is not necessary for the ripening process. *E8* gene expression has been reported to be ethylene regulated (Giovanonni et al. 1989), but our results suggest that *E8* is ethylene independent (Theologis et al. 1993). Since *E8* is encoded by a multigene family (Deikman and Fischer 1988), the possibility exists that the cloned *E8* gene is ethylene regulated but its mRNA is masked by the non-ethylene-regulated *E8* genes. It has been suggested that *J49* and *E17* gene expression is regulated by changes in $C_2H_4$ sensitivity (Lincoln and Fischer 1988b). Results with antisense fruit show that *J49* and *E17* are ethylene independent and their expression patterns vary depending on the ripening stage of the fruit (Theologis et al. 1993).

## ETHYLENE SIGNAL TRANSDUCTION PATHWAY(S) DURING FRUIT RIPENING

The analysis of ACC synthase antisense fruits reveals that at least two signal transduction pathways are operational during tomato fruit ripening. The ethylene-independent (developmental) pathway is responsible for transcriptional activation (or repression) of genes such as ACC oxidase, *PG, D2, E17, J49,* and *E8-X.* On the other hand, the ethylene-dependent pathway is responsible for the transcriptional and posttranscriptional regulation of genes involved in lycopene and aroma biosynthesis, respiratory metabolism, ACC synthase gene expression, and translation of genes such as *PG* and cellulase (Theologis et al. 1993).

Ethylene appears to have a dual role in fruit senescence. First, it activates transcription of yet-to-be-identified genes whose products are unstable but required for fruit senescence. Second, it regulates the translation of developmentally regulated mRNA such as *PG*. The possibility exists that a large number of genes encoding secretory proteins responsible for cell wall degradation are regulated by ethylene posttranscriptionally (translational or posttranslational control). It remains to be determined which are the hormones and developmental signals responsible for the activation of the ACC synthase genes that will lead to the first $C_2H_4$ production responsible for activation of its signal transduction pathway that leads to fruit ripening.

## ETHYLENE PERCEPTION

Although the pathway of ethylene biosynthesis has been elucidated, the mechanisms of ethylene detection and signal transduction have remained a mystery until very recently (Kieber and Ecker 1993). It has been postulated that since ethylene is an olefin, its receptor may be either a $Zn^{++}$- or a $Cu^+$-containing metalloprotein (Burg and Burg 1967; Sisler 1991). The suggestion is attractive in view of the fact that compounds such as CO, which are known to complex with transition metals, have ethylene-like effects on plants (Burg and Burg 1967; Sisler 1991). It is of great interest, therefore, that tomato seedlings deficient in $Zn^{++}$ do not respond to ethylene (Burg and Burg 1967). However, attempts to isolate the putative metalloprotein receptor for ethylene have been frustrating and unsuccessful (Hall et al. 1990).

## ETHYLENE-MEDIATED TRIPLE RESPONSE OF *ARABIDOPSIS*

Exposure of etiolated seedlings to ethylene brings about morphological changes that are collectively known as the triple response (Knight et al. 1910a). This response was first characterized in peas by Neljubov (1901) and consists of a shortening and radial swelling of the epicotyl and a loss of normal gravitropism (diageotropism). The triple response was used as a bioassay for ethylene during the subsequent half-century (Knight et al. 1910b). Characteristics of the ethylene-mediated triple response in *Arabidopsis* include exaggerated tightening of the apical hook, radial swelling of the hypocotyl, and inhibition of root and hypocotyl elongation (Fig. 4). Due to its high reproducibility, ease of screening large numbers of individuals, and the early stage of development at which the screen is carried out, the triple response phenotype provides a facile

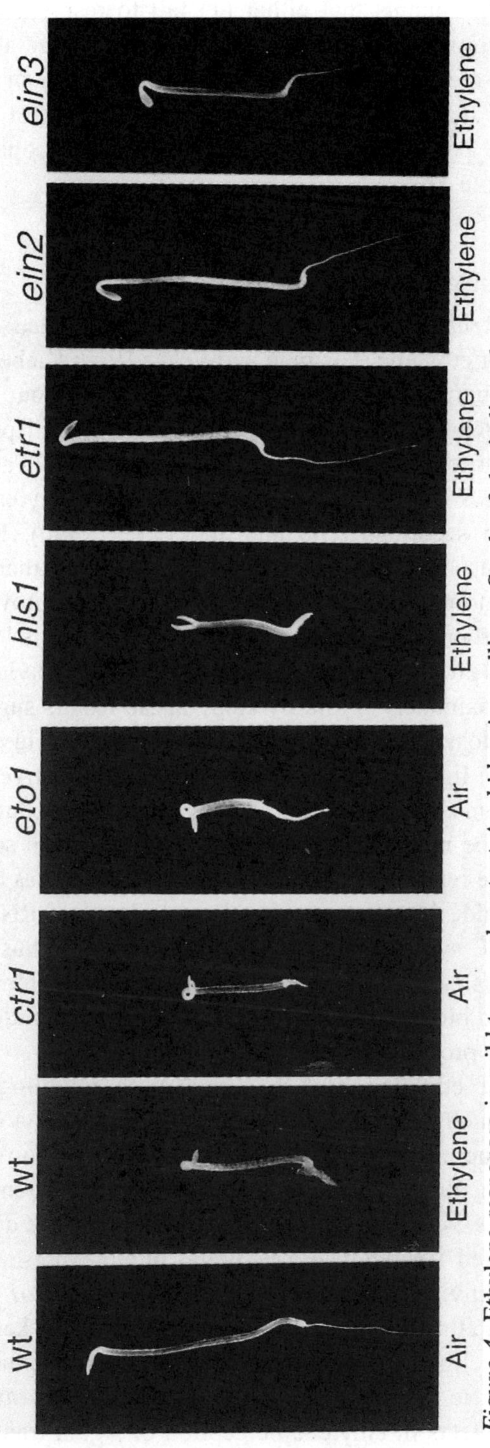


*Figure 4*  Ethylene responses in wild-type and mutant *Arabidopsis* seedlings. Seeds of the indicated genotypes were germinated and grown for 3 days in the dark in either air or air containing 10 ppm ethylene as indicated.

means to identify mutants that either (1) fail to respond to exogenous ethylene or (2) constitutively display the response in the absence of the hormone. Such screens have allowed the identification of a number of genes that are likely to be involved in the control of ethylene biosynthesis, the perception of ethylene, and/or the propagation of its stimulus (see Table 1).

## GENES THAT REGULATE ETHYLENE BIOSYNTHESIS

A number of *Arabidopsis* mutants have been isolated that produce elevated levels of ethylene (Guzman and Ecker 1990; Kieber et al. 1993). The *et*hylene *o*verproduction (*eto*) mutants have been identified by screening mutagenized $M_2$ seedlings for plants that display the triple response in the absence of exogenously added ethylene (Table 1; Fig. 4). The recessive *eto1* mutant produces 10-fold more ethylene than wild-type dark-grown seedlings (Guzman and Ecker 1990). Recently, two dominant mutants have been isolated, *eto2* and *eto3*, that produce 20- and 100-fold more ethylene than wild-type, dark-grown seedlings, respectively (Kieber et al. 1993). Interestingly, the rosette leaves of these mutants do not produce highly elevated levels of ethylene, the most being only approximately 2-fold by *eto3*. These results suggest that perhaps ethylene biosynthesis is regulated independently in seedlings and adult plants or in light- and dark-grown plants. Alternatively, a negative feedback mechanism may repress excess ethylene production in adult Eto⁻ mutants. The recessive nature of the *eto1* mutation suggests that it acts as a negative regulator of ethylene biosynthesis. It has been observed that cyclohexamide increases the steady-state level of RNA for all five *Arabidopsis* ACC synthase genes (Liang et al. 1992). Thus, it is possible that ETO1 acts as a short-lived negative regulator of ethylene biosynthesis, and blocking its synthesis with cyclohexamide results in increased ethylene production.

Treatment of etiolated Eto⁻ seedlings with inhibitors of ethylene biosynthesis, such as aminoethoxyvinylglycine (AVG), or with antagonists of the ethylene receptor such as *trans*-cyclo-octene or Ag⁺, results in abrogation of the constitutive triple response phenotype (Kieber et al. 1993). It has not been possible to directly quantitate ACC levels in etiolated *Arabidopsis* seedlings, but the presence of normal ACC oxidase activity suggests that the alteration in *eto1* occurs before the final step in ethylene biosynthesis (Guzman and Ecker 1990). Interestingly, mature light-grown Eto⁻ plants produce amounts of ethylene that are similar to wild-type plants and exhibit a normal phenotype. Mutants with defects in ethylene perception or signal transduction show

*Table 1*  *Arabidopsis* ethylene mutants

| Mutant | Phenotype | Chromosome | Comments | References |
|---|---|---|---|---|
| *etr* | insensitive | 1, bottom | dominant; reduced ethylene binding, gene cloned | Bleecker et al. (1988); Chang et al. (1993) |
| *ein1* | insensitive | 1, bottom | dominant; allelic to *etr* | Guzman and Ecker (1990) |
| *ein2* | insensitive | 5, top | recessive; tolerant to pathogens, gene cloned | G. Roman and J.R. Ecker (unpubl.) |
| *ein3* | insensitive | 3, top | recessive; "weak" phenotype; gene cloned | M. Rothenberg and J.R. Ecker (unpubl.) |
| *ein4* | insensitive | ? | dominant; not allelic to *etr1* | G. Roman and J.R. Ecker (unpubl.) |
| *ein5* | insensitive | 1, middle | recessive; possibly allelic to *ain1* | G. Roman and J.R. Ecker (unpubl.) |
| *ain1* | insensitive | 1, middle | recessive; "weak "phenotype | Van der Straeten et al. (1993) |
| *eti* | insensitive | ? | 5 isolates; poorly characterized genetically | Harpham et al. (1991) |
| *eir1* | insensitive | 5, bottom | recessive; ethylene-insensitive root | G. Roman and J.R. Ecker (unpubl.) |
| *eto1* | constitutive | 3, bottom | recessive; ethylene overproducer | Guzman and Ecker (1990) |
| *eto2* | constitutive | 5, bottom | dominant; ethylene overproducer | Kieber et al. (1993) |
| *eto3* | constitutive | 3, bottom | dominant; ethylene overproducer | Kieber et al. (1993) |
| *ctr1* | constitutive | 5, top | recessive; gene cloned | Kieber et al. (1993) |
| *hls1* | hookless | 4, bottom | recessive; gene cloned | Guzman and Ecker 1990; A. Lehman and J.R. Ecker (unpubl.) |

increased ethylene production relative to wild-type plants (Guzman and Ecker 1990). The increase in ethylene production in these mutants suggests that auto-inhibition of ethylene biosynthesis may be affected by the defect in ethylene perception (Bleecker et al. 1988).

## GENES THAT CONTROL ETHYLENE PERCEPTION OR SIGNAL TRANSDUCTION

A number of *Arabidopsis* mutants that show varying degrees of insensitivity to ethylene have been isolated (Table 1). Ethylene-resistant (*etr*) (Bleecker et al. 1988) , -insensitive (*ein*) (Guzman and Ecker 1990) or *eti* (Harpham et al. 1991), and ACC-insensitive (*ain*) (Van der Straeten et al. 1993) plants show either complete deficiency or reduction in the magnitude of the ethylene-mediated triple response. Mutant (tall) seedlings are readily identified protruding above the "lawn" of wild-type (short) seedlings when mutagenized populations are plated in the dark in the presence of ethylene (Fig. 4). The dominant *etr1* mutant (also previously called *etr* and *ein1*) is inherited as a single-gene, dominant mutation (Bleecker et al. 1988; Guzman and Ecker 1990). *etr1* was selected on the basis of its insensitivity to ethylene-mediated inhibition of hypocotyl elongation in etiolated seedlings. Further analysis showed that this mutant is defective in a number of other ethylene responses, including promotion of seed germination, enhancement of peroxidase activity, acceleration of senescence of detached leaves, and negative feedback of ethylene biosynthesis (Bleecker et al. 1988; Guzman and Ecker 1990). Furthermore, ethylene treatment of *etr1* plants does not cause expression of the ethylene-regulated basic chitinase gene (Samac et al. 1990). Notably, the *etr1* mutant was shown to bind only one-fifth the amount of ethylene bound by wild-type plants. The pleiotropic effects of *etr1* suggest that the wild-type gene may encode an ethylene receptor or act at an early step in the signal transduction pathway (Bleecker et al. 1988). The *ETR1* gene has been isolated by positional cloning (Chang et al. 1993). The predicted translation product of *ETR1* showed striking homology with a family of bacterial histidine protein kinases and a yeast gene called SLN1 (Ota and Varshavsky 1993). Four dominant *ETR1* alleles were sequenced and all of the mutations resulted from amino acid substitutions in either of two predicted transmembrane domains. Limited gene dosage experiments suggest that *etr1* alleles may result from gain-of-function mutations (i.e., constitutive kinase activity). The absence of recessive (loss-of-function) alleles may indicate redundancy of ETR1 function. In support of this hypothesis, several *ETR1* homologous genes have been isolated (Chang et al. 1993; C. Chang and E. Meyerowitz, pers. comm.). Alternatively, null mutations in the *ETR1* gene may not allow for plant viability.

A second ethylene-insensitive mutation called *ein2* is recessive and not allelic to *etr1*. Similar to *etr1*, strong alleles of *ein2* are pleiotropic and lack all known ethylene responses (Guzman and Ecker 1990; Rothenberg and Ecker 1993; Kieber et al. 1993; Lawton et al. 1994; G. Roman and J.R. Ecker, unpubl.). Interestingly, alleles of *ein2*, but not *etr1* or another mutant called *ein3*, demonstrated increased tolerance to virulent bacterial pathogens (Bent et al. 1992; see below). Recently, the *EIN2* gene has been isolated by positional cloning (G. Roman and J.R. Ecker, in prep.). Analysis of the *EIN2* protein product may provide clues about its normal requirement for both ethylene and pathogen sensitivity.

Additional Ein$^-$ mutants have been identified that show reduced sensitivity to ethylene in the seedling triple response assay. A representative member of this class of "weak" insensitive mutants is the recessive mutation *ein3* (Fig. 4) (Kieber et al. 1993). The *EIN3* gene has been cloned, and although DNA sequence analysis has revealed that at least one allele is a null mutation, all of the ethylene phenotypes of *ein3* mutants are less pronounced than that of either *etr1* or *ein2* (M. Rothenberg and J.R. Ecker, unpubl.). Low-stringency hybridization experiments indicate that there are at least three other sequences that are closely related to *EIN3* in the *Arabidopsis* genome (M. Rothenberg and J.R. Ecker, unpubl.). These additional genes may represent a functional redundancy of *EIN3*, which could account for the "leaky" ethylene-insensitive phenotype. Furthermore, several novel Ein$^-$ mutants have been identified (*ein5*, *ein6*, and *ein7*) that also confer similar levels of insensitivity to ethylene (G. Roman et al., unpubl.).

An additional recessive ethylene-insensitive mutant called *ain1* was identified using the immediate precursor of ethylene, ACC (Van der Straeten et al. 1993). Similar to *ein3*, *ain1* showed greater sensitivity to ethylene in the triple response assay than either *etr1* or *ein2*. However, these mutations are not allelic, since *ain1* has been mapped to the middle of chromosome 1 (D. Van der Straeten, pers. comm.), whereas *ein3* has been mapped to chromosome 3 (M. Rothenberg and J.R. Ecker, unpubl.).

In another study, five ethylene-insensitive seedlings were isolated that were referred to as *eti* (Harpham et al. 1991). These plants have been only partially characterized genetically, and so it is unclear whether they represent independent or novel loci. Like the other ethylene-insensitive mutants, etiolated Eti$^-$ seedlings lack the ethylene-promoted apical hook curvature response: a component of the etiolated seedling triple response. As Haberlandt first proposed (quoted in Darwin 1869), the apical hook of dicotyledonous seedlings may serve to protect the delicate apical meristem during emergence through the soil. Ethylene is produced at high levels in etiolated seedlings in response to physical impedance and causes inhibition of normal elongation of etiolated epicotyl or hypocotyl cells, with more radial enlargement of these same cells (Goeshel et al. 1966). The change in cell shape is accompanied by (or possibly results from) a dramatic change in the orientation of microtubules (Apelbaum and Burg 1971; Eisinger 1983). It has been suggested that this morphological response to ethylene is a mechanism by which seedlings circumvent mechanical impediments in soil during emergence (Goeschl et al. 1966 ). The ethylene-induced radial expansion of the seedling stem may provide the plant with greater ability to penetrate compact soil. In support of these suggestions, the ability of the *eti* mutant seedlings to emerge through compacted sand was reported to be directly proportional to their sensitivity to ethylene (Harpham et al. 1991).

Although the identification of all of the *Arabidopsis* ethylene-insensitive mutants has been based on a phenotype displayed by etiolated seedlings, these mutations also have effects on all known adult plant responses to ethylene. Such results suggest that the ethylene signal perception and/or transduction pathways in seedlings and adult plants must share at least some common components in *Arabidopsis*. Recent studies indicate that this may also be true for other plants, including those that undergo an ethylene-mediated climacteric such as tomato. A number of tomato mutants that are affected in the ripening process are known (Abeles et al. 1992). The *Never ripe* (*Nr*), *ripening inhibitor* (*rin*), and *non-ripening* (*nor*) mutants are delayed in fruit ripening. The *rin* mutant fails to display the burst of ethylene production normally observed during fruit ripening but shows normal induction of several genes upon treatment with exogenous ethylene (Herner and Sink 1973; Lincoln and Fischer 1988a). However, treatment of *rin* fruits with ethylene does not result in fruits that are fully ripened (Lincoln and Fischer 1988a). The partially dominant *Nr* mutation shows pleiotropic effects on plant development (Lanahan et al. 1994). Not only is fruit ripening blocked, but flower and petal abscission and epinasty are also affected. Furthermore, *Nr* mutants fail to display the seedling triple response in the presence of ethylene, but *rin* and *nor* seedlings show a normal triple response (Lanahan et al. 1994). These results suggest that *Nr* may generally affect ethylene perception, but that *rin* and *nor* may affect ethylene sensitivity specifically during fruit ripening. Thus, as with *Arabidopsis*, a single gene mutation in tomato has profound effects on both seedling (triple response) and adult (abscission, senescence, and fruit ripening) ethylene responses. The semidominant *Nr* mutant may correspond to one of the previously identified *Arabidopsis* mutants; possibly *etr1* (Chang et al.

1993), *ein4*, or *ein7* (G. Roman and J.R. Ecker, unpubl.).

As described above, the ease with which millions of plants can be tested for ethylene responsiveness using the seedling triple response screen should provide a facile means for the identification of new tomato fruit-ripening mutants and additional alleles of existing mutants such as *Nr*. In principle, this assay may be used to identify similar mutants in any plant whose fruits undergo ethylene-mediated climacteric ripening or whose flowers show ethylene-induced petal abscission and where it is possible to generate large quantities of mutagenized seeds.

## MUTANTS THAT DISPLAY CONSTITUTIVE ETHYLENE RESPONSES

A number of mutants have been identified that display the triple response in the absence of exogenously added ethylene, one class of which are the ethylene-overproducing mutants discussed above. A second class constitutively display the triple response even in the presence of inhibitors of ethylene biosynthesis and binding, suggesting that these mutants affect ethylene signal transduction (Fig. 4) (Kieber et al. 1993). All of these are recessive and fall into a single complementation group called *ctr1* (*constitutive triple response*). Thus, at least one component of the ethylene response pathway may be under negative control because loss of function of the *CTR1* gene results in constitutive activation of all known ethylene responses.

Unlike the Eto⁻ mutants, *ctr1* produces less ethylene than the wild type, and its phenotype cannot be reverted by inhibitors of ethylene biosynthesis (AVG) or action (Ag⁺ or *trans*-cyclo-octene). Adult *ctr1* plants show an altered phenotype, with their growth habit being compact and epinastic, resembling wild-type plants grown in 10 µl ethylene per liter of air (Kieber et al. 1993). The *ctr1* mutant allele is transmitted at a reduced frequency relative to the wild-type allele, and reciprocal back-crosses revealed that this is due to a defect in female gametophytes (Kieber and Ecker 1994). The *ctr1* mutation has dramatic effects on the morphology and development of seedlings and adult plants. Etiolated *ctr1* seedlings take longer to open the apical hook and expand their cotyledons when shifted to light than do wild-type seedlings; this phenotype is opposite that of the Ein⁻ mutants, where the apical hook opens in complete darkness. *ctr1* rosette leaves are smaller, the root system is much less extensive, the plants bolt later, and the inflorescence is much more compact than that of wild-type plants. Epidermal cells from *ctr1* leaves are 5-fold smaller than those from wild-type plants. This reduction in cell size accounts for at least part of the decrease in leaf size observed in *ctr1* and may underlie some of the other phenotypes such as the shortened hypocotyl, compacted inflorescence, and reduced root system. Consistent with these results, ethylene has been shown to inhibit cell elongation in other systems, perhaps due to a reorientation of the cytoskeleton and/or cell wall (for review, see Eisinger 1983). However, ethylene has been shown to inhibit DNA synthesis and subsequent cell division in etiolated seedlings (Apelbaum and Burg 1972). Thus, a reduction in cell number may also contribute to the Ctr⁻ phenotype. Both *etr1* and *ein2* have slightly larger rosette leaves than wild-type plants, which suggests that they may have larger cells and/or more cells than wild-type plants, perhaps due to a failure to respond to a basal level of ethylene (Bleecker et al. 1988; Guzman and Ecker 1990). Ctr⁻ can be phenocopied by growth of wild-type plants in ethylene, suggesting that *ctr1* constitutively displays both seedling and adult ethylene responses (Kieber et al. 1993).

The *CTR1* gene has been cloned (Kieber et al. 1993). The carboxyl half of this 92-kD protein has all the hallmark features of a serine/threonine protein kinase. The highest similarity obtained in searches of the protein databases with CTR1 is to the Raf family of protein kinases. Raf was originally identified as a cellular homolog of v-*raf*, the transforming gene from an avian retrovirus (Rapp et al. 1988). In humans there are three Raf genes, the best characterized of which is *Raf-1*. The transduction of a number of external regulatory signals including mitogens and growth hormones is controlled by *Raf-1* (Rapp 1991). A *Drosophila* Raf homolog, Draf, which was identified genetically by the *polehole* mutation, has been shown to act downstream from *torso* in the signal transduction pathways leading to the proper development of the posterior and anterior ends of larvae (Brand and Perrimon 1994). The Draf gene has also been shown to act in the signal transduction pathway leading to development of the R7 photosystem in the *Drosophila* eye (Dickson et al. 1992). Recently, a mutation in a Raf homolog in *Caenorhabditis elegans* (*lin*45) has been shown to be involved in vulva development, acting downstream from the *let*-23 gene product (Han et al. 1993).

Two of the *ctr1* alleles resulted from single amino acid changes. Both substitutions were in residues that are invariant or nearly invariant in all known protein kinases (Hanks et al. 1988; Hanks and Quinn 1991). *ctr1-4* results in a Glu→Lys change at the position corresponding to Glu-91. This residue is involved in an ionic interaction with Lys-72, and thus alteration to an amino acid with an opposite charge would be expected to disrupt the structure of the protein. The change in *ctr1-1* is an Asp→Glu change at the position corresponding to Asp-184. This residue is involved in chelating Mg⁺⁺, as well as in hydrogen bonding with both Glu-

91 and Asn-171. This hydrogen bonding is important for keeping the two lobes of the catalytic domain together (see Taylor et al. 1992). Therefore, it is not surprising that even a conservative change to a glutamic acid residue would disrupt the kinase activity of CTR1. A systematic mutagenesis of yeast cAPK (Gibbs and Zoller 1991) has shown that when either of these residues is changed to an uncharged alanine, catalytic activity is either very strongly reduced (Glu-91) or essentially abolished (Asp-184).

The *CTR1* gene may represent a widespread component of ethylene signal transduction in plants, as similar DNA sequences were detected in a number of plant species, including tomato, tobacco, carrot, beet, and monocots such as corn and rice (Kieber and Ecker 1994).

## GENES THAT CONTROL TISSUE/CELL-SPECIFIC RESPONSES TO ETHYLENE

### Hypocotyl Hook Mutants

The hypocotyl hook can be considered as a standing wave of plant growth (Silk and Erickson 1978). Cells, produced at the apical meristem, appear to "flow" through the hook as they elongate. At different points in its passage through the hook, each cell must be capable of accelerating and decelerating its rate of elongation. A complex pattern of cell "movement" (elongation) is needed to establish and maintain this structure where cell "passage" through the hook is occurring over a relatively short period of time (hours). Ethylene is involved in the development of the etiolated seedling apical hook (Kang et al. 1967). Application of exogenous ethylene results in exaggeration of the curvature in the hook region of the hypocotyl (Fig. 4). Furthermore, mutations or chemicals that alter sensitivity to ethylene have reduced curvature in the hook region (Fig. 4 and data not shown).

Similar to ethylene-mediated inhibition of hypocotyl and root growth in etiolated seedlings, cell elongation in the apical hook region is also affected by ethylene. However, in this region of the etiolated seedling, cells in the hypocotyl undergo a differential growth response to ethylene; cells on the inside of the apical hook are more inhibited in growth in response to ethylene treatment than the cells on the outside. One can speculate that a "gradient" of ethylene responsiveness of cells in the hypocotyl may result in exaggeration of the curvature of the hook in the presence of ethylene (Fig. 4). Formation and maintenance of the apical hook are equally dependent on a second hormone, auxin. Inhibition of auxin transport using drugs such as N-1-naphthylphthalamic acid (NPA) or 2,3,5-tri-iodobenzoic acid (TIBA) completely blocks hook formation in *Arabi-*

*dopsis* (A. Lehman and J.R. Ecker, unpubl.). In addition, germination of wild-type seedlings on medium with 10 µм 2,4-dichlorophenoxyacetic acid (2,4-D), a synthetic auxin, prevents apical hook formation (A. Lehman and J.R. Ecker, unpubl.).

As described above, there is a complex interplay between ethylene and auxin in the establishment and maintenance of this differential growth process. Auxin can stimulate transcription of the ethylene biosynthetic gene, ACC synthase, and it is well established that transport of auxin is greatly affected by ethylene (Burg and Burg 1967). However, it is known that many of the effects of high-level auxin treatment on plant development are not mediated by auxin induction of ethylene (Romano et al. 1993). Mutations that would affect ethylene's ability to control auxin transport, as well as those affecting the hypothetical gradient of ethylene or auxin perception of cells in the hypocotyl hook region, may be identified in screens for "hook-affected" phenotypes. A number of new or existing ethylene and auxin mutants have been identified using this assay. The *hookless3* (*hls3*) mutation results in 6-fold overproduction of auxin and loss of the apical hook (A. Lehman and J.R. Ecker, unpubl.). Mutations that result in increased resistance to auxin (such as *aux1*) or ethylene (such as *etr1*, *ein2*) also diminish hook formation or maintenance (MacDonald et al. 1983; A. Lehman and J.R. Ecker, unpubl.). The recessive *hookless1* (*hls1*) mutation completely abolishes apical hook development (Fig. 4) (Guzman and Ecker 1990). The *HLS1* gene has been cloned (A. Lehman and J.R. Ecker, in prep.). Interestingly, certain alleles of *hls1* (*hls1-2*) can be partially complemented in hook formation by the addition of ethylene. Its primary sequence may yield clues as to the role of ethylene in apical hook development and, more generally, may contribute to our understanding of differential cell growth processes in plants.

### Root Mutants

A second example of a "tissue"-specific mutation is the ethylene insensitive root (*eir1*) mutant. The *eir1* mutation affects ethylene sensitivity specifically in the root (G. Roman and J.R. Ecker, in prep.). Unlike the *aux1* mutant, which shows resistance to auxin and ethylene in the root elongation assay (Pickett et al. 1990), *eir1* shows normal sensitivity to high levels of auxin. However, *eir1* is similar to *aux1*, in that it demonstrates a root gravitropism defect. Further analysis of the *eir1* mutation should provide significant information on the enigmatic role of ethylene in the perception of gravity (Wheeler and Salisbury 1980).

## CROSS-TALK BETWEEN ETHYLENE, OTHER HORMONES, AND LIGHT

### Multihormone Resistance

Ethylene is involved in numerous developmental and environmental responses in higher plants (Abeles et al. 1992). Many of these responses are likely the result of a complex interplay between various hormonal and nonhormonal signals. For example, the triple response involves signals from ethylene, auxin, and light, as well as intrinsic developmental control (A. Lehman and J.R. Ecker, unpubl.). How the integration of numerous inputs is accomplished in plants is unclear. One possibility is that multiple signals are funneled through a single downstream target and then branch again, as seems to be the emerging theme for the Raf signal pathway. Interestingly, the *Arabidopsis aux1 axr1* and *axr2* mutants, which were isolated in a screen for resistance to growth-inhibiting levels of auxin, also show increased resistance to other plant hormones, including abscisic acid, and ethylene (Pickett et al. 1990; Wilson et al. 1990). The same is true for all of the ethylene-insensitive mutants in response to exogenously applied cytokinin; they show resistance to cytokinin-mediated inhibition of root elongation (G. Roman and J.R. Ecker, unpubl.). However, recent studies have revealed that the inhibition of cell elongation in wild-type seedlings by cytokinin is, in fact, mediated by ethylene. The "cytokinin" effect (inhibition of root and hypocotyl elongation and cotyledon expansion) can be completely suppressed by blocking ethylene biosynthesis or perception with AVG or Ag$^+$, respectively (G. Roman and J.R. Ecker, unpubl.). Furthermore, a cytokinin-resistant *Arabidopsis* mutant called *ckr1* (Su and Howell 1992) was found to be an allele of *ein2* (G. Roman et al., in prep.). Thus, the cytokinin resistance of Ein$^-$ mutants is mediated by loss of sensitivity to ethylene.

### Photomorphogenic Phenotypes

Maintenance of the hypocotyl hook is suppressed by red (Rubinstein 1971) and blue light (Liscum and Hangarter 1993). Cells on the inner side of the hook begin to elongate at a greater rate relative to those on the outer side, which results in hook opening (Silk and Erickson 1978). As described above, etiolated *ctr1* seedlings show a drastic reduction in rapidity of light-mediated de-etiolation responses: cotyledon expansion and hook opening. One might expect that mutants which demonstrate constitutive light phenotypes in the absence of light (*cop* and *det*) may overlap with those of the apical hookless class (Chory et al. 1989; Deng et al. 1991). As with *hls1* and *hls2*, etiolated *cop2* and *cop3* seedlings have expanded cotyledons and lack a hypocotyl hook (Hou et al. 1993).

Unlike the other Cop⁻ mutants, *cop3* and *cop2* seedlings showed normal (1) regulation of light-regulated genes, (2) elongation of the hypocotyl and root, and (3) chloroplast development. These results suggest that these mutations (*cop2* and *cop3*) may effect ethylene or auxin regulation of cell expansion in the apical region of the hypocotyl. In support of this hypothesis, genetic complementation studies indicate that *cop3* is an allele of *hls1* (A. Lehman and J.R. Ecker, unpubl.). In regard to regulation of differential growth in the apical hook region by both light and ethylene, it will be interesting to examine the effects of combinations of mutations that cause constitutive suppression of the hook (such as *det1* or *cop1*) with those that cause constitutive hook formation (such as *ctr1* or *eto1*). Such studies may allow positioning of the point(s) of intersection between ethylene and light signaling pathways in the control of skoto-/ photo-morphogenic development.

## ROLE OF ETHYLENE IN PLANT DISEASE TOLERANCE

Ethylene has been implicated in the response of plants to pathogen attack (see, e.g. Ecker and Davis 1987). The response of *Arabidopsis* ethylene-insensitive and overproducer mutants to the pathogen *Pseudomonas syringae* pv. *tomato* was evaluated (Bent et al. 1992). When plants are challenged with a pathogen to which they are resistant (avirulent), they display a patch of localized cell death at the site of infection, known as the hypersensitive response (Keen 1990). This response effectively isolates the infection, preventing further damage to the plant. Wild-type, *etr1*, and *ein2* mutants displayed a normal hypersensitive response when challenged with an avirulent *P. syringae* strain, suggesting that ethylene is not critical for this process. Upon infection with several strains of virulent bacteria, wild-type, *etr1*, and *ein3* plants showed a number of disease symptoms, including chlorosis and the presence of water-soaked lesions. However, these symptoms were significantly abated in *ein2* alleles. The differing reactions of *etr1* and *ein3* relative to *ein2* suggest that only the EIN2 gene product is involved in the development of disease symptoms. These results suggest that development of symptoms may be mediated through ethylene and that the ethylene response pathway may branch after the *EIN2* gene product. Alternatively, the difference in symptoms observed between the *ein2* and the other ethylene-insensitive mutants may be due to leakiness of the *ein1* and *ein3* alleles. The disease tolerance of the *ein2* mutant suggests that EIN2 plays a dual role in plants, mediating both ethylene sensitivity and pathogen-induced damage (Bent et al. 1992).

Many plants can respond to pathogen infection by induction of broad-spectrum resistance. This phenomenon is known as systemic acquired resistance (SAR) (Ross 1961). Inducers of SAR in *Arabidopsis* include pathogens, salicylic acid, and ethephon (an ethylene-releasing compound). It has been suggested that ethylene may act as a signal involved in salicylic acid-mediated SAR (Raz and Fluhr 1992). The role of ethylene in SAR has recently been clarified using mutants that are insensitive to ethylene (Bleecker et al. 1988; Guzman and Ecker 1990; Kieber et al. 1993). The pattern of gene expression in ethylene-insensitive *Arabidopsis* mutants clearly demonstrated that chemical breakdown products of ethephon (hydrochloric and phosphoric acids) and not ethylene were responsible for induction of SAR (Lawton et al. 1994), although ethylene was found to potentiate the effect of salicylic acid on PR gene induction. In light of these results, previous experiments in which ethephon has been used as an ethylene source must be reevaluated.

Ethylene has been shown to elevate the transcription and/or steady-state level of mRNA for a number of "plant defense genes," including chitinase, glucanase pathogen-related protein PR1b, chalcone synthase, hydroxyproline-rich glycoproteins, and ripening-related genes (see Eyal et al. 1993 and reference therein). The application of exogenous ethylene fails to induce ethylene-regulated genes in *etr1*, *ein2*, and *ein3* (Samac et al. 1990; Lawton et al. 1994, M. Rothenberg and J.R. Ecker, unpubl.), and in *ctr1*, ethylene-regulated genes are expressed constitutively at a high level (Kieber et al. 1993). The minimal DNA sequences that are sufficient and necessary for ethylene regulation of gene expression have been defined in the promoter of the PR1b gene (Meller et al. 1993). Several *cis* elements, including the well-known G-box motif (Giuliano et al. 1988), were found to be essential for ethylene responsiveness. Several DNA-binding proteins have been detected that interact with these elements; however, none of these has been purified (Meller et al. 1993).

## A GENETIC MODEL FOR THE ETHYLENE RESPONSE PATHWAY

Epistasis analysis is a powerful tool for ordering gene products that act in a given pathway (Avery and Wasserman 1992). A genetic model for the ethylene signal transduction pathway in *Arabidopsis* has been devised that is based on epistasis relationships between the various mutants (Kieber et al. 1993; G. Roman et al., in prep.). *etr1* is proposed to mediate early steps in the signal transduction pathway. The *etr1* mutation is epistatic to the Eto$^-$ mutations; the triple response phenotype of Eto$^-$ mutants is suppressed by the ethylene insensitivity of *etr1*. However, high-level ethylene production is not suppressed in the double mutants

(J. Kieber and J.R. Ecker, unpubl.). The recessive nature of the *eto1* mutation suggests that ethylene biosynthesis is regulated by at least one gene that acts in a negative manner. ETR2 and EIN2 are placed in the same pathway rather than in separate pathways because the effects of the *etr1* and *ein2* mutations are not additive (G. Roman et al., in prep.). However, the precise position of *ein2* in this pathway has not been determined. The *ctr1 etr1* double mutant shows a constitutive triple response phenotype, and the *ctr1 ein3* double mutant is insensitive to ethylene (Kieber et al. 1993). These results suggest that CTR1 acts downstream from ETR1, but upstream of EIN3. The *hls1* mutation is epistatic to *eto1 eto2* as well as the *ctr1* mutation, as expected, since mutations that abolish hook formation would be expected to be epistatic to mutations that cause constitutive exaggeration of the apical hook (J. Kieber and J.R. Ecker, unpubl.). Therefore, the assigned order of gene action is (*ETO1 ETO2 ETO3*), *ETR1*, *CTR1*, *EIN3*, and *HLS1*. Gene orders based on analysis of these and other double mutant combinations are summarized in Figure 5.

## A BIOCHEMICAL MODEL FOR THE ETHYLENE-SENSING APPARATUS

Signal perception of ethylene most likely begins at the plasma membrane and ends in alterations in gene expression in the nucleus. The first step in ethylene perception is presumably binding to a receptor molecule. Ethylene-binding components that fit the pharmacological criteria for authentic receptors have been identified in several plant species, including *Arabidopsis* (Sisler 1991). These binding components show high affinity and saturable binding with a $K_d$ that is consistent with physiologically active concentrations of ethylene. It appears that two classes of binding proteins are present in a variety of plant species: one with a relatively low rate constant of association and dissociation and one with a high rate. Inhibitors of ethylene action such as *trans*-cyclo-octene have been shown to inhibit binding of ethylene to these proteins. However, it is unclear if any of these binding proteins are authentic receptors.

It is possible that one of the genetically identified genes discussed above encodes an ethylene receptor. The most likely candidate is *ETR1*, which acts early in the signaling pathway and, in mutant form, demonstrates pleiotropic effects on ethylene physiology (Bleecker et al. 1988; Guzman and Ecker 1990; Chang et al. 1993). The ETR1 gene product also shows strong similarity to bacterial two-component histidine kinase "sensors." Interestingly, *ETR1* adult plants showed an 80% reduction in the amount of ethylene that they can bind, in a competitive bind-

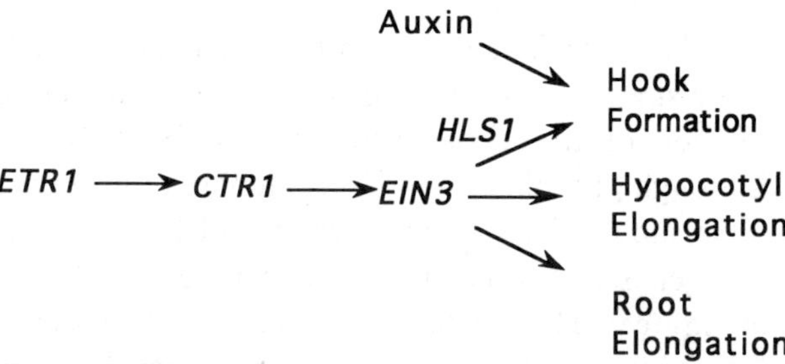


*Figure 5* Genetic model of interactions among components of the ethylene signal transduction pathway. This model shows the predicted order in which the various gene products act, based on the epistatic relationships among the mutants. The seedling ethylene responses are indicated on the right.

ing assay. The significance of this result is somewhat diminished by the finding that at least one allele of *etr1* (*etr1-3*, also previously called *ein1-1*) has been shown to have increased ethylene production (Guzman and Ecker 1990). Similarly, two partially dominant isolates, *eti5* and *eti8*, displayed decreased ethylene binding and increased ethylene production (Sanders et al. 1991). Most of the difference in binding in the Eti⁻ mutants could be accounted for by the increased level of ethylene production. However, it is unclear whether the semidominant *eti* loci are different from each other or from *etr1*.

The *ctr1* mutation is recessive and so likely represents a loss of function. This conclusion is supported by sequence analysis of several *ctr1* alleles, some of which are most likely null mutations (Kieber et al. 1993). This suggests that the wild-type *CTR1* gene product acts as a negative regulator of the ethylene response pathway, as its loss results in plants that constitutively display ethylene responses. This same kind of reasoning leads to the conclusion that the *EIN2* and *EIN3* gene products act as positive regulators in this pathway. The nature of action of the *ETR1* gene product is not as straightforward, as *ETR1* alleles are dominant. If the *ETR1* mutation results in a gain of function, then the wild-type protein may act as a negative regulator of ethylene action. However, if the mutation is a dominant negative, it would suggest that the wild-type protein acts as a positive regulator. Alternatively, if *etr1* is neomorphic, the wild-type *ETR1* gene product may not normally function in the ethylene signal transduction pathway.

The amino terminus of CTR1 does show significant, albeit weak, similarity to the amino terminus of Raf, although it is unlikely that CTR1 binds to the same ligands as Raf-1. Raf-1 is involved in the transduction of numerous signals in human cells, in *C. elegans*, and in multiple developmental programs in *Drosophila* (Bar-Sagi et al. 1993). Control of Raf-

1 is exerted by phosphorylation of its amino terminus by numerous upstream regulators (Van Aelst et al. 1993; Warne et al. 1993; Zhang et al. 1993). In plants, regulation of the kinase activity of CTR1 may be analogously regulated through phosphorylation by a number of upstream activators/repressors. One model that incorporates this idea as well as other ethylene response genes and what has been learned about their genetics is presented in Figure 6. This model proposes that in the absence of ethylene, the kinase activity of CTR1 may subsequently phosphorylate several genetically downstream targets (such as EIN3) and may include targets that control cell elongation or division. The *EIN3* gene product may then activate a number of terminal ethylene-regulated genes. When ethylene is present, it binds to a receptor, possibly the *ETR1* gene product. This ethylene/receptor complex may then inactivate CTR1, either

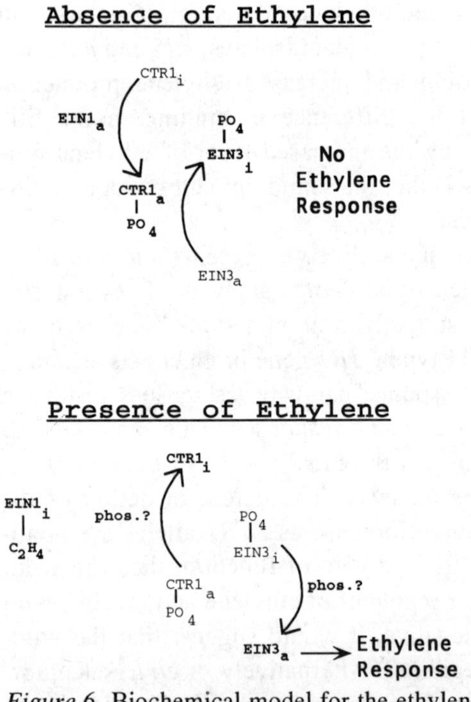

*Figure 6* Biochemical model for the ethylene sensing apparatus. In the absence of ethylene, the kinase activity of CTR1 is active and phosphorylates genetically downstream gene products (possibly EIN3). EIN3 may then activate a number of terminal ethylene-regulated genes. In the presence of ethylene, it binds to a receptor, possibly the ETR1 gene product. The ethylene-receptor complex then inactivates CTR1, either directly or indirectly, and possibly via phosphorylation. The inactive CTR1 no longer phosphorylates the EIN3 gene product, which is then activated, possibly via dephosphorylation by a protein phosphatase.

directly or indirectly, and possibly via phosphorylation (as in the case with several Raf proteins). In its inactive form, CTR1 can no longer phosphorylate EIN3 and other targets. Dephosphorylation of EIN3 by a protein phosphatase would produce the active form of the protein. This model is consistent with what is known about these genes, but is at this point purely speculative. Recently, Raz and Fluhr (1993) demonstrated that ethylene application induces very rapid and transient protein phosphorylation in tobacco leaves. Furthermore, this effect was abolished in the presence of protein kinase inhibitors. Reciprocally, treatment of excised tobacco leaves with inhibitors of type 1 and 2A protein phosphatases caused increased phosphorylation and accumulation of pathogenesis-related proteins. Earlier pharmacological experiments by this group revealed that calcium is necessarily involved in the ethylene-mediated pathogenesis response in tobacco, as exemplified by the induction of the chitinase gene (Raz and Fluhr 1992). Taken together, the results of biochemical and pharmacological studies in tobacco support the conclusion that several ethylene-evoked responses in plants are mediated by phosphorylated intermediates and suggest that at least some of these processes have a requirement for calcium. Cloning of additional genes involved in this pathway and biochemical analysis of the products should increase our understanding of the molecular basis for ethylene signal transduction.

## CONCLUSIONS AND PERSPECTIVES

Hormones play a central role in the regulation of plant growth and development. The use of tomato mutants constructed by reverse genetic techniques has made it possible to better understand the ethylene-mediated signal transduction pathway that leads to tomato fruit ripening (Sheehy et al. 1988; Smith et al. 1988; Hamilton et al. 1990; Klee et al. 1991; Oeller et al. 1991; Penarrubia et al. 1992). The components of the pathway will be further elucidated by cloning the *Arabidopsis* mutant genes (Kieber et al. 1993; Chang et al. 1993). Molecular genetic studies in *Arabidopsis* are beginning to unravel the biochemical processes controlling ethylene biosynthesis, perception, and signal transduction. The *CTR1* and *ETR1* genes have been cloned, and conceptual translation of the DNA sequences revealed that both encode protein kinases. The *CTR1* gene encodes a putative serine/threonine protein kinase that is most closely related to the Raf family of protein kinases. The *ETR1* gene encodes a predicted protein kinase that is most closely related to bacterial histidine kinases. These results, along with genetic epistasis studies, are consistent with a model in which CTR1 and ETR1 act in a multistep signal transduction pathway through phosphorylation of proteins in a cascade, possibly including the *EIN2* and *EIN3* gene products. Thus, the identification of the first components of the ethylene signal apparatus in *Arabidopsis* ends a 65-year-old view that ethylene mediates its effects by directly altering membrane permeability (Blackman and Parija 1928; Solomos and Laties 1973). Further analysis of the *ETR1*, *CTR1*, *EIN2*, *EIN3*, and *HLS1* gene products, in concert with the identification of additional genes that interact with these, will provide further insight into the molecular nature of the signal transduction pathway and the biochemical interactions between these proteins.

It is likely that additional loci remain to be detected, since there is no evidence that the mutant screens are saturated, and only a limited attempt has been made to recover mutants that are weak, lethal, or infertile. Suppressor analysis of the existing mutants may also identify novel loci. Genetic, molecular, biochemical, and physiological studies of ethylene response mutants will define more precisely the roles of these genes in the ethylene action pathway.

## ACKNOWLEDGMENTS

We dedicate this chapter to our teacher and mentor, Dr. Ronald W. Davis (Department of Biochemistry, Stanford University). We thank the members of our laboratories for useful discussions, in particular, and Dr. Xiaowu Liang for providing the unpublished data shown in Figure 2; Gregg Roman, Anne Lehman, and Drs. Madge Rothenberg and Joseph Kieber for providing results prior to publication; Drs. Elliot Meyerowitz, Alexander Varshavsky, and John Ryals for providing us with preprints of papers in press; Ms. Barbara Alonso for typing the manuscript, and Dr. Ron Wells for editing the chapter. The generous support of the U.S. Department of Energy (DE-FR02-93ER20104) and the National Science Foundation (IBN-92-05342) to J.R.E., and of the National Science Foundation (DCB-8645952, -8819129, -8916286, and MCB-9316475) and the U.S. Department of Agriculture (5335-21430-002-00D; 5335-21430-003-00D) to A.T. is gratefully acknowledged.

## REFERENCES

Abeles, F.B., P.W. Morgan, and M.E. Saltveit, Jr. 1992. *Ethylene in plant biology.* Academic Press, New York.

Apelbaum, A. and S.P. Burg. 1971. Altered cell microfibrillar orientation in ethylene-treated *Pisum sativum* stems. *Plant Physiol.* **48:** 648–652.

_______. 1972. Effects of ethylene on cell division and deoxyribonucleic acid synthesis in *Pisum sativum. Plant Physiol.* **50:** 117–124.

Avery, L. and S. Wasserman. 1992. Ordering gene function: Interpretations of epistasis in regulatory hierarchies. *Trends Genet.* **8:** 312–316.

Bailey, B.A., A. Avni, N. Li, A.K. Mattoo, and J.M. Anderson. 1992. Nucleotide sequence of the *Nicotiana tabacum* cv Xanthi gene encoding 1-aminocyclopropane-1-carboxylate synthase. *Plant Physiol.* **100:** 1615–1616.

Bar-Sagi, D., D. Bowtell, J. Downward, G. Rubin, J. Schlessinger, R. Weinberg, and M. Wigler. 1993. Forging a pathway to the nucleus. *Science* **260:** 1588–1560.

Bent, A.J., R.W. Innes, J.R. Ecker, and B.J. Staskawicz. 1992. Disease development in ethylene-insensitive *Arabidopsis thaliana* infected with virulent and avirulent *Pseudomonas* and *Xanthomonas* pathogens. *Mol. Plant Microbe Interact.* **5:** 372–378.

Biale, J.B. 1960. The postharvest biochemistry of tropical and subtropical fruits. *Adv. Food Res.* **10:** 293–354.

–––––––. 1964. Growth, maturation and senescence in fruits. *Science* **146:** 880–888.

Biale, J.B. and R.E. Young. 1981. Respiration and ripening in fruits—Retrospect and prospect. In *Recent advances in the biochemistry of fruits and vegetables* (ed. J. Friend and M.J.C. Rhodes), pp. 1–39. Academic Press, London.

Blackman, F.F. and P. Parija. 1928. Analytical studies in plant respiration. I. The respiration of a population of senescent respiring apples. *Proc. R. Soc. Lond. B Biol. Sci.* **103:** 422–445.

Blanpied, G.D. 1985. Introduction to the symposium. *HortScience* **20:** 40–41.

Bleecker, A.B., M.A. Estelle, C. Somerville, and H. Kende. 1988. Insensitivity to ethylene conferred by a dominant mutation in *Arabidopsis thaliana*. *Science* **241:** 1086–1089.

Botella, J.R., J.M. Arteca, C.D. Schlagenhaufer, R.N. Arteca, and A.T. Phillips. 1992. Identification and characterization of a full length cDNA encoding for an auxin-induced 1-aminocyclopropane-1-carboxylate synthase from etiolated mung bean hypocotyl segments and expression of its mRNA in response to indole-3-acetic acid. *Plant Mol. Biol.* **20:** 425–436.

Botella, J.R., C.D. Schlagnhaufer, J.M. Arteca, R.N. Arteca, and A.T. Phillips. 1993. Identification of two new members of the 1-aminocyclopropane-1-carboxylate synthase-encoding multigene family in mung bean. *Gene* **123:** 249–253.

Brand, A.H. and N. Perrimon. 1994. Raf acts downstream of the EGF receptor to determine dorsoventral polarity during *Drosophila* oogenesis. *Genes Dev.* **8:** 629–639.

Burg, S.P. 1962. The physiology of ethylene formation. *Annu. Rev. Plant Physiol.* **13:** 265–302.

Burg, S.P. and E.A. Burg. 1962. Role of ethylene in fruit ripening. *Plant Physiol.* **37:** 179–189.

–––––––. 1965. Ethylene action and the ripening of fruits. *Science* **148:** 1190–1196.

–––––––. 1966. Fruit storage at subatmospheric pressures. *Science* **153:** 314–315.

–––––––. 1967. Molecular requirements for the biological activity of ethylene. *Plant Physiol.* **42:** 144–152.

Buse, E.L. and G.G. Laties. 1993. Ethylene-mediated post-transcriptional regulation in ripening avocado mesocarp discs. *Plant Physiol.* **102:** 417–423.

Callahan, A.M., P.H. Morgens, P. Wright, and K.E. Nichols. 1992. Comparison of Pch313 (pTOM13 homolog) RNA accumulation during fruit softening and wounding of two phenotypically different peach cultivars. *Plant Physiol.* **100:** 482–488.

Chang, C., S.F. Kwok, A.B. Bleecker, and E.M. Meyerowitz. 1993. *Arabidopsis* ethylene-response gene *ETR1*: Similarity of product to two-component regulators. *Science* **262:** 539–544.

Chory, J., C. Peto, R. Feinbaum, L. Pratt, and F. Ausubel. 1989. *Arabidopsis thaliana* mutant that develops as a light-grown plant in the absence of light. *Cell* **58**: 991–999.

Christoffersen, R.E. and G.G. Laties. 1982. Ethylene regulation of gene expression in carrots. *Proc. Natl. Acad. Sci* **79**: 4060–4063.

Darwin, C. 1869. *The power of movement in plants.* Appleton, New York.

Deikman, J. and R.L. Fischer. 1988. Interaction of a DNA binding factor with the 5′-flanking region of an ethylene-responsive fruit ripening gene from tomato. *EMBO J.* **7**: 3315–3320.

Deng, X.-W., T. Caspar, and P.H. Quail. 1991. *cop1*: A regulatory locus involved in light-controlled development and gene expression in *Arabidopsis. Genes Dev.* **5**: 1172–1182.

Denny, F.E. 1924. Hastening the coloration of lemons. *J. Agric. Res.* **27**: 757–769.

Dickson, B., F. Sprenger, D. Morrison, and E. Hafen. 1992. Raf functions downstream of Ras1 in the Sevenless signal transduction pathway. *Nature* **360**: 600–603.

Dong, J.G., J.C. Fernandez-Maculet, and S.F. Yang. 1992a. Purification and characterization of 1-aminocyclopropane-1-carboxylate oxidase from apple fruit. *Proc. Natl. Acad. Sci.* **89**: 9789–9793.

Dong, J.G., D. Olson, A.L. Silverstone, and S.F. Yang. 1992b. Sequence of a cDNA coding for 1-aminocyclopropane-1-carboxylate oxidase homolog from apple fruit. *Plant Physiol.* **98**: 1530–1531.

Dong, J.G., W.T. Kim, W.K. Yip, G.A. Thompson, L. Li, A.B. Bennett, and S.F. Yang. 1991. Cloning of a cDNA encoding 1-aminocyclopropane-1-carboxylate synthase and expression of its mRNA in ripening apple fruit. *Planta* **185**: 38–45.

Ecker, J.R. and R.W. Davis. 1987. Plant defense genes are regulated by ethylene. *Proc. Natl. Acad. Sci.* **84**: 5202–5206.

Eisinger, W. 1983. Regulation of pea internode expansion by ethylene. *Annu. Rev. Plant Physiol.* **34**: 225–240.

Eyal, Y., Y. Meller, S. Lev-Yadur, and R. Fluhr. 1993. A basic type PR-1 promoter directs ethylene responsiveness vascular and abscission zone-specific expression. *Plant J.* **4**: 225–234.

Gibbs, C.S. and M.J. Zoller. 1991. Rational scanning mutagenesis of a protein kinase identifies functional regions involved in catalysis and substrate interactions. *J. Biol. Chem.* **266**: 8923–8931.

Giovannoni, J.J., D. DellaPenna, A.B. Bennett, and R.L. Fischer. 1989. Expression of a chimeric polygalacturonase gene in transgenic *rin* (ripening inhibitor) tomato fruit results in polyuronide degradation but not fruit softening. *Plant Cell* **1**: 53–63.

Giovannoni, J.J., D. DellaPenna, C.C. Lashbrook, A.B. Bennett, and R.L. Fischer. 1990. Expression of a chimeric polygalacturonase gene in transgenic *rin* (ripening inhibitor) tomato fruit. In *Horticultural biotechnology* (ed. A.B. Bennett and S.D. O'Neill), pp. 217–227. Wiley-Liss, New York.

Giuliano, G., E. Pichersky, V.S. Malik, M.P. Timko, P.A. Scolnik, and A.R. Cashmore. 1988. An evolutionarily conserved protein binding sequence upstream of a plant light-regulated gene. *Proc. Natl. Acad. Sci.* **85**: 7089–7093.

Goeschl, J.D., L. Rappaport, and H.K. Pratt. 1966. Ethylene as a factor regulating the growth of pea epicotyls subjected to physical stress. *Plant Physiol.* **41**: 877–884.

Gray, J., S. Pictor, J. Shabbeer, W. Schuch, and D. Grierson. 1992. Molecular biology of fruit ripening and its manipulation with antisense genes. *Plant Mol. Biol.* **19**: 69–87.

Guzman, P. and J.R. Ecker. 1990. Exploiting the triple response of *Arabidopsis* to identify ethylene-related mutants. *Plant Cell* **2**: 513–523.

Hall, M.A., C.P.K. Connern, N.V.J. Harpman, K. Ishizawa, G. Roveda-Hoyos, I. Raskin, I.O. Sanders, R. Smith, and C.K. Woods. 1990. Ethylene: Receptors and action. *Symp.*

*Soc. Exp. Biol.* **44:** 87–110.

Hamilton, A.J., M. Bouzayen, and D. Grierson. 1991. Identification of a tomato gene for the ethylene-forming enzyme by expression in yeast. *Proc. Natl. Acad. Sci.* **88:** 7434–7437.

Hamilton, A.J., G.W. Lycett, and D. Grierson. 1990. Antisense gene that inhibits synthesis of the hormone ethylene in transgenic plants. *Nature* **346:** 284–287.

Han, M., A. Golden, Y. Han, and P.W. Sternberg. 1993. *C. elegans lin-45 raf* gene participates in *let-60 ras*-stimulated vulval differentiation. *Nature* **363:** 133–140.

Hanks, S.K. and A. Quinn. 1991. Protein kinase catalytic domain sequence database: Identification of conserved features of primary structure and classification of family members. *Methods Enzymol.* **200:** 38–81.

Hanks, S.K., A.M. Quinn, and T. Hunter. 1988. The protein kinase family: Conserved features and deduced phylogeny of the catalytic domains. *Science* **241:** 42–52.

Harpham, N.V.J., A.W. Berry, E.M. Knee, G. Roveda-Hoyos, I. Raskin, I.O. Sanders, C.K. Smith, C.K. Wood, and M.A. Hall. 1991. The effect of ethylene on the growth and development of wild type and mutant *Arabidopsis thaliana* (L.) Heynh. *Ann. Bot.* **68:** 55–62.

Herner, R.C. and K.C. Sink, Jr.. 1973. Ethylene production and respiratory behavior of the *rin* tomato mutant. *Plant Physiol.* **52:** 38–42.

Hou, Y., A.G. von Arnim, and X.-W. Deng. 1993. A new class of *Arabidopsis* constitutive photomorphogenic genes involved in regulating cotyledon development. *Plant Cell* **5:** 329–339.

Huang, P.L., J.E. Parks, W.E. Rottmann, and A. Theologis. 1991. Two genes encoding 1-aminocyclopropane-1-carboxylate synthase in zucchini (*Cucurbita pepo*) are clustered and similar but differentially regulated. *Proc. Natl. Acad. Sci.* **88:** 7021–7025.

Hulme, A.C. 1954. Studies in the nitrogen metabolism of apple fruits. The climacteric rise in respiration in relation to changes in the equilibrium between protein synthesis and breakdown. *J. Exp. Bot.* **5:** 159–172.

Inouye, M. 1988. Antisense RNA: Its functions and applications in gene regulation—A review. *Gene* **72:** 25–34.

Jackson, M.B. 1985. Ethylene and responses of plants to soil waterlogging and submergence. *Annu. Rev. Plant Physiol.* **36:** 145–174.

Kang, B.G., C.S. Yocum, S.P. Burg, and P.M. Ray. 1967. Ethylene and carbon dioxide: Mediation of hook opening response. *Science* **156:** 958–959.

Keen, N.T. 1990. Gene-for-gene complementarity in plant-pathogen interactions. *Annu. Rev. Genet.* **24:** 447–463.

Kende, H. 1987. Studies on internodal growth using deepwater rice. In *Physiology of cell expansion during plant growth* (ed. D.J. Cosgrove and D.P. Knievel), pp. 227–238. American Society of Plant Physiologists, Rockville, Maryland.

———. 1989. Enzymes of ethylene biosynthesis. *Plant Physiol.* **91:** 1–4.

———. 1993. Ethylene biosynthesis. *Annu. Rev. Plant Physiol. Plant Mol. Biol.* **44:** 283–307.

Kidd, F. and C. West. 1930. Physiology of fruit. I. Changes in the respiratory activity of apples during their senescence at different temperatures. *Proc. R. Soc. Lond. B Biol. Sci.* **106:** 93–109.

Kieber, J.J. and J.R. Ecker. 1993. Ethylene, it's not just for ripening anymore. *Trends Genet.* **9:** 356–362.

———. 1994. Molecular and genetic analysis of the constitutive ethylene response mutation *ctr1. NATO ASI Ser. Ser. H Cell Biol.* **81:** 193–202.

Kieber, J.J., M. Rothenberg, G. Roman, K.A. Feldmann, and J.R. Ecker. 1993. *CTR1*, a negative regulator of the ethylene response pathway in *Arabidopsis*, encodes a member of the Raf family of protein kinases. *Cell* **72**: 427–441.

Kim, W.T., A. Silverstone, W.K. Yip, J.G. Dong, and S.F. Yang. 1992. Induction of 1-aminocyclopropane-1-carboxylate synthase mRNA by auxin in mung bean hypocotyls and cultured apple shoots. *Plant Physiol.* **98**: 465–471.

Klee, H. and M. Estelle. 1991. Molecular genetic approaches to plant hormone biology. *Annu. Rev. Plant Physiol. Plant Mol. Biol.* **42**: 529–551.

Klee, H.J., M.B. Hayford, K.A. Kretzmer, G.F. Barry, and G.M. Kishore. 1991. Control of ethylene synthesis by expression of a bacterial enzyme in transgenic tomato plants. *Plant Cell* **3**: 1187–1193.

Knight, L.I., R.C. Rose, and W. Crocker. 1910a. Effects of various gases and vapors upon etiolated seedlings of the sweet pea. *Science* **31**: 635–636.

————. 1910b. A new method of detecting traces of illuminating gas. *Science* **31**: 636.

Lanahan, M.B., H.S. Yen, J.J. Giovannoni, and H.J. Klee. 1994. The *Never ripe* mutation blocks ethylene perception in tomato. *Plant Cell* **6**: 521–530.

Lawton, K.A., S.S. Potter, S. Uknes, and J. Ryals. 1994. Acquired resistance signal transduction in Arabidopsis is ethylene independent. *Plant Cell* **6**: 581–588.

Liang, X.W., S. Abel, J.A. Keller, N.F. Shen, and A. Theologis. 1992. The 1-aminocyclopropane-1-carboxylate synthase gene family of *Arabidopsis thaliana*. *Proc. Natl. Acad. Sci.* **89**: 11046–11050.

Lincoln, J.E. and R.L. Fischer. 1988a. Regulation of gene expression by ethylene in wild-type and *rin* tomato (*Lycopersicon esculentium*) fruit. *Plant Physiol.* **88**: 370–374.

————. 1988b. Diverse mechanisms for the regulation of ethylene-inducible gene expression. *Mol. Gen. Genet.* **212**: 71–75.

Lincoln, J.E., A.D. Campbell, J. Oetiker, W.H. Rottmann, P.W. Oeller, N.F. Shen, and A. Theologis. 1993. LE-ACS4, a fruit ripening and wound-induced 1-aminocyclopropane-1-carboxylate synthase gene of tomato (*Lycopersicon esculentum*). Expression in *Escherichia coli*, structural, characterization, expression characteristics, and phylogenetic analysis. *J. Biol. Chem.* **268**: 19422–19430.

Liscum, E. and R.P. Hangarter. 1993. Light-stimulated apical hook opening in wild-type *Arabidopsis thaliana* seedlings. *Plant Physiol.* **101**: 567–572.

MacDonald, I.R., D.C. Gordon, J.W. Hart, and E.P. Maher. 1983. The positive hook: The role of gravity in the formation and opening of the apical hook. *Planta* **158**: 76–81.

McGarvey, D.J., R. Sirevug, and R.E. Christoffersen. 1992. Ripening-related gene from avocado fruit. *Plant Physiol.* **98**: 554–559.

McGarvey, D.J., H. Yu, and R.E. Christoffersen. 1990. Nucleotide sequence of a ripening-related cDNA from avocado fruit. *Plant Mol. Biol.* **15**: 165–167.

Meller, Y., G. Sessa, Y. Eyal, and R. Fluhr. 1993. DNA-protein interactions on a *cis*-DNA element essential for ethylene regulation. *Plant Mol. Biol.* **23**: 453–463.

Nadeau, J.A., X.S. Zhang, H. Nair, and S.D. O'Neill. 1993. Temporal and spatial regulation of 1-aminocyclopropane-1-carboxylate oxidase in the pollination-induced senescence of orchid flowers. *Plant Physiol.* **103**: 31–39.

Nakagawa, N., H. Mori, K. Yamazaki, and H. Imaseki. 1991. Cloning of a complementary DNA for auxin-induced 1-aminocyclopropane-1-carboxylate synthase and differential expression of the gene by auxin and wounding. *Plant Cell Physiol.* **32**: 1153–1163.

Nakajima, N., H. Mori, K. Yamazaki, and H. Imaseki. 1990. Molecular cloning and sequence of a complementary DNA encoding 1-aminocyclopropane-1-carboxylate synthase induced by tissue wounding. *Plant Cell Physiol.* **31**: 1021–1029.

Neljubov, D. 1901. Öber die horizontale Nutation der Stengel von *Pisum sativum* und einiger anderer. *Pflanzen. Beih. Bot. Zentralbl.* **10:** 128–239.

Oeller, P.W., L.M. Wong, L.P. Taylor, D.A. Pike, and A. Theologis. 1991. Reversible inhibition of tomato fruit senescence by antisense RNA. *Science* **254:** 437–439.

Olson, D.C., J.A. White, L. Edelman, R.N. Harkins, and H. Kende. 1991. Differential expression of two genes for 1-aminocyclopropane-1-carboxylate synthase in tomato fruits. *Proc. Natl. Acad. Sci.* **88:** 5340–5344.

Ota, I. and A. Varshavsky. 1993. A yeast protein similar to bacterial 2-component regulators. *Science* **262:** 566–569.

Park, K.Y., A. Drory, and W.R. Woodson. 1992. Molecular cloning of an 1-aminocyclopropane-1-carboxylate synthase from senescing carnation flower petals. *Plant Mol. Biol.* **18:** 377–386.

Peleman, J., W. Boerjan, G. Engler, J. Seurinck, J. Botterman, T. Alliote, M. Van Montagu, and D. Inze. 1989. Strong cellular preference in the expression of a housekeeping gene of *Arabidopsis thaliana* encoding S-adenosylmethionine synthetase. *Plant Cell* **1:** 81–93.

Penarrubia, L., M. Aguilar, L. Margossian, and R.L. Fischer. 1992. An antisense gene stimulates ethylene hormone production during tomato fruit ripening. *Plant Cell* **4:** 681–687.

Pickett, F.B., A.K. Wilson, and M. Estelle. 1990. The *aux1* mutation of *Arabidopsis* confers both auxin and ethylene resistance. *Plant Physiol.* **94:** 1462–1466.

Picton, S., S.L. Barton, M. Bouzayen, A.J. Hamilton, and D. Grierson. 1993. Altered fruit ripening and leaf senescence in tomatoes expressing an antisense ethylene-forming enzyme transgene. *Plant J.* **3:** 469–481.

Rapp, U.R. 1991. Role of *Raf*-1 serine/threonine protein kinase in growth factor signal transduction. *Oncogene* **6:** 495–500.

Rapp, U.R., G. Heidecker, M. Huleihel, J.L. Cleveland, W.C. Choi, T. Pawson, J.N. Ihle, and W.B. Anderson. 1988. *raf* family serine/threonine protein kinases in mitogen signal transduction. *Cold Spring Harbor Symp. Quant. Biol.* **53:** 173–182.

Raz, V. and R. Fluhr. 1992. Calcium requirement for ethylene-dependent responses. *Plant Cell* **4:** 1123–1130.

———. 1993. Ethylene signal is transduced via protein phosphorylation events in plants. *Plant Cell* **5:** 523–530.

Rodrigues-Pousada, R.A., R. De Rycke, A. Dedonder, W. Van Caeneghem, G. Engler, M. Van Montagu, and D. Van der Straeten. 1993. The *Arabidopsis* 1-aminocyclopropane-1-carboxylate synthase gene 1 is expressed during early development. *Plant Cell* **5:** 897–911.

Romano, C.P., M.L. Cooper, and H.J. Klee. 1993. Uncoupling auxin and ethylene effects in transgenic tobacco and *Arabidopsis* plants. *Plant Cell* **5:** 181–189.

Ross, A.F. 1961. Localized acquired resistance to plant virus infections hypersensitive hosts. *Virology* **14:** 329–339.

Ross, G.S., M.L. Knighton, and M. Lay-Yee. 1992. An ethylene-related cDNA from ripening apples. *Plant Mol. Biol.* **19:** 213–238.

Rothenberg, M. and J.R. Ecker. 1993. Mutant analysis as an experimental approach toward understanding plant hormone action. *Semin. Dev. Biol. Plant Dev. Genet.* **4:** 3–13.

Rottmann, W.E., G.F. Peter, P.W. Oeller, J.A. Keller, N.F. Shen, B.P. Nagy, L.P. Taylor, A.D. Campbell, and A. Theologis. 1991. 1-Aminocyclopropane-1-carboxylate synthase in tomato is encoded by a multigene family whose transcription is induced during fruit and floral senescence. *J. Mol. Biol.* **222:** 937–961.

Rubinstein, B. 1971. The role of various regions of the bean hypocotyl on red light-induced hook opening. *Plant Physiol.* **48:** 183–186.

Samac, D.A., C.M. Hironaka, P.E. Yallaly, and D.M. Shah. 1990. Isolation and characterization of the genes encoding basic and acidic chitinase in *Arabidopsis thaliana. Plant Physiol.* **93:** 907–914.

Sanders, I.O., N.V.J. Harpham, I. Raskin, A.R. Smith, and M.A. Hall. 1991. Ethylene binding in wild type and mutant *Arabidopsis thaliana* (L.) Heynh. *Am. J. Bot.* **68:** 97–104.

Sato, T. and A. Theologis. 1989. Cloning the mRNA encoding 1-aminocyclopropane-1-carboxylate synthase, the key enzyme for ethylene biosynthesis in plants. *Proc. Natl. Acad. Sci.* **86:** 6621–6625.

Sato, T., P.W. Oeller, and A. Theologis. 1991. The 1-aminocyclopropane-1-carboxylate synthase of *Curcurbita*. Purification, properties, expression in *E. coli* and primary structure determination by DNA sequence analysis. *J. Biol. Chem.* **266:** 3752–3759.

Sheehy, R.E., M. Kramer, and W.R. Hiatt. 1988. Reduction of polygalacturonase activity in tomato fruit by antisense RNA. *Proc. Natl. Acad. Sci.* **85:** 8805–8809.

Silk, W.K. and R.O. Erickson. 1978. Kinematics of hypocotyl curvature. *Am. J. Bot.* **65:** 310–319.

Sisler, E.C. 1991. Ethylene-binding components in plants. In *The plant hormone ethylene* (ed. A.K. Matoo and J.C. Suttle), pp. 81–99. CRC Press, Boca Raton, Florida.

Slater, A., M.J. Maunders, K. Edwards, W. Schuch, and D. Grierson. 1985. Isolation and characterization of cDNA clones for tomato polygalacturonase and other ripening-related proteins. *Plant Mol. Biol.* **5:** 137–147.

Smith, C.J.S., C.F. Watson, J. Ray, C.R. Bird, P.C. Morriss, W. Schuch, and D. Grierson. 1988. Antisense RNA inhibition of polygalacturonase gene expression in transgenic tomatoes. *Nature* **334:** 724–726.

Snyder, S.H. 1992. Nitric oxide: First in a new class of neurotransmitters. *Science* **257:** 494–496.

Solomos, T. and G.G. Laties. 1973. Cellular organization and fruit ripening. *Nature* **245:** 390–392.

Spanu, P., D. Reinhardt, and T. Boller. 1991. Analysis and cloning of the ethylene-forming enzyme from tomato by functional expression of its mRNA in *Xenopus laevis* oocytes. *EMBO J.* **10:** 2007–2013.

Su, W. and S.H. Howell. 1992. A single genetic locus, *ckr1*, defines *Arabidopsis* mutants in which root growth is resistant to low concentrations. *Plant Physiol.* **99:** 1569–1574.

Taylor, S.S., D.R. Knighton, J. Zheng, L.F. Ten Eyck, and J.M. Sowadski. 1992. Structural framework for the protein kinase family. *Annu. Rev. Cell Biol.* **8:** 429–462.

Theologis, A. 1992. One rotten apple spoils the whole bushel: The role of ethylene in fruit ripening. *Cell* **70:** 181–184.

———. 1993. What a gas! *Curr. Biol.* **3:** 369–371.

Theologis, A., P.W. Oeller, L.M. Wong, W.H. Rottmann, and D.M. Gantz. 1993. Use of a tomato mutant constructed with reverse genetics to study fruit ripening, a complex developmental process. *Dev. Genet.* **14:** 282–295.

Tigchelaar, E.C., W.B. McGlassen, and R.W. Buescher. 1978. Genetic regulation of tomato fruit ripening. *HortScience* **13:** 508–513.

Van Aelst, L., M. Barr, S. Marcus, A. Polverino, and M. Wigler. 1993. Complex formation between RAS and RAF and other protein kinases. *Proc. Natl. Acad. Sci.* **90:** 6213–6217.

Van der Straeten, D., L. Van Wiemeersch, H.M. Goodman, and M. Van Montagu. 1990. Cloning and sequence of two different cDNAs encoding 1-aminocyclopropane-1-carboxylate synthase in tomato. *Proc. Natl. Acad. Sci.* **87:** 4859–4863.

Van der Straeten, D., A. Djudzman, W. Vancaeneghem, J. Smalle, and M. Van Montagu. 1993. Genetic and physiological analysis of a new locus in *Arabidopsis* that confers resistance to 1-aminocyclopropane-1-carboxylic acid and ethylene and specifically affects the ethylene signal- transduction pathway. *Plant Physiol.* **102:** 401–408.

Van der Straeten, D.V., R.AR. Pousada, R. Villarroel, S. Handey, H.M. Goodman, and M. Van Montagu. 1992. Cloning, genetic mapping, and expression analysis of an *Arabidopsis thaliana* gene that encodes 1-aminocyclopropane-1-carboxylate synthase. *Proc. Natl. Acad. Sci.* **89:** 9969–9973.

Varner, J.E. 1961. Biochemistry of senescence. *Annu. Rev. Plant Physiol.* **12:** 245–264.

Verma, A., D.J. Hirsch, C.E. Glatt, G.V. Ronnett, and S.H. Snyder. 1993. Carbon monoxide: A putative neural messenger. *Science* **259:** 381–384.

Wang, H. and W.R. Woodson. 1991. A flower senescence-related mRNA from carnation shares sequence similarity with fruit ripening-related mRNAs involved in ethylene biosynthesis. *Plant Physiol.* **96:** 1000–1001.

Warne, P. H., P.R. Viciana, and J. Downward. 1993. Direct interaction of Ras and the amino-terminal region of Raf-1 *in vitro. Nature* **364:** 352–355.

Wheeler, R.M. and F.B. Salisbury. 1980. Gravitropism in plant stems may require ethylene. *Science* **209:** 1126–1128.

Wilson, A.K., F.B. Pickett, J.C. Turner, and M. Estelle. 1990. A dominant mutation in *Arabidopsis* confers resistance to auxin, ethylene, and abscisic acid. *Mol. Gen. Genet.* **222:** 377–383.

Woodson, W.R., K.Y. Park, A. Drory, P.B. Larsen, and H. Wang. 1992. Expression of ethylene biosynthesis pathway transcripts in senescing carnation flowers. *Plant Physiol.* **99:** 526–532.

Woolf, T.M., D.A. Melton, and C.G.B. Jennings. 1992. Specificity of antisense oligonucleotides *in vivo. Proc. Natl. Acad. Sci.* **89:** 7305–7309.

Yang, S.F. 1987. The role of ethylene and ethylene synthesis in fruit ripening. In *Plant senescence: Its biochemistry and physiology* (ed. W.W. Thomson et al.), pp. 156–166. American Soceity of Plant Physiologists, Rockville, Maryland.

Yang, S.F. and N.E. Hoffman. 1984. Ethylene biosynthesis and its regulation in higher plants. *Annu. Rev. Plant Physiol.* **35:** 155–189.

Yip, W.K., T. Moore, and S.F. Yang. 1992. Differential accumulation of transcripts for four tomato 1-aminocyclopropane-1-carboxylate synthase homologs under various conditions. *Proc. Natl. Acad. Sci.* **89:** 2475–2479.

Yip, W.K., J.G. Dong, J.W. Kenny, G.A. Thompson, and S.F. Yang. 1990. Characterization and sequencing of the active site of 1-aminocyclopropane-1-carboxylate synthase. *Proc. Natl. Acad. Sci.* **87:** 7930–7934.

Zarembinski, T. and A. Theologis. 1993. Anaerobiosis and plant growth hormones induce two genes encoding 1-aminocylopropane-1-carboxylate synthase in rice (*Oryza sativa* L.). *Mol. Biol. Cell* **4:** 363–373.

Zhang, X., J. Settleman, J.M. Kyriakis, E. Takeuchi-Suzuki, S.J. Elledge, M.S. Marshall, J.T. Bruder, U.R. Rapp, and J. Avruch. 1993. Normal and oncogenic p21$^{ras}$ proteins bind to the amino-terminal regulatory domain of a c-Raf-1. *Nature* **364:** 308–313.

# 20

# Gibberellin and Abscisic Acid Biosynthesis and Response

**Ruth R. Finkelstein**
Department of Biological Sciences
University of California
Santa Barbara, California 93106

**Jan A.D. Zeevaart**
MSU-DOE Plant Research Laboratory
Michigan State University
East Lansing, Michigan 48824

## INTRODUCTION

Evidence for the roles of hormones in plant growth and development has been classically obtained by (1) applying the pertinent hormone and afterwards observing the resulting responses, and (2) correlating endogenous hormone levels with physiological effects. More recently, hormone-deficient mutants have been used for establishing the roles of a particular hormone.

To elicit a response, a certain hormone concentration and a receptor are both needed. Hormone response mutants are supposedly impaired either in the receptor or at a step in the transduction pathway. These mutants are phenotypically similar to hormone-deficient mutants but still contain the hormone and do not respond to that hormone, whether endogenous or applied. Biosynthetic and response mutants for both gibberellins (GA) and abscisic acid (ABA) are known in a number of species, including *Arabidopsis*.

In this chapter, we describe various physiological functions of GA and ABA in *Arabidopsis*. In addition, we discuss how mutants for both hormones have been exploited to facilitate further understanding of GA and ABA biosynthesis and action.

## Physiological Effects of Gibberellins

The most striking effect of applied GA is an increase in stem length, especially in dwarf mutants of maize, pea, and rice (Phinney 1984), and

*Arabidopsis*
© 1994 Cold Spring Harbor Laboratory Press 0-87969-428-9/94 $5 + .00

rosette plants (Zeevaart 1983). The cellular basis for stem elongation is a combination of cell division and cell elongation (Talón et al. 1991). In addition, GAs are involved in flower formation (Zeevaart 1983), male sex expression, fruit set and growth (Pharis and King 1985), and seed germination (Graebe 1987). GAs also induce the synthesis of a number of hydrolytic enzymes that mobilize the reserves to nourish the sprouting embryo (Jacobsen and Chandler 1987; Rodriguez et al. 1987).

## Physiological Effects of ABA

ABA regulates many physiological and developmental processes, such as stomatal closure, embryo dormancy, seed germination, and adaptation to environmental stresses (e.g., drought, chilling, and salt) (Zeevaart and Creelman 1988; Hetherington and Quatrano 1991). The physiological responses to ABA range from very rapid to long term, implying that ABA has different modes of action. The rapid responses, such as stomatal closure, occur at the plasmalemma and are induced only by (+)-ABA, whereas long-term effects can be induced by both (+)- and (–)-ABA (Zeevaart and Creelman 1988; Walker-Simmons et al. 1992). Historically, ABA has been thought of as a growth inhibitor. However, in ABA-deficient plants it can also *promote* growth by reducing transpiration and establishing turgor (Zeevaart and Creelman 1988). Furthermore, in water-stressed plants, ABA is required for continued root growth while inhibiting shoot growth (Saab et al. 1990).

ABA and GAs are traditionally considered as acting antagonistically. The antagonism between GA and ABA is most clearly demonstrated in aleurone cells of many cereal species in which ABA negates GA induction of several hydrolytic enzymes (Jacobsen and Chandler 1987). However, ABA-induced transcription of genes thought to be involved in ameliorating various stresses is not counteracted by GA.

## BIOSYNTHETIC MUTANTS

The initial selection of mutants in *Arabidopsis* for GA- or ABA-deficiency was based on the hypothesis that GA and ABA interact in controlling seed dormancy: ABA would induce seed dormancy and inhibit germination, whereas GA would be required to break dormancy, resulting in promotion of germination (Koornneef et al. 1985a; Koornneef and Karssen, this volume). Thus, the strategy was to look for dormant mutants among germinating seeds, and for germinating mutants among a population of dormant seeds. Selection at the seed level made it possible to screen large populations in petri dishes.

## Gibberellin Biosynthetic Mutants

To isolate nongerminating, GA-responsive mutants, $M_2$ seeds of Landsberg *erecta* (L*er*) were screened for seeds that initially failed to germinate, but did germinate after transfer to a medium containing GA. The germinating seeds developed into dwarfed plants to which the wild-type phenotype could be restored completely by spraying with GA (Fig. 1) (Koornneef and van der Veen 1980). These nongerminating dwarfs grow as rosettes with dark green leaves, or they form short stems and sterile flowers with poorly developed petals and stamens. In addition, germinating mutants with short stems were also found (Koornneef and van der Veen 1980). These mutants are apparently leaky; their phenotypes indicate that the GA requirement for seed germination is lower than that for stem growth and flower development. All the GA-responsive mutants are monogenic recessive to wild-type (L*er*), and the gene symbol *ga* was assigned to this class of mutants. It was found that the 56 GA-responsive mutants represent five different loci located on three of the five *Arabidopsis* chromosomes. The nongerminating mutants at the *GA1*, *GA2*, and *GA3* loci have a similar morphology and grow as rosettes after germination has been induced by GA application. Several alleles of *GA1* have been isolated, some of which are leaky (Koornneef et al. 1983). The available *ga4* and *ga5* mutants are also leaky, permitting germination without added GA but resulting in a semidwarf phenotype: Their stems are approximately half the length of those of wild-type (L*er*) and they flower and produce seeds without addition of GA (Koornneef and van der Veen 1980; Talón et al. 1990a).

### *Characterization of the* ga1, ga2, *and* ga3 *Mutants*

The GA-deficient mutants *ga1*, *ga2*, and *ga3* have been characterized by several different approaches. First, the endogenous GAs present in these mutants have been quantified by gas chromatography-selected ion monitoring (GC-SIM) using $^2$H- or $^{13}$C-labeled GAs as internal standards. In comparison with the GA levels in wild-type plants, those of the three mutants are very low. This indicates that the lesions are early in the GA biosynthetic pathway, prior to $GA_{12}$ (Fig. 2). The *ga1-3* allele has a large deletion encompassing much of the gene (Koornneef et al. 1983; Sun et al. 1992). However, *ga1-3* plants still contain small amounts of GAs (Zeevaart and Talón 1992), suggesting that the *ga1-3* gene product may have some residual enzymatic function, or that another locus can provide a limited amount of the required enzymatic activity.

Second, application of various GAs and GA precursors has established that *ga1* and *ga2* plants respond with normal growth and develop-

*Figure 1*  Plants of the dwarf *ga1-2* mutant, initially grown under short days. After transfer to long days, the plant on the right was treated with GA. Photograph taken 24 days after GA treatment started.

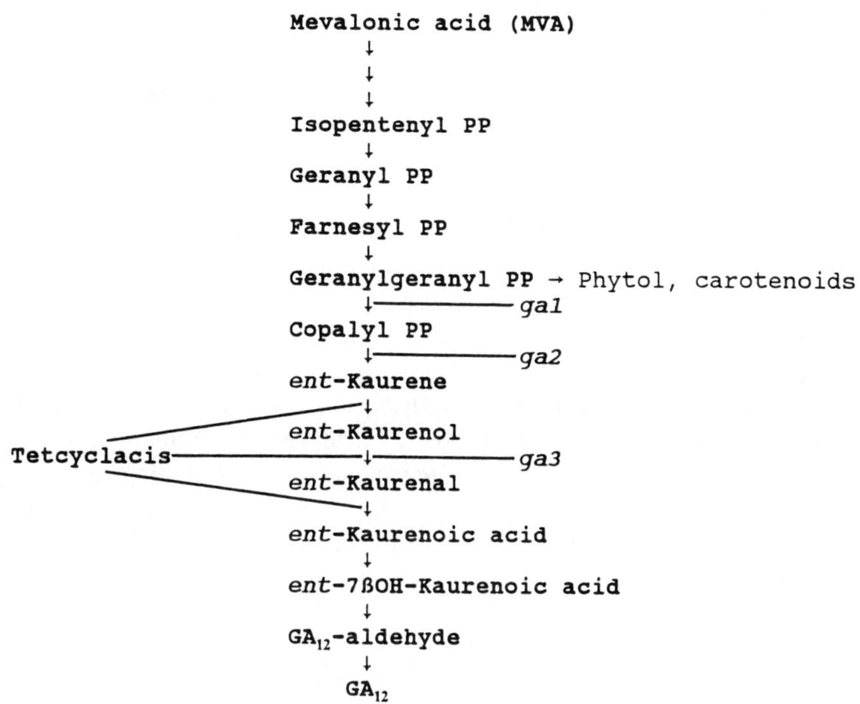



*Figure 2* Early steps in the GA biosynthetic pathway with sites of inhibition by the growth retardant tetcyclacis and the proposed steps blocked by the *ga1*, *ga2*, and *ga3* mutants indicated.

ment (leaf expansion, stem elongation, flowering, and fruiting) to application of *ent*-kaurene and all later intermediates of the GA biosynthetic pathway. On the basis of these results, it can be concluded that the *ga1* and *ga2* mutants are blocked prior to *ent*-kaurene. Since these mutants have a normal chlorophyll and carotenoid content, the biosynthetic pathway to geranylgeranyl PP can be assumed to function (Fig. 2). The most logical interpretation of these observations is, therefore, that both mutants are impaired in *ent*-kaurene synthetase activity, one lacking **A** activity (geranylgeranyl PP→copalyl PP) and the other **B** activity (copalyl PP→*ent*-kaurene). Indeed, it has been shown recently that *Escherichia coli* cells that express both the *Arabidopsis GA1* gene and the *Erwinia uredovora* gene encoding geranylgeranyl PP synthase accumulate copalyl PP (Kamiya et al. 1994). This result establishes that the *GA1* gene encodes *ent*-kaurene synthetase A, and by the process of elimination it follows that the *GA2* gene encodes *ent*-kaurene synthetase B, although a regulatory function for this gene cannot be ruled out at the present time. Plants of the *ga3* mutant do not respond to *ent*-kaurene, and only slightly to *ent*-kaurenol, but they show a growth response following application of *ent*-kaurenal and *ent*-kaurenoic acid. This means that the lesion in the *ga3* mutant is probably in the oxidation of *ent*-kaurenol to *ent*-kaurenal (Fig. 2). On the basis of our current understanding of GA biosynthesis (Graebe 1987), the *GA3* locus presumably encodes a cyto-chrome P-450 monooxygenase.

Third, the ability of wild-type (L*er*) and various *ga* mutants to ac-cumulate *ent*-kaurene following application of tetcyclacis, an inhibitor of GA biosynthesis, was measured by isotopic dilution of added *ent*-[$^{14}$C]kaurene and GC-SIM. Tetcyclacis inhibits oxidation of *ent*-kaurene to *ent*-kaurenoic acid (Fig. 2), so that *ent*-kaurene accumulates, presumably at the rate of its biosynthesis (Grosselindemann et al. 1991; Zeevaart and Gage 1993). Following tetcyclacis treatment, no *ent*-kaurene accumulated in *ga1* and *ga2* plants, thus confirming that both mutants are impaired in synthesis of *ent*-kaurene (see above). Untreated wild-type and *ga3* plants contained approximately 50 pmole and 3000 pmole of *ent*-kaurene per gram dry weight, respectively (J.A.D. Zeevaart, unpubl.). Following tetcyclacis treatment, the *ent*-kaurene content in-creased 50-fold in wild-type plants, but remained unchanged in *ga3* plants. Considering that *ga3* plants do not respond to applied *ent*-kaurenol (see above), it is to be expected that *ent*-kaurenol accumulates as well in this mutant.

Counterparts of the *ga1* and *ga2* mutants in *Arabidopsis* are known in a number of plants: in tomato, *gib-1* and *gib-3* are blocked in *ent*-kaurene synthetase **A** and **B** activities, respectively (Bensen and Zeevaart 1990; Koornneef et al. 1990); the *d5* mutant of maize is altered in *ent*-kaurene synthetase B (Hedden and Phinney 1979), and the *ls* mutant of pea is blocked in *ent*-kaurene synthetase A (Reid and Ross 1993). Other dwarf mutants that are impaired in the GA biosynthetic pathway prior to *ent*-kaurene, most probably at *ent*-kaurene synthetase, are *ga1* in barley (Boother et al. 1991) and the EMS-141 mutant of *Thlaspi arvense* (Metzger and Hassebrock 1990). There are no reported mutations in other species that are analogous to the *ga3* mutation in *Arabidopsis*.

*Characterization of the* ga4 *and* ga5 *Mutants*

*Arabidopsis* shoots contain an unusually large number of GAs that can be arranged in three parallel pathways: (1) the early 13-hydroxylation pathway (GA$_{53}$, GA$_{44}$, GA$_{19}$, GA$_{17}$, GA$_{20}$, GA$_1$, GA$_{29}$, and GA$_8$); (2) the non-3,13-hydroxylation pathway (GA$_{12}$, GA$_{15}$, GA$_{24}$, GA$_{25}$, GA$_9$, and GA$_{51}$); and (3) the early-3-hydroxylation pathway (GA$_{37}$, GA$_{27}$, GA$_{36}$, GA$_{13}$, GA$_4$, and GA$_{34}$) (Fig. 3) (Talón et al. 1990a). The three

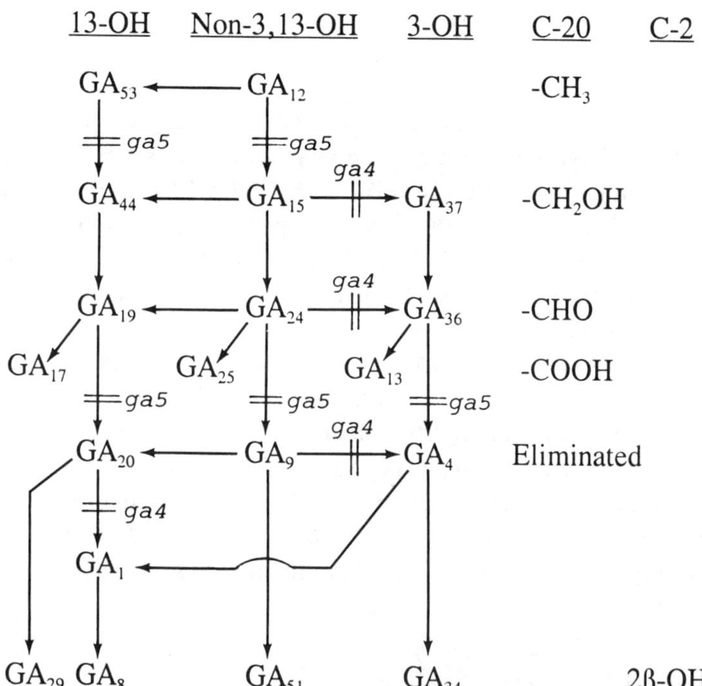



*Figure 3*  Possible biosynthetic pathways in *Arabidopsis thaliana* after $GA_{12}$ with the proposed positions of the steps blocked in the *ga4* and *ga5* mutants. All $C_{19}$-gibberellin conversions have been demonstrated in feeding experiments through identification of the products by combined gas chromatography-mass spectrometry (Zeevaart and Talón 1992). Numbering system of *ent*-gibberellane skeleton is indicated on top. Inactivation of GAs occurs by 2β-hydroxylation: $GA_{29}$, $GA_8$, $GA_{51}$, and $GA_{34}$. GAs are also inactivated by oxidation of C-20 to -COOH: $GA_{17}$, $GA_{25}$, and $GA_{13}$.

pathways are connected, so that hydroxylations can occur at different levels of oxidation at C-20. For example, applied $GA_9$ is not only converted to $GA_{51}$ (2β-OH), but also to $GA_{20}$ (13-OH) and $GA_4$ (3β-OH) (Zeevaart and Talón 1992).

Quantification of GAs by GC-SIM indicates that the *ga4* and *ga5* mutants are leaky, suggesting that the genes may be expressed at a lower level than in wild type or may encode altered proteins with partially impaired catalytic activities (Talón et al. 1990a). In comparison with wild-type, *ga4* plants had reduced levels of 3-hydroxy-GAs ($GA_{37}$, $GA_{36}$, $GA_4$, $GA_{13}$, and $GA_{34}$) and 3,13-dihydroxy-GAs ($GA_1$ and $GA_8$) and accumulated 13-hydroxy-GAs ($GA_{44}$, $GA_{19}$, $GA_{17}$, and $GA_{20}$) and non-3,13-hydroxy-GAs ($GA_{15}$, $GA_{24}$, $GA_{25}$, $GA_9$, and $GA_{51}$). Thus, the *ga4* mutation blocks 3β-hydroxylation in all three pathways, suggesting that these reactions are catalyzed by a single enzyme with low substrate specificity, as is common for GA hydroxylases (Graebe 1987).

The *ga5* mutant had reduced levels of $C_{19}$-GAs ($GA_9$, $GA_{51}$, $GA_4$, $GA_{34}$, $GA_{20}$, $GA_1$, $GA_{29}$, and $GA_8$), independent of the degree of oxidation and position of hydroxyl groups. The GAs with C-20=-$CH_3$ ($GA_{12}$ and $GA_{53}$) accumulated, as did the GAs with C-20=-CHO ($GA_{24}$ and $GA_{36}$). On the basis of these results, it was proposed that the *ga5* mutation decreases the elimination of C-20 at the aldehyde level and possibly also the first oxidation step at C-20 (Talón et al. 1990a). Results obtained with purified enzymes from spinach suggest that a single enzyme catalyzes C-20 hydroxylation of $GA_{53}$ and C-20 elimination from $GA_{19}$ (Zeevaart et al. 1990). This would explain why the GA biosynthetic pathway appears to be blocked at two different sites by the *ga5* mutation.

The *ga4* mutant in *Arabidopsis* is analogous to the *d1* mutant in maize (Phinney 1984), the *le* mutant in pea (Reid and Ross 1993), the *l* mutant in *Lathyrus* (Ross et al. 1990), and the *dy* mutant in rice (Takahashi and Kobayashi 1990). No counterpart of the *ga5* mutation has been described so far in other species.

*Effectiveness of Different GAs*

When $GA_1$, $GA_4$, and $GA_9$ were applied to *ga4* plants, $GA_4$ restored the wild-type phenotype, $GA_9$ had very low activity, and $GA_1$ caused an intermediate response (Talón et al. 1990a). In contrast, $GA_4$ and $GA_9$ were equally effective when applied to *ga1* plants (Zeevaart and Talón 1992). This difference in effectiveness of $GA_9$ is due to the lesion in 3β-hydroxylation in *ga4* plants, whereas in *ga1* plants, 3β-hydroxylation can proceed as in normal plants. Thus, as has been shown for maize (Phinney 1984) and pea (Reid and Ross 1993), 3β-hydroxylation is essential for GAs to be active per se. This has been further substantiated by the use of the growth retardant BX-112 (calcium 3,5-dioxo-4-propionylcyclohexanecarboxylate), which inhibits α-ketoglutarate-dependent dioxygenases. In the case of GA biosynthesis, BX-112 is particularly effective in blocking 3β-hydroxylation. Application of BX-112 to wild-type plants resulted in reduced stem growth. This growth inhibition could be overcome by

3-hydroxy-GAs ($GA_4$, $GA_{36}$, and to a lesser extent by $GA_1$), but not by non-3-hydroxy-GAs ($GA_9$, $GA_{19}$, $GA_{20}$) (Zeevaart and Talón 1992). In wild-type plants treated with BX-112, the levels of the 3-hydroxy-GAs ($GA_4$ and $GA_1$) were much reduced, whereas the non-3-hydroxylated GAs ($GA_9$, $GA_{20}$, and $GA_{19}$) accumulated. Thus, treatment of *Arabidopsis* with BX-112 is a chemical means by which the *ga4* mutation can be simulated.

Since $GA_4$ is more active than $GA_1$ in causing stem elongation in *Arabidopsis*, it was suggested that $GA_4$ is active per se (Talón et al. 1990a; Zeevaart and Talón 1992). However, $GA_4$ applied to *Arabidopsis* is 13-hydroxylated to $GA_1$ (Zeevaart and Talón 1992; Kobayashi et al. 1993). It is possible, therefore, that the biological activity of $GA_4$ is due to its conversion to $GA_1$, although this does not explain why the precursor $GA_4$ is more active than its product $GA_1$. Thus, whereas 3β-hydroxylation is a prerequisite for GA activity in *Arabidopsis*, the requirement for 13-hydroxylation is uncertain. This question can be resolved only with the aid of a mutant that is blocked in 13-hydroxylation or with an inhibitor that specifically blocks this step (Zeevaart and Talón 1992).

## ABA Biosynthetic Mutants

Nongerminating seeds of the *ga1* mutant were mutagenized to obtain revertants; i.e., mutants that germinate in the absence of GA. In addition to reduced seed dormancy, these mutants are characterized by reduced stem length and leaf size, increased transpiration, wilting, and a lowered ABA content (Fig. 4) (Koornneef et al. 1982; Rock and Zeevaart 1991). Regular spraying with ABA or with the ABA analog LAB 173711 (Koornneef et al. 1982; Jung and Grossmann 1985; J.A.D. Zeevaart, unpubl.) restores the normal phenotype. The revertant mutants are homozygous recessive for the *ABA* locus, which is involved in ABA biosynthesis. Three alleles of this locus, listed in order of increasing phenotypic severity (reduced growth and increased transpiration), were isolated: *aba-3*, *aba-1*, and *aba-4* (Koornneef et al. 1982). Both leaves and seeds of these mutants are deficient in ABA. Although the revertant selection was successful in *Arabidopsis*, it did not work in tomato: The double mutant *sit/sit gib-1/gib-1* is ABA- and GA-deficient, but still requires GA for germination (Koornneef et al. 1985a). In contrast, there is no GA requirement per se for seed germination in *Arabidopsis*, since genotypes that combine GA- and ABA-deficiency readily germinate.

The biosynthetic pathway of ABA, a sesquiterpenoid ($C_{15}H_{20}O_4$), has only recently been elucidated. Originally, two pathways were pro-

*Figure 4* Wild-type Landsberg *erecta* plant (*left*) and *aba-1* mutant plant (*right*) with droopy stem.

posed: (1) the direct pathway from mevalonic acid via farnesyl pyrophosphate and (2) the indirect pathway from xanthophylls with the cleavage product xanthoxin as intermediate. $^{18}O_2$ labeling experiments with water-stressed leaves have shown that the side-chain carboxyl group of ABA becomes highly enriched in $^{18}O$, whereas little $^{18}O$ is incorporated in the oxygens on the ring, indicating that there is a large ABA precursor pool that already contains oxygens on the ring (Zeevaart and Creelman 1988). When water-stressed leaves of wild type and the three *aba* mutants were incubated in an atmosphere containing $^{18}O_2$, little [$^{18}O$]ABA and its catabolites, phaseic acid and ABA-glucose ester, were produced. This establishes that the low ABA levels in the mutants are due to low rates of biosynthesis and not to a high rate of ABA degradation. Furthermore, ABA from the *aba-1* mutant, and even more so from *aba-4* plants, showed high $^{18}O$ incorporation into the ring-attached oxygens (Rock and Zeevaart 1991). This means that the *aba* mutant has a small ABA precursor pool that presumably consists of xanthophylls. Quantification of the carotenoids from wild-type and mutant leaves established that the *aba* mutation causes deficiency of the epoxy-carotenoids violaxanthin and neoxanthin, and an accumulation of their precursor zeaxanthin (Duckham et al. 1991; Rock and Zeevaart 1991). Roots of the *aba* mutant also had reduced epoxy-carotenoid levels compared to wild-type (Parry and Horgan 1992). Thus, the *ABA* locus affects an enzyme that functions in the epoxidation of zeaxanthin via antheraxanthin to violaxanthin (Rock and Zeevaart 1991). These results obtained with the *aba* mutant demonstrate that violaxanthin and neoxanthin are essential for ABA biosynthesis and therefore provide conclusive proof for the indirect pathway of ABA biosynthesis in higher plants (Fig. 5).

The *aba* mutant also provides a means to study the function of xanthophylls in photosynthesis. In vivo chlorophyll fluorescence of the *aba* mutant was lower than that of wild-type (Rock et al. 1992). This result is as expected, since zeaxanthin, which accumulates in *aba* leaves, is known to be involved in photoprotective energy dissipation in plants under excess light (Demmig-Adams and Adams 1992). Thylakoids isolated from *aba* leaves had a lower rate of oxygen evolution than thylakoids from wild-type. In addition, the ultrastructure of chloroplasts of *aba* leaves is abnormal: The mutant has fewer thylakoid lamellae per granum stack than wild-type chloroplasts, but there are more grana per chloroplast than in wild-type (Rock et al. 1992).

No ABA biosynthetic mutants analogous to the *aba* mutant in *Arabidopsis* are known in other species. All other characterized ABA-deficient mutants are impaired either in carotenoid biosynthesis or in the final step in ABA biosynthesis from ABA-aldehyde to ABA (Fig. 5): *flacca* and *sitiens* in tomato (Sindhu and Walton 1988), *droopy* in potato (Duckham et al. 1989), the *Az34* mutant in barley (Walker-Simmons et

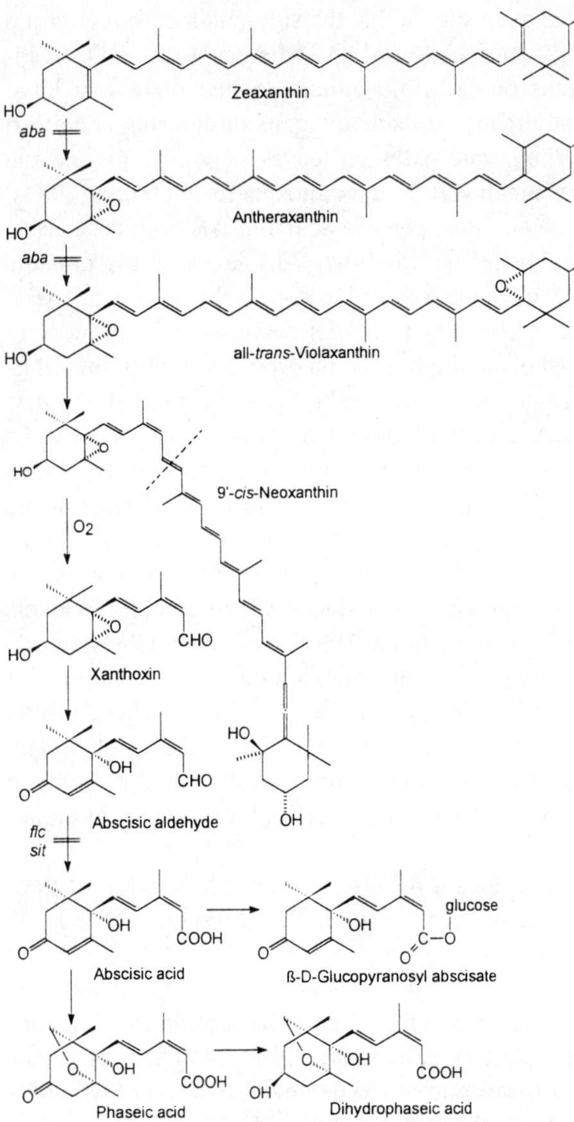

*Figure 5* Proposed biosynthetic pathway of ABA after zeaxanthin with positions of the steps blocked in the *aba* mutant of *Arabidopsis* and the *flacca* and *sitiens* mutants of tomato. Also indicated are the major inactivation products of ABA: phaseic acid, dihydrophaseic acid, and the glucose ester of ABA, β-D-glucopyranosyl abscisate.

al. 1989), and the *CKR1* mutant in *Nicotiana plumbaginifolia* (Parry et al. 1991). Considering that mutants for only two steps in the pathway are available, mutations at other points would add to our understanding of the pathway and its regulation.

**RESPONSE MUTANTS**

Hormone-response mutants have traditionally been defined as individuals that resemble mutants with defects in hormone biosynthesis, yet cannot be restored to a wild-type phenotype by addition of the relevant hormone. In general, the classification of hormone-response mutant is reserved for those showing highly pleiotropic phenotypes. Assuming that the myriad of hormone-regulated responses are controlled by branching regulatory pathways originating from one to a few types of recognition events, such criteria should skew the mutant collection toward those affecting early steps in the signal transduction pathway(s).

## GA Response Mutants

Two loci affecting GA response in *Arabidopsis* have been described: *gibberellin insensitive* (*GAI*) (Koornneef et al. 1985b) and *spindly* (*SPY*) (Jacobsen and Olszewski 1993a), so named because the *spy* mutants display the extreme elongation characteristic of plants exposed to saturating concentrations of GA. Interestingly, mutations at these loci can produce opposite phenotypes. The first gibberellin-insensitive (*gai*, originally designated *Gai*) mutant isolated resembles the gibberellin-deficient (*ga*) mutants, in that *gai* mutant plants are dark green dwarfs with reduced apical dominance and reduced seed germination. The *gai* mutant differs from the severe GA-deficient mutants in that *gai* floral organs and fertility are normal. However, leaky *ga* mutants also have normal flowers and fertility, raising the possibility that the *gai* allele might also be leaky. The most significant phenotypic difference between GA-insensitive and GA-deficient mutants is that the normal phenotype cannot be restored by application of GA (Table 1). Recently, four irradiation-induced derivative alleles of the *gai* mutation were isolated as suppressors of the dwarf phenotype (Peng and Harberd 1993). The screen was based on the fact that the original *gai* mutant acted semi-dominantly, so loss of one *gai* allele should partially rescue the dwarf phenotype. Further genetic analysis of these mutants indicated that all four were recessive intragenic suppressors. Three of these produce a wild-type phenotype when homozygous; the fourth has not been obtained as a homozygote because its transmission through pollen is severely reduced. The simplest interpretation of these results is that the original *gai* mutant is a gain of function, creating a gene product that interferes with GA response, and that the *GAI* locus may not normally be involved in GA response.

The *spy* mutants were isolated as plants whose germination and vegetative growth were resistant to inhibition by the GA biosynthesis in-

*Table 1* Characteristics of GA-deficient and GA-insensitive mutants

| Characteristic | WT | ga | gai | spy | References |
|---|---|---|---|---|---|
| Stem elongation | + | – | – | ++ | 1, 2, 3 |
| Apical dominance | + | – | – | + | 1, 2 |
| Germination | | | | | |
|  darkness | – | – | – | | 2 |
|  light | + | – | + | | 2 |
|  light + GA | ++ | ++ | + | | 2 |
|  cold | – | – | – | | 2, 4 |
|  cold + GA | ++ | ++ | ± | | 2, 4 |
|  light + cold | ++ | – | ++ | | 2, 4 |
|  light + cold + GA | ++ | ++ | ++ | | 2, 4 |
|  paclobutrazol | – | – | | + | 1 |
| Flowering | | | | | |
|  short days | + | –[a] | ± | | 5 |
|  short days + cold | ++ | –[a] | + | | 5 |
|  short days + GA | ++ | +[a] | ± | | 5 |
|  continuous light | ++ | +[a] | +± | +++ | 1, 5 |
| Male fertility | + | – | ±[b] | – | 1, 6 |

References: (1) Jacobsen and Olszewski 1993a; (2) Koornneef et al. 1985b; (3) Koornneef and van der Veen 1980; (4) Karssen and Laçka 1985; (5) Wilson et al. 1992; (6) Wilson 1993.
[a]Defect observed only in severe alleles.
[b]Defect observed only in GA-deficient background.

hibitor paclobutrazol (Jacobsen and Olszewski 1993a). In the absence of paclobutrazol, they resemble wild-type plants that have been repeatedly treated with GA (Fig. 6). The *spy* mutants have increased stem elongation, pale green leaves, early flowering, increased parthenocarpy, and partial male sterility. These characteristics, and the *spy* mutants' resistance to growth inhibition by GA biosynthesis inhibitors, resemble the "slender" mutants of pea (Potts et al. 1985), barley (Lanahan and Ho 1988), and tomato (Jones 1987). The recessive nature of all the slender mutations (all except pea are monogenic mutants) has led to the suggestion that the wild-type gene products are repressors of GA signal transduction.

A relatively trivial cause of a GA nonresponsive phenotype would be a mutation causing a high rate of GA turnover or inactivation. However, this does not appear to be the case in either the *gai* or *spy* mutants. Measurements of endogenous GAs in *gai* mutant plants showed decreased levels of $C_{20}$ precursors and greatly increased levels of $C_{19}$-GAs, with the bioactive $GA_1$ and $GA_4$ and the inactivated $GA_8$ and $GA_{34}$ most abundant (Talón et al. 1990b). $GA_1$, the GA active in regulating shoot growth of maize, rice, and pea (Phinney 1984), shows the

*Figure 6*   *spy-1* mutant and Columbia (Col) wild-type plants photographed 18 days after germination under long photoperiods. (Reprinted, with permission, from Jacobsen and Olszewski 1993a.)

greatest increase; *gai* mutants accumulate approximately 30-fold more $GA_1$ than wild-type plants (Talón et al. 1990b). Similar results were obtained for the maize (Fujioka et al. 1988) and wheat (Appleford and Lenton 1991) non-GA-responding dwarfs (*D8* and *Rht3*, respectively), indicating that these mutants may reflect some sort of feedback regulation of GA synthesis.

The *spy* mutations are unlikely to alter GA metabolism as their primary effect because they suppress the effects of GA deficiency conferred by genetic or chemical blocks at more than one step in the biosynthetic pathway (Jacobsen and Olszewski 1993a). Furthermore, the effects of the *spy* mutations and applied GA are additive, consistent with a normal GA sensitivity superimposed on GA-independent activation of responses that usually require GA. Although endogenous GA levels have not been measured in the *spy* mutants, the slender mutants of pea and tomato have reduced endogenous GA levels (Potts et al. 1985; Jones 1987), again consistent with feedback regulation of GA synthesis.

## Role of GA in Germination

The role of GA in seed germination has been studied extensively in *Arabidopsis* by comparing wild-type, GA-deficient, and GA-insensitive lines under various environmental conditions (Koornneef and van der Veen 1980; Karssen and Laçka 1985; Koornneef et al. 1985b). In addition to hormonal regulation, seed germination is affected by a variety of extrinsic factors, including environmental conditions during seed development, time in storage, cold treatments, and light conditions during germination (for review, see Koornneef and Karssen, this volume). Therefore, GA sensitivity has been compared for seeds in either light or darkness, with or without cold treatment during imbibition, using seed lots grown, harvested, and stored under similar conditions. Several studies have indicated that light-induction of germination is at least partly dependent on light-induced GA synthesis (Karssen and Laçka 1985; Nambara et al. 1991), but this may not account for the entire light-induced increase in germination. Exposure to light reduces the amount of exogenous GA required to induce germination of both wild-type and GA-deficient seeds, implying that light may also increase GA sensitivity (Koornneef et al. 1985b). However, since none of the GA-deficient mutants completely lacks endogenous GAs, their light-induced reduction in requirement for exogenous GA may partly reflect residual light-induced GA synthesis.

In contrast to light treatment, cold treatment alone (i.e., followed by incubation in darkness) is not sufficient to induce germination of any genotype. Cold treatment results in comparably increased sensitivity to applied GA for both wild-type and *ga1* mutant seeds (Karssen and Laçka 1985), but germination of cold-treated *gai* seeds is negligible in darkness, regardless of whether GA (even up to 1 mM) is added (Koornneef et al. 1985b). However, deeply dormant *gai* or *gai ga1* seeds require preincubation in cold for light or GA induction of germination to be effective (Table 1), showing that *gai* mutants perceive chilling (Derkx 1993). Derkx concluded that GA synthesis and sensitivity are essential for germination but are not the primary factors controlling dormancy.

## Role of GA in Flowering

GAs have been shown to induce flowering, or at least stem elongation, of some plants, especially rosette plants (for review, see Zeevaart 1983).

The GA synthesis and response mutants of *Arabidopsis* provide an opportunity to test the role of endogenous GAs in floral induction. All races of *Arabidopsis* characterized to date have a quantitative requirement for long days in that they flower earlier under long-day conditions but will flower eventually under short days (Brown and Klein 1971; Napp-Zinn 1985). A role for GAs in regulating flowering of *Arabidopsis* was suggested by the observation that exogenous GA can accelerate flowering under short days (Langridge 1957). Consistent with this, the constitutive GA response *spy* mutants flower earlier than wild type (Jacobsen and Olszewski 1993a).

Studies of late-flowering mutants led Martínez-Zapater and Somerville (1990) to propose a model in which flowering can be induced by at least three independent environmental stimuli: cold, far-red light, and blue light or assimilate availability. The role of GA in mediating cold and photoperiodic effects on floral induction has been tested by comparing flowering of GA-deficient and GA-insensitive plants grown under short days or continuous light, with or without a cold treatment (Wilson et al. 1992). These studies did not attempt to distinguish between far-red and blue light effects. Wilson et al. (1992) found that a severely GA-deficient mutant (*ga1-3*) failed to flower in short days and flowered more slowly than wild-type plants in continuous light, indicating that light could not completely override the requirement for GA (Table 1). Long photoperiods have been shown to stimulate GA synthesis in other rosette plants (Talón et al. 1991), and it is likely that this is also true in *Arabidopsis*.

Chilling, the other major environmental stimulus of flowering described by Martínez-Zapater and Somerville (1990), also appears to require a minimal amount of GA to be effective, since cold treatment triggers flowering of wild type, but not of the severely GA-deficient mutant (*ga1-3*) in short days. Cold induction of flowering is also slightly impaired in the *gai* mutant, but given the wild-type phenotype of the *gai*-derivative alleles, it is not clear whether *GAI* has a normal role in flowering.

The importance of GA in mediating floral induction by cold and light has also been tested in mutants defective in response to some of these environmental stimuli. The late-flowering *fca* mutant remains responsive to GA, far-red light, and cold (Bagnall 1992). The fact that applied GAs can circumvent the *fca* defect could be explained by GA action through any of the environmentally induced signaling pathways, provided that GA response functions at a later step than *FCA*. Thus, although experiments point toward a role for GAs in floral induction, their role in mediating responses to specific environmental signals remains ambiguous.

## ABA Response Mutants

Mutations at three loci affecting ABA response (*ABI*, for *ABA-insensitive*) have been identified and extensively characterized (Koornneef et al. 1984, 1989; Finkelstein and Somerville 1990). $M_2$ seeds of Landsberg *erecta* were germinated on a medium containing 10 μM (±)-ABA to select mutants with reduced sensitivity to ABA inhibition of seed germination. Phenotypically, these mutants resemble the *aba* mutants in exhibiting reduced seed dormancy. However, all have at least wild-type levels of endogenous ABA, indicating that their insensitivity is not due to increased turnover or inactivation (Koornneef et al. 1984). None of the mutants has lost ABA inducibility for all responses tested, suggesting that none affects a putative "universal" ABA receptor (Finkelstein and Somerville 1990).

### *Physiological Characterization of the* abi *Mutants*

The failure of all *abi* mutants to become dormant, and their decreased sensitivity to exogenous ABA at seed maturity, imply that all of the *ABI* gene products are required in maturing and mature seeds. However, several studies have indicated that there are more processes in maturing seeds requiring *ABI3* than *ABI1* or *ABI2* action. A weak mutant allele of *ABI3* (*abi3-1*) produces a slight reduction in storage protein and lipid accumulation (2- and 3-fold, respectively) (Finkelstein and Somerville 1990) and a 10-fold decrease in mRNA levels for the *Arabidopsis* homolog of the ABA-responsive cereal *Em* gene, a late embryogenesis-abundant transcript (Finkelstein 1993). Double mutants combining the *abi3-1* allele with ABA deficiency (the *aba-1* mutant) produce seeds that fail to lose chlorophyll, accumulate storage proteins, or attain desiccation tolerance (Koornneef et al. 1989). Although applied ABA cannot restore dormancy induction to ABA synthesis or response mutants (Karssen et al. 1983), it can reverse the *aba abi3-1* effects on desiccation tolerance and seed protein accumulation (Meurs et al. 1992). One possible explanation for this observation is that induction of desiccation tolerance and storage reserve accumulation requires much less ABA than dormancy induction. However, it is worth noting that even the most severe mutations in ABA biosynthesis, resulting in a greater than 27-fold reduction of seed ABA levels (Koornneef et al. 1982), do not produce a "green seed" phenotype or reduce storage reserve accumulation. Recently, new alleles of *ABI3* (*abi3-3* and *abi3-4*) have been isolated which do produce a green seed phenotype, even in plants with wild-type ABA biosynthetic capacity (Giraudat et al. 1992; Nambara et al. 1992). Taken together, these results suggest that *ABI3* is required for processes in seed development which can respond to, but do not require, high endogenous ABA. It is still not clear whether *ABI3* mediates responses to ABA, per se, or is involved in an overlapping regulatory pathway. In contrast to the results with *abi3* mutants, no available alleles of *ABI1* or *ABI2* affect storage reserve synthesis or desiccation tolerance, whether or not ABA biosynthesis is altered by the presence of the *aba-1* allele (Koornneef et al. 1989; Finkelstein and Somerville 1990).

The relative importance of the *ABI* loci changes after germination. Seedling growth of *abi3* plants is slightly less sensitive to ABA inhibition than wild type, but *abi1* and *abi2* seedling growth is not inhibited by ABA at concentrations resulting in 50–70% inhibition of wild-type seedling growth (Finkelstein and Somerville 1990). This difference is apparent in both shoot and root growth (R.R. Finkelstein, unpubl.), even at the level of individual root hair elongation (Table 2) (Schnall and Quatrano 1992).

In young plantlets, the differences between *abi3* versus *abi1* and *abi2* mutants become more pronounced. Comparison of ABA-induced changes in protein synthesis and proline accumulation shows that *abi3* plants closely resemble wild type, but *abi1* and *abi2* show reduced response to ABA (Finkelstein and Somerville 1990). Consistent with this, Gosti et al. (1993) have identified cDNAs corresponding to genes induced by either ABA or drought stress in wild-type plants, but not in *abi1* and *abi2* mutants. Interestingly, drought-stressed *abi1* and *aba* plants accumulate both ABA and proline, but proline accumulation is not proportional to endogenous ABA levels or sensitivity (Gottlieb and Bray 1991). Similar results were obtained in experiments with osmotically stressed barley and tomato, whose ABA biosynthesis was impaired either chemically or genetically (Stewart and Voetberg 1987). These studies indicate that, during drought stress, proline accumulation is not directly tied to ABA accumulation or response.

Another water-stress response, stomatal closure, is dramatically reduced in only the *abi1* and *abi2* mutants, resulting in a "wilty" phenotype similar to that of the ABA-deficient mutant (Koornneef et al. 1984). Surprisingly, the *abi1* and *abi2* mutants differ from the *aba* mutant in that exogenous ABA actually intensifies, rather than reverses, withering of stems and siliques due to long-term water stress (Koornneef et al. 1984). One possible explanation for this is that the mutants are capable of recognizing the ABA treatment and may respond nonproductively by increasing turnover, decreasing synthesis, or altering redistribution of ABA to the relevant location, i.e., the guard cells. Testing this hypothesis would require assaying ABA distribution at the cellular or subcellular level.

*Table 2* Characteristics of ABA-deficient and ABA-insensitive mutants

| Characteristic | WT | *aba* | *abi1* | *abi2* | *abi3* | References |
|---|---|---|---|---|---|---|
| Peak seed ABA level (ng/g FW) | 169 | <10 | 348 | 473 | 279 | 1 |
| Storage protein accumulation | + | + | + | + | − | 2, 3 |
| Storage lipid accumulation | + | + | + | + | − | 2 |
| *Lea* expression[a] | + | + | + | + | − or + | 4, 5 |
| Desiccation tolerance | + | + | + | + | −[b] | 3, 5 |
| Dormancy | + | − | − | − | − | 1, 6 |
| ABA inhibition of | | | | | | |
|   germination | + | ± | − | − | − | 1 |
|   seedling growth | + | − | − | ± | | 2 |
|   root hair elongation | + | − | − | + | | 7 |
| Stomatal regulation | + | − | − | − | + | 1, 6 |
| Proline accumulation | | | | | | |
|   ABA-induced | + | − | − | + | | 2 |
|   drought-induced | + | + | + | | | 8 |
| Cold tolerance: cold-induced | + | − | + | + | + | 9, 10 |
| *Cor/lti* gene expression | | | | | | |
|   ABA-induced | + | + | − | + | + | 9, 11–13 |
|   cold-induced[c] | + | − or + | + or − | + | + | 9, 11–13 |

References: (1) Koornneef et al. 1984; (2) Finkelstein and Somerville 1990; (3) Nambara et al. 1992; (4) Finkelstein 1993; (5) Ooms et al. 1993; (6) Koornneef et al. 1982; (7) Schnall and Quatrano 1992; (8) Gottlieb and Bray 1991; (9) Gilmour and Thomashow 1991; (10) Heino et al. 1990; (11) Lang and Palva 1992; (12) Nordin et al. 1991; (13) Nordin et al. 1993.
[a]*ABI3* regulation varies among *Lea* genes.
[b]Defect observed only in severe alleles.
[c]Cold-inducibility varies among genes.

ABA may also play a role in cold tolerance, since ABA can substitute for cold treatment in inducing both cold hardening and expression of some cold-induced genes (for reviews, see Guy 1990; Thomashow, this volume). Consistent with this, the *aba* mutant does not harden as well as wild type in response to cold treatment (Heino et al. 1990; Gilmour and Thomashow 1991). In contrast, the *abi* mutants are not impaired in cold-induced hardening (Gilmour and Thomashow 1991); they have not been tested for ABA-induced hardening. Comparison of ABA and cold induction of the cold-regulated genes in the various ABA synthesis and response backgrounds indicates that ABA and cold probably regulate some, but not all, of these genes through separate mechanisms (Gilmour and Thomashow 1991; Nordin et al. 1991, 1993; Lang and Palva 1992).

Although most of the examples in the preceding paragraphs indicate that mutations in *ABI1* and *ABI2* have similar effects, some distinctions have been observed. ABA regulation of the cold-induced genes is impaired only in the *abi1* mutant (Gilmour and Thomashow 1991; Nordin et al. 1991). Similarly, germination of wild-type and *abi2* seeds is inhibited by the GA biosynthesis inhibitor uniconazol, but *abi1* and *abi3* are unaffected (Nambara et al. 1991). Either of these observations could be explained by greater leakiness of the *abi2* mutation. Alternatively, the differences in sensitivity could reflect action through multiple ABA response pathways.

## Genetic Analysis of the abi Mutants

A standard approach to determining whether mutations affect the same or parallel response pathways is to construct double mutants and look for additive or epistatic interactions. Pairwise combinations of mutations at all three *ABI* loci have been tested for ABA sensitivity of germination inhibition, the only readily quantified response affected in all three mutant classes (Finkelstein and Somerville 1990). The *abi1 abi2* digenic mutant showed the same degree of ABA sensitivity as *abi1*, the more severe of the two monogenic parents. However, the *abi1 abi3-1* digenic mutant was more than 100-fold less sensitive to ABA than *abi1* or *abi3-1* monogenic mutants, yet still did not show a green seed phenotype. It was not initially possible to obtain an *abi2 abi3-1* double homozygote, but the *abi2/+ abi3-1/abi3-1* seeds showed a similarly drastic decrease in ABA sensitivity and again no green seed phenotype. We have recently obtained *abi2 abi3-1* homozygotes and found that seed viability declines rapidly in this genotype, even though the seeds do lose chlorophyll and undergo desiccation (R.R. Finkelstein, unpubl.). The triple mutant has not yet been constructed. Although the presumed leakiness of all three mutant alleles in this study makes it difficult to say which loci are acting in the same pathway, it is clear that there are two phenotypic classes of double mutants, consistent with gene action in at least two pathways. Furthermore, the fact that combination of a weak allele of *ABI3 abi3-1* with either *abi1* or *abi2* does not produce the green seed phenotype observed for severe alleles of *abi3* suggests that *ABI3* is acting in a separate pathway from *ABI1* and *ABI2*. This is also consistent with the observation that combining the *aba-1* allele with an *abi1* mutation does not produce the green seed phenotype of the *aba-1 abi3-1* digenic mutant (Koornneef et al. 1989).

Two additional interesting observations emerged from the double mutant analysis. First, the initial failure to obtain *abi2 abi3-1* homozygotes provided the first evidence of a role for ABA response in gamete development or transmission. Although we have since obtained the digenic mutant, it is grossly underrepresented in a segregating population derived from self-fertilization of *abi2/+ abi3/abi3* individuals. An effect on gamete transmission is not obvious in either the *abi2* or *abi3* monogenic mutants, again consistent with action of the *ABI2* and *ABI3* gene products in separate pathways with overlapping function (R.R. Finkelstein, unpubl.). Second, the greatly decreased ABA sensitivity of the *abi2/+ abi3/abi3* seeds indicated that the *abi2* mutation was at least partially dominant. Subsequent studies have shown that, for both the *abi1* and *abi2* mutants, the degree of dominance varies among ABA-inducible responses, ranging from recessive to nearly fully dominant (Finkelstein 1994b). Interestingly, the sensitivity of mature seeds to applied ABA appears to be partially maternally controlled. A partially dominant loss of sensitivity could reflect control by either rate-limiting levels of the *ABI* gene products or inactivation of an *abi*-containing complex by subunit poisoning. In contrast, all *abi3* mutant alleles isolated to date are recessive, and their effects could be explained by a simple loss of function. Molecular studies of the products of these loci should make it possible to test these hypotheses.

Finally, the studies summarized above indicate that there are at least two independent pathways of ABA responses with overlapping effects. Mutations at three loci do not saturate these or any additional pathways. Several research groups are currently screening for new mutants, which will help construct a more complete picture of ABA responses. We have identified at least four new loci affecting ABA sensitivity of germination inhibition (Finkelstein and Doyle 1993), two of which appear physiologically and genetically similar to the *abi3* mutants (Finkelstein 1994a).

## MOLECULAR APPROACHES

The studies summarized in the previous sections have provided the basis for models of gene action that could be tested with molecular probes for the relevant loci. All of the hormone response loci share the general characteristic that the biochemical nature of the gene product cannot be deduced from the mutant phenotype. Although it is theoretically possible to purify and microsequence the relevant proteins for the loci encoding hormone biosynthetic enzymes, such enzymes are likely to be of extremely low abundance. A biochemical approach is particularly impractical when working with an organism of such small size as *Arabidopsis*. Therefore, a genetically based cloning strategy is the approach of choice in *Arabidopsis*. Three types of strategies are currently being employed: chromosome walking, genomic subtraction, and insertional mutagenesis.

Chromosome walking is a labor-intensive method that can target any gene defined by mutation, regardless of the nature of the mutation. All of the original *abi* mutations were produced by ethylmethane sulfonate

(EMS) mutagenesis (Koornneef et al. 1984, 1985b), a technique unlikely to produce significant local structural changes in the genome. All have been or are targets of chromosome walks. To date, two of these loci have been cloned and sequenced: *ABI3* (Giraudat et al. 1992) and *ABI1* (Leung et al. 1994; Meyer et al. 1994). Giraudat et al. (1992) cloned the *ABI3* locus in a one-step walk in the overlapping cosmid library being used to construct a physical map of the *Arabidopsis* genome (Nam et al. 1989). Sequence analysis of a cDNA clone homologous to the cosmid subclone capable of complementing the *abi3-1* mutation showed that the *ABI3* gene has regions of strong homology with the *Vp1* gene of maize (Giraudat et al. 1992). These loci are similar in that mutations in both have pleiotropic effects on seed development, such as loss of ABA inhibition of germination and reduced storage reserve and late embryogenesis-abundant protein accumulation. Not all aspects of the mutant phenotypes are identical or can be easily explained as defects in ABA response. Consistent with this, the *Vp1* gene product appears to be a transcriptional activator that acts synergistically with ABA (McCarty et al. 1991). Although it has not been shown to bind DNA directly, promoter deletion analyses have indicated that *Vp1* action is dependent on the presence of a sequence other than the ABA response element conserved among many ABA-regulated genes (Hattori et al. 1992).

The *ABI3* and *Vp1* loci differ in that the *vp1* mutants not only fail to complete maturation, but actually germinate precociously (for review, see Koornneef 1986). The precocious germination could reflect a minor difference in seed environment such as the relative humidities of *Arabidopsis* siliques in cool, dry air versus corn husks in warm, humid air. Alternatively, *Arabidopsis* may have an additional level of germination control since, unlike maize, *Arabidopsis* seeds are normally truly dormant (for review, see Koornneef and Karssen, this volume). Therefore, the specifics of *ABI3* and *Vp1* function should provide interesting evolutionary and physiological comparisons.

The *ABI1* gene was cloned independently by two groups using yeast artificial chromosome libraries for their walks (Leung et al. 1994; Meyer et al. 1994). Because the *abi1-1* allele is dominant, functional evidence of the relevant gene required the demonstration that mutant DNA conferred the *abi1* phenotype on transformed plants. Sequence analysis of the relevant cDNA showed this to encode a novel protein, with domain homologies suggesting that it is a $Ca^{++}$-modulated protein phosphatase. This raises the intriguing possibility that the ABI1 protein integrates ABA and $Ca^{++}$ regulation of phosphorylation-regulated signaling pathways. The lack of any loss-of-function (recessive) alleles of *ABI1* leaves the "normal" physiological function of the ABI1 protein in question.

Genomic subtraction is dependent on availability of a deletion allele, creating a basis for enrichment of the desired sequence by repeated rounds of subtractive hybridization (Straus and Ausubel 1990). A major limitation of this technique is the decreasing efficiency of subtraction with increasing genome size. The fast-neutron-generated *ga1-3* allele, previously shown by elegant intragenic recombination studies to be a deletion allele (Koornneef et al. 1983), provided an opportunity for Sun et al. (1992) to test whether this technique was feasible in *Arabidopsis*. By enriching for sequences corresponding to a 5-kb deletion in the *ga1-3* allele, the *GA1* locus was cloned and subsequently shown to encode a 2.8-kb transcript of extremely low abundance. This gene is expressed at slightly higher levels in immature seeds than in vegetative tissue, consistent with the higher levels of GA synthesis in developing seeds, but still comprises only $5 \times 10^{-4}\%$ of the sequences in a seed cDNA library. Previous biochemical studies of the *ga1* mutants had shown that the locus was required for the first committed step in GA biosynthesis (Barendse et al. 1986; Zeevaart 1986), a likely control point. Further studies of *GA1* expression should permit testing of this regulatory role and analysis of environmental and developmental regulation of GA biosynthesis.

Finally, many genes of unknown biochemical function have been cloned by insertional mutagenesis. Insertional mutagenesis creates a mutation "tagged" with a known sequence, making it possible to clone the affected locus directly. A large collection of potential mutants has already been generated by *Agrobacterium*-mediated transformation of *Arabidopsis* seeds (Feldmann 1991). Jacobsen and Olszewski (1993b) recently reported isolation of a tagged allele of the *SPY* locus; cloning is in progress. We have screened pools comprising more than 12,000 independent transformant lines and identified 8 new ABA-insensitive mutant lines (Finkelstein and Doyle 1993), representing at least two new *ABI* loci. Genetic analysis is under way to determine whether any of these mutations are due to T-DNA insertions.

## CONCLUSIONS AND PERSPECTIVES

Studies of the GA and ABA biosynthesis and response mutants of *Arabidopsis* have complemented similar studies in other species. The hormone-deficient mutants have been valuable in providing confirmations of proposed biosynthetic pathways, such as the "indirect" pathway of ABA synthesis (Rock and Zeevaart 1991). In addition, the collection of GA-deficient mutants blocked in various steps makes it possible to analyze the effectiveness of different applied GAs under conditions when their metabolism to other GAs is inhibited. Whereas the *GA1*, *GA2*, and *GA4* loci correspond to mutationally defined loci in other species, the *GA3* and *GA5* loci control steps that were not previously accessible (Phinney 1984; Reid and Ross 1993).

Although the number of available mutants is still far from saturating the biosynthesis pathways, any mutants deficient in bioactive hormones are useful for studying the roles of those hormones in physiological and developmental processes. Such studies have provided further evidence that endogenous GAs mediate some aspects of environmental regulation of germination and flowering. Similarly, studies of ABA-deficient mutants have confirmed the role of endogenous ABA in dormancy induction and stomatal regulation. However, they have also shown that many responses induced by either environmental stresses (e.g., drought, cold, or salinity) or exogenous ABA probably do not require endogenous ABA to mediate response to the environmental cues. Finally, the range of phenotypes displayed by different alleles can be explained most easily by differing threshold sensitivities of hormonally regulated processes. For example, floral differentiation appears to have the lowest requirement for GA, with germination, leaf expansion, and stem elongation requiring progressively more.

The hormone-response mutants provide a means of dissecting the signal transduction pathways used by each hormone. Studies of GA response have been limited by the availability, until recently, of only a single mutation affecting response. In contrast, *Arabidopsis* is the only flowering plant for which multiple ABA response loci have been identified. These loci provide an opportunity to test the legitimacy of classifying ABA responses according to their kinetics. Mutations in two of the *ABI* loci disrupt both "fast" and "slow" ABA responses (e.g., stomatal closure and dormancy induction, respectively), indicating that these subsets of ABA responses share common transduction elements (Koornneef et al. 1984) even though they differ in specificity for (+)- versus (−)-ABA (Zeevaart and Creelman 1988; Walker-Simmons et al. 1992). Genetic studies have shown that the three best-characterized *ABI* loci operate in at least two pathways with overlapping effects (Finkelstein and Somerville 1990). This is consistent with the view, emerging from studies in animals and fungi, that signaling pathways have a high degree of redundancy (for review, see Hoffman 1991). A further implication of this finding is that defects in any single locus may not produce a dramatic phenotype because an alternative pathway could still be functional. For these reasons, identification and genetic analysis of new response mutants should be valuable.

Finally, all of the mutations discussed above provide opportunities to clone the affected loci using genetically based strategies. With such genes in hand, it will be possible to determine the biochemical nature of elements of the response pathways and study their interactions as well as the developmental and environmental regulation of both hormone response and biosynthetic loci. In addition, the *Arabidopsis* genes can be used to isolate homologous sequences from crop species for eventual construction of transgenic plants with conditionally altered hormone synthesis or response. Given the importance of GA and ABA in regulating such basic processes as seed development, germination, transpiration, and flowering, these could have important biotechnological applications.

## ACKNOWLEDGMENTS

We thank M. Koornneef, T. Lynch, and R. Wilson for critical reading of the manuscript. N.E. Olszewski kindly provided Figure 6. Our research discussed in this chapter has been supported by the National Science Foundation grants DCB-9105241 (R.R.F.) and IBN-9118377 (J.A.D.Z.), and by the U.S. Department of Energy grant DE-FG02-90ER20021 (J.A.D.Z.).

## REFERENCES

Appleford, N.E.J. and J.R. Lenton. 1991. Gibberellins and leaf expansion in near-isogenic wheat lines containing *Rht1* and *Rht3* dwarfing alleles. *Planta* **183:** 229–236.

Bagnall, D.J. 1992. Control of flowering in *Arabidopsis thaliana* by light, vernalisation and gibberellins. *Aust. J. Plant Physiol.* **19:** 401–409.

Barendse, G.W.M., J. Kepczynski, C.M. Karssen, and M. Koornneef. 1986. The role of endogenous gibberellins during fruit and seed development: Studies on gibberellin-deficient genotypes of *Arabidopsis thaliana*. *Physiol. Plant.* **67:** 315–319.

Bensen, R.J. and J.A.D. Zeevaart. 1990. Comparison of *ent*-kaurene synthetase A and B activities in cell-free extracts from young tomato fruits of wild-type and *gib-1, gib-2,* and *gib-3* tomato plants. *J. Plant Growth Regul.* **9:** 237–242.

Boother, G.M., M.D. Gale, P. Gaskin, J. MacMillan, and V.M. Sponsel. 1991. Gibberellins in shoots of *Hordeum vulgare*. A comparison between cv. Triumph and two dwarf mutants which differ in their response to gibberellin. *Physiol. Plant.* **81:** 385–392.

Brown, J.A.M. and W.H. Klein. 1971. Photomorphogenesis in *Arabidopsis thaliana* (L.) Heynh: Threshold intensities and blue-far-red synergism in floral induction. *Plant Physiol.* **47:** 393–399.

Demmig-Adams, B. and W.W. Adams III. 1992. Photoprotection and other responses of plants to high light stress. *Annu. Rev. Plant Physiol. Plant Mol. Biol.* **43:** 599–626.

Derkx, M.P.M. 1993. "Regulation of seasonal patterns in seed dormancy." Ph.D. thesis, Agricultural University, Wageningen, The Netherlands.

Duckham, S.C., R.S.T. Linforth, and I.B. Taylor. 1991. Abscisic-acid-deficient mutants at the *aba* locus of *Arabidopsis thaliana* are impaired in the epoxidation of zeaxanthin.

*Plant Cell Environ.* **14:** 601–606.

Duckham, S.C., I.B. Taylor, R.S.T. Linforth, R.J. Al-Naieb, B.A. Marples, and W.R. Bowman. 1989. The metabolism of *cis*-ABA-aldehyde by the wilty mutants of potato, pea, and *Arabidopsis thaliana. J. Exp. Bot.* **40:** 901–905.

Feldmann, K.A. 1991. T-DNA insertion mutagenesis in *Arabidopsis*: Mutational spectrum. *Plant J.* **1:** 71–82.

Finkelstein, R.R. 1993. Abscisic acid-insensitive mutations provide evidence for stage-specific signal pathways regulating expression of an *Arabidopsis* late embryogenesis-abundant gene. *Mol. Gen. Genet.* **238:** 401–408.

———. 1994a. Mutations at two new *Arabidopsis* ABA response loci are similar to *abi3* mutations. *Plant J.* (in press).

———. 1994b. Maternal effects govern variable dominance of two ABA response mutations in *Arabidopsis thaliana. Plant. Physiol.* (in press).

Finkelstein, R.R. and M.P. Doyle. 1993. Molecular genetic analysis of abscisic acid signal transduction in *Arabidopsis*. In *Abstracts from the 5th International Conference on* Arabidopsis *Research*, Columbus, Ohio, p. 46. Ohio State University, Columbus.

Finkelstein, R.R. and C.R. Somerville. 1990. Three classes of abscisic acid (ABA)-insensitive mutations of *Arabidopsis* define genes that control overlapping subsets of ABA responses. *Plant Physiol.* **94:** 1172–1179.

Fujioka, S., H. Yamane, C.R. Spray, M. Katsumi, B.O. Phinney, P. Gaskin, J. MacMillan, and N. Takahashi. 1988. The dominant non-gibberellin-responding dwarf mutant (*D8*) of maize accumulates native gibberellins. *Proc. Natl. Acad. Sci.* **85:** 9031–9035.

Gilmour, S.J. and M.F. Thomashow. 1991. Cold acclimation and cold-regulated gene expression in ABA mutants of *Arabidopsis thaliana. Plant Mol. Biol.* **17:** 1233–1240.

Giraudat, J., B. Hauge, C. Valon, J. Smalle, F. Parcy, and H.M. Goodman. 1992. Isolation of the *Arabidopsis ABI3* gene by positional cloning. *Plant Cell* **4:** 1251–1261.

Gosti, F., N. Bertauche, M. Bouvier, N. Vartanian, and J. Giraudat. 1993. Characterization of genes differentially expressed in *Arabidopsis* roots upon progressive drought stress. In *Abstracts from the 5th International Conference on* Arabidopsis *Research*, Columbus, Ohio, p. 21. Ohio State University, Columbus.

Gottlieb, M.L. and E.A. Bray. 1991. The induction of free proline accumulation by endogenous ABA in *Arabidopsis thaliana* during drought. *Plant Physiol.* (suppl.) **96:** S-21.

Graebe, J.E. 1987. Gibberellin biosynthesis and control. *Annu. Rev. Plant Physiol.* **38:** 419–465.

Grosselindemann, E., J.E. Graebe, D. Stöckl, and P. Hedden. 1991. *ent*-Kaurene biosynthesis in germinating barley (*Hordeum vulgare* L., cv Himalaya) caryopses and its relation to α-amylase production. *Plant Physiol.* **96:** 1099–1104.

Guy, C. 1990. Cold acclimation and freezing stress tolerance: Role of protein metabolism. *Annu. Rev. Plant Physiol. Plant Mol. Biol.* **41:** 187–223.

Hattori, T., V. Vasil, L. Rosenkrans, L.C. Hannah, D.R. McCarty, and I.K. Vasil. 1992. The *viviparous-1* gene and abscisic acid activate the *C1* regulatory gene for anthocyanin biosynthesis during seed maturation in maize. *Genes Dev.* **6:** 609–618.

Hedden, P. and B.O. Phinney. 1979. Comparison of *ent*-kaurene and *ent*-isokaurene synthesis in cell-free systems from etiolated shoots of normal and *dwarf-5* maize seedlings. *Phytochemistry* **18:** 1475–1479.

Heino, P., G. Sandman, V. Lang, K. Nordin, and E.T. Palva. 1990. Abscisic acid deficiency prevents development of freezing tolerance in *Arabidopsis thaliana* (L.) Heynh. *Theor. Appl. Genet.* **79:** 801–806.

Hetherington, A.M. and R.S. Quatrano. 1991. Mechanisms of action of abscisic acid at the cellular level. *New Phytol.* **119:** 9–32.

Hoffman, F.M. 1991. *Drosophila abl* and genetic redundancy in signal transduction. *Trends Genet.* **7:** 351–355.

Jacobsen, J.V. and P.M. Chandler. 1987. Gibberellin and abscisic acid in germinating cereals. In *Plant hormones and their role in plant growth and development* (ed. P.J. Davies), pp. 164–193. Martinus Nijhoff, Dordrecht, The Netherlands.

Jacobsen, S.E. and N.E. Olszewski. 1993a. Mutations at the *SPINDLY* locus of *Arabidopsis* alter gibberellin signal transduction. *Plant Cell* **5:** 887–896.

————. 1993b. Mutations at the *SPINDLY* locus of *Arabidopsis thaliana* alter gibberellin signal transduction. In *Abstracts from the 5th International Conference on Arabidopsis Research*, Columbus, Ohio, p. 12. Ohio State University, Columbus.

Jones, M.G. 1987. Gibberellins and the *procera* mutant of tomato. *Planta* **172:** 280–284.

Jung, J. and K. Grossmann. 1985. Effectiveness of new terpenoid derivatives, abscisic acid and its methyl ester on transpiration and leaf senescence of barley. *J. Plant Physiol.* **121:** 361–367.

Kamiya, Y., T. Saito, and T.-P. Sun. 1994. Function of the *GA1* gene in *Arabidopsis*. *Plant Cell Physiol.* (suppl.) **35:** s77.

Karssen, C.M. and E. Laçka. 1985. A revision of the hormone balance theory of seed dormancy: Studies on gibberellin and/or abscisic acid-deficient mutants of *Arabidopsis thaliana*. In *Plant growth substances 1985* (ed. M. Bopp), pp. 315–323. Springer Verlag, Heidelberg.

Karssen, C.M., D.L.C. Brinkhorst-van der Swan, A.E. Breekland, and M. Koornneef. 1983. Induction of dormancy during seed development by endogenous abscisic acid: studies of abscisic acid deficient genotypes of *Arabidopsis thaliana* (L.) Heynh. *Planta* **157:** 158–165.

Kobayashi, M., P. Gaskin, C.R. Spray, Y. Suzuki, B.O. Phinney, and J. MacMillan. 1993. Metabolism and biological activity of gibberellin $A_4$ in vegetative shoots of *Zea mays*, *Oryza sativa*, and *Arabidopsis thaliana*. *Plant Physiol.* **102:** 379–386.

Koornneef, M. 1986. Genetic aspects of abscisic acid. In *A genetic approach to plant biochemistry* (ed. A.D. Blonstein and P.J. King), pp. 35–54. Springer Verlag, New York.

Koornneef, M. and J.H. van der Veen. 1980. Induction and analysis of gibberellin sensitive mutants in *Arabidopsis thaliana* (L.) Heynh. *Theor. Appl. Genet.* **58:** 257–263.

Koornneef, M., G. Reuling, and C.M. Karssen. 1984. The isolation and characterization of abscisic acid-insensitive mutants of *Arabidopsis thaliana*. *Physiol. Plant.* **61:** 377–383.

Koornneef, M., C.J. Hanhart, H.W.M. Hilhorst, and C.M. Karssen. 1989. *In vivo* inhibition of seed development and reserve protein accumulation in recombinants of abscisic acid biosynthesis and responsiveness mutants in *Arabidopsis thaliana*. *Plant Physiol.* **90:** 463–469.

Koornneef, M., M.L. Jorna, D.L.C. Brinkhorst-van der Swan, and C.M. Karssen. 1982. The isolation of abscisic acid (ABA)-deficient mutants by selection of induced revertants in non-germinating gibberellin sensitive lines of *Arabidopsis thaliana* (L.) Heynh. *Theor. Appl. Genet.* **61:** 385–393.

Koornneef, M., J. van Eden, C.J. Hanhart, and A.M.M. de Jongh. 1983. Genetic fine-structure of the *GA-1* locus in the higher plant *Arabidopsis thaliana* (L.) Heynh. *Genet. Res.* **41:** 57–68.

Koornneef, M., T.D.G. Bosma, C.J. Hanhart, J.H. van der Veen, and J.A.D. Zeevaart. 1990. The isolation and characterization of gibberellin-deficient mutants in tomato. *Theor. Appl. Genet.* **80:** 852–857.

Koornneef, M., J.W. Cone, C.M. Karssen, R.E. Kendrick, J.H. van der Veen, and J.A.D. Zeevaart. 1985a. Plant hormone and photoreceptor mutants in *Arabidopsis* and tomato. *UCLA Symp. Mol. Cell. Biol. New Ser.* **35:** 103–114.

Koornneef, M., A. Elgersma, C.J. Hanhart, E.P. van Loenen-Martinet, L. van Rijn, and J.A.D. Zeevaart. 1985b. A gibberellin insensitive mutant of *Arabidopsis thaliana*. *Physiol. Plant.* **65:** 33–39.

Lanahan, M.B. and T.-H.D. Ho. 1988. Slender barley: A constitutive gibberellin-response mutant. *Planta* **175:** 107–114.

Lang, V. and E.T. Palva. 1992. The expression of a *rab*-related gene, *rab18*, is induced by abscisic acid during the cold acclimation process of *Arabidopsis thaliana* (L.) Heynh. *Plant Mol. Biol.* **20:** 951–962.

Langridge, J. 1957. Effect of day-length and gibberellic acid on the flowering of *Arabidopsis*. *Nature* **180:** 36–37.

Leung, J., M. Bouvier-Durand, P.-C. Morris, D. Guerrier, F. Chefdor, and J. Giraudat. 1994. *Arabidopsis* ABA response gene *ABI1:* Features of a calcium-modulated protein phosphatase. *Science* **264:** 1448–1452.

Martínez-Zapater, J.M. and C.R. Somerville. 1990. Effect of light quality and vernalization on late flowering mutants of *Arabidopsis thaliana*. *Plant Physiol.* **92:** 770–776.

McCarty, D.R., T. Hattori, C.B. Carson, V. Vasil, M. Lazar, and I.K. Vasil. 1991. The *Viviparous-1* developmental gene of maize encodes a novel transcriptional activator. *Cell* **66:** 895–905.

Metzger, J.D. and A.T. Hassebrock. 1990. Selection and characterization of a gibberellin-deficient mutant of *Thlaspi arvense* L. *Plant Physiol.* **94:** 1655–1662.

Meurs, C., A.S. Basra, C.M. Karssen, and L.C. van Loon. 1992. Role of abscisic acid in the induction of desiccation tolerance in developing seeds of *Arabidopsis thaliana*. *Plant Physiol.* **98:** 1484–1493.

Meyer, K., M.P. Leube, and E. Grill. 1994. A protein phosphatase 2C involved in ABA signal transduction in *Arabidopsis thaliana*. *Science* **264:** 1452–1455.

Nam, H.-G., J. Giraudat, B. den Boer, F. Moonan, W.D.B. Loos, B.M. Hauge, and H.W. Goodman. 1989. Restriction fragment length polymorphism linkage map of *Arabidopsis thaliana*. *Plant Cell* **1:** 699–705.

Nambara, E., T. Akazawa, and P. McCourt. 1991. Effects of the gibberellin biosynthetic inhibitor uniconazol on mutants of *Arabidopsis*. *Plant Physiol.* **97:** 736–738.

Nambara, E., S. Naito, and P. McCourt. 1992. A mutant of *Arabidopsis* which is defective in seed development and storage protein accumulation is a new *abi3* allele. *Plant J.* **2:** 435–441.

Napp-Zinn, K. 1985. *Arabidopsis thaliana*. In *Handbook of flowering* (ed. A.H. Halevy), vol.1, pp. 492–503. CRC Press, Boca Raton, Florida.

Nordin, K., P. Heino, and E. T. Palva. 1991. Separate signal pathways regulate the expression of a low-temperature-induced gene in *Arabidopsis thaliana* (L.) Heynh. *Plant Mol. Biol.* **16:** 1061–1071.

Nordin, K., T. Vahala, and E.T. Palva. 1993. Differential expression of two related, low-temperature-induced genes in *Arabidopsis thaliana* (L.) Heynh. *Plant Mol. Biol.* **21:** 641–653.

Ooms, J.J.J., K.M. Léon-Kloosterziel, D. Bartels, M. Koornneef, and C.M. Karssen. 1993. Acquisition of desiccation tolerance and longevity in seeds of *Arabidopsis thaliana*. A comparative study using abscisic acid-insensitive *abi3* mutants. *Plant Physiol.* **102:** 1185–1191.

Parry, A.D. and R. Horgan. 1992. Abscisic acid biosynthesis in roots. I. The identification of potential abscisic acid precursors, and other carotenoids. *Planta* **187:** 185–191.

Parry, A.D., A.D. Blonstein, M.J. Babiano, P.J. King, and R. Horgan. 1991. Abscisic-acid metabolism in a wilty mutant of *Nicotiana plumbaginifolia*. *Planta* **183:** 237–243.

Peng, J. and N.P. Harberd. 1993. Derivative alleles of the *Arabidopsis* gibberellin-insensitive (*gai*) mutation confer a wild-type phenotype. *Plant Cell* **5:** 351–360.

Pharis, R.P. and R.W. King. 1985. Gibberellins and reproductive development in seed plants. *Annu. Rev. Plant Physiol.* **36:** 517–568.

Phinney, B.O. 1984. Gibberellin $A_1$, dwarfism and the control of shoot elongation in higher plants. In *The biosynthesis and metabolism of plant hormones* (ed. A. Crozier and J.R. Hillman), pp. 17–41, Cambridge University Press, Cambridge, United Kingdom.

Potts, W.C., J.B. Reid, and I.C. Murfet. 1985. Internode length in *Pisum*. Gibberellins and the slender phenotype. *Physiol. Plant.* **63:** 357–364.

Reid, J.B. and J.J. Ross. 1993. A mutant-based approach, using *Pisum sativum*, to understanding plant growth. *Int. J. Plant Sci.* **154:** 22–34.

Rock, C.D. and J.A.D. Zeevaart. 1991. The *aba* mutant of *Arabidopsis thaliana* is impaired in epoxy-carotenoid biosynthesis. *Proc. Natl. Acad. Sci.* **88:** 7496–7499.

Rock, C.D., N.R. Bowlby, S. Hoffmann-Benning, and J.A.D. Zeevaart. 1992. The *aba* mutant of *Arabidopsis thaliana* (L.) Heynh. has reduced chlorophyll fluorescence yields and reduced thylakoid stacking. *Plant Physiol.* **100:** 1796–1801.

Rodriguez, D., J. Dommes, and D.H. Northcote. 1987. Effect of abscisic and gibberellic acids on malate synthase transcripts in germinating castor bean seeds. *Plant Mol. Biol.* **9:** 227–235.

Ross, J.J., N.W. Davies, J.B. Reid, and I.C. Murfet. 1990. Internode length in *Lathyrus odoratus*. Effects of mutants *l* and *lb* on gibberellin metabolism and levels. *Physiol. Plant.* **79:** 453–458.

Saab, I.N., R.E. Sharp, J. Pritchard, and G.S. Voetberg. 1990. Increased endogenous abscisic acid maintains primary root growth and inhibits shoot growth of maize seedlings at low water potential. *Plant Physiol.* **93:** 1329–1336.

Schnall, J.A. and R.S. Quatrano. 1992. Abscisic acid elicits the water-stress response in root hairs of *Arabidopsis thaliana*. *Plant Physiol.* **100:** 216–218.

Sindhu, R.K. and D.C. Walton. 1988. Xanthoxin metabolism in cell-free preparations from wild type and wilty mutants of tomato. *Plant Physiol.* **88:** 178–182.

Stewart, C.R. and G. Voetberg. 1987. Abscisic acid accumulation is not required for proline accumulation in wilted leaves. *Plant Physiol.* **83:** 747–749.

Straus, D. and F. Ausubel. 1990. Genomic subtraction for cloning DNA corresponding to deletion mutations. *Proc. Natl. Acad. Sci.* **87:** 1889–1893.

Sun, T.-P., H.M. Goodman, and F.M. Ausubel. 1992. Cloning the *Arabidopsis GA1* locus by genomic subtraction. *Plant Cell* **4:** 119–128.

Takahashi, N. and M. Kobayashi. 1990. Organ-specific gibberellins in rice: Roles and biosynthesis. In *Gibberellins* (ed. N. Takahashi et al.), pp. 9–21. Springer Verlag, New York.

Talón, M., M. Koornneef, and J.A.D. Zeevaart. 1990a. Endogenous gibberellins in *Arabidopsis thaliana* and possible steps blocked in the biosynthetic pathways of the semidwarf *ga4* and *ga5* mutants. *Proc. Natl. Acad. Sci.* **87:** 7983–7987.

––––––. 1990b. Accumulation of $C_{19}$-gibberellins in the gibberellin-insensitive dwarf mutant *gai* of *Arabidopsis thaliana* (L.) Heynh. *Planta* **182:** 501–505.

Talón, M., F.R. Tadeo, and J.A.D. Zeevaart. 1991. Cellular changes induced by exogenous and endogenous gibberellins in shoot tips of the long-day plant *Silene armeria*. *Planta* **185:** 487–493.

Walker-Simmons, M., D.A. Kudrna, and R.L. Warner. 1989. Reduced accumulation of

ABA during water stress in a molybdenum cofactor mutant of barley. *Plant Physiol.* **90:** 728–733.

Walker-Simmons, M.K., R.J. Anderberg, P.A. Rose, and S.R. Abrams. 1992. Optically pure abscisic acid analogs—Tools for relating germination inhibition and gene expression in wheat embryos. *Plant Physiol.* **99:** 501–507.

Wilson, R.N. 1993. "Gibberellin action in *Arabidopsis thaliana*." Ph.D. thesis, Michigan State University, East Lansing.

Wilson, R.N., J.W. Heckman, and C.R. Somerville. 1992. Gibberellin is required for flowering in *Arabidopsis thaliana* under short days. *Plant Physiol.* **100:** 403–408.

Zeevaart, J.A.D. 1983. Gibberellins and flowering. In *The biochemistry and physiology of gibberellins* (ed. A. Crozier), vol. 2, pp. 333–374. Praeger, New York.

———. 1986. In *Plant research '86* (Annual Report of the MSU-DOE Plant Research Laboratory), pp. 130–131. Michigan State University, East Lansing.

Zeevaart, J.A.D. and R.A. Creelman. 1988. Metabolism and physiology of abscisic acid. *Annu. Rev. Plant Physiol. Plant Mol. Biol.* **39:** 439–473.

Zeevaart, J.A.D. and D.A. Gage. 1993. *ent*-Kaurene biosynthesis is enhanced by long photoperiods in the long-day plants *Spinacia oleracea* L. and *Agrostemma githago* L. *Plant Physiol.* **101:** 25–29.

Zeevaart, J.A.D. and M. Talón. 1992. Gibberellin mutants in *Arabidopsis thaliana*. In *Progress in plant growth regulation* (ed. C.M. Karssen et al.), pp. 34–42. Kluwer, Dordrecht, The Netherlands.

Zeevaart, J.A.D., M. Talón, and T.M. Wilson. 1990. Stem growth and gibberellin metabolism in spinach in relation to photoperiod. In *Gibberellins* (ed. N. Takahashi et al.), pp. 273–279. Springer-Verlag, New York.

# 21
# Auxin and Cytokinin in *Arabidopsis*

**Mark Estelle**
Department of Biology
Indiana University
Bloomington, Indiana 47405

**Harry J. Klee**
Monsanto Company
St. Louis, Missouri 63198

The phytohormones auxin and cytokinin are believed to play a critical role in virtually every aspect of plant growth and development (Evans 1984; Davies 1987). At the cellular level, auxin acts by altering both cell elongation and cell division, whereas cytokinin appears to act primarily by stimulating cell division. Perhaps the most striking demonstration of the importance of these two compounds is their effect on cultured plant cells. In general, growth of cells in culture, either as callus on solid medium or as a cell suspension in liquid medium, requires exogenous cytokinin and auxin (Davies 1987). In addition, auxin and cytokinin have dramatic effects on in vitro organogenesis. Skoog and Miller (1957) were the first to demonstrate that increasing the ratio of cytokinin to auxin in the growth medium promoted shoot development from callus tissue, and decreasing this ratio promoted root development.

In the intact plant, the effects of auxin and cytokinin are exceedingly diverse (Davies 1987), and we do not attempt a comprehensive review in this article. One example of an auxin response is the rapid stimulation of cell elongation in excised stem segments (Brummell and Hall 1987). Numerous studies have shown that auxin-induced cell elongation is correlated with two biochemical responses: activation of a plasma membrane-localized proton-pumping ATPase and consequent acidification of the cell wall space (Brummell and Hall 1987) and the stimulation of transcription of specific genes (Key 1989). The exact function of these two responses in auxin-regulated growth is unclear, but there is increasing evidence that both play an important role (Estelle 1992).

Tropistic responses are also thought to be mediated by auxin (Feldman 1985; Kaufman and Song 1987). According to the Cholodny and Went hypothesis, tropic responses are due to differential growth on the two sides of a responding organ. It has been suggested that differential growth is caused by the asymmetric distribution of auxin across the organ. In the case of root gravitropism, the existence of an auxin gradient across a gravitropically responding root is controversial (Feldman 1985). However, genetic evidence in *Arabidopsis* (described below) confirms that auxin plays a major role in gravitropic response.

Both auxin and cytokinin appear to play a pivotal role in determining the overall architecture of a plant through a process called apical dominance. Physiological studies indicate that auxin, transported down from the apex of the plant, acts to inhibit the growth of axillary meristems (Tamas 1987). Cytokinin, on the other hand, acts to promote the growth of these meristems. Thus, the activity of the axillary meristems, and consequently the bushiness of the plant, may be regulated by the competing effects of these two hormones. Experiments with transgenic plants, described below, provide additional support for this view.

Despite the wealth of information on the physiological role of auxin and cytokinin, much remains to be learned. For example, we know almost nothing about the mode of action of either hormone. *Arabidopsis* has significant advantages as well as disadvantages as a model system to study phytohormones. As is the case for many other biochemical processes, the major advantage of *Arabidopsis* is the availability of mutants affected in metabolism or perception of the various phytohormones. The disadvantage of *Arabidopsis* is the lack of information on auxin and cytokinin levels and metabolism. Furthermore, almost no information is available concerning hormone levels in specific tissues or variation through development. Although some information is becoming available for auxins, virtually nothing is known about cytokinins. As analytical techniques become simpler and more sensitive and more researchers enter the field, this is likely to change rapidly.

In this review, we summarize what is known about auxin and cytokinin in *Arabidopsis*. A number of mutants that are affected in aspects of hormone perception are described. In addition, transgenic plants engineered to have altered hormone levels are described. We hope that this summary will provide the reader with sufficient information about these hormones to predict what a mutation leading to an auxin or cytokinin alteration would look like.

## EFFECTS OF AUXIN TRANSGENES ON *ARABIDOPSIS* MORPHOLOGY

Although the endogenous pathway for auxin synthesis is not well defined, bacterial genes that alter auxin's metabolism are available.

Endogenous pools of the active auxin, indole-3-acetic acid (IAA), can be increased by expression of the *Agrobacterium tumefaciens* tryptophan monooxygenase (iaaM) gene product (Klee et al. 1987; Sitbon et al. 1992). Conversely, the free IAA pool can be reduced by expression of the IAA-lysine synthetase gene product (iaaL) from *Pseudomonas savastanoi* (Romano et al. 1991).

The effects of auxin increase or depletion on *Arabidopsis* morphology are similar to those obtained in other dicots. The most obvious effects of altered auxin are in apical dominance and leaf morphology (Fig. 1). In the case of apical dominance, increased auxin results in suppression of secondary inflorescences, and auxin depletion has the opposite effect with more rapid release of the secondary inflorescences. In the case of leaf morphology, plants with increased IAA content have narrow epinastic leaves relative to controls. The abnormal shape is due to greater expansion of adaxial than abaxial cells. The results indicate that altered auxin has a profound effect on the normal pattern of cell expansion in developing leaves.

The elevated auxin in transgenic iaaM plants also causes elongated hypocotyls. The elongation resembles that of the *hy* mutants and is light independent. Histological analysis indicates that the elongation is due to cell elongation in the transgenic cells and does not seem to be due to an increase in cell number. When the iaaM line was crossed to the *hy6* mutant, there was no additional hypocotyl elongation observed. The effect of iaaM on the other classes of *hy* mutants has not yet been examined.

Also consistent with the results obtained in other dicots (Klee et al. 1987; Romano et al. 1991), altered auxins had no effect on cell identity; i.e., within the limits of the experimental approach, auxin does not appear to be involved in determination of tissue or organ identity. Rather, auxin seems to be involved in regulating cell expansion and possibly the rate of cell division. It should be noted that tissue-specific gene expression that significantly alters auxin distribution rather than content may still have a role in cell identity.

## INTERACTIONS BETWEEN TRANSGENIC AND MUTANT PLANTS

Where *Arabidopsis* has a major advantage over many other organisms for the study of hormone biology is in the wealth of mutants available. This is particularly advantageous in hormone research because of the opportunity to sort out the complex interactions between the various phytohormones. For example, there are mutants that are insensitive to all of the major classes of hormones (see below). Furthermore, there are

*Figure 1* The effects of auxin and ethylene overproduction in transgenic *Arabidopsis* plants. A nontransgenic control plant is in the middle. The plant on the left contains the *iaaM* gene, which leads to increased auxin and ethylene. The plant on the right contains, in addition, a gene encoding ACC deaminase. This gene reduces ethylene synthesis to wild-type levels. This plant is phenotypically indistinguishable from the plant on the left, indicating that the abnormal leaf morphology and increased apical dominance are caused by elevated auxin alone. (Reprinted, with permission, from Romano et al. 1993.)

GA-deficient and ABA-deficient mutants. The availability of the bacterial genes capable of altering endogenous hormone levels such as iaaM, iaaL, and isopentenyl transferase plus a facile transformation system results in an abundance of genetic tools for analysis of hormone effects either in isolation or in combination. There are few plant species in which such a thorough analysis of hormone action can be performed.

The complex interaction of phytohormones is well illustrated in the case of auxin and ethylene. Exogenous application of auxin is known to lead to a rapid increase in ethylene synthesis due to induction of ACC synthase gene expression. In the same way, endogenous increases in auxin synthesis in transgenic plants also cause increased ethylene synthesis. The magnitude of the increase can vary from two- to threefold for *Arabidopsis* to five- to tenfold for tobacco and petunia (Romano et al. 1993). Since ethylene is a potent phytohormone that causes morphological changes at concentrations as low as 0.1 ppm (Reid 1987), it has historically been difficult to dissociate auxin from ethylene effects. Thus, until recently, it has not been possible to determine unequivocally whether the effects of transgenes on apical dominance and leaf morphology are auxin effects or merely a consequence of auxin-induced ethylene. To address this question, transgenic auxin-overproducing plants were crossed to ethylene-insensitive mutants, *ein1* and *ein2* (Romano et al. 1993). Plants containing the transgene and homozygous for the ethylene-insensitivity loci were phenotypically indistinguishable from those containing the transgene alone. This result was confirmed independently by crossing the transgenic plants to a second transgenic line expressing ACC deaminase. The gene in this latter line degrades ACC, the immediate precursor of ethylene. These plants synthesize low levels of ethylene and the crossed lines containing the two genes produce elevated auxin and less than wild-type ethylene. Again, as illustrated in Figure 1, plants are phenotypically indistinguishable from those containing the iaaM gene alone. Taken together, the results unequivocally demonstrate the effects of auxin alteration on development. They further eliminate any question of a possible role for ethylene in control of apical dominance, a role that has been postulated for a number of years (Cline 1991).

The transgenic auxin-overproducing *Arabidopsis* plants have also been combined with the auxin-resistant *axr1* mutation. In this case, *axr1* is epistatic to iaaM; i.e., the effects of auxin overproduction from iaaM are abolished despite a measured increase in IAA (C. Romano et al., in prep.). This result indicates that *axr1* is an auxin response mutation and not altered in auxin uptake, transport, or metabolism. It further confirms that the phenotypic effects exhibited by iaaM lines are, indeed, due to auxin.

It has been suggested that the *pinform* mutation in *Arabidopsis* is related to a deficiency in auxin transport (Okada et al.1991) (see also below). If this is the case, crossing *pin* with an auxin-overproducing or -underproducing line might have effects on the phenotype of the mutant. We introduced the *iaaM* and *iaaL* genes independently into the *pin* background. Neither gene had any effect on the morphology of the *pin* inflorescence, nor did the *pin* mutation have any effect on the vegetative morphology of the transgene-containing plants. Although this experiment does not rule out a role for auxin transport in the *pinform* inflorescences, it does suggest that the phenotype is not simply the consequence of too much or too little IAA in some tissue. To completely rule out auxin accumulation as a cause for the *pin* phenotype, it would be interesting to introduce it into an *axr1* background. This experiment has not yet been done.

## AUXIN-BINDING PROTEINS

Identification and isolation of the auxin "receptor" protein has been the focus of much work over the years. Despite a great deal of effort, no convincing candidate proteins have been identified. One promising approach to identification of auxin receptor proteins has utilized photoaffinity labeling to tag proteins that bind with auxin with a high affinity and degree of specificity. Several groups have identified proteins that bind auxins with a high affinity. One of these proteins, the Zm-ERabp1 from maize, has been used to clone an *Arabidopsis* homolog (Palme et al.1992; Shimomura et al. 1993). As is true in maize, the At-ERabp is approximately 22 kD and contains an amino-terminal hydrophobic signal sequence and a carboxy-terminal KDEL sequence, suggesting that the protein may be localized to the endoplasmic reticulum. There is a single copy of the gene in *Arabidopsis* that maps near the top of chromosome 4. Beyond its ability to bind auxin, nothing is known about the function of this protein at present.

## AUXIN-REGULATED GENES

Auxins are known to rapidly induce gene expression in a number of plant species (Key 1989). Auxin-regulated genes have been identified in soybean (Walker and Key 1982; Hagen et al. 1984; McClure et al. 1989), pea (Theologis et al. 1985), tobacco (Takahashi et al. 1989), and mung bean (Yamamoto et al. 1992). In some instances, auxin has been shown to stimulate transcription within several minutes, suggesting that gene expression is a primary response to hormone treatment (Theologis 1986;

McClure et al. 1989; Ballas et al. 1993). Members of two families of auxin-regulated genes have been identified in *Arabidopsis*. Conner et al. (1990) used the soybean gene *GmAux22* to recover two related genes from *Arabidopsis*. More recently, Oeller et al. (1993) have shown that these genes define a large multigene family with members present in mung bean, soybean, *Arabidopsis*, and pea. The pea genes (*PS-IAA4/5* and *PS-IAA6*) are the best-characterized members of this family. They encode short-lived nuclear proteins with structural similarity to the Arc family of prokaryotic repressor proteins (Abel et al. 1994). These workers suggest that the PS-IAA4 and PS-IAA5 proteins act as activators or repressors of genes required for auxin responses. This important work provides the first specific information on the function of auxin-regulated genes.

Several members of the SAUR (small auxin up-regulated RNAs) gene family have also been identified in *Arabidopsis* (Gil et al. 1994; T. Guilfoyle, pers. comm.). At least one of these genes, *SAUR-AC1,* is regulated by auxin. The *SAUR-AC1* gene has been used to examine auxin responses in a number of auxin-resistant and gravity-response mutants of *Arabidopsis* (Gil et al. 1994; C. Timpte and M.A. Estelle, unpubl.). These results showed that the gravitropic mutants *mg20* and *mg421*, as well as the auxin-resistant mutants *axr1, aux1,* and *axr2*, are deficient in auxin-regulated *SAUR-AC1* expression (see below).

## CYTOKININS AND TRANSGENIC PLANTS

Cytokinin levels have been manipulated in transgenic plants using the *Agrobacterium* isopentenyl transferase gene (*ipt*) (for review, see Klee 1994). The ipt enzyme catalyzes the condensation of isopentenyl pyrophosphate and adenosine monophosphate to synthesize the biologically active cytokinin, isopentenyl adenosine-5′-monophosphate (IPA). This IPA is then rapidly converted by endogenous activities to the more active zeatin cytokinins. The *ipt* gene has proven to be difficult to work with experimentally, because even very low levels of expression suppress root growth. Thus, it has been very difficult to regenerate transgenic plants. In tobacco, overexpression of *ipt* leads to reductions in stature and apical dominance. There is also some quantitative reduction in root growth. Very little work has been done in the area of cytokinins in *Arabidopsis*. There are no published reports on the levels or occurrence of the various cytokinins in *Arabidopsis*. Indeed, even the identities of some of the cytokinin-related molecules in *Arabidopsis* remain to be determined (Nicander et al. 1993). Only one report of plants with altered cytokinins has been published (Medford et al. 1989). In this case, a chimeric hsp70 promoter/*ipt* gene was introduced. Although expression of the gene was demonstrated, no quantitation of endogenous cytokinins has been done. Phenotypically, transgenic plants had reduced amounts of xylem when grown in short days. The plants also exhibited reduced apical dominance. As in tobacco, they also showed reduced root growth rates and a corresponding decrease in the size of the root elongation zone.

## GENETIC ANALYSES

One approach to the identification of genes involved in hormone action is the isolation and characterization of mutants deficient in hormone response. These mutants are most easily identified by screening for resistance to the effects of exogenous hormone. Both auxin and cytokinin inhibit root growth at very low concentrations (Estelle and Somerville 1987; Su and Howell 1992). Thus, response mutants can be recovered by screening large populations of $M_2$ seedlings for plants that grow roots on growth-inhibiting concentrations of hormone. This strategy has been used to recover both auxin- and cytokinin-resistant mutants. Although this approach has been used successfully to identify a number of interesting genes, several classes of mutants may not be recovered by screening for root resistance. For example, mutations that specifically affect auxin response in structures other than the seedling root would not be identified. In addition, mutations in genes that are essential for viability may not be recovered in such a screen.

## AUXIN-RESISTANT MUTANTS

Table 1 lists all of the auxin-resistance loci that have been identified. Mutations in these genes were recovered by screening seedlings for resistance to either IAA, 2,4-dichlorophenoxyacetic acid (2,4-D), or 2-naphthalene acetic acid (NAA). In addition to hormone resistance, mutations in each gene cause characteristic defects in plant growth and development (see below). However, the mutants also share some characteristics. First, each mutant is resistant to all active auxins examined. Thus, mutants that are recovered in a screen for 2,4-D resistance are also resistant to IAA. This result indicates that none of the mutants is affected in functions that are specifically involved in the herbicidal action of 2,4-D. Second, mutations in each of these genes result in a defect in gravitropism. The gravitropic defect can be relatively minor as in *axr1* and *axr4*, or quite dramatic as in *axr2*, *axr3*, *aux1*, and *dwf*. These results provide strong support for a role for auxin in gravitropism. Third, each of

*Table 1* Auxin-resistant mutants of *Arabidopsis*

| Gene designation | Number of alleles | Genetic behavior |
| --- | --- | --- |
| *axr1* | 8 | recessive |
| *axr2* | 2 | dominant |
| *axr3* | 2 | semidominant |
| *axr4* | 2 | recessive |
| *aux1* | at least 9 | recessive |
| *dwf* | 1 | dominant |

the auxin-resistant mutants is resistant to at least one additional plant hormone. The pattern of cross-resistance is illustrated in Table 2. The sensitivity of the *dwf* mutant to hormones other than auxin has not been reported. There are several possible explanations for this cross-resistance. One possibility is that the auxin-resistance genes encode proteins required for transduction of more than one hormone signal. Alternatively, sensitivity to one hormone may be regulated by the action of a second hormone. It is also worth noting that neither ethylene-resistant nor ABA-resistant *Arabidopsis* mutants are resistant to additional plant hormones. Thus, the cross-resistance of the auxin-resistant mutants may reflect a unique role for auxin in plant hormone action.

## AXR1

The *AXR1* gene is defined by eight recessive mutations that were recovered by screening for mutants with reduced sensitivity to the auxins IAA, 2,4-D, or NAA (Estelle and Somerville 1987; Lincoln et al. 1990). Subsequent studies have shown that the *axr1* mutants are also less sensitive to cytokinin and ethylene than wild-type seedlings (C. Lincoln et al., in prep.). In addition to hormone resistance, the *axr1* mutants display a number of morphological defects (Lincoln et al. 1990). These include a

*Table 2* Hormone cross-resistance in the auxin-resistant mutants

| Mutant | IAA | Ethylene | Cytokinin | ABA | α-Methyl trp |
| --- | --- | --- | --- | --- | --- |
| *axr1* | ++ | + | + | – | – |
| *axr2* | +++ | +++ | ? | ++ | – |
| *axr3* | ++++ | ++ | (s) | – | – |
| *axr4* | + | – | – | – | – |
| *aux1* | ++ | +++ | + | – | – |

+ Indicates resistance, – normal sensitivity, and (s) increased sensitivity. The compound α-methyl tryptophan is a toxic tryptophan analog that inhibits tryptophan biosynthesis.

reduction in both hypocotyl and stem elongation, an increase in the number of shoot lateral branches, leaf wrinkling, poorly developed flowers, a reduction in root gravitropism, an increase in root length, and a decrease in root branching. Although changes in ethylene and/or cytokinin sensitivity may play a role in the *axr1* phenotype, a reduction in auxin sensitivity is sufficient to explain most aspects of the phenotype. It is noteworthy that neither cytokinin-resistant (*ckr1*, Su and Howell 1992) or ethylene-resistant (*etr*, Bleecker et al. 1988; *ein1*, *ein2*, Guzman and Ecker 1990) mutants of *Arabidopsis* have any of the defects displayed by the *axr1* mutants.

To determine if the *AXR1* gene is required for an early step in auxin action, it is important to compare rapid auxin responses in mutant and wild-type plants. Mike Evans and collaborators have analyzed rapid changes in root elongation using a video digitizer system. They have found that auxin inhibits root growth in wild-type *Arabidopsis* seedlings with a lag period as short as 11 minutes (M. Evans, pers. comm.). When *axr1* seedlings were exposed to auxin, the kinetics of the response was the same, but the magnitude of the inhibition was significantly less. These results show that the *axr1* mutants are deficient in a very rapid auxin growth response. In addition, they rule out the possibility that resistance is due to a difference in time-dependent adaptation to auxin. To learn more about the role of the *AXR1* gene in rapid hormone responses, C. Timpte and M. Estelle (unpubl.) have examined expression of the *SAUR* gene family in wild-type and mutant *Arabidopsis* plants. In soybean, transcription of the *SAUR* genes is stimulated very rapidly in response to auxin treatment. Recently a *SAUR* gene in *Arabidopsis* called *SAUR-AC1* has been isolated (Gil et al. 1994). We have shown that *SAUR-AC1* transcripts accumulate in response to auxin treatment in wild-type hypocotyls, rosettes, stems, and roots. In contrast, we find that auxin induction of *SAUR-AC1* is significantly reduced in *axr1* plants compared to wild type. This effect is seen in mutant roots, leaves, and stems and is most dramatic in the strong *axr1-12* mutant. These results confirm that the *AXR1* gene is required very early in an auxin response pathway.

The *AXR1* gene has been cloned using a map-based strategy (Leyser et al. 1993). The gene encodes a novel protein of 540 amino acids that is related to ubiquitin-activating enzyme (E1). This enzyme catalyzes the first step in the biosynthesis of ubiquitin-protein conjugates (Finley and Chau 1991). E1 enzymes have been characterized in a variety of eukaryotic species and found to be highly conserved. Two E1 genes from *Arabidopsis* have recently been cloned and sequenced. The proteins encoded by these genes are similar to all other E1 proteins that have been described and not identical to AXR1, which is a distant relative to the E1s (P. Hatfield and R. Vierstra, pers. comm.). Although AXR1 may have E1 activity, a comparison of the two proteins suggests that this is unlikely. AXR1 is approximately half the size of conventional E1 proteins and lacks several residues and domains thought to be important for E1 function (Leyser et al. 1993). It is possible that AXR1 acts to regulate ubiquitin conjugation, perhaps to a specific protein target. The suggestion that AXR1 and the E1 enzymes are functionally related is supported by analysis of the *axr1-3* mutation. This mutation is a weak allele and substitutes tyrosine at Cys-154, a residue that is conserved between AXR1 and all E1 enzymes characterized. In the wheat enzyme, changing this cysteine to serine results in a dramatic decrease in enzyme activity (Hatfield and Vierstra 1992). These results suggest that this cysteine residue has an important, and possibly similar, function in both AXR1 and the E1 enzymes.

The role of the ubiquitin pathway in cellular regulation is currently an area of active investigation. One well-known function of ubiquitin is to mark proteins for degradation (Finley and Chau 1991). Several targets of ubiquitin-mediated proteolysis have been identified, including damaged or defective proteins and short-lived regulatory proteins such as the plant photoreceptor phytochrome, cyclins, and the yeast MATα2 repressor (Finley and Chau 1991). Genetic studies in yeast indicate that the ubiquitin pathway is involved in many different cellular processes, including DNA repair, cell cycle control, and peroxisome biogenesis (Jentsch 1992). The precise role of ubiquitination in each of these processes is unknown. Several instances of stable ubiquitin conjugates have been described, suggesting that ubiquitination may have functions other than targeting proteins for degradation (Finley and Chau 1991).

The role of the ubiquitin pathway in auxin action is unknown. However, studies of auxin-regulated gene expression suggest that changes in protein stability may have an important role in auxin response. Auxin treatment of plant tissues results in the rapid accumulation of several classes of auxin-regulated RNA species (Key 1989). Two well characterized families of auxin-regulated genes are the *SAUR* genes from soybean (McClure et al. 1989; Gee et al. 1991) and *Arabidopsis* (Y. Liu and P. Green, pers. comm.) and the *PS-IAA4/5* genes from pea (Theologis 1986). Both *SAUR* and *PS-IAA4/5* transcripts accumulate in response to auxin as well as the protein synthesis inhibitor cycloheximide (Theologis 1986; McClure et al. 1989; Ballas et al. 1993). These results suggest that in the absence of auxin, expression of the auxin-regulated genes is repressed by the action of a short-lived regulatory protein (Theologis 1986). It is possible that auxin acts to relieve this repression by stimulating ubiquitin-mediated degradation of the putative repressor through the action of AXR1 (Leyser et al. 1993; Abel et al. 1994). Alternatively, AXR1 may regulate the ubiquitination of some other protein required for auxin action, such as an auxin receptor. Studies in vertebrate cells indicate that several membrane receptors are ubiquitinated in response to ligand binding (Cenciarelli et al. 1992; Paolini and Kinet 1993). In the case of the IgE receptor, ubiquitination is rapidly reversible upon ligand dissociation, suggesting that ubiquitination may have a regulatory function (Paolini and Kinet 1993).

To identify loci that interact with the *axr1* gene, screens for mutations that suppress the *axr1* phenotype have been performed (C. Lincoln and M. Estelle, unpubl.). The progeny of approximately 30,000 EMS-mutagenized *axr1-3* plants were screened and 11 suppressor lines were recovered (C. Lincoln and M. Estelle, unpubl.). One of these lines has been analyzed in some detail. Suppression is due to a single recessive mutation called *suppressor* of *auxin resistance 1* or *sar1*. The *sar1* mutation significantly increases the auxin sensitivity of *axr1* plants but has no effect on auxin sensitivity in wild-type plants. Interestingly, the *sar1* mutation also confers a new phenotype, both in the presence and absence of the *axr1* mutation. Homozygous *sar1* plants flower much earlier than wild type and are dwarfed relative to wild-type plants.

### AXR2

The *AXR2* gene was originally defined by a single dominant mutation (Wilson et al. 1990). This mutation, called *axr2-1*, was recovered from more than 500,000 $M_2$ seedlings screened for auxin resistance. The phenotype of the *axr2-1* mutant is described in detail by Wilson et al. (1990). Briefly, *axr2-1* plants are extreme dwarfs and display defects in both root and shoot gravitropism. Dark-grown mutant seedlings have very short agravitropic hypocotyls and agravitropic roots. Mature *axr2-1* plants are also severely dwarfed relative to wild-type plants, and mutant inflorescences appear to be deficient in gravitropic response. To determine if the dwarf phenotype of *axr2-1* plants is due to a reduction in cell number or a decrease in cell elongation, longitudinal sections of mutant and wild-type stems were examined using light and scanning electron microscopy. The results suggest that most aspects of the *axr2-1* phenotype are due to a dramatic decrease in cell elongation (Timpte et al. 1992).

Like the *axr1* mutations, the *axr2-1* mutation has a pronounced effect on SAUR-AC1 gene expression. Gil et al. (1994) found that etiolated *axr2-1* seedlings are deficient in auxin regulation of SAUR-AC1 expression. C. Timpte and M. Estelle (unpubl.) have extended this analysis and found that no SAUR-AC1 transcript was detected before or after auxin treatment in *axr2-1* hypocotyl, rosette leaf, and stem tissue. These results indicate that the *axr2-1* mutation disrupts auxin action at an early step, probably in perception or transduction of the auxin signal.

Because the *axr2-1* mutation is dominant and extremely rare, it is probably a gain-of-function mutation. This conclusion is supported by dosage studies in which plants with various ratios of wild-type to mutant *AXR2* genes were constructed and compared (A. Wilson et al., in prep.). These plants included the diploids +/+, +/*axr2-1*, and *axr2-1/axr2-1*, and the triploids +/+/+, and *axr2-1/+/+*. In general, the mutant phenotype was not ameliorated by increasing the ratio of wild-type to mutant genes, suggesting that *axr2-1* is a neomorphic mutation. Because *axr2-1* is a gain-of-function mutation, it is difficult to infer function of the wild-type gene. Thus, it is possible that *AXR2* does not normally function in hormone action. To identify recessive mutations in the *AXR2* gene, screens for revertants of the dominant *axr2-1* phenotype were performed. Three revertants were recovered from approximately 250,000 $M_2$ seedlings (A.K. Wilson and M. Estelle, unpubl.). Two of these were completely wild type in appearance. Genetic analysis indicated that these two lines no longer carried the *axr2-1* mutation and may be true revertants. Genetic analysis of the third revertant line showed that this line carries a second mutation in the *axr2-1* gene that modifies but does not completely revert the *axr2-1* phenotype. Since it is not practical to genetically separate the original *axr2-1* mutation from the new mutation, we have called this revertant line *axr2-1-r3*. Revertant seedlings have an intermediate level of auxin sensitivity and display a dominant phenotype that is distinct from the *axr2-1* phenotype. Unlike *axr2-1* plants, both the roots and primary stems of *axr2-1-r3* plants are gravitropic. However, *axr2-1-r3* plants have shorter inflorescences than wild-type plants. It is likely that in *axr2-1-r3*, the *axr2* gain of function is modified but not eliminated.

### AXR3

The *AXR3* gene is defined by two semidominant mutants (O.H.M. Leyser et al., unpubl.). These mutants have a number of features that set them apart from the other auxin-resistant mutants. First, they display an extreme lack of auxin sensitivity. They are approximately 1000-fold less sensitive to IAA than wild-type plants. Second, cytokinin acts to stimulate growth of mutant roots at concentrations that cause a 70% inhibition of wild-type roots. Third, they have a phenotype that is qualitatively opposite to the *axr1* mutants. Mutant plants always produce one un-branched inflorescence only. This is in contrast to wild-type plants that typically produce from two to five inflorescences and *axr1* plants that produce highly branched inflorescences. The roots of *axr3* plants are particularly unusual. Whereas 18-day-old wild-type seedlings have one or two main roots growing from the base of the hypocotyl and no roots growing from the hypocotyl itself, mutant seedlings of the same age have an average of eight roots growing from the hypocotyl. These roots are agravitropic, extremely twisted, and lack root hairs.

The phenotype of the *axr3* mutant has paradoxical aspects. The proliferation of roots and the pronounced apical dominance are best explained by an increase in either auxin sensitivity or auxin level. Since the ratio of cytokinin to auxin is important in many developmental processes, cytokinin stimulation of root growth is also consistent with an increase in auxin level or auxin sensitivity. However, the roots of *axr3* seedlings are almost insensitive to exogenous IAA (1000-fold less sensitive than wild-type). This may be because the mutation causes a constitutive auxin response, and as a result, mutant tissues are unable to respond to changes in exogenous auxin. Alternatively, if auxin levels are very high, additional auxin would not cause significant changes in the auxin pool and thus not produce changes in growth. Both of these models are consistent with the semidominant nature of the *axr3* mutation. A preliminary analysis of IAA levels in *axr3* seedlings indicates that auxin levels are unchanged in the mutant (J. Turner and M. Estelle, unpubl.). Thus, the *axr3* mutation may confer a constitutive auxin response.

### *AXR4* and *AUX1*

The *aux1* and *axr4* mutants have less pleiotropic phenotypes than any of the other auxin-resistant mutations that have been described. Both *axr4* and *aux1* plants are normal in appearance except for defects in root gravitropism. Thus, these two genes may be specifically involved in hormonal regulation of gravitropism.

The original *aux1-1* mutant was recovered by Maher and Martindale (1980). Since then, at least nine independent recessive mutations in the *AUX1* gene have been recovered. A detailed physiological analysis of these alleles indicates that all of the mutants display a similar reduction in auxin and ethylene sensitivity, as well as a dramatic reduction in root gravitropism (Pickett et al. 1990; F.B. Pickett and M. Estelle, unpubl.). In addition, Okada and Shimura (1990) have shown that the roots of *aux1* seedlings are deficient in touch-induced rotation of root tips. The phenotype of the *aux1* mutants suggests that this gene functions primarily in the root. However, the phenotype of *aux1 axr1* double mutants indicates that *AUX1* also functions in aerial structures. The two mutations are additive in their effects on hormone sensitivity, suggesting that each mutation confers resistance by a different mechanism. In the aerial part of the plant, however, the *aux1-7* mutation acts to suppress the strong *axr1-12* mutation, indicating the *AXR1* and *AUX1* genes interact during plant development, either directly or indirectly (C. Lincoln et al., unpubl.). Since the *AXR1* gene has been cloned and the *AUX1* gene may soon be cloned as well (see below), the biochemical nature of this interaction may soon be apparent.

An *aux1* allele has been recovered from T-DNA insertion lines by M. Bennett and K. Feldmann (pers. comm.). Genetic studies indicate that the mutation is due to a T-DNA insertion in this line. Efforts to clone the *AUX1* gene are currently under way (M. Bennett, pers. comm.).

Two alleles of the *AXR4* gene have been recovered (L. Hobbie and M. Estelle, unpubl.). Like the *aux1* mutants, the *axr4* mutants are normal in appearance except for a reduction in root gravitropism. However, unlike *aux1*, the *axr4* mutants are not resistant to ethylene. One of the *axr4* mutations also appears to be due to a T-DNA insertion, so molecular information on the nature of the *AXR4* gene should also be available soon.

## CYTOKININ-RESISTANT MUTANTS

As noted above, several mutants that were isolated in screens for auxin resistance also have altered sensitivity to cytokinin. The *axr1* and *aux1* mutations confer decreased sensitivity to cytokinin (C. Lincoln et al., unpubl.), whereas the growth of *axr3* roots is stimulated by cytokinin (O. Leyser and M. Estelle, unpubl.). This cross-resistance is particularly interesting because auxin and cytokinin cooperate to regulate many aspects of plant development. Growth processes such as apical dominance appear to be regulated by the ratio of auxin to cytokinin rather than the absolute concentration of either hormone. It is possible that auxin-cytokinin cross-resistance in *aux1* and *axr1* plants is related to this interaction.

Mutants that are specifically resistant to cytokinin have also been identified (Su and Howell 1992). When wild-type seedlings are exposed to cytokinin, the roots are shorter than in wild-type seedlings, but the root hairs are longer. Cytokinin-resistant mutants were identified by screening for seedlings that did not display this characteristic response on 2.5 $\mu$M benzyladenine. A total of nine mutants were recovered from 1.4 x $10^5$ $M_2$ seedlings. Five of these lines were characterized further, and each was found to carry a single recessive mutation at the same locus, designated *CKR1*. In the absence of exogenous cytokinin, the *ckr* mutants have longer roots than wild type but shorter root hairs, suggesting that endogenous cytokinin acts to inhibit root growth and stimulate root hair elongation. Initial studies showed that the mutants are cytokinin resistant with respect to inhibition of root elongation and stimulation of root hair elongation (Su and Howell 1992). However, recent experiments have shown that the *ckr* mutants are affected in other cytokinin responses. For example, the mutants have an attenuated response to infection by *Agrobacterium* strains (W. Su and S.H. Howell, pers. comm.). Furthermore, *ckr* seedlings display additional defects when grown in low light. Under these conditions, the *ckr* mutants are resistant to cytokinin inhibition of hypocotyl elongation and have smaller leaves and reduced chlorophyll levels (Su and Howell 1993). These observations are consistent with proposed interactions between light and cytokinin.

Interactions between light and cytokinin are also suggested by studies of mutants that display constitutive photomorphogenesis. Mutants of this type include the *de-etiolated* (*det*) mutants (Chory 1993) and a mutant called *amp1* (Chaudhury et al. 1993). Chory et al. (1994) have shown that dark-grown wild-type seedlings have a similar phenotype to the *det* mutants when grown on medium containing cytokinin. Cytokinin levels are not dramatically altered in the *det1* and *det2* mutants. However, the behavior of mutant tissue in culture, as well as the results of a detached leaf senescence assay, indicate that both mutants have an altered response to cytokinin. These results suggest that some aspects of the *det* phenotype may be related to a change in cytokinin sensitivity. In contrast, the *amp1* mutant has 6-fold higher cytokinin levels compared to wild-type plants. In addition to its effects on photomorphogenesis, the *amp1* mutation causes early flowering and altered phyllotaxy. Further studies, including the analysis of various double mutant combinations, may lead to additional insights into how cytokinin and light interact during photomorphogenesis.

## AUXIN BIOSYNTHESIS

The biosynthesis of IAA by bacterial-encoded enzymes such as iaaM and iaaH is well understood (Weiler and Schroder 1987). In contrast, the IAA biosynthetic pathway is not known for any plant species (Cohen and Bialek 1984). Experiments in which radiolabeled tryptophan, fed to plant tissues, resulted in the production of radiolabeled IAA suggested that IAA is synthesized from tryptophan (Cohen and Bialek 1984). However, experiments utilizing tryptophan auxotrophs of *Arabidopsis* indicate that there may be at least two pathways of IAA biosynthesis, one in which tryptophan is an intermediate, and a second which is tryptophan-independent. The *trp2* mutant is deficient in tryptophan synthase subunit B (TSB) activity and is a tryptophan auxotroph when grown in high light (Last and Fink 1988; Last et al. 1991). In addition, the mutant accumulates high levels of indole, the substrate of the TSB activity. GC-MS analysis of IAA and IAA metabolites indicates that the levels of free IAA are similar in wild-type and *trp2* plants, but the levels of IAA conjugates are up to 25-fold higher in mutant tissue (Normanly et al. 1993). These workers suggest that IAA conjugates are higher in mutant tissues because there is an increase in IAA biosynthesis through a pathway that diverges from the tryptophan pathway prior to the TSB step. Similar results have been obtained in maize plants deficient in TSB activity (Wright et al. 1992). Furthermore, feeding studies in *Arabidopsis* confirm that IAA can be synthesized from tryptophan or from indole (Normanly et al. 1993). The relative importance of these two pathways remains to be determined. However, these results do provide a partial explanation for the lack of auxin biosynthetic mutants in any plant species.

## HORMONE AUTONOMOUS CELL LINES

Chris Town and colleagues have developed an alternative approach to the identification of genes involved in either auxin action or auxin biosynthesis. Treatment of seeds or young seedlings with γ-irradiation occasionally results in the development of tumors on the hypocotyl or apical region of the developing plant. Some of these tumors can be propagated on hormone-free culture medium after excision from the plant (Persinger and Town 1991). Campbell and Town (1991) have analyzed five such tumors in detail. Three of the tumor lines have levels of auxin and auxin conjugates similar to normal tissues, suggesting that hormone autonomy is due to changes in hormone signal transduction pathways. A fourth line has increased levels of auxin conjugates, and a derivative of this line has increased levels of auxin conjugates as well as an increase in free IAA levels. In the case of the latter two lines, hormone autonomy may be due to a change in auxin biosynthesis or metabolism. Since plants cannot be regenerated from these lines, a conventional genetic analysis has not been possible. Thus, tumorigenesis may be caused by an epigenetic change, a single mutation, or several mutations. If a single mutation is responsible, it is not clear at present if the affected gene can be cloned. Despite these complications, it is likely that these tumor lines will provide important information on how plant cells regulate cell division, particularly as the genes involved in hormone action become available through other approaches.

Another promising strategy for identifying interesting auxin-related genes has recently been developed in *N. tabacum* by Hayashi et al. (1992). Briefly, this approach involved transformation of *N. tabacum* SR1 protoplasts with a T-DNA-derived vector that contains multiple enhancer sequences. The enhancers should stimulate transcription in genes adjacent to the integrated T-DNA. Normally, *N. tabacum* protoplasts require auxin for cell division. By selecting for transformants in auxin-free medium, Hayashi et al. recovered 12 lines that were auxin independent. Plants could be regenerated from 11 of these lines and one, called *axi* 159, was analyzed in detail. Protoplasts derived from this plant had the same auxin-independent growth characteristics as the original transformed line. Genetic and molecular studies indicated that auxin-independent growth in *axi* 159 was caused by the presence of a T-DNA insert adjacent to a gene that encodes a highly basic protein, 570 amino acids in length. When this gene was placed under control of the 35S promoter and introduced into wild-type protoplasts, it again conferred auxin-independent growth. At present, the function of the *axi* 159 gene is unknown. However, continued analysis of the *axi* 159 line, as well as the other auxin-independent lines, will certainly provide new information on auxin action and metabolism. In addition, it is very likely that a similar approach can be utilized in *Arabidopsis*.

### AUXIN TRANSPORT

Polar transport of IAA through plant tissues has been well documented in a number of plant species (Goldsmith 1977). The primary sites of IAA biosynthesis are thought to be the apical meristem and young developing leaves. However, there is also evidence that IAA is synthesized in the root meristem and may be synthesized in a variety of additional tissues. In the aerial part of the plant, IAA moves from apical tissues or organs toward the basal part of the plant. In the roots, the situation is less clear. Depending on the study and the plant species, there is evidence for both acropetal (toward the apex of the root) and/or basipetal IAA transport (Goldsmith 1977).

Physiological studies suggest that the auxin transport system may play an important role in the regulation of many different developmental processes, including stem elongation, apical dominance, leaf abscission, and tropisms. Many of these studies involve the use of specific inhibitors of polar auxin transport such as 2,3,5-triiodobenzoic acid (TIBA) or N-1-naphthylphthalamic acid (NPA). Both of these compounds have been shown to specifically inhibit auxin efflux (see below), and neither affects movement of IAA through the vascular system. Thus, disruption of a developmental process by NPA or TIBA suggests a role for the polar transport system. For example, in the case of apical dominance, IAA produced in the apical meristem is thought to inhibit the growth of axillary meristems. In classic experiments, application of NPA or TIBA just below the apex of plants resulted in the release of dormant axillary buds (Tamas 1987). This result suggests that IAA is transported to the dormant meristem via the polar transport system. Similar studies have implicated the polar transport system in the regulation of phototropism and gravitropism. Curvature of an elongating organ in response to light or gravity is due to differential growth on the two sides of the organ. It has been proposed that this differential growth is caused by an asymmetric distribution of auxin produced by polar transport of auxin.

Some progress has been made in the biochemical characterization of the auxin transport system. Transport is believed to be mediated by an auxin influx carrier located in the plasma membrane (Edwards and Goldsmith 1980; Hertel et al. 1983) and an efflux carrier located preferentially on the basal side of transporting cells (Depta et al. 1983). A specific influx carrier has been proposed because the transport of IAA into stem segments and purified membrane vesicles of *Cucurbita pepo* is partially saturable (Depta et al. 1983). The auxin efflux carrier has been defined by inhibitor studies. NPA and TIBA, as well as a host of other transport inhibitors, inhibit polar IAA transport in vivo (Thomson et al. 1973; Katekar and Geissler 1980) and IAA efflux from purified membrane vesicles (Hertel et al. 1983). In addition, radiolabeled NPA binds specifically to a membrane-associated binding activity in a variety of plant species (Trillmich and Michalke 1979; Muday et al. 1993). Zettl et al. (1992) have used an azido derivative of NPA to identify an NPA-binding protein in maize. This sensitive assay may permit the facile purification of the NPA-binding protein. At present, neither the NPA-binding protein nor any other component of the auxin transport system has been purified.

Several groups have begun to use *Arabidopsis* to study auxin transport. The *pin1* mutant was originally identified because of abnormalities in the inflorescence. In its most extreme form, *pin* plants completely lack flower buds. Okada et al. (1991) have shown that *pin1* stem explants are deficient in auxin transport. Furthermore, they show that treatment of *Arabidopsis* plants with inhibitors of auxin transport produces a pin-formed phenotype, confirming that a reduction in auxin transport, and possibly accumulation of auxin in the meristem, results in meristem defects. These effects are also apparent in embryos in the work of Liu et al. (1993), who observed fused cotyledons in the *pin1-1* mutant. It is possible that the *PIN1* gene encodes a component of the auxin transport system. Alternatively, the auxin transport defect could be a secondary effect of an alteration in meristem or inflorescence structure.

A number of mutants that are resistant to the growth-inhibiting properties of auxin transport inhibitors have been isolated (M. Ruegger et al., unpubl.). These mutants define at least five genes. One class of mutants, the *cmr1* mutants, were recovered in a screen for resistance to the compound chlorfurenol-methyl (CM). This compound is not believed to inhibit auxin transport itself (Thomson and Leopold 1974). Rather, it is thought that CM is hydrolyzed by plant tissues to form the active compound, chlorfurenol (CFl). The *cmr1* mutants are resistant to CM but not to CFl or to other auxin transport inhibitors (M. Ruegger and M. Estelle, unpubl.). This result suggests that the *Cmr1* gene may encode an esterase that converts CM into CFl. The other transport inhibitor-resistant mutants are resistant to a number of auxin transport inhibitors. The physiological and genetic characterization of these mutants is in progress.

## CONCLUDING REMARKS

The studies described in this review provide new insight into the physiological role of auxin and cytokinin during plant growth as well as the molecular mechanism of hormone action. However, it is clear that much remains to be learned. By manipulating hormone levels in transgenic plants in a cell- and tissue-specific fashion, it should be possible to obtain much more detailed information about the role of auxin and cytokinin in specific aspects of plant development. In addition, it is likely that novel genetic approaches to the study of auxin and cytokinin metabolism will help us understand how the plant regulates hormone levels. Before *Arabidopsis* can achieve its full potential as a model for hormone biology, much more analytical work needs to be performed. Finally, new genetic approaches to the problem of hormone action need to be developed. It is likely that many of the genes required for auxin and cytokinin action are essential for plant viability. Mutations in these genes may not be recovered in screens for hormone resistance. Thus, it will continue to be important to design and perform new screens for mutants deficient in hormone action.

## REFERENCES

Abel, S., P.W. Oeller, and A. Theologis. 1994. Early auxin-induced genes encode short-lived nuclear proteins. *Proc. Natl. Acad. Sci.* **91:** 326–330.

Ballas, N., L.-M. Wong, and A. Theologis. 1993. Identification of the auxin-responsive element *AuxRE*, in the primary indoleacetic acid-inducible gene, *PS-IAA4/5*, of pea. *J. Mol. Biol.* **233:** 580–596.

Bleecker, A.B., M.A. Estelle, C.R. Somerville, and H. Kende. 1988. Insensitivity to ethylene conferred by a dominant mutation in *Arabidopsis thaliana. Science* **241:**

1086–1089.

Brummell, D.A. and J.L. Hall. 1987. Rapid cellular responses to auxin and the regulation of growth. *Plant Cell Environ.* **10:** 523–543.

Campbell, B.R. and C.D. Town. 1991. Physiology of hormone autonomous tissue lines derived from radiation-induced tumors of *Arabidopsis thaliana. Plant Physiol.* **97:** 1166–1173.

Cenciarelli, C., D. Hou, K.-C. Hsu, B.L. Rellahan, D.L. Wiest, H.L. Smith, V.A. Fried, and A.M. Weissman. 1992. Activation-induced ubiquitination of the T cell antigen receptor. *Science* **257:** 795–797.

Chaudhury, A., S. Letham, S. Craig, and E.S. Dennis. 1993. *amp1*—A mutant with high cytokinin levels and altered embryonic pattern, faster vegetative growth, constitutive photomorphogenesis, and precocious flowering. *Plant J.* **4:** 907–916.

Chory, J. 1993. Out of darkness: Mutants reveal pathways controlling light-regulated development in plants. *Trends Genet.* **9:** 167–172.

Chory, J., D. Reinecke, S. Sim, T. Washburn, and M. Brenner. 1994. A role for cytokinins in de-etiolation in *Arabidopsis. Plant Physiol.* **104:** 339–347.

Cline, M. 1991. Apical dominance. *Bot. Rev.* **57:** 318–358.

Cohen, J.D. and K. Bialek. 1984. The biosynthesis of indole-3-acetic acid in higher plants. In *The biosynthesis and metabolism of plant hormones* (ed. A. Crozier and J.R. Hillman), pp. 165-181. Cambridge University Press, Cambridge, United Kingdom.

Conner, T., V. Goekjian, P. La Fayette, and J. Key. 1990. Structure and expression of two auxin-inducible genes from *Arabidopsis. Plant. Mol. Biol.* **15:** 623–632.

Davies, P.J., ed. 1987. The plant hormones: Their nature, occurrence, and functions. In *Plant hormones and their role in plant growth and development*, pp. 1–11. Martinus Nijhoff, Dordrecht, The Netherlands.

Depta, H., K.H. Eisele, and R. Hertel. 1983. Specific inhibitors of auxin transport: Action in tissue segments and *in vitro* binding to membranes from maize coleoptiles. *Plant Sci. Lett.* **31:** 181–192.

Edwards, K.L. and M.H.M. Goldsmith. 1980. pH-dependent accumulation of indoleacetic acid by corn coleoptile sections. *Planta* **147:** 457–466.

Estelle, M.A. 1992. The plant hormone auxin: Insight in sight. *BioEssays* **14:** 439–444.

Estelle, M.A. and C.R. Somerville. 1987. Auxin resistant mutants of *Arabidopsis* with an altered morphology. *Mol. Gen. Genet.* **206:** 200–206.

Evans, M. L., 1984. Functions of hormones at the cellular level of organization. In *Encyclopedia of plant physiology: Hormonal regulation of development II* (ed. T.K. Scott), vol. 10, pp. 23–79. Springer-Verlag, Berlin.

Feldman, L.J. 1985. Root gravitropism. *Physiol. Plant.* **65:** 341–344.

Finley, D. and Y. Chau. 1991. Ubiquitination. *Annu. Rev. Cell Biol.* **7:** 25–69.

Gee, M.A., G. Hagen, and T.J. Guilfoyle. 1991. Tissue-specific and organ-specific expression of soybean auxin-responsive transcripts. *Plant Cell* **3:** 419–430.

Gil, P., Y. Liu, V. Orbovic, E. Verkamp, K. Poff, and P.J. Green. 1994. Characterization of the auxin-inducible *SAUR-AC1* gene for use as a molecular tool in *Arabidopsis. Plant Physiol.* **104:** 777–784.

Goldsmith, M.H.M. 1977. The polar transport of auxin. *Annu. Rev. Plant Physiol.* **28:** 439–478.

Guzman, P. and J.R. Ecker. 1990. Exploiting the triple response of *Arabidopsis* to identify ethylene-related mutants. *Plant Cell* **2:** 513–523.

Hagen, G., A. Kleinschmidt, and T. Guilfoyle. 1984. Auxin-regulated gene expression in intact soybean hypocotyl and excised hypocotyl sections. *Planta* **162:** 147–153.

Hatfield, P.M. and R.D. Vierstra. 1992. Multiple forms of ubiquitin-activating enzyme E1 from wheat. *J. Biol. Chem.* **267:** 14799–14803.

Hayashi, H., I. Czaja, H. Lubenow, J. Schell, and R. Walden. 1992. Activation of a plant gene by T-DNA tagging: Auxin-independent growth *in vitro. Science* **245:** 1350–1353.

Hertel, R., T. Lomax, and W.R. Briggs. 1983. Auxin transport in membrane vessicles from *Cucurbita pepo. Planta* **157:** 193–201.

Jentsch S. 1992. Ubiquitin-dependent protein degradation: A cellular perspective. *Trends Cell Biol.* **2:** 98–103.

Katekar, G.F. and A.E. Geissler. 1980. Auxin transport inhibitors: IV. Evidence of a common mode of action for a proposed class of auxin transport inhibitors: The phytotropins. *Plant Physiol.* **66:** 1190–1195.

Kaufman, P.B. and I. Song. 1987. Hormones and orientation of growth. In *Plant hormones and their role in plant growth and development* (ed. P.J. Davies), pp. 375–392. Martinus Nijhoff, Dordrecht, The Netherlands.

Key, J.L. 1989. Modulation of gene expression by auxin. *BioEssays* **11:** 52–58.

Klee, H. 1994. Transgenic plants and cytokinin biology. In *Cytokinins: Chemistry, activity and function* (ed. D.W. Mok and M. Mok), pp. 289–293. CRC Press, Boca Raton, Florida.

Klee, H.J., R.B. Horsch, M.A. Hinchee, M.B. Hein, and N.L. Hoffmann. 1987. The effects of overproduction of two *Agrobacterium tumefaciens* T-DNA auxin biosynthetic gene products in transgenic petunia plants. *Genes Dev.* **1:** 86–96.

Last, R.L. and G.R. Fink. 1988. Tryptophan-requiring mutants of the plant *Arabidopsis thaliana. Science* **240:** 305–310.

Last, R.L., P.H. Bissinger, D.J. Mahoney, E.R. Radwanski, and G.R. Fink. 1991. Tryptophan mutants in *Arabidopsis*: The consequences of duplicated tryptophan synthase B genes. *Plant Cell* **3:** 345–358.

Leyser, H.M.O., C. Lincoln, C. Timple, D. Lammer, J. Turner, and M. Estelle. 1993. *Arabidopsis* auxin resistance gene *AXR1* encodes a protein related to ubiquitin-activating enzyme E1. *Nature* **364:** 161–164.

Lincoln, C., J.H. Britton, and M. Estelle. 1990. Growth and development of the *axr1* mutants of *Arabidopsis. Plant Cell* **2:** 1071–1080.

Liu, C.-M., Z. Xu, and N.-H. Chua. 1993. Auxin polar transport is essential for the establishment of bilateral symmetry during early plant embryogenesis. *Plant Cell* **5:** 621–630.

Maher, E.P. and S.J.B. Martindale. 1980. Mutants of *Arabidopsis* with altered responses to auxins and gravity. *Biochem. Genet.* **18:** 1041–1053.

McClure, B.A., G. Hagen, C.S. Brown, M.A. Gee, and T.J. Guilfoyle. 1989. Transcription, organization and sequence of an auxin-induced gene cluster in soybean. *Plant Cell* **1:** 229–239.

Medford, J.I., R. Horgan, Z. El-Sawi, and H.J. Klee. 1989. Alterations of endogenous cytokinins in transgenic plants using a chimeric isopentenyl transferase gene. *Plant Cell* **4:** 403–413.

Muday, G., S. Brunn, P. Haworth, and M. Subramanian. 1993. Evidence for a single naphthylphthalamic acid binding site on the zucchini plasma membrane. *Plant Physiol.* **103:** 449–456.

Nicander, B., U. Ståhl, P. Bjorkman, and E. Tilberg. 1993. Immunoaffinity copurification of cytokinins and analysis by high-performance liquid chromatography with ultraviolet-spectrum detection. *Planta* **189:** 312–320.

Normanly, J., J.D. Cohen, and G.R. Fink. 1993. *Arabidopsis thaliana* auxotrophs reveal a tryptophan-independent biosynthetic pathway for indole-3-acetic acid. *Proc. Natl. Acad. Sci.* **90:** 10355–10359.

Oeller, P.W., J.A. Keller, J.E. Parks, J.E. Silbert, and A. Theologis. 1993. Structural characterization of the early indoleacetic acid-inducible genes, *PS-IAA4/5* and *PS-IAA6*, of pea (*Pisum sativum* L.). *J. Mol. Biol.* **233**: 789–798.

Okada, K. and Y. Shimura. 1990. Aspects of recent developments in mutational studies of plant signalling pathways. *Cell* **70**: 369–372.

Okada, K., J. Ueda, M.K. Komaki, C.J. Bell, and Y. Shimura. 1991. Requirements of the auxin polar transport system in early stages of *Arabidopsis* floral bud formation. *Plant Cell* **3**: 677–684.

Palme, K., T. Hesse, N. Campos, C. Garbers, M. Yanofsky, and J. Schell. 1992. Molecular analysis of an auxin binding protein gene located on chromosome 4 of *Arabidopsis*. *Plant Cell* **4**: 193–201.

Paolini, R. and J.-P. Kinet. 1993. Cell surface control of the multiubiquitination and deubiquitination of high-affinity immunoglobulin E receptors. *EMBO J.* **12**: 779–786.

Persinger, S. and C.D. Town. 1991. Isolation and characterization of hormone autonomous tumours of *Arabidopsis thaliana*. *J. Exp. Bot.* **42**: 1363–1370.

Pickett, F.B., A.K.Wilson, and M. Estelle. 1990. The *aux1* mutation of *Arabidopsis* confers both auxin and ethylene resistance. *Plant Physiol.* **94**: 1462–1466.

Reid, M.S. 1987. Ethylene in plant growth, development, and senesence. In *Plant hormones and their role in plant growth and development* (ed. P.J. Davies), pp. 257–279. Martinus Nijhoff, Dordrecht, The Netherlands.

Romano, C.P., M.L. Cooper, and H.J. Klee. 1993. Uncoupling auxin and ethylene effects in transgenic tobacco and *Arabidopsis* plants. *Plant Cell* **5**: 181–189.

Romano, C.P., M.B. Hein, and H.J. Klee. 1991. Inactivation of auxin in tobacco transformed with the indoleacetic acid-lysine synthetase gene of *Pseudomonas savastanoi*. *Genes Dev.* **5**: 438–446.

Shimomura, S., W. Liu, N. Inohara, S. Watanabe, and M. Futai. 1993. Structure of the gene for an auxin-binding protein and a gene for 7SL RNA from *Arabidopsis thaliana*. *Plant Cell Physiol.* **34**: 633–637.

Sitbon, F., S. Hennion, B. Sundberg, C.H.A. Little, O. Olsson, and G. Sandberg. 1992. Transgenic tobacco co-expressing the *A. tumefaciens iaaM* and *iaaH* genes display altered growth and IAA metabolism. *Plant Physiol.* **99**: 1062–1069.

Skoog, F. and C.O. Miller. 1957. Chemical regulation of growth and organ formation in plant tissues cultured in vitro. *Symp. Soc. Exp. Biol.* **11**: 188–231.

Su, W. and S.H. Howell. 1992. A single genetic locus, *Ckr1*, defines *Arabidopsis* mutants in which root growth is resistant to low concentrations of cytokinin. *Plant Physiol.* **99**: 1569–1574.

———. 1993. Properties of cytokinin resistant mutant, *ckr1*, indicate interaction of cytokinin and light on photomorphogenesis. In *Abstracts from the 5th International Conference on* Arabidopsis *Research*, Columbus, Ohio, p. 62. University of Ohio, Columbus.

Takahashi, Y. and T. Nagata. 1992. *parB*: An auxin regulated gene encoding glutathione-S-transferase. *Proc. Natl. Acad. Sci.* **89**: 56–59.

Tamas, I.A. 1987. Hormonal regulation of apical dominance. In *Plant hormones and their role in plant growth and development* (ed. P.J. Davies), pp. 393–410. Martinus Nijhoff, Dordrecht, The Netherlands.

Theologis, A. 1986. Rapid gene regulation by auxin. *Annu. Rev. Plant Physiol.* **37**: 407–438.

Theologis, A., T.V. Huynh, and R.W. Davis. 1985. Rapid induction of specific mRNAs by auxin in pea epicotyl tissue. *J. Mol. Biol.* **183**: 53–68.

Thomson, K.-S. and A.C. Leopold. 1974. *In vitro* binding of morphactins and 1-N-naphthylphthalamic acid in corn coleoptiles and their effects on auxin transport. *Planta* **115:** 259–270.

Thomson, K.-S., R. Hertel, S. Muller, and J.E. Tavares. 1973. 1-N-naphthylphthalamic acid and 2,3,5-triiodobenzoic acid. *In vitro* binding to particulate cell fractions and action on auxin transport in corn coleoptiles. *Planta* **109:** 337–352.

Timpte, C., A.K. Wilson, and M. Estelle. 1992. Effects of the *axr2* mutation of *Arabidopsis* on cell shape in hypocotyl and inflorescence. *Planta* **188:** 271–278.

Trillmich, K. and W. Michalke. 1979. Kinetic characterization of 1-N-naphthylphthalamic acid binding sites from maize coleoptile homogenates. *Planta* **145:** 119–127.

Walker, J. and J. Key. 1982. Isolation of cloned cDNAs to auxin-responsive poly(A)$^+$ RNAs of elongating soybean hypocotyl. *Proc. Natl. Acad. Sci.* **79:** 7185–7189.

Weiler, E. and J. Schroder. 1987. Hormone genes and crown gall disease. *Trends Biol. Sci.* **12:** 271–275.

Wilson, A.K., F.B. Pickett, J.C. Turner, and M. Estelle. 1990. A dominant mutation in *Arabidopsis* confers resistance to auxin, ethylene and abscisic acid. *Mol. Gen. Genet.* **222:** 377–383.

Wright, A.D., M.B. Sampson, M.G. Neuffer, L. Michakzuk, J.P. Slovin, and J.D. Cohen. 1992. Indole-3-acetic acid biosynthesis in the *orange pericarp* maize mutant, a tryptophan auxotroph. *Science* **254:** 998–1000.

Yamamoto, K.T., H. Mori, and H. Imaseki. 1992. cDNA cloning of indole-3-acetic acid-regulated genes: *Aux22* and *SAUR* from mung bean hypocotyl tissue. *Plant Cell Physiol.* **33:** 93–97.

Zettl, R., J. Feldwisch, W. Boland, J. Schell, and K. Palme. 1992. 5′ Azido-[3,6-$^3$H$_2$]-1-naphthylphthalamic acid, a photoactivatable probe for naphthylphthalamic acid receptor proteins from higher plants: Identification of a 23-kDa protein from maize coleoptile plasma membranes. *Proc. Natl. Acad. Sci.* **89:** 480–484.

# 22
# Light Signal Transduction and the Control of Seedling Development

**Joanne Chory and Ronald E. Susek**
Plant Biology Laboratory
The Salk Institute for Biological Studies
San Diego, California 92186-5800

Numerous environmental factors, including temperature, touch, water, gravity, and light, influence plant development. For the emerging seedling whose development is governed by strategies that maximize photosynthetic capacity, light serves an especially important role. Light stimulates chloroplast biogenesis, leaf meristem differentiation, and induction of the coordinate expression of many light-regulated nuclear and chloroplast-encoded genes. This light-dependent development of plants, called photomorphogenesis, is controlled by the combined action of several photoreceptor systems within the plant. The biochemistry and molecular biology of the signal transduction and developmental pathways that lead to leaf and chloroplast biogenesis are largely unknown.

Although very little is known about the mechanisms of light signal transduction in plants, several approaches are beginning to yield new insight into the number and complexity of the pathways involved. This chapter highlights advances in three separate areas: (1) the biochemical analysis of the photoreceptors, (2) the identification of the *cis*-acting sequences and *trans*-acting factors that regulate the downstream light-regulated genes, and (3) the genetic dissection of the photoreceptor action pathways and signaling pathways that control chloroplast development. In particular, we focus on recent molecular, biochemical, and genetic studies that have aided in the dissection of the red and blue light signal transduction pathways that affect leaf and chloroplast development and the expression of nuclear light-regulated genes encoding chloroplast-destined proteins in dicots. We also discuss various mutations that affect nuclear control of chloroplast development and chloroplast signaling back to genes in the nuclear compartment. Although most of what has been learned about the biochemistry of the photoreceptors and *trans*-acting factors has been learned from studies in other plants, many of the experimental results appear to hold true in *Arabidopsis* as well. In con-

*Arabidopsis*
© 1994 Cold Spring Harbor Laboratory Press 0-87969-428-9/94 $5 + .00

trast, most of the genetic studies have been performed in *Arabidopsis*, but it is likely that similar mutants could be isolated in other dicots.

Other recent reviews cover light-regulated gene expression and chloroplast development in organisms other than *Arabidopsis*. We refer interested readers to these reviews on phytochrome (Quail 1991), chloroplast development in C3 monocots (Mullet 1988), the mechanisms of plastid gene regulation (Gruissem 1989), the interactions of the nuclear and chloroplast compartments in the maintenance of the photosynthetic state (Taylor 1989; Susek and Chory 1992), and the regulated expression of nuclear light-regulated genes (Gilmartin et al. 1990; Katagiri and Chua 1992).

## CHLOROPLAST DEVELOPMENT

### A Brief Review

Chloroplast development involves the temporally regulated biosynthesis of components of the photosynthetic apparatus and the carbon reduction cycle and requires the coordinated expression of both nuclear and chloroplast genes (Leech 1976; Kirk and Tilney-Bassett 1978; Mullet 1988). In dicots, the initial stages of chloroplast differentiation occur during the formation of mesophyll cells from undifferentiated meristematic cells. This suggests that chloroplast development is intimately connected to the determination of mesophyll cell fate and leaf development (Dean and Leech 1982; Dale 1988).

Figure 1 outlines the stages of coordinate chloroplast and leaf development in light-grown plants. Red or blue light signals are required to initiate this process; however, high fluence rates of white light (including both red and blue light) are required for the build-up and maintenance phases (Kirk and Tilney-Bassett 1978). In dicots, most leaf cell divisions are completed by the end of the first stage of development (Steeves and Sussex 1989). The second stage of chloroplast development, the chloroplast "build-up" phase (for terminology, see Mullet 1988), is characterized by increases in the number of chloroplasts per cell and chloroplast volume until mature chloroplasts occupy a large part of the total mesophyll cell volume. Leaf growth continues but consists almost entirely of cell expansion. The chloroplast genes and those nuclear light-regulated genes that encode chloroplast-destined proteins are expressed at very high levels (up to 100 times their levels in dark-grown *Arabidopsis* seedlings). The third stage of chloroplast development is a maintenance phase. Photosynthetic capacity of the leaf is maintained at a steady-state level, determined by the ambient light, temperature, and water conditions. During this phase, the light-regulated genes are moder-

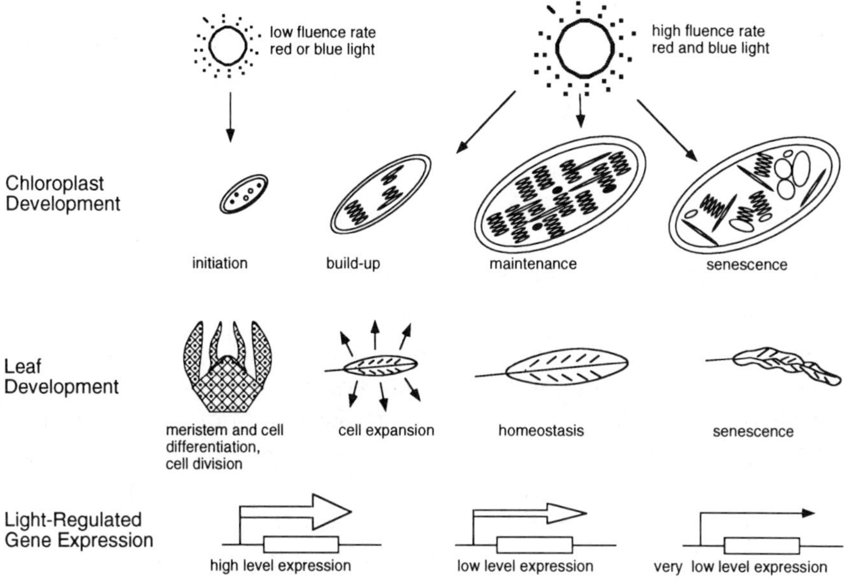


*Figure 1* Stages of chloroplast development in dicotyledonous plants. The four stages of chloroplast development are as defined in the text and are shown in relation to leaf development and expression of a subset of light-regulated nuclear and chloroplast genes that code for chloroplast-localized proteins.

ately highly expressed, but not as highly as during the build-up phase. During the final stage, leaves and chloroplasts senesce. Both chloroplast volume and number per cell decline, and expression of the light-regulated genes falls to about 10% of their maximum during early seedling development.

In contrast, if plants are germinated in the dark, plastids in the cotyledons initiate chloroplast development but arrest as etioplasts, characterized by their irregular shapes and central paracrystalline prolamellar bodies (Whatley 1974, 1977). In dicotyledons, there is little or no expression of the chloroplast genes and nuclear genes encoding chloroplast proteins in the dark. After etiolated plants are transferred to the light, etioplasts develop rapidly into chloroplasts and the light-regulated genes are transcribed at high levels (Mullet 1988; Gruissem 1989). Furthermore, plastids in non-photosynthetic cells differentiate into other distinct plastid types (Thomson and Whatley 1980). Thus, not only light, but also nutritional state, temperature, and intrinsic cell specificity factors all contribute to the determination of the plastid's developmental fate (Possingham 1980). Since chloroplasts retain the ability to adjust their structure and photosynthetic capacity in response to changing environmental conditions (Possingham 1980; Chow et al. 1991), as well as the potential to redifferentiate into other types of plastids (Thomson and Whatley 1980), their development is both a dynamic and a plastic process. How plants control the course of plastid development is not known, yet recent advances in *Arabidopsis* genetics promise to elucidate some of the mechanisms.

## Genetic Control of Chloroplast Development

Some of the earliest events in chloroplast biogenesis are the (1) activation of plastid transcription and (2) increases in chloroplast number and DNA content (Leech et al. 1981; Klein and Mullet 1990). Several new *Arabidopsis* mutants may help to explain the control of plastid division. Cell size normally appears to control chloroplast number (Pyke and Leech 1987). Pyke and Leech (1991, 1992) have identified *Arabidopsis* mutations that have altered this control and result in mutants with altered numbers of chloroplasts per mesophyll cell. The *arc1* mutant (*a*ccumulation and *r*eplication of *c*hloroplast) shows an increase in chloroplast number per cell. *arc1* also exhibits a compensatory decrease in the average size of a chloroplast when compared to wild type. The *arc2* and *arc3* mutants show the reverse; a decrease in chloroplast number is coupled to an increase in chloroplast size. Thus, the mutations appear to alter the control of chloroplast division, yet the cell retains the capacity to balance total chloroplast volume or photosynthetic capacity against the total cell volume.

Observations using *Arabidopsis fad7* (formerly *fadD*) mutants are also consistent with this notion of balancing chloroplast size and cell size. Mutations in *fad7* cause a decrease in $C_{18:3}$ and $C_{16:3}$ fatty acid accumulation due to a defect in an n-3 *f*atty *a*cid *d*esaturase. *fad7* also exhibits a decrease in chloroplast cross-sectional area and thylakoid area per chloroplast that is compensated for by an increase in the number of chloroplasts per cell (McCourt et al. 1987). This suggests that lipids play a role in the development of chloroplasts.

Other *Arabidopsis* mutations affect later stages of chloroplast development (Fig. 2). *det* (*de-et*iolated) and *cop* (*c*onstitutively *p*hotomorphogenic) mutations affect etioplast development and etiolated plant morphology (for review, see Chory 1993). *det1*, *cop1*, and *cop9* mutants are described in detail in a later section. It is of importance here that the plastids of these mutants grown in the dark fail to arrest in the etioplast stage. Whereas wild-type plants develop etioplasts in the dark, *det1*, *cop1*, and *cop9* mutants develop chloroplasts (Chory et al. 1989a; Deng et al. 1991; Wei and Deng 1992). In contrast, *gun* (*g*enomes *un*coupled)

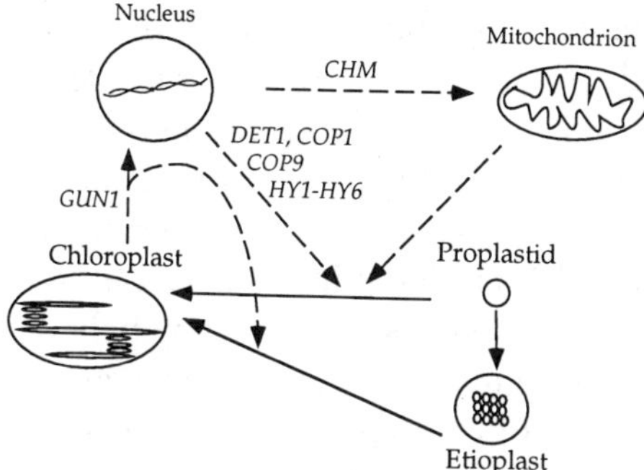


*Figure 2* Model of an intracellular regulatory network controlling chloroplast development. In light-grown plants, plastids differentiate into chloroplasts. In dark-grown plants, differentiating proplastids arrest as etioplasts. In etiolated plants transferred to the light, the etioplasts differentiate into chloroplasts. There is genetic evidence that the nucleus (DET1-COP1, COP9, HY1-HY6) and mitochondria (CHM) regulate chloroplast development. GUN1 plays a role in chloroplast signaling to the nucleus and also in the regulation of the etioplast to chloroplast transition. Developmental transitions are indicated by solid arrows; regulatory interactions by dashed arrows.

mutants develop normally in the dark but are defective in the escape from the etiolated state (Susek et al. 1993). The *gun* mutants were originally identified in a screen that required uncoupling of nuclear *CAB* (gene for chlorophyll *a/b* binding protein) gene expression from its normal dependence on chloroplast function (Susek et al. 1993). Thus, *gun* mutations interrupt a signal transduction pathway from the chloroplast to the nucleus (Fig. 2), but they also impede the transition from etioplast to chloroplast. Another class of *Arabidopsis* mutations affecting later stages of chloroplast development includes the *hy* or long *hy*pocotyl mutants (for details, see following sections; Koornneef et al. 1980; Chory et al. 1989b). Several of the *hy* mutants are deficient in the red/far-red light photoreceptor, phytochrome. Plastids from these phytochrome-deficient *hy* mutants differentiate into chloroplasts; however, they appear arrested in a state of partial development (Chory et al. 1989b). The *hy* mutant phenotypes suggest that the phytochrome signal transduction pathway is not needed to initiate chloroplast biogenesis but is required to complete chloroplast development.

A variety of pigmentation mutants may serve as additional tools for unraveling the control of chloroplast biogenesis. For example, mutations of the nuclear *CHM* (*ch*loroplast *m*utator) gene cause cytoplasmically inherited, somatic clonal variegation (Rédei 1973). The recent demonstration that *chm* mutants suffer mitochondrial genome mutations, but no obvious chloroplast genome mutations, suggests a regulatory role for mitochondria in early chloroplast biogenesis (Fig. 2) (Martínez-Zapater et al. 1992). This interesting mutant is discussed more fully by Palmer et al. (this volume). Another pigmentation mutant with a similar variegation phenotype is *im* or *im*mutans (Rédei, 1967). The primary defect in *im* is not known, although the variegation phenotype is modulated by light and temperature. The sharp boundaries of pigmentation in both *chm* and *im* demonstrate the cell-autonomous nature of chloroplast development. The *chlorina1 (ch1)* pigmentation mutant is yellow/green in color, lacks chlorophyll *b*, and grows slowly (Murray and Kohorn 1991). Although *ch1* plants accumulate normal levels of LHCPII or *CAB* mRNA (light-harvesting chlorophyll protein complex II, the apoproteins of which are encoded by *CAB* genes), they accumulate no LHCPII protein. The normal thylakoid grana stacking in *ch1* chloroplasts was unexpected because the mutant lacks LHCPII, yet LHCPII has been proposed to play a critical role in thylakoid membrane appression (for review, see Anderson 1986). Further analysis of *ch1* will be helpful in understanding the mechanism of thylakoid stacking, the biosynthesis of and function of chlorophyll *b*, and stabilization of LHCPII by chlorophyll *b*. Another pigmentation gene, *chlorata,* was recently T-DNA tagged and cloned (Koncz et al. 1990). It is a light-regulated nuclear gene encoding a chloroplast stromal protein of unknown function. Other pigmentation mutants of *Arabidopsis* have been reported (see, e.g., Hirono and Rédei 1963); however, they are less well characterized.

Chloroplast metabolic mutants are helping to elucidate the regulation of plastid development (for review, see Somerville 1986). There are *Arabidopsis* metabolic mutants affecting photorespiration, Rubisco activation, and starch accumulation (Caspar et al. 1985; Salvucci et al. 1986; Somerville 1986). One mutant that requires a high $CO_2$ atmosphere demonstrates a novel role for $CO_2$ in chloroplast development (Artus and Somerville 1988). The *fad6* (formerly *fadC*) mutant is defective in the n-6 fatty acid desaturase for $C_{16:1}$ and $C_{18:1}$ and is defective in polyunsaturated fatty acid accumulation (Hugly et al. 1989). *fad6* mutant chloroplasts have less thylakoid grana stacking than wild-type plants, suggesting a role for polyunsaturated lipids in chloroplast membrane elaboration.

Further characterization of these and new chloroplast developmental and metabolic mutants will help elucidate the control of chloroplast biogenesis and plasticity. In the remaining sections of this chapter, we focus on the role of light and its effects on the initiation and progression of plastid development.

## PHYTOCHROME: BIOCHEMISTRY AND MUTANTS

Most efforts to study light-controlled seedling development have been directed toward the study of the red/far-red light photoreceptor, phytochrome. Over the last 40 years, various physiological studies have demonstrated phytochrome involvement in almost every stage of plant development, including germination (Borthwick et al. 1952b), photomorphogenesis (Parker et al. 1949; Dale 1988), and floral induction (Borthwick et al. 1952a). In the few cases where molecular data are available, phytochrome regulates these developmental responses by modulating gene expression (Silverthorne and Tobin 1987; Gilmartin et al. 1990).

### Biochemistry and Regulation

Phytochromes are soluble chromoproteins that exist as dimers of two 120–127-kD polypeptides (depending on the plant and the phytochrome type), each with a covalently attached linear tetrapyrrole chromophore (for review, see Lagarias 1985; Furuya 1989). The pigment-protein can undergo a reversible photoinduced conversion between two spectrally and biochemically distinct forms, a red-absorbing form, Pr ($A_{max}$ = 666–668 nm), and a far-red-absorbing form, Pfr ($A_{max}$ = 730–734 nm) (Vierstra and Quail 1983a,b; Lagarias et al. 1987a; Smith and Cyr 1988). Pfr is generally considered to be the physiologically active form because low fluences of red light, but not far-red light, convert Pr to Pfr and activate phytochrome responses, such as the initiation of chloroplast development and transcriptional activation of several nuclear genes encoding chloroplast proteins (Silverthorne and Tobin 1987; Karlin-Neumann et al. 1988). However, two exceptions to the "Pfr is active form" rule have recently been reported in the literature (Liscum and Hangarter 1993; Shinomura et al. 1994). These are discussed in detail below.

Recent molecular, spectrophotometric, and biochemical evidence indicates that phytochromes exist in multiple forms within a plant (Abe et al. 1985, 1989; Shimazaki and Pratt 1985; Tokuhisa et al. 1985; Tokuhisa and Quail 1987; Sharrock et al. 1988). *Arabidopsis* expresses at least five apoprotein genes that encode distinct phytochromes (Sharrock and Quail 1989; R. Sharrock, pers. comm.). It is possible that different phytochromes act at different times or places during development to control the many downstream red-light-regulated responses. Presently, however, most of our knowledge of the biochemical properties of phytochrome comes from studies on the phytochrome that accumulates to high levels in dark-grown (etiolated) seedlings; therefore, much of what follows is based on studies of this particular phytochrome type (herein called PHYA, gene designation, *PHYA*). It should be noted here that in the older literature, this phytochrome has been referred to as "phytochrome," the "light-labile" phytochrome, or the "type I" phytochrome.

The inactive Pr form of *Avena* (oat) PHYA accumulates to extremely high levels in seedlings maintained in absolute darkness. If these etiolated seedlings are exposed to red light, Pr is converted to Pfr, which is rapidly degraded (Quail et al. 1973; Pratt et al. 1974). This rapid degradation of Pfr in the light is at least partially ubiquitin-mediated (Shanklin et al. 1987). In addition, Pfr autoregulates the accumulation of its own mRNA, largely by repressing *PHYA* transcription in both *Avena* (Colbert et al. 1983; Lissemore and Quail 1988; Bruce et al. 1991) and rice (Kay et al. 1989a). Thus, in *Avena*, PHYA levels are tightly controlled by a light-mediated mechanism that regulates *PHYA* transcription, mRNA turnover (for discussion, see Colbert 1988), and the differential degradation of Pr and Pfr (Shanklin et al. 1987). Consequently, the level of PHYA in the light is only 1–3% of its level in etiolated tissue. In plants returned to the dark, new Pr is synthesized and PHYA accumulates to its original high levels. Studies in *Arabidopsis*, although lacking in details, indicate  that PHYA accumulates to high levels in etiolated seedlings and decreases upon exposure to either red or white light in dicots as well (Chory et al. 1989b; Parks et al. 1989; Somers et al. 1991).

The role of PHYA in plant development is beginning to be elucidated through the study of mutants (see below) and by molecular and biochemical analyses. Although there are few available data on the spatial or temporal expression of *PHYA* in light-grown plants, immunocytological studies indicate that PHYA accumulates in coleoptiles and roots of etiolated *Avena* seedlings (McCurdy and Pratt 1986; Pratt 1986). More recently, studies with rice show that *PHYA* mRNA accumulates at approximately twofold higher levels in leaves than in roots of etiolated rice seedlings (Kay et al. 1989a). Surprisingly, the opposite was true in green rice seedlings, where *PHYA* mRNA accumulated to much higher levels in roots than in photosynthesizing leaves (Kay et al. 1989a).

*PHYA* genes have been ectopically expressed in an effort to elucidate the function of this phytochrome in green plants. In these experiments, either the rice (Kay et al. 1989b) or *Avena PHYA* (Boylan and Quail 1989; Keller et al. 1989) gene was overexpressed under the control of the highly and constitutively expressed cauliflower mosaic virus 35S promoter using either transgenic tomato (Boylan and Quail 1989), *Arabidopsis* (Boylan and Quail 1991), or tobacco (Kay et al. 1989b; Keller et al. 1989). Transgenic tobacco and tomato plants were small, dark-green dwarfs (Boylan and Quail 1989; Keller et al. 1989). In transgenic

*Arabidopsis*, ectopic expression of *Avena PHYA* resulted in shortened hypocotyls but normal pigmentation and stature when compared with the wild type (Boylan and Quail 1991). In one study, the circadian expression of a downstream light-regulated gene was affected by phytochrome overexpression (Kay et al. 1989b). Fluence rate response experiments, using hypocotyl growth inhibition as a parameter, indicate that seedlings overexpressing phytochrome A are more sensitive than the wild type to low fluences of both red and far-red light (McCormac et al. 1992, 1993). These observations point to roles for PHYA in hypocotyl growth inhibition, cell expansion, and regulated gene expression; however, the elucidation of the normal spatial distribution of PHYA is required before we can fully understand the phenotypes observed when the protein is produced ectopically at high levels.

The PHYA-overexpressing transgenic plants have been useful in determining important functional domains of the phytochrome apoprotein. Recent studies have shown the requirement for both the 6-kD amino-terminal domain and the last 35 amino acids of the carboxy-terminal domain (Cherry et al. 1992, 1993). Furthermore, the first 10 serine residues at the amino terminus play a role in biological activity, since mutation of these serines to alanines results in a phytochrome with increased red-light responsiveness (Stockhaus et al. 1992). Thus, it appears that the entire apoprotein is required for full biological activity, even though only the amino-terminal 60 kD is required for photoreversibility (see, e.g., Cherry et al. 1993).

The other phytochrome types are less well characterized (Abe et al. 1985; Shimazaki and Pratt 1985; Tokuhisa et al. 1985; Tokuhisa and Quail 1987; Furuya 1989). A purified light-stable "type II" phytochrome from pea has slightly different spectral properties from PHYA, with the absorption maximum for Pr shifted approximately 15 nm toward the blue (Abe et al. 1989). A gene (*PHYB*) with homology to *PHYA* and amino acid sequence similarity to pea type II has been cloned from *Arabidopsis* (Sharrock and Quail 1989). Unlike *PHYA*, the *Arabidopsis PHYB* mRNA (Sharrock and Quail 1989) and protein (Somers et al. 1991) do not appear to be regulated by light. It has been proposed that the type II phytochrome controls end-of-day responses (Furuya 1989), such as far-red stimulation of stem elongation; however, a role for PHYA or PHYC, -D, or -E in these responses has not been excluded, since there are measurable amounts of these proteins present in the experimental plants. As with PHYA, ectopic expression of PHYB in *Arabidopsis* causes a short hypocotyl phenotype (Wagner et al. 1991). However, in contrast to PHYA-overexpressing lines, the PHYB overexpressors have an increased sensitivity to red light, with no differences from the wild type observed in far-red light (McCormac et al. 1993). Together, these ectopic expression studies suggest that both PHYA and PHYB play a role in the inhibition of hypocotyl growth rate responses. However, it is unlikely that PHYA alone plays a significant role in hypocotyl growth inhibition because null mutations in *PHYB* result in seedlings with extremely elongated hypocotyls in white light (see below) (Reed et al. 1994).

A third phytochrome apoprotein gene (*PHYC*) has been cloned from *Arabidopsis*. As for *PHYB*, both dark- and light-grown *Arabidopsis* seedlings express the *PHYC* mRNA and protein at low constitutive levels (Sharrock and Quail 1989; Somers et al. 1991). By definition, then, PHYC is a light-stable or "type II" phytochrome. It still remains to be shown that PHYC has an associated phytochrome spectral activity. The biological functions of *PHYC* and the two additional unsequenced putative *PHY* genes (*PHYD* and *PHYE*) from *Arabidopsis* are currently unknown (Sharrock and Quail 1989).

The primary mode of phytochrome action in the regulation of developmental or physiological responses is unknown, although phosphorylation may be involved (Wong et al. 1986; Lagarias et al. 1987b; Wong and Lagarias 1989; McMichael and Lagarias 1990). *Avena* PHYA (amino acids 403–415) shares some sequence similarity with the catalytic domains of known protein kinases (Lagarias et al. 1987b) and with bacterial sensing proteins of the *NTRB* class (these are aspartate phosphotransferases; Schneider-Poetsch et al. 1991). Highly purified phytochrome preparations contain a polycation-stimulated protein kinase activity (Wong et al. 1986). This suggests that protein kinase activity is a crucial functional property of phytochrome and is supported by the observation that phytochrome is labeled with ATP analogs, indicating the presence of an ATP-binding site (Wong and Lagarias 1989). However, the specific activity of the polycation-stimulated kinase found in purified phytochrome preparations is low, raising the possibility that this is a copurifying activity that cannot be visualized by staining denaturing protein gels. Two recent reports claimed that highly purified phytochrome did not possess a kinase activity, but rather that a kinase copurifies with a subpopulation of the phytochrome molecules (Grimm et al. 1989; Kim et al. 1989). These results suggest that a protein kinase may copurify with phytochrome owing to specific interactions with phytochrome. Alternatively, there may be a subset of "activated" phytochromes with protein kinase activity.

Regardless of whether or not the kinase activity is intrinsic to phytochrome, phytochrome is a good substrate for the polycation-induced kinase. Phosphopeptide mapping of the *Avena* phytochrome phosphorylated by either mammalian kinases or the phytochrome-associated polycation-stimulated kinase indicates that there are two phosphorylation sites on phytochrome, and that Pr and Pfr are differentially phosphorylated at the two sites (McMichael and Lagarias 1990). Although additional work needs to be done to understand the biological significance of these observations, it is likely that light-induced conformational changes in phytochrome participate in the specific interactions of Pfr with regulatory molecules or components of the signal transduction pathway.

**Mutants**

To define the precise biological role for each phytochrome, it will be necessary to identify null mutations in individual phytochrome genes. Probable null mutations in the *Arabidopsis PHYA* and *PHYB* genes have been identified from a class of photomorphogenetic mutants that exhibit a reduced or delayed greening response to far-red or white light (Dehesh et al. 1993; Reed et al. 1993, 1994; Whitelam et al. 1993). This phenotypic class is characterized by a long hypocotyl because mutants are defective in the light-controlled response of stem growth inhibition. A large number of long hypocotyl mutant alleles defining at least thirteen complementation groups have been identified in *Arabidopsis* (Table 1) (Rédei and Horono 1964; Koornneef et al. 1980; Chory et al. 1989b; Liscum and Hangarter 1991; Nagatani et al. 1993; Parks and Quail 1993; Whitelam et al. 1993). The mutants were isolated as showing reduced responses to continuous white, blue, or far-red light.

Mutations in three separate *Arabidopsis* loci, *hy1*, *hy2*, or *hy6* (long *h*ypocotyl), reduce red- and far-red-light responsiveness and result in dark-grown seedlings with little or no detectable phytochrome activity (Koornneef et al. 1980; Chory et al. 1989b; Parks et al. 1989). These lesions likely affect either the synthesis or attachment of the bilitriene chromophore of phytochrome, since these mutants contain both the major light-labile and light-stable phytochrome apoproteins and the long hypocotyl phenotypes can be suppressed by feeding plants with an intermediate in the chromophore biosynthetic pathway (Parks and Quail 1991). Since the two major phytochromes (and perhaps all phytochromes) share the same chromophore, strong mutant alleles at these loci may be severely deficient in all phytochrome activities, despite showing slight phytochrome responses (Chory et al. 1989b; Parks et al. 1989). The existence of these mutants suggests that very few or no molecules of spectrally active phytochromes are required for plant viability. Although the precise biochemical lesions in these mutants are not known, it is possible that the mutations affect early enzymes in the protoporphyrin biosynthetic pathway rather than late enzymes specific

*Table 1* Summary of reduced light response mutants

| Genotype | Light deficiency | Defect |
| --- | --- | --- |
| Wild type | — | — |
| *hy1* | red, far-red | phytochrome chromophore biosynthesis |
| *hy2* | red, far-red | phytochrome chromophore biosynthesis |
| *hy3* | red | phytochrome B |
| *hy4* | blue | blue-light photoreceptor? |
| *hy5* | red, far-red, blue | signal transduction? |
| *hy6* | red, far-red | phytochrome chromophore biosynthesis |
| *hy8, fre1, fhy2* | far-red | phytochrome A |
| *fhy1* | far-red | signal transduction? |
| *fhy3* | far-red | signal transduction? |
| *blu1* | blue | unknown; affects blue-light-induced hypocotyl growth inhibition |
| *blu2* | blue | unknown; affects blue-light-induced hypocotyl growth inhibition |
| *blu3* | blue | unknown; affects blue-light-induced hypocotyl growth inhibition |
| JK224 | blue | phototropism; affects signal perception or early transduction |
| JK218 | blue | phototropism; signal transduction |

for phytochrome chromophore biosynthesis. If this were the case, chlorophyll and heme synthesis might also be affected in the mutants, making it difficult to interpret the pleiotropic mutant phenotypes.

In addition to the long hypocotyl for which they were selected, severe alleles of *hy1*, *hy2*, and *hy6* have a pale yellow color, fewer leaves, and increased apical dominance, and flower prematurely when compared to wild-type plants (Chory et al. 1989b; Chory 1992). Molecular, biochemical, and ultrastructural studies suggest that these mutants do not complete the normal leaf and chloroplast developmental programs when grown in white light (Chory et al. 1989b). Etiolated *hy1*, *hy2*, and *hy6* seedlings subjected to a pulse of red light show considerably less accumulation of *CAB* mRNAs than etiolated wild-type seedlings treated in the same way (Chory et al. 1989b). No differences in *CAB* mRNA accumulation were observed in the mutants when grown under high photon fluence rate white light (Chory et al. 1989b). An explanation for these results is that when phytochrome activity is deficient, the blue-light signal transduction pathways are sufficient to initiate the early events of leaf and chloroplast development (see below). Thus, phytochromes may have a redundant role with blue-light receptors in the initiation of chloroplast development, but they may play a unique role in the degree of modulation of chloroplast development.

In contrast to the *hy1*, *hy2*, and *hy6* mutants, the precise biochemical lesions in the red-light-insensitive *hy3* plants are known (Table 1) (Reed et al. 1993). *HY3* encodes the type B phytochrome apoprotein. This was shown both genetically (*hy3* mutations are tightly linked to the *PHYB* gene) and by molecular analysis of five independent *hy3* alleles (Reed et al. 1993). Three of the alleles have nonsense mutations in the first exon of *PHYB* and are probably null mutants. The *hy3* null mutants have a striking phenotype with elongated hypocotyls, petioles, root hairs, and flowering stem. It is notable that the most severe *hy1* allele (*hy1* is likely to be extremely deficient in all phytochrome types) has hypocotyls that are no longer than the *hy3* null mutants. Therefore, PHYB appears to be the major phytochrome controlling cell elongation in *Arabidopsis*. Plants homozygous for *hy3* mutations are also pale green, accumulating less chlorophyll and having fewer chloroplasts per mesophyll cell than wild-type plants. Thus, PHYB acts in numerous tissues throughout development to affect growth and physiological processes in *Arabidopsis* (Reed et al. 1993).

One very exciting use for the *hy1*, *hy2*, *hy3*, and *hy6* mutants has been in showing that the Pr form of PHYB plays a physiologically active role in *Arabidopsis* development (Liscum and Hangarter 1993; Shinomura et al. 1994). Physiological studies performed in wild-type plants during the past 40 years have indicated that the Pr form of phytochromes is inactive and that formation of the active Pfr form initiates the events that lead to altered growth responses. However, these studies have been performed largely on etiolated seedlings given pulses of red or far-red light and, thus, largely reflect the activity of PHYA (the light-labile phytochrome). The existence of *hy3* null mutants has allowed recent physiological studies in which the Pr form of phytochrome B was shown to control shoot gravitropism (Liscum and Hangarter 1993) and germination in continuous far-red light (Shinomura et al. 1994). This is in contrast to PHYA in the Pr form, which appears to be biologically inactive in all responses studied to date (Furuya 1989).

Other *Arabidopsis* mutations lie in the phytochrome A apoprotein gene (Dehesh et al. 1993; Nagatani et al. 1993; Parks and Quail 1993; Whitelam et al. 1993; Reed et al. 1994). These mutants, called *hy8* (long *hy*pocotyl-8), *fre1* (*f*ar-*re*d *e*longated), or *fhy2* (*f*ar-red *hy*pocotyl-2), are characterized by insensitivity to inhibition of hypocotyl elongation by continuous far-red light, a response that had previously been attributed to the light-labile phytochrome, PHYA (Smith and Whitelam 1990). The *hy8*, *fre1*, and *fhy2* mutations are allelic, since all alleles examined to date have mutations in the phytochrome A gene that maps to chromosome 1 (Table 1) (Dehesh et al. 1993; Whitelam et al. 1993; Reed et al.

1994). The *hy8*, *fre1*, and *fhy2* mutants appear completely normal under continuous white or red light, suggesting that PHYA is less important than PHYB under normal growth conditions. PHYA may play a highly specialized role in allowing seedlings to germinate in extreme shaded environments that are highly enriched in far-red light.

Mutants that are null for both phytochrome A and B function have revealed that these two phytochromes have overlapping but distinct functions in *Arabidopsis* development (Reed et al. 1994). For instance, the *phyA* mutant is deficient in inhibition of hypocotyl elongation, but this phenotype is only visible at low light fluences, or in the *phyB* mutant background. In continuous red light, the *phyA phyB* double mutant, but neither single mutant, has poorly developed cotyledons. The double mutant is also defective in red-light induction of *CAB* gene expression and potentiation of chlorophyll induction. Although the *phyA* mutant has a deficiency in germination in far-red light, the *phyB* mutant germinates well in far-red light but has a slight germination defect in the dark. Moreover, the germination defect caused by the *phyA* mutation in far-red light can be suppressed by a *phyB* mutation, suggesting that PHYB can have both stimulatory and inhibitory effects on germination. Finally, the *phyA phyB* double mutant flowers early and has a slight defect in sensing an inductive light treatment in the middle of the night, an additive response of the two single mutant parents. These studies show that PHYA and PHYB have complementary functions in controlling germination, seedling development, and flowering. It will be interesting to discover whether the two phytochromes affect these processes through different effects on the same signal transduction systems or through distinct signal transduction systems.

Similar reduced red-light response mutants have been described in other plant species. For example, mutants with phenotypes similar to *hy3* include the *lh* (*l*ong *h*ypocotyl) mutant of cucumber (Lopez-Juez et al. 1992), the *ein* (*e*longated *in*ternode) mutant of *Brassica* (Devlin et al. 1992), and *ma3*[R] allele of sorghum (Childs et al. 1992); however, unlike *hy3*, the molecular lesions in these mutants are not known. A tomato mutant (*au*rea) may be a phytochrome chromophore mutant, similar to the *hy1*, *hy2*, and *hy6* mutants of *Arabidopsis* (Koornneef et al. 1985; Parks et al. 1987; Sharrock et al. 1988).

## BLUE-LIGHT RECEPTORS AND MUTANTS

### Biochemistry

Blue light also plays a role in chloroplast development and gene expression. Very little is known about this photoreceptor(s) or its signal transduction pathway (Senger and Schmidt 1986). Blue-light signal transduction is likely to be complex because "pure" blue-light responses are relatively rare. More common are the blue-light responses that involve a coaction of a blue-light receptor with phytochrome (Oelmüller and Mohr 1985; Mohr 1986). Specifically, it is thought that the blue-light receptor establishes responsiveness to Pfr, thereby modulating the rate of the downstream response (e.g., pigment synthesis, chloroplast development).

Very recently, new data concerning the nature of the blue-light photoreceptor and the signal transduction pathway have renewed excitement in the blue-light field. Molecular analysis of the *Arabidopsis HY4* locus, defined by mutations that make seedlings insensitive to blue light, has revealed that this locus encodes a flavoprotein with homology to photolyases and is thus likely to encode a blue-light photoreceptor that plays a role in hypocotyl growth inhibition responses (Ahmad and Cashmore 1993). Studies in pea indicate that the blue-light receptor involved in phototropism and activation of a membrane-associated GTP-binding protein is likely to be a plasma membrane-localized flavoprotein (Short et al. 1992; Warpeha et al. 1992). Blue-light-mediated phototropism involves phosphorylation of a 120-kD membrane protein (Gallagher et al. 1988; Short and Briggs 1990). This phosphorylation reaction appears to be ubiquitous in higher plants, since crude microsomal membrane preparations from stem or coleoptile segments of dark-grown pea, sunflower, zucchini, *Arabidopsis*, tomato, maize, barley, wheat, oat, and sorghum all contain a 114–130-kD protein that is phosphorylated upon illumination by blue light (Reymond et al. 1992a).

Physiologically significant fluences of blue light also activate a GTPase activity in the plasma membranes of the apical buds of etiolated peas (Warpeha et al. 1991). The activity of this G protein is specifically regulated by low fluences of blue light in the range required for the activation of *CAB* genes. The activity of the proposed flavoprotein blue-light receptor may be coupled to a G protein, since quenchers of flavin-excited states inhibit the blue-light-enhanced binding of GTPγS to a protein in the plasma membrane (Warpeha et al. 1991, 1992).

## Mutants

*Arabidopsis* is the only higher plant for which blue-light response mutants have been identified (Table 1). These mutants fall into two classes: those that affect hypocotyl growth rate inhibition and those that affect phototropic curvature. *hy4* mutants show reduced inhibition of hypocotyl elongation in blue light but normal phytochrome levels and responses (Koornneef et al. 1980). As mentioned briefly above, *HY4* has been cloned (Ahmad and Cashmore 1993). The sequence of the *HY4* gene reveals homologies with flavin-containing photolyases, suggesting that HY4 is a blue-light photoreceptor controlling hypocotyl growth inhibition responses in the developing seedling. An additional three loci, *blu1*, *blu2*, and *blu3* (*blue light uninhibited*) were identified (Liscum and Hangarter 1991). In contrast to *hy4* alleles, mutations in either *blu1*, *blu2*, or *blu3* affect hypocotyl elongation responses in blue light but not in white light. To date, phenotypic analyses of mutant plants indicate that *HY4* and the *BLU* genes have little to do with growth of adult plants (Liscum and Hangarter 1991). This is in contrast to the phytochrome-deficient *hy1*, *hy2*, *hy3*, and *hy6* mutations that show pleiotropic effects in adult plants (Chory et al. 1989b; Reed et al. 1993).

The other class of *Arabidopsis* blue-light response mutants are phototropism mutants (see Poff et al., this volume). Among these mutants are a null phototropism line (JK218) and a mutant line (JK224) which requires 20- to 30-fold higher blue-light levels for both the threshold and saturation of first positive phototropic curvature (Khurana et al. 1989). These lines show normal hypocotyl elongation responses in blue light (Liscum et al. 1992). As discussed above, blue light mediates a very rapid phosphorylation of a 120-kD membrane-associated protein in pea epicotyls. The mutation in JK224 plants dramatically reduces the level of phosphorylation of this 120-kD membrane protein (Reymond et al. 1992b). This suggests that JK224 has a deficiency prior to the phosphorylation event, possibly in the photoreceptor itself or in an early component of the signal transduction chain. JK218, in contrast, shows nearly normal protein phosphorylation. This suggests the JK218 deficiency is after the phosphorylation event. These collaborative experiments show the power of combining "big" seedling (pea) biochemistry with *Arabidopsis* genetics.


## SIGNAL TRANSDUCTION: MOLECULAR, BIOCHEMICAL, AND GENETIC STUDIES

### Biochemistry

Little is known about the biochemistry of signal transduction from the photoreceptors, although recent studies implicate a role for G proteins, calmodulin, and $Ca^{++}$ (Romero et al. 1991; Shacklock et al. 1992; Neuhaus et al. 1993; Romero and Lam 1993). As discussed in the two previous sections, phosphorylation events are also likely to be involved in early events associated with both red- and blue-light perception. Phosphorylation/dephosphorylation reactions also seem to regulate binding of the *trans*-acting factors to the downstream light-regulated promoters. For instance, phosphorylation of either the *Arabidopsis* G-box-binding factor or factor 3AF3 (see below) stimulates DNA binding (Klimczak et al. 1992; Sarokin and Chua 1992). Phosphorylation prevents the in vitro binding of a second factor to an AT-rich element in the promoters of the downstream genes (Datta and Cashmore 1989). Therefore, a phosphorylation cascade appears to be involved in signal transduction from the photoreceptors.

Calcium may function in the light signal transduction pathway. Transient increases in the levels of free cytosolic calcium appear to modulate phytochrome-mediated protoplast swelling in monocots (Shacklock et al. 1992). Since the changes in protoplast volume were similar either when $Ca^{++}$ was released from caged probes or by red-light induction, the authors concluded that red-light-mediated effects can be transduced through $Ca^{++}$. It is still not known if photorelease of $Ca^{++}$ from its caged probe inside protoplasts influences the expression of phytochrome-regulated genes. Studies with calmodulin antagonists provide additional evidence for the role of calcium or calmodulin in phytochrome signal transduction (Lam et al. 1989; Neuhaus et al. 1993). Of particular note is a recent study in which single cell assays were developed to visualize phytochrome responses (Neuhaus et al. 1993). These authors injected $Ca^{++}$ and activated calmodulin into individual cells of phytochrome-deficient tomato mutants and showed that they could activate a *CAB* promoter and stimulate the synthesis and assembly of some of the photosynthetic complexes in the chloroplast.

Trimeric G proteins have also been implicated in phytochrome signal transduction (Romero et al. 1991; Neuhaus et al. 1993; Romero and Lam 1993). It is not yet clear if the G protein implicated in phytochrome signaling is related to the one involved in blue-light signal transduction. However, it is notable that only one G-protein α-subunit gene has been cloned from *Arabidopsis* (Ma et al. 1990), and genetic studies indicate that phytochrome and blue-light signal transduction pathways are likely to converge on common intermediates (see below) (Chory 1992, 1993).


**The End of the Signal Transduction Pathways: *trans*-Acting
Factors That Bind *cis*-Elements of Light-regulated
Nuclear Genes**

Light-regulated nuclear genes can be divided into two broad classes: those that are positively regulated by light and those that are negatively regulated by light (Silverthorne and Tobin 1987). These classes of genes are expressed in different cell types within the plant, and they share common *cis*-regulatory sequence motifs (Gilmartin et al. 1990; Schindler and

Cashmore 1990). Many recent papers have described the light-regulated *cis*-elements and the factors that bind them (for review, see Gilmartin et al. 1990; Katagiri and Chua 1992; Li et al. 1993). We do not review this vast literature here, but instead briefly illustrate the potential complexities of the signal transduction pathways involved. The expression of these genes is phytochrome-regulated (Silverthorne and Tobin 1987), yet also modulated by other photoreceptors, including blue-light and UV-B photoreceptor(s) (Fluhr and Chua 1986; Schulze-Lefert et al. 1989). Furthermore, the responsiveness of a particular gene to a given wavelength of light varies with the light intensity, as well as with the developmental state of the plant (Bennett et al. 1984; Glick et al. 1986; Ehmann et al. 1991). Together, these observations raise the question of whether the signal transduction pathways for the different photoreceptors are parallel or converge to act on the same *cis*-acting regulatory elements in the promoters of these genes.

Of all the light-regulated genes, most information is available on nuclear genes that encode chloroplast-destined proteins. The best studied of these are the chlorophyll *a/b* binding protein genes (*CAB*) and the genes for the ribulose bisphosphate carboxylase/oxygenase small subunit (*RBCS*). Both are positively regulated by light and show tight cell-type specificity, being expressed exclusively in chloroplast-containing cells (see, e.g., Fluhr et al. 1986). As little as 300–400 bp of upstream DNA is needed for light- and tissue-specific expression of *CAB* and *RBCS* genes. The pea *RBCS* 3A promoter has six sequences (Boxes II or III, and their derivatives) with homology to the SV40-core enhancer region that confer both positive and negative regulation on reporter genes (Kuhlemeir et al. 1987, 1988). Box II is bound by a factor called GT-1. A conserved "G-box" motif, similar to Myc-binding motifs, is found upstream of several light-regulated plant genes (Guilliano et al. 1988; Donald and Cashmore 1990). The G-box-binding factor appears to be a general transcription factor that works in concert with tissue-specific factors to mediate the transcription of a variety of differentially regulated genes (Guilliano et al. 1988; Schindler and Cashmore 1990). AT-1 binds an AT-rich sequence in *CAB* and *RBCS* promoters (Lam et al. 1990; Schindler and Cashmore 1990). Phosphorylation prevents the in vitro binding of AT-1 to this sequence (Datta and Cashmore 1989). Multiple GATA elements (ATGATAAGG) are necessary, but not sufficient, for *CAB* gene induction (Schindler et al. 1992a; Sun et al. 1993). They may confer leaf-specific expression (Lam and Chua 1989). Red light increases the in vitro binding of a factor, LRF-1, to a GATA sequence in the *Lemna gibba* *RBCS* promoter, suggesting a specific role for LRF-1 and GATA elements in a phytochrome signal transduction chain (Buzby et al. 1990).

Last, a factor, CA-1, that binds an ACGT motif in the *Arabidopsis CAB1* promoter and whose binding activity is missing in *det1* mutants (see below) may be a transcriptional repressor of *CAB* expression in the dark (Sun et al. 1993).

Other light-activated nuclear genes not involved in chloroplast functions are expressed in cells that do not contain chloroplasts (Schmelzer et al. 1988; Chory and Peto 1990). These include the anthocyanin biosynthetic genes, the best characterized of which encodes chalcone synthase (*CHS*) (Lamb et al. 1989). Depending on the plant species, *CHS* is regulated by either ultraviolet, blue, or red light alone, or various combinations of these wavelengths (Rabino and Mancinelli 1986; Lipphardt et al. 1988). The *CHS* promoters of bean and parsley contain both positively and negatively acting light regulatory elements. Some of the elements defined for *CAB* and *RBCS* (e.g., Boxes II and III, G box) are also implicated in light- and tissue-specific *CHS* expression (Schulze-Lefert et al. 1989; Staiger et al. 1989; Lawton et al. 1991). One of these factors, SBF, binds only when phosphorylated (Harrison et al. 1991).

Finally, a small group of nuclear genes is negatively regulated by light (Bruce et al. 1991; Okubara and Tobin 1991). The best studied of these is *PHYA* (Colbert 1988). An element important for negative phytochrome regulation was identified in the oat *PHYA* promoter (Bruce et al. 1991). In addition, *PHYA* expression is positively regulated by a factor, GT-2 (Kay et al. 1989a; Bruce and Quail 1990; Dehesh et al. 1990, 1992).

Genes that encode proteins that bind to Box II (Gilmartin et al. 1992; Perisic and Lam, 1992), 3AF1 (Lam et al. 1990), G box (Oeda et al. 1991; Weisshaar et al. 1991; Schindler et al. 1992a,b), and GT-2 (Dehesh et al. 1992) have been cloned. These genes are members of small gene families. The proteins are members of the bZIP (G box), HLH (GT-1, GT-2), or Zn-finger (3AF1) transcription factor subgroups. At least one member of the G-box-binding protein family is differentially expressed in light- and dark-grown plants, implicating this factor in light-regulated gene expression (Weisshaar et al. 1991; Schindler et al. 1992a). Future studies to identify the spatial expression patterns of the various family members and to block the expression of specific factors may help to elucidate the function of the various transcription factors in photoregulated signal transduction pathways.

**Mutants**

Photomorphogenesis is a complex process, involving a variety of photoreceptors and potentially a large number of signal transduction pathways. One approach to dissect the complex regulatory circuitry is to isolate and characterize developmental mutants that affect some or all of the downstream light-regulated responses. One predicted class of such mutations would result in constitutive elaboration of the light response, i.e., mutants would show many characteristics of light-grown plants even when grown in complete darkness.

Using such a screen, we have identified four *Arabidopsis* loci, mutations that result in dark-grown seedlings with the morphology of light-grown seedlings (Chory et al. 1989a, 1991; Chory 1991; Cabrera y Poch et al. 1993; H. Cabrera y Poch and J. Chory, unpubl.). These mutants were designated *det*, because they are *de-et*iolated in the dark, instead of having the usual etiolated seedling morphology. Similar mutations in three additional genes, *cop1*, *cop4*, and *cop9* (*c*onstitutively *p*hotomorphogenic) have been identified as well (Deng et al. 1991; Wei and Deng 1992; Hou et al. 1993). In addition, there are several other loci which, when mutated, alter etiolated seedling morphology. These include *cop2* and *cop3* (Hou et al. 1993), *axr1* and *axr2* (*a*uxin-resistant) (Estelle and Somerville 1987; Timpte et al. 1992), and *hls* (*h*ook*l*ess) (Guzman and Ecker 1990).

These mutants share several common morphological and gene expression characteristics when grown in the dark (Table 2). These include short hypocotyls, expanded cotyledons, developed leaves, and elevated expression of several light-regulated nuclear and chloroplast genes, including *CAB, RBCS,* and *CHS.* To date, there are three *Arabidopsis* genes (*DET1, COP1,* and *COP9*) and one pea gene *LIP1* (*l*ight-*i*ndependent *p*hotomorphogenesis) that have been mutated to give such phenotypes (Chory et al. 1989a; Deng et al. 1991; Frances et al. 1992; Wei and Deng 1992). In addition, three other loci, *DET2, DET3,* and *COP4,* have been identified (Chory et al. 1991; Cabrera y Poch et al. 1993; Hou et al. 1993). Dark-grown *det2* mutants exhibit morphological and gene expression changes in the absence of chloroplast development (Chory et al. 1991). The *det3* mutant develops multiple leaves in the dark; however, this mutant does not develop chloroplasts or show elevated expression of nuclear- and chloroplast-encoded light-regulated mRNAs, suggesting that DET3 acts on a pathway that affects only leaf development (Cabrera y Poch et al. 1993). A dark-grown *cop4* mutant has partially expanded cotyledons and slightly increased expression of photoregulated nuclear mRNAs. Moreover, this mutant exhibits shoot and root agravitropism in both the dark and light (Hou et al. 1993). Since all mutant *det* and *cop* alleles identified to date are recessive and pleiotropic, these genes have been proposed to play a negative regulatory role in the control of de-etiolation in *Arabidopsis* (Chory et al. 1989a). The possible existence of these negative regulators implies that de-

*Table 2* Comparison of de-etiolated mutants

| Phenotypic traits | det1 | det2 | det3 | cop1 | cop4 | cop9 | WT[a] |
|---|---|---|---|---|---|---|---|
| **Dark** | | | | | | | |
| cotyledon development | yes | yes | yes | yes | yes | yes | no |
| leaf development | yes | partial | yes | yes | no | no (lethal) | no |
| hypocotyl | short[b] | short | short | short[b] | long | short | long |
| chloroplast development | partial | none | none | partial | none | partial | none |
| levels of light-specific mRNAs (in dark) | high | high | low | high | medium | high | low |
| **Light** | | | | | | | |
| height | short | short | short | very short | n.d. | lethal | tall |
| color | pale green | dark green | normal | normal | n.d. | n.d. | normal |
| flowering time | normal | late | normal | n.d. | n.d. | n.a. | normal |
| fertility | reduced | male-sterile | reduced | very reduced | n.d. | n.a. | normal |
| senescence | normal | delayed | normal | n.d. | n.d. | n.a. | normal |
| green roots | yes | no | no | yes | n.d. | yes | no |
| apical dominance | reduced | reduced | very reduced | normal | n.d. | n.a. | normal |
| ectopic expression of light-regulated genes | yes | no | no | n.d. | n.d. | n.d. | no |
| shoot/root gravitropism | normal | normal | normal | normal | defective | normal | normal |

n.a. indicates not applicable. n.d. indicates no data available.
[a]Wild type is *Arabidopsis*, ecotype Columbia.
[b]Some alleles have long hypocotyls.

etiolation is neither a simple nor direct series of positive regulatory events leading from light perception to gene induction and other light-dependent responses.

Negative regulators may also integrate developmental (spatial and temporal) information with photomorphogenetic cues. In this respect, it is noteworthy that the *det1* mutant phenotype includes a number of developmental aberrations. For example, in the roots of *det1* plants, proplastids differentiate into chloroplasts rather than into amyloplasts as in wild-type roots (Chory and Peto 1990). Several chloroplast and nuclear genes are expressed at higher levels in *det1* roots than in wild-type roots. Furthermore, expression of the *CHS* promoter is restricted to epidermal and vascular tissues in wild-type leaves, whereas in *det1* leaves the *CHS* promoter is active in all cell types (Chory and Peto 1990). On the basis of these studies, we predict that *DET1* acts negatively at the first stage of chloroplast development, i.e., in the establishment of the chloroplast developmental program in a given cell type. Likewise, *COP1* may also repress chloroplast development in *Arabidopsis* roots, since *cop1* mutants also have green chloroplast-containing roots (Deng and Quail 1992).

*det2* mutations, in contrast, appear to affect photoperiodic responses in light-grown plants, causing delayed flowering, reduced dark adaptation of *CAB* and *RBCS* gene expression, and delayed leaf and chloroplast senescence (Chory et al. 1991). Given the recessive nature of the *det2* mutations and phenotype of the mutants, one prediction is that wild-type *DET2* activity is high prior to light-induced morphogenesis, low during chloroplast differentiation and build-up (Fig. 1), and relatively high during the later stages of vegetative growth.

Efforts are currently under way to clone several of these regulatory genes. The *COP1* locus has recently been cloned using a T-DNA tagged line (Deng et al. 1992). The DNA sequence shows that *COP1* has a basic domain (possibly indicating a DNA-binding domain) and a limited homology with a protein-protein contact domain of the *TUP1* protein of yeast. *TUP1* encodes a general transcriptional repressor that regulates several different pathways (Keleher et al. 1992). These homologies suggest that *COP1* may function as a transcriptional repressor, playing a role at the end of a photoreceptor signal transduction pathway.

Similar mutants have been identified in other organisms, such as the fern *Ceratopteris richardii* (Chasan 1992), pea (*lip* or *light-independent photomorphogenesis* mutant; Frances et al. 1992), and *Aspergillus nidulans* (the *vel*A or *vel*vet mutant; Mooney and Yager 1990).

A reduced-response signal transduction mutant may be defined by mutations at the *HY5* locus (Koornneef et al. 1980). Homozygous recessive mutations in *HY5* result in seedlings that are primarily deficient in red- and far-red-light-controlled hypocotyl growth inhibition responses (Table 1). To a lesser degree, *hy5* is also insensitive to blue light. *hy5* seedlings have normal levels of phytochrome and therefore are probably not photoreceptor mutants. In addition, *hy5* mutations have been isolated in a screen designed to find extragenic suppressors of *det1*, suggesting that *HY5* may act in a manner opposite to *DET1* (A. Pepper and J. Chory, unpubl.).

Two reduced response signal transduction mutants that might be specific for a phytochrome A signal transduction pathway were recently described (Whitelam et al. 1993). Like the phytochrome A mutants, these mutants, called *fhy1* and *fhy3*, were isolated after a screen in continuous far-red light. Mutations in *fhy1* and *fhy3* complement a phytochrome A mutation and each other, thus defining two additional loci involved in the inhibition of hypocotyl elongation in far-red light (Whitelam et al. 1993). The *fhy1* and *fhy3* mutations show their effects specifically in far-red light, and as such, may define signal transduction components on a phytochrome A signal transduction pathway.

The *det*, *cop*, and *hy* mutations are pleiotropic, affecting a number of different downstream light-regulated responses. To identify regulatory components that define downstream branches in the light-regulated signal transduction pathway, we have isolated mutants specifically affected in only one light-regulated response (Li et al. 1994). The screen was designed to isolate mutants that overexpressed *CAB* genes in the dark, without causing a change in the morphology of the etiolated seedling. Using a transgenic line containing a T-DNA construct with two *CAB3* promoter-reporter fusions, recessive mutations defining three genes were identified. The new mutants showed aberrant expression of both *CAB3* promoters and were designated *doc*, for *d*ark *o*verexpression of *CAB*. The *doc* mutants fall into two phenotypic classes. Mutants of the first class have both elevated *CAB* and *RBCS* (the small subunit of RuBP carboxylase) transcript levels in the dark, whereas mutants of the second class specifically derepress the expression of *CAB* genes in the dark but do not affect the expression of *RBCS* genes. The phenotypes of *doc* mutants suggest that morphological changes can be genetically separated from changes in *CAB* gene expression. Moreover, the regulation of *CAB* gene expression can be further separated from the regulation of *RBCS* gene expression.

## GENETIC INTERACTIONS CONTROLLING PHOTOMORPHOGENESIS

Analyses of *det*, *cop1*, and *hy* single mutant lines suggest that the *DET1*, *DET2*, and *COP1* gene products are required to couple red- and blue-light signals to the downstream developmental and gene expression responses in *Arabidopsis*. In a variety of other genetic systems, epistasis tests have revealed functional relationships between mutationally defined genes (Ferguson et al. 1987; Ambros 1989; Chant and Herskowitz 1991). Phenotypic analysis of double mutant lines should thus provide a genetic framework for understanding the interactions between the photomorphogenetic genes of *Arabidopsis*. We have constructed double mutants between the most severe alleles of existing *hy* and *det* mutants (Chory 1992; Cabrera y Poch et al. 1993). The results to date suggest a hierarchical regulatory network among these genes (Fig. 3). Phenotypes of the *hy1 det1*, *hy2 det1*, *hy3 det1*, and *hy6 det1* double mutants resemble that of *det1*, indicating that *det1* is epistatic to *hy1*, *hy2*, *hy3*, and *hy6* (Chory 1992). Likewise, *det2* is epistatic to *hy1*, *hy2*, *hy3*, and *hy6* (Chory 1992). Since both *det1* and *det2* contain wild-type levels of phytochrome activity, these results are consistent with a model where formation of the active form of phytochrome normally results in a decrease in DET1 and DET2 activities. This in turn derepresses the downstream light-regulated responses. In addition, *det1* and *det2* are epistatic to *hy4*, suggesting that blue-light signals are also involved in decreasing DET1 and DET2 activities (Chory 1992). The phenotype of *det1 det2* double mutants is additive, suggesting either that DET1 and DET2 do not lie within a common signal transduction pathway or that they act in separate branches of a common signal transduction pathway. In a different study, it was shown that the *cop1* mutation is epistatic to *hy1*, implying that COP1 may also be a transduction element on a phytochrome signal transduction pathway (Deng and Quail 1992).

In contrast to the other *hy det* double mutant combinations, the *det1 hy5* and *det2 hy5* double mutants showed features of both single mutant parents when grown in either the light or dark (Chory 1992). These results are consistent with a model in which HY5 acts in a separate signal transduction pathway from DET1 and DET2 (Chory 1992). It should be noted here, however, that it is unlikely that the *det1* and *hy5* alleles used in these studies are nulls; therefore, we cannot exclude models in which *det1* and *hy5* act on a single pathway (Fig. 3). This latter model is supported by our observation that new *hy5* alleles were isolated in a screen designed to identify suppressors of *det1* (A. Pepper and J. Chory, unpubl.).

Double mutant combinations between blue-light-deficient and red-light-deficient long hypocotyl mutants support the notion that the blue- and red-light signal transduction pathways are at least partially independent. In one study, double mutants were made between *hy4* and the phytochrome-deficient *hy* mutants, *hy1*, *hy2*, and *hy3* (Koornneef et al.

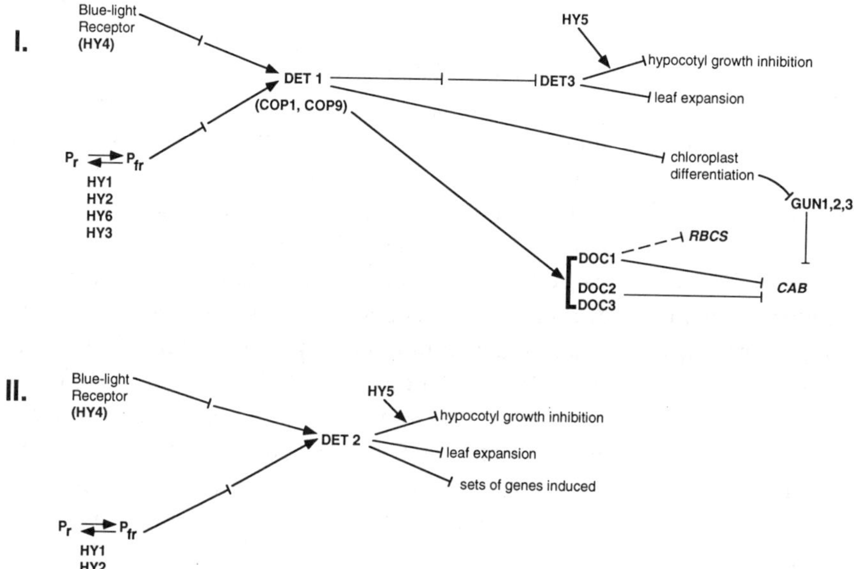

*Figure 3* Genetic control of photomorphogenesis in *Arabidopsis*. Models for the functional relationships among the *DET*, *DOC*, and *HY* genes. The phenotypes of double mutant lines suggest a hierarchical regulatory network among genes controlling the downstream light-regulated responses (inhibition of hypocotyl elongation, and promotion of leaf expansion, chloroplast differentiation, and gene expression). The models are formal and make no prediction as to the precise molecular nature of the proposed interactions among genes or gene products. (*I*) *DET1*, *COP1*, and *COP9* are negative regulators of a large number of de-etiolation responses. Formation of Pfr results in a decrease in activity of DET1 (or COP1, COP9), leading to derepression of the downstream light-regulated morphological and gene expression responses. *DET1* activity is influenced similarly by blue-light signals (*hy4*). *hy5* suppresses the phenotype of *det1* mutant seedlings, suggesting that HY5 may act downstream from DET1. DET3 acts downstream from DET1 in the regulation of morphological changes in the seedling; however, DET3 acts on a separate branch of the pathway from DOC1, 2, and 3. The *DOC* gene products control the expression of the downstream light-regulated nuclear genes. Last, GUN1, 2, and 3 appear to be components of an intracellular signaling pathway that regulates the expression of nuclear light-regulated genes in response to the functional state of the chloroplast. (*II*) DET2 is a negative regulator that acts downstream from both phytochrome and a blue-light receptor to control light-regulated seedling development. The additive phenotypes of *det2 hy5*, *det2 det1*, and *det2 det3* suggest that DET2 does not lie within a DET1 transduction pathway or that DET2 acts in a separate branch of the pathway from the DET1 branch.

1980). In each case, the double mutant had a longer hypocotyl than either single mutant, suggesting that phytochrome does not act through the blue-light pathway defined by *hy4* or vice versa. Furthermore, plants homozygous for *blu1* and *hy6* had very long hypocotyls and closed cotyledons characteristic of an etiolated seedling (Liscum and Hangarter 1991). This result suggests that blue-light and phytochrome-mediated responses together may account for most or all of the de-etiolation responses to white light.

To better define downstream branches of the phototransduction pathways, double mutants between *det1* and *det3* and *det2* and *det3* were made (Cabrera y Poch et al. 1993). The *det2 det3* double mutant has an additive phenotype of both parents, suggesting that DET2 and DET3 do not lie within a common signal transduction pathway (Fig. 3). In contrast, *det1* is epistatic to *det3*, indicating that DET1 acts before DET3 in the control of light-regulated development (Fig. 3). In addition, double mutant studies of *det1 doc1* and *det3 doc2* suggest that the *DOC* gene products are downstream from DET1 and on a different branch from DET3 in the light-regulated signal transduction pathway (Fig. 3) (Li et al. 1994).

## CONCLUSIONS AND PERSPECTIVES

Recent reports suggest roles for protein kinases/phosphatases, calcium, and G proteins in the red- and blue-light signal transduction pathways in higher plants. Although it is still unclear whether phytochromes or the blue-light photoreceptors have intrinsic kinase activities, the data show that phosphorylation events are associated with very early events of the signal transduction pathway(s). Phosphorylation/dephosphorylation events also seem to regulate the expression of genes associated with chloroplast development; however, in immunocytochemical studies, no nuclear localization of phytochrome has ever been observed. Therefore, it seems likely that additional molecules act to transduce the signal from the photoreceptors to regulate gene expression and other downstream events. The challenges now are to confirm the role of phosphorylation and to identify the other signal transduction components.

The number and sequence of events after light excitation of the photoreceptors are unknown. Furthermore, the number of photoreceptors involved in any one particular response in a given cell type is also not known. To understand the genetic control of seedling development in response to light, additional genes that direct the downstream light-regulated responses need to be identified; in particular, ones in which only a subset of the downstream responses are affected. The genic interactions between the various mutations will need to be examined to define the number and complexity of the pathway(s) involved in light-regulated seedling development. With mutations at just a few loci, it already appears that there are two to three distinct pathways that affect downstream light-regulated genes. Finally, cloning the genes defined by the various mutations will help unravel the molecular mechanisms of light-mediated seedling development.

## ACKNOWLEDGMENT

We thank our colleagues for communicating results prior to publication.

## REFERENCES

Abe, H., K. Takio, K. Titani, and M. Furuya. 1989. Amino-terminal amino acid sequences of pea phytochrome II fragments obtained by limited proteolysis. *Plant Cell Physiol.* **30:** 1089–1097.

Abe, H., K.T. Yamamoto, A. Nagatani, and M. Furuya. 1985. Characterization of green tissue-specific phytochrome isolated immunochemically from pea seedlings. *Plant Cell Physiol.* **26:** 1387–1399.

Ahmad, M. and A.R. Cashmore. 1993. *HY4* gene of *A. thaliana* encodes a protein with characteristics of a blue-light photoreceptor. *Nature* **366:** 162–166.

Ambros, V. 1989. A hierarchy of regulatory genes controls a larva-to-adult development switch in *C. elegans. Cell* **57:** 49–57.

Anderson, J.M. 1986. Photoregulation of the composition, function, and structure of thylakoid membranes. *Annu. Rev. Plant Physiol.* **37:** 93–136.

Artus, N. and C. Somerville,. 1988. A mutant of *Arabidopsis thaliana* that exhibits chlorosis in air but not in atmospheres enriched in $CO_2$. *Plant Physiol.* **87:** 83–88.

Bennett, J., G.I. Jenkins, and M.R. Hartley. 1984. Differential regulation of the accumulation of the light-harvesting chlorophyll a/b complex and ribulose bisphosphate carboxylase/oxygenase in greening pea leaves. *J. Cell. Biochem.* **25:** 1–13.

Borthwick, H.A., S.B. Hendricks, and M.W. Parker. 1952a. The reaction controlling floral initiation. *Proc. Natl. Acad. Sci.* **38:** 929–934.

Borthwick, H.A., S.B. Hendricks, M.W. Parker, E.H. Toole, and V.K. Toole. 1952b. A reversible photoreaction controlling seed germination. *Proc. Natl. Acad. Sci.* **38:** 662–666.

Boylan, M.T. and P.H. Quail. 1989. Oat phytochrome is biologically active in transgenic tomatoes. *Plant Cell* **1:** 765–773.

———. 1991. Phytochrome A overexpression inhibits hypocotyl elongation in transgenic *Arabidopsis. Proc. Natl. Acad. Sci.* **88:** 10806–10810.

Bruce, W.B. and P.H. Quail. 1990. *cis*-Acting elements involved in photoregulation of an oat phytochrome promoter in rice. *Plant Cell* **2:** 1081–1089.

Bruce, W.B., X.D. Deng, and P.H. Quail. 1991. A negatively acting DNA sequence element mediates phytochrome-directed repression of *phy*A gene transcription. *EMBO J.* **10:** 3015–3024.

Buzby, J.S., T. Tamada, and E.M. Tobin. 1990. A light-regulated DNA-binding activity interacts with a conserved region of a *Lemna gibba rbc*S promoter. *Plant Cell* **2:** 805–814.

Cabrera y Poch, H., C.A. Peto, and J. Chory. 1993. A mutation in the *Arabidopsis DET3* gene uncouples photoregulated leaf development from gene expression and chloroplast biogenesis. *Plant J.* **4:** 671–682.

Caspar, T., S. Huber, and C. Somerville. 1985. Alterations in growth, photosynthesis, and respiration in a starchless mutant of *Arabidopsis thaliana* (L.) deficient in chloroplast phosophoglucomutase activity. *Plant Physiol.* **79:** 11–17.

Chant, J. and I. Herskowitz. 1991. Genetic control of bud site selection in yeast by a set of gene products that constitute a morphogenetic pathway. *Cell* **65:** 1203–1212.

Chasan, R. 1992. *Ceratopteris*: A model plant for the 90s. *Plant Cell* **4:** 113–115.

Cherry, J.R., D. Hondred, J.M. Walker, and R.D. Vierstra. 1992. Phytochrome requires the 6-kDa N-terminal domain for full biological activity. *Proc. Natl. Acad. Sci.* **89:** 5039–5043.

Cherry, J.R., D. Hondred, J.M. Walker, J.M. Keller, H.P. Hershey, and R. Vierstra. 1993. Carboxy-terminal deletion analysis of oat phytochrome A reveals the presence of separate domains required for structure and biological activity. *Plant Cell* **5:** 565–575.

Childs, K.L., M.M. Cordonnier-Pratt, L.H. Pratt, and P.W. Morgan. 1992. Genetic regulation of development in *Sorghum bicolor*. VII. *ma*3R flowering mutant lacks a phytochrome that predominates in green tissue. *Plant Physiol.* **99:** 765–770.

Chory, J. 1991. Light signals in leaf and chloroplast development: Photoreceptors and downstream responses in search of a transduction pathway. *New Biol.* **3:** 538–548.

———. 1992. A genetic model for light-regulated seedling development in *Arabidopsis*. *Development* **115:** 337–354.

———. 1993. Out of darkness: Mutants reveal pathways controlling light-regulated development in plants. *Trends Genet.* **9:** 167–172.

Chory, J. and C. Peto. 1990. Mutations in the *DET1* gene affect cell-type-specific expression of light-regulated genes and chloroplast development in *Arabidopsis*. *Proc. Natl. Acad. Sci.* **87:** 8776–8780.

Chory, J., P. Nagpal, and C.A. Peto. 1991. Phenotypic and genetic analysis of *det2*, a new mutant that affects light-regulated seedling development in *Arabidopsis*. *Plant Cell* **3:** 445–459.

Chory, J., C. Peto, R. Feinbaum, L. Pratt, and F. Ausubel. 1989a. *Arabidopsis thaliana* mutant that develops as a light-grown plant in the absence of light. *Cell* **58:** 991–999.

Chory, J., C.A. Peto, M. Ashbaugh, R. Saganich, L. Pratt, and F. Ausubel. 1989b. Different roles for phytochrome in etiolated and green plants deduced from characterization of *Arabidopsis thaliana* mutants. *Plant Cell* **1:** 867–880.

Chow, W., C. Miller, and J. Anderson. 1991. Surface charges, the heterogeneous lateral distribution of the two photosystems, and thylakoid stacking. *Biochim. Biophys. Acta* **1057:** 69–77.

Colbert, J.T. 1988. Molecular biology of phytochrome. *Plant Cell Environ.* **11:** 305–318.

Colbert, J.T., H.P. Hershey, and P.H. Quail. 1983. Autoregulatory control of translatable phytochrome mRNA levels. *Proc. Natl. Acad. Sci.* **80:** 2248–2252.

Dale, J.E. 1988. The control of leaf expansion. *Annu. Rev. Plant Physiol. Plant. Mol. Biol.* **39:** 267–295.

Datta, N. and A.R. Cashmore. 1989. Binding of a pea nuclear protein to promoters of certain photoregulated genes is modulated by phosphorylation. *Plant Cell* **1:** 1069–1077.

Dean, C. and R.M. Leech. 1982. Genome expression during normal leaf development. *Plant Physiol.* **69:** 904–910.

Dehesh, K., W.B. Bruce, and P.H. Quail. 1990. A *trans*-acting factor that binds to a GT-motif in a phytochrome gene promoter. *Science* **250:** 1397–1399.

Dehesh, K., H. Hung, J. Tepperman, and P.H. Quail. 1992. GT-2: A transcription factor with twin autonomous DNA-binding domains of closely related but different target sequence specificity. *EMBO J.* **11**: 4131–4144.

Dehesh, K., C. Franci, B.M. Parks, K.A. Seeley, T.W. Short, J.M. Tepperman, and P.H. Quail. 1993. *Arabidopsis HY8* locus encodes phytochrome A. *Plant Cell* **5**: 1081–1088.

Deng, X.W. and P.H. Quail. 1992. Genetic and phenotypic characterization of *cop1* mutants of *Arabidopsis*. *Plant J.* **2**: 83–95.

Deng, X.W., T. Caspar, and P.H. Quail. 1991. *cop1*: A regulatory locus involved in light-controlled development and gene expression in *Arabidopsis*. *Genes Dev.* **5**: 1172–1182.

Deng, X.W., M. Matsui, N. Wei, D. Wagner, A.M. Chu, K.A. Feldmann, and P.H. Quail. 1992. *COP1*, an *Arabidopsis* regulatory gene, encodes a protein with both a Zn-binding motif and a GB-protein homologous domain. *Cell* **71**: 791–801.

Devlin, P.F., S.B. Rood, D.E. Somers, P.H. Quail, and G.C. Whitelam. 1992. Photophysiology of the *elongated internode* (*ein*) mutant of *Brassica rapa*: The *ein* mutant lacks a detectable phytochrome B-like polypeptide. *Plant Physiol.* **100**: 1442–1447.

Donald, R.G.K. and A.R. Cashmore. 1990. Mutation of either G box or I box sequences profoundly affects expression from the *Arabidopsis rbc*S1-A promoter. *EMBO J.* **9**: 1717–1726.

Ehmann, B., B. Ocker, and E. Schafer. 1991. Development- and light-dependent regulation of the expression of two different chalcone synthase transcripts in mustard cotyledons. *Planta* **183**: 416–422.

Estelle, M. and C. Somerville. 1987. Auxin-resistant mutants of *Arabidopsis thaliana* with an altered morphology. *Mol. Gen. Genet.* **206**: 200–206.

Ferguson, E.L., P.W. Sternberg, and H.R. Horwitz. 1987. A genetic pathway for the specification of the vulval cell lineages of *C. elegans*. *Nature* **326**: 259–267.

Fluhr, R. and N.H. Chua. 1986. Developmental regulation of two genes encoding ribulose-bisphosphate carboxylase small subunit in pea and transgenic petunia plants: Phytochrome response and blue-light induction. *Proc. Natl. Acad. Sci.* **83**: 2358–2362.

Fluhr, R., C. Kuhlemeier, F. Nagy, and N.H. Chua. 1986. Organ-specific and light-induced expression of plant genes. *Science* **232**: 1106–1112.

Frances, S., M.J. White, M.D. Edgerton, A.M. Jones, R.C. Elliot, and W.F. Thompson. 1992. Initial characterization of a pea mutant with light-independent photomorphogenesis. *Plant Cell* **4**: 1519–1530.

Furuya, M. 1989. Molecular properties and biogenesis of phytochrome I and II. *Adv. Biophys.* **25**: 133–167.

Gallagher, S., T.W. Short, P.M. Ray, L.H. Pratt, and W.R. Briggs. 1988. Light-mediated changes in two proteins found associated with plasma membrane fractions from pea stem sections. *Proc. Natl. Acad. Sci.* **85**: 8003–8007.

Gilmartin, P.M., L. Sarokin, J. Memelink, and N.H. Chua. 1990. Molecular light switches for plant genes. *Plant Cell* **2**: 369–378.

Gilmartin, P.M., J. Memelink, K. Hiratsuka, S.A. Kay, and N.H. Chua. 1992. Characterization of a gene encoding a DNA binding protein with specificity for a light-responsive element. *Plant Cell* **4**: 839–849.

Glick, R.E., S.W. McCauley, W. Gruissem, and A. Melis. 1986. Light quality regulates expression of chloroplast genes and assembly of photosynthetic membrane complexes. *Proc. Natl. Acad. Sci.* **83**: 4287–4291.

Grimm R., D. Gast, and W. Rudiger. 1989. Characterization of a protein-kinase activity associated with phytochrome from etiolated oat (*Avena sativa* L) seedlings. *Planta* **178**: 199–206.

Gruissem, W. 1989. Chloroplast gene expression: How plants turn their plastids on. *Cell* **56:** 161–170.

Guilliano, G., E. Pichersky, U.S. Malik, M.P. Timko, P.A. Scolnik, and A.R. Cashmore. 1988. An evolutionarily conserved protein binding sequence upstream of a plant light-regulated gene. *Proc. Natl. Acad. Sci.* **85:** 7089–7093.

Guzman, P. and J. Ecker. 1990. Exploiting the triple response of *Arabidopsis* to identify ethylene-related mutants. *Plant Cell* **2:** 513–523.

Harrison, M.J., M.A. Lawton, C.J. Lamb, and R.A. Dixon. 1991. Characterization of a nuclear protein that binds to three elements within the silencer region of a bean chalcone synthase gene promoter. *Proc. Natl. Acad. Sci.* **88:** 2515–2519.

Hirono, Y. and G.P. Rédei. 1963. Multiple alleleic control of chlorophyll b level in *Arabidopsis thaliana. Nature* **197:** 1324–1325.

Hou, Y., A.G. von Arnim, and X.-W. Deng. 1993. A new class of *Arabidopsis* constitutive photomorphogenic genes involved in regulating cotyledon development. *Plant Cell* **5:** 329–339.

Hugly, S., L. Kunst, J. Browse, and C. Somerville. 1989. Enhanced thermal tolerance of photosynthesis and altered chloroplast ultrastructure in a mutant of *Arabidopsis* deficient in lipid desaturation. *Plant Physiol.* **90:** 1134–1142.

Karlin-Neumann, G.A, L. Sun, and E.M. Tobin. 1988. Expression of light-harvesting chlorophyll a/b protein genes is phytochrome-regulated in etiolated *Arabidopsis thaliana* seedlings. *Plant Physiol.* **88:** 1323–1331.

Katagiri, F. and N.H. Chua. 1992. Plant transcription factors: Present knowledge and future challenges. *Trends Genet.* **8:** 22–27.

Kay, S.A., B. Keith, K. Shinozaki, M.-L. Chye, and N.H. Chua. 1989a. The rice phytochrome genes: Structure, autoregulated expression, and binding of GT-1 to a conserved site in the 5′ upstream region. *Plant Cell* **1:** 351–360.

Kay, S.A., A. Nagatani, B. Keith, M. Deak, M. Furuya, and N.H. Chua. 1989b. Rice phytochrome is biologically active in transgenic tobacco. *Plant Cell* **1:** 775–782.

Keleher, C.A., M.J. Redd, J. Schmultz, M. Carlson, and A.D. Johnson. 1992. Ssn6-Tup1 is a general repressor of transcription in yeast. *Cell* **68:** 709–719.

Keller, J.M., J. Shanklin, J.D. Vierstra, and H.P. Hershey 1989. Expression of a functional monocotyledonous phytochrome in transgenic tobacco. *EMBO J.* **8:** 1005–1012.

Khurana, J.P., Z. Ren, B. Steinitz, B. Parks, T.R. Best, and K.L. Poff. 1989. Mutants of *Arabidopsis thaliana* with decreased amplitude in their phototropic response. *Plant Physiol.* **91:** 685–689.

Kim, I.-S., U. Bai, and P.S. Song. 1989. A purified 124-kDa oat phytochrome does not possess a protein kinase activity. *Photochem. Photobiol.* **49:** 319–323.

Kirk, J.T.O. and R.A.E. Tilney-Bassett. 1978. *The plastids: Their chemistry, structure, growth, and inheritance.* Elsevier/North-Holland Biomedical, Amsterdam.

Klein, R.R. and J.E. Mullet. 1990. Light-induced transcription of chloroplast genes. *psbA* transcription is differentially enhanced in illuminated barley. *J. Biol. Chem.* **265:** 1895–1902.

Klimczak, L.J., U. Schindler, and A.R. Cashmore. 1992. DNA binding activity of the *Arabidopsis* G-box binding factor GBF1 is stimulated by phosphorylation by casein kinase II from broccoli. *Plant Cell* **4:** 87–98.

Koncz, C., R. Mayerhofer, Z. Koncz-Kalman, C. Nawrath, B. Reiss, G.P. Rédei, and J. Schell. 1990. Isolation of a gene encoding a novel chloroplast protein by T-DNA tagging in *Arabidopsis thaliana. EMBO J.* **9:** 1337–1346.

Koornneef, M., E. Rolff, and C.J.P. Spruit. 1980. Genetic control of light-inhibited hypocotyl elongation in *Arabidopsis thaliana* (L.) Heynh. *Z. Pflanzenphysiol.* **100S:**

147–160.

Koornneef, M., J.W. Cone, R.G.E. Kekens, G. O-Herne-Robers, C.J.P. Spruit, and R.E. Kendrick. 1985. Photomorphogenic responses of long hypocotyl mutants of tomato. *J. Plant Physiol.* **120:** 153–165.

Kuhlemeir, C., R. Fluhr, P.J. Green, and N.H. Chua. 1987. Sequences in the pea *rbc*S-3A gene have homology to constitutive mammalian enhancers but function as negative regulatory elements. *Genes Dev.* **1:** 247–255.

Kuhlemeier, C., M. Cuozzo, P.J. Green, E. Goyvaerts, K. Ward, and N.H. Chua. 1988. Localization and conditional redundancy of regulatory elements in *rbc*S-3A, a pea gene encoding the small subunit of ribulose-bisphosphate carboxylase. *Proc. Natl. Acad. Sci.* **85:** 4662–4666.

Lagarias, J.C. 1985. Progress in the molecular analysis of phytochrome. *Photochem. Photobiol.* **42:** 811–820.

Lagarias, J.C., J.M. Kelly, K.L. Cyr, and W.O. Smith. 1987a. Comparative photochemical analysis of highly purified 124 kilodalton oat and rye phytochromes *in vitro. Photochem. Photobiol.* **46:** 5–13.

Lagarias, J.C., Y.S. Wong, T.R. Berkelman, D.G. Kidd, and R.W. McMichael. 1987b. Structure-function studies on *Avena* phytochrome. In *Phytochrome and photoregulation in plants* (ed. M. Furuya), pp. 51–61. Academic Press, Tokyo.

Lam, E. and N.H. Chua. 1989. ASF-2: A factor that binds to the cauliflower mosaic virus 35S promoter and a conserved GATA motif in *cab* promoters. *Plant Cell* **1:** 1147–1156.

Lam, E., M. Benedyk, and N.H. Chua. 1989. Characterization of phytochrome-regulated gene expression in a photoautotrophic cell suspension: Possible role for calmodulin. *Mol. Cell. Biol.* **9:** 4819–4823.

Lam, E., Y. Kano-Murakami, P. Gilmartin, B. Niner, and N.H. Chua. 1990. A metal-dependent DNA-binding protein interacts with a constitutive element of a light-responsive promoter. *Plant Cell* **2:** 857–866.

Lamb, C.J., M.A. Lawton, M. Dron, and R.A. Dixon. 1989. Signals and transduction mechanisms for activation of plant defenses against microbial attack. *Cell* **56:** 215–224.

Lawton, M.A,. S.M. Dean, M. Dron, J.M. Kooter, K.M. Kragh, M.J. Harrison, L. Yu, L. Tanguay, R.A. Dixon, and C.J. Lamb. 1991. Silencer region of a chalcone synthase promoter contains multiple binding sides for a factor, SBF-1, closely related to GT-1. *Plant Mol. Biol.* **16:** 235–249.

Leech, R.M. 1976. Plastid development in isolated etiochloroplasts and isolated etioplasts. In *Perspective in experimental biology* (ed. N. Sunderland), pp. 145–162. Permagon, New York.

Leech, R., W. Thompson, and K. Platt-Aloia. 1981. Observations on the mechanism of chloroplast division in higher plants. *New Phytol.* **87:** 1–9.

Li, H.-M., L. Altschmied, and J. Chory. 1994. *Arabidopsis* mutants define downstream branches in the phototransduction pathway. *Genes. Dev.* **8:** 339–349.

Li, H.-M., T. Washburn, and J. Chory. 1993. Regulation of gene expression by light. *Curr. Opin. Cell Biol.* **5:** 455–460.

Lipphardt, S., R. Brettschneider, F. Kreuzaler, J. Schell, and J.L. Dangl. 1988. UV-inducible transient expression in parsley protoplasts identifies regulatory *cis*-elements of a chimeric *Antirrhinum majus* chalcone synthase gene. *EMBO J.* **7:** 4027–4033.

Liscum, E. and R.P. Hangarter. 1991. *Arabidopsis* mutants lacking blue light-dependent inhibition of hypocotyl elongation. *Plant Cell* **3:** 685–694.

———. 1993. Genetic evidence that the red-absorbing form of phytochrome B modulates gravitropism in *Arabidopsis thaliana. Plant Physiol.* **103:** 15–19.

Liscum, E., J.C. Young, K.L. Poff, and R.P. Hangarter. 1992. Genetic separation of phototropism and blue light inhibition of stem elongation. *Plant Physiol.* **100:** 267–271.

Lissemore, J.L. and P.H. Quail. 1988. Rapid transcriptional regulation by phytochrome of the genes for phytochrome and chlorophyll a/b-binding protein in *Avena sativa. Mol. Cell. Biol.* **8:** 4840–4850.

Lopez-Juez, E., A. Nagatani, K.I. Tomizawa, M. Deak, R. Kern, R.E. Kendrick, and M. Furuya. 1992. The cucumber long hypocotyl mutant lacks a light-stable PHYB-like phytochrome. *Plant Cell* **4:** 241–251.

Ma, H., M.F. Yanofsky, and E.M. Meyerowitz. 1990. Molecular cloning and characterization of *GPA*1, a G protein α subunit gene from *Arabidopsis thaliana. Proc. Natl. Acad. Sci.* **87:** 3821–3825.

Martínez-Zapater, J.M., P. Gil, J. Capel, and C.R. Somerville. 1992. Mutations at the *Arabidopsis CHM* locus promote rearrangements of the mitochondrial genome. *Plant Cell* **4:** 889–899.

McCormac, A., G. Whitelam, and H. Smith. 1992. Light-grown plants of transgenic tobacco expressing an introduced oat phytochrome A gene under the control of a constitutive viral promoter exhibit persistent growth inhibition by far-red light. *Planta* **188:** 173–181.

McCormac, A., D. Wagner, M.T. Boylan, P.H. Quail, H. Smith, and G. Whitelam. 1993. Photoresponses of transgenic *Arabidopsis* seedlings expressing introduced phytochrome B-encoding cDNAs: Evidence that phytochrome A and phytochrome B have distinct photoregulatory functions. *Plant J.* **4:** 19–27.

McCourt, P., L. Kunst, J. Browse, and C. Somerville. 1987. The effects of reduced amounts of lipid unsaturation on chloroplast ultrastructure and photosynthesis in a mutant of *Arabidopsis. Plant Physiol.* **84:** 353–360.

McCurdy, D.W. and L.H. Pratt. 1986. Immunogold electron microscopy of phytochrome of *Avena*: Identification of intracellular sites responsible for phytochrome sequestering and enhanced pelletability. *J. Cell Biol.* **103:** 2541–2550.

McMichael, R.W. and J.C. Lagarias. 1990. Phosphopeptide mapping of *Avena* phytochrome phosphorylated by protein kinases in vitro. *Biochemistry* **29:** 3872–3878.

Mohr, H. 1986. Coaction between pigment systems. In *Photomorphogenesis in plants* (ed. R. Kendrick and G. Kronenberg), pp. 547–564. Martinus-Nijhoff, Dordrecht, The Netherlands.

Mooney, J.L. and L.N. Yager. 1990. Light is required for conidiation in *Aspergillus nidulans. Genes Dev.* **4:** 1473–1482.

Mullet, J.E. 1988. Chloroplast development and gene expression. *Annu. Rev. Plant Physiol.* **39:** 475–502.

Murray, D. and B. Kohorn. 1991. Chloroplasts of *Arabidopsis thaliana* homozygous for the *ch-1* locus lack chlorophyll b, lack stable LHCPII and have stacked thylakoids. *Plant Mol. Biol.* **16:** 71–79.

Nagatani, A., J.W. Reed, and J. Chory. 1993. Isolation and initial characterization of *Arabidopsis* mutants that are deficient in phytochrome A. *Plant Physiol.* **102:** 269–277.

Neuhaus, G., C. Bowler, R. Kern, and N.-H. Chua. 1993. Calcium/calmodulin-dependent and -independent phytochrome signal transduction pathways. *Cell* **73:** 937–952.

Oeda, K., J. Salinas, and N.H. Chua. 1991. A tobacco bZip transcription activator (TAF-1) binds to a G-box-like motif conserved in plant genes. *EMBO J.* **10:** 1793–1802.

Oelmüller, R. and H. Mohr. 1985. Mode of coaction between blue/UV light and light absorbed by phytochrome in light-mediated anthocyanin formation in the milo (*Sorghum valgare* Pers.) seedling. *Proc. Natl. Acad. Sci.* **82:** 6124–6128.

Okubara, P.A. and E.M. Tobin. 1991. Isolation and characterization of three genes negatively regulated by phytochrome action in *Lemna gibba. Plant Physiol.* **96:** 1237–1245.

Parker, M.W., S.B. Hendricks, H.A. Borthwick, and F.W. Went. 1949. Spectural sensitivities for leaf and stem growth of etiolated pea seedlings and their similarity to action spectra for photoperiodism. *Am. J. Bot.* **36:** 194–204.

Parks, B.M. and P.H. Quail. 1991. Phytochrome-deficient *hy1* and *hy2* long hypocotyl mutants of *Arabidopsis* are defective in phytochrome chromophore biosynthesis. *Plant Cell* **3:** 1177–1186.

————. 1993. *hy8,* a new class of *Arabidopsis* long hypocotyl mutants deficient in functional phytochrome A. *Plant Cell* **5:** 39–48.

Parks, B.M., J. Shanklin, M. Koornneef, R.E. Kendrick, and P.H. Quail. 1989. Immunochemically detectable phytochrome is present at normal levels but is photochemically nonfunctional in the *hy1* and *hy2* long hypocotyl mutants of *Arabidopsis. Plant Mol. Biol.* **12:** 425–437.

Parks, B.M., A.M. Jones, P. Adamse, M. Koornneef, R.E. Kendrick, and P.H. Quail. 1987. The *aurea* mutant of tomato is deficient in spectrophotometrically and immunochemically detectable phytochrome. *Plant Mol. Biol.* **9:** 97–107.

Perisic, O. and E. Lam. 1992. A tobacco DNA binding protein that interacts with a light-responsive box II element. *Plant Cell* **4:** 831–838.

Possingham, J. 1980. Plastid replication and development in the life cycle of higher plants. *Annu. Rev. Plant Physiol.* **31:** 113–129.

Pratt, L.H. 1986. Localization within the plant. In *Photomorphogenesis in plants* (ed. R. Kendrick and G. Kronenberg), pp. 61–81. Martinus-Nijhoff, Dordrecht, The Netherlands.

Pratt, L.H., G.H. Kidd, and R.A. Coleman. 1974. An immunochemical characterization of the phytochrome destruction reaction. *Biochim. Biophys. Acta* **365:** 93–107.

Pyke, K. and R. Leech. 1987. The control of chloroplast number in wheat mesophyll cells. *Planta* **170:** 416–420.

————. 1991. Rapid image analysis screening procedure for identifying chloroplast number mutants in mesophyll cells of *Arabidopsis thaliana* (L.) Heynh. *Plant Physiol.* **96:** 1193–1195.

————. 1992. Chloroplast division and expansion is radically altered by nuclear mutations in *Arabidopsis thaliana. Plant Physiol.* **99:** 1005–1008.

Quail, P.H. 1991. Phytochrome: A light-activated molecular switch that regulates plant gene expression. *Annu. Rev. Genet.* **25:** 389–409.

Quail, P.H., E. Schafer, and D. Marme. 1973. Turnover of phytochrome in pumpkin cotyledons. *Plant Physiol.* **52:** 128–131.

Rabino, I. and A.L. Mancinelli. 1986. Light, temperature, and anthocyanin production. *Plant Physiol.* **81:** 922–924.

Rédei, G.P. 1967. Biochemical aspects of a genetically determined variegation in Arabidopsis. *Genetics* **56:** 431–443.

————. 1973. Extra-chromosomal mutability determined by a nuclear gene locus in *Arabidopsis. Mutat. Res.* **18:** 149–162.

Rédei, G.P. and Y. Hirono. 1964. Linkage studies. *Arabidopsis Inf. Serv.* **1:** 9–10.

Reed, J.W., A. Nagatani, T.D. Elich, M. Fagan, and J. Chory. 1994. PHYA and PHYB have overlapping but distinct functions in *Arabidopsis* development. *Plant Physiol.* **104:** 1139–1149.

Reed, J.W., P. Nagpal, D.S. Poole, M. Furuya, and J. Chory. 1993. Mutations in the gene for the red/far-red light receptor phytochrome B alter cell elongation and physiological responses throughout *Arabidopsis* development. *Plant Cell* **5:** 147–157.

Reymond, P., T.W. Short, and W.R. Briggs. 1992a. Blue light activates a specific protein kinase in higher plants. *Plant Physiol.* **100:** 655–661.

Reymond, P., T.W. Short, W.R. Briggs, and K.L. Poff. 1992b. Light-induced phosphorylation of a membrane protein plays an early role in signal transduction for phototropism in *Arabidopsis thaliana. Proc. Natl. Acad. Sci.* **89:** 4718–4721.

Romero, L.C. and E. Lam. 1993. Guanine nucleotide binding protein involvement in early steps of phytochrome-regulated gene expression. *Proc. Natl. Acad. Sci.* **90:** 1465–1469.

Romero, L.C., D. Sommer, C. Gotor, and P.-S. Song. 1991. G-proteins in etiolated *Avena* seedlings: Possible phytochrome regulation. *FEBS Lett.* **282:** 341–346.

Salvucci, M., A.J. Portis, and W. Ogren. 1986. Light and $CO_2$ response of ribulose-1,5-bisphosphate carboxylase/oxygenase activation in *Arabidopsis* leaves. *Plant Physiol.* **80:** 655–659.

Sarokin, L. and N.H. Chua. 1992. Binding sites for the two novel phosphoproteins, 3AF5 and 3AF3, are required for *rbc*S-3A expression. *Plant Cell* **4:** 473–483.

Schindler, U. and A.R. Cashmore. 1990. Photoregulated gene expression may involve ubiquitous DNA binding proteins. *EMBO J.* **9:** 3415–3427.

Schindler, U., A.E. Menkens, H. Beckmann, J. Ecker, and A.R. Cashmore. 1992a. Heterodimerization between light-regulated and ubiquitously expressed *Arabidopsis* GBF bZIP proteins. *EMBO J.* **11:** 1261–1273.

Schindler, U., W. Terzaghi, H. Beckmann, T. Kadesch, and A.R. Cashmore. 1992b. DNA binding site preferences and transcriptional activation properties of the *Arabidopsis* transcription factor GBF1. *EMBO J.* **11:** 1275–1289.

Schmelzer, E., W. Jahnen, and K. Hahlbrock. 1988. *In situ* localization of light-induced chalcone synthase mRNA, chalcone synthase, and flavonoid end products in epidermal cells of parsley leaves. *Proc. Natl. Acad. Sci.* **85:** 2989–2993.

Schneider-Poetsch, H.A.W., B. Braun, S. Marx, and A. Schaumburg. 1991. Phytochromes and bacterial sensor proteins are related by structural and functional homologies: Hypothesis on phytochrome-mediated signal transduction. *FEBS Lett.* **281:** 245–249.

Schulze-Lefert, P., M. Becker-Andre, W. Schulz, K. Hahlbrock, and J.L. Dangl. 1989. Functional architecture of the light-responsive chalcone synthase promoter from parsley. *Plant Cell* **1:** 707–714.

Senger, H. and W. Schmidt. 1986. Cryptochrome and UV receptors. In *Photomorphogenesis in plants* (ed. R. Kendrick and G. Kronenberg), pp. 137–158. Martinus-Nijhoff, Dordrecht, The Netherlands.

Shacklock, P.S., N.D. Read, and A.J. Trewavas. 1992. Cytosolic free calcium mediates red light-induced photomorphogenesis. *Nature* **358:** 753–755.

Shanklin, J., M. Jabben, and R.D. Vierstra. 1987. Red light-induced formation of ubiquitin-phytochrome conjugates: Identification of possible intermediates of phytochrome degradiation. *Proc. Natl. Acad. Sci.* **84:** 359–363.

Sharrock, R.A. and P.H. Quail. 1989. Novel phytochrome sequences in *Arabidopsis thaliana*: Structure, evolution, and differential expression of a plant regulatory photoreceptor family. *Genes Dev.* **3:** 1745–1757.

Sharrock, R.A., B.M. Parks, M. Koornneef, and P.H. Quail. 1988. Molecular analysis of the phytochrome deficiency in an *aurea* mutant of tomato. *Mol. Gen. Genet.* **213:** 9–14.

Shimazaki, Y. and L.H. Pratt. 1985. Immunochemical detection with rabbit polyclonal and mouse monoclonal antibodies of different pools of phytochrome from etiolated and green *Avena* shoots. *Planta* **164:** 333–344.

Shinomura, T., A. Nagatani, J. Chory, and M. Furuya. 1994. The induction of seed germination in *Arabidopsis thaliana* is regulated principally by phytochrome B and secondarily by phytochrome A. *Plant Physiol.* **104:** 363–371.

Short, T.W. and W.R. Briggs. 1990. Characterization of a rapid, blue light-mediated change in detectable phosphorylation of a plasma membrane protein from etiolated pea (*Pisum sativum*) seedlings. *Plant Physiol.* **92:** 179–185.

Short, T.W., M. Porst, and W.R. Briggs. 1992. A photoreceptor system regulating *in vivo* and *in vitro* phosphorylation of a pea plasma membrane protein. *Photochem. Photobiol.* **55:** 773–781.

Silverthorne, J. and E.M. Tobin. 1987. Phytochrome regulation of nuclear gene expression. *BioEssays* **7:** 18–22.

Smith, H. and G. Whitelam. 1990. Phytochrome, a family of photoreceptors with multiple physiological roles. *Plant Cell Environ.* **13:** 695–707.

Smith, W.O and K.L. Cyr. 1988. Modifications of sulfhydryl groups on phytochrome and their influence on physicochemical differences between the red- and far-red-absorbing forms. *Plant Physiol.* **87:** 195–200.

Somers, D.E., R.A. Sharrock, J.M. Tepperman, and P.H. Quail. 1991. The *hy3* long hypocotyl mutant of *Arabidopsis* is deficient in phytochrome B. *Plant Cell* **3:** 1263–1274.

Somerville, C. 1986. Analysis of photosynthesis with mutants of higher plants and algae. *Annu. Rev. Plant Physiol.* **37:** 467–507.

Staiger, D., H. Kaulen, and J. Schell. 1989. A CACGTG motif of the *Antirrhinum majus* chalcone synthase promoter is recognized by an evolutionarily conserved nuclear protein. *Proc. Natl. Acad. Sci.* **86:** 6930–6934.

Steeves, T.A. and I.M. Sussex. 1989. *Patterns in plant development,* 2nd edition, pp. 110–123. Cambridge University Press, Cambridge, United Kingdom.

Stockhaus, J., A. Nagatani, U. Halfter, S. Kay, M. Furuya, and N.-H. Chua. 1992. Serine-to-alanine substitutions at the amino-terminal region of phytochrome A result in an increase in biological activity. *Genes Dev.* **6:** 2364–2372.

Sun, L., R. Doxsee, E. Harel, and E.M. Tobin. 1993. CA-1, a novel phosphoprotein, interacts with the promoter of the *cab*140 gene in *Arabidopsis* and is not detectable in the *det*1 mutant. *Plant Cell* **5:** 109–121.

Susek, R.E. and J. Chory. 1992. A tale of two genomes: Role of a chloroplast signal in coordinating nuclear and plastid genome expression. *Aust. J. Plant Physiol.* **19:** 387–399.

Susek, R.E., F.M. Ausubel, and J. Chory. 1993. Signal transduction mutants of *Arabidopsis* uncouple nuclear *CAB* and *RBC*S gene expression from chloroplast development. *Cell* **74:** 787–799.

Taylor, W.C. 1989. Regulatory interactions between nuclear and plastid genomes. *Annu. Rev. Plant Physiol. Plant Mol. Biol.* **40:** 211–233.

Thomson, W.W. and J.M. Whatley. 1980. Development of nongreen plastids. *Annu. Rev. Plant Physiol.* **31:** 375–394.

Timpte, C., A. Wilson, and M. Estelle. 1992. Effects of the *axr2* mutation of *Arabidopsis* on cell shape in hypocotyl and inflorescence. *Planta* **188:** 271–288.

Tokuhisa, J.G. and P.H. Quail. 1987. The levels of two distinct species of phytochrome are regulated differently during germination of *Avena sativa* L. *Planta* **172:** 371–377.

Tokuhisa, J.G., S.M. Daniel, and P.H. Quail. 1985. Phytochrome in green tissue: Spectral and immunochemical evidence for two distinct molecular species of phytochrome in light-grown *Avena sativa* L. *Planta* **164:** 321–332.

Vierstra, R.D. and P.H. Quail. 1983a. Purification and initial characterization of 123-kilodalton phytochrome from *Avena. Biochemistry* **22:** 2498–2505.

————. 1983b. Photochemistry of 124 kilodalton *Avena* phytochrome in vitro. *Plant Physiol.* **82:** 264–267.

Wagner, D., J.M. Tepperman, and P.H. Quail. 1991. Overexpression of phytochrome B induces a short hypocotyl phenotype in transgenic *Arabidopsis*. *Plant Cell* **3:** 1275–1288.

Warpeha, K.M.F., L.S. Kaufmann, and W.R. Briggs. 1992. A flavoprotein may mediate the blue light-activated binding of guanosine 5′-triphosphate to isolated plasma membranes of *Pisum sativum* L. *Photochem. Photobiol.* **55:** 595–603.

Warpeha, K.M.F., H.E. Hamm, M.M. Rasenick, and L.S. Kaufmann. 1991. A blue-light-activated GTP-binding protein in the plasma membranes of etiolated peas. *Proc. Natl. Acad. Sci.* **88:** 8925–8929.

Wei, N. and X.-W. Deng. 1992. *COP9*: A new genetic locus involved in light-regulated development and gene expression in *Arabidopsis*. *Plant Cell* **4:** 1507–1518.

Weisshaar, B., G. Armstrong, A. Block, O. da Costa e Silva, and K. Hahlbrock. 1991. Light-inducible and constitutively expressed DNA-binding proteins recognizing a plant promoter element with functional relevance in light responsiveness. *EMBO J.* **10:** 1777–1786.

Whatley, J. 1974. Chloroplast development in primary leaves of *Phaseolus vulgaris*. *New Phytol.* **73:** 1097–1110.

————. 1977. Variations in the basic pathway of chloroplast development. *New Phytol.* **78:** 407–420.

Whitelam, G., E. Johnson, J. Peng, P. Carol, M.L. Anderson, J.S. Cowl, and N.P. Harberd. 1993. Phytochrome A *null* mutants of *Arabidopsis* display a wild-type phenotype in white light. *Plant Cell* **5:** 757–768.

Wong, Y.S. and J.C. Lagarias. 1989. Affinity labeling of *Avena* phytochrome with ATP analogs. *Proc. Natl. Acad. Sci.* **86:** 3469–3473.

Wong, Y.-S., H.-C. Cheng, D.A. Walsh, and J.C. Lagarias. 1986. Phosphorylation of *Avena* phytochrome in vitro as a probe of light-induced conformational changes. *J. Biol. Chem.* **261:** 12089–12097.

# 23

# Circadian Rhythms in
# *Arabidopsis thaliana*


**C. Robertson McClung**
Department of Biological Sciences
Dartmouth College
Hanover, New Hampshire 03755

**Steve A. Kay**
Department of Biology
University of Virginia
Charlottesville, Virginia 22901


## INTRODUCTION

Much of an organism's physiology, biochemistry, and behavior is temporally organized with respect to the environmental oscillation of day and night. Thus, most organisms express diurnal rhythms. When deprived of environmental time cues, such as light-dark cycles or temperature cycles, many of these rhythms persist, indicating that organisms have the capacity to measure time and to use this time information to temporally regulate aspects of their biology. Circadian rhythms, that subset of endogenous rhythms with periods of approximately 24 hours, have been known for many years, and a great deal of information has been gathered concerning the rhythmic processes of many organisms (Pittendrigh 1981b; Edmunds 1988). Circadian rhythms share certain characteristics. They persist in the absence of environmental time cues. The period of the free-running rhythm is approximately, although not exactly, 24 hours. The deviation of the free-running period from exactly 24 hours has been termed the "strongest single piece of evidence that the overt rhythm is under the control of an endogenous timing mechanism" (Feldman 1982). The period of the rhythm is temperature-compensated; that is, the period remains relatively constant over the range of physiologically relevant temperatures. Note, however, that temperature compensation applies to organisms maintained at one (of several possible) constant temperature and does not imply that temperature changes or cycles are without effect on the clock; indeed, temperature cues and light cues are two of the most potent stimuli that can shift the phase of the clock. The primary effect of the environmental time cues associated with the diurnal cycle is to entrain the endogenous timing system to a period of 24 hours (Pittendrigh 1981a), which precisely corresponds to the environmental period resulting from the rotation of the earth on its axis. This entrainment ensures that the endogenous circadian system and the rhythms it controls will be in adaptive relationships with the day-night cycle.

A voluminous literature describing rhythmic phenomena in many organisms and addressing theoretical considerations associated with these rhythms has been generated in the more than 200 years since the scientific investigation of circadian rhythms was initiated by de Mairan (1729), who showed that the rhythmic leaf movement of the sensitive plant (probably *Mimosa pudica*) persisted under constant temperature in the absence of a light-dark cycle. However, the molecular basis by which any organism keeps time is apparently complex and remains quite mysterious. Over the last several years, using a combined approach of genetic and molecular biological experiments, considerable progress has been made toward the elucidation of the mechanism of biological timekeeping. The rationale behind the genetic approach is straightforward: Components of the circadian clock include macromolecules that are encoded by genes (or that are synthesized by gene products), and it should be possible to identify at least some of those genes (and their products) through the isolation and characterization of mutants which show altered clock function. Molecular biological approaches then allow the determination not only of the structure and function, but also of the spatial and temporal distribution of the products of those genes which, when mutated, confer the phenotype of altered clock function. Ideally, genetic and molecular biological approaches can allow access to organizational levels of clocks which are many steps removed from the overt expression of clock function, the phenomenon originally observed to be rhythmic.

To date, most progress toward understanding the molecular basis of biological clocks has been made in two model systems, *Drosophila melanogaster* and *Neurospora crassa* (Rosbash and Hall 1989; Dunlap 1993; Dunlap et al. 1993; Jackson 1993). In both systems, mutant screens have identified sets of loci which, when mutated, confer altered clock properties. Particularly interesting are allelic series of mutations identified at the *period* (*per*) locus of *Drosophila* and at the *frequency* (*frq*) locus of *Neurospora*. These alleles include semidominant variants that confer periods longer or shorter than wild type, and recessive variants that confer arrhythmicity. Both *per* (Bargiello and Young 1984; Reddy et al. 1984) and *frq* (McClung et al. 1989) have been cloned, although their biochemical functions remain unclear. It is noteworthy that both *per*

(Hardin et al. 1990) and *frq* (Dunlap 1993) are themselves regulated by the clock at the level of mRNA accumulation and, at least for *per*, this circadian regulation of mRNA abundance is at the level of transcription (Hardin et al. 1992). *per* or *frq* protein function is necessary for these oscillations of *per* or *frq* mRNA (Zwiebel et al. 1991; Dunlap 1993). *per* protein levels also show circadian oscillation (Siwicki et al. 1988; Zerr et al. 1990); it is not yet known if *frq* protein levels oscillate in *Neurospora*. In adult flies, *per* protein is predominantly nuclear (Liu et al. 1992). The *per* protein shows similarity to a family of transcription factors, although *per* itself does not contain any known DNA-binding motifs, and *per* has not yet been shown to bind DNA (Hardin et al. 1992; Takahashi 1992). It is currently hypothesized that *per* protein acts as a dominant negative regulator of transcription through the formation of heterodimers with DNA-binding proteins (Huang et al. 1993).

## Characteristics of Model Organisms for the Analysis of Circadian Rhythms

As described above, much of what is known of the molecular basis of circadian rhythmicity stems from research in *Drosophila* and *Neurospora*. Much of the rest of this volume addresses the utility of *Arabidopsis thaliana* as a model system in which to study most problems in higher plants. In this chapter, we present the case for the utility of *Arabidopsis* as a system in which the confluence of genetic and molecular biological approaches may contribute to our understanding of the biological clock.

## Plants as Model Systems in the Study of Circadian Rhythms

Prior to 1900, plants were the primary system for the study of circadian rhythms, and plant research generated the historical milestones of circadian rhythm research (for discussions, see Moore-Ede et al. 1982; Sweeney 1987). The earliest known writings on what are now known to be circadian rhythms date from the 4th century B.C., when Androsthenes, in descriptions of the marches of Alexander the Great, described the diurnal leaf movements of the tamarind (*Tamarindus indicus*). The first experimental test of the assumption that these rhythmic leaf movements were direct responses to the environment was the demonstration, in 1729, that the leaf movements of the "sensitive" heliotrope (probably *Mimosa pudica*) persisted in the absence of a light-dark cycle (de Mairan 1729). Some 30 years later, variations in environmental temperature were also eliminated as cues for rhythmic leaf movements (Duhamel duMonceau 1759). Thus, plant research provided the experimental evidence of the endogenous nature of circadian rhythms. The deviation of the endogenous period from exactly 24 hours was first described for the free-running period of the leaf movements of *M. pudica* (de Candolle 1832). Circadian rhythms have been described in diverse species, from unicells to multicellular higher plants, and many, if not most, plant processes exhibit circadian regulation. Circadian rhythms have been described in enzyme synthesis and activity, photosynthetic capacity, phototaxis, cell division, stomatal opening, flower opening, and odor production (Cumming and Wagner 1968; Sweeney 1979, 1987). The ubiquity of circadian regulation in plants may reflect the overwhelming dependence of plants on the environmental cycling of light and dark.

## EXPRESSION OF THE BIOLOGICAL CLOCK IN *ARABIDOPSIS*

Until recently, *Arabidopsis* had not been examined for circadian rhythms. A single report in the literature described an annual dormancy/nondormancy cycle of seeds exposed to natural seasonal temperature changes (Baskin and Baskin 1983). However, several groups have now demonstrated that *Arabidopsis* shows endogenous circadian rhythms in the expression of many genes, most notably the chlorophyll *a/b* binding protein genes (*CAB*; see below). Quite recently, a rhythm in movement of the leaves has been described for young *Arabidopsis* (see below).

### Physiological Rhythms: Rhythmic Leaf Movement in *Arabidopsis*

Leaf movement rhythms offer a number of advantages as a system for experimental analysis of circadian rhythm. The assay is simple, inexpensive, and readily automated. In addition, the assay is nonintrusive and nondestructive. Therefore, the recent description of rhythmic leaf movement (Engelmann et al. 1992) is quite exciting and may prove extremely useful. The authors coupled a video camera to a computer via a digitizer and used the program OXALIS for data acquisition and the program OXALDIFI for period-length determination (Schuster and Engelmann 1990). Rhythmic movements of the expanded cotyledons were detected in very young plants. In slightly older plants, rhythmic movements of the first pair of secondary leaves were detected. Cotyledon movements showed a period length of approximately 22 hours, and the secondary leaves showed a period of 24.5–26 hours (Engelmann et al.

1992). More recent experiments have indicated that these leaf movement rhythms are temperature-compensated (W. Engelmann, pers. comm.). This leaf movement rhythm appears to be robust in many different lines of *Arabidopsis*, which should allow one to study the effect of different mutant backgrounds on the leaf movement rhythm (I. Carré and S.A. Kay, unpubl.). These studies are significant because the leaf movement rhythm offers a second means to monitor the circadian oscillator that is independent of the more laborious (and destructive) assays based on direct monitoring of gene expression, although the in vivo luciferase assay of *CAB-luc* fusion-bearing transgenics described by Millar and colleagues (Millar et al. 1992a,b) and discussed below also avoids the problems of destructive assays of gene expression. Optimization of the measurement of the leaf movement rhythm will permit the investigation of many of the classical parameters of clock biology, including the effects on the phase and period of the clock of various treatments, such as light pulses and chronic exposure to light of different intensities, to light of different wavelengths, and to chemical and hormonal treatments of short-term and chronic durations.

## Rhythms in Gene Expression

### Transcriptional Regulation of CAB Genes

Numerous studies have indicated that photosynthetic capacity oscillates in a circadian fashion in many plants (Sweeney 1987). Therefore, the search for plant genes whose expression is regulated by the circadian clock has concentrated on an examination of genes involved in photosynthetic light harvesting or carbon metabolism. The expression of many of these genes was already known to be regulated by light, and the discovery of circadian regulation of gene expression resulted from a logical extension of the experiments on light regulation. A number of reports now exist concerning the circadian regulation of *CAB* genes; these have been reviewed recently (Kay 1993; Kay and Millar 1993; McClung 1993; Piechulla 1993). Both common and contrasting features emerge from these data. In light/dark cycles, *CAB* mRNA has been observed to fluctuate in all species tested, with a number of qualitative and quantitative differences. In general, *CAB* mRNA levels peak at approximately Zeitgeber time (ZT) 3 to ZT8, with minima occurring in the dark period around ZT10 to ZT16 (Fig. 1). Times are expressed in hours as ZT, which is simply the time since the onset of illumination. Zeitgeber ("time giver" in German) is a term commonly used in chronobiology for the environmental signals that reset the circadian clock. ZT is used to allow comparison of experiments conducted under different light-dark regimes.

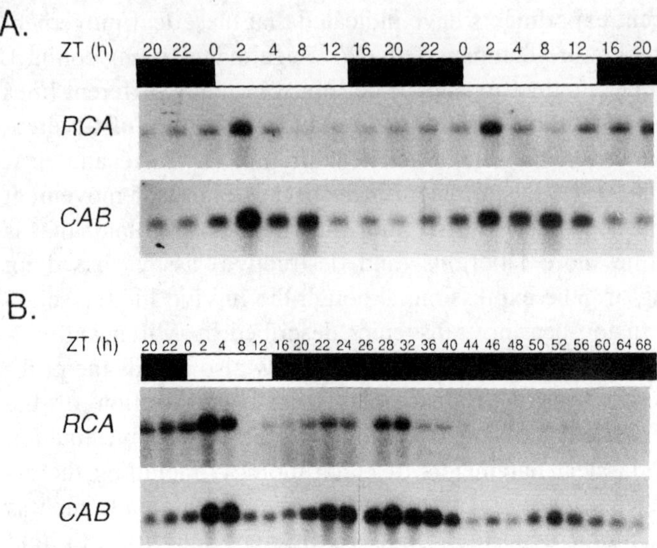

*Figure 1* Northern blots showing circadian regulation of *RCA* and *CAB* mRNA accumulation. Northern blots of total RNA from 5-week-old *Arabidopsis* were hybridized with a probe to the coding sequence of the *CAB1* gene (which detects *CAB1*, *CAB2*, and *CAB3* transcripts), then stripped and rehybridized with an *RCA* gene probe. (*A*) RNA prepared from plants that were grown in a 14:10 photoperiod. (*B*) RNA prepared from plants that were grown in a 14:10 photoperiod and then transferred to continuous darkness. Bars above blots indicate the light regime, where the filled areas indicate dark, and the open areas indicate light. Zeitgeber time is indicated above the bars. (Reprinted, with permission, from Pilgrim and McClung 1993.)

The rise in *CAB* transcript abundance begins several hours before dawn, indicating the light-independence of this expression pattern (Fig. 1). Rhythmicity persists when green plants are transferred to constant light (LL) or constant darkness (DD). Rapid reduction of peak levels is observed in DD (Fig. 1), although the rate of damping varies among species. The majority of the analyses mentioned above employed probes detecting the sum of all mRNAs from all genes of the PSII *CAB* type 1 family (Green et al. 1991). This approach cannot discern any differences in regulation among family members, which may be considerable. For example, in *Arabidopsis*, the steady-state levels of both the *CAB2* and *CAB3* transcripts exhibit a dramatic circadian fluctuation that is not evident in the abundance of *CAB1* mRNA (Millar and Kay 1991). However, in vitro run-on transcription assays revealed that the transcription rate of the *CAB1* gene does indeed follow a circadian rhythm. Thus, a posttranscriptional mechanism, such as high mRNA stability, results in a constitutive mRNA level despite circadian transcription of the *CAB1* gene. A number of other investigators have utilized in vitro nuclear run-on assays to demonstrate that the clock does indeed act at the level of transcription in regulating *CAB* gene expression in several species (for review, see Kay and Millar 1993; McClung 1993; Piechulla 1993).

The identification of *cis*-acting elements of *CAB* promoters from wheat (Nagy et al. 1988) and *Arabidopsis* (Millar and Kay 1991; Millar et al. 1992a,b) now permits molecular analyses of the mechanisms mediating circadian-responsive transcription of specific genes. For the wheat *Cab-1* gene, this initially was shown by transferring the entire wheat gene to tobacco and demonstrating that *Cab-1* cycled in the transgenic tobacco background (Nagy et al. 1988). This demonstrates that the phytochrome and circadian modes of regulation have been conserved among monocots and dicots, which diverged about 100 million years ago. Furthermore, when a fragment of the wheat *Cab-1* gene extending from −1800 to +31 is fused to the bacterial chloramphenicol acetyltransferase (CAT) gene and transferred to tobacco, the bacterial CAT RNA exhibits circadian rhythmicity in the transgenic plants. Thus, the 5′ upstream region of the wheat *Cab-1* gene contains a *cis*-acting element that confers circadian regulation. If a perfect copy of the wheat *Cab-1* mRNA is expressed in tobacco under the control of a constitutive enhancer, no circadian or diurnal fluctuations are observed (Nagy et al. 1988). This demonstrates that the clock does not alter *Cab-1* mRNA stability, but acts at the level of transcription. The wheat *Cab-1* circadian enhancer element has been localized between −357 and −90, by analyzing *Cab1-GUS* fusions in transgenic tobacco (Fejes et al. 1990). The 5′ border of this element lies at or downstream from −211, although loss in peak expression level occurs in this construct, implying that positive elements not necessary for circadian regulation have been removed. In the case of the *Arabidopsis CAB2* promoter (Fig. 2), −321 to +1 is sufficient to mediate high expression levels that are circadian- and light-regulated in both transgenic tobacco (Millar and Kay 1991; Millar et al. 1992a) and *Arabidopsis* (Millar et al. 1992b). Preliminary data suggest that the circadian-responsive element lies between −111 and −33 and can confer circadian regulation on a heterologous promoter containing an element for the general positive factor bZIP factor ASF-1 (S.A. Kay and S. Anderson, unpubl.). Obviously, the identification of the protein factors that interact with these *cis*-acting elements will be the next step toward defining the signal transduction pathway that lies between the circadian oscillator and the genomic target. Indeed, Borello et al. (1993) have shown that the abundance of the tomato protein IBF-2a that binds the *Ic* and *In* boxes of the circadian clock-responsive *CAB* and *NIA* (nitrate

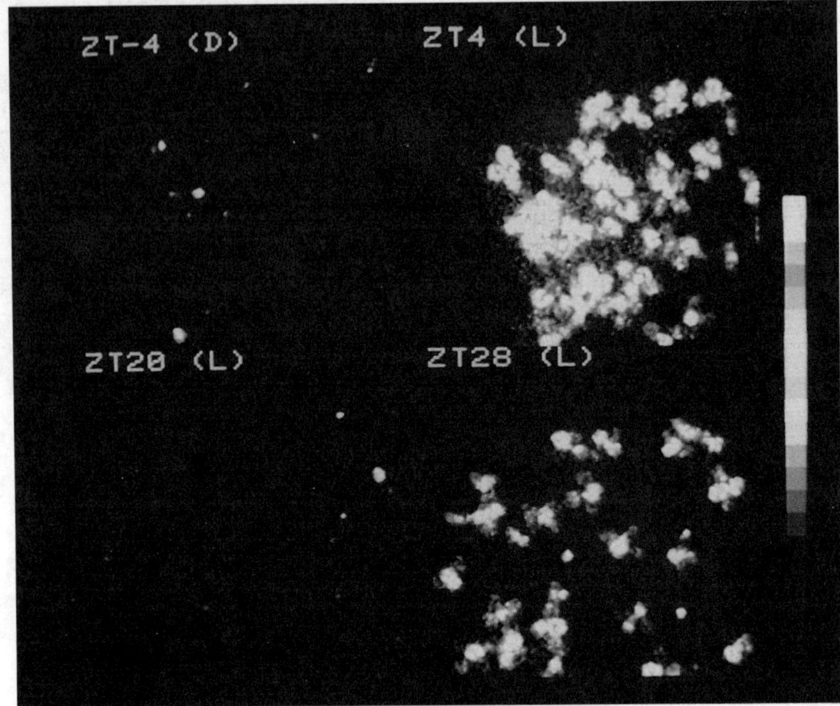


*Figure 2* Images of luminescence from transgenic *Arabidopsis* seedlings, stably transformed with *CAB2::Luc* fusions, that were grown in a light-dark (LD) cycle and transferred from LD into continuous light (LL). Seedlings growing in tissue culture dishes were imaged in a low-light video system (Millar et al. 1992a,b) at the time indicated. The upper panels show seedlings growing in LD imaged at ZT-4 (ZT20), in the night (dark), and at ZT4, in the day (light). The lower panels show seedlings transferred from LD into LL and imaged at ZT20 (subjective night) and ZT28 (subjective day). The gray scale at the right indicates intensity of luminescence from dark gray (lowest) to white (highest).

reductase; see below) genes fluctuates in diurnal and in circadian fashion, making IBF-2a a prime candidate as a circadian clock-regulated transcription factor.

## Circadian Regulation of Other Photosynthetic Genes

As indicated by Kloppstech's initial study (1985), circadian regulation of gene expression is not restricted to the *CAB* gene families of higher plants. A number of other nuclear genes encoding products involved in photosynthesis, as well as several nonphotosynthetic genes, exhibit circadian oscillations in expression. *Arabidopsis* genes whose expression is regulated by the circadian clock are summarized in Table 1. For exam-

*Table 1* **Arabidopsis** genes whose expression is regulated by the circadian clock

| Gene designation (gene encodes) | Assay | Level of regulation | References |
| --- | --- | --- | --- |
| *CAB1* (chlorophyll *a/b*-binding protein) | S1 protection, in vitro run-on, gene fusions | transcriptional, posttranscriptional | Millar and Kay (1991); Millar et al. (1992a) |
| *CAB2*, *CAB3* (chlorophyll *a/b*-binding protein) | S1 protection, in vitro run-on, gene fusions | transcriptional | Millar and Kay (1991); Millar et al. (1992a,b) |
| *CCR1*, *CCR2* (glycine-rich putative RNA-binding protein) | Northern | not determined | Carpenter et al. (1994) |
| *FEDA* (ferredoxin) | Northern, slot blots | not determined | E.A. Pease and C.R. McClung (unpubl.) |
| *RBCS* (Rubisco small subunit) | Northern, slot blots, in vitro run-on | not determined | Pilgrim and McClung (1993) |
| *RCA* (Rubisco activase) | Northern, slot blots, in vitro run-on | transcriptional | Pilgrim and McClung (1993) |
| *CAT1* (catalase) | Northern, slot blots | not determined | H.H. Zhong and C.R. McClung (unpubl.) |
| *CAT2* (catalase) | Northern, slot blots | not determined | Zhong et al. (1994) |
| *NIA1* (nitrate reductase apoprotein) | Northern | not determined | Cheng et al. (1991) |
| *NIA2* (nitrate reductase apoprotein) | Northern, slot blots, in vitro run-on | posttranscriptional | Cheng et al. (1991); Pilgrim et al. (1993) |

ple, the *Arabidopsis FEDA* gene, encoding ferredoxin A (Somers et al. 1990; Vorst et al. 1990; Caspar and Quail 1993), shows a robust circadian rhythm in DD and in LL (E.A. Pease and C.R. McClung, unpubl.). Recently, Adamska et al. (1991) showed that, in pea, the nuclear-encoded Rieske Fe-S protein, as well as the early light-inducible protein (ELIP) and the *CAB*-encoded chlorophyll *a/b*-binding protein (LHCP) oscillated in circadian fashion, although they did not investigate mRNA levels. Genes encoding products required for photosynthetic electron transport (PSAD) and oxygen evolution (OEE1) are transcriptionally regulated by the circadian clock in tomato (Giuliano et al. 1988).

In addition to photosynthetic light harvesting, the process of photosynthetic carbon assimilation is also regulated by the circadian clock. In *Arabidopsis*, transcript abundance for the Rubisco small subunit genes (*RBCS*), but not for the large subunit gene (*RBCL*), exhibits circadian oscillation (Pilgrim and McClung 1993). The *Arabidopsis RBCS* gene family is small (4 members; Krebbers et al. 1988), offering an advantageous system in which to investigate the regulation of an entire gene family in detail. A number of studies have demonstrated that the *RBCS* gene family is regulated by the circadian clock in other species, such as pea (Kloppstech 1985; Spiller et al. 1987; Otto et al. 1988). However, circadian regulation of *RBCS* mRNA was not detected in tobacco (Paulsen and Bogorad 1988), tomato (Giuliano et al. 1988; Piechulla 1989), or wheat (Nagy et al. 1988). As in *Arabidopsis*, transcript abundance for *RBCL* does not oscillate in a circadian fashion in tomato (Piechulla 1988), and abundance of large Rubisco subunit protein does not oscillate in pea (Adamska et al. 1991). That *RBSC* but not *RBCL* is clock-regulated should not be generalized to imply that circadian control is restricted to nuclear-encoded genes. Paulsen and Bogorad (1988) did not observe circadian regulation of transcript abundance for two tobacco chloroplast-encoded genes, *psbA* and *psbE*, which encode products required for photosynthetic electron transport. However, circadian regulation of transcript abundance has been described in *Chlamydomonas* for three chloroplast-encoded genes, *atpA*, *atpB*, and *tufA*, out of seven genes surveyed (Salvador et al. 1993). Similarly, not all nuclear-encoded genes with photosynthetic products are clock-controlled, as is indicated by the results which show that the *RBCS* transcript abundance oscillates in some but not in all species examined. In addition, in *Arabidopsis*, transcripts of the nuclear genes encoding the chloroplast chaperonins, *CPN60α* and *CPN60β*, oscillate in a diurnal but not in a circadian fashion (Pilgrim and McClung 1993). Why some nuclear-encoded and some plastid-encoded genes should be clock-regulated whereas other genes with related functions are not remains unclear.

Rubisco activity is regulated in the plant. Key to the disc ery that the catalytic activity of Rubisco in the plant requires enzyma activation by Rubisco activase (Salvucci et al. 1985) was the identification and characterization of an *Arabidopsis* mutant, *rca*, in which Rubisco was nonactivatable in vivo (Somerville et al. 1982). Rubisco activase is phylogenetically widespread, and control of Rubisco by Rubisco activase may be ubiquitous among eukaryotic photosynthetic organisms (Salvucci et al. 1987). In *Arabidopsis*, Rubisco activase is encoded by a single, nuclear gene, *RCA* (Werneke et al. 1988, 1989). *RCA* mRNA oscillates in *Arabidopsis* grown in a light-dark cycle, and this oscillation persists in plants transferred to continuous light or to continuous dark (see Fig. 1) (Pilgrim and McClung 1993). Similarly, *RCA* expression is regulated by the circadian clock in tomato (Martino-Catt and Ort 1992) and in apple (Watillon et al. 1993). Nuclear run-on experiments show that this circadian regulation of *RCA* expression includes a transcriptional component in both *Arabidopsis* and tomato (Martino-Catt and Ort 1992; Pilgrim and McClung 1993).

## Circadian Regulation of Genes Encoding Catalase

In *Arabidopsis*, catalase is encoded by at least two different genes (Chevalier et al. 1992; H.H. Zhong et al., unpubl.). *CAT2* is expressed during germination and in the photosynthetically competent plant, where catalase is required for photorespiration (Zhong et al. 1994). In mature plants, *CAT2* mRNA accumulation shows robust circadian oscillations in DD and in LL, which persist for at least 5 days in LL; the period length of this oscillation is less than 24 hours (Zhong et al. 1994). It is not yet known whether these circadian oscillations are also detectable at the levels of catalase protein and activity. Not surprisingly, the phase of the accumulation of *Arabidopsis CAT2* transcript is during the subjective day with maximal mRNA abundance at or shortly after dawn, presumably reflecting catalase-mediated detoxification of $H_2O_2$ generated during the photorespiratory oxidation of glycolate in the peroxisome (Ogren 1984). mRNA accumulation of a second *Arabidopsis* catalase gene, *CAT1*, also shows circadian oscillations (H.H. Zhong and C.R. McClung, unpubl.). *CAT1* is most abundantly expressed in roots and stems and is only weakly expressed in leaves and, therefore, probably is not involved in photorespiration. It is interesting that the phase of maximal *CAT1* mRNA abundance is late in the day, approximately 9 hours after dawn (H.H. Zhong and C.R. McClung, unpubl.). That *CAT1* and *CAT2* mRNAs reach maximal abundance some 9 hours out of phase with each other suggests that the molecular elements mediating clock control of the two *CAT* genes may differ.

Rhythmicity of catalase gene expression had also been observed in the C4 plant, maize. In maize, a family of three unlinked genes, designated *Cat1*, *Cat2*, and *Cat3*, encodes three biochemically distinct catalase isozymes. The three *Cat* genes and corresponding CAT isozymes exhibit distinct spatial patterns of expression and also respond differently to light (Scandalios 1990). Redinbaugh et al. (1990) observed diurnal fluctuations in *Cat3* mRNA abundance which persisted in plants transferred to constant dark or to constant light and, hence, represent a circadian rhythm. As determined by nuclear run-on analysis, most of the variation in *Cat3* mRNA abundance could be attributed to changes in transcription. *Cat3* was the only catalase gene that was regulated by the circadian clock; no circadian fluctuations were observed in mRNA abundance of *Cat1* or *Cat2*. The function of CAT3 and, hence, the functional significance of this oscillation in *Cat3* expression are not known. The circadian regulation of *Cat3* expression, like that of *Arabidopsis CAT1*, differs from the examples of photosynthetic and photorespiratory genes described above, in that *Cat3* mRNA abundance peaked late in the light period (ZT 10; i.e., 10 hr after dawn), whereas the photosynthetic and photorespiratory genes showed peaks of mRNA abundance earlier in the light period. In further contrast to results obtained in studies on *CAB* expression (Nagy et al. 1988), pulses of red or far-red light had no effect on the fluctuations in *Cat3* mRNA abundance, indicating that phytochrome does not play a role in regulation of *Cat3* expression (Redinbaugh et al. 1990). It is also noteworthy that, when seeds were imbibed, germinated, and grown in constant conditions (either constant light or constant dark), *Cat3* showed constitutively high and uniform expression (Acevedo et al. 1991). Thus, expression of the circadian rhythm in *Cat3* expression required the entraining signal of a light-dark cycle (it is not known how many light-dark cycles would be sufficient). This is in marked contrast to *CAB* expression, which showed circadian oscillations in wheat seedlings grown in continuous light (Nagy et al. 1988). It is also significant that, in the absence of a clock-entraining stimulus, maize *Cat3* expression is maximal. One simple interpretation is that circadian regulation is mediated by a cyclic repression of *Cat3* expression (Acevedo et al. 1991).

## Circadian Regulation of Nitrogen Assimilation Genes

The assimilation of nitrate entails reduction of nitrate to ammonia via nitrate and nitrite reductases. Nitrate reductase (NR) is probably the key regulated step in this pathway. NR activity is inducible by nitrate (Tang and Wu 1957), with induction reflecting de novo enzyme synthesis, rather than activation of previously synthesized enzyme (Somers et al. 1983; Remmler and Campbell 1986). The cloning of genes encoding the NR apoprotein allowed the demonstration that NR apoprotein synthesis in response to nitrate correlates with increased mRNA accumulation (Cheng et al. 1986; Crawford et al. 1986, 1988). NR activity is also light-inducible (Duke and Duke 1984) and responds to red light, mediated by phytochrome (Duke and Duke 1984; Rajasekhar et al. 1988; Melzer et al. 1989), and to blue light, mediated by a blue-light receptor (Melzer et al. 1989). In etiolated *Arabidopsis*, *NIA2* mRNA accumulates (a sevenfold induction in response to a 5-min pulse of red light) in a phytochrome-mediated very low fluence response; that is, the induction of *NIA2* mRNA accumulation by red light is not fully reversible by far-red light, and far-red light alone results in a substantial (threefold over the etiolated control) induction (Pilgrim et al. 1993). Curiously, supplementation of dark-adapted green *Arabidopsis* with photosynthetically produced sugars obviates the light requirement for *Arabidopsis NIA1* gene expression, suggesting that light effects on *NIA1* expression in green plants may be indirect, mediated through effects on photosynthesis (Cheng et al. 1992). Thus, it appears that *NIA2* expression responds directly to red and far-red light (a 5-min pulse of red light applied to etiolated seedlings does not induce photosynthetic competence), whereas the response of *NIA1* is indirect. However, differences in the two experimental procedures (etiolated seedlings vs. dark-adapted green plants) complicate the comparison of *NIA1* and *NIA2* regulation. Expression of nitrate reductase is also influenced by phytohormones; barley nitrate reductase gene transcription is regulated by cytokinin-abscisic acid interactions (Lu et al. 1992).

Circadian oscillations in NR activity were first described almost 20 years ago by Cohen and Cumming (1974). Diurnal oscillations in NR activity are widespread and have been described in wheat, oat, and barley (Lillo 1984; Lillo and Henriksen 1984). In barley, these oscillations in NR activity persist in constant light and, hence, represent true circadian rhythms (Lillo 1991). In tobacco and in maize, NR apoprotein mRNA abundance, apoprotein level, and NR activity oscillate diurnally (Galangau et al. 1988; Lillo and Ruoff 1989). These diurnal oscillations persist in plants transferred to continuous light and, hence, are true circadian rhythms (Deng et al. 1989, 1990; Lillo and Ruoff 1989). In tobacco transferred to continuous dark, the mRNA oscillations persisted for two cycles, but the apoprotein and activity levels declined steadily and did not oscillate from the time of transfer into the dark (Deng et al. 1990). In *Arabidopsis*, there is circadian regulation of *NIA1* and *NIA2* mRNA abundance in plants transferred either to continuous light (Cheng et al. 1991; Pilgrim et al. 1993) or to continuous dark (Pilgrim et al.

1993). This regulation of *NIA2* expression does not occur at the level of transcription, as determined by nuclear run-off experiments (Pilgrim et al. 1993), suggesting that circadian regulation is occurring posttranscriptionally, possibly through changes in *NIA2* mRNA stability. Circadian oscillations in *NIA2* mRNA abundance are correlated to oscillations in NR activity in plants transferred either to continuous light or to continuous dark (Pilgrim et al. 1993). In *Arabidopsis*, tobacco, and maize, the amplitude of the NR activity oscillations was approximately twofold, in contrast to the much greater oscillation seen in mRNA abundance (7- to 20-fold in *Arabidopsis* [Pilgrim et al. 1993] and 50- to 100-fold in tobacco [Galangau et al. 1988; Deng et al. 1990] and in maize [Lillo and Ruoff 1988]). These observations, coupled with the persistence of oscillations in mRNA abundance but not in apoprotein level or in activity in tobacco transferred to continuous dark, indicate that posttranscriptional regulation of NR expression must also occur at levels in addition to changes in mRNA stability.

## Circadian Regulation of Other Genes

To date, efforts to identify plant genes regulated by the circadian clock have centered on processes with obvious links to the light-dark diurnal cycle, such as photosynthetic light-harvesting and carbon assimilation or the energy-intensive assimilation of nitrate. It is highly likely that circadian regulation of gene expression will be found in other aspects of plant biology. Recently, Carpenter et al. (1994) have described circadian oscillations in mRNA abundance of two *Arabidopsis* genes (*CCR1* and *CCR2*) that encode glycine-rich proteins with putative RNA-binding domains. The function of these proteins is not known. *CCR1* and *CCR2* are highly and differentially regulated, and transcript levels respond to cold stress, to abscisic acid treatment, and to drought (transcripts of the two genes respond in opposite fashion to drought). Undoubtedly, the list of circadian-clock-regulated *Arabidopsis* genes will continue to grow as more investigators become aware of the widespread nature of clock control of gene expression and begin to look for evidence of it in their own studies.

## ISOLATION OF MUTANTS AFFECTED IN CIRCADIAN REGULATION

A substantial limitation to past genetic approaches for analysis of circadian rhythms has been the lack of a selectable clock phenotype. Mutants described to date were isolated by screening large numbers of mutants for those displaying an altered clock phenotype. Lack of a selectable clock phenotype precludes the identification of clock mutants by direct selection and makes reversion analysis of identified mutations difficult due to the large numbers of mutagenized lines which would need to be screened. Analysis of both intra- and intergenic suppressors of clock mutations would be informative, providing information on functional interactions among domains of putative clock proteins and possibly identifying new clock components and providing insight into interactions among clock components.

Is it possible to develop a screen of *Arabidopsis* for mutants displaying clock phenotypes? The small genome size of *Arabidopsis* and the small growth habit allow the screening of sufficient numbers of mutagenized plants to detect mutant phenotypes at reasonable frequencies. Screening mutagenized *Arabidopsis* has allowed the identification of mutants (e.g., mutants that fail to synthesize starch and mutants showing altered fatty acid profiles) at frequencies of 1 in 2000 plants (Estelle and Somerville 1986), which is similar to the frequency with which clock mutants were found by brute force screening of mutagenized strains of *Neurospora* (Feldman 1982). Previous mutant hunts in *Neurospora* and *Drosophila* have chiefly yielded mutants with altered period (Feldman 1982; Hall 1990; Jackson 1993). By analogy, in *Arabidopsis* one would anticipate classes of mutants in which clock-controlled markers are expressed, but in which the periodicity of expression is either longer or shorter than in wild type. A short-period mutant would express a clock-controlled marker earlier after transfer to DD (or to LL) than in wild type. A long-period variant would express the marker in DD (or LL) after the wild type had stopped expression. Thus, both long- and short-period mutants would be identifiable.

The most advanced effort relying on a gene fusion strategy to screen for clock mutants is that of Kay and colleagues, who have made transgenic tobacco and *Arabidopsis* lines in which luciferase expression is controlled by the clock using an *Arabidopsis CAB-luc* fusion construct (Millar et al. 1992a,b). Empirically, the firefly luciferase activity is sufficiently unstable that clear oscillations in light production can be observed (Fig. 2). Transgenic *Arabidopsis* seedlings bearing the *CAB-luc* construct exhibit rhythmic bioluminescence for at least five cycles in constant light (Millar et al. 1992b). This rhythmic bioluminescence constitutes a novel circadian phenotype that is currently being used to screen for aberrant patterns in *CAB-luc* drive bioluminescence. A subset of these mutants should be altered in components that constitute the oscillator and the signal transduction chain from the oscillator to the *CAB* promoter. A major advantage of this phenotype in the search for clock mutants is the dynamic nature of the screen: Subtle changes in period length and phase can be readily detected many times within a single life-cycle, enabling the elimination of much of the background noise in the screen.

## INPUTS TO AND OUTPUTS FROM THE CIRCADIAN CLOCK

By definition, circadian rhythms persist in plants transferred into constant conditions, in the absence of external time cues. However, outside the laboratory, plants are entrained to the natural cycles that result from the rotation of the earth on its axis. What are the signals that are capable of providing time information to the circadian clock, and what is known of the signal transduction pathways by which this sidereal time information is used to reset the circadian clock? Figure 3 summarizes what little is known of clock inputs. At this point, the best-characterized input to the circadian clock is light. Recently, however, cyclic heat shocks or temperature changes have been suggested to synchronize the circadian clock of barley (Kloppstech et al. 1991; Beator et al. 1992). Daily heat shocks applied to etiolated barley induced photomorphogenetic responses, including shortening of the primary leaves and coleoptiles to the lengths observed in light-grown seedlings. In addition, the transcript levels of several light-induced genes, including *CAB* and *RBCS*, were induced in response to the heat treatment. Interestingly, the elevated *CAB* and *RBCS* mRNAs showed circadian oscillations in abundance. It is intriguing to speculate that the cyclic heat shock treatments initiated or synchronized the circadian clock in the treated seedlings. However, an alternate explanation is that the circadian clock was already synchronized and running, but the levels of the light-inducible genes were too low to allow the reliable observation of transcript oscillations. The later notion is supported by the observation that basal *CAB* gene transcription is regulated in a circadian manner in etiolated tobacco seedlings prior to any detectable exposure to light (Millar et al. 1992a). The heat shock treatments employed in barley might have had the primary effect of elevating expression, allowing the detection of transcript oscillations. Certainly, the potential role of temperature treatments in influencing the circadian clock warrants further investigation. However, it should be noted that although, in *Arabidopsis*, heat shock can produce a phenocopy of a *det* or *cop* phenotype, the required heat shock is extreme and the resulting plants are sickly (S.A. Kay, unpubl.).

Studies of light perception in plants are well-advanced, and the details of these studies are reviewed elsewhere in this volume. However, although a great deal is known about the earliest events in the perception of light, very little is known of the subsequent events in signal transduc-

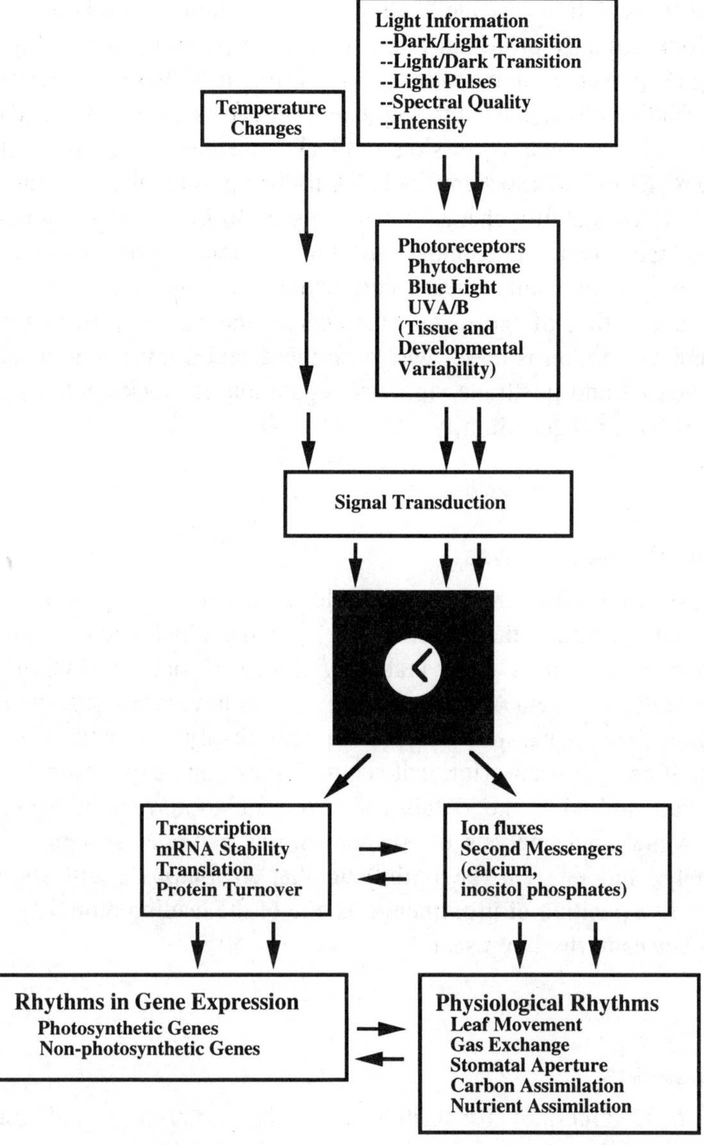


*Figure 3* Inputs to and outputs from the circadian clock. The clock is metaphorically shown as a black box, consistent with our nearly total ignorance of the elements necessary to constitute a circadian oscillator. The best-characterized stimuli known to convey sidereal time information to the clock are light and temperature. Not shown is the very likely possibility that the phase of the clock may affect the input pathways and the sensitivity of the clock itself to input. That is, there are likely to be loops in which clock outputs (for example, gene expression) feed back upon elements of the input pathways and upon the circadian clock itself.

tion. Within the context of circadian rhythms, almost nothing is known of the signal transduction cascade leading from the photoreceptors to the circadian clock. Equally profound is our ignorance (summarized in Fig. 3) of the signal transduction cascade by which time information generated by the circadian clock is used to regulate the expression of "the hands of the clock," such as gene expression and leaf movement. Regulation of *Arabidopsis CAB* gene expression includes transcriptional and posttranscriptional (mRNA stability changes) mechanisms. In *Bryophyllum*, posttranslational mechanisms (phosphorylation) are responsible for circadian oscillations in phosphoenolpyruvate carboxylase activity (Carter et al. 1991). The unraveling of the distal elements of the pathway from the clock to gene expression is under way with the detailed investigation of the transcriptional and posttranscriptional regulation of clock-controlled genes such as the *CAB* gene family.

## CONCLUSIONS AND PERSPECTIVES

To date, most of the effort in circadian biology of *Arabidopsis* has been descriptive, demonstrating the presence of a circadian clock and defining rhythmic phenomena. It is clear that *Arabidopsis* is rich in rhythms. Rhythms in leaf movement and in gene expression have been observed, and rhythmic gene expression has proven remarkably common. Both transcriptional and posttranscriptional regulation of gene expression has been observed, although the details of this regulation remain to be elucidated. Mutational analysis of *Arabidopsis* as a model system has proven timely, and one can be optimistic that *Arabidopsis* will soon return plants to a position of prominence as one of the leading model systems for advances in rhythm research.

## ACKNOWLEDGMENTS

We thank M.L. Guerinot for criticism of the manuscript and for numerous helpful discussions. Work in the McClung laboratory is supported by grants from the National Science Foundation (DMB-9005345) and the American Cancer Society (JFRA-253). Work in the Kay laboratory on circadian rhythms and phytochrome is supported by grants to S.A.K. from the National Science Foundation (MCB-9216399), the National Institutes of Health (GM-43884), the Jeffress Trust, and The National Science Foundation for Biological Timing. S.A.K. is supported by an award from the W.M. Keck Foundation.

## REFERENCES


Acevedo, A., J.D. Williamson, and J.G. Scandalios. 1991. Photoregulation of the *Cat2* and *Cat3* catalase genes in pigmented and pigment-deficient maize: The circadian regulation of *Cat3* is superimposed on its quasi-constitutive expression in maize leaves. *Genetics* **127**: 601–607.

Adamska, I., B. Scheel, and K. Kloppstech. 1991. Circadian oscillations of nuclear-encoded chloroplast proteins in pea (*Pisum sativum*). *Plant Mol. Biol.* **17**: 1055–1065.

Bargiello, T.A. and M.W. Young. 1984. Molecular genetics of a biological clock in *Drosophila. Proc. Natl. Acad. Sci.* **81**: 2142–2146.

Baskin, J.M. and C.C. Baskin. 1983. Seasonal changes in the germination responses of buried seeds of *Arabidopsis thaliana* and ecological interpretation. *Bot. Gaz.* **144**: 540–543.

Beator, J., E. Pötter, and K. Kloppstech. 1992. The effect of heat shock on morphogenesis in barley. *Plant Physiol.* **100**: 1780–1786.

Borello, U., E. Ceccarelli, and G. Giuliano. 1993. Constitutive, light-responsive and circadian clock-responsive factors compete for the different *I* box elements in plant light-regulated promoters. *Plant J.* **4**: 611–619.

Carpenter, C.D., J.A. Kreps, and A.E. Simon. 1994. Genes encoding glycine-rich *Arabidopsis thaliana* proteins with RNA-binding motifs are influenced by cold treatment and an endogenous circadian rhythm. *Plant Physiol.* **104**: 1015–1025.

Carter, P.J., H.G. Nimmo, C.A. Fewson, and M.B. Wilkins. 1991. Circadian rhythms in the activity of a plant protein kinase. *EMBO J.* **10**: 2063–2068.

Caspar, T. and P.H. Quail. 1993. Promoter and leader regions involved in the expression of the *Arabidopsis* ferredoxin A gene. *Plant J.* **3**: 161–174.

Cheng, C., G.N. Acedo, M. Cristinsin, and M.A. Conkling. 1992. Sucrose mimics the light induction of *Arabidopsis* nitrate reductase gene transcription. *Proc. Natl. Acad. Sci.* **89**: 1861–1864.

Cheng, C.-L., J. Dewdney, A. Kleinhofs, and H.M. Goodman. 1986. Cloning and nitrate induction of nitrate reductase mRNA. *Proc. Natl. Acad. Sci.* **83**: 6825–6828.

Cheng, C.-L., G.N. Acedo, J. Dewdney, H.M. Goodman, and M.A. Conkling. 1991. Differential expression of the two *Arabidopsis* nitrate reductase genes. *Plant Physiol.* **96**: 275–279.

Chevalier, C., J. Yamaguchi, and P. McCourt. 1992. Nucleotide sequence of a cDNA for catalase from *Arabidopsis thaliana. Plant Physiol.* **99**: 1726–1728.

Cohen, A.S. and B.G. Cumming. 1974. Endogenous rhythmic activity of nitrate reductase in a selection of *Chenopodium rubrum. Can. J. Bot.* **52**: 2351–2360.

Crawford, N.M., W.H. Campbell, and R.W. Davis. 1986. Nitrate reductase from squash: cDNA cloning and nitrate regulation. *Proc. Natl. Acad. Sci.* **83**: 8073–8076.

Crawford, N.M., M. Smith, D. Bellissimo, and R.W. Davis. 1988. Sequence and nitrate regulation of the *Arabidopsis thaliana* mRNA encoding nitrate reductase, a metalloflavoprotein with three functional domains. *Proc. Natl. Acad. Sci.* **85**: 5006–5010.

Cumming, B.G. and E. Wagner. 1968. Rhythmic processes in plants. *Annu. Rev. Plant Physiol.* **19**: 381–416.

de Candolle, A.P. 1832. *Physiologie vegetale*. Bechet Jeune, Paris.

de Mairan, J. 1729. Observation botanique. *Histoire de L'Academie Royale des Sciences*, pp. 35–36.

Deng, M.D., T. Moureaux, and T. Lamaze. 1989. Diurnal and circadian fluctuation of malate levels and its close relationship to nitrate reductase activity in tobacco leaves. *Plant Sci.* **65**: 191–197.

Deng, M.D., T. Moureaux, M.-T. Leydecker, and M. Caboche. 1990. Nitrate-reductase expression is under control of a circadian rhythm and is light inducible in *Nicotiana tabacum* leaves. *Planta* **180:** 257–261.

Duhamel duMonceau, H.L. 1759. *La physique des arbres.* H.L. Guerin and L.F. Delatour, Paris.

Duke, S.H. and S.O. Duke. 1984. Light control of extractable nitrate reductase activity in higher plants. *Physiol. Plant.* **62:** 485–493.

Dunlap, J.C. 1993. Genetic analysis of circadian clocks. *Annu. Rev. Physiol.* **55:** 683–728.

Dunlap, J.C., Q. Liu, K.A. Johnson, and J.J. Loros. 1993. Genetic and molecular dissection of the *Neurospora* clock. In *The molecular genetics of biological rhythms* (ed. M.W. Young), pp. 37–54. Marcel Dekker, New York.

Edmunds, L.N. 1988. *Cellular and molecular bases of biological clocks.* Springer-Verlag, New York.

Engelmann, W., K. Simon, and C.J. Phen. 1992. Leaf movement rhythm in *Arabidopsis thaliana. Z. Naturforsch.* **47c:** 925–928.

Estelle, M.A. and C.R. Somerville. 1986. The mutants of *Arabidopsis. Trends Genet.* **2:** 89–93.

Fejes, E., A. Pay, I. Kanevsky, M. Szell, E. Adam, S. Kay, and F. Nagy. 1990. A 268 bp upstream sequence mediates the circadian clock-regulated transcription of the wheat *Cab*-1 gene in transgenic plants. *Plant Mol. Biol.* **15:** 921–932.

Feldman, J.F. 1982. Genetic approaches to circadian clocks. *Annu. Rev. Plant Physiol.* **33:** 583–608.

Galangau, F., F. Daniel-Vedele, T. Moureaux, M.-F. Dorbe, M.-T. Leydecker, and M. Caboche. 1988. Expression of leaf nitrate reductase genes from tomato and tobacco in relation to light-dark regimes and nitrate supply. *Plant Physiol.* **88:** 383–388.

Giuliano, G., N.E. Hoffman, K. Ko, P.A. Scolnick, and A.R. Cashmore. 1988. A light-entrained circadian clock controls transcription of several plant genes. *EMBO J.* **7:** 3635–3642.

Green, B.R., E. Pichersky, and K. Kloppstech. 1991. Chlorophyll a/b-binding proteins: An extended family. *Trends Biochem. Sci.* **16:** 81–86.

Hall, J.C. 1990. Genetics of circadian rhythms. *Annu. Rev. Genet.* **24:** 659–697.

Hardin, P.E., J.C. Hall, and M. Rosbash. 1992. Circadian oscillations in period gene mRNA levels are transcriptionally regulated. *Proc. Natl. Acad. Sci.* **89:** 11711–11715.

––––––––. 1990. Feedback of the *Drosophila period* gene product on circadian cycling of its messenger RNA levels. *Nature* **343:** 536–540.

Huang, Z.J., I. Edery, and M. Rosbash. 1993. PAS is a dimerization domain common to *Drosophila period* and several transcription factors. *Nature* **364:** 259–262.

Jackson, F.R. 1993. Circadian rhythm mutants of *Drosophila.* In *The molecular genetics of biological rhythms* (ed. M.W. Young), pp. 91–121. Marcel Dekker, New York.

Kay, S.A. 1993. Shedding light on circadian regulated *cab* gene transcription in higher plants. *Semin. Cell Biol.* **4:** 81–86.

Kay, S.A. and A.J. Millar. 1993. Circadian-regulated *cab* gene transcription in higher plants. In *The molecular genetics of biological rhythms* (ed. M.W. Young), pp. 73–89. Marcel Dekker, New York.

Kloppstech, K. 1985. Diurnal and circadian rhythmicity in the expression of light-induced nuclear messenger RNAs. *Planta* **165:** 502–506.

Kloppstech, K., B. Otto, and W. Sierralta. 1991. Cyclic temperature treatments of dark-grown pea seedlings induce a rise in specific transcript levels of light-regulated genes related to photomorphogenesis. *Mol. Gen. Genet.* **225:** 468–473.

Krebbers, E., J. Seurinck, L. Herdies, A.R. Cashmore, and M.P. Timko. 1988. Four genes in two diverged subfamilies encode the ribulose-1,5-bisphosphate carboxylase small subunit polypeptides of *Arabidopsis thaliana. Plant Mol. Biol.* **11:** 745–759.

Lillo, C. 1984. Diurnal variations of nitrite reductase, glutamine synthetase, glutamate synthase, alanine aminotransferase and aspartate aminotransferase in barley leaves. *Physiol. Plant.* **61:** 214–218.

———. 1991. Diurnal variations of corn leaf nitrate reductase: An experimental distinction between transcriptional and post-transcriptional control. *Plant Sci.* **73:** 149–154.

Lillo, C. and A. Henriksen. 1984. Comparative studies of diurnal variations of nitrate reductase activity in wheat, oat and barley. *Physiol. Plant.* **62:** 89–94.

Lillo, C. and P. Ruoff. 1989. An unusually rapid light-induced nitrate reductase mRNA pulse and circadian oscillations. *Naturwissenschaften* **76:** 526–528.

Liu, X., L.J. Zweibel, D. Hinton, S. Benzer, J.C. Hall, and M. Rosbash. 1992. The *period* gene encodes a predominantly nuclear protein in adult *Drosophila. J. Neurosci.* **12:** 2735–2744.

Lu, J.L., J.R. Ertl, and C.M. Chen. 1992. Transcriptional regulation of nitrate reductase mRNA levels by cytokinin-abscisic acid interactions in etiolated barley leaves. *Plant Physiol.* **98:** 1255–1260.

Martino-Catt, S. and D.R. Ort. 1992. Low temperature interrupts circadian regulation of transcriptonal activity in chilling-sensitive plants. *Proc. Natl. Acad. Sci.* **89:** 3731–3735.

McClung, C.R. 1993. The higher plant, *Arabidopsis thaliana,* as a model system for the molecular analysis of circadian rhythms. In *The molecular genetics of biological rhythms* (ed. M. Young), pp. 1–35. Marcel Dekker, New York.

McClung, C.R., B.A. Fox, and J.C. Dunlap. 1989. The *Neurospora* clock gene *frequency* shares a sequence element with the *Drosophila* clock gene *period. Nature* **339:** 558–562.

Melzer, J.M., A. Kleinhofs, and R.L. Warner. 1989. Nitrate reductase regulation: Effects of nitrate and light on nitrate reductase mRNA accumulation. *Mol. Gen. Genet.* **217:** 341–346.

Millar, A.J. and S.A. Kay. 1991. Circadian control of *cab* gene transcription and mRNA accumulation in *Arabidopsis. Plant Cell* **3:** 541–550.

Millar, A.J., S.R. Short, N.-H. Chua, and S.A. Kay. 1992a. A novel circadian phenotype based on firefly luciferase expression in transgenic plants. *Plant Cell* **4:** 1075–1087.

Millar, A.J., S.R. Short, K. Hiratsuka, N.-H. Chua, and S.A. Kay. 1992b. Firefly luciferase as a reporter of regulated gene expression in higher plants. *Plant Mol. Biol. Rep.* **10:** 324–337.

Moore-Ede, M.C., F.M. Sulzman, and C.A. Fuller. 1982. *The clocks that time us.* Harvard University Press, Cambridge, Massachusetts.

Nagy, F., S.A. Kay, and N.-H. Chua. 1988. A circadian clock regulates transcription of the wheat *Cab-1* gene. *Genes Dev.* **2:** 376–382.

Ogren, W.L. 1984. Photorespiration: Pathways, regulation, and modification. *Annu. Rev. Plant Physiol.* **35:** 415–442.

Otto, B., B. Grimm, P. Ottersbach, and K. Kloppstech. 1988. Circadian control of the accumulation of mRNAs for light- and heat-inducible chloroplast proteins in pea (*Pisum sativum* L.). *Plant Physiol.* **88:** 21–25.

Paulsen, H. and L. Bogorad. 1988. Diurnal and circadian rhythms in the accumulation and synthesis of mRNA for the light-harvesting chlorophyll a/b-binding protein in tobacco. *Plant Physiol.* **88:** 1104–1109.

Piechulla, B. 1988. Plastid and nuclear mRNA fluctuations in tomato leaves—Diurnal and circadian rhythms during extended dark and light periods. *Plant Mol. Biol.* **11:** 345–353.

———. 1989. Changes of the diurnal and circadian (endogenous) mRNA oscillations of the chlorophyll a/b binding protein in tomato leaves during altered day/night (light/dark) regimes. *Plant Mol. Biol.* **12:** 317–327.

———. 1993. "Circadian clock" directs the expression of plant genes. *Plant Mol. Biol.* **22:** 533–542.

Pilgrim, M.L. and C.R. McClung. 1993. Differential involvement of the circadian clock in the expression of genes required for ribulose-1,5-bisphosphate carboxylase/oxygenase synthesis, assembly, and activation in *Arabidopsis thaliana. Plant Physiol.* **103:** 553–564.

Pilgrim, M.L., T. Caspar, P.H. Quail, and C.R. McClung. 1993. Circadian and light-regulated expression of nitrate reductase in *Arabidopsis. Plant Mol. Biol.* **23:** 349–364.

Pittendrigh, C.S. 1981a. Circadian rhythms: Entrainment. In *Handbook of behavioral neurobiology: Biological rhythms* (ed. J. Aschoff), pp. 95–124. Plenum Press, New York.

———. 1981b. Circadian rhythms: General perspective. In *Handbook of behavioral neurobiology: Biological rhythms* (ed. J. Aschoff), pp. 57–80. Plenum Press, New York.

Rajasekhar, V.K., G. Gowri, and W.H. Campbell. 1988. Phytochrome-mediated light regulation of nitrate reductase expression in squash cotyledons. *Plant Physiol.* **88:** 242–244.

Reddy, P., W.A. Zehring, D.A. Wheeler, V. Pirotta, C. Hadfield, J.C. Hall, and M. Rosbash. 1984. Molecular analysis of the *period* locus in *Drosophila melanogaster* and identification of a transcript involved in biological rhythms. *Cell* **38:** 701–710.

Redinbaugh, M.G., M. Sabre, and J.G. Scandalios. 1990. Expression of the maize *Cat3* catalase gene is under the influence of a circadian rhythm. *Proc. Natl. Acad. Sci.* **87:** 6853–6857.

Remmler, J.L. and W.H. Campbell. 1986. Regulation of corn leaf nitrate reductase. II. Synthesis and turnover of the enzyme's activity and protein. *Plant Physiol.* **80:** 442–447.

Rosbash, M. and J.C. Hall. 1989. The molecular biology of circadian rhythms. *Neuron* **3:** 387–398.

Salvador, M.L., U. Klein, and L. Bogorad. 1993. Light-regulated and endogenous fluctuations of chloroplast transcript levels in *Chlamydomonas.* Regulation by transcription and RNA degradation. *Plant J.* **3:** 213–219.

Salvucci, M.E., A.R. Portis, and W.L. Ogren. 1985. A soluble chloroplast protein catalyzes ribulosebisphosphate carboxylase/oxygenase activation in vivo. *Photosynth. Res.* **7:** 193–201.

Salvucci, M.E., J.M. Werneke, W.L. Ogren, and A.R. Portis, Jr. 1987. Purification and species distribution of RUBISCO activase. *Plant Physiol.* **84:** 930–936.

Scandalios, J.G. 1990. Response in plant antioxidant defense genes to environmental stress. *Adv. Genet.* **28:** 1–41.

Schuster, J. and W. Engelmann. 1990. Recording of rhythms in organisms using video-digitizing. In *Chronobiology: Its role in clinical medicine, general biology, and agriculture* (ed. D.K. Hayes et al.), part B, pp. 389–396. Wiley-Liss, New York.

Siwicki, K.K., C. Eastman, G. Petersen, M. Rosbash, and J.C. Hall. 1988. Antibodies to the *period* gene product of *Drosophila* reveal diverse tissue distribution and rhythmic changes in the visual system. *Neuron* **1:** 141–150.

Somers, D.A., T.-M. Kuo, A. Kleinhofs, R.L. Warner, and A. Oaks. 1983. Synthesis and degradation of barley nitrate reductase. *Plant Physiol.* **72:** 949–952.

Somers, D.E., T. Caspar, and P.H. Quail. 1990. Isolation and characterization of a ferredoxin gene from *Arabidopsis thaliana. Plant Physiol.* **93:** 572–577.

Somerville, C.R., A.R. Portis, Jr., and W.L. Ogren. 1982. A mutant of *Arabidopsis thaliana* which lacks activation of RuBP carboxylase *in vivo. Plant Physiol.* **70:** 381–387.

Spiller, S.C., L.S. Kaufman, W.F. Thompson, and W.R. Briggs. 1987. Specific mRNA and rRNA levels in greening pea leaves during recovery from iron stress. *Plant Physiol.* **84:** 409–414.

Sweeney, B.M. 1979. Endogenous rhythms in the movements of plants. In *Encyclopedia of plant physiology* (ed. W. Haupt and M.E. Feinleib), pp. 71–93. Springer-Verlag, New York.

———. 1987. *Rhythmic phenomena in plants.* Academic Press, New York.

Takahashi, J.S. 1992. Circadian clock genes are ticking. *Science* **258:** 238–240.

Tang, P.S. and H.Y. Wu. 1957. Adaptive formation of nitrate reductase in rice seedlings. *Nature* **179:** 1355–1356.

Vorst, O., F. van Dam, R. Oosrethoff-Teertstra, S. Smeekens, and P. Weisbeek. 1990. Tissue-specific expression directed by an *Arabidopsis thaliana* pre-ferredoxin promoter in transgenic tobacco plants. *Plant Mol. Biol.* **14:** 491–499.

Watillon, B., R. Kettman, P. Boxus, and A. Burny. 1993. Developmental and circadian pattern of rubisco activase mRNA accumulation in apple plants. *Plant Mol. Biol.* **23:** 501–509.

Werneke, J.M., J.M. Chatfield, and W.L. Ogren. 1989. Alternative mRNA splicing generates the two ribulosebisphosphate carboxylase/oxygenase activase polypeptides in spinach and *Arabidopsis. Plant Cell* **1:** 815–825.

Werneke, J.M., R.E. Zielinski, and W.L. Ogren. 1988. Structure and expression of spinach leaf cDNA encoding ribulosebisphosphate carboxylase/oxygenase activase. *Proc. Natl. Acad. Sci.* **85:** 787–791.

Zerr, D.M., J.C. Hall, M. Rosbash, and K.K. Siwicki. 1990. Circadian fluctuations of period protein immunoreactivity in the CNS and the visual system of *Drosophila. J. Neurosci.* **10:** 2749–2762.

Zhong, H.H., J.C. Young, E.A. Pease, R.P. Hangarter, and C.R. McClung. 1994. Interactions between light and the circadian clock in the regulation of *CAT2* expression in *Arabidopsis. Plant Physiol.* **104:** 889–898.

Zweibel, L.J., P.E. Hardin, X. Liu, J.C. Hall, and M. Rosbash. 1991. A post-transcriptional mechanism contributes to circadian cycling of a per-β-galactosidase fusion protein. *Proc. Natl. Acad. Sci.* **88:** 3882–3886.

# 24

# The Physiology of Tropisms

**Kenneth L. Poff, Abdul-Kader Janoudi,**
**Elizabeth S. Rosen, and Vladimir Orbović**
MSU-DOE Plant Research Laboratory
Michigan State University
East Lansing, Michigan 48824

**Radomir Konjević**
Institute of Botany, Faculty of Science and
Institute for Biological Research
University of Belgrade, Yugoslavia

**Marie-Claude Fortin**
Center for Land and Biological Resources Research
Agriculture Canada, Vancouver
British Columbia, Canada V6T-1X2

**Tom K. Scott**
Department of Biology
University of North Carolina
Chapel Hill, North Carolina 27599


Since the historic work of Darwin (1896), one of the driving forces in the field of plant physiology has been the study of tropisms. Tropisms are directed growth responses to environmental stimuli such as light, gravity, temperature (heat), and water. Thus, the tropism by the plant is an integrated response to a number of pieces of information concerning the status of the plant's environment. Tropisms have been widely studied, largely because they are obvious outward evidence of the processing of information by the plant. Despite the great interest in tropisms by plant biologists during the past century, progress in this area has not been rapid. This can be partially attributed to the fact that many studies have been limited to observations of the final growth response of the organism as a consequence of the manipulation of the environmental stimulus. It seems clear that an understanding of the steps between the stimulus and the final growth response will require new approaches to the problem.

In recognition of this requirement for new approaches to understanding tropisms, and in recognition of the promise of *Arabidopsis thaliana* for genetics and molecular genetics, this plant was adapted as a model

*Arabidopsis*
© 1994 Cold Spring Harbor Laboratory Press  0-87969-428-9/94 $5 + .00

system for sensory physiology study slightly more than a decade ago. Since that time, research efforts have been directed toward the biophysical definition of tropisms in this model plant system and the development of genetic systems for their study. In this chapter, our goal is to describe the present state of information concerning tropisms in *A. thaliana* and to underscore the many problems that remain to be solved. In many cases, because there is inadequate information available for *A. thaliana*, we include information from other plant systems. It is our hope that this chapter will stimulate the application of many exciting new technologies to a true understanding of the ancient question of how plants orient their growth with respect to the conditions of their environment.

## PHOTOTROPISM

Phototropism is the growth response whereby a plant orients with respect to direction of light propagation. In *A. thaliana*, this is manifested by curvature of the hypocotyl and flowering stalk toward the light source. Phototropism may also be involved along with gravitropism and epinasty in orienting the leaves such that the upper surface of the leaf faces the light source. Ultimately, we want to know the sequence of reactions from photoreception to the growth response, and the mechanism whereby the plant detects, via photoreception, the direction from which the light is incident. Research has been concentrated at the two ends of the sensory transduction sequence, photoreception and the growth response.

### Identity of the Photoreceptor Pigment

Action spectra for phototropism in *Avena* (Thimann and Curry 1961) and in the fungus, *Phycomyces* (Curry and Gruen 1959; Delbrück and Shropshire 1960), are quite similar. In general, the spectra show peaks in the blue region, a peak at 370 nm, little action at wavelengths higher than 500 nm (Fig. 1), and a peak in the UV range at 280 nm (Dennison 1979). On the basis of such spectra, the responsible pigment has been referred to as the blue-light photoreceptor pigment (see, e.g., Schmidt 1984). The consensus in the literature has been that these spectra most closely match the absorption spectrum of a flavoprotein in a relatively hydrophobic environment (Dennison 1979). The 280-nm band is thought to be a consequence of absorption by the protein, and the 370-nm band and those in the blue are thought to result from absorption by the flavin moiety (Dennison 1979). Although an action spectrum has not been measured for phototropism in *A. thaliana*, there is no particular reason to expect an action spectrum to differ from those measured in other plants.

Many efforts have been directed toward the identification of the blue-

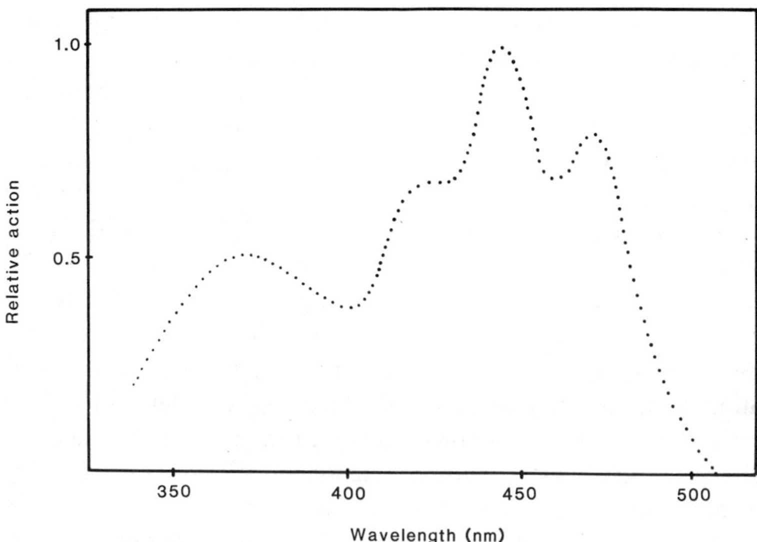

*Figure 1*  Action spectrum for phototropism in *Avena*. (Redrawn from Thimann and Curry 1961.)

light photoreceptor pigment involved in phototropism (for review, see Pohl and Russo 1984). Thus far, all have been notably unsuccessful. However, these unsuccessful attempts underscored the inability to demonstrate that any potential assay was in fact related to the physiological photoreceptor pigment and not some other pigment with similar absorption characteristics but uninvolved in phototropism. This situation gave rise to efforts to identify mutants in *Phycomyces* and *A. thaliana* with alterations in the photoreceptor pigment for phototropism.

A significant percentage of the unsuccessful efforts to identify the photoreceptor pigment have been based on the assumption that the response is controlled by a single pigment. Although this was a reasonable initial assumption, it has proven to be incorrect for *Phycomyces* and *A. thaliana*, the two organisms in which the assumption has been rigorously tested. Galland and Lipson (1985a,b, 1987) reached the conclusion that multiple photoreceptor pigments are involved in *Phycomyces* phototropism, based on their comparative analysis of action spectra for the wild type and several mutants. The same conclusion has been reached for *A. thaliana*, based on an analysis of the wavelength dependence for the fine structure of the fluence response relationship (Konjević et al. 1989b). The *A. thaliana* mutant, JK224, exhibits an apparent alteration in one photoreceptor pigment while a second pigment appears unchanged (Konjević et al. 1992). Previous efforts had been directed toward identification of *the* photoreceptor pigment. Recognition that tropisms are mediated by multiple photoreceptor pigments should greatly assist toward real progress in the identification of these pigments.

Recently, Briggs and co-workers have made what appears to be major progress toward identification of one of the photoreceptor pigments. An approximately 120-kD protein associated with the plasma membrane is phosphorylated during exposure to blue light (Gallagher et al. 1988). Although first observed in pea, phosphorylation of this protein is also seen in a number of other plants, including *A. thaliana* (Reymond et al. 1992). Phosphorylation is seen at approximately wild-type levels in JK218, an *A. thaliana* mutant that exhibits no phototropism. This demonstrates that phosphorylation is not a consequence of phototropism. However, phosphorylation is seen at approximately 5% of the wild-type level in JK224, which is thought to be a photoreceptor pigment mutant (see above). Thus, the blue-light-induced phosphorylation of a 120-kD protein appears to be closely connected to one specific blue-light photoreceptor pigment that mediates phototropism in *A. thaliana* (Reymond et al. 1992). This protein behaves as a kinase that phosphorylates itself (Short and Briggs 1990). The blue-light-absorbing chromophore appears to be specifically associated with the kinase in solution, but it is not yet certain that the chromophore migrates with the kinase in an SDS gel.

The next obvious step is to clone the gene for the 120-kD protein. If transforming the mutant, JK224, with the wild-type gene restores wild-type levels of blue-light-induced phosphorylation and wild-type phototropism, it can then be concluded that the 120-kD protein is required for phototropism, and the chromophore associated with this autophosphorylatable protein is likely to be one of the phototropism photoreceptor pigments. The remaining challenge will then be to identify the other phototropism photoreceptor pigment(s).

The recent identification of a gene coding for a blue-light photoreceptor pigment involved in the blue-light suppression of hypocotyl elongation in *A. thaliana* (Ahmad and Cashmore 1993) is of considerable interest. Although the phototropism photoreceptor pigments for phototropism appear to be distinct from the hypocotyl suppression photoreceptor pigments (Liscum et al. 1992), they may belong to the same family of related pigments. Thus, we may finally be approaching the elusive blue-light photoreceptor pigments in phototropism.

## Fluence Response Relationship and Adaptation

Fluence response relationships for phototropism have been measured for several plant species (for review, see Pohl and Russo 1984). However, most studies have been descriptive in nature with only a few (see, e.g., Zimmermann and Briggs 1963a,b; Iino 1987; Janoudi and Poff 1991) attempting to characterize the factors contributing to the complexity of the fluence response relationships.

The basic features of a typical fluence response relationship include a bell-shaped curve describing what is known as first positive phototropism, an indifferent zone where little or no measurable response is observed, and an ascending curve describing second positive phototropism in response to prolonged exposures to blue light (Fig. 2). It is important for two reasons to understand the basis of this complexity. First, the complexity has been a major impediment to an understanding of phototropism for a number of decades and has led to a study of only first positive phototropism (see, e.g., Steinitz and Poff 1986), with second positive phototropism, the response under field conditions, being virtually ignored. Second, recent efforts to understand this complexity have identified elements in phototropism that can now be specifically studied (Janoudi et al. 1992). The major factor contributing to the complexity of the phototropic response is thought to be photosensory adaptation (Iino 1987; Galland 1991; Janoudi and Poff 1991, 1992; Janoudi et al. 1992). We analyze each of the components of the fluence response relationship and discuss the possible underlying mechanisms with particular emphasis on the process of adaptation in phototropism.

## First Positive Phototropism

Phototropic curvature in response to short irradiations at low fluences is referred to as first positive phototropism. In general, above a specific threshold fluence, curvature increases to a maximum with increasing fluence of blue light and then decreases almost symmetrically to reach the zone of indifference (Fig. 2). In the ascending arm of first positive phototropism, curvature appears to be the result of a simple photochemical reaction dependent on the number of unilateral quanta received by the plant. Thus, the Bunsen-Roscoe law of reciprocity (Bunsen and Roscoe 1863) is valid in the ascending arm of first positive phototropism (Janoudi and Poff 1990). This law states that there is a reciprocal relationship between the fluence rate of the light and the time of irradiation, such that a constant response is observed for a particular fluence. Fluence is the product of fluence rate (referred to as intensity in the older literature) multiplied by the time of irradiation (Janoudi and Poff 1990). Therefore, in modern terms, fluence translates into the number of quanta. Since curvature in the ascending arm of first positive phototropism depends on fluence (regardless of the fluence rate used), the plant is act-

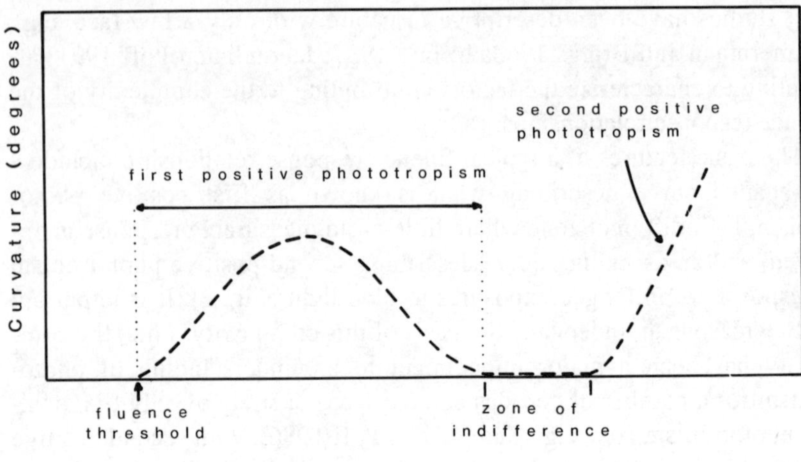


*Figure 2* Diagrammatic fluence response relationship for phototropism.

ing as a quantum counter. Recent studies (Galland and Lipson 1987; Konjevic et al. 1989b, 1992) have shown that reciprocity fails at high fluence rates. This failure has been attributed to the involvement of multiple photoreceptor pigments (see discussion of multiple pigments above). Reciprocity also fails in second positive phototropism, but the explanation for this failure in second positive phototropism is the time requirement of adaptation.

One would expect the fluence response relationship for a simple photochemical reaction to rise from a fluence threshold to a plateau at saturation. Thus, whereas curvature in the ascending arm of first positive phototropism may be accounted for by a relatively simple photochemical reaction, some additional factor or factors must be invoked to account for the descending arm. Several possibilities could account for this decrease in curvature. One possibility is that the photoreceptor pigment is bleached by light and that this bleaching increases with increasing fluence. Bleaching the photoreceptor pigment would decrease the pool of available active pigment and thus decrease perception of the stimulus at higher fluences.

It is known that blue light causes desensitization, which is a decrease or loss of responsiveness to a subsequent exposure to blue light (Janoudi and Poff 1991). Thus, a preirradiation with blue light leads to a decrease in the magnitude of the response, with no apparent change in sensitivity (shift in fluence threshold) that would indicate a decreased pool of available pigment (Janoudi and Poff 1991). In contrast, maize exhibits a change in sensitivity following a preirradiation with blue light (Iino

1988). Therefore, bleaching of the photoreceptor pigment could be the basis for the descending arm in first positive phototropism in maize but not in *A. thaliana*. However, since the shape of the fluence response relationship is essentially identical in a wide variety of organisms, one expects to find the same basis for its characteristics in all of those organisms. Therefore, bleaching is not the most probable explanation for the descending arm of first positive phototropism.

A second possible explanation for the descending arm in first positive phototropism is based on the premise that curvature is the consequence of the difference in growth rate on the shaded and lighted sides of the shoot. At high light levels (a fluence on the descending arm), sufficient light may penetrate the shoot for photoinduction of phototropism on the shaded side. This photoinduction would induce curvature in the opposite direction from photoinduction on the lighted side. Thus, the difference in photoinduction, and therefore difference in growth rate, on the two sides would decrease, leading to a decrease in curvature toward the light source (Poff 1983). However, the fluence range for first positive phototropism (the width of the first positive curve from the fluence threshold to the zone of indifference) would be expected to be related to the optical density of the responding organ. In fact, this fluence range is the same for a wide variety of material from the shoot of mung bean (Konjević et al. 1989a) to that of *A. thaliana* (Steinitz and Poff 1986; Khurana and Poff 1989). A demonstration that these plants in fact have significantly different optical cross sections would constitute evidence that this possible explanation for the descending arm in first positive phototropism is incorrect.

A third possible explanation for the descending arm in first positive phototropism is that the level of some component of the signal transduction chain, downstream from the photoreceptor, is decreasing and becoming limited in response to irradiation with blue light. This hypothesis is consistent with the effector adaptation observed in *A. thaliana* (Janoudi and Poff 1991).

*The Zone of Indifference*

This refers to a region of the fluence response relationship where minimal or no phototropic curvature is observed. The fluence range for this region is dependent on the fluence rate of the light used to induce curvature and increases with increasing fluence rate. The zone of indifference may be the consequence of a time requirement for the initiation of second positive phototropism (see below). During this time, recovery from blue-light-induced desensitization could occur. Thus, following an irradiation, the plant enters a refractory period during which it is insensitive to further stimulation (Galland 1991) and does not respond phototropically until recovery occurs. The time requirement permits that recovery. It then follows that the range of fluence over which the plant is indifferent would be a function of fluence rate, since the time interval, not the fluence, is the critical element.

## Second Positive Phototropism

Phototropism in response to long irradiations is referred to as second positive phototropism. The response in second positive phototropism is a function of the light stimulus duration and of fluence rate. A minimum time of irradiation, a time threshold, is required before second positive phototropism is exhibited (Janoudi and Poff 1990). This time threshold varies between species but ranges between 4 minutes and 60 minutes (Briggs 1960; Dennison 1979; Pohl and Russo 1984; Janoudi and Poff 1990; Janoudi et al. 1992). The time threshold for second positive phototropism also varies depending on the irradiation history of the plant. When etiolated *A. thaliana* and tobacco seedlings are preirradiated with red light 2 hours before stimulation with phototropic light, the time threshold for second positive phototropism is decreased by a factor of four (Janoudi et al. 1992).

For the second positive response to be exhibited, the fluence threshold *and* the time threshold must be exceeded (Janoudi and Poff 1990). Thus, first positive phototropism exhibits *only* a fluence threshold; second positive phototropism exhibits *both* a fluence threshold *and* a time threshold. Because the limiting fluence threshold is the same for first and second positive phototropism (Janoudi and Poff 1990), it is unnecessary to postulate different photoreceptor pigments for the two responses. Thus, although the evidence indicates multiple photoreceptor pigments in phototropism (see above), it is unlikely that first and second positive phototropism are mediated by different pigments.

## Adaptation and the Fluence Response Relationship

Adaptation in phototropism is well documented (Blaauw and Blaauw-Jansen 1970; Shropshire 1979; Iino 1987, 1988; Galland 1991; Janoudi and Poff 1991, 1992). Upon exposure to light, a plant's sensitivity and/or responsiveness to a subsequent exposure to light may change (Galland 1991; Janoudi and Poff 1991). Thus, the plant adapts in response to the initial light exposure. Two types of adaptation have been defined by Galland (1991). In one type, sensor adaptation, the range of the sensor, i.e., its sensitivity, is modulated. The other type, effector adaptation, involves a modulation in the magnitude of the response. *A. thaliana* exhibits effector adaptation in phototropism, whereas maize exhibits sensor adaptation in phototropism. Immediately following an initial exposure to blue light from above, *A. thaliana* loses its phototropic capacity to respond to unilateral light. Following a nonresponsive period, this capacity is slowly regained. Even more slowly, the responsiveness of the plant increases to a level greater than the responsiveness before the initial exposure to light. Thus, in *A. thaliana* phototropism, the process of adaptation consists of desensitization, a refractory period, a period of recovery, and enhancement of phototropic responsiveness (Janoudi and Poff 1991).

Desensitization is induced by blue light in *A. thaliana* (Janoudi and Poff 1991) but by either blue or red light in maize (Iino 1988). In contrast, enhancement in both plants appears to be a blue and red light response which is mediated by phytochrome (Chon and Briggs 1966; Janoudi et al. 1992). Which phytochrome mediates enhancement and the mechanism of this mediation are both unknown.

Adaptation introduces complexities in the fluence response relationship. This is due in part to the fluence dependence of adaptation. The degree of desensitization, the time required for recovery, and the level of curvature enhancement are directly proportional to the fluence of blue light received by the plant (Janoudi and Poff 1991). Fluences of blue light that induce phototropism overlap with fluences that can induce desensitization. Concurrently, similar fluences of blue light can also induce enhancement (Janoudi and Poff 1991, 1992). Thus, blue light is perceived by the plant as a signal that induces and enhances phototropism and may simultaneously cause desensitization, which lowers the plant's responsiveness to blue light. Therefore, the net response to a given phototropic stimulus is a function of each component of adaptation, and these components exhibit different time constants.

### Components of Adaptation

Among the various components of adaptation, desensitization is the most rapid, occurring within 2 seconds following an exposure to blue light (Janoudi and Poff 1991). Recovery from desensitization requires several minutes, and enhancement requires up to 2 hours for completion. Consequently, in first positive phototropism, where exposure times are relatively short, a plant can respond to the curvature inducing, and also desensitizing, light stimulus. Recovery and enhancement, however, typically require longer times and are not likely to have an appreciable effect during these short irradiations. Thus, the shape of the fluence response curve in first positive phototropism reflects the net result of the rapid components, curvature induction and desensitization, in phototropism.

The zone of indifference is a consequence of the time threshold for second positive phototropism. This establishes a period during which the plant is receiving light but is not responding with phototropic curvature. It is not known what factors set the time threshold. However, the time threshold can be decreased, although not eliminated, by preirradiation with red light. As expected, red-light preirradiation also decreases the fluence range of the indifferent zone (Janoudi et al. 1992) and decreases the time requirement for recovery from desensitization (Janoudi and Poff 1993). However, the time threshold is always longer than the time requirement for recovery. Thus, factors other than recovery from desensitization must be involved in setting the time threshold.

In contrast with first positive phototropism, during second positive phototropism, plants are exposed to irradiations sufficiently long to allow enhancement to occur. Because the degree of enhancement is time-dependent, the magnitude of phototropic curvature is also time-dependent. Thus, at a given fluence, a plant has more time for enhancement and curves more when irradiated at a low fluence rate than when irradiated at a higher fluence rate.

In summary, the complexity of the fluence response relationship for phototropism is a consequence of the process of adaptation in which a plant's sensitivity and/or responsiveness to blue light is altered. This alteration results in differences in the magnitude of phototropic curvature depending on the irradiation history of the plant and the total fluence and fluence rate of the light stimulus. With increasing knowledge of the specific elements in adaptation, efforts can now be specifically directed toward their genetic manipulation and molecular identification.

## Elements of the Transduction Chain

Not one element in the phototropism transduction chain is known with certainty. Much of the effort in this arena has centered on the Cholodny-Went theory, which is based on the independent work of Cholodny (1927) and Went (1928). In this theory, tropistic curvature is a consequence of the movement of a growth substance (auxin) from one side of the organ to the other side. An increase in auxin concentration leads to an increased cellular elongation, whereas a decrease in auxin concentration leads to a decreased cellular elongation. This differential in cellular elongation on the two sides of the organ results in curvature of the organ. In the case of phototropism, the auxin is proposed to move from the lighted side to the shaded side. This results in increased cellular elongation on the shaded side, decreased cellular elongation on the lighted side, and thus, curvature toward the source of light. In photo- and gravitropism, the unequal distribution is proposed to result from a transverse polarization of the cells, which results in lateral transport of auxin (Went and Thimann 1937). Considerable controversy continues to surround the Cholodny-Went theory. This controversy has been best summarized recently in a multi-author forum (Trewavas et al. 1992).

Alternatives to the Cholodny-Went theory are the theory of Boysen Jensen (1928) and the theory of Blaauw (1918) and Paál (1919). The Boysen Jensen theory predicts an increased rate of cellular elongation on the shaded side of the organ in phototropism with no change on the lighted side (Boysen Jensen 1928). The Blaauw/Paál theory predicts a general decrease in cellular elongation with a greater decrease on the lighted side than on the shaded side (Blaauw 1918; Paál 1919). Thus, as has been discussed by Pohl and Russo (1984), these theories should be easily distinguished by careful measurements of growth rate on the two sides of the curving organ. The most convincing evidence supporting the Cholodny-Went theory consists of such data showing that growth on the lighted side of a corn coleoptile decreases during curvature while that on the shaded side increases (Iino and Briggs 1984). This follows the predictions of the Cholodny-Went theory, but not of the alternate theories. Similar experiments result in the same basic conclusions with *A. thaliana* (Orbović and Poff 1993). In addition, direct measurements of auxin also support the Cholodny-Went theory (Gardner et al. 1974; Iino 1991). Such direct measurements would be technically quite difficult to make in the hypocotyl of *A. thaliana* because of its small size. An indirect indication of auxin redistribution is the report that a small auxin up-regulated RNA (SAUR) is differentially induced during phototropism of transgenic tobacco seedlings (Li et al. 1991).

A number of papers have argued recently against the Cholodny-Went theory. These arguments have been based on evidence or proposals that growth regulators other than auxin are involved in tropisms (Bruinsma et al. 1975; Hasegawa and Togo 1989), that the concentration of auxin on the lighted side of the shoot is the same as that on the shaded side (Togo and Hasegawa 1991; Hasegawa and Yamada 1992), and/or that the difference in auxin concentration is insufficient to account for the observed changes in growth rate on the two sides.

There is evidence that cytosolic $[Ca^{++}]$ may also play a role in the transduction chain for phototropism. Gehring et al. (1990) reported that cells on the shaded side of unilaterally irradiated maize coleoptile tips showed rapid changes in cytosolic $[Ca^{++}]$ and pH. There is also some evidence for a role of $Ca^{++}$ in gravitropism (see below). A major challenge for those interested in tropisms is to devise approaches with which one can distinguish between fortuitous and causal connections between fluctuations in $Ca^{++}$ concentrations and tropistic responses.

Insight into an initial signal transduction step in phototropism may have been provided by the demonstration that the 120-kD protein, which is phosphorylated by blue light, is related to phototropism in *A. thaliana* (Reymond et al. 1992; see above). The protein could conceivably be phosphorylated by blue light without being involved in any of the physiologically relevant steps in phototropism; phosphorylation of the protein could be the initial step in the induction of phototropism; or phosphorylation could be involved in desensitization rather than the induction of phototropism. At present, there are no data in direct support of any one of these possibilities. Determining the role of the autophosphorylation will be a challenge.

A second blue-light-activated protein (a G protein) has been isolated from the plasma membrane of etiolated pea buds (Warpeha et al. 1992). This might be related to phototropism or to another blue-light physiological response. Items such as this can best be resolved through the judicious use of mutants. Ideally, this would involve the use of tropism mutants and also mutants with alterations in specific components in the G-protein response mechanism. Although a number of phototropism mutants have been described previously (see below), mutants with alterations in possible transduction elements such as the G-protein response mechanism or $Ca^{++}$-transport elements will be more difficult to obtain.

The phototropism mutants thus far described fall into several categories: amplitude mutants, null mutants, and ostensible photoreceptor pigment mutants. The amplitude mutants exhibit a fluence response relationship which is the same as that of the wild-type parent; only the amount of response is diminished (Khurana et al. 1989). A single null mutant has been identified. This mutant, JK218, exhibits no curvature following irradiations up to 2 hours in length (Khurana and Poff 1989). Perhaps the most interesting mutant described thus far has been JK224 (Khurana and Poff 1989). This strain exhibits normal first and second positive phototropism, but the fluence threshold for first positive phototropism is about 30 times higher than that of the wild-type parent. This is the expected phenotype for a mutant with an alteration in the photoreceptor pigment (Khurana and Poff 1989). This strain has been used for showing the relationship between phosphorylation and phototropism (see above, Identity of the Photoreceptor Pigment).

### Mechanism for Measuring Light Direction

A plant detects the direction from which the light is incident and grows toward that light. This is accomplished in the absence of any specialized photoreceptor organelle such as an eye. It is likely that the plant has a mechanism for establishing a gradient in the quantum concentration. This light gradient would be translated into a difference in concentration of some photoproduct on the two sides of the plant shoot, and this could result in curvature due to unequal growth rates (cellular elongation rates) of the two sides. Two models have been proposed for establishing a light gradient. One is based on refraction; the second is based on screening.

In the refraction model, light is refracted at the air/organism interface and thereby focused onto the distal side of the organism. Thus, a higher quantum density is established on the distal side than on the proximal side of the organism. There is considerable evidence to support this model for the very small and rather transparent organs supporting the sporangia of some phototropic fungi. For example, if the sporangiophore of *Phycomyces blakesleeanus* is submerged in oil with an optical index of refraction close to that of the cytoplasm, its ability to grow toward a unilateral light source is lost (Banbury 1959). Similar experiments have not been successful with plant shoots.

In the screening model, a difference in quantum density is established across the organ by screening of light within the organ. Thus, a higher quantum density is established on the proximal side than on the distal side of the organism. The screening is a consequence of scattering and of absorption. This appears to be the mechanism whereby plants establish a light gradient. The evidence for this has been obtained by manipulating the absorption component of screening. Use of inhibitors and mutants decreasing the visible-absorbing carotenoids has been shown to cause a decrease in the amplitude of phototropic curvature in corn (Vierstra and Poff 1981; Piening and Poff 1988). These treatments are specific for phototropism. They have no effect on the threshold for phototropism, and they affect the amplitude of curvature only if phototropism is induced by wavelengths that are absorbed by the carotenoids. Therefore, it has been concluded that the carotenoids do not function as phototropism photoreceptor pigments, but rather as screening pigments (Vierstra and Poff 1981; Piening and Poff 1988). *A. thaliana* presents a special case because of the small size of the etiolated hypocotyl. The carotenoid functions as a screening pigment but the scattering- and absorption-based screening is barely enough to establish a light gradient across the hypocotyl (K.L. Poff, unpubl.). Thus, the *A. thaliana* hypocotyl may be the smallest material capable of using the screening mechanism for measuring light direction.

## GRAVITROPISM

Gravitropism is the growth response whereby a plant orients with respect to the gravity vector. This can be observed in *A. thaliana* as an upward curvature of the hypocotyl and flowering stalk and a downward curvature of the root. Considering the large amount of time that has been invested in the study of gravitropism and the remarkable number of papers with gravitropism as their major topic, an incredibly small amount of definitive information is available concerning the mechanism for this sensory response. As is the case with phototropism, research has been concentrated at the two ends of the transduction sequence, graviperception and the growth response.

The only mechanism thus far conceived by which gravity can be measured is its effect on a mass, lending that mass the attribute of weight. Perception of gravity by a plant is usually divided into several steps, the first of which is susception, the exertion of weight or motion of some mass acted upon by gravity. Perception is considered to have occurred after susception has altered some biochemical/physiological step. The transduction pathway includes all of the processes from susception to the final growth response.

Unfortunately, gravity as an environmental stimulus is extremely difficult to study because every mass is acted upon by the all-pervasive $1g$ of earth's gravity. Thus, the experimenter can manipulate the direction of the gravitational vector and subject the organism to a gravitational acceleration above $1g$ through centrifugation, but the minimum gravitational stimulus is $1g$ in an earth-based experiment. To avoid this constraint, the clinostat is used in an earth-based experiment to slowly rotate the organism such that the organism experiences a constantly changing gravitational direction. The clinostat does not eliminate the $1g$ force of gravity. It only eliminates its constant directionality (Brown et al. 1976). Recently, access to a micro-$g$ environment using parabolic flights (Volkmann et al. 1986) and space capsules has become available (Halstead and Dutcher 1987). Unfortunately, such access is limited at present. In addition, every mass in the organism is acted upon by gravity; there is no specificity at this level. A number of reviews are available on gravitational responses by plants and should be consulted for further information (Audus 1975; Volkmann and Sievers 1979; Feldman 1985; Pickard 1985; Bjorkman 1988; Sack 1991).

## Susception

Two major candidates have been proposed as the susceptor in gravitropism. The most widely accepted candidate has been the amyloplast. According to the starch-statolith model, these starch-containing plastids physically respond to gravity through sedimentation. This sedimentation activates some cellular response mechanism which initiates the signal transduction chain leading to gravitropic curvature

(Sack 1991). Considerable evidence has been advanced supporting the starch-statolith model (Haberlandt 1905 as reviewed in Audus 1975; Heathcote 1981; Hillman and Wilkins 1982), but at least much of this evidence is correlative (see, e.g., Hillman and Wilkins 1982).

Regrettably, the use of starchless or starch-deficient mutants to determine the role of plastid density and plastid sedimentation in gravitropism has not yet provided definitive evidence for this model. A starch-deficient tobacco mutant was reported to be reduced in gravitropic responsiveness in roots (Kiss and Sack 1989) and hypocotyls (Kiss and Sack 1990). The use of starchless *A. thaliana* mutants has proven more controversial. In a starchless *A. thaliana* mutant altered in phosphoglucomutase, Caspar and Pickard (1989) reported that gravitropism was virtually unimpaired despite the lack of detectable amyloplast movement during a 2-hour period of inversion. However, further work with the same mutants showed a small reduction in sensitivity (higher threshold) to gravity corresponding to the reduced plastid density (Kiss et al. 1989). On the basis of these data, Kiss et al. (1989) suggested that the starchless plastids could still function as statoliths but with a decreased sensitivity to gravity. This possibility that the amyloplast itself could function as a susceptor in the absence of the dense starch grain complicates the interpretation of these mutant studies.

An alternative model of gravity perception proposes that the gravity-directed distribution of the weight of the protoplasm could serve as a susceptor. A version of this was originally proposed by Czapek (cited in Wayne et al. 1990), who suggested that the weight of the protoplasm on the lower cell membrane could serve as a mechanism for gravity detection. A more recent model has been proposed by Wayne et al. (1990) based on experimental data from a characean alga that contains no visible statoliths. They present calculations to show that the movement of the plasma membrane induced by protoplasm weight redistribution could produce sufficient potential energy to open ion channels in the plasma membrane. Although this argument is convincing for the mass of the 4-cm- to 6-cm-long cell used by Wayne et al. (1990), it remains to be demonstrated that protoplasm settling in a considerably smaller higher plant cell would produce a detectable change in potential energy.

One experimental approach that could distinguish between the two models would be the use of plants with increased cytoplasmic density. If, in fact, the function of the starch-containing amyloplasts is to increase the signal-to-noise ratio in gravity detection by increasing the total density of the protoplast (Wayne et al. 1990), then any increase in cytoplasmic density should increase gravitropic sensitivity. In contrast, in the starch-statolith model, increasing the density of the cytoplasm should increase the buoyancy of the amyloplasts, and the decrease in their settling would decrease the sensitivity to gravity.

The weight or movement of the cellular component functioning as the gravitropic susceptor must itself be detected by an additional cellular component. Components that have been suggested to fill this role include the endoplasmic reticulum (Juniper 1976), the cytoskeleton (Bjorkman 1988), and stretch-activated ion channels in the plasma membrane (Wayne et al. 1990). It was suggested that the amyloplasts settled onto the endoplasmic reticulum, due largely to the spatial proximity between the endoplasmic reticulum and the amyloplasts and to the disruption of this orientation in an agravitropic mutant (Olsen and Iversen 1980) and in a wild-type plant following prolonged exposure on a clinostat (Hensel and Sievers 1980). More recently, it has been noted that the largely polar orientation of organelles in many plant cells leads to the possibility that the weight of the susceptor is detected by the cytoskeleton (Sack 1991). The cytoskeleton could detect movement of the susceptor either through a shearing effect or through a pulling of a susceptor attached directly to the cytoskeleton (Bjorkman 1988). Disruption of the cytoskeleton has shown that structural polarity of gravity-sensing cells is necessary for gravitropism (Hensel and Sievers 1980). However, no evidence is available to determine if the cytoskeleton is directly involved in reception or transduction of the gravitropic signal, or if the cytoskeleton serves to maintain cellular polarity, which is necessary for gravity detection or response.

### Site of Perception

In higher plant systems, the root tip has been shown to be the site of gravity sensing in many classical experiments (for review, see Jackson and Barlow 1981; Poff and Martin 1989). Many authors (see, e.g., Juniper et al. 1966; Pilet 1982) have extrapolated from correlative evidence that the root cap is the site of perception. Some authors (see, e.g., Konings 1968) have even ascribed this function to the columella cells within the root cap. However, the only data specifically directed toward the site of perception ascribe this function to the root tip, which includes the root cap in addition to the zone of division, and possibly part of the growing zone (Poff and Martin 1989).

In hypocotyls and coleoptiles, there is little convincing evidence for a specific region along the shoot axis that is required for gravity sensing. The region of maximum sensitivity has been directly shown to be present in the most strongly growing region of the elongation zone (Sack 1991). Less direct studies have shown that the surgical removal of the tips of stems renders the stems less responsive to gravity (MacDonald and Hart 1985). However, surgical removal of a segment of the plant, thereby decreasing the gravitropic response, shows only that the segment is required for the response and does not show that perception occurs in that segment. Consistent with this critique is the observation that the gravitropic response of the de-tipped shoots has frequently been restored by the application of auxin (Sliwinski and Salisbury 1984). If perception had been diminished by de-tipping, then one would have to argue that the application of auxin restored perception. It seems far more likely that auxin is involved in signal transduction than in perception.

### Elements of the Transduction Chain

As gravitropism is the result of differential growth on opposite sides of a responding organ, it is not surprising that a large number of growth regulators have been associated with this response. These include abscisic acid and auxin (Pickard 1985). There is considerable evidence that auxin has a role in gravitropism (Evans 1991; Trewavas et al. 1992), although there are discrepancies between the timing of auxin asymmetry development and the gravitropic curvature (Firn and Digby 1980). The role of auxins in gravitropism has received strong support from studies of *A. thaliana* mutants. Mutations at the *aux1* locus confer both resistance to exogenous auxin and agravitropism (Mirza et al. 1984). Molecular approaches have yielded additional support for the role of auxin in gravitropism. Auxin-regulated mRNAs, or SAURs, are more abundant in the side of the gravitropically responding organ that is expected to contain the higher auxin concentration (Li et al. 1991). Together with the observation that a SAUR is found in reduced amounts in a gravitropism-minus *A. thaliana* mutant (Gil et al. 1994), this supports some role of auxin in gravitropism.

Calcium has also been implicated in gravitropism (Lee et al. 1983). Studies that used fluorescent cation indicators to detect free cytosolic calcium have shown rapid (within about 3 min) increases in calcium concentration in the lower side of a maize coleoptile following a 90° alteration in gravity vector (Gehring et al. 1990). Alterations of calcium levels have been shown in gravistimulated oat coleoptiles, but the changes are less rapid than those reported in maize. The differences in kinetics could be due to differences in the methods used to detect calcium (Slocum and Roux 1983). In addition, calmodulin, a protein that binds and regulates intracellular calcium, has been suggested to be involved in gravitropism. The use of calmodulin inhibitors has been shown to inhibit gravitropism without alteration of growth rate (Bjorkman and Leopold 1987a). However, calcium plays a central role in the regulation of a number of biochemical and physiological processes. It would seem surprising if calcium were not involved in gravitropism, but also most surprising if calmodulin inhibitors were specific for gravitropism.

Several investigators have found changes in the electrical properties of plants and have suggested that these changes are associated with gravitropism (see, e.g., Bjorkman and Leopold 1987b; Ishikawa and Evans 1990; Imagawa et al. 1991). Thus far, we are not aware of data proving that these electrical properties are linked to gravitropism. The correlative evidence is consistent with gravity-induced changes in these properties, but this does not constitute evidence that there is an association with the specific, gravity-sensing system, gravitropism. New approaches are required to establish causal relationships with basic physiological/cellular mechanisms.

Mutants of *A. thaliana* with alterations in their gravitropic responses (see, e.g., Bullen et al. 1990) may prove to be the tools with which gravitropism can finally be dissected. Such mutants offer the potential for establishing the involvement of a particular component in gravitropism, for establishing the relationship between gravitropism and other tropisms, and ultimately for the identification of the molecular components in the signal transduction pathway.

In summary, it appears likely that gravity is perceived by the plant through the weight of some susceptor (possibly the starch-containing amyloplast or possibly the entire cell). The weight of the susceptor is detected by its effect on some cellular component (endoplasmic reticulum, cytoskeleton, stretch-activated channel, etc.). The transduction chain may involve changes in the electrical properties of the cells and uses one or more plant growth regulators to modulate growth rate in the growing zone of the root or shoot.

## THERMOTROPISM

Thermotropism is the growth response whereby a plant orients to temperature. The very existence of thermotropism has been questioned in the past. Although the first experiments on thermotropism date back to 1884 (Hooker 1914), the literature on this response has been contradictory, with many reports of positive and negative thermotropism, and even reports of the absence of any thermotropic response (for review, see Aletsee 1962). Recently, corn roots were shown to exhibit thermotropism when grown in a horizontal thermal gradient perpendicular to the vertical roots (Fortin and Poff 1990). At lower temperatures (9°C–21°C), the roots are positively thermotropic, and the amplitude of the response increases with increasingly steep thermal gradient (Fortin and Poff 1991). There is no thermotropism at temperatures ranging from 24ºC to 30ºC, depending on the gradient strength. At high temperatures (31ºC, 39ºC, and 42ºC) and relatively steep gradients, the growth is negatively thermotropic (Fortin and Poff 1991). These results provide a strong basis for reevaluating the contradictory reports in the early literature.

Thermosensing can be difficult to prove. It is not easy to distinguish between a thermal effect on general metabolism and a more specific sensing of temperature, because heat affects the rate of every chemical reaction in the plant. Fortunately, it has been easy to distinguish between the two effects in corn. Fortin and Poff (1991) demonstrated that over the temperature range in which corn roots are positively thermotropic, their growth rate increases with increasing temperature (Fortin and Poff 1991). If the directed growth response were simply the consequence of temperature on general metabolism, the root would be expected to be negatively thermotropic, because the growth rate would be expected to be greater on the warmer side of the root. Thus, the observation that the roots are positively thermotropic at temperatures that increase the growth rate argues for a relatively specific sensing of temperature.

To our knowledge, there has been only one attempt to observe thermotropism in *A. thaliana*. Fortin (1989) was unable to measure any significant thermotropic response of hypocotyls or roots. However, this was tested at only one temperature (21ºC) and one thermal gradient strength (4.2ºC cm$^{-1}$). Interestingly, the response of a gravity-insensitive mutant was slightly but significantly higher than that of the wild type. This implies that root behavior by *A. thaliana* in a temperature gradient is not independent of gravity sensing. This illustrates how profitable it may be to look for thermotropism in *A. thaliana* at a number of temperatures and gradient strengths. It would be particularly advantageous to describe thermotropism in *A. thaliana* because of the relative ease of mutant screening and the value of mutants for any investigations of the thermosensory transduction chain.


## HYDROTROPISM

Hydrotropism is the growth response whereby a plant orients to water. Although described early in the history of plant physiology by Darwin (1896) and Hooker (1914), hydrotropism, like thermotropism, has received little attention because of technical and conceptual difficulties. It is difficult at best to imagine studying the sensory response to the most prevalent compound in the plant's internal (and frequently external) environment. Moreover, this is a compound that cannot be eliminated while maintaining physiological conditions.

The more recent literature has concentrated on demonstrating the existence of the phenomenon. It is clear that corn roots exhibit curvature in response to a differential in relative humidity (Takahashi and Scott 1993). In corn, the orientation of the root is a function of its hydrotropic and gravitropic responses (Takahashi and Scott 1991). In addition, the agravitropic pea mutant, ageotropum, exhibits normal hydrotropism (Takahashi et al. 1992). However, the tools are not yet available to demonstrate that hydrotropism and thermotropism are separate responses. Ideally, this would be accomplished with a mutant lacking only one of the responses.

The site of perception and elements in the transduction chain are open problems at this time. Surgical removal of the root cap renders the root nonhydrotropic, demonstrating that the root cap is required for hydrotropism (Takahashi and Scott 1993). As discussed above for gravitropism, such data are not sufficient to conclude that the root cap is the site of perception of water.

## CONCLUSIONS AND PERSPECTIVES

In this chapter, we have attempted to present a realistic assessment of the present state of knowledge on plant tropisms. In most instances, this knowledge is quite meager or is supposition with little support. The tools of genetics and molecular biology show great promise for considerable advances in our understanding of tropisms in *A. thaliana*, if based on an understanding of the physiology and biophysics. It is now abundantly clear that the apparent simplicity of tropisms has been a fantasy. In fact, every model system appears more complex following serious study. The available evidence indicates that tropisms in *A. thaliana* are controlled by a complex network leading from a number of sensory inputs to a differential growth response (Fig. 3). We are confident that great strides will be made toward a thorough understanding if the true physiological complexities are sufficiently well described to permit careful definition of the experimental system. For example, a mutant screen will always give exactly the mutants being screened for, although these are not necessarily the mutants for which the screen was ostensibly designed. Knowledge of the physiology is a prerequisite for intelligent use of genetics and molecular genetics, and for the eventual understanding of tropisms at the molecular level. If this chapter succeeds in stimulating the application of new approaches to the exciting questions in tropisms, we will have achieved our objectives.

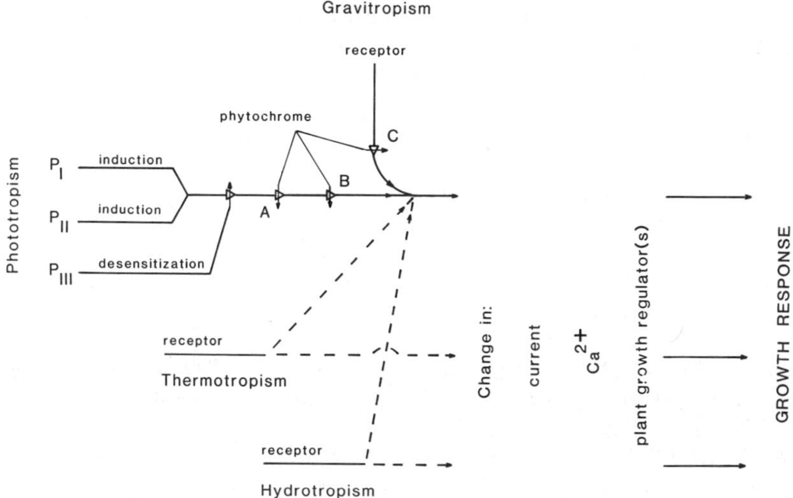


*Figure 3* Summary network for plant tropisms. (*A*) Amplifier representing enhancement in phototropism; (*B*) amplifier representing the timer for the time threshold in second positive phototropism; (*C*) amplifier representing enhancement in gravitropism; dashed lines represent uncertainty.

## ACKNOWLEDGMENTS

The preparation of this chapter was supported by the U. S. Department of Energy under grant number DE-FG02-90-ER20021 and by the U. S. National Aeronautics and Space Administration under grant numbers NAGW-882 to K.L.P., 61-2750 to E.S.R., and NAGW-1842 to T.K.S.

## REFERENCES


Ahmad, M. and A.R. Cashmore. 1993. Hy4 gene of *A. thaliana* encodes a protein with characteristics of a blue-light photoreceptor. *Nature* **366:** 162–166.

Aletsee, L. 1962. Thermotropismus. In *Encyclopedia of plant physiology* (ed. W. Ruhland), vol. 17, pp. 1–14. Springer-Verlag, Berlin.

Audus, L.J. 1975. Geotropism in roots. In *The development and function of roots* (ed. J.C. Torrey and D.T. Clarkson), pp. 327–363. Academic Press, London.

Banbury, G.H. 1959. Phototropism of lower plants. In *Encyclopedia of plant physiology* (ed. W. Ruhland), vol. 17, pp. 530–578. Springer-Verlag, Berlin.

Bjorkman, T. 1988. Perception of gravity by plants. *Adv. Bot. Res.* **15:** 1–41.

Bjorkman, T. and A.C. Leopold. 1987a. Effect of inhibitors of auxin transport and of calmodulin on a gravisensing-dependent current in maize roots. *Plant Physiol.* **84:** 847–850.

———. 1987b. An electric current associated with gravity sensing in maize roots. *Plant Physiol.* **84:** 841–846.

Blaauw, A.H. 1918. Licht und Wachstum III. *Meded. Landbouwhogesch. Wageningen* **15:** 89–204.

Blaauw, O.H. and G. Blaauw-Jansen. 1970. The phototropic responses of *Avena* coleoptiles. *Acta Bot. Neerl.* **19**: 755–763.

Boysen Jensen, P. 1928. Die phototropishe Induktion in der Spitze der *Avena* coleoptile. *Planta* **5**: 464–477.

Briggs, W.R. 1960. Light dosage and phototropic responses of corn and oat coleoptiles. *Plant Physiol.* **35**: 951–962.

Brown, A.H., A.O. Dahl, and D.K. Chapman. 1976. Morphology of *Arabidopsis* grown under chronic centrifugation and on the clinostat. *Plant Physiol.* **57**: 358–364.

Bruinsma, J., C.M. Karssen, M. Benschop, and J.B. Van Dort. 1975. Hormonal regulation of phototropism in the light-grown sunflower seedling, *Helianthus annuus* L.: Immobility of endogenous indoleacetic acid and inhibition of hypocotyl growth by illuminated cotyledons. *J. Exp. Bot.* **26**: 411–418.

Bullen, B.L., T.R. Best, M.M. Gregg, S.-E. Barsl, and K.L. Poff. 1990. A direct screening procedure for gravitropism mutants in *Arabidopsis thaliana* (L.) Heynh. *Plant Physiol.* **93**: 525–531.

Bunsen, R. and H.E. Roscoe. 1863. Photochemical researches. V. On the direct measurement of the chemical action of sunlight. *Philos. Trans. R. Soc. Lond. B Biol. Sci.* **153**: 139–160.

Caspar, T. and B.G. Pickard. 1989. Gravitropism in a starchless mutant of *Arabidopsis:* Implications for the starch-statolith theory of gravity sensing. *Planta* **177**: 185–197.

Cholodny, N. 1927. Wuchshormone und Tropismen bei den Pflanzen. *Biol. Zentralbl.* **47**: 604–626.

Chon, H.P. and W.R. Briggs. 1966. Effect of red light on the phototropic sensitivity of corn coleoptiles. *Plant Physiol.* **41**: 1715–1724.

Curry, G.M. and H.E. Gruen. 1959. Action spectra for the positive and negative phototropism of *Phycomyces* sporangiophores. *Proc. Natl. Acad. Sci.* **45**: 797–804.

Darwin, C. 1896. *The power of movements in plants.* Appleton, New York.

Delbrück, M. and W. Shropshire, Jr. 1960. Action and transmission spectra of *Phycomyces. Plant Physiol.* **35**: 194–204.

Dennison, D. 1979. Phototropism. In *Encyclopedia of plant physiology* (ed. W. Haupt and M. Feinleib), vol. 7, pp. 506–566. Springer-Verlag, Berlin.

Evans, M.L. 1991. Gravitropism: Interaction of sensitivity modulation and effector redistribution. *Plant Physiol.* **95**: 1–5.

Feldman, L.J. 1985. Root gravitropism. *Physiol. Plant* **65**: 341–344.

Firn, R.D. and J. Digby. 1980. The establishment of tropic curvatures in plants. *Annu. Rev. Plant Physiol.* **31**: 131–148.

Fortin, M.-C. 1989. "Developmental and tropic responses of *Zea mays* (L.) and *Arabidopsis thaliana* (L.) Heynh. to temperature conditions simulating conservation tillage." Ph.D. thesis, Michigan State University, East Lansing.

Fortin, M.-C. and K.L. Poff. 1990. Temperature sensing by primary roots of maize. *Plant Physiol.* **94**: 367–369.

———. 1991. Characterization of thermotropism in primary roots of maize: Dependence on temperature and temperature gradient, and interaction with gravitropism. *Planta* **184**: 410–414.

Gallagher, S., T.W. Short, P.M. Ray, L.H. Pratt and W.R. Briggs. 1988. Light-mediated changes in two proteins found associated with plasma membrane fractions from pea stem sections. *Proc. Natl. Acad. Sci.* **85**: 8003–8007.

Gardner, G., S. Shaw, and M.B. Wilkins. 1974. IAA transport during the phototropic responses of intact *Zea* and *Avena* coleoptiles. *Planta* **121**: 237–251.

Galland, P. 1991. Photosensory adaptation in aneural organisms. *Photochem. Photobiol.*

**54:** 1119–1134.

Galland, P. and E.D. Lipson. 1985a. Action spectra for phototropic balance in *Phycomyces blakesleeanus* dependence on reference wavelength and intensity range. *Photochem. Photobiol.* **41:** 323–329.

———. 1985b. Modified action spectra of photogeotropic equilibrium in *Phycomyces blakesleeanus* mutants with defects in genes *madA*, *madC*, and *madH*. *Photochem. Photobiol.* **41:** 331–335.

———. 1987. Blue-light reception in *Phycomyces* phototropism: Evidence for two photosystems operating in low- and high-intensity ranges. *Proc. Natl. Acad. Sci.* **84:** 104–108.

Gehring, C.A., D.A. Williams, S.H. Cody, and R.W. Parish. 1990. Phototropism and geotropism in maize coleoptiles are spatially correlated with increases in cytosolic free calcium. *Nature* **345:** 528–530.

Gil, P., Y. Liu, V. Orbović, E. Verkanp, K.L. Poff, and P.J. Green. Characterization of the auxin-inducible *SAUR-AC1* gene for use as a molecular genetic tool in *Arabidopsis*. *Plant Physiol.* **104:** 777–784.

Halstead, T.W. and F.R. Dutcher. 1987. Plants in space. *Annu. Rev. Plant Physiol.* **38:** 317–345.

Hasegawa, K. and S. Togo. 1989. Phototropism in hypocotyls of radish. VII. Involvement of the growth inhibitors, raphanusol A and B in phototropism of radish hypocotyls. *J. Plant Physiol.* **135:** 110–113.

Hasegawa, K. and K. Yamada. 1992. Even distribution of endogenous indole-3-acetic acid in phototropism of pea epicotyls. *J. Plant Physiol.* **139:** 455–459.

Heathcote, D.G. 1981. The geotropic reaction and statolith movements following geostimulation of mung bean hypocotyls. *Plant Cell Environ.* **4:** 131–140.

Hensel, W. and A. Sievers. 1980. Effects of prolonged omnilateral gravistimulation on the ultrastructure of statocytes and on the graviresponse of roots. *Planta* **150:** 338–346.

Hillman, S.K. and M.B. Wilkins. 1982. Gravity perception in decapped roots of *Zea mays*. *Planta* **155:** 267–271.

Hooker, J.D.J. 1914. Thermotropism in roots. *Plant World* **17:** 135–153.

Imagawa, K., K. Toko, S. Ezaki, K. Hayashi, and K. Yamafuji. 1991. Electrical potentials during gravitropism in bean epicotyls. *Plant Physiol.* **97:** 193–196.

Iino, M. 1987. Kinetic modelling of phototropism in maize coleoptiles. *Planta* **171:** 110–126.

———. 1988. Desensitization by red and blue light of phototropism in maize coleoptiles. *Planta* **176:** 183–188.

———. 1991. Mediation of tropisms by lateral translocation of endogenous indole-3-acetic acid in maize coleoptiles. *Plant Cell Environ.* **14:** 279–286.

Iino, M. and W.R. Briggs. 1984. Growth distribution during first positive phototropic curvature of maize coleoptiles. *Plant Cell Environ.* **7:** 97–104.

Ishikawa, H. and M.L. Evans. 1990. Electrotropism of maize roots. Role of the root cap and relationship to gravitropism. *Plant Physiol.* **94:** 913–918.

Jackson, M.B. and P.W. Barlow. 1981. Root geotropism and the role of growth regulators from the cap: A re-examination. *Plant Cell Environ.* **4:** 107–123.

Janoudi, A.-K. and K.L. Poff. 1990. A common fluence threshold for first positive and second positive phototropism in *Arabidopsis thaliana*. *Plant Physiol.* **94:** 1605–1608.

———. 1991. Characterization of adaptation in phototropism of *Arabidopsis thaliana*. *Plant Physiol.* **95:** 517–521.

———. 1992. Action spectrum for enhancement of phototropism by *Arabidopsis thaliana* seedlings. *Photochem. Photobiol.* **56:** 655–659.

————. 1993. Desensitization and recovery in phototropism or *Arabidopsis thaliana*. *Plant Physiol.* **101:** 1175–1180.

Janoudi, A.-K., R. Konjević, P. Apel, and K.L. Poff. 1992. Time threshold for second positive phototropism is decreased by a preirradiation with red light. *Plant Physiol.* **99:** 1422–1425.

Juniper, B.E. 1976. Geotropism. *Annu. Rev. Plant Physiol.* **27:** 385–406.

Juniper, B.E., S. Groves, B. Landau-Schachar, and L.J. Audus. 1966. Root cap and the perception of gravity. *Nature* **209:** 93–94.

Khurana, J.P and K.L. Poff. 1989. Mutants of *Arabidopsis thaliana* with altered phototropism. *Planta* **178:** 400–406.

Khurana, J.P., Z. Ren, B. Steinitz, B. Parks, T.R. Best, and K.L. Poff. 1989. Mutants of *Arabidopsis thaliana* with decreased amplitude in their phototropic response. *Plant Physiol.* **91:** 685–689.

Kiss, J.Z. and F.D. Sack. 1989. Reduced gravitropic sensitivity in roots of a starch-deficient mutant of *Nicotiana sylvestris*. *Planta* **180:** 123–130.

————. 1990. Severely reduced gravitropism in dark-grown hypocotyls of a starch-deficient mutant of *Nicotiana sylvestris*. *Plant Physiol.* **94:** 1867–1873.

Kiss, J.Z., R. Hertel, and F.D. Sack. 1989. Amyloplasts are necessary for full gravitropic sensitivity in roots of *Arabidopsis thaliana*. *Planta* **177:** 198–206.

Konings, H. 1968. Significance of the root cap for geotropism. *Acta Bot. Neerl.* **17:** 203–221.

Konjević, R., D. Grubisić, and M. Nesković. 1989a. Growth retardant-induced changes in phototropic reaction of *Vigna radiata* seedlings. *Plant Physiol.* **89:** 1085–1087.

Konjević, R., J.P. Khurana, and K.L. Poff. 1992. Analysis of multiple photoreceptor pigments for phototropism in a mutant of *Arabidopsis thaliana*. *Photochem. Photobiol.* **55:** 789–792.

Konjević, R., B. Steinitz, and K.L. Poff. 1989b. Dependence of the phototropic response of *Arabidopsis thaliana* on fluence rate and wavelength. *Proc. Natl. Acad. Sci.* **86:** 9876–9880.

Lee, J.S., T.J. Mulkey, and M.L. Evans. 1983. Reversible loss of gravitropic sensitivity in maize roots after tip application of calcium chelators. *Science* **220:** 1375–1376.

Li, Y., G. Hagen, and T.J. Guilfoyle. 1991. An auxin-responsive promoter is differentially induced by auxin gradients during tropisms. *Plant Cell* **3:** 1167–1175.

Liscum, E., J.C. Young, K.L. Poff, and R.P. Hangarter. 1992. Genetic separation of phototropism and blue light inhibition of stem elongation. *Plant Physiol.* **100:** 267–271.

MacDonald, I.R. and J.W. Hart. 1985. The role of the apex in normal and tropic growth of sunflower hypocotyls. *Planta* **163:** 549–553.

Mirza, J.I., G.M. Olsen, T.-H. Iversen, and E.P. Maher. 1984. The growth and gravitropic responses of wild-type and auxin-resistant mutants of *Arabidopsis thaliana*. *Physiol. Plant.* **60:** 516–522.

Olsen, G.M. and T.-H. Iversen. 1980. Ultrastructure and movements of cell structures in normal pea and an ageotropic mutant. *Physiol. Plant.* **50:** 275–284.

Orbović, V. and K.L. Poff. 1993. Growth distribution during phototropism of *Arabidopsis thaliana* seedlings. *Plant Physiol.* **103:** 157–163.

Paál, A. 1919. Uber Phototropishe Rezleintungen. *Jahrb. Wiss. Bot.* **58:** 406–458.

Pickard. B.G. 1985. Roles of hormones, protons and calcium in geotropism. In *Encyclopedia of plant physiology* (ed. R. Pharis and D. Reid), vol. 11, pp. 193–281. Springer-Verlag, Berlin.

Piening, C.J. and K.L. Poff. 1988. Mechanism of detecting light direction in first positive phototropism in *Zea mays* L. *Plant Cell Environ.* **11:** 143–146.

Pilet, P.-E. 1982. Importance of the cap cells in maize root gravireaction. *Planta* **156:** 95–96.

Poff, K.L. 1983. Perception of a unilateral light stimulus. *Philos. Trans. R. Soc. Lond. B. Biol. Sci.* **303:** 479–487.

Poff, K.L. and H.V. Martin. 1989. Site of graviperception in roots: A re-examination. *Physiol. Plant.* **76:** 451–455.

Pohl, U. and V.E.A. Russo. 1984. Phototropism. In *Membranes and sensory transduction* (ed. G. Colombetti and F. Lenci), pp. 231–329. Plenum Press, New York.

Reymond, P., T.W. Short, W.R. Briggs, and K.L. Poff. 1992. Light-induced phosphorylation of a membrane protein plays an early role in signal transduction for phototropism in *Arabidopsis thaliana. Proc. Natl. Acad. Sci.* **89:** 4718–4721.

Sack, F.D. 1991. Plant gravity sensing. *Int. Rev. Cytol.* **127:** 193–252.

Schmidt, W. 1984. Bluelight physiology. *BioScience* **34:** 698–704.

Short, T.W. and W.R. Briggs. 1990. Characterization of a rapid, blue light-mediated change in detectable phosphorylation of a plasma membrane protein from etiolated pea (*Pisum sativum* L.) seedlings. *Plant Physiol.* **92:** 179–185.

Shropshire, W., Jr. 1979. Stimulus perception In *Encyclopedia of plant physiology* (ed. W. Haupt and M. Feinleib), vol. 7, pp. 10–41. Springer-Verlag, Berlin.

Sliwinski, J.E. and F.B. Salisbury. 1984. Gravitropism in higher plant shoots. III. Cell dimensions during gravitropic bending; perception of gravity. *Plant Physiol.* **76:** 1000–1008.

Slocum, R.D. and S.J. Roux. 1983. Cellular and subcellular localization of calcium in gravistimulated oat coleoptiles and its possible significance in the establishment of tropic curvature. *Planta* **157:** 481–492.

Steinitz, B. and K.L. Poff. 1986. A single positive phototropic response induced with pulsed light in hypocotyls of *Arabidopsis thaliana* seedlings. *Planta* **168:** 305–315.

Takahashi, H. and T.K. Scott. 1991. Hydrotropism and its interaction with gravitropism in maize roots. *Plant Physiol.* **96:** 558–564.

————. 1993. Intensity of hydrostimulation for the induction of root hydrotropism and its sensing by the root cap. *Plant Cell Environ.* **161:** 99–103.

Takahashi, H., C.S. Brown, T.W. Dreschel, and T.K. Scott. 1992. Hydrotropism in pea roots in a porous-tube water delivery system. *HortScience* **27:** 430–432.

Thimann, K.V. and G.M. Curry. 1961. *Phototropism. In light and life* (ed. W.D. McElroy and B. Glass), pp. 646–672. Johns Hopkins University Press, Baltimore, Maryland.

Togo, S. and K. Hasegawa. 1991. Phototropic stimulation does not induce unequal distribution of indole-3-acetic acid in maize coleoptiles. *Physiol. Plant.* **81:** 555–557.

Trewavas, T., W.R. Briggs, J. Bruinsma, M.L. Evans, R. Firn, R. Hertel, M. Iino, A.M. Jones, A.C. Leopold, P.E. Pilet, K.L. Poff, S.J. Roux, F.B. Salisbury, T.K. Scott, A. Sievers, H.E. Zieschaug, and R. Wayne. 1992. Forum. What remains of the Cholodny-Went theory? *Plant Cell Environ.* **15:** 759–794.

Vierstra, R. and K.L. Poff. 1981. Role of carotenoids in the phototropic response of corn seedlings. *Plant Physiol.* **68:** 798–801.

Volkmann, D. and A. Sievers. 1979. Graviperception in multicellular organs. In *Encyclopedia of plant physiology* (ed. W. Haupt and M. Feinleib), vol. 7, pp. 574–600. Springer-Verlag, Berlin.

Volkmann, D., H.M. Behrens, and S. Sievers. 1986. Development and gravity sensing of cress roots under microgravity. *Naturwissenschaften* **73:** 438–441.

Warpeha, K.M.F., L.S. Kaufman, and W.R. Briggs. 1992. A flavoprotein may mediate the blue light-activated binding of guanosine 5'-triphosphate to isolate plasma membranes of *Pisum sativum* L. *Photochem. Photobiol.* **55:** 595–603.

Wayne, R., M.P. Staves, and A.C. Leopold. 1990. Gravity-dependent polarity of cytoplasmic streaming in *Nitellopsis*. *Protoplasma* **155:** 43–57.

Went, F.W. 1928. Wuchsstoff und Wachstum. *Recl. Trav. Bot. Neerl.* **25:** 1–116.

Went. F.W. and K.V. Thimann. 1937. *Phytohormones*. Macmillan, New York.

Zimmerman, B.K. and W.R. Briggs. 1963a. Phototropic dosage-response curves for oat coleoptiles. *Plant Physiol.* **38:** 237–247.

———. 1963b. A kinetic model for phototropic responses of oat coleoptiles. *Plant Physiol.* **38:** 253–261.

# 25
# Modulation of Root Growth by Physical Stimuli


**Kiyotaka Okada**[1] **and Yoshiro Shimura**[1,2]
[1]Division 1 of Gene Expression and Regulation
National Institute for Basic Biology
Okazaki 444, Japan
[2]Department of Biophysics
Faculty of Science
Kyoto University
Kyoto 606, Japan


Roots and shoots of higher plants change their growth patterns in response to physical stimuli such as gravity, light, humidity, and obstacles. One of the earliest studies of the stimulus-response interactions in young seedlings was done by Charles Darwin, who was interested in the circular movement of hypocotyls and roots, and who examined effects of light, gravity, and touching stimulation using experimental devices designed by himself (1880). Since the pioneering studies of Darwin, much information on the mechanism of root behavior triggered by several kinds of physical stimuli has been accumulated from physiological as well as anatomical analyses of roots (for review, see Wilkins 1966; Juniper 1976; Jackson and Barlow 1981; Evans 1991). Extensive studies on the genetic network of stimulus-response interactions in root systems have started recently with the isolation and analysis of mutants of aberrant morphology or of abnormal growth pattern. We have used *Arabidopsis thaliana* for the molecular genetic approach to elucidate the genetic network ruling the signaling pathways in roots, because roots of young seedlings show a variety of responses to several physical and chemical stimuli, and the structure of *Arabidopsis* roots is simpler than that of other plants (Schiefelbein and Benfey 1991; Dolan et al. 1993). To analyze the responses and to obtain mutant plants that show aberrant responses, we have developed a series of experimental procedures using agar plates (Fig. 1) (Okada and Shimura 1992a). The agar medium contains 1.5% agar and 0.5x *Arabidopsis* mineral nutrient solution. Young seedlings grow well on the surface of agar medium, because sufficient

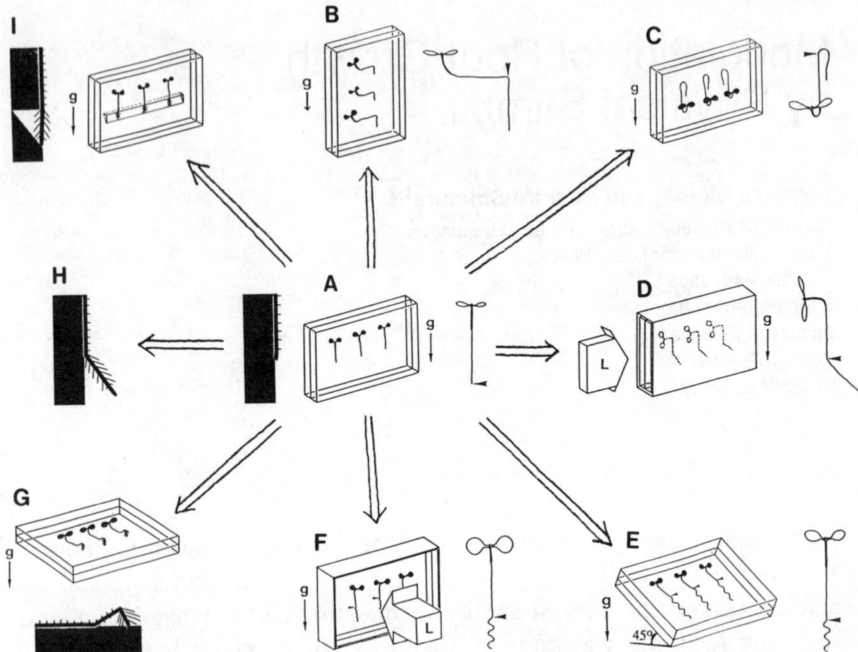

*Figure 1* Agar plate systems for analysis of root and root-hair growth. Seeds of *Arabidopsis* are sterilized with liquid bleach and sown on the surface of agar medium containing 1.5% agar and mineral nutrient solution. Roots of wild-type plants grow straight downward on the surface of agar plates in the vertical position (*A*). Gravitropic responses are examined by changing the position of the plates to the side position (*B*) or to the inverted position (*C*), where roots of wild-type plants bend 90° or 180°, respectively. Phototropic responses are tested by covering the plates with black opaque sheets and illuminating with light from one side (*D*). Positions *E* and *F* are used to examine responses to obstacle-touching stimulus. In the angled position (*E*), the plate is inclined at 45°. In position F, the plate is illuminated from the front. Panels *G*, *H*, and *I* show three different cases in which the structures of root hairs are changed. The dark rectangles indicate cross sections of agar medium. When the plate is set horizontally (*G*), the top region of the root bends downward and lifts up from the agar surface because the root cannot penetrate the hard agar medium. Panel *H* indicates a root accidentally separated from the agar surface. Panel *I* shows an agar plate with a trough. Arrowheads on the roots of seedlings in panels *A–F* indicate the position of the root top when the agar plate is changed from the vertical position to the position of each panel. (g) Gravity, (L) light.

amounts of nutrients, water, and humidity are constantly available. We can control the conditions of physical stimuli by simply changing the position of the plates.

## STRUCTURE OF ROOTS OF WILD-TYPE SEEDLINGS GROWING ON VERTICAL AGAR PLATES

Major roots of *Arabidopsis* seedlings have a simple structure. The diameter of a root is constant (~0.12 mm), indicating that cells divide and elongate mainly in the direction parallel to the axis of root. The growing roots are structurally separated into three regions: apical, middle, and basal. The apical region from 0 to about 0.25 mm from the extreme tip of roots includes the root tip and most of the cell-division zone. The middle region from 0.25 to 0.7 mm from the root tip includes the elongation zone. When roots grow, the apical region slides on the agar surface as the middle region elongates. The speed of growth is estimated to be 200 ± 30 mm per hour (Okada and Shimura 1992b). Although the basal region remains stationary on the agar surface, cell differentiation, such as formation of root hairs and vascular bundle systems, is observed at the basal region. Files of the epidermal cells become visible in the basal region. Root hairs grow perpendicular to the root surface and finally reach about 0.4 mm in length after 15–20 hours. However, as described below, length and direction of root hairs change in response to several physical and chemical stimuli.

## RESPONSE TO GRAVITY

It is strong evidence of positive gravitropism that roots of wild-type plants grow straight downward on a vertical agar plate. To examine how straight roots grow, we measured angles formed between the direction of roots and the direction of gravity. The statistical analysis showed that the angle was distributed in a 30° range from straight downward (Okada and Shimura 1992b). When the agar plate is rotated 90° or 180° (Fig. 1B,C), roots bend to the new direction of gravity (Fig. 2A). The apical region of roots starts to bend downward within 1 hour after the position change, and completes bending in about 8 hours (Sakamoto et al. 1993). When *Arabidopsis* roots change their growth direction on agar plates, two different mechanisms are known; one is differential elongation, and the other is twisting. The former means that the cells located at one side of the root elongate more than those at the other side (schematically shown in Fig. 2F), and the latter means that the cells of the elongation zone rotate the apical part (see Fig. 4C,D). Gravity-induced curvature occurs by differential elongation. This is shown in the enlarged photograph of a bending region where the files of epidermal cells are parallel to the axis of the root (Fig. 2B).

We used the agar-plate system for screening mutants defective in gravitropic responses. Seven mutants were isolated from about 3000

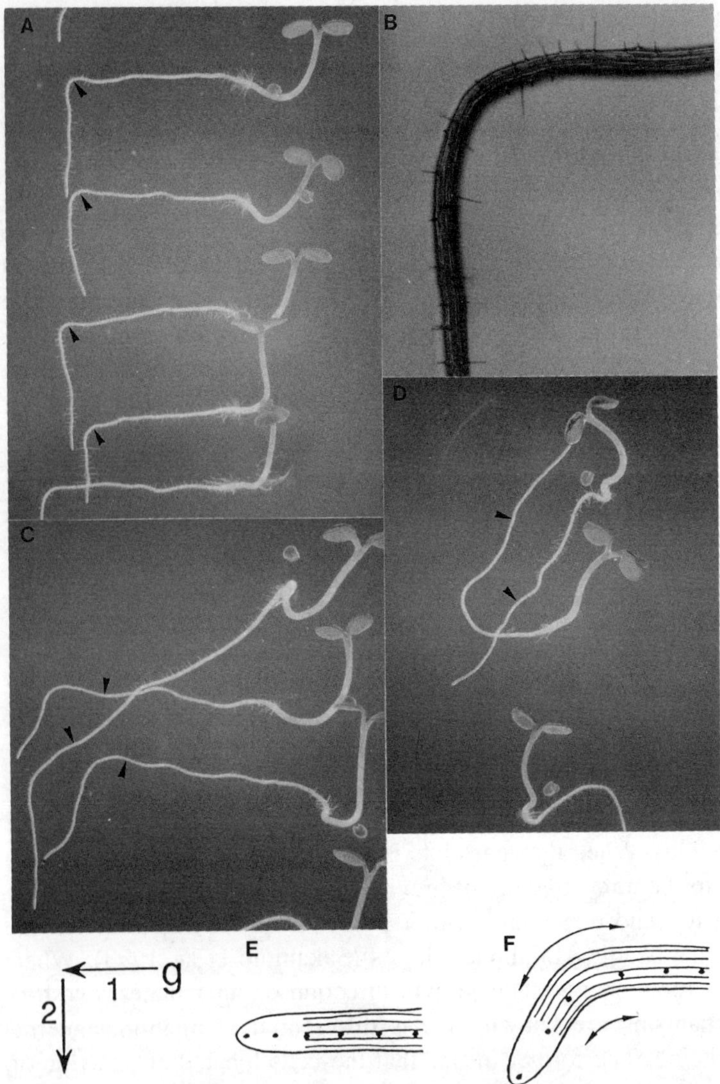

*Figure 2* Gravitropic responses of wild type and mutant plants. Plants were grown on agar plates which shifted from the vertical position (see Fig. 1A) to the side position (see Fig. 1B). (*A*) Wild-type plants. (*B*) Bending region of a wild-type root. (*C*) A reduced-response type mutant, *aux1-31*. (*D*) A no-response type mutant, *aux1-33*. (*E,F*) Schematic demonstration of the root tips before (*E*) and after (*F*) the position change. Arrowheads indicate the position of the root tip when the position of the agar plate is changed. Arrows 1 and 2 show the direction of gravity following the position change.

EMS-mutagenized $M_2$ seeds and classified from their phenotypes into two types: reduced-response type and no-response type (Fig. 2C,D). Roots of the former type grow downward but deviate from a vertical line by about 60–90° (Okada and Shimura 1992b). When the agar plates are shifted to the side position, the roots start to bend at about 7 hours after the position change, but do not bend completely downward even after 31 hours (Okada and Shimura 1992b). On the other hand, the mutants of the no-response type do not show a gravitropic response at all. Distribution of root direction is widely dispersed to almost all angles on the vertical agar surface (Okada and Shimura 1992b).

When the agar plates are shifted to the inverted position (Fig. 1C), roots of wild-type plants make hairpin-like curves with a radius of 0.3–0.6 mm. The reduced-response mutants also form hairpin structures with a larger radius of 1–2 mm. However, the no-response mutants often continue to grow upward. Therefore, the radius of the hairpin curves provides a good parameter to evaluate the degree of gravitropic response. Although both types of mutants show an aberrant gravitropic response of roots, the gravitropic response of hypocotyls looks normal. In contrast, several mutants with aberrant gravitropic response in hypocotyls show normal gravitropism in roots (Bullen et al. 1990). However, one of the photomorphogenesis mutants, *cop4*, has been shown to be defective in both root and hypocotyl gravitropic responses (Hou et al. 1993). These results indicate that some components of the pathways of root and hypocotyl gravitropic responses are common, but some others are not.

Genetic complementation experiments revealed that all mutants contained single recessive mutations and that one mutant of the reduced-response type and six mutants of the no-response type were allelic to a previously reported agravitropic mutant, *aux1* (Mirza et al. 1984; Mirza 1987; Pickett et al. 1990), and one mutant of the reduced-response type was allelic to another agravitropic mutant, *agr1* (Bell and Maher 1990; Maher and Bell 1990). Bell and Maher (1990) reported three *agr1* mutants: one no-response type and two reduced-response type. Therefore, mutants in both the *aux1* and *agr1* genes generate mutants of either no-response type or reduced-response type. When a mutant of reduced-response type is crossed with a mutant of the same locus but of the no-response type, the $F_1$ progeny plants show reduced response (Okada and Shimura 1992b). This indicates that mutants of the reduced-response type carry a less severe genetic lesion than mutants of the no-response type.

It has been widely accepted that the gravistimulus is perceived by cells at the root cap containing a number of amyloplasts that may function as statoliths. When the direction of growth of a root exceeds a threshold value of the deviation angle between the orientation of the root and the direction of gravity, a mechanism to correct the direction of root growth is triggered. The *aux1* and *agr1* genes may have a function associated with determining the threshold value, because mutations of the two genes are considered to raise the value required to trigger the correction mechanism. A second possible interpretation is that the two genes are involved in the cell-to-cell signal transfer mechanism, which works to deliver the gravistimulated signal from root cap cells to cells at the elongation zone, where the bending response occurs. Mutations of the genes may cause reduction of the level of signal that arrives at the target cells. A third possible interpretation is that the genes regulate the cell elongation rate in the elongation zone, and the mutations may disrupt the concerted cell growth. However, the last interpretation is not likely, because the mutants show normal phototropic curvature, as described below.

In addition to *aux1*, *agr1*, and *cop4*, three other genes are reported to be involved in the root gravitropic response (*dwf*, Mirza et al. 1984; *axr1*, Estelle and Somerville 1987; Lincoln et al. 1990; *axr2*, Wilson et al. 1990; Timpte et al. 1992). The *aux1*, *dwf*, *axr1*, and *axr2* mutants show resistance to an excess amount of auxin, suggesting the involvement of auxin in the process of gravitropic response.

### RESPONSE TO INCOMING LIGHT

Phototropic responses of seedlings were tested by covering agar plates in the upright position with a black plastic sheet and illuminating them with light from one side (Fig. 1D). Usually white fluorescent lamps are used, but for illumination with red or blue light, plastic transparent sheets of red or blue color are placed between the lamps and agar plates. The side illumination causes hypocotyls to bend toward light and roots to grow in a direction about 45° away from light (Fig. 3A,C). The growing direction of roots appears to fit the sum of the vertical vector of gravity and the horizontal vector of light, suggesting that positive gravitropism and negative phototropism of roots work additively (Fig. 3G). It has been shown that the differential elongation mechanism (Fig. 2F) is working in the photoinduced bending of roots, because the files of epidermal cells are kept parallel to the root axis (Fig. 3B). Differing from the negative phototropic response of roots, the positive response of hypocotyls seems to dominate the negative gravitropic response (Fig. 3C). The positive phototropic bending of hypocotyls is known to be induced by blue and green light (Steinitz et al. 1985). We showed that the phototropic bending of roots is also induced by blue light (Fig. 3D), and not by red (Fig.

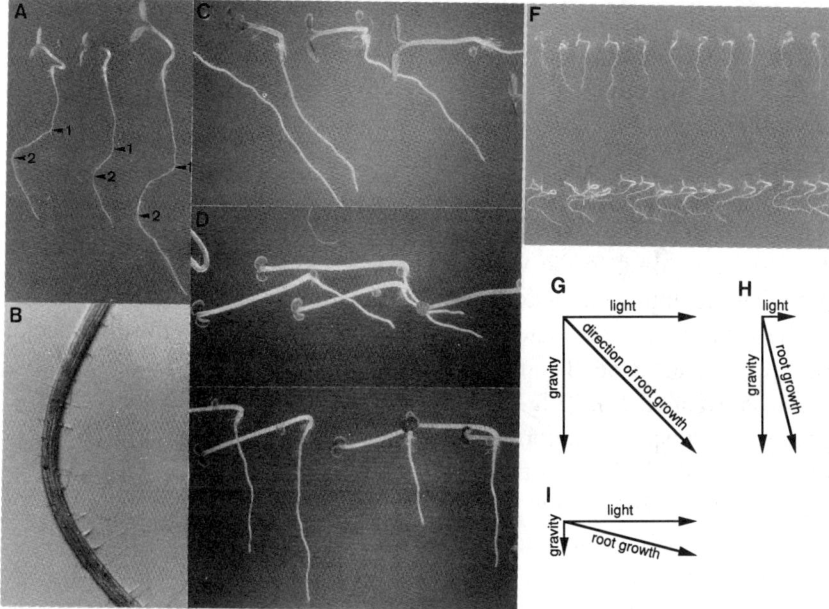

*Figure 3*  Phototropic responses of wild-type and mutant plants. (*A*) Wild-type plants grown on agar plates in the vertical position and illuminated from the side (see Fig. 1D). Arrowheads 1 and 2 indicate the positions of root tips when illuminated with white light from the right and from the left, respectively. (*B*) Enlarged picture of the curving region of a root in *A*. (*C–E*) Wild-type seedlings illuminated with white (*C*), blue (*D*), or red (*E*) light from the left. (*F*) Seedlings of a photoresponse-defective mutant, *rpt2-1* (top), and a graviresponse-defective mutant, *aux1-31* (*bottom*) illuminated with white light from the right, and then from the left. (*G–I*) The vector hypothesis. The direction of root growth fits the sum of the two vectors representing the incoming light and gravity. *G* represents the direction of root growth of wild type, which has normal phototropic and gravitropic responses. When the phototropic response or the gravitropic response is weakened by mutation or by chemical compounds, roots grow to the direction close to the vertical line (*H*) or to the horizontal line (*I*), respectively.

3E) or green light (data not shown). It is not known, however, how the direction of incoming light is recognized by roots or whether the same recognition system works in hypocotyls and in roots.

Root phototropism mutants were screened using the agar plate system. In our first screening, two mutants designated *rpt1-1* and *rpt2-1* (root phototropism) were isolated (Fig. 3F) (Okada and Shimura 1992b). Recently we isolated the third mutant, *rpt3-1*. The direction of root growth was straight downward when illuminated from one side, indicating that the mutants do not have a phototropic response but do have a normal gravitropic response (Fig. 3H). Hypocotyls of the mutants show normal positive phototropism, suggesting that the *rpt1*, *rpt2*, and *rpt3* genes regulate phototropic response of roots but not of hypocotyls.

To examine whether the root graviresponse genes, *aux1* and *agr1*, would have any functions in root phototropic response, seedlings of the mutants were illuminated from one side. The path of roots of the mutants, both of the no-response type mutants and the reduced-response type mutants, were nearly horizontal (Fig. 3F) (Okada and Shimura 1992b). According to the vector hypothesis, the direction of the mutant roots can be explained as follows: The vector of gravity is shortened due to the reduced or nonexistent graviresponse activity, whereas the vector of light is unaffected. Therefore, the sum of the two vectors comes close to the horizontal line (Fig. 3I). It can be concluded, therefore, that the phototropic response is normal in the mutants and that the *aux1* and *agr1* genes are not involved in the phototropic response of roots. Because the phototropic response and the gravitropic response are independent of each other, we can postulate that direction of growth of roots will be completely random if both of the two responses are abolished. We have made a double mutant carrying both the *rpt1-1* and *aux1-33* mutations. As expected, roots of the double mutant strain grow in almost all directions.

Phenotypic analyses of the mutants suggest that the mechanisms controlling the phototropic response of roots are separated, at least partly, from that of the gravitropic response of roots or from that of the phototropic response of hypocotyls. In contrast, several genes are reported to control both the phototropic and the gravitropic responses of hypocotyls (Khurana et al. 1989).

### RESPONSE TO TOUCHING STIMULUS

When root tips encounter obstacles in soil, they avoid the obstacles by changing the direction of their growth. We have developed an assay system to provide obstacle-touching stimulus continuously to the root tips of young seedlings. When seedlings are incubated on agar plates set at an angle of 45º from the vertical and illuminated from above (the angled position, Fig. 1E), roots bend downward due to positive gravitropic response and negative phototropic response and thereby encounter the agar surface. Because the roots cannot penetrate the agar medium used, they experience a touching stimulus. Following a positional shift of the agar plates from the vertical to the angled position, roots of wild-type seedlings began to grow in a waving pattern (Fig. 4A). If the plates were returned to the vertical position, the wavy growth came to an immediate halt and straight growth resumed, showing that a continuous obstacle-

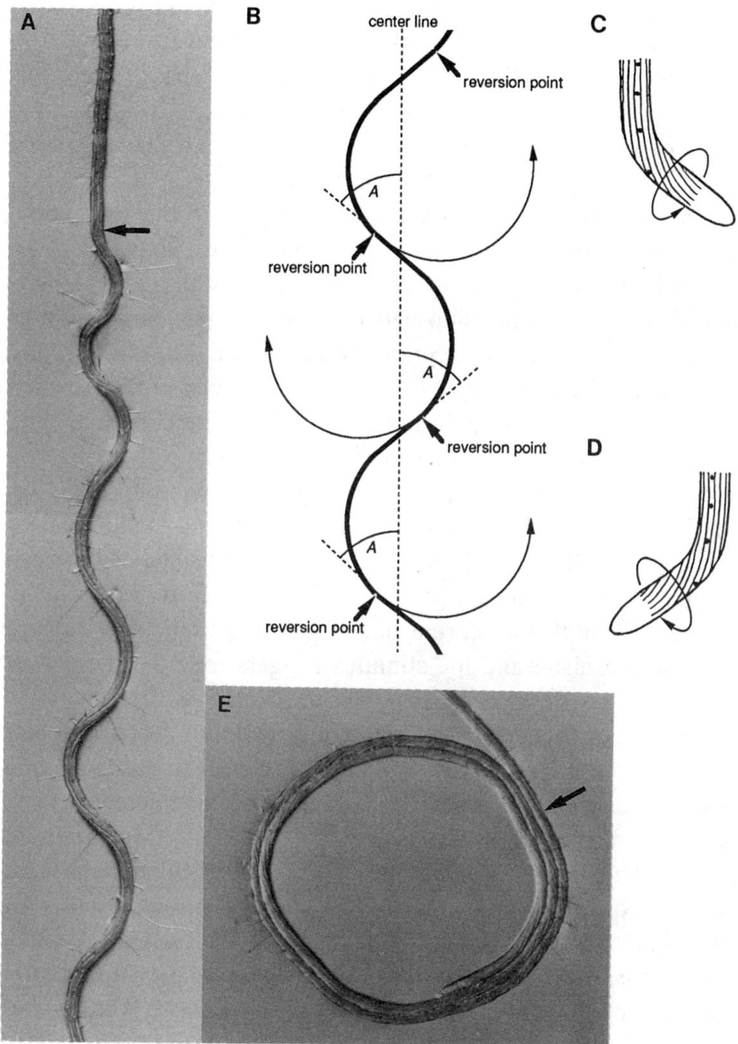



*Figure 4* Wavy growth of roots on the angled agar plates. (*A*) Root of wild-type plant. (*B*) Model of the wavy growth. The thick waving line indicates the path of a root. The thin solid circular lines show the path of a root assumed if rotation of the root tip is not reverted. Root changes the direction of root-tip rotation at the "reversion points" where tangential lines cross the center line of waves at an angle A. (*C,D*) Schematic demonstration of the root tips on the angled agar plates. Right-handed or left-handed rotation of a root tip causes clockwise or counterclockwise curvature, respectively. (*E*) Root of the *aux1-33* mutant. Arrows in *A* and *E* indicate the position of the root tip when the agar plate was shifted from the vertical position to the angled position.

touching stimulus is required for the wavy growth. Microscopic observation of the wavy roots showed that the files of epidermal cells were arranged in a twisted pattern around the root axis, whereas in roots showing straight vertical growth, the files of epidermal cells were arranged in parallel along the root axis. These observations suggest that the root tips of wavy roots are rotating. Direct proof of root-tip rotation was obtained from a series of photographs taken every hour of a root that had been marked with carbon grains (Okada and Shimura 1990). The carbon grains put on the root tip moved from side to side as the root grew. Detailed observation showed that the apical region of the root was rotating and that cells in the elongation zone were responsible for making it rotate. There is a close relationship between the direction of rotation of the root tip and the direction of curvature of the root on the surface of the agar: Left-handed twisting of the tip was observed during clockwise curvature and right-handed twisting during counterclockwise curvature (Fig. 4C,D). Reversion of rotation occurs periodically every 5–6 hours. As schematically shown in Figure 4B, roots change the direction of root-tip rotation at fixed points where tangential lines cross the center line of waves at an angle of 35–45°. The wavy growth of roots can also be observed on agar plates placed upright and illuminated from the front by blue light (Fig. 1F). In this case, root tips are pressed to the agar surface by negative phototropism, and the stimulus triggers rotation of the root tips.

To examine genes that regulate the wavy growth, we tested the root growth pattern of the previously isolated gravitropism mutants and phototropism mutants on the angled agar plates and then screened new mutants which formed aberrant waves or which could not form waves at all. Isolated mutants were divided into six complementation groups, designated *wav1* to *wav6* (wavy growth) (Okada and Shimura 1990). All of the mutants have a single, recessive mutation in the nuclear genome. The root path of the *wav1-1* mutant is not wavy, because root-tip rotation is not induced on an agar plate in the angled position (Okada and Shimura 1990). The *wav1-1* mutant is considered to be less sensitive to the touching stimulus, because the mutants often induce the root-tip rotation on an agar plate set horizontal. The *wav1* gene appears to be the *rpt1* gene (root phototropism 1, see above), because the two different phenotypes, no waves and no phototropism, were not separated after three backcrosses. The other mutants, *wav2-1* and *wav3-1*, have waves of a shorter pitch due to a higher rate of rotation (Okada and Shimura 1990). The growth rate and the timing of reversion of rotation of the two mutants are almost the same as those of wild type. Gravitropic and phototropic responses also appear to be normal. However, when the agar plates are put upside down (Fig. 1C), roots of the two mutants show hairpin-like forms of smaller radius (0.1–0.2 mm) than that of wild type (0.3–0.6 mm). These results indicate that the *wav2* and *wav3* genes are involved in the touching-induced root-tip twisting and the bending mechanism of differential elongation. It is also supposed that the sensitivity to the touching stimulus is increased in the two mutants, because root-tip rotation is often induced when they are grown on agar plates in the vertical position. The *wav4* mutants, *wav4-1* and *wav4-2*, show rectangular waves, possibly owing to irregular timing of reversion of rotation (Okada and Shimura 1990). Gravitropic and phototropic responses of the *wav4* mutants appear normal.

Mutants of two other alleles, *wav5* and *wav6*, show no waves and reduced gravitropism (Okada and Shimura 1990). Genetic analysis revealed that the *wav5* mutants and the *wav6* mutants are allelic to the *aux1* mutants and the *agr1* mutants, respectively. All of the examined graviresponse mutants except two *aux1* mutants, *aux1-33* and *aux1-66*, show no or few waves on the inclined agar plates. The *aux1-33* mutant, however, forms a circular path of clockwise curve (Fig. 4E), because the root tips rotate in one direction and are unable to revert to the other direction. To demonstrate the unidirectional rotation of the *aux1-33* mutant, carbon grains were put on the root. Following root-tip rotation, the carbon grains were conveyed from one side to the other side of the root (Fig. 5). The *aux1-66* mutant forms circular paths of either clockwise or counterclockwise curve. Thus, in this mutant, the periodic reversion of the twisting direction is abolished. However, reversion of the twisting direction can take place spontaneously in the *aux1-66* mutant.

Our results suggest the existence of a signal transduction mechanism, beginning with perception of the touching stimulus at the root tip, followed by signal transfer from the root tip to the elongation zone, induction of rotation, and the periodic reversion of rotation that results in wavy growth. In this model, the genetic defect in the *wav2* and *wav3* mutants may be in signal processing or in the process of induction of rotation. The *wav4* mutants may have a defect in the process of timing adjustment for the reversion of rotation.

Reversion of the twisting direction seems to be dependent on correct gravity sensing and responsiveness. As shown in Figure 4B, if a wild-type root does not revert the twisting direction at a point of reversion and continues to rotate in one direction, the root will follow a circular path and finally grow upward. However, the root continues to grow downward if the twisting direction reverts. Therefore, the reversion would be a mechanism to ensure that roots elongate along the direction of gravity. This hypothesis is supported by our observation that all of the

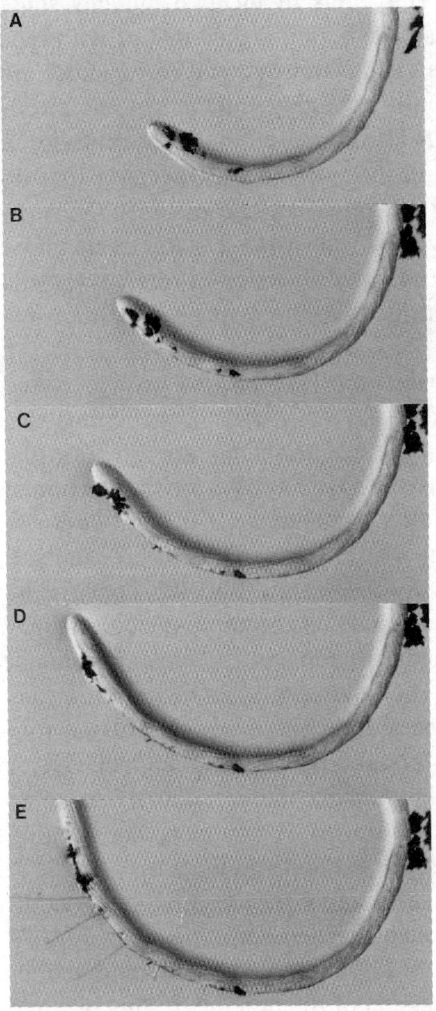

*Figure 5* Movement of carbon grains on a twisting root of the *aux1-33* mutant. The plant was incubated on an agar plate in the angled position. Photographs were taken at 0 (*A*), 2.5 (*B*), 5 (*C*), 7 (*D*), and 22 hr (*E*) after carbon grains were put on the root.

graviresponse-defective mutants do not show periodical reversion of rotation.

The touching-induced root-tip rotation is closely related to the ability to evade obstacles in the root path. When small sand grains are put in the path of roots growing on a vertical agar plate, roots of the *wav* mutants are often trapped on and unable to pass the obstacles, whereas wild-type roots easily pass them.

## STRESS-DEPENDENT ROOT-HAIR GROWTH

Normal root hairs of wild-type plants are about 0.4 mm in length and are formed perpendicular to the root surface, as usually observed on roots growing on vertically oriented agar plates. We noticed that the pattern of root-hair growth is changed by several physical or chemical stimuli. When the agar plates are shifted to the horizontal position (Fig. 1G), roots bend downward and try to penetrate into the agar medium. However, if the agar medium is hard, the roots lift themselves up as they continue to grow (Fig. 6A–C). Root hairs formed in the vicinity of the bend are long, up to 0.8 mm, and grow obliquely to the root surface. The long and tilted root hairs are formed under several unusual growth conditions as schematically shown in Figure 1G, H, and I; namely, when root tips do not touch the agar surface (Fig. 1H) or when roots grow over troughs cut into agar plates (Figs. 1I, 6D). Long but not angled root hairs are observed on the waving roots growing on inclined agar plates (see Fig. 4A). In addition, quality of light influences the root-hair development. Root hairs develop well when illuminated by white light, but not in red or blue light (Fig. 3C–E). It is interesting that a phytochrome B-defective mutant, *hy3*, forms long root hairs in white light (Reed et al. 1993). Finally, long and tilted root hairs are formed when wild-type seedlings are transferred to agar plates containing chemical compounds that have effects to decrease root growth rate: high concentration of auxin (indole-3-acetic acid or 2,4-D at 30 μM) (Fig. 6G), cytokinin (benzyladenine at 30 μM; see also Su and Howell 1992), or auxin transport inhibitor (9-hydroxyfluorene-9-carboxylic acid at 40 μM). These results suggest that longer and/or tilted root hairs are formed when roots receive physical or chemical stress. Thus, root-hair development is under the control of two distinct pathways: a stress-induced pathway, and a stress-independent and developmentally regulated pathway. Although the latter pathway has been extensively studied (Schiefelbein and Somerville 1990), the former pathway is not well understood.

As the first step to analyze the genetic system regulating the stress-dependent root-hair growth, we have started to examine root hairs of the previously isolated gravitropic, phototropic, and waving mutants. Although the study is not finished yet, we have made several interesting observations. First, on an angled agar plate, the *wav1-1* mutant forms root hairs longer than those formed on a vertical agar plate (Fig. 6E), although twisting of root tips is not induced. The result suggests that the two responses induced by the touching stimulus, twisting of root tips and formation of long root hairs, are genetically separated and that the *wav1-1* mutation blocks the signaling pathway that induces twisting but does not block another pathway that leads to long root hairs. An alternative in-

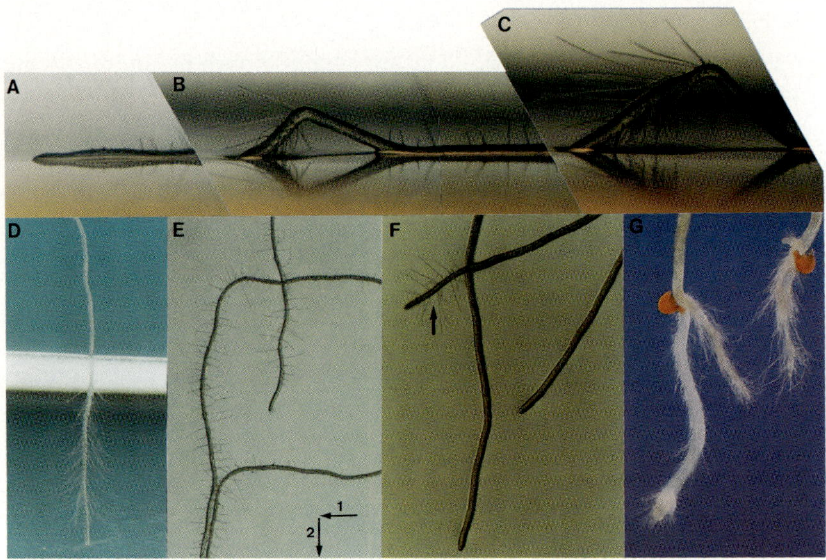



*Figure 6* Root-hair growth patterns of wild-type and mutant plants. (*A–C*) Wild-type plants were grown on an agar plate shifted from the vertical to horizontal position (see Fig. 1G). Photographs were taken at 0 (*A*), 24 (*B*), and 32 hr (*C*) after the shift. (*D*) Wild-type root growing over a trough (see Fig. 1I). A trough is the lower half of the picture. (*E*) Roots of the *wav1-1* mutant on an agar plate shifted from the vertical to the side position. Arrows 1 and 2 indicate the directions of gravity before and after the position shift, respectively. (*F*) Roots of the *aux1-33* mutant on an agar plate placed vertically. A part of a root indicated by an arrow was accidentally separated from the agar surface. (*G*) Roots of wild-type plant on an agar medium containing auxin.

terpretation is that the signal(s) required for triggering the twisting is stronger than that for induction of long root hairs and that the *wav1-1* mutation results in decreasing the strength of the signal(s) to a level that is too weak to trigger twisting, but is strong enough to induce long root hairs. Second, the *aux1* mutants have no or fewer root hairs on the vertical agar plates. However, roots that do not touch agar medium have long and tilted root hairs (Fig. 6F). The result indicates that the *aux1* mutant retains the stress-induced pathway of root hair development, although the stress-independent, developmentally regulated pathway may be abolished. We are continuing to screen and characterize new mutants defective in stress-induced root-hair development.

## GENETIC NETWORKS OF STIMULUS-RESPONSE INTERACTIONS IN ROOTS

The growth pattern of roots and root hairs is largely modulated by different kinds of physical stimuli. When grown on an agar surface, roots show clear and quick responses to gravity, light, and obstacle-touching. However, when roots do not touch the agar surface, they bend slowly and incompletely and do not show the waving response. The form of root hairs is also changed when roots are separated from the agar surface. Therefore, the condition of not-touching to some wet solid material induces signal(s) that may repress the response systems for gravity, light, or touching stimulus. It will be interesting to examine the molecular nature of the signal(s) recognized by roots when touched or not touched to the agar surface, and how the signal(s) represses gravitropic, phototropic, and obstacle-touching responses.

As listed in Table 1, some mutants show genetic defects in more than two different stimulus-response interactions. For example, the *aux1* mutants show abnormal responses to gravity and touching, as well as aberrant root-hair formation, and the *rpt1* mutants have defects in phototropic and touching responses. Figure 7 shows the genetic interrelationships of root responses to physical stimuli based on the pleiotropic phenotype of the mutants. Further genetic analysis of the mutants will clarify molecular function(s) of the genes involved in the responses.

## PHYSIOLOGICAL AND BIOCHEMICAL ANALYSES OF *ARABIDOPSIS* ROOTS

Physiological studies of root tropisms have been done mainly using maize and plants other than *Arabidopsis*, probably because *Arabidopsis* roots are too small to perform microsurgery or to collect enough material for further experiments. However, biochemical and physiological studies of *Arabidopsis* roots would definitely be necessary to examine whether models or hypotheses based on research with other plants are also valid with *Arabidopsis* and whether there is anything new and unique in *Arabidopsis* systems. We have tested several kinds of chemical compounds that may affect tropic responses and found that gravitropic, phototropic, and touching responses are greatly reduced by some compounds known as inhibitors of auxin polar transport.

Roots of wild-type seedlings show a reduced response to gravity when naphthylphthalamic acid (NPA) or 9-hydroxyfluorene-9-carboxylic acid (HFCA) is added to the agar medium at a low concentration (4 μM). When the compound is added at a high concentration (40 μM), roots show no response to gravistimulus. As already mentioned, weaker alleles of gravitropism mutants, *aux1* or *agr1*, show reduced responses, whereas stronger alleles show no response. The parallel relationships between the degree of mutation and the dose of inhibitors suggests that both *aux1* and

*Table 1*  List of mutants

| Strain | gravitropic response | phototropic response | touching-induced waving | root hairs on roots growing on vertical agar plates | stress-induction of longer and tilted root hairs |
| --- | --- | --- | --- | --- | --- |
| | | | Phenotypes | | |
| Wild type | normal | normal | normal waves | normal | normal |
| *aux1-31* (*wav5-31*) | reduced response | normal[a] | no waves | few | normal |
| *aux1-33* (*wav5-33*) | no response | normal[a] | counterclockwise circles | no or very few | normal |
| *aux1-66* (*wav5-66*) | no response | normal[a] | clockwise or counter-clockwise circles | short, few | normal |
| *agr1-52* (*wav6-52*) | reduced response | normal[a] | no waves | few | reduced? |
| *rpt1-1* (*wav1-1*) | normal | no response | no or less waves | normal | normal |
| *rpt2-1* | normal | no response | normal | normal | normal |
| *wav2-1* | normal[b] | normal | short pitch | short, few | normal |
| *wav3-1* | normal[b] | normal | short pitch | short, few | normal |
| *wav4-1* | normal | normal | rectangular waves | short, no or very few | normal |

[a]The response appears to be hypersensitive, because of no or reduced gravitropic response.
[b]Radius of the curvature is smaller than wild type.

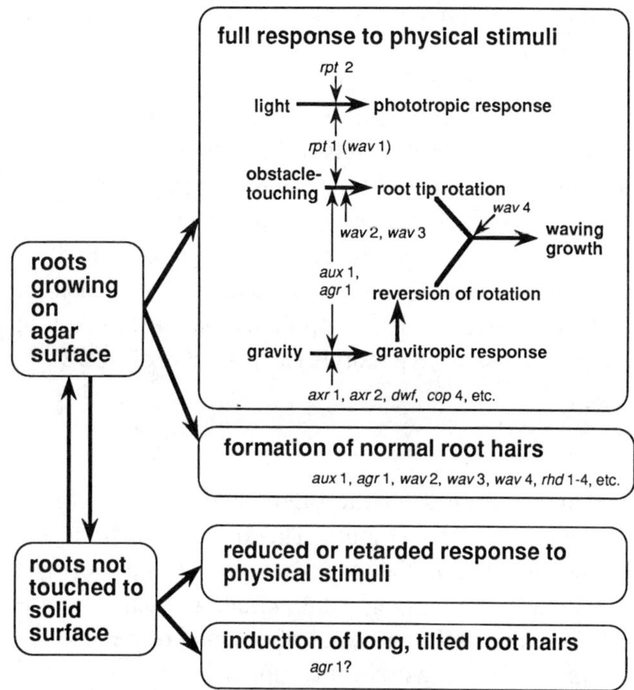


*Figure 7* Genetic regulatory network of root responses.

*agr1* gene products are involved in the auxin transport system. However, the function of the *aux1* and *agr1* genes may be separable, because the *aux1* mutants are resistant to auxin and the *agr1* mutants are not.

Phototropic responsiveness is also repressed by the auxin transport inhibitors at a high concentration (40 μM). Interestingly, however, roots show an increased response to incoming light at a low concentration (4 μM). It could be interpreted that the chemicals have a two-phase effect on phototropic response of wild-type roots; namely, promotion or repression of the response at a low or high concentration, respectively. A similar biphasic effect is known for auxin, which promotes or represses tissue growth at a low or high concentration, respectively. However, an alternative interpretation is that the gravitropic response is more sensitive to the drug than is the phototropic response. Thus, the gravitropic response is weakened, whereas the phototropic response remains almost intact at 4 μM of the drug. Because the path of roots fits to the sum of the two vectors, gravitropism and phototropism, weakened gravitropism causes roots to grow at smaller angles to the horizontal line (see Fig. 3I). The auxin transport inhibitors also affect the touching-induced waving growth. These results suggest that the auxin transport system(s) plays an important role at some common steps of the signaling pathways of the three stimulus-response interactions.

As an approach for biochemical analysis of *Arabidopsis* roots, proteins synthesized or modified by physical stimuli have been studied using two-dimensional polyacrylamide gels (Sakamoto et al. 1993). Several new protein spots appear after 24 hours of continuous rocking on a rocking shaker, or after obstacle-touching stimulation. Four of the protein spots are commonly enhanced by continuous rocking and by touching stimulation. However, the function and induction mechanism of the proteins remain to be solved. A similar approach has been taken to analyze the phototropic response of hypocotyls (Reymond et al. 1992).

## OUTLOOK

There are three different approaches for the future study of root-growth responses to physical and chemical stimuli. The first is isolation and characterization of genes involved in the responses by means of new and efficient procedures such as gene tagging, subtraction, and chromosome walking. The second is isolation of new types of mutants or new alleles of known genes. New screening ideas will undoubtedly help obtain new classes of mutants. Examination of the whole spectrum of mutations in root stimulus-response interactions will help us to identify the genetic regulatory networks of signaling pathways. The third is integration of the accumulated data of genetic, physiological, developmental, and biochemical studies on root responses. By combining the results from the three different approaches, we will be able to clarify the regulatory networks of the unique signaling pathways in plants.

## ACKNOWLEDGMENTS

We thank present and past members of our laboratory who worked with us in the root project, in particular, Drs. Callum J. Bell, Sumie Ishiguro, Nobuyoshi Mochizuki, Tokitaka Oyama, Koji Sakamoto, Hideaki Shiraishi, and Takuji Wada.

## REFERENCES

Bell, C.J. and E.P. Maher. 1990. Mutants of *Arabidopsis thaliana* with abnormal gravitropic responses. *Mol. Gen. Genet.* **220:** 289–293.

Bullen, B.L., T.R. Best, M.M. Gregg, S.-E. Barsel, and K.L. Poff. 1990. A direct screening procedure for gravitropism mutants in *Arabidopsis thaliana* (L.) Heynh. *Plant Physiol.* **93:** 525–531.

Darwin, C. 1880. *Power of movement in plants.* John Murray, London.

Dolan, L., K. Janmaat, V. Willemsen, P. Linstead, S. Poethig, K. Roberts, and B. Scheres. 1993. Cellular organization of the *Arabidopsis thaliana* root. *Development* **119:** 71–84.

Estelle, M.A. and C. Somerville. 1987. Auxin-resistant mutants of *Arabidopsis thaliana* with an altered morphology. *Mol. Gen. Genet.* **206:** 200–206.

Evans, M.L. 1991. Gravitropism: Interaction of sensitivity modulation and effector redistribution. *Plant Physiol.* **95:** 1–5.

Hou, Y., A.G. von Arnim, and X.-W. Deng. 1993. A new class of *Arabidopsis* constitutive photomorphogenic genes involved in regulation cotyledon development. *Plant Cell* **5:** 329–339.

Jackson, M.B. and P.W. Barlow. 1981. Root gravitropism and the role of growth regulators from the cap: A re-examination. *Plant Cell Environ.* **4:** 107–123.

Juniper, B.E. 1976. Geotropism. *Annu. Rev. Plant Physiol.* **27:** 385–406.

Khurana, J.P. and K.L. Poff. 1989. Mutants of *Arabidopsis thaliana* with altered phototropism. *Planta* **178:** 400–406.

Lincoln, C., J.H. Britton, and M. Estelle. 1990. Growth and development of the *axr*1 mutants of *Arabidopsis*. *Plant Cell* **2:** 1071–1080.

Maher, E.P. and C.J. Bell. 1990. Abnormal responses to gravity and auxin in mutants of *Arabidopsis thaliana*. *Plant Sci.* **66:** 131–138.

Mirza, J.I. 1987. The effects of light and gravity on the horizontal curvature of roots of gravitropic and agravitropic *Arabidopsis thaliana* L. *Plant Physiol.* **83:** 118–120.

Mirza, J.I., G.M. Olsen, T.-H. Iversen, and E.P. Maher. 1984. The growth and gravitropic responses of wild-type and auxin-resistant mutants of *Arabidopsis thaliana*. *Physiol. Plant.* **60:** 516–522.

Okada, K. and Y. Shimura. 1990. Reversible root tip rotation in *Arabidopsis* seedlings induced by obstacle-touching stimulus. *Science* **250:** 274–276.

———. 1992a. Aspects of recent developments in mutational studies of plant signaling pathways. *Cell* **70:** 369–372.

———. 1992b. Mutational analysis of root gravitropism and phototropism of *Arabidopsis thaliana* seedlings. *Aust. J. Plant Physiol.* **19:** 439–448.

Pickett, F.B., A.K. Wilson, and M. Estelle. 1990. The *aux*1 mutation of *Arabidopsis* confers both auxin and ethylene resistance. *Plant Physiol.* **94:** 1462–1466.

Reed, J.W., P. Nagpal, D.S. Poole, M. Furuya, and J. Chory. 1993. Mutations in the gene for the red/far-red light receptor phytochrome B alter cell elongation and physiological responses throughout *Arabidopsis* development. *Plant Cell* **5:** 147–157.

Reymond, P., T.W. Short, W.R. Briggs, and K.L. Poff. 1992. Light-induced phosphorylation of a membrane protein plays an early role in signal transduction for phototropism in *Arabidopsis thaliana*. *Proc. Natl. Acad. Sci.* **89:** 4718–4721.

Sakamoto, K., H. Shiraishi, K. Okada, and Y. Shimura. 1993. Proteins induced by physical stimuli in root tips of *Arabidopsis thaliana* seedlings. *Plant Cell Physiol.* **34:** 297–304.

Schiefelbein, J.W. and P.N. Benfey. 1991. The development of plant roots: New approaches to underground problems. *Plant Cell* **3:** 1147–1154.

Schiefelbein, J.W. and C. Somerville. 1990. Genetic control of root hair development in *Arabidopsis thaliana*. *Plant Cell* **2:** 235–243.

Steinitz, B., Z. Ren, and K.L. Poff. 1985. Blue and green light-induced phototropism in *Arabidopsis thaliana* and *Lactuca sativa* L. seedlings. *Plant Physiol.* **77:** 248–251.

Su, W. and S.H. Howell. 1992. A single genetic locus, Ckr1, defines *Arabidopsis* mutants in which root growth is resistant to low concentration of cytokinin. *Plant Physiol.* **99:** 1569–1574.

Timpte, C.S., A.K. Wilson, and M. Estelle. 1992. Effects of the *axr*2 mutation of *Arabidopsis* on cell shape in hypocotyl and inflorescence. *Planta* **188:** 271–278.

Wilkins, M.B. 1966. Geotropism. *Annu. Rev. Plant Physiol.* **17:** 379–408.

Wilson, A.K., F.B. Pickett, J.C. Turner, and M. Estelle. 1990. Dominant mutation in *Arabidopsis* confers resistance to auxin, ethylene and abscisic acid. *Mol. Gen. Genet.* **222:** 377–383.

# 26

# Interactions between *Arabidopsis thaliana* and Viruses

**Anne E. Simon**

Department of Biochemistry and Molecular Biology
University of Massachusetts
Amherst, Massachusetts 01003

Interactions between plant viruses and their hosts have been intensively studied since tobacco mosaic virus (TMV) was first characterized early this century. Advancements in nucleic acid technology in the 1980s provided the tools that led to the complete sequence determination of scores of both RNA and DNA viruses. The ability to synthesize infectious transcripts of RNA viruses in vitro, coupled with the development of protoplast inoculation methods (Takebe 1977), led to the identification of sequences involved in symptom expression (Neeleman et al. 1991; Rodriguez-Cerezo et al. 1991; Shintaku et al. 1992), replication (David et al. 1992), and systemic movement (Meshi et al. 1987; Hacker et al. 1992). The availability of in vitro systems for the complete replication of viral and subviral genomes expanded our understanding of the role that viral determinants play in virus replication (Hayes and Buck 1990; Hayes et al. 1992). The ability to transform plants with specific viral open reading frames revolutionized our understanding of resistance mechanisms and is providing new approaches to controlling virus infections (Beachy et al. 1990; Braun and Hemenway 1992; MacFarlane and Davies 1992; Anderson et al. 1992).

Although our knowledge of the structure and function of viral nucleic acids has proceeded at a rapid pace, much less is known about how the plant responds at the molecular level to virus invasion. To date, no plant genes specifying resistance to any viral pathogen have been isolated, and little is known about molecular events that contribute to disease symptoms. The attributes that make *A. thaliana* an excellent system for molecular genetic studies contribute to its potential as a model system for the elucidation of the plant side of plant/virus interactions. In this chapter, an overview of the field of plant/virus interactions is presented, followed by a summary of the progress that has been made in developing *Arabidopsis* as a host system.

## OVERVIEW OF PLANT VIROLOGY

### Structure of Viral Nucleic Acids

Plant viruses, like animal viruses, consist of nucleic acid (either DNA or RNA) encapsidated in a protein or lipoprotein coat. Most plant viruses have small genomes ranging from 2,500 to 10,000 nucleotides with most between 4,000 and 7,000 nucleotides. Plant viruses can be either single-stranded or double-stranded and, unlike most animal viruses, can frequently have multiple genomic segments generally encapsided in separate viral particles. Coding strands of single-stranded RNA viruses are referred to as the positive or plus strand. A virus is considered to be a plus or minus strand virus, depending on the encapsidated strand. Since the vast majority of plant viruses have positive, single-stranded RNA genomes, this review is mainly concerned with this type of virus.

Although single-stranded RNA viruses exhibit wide variation in symptom production, capsid morphology, host range, and genetic organization, similarities in the sequence of nonstructural genes and other common features have led to the grouping of these viruses into two or possibly three supergroups (Goldback and Wellink 1988). Sindbis-like viruses (related to animal alphaviruses) have single or multiple genomes and share such features as 5′ cap structures and translation of structural and/or nonstructural proteins from subgenomic RNAs. Picorna-like viruses (related to animal picornaviruses) have one or two genomic RNAs, are polyadenylated at the 3′ end, contain a small virus-encoded protein (Vpg) at the 5′ end, and express their genome through the production of polyproteins. Cleavage of the polyproteins by virus-encoded proteases results in the production of functional, smaller proteins. The plant carmoviruses may belong to a third supergroup based on their small genome size (4 kb) and the presence of only one of three conserved domains in the presumptive RNA-dependent RNA polymerase open reading frame.

Many plant RNA viruses are associated with subviral RNAs in addition to their genomic RNA(s). Subviral RNAs known as subgenomic RNAs are templates for the synthesis of structural or nonstructural proteins and are derived from a continuous segment of 3′ terminal genomic RNA. Defective interfering (DI) RNAs are generally noncoding molecules derived exclusively or nearly exclusively from several noncontiguous segments of the genomic RNAs. Satellite (sat-) RNAs share little or no sequence similarity with plant viruses but, along with DI RNAs, require a specific helper virus for multiplication in plants.

## Plant Virus Movement

Unlike animal viruses, there are no known receptors for plant viruses on the surface of cells (Matthews 1991). In general, viruses are naturally introduced into plant cells with the aid of a specific virus vector that might be an insect, nematode, or fungus. Many viruses also can be mechanically inoculated by rubbing leaves of young plants with an inoculum containing viral particles or nucleic acids and an abrasive such as diatomaceous earth (Celite).

Virus spread from the initially inoculated cell is thought to occur in two separate phases (for review, see Citovsky and Zambryski 1991; Maule 1991; Deom et al. 1992). Short-distance or cell-to-cell spread is mediated by plasmodesmata, the channels that join all plant cells by forming a continuous symplastic connection (Robards and Lucas 1990). Long-distance spread (e.g., leaf-to-leaf) is a poorly understood process involving transport of the virus through the vascular system.

Cell-to-cell transport of plant viruses is distinctly different from that of animal viruses; animal viruses usually move from cell to cell by fusing with the cell membrane or by receptor-mediated endocytosis, processes not possible in plant cells due to rigid cell walls. Movement through plasmodesmata requires the presence of a unique plant viral product(s) known as the movement protein(s) (Melcher 1990). Several recent reports have indicated that movement proteins participate in short-distance virus spread in two ways: by intercalating into secondary (branched) plasmodesmata in nonvascular cells, thereby increasing the diameter of the pore through which the virus must travel (Wolf et al. 1989; Ding et al. 1992) or by acting as nonspecific, single-stranded nucleic acid-binding proteins that coat and extend viral nucleic acids, producing an elongated ribonucleoprotein complex of a size suitable for transport through plasmodesmata (Citovsky et al. 1992). Besides the movement protein, viral coat protein is also frequently required for short-distance and/or long-distance spread (Dawson et al. 1988; Allison et al. 1990; Saito et al. 1990; Chapman et al. 1992; Hacker et al. 1992). Whether the movement and coat proteins interact is unclear; a hybrid virus containing an unrelated movement protein was still able to systemically infect its host, suggesting that, in this case, the movement protein did not need to interact with specific sequence or structural elements from the normally associated coat protein or viral genome (DeJong and Ahlquist 1992).

Despite these recent advances in our understanding of plant virus movement, many questions still remain. Since viral products involved in movement have only been analyzed in detail for a few viral systems (most notably TMV, cauliflower mosaic virus [CaMV], and cowpea

mosaic virus), it is unclear if plant viruses in general have unique movement mechanisms or whether universal parameters apply. Also unresolved is the form by which plant viruses move. Viruses may move as
viral particles, or as complexes between genomic nucleic acids and coat
protein and/or movement protein, or as some combination of particles
and nucleoprotein complexes, dependent on whether the virus is moving
cell-to-cell or long distances. Other unanswered questions concern the
role of the host in virus spread. Since movement proteins appear to bind
nonspecifically to single-stranded nucleic acids, and since it is unlikely
that cellular nucleic acids move freely between cells, host factors are
probably involved which limit cell-to-cell spread to the virus.

## Symptom Induction by Plant Viruses

Although a wealth of information exists on the effects of plant virus infection on host metabolism (for review, see Fraser 1987; van Loon
1987), little is known about how the virus triggers these metabolic
changes. Susceptible interactions between plants and viruses can result in
a variety of visible symptoms ranging from mild stunting to overall
necrosis. In some plant/virus interactions, normal levels of viral replication result in no detectable symptoms (Li and Simon 1990), a condition
known as tolerance. Complicating the analysis of susceptible plant/virus
interactions is the influence of such factors as environmental conditions
(e.g., light intensity, temperature, nutrition, and day length), physiological age of the inoculated tissue, and the presence of subviral RNAs.

DI RNAs, determined to be associated with an increasing number of
plant viruses, can modulate the symptoms expressed by their helper
viruses (for review, see Roux et al. 1991). DI RNAs can attenuate
(Hillman et al. 1987; Burgyan et al. 1989), intensify (Li et al. 1989), or
have no effect on symptoms (White et al. 1992). Sat-RNAs can also
dramatically influence the symptoms expressed by satellite-free helper
viruses (for review, see Collmer and Howell 1992). The symptom
modulation ability of sat-RNAs has been mapped to specific domains
(Simon et al. 1988) and even to single bases (Masuta and Takanami
1989; Jaegle et al. 1990; Sleat and Palukaitis 1990). In general, DI RNAs
and sat-RNAs that decrease the normal symptoms produced by their
helper viruses interfere either directly or indirectly with the replication of
the helper virus genome (Roux et al. 1991; Collmer and Howell 1992).
Whether the subviral RNAs that intensify symptoms do so by interaction
with the helper virus and/or some undetermined host factor is still unknown.

Attempts to elucidate the molecular events that lead to symptom formation have mainly centered on determining the role that viral nucleic acids and proteins play in development of macroscopic disease symptoms. By generating transgenic plants expressing the gene VI product of CaMV, Baughman et al. (1988) were able to demonstrate that virus-like symptoms could be induced in a plant that is not a normal host. A second, very fruitful approach to identifying viral determinants of symptom production has been the use of "reverse genetics." Viral nucleic acids are mutagenized in vitro; then inoculated on plants either directly in the case of DNA viruses, or following the production of in vitro synthesized transcripts for RNA viruses; and the effects of the mutation on symptom expression are determined. Using this approach, the coat proteins of a number of viruses have been implicated in disease production (Dawson et al. 1988; Stratford and Covey 1989; Petty and Jackson 1990; Heaton et al. 1991; Neeleman et al. 1991; Chapman et al. 1992; Shintaku et al. 1992).

Although much of the research on viral symptom determinants has focused on the effects of altering specific viral products, nucleotide sequence changes that do not affect coding capacity can also contribute to symptom production (Petty et al. 1990; Boulton et al. 1991; Rodriguez-Cerezo et al. 1991). The symptoms induced by many viroids (small circular RNAs that lack mRNA activity) are also similar to those of the larger, more complex plant viruses. These results suggest that viral and viroid nucleic acids, either the primary sequence or secondary structure, in addition to specific viral products, might interact with a host component(s) leading to a cascade of events resulting in disease. Although such host components have not yet been identified, intriguing sequence complementarity between viroids and plant 7S RNA (Haas et al. 1988), and between one sat-RNA and $tRNA^{Glu}$ (Masuta et al. 1992), have been noted.

To date, little is known about the changes in plant gene expression that follow the successful invasion of a plant by a virus or viroid. Unlike animal virus infections, the overall population of $poly(A)^+$ RNA in infected plants does not appear to change substantially from levels in uninfected plants (Matthews 1991). Attempts to identify specific proteins induced in susceptible interactions using a molecular genetic approach have been few and have so far failed to clarify the process of disease induction (Stratford and Covey 1988; Saunders et al. 1989).


**Resistance to Plant Viruses**

Much research on plant viruses is driven by a need to protect plants against virus infections. Viruses and other pathogens are responsible for nearly 10 billion dollars in crop damages each year in the United States (Agrios 1988). Most plants, however, are resistant to most viruses (Matthews 1991; White and Antoniw 1991). According to Fraser (1990), if all members of a species permit no virus accumulation, the resistance is termed *non-host*. Genetic resistance (also referred to as *cultivar resistance*) occurs if some members of a plant species are resistant and others are susceptible. For both non-host and genetic resistance, a plant is said to be *immune* if no virus replication occurs even in initially inoculated cells. This is generally assayed by determining if the virus can replicate in isolated protoplasts. If the virus can replicate in protoplasts, and the virus is localized to the initially inoculated cells or a few surrounding cells in inoculated tissue, a *subliminal* infection is said to have occurred. Subliminal infections are often accompanied by a hypersensitive response at the site of infection involving rapid cell death resulting in localized necrotic lesions. The appearance of local necrotic lesions is usually correlated with restriction of virus spread (Fraser 1987) and the induction of systemic acquired resistance (see below).

### Genetic Resistance

Genetic resistance to viruses is mainly specified by single dominant genes. However, it should be noted that resistance to viruses has been studied nearly exclusively in crop plants that have been specifically bred to have resistance under control of single genes. There are probably several different biochemical mechanisms involved in genetic resistance of plants to virus infections; these include restriction of viral spread (Holmes 1938) or inhibition of viral multiplication at the level of transcription, translation, or assembly (Wyatt and Kuhn 1979; Fraser and Gerwitz 1980; Maule et al. 1980). Despite progress made in recent years defining viral genes involved in genetic resistance, virtually nothing is known about plant resistance genes at the molecular level.

The best-studied loci involved in genetic resistance to plant viruses are the tobacco *N* and *N'* genes, which confer resistance to TMV. The *N* gene is a single dominant gene that localizes all strains of TMV to small necrotic lesions at the inoculation site. Despite many years of work by laboratories throughout the world, the *N* gene remains to be cloned, and the mechanism by which its expression results in resistance to TMV has not been identified. One suggestion is that the *N* gene is involved in the activation of an antiviral factor (AVF) that can affect virus replication, analogous to the interferon system in mammals (Antignus et al. 1977; Sela et al. 1978). In TMV-infected protoplasts, the presence of the *N* gene is also correlated with the release of an inhibitor of virus replication into the media (IVR; Loebenstein and Gera 1981; Spiegel et al. 1989).

The *N'* gene is a second dominant gene conferring hypersensitivity to TMV. Unlike the *N* gene, the *N'* gene localizes many, but not all, strains of TMV (Valleau 1952; Fraser 1983). Several more recent reports have strongly implicated the coat protein in eliciting the hypersensitive reaction in N' plants although the coat protein was not involved in development of systemic symptoms (Saito et al. 1987; Knorr and Dawson 1988; Culver and Dawson 1989). As is the case for the *N* gene, nothing is known about how the product of the *N'* gene might be involved in this resistance response.

In tomato, two dominant genetic loci, *Tm-1*, and *Tm-2* and the allele of *Tm-2*, *Tm-2²*, have been recognized as conferring resistance to some strains (*Tm-1* and *Tm-2*) or all natural strains of TMV (*Tm-2²*) tested (Pelham 1966; Hall 1980). The resistance specified by *Tm-1* is at the level of virus replication (Watanabe et al. 1987) and resistance-breaking strains of TMV contain alterations in the putative replicase open reading frame (Meshi et al. 1988). Resistance conferred by *Tm-2* or *Tm-2²* is expressed only in whole plants or in leaf disks but not in isolated protoplasts (Motoyoshi and Oshima 1975, 1977; Stobbs and MacNeill 1980). A strain of TMV that could overcome the resistance specified by *Tm-2* had alterations in the movement protein. This implied that resistance in *Tm-2* tomatoes was due to an inhibition in virus cell-to-cell spread (Meshi et al. 1989).

## Systemic Acquired Resistance

Hypersensitive responses to viruses and other pathogens can be associated with the phenomenon of systemic acquired resistance. Systemic acquired resistance has been defined recently as the increased resistance to further pathogen or insect attack that results from a localized interaction with viral, bacterial, or fungal pathogens or invertebrate pests (Enyedi et al. 1992). The elicitation of systemic acquired resistance is correlated with the appearance of pathogenesis-related (PR) proteins, a group of mostly extracellular proteins, some of which have antimicrobial activity (Bol et al. 1990; Linthorst 1991). Although transgenic tobacco constitutively expressing one or a few PR proteins did not exhibit increased resistance to virus infection (Cutt et al. 1989; Linthorst et al. 1989), the correlation of PR protein expression with the hypersensitive reaction and systemic acquired resistance suggests their involvement in restriction of viral spread. Recent work on the elucidation of the signal transduction pathway between the onset of virus infection and the expression of these proteins has demonstrated a correlation between increases in endogenous salicylic acid levels and PR protein gene expression (Malamy et al. 1990;

Métraux et al. 1990). However, the mechanism by which PR proteins might be involved in viral resistance is unknown.

### *ARABIDOPSIS*/VIRUS INTERACTIONS

Since most work in plant virology is being driven by agronomic considerations, only recently has *Arabidopsis* been used as a host to study plant/virus interactions. The utility of *Arabidopsis* as a model system has not gone unnoticed, however, and several viruses previously found to be pathogenic on crucifers have also been found to infect *Arabidopsis*. Studies using *Arabidopsis* as a host are still in the early stages of development. As outlined below, such studies have mainly concentrated on identifying resistant and susceptible ecotypes as a prelude to the isolation of host factors that affect virus replication and spread.

### Beet Curly Top Virus

BCTV is a single-stranded DNA virus with a broad host-range that includes a number of crucifers including *Arabidopsis* (Davis 1992). As with other leafhopper-transmitted geminiviruses, BCTV is not infectious when mechanically inoculated on plants but must be introduced using a technique known as agroinfection (Briddon et al. 1989). In agroinfection, the viral genome (usually a head-to-tail dimer) is incorporated into the Ti plasmid of *Agrobacterium tumefaciens*. After infection of the plant with *Agrobacteria* and transfer of the T-DNA, transcription of viral products and replication of the viral genome occur, leading to the production of movement-competent virus (Grimsley et al. 1986). *Arabidopsis* plants have been successfully agroinfected with BCTV by inoculating plants just prior to bolting with several drops of an overnight *Agrobacterium* culture on the middle of the rosette and pricking the crown a few times with an insect pin (K. Davis et al., unpubl.).

Two strains of BCTV (Logan and CFH) are virulent on *Arabidopsis* ecotype Columbia (Col-0). Symptoms appeared 2 weeks following inoculation and were characterized by curled leaves and stunted inflorescences. Flower structures and siliques also developed abnormally. In infected Col-0 plants, symptom development was correlated with the accumulation of viral genomic DNA. Viral DNA was detectable by 10 days following inoculation and continued to accumulate over the next 4 weeks. By 28 days postinoculation, the level of BCTV CFH was about five times higher than that observed for BCTV Logan. Symptoms produced by BCTV CFH were more severe than those produced by BCTV Logan on a different *Arabidopsis* ecotype, with infected plants frequently dying within 4 weeks of inoculation. Symptom severity in these plants was correlated with an increased accumulation of viral DNA; 12-fold more viral DNA accumulated in CFH-infected plants than in Logan-infected plants (K. Davis et al., unpubl.).

Of the 40 ecotypes screened for susceptibility to BCTV, most displayed symptoms similar to those of Col-0, whereas 2 ecotypes were asymptomatic when inoculated with Logan but not CFH. Analysis of viral DNA levels in asymptomatic plants revealed that plants inoculated with Logan did not accumulate detectable levels of viral DNA. $F_1$ progeny of crosses between the asymptomatic ecotypes and Col-0 were all susceptible to BCTV, suggesting that resistance is a recessive trait (K. Davis et al., unpubl.).

## Cauliflower Mosaic Virus

CaMV, a double-stranded DNA virus, was one of the first characterized pathogens of *Arabidopsis* (Balàzs and Lebeurier 1981). CaMV can be mechanically inoculated onto *Arabidopsis* by dipping a cotton-tipped applicator or plastic spatula into an inoculum containing Celite and either cell sap from infected plants or virion preparations, and then rubbing the inoculum lightly on several leaves of 2- to 3-week-old plants. Symptoms produced by CaMV infection are dependent on the strain of the virus and ecotype of *Arabidopsis* (Balàzs and Lebeurier 1981; Melcher 1989; Leisner and Howell 1992). In general, symptoms included vein clearing and chlorotic spots on rosette leaves, vein clearing and chlorosis on cauline leaves, mottled and chlorotic stems, and yellow to brown siliques that arose from chlorotic stem sectors. Occasionally, the chlorosis on rosette leaves proceeded to necrosis leading to the death of the plant. In a study using 23 ecotypes and three virus strains, Leisner and Howell (1992) found a correlation between the rate of development of an ecotype, as assayed by time of bolting, and the nature of the systemic symptoms. In general, when inoculated with any of the three virus strains, ecotypes with early bolting times only displayed symptoms on the flower stalks and cauline leaves, whereas late-bolting plants also exhibited symptoms on the rosette leaves.

Whole-plant hybridizations (also known as plant skeleton hybridizations) were used to examine the pattern of movement of CaMV in ecotypes of *Arabidopsis* with differing developmental rates (Leisner et al. 1993). This variation of in situ hybridization was originally developed by Melcher et al. (1981) and modified for work with CaMV in turnip leaves by Leisner et al. (1992) and *Arabidopsis* (Leisner et al. 1993). Briefly, infected plants were treated in EtOH until leaves turned white and were then incubated in a solution containing sodium azide, sodium dodecyl sulfate, and proteinase K that released the DNA from viral particles. As with DNA immobilized on a membrane, plants were subjected to prehybridization, hybridization, and washing steps. The treated plants were then dried and exposed to X-ray film (Fig. 1). Using this technique, Leisner et al. (1993) determined that CaMV moved into the petioles of inoculated leaves of ecotype Col-0 by 14 days postinoculation and was detectable in systemic leaves 4 days later. By 26 days postinoculation, all plant tissues accessible to CaMV had detectable levels of virus.

The ability of CaMV to move in *Arabidopsis*, as detected by plant skeleton hybridizations, was correlated with the developmental rate of the plant; plants subjected to slow growth conditions (short day length) allowed more widespread virus movement (Leisner et al. 1993). The authors suggested that tissue maturation leads to a change in the sink-source relationship and a corresponding change in accessibility to virus import. Since the apparent resistance to virus systemic infection by fast-developing plants was reversible by altering the growth conditions, the authors have coined the term "developmental resistance" to differentiate between resistance to systemic spread exhibited by fast-developing ecotypes and resistance that is not based on developmental or growth conditions.

Three *Arabidopsis* ecotypes were exceptions to this trend, having no detectable symptoms even though they differed substantially from each other in developmental rate (Leisner and Howell 1992). Using plant skeleton hybridization, the authors demonstrated that one of these ecotypes, En-2, did not support systemic movement of CaMV but did support localized cell-to-cell movement (Leisner et al. 1993). Analysis of $F_2$ progeny from crosses between ecotypes En-2 and Col-0 revealed that En-2 plants contain a single dominant trait that correlates with resistance to CaMV.

## Tobacco Mosaic Virus

TMV is a widely studied single-stranded RNA virus that can be applied mechanically to its hosts. A strain of TMV that is infectious on crucifers (TMV-Cg; Li et al. 1983) was found to also be a pathogen of *Arabidopsis*, as assayed by accumulation of viral coat protein (Ishikawa et al. 1992). Whereas the more widely studied tomato strain of TMV multiplied at low levels in *Arabidopsis*, the TMV-Cg strain reached levels of about 6 mg/g tissue by 14 days postinoculation in five different ecotypes including Col-0. Col-0 plants infected at an early age (15–20 days after sowing) exhibited mild disease symptoms, including stunting

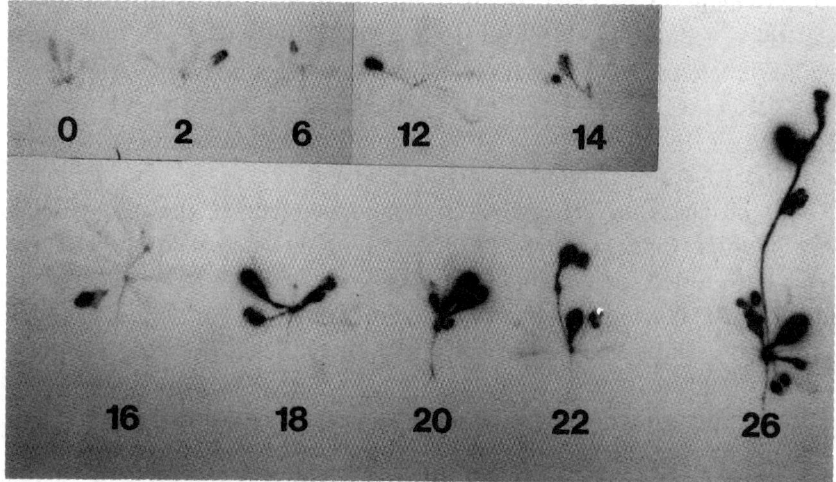



*Figure 1* Time course of virus movement during systemic infection. *A. thaliana* (Col-0 ecotype) was inoculated on one of the rosette leaves with CaMV CM4-184, and virus movement was visualized by plant skeleton hybridization. Numbers indicate days after inoculation. (Reprinted, with permission, from Leisner et al. 1993 [American Society of Plant Physiologists].)

and twisting of stems and dark-greening of rosette leaves. As is the case for many plant/virus interactions (Matthews 1991), older plants (>3 weeks after sowing) displayed weaker or undetectable symptoms.

Ishikawa et al. (1992) screened for mutants of Col-0 with a decreased ability to accumulate TMV-Cg coat protein. Of 6000 $M_2$ plants derived from ethyl methanesulfonate (EMS)-treated seeds, two independent plants were reproducibly found to accumulate only a fraction of the normal level of coat protein. Crosses between the two plants and to the parental Col-0 ecotype revealed that each plant had a monogenic recessive Mendelian trait that was allelic. Plants homozygous recessive at the locus, coined *TOM-1* (*To*bamovirus *m*ultiplication) were still normally susceptible to two other viruses that infect *Arabidopsis*, turnip crinkle virus (TCV) (see below) and turnip yellow mosaic virus, suggesting that these specific mutations at the *TOM-1* locus only influence the coat protein accumulation of TMV. Still undetermined is whether the *tom-1* lines affect virus replication, translation, or movement.

## Turnip Crinkle Virus

TCV is a single-stranded RNA virus that infects a large number of cruciferous and noncruciferous hosts (Broadbent and Heathcote 1958).

TCV is naturally associated with a number of subviral RNAs, including sat-RNAs (Simon and Howell 1986), DI RNAs (Li et al. 1989), and chimeric RNAs derived from recombination events between various subviral RNAs (Cascone et al. 1990) and between the subviral and genomic RNAs (Zhang et al. 1991). TCV genomic RNA, in combination with subviral RNAs containing the 3′ untranslated end of the viral genomic RNA, produces severe symptoms on a number of hosts including *Arabidopsis* (Li and Simon 1990). Twenty-eight ecotypes of *Arabidopsis* have been tested for susceptibility to TCV (Li and Simon 1990; Simon et al. 1992). The symptoms produced by TCV-infected *Arabidopsis* depend greatly on the presence or absence of virulent subviral RNAs. Twenty-seven of 28 ecotypes inoculated with TCV without virulent subviral RNAs displayed generally mild symptoms such as stunted inflorescence and chlorosis of rosette and cauline leaves. The same ecotypes inoculated with TCV containing at least one virulent subviral RNA failed to bolt and were killed 18 days postinoculation by a rapidly spreading necrosis (Fig. 2). Viral genomic RNA was detected by Northern hybridization by 3 days postinoculation and reached maximal levels 4 days later (Simon et al. 1992).

Ecotype Dijon (Di-0) was highly resistant to TCV infection (Simon et al. 1992), accumulating little detectable viral RNA until late in infection. Nearly all TCV-infected Di-0 plants were symptomless 3 weeks postinoculation, with about 25% of the plants displaying localized necrotic lesions on inoculated leaves (Simon et al. 1992). In addition, a few plants exhibited mild symptoms such as stunted rosette leaves and inflorescence, curled bolts and siliques, and early desiccation. The number of Di-0 plants exhibiting symptoms after virus infection varied depending on the growth environment. Conditions conducive for rapid growth of Di-0 (long day length, high light intensity) resulted in the highest percentage of symptomless plants (>90–95%), with the remaining plants expressing very mild symptoms (Simon et al. 1992; J. Lew and A. Simon, unpubl.). However, nearly 25% of TCV-inoculated Di-0 plants grown under conditions that promoted slow growth (8 hr light per 24-hr cycle) were killed by a rapidly spreading systemic necrosis similar to that of susceptible ecotypes. Interestingly, susceptible Col-0 plants grown under conditions promoting slow growth were less susceptible to TCV infection as assayed by amount of virus accumulating 5 days postinoculation and visible symptoms 9 days postinoculation (Simon et al. 1992). These results differed substantially from the results using CaMV, where a correlation between slow plant growth and a more susceptible phenotype was found. The resistance to TCV exhibited by Di-0 did not extend to a second, closely related virus (cardamine chlorotic

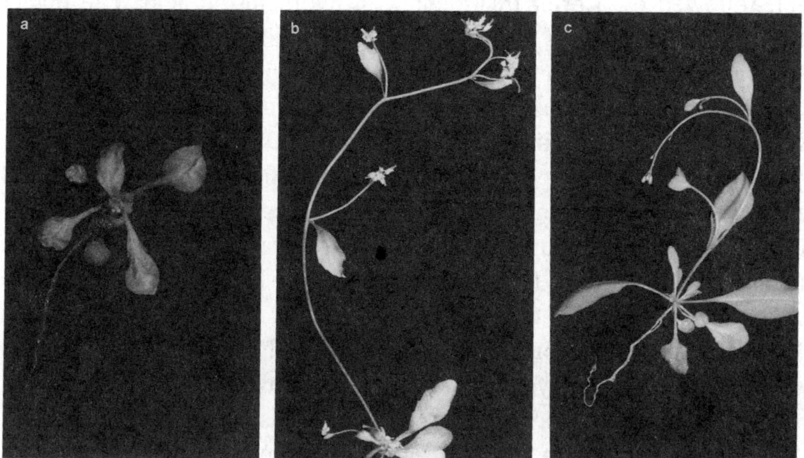

*Figure 2* Symptoms of *Arabidopsis* plants infected with TCV. Col-0 (*a*) or Di-0 (*b*) ecotypes 2 weeks postinoculation with TCV containing the virulent satellite RNA, sat-RNA C. Most Di-0 plants remained symptomless. The stem of the plant shown displayed mild symptoms of curling, especially at the tip. Col-0 plant shown in *c* was mock-infected. The stem was purposely curled to fit the photograph.

fleck; J. Hoe and A. Simon, unpubl.) or to CaMV. This suggests that, as with TMV, resistance is virus-specific and does not involve a general inability to support virus accumulation.

As described above, resistance can be at the cellular level (inhibition of virus replication) or at the level of systemic infection (inhibition of cell-to-cell or long-distance movement). To investigate the basis of resistance of Di-0 to TCV, a system was developed for the replication of TCV in *Arabidopsis* protoplasts. Protoplasts were prepared from axenic callus cultures of either resistant Di-0 or susceptible Col-0 leaves and infected with the virulent form of TCV in a solution containing 40% polyethylene glycol (Simon et al. 1992). TCV genomic RNA replicated at a similar rate in protoplasts from either ecotype, indicating that inhibition of virus replication was not the cause of the resistance phenotype.

To determine if the ability of TCV to spread cell-to-cell or long distances was inhibited in Di-0, the whole-plant hybridization procedure used in the analysis of DNA virus movement was adapted for the detection of RNA viruses (Simon et al. 1992). Since this publication, the protocol has been further refined to detect lower RNA levels and increase the signal-to-noise ratio (C. Kong and A. Simon, unpubl.). Briefly, infected *Arabidopsis* plants are incubated in EtOH under conditions that do not result in degradation of plant RNAs. The plants are then incubated in a buffer containing SDS, sodium azide, pronase A, and EDTA overnight, followed by a brief treatment with hydrochloric acid. The plants are baked under vacuum and then treated in a similar fashion to RNA-bound membranes, with prehybridization, hybridization, and washing steps (Fig. 3). Results using whole-plant hybridizations indicated that TCV moved cell-to-cell in inoculated leaves of Di-0 and migrated to the opposite leaf by 7 days postinoculation. In Col-0, virus was detected in the vascular tissue of uninoculated younger leaves by 7 days postinoculation and in numerous younger leaves soon afterward. No such signal was found in Di-0 vascular tissue, suggesting that long-distance systemic movement is inhibited in this ecotype.

S. Uknes et al. (in prep.) investigated whether inoculation of Di-0 with TCV produces a systemic acquired response. Under their growth conditions, inoculated leaves of all Di-0 plants displayed necrotic lesions 2–5 days postinoculation. As described above, hypersensitive responses are frequently associated with systemic acquired resistance that is correlated with the induction of salicylic acid levels and the appearance of PR proteins. The authors determined that 7 days postinoculation with TCV, salicylic acid levels and transcripts of several PR proteins increased in non-inoculated leaves. TCV-inoculated Di-0 plants were also tested for acquired resistance against normally virulent bacteria (*Pseudomonas syringae* pv *tomato* DC3000), fungi (*Peronospora parasitica*), and further TCV infection. Plants inoculated with TCV 7 days before being challenged accumulated tenfold less *P. syringae*, did not exhibit normal downy mildew disease symptoms when infected with *P. parasitica*, and produced fewer lesions on the leaves subjected to further inoculation with TCV. These results extended the authors' previous findings to include interaction with a resistant pathogen, as well as treatment with the immunization compound 2,6-dichloroisonicotinic acid (INA; Uknes et al. 1992), as inducers of systemic acquired resistance in *Arabidopsis*.

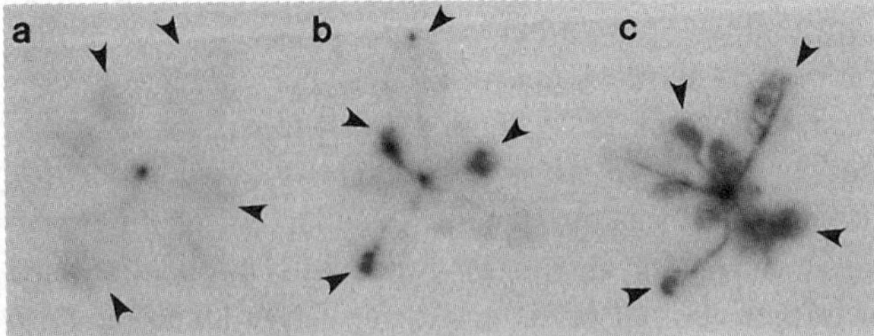

*Figure 3* Plant skeleton hybridization to *Arabidopsis* ecotypes Col-0 (*a,c*) or Di-0 (*b*). Plant *a* was mock-infected with buffer, and plants *b* and *c* were inoculated with TCV and sat-RNA C, 6 days prior to hybridization. Arrowheads denote inoculated leaves.

## PROSPECTS FOR FUTURE RESEARCH

One of the most important problems in the field of plant/virus interactions is the identity of host factors that, by their presence or absence, specify resistance or susceptibility to viruses. The wealth of natural genetic diversity, coupled with the ease of isolating genes, positions *Arabidopsis* as an important tool in the effort to understand the role the plant plays in virus replication, movement, and symptom production. Whether the information gained using *Arabidopsis* will be applicable to more agronomically important virus/crop combinations is not known. In the area of systemic acquired resistance, the responses of *Arabidopsis* are very similar to the well-characterized responses of tobacco, suggesting that *Arabidopsis* will be an important system for the further elucidation of the signals that mediate acquired resistance. Furthermore, studies on the relationship between plant age/developmental rate and virus susceptibility should continue to result in new insights on the poorly understood area of long-distance virus movement, which may be applicable to a wide range of plant/virus combinations.

Additional areas where *Arabidopsis* should prove useful include the analysis of gene expression in susceptible interactions and the relationship between virus amplification and symptom induction. Most research has been concerned with the identification of virus-induced products in resistant interactions (e.g., PR proteins). However, understanding susceptible interactions may lead to the ability to block early stages of virus infection, possibly resulting in new strategies in virus control. In the area of disease production, the coat proteins of a number of viruses have been implicated in symptoms, yet nothing is known about plant factors that interact with this protein. With the increasing number of researchers using *Arabidopsis* as a host for virus studies, the role that the plant plays in plant/virus interactions may soon be clarified.

## ACKNOWLEDGMENTS

I am grateful to S.H. Howell, S. Uknes, and K.R. Davis for providing unpublished results, and to J. Krebs for helpful comments on the manuscript. Research in my laboratory is supported by the National Science Foundation (MCB-9315948 and DMB-9105890).

## REFERENCES

Agrios, G.N. 1988. *Plant pathology*, 3rd edition. Academic Press, San Diego.
Allison, R.F., C. Thompson, and P. Ahlquist. 1990. Regeneration of a functional RNA virus genome by recombination between deletion mutants, and requirement for cowpea

chlorotic mottle virus 3a and coat genes for systemic infection. *Proc. Natl. Acad. Sci.* **87:** 1820–1824.

Anderson, J.M., P. Palukaitis, and M. Zaitlin. 1992. A defective replicase gene induces resistance to cucumber mosaic virus in transgenic tobacco plants. *Proc. Natl. Acad. Sci.* **89:** 8759–8763.

Antignus, Y., I. Sela, and I. Harpaz. 1977. Further studies on the biology of antiviral factor (AVF) from virus infected plants and its association with the *N*-gene of *Nicotiana* species. *J. Gen. Virol.* **35:** 107–116.

Balàzs, E. and G. Lebeurier. 1981. *Arabidopsis* is a host of cauliflower mosaic virus. *Arabidopsis Inf. Serv.* **18:** 130–134.

Baughman, G.A., J.D. Jacobs, and S.H. Howell. 1988. Cauliflower mosaic virus gene VI produces a symptomatic phenotype in transgenic tobacco plants. *Proc. Natl. Acad. Sci.* **85:** 733–737.

Beachy, R.N., S. Loesch-Fries, and N.E. Tumer. 1990. Coat protein-mediated resistance against virus infection. *Annu. Rev. Phytopathol.* **28:** 451–474.

Bol, J.F., H.J.M. Linthorst, and B.J.C. Cornelissen. 1990. Plant pathogenesis-related proteins induced by virus infection. *Annu. Rev. Phytopathol.* **28:** 113–138.

Boulton, M.I., D.I. King, J. Donson, and J.W. Davies. 1991. Point substitutions in a promoter-like region and the V1 gene affect the host range and symptoms of maize streak virus. *Virology* **183:** 114–121.

Braun, C.J. and C.L. Hemenway. 1992. Expression of amino-terminal portions of full-length viral replicase genes in transgenic plants confers resistance to potato virus X infection. *Plant Cell* **4:** 735–744.

Briddon, R.W., J. Watts, P.G. Markham, and J. Stanley. 1989. The coat protein of beet curly top virus is essential for infectivity. *Virology* **172:** 628–633.

Broadbent, L. and G.D. Heathcote. 1958. Properties and host range of turnip crinkle, rosette and yellow mosaic viruses. *Ann. Appl. Biol.* **46:** 585–592.

Burgyan, J., F. Grieco, and M. Russo. 1989. A defective interfering RNA molecule in cymbidium ringspot virus infections. *J. Gen. Virol.* **70:** 235–239.

Cascone, P.J., C.D. Carpenter, X.H. Li, and A.E. Simon. 1990. Recombination between satellite RNAs of turnip crinkle virus. *EMBO J.* **9:** 1709–1715.

Chapman, S., G. Hills, J. Watts, and D. Baulcombe. 1992. Mutational analysis of the coat protein gene of potato virus X: Effects on virion morphology and viral pathogenicity. *Virology* **191:** 223–230.

Citovsky, V. and P. Zambryski. 1991. How do plant virus nucleic acids move through intercellular connections? *BioEssays* **13:** 373–379.

Citovsky, V., M.L. Wong, A.L. Shaw, B.V. Venkataram Prasad, and P. Zambryski. 1992. Visualization and characterization of tobacco mosaic virus movement protein binding to single-stranded nucleic acids. *Plant Cell* **4:** 397–411.

Collmer, C.W. and S.H. Howell. 1992. Role of satellite RNA in the expression of symptoms caused by plant viruses. *Annu. Rev. Phytopathol.* **30:** 419–442.

Culver, J. and W.O. Dawson. 1989. Tobacco mosaic virus coat protein: An elicitor of the hypersensitive reaction but not required for the development of mosaic symptoms in *N. sylvestris. Virology* **173:** 755–758.

Cutt, J.R., M.H. Harpster, D.C. Dixon, J.P. Carr, P. Dunsmuir, and D.F. Klessig. 1989. Disease response to tobacco mosaic virus in transgenic tobacco plants that constitutively express the pathogenesis-related *PR1b* gene. *Virology* **173:** 89–97.

David, C., R. Gargouri-Bouzid, and A-L. Haenni. 1992. RNA replication of plant viruses containing an RNA genome. *Prog. Nucleic Acid Res.* **42:** 157–227.

Davis, K.R. 1992. *Arabidopsis thaliana* as a model host for studying plant-pathogen interactions. In *Molecular signals in plant-microbe communications* (ed. D.P.S. Verma), pp. 1–4. CRC Press, Boca Raton, Florida.

Dawson, W.O., P. Bubrick, and G.L. Grantham. 1988. Modification of the tobacco mosaic virus coat protein gene affecting replication movement and symptomatology. *Phytopathology* **78:** 783–789.

De Jong, W. and P. Ahlquist. 1992. A hybrid plant RNA virus made by transferring the noncapsid movement protein from a rod-shaped to an icosahedral virus is competent for systemic infection. *Proc. Natl. Acad. Sci.* **89:** 6808–6812.

Deom, C.M., M. Lapidot, and R.N. Beachy. 1992. Plant virus movement proteins. *Cell* **69:** 221–224.

Ding, B., J.S. Haudenshield, R.J. Hull, S. Wolf, R.N. Beachy, and W.J. Lucas. 1992. Secondary plasmodesmata are specific sites of localization of the tobacco mosaic virus movement protein in transgenic tobacco plants. *Plant Cell* **4:** 915–928.

Enyedi, A.J., N. Yalpani, P. Silverman, and I. Raskin. 1992. Signal molecules in systemic plant resistance to pathogens and pests. *Cell* **70:** 879–886.

Fraser, R.S.S. 1983. Varying effectiveness of the *N'* gene for resistance to tobacco mosaic virus in tobacco infected with virus strains differing in coat protein properties. *Physiol. Plant Pathol.* **22:** 109–119.

———. 1987. *Biochemistry of virus-infected plants*. Research Studies Press, Letchworth, United Kingdom.

———. 1990. The genetics of resistance to plant viruses. *Annu. Rev. Phytophathol.* **28:** 179–200.

Fraser, R.S.S. and A. Gerwitz. 1980. Tobacco mosaic virus infection does not alter the polyadenylated messenger RNA content of tobacco leaves. *J. Gen. Virol.* **46:** 139–148.

Goldback, R. and J. Wellink. 1988. Evolution of plus-strand RNA viruses. *Intervirology* **29:** 260–267.

Grimsley, N., B. Hohn, T. Hohn, and R. Walden. 1986. "Agroinfection," an alternative route for viral infection of plants by using the Ti plasmid. *Proc. Natl. Acad. Sci.* **83:** 3282–3286.

Haas, B., A. Klanner, K. Ramm, and H.L. Sanger. 1988. The 7s RNA from tomato leaf tissue resembles a signal recognition particle RNA and exhibits a remarkable sequence complementarity to viroids. *EMBO J.* **7:** 4063–4074.

Hacker, D.L., I.T.D. Petty, N. Wei, and T.J. Morris. 1992. Turnip crinkle virus genes required for RNA replication and virus movement. *Virology* **186:** 1–8.

Hall, T.J. 1980. Resistance at the *Tm-2* locus in tomato to tomato mosaic virus. *Euphytica* **29:** 189–197.

Hayes, R.J. and K.W. Buck. 1990. Complete replication of a eukaryotic virus RNA in vitro by a purified RNA-dependent RNA polymerase. *Cell* **63:** 363–368.

Hayes, R.J., D. Tousch, M. Jacquemond, V.C. Pereira, K.W. Buck, and M. Tepfer. 1992. Complete replication of a satellite RNA in vitro by a purified RNA-dependent RNA polymerase. *J. Gen. Virol.* **73:** 1597–1600.

Heaton, L.A., T.C. Lee, N. Wei, and T.J. Morris. 1991. Point mutations in the turnip crinkle virus capsid protein affect the symptoms expressed by *Nicotiana benthamiana*. *Virology* **183:** 143–150.

Hillman, B.I., J.C. Carrington, and T.J. Morris. 1987. A defective interfering RNA that contains a mosaic of a plant virus genome. *Cell* **51:** 427–433.

Holmes, F.O. 1938. Inheritance of resistance to tobacco mosaic disease in tobacco. *Phytopathology* **28:** 553–561.

Ishikawa, M., F. Obata, T. Kumagai, and T. Ohno. 1992. Isolation of mutants of *Arabidopsis thaliana* in which accumulation of tobacco mosaic virus coat protein is reduced to low levels. *Mol. Gen. Genet.* **230**: 33–38.

Jaegle, M., M. Devic, M. Longstall, and D. Baulcombe. 1990. Cucumber mosaic virus satellite RNA (Y strain): Analysis of sequences which affect yellow mosaic symptoms on tobacco. *J. Gen. Virol.* **71**: 1905–1912.

Knorr, D.A. and W.O. Dawson. 1988. A point mutation in the tobacco mosaic virus capsid protein gene induces hypersensitivity in *N. sylvestris. Proc. Natl. Acad. Sci.* **84**: 2363–2367.

Leisner, S.M. and S.H. Howell. 1992. Symptom variation in different *Arabidopsis thaliana* ecotypes produced by cauliflower mosaic virus. *Phytopathology* **82**: 1042–1046.

Leisner, S.M., R. Turgeon, and S.H. Howell. 1992. Long distance movement of cauliflower mosaic virus in infected turnip plants. *Mol. Plant-Microbe Interact.* **5**: 41–47.

———. 1993. Effects of host plant development and genetic determinants on the long-distance movement of cauliflower mosaic virus in *Arabidopsis. Plant Cell* **5**: 191–202.

Li, X.H. and A.E. Simon. 1990. Symptom intensification on cruciferous hosts by the virulent satellite RNA of turnip crinkle virus. *Phytopathology* **80**: 238–242.

Li, X.H., L.A. Heaton, T.J. Morris, and A.E. Simon. 1989. Turnip crinkle virus defective interfering RNAs intensity viral symptoms and are generated *de novo. Proc. Natl. Acad. Sci.* **86**: 9173–9177.

Li, Z.-Y., I. Uyeda, and E. Shikata. 1983. Crucifer strain of tobacco mosaic virus isolated from garlic. *Mem. Fac. Agric. Hokkaido Univ.* **13**: 542–549.

Linthorst, H.J.M. 1991. Pathogenesis-related proteins of plants. *Crit. Rev. Plant Sci.* **10**: 123–150.

Linthorst, H.J.M., R.L.J. Meuwissen, S. Kauffmann, and J.F. Bol. 1989. Constitutive expression of pathogenesis-related proteins PR-1, GRP, and PR-S in tobacco has no effect on virus infection. *Plant Cell* **1**: 285–291.

Loebenstein, G. and A. Gera. 1981. Inhibitor of virus replication released from tobacco mosaic virus infected protoplasts of a local lesion-responding cultivar. *Virology* **114**: 132–139.

MacFarlane, S.A. and J.W. Davies. 1992. Plants transformed with a region of the 201-kilodalton replicase gene from pea early browning virus RNA1 are resistant to virus infection. *Proc. Natl. Acad. Sci.* **89**: 5829–5833.

Malamy, J., J.P. Carr, D.R. Klessig, and I. Raskin. 1990. Salicylic acid: A likely endogenous signal in the resistance response of tobacco to viral infection. *Science* **250**: 1002–1004.

Masuta, C. and Y. Takanami. 1989. Determination of sequence and structural requirements for pathogenicity of a cucumber mosaic virus satellite RNA (Y-satRNA). *Plant Cell* **1**: 1165–1173.

Masuta, C., S. Kuwata, T. Matzuzaki, Y. Takanami, and A. Koiwai. 1992. A plant virus satellite RNA exhibits a significant sequence complementarity to a chloroplast tRNA. *Nucleic Acids Res.* **20**: 2885.

Matthews, R.E.F. 1991. *Plant virology*, 3rd edition. Academic Press, San Diego.

Maule, A.J. 1991. Virus movement in infected plants. *Crit. Rev. Plant Sci.* **9**: 457–473.

Maule, A.J., M.I. Bouton, and K.R. Wood. 1980. Resistance of cucumber protoplasts to cucumber mosaic virus: A comparative study. *J. Gen. Virol.* **51**: 271–279.

Melcher, U. 1989. Symptoms of cauliflower mosaic virus infection in *Arabidopsis thaliana* and turnip. *Bot. Gaz.* **150**: 139–147.

———. 1990. Similarities between putative transport proteins of plant viruses. *J. Gen. Virol.* **71**: 1009–1018.

Melcher, U., C.O. Gardner, and R.C. Essenberg. 1981. Clones of cauliflower mosaic virus identified by molecular hybridization in turnip leaves. *Plant Mol. Biol.* **1:** 63–73.

Meshi, T., F. Motoyoshi, A. Adachi, Y. Watanabe, N. Takamatsu, and Y. Okada. 1988. Two concomitant base substitutions in the putative replicase genes of tobacco mosaic virus confer the ability to overcome the effects of a tomato resistance gene, *Tm-1*. *EMBO J.* **7:** 1575–1581.

Meshi, T., F. Motoyoshi, T. Maeda, S. Yoshiwoka, H. Watanabe, and Y. Okada. 1989. Mutations in the tobacco mosaic virus 30-kd protein gene overcome *Tm-2* resistance in tomato. *Plant Cell* **1:** 515–522.

Meshi, T., Y, Watanabe, T. Saito, A. Sugimoto, T. Maeda, and Y. Okada. 1987. Function of the 30 kD protein to tobacco mosaic virus: Involvement in cell-to-cell movement and dispensability for replication. *EMBO J.* **6:** 2557–2593.

Métraux, J.P., H. Signer, J. Ryals, E. Ward, M. Wyss-Benz, J. Gaudin, K. Raschdorf, E. Schmid, W. Blum, and B. Inverardi. 1990. Increase in salicylic acid at the onset of systemic acquired resistance in cucumber. *Science* **250:** 1004–1006.

Motoyoshi, R. and N. Oshima. 1975. Infection with tobacco mosaic virus of leaf mesophyll protoplasts from susceptible and resistant lines of tomato. *J. Gen. Virol.* **29:** 81–91.

———. 1977. Expression of genetically controlled resistance to tobacco mosaic virus infection in isolated tomato leaf mesophyll protoplasts. *J. Gen. Virol.* **34:** 499–506.

Neeleman, L., A.C. van der Kuyl, and J.R. Bol. 1991. Role of alfalfa mosaic virus coat protein gene in symptom formation. *Virology* **181:** 687–693.

Pelham, J. 1966. Resistance in tomato to tobacco mosaic virus. *Euphytica* **15:** 258–267.

Petty, I.T.D. and A.O. Jackson. 1990. Mutational analysis of barley stripe mosaic virus RNA β. *Virology* **179:** 712–718.

Petty, I.T.D., M.C. Edwards, and A.O. Jackson. 1990. Systemic movement of an RNA plant virus determined by a point substitution in a 5′ leader sequence. *Proc. Natl. Acad. Sci.* **87:** 8894–8897.

Robards, A.W. and W.J. Lucas. 1990. Plasmodesmata. *Annu. Rev. Plant Physiol. Plant Mol. Biol.* **41:** 369–419.

Rodriguez-Cerezo, E., P.G. Klein, and J.G. Shaw. 1991. A determinant of disease symptom severity is located in the 3′-terminal noncoding region of the RNA of a plant virus. *Proc. Natl. Acad. Sci.* **88:** 9863–9867.

Roux, L., A.E. Simon, and J.J. Holland. 1991. Effects of defective interfering viruses on virus replication and pathogenesis in vitro and in vivo. *Adv. Virus Res.* **40:** 181–211.

Saito, T., K. Yamanaka, and Y. Okada. 1990. Long-distance movement and viral assembly of tobacco mosaic virus mutants. *Virology* **176:** 329–336.

Saito, T., T. Meshi, N. Takanatsu, and Y. Okada. 1987. Coat protein gene sequence of tobacco mosaic virus encodes a host response determinant. *Proc. Natl. Acad. Sci.* **84:** 6074–6077.

Saunders, K., A.P. Lucy, and S.N. Covey. 1989. Chracterization of cDNA clones of host RNAs isolated from cauliflower mosaic virus-infected turnip leaves. *Physiol. Mol. Plant Pathol.* **35:** 339–346.

Sela, I., A. Hauschner, and R. Mozes. 1978. The mechanisms of stimulation of the antiviral factor (AVF) in *Nicotiana* leaves. The involvement of phosphorylation and the role of the *N*-gene. *Virology* **89:** 1–6.

Shintaku, M.H., L. Zhang, and P. Palukaitis. 1992. A single amino acid substitution in the coat protein of cucumber mosaic virus induces chlorosis in tobacco. *Plant Cell* **4:** 751–757.

Simon, A.E. and S.H. Howell. 1986. The virulent satellite RNA of turnip crinkle virus has a major domain homologous to the 3′-end of the helper virus genome. *EMBO J.* **5**: 3423–3428.

Simon, A.E., H. Engel, R.P. Johnson, and S.H. Howell. 1988. Identification of regions affecting virulence, RNA processing and infectivity in the virulent satellite of turnip crinkle virus. *EMBO J.* **7**: 2645–2651.

Simon, A.E., X.H. Li, J.E. Lew, R. Stange, C. Zhang, M. Polacco, and C.D. Carpenter. 1992. Susceptibility and resistance of *Arabidopsis thaliana* to turnip crinkle virus. *Mol. Plant-Microbe Interact.* **5**: 496–503.

Sleat, D.E. and P. Palukaitis. 1990. Site-directed mutagenesis of a plant viral satellite RNA changes its phenotype from ameliorative to necrogenic. *Proc. Natl. Acad. Sci.* **87**: 2946–2950.

Spiegel, S., A. Gera, R. Salomon, P. Ahl, S. Harlap, and G. Loebenstein. 1989. Recovery of an inhibitor of virus replication from the intercellular fluid of hypersensitive tobacco infected with tobacco mosaic virus and from uninfected induced-resistant tissue. *Phytopathology* **79**: 258–262.

Stratford, R. and S.N. Covey. 1988. Changes in turnip leaf messenger RNA populations during systemic infection by severe and mild strains of cauliflower mosaic virus. *Mol. Plant-Microbe Interact.* **1**: 243–249.

———. 1989. Segregation of cauliflower mosaic virus symptoms genetic determinants. *Virology* **172**: 451–459.

Stobbs, L.W. and B.H. MacNeill. 1980. Response to tobacco mosaic virus of a tomato cultivar homozygous for gene *Tm-2*. *Can. J. Plant Pathol.* **2**: 5–11.

Takebe, I. 1977. Protoplasts in the study of plant virus replication. *Compr. Virol.* **11**: 237–283.

Uknes, S., B. Mauch-Mani, M. Moyer, S. Potter, S. Williams, S. Dincher, D. Chandler, A. Slusarenko, E. Ward, and J. Ryals. 1992. Acquired resistance in *Arabidopsis*. *Plant Cell* **4**: 645–656.

Valleau, W.D. 1952. Breeding tobacco for disease resistance. *Econ. Bot.* **6**: 69–102.

van Loon, L.C. 1987. Disease induction by plant viruses. *Adv. Virus Res.* **33**: 205–255.

Watanabe, Y., N. Kishibayashi, F. Motoyoshi, and Y. Okada. 1987. Characterization of *Tm-1* gene action of replication of common isolates and a resistance-breaking isolate of TMV. *Virology* **161**: 527–532.

White, K.A., J.B. Bancroft, and G.A. Mackie. 1992. Coding capacity determines in vivo accumulation of a defective RNA of clover yellow mosaic virus. *J. Virol.* **66**: 3069–3076.

White, R.F. and J.F. Antoniw. 1991. Virus-induced resistance responses in plants. *Crit. Rev. Plant Sci.* **9**: 443–455.

Wolf, S., C.M. Deom, R.N. Beachy, and W.J. Lucas. 1989. Movement protein of tobacco mosaic virus modifies plasmodesmatal size exclusion limit. *Science* **246**: 377–379.

Wyatt, S.D. and C.W. Kuhn. 1979. Replication and properties of cowpea chlorotic mottle virus in resistant cowpeas. *Phytopathology* **69**: 125–129.

Zhang, C., P.J. Cascone, and A.E. Simon. 1991. Recombination between satellite and genomic RNAs of turnip crinkle virus. *Virology* **184**: 791–794.

# 27
# Microbial Pathogenesis of *Arabidopsis*


**Ian Crute**
Plant Pathology and Weed Science Department
Horticulture Research International
Wellesbourne, Warwick CV35 9EF, United Kingdom

**Jim Beynon**
Department of Biological Sciences
Wye College, University of London
Ashford, Kent, TN25 5AH, United Kingdom

**Jeff Dangl**
Max-Delbrück Laboratory in the Max-Planck Society
D-52789 Köln, Germany

**Eric Holub**
Plant Pathology and Weed Science Department
Horticulture Research International, East Malling
West Malling, Kent ME19 6BJ, United Kingdom

**Brigitte Mauch-Mani and Alan Slusarenko**
Institut für Pflanzenbiologie, Cytologie, Universität Zürich
CH-8008, Zürich, Switzerland

**Brian Staskawicz**
Department of Plant Pathology
University of California, Berkeley, California 94720

**Fred Ausubel**
Department of Molecular Biology
Massachusetts General Hospital
Boston, Massachusetts 02114


## A PRIMER ON PLANT PATHOGENESIS

In common with all other terrestrial angiosperms, *Arabidopsis* provides ecological niches for an array of microorganisms. They may inhabit the aerial parts of the plant or be confined to the roots; they may live within the plant or on its outer surfaces; and they may have detrimental, beneficial, or neutral effects. In this chapter we consider the relationships between *Arabidopsis* and its pathogens; that is, microorganisms causing


*Arabidopsis*
© 1994 Cold Spring Harbor Laboratory Press 0-87969-428-9/94 $5 + .00

overt symptoms of disease. Plant pathogens are in no sense a biologically homogeneous assemblage of organisms, and their diversity is well illustrated by the fungi and bacteria, listed in Tables 1 and 2, respectively, capable of parasitic growth on *Arabidopsis*. Despite this biological diversity, there is well-founded optimism that studies of *Arabidopsis* as a host to a variety of pathogens will facilitate a deeper understanding of common processes in microbial pathogenesis of plants.

## Modes of Parasitism

Microbial plant parasites obtain their nutrients either "biotrophically" from living cells or "necrotrophically" from cells which they have killed (Lewis 1973). Examples of the former include fungi, such as powdery and downy mildews, and mollicutes, such as mycoplasma-like organisms which are obligately biotrophic and have not thus far been cultured on any synthetic medium. These obligate biotrophs invade and extensively colonize susceptible plants in such a way that host cells either remain alive or die only after the pathogen has grown on to exploit other living cells.

*Table 1* Eukaryotic microorganisms recorded as pathogenic on *Arabidopsis*

| Microorganism | References |
|---|---|
| Plasmodiophoromycete | |
| *Plasmodiophora brassicae* | Naumov (1925); Mithen and Magrath (1992) |
| Oomycetes | |
| *Peronospora parasitica* | Koch and Slusarenko (1990a); Holub et al. (1991, 1994) |
| *Albugo candida* | Holub et al. (1991, 1993) |
| *Pythium* spp. | Mauch-Mani et al. (1993) |
| Ascomycetes | |
| *Alternaria brassicae* | Berger (1965) |
| *Sclerotinia sclerotiorum* | Morgan (1971); Dickman and Mitra (1992) |
| *Leptosphaeria maculans* (*Phoma lingam*) | Sjödin and Glimelius (1988) |
| *Erysiphe cruciferarum* | Koch and Slusarenko (1990b) |
| *Botrytis cinerea* | Koch and Slusarenko (1990b) |
| *Fusarium oxysporum* | Mauch-Mani and Slusarenko (1993b, 1994) |
| *Cladosporium* spp. | Mauch-Mani and Slusarenko (1993b) |
| Basidiomycetes | |
| *Thanatephorus cucumeris* (*Rhizoctonia solani*) | Koch and Slusarenko (1990b) |
| *Puccinia thlaspeos* | Brandenburger (1985) |

*Table 2* Prokaryotic microorganisms recorded as pathogenic on *Arabidopsis*

| Microorganism | References |
| --- | --- |
| **Mollicutes** | |
| *Spiroplasma citri* | Fletcher and Eastman (1992) |
| Beet leafhopper transmitted virescence MLO | Golino et al. (1988) |
| **Bacteria** | |
| *Xanthomonas campestris* pv. *campestris* | Tsuji and Somerville (1988, 1992); Simpson and Johnson (1990); Davis et al. (1991); Tsuji et al. (1991); Parker et al. (1993a) |
| *Xanthomonas campestris* pv. *amoraciae* | Davis et al. (1991) |
| *Pseudomonas syringae* pv. *tomato* | Davis et al. (1991); Whalen et al. (1991) |
| *Pseudomonas syringae* pv. *maculicola* | Davis et al. (1991); Debener et al. (1991) |
| *Pseudomonas syringae* pv. *pisi* | Davis et al. (1991) |
| *Pseudomonas cichorii* | Davis et al. (1991) |

Many of the most destructive plant pathogenic fungi (e.g., *Botrytis* spp. and *Pythium* spp.) and bacteria (e.g., *Xanthomonas* spp. and *Erwinia* spp.) have a necrotrophic mode of nutrition. Some necrotrophic parasites can also live as saprophytes on plant debris or healthy plant surfaces, whereas others are *facultative* obligate parasites. Infections by necrotrophic pathogens are frequently observed as visible lesions composed of dead tissue. Although necrotrophic pathogens clearly employ toxins and cell-wall-degrading enzymes to kill plant cells, the molecular mechanisms that distinguish biotrophic and necrotrophic lifestyles remain elusive.

Variation in modes of parasitism, together with the array of ecological niches in the plant kingdom, has resulted in parasite adaptation to particular plant species. *Arabidopsis*, like all other angiosperms, is not parasitized by the vast majority of potentially parasitic microorganisms with which it comes into contact; it is a "non-host" to these pathogens of other plant species. Although essentially nothing is known about the molecular basis of host specificity, it is generally the case that obligate biotrophs have more limited host ranges than necrotrophic pathogens. For example, *Arabidopsis* is a host for several biotrophic pathogens such as *Peronospora parasitica* and *Albugo candida* that are only capable of parasitizing members of the Cruciferae. In contrast, necrotrophic *Arabidopsis* pathogens such as *Verticillium* spp. and *Pythium* spp., as well as some pathovars of *Pseudomonas syringae*, have wider host ranges that encompass numerous plant species taxonomically remote from the Cruciferae.

## Natural Variation in Resistance to a Particular Pathogen

Whether or not infection by a particular pathogen on a particular host leads to disease depends on many factors, including the activation of host defenses and the production of appropriate virulence factors by the path-ogen. The outcome of a particular plant/pathogen interaction is described using the terms pathogenicity, resistance, virulence, avirulence, com-patibility, and incompatibility. "Pathogenicity" is used to describe the qualitative capability of a microorganism to be a pathogen. This is in contrast to "virulence," which is a quantitative descriptor of the degree to which a particular parasite invades and colonizes a particular host and causes disease. "Resistance" is defined as the ability of a plant to impede invasion and colonization by the parasite. As an example, *P. syringae* pathovar *phaseolicola* strain NPS3121 is a well-studied bacterial patho-gen that causes severe disease symptoms on particular cultivars of com-mon bean. In this case, where disease is the consequence of the interac-tion between NPS3121 and a particular bean cultivar, the pathogen is said to be "virulent," the host "susceptible," and the interaction "com-patible." In contrast, other cultivars of beans are resistant to NPS3121 and display a hypersensitive response (HR) described in detail below. In this latter case, where a particular bean cultivar mounts an effective defense response, NPS3121 is said to be "avirulent," the bean cultivar "resistant," and the interaction "incompatible." Likewise, different iso-lates of *P. syringae* pv. *phaseolicola* vary with respect to their virulence or avirulence on a range of different bean cultivars. As described in detail below, this latter naturally occurring variation in resistance and virulence in many cases can best be described in terms of gene-for-gene interactions between specific pathogen and host genes (Flor 1971).

Although NPS3121 is a highly successful pathogen of bean, it does not cause disease in any of a large number of *Arabidopsis* ecotypes that have been tested. Moreover, NPS3121 does not elicit a strong defense response in *Arabidopsis*. This type of resistance is referred to as "non-host resistance" and NPS3121 is consequently a "non-pathogen" of *Arabidopsis*. The type of species specificity exhibited by NPS3121 is generally described in terms of "host range" and NPS3121 is said to be "non-pathogenic," rather than "avirulent," with respect to its inability to cause disease on *Arabidopsis*. It is not clear whether the status of non-host reflects the absence of some essential character required for parasitism or the presence of some character that inhibits parasitism (it could be both).

## Gene-for-Gene Relationships in Host/Pathogen Interactions

Variation in symptomology is observed when a collection of host variants (or cultivars), differing in resistance, is tested in combination with a set of pathogen isolates differing in virulence. This variation in the interaction phenotype (symptoms) can take two extreme forms. In one case, the relative resistance of host variants is observed to be independent of pathogen variation and the relative virulence of pathogen isolates is observed to be independent of host variation. This type of variation for resistance is called race nonspecific, and the variation for virulence is called host nonspecific. Alternatively, the relative resistance of different hosts depends on the particular pathogen isolate, and the relative virulence of pathogen isolates depends on the particular host variant. That is, the degree of host resistance is specific to a particular pathogen isolate and the degree of virulence is specific to a particular host ecotype. Under these circumstances, pathogen isolates may be readily identified as distinct pathotypes (races) by their interactive virulence on a range of differentially resistant host genotypes. Genetic analyses of these specific interactions between pathotypes and differential host genotypes have been conducted in numerous host/parasite associations, and a gene-for-gene relationship provides the simplest and most parsimonious explanation of the data (Person 1959; Flor 1971; Keen 1982, 1990; Crute 1985).

In a gene-for-gene relationship, for each plant gene that affects specific resistance (defined as a *resistance* [*R*] gene) there appears to be a matching pathogen gene regulating specific virulence (defined as an *avirulence* [*avr*] gene). In the simplest molecular formulation of this gene-for-gene model, a specific *R* gene encodes a receptor for a specific signal generated by a specific *avr* gene. Incompatibility (host resistance and pathogen avirulence) thus requires a specific matched pair consisting of a host *R* gene and a pathogen *avr* gene. It appears likely that the specific gene-for-gene controlled recognition event is in some way transduced to activate a set of nonspecific plant response genes that bring about the resistant condition (Lamb et al. 1989).

Although several *avr* genes have been cloned from bacteria (Keen and Staskawicz 1988) and fungi (van den Ackerveken et al. 1992), the nature of the *avr*-generated signal has only been defined in a limited number of cases. In the case of the fungal pathogen *Cladosporium fulvum*, the product of the *avr9* gene has been shown to be processed into a 28-amino-acid polypeptide, which in purified form elicits an HR in tomato cultivars carrying the *Cf9* resistance gene (van den Ackerveken et al. 1992). In contrast to the *C. fulvum avr9* gene product, there is no direct evidence that the product of any bacterial *avr* gene is either exported from or displayed on the bacterial cell surface where it could interact with the product of a plant-encoded *R* gene. It is possible that many bacterial *avr* genes encode enzymes that are involved in the synthesis of signal molecules. In support of this view, the *P. syringae* pv. *glycinea avrD* locus appears to encode an enzyme(s) involved in the synthesis of low-molecular-weight syringolides that act as HR elicitors to soybean cultivars carrying the resistance gene *Rpg4* (Keen et al. 1990; Smith et al. 1993).

Compared to *avr* genes, relatively few *R* genes have been cloned from pathosystems demonstrably under gene-for-gene control. Martin et al. (1993) cloned the tomato *PTO* gene conferring resistance to *P. syringae* strains expressing *avrPto*. The inferred product of the *PTO* gene appears to be similar to proteins with serine-threonine protein kinase activity, and a role in a signal transduction pathway is therefore suggested. Recently, resistance genes that conform to the gene-for-gene model have also been cloned from *Arabidopsis* (*RPS2:* Bent et al. 1994; Mindrinos et al. 1994b), tobacco (*N:* Whitham et al. 1994), and tomato (*Cf9:* J. Jones, pers. comm.). *RPS2* confers resistance to the *P. syringae* strains expressing the avirulence gene *avrRpt2*. *N* confers resistance to tobacco mosaic virus, and *Cf9* confers resistance to strains of the fungal pathogen *C. fulvum* expressing the avirulence gene *avr9*. The proteins encoded by all three of these genes share no similarity to PTO protein. Rather, they all contain a prominent leucine-rich tandem repeat region. In addition, the RPS2 and N proteins contain a nucleotide-binding site. The Cf9 protein has the structure expected of a trans membrane receptor, whereas the RPS2 and N proteins are probably localized in the cytoplasm. Remarkably, the RPS2 and N proteins share approximately 25% identity and 50% similarity dispersed over the entire molecules, suggesting a common underlying mechanism at least for these two plant resistance genes.

### The Plant Defense Response

Some aspects of plant defense against a potential parasite may be based on structural characteristics of the plant or constitutively expressed defenses (Crute et al. 1985; Hahlbrock and Scheel 1989; Lamb et al. 1989; Dixon and Harrison 1990). A successful parasite needs to have the capacity to breach such defenses to effect invasion. However, plants also activate a range of responses to attempted invasion by microorganisms whether or not they are pathogens of the species. Rapid alterations in plant gene expression occur following contact between microorganism and plant; synthesis of a battery of new proteins is sequentially activated. These include degradative enzymes, biosynthetic enzymes for the production of toxic secondary metabolites and for the reinforcement of cell walls, enzymes involved in signal transduction, and proteins of unknown function (Hahlbrock and Scheel 1989; Lamb et al. 1989; Dixon and Harrison 1990). These changes in plant gene expression are presumed to occur in response to signals (elicitors) from the invader, such as those encoded by a pathogen *avr* gene. A microorganism may be a successful pathogen because it fails to deliver signals that activate the plant's response; alternatively, success as a parasite may result from attributes that overwhelm the plant's defenses or render them ineffective.

Plants employ several specific strategies actively to defend themselves against potentially pathogenic microorganisms. Plant defense responses can be localized in the near vicinity of the infection or can be systemic. The most extensively studied plant defense response is the so-called hypersensitive response (HR), which is characterized by rapid necrosis of a variable number of cells near the infection site. Although the HR involves tissue damage, the spread of the pathogen is limited. Many studies have shown that an HR is accompanied by biochemical changes both at the site of infection and more remotely in the plant (Malamy and Klessig 1992). At the infection site, the HR is correlated with activation of specific defense-related genes, accumulation of low-molecular-weight antimicrobial compounds (phytoalexins), and alterations of the plant cell wall. Another local response associated with the HR is the "oxidative burst," NADPH-dependent production of active oxygen species such as $O_2^-$ and $H_2O_2$. Active oxygen species may be directly toxic to invading microorganisms, may function as signal molecules that trigger HR, or may be directly involved in the rapid necrosis of plant tissue. The oxidative burst also supplies $H_2O_2$, the substrate for the peroxidatic activities involved in lignification and cross-linking of hydroxyproline-rich glycoproteins.

In addition to the HR elicited at the infection site, infection by an avirulent pathogen is often associated with increased resistance throughout the plant to subsequent infection by a broad spectrum of normally virulent pathogens (Kuć 1982). This phenomenon is termed systemic acquired resistance (SAR; Chester 1933; Ross 1961). Expression of a number of defense-related genes, in particular the pathogenesis-related (PR) genes, correlates with the establishment of SAR (Ward et al. 1991), and the secondary metabolite salicylic acid is a necessary signal for SAR (Gaffney et al. 1993).

No experimental studies have identified the cause of the HR; it is unclear whether it is caused by pathogen toxicity or by the activation of a suicide program that is generated by the plant in response to attack. Furthermore, it is unknown whether the HR is obligatorily coupled to the diverse biochemical responses that occur upon infection. In general, however, the HR is only triggered by avirulent pathogens or, in some cases, by non-pathogens of the host. In the former case, the HR is frequently the hallmark of a gene-for-gene interaction involving a pathogen *avr* gene and a host resistance gene. In the case of bacterial pathogens on non-hosts, a pathogen-encoded and exported protein called harpin can be purified from a bacterial culture and shown to be sufficient to elicit an HR-like response. Since metabolic inhibitors block the apparent HR caused by purified harpin, it has been argued (Wei et al. 1992; He et al. 1993) that the HR is not caused by a direct toxic effect of harpin, but rather is caused by the triggering of a plant cell death program.

Although the HR appears to be specifically activated by avirulent or non-pathogens, a variety of defense responses are activated by both virulent and avirulent pathogens. These include the reinforcement of cell walls by (1) lignification, suberization, callose biosynthesis, and cross-linking of hydroxyproline-rich glycoproteins; (2) production of chitinases and glucanases; and (3) synthesis of a variety of species-specific flavonoid, courmarin, terpenoid, or other aromatic phytoalexins.

Over the course of the past few years, a variety of pathogen-induced genes that are associated with many aspects of the plant defense response discussed above have been cloned and characterized. A large body of work has shown that the induction of these pathogen-induced genes is often correlated with disease resistance in specific plant/microbe interactions (Collinge and Slusarenko 1987; Hahlbrock and Scheel 1989; Lamb et al. 1989; Dixon and Harrison 1990; Dixon and Lamb 1990) and that many of them also have roles during normal development. The promoter elements required for expression of a number of these genes have been identified and studied (Lois et al. 1989; Ohl et al. 1990; Ohshima et al. 1990; Van de Rhee et al. 1990; Wingender et al. 1990; Broglie et al. 1991; Meier et al. 1991; Samac and Shah 1991). One of the most important conclusions from studies on pathogen-induced genes is that the major difference between the defense responses elicited by avirulent and virulent pathogens lies in the kinetics and/or the magnitude of activation of these genes. In incompatible interactions, the defense responses are either activated more rapidly and/or they are activated to higher levels than in compatible interactions. However, despite a large effort by many laboratories, because no mutant plants have been identified that are defective in any pathogen-induced gene, it has been difficult to demonstrate the significance of particular defense-related genes in conferring resistance. In addition, despite the extensive analysis of the promoters of defense-related genes, very little is known at the molecular level about the signal transduction pathways that lead to their activation. At a more global level, unambiguous evidence has not been provided that "defense-related responses" observed following invasion by a potential parasite are themselves a cause rather than a consequence of resistance to the primary invader. The genetic dissection of the *Arabidopsis* defense response should provide some answers to these basic questions.

## IDENTIFICATION OF *ARABIDOPSIS* PATHOGENS

### Bacterial Pathogens

There is only one report of the isolation of a naturally occurring bacterial pathogen on *Arabidopsis* (Tsuji and Somerville 1992). However, by screening collections of known *Brassica* pathogens, a variety of bacterial strains have been identified, including many different *Pseudomonas* and *Xanthomonas* species, that infect *Arabidopsis* (Table 2).

Most of the effort in developing model systems to study bacterial pathogenesis of *Arabidopsis* has been focused on the characterization of *Xanthomonas campestris* pv. *campestris,* the causal agent of black rot on crucifers, and on two *P. syringae* pathovars: pv. *maculicola* and pv. *tomato,* the causal agents of bacterial speck disease on crucifers and tomato, respectively. The notation "pv." refers to "pathovar," subspecific variants that are infectious on certain host plants. For example, pathovar *maculicola* strains originate from cruciferous plants and the host range of many such strains is limited to species in the family Crucifereae. Pathovar designations are not necessarily indicative of the phylogenetic relationship between different *P. syringae* isolates. For example, *P. syringae* pv. *maculicola* (*Psm*) and *P. syringae* pv. *tomato* (*Pst*) appear to be very similar. Several *P. syringae* and *X. campestris* strains have been shown to be highly virulent on *Arabidopsis*, including *Psm* strains ES4326 (Dong et al. 1991) and M4 (Debener et al. 1991), *Pst* strain DC3000 (Whalen et al. 1991), and *X. campestris* pv. *campestris* (*Xcc*) strains 2D520 and 8004 (Tsuji and Somerville 1988; Tsuji et al. 1991; Parker et al. 1993a). These strains proliferate in *Arabidopsis* leaves and elicit chlorotic water-soaked lesions. When infiltrated into *Arabidopsis* leaves at a dose of approximately $10^4$ cells per $cm^2$, *P. syringae* strains multiply about $10^4$-fold and *X. campestris* strains multiply about $10^3$-fold.

### Fungal Pathogens

In contrast to the lack of any references in early literature to bacterial pathogens, reports of fungal pathogens on *Arabidopsis* are to be found at least as far back as 1861 (for references, see Koch and Slusarenko 1990a). However, the historical record is complicated by early synonymy for *Arabidopsis*; usually a simple record of the occurrence of a disease is all that can be gained from early reports. Hence, *Arabidopsis* is reported as a host for *Peronospora parasitica* (downy mildew; Gäumann 1918), *Plasmodiophora brassicae* (clubroot; Naumov 1925), *Albugo candida* (white blister; Jørstad 1964), *Alternaria brassicae* (dark leaf-spot; Berger 1965), *Sclerotinia sclerotiorum* (white mold or soft rot; Morgan 1971; Dickman and Mitra 1992), *Puccinia thlaspeos* (rust; Brandenburger 1985), and *Phoma lingam* (anamorph of *Leptosphaeria maculans*) (blackleg; Sjödin and Glimelius 1988). To this list, Koch and Slusarenko (1990b), Mauch-Mani et al. (1993), and Mauch-Mani and Slusarenko (1993a,b, 1994) have added *Erysiphe cruciferarum* (powdery mildew), *Rhizoctonia solani* (sclerotial state of *Thanatephorus cucumeris*) (damping-off or wire-stem), *Botrytis cinerea* (gray mold), two species of *Pythium* (damping-off), *Cladosporium* sp. (leaf mold or leaf spot), and two morphologically distinct isolates of *Fusarium oxysporum* (vascular wilt). Eukaryotic microorganisms reported as pathogenic on *Arabidopsis* are detailed in Table 1. In addition, with the exception of *Peronospora parasitica*, which is described in detail below, the Appendix to this chapter contains a short description of a variety of *Arabidopsis* fungal pathogens. The detailed molecular-genetic analysis of the diversity of these fungi in their mode of infection should provide important insights into the molecular mechanism of pathogenesis and host defense.

Most of the published work on *Arabidopsis* fungal pathogens has been carried out with *Peronospora parasitica*, causing the disease downy mildew (Koch and Slusarenko 1990a; Holub et al. 1993, 1994; Parker et al. 1993b). *P. parasitica* is an obligately biotrophic member of the Oomycete family Peronosporaceae, mycelial plant pathogens that are diploid in their vegetative phase. *P. parasitica* is exclusively parasitic on members of the Crucifereae and exists as numerous host-adapted forms, each specialized to particular subfamily taxa (assemblages of genera, species, or genotypes, including cultivated ones) (Channon 1981; Sherriff and Lucas 1990).

Long-term survival of *P. parasitica* is by means of sexually produced oospores, which remain viable in soil and debris. These oospores germinate by means of a germ tube and probably initiate infections by penetrating between rhizodermal cells. After penetration, the pathogen produces a mycelium that ramifies intercellularly throughout host tissues. As the hyphae grow, large numbers of pear-shaped haustoria are produced along their length, primarily within cortical and mesophyll cells. It is believed that these structures, which invaginate the host plasmalemma, are involved in the extraction of nutrients. In a compatible host genotype, once hyphae have reached a particular density within a region of the host tissue, conidiophores (asexual sporophores) are differentiated and emerge from the stomata.

Macroscopic symptoms of downy mildew on mature plants of *Arabidopsis* are discrete, pale green patches on rosette leaves which eventually turn chlorotic. When humidity is high, and often before any other symptoms are evident, asexual sporulation is seen as a downy, white covering over the affected foliage which is most profuse on the lower surface. Sporulation may also be observed over the length of the flowering stem when infected plants bolt. When small seedlings are infected, they are stunted and pale green by comparison with healthy individuals; profuse sporulation occurs on young seedlings. Although infection at the seedling stage can be lethal, it is not invariably so, and older plants will grow and reproduce despite extensive colonization by the pathogen. Isolates of *P. parasitica* derived from *Arabidopsis* are, so far as is known, only capable of parasitizing *Arabidopsis suecica* in addition to *Arabidopsis thaliana*, although extensive cross-inoculation studies have yet to be conducted (E.B. Holub, unpubl.). Koch and Slusarenko (1990a) and Mauch-Mani and Slusarenko (1993a) illustrate most stages of the life cycle of *P. parasitica* photographically.

## Natural Variation in *Arabidopsis*/Pathogen Interactions

Natural variation among *Arabidopsis* ecotypes, characteristic of gene-for-gene relationships, has been demonstrated for the bacterial pathogens *P. syringae* (Debener et al. 1991) and *X. campestris* (Tsuji et al. 1991), and for the fungal pathogens *P. parasitica* (Koch and Slusarenko 1990a; Parker et al. 1993b; Holub et al. 1994) and *Erisiphe cruciferarum* (D. Hall et al.; L. Adam and S. Somerville; both pers. comm.). In the case of *P. syringae*, Debener et al. (1991) demonstrated that *Psm* $M_2$ was avirulent on certain ecotypes (Oy-0 and Col-0) but virulent and caused disease on another (Nd-0), whereas *Psm* $M_4$ was virulent and caused disease on most ecotypes of *Arabidopsis*. Segregation analysis of a cross between ecotype Nd-0 (susceptible to *Psm* $M_2$) and ecotype Col-0 (resistant to *Psm* $M_2$) led to the identification of an *Arabidopsis* resistance gene, *RPM1*. The dominant allele of *RPM1* that specified resistance to *Psm* $M_2$ segregated at a single locus and mapped to chromosome 3.

The differential response of *Psm* $M_2$ and *Psm* $M_4$ on single *Arabidopsis* ecotypes such as Nd-0 and Col-0 permitted the molecular cloning of an *avr* gene, *avrRpm1*, that interacts with *RPM1* (Debener et al. 1991). Using methods developed previously to clone other bacterial *avr* genes (Staskawicz et al. 1984), Debener et al. (1991) used DNA from the avirulent strain *Psm* $M_2$ to construct a library in a broad host-range cosmid vector. Individual clones from the library were transferred into the virulent strain *Psm* $M_4$ and tested for their ability to elicit an HR on *Arabidopsis* ecotype Oy-0 in common with *Psm* $M_2$. Importantly, resistance to *Psm* $M_4$ carrying the cloned *avrRpm1* gene cosegregated with resistance to *Psm* $M_2$. These experiments established that *RPM1* and *avrRpm1* conform to a classic gene-for-gene relationship (Debener et al. 1991).

In the case of *X. campestris*, *Xcc* strain 2D520 exhibited differential virulence on different *Arabidopsis* ecotypes (Tsuji and Somerville 1988). Resistant ecotypes (Col-0) remained asymptomatic following infiltration, whereas susceptible ecotypes (Pr-0) developed a spreading chlorosis after 3–4 days (Tsuji and Somerville 1988; Tsuji et al. 1991). Interestingly, bacterial multiplication in both the resistant and the susceptible ecotypes was similar, suggesting a tolerance in Col-0 to bacterial multiplication. Genetic analyses with crosses between Col-0 and Pr-0 indicated that a dominant allele of a single nuclear gene, *RXC1*, governs tolerance to *Xcc* 2D520 (Tsuji et al. 1991). *RXC1* has been mapped to chromosome 2 (S. Somerville, pers. comm.).

Koch and Slusarenko (1990a) provided the first evidence for phenotypic variation among ecotypes of *Arabidopsis* for response to *P. parasitica*. They demonstrated that whereas ecotype Wei-0 supported extensive mycelial development and copious sporulation of a particular isolate (subsequently referred to as *Wela1*), the development of the pathogen on ecotype Rld-1 was restricted and associated with a hypersensitive response of those epidermal cells penetrated by haustoria. Subsequently, development of *Wela1* in Col-0 was also shown to be partially restricted, although to a lesser degree than in Rld-1, and ecotype Ler-0 was shown to be fully susceptible.

Extensive reciprocal phenotypic variation has subsequently been revealed by studying interactions between numerous ecotypes of *Arabidopsis* and several isolates of *P. parasitica* (Holub et al. 1993, 1994; Dangl et al. 1992; Crute et al. 1993). The considerable variation observed among interaction phenotypes has been described in terms of the timing and degree of sporulation and the characteristics of the host response. Table 3 provides a summary of the range of phenotypes observed in one study among eleven ecotypes of *Arabidopsis* following inoculation with seven isolates of *P. parasitica*. Genetic analysis of these interactions led to the identification of at least eight putative resistance loci (*RPP1*, *RPP2*, *RPP3*, *RPP4*, *RPP7*, *RPP8*, *RPP9*, and *RPP10*)

(Table 3) (Holub et al. 1994), and additional studies designating additional *RPP* loci are under way in several laboratories (Crute et al. 1993; Mauch-Mani et al. 1993; Parker et al. 1993b).

One advantage of the *Arabidopsis/P. parasitica* interaction is the ability to readily discern quite subtle differences among different interaction phenotypes. This will allow studies to be undertaken on phenomena such as genetic epistasis and additivity between *RPP* loci. For example, segregation of *Arabidopsis/P. parasitica* interaction phenotypes among $F_3$ progeny from a single cross (Col-*gl1* – Nd-0) made it possible to determine the $F_2$ genotype with respect to alleles as five loci conditioning a range of interaction phenotypes (Holub et al. 1993). Hence, *RPP1.1* from Nd-0 was associated with the occurrence of necrotic pits and no sporulation, whereas *RPP2.1*, *RPP4.1*, *RPP6.1*, and *RPP7.1* (all from Col-*gl1*) were associated with the occurrence of necrotic flecks and, in some cases, delayed sporulation in comparison with a fully susceptible phenotype (Table 3). A summary of the gene-for-gene model currently explaining all the available phenotypic and genotypic data is also provided in Table 3. Details of the *Arabidopsis* loci involved in the specific recognition of *P. parasitica* are provided in Table 4. Table 4 also shows the map positions of several *RPP* loci (Tör et al. 1994). It is apparent from these analyses that the number of loci involved in the genotype-specific recognition of *P. parasitica* by *Arabidopsis* may be very large.

What is currently missing in the *Arabidopsis/P. parasitica* system is a genetic analysis of *P. parasitica*. For example, it is important to demonstrate that a single *avr* locus is involved in matching each of the *RPP* loci. Some isolates of *P. parasitica* from crucifers other than *Arabidopsis* are known to be heterothallic; they exist as two mating types, and sexual sporulation occurs when both types are intimately associated in the same tissue (Sherriff and Lucas 1990). All of the *P. parasitica* isolates from *Arabidopsis* so far studied appear to be homothallic, but the prospect of carrying out genetic analyses following selfing of $F_1$ progeny is being investigated (I. Crute and E. Holub, unpubl.).

## *ARABIDOPSIS* PATHOGEN-INDUCED GENES

As described in the first section of this chapter, the plant defense response is characterized by the activation of a variety of defense-related genes. In this section, we describe various methods that have been used to identify *Arabidopsis* defense-related genes and to study their expression patterns both near the site of an infection and during the SAR response.

*Table 3*  Variation for interaction phenotype observed among eleven ecotypes of *Arabidopsis* following inoculation with seven isolates of *Peronospora parasitica* together with a summary of host and parasite loci and alleles postulated from genetic analyses to explain the genotype specificity associated with each interaction phenotype

|            | *P. parasitica* isolate[a] | | | | | | |
|------------|-------|-------|-------|-------|-------|-------|-------|
| *ATR*[b]   | *Emoy2* | *Cala2* | *Hiks1* | *Emwa1* | *Noks1* | *Wela3* | *Cand3* |
| 1  | 1 | 0 | 2 | 0 | 0 | 0 | 0 |
| 2  | 0 | 1 | 0 | 0 | 0 | 0 | 0 |
| 3  | 0 | 1 | 0 | 0 | 0 | ? | 0 |
| 4  | 1 | 0 | 0 | 2 | 0 | 0 | ? |
| 5  | 0 | 0 | 0 | 0 | 1 | 0 | 0 |
| 6  | 0 | 0 | 0 | 0 | 0 | 1 | ? |
| 7  | 0 | 0 | 1 | 0 | 0 | 0 | ? |
| 8  | 1 | 0 | 0 | ? | 0 | 0 | ? |
| 9  | ? | 0 | 1 | ? | ? | 0 | ? |
| 10 | ? | 1 | ? | 0 | ? | ? | ? |

*Arabidopsis* *RPP*[c]

| Ecotype | 1 | 2 | 3 | 4 | 5 | 6 | 7 | 8 | 9 | 10 | | | | | | | |
|---|---|---|---|---|---|---|---|---|---|---|---|---|---|---|---|---|---|
| Nd-0 | 1 | 0 | 0 | 0 | 0 | 0 | 0 | 0 | 0 | 0 | PN[d] | EH | PN | DH | EH | DM | EH |
| Col-gl1 | 0 | 1 | 0 | 1 | 0 | 1 | 1 | 0 | 0 | 0 | FDL | FR | FR | FR | EH | FR | FN |
| Ler-o | 0 | 0 | 0 | 2 | 1 | 0 | 2 | 1 | 0 | 0 | FN | EH | FN | FN | CN | EH | FN |
| Oy-0 | 0 | 3 | 1 | 0 | 2 | ? | 0 | 0 | 0 | 0 | EH | FN | DH | DH | FN | FN | EH |
| Wei-0 | 0 | 0 | 0 | 0 | 0 | 0 | 0 | 0 | 1 | ? | DL | DL | DL | DL | FDL | EH | DL |
| Ws-0 | ? | 0 | 0 | 0 | 0 | 0 | 0 | 0 | 0 | 1 | PN | CN | PN | DH | PN | FN | CN |
| Ksk-1 | 0 | 2 | 0 | 0 | 0 | 0 | 0 | ? | 0 | ? | FN | FN | EH | FN | EH | FN | EH |
| Ema-1 | 2 | 0 | 0 | 0 | ? | 0 | 0 | 0 | 0 | 0 | PN | EH | PN | DH | FN | DH | EH |
| Cnt-1 | 3 | 0 | 0 | ? | ? | ? | ? | ? | ? | 0 | PN | EH | PN | FN | FN | FN | EH |
| Rld-1 | 0 | ? | ? | ? | 0 | ? | ? | ? | ? | ? | FN | FN | FN | FN | FDM | FN | FN |
| Tsu-0 | 0 | 0 | 0 | 0 | 0 | 0 | ? | 0 | 0 | ? | EM | EM | EM | EM | EH | FN | EM |

Data from Holub et al. (1994).

[a]Each isolate was derived from an oospore population. *Noks1* was derived from an oospore population of *Noco2* provided by J. Parker (Nowrick, UK) and *Wela3* was derived from an oospore population of *Wela1* provided by A. Slusarenko (Zurich, Switzerland).

[b]Locus abbreviation for "*Arabidopsis* recognize." Rows to the right represent the nine predicted loci which correspond pairwise with *RPP* loci. The allele designation "0" represents the allele associated with full susceptibility or lack of recognition. Note that allele numbers at *ATR* loci do not correspond directly with allele numbers at *RPP* loci. For example, allele *ATR.1.1* from *Emoy2* is probably recognized by alleles *RPP1.1*, *RPP1.2*, and *RPP1.3*, and allele *ATR2.1* from *Cala2* is probably recognized by *RPP2.1*, *RPP2.2*, and *RPP2.3*. See following footnote for explanation of question marks.

[c]Locus abbreviation for "recognition of *P. parasitica*." Columns below represent the nine loci which have been identified and alleles at each locus are designated for each ecotype. Alleles from different sources for which there is no evidence of recombination are assigned to the same locus. For example, three alleles (1–3) have been designated at locus *RPP1*. The designation "0" at each locus represents the allele associated with full susceptibility (EH). A question mark indicates that there is no genetic evidence to postulate either the presence or absence of a functional allele at the locus; an allele at a different locus or loci may need to be postulated in the future. No gene pairs have been postulated to explain EM, DM, or DH phenotypes.

[d]Interaction phenotypes were characterized using an assessment of host and parasite characteristics: the emergence of sporangiophores of *P. parasitica* as early (E) 3 days after inoculation (dai) or delayed (D) >4 dai; the intensity of asexual sporulation as heavy (H; >20 sporangiophores per cotyledon), moderate (M; 10–20 sporangiophores), light (L; <10 sporangiophores), rare (R; <5 sporangiophores on <10% of inoculated seedlings), or none (N); and the type of response by *Arabidopsis* as minute necrotic flecks (F) evident 7 dai, flecks clearly visible 3 dai which form necrotic cavities (C) 7 dai, or necrotic pits (P) observed as early as 3 dai.

*Table 4* Summary of map positions and interaction phenotypes for named loci associated with the genotype-specific recognition of *Peronospora parasitica* and *Albugo candida* in *Arabidopsis*

| Locus/Allele | Origin | Phenotype | Diagnostic isolate(s) | Chromosome location |
|---|---|---|---|---|
| *Peronospora parasitica* | | | | |
| *RPP1.1* | Nd-0 | PN | *Emoy2, Hiks1* | chromosome 3; interval M249-OPC12$_{250}$ |
| *RPP1.2* | Ema-1 | PN | *Emoy2, Hiks1* | chromosome 3; cosegregation with *RPP1.1* |
| *RPP1.3* | Cnt-1 | PN | *Emoy2, Hiks1* | chromosome 3; cosegregation with *RPP1.1* |
| *RPP2.1* | Col-*gl1* | FR | *Cala2* | chromosome 4; interval M557-M600 |
| *RPP2.2* | Ksk-1 | FR | *Cala2* | chromosome 4; interval *ag*-B9 |
| *RPP2.3* | Oy-0 | FR | *Cala2* | chromosome 4; cosegregation with *RPP2.1* |
| *RPP3.1* | Oy-0 | FR | *Cala2* | to be determined |
| *RPP4.1* | Col-*gl1* | FDL | *Emoy2, Emwa1* | chromosome 4; interval M557-M326 |
| *RPP4.2* | Ler-0 | FDL | *Emoy2, Emwa1* | chromosome 4; cosegregation with *RPP4.1* |
| *RPP5.1* | Ler-0 | FN | *Noco2* | chromosome 4; interval M226-g3845 |
| *RPP5.2* | Oy-0 | FN | *Noks1* | cosegregation with *RPP5.1* |
| *RPP6.1* | Col-0 | FR | *Wela2* | chromosome 1? |
| *RPP7.1* | Col-*gl1* | FR | *Hiks1* | chromosome 1 |
| *RPP7.2* | Ler-0 | FN | *Hiks1* | cosegregation with *RPP7.1* |
| *RPP8.1* | Ler-0 | FN | *Emoy2* | chromosome 5; near tt3 |
| *RPP9.1* | Wei-0 | DL | *Hiks1* | chromosome 1; cosegregation with *RAC1.1* |
| *RPP9.1* | Ws-0 | CN (PN) | *Cala2* | chromosome 3; interval M249-*gl1* |
| *RPP11.1* | Rld-1 | FN | *Wela1* | chromosome 3; near M249 and g2534 |
| *RPP12.1* | Ws-0 | FN | *Wela1* | chromosome 4? |
| *Albugo candida* | | | | |
| *RAC1.1* | Ksk-1 | NF | *Acem1* | chromosome 1; interval M335-M253 |

(n.0) Allele at locus in conditioning "full" susceptibility. (n.1 to n.x.) Alleles at locus n conditioning any recognizable deviation from "full" susceptibility; the decimal point reflects either a different specificity or a different origin. (PN) No asexual sporulation; necrotic pit response. (FN) No asexual sporulation; necrotic fleck response. (CN) No asexual sporulation; necrotic cavity response (intermediate between pit and fleck). (FDL) Delayed, light asexual sporulation; necrotic fleck response. (FR) Rare sporophore produced; necrotic fleck response.

## Identification of *Arabidopsis* Pathogen-induced Genes

To lay the groundwork for a genetic analysis of *Arabidopsis* pathogen-induced genes, RNA blot analysis has been used to monitor the activation of a variety of *Arabidopsis* genes by pathogen infection. There have been two overall goals for these experiments. One goal was to determine when and where specific genes are activated during an infection, and whether this activation occurs during virulent interactions, the HR, or as part of the SAR response. A second goal has been to identify genes that are highly and specifically activated in the hope that their promoters could be fused to reporter genes and then used in genetic selections designed to identify signal transduction mutants.

Several *Arabidopsis* pathogen-induced genes were identified by their homology with defense-regulated genes in other plants. *Arabidopsis* genes identified in this way include those corresponding to tobacco pathogenesis-related (PR) proteins and to parsley elicitor induced (ELI) cDNAs (Davis and Ausubel 1989; Samac et al. 1990; Dong et al. 1991; Metzler et al. 1991; Verberg and Huynh 1991; Kiedrowski et al. 1992; Uknes et al. 1992; Kawalleck et al. 1993). The PR proteins and ELIs are both heterogeneous collections that include genes encoding both known and unknown biochemical functions. Other *Arabidopsis* genes cloned on the basis of homology to known pathogen-induced genes include those encoding chalcone synthase (*CHS*; Feinbaum and Ausubel 1988), phenylalanine ammonia lyase (*PAL1*; Ohl et al. 1990; Davis et al. 1991), superoxide dismutase (*SOD1*; Hindges and Slusarenko 1991), and lipoxygenase (*LOX1*; Melan et al. 1993). PAL is the first committed enzyme required for general phenylpropanoid biosynthesis and CHS is the first committed step in flavonoid biosynthesis. Glucanases are lytic enzymes that hydrolyze fungal cell walls, superoxide dismutase is most likely involved in protecting plant cells from the deleterious effects of an oxidative burst, and the physiological function of lipoxygenase in the defense response is not clear.

Several *Arabidopsis* pathogen-induced genes have been identified by differential screening of *Arabidopsis* cDNA libraries constructed before and after pathogen attack. These genes include one encoding a glutathione transferase (*GST1*) and two different genes called *PIG2* and *PIG18* (pathogen-induced-gene) encoding calmodulin-like proteins (G.-L. Yu and F. Ausubel, unpubl.). In addition to these genes, Yu and Ausubel (unpubl.) showed that the previously identified *Arabidopsis* calmodulin-encoding gene *TCH3* and the G protein encoding gene *GPA1* are also pathogen-inducible. The calmodulins and G protein, which have not previously been shown to be pathogen-inducible in other plants, are presumably involved in signal transduction. A likely role for glutathione transferase is the detoxification of hydroxyalkenals, toxic by-products of lipid peroxidation generated in an oxidative burst (Prohaska 1980; Morgenstern and de Pierre 1983; Alin et al. 1985; Morgenstern et al. 1989; Dudler et al. 1991).

Finally, based on the notion that increased flux through the phenyl-propanoid pathway would increase the need for phenylalanine, mRNA corresponding to one of the two genes encoding 3-deoxy-D-arabino-heptulosonate 7-phosphate synthase (*DHS1*), which catalyzes the first committed step in aromatic amino acid biosynthesis, was shown to accumulate during pathogen attack (Keith et al. 1991). Similarly, *ASA1* mRNA, which encodes anthranilate synthase, also accumulates following infection with a bacterial pathogen (Niyogi and Fink 1992). Anthranilate is a precursor of the *Arabidopsis* indole-based phytoalexin, camalexin, which also accumulates following pathogen attack (Tsuji et al. 1993).

## Activation of *Arabidopsis* Pathogen-induced Genes

The results of a large number of RNA blot experiments are summarized in Table 5. Three major conclusions arise from this work. First, different *Arabidopsis* defense-related genes display markedly different patterns of mRNA accumulation after bacterial inoculation (Davis et al. 1991; Dong et al. 1991; Keith et al. 1991; Kiedrowski et al. 1992). Second, at the very least, it appears that different signal transduction pathways are involved in activating genes during a compatible interaction, during an HR, and as part of the SAR response. For example, *PR1*, *PAL1*, *DHS1*, *GST1*, *LOX1*, *TCH3*, and *ELI3* appear to be activated more strongly by avirulent strains that express an *avr* gene than by virulent strains that elicit disease symptoms (Davis et al. 1991; Dong et al. 1991; Keith et al. 1991; Kiedrowski et al. 1992; Greenberg et al. 1994). Third, if a defense-related gene is a member of a small gene family, then typically it is the only gene in the family that is activated by pathogen attack (mRNA corresponding to *DHS2*, *GST2*, *GST3*, *BGL1*, and *BGL3* does not increase following infection with virulent or avirulent strains; Keith et al. 1991; X. Dong et al., unpubl.). Together, these observations are consistent with a battery of observations made in other plant systems (Collinge and Slusarenko 1987; Lamb et al. 1989; Dixon and Harrison 1990; Dixon and Lamb 1990).

Ethylene has been implicated as a signal molecule in the activation of various plant defense-related genes. Ethylene is synthesized as a result of wounding and pathogen invasion (Yang and Hoffman 1984), and β-glucanase and chitinase genes in several plants including *Arabidopsis* have been shown to be induced by ethylene (Samac et al. 1990). The

*Table 5* Accumulation of mRNA corresponding to *Arabidopsis* genes following infection with virulent or avirulent strains of *P. syringae*

| Gene | mRNA accumulation elicited by virulent strains that elicit disease symptoms[a] | | | | | mRNA accumulation elicited by avirulent strains that elicit a hypersensitive response[a] | | | | | Systemic induction in plants with necrotic lesions |
| | accumulation (hr) | | | | fold induction[b] | accumulation (hr) | | | | fold enhanced induction by avirulent strain[c] | |
| | 6 | 12 | 24 | 48 | | 6 | 12 | 24 | 48 | | |
|---|---|---|---|---|---|---|---|---|---|---|---|
| *CHS*[d] | − | − | − | − | 1 | − | − | − | − | 1 | n.d. |
| *PAL1*[d] | ± | − | − | − | 1–2 | + | − | − | − | 2–4 | − |
| *DHS1*[e] | − | − | + | ± | ~3 | + | − | − | − | ~3 | + |
| *GST1*[f,g] | ± | − | − | − | 1–2 | ± | + | + | − | 2–3 | + |
| *ELI3*[h] | − | − | ± | + | 50–100 | + | + | ± | − | 25–50 | n.d. |
| *LOX1*[i] | − | − | − | + | 5 | − | − | + | + | 2–4 | − |
| *TCH3*[g] | + | + | + | − | 2–5 | + | + | + | − | 2–4 | n.d. |
| *GPA1*[g] | + | + | + | + | 1–2 | + | + | + | + | ~1 | n.d. |
| *PIG2*[g] | − | + | + | + | 5–10 | − | + | + | + | ~1 | n.d. |
| *PIG18*[g] | + | + | + | + | 5–10 | + | + | + | + | ~1 | n.d. |
| *SOD1*[g] | + | + | + | + | 2–3 | + | + | + | + | ~1 | n.d. |
| *ASA1*[j] | − | + | + | + | 3–5 | − | + | + | + | 3–5 | n.d. |
| *PR1*[f] | − | − | − | + | 50–500 | − | − | + | + | 2–10 | + |
| *BGL2*[d] | − | − | − | + | 10–20 | − | − | − | + | ~1 | + |
| *PR5*[j] | − | − | − | + | 30 | − | − | − | + | ~1 | + |

n.d. indicates not determined.

[a]In comparison to mock-infected control infiltrated with $MgCl_2$, ± means that induction was slightly higher than the mock-infected control.
[b]Fold induction by virulent strain in comparison to mock-infected control.
[c]Fold induction by avirulent strain in comparison to virulent strain.
[d]Dong et al. (1991); BGL2 is also known as PR2.
[e]Keith et al. (1991).
[f]Greenberg et al. (1994).
[g]G.L. Yu and F. Ausubel (unpubl.).
[h]Kiedrowski et al. (1992).
[i]Melan et al. (1993).
[j]Niyogi and Fink (1992); J. Glazebrook and F. Ausubel (unpubl.).

*Arabidopsis BGL2* and *BGL3* genes (X. Dong and F. Ausubel, unpubl.) are also induced by ethylene, as are the *Arabidopsis GST1* and *GST2* genes (Zhou and Goldsbrough 1993; G.-L Yu and F. Ausubel, unpubl.). Interestingly, in both of these cases, at least two genes in these small gene families are ethylene-inducible, whereas only one of the genes is pathogen-inducible. Importantly, *BGL2* and *GST1* are still pathogen-inducible in the ethylene response mutant *etr1*, suggesting that ethylene does not serve as a secondary messenger in a pathogen-mediated signal transduction pathway (X. Dong et al., unpubl.). On the other hand, Bent et al. (1992) have shown that another ethylene response mutant, *ein2*, shows less severe symptoms than wild type when infiltrated with *Pst* DC3000, even though the *in planta* bacterial growth remained the same as in wild type. This latter result suggests that although ethylene may not be directly involved in the activation of certain pathogen-induced genes during pathogenesis, it may play a role in the development of host symptoms in response to pathogens.

## *ARABIDOPSIS* DEFENSE-RELATED MUTANTS

As described in the first section of this chapter, the host defense response to pathogen attack is multifaceted and involves both local and systemic events. The identification of pathogens for *Arabidopsis* now provides an opportunity to apply molecular-genetic techniques to help elucidate the roles of various defense-related responses in conferring resistance. In this section, we discuss the use of *Arabidopsis* to identify signal transduction mutants in the pathways leading to an HR. We also describe the isolation of *Arabidopsis* mutants that synthesize decreased levels of camalexin, an *Arabidopsis* phytoalexin. One of the major goals of these mutant searches is to clone the corresponding wild-type genes. This has been accomplished in the case of the *Arabidopsis RPS2* gene, a resistance gene corresponding to the *P. syringae avr* gene *avrRpt2* (Bent et al. 1994; Mindrinos et al. 1994b).

### *Arabidopsis* Resistance Gene Mutants

As discussed above, most of our knowledge concerning the genetic control of disease resistance comes from studies that have exploited natural genetic variation in both the host and the pathogen. These studies led to the formulation of the gene-for-gene hypothesis and subsequent speculation that the resistance phenotype is the consequence of a direct interaction between the primary protein products of the resistance gene and the avirulence gene. It has been argued that other gene products in a signal transduction pathway are not required for the expression of resistance (Ellingboe 1981). This hypothesis is currently being tested by isolating and characterizing disease-susceptible mutants and determining if all of these mutants map at the disease-resistance locus (as predicted by the gene-for-gene hypothesis) or whether additional loci can be identified.

The underlying strategy is to utilize a pair of isogenic bacterial strains which differ only in the expression of a single *avr* gene, and to screen mutagenized $M_2$ seed of a normally resistant *Arabidopsis* ecotype for disease-susceptible mutants. A genetically uncharacterized pathogen might contain more than one avirulence determinant recognized by the host. In this latter case, both corresponding resistance genes would have to be mutated to observe a susceptible phenotype. The availability of isogenic pathogen strains containing single *avr* genes greatly increases the probability that mutants will be identified.

The isolation of *Arabidopsis* resistance gene mutants has been pursued with the *avr* genes *avrRpm1*, *avrB*, and *avrRpt2*, all of which generate signals that lead to the elicitation of an HR response in particular *Arabidopsis* ecotypes as illustrated in Table 6. As described above, *avrRpm1* was identifed as the *avr* gene that interacts with the resistance gene *RPM1* on the basis of natural variation in virulence between different *P. syringae* pv. *maculicola* strains on different *Arabidopsis* ecotypes (Debener et al. 1991).

The avirulence gene *avrB* was first identified in *P. syringae* pv. *glycinea* (Staskawicz et al. 1987) but was shown to elicit an HR in *Arabidopsis* leaves when transferred to virulent strains such as *Pst* DC3000 or *Psm* ES4326. As was the case with *RPM1*, a dominant resistance gene corresponding to *avrB*, *RPS3*, was also identified by the genetic analysis of natural variation in response to *Pst* DC3000 carrying *avrB*. Interestingly, *avrB* is recognized by the same ecotypes as *avrRpm1*, and *RPS3* maps to the same position on chromosome 3 as *RPM1* (Innes et al. 1993b). Recently, Bisgrove et al. (1994) isolated 12 *Arabidopsis* mutants that are susceptible to *Pst* DC3000 carrying *avrB*. All 12 mutants were also susceptible to *Pst* DC3000 carrying *avrRpm1*. It thus appears that *RPM1* and *RPS3* are the same gene. This was an unexpected result, since *avrB* and *avrRpm1* share no sequence homology.

The third *avr* gene for which a corresponding resistance gene has been identified, *avrRpt2*, was cloned from the avirulent strain *Pst* JL1065 (Dong et al. 1991; Whalen et al. 1991; Innes et al. 1993a). In contrast to the case with *avrRpm1* and *avrB*, instead of initially relying on the natural variation among *Arabidopsis* ecotypes to identify a resistance gene corresponding to *avrRpt2*, mutant plants were isolated, as described below, that did not respond with an HR following inoculation with *Psm*

*Table 6* Responses of *Arabidopsis* ecotypes to isolates of *Pseudomonas syringae* pv. *tomato* with and without *avr* genes

| Pathogen (*avr* gene) | *Arabidopsis* ecotypes | | | | |
|---|---|---|---|---|---|
| | Col-0 | Nd-0 | Bla-2 | Mt-0 | Wü-0 |
| *Pst*DC3000[a] | S | S | S | S | S |
| *Pst*DC3000 (*avrRpt2*) | R | R | R | S | S |
| *Pst*DC3000 (*avrRpm1*) | R | S | S | S | R |
| *Pst*DC3000 (*avrB*) | R | S | S | S | R |

(S) Susceptible phenotype (development of chlorotic lesions and prolific bacterial growth). (R) Resistant phenotype (hypersensitive reaction and inhibited bacterial growth).

[a]*Psm*ES4326 containing the above avirulence genes has the same phenotype as *Pst*DC3000.

ES4326 or *Pst* DC3000 carrying *avrRpt2*. Genetic characterization of these mutants showed that the mutations resulting in loss of resistance all mapped at or very close to a locus called *RPS2* located on the bottom of chromosome 4 (Kunkel et al. 1993; Yu et al. 1993; Mindrinos et al. 1994a). Subsequent to defining *RPS2* by mutant analysis, the *Arabidopsis* ecotype Wü-0 was found to be unable to recognize *avrRpt2*; allelism tests using *rps2* mutants showed that Wü-0 also carries a recessive *rps2* allele (Kunkel et al. 1993). *RPS2* differs from *RPM1* and *RPS3* in that it is incompletely dominant. Heterozygous *RPS2/rps2* plants respond with an HR, but the response is slower to appear and requires a higher inoculum of avirulent bacteria to elicit it than in the case of *RPS2/RPS2* homozygous plants. This intermediate phenotype is reflected in the fact that strains expressing *avrRpt2* grow almost as well in *RPS2/rps2* heterozygous plants as in *rps2/rps2* homozygous plants (Kunkel et al. 1993; Yu et al. 1993).

Several methods have been used to screen *Arabidopsis* $M_2$ plants for mutants that do not mount an HR in response to a particular *avr* gene. All of these screens have made use of a normally virulent *P. syringae* strain carrying a single cloned *avr* gene. In the first method, $M_2$ plants are hand-inoculated one by one with a high inoculum and examined for the lack of an HR (Yu et al. 1993). In the second method, $M_2$ plants are dipped in a bacterial suspension in the presence of a wetting agent (Silwet L-77). In this latter case, resistant plants show no symptoms, whereas resistance-gene mutants develop disease symptoms (Kunkel et al. 1993; Bisgrove et al. 1994). Susceptible mutants can be recovered using this dipping method because the pathogen does not cause a systemic infection and susceptible mutants survive to produce seed.

A third method for identifying susceptible mutants takes advantage of the observation that no symptoms appear on mature *Arabidopsis* plants that have been infiltrated with the bean pathogen, *P. syringae* pv.

*phaseolicola* (*Psp*). In contrast, *Psp* carrying *avrRpt2*, *avrRpm1*, or *avrB* elicits a strong HR.

Interestingly, Yu et al. (1993) observed that if *Psp* carrying an *avr* gene is vacuum infiltrated into young seedlings growing on petri plates, the seedlings undergo a systemic HR and become necrotic, whereas seedlings infiltrated with *Psp* mostly survive. Similarly, mutant plants that fail to mount an HR also survive when they are infiltrated with *Psp* carrying an *avr* gene. It is anticipated that this method will permit saturation mutagenesis leading to the dissection of the putative defense pathway(s) controlling disease resistance in *Arabidopsis*.

Using these methods, mutants have been isolated that fail to mount an HR in response to *avrRpt2* (Kunkel et al. 1993; Yu et al. 1993; Mindrinos et al. 1994a) and *avrB* (Bisgrove et al. 1994). Among 12 mutants that do not respond to *avrB*, all 12 also failed to respond to *avrRpm1* (as described above), and genetic analysis showed that 4 of these mutants carry mutations at the *RPS3* locus. Among 8 mutants that do not respond to *avrRpt2*, 7 are *RPS2* alleles (Kunkel et al. 1993; Yu et al. 1993; Mindrinos et al. 1994a). The single mutant that is not an *rps2* mutant also fails to mount an HR in response to *avrB* and *avrRpm1* and may correspond to a downstream component of the signal transduction pathway leading to the HR (K. Century and B. Staskawicz, unpubl.). Recently, a map-based cloning strategy has been used to clone the *RPS2* gene (B. Staskawicz and F. Ausubel, unpubl.). Fine-structure restriction fragment length polymorphism (RFLP) mapping showed that the *RPS2* gene was located in a 30-kb region on the bottom of chromosome 4 near the RFLP markers M600 and PG11. The *RPS2* gene was identified in this 30-kb region by comparing the DNA sequences of five cDNA clones encoded in the 30-kb region with the DNA sequences of polymerase chain reaction products amplified using RNA isolated from four different *rps2* mutants. The *RPS2* gene encodes a 105-kD protein containing a variety of identifiable motifs including a leucine zipper, a nucleotide-binding site, and 14 imperfect leucine-rich repeats (Bent et al. 1994; Mindrinos et al. 1994b).

## Does Defense Gene Activation Depend on Specific Gene-for-Gene Triggers?

In other plant-pathogen systems, a variety of experiments have correlated the induction of different host-defense-related genes with the elicitation of an HR (Lamb et al. 1989; Bowles 1990). Fewer experiments have clearly correlated the activation of these genes with specific *avr-*

resistance gene interactions. Isogenic mutants of the pathogen, differing only in the presence of a particular *avr* gene, as well as isogenic mutants or segregating populations of the host, differing only in the presence of the corresponding resistance gene, are required to carry out a definitive experiment which shows that an *avr*-generated signal can lead to the activation of defense genes. The isolation of *Arabidopsis* resistance gene mutants, described earlier in this chapter, and *Arabidopsis* populations segregating only for the presence of a resistance gene, can be used to definitively test the role of resistance genes in the induction of defense-related genes.

As illustrated in Table 5, enhanced mRNA accumulation corresponding to several *Arabidopsis* defense-related genes occurs following infiltration of avirulent compared to virulent bacterial pathogens. This result suggested that a signal transduction pathway that is dependent on a resistance gene may be involved in defense gene activation during an incompatible interaction. To test the role of *avrRpt2* and *RPS2* in activation of these defense-related genes, G.-L. Yu and F. Ausubel (unpubl.) monitored the accumulation of *PAL1* and *GST1* mRNA in wild-type and *rps2* plants. Importantly, the difference in mRNA induction kinetics resulting from infection of *Psm* ES4326 compared to *Psm* ES4326/*avrRpt2* was not observed in the *rps* mutant. These results indicate that the enhancement of *PAL1* and *GST1* mRNA levels in infected leaves was dependent on a signal transduced from the resistance gene *RPS2*.

A different approach was taken by Kiedrowski et al. (1992) in the analysis of pathogen-induced expression of *ELI3*. Two closely linked *ELI3* genes encode proteins with approximately 50% similarity to cinnamyl-alcohol dehydrogenase from *Arabidopsis* (S. Kiedrowski and J. Dangl, unpubl.). Using $F_3$ families selected for their genotypes as either *RPM1/RPM1* or *rpm1/rpm1*, but freely segregating at all other loci, Kiedrowski et al. (1992) demonstrated that rapid and massive differential accumulation of *ELI3* mRNA was dependent on the *RPM1* resistance gene after infiltration with *Psm* $M_2$ or *Psm* $M_4$ carrying the *avrRpm1* gene from *Psm* $M_2$.

## Programmed Cell Death and Intercellular Signaling as Part of the Plant Defense Mechanism

The products of resistance genes are thought to be receptors that are located at the top of a signal transduction cascade leading to the induction of the HR as well as a variety of associated defense responses, including the activation of pathogen-induced genes, the accumulation of phytoalexins, and the initiation of SAR. What has not been clear from a variety of physiological and biochemical studies, however, is whether the various components of the defense response associated with the HR are part of a single genetically programmed event or the consequence of multiply activated parallel signal transduction pathways. It was also not clear in the case of the HR whether the observed necrotic symptoms are due to the activation of a programmed cell death pathway or a consequence of pathogen toxicity.

In maize, a variety of mutants have been identified that form necrotic lesions in the absence of any pathogens (Walbot 1991). It is generally thought that these mutants identify genes involved in symptom development and possibly the elaboration of disease resistance. Unfortunately, these so-called lesion mimic (*les*) mutants have not been thoroughly tested to determine if the lesions resemble bona fide pathogen-induced lesions. However, progress in this regard has recently been made with *Arabidopsis* mutants called *acd* for *a*ccelerated *c*ell *d*eath (Greenberg and Ausubel 1993; Greenberg et al. 1994) and *lsd* for *l*esions *s*imulating *d*isease (Dietrich et al. 1994). For both *acd* and *lsd* mutants, uninfected sterile plants form lesions in the absence of any stimulus. The lesion tissue accumulates the antimicrobial defense compound camalexin, defense-related cell wall modifications, and defense-related gene transcripts (Dietrich et al. 1994; Greenberg et al. 1994). Moreover, the *acd2* mutants with lesions exhibit an SAR response typical of plants that have been infected with a pathogen that elicits an HR; *acd2* plants with lesions show high levels of salicylic acid and high levels of SAR-associated gene transcripts (Greenberg et al. 1994). Finally, *acd* and *lsd* mutants were shown to exhibit high-level resistance to *Pseudomonas* and *Peronospora parasitica* (Dietrich et al. 1994; Greenberg et al. 1994; E. Holub, unpubl.). These observations suggest that *acd*- and *lsd*-encoded products negatively regulate the HR, since the mutations are recessive (Dietrich et al. 1994; Greenberg et al. 1994).

The observation that the HR and SAR can be activated in the absence of any pathogens in *acd2* and *lsd* mutants suggests a model for how these pathways are activated. The simplest model is that the HR and SAR are triggered by a plant-derived product(s), the synthesis of which does not strictly depend on pathogen infection. It had previously been hypothesized that SAR was the result of a plant-derived systemic signal, since in the case of virus-induced SAR, no virus was detected in the uninoculated leaves (see, e.g., Dempsey et al. 1993). The observation that SAR is activated in the *acd* and *lsd* mutants in the absence of any pathogen definitively establishes that the systemic signal is derived from the plant.

### *Arabidopsis* Phytoalexin Mutants

A large number of plant defense responses have been identified, mainly on the basis of observations that they are induced in response to pathogen attack. Relatively little is known concerning which of these responses actually contribute to resistance against particular pathogens. As pointed out in the first section of this paper, this is partly because no plant mutants with defects in individual components of the defense response have been isolated, making it difficult to study the effectiveness of particular defense responses in vivo. The development of *Arabidopsis* as a model for studying plant/pathogen interactions has made a genetic approach to this problem feasible.

One defense response that has been studied extensively is phytoalexin synthesis. Phytoalexins are small molecules with antimicrobial activity that are synthesized by plants in response to pathogen attack. As such, they seem likely to play an important role in plant defense. Only one phytoalexin, 3-thiazol-2′-yl-indole, called camalexin (Browne et al. 1991), has been detected in *Arabidopsis* (Tsuji et al. 1992). Camalexin was found to accumulate in *Arabidopsis* leaves following infection with *P. syringae* strains. Camalexin accumulated to similar high levels in response to either the virulent bacterial strain *Psm* ES4326 or an isogenic avirulent strain. However, camalexin accumulated more slowly in response to a different virulent strain, *Pst* DC3000, than in response to an isogenic avirulent strain. Thus, camalexin accumulation appears to be affected by a variety of pathogen characteristics, including the presence of avirulence genes (Glazebrook and Ausubel 1994).

A biochemical screen was used to isolate three mutants of *A. thaliana* ecotype Columbia that synthesized less camalexin than wild-type plants in response to infection by *P. syringae* (Glazebrook and Ausubel 1994). These were labeled *pad* for phytoalexin *d*eficient. The *pad1* and *pad2* mutants accumulated camalexin to 30% and 10%, respectively, of the level in wild-type plants. No camalexin was detected in the *pad3* mutant. The mutations *pad1*, *pad2*, and *pad3* were found to be recessive alleles of three different genes. *pad1* and *pad2* were mapped to chromosome 4 and *pad3* was mapped to chromosome 3.

Infection of *pad* mutant plants with several isogenic pairs of virulent and avirulent *P. syringae* strains revealed that the growth of avirulent strains relative to isogenic virulent strains was reduced to similar extents in both wild-type and *pad* mutant plants. This result strongly suggests that in *A. thaliana*, phytoalexin biosynthesis is not required for resistance to avirulent *P. syringae* pathogens. Two of the *pad* mutants, but not the third, displayed enhanced susceptibility to virulent *P. syringae* pathogens. Therefore, it is possible that camalexin serves to limit the growth of virulent bacteria, but it will not be possible to draw a definitive conclusion until the variation among the *pad* mutants has been explained.

These phytoalexin-deficient mutants can now be used to assess the importance of camalexin in defense against other bacterial and fungal pathogens. The fact that it was possible to use *Arabidopsis* to isolate mutations affecting a single component of the defense response and to use them effectively to study the importance of that component in combating a particular pathogen demonstrates the power of *Arabidopsis* for studying plant defense responses. Similar approaches could be used to study other aspects of the defense response, such as synthesis of pathogenesis-related proteins, glucanases, and chitinases.

## CONCLUDING REMARKS

Remarkable progress has been achieved in the few years since *Arabidopsis* began to be exploited as a model for addressing important questions about the microbial pathogenesis of plants. The unique benefits of biology and technology provided by *Arabidopsis* as a tool for research in plant science are well-illustrated by the rapidity with which it has been possible to elucidate the genetic control of host/pathogen interactions in several pathosystems, the relative ease with which interesting mutants have been identified, and the fact that several loci of potential importance in pathogen recognition and plant response have already been located in the genome. It is too early to claim that any major new insights have emerged from research on pathogenesis of *Arabidopsis*. However, there is a high level of cooperation among research groups interested in the pathology of *Arabidopsis* and every reason to be optimistic that significant advances in knowledge will not be long in coming. In particular, the isolation and molecular characterization of several alleles involved in genotype-specific recognition phenomena can be anticipated. This may rapidly lead to an understanding of the molecular basis for gene-for-gene specificity, the isolation of homologous alleles in other plant species, and the prospect of creating new sequences with novel recognition capability. It is premature to predict how these advances may be exploited to the practical benefit of disease control in crops, but there will inevitably be insights into host/parasite coevolution within the context of agricultural and natural ecosystems. There are increasing opportunities to influence and manipulate the resistance of plants to microbial parasites through various applications of transformation technology. It is now clear that *Arabidopsis* is a host to a more complete array of biologically diverse microbial parasites than the plant species used hitherto to evaluate the efficacy of transgenic technology. A more prominent role for *Arabidopsis* and its parasites in this context can be persuasively argued.

# APPENDIX

## *Arabidopsis* **Fungal Pathogens**

In terms of economic loss to agriculture, and perhaps also in natural ecosystems, the most important plant pathogens are fungi. The fungi are a taxonomically diverse assemblage; Oomycetes and Plasmodiophoromycetes are members of separate phyla and are evolutionarily remote from "true" fungi: Ascomycetes, Basidiomycetes, and Deuteromycetes. In the preceding sections of this chapter, we have presented a detailed discussion of the biology and pathophysiology of *Peronospora parasitica/Arabidopsis* interactions. In this Appendix, we discuss several other fungi which have also been shown to be *Arabidopsis* pathogens.

### *Albugo candida* (White Blister/White Rust)

*Albugo candida* is phylogenetically close to the downy mildews (*Peronosporales*) and is an obligately biotrophic parasite exclusive to members of the Crucifereae. In common with *P. parasitica*, *A. candida* also exists as numerous host-adapted forms, each specialized to particular subfamily taxa. *A. candida* sporangia are air-borne and dispersed in a desiccated condition that appears tolerant of low humidity. When sporangia rehydrate in water, they release numerous motile, biflagellate zoospores that migrate to stomatal pores where they encyst. The cyst produces a germ tube that enters the leaf through the stomate; thereafter, an extensive intercellular mycelium develops with globular haustoria inserted into host cells along the length of the hyphae. *A. candida* also produces long-lived, thick-walled oospores which germinate to produce zoospores that are presumed to infect roots or aerial parts of the host after being dispersed by rain-splash. It is probable that some isolates of *A. candida* are heterothallic, but all isolates from *Arabidopsis* studied so far appear to be homothallic.

Infection of the foliage of *Arabidopsis* with *A. candida*, as with other crucifers, is characterized by the occurrence of pale-green patches on the upper surface which eventually turn chlorotic and may develop red borders. On the under surface, the fungus produces white pustules or "blisters" similar in appearance to the "true" rusts (this disease is sometimes referred to as white rust). Within the pustule, beneath the host epidermis, the fungus produces a mass of unbranched sporangiophores on which develop linear chains of sporangia. When the pustule is mature, the epidermis ruptures to release a mass of sporangia as a dry, white powder. *A. candida* can become systemic in its host and, under these circumstances, may cause malformation of stems and phyllody of floral parts (referred to as "stag-heads"), presumably due to a disturbance of normal hormone balances.

Isolate-specific interaction phenotypes have been examined for more than 30 ecotypes of *Arabidopsis* following inoculation of seedlings at the cotyledon stage with two different isolates of *A. candida* (Acem1 and Acks1) (Crute et al. 1993; E.B. Holub, unpubl.). The observed phenotypes have been placed into four categories ranging from extensive pustule production without early occurrence of chlorosis or necrosis to the occurrence of necrotic flecks in the absence of chlorosis or pustules. Segregation among $F_2$ progeny of the cross Wei-0 x Ksk-1 inoculated with isolate Acem1 indicated that a dominant allele at a single locus was associated with the resistant phenotype characteristic of Ksk-1 (no sporulation). Wei-0 is susceptible to Acem1 (E.B. Holub, unpubl.). Resistance to isolate Acem1 resides at the *RAC1* locus on chromosome 1 in the Ksk-1 ecotype (E. Brose et al., unpubl.).

Interestingly, studies of natural *Arabidopsis* populations (Holub et al. 1994; E.B. Holub and I.R. Crute, unpubl.) revealed that areas of foliage bearing pustules of *A. candida* are frequently co-infected by *P. parasitica*. The two organisms exist in an intimate association, with oospores of both evident within the infected tissue. Prior infection by one organism (probably *A. candida*) predisposes tissue to infection by the other. Similar observations have been recorded on *Capsella bursa-pastoris* (Sansome and Sansome 1974), *Brassica rapa* (synonym: *B. campestris*) (Chaurasia et al. 1982), and *B. juncea* (Bains and Jhooty 1985). A comparison between alleles involved in the specific recognition of *A. candida* and *P. parasitica* will be of particular importance for a full appreciation of the molecular basis for parasite recognition.

## *Erysiphe cruciferarum* (Powdery Mildew)

The powdery mildews (Erysiphaceae), a group of obligately biotrophic Ascomycetes, are airborne pathogens that spread by means of short-lived conidia produced in chains at the surface of infected plants. These conidia will germinate in the absence of free water provided humidity is high and, as a consequence, there are fewer environmental constraints on epidemic development than with most other plant diseases. The powdery mildews produce a haploid mycelium on infected plant surfaces and place haustoria into the epidermal cells beneath by which they extract nutrients. Powdery mildew fungi are frequently heterothallic with sexual spores being formed only when clones of opposite sexual compatibility type infect the same tissue.

Genera and species in the Erysiphaceae are parasites of numerous plant families and cause many diseases of economic significance on crops. Powdery mildew fungi exhibit a high degree of host specificity. *Erysiphe cruciferarum* is the species of powdery mildew that parasitizes members of the Crucifereae. The characteristic symptom of powdery mildew infection is a white mat of mycelium and conidial chains over the affected aerial plant surfaces. This degree of colonization has been observed in natural infections of *Arabidopsis* ecotype Col-0 (K. Michelson and E.B. Holub, unpubl.). Isolates of *E. cruciferarum* obtained from *Brassica napus* have also been shown to grow and reproduce on *Arabidopsis* (D. Hall et al., pers. comm.). The sexual stage of *E. cruciferarum* has not been observed on *Arabidopsis* and there is, as yet, no information on sexual compatibility of isolates.

Among 90 ecotypes of *Arabidopsis*, a range of interaction phenotypes have been identified following inoculation with two isolates of *E. cruciferarum* (D. Hall et al.; L. Adam and S. Somerville; both pers. comm.). In the case of the UK strain (from *B. napus*), 50 *Arabidopsis* ecotypes could be allocated to at least eight categories using the extent of mycelial growth, sporulation, and the elicitation of an HR as criteria (D. Hall et al., pers. comm.). Segregation in the $F_2$ generation among progeny from crosses between the resistant ecotypes Ms-0 and Can-0 and the fully susceptible ecotype Col-0 suggest that resistance to the UK race is controlled by a dominant allele at a single locus in Ms-0 and by two dominant alleles at two independent loci in Can-0.

In the case of the USC strain, a survey of 50 *Arabidopsis* ecotypes showed that most were susceptible (L. Adam and S. Somerville, pers. comm.). Six resistant ecotypes were identified. The resistance phenotypes in Wa-0 and Kas-1 were characterized by hypersensitive necrosis as evidenced by the rapid (24–48 hr) appearance of small necrotic flecks at the inoculation site. Genetic analyses of these two ecotypes were consistent with the hypothesis that resistance is determined by a semidominant allele at a single, nuclear gene. Resistance in Su-0 and Stw-0 was characterized by a reduction in fungal growth without any other visible reaction. Disease resistance was determined by a semidominant allele in both ecotypes; genetic analysis indicated that a single locus controlled the resistance phenotype. Sl-0 and Te-0 showed no fungal growth 7 days after inoculation. With both ecotypes, the $F_1$ were suceptible to powdery mildew attack. In the $F_2$ population, the segregation ratios were 9 susceptible:6 intermediate:1 resistant (2-gene model) and 3 susceptible:1 resistant (1-gene model), respectively, for Sl-0 and Te-0. Examples of dominant suceptiblility alleles and of 2-gene mechanisms of resistance are uncommon, and it will be interesting to compare these resistance genes with dominant resistance genes. Between the US and UK laboratories, one ecotype, Kas-1, was found to be resistant to both the UK and USC races of powdery mildew, raising the question of whether the same resistance allele conditions resistance to both races.

## *Plasmodiophora brassicae* (Clubroot)

*Plasmodiophora brassicae* is an intracellular obligate parasite of crucifers belonging to the class Plasmodiophoromycete within the Myxomycota (a phylum of the kingdom Protista). Infection by *P. brassicae* results when long-lived haploid resting spores germinate to release a motile biflagellate zoospore which encysts at the root surface and inserts its content into a root-hair or rhizodermal cell. After infection, a multinucleate myxamoeboid plasmodium develops (the so-called primary plasmodium) capable of migrating from cell to cell throughout the root (Mithen and Magrath 1992). Subsequently, large secondary plasmodia develop, possibly by fusion of several myxamoebae. Haploid resting spores are differentiated from secondary plasmodia possibly after karyogamy and meiosis.

*P. brassicae* causes hypertrophy and hyperplasia in the cortex of roots of plant species in the family Crucifereae including *Arabidopsis* (Koch et al. 1991a). Infection results in the formation of a gall, which, when it finally decays, releases an enormous resting spore population into the surrounding soil. The disease, clubroot, imposes a significant economic constraint on the production of crucifer crops in infested areas and is characterized by severe plant growth reduction as a consequence of nutrient and water stress resulting from root malformation and galling.

*P. brassicae* is known to exist as pathotypic variants identified by the ability or inability to differentially incite galling in different cultivars of *Brassica rapa* and *B. napus* carrying genes for resistance (Crute 1986); a gene-for-gene relationship has been proposed. Up to now, however, no variation in the occurrence of gall formation has been observed among ecotypes of *Arabidopsis* in response to inoculation by *P. brassicae* (Koch et al. 1991a; R. Mithen, pers. comm.), although gall formation on some ecotypes occurs more slowly than on others and some variation in the size of galls produced has also been observed.

There has been a report that indicates the possibility of horizontal transfer of DNA from cruciferous hosts to *P. brassicae* associated particularly with the intimate relationship between the secondary plasmodium and the host nucleus prior to sporogenesis (Bryngelsson et al. 1988; Mithen and Magrath 1992). *Arabidopsis* provides the opportunity to validate this observation and gain an appreciation of its importance in pathogenesis and host/parasite co-evolution.

## *Leptosphaeria maculans* (Canker/Blackleg)

The necrotrophic Ascomycete *Leptosphaeria maculans* (order Dothideales) is a common heterothallic pathogen of wild and cultivated crucifers and is of considerable economic importance in *Brassica* crops. Infections are initiated from splash-dispersed ascospores produced when individuals of opposite sexual compatibility type infect the same host tissue. The pathogen can cause damping-off of seedlings and, on older plants, leaf lesions and debilitating root rot and stem canker. Differential interactions between isolates of *L. maculans* and genotypes of *Brassica* spp. have been reported, but these have not yet been subjected to detailed genetic studies (Hill and Williams 1988). Isolates of *L. maculans* from *Brassica* spp. are known to fall into two sexually incompatible groups on the basis of their pathogenicity; these have been referred to as "aggressive" and "nonaggressive" (Koch et al. 1991b). Analyses of RFLPs between numerous isolates of each group suggest that they are phylogenetically distinct. Sjödin and Glimelius (1988) report the occurrence of *L. maculans* on *Arabidopsis*, but no details of symptoms are provided.

Hill and Williams (1988) and Koch et al. (1991b) have described the many advantages that *L. maculans* has as a pathogen worthy of detailed genetic investigation. These include the ease of axenic culture and asexual inoculum production; the ease of sexual genetic studies; the availability of considerable morphological, physiological, and pathogenic variation; and the ability to conduct genetic transformation. There are, as yet, no reports of variation among ecotypes or mutants of *Arabidopsis* for response to different isolates of *L. maculans*.

## *Rhizoctonia solani* (Wire Stem/Damping-off)

The common soil-inhabiting fungus, *Rhizoctonia solani*, is the anamorph (imperfect state) of the Basidiomycete *Thanatephorus cucumeris* (class Hymenomycetes, order Tulasnellales). *R. solani* survives in soil as sclerotia (decay-resistant assemblages of melanized cells); these structures are the infectious propagules of the pathogen that germinate to infect new host tissue in their vicinity. A new crop of sclerotia is produced in infected, rotting tissue and subsequently released to the soil.

*R. solani* is a necrotrophic pathogen that causes root and stem rots, foliage blights of a wide range of plant species, and, in particular, results in the death (damping-off) of young seedlings before or soon after emergence. In cultivated crucifers, this pathogen causes a disease referred to as wire stem when it causes their collapse by girdling the stem of young plants at soil level. Koch and Slusarenko (1990b) reported natural infection of *Arabidopsis* in an experimental glasshouse where the pathogen caused rapid death of seedlings of several ecotypes (Rld-1, Wei-0, and Ler-0).

*R. solani* exists as four main anastomosis groups (AG1–4) (assemblages of isolates whose hyphae are capable of fusing with one another) that differ for various ecological and physiological characters including pathogenicity and host adaptation. Although the AG identities of the isolates recovered from *Arabidopsis* have not been determined, isolates from cruciferous hosts have previously been shown to belong to AG1 and AG2 (Ogoshi 1978; Adams 1988). Because *R. solani* is one of the few necrotrophic fungal pathogens that exhibits a degree of isolate-specific host adaptation, it would be worthwhile determining if isolates from all AG groups are potential pathogens of *Arabidopsis* and, similarly, if there is variation among ecotypes for susceptibility to isolates from different AG groups.

### Botrytis cinerea (Gray Mold)

*Botrytis cinerea* is a highly variable species aggregate that may correspond to several teleomorphs of the Ascomycete genus *Botrytinia* (order Helotiales). *B. cinerea* is a ubiquitous, necrotrophic pathogen exhibiting variation for virulence among isolates but not characterized by any known host specialization. In addition to being parasitic, *B. cinerea* is capable of saprophytic growth and primarily survives on infected plant debris. The main means of dispersal is via the airborne conidia that are produced in large numbers.

*B. cinerea* causes a brown soft-rot of foliage, fruits, and flowers, accompanied by profuse production of gray conidiophores under conditions of high humidity. The fungus is generally considered only to be able to establish initially on senescent or damaged tissue or on sugary substrates such as fruits; however, following establishment, *B. cinerea* is capable of killing and invading healthy tissue of many plant species. Koch and Slusarenko (1990b) reported the natural infection of *Arabidopsis* at the rosette stage and noted that the infection progressed after flowering.

### Verticillium spp. (Vascular Wilt)

The root-infecting vascular pathogens, *Verticillium dahliae* and *V. alboatrum*, are imperfect fungi (Deuteromycetes) for which no teleomorph (sexual stage) is known, although they are often assumed to be closely allied to the Ascomycete order Hypocreales. Populations of *Verticillium* spp. reside in soil as microsclerotia (*V. dahliae*) or black resting hyphae (*V. alboatrum*). Susceptible host plants are invaded through the roots, and inoculum is returned to the soil with plant debris. These fungi have a wide host range and cause damaging diseases of numerous annual and perennial crops known collectively as vascular wilts. Recently, analyses of RFLPs among a range of isolates of both species have clearly established the existence of several distinct subspecific groupings (Carder and Barbara 1991; Okoli et al. 1993).

The classic symptom of infection by *Verticillium* spp. is wilting or reduced vigor and premature foliar senescence and necrosis. However, *Verticillium* spp. infect and colonize many host species, particularly herbaceous weeds, without provoking overt symptoms. Variation within crop species for response to *Verticillium* spp. has frequently been exploited in plant breeding programs, and quantitative variation among fungal isolates with respect to pathogenicity on hosts of differing resistance is well known. However, the genetic basis of this variation is not well elucidated, although there is no evidence for specific genotype X genotype interactions indicative of a gene-for-gene relationship. Following root inoculation of several ecotypes of *Arabidopsis* at the rosette stage with both *V. dahliae* and *V. alboatrum*, systemic colonization was confirmed by reisolation from the flowering stem several weeks later (D.J. Barbara, pers. comm.). Obvious macroscopic symptoms of infection were not evident, however.


## *Pythium* spp. (Damping-off)

Species of *Pythium* comprise a large genus of soil-inhabiting, necrotrophic, Oomycetes (order Peronosporales; family Pythiaceae), several of which are plant pathogenic and primarily cause pre- and postemergence damping-off of seedlings and subclinical root necroses. *Pythium* spp. usually survive in soil as oospores which germinate to produce infective motile biflagellate zoospores that encyst on the surface of roots. An extensive coenocytic mycelium rapidly colonizes the root from the site of infection, causing necrosis and death. Further generations of zoospores are typically produced during the growth of the vegetative mycelium. The production of oospores following meiosis in the sexual organs (antheridia and oogonia) is dependent on whether the species is homothallic or heterothallic. Poor emergence after sowing or rapid collapse of newly emerged seedlings are the main symptoms of infection by *Pythium* spp.

*Pythium* spp. are generally considered to be polyphagus, and there is little evidence that they express any marked host adaptation. Two as-yet-unidentified species of *Pythium*, differentiated by the morphology of their oospores, have been isolated from *Arabidopsis* seedlings suffering severe damping-off (Mauch-Mani et al. 1993). Reinfection of *Arabidopsis* seedlings with isolates of both species established their pathogenicity.

## *Penicillium olsonii* (Blue Mold)

There are several plant pathogenic species in the predominantly saprophytic ascomycetous genus *Penicillium* (order Eurotiales), although most cause storage rots of fruits and bulbs rather than growing plants. *P. olsonii* was identified as a pathogen of *Arabidopsis* when it occurred as a chance contaminant on glasshouse-grown plants (W. Schäfer, pers. comm.). The fungus was transformed with the *uidA* reporter under the control of a promoter from the related fungus *Aspergillus nidulans*. Conventional GUS staining allowed the development of the fungus through the plant to be monitored after sterile leaves were inoculated with conidia. Growth was slow and sparse through the leaf, and only a few hyphae entered the peduncle within the xylem. Thereafter, flowers were rapidly colonized and profuse sporulation occurred over the surface of the flower. More extensive mycelial development then took place as the fungus grew back into older tissues as they seemingly became more susceptible. Plant death finally resulted from this heavy infection. Three ecotypes of *Arabidopsis* apparently resistant to *P. olsonii* have been identified, and this resistance is expressed in $F_1$ individuals from crosses with a susceptible ecotype. *P. olsonii* can be readily transformed, making it a potentially attractive model system.

## *Sclerotinia sclerotiorum* (White Mold or Soft Rot)

The soil-borne ascomycetous fungus *Sclerotinia sclerotiorum* (order Helotiales) has one of the widest host ranges of all fungal pathogens, causing a wet rot of many herbaceous species. The fungus survives in soil as resistant sclerotia which may germinate to give rise to vegetative hyphae or cup-shaped apothecia (sexual structures) in which ascospores are produced. The pathogenesis of plants by *S. sclerotiorum* has been the subject of much study and, in particular, the role of degradative and lytic enzymes in that pathogenesis. Reports that oxalic acid plays a role in the process of plant infection have been confirmed by Dickman and Mitra (1992), who utilized *Arabidopsis* as an experimental host. Bacteria capable of degrading oxalic acid have been shown by Dickman and Mitra (1992) to inhibit infection of *Arabidopsis* by *S. sclerotiorum*. The isolation of genes involved in the degradation process, and their expression in transgenic *Arabidopsis*, may provide evidence of their value as transgenes to provide enhanced resistance to this important pathogen.

### *Fusarium oxysporum* (Vascular Wilt)

Worldwide in occurrence, *Fusarium oxysporum* Schlechtend.: Fr. is the most frequent of all Fusaria found in the soil (40–70% of all Fusaria isolated). The teleomorphic (perfect) form of *F. oxysporum* is unknown. Various forms are capable of causing tracheomycoses (leading to vascular wilt diseases) and damping-off in a variety of annual vegetables, perennials, and weeds. In axenic culture, *F. oxysporum* isolates can show a large variation in morphology. However, the pathogenicity remains relatively constant and seems to be independent of the morphological variation.

Different isolates of *F. oxysporum* often show host specialization, e.g., *F. oxysporum* f.sp. *lycopersici* on tomatoes and *F.o.* f.sp. *conglutinans* on cabbage. In addition, it is sometimes possible to distinguish between physiologically specialized forms (races) which attack only certain cultivars of a host. In cabbage (*Brassica oleracea* var. *capitata*), for example, cultivars with single dominant genes controlling high degrees of resistance against specific races of *F. oxysporum*, as well as cultivars displaying a multigenic type of resistance, have been known for many years. Recently, *F. oxysporum* has been genetically transformed (Kistler and Benny 1988), making it a good subject for molecular analysis.

Mauch-Mani and Slusarenko (1994) isolated *F. oxysporum* from naturally infected *Arabidopsis* (ecotype Landsberg *erecta*) plants. This isolate varied in virulence on different ecotypes and was capable of immunizing *Arabidopsis* against a subsequent infection by *Peronospora parasitica*.

## REFERENCES

Adams, G.C. 1988. *Thanatephorus cucumeris* (*Rhizoctonia solani*), a species complex of wide host range. In *Advances in plant pathology: Genetics of plant pathogenic fungi* (ed. D.S. Ingram et al.), vol. 6, pp. 535–552. Academic Press, London.

Alin, P., U.H. Danielson, and B. Mannervik. 1985. 4-Hydroxylalkenals are substrates for glutathione transferase. *FEBS Lett.* **179:** 267–270.

Bains, S.S. and J.S. Jhooty. 1985. Association of *Peronospora parasitica* with *Albugo candida* on *Brassica juncea* leaves. *Phytopathol. Z.* **112:** 28–31.

Bent, A.F., R.W. Innes, J.R. Ecker, and B.J. Staskawicz. 1992. Disease development in ethylene-insensitive *Arabidopsis thaliana* infected with virulent and avirulent *Pseudomonas* and *Xanthomonas* pathogens. *Mol. Plant-Microbe Interact.* **5:** 372–378.

Bent, A.F., B.N. Kunkel, D. Dahlbeck, K.L. Brown, R. Schmidt, J. Giraudat, J. Leung, and B.J. Staskawicz. 1994. *RPS2* of *Arabidopsis thaliana* represents a new class of plant disease resistance genes. *Science* (in press).

Berger, B. 1965. Epidemic appearance of *Alternaria brassicae* on *Arabidopsis* growing in greenhouses. *Arabidopsis Inf. Serv.* **2:** 32–33.

Bisgrove, S.R., M.T. Simonich, N.M. Smith, A. Sattler, and R.W. Innes. 1994. A disease resistance gene in *Arabidopsis* with specificity for two different pathogen avirulence genes. *Plant Cell* **6:** 927–933.

Bowles, D.J. 1990. Defense-related proteins in higher plants. *Annu. Rev. Biochem.* **59:** 873–907.

Brandenburger, W. 1985. *Parasitische Pilze an Gefässpflanzen in Europa.* Gustav Fischer, New York.

Broglie, K., I. Chet, M. Holliday, R. Cressman, P. Biddle, S. Knowlton, C. J. Mauvis, and R. Broglie. 1991. Transgenic plants with enhanced resistance to the fungal pathogen *Rhizoctonia solani. Science* **254:** 1194–1197.

Browne, L.M., K.L. Conn, W.A. Ayer, and J.P. Tewari. 1991. The camalexins: New phytoalexins produced in the leaves of *Camelina sativa* (Cruciferae). *Tetrahedron* **47:** 3909–3914.

Bryngelsson, T., M. Gustaffson, B. Green, and C. Lind. 1988. Uptake of host DNA by the parasitic fungus *Plasmodiophora brassicae. Physiol. Mol. Plant Pathol.* **33:** 163–171.

Carder, J.H. and D.J. Barbara. 1991. Molecular variation and restriction fragment length polymorphisms (RFLPs) within and between six species of *Verticillium. Mycol. Res.* **95:** 935–942.

Channon, A.G. 1981. Downy mildew of brassicas. In *The downy mildews* (ed. D.M. Spencer), pp. 381–339. Academic Press, London.

Chaurasia, S.N.P., U.P. Singh, and H.B. Singh. 1982. Preferential parasitism of *Peronospora parasitica* on galls of *Brassica campestris* caused by *Albugo cruciferarum. Trans. Br. Mycol. Soc.* **78:** 379–381.

Chester, K.S. 1933. The problem of acquired physiological immunity in plants. *Q. Rev. Biol.* **8:** 275–324.

Collinge, D.B. and A.J. Slusarenko. 1987. Plant gene expression in response to pathogens. *Plant Mol. Biol.* **9:** 389–410.

Crute, I.R. 1985. The genetic bases of relationships between microbial parasites and their hosts. In *Mechanisms of resistance to plant disease* (ed. R.S.S. Fraser), pp. 80–143. Martinus Nijhoff/Kluwer Academic, Dordrecht, The Netherlands.

———. 1986. The relationship between *Plasmodiophora brassicae* and its hosts: The application of concepts relating to variation in inter-organismal associations. In *Ad-*

*vances in plant pathology* (ed. D.S. Ingram and P.H. Williams), vol. 5, pp. 1–52. Academic Press, London.

Crute, I.R., P.J.G.M. De Wit, and M. Wade. 1985. Mechanisms by which genetically controlled resistance and virulence influence host colonisation by fungal and bacterial parasites. In *Mechanisms of resistance to plant disease* (ed R.S.S. Fraser), pp. 197–309. Martinus Nijhoff/Kluwer Academic, Dordrecht, The Netherlands.

Crute, I.R., E.B. Holub, M. Tör, E. Brose, and J.L. Beynon. 1993. The identification and mapping of loci in *Arabidopsis thaliana* for recognition of the fungal pathogens: *Peronospora parasitica* (downy mildew) and *Albugo candida* (white blister). In *Advances in molecular genetics of plant-microbe interactions* (ed. E.W. Nester and D.P.S. Verma), vol. 2, pp. 437–444. Kluwer Academic, Dordrecht, The Netherlands.

Dangl, J.L., E.B. Holub, T. Debener, H. Lehnackers, C. Ritter, and I.R. Crute. 1992. Genetic definition of *Arabidopsis* loci involved in plant-pathogen interactions. In *Methods in Arabidopsis research* (ed. C. Koncz et al.), pp. 393–418. World Scientific, Singapore.

Davis, K.R. and F.M. Ausubel. 1989. Characterization of elicitor-induced defense responses in suspension-cultured cells of *Arabidopsis*. *Mol. Plant-Microbe Interact.* **2:** 363–368.

Davis, K.R., E. Schott, and F.M. Ausubel. 1991. Virulence of selected phytopathogenic pseudomonads in *Arabidopsis thaliana*. *Mol. Plant-Microbe Interact.* **4:** 477–488.

Debener, T., H. Lehnackers, M. Arnold, and J. Dangl. 1991. Identification and molecular cloning of a single *Arabidopsis* locus conferring resistance against a phytopathogenic *Pseudomonas* isolate. *Plant J.* **1:** 289–302.

Dempsey, D.A., K.K. Wobbe, and D.F. Klessig. 1993. Resistance and susceptible responses of *Arabidopsis thaliana* to turnip crinkle virus. *Phytopathology* **83:** 1021–1029.

Dickman, M.B. and A. Mitra. 1992. *Arabidopsis thaliana* as a model for studying *Sclerotinia sclerotiorum* pathogenesis. *Physiol. Mol. Plant Pathol.* **41:** 255–263.

Dietrich, R.A., T.P. Delaney, S.J. Uknes, E.R. Ward, J.A. Ryals, and J.L. Dangl. 1994. *Arabidopsis* mutants simulating disease resistance response. *Cell* **77:** 565–577.

Dixon, R.A. and M.J. Harrison. 1990. Activation, structure and organization of genes involved in microbial defense in plants. *Adv. Genet.* **28:** 165–234.

Dixon, R.A. and C.J. Lamb. 1990. Molecular communication in interactions between plants and microbial pathogens. *Annu. Rev. Plant Physiol. Plant Mol. Biol.* **41:** 339–367.

Dong, X., M. Mindrinos, K.R. Davis, and F.M. Ausubel. 1991. Induction of *Arabidopsis thaliana* defense genes by virulent and avirulent *Pseudomonas syringae* strains and by a cloned avirulence gene. *Plant Cell* **3:** 61–72.

Dudler, R., C. Hertig, G. Rebmann, J. Bull, and F. Mauch. 1991. A pathogen-induced wheat gene encodes a protein homologous to glutathione-S-transferases. *Mol. Plant-Microbe Interact.* **4:** 14–18.

Ellingboe, A.H. 1981. Changing concepts in host-pathogen genetics. *Annu. Rev. Phytopathol.* **19:** 125–143.

Feinbaum, R.L. and F.M. Ausubel. 1988. Transcriptional regulation of the *Arabidopsis thaliana* chalcone synthase gene. *Mol. Cell. Biol.* **8:** 1985–1992.

Fletcher, J. and C.E. Eastman. 1992. *Arabidopsis thaliana* as an experimental host of the mollicute *Spiroplasma citri*. *Plant Dis.* **76:** 862–864.

Flor, H.H. 1971. Current status of gene-for-gene concept. *Annu. Rev. Phytopathol.* **9:** 275–296.

Gaffney, T., L. Friedrich, B. Vernooij, D. Negrotto, G. Nye, S. Uknes, E. Ward, H.

Kessmann, and J. Ryals. 1993. Requirement of salicylic acid for the induction of systemic acquired resistance. *Science* **261:** 754–756.

Gäumann, E. 1918. Ueber die Formen der *Peronospora parasitica* (Pers.) Fries. *Beih. Bot. Zentralbl.* **35**(1. Abt): 395–533.

Golino, D.A., M. Shaw, and L. Rappaport. 1988. Infection of *Arabidopsis thaliana* (L.) Heyhn. with a mycoplasma-like organism, the beet leafhopper transmitted virescence agent. *Arabidopsis Inf. Serv.* **26:** 9–14.

Glazebrook, J. and F.M. Ausubel. 1994. Isolation of phytoalexin-deficient mutants of *Arabidopsis thaliana* and characterization of their interactions with bacterial pathogens. *Proc. Natl. Acad. Sci.* **91:** 8955–8959.

Greenberg, J. and F.M. Ausubel. 1993. *Arabidopsis* mutants compromised for the control of cellular damage during pathogenesis and aging. *Plant J.* **4:** 327–341.

Greenberg, J.T., A. Guo, D.F. Klessig, and F.M. Ausubel. 1994. Programmed cell death in plants: A pathogen-triggered response activated coordinately with multiple defense functions. *Cell* **77:** 551–563.

Hahlbrock, K. and D. Scheel. 1989. Physiology and molecular biology of phenylpropanoid metabolism. *Annu. Rev. Plant Physiol. Plant Mol. Biol.* **40:** 347–369.

He, S.H., H.-C. Huang, and A. Collmer. 1993. *Pseudomonas syringae* pv. *syringae* Harpin: A protein that is secreted via the Hrp pathway and elicits the hypersensitive response in plants. *Cell* **73:** 1255–1266.

Hill, C.B. and P.H. Williams. 1988. *Leptosphaeria maculans*, cause of blackleg of crucifers. In *Advances in plant pathology: Genetics of plant pathogenic fungi* (ed. D.S. Ingram et al.), vol. 6, pp. 169–174. Academic Press, London.

Hindges, R. and A. Slusarenko. 1991. cDNA and derived amino acid sequence of a cytosolic Cu,Zn superoxide dismutase from *Arabidopsis thaliana* (L.) Heyhn. *Plant Mol. Biol.* **17:** 1–3.

Holub, E.B., J.L. Beynon, and I.R. Crute. 1994. Phenotypic and genotypic characterization of interactions between isolates of *Peronospora parasitica* and accessions of *Arabidopsis thaliana*. *Mol. Plant-Microbe Interact.* **7:** 223–239.

Holub, E.B., P.H. Williams, and I.R. Crute. 1991. Natural infection of *Arabidopsis thaliana* by *Albugo candida* and *Peronospora parasitica*. *Phytopathology* **81:** 1226.

Holub, E., I. Crute, E. Brose, and J. Beynon. 1993. Identification and mapping of loci in *Arabidopsis* for resistance to downy mildew and white blister. In Arabidopsis *as a model in plant-pathogen interactions* (ed. K.R. Davis and R. Hammerschmidt), pp. 21–35. American Phytopathological Society Press, St. Paul, Minnesota.

Innes, R.W., A.F. Bent, B.N. Kunkel, S.R. Bisgrove, and B.J. Staskawicz. 1993a. Molecular analysis of avirulence gene *avrRpt2* and identification of a putative regulatory sequence common to all known *Pseudomonas syringae* avirulence genes. *J. Bacteriol.* **175:** 4859–4869.

Innes, R.W., S.R. Bisgrove, N.M. Smith, A.F. Bent, B.J. Staskawicz, and Y.-C. Liu. 1993b. Identification of a disease resistance locus in *Arabidopsis* that is functionally homologous to the *RPG1* locus of soybean. *Plant J.* **4:** 813–820.

JØrstad, I. 1964. The phycomycetous genera *Albugo*, *Bremia*, *Plasmopara* and *Pseudoperonospora* in Norway with an appendix containing unpublished finds of *Peronospora*. *NYTT Mag. Bot.* **11:** 47–82.

Kawalleck, P., H. Keller, K. Hahlbrock, D. Scheel, and I.E. Somssich. 1993. A pathogen-responsive gene of parsley encodes tyrosine decarboxylase. *J. Biol. Chem.* **268:** 2189–2194.

Keen, N.T. 1982. Specific recognition in gene-for-gene host-parasite systems. In *Advances in plant pathology* (ed. D.S. Ingram and P.H. Williams), vol. 1, pp. 35–82. Aca-

744   I. Crute et al.

demic Press, London.

———. 1990. Gene-for-gene complementarity in plant-pathogen interactions. *Annu. Rev. Genet.* **24:** 447–463.

Keen, N.T. and B.J. Staskawicz. 1988. Host range determinants in plant pathogens and symbionts. *Annu. Rev. Microbiol.* **42:** 421–440.

Keen, N.T., D. Tamaki, D. Kobayashi, M. Gerhold, H. Stayton, S. Shen, J. Gold, H. Lorang, D. Thordal-Christensen, D. Dahlbeck, and B. Staskawicz. 1990. Bacteria expressing avirulence gene *D* produce a specific elicitor of the soybean hypersensitive reaction. *Mol. Plant-Microbe Interact.* **3:** 112–121.

Keith, B., X. Dong, F.M. Ausubel, and G.R. Fink. 1991. Differential induction of 3-deoxy-D-*arabino*-heptulosonate 7-phosphate synthase gene in *Arabidopsis thaliana* by wounding and pathogen attack. *Proc. Natl. Acad. Sci.* **88:** 8821–8825.

Kiedrowski, S., P. Kawalleck, K. Hahlbrock, I.E. Somssich, and J.L. Dangl. 1992. Rapid activation of a novel plant defense gene is strictly dependent on the *Arabidopsis RPM1* disease resistance locus. *EMBO J.* **11:** 4677–4684.

Kistler, H.C. and U.K. Benny. 1988. Genetic transformation of the fungal plant wilt pathogen, *Fusarium oxysporum. Curr. Genet.* **13:** 145–149.

Koch, E. and A.J. Slusarenko. 1990a. *Arabidopsis* is susceptible to infection by a downy mildew fungus. *Plant Cell* **2:** 437–445.

———. 1990b. Fungal pathogens of *Arabidopsis thaliana* (L.) Heyhn. *Bot. Helv.* **100:** 257–268.

Koch, E., R. Cox, and P.H. Williams. 1991a. Infection of *Arabidopsis thaliana* by *Plasmodiophora brassicae. J. Phytopathol.* **132:** 99–104.

Koch, E., K. Song, T.C. Osborne, and P.H. Williams. 1991b. Relationship between pathogenicity and phylogeny based on restriction fragment length polymorphisms in *Leptosphaeria maculans. Mol. Plant-Microbe Interact.* **4:** 341–349.

Kuč, J. 1982. Induced immunity to plant disease. *BioScience* **32:** 854–860.

Kunkel, B.N., A.F. Bent, D. Dahlbeck, R.W. Innes, and B.J. Staskawicz. 1993. *RPS2*, an *Arabidopsis* disease resistance locus specifying recognition of *Pseudomonas syringae* strains expressing the avirulence gene *avrRpt2. Plant Cell* **5:** 865–875.

Lamb, C.J., M.A. Lawton, M. Dron, and R.A. Dixon. 1989. Signals and transduction mechanisms for activation of plant defences against microbial attack. *Cell* **56:** 215–224.

Lewis, D.H. 1973. Fungal nutrition and the origin of biotrophy. *Biol. Rev.* **48:** 261–278.

Lois, R., A. Dietrich, K. Hahlbrock, and W. Schulz. 1989. A phenylalanine ammonia-lyase gene from parsley: Structure, regulation and identification of elicitor and light responsive *cis*-acting elements. *EMBO J.* **8:** 1641–1648.

Malamy, J. and D.F. Klessig. 1992. Salicylic acid and plant disease resistance. *Plant J.* **2:** 643–654.

Martin, G.B., S.H. Brommonschenkel, J. Chunwongse, A. Frary, M.W. Ganal, R. Spivey, T. Wu, E.D. Earle, and S.D. Tanksley. 1993. Map-based cloning of a protein kinase gene conferring disease resistance in tomato. *Science* **262:** 1432–1436.

Mauch-Mani, B. and A.J. Slusarenko. 1993a. Downy mildew of *Arabidopsis thaliana*. In Arabidopsis: *An atlas of morphology and development* (ed. J. Bowman). Springer Verlag, New York. (In press.)

———. 1993b. *Arabidopsis* as a model host for studying plant-pathogen interactions. *Trends Microbiol.* **1:** 265–270.

———. 1994. Systemic acquired resistance in *Arabidopsis thaliana* induced by a predisposing infection with a pathogenic isolate of *Fusarium oxysporum. Mol. Plant-Microbe Interact.* **7:** 378–383.

Mauch-Mani, B., K.P.C. Croft, and A.J. Slusarenko. 1993. The genetic basis of resistance of *Arabidopsis thaliana* L. Heyhn to *Peronospora parasitica*. In Arabidopsis *as a model in plant-pathogen interactions* (ed K.R. Davis and R. Hammerschmidt), pp. 5–20. American Phytopathological Society Press, St. Paul, Minnesota.

Meier, I., K. Hahlbrock, and E. Somssich. 1991. Elicitor-inducible and constitutive *in vivo* DNA footprints indicate novel *cis*-acting elements in the promoter of a parsley gene encoding pathogenesis-related proteins. *Plant Cell* **3**: 309–315.

Melan, M.A., X. Dong, M.E. Endara, K.R. Davis, F.M. Ausubel, and T.K. Peterman. 1993. An *Arabidopsis thaliana* lipoxygenase gene can be induced by pathogens, abscisic acid, and methyl jasmonate. *Plant Physiol.* **101**: 441–450.

Metzler, M.C., J.R. Cutt, and D.F. Klessig. 1991. Isolation and characterisation of a gene encoding a PR-1-like protein from *Arabidopsis thaliana*. *Plant Physiol.* **96**: 346–348.

Mindrinos, M., F. Katagiri, J. Glazebrook, and A.M. Ausubel. 1994a. Identification and characterization of an *Arabidopsis* ecotype which fails to mount a hypersensitive response when infiltrated with *Pseudomonas syringae* strains carrying *avrRpt2*. In *Advances in molecular genetics of plant-microbe interactions* (ed. M. Daniels), vol. 3. Kluwer Academic, Dordrecht, The Netherlands. (In press.)

Mindrinos, M., F. Katagiri, G.-L. Yu, and F.M. Ausubel. 1994b. A new class of plant disease resistance gene: The *Arabidopsis RPS2* gene encodes a protein containing a nucleotide binding site and leucine rich repeats. *Cell* (in press).

Mithen, R. and R.A. Magrath. 1992. A contribution to the life history of *Plasmodophora brassicae*: Secondary plasmodia development in root galls of *Arabidopsis thaliana*. *Mycol. Res.* **96**: 877–885.

Morgan, O.D. 1971. A study of four weed hosts of *Sclerotinia* sp. in alfalfa fields. *Plant Dis. Rep.* **55**: 1087–1089.

Morgenstern, R. and J.W. de Pierre. 1983. Microsomal glutathione transferase: Purification in unactivated form and further characterization of the activation process, substrate specificity and amino acid composition. *Eur. J. Biochem.* **134**: 591–597.

Morgenstern, R., G. Lundquist, H. Jornvall, and J.W. de Pierre. 1989. Activation of rat liver microsomal glutathione transferase by limited proteolysis. *Biochem. J.* **260**: 577–582.

Naumov, N.A. 1925. Contribution to the study of club root of cabbage (in Russian). *Morib. Plant* **14**: 49–73.

Niyogi, K.K. and G.R. Fink. 1992. Two anthranilate synthase genes in *Arabidopsis*: Defense-related regulation of the tryptophan pathway. *Plant Cell* **4**: 721–733.

Ogoshi, A. 1978. Studies on the grouping of *Rhizoctonia solani* Kuhn with hyphal anastomosis and the perfect stages of groups. *Bull. Natl. Inst. Agric. Sci. Ser. C* **30**: 1–163.

Ohl, S., S. Hedrick, J. Chory, and C.J. Lamb. 1990. Function properties of a phenylalanine ammonia-lyase promoter from *Arabidopsis*. *Plant Cell* **2**: 837–848.

Ohshima, M., H. Itoh, M. Matsuoka, T. Murakami, and Y. Ohashi. 1990. Analysis of stress-induced or salicylic acid-induced expression of the pathogenesis-related 1a protein gene in transgenic tobacco. *Plant Cell* **2**: 95–106.

Okoli, C.A.N., J.H. Carder, and D.J. Barbara. 1993. Molecular variation and sub-specific groupings within *Verticillium dahliae*. *Mycol. Res.* **97**: 233–239.

Parker, J.E., C.E. Barber, F. Mi-Jiao, and M.J. Daniels. 1993a. Interaction of *Xanthomonas campestris* with *Arabidopsis thaliana* characterization of a gene from *X. campestris* pv. *raphani* that confers avirulence on most *A. thaliana* accessions. *Mol. Plant-Microbe Interact.* **6**: 216–224.

Parker, J.E., V. Szabo, B.J. Staskawicz, C. Lister, C. Dean, M.J. Daniels, and J.D.G. Jones. 1993b. Phenotypic characterisation and molecular mapping of the *Arabidopsis thaliana* locus, *RPP5*, determining disease resistance to *Peronospora parasitica*. *Plant J.* **4:** 821–831.

Person, C. 1959. Gene-for-gene relationships in host:parasite systems. *Can. J. Bot.* **37:** 1101–1130.

Prohaska, J.R. 1980. The glutathione peroxidase activity of glutathione S-transferases. *Biochim. Biophys. Acta* **611:** 87–98.

Ross, R.F. 1961. Systemic acquired resistance induced by localized virus infections in plants. *Virology* **14:** 340–358.

Samac, D.A. and D.M. Shah. 1991. Developmental and pathogen-induced activation of the *Arabidopsis* acidic chitinase promoter. *Plant Cell* **3:** 1063–1072.

Samac, D., C.M. Huribaja, P.E. Yallaly, and D.M. Shah. 1990. Isolation and characterization of the genes encoding basic and acidic chitinase in *Arabidopsis thaliana*. *Plant Physiol.* **93:** 907–914.

Sansome, E. and F.W. Sansome. 1974. Cytology and life-history of *Peronospora parasitica* on *Capsella bursa-pastoris* and of *Albugo candida* on *C. bursa-pastoris* and *Lunaria annua*. *Trans. Br. Mycol. Soc.* **62:** 323–332.

Sherriff, C. and J.A. Lucas. 1990. A study of the host range of isolates of *Peronospora parasitica* (Pers. ex Fr.) Fr. from *Brassica* crop species. *Plant Pathol.* **39:** 77–91.

Simpson, R.B. and L.J. Johnson. 1990. *Arabidopsis thaliana* as a host for *Xanthomonas campestris* pv. *campestris*. *Mol. Plant-Microbe Interact.* **3:** 233–237.

Sjödin, C. and K. Glimelius. 1988. Screening for resistance to blackleg *Phoma lingam*. *J. Phytopathol.* **123:** 322–332.

Smith, M., E. Mazzola, J. Sims, S. Midland, and N. Keen. 1993. The syringolides: Bacterial C-glycosyl lipids that trigger plant disease resistance. *Tetrahedron Lett.* **34:** 223–226.

Staskawicz, B.J., D. Dahlbeck, and N.T. Keen. 1984. Cloned avirulence gene of *Pseudomonas syringae* pv. *glycinea* determines race-specific incompatibility on *Glycine max* (L.) *Merr. Proc. Natl. Acad. Sci.* **81:** 6024–6028.

Staskawicz, B.J., D. Dahlbeck, N. Keen, and C. Napoli. 1987. Molecular characterization of cloned avirulence genes from race 0 and race 1 of *Pseudomonas syringae* pv. *glycinea*. *J. Bacteriol.* **169:** 5789–5794.

Tör, M., E.B. Holub, E. Brose, R. Musker, N. Gunn, C. Can, I.R. Crute, and J.L. Beynon. 1994. Map positions of four loci in *Arabidopsis thaliana* associated with isolate specific recognition of *Peronospora parasitica* (downy mildew). *Mol. Plant-Microbe Interact.* **7:** 214–222.

Tsuji, J. and S.C. Somerville. 1988. *Xanthomonas campestris* pv. *campestris* induced chlorosis in *Arabidopsis*. *Arabidopsis Inf. Serv.* **26:** 1–8.

————. 1992. First report of natural infection of *Arabidopsis thaliana* by *Xanthomonas campestris* pv. *campestris*. *Plant Dis.* **76:** 539.

Tsuji, J., S.C. Somerville, and R.M. Hammerschmidt. 1991. Identification of a gene in *Arabidopsis thaliana* that controls resistance to *Xanthomonas campestris* pv. *campestris*. *Physiol. Mol. Plant Pathol.* **38:** 57–65.

Tsuji, J., E.P. Jackson, D.A. Gage, R. Hammerschmidt, and S.C. Sommerville. 1992. Phytoalexin accumulation in *Arabidopsis thaliana* during the hypersensitive reaction to *Pseudomonas syringae* pv. *syringae*. *Plant Physiol.* **98:** 1304–1309.

Tsuji, J., M. Zook, S.C. Somerville, R.L. Last, and R. Hammerschmidt. 1993. Evidence that tryptophan is not a direct biosynthetic intermediate of camalexin in *Arabidopsis thaliana*. *Physiol. Mol. Plant Pathol.* **43:** 221–229.

Uknes, S., B. Mauch-Mani, M. Moyen, S. Potter, S. Williams, S. Dincher, D. Chandler, A. Slusarenko, E. Ward, and J. Ryals. 1992. Acquired resistance in *Arabidopsis*. *Plant*

*Cell* **4:** 645–656.

Van de Rhee, M.D., J.A.L. Van Kan, M.T. Gonzalez-Jaen, and J.F. Bol. 1990. Analysis of regulatory elements involved in the induction of two tobacco genes by salicylate treatment and virus infection. *Plant Cell* **2:** 357–366.

van den Ackerveken, G.F.J.M., J.A.L. van Kan, and P.J.G.M. De Wit. 1992. Molecular analysis of the avirulence gene *avr9* of the fungal tomato pathogen *Cladosporium fulvum* fully supports the gene-for-gene hypothesis. *Plant J.* **2:** 359–366.

Verberg, J.G. and Q.K. Huynh. 1991. Purification and characterization of an antifungal chitinase from *Arabidopsis thaliana*. *Plant Physiol.* **95:** 450–455.

Walbot, V. 1991. Maize mutants for the 21st century. *Plant Cell* **3:** 851–856.

Ward, E.R., S.J. Uknes, S.C. Williams, S.S. Dincher, D.L. Wiederhold, D.C. Alexander, P. Ahl-Goy, J.P. Métraux, and J.A. Ryals. 1991. Coordinate gene activity in response to agents that induce systemic acquired resistance. *Plant Cell* **3:** 1085–1094.

Wei, Z.-M., R.J. Larby, C.H. Zumoff, D.W. Bauer, S.Y. He, A. Collmer, and S.V. Beer. 1992. Harpin, elicitor of the hypersensitive response produced by the plant pathogen *Erwinia amylovora*. *Science* **257:** 85–88.

Whalen, M.C., R.W. Innes, A.F. Bent, and B.J. Staskawicz. 1991. Identification of *Pseudomonas syringae* pathogens of *Arabidopsis* and a bacterial locus determining avirulence on both *Arabidopsis* and soybean. *Plant Cell* **3:** 49–59.

Whitham, S., S.P. Dinesh-Kumar, D. Choi, R. Hehl, C. Corr, and B. Baker. 1994. The product of the tobacco mosaic virus resistance gene *N:* Similarity to toll and the interleukin-1 receptor. *Cell* (in press).

Wingender, R., H. Rogrig, C. Horicke, and J. Schell. 1990. *Cis*-regulatory elements involved in ultraviolet light regulation and plant defense. *Plant Cell* **2:** 1019–1026.

Yang, S.F. and N.E. Hoffman. 1984. Ethylene biosynthesis and its regulation in higher plants. *Annu. Rev. Plant Physiol.* **35:** 155–189.

Yu, G.-L., F. Katagiri, and F.M. Ausubel. 1993. *Arabidopsis* mutations at the *RPS2* locus result in loss of resistance to *Pseudomonas syringae* strains expressing the avirulence gene *avrRpt2*. *Mol. Plant-Microbe Interact.* **6:** 434–443.

Zhou J. and P.B. Goldsbrough. 1993. An *Arabidopsis* gene with homology to glutathione S-transferases. *Plant Mol. Biol.* **22:** 517–523.

# 28
# Plant-parasitic Nematodes

**Peter C. Sijmons**
MOGEN Int. nv
Leiden, 2333 CB
The Netherlands

**Nicola von Mende**
Entomology and Nematology Department
AFRC IACR, Rothamsted Experimental Station
Harpenden, Herts, AL5 2JQ United Kingdom

**Florian M.W. Grundler**
Institute of Phytopathology
University of Kiel
D-2300 Kiel, Germany

Plant-parasitic nematodes are a major problem in modern agriculture and cause diseases in all crops of economic importance. Worldwide, yield losses are estimated to run into tens of billions of dollars per year. In general, symptoms of nematode infestation in the field are patches of poorly growing plants, often accompanied by wilting. In roots, symptoms depend on the nematode species and may involve galling, stunted root growth, and necrosis. To prevent the buildup of unacceptably high population densities, farmers tend to use integrated pest management systems, which combine the use of nematicides, crop rotation, and, if available, resistant cultivars. Nematicides are the most expensive and toxic pesticides that are in widespread use in modern agriculture and evoke increasing opposition for environmental reasons. This is illustrated by the drastic reductions in the use of nematicides that are proposed by the European Community for the next decade. A large-scale ban on nematicides will further increase the agronomic problems that result from these pests. Since only a limited number of resistance traits are available for use in breeding programs, genetic engineering of nematode resistance may provide crucial alternatives for germ plasm improvement.

Compared to fungal and bacterial phytopathology, our knowledge of plant/nematode interactions is limited and progressing slowly (Sijmons et al. 1994). The long and complex life cycle of these obligate parasites poses a major limitation for experimental approaches under laboratory conditions. There are no protocols available for axenic growth, and only a few species have been reported to grow on undifferentiated callus tissue. The majority will only develop on complete root systems, preferably under standard soil conditions, although monoxenic culture protocols (i.e., one nematode species inhabiting one host species) have been described for some of the economically important nematode species. As a result of these practical constraints, parasitic nematodes have rarely become the object of choice for molecular studies on plant/pathogen interactions. Research funding in nematology is often aimed at specific local problems, thus further diluting progress on fundamental issues. A few interactions are being studied more intensively because of their widespread importance in agriculture, e.g., *Meloidogyne incognita* on tomato, *Globodera rostochiensis* on potato, *Heterodera schachtii* on sugarbeet, and *H. glycines* on soybean, but these plant species all have their particular drawbacks for molecular approaches. Because the pathogen is relatively difficult to handle under laboratory conditions, it is especially important to minimize all other aspects that may complicate experimentation. Thus, the need for a model host plant to study plant/nematode interactions is undisputed; for the reasons emphasized in this book, *Arabidopsis* is a prime candidate.

The foregoing may intimidate new researchers in this field, but several aspects make parasitic nematodes unique pathogens and a fascinating research subject. We highlight in this chapter the most intriguing phenomena, as well as the large gaps in our knowledge of the parasitic interaction. We hope that both these aspects and the recent availability of *Arabidopsis* as a host plant will attract new research groups.

## BIOLOGY OF NEMATODES AND INTERACTION WITH *ARABIDOPSIS*

### Life Cycles

All root-parasitic nematodes are obligate biotrophs and modify living root cells according to their parasitic habits and nutritional demands (Zuckerman and Rohde 1981; Dropkin 1989). During a life cycle, nematodes pass through several developmental stages: the egg, four juvenile stages, and adult. Depending on their feeding behavior, plant-parasitic nematodes can be divided into two groups, the migratory and the sedentary nematodes. The latter become immobile once they have selected a proper cell on which they start feeding. A common feature of all plant-parasitic nematodes is the mouth stylet that they use for feeding. This discriminates them from free-living soil nematodes such as *Caenorhabditis elegans*, which feeds through uptake of cell debris or bacteria.

Sedentary nematodes, the cyst-forming species (e.g., *Heterodera schachtii*) and the root-knot nematodes (e.g., *Meloidogyne incognita*), have evolved highly sophisticated relationships with their hosts. They pass through distinct free-living and parasitic phases (Fig. 1). Only the second-stage juvenile is able to infect a host plant. After initiation of a permanent feeding site within the vascular tissue (described below in more detail), the juveniles molt three times to become adults. Females of cyst-forming nematodes are initially endoparasitic. As they mature, they become spherical and break through the root cortex, with the head still attached to the feeding site. The body wall hardens and tans to form the typical brown cysts in which hundreds of second-stage juveniles, each enclosed in an eggshell, can remain dormant for several years. Males finish their development earlier than females. The adult males emerge from the cast cuticles of earlier stages and move toward females for mating, attracted by pheromones. In general, cyst-forming nematodes reproduce sexually. Fertilized eggs develop into first-stage juveniles, which molt once within the eggshell. A generation can last between 5 and 8 weeks, depending on temperature. Several species of cyst-forming nematodes go through a diapause, which can only be broken by a cold period (~6 weeks at 4ºC).

Like cyst-forming nematodes, females of root-knot nematodes increase in size during development at well-established feeding sites. However, the cuticle does not tan and harden, but instead, surrounding plant tissue is induced to form galls, which eventually embed the females. Several hundred eggs may be deposited into gelatinous material secreted by the females to produce the so-called egg masses. These are exposed on the root surface. Second-stage juveniles hatch, infect host plants at the

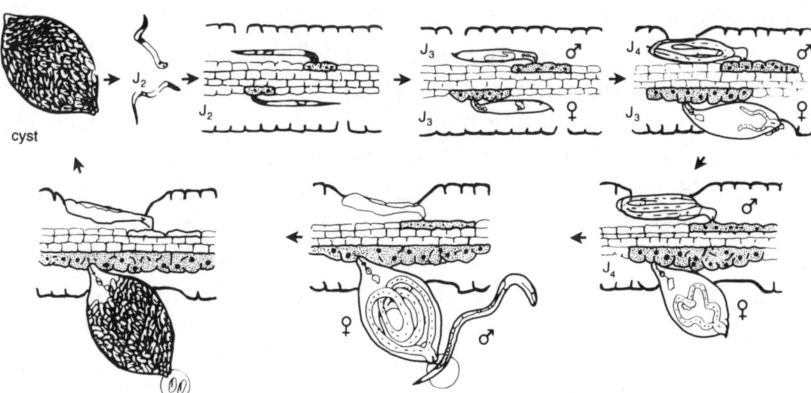


*Figure 1*  Schematic representation of the life cycle of a cyst nematode (*H. schachtii*). The development of both a male and a female is shown. (J₂) Second-stage juvenile; (J₃) third-stage juvenile; (J₄) fourth-stage juvenile. The syncytium is shaded. (Drawing by Grundler and VanderLee.)

root tips, and initiate the permanent feeding site within the vascular cylinder. After feeding for about 2 weeks, the juveniles quickly molt three times (to the adult stage), during which process feeding is interrupted. Only the adult females recommence feeding. Reproduction in most *Meloidogyne* species is usually by parthenogenesis, although males are commonly found.

Migratory nematodes either feed on epidermal and meristematic cells, or they invade the roots and feed on the cortical tissue. In the case of *Pratylenchus* species (migratory endoparasites), the adult female, male, and all juvenile stages except the first (which occurs only within the eggshell) are able to invade roots. Usually they feed on and migrate through cortical cells, causing extensive destruction. Single eggs are deposited in root tissue or in the soil. Depending on the temperature, a life cycle can last between 4 and 8 weeks.

## Infection Processes

Although *Arabidopsis* can successfully be inoculated with *H. schachtii* and *M. incognita* in sand/soil mixtures (J. Klap and P.C. Sijmons, unpubl.), agar culture techniques are more suitable for detailed studies of feeding structure development and plant/nematode interactions. Two factors are important for monoxenic culture: the medium composition and the agar purity. Their influence on nematode reproduction cannot be ascribed only to their effects on root growth vigor, as T-DNA *Arabidopsis* mutants with strongly inhibited root growth also allowed completion of the nematode life cycle. High infection rates have been achieved for a number of nematode species (Sijmons et al. 1991) with a medium composition that was optimized for hydroponic culture (Sijmons and Bienfait 1983) in combination with tissue-culture-quality agar (Table 1). The agar layer should not be made too thick as this dilutes the attracting agents and decreases the infection rate. Depending on the purpose of the experiment, *Arabidopsis* seedlings can be grown on slightly tilted agar plates (resulting in parallel root growth toward the base of the plate) or in small agar droplets in petri dishes. The latter method allows easy identification of individual plant lines. Microtiter plates are less suitable because of stress phenomena that are observed in the aerial plant parts. Feeding structures and the attached nematode can be isolated from such plates for further analysis with a minimum of disturbance.

The thin transparent roots of *Arabidopsis* provide good conditions to observe nematode parasitic behavior and plant responses in vivo with light microscopic techniques. In vivo observations give valuable information about the dynamic processes during nematode development

*Table 1*  Nematode species on *Arabidopsis* (L*er*) in monoxenic conditions

| Nematode | Last host | Life cycle | Comments |
|---|---|---|---|
| *Heterodera schachtii* | mustard | 6 weeks | necrosis at invasion and feeding site |
| *H. trifolii* | clover | 8 weeks | — |
| *H. cajani* | cowpea | 8 weeks | — |
| *H. goettingiana* | field beans | not complete | $J_2$ invade roots but destroy the tissue |
| *Globodera rostochiensis* | potato cv. Desiree | not complete | strong necrosis |
| *Meloidogyne incognita* | tomato cv. Pixie | 4–5 weeks | little or no necrosis |
| *M. arenaria* | tomato cv. Pixie | 6 weeks | little or no necrosis |
| *Pratylenchus penetrans* | carrots/maize | 6 weeks | necrosis |

and complement the data from serial sections or molecular analysis. Using proper equipment, roots and nematodes within root tissue can be examined over a period of several weeks in special observation chambers that allow root growth between microscope coverslips (a complete observation chamber set can be obtained from F.M.W. Grundler).

## In Vivo Observations with H. schachtii

The behavior of *H. schachtii* and its development inside roots of *Arabidopsis* correspond closely with those in other cruciferous plants, such as *Brassica napus* and *Brassica campestris* (Wyss and Zunke 1986). However, more detailed information is now available from in vivo observations with *Arabidopsis* roots (Wyss 1992). Infective juveniles ($J_2$) of *H. schachtii* invade the root mostly in the elongation zone behind the root tip. The rhizodermal cell walls are perforated with the stylet in a specific sequence. After penetration of the root, the juveniles orient themselves directly toward the vascular cylinder. With destructive stylet thrustings, they migrate through the cortex. Within the stele, a single parenchymatous cell, normally in close proximity to xylem elements, is selected as the initial feeding site. A perforation is made in the wall of this cell by gentle stylet thrustings. The tip of the stylet is then inserted into the cell, where it remains for about 7 hours. During this period, no activity of the nematode's median bulb (i.e., the pump organ for food uptake) can be observed, but changes in the optical density of the three esophageal secretory glands become visible. This period can be interpreted as a phase of transition from the infective-migratory to the parasitic-sedentary stage, because afterward, locomotion is no longer possible. The nematode then starts feeding with coordinated contractions of the median bulb. The behavior of the $J_2$ during feeding is composed of three repeating phases: Phase I is the period of food uptake during which nutrients are withdrawn from the feeding site by pumping of the median bulb; phase II is characterized by retraction and subsequent reinsertion of the stylet; and phase III is when secretory granules of the dorsal gland cells seem to become mobilized. During female development, the duration of phase I increases continuously, whereas the duration of phase II remains steady and that of phase III decreases. At the feeding site within the cytoplasm, a modified zone surrounding the stylet tip can be observed. Eventually, feeding tubes, which are formed from secretions released out of the stylet orifice during "salivation," become visible. At the end of the $J_2$ stage, female and male juveniles can already be distinguished by the size of the dorsal gland (larger in females) and the genital primordia, which are more rounded and less elongate in females.

Depending on the position of the nematode body and the syncytium in the root, more or less detailed studies of the processes within the nematode and the plant during the entire development procedure can be conducted. In later stages, observations may become problematic, as the steady enlargement of the syncytium and the hyperplastic response of the surrounding tissue cause difficulties in focusing on details.

### *In Vivo Observations with* M. incognita

Infective juveniles of *M. incognita* generally enter the root in the elongation zone, where two types of root invasion can be observed: Usually the $J_2$ penetrates a rhizodermal cell by perforation of the cell wall, but sometimes invasion is by entering the intercellular space between two rhizodermal cells. The penetration site has to be conditioned by intensive lip rubbings and stylet thrustings aimed at a single point of the root surface. As soon as the juveniles move into the cortex, they orient themselves toward the root tip. After intracellular invasion, they first have to enter the intercellular space between subepidermal cortical cells. Within the intercellular space, they migrate into the root-tip meristem. At high magnification, fluid droplets of unknown origin can be observed, and these are probably introduced into the intercellular space by the nematodes. It is most probable that these droplets have enzymatic activity and are produced by the nematodes to facilitate intercellular migration. When the juveniles reach the meristem, they turn around and enter the developing vascular cylinder, within which they migrate toward the base of the root. Migration stops as soon as the differentiation zone is reached, where the protoxylem is already differentiated. The indirect invasion of the central cylinder via the root-tip meristem seems to be obligate, as the juveniles are probably not able to penetrate the endodermis. Within the early central cylinder they begin a specific pattern of behavior that consists of gentle stylet thrusts directed at the xylem parenchymatous cells surrounding their heads, stylet protrusions into the cells, and pumping activity of the median bulb. The juveniles feed from the giant cells by inserting their stylet into one cell after another and withdrawing food with the aid of the median bulb. During the following days, the giant cells and the root tissue in which they are embedded proliferate permanently. The formation of galling tissue prevents in vivo observation for longer than 4–5 days, as details within the root can no longer be discerned. The invasion process, selection of feeding site, and induction of the giant cells are documented on videotape (available from F.M.W. Grundler) and are described in detail by Wyss et al. (1992).


**Root Responses to Nematode Parasitism**

H. schachtii

Nematode secretions that are injected into the initial feeding cell trigger dramatic cytological alterations. The selected cell undergoes a phase of redifferentiation. First indications of this process are the acceleration of cytoplasmic streaming and the enlargement of the nucleus, which can be observed a few hours after the juveniles have commenced feeding. Further redifferentiation occurs by fragmentation of the large central vacuole to give several small vacuoles which are dispersed over the cytoplasm. At the same time, the cell walls adjoining neighboring parenchymatic cells are successively dissolved and the protoplasts of the cells form a syncytium. The cell walls of the syncytium undergo striking structural changes. They become thickened and seem to enclose the syncytium, separating it from the surrounding tissue. The cell walls also become invaginated, resulting in enlargement of the surface area, mainly on the surfaces toward xylem vessels. The syncytium expands continuously along the central cylinder during the development of the nematode and, in the case of females, reaches dimensions of more than 10 mm (Fig. 2, a–c). As a response to the expansion and hypertrophy of the syncytium, the cortex produces new tissue by division of those cells that ensheath the thickened vascular cylinder. At the same time, differentiation of xylem elements is suppressed in the region of the syncytium (Golinowski and Grundler 1992).

Whereas females feed from very large syncytia, males are associated with and feed from much smaller syncytia. Investigations with different

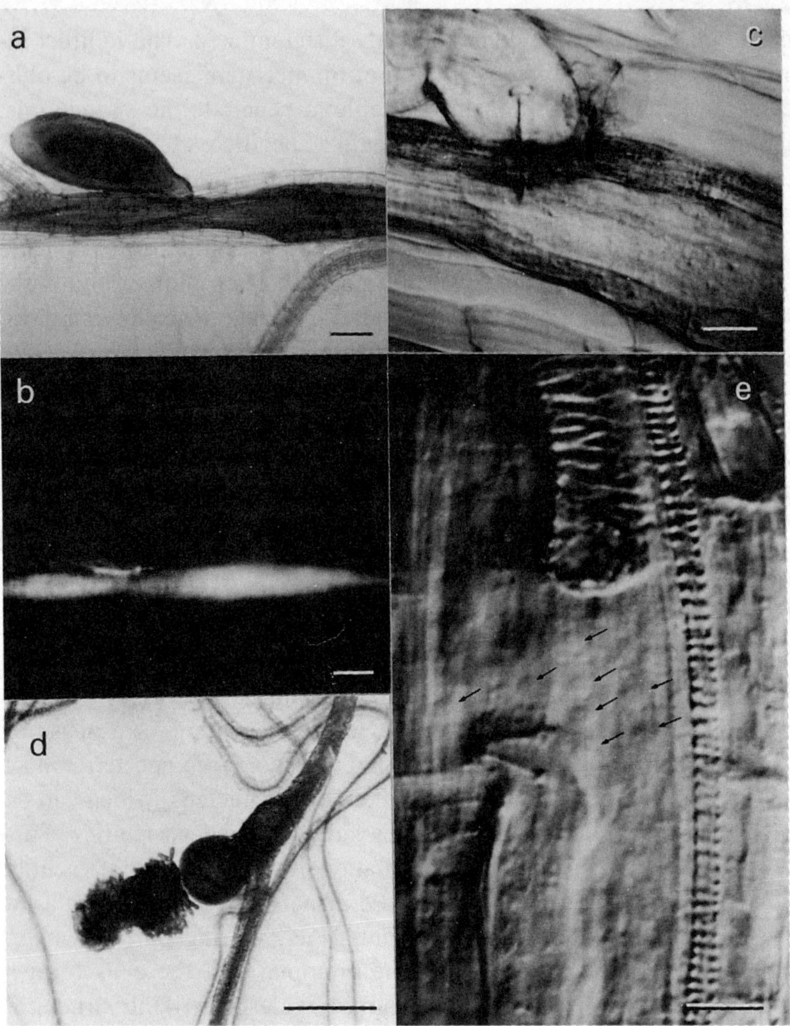

*Figure 2*  Monoxenic cultures of plant-parasitic nematodes on roots of *Arabidopsis thaliana* ecotype Landsberg *erecta*. (*a*) Female $J_4$ (*H. schachtii*) just before the last molt, anchored in a large syncytium. For comparison, uninfected root tissue can be seen in the same picture. Bar, 200 μm. (*b*) Autofluorescence of the root in panel *a*. The fluoresence indicates the size of the syncytium and is a secondary effect of the in vitro culture conditions; in the presence of light, chloroplasts can develop inside syncytial cells. Bar, 200 μm. (*c*) $J_2$ female of *H. schachtii* with stylet inserted in the developing syncytium, prior to the molt. Bar, 10 μm. (*d*) Mature egg-producing female of *M. incognita*. A gall is visible at the site of infection. Bar, 1 mm. (*e*) *M. incognita* $J_2$, sedentary within the vascular cylinder, feeding from multinucleate giant cells. The photograph was taken 3 days after the selection of the initial feeding cell. Bar, 10 μm. (Reprinted, with permission of Blackwell Scientific Publications Ltd., from Sijmons et al. 1991.)

cruciferous host plants suggest that the host influences the sexual differentiation of the nematode. The infective and sexually undifferentiated juveniles enter the root and find either favorable or unfavorable conditions. Favorable conditions lead to the formation of large syncytia and thereby allow the development of females, whereas unfavorable conditions trigger the differentiation of males, which are able to develop within a short time and with a very limited food supply. Unfavorable conditions can be produced experimentally by reduced nutrient supply, advanced age of the host plant at the date of inoculation, or use of plants with resistance to *H. schachtii*, as was demonstrated with oil seed radish (Grundler et al. 1991). Nothing is known as yet about the mechanisms of sex determination in cyst nematodes or in root-knot nematodes, where similar processes have been described (Papadopoulou and Triantaphyllou 1982).

## M. incognita

Juveniles of root-knot nematodes enter the root without causing any obvious damage. With the aid of secretions—most probably a set of enzymes—they are able to widen the intercellular space, through which they move toward the central cylinder (see above, In Vivo Observations). Within the xylem parenchymatous tissue, they stop migration and induce their permanent feeding site, a system of giant cells (Fig. 2d). Each giant cell is initiated by the injection of salivary secretions, which at this stage are of a different nature from those secreted during migration (Hussey 1989). These secretions probably originate from the dorsal esophageal gland. The secretions trigger dramatic cellular responses, viz., nuclear division without cytokinesis, leading to hypertrophied multinucleate cells (Fig. 2e). Like syncytia, giant cells have thickened and invaginated cell walls abutting the xylem vessels. The formation of giant cells is accompanied by galling of the cortical tissue enclosing the feeding site and the growing nematode. The gall size depends on the number of nematodes enclosed and the root-knot nematode species. After completion of its life cycle, the nematode dies and the giant cells and gall tissue degenerate.

## Pratylenchus penetrans

Migratory nematodes such as *P. penetrans* do not induce specific feeding structures within host plants. The roots are invaded by piercing epidermal cells with the stylet. For feeding, the stylet is carefully inserted through the cell wall into the cytoplasm. Before nutrients are withdrawn through the lumen of the stylet, secretory fluids, originating from the dorsal esophageal gland cell of the nematode, are injected into the host cell. Around the tip of the inserted stylet, the cytoplasm is modified and becomes distinctly condensed. After food withdrawal, the nematode migrates to the next cell. This basic pattern of events during feeding is common to all migratory species of the order Tylenchida (Wyss 1988). The extensive migratory activity of *P. penetrans* is primarily within the cortical tissue and leads to the development of expanded necrotic regions in the roots that are typical, but which produce nonspecific symptoms in top growth.

## SCREENING AND GENETICS

For screening purposes, a high degree of standardization of infection rate and plant/nematode development is essential; however, the varying infection strategies of the nematode species have to be considered. Preliminary screening results indicated that certain conditions (e.g., medium composition, age of host plant, size of growing vessel) influence the success of the host/parasite interaction. Under in vitro conditions, most factors can be standardized easily and screening can be done visually. Highly standardized procedures are now being used both to analyze *Arabidopsis* ecotypes with different populations of *H. schachtii* and to screen ethyl methanesulfonate (EMS)-mutated plants with *H. schachtii* or *M. incognita*. Screening procedures can also be performed under more natural conditions in soil, although in the case of *H. schachtii*, a quantitative evaluation of infection will be tedious and time consuming because mature cysts can become detached from the roots. Root-knot infection can be easily scored by a galling index, as is done in conventional resistance breeding programs. Screening of random EMS mutants has recently been initiated in several institutes. In a screening with 5000 plants infected with *M. incognita*, one mutant line was identified where root penetration was affected (A. Niebel and I. Vercauteren, pers. comm.). The mutation inherits as a single locus after backcrossing with a wild type. A similar screen with *H. schachtii* did not result in stable plant lines in which the infection process was disturbed (A. Munch and F.M.W. Grundler, unpubl.).

### Ecotype Screening

For several agronomic crops (potato, tomato, soybean, sugarbeet), resistance reactions have been identified against sedentary parasitic nematodes, ranging from recessive polygenic to single dominant *R* genes (Trudgill 1991). The resistance reactions usually involve necrotic reactions around the infection or feeding site and inhibit the normal development of a feeding structure. Among the wide range of *Arabidopsis* ecotypes that are available, resistance genes that coevolved naturally as a defense against nematode infections may be present in specific populations. However, in a preliminary screening of 74 *Arabidopsis* ecotypes with *H. schachtii*, all were susceptible, although differences were found in the number of females developing in the roots (Sijmons et al. 1991). Different *H. schachtii* populations were also infectious without exception (F.M.W. Grundler, unpubl.). This is in sharp contrast to bacterial and fungal strains that can infect *Arabidopsis*. Resistant ecotypes are found with a relatively high frequency (Koch and Slusarenko 1990; Debener et al. 1991). Similar screens must be done using an incompatible (but related) nematode species to detect a possible breakage of resistance.

### Hormone Mutants

Several *Arabidopsis* hormone mutants have been screened to examine the influence of phytohormones on the development of plant-parasitic nematodes. Of particular interest is the effect of hormones on the initiation and maintenance of the feeding site of cyst-forming and root-knot nematodes. It has been shown repeatedly for these nematode species that hormonal differences do occur between infected and noninfected plant material, such as altered levels of auxins, cytokinin, ethylene, and gibberellin in galls produced by *M. incognita* (Yu and Viglierchio 1964; Bird and Loveys 1980; Glatzer et al. 1986). Since the formation of feeding structures impairs transport functions and affects both the acropetal and basipetal hormone movement (Dorhout et al. 1988, 1991), it is difficult to establish a causal relationship. Analysis of characterized mutants should provide more direct evidence for the involvement of phytohormones.

In a preliminary test, 14 different hormone mutants have been tested and compared with ecotype Landsberg as a control. The mutants were defective in either the biosynthesis or response to auxin (Mirza et al. 1984; Estelle and Somerville 1987), abscisic acid (Koornneef et al. 1982, 1984), gibberellins (Koornneef and van der Veen 1980; Koornneef et al. 1985), and ethylene (Bleecker et al. 1988). Differences in the development of *H. schachtii* or *M. incognita* were expected, especially in the hormone receptor mutants. Surprisingly, both species developed well on all mutants and reached maturity by producing cysts or egg sacs filled with embryonated eggs. Additional observations were made on invasion behavior, galling, and syncytium formation, as well as lateral root development. Relatively early invasion was found with the ethylene mutant, possibly caused by the vigorous root growth of this mutant. No differences were found in gall and syncytium formation in any of the mutant plant lines tested. The most striking differences were in the formation of lateral roots after infection with *H. schachtii*. On Landsberg, an average of 2.5 lateral roots were found per feeding site. On the *axr-1* mutants, the formation of lateral roots was clearly inhibited (not in *aux-1*), but the number of lateral roots of gibberellin mutant *ga-5* was above average, with 4.5 per feeding site. Similar studies on feeding structure development have now been initiated in root-form mutants.


## T-DNA Tagging

The (de)regulation of gene activity in cells comprising the feeding structure is one of the most fascinating aspects of the biology of parasitic nematodes. The in vivo analysis described above clearly shows that glandular contents are injected into the host cell through the stylet, although no direct immunological evidence has yet been published that proteins of nematode origin are present in the cytoplasm of root cells. An intriguing possibility is that they interact directly or indirectly with the host genome, resulting in the reprogramming of the pattern of gene expression required for completion of the feeding structure.

In an attempt to analyze gene activity in syncytia, *Arabidopsis* plants that were transgenic for a range of promoter/*GUS* constructs—commonly available from specialists working with *Arabidopsis*—were infected with *H. schachtii*. Most promoters were selected for their activity in storage or root vascular tissue. Almost without exception, the promoter activities were down-regulated in syncytial cells, and *GUS* expression could no longer be detected (Goddijn et al. 1993). The reverse situation (a blue syncytium amid unstained tissue) was also found in later studies (Cramer 1992; Taylor et al. 1992; Goddijn et al. 1993), indicating that the lack of *GUS* activity is not the result of a decreased substrate permeability or mRNA stability. Even promoters with strong activity in the vascular cylinder, such as CaMV-35S (Benfey et al. 1989), are silenced during development of the feeding structure. This unexpected finding was further substantiated with analysis of transgenic *Arabidopsis* plants carrying a promoterless *GUS* gene (Topping et al. 1991). This type of T-DNA tagging technique allows identification of upstream regulatory sequences involved in a particular pattern of expression. Usually, tagged sequences that gave rise to *GUS* expression in noninfected root vascular tissue stopped expressing the *GUS* gene during syncytium development after infection with *H. schachtii* (Goddijn et al. 1993). Apparently, a large part of reprogramming the plant expression pattern involves the down-regulation of those genes that are not essential for the formation and development of the feeding structure, possibly allowing the full machinery of protein synthesis to support the nutritional requirements of the nematode. Two reports were recently published on promoters that are active inside giant cells after infection of transgenic tobacco plants with *M. incognita*. Both promoters regulate genes that can be interpreted as necessary for high metabolic activity; the first is the *TobRB7* gene (Taylor et al. 1992), coding for a membrane channel protein that is normally synthesized only during development of the vascular cylinder (Yamamoto et al. 1991), and the second codes for hydroxymethyl-glutaryl CoA reductase, a key enzyme for terpenoid and sterol synthesis (Cramer 1992).

Through the use of the T-DNA tagging techniques, inserting promoterless *GUS* constructs into the plant genome and inverted polymerase chain reaction (PCR) isolation of sequences that give rise to deviant *GUS* expression inside the nematode's feeding structure, the mechanism of reprogramming may be elucidated one step further.


## ENGINEERING NEMATODE RESISTANCE: *ARABIDOPSIS*, THE TEST CASE

A major challenge for genetic engineering will be to develop genetic cassettes that will, after transfer to the genome of the target crop, render that crop resistant to nematode damage. For agricultural purposes, complete resistance may not be necessary as long as an increase in nematode population densities can be prevented. Provided the target crop has some degree of tolerance, infection can still be allowed when the engineered trait inhibits completion of the nematode's life cycle, i.e., decreases the amount of offspring. Ideally, the genetic cassette should be applicable in a wide range of crop plants and effective against different nematode families.

In theory, there are several stages where one can attempt to disrupt the life cycle of the nematode, ranging from interference with egg hatching to interference with the later parasitic stages. Several institutes are developing biological control methods, e.g., through the use of nematode-pathogenic microorganisms or synthetic hatching factors, but at present no data demonstrate the durability of such methods on an agronomic scale. We consider only those approaches that would involve genetic engineering of the host plant. Although some may not be realistic with the present state of knowledge, all possible strategies are discussed in relation to the chronology of the life cycle.

**Hatching Factors**

The eggs of certain species of cyst nematodes require stimulation by host root diffusate before they can hatch; in other species, the presence of root diffusate enhances the rate of hatch. Once the biologically most active hatching factors are identified, attempts can be made to alter root metabolism in such a way that it produces less of that particular compound. There are several drawbacks to this approach. One is that, usually, part of the egg population will hatch even in the absence of host roots (Perry 1989; Dennijs and Lock 1992), and the use of such plants would quickly select for that response. Second, it has proved extremely difficult to identify specific hatching factors (Masamune et al. 1982; Atkinson et al. 1987) and, as these will be different for every plant/nematode combination, this strategy will be a formidable task. Third, even when the hatching factor is identified, it may be questionable whether root metabolism can be altered to such an extent that the leached quantities remain below hatching levels, especially because these factors are active even at extreme dilutions.

**Recognition Processes**

The specific site of entry into the root and the behavior of the migrating nematode through the root tissue indicate that the nematode is able to recognize particular root cells. At the next stage, the behavior of the nematode alters and becomes less destructive when the initial feeding cell is selected. Again, a recognition process must be involved. The molecular mechanism must be similar over a wide range of plant species, since most nematodes are able to find these cell types in a number of plant species. For example, the potato cyst nematode *Globodera rostochiensis* is able to induce a feeding structure inside *Arabidopsis* roots even though development to the adult stage does not occur (Holz et al. 1992). At present, there is no clue as to what host compounds are involved in the recognition processes, but screening random *Arabidopsis* mutants may provide plant lines that show abortion of the infection process at these early stages. The simple architecture of *Arabidopsis* roots (Schiefelbein and Benfey 1991) will help to identify destined feeding cells at the anatomic level. Analysis of the molecular mechanisms underlying this recognition may provide tools to interfere with this stage of the interaction.

Recognition processes also play a role on the plant side; several cultivars have emerged from breeding programs that show a hypersensitive response after nematode invasion. Necrotic tissue can be seen after mechanical wounding but can also be seen around developing feeding structures, which become encapsulated and are no longer able to expand fully (Trudgill 1991). This type of plant response has not yet been identified in *Arabidopsis* ecotypes after screening with *H. schachtii*.


### Anti-nematode Proteins

One obvious way to engineer nematode resistance would be to make the plant produce a nematicidal protein inside the feeding structure. Direct contact with or ingestion of such a protein with the host cytoplasm should then inhibit normal development of the nematode. Possible examples include such proteins as collagenases, proteinase inhibitors, or *Bacillus thuringiensis* toxins. Preferably, the protein should not have any effect on plant cells, otherwise the expression of the gene coding for that protein must be tightly regulated and only be transcribed by cells comprising the feeding structure. Of course, a crucial factor for this approach is the choice of the anti-nematode protein. Enzymes that act on the surface of the nematode may be screened in vitro, although alterations in the surface structure after molting may decrease the predictability of effects identified in such screening experiments. Proteins that interfere with the digestive tract of the nematode cannot be assayed directly because sedentary nematodes are obligate parasites and do not feed on artificial diets. Expression of a cowpea trypsin inhibitor in potato roots induced a shift in the female/male ratio when infected with *Globodera*, indicating the feasibility of such an approach (Atkinson and Koritsas 1993). An alternative screening method may be to inject the protein directly into the feeding structure. The clearly visible feeding structures in *Arabidopsis* roots are excellent for micromanipulation techniques. The feasibility of this technique is being assessed (A. Bückenhoff and F.M.W. Grundler, unpubl.).

Another possible approach is to synthesize antibodies inside the feeding cells that are directed against one of the nematode saliva proteins (Hiatt et al. 1989; Hussey et al. 1990). If these saliva proteins are indeed involved in induction of the feeding structure or the feeding process itself, then the local production of sufficient antibodies may interfere with nematode development.


### Anti-feeding Structure Proteins

Finally, a target for engineering nematode resistance would be the cells of the feeding structure. During the parasitic stages of their life cycles, sedentary nematodes are completely dependent on this structure, and even a small disturbance of the feeding cells may have profound effects on the  multiplication rate of  the nematode. In this scenario, the gene products must be directed against plant cells, and extremely precise gene regulation would be required. The use of phytotoxic genes would mean that any promoter activity outside the feeding structure would affect plant cell metabolism and, eventually, the agronomic value of crop plants. Therefore, either an extremely specific promoter must be used which can then be fused to a range of phytotoxic genes coding for such proteins as DNase, RNase (Mariani et al. 1990), ribosome-inactivating protein  (Logemann et al. 1992), etc., or the phytotoxicity of the gene product must be limited such that some leakage of promoter activity outside the feeding structure can be tolerated. Examples of genes in the latter category may be in the area of *onc* genes, proteinase, or anti-sense constructs. The isolation of a syncytium-specific promoter is anticipated in the work of Gurr et al. (1991), where a gene was identified that is specific for the compatible interaction between a potato and the cyst nematode *G. rostochiensis*. Recent work of Taylor et al. (1992) is even more  promising; in deletion studies of the *TobRB7*  gene promoter, a 1.5-kb 5′deletion resulted in a 300-bp regulatory sequence that was, after fusion to *GUS*, no longer active in vascular root tissue (Yamamoto et al. 1991) but gave high *GUS* expression in giant cells after infection with *M. incognita*. Thus, 5′deletion of this promoter made it highly specific for the feeding structure, and it may therefore be possible to use it to express phytotoxic genes without detrimental side effects in other plant parts.

## Extrapolation of *Arabidopsis*/Nematode Research to Crop Plants

The essence of a model plant is that the results generated are applicable to the more complex species that it is replacing. It is more than likely that this is the case with studies on plant/nematode interactions. The wide host ranges of many nematode species indicate host/parasite interaction mechanisms that must be very similar over a large range of plant species. Indeed, the whole infection process and root responses show similarities between cyst and root-knot nematodes, irrespective of the plant species. Furthermore, despite the morphological differences between giant cells and syncytia, there are similarities at the level of gene regulation; histochemical analysis of transgenic *Arabidopsis* and tobacco plants, carrying the same promoter-*GUS* fusions, showed similar changes in *GUS* expression after infection with either *M. incognita* or *H. schachtii* (Goddijn et al. 1993). In conclusion, it is probably safe to assume that the results obtained through studies of *Arabidopsis*/nematode interactions can be extrapolated directly to major crop plants.

**CONCLUSIONS AND PERSPECTIVES**

In the coming years, there will undoubtedly be several attempts to engineer nematode resistance by one of the methods mentioned above. The obligate parasitic life cycle of sedentary nematodes implies that there are no easy ways to screen and compare different approaches; the ultimate test to check the effectiveness of a certain promoter-gene combination can only be done after infection of appropriately transformed plants. *Arabidopsis* transformation has become a routine technology but, more importantly, the effect of toxic gene constructs on feeding structures can be observed in vivo at the cellular level in this species and can be tested relatively quickly. Once the choice of strategies has been narrowed to acceptable numbers, field testing of agronomically important transformed crops will filter out the best approaches.

**REFERENCES**

Atkinson, H.J. and V.M. Koritsas. 1993. Development of a novel strategy for nematode control based on proteinase inhibition. In *Abstracts from the 6th International Congress on Plant Pathology*, Montreal, Canada, p.192. National Research Council of Canada, Ottawa.

Atkinson, H.J., M. Fowler, and R.E. Isaac. 1987. Partial purification of hatching activity for *Globodera rostochiensis* from potato root diffusate. *Ann. Appl. Biol.* **110:** 115–125.

Benfey, P.N., L. Ren, and N. Chua. 1989. The CaMV 35S enhancer contains at least two domains which can confer different developmental and tissue-specific expression patterns. *EMBO J.* **8:** 2195–2202.

Bird, A.F. and B.R. Loveys. 1980. The involvement of cytokinins in a host-parasite relationship between tomato (*Lycopersicon esculentum*) and a nematode (*Meloidogyne javanicum*). *Parasitology* **80:** 497–505.

Bleecker, A.B., M.A. Estelle, C. Somerville, and H. Kende. 1988. Insensitivity to ethylene conferred by a dominant mutation in *Arabidopsis thaliana*. *Science* **241:** 1086–1089.

Cramer, C.L. 1992. Regulation of defense-related gene expression during plant-pathogen interaction. *J. Nematol.* **24:** 586–587.

Debener, T., H. Lehnackers, M. Arnold, and J.L. Dangl. 1991. Identification and molecular mapping of a single *Arabidopsis thaliana* locus determining resistance to a phytopathogenic *Pseudomonas syringae* isolate. *Plant J.* **1:** 289–302.

Dennijs, L.J.M. and C.A.M. Lock. 1992. Differential hatching of the potato cyst nematodes *Globodera rostochiensis* and *G. pallida* in root diffusates and water of differing ionic composition. *Neth. J. Plant Pathol.* **98:** 117–128.

Dorhout, R., F.J. Gommers, and C. Kolloffel. 1991. Water transport through tomato roots infected with *Meloidogyne incognita*. *Phytopathology* **81:** 379–385.

Dorhout, R., C. Kolloffel, and F.J. Gommers. 1988. Transport of an apoplastic fluorescent dye to feeding sites induced in tomato roots by *Meloidogyne incognita*. *Phytopathology* **78:** 1421–1424.

Dropkin, V.H. 1989. *Introduction to plant nematology*, 2nd edition. Wiley, New York.

Estelle, M.A. and C. Somerville. 1987. Auxin-resistant mutants of *Arabidopsis thaliana* with an altered root morphology. *Mol. Gen. Genet.* **206:** 200–206.

Glatzer, I., E. Epstein, D. Orion, and A. Apelbaum. 1986. Interactions between auxin and ethylene in root-knot nematode (*Meloidogyne javanica*) infected tomato roots. *Physiol. Mol. Plant Pathol.* **28:** 171–179.

Goddijn, O.J.M., K. Lindsey, F.M. van der Lee, J.C. Klap, and P.C. Sijmons. 1993. Differential gene expression in nematode-induced feeding structures of transgenic plants harbouring promoter-*gus*A fusion constructs. *Plant J.* **4:** 863–873.

Golinowski, W. and F.M.W. Grundler. 1992. The structure of feeding sites (syncytia) of *Heterodera schachtii* in roots, stems and leaves of *Arabidopsis thaliana*. *Nematologica* **38:** 413. (Abstr.)

Grundler, F., M. Betka, and U. Wyss. 1991. Influence of changes in the nurse cell system (syncytium) on sex determination and development of the cyst nematode *Heterodera schachtii*—Total amounts of proteins and amino acids. *Phytopathology* **81:** 70–74.

Gurr, S.J., M.J. Mcpherson, C. Scollan, H.J. Atkinson, and D.J. Bowles. 1991. Gene expression in nematode-infected plant roots. *Mol. Gen. Genet.* **226:** 361–366.

Hiatt, A., R. Cafferkey, and K. Bowdish. 1989. Production of antibodies in transgenic plants. *Nature* **342:** 76–78.

Holz, G., W. Selig, and F.M.W. Grundler. 1992. *Globodera pallida* in *Arabidopsis thaliana*: Nematode behaviour and plant response in an incompatible system. *Nematologica* **38:** 416. (Abstr.)

Hussey, R.S. 1989. Disease-inducing secretions of plant-parasitic nematodes. *Annu. Rev. Phytopathol.* **27:** 123–141.

Hussey, R.S., O.R. Paguio, and F. Seabury. 1990. Localization and purification of a secretory protein from the esophageal glands of *Meloidogyne incognita* with a monoclonal antibody. *Phytopathology* **80:** 709–714.

Koch, E. and A. Slusarenko. 1990. *Arabidopsis* is susceptible by a downy mildew fungus. *Plant Cell* **2:** 437–445.

Koornneef, M. and J.H. van der Veen. 1980. Induction and analysis of gibberellin sensitive mutants in *Arabidopsis thaliana* (L.) Heynh. *Theor. Appl. Genet.* **58:** 257–263.

Koornneef, M., G. Reuling, and C.M. Karssen. 1984. The isolation and characterization of abscisic acid-insensitive mutants of *Arabidopsis thaliana*. *Physiol. Plant.* **61:** 377–384.

Koornneef, M., M.L. Jorna, A. Brinkhorst-van der Swan, and C.M. Karssen. 1982. The isolation of abscisic acid (ABA) deficient mutants by selection of induced revertants in non-germinating gibberellin sensitive lines of *Arabidopsis thaliana* (L.) Heynh. *Theor. Appl. Genet.* **61:** 385–393.

Koornneef, M., A. Elgersma, C.J. Hanhart, E.P. Van Loenen-Martinet, L. Van Rijn, and J.A.D. Zeevaart. 1985. A gibberellin insenstitive mutant of *Arabidopsis thaliana*. *Physiol. Plant.* **65:** 33–39.

Logemann, J., G. Jach, H. Tommerup, J. Mundy, and J. Schell. 1992. Expression of a barley ribosome-inactivating protein leads to increased fungal protection in transgenic tobacco plants. *Bio/Technology* **10:** 305–308.

Mariani, C., M. De Beuckeleer, J. Truettner, J. Leemans, and R.B. Goldberg. 1990. Induction of male sterility in plants by a chimaeric ribonuclease gene. *Nature* **347:** 737–741.

Masamune, T., M. Anetai, M. Takasugi, and N. Katsui. 1982. Isolation of a natural hatching stimulus for glycinoclepin A, for the soybean cyst nematode. *Nature* **297:** 495–496.

Mirza, J.I., G.M. Olsen, T. Iversen, and E.P. Maher. 1984. The growth and gravitropic responses of wild-type and auxin resistant mutants of *Arabidopsis thaliana*. *Physiol. Plant.* **60:** 516–522.

Papadopoulou, J. and A.C. Triantaphyllou. 1982. Sex differentiation in *Meloidogyne incognita* and anatomical evidence for sex reversal. *J. Nematol.* **14:** 549–566.

Perry, R.N. 1989. Dormancy and hatching of nematode eggs. *Parasitol. Today* **5:** 377–383.

Schiefelbein, J.W. and P.N. Benfey. 1991. The development of plant roots—New approaches to underground problems. *Plant Cell* **3:** 1147–1154.

Sijmons, P.C. and H.F. Bienfait. 1983. Source of electrons for extracellular Fe(III) reduction in iron-deficient bean roots. *Physiol. Plant.* **59:** 409–415.

Sijmons, P.C., H.J. Atkinson, and U. Wyss. 1994. Host cell responses to plant-parasitic nematodes. *Annu. Rev. Phytopathol.* **32:** 235–259.

Sijmons, P.C., F.M.W. Grundler, N. Vonmende, P.R. Burrows, and U. Wyss. 1991. *Arabidopsis thaliana* as a new model host for plant-parasitic nematodes. *Plant J.* **1:** 245–254.

Taylor, C.G., W. Song, C.H. Opperman, and M.A. Conkling. 1992. Characterization of a nematode-responsive plant gene promoter. *J. Nematol.* **24:** 621. (Abstr.)

Topping, J.F., W.B. Wei, and K. Lindsey. 1991. Functional tagging of regulatory elements in the plant genome. *Development* **112:** 1009–1019.

Trudgill, D.L. 1991. Resistance to and tolerance of plant parasitic nematodes in plants. *Annu. Rev. Phytopathol.* **29:** 167–192.

Wyss, U. 1988. Pathogenesis and host-parasite specificity in nematodes. In *Experimental and conceptual plant pathology* (ed. R.S. Singh et al.), pp. 417–432. Oxford & IBH Publishing, New Delhi, India.

———. 1992. Observations on the feeding behaviour of *Heterodera schachtii* throughout development, including events during moulting. *Fundam. Appl. Nematol.* **15:** 75–89.

Wyss, U. and U. Zunke. 1986. Observations on the behaviour of second-stage juveniles of *Heterodera schachtii* inside host roots. *Rev. Nematol.* **9:** 153–165.

Wyss, U., F.M.W. Grundler, and A. Munch. 1992. The parasitic behaviour of 2nd-stage juveniles of *Meloidogyne incognita* in roots of *Arabidopsis thaliana*. *Nematologica* **38:** 98–111.

Yamamoto, Y.T., C.G. Taylor, G.N. Acedo, C.L. Cheng, and M.A. Conkling. 1991. Characterization of *cis*-acting sequences regulating root-specific gene expression in tobacco. *Plant Cell* **3:** 371–382.

Yu, P.K. and D.R. Viglierchio. 1964. Plant growth substances and parasitic nematodes. I. Root-knot nematodes and tomato. *Exp. Parasitol.* **15:** 242–248.

Zuckerman, B.M. and R.A. Rohde, eds. 1981. *Plant parasitic nematodes*, vol. III. Academic Press, London.

# 29

# Environmental Stress and Gene Regulation

Christine J. Daugherty,[1] Michael F. Rooney,[1] Anna-Lisa Paul,[1]
Nick de Vetten,[1] Miguel A. Vega-Palas,[1] Guihua Lu,[1]
William B. Gurley,[2] and Robert J. Ferl[1]

[1]Horticultural Sciences Department
[2]Department of Microbiology
Program in Plant Molecular and Cellular Biology
University of Florida
Gainesville, Florida 32611

The many facets of plant responses to environmental stress may include morphological and ultrastructural changes, alterations in metabolic pathways, and transcriptional regulation of genes. Despite a great deal of research on plant stress, the mechanisms by which plants sense and respond to many environmental changes are largely unknown, particularly at the gene level. In addition, although considerable information concerning the responses of plants to different types of stress has been published in other species, relatively few studies have been conducted with *Arabidopsis*.

In this chapter, we discuss the processes involved in the environmental signal transduction chain, linking a given environmental stimulus to gene expression and regulation. A multitude of genes, their sequences, and their subsequent protein profiles produced have been reported in numerous types of plants under many environmental stresses. Therefore, discussion of what is known about stresses in other plants precedes discussions concerning *Arabidopsis*. Many of the genes and proteins described in the literature have been named according to an extraction procedure, or according to the particular stress (e.g., cold, salt, water) they were associated with when described. It is becoming evident that many of these genes and proteins are induced in response to more than one stress, and that those expressed under different stresses are similar in sequence or function. Therefore, the nomenclature of many of the genes presented here is different because of historical context, even if they share an evolutionary base or function.

*Arabidopsis*
© 1994 Cold Spring Harbor Laboratory Press 0-87969-428-9/94 $5 + .00

In this chapter, we review UV light, water, heat, and hypoxia stress in relation to gene regulation. The response to UV light stress and chalcone synthase regulation is the topic of section one. Section two discusses water stress, and its relationship to other environmental stresses and to stages of plant development that involve water stress. In section two, responses to abscisic acid (ABA), a hormone associated with the desiccation process, are also discussed in the context of environmental gene regulation. Section three emphasizes high-temperature stress, focusing on heat shock proteins. Hypoxia and the regulation of alcohol dehydrogenase are presented in section four. Section five focuses on the G-box motif, a conserved DNA sequence found in many stress-related genes, and the role it may play in gene regulation.

## UV LIGHT STRESS

### Response to UV Light Stress

High intensity and UV light stress affect the phenylpropanoid and flavonoid pathways in plants, causing the induction and subsequent accumulation of UV-absorbing flavonoids (for review, see Hahlbrock and Scheel 1989). Flavonoids are thought to be synthesized as UV protective pigments in plants (Schmelzer et al. 1988). The first step of the flavonoid-specific branch of the phenylpropanoid pathway is catalyzed by chalcone synthase (CHS), the most studied enzyme in the pathway (Hahlbrock and Grisebach 1979). CHS activity increases in response to UV light, and in parsley (*Petroselinum hortense*), the accumulation of *Chs* mRNA is controlled at the transcriptional level (Chappell and Hahlbrock 1984).

The *Chs* gene has been cloned from *Arabidopsis* and other species such as *P. hortense*, *Phaseolus vulgaris*, *Zea mays*, *Petunia hybrida*, and *Antirrhinum majus* (snapdragon), but perhaps the most studied *Chs* system is from cultured parsley cells (Reimold et al. 1983; Ryder et al. 1984; Koes et al. 1986; Sommer and Saedler 1986; Wienand et al. 1986). Parsley displays three photoreceptors that regulate the expression of *Chs*: (1) a UV-B receptor, (2) a blue-light receptor, and (3) phytochrome (Bruns et al. 1986). When the photoreceptors are exposed to blue and red light alone, a small concentration of *Chs* mRNA is produced, but when UV light is added, *Chs* mRNA increases over the blue and red light levels alone (Ohl et al. 1989). UV-light-irradiated cultured parsley cells initiate transcription of *Chs* (Chappell and Hahlbrock 1984), but blue and red light, acting via the blue-light receptor and phytochrome, respectively, modulate *Chs* expression (Duell-Pfaff and Wellmann 1982).

Within the parsley *Chs* gene promoter, there are two *cis*-acting units that appear to be responsive to UV-containing white light (Schulze-Lefert et al. 1989a). Each unit, approximately 50 nucleotides in size, contains a set of two light-induced in vivo footprints (Schulze-Lefert et al. 1989a,b). Boxes I and II together form one *cis*-acting unit (designated Unit I). Replacement of either of these boxes with mutant sequences leads to complete loss of light responsiveness, suggesting that both are necessary components of the minimal UV-light-responsive promoter. The second unit (Unit II), composed of boxes III and IV, enhances UV-light-dependent promoter activation only slightly: twofold activation for Unit II versus tenfold for Unit I (Schulze-Lefert et al. 1989a,b).

Closer examination by in vivo footprinting indicated that Unit I was footprinted a few hours after the onset of UV irradiation. A detailed functional analysis of Box II identified it as a G box, containing the heptameric asymmetric sequence 5'-ACGTGGC-3' that is crucial for activity of the *Chs* minimal UV-light-responsive promoter (Block et al. 1990). To further characterize the Unit I *cis*-acting element, various constructs were tested containing the Unit I as monomer, dimer, and tetramer, either in the original or reverse orientation. The analyses showed that promoter activity increased with increasing number of copies of Unit I after exposure to UV light. Furthermore, one copy of Unit I in either orientation conferred UV-light responsiveness, suggesting that Unit I has properties of a UV-light-dependent enhancer element (Weisshaar et al. 1991).

Transcription factors that bind to gene promoters have been isolated from *Arabidopsis* and other plant species. One type of nuclear factor, CG-1, was found in *Arabidopsis*, *Nicotiana tabacum*, *A. majus*, and *P. hybrida* (Staiger et al. 1989). CG-1 isolated from tobacco seedlings specifically recognized and bound to an oligonucleotide with the core G box 5'-CACGTG-3' motif, as demonstrated by competition assays, and was possibly involved, but not required, for the UV-light regulation of the *Chs* gene (Staiger et al. 1989).

Other putative G-box-binding factors are the common plant regulatory factors (CPRFs). CPRF-1,2,3 were isolated by their ability to bind to the G-box sequence located in the UV-light-responsive *Chs* promoter in parsley. Expression analysis of these clones showed that during UV illumination, the levels of CPRF-2 and CPRF-3 mRNA decreased slightly or were not influenced. In contrast to CPRF-2 and CPRF-3, CPRF-1 mRNA began to accumulate shortly after the onset of irradiation. Maximal accumulation of CPRF-1 mRNA coincided with the steepest increase in *Chs* mRNA. These results suggest that CPRF-1 is involved in *Chs* gene activation in response to UV light (Weisshaar et al. 1991).

## UV Light Stress in *Arabidopsis*

In *Arabidopsis*, high intensity and UV light stress also affect the flavonoid pathway and the *Chs* gene. The *Arabidopsis Chs* gene was cloned using a parsley *Chs* cDNA clone (Feinbaum and Ausubel 1988). *Arabidopsis Chs* is encoded by a single-copy gene with one intron, and its protein sequence is 90% similar to the parsley gene. When *Arabidopsis* plants were exposed to continuous high-intensity light, 50-fold increases in CHS enzyme activity occurred due to a 25-fold increase of *Chs* steady-state mRNA. Nuclear run-on experiments showed that the increase in mRNA concentration was due to an increased transcription rate of the *Chs* gene (Feinbaum and Ausubel 1988).

The effect of wavelengths on the induction of *Chs* was tested by ir-radiating *Arabidopsis* with various wavelengths of light: white, blue, red, and UV (Feinbaum et al. 1991). Plants exposed to white, blue, or UV light exhibited increased *Chs* mRNA levels, but red light showed little effect on mRNA concentrations. Further evidence that wavelength played a role in the induction of genes was seen in the expression of four flavonoid genes, phenylalanine ammonia-lyase (*Pal1*), chalcone synthase (*Chs*), chalcone isomerase (*Chi*), and dihydroflavonol reductase (*Dfr*). When dark-grown *Arabidopsis* seedlings were exposed to both UV-B (310 nm) and blue light, the levels of expression were greater with UV-B light treatment than with blue light treatment (Kubasek et al. 1992).

When *Arabidopsis* was exposed to UV-B light, the sequence of mes-sage induction was the same in the phenylpropanoid and flavonoid path-ways: i.e., it started with *Pal1* and was followed by *Chs*, *Chi*, and *Dfr* (Kubasek et al. 1992). Both the *Arabidopsis Pal1* and 3-hydroxy-3-methylglutaryl coenzyme A reductase (*Hmg-CoA*) mRNA levels in-creased after exposure to UV-C light (290 nm and below), suggesting that carotenoids were indicators of UV-C-induced stress in *Arabidopsis* (Campos et al. 1991).

General deletion studies on the *Arabidopsis Chs* promoter showed that the region between 523 bp and 17 bp upstream of the transcription start site was necessary for constitutive levels of *Chs* mRNA (Feinbaum et al. 1991). Within this region there were two sequences, 5′-CACGTG-3′ located at −442 and 5′-CACCTG-3′ at −285, that were similar to the G-box element in the *Chs* gene of snapdragon and parsley. Other deletion studies using the *Arabidopsis Pal1* gene showed that 290 bp of the proximal region of the promoter was needed for normal tissue-specific expression of *Pal1*, but 540 bp was required for a response to light stress (Ohl et al. 1990).

The GC-1 nuclear factor was shown to be present in *Arabidopsis* and may be involved in UV-light regulation of the *Chs* gene (Staiger et al.

1989). Footprinting studies of the *Arabidopsis Chs* gene promoter have not yet been carried out, but it is expected that multiple regions responsible for UV-light induction of *Chs* will be present, with one of the regions being the G box as in the *Chs* genes of snapdragon and parsley (Schulze-Lefert et al. 1989a,b; Staiger et al. 1989).

## WATER STRESS

### Response to Water Stress

Plants, like other living organisms, need to take up water from the surrounding environment to fulfill their physiological requirements. This process depends on environmental conditions and is largely influenced by different stresses such as drought, salt, and cold. Under these environmental stresses, plants ultimately suffer from water stress, defined as a deficit or lack of water. The resulting negative water potential can lead to reduced water uptake, alterations in transpiration, turgor loss, and alterations in mineral nutrition (Levitt 1980; Paternak 1987; Neumann et al. 1988; Skriver and Mundy 1990). The cellular responses to the stresses mentioned above include modifications of the lipid composition of the cytoplasmic membrane (Hirayama and Miara 1987; Steponkus et al. 1988; Skriver and Mundy 1990) and synthesis of osmoprotectant solutes (Levitt 1980; Hanson and Hitz 1982; Binzel et al. 1987). Specific gene expression is induced in response to drought, salt, and cold stress, and a significant number of these genes belong to the same gene family (Close et al. 1989; Godoy et al. 1990; Skriver and Mundy 1990; Galau and Close 1992; Gilmour et al. 1992). In addition, the protein products of these stress-induced genes can be found in response to more than one type of stress (Skriver and Mundy 1990; Skriver et al. 1991; Guy et al. 1992). In some cases, it has been demonstrated that the exposure of plants to one stress results in acclimation to another stress (Chen et al. 1977; Siminovitch and Cloutier 1982). These data indicate that plants exposed to several different stresses have common responses which may play a similar role in the acclimation to cold, water, and salt stress.

### Relationship between Plant Development, ABA, and Stress

The natural process of seed development in plants involves physiological and biochemical changes that may be similar to responses observed in plants exposed to stress (Quatrano 1986). During the late part of embryogenesis, the seed undergoes desiccation, allowing the embryo to withstand dehydration for an extended period of time (Kermode and Bewley

1987). The storage proteins and the LEA (*late* embryogenesis *a*bundant) proteins are also synthesized at this time (Casey and Domoney 1987; Baker et al. 1988; Dure et al. 1989). Several of the *lea* genes are also induced under drought or osmotic stress (Hughes and Galau 1991). Other genes similar to *lea* genes can be induced by drought (Raynal et al. 1990), salt (Godoy et al. 1990), or cold stress (Gilmour et al. 1992).

A common factor involved in the response to drought, salt, and cold stress, as well as in the development and dormancy of seeds, is ABA. The level of ABA increases in plants during seed development and under many environmental stresses (Chen et al. 1983; Kermode and Bewley 1987; Skriver and Mundy 1990). It has been demonstrated that treatment with ABA makes many plants more resistant to drought (Bartels et al. 1990), salt (LaRosa et al. 1987), and cold stress (Chen and Gusta 1983). In addition, levels of LEA proteins, which are normally low in germinated seedlings, increase after treatment with ABA (Mundy and Chua 1988). Several ABA-responsive genes (*rab* genes) are also induced in response to drought (Pla et al. 1991, 1993), salt (Yamagushi-Shinozaki et al. 1989), and cold stress (Hahn and Walbot 1989), as well as during embryogenesis (Vilardell et al. 1990). Some of the *rab* genes are homologous to *lea*, *dhn* (dehydrin genes), salt, and cold stress genes (Pla et al. 1991; Galau and Close 1992; Gilmour et al. 1992; King et al. 1992). Several *lea* genes and other genes associated with drought, cold, or salt stress are also inducible in response to exogenous ABA (Close et al. 1989; Marcotte et al. 1989; Godoy et al. 1990; Gilmour et al. 1992).

It appears that ABA is involved in the response of plants to environmental stresses and in the development of the seed, but the mechanism of action is not entirely clear. Current models envision the action of turgor pressure sensors, followed by increases in ABA levels, and ultimately, the induction of ABA-regulated genes (Skriver and Mundy 1990).

### Relationship of Water Stress and ABA in *Arabidopsis*

Six genes that are induced in response to water and cold stress have been isolated from *Arabidopsis*, and the nucleotide sequences of four have been determined (Hajela et al. 1990; Kurkela and Frank 1990; Nordin et al. 1991; Gilmour et al. 1992). One of these genes, *Iti140*, codes for a 140-kD protein that contains a perfect repetition of 21 amino acids and has no significant similarity to any other known proteins (Nordin et al. 1991). Two other genes, *kin1* and *cor6.6*, code for proteins of 6.5 kD and 6.6 kD, respectively. These two proteins are 94% similar and show similarities with several small antifreeze proteins of Arctic flounders (Kurkela and Frank 1990; Gilmour et al. 1992). In addition, it has been shown that the COR6.6 protein has a 30% similarity to a LEA protein from *Brassica napus* (Kurkela and Frank 1990; Gilmour et al. 1992). The fourth gene, *cor47*, codes for a 47-kD protein that contains lysine and serine clusters. Both of these clusters have been found in LEA (Gilmour et al. 1992), RAB (Mundy and Chua 1988), and DHN proteins (Close et al. 1989), as well as in salt-induced proteins (Godoy et al. 1990). Unlike *lea* genes, the *cor47* and *cor6.6* genes are not abundant in late stages of seed development (Gilmour et al. 1992).

In *Arabidopsis*, as in other plants, ABA has been shown to be involved in the responses to cold and drought stress (Heino et al. 1990; Schnall and Quatrano 1992), as well as in the development of the seed (Koornneef et al. 1989; Finkelstein and Somerville 1990; Meurs et al. 1992). The six genes mentioned above are also inducible by exogenous ABA (Hajela et al. 1990; Kurkela and Frank 1990; Nordin et al. 1991; Gilmour et al. 1992). Induction of *Iti140* by water stress has been shown to involve ABA (Nordin et al. 1991). When exposed to water stress, both the ABA-insensitive mutant *abi-1* and the wild-type strain in the presence of fluridone (an ABA biosynthesis inhibitor) showed decreased expression of the *Iti140* gene. The induction of *Iti140* by cold stress does not seem to be mediated by ABA. However, induction of *Iti140* by exogenous ABA and cold stress is not cumulative, suggesting that both activation pathways converge (Nordin et al. 1991). Gene expression in response to water stress appears to be similar both in *Arabidopsis* and in other higher plants.

## HEAT SHOCK

### Heat Shock Response

Plants, like most other organisms, respond to an elevation in temperature by greatly reducing translation of most proteins and by induction of heat shock (*HS*) genes. The heat shock response can also be triggered by other stresses, such as exposure to arsenite, ethanol, amino acid analogs, glucose starvation, and uncouplers of oxidative phosphorylation. The *HS* genes encode proteins (hsps) that are highly conserved among prokaryotic and eukaryotic organisms. These proteins comprise five major classes: hsp110, hsp90 (80–95 kD), hsp70 (63–78 kD), hsp60 (53–62 kD), and the low-molecular-weight (LMW) hsps (14–30 kD) (Neumann et al. 1989; Vierling, 1991). The hsp90, -70, and -60 classes are the best characterized and act to stabilize protein conformation. These proteins are thought to aid in the proper folding of newly synthesized nascent peptides, transport across membranes, assembly of oligomeric complexes, and maintenance of steroid receptor conformation. Very little is known regarding the action of members of the hsp110 and LMW hsp classes. The *hsp104* (hsp110 class) gene has been shown to be essential for acquisition of thermal tolerance in yeast (Sanchez and Lindquist 1990). Although the precise mechanisms are not known, the ability to survive during high-temperature stress as well as during normal growth conditions is correlated with the presence of certain classes of hsps.

**Regulation of the Heat Shock Response**

The organism's response to thermal stress is modulated at both the transcriptional and translational levels. Transcriptional induction of *HS* genes is mediated by the binding of the heat shock transcription factor (HSF) to DNA sequences in the promoter known as heat shock elements (HSEs). The HSE was originally defined as the 14-bp sequence 5'-CTnGAAnnTTCnAG-3' from studies conducted with *Drosophila HS* genes (Mirault et al. 1982; Pelham 1982; Pelham and Bienz 1982). More recent analyses have shown the HSE to comprise a 5-bp cluster (5'-nGAAn-3') of alternating orientations (Amin et al. 1988; Xiao and Lis 1988). These core repeats have been shown to be essential in plant *HS* genes. The optimal bases for the nonconserved positions are an A preceding the GAA core and a G following (Czarnecka et al. 1989; Barros et al. 1992). A single HSE is often hard to delineate, but usually comprises three or more 5-bp repeats with imperfect trinucleotide core repeats usually present (Gurley and Key 1991). In most cases, the *HS* promoter contains at least two clusters of trinucleotide repeats located between the TATA motif and position –120. Although most *HS* genes contain a TATA, some, like petunia *hsp70* (Winter et al. 1988) and human *hsp70*, do not (Greene and Kingston 1990).

**Heat Shock Transcription Factor**

There appear to be two components to induction of transcription by the heat shock transcription factor (HSF): HSF binding to the HSE, and activation of the HSF protein to a state capable of inducing transcription. In *Saccharomyces cerevisiae*, HSF binding to the promoter occurs at normal and heat shock temperatures (Sorger et al. 1987; Jakobsen and Pelham 1988; Wu et al. 1990). Stress-induced activation of prebound HSF results in transcription of *HS* genes. In higher eukaryotes, both DNA binding and transcriptional activation of the HSF protein are induced by heat shock. Vertebrate and insect cells show a 10- to 30-fold increase in DNA-binding activity in extracts prepared from heat shocked cells (Kingston et al. 1987; Zimarino and Wu 1987; Mosser et al. 1990; Wu et al. 1990; Zimarino et al. 1990). A similar induction in binding activity has been shown in tomato and soybean extracts, indicating that plants must use a mode of induction similar to animal cells (Czarnecka et al. 1990; Scharf et al. 1990). Evidence that transcriptional activation of the HSF is also required in higher eukaryotes is seen in the uncoupling of DNA binding from transcriptional activation by the treatment of HeLa cells with salicylate (Jurivich et al. 1992).

In animal cells, HSF binding is reversible in vivo and not dependent on protein synthesis at high temperatures. Protein synthesis is required, however, at intermediate temperatures which range from 2°C to 7°C below the optimum for heat shock induction (Zimarino et al. 1990). It is not known if a similar requirement for protein synthesis at intermediate temperatures exists for plants.


**Expression of the HSF Gene**

HSF cDNAs have been cloned from a variety of organisms including yeast, *Drosophila*, mice, humans, and tomato (Wiederrecht et al. 1988; Clos et al. 1990; Scharf et al. 1990; Jakobsen and Pelham 1991; Rabindran et al. 1991; Sarge et al. 1991; Schuetz et al. 1991). The DNA-binding domains and oligomerization domains are conserved among all eukaryotes, whereas other regions of the protein show very little similarity in amino acid sequences. A single HSF has been cloned from *Drosophila*, and multiple HSFs have been cloned from HeLa, mouse, and tomato. Although constitutively expressed HSFs are present in all of these organisms, tomato is unique in having both constitutive and heat-shock-inducible HSFs (Scharf et al. 1990). Presumably, the constitutive HSFs are utilized in the initial stages of the heat shock response and, under extended heat shock conditions, the inducible forms may assist or replace the constitutive HSFs. It may be that heat-shock-inducible HSFs delay the attenuation of *HS* gene transcription, normally seen 2–4 hours after heat shock, by enlarging the pool of unrepressed HSF. Some results suggest that *Arabidopsis* may also contain multiple HSFs (E. Czarnecka et al., unpubl.). It is not known if any of the *Arabidopsis* HSF genes are heat inducible.


**Negative Regulation of HSF by hsp70**

Evidence is accumulating that the heat shock response is negatively regulated by hsp70. This possibility was first suggested because of the high degree of conservation between the DnaK protein of *Escherichia coli* and hsp70 of eukaryotes. DnaK proteins, along with DnaJ and GrpE proteins, are involved in negatively modulating the activity of the heat-shock-induced sigma factor, $\sigma^{32}$, in *E. coli* (Straus et al. 1989; Tilly et al. 1989; Straus 1990). Recent evidence has shown that hsp70 binds to human HSF and is thought to maintain the repressed state at normal temperatures (Abravaya et al. 1992). Presumably, the interaction of hsp70 with HSF prevents hexamerization of HSF, which is necessary for DNA binding, and may also mask the transcriptional activation domain (for review, see Czarnecka-Verner et al. 1994). After a heat shock, the cellular pool of hsp70 available for repression of HSF diminishes as hsp70 binds to proteins denatured from the stress. It has also been postulated that heat stress and other agents that activate the heat shock response lower cellular ATP levels (Beckmann et al. 1990). Since hsp70 requires ATP to disengage from protein-protein interactions (Pelham 1990), it would become locked in its associations with other proteins, thereby contributing to the decrease in levels of free hsp70. This hypothetical mechanism for transcriptional regulation, based on levels of free hsp70, would be responsive to stress in two ways: harsh environmental conditions are assessed by monitoring levels of denatured proteins, and metabolic stress is assessed by its sensitivity to low ATP concentrations.

### Heat Shock Response in *Arabidopsis*

The response to heat shock in *Arabidopsis* has been characterized at both the transcriptional and translational levels. The optimal temperature for this alpine plant is lower than most crop species and as a result, *Arabidopsis* has been called a "heat sensitive genotype" (Binelli and Mascarenhas 1990). When the promoter of the *hsp18.2* gene was fused to a *GUS* reporter gene, the optimal temperature for expression was 35°C in transgenic *A. thaliana* ecotype Columbia (Takahashi et al. 1992). No *GUS* activity was observed at 40°C, which is the optimum for soybean, but some induction of HS mRNA does occur at this temperature. Apparently, transcription is less sensitive to heat than protein synthesis or protein stability. An earlier study using the Enkheim race indicated an optimum for the heat shock response at 37°C and also found a greatly reduced response at 40°C (Klyueva and Samokhvalov 1991). The optimum for heat shock protein synthesis was also found to be 37°C (Wu et al. 1988). Differences in the optimum temperature for the heat shock response (34–39°C) have been attributed to differences in age of the plants, tissues and organs assayed, experimental methods, and varieties.

Regardless of reported differences in the precise temperature of the optimum, all of these studies indicated that exposure to 40°C temperatures constitutes a severe shock to the species. The sensitive nature of *Arabidopsis* was further illustrated in experiments where seedlings were exposed to 42°C for 2 hours, followed by a return to 23°C (Binelli and Mascarenhas 1990). No damage was immediately visible, but all plants were dead after 96 hours. Exposure to 37°C for 2 hours and return to 23°C had no harmful effect. Soybean seedlings were able to survive both treatments with no harmful effects.

## Hsp Induction: Classes and Timing

The expression of hsps in *Arabidopsis* is similar to that seen in other plants. The first hsps to appear are hsp110, hsp90, hsp76, and hsp22, which are induced within 30 minutes of heat shock; hsps19 and 27 are expressed by 60 minutes (Wu et al. 1988). Small amounts of hsp70 and hsp90 class proteins are present at normal growth temperatures, but levels are increased at 37°C (Klyueva and Samokhvalov 1991). For example, the hsp70s represent 5.6% of total protein synthesis at 23°C and 14.3% at 37°C. After 2 hours, the high-molecular-weight hsps represent 35% of total cellular synthesis and the LMW hsps 18%. Unlike *Drosophila* and other organisms, a substantial amount of synthesis of non-hsp proteins (47%) can occur during heat shock in *Arabidopsis*. This phenomenon is most prominent at intermediate temperatures between 31°C and 34°C, where both hsps and normal proteins are synthesized.

In many plants, the LMW hsps are the most abundantly expressed class of hsp (Kimpel and Key 1985; Nagao et al. 1986). In *Arabidopsis*, LMW hsp synthesis occurs later than higher-molecular-weight classes, and final levels of accumulation are not as great (Wu et al. 1988). In one study, LMW hsps were much less abundant than the higher-molecular-weight hsps (Binelli and Mascarenhas 1990). This is in contrast to soybean, pea, and maize, where the LMW hsps predominate. Although LMW hsps are not detected until mid-heat shock, transcriptional expression of the *Arabidopsis hsp18.2* promoter fused to *GUS* was evident at low levels at 22°C in the vascular tissues of leaves and sepals, and at the tips of carpels (Takahashi et al. 1992). Surprisingly, some expression was also seen in heat-treated pollen tubes, although other studies have indicated that there is almost no heat shock induction in germinating pollen tubes of other plants (Xiao and Mascarenhas 1985; van Herpen et al. 1989; Dupuis and Dumas 1990). A number of *Arabidopsis* cDNAs for LMW hsps have been cloned. Those localized to the cytoplasm include hsp17.6 (Helm et al. 1989), hsp17.6-II (Bartling et al. 1992), hsp17.4 and hsp18.2 (Takahashi and Komeda 1989). Nuclear-encoded hsp21 of the chloroplast has also been cloned as a cDNA (Chen and Vierling 1991).

The chloroplast hsps are conserved among plants and contain a unique structural feature (conserved region III) that consists of an amphipathic α-helix with a cluster of methiones on one face. This region, called a "methione bristle," is thought to bind signal peptides by hydrophobic interactions (Berstein et al. 1989; Chen and Vierling 1991).

The hsp70 proteins are expressed under normal growth conditions and show enhanced expression during heat shock. Often, certain members within the group are expressed during specific stages of embryogenesis or development (Zimmerman and Cohill 1991). Those genes that are regulated by conditions other than heat shock are termed hsp70 cognates (hsc70). In many organisms, the coding region of hsc70 is interrupted by introns, whereas hsp70 genes are not. In *Arabidopsis*, there are least 12 hsp70-related polypeptides, 4 of which are strongly induced by heat shock, whereas the rest are constitutively expressed (Wu et al. 1988). All three of the hsp70-related genes that have been cloned in *Arabidopsis* appear to be cognates, based on the presence of an intron at amino acid 72. The hsc70 genes in *Drosophila* and rat also have an intron at this position. In plants, the distinction between hsc70 and hsp70 genes is blurred with respect to expression since hsp70-like genes that contain an intron are sometimes heat shock inducible. Of the three cloned hsp70-related genes, hsp70-2 is apparently not expressed, and hsp70-3 is very weakly expressed after heat shock in leaves only. The third gene, hsp70-1, is constitutively expressed in leaves and shows a 4- to 5-fold increase in expression after heat shock. A similar situation exists in maize, where the cloned hsp70 gene is heat shock inducible, but contains an intron (Winter et al. 1988).

A single gene encodes hsp60 in *Arabidopsis*. This gene is developmentally regulated during seed germination and is also heat shock inducible (Prasad and Stewart 1992). Transcripts to hsp60 are present in dry seed embryos, increase 15-fold during the first 24 hours of imbibition, and then decline to low levels after 120 hours. In plants, hsp60 proteins are known to be involved in macromolecular assembly of protein complexes in mitochondria and chloroplasts. Hsp60 is one of two plant homologs to the GroEL protein of *E. coli* (the other is the Rubisco subunit binding protein, RBP or plastid Cpn60) (Ellis 1990). In *E. coli*, GroEL and GroES participate in bacteriophage head protein assembly (Georgopoulos et al. 1973).

The hsp90 family of proteins (80–95 kD) form soluble dimers in animal cells. They have been found in association with various proteins and are thought to play a critical role in signal transduction by interacting with steroid receptors (Picard et al. 1990). The *hsp81-1* and *hsp81-2* genes from *Arabidopsis* have been cloned and are members of the hsp90 family (Takahashi 1992). Hsp81-1 has very low basal expression, but is strongly induced by heat shock. Hsp81-2 is constitutively expressed at 22°C and shows moderately increased expression at heat shock temperatures. The function of these proteins in plants has yet to be determined; however, in yeast, disruption of the two copies of hsp82 is lethal (Borkovich et al. 1989).

### Related Gene Families

Two other gene families that have been associated with heat stress are ubiquitin and calmodulin-related touch (*TCH*) genes. The ubiquitin gene family in *Arabidopsis* comprises approximately 11 members that show differential patterns of expression (Burke et al. 1988). As a whole, expression is most prominent in young green plants and in immature tissues from older plants. The ubiquitin protein covalently attaches to a variety of proteins and is thought to selectively target them for degradation. Although some polyubiquitin mRNAs increase in abundance after heat shock in maize and sunflower (Christensen and Quail 1989; Binet et al. 1991), there seems to be only slight accumulation of ubiquitin transcripts in heat-stressed *Arabidopsis* (Burke et al. 1988). Although most ubiquitin mRNAs are decreased by heat shock, levels of one of four ubiquitin transcripts was increased, but then only in flowers. In the case of *TCH* genes, three of four genes investigated showed clear induction after heat shock (Braam 1992). These genes are induced by a variety of stimuli including touch, rain, wind, wounding, darkness, and calcium treatment. The heat-induced stimulation is thought to be due to an increase in intracellular calcium rather than thermal stress directly, since hsp70 is not induced by $Ca^{++}$, and heat-shock-induced release of $Ca^{++}$ is known to occur.

## HYPOXIC STRESS

### Response to Hypoxia

Hypoxia is a common environmental stress imposed upon plants in naturally oxygen-poor soils and during flooding of normally aerated soils. Higher plants tend to deal with the environmental stress of hypoxia in two ways: (1) by altering structural morphology to maximize the use of any available oxygen and (2) by adopting alternate metabolic pathways that can function with limited oxygen.

Gross morphological changes in response to flooding include induction of adventitious roots (Wenkert et al. 1981) and an increase in aerenchymatous tissue (Drew et al. 1981). Both mechanisms function to increase oxygen availability to the waterlogged roots. These responses are thought to be mediated via plant growth regulators, especially ethylene (Drew et al. 1979; Jackson et al. 1981). Spatial concentrations of hormones in the shoot near the soil surface may also contribute to the initiation of adventitious roots (Jackson 1984).

Cellular alterations that occur under hypoxic conditions may include changes in membrane properties, mitochondrial ultrastructure, and metabolic energy balance (Crawford 1978; Aldrich et al. 1985; Vartapetian et al. 1987). Under limited oxygen, root respiratory processes are immediately affected. There are several metabolic strategies by which plants respond to the problems of anaerobic stress (Crawford 1978): (1) by controlling the rate of glycolytic metabolism, (2) by diversifying the end products of glycolysis, (3) by providing an adequate store of carbohydrates, and (4) by coupling metabolic pathways to facilitate proton disposal and provide additional ATP. The utilization of each of these strategies is mediated by a suite of enzymes, and their regulators, in response to hypoxia.

During hypoxic stress, the Krebs cycle is usually suppressed and alcoholic fermentation is accelerated, leading to the accumulation of pyruvate. Two enzymes act on pyruvate in anaerobic metabolism: lactose dehydrogenase (LDH), which catalyzes the formation of lactate from pyruvate (Hoffman et al. 1986; Hoffman and Hanson 1986; Good and Paetkau 1992), and pyruvate decarboxylase (PDC) (Laszlo and St. Lawrence 1983; Kimmer 1987; Kelley 1989). PDC is an important branch point in anaerobic metabolism. In this reaction, pyruvate is converted to acetaldehyde by pyruvate decarboxylase, the accumulating $NADH_2$ is oxidized, and acetaldehyde is reduced by alcohol dehydrogenase to ethanol. This regenerates the $NAD^+$ needed to sustain glycolysis and so maintains a slow, but steady, output of ATP. Because the yield of ATP is low, glycolysis must be accelerated (Pasteur effect) to support cellular needs. The acceleration leads to rapid exhaustion of carbohydrate reserves and results in the generation of a toxic end product (ethanol). Plants adapted to hypoxic environments often fail to show the acceleration of glycolysis that is typically associated with the hypoxic response (Crawford 1978). In addition, if ethanol is formed at all, the accumulation is below concentrations typically produced from intolerant species under similar conditions. The key enzymes in the adaptive pathways appear to be those involved in the latter stages of alcoholic fermentation: pyruvate decarboxylase, lactate dehydrogenase, and alcohol dehydrogenase (ADH) (Kennedy et al. 1992). Of these enzymes, ADH is the best characterized. ADH is the principal terminal enzyme of alcoholic fermentation in plants and is responsible for recycling $NAD^+$ during hypoxia. ADH is crucial for the survival of many plants intolerant of flooding during hypoxic stress (Harberd and Edwards 1982; Roberts et al. 1984).

## Hypoxic Stress in *Arabidopsis*

Many of the morphological alterations (aerenchymatous tissue and adventitious roots) that occur in other plants under hypoxic stress have not been characterized in *Arabidopsis*. However, alcoholic fermentation, the reduction of acetaldehyde to ethanol by ADH, and the ADH enzyme have been well studied in *Arabidopsis*. Whereas most plants contain two or three genes that code for ADH, *Arabidopsis* is unusual in that it has only a single gene encoding ADH. There are three naturally occurring alleles of *Arabidopsis* ADH, which have been described as the electrophoretic variants A, F, and S (Dolferus and Jacobs 1984). The *Adh* gene is normally expressed in the seed and seedling, but expression quickly diminishes after germination. ADH is virtually undetectable in the seedling after 10 days (Dolferus and Jacobs 1991). In mature *Arabidopsis*, ADH is found in root tips and pollen, and is absent in leaves. However, hypoxia increases *Adh* mRNA levels by 10- to 15-fold in the roots of *Arabidopsis* plants (Dolferus et al. 1985; Jacobs et al. 1988). *Adh* is expressed constitutively in some cultured cell lines (Ferl and Laughner 1989), but the tissue source for the cell line and the levels of 2,4-D in the culture medium are important factors that influence the expression of *Adh* (Dolferus and Jacobs 1991).

## *Adh* Genes

Comparisons of the *Adh* genes from *Arabidopsis* and a variety of plants show that the coding sequence is highly conserved across species (Walker et al. 1987b). The coding region of the *Arabidopsis Adh* gene has 73% similarity to maize *Adh1*, another well-characterized *Adh* gene. The *Arabidopsis Adh* gene contains six short introns that coincide with the positions of six of the nine introns of the maize *Adh1* and *Adh2* genes (Chang and Meyerowitz 1986). The physicochemical properties of *Arabidopsis Adh* and maize *Adh1* are very similar, and antibodies for maize ADH recognize *Arabidopsis* ADH. In addition, *Arabidopsis* ADH and maize ADH are able to form functional heterodimers (Dolferus and Jacobs 1991). Clearly, the *Adh* gene products from these two very different plants are highly conserved.

Even though the *Arabidopsis Adh* and maize *Adh1* genes respond to hypoxia in a similar fashion, the regulatory portions of the respective

*Adh* genes have only a few short sequence motifs that are conserved among species. Because of the paucity of conserved sequence within the promoters, it was presumed that the conserved motifs were probably important for the anaerobic regulation of each gene (Walker et al. 1987b).

### *Adh* Promoter Motifs

Analyses of the promoter region of *Arabidopsis Adh* indicate that there are several motifs which might function as regulatory elements. Several of these motifs have been shown to be associated with DNA-binding factors in vivo (Ferl and Laughner 1989). The two footprinted sequence motifs known to have functional significance in other plant promoters are a 4C motif (the 4C box) located at −145 and the G box (5′-CCACGTGG-3′) centered around −214. The 4C box is the most likely candidate for an anaerobic regulatory element in *Arabidopsis*.

Mutagenesis and deletions of the maize promoter defined regions of the promoter that were necessary for a positive response to hypoxic stress (Ellis et al. 1987; Howard et al. 1987; Lee et al. 1987). The minimum promoter required for hypoxic induction was a region that extended 5′ to the *Pst*I site at approximately −140. It was later shown that a motif within this region between −140 and −99 was absolutely required for anaerobic induction. This 41-bp sequence was dubbed the *a*naerobic *r*esponse *e*lement (ARE) (Walker et al. 1987a). A synopsis of the functional analyses of the *Arabidopsis* and maize *Adh* promoter is shown in Figure 1.

The ARE in maize is composed of two subelements positioned between −133 and −124 (ARE I), and between −113 and −99 (ARE II).

*Figure 1* Functional analyses of the *Arabidopsis Adh* and the maize Adh1 promoters. (*A*) Sections of the 5′-flanking regions of both *Adh* genes. The ARE motifs are highlighted with lines showing the orientation of the motif. Open boxes define regions that have had mutations introduced in vitro for transient gene expression assays. Specific mutations are indicated within the boxes below the corresponding sequence. The results of the transient gene expression assays are reflected by the filled boxes below each mutated site. The circles located above the sequences indicate the specific bases involved in protein interactions in vivo. Filled circles denote enhanced interactions and open circles denote protected sites. (*B*) Deletion analyses of the *Arabidopsis* and maize *Adh* promoters. The dark line represents the promoter of each gene. Sites of in vivo protein interactions are denoted by ovals on the promoter lines. Positions of the ARE motifs are shown within the open rectangles. Stippled bars beneath each promoter show the deletions with the relative activities of each deletion in transient assays.

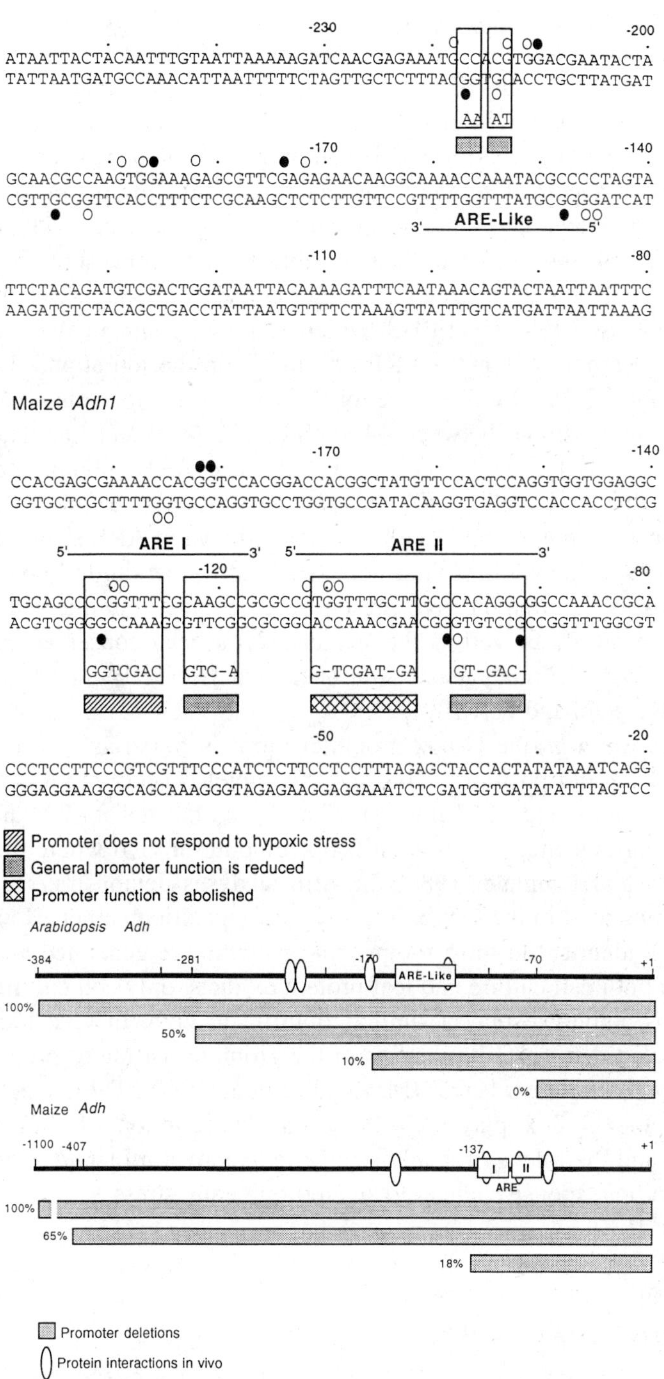



*Figure 1 (See facing page for legend.)*

Mutations within ARE I reduce the level of expression in response to anaerobic stress by 75%, suggesting that the motif 3′-CCGGTTT-5′ in some way contributes to the anaerobic response (Walker et al. 1987a). Mutation within the 3′-GGTTT-5′ motif in the ARE II subunit completely obliterates promoter function, so it is difficult to evaluate the specific importance of ARE II to the detection of hypoxia. However, it is essential to have a pair of subunits for anaerobic induction in transient gene expression assays. The subunits appear to function cooperatively, as increasing the spacing between them diminishes anaerobic induction (Olive et al. 1990). The ARE I mutation described above disrupts a group of 4Cs as well as the 5′-GGTTT-3′ which is found again in ARE II. *Arabidopsis* does not contain an ARE-like motif on the top strand, but the sequence 5′-TTGGTTT-3′ can be found just 5′ of the footprinted 4C box, on the bottom strand between –150 and –160 (Ferl and Laughner 1989; Dolferus and Jacobs 1991). Since the maize ARE subunits can function in either orientation, the *Arabidopsis* inverted 5′-GGTTT-3′ motif adjacent to the 4C box can be considered a valid ARE sequence (Olive et al. 1990). In addition, since both ARE motifs are similarly positioned and display similar in vivo protein interactions, it is likely that function of the motif, as well as the sequence, has been conserved between *Arabidopsis Adh* and maize *Adh* (Ferl and Nick 1987; Ferl and Laughner 1989; Paul and Ferl 1991).

In *Arabidopsis Adh*, the G-box promoter motif is found as a perfect dyad centered around position –214, and incomplete G-box motifs can be found centered around –187 and –310. The G box located at –214 displays prominent evidence of a protein factor binding in vivo where *Adh* is active (Ferl and Laughner 1989). In vitro analyses demonstrate that isolated proteins bind to the G box in a sequence-specific manner. However, although identical in vitro footprinting patterns are generated with extracts from both cell culture and leaf protein extracts, only cell cultures (where *Adh* is active) display a similar footprint in vivo. In vivo footprinting revealed that the G-box region of the promoter is free of protein interactions in *Arabidopsis* leaves (McKendree et al. 1990). These results suggest that the G box plays a role in the transcription of *Adh* in *Arabidopsis*, and the role is probably as a general activator of *Adh* rather than as a specific response element to an environmental stress.

## THE G-BOX MOTIF

### A Conserved *cis*-Element, the G-box Motif

Generally, a given promoter contains several different sequence motifs that may also reside in other promoters regulated by diverse stimuli (for review, see Dynan 1989). Responses of eukaryotic cells to external stimuli are at least in part modulated by changes in gene expression at the level of promoter activity, through the combination of *cis*-acting elements and their respective suite of *trans*-acting factors. One highly conserved DNA sequence has been identified in the promoter region of plant genes exhibiting regulation by a wide variety of environmental signals (Table 1). This sequence, referred to as the G box, has a hexameric core with internal dyad symmetry, 5'-CACGTG-3', and was first identified as a highly conserved promoter element encoding the small subunit of ribulose bisphosphate carboxylase (*RbcS*) (Giuliano et al. 1988). Subsequently, sequences similar to the G box were found in a number of other, mainly inducible, genes (see Table 1). Although discussion here is limited to the role of this particular sequence in plant gene promoters, it should be noted that similar sequences are found in at least 73 different promoters, including yeast, viral, and mammalian promoters (Chodosh et al. 1989; Donald et al. 1990; Williams et al. 1992).

For many of the genes described in Table 1, functional analyses using transient gene expression systems and transgenic plants have identified regions in the promoter, in addition to the G box, that appear to contain sequences necessary for the specific induction of the gene. In the case of the chalcone synthase (*Chs*) promoter, in vivo footprinting and transient

*Table 1*  Partial list of G-box elements in plant genes responding to  different signals

| Signal | Plant species | Gene | Sequence |
|---|---|---|---|
| Low oxygen | *Arabidopsis* | *Adh* | cCACGTGG |
| Heat shock | *Arabidopsis* | *hsp70* | gCACGTGG |
| UV light | *Arabidopsis* | *Chs* | cCACGTGG |
|  | petunia | *Chs* | aCACGTGG |
|  | parsley | *Chs* | CCACGTGG |
| Auxin | *Arabidopsis* | *Dpr* | tgACGTGG |
|  | wheat | *Em1* | aCACGTGG |
| ABA | rice | *Rab16* | gtACGTGG |
| Ethylene | *Arabidopsis* | *AT-ACC1* | tCACGTGG |
| Light | *Arabidopsis* | *RbcS-1A* | CCACGTGG |
|  | tomato | *RbcS* | gCACGTGG |
|  | *Arabidopsis* | *cab1* | atACGTGG |
|  | *Arabidopsis* | *Acp* | gCACGTGG |
| Development | *Arabidopsis* | *at2s* | acACGTGG |
|  | *Arabidopsis* | *Als* | tCACGTGG |
|  | *Arabidopsis* | *EPSP* | aCACGTGG |
| Wounding | potato | *PI-II* | tCACGTGG |

gene expression analysis in parsley protoplasts revealed two *cis*-acting units that are important for UV-light-regulated promoter activity (Schulze-Lefert et al. 1989a,b). One unit was highly conserved among light-related genes, and the other unit was not conserved. The highly conserved footprinted region of the parsley *Chs* gene promoter is the G box (Schulze-Lefert et al. 1989a). A detailed function analysis of the G-box motif showed that it was crucial for responsiveness of the *Chs* gene to UV-light irradiation (Block et al. 1990)

The ABA-responsive *Em* gene promoter also has a unit of two *cis*-acting elements required for promoter activity analogous to the UV-light-inducible *Chs* promoter. The two elements within the promoter were identified on the basis of similarity with other ABA-regulated promoters (Marcotte et al. 1989). One of the elements, referred to as *Em1* (5′-ACACGTGG-3′), matches the G-box core sequence. To determine the function of the G box in the minimal *Em* promoter, transient gene expression assays were performed with reporter gene constructs (Guiltinan et al. 1990). Substitution of bases within the G-box sequence almost eliminated the ability of the ABA response element to activate gene expression in an ABA-dependent manner.

These data suggest that the G box works together with other *cis*-acting elements for the activation of a given promoter in response to a stimulus. In the simplest situation, the activity of these promoters could be influenced by protein-protein interaction: a G-box-binding protein with a protein binding to the surrounding *cis*-acting element. Elimination of one of the protein-binding sites would affect protein-protein interaction and in turn could destabilize the protein complex and/or interaction with RNA polymerase. This model would suggest that the spacing and placement of the G-box elements are major determinants of promoter activity in stress-related genes.

### G-box Motif in *Arabidopsis*

In *Arabidopsis*, an extended palindromic G box (5′-CCACGTGG-3′) motif is located in the *Adh* promoter (Ferl and Laughner 1989; DeLisle and Ferl 1990; McKendree et al. 1990; McKendree and Ferl 1992). Besides the hypoxic-inducible *Adh* gene, at least 10 different types of gene promoters in *Arabidopsis* contain the G box or the core motif (5′-CACGTG-3′). These genes include the light-regulated *RbcS* gene (Krebbers et al. 1988a), UV-light and fungal-pathogen-inducible chalcone synthase gene (*Chs*) (Feinbaum and Ausubel 1988), ethylene-induced 1-aminocyclopropane-1-carboxylate synthase (*AT-ACC1*) (Van der Straeten et al. 1992), light-regulated chlorophyll *a/b* binding protein gene (*cab1*) (Ha and An 1988), heat shock protein genes (*HS*) (Wu et al. 1988; Takahashi and Komeda 1989), development-regulated 2S albumin genes (*at2S*) (Krebbers et al. 1988b), mutant acetolactate synthase gene (*Als*) (Sathasivan et al. 1990), 5-endolpyruvylshikimate-3-phosphate synthase gene (*EPSP*) (Klee et al. 1987), light-regulated acyl-carrier protein gene (*Acp*) (Post-Beittenmiller et al. 1989), light-regulated pre-ferredoxin gene (Vorst et al. 1990), and an auxin-regulated gene (*Dpr*) (Alliotte et al. 1989). Although all of these genes have a G-box element in the promoters, they encode diverse proteins and are mediated by unrelated environmental cues.

In the *Arabidopsis RbcS-1A* promoter, the G box forms a transcriptional unit with another *cis*-acting element. Deletion analysis of the *RbcS-1A* defined a 196-bp region containing what were referred to as the GT box and I boxes, in addition to the G box, which were capable of conferring light-regulated and tissue-specific expression (Donald and Cashmore 1990). Deletion and mutations of these sequences and the subsequent expression analysis showed that both G-box and I-box mutations substantially reduced expression of the reporter genes (Donald and Cashmore 1990; Donald et al. 1990).

In the *Arabidopsis Adh* promoter, deletion of the G box dramatically reduces activity of the *Adh* promoter. Mutations within the G box that result in more than 60% reduction in activity also disrupt G-box-factor binding (McKendree and Ferl 1992). In *Arabidopsis*, the G box is a *cis*-acting element whose role is more general in promoter activity.

**Protein Interactions with the G Box: G-box-binding Factors**

Nuclear proteins binding to the G box have been found in *Arabidopsis* (Staiger et al. 1989) and a variety of mono- and dicotyledonous plants, including wheat (Guiltinan et al. 1990), maize (De Vetten et al. 1992), tobacco (Staiger et al. 1989), parsley (Armstrong et al. 1992), snapdragon, petunia, tomato, and pea (Giuliano et al. 1988). Interestingly, no differences or subtle differences in the complexes formed between the G box and nuclear protein were found upon subjection of plants or cells to various environmental stimuli (Guiliano et al. 1988; Staiger et al. 1989; Guiltinan et al. 1990; Armstrong et al. 1992). This suggests that transcriptional activation of these genes by the different environmental signals does not require de novo synthesis or major modification of G-box-binding activity. It is possible that in vivo differences in the G-box binding in response to environmental stimuli are the result of influences other than DNA-binding characteristics. For example, a change in chromatin structure has been suggested to account for differential G-box binding in vivo (McKendree et al. 1990). Nuclear proteins binding to the *Chs* G-box motif were isolated from tobacco seedlings using differential sequence-specific DNA affinity chromatography employing wild-type and mutated binding sites (Staiger et al. 1991). SDS polyacrylamide gel electrophoresis revealed polypeptides of 20, 30, and 40 kD that were highly enriched in the purified fraction. These polypeptides were also identified by DNA-protein crosslinking, suggesting the presence of multiple polypeptides binding to the *Chs* G box in tobacco (Staiger et al. 1991).

The multiplicity of *G-box-binding factors* (GBF) has been confirmed by molecular cloning studies. To date, several cDNA clones have been isolated that encode proteins which specifically interact with sequences containing the G-box core motif; these include *Arabidopsis* GBF-1,2,3 (Schindler et al. 1992a,b), wheat EmBP-1 (Guiltinan et al. 1990), tobacco TAF-1 (Oeda et al. 1991), and parsley CPRF-1,2,3 (Weisshaar et al. 1991). The deduced amino acid sequences of the clones reveal that they possess the characteristic features of the basic leucine zipper (bZIP) class of *trans*-acting factors. The bZIP proteins are characterized by their bipartite DNA-binding domain, which consists of a region enriched in basic amino acids and a leucine zipper that can mediate protein-protein interactions (Abel and Maniatis 1989). Amino acid sequence comparison between the cloned GBFs shows a high degree of similarity only in the basic domain, which is involved in DNA binding (Fig. 2).

The amino-terminal domains of the parsley CPRFs and wheat EMBP-1 are rich in proline residues. A number of transcription activators identified in mammalian cells also contain proline-rich regions (Mitchell and Tjian 1989). For GBFs, it was shown that the amino-terminal portion can stimulate transcription in plant protoplasts (Schindler et al. 1992c). Tobacco TAF-1 was also shown to function as a *trans*-activator in vivo (Oeda et al. 1991). However, instead of a proline-rich region, the tobacco transcription factor contains an acidic amino-terminal region that may serve as an acidic transcription activation domain.

### G-box-binding Factors in *Arabidopsis*

Multiple proteins with distinct G-box DNA-binding specificity have been identified in *Arabidopsis* extracts (DeLisle and Ferl 1990; McKendree et al. 1990). By UV-cross-linking and southwestern experiments, several proteins of 18–100 kD molecular mass have been shown to interact with the *Arabidopsis Adh* G box (DeLisle and Ferl 1990; Lu et al. 1992).

Four *Arabidopsis* cDNA clones encoding G-box or related binding proteins were isolated and referred to as GBF-l, GBF-2, GBF-3, and TGA-1 (Schindler et al. 1992a,b). The cDNA clones encode bZIP type

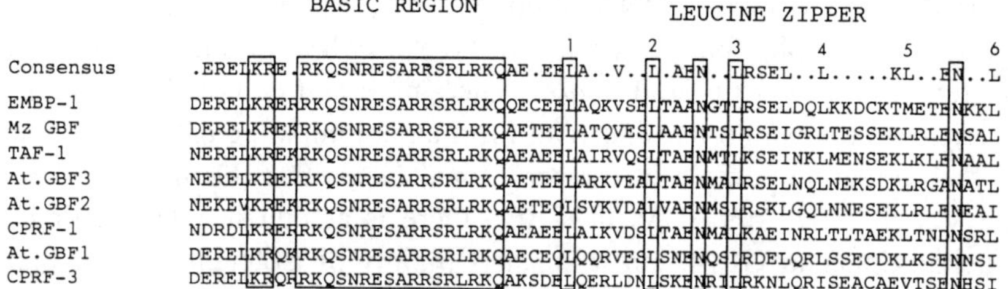


*Figure 2* Comparison of amino acid sequence of the basic domains and leucine repeats of plant GBF proteins. Conserved amino acids are boxed. Amino acids identical in at least 60% of the sequences are indicated on top as consensus sequence. Repeated leucine residues are numbered.

proteins, and the GBF proteins have an amino-terminal proline-rich domain. Two additional bZIP proteins, the HBP-lb homolog (Kawata et al. 1992) and FosF21 (Aeschbacher et al. 1991), have been found in *Arabidopsis*, but apparently they do not interact with the G box. The basic domains of the three *Arabidopsis* GBFs and other plant GBFs are conserved except in the leucine zipper region (Fig. 2). The proline-rich domain of *Arabidopsis* GBF-1, when fused to a heterologous DNA-binding domain, stimulates transcription in both plant protoplasts and mammalian cells (Schindler et al. 1992c). The three GBFs bind to the G-box motif as a homodimer or heterodimers with distinct binding properties. GBF-1-binding activity is stimulated by phosphorylation (Klimczak et al. 1992). GBF-l and GBF-2 are expressed in light- and dark-grown leaves as well as in roots, but GBF-3 is only expressed in dark-grown tissue. The GBF proteins possess both distinct DNA-binding specificity and expression characteristics, suggesting the proteins may individually mediate distinct subclasses of expression properties assigned to the G box (Schindler et al. 1992b,c; Williams et al. 1992).

## Proteins Associated with G-box Factors

Screening with oligonucleotides to the G box is a direct method for isolating DNA-binding factors that may bind to the motif. However, proteins with indirect DNA-binding activity (i.e., those proteins that do not bind to the DNA, but to the factors associated with the G-box motif) cannot be isolated by direct oligonucleotide screening. To recover cDNAs for proteins involved in G-box-protein complexes, monoclonal antibodies of the G-box-binding complex were used as a screening tool (De Vetten et al. 1992; Lu et al. 1992). By this technique, several different cDNA clones, referred to as GF14, have been isolated from

*Arabidopsis* and maize cDNA libraries. The GF14 is apparently part of the G-box complex but does not bind to the G box in and of itself (De Vetten et al. 1992; Lu et al. 1992). The amino and carboxyl termini are acidic, and the amino terminus can form an amphipathic α helix, similar to the activation domain of transcriptional factor Fos (Chiu et al. 1988; De Vetten et al. 1992). As in GBFs, there is a sequence resembling a leucine zipper in GF14, but it consists mainly of isoleucine rather than leucine and is not preceded by a basic domain, which is present in bZIP proteins (Lu et al. 1992).

GF14 may be operationally similar to Fos (Chiu et al. 1988; Abel and Maniatis 1989), possessing a dimerization zipper and acidic activation domain but being able to bind DNA only in the presence of heterologous partner protein, such as the *Arabidopsis* GBFs or the 69-kD proteins detected by southwestern assays (Lu et al. 1992). A search of the Gen-Bank database found no similarity with any proteins known to interact with DNA or participate in protein/DNA complexes. Interestingly, GF14 has more than 60% identity with a small class of proteins from mammalian brains identified as kinase-dependent activators of the tryptophan and tyrosine hydroxylase, termed 14-3-3 proteins (Ichimura et al. 1988), and inhibitors of protein kinase C (KCIP) (Aitken et al. 1991) (Fig. 3). The activation domain of 14-3-3 protein for tyrosine and tryptophan hydroxylase appears to be in the carboxy-terminal acidic region (Ichimura et al. 1988). The acidic carboxy-terminal region is conserved in GF14s and may have a transcriptional activating role in gene expression. In addition, GF14s contain putative protein kinase A and C recognition sites (De Vetten et al. 1992; Lu et al. 1992), and these sites are thought to be involved in the inhibitory sequences in the regulatory domains of protein kinase C in mammals (Aitken et al. 1991).

Other functions of the GF14/14-3-3 proteins in several systems have been addressed. Human 14-3-3 protein binds $Ca^{++}$ with high affinity and mediates phospholipolysis (Zupan et al. 1992). Members of the 14-3-3 family, termed Exo 1 and Exo 2 from brain cytosol, stimulate $Ca^{++}$-dependent exocytosis in permeabilized adrenal chromaffin (Morgan and Burgoyne 1992). This function was thought to be related to the stimulatory effect of 14-3-3 protein on the protein kinase C (Isobe et al. 1992). The *Drosophila* 14-3-3 protein homolog may be related to embryogenesis (Swanson and Ganguly 1992), and yeast 14-3-3 protein has a function in growth regulation (Van Heusden et al. 1992). 14-3-3 homologs are involved in secretions from the pituitary gland of *Xenopus* during background adaptation (Martens et al. 1992) and contribute to maintaining $Ca^{++}$-dependent noradrenaline secretion in bovine chromaffin cells (Wu et al. 1992). A 14-3-3 homolog from barley was induced by

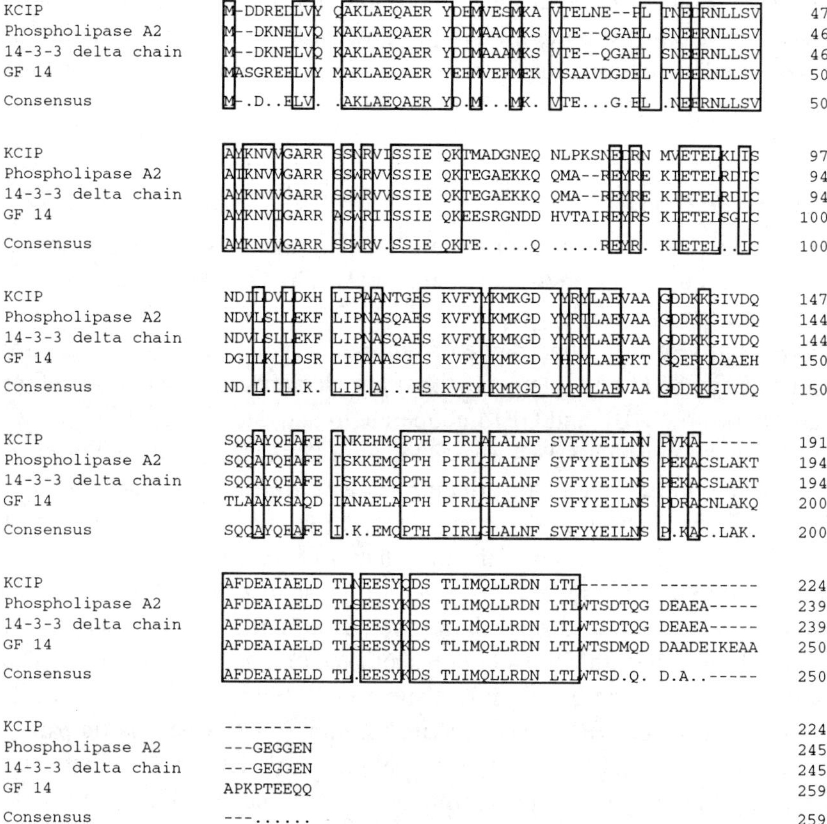

*Figure 3*  Alignment of the amino acid sequences (single-letter notation) of *Arabidopsis* GF14 (Lu et al. 1992) with 14-3-3 protein, KCIP, and phospholipase A2. 14-3-3 delta chain activates protein kinase C (Isobe et al. 1992); KCIP is an inhibitor of protein kinase C (Aitken et al. 1991); and phospholipase A2 binds Ca$^{++}$ with high affinity (Zupan et al. 1992). Residues identical to *Arabidopsis* GF14 are boxed.

inoculation with an avirulent isolate of *Erysiphe graminis* (Brandt et al. 1992). Although the pea 14-3-3 protein has an inhibitory effect on protein kinase C (Hirsch et al. 1992), recent observations indicate the GF14 homolog from *Arabidopsis* binds Ca$^{++}$ and activates the mammalian tryptophan hydroxylase and protein kinase C (G. Lu et al., unpubl.). These results indicate that GF14/14-3-3 proteins are a family of highly conserved proteins which serve a broad set of perhaps interrelated functions involved in Ca$^{++}$ and protein kinase.

The physical association of these regulatory proteins with the GBF activity may present a means for the signal transduction pathway to communicate with the G box. Since G-box-binding factors are known to be phosphorylated by a protein kinase (Klimczak et al. 1992), regulation of kinase-mediated activation would be a logical modulation point in the pathway.

### Summary of G Box

In summary, the G-box element is present in many types of genes that respond to numerous environmental signals. The G-box element, G-box-binding activity, GBFs, and GF14s (14-3-3 proteins) exist in *Arabidopsis*, but also in a wide range of other eukaryotic organisms. It is not known how the G box interacts with other elements to enhance promoter activity or how GBF and GF14 cooperate to activate transcription. However, it is known that GF14 associates with the G-box-protein complex, is a protein kinase regulator protein, has $Ca^{++}$-binding affinity, and that GBF activity is stimulated by phosphorylation. These facts suggest that GF14 and GBFs may play an important role in environmental signal transduction. In the previous sections, we presented several environmental stresses and showed how these stresses affected gene regulation. In many of the genes, there is a G-box sequence or a core G-box motif. The flanking sequence of the G box, distinct GBF-binding activity, and different isoforms of GF14 may account for conserved G-box participation in the diverse signal transduction response to a variety of environmental stresses. However, it should be remembered that the G box is not the sole determinant of environmental activation of genes.

### CONCLUSIONS AND PERSPECTIVES

In *Arabidopsis*, as in other species, the relationship between stress and gene regulation involves a complex set of components beginning with the environmental signal and ending in production of proteins that lead to survival. Numerous genes that are influenced by diverse environmental stresses contain a common motif, the G box. The G-box motif, G-box-binding factors, and the GF14 proteins associated with the binding factors may be involved in the environmental stress signal transduction pathway. However, in each case, there appear to be separate stress-specific signals mediated by unique promoter elements.

Several of the genes discussed in this chapter (e.g., *lea, rab, dhn, cor*) have been identified and sequenced, but the roles these genes play in plants are unclear. For example, it has been suggested that certain ABA-responsive genes encode RNA regulatory proteins (Skriver and Mundy 1990) and a salt-induced gene encodes a lipid transfer protein (Torres-

Shumann et al. 1992). Little is known about the mechanisms that regulate the expression of many of these genes. In addition, many of the genes reviewed in this chapter are named not on an evolutionary or functional basis, but from isolation procedures or the type of stress that induced them. With numerous names for the genes, it is becoming evident that a nomenclature based on evolutionary relationships and biological functions would be helpful. In addition, the proper nomenclature could be used to classify respective genes and their functions. Finally, many environmental stresses, such as ionic toxicity from heavy metals, air pollutants, and acid rain, have been researched in other plants, but little or no data exist for *Arabidopsis*. Identification of the genes and transcription factors associated with these stresses and the ways in which they are regulated in *Arabidopsis* is needed. Ultimately, knowledge of how genes are regulated in response to environmental conditions may aid in our understanding of the connection between environmental stress signals and evolutionary continuity of plant genes.

## ACKNOWLEDGMENTS

Support for the authors was provided by the National Institutes of Health (grants GM-39732 and GM-40061), U.S. Department of Agricultural-National Research Initiative (grants 93-37304-9608 and 91-37301-6372), and the Spanish Ministry for Education and Sciences.

## REFERENCES

Abel, T. and T. Maniatis. 1989. Action of leucine zipper. *Nature* **341:** 24–25.

Abravaya, K., M.P. Myers, S.P. Murphy, and R.I. Morimoto. 1992. The human heat shock protein hsp70 interacts with HSF, the transcription factor that regulates heat shock gene expression. *Genes Dev.* **6:** 1153–1164.

Aeschbacher, R.A., M. Schrott, I. Potrykus, and M.W. Saul. 1991. Isolation and molecular characterization of *FosF21*, an *Arabidopsis thaliana* gene which shows characteristics of a b-Zip class transcription factor. *Plant J.* **1:** 303–316.

Aitken, A., C.A. Ellis, A. Harris, L.A. Sellers, and A. Toker. 1991. Kinase and neurotransmitters. *Nature* **344:** 594.

Aldrich, H.C., R.J. Ferl, M.H. Hils, and D.E. Akin. 1985. Ultrastructure correlates of anaerobic stress in corn roots. *Tissue Cell* **17:** 341–348.

Alliotte, T., C. Tire, G. Engler, J. Peleman, A. Caplan, M.V. Montagu, and D. Inze. 1989. An auxin-regulated gene of *Arabidopsis thaliana* encoding a DNA-binding protein. *Plant Physiol.* **89:** 743–752.

Amin, J., J. Ananthan, and R. Voellmy. 1988. Key features of heat shock regulatory elements. *Mol. Cell. Biol.* **8:** 3761–3769.

Armstrong, A.M., B. Weisshaar, and K. Hahlbrock. 1992. Homodimeric and heterodimeric leucine zipper proteins and nuclear factors from parsley recognize diverse promoter elements with ACGT cores. *Plant Cell* **4:** 525–537.

Baker, J., C. Steele, and L. Dure III. 1988. Sequence and characterization of 6 LEA proteins and their genes from cotton. *Plant Mol. Biol.* **11:** 277–291.

Barros, M.D., E. Czarnecka, and W.B. Gurley. 1992. Mutational analysis of a plant heat shock element. *Plant Mol. Biol.* **19:** 665–675.

Bartels, D., K. Scheneider, G. Terstappen, D. Piatkowski, and F. Salamini. 1990. Molecular cloning of abscisic acid-modulation genes which are induced during desiccation of the resurrection plant *Craterostigma plantagineum*. *Planta* **181:** 27–34.

Bartling, B., H. Bütler, K. Liebeton, and E.W. Weiler. 1992. An *Arabidopsis thaliana* cDNA clone encoding a 17.6 kDa class II heat shock protein. *Plant Mol. Biol.* **18:** 1007–1008.

Beckmann, R.P., L.A. Mizzen, and W.J. Welch. 1990. Interactions of hsp70 with newly synthesized proteins: Implications for protein folding and assembly. *Science* **248:** 850–854.

Berstein, H.D., M.A. Poritz, K. Strub, P.J. Hoben, S. Brenner, and P. Walter. 1989. Model for signal sequence recognition from amino acid sequence of 54K subunit of signal recognition particle. *Nature* **340:** 482–486.

Binelli, G. and J.P. Mascarenhas. 1990. *Arabidopsis*: Sensitivity of growth to high temperature. *Dev. Genet.* **11:** 294–298.

Binet, M., J. Weil, and L. Tessier. 1991. Structure and expression of sunflower ubiquitin genes. *Plant Mol. Biol.* **17:** 395–407.

Binzel, M.L., P.M. Hasegawa, D. Rhodes, S. Handa, A.K. Handa, and R.A. Bressan. 1987. Solute accumulation of tobacco cells adapted to NaCl. *Plant Physiol.* **84:** 1408–1415.

Block, A., J.L. Dangl, K. Halhbrock, and P. Schulze-Lefert. 1990. Functional borders, genetic fine structure, and distance requirements of *cis*-elements mediating light responsiveness of the parsley chalcone synthase promoter. *Proc. Natl. Acad. Sci.* **87:** 5387–5391.

Borkovich, K.A., F.W. Farrelly, D.B. Finkelstein, J. Taulien, and S. Lindquist. 1989. Hsp82 is an essential protein that is required in higher concentrations for growth of cells at higher temperatures. *Mol. Cell. Biol.* **9:** 3919–3930.

Braam, J. 1992. Regulated expression of the calmodulin-related *TCH* genes in cultured *Arabidopsis* cells: Induction by calcium and heat shock. *Proc. Natl. Acad. Sci.* **89:** 3213–3216.

Brandt, J., H. Thordal-Christensen, K. Vad, P.L. Gregersen, and D.B. Collinge. 1992. A pathogen-induced gene of barley encodes a protein show high similarity to a protein kinase regulator. *Plant J.* **2:** 815–820.

Bruns, B., K. Hahlbrock, and E. Schafer. 1986. Fluence dependence of the ultraviolet-light-induced accumulation of chalcone synthase mRNA and effects of blue and far-red light in cultured parsley cells. *Planta* **169:** 393–398.

Burke, T.J., J. Callis, and R.D. Vierstra. 1988. Characterization of a polyubiquitin gene from *Arabidopsis thaliana*. *Mol. Gen. Genet.* **213:** 435–443.

Campos, J.L., X. Figueras, M.T. Pinol, A. Boronat, and A.F. Tiburcio. 1991. Carotenoid and conjugated polyamine levels as indicators of ultraviolet-C induced stress in *Arabidopsis thaliana*. *Photochem. Photobiol.* **53:** 689–693.

Casey, R. and C. Domoney. 1987. The structure of plant storage protein genes. *Plant Mol. Biol. Rep.* **5:** 261–281.

Chang, C. and E.M. Meyerowitz. 1986. Molecular cloning and DNA sequence of the *Arabidopsis thaliana* alcohol dehydrogenase gene. *Proc. Natl. Acad. Sci.* **83:** 1408–1412.

Chappell, J. and K. Hahlbrock. 1984. Transcription of plant defense genes in response to

UV light or fungal elicitor. *Nature* **311:** 76–78.

Chen, H.H., P.H. Li, and M.L. Brenner. 1983. Involvement of abscisic acid in potato cold acclimation. *Plant Physiol.* **71:** 362–365.

Chen, P.M., P.H. Li, and M.J. Burke. 1977. Induction of frost hardiness in stem cortical tissues of *Cornus stolonifera* Michx. by water stress. *Plant Physiol.* **59:** 236–239.

Chen, Q. and E. Vierling. 1991. Analysis of conserved domains identifies a unique structural feature of a chloroplast heat shock protein. *Mol. Gen. Genet.* **226:** 425–431.

Chen, T.H.H. and L.V. Gusta. 1983. Abscisic acid induced freezing resistance in cultured plant cells. *Plant Physiol.* **73:** 71–75.

Chiu, R., W.J. Boyle, J. Meek, T. Smeal, T. Hunter, and M. Karin. 1988. The C-Fos protein interacts with c-Jun/AP-l to stimulate transcription of AP-1 responsive genes. *Cell* **54:** 541–552.

Chodosh, L.A., S. Buratowski, and P.A. Sharp. 1989. A yeast protein possesses the DNA-binding properties of the adenovirus major late transcription factor. *Mol. Cell. Biol.* **9:** 820–822.

Christensen, A.H. and P.H. Quail. 1989. Sequence analysis and transcriptional regulation by heat shock of polyubiquitin transcripts from maize. *Plant Mol. Biol.* **12:** 619–632.

Clos, J., J.T. Westwood, P.B. Becker, S. Wilson, K. Lambert, and C. Wu. 1990. Molecular cloning and expression of a hexameric *Drosophila* heat shock factor subject to negative regulation. *Cell* **63:** 1085–1097.

Close, T.J., A.A. Kortt, and P.M. Chandler. 1989. A cDNA-based comparison of dehydration-induced proteins (dehydrins) in barley and corn. *Plant Mol. Biol.* **13:** 95–108.

Crawford, R.M.M. 1978. Metabolic adaptations to anoxia. In *Plant life in anaerobic environments* (ed. D.D. Hook and R.M.M. Crawford), pp. 119–137. Ann Arbor Scientific, Ann Arbor, Michigan.

Czarnecka, E., P.C. Fox, and W.B. Gurley. 1990. *In vitro* interaction of nuclear proteins with the promoter of soybean heat shock gene *Gmhsp17.5E*. *Plant Physiol.* **94:** 935–943.

Czarnecka, E., J.L. Key, and W.B. Gurley. 1989. Regulatory domains of the *Gmhsp17.5E* heat shock promoter of soybean: Mutational analysis. *Mol. Cell. Biol.* **9:** 3457–3463.

Czarnecka-Verner, E., M.D. Barros, and W.B. Gurley. 1994. Regulation of plant heat shock gene expression. In *Stress-induced gene expression* (ed. A.S. Basra), pp. 131–161. Harwood Academic, Switzerland.

DeLisle, A.J. and R.J. Ferl. 1990. Characterization of the *Arabidopsis Adh* G-box binding factor. *Plant Cell* **2:** 547–557.

De Vetten, N.C., G. Lu, and R.J. Ferl. 1992. A maize protein associated with the G-box binding complex has homology to brain regulatory proteins. *Plant Cell* **4:** 1295–1307.

Dolferus, R. and M. Jacobs. 1984. Polymorphism of ADH in *Arabidopsis thaliana* (L.) Heynh: Genetic and biochemical characterization. *Biochem. Genet.* **22:** 817–838.

———. 1991. Another ADH system—From *Arabidopsis thaliana*: An overview. *Maydica* **36:** 169–187.

Dolferus, R., G. Marbaix, and M. Jacobs. 1985. ADH in *Arabidopsis*: Analysis of the induction phenomenon in plantlets and tissue cultures. *Mol. Gen. Genet.* **199:** 256–302.

Donald, R.G.K. and A.R. Cashmore. 1990. Mutation of either G box or I box sequences profoundly affects expression from the *Arabidopsis rbcS-1 A* promoter. *EMBO J.* **9:** 1717–1726.

Donald, R.G.K., U. Schindler, A. Batschauer, and A.R. Cashmore. 1990. The plant G box promoter sequence activates transcription in *Saccharomyces cerevisiae* and is bound *in vitro* by a yeast activity similar to GBF, the plant G box binding factor. *EMBO J.* **9:**

1727–1735.

Drew, M.C., M.B. Jackson, and S.C. Giffard. 1979. Ethylene-promoted adventitious rooting and development of cortical air spaces (aerenchyma) in roots may be adaptive responses to flooding in *Zea mays* L. *Planta* **147:** 83–86.

Drew, M.C., M.B. Jackson, S.C. Giffard, and R. Campbell. 1981. Inhibition by silver ions of gas space (aerenchyma) formation in adventitious roots of *Zea mays* L. subjected to exogenous ethylene or to oxygen deficiency. *Planta* **153:** 217–224.

Duell-Pfaff, N. and E. Wellmann. 1982. Involvement of phytochrome and a blue light photoreceptor in UV-B induced flavonoid synthesis in parsley (*Petroselinum hortense* Hoffm.) cell suspension cultures. *Planta* **156:** 213–217.

Dupuis, I. and C. Dumas. 1990. Influence of temperature stress on *in vitro* fertilization and heat shock protein synthesis in maize (*Zea mays* L.) reproductive tissues. *Plant Physiol.* **94:** 665–670.

Dure L., III, M. Crouch, J. Harada, T.D. Ho, J. Mundy, R. Quatrano, T. Thomas, and Z.R. Sung. 1989. Common amino acid sequence domains among LEA proteins of higher plants. *Plant Mol. Biol.* **12:** 475–486.

Dynan, W.S. 1989. Modularity in promoters and enhancers. *Cell* **58:** 1–4.

Ellis, R.J. 1990. Molecular chaperones: The plant connection. *Science* **250:** 954–959.

Ellis, J.G., D.J. Llewellyn, E.S. Dennis, and W.J. Peacock. 1987. Maize promoter sequences control anaerobic regulation: Addition of upstream promoter elements from constitutive genes is necessary for expression in tobacco. *EMBO J.* **6:** 11–16.

Feinbaum, R.L. and F.M. Ausubel. 1988. Transcriptional regulation of *Arabidopsis thaliana* chalcone synthase gene. *Mol. Cell. Biol.* **8:** 1985–1992.

Feinbaum, R.L., G. Storz, and F.M. Ausubel. 1991. High intensity and blue light regulated expression of chimeric chalcone synthase genes in transgenic *Arabidopsis thaliana* plants. *Mol. Gen. Genet.* **226:** 449–456.

Ferl, R.J. and B.H. Laughner. 1989. *In vivo* detection of regulatory factor binding sites of *Arabidopsis thaliana Adh. Plant Mol. Biol.* **12:** 357–366.

Ferl, R.J. and H.S. Nick. 1987. *In vivo* detection of regulatory factor binding sites in the 5′ flanking region of maize *Adh1. J. Biol. Chem.* **262:** 7947–7950.

Finkelstein, R.R. and C. Somerville. 1990. Three classes of abscisic acid (ABA)-insensitive mutations of *Arabidopsis* define genes that control overlapping subsets of ABA responses. *Plant Physiol.* **94:** 1172–1179.

Galau, G.A. and T.J. Close. 1992. Sequences of the cotton 2 LEA/RAB/Dehydrin proteins encoded by *lea3* cDNAs. *Plant Physiol.* **98:** 1523–1525.

Georgopoulos, C.P., R.N. Hendrix, S.R. Casjens, and A.D. Kaiser. 1973. Host participation in bacteriophage K head assembly. *J. Mol. Biol.* **76:** 45–60.

Gilmour, S.J., N.N. Artus, and M.F. Thomashow. 1992. cDNA sequence analysis and expression of two cold-regulated genes of *Arabidopsis thaliana. Plant Mol. Biol.* **18:** 13–21.

Giuliano, G., E. Pichersky, V.S. Malik, M.P. Timko, P.A. Scolnik, and A.R. Cashmore. 1988. An evolutionarily conserved protein binding sequence upstream of a plant light-regulated gene. *Proc. Natl. Acad. Sci.* **85:** 7089–7093.

Godoy, J.A., J.M. Pardo, and J.A. Pintor-Toro. 1990. A tomato cDNA inducible by salt stress and abscisic acid: Nucleotide sequence and expression pattern. *Plant Mol. Biol.* **15:** 695–705.

Good, A.G. and D.H. Paetkau. 1992. Identification and characterization of a hypoxically induced maize lactate dehydrogenase gene. *Plant Mol. Biol.* **19:** 693–697.

Greene, J.M. and R.E. Kingston. 1990. TATA-dependent and TATA-independent function of the basal and heat shock elements of a human *hsp70* promoter. *Mol. Cell. Biol.*

**10:** 1319–1328.

Guiltinan, M.J., W.R. Marcotte, and R.S. Quatrano. 1990. A plant leucine zipper protein that recognizes an abscisic acid response element. *Science* **250:** 267–271.

Gurley, W.B. and J.L. Key. 1991. Transcriptional regulation of the heat-shock response: A plant perspective. *Biochemistry* **30:** 1–12.

Guy, C., D. Haskell, L. Niven, P. Klein, and C. Smelser. 1992. Hydration-state-responsive proteins link cold and drought stress in spinach. *Planta* **188:** 265–270.

Ha, S.-B. and G. An. 1988. Identification of upstream regulatory elements involved in the developmental expression of the *Arabidopsis thaliana cab1* gene. *Proc. Natl. Acad. Sci.* **85:** 8017–8021.

Hahlbrock, K. and H. Grisebach. 1979. Enzymatic controls in the biosynthesis of lignin and flavonoids. *Annu. Rev. Plant Physiol.* **30:** 105–130.

Hahlbrock, K. and D. Scheel. 1989. Physiology and molecular biology of phenyl-propanoid metabolism. *Annu. Rev. Plant Physiol. Plant Mol. Biol.* **40:** 347–369.

Hahn, M. and V. Walbot. 1989. Effects of cold-treatment on protein synthesis and mRNA levels in rice leaves. *Plant Physiol.* **91:** 930–938.

Hajela, R.K., D.P. Horvath, S.J. Gilmour, and M.F. Thomashow. 1990. Molecular cloning and expression of *cor* (cold-regulated) genes in *Arabidopsis thaliana*. *Plant Physiol.* **93:** 1246–1252.

Hanson, A.D. and W.D. Hitz. 1982. Metabolic responses of mesophytes to plant water deficits. *Annu. Rev. Plant Physiol.* **33:** 163–203.

Harberd, N.P. and K.J.R. Edwards. 1982. The effect of a mutation causing alcohol dehydrogenase deficiency on flooding tolerance in barley. *New Phytol.* **90:** 631–644.

Heino, P., G. Sandman, V. Lang, K. Nordin, and E.T. Palva. 1990. Abscisic acid deficiency prevents development of freezing tolerance in *Arabidopsis thaliana* (L.) Heynh. *Theor. Appl. Genet.* **79:** 801–806.

Helm, K.W., N.S. Petersen, and R.H. Abernethy. 1989. Heat shock response of germinating embryos of wheat. Effects of imbibition time and seed vigor. *Plant Physiol.* **90:** 598–605.

Hirayama, O. and M. Miara. 1987. Characterization of membrane lipids of higher plants differing in salt tolerance. *Agric. Biol. Chem.* **51:** 3215–3221.

Hirsch, S., A. Aitken, U. Bertsch, and J. Soll. 1992. A plant homologue to mammalian brain 14-3-3 protein and protein kinase C inhibitor. *FEBS Lett.* **296:** 222–224.

Hoffman, N.E. and A.D. Hanson. 1986. Purification and properties of hypoxically induced lactate dehydrogenase from barley roots. *Plant Physiol.* **82:** 664–670.

Hoffman, N.E., A.F. Bent, and A.D. Hanson. 1986. Induction of lactate dehydrogenase isozymes by oxygen deficit in barley root tissue. *Plant Physiol.* **82:** 658–663.

Howard E.A., J.C. Walker, E.S. Dennis, and W.J. Peacock. 1987. Regulated expression of an *alcohol dehydrogenase-1* chimeric gene introduced into maize protoplasts. *Planta* **170:** 535–540.

Hughes, D.W. and G.A. Galau. 1991. Developmental and environmental induction of *Lea* and *LeaA* mRNAs and the post abscission program during embryo culture. *Plant Cell* **3:** 605–618.

Ichimura, T., T. Isobe, T. Okuyama, N. Takahashi, K. Araki, R. Kuwano, and Y. Takahashi. 1988. Molecular cloning of cDNA coding brain-specific 14-3-3 protein, a protein kinase-dependent activator of tyrosine and tryptophan hydroxylase. *Proc. Natl. Acad. Sci.* **85:** 7084–7088.

Isobe, T., Y. Hiyane, T. Ichimura, T. Okuyama, N. Takahashi, S. Nakajo, and K. Nakaya. 1992. Activation of protein kinase C by the 14-3-3 proteins homologous with Exo1 protein that stimulates calcium dependent exocytosis. *FEBS Lett.* **308:** 121–124.

Jackson, M.B. 1984. Regulation of root growth and morphology by ethylene and other externally applied growth substances. In *Growth regulators in root development* (ed. M.B. Jackson and A.D. Stead), pp. 103–116. British Plant Growth Regulator Group, London.

Jackson, M.B., M.C. Drew, and S.C. Giffard. 1981. Effects of applying ethylene to the root system of *Zea mays* on growth and nutrient concentration in relation to flooding tolerance. *Physiol. Plant* **52:** 23–28.

Jacobs, M., R. Dolferus, and D. van den Bossche. 1988. Isolation and biochemical analysis of ethyl methane sulfonate-induced ADH null mutants of *Arabidopsis thaliana*. *Biochem. Genet.* **26:** 105–122.

Jakobsen, B.K. and H.R.B. Pelham. 1988. Constitutive binding of yeast heat shock factor to DNA *in vivo*. *Mol. Cell. Biol.* **8:** 5040–5042.

———. 1991. A conserved heptapeptide restrains the activity of the yeast heat shock transcription factor. *EMBO J.* **10:** 369–375.

Jurivich, D.A., L. Sistonen, R.A. Kroes, and R.I. Morimoto. 1992. Effect of sodium salicylate on the human heat shock response. *Science* **255:** 1243–1245.

Kawata, T., T. Imada, H. Shiraishi, K. Okada, Y. Shimura, and M. Iwabuchi. 1992. A cDNA clone encoding HBP-1b homologue in *Arabidopsis thaliana*. *Nucleic Acids Res.* **20:** 1141.

Kelley, P.M. 1989. Maize pyruvate decarboxylase mRNA is induced anaerobically. *Plant Mol. Biol.* **13:** 213–222.

Kennedy, R.A., M.E. Rumpho, and T.C. Fox. 1992. Anaerobic metabolism in plants. *Plant Physiol.* **100:** 1–6.

Kermode, A.R. and J.D. Bewley. 1987. Regulatory processes involved in the switch from seed development to germination: Possible roles for desiccation and ABA. In *Drought resistance in plants: Physiological and genetic aspects* (ed. L. Monti and E. Porceddu), pp. 59–76. EEC, Brussels.

Kimmer, T.W. 1987. Alcohol dehydrogenase and pyruvate decarboxylase activity in leaves and roots of eastern cottonwood (*Populus deltoides* Bartr.) and soybean (*Glycine max* L.). *Plant Physiol.* **84:** 1210–1213.

Kimpel, J.A. and J.L. Key. 1985. Heat shock in plants. *Trends Biochem. Sci.* **10:** 353–357.

King, S.W., C.P. Joshi, and H.T. Nguyen. 1992. DNA sequence of an ABA-responsive gene (*rab 15*) from water-stressed wheat roots. *Plant Mol. Biol.* **18:** 119–121.

Kingston, R.E., T.J. Schuetz, and Z. Larin. 1987. Heat-inducible human factor that binds to a human hsp70 promoter. *Mol. Cell. Biol.* **7:** 1530–1534.

Klee, H.J., Y.M. Muskopf, and C.S. Gasser. 1987. Cloning of an *Arabidopsis thaliana* gene encoding 5-endolpyruvylshikimate-3 phosphate synthase: Sequence analysis and manipulation to obtain glyphosate-tolerant plants. *Mol. Gen. Genet.* **210:** 437–442.

Klimczak, L.J., U. Schindler, and A.R. Cashmore. 1992. DNA binding activity of the *Arabidopsis* G-box binding GBF 1 is stimulated by phosphorylation by casein kinase II from broccoli. *Plant Cell* **4:** 87–96.

Klyueva, N.Y. and I.M. Samokhvalov. 1991. Synthesis of heat shock proteins in leaves of *Arabidopsis thaliana*. *Sov. Plant Physiol.* **37:** 564–571.

Koes, R.E., C.E. Spelt, H.J. Reif, P.J.M. van den Elzen, E. Veltkamp, and J.N.M. Mol. 1986. Floral tissue of *Petunia hybrida* (V30) expresses only one member of the chalcone synthase multigene family. *Nucleic Acids Res.* **14:** 5229–5239.

Koornneef, M., C.J. Hanhart, H.W. Hilhorst, and C.M. Karssen. 1989. *In vivo* inhibition of seed development and reserve protein accumulation in recombinants of abscisic acid biosynthesis and responsiveness mutants in *Arabidopsis thaliana*. *Plant Physiol.* **90:**

463–469.

Krebbers, E., J. Seurinck, L. Herdies, A.R. Cashmore, and M.P. Timko. 1988a. Four genes in 2 diverged subfamilies encode the ribulose 1,5-bisphosphate carboxylase small subunit polypeptides of *Arabidopsis thaliana. Plant Mol. Biol.* **11:** 745–759.

Krebbers, E., L. Herdies, A. DeClercq, J. Seurinck, J. Leemans, J. Van Damme, M. Segura, G. Gheysen, M. Van Montagu, and J. Vandekerckhove. 1988b. Determination of the processing sites of an *Arabidopsis* 2S albumin and characterization of the complete gene family. *Plant Physiol.* **87:** 859–866.

Kubasek, W.L., B.W. Shirley, A. McKillop, H.M. Goodman, W. Briggs, and F.M. Ausubel. 1992. Regulation of flavonoid biosynthetic genes in germinating *Arabidopsis* seedlings. *Plant Cell* **4:** 1229–1236.

Kurkela, S. and M. Frank. 1990. Cloning and characterization of a cold- and ABA-inducible *Arabidopsis* gene. *Plant Mol. Biol.* **15:** 137–144.

LaRosa, P.C., P.M. Hasegawa, D. Rhodes, J.M. Clithero, A. Watad, and R.A. Bressan. 1987. Abscisic acid stimulated osmotic adjustment and its involvement in adaptation of tobacco cells to NaCl. *Plant Physiol.* **85:** 174–181.

Laszlo, A. and P. St. Lawrence. 1983. Parallel induction and synthesis of PDC and ADH in anoxic maize roots. *Mol. Gen. Genet.* **192:** 110–117.

Lee, L., C. Fenoll, and J.L. Bennetzen. 1987. Construction and homologous expression of a maize based *Nco*I cassette vector. *Plant Physiol.* **85:** 327–330.

Levitt, J. 1980. *Responses of plants to environmental stress. Chilling, freezing, and high temperature stresses*, 2nd edition. Academic Press, New York.

Lu, G., A.J. DeLisle, N. C. de Vetten, and R.J. Ferl. 1992. Brain proteins in plants: An *Arabidopsis* homologue to neurotransmitter pathway activators is part of a DNA binding complex. *Proc. Natl. Acad. Sci.* **89:** 11490–11494.

Marcotte, W.R., S.H. Russell, Jr., and R.S. Quatrano. 1989. Abscisic acid responsive sequence from the *Em* gene of wheat. *Plant Cell* **1:** 969–976.

Martens, G.J.M., P.A. Piosik, and E.H.J. Danen. 1992. Evolutionary conservation of the 14-3-3 protein. *Biochem. Biophys. Res. Commun.* **184:** 1456–1459.

McKendree, W.L. and R.J. Ferl. 1992. Functional elements of the *Arabidopsis Adh* promoter include the G-box. *Plant Mol. Biol.* **19:** 859–862.

McKendree, W.L., A.-L. Paul, A.J. DeLisle, and R.J. Ferl. 1990. *In vivo* and *in vitro* characterization of protein interactions with the dyad G-box of the *Arabidopsis Adh* gene. *Plant Cell* **2:** 207–214.

Meurs, C., A.S. Basra, C.M. Karssen, and L.C. van Loon. 1992. Role of abscisic acid in the induction of desiccation tolerance in developing seeds of *Arabidopsis thaliana. Plant Physiol.* **98:** 1484–1493.

Mirault, M.E., R. Southgate, and E. Delwart. 1982. Regulation of heat-shock genes: A DNA sequence upstream of *Drosophila* hsp70 genes is essential for their induction in monkey cells. *EMBO J.* **1:** 1279–1285.

Mitchell, P.J. and R. Tjian. 1989. Transcriptional regulation in mammalian cells by sequence-specific DNA binding proteins. *Science* **245:** 371–378.

Morgan, A. and R.D. Burgoyne. 1992. Exo1 and Exo2 proteins stimulate calcium-dependent exocytosis in permeabilized adrenal chromaffin cells. *Nature* **355:** 833–836.

Mosser, D.D., P.T. Kotzbauer, K.D. Sarge, and R.I. Morimoto. 1990. *In vitro* activation of heat shock transcription factor DNA-binding by calcium and biochemical conditions that affect protein conformation. *Proc. Natl. Acad. Sci.* **87:** 3748–3752.

Mundy, J. and N.-H. Chua. 1988. Abscisic acid and water-stress induce the expression of a novel rice gene. *EMBO J.* **7:** 2279–2286.

Nagao, R.T., J.A. Kimpel, E. Vierling, and J.L. Key. 1986. The heat shock response: A comparative analysis. *Oxf. Surv. Plant Mol. Cell Biol.* **3**: 384–438.

Neumann, D., L. Nover, B. Parthier, R. Rieger, K.-D. Scharf, R. Wollgiehn, and U.Z. Nieden. 1989. Heat shock and other stress response systems of plants. *Biol. Zentralbl.* **108**: 1–156.

Neumann, P.M., E.V. Volkenburg, and R.E. Cleland. 1988. Salinity stress inhibits bean leaf expansion by reducing turgor, not wall extensibility. *Plant Physiol.* **88**: 233–237.

Nordin, K., P. Heino, and E.T. Palva. 1991. Separate signal pathways regulate the expression of a low-temperature-induced gene in *Arabidopsis thaliana* (L.) Heynh. *Plant Mol. Biol.* **16**: 1061–1071.

Oeda, K., J. Salinas, and N.-H. Chua. 1991. A tobacco bZIP transcription factor (TAF-1) binds to a G-box-like motif conserved in plant genes. *EMBO J.* **10**: 1793–1802.

Ohl, S., K. Hahlbrock, and E. Schafer. 1989. A stable blue-light-derived signal modulates ultraviolet-light-induced activation of the chalcone-synthase gene in cultured parsley cells. *Planta* **177**: 228–236.

Ohl, S., S.A. Hedrick, J. Chory, and C.J. Lamb. 1990. Functional properties of a phenylalanine ammonia-lyase promoter from *Arabidopsis*. *Plant Cell* **2**: 837–848.

Olive, M.R., J.C. Walker, K. Singh, E.S. Dennis, and J.W. Peacock. 1990. Functional properties of the anaerobic response element of the maize gene. *Plant Mol. Biol.* **15**: 593–604.

Paternak, D. 1987. Salt tolerance and crop production—A comprehensive approach. *Annu. Rev. Phytopathol.* **25**: 271–291.

Paul, A.-L. and R.J. Ferl. 1991. *In vivo* footprinting reveals unique *cis*-elements and different modes of hypoxic induction in maize *Adh1* and *Adh2*. *Plant Cell* **3**: 159–168.

Pelham, H.R.B. 1982. A regulatory upstream promoter element in the *Drosophila hsp70* heat-shock gene. *Cell* **30**: 517–528.

———. 1990. Functions of the hsp70 protein family: An overview. In *Stress proteins in biology and medicine* (ed. R.I. Morimoto et al.), pp. 287–299. Cold Spring Harbor Laboratory Press, Cold Spring Harbor, New York.

Pelham, H.R.B. and M. Bienz. 1982. A synthetic heat-shock promoter element confers heat-inducibility on the herpes simplex virus thymidine kinase gene. *EMBO J.* **1**: 1473–1477.

Picard, D., B. Khursheed, M.J. Garabedian, M.G. Fortin, S. Lindquist, and K.R. Yamamoto. 1990. Reduced levels of hsp90 compromise steroid receptor action *in vivo*. *Nature* **348**: 166–168.

Pla, M., J. Gomez, A. Goday, and M. Pages. 1991. Regulation of the abscisic acid-responsive gene *rab28* in maize viviparous mutants. *Mol. Gen. Genet.* **230**: 394–400.

Pla, M., J. Vilardell, M.J. Guiltinana, W.R. Marcotte, M.F. Niogret, R.S. Quatrano, M. Pages. 1993. The *cis*-regulatory element CCACGTGG is involved in ABA and water-stress responses of the maize gene *rab28*. *Plant Mol. Biol.* **21**: 259–266.

Post-Beittenmiller, M.A., A. Hlousek-Radojcic, and J.B. Ohlrogge. 1989. DNA sequence of a genomic clone encoding an *Arabidopsis* acyl carrier protein (ACP). *Nucleic Acids Res.* **17**: 1777.

Prasad, T.K. and C.R. Stewart. 1992. cDNA clones encoding *Arabidopsis thaliana* and Z. *mays* mitochondrial chaperonin HSP60 and gene expression during seed germination and heat shock. *Plant Mol. Biol.* **18**: 873–885.

Quatrano, R.S. 1986. Regulation of gene expression by abscisic acid during angiosperm embryo development. *Oxf. Surv. Plant Mol. Cell. Biol.* **3**: 467–477.

Rabindran, S.K., G. Giorgi, J. Clos, and C. Wu. 1991. Molecular cloning and expression of a human heat shock factor, HSF1. *Proc. Natl. Acad. Sci.* **88**: 6906–6910.

Raynal, M., P. Gaubier, F. Grellet, and M. Delseny. 1990. Nucleotide sequence of a radish cDNA clone coding for a late embryogenesis abundant (LEA) protein. *Nucleic Acids Res.* **18:** 6132.

Reimold, U., M. Kroger, F. Kreuzaler, and K. Hahlbrock. 1983. Coding and 3′ non-coding nucleotide sequence of chalcone synthase mRNA and assignment of amino acid sequence of the enzyme. *EMBO J.* **2:** 1801–1805.

Roberts, J.K.M., J. Callis, D. Wemmer, V. Walbot, and O. Jardetzy. 1984. Mechanism of cytoplasmic pH regulation in hypoxic maize root tips and its role in survival under hypoxia. *Proc. Natl. Acad. Sci.* **81:** 3379–3383.

Ryder, T.B., C.L. Cramer, J.N. Bell, M.P. Robbins, R.A. Dixon, and C.J. Lamb. 1984. Elicitor rapidly induces chalcone synthase mRNA in *Phaseolus vulgaris* cells at the onset of the phytoalexin defense response. *Proc. Natl. Acad. Sci.* **81:** 5724–5728.

Sanchez, Y. and S.L. Lindquist. 1990. HSP104 required for induced thermotolerance. *Science* **248:** 1112–1115.

Sarge, K.D., V. Zimarino, K. Holm, C. Wu, and R.I. Morimoto. 1991. Cloning and characterization of two mouse heat shock factors with distinct and constitutive DNA-binding ability. *Genes Dev.* **5:** 1902–1911.

Sathasivan, K., G.M. Haughn, and N. Murai. 1990. Nucleotide sequence of a mutant acetolactate synthase gene from imidazolinone-resistant *Arabidopsis thaliana* var. Columbia. *Nucleic Acids Res.* **18:** 2188.

Scharf, K.-D., S. Rose, W. Zott, F. Schöffl, and L. Nover. 1990. Three tomato genes code for heat stress transcription factors with a region of remarkable homology to the DNA-binding domain of the yeast HSF. *EMBO J.* **9:** 4495–4501.

Schindler, U., H. Beckmann, and A.R. Cashmore. 1992a. TGA1 and G-box factors: Two distinct classes of *Arabidopsis* leucine zipper proteins compete for the G-box-like element TGACGTGG. *Plant Cell* **4:** 1309–1319.

Schindler, U., A.E. Menkens, H. Beckmann, J.R. Ecker, and A.R. Cashmore. 1992b. Heterodimerization between light-regulated and ubiquitous expressed *Arabidopsis* GBF bZIP proteins. *EMBO J.* **11:** 1261–1273.

Schindler, U., W. Terzaghi, H. Beckmann, T. Kadesch, and A.R. Cashmore. 1992c. DNA binding site preferences and transcriptional activation properties of the *Arabidopsis* transcription factor GBF 1. *EMBO J.* **11:** 1275–1289.

Schmelzer, E., W. Jahnen, and K. Hahlbrock. 1988. *In situ* localization of light induced chalcone synthase mRNA, chalcone synthase, and flavonoid end products in epidermal cells of parsley leaves. *Proc. Natl. Acad. Sci.* **85:** 2989–2993.

Schnall, J.A. and R.S. Quatrano. 1992. Abscisic acid elicits the water-stress response in root hairs of *Arabidopsis thaliana*. *Plant Physiol.* **100:** 216–218.

Schuetz, T.J., G.J. Gallo, L. Sheldon, P. Tempst, and R.E. Kingston. 1991. Isolation of a cDNA for HSF2: Evidence for two heat shock factor genes in humans. *Proc. Natl. Acad. Sci.* **88:** 6911–6915.

Schulze-Lefert, P., M. Becker-Andre, W. Schulz, K. Hahlbrock, and J.L. Dangle. 1989a. Functional architecture of the light-responsive chalcone synthase promoter from parsley. *Plant Cell* **1:** 707–714.

Schulze-Lefert, P., J.L. Dangl, M. Becker-Andre, K. Hahlbrock, and W. Schulz. 1989b. Inducible *in vivo* DNA footprints define sequences necessary for UV-light activation of the parsley chalcone synthase gene. *EMBO J.* **8:** 651–656.

Siminovitch, D. and Y. Cloutier. 1982. Twenty-four-hour induction of freezing and drought tolerance in plumules of winter rye seedlings by desiccation stress at room temperature in the dark. *Plant Physiol.* **69:** 250–255.

Skriver, K. and J. Mundy. 1990. Gene expression in response to abscisic acid and osmotic stress. *Plant Cell* **2:** 503–512.

Skriver, K., F.L. Olsen, J.C. Rogers, and J. Mundy. 1991. *Cis*-acting DNA elements responsive to gibberellin and its antagonist abscisic acid. *Proc. Natl. Acad. Sci.* **88:** 7266–7270.

Sommer, H. and H. Saedler. 1986. Structure of the chalcone synthase gene of *Antirrhinum majus. Mol. Gen. Genet.* **202:** 429–434.

Sorger, P.K., M.J. Lewis, and H.R.B. Pelham. 1987. Heat shock factor is regulated differently in yeast and HeLa cells. *Nature* **329:** 81–84.

Staiger, D., H. Kaulen, and J. Schell. 1989. A CACGTG motif of the *Antirrhinum majus* chalcone synthase promoter is recognized by an evolutionarily conserved nuclear protein. *Proc. Natl. Acad. Sci.* **86:** 6930–6934.

Staiger, D., F. Becker, J. Schell, C. Koncz, and K. Palme. 1991. Purification of tobacco nuclear proteins binding to a CACGTG motif of the chalcone synthase promoter by DNA affinity chromatography. *Eur. J. Biochem.* **199:** 519–527.

Steponkus, P.L., M. Uemura, R.A. Balsamo, T. Arvinte, and D.V. Linch. 1988. Transformation of the cryobehavior of rye protoplasts by modification of the plasma membrane lipid composition. *Proc. Natl. Acad. Sci.* **85:** 9026–9030.

Straus, D.B. 1990. DnaK, DnaJ, and GrpE heat shock proteins negatively regulate heat shock gene expression controlling the synthesis and stability of $\sigma^{35}$. *Genes Dev.* **4:** 2202–2209.

Straus, D.B., W.A. Walter, and C.A. Gross. 1989. The activity of $\sigma^{35}$ is reduced under conditions of excess heat shock protein production in *Escherichia coli. Genes Dev.* **3:** 2003–2010.

Swanson, K.D. and R. Ganguly. 1992. Characterization of a *Drosophila melanogaster* gene similar to the mammalian genes encoding the tyrosine/tryptophan hydroxylase activator and protein kinase C inhibitor proteins. *Gene* **113:** 183–190.

Takahashi, T. 1992. Isolation and analysis of the expression of two genes for the 81-kilodalton heat shock proteins from *Arabidopsis. Plant Physiol.* **99:** 383–390.

Takahashi, T. and Y. Komeda. 1989. Characterization of two genes encoding small heat-shock proteins in *Arabidopsis thaliana. Mol. Gen. Genet.* **219:** 365–372.

Takahashi, T., S. Naito, and Y. Komeda. 1992. The *Arabidopsis HSP18.2* promoter/*GUS* gene fusion in transgenic *Arabidopsis* plants: A powerful tool for the isolation of the heat-shock response. *Plant J.* **2:** 751–761.

Tilly, K., J. Spence, and C. Georgopoulos. 1989. Modulation of stability of the *Escherichia coli* heat shock regulatory factor $\sigma^{35}$. *J. Bacteriol.* **171:** 1585–1589.

Torres-Shumann, S., J.A. Godoy, and J.A. Pintor-Toro. 1992. A probable lipid transfer protein gene is induced by NaCl in stems of tomato plants. *Plant Mol. Biol.* **18:** 749–757.

Van der Straeten, D., R.A. Rodrigues-Pousada, R. Villarroel, S. Hanley, H.M. Goodman, and M. Van Montagu. 1992. Cloning, genetic mapping, and expression analysis of an *Arabidopsis thaliana* gene that encodes 1-aminocyclopropane-1-carboxylate synthase. *Proc. Natl. Acad. Sci.* **89:** 9969–9973.

van Herpen, M.M.A., W.H. Reijnen, J.A.M. Schrauwen, P.F.M. de Groot, J.W.H. Jager, and G.J. Wullems. 1989. Heat shock proteins and survival of germinating pollen of *Lilium longiflorum* and *Nicotiana tabacum. Plant Physiol.* **134:** 345–351.

Van Heusden, G.P.H., T.J. Wenzil, E.L. Laqendi, H.Y. de Steensma, and J.A. van den Berg. 1992. Characterization of the yeast BMH 1 gene encoding a putative protein homologous to mammalian protein kinase II activators and protein kinase inhibitors. *FEBS Lett.* **302:** 145–150.

Vartapetian, B.B., H.H. Snkhchian, and I.P. Generozova. 1987. Mitochondrial fine structure in imbibing seeds and seedlings of *Zea mays* L. under anoxia. In *Plant life in aquatic and amphibious habitats* (ed R.M.M. Crawford), pp. 205–223. Blackwell Scientific, Oxford, United Kingdom.

Vierling, E. 1991. Roles of heat shock proteins in plants. *Annu. Rev. Plant Physiol. Plant Mol. Biol.* **42:** 579–620.

Vilardell, J., A. Goday, M.A. Freire, M. Torrent, M.C. Martinez, J.M. Torne, and M. Pages. 1990. Gene sequence, development expression, and protein phosphorylation of RAB-17 in maize. *Plant Mol. Biol.* **14:** 423–432.

Vorst, O., F. Van Dam, R. Oosterhoff-Teetstra, and P. Weisbeek. 1990. Tissue-specific expression directed by an *Arabidopsis thaliana* pre-ferredoxin promoter in transgenic tobacco plants. *Plant Mol. Biol.* **14:** 491–499.

Walker, J.C., E.A. Howard., E.S. Dennis, and W.J. Peacock. 1987a. DNA sequences required for anaerobic expression of the maize *alcohol dehydrogenase 1* gene. *Proc. Natl. Acad. Sci.* **84:** 6624–6628.

Walker, J.C., D.J. Llewellyn, L.E. Mitchell, and E.S. Dennis. 1987b. Anaerobically regulated gene expression: Molecular adaptations of plants for survival under flooded conditions. *Oxf. Surv. Plant Mol. Cell Biol.* **4:** 71–93.

Weisshaar, B., G.A. Armstrong, A. Block, O. de Costa e Silva, and K. Hahlbrock. 1991. Light-inducible and constitutively expressed DNA-binding proteins recognizing a plant element with functional relevance in light responsiveness. *EMBO J.* **10:** 1777–1786.

Wenkert, W., N.R. Fausey, and H.D. Watters. 1981. Flooding responses in *Zea mays* L. *Plant Soil* **62:** 351–366.

Wiederrecht, G., D. Seto, and C.S. Parker. 1988. Isolation of the gene encoding the *S. cerevisiae* heat shock transcription factor. *Cell* **54:** 841–853.

Wienand, U., U. Weydemann, U. Niesbach-Klosgen, P.A. Peterson, and H. Saedler. 1986. Molecular cloning of the *c2* locus of *Zea mays*, the gene coding for chalcone synthase. *Mol. Gen. Genet.* **203:** 202–207.

Williams, M.E., R. Foster, and N.-H. Chua. 1992. Sequences flanking the hexameric G-box core CACGTG affect the specificity of protein binding. *Plant Cell* **4:** 485–496.

Winter, J., R. Wright, N. Duck, C. Gasser, C. Fraley, and D. Shah. 1988. The inhibition of petunia hsp70 mRNA processing during $CdCl_2$. *Mol. Gen. Genet.* **211:** 315–319.

Wu, C.H., T. Caspar, J. Browse, S. Lindquist, and C. Somerville. 1988. Characterization of an HSP70 cognate gene family in *Arabidopsis*. *Plant Physiol.* **88:** 731–740.

Wu, C., V. Zimarino, C. Tsai, B. Walker, and S. Wilson. 1990. Transcriptional regulation of heat shock genes. In *Stress proteins in biology and medicine* (ed. R.I. Morimoto et al.), pp. 429–442. Cold Spring Harbor Press, Cold Spring Harbor, New York.

Wu, Y.N., N.-D. Vu, and P.D. Wagner. 1992. Anti-(14-3-3 protein) antibody inhibits stimulation of noradrenaline (norepinephrine) secretion by chromaffin-cell cytosolic proteins. *Biochem. J.* **285:** 697–700.

Xiao, C.M. and J.P. Mascarenhas. 1985. High temperature-induced thermotolerance in pollen tubes of *Tradescantia* and heat-shock proteins. *Plant Physiol.* **78:** 887–890.

Xiao, H. and J.T. Lis. 1988. Germline transformation used to define key features of heat-shock response elements. *Science* **239:** 1139–1141.

Yamaguchi-Shinozaki, K., J. Mundy, and N.-H. Chua. 1989. Four tightly linked *rab* genes are differentially expressed in rice. *Plant Mol. Biol.* **14:** 29–39.

Zimarino, V. and C. Wu. 1987. Induction of sequence specific binding of *Drosophila* heat shock activator protein synthesis. *Nature* **327:** 727–730.

Zimarino, V., C. Tsai, and C. Wu. 1990. Complex modes of heat shock factor activation. *Mol. Cell. Biol.* **10:** 752–759.

Zimmerman, J.L. and P.R. Cohill. 1991. Heat shock and thermotolerance in plant and animal embryogenesis. *New Biol.* **3:** 641–650.

Zupan, L.A., D. L. Steffens, C.A. Berry, M.L. Landt, and R.W. Gross. 1992. Cloning and expression of human 14-3-3 protein mediating phospholipolysis. *J. Biol. Chem.* **267:** 8707–8710.

# 30

## *Arabidopsis thaliana* as a Model for Studying Mechanisms of Plant Cold Tolerance

**Michael F. Thomashow**
Department of Crop and Soil Sciences
Department of Microbiology
Michigan State University
East Lansing, Michigan 48824

Plants vary widely in their ability to tolerate cold temperatures. Many species from tropical and subtropical regions are chilling-sensitive (Lyons 1973; Levitt 1980). Such plants suffer a wide range of damage, ranging from loss of vigor to death, when temperatures drop below about 10°C. In contrast, plants from temperate regions are not only chilling-tolerant, but many can survive freezing at high subzero temperatures (Sakai and Larcher 1987). Moreover, many chilling-tolerant plants increase in freezing tolerance in response to low nonfreezing temperature, a process known as cold acclimation (Levitt 1980; Sakai and Larcher 1987). Nonacclimated wheat and rye, for example, are killed at about −5°C, but after cold acclimation, they can survive temperatures as low as about −20°C and −30°C, respectively.

A major challenge for plant scientists interested in low-temperature biology is to determine the molecular bases for plant chilling and freezing tolerance. Why are certain plants such as rice and cucumber sensitive to chilling temperatures whereas plants such as wheat and rye are not? Why is it that the common potato, *Solanum tuberosum*, cannot cold-acclimate although its close tuber-bearing relative, *Solanum commersonii*, can? What is the molecular basis for the enhancement of freezing tolerance that occurs with cold acclimation? Considerable effort has been directed toward addressing these and related questions, and much has been learned. However, our understanding of the underlying bases for chilling sensitivity and freezing tolerance remains at an elementary level. We cannot yet answer the questions posed above, for instance, or rationally design from first principles a more chilling-tolerant rice plant or engineer a freezing-tolerant tomato.

*A. thaliana* is a chilling-tolerant plant that can be found throughout the northern regions of North America, Europe, and Asia (Rédei 1970). In nature, seeds typically germinate at the end of summer or early fall and produce rosette plants which overwinter at subzero temperatures. In early spring, the rosettes flower and shed seed before many other plant species have become established. Given this natural history and habitat, it would seem that *Arabidopsis* would have the potential to serve as a model plant to determine basic mechanisms of cold tolerance. Accordingly, studies on *Arabidopsis* chilling tolerance and cold acclimation were initiated several years ago in a number of laboratories. Here, I summarize the progress that has been made in these areas, emphasizing the work on cold acclimation. For readers who are unfamiliar with the topics of chilling tolerance and cold acclimation, I present pertinent background information. For a more rounded appreciation of chilling tolerance, I recommend the review articles by Lyons (1973), Graham and Patterson (1982), Markhart (1986), and Lynch (1990). For further information on the biochemistry and genetics of cold acclimation, I recommend the review articles by Guy (1990) and Thomashow (1990, 1993), and for cogent presentations of the roles of membranes in cold acclimation, I recommend two articles by Steponkus and colleagues (Steponkus 1984; Steponkus et al. 1993). Finally, the treatise by Levitt (1980) on environmental stress is a classic "must read" for those seriously interested in the subject of low-temperature tolerance.

## CHILLING TOLERANCE

### Background

Chilling injury is a complex phenomenon. It can take different forms—loss of vigor, wilting, chlorosis, sterility, and/or death—in different plant species and is associated with numerous cellular and metabolic dysfunctions, including impaired photosynthesis, altered respiration, cessation of protoplasmic streaming, and changes in membrane permeability (Levitt 1980; Lynch 1990). The molecular basis for this complexity is uncertain. It might indicate that there are numerous potential cold-sensitive "sites" in plants. Alternatively, there may be a primary cold-sensitive site which, when impaired, leads to a cascade of events resulting in various forms of dysfunction and injury. Lyons and Raison (Lyons 1973; Raison 1973) suggested that the latter is the case and that the primary site of damage is the plant cell membrane.

The Lyons and Raison "membrane hypothesis" has been the most widely acknowledged, debated, and tested hypothesis on chilling sensitivity. The basic tenet of the proposal is that low temperature causes a deleterious change in the physical state of one or more cellular membrane systems in chilling-sensitive plants and that this has various primary effects ranging from changes in membrane semipermeability to modifications in the activities of membrane-associated enzymes. These primary effects, in turn, are envisioned to have numerous potential secondary effects, such as the cellular and metabolic dysfunctions mentioned above. In the original version of the hypothesis, Lyons and Raison suggested that the deleterious change in membrane state was a liquid crystalline-to-gel-phase transition in bulk lipids. Subsequent studies, however, indicated that this is not a likely mechanism: Phase transitions in bulk membrane lipids would not be expected to occur above 0°C, even in chilling-sensitive plants, due to the high level of unsaturation of most plant fatty acids (see Lynch 1990).

In a refinement of the membrane hypothesis, Murata and colleagues (Murata et al. 1982; Murata 1983) suggested that it is the degree of fatty acid unsaturation in a specific phospholipid, phosphatidylglycerol, that is critical to chilling tolerance. In a test of this hypothesis, Murata et al. (1992) altered the level of phosphatidylglycerol fatty acid unsaturation in chilling-sensitive tobacco by transforming it with the glycerol-3-phosphate acyltransferase gene from either squash, a plant that is more chilling-sensitive than tobacco, or chilling-tolerant *Arabidopsis*. Those transgenic tobacco plants that received the glycerol-3-phosphate acyltransferase gene from squash had lower levels of *cis*-unsaturated fatty acids in phosphatidylglycerol than did wild-type tobacco and were found to be more chilling-sensitive than tobacco (measured by photoinhibition of photosynthesis and visible damage of the whole plant). In contrast, those transgenic plants carrying the glycerol-3-phosphate acyltransferase gene from *Arabidopsis* had greater levels of *cis*-unsaturated fatty acids in phosphatidylglycerol than did wild-type tobacco and displayed increased levels of chilling tolerance.

The results of Murata et al. (1992) provide clear evidence that the levels of fatty acid unsaturation in phosphatidylglycerol can affect the chilling tolerance of plants. The mechanism that underlies this relationship must now be determined. It also remains to be determined whether the low-temperature sensitivity of chilling-sensitive crops is generally attributable to a deleterious membrane lipid composition.

## Chilling Tolerance of *Arabidopsis*

### Studies on Desaturase Mutants

The availability of a comprehensive collection of *Arabidopsis* mutants with defects in membrane lipid fatty acid unsaturation (see Browse and

Somerville 1991) provides a unique opportunity to determine the role of membrane composition in chilling tolerance. Indeed, the results that Miquel et al. (1993) have obtained with the *fad2* mutants indicate that *Arabidopsis* requires polyunsaturated lipids for low-temperature survival. *FAD2* is believed to encode the structural gene for the microsomal 18:1 desaturase, the enzyme responsible for producing most of the polyunsaturated lipids in extrachloroplast membranes (Miquel and Browse 1992). Inactivation of this locus results in dramatically reduced levels of 18:2 and 18:3 fatty acids in phosphatidylcholine, the most abundant extrachloroplast lipid. An effect of this alteration in membrane composition is chilling sensitivity. Whereas wild-type plants grow and develop normally at 6°C, the *fad2-2* mutant dies. The process, however, is slow. After 3 days at 6°C, there is no discernible difference between the wild-type and mutant plants. Then, at about 10 days, the leaves of the mutant begin to deteriorate, first displaying small necrotic lesions followed by extensive production of anthocyanins. Finally, upon continued incubation at low temperature, the mutant plants die. The *fad2-2* mutant also displays chilling sensitivity at 12°C, but the phenotype is more subtle: The stems do not elongate during reproductive growth. Normal flowers and viable seed are produced, however, thus leading to the visually striking situation of having normal flowers and siliques develop on stems that are no more that 15 mm in length.

The results with the *fad2* mutants support the hypothesis that the level of fatty acid unsaturation can be an important factor in chilling tolerance. Furthermore, the data indicate that specific changes in fatty acid unsaturation can lead to specific membrane-associated dysfunctions at low temperature; e.g., at 12°C, the *fad2-2* mutants were only affected in stem elongation during reproductive growth. The results obtained with other fatty acid desaturase mutants are consistent with this notion. At least eight *Arabidopsis* lipid mutants show reduced levels of fatty acid unsaturation (Browse and Somerville 1991), but only three, the *fadB*, *fadC*, and *fab1* mutants, appear to be sensitive to chilling temperatures (Hugly and Somerville 1992; J. Wu and J. Browse, unpubl.). Furthermore, in these cases, the phenotype is different from that of the *fad2* mutants. For instance, the *fadC* mutant, which has decreased levels of unsaturated fatty acids in all lipids, turns chlorotic at low temperature (5°C) but continues to grow and complete its life cycle (Hugly and Somerville 1992).

### Genes Required at Low Temperature

*Arabidopsis* mutants have been isolated that are indistinguishable from the wild type at normal growth temperatures (about 22°C), but which turn chlorotic, leak ions, and die following several days of exposure to temperatures in the range of 10°C–13°C (Hugly et al. 1990; C. Somerville, pers. comm.). The mutations suffered by these chilling-sensitive mutants map to a number of different loci (Hugly et al. 1990; C. Schneider, unpubl.). However, only those mutants with mutations at the *chs1* locus, which maps to the top of chromosome 1, have been examined in any detail (Hugly and Somerville 1992; Schneider et al. 1994). Recent evidence indicates that at low temperature, *chs1* mutants are deficient in the accumulation of chloroplast proteins (Schneider et al. 1994). The nature of the biochemical defect in the mutant is not yet known. However, several independent mutant alleles of *chs1* have been isolated, suggesting that the mutant phenotype results from a loss of function of a gene product that is only required for survival at low temperature. The eventual identification of the function of the *chs1* gene may provide insight into the mechanism of chilling tolerance.

## COLD ACCLIMATION

### Background

*Changes in Membrane Cryobehavior*

Immediately after a killing freeze-thaw cycle, plant tissues typically appear flaccid and water-soaked, suggesting that cell lysis has occurred. Indeed, some 80 years ago, Maximov (1912) proposed that disruption of cellular membranes, particularly the plasma membrane, was the primary cause of freezing injury in plants. Subsequent studies have supported this notion and have indicated that membrane damage results largely from the severe dehydration that occurs during a freeze-thaw cycle (see Steponkus 1984). As the temperature drops below freezing, ice generally forms in the extracellular spaces (the freezing point of the apoplastic fluid is higher than that of intracellular water). Because the chemical potential of ice is lower than that of water at a given temperature, the intracellular water moves out of the cell. The dehydration and cell contraction that occurs is quite extensive: Plant cells frozen to –10°C lose more than 90% of their osmotically active water. Upon thawing, the process is reversed. The chemical potential of the water inside the cell is lower than outside and thus water moves back into the cell. If the cell is to survive the freeze-thaw cycle, the plasma membrane must be able to withstand the major efflux and influx of water.

Steponkus and colleagues (see Steponkus 1984) conducted an elegant series of experiments indicating that "expansion-induced lysis" is the major form of injury that occurs in cells of nonacclimated plants frozen between –3°C and –7°C. When protoplasts are isolated from the leaves of nonacclimated rye plants and frozen to −5°C, they dehydrate, and as they shrink, endocytotic vesicles bud off from the plasma membrane. When the protoplasts are thawed, the vesicular material is not reincorporated into the plasma membranes. Consequently, rehydration results in an intolerable osmotic pressure, and the cells burst. The cryobehavior of protoplasts from the leaves of cold-acclimated rye plants, however, is quite different. When frozen, such protoplasts also dehydrate and shrink, but instead of forming endocytotic vesicles, the plasma membranes form exocytotic extrusions. Unlike the endocytotic vesicles, the membrane extrusions remain in association with the plasma membrane and are reincorporated into it during rehydration. Thus, the protoplasts are able to swell to their original size and do not lyse. Whether the plasma membranes from whole plant cells behave in precisely the same way as those from protoplasts has been questioned (see, e.g., Pearce and Willison 1985; Singh et al. 1987), but even in whole plants, it appears that nonacclimated cells suffer expansion-induced lysis and acclimated cells do not.

Expansion-induced lysis is not the only form of damage that membranes suffer in response to freezing (see Steponkus and Webb 1992). As temperatures drop below −10°C, and dehydration becomes more severe, injury is manifested as a loss of osmotic responsiveness. The membranes from nonacclimated cells frozen to −10°C form lateral phase separations, aparticulate lamellae, and lamellar-to-hexagonal-II phase transitions. Similar membrane damage is suffered by such cells subjected to equivalent degrees of dehydration induced by suspension in hypertonic solutions at 0°C, indicating that the injury is due to dehydration rather than low temperature per se. In cold-acclimated cells, these forms of injury are greatly reduced: Whereas osmotic nonresponsiveness occurs when protoplasts from nonacclimated rye leaves are frozen to −10°C, protoplasts isolated from acclimated leaves can be frozen to −10°C without injury (Gordon-Kamm and Steponkus 1984). Protoplasts from acclimated rye leaves do become osmotically nonresponsive when frozen to about −25°C to −30°C, but even when frozen to −35°C, the plasma membranes do not undergo lamellar-to-hexagonal-II phase transitions (Gordon-Kamm and Steponkus 1984). Instead, they form "fracture-jump lesions" (Fujikawa and Steponkus 1990).

What accounts for the difference in cryobehavior between plasma membranes from cold-acclimated and nonacclimated plants? The work of Steponkus et al. (1988) indicates that at least some of the difference is due to the phospholipid composition of the membranes. In particular, Steponkus and colleagues found that they could eliminate expansion-induced lysis in nonacclimated protoplasts by elevating the levels of mono- and di-unsaturated fatty acids in phosphatidylcholine (the levels of mono- and di-unsaturated molecular species of phosphatidylcholine normally increase during cold acclimation); the fatty acid modifications were accomplished by fusing the protoplasts with liposomes composed of different phospholipids. Significantly, the plasma membranes of the modified protoplasts formed extracellular extrusions instead of endocytotic vesicles upon moderate dehydration. Whether changes in the levels of fatty acid unsaturation in phosphatidylcholine is the only plasma membrane alteration that can abolish expansion-induced lysis, and whether this alteration accounts for cold acclimation-induced freezing tolerance in a wide range of plant species, remain to be determined. Moreover, the enrichment of mono- and di-unsaturated molecular species of phosphatidylcholine does not impart freezing tolerance to temperatures of $-10^{\circ}C$, the point where dehydration is severe enough to cause lateral phase separations, lamellar-to-hexagonal-II phase transitions, and osmotic unresponsiveness. Thus, the freezing tolerance of plasma membranes appears to involve multiple alterations, each contributing to a different aspect of freezing tolerance.

## Biochemistry

A number of biochemical changes occur in addition to membrane alterations during cold acclimation. Common examples of such changes include the appearance of new isozymes, increased sugar and soluble protein content, and increased levels of proline and other organic acids (see Levitt 1980; Sakai and Larcher 1987; Guy 1990). The precise role that each of these changes has in the cold acclimation process, however, is not certain. Some may contribute directly to the freezing tolerance of acclimated cells. Indeed, proline and various simple sugars can act as cryoprotectants (Rudolph and Crowe 1985; Carpenter and Crowe 1988). Moreover, Volger and Heber (1975) and Hincha and colleagues (Hincha et al. 1989, 1990) have reported that certain soluble polypeptides that accumulate during cold acclimation in spinach and cabbage have cryoprotective properties: The proteins have been estimated to be about $10^4$ times more effective than sucrose (on a molar basis) in protecting isolated thylakoids against freeze-thaw damage in vitro. Whether these polypeptides have important cryoprotective roles in vivo and contribute significantly to freezing tolerance remains to be determined.

## Genetics

It has been long established that the ability to cold-acclimate is a quantitative genetic trait (see Thomashow 1990). In addition, Stone et al.

(1993) have reported that the genes involved in cold acclimation of two tuber-bearing *Solanum* species, *S. commersonii* and *S. cardiophyllum*, are different from those that impart freezing tolerance in nonacclimated plants. However, the number of genes involved in cold acclimation and freezing tolerance is not known for any plant species, and no individual genes that are certain to have direct roles in cold acclimation or freezing tolerance have been identified.

Cold acclimation is associated with changes in gene expression in a wide range of plants, including *Arabidopsis* (see Thomashow 1993). The alterations in gene expression include the appearance of novel transcripts and polypeptides and increased or decreased levels of others. The number of genes affected does not appear to be as extensive as that affected by heat shock (Vierling 1991) and anaerobic stress (Sachs and Ho 1986): Typically, in vitro translation patterns of RNA isolated from cold-acclimated and nonacclimated plants only differ by some 10–20 polypeptides out of a total display of a few hundred polypeptides on two-dimensional SDS-PAGE (see Guy 1990; Thomashow 1993). However, these results only reflect the changes in the most abundantly expressed genes. If the frequency observed is typical for all gene classes, it would mean that the expression of hundreds of genes is altered with cold acclimation.

Studies on cold-regulated gene expression are now focused on two main issues. One is the mechanism of low-temperature gene regulation; that is, how do plants sense low temperature and relay this information into altered gene expression? The other major issue regards the function of the cold-regulated genes. In particular, do these genes have critical roles in freezing tolerance or some other aspect of low-temperature survival?


### Freezing Tolerance of *Arabidopsis*

#### Arabidopsis *Can Cold-acclimate*

*Arabidopsis* increases in freezing tolerance in response to low nonfreezing temperatures (Gilmour et al. 1988; Kurkela et al. 1988). This can be dramatically demonstrated by whole-plant freeze tests. Plants that are grown at 22°C and subjected to a 3-day freeze at –5°C turn flaccid and water-soaked immediately after thawing and die within a few days (Fig. 1). In contrast, plants that are cold-acclimated at 4°C for 3 days suffer no obvious damage upon freezing to –5°C (Fig. 1).

Electrolyte leakage tests indicate that *Arabidopsis* alters freezing tolerance quickly in response to temperature (Fig. 2). In this test, leaves are removed from plants, frozen to a given temperature, and thawed; then

*Figure 1*    Response of cold-acclimated (*left*) and nonacclimated (*right*) *Arabidopsis* to freezing. Plants that had been either cold-acclimated (3 days at 3ºC) or nonacclimated (22ºC) were frozen at –5ºC for 3 days, returned to normal growth temperature (22ºC) for 4 days, and then photographed (N.N. Artus, unpubl.).

the extent of cell damage is estimated by measuring the release of ions (using a conductivity meter). Leakage of 50% of the total electrolytes is generally designated as the $LT_{50}$ (temperature that gives 50% killing). Using this method, increased freezing tolerance can be detected in *Arabidopsis* plants that have been cold-acclimated for as little as 24 hours. In addition, it can be demonstrated that *Arabidopsis* "deacclimates" quickly upon return to normal growth temperatures: Within 3 days, the level of freezing tolerance in cold-acclimated plants returns to that of nonacclimated plants (Fig. 2). The maximum freezing tolerance obtained with *Arabidopsis*, estimated using the electrolyte leakage test, has been about –10ºC (Gilmour et al. 1988; Heino et al. 1990), even upon prolonged exposure (30 days) of the plants to low temperature (S.J. Gilmour, unpubl.). However, Laroche et al. (1992), estimating cell damage by measuring uptake of fluorescein diacetate and neutral red, concluded that long-term incubation of *Arabidopsis* at low temperature (about 30 days) results in plants that have an $LT_{50}$ of –17ºC.

*Genotypic Variation in* Arabidopsis *Freezing Tolerance*

All the studies published to date on *Arabidopsis* freezing tolerance have been with either the Columbia or Landsberg ecotypes. An important question is whether there is significant variation in the level of freezing

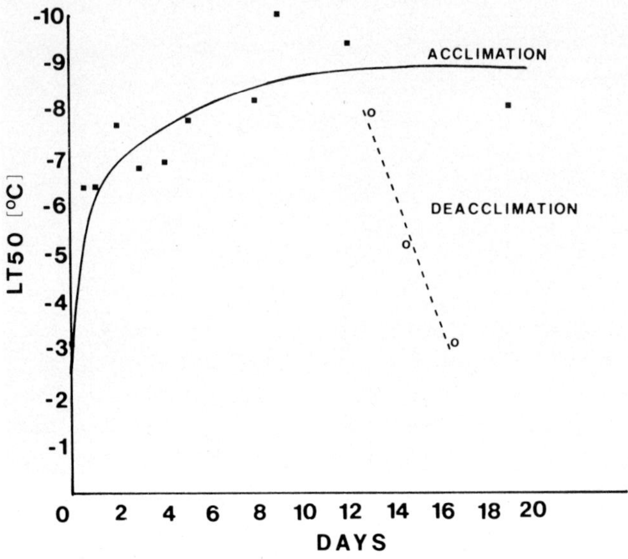



*Figure 2* Freezing tolerance of *Arabidopsis*. Plants grown at 22°C for 2–3 weeks were transferred to 4°C to cold-acclimate (*solid line*). Plants that had been cold-acclimated for 13 days were returned to 22°C for "deacclimation" (*dashed line*). At various times, freezing tolerance was estimated using the electrolyte leakage test as described previously (Gilmour et al. 1988). (Reprinted, with permission of John Wiley & Sons, Inc., from Thomashow et al. 1990.)

tolerance among *Arabidopsis* ecotypes. If so, one could potentially map freezing tolerance loci and use map-based cloning strategies to isolate genes with roles in cold acclimation. Toward this end, S.J. Gilmour (unpubl.) compared the freezing tolerances of eight ecotypes that had been isolated from a variety of cold and warm habitats: Pa-1, Ct-1, Est-O, Kn-O, St-O, Cvi-O, Pi-O, and Kas-1 (obtained from A.R. Kranz). Differences in $LT_{50}$ values were observed, but they were small (the maximum difference was only about 3°C). These data suggest that there may not be much natural variation in the ability of *Arabidopsis* ecotypes to cold-acclimate. However, these results must be considered preliminary, as only one condition for cold acclimation (5 days at 4°C) and one assay for freezing tolerance (electrolyte leakage test) were used in the survey. In addition, a greater array of ecotypes must be examined before the issue of genetic variability in freezing tolerance can be considered closed.

## Freezing Tolerance Mutants

Another potential approach to identifying genes involved in cold acclimation would be to isolate and characterize mutants that are either deficient or enhanced in freezing tolerance. At this point, there is only one preliminary report in this area (Artus and Thomashow 1992), but extensive screens for such mutants are now in progress in a number of laboratories. The approach, however, has potential limitations. In particular, one can conceive of a mutation that might result in a generally less robust plant, which in turn might have an indirect negative effect on the ability of the plant to cold-acclimate. Such mutations would not be of primary interest, as they would not be a part of the cold acclimation mechanism per se. Thus, particularly in the isolation of freezing-deficient mutants, it will be necessary to establish criteria by which "noninformative" mutations can be distinguished from those that affect the cold acclimation mechanism.

The *Arabidopsis* desaturase mutants also have the potential to yield important information on the role of fatty acid unsaturation in freezing tolerance. As discussed earlier, there is strong evidence that the increase in freezing tolerance that occurs with cold acclimation involves increased levels of fatty acid unsaturation in the phospholipids of plasma membranes (Steponkus et al. 1988). Thus, it would be of interest to determine whether the *Arabidopsis fad2* mutants (discussed above) are affected in freezing tolerance.


## Role of ABA in Cold Acclimation

Two lines of evidence suggest that the phytohormone ABA may have an important role in the cold acclimation process. First, ABA treatment at normal growth temperatures can enhance the freezing tolerance of a wide range of plants (Chen et al. 1979, 1983; Mohapatra et al. 1988), including *Arabidopsis* (Lång et al. 1989). Second, ABA levels increase at least transiently in a number of plants (Chen and Gusta 1983; Chen et al. 1983; Lalk and Dörffling 1985; Guy and Haskell 1988) in response to low temperature. It has therefore been hypothesized that ABA levels normally increase in plants during cold acclimation and that this induces changes that enhance freezing tolerance.

The role of ABA in cold acclimation has been assessed by examining the abilities of the *Arabidopsis* ABA mutants (Koornneef et al. 1982, 1984) to cold-acclimate. The ABA mutants fall into two classes: ABA-deficient, designated *aba*, and ABA-insensitive, designated *abi*. The *aba* mutants have abnormally low levels of ABA in both seeds and leaves of whole plants (Koornneef et al. 1982; Rock and Zeevaart 1991), whereas the *abi* mutants are impaired or deficient in various responses that are regulated by ABA (Koornneef et al. 1984; Finkelstein and Somerville 1990). If ABA has a central role in *Arabidopsis* cold acclimation, certain of the ABA mutants might be impaired in their ability to increase in freezing tolerance in response to low temperature. This is not the case with the *abi1*, *abi2*, and *abi3* mutants: No difference in freezing tolerance was detected between nonacclimated and acclimated wild-type and mutant plants (Heino et al. 1990; Gilmour and Thomashow 1991). In contrast, the *aba1* and *aba4* mutants were severely impaired in their ability to cold-acclimate (Heino et al. 1990; Gilmour and Thomashow 1991). Exogenous application of ABA could suppress the impaired cold-acclimation phenotype (Heino et al. 1990), as would be expected if it was due to decreased levels of ABA.

The results with the *aba* mutants clearly indicate that abnormally low levels of ABA severely impair the ability of *Arabidopsis* to cold-acclimate. However, it remains uncertain whether ABA has a specific role in cold acclimation and the enhancement of freezing tolerance. ABA has been demonstrated, or implicated, to have roles in a variety of plant developmental and physiological processes (Davies and Jones 1991). Thus, it is not surprising that the *aba* mutations have pleiotropic effects (Koornneef et al. 1982): Among other things, they affect seed dormancy and water relations. In addition, the *Arabidopsis aba* mutants, like ABA-deficient mutants of other plant species (Quarrie 1987), have an "ion leaky" phenotype: Nonfrozen leaves from nonacclimated or cold-acclimated wild-type *Arabidopsis* plants release less than 10% of their ions when incubated in distilled water, whereas nonfrozen leaves of non-acclimated and cold-acclimated plants carrying the *aba4* mutation release about 20% and 40% of their ions, respectively (Gilmour and Thomashow 1991). Interestingly, this leakiness is markedly suppressed in the *Arabidopsis* mutants if the leaves are suspended in 0.5 M sorbitol instead of distilled water (S.J. Gilmour, unpubl.). One possibility thus raised is that the *aba* mutants may be more sensitive to osmotic stress than are wild-type plants. If true, this could account, at least in part, for the decreased freezing tolerance of the ABA-deficient mutants.

## Cold-regulated Genes of *Arabidopsis*

Several *Arabidopsis* cold-regulated genes have been identified (Table 1). Most of these, designated either *COR* (*cold-regulated*), *LTI* (*low-temperature-induced*), or *KIN* (*kykna-indusoitu*; Finnish for cold-induced), were isolated by differential hybridization screens of cDNA or genomic libraries. Of these, six have been shown to be members of gene pairs: *KIN1* and *KIN2* (*COR6.6*) (Kurkela and Borg-Franck 1992); *COR15a* and *COR15b* (Wilhelm and Thomashow 1993); and *LTI78* (*RD29A, COR78*), and *LTI65* (*RD29B*) (Nordin et al. 1993; Yamaguchi-

*Table 1  Arabidopsis* cold-regulated genes

| Gene | Other designations | Polypeptide[a] (kD) | GenBank accession number | Reference |
|---|---|---|---|---|
| *KIN1* | | 6.5 | X51474 | Kurkela and Franck (1990) |
| *KIN2* | | 6.6 | X62281 | Kurkela and Borg-Franck (1992) |
| | *COR6.6* | 6.6 | X55053 | Gilmour et al. (1992) |
| *COR15a* | *COR15* | 15 | U01377 | Lin and Thomashow (1992a) |
| *COR15b* | | 15 | L24070 | Wilhelm and Thomashow (1993) |
| *RAB18* | | 18 | X68042 | Lång and Palva (1992) |
| *COR47* | | 47 | X59814 | Gilmour et al. (1992) |
| *LTI65* | | 65 | X67670 | Nordin et al. (1993) |
| | *RD29B* | 65 | D13044 | Yamaguchi-Shinozaki and Shinozaki (1993) |
| *LTI78* | *LTI140* | 78 | X67671 | Nordin et al. (1993) |
| | *RD29A* | 78 | D13044 | Yamaguchi-Shinozaki and Shinozaki (1993) |
| | *COR78* | 78 | L22568 | Horvath et al. (1993) |
| | *COR160* | 78 | | Thomashow et al. (1992) |
| *ADH* | | 41 | M12196 | Jarillo et al. (1993) |

[a]Molecular masses of all polypeptides except COR47 are deduced from DNA sequence analysis of the corresponding gene. The molecular mass of COR47 was estimated from the migration of the polypeptide on SDS-PAGE (Lin et al. 1990).

Shinozaki and Shinozaki 1993). The members of each gene pair have a high degree of nucleic acid sequence identity, are deduced to encode very similar polypeptides, and are physically linked on the chromosome in tandem array. Presumably, each gene pair arose from an independent gene duplication event followed by fixation in the population.

## Regulation

*Low Temperature.* The molecular basis of cold-regulated gene expression is not well understood in any organism. Outside of plants, the most progress has been made with the bacterium *Escherichia coli.* A "cold-shock" transcription factor, encoded by *cspA*, has been identified and shown to increase dramatically in level in response to low temperature (Goldstein et al. 1990; Tanabe et al. 1992). Increased levels of the factor appear to be responsible for the induction of a number of cold-shock proteins (La Teana et al. 1991; Jones et al. 1992). It is not yet known, however, how low temperature brings about increased levels of *cspA* expression.

The transcript levels for most of the *Arabidopsis* cold-regulated genes listed in Table 1 increase dramatically in response to low temperature, remain high for as long as the plants are kept at cold temperatures, and rapidly return to low basal levels upon return to normal growth temperatures (see Fig. 3) (Hajela et al. 1990; Kurkela and Borg-Franck 1992; Lång and Palva 1992; Nordin et al. 1993; Jarillo et al. 1993; Wilhelm and Thomashow 1993; Yamaguchi-Shinozaki and Shinozaki 1993). Similar results have been obtained with cold-regulated genes from alfalfa (Mohapatra et al. 1989), barley (Cattivelli and Bartels 1990; Dunn et al. 1991), spinach (Neven et al. 1993), and wheat (Guo et al. 1992; Houde et al. 1992). However, there are exceptions to this general pattern. Most notably, the *Arabidopsis ADH* gene is only transiently expressed in response to low temperature (Jarillo et al. 1993): High transcript levels are present in plants (leaves) cold-treated for 12 and 24 hours, but not in plants cold-treated for 4 days. In addition, the relative responses of *LTI65* and *RAB18* to low temperature, in comparison to other cold-regulated genes, are weak: The transcript level of *LTI65*, for instance, is only about one-tenth that of *LTI78* in cold-acclimated plants (Nordin et al. 1993).

Efforts in a number of laboratories are now being directed at deter-

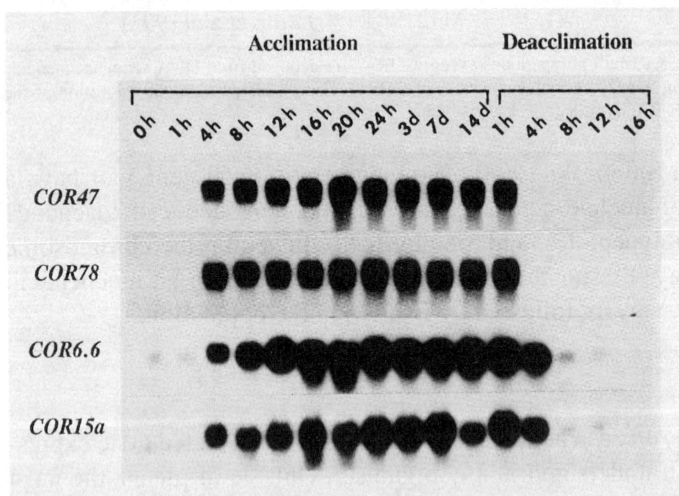


*Figure 3* *Arabidopsis COR* transcript levels during cold acclimation and deacclimation. Total RNA (5 μg) isolated from *Arabidopsis* leaves and stems was fractionated on formaldehyde agarose gels, transferred to membranes, and hybridized with $^{32}$P-labeled cDNA inserts representing the indicated *COR* genes. Acclimation was for the indicated time at 5°C and deacclimation was at 22°C. The plants used in the deacclimation experiment had been cold-acclimated for 3 days. (Adapted, with permission, from Hajela et al. 1990 [copyright held by the American Society of Plant Physiologists].)

mining the mechanism(s) responsible for the accumulation of cold-regulated gene transcripts in response to low temperature. An initial question was whether any of the *Arabidopsis* cold-regulated genes had a promoter that was induced by low temperature. The results of nuclear run-on experiments indicated that at least one of the *COR15* genes did (Hajela et al. 1990). The results of promoter fusion experiments have been consistent with this notion (Baker and Thomashow 1992; Baker et al. 1994). Specifically, it has been shown that a chimeric gene containing the 5′ region of *COR15a* (–900 to +78 relative to the site of transcription initiation) fused to the *GUS* reporter gene (Jefferson 1987) is cold-regulated in transgenic *Arabidopsis* (Fig. 4). Additional experiments have indicated that the region between –305 and +78 can impart cold-regulated gene expression whereas the region between positions –129 and +78 cannot (S.S. Baker et al., in prep.). Thus, a *cis*-acting element involved in cold-regulated gene expression appears to be located between positions –305 and –129.

Gene fusion experiments indicate that the 5′ region of *COR78* (*RD29A*, *LTI78*) also has an element(s) that can impart cold-regulated gene expression. In particular, fusion of *COR78* sequences between –808 and +5 (Horvath et al. 1993) or *RD29A* sequences between –880 and +81 (Yamaguchi-Shinozaki and Shinozaki 1993) to *GUS* has yielded chimeric genes that are cold-regulated in transgenic *Arabidopsis*. A comparison of the 5′ regions of *COR15a* with *COR78* indicates the presence of a 5-bp sequence, CCGAC, that is repeated three times in the former upstream region and four times in the latter. Interestingly, the sequence is present between positions –305 and –129 of *COR15a*, the region that appears to have a cold-regulatory element. Whether this sequence is part of a *cis*-acting cold-regulatory element remains to be determined.

*ABA.* In view of the lines of evidence indicating a role for ABA in cold acclimation, Chen et al. (1983) suggested that low temperature might bring about increased levels of ABA, which in turn might trigger changes in gene expression that result in increased freezing tolerance. If this is true, then one would expect that cold-regulated genes would be responsive to ABA. Indeed, with the exception of *ADH* (Jarillo et al. 1993), all of the *Arabidopsis* cold-regulated genes listed in Table 1 have been shown to be ABA-responsive; i.e., exogenous application of ABA leads to increased transcript levels for these genes (Hajela et al. 1990; Kurkela and Borg-Franck 1992; Lång and Palva 1992; Nordin et al. 1993; Wilhelm and Thomashow 1993; Yamaguchi-Shinozaki and Shinozaki 1993). Moreover, the 5′ promoter regions of *COR15a*, *COR15b*, *COR78*, *LTI65*, and *RAB18* have potential ABA-responsive elements

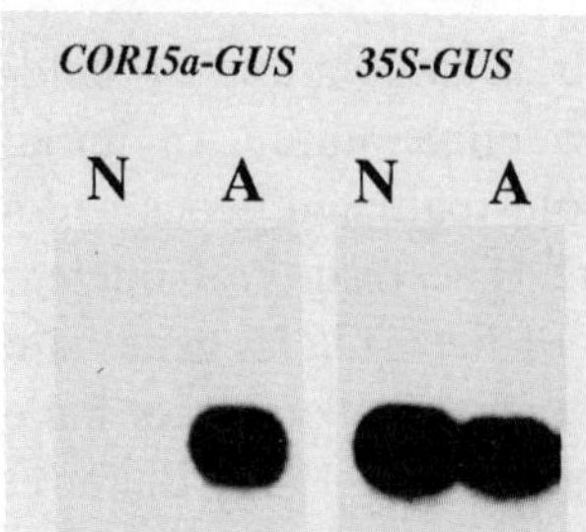



*Figure 4*  Evidence that *COR15a* has a cold-regulated promoter. The 5′ region of *COR15a* (from position −900 to +78 relative to the site of transcription initiation) was fused to the *GUS* reporter gene in pBI101.2 (Jefferson 1987) creating *COR15-GUS*. The CaMV *35S-GUS* gene carried on pBI121 (Jefferson 1987) served as a control; the CaMV *35S* promoter is not cold-regulated. Both genes were transformed into *Arabidopsis* RLD by the root transformation procedure (Valvekens et al. 1988). Expression of the two genes was determined by Northern analysis. Total RNA (20 μg) was isolated from the leaves and stems of cold-acclimated (A) and nonacclimated (N) transgenic plants carrying either *COR15a-GUS* or CaMV *35S-GUS*. The RNA was fractionated on formaldehyde agarose gels, transferred to membranes, and hybridized with a $^{32}$P-labeled fragment from the *GUS* gene (S.S. Baker, unpubl.).

(ABREs) (Baker and Thomashow 1992; Lång and Palva 1992; Horvath et al. 1993; Nordin et al. 1993; Wilhelm and Thomashow 1993; Yamaguchi-Shinozaki and Shinozaki 1993; Baker et al. 1994), *cis*-acting DNA sequences that are involved in the ABA-regulated expression of a number of genes (Guiltinan et al. 1990; Mundy et al. 1990).

The question raised by these observations is whether ABA normally mediates the changes in gene expression that occur during cold acclimation. To address this question, Nordin et al. (1991) and Gilmour and Thomashow (1991) examined the ABA- and low-temperature-regulated expression of cold-regulated genes in the *abi* mutants of *Arabidopsis*. The results indicated that ABA-regulated expression of *LTI78*, *COR47*, and *COR6.6* is severely impaired in the *abi1* mutant, but that cold-regulated expression of these genes is unaffected. Thus, cold-regulated expression of *LTI78*, *COR47*, and *COR6.6* does not appear to require the action of ABA. In contrast, both ABA- and cold-regulated expression of *LTI65* (Nordin et al. 1993) and *RAB18* (Lång and Palva 1992) are impaired in the *abi1* mutant. Therefore, it appears that the low-temperature-regulated expression of certain cold-regulated genes can occur independently of ABA action, whereas low-temperature-regulated expression of others requires this hormone. Proof of this hypothesis will require a detailed analysis of the *cis*-acting elements and *trans*-acting factors involved in the cold- and ABA-regulated expression of these genes.

*Drought.* The transcript levels for most of the cold-regulated genes listed in Table 1 have been shown to increase when plants are drought-stressed (Hajela et al. 1990; Kurkela and Borg-Franck 1992; Lång and Palva 1992; Yamaguchi-Shinozaki and Shinozaki 1993). On first consideration, this result might seem expected, since these genes are subject to ABA regulation and since ABA levels generally increase in response to drought (Zeevaart and Creelman 1988). However, not all genes that are ABA-induced are responsive to drought. For instance, the transcript levels for both *KIN1* (Kurkela and Borg-Franck 1992) and *COR15b* (Wilhelm and Thomashow 1993) increase in response to exogenous application of ABA, but not in response to drought. Moreover, the results of Nordin et al. (1991) suggest that ABA- and drought-regulated expression of *COR78* are through independent pathways: Whereas ABA-regulated expression of *Arabidopsis COR78* is nearly abolished in the *abi1* mutant, drought-regulated expression of the gene is only slightly impaired. The fact that *KIN1* (Kurkela and Borg-Franck 1992) and *COR15b* (Wilhelm and Thomashow 1993) transcripts accumulate in response to low temperature, but not drought, suggests that cold- and drought-regulated gene expression can also proceed through independent pathways.

### Function

Except for *ADH*, the activities and functions of *Arabidopsis* cold-regulated genes are unknown. Presumably some have roles in low-temperature survival. Certain genes may be involved in chilling tolerance, bringing about structural or metabolic changes required for life at low, nonfreezing temperatures. Others may have direct roles in the enhancement of freezing tolerance. Studies directed at determining the functions of the cold-regulated genes are only now moving out of the introductory phase. However, many of the initial results are interesting and suggest hypotheses that can now be tested.

*Role in Cryoprotection.* As mentioned earlier, Heber, Hincha, and colleagues (Volger and Heber 1975; Hincha et al. 1989, 1990) have proposed that certain polypeptides encoded by cold-regulated genes of spinach and cabbage act as cryoprotectants during a freeze-thaw cycle. This hypothesis is based on the finding that cold-acclimated spinach and cabbage, but not nonacclimated plants, synthesize polypeptides that are highly active in protecting isolated thylakoids against freeze-thaw inactivation in vitro. At this point, however, there is no direct evidence to indicate that these polypeptides have significant cryoprotective roles in vivo. Indeed, none of the putative cryoprotective polypeptides has been identified with certainty, and the genes encoding them have not been isolated.

In an exploratory study, Lin et al. (1990) sought to determine whether *Arabidopsis* might have polypeptides that are analogous to the putative cryoprotective polypeptides of spinach and cabbage. In particular, Lin at al. (1990) asked whether any of the polypeptides that were synthesized during cold acclimation in *Arabidopsis* remained soluble upon boiling. This question was prompted by the procedures that had been used to enrich for the spinach and cabbage polypeptides. Specifically, these procedures indicated that the polypeptides were very hydrophilic and that they probably did not coagulate upon heating to near boiling (Volger and Heber 1975). Indeed, Lin et al. (1990) found that transcripts for a number of "boiling-soluble" polypeptides accumulated in *Arabidopsis* during cold acclimation (Fig. 5). Subsequent studies established that the boiling-soluble polypeptides with apparent molecular masses of 160, 47, and 15 kD correspond to the polypeptides encoded by *COR78*, *COR47*, and *COR15a*, respectively (Thomashow et al. 1992; Horvath et al. 1993), and that *COR6.6* also encodes a boiling-soluble polypeptide (S.J. Gilmour, unpubl.). In addition, DNA sequence analysis has established that all of the cold-regulated genes listed in Table 1, with the exception of *ADH*, encode very hydrophilic polypeptides (Kurkela and Franck 1990; Lång and Palva 1992; Gilmour et al. 1992; Lin and Thomashow 1992a; Kurkela and Borg-Franck 1992; Nordin et al. 1993; Wilhelm and Thomashow 1993; Yamaguchi-Shinozaki and Shinozaki 1993). Thus, the polypeptides encoded by certain cold-regulated genes of *Arabidopsis* have three distinctive properties in common with the putative cryoprotective polypeptides of spinach and cabbage: synthesis in response to low temperature, hydrophilicity, and boiling-solubility. Furthermore, there is evidence to suggest that the COR15a polypeptide might act as a cryoprotectant.

The *COR15a* gene encodes a 15-kD polypeptide, COR15a, that is targeted to chloroplasts (Lin and Thomashow 1992a). Lin and Thomashow (1992b) found that COR15a has potent cryoprotective activity in an in vitro assay: On a molar basis, the polypeptide was nearly $10^9$ times more effective than sucrose and about $10^2$ times more effective than bovine serum albumin (BSA) in protecting lactate dehydrogenase from freeze-thaw inactivation. Moreover, Uemura et al. (1993) found that COR15am (the mature processed form of the COR15a polypeptide), as well as the COR6.6 polypeptide, strongly inhibits freeze-induced fusion of small unilamellar vesicles of dioleoylphosphatidylcholine (BSA did not have his activity). Both COR15am and COR6.6, however, were

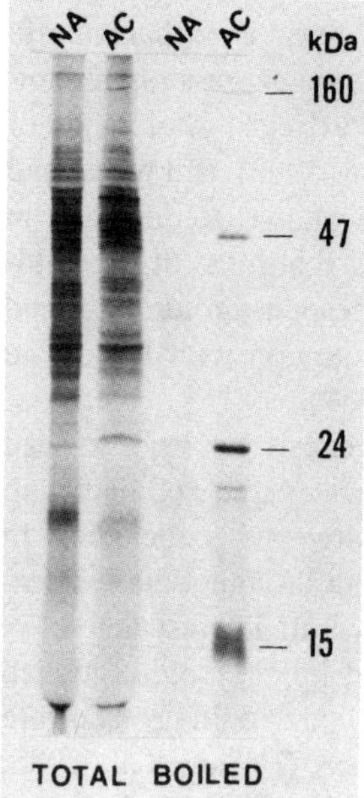


*Figure 5* Accumulation of transcripts encoding boiling-stable polypeptides in cold-acclimated *Arabidopsis* plants. Poly(A)$^+$ RNA isolated from cold-acclimated (AC) and nonacclimated (NA) *Arabidopsis* was translated in vitro, and the polypeptide products were either fractionated directly by SDS-PAGE (TOTAL) or were first boiled, sedimented by centrifugation in an Eppendorf microfuge (15 min), and the soluble material fractionated by SDS-PAGE (BOILED). (Reprinted, with permission, from Lin et al. 1990 [copyright held by the American Society of Plant Physiologists].)

also found to promote leakage of molecules from small unilamellar vesicles of dioleoylphosphatidylcholine.

The preliminary indications that COR15a and COR6.6 can influence membrane cryobehavior is exciting, given the central role of membranes in freezing tolerance. Indeed, the possibility that COR polypeptides prevent membrane fusion indicates precisely the kind of activity that one might predict for a protein that protects membranes against freezing damage. On the other hand, an ability to promote membrane leakiness would not seem to be a desirable characteristic for a cryoprotectant. Thus, whether the COR polypeptides have properties that might help stabilize membranes during a freeze-thaw cycle is unclear. Furthermore, it must be stressed that there is no direct evidence to indicate that either COR15am or COR6.6 has a cryoprotective role in vivo.

*Role in Dehydration Tolerance.* Given that tolerance to freezing must include tolerance to dehydration stress, it seems reasonable to speculate that freezing and drought tolerance might have certain mechanisms in common and that such mechanisms might include the expression of common genes. In this regard, it is of interest that the freezing tolerance of cabbage (Cox and Levitt 1976), spinach (Guy et al. 1992), and wheat and rye (Siminovitch and Cloutier 1983) has been shown to increase in response to drought. Furthermore, it is intriguing that most of the cold-regulated genes listed in Table 1 are induced in response to drought and that a number of these have characteristics in common with LEA (*late embryogenesis abundant*) proteins (Dure et al. 1989).

LEA proteins are synthesized late in embryogenesis just prior to seed desiccation and in plant seedlings in response to water stress (Galau et al. 1986). These and other observations led Dure and colleagues to hypothesize that LEA proteins might have roles in drought and desiccation tolerance (Baker et al. 1988). Therefore, it is of interest that LEA proteins, like most of the polypeptides encoded by the cold-regulated genes, are very hydrophilic, remain soluble upon boiling, and are synthesized in response to ABA and water stress (Baker et al. 1988; Mundy and Chua 1988; Close et al. 1989; Dure et al. 1989; Jacobsen and Shaw 1989). Furthermore, two of the cold-induced proteins from *Arabidopsis*, COR47 (Gilmour et al. 1992) and RAB18 (Lång and Palva 1992), have amino acid sequence homology with group II LEA proteins: They have a lysine-rich repeat that defines the group II class of LEA proteins as well as a serine cluster that is found in most LEA proteins (Dure et al. 1989; Close et al. 1993). Taken together, these observations encourage the notion that there is a genetic link between freezing and drought tolerance and suggest that certain *COR* and *LEA* genes might be a part of that link.

*Role as Antifreeze Proteins.* Certain fish and insects that cold-acclimate synthesize "antifreeze" proteins (Zachariassen 1985; Davies and Hew 1990). A hallmark characteristic of these proteins is their thermal hysteresis activity: They depress the freezing point of a solution more than they lower the melting temperature. In addition, the proteins can affect the shape of ice crystals and inhibit ice recrystallization.

Duman and colleagues (Urrutia et al. 1992) and Griffith and colleagues (Griffith et al. 1992; Hon et al. 1994) have recently established that plants also synthesize antifreeze proteins, although the degree of thermal hysteresis activity detected in protein extracts has been rather low (~ 0.2–0.4°C). The genes encoding these proteins have not yet been isolated. However, Kurkela and Franck (1990) have suggested that the polypeptide encoded by the *Arabidopsis KIN1* gene might be an antifreeze protein. This hypothesis was based on amino acid sequence similarities noted between KIN1 and the protein B antifreeze from winter flounder (Pickett et al. 1984); alignment of KIN1 with protein B gave 28% amino acid sequence identity and 41% amino acid sequence similarity.

Whether the KIN1 polypeptide has any of the functional properties associated with antifreeze proteins is not yet known. However, it would appear that the closely related COR6.6 polypeptide has little if any thermal hysteresis activity. In particular, S.J. Gilmour (unpubl.) expressed the *COR6.6 (KIN2)* gene in *E. coli* and purified the recombinant COR6.6 protein to near homogeneity, and both J.G. Duman and P.L. Steponkus (unpubl.) assayed the preparation for thermal hysteresis activity. None was detected. It should be noted, however, that in constructing the recombinant *COR6.6* gene, a mutation was introduced into the polypeptide: The second amino acid was changed from a serine to an alanine. Although it seems unlikely that this difference would abolish all thermal hysteresis activity of the polypeptide, it remains a formal possibility. More importantly, however, it is unknown whether COR6.6 is modified in vivo. If it is, such modification might affect the activity of the polypeptide.

*Role of* ADH. *ADH* is unique among the cloned cold-regulated genes of *Arabidopsis* in that its function is known: It encodes alcohol dehydrogenase. The role that the *ADH* has at low temperature, however, is uncertain. Jarillo et al. (1993) have shown that the transcript levels for *ADH* increase dramatically in response to low temperature. However, the increase is only transient. Moreover, the freezing tolerance of the *adh* null mutant R002 was found to be indistinguishable from that of wild-type *Arabidopsis* before or after exposure to low nonfreezing temperature (Jarillo et al. 1993). Thus, the *ADH* gene is not essential for cold acclimation. Whether the *adh* null mutant has a chilling-sensitive phenotype, however, was not specifically addressed in the study by Jarillo et al. (1993).

### *Conservation of* Arabidopsis *Cold-regulated Genes*

Information is sketchy on whether there are homologs of the *Arabidopsis* cold-regulated genes in other plant species. Genes related to *Arabidopsis LTI78* and *LTI65* have yet to be described in any other plants. Cold-regulated genes related to *Arabidopsis COR15a/COR15b* and *KIN1/KIN2* have been isolated from *Brassica napus* by Singh and colleagues (Orr et al. 1992; Weretilnyk et al. 1992), but whether related genes are present outside of the crucifer family is not known. The *ADH* gene, of course, is highly conserved among plant species, and in rice and corn it is cold-regulated (Christie et al. 1991), but whether *ADH* is subject to cold regulation in a wide range of chilling-sensitive and chilling-tolerant plants is not known.

## CONCLUSIONS AND PERSPECTIVES

A few years ago, it was uncertain whether *Arabidopsis* would serve as a good model to study the molecular basis of plant cold tolerance. Clearly, as is evident from the results presented here, it does, especially in situations where genetic approaches can be taken. Indeed, studies with the *Arabidopsis fad* mutants have already provided significant new insight into the effects that alterations in fatty acid unsaturation can have on chilling tolerance. Similarly, studies with the *abi* and *aba* mutants have indicated that low temperature and ABA regulate certain cold-regulated genes through independent pathways. Continued analysis of the *fad* mutants, along with more detailed studies on chilling-sensitive mutants such as *chs1*, should add greatly to our understanding of the factors that are critical for chilling tolerance. Likewise, an analysis of mutants affected in freezing tolerance, along with studies on the function and regulation of cold-regulated genes, should add significantly to our understanding of the molecular basis of cold acclimation. Over the past few years, it has been exciting to see the establishment of *Arabidopsis* as a model organism to study mechanisms of low-temperature tolerance. Over the next few years, it should be even more exciting to see the contributions that *Arabidopsis* makes to our understanding of cold-tolerance mechanisms.

### Note Added in Proof

Yamaguchi-Shinozaki and Shinozaki (1994) have demonstrated that a CCGAC-containing sequence, TACCGACAT, of the *Arabidopsis RD29A* gene is part of a *cis*-acting element involved in both cold- and drought-regulated gene expression.

### ACKNOWLEDGMENTS

I am grateful to Sarah Gilmour, Stokes Baker, and Nancy Artus for their help in preparing the figures for this paper and their critical readings of the manuscript. I thank Chris Somerville for ongoing discussions regarding *Arabidopsis* cold tolerance, unpublished results, and specific suggestions to improve the manuscript. I also thank Carrie Schneider and my collaborators Peter Steponkus and Jack Duman for unpublished results. Research from the author's laboratory was funded by grants from the USDA National Research Initiative Competitive Grants Program, the National Science Foundation, the Michigan Agriculture Experiment Station, and the Michigan Research Excellence Fund.


## REFERENCES

Artus, N.N. and M.F. Thomashow. 1992. A cold acclimation mutant of *Arabidopsis*. *Plant Physiol.* (suppl.) **99:** 123. (Abstr. 735.)

Baker, J., C. Steele, and L. Dure III. 1988. Sequence and characterization of 6 Lea proteins and their genes from cotton. *Plant Mol. Biol.* **11:** 277–291.

Baker, S.S. and M.F. Thomashow. 1992. Transcriptional regulation of *cor15*, a cold-regulated gene of *Arabidopsis thaliana*. *Plant Physiol.* (suppl.) **99:** 33. (Abstr. 194.)

Baker, S.S., K.S. Wilhelm, and M.F. Thomashow. 1994. The 5'-region of *Arabidopsis thaliana cor15a* has *cis*-acting elements that confer cold-, drought-, and ABA-regulated gene expression. *Plant Mol. Biol.* **24:** 701–713.

Browse, J. and C. Somerville. 1991. Glycerolipid synthesis: Biochemistry and regulation. *Annu. Rev. Plant. Physiol. Plant Mol. Biol.* **42:** 467–506.

Carpenter, J.F. and J.H. Crowe. 1988. The mechanism of cryoprotection of proteins by solutes. *Cryobiology* **25:** 244–255.

Cattivelli, L. and D. Bartels. 1990. Molecular cloning and characterization of cold-regulated genes in barley. *Plant Physiol.* **93:** 1504–1510.

Chen, T.H.H. and L.V. Gusta. 1983. Abscisic acid-induced freezing resistance in cultured plant cells. *Plant Physiol.* **73:** 71–75.

Chen, T.H.H., P. Gavinlertvatana, and P.H. Li. 1979. Cold acclimation of stem-cultured plants and leaf callus of *Solanum* species. *Bot. Gaz.* **140:** 142–147.

Chen, H.-H., P.H. Li, and M.L. Brenner. 1983. Involvement of abscisic acid in potato cold acclimation. *Plant Physiol.* **71:** 362–365.

Christie, P.J., H. Matthias, and V. Walbot. 1991. Low-temperature accumulation of alcohol dehydrogenase-1 mRNA and protein activity in maize and rice seedlings. *Plant Physiol.* **95:** 699–706.

Close, T.J., A.A. Kortt, and P.M. Chandler. 1989. A cDNA-based comparison of dehydration-induced proteins (dehydrins) in barley and corn. *Plant Mol. Biol.* **13:** 95–108.

Close, T.J., R.D. Fenton, A. Yang, R. Asghar, D.A. DeMason, D.E. Crone, N.C. Meyer, and F. Moonan. 1993. Dehydrin: The protein. In *Plant responses to cellular dehydration during environmental stress* (ed. T.J. Close et al.), pp. 104–118. American Society for Plant Physiologists, Rockville, Maryland.

Cox, W. and J. Levitt. 1976. Interrelations between environmental factors and freezing resistance of cabbage leaves. *Plant Physiol.* **57:** 553–555.

Davies, P.L. and C.L. Hew. 1990. Biochemistry of fish antifreeze proteins. *FASEB J.* **4:** 2460–2468.

Davies, W.J. and H.G. Jones. 1991. *Abscisic acid: Physiology and biochemistry*. BIOS Scientific Publishers, Oxford, United Kingdom.

Dunn, M.A., M.A. Hughes, L. Zhang, R.S. Pearce, A.S. Quigley, and P.L. Jack. 1991. Nucleotide sequence and molecular analysis of the low temperature induced cereal gene, *BLT4. Mol. Gen. Genet.* **229:** 389–394.

Dure, L., III, M. Crouch, J. Harada, T-H.D. Ho., J. Mundy, R. Quatrano, T. Thomas, and Z.R. Sung. 1989. Common amino acid sequence domains among the LEA proteins of higher plants. *Plant Mol. Biol.* **12:** 475–486.

Finkelstein, R.R. and C.R. Somerville. 1990. Three classes of abscisic acid (ABA)-insensitive mutations of *Arabidopsis* define genes that control overlapping subsets of ABA responses. *Plant Physiol.* **94:** 1172–1179.

Fujikawa, S. and P.L. Steponkus. 1990. Freeze-induced alterations in the ultrastructure of the plasma membrane of rye protoplasts isolated from cold-acclimated leaves. *Cryobiology* **27:** 665–666.

Galau, G.A., D.W. Hughes, and L. Dure III. 1986. Abscisic acid induction of cloned late embryogenesis abundant (Lea) mRNAs. *Plant Mol. Biol.* **7:** 155–170.

Gilmour, S.J. and M.F. Thomashow. 1991. Cold acclimation and cold-regulated gene expression in ABA mutants of *Arabidopsis thaliana. Plant Mol. Biol.* **17:** 1233–1240.

Gilmour, S.J., N.N. Artus, and M.F. Thomashow. 1992. cDNA sequence analysis and expression of two cold-regulated genes of *Arabidopsis thaliana. Plant Mol. Biol.* **18:** 13–21.

Gilmour, S.J., R.K. Hajela, and M.F. Thomashow. 1988. Cold acclimation in *Arabidopsis thaliana. Plant Physiol.* **87:** 745–750.

Goldstein, J., N.S. Pollitt, and M. Inouye. 1990. Major cold shock protein of *Escherichia coli. Proc. Natl. Acad. Sci.* **87:** 283–287.

Gordon-Kamm, W.J. and P.L. Steponkus. 1984. Lamellar-to-hexagonal-II phase transitions in the plasma membrane of isolated protoplasts after freeze-induced dehydration. *Proc. Natl. Acad. Sci.* **81:** 6373–6377.

Graham, D. and B.D. Patterson. 1982. Responses of plants to low, nonfreezing temperature: Proteins, metabolism and acclimation. *Annu. Rev. Plant Physiol.* **33:** 347–372.

Griffith, M., P. Ala, D.S.C. Yang, W.-C. Hon, and B.A. Moffatt. 1992. Antifreeze protein produced endogenously in winter rye leaves. *Plant Physiol.* **100:** 593–596.

Guiltinan, M.J., W.R. Marcotte, Jr., and R.S. Quatrano. 1990. A plant leucine zipper protein that recognizes an abscisic acid response element. *Science* **250:** 267–271.

Guo, W., R.W. Ward, and M.F. Thomashow. 1992. Characterization of a cold-regulated wheat gene related to *Arabidopsis cor47. Plant Physiol.* **100:** 1–8.

Guy, C.L. 1990. Cold acclimation and freezing stress tolerance: Role of protein metabolism. *Annu. Rev. Plant Physiol. Plant Mol. Biol.* **41:** 187–223.

Guy, C.L. and D. Haskell. 1988. Detection of polypeptides associated with the cold acclimation process in spinach. *Electrophoresis* **9:** 787–796.

Guy, C.L., D. Haskell, L. Neven, P. Klein, and C. Smelser. 1992. Hydration-state-responsive protein link cold and drought stress in spinach. *Planta* **188:** 265–270.

Hajela, R.K., D.P. Horvath, S.J. Gilmour, and M.F. Thomashow. 1990. Molecular cloning and expression of *cor* (cold-regulated) genes in *Arabidopsis thaliana. Plant Physiol.* **93:** 1246–1252.

Heino, P., G. Sandman, V. Lång, K. Nordin, and E.T. Palva. 1990. Abscisic acid deficiency prevents development of freezing tolerance in *Arabidopsis thaliana* (L) Heynh. *Theor. Appl. Genet.* **79:** 801–806.

Hincha, D.K., U. Heber, and J.M. Schmitt. 1989. Freezing ruptures thylakoid membranes in leaves, and rupture can be prevented *in vitro* by cryoprotective proteins. *Plant Physiol. Biochem.* **27:** 795–801.

———. 1990. Proteins from frost-hardy leaves protect thylakoids against mechanical freeze-thaw damage *in vitro. Planta* **180:** 416–419.

Hon, W.-C., M. Griffith, P. Chong, and D.S.C. Yang. 1994. Extraction and isolation of antifreeze proteins from winter rye (*Secale cereale* L.) leaves. *Plant Physiol.* **104:** 971–980.

Horvath, D.P., B.K. McLarney, and M.F. Thomashow. 1993. Regulation of *Arabidopsis thaliana* L. (Heyn) *cor78* in response to low temperature. *Plant Physiol.* **103:** 1047–1053.

Houde, M., J. Danyluk, J.-F. Laliberte, E. Rassart, R.S. Dhindsa, and F. Sarhan. 1992. Cloning, characterization, and expression of a cDNA encoding a 50-kilodalton protein specifically induced by cold acclimation in wheat. *Plant Physiol.* **99:** 1381–1387.

Hugly, S. and C. Somerville. 1992. A role of membrane lipid polyunsaturation in chloroplast biogenesis at low temperature. *Plant Physiol.* **99:** 197–202.

Hugly, S., P. McCourt, J. Browse, G.W. Patterson, and C.R. Somerville. 1990. A chilling sensitive mutant of *Arabidopsis* with altered steryl-ester metabolism. *Plant Physiol.* **93:** 1053–1062.

Jacobsen, J.V. and D.C. Shaw. 1989. Heat-stable proteins and abscisic acid action in barley aleurone cells. *Plant Physiol.* **91:** 1520–1526.

Jarillo, J.A., A. Leyva, J. Salinas, and J.M. Martínez-Zapater. 1993. Low temperature induces the accumulation of alcohol dehydrogenase mRNA in *Arabidopsis thaliana*, a chilling-tolerant plant. *Plant Physiol.* **101:** 833–837.

Jefferson, R.A. 1987. Assaying dimeric genes in plants: The GUS gene fusion system. *Plant Mol. Biol. Rep.* **5:** 387–405.

Jones, P.G., R. Krah, S.R. Tafuri, and A.P. Wolffe. 1992. DNA gyrase, CS7.4, and the cold shock response in *Escherichia coli. J. Bacteriol.* **174:** 5798–5802.

Koornneef, M., G. Reuling, and C.M. Karssen. 1984. The isolation and characterization of abscisic acid-insensitive mutants of *Arabidopsis thaliana. Physiol. Plant.* **61:** 377–383.

Koornneef, M., M.L. Jorna, D.L.C. Brinkhorst-van der Swan, and C.M. Karssen. 1982. The isolation of abscisic acid (ABA) deficient mutants by selection of induced revertants in non-germinating gibberellin sensitive lines of *Arabidopsis thaliana* (L.) Heynh. *Theor. Appl. Genet.* **61:** 385–393.

Kurkela, S. and M. Borg-Franck. 1992. Structure and expression of *kin2*, one of two cold- and ABA-induced genes of *Arabidopsis thaliana. Plant Mol. Biol.* **19:** 689–692.

Kurkela, S. and M. Franck. 1990. Cloning and characterization of a cold- and ABA-inducible *Arabidopsis* gene. *Plant Mol. Biol.* **15:** 137–144.

Kurkela, S., M. Franck, P. Heino, V. Lång, and E.T. Palva. 1988. Cold induced gene expression in *Arabidopsis thaliana* L. *Plant Cell Rep.* **7:** 495–498.

Lalk, I. and K. Dörffling. 1985. Hardening, abscisic acid, proline and freezing resistance in two winter wheat varieties. *Physiol. Plant.* **63:** 287–292.

Lång, V. and E.T. Palva. 1992. The expression of a *rab*-related gene, *rab18*, is induced by abscisic acid during the cold acclimation process of *Arabidopsis thaliana* (L.) Heynh. *Plant Mol. Biol.* **20:** 951–962.

Lång, V., P. Heino, and E.T. Palva. 1989. Low temperature acclimation and treatment with exogenous abscisic acid induce common polypeptides in *Arabidopsis thaliana* (L.) Heynh. *Theor. Appl. Genet.* **77:** 729–734.

Laroche, A., X.-M. Geng, and J. Singh. 1992. Differentiation of freezing tolerance and vernalization responses in Cruciferae exposed to low temperature. *Plant Cell Environ.* **15:** 439–445.

La Teana, A., A. Brandi, M. Falconi, R. Spurio, C.L. Pon, and C.O. Gualerzi. 1991. Identification of a cold shock transcriptional enhancer of the *Escherichia coli* gene encoding nucleoid protein H-NS. *Proc. Natl. Acad. Sci.* **88:** 10907–10911.

Levitt, J. 1980. *Responses of plants to environmental stress. Chilling, freezing, and high temperature stresses*, 2nd edition. Academic Press, New York.

Lin, C. and M.F. Thomashow. 1992a. DNA sequence analysis of a complementary DNA for cold-regulated *Arabidopsis* gene *cor15* and characterization of the COR15 polypeptide. *Plant Physiol.* **99:** 519–525.

————. 1992b. A cold-regulated *Arabidopsis* gene encodes a polypeptide having potent cryoprotective activity. *Biochem. Biophys. Res. Commun.* **183:** 1103–1108.

Lin, C., W.W. Guo, E. Everson, and M.F. Thomashow. 1990. Cold acclimation in *Arabidopsis* and wheat. A response associated with expression of related genes encoding "boiling-stable" polypeptides. *Plant Physiol.* **94:** 1078–1083.

Lyons, J.M. 1973. Chilling injury in plants. *Annu. Rev. Plant Physiol.* **24:** 445–466.

Lynch, D.V. 1990. Chilling injury in plants: The relevance of membrane lipids. In *Environmental injury to plants* (ed. F. Katterman), pp. 17–34. Academic Press, San Diego, California.

Markhart, A.H., III 1986. Chilling injury: A review of possible causes. *Hortic. Sci.* **2:** 1329–1333.

Maximov, N.A. 1912. Chemische schulzmittel der pflanzen gegan erfrieren. *Ber. Dtsch. Bot. Ges.* **30:** 52–65.

Miquel, M. and J. Browse. 1992. *Arabidopsis* mutants deficient in polyunsaturated fatty acid synthesis. *J. Biol. Chem.* **267:** 1502–1509.

Miquel, M., D. James, Jr., H. Dooner, and J. Browse. 1993. *Arabidopsis* requires polyunsaturated lipids for low-temperature survival. *Proc. Natl. Acad. Sci.* **90:** 6208–6212.

Mohapatra, S.S., R.J. Poole, and R.S. Dhindsa. 1988. Abscisic acid-regulated gene expression in relation to freezing tolerance in alfalfa. *Plant Physiol.* **87:** 468–473.

Mohapatra, S.S., L. Wolfraim, R.J. Poole, and R.S. Dhindsa. 1989. Molecular cloning and relation to freezing tolerance of cold-acclimation-specific genes of alfalfa. *Plant Physiol.* **89:** 375–380.

Mundy, J. and N.-H. Chua. 1988. Abscisic acid and water stress induce the expression of a novel rice gene. *EMBO J.* **7:** 2279–2286.

Mundy, J., K. Yamaguchi-Shinozaki, and N.-H. Chua. 1990. Nuclear proteins bind conserved elements in the abscisic acid-responsive promoter of rice *rab* gene. *Proc. Natl. Acad. Sci.* **87:** 1406–1410.

Murata, N. 1983. Molecular species composition of phosphatidylglycerols from chilling-sensitive and chilling-resistant plants. *Plant Cell Physiol.* **24:** 81–86.

Murata, N., N. Sato, N. Takahashi, and Y. Hamazaki. 1982. Compositions and positional distributions of fatty acids in phospholipids from leaves of chilling-sensitive and chilling-resistant plants. *Plant Cell Physiol.* **23:** 1071–1079.

Murata, N., O. Ishizaki-Nishizawa, S. Higashi, H. Hayashi, Y. Tasaka, and I. Nishida. 1992. Genetically engineered alteration in the chilling sensitivity of plants. *Nature* **356:** 710–713.

Neven, L.G., D.W. Haskell, A. Hofig, Q.-B. Li, and C.L. Guy. 1993. Characterization of a spinach gene responsive to low temperature and water stress. *Plant Mol. Biol.* **21:** 291–305.

Nordin, K., P. Heino, and E.T. Palva. 1991. Separate signal pathways regulate the expression of a low-temperature-induced gene in *Arabidopsis thaliana* (L.) Heynh. *Plant Mol. Biol.* **16:** 1061–1071.

Nordin, K., T. Vahala, and E.T. Palva. 1993. Differential expression of two related, low-temperature-induced genes in *Arabidopsis thaliana* (L.) Heynh. *Plant Mol. Biol.* **21:** 641–653.

Orr, W., B. Lu, T.C. White, L.S. Robert, and J. Singh. 1992. Complementary DNA sequence of a low temperature-induced *Brassica napus* gene with homology to the *Arabidopsis thaliana kin1* gene. *Plant Physiol.* **98:** 1532–1534.

Pearce, R.S. and J.H.M. Willison. 1985. Wheat tissues freeze-etched during exposure to extracellular freezing: Distribution of ice. *Planta* **163:** 295–303.

Pickett, M., G. Scott, P. Davies, N. Wang, S. Joshi, and C. Few. 1984. Sequence of an antifreeze protein precursor. *Eur. J. Biochem.* **143:** 35–38.

Quarrie, S.A. 1987. Use of genotypes differing in endogenous abscisic acid levels in studies of physiology and development. In *Hormone action in plant development: A critical appraisal* (ed. G.V. Hoad et al.), pp. 89–105. Butterworths, London.

Raison, J.K. 1973. The influence of temperature-induced phase changes on kinetics of respiratory and other membrane-associated enzymes. *J. Bioenerg.* **4:** 285–309.

Rédei, G.P. 1970. *Arabidopsis thaliana* (L.) Heynh. A review of the genetics and biology. *Bibliogr. Genet.* **20:** 1–151.

Rock, C.D. and J.A.D. Zeevaart. 1991. The *aba* mutant of *Arabidopsis thaliana* is impaired in epoxy-carotenoid biosynthesis. *Proc. Natl. Acad. Sci.* **88:** 7496–7499.

Rudolph, A.S. and J.H. Crowe. 1985. Membrane stabilization during freezing: The role of two natural cryoprotectants, trehalose and proline. *Cryobiology* **22:** 367–377.

Sachs, M.M. and T.-H.D. Ho. 1986. Alterations of gene expression during environmental stress in plants. *Annu. Rev. Plant. Physiol.* **37:** 363–376.

Sakai, A. and W. Larcher. 1987. *Frost survival of plants: Responses and adaptations to freezing stress.* Springer-Verlag, New York.

Schneider, J.C., E. Nielsen, and C. Somerville. 1994. A chilling-sensitive mutant of *Arabidopsis* is deficient in chloroplast protein accumulation at low temperature. *Plant Cell Environ.* (in press).

Siminovitch, D. and Y. Cloutier. 1983. Drought and freezing tolerance and adaptation in plants: Some evidence for near equivalences. *Cryobiology* **20:** 487–503.

Singh, J., B. Iu, and A.M. Johnson-Flanagan. 1987. Membrane alterations in winter rye and *Brassica napus* cells during lethal freezing and plasmolysis. *Plant Cell Environ.* **10:** 163–168.

Steponkus, P.L. 1984. Role of the plasma membrane in freezing injury and cold acclimation. *Annu. Rev. Plant Physiol.* **35:** 543–584.

Steponkus, P.L. and M.S. Webb. 1992. Freeze-induced dehydration and membrane destabilization in plants. In *Water and life: Comparative analysis of water relationships at the organismic, cellular and molecular level* (ed. G. Somero and B. Osmond), pp. 338–362. Springer Verlag, Berlin.

Steponkus, P.L., M. Uemura, and M.S. Webb. 1993. A contrast of the cryostability of the plasma membrane of winter rye and spring oat. Two species that widely differ in their freezing tolerance and plasma membrane lipid composition. In *Advances in low-temperature biology* (ed. P.L. Steponkus), vol. 2, pp. 211–312. JAI Press, London.

Steponkus, P.L., M. Uemura, R.A. Balsamo, T. Arvinte, and D.V. Lynch. 1988. Transformation of the cryobehavior of rye protoplasts by modification of the plasma membrane lipid composition. *Proc. Natl. Acad. Sci.* **85:** 9026–9030.

Stone, J., J.P. Palta, J.B. Bamberg, L.S. Weiss, and J.F. Harbage. 1993. Inheritance of freezing resistance in tuber-bearing *Solanum* species: Evidence for independent genetic control of nonacclimated freezing tolerance and cold acclimation capacity. *Proc. Natl. Acad. Sci.* **90:** 7869–7873.

Tanabe, H., J. Goldstein, M. Yang, and M. Inouye. 1992. Identification of the promoter region of the *Escherichia coli* major cold shock gene, *cspA. J. Bacteriol.* **174:** 3867–3873.

Thomashow, M.F. 1990. Molecular genetics of cold acclimation in higher plants. *Adv.*

*Genet.* **28:** 99–131.

————. 1993. Genes induced during cold acclimation in higher plants. In *Advances in low temperature biology* (ed. P.L. Steponkus), vol. 2, pp. 183–210. JAI Press, London.

Thomashow, M.F., S.J. Gilmour, and C. Lin. 1992. Cold-regulated genes of *Arabidopsis thaliana*. In *Advances in plant cold hardiness* (ed. P.H. Li et al.), pp. 32–44. CRC Press, Boca Raton, Florida.

Thomashow, M.F., S.J. Gilmour, R. Hajela, D. Horvath, C. Lin, and W. Guo. 1990. Studies on cold acclimation in *Arabidopsis thaliana*. In *Horticultural biotechnology* (ed. A.B. Bennett and S.D. O'Neill), pp. 305–314. Wiley-Liss, New York.

Uemura, M., S.J. Gilmour, M.F. Thomashow, and P.L. Steponkus. 1993. Effect of COR proteins on the freeze-induced fusion of liposomes. *Plant Physiol.* (suppl.) **102:** 35. (Abstr. 177.)

Urrutia, M.E., J.G. Duman, and C.A. Knight. 1992. Plant thermal hysteresis proteins. *Biochim. Biophys. Acta* **1121:** 199–206.

Valvekens, D.M., M. Van Montagu, and M. Van Lijsebettens. 1988. *Agrobacterium tumefaciens*-mediated transformation of *Arabidopsis thaliana* root explants by using kanamycin selection. *Proc. Natl. Acad. Sci.* **85:** 5536–5540.

Vierling, E. 1991. The roles of heat shock proteins in plants. *Annu. Rev. Plant Physiol. Plant Mol. Biol.* **42:** 579–620.

Volger, H.G. and U. Heber. 1975. Cryoprotective leaf proteins. *Biochim. Biophys. Acta* **412:** 335–349.

Weretilnyk, E., W. Orr, T.C. White, B. Iu, and J. Singh. 1992. Characterization of three related low temperature regulated cDNAs from winter *Brassica napus*. *Plant Physiol.* **101:** 171–177.

Wilhelm, K.S. and M.F. Thomashow. 1993. *Arabidopsis thaliana cor15b*, an apparent homologue of *cor15a*, is strongly responsive to cold and ABA, but not drought. *Plant Mol. Biol.* **23:** 1073–1077.

Yamaguchi-Shinozaki, K. and K. Shinozaki. 1993. Characterization of the expression of a desiccation-responsive *rd29* gene of *Arabidopsis thaliana* and analysis of its promoter in transgenic plants. *Mol. Gen. Genet.* **236:** 331–340.

————. 1994. A novel *cis*-acting element in an *Arabidopsis* gene is involved in responsiveness to drought, low-temperature, or high-salt stress. *Plant Cell* **6:** 251–264.

Zachariassen, K.E. 1985. Physiology of cold tolerance in insects. *Physiol. Rev.* **65:** 795–832.

Zeevaart, J.A.D. and R.A. Creelman. 1988. Metabolism and physiology of abscisic acid. *Annu. Rev. Plant Physiol. Plant Mol. Biol.* **39:** 439–473.

# 31

# Molecular Genetics of Amino Acid, Nucleotide, and Vitamin Biosynthesis

**Alan B. Rose and Robert L. Last**
Plant Molecular Biology Program
Boyce Thompson Institute for Plant Research
and Section of Genetics and Development
Cornell University
Ithaca, New York 14853-1801

The study of plant intermediary metabolism has recently been energized by recruitment of the techniques of genetics and molecular biology. This is an interesting field of study in plant biology because there are a number of special uses that plants make of amino acids, nucleotides, and their biosynthetic intermediates, in addition to their role in protein and nucleic acid biosynthesis. These include synthesis of secondary metabolites, such as flavonoids and the hormones auxin, cytokinin, and ethylene; the structural lignins; and the nitrogen storage compounds known as ureides. A deeper understanding of the regulation of these primary metabolic pathways should provide strategies for manipulating such agronomic traits as nutritional quality, growth rate and morphology, and environmental stress susceptibility. Studies of the regulation of metabolic pathways and coregulation of primary and secondary metabolism should lead to insights about the molecular genetic mechanisms governing complex processes in plants.

This is not an exhaustive literature review. Instead, we focus on some pathways that are of particular interest to plant biologists and about which a reasonable amount of biochemical, molecular, or genetic information is available. Whenever possible, work on *Arabidopsis thaliana* is highlighted, or mention is made of important questions that could be answered using *Arabidopsis*. We encourage interested readers to consult the recent review articles cited throughout the text to obtain more details.

## AMMONIUM ASSIMILATION INTO AMINO ACIDS

### Genetic Analysis of the GS-GOGAT Cycle

The nitrogen atoms incorporated into amino acids, nucleotides, and vitamins are ultimately derived from ammonium. This is generated by a variety of mechanisms in flowering plants, including reduction of nitrate assimilated from soil (see Crawford, this volume), photorespiration, catabolism of nitrogen-containing compounds, and symbiotic nitrogen fixation in legumes. The relative contribution of each source depends on the developmental stage and growth conditions of the plant. For example, mobilization of nitrogenous seed storage reserves is important during seed germination, and appreciable catabolism of amino acids and ureides is observed upon transition from dormancy to active growth conditions for perennial plants. Massive amounts of ammonium are liberated in the mitochondria of $C_3$ plants by photorespiratory oxidative decarboxylation of glycine. It is estimated that, at a given moment, tenfold more ammonium is being generated by photorespiration than by primary nitrogen assimilation (Keys et al. 1978), indicating the importance of coordinating ammonium assimilation with photorespiration.

The primary mechanism for the incorporation of ammonium into amino acids, which subsequently serve as the source of nitrogen atoms for other molecules, is the conversion of glutamate and ammonium into glutamine (Fig. 1, reaction 1) by glutamine synthetase (GS). The enzyme glutamate synthase (GOGAT; L-glutamate:NAD(P)$^+$ oxidoreductase [*trans*-aminating] or L-glutamate:ferredoxin oxidoreductase [*trans*-aminating]) then transfers the glutamine amide group to α-ketoglutarate to yield two molecules of glutamate (Fig. 1, reaction 2). There are chloroplast- and cytosol-localized isoenzymes of GS and GOGAT. The chloroplast isoforms are thought to be important in the reincorporation of ammonium released during photorespiration, whereas the cytosolic polypeptides are proposed to function in primary ammonium assimilation, reassimilation during amino acid catabolism, and phenylpropanoid biosynthesis (Miflin and Lea 1980). The isoforms of GOGAT are also differentiated by the reductants they utilize; the chloroplastic forms require ferredoxin (Fd-GOGAT), but the cytoplasmic isoenzymes are NADH-dependent (NADH-GOGAT).

The importance of the GS-GOGAT pathway in plants was demonstrated by $^{15}$N-labeling studies and experiments utilizing GS inhibitors (for review, see Lea et al. 1990). Analysis of photorespiratory mutants reinforced the in vivo relevance of these biochemical results. *Arabidopsis* (Somerville and Ogren 1980) and barley (*Hordeum vulgare*) (Kendall et al. 1986) mutants defective in Fd-GOGAT and barley chloroplastic glutamine synthetase (*gs2*) mutants (Blackwell et al. 1987; Wallsgrove et al. 1987) are unable to grow under photorespiratory growth conditions (atmospheric $CO_2$ and $O_2$), presumably due to an inability to reassimilate photorespiratory ammonium released in the mitochondrion (Givan et al. 1988). As predicted for the GS-GOGAT as-

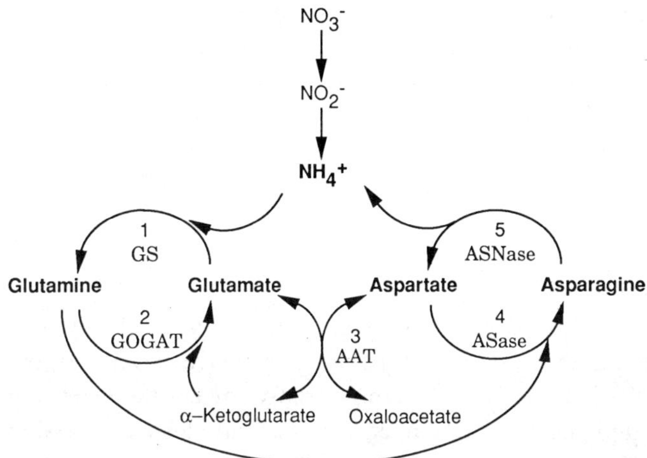


*Figure 1*    Nitrogen assimilatory amino acid biosynthesis. Enzymes: (1) glutamine sythetase (GS, EC 6.3.1.2); (2) glutamate synthase (GOGAT, NADH enzyme: EC 1.4.1.14, NADPH enzyme: EC 1.4.1.13, ferredoxin enzyme: EC 1.4.7.1); (3) aspartate aminotransferase (AAT, EC 2.6.1.10); (4) asparagine synthetase (ASase, EC 6.3.5.4); (5) asparaginase (ASNase EC 3.5.1.1).

similatory pathway, *gs2* mutants accumulate very high levels of ammonium (Blackwell et al. 1987; Wallsgrove et al. 1987) when grown under photorespiratory conditions, whereas Fd-GOGAT mutants accumulate lower concentrations (Somerville and Ogren 1980; Kendall et al. 1986). Consistent with the hypothesis that the GS reaction precedes that of GOGAT, the barley GS mutation is epistatic to that in Fd-GOGAT: The double mutant accumulates the very high levels of ammonium characteristic of the *gs2* mutant, which is defective in chloroplastic GS (Blackwell et al. 1988).

The viability of the chloroplastic *gs2* and Fd-GOGAT mutants under non-photorespiratory growth conditions indicates that the plastidic forms of GS and GOGAT are not essential for primary ammonium assimilation. Based on growth characteristics of these mutants under high $CO_2$ or low $O_2$, it appears that the cytosolic GS isoenzymes and NADH-dependent GOGAT provide sufficient assimilation of soil ammonium and nitrate for normal growth and development. In a [15]N-labeling study, Joy and coworkers (1992) demonstrated that despite normal growth, the barley photorespiratory mutants have significantly altered pools and labeling patterns of amino acids under non-photorespiratory conditions. These results suggest that, under the experimental conditions employed, barley normally has an ammonium assimilation capacity beyond that required.

This point deserves further attention, considering the importance of nitrogen assimilation in plant nutrition. However, unlike barley, no *gs2* photorespiratory mutant is yet available in *Arabidopsis*. Furthermore, the existence of multiple cytosolic GS genes in all plants means that regulatory mutants or novel approaches to inactivation of the various isoenzymes will be required (Bennett and Cullimore 1992; Last 1993).

## Molecular Biology of Glutamine Synthetase

Consistent with its role as committing enzyme in the pathway, the molecular biology of GS has received the most extensive scrutiny of the ammonium assimilation enzymes. We only briefly review the extensive literature on this topic because a number of review articles are available that describe the structure and expression of GS genes (Lea et al. 1990; Coruzzi 1991; Cullimore et al. 1992; Last 1993). In all plants examined, the distinct organ-specific and developmentally regulated isoenzymes of GS are encoded by small families of nuclear genes that produce chloroplast- and cytosol-localized isoenzymes. The chloroplastic isoform is encoded by a single gene (*GS2*) in a number of plant species (Lightfoot et al. 1988; Tingey et al. 1988; Freeman et al. 1990). Consistent with a role in reassimilating ammonium released during photorespiration, *GS2* expression is induced by light and during growth under photorespiratory conditions (Tingey et al. 1988; Edwards and Coruzzi 1989; Cock et al. 1991, 1992). In contrast, cytosolic GS is encoded by three expressed genes in pea (Walker and Coruzzi 1989) and bean (*Phaseolus vulgaris*) (Cock et al. 1991). These genes are expressed in nonphotosynthetic tissues (Forde et al. 1989; Edwards et al. 1990; Brears et al. 1991), as expected if their functions are distinct from the chloroplastic isoenzymes.

The family of GS genes in *Arabidopsis* follows this pattern (Peterman and Goodman 1991). Four classes of cDNA clones were characterized, one with a presumed chloroplast target sequence. The 1.6-kb mRNA for this presumptive plastidic isoenzyme is most highly expressed in leaf tissue, and it appears to be light-regulated because it is expressed at higher levels in green tissue than in etiolated seedlings. The other three cDNAs hybridize to 1.4-kb transcripts whose abundances vary in a tissue-specific manner and are higher in root or seeds than in leaves. These shorter mRNAs presumably encode cytosolic GS polypeptides. Although no corresponding cDNA clone was isolated, genomic Southern blot hybridization analysis suggests the existence of a fifth GS gene in *Arabidopsis*. This gene might represent a pseudogene or, alternatively, a gene that is expressed at low levels or in a limited number of cells.

There are some surprising features of the *Arabidopsis* GS gene family. First, the organization of GS genes is as complex in this crucifer as in legumes, despite vast differences in total genome sizes and the lack of symbiotic nitrogen fixation. Second, no GS-deficient photorespiratory mutant is reported in *Arabidopsis*, even though only a single chloroplastic GS gene was identified.

## Molecular Biology of Glutamate Synthase

Fd-GOGAT clones have been isolated from maize (Sakakibara et al. 1991), tobacco (Zehnacker et al. 1992), and *Arabidopsis* (K. Coshigano and G. Coruzzi, pers. comm.). The mRNAs for either the maize or tobacco Fd-GOGAT enzymes are increased coordinately with the respective chloroplast GS transcripts during greening of etiolated tissue (Sakakibara et al. 1992; Zehnacker et al. 1992). This is consistent with the combined roles of these proteins in photorespiratory ammonium assimilation. The availablity of a large number of *Arabidopsis* Fd-GOGAT-defective *gluS* mutants (Somerville and Ogren 1980; K. Coshigano and G. Coruzzi, pers. comm.) should allow an in-depth molecular genetic analysis of this enzyme. For example, some of these mutants might be found to be defective in *cis*- or *trans*-acting regulatory loci. Sequence analysis of mutant GOGAT alleles would provide information about catalytically important amino acids.

## Aspartate Aminotransferase

*trans*-Amination reactions are crucial for amino acid biosynthesis and catabolism and are an important means of transferring nitrogen atoms from amino acids into other compounds. Many amino acids are created by *trans*-amination of an oxoacid by an amino acid to form a new amino acid and oxoacid pair. Aspartate aminotransferase (AAT) plays a central role in plant nitrogen assimilation because it is the link between glutamate, the product of the GS-GOGAT pathway, and aspartate (Fig. 1, reaction 3). Aspartate in turn serves as precursor to the amino acids asparagine (Fig. 1, step 4), threonine, lysine, isoleucine, and methionine (Fig. 2); the C1 donor *S*-adenosylmethionine; and the hormone ethylene. Plant AAT, which functions as a homodimer, exists in a variety of isoenzymic forms. At least one of these forms is cytosolic, whereas other activities are found in chloroplasts, glyoxosomes, mitochondria, and peroxisomes (Wightman and Forest 1978; Givan 1980). The diversity of subcellular locations, combined with its central part in amino acid biosynthesis, makes this an attractive enzyme for studies of the role of cell biology in nitrogen assimilation and amino acid biosynthesis.

The regulation of AAT is of special significance in temperate legumes, which transport much of their symbiotically fixed nitrogen as aspartate and asparagine. Two immunologically distinct isoforms (AAT1 and AAT2) are found in alfalfa (*Medicago sativa*) (Griffith and Vance 1989; Farnham et al. 1990). AAT1 and AAT2 mRNAs are both induced during nodule development (Udvardi and Kahn 1991; Gantt et al. 1992), supporting the proposed role of this enzyme in nodule nitrogen assimilation. The observation that AAT2 mRNA levels are higher in alfalfa lines that make nitrogen-fixing (effective) nodules than in ineffective mutants suggests that the synthesis of AAT2 mRNA might be regulated by nitrogen fixation (Gantt et al. 1992), perhaps in response to substrate availability.

An alfalfa *AAT1* cDNA was isolated by suppression of an *Escherichia coli aspC* mutation (Udvardi and Kahn 1991), and a carrot *AAT1* cDNA was obtained by polymerase chain reaction (Turano et al. 1992). Alfalfa

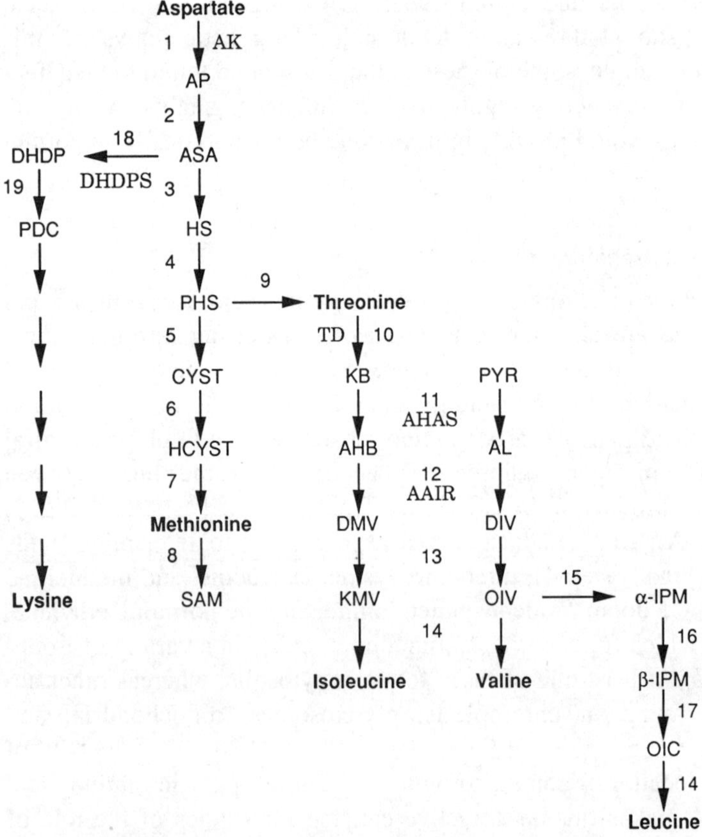


*Figure 2 (See facing page for legend.)*

*AAT2* cDNAs, cloned by immunological screening, were also shown to suppress an *E. coli aspC* mutation (Gantt et al. 1992). Whereas the *AAT2* cDNA coding region has an amino-terminal extension that is thought to encode a plastid target sequence, the subcellular locations of the proteins encoded by the *AAT1* cDNAs are less clear. The size of the gene families encoding these isoforms is not yet known.

Four classes of *Arabidopsis* cDNAs that represent four different genes (*ASP1–ASP4*) were recently isolated (C. Shultz and G. Coruzzi 1992; and pers. comm.). DNA sequence analysis suggests that *ASP2* and *ASP4* might encode mitochondrial and plastidic isoenzymes, respectively. RNA blot hybridization indicates that *ASP3* mRNA is dramatically increased in dark-adapted rosette-stage leaves. This type of "photophobic" regulation resembles the regulation of the asparagine synthetase genes of pea, suggesting coordinate regulation of aspartate and asparagine synthesis (see below).

*Arabidopsis* has two major AAT isoenzymes that are visible by activity staining of proteins fractionated by nondenaturing gel electrophoresis (C. Shultz and G. Coruzzi, pers. comm.). Several mutants with altered activity of the isoenzymes were identified by screening ex-

*Figure 2* Biosynthesis of the aspartate-derived amino acids lysine, threonine, isoleucine, and methionine. Intermediates: AP, aspartate 3-phosphate; ASA, aspartate-3-semialdehyde; HS, homoserine; PHS, *O*-phosphohomoserine; CYST, cystathionine; HCYST, homocysteine; SAM, *S*-adenosylmethionine; DHDP, 2,3-dihydrodipicolinate; PDC, piperidine-2,6-dicarboxylate; KB, $\alpha$-ketobutyrate; AHB, $\alpha$-aceto-$\alpha$-hydroxybutyrate; DMV, $\alpha,\beta$-dihydroxy-$\beta$-methylvalerate; KMV, $\alpha$-keto-$\beta$-methylvalerate; PYR, pyruvate; AL, 2-acetolactate; DIV, 2,3-dihydroxyisovalerate; OIV, 2-oxoisovalerate; $\alpha$-IPM, $\alpha$-isopropylmalate; $\beta$-IPM, $\beta$-isopropylmalate; OIC, 2-oxoisocaproate. Enzymes: (1) aspartate kinase (AK, EC 2.7.2.4); (2) aspartate-semialdehyde dehydrogenase (EC 1.2.1.11); (3) homoserine dehydrogenase (EC 1.1.1.3); (4) homoserine kinase (EC 2.7.1.39); (5) cystathionine $\gamma$-synthase (EC 4.2.99.9); (6) cystathionine $\beta$-lyase (EC 4.4.1.8); (7) methionine synthase (E.C. 2.1.1.13); (8) methionine adenosyltransferase (EC 2.5.1.6); (9) threonine synthase (EC 4.2.99.2); (10) threonine dehydratase TD (EC 4.2.1.16); (11) acetohydroxy acid synthase (AHAS, EC 4.1.3.18); (12) acetohydroxy acid isomeroreductase (AAIR, EC 1.1.1.86); (13) dihydroxyacid dehydratase (EC 4.2.1.9); (14) branched-chain amino acid glutamate transaminase (EC 2.6.1.42); (15) $\alpha$-isopropylmalate synthase (EC 4.1.3.12); (16) 3-isopropylmalate dehydratase (EC 4.2.1.33); (17) $\beta$-isopropylmalate dehydrogenase (EC 1.1.1.85); (18) dihydrodipicolinate synthase (DHDPS, EC 4.2.1.52); (19) dihydrodipicolinate reductase (EC 1.3.1.26). Abbreviations for enzymes discussed in the text are shown beside the reactions they catalyze. The steps leading from piperidine-2,6-dicarboxylate to lysine were excluded due to uncertain data for this pathway in plants (Bryan 1990).

tracts from individual $M_2$ plants. Identification of mutants altered in a single AAT isoenzyme, and the availability of the structural genes, would permit critical analysis of their roles in nitrogen assimilation and redistribution during *Arabidopsis* development.

## Asparagine Synthesis and Degradation

Asparagine is a major nitrogen storage and transport molecule in many flowering plants, perhaps due to its high nitrogen-to-carbon ratio, high solubility, and chemical stability. These properties are presumably important during seed germination, when large amounts of nitrogen are mobilized and transported to the growing axes, and during dark growth, when carbon skeletons are limiting. Although several pathways were proposed for asparagine synthesis (for review, see Lea et al. 1990), the asparagine synthetase (ASase; Fig. 1, reaction 4)-mediated *trans*-amidation of aspartate by glutamine is favored as a general mechanism (Joy et al. 1983; Sieciechowicz et al. 1988). Now that ASase clones are available (Tsai and Coruzzi 1990, 1991), it should be possible to make transgenic plants with reduced ASase activity and to test the in vivo importance of these other pathways of asparagine synthesis.

Regulation of the expression of the two ASase genes of pea is consistent with the proposed role in asparagine synthesis (Tsai and Coruzzi 1990, 1991). The mRNA levels are increased during nitrogen fixation and seed germination, as expected for an enzyme involved in mobilization of ammonium generated by nitrogen fixation and catabolism of cotyledonary protein reserves. This regulatory behavior is similar to that previously observed for cytosolic GS.

The observed diurnal fluctuations in free asparagine levels appear to result from changes in levels of asparagine synthetic and catabolic enzymes. When pea plants are grown under 16 hours of daylight, the highest concentrations of asparagine are found in tissue harvested during the dark period (Sieciechowicz et al. 1985). This is correlated with increased asparagine synthetase gene transcription and mRNA accumulation during the dark phase (Tsai and Coruzzi 1990, 1991). In fact, transcription of both *AS* genes of pea is rapidly repressed by light (Tsai and Corruzzi 1991). Thus, increased expression of pea ASase gene expression, and presumably enzyme activity, can account for higher asparagine levels during the dark growth period. The drop in soluble asparagine in the light is also correlated with a rapid increase in asparaginase (ASNase; Fig. 1, reaction 5) activity (Sieciechowicz et al. 1985). This catabolic enzyme removes the asparagine amide group to form aspartate and ammonium, allowing the mobilization of ammonium that accumulates in asparagine during the dark growth period. Thus, pea leaf asparagine levels appear to be governed by the opposing forces of synthesis and degradation, predominant during dark and light growth, respectively. With the availability of lupine ASNase clones (Lough et al. 1992), it is now possible to study the genetic regulation of both asparagine synthetic and catabolic enzymes. One approach is to reduce ASase activity (and thus asparagine levels) in transgenic plants and to monitor changes in the gene expression and activity of ASNase or, conversely, to assay the effects of reduced asparaginase on ASase regulation.

## ASPARTATE-DERIVED AMINO ACIDS

Aspartic acid plays a central role in amino acid biosynthesis because it is a precursor to four amino acids in addition to asparagine (Fig. 1, reaction 4): lysine, threonine, methionine, and isoleucine (Fig. 2). Although not yet intensively studied in *Arabidopsis*, the synthesis of aspartate-derived amino acids continues to be the focus of considerable attention in other plants because cereals and legumes are deficient in lysine and methionine, respectively. Strategies for increasing the biosynthesis of these nutritionally important amino acids in crop plants have been adopted based on current ideas about the regulation of the pathway. Although there is evidence for biochemical regulation at several steps in plants (Bryan 1980; Wallsgrove and Mazelis 1981; Giovanelli et al. 1989a; Ghislain et al. 1990; Turano et al. 1990), we focus on two enzymes for which in vivo data indicate their importance: aspartate kinase (Fig. 2, reaction 1) and dihydrodipicolinate synthase (DHDPS; Fig. 2, reaction 18).

Addition of lysine plus threonine to the culture medium is toxic to plants and cell cultures, and this growth inhibition is reversed by methionine (Bright et al. 1978; Hibberd et al. 1980). The growth repression appears to result from inhibition of aspartate kinase isoenzymes by lysine and/or threonine, leading to a methionine deficiency (see Fig. 2). This observation led to selection of plant mutants that were resistant to this inhibitory mixture of amino acids, caused by dominant mutations that render an aspartate kinase isoenzyme less sensitive to lysine inhibition (Hibberd et al. 1980; Bright et al. 1982; Rognes et al. 1983; Arruda et al. 1984; Frankard et al. 1991). Genetic and biochemical analysis of lysine-plus-threonine-resistant mutants in both barley (Bright et al. 1982; Rognes et al. 1983; Arruda et al. 1984) and maize (Hibberd et al. 1980; Hibberd and Green 1982; Diedrick et al. 1990) revealed two genetically unlinked aspartate kinase loci capable of mutating to resistance. Although it was hoped that lysine-plus-threonine-resistant mutants would accumulate lysine, increases only in soluble threonine levels were observed. As was the case for the lysine-resistant aspartate kinase mutants, expression of a feedback-desensitized mutant aspartate kinase gene from *E. coli* in transgenic tobacco also led to threonine overproduction (Shaul and Galili 1992) without the desired lysine accumulation. These results suggest operation of an additional mechanism for regulating lysine levels in vivo.

Alteration of DHDPS activity was shown to enhance plant lysine content. A high-lysine mutant of *Nicotiana sylvestris* was identified by selecting for growth of protoplasts on the toxic lysine analog S-(2-aminoethyl)L-cysteine (Negrutiu et al. 1984). This dominant mutant was shown to have a feedback-insensitive form of DHDPS. Similarly, increased lysine levels were observed in transgenic tobacco (Shaul and Galili 1992) and potato (Perl et al. 1992) expressing *E. coli* DHDPS. These results supported biochemical results in duckweed (*Lemna paucicostata*) that indicated the importance of end-product inhibition of the enzyme DHDPS in regulating in vivo lysine biosynthesis (Giovanelli et al. 1989b). These studies also suggested that simultaneous genetic manipulation of both aspartate kinase and DHDPS might lead to lysine overproduction in agronomically important species.

Double mutant analysis confirmed this as a potentially useful strategy. A double mutant heterozygous for both lysine-insensitive aspartate kinase and DHDPS was described in *N. sylvestris* (Frankard et al. 1992). As predicted, these plants accumulate very high concentrations of soluble lysine in rosette leaves (5- to 15-fold higher than the wild type and 1.5- to 2.5-fold higher than the DHDPS single mutant). The observation that threonine accumulation characteristic of the single aspartate kinase mutant is suppressed in the double mutant reinforces the importance of DHDPS in the regulation of flux through this pathway.

Phenotypic peculiarities of the lysine-overproducing double mutant tobacco have implications for engineering high-lysine plants. Unlike either single mutant parent, the double mutant heterozygote plants had extreme morphological abnormalities, failed to produce flower stalks, and thus were sterile (Frankard et al. 1992). Similar, but less severe, growth changes were reported for the highest-lysine-accumulating tobacco plants that expressed the *E. coli* DHDPS gene (Shaul and Galili 1992). Another observation was that leaf-soluble lysine accumulation was developmentally regulated in the doubly heterozygous mutant (Frankard et al. 1992). Significantly increased lysine levels were observed in adult plants only within a relatively narrow time frame. If the morphological abnormalities are found in genetically engineered high-lysine varieties of other plant species, strategies for reversing the deleterious effects will be needed. For example, if free lysine is directly causing the abnormalities, diversion of the amino acid into less detrimental forms such as vegetative or seed storage proteins might return plant growth to normal. Recently, clones for maize DHDPS (Frisch et al. 1991) and the *Arabidopsis* (Ghislain et al. 1992), carrot (Matthews et al. 1992), and maize (Muehlbauer et al. 1992) aspartate kinase-homoserine dehydrogenase bifunctional enzymes were reported. It should now be possible to isolate genes for feedback-insensitive plant enzymes and to introduce them into transgenic plants under promoters that confer appropriate quantitative and tissue-specific expression patterns.

It is of interest to identify the means by which lysine is affecting plant growth and development and to dissect the mechanisms that regulate lysine accumulation during development. For example, unlinked enhancer or suppressor mutations could be sought to increase or reduce the morphological defects of a lysine overproducer, respectively.

## BRANCHED-CHAIN AMINO ACIDS

The isoleucine, valine, and leucine pathways have received a considerable amount of attention in *Arabidopsis* and other plants. The committing enzyme, acetohydroxy acid synthase (AHAS or ALS; Fig. 2, reaction 11), is the target of a variety of commercial herbicides, including sulfonylureas (Ray 1984), imidazolinones (Shaner et al. 1984; Muhitch et al. 1987), and triazolopyrimidine sulfonanilides (Subramanian et al. 1991). Dominant *Arabidopsis* mutations that confer herbicide resistance were identified in the single-copy *CSR1* AHAS structural gene. The *csr1* allele (originally identified in line GH50) confers resistance to sulfonylureas (Haughn and Somerville 1986; Haughn et al. 1988), and *csr1-2* (originally isolated in line GH90) causes imidazolinone tolerance (Haughn and Somerville 1990; Sathasivan et al. 1990). Another mutation, which causes resistance to triazolopyrimidine and sulfonylurea herbicides, was originally called *csr1-2* (Mourad and King 1992), and renamed *csr1-3* (Mourad et. al. 1993). The *Arabidopsis csr1* and *csr1-1* mutant alleles confer herbicide resistance in transgenic tobacco (Haughn et al. 1988; Gabard et al. 1989; Sathasivan et al. 1991), validating their use as selectable markers for plant transformation. A cDNA for *Arabidopsis* acetohydroxy acid isomeroreductase, the enzyme that catalyzes the next step in the branched chain amino acid biosynthetic pathway (AAIR; Fig. 2, reaction 12) was also described previously (Curien et al. 1993).

In plants, AHAS is subject to end-product inhibition by leucine and valine (Borstlap and Vernooy-Gerritsen 1985; Relton et al. 1986). As a result, addition of valine to the growth medium causes isoleucine-reversible plant growth restriction, presumably because inhibition of AHAS causes isoleucine starvation. Mutants with altered response to these branched-chain amino acids were identified. An *Arabidopsis* mutant (405-A-154) that is hypersensitive to valine, or a mixture of valine and leucine, was described previously (Acedo and Rédei 1976). Several valine-resistant mutants from tobacco-cell cultures were genetically and biochemically characterized (Bourgin 1978; Bourgin et al. 1985). In one class of tobacco mutants (referred to as Val$^r$-2, Val$^r$-3, Val$^r$-4, and Val$^r$-5 in Borstlap et al. [1985] and in Bourgin et al. [1985]), valine resistance was transmitted as a digenic recessive trait. These *vr2*, *vr3* double mutants are altered in uptake of neutral and acidic amino acids, including leucine and valine (Borstlap et al. 1985; Bourgin et al. 1985). In *Arabidopsis*, a T-DNA-induced valine-resistant mutant was shown to have a mutation at a locus that cross-hybridizes with the neutral amino acid carrier system II gene, *NAT1* (Hsu et al. 1993; D. Bush, pers. comm.). Analysis of this and related mutants will provide new insights into the roles of the various amino acid transporter systems in *Arabidopsis*.

In contrast, valine-resistant tobacco that has dominant *Vr1* mutations produces feedback-insensitive AHAS enzyme (Bourgin et al. 1985; Relton et al. 1986). Overexpression of the *Arabidopsis* AHAS activity also causes valine resistance in transgenic tobacco (Tourneur et al. 1993). The amino acid resistance of plants with altered AHAS activity is consistent with a role for AHAS as a regulatory enzyme in valine and leucine biosynthesis (Borstlap and Vernooy-Gerritsen 1985). Surprisingly, *Arabidopsis* AHAS enzyme purified from *E. coli* was insensitive to inhibition by valine and leucine (Singh et al. 1992). It appears that endproduct inhibition of the *Arabidopsis* enzyme requires other plant factors.

AHAS does not appear to play a role in regulation of plant isoleucine biosynthesis (Borstlap and Vernooy-Gerritsen 1985), whereas threonine deaminase (TD; Fig. 2, step 10) is a good candidate for controlling isoleucine production. TD is inhibited by isoleucine, and dominant *omr1*$^-$ *Arabidopsis* mutations were identified that cause resistance to the toxic effects of L-*O*-methylthreonine (G. Mourad and J. King, in prep.). The resistant mutant produces TD activity with a $K_i$ for isoleucine 50 times higher than that of the wild-type enzyme. The mutant accumulates 20-fold higher soluble isoleucine than the wild type, as a result of the feedback-insensitive TD activity. This result is consistent with the hypothesis that the analog is toxic because it competes with isoleucine for incorporation into proteins. The mutation maps to chromosome III, linked to *hy2*.

## AROMATIC AMINO ACID BIOSYNTHESIS

In addition to phenylalanine, tryptophan, and tyrosine, the aromatic amino acid biosynthetic pathways in plants are responsible for synthesis of a variety of important non-amino acid products, including vitamins and a great diversity of secondary metabolites (Gilchrist and Kosuge 1980; Floss 1986). Many functions are ascribed to these compounds, including regulation of growth processes (the auxin indole-3-acetic acid), protection from abiotic environmental stresses such as UV-B (flavonols, flavonones, and cinnamate derivatives), attraction of pollinating insects (flavonoid pigments), protection from herbivores (alkaloids), enhancement and inhibition of microbial growth and gene expression (phytoalexins, flavonoids, and cinnamate derivatives), and cell wall structural components (lignins). There is significant carbon flux through these secondary pathways and, as a result, up to 25% of fixed carbon is thought to flow through the prechorismate pathway (Singh et al. 1991). The production of some aromatic secondary compounds is influenced by environmental or developmental signals, suggesting that the synthesis of aromatic amino acids or precursor molecules might be subject to complex regulation. An important long-term goal in the study of both aromatic amino acid and secondary metabolism in plants is to manipulate the synthesis of these important compounds at will. This would allow the introduction of desirable agronomic traits such as pest resistance, synthesis of novel organic compounds, and altered plant development.

The prechorismate (Fig. 3, reactions 1–7) and tryptophan (reactions 8–12) pathways appear to be conserved among bacteria, fungi, and plants (Bryan 1990; Singh et al. 1991; Last 1993). In contrast, there is increasing biochemical evidence that arogenate is the final common intermediate in tyrosine and phenylalanine biosynthesis (reactions 14–16; for review, see Singh et al. 1991). This differs from the pathway in commonly studied microbes such as *E. coli* and bakers yeast, where prephenate is the final shared intermediate. The lack of phenylalanine- or tyrosine-requiring mutants prevents critical evaluation of the relative importance of these two pathways in plants in vivo.

### Prechorismate Pathway

In the pathway common to the three aromatic amino acids (the shikimate, or prechorismate, pathway), phosphoenolpyruvate, derived from glycolysis, and erythrose-4-phosphate, from the pentose phosphate and photosynthetic carbon reduction pathways, are precursors for the synthesis of chorismate (Fig. 3). Chorismate is then converted to phenylalanine or

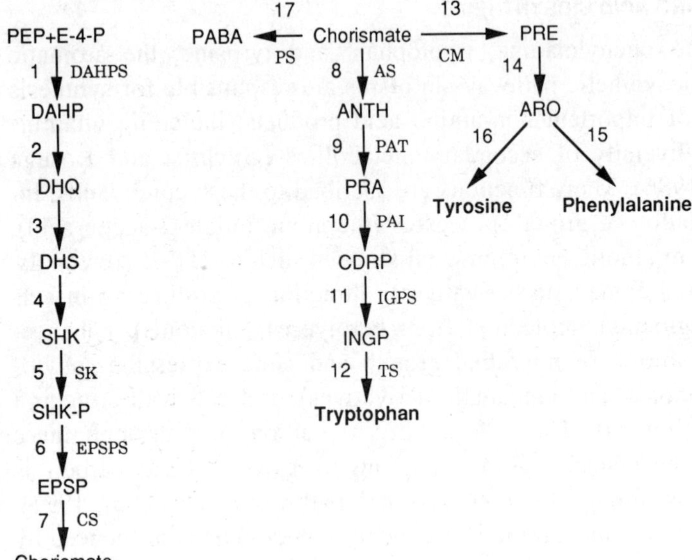


*Figure 3* Biosynthesis of the aromatic amino acids. Intermediates: PEP, phosphoenolpyruvate; E-4-P, erythrose-4-phosphate; DAHP, 3-deoxy-D-*arabino*-heptulosonate 7-phosphate; DHQ, 3-dehydroquinate; DHS, 3-dehydroshikimate; SHK, shikimate; SHK-P, shikimate-3-phosphate; EPSP, 5-enolpyruvylshikimate 3-phosphate; ANTH, anthranilate; PRA, phosphoribosyl anthranilate; CDRP, 1-(0-carboxyphenylamino)-1-deoxyribulose-5-phosphate; INGP, indole-3-glycerol phosphate; PRE, prephenate; ARO, arogenate; PABA, *p*-aminobenzoic acid. Enzymes: (1) deoxy-D-*arabino*-D-heptulosonate 7-phosphate synthase (DAHPS, EC 4.1.2.15); (2) 3-dehydroquinate synthase (EC 4.6.1.3); (3) 3-dehydroquinate dehydratase (EC 4.2.1.10); (4) shikimate dehydrogenase (EC 1.1.1.25); (5) shikimate kinase (SK, EC 2.7.1.71); (6) 5-*enol*pyruvylshikimate 3-phosphate synthase (EPSPS, EC 2.5.1.19); (7) chorismate synthase (CS, EC 4.6.1.4); (8) anthranilate synthase (AS, EC 4.1.3.27); (9) phosphoribosylanthranilate transferase (PAT, EC 2.4.2.18); (10) phosphoribosylanthranilate isomerase (PAI, EC 5.3.1.24); (11) indole-3-glycerol-phosphate synthase (IGPS, EC 4.1.1.48); (12) tryptophan synthase (TS, EC 4.2.1.20); (13) chorismate mutase (CM, EC 5.4.99.5); (14) prephenate aminotransferase; (15) arogenate dehydratase (E.C. 4.2.1.91); (16) arogenate dehydrogenase (E.C. 1.3.1.43); (17) *p*-amino benzoate synthase (PS). Abbreviations for enzymes discussed in the text are shown beside the reactions they catalyze.

tyrosine (via chorismate mutase, reaction 13), tryptophan (via anthranilate synthase, reaction 8), or *p*-aminobenzoic acid (PABA; via PABA synthase, reaction 17), a component of the vitamin folic acid.

## DAHP Synthase

The committing step of aromatic amino acid biosynthesis is catalyzed by the enzyme 3-deoxy-D-*arabino*-heptulosonate 7-phosphate synthase (DAHPS; Fig. 3, reaction 1). Intriguingly, two classes of enzymes capable of carrying out the synthesis of DAHP are known: a plastid-localized $Mn^{++}$ enzyme and a non-plastidic form that is activated by $Co^{++}$ and $Mg^{++}$ (Morris et al. 1989; Bryan 1990; Singh et al. 1991). Chorismate mutase (Fig. 3, reaction 13) is the only other enzyme of aromatic amino acid biosynthesis for which a cytosolic form is characterized (d'Amato et al. 1984). The significance of these non-plastidic activities in chorismate biosynthesis remains unclear. For example, the $Co^{++}/Mg^{++}$ DAHPS has a broad substrate specificity, with a variety of sugars able to efficiently substitute for erythrose-4-phosphate (Doong et al. 1992). Thus, the importance of this isoenzyme in aromatic amino acid biosynthesis is questionable. No molecular or genetic analysis of the $Co^{++}/Mg^{++}$ DAHPS or cytosolic chorismate mutase protein has been reported.

In contrast, the molecular biology of the $Mn^{++}$ DAHPS enzyme is being pursued in a number of plants, including *Arabidopsis*. cDNAs for two potato $Mn^{++}$ isoenzymes were reported (Dyer et al. 1990; Zhao and Herrmann 1992), as were two in *Arabidopsis* (Keith et al. 1991), one from tobacco (Wang et al. 1991), and a tomato (*Lycopersicon esculentum*) gene (K. Herrmann, pers. comm.). The protein abundance and enzyme activity of $Mn^{++}$ DAHPS are highly regulated (Suzich et al. 1985; Dyer et al. 1989; McCue and Conn 1989, 1990), consistent with the hypothesis that this enzyme plays a key role in primary and secondary aromatic metabolism. For example, synthesis of the $Mn^{++}$-activated enzyme is induced in response to inhibition of the prechorismate pathway enzyme 5-*enol*pyruvylshikimate-3-phosphate synthase (EPSPS; Fig. 3, reaction 6) by the herbicide glyphosate (Pinto et al. 1988). The mRNA level and enzyme activity of $Mn^{++}$ DAHPS are also induced by environmental stresses that cause increased demand for aromatic secondary products: wounding (Dyer et al. 1989; Keith et al. 1991; Muday and Herrmann 1992), high light intensity (McCue and Conn 1990), fungal elicitor treatment (McCue and Conn 1989; Henstrand et al. 1992), and bacterial pathogenesis (Keith et al. 1991). Similar stress-induction kinetics are seen for DAHPS and phenylalanine ammonia lyase, the committing enzyme in secondary aromatic biosynthesis. These observations suggest that synthesis of these key primary and secondary metabolic enzymes is coregulated (Dyer et al. 1989; Keith et al. 1991; Henstrand et al. 1992). It is of great interest to understand the mechanism underlying this regulatory network.

There is evidence for differential regulation of duplicate DAHPS genes in plants. In *Arabidopsis*, expression of only one (*DHS1*) of two cloned genes is induced by wounding and bacterial pathogenesis (Keith et al. 1991). A similar situation is seen in potato, where DAHPS mRNA and enzyme encoded by the *shkA* gene are stress-induced, whereas *shkB* is constitutively expressed (Muday and Herrmann 1992; J. Zhao and K. Herrmann, pers. comm.). The similar response of these cruciferous and solanaceous plants suggests the possibility that this is a widespread regulatory strategy. It is possible that the *DHS1/shkA* genes have the specialized role of production of aromatic secondary compounds and that *DHS2/shkB* are "housekeeping" genes for protein synthesis.

## Other Shikimate Pathway Genes

Genes for the enzymes that catalyze the final three steps in chorismate biosynthesis are available. The molecular biology of EPSP synthase (Fig. 3, reaction 6) was vigorously pursued because it is the target for the herbicide glyphosate (Amrhein et al. 1980; Kishore and Shah 1988; Mazur and Falco 1989). Two nonallelic genes are detected in *Arabidopsis* (Klee et al. 1987; H. Klee, pers. comm.), and one or two genes are found in petunia and tomato (Gasser et al. 1988). Transgenic plants resistant to high levels of glyphosate were generated by overexpressing plant EPSPS (Kishore and Shah 1988) or by introducing EPSPS genes isolated from microorganisms that can grow in extremely high herbicide concentrations (Barry et al. 1992). cDNA clones for a single shikimate kinase gene (Fig. 3, reaction 5) were isolated from tomato (Schmid et al. 1992), and chorismate synthase (Fig. 3, reaction 7) cDNAs were reported from *Cordydalis sempervirens* (Schaller et al. 1991) and tomato (J. Schmid and N. Amrhein, pers. comm.).

## Subcellular Localization of Aromatic
## Amino Acid Biosynthesis

Many aromatic amino acid biosynthetic enzymes are found in the plastid (Mousdale and Coggins 1985), and tryptophan biosynthesis is thought to occur only in the chloroplast (Bryan 1990; Singh et al. 1991). In agreement with a plastidic localization, all of the genes of the prechorismate and tryptophan pathways that have been isolated encode proteins with amino-terminal peptides that resemble plastid target sequences. In-vitro-synthesized potato *shkA* and *shkB* gene products (J. Zhao and K. Herrmann, pers. comm.) and the tomato shikimate kinase precursor (Schmid et al. 1992) are translocated into isolated chloroplasts and cleaved by chloroplast peptidase, verifying that these amino-terminal sequences function in plastid targeting.

## Molecular Genetics of *Arabidopsis* Tryptophan Biosynthesis

A molecular genetic analysis is providing a detailed understanding of the *Arabidopsis* tryptophan pathway, which produces many indolic secondary compounds as well as the amino acid. Analysis of tryptophan-requiring mutants and tryptophan biosynthetic enzyme genes indicates that the plant pathway is identical to that found in microorganisms. Tryptophan-requiring (auxotrophic) *Arabidopsis* mutants were the first whole-plant amino acid auxotrophs isolated (Last and Fink 1988; Last et al. 1991); these mutants show that at least some amino acid-requiring mutants are viable. The first tryptophan auxotrophs were isolated by selection for 5-methylanthranilate (5MA) resistance (Last and Fink 1988). This selection works presumably because 5MA is converted to the toxic 5-methyltryptophan (5MT) by the same enzymes that synthesize tryptophan from anthranilate. This approach yielded *trp1⁻* mutations, defective in phosphoribosylanthranilate transferase (PAT; Fig. 3, reaction 9) (Last and Fink 1988; Rose et al. 1992), *trp2⁻* mutations altered in the tryptophan synthase β subunit (TSβ; Fig. 3, reaction 12) (Last et al. 1991), and *trp3⁻* mutants that are affected in the α subunit of tryptophan synthase (TSα; Fig. 3, reaction 12; E. Radwanski and R. Last, unpubl.). The analog 6-methylanthranilate (6MA) is especially effective for selection of *Arabidopsis* tryptophan pathway mutants (J. Li and R. Last, unpubl.). Inhibition of *Arabidopsis* growth by this anthranilate analog agrees with a previous report of herbicidal activity on a variety of plant species (Thomas 1984). Other strategies that led to identification of defects in specific enzymes are described in our treatment of the individual enzymes.

Cloned genes are available for all enzymes of tryptophan biosynthesis in *Arabidopsis*, and many have been genetically mapped (Table 1). Heterologous hybridization probes from *Saccharomyces cerevisiae* were used to isolate the first anthranilate synthase α subunit (Niyogi and Fink 1992) and tryptophan synthase β subunit (Berlyn et al. 1989) clones. cDNA clones for all enzymes in the tryptophan pathway are biologically active in *E. coli*, which enabled direct cDNA cloning of many of the genes by suppression of *E. coli* mutations (Elledge et al. 1991; Rose et al. 1992; Niyogi et al. 1993; J. Li and R. Last; E. Radwanski and R. Last; both in prep.).

The ability to express functional *Arabidopsis* enzymes and in-vitro-generated mutant polypeptides in microbes will facilitate their enzymological characterization. Transformation of an *E. coli* or yeast null mutant with mutant cDNA alleles of tryptophan pathway structural genes should become a powerful approach to structure-function studies of the

*Table 1*  Cloned genes of the tryptophan pathway

| Enzyme | Gene name | Method of isolation[a] | Location | Reference[b] |
|---|---|---|---|---|
| Anthranilate synthase | | | | |
| α subunit | *ASA1* | Hyb. | Chr. 5 near RFLP 3837 | 1 |
| | *ASA2* | Hyb. | Chr. 2 between *er* and RFLP 21502 | 1 |
| β subunit | *ASB1* | Sup. | Chr. 1, 14.9 cM from *GAPB* and 21.6 cM from *chl1* | 2 |
| | *ASB2* | Hyb. | Chr. 5 near *lfy* | 2 |
| | *ASB2* | Hyb. | | 2 |
| Phosphoribosylanthranilate transferase | *PAT1* | Sup. | Chr. 5 between RFLPs pCIT1243 and pCIT116 | 3 |
| Phosphoribosylanthranilate isomerase | *PAI1* | Sup. | Chr. 1 near RFLP m488 | 4 |
| | *PAI2* | Sup. | Chr. 5 near *ASA1* | 4 |
| | *PAI3* | Sup. | | 4 |
| Indole-3-glycerolphosphate synthase | *IGS1* | Sup. | | 4 |
| Tryptophan synthase | | | | |
| α subunit | *TSA1* | Sup. | | 5 |
| β subunit | *TSB1* | Hyb. | Chr. 5 near RFLP 4130 | 6 |
| | *TSB2* | Hyb. | Chr. 4 near RFLP λAt272 | 6 |

[a](Hyb.) Hybridization with heterologous probe; (Sup.) suppression of *E. coli* mutation.
[b](1) Niyoka and Fink 1992; (2) Niyogi et al. 1993; (3) Rose et al. 1992; (4) J. Li and R. Last, in prep.; (5) E. Radwanski et al., in prep.; (6) Last et al. 1991.

plant enzymes. Alternatively, direct mutant selections in microorganisms that express the plant cDNA will allow identification of alleles that confer desirable properties such as inhibitor resistance.

This approach was used to facilitate genetic engineering of 5-MT-resistant transgenic *Arabidopsis* that overproduce tryptophan (Niyogi 1993). Anthranilate synthase (AS) mutations that were predicted to cause reduced feedback inhibition by tryptophan were created by site-directed mutagenesis of cDNAs encoding *Arabidopsis* AS α subunit. Prior to introduction of these mutant alleles into transgenic *Arabidopsis*, they were tested for biological activity in *S. cerevisiae*. Only one of three amino acid changes conferred 5-MT-resistance to yeast. When this mutant AS subunit was expressed from the 35S cauliflower mosaic virus promoter, the transgenic *Arabidopsis* plants were resistant to 5-MT and accumulated approximately 50% higher soluble tryptophan than vector transformants.

## AS

AS is the committing enzyme of the tryptophan pathway, converting chorismate to the aromatic compound anthranilate (for reviews of this and other microorganismal pathway enzymes, see Hütter et al. 1986; Crawford 1989). It is likely that this enzyme catalyzes an important regulatory step in the pathway, because plant AS activity is inhibited by micromolar concentrations of tryptophan (Kreps and Town 1992). The larger α subunit is responsible for aromatization of chorismate and acts in concert with the β subunit to transfer an amino group from glutamine. In high ammonium concentrations, the α subunit alone can produce anthranilate from chorismate. It would be interesting to test whether Fd-GOGAT mutations, which cause ammonium accumulation, suppress *trp4* mutations in the AS β subunit gene *ASB1* (Niyogi et al. 1993).

Two nonallelic *Arabidopsis* genes for the α subunit (*ASA1* and *ASA2*) were characterized previously (Niyogi and Fink 1992); these share approximately 65% amino acid identity. Despite the strong amino acid conservation, mRNA accumulation patterns from the two α subunit genes are dramatically different, suggesting the possibility of unique roles for the two isoenzymes. *ASA1* mRNA varies among tissue types, with high levels in leaf and root and low or undetectable levels in stem, flower, green silique, and dry seed. In contrast, *ASA2* mRNA is present at low or undetectable levels in all tissues examined. This pattern of two genes that produce transcripts of dramatically different abundance is similar to that observed for the two highly conserved genes encoding TSβ, the final step in the pathway (Last et al. 1991; Pruitt and Last 1993).

Another feature of regulation of the two *ASA* genes is that, unlike the "unregulated" *ASA2* gene, *ASA1* leaf mRNA increases in response to wounding and infiltration of both virulent and avirulent strains of the bacterial pathogen *Pseudomonas syringae* (Niyogi and Fink 1992). This response resembles the induction of the mRNA for *DHS1* described above. Although gene-specific probes were not used, it was demonstrated that AS β subunit mRNA is also induced by bacterial pathogens (Niyogi et al. 1993). The co-induction of mRNA accumulation for the committing enzymes in both the prechorismate and tryptophan branches of the pathway raises the possibility of coordinate genetic regulation of these important enzymes. The increased *ASA* and *ASB* mRNA might reflect an enhanced need for tryptophan or pathway intermediates as precursors to indolic secondary metabolites because crucifers, including *Arabidopsis*, produce indolic glucosinolates (Hogge et al. 1988) and phytoalexins (Tsuji et al. 1992), compounds that are postulated to function in plant defense. Now that *Arabidopsis* AS β subunit clones (*ASB1*, *ASB2*, and *ASB3*) are available (Niyogi et al. 1993), it will be interesting to compare the organization and regulation of expression of the individual genes for these two subunits.

Two classes of AS activity mutants were identified in *Arabidopsis*. A dominant *amt1⁻* mutant, identified as resistant to the toxic analog α-methyltryptophan, has altered AS activity (Kreps and Town 1992). Extracts from *amt1⁻* mutant plants contain an AS activity that is less sensitive to end-product inhibition by tryptophan. Leaves and callus of *amt1⁻* contain increased pools of free tryptophan, consistent with altered biochemical regulation of AS activity. Genetic mapping indicates that the *amt1⁻* mutation is linked to the AS α subunit gene *ASA1* (J. Kreps and C. Town, pers. comm.). AS β subunit-deficient mutants (*trp4⁻*) were isolated in a screen for phenotypic revertants of the blue fluorescence phenotype associated with anthranilate accumulation in the PAT-deficient *trp1-100* mutant (Niyogi et al. 1993). The rationale for this screen was that defects in activities earlier in the tryptophan pathway, potentially including enzymes of the prechorismate pathway, would cause reduced synthesis and accumulation of anthranilate (Fig. 3, reaction 9). The *trp4 trp1-100* double mutant is a tryptophan-requiring auxotroph, despite the fact that either single mutant is able to grow in the absence of tryptophan. Thus, it appears that the two leaky mutations combine to yield a more severely affected "synthetic" double mutant phenotype. Additive phenotypes are also observed for tryptophan auxotrophic mutations: *trp1-1 trp2-1* and *trp1-1 trp3-1* double mutants are more severely affected than any of the single mutant plants (R. Last, unpubl.). These observations suggest that it might be possible to identify plants bearing leaky mutations throughout the prechorismate and tryptophan pathways by screening for second-site mutations that enhance the severity of prototrophic mutations such as *trp1-100* or *trp4*. This screen, and identification of other suppressors of *trp1* blue fluorescence, could open up the prechorismate pathway to genetic analysis.

## *PAT*

The PAT enzyme catalyzes the condensation of anthranilate and phosphoribosylpyrophosphate (PRPP) to yield phosphoribosylanthranilate and pyrophosphate (Fig. 3, reaction 9). *Arabidopsis* PAT-deficient *trp1* mutants have a diverse group of phenotypes, including blue fluorescence under UV light, caused by accumulation of anthranilate compounds, and resistance to 5- and 6-methylanthranilate (Last and Fink 1988; Rose et al. 1992; J. Li and R. Last, unpubl.). Although the first *trp1* mutant (*trp1-1*) was identified by 5MA resistance (Last and Fink 1988), many alleles were subsequently obtained by direct screening for blue-fluorescent plants (Rose et al. 1992; A. Rose and R. Last, in prep.). The phenotypes of *trp1* mutants range from tryptophan-requiring mutants with severe morphological defects (*trp1-1*, for example) and reduced PAT protein accumulation to prototrophic mutants with normal morphology (including *trp1-100*).

The *trp1-1* and *trp1-100* mutations were demonstrated to be alleles of the single-copy *Arabidopsis* PAT structural gene (*PAT1*) by mapping and complementation studies (Rose et al. 1992). The intense blue fluorescence of the *trp1* mutants indicates that the *PAT1* gene might be useful as a selectable marker or reporter of gene expression. Unlike other plant reporter genes, it should be possible to screen and select for both increased and decreased PAT activity. Expression of functional PAT protein can be selected (tryptophan prototrophy), screened (loss of blue fluorescence), counterselected (resistance to methylanthranilate), and counterscreened (blue fluorescence, tryptophan requirement, and bushy phenotype).

## *Phosphoribosylanthranilate Isomerase*

cDNA clones for phosphoribosylanthranilate isomerase (PAI; Fig. 3 reaction 10) were identified by suppression of *E. coli trpC* mutations (J. Li and R. Last, in prep.). There are three nonallelic *PAI* genes detectable by high-stringency genomic Southern analysis. cDNA clones for two of these genes (*PAI1* and *PAI2*) were found, indicating that they are transcribed. A transcript-specific hybridization probe demonstrated that *PAI3* is also transcribed. The presumptive transcribed regions of all three genes are remarkably well conserved. For example, there is only a single conservative amino acid difference between *PAI1* and *PAI2*. Triplicate genes might preclude identification of PAI-deficient mutants of *Arabidopsis* by traditional genetic screens or selections.

### Tryptophan Synthase

Tryptophan synthase (TS; Fig. 3 reaction 12) catalyzes the conversion of indole-3-glycerol phosphate to tryptophan (Miles 1991). Of all the microbial tryptophan biosynthetic enzymes, the structure, function, and genetic regulation of TS is particularly well understood (Yanofsky and Crawford 1987; Miles 1991). In bacteria, and apparently in plants, the enzyme is a heterotetramer of structure $\alpha\beta\beta\alpha$, and each subunit can catalyze a half reaction, albeit less efficiently than in the active complex:

$\alpha$-subunit activity:

Indole-3-glycerol phosphate $\longleftrightarrow$ Indole + Glyceraldehyde-3-phosphate

$\beta_2$-subunit activity:

Indole + L-Serine $\rightarrow$ L-Tryptophan + $H_2O$

In *Arabidopsis*, there are two unlinked genes (*TSB1* and *TSB2*) that encode very similar $\beta$ subunit proteins (Berlyn et al. 1989; Last et al. 1991). *TSB1* produces the majority of mRNA in leaf and other tissues, whereas *TSB2* mRNA is present at a constant low level in the tissues examined (Last et al. 1991; Pruitt and Last 1993). These mRNA expression results were extended by *TSB* promoter-*GUS* fusion studies (Pruitt and Last 1993), which demonstrated higher *GUS* activity in plants transformed with the *TSB1-GUS* fusion constructs. The *TSB1* promoter specifies developmentally regulated *GUS* reporter gene expression in stem, vascular tissues, root tips, mesophyll cells, and cells located at the base and tip of the anther filament. *TSB2-GUS* activity is apparent later in development in a less complex pattern, which overlaps that of *TSB1*.

The observation that recessive *trp2*[−] mutations in the more highly expressed *TSB1* gene lead to a tryptophan auxotrophy (Last et al. 1991) is consistent with the differences in mRNA abundances: The *TSB2* gene presumably makes a modest contribution to total tryptophan synthesis. This is an example where genetic redundancy does not preclude isolation of amino acid-requiring mutants. In contrast, two different maize TS$\beta$ subunit genes (*ORP1* and *ORP2*) must be mutant for a tryptophan requirement (Wright et al. 1992).

*Arabidopsis* mutants that are deficient in TS$\beta$ activity are obtained by selection for 5-fluoroindole resistance. The basis for this selection is that TSβ converts indole to tryptophan and can convert exogenously supplied 5-fluoroindole to the toxic analog 5-fluorotryptophan. A range of variants were obtained by this method, including tryptophan-requiring *trp2⁻* alleles and prototrophic Trp⁺ mutants (A. Barczak et al., in prep.). A number of these mutants have altered accumulation of TSβ mRNA or protein. As described for the *PAT1* gene, the variety of selectable and screenable phenotypes associated with TSβ-deficient *Arabidopsis* mutants makes this an attractive reporter gene and selectable marker. The availabilty of a spectrum of *trp2* mutant alleles, and the ability to express the plant enzyme in *E. coli*, should permit structure-function studies with this interesting protein.

Tryptophan synthase α subunit cDNA and *TSA1* gene were cloned by suppression of an *E. coli trpA* mutation (E. Radwanski and R. Last, in prep.). It was demonstrated that the plant TSα subunit is not biologically active unless the plant TSβ subunit is also expressed in *E. coli*. This result suggests that the plant α subunit does not form a productive interaction with the bacterial β protein. This hypothesis is reasonable because it is known that the rate of the bacterial α subunit reaction is enhanced 150-fold following binding of serine to the adjacent β subunit pyridoxal phosphate (Anderson et al. 1991). This approach was also used to identify *Arabidopsis* AS β subunit cDNAs by screening for clones that rendered an *E. coli* strain containing the plant AS α able to utilize glutamine as amino donor (Niyogi et al. 1993).

### Can This Pathway Be Studied in Other Plants?

The feasibility of using 5MA to select for mutants in the tryptophan pathway in other plants and unicellular photosynthetic eukaryotes is supported by work on the green alga *Chlamydomonas reinhardtii* (Dutcher et al. 1992). Several of the 5MA-resistant mutants have phenotypes suggestive of defects in the tryptophan biosynthetic pathway, including a mutant that cannot convert anthranilate to indole-3-glycerol phosphate. Surprisingly, none of these mutants has a strict tryptophan requirement, consistent with genetic redundancy or biochemically leaky mutations. Tryptophan auxotrophs would not be recovered if this alga is unable to efficiently utilize tryptophan from the growth medium. Blue fluorescence and 5-fluoroindole resistance should also be useful phenotypes for isolating plant tryptophan pathway mutants. For example, the maize tryptophan synthase-deficient *orp1 orp2* double mutant is blue fluorescent and 5-fluoroindole-tolerant (Wright et al. 1992).

### PROLINE BIOSYNTHESIS

There is great interest in the regulation of proline biosynthesis in plants because soluble proline accumulates in response to a variety of stress conditions, including drought and salinity (Hanson and Hitz 1982), and this is proposed to constitute a protective response. Plants appear to produce the imino acid proline by two mechanisms: from glutamate and by catabolism of arginine and ornithine (Thompson 1980; Bryan 1990). It is not possible to evaluate the relative importance of these two pathways in vivo because no plant mutants auxotrophic for proline due to a biosynthetic defect are reported. In addition to a possible role in stress response, the glutamate-derived pathway was proposed to be important in regulating de novo purine biosynthesis in plants (Kohl et al. 1988). This is because two molecules of $NADP^+$ are generated for each molecule of proline. $NADP^+$, in turn, stimulates the pentose phosphate pathway to produce the purine precursor ribose-5-phosphate. Thus, proline and nucleotide biosynthesis might be linked in plants. This inter-relationship is thought to be especially important under situations that favor synthesis of the ureides allantoin and allanoate, such as in nitrogen-fixing nodules of tropical legumes (Schubert and Boland 1990). This is because purines are precursors to these abundant nitrogen storage and transport molecules.

Analog-resistant mutants that overproduce soluble proline are available in carrot (Widholm 1976; Cella et al. 1982) and barley (Kueh and Bright 1981). Growth of the barley mutants is slightly less sensitive to the inhibitory effects of NaCl (Kueh and Bright 1982), as expected if soluble proline is involved in salinity tolerance. Recent progress on the molecular biology of glutamate-derived proline biosynthesis in *Arabidopsis* and other plants should provide new approaches to test the importance of proline accumulation in stress response.

The existence of the enzymes of the glutamate-derived proline biosynthetic pathway in plants was confirmed by molecular cloning experiments. Genes for the enzyme $\Delta^1$-pyrroline-5-carboxylate reductase (P5CR), which catalyzes the final step in proline biosynthesis (Fig. 4, reaction 4), were identified from a number of species. The first clones were obtained by functional suppression of an *E. coli proC* mutation using a soybean cDNA library (Delauney and Verma 1990), and clones were later isolated from pea (Williamson and Slocum 1992) and *Arabidopsis* (Verbruggen et al. 1993). *Arabidopsis* P5CR mRNA is regulated in a tissue-specific manner, with the highest concentrations in seeds, roots, and flowers and lowest in stem and leaf tissue. The *Arabidopsis* gene shows a rapid accumulation of P5CR mRNA in response to NaCl stress, as was previously reported in soybean and pea

*Figure 4* Glutamate-derived proline biosynthesis. Intermediates: G5P, glutamyl-5-phosphate; G5SA, glutamyl-5-semialdehyde; P5C, $\Delta^1$-pyrroline-5-carboxylate. Enzymes: (1) the bifunctional $\Delta^1$-pyrroline-5-carboxylate synthetase (P5CS) found in mothbean (Hu et al. 1992); (2) γ-glutamyl kinase (E.C. 2.7.2.11) activity; (3) γ-glutamyl phosphate reductase activity (also called glutamic-γ-semialdehyde dehydrogenase) (EC 1.2.1.41); (4) $\Delta^1$-pyrroline-5-carboxylate reductase (P5CR, E.C. 1.5.1.2).

plants. All of the P5CR clones characterized thus far encode proteins without obvious organelle target sequences, despite the fact that both chloroplast-localized and cytosolic P5CR were detected in leaves of pea (Rayapati et al. 1989) and soybean (Szoke et al. 1992). Verbruggen and co-workers (1993) observed hybridization of an *Arabidopsis* P5CR cDNA clone to multiple genomic sites; perhaps these represent genes for both cytosolic and chloroplastic isoenzymes.

cDNA clones for γ-glutamyl kinase (Fig. 4, reaction 2), the committing enzyme in glutamate-derived proline biosynthesis, were recently isolated by suppression of the *E. coli proB* mutation with a mothbean (*Vigna aconitifolia*) expression library (Hu et al. 1992). Surprisingly, the plant cDNAs encode activities for both γ-glutamyl kinase and the next enzyme in the pathway, γ-glutamyl phosphate reductase (Fig. 4, reaction 3). Thus, the committing reaction in this pathway is encoded by a bifunctional enzyme, $\Delta^1$-pyrroline-5-carboxylate synthetase (P5CS; Fig. 4, reaction 1). Two reported properties of the plant P5CS are of considerable interest. The γ-glutamyl kinase activity of the plant protein synthesized in *E. coli* is 30 times less sensitive to feedback inhibition by proline than the *E. coli* enzyme. This property might promote high-level proline synthesis in nodules and under stress conditions. In addition, as previously seen for P5CR, the concentration of P5CS mRNA increases in response to NaCl treatment. Thus, salt-induced gene expression is observed for the entire biosynthetic pathway. With the molecular dissection of this pathway, it is now possible to directly test whether genetic engineering of high-level proline accumulation will confer stress tolerance to plants. Similarly, it should be possible to determine the relative importance of this pathway in proline biosynthesis compared to catabolism of arginine and ornithine.

## NUCLEOTIDES

There are two kinds of pathways for the synthesis of nucleotides, the de novo and the salvage routes. The de novo pathways, illustrated in Figures 5 and 6, catalyze the multistep synthesis of nucleotides from simpler molecules such as PRPP, amino acids, and tetrahydrofolate. These are energetically expensive pathways. For example, ATP or GTP hydrolysis occurs in 5 of the 12 steps in the de novo pathway between PRPP and AMP (Fig. 5). The enzymes that constitute the salvage pathways (Fig. 7) recycle the purine and pyrimidine bases released by the degradation of nucleic acids (or supplied exogenously) back into nucleotides. In contrast to the energy requirements of the de novo pathway, AMP is created from PRPP and free adenine by a single salvage reaction (Fig. 7) that does not require ATP.

The available biochemical and molecular evidence suggests that plants utilize the same de novo and salvage reactions as those defined in microbes. Enzyme activities for various steps throughout these pathways have been found in a wide range of plant species, as recently reviewed by Wagner and Backer (1992). Furthermore, several plant cDNAs have been isolated by suppression of bacterial or yeast nucleotide auxotrophic mutations, suggesting functional conservation of these pathways between kingdoms. However, it will not be possible to state firmly that the pathways shown in Figures 5–7 exist in *Arabidopsis* until biochemical assays are performed. Given the biochemical diversity of plants, it is possible that some aspects of the pathways illustrated in Figures 5–7 will be inaccurate for some plant species, including *Arabidopsis*.

Very few plant mutants defective in nucleotide biosynthesis were isolated. Mutations that completely block synthesis of any nucleotide are almost certainly lethal, because nucleotides and nucleotide derivatives are abundant molecules with essential functions in a multitude of biochemical processes. Exogenously added purine or pyrimidine bases would need to be transported to the appropriate location for synthesis of all of the necessary nucleotides and derivatives, and therefore may not be able to rescue whole plants that are nucleotide auxotrophs. This may explain why the only reported plant adenine auxotroph is a cell line, rather than a fertile plant (King et al. 1980).

Two *Arabidopsis* mutants are known to have perturbed nucleotide metabolism. One is defective in an adenine salvage enzyme, and the other is deficient in a cofactor required by several enzymes, including one involved in purine degradation. Considerable progress has recently been made in isolating *Arabidopsis* genes encoding enzymes of nucleotide synthesis. These mutants and genes, and relevant examples from other plant species, are described in the following section.

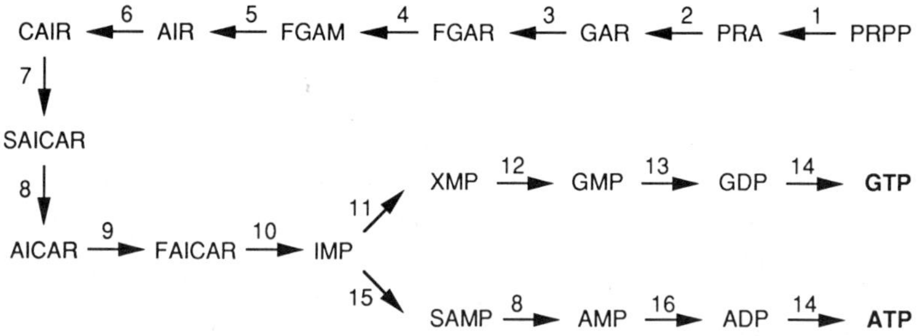

CAIR ←⁶— AIR ←⁵— FGAM ←⁴— FGAR ←³— GAR ←²— PRA ←¹— PRPP

7 ↓

SAICAR

8 ↓

AICAR —⁹→ FAICAR —¹⁰→ IMP

XMP —¹²→ GMP —¹³→ GDP —¹⁴→ GTP

11 ↗ (from IMP)

15 ↘ SAMP —⁸→ AMP —¹⁶→ ADP —¹⁴→ ATP

*Figure 5*  De novo purine nucleotide biosynthesis. Intermediates: PRPP, phosphoribosylpyrophosphate; PRA, 5-phosphoribosylamine; GAR, glycinamide ribotide; FGAR, *N*-formylglycinamide ribotide; FGAM, *N*-formylglycinamidine ribotide; AIR, 5-aminoimidazole ribotide; CAIR, carboxyaminoimidazole ribotide; SAICAR, succinoaminoimidazole carboxyamide ribotide; AICAR, aminoimidazole carboxyamide ribotide; FAICAR, formylaminoimidazole carboxyamide ribotide; IMP, inosine monophosphate; XMP, xanthosine monophosphate; GMP, guanosine monophosphate; GDP, guanosine diphosphate; GTP, guanosine triphosphate; SAMP, succinoadenosine monophosphate; AMP, adenosine monophosphate; ADP, adenosine diphosphate; ATP, adenosine triphosphate. Enzymes: (1) amidophosphoribosyl transferase (E.C. 2.4.2.14); (2) GAR synthetase (E.C. 6.3.4.13); (3) GAR formyltransferase (E.C. 2.1.2.2); (4) FGAM synthetase (E.C. 6.3.5.3); (5) AIR synthetase (E.C. 6.3.3.1); (6) AIR carboxylase (E.C. 4.1.1.21); (7) SAICAR synthetase (E.C. 6.3.2.6); (8) adenylosuccinate lyase (E.C. 4.3.2.2); (9) AICAR formyltransferase (E.C. 2.1.2.3); (10) IMP cyclohydrolase (E.C. 3.5.4.10); (11) IMP dehydrogenase (E.C. 1.1.1.205); (12) GMP synthetase (E.C. 6.3.5.2); (13) GMP kinase (E.C. 2.7.4.8); (14) nucleoside diphosphate kinase (E.C. 2.7.4.6); (15) SAMP synthetase (E.C. 6.3.4.4); (16) AMP kinase (E.C. 2.7.4.3).

## Purines

The *apt1* mutant is the most thoroughly characterized *Arabidopsis* mutant defective in nucleotide metabolism (Moffatt and Somerville 1988). These plants were isolated using a selection for resistance to 2,6-diaminopurine, an adenine analog that is converted by salvage pathway enzymes into a toxic nucleotide. Biochemical analysis showed that the *apt1* mutants lack adenine phosphoribosyl transferase (APRT) (Moffatt and Somerville 1988), an enzyme that uses PRPP to convert adenine into AMP (Fig. 7). The *apt1* mutants are also unable to convert the cytokinin benzyladenine into its nucleotide form, benzyladenine monophosphate (Moffatt et al. 1991). In vitro, the *Arabidopsis* APRT can also convert the cytokinins benzyladenine and isopentenyladenine into nucleotides, albeit much less efficiently than adenine: the $K_m$ for benzyladenine is 730 μM, compared with 4.5 μM for adenine (D. Lee and B. Moffatt, pers.

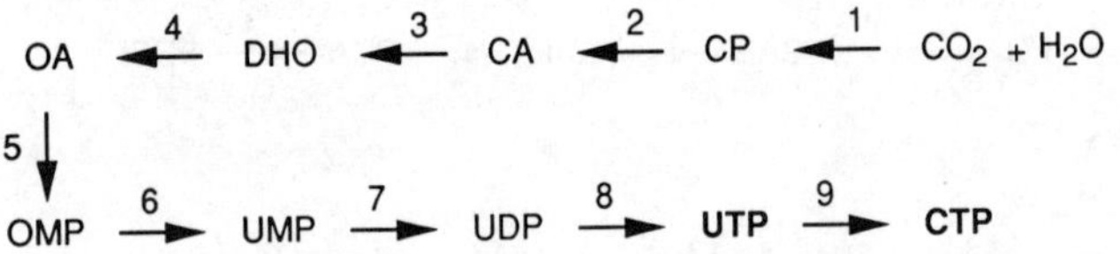

$$\text{OA} \xleftarrow{\ 4\ } \text{DHO} \xleftarrow{\ 3\ } \text{CA} \xleftarrow{\ 2\ } \text{CP} \xleftarrow{\ 1\ } \text{CO}_2 + \text{H}_2\text{O}$$

$$5 \downarrow$$

$$\text{OMP} \xrightarrow{\ 6\ } \text{UMP} \xrightarrow{\ 7\ } \text{UDP} \xrightarrow{\ 8\ } \text{UTP} \xrightarrow{\ 9\ } \text{CTP}$$

*Figure 6* De novo pyrimidine nucleotide biosynthesis. Intermediates: CP, carbamoyl phosphate; CA, carbamoyl aspartate; DHO, dihydro-orotate; OA, orotic acid (orotate); OMP, orotate monophosphate; UMP, uridine monophosphate; UDP, uridine diphosphate; UTP, uridine triphosphate; CTP, cytidine triphosphate. Enzymes: (1) carbamoyl phosphate synthase (E.C. 6.3.5.5); (2) aspartate carbamoyltransferase (E.C. 2.1.3.2); (3) dihydro-orotate synthase (E.C. 3.5.2.3); (4) dihydro-orotate dehydrogenase (E.C. 1.3.99.11); (5) orotate phosphoribosyltransferase (E.C. 2.4.2.10); (6) orotidine monophosphate decarboxylase (E.C. 4.1.1.23); (7) nucleoside monophosphate kinase (E.C. 2.7.4.4); (8) nucleoside diphosphate kinase (E.C. 2.7.4.6); (9) CTP synthetase (E.C. 6.3.4.2).

comm.). The biological significance of cytokinin nucleotides remains unclear, as reviewed by McGaw (1987).

An unexpected phenotype is that the *apt1* mutants are male sterile due to abnormal pollen production (Moffatt and Somerville 1988). Visible defects in anther development appear just after the pollen mother cells undergo meiosis (Regan and Moffatt 1990), suggesting that sufficient nucleotides are present for the S phase that precedes meiosis I. The arrest of development could be caused either by the lack of a needed compound normally formed by APRT, or by an increase to inhibitory levels of a metabolite on which APRT usually acts. For example, the conversion of

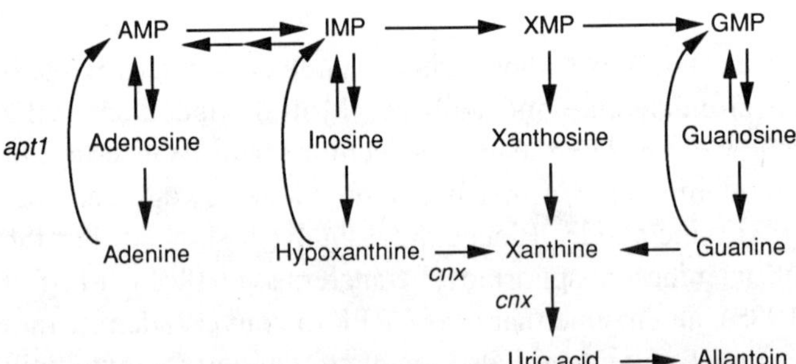

*Figure 7* Purine salvage pathways. Intermediates: AMP, adenosine monophosphate; IMP, inosine monophosphate; XMP, xanthosine monophosphate; GMP, guanosine monophosphate. The enzymes defective in the *apt1* (adenine phosphoribosyltransferase, E.C. 2.4.2.7) and *cnx* (xanthine oxidase, E.C. 1.1.3.22, also called xanthine dehydrogenase) mutants of *Arabidopsis* are indicated.

cytokinins into their nucleotide forms might be important in the regulation of pollen development. Alternatively, cells lacking APRT activity may accumulate a compound such as methylthioadenine, which, as a byproduct of polyamine and ethylene synthesis, might be produced in relatively large amounts during flower development (Moffatt et al. 1991).

Both cDNA and genomic clones for *Arabidopsis* APRT have been isolated (Moffatt et al. 1992), and it was demonstrated that expression of the cDNA clone reverses the *apt1* infertility phenotype (S. Regan and B. Moffatt, pers. comm.). The APRT was purified from *Arabidopsis* and some of the characteristics of this enzyme were analyzed (Lee and Moffatt 1993). The native enzyme is a homodimer composed of two 27-kD subunits, in agreement with the size predicted from the cDNA sequence. The organs with the highest APRT-specific activity in wild-type plants are the flowers and roots, which contain roughly three times as much APRT activity as leaves.

A mutant defective in the initial steps of purine degradation was isolated. The *Arabidopsis cnx* mutant has reduced levels of xanthine oxidase (also called xanthine dehydrogenase), a protein that catalyzes the conversion of hypoxanthine to xanthine, and xanthine to uric acid (see Fig. 7). The reason for this enzyme defect is that the mutants lack a molybdopterin cofactor that is required for the activity of most molybdenum-containing enzymes, including xanthine oxidase and nitrate reductase (Braaksma and Feenstra 1982). These plants, isolated from a screen for resistance to chlorate, were first recognized as nitrate reductase mutants (see Crawford, this volume). Allopurinol, a specific and potent inhibitor of xanthine oxidase, may provide a means of determining the phenotypic consequences of reduced xanthine oxidase activity alone.

The only reported plant that has abnormal de novo purine synthesis is the Ad1 cell line of *Datura innoxia* (King et al. 1980). These cells can grow only when adenine, adenosine, inosine, or aminoimidazolecar boxamide ribotide (AICAR) is supplied in the medium (Mitsui and Ashihara 1988), suggesting a block in de novo purine synthesis at a step before AICAR formation (Fig. 5, reaction 8). The AICAR released as a by-product of histidine biosynthesis may be unable to meet the demand for purines in these cells.

### Genes of De Novo Purine Synthesis

Preliminary reports indicate that plant cDNAs encoding all except the fourth and tenth steps of the de novo purine pathway between PRPP and IMP (Fig. 5, reactions 1–3, 5–9) have been isolated by suppression of yeast or *E. coli* mutations. Several groups constructed plant cDNA libraries in an expression vector and transformed these libraries into microbial mutants defective in a known step. *Arabidopsis* cDNAs that provide functional suppression of the *E. coli purD* and *purN* mutations (impaired in reactions 2 and 3, respectively, Fig. 5; K. Schnorr, pers. comm.), the *E. coli purM*, *purC*, *purB*, and *purH* mutations (reactions 5, 7–9, Fig. 5; Senecoff and Meagher 1993), and the yeast *ade2* mutation (defective in reaction 6; Minet et al. 1992) were isolated in this way. The *E. coli purF*, *purE*, and *purC* mutations (lacking reactions 1, 6, and 7, respectively) have also been suppressed by cDNAs from soybean and mothbean libraries (D. Verma, pers. comm.).

The sequence of the *Arabidopsis* cDNAs isolated by suppression of the *E. coli purD* (reaction 2, K. Schnorr, pers. comm.) and *purM* (reaction 5, Senecoff and Meagher 1993) mutations were determined. They encode proteins that resemble the *E. coli* glycinamide ribotide synthetase and aminoimidazole ribotide synthetase, respectively, suggesting that the isolated genes encode the analogous *Arabidopsis* genes, rather than causing an artifactual suppression. Preliminary reports suggest that most of the enzymes of nucleotide metabolism in *Arabidopsis* are encoded by unique genes and catalyze single reactions.


### Pyrimidines

No *Arabidopsis* mutants defective in pyrimidine metabolism are described, but all steps of the de novo pyrimidine pathway to uridylate (Fig. 6, reactions 1–6) are reported to be represented by cloned cDNAs. *Arabidopsis* cDNAs capable of suppressing the yeast *ura1* (defective in reaction 4, Fig. 6), *ura2* (reactions 1 and 2), *ura4* (reaction 3), and *ura5 ura10* double (reaction 5 is completely eliminated only in strains containing both mutations) mutations were reported (Minet et al. 1992). The cDNA that suppresses the yeast *ura5 ura10* double mutations also complements a *ura3* (lacks reaction 6, Fig. 6) mutation (Minet et al. 1992), providing support for biochemical evidence that plants contain a bifunctional protein with both orotate phosphoribosyltransferase and orotidine monophosphate decarboxylase activity (Walther et al. 1984; Doremus 1986). Tobacco cell culture mutants defective in the orotate phosphoribosyltransferase activity were obtained by selection for resistance to the toxic compound 5-fluoro-orotate (Santoso and Thornburg 1992). This inhibitor is widely used in yeast research to select against cells with a functional *URA3* or *URA5* gene (Boeke et al. 1984). A well-defined counterselectable marker analogous to the yeast *URA3* gene would prove quite useful in *Arabidopsis* research.

## VITAMINS

Plants synthesize the vitamin cofactors that animals require in their diets. *Arabidopsis* mutants that require either thiamine or biotin for normal growth and development were described, as reviewed by Rédei (1970, 1975) and Rédei and Koncz (1992). Vitamin auxotrophs were also identified in other plant species, including the thiamine mutants of pea, barley, and tomato (Langridge and Brock 1961; Kumar and Sharma 1987), pyridoxine mutants of pea (Kumar 1988), and the pantothenate-requiring cell line of *D. innoxia* (King et al. 1980; Sahi et al. 1988). It might be possible to find *Arabidopsis* mutants defective in the synthesis of these or other vitamins if a suitable screen or selection can be devised.

The biochemical analysis of vitamin synthesis is extremely difficult, in large part because of the minuscule amounts of product. As a result, the vitamin pathways are not completely understood even in *E. coli*. Microbial mutants have been useful in identifying some of the steps of selected pathways, but more biochemical data will be needed before these pathways can be fully defined. The assumption that plants and bacteria utilize the same pathways is supported by the limited evidence available.

### Thiamine

The isolation of thiamine-requiring mutants is the only published successful screen for *Arabidopsis* auxotrophs performed by looking for unhealthy plants that could be rescued by a rich mixture of nutrients (Feenstra 1964; Li et al. 1967; Li and Rédei 1969). In this way, more than 200 thiamine-requiring mutants in five complementation groups were identified (Langridge 1955; Rédei 1975; Koornneef and Hanhart 1981), the first one reported in 1955. The failure to find other types of auxotrophs by this approach suggests that thiamine is unusual in its ability to be taken up or transported through plants, or that the consequences of a lack of thiamine are less severe or more easily reversed than for other metabolites (Li et al. 1967). Alternatively, the enzymes of other pathways may be encoded by duplicate genes.

Thiamine is synthesized by the production of pyrimidine pyrophosphate and thiazole monophosphate, which are joined to form thiamine monophosphate. This compound is ultimately converted into the coenzyme form, thiamine pyrophosphate, as outlined in Figure 8. The formation of the pyrimidine (4-amino-5-hydroxymethyl-2-methyl pyrimidine) is unusual in that it is derived from an intermediate of purine biosynthesis, 5-aminoimidazole ribotide (AIR). Although less is known about the origin of 4-methyl-5-($\beta$-hydroxyethyl) thiazole (Brown and

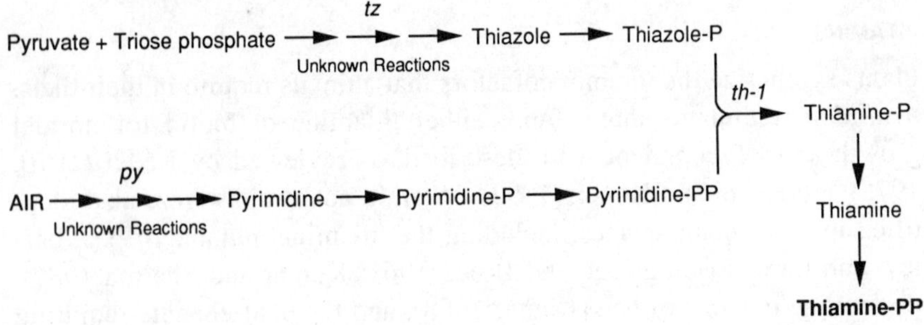


*Figure 8*   Biosynthesis of thiamine pyrophosphate. Abbreviations: AIR, 5-aminoimidazole ribotide; pyrimidine, 4-amino-5-hydroxymethyl-2-methyl pyrimidine; thiazole, 4-methyl-5-(β-hydroxyethyl) thiazole; -P, monophosphate; -PP, pyrophosphate. The steps affected by the *tz*, *py*, and *th-1* mutants of *Arabidopsis* are indicated.

Williamson 1987), labeling studies implicated tyrosine, cysteine, and a pentulose. A similar thiazole ring is found in the *Arabidopsis* phytoalexin camalexin (Tsuji et al. 1992). Analysis of camalexin synthesis in *tz* mutants should reveal if the thiazole moieties are derived from the same biosynthetic pathway.

Pathway intermediate feeding studies were useful in determining where in the thiamine pathway the mutants are blocked (Li and Rédei 1969). The *py* mutants can grow normally if supplemented with the pyrimidine, indicating a block in the conversion of AIR to the un-phosphorylated pyrimidine (Fig. 8). Similarly, the *tz* mutants must be unable to synthesize the thiazole molecule because exogenous thiazole reverses the deficiency. The *th-1*, *th-2*, and *th-3* mutants are rescued only with thiamine and not pyrimidine or thiazole, consistent with blocks in steps later in the pathway. In fact, it was demonstrated that the *th-1* mutant lacks thiamine phosphate pyrophosphorylase (E.C. 2.5.1.3, Komeda et al. 1988), the enzyme that joins the pyrimidine and thiazole moieties. The *th-2* and *th-3* mutants could be defective in the kinases that phosphorylate pyrimidine or thiazole, or in thiamine phosphate phos-phorylase activity.

## Biotin

Why were no other auxotrophs isolated from the screen that yielded the thiamine mutants? One possible explanation is that other auxotrophic mutations may be embryo-lethal, so that homozygous mutant seeds either develop improperly or fail to germinate. Evidence that some auxotrophs fail to produce viable seed comes from screening a collection of

*Arabidopsis* embryo-lethal mutants for embryos that could be rescued in culture by a mixture of amino acids, nucleosides, and vitamins (Baus et al. 1986). Analysis of one such mutant (*bio1*) revealed that supplementation with biotin alone allowed the embryos or homozygous seeds to grow (Schneider et al. 1989).

The *bio1* mutants could also be rescued by two of the intermediates of biotin biosynthesis, desthiobiotin (DTB) (Schneider et al. 1989) and 7,8-diaminopelargonic acid (DAPA) (Shellhammer and Meinke 1990), but not by 7-keto-8-aminopelargonic acid (KAPA) (Shellhammer 1991) (Fig. 9). These results place the *bio1* lesion at the DAPA synthase step and suggest that the last two steps of the plant biotin pathway are the same as in bacteria (Eisenberg 1987).

One quarter of the seeds produced by self-fertilization of heterozygous (*bio1/BIO1*) plants are inviable under normal growth conditions. The biotin concentration in these *bio1/bio1* embryos is very low (Shellhammer and Meinke 1990) and is presumably insufficient to allow normal seed development and germination. This indicates that very little biotin is normally transported into the developing *bio1/bio1* embryos from the heterozygous mother plant. However, if the heterozygous plants are regularly watered with a biotin solution, the homozygous mutant seeds they produce are viable and can be grown to maturity (Schneider et al. 1989). This result suggests that there is some transport of biotin to the developing seed.

No reports describing the cloning of plant genes for thiamine or biotin biosynthesis have appeared. The *Arabidopsis th-1* and *bio1* mutants are blocked in steps analogous to the *E. coli thiB* and *bioA* mutants, respectively. It may be possible to isolate the wild-type *Arabidopsis* genes defective in the mutants by screening a cDNA library for clones able to suppress these bacterial mutants. The *E. coli thiA* and *thiC* mutants are blocked in pyrimidine and thiazole synthesis, respectively, as is the case for the *Arabidopsis py* and *tz* mutants. However, cloning by suppression may not succeed because it remains unclear which precise reactions are defective in these mutants. A further obstacle is that vitamin biosynthetic enzymes are likely to be encoded by low-abundance mRNAs, which may be underrepresented in a cDNA library.

Unknown steps                                      *bio1*

→ ━▶ ━▶ PCA ━▶ KAPA ━▶ DAPA ━▶ DTB ━▶ biotin

*Figure 9* Biotin biosynthetic pathway. Intermediates: PCA, pimeloyl- coenzyme A; KAPA, 7-keto-8-amino pelargonic acid; DAPA, 7,8-diaminopelargonic acid; DTB, desthiobiotin. The reaction blocked in the *Arabidopsis bio1* mutant (DAPA synthase) is indicated.

## CONCLUSIONS AND PERSPECTIVES

The techniques of molecular biology and genetics have only recently been combined with biochemistry in the study of the diverse and complex metabolism of plants. This promising approach has yielded a fairly complete description of the ammonium assimilation and tryptophan biosynthetic pathways in *Arabidopsis* and is providing important inroads for the study of how other amino acids and nucleotides are synthesized. *Arabidopsis* is an excellent choice for the study of plant metabolism because of the molecular and genetic advantages that it offers.

The similarity of most plant primary metabolic pathways to those in microorganisms has facilitated the isolation of many *Arabidopsis* biosynthetic genes, either by cross-hybridization, suppression of bacterial or yeast mutations, or EST analysis. The cloned genes are valuable tools for the study of transcriptional regulation and can be used to perturb the normal activity of a pathway. Much of what is known about gene regulation was first learned from studying microbial metabolic pathways. Plants have an even greater potential than microorganisms to provide interesting examples of metabolic gene regulation, because plants are complex multicellular organisms that respond to a broad range of stimuli.

Important directions for future research in plant metabolism include detailed analysis of several amino acid pathways. For example, cysteine biosynthesis is an exciting area for genetic and molecular dissection because this process bridges inorganic nutrition (sulfate uptake and reduction) and amino acid biosynthesis (for a recent review, see Schmidt and Jäger 1992). The pathways of cysteine biosynthesis are important for plant environmental stress responses, including detoxification of the air pollutant sulfur dioxide. A cDNA for spinach cysteine synthase was identified by oligonucleotide hybridization screening (Saito et al. 1992), opening up new approaches to this pathway. As a precursor to phytochelatins, the synthesis of glutathione from cysteine plays a role in the detoxification of heavy metals. Glutathione is also a component of the plant oxidative stress response. Amino acid catabolism is another neglected area of amino acid metabolism that would benefit from creative molecular genetic approaches.

The transport of amino acids, nucleotides, and pathway intermediates is another field that is rather poorly understood. Intracellular trafficking presumably plays an important role in many aspects of plant metabolism. For example, the *Arabidopsis dct* (*dicarboxylate transport*) photorespiratory mutant is defective in transport of glutamate and aspartate into and out of the chloroplast (Somerville and Ogren 1983). The properties of this mutant suggest that the primary in vivo role for the dicarboxylate transporter is the transport of 2-oxoglutarate and glutamate across the chloroplast envelope. There is good evidence for the importance of transport throughout the plant of the amino acids central to nitrogen metabolism, such as glutamine and asparagine. In contrast, the physiological significance of long-distance movement of lower-abundance amino acids is less clear. Isolation of cDNAs that encode the plasma membrane neutral system I (Frommer et al. 1993) and *Arabidopsis* neutral system II (Hsu et al. 1993) amino acid transporters provide new approaches to studying amino acid translocation. Plant mutants deficient in amino acid or nucleotide biosynthesis or transport, and cloned genes for the affected proteins, should be useful tools in studying this and other aspects of amino acid metabolism.

## ACKNOWLEDGMENTS

The authors thank Jianmin Zhao for helpful comments, and Mary Westlake for assistance with the manuscript. Work in the laboratory of the authors was supported by a grant from the Biotechnology Research and Development Corporation, National Institutes of Health grant GM-43134, and National Science Foundation Presidential Young Investigator Award DMB-9058134 to R.L.L.

## REFERENCES

Acedo, G.N. and G.P. Rédei. 1976. Regulation of amino acid metabolism in *Arabidopsis*. *Arabidopsis Inf. Serv.* **13:** 120–124.

Amrhein, H., B. Teus, P. Gehrke, and H.C. Steinrucken. 1980. The site of inhibition of shikimate pathway by glyphosate. *Plant Physiol.* **66:** 830–834.

Anderson, K.S., E.W. Miles, and K.A. Johnson. 1991. Serine modulates substrate channeling in tryptophan synthase. *J. Biol. Chem.* **266:** 8020–8030.

Arruda, P., S.W.J. Bright, J.S.H. Kueh, P.J. Lea, and S.E. Rognes. 1984. Regulation of aspartate kinase isoenzymes in barley mutants resistant to lysine plus threonine. Construction and analysis of combinations of the *Ltla*, *Ltlb*, and *Lt2* mutant genes. *Plant Physiol.* **76:** 442–446.

Barry, G., G. Kishore, S. Padgette, M. Taylor, K. Kolacz, M. Weldon, D. Re, D. Eicholtz, K. Fincher, and L. Hallas. 1992. Inhibitors of amino acid biosynthesis: Strategies for imparting glyphosate tolerance to crop plants. *Curr. Top. Plant Physiol. Am. Soc. Plant Physiol. Ser.* **7:** 139–145

Baus, A.D., L. Franzman, and D.W. Meinke. 1986. Growth *in vitro* of arrested embryos from lethal mutants of *Arabidopsis thaliana*. *Theor. Appl. Genet.* **72:** 577–586.

Bennett, M.J. and J.V. Cullimore. 1992. Selective cleavage of closely-related mRNAs by synthetic ribozymes. *Nucleic Acids Res.* **20:** 831–837.

Berlyn, M.B., R.L. Last, and G.R. Fink. 1989. A gene encoding the tryptophan synthase β subunit of *Arabidopsis thaliana*. *Proc. Natl. Acad. Sci.* **86:** 4604–4608.

Blackwell, R.D., A.J.S. Murray, and P.J. Lea. 1987. Inhibition of photosynthesis in barley with decreased levels of chloroplastic glutamine synthetase activity. *J. Exp. Bot.* **38:**

1799–1809.

Blackwell, R.D., A.J.S. Murray, P.J. Lea, and K.W. Joy. 1988. Photorespiratory amino donors, sucrose synthesis and the induction of $CO_2$ fixation in barley deficient in glutamine synthetase and/or glutamate synthase. *J. Exp. Bot.* **39:** 845–858.

Boeke, J.D., F. LaCroute, and G.R. Fink. 1984. A positive selection for mutants lacking orotidine-5′-phosphate decarboxylase activity in yeast: 5-fluoro-orotic acid resistance. *Mol. Gen. Genet.* **197:** 345–346.

Borstlap, A.C. and A.C. Vernooy-Gerritsen. 1985. In-vivo control of valine and leucine synthesis in the duckweed *Spirodela polyrhiza. Planta* **164:** 129–134.

Borstlap, A.C., J. Schuurmans, and J.P. Bourgin. 1985. Amino-acid-transport mutant of *Nicotiana tabacum* L. *Planta* **166:** 141–144.

Bourgin, J.-P. 1978. Valine-resistant plants from in vitro selected tobacco cells. *Mol. Gen. Genet.* **161:** 225–230.

Bourgin, J.P., J. Goujaud, C. Missonier, and C. Pethe. 1985. Valine-resistance, a potential marker in plant cell genetics. I. Distinction between two types of valine-resistant tobacco mutants isolated from protoplast-derived cells. *Genetics* **109:** 393–407.

Braaksma, F.J. and W.J. Feenstra. 1982. Isolation and characterization of nitrate reductase-deficient mutants of *Arabidopsis thaliana. Theor. Appl. Genet.* **64:** 83–90.

Brears, T., E.L. Walker, and G.M. Coruzzi. 1991. A promoter sequence involved in cell-specific expression of the pea glutamine synthetase *GS3A* gene in organs of transgenic tobacco and alfalfa. *Plant J.* **1:** 235–244.

Bright, S.W.J., E.A. Wood, and B.J. Miflin. 1978. The effect of aspartate-derived amino acids (lysine, threonine, methionine) on the growth of excised embryos of wheat and barley. *Planta* **139:** 113–117.

Bright, S.W.J., J.S.H. Kueh, J. Franklin, S.E. Rognes, and B.J. Miflin. 1982. Two genes for threonine accumulation in barley seeds. *Nature* **299:** 278–279.

Brown, G.M. and J.M. Williamson. 1987. Biosynthesis of folic acid, riboflavin, thiamine, and pantothenic acid. In Escherichia coli *and* Salmonella typhimurium: *Cellular and molecular biology* (ed. F.C. Neidhardt et al.), vol. 1, pp. 521–538. American Society for Microbiology, Washington, D.C.

Bryan, J.K. 1980. Synthesis of the aspartate family and branched chain amino acids. In *The biochemistry of plants* (ed. B.J. Miflin), vol. 5, pp. 403–452. Academic Press, New York.

———. 1990. Advances in the biochemistry of amino acid biosynthesis. In *The biochemistry of plants* (ed. B.J. Miflin and P.J. Lea), vol. 16, pp. 161–195. Academic Press, New York.

Cella, R., B. Parisi, and E. Nielsen. 1982. Characterization of a carrot cell line resistant to azetidine-2-carboxylic acid. *Plant Sci. Lett.* **24:** 125–135.

Cock, J.M., P. Hemon, and J.V. Cullimore. 1992. Characterization of the gene encoding the plastid-located glutamine synthetase of *Phaseolus vulgaris* regulation of beta glucuronidase gene fusions in transgenic tobacco. *Plant Mol. Biol.* **18:** 1141–1149.

Cock, J.M., I.W. Brock, A.T. Watson, R. Swarup, A.P. Morby, and J.V. Cullimore. 1991. Regulation of glutamine synthetase genes in leaves of *Phaseolus vulgaris. Plant Mol. Biol.* **17:** 761–771.

Coruzzi, G.M. 1991. Molecular approaches to the study of amino acid biosynthesis in plants. *Plant Sci.* **74:** 145–155.

Crawford, I.P. 1989. Evolution of a biosynthetic pathway: The tryptophan paradigm. *Annu. Rev. Microbiol.* **43:** 567–600.

Cullimore, J.V., J.M. Cock, T.J. Daniell, R. Swarup, and M.J. Bennett. 1992. Inducibility of the glutamine synthetase gene family of *Phaseolus vulgaris* L. In *Biochemistry and molecular biology of inducible enzymes and proteins in higher plants* (ed. J.L. Wray), pp. 79–95. Cambridge University Press, Cambridge, United Kingdom.

Curien, G., R. Dumas, and R. Douce. 1993. Nucleotide sequence and characterization of a cDNA encoding the acetohydroxy acid isomeroreductase from *Arabidopsis thaliana*. *Plant Mol. Biol.* **21:** 717–722.

d'Amato, T.A., R.J. Ganson, C.G. Gaines, and R.A. Jensen. 1984. Subcellular localization of chorismate-mutase isoenzymes in protoplasts from mesophyll and suspension-cultured cells of *Nicotiana silvestris*. *Planta* **162:** 104–108.

Delauney, A.J. and D.P.S. Verma. 1990. A soybean gene encoding $D^1$-pyrroline-5-carboxylate reductase was isolated by functional complementation in *Escherichia coli* and is found to be osmoregulated. *Mol. Gen. Genet.* **221:** 299–305.

Diedrick, T.J., D.A. Frisch, and B.G. Gegenbach. 1990. Tissue culture isolation of a second mutant locus for increased threonine accumulation in maize. *Theor. Appl. Genet.* **79:** 209–215.

Doong, R.L., J.E. Gander, R.J. Ganson, and R.A. Jensen. 1992. The cytosolic isoenzyme of 3-deoxy-D-arabino-heptulosonate-7-phosphate synthase in *Spinacia oleracea* and other higher plants: Extreme substrate ambiguity and other properties. *Physiol. Plant.* **84:** 351–360.

Doremus, H.D. 1986. Organization of the pathway of *de novo* pyrimidine nucleotide biosynthesis in pea (*Pisum sativum* L. cv Progress No. 9) leaves. *Arch. Biochem. Biophys.* **250:** 112–119.

Dutcher, S.K., R.E. Galloway, W.R. Barclay, and G. Poortinga. 1992. Tryptophan analog resistant mutations in *Chlamydomonas reinhardtii*. *Genetics* **131:** 593–607.

Dyer, W.E., J.M. Henstrand, A.K. Handa, and K.M. Herrmann. 1989. Wounding induces the first enzyme of the shikimate pathway in Solanaceae. *Proc. Natl. Acad. Sci.* **86:** 7370–7373.

Dyer, W.E., L.M. Weaver, J. Zhao, D.N. Kuhn, S.C. Weller, and K.M. Herrmann. 1990. A cDNA encoding 3-deoxy-D-arabino-heptulosonate 7-phosphate synthase from *Solanum tuberosum* L. *J. Biol. Chem.* **265:** 1608–1614.

Edwards, J.W. and G.M. Coruzzi. 1989. Photorespiration and light act in concert to regulate the expression of the nuclear gene for chloroplast glutamine synthetase. *Plant Cell* **1:** 241–248.

Edwards, J.W., E.L. Walker, and G.M. Coruzzi. 1990. Cell-specific expression in transgenic plants reveals nonoverlapping roles for chloroplast and cytosolic glutamine synthetase. *Proc. Natl. Acad. Sci.* **87:** 3459–3463.

Eisenberg, M. 1987. Biosynthesis of biotin and lipoic acid. In Escherichia coli *and* Salmonella typhimurium: *Cellular and molecular biology* (ed. F.C. Neidhardt et al.), vol. 1, pp. 544–550. American Society for Microbiology, Washington, D.C.

Elledge, S.J., J.T. Mulligan, S.W. Ramer, M. Spottswood, and R.W. Davis. 1991. λYES: A multifunctional cDNA expression vector for the isolation of genes by complementation of yeast and *Escherichia coli* mutations. *Proc. Natl. Acad. Sci.* **88:** 1731–1735.

Farnham, M.W., S.S. Miller, S.M. Griffith, and C.P. Vance. 1990. Aspartate aminotransferase in alfalfa root nodules. II. Immunological distinction between two forms of the enzyme. *Plant Physiol.* **93:** 603–610.

Feenstra, W.J. 1964. Isolation of nutritional mutants in *Arabidopsis thaliana*. *Genetica* **35:** 259–269.

Floss, H.G. 1986. The shikimate pathway: An overview. *Recent Adv. Phytochem.* **20:** 13–55.

Forde, B.G., H.M. Day, J.F. Turton, S. Wen-jun, J.V. Cullimore, and J.E. Oliver. 1989. Two glutamine synthetase genes from *Phaseolus vulgaris* L. display contrasting developmental and spatial patterns of expression in transgenic *Lotus corniculatus* plants. *Plant Cell* **1**: 391–401.

Frankard, V., M. Ghislain, and M. Jacobs. 1992. Two feedback-insensitive enzymes of the aspartate pathway in *Nicotiana sylvestris*. *Plant Physiol.* **99**: 1285–1293.

Frankard, V., M. Ghislain, I. Negrutiu, and M. Jacobs. 1991. High threonine producer mutant of *Nicotiana sylvestris* (Spegg. and Comes). *Theor. Appl. Genet.* **82**: 273–282.

Freeman, J., A.J. Maquez, R.M. Wallsgrove, R. Saarelainen, and B.G. Forde. 1990. Molecular analysis of barley mutants deficient in chloroplast glutamine synthetase. *Plant Mol. Biol.* **14**: 297–311.

Frisch, D.A., A.M. Tommey, B.G. Gegenbach, and D.A. Somers. 1991. Direct selection of a maize cDNA for dihydropicolinate synthase in an *Escherichia coli dapA⁻* auxotroph. *Mol. Gen. Genet.* **228**: 287–293.

Frommer, W.B., S. Hummel, and J.W. Riesmeier. 1993. Expression cloning of a cDNA encoding a broad specificity amino acid permease from *Arabidopsis thaliana*. *Proc. Natl. Acad. Sci.* **90**: 5944–5948.

Gabard, J.M., P.J. Charest, V.N. Iyer, and B.L. Miki. 1989. Cross-resistance to short residual sulfonylurea herbicides in transgenic tobacco plants. *Plant Physiol.* **91**: 574–580.

Gantt, J.S., R.J. Larson, M.W. Farnham, S.M. Pathirana, S.S. Miller, and C.P. Vance. 1992. Aspartate aminotransferase in effective and ineffective alfalfa nodules. *Plant Physiol.* **98**: 868–878.

Gasser, C.S., J.A. Winter, C.M. Hironaka, and D.M. Shah. 1988. Structure, expression, and evolution of the 5-enolpyruvylshikimate-3-phosphate synthase genes of petunia and tomato. *J. Biol. Chem.* **263**: 4280–4289.

Ghislain, M., V. Frankard, and M. Jacobs. 1990. Dihydropicolinate synthase of *Nicotiana sylvestris*, a chloroplast-localized enzyme of the lysine pathway. *Planta* **180**: 480–486.

Ghislain, M., V. Frankard, M. Vauterin, B. Matthews, and M. Jacobs. 1992. Molecular analysis of key enzymes of the aspartate-derived amino acid pathway. *Curr. Top. Plant Physiol. Am. Soc. Plant Physiol. Ser.* **7**: 326–327.

Gilchrist, D.G. and T. Kosuge. 1980. Aromatic amino acid biosynthesis and its regulation. In *The biochemistry of plants* (ed. B.J. Miflin), vol. 5, pp. 507–531. Academic Press, New York.

Giovanelli, J., S.H. Mudd, and A.H. Datko. 1989a. Aspartokinase of *Lemna paucicostata* Hegelm. 6746. *Plant Physiol.* **90**: 1577–1583.

———. 1989b. Regulatory structure of the biosynthetic pathway for the aspartate family of amino acids in *Lemna paucicostata* Hegelm. 6746, with special reference to the role of aspartokinase. *Plant Physiol.* **90**: 1584–1599.

Givan, C.V. 1980. Aminotransferases in higher plants. In *The biochemistry of plants* (ed. B.J. Miflin), vol. 5, pp. 329–357. Academic Press, New York.

Givan, C.V., K.W. Joy, and L.A. Kleczkowski. 1988. A decade of photorespiratory nitrogen cycling. *Trends Biochem.* **13**: 433–437.

Griffith, S.M. and C.P. Vance. 1989. Aspartate aminotransferase in alfalfa root nodules. I. Purification and partial characterization. *Plant Physiol.* **90**: 1622–1629.

Hanson, A.D. and W.D. Hitz. 1982. Metabolic responses of mesophytes to plant water deficits. *Annu. Rev. Plant Physiol.* **33**: 163–203.

Haughn, G.W. and C.R. Somerville. 1986. Sulfonylurea resistant mutants of *Arabidopsis thaliana*. *Mol. Gen. Genet.* **204**: 430–434.

———. 1990. A mutation causing imidazolinone resistance maps to the *csr1* locus of *Arabidopsis thaliana*. *Plant Physiol.* **92**: 1081–1085.

Haughn, G.W., J. Smith, B. Mazur, and C. Somerville. 1988. Transformation with a mutant *Arabidopsis* acetolactate synthase gene renders tobacco resistant to sulfonylurea herbicides. *Mol. Gen. Genet.* **211**: 266–271.

Henstrand, J.M., K.F. McCue, K. Brink, A.K. Handa, K.M. Herrmann, and E.E. Conn. 1992. Light and fungal elicitor induce 3-deoxy-D-arabino-heptulosonate 7-phosphate synthase mRNA in suspension cultured cells of parsley. *Plant Physiol.* **98**: 761–763.

Hibberd, K.A. and C.E. Green. 1982. Inheritance and expression of lysine plus threonine resistance in maize tissue culture. *Proc. Natl. Acad. Sci.* **79**: 559–563.

Hibberd, K.A., T. Walter, C.E. Green, and B.G. Gengenbach. 1980. Selection and characterization of a feedback-insensitive tissue culture of maize. *Planta* **148**: 183–187.

Hogge, L.R., D.W. Reed, E.W. Underholl, and G.W. Haughn. 1988. HPLC separation of glucosinolates from leaves and seeds of *Arabidopsis thaliana* and their identification using thermospray liquid chromatography/mass spectroscopy. *J. Chromatogr. Sci.* **26**: 551–556.

Hsu, L.-C., T.-J. Chiou, L. Chen, and D.R. Bush. 1993. Cloning a plant amino acid transporter by functional complementation of a yeast amino acid transport mutant. *Proc. Natl. Acad. Sci.* **90**: 7441–7445.

Hu, C.-A., A.J. Delauney, and D.P.S. Verma. 1992. A bifunctional enzyme ($\Delta^1$-pyrroline-5-carboxylate synthetase) catalyzes the first two steps in proline biosynthesis in plants. *Proc. Natl. Acad. Sci.* **89**: 9354–9358.

Hütter, R., P. Niederberger, and J.A. DeMoss. 1986. Tryptophan biosynthetic genes in eukaryotic microorganisms. *Annu. Rev. Microbiol.* **40**: 55–77.

Joy, K.W., R.D. Blackwell, and P.J. Lea. 1992. Assimilation of nitrogen in mutants lacking enzymes of the glutamate synthase cycle. *J. Exp. Bot.* **43**: 139–145.

Joy, K.W., R.J. Ireland, and P.J. Lea. 1983. Asparagine synthesis in pea leaves, and the occurrence of an asparagine synthetase inhibitor. *Plant Physiol.* **73**: 165–168.

Keith, B., X. Dong, F.M. Ausubel, and G.R. Fink. 1991. Differential induction of 3-deoxy-D-arabino-heptulosonate 7-phosphate synthase genes in *Arabidopsis thaliana* by wounding and pathogenic attack. *Proc. Natl. Acad. Sci.* **88**: 8821–8825.

Kendall, A.C., R.M. Wallsgrove, N.P. Hall, J.C. Turner, and P.J. Lea. 1986. Carbon and nitrogen metabolism in barley (*Hordeum vulgare* L.) mutants lacking ferredoxin-dependent glutamate synthase. *Planta* **168**: 316–323.

Keys, A.J., I.F. Bird, M.J. Cornelius, P.J. Lea, R.M. Wallsgrove, and B.J. Miflin. 1978. Photorespiratory nitrogen cycle. *Nature* **275**: 741–743.

King, J., R.B. Horsch, and A.D. Savage. 1980. Partial characterization of two stable auxotrophic cell strains of *Datura innoxia* Mill. *Planta* **149**: 480–484.

Kishore, G.M. and D.M. Shah. 1988. Amino acid biosynthesis inhibitors as herbicides. *Annu. Rev. Biochem.* **57**: 627–663.

Klee, H.J., Y.M. Muskopf, and C.S. Gasser. 1987. Cloning of an *Arabidopsis thaliana* gene encoding 5-enolpyruvylshikimate-3-phosphate synthase: Sequence analysis and manipulation to obtain glyphosate-tolerant plants. *Mol. Gen. Genet.* **210**: 437–442.

Kohl, D.H., K.R. Schubert, M.B. Carter, and C.H. Hagedorn. 1988. Proline metabolism in $N_2$-fixing root nodules: Energy transfer and regulation of purine synthesis. *Proc. Natl. Acad. Sci.* **85**: 2036–2040.

Komeda, Y., M. Tanaka, and T. Nishimune. 1988. A *th-1* mutant of *Arabidopsis thaliana* is defective for a thiamin-phosphate-synthesizing enzyme: Thiamin phosphate pyrophosphorylase. *Plant Physiol.* **88**: 248–250.

Koornneef, M. and C.J. Hanhart. 1981. A new thiamine locus in *Arabidopsis*. *Arabidopsis Inf. Serv.* **18**: 52–58.

Kreps, J.A. and C.D. Town. 1992. Isolation and characterization of a mutant of *Arabidopsis thaliana* resistant to α-methyltryptophan. *Plant Physiol.* **99**: 269–275.

Kueh, J.S.H. and S.W.J. Bright. 1981. Proline accumulation in a barley mutant resistant to *trans*-4-hydroxy-L-proline. *Planta* **153:** 166–171.

———. 1982. Biochemical and genetical analysis of three proline-accumulating barley mutants. *Plant Sci. Lett.* **27:** 233–241.

Kumar, S. 1988. Recessive monogenic mutation in grain pea (*Pisum sativum*) that causes pyridoxine requirement for growth and seed production. *J. Biosci.* **13:** 415–418.

Kumar, S. and S.B. Sharma. 1987. Selectable markers suitable for genetic manipulations in crop plants: Development of thiamin auxotrophs in grain pea and barley. In *Biotechnology in agriculture* (ed. S. Natesh et al.), pp. 245–249. Oxford and IBH, New Delhi, India.

Langridge, J. 1955. Biochemical mutations in the crucifer *Arabidopsis thaliana* (L.) Heynh. *Nature* **176:** 260–261.

Langridge, J. and R.D. Brock. 1961. A thiamin-requiring mutant of the tomato. *Aust. J. Biol. Sci.* **14:** 66–69.

Last, R.L. 1993. The genetics of nitrogen assimilation and amino acid biosynthesis in plants: Progress and prospects. *Int. Rev. Cytol.* **143:** 297–330.

Last, R.L. and G.R. Fink. 1988. Tryptophan-requiring mutants of the plant *Arabidopsis thaliana*. *Science* **240:** 305–310.

Last, R.L., P.H. Bissinger, D.J. Mahoney, E.R. Radwanski, and G.R. Fink. 1991. Tryptophan mutants in *Arabidopsis*: The consequences of duplicated tryptophan synthase β genes. *Plant Cell* **3:** 345–358.

Lea, P.J., S.A. Robinson, and G.R. Stewart. 1990. The enzymology and metabolism of glutamine, glutamate, and asparagine. In *The biochemistry of plants* (ed. B.J. Miflin and P.J. Lea), vol. 16, pp. 121–159. Academic Press, New York.

Lee, D. and B.A. Moffatt. 1993. Purification and characterization of adenine phosphoribosyltransferase from *Arabidopsis thaliana*. *Physiol. Plant.* **87:** 483–492.

Li, S.L. and G.P. Rédei. 1969. Thiamine mutants of the crucifer, *Arabidopsis*. *Biochem. Genet.* **3:** 163–170.

Li, S.L., G.P. Rédei, and C.S. Gowans. 1967. A phylogenetic comparison of mutation spectra. *Mol. Gen. Genet.* **100:** 77–83.

Lightfoot, D.A., N.K. Green, and J.V. Cullimore. 1988. The chloroplast-located glutamine synthetase of *Phaseolus vulgaris* L.: Nucleotide sequence, expression in different organs and uptake into isolated chloroplasts. *Plant Mol. Biol.* **11:** 191–202.

Lough, T.J., B.D. Reddington, M.R. Grant, D.F. Hill, P.H.S. Reynolds, and K.J.F. Farnden. 1992. The isolation and characterization of a cDNA clone encoding L-asparaginase from developing seeds of lupin (*Lupinus arboreus*). *Plant Mol. Biol.* **19:** 391–399.

Matthews, B.F., J.M. Weisemann, K.M. Lewin, G.J. Wadsworth, and J.S. Gebhardt. 1992. Cloning and analysis of cDNAs encoding bifunctional aspartate kinase-homoserine dehydrogenase activities in carrot and soybean. *Curr. Top. Plant Physiol. Am. Soc. Plant Physiol. Ser.* **7:** 294–295.

Mazur, B. and S.C. Falco. 1989. The development of herbicide resistant crops. *Annu Rev. Plant Physiol.* **40:** 441–470.

McCue, K.F. and E.E. Conn. 1989. Induction of 3-deoxy-D-*arabino*-hepulosonate-7-phosphate synthase activity by fungal elicitor in cultures of *Petroselinum crispum*. *Proc. Natl. Acad. Sci.* **86:** 7374–7377.

———. 1990. Induction of shikimic acid pathway enzymes by light in suspension cultured cells of parsley (*Petroselinum crispum*). *Plant Physiol.* **94:** 507–510.

McGaw, B.A. 1987. Cytokinin biosynthesis and metabolism. In *Plant hormones and their role in plant growth and development* (ed. P.J. Davies), pp. 76–93. Martinus Nijhoff,

Boston.

Miflin, B.J. and P.J. Lea. 1980. Ammonium assimilation. In *The biochemistry of plants* (ed. B.J. Miflin), pp. 169–202. Academic Press, New York.

Miles, E.W. 1991. Structural basis for catalysis by tryptophan synthase. *Adv. Enzymol. Relat. Areas Mol. Biol.* **64:** pp. 93–172.

Minet, M., M.-E. Dufour, and F. Lacroute. 1992. Complementation of *Saccharomyces cerevisiae* auxotrophic mutants by *Arabidopsis thaliana* cDNAs. *Plant J.* **2:** 417–422.

Mitsui, K. and H. Ashihara. 1988. Nucleotide pools and purine metabolism in tissue cultures of wild-type *Datura innoxia* and adenine-requiring auxotroph. *Plant Cell Physiol.* **29:** 1177–1183.

Moffatt, B. and C. Somerville. 1988. Positive selection for male-sterile mutants of *Arabidopsis* lacking adenine phosphoribosyltransferase activity. *Plant Physiol.* **86:** 1150–1154.

Moffatt, B., C. Pethe, and M. Laloue. 1991. Metabolism of benzyladenine is impaired in a mutant of *Arabidopsis thaliana* lacking adenine phosphoribosyltransferase activity. *Plant Physiol.* **95:** 900–908.

Moffatt, B.A., E.A. McWhinnie, W.E. Burkhart, J.J. Pasternak, and S.J. Rothstein. 1992. A complete cDNA for adenine phosphoribosyltransferase from *Arabidopsis thaliana*. *Plant Mol. Biol.* **18:** 653–662.

Morris, P.F., R.-L. Doong, and R.A. Jensen. 1989. Evidence from *Solanum tuberosum* in support of the dual-pathway hypothesis of aromatic biosynthesis. *Plant Physiol.* **89:** 10–14.

Mourad, G. and J. King. 1992. Effect of four classes of herbicides on growth and acetolactate-synthase activity in several variants of *Arabidopsis thaliana*. *Planta* **188:** 491–497.

Mourad, G., B. Pandey, and J. King. 1993. Isolation and characterization of a triazolopyrimidine-resistant mutant of *Arabidopsis*. *J. Heredity* **84:** 91–96.

Mousdale, D.M. and J.R. Coggins. 1985. Subcellular localization of the common shikimate-pathway enzymes in *Pisum sativum* L. *Planta* **163:** 241–249.

Muday, G.K. and K.M. Herrmann. 1992. Wounding induces one of two isoenzymes of 3-deoxy-D-arabino-heptulosonate 7-phosphate synthase in *Solanum tuberosum* L. *Plant Physiol.* **98:** 496–500.

Muehlbauer, G.J., D.A. Somers, B.F. Matthews, and B.G. Gegenbach. 1992. Isolation and characterization of a maize aspartate kinase-homoserine dehydrogenase cDNA clone. *Curr. Top. Plant Physiol. Am. Soc. Plant Physiol. Ser.* **7:** 324–325.

Muhitch, M.J., D.L. Shaner, and M.A. Stidham. 1987. Imidazolinones and acetohydroxyacid synthase from higher plants. *Plant Physiol.* **83:** 451–456.

Negrutiu, I., A. Cattoir-Reynearts, I. Verbruggen, and M. Jacobs. 1984. Lysine overproducer mutant with an altered dihydrodipicolinate synthase from protoplast culture of *Nicotiana sylvestris* (Spegazzini and Comes). *Theor. Appl. Genet.* **68:** 11–20.

Niyogi, K.K. 1993. Ph.D. thesis, Massachusetts Institute of Technology, Cambridge.

Niyogi, K.K. and G.R. Fink. 1992. Two anthranilate synthase genes in *Arabidopsis*: Defense-related regulation of the tryptophan pathway. *Plant Cell* **4:** 721–733.

Niyogi, K.K., R.L. Last, G.R. Fink, and B. Keith. 1993. Suppressors of *trp1* fluorescence identify a new *Arabidopsis* gene, *TRP4*, encoding the anthranilate synthase β subunit. *Plant Cell* **5:** 1011–1027.

Perl, A., O. Shaul, and G. Galili. 1992. Regulation of lysine synthesis in transgenic potato plants expressing a bacterial dihydrodipicolinate synthase in their chloroplasts. *Plant Mol. Biol.* **19:** 815–823.

Peterman, T.K. and H.M. Goodman. 1991. The glutamine synthetase gene family of

*Arabidopsis thaliana* light-regulation and differential expression in leaves, roots and seeds. *Mol. Gen. Genet.* **230:** 145–154.

Pinto, J.E.B.P., W.E. Dyer, S.C. Weller, and K.M. Herrmann. 1988. Glyphosate induces 3-deoxy-D-arabino-hepulosonate-7-phosphate synthase in potato (*Solanum tuberosum* L.) cells grown in suspension culture. *Plant Physiol.* **87:** 891–893.

Pruitt, K.P. and R.L. Last. 1993. Expression patterns of duplicate tryptophan synthase β genes in *Arabidopsis thaliana. Plant Physiol.* **102:** 1019–1026.

Ray, T.B. 1984. Site of action of chlorsulfuron. Inhibition of valine and isoleucine biosynthesis in plants. *Plant Physiol.* **75:** 827–831.

Rayapati, P.J., C.R. Stewart, and E. Hack. 1989. Pyrroline-5-carboxylate reductase is in pea leaf chloroplasts. *Plant Physiol.* **91:** 581–586.

Rédei, G.P. 1970. *Arabidopsis thaliana* (L.) Heynh. A review of the genetics and biology. *Bibliogr. Genet.* **20:** 1–151.

————. 1975. *Arabidopsis* as a genetic tool. *Annu. Rev. Genet.* **9:** 111–127.

Rédei, G.P. and C. Koncz. 1992. Classical mutagenesis. In *Methods in* Arabidopsis *research* (ed. C. Koncz et al.), pp. 16–82. World Scientific, Singapore.

Regan, S.M. and B.A. Moffatt. 1990. Cytochemical analysis of pollen development in wild-type *Arabidopsis* and a male-sterile mutant. *Plant Cell* **2:** 877–889.

Relton, J.M., R.M. Wallsgrove, J.-P. Bourgin, and S.W.J. Bright. 1986, Altered feedback sensitivity of acetohydroxyacid synthase from valine-resistant mutants of tobacco (*Nicotiana tabacum* L.). *Planta* **169:** 46–50.

Rognes, S.E., S.W.J. Bright, and B.J. Miflin. 1983. Feedback-insensitive aspartate kinase isoenzymes in barley mutants resistant to lysine plus threonine. *Planta* **157:** 32–38.

Rose, A.B., A.L. Casselman, and R.L. Last. 1992. A phosphoribosylanthranilate transferase gene is defective in blue fluorescent *Arabidopsis thaliana* tryptophan mutants. *Plant Physiol.* **100:** 582–592.

Sahi, S.V., P.K. Saxena, G.D. Abrams, and J. King. 1988. Identification of the biochemical lesion in a pantothenate-requiring auxotroph of *Datura innoxia* P. Mill. *J. Plant Physiol.* **133:** 277–280.

Saito, K., N. Miura, M. Yamazaki, H. Hirano, and I. Murakoshi. 1992. Molecular cloning and bacterial expression of cDNA encoding a plant cysteine synthase. *Proc. Natl. Acad. Sci.* **89:** 8078–8082.

Sakakibara, H., M. Watanabe, H. Toshiharu, and T. Sugiyama. 1991. Molecular cloning and characterization of complementary DNA encoding for ferredoxin-dependent glutamate synthase in maize leaf. *J. Biol. Chem.* **266:** 2028–2035.

Sakakibara, H., S. Kawabata, H. Takahashi, T. Hase, and T. Sugiyama. 1992. Molecular cloning of the family of glutamine synthetase genes from maize expression of genes for glutamine synthetase and ferredoxin- dependent glutamate synthase in photosynthetic and non-photosynthetic tissues. *Plant Cell Physiol.* **33:** 49–58.

Santoso, D. and R.W. Thornburg. 1992, Isolation and characterization of UMP synthase mutants from haploid cell suspensions of *Nicotiana tabacum. Plant Physiol.* **99:** 1216–1225.

Sathasivan, K., G.W. Haughn, and N. Murai. 1990. Nucleotide sequence of a mutant acetolactate synthase gene from an imidazolinone-resistant *Arabidopsis thaliana* Columbia. *Nucleic Acids Res.* **18:** 2188.

————. 1991, Molecular basis of imidazolinone herbicide resistance in *Arabidopsis thaliana* var Columbia. *Plant Physiol.* **97:** 1044–1050.

Schaller, A., J. Schmid, U. Leibinger, and N. Amrhein. 1991. Molecular cloning and analysis of a cDNA coding for chorismate synthase from the higher plant *Corydalis sempervirens* Pers. *J. Biol. Chem.* **266:** 21434–21438.

Schmid, J., A. Schaller, U. Leibinger, W. Boll, and N. Amrhein. 1992. The *in vitro* synthesized tomato shikimate kinase precursor is enzymatically active and is imported and processed to the mature enzyme by chloroplasts. *Plant J.* **2:** 375–383.

Schmidt, A. and K. Jäger. 1992. Open questions about sulfur metabolism in plants. *Annu. Rev. Plant Physiol. Plant Mol. Biol.* **43:** 325–349.

Schneider, T., R. Dinkins, K. Robinson, J. Shellhammer, and D.W. Meinke. 1989. An embryo-lethal mutant of *Arabidopsis thaliana* is a biotin auxotroph. *Dev. Biol.* **131:** 161–167.

Schubert, K.R. and M.J. Boland. 1990. The ureides. In *The biochemistry of plants* (ed. B.J. Miflin and P.J. Lea), vol. 16, pp. 197-282. Academic Press, New York.

Schultz, C.J. and G.M. Coruzzi. 1992. Towards an understanding of the role of the multiple isoenzymes for aspartate aminotransferase in *Arabidopsis thaliana. Curr. Top. Plant Physiol. Am. Soc. Plant Physiol. Ser.* **7:** 318–319.

Senecoff, J.F. and R.B. Meagher. 1993. Isolating the *Arabidopsis thaliana* genes for *de novo* purine synthesis by functional complementation of *Escherichia coli* mutants. I. 5′-Phosphoribosyl-5-aminoimidazole synthetase. *Plant Physiol.* **102:** 387–399.

Shaner, D.L., P.C. Anderson, and M.A. Stidham. 1984. Imidazolinones: Potent inhibitors of acetohydroxyacid synthase. *Plant Physiol.* **76:** 545–546.

Shaul, O. and G. Galili. 1992. Increased lysine synthesis in tobacco plants that express high levels of bacterial dihydrodipicolinate synthase in their chloroplasts. *Plant J.* **2:** 203–209.

Shellhammer, A.J. 1991. "Analysis of a biotin auxotroph of *Arabidopsis thaliana.*" Ph.D. thesis, Oklahoma State University, Stillwater.

Shellhammer, J. and D. Meinke. 1990. Arrested embryos from the *bio1* auxotroph of *Arabidopsis thaliana* contain reduced levels of biotin. *Plant Physiol.* **93:** 1162–1167.

Sieciechowicz, K., R.J. Ireland, and K.W. Joy. 1985. Diurnal variation of asparaginase in developing pea leaves. *Plant Physiol.* **77:** 506–508.

Sieciechowicz, K.A., K.W. Joy, and R.J. Ireland. 1988. The metabolism of asparagine in plants. *Phytochemistry* **27:** 663–671.

Singh, B.K., D.L. Siehl, and J.A. Connelly. 1991. Shikimate pathway: Why does it mean so much to so many? *Oxf. Surv. Plant Mol. Cell Biol.* **7:** 143–186.

Singh, B., I. Szamosi, J.M. Hand, and R. Misra. 1992. *Arabidopsis* acetohydroxyacid synthase expressed in *Escherichia coli* is insensitive to the feedback inhibitors. *Plant Physiol.* **99:** 812–816.

Somerville, C.R. and W.L. Ogren. 1980. Inhibition of photosynthesis in *Arabidopsis* mutants lacking leaf glutamate synthase activity. *Nature* **286:** 257–259.

———. 1983. An *Arabidopsis thaliana* mutant defective in dicarboxylate transport. *Proc. Natl. Acad. Sci.* **80:** 1290–1294.

Subramanian, M.V., V. Loney-Gallant, J.M. Dias, and L.C. Mireles. 1991. Acetolactate synthase inhibiting herbicides bind to the regulatory site. *Plant Physiol.* **96:** 310–313.

Suzich, J.A., J.F.D. Dean, and K.M. Herrmann. 1985. 3-Deoxy-D-*arabino*-heptulosonate 7-phosphate synthase from carrot root (*Daucus carota*) is a hysteretic enzyme. *Plant Physiol.* **79:** 765–770.

Szoke, A., G.-H. Miao, Z. Hong, and D.P.S. Verma. 1992. Subcellular location of $\Delta^1$-pyrroline-5-carboxylate reductase in root/nodule and leaf of soybean. *Plant Physiol.* **99:** 1642–1649.

Thomas, G.J. 1984. Herbicidal activity of 6-methylanthranilic acid and analogues. *J. Agric. Food Chem.* **32:** 747–749.

Thompson, J.F. 1980. Arginine synthesis, proline synthesis and related processes. In *The biochemistry of plants* (ed. B.J. Miflin), vol. 5, pp. 375–402. Academic Press, New

York.

Tingey, S.V., F.-Y. Tsai, J.W. Edwards, E.L. Walker, and G.M. Coruzzi. 1988. Chloroplast and cytosolic glutamine synthetase are encoded by homologous nuclear genes which are differentially expressed *in vivo. J. Biol. Chem.* **263**: 9651–9657.

Tourneur, C., L. Jouanin, and H. Vaucheret. 1993. Over-expression of acetolactate synthase confers resistance to valine in transgenic tobacco. *Plant Sci.* **88**: 159–168.

Tsai, F.-Y. and G.M. Coruzzi. 1990. Dark-induced and organ-specific expression of two asparagine synthetase genes in *Pisum sativum. EMBO J.* **9**: 323–332.

————. 1991. Light represses the transcription of plant asparagine synthetase genes in photosynthetic and non-photosynthetic organs of plants. *Mol. Cell. Biol.* **11**: 4966–4972.

Tsuji, J., E.P. Jackson, D.A. Gage, R. Hammerschmidt, and S.C. Somerville. 1992. Phytoalexin accumulation in *Arabidopsis thaliana* during the hypersensitive reaction to *Pseudomonas syringae* pv. *syringae. Plant Physiol.* **98**: 1304–1309.

Turano, F.J., R.L. Jordan, and B.F. Matthews. 1990. Immunological characterization of in vitro forms of homoserine dehydrogenase from carrot suspension cultures. *Plant Physiol.* **92**: 395–400.

Turano, F.J., J.M. Weisemann, and B.F. Matthews. 1992. Identification and expression of a cDNA clone encoding aspartate aminotransferase in carrot. *Plant Physiol.* **100**: 374–381.

Udvardi, M.K. and M.L. Kahn. 1991. Isolation and analysis of a cDNA clone that encodes an alfalfa (*Medicago sativa*) aspartate aminotransferase. *Mol. Gen. Genet.* **231**: 97–105.

Verbruggen, N., R. Villarroel, and M. Van Montague. 1993. Osmoregulation of a pyrroline-5-carboxylate reductase-encoding gene in *Arabidopsis thaliana. Plant Physiol.* **103**: 771–781.

Wagner, K.G. and A.I. Backer. 1992. Dynamics of nucleotides in plants studied on a cellular basis. *Int. Rev. Cytol.* **134**: 1–84.

Walker, E.L. and G.M. Coruzzi. 1989. Developmentally regulated expression of the gene family for cytosolic glutamine synthetase in *Pisum sativum. Plant Physiol.* **91**: 702–708.

Wallsgrove, R.M. and M. Mazelis. 1981. Spinach leaf dihydrodipicolinate synthase: Partial purification and characterization. *Phytochemistry* **20**: 2651–2655.

Wallsgrove, R.M., J.C. Turner, N.P. Hall, A.C. Kendall, and S.W.J. Bright. 1987. Barley mutants lacking chloroplast glutamine synthetase—Biochemical and genetic analysis. *Plant Physiol.* **83**: 155–158.

Walther, R., K. Wald, K. Glund, and A. Tewes. 1984. Evidence that a single polypeptide catalyses the two step conversion of orotate to UMP in cells from a tomato suspension culture. *J. Plant Physiol.* **116**: 301–311.

Wang, J., K.M. Herrmann, S.C. Weller, and P.B. Goldsbrough. 1991. Cloning and nucleotide sequence of a complementary DNA encoding 3-deoxy-Δ-*arabino*-hepulosonate-7-phosphate synthase from tobacco. *Plant Physiol.* **97**: 847–848.

Widholm, J. 1976. Selection and characterization of cultured carrot and tobacco cells resistant to lysine, methionine, and proline analogs. *Can. J. Bot.* **54**: 1523–1529.

Wightman, F. and J.C. Forest. 1978. Properties of plant aminotransferases. *Phytochemistry* **17**: 1455–1471.

Williamson, C.L. and R.D. Slocum. 1992. Molecular cloning and evidence for osmoregulation of the $\Delta^1$-pyrroline-5-carboxylate reductase gene in pea (*Pisum sativum* L.). *Plant Physiol.* **100**: 1464–1470.

Wright, A.D., C.A. Moehlenkamp, G.H. Perrot, M.G. Neuffer, and K.C. Cone. 1992. The maize auxotrophic mutant *orange pericarp* is defective in duplicate genes for tryptophan synthase β. *Plant Cell* **4:** 711–719.

Yanofsky, C. and I.P. Crawford. 1987. The tryptophan operon. In Escherichia coli *and* Salmonella typhimurium: *Cellular and molecular biology* (ed. F.C. Neidhardt et al.), vol. 2, pp. 1453–1472. American Society for Microbiology, Washington, D.C.

Zehnacker, C., T.W. Becker, A. Suzuki, E. Carrayol, M. Caboche, and B. Hirel. 1992. Purification and properties of tobacco ferredoxin-dependent glutamate synthase and isolation of corresponding cDNA clones. Light-inducibility and organ-specificity of gene transcription and protein expression. *Planta* **187:** 266–274.

Zhao, J. and K.M. Herrmann. 1992. Cloning and sequencing of a second cDNA encoding 3-deoxy-D-*arabino*-heptulosonate 7-phosphate synthase from *Solanum tuberosum* L. *Plant Physiol.* **100:** 1075–1076.

# 32
# Glycerolipids

**John Browse**
Institute of Biological Chemistry
Washington State University
Pullman, Washington 99164-6340

**Chris R. Somerville**
Carnegie Institution of Washington
Plant Biology Department
Stanford, California 94305-4101

The membranes of a typical plant cell collectively contain several hundred chemically distinct lipids. The functional significance of this chemical complexity and the mechanisms that regulate the amount of each constituent are not known. These problems may be subdivided into a wide range of more specific questions that direct much of the current research on plant and animal lipids. Most research on lipids in *Arabidopsis* has been focused on problems in three major areas: the biosynthesis of membrane glycerolipids, the roles of these lipids in the structure and function of membranes, and the synthesis of storage lipids (triacylglycerols) in seeds. Relatively little work has been published concerning the role of lipid derivatives, such as jasmonate (Farmer and Ryan 1990) and phosphoinositides (Blatt et al. 1990; Cote and Crain 1993), as components of signal transduction pathways. Similarly, in *Arabidopsis*, there appears to have been very little work on sterols (Patterson et al. 1993), sphingolipids, or chlorophylls, which are all important membrane components (Somerville 1986). The various functions of acyl lipids—involving fatty acids, which are produced from the 2-carbon precursor acetate—and the roles of isoprene lipids, produced from a 5-carbon precursor which itself is derived from acetate, are listed in Table 1 along with references to appropriate reviews. The role of fatty acid derivatives in the formation of cutin and wax are described elsewhere in this volume.

Although *Arabidopsis* contains only a small subset of the fatty acids found in various plant species (Hilditch and Williams 1964; Harwood 1980), the lipid composition of *Arabidopsis* is similar to that of the major crop and horticultural species. Thus, *Arabidopsis* is a good model for understanding the ubiquitous aspects of lipid metabolism and function. As

*Table 1* The functions of lipid molecules in higher plants

| Function | Lipid types involved |
| --- | --- |
| 1.  Membrane structural components | glycerolipids<br>sphingolipids<br>sterols |
| 2.  Storage compounds | triacylglycerols<br>waxes |
| 3.  Catalysts in biochemical and biophysical reactions | chlorophyll and other pigments, ubiquinone, plastiquinone |
| 4.  Photoprotection | carotenoids (xanthophyll cycle) |
| 5.  Protection of membranes against free radical damage | trocopherols |
| 6.  Waterproofing and surface protection cutin, suberin, surface waxes | long chain and very long chain fatty acids and their derivatives, tri-terpenes |
| 7.  Protein modification | |
|     a.  acylation | mainly 14:0 and 16:0 fatty acids |
|     b.  prenylation | |
|     c.  glycosylation | dolichol? |
|     d.  phosphatidylinositol anchors? | |
| 8.  Signaling | |
|     a.  internal | abscisic acid, giberellins, kinetin<br>synthesis of jasmonate from 18.3<br>inositol phosphates<br>diacylglycerols? |
|     b.  external | jasmonate<br>volatile insect attractants |
| 9.  Defense and anti-feeding compounds | essential oils, latex components (rubber, etc.), resin components (terpenes) |

References: Porter and Spurgeon (1981); Stumpf and Conn (1980, 1987); Goodwin and Mercer (1983); Taiz and Zeiger (1991); Salisbury and Ross (1992); Moore (1993).

with most areas of biological research, the use of mutants has been an essential tool. The mutants were isolated by direct screening of mutant populations for alterations in leaf or seed fatty acid composition using gas chromatography (Browse et al. 1985; James and Dooner 1990; Lemieux et al. 1990). Subsequent investigations of the mutants have contributed to developments in four main areas. They have allowed novel insights into the biochemistry and regulation of membrane glycerolipid synthesis. They have been important in this respect because many of the enzymes involved are integral membrane proteins that have been recalcitrant to characterization by traditional biochemical approaches. The mutants have also been used as markers to facilitate the cloning of genes that encode enzymes of lipid metabolism by chromosome walking (Arondel et al. 1992) and gene tagging (Yadav et al. 1993; Okuley et al. 1994). The availability of a series of plant lines with specific changes in lipid composition has provided a new resource for studying the role of lipids in the structure and function of plant membranes. Finally, mutations affecting seed lipid metabolism have provided tests of the consequences of modifying the fatty acid composition of seed oils.

To set the context for studies of lipid metabolism and function, it is useful to consider the overall composition of leaves and other tissues. Despite the fact that each $cm^2$ of *Arabidopsis* leaf contains approximately 400 $cm^2$ of membrane, lipids are a minor proportion of leaf dry matter (Fig. 1). This reflects the fact that if a leaf mesophyll cell were expanded a million times (to the size of a large auditorium), the membranes would still be less than 1 cm thick. Despite their slight dimensions, however, the lipid membranes are the barriers that delineate the cell and its compartments, and they form the sites where many essential processes occur, including the light harvesting and electron transport reactions of photosynthesis. Not unexpectedly, many of the other lipid components from plant leaves are chloroplast membrane components involved in photosynthesis (Fig. 1).

Table 2 shows the overall membrane glycerolipid compositions of different tissues of *Arabidopsis* and the compositions of chloroplast and extrachloroplast membranes from leaf mesophyll protoplasts. No other data are available on the lipid compositions of individual organelles and membranes from *Arabidopsis*, but these are likely to be similar to the compositions reported for other angiosperms (Harwood 1980; Douce

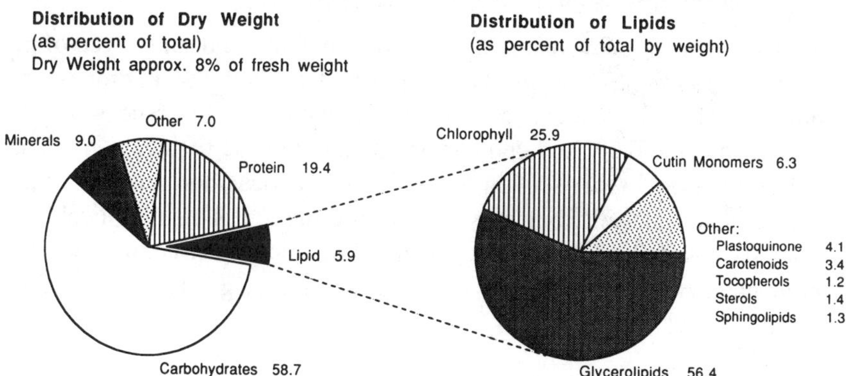


*Figure 1* Approximate data for the distribution of dry matter in leaf tissues of *Arabidopsis*. Some figures were extrapolated from results obtained with other plants.

*Table 2* The membrane glycerolipid compositions of different organs and subcellular fractions of *Arabidopsis*

| | | | Source of lipid extract | | |
|---|---|---|---|---|---|
| Lipid | leaves | chloroplasts | extrachloroplast | roots | mature seeds |
| PC | 21.8 | 12.0 | 47.8 | 45.4 | 48.1 |
| PE | 10.0 | — | 36.5 | 27.5 | 22.1 |
| PI + PS | 3.0 | — | 10.9 | 12.9 | 18.9 |
| SL | 2.8 | 3.9 | — | — | — |
| DGD | 15.2 | 20.9 | — | 2.0 | 3.3 |
| PG | 8.1 | 9.5 | 4.4 | 3.8 | 4.6 |
| MGD | 39.0 | 53.7 | — | 3.4 | 3.0 |
| Total lipid μg/g F.W. | 3.2 | 2.3 | 0.9 | 1.5 | — |

Data are expressed as mole %.

1985; Larsson et al. 1990). The fatty acid compositions of individual lipids of *Arabidopsis* leaf tissue are shown in Table 3. Phosphatidylcholine (PC) is present in substantial amounts in leaves, roots, and seeds and is also found in chloroplasts (where it is largely restricted to the outer leaflet of the outer envelope) and in extrachloroplast membranes. The fatty acid composition of PC from these different sources is similar (Miquel and Browse 1992), suggesting that the data in Table 3 may be a good guide to the fatty acid profiles of the major lipids from different tissues and cell types.

## BIOCHEMISTRY OF MEMBRANE GLYCEROLIPID SYNTHESIS

In many respects, the reactions involved in fatty acid and glycerolipid synthesis in plants are similar to those described for microbes and animals. However, there are many fundamental differences in the detailed biochemistry and cellular organization of the pathways in plants and non-plants. In this section, we specifically focus on membrane lipid synthesis in leaf cells, but most of the information is also applicable to other tissues and organs of the plant. The chemical structures of lipid molecules can be found in other texts (see, e.g., Harwood 1980; Gunstone et al. 1986).

### Fatty Acid Synthesis

The synthesis of malonyl-CoA by acetyl-CoA carboxylase is the first committed step in fatty acid synthesis, and in animals it is well estab-

*Table 3* The fatty acid compositions of the major leaf glycerolipids from wild-type *Arabidopsis*

| | Lipid | | | | | | |
|---|---|---|---|---|---|---|---|
| Fatty acid | PC | PE | PI | SL | DGD | PG | MGD |
| 16:0 | 20.6 | 31.2 | 43.5 | 43.2 | 13.6 | 20.7 | 1.5 |
| 16:1 | 0.6 | — | — | — | 0.3 | 33.5[a] | 1.5 |
| 16:2 | — | — | — | — | 0.6 | — | 1.3 |
| 16:3 | — | — | — | — | 2.1 | — | 30.6 |
| 18:0 | 2.7 | 3.4 | 5.2 | 3.7 | 1.1 | 1.8 | 0.2 |
| 18:1 | 4.4 | 3.3 | 4.3 | 5.3 | 1.3 | 6.0 | 1.5 |
| 18:2 | 38.8 | 43.0 | 27.0 | 10.4 | 5.0 | 12.5 | 3.4 |
| 18:3 | 32.1 | 18.7 | 20.0 | 37.4 | 75.9 | 25.6 | 60.0 |

Data are expressed as mole %.
[a]16:1 Δ3 *trans*.

lished that acetyl-CoA carboxylase is the rate-limiting step for fatty acid synthesis. Recent results suggest that acetyl-CoA carboxylase is also the most highly regulated enzyme in plant fatty acid synthesis (Post-Beitenmiller et al. 1991). The recent cloning of cDNAs encoding the plastid acetyl-CoA carboxylase (Konishi and Sasaki 1994) will allow more direct investigation of the processes involved in regulation of this enzyme.

The de novo synthesis of 16-carbon and 18-carbon fatty acids takes place almost exclusively in the plastids (Ohlrogge et al. 1991, 1993). Subsequent elongation steps (to 20 carbons or longer) that provide long-chain fatty acids for seed triacylglycerols and other specialized purposes (Table 1) are believed to occur in the endoplasmic reticulum. Synthesis of an 18-carbon fatty acid from acetyl-CoA and malonyl-CoA precursors (Fig. 2) is catalyzed by seven enzymes (Browse and Somerville 1991; Ohlrogge et al. 1993). In addition, acyl carrier protein (ACP) is a required cofactor to which the 32 intermediates in the pathway are attached as thioesters. The first step in the pathway is the transfer of malonate from Coenzyme A to ACP. Three condensing enzymes then utilize malonyl-ACP as the 2-carbon donor for elongation of the growing acyl chain. The first condensation reaction is now believed to be between malonyl-ACP and acetyl-CoA by the action of a short-chain 3-ketoacyl-ACP synthase III (Jaworski et al. 1989). Subsequent condensations are between malonyl-ACP and acyl-ACP intermediates and are catalyzed by the 3-ketoacyl-ACP synthase I. The final 2-carbon elongation occurring in plastids is from 16:0 to 18:0 and requires 3-ketoacyl-ACP synthase II. After each condensation, the 3-ketoacyl-ACP intermediates are reduced (by 3-ketoacyl-ACP reductase), dehydrated (by 3-hydroxy-acyl-ACP de-

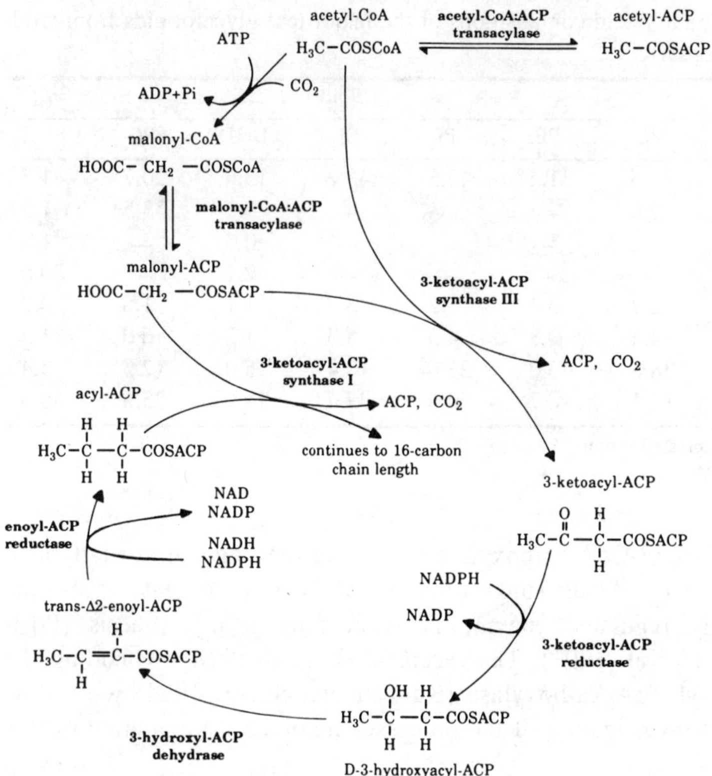

*Figure 2* The pathway of fatty acid synthesis in plants.

hydratase), and reduced again (by enoyl-ACP reductase) to yield the saturated acyl-ACP intermediates (Fig. 2). Unlike the complex multienzyme polypeptides that catalyze fatty acid synthesis in animals and yeast, each of the component enzymes of the plant pathway can be isolated separately. In this and other respects, the synthase system is similar to the Type II fatty acid synthase of *Escherichia coli* rather than the multifunctional Type I synthases of other eukaryotes.

### Attachment of the Fatty Acids and Head Group to the Glycerol Backbone

The first steps of glycerolipid synthesis are two acylation reactions that transfer fatty acids to glycerol-3-P to form phosphatidic acid (PA) (Fig. 3). Diacylglycerol (DAG) is produced from PA by a specific phosphatase, whereas a nucleotide-activated form of DAG (CDP-DAG) is produced from the reaction of PA with CTP. The energy to drive attachment

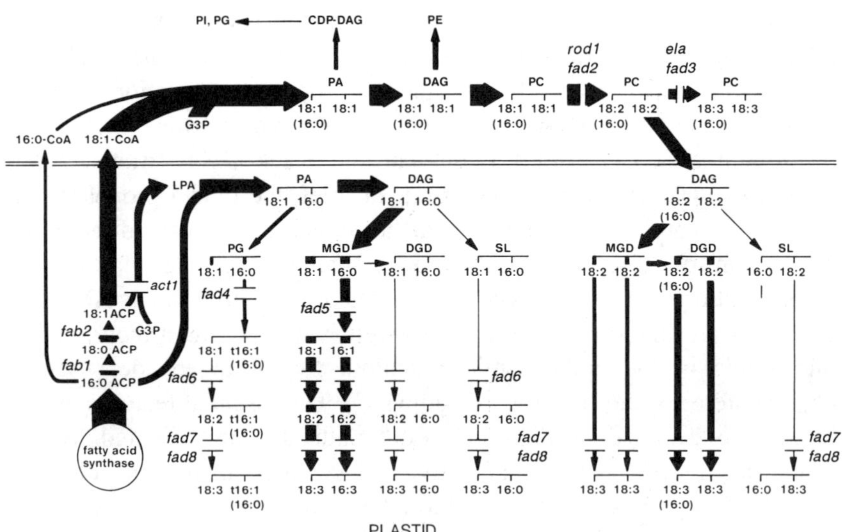



*Figure 3*  An abbreviated diagram of the two-pathway scheme of membrane glycerolipid synthesis in *Arabidopsis* leaves. See text for details. Widths of the lines show the relative fluxes through different reactions. The breaks indicate the putative enzyme deficiencies in various mutants (see Table 4). No attempt has been made to accurately represent the various lipid species that are transferred between the endoplasmic reticulum and the plastid.

of the polar head group during de novo glycerolipid synthesis is provided by nucleotide activation. When DAG is the lipid substrate, it is the head group that is activated. Thus, CDP-choline, CDP-ethanolamine, and CDP-methylethanolamine can be substrates for phospholipid synthesis, and UDP-galactose and UDP-sulfoquinovose are substrates for monogal-actosyldiacylglycerol (MGD) and sulfoquinovo-syldiacylglycerol (SL) synthesis, respectively. Conversely, phospatidylinositol (PI), phospha-tidylserine (PS), and phosphatidylglycerol phosphate (the precursor of phosphatidylglycerol [PG]) are formed by reactions of myo-inositol, serine, or glycerol-3-P with CDP-DAG.

Although the synthetic pathways are presented here as a linear series of simple enzymatic steps (see below and Fig. 3), the actual biochemistry involved is complicated by the possibilities of head group modification (for example, the synthesis of PC from phosphatidylmethylethanolamine by two rounds of methylation) and head group exchange. The details of the reactions involved and the synthetic routes that probably operate in higher plants can be found in recent reviews (Browse and Somerville 1991; Joyard et al. 1993; Kinney 1993).

## Two Pathways for Glycerolipid Synthesis

In broad terms, we can consider two major pathways by which the acyl-ACP products of plastid fatty acid synthesis are utilized in plant cells for the biosynthesis of glycerolipids and the associated production of polyunsaturated fatty acids. The evidence for the "two pathway" model has been summarized elsewhere (Roughan and Slack 1982; Browse et al. 1986b; Browse and Somerville 1991). In brief, the model proposes that fatty acids synthesized de novo in plastids either may be used directly for production of chloroplast lipids by a pathway in the plastid (the "prokaryotic pathway") or may be exported to the cytoplasm as free fatty acids that are first converted to CoA esters, which are then incorporated into lipids in the endoplasmic reticulum by an independent set of acyltransferases (the "eukaryotic pathway"). The essential features of this model are depicted in Figure 3. Both pathways are initiated by the synthesis of 16:0-ACP by the fatty acid synthase in the plastid. This 16:0-ACP may be elongated to 18:0-ACP and then desaturated to 18:1-ACP by a soluble desaturase (Shanklin and Somerville 1990) so that 16:0-ACP and 18:1-ACP are the primary products of plastid fatty acid synthesis. These thioesters may be used within the plastid for the synthesis of PA, or they may be hydrolyzed to free fatty acids which move through the plastid envelope to be converted to CoA thioesters in the outer envelope membrane by acyl-CoA synthetase.

Because of the substrate specificities of the plastid acyltransferases (Frentzen 1993), the PA made by the prokaryotic pathway has 16:0 at the *sn*-2 position and, in most cases, 18:1 at the *sn*-1 position. This PA is used for the synthesis of PG or is converted to DAG by a PA-phosphatase located in the inner plastid envelope. This DAG pool can act as a precursor for the synthesis of the other major plastid membrane lipids MGD, digalactosyldiacylglycerol (DGD), and SL (Joyard et al. 1993). Acyl groups exported from the plastid are used for the synthesis of PA, mainly in the endoplasmic reticulum. In contrast to the plastid isozymes, the acyltransferases of the endoplasmic reticulum produce PA highly enriched with 18-carbon fatty acids at the *sn*-2 position; 16:0, when present, is confined to the *sn*-1 position. This PA gives rise to the phospholipids such as PC, phosphatidylethanolamine (PE), and PI, which are characteristic of the various extrachloroplast membranes. In addition, however, the DAG moiety of PC is returned to the chloroplast envelope where it enters the DAG pool and contributes to the synthesis of plastid lipids (Fig. 3). Evidence from several *Arabidopsis* mutants indicates that lipid exchange between the endoplasmic reticulum and the chloroplast is reversible to some extent (Miquel and Browse 1992; Browse et al. 1993a).

In many species of higher plants, PG is the only product of the prokaryotic pathway, and the remaining chloroplast lipids are synthesized entirely by the eukaryotic pathway. In other species, including *Arabidopsis*, in which both pathways contribute about equally to the synthesis of MGD, DGD, and SL (Browse and Somerville 1991), the leaf lipids characteristically contain substantial amounts of hexadecatrienoic acid (16:3), which is found only in MGD and DGD molecules produced by the prokaryotic pathway. These plants have been termed "16:3 plants" to distinguish them from the other angiosperms (18:3 plants) whose galactolipids contain predominantly linolenate. The contribution of the eukaryotic pathway to MGD, DGD, and SL synthesis is reduced in lower plants, and in many green algae the chloroplast is almost entirely autonomous with respect to membrane lipid synthesis.

## The Problem of Lipid Transfer

The exchange of lipids between the endoplasmic reticulum and chloroplasts of leaf cells is well documented, and lipid exchange to mitochondria and other sites is assumed (Browse and Somerville 1991). However, essentially nothing is known about the mechanism of this process. Membrane flow to the plasma membrane and the tonoplast is known to occur in plant cells, but vesicular transport does not seem to be involved in transfer to the chloroplast. A class of soluble proteins characterized (from in vitro experiments) as lipid transfer proteins (LTPs) had been considered to be the intracellular transporters, but biochemical and immunohistochemical evidence has made it clear that these proteins are extracellular and therefore cannot fulfill their proposed role (Bernhard et al. 1991; Sterk et al. 1991; Thoma et al. 1993). At least three different LTP genes have been identified by expressed sequence tag (EST) sequencing, and the pattern of expression of one of these has been examined in some detail using both promoter-*GUS* fusions and in situ hybridization (Thoma et al. 1994). The *LTP-1* gene appears to be expressed only in epidermal cells and is most strongly expressed in tissues such as the stigma, which are known to secrete lipophilic materials. Thus, the available evidence is consistent with a possible role in some secretory process. Transgenic plants in which the amount of the LTP-1 product has been strongly reduced by expression of an antisense construct have no morphological phenotype, no apparent defect in cutin or wax composition, and no growth defect, but are very late flowering (S. Thoma and C. Somerville, unpubl.). This raises the possibility that one of the late-flowering mutants could have a defect in the *LTP-1* gene. At present, it is not clear whether other, intracellular, lipid transfer proteins may exist or whether it will be necessary to invoke other mechanisms for the transfer of membrane lipids (Kader 1993).

### Membrane Desaturases

In all plant tissues, the major glycerolipids are first synthesized using only 16:0 and 18:1 acyl groups. Subsequent desaturation of the lipids to the highly unsaturated forms typical of the membranes of plant cells is carried out by membrane-bound desaturases of the chloroplast and the endoplasmic reticulum (Browse and Somerville 1991; Heinz 1993). Investigation of these desaturases by traditional biochemical approaches has been limited because solubilizing and purifying them has proven very difficult. To date, the only glycerolipid desaturase purified to homogeneity is the 16:1/18:1 desaturase from spinach chloroplasts (E. Heinz, unpubl.). Understanding of the mechanisms and regulation of the chloroplast and endoplasmic reticulum desaturases has benefited considerably from the characterization of seven classes of *Arabidopsis* mutants, each one deficient in a specific desaturation step (Browse and Somerville 1991; Somerville and Browse 1991). The mutations that define four of these loci were originally called *fadA*, *fadB*, *fadC*, and *fadD*, but these have now been renamed *fad4*, *fad5*, *fad6*, and *fad7*, respectively.

Two of the chloroplast desaturases are highly substrate specific. The *FAD4* (*FADA*) gene product controls a Δ3 desaturase that inserts a *trans* double bond into 16:0 esterified to position *sn*-2 of PG, whereas the *FAD5* (*FADB*) gene product is responsible for the synthesis of Δ7 16:1 on MGD and possibly DGD (Browse et al. 1985; Kunst et al. 1989a). In contrast, the other two chloroplast desaturases act on acyl chains with no apparent specificity for the length of the fatty acid chain (16- or 18-carbon), its point of attachment to the glycerol backbone (*sn*-1 or *sn*-2), or the nature of the lipid head group. The 16:1/18:1 desaturase is encoded by the *FAD6* (*FADC*) gene (Browse et al. 1989), whereas two 16:2/18:2 isozymes are encoded by *FAD7* (*FADD*) and *FAD8* (*FADE*) (Browse et al. 1986a; S. Hugly et al., unpubl.). The endoplasmic reticulum 18:1 (*FAD2*) and 18:2 (*FAD3*) desaturases (Miquel and Browse 1992; Browse et al. 1993a) also act on fatty acids at both the *sn*-1 and *sn*-2 positions of the molecule. They have been characterized as PC desaturases, but it is possible that they also act on other phospholipids.

The *fab1* and *fab2* mutants of *Arabidopsis* are characterized by increased levels of 16:0 or 18:0, respectively, in seed and leaf tissues (James and Dooner 1990; J. Wu et al., unpubl.). In the *fab1* mutant, the biochemical defect appears to be a reduction in the activity of the condensing enzyme (3-ketoacyl-ACP synthase II) responsible for the elongation of 16:0 to 18:0. In *fab2*, it is assumed that 18:0-ACP desaturase activity is reduced. In both mutants, the saturated fatty acids are incorporated into all the major membrane glycerolipids (although MGD contains relatively low proportions). Both the mutants appear to be leaky, so that the changes in overall membrane fatty acid composition are moderately small. Nevertheless, the changes do have profound effects on the growth and development of these plants (see below).

## Regulation of Membrane Lipid Synthesis

Because each of the prokaryotic and eukaryotic pathways has its own set of associated desaturases, the mutational block of any single desaturation step still allows unsaturated fatty acid synthesis by the other pathway. Furthermore, several of the mutants isolated show a significant shift in the balance of lipid synthesis between the two pathways. The *fad5* and *fad6* mutants show decreased synthesis via the prokaryotic pathway (by 20% and 30%, respectively, relative to wild type) and a corresponding increase in chloroplast lipid synthesis by the eukaryotic pathway (Browse et al. 1989; Kunst et al. 1989a). The *fad2* mutants show increased activity of the prokaryotic pathway and a decreased flux from PC to the chloroplast lipids (Miquel and Browse 1992). The implication from these results is that the pathways of lipid synthesis can be regulated in order to ameliorate to some extent the consequences of the desaturase defects on the fatty acid composition of cell membranes.

A more extensive demonstration of the regulation underlying lipid metabolism in leaf cells comes from the *act1* mutant (Kunst et al. 1988). In wild-type *Arabidopsis*, approximately half of the chloroplast glycerolipids are derived from the prokaryotic pathway (Fig. 1). However, the *act1* mutant is deficient in the chloroplast acyl-ACP:glycerol-3-P acyltransferase—the first enzyme of the prokaryotic pathway. The blockage does not result in the accumulation of precursors (18:1- and 16:0-ACP) upstream of the enzyme deficiency but causes a redirection of lipid metabolism so that the eukaryotic pathway predominates in the mutant. Somewhat unexpectedly, this redirection has little effect on the amount of each lipid that accumulates or on the total glycerolipid content of the tissue. The enhanced flux through the eukaryotic pathway in the mutant is accompanied by a change in the proportion of individual lipids synthesized by this pathway. Thus, the amount of lipids such as PE and PI in the extrachloroplast membranes of the mutant is similar to that in the wild type. However, to compensate for the loss of the prokaryotic pathway, the amount of PC synthesized must be increased about twofold to provide the DAG moieties required for normal levels of glycerolipid synthesis by the chloroplast. Nevertheless, the amount of PC that accumulates is apparently only that required for synthesis of the extrachloroplast membranes, since the level of this lipid in the mutant is only slightly higher than in the wild type. Clearly, the synthesis of PC and export from the microsomal membranes are regulated in concert to meet the requirements for lipid synthesis by chloroplasts. Although it is not currently possible to infer the mechanisms, it is apparent that membrane biosynthesis by the chloroplast and the microsomal membranes is closely coordinated.

### Further Metabolism of Membrane Glycerolipids

Very little is known about the cell biology of lipid turnover and degradation. The enzymes involved, including phospholipases, galactolipases, and lipoxygenases have been thoroughly investigated for a number of years (Galliard 1980; Vick 1993; Wang 1993). However, it is not at all clear how these and other enzymes are regulated during the senescence of an organ, for example, a leaf, even though it is obvious that the whole process must be controlled to bring about the orderly dismantling of cells and organelles and to facilitate the export of those lipid breakdown products that can be used as precursors elsewhere in the plant.

Two more specific examples of the further metabolism of membrane lipids involve cellular signaling processes. The presence of phosphatidylinositol 4,5-bisphosphate in higher plants has been established (Cote and Crain 1993), and by analogy with the well-studied phosphoinositide system in animal cells (Majerus 1992), it is assumed that this compound may act as a precursor for signal molecules (DAG and inositol [1,4,5] trisphosphate, $IP_3$). The action of $IP_3$ in releasing $Ca^{++}$ into the cytoplasm (through specific channels in the tonoplast and possibly other membranes) and thereby regulating cellular processes has been demonstrated in several plant systems (Alexandre and Lassalles 1992; Cote and Crain 1993). On the other hand, it is not at all certain that the DAG arm of the phosphoinositide signaling pathway is present in plants (Cote and Crain 1993).

Jasmonate is one of several lipid-derived plant growth regulators (Table 1). Recent investigations of jasmonate action have revealed that it is able, at low concentrations, to strongly induce specific genes, including proteinase inhibitors and other plant defense genes (Farmer and Ryan 1990; Gundlach et al. 1992) and vegetative storage proteins (Mason and Mullet 1990). The structure and biosynthesis of jasmonate have intrigued plant biologists because of the parallels to some eicosanoids that are central to inflammatory responses and other physiological processes in mammals (Creelman et al. 1992). In plants, jasmonate is synthesized from linolenate (which is presumably released from membrane lipids by a phospholipase $A_2$) by a pathway that starts with the action of lipoxygenase on 18:3 (Fig. 4). The proposed pathway has not been characterized at the level of purified enzymes. Several lipoxygenase isozymes have been purified and thoroughly studied (Vick 1993), but they typically have high activities. It is uncertain that these lipoxygenases would be the enzymes operating in a pathway that synthesizes about 3 nmole/g fresh weight of product in a strongly induced plant (Creelman et al. 1992). Fortunately, both the biochemistry and cell biology of jasmonate synthesis are now being given more attention, so that we can look forward to having a more complete picture of jasmonate biology.

## PHYSIOLOGY AND CELL BIOLOGY OF MEMBRANE LIPID COMPOSITION

The central issue with respect to the role of glycerolipids in membranes is framed by the observations that each membrane of the cell has a characteristic and distinct complement of glycerolipid types, and that within a single membrane, each class of lipids has a distinct fatty acid composition (Browse et al. 1993b). This diversity implies that differences in lipid structure are important for membrane function. However, despite considerable effort, the details of this structure/function relationship have remained elusive; in large part, because there have not been instructive biological examples available in which changes in lipid composition have been clearly shown to have specific effects on membrane processes. The *Arabidopsis* lipid mutants have provided such examples, and an understanding of the mutant phenotypes at a mechanistic level will undoubtedly advance the field considerably.

### Significance of Chloroplast Glycerolipid Composition

Chloroplast membranes have a characteristic and unusual fatty acid composition. Typically, 18:3, or a combination of 18:3 and 16:3, fatty acids account for approximately 67% of all the thylakoid membrane fatty acids and more than 90% of the fatty acids of MGD, the most abundant chloroplast lipid. The atypical fatty acid $\Delta$3,*trans*-hexadecenoate (*trans*-16:1) is present as a component of the major thylakoid phospholipid, PG. The fact that these and other characteristics of chloroplast lipids are common to most or all higher plant species suggests that the lipid fatty acid composition is important for maintaining photosynthetic function. Fur-

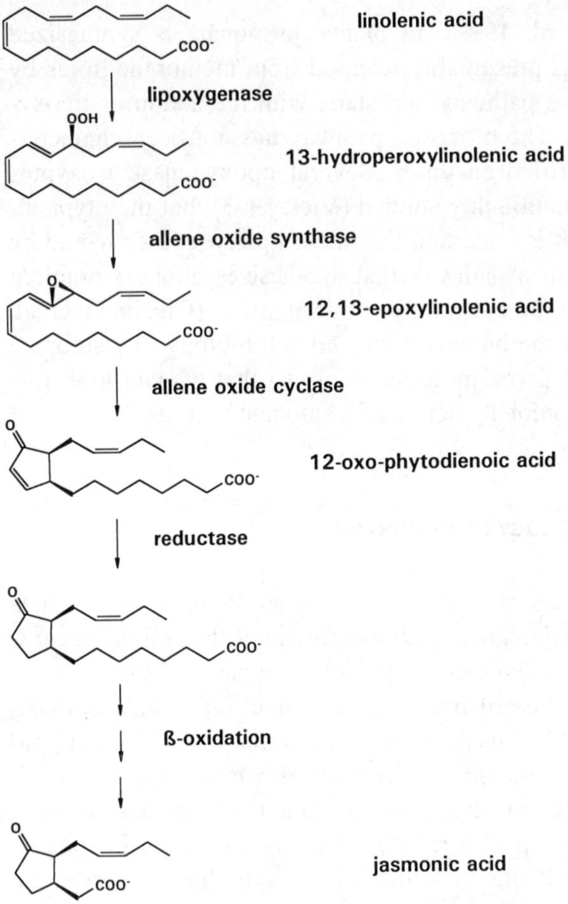

*Figure 4* The pathway for conversion of linolenate to jasmonate.

thermore, the thylakoid membrane is the site of light absorption and oxygen production. The free radicals that are by-products of these reactions will stimulate oxidation of the polyunsaturated fatty acids. Since this might be expected to mediate against a high degree of unsaturation, it has been inferred that there is a strong selective advantage to having such high levels of trienoic fatty acids in thylakoid membranes.

Many different approaches have been used in attempts to understand the significance of membrane fatty acid composition to photosynthetic function. These include the correlation of events during chloroplast development; physical studies of model membrane systems; reconstitution of photosynthetic components with lipid mixtures; alteration of lipids in situ by heat stress, lipase treatment, or hydrogenation of unsaturated fatty acids; and alteration of lipids in vivo by chemical inhibitors or molecular genetics techniques (Quinn et al. 1989; Murata et al. 1992; Wolter et al.

1992; Browse et al. 1993b). However, in general, these approaches have not been successful in establishing unequivocal relationships between membrane form and function. Indeed, some of the relationships that have been proposed are now untenable in view of the results we have obtained from *Arabidopsis* mutants (see below).

Table 4 shows the leaf fatty acid compositions of eight classes of mutants deficient in (nuclear encoded) chloroplast enzymes involved in lipid synthesis. Because chloroplast membranes account for 75% of the total leaf glycerolipids (Table 2), the data in Table 4 are sufficient to illustrate the overall change in chloroplast membrane fatty acid composition in each mutant line. The phenotypes of each mutant are summarized in Table 5.

In the *fad7* (*fadD*) mutants, there is a considerable decrease in the amount of both 16:3 and 18:3 fatty acids associated with a concomitant increase in the levels of 16:2 and 18:2. The photosynthetic characteristics of the *Arabidopsis fad7* mutant have been examined in detail with particular attention to the proposals that have been made regarding the importance of trienoic fatty acids to the photosynthetic competence of chloroplast membranes. The average number of double bonds per glycerolipid

*Table 4*  The overall fatty acid compositions of leaf lipids from *Arabidopsis* mutants

| Mutant line | \multicolumn{9}{c}{Fatty acid} | | | | | | | | |
|---|---|---|---|---|---|---|---|---|---|
| | 16:0 | 16:1c | 16:1t | 16:2 | 16:3 | 18:0 | 18:1 | 18:2 | 18:3 |
| Wild type | 1 5 | t r | 3 | t r | 1 4 | 1 | 3 | 1 4 | 4 8 |
| *fab1* | **2 3** | 1 | 4 | t r | 1 7 | 1 | 3 | 1 1 | 3 9 |
| *fab2* | 1 4 | t r | 2 | t r | 6 | **14** | 3 | 1 8 | 4 2 |
| *act1* | 1 0 | t r | 2 | t r | **1** | 1 | 8 | 2 3 | 5 4 |
| *fad4* | 1 8 | t r | **0** | t r | 1 2 | 1 | 3 | 1 9 | 4 7 |
| *fad5* | **2 4** | 1 | 3 | t r | **t r** | 1 | 3 | 1 7 | 5 0 |
| *fad6* | 1 4 | **1 1** | 4 | t r | **t r** | 1 | **1 6** | 1 7 | **3 7** |
| *fad7*[a] | 1 7 | 3 | 3 | **6** | **2** | 1 | 9 | **3 9** | **1 9** |
| *fad2* | 1 2 | 1 | 3 | t r | 1 7 | 1 | **2 1** | **4** | 4 1 |
| *fad3* | 1 5 | t r | 3 | t r | 1 6 | 1 | 3 | **2 1** | **4 1** |
| Multiple mutants | | | | | | | | | |
| *fad7 fad8*[b] | 1 5 | 1 | 2 | 1 0 | 0 | 1 | 4 | 4 9 | 1 7 |
| *fad2 fad6* | 1 1 | 1 6 | 4 | 1 | 0 | 1 | 60 | 5[c] | 0 |
| *fad3 fad7 fad8* | 1 5 | 1 | 2 | 9 | 0 | 1 | 6 | 6 5 | 0 |

Data are mole %.
[a]Plants grown at 28°C.
[b]The fatty acid composition of *fad8* alone is indistinguishable from wild type.
[c]Probably the $\Delta 9$, $\Delta 15$ isomer.

*Table 5* Responses of *Arabidopsis* lipid mutants to high and low temperature relative to the performance of wild-type plants

| Mutant line | Observed phenotypes |
| --- | --- |
| *fad4* | None detected |
| *fad5* | Enhanced growth rate and thermotolerance of photosynthetic electron transport at high temperatures. Leaf chlorosis, reduced growth rate, and impaired chloroplast development at low temperature. |
| *fad6* | Enhanced thermotolerance of photosynthetic electron transport (but not enhanced growth) at high temperatures. Leaf chlorosis, reduced growth rate and impaired chloroplast development at low temperature. |
| *fad7* | Reduced chloroplast size and altered chloroplast ultrastructure at temperatures (>25°C) at which fatty acid composition is affected. |
| *act1* | Enhanced growth and slightly enhanced thermostability of photo-synthetic processes at high temperatures. |
| *fad2* | Greatly reduced stem elongation at 12°C. Death of plants at 6°C. |
| *fad3* | None detected. |
| *fab1* | Death of plants after prolonged exposure to 2°C. |
| *fab2* | Dwarf phenotype at 22°C is ameliorated at temperatures above 35°C. |

molecule is reduced from 4.4 in the wild type to 3.4 in the mutant, and this reflects a change in the degree of unsaturation of all the chloroplast glycerolipids (Browse et al. 1986a). Despite this change, fluorescence polarization measurements indicated that the viscosity of thylakoid membranes from the mutant was almost unaltered compared with thylakoids from wild-type plants. Furthermore, there was little or no effect on a range of photosynthetic parameters, including net $CO_2$ exchange of whole plants, the photosystem I, photosystem II, whole-chain electron transport rates of thylakoid preparations, and the primary photochemistry as revealed by both room temperature and 77°K fluorescence analysis (McCourt et al. 1987). Instead, the most pronounced change in the mutants is a 45% reduction in the cross-sectional area of chloroplasts relative to wild-type plants with a corresponding increase in the number of chloroplasts so that the total amount of membrane is unchanged. At present, we do not understand how the *fad7* mutations exert this effect, but the observation highlights the lack of knowledge of the factors that determine organelle size and shape. The effects of the *fad7* mutations are not due simply to a reduction in the gross level of unsaturation, because the greater reduction in chloroplast membrane unsaturation in the *fad6* mutants provokes quite different effects on chloroplast ultrastructure

(Hugly et al. 1989). These observations and work on the *fad4* and *act1* mutants (McCourt et al. 1985; Kunst et al. 1989b) indicate that photosynthetic processes are somewhat insensitive to at least some changes in thylakoid fatty acid composition. In the *act1* and *fad7* mutants, the main effects of alterations in lipid biosynthesis appear to be on chloroplast ultrastructure (McCourt et al. 1987; Kunst et al. 1989b).

Studies of two other mutants, *fad5* and *fad6*, have revealed more direct effects of lipid composition on the function of chloroplast membranes (Table 5). Leaf tissue of the *fad6* (*fadC*) mutant (Table 4) contains reduced levels of both 18:3 and 16:3 fatty acids and increased levels of the 18:1 and 16:1 precursors, due to a deficiency in chloroplast desaturation (Browse et al. 1989). The fatty acid composition of the *fad5* mutant is consistent with the loss of function of the MGD-specific 16:0 desaturase required for the synthesis of 16:3 in MGD (Kunst et al. 1989a). In the *fad6* mutant, the substantial decrease in lipid unsaturation is accompanied by reduced chloroplast membrane synthesis and a slight decrease in photosynthesis in plants grown and assayed at 22°C (Hugly et al. 1989). However, on the basis of two different criteria, thylakoid membranes from *fad6* plants showed increased stability to thermal disruption. First, the temperature at which fluorescence yield enhancement occurred in the mutant was increased 3°C relative to wild type, from 42°C to 45°C. Second, when thylakoid preparations were heat-treated prior to the measurement of photosynthetic electron transport, membranes from the mutant consistently retained higher relative rates than membranes from wild-type plants. The magnitude of the effects observed in the *fad6* mutant is comparable with observations on natural populations of adapted and nonadapted plants (Downton et al. 1984) and consistent with the effects of catalytic hydrogenation on the thermal properties of thylakoid preparations (Quinn et al. 1989). Qualitatively similar, but smaller, effects were later demonstrated for the *fad5* mutant (Kunst et al. 1989c).

The high-temperature tolerance of photosynthesis in the *fad6* and *fad5* mutants is correlated with an increased sensitivity of these two lines to chilling temperatures (Table 5). The wild type and the mutant lines *fad4*, *fad7*, and *act1*, are indistinguishable from one another in growth and appearance when plants are kept at 6°C. In contrast, the *fad5* and *fad6* mutants become chlorotic at this temperature and show a 20–30% reduction in growth rate relative to the wild type (Hugly and Somerville 1992). Ultrastructural analysis of chloroplasts from plants grown at low temperature indicated a major reduction in the amount of both appressed and nonappressed thylakoids, and this was associated with simultaneous reductions in amounts of chlorophyll, lipid, and protein. Taken together, the results suggest that the changes in fatty acid composition in the *fad5* and *fad6* mutants impair chloroplast development at low temperatures. These observations notwithstanding, both the *fad5* and *fad6* mutants continue to grow and complete their life cycles normally at 6°C.

The increased 16:0 content in leaves of *fab1* plants includes the synthesis of molecular species of chloroplast PG that contain only 16:0 + 16:1 *trans* + 18:0. Such molecular species account for 42% of the PG in leaves of the mutant. Such "high melting point" molecular species of PG have been correlated with chilling sensitivity of plants, including transgenic *Arabidopsis* whose chloroplast lipid composition was altered by the expression of a modified *E. coli plsB* gene (Wolter et al. 1992). Experiments with transgenic tobacco have also been interpreted as supporting a major role for high-melting-point PG species in determining plant chilling sensitivity (Murata et al. 1992). However, the *fab1* mutants are able to grow and complete their life cycles normally at 10°C. They are also unaffected (when compared with wild-type controls) by more severe chilling treatments that quickly lead to the death of cucumber and other chilling-sensitive plants. These treatments include exposure to 2°C for up to 7 days and freezing to –2°C for 24 hours. When grown continuously at 2°C, *fab1* plants eventually die, whereas wild-type *Arabidopsis* flower and produce seed at this temperature. The implication of these results is that moderately high levels of disaturated PG are a minor component of plant chilling sensitivity, although they may compromise long-term survival of plants at low temperatures. The different findings of Wolter et al. (1992) may be a result of the higher levels of high-melting-point PG (>50%) in their plants. However, these genetically altered plants did not grow as well as wild type at normal temperatures; this observation raises the possibility that some other aspect of *plsB* expression is contributing to the low-temperature phenotype.

## Mutants Deficient in Desaturation of
## Extrachloroplast Membranes

The *fad2* mutants of *Arabidopsis* are deficient in activity of the microsomal 18:1 desaturase responsible for production of polyunsaturated lipids on the eukaryotic pathway of lipid synthesis (Miquel and Browse 1992). Because the 16:1/18:1 desaturase (the *FAD6* gene product) is able to desaturate all the major chloroplast glycerolipids (Browse et al. 1989), the biochemical effect of the *fad2* lesion is largely confined to the extrachloroplast membranes of the cell. Furthermore, exchange of lipids between the eukaryotic pathway and the prokaryotic pathway leads to the appearance of substantial amounts of 18:3 (but not 18:2) in the extrachloroplast membranes of leaves of *fad2* mutant plants. Thus, PC, the most abundant extrachloroplast lipid, contains 2% 18:2 and 14% 18:3 in leaves of *fad2* plants, compared with wild-type levels of 33% and 40%, respectively. In tissues such as roots that lack chloroplasts, the effect of the mutation is greater. Roots of *fad2* plants contained 7–9% 18:2+18:3 compared with 65% in wild type.

When grown at 22°C, *fad2* plants are similar in growth and appearance to the wild type. However, these mutants show a dramatic chilling-sensitive phenotype when plants are transferred to 6°C (Miquel et al. 1993). At this temperature, wild-type *Arabidopsis* continued to grow and develop normally, but leaves of three allelic *fad2* mutants gradually deteriorated, displaying patches of necrosis and extensive accumulation of anthocyanins. Death of the leaves and of the whole plants eventually followed. Deterioration of the mutant plants was not associated with any further decrease in membrane unsaturation induced by low temperature. The observation that three independent alleles at the *fad2* locus all showed the same symptoms at 6°C indicates that the biochemical defect in polyunsaturated lipid synthesis, rather than any other mutation, is the cause of the chilling-sensitive phenotype. Therefore, even though the lesions at the *fad2* locus are largely irrelevant to growth and survival of plants at normal temperatures, the results demonstrate that polyunsaturated membranes are essential for maintaining cellular function and plant viability at low temperatures. Although the deficiency in polyunsaturated lipids has severe consequences, several lines of evidence indicate that the change(s) in membrane function in *fad2* plants at 6°C only begins to compromise cell viability after at least several days. For example, appearance of visual symptoms occurred gradually in *fad2* plants over a period of 6–10 days at low temperature. Furthermore, *fad2* plants returned to 22°C after 3 days exposure at 6°C recovered fully and showed no visible signs of tissue damage. Even plants held for several weeks at 6°C were sometimes able to produce new leaves, flower, and set seed after being returned to 22°C.

When plants grown at 22°C are transferred to intermediate temperatures (10–12°C), the *fad2* plants exhibit another unusual phenotype. In contrast with the usual extensive stem elongation of the wild type, the mutants produce flowers and, in due course, viable seeds on stems that average no more than 15 mm in length (Miquel et al. 1993). In other respects, however, the mutant plants appear to be healthy. Because flower and silique sizes are not markedly altered in the mutants, these findings point to a role for polyunsaturated membrane glycerolipids in ensuring correct cellular responses to tissue-specific growth signals.

The hypothesis that an acute loss of membrane integrity at low temperatures, as a result of a phase change within the lipid bilayer, constitutes the primary event leading to injury and death of chilling-sensitive plants has been a dominant theme in considering the functional relevance of glycerolipid unsaturation in higher plants (Lyons 1973; Murata et al. 1992). It is highly unlikely that any such major disruption of membrane structure is involved in producing the phenotypes described here for the *fad2* mutant. Liquid-crystalline-phase to gel-phase transition temperatures for 18:1/18:1- and 16:0/18:1-phosphatidylcholines (the most abundant diunsaturated and monounsaturated lipid species in extrachloroplast membranes of *fad2* plants) are $-17^{\circ}C$ and $-3^{\circ}C$, respectively (Marsh 1990). In contrast, the reduction in stem elongation was observed in *fad2* plants at $12^{\circ}C$, apparently in the absence of any major disruption of plant metabolism. Furthermore, the onset of damage at $6^{\circ}C$ was gradual and chronic. These observations are clearly not consistent with a gross disruption of membrane structure but, instead, suggest a limited defect in membrane function.

Whereas most of the lipid mutants grow and develop normally at $22^{\circ}C$, plants of the *fab2* mutant line exhibit an extreme dwarf phenotype (Lightner et al. 1994). Following mutagenesis of the *fab2* line, a suppressor mutation provisionally designated *shs* (*s*uppressor of *h*igh *s*tearate) was identified that resulted in reversion of leaf fatty acid composition and plant size to wild type. This indicates that in *fab2*, both the high 18:0 content and the dwarf phenotype are caused by the same mutation. Leaf number, leaf shape, and the rate of leaf production appear to be relatively normal in the mutant, suggesting that pattern formation processes are unperturbed. Many cell types in the mutant, including leaf mesophyll and epidermal cells, fail to expand, whereas others, including stomatal guard cells and the leaf trichomes, develop to the normal wild-type sizes. The mechanisms through which physical parameters determine the function of biological membranes are poorly understood. However, with respect to the fatty acid components of membrane glycerolipids, it is generally accepted that three interrelated features are of primary importance: the phase behavior of the lipid, the viscosity characteristic contributed to the lipid bilayer, and the shape of the molecule (Browse et al. 1993b). These characteristics are influenced by the identity of the lipid head group but, in all cases, the thermodynamic behavior is such that the physical effects of increased stearate in membrane lipids would be expected to be offset by increased temperature. Growth of *fab2* plants at $36^{\circ}C$ does not alter leaf fatty acid composition, but cell expansion and leaf size more closely resemble wild type at this temperature (J. Lightner and J. Browse, unpubl.). Such findings could be consistent with high temperature alleviating the effect of saturated fatty acids on membrane fluidity.

## More Extreme Phenotypes: Generation of
## Multiple-mutant Lines

With only one exception, all the *Arabidopsis* mutants deficient in lipid biosynthesis have been isolated as healthy plants that are not readily distinguished from the wild type when grown at 22°C. In line with this observation, relatively small changes have been found in the photosynthetic characteristics, growth, or development of these plants, except at extreme temperatures. However, in all of these mutants, the level of polyunsaturated fatty acids is at least 60–80% of that found in the wild type (Table 4), because each pathway of lipid synthesis has an associated set of desaturases (Fig. 3). To expand the range of mutant phenotypes available, it was necessary to generate double mutants with more extensive alterations in the fatty acid composition of their thylakoid membranes. Some of the double mutants produced show only a small additional change in chloroplast lipid composition. However, in some cases, it has been possible to produce very substantial alterations.

The most extreme fatty acid composition obtained to date is that for the *fad2 fad6* double mutant. The *fad6* mutant is deficient in the synthesis of polyunsaturated fatty acids by the prokaryotic pathway, but leaves of *fad6* plants still contain 52% 18:2+18:3+16:3 (Table 4), compared with 77% in the wild type. The *fad2* mutant is deficient in activity of the endoplasmic reticulum 18:1 desaturase, but chloroplast lipids derived from the eukaryotic pathway in this mutant are desaturated by the *FAD6* gene product. This results in *fad2* plants containing 63% polyunsaturated fatty acids in their leaf lipids. Analysis of 146 $F_2$ plants derived from a cross between the two mutants revealed a large deviation from the expected Mendelian ratio, and no double mutants were recovered. Instead, plants that were homozygous for *fad6* and heterozygous for *fad2* were kept. Individual seeds from these plants were analyzed to show that they were segregating 1:2:1 at the *fad2* locus. Subsequent germination tests revealed that only 40% of the homozygous *fad2 fad6* seeds germinated and that those seedlings produced were not capable of autotrophic growth. Although other possibilities exist, the simplest explanation for these results is that chloroplasts of the double mutant are unable to carry out sufficient photosynthesis to sustain the plant. The double mutants can be grown on sucrose-supplemented media and, under these conditions, they contain less than 6% polyunsaturated fatty acids. The *fad2 fad6* plants contain less chlorophyll than wild-type controls, but their growth and organ development are remarkably normal. These observations indicate that the vast majority of receptor-mediated and transport-related membrane functions required to sustain an organism and to induce proper development are well supported in this double mutant that contains almost no polyunsaturated membrane lipids.

There are three gene products in *Arabidopsis* that mediate the synthesis of trienoic fatty acids. The *fad7* and *fad8* genes encode chloroplast isozymes, and the *fad3* gene product is the endoplasmic reticulum desaturase. We have recently succeeded in generating triple mutants *fad3 fad7 fad8* that contain no detectable 18:3 in their leaf lipids. Preliminary observations indicate that these plants are capable of robust autotrophic growth under our normal growth conditions (22°C; 150 μmole quanta/m$^2$/s constant illumination). Their development matches the wild type and they produce abundant flowers that appear normal. However, the triple mutants do not produce seeds because they are male sterile. This may again be a striking example of a change in membrane composition having a very specific effect on an essential step in the plant's life cycle, while producing no overall phenotype that could be attributed to a general defect in membrane function. We do not yet know whether the absence of trienoic fatty acids in the membranes of these mutants has any consequences for growth at reduced or elevated temperatures.

It is well accepted that molecular shape and the degree of acyl unsaturation must be important factors in determining the properties of membrane lipids. However, with a few exceptions, the field of membrane biology has lacked examples of how specific changes in lipid composition can influence membrane, cell, and organism function. Most of the *Arabidopsis* lipid mutants exhibit decreased levels of membrane unsaturation (Table 4). In many of the mutants, the changes in unsaturation are accompanied by compromised growth or plant function at low temperatures and, in some cases, by small increases in high-temperature tolerance. Nevertheless, we do not interpret our results as supporting a general role for lipid unsaturation in maintaining overall membrane integrity. The diverse range of phenotypes that we have observed among the different mutants (Table 5) is simply not consistent with a gross disruption of membrane structure by low temperature. Rather, each class of mutants displays a set of symptoms that suggest specific and limited defects in membrane function. The series of *Arabidopsis* lipid mutants provides one means to investigate these defects further. Furthermore, recent successes in cloning genes encoding desaturases and other enzymes of lipid metabolism indicate that molecular tools will soon be available to assist in the dissection of the mutants. Genes have recently been isolated for *act1* (Nishida et al. 1993), *fad1* (stearoyl-ACP desaturase), *fad2*, *fad3*, *fad6*, *fad7*, and *fad8* (Arondel et al. 1992; Iba et al. 1993; Yadav et al. 1993; Okuley et al. 1994; J. Shanklin et al., unpubl.).

## SEED LIPID SYNTHESIS

Lipids in the form of triacylglycerols are widely found as a major energy reserve in seeds and fruits. Vegetable oils from seed crops constitute one of the world's most important plant commodities. The major use of these oils is for human consumption, but a significant proportion find use in manufacturing industries, particularly in the production of detergents, coatings, plastics, and specialty lubricants. For both food and industrial applications, it is the fatty acid composition of the oil which determines its usefulness and, therefore, its commercial value.

The biochemistry of storage lipid synthesis in developing oilseeds is less well understood than lipid synthesis in leaves. Some of the steps involved in triacylglycerol synthesis and many aspects of subcellular compartmentation are the same (Stymne and Stobart 1987; Browse and Somerville 1991; Miquel and Browse 1994). However, it is useful to consider seed lipid biosynthesis in terms of a quite different scheme, shown in Figure 5. This scheme was developed using results from many oilseed species (Slack and Browse 1984; Stymne and Stobart 1987) but is drawn to describe the metabolism of developing *Arabidopsis* seeds. *Arabidopsis* seed lipids contain substantial proportions of both unsaturated 18-carbon fatty acids (30% 18:2, 20% 18:3) and long-chain fatty acids (22% 20:1) derived from 18:1 (Table 3). As a result, *Arabidopsis* is a good model for the biochemistry of both 18:2/18:3-rich oilseeds and those species containing longer fatty acids.

In seeds, as in leaves, 16:0-ACP and 18:1-ACP are the major products of plastid fatty acid synthesis (1) and 18:0-ACP desaturase activity (2). (Numbers in parentheses refer to steps in Fig. 5.) 18:0-ACP is an intermediate but is normally found only as a minor component in triacylglycerols and other plant lipids. Presumably in most plants, the thioesterase involved in acyl-ACP hydrolysis is specific for 16:0 and 18:1. By analogy, with leaf chloroplasts, the free fatty acid products probably move through the plastid envelope and are converted to acyl-CoA thioesters on the outer envelope. Certainly, acyl-CoAs are the primary substrates for subsequent reactions in other cellular compartments (Stymne and Stobart 1987). They are used for the synthesis of PC (4) which is the major substrate for 18:1 and presumably 18:2 desaturation (5,6) by microsomal enzymes (Browse and Somerville 1991). The synthesis of PC from DAG by choline phosphotransferase (7) is reversible (Slack et al. 1985), so that PC can be a direct precursor of the highly unsaturated species of DAG used for triacylglycerol (TAG) synthesis. However, the acyl-CoA pool in the seed is a much more complex collection of fatty acyl groups than that in leaf mesophyll cells. Exchange of 18:1 from CoA with the fatty acids at position *sn*-2 of PC (8) provides in-

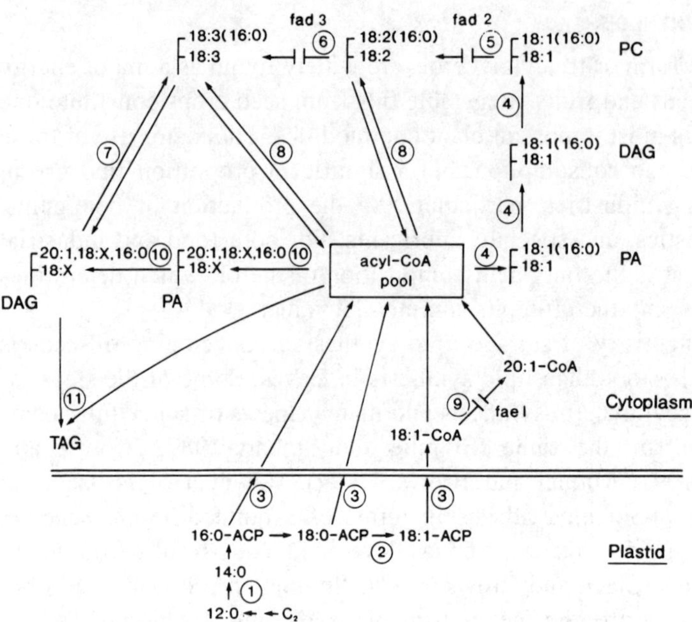

*Figure 5* A simplified scheme for triacylglycerol synthesis in developing *Arabidopsis* seeds. See text for details.

puts of 18:2 and 18:3 (Stymne et al. 1983). Elongation (9) of 18:1-CoA to 20:1-CoA and 22:1-CoA also occurs in *Arabidopsis*, as it does in rapeseed and other cruciferous oilseeds (Stumpf and Pollard 1983). Obviously, synthesis of DAG (10) via PA may also involve these other components of the acyl-CoA pool (Stymne and Stobart 1987), as does the final acylation (11) of DAG to form TAG (Sun et al. 1988). Thus, the pool of DAG used for TAG synthesis is fed by both phosphatidic acid phosphatase and choline phosphotransferase. The evidence for many of the steps in Figure 5 is incomplete, but it is a good working model for considering biochemical and genetic evidence about oilseed lipid synthesis and the possibilities to modify the fatty acid composition of seed oils.

## Mutants Altered in Seed Fatty Acid Composition

Additional information on many aspects of seed lipid synthesis in *Arabidopsis* has been obtained by the isolation and characterization of mutants with alterations in seed fatty acid composition (James and Dooner 1990; Lemieux et al. 1990). Single gene mutations were found that substantially blocked the desaturation of 18:1 to 18:2 or of 18:2 to

18:3 (Fig. 5). These mutants with strong fatty acid phenotypes are allelic to the *fad2* and *fad3* mutants, respectively, which also affect leaf and root fatty acid compositions. The identification of constitutively expressed desaturase genes in *Arabidopsis* contrasts with results for other oilseeds, which indicate that, in many cases, changes in seed oil compositions are not accompanied by changes in the fatty acid composition of leaf membrane lipids (Ohlrogge et al. 1991).

In addition to the *fad2* mutants, a number of other mutants whose seed contained increased levels of 18:1 and correspondingly decreased levels of 18:2 and 18:3 were isolated (Lemieux et al. 1990). One of these lines was shown by complementation tests to carry a recessive nuclear mutation at a locus designated *rod1* which was distinct from the *fad2* locus. It is not apparent where in the pathway of desaturation the *rod1* mutants exert an effect. Obviously, genes regulating the production or activity of the *fad2* desaturase are possibilities. In addition, however, Figure 5 shows that synthesis of the desaturase substrate (18:1-PC) can be affected by the activities of both choline phosphotransferase (reactions 7, 4) and acyl-CoA:lyso-PC acyltransferase (reaction 8). Mutations leading to a decrease in either of these activities are predicted to reduce the proportion of polyunsaturated fatty acids in the seed oil without necessarily reducing overall TAG synthesis.

One exceptional mutant had increased levels of 18:3 and decreased levels of 18:2 and 18:1 (Lemieux et al. 1990). The mutant cannot be explained in terms of a deficiency in any known step of fatty acid desaturation. Nevertheless, genetic analysis indicates that it is heritable and due to a single recessive mutation at a locus provisionally designated *ela* (*e*nhanced *l*inolenate *a*ccumulation). The phenotype exhibits a relatively subtle change in seed fatty acid composition and has no detectable effects on the lipid composition of leaves and roots. The phenotype of the *ela* mutant may be caused by an altered response of linolenic acid synthesis to the standard growth conditions used. Our inability to propose a mechanistic explanation for the mutant emphasizes the fact that we do not understand what it is that normally prevents complete desaturation of fatty acids to linolenate. One possibility is that the degree of desaturation is limited by competition between the 18:2 desaturase and enzymes such as choline phosphotransferase which participate in conversion of PC to triacylglycerols. If this is the case, a leaky mutation in the choline phosphotransferase might be expected to result in an increase in the proportion of linolenate by simply decreasing the rate of flux of lipid from PC to TAG.

*Arabidopsis* seed triacylglycerols contain about 20% of very long chain fatty acids (i.e., ~ 17% 20:1 and 3% 22:1). More than 10 mutants deficient in the elongation of 18:1 have been isolated (Kunst et al. 1992). The most extreme mutant of this class had only about 0.2% 20:1 in its seed lipids due to a single codominant nuclear mutation in a gene designated *fae1*. The codominant nature of the *fae1* mutation indicates that the level of its expression is rate-limiting to elongation. Recessive mutations at two other loci, *fae2* and *fae3*, resulted in intermediate levels of 20:1. It is not apparent how these gene products are involved in elongation. These could be leaky mutations affecting other steps in elongation, or they could reflect the existence of multiple genes for some of the enzymatic steps. The elongation of 18:1 involves four reactions (besides those required for synthesis of malonyl-CoA): condensation of 18:1-CoA with malonyl-CoA, and the subsequent reduction, dehydration, and reduction steps to give 20:1. If each step is catalyzed by a separate enzyme, then, in principle, each of them might be a target for mutation. Alternatively, the synthesis of malonyl-CoA might be affected. None of the *fae* mutants has been shown to affect leaf lipid metabolism. In particular, none shows an eceriferum phenotype, suggesting that the seed elongase system is distinct from the elongation pathway that gives rise to long-chain wax precursors in the epidermal cells of leaves and stems. The leaky *abi3-1* mutation also reduces the proportion of long-chain fatty acids (Finkelstein and Somerville 1990). This finding is interpreted as evidence that the elongation pathway is under control of the *abi3* transcription factor.

One of the screening methods used to identify *Arabidopsis* seed mutants was designed to permit the recovery of any mutants that accumulated short-chain fatty acids (James and Dooner 1990). Although no such mutants were found in approximately 1800 families, several novel mutants that accumulate high levels of saturated fatty acids were recovered. A recessive mutation at a locus designated *fab1* had a two- to threefold increase in the amount of 16:0 (to about 19% of total fatty acids), a 50% decrease in 18:1, an increase in 16:1, and an increase in 20:0. This mutant could be a leaky mutation in an enzyme involved in the elongation of 16:0 to 18:0, most probably 3-ketoacyl-ACP synthase II, which is thought to be the only enzyme specific for this elongation step. As discussed earlier, the *fab1* mutation affects leaf lipid composition, although the mutant was first isolated on the basis of its seed fatty acid composition. The *fab2* mutant that exhibits an extreme dwarf phenotype was also isolated as a seed mutant (James and Dooner 1990).

## CONCLUSION AND PERSPECTIVES

*Arabidopsis* has proven to be a useful and versatile experimental organism for genetic and biochemical studies of the biosynthesis and function of fatty acids and glycerolipids. However, a relatively small proportion of the enzymes thought to be involved in catalyzing or regulating lipid biosynthesis has been marked by mutation. Because many of the enzymes are probably indispensable, future progress in this area will be dependent on implementing novel strategies for mutant isolation or gene identification. Some of the currently available mutants, such as *act1*, demonstrate that there must be powerful regulation of lipid biosynthesis in higher plants. Unraveling the mechanisms involved in this regulation is another task well suited to the tools available in *Arabidopsis*. Two other subjects barely touched upon in this review are the degradation of membrane lipids and the β-oxidation of fatty acids. An indication that these processes can be usefully studied in *Arabidopsis* may be inferred from the observation that a number of the ESTs which have been produced by the large-scale cDNA sequencing projects show significant sequence homology with enzymes of β-oxidation in other eukaryotes. In addition, the approaches that have yielded so much information about glycerolipid metabolism can no doubt also be profitably applied to studying the biochemistry of the other lipid classes (Fig. 1).

Because of the importance of plant oils as abundant sources of industrial materials, there is continuing interest in the use of genetic technology to produce novel fatty acids and related materials in plants. The demonstration of thermoplastic production in *Arabidopsis* is an example of the long-term possibilities in this respect (Poirier et al. 1992). In this instance, granules of polyhydroxybutyrate (PHB) were accumulated by transgenic plants that expressed several genes for PHB synthesis from *Alcaligenes eutrophus*. The collection of *Arabidopsis* mutants and transgenic plants with various alterations in the pathways of fatty acid and lipid biosynthesis provides many opportunities to explore the consequences of other metabolic diversions. Thus, it seems likely that this small weed will indefinitely remain a central component of both basic and applied research in this area of plant biology.

### ACKNOWLEDGMENTS

We are grateful to many of our colleagues who have provided ideas and perspectives on the work discussed in this review, including preprints of their publications. We also thank the Department of Energy, the National Science Foundation, the U.S. Department of Agriculture, Procter and Gamble, Monsanto, and Pioneer Hi-Bred International for support of research in our laboratories. (This is Carnegie Institution of Washington Department of Plant Biology publication #1227.)

## REFERENCES

Alexandre, J. and J.P. Lassalles. 1992. Intracellular $Ca^{2+}$ release by $InsP_3$ in plants and effects of buffers on $Ca^{2+}$ diffusion. *Philos. Trans. R. Soc. Lond. B* **338:** 53–61.

Arondel, V., B. Lemieux, I. Hwang, S. Gibson, H.M. Goodman, and C.R. Somerville. 1992. Map-based cloning of a gene controlling omega-3 fatty acid desaturation in *Arabidopsis. Science* **258:** 1353–1355.

Bernhard, W.R., S. Thoma, J. Botella, and C.R. Somerville. 1991. Isolation of a cDNA clone for spinach lipid transfer protein and evidence that the protein is synthesized by the secretory pathway. *Plant Physiol.* **95:** 164–170.

Blatt, M.R., G. Thiel, and D.R. Trentham. 1990. Reversible inactivation of $K^+$ channels of *Vicia* stomatal guard cells following the photolysis of caged inositol 1,4,5-triphosphate. *Nature* **346:** 766–769.

Browse, J. and C. Somerville. 1991. Glycerolipid metabolism: Biochemistry and regulation. *Annu. Rev. Plant Physiol. Plant Mol. Biol.* **42:** 467–506.

Browse, J.A., P.J. McCourt, and C.R. Somerville. 1985. A mutant of *Arabidopsis* lacking a chloroplast-specific lipid. *Science* **227:** 763–765.

———. 1986a. A mutant of *Arabidopsis* deficient in $C_{18:3}$ and $C_{16:3}$ leaf lipids. *Plant Physiol.* **81:** 751–756.

Browse, J., M. McConn, D. James, and M. Miquel. 1993a. Mutants of *Arabidopsis* deficient in the synthesis of α-linolenate. Biochemical and genetic characterization of the endoplasmic reticulum linoleoyl desaturase. *J. Biol. Chem.* **268:** 16345–16351.

Browse, J.A., M. Miquel, M. McConn, and J. Wu. 1993b. *Arabidopsis* mutants and genetic approaches to the control of lipid composition. In *The temperature adaptation of biological membranes* (ed. A. Cossins), pp. 141–154. Portland Press, London.

Browse, J.A., N. Warwick, C.R. Somerville, and C.R. Slack. 1986b. Fluxes through the prokaryotic and eukaryotic pathways of lipid synthesis in the 16:3 plant *Arabidopsis thaliana. Biochem. J.* **235:** 25–31.

Browse, J., L. Kunst, S. Anderson, S. Hugly, and C.R. Somerville. 1989. A mutant of *Arabidopsis* deficient in the chloroplast 16:1/18:1 desaturase. *Plant Physiol.* **90:** 522–529.

Cote, G.G. and R.C. Crain. 1993. Biochemistry of phosphoinositides. *Annu. Rev. Plant Physiol. Plant Mol. Biol.* **44:** 333–356.

Creelman, R.A., M.L. Tierney, and J.E. Mullet. 1992. Jasmonic acid/methyl jasmonate accumulate in wounded soybean hypocotyls and modulate wound gene expression. *Proc. Natl. Acad. Sci.* **89:** 4938–4941.

Douce, R. 1985. *Mitochondria in higher plants. Structure function and biogenesis.* Academic Press, Orlando, Florida.

Downton, W.J.S., J.A. Berry, and J.R. Seemann. 1984. Tolerance of photosynthesis to high temperature in desert plants. *Plant Physiol.* **74:** 786–790.

Farmer, E.E. and C.A. Ryan. 1990. Interplant communication: Airborne methyl jasmonate induces synthesis of proteinase inhibitors in plant leaves. *Proc. Natl Acad. Sci.* **87:** 7713–7716.

Finkelstein, R. and C.R. Somerville. 1990. Three classes of ABA-insensitive mutants of *Arabidopsis* define genes which control overlapping subsets of ABA responses. *Plant Physiol.* **94:** 1172–1179.

Frentzen, M. 1993. Acyltransferases and triacylglycerols. In *Lipid metabolism in plants* (ed. T.S. Moore, Jr.), pp. 195–231. CRC Press, Boca Raton, Florida.

Galliard, T. 1980. Degradation of acyl lipids: Hydrolytic and oxidative enzymes. In *The biochemistry of plants: A comprehensive treatise* (ed. P.K. Stumpf and E.E. Conn), vol.

4, pp. 85–116. Academic Press, New York.

Goodwin, T.W. and E.I. Mercer. 1983. *Introduction to plant biochemistry.* Pergamon Press, Oxford, United Kingdom.

Gundlach, H., M.J. Müller, T.M. Kutchan, and M.H. Zenk. 1992. Jasmonic acid is a signal transducer in elicitor induced plant cell cultures. *Proc. Natl. Acad. Sci.* **89:** 2389–2393.

Gunstone, F.D., J.L. Harwood, and F.B. Padley, eds. 1986. *The lipid handbook.* Chapman Hall, London.

Harwood, J.L. 1980. Plant acyl lipids: Distribution and analysis. In *The biochemistry of plants: A comprehensive treatise* (ed. P.K. Stumpf and E.E. Conn), vol. 4, pp. 1–55. Academic Press, New York.

Heinz, E. 1993. Biosynthesis of polyunsaturated fatty acids. In *Lipid metabolism in plants* (ed. T.S. Moore, Jr.), pp. 33–90. CRC Press, Boca Raton, Florida.

Hilditch, T.P. and P.N. Williams. 1964. *The chemical constituents of natural fats*, 4th edition. Chapman and Hall, London.

Hugly, S. and C. Somerville. 1992. A role for membrane lipid polyunsaturation in chloroplast biogenesis at low temperature. *Plant Physiol.* **99:** 197–202.

Hugly, S., L. Kunst, J. Browse, and C. Somerville. 1989. Enhanced thermal tolerance and altered chloroplast ultrastructure in a mutant of *Arabidopsis* deficient in lipid desaturation. *Plant Physiol.* **90:** 1134–1142.

Iba, K., S. Gibson, T. Nishiuchi, T. Fuse, M. Nishimura, V. Arondel, S. Hugly, and C. Somerville. 1993. A gene encoding a chloroplast omega-3 fatty acid desaturase complements alterations in fatty acid desaturation and chloroplast copy number of the *fad7* mutant of *Arabidopsis thaliana. J. Biol. Chem.* **268:** 24099–24105.

James, D.W. and H.K. Dooner. 1990. Isolation of EMS-induced mutants in *Arabidopsis* altered in seed fatty acid composition. *Theor. Appl. Genet.* **80:** 241–245.

Jaworski, J.G., R.C. Clough, and S.R. Barnum. 1989. A cerulenin insensitive short chain 3-ketoacyl-acyl carrier protein synthetase in *Spinacia oleracea* leaves. *Plant Physiol.* **90:** 41–44.

Joyard, J., M.A. Block, A. Malherbe, E. Marachal, and R. Douce. 1993. Origin and synthesis of galactolipid and sulfolipid headgroups. In *Lipid metabolism in plants* (ed. T.S. Moore, Jr.), pp. 231–258. CRC Press, Boca Raton, Florida.

Kader, J.-C. 1993. Lipid transport in plants. In *Lipid metabolism in plants* (ed. T.S. Moore, Jr.), pp. 309–338. CRC Press, Boca Raton, Florida.

Kinney, A.J. 1993. Phospholipid headgroups. In *Lipid metabolism in plants* (ed. T.S. Moore, Jr.), pp. 259–284. CRC Press, Boca Raton, Florida.

Konishi, T. and Y. Sasaki. 1994. Compartmentalization of two forms of acetyl-CoA carboxylase in plants and the origin of their tolerance toward herbicides. *Proc. Natl. Acad. Sci.* **91:** 3598–3601.

Kunst, L., J. Browse, and C. Somerville. 1988. Altered regulation of lipid biosynthesis in a mutant of *Arabidopsis* deficient in chloroplast glycerol phosphate acyltransferase activity. *Proc. Natl. Acad. Sci.* **85:** 4143–4147.

———. 1989a. A mutant of *Arabidopsis* deficient in desaturation of palmitic acid in leaf lipids. *Plant Physiol.* **90:** 943–947.

———. 1989b. Altered chloroplast structure and function in a mutant of *Arabidopsis* deficient in plastid glycerol-3-phosphate acyltransferase activity. *Plant Physiol.* **90:** 846–853.

———. 1989c. Enhanced thermal tolerance in a mutant of *Arabidopsis* deficient in palmitic acid unsaturation. *Plant Physiol.* **91:** 401–408.

Kunst, L., D.C. Taylor, and E.W. Underhill. 1992. Fatty acid elongation in developing seeds of *Arabidopsis thaliana*. *Plant Physiol. Biochem.* **30:** 425–434.

Larsson, C., J.M. Moller, and S. Widell. 1990. Introduction to the plant plasma membrane. In *The plant plasma membrane: Structure function and molecular biology* (ed. C. Larsson and I.M. Moller), pp. 1–15. Springer Verlag, Berlin.

Lemieux, B., M. Miquel, C. Somerville, and J. Browse. 1990. Mutants of *Arabidopsis* with alterations in seed lipid fatty acid composition. *Theor. Appl. Genet.* **80:** 234–240.

Lightner, J., D.W. James, Jr., H.K. Dooner, and J. Browe. 1994. Altered body morphology is caused by increased stearate levels in a mutant of *Arabidopsis*. *Plant J.* (in press).

Lyons, J.M. 1973. Chilling injury in plants. *Annu. Rev. Plant Physiol.* **24:** 445–466.

Majerus, P.W. 1992. Inositol phosphate biochemistry. *Annu. Rev. Biochem.* **61:** 225–250.

Marsh, D. 1990. *CRC handbook of lipid bilayers.* CRC Press, Boca Raton, Florida.

Mason, H.S. and J.E. Mullet. 1990. Expression of two soybean vegetative storage protein genes during development and in response to water deficit, wounding, and jasmonic acid. *Plant Cell* **2:** 569–579.

McCourt, P.J., L. Kunst, J. Browse, and C.R. Somerville. 1987. The effects of reduced amounts of lipid unsaturation on chloroplast ultrastructure and photosynthesis in a mutant of *Arabidopsis*. *Plant Physiol.* **84:** 353–360.

McCourt, P.J., J.A. Browse, J. Watson, C. Arntzen, and C.R. Somerville. 1985. Analysis of photosynthetic antenna function in a mutant of *Arabidopsis thaliana* (L.) lacking *trans*-hexadecenoic acid. *Plant Physiol.* **78:** 853–858.

Miquel, M. and J. Browse. 1992. *Arabidopsis* mutants deficient in polyunsaturated fatty acid synthesis. Biochemical and genetic characterization of a plant oleoyl-phosphatidylcholine desaturase. *J. Biol. Chem.* **267:** 1502–1509.

———. 1994. Lipid biosynthesis in developing seeds. In *Seed development and germination* (ed. G. Galili et al.). Marcel Dekker, New York. (In press.)

Miquel, M., D. James, H. Dooner, and J. Browse. 1993. *Arabidopsis* requires polyunsaturated lipids for low-temperature survival. *Proc. Natl. Acad. Sci.* **90:** 6208–6212.

Moore, T.S., ed. 1993. *Lipid metabolism in plants.* CRC Press, Boca Raton, Florida.

Murata, N., O. Ishizaki-Nishizawa, S. Higashi, H., Hayashi, Y. Tasaka, and I. Nishida. 1992. Genetically engineered alteration in the chilling sensitivity of plants. *Nature* **356:** 710–713.

Nishida, I., Y. Tasaka, H. Shiraishi, and N. Murata. 1993. The gene and the RNA for the precursor to the plastid-located glycerol-3-phosphate acyltransferase of *Arabidopsis thaliana*. *Plant Mol. Biol.* **21:** 267–277.

Ohlrogge, J.B., J. Browse, and C.R. Somerville. 1991. The genetics of plant lipids. *Biochim. Biophys. Acta* **1082:** 1–26.

Ohlrogge, J.B., J.G. Jaworski, and D. Post-Beittenmiller. 1993. De novo fatty acid biosynthesis. In *Lipid metabolism in plants* (ed. T.S. Moore, Jr.), pp. 3–32. CRC Press, Boca Raton, Florida.

Okuley, J., J. Lightner, K. Feldmann, N. Yadav, E. Lark, and J. Browse. 1994. *Arabidopsis FAD2* gene encodes the enzyme that is essential for polyunsaturated lipid synthesis. *Plant Cell* **6:** 147–158.

Patterson, G.W., S. Hugly, and D. Harrison. 1993. Sterols and phytyl esters of *Arabidopsis thaliana* under normal and chilling temperatures. *Phytochemistry* **33:** 1381–1383.

Poirier, Y.P., D.E. Dennis, K. Klomparens, and C.R. Somerville. 1992. Production of polyhydroxybutyrate, a biodegradable thermoplastic, in higher plants. *Science* **256:** 520–523.

Porter, J.W. and S.L. Spurgeon, eds. 1981. *Biosynthesis of isoprenoid compounds.* Wiley, New York.

Post-Beitenmiller, D., J.G. Jaworski, and J.B. Ohlrogge. 1991. *In vivo* pools of free and acylated acyl carrier proteins in spinach: Evidence for sites of regulation of fatty acid biosynthesis. *J. Biol. Chem.* **266:** 1858–1865.

Quinn, P.J., F. Joo, and L. Vigh. 1989. The role of unsaturated lipids in membrane structure and stability. *Prog. Biophys. Mol. Biol.* **53:** 71–103.

Roughan, P.G. and C.R. Slack. 1982. Cellular organization of glycerolipid metabolism. *Annu. Rev. Plant Physiol.* **33:** 97–123.

Salisbury, F.B. and C.W. Ross. 1992. *Plant physiology*, 4th edition. Wadsworth, Belmont, California.

Shanklin, J. and C.R. Somerville. 1990. The cDNA clones for stearoyl-ACP desaturase from higher plants are not homologous to yeast or mammalian genes encoding stearoyl-CoA desaturase. *Proc. Natl. Acad. Sci.* **90:** 2486–2490.

Slack, C.R. and J.A. Browse. 1984. Lipid synthesis during seed development. In *Seed physiology* (ed. D. Murray), vol. 1, pp. 209–244. Academic Press, Sydney.

Slack, C.R., P.G. Roughan, J.A. Browse, and S.E. Gardiner. 1985. Some properties of cholinephosphotransferase from developing safflower cotyledons. *Biochim. Biophys. Acta* **833:** 438–448.

Somerville, C.R. 1986. Analysis of photosynthesis with mutants of higher plants and algae. *Annu. Rev. Plant Physiol.* **37:** 467–507.

Somerville, C. and J. Browse. 1991. Plant lipids: Metabolism mutants and membranes. *Science* **252:** 80–87.

Sterk, P., H. Booij, G.A. Schellekens, A. Van Kammen, and S.C. De Vries. 1991. Cell-specific expression of the carrot EP2 lipid transfer protein. *Plant Cell* **3:** 907–921.

Stumpf, P.K. and E.E. Conn. 1980. *The biochemistry of plants. A comprehensive treatise*, vol. 4. Academic Press, New York.

———. 1987. *The biochemistry of plants. A comprehensive treatise*, vol. 9. Academic Press, New York.

Stumpf, P.K. and M.R. Pollard. 1983. Pathways of fatty acid biosynthesis in higher plants with particular reference to developing rapeseed. In *High and low erucic acid rapeseed oils* (ed. J.K.G. Kramer et al.), pp. 131–141. Academic Press, Toronto.

Stymne, S. and A.K. Stobart. 1987. Triacylglycerol biosynthesis. In *The biochemistry of plants. A comprehensive treatise* (ed. P.K. Stumpf and E.E. Conn), vol. 9, pp. 175–214. Academic Press, New York.

Stymne, S., A.K. Stobart, and G. Glad. 1983. The role of the acyl-CoA pool in the synthesis of polyunsaturated 18-carbon fatty acids and triacylglycerol production in the microsomes of developing safflower seeds. *Biochim. Biophys. Acta* **752:** 198–208.

Sun, C., Y.-Z. Cao, and A.H.C. Huang. 1988. Acyl coenzyme-A preference of the glycerol phosphate pathway in the microsomes from the maturing seeds of palm, maize and rapeseed. *Plant Physiol.* **88:** 56–60.

Taiz, L. and E. Zeiger. 1991. *Plant physiology*, Benjamin/Cummings, Redwood City, California.

Thoma, S., Y. Kaneko, and C.R. Somerville. 1993. The non-specific lipid transfer protein from *Arabidopsis* is a cell wall protein. *Plant J.* **3:** 427–437.

Thoma, S., U. Hecht, A. Kippers, J. Botella, S. De Vries, and C.R. Somerville. 1994. Tissue-specific expression of a gene encoding a cell-wall-localized lipid transfer protein from *Arabidopsis. Plant Physiol.* **105:** 35–45.

Vick, B.A. 1993. Oxygenated fatty acids of the lipoxygenase pathway. In *Lipid metabolism in plants* (ed. T.S. Moore, Jr.), pp. 167–194. CRC Press, Boca Raton, Florida.

Wang, X. 1993. Phospholipases. In *Lipid metabolism in plants* (ed. T.S. Moore, Jr.), pp. 505–527. CRC Press, Boca Raton, Florida.

Wolter, F.P., R. Schmidt, and E. Heinz. 1992. Chilling sensitivity of *Arabidopsis thaliana* with genetically engineered membrane lipids. *EMBO J.* **11:** 4685–4692.

Yadav, N.S., A. Wierzbicki, M. Aegerter, C.S. Caster, L. Perez-Grau, A.J. Kinney, W.D. Hits, R. Booth, Jr., B. Schweiger, K.L. Stecca, S.M. Allen, M. Blackwell, R.S. Reiter, T.J. Carlson, S.H. Russell, K.A. Feldmann, J. Pierce, and I. Browse. 1993. Cloning of higher plant omega-3 fatty acid desaturases. *Plant Physiol.* **103:** 467–476.

# 33

# Genetic Dissection of the Biosynthesis, Degradation, and Biological Functions of Starch

**Timothy Caspar**
Central Research and Development
E.I. DuPont de Nemours & Company
Wilmington, Delaware 19880-0402

Starch is a polymer composed of glucose residues linked through a backbone of $\alpha$-(1,4)-glycosidic bonds with branch points in the backbone formed by $\alpha$-(1,6)-glycosidic bonds. Starch can be categorized as amylose, which typically contains few if any branch points, and amylopectin, which consists of short linear chains (about 20 residues in length on average) linked together into a branched structure (for a review of starch structure, see Kainuma 1988). In addition to differences in degree of branching, starches from different sources differ in the organization of the branch points, the lengths of the backbone chains, and the presence of small amounts of covalently or noncovalently bound protein, lipid, and phosphate. Starch accumulates in insoluble, partially crystalline particles called starch grains, which range in size from less than 1 μm in green tissues to more than 100 μm in diameter in potato tubers. Starch is made only inside plastids (chloroplasts and amyloplasts), and due to its insoluble nature, remains inside the plastid except in specialized cases such as cereal endosperm, where cellular compartments degrade as the endosperm matures.

Starch is synthesized by all higher plants and accumulates in large amounts in most plants. Transitory starch plays a key role in photosynthetic carbon metabolism, where it is synthesized inside chloroplasts during photosynthesis and then degraded to supply energy and carbon skeletons for metabolism during the subsequent dark period. Reserve starch is synthesized in amyloplasts in storage tissues (seeds, tubers, etc.) where it accumulates during one developmental phase (e.g., seed development) and then is degraded during a subsequent phase (e.g., seed germination).

Starch metabolism is a quantitatively important pathway; many plants incorporate much of their newly fixed carbon initially into starch. For ex-

*Arabidopsis*
© 1994 Cold Spring Harbor Laboratory Press 0-87969-428-9/94 $5 + .00

ample, about 30% of the carbon in an *Arabidopsis* plant was initially incorporated into the starch pool (calculated from data in Caspar et al. 1985). Because of the high flux through this pathway, and the importance of starch as an agricultural commodity (about half of the calories consumed by humans world wide come directly from starch), much study has been devoted to starch metabolism. In nonphotosynthesizing sink tissues, starch is synthesized inside amyloplasts from imported photosynthate. In leaves, starch which is made in the chloroplast and sucrose which is made in the cytosol are usually considered to be alternate sinks for newly fixed carbon. Current models of the complex regulation of partitioning of carbon between starch and sucrose suggest that in many plants, photosynthate is channeled into starch synthesis when the capacity for sucrose synthesis or utilization is exceeded by the rate of photosynthesis (for review, see Stitt and Quick 1989).

Despite the large amount of study directed toward starch, many questions have been difficult to address solely by traditional biochemical and physiological approaches. These questions include many details regarding the synthesis of different chemical forms of the starch molecule, as well as the physical forms of the granule, the regulation of starch synthetic and degradative pathways which control the accumulation of starch, and the roles starch plays in the physiology of such processes as photosynthesis and gravitropism. This difficulty is due at least in part to the complex and poorly understood structure of the native starch grain, the lack of useful inhibitors of starch synthesis and degradation, the presence of multiple enzymes with similar activities for many of the metabolic steps, and the difficulty of reconstituting in vitro a system capable of synthesizing normal starch grains. For these reasons, genetic approaches using mutants with alterations in starch metabolism have been important in the field. Specifically, mutants with lesions in pathways of starch synthesis and degradation allow identification and analysis of the steps in these pathways. Through detailed study of mutants, one can address such issues as regulation of the individual steps and the pathways as a whole, the role of starch in various physiological and biochemical processes, and the identification, cloning, and analysis of the genes in these pathways.

The history of the study of starch mutants can be traced to the very beginning of the science of genetics. One of the original seven traits used by Mendel in his studies of garden peas, rough versus smooth seeds, has recently been shown to be caused by a mutation in starch branching enzyme (Bhattacharyya et al. 1990). Other mutations that affect the starch content of corn and other cereal seeds, potato, and *Chlamydomonas*, in addition to pea, have been isolated and characterized (Shannon and Garwood 1984; Delrue et al. 1992; Visser and Jacobsen 1993). Although studies of these mutants have greatly added to our knowledge of starch metabolism, their utility has been limited in several ways. In most cases, they are relatively specific for the seed and other reproductive tissues and do not affect starch metabolism in the vegetative part of the plant (or have only recently been shown to affect the vegetative parts of the plant; Smith et al. 1990). Furthermore, most of these mutants have been identified in species that are relatively difficult to study at the molecular genetics level. Because of this, mutants that affect starch metabolism in the vegetative parts of *Arabidopsis* have been a useful adjunct to the seed, tuber, and algal starch mutants. The rest of this review is devoted to a description of the isolation and characterization of mutants of *Arabidopsis* with lesions in starch metabolism and how these mutants have been useful in understanding starch metabolism and its role in the plant. Brief mention is also made of similar mutants that have been isolated in other species.

## ISOLATION OF STARCH MUTANTS

Starch mutants have been identified both by their morphological effects on seed development and by direct assays for starch content. For example, numerous mutations that affect seed starch content in corn, other cereals, and pea have been identified because the decreased starch content of these seeds causes the seeds to shrink during the final stages of maturation, producing wrinkled or shrunken phenotypes. Similarly, some mutations that affect seed starch have been identified by their effect on the overall appearance of the endosperm (e.g., *waxy* and *dull* mutants).

Starch may also be directly assayed by an iodine/potassium iodide mixture (hereafter termed iodine stain) which intensely stains starch and starch-containing plant tissues. The absorption of the iodine-starch complex is affected by the structure of the starch. Mutants that have altered types of starch have been identified by their altered staining characteristics (e.g., *waxy* mutants which lack amylose stain red with iodine as opposed to the wild type, which stains blue).

Iodine staining of starch has also been exploited to screen directly for *Arabidopsis* mutants that affect leaf starch accumulation. Leaves from individually marked, mutagenized *Arabidopsis* plants were removed from plants (at least 8 hours after the beginning of the light period), extracted with ethanol to remove chlorophyll, and then stained with iodine (Caspar et al. 1985). Leaves from wild-type plants accumulate large quantities of starch (1–80 mg/g fresh weight, depending on the light treatments and other growth conditions) and thus stain a dark blue-black. Starchless mutants[1] were identified by their light yellow or orange color when stained with iodine (Fig. 1). Mutants with reduced levels of starch stain a light gray or brown. Using this screen, mutants in five loci that reduce starch accumulation have been isolated in *Arabidopsis* (Table 1). Mutants in tobacco (Hanson and McHale 1988) and *Chlamydomonas* (Ball et al. 1991; Delrue et al. 1992) have also been isolated by similar procedures. Finally, mutants with reduced starch have also been isolated in *Clarkia* by direct screens for reduced activity of phosphogluco-isomerase (PGI), a starch biosynthetic enzyme (Jones et al. 1986).

A modification of the iodine leaf screen has been used to identify mutants that affect starch degradation (Caspar et al. 1991). In this screen, *Arabidopsis* plants were transferred to darkness for at least 12 hours before leaves were stained for starch. During this dark treatment, the wild type completely degrades the starch in its leaves, and thus the leaves do not stain with iodine. Mutants in three loci that are unable to degrade starch were identified by the retention of their intense iodine staining after the dark treatment (Fig. 1; Table 1).

## BIOCHEMICAL CHARACTERIZATION OF STARCH MUTANTS

### Starch Metabolic Pathways

Starch is synthesized in photosynthesizing chloroplasts from the Calvin cycle intermediate fructose-6-phosphate by a short pathway (Preiss 1988; Beck and Ziegler 1989). In nonphotosynthetic plastids, the same pathway is utilized, but the carbon is imported from the cytosol, probably as a hexose-phosphate. Figure 2 shows the presumed pathway of starch biosynthesis in *Arabidopsis*. The enzymatic deficiencies have been identified in three of the five *Arabidopsis* mutants that are partially or completely starch deficient. Starch is degraded by the combined actions of α- and β-amylase, phosphorylase, debranching enzyme, and disproportion-ating enzyme to produce smaller oligosaccharides and sugars (Steup 1988). These compounds are then further degraded by a variety of enzymes to produce compounds that can be exported from the plastid. Because of the multiple isozymes that exist for many of these enzymes, the fact that many of these enzymes utilize similar substrates and produce

---

[1]The "starchless" phenotypic designation is operationally defined to refer to those mutants which when grown in normal conditions do not accumulate sufficient starch to detect with iodine staining or standard biochemical assays. Subsequent, more sensitive assays of these mutants have shown that they can accumulate very small amounts of starch (<1% of that in the wild type) in certain cell types or in specialized growth conditions. Nevertheless, the starchless designation has been retained to indicate the essentially complete absence of starch in most tissues when grown in normal growth conditions and to emphasize the clear difference between this phenotype and the "reduced starch" phenotype, which is characterized by easily detectable but reduced starch levels.

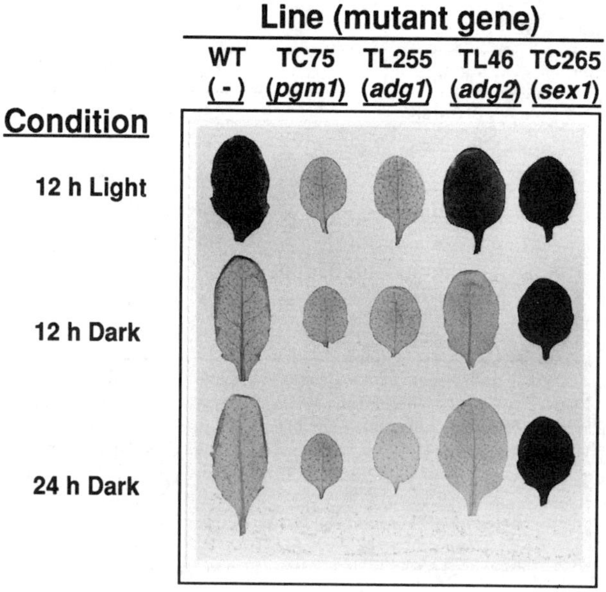

*Figure 1* Iodine-stained leaves of *Arabidopsis* wild type and starch mutants. Leaves from plants grown in a 12-hr photoperiod were sampled at the end of the light period, end of the dark period, and after an additional 12 hr of darkness (24 hr darkness total), extracted with ethanol, and then stained with iodine (20 g/l KI, 2 g/l I$_2$). After 12 hr of light, the starchless TC75 and TL255 mutants completely lack starch staining and the reduced starch mutant TL46 has only somewhat lighter staining than the wild type (the difference is difficult to detect in black and white photographs). The wild type almost completely degrades its starch within the first 12 hr of darkness, whereas the starch degradation mutant TC265 retains all of its starch even after 24 hr of darkness.

similar products, and the likelihood that the pathway may vary in different species and in tissues within a given species, it has not been possible to determine a consensus pathway for starch degradation. Moreover, with the exception of some storage tissues, little is known about the regulation of these degradative pathways.

## Phosphoglucomutase Mutants

The first class of starchless mutants identified contains a lesion affecting phosphoglucomutase (PGM), which catalyzes the reversible conversion of glucose-6-phosphate and glucose-1-phosphate (Caspar et al. 1985). *Arabidopsis* contains three isozymes of PGM, one located in the plastid and two in the cytosol. The best-characterized PGM mutant, TC7, completely lacks activity of the plastid PGM isozyme, whereas the cytosolic

*Table 1*  *Arabidopsis* starch mutants

| Mutant gene | Original line[a] | Phenotype | Enzymatic deficiency |
| --- | --- | --- | --- |
| *pgm1*[b] | TC7 | starchless | plastid PGM |
| *adg1* | TL25 | starchless | ADPG PPase |
| *adg2* | TL46 | reduced starch | ADPG PPase |
| *stf1* | TL14 | starchless | unknown |
| [c] | TC135 | reduced starch | unknown |
| *sex1* | TC26 | no starch degradation | unknown |
| [c] | TL50 | no starch degradation | unknown |
| [c] | TL54 | no starch degradation | unknown |

[a]The lines listed are the original designation of the lines which have been studied most extensively. In all cases, these lines carry the first allele of the series. In many cases, other mutant alleles for these genes have also been identified. Lines that have been backcrossed with the wild type and reselected have been given new names. TC75, TL255, and TC265 were derived from TC7, TL25, and TC26, respectively, by five rounds of backcrossing. TL46BC1 was obtained from TL46 after one backcross to the wild type.

[b]The *pgm1* gene was originally named *pgmP* but has been renamed to conform with the conventions for naming *Arabidopsis* genes.

[c]The genes mutated in TC135, TL50, and TL54 have not been named, pending further characterization.

isozymes are not affected (Caspar et al. 1985; Caspar and Pickard 1989; Sicher and Kremer 1992). In addition to TC7, at least four other independent mutants have been isolated which are also starchless and have no detectable activity of the plastid PGM isozyme. All of these mutations are located in a single complementation group which identifies the *pgm1* gene, with *pgm1-1* being derived from TC7. (This gene was originally termed *pgmP*, but the name has been changed to *pgm1* to conform with the conventions for naming *Arabidopsis* genes.) The *pgm1* gene has been mapped to the bottom of chromosome 5 (T. Caspar and C.R. Somerville, unpubl.). Lines containing *pgm1-1* (e.g., TC7 and TC75) are essentially starchless. For example, in biochemical assays, only background levels of starch were detected in leaves or stems grown in ambient conditions (Caspar et al. 1985). However, in 1% $CO_2$, where the rate of photosynthetic carbon fixation is greatly increased because of the suppression of photorespiration, very small amounts of starch (less than 0.1% of that found in the wild type) were observed in leaves by electron microscopy (T. Caspar and C.R. Somerville, unpubl.). In light and electron microscopic examinations of seedling hypocotyls and root caps (which normally contain large amounts of starch in the wild type), no starch was detected by three separate groups (Caspar and Pickard 1989; Moore 1989; Sack and Kiss 1989). However, one group has reported significant quantities of starch (about 14% of that in the wild type) in seedling root caps of TC75 (Sæther and Iversen 1991). The cause for this dis-

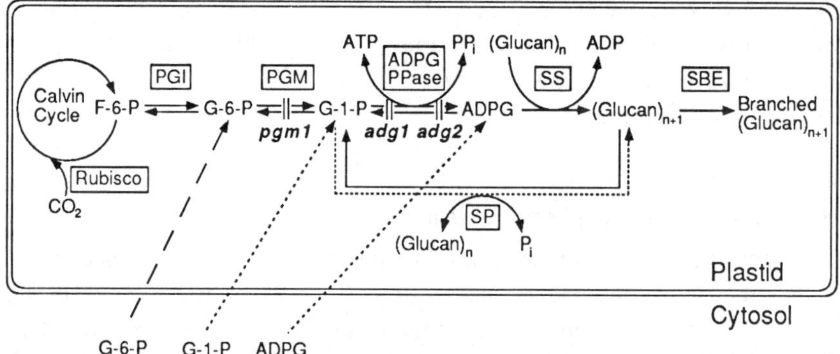


*Figure 2* Pathway of starch biosynthesis in *Arabidopsis* chloroplasts and amyloplasts. Two alternate sources of carbon for starch biosynthesis, the Calvin cycle in photosynthesizing chloroplasts and imported sugars from the cytosol in nonphotosynthetic amyloplasts, are shown. The sites of the known lesions in the *Arabidopsis* starch mutants are indicated by breaks (⊣⊢) labeled with the name of the gene. Solid lines indicate the presumed pathways. Dotted lines indicate pathways that have been suggested in other species but are unlikely to function in *Arabidopsis* (see text for details). The dashed line indicates a likely pathway of carbohydrate into the amyloplast, but direct evidence for this path in *Arabidopsis* does not exist. The starch synthase and starch branching steps have been simplified for the sake of clarity. There are likely to be multiple isozymes for each step, and these isozymes may have specialized roles in the synthesis of linear versus branched starch molecules. Enzymes are boxed. Abbreviations used for enzymes: (Rubisco) ribulose bisphosphate carboxylase oxygenase; (PGI) phosphoglucoisomerase; (PGM) phosphoglucomutase; (ADPG PPase) ADP glucose pyrophosphorylase; (SS) starch synthase; (SBE) starch branching enzyme; (SP) starch phosphorylase. Abbreviations used for metabolites: (F-6-P) fructose-6-phosphate; (G-6-P) glucose-6-phosphate; (G-1-P) glucose-1-phosphate; (ADPG) ADP glucose. $(Glucan)_n$ is a starch molecule containing $n$ glucose residues.

crepancy is unclear; the four groups that studied root caps used similar growth conditions and analysis techniques. It is possible that the lesion in TC7 is conditional on an unidentified environmental variable. However, one likely environmental variable, temperature, cannot explain the variation in reported phenotype. We have grown TC75 seedlings in the same conditions as used by Sæther and Iversen (1991), but at 15° and 27°C, in addition to the standard 23°C, and all were essentially starchless (T. Caspar and T. Bourett, unpubl.). Only 1 plastid containing a single small starch grain was found among 206 plastids examined by electron microscopy in root caps of TC75, whereas almost every plastid in the wild type contained numerous large starch grains. Nevertheless, because very small amounts of starch were observed here as well as in high-$CO_2$-grown leaves, it is possible that the lesion may be leaky or that alternative pathways to starch biosynthesis that bypass plastid PGM may be able to produce very small amounts of starch.

### ADP Glucose Pyrophosphorylase Mutants

A second starchless mutant class is caused by a complete absence of ADP glucose pyrophosphorylase (ADPG PPase) activity (Lin et al. 1988a). Two independent, allelic mutations have been found which identify the *adg1* gene (also listed as *adg* in AAtDB version 1.3, an *Arabidopsis thaliana* database version 1.3) (T. Caspar, unpubl.). *adg1* mutants do not contain detectable starch in leaves or roots when grown in ambient conditions (Lin et al. 1988a, F. Sack, pers. comm.). However, when they are grown on sucrose (1%) in elevated $CO_2$ (2%), *adg1* mutants can accumulate up to about 1% of the starch in the wild type (J. Lightner and T. Caspar, unpubl.). The ADPG PPase enzyme is a heterotetramer, and immunoblotting analyses using antibodies prepared against the spinach enzyme indicate that the mutant lacks both subunits (Lin et al. 1988a). Interestingly, the mutation does not affect the enzyme activity in a dosage-dependent manner, suggesting *adg1* may have a regulatory function or that the *adg1* gene may be feedback-regulated. Heterozygotes containing one null and one wild-type allele have the same ADPG PPase activity as the wild type (Lin et al. 1988a). In contrast, mutations at a second locus that affects ADPG PPase (*adg2*; see below) and at *pgm1* show a more conventional response where the heterozygote contains enzyme activity approximately intermediate between that of the wild type and the homozygous mutant (Lin et al. 1988b; T. Caspar and C.R. Somerville, unpubl.).

A second locus that affects ADPG PPase activity contains reduced amounts of starch and about 6% of the ADPG PPase activity of the wild type (Lin et al. 1988b). Only two mutants have been identified with this phenotype, and they were isolated from a single $M_2$ population and may represent siblings. These two allelic mutants complement mutations in the *adg1* locus and thus define a second locus affecting ADPG PPase, the *adg2* gene. The ADPG PPase in the mutant lacks the larger subunit, but a homotetrameric form of the enzyme accumulates which contains only the smaller subunits (Lin et al. 1988b; Li and Preiss 1992). This altered form of the enzyme is catalytically active, and the mutant accumulates about 40% as much starch as the wild type (Lin et al. 1988b). ADPG PPase is an important control point in starch biosynthesis, being allosterically activated by 3-phosphoglycerate and inhibited by inorganic phosphate.

This allosteric regulation is retained in the homotetrameric mutant form of the enzyme; however, the enzyme from the mutant required 30-fold higher levels of 3-phosphoglycerate for half-maximal activation and was five times more sensitive to inhibition by inorganic phosphate (Li and Preiss 1992). Clearly then, the small subunit must contain sites that are able to interact with and respond to these regulators without input from the large subunit. However, both subunits are required for a fully functional enzyme. These results suggest that the large and small subunits, which have sequence similarity and appear to have evolved from the single subunit that forms the prokaryotic ADPG PPase enzyme (Smith-White and Preiss 1992), have only partially diverged and, in the process, have achieved only an incomplete specialization of function.

**Starch Degradation Mutants**

A mutant class termed *sex1*, for *s*tarch *ex*cess,[2] accumulates more starch than the wild type because of an apparent lesion in starch degradation (Caspar et al. 1991). Two independent alleles at this locus with similar phenotypes have been isolated. The gene has been mapped to the top of chromosome 1 (Caspar et al. 1991). In this mutant, starch accumulates to high levels not only in leaves, but also in the seed, anther, sepal, root cap, and the body of the root, indicating that its function is involved with starch degradation throughout the plant. The enzymatic lesion in this mutant has not been identified. The activities of the starch degradative enzymes α- and β-amylase, phosphorylase, debranching enzyme, and disproportionating enzyme were all normal in this mutant, as were the activities of all the isozymes of amylase and phosphorylase identified by activity staining following separation on native PAGE (Caspar et al. 1991). There are several possible explanations for this result. It is possible that an enzyme not known to be required for starch degradation might be affected in the mutant. For example, there might be a unique enzyme required to cleave particular residues of the starch polymer. Alternatively, a later step in the degradation pathway, for example, export of carbon from the plastid, might be affected, and this lesion might inhibit net starch degradation through a feedback regulatory system or by promoting resynthesis of starch due to the accumulation of carbohydrate inside the plastid. The lesion might also affect an enzyme in a way that is

[2]The *sex1* gene was originally named *sop1*, for *s*tarch *o*ver-producer, and is referred to by this name in some early publications. As more was learned about the mutant, the name was changed to *sex1* to better describe the phenotype. Due to a clerical error, both the *sop1* and *sex1* names appear in the primary description of this mutant (Caspar et al. 1991), and both the *sex1* and *sop1* designations appear in AAtDB v. 1.3.

not detectable by in vitro assays. For example, the enzyme might be present and active, but localized to the wrong subcellular compartment or altered such that it is inactive in the conditions inside the cell but is active in vitro. Finally, the mutation might affect a step that is involved in regulating the pathway. Given the limited knowledge of the pathways and regulation of starch degradation (Steup 1988), the eventual elucidation of the lesion in the *sex1* mutant may significantly add to our knowledge of these processes.

### Other Starch Mutants

In addition to the four well-described starch mutants discussed above, several poorly characterized mutants have also been identified. A starchless mutant TL14 contains a mutation in the the *stf1* (for *starch free*) gene (Caspar et al. 1989), and a starch-deficient mutant TC135 contains a mutation in an unnamed gene (Caspar et al. 1985). These have normal activities of the biosynthetic enzymes PGI, PGM, ADPG PPase, and starch synthase (T. Caspar et al., unpubl.). Two additional mutants that appear to affect starch degradation, TL50 and TL54, have also been identified but not well-characterized (Caspar et al. 1991). These mutations are nonallelic to *sex1* and to each other, and their genes have not been named, pending further characterization. In some of these poorly characterized mutants, the effect on starch metabolism may only be a secondary effect of the mutation. For example, TC135 is also deficient in chlorophyll (T. Caspar and C.R. Somerville, unpubl.), suggesting its starch deficiency may be a secondary effect of a lesion affecting photosynthesis.

### Comparison to Other Species

Some of the starch mutants that have been identified in other species are similar to the *Arabidopsis* starch mutants. For example, a starch-deficient mutant that affects PGM has been isolated in tobacco. The mutation alters the kinetic properties of the plastid enzyme, suggesting it lies in the structural gene for the enzyme (Hanson and McHale 1988). The endosperm starch-deficient corn mutants, *sh2* and *bt2*, have very low activities of ADPG PPase and specifically affect individual subunits of the enzyme (Bhave et al. 1990; Preiss et al. 1990). The *sh2* mutant is similar to *adg2* in that both lack the large subunit of the enzyme and both have a small (~5%) residual enzyme activity. The *bt2* mutant differs from the *adg1 Arabidopsis* mutant in that *adg1* has no ADPG PPase activity and lacks both subunits of the ADPG PPase enzyme, whereas *bt2* has about

3% of normal ADPG PPase activity and accumulates the large subunit. A potato transgenic line containing an antisense construct for the potato small subunit of ADPG PPase is also similar to the *bt2* mutant in that ADPG PPase activity and starch accumulation are greatly inhibited (Müller-Röber et al. 1992). The *Chlamydomonas* low-starch mutant *st-1-1* is similar to the *Arabidopsis adg2* mutant in that it has altered allosteric regulation and a low specific activity; however, the subunit structure of the enzyme in this *Chlamydomonas* mutant has not been determined (Ball et al. 1991). The *rb* pea mutant has also been shown to be starch deficient due to a reduced activity of ADPG PPase (Smith et al. 1989). No mutants with lesions in starch degradation similar to the *Arabidopsis sex1* mutants have been identified in other species.

## PHYSIOLOGICAL AND BIOCHEMICAL STUDIES OF STARCH MUTANTS

### Role of Starch in Growth, Respiration, and Photosynthesis

One of the most striking features of the *Arabidopsis pgm1*, *adg1*, and *sex1* starch mutations is their effect on overall plant growth (Caspar et al. 1985, 1991; Lin et al. 1988a; Schulze et al. 1991). When grown in continuous light, the mutants have growth rates indistinguishable from the wild type. On the other hand, when the mutants are grown in a light-dark photoperiod, they grow much more slowly than the wild type. Taken together, these results indicate there is a photoperiod-conditional requirement for starch synthesis and turnover. Starch synthesis, accumulation, and turnover are completely dispensable for growth and development in continuous light; however, in alternating photoperiods, the storage of carbon in starch is critical to efficient growth. This conditional requirement for starch has facilitated the isolation of these *Arabidopsis* mutants and similar mutants in other species, since mutants can be identified by screening populations grown in continuous light where they grow well. Furthermore, the conditional growth pattern has greatly simplified their study, since mutants grown in continuous light are developmentally equivalent to the wild type. However, by switching the light conditions, the effects of the mutations can be readily seen and studied.

The photoperiod-conditional growth of the starch mutants is most easily understood by considering the turnover of the starch pool. During growth in continuous light, the fully developed leaves of the wild type must reach a steady-state level where no further net synthesis occurs (Fig. 3a). In the starch degradation mutants, a similar steady-state condition is reached. In the starchless mutants, there is also no net synthesis

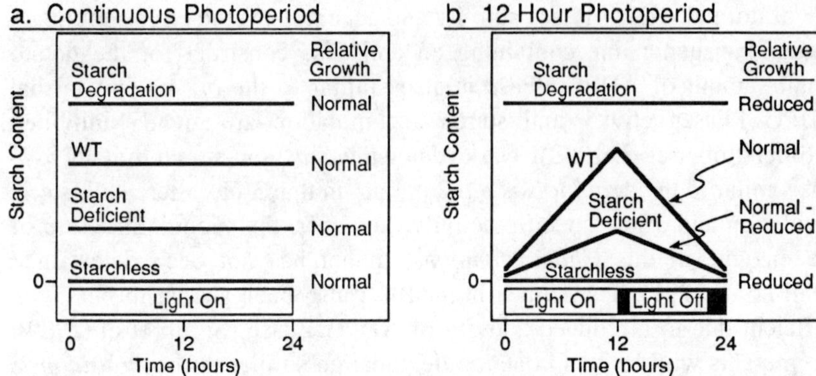



*Figure 3* Starch levels and relative growth rates in wild type and starch mutants in (*a*) continuous light and (*b*) a 12-hr photoperiod. Starch levels are indicated schematically for all lines and the growth rates are indicated relative to that of the wild type (data from Caspar et al. 1985, 1991; Lin et al. 1988a,b; Schulze et al. 1991). The starchless mutants have lesions in the *pgm1* and *adg1* genes; the starch-deficient mutant in *adg2*; and the starch degradation mutants in *sex1*. The growth of the starch-deficient *adg2* mutant in alternating photoperiod varies depending on the irradiance, being normal in low light and reduced in high light (see text for details).

because of the lesion in biosynthesis. In each case, there is no net starch synthesis, but the size of the starch pool at steady state ranges from empty in the starchless mutants to higher than the wild type in the starch degradation mutants. Since all of these plants have indistinguishable growth rates, the absolute size of the starch pool does not affect growth in these conditions.

The situation is quite different when plants are grown in an alternating photoperiod (Fig. 3b). Here, the wild type accumulates starch in the light and breaks it down during the dark period, such that a constant cycling of the starch occurs. In contrast, in the starchless mutants no synthesis and thus no cycling can occur. Similarly, in the starch breakdown mutants, no degradation occurs in the dark and thus the starch pool reaches a steady-state level and no net synthesis occurs. Thus, these two classes of mutants that have different sizes of starch pools are functionally similar in lacking a cycling of carbon in and out of the starch pool, but are quite distinct from the wild type. In the absence of the daily flux of carbon in and out of the starch pool, these two classes grow at similar rates but much more slowly than the wild type. Clearly, the absolute size of the starch pool is not an important criterion determining growth rates in these conditions. Rather, a flux of carbon through the starch pool is necessary for normal growth.

The *adg2* mutant, which accumulates about 40% as much starch as the wild type, grows at the same rate as the wild type in continuous light and a 12-hour photoperiod of moderate irradiance (120 $\mu$mole/m$^2$/s) (Lin et al. 1988b). However, in a high light (600 $\mu$mole/m$^2$/s), 14-hour photoperiod growth is only about half that of the wild type (Schulze et al. 1991). Apparently, normal rates of starch turnover are critical to support growth in high light. In contrast, in lower light, a limited capacity for starch turnover is sufficient to support normal growth.

The basis for the slow growth rates of the starch mutants when grown in a light/dark photoperiod has been investigated by measuring both metabolite levels and rates of photosynthesis and dark respiration. Interpretation of these results is complicated by the different growth conditions and experimental protocols used by various groups. The effects of the mutations on the regulation of the pathways of carbohydrate metabolism and the subsequent effects on growth have complex interactions with the plant's environment. Photoperiod, irradiance, $CO_2$ concentration, nutrient levels, plant age, and the use of intact plants versus detached leaves are among the variables that have been shown to, or are likely to, affect growth or metabolism in the mutants. Because of this, care must be taken in comparing results from different studies. Whenever possible, I have avoided using results obtained with one set of conditions to interpret results from a different set of conditions. An example of the effect different growth and measurement conditions can have on the results is that of soluble sugar accumulation. The *pgm1* mutant accumulated sugar about eight times as rapidly as the wild type when grown in air at 200 $\mu$mole/m$^2$/s in a 12-hour photoperiod (Caspar et al. 1985). In contrast, when the plants were grown in air at 600 $\mu$mole/m$^2$/s on an 18-hour photoperiod, the rate of sucrose synthesis in the *pgm1* and *adg2* mutants measured in detached leaf discs in saturating $CO_2$ was only about 60% of the wild-type rate (Neuhaus and Stitt 1990). The synthesis of soluble sugars in the mutant thus may be either increased or decreased when plants are grown in a light/dark photoperiod (see, e.g. Caspar et al. 1985, 1991; Neuhaus and Stitt 1990; Schulze et al. 1991; Sicher and Kremer 1992). It is not clear why environmental conditions have such a strong effect on the rate of sugar accumulation by the mutants relative to the wild type.

Under conditions where the *pgm1* mutant accumulated high levels of soluble sugars, the rate of respiration in the mutant was much higher than the wild type at the beginning of the dark period and then dropped back to the level of the wild type during the remainder of the dark period (Caspar et al. 1985). In contrast, the rate of respiration in the wild type was relatively constant during the entire dark period. The very high rate of respiration in the mutant at the beginning of the dark period was attributed to the elevated sugar levels in the mutant (Caspar et al. 1985), indicating, as has been suggested by others (see, e.g., Azcon-Bieto and Osmond 1983), that respiration is controlled by substrate availability rather than demand for ATP and NADH. In agreement with the stable respiratory rates in the wild type during the course of the dark period, the starch pool was gradually degraded such that by the end of the dark period there was still starch available to support respiration. In contrast, the sugar levels in the *pgm1* mutant (as well as the *sex1* mutant; Caspar et al. 1991) were exhausted in the mutant before the end of the dark period (Caspar et al. 1985), indicating that at the end of the dark period, respiration must be supported by another source.

Taken together, these respiration and carbohydrate pool measurements suggest that one of the roles of starch is to serve as a storage form of carbohydrate that is not rapidly metabolized during the dark. Rather, the plant is able to regulate its rate of starch degradation to have carbohydrate available throughout the dark period. The starch-deficient *adg2* mutant also provides good evidence of the ability of the plant to monitor the size of the starch pool and to regulate starch degradation accordingly. Because of the deficiency of ADPG PPase in this mutant, only about 40% as much starch as in the wild type is accumulated during a 12-hour photoperiod. During the dark period, the rate of starch degradation is reduced in the mutant in proportion to the size of the starch pool, such that the degradation proceeds at a steady rate throughout the dark period at about 40% of the rate of the wild type. It is unknown both how the plant is able to measure the size of the starch pool and how it then regulates the rate of starch degradation to meter out the starch during the entire dark period.

In addition to the effect on respiration, photosynthesis is also affected in the *pgm1* mutant when it is grown in a light/dark photoperiod. Intact mutant plants grown in a 12-hour photoperiod had a 30% lower photosynthetic rate than the wild type when measured in ambient conditions (Caspar et al. 1985). Similar reductions were observed for both the *pgm1* and *adg2* mutants grown in an alternating photoperiod when measurements were made on detached leaves in saturating $CO_2$ (Sivak and Rowell 1988; Neuhaus and Stitt 1990; Sicher and Kremer 1992). In contrast, when plants were grown in continuous light, the photosynthetic rate was indistinguishable for the wild type and mutant (Caspar et al. 1985). It is not known whether this decreased photosynthetic rate is directly due to the lack of starch synthesis or whether it is a secondary effect of the increased accumulation of sugar, increased dark respiration, or other metabolic alterations.

In addition to the effects on photosynthesis in ambient conditions, Sivak and Rowell (1988) have shown that fluorescence oscillations in the *pgm1* mutant can be induced by increases in $[CO_2]$ or decreases in $[O_2]$. These responses could be reduced by feeding inorganic phosphate to the mutant and could be mimicked in the wild type by feeding phosphate chelators. These results suggest that the mutant is impaired in its ability to adapt to conditions where increased photosynthesis requires rapid recycling of phosphate. Starch metabolism apparently plays an important role in facilitating the rapid recycling of phosphate during periods of transient increases in photosynthesis. It is unclear from these results, however, whether phosphate limitation plays a role in limiting photosynthesis and growth of the mutant in steady-state, ambient conditions.

## Mutants as Tools to Determine the Pathway and Control Points of Starch Synthesis

Mutants are powerful tools for analyzing biochemical pathways. Although the pathway for starch biosynthesis is generally accepted to utilize plastid forms of PGM, ADPG PPase, and starch synthase to synthesize the backbone of the starch molecule, reports have suggested that other routes are utilized (see dotted lines in Fig. 2). For example, it has been shown that ADP glucose can be transported into plastids (Pozueta-Romero et al. 1991), suggesting that cytosolically synthesized ADP glucose might be an important substrate for starch biosynthesis. Similarly, it has been suggested that glucose-1-phosphate is the form of carbohydrate imported into wheat amyloplasts (Tyson and ap Rees 1988). Finally, starch phosphorylase has been implicated as an important starch biosynthetic enzyme in some nonphotosynthesizing tissues (see, e.g., Obata-Sasamoto and Suzuki 1979). However, none of these pathways can be quantitatively important in *Arabidopsis*, since lesions in PGM and ADPG PPase produce starchless phenotypes in both chloroplast- and amyloplast-containing tissues. The effects of the *pgm1* and *adg1* mutations in nonphotosynthetic tissue indicate that the carbohydrate imported into amyloplasts must be a compound which lies before these steps in the biosynthetic pathway (e.g., glucose-6-phosphate or triose phosphate). More generally, since the *pgm1*, *adg1*, and *sex1* mutations all affect starch metabolism in both photosynthetic and nonphotosynthetic tissues, and in both vegetative and reproductive tissues, the pathways of starch synthesis and degradation are similar throughout the plant and there is no evidence of tissue-specific specialization of these steps.

The control of a metabolic pathway is often distributed among several steps, such that a change in the activity of any of these enzymes results in a change in the flux through the pathway. The degree of control over the pathway for any enzyme can be described by the flux-control coefficient (Kacser and Porteous 1987), which is defined such that the sum of all the flux-control coefficients for a given pathway is 1, a higher flux-control coefficient for a given enzyme signifies it has greater control over the pathway, and negative flux control coefficients can exist. Mutants with decreased activity for specific enzymes can be utilized to determine flux-control coefficients. Using the *Arabidopsis* PGM and ADPG PPase (Neuhaus and Stitt 1990), the pea starch branching enzyme (Smith et al. 1990), and the *Clarkia* PGI (Kruckeberg et al. 1989) starch mutants and heterozygotes derived from them, flux-control coefficients for the starch biosynthetic pathway have been determined. As emphasized by these authors, these values are only approximations because the limited spectrum of altered enzyme activities available requires extensive extrapolation to estimate the situation in the wild type. Moreover, the measurements of rates of starch synthesis were made for short times (20–25 min) in saturating $CO_2$, which may produce different results than in steady-state, ambient conditions. Nevertheless, these measurements provide the first quantitative information on the distribution of control within the starch biosynthetic pathway. In light-limited conditions (75 $\mu$mole/m$^2$/s for *Arabidopsis* ) only ADPG PPase has significant control over the pathway (flux-control coefficient = 0.28) whereas the other enzymes have no control (flux-control coefficients $\leq$0.02). These results confirm the importance of ADPG PPase in regulating the pathway but also indicate that, in addition to ADPG PPase, PGI, PGM, and starch branching enzyme, other steps in the pathway to starch must be able to exert substantial positive control over the pathway. In light-saturating conditions (600 $\mu$mole/m$^2$/s for *Arabidopsis*), where there is a higher flux through the pathway, control is distributed among all four of these enzymes (flux-control coefficients: 0.35 for PGI; 0.21 for PGM; 0.64 for ADPG PPase; and 0.13 for starch branching enzyme).

In addition to estimations of the flux-control coefficients for the wild type, the study of these mutants also demonstrated that despite its relatively low control coefficient in high light, PGM exists in less than twofold excess. That is, in lines containing half of the normal PGM activity, the control coefficient rises to nearly unity, indicating that the flux through the pathway is almost directly proportional to the activity of this enzyme (Neuhaus and Stitt 1990). Furthermore, comparison of the activities of all four of these enzymes measured in vitro in the wild type and various mutants with the observed rates of starch synthesis in vivo, indicates that none of the enzymes operate at their full potential in vivo. For example, ADPG PPase activity utilized in vivo is less than 20% of the activity measured in vitro, presumably due to limitation by substrate availability or allosteric inhibition (Neuhaus and Stitt 1990). As pointed out by these authors, these types of results have important implications in interpreting the significance of developmentally or environmentally induced changes in enzyme activities and in directing efforts to alter the flux through these pathways.

## Secondary Effects of Starch Mutations on Other Enzymes

In addition to affecting photosynthesis and respiration, the mutations in starch metabolism affect the regulation of other enzymes involved in carbohydrate metabolism. The largest effect of the starch mutations is on a specific isozyme of β-amylase. The starch mutants, when grown in a 12-hour photoperiod, have up to a 40-fold increase in activity of this enzyme (Caspar et al. 1989), due to an elevated level of the protein (Monroe and Preiss 1990). Since this elevated expression is caused by mutations in at least four different genes, including those with starchless and starch degradation phenotypes, the increase in expression of this β-amylase must be a secondary effect caused by altered carbohydrate metabolism in the mutants. The similar effects observed for the regulation of this enzyme in these two different classes of mutants emphasize that the critical feature which determines the secondary responses of these mutations is not the size of the starch pool, but rather the flux through it. In addition, in agreement with the overall apparent functional equivalence of the mutants with the wild type when grown in a continuous photoperiod, the mutants show no increase in expression of the β-amylase when grown in this condition. Surprisingly, the β-amylase that is affected is located outside the chloroplast. Since starch is confined to the plastid, this β-amylase has access to no known substrate. The fact that its expression is highly regulated by alterations in carbohydrate metabolism not only demonstrates the presence of an active regulatory system for this gene, but also suggests that it has an important function.

## Role of Starch in Gravitropism

Since the turn of the century, the predominant theory for the mechanism by which plants perceive gravity has involved starch-filled plastids, termed statoliths. These statoliths, by virtue of their large size and density, are thought to indicate the direction of gravity by sedimenting through the cell or by remaining stationary but pushing or pulling on other cellular components. Gravitropism of the starchless mutants has been intensely studied because of the possibility that they may allow a decisive test of the starch statolith theory. The *pgm1-1* mutant TC7 (or the five-times back-crossed derivative TC75) have been used for the most thorough investigations, but similar results have been obtained using other *pgm1* alleles and the other starchless mutant *adg1* (Caspar and Pickard 1989). Electron microscopic examinations by three independent groups have each shown that the *pgm1-1* mutant has no starch in either the root cap or hypocotyl (Caspar and Pickard 1989; Kiss et al. 1989; Moore 1989; Sack and Kiss 1989). However, as mentioned earlier, a fourth group, using essentially identical growth conditions and genetic materials, concluded that significant starch is present in the root cap of the *pgm1* mutant (Sæther and Iversen 1991). Although the reason for these inconsistent results has not been determined (see also above), the reproduction of the starchless results by three laboratories supports the conclusion that the *pgm1* mutant is in fact starchless, at least in the conditions used by these groups. The lack of starch in the *pgm1* mutant does not eliminate gravitropic responsiveness as predicted by a literal interpretation of the starch statolith theory. Rather, the roots, hypocotyls, and flower stalks of the mutant when grown in the light still retain significant, although reduced, graviresponsiveness relative to the wild type (Caspar and Pickard 1989; Kiss et al. 1989; Moore 1989; Sæther and Iversen 1991). Clearly, starch is not absolutely required for graviresponsiveness in light-grown *Arabidopsis*. The reduced graviresponsiveness of the starchless mutant may, however, indicate that starch is important for gravity perception. Starch-free plastids may be capable of receiving the gravitational signals, but starch improves their sensitivity or efficiency. Alternatively, the starch may be involved in later phases of the gravitational response through its role as a source of metabolic energy.

In contrast to the case for light-grown seedlings, when the starchless mutant is grown in the dark, the gravitropism is almost completely eliminated (Caspar and Pickard 1989). This phenomenon has not been well studied, but it may be similar to the well-known stimulatory effect of light on gravitropism in some other species (e.g., pea; Britz and Galston 1982). Further study of the gravitropism in the starch mutants is needed to resolve the role of starch in gravitropism in both light- and dark-grown plants.

### Other Uses of Starch Mutants

In addition to its roles in energy storage and gravitropism, starch may play a role in other aspects of plant physiology and development. These roles can be tested using the starchless mutants. Furthermore, if the starchless mutants indicate that starch has a role in a particular process, then the starch degradation mutants can be used to determine whether the process requires active flux through the starch pool or only the presence of starch per se. Since the starchless mutants are generally normal when grown in continuous light, it seems likely that starch is not absolutely required for any critical functions. Starch may, however, have effects too small to be noticed by casual observation or in highly optimized laboratory growth conditions.

The processes that determine why some nonphotosynthetic tissues accumulate starch and others do not are not well understood. In addition to its effect on the degradation of transitory starch in photosynthetic tissues, the *sex1* mutant also causes the accumulation of starch in many tissues (e.g., seed coats, anthers, sepals, and the bodies of roots) where the wild type accumulates little or none (Caspar et al. 1991). This shows that the lack of starch accumulation in these tissues in the wild type must be caused, at least in part, by the rapid degradation of starch, rather than simply by the lack of its synthesis.

From a practical standpoint, the presence of starch often complicates experimental protocols. The starchless mutants can simplify these types of experiments. For example, intact, starch-filled chloroplasts are more difficult to isolate than starch-free chloroplasts, since the starch grains may rupture the envelope during centrifugations. For this reason, plants are often dark-adapted prior to chloroplast isolation to reduce the starch content. The use of starchless mutants as the starting material removes the requirement for this potentially artifactual dark-adaptation. Similarly, nucleic acid isolations are sometimes improved by reducing the starch content by dark-adaptation. Finally, microscopic observations of some plant tissues (e.g., developing seeds) may be obscured by starch accumulation. Again, use of the mutants would reduce this problem.

## CONCLUSIONS AND PERSPECTIVES

The three starchless and starch-deficient mutants, two affecting ADPG PPase and one affecting PGM, represent only two of the five enzymatic steps between the Calvin cycle and starch. Of the approximately 10,000 mutagenized *Arabidopsis* plants screened, no mutations have been identified yet in PGI, starch synthase, or starch branching enzyme. For PGI, this is probably because the product of this step, glucose-6-phosphate, also serves as the starting point for other pathways such as inositol biosynthesis and the oxidative pentose phosphate cycle and, thus, null PGI mutants are likely to be lethal. In agreement with this, leaky, but not null, mutants of chloroplast PGI have been identified by direct enzyme activity screens in *Clarkia* (Jones et al. 1986). For starch synthase and starch branching enzyme, the lack of mutants is probably not due to the deleterious effects of these mutations, since these enzymes lie after PGM and ADPG PPase in the pathway, and null mutations in PGM and ADPG PPase are not strongly deleterious. It is possible that the genes for starch synthase and starch branching enzyme have low mutabilities and that further screening may identify these mutants. More likely, because of multiple copies of the genes for these enzymes, mutations affecting any one copy will only partially reduce starch accumulation (see, e.g., Smith et al. 1990), although they may have substantial effects on starch structure. The leaf-staining protocol used to isolate the present collection of *Arabidopsis* starch mutants is not well-suited to identifying mutants with only a partial loss of starch or an altered starch structure. Both subtle microclimate differences and the indirect effects of mutations that reduce plant vigor combine to create a high background of plants with reduced starch in the leaf screen. Improved assays that are less sensitive to genetic and environmental noise and are better suited to identifying alterations in starch structure may be required to allow the isolation of other lesions in the pathway.

Mutations in three genes that affect the starch degradation pathway have been identified, but only one has been characterized in any detail (Caspar et al. 1991). These mutations were identified in a relatively small screen, and the fact that only a single mutant allele has been identified for two of the three genes suggests further screening could identify new loci. Given the limited knowledge of the starch degradation pathway, mutations of this type may be of significant use.

In addition to mutations that affect starch biosynthesis and degradation per se, mutations that affect the regulation of these processes would also be quite useful tools, since the biochemical and genetic mechanisms that control this regulation are largely unknown. For example, although much is known about the regulation of starch and sucrose biosynthesis in photosynthetic tissue, the mechanisms that place a limit on starch accumulation in plants growing in continuous light (Lin et al 1988b; Caspar et al. 1991) or at the end of a natural photoperiod (Li et al. 1992) or after a switch in photoperiods (Chatterton and Silvius 1979; Li et al. 1992) are unknown. Moreover, little is known about the mechanisms that determine the spatial distribution of starch accumulation. For example, the root cap of all higher plants accumulates large amounts of starch, but in many plants, including *Arabidopsis*, the body of the root, including cells only a layer or two away from the cap, accumulate almost no starch. At a finer level, the structure of starch grains is highly organized, conserved, and species-specific, but the processes that govern this organization are largely unknown.

Partitioning between starch and sucrose in photosynthesizing tissues may be a key control point limiting the capacity of plants to fix carbon. Starch mutants in *Arabidopsis* have added to our knowledge of the processes that regulate this partitioning and have pointed out potentially fruitful avenues for future efforts at modifying carbohydrate partitioning and photosynthesis (see, e.g., Neuhaus and Stitt 1990). The engineering of plants with improved photosynthetic characteristics, if successful, would have a major impact on agricultural productivity. In addition to its effects on partitioning in photosynthetic tissues, starch is an important agronomic commodity in its own right. The ability to increase starch accumulation in crop plants or to alter its properties has a great potential value. Analysis of starch mutants should aid in the analysis of genes involved in starch accumulation. This type of analysis should greatly facilitate genetic engineering approaches to modifying starch quantity and structure (for review, see Visser and Jacobsen 1993).

Notwithstanding the agronomic importance of starch, it may become possible to engineer plants to accumulate other higher value products in place of starch. To achieve this efficiently, carbon will have to be diverted from starch into the novel product. In addition to a good understanding of the pathways of starch biosynthesis and degradation and their regulation, knowledge of the effects of this diversion of carbon away from starch will be necessary. For example, analysis of the starch mutants indicates that plants in which all carbon is diverted away from starch in the photosynthetic tissues will grow slowly in natural conditions and thus be poor crops. On the other hand, the capacity to accumulate starch in reduced quantities, as in the leaky *adg2* mutant, significantly improves the growth rate, indicating it may be feasible to partially divert carbon from starch to other products.

The starch mutants have allowed functional analyses of the roles of several genes in the starch biosynthetic pathway, the identification of novel genes involved in starch degradation, and an analysis of the role of starch in photosynthesis, respiration, gravitropism, and other areas of plant growth and development. In the future, these mutants should become a greater resource for cloning genes involved in these pathways as technologies for mutant-based cloning improve. The starch mutants, combined with the increased use of plants genetically engineered to over- or underexpress particular enzyme activities (see, e.g., Visser and Jacobsen 1993), should continue to be useful tools for studying starch metabolism and its role in the plant. New mutants will allow other, uncharacterized steps to be analyzed and, more importantly, may aid in the identification of regulatory mechanisms that control the starch metabolic pathways and their integration into the overall metabolism of the plant.

## Note Added In Proof

Chloroplasts isolated from the starch degradation mutant TC265 have recently been shown to be deficient in their ability to take up glucose (Trethewey and ap Rees 1994), suggesting the lesion in this mutant affects a sugar transporter required for export of starch degradation products from the plastid.

## ACKNOWLEDGMENTS

I thank Tim Bourett for assistance with microscopy and Natalie Hubbard, Ellen Johnson, Richard Trethewey, Pablo Scolnik, David Patton, and Tim Bourett for their critical reading of this manuscript.

## REFERENCES

Azcon-Bieto, J. and C.B. Osmond. 1983. Relationship between photosynthesis and respiration. The effect of carbohydrate status on the rate of $CO_2$ production by respiration in darkened and illuminated wheat leaves. *Plant Physiol.* **71:** 574–581.

Ball, S., T. Marianne, L. Dirick, M. Fresnoy, B. Delrue, and A. Decq. 1991. A *Chlamydomonas reinhardtii* low-starch mutant is defective for 3-phosphoglycerate activation and orthophosphate inhibition of ADP-glucose pyrophosphorylase. *Planta* **185:** 17–26.

Beck, E. and P. Ziegler. 1989. Biosynthesis and degradation of starch in higher plants. *Annu. Rev. Plant Physiol. Plant Mol. Biol.* **40:** 95–117.

Bhattacharyya, M.K., A.M. Smith, T.H.N. Ellis, C. Hedley, and C. Martin. 1990. The wrinkled-seed character of pea described by Mendel is caused by a transposon-like insertion in a gene encoding starch-branching enzyme. *Cell* **60:** 115–121.

Bhave, M.R., S. Lawrence, C. Barton, and L.C. Hannah. 1990. Identification and molecular characterization of *Shrunken-2* cDNA clones of maize. *Plant Cell* **2:** 581–588.

Britz, S.J. and A.W. Galston. 1982. Light-enhanced perception of gravity in stems of intact pea seedlings. *Planta* **154:** 189–192.

Caspar, T. and B.G. Pickard. 1989. Gravitropism in a starchless mutant of *Arabidopsis* : Implications for the starch-statolith theory of gravity sensing. *Planta* **177:** 185–197.

Caspar, T., S.C. Huber, and C.R. Somerville. 1985. Alterations in growth, photosynthesis, and respiration in a starchless mutant of *Arabidopsis thaliana* (L.) deficient in chloroplast phosphoglucomutase activity. *Plant Physiol.* **79:** 11–17.

Caspar, T., T.-P. Lin, G. Kakefuda, L. Benbow, J. Preiss, and C.R. Somerville. 1991. Mutants of *Arabidopsis* with altered regulation of starch degradation. *Plant Physiol.* **95:** 1181–1188.

Caspar, T., T.-P. Lin, J. Monroe, W. Bernhard, S. Spilatro, J. Preiss, and C.R. Somerville. 1989. Altered regulation of ß-amylase activity in mutants of *Arabidopsis* with lesions in starch metabolism. *Proc. Natl. Acad. Sci.* **86:** 5830–5833.

Chatterton, N.J. and J.E. Silvius. 1979. Photosynthate partitioning into starch in soybean leaves. I. Effects of photoperiod versus photosynthetic period duration. *Plant Physiol.* **64:** 749–753.

Delrue, B., T. Fontaine, F. Routier, A. Decq, J.-M. Wieruszeski, N.V.D. Koornhuyse, M.-L. Maddelein, B. Fournet, and S. Ball. 1992. Waxy *Chlamydomonas reinhardtii:* Monocellular algal mutants defective in amylose biosynthesis and granule-bound starch synthase activity accumulate a structurally modified amylopectin. *J. Bacteriol.* **174:** 3612–3620.

Hanson, K.R. and N.A. McHale. 1988. A starchless mutant of *Nicotiana sylvestris* containing a modified plastid phosphoglucomutase. *Plant Physiol.* **88:** 838–844.

Jones, T.W.A., L.D. Gottlieb, and E. Pichersky. 1986. Reduced enzyme activity and starch level in an induced mutant of chloroplast phosphoglucose isomerase. *Plant Physiol.* **81:** 367–371.

Kacser, H. and J. Porteous. 1987. Control of metabolism: What do we have to measure? *Trends Biochem. Sci.* **12:** 5–14.

Kainuma, K. 1988. Structure and chemistry of the starch granule. In *The biochemistry of plants* (ed. J. Preiss), pp. 141–180. Academic Press, San Diego.

Kiss, J.Z., R. Hertel, and F.D. Sack. 1989. Amyloplasts are necessary for full gravitropic sensitivity in roots of *Arabidopsis thaliana. Planta* **177:** 198–206.

Kruckeberg, A.L., H.E. Neuhaus, R. Feil, L.D. Gottlieb, and M. Stitt. 1989. Decreased-activity mutants of phosphoglucose isomerase in the cytosol and chloroplast of *Clarkia xantiana:* Impact on mass-action ratios and fluxes to sucrose and starch, estimation of flux control coefficients and elasticity coefficients. *Biochem. J.* **261:** 457–467.

Li, B., D.R. Geiger, and W.J. Shieh. 1992. Evidence for circadian regulation of starch and sucrose synthesis in sugar beet leaves. *Plant Physiol.* **99:** 1393–1399.

Li, L. and J. Preiss. 1992. Characterization of ADPglucose pyrophosphorylase from a starch-deficient mutant of *Arabidopsis thaliana* (L.). *Carbohydr. Res.* **227:** 227–239.

Lin, T.-P., T. Caspar, C.R. Somerville, and J. Preiss. 1988a. Isolation and characterization of starchless mutant of *Arabidopsis thaliana* (L.) Heynh lacking ADPglucose pyrophosphorylase activity. *Plant Physiol.* **86:** 1131–1135.

————. 1988b. A starch deficient mutant of *Arabidopsis thaliana* with low ADPglucose pyrophosphorylase activity lacks one of the two subunits of the enzyme. *Plant Physiol.* **88:** 1175–1181.

Monroe, J. and J. Preiss. 1990. Purification of a β-amylase that accumulates in *Arabidopsis thaliana* mutants defective in starch metabolism. *Plant Physiol.* **94:** 1033–1039.

Moore, R. 1989. Root graviresponsiveness and cellular differentiation in wild-type and a starchless mutant of *Arabidopsis thaliana. Ann. Bot.* **64:** 271–277.

Müller-Röber, B., U. Sonnewald, and L. Wilmitzer. 1992. Inhibition of the ADP-glucose pyrophosphorylase in transgenic potatoes leads to sugar-storing tubers and influences tuber formation and expression of tuber storage protein genes. *EMBO J.* **11:** 1229–1238.

Neuhaus, H.E. and M. Stitt. 1990. Control analysis of photosynthate partitioning: Impact of reduced activity of ADP-glucose pyrophosphorylase or plastid phosphoglucomutase on the fluxes to starch and sucrose in *Arabidopsis thaliana* (L.) Heynh. *Planta* **182:** 445–454.

Obata-Sasamoto, H. and H. Suzuki. 1979. Activities of enzymes relating to starch synthesis and endogenous levels of growth regulators in potato stolon tips during tuberization. *Physiol. Plant.* **45:** 320–324.

Pozueta-Romero, J., M. Frehner, A.M. Viale, and T. Akazawa. 1991. Direct transport of ADPglucose by adenylate transporter is linked to starch biosynthesis in amyloplasts. *Proc. Natl. Acad. Sci.* **88:** 5769–5773.

Preiss, J. 1988. Biosynthesis of starch and its regulation. In *The biochemistry of plants* (ed. J. Preiss), pp. 181–254. Academic Press, San Diego.

Preiss, J., S. Danner, P.S. Summers, M. Morell, C.R. Barton, L. Yang, and M. Nieder. 1990. Molecular characterization of the *Brittle-2* gene effect on maize endosperm ADPglucose pyrophosphorylase subunits. *Plant Physiol.* **92:** 881–885.

Sack, F.D. and J.Z. Kiss. 1989. Rootcap structure in wild-type and in a starchless mutant of *Arabidopsis. Am. J. Bot.* **76:** 454–464.

Sæther, N. and T.-H. Iversen. 1991. Gravitropism and starch statoliths in an *Arabidopsis* mutant. *Planta* **184:** 491–497.

Schulze, W., M. Stitt, E.-D. Schulze, H.E. Neuhaus, and K. Fichtner. 1991. A quantification of the significance of assimilatory starch for growth of *Arabidopsis thaliana* L. Heynh. *Plant Physiol.* **95:** 890–895.

Shannon, J.C. and D.L. Garwood. 1984. Genetics and physiology of starch development. In *Starch chemistry and technology,* 2nd edition (ed. R.L. Whistler et al.), pp. 25–86. Academic Press, San Diego.

Sicher, R.C. and D.F. Kremer. 1992. Control of carbohydrate metabolism in a starchless mutant of *Arabidopsis thaliana. Physiol. Plant.* **85:** 446–452.

Sivak, M.N. and J. Rowell. 1988. Phosphate limits photosynthesis in a starchless mutant of *Arabidopsis thaliana* deficient in chloroplast phosphoglucomutase activity. *Plant Physiol. Biochem.* **26:** 493–501.

Smith, A.M., M. Bettey, and I.D. Bedford. 1989. Evidence that the *rb* locus alters the starch content of developing pea embryos through an effect on ADP glucose pyrophosphorylase. *Plant Physiol.* **89:** 1279–1284.

Smith, A.M., H.E. Neuhaus, and M. Stitt. 1990. The impact of decreased activity of starch-branching enzyme on photosynthetic starch synthesis in leaves of wrinkled-seeded peas. *Planta* **181:** 310–315.

Smith-White, B.J. and J. Preiss. 1992. Comparison of proteins of ADP-glucose pyrophosphorylase from diverse sources. *J. Mol. Evol.* **34:** 449–464.

Steup, M. 1988. Starch degradation. In *The biochemistry of plants* (ed. J. Preiss), pp. 255–296. Academic Press, San Diego.

Stitt, M and W.P. Quick. 1989. Photosynthetic carbon partitioning: Its regulation and possibilities for manipulation. *Physiol. Plant.* **77:** 633–641.

Trethewey, R.N. and T. ap Rees. 1994. A mutant of *Arabidopsis thaliana* lacking the ability to transport glucose across the chloroplast envelope. *Biochem. J.* (in press).

Tyson, R. H. and T. ap Rees. 1988. Starch synthesis by isolated amyloplasts from wheat endosperm. *Planta* **175:** 33–38.

Visser, R.G.F. and E. Jacobsen. 1993. Towards modifying plants for altered starch content and composition. *Trends Biotechnol.* **11:** 63–68.

# 34
# Photosynthesis

**William L. Ogren**

Photosynthesis Research Unit
Agricultural Research Service
United States Department of Agriculture
Urbana, Illinois 61801-3838

*Arabidopsis* photosynthesis has not been studied in a systematic manner; rather, many investigators have applied their particular expertise to *Arabidopsis* because of the oft-stated advantages in genetic and molecular studies. Consequently, the information available on *Arabidopsis* photosynthesis is very detailed in a few areas and totally absent in many others. In this review, I have considered photosynthesis from a functional perspective, referencing those papers that provide information on the biophysics, biochemistry, and physiology of the energy transduction and $CO_2$ assimilation processes. This account, which cites most of the individual papers on *Arabidopsis* photosynthesis, can therefore serve as a starting place for individuals interested in pursuing either the molecular mechanisms or genetics of this topic.

## THYLAKOID MEMBRANES

### Chlorophyll

The earliest *Arabidopsis* mutants were identified on the basis of a visual phenotype, and plants deficient in chlorophyll *b* were among the first characterized (Röbbelen 1957). Similar mutants were described by Hirono and Rédei (1963) and Kranz (1973). The chlorophyll *b* mutants grew more slowly than wild type and had reduced levels of accessory pigments, but no measurements of any photosynthetic parameter were made. In addition to plants deficient in chlorophyll *b*, several mutants with temperature-sensitive alterations in chlorophyll content have been selected (Markwell and Osterman 1992). In all cases, the specific biochemical defect leading to the absence of chlorophyll *b* or to alterations in chlorophyll content is unknown.

More recent analyses of *Arabidopsis* chlorophyll *b* mutants found that such plants lack the photosystem II (PSII) light-harvesting chlorophyll-

*Arabidopsis*
© 1994 Cold Spring Harbor Laboratory Press 0-87969-428-9/94 $5 + .00

protein (LHCPII) complexes (Rühle et al. 1983; Murray and Kohorn 1991), even though normal levels of translatable LHCPII mRNA were found (Murray and Kohorn 1991). Thus, as is the case for other species, LHCPII is not stable in the absence of chlorophyll *b*. In an *Arabidopsis* mutant lacking Δ3-*trans*-hexadecanoic acid, LHCPII oligomer, but not the LHCPII monomer, was found to be labile to SDS (McCourt et al. 1985).

A cDNA for one enzyme in the path of chlorophyll biosynthesis, NADPH-protochlorophyllide oxidoreductase, has been isolated and sequenced (Benli et al. 1991). This enzyme catalyzes the light-dependent reduction of protochlorophyllide to chlorophyllide. It is present in etiolated seedlings but rapidly declines to only trace amounts after 8 hours of illumination, even though chlorophyll accumulation is reaching its maximum level at this time. *Arabidopsis* is similar to other higher plants in this puzzling response.

## Carotenoids

A primary function of carotenoids and other accessory pigments is to protect the photosynthetic apparatus from photooxidation during times of high light or other environmental stresses. Exposing *Arabidopsis* to ultraviolet irradiation (UV-C, 10–16 W m$^{-2}$ at 254 nm) stimulated the level of total carotenoids and led to a threefold increase in the carotenoid/chlorophyll ratio (Campos et al. 1991). Additionally, the mRNA level for 3-hydroxy-3-methylglutaryl coenzyme A reductase, an enzyme in the carotenoid biosynthetic pathway, was dramatically increased. It was suggested that the stimulation in carotenoid synthesis was a protective response to reduce UV damage, and that enhanced carotenoid levels may be used as an indicator of UV-induced stress in plants. Photosynthesis was not measured in the irradiated plants, but UV treatment reduced dry weight by about 50%.

The plant growth regulator abscisic acid (ABA) has been suggested to arise from the breakdown of one or several xanthophylls, a family of oxygenated carotenoids. Alternatively, ABA could be synthesized more directly from mevalonate through the isopentenyl pyrophosphate pathway (Milborrow 1974). The identification of *Arabidopsis* mutants which are deficient in ABA while accumulating the xanthophylls zeaxanthin (Rock and Zeevaart 1991), violaxanthin, and 9'-*cis*-neoxanthin (Duckham et al. 1991) has firmly established carotenoids as the more immediate biosynthetic precursors of ABA. The accumulation of zeaxanthin in ABA mutants also quenches chlorophyll fluorescence, suggesting that some carotenoids may fulfill a regulatory role in photosystem II activity (Rock et al. 1992).

## Lipids

Thylakoid membranes are rich in lipid, including fatty acids not found elsewhere. The isolation of several *Arabidopsis* mutant lines deficient in specific thylakoid fatty acids has permitted the first definitive analyses of the function or requirement for these lipids in the photosynthetic process. The first such mutant characterized (Browse et al. 1985) lacked the chloroplast-specific lipid $\Delta 3$-*trans*-hexadecanoic acid, proposed to be involved in the formation of oligomeric LHCP from monomeric LHCP (Dubacq and Trémolières 1983). The rates of electron transport through photosystems I and II in thylakoid membranes isolated from wild-type and mutant leaves were identical over a wide range of light intensities, demonstrating that this lipid fulfills no requisite role in the light-harvesting process. Additionally, fluorescence measurements on leaves of intact plants over an extended temperature range (25°C–55°C) and on isolated thylakoid membranes at several NaCl concentrations indicated that the absence of this fatty acid had no effect on the efficiency of energy transfer from LHCP to the photosynthetic reaction center (McCourt et al. 1985). Analysis of the fatty acid content of wild-type and mutant leaves showed that the unsaturated $C_{16:1}$ hexadecanoic acid was replaced by saturated $C_{16:0}$ hexadecanoic acid. Thus, the mutant is deficient in $C_{16:0}$ desaturase activity.

Mutations leading to deficiencies in other thylakoid membrane fatty acids showed more dramatic effects on photosynthesis. A mutant deficient in $\omega 9$ desaturase activity, which prevents the oxidation of palmitic acid ($C_{16:0}$) to $C_{16:1}$ at position 2 in monogalactosyl-diacylglycerol, was similar to wild type in growth and photosynthetic activities at temperatures between 10°C and 28°C but showed greater growth and membrane stability at higher temperatures (Kunst et al. 1989a). Enhanced membrane thermal stability was also found in a mutant lacking the desaturase that converts $C_{16}$ and $C_{18}$ fatty acids containing a single double bond to higher degrees of unsaturation (Hugly et al. 1989). Conversely, mutants with higher degrees of membrane lipid saturation than wild type became chlorotic when grown at 5°C, whereas the wild type remained green (Hugly and Somerville 1992). These observations provide new support to the concept, based on comparisons of lipid composition in plants from different temperature environments, that the photosynthetic apparatus adapts to temperature changes by altering the fatty acid composition of membrane lipid.

Photosynthesis rates per leaf weight in the mutant lacking $C_{16:1}$ fatty acid in chloroplast membrane galactolipid were similar to wild type, indicating that the mutation had no effect on the level of photosynthetic carbon reduction pathway enzymes (Kunst et al. 1989a). Photosynthetic

$CO_2$ fixation per milligram of chlorophyll, however, was greater in the mutant than in wild type, which corresponded to a reduced level of chlorophyll *b* and LHCP in the mutant. The rates of electron transfer through PSI and PSII individually were identical, but whole-chain electron transfer in the mutant was slightly reduced, so the efficiency of energy transfer between the two photosystems was reduced in the mutant. There was a greater reduction in antenna chlorophyll than in reaction center chlorophyll, so the rate of electron transport in mutants incapable of synthesizing dienoic thylakoid lipids was higher per milligram chlorophyll than wild type (Hugly et al. 1989). In a mutant incapable of desaturating dienoic lipids to their trienoic analog, $CO_2$ fixation rates were identical on a chlorophyll basis but less per unit leaf area due to reduced chlorophyll content (McCourt et al. 1987).

A mutant missing the chloroplast enzyme glycerol-3-phosphate acyltransferase, which consequently produced chloroplast lipid with a much reduced hexadectrienoic ($C_{16:3}$) content (Kunst et al. 1989b), showed increased thermal tolerance with slightly higher rates of electron transport per milligram of chlorophyll. This mutant also had reduced amounts of appressed membrane, the site of photosystem II. The ratio of the two photosystems did not appear to be altered, but the change in membrane structure arising from the mutation reduced the amount of energy spillover from PSII to PSI.

Although alterations in photosynthetic $CO_2$ fixation, $O_2$ evolution, electron transport activity, and fluorescence were observed in some of the lipid-deficient mutants, the primary phenotypical change in several strains was less thylakoid membrane and reduced chloroplast size (McCourt et al. 1987; Hugly et al. 1989; Hugly and Somerville 1992). Thus, the fatty acids are very important in proper chloroplast biogenesis. *Arabidopsis* mutants with the phenotype of altered chloroplast size and number per cell have also been found by Pyke and Leech (1991, 1992). The genetic basis for the differences observed in these mutants is not known, but it was suggested that these plants may represent mutations in the pathways of chloroplast development or division.

### Proteins

*Cab Proteins*

Most of the chlorophyll in plants is used for light harvesting, absorbing photons, and transferring the light energy to the few chlorophyll molecules involved in biochemistry at the photochemical reaction centers. The antenna chlorophyll molecules are bound to protein and inserted into thylakoid membranes to form the LHCP complexes. The proteins in these complexes, chlorophyll *a/b*-binding (Cab) proteins, are encoded by a family of up to 16 nuclear genes in some species (Leutwiler et al. 1986). In *Arabidopsis*, three LHCP genes for PSII encoding an identical deduced amino acid sequence for a 28.2-kD protein were initially found, clustered on one 11-kb genomic fragment (Leutwiler et al. 1986). Subsequently, one LHCP gene similar to PSI *cab* genes from other species and one Cab-like protein of unknown function have also been identified (Zhang et al. 1991). The nucleotide sequence homology of the three genes described by Leutwiler et al. (1986) is 96%.

Two of the LHCPII genes show very high homology in the noncoding region as well, such that the two gene products cannot be readily distinguished (Leutwiler et al. 1986). However, considerable divergence was found in the third gene. In white light, both types of mRNA are produced in about equal amount (Karlin-Neumann et al. 1988), whereas in etiolated seedlings, red light preferentially induced expression of the divergent gene. Thus, there is differential expression of the LHCPII genes, but both are under the control of phytochrome (Karlin-Neumann et al. 1988; Sun and Tobin 1990).

A second locus, encoding two LHCPII genes, was subsequently found by McGrath et al. (1992). Southern immunoblots of *Arabidopsis* genomic DNA from the Columbia, Landsberg, and Niederzenz ecotypes, probed with the *Arabidopsis* LHCPII and tomato LHCPI and LHCPII *cab* genes, indicated that the *Arabidopsis* genome contains one or more loci for at least one copy of the five LHCPI and three LHCPII types of *cab* genes found in tomato. One of the *Arabidopsis* LHCPI genes has been isolated and sequenced (Jensen et al. 1992). Additional *cab* gene sequences may be found on the list of dbEST homologies.

Biochemical characterization of the major light-harvesting chlorophyll proteins from *Arabidopsis* thylakoids identified two distinct polypeptides, with molecular masses of 28 kD and 25 kD (Morishige and Thornber 1991). Sequence analysis of the two polypeptides and comparison with the gene nucleotide sequences showed that the larger polypeptide was encoded by one or more of the three LHCPII *cab* genes sequenced (Leutwiler et al. 1986), whereas synthesis of the distinctly different 25-kD polypeptide was directed by a gene that was not isolated.

Since *cab* genes are nuclear-encoded but their protein products are located in the chloroplast, their expression must be coordinately expressed with chloroplast gene expression. Thus, in normal plants, preventing chloroplast formation suppresses the expression of nuclear genes for chloroplast enzymes. Recently, three nuclear genes were identified in *Arabidopsis* which uncoupled nuclear gene expression from chloroplast development (Susek et al. 1993), demonstrating the existence of inter-genomic communication in the development of the photosynthetic apparatus.

Promoters of the three *Arabidopsis cab* genes described by Leutwiler et al. (1986) were equally expressed in green leaves of transformed and regenerated tobacco plants, but only weakly expressed in roots, stems, and senescing leaves (An 1987). About 1000 bp of the promoter region was sufficient for light regulation of expression and tissue specificity (An 1987; Brusslan and Tobin 1992). Several *cis*-acting elements in the promoter have been identified (Ha and An 1988; Mitra et al. 1989), including domains involved in regulating light response, tissue specificity, and maximum expression. Increases in *cab* expression were observed in the dark, suggesting the occurrence of developmental regulation, and in the light under phytochrome control (Brusslan and Tobin 1992). Additionally, transcription of two of the three *cab* genes described by Leutwiler et al. (1986) was found to be regulated by the circadian clock, although the steady-state level of all three mRNAs was similar (Millar and Kay 1991). This finding led to the suggestion that there is posttranscriptional regulation of the third *cab* gene product.

## *Other Proteins*

The gene for a 10-kD polypeptide component of PSII in *Arabidopsis* was isolated and sequenced (Gil-Gómez et al. 1991). Expression of the gene was strongly regulated by light. By comparison to light-regulated genes encoding chloroplast polypeptides from other species, several putative *cis*-elements were identified.

Two *Arabidopsis* genes coding for γ subunits of chloroplast ATP synthase were cloned and found to have an amino acid sequence homologous to the spinach sequences (Inohara et al. 1991). In contrast to the spinach γ subunit genes, which contained two introns, the *Arabidopsis* genes contained none.

Delayed luminescence was used to characterize a series of plants with allelic mutations at the *flavoridis* locus (Usmanova et al. 1991). In addition to 50% or greater reduction in delayed luminescence following a flash of light after dark adaptation, these mutants are characterized by reduced amounts of chlorophyll *a* and *b*, a reduced number of chloroplasts per cell, smaller chloroplasts, a decreased amount of thylakoid membrane, and reduced thermal tolerance. The nature of the mutations is not known, although the phenotype is reminiscent of plants described above with altered fatty acid composition of the chloroplast membrane lipids.

## STROMAL METABOLISM

### Rubisco Regulation

An important new regulatory enzyme of photosynthetic $CO_2$ fixation, ribulose bisphosphate carboxylase/oxygenase (Rubisco) activase, was discovered by characterizing an *Arabidopsis* mutant deficient in this activity (Salvucci et al. 1985). The mutant strain, which grows only at elevated $CO_2$ concentrations, was isolated in a screening procedure designed to select mutants defective in enzymes of the photorespiratory pathway (Somerville et al. 1982). Physiologically, the *rca* mutant strain fixed $CO_2$ at a very low rate, and the carboxylation substrate, ribulose bisphosphate (RuBP), accumulated to a concentration tenfold higher than in wild-type plants. The amounts and kinetics of Rubisco were identical in both wild-type and mutant plants. A more detailed examination of Rubisco activity found that the enzyme, which must be carbamylated to be active (Lorimer and Miziorko 1980), was not activated in the light.

The first evidence that a protein factor was necessary for Rubisco activation in vivo came from two-dimensional gel analysis of mutant and wild-type chloroplast extracts (Salvucci et al. 1985). Two polypeptides, of 42 kD and 46 kD molecular mass, were found to be missing in the mutant. Stimulation of Rubisco activation by stromal extracts from both *Arabidopsis* and spinach was then observed (Salvucci et al. 1985), followed by the purification of the Rubisco activase enzyme (Salvucci et al. 1987; Robinson et al. 1989).

*Arabidopsis* Rubisco activase cDNA (Werneke and Ogren 1989) and genomic DNA (Orozco et al. 1993) have been cloned and sequenced, and the amino acid sequence is highly homologous to that of the enzyme from other species (Wang et al. 1992). The two *Arabidopsis* Rubisco activase polypeptides arise from alternative splicing of the same or very similar pre-mRNAs (Werneke et al. 1989). The *Arabidopsis* gene contains six introns, and the alternative splicing occurs in the intron nearest the 3' end of the mRNA. The larger polypeptide is synthesized when the entire intron is removed. The alternatively spliced mRNA contains 11 additional nucleotides at the 5' splice junction, causing a shift in reading frame that leads to a termination signal within the 3' exon. Alternative splicing has also been shown to occur in spinach (Werneke et al. 1989) and barley (Rundle and Zielinski 1991). Genomic sequence comparison of wild type and the Rubisco activase null strain determined that the mutation was caused by a G to A transition at the 5' splice junction of intron 3, leading to three aberrant mRNAs (Orozco et al. 1993). These mRNAs do not seem to lead to the synthesis of any stable polypeptide products, as Western immunoblot analysis indicated the total absence of cross-reacting material in the mutant (Salvucci et al. 1985; Werneke et al. 1988).

*Arabidopsis* has played a central role in probing the still unknown mechanism of light regulation of Rubisco activity and photosynthetic $CO_2$ fixation. The response of *Arabidopsis* photosynthesis to changes in light intensity is similar to that of other species, saturating at 600 µmole photons $m^{-2}$ $s^{-1}$ for plants grown at 300 µmole photons (Salvucci et al. 1986). At 2% $O_2$, the rates of photosynthesis in wild-type and Rubisco-activase-deficient plants were very closely correlated with the state of Rubisco activation, except in the dark. In the dark, Rubisco activation state was the same in both wild type and mutant, increasing with higher ambient $CO_2$ concentrations (Salvucci et al. 1986). Thus, Rubisco activation state in the dark is determined by the ambient $CO_2$ concentration and is not regulated by the Rubisco activase mechanism. In the light, $CO_2$ concentration has very little effect on Rubisco activation state.

When plants were illuminated at saturating light intensity, Rubisco became fully activated in the wild type, whereas in the *rca* mutant, activity declined to a very low level (Somerville et al. 1982). The reduction of activity in the mutant was presumably due to the accumulation of RuBP, a potent inhibitor of Rubisco activation (Jordan and Chollet 1983). Using *Arabidopsis*, Brooks and Portis (1988) showed that the amount of RuBP bound to Rubisco was inversely correlated with Rubisco activation state. Thus, the current concept of regulation of Rubisco activation state is that, upon illumination of a leaf, RuBP is synthesized and binds to Rubisco (Salvucci 1989; Portis 1992). At higher light intensities, Rubisco activase removes RuBP from Rubisco by a mechanism not yet determined and promotes full activation of Rubisco. At low light, where full Rubisco activation would lead to an imbalance between RuBP synthesis and regeneration, the system acts to remove only some of the RuBP from Rubisco, leading to partial Rubisco activation (Brooks and Portis 1988).

There is some degree of species-specificity in the interaction of Rubisco with Rubisco activase. A limited survey of higher plant species has identified two categories of plants with respect to this interaction, the Solanaceae and the non-Solanaceae. Rubisco activase isolated from non-Solanaceae species, including *Arabidopsis* and *Chlamydomonas*, effectively activated Rubisco from other non-Solanaceae plants, but not from Solanaceae. Conversely, Rubisco activase from tobacco and petunia almost fully activated Rubisco from other Solanaceae plants, but functioned poorly with non-Solanaceae Rubisco. Sequence analysis of Rubisco activase cDNA did not reveal any regions of the enzyme that were notably different within the Solanaceae and non-Solanaceae enzymes (Wang et al. 1992).

Many species synthesize considerable quantities of 2-carboxyarabinitol-1-phosphate (CA1P), a naturally occurring inhibitor of Rubisco, in the dark (Gutteridge et al. 1986). CA1P, when present, is removed upon illumination by the action of Rubisco activase (Robinson and Portis 1988). *Arabidopsis* synthesizes only small quantities of CA1P (Moore et al. 1991), so it is not a significant factor in this species.

Leaf senescence in *Arabidopsis* is initiated shortly after the leaf reaches full expansion and is first marked by a decline in photosynthesis rate (Hensel et al. 1993). Photosynthesis ceases gradually over about 10 days, and the loss in $CO_2$ assimilation activity is accompanied by proportional losses in Rubisco activity and leaf soluble protein. The kinetics of chlorophyll loss is slightly different, with the onset of chlorosis occurring about 4 days after the decline in photosynthesis begins. Unlike previous species examined, *Arabidopsis* senescence did not seem to be related to the beginning of reproductive development. Rather, Hensel et al. (1993) postulate that senescence was triggered by unspecified maturation signals and age-related changes in metabolites of photosynthesis. The decline in photosynthetic rate and associated enzymes is accompanied by declines in the expression of CAB and Rubisco small subunit genes and by increases in the expression of other genes, including one clone with homology to cysteine proteinase.

### Stromal Proteins

Sequence information is available for several *Arabidopsis* enzymes of the photosynthetic carbon cycle and closely related metabolism. Four genes have been identified for the small subunit of Rubisco (Krebbers et al. 1988). Three of the genes were found in tandem in a single 8-kb region of the genome, and the fourth gene was completely unlinked from the others. From a high degree of nucleotide homology in both the coding region and introns of the three genes in tandem, and the relative divergence in sequence of the fourth gene, it was suggested that the *rbcS* genes in *Arabidopsis* constitute two gene families. The four genes, which are similar in structure to *rbcS* genes from other plants, were all expressed and were considered to represent the entire gene family (Krebbers et al. 1988).

Nucleotide sequences have been determined for cDNAs for three other enzymes of the photosynthetic carbon reduction cycle: phosphoribulokinase (Horsnell and Raines 1991a), fructose-1,6-bisphosphatase (Horsnell and Raines 1991b), and glyceraldehyde-3-phosphate dehydrogenase (GAPDH) (Shih et al. 1991, 1992). These three enzymes are activated in light by the ferredoxin/thioredoxin system (Buchanan 1980), and the sequences were found to be similar to those from other species. Three GAPDH genes were found by genomic analysis. Two of the genes, closely related, encode chloroplast enzymes, whereas the third, much less similar, encodes a cytosolic enzyme. This observation is consistent with the suggestion that the chloroplast and cytosolic enzymes evolved from different lineages that diverged earlier than the prokaryotes and eukaryotes diverged (Shih et al. 1991, 1992). The expression of chloroplast GAPDH mRNA was found to be stimulated and absorbed by both blue-light photoreceptors and phytochrome (Dewdney et al. 1993), but not by heat shock, anaerobiosis, or exogenous sucrose (Yang et al. 1993). In contrast, cytosolic GAPDH expression was increased by heat shock and anaerobiosis (Yang et al. 1993).

Nucleotide sequences for genes that may be related to photosynthetic activity include chloroplast ascorbic acid peroxidase (Kubo et al. 1992), which may protect the photosynthetic apparatus by scavenging oxygen radicals; enolase (Van Der Straeten et al. 1991), which is thought to participate in chloroplast glycolysis; and carbonic anhydrase (Raines et al. 1992), which may function to maintain the supply of $CO_2$ to Rubisco. All three enzymes are nuclear encoded, and there was evidence for only a single copy of the genes for ascorbic acid peroxidase and enolase. *Arabidopsis* enolase did not possess a transit peptide and, therefore, is only in the cytosol. This contrasts with all other species examined, which possess a chloroplastic enolase, and indicates that *Arabidopsis* chloroplasts are distinct in not containing a complete set of glycolytic pathway enzymes (Van Der Straeten et al. 1991). Carbonic anhydrase mRNA expression was increased when *Arabidopsis* plants were grown at elevated $CO_2$ (Raines et al. 1992). As mentioned above, nucleotide sequences for *Arabidopsis cab* genes (Leutwiler et al. 1986; Zhang et al. 1991) and Rubisco activase (Werneke and Ogren 1989; Orozco et al. 1993) have also been determined. Several other full or partial sequences homologous to chloroplast and photosynthesis enzymes are found in the dbEST database.

Chlorophyll fluorescence, an indication of ATP demand during photosynthesis, was more susceptible to $CO_2$- and $O_2$-induced oscillations in the phosphoglucomutase mutant strain than in wild type (Sivak and Rowell 1988). In addition, photosynthesis rate in the mutant reached light saturation at much lower intensities. The interpretation of these observations was that photosynthesis in the *pgm1* mutant is more limited than in wild type by the availability of orthophosphate (Sivak and Rowell 1988). Biochemical analysis showed that the concentration of hexose phosphates was considerably higher in the wild type than in the *pgm1* mutant (Sicher and Kremer 1992), an observation consistent with phosphate limitation of photosynthesis in the mutant. A second *Arabidopsis* mutant defective in starch metabolism, containing only about 5% as much ADP-glucose pyrophosphorylase as wild type, has also been described (Lin et al. 1988). The activity of all other enzymes examined, including Rubisco, was present in wild-type amounts.

White light was found to activate a $K^+$ channel in *Arabidopsis* leaves, but this effect was not observed in leaves treated with a photosynthesis inhibitor in albino leaves (Spalding and Goldsmith 1993). The similar $K^+$ channel was shown by patch clamp to be activated by ATP when applied to the cytoplasmic side of the patch. On the basis of these observations, it was proposed that $K^+$ channels in the plasma membrane were activated by ATP produced during photosynthesis and that the $K^+$ influx modulates photosynthetic $CO_2$ fixation by maintaining the proper cytoplasmic pH.

### Photorespiration

A series of *Arabidopsis* mutants deficient in various enzymes of the photorespiratory pathway permitted definitive conclusions to be made on previously controversial steps in the process. Selection of the mutants was based on the mutual competition at Rubisco by the gaseous substrates $CO_2$, which initiates the photosynthetic carbon reduction cycle, and $O_2$, which initiates the photorespiratory carbon oxidation cycle (Somerville and Ogren 1982a). At the elevated $CO_2$ concentration of 1% used in the experiments (standard atmospheric concentration is about 0.035%), oxygenase activity is suppressed and carbon enters the photorespiratory cycle at a greatly reduced rate, if at all. Under these conditions, enzymes used solely for photorespiration are not needed. In the same plants in air, however, carbon enters the pathway, and if it cannot be processed to the photorespiratory products $CO_2$, $NH_3$, and glycerate, plant senescence occurs. Thus, the selected phenotype was plant growth at 1% $CO_2$ but chlorosis in air.

Activity of six enzymes in the pathway was found to be missing in the various mutants characterized (Somerville and Ogren 1982b). Enzymes affected in the mutants included phosphoglycolate phosphatase (Somerville and Ogren 1979), which verified that RuBP oxygenase activity was the sole source of photorespiratory glycolate. Analysis of mutants deficient in serine-glyoxyolate aminotransferase (Somerville and Ogren 1980a) and two mitochondrial enzymes, serine transhydroxymethylase (Somerville and Ogren 1981; Somerville and Somerville 1983) and glycine decarboxylase (Somerville and Ogren 1982c), established that glycine oxidation was the primary, if not only, source of photorespiratory $CO_2$. Two additional mutants were found that interfered with photorespiratory ammonia recycling: glutamate synthase (Somerville and Ogren 1980b) and a chloroplast membrane protein that transports oxoglutarate into the chloroplast and glutamate out of the chloroplast (Somerville and Ogren 1983; Somerville and Somerville 1985). The existence of these mutants confirmed the validity of the hypothesis by Keys et al. (1978) that photorespiratory $NH_3$, released by glycine oxidation in the mitochondria, is reassimilated into amino acids by the glutamine synthetase/glutamate synthase pathway in the chloroplast. Several mutants that possessed the high-$CO_2$-requiring phenotype, yet had no identifiable lesion in the photorespiratory pathway, were also recovered (Artus and Somerville 1988).

The biochemical and physiological analyses required to characterize the photorespiration mutants provided considerable background information on the details of *Arabidopsis* photosynthesis, including techniques for isolating photosynthetically active protoplasts and chloroplasts (Somerville et al. 1981). The kinetics of *Arabidopsis* gas exchange is typical for $C_3$ plants, including $O_2$ inhibition of $CO_2$ fixation, $O_2$ stimulation of $CO_2$ evolution in the light, and an oxygen-dependent postillumination burst of photorespiratory $CO_2$ (Somerville and Ogren 1979, 1980a, 1981). The synthesis and distribution of photosynthetic products in the presence of $^{14}CO_2$ are also those expected of a $C_3$ plant (Somerville and Ogren 1979, 1980a,b, 1981).

The kinetics of photosynthetic fixation in the photorespiration mutants at 21%, but not 2%, $O_2$ showed normal $CO_2$ uptake for about 5 minutes followed by a decline to lower rates. A more detailed examination of these plants revealed that, in all cases except the phosphoglycolate phosphatase mutant, the decline in photosynthesis was correlated with a decline in Rubisco activation state (Chastain and Ogren 1985). A similar reduction in Rubisco activation state occurred in the presence of aminoacetonitrile, an inhibitor of glycine decarboxylation. Examination of the photorespiratory pathway (Somerville and Ogren 1982b) indicates that inhibiting these reactions leads, in all cases except for the phosphatase mutant, to the accumulation of glyoxylate, since all the photorespiratory $NH_3$ becomes tied up in amino acids. Thus, it was suggested that glyoxylate may act to regulate Rubisco activation state in vivo (Chastain and Ogren 1985). Subsequently, it was found that glyoxylate accumulated to about threefold higher amounts in the mutants than in wild-type plants and that this concentration was sufficient to cause Rubisco deactivation in isolated chloroplasts (Chastain and Ogren 1989). Glyoxylate-mediated deactivation of Rubisco has also been demonstrated in lysed and reconstituted chloroplasts, but the mechanism has not yet been determined (Campbell and Ogren 1990).

**CONCLUSIONS AND PERSPECTIVES**

The data currently in the literature indicate that *Arabidopsis* is in every respect a typical $C_3$ plant. Because of its small size, it is not the plant of choice for general analysis of photosynthetic processes. However, techniques have been developed that allow for plant growth analysis; measurement of photosynthetic, photorespiratory, and respiratory $CO_2$ gas exchange; the isolation of active protoplasts and intact chloroplasts for $^{14}CO_2$ feeding and $O_2$ evolution studies; and the isolation of thylakoid membranes and soluble enzymes. Thus, the full range of standard biochemical and physiological tools is available to complement genetic studies.

In little more than a decade, *Arabidopsis* research has made significant contributions to our understanding of the photosynthetic process. The discovery of Rubisco activase was a major breakthrough and, because Rubisco activation can proceed spontaneously in vitro, is a component of regulation that likely would have eluded our notice for many years. The isolation of mutant plants with altered thylakoid membrane fatty acid composition has begun to sort out the various requisite functions of the many lipids present. Integration and regulation of photosynthetic $CO_2$ uptake by the end products starch and sucrose are now being definitively addressed by the analysis of mutants with defects in these pathways. Long-standing controversies in photorespiration and abscisic acid biosynthesis have been resolved with suitable *Arabidopsis* mutants.

Comparable advances will continue to be made in the future. Several areas amenable to analysis through the induced mutation approach, including proteins of chlorophyll antenna and reaction centers, electron transfer components, ATP synthase, enzymes of the photosynthetic carbon reduction cycle, and other participants in the photosynthetic process, remain to be exploited. *Arabidopsis* plants can be transformed with *Agrobacterium tumefaciens*. Two related approaches which hold considerable potential are manipulation of enzyme levels by antisense and overexpressed mRNAs. One breakthrough that would greatly stimulate progress in photosynthesis research is a method to grow *Arabidopsis* photosynthesis null mutants, plants which contain normal chloroplasts yet are incapable of photosynthesis because a single requisite enzyme for the $CO_2$-fixing process is missing. Such a method, particularly if used in conjunction with insertion mutagenesis, would allow the identification of essential regulatory and functional components of the photosynthetic process which are still unknown because when they are absent, the plant is not viable. Devising a procedure for selecting and maintaining such mutants would herald an exciting new era in photosynthesis research.

## REFERENCES

An, G. 1987. Integrated regulation of the photosynthetic gene family from *Arabidopsis thaliana* in transformed tobacco cells. *Mol. Gen. Genet.* **207**: 210–216.

Artus, N.N. and C. Somerville. 1988. A mutant of *Arabidopsis thaliana* that exhibits chlorosis in air but not in atmospheres enriched in $CO_2$. *Plant Physiol.* **87**: 83–88.

Benli, M., R. Schulz, and K. Apel. 1991. Effect of light on the NADPH-protochlorophyllide oxidoreductase of *Arabidopsis thaliana*. *Plant Mol. Biol.* **16**: 615–625.

Brooks, A. and A.R. Portis, Jr. 1988. Protein-bound ribulose bisphosphate correlates with deactivation of ribulose bisphosphate carboxylase in leaves. *Plant Physiol.* **87**: 244–249.

Browse, J., P. McCourt, and C.R. Somerville. 1985. A mutant of *Arabidopsis* lacking a chloroplast-specific lipid. *Science* **227**: 763–765.

Brusslan, J.A. and E.M. Tobin. 1992. Light-independent developmental regulation of *cab* gene expression in *Arabidopsis thaliana* seedlings. *Proc. Natl. Acad. Sci.* **89**: 7791–7795.

Buchanan, B.B. 1980. Role of light in the regulation of chloroplast enzymes. *Annu. Rev. Plant Physiol.* **31**: 341–374.

Campbell, W.J. and W.L. Ogren. 1990. Glyoxylate inhibition of ribulosebisphosphate carboxylase/oxygenase activation in intact, lysed, and reconstituted chloroplasts. *Photosynth. Res.* **23**: 257–268.

Campos, J.L., X. Figueras, M.T. Piñol, A. Boronat, and A.F. Tiburcio. 1991. Carotenoid and conjugated polyamine levels as indicators of ultraviolet-C induced stress in *Arabidopsis thaliana*. *Photochem. Photobiol.* **53**: 689–693.

Chastain, C.J. and W.L. Ogren. 1985. Photorespiration-induced reduction of ribulosebisphosphate carboxylase activation level. *Plant Physiol.* **77**: 851–856.

———. 1989. Glyoxylate inhibition of ribulosebisphosphate carboxylase/oxygenase activation state *in vivo*. *Plant Cell Physiol.* **30**: 937–944.

Dewdney, J., T.R. Conley, M.-C. Shih, and H.M. Goodman. 1993. Effects of blue and red light on expression of nuclear genes encoding chloroplast glyceraldehyde-3-phosphate dehydrogenase of *Arabidopsis thaliana*. *Plant Physiol.* **103**: 115–1121.

Dubacq, J.-P. and A. Trémolières. 1983. Occurrence and function of phosphatidyl-glycerol containing Δ3-*trans*-hexadecenoic acid in photosynthetic lamellae. *Physiol. Vég.* **21**: 293–312.

Duckham, S.C., R.S.T. Linforth, and I.B. Taylor. 1991. Abscisic-acid-deficient mutants at the *aba* gene locus of *Arabidopsis thaliana* are impaired in the epoxidation of zeaxanthin. *Plant Cell Environ.* **14**: 601–606.

Gil-Gómez, G., P.F. Marrero, D. Haro, J. Ayté, and F.G. Hegardt. 1991. Characterization of the gene encoding the 10 Kda polypeptide of photosystem II from *Arabidopsis thaliana*. *Plant Mol. Biol.* **17**: 517–522.

Gutteridge, S., M.A.J. Parry, S. Burton, A.J. Keys, A. Mudd, J. Feeney, J.C. Servaites, and J. Pierce. 1986. A nocturnal inhibitor of carboxylation in leaves. *Nature* **324**: 274–276.

Ha, S.-B. and G. An. 1988. Identification of upstream regulatory elements involved in the developmental expression of the *Arabidopsis thaliana cab1* gene. *Proc. Natl. Acad. Sci.* **85**: 8017–8021.

Hensel, L.L., Grbic, D.A. Baumgarten, and A.B. Bleecker. 1993. Developmental and age-related processes that influence the longevity and senescence of photosynthetic tissues in *Arabidopsis*. *Plant Cell* **5**: 553–564.

Hirono, Y. and G.P. Rédei. 1963. Multiple allelic control of chlorophyll *b* level in *Arabidopsis thaliana. Nature* **197:** 1324–1325.

Horsnell, P.R. and C.A. Raines. 1991a. Nucleotide sequence of a cDNA clone encoding chloroplast phosphoribulokinase from *Arabidopsis thaliana. Plant Mol. Biol.* **17:** 183–184.

————. 1991b. Nucleotide sequence of a cDNA clone encoding chloroplast fructose-1,6-bisphosphatase from *Arabidopsis thaliana. Plant Mol. Biol.* **17:** 185–186.

Hugly, S. and C. Somerville. 1992. A role for membrane lipid polyunsaturation in chloroplast biogenesis at low temperature. *Plant Physiol.* **99:** 197–202.

Hugly, S., L. Kunst, J. Browse, and C. Somerville. 1989. Enhanced thermal tolerance of photosynthesis and altered chloroplast ultrastructure in a mutant of *Arabidopsis* deficient in lipid desaturation. *Plant Physiol.* **90:** 1134–1142.

Inohara, N., A. Iwamoto, Y. Moriyama, S. Shimomura, M. Maeda, and M. Futai. 1991. Two genes, *atpC1* and *atpC2*, for the subunit of *Arabidopsis thaliana* chloroplast ATP synthase. *J. Biol. Chem.* **266:** 7333–7338.

Jensen, P.E., M. Kristensen, T. Hoff, J. Lehmbeck, B.M. Stummann, and K.W. Hennigsen. 1992. Identification of a single-copy gene encoding a type I chlorophyll *a/b*-binding polypeptide of photosystem I in *Arabidopsis thaliana. Physiol. Plant.* **84:** 561–567.

Jordan, D.B. and R. Chollet. 1983. Inhibition of ribulose bisphosphate carboxylase by substrate ribulose 1,5-bisphosphate. *J. Biol. Chem.* **258:** 13752–13758.

Karlin-Neumann, G.A., L. Sun, and E.M. Tobin. 1988. Expression of light-harvesting chlorophyll *a/b*-protein genes is phytochrome-regulated in etiolated *Arabidopsis thaliana* seedlings. *Plant Physiol.* **88:** 1323–1331.

Keys, A.J., I.F. Bird, M.J. Cornelius, P.J. Lea, R.M. Wallsgrove, and B.J. Miflin. 1978. Photorespiratory nitrogen cycle. *Nature* **275:** 741–743.

Kranz, A.R. 1973. Monogen kontrollierte Langzeit-Transformationen der Chlorophylle in Blättern von *Arabidopsis thaliana* (L.) Heynh. *Z. Pflanzenphysiol.* **70:** 333–349.

Krebbers, E., J. Scurinck, L. Heries, A.R. Cashmore, and M.P. Timpko. 1988. Four genes in two diverged subfamilies encode the ribulosebisphosphate-1,5-carboxylase small subunit polypeptides in *Arabidopsis thaliana. Plant Mol. Biol.* **11:** 745–759.

Kubo, A., H. Saji, K. Tanaka, K. Tanaka, and N. Kondo. 1992. Cloning and sequencing of a cDNA encoding ascorbate peroxidase from *Arabidopsis thaliana. Plant Mol. Biol.* **18:** 691–701.

Kunst, L., J. Browse, and C.R. Somerville. 1989a. Altered chloroplast structure and function in a mutant of *Arabidopsis* deficient in plastid glycerol-3-phosphate acyltransferase activity. *Plant Physiol.* **90:** 846–853.

————. 1989b. Enhanced thermal tolerance in a mutant of *Arabidopsis* deficient in palmitic acid unsaturation. *Plant Physiol.* **91:** 401–408.

Leutwiler, L.S., E.M. Meyerowitz, and E.M. Tobin. 1986. Structure and expression of three light-harvesting chlorophyll *a/b*-binding protein genes in *Arabidopsis thaliana. Nucleic Acids Res.* **14:** 4051–4064.

Lin, T.-P., T. Caspar, C.R. Somerville, and J. Preiss. 1988. A starch deficient mutant of *Arabidopsis thaliana* with low ADPglucose pyrophosphorylase activity lacks one of the two subunits of the enzyme. *Plant Physiol.* **88:** 1175–1181.

Lorimer, G.H. and H.M. Miziorko. 1980. Carbamate formation on the ε-amino group of a lysyl residue as the basis for the activation of ribulosebisphosphate carboxylase by $CO_2$ and $Mg^{2+}$. *Biochemistry* **19:** 5321–5328.

Markwell, J. and J.C. Osterman. 1992. Occurrence of temperature-sensitive phenotypic plasticity in chlorophyll-deficient mutants of *Arabidopsis thaliana*. *Plant Physiol.* **98:** 392–394.

McCourt, P., L. Kunst, J. Browse, and C.R. Somerville. 1987. The effects of reduced amounts of lipid unsaturation on chloroplast ultrastructure and photosynthesis in a mutant of *Arabidopsis*. *Plant Physiol.* **84:** 353–360.

McCourt, P., J. Browse, J. Watson, C.J. Arntzen, and C.R. Somerville. 1985. Analysis of photosynthetic antenna function in a mutant of *Arabidopsis thaliana*(L.) lacking *trans*-hexadecenoic acid. *Plant Physiol.* **78:** 853–858.

McGrath, J.M., W.B. Terzaghi, P. Sridhar, A.R. Cashmore, and E. Pichersky. 1992. Sequence of the fourth and fifth photosystem II type I chlorophyll *a/b*-binding protein genes of *Arabidopsis thaliana* and evidence for the presence of a full complement of the extended CAB gene family. *Plant Mol. Biol.* **19:** 725–733.

Milborrow, B.V. 1974. The chemistry and physiology of abscisic acid. *Annu. Rev. Plant Physiol.* **25:** 259–307.

Millar, A.J. and S.A. Kay. 1991. Circadian control of *cab* gene transcription and mRNA accumulation in *Arabidopsis*. *Plant Cell* **3:** 541–550.

Mitra, A., H.K. Choi, and G. An. 1989. Structural and functional analyses of *Arabidopsis thaliana* chlorophyll *a/b*-binding protein (*cab*) promoters. *Plant Mol. Biol.* **12:** 169–179.

Moore, B.D., J. Kobza, and J.R. Seemann. 1991. Measurement of 2-carboxyarabinitol 1-phosphate in plant leaves by isotope dilution. *Plant Physiol.* **96:** 208–213.

Morishige, D.T. and J.P. Thornber. 1991. Correlation of apoproteins with the genes of the major chlorophyll *a/b* binding protein of photosystem II in *Arabidopsis thaliana*. *FEBS Lett.* **293:** 183–187.

Murray, D.L. and B.D. Kohorn. 1991. Chloroplasts of *Arabidopsis thaliana* homozygous for the ch-1 locus lack chlorophyll *b*, lack stable LHCPII and have stacked thylakoids. *Plant Mol. Biol.* **16:** 71–79.

Orozco, B.M., C.R. McClung, J.M. Werneke, and W.L. Ogren. 1993. The molecular basis of the rubisco activase mutation in *Arabidopsis thaliana* is a guanosine to adenine transition at the 5′ splice junction of intron 3. *Plant Physiol.* **102:** 227–232.

Portis, A.R., Jr. 1992. Regulation of ribulose 1,5-bisphosphate carboxylase/oxygenase activity. *Annu. Rev. Plant Physiol. Plant Mol. Biol.* **43:** 415–437.

Pyke, K.A. and R.M. Leech. 1991. Rapid image analysis screening procedure for identifying chloroplast number mutants in mesophyll cells of *Arabidopsis thaliana* (L.) Heynh. *Plant Physiol.* **96:** 1193–1195.

———. 1992. Chloroplast division and expansion is radically altered by nuclear mutations in *Arabidopsis thaliana*. *Plant Physiol.* **99:** 1005–1008.

Raines, C.A., P.R. Horsnell, C. Holder, and J.C. Lloyd. 1992. *Arabidopsis thaliana* carbonic anhydrase: cDNA sequence and effect of $CO_2$ on mRNA levels. *Plant Mol. Biol.* **20:** 1143–1148.

Robinson, S.P. and A.R. Portis, Jr. 1988. Release of the nocturnal inhibitor, carboxyarabinitol-1-phosphate, from ribulose bisphosphate carboxylase/oxygenase by rubisco activase. *FEBS Lett.* **2:** 413–416.

Robinson, S.P., V.J. Streusand, J.M. Chatfield, and A.R. Portis, Jr. 1989. Purification and assay of rubisco activase from leaves. *Plant Physiol.* **88:** 1008–1014.

Röbbelen, G. 1957. Eine Blattfarbmutante ohne Chlorophyll *b* von *Arabidopsis thaliana* (L.) Heynh. *Naturwissenschaften* **44:** 288–289.

Rock, C.D. and J.A.D. Zeevaart. 1991. The *aba* mutant of *Arabidopsis thaliana* is impaired in epoxy-carotenoid biosynthesis. *Proc. Natl. Acad. Sci.* **88:** 7496–7499.

Rock, C.D., N.R. Bowlby, S. Hoffmann-Benning, and J.A.D. Zeevaart. 1992. The *aba* mutant of *Arabidopsis thaliana* (L.) Heynh. has reduced chlorophyll fluorescence yields and reduced thylakoid stacking. *Plant Physiol.* **100:** 1796–1801.

Rühle, H., H.H. Reiländer, K.-D. Otto, and A. Wild. 1983. Chlorophyll-protein-complexes of thylakoids of wild type and chlorophyll *b* mutants of *Arabidopsis thaliana. Photosynth. Res.* **4:** 301–305.

Rundle, S.J. and R.E. Zielinski. 1991. Organization and expression of two tandemly oriented genes encoding ribulosebisphosphate carboxylase/oxygenase activase in barley. *J. Biol. Chem.* **266:** 4677–4685.

Salvucci, M.E. 1989. Regulation of rubisco in vivo. *Physiol. Plant.* **77:** 164–171.

Salvucci, M.E., A.R. Portis, Jr., and W.L. Ogren. 1985. A soluble chloroplast protein catalyzes ribulosebisphosphate carboxylase/oxygenase activation *in vivo. Photosynth. Res.* **7:** 193–201.

————. 1986. Light and $CO_2$ response of ribulose-1,5-bisphosphate carboxylase/oxygenase activation in *Arabidopsis* leaves. *Plant Physiol.* **80:** 655–659.

Salvucci, M.E., J.M. Werneke, W.L. Ogren, and A.R. Portis, Jr. 1987. Purification and species distribution of rubisco activase. *Plant Physiol.* **84:** 930–936.

Shih, M.-C., P. Heinrich, and H.M. Goodman. 1991. Cloning and chromosomal mapping of nuclear genes encoding chloroplast and cytosolic glyceraldehyde-3-phosphate-dehydrogenase from *Arabidopsis thaliana. Gene* **104:** 133–138.

————. 1992. Cloning and chromosomal mapping of nuclear genes encoding chloroplast and cytosolic glyceraldehyde-3-phosphate-dehydrogenase from *Arabidopsis thaliana. Gene* **119:** 317–319.

Sicher, R.C. and D.F. Kremer. 1992. Control of carbohydrate metabolism in a starchless mutant of *Arabidopsis thaliana. Physiol. Plant.* **85:** 446–452.

Sivak, M.N. and J. Rowell. 1988. Phosphate limits photosynthesis in a starchless mutant of *Arabidopsis thaliana* deficient in chloroplast phosphoglucomutase activity. *Plant Physiol. Biochem.* **26:** 493–501.

Somerville, C.R. and W.L. Ogren. 1979. A phosphoglycolate phosphatase-deficient mutant in Arabidopsis. *Nature* **280:** 833–836.

————. 1980a. Photorespiration mutants of *Arabidopsis thaliana* deficient in serine-glyoxylate aminotransferase activity. *Proc. Natl. Acad. Sci.* **77:** 2684–2687.

————. 1980b. Inhibition of photosynthesis in mutants of *Arabidopsis* lacking glutamate synthase activity. *Nature* **286:** 257–259.

————. 1981. Photorespiration-deficient mutants of *Arabidopsis thaliana* lacking mitochondrial serine transhydroxymethylase activity. *Plant Physiol.* **67:** 666–671.

————. 1982a. Isolation of photorespiration mutants in *Arabidopsis thaliana.* In *Chloroplast molecular biology* (ed. M. Edelman et al.), pp. 129–138. Elsevier/North-Holland, Amsterdam.

————. 1982b. Genetic modification of photorespiration. *Trends Biochem. Sci.* **7:** 171–174.

————. 1982c. Mutants of the cruciferous plant *Arabidopsis thaliana* lacking glycine decarboxylase activity. *Biochem. J.* **202:** 373–380.

Somerville, C.R., A.R. Portis, and W.L. Ogren. 1982. A mutant of *Arabidopsis thaliana* which lacks activation of RuBP carboxylase *in vivo. Plant Physiol.* **70:** 381–387.

Somerville, C.R., S.C. Somerville, and W.L. Ogren. 1981. Isolation of photosynthetically active protoplasts and chloroplasts from *Arabidopsis thaliana*. *Plant Sci. Lett.* **21**: 89–96.

Somerville, S.C. and W.L. Ogren. 1983. An *Arabidopsis thaliana* mutant defective in chloroplast dicarboxylate transport. *Proc. Natl. Acad. Sci.* **80**: 1290–1294.

Somerville, S.C. and C.R. Somerville. 1983. Effect of oxygen and carbon dioxide on photorespiratory flux determined from glycine accumulation in a mutant of *Arabidopsis thaliana*. *J. Exp. Bot.* **34**: 415–424.

————. 1985. A mutant of *Arabidopsis* deficient in chloroplast dicarboxylate transport is missing an envelope protein. *Plant Sci. Lett.* **37**: 217–220.

Spalding, E.P. and M.H.M. Goldsmith. 1993. Activation of $K^+$ channels in the plasma membrane of *Arabidopsis* by ATP produced photosynthetically. *Plant Cell* **5**: 477–484.

Sun, L. and E.M. Tobin. 1990. Phytochrome-regulated expression of genes encoding light-harvesting chlorophyll *a/b*-protein in two long hypocotyl mutants and wild type plants of *Arabidopsis thaliana*. *Photochem. Photobiol.* **52**: 51–56.

Susek, R.E., F.M. Ausubel, and J. Chory. 1993. Signal transduction mutants of *Arabidopsis* uncouple nuclear *CAB* and *RBCS* gene expression from chloroplast development. *Cell* **74**: 787–799.

Usmanova, O.V., T.V. Veselova, V.A. Veselovskii, and P.D. Usmanov. 1991. Functioning of the photosynthetic apparatus in allelic mutants of *Arabidopsis thaliana* (L.) Heynh. *Sov. Plant Physiol.* **38**: 470–474.

Van Der Straeten, D., R.A. Rodrigues-Pousada, H.M. Goodman, and M. Van Montagu. 1991. Plant enolase: Gene structure, expression, and evolution. *Plant Cell* **3**: 719–735.

Wang, Z.-Y., G.W. Snyder, B.D. Esau, A.R. Portis, Jr., and W.L. Ogren. 1992. Species-dependent variation in the interaction of substrate-bound ribulose-1,5-bisphosphate carboxylase/oxygenase (rubisco) and rubisco activase. *Plant Physiol.* **100**: 1858–1862.

Werneke, J.M. and W.L. Ogren. 1989. Structure of an *Arabidopsis thaliana* cDNA encoding rubisco activase. *Nucleic Acids Res.* **17**: 2871.

Werneke, J.M., J.M. Chatfield, and W.L. Ogren. 1989. Alternative mRNA splicing generates the two ribulose bisphosphate carboxylase/oxygenase activase polypeptides in spinach and *Arabidopsis*. *Plant Cell* **1**: 815–825.

Werneke, J.M., R.E. Zielinski, and W.L. Ogren. 1988. Structure and expression of spinach leaf cDNA encoding rubisco activase. *Proc. Natl. Acad. Sci.* **85**: 787–791.

Yang, Y., H.-B. Kwon, H.-P. Peng, and M.-C. Shih. 1993. Stress responses and metabolic regulation of glyceraldehyde-3-phosphate dehydrogenase genes in *Arabidopsis*. *Plant Physiol.* **101**: 209–216.

Zhang, H., S. Hanley, and H.M. Goodman. 1991. Isolation, characterization, and chromosomal location of a new *cab* gene from *Arabidopsis thaliana*. *Plant Physiol.* **96**: 1387–1388.

# 35

# Structure, Synthesis, and Function of the Plant Cell Wall

**Wolf-Dieter Reiter**

Department of Molecular and Cell Biology
and Institute of Materials Science
University of Connecticut, U-125
Storrs, Connecticut 06269

The cell walls of higher plants form a contiguous scaffold surrounding the individual protoplasts within the plant body. In addition to providing mechanical support for the cells, plant cell walls have a pivotal role in growth and development, in cell-cell recognition and interaction, and in defense responses (for recent reviews, see Roberts 1990; Levy and Staehelin 1992; Carpita and Gibeaut 1993). In contrast to animal systems, cell migrations make little contribution to plant development; instead, the ordered formation of plant tissues and organs relies solely on the combination of two processes: the division of cells in specific planes, and the subsequent three-dimensional expansion of the cells in specific dimensions. This latter process of extension growth is primarily a function of the mechanical properties of the walls and the synthesis of new cell wall material. Since all cells within a growing tissue or organ are held together via their middle lamella, extension growth needs to be tightly coordinated between the various cell types in growing parts of the plant body. Since different cell types within a growth zone are distinguished not only by their shape and size, but also by the thickness of their walls, a very complex network of regulatory mechanisms is required to ensure the ordered formation of plant tissues and organs.

Cell wall formation in higher plants essentially takes place in three stages: cell plate formation, extension growth, and secondary growth. The cell plate is a structure formed during the final stages of cell division by the fusion of Golgi-derived vesicles containing cell wall material. Outward growth of the cell plate from the center of the dividing cell connects it to the mother cell wall; the membranes of the cell plate fuse with the plasma membrane of the mother cell completing cell separation. The newly formed daughter cells then undergo extension growth by deposition of new cell wall material throughout the existing wall area. This



"

mode of cell expansion is unique to plant cell walls and is quite distinct from the tip-growth mechanism found in other wall-forming organisms, such as filamentous fungi, where new cell wall material is inserted at localized growth zones pushing the existing wall apart. Since the various components of the growing plant cell wall are covalently and non-covalently cross-linked (Fry 1986), it is generally believed that synthesis of new cell wall material is accompanied by biochemical "loosening" events that permit a continuous reorganization of fibrillar elements within the existing wall structure (Cleland 1981; Taiz 1984; Cosgrove 1993). Most workers believe that cleavage of cell wall polysaccharides is required to maintain an extensible wall structure; however, this view has recently been challenged by the isolation of cell wall proteins that induce cell wall extension but lack detectable glycanase or transglycosylase activities (McQueen-Mason et al. 1992, 1993).

The cell wall formed during extension growth is called the primary cell wall; its thickness is usually 100–200 nm, although some cell types have much thicker primary walls. Since these walls are actually composed of two parts separated by the middle lamella, the wall synthesized by a meristematic cell may only be about 50 nm thick, permitting the deposition of no more than four layers of cellulose microfibrils (see Fig. 1) (McCann and Roberts 1991). The dimensions of cell wall extension appear to be determined by the orientation of cellulose microfibrils, which wrap around cells essentially forming a cocoon restricting cell extension. A role of microfibril orientation in determining cell shape is primarily inferred from electron microscopic observations where the walls of isodiametric cells are characterized by a random orientation of cellulose microfibrils and unidimensionally extending cells are characterized by microfibrils oriented essentially transverse to the dimension of extension (see Fig. 2) (McCann et al. 1990). Extension growth appears to be terminated by a reduction in the yielding properties of the primary wall accompanied by a cessation of the synthesis of cell wall material; however, the exact mechanisms governing extension growth and its termination are poorly understood. Some cell types continue the synthesis of cell wall material after reaching their final size; the resulting thickened wall structure is termed a secondary wall, which differs from the primary wall in its polysaccharide composition. Pronounced secondary wall structures are formed by cell types that have mechanical support functions within the plant body. In many cases, protoplasts synthesizing secondary walls undergo programmed cell death, leaving the wall as the only functional element. This type of developmental pathway is exemplified by the water-conducting vessel elements of the xylem, which are characterized by spiral-shaped or reticulate wall thick-

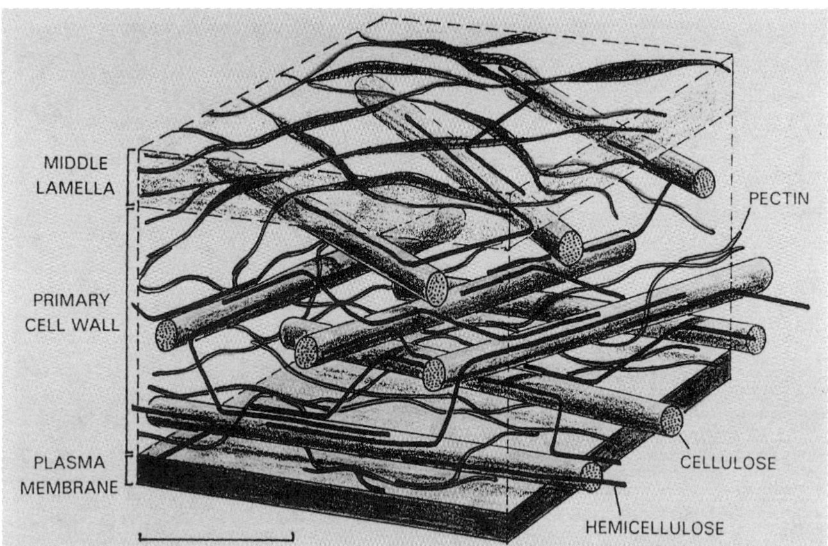


*Figure 1* Model of the primary cell wall of an isodiametric cell. Several layers of cellulose microfibrils coated and cross-linked by hemicelluloses are embedded in a matrix of pectic polysaccharides. Bar, 50 nm. (Reprinted, with permission, from McCann and Roberts 1991.)

enings that prevent them from collapsing under the pressure of the surrounding cells (Esau 1977).

Plant cell walls are primarily composed of polysaccharides, which are classified as cellulose, hemicelluloses, and pectins, based on their solubility properties and chemical composition. Furthermore, they contain structural proteins classified according to their amino acid composition as hydroxyproline-rich glycoproteins (HRGPs), proline-rich proteins (PRPs), and glycine-rich proteins (GRPs). Many secondary cell walls are reinforced by lignin, a three-dimensional network of cross-linked phenylpropanoid residues. Furthermore, walls of specific cell types (e.g., epidermis, cork cells, and endodermis) are impregnated by hydrophobic components such as cutin, a polyester composed of hydroxy-fatty acids, and suberin, a copolymer of phenylpropanoid and fatty acid residues. Reviews on the biochemistry and function of cutin and suberin are available (Kolattukudy 1980, 1984; von Wettstein-Knowles 1993).

In addition to the structural wall components listed above, plant cell walls contain arabinose- and galactose-rich proteoglycans (so-called arabinogalactan proteins; AGPs) and a large number of nonstructural proteins such as glycanases and peroxidases, many of which appear to be involved in cell wall modification events. These proteins, glycoproteins,

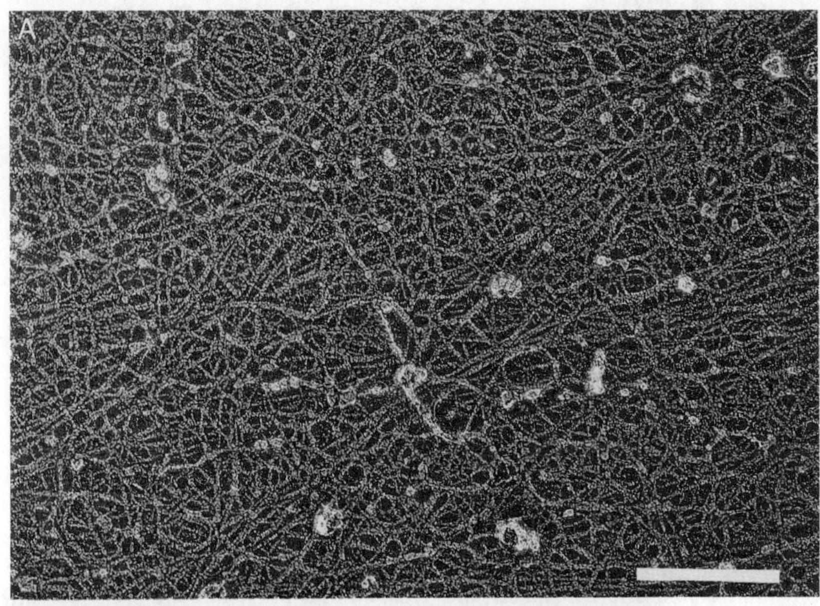

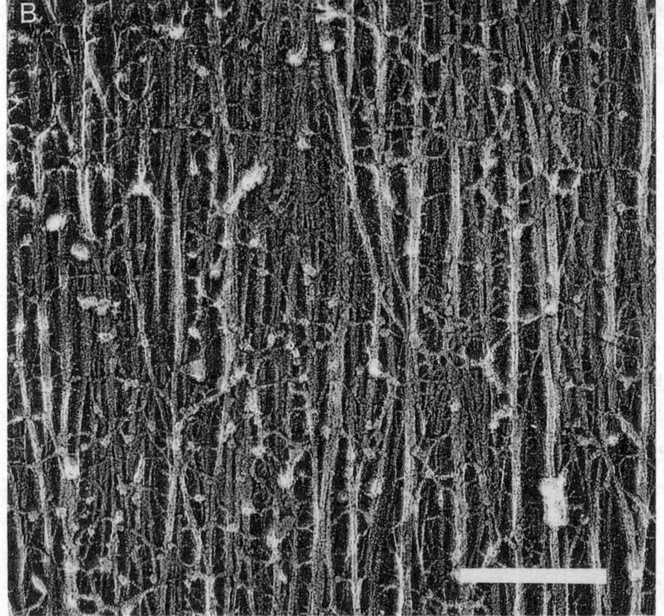

*Figure 2* Electron micrographs of the wall from an isodiametric onion cell (*A*) and an elongating carrot suspension cell (*B*) prepared by the fast-freeze, deep-etch, rotary-shadowing technique (McCann et al. 1990). The main fibrillar elements are cellulose microfibrils cross-linked by hemicelluloses. The parallel orientation of the microfibrils in *B* is perpendicular to the long axis of the cell. Bars, 200 nm. Courtesy of K. Roberts.

and proteoglycans are considered apoplastic macromolecules but not integral parts of the cell wall. Arabinogalactan proteins have been reviewed by Fincher and Stone (1983) and Showalter and Varner (1989); nonstructural cell wall proteins are included in a recent review by Cassab and Varner (1988).

The literature on the structure, composition, mechanical properties, and digestibility of plant cell walls is immense, partly due to the generally recognized importance of cell walls for various aspects of plant development and partly because of the importance of plant cell walls for applied fields. The vast amount of descriptive literature on plant cell walls contrasts sharply with the current knowledge about the biosynthesis of cell wall material and its regulation with regard to cell cycle, extension growth, secondary growth, and environmental parameters. This lack of knowledge is particularly evident in the case of cell wall polysaccharides, where not a single biosynthetic enzyme has been purified to homogeneity and not a single gene for any such enzyme has been cloned. The primary reason for this gap in our understanding of cell wall synthesis is the difficulty of addressing this issue by conventional biochemical approaches. To understand the function of individual cell wall components and to dissect the mechanisms involved in regulating cell wall synthesis, a genetic approach is clearly required. Although a few mutants with changes in cellulose content and lignin composition have been identified in grasses (Bucholtz et al. 1980; Grand et al. 1985; Lapierre et al. 1988; Kokubo et al. 1989, 1991; Pillonel et al. 1991), genetic work in these plants is hampered by long generation times and complex genomes. Recently, several groups have begun to use the powerful tool of *Arabidopsis* genetics to address cell wall-related issues. The purpose of this chapter is to summarize our current knowledge on plant cell walls, to delineate the biological questions connected with cell wall synthesis and structure, and to outline the ways in which *Arabidopsis* genetics can be used to get a better understanding of the role of cell walls for various aspects of plant biology.

## STRUCTURE AND SYNTHESIS OF CELL WALL POLYSACCHARIDES

### General Background

The predominant structural components of plant cell walls are polysaccharides, which account for about 90% of the dry weight of primary walls. Cell wall polysaccharides are classified as cellulose, hemicelluloses, and pectins, based on their solubility properties and chemical composition. The most abundant monosaccharide components of plant cell walls are D-glucose, D-galactose, D-mannose, D-xylose, L-arabinose,

L-rhamnose (6-deoxy-L-mannose), L-fucose (6-deoxy-L-galactose), D-galacturonic acid, and D-glucuronic acid. Some of these monomeric cell wall components may be modified by *O*-acetylation or *O*-methylation. Other monosaccharides such as apiose, aceric acid, 3-deoxy-D-*manno*-octulosonic acid (KDO), and 3-deoxy-D-*lyxo*-2-heptulosaric acid (DHA) are found in trace amounts derived primarily or exclusively from the pectic component RG II (York et al. 1985; Stevenson et al. 1988a). All of these monosaccharides are classified as either D- or L-sugars, depending on the stereochemistry of the last chiral carbon atom. Typically, only one member of an enantiomeric pair (i.e., D or L) is found in cell wall components. The nomenclature of cell wall polysaccharides primarily reflects the sugar residues of which they are composed. Arabinans, for example, are polymers solely composed of L-arabinose monomers, whereas arabinogalactans contain both L-arabinose and D-galactose residues. In aqueous solutions, most of the free monomeric units of cell wall polysaccharides exist as a mixture of furanose (i.e., five-membered) rings, pyranose (i.e., six-membered) rings, and linear forms; however, in cell wall polysaccharides, they are locked into either the furanose or the pyranose form. In most cases, sugars within cell wall polymers exist as pyranoses; the exceptions are apiose and arabinose, which are found exclusively (apiose) or predominantly (arabinose) in their furanose forms. The hydroxyl group at C1 of a circularized monosaccharide can exist in two different stereochemical configurations ($\alpha$ and $\beta$), $\alpha$ indicating the same configuration as the last chiral carbon atom, and $\beta$ indicating the opposite configuration (Fig. 3). Considering the stereochemical complexity of monosaccharides and the possibility that each of these building blocks can be linked to any hydroxyl on any other monosaccharide, the potential for structural diversity in polysaccharides is enormous. As a consequence, polysaccharides are not only important as structural components, but they also have a considerable potential as informational molecules (i.e., receptors or receptor ligands). The following sections summarize what is known about the synthesis, structure, and function of the three major classes of cell wall polysaccharides.

## Cellulose

Cellulose is composed of a linear chain of $\beta(1\rightarrow4)$-linked D-glucose molecules. All of its hydroxyl groups are involved in hydrogen bond interactions either within a single chain or between chains leading to a water-insoluble semicrystalline structure. In the walls of higher plants, cellulose forms microfibrils approximately 5–12 nm in diameter that are covered by a sheath of hemicelluloses (Roberts 1990). Cellulose ac-α-D-glucopyranose          β-D-xylopyranose

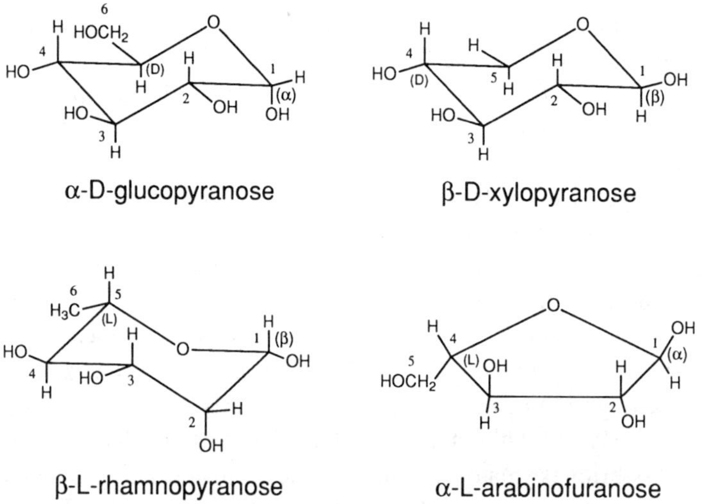

β-L-rhamnopyranose         α-L-arabinofuranose

*Figure 3*  Structures of some monosaccharides commonly found in cell wall polysaccharides. The free monomers can exist in one of four different circular forms: α-furanose, β-furanose, α-pyranose, and β-pyranose. The numbering of carbon atoms, and the stereochemical configurations at carbon atom 1 (α or β) and the last chiral carbon atom (D or L) are indicated.

counts for one-fourth to one-third of total cell wall polysaccharides in primary walls. In secondary walls, the cellulose content tends to be higher, reaching a value of about 94% in cotton seed hairs (Bacic et al. 1988). Due to its tendency to form strong fibrillar structures, cellulose is considered the major load-bearing element of plant cell walls.

## Hemicelluloses

Hemicelluloses account for about one-fourth of the dry mass of primary cell walls and are characterized by their ability to associate tightly with cellulose microfibrils via hydrogen bonding. Plant cell walls contain two major classes of hemicelluloses, xyloglucans and xylans. Xyloglucans represent the most abundant hemicellulose in the primary wall of dicots, whereas xylans represent the predominant hemicellulose in the primary wall of grasses and in the secondary wall of higher plants in general (Bacic et al. 1988).

All xyloglucans possess a linear backbone of β(1→4)-linked D-glucose residues which is structurally identical to cellulose. This stereochemical similarity explains the tight binding of xyloglucans to cellulose microfibrils both in vivo and in vitro (Hayashi and Maclachlan 1984; Hayashi 1989). α-D-Xylosyl residues are attached to the 6 position of most of the glucose units; furthermore, some of these xylose residues are modified at position 2 by the attachment of β-D-galactosyl or α-L-fucosyl-(1→2)-β-D-galactosyl units (Fig. 4). Since xyloglucans bind to cellulose microfibrils but have no strong affinity for each other, their main function appears to be the coating of cellulose microfibrils, thereby preventing their aggregation (Hayashi et al. 1987). Electron microscopic evidence indicates that xyloglucan molecules form cross-bridges between cellulose microfibrils, thereby establishing a cellulose-xyloglucan network contributing to the mechanical strength of the wall (McCann et al. 1990). This observation is of particular interest, since xyloglucans are believed to undergo limited endoglycolytic cleavage during auxin-stimulated cell elongation (Gilkes and Hall 1977; Terry and Jones 1981; Taiz 1984; Fry 1989). This suggests that the partial cleavage of xyloglucan cross-bridges may be one of the wall-loosening events believed to lead to cell expansion by increasing cell wall extensibility. Alternatively, transglycosylation reactions between xyloglucan molecules may lead to a reconstruction of the wall, permitting continuous extension growth (Smith and Fry 1991; Fry et al. 1992; Nishitani and Tominaga 1992; de Silva et al. 1993; Fanutti et al. 1993). This hypothesis is not supported by measurements on the extensibility of heat-inactivated cucumber hypocotyls in the presence or absence of xyloglucan endotransglycosylase (XET) (McQueen-Mason et al. 1993); however, this in vitro system may not accurately reflect extension mechanisms in the intact plant.

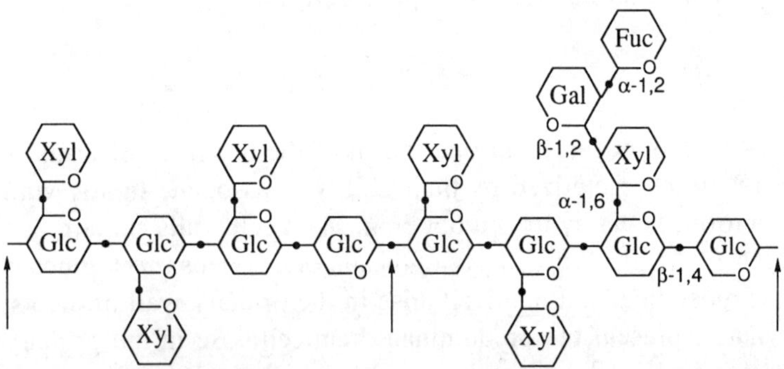



heptasaccharide          nonasaccharide

*Figure 4* Structure of xyloglucan from dicotyledonous plants consisting of heptasaccharide (XG7) and nonasaccharide (XG9) blocks. Some of these repeat units are modified by xylosyl and arabinosyl groups (Hisamatsu et al. 1992). Arrows indicate cleavage sites for endoglucanases. (Glc) D-glucose; (Xyl) D-xylose; (Gal) D-galactose; (Fuc) L-fucose; (—●—) O-glycosidic bond.

Two genes (*Meri-5* and *TCH4*) with a high degree of sequence similarity to nasturtium seed XET (de Silva et al. 1993) have been cloned from *Arabidopsis*. The *Meri-5* gene was isolated on the basis of its strong expression in the shoot apical meristem (Medford et al. 1991); the *TCH4* gene was isolated on the basis of its inducibility by mechanical stimuli such as touch, rain, and wind (Braam and Davis 1990; Braam 1992). *TCH4* is 89% identical with *Meri-5* over 112 amino acids, and both *TCH4* and *Meri-5* are about 50% identical over 111 amino acids with the 31-kD nasturtium XET (J. Braam, pers. comm.). These findings suggest a role of the *TCH4* and *Meri-5* gene products in regulating extension growth; however, the precise functions of these two genes remain to be determined.

It has been shown in elongation assays that specific oligosaccharides derived from xyloglucan by endoglucanase action inhibit auxin- or gibberellin-stimulated cell extension in nanomolar concentrations (York et al. 1984; McDougall and Fry 1988; Warneck and Seitz 1993). These observations have led to the idea that oligosaccharides can act as potent messenger molecules (Albersheim et al. 1983; Ryan 1987; Ryan and Farmer 1991) and suggest the existence of a feedback loop preventing excessive xyloglucan degradation during auxin- or gibberellin-induced extension growth.

The second major group of hemicelluloses, the xylans, are composed of a backbone of $\beta(1{\rightarrow}4)$-linked xylose residues containing short side chains, often consisting of single arabinosyl or 4-*O*-methyl-glucuronosyl residues (Darvill et al. 1980; McNeil et al. 1984). They possess a lower affinity to cellulose than the xyloglucans (Hayashi et al. 1987). Although xylans constitute the major hemicellulosic component in the primary cell wall of grasses, they account for only about 5% of the primary cell wall material in dicots.


## Pectins

Pectic polysaccharides can be defined as cell wall components rich in galacturonic acid; they are at least partially extractable by hot water or by aqueous solutions containing metal-chelating agents. Pectins account for about one-third of the primary cell wall polymers of dicots (Darvill et al. 1980; Fry 1988) and comprise a number of chemically distinct polysaccharides, including homogalacturonans and rhamnogalacturonans (Aspinall 1980). Circumstantial evidence suggests that pectins may be composed of structurally distinct polysaccharide units covalently attached to each other (Aspinall 1980). In blocks described as "rhamnogalacturonan-I" (RG I), neutral side chains mainly consisting of

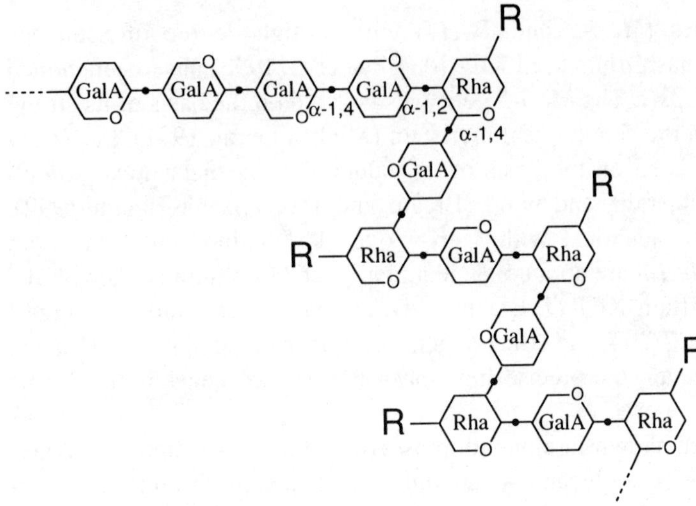

*Figure 5* Schematic model of the proposed structure for a segment of RG I linked to a stretch of polygalacturonic acid residues (Moore et al. 1991). (GalA) D-Galacturonic acid; (Rha) L-rhamnose; (R) potential side chain primarily consisting of D-galactose and L-arabinose residues; (–•–) *O*-glycosidic bond.

arabinose and galactose residues are attached to "backbone" rhamnose residues (Fig. 5) (McNeil et al. 1980; Thomas et al. 1989). "Rhamnogalacturonan-II" (RG II) is a highly branched pectic polysaccharide consisting of at least 12 different monosaccharides interconnected by a variety of glycosidic linkages (Darvill et al. 1978; Stevenson et al. 1988b). Despite its complex structure, rhamnogalacturonan-II appears to be evolutionarily conserved (Stevenson et al. 1988b). The presence of negatively charged oligogalacturonate blocks within pectic polysaccharides is believed to permit the lateral association of pectin molecules via $Ca^{++}$ bridges leading to the formation of a network of noncovalently linked polymers (Grant et al. 1973; Jarvis 1984). A substantial portion of the galacturonic acid residues in pectin is methyl-esterified, thereby eliminating negative charges and reducing the potential for $Ca^{++}$-mediated cross-linking (Bacic et al. 1988). Mainly on the basis of the solution properties of isolated pectins, these molecules are often assumed to form an amorphous gel-like matrix into which fibril-forming polysaccharides like cellulose are embedded (Jarvis 1984). This view has recently been challenged by structure-preserving electron microscopic studies indicating that pectins may form an independent fibrillar network (McCann et al. 1990) which determines wall porosity (Baron-Epel et al. 1988). A structural role for pectins beyond simple gel-forming properties is also suggested by the recent isolation of a tomato cell line adapted to growth in the presence of an inhibitor of cellulose biosynthesis. The walls of this cell line are almost entirely composed of pectic material, indicating that pectins alone can form a wall structure sufficient for cell growth and structural integrity under tissue culture conditions (Shedletzky et al. 1990, 1992).


## Biosynthesis of Cell Wall Polysaccharides

### *Precursor Synthesis*

The polysaccharides of plant cell walls are generated from NDP-sugars, which are synthesized by de novo pathways and by salvage pathways (for review, see Feingold and Avigad 1980; Feingold and Barber 1990). Most of the NDP-sugars are derived from UDP-glucose by a series of nucleotide sugar interconversion pathways. A separate pathway leads from GDP-D-mannose (formed from GTP and mannose-1-phosphate) to GDP-L-fucose. Monosaccharides released from the wall by glycanases during turnover processes are reutilized by the combined action of monosaccharide kinases and nucleotidyl transferases. Most of the NDP-sugars in plant cells are primarily or exclusively used for the synthesis of cell wall polysaccharides, making the NDP-sugar interconversion pathways good candidates for the regulation of cell wall synthesis. Little is known about the enzymes involved in these pathways, and none of their genes has been cloned from higher plants. Consequently, virtually nothing is known about regulatory mechanisms governing the synthesis of these polysaccharide precursors.


### *Polymer Synthesis*

Cellulose is believed to be synthesized at the plasma membrane by rosette-shaped complexes that are readily visualized by electron microscopy (Delmer 1987). The orientation of cellulose microfibrils reflects the orientation of cytoskeletal elements during cell wall synthesis (Heath and Seagull 1982; Lloyd 1984). One model for the ordered deposition of cellulose microfibrils assumes that the rosettes are pushed backward by the nascent cellulose chains and follow a path defined by plasma-membrane-associated microtubules which restrict lateral movement of the synthase complexes (Mueller and Brown 1982; Staehelin and Giddings 1982). The enzymology of cellulose synthesis is poorly understood, primarily due to problems with establishing cellulose-synthesizing in vitro systems. Attempts to clone the gene(s) for plant cellulose synthase(s) have been unsuccessful.

Cell wall polysaccharides other than cellulose are synthesized at the endomembrane system (primarily the Golgi apparatus; for review, see Delmer and Stone 1988). The assembly of some pectic and hemicellulosic cell wall polymers has been investigated by immunological approaches. From these studies it appears that individual Golgi stacks can process glycoproteins and complex polysaccharides simultaneously (Moore et al. 1991). Very little is known about the enzymology of pectin and hemicellulose synthesis. Purification of polysaccharide synthases has not progressed beyond crude extract systems, and it is far from clear whether lipid intermediates are involved in sugar transport and assembly and what primers are used to initiate polysaccharide synthesis. Primarily due to these difficulties with conventional biochemical approaches, no genes directly involved in polysaccharide synthesis have been cloned to date.

## CELL WALL PROTEINS

### Overview

The cell wall space contains a large number of proteins, some of which have a catalytic function (e.g., peroxidases, methylesterases, glycosidases, and transglycosylases). These enzymatic cell wall proteins appear to be involved in cell wall modification and turnover processes that are important to maintain a dynamic wall structure during plant growth and development. Other cell wall proteins have no demonstrable catalytic activity and are characterized by a high abundance of a single amino acid, repetitive sequence motifs, and a tendency to become insolubilized in the wall. These so-called structural cell wall proteins are classified according to their amino acid composition as hydroxyproline-rich glycoproteins (HRGPs), proline-rich proteins (PRPs), and glycine-rich proteins (GRPs). In most plants studied to date, each class of structural cell wall protein is encoded by a family of related genes, each showing specific patterns of expression with regard to cell type, stage of development, and environmental parameters, including wounding and pathogen attack. Considering these highly specific expression patterns and the low abundance of structural cell wall proteins in some cell types, it appears that these proteins play specific roles for the development and properties of certain cell types; however, the precise function of individual cell wall proteins remains to be determined. The following sections summarize what is known about structural cell wall proteins in higher plants. A comprehensive review on both structural and enzymatic cell wall proteins has been published by Cassab and Varner (1988). The literature on structural cell wall proteins has recently been reviewed by Showalter (1993) and Keller (1993).

## HRGPs

At least three different classes of HRGPs have been described in plants: the wall-bound HRGPs or "extensins," the arabinogalactan proteins (AGPs) representing arabinose and galactose-rich apoplastic proteoglycans with a protein content of 2–10%, and certain Solanaceous lectins. A comprehensive review covering all three classes of HRGPs has recently been published (Showalter and Varner 1989). The term HRGPs is mostly used for the extensins, a class of highly basic cell wall proteins characterized by multiple copies of the sequence Ser(Hyp)$_4$. Most of the hydroxyproline residues in extensins carry a linear chain of up to four arabinose residues leading to an overall carbohydrate content of 40–60% (Showalter and Varner 1989). Although the Ser(Hyp)$_4$ repeat is considered a diagnostic feature of extensin structure, HRGPs have been isolated where this repeat is rare (THRGP from maize; Stiefel et al. 1988; Kieliszewski et al. 1990) or absent (HRGP from sugar beet; Li et al. 1990). Monomeric extensins are water-soluble (see, e.g., Smith et al. 1984), but only very small amounts of extensins are extractable from most wall preparations (Lamport 1977; Wilson and Fry 1986), indicating that they become rapidly insolubilized in the wall (Cooper and Varner 1983). The complete removal of cell wall polysaccharides leaves an insoluble network of extensin molecules (Mort and Lamport 1977), suggesting that the extensin monomers are cross-linked to each other. The observation of an intramolecular "isodityrosine" cross-link between two tyrosine residues (Epstein and Lamport 1984) suggests that the same type of cross-link exists between different extensin molecules; however, the precise type of connection between extensin monomers has not been established. The oligoarabinosyl side chains of the Hyp residues appear to stabilize a polyproline II conformation of the extensin molecules (van Holst and Varner 1984), resulting in an extended structure that can be visualized by electron microscopy as kinked rods approximately 80 nm in length (Stafstrom and Staehelin 1986; Heckman et al. 1988). On the basis of the capability of extensins to form insoluble networks, it has been proposed that the cross-linking of extensin molecules restricts cell wall extensibility and is therefore involved in the termination of extension growth (Carpita and Gibeaut 1993). A wall model proposed by Lamport (1986) envisions a network of extension molecules that is interwoven with the cellulose-hemicellulose network (so-called warp-weft model); however, direct evidence in support of this model and for the proposed role of extensin in terminating extension growth has not been obtained.

The expression of extensin genes and the accumulation of extensins within the wall has been studied in carrot (Chen and Varner 1985; Stafstrom and Staehelin 1986, 1987, 1988; Swords and Staehelin 1993), soybean (Cassab and Varner 1987; Ye and Varner 1991), bean (Sheng et al. 1993), maize (Ruiz-Avila et al. 1992), tomato (Ye et al. 1991; Showalter et al. 1992), tobacco and petunia (Ye et al. 1991). Due to the presence of gene families encoding extensins, and plant-to-plant differences in expression patterns, it is difficult to draw general conclusions from these data; however, it appears that extensin genes are preferentially expressed in dividing cells and cells that are subject to mechanical stress during growth such as developing vascular elements. More specialized roles for extensins in certain cell types are indicated by the abundance of extensins in sclerified tissues of soybean seed coats (Cassab and Varner 1987), the expression of a specific extensin gene in a small set of cells in newly formed lateral roots (Keller and Lamb 1989), and the observation that some flower- or pistil-specific genes have extensin-like structures (Chen et al. 1992; de S. Goldman et al. 1992; Wu et al. 1993). Furthermore, it is well established that expression of extensin genes is induced around wound and infection sites (Corbin et al. 1987; Showalter et al. 1991; Tagu et al. 1992; Zhou et al. 1992), presumably to strengthen the wall by providing an impenetrable barrier consisting of cross-linked extensin molecules; alternatively, extensin may participate in defense responses by agglutinating invading bacteria (Leach et al. 1982).

## PRPs

Proline-rich cell wall proteins (PRPs; also referred to as repetitive proline-rich proteins, RPRPs) are characterized by repetitive proline-rich motifs but lack the Ser(Hyp)$_4$ repeat that characterizes extensins. The repeat sequences of PRPs vary considerably between different plant species and between PRP genes within a plant, the common denominator being the presence of two consecutive proline residues. Genes encoding PRPs have been cloned from soybean (Hong et al. 1987, 1989, 1990; Averyhart-Fullard et al. 1988; Datta et al. 1989; Kleis-San Francisco and Tierney 1990), bean (Sheng et al. 1991), carrot (Chen and Varner 1985; Tierney et al. 1988), maize (Josè-Estanyol et al. 1992), and *Arabidopsis*, where four PRP genes are known (Fowler et al. 1993; Fowler and Tierney 1993). About half of the proline residues in soybean PRPs are hydroxylated (Datta et al. 1989; Kleis-San Francisco and Tierney 1990), and the same may be true for other PRPs whose amino acid sequence has only been deduced from cDNA clones. The recent discovery that some hydroxyproline residues in Douglas fir PRPs show arabinosylation patterns similar to extensins (Kieliszewski et al. 1992), and the recognition that PRPs and extensins share sequence motifs other than the clustered proline or hydroxyproline residues (Kieliszewski et al. 1992 and references therein), suggest that the distinction between PRPs and extensins is becoming diffuse, and that the two classes of proteins may be considered members of the same family (Li et al. 1990). Like the extensins and the GRPs, expression of PRP genes is regulated with regard to development and cell type, and in response to wounding and pathogen attack (Fowler and Tierney 1993; Suzuki et al. 1993; Sheng et al. 1993 and references therein). A class of PRPs termed nodulins are specifically synthesized during nodule formation in legumes (Franssen et al. 1987; Scheres et al. 1990), indicating an important role of PRPs in establishing symbiotic relationships that require restructuring of the wall and the synthesis of new cell wall material. Immunolocalization studies of PRPs in several dicotyledonous species indicate preferential deposition in tissues that become lignified during development (Ye and Varner 1991; Ye et al. 1991). Although these data suggest that PRPs may be directly or indirectly involved in strengthening the wall, their precise function(s) remains to be established.

## GRPs

There are several instances where glycine rather than hydroxyproline is the most abundant amino acid in plant cell walls (Varner and Cassab 1986), indicating a high abundance of glycine-rich proteins in these walls. Genes encoding glycine-rich proteins have been cloned from several plants; however, not all of the corresponding gene products are targeted to the wall. The best-studied glycine-rich cell wall proteins are those from petunia (Condit and Meagher 1986, 1987; Condit et al. 1990) and bean (Keller et al. 1988, 1989; Ryser and Keller 1992), which have a $(Gly-X)_n$ structure where X is frequently glycine. This structural pattern is believed to lead to extended conformations with antiparallel strands forming a $\beta$-pleated sheet (Condit and Meagher 1986; Keller et al. 1988). Although there appear to be species-specific differences in the cell type specificity of GRPs, the available data indicate preferential accumulation of GRPs in vascular tissues (Keller et al. 1989; Condit et al. 1990; Ye and Varner 1991; Ye et al. 1991). This expression pattern suggests some connection between GRP deposition and lignification processes, but a detailed immunolocalization study by Ryser and Keller (1992) indicates that GRPs are most abundant in the non-lignified primary walls of xylem elements rather than in the secondary walls that undergo lignification. Interestingly, synthesis of GRPs was not observed in living protoxylem cells but in the neighboring xylem parenchyma cells, which have essentially GRP-free walls. This finding suggests that the GRPs synthesized in the parenchyma cells are targeted to the wall of the neighboring xylem elements.

Six genes encoding GRPs have been described for *Arabidopsis* by de Oliveira et al. (1990) and Quigley et al. (1991). Three of these genes (atGRP-1, atGRP-5, and the GRP gene isolated by Quigley et al.) encode polypeptides that conform to the $(Gly-X)_n$ structure characteristic of glycine-rich cell wall proteins found in other plants and contain a putative signal peptide for secretion, with the possible exception of atGRP-1, where the amino-terminal sequence is not known. Organ-specific differences in expression were described for all three putative cell wall GRP genes of *Arabidopsis*, and atGRP-1 and atGRP-5 are regulated in response to stress conditions (de Oliveira et al. 1990). Although it is clear from the available data that GRP genes show complex expression patterns, the roles of GRPs for wall structure and plant development are far from clear.

## LIGNIN

Lignin is a three-dimensional polymer of phenylpropanoid alcohols ("monolignols") that imparts mechanical rigidity to plant tissues specialized in solute conductance or mechanical support (for a comprehensive review on lignin synthesis and structure, see Lewis and Yamamoto 1990). Furthermore, lignin formation is observed at sites of wounding or pathogen attack, apparently in an effort to strengthen the wall at these sites of damage (Vance et al. 1980; Ride 1983). Monomeric precursors to lignin formation are *p*-coumaryl alcohol, coniferyl alcohol, and sinapyl alcohol, which are derived from *p*-coumaric acid by hydroxylations, *O*-methylations, and carboxy reductions (see Fig. 6A) (Grisebach 1981; Higuchi 1985). The substituted phenyl moieties corresponding to these three monolignols are termed *p*-hydroxyphenyl, guaiacyl, and syringyl units (Fig. 6B). The relative amounts of these moieties within lignin show some plant and cell type specificity; in general, gymnosperms are characterized by lignins almost only containing guaiacyl units, whereas both guaiacyl and syringyl units are present in angiosperm lignins. The most complex type of lignin is found in grasses, where *p*-hydroxyphenyl, guaiacyl, and syringyl units are present (Lewis and Yamamoto 1990). Lignin is thought to be formed extracellularly from the monolignols by a free-radical polymerization presumably catalyzed by wall-resident peroxidases or oxidases (Bao et al. 1993), resulting in a highly cross-linked structure that confers rigidity on secondary cell walls. During the polymerization process, lignin becomes cross-linked to cell wall polysaccharides, possibly via hydroxycinnamic acids that have been found esterified to polysaccharides (Fry 1983, 1987; Meyer et al. 1991). The regulation of lignification is poorly understood; in particular, it is far from clear how monolignols are transported to their apoplastic destina-

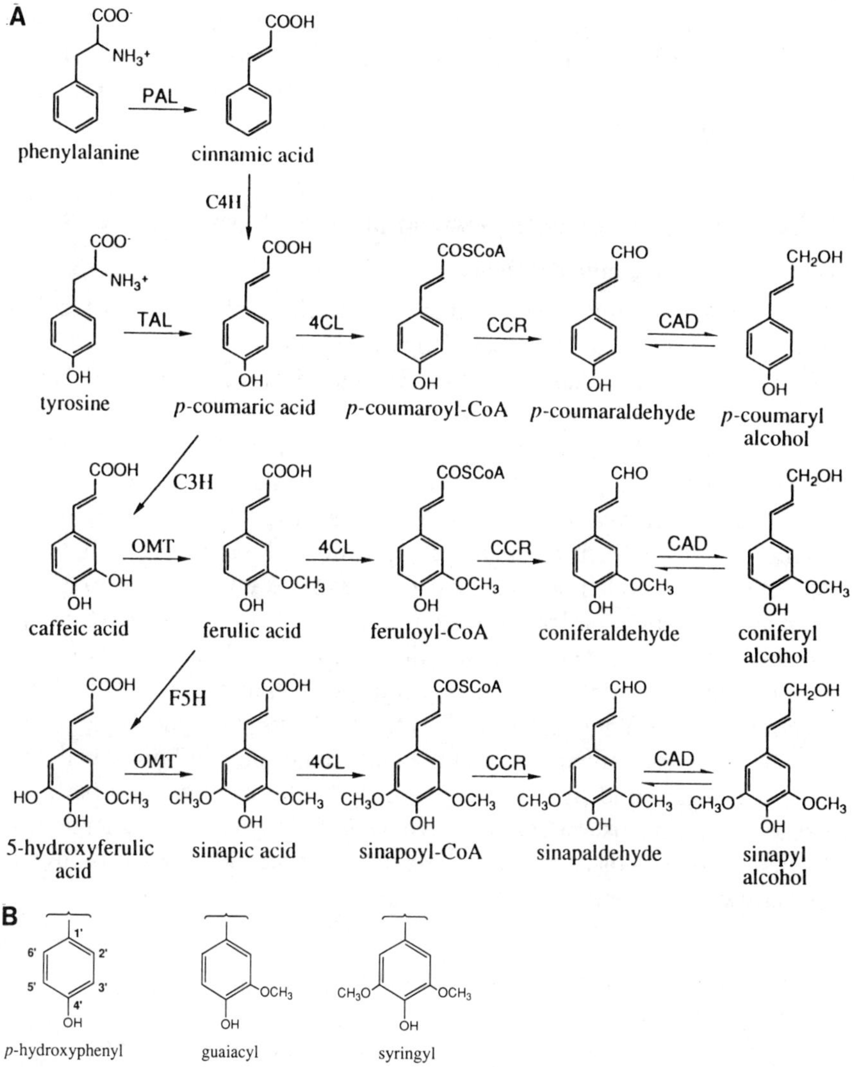

*Figure 6*  (*A*) Biosynthesis of monolignols. Abbreviations for enzymes: (C3H) Coumarate-3-hydroxylase; (C4H) cinnamate-4-hydroxylase; (CAD) hydroxycinnamyl alcohol dehydrogenase; (CCR) hydroxycinnamoyl coenzyme A reductase; (4CL) 4-hydroxycinnamate coenzyme A ligase; (F5H) ferulate-5-hydroxylase; (OMT) caffeate/5-hydroxyferulate *O*-methyltransferase; (PAL) phenylalanine ammonia lyase; (TAL) tyrosine ammonia lyase. (*B*) Structures of substituted phenyl moieties found in lignins.

tion, and what nucleation sites, if any, may trigger the lignification process.

Lignification in *Arabidopsis* has been studied by Dharmawardhana et al. (1992), who indicate that the developmental pattern of lignification in *Arabidopsis* is similar to that in other higher plants. Furthermore, *Arabidopsis* contains guaiacyl/syringyl lignin similar to that of other herbaceous angiosperms. From these data, it can be concluded that *Arabidopsis* can be used as a model system to address lignification-related biological questions.

## MUTANTS IN POLYSACCHARIDE COMPOSITION

### Culm Strength Mutants of Grasses

"Brittle" mutants with decreases in the mechanical strength of culms have been described in maize (Briggs and Robert 1966), rice (Jones 1933; Nagao and Takahashi 1963; Takahashi et al. 1968), and barley (Takahashi et al. 1953, 1966; Kokubo et al. 1989, 1991). Furthermore, barley plants with increased culm stiffness have been described by Hozyo and Oda (1964, 1965). These wall strength phenotypes are observed in non-elongating parts of the plants, suggesting that they reflect changes in the structure and/or composition of secondary cell walls. Studies on the wall thickness and carbohydrate composition of brittle mutants of barley indicate an approximately twofold decrease in the breaking force correlated with thinner walls and an approximately fivefold decrease in the amount of cellulose (Kokubo et al. 1989, 1991). The mutants were not significantly different from wild type in the length of their cellulose molecules or in the amounts of non-cellulosic cell wall polysaccharides and lignin, suggesting that culm brittleness in barley is caused by a decrease in cellulose synthesis. It is not known whether this reduction in cellulose deposition is specific for secondary walls or affects primary walls as well.

### Mutants of *Arabidopsis* with Changes in Cell Wall Polysaccharide Composition

*Overview*

To isolate *Arabidopsis* mutants affected in the synthesis of cell wall polysaccharides, four different approaches have been taken over the last few years: (1) the isolation of mutants resistant to inhibitors of cell wall synthesis, (2) the isolation of mutants with visible phenotypes predicted to arise from perturbations in cell wall synthesis, (3) the isolation of mutants nonreactive with antibodies against carbohydrate epitopes, and

(4) the isolation of mutants altered in the monosaccharide composition of cell wall material.

*Mutants Resistant to Cell Wall Synthesis Inhibitors*

Heim et al. (1989, 1990a,b) isolated *Arabidopsis* plants resistant to the herbicide isoxaben, which has a mode of action very similar to 2,6-dichlorobenzonitrile (DCB), an inhibitor of cellulose synthesis (Delmer 1987). Like DCB, isoxaben does not completely eliminate cellulose synthesis but reduces it by about 80%. Approximately 50 mutant lines with various degrees of isoxaben resistance were recovered from a population of about 1 million $M_2$ plants. Genetic analyses of three highly resistant lines revealed two loci involved in isoxaben resistance, *ixrA* (two known alleles: *ixrA1* and *ixrA2*) and *ixrB1* (one known allele). Although both loci map to chromosome 5, they are sufficiently separated to appear unlinked in genetic crosses (Heim et al. 1990a). The *ixrA* mutations confer a higher level of resistance on the mutant plants than the *ixrB* mutation (300-fold for *ixrA1*, 90-fold for *ixrA2*, and 28-fold for *ixrB1*). Plants heterozygous for the *ixr* mutations show only weak levels of resistance (3-fold for *ixrA1*, 6-fold for *ixrA2*, and 2-fold for *ixrB1*), indicating that the resistant phenotype is essentially recessive with a small semidominant component. *ixrA ixrB* double mutants were not recognizable by an increased level of resistance in $F_2$ populations segregating for both loci (Heim et al. 1990a), suggesting that the two mutations have no additive effect and may affect the same metabolic pathway. It is not known whether isoxaben interacts directly with cellulose synthase or some other component connected with cellulose synthesis, making it difficult to arrive at conclusions concerning the gene products encoded by the *IXRA* and *IXRB* loci. Although the essentially recessive nature of the *ixr* mutations speaks against the idea that they represent gain-of-function mutations in cellulose synthase components, it is conceivable that multisubunit synthase molecules containing both isoxaben-sensitive and isoxaben-resistant components are nonfunctional in the presence of the herbicide, leading to dominance of the wild-type allele. Because of this possibility, the *ixr* mutants may be very useful to clone genes encoding cellulose synthase components, a task that has frustrated workers using conventional biochemical techniques.

*Cell Wall Mutants Identified by Screening for
Visible Phenotypes*

Since the precise function of most cell wall polysaccharides is not known, it is difficult to make predictions about visible phenotypes associated with alterations in cell wall composition and structure. One exception to this rule is mutants deficient in cellulose synthesis, which are expected to be lethal in a homozygous state but could be recovered as heterozygotes segregating for lethals or by screening for conditional lethals such as temperature-sensitive mutants. Since the orientation of cellulose microfibrils is believed to determine the dimensions of extension growth, perturbations in cellulose synthesis or in its ordered deposition should cause all cells in a growth zone to assume an isodiametric shape and possibly become enlarged due to a weakened wall structure. In agreement with this prediction, plants exposed to the cellulose synthesis inhibitors DCB and isoxaben exhibit a lateral swelling of their root tips (Potikha and Delmer 1993). In an effort to isolate *Arabidopsis* mutants with altered cell walls, R. Williamson et al. (pers. comm.) screened about 200,000 chemically mutagenized plants for temperature-sensitive radial swelling of their roots and obtained several plants whose root tips were almost indistinguishable from wild type at $\leq 21^\circ$C but doubled their diameter within 48 hours of transfer to $31^\circ$C (for a description of the first of these mutants, see Baskin et al. 1992). These mutants are currently being screened for alterations in the incorporation of radiolabeled glucose into cellulose, hemicelluloses, and pectins. One mutant line homozygous for a single gene mutation synthesizes cellulose at near wild-type rates at $21^\circ$C but shows a reduction in cellulose synthesis by approximately 70% within 24 hours of transfer to $31^\circ$C. The mutation is specific for cellulose because incorporation of radiolabel into other cell wall polysaccharides is not comparably inhibited (R. Williamson et al., in prep.). The degree of inhibition of cellulose synthesis in the mutant is very similar to that observed in plants grown in the presence of the cellulose synthesis inhibitors DCB and isoxaben, suggesting that the two herbicides and the temperature-sensitive mutation affect the same biosynthetic pathway.

A different approach to isolate mutants in cellulose synthesis or deposition was taken by Potikha and Delmer (1993). They screened mutagenized $M_2$ seedlings for alterations in shape and birefringence of leaf trichomes and xylem elements. Birefringence of cell walls is thought to be caused by the ordered deposition of cellulose microfibrils and is reduced in plants grown in the presence of the cellulose synthesis inhibitor DCB (Potikha and Delmer 1993). One of the most interesting mutants obtained in this screen appears to be impaired in the synthesis of cellulose ($\beta$-$(1{\rightarrow}4)$-D-glucan) and the structurally related callose ($\beta$-$(1{\rightarrow}3)$-D-glucan) which is found in the walls of some specialized cell types and is rapidly synthesized in response to wounding (Delmer 1987). This $\beta$-D-glucan mutant is characterized by a total lack of birefringence in the trichomes, reduced birefringence in most other cell walls including xylem elements, and an abnormal pattern of callose synthesis in response to wounding. The mutation is dominant, leads to an approximately 40% reduction in the amount of cellulose in leaf material, but does not seriously affect plant viability and seed-set. The alterations in both cellulose and callose synthesis in this mutant provide genetic evidence for some link in the synthesis of the two β-D-glucans confirming biochemical data on cellulose and callose synthesis (Delmer 1987).

## Mutant Screening by Immunological Approaches

The use of antibodies directed against cell wall epitopes should be a promising approach to identify cell wall mutants, especially when used in combination with the tissue printing technique that allows the detection of cell wall epitopes on the level of individual cells (Cassab and Varner 1987). The usefulness of immunological screening procedures is currently limited by the number of antibodies directed against specific polysaccharide epitopes and by the small stem diameter of *Arabidopsis* plants, which complicates the use of tissue printing procedures; however, the first limitation is likely to be overcome as more experience is gained with the production of polysaccharide-specific antibodies. The recent identification of *Arabidopsis* mutants nonreactive with antibodies against carbohydrate epitopes in glycoproteins (von Schaewen et al. 1993; K. Roberts, pers. comm.) indicates the general usefulness of immunological procedures for mutant isolation.

## Isolation of Polysaccharide Mutants by Direct Biochemical Screening

Since cell wall polysaccharides vary widely in their monosaccharide composition (e.g., rhamnose is almost only found in pectins, and xylose is primarily found in hemicelluloses), many changes in the composition of cell wall polysaccharides are reflected by changes in the relative amounts of monosaccharides in hydrolyzed cell wall material. Quantitation procedures for these monosaccharides should therefore be useful to screen for mutants which are partially or completely deficient in cell wall polysaccharides, or which synthesize cell wall components in abnormal ratios due to regulatory defects. Since the monosaccharide composition of cell wall material shows a considerable plant-to-plant variation in wild-type *Arabidopsis*, such a screening protocol would primarily identify mutants with relatively major changes in monosaccharide composition. A gas-chromatographic quantitation procedure for neutral monosaccharides released from leaf-derived cell wall material has recently been used to screen more than 5000 chemically mutagenized *Arabidopsis* plants for unusual monosaccharide composition values, leading to the identification of 38 mutant lines representing more than ten complementation groups (Reiter et al. 1993).

On the basis of the observed alterations in monosaccharide composition, these mutant lines can essentially be placed in one of the following categories: (1) lines with complex changes in the relative amounts of several monosaccharides, (2) lines with a substantial reduction in the amount of a single monosaccharide, and (3) lines completely deficient in a specific monosaccharide. Most of the mutants in category 1 show alterations in growth habit, morphology, and physiology that were genetically inseparable from the changes in cell wall composition. Visible phenotypes associated with altered cell wall compositions include extremely stunted morphologies, distortions in leaf shape and size, strongly reduced fertility, and changes in the control of flowering. Most of the mutants in category 2 show up to 50% reductions in the amounts of cell wall-derived arabinose (three loci) or fucose (two loci) but are morphologically indistinguishable from wild-type plants. The complete absence of a monosaccharide was only observed in the case of fucose, where nine allelic lines at a locus designated *mur1* were obtained. Plants carrying tight *mur1* alleles have almost undetectable levels of fucose in shoot organs (less than 2% of the wild-type level), but the amount of fucose in the roots of the mutants is only reduced by about 40%.

*Mur1* plants grown in the presence of L-fucose contain the wild-type level of fucose in their cell wall material, indicating that they are deficient in the de novo synthesis of fucose. A simple explanation for the differences in fucose deficiency between shoots and roots would be the presence of at least two genes encoding isofunctional enzymes in the fucose biosynthetic pathway. One of these genes would correspond to the *MUR1* locus and be expressed throughout the plant, whereas the other one(s) would be expressed only in the root or part(s) thereof.

The *mur1* mutation causes a slightly dwarfed growth habit of the mutant plants and a decreased apical dominance, which does not appear to be due to an auxin or gibberellin deficiency. *Mur1* plants grown in the presence of fucose are morphologically indistinguishable from wild-type plants, indicating that the fucose deficiency is directly responsible for the change in growth habit (Reiter et al. 1993).

Fucose is a component of several cell wall polysaccharides including xyloglucan, the major hemicellulose in the primary walls of dicots. The virtual absence of fucose from xyloglucan in the *mur1* plants has some implications for the "oligosaccharin" hypothesis concerning the biological activity of xyloglucan-derived oligomers (York et al. 1984;

McDougall and Fry 1988; Augur et al. 1992; Warneck and Seitz 1993). According to this hypothesis, auxin-mediated extension growth is regulated on the level of endoglucanases that cleave xyloglucan cross-links between cellulose microfibrils. One of the degradation products, a fucose-containing nonamer (XG9), has been shown to be a potent inhibitor of auxin-mediated extension growth in a bioassay system, the biological activity depending on the presence of a terminal fucose residue within XG9 (McDougall and Fry 1989). The auxin-antagonistic effect of XG9 is believed to constitute a feedback loop preventing excessive extension growth. Since fucose-containing XG9 should be absent or strongly reduced in the mutant, the oligosaccharin hypothesis in its current form would predict excessive extension growth due to the absence of an auxin antagonist, which is not in agreement with the dwarfed growth habit of the *mur1* plants.

Elongating stem segments of *mur1* plants show a more than twofold decrease in the force required to break their primary walls upon longitudinal stretching. This decreased wall strength is not due to differences in stem diameter, in the amount of cellulose, or in the amount of neutral cell wall polysaccharides, indicating that the reduction in overall wall strength is not caused by a decrease in wall synthesis but reflects changes in the intrinsic mechanical properties of the walls. Since the *mur1* mutation affects the structure of xyloglucan, which noncovalently cross-links cellulose microfibrils, it is possible that the decreased wall strength in the *mur1* plants is due to disturbances in the cellulose-xyloglucan network. Computer simulations of xyloglucan conformations indicate that the fucosylated side chain stabilizes an extended structure that can favorably interact with cellulose microfibrils (Levy et al. 1991), supporting the idea that *mur1*-derived xyloglucan may be impaired in its ability to coat and cross-link cellulose microfibrils.

## MUTANTS IN LIGNIN COMPOSITION

### Brown-midrib Mutants of Grasses

A number of "brown-midrib" mutants have been described in the graminaceous monocots sorghum (Porter et al. 1978; Bucholtz et al. 1980), pearl millet (Cherney et al. 1990), and maize (four known loci; Kuc and Nelson 1964; Kuc et al. 1968). This class of mutants is of particular interest for workers concerned with forage digestibility by ruminants, since the brown-midrib phenotype correlates with a reduction in lignin content and a change in lignin structure that improves the enzymatic degradability of cell wall polysaccharides (Cherney et al. 1990 and references therein). The visible phenotype of these mutants is directly caused by an unusually dark appearance of their lignins (Kuc and Nelson 1964). The biochemical alterations leading to the brown-midrib phenotype have been investigated in the *bm-3* mutation of maize (Grand et al. 1985; Lapierre et al. 1988) and the BMR-6 line of sorghum (Bucholtz et al. 1980; Pillonel et al. 1991). These studies indicate a partial block in the conversion of 5-hydroxyferulate to sinapate in the *bm-3* mutant (catalyzed by caffeate/5-hydroxy-ferulate *O*-methyltransferase; OMT) leading to the incorporation of 5-hydroxyguaiacyl units into lignin (Lapierre et al. 1988). A decrease of OMT activity was also seen in the BMR-6 line of sorghum; however, the change in lignin structure in this mutant appeared to be primarily due to a decreased activity of hydroxycinnamyl alcohol dehydrogenase (CAD), leading to an increased incorporation of cinnamaldehydes into the polymer (Bucholtz et al. 1980; Pillonel et al. 1991). These biochemical data suggest a low degree of substrate specificity during the final steps of lignin formation, leading to the incorporation of biosynthetic intermediates into the lignins of the mutants. Although the chemical basis for the brown appearance of the mutant lignins is not known, it is likely that the presence of highly reactive 5-hydroxyguaiacyl and/or cinnamaldehyde groups in a peroxidase-containing environment leads to the formation of colored polymerization products. The availability of brown-midrib mutants opens the possibility to clone genes involved in lignin formation by a genetic approach, particularly in the case of maize, where powerful transposon-tagging strategies are available; however, cloning of genes connected with brown-midrib phenotypes has not been reported to date.

### Lignin Mutants of *Arabidopsis*

Recently, a mutant of *Arabidopsis* has been described that lacks syringyl units in its lignin due to an inability to convert ferulate into sinapate (Chapple et al. 1992). This change in lignin composition does not lead to readily observable alterations in the growth habit of the plants, suggesting that the presence of syringyl units in angiosperm lignin is not essential for the structural integrity of the plant but plays a more subtle role that remains to be elucidated. One T-DNA-tagged allele of this mutation is available, leading to the cloning of the affected gene. Preliminary sequence data indicate a high degree of sequence similarity with P-450-dependent hydroxylases (C. Chapple, unpubl.; Chapple et al., this volume), suggesting that the mutated gene encodes ferulate-5-hydroxylase. This is in agreement with biochemical data on the mutant (Chapple et al. 1992).

Another *Arabidopsis* mutant affected in lignin synthesis has been isolated by T. Potikha and D. Delmer (pers. comm.). This mutant fails to synthesize lignin in response to wounding and is inviable at elevated temperatures. The biochemical or regulatory defect in this mutant line remains to be established.


**CONCLUSIONS AND PERSPECTIVES**

Most of the literature on plant cell walls is largely descriptive, dealing with issues like cell wall composition, ultrastructure, and enzymatic digestibility. In contrast, very little information is available on the enzymology and molecular biology of cell wall synthesis, primarily due to difficulties with conventional biochemical approaches. A considerable amount of information is available on the molecular biology of structural cell wall proteins, but the currently available data on gene expression patterns and protein localization do not directly address questions concerning their function within the wall.

To obtain direct evidence on the roles of the various cell wall components, it is necessary to use genetic approaches; i.e., to eliminate specific cell wall components in intact plants and to evaluate these plants for changes in development and physiology. The availability of genes for structural cell wall proteins allows the use of reverse genetic approaches such as the introduction of antisense constructs; however, this approach suffers from the inability to completely eliminate gene function and from difficulties in targeting one specific member of a multigene family. An alternative approach is the screening of mutagenized plant populations for mutants lacking an epitope that is specific for a certain cell wall protein. To carry out this type of mutant screen, the use of *Arabidopsis* as a genetic system is of crucial importance, since no other plant allows the efficient and rapid screening of sufficiently large numbers of individuals to identify deletion or nonsense mutations that would lead to null alleles of genes encoding structural cell wall proteins. Once such single-gene mutants have been obtained, they could be used to construct multiply mutant lines deficient in two or more members of a cell wall protein family until a whole class of cell wall proteins is rendered nonfunctional, ultimately allowing an evaluation of the roles of these proteins, even in the case of functional redundancy. A similar point can be made for cell wall polysaccharides and lignin components where *Arabidopsis* mutants have already been obtained and are being analyzed for changes in their development and physiology. The isolation of polysaccharide and lignin mutants in *Arabidopsis* has the additional advantage that these lines can be used to clone genes involved in the synthesis of cell wall polysaccharides and lignin components that are difficult to obtain by other means. In summary, it can be concluded that the use of *Arabidopsis* genetics will be of considerable importance to address the complex biological questions concerning the mechanisms of cell wall assembly and the way in which the synthesis and the properties of cell walls are being regulated.

## ACKNOWLEDGMENTS

I thank J. Braam, C. Chapple, D. Delmer, T. Potikha, K. Roberts, and R. Williamson for the communication of unpublished results. Special thanks go to C. Chapple for his help in generating an outline for this chapter, to K. Roberts for providing materials for figures, and to N. Carpita and the editors for their helpful comments and suggestions.

## REFERENCES

Albersheim, P., A.G. Darvill, M. McNeil, B. Valent, J.K. Sharp, E.A. Nothnagel, K.R. Davis, N. Yamazaki, D.J. Gollin, W.S. York, W.F. Dudman, J.E. Darvill, and A. Dell. 1983. Oligosaccharins: Naturally occurring carbohydrate with biological regulatory functions. In *Structure and function of plant genomes* (ed. O. Ciferri and L. Dure III), pp. 293–312. Plenum Press, New York.

Aspinall, G.O. 1980. Chemistry of cell wall polysaccharides. In *The biochemistry of plants: A comprehensive treatise* (ed. P.K. Stumpf and E.E. Conn), vol. 3, pp. 473–500. Academic Press, New York.

Augur, C., L. Yu, K. Sakai, T. Ogawa, P. Sinaÿ, A.G. Darvill, and P. Albersheim. 1992. Further studies of the ability of xyloglucan oligosaccharides to inhibit auxin-stimulated growth. *Plant Physiol.* **99:** 180–185.

Averyhart-Fullard, V., K. Datta, and A. Marcus. 1988. A new hydroxyproline-rich protein in the soybean cell wall. *Proc. Natl. Acad. Sci.* **85:** 1082–1085.

Bacic, A., P.J. Harris, and B.A. Stone. 1988. Structure and function of plant cell walls. In *The biochemistry of plants: A comprehensive treatise* (ed. P.K. Stumpf and E.E. Conn), vol. 14, pp. 297–371. Academic Press, New York.

Bao, W., D.M. O'Malley, R. Whetten, and R.R. Sederoff. 1993. A laccase associated with lignification in loblolly pine xylem. *Science* **260:** 672–674.

Baron-Epel, O., P.K. Gharyal, and M. Schindler. 1988. Pectins as mediators of wall porosity in soybean cells. *Planta* **175:** 389–395.

Baskin, T.I., A.S. Betzner, R. Hoggart, A. Cork, and R.E. Williamson. 1992. Root morphology mutants in *Arabidopsis thaliana*. *Aust. J. Plant Physiol.* **19:** 427–437.

Braam, J. 1992. Regulated expression of the calmodulin-related TCH genes in cultured *Arabidopsis* cells: Induction by calcium and heat shock. *Proc. Natl. Acad. Sci.* **89:** 3213–3216.

Braam, J. and R.W. Davis. 1990. Rain-, wind-, and touch-induced expression of calmodulin and calmodulin-related genes in *Arabidopsis*. *Cell* **60:** 357–364.

Briggs, R.W. and W. Robert. 1966. Recognition and classification of some genetic traits in maize. *J. Hered.* **57:** 35–42.

Bucholtz, D.L., R.P. Cantrell, J.D. Axtell, and V.L. Lechtenberg. 1980. Lignin biochemistry of normal and brown midrib mutant *Sorghum*. *J. Agric. Food Chem.* **28:** 1239–1241.

Carpita, N.C. and D.M. Gibeaut. 1993. Structural models of primary cell walls in flowering plants: Consistency of molecular structure with the physical properties of the walls during growth. *Plant J.* **3:** 1–30.

Cassab, G.I. and J.E. Varner. 1987. Immunocytolocalization of extensin in developing soybean seed coats by immunogold-silver staining and by tissue printing on nitrocellulose paper. *J. Cell Biol.* **105:** 2581–2588.

————. 1988. Cell wall proteins. *Annu. Rev. Plant Physiol. Plant Mol. Biol.* **39:** 321–353.

Chapple, C.C.S., T. Vogt, B.E. Ellis, and C.R. Somerville. 1992. An *Arabidopsis* mutant defective in the general phenylpropanoid pathway. *Plant Cell* **4:** 1413–1424.

Chen, C.-G., E.C. Cornish, and A.E. Clarke. 1992. Specific expression of an extensin-like gene in the style of *Nicotiana alata*. *Plant Cell* **4:** 1053–1062.

Chen, J. and J.E. Varner. 1985. Isolation and characterization of cDNA clones for carrot extensin and a proline-rich 33-kDa protein. *Proc. Natl. Acad. Sci.* **82:** 4399–4403.

Cherney, D.J.R., J.A. Patterson, and K.D. Johnson. 1990. Digestibility and feeding value of pearl millet as influenced by the brown-midrib, low-lignin trait. *J. Anim. Sci.* **68:** 4345–4351.

Cleland, R. 1981. Wall extensibility: Hormones and wall extension. In *Plant carbohydrates. II. Extracellular carbohydrates (Encyclopedia of plant physiology*, vol. 13B) (ed. W. Tanner and F.A. Loewus), pp. 255–276. Springer Verlag, Berlin.

Condit, C.M. and R.B. Meagher. 1986. A gene encoding a novel glycine-rich structural protein of petunia. *Nature* **323:** 178–181.

————. 1987. Expression of a gene encoding a glycine-rich protein in petunia. *Mol. Cell. Biol.* **7:** 4273–4279.

Condit, C.M., B.G. McLean, and R.B. Meagher. 1990. Characterization of the expression of the petunia glycine-rich protein-1 gene product. *Plant Physiol.* **93:** 596–602.

Cooper, J.B. and J.E. Varner. 1983. Insolubilization of hydroxyproline-rich glycoprotein in aerated carrot root slices. *Biochem. Biophys. Res. Commun.* **112:** 161–167.

Corbin, D.R., N. Sauer, and C.J. Lamb. 1987. Differential regulation of a hydroxyproline-rich glycoprotein gene family in wounded and infected plants. *Mol. Cell. Biol.* **7:** 4337–4344.

Cosgrove, D.J. 1993. How do plant cell walls extend? *Plant Physiol.* **102:** 1–6.

Darvill, A.G., M. McNeil, and P. Albersheim. 1978. Structure of plant cell walls. VIII. A new pectic polysaccharide. *Plant Physiol.* **62:** 418–422.

Darvill, A., M. McNeil, P. Albersheim, and D.P. Delmer. 1980. The primary cell walls of flowering plants. In *The biochemistry of plants: A comprehensive treatise* (ed. P.K. Stumpf and E.E. Conn), vol. 1, pp. 91–162. Academic Press, New York.

Datta, K., A. Schmidt, and A. Marcus. 1989. Characterization of two soybean repetitive proline-rich proteins and a cognate cDNA from germinating axes. *Plant Cell* **1:** 945–952.

Delmer, D.P. 1987. Cellulose biosynthesis. *Annu. Rev. Plant Physiol.* **38:** 259–290.

Delmer, D.P. and B.A. Stone. 1988. Biosynthesis of plant cell walls. In *The biochemistry of plants: A comprehensive treatise* (ed. P.K. Stumpf and E.E. Conn), vol. 14, pp. 373–420. Academic Press, New York.

de Oliveira, D.E., J. Seurinck, M. van Montagu, and J. Botterman. 1990. Differential expression of five *Arabidopsis* genes encoding glycine-rich proteins. *Plant Cell* **2:** 427–436.

de S. Goldman, M.H., M. Pezzotti, J. Seurinck, and C. Mariani. 1992. Developmental expression of tobacco pistil-specific genes encoding novel extensin-like proteins. *Plant Cell* **4:** 1041–1051.

de Silva, J., C.D. Jarman, D.A. Arrowsmith, M.S. Stronach, S. Chengappa, C. Sidebottom, and J.S.G. Reid. 1993. Molecular characterization of a xyloglucan-specific *endo*-$(1\rightarrow4)$-β-D-glucanase (xyloglucan *endo*-transglycosylase) from nasturtium seeds. *Plant J.* **3:** 701–711.

Dharmawardhana, D.P., B.E. Ellis, and J.E. Carlson. 1992. Characterization of vascular lignification in *Arabidopsis thaliana*. *Can J. Bot.* **70:** 2238–2244.

Epstein, L. and D.T.A. Lamport. 1984. An intramolecular linkage involving isodityrosine in extensin. *Phytochemistry* **23:** 1241–1246.

Esau, K. 1977. *Anatomy of seed plants.* 2nd edition. Wiley, New York.

Fanutti, C., M.J. Gidley, and J.S.G. Reid. 1993. Action of a pure xyloglucan *endo*-transglycosylase (formerly called xyloglucan-specific *endo*-$(1\rightarrow4)$-β-D-glucanase) from the cotyledons of germinated nasturtium seeds. *Plant J.* **3:** 691–700.

Feingold, D.S. and G. Avigad. 1980. Sugar nucleotide transformations in plants. In *The biochemistry of plants: A comprehensive treatise* (ed. P.K. Stumpf and E.E. Conn), vol. 3, pp. 101–170. Academic Press, New York.

Feingold, D.S. and G.A. Barber. 1990. Nucleotide sugars. In *Methods in plant biochemistry* (ed. P.M. Dey and J.B. Harborne), vol. 2, pp. 39–78. Academic Press, London.

Fincher, G.B. and B.A. Stone. 1983. Arabinogalactan-proteins: Structure, biosynthesis, and function. *Annu. Rev. Plant Physiol.* **34:** 47–70.

Fowler, T.J. and M.L. Tierney. 1993. Isolation and differential expression of proline-rich cell wall protein genes in *Arabidopsis*. In *Abstracts from the 5th International Conference on* Arabidopsis *Research*, Columbus, Ohio (ed. R. Hangarter et al.), p. 155. Ohio State University, Columbus.

Fowler, T.J., J.C. Young, R.P. Hangarter, and M.L. Tierney. 1993. Light modulation of *Arabidopsis* PRP gene expression during seedling growth. In *Abstracts from the 5th International Conference on* Arabidopsis *Research*, Columbus, Ohio (ed. R. Hangarter et al.), p. 57. Ohio State University, Columbus.

Franssen, H.J., J.-P. Nap, T. Gloudemans, W. Stiekema, H. van Dam, F. Govers, J. Louwerse, A. van Kammen, and T. Bisseling. 1987. Characterization of cDNA for nodulin-75 of soybean: A gene product involved in early stages of root nodule development. *Proc. Natl. Acad. Sci.* **84:** 4495–4499.

Fry, S.C. 1983. Feruloylated pectins from the primary cell wall: Their structure and possible functions. *Planta* **157:** 111–123.

———. 1986. Cross-linking of matrix polymers in the growing cell walls of angiosperms. *Annu. Rev. Plant Physiol.* **37:** 165–186.

———. 1987. Intracellular feruloylation of pectic polysaccharides. *Planta* **171:** 205–211.

———. 1988. *The growing plant cell wall: Chemical and metabolic analysis.* Wiley, New York.

———. 1989. The structure and functions of xyloglucan. *J. Exp. Bot.* **40:** 1–11.

Fry, S.C., R.C. Smith, K.F. Renwick, D.J. Marten, S.K. Hodge, and K.J. Matthews. 1992. Xyloglucan endotransglycosylase, a new wall-loosening enzyme activity from plants. *Biochem J.* **282:** 821–828.

Gilkes, N.R. and M.A. Hall. 1977. The hormonal control of cell wall turnover in *Pisum sativum* L. *New Phytol.* **78:** 1–15.

Grand, C., P. Parmentier, A. Boudet, and A.M. Boudet. 1985. Comparison of lignins and of enzymes involved in lignification, in normal and brown midrib (*bm3*) mutant corn seedlings. *Physiol. Veg.* **23:** 905–911.

Grant, G.T., E.R. Morris, D.A. Reese, P.J.C. Smith, and D. Thom. 1973. Biological interactions between polysaccharides and divalent cations: The egg-box model. *FEBS Lett.* **32:** 195–198.

Grisebach, H. 1981. Lignins. In *The biochemistry of plants: A comprehensive treatise* (ed. P.K. Stumpf and E.E. Conn), vol 7, pp. 457–478. Academic Press, New York.

Hayashi, T. 1989. Xyloglucans in the primary cell wall. *Annu. Rev. Plant Physiol. Plant Mol. Biol.* **40:** 139–168.

Hayashi, T. and G. Maclachlan. 1984. Pea xyloglucan and cellulose. I. Macromolecular organization. *Plant Physiol.* **75:** 596–604.

Hayashi, T., M.P.F. Marsden, and D.P. Delmer. 1987. Pea xyloglucan and cellulose. V. Xyloglucan-cellulose interactions *in vitro* and *in vivo*. *Plant Physiol.* **83:** 384–389.

Heath, I.B. and R.W. Seagull. 1982. Oriented cellulose fibrils and the cytoskeleton. In *The cytoskeleton in plant growth and development* (ed. C.W. Lloyd), pp. 163–182. Academic Press, London.

Heckman, J.W., Jr., B.T. Terhune, and D.T.A. Lamport. 1988. Characterization of native and modified extensin monomers and oligomers by electron microscropy and gel filtration. *Plant Physiol.* **86:** 848–856.

Heim, D.R., J.L. Roberts, P.D. Pike, and I.M. Larrinua. 1989. Mutation of a locus of *Arabidopsis thaliana* confers resistance to the herbicide isoxaben. *Plant Physiol.* **90:** 146–150.

———. 1990a. A second locus, *ixr* B1 in *Arabidopsis thaliana*, that confers resistance to the herbicide isoxaben. *Plant Physiol.* **92:** 858–861.

Heim, D.R., J.R. Skomp, E.E. Tschobold, and I.M. Larrinua. 1990b. Isoxaben inhibits the synthesis of acid insoluble cell wall materials in *Arabidopsis thaliana*. *Plant Physiol.* **93:** 695–700.

Higuchi, T. 1985. Biosynthesis of lignin. In *Biosynthesis and biodegradation of wood components* (ed. T. Higushi), pp. 141–162. Academic Press, Orlando, Florida.

Hisamatsu, M., W.S. York, A.G. Darvill, and P. Albersheim. 1992. Characterization of seven xyloglucan oligosaccharides containing from seventeen to twenty glycosyl residues. *Carbohydr. Res.* **227:** 45–71.

Hong, J.C., R.T. Nagao, and J.L. Key. 1987. Characterization and sequence analysis of a developmentally regulated putative cell wall protein gene isolated from soybean. *J. Biol. Chem.* **262:** 8367–8376.

———. 1989. Developmentally regulated expression of soybean proline-rich cell wall protein genes. *Plant Cell* **1:** 937–943.

———. 1990. Characterization of a proline-rich cell wall protein gene family of soybean. A comparative analysis. *J. Biol. Chem.* **265:** 2470–2475.

Hozyo, Y. and K. Oda. 1964. Studies on the stiffness of culms in barley plants (*Hordeum sativum*). 2. The development of physical properties in culms. *Jpn. J. Crop Sci.* **33:** 259–262.

———. 1965. Studies on the stiffness of culms in barley plants (*Hordeum sativum*). 10. Bending rigidity of culms. *Jpn. J. Crop Sci.* **34:** 163–170.

Jarvis, M.C. 1984. Structure and properties of pectin gels in plant cell walls. *Plant Cell Environ.* **7:** 153–164.

Jones, J.W. 1933. Inheritance of characters in rice. *J. Agric. Res.* **47:** 771–782.

Josè-Estanyol, M., L. Ruiz-Avila, and P. Puigdomènech. 1992. A maize embryo-specific gene encodes a proline-rich and hydrophobic protein. *Plant Cell* **4:** 413–423.

Keller, B. 1993. Structural cell wall proteins. *Plant Physiol.* **101:** 1127–1130.

Keller, B. and C.J. Lamb. 1989. Specific expression of a novel cell wall hydroxyproline-rich glycoprotein gene in lateral root initiation. *Genes Dev.* **3:** 1639–1646.

Keller, B., N. Sauer, and C.J. Lamb. 1988. Glycine-rich cell wall proteins in bean: Gene structure and association of the protein with the vascular system. *EMBO J.* **7:** 3625–3633.

Keller, B., M.D. Templeton, and C.J. Lamb. 1989. Specific localization of a plant cell wall glycine-rich protein in protoxylem cells of the vascular system. *Proc. Natl. Acad. Sci.* **86:** 1529–1533.

Kieliszewski, M.J., J.F. Leykam, and D.T.A. Lamport. 1990. Structure of the threonine-rich extensin from *Zea mays*. *Plant Physiol.* **92:** 316–326.

Kieliszewski, M., R. de Zacks, J.F. Leykam, and D.T.A. Lamport. 1992. A repetitive proline-rich protein from the gymnosperm Douglas fir is a hydroxyproline-rich glycoprotein. *Plant Physiol.* **98:** 919–926.

Kleis-San Francisco, S.M. and M.L. Tierney. 1990. Isolation and characterization of a proline-rich cell wall protein from soybean seedlings. *Plant Physiol.* **94:** 1897–1902.

Kokubo, A., S. Kuraishi, and N. Sakurai. 1989. Culm strength of barley. Correlation among maximum bending stress, cell wall dimensions, and cellulose content. *Plant Physiol.* **91:** 876–882.

Kokubo, A., N. Sakurai, S. Kuraishi, and K. Takeda. 1991. Culm brittleness of barley (*Hordeum vulgare* L.) mutants is caused by smaller number of cellulose molecules in cell wall. *Plant Physiol.* **97:** 509–514.

Kolattukudy, P.E. 1980. Cutin, suberin, and waxes. In *The biochemistry of plants: A comprehensive treatise* (ed. P.K. Stumpf and E.E. Conn), vol. 4, pp. 571–645. Academic Press, New York.

———. 1984. Biochemistry and function of cutin and suberin. *Can. J. Bot.* **62:** 2918–2933.

Kuc, J. and O.E. Nelson. 1964. The abnormal lignins produced by the brown-midrib mutants of maize. I. The brown-midrib-1 mutant. *Arch. Biochem. Biophys.* **105:** 103–113.

Kuc, J., O.E. Nelson, and P. Flanagan. 1968. Degradation of abnormal lignins in the brown-midrib mutants and double mutants of maize. *Phytochemistry* **7:** 1435–1436.

Lamport, D.T.A. 1977. Structure, biosynthesis and significance of cell wall glycoproteins. *Recent Adv. Phytochem.* **11:** 79–115.

———. 1986. The primary cell wall: A new model. In *Cellulose: Structure, modification and hydrolysis* (ed. R.A. Young and R.M. Rowell), pp. 77–90. John Wiley, New York.

Lapierre, C., M.-T. Tollier, and B. Monties. 1988. Occurrence of additional monomeric units in the lignins from internodes of a *brown-midrib* mutant of maize bm3. *C.R. Acad. Sci. Ser. III* **307:** 723–728.

Leach, J.E., M.A. Contrell, and L. Sequeira. 1982. A hydroxyproline-rich bacterial agglutinin from potato: Its localization by immunofluorescence. *Physiol. Plant Pathol.* **21:** 319–325.

Levy, S. and L.A. Staehelin. 1992. Synthesis, assembly and function of plant cell wall macromolecules. *Curr. Opin. Cell Biol.* **4:** 856–862.

Levy, S., W.S. York, R. Stuike-Prill, B. Meyer, and L.A. Staehelin. 1991. Simulations of the static and dynamic molecular conformations of xyloglucan. The role of the fucosylated sidechain in surface-specific sidechain folding. *Plant J.* **1:** 195–215.

Lewis, N.G. and E. Yamamoto. 1990. Lignin: Occurrence, biogenesis and biodegradation. *Annu. Rev. Plant Physiol. Plant Mol. Biol.* **41:** 455–496.

Li, X., M. Kieliszewski, and D.T.A. Lamport. 1990. A chenopod extensin lacks repetitive tetrahydroxyproline blocks. *Plant Physiol.* **92:** 327–333.

Lloyd, C.W. 1984. Toward a dynamic helical model for the influence of microtubules on wall patterns in plants. *Int. Rev. Cytol.* **86:** 1–51.

McCann, M.C. and K. Roberts. 1991. Architecture of the primary cell wall. In *The cyto-*

*skeletal basis of plant growth and form* (ed. C.W. Lloyd), pp. 109–129. Academic Press, London.

McCann, M.C., B. Wells, and K. Roberts. 1990. Direct visualization of cross-links in the primary plant cell wall. *J. Cell Sci.* **96:** 323–334.

McDougall, G.J. and S.C. Fry. 1988. Inhibition of auxin-stimulated growth of pea stem segments by a specific nonasaccharide of xyloglucan. *Planta* **175:** 412–416.

———. 1989. Structure-activity relationships for xyloglucan oligosaccharides with antiauxin activity. *Plant Physiol.* **89:** 883–887.

McNeil, M., A.G. Darvill, and P. Albersheim. 1980. Structure of plant cell walls. X. Rhamnogalacturonan-I, a structurally complex pectic polysaccharide in the walls of suspension-cultured sycamore cells. *Plant Physiol.* **66:** 1128–1134.

McNeil, M., A.G. Darvill, S.C. Fry, and P. Albersheim. 1984. Structure and function of the primary cell walls of plants. *Annu. Rev. Biochem.* **53:** 625–663.

McQueen-Mason, S., D.M. Durachko, and D.J. Cosgrove. 1992. Two endogenous proteins that induce cell wall extension in plants. *Plant Cell* **4:** 1425–1433.

McQueen-Mason, S.J., S.C. Fry, D.M. Durachko, and D.J. Cosgrove. 1993. The relationship between xyloglucan endotransglycosylase and in-vitro cell wall extension in cucumber hypocotyls. *Planta* **190:** 327–331.

Medford, J.I., J.S. Elmer, and H.J. Klee. 1991. Molecular cloning and characterization of genes expressed in shoot apical meristems. *Plant Cell* **3:** 359–370.

Moore, P.J., K.M.M. Swords, M.A. Lynch, and L.A. Staehelin. 1991. Spatial organization of the assembly pathways of glycoproteins and complex polysaccharides in the Golgi apparatus of plants. *J. Cell Biol.* **112:** 589–602.

Mort, A.J. and D.T.A. Lamport. 1977. Anhydrous hydrogen fluoride deglycosylates glycoproteins. *Anal. Biochem.* **82:** 289–309.

Meyer, K., A. Kohler, and H. Kauss. 1991. Biosynthesis of ferulic acid esters of plant cell wall polysaccharides in endomembranes from parsley cells. *FEBS Lett.* **290:** 209–212.

Mueller, S.C. and R.M. Brown, Jr. 1982. The control of cellulose microfibril deposition in the cell wall of higher plants. *Planta* **154:** 489–515.

Nagao, S. and M. Takahashi. 1963. Trial construction of twelve linkage groups in Japanese rice. Genetical studies on rice plant XXVII. *J. Fac. Agric. Hokkaido University* **53:** 72–130.

Nishitani, K. and R. Tominaga. 1992. Endo-xyloglucan transferase, a novel class of glycosyltransferase that catalyzes transfer of a segment of xyloglucan molecule to another xyloglucan molecule. *J. Biol. Chem.* **267:** 21058–21064.

Pillonel, C., M.M. Mulder, J.J. Boon, B. Forster, and A. Binder. 1991. Involvement of cinnamyl-alcohol dehydrogenase in the control of lignin formation in *Sorghum bicolor* L. Moench. *Planta* **185:** 538–544.

Porter, K.S., J.D. Axtell, V.L. Lechtenberg, and V.F. Colenbrander. 1978. Phenotype, fiber composition, and in vitro dry matter disappearance of chemically induced brown midrib (bmr) mutants of *Sorghum*. *Crop Sci.* **18:** 205–208.

Potikha, T. and D.P. Delmer. 1993. A search for *Arabidopsis* mutants impaired in cell wall biosynthesis. *J. Cell. Biochem. Suppl.* **17A:** 16.

Quigley, F., M.-L. Villiot, and R. Mache. 1991. Nucleotide sequence and expression of a novel glycine-rich protein gene from *Arabidopsis thaliana*. *Plant Mol. Biol.* **17:** 949–952.

Reiter, W.-D., C.C.S. Chapple, and C.R. Somerville. 1993. Altered growth and cell walls in a fucose-deficient mutant of *Arabidopsis*. *Science* **261:** 1032–1035.

Ride, J.P. 1983. Cell walls and structural barriers in defense. In *Biochemical plant pathology* (ed. J.A. Callow), pp. 215–236. Wiley, London.

Roberts, K. 1990. Structures at the plant cell surface. *Curr. Opin. Cell Biol.* **2:** 920–928.

Ruiz-Avila, L., S.R. Burgess, V. Stiefel, M.D. Ludevid, and P. Puigdomènech. 1992. Accumulation of cell wall hydroxyproline-rich glycoprotein mRNA is an early event in maize embryo cell differentiation. *Proc. Natl. Acad. Sci.* **89:** 2414–2418.

Ryan, C.A. 1987. Oligosaccharide signalling in plants. *Annu. Rev. Cell Biol.* **3:** 295–317.

Ryan, C.A. and E.E. Farmer. 1991. Oligosaccharide signals in plants: A current assessment. *Annu. Rev. Plant Physiol. Plant Mol. Biol.* **42:** 651–674.

Ryser, U. and B. Keller. 1992. Ultrastructural localization of a bean glycine-rich protein in unlignified primary walls of protoxylem cells. *Plant Cell* **4:** 773–783.

Scheres, B., C. van de Wiel, A. Zalensky, B. Horvath, H. Spaink, H. van Eck, F. Zwartkruis, A.-M. Wolters, T. Gloudemans, A. van Kammen, and T. Bisseling. 1990. The ENOD12 gene product is involved in the infection process during the pea-*Rhizobium* interaction. *Cell* **60:** 281–294.

Shedletzky, E., M. Shmuel, D.P. Delmer, and D.T.A. Lamport. 1990. Adaptation and growth of tomato cells on the herbicide 2,6-dichlorobenzonitrile leads to production of unique cell walls virtually lacking a cellulose-xyloglucan network. *Plant Physiol.* **94:** 980–987.

Shedletzky, E., M. Shmuel, T. Trainin, S. Kalman, and D. Delmer. 1992. Cell wall structure in cells adapted to growth on the cellulose-synthesis inhibitor 2,6-dichlorobenzonitrile. *Plant Physiol.* **100:** 120–130.

Sheng, J., R. D'Ovidio, and M.C. Mehdy. 1991. Negative and positive regulation of a novel proline-rich protein mRNA by fungal elicitor and wounding. *Plant J.* **1:** 345–354.

Sheng, J., J. Jeong, and M.C. Mehdy. 1993. Developmental regulation and phytochrome-mediated induction of mRNAs encoding a proline-rich protein, glycine-rich proteins, and hydroxyproline-rich glycoproteins in *Phaseolus vulgaris* L. *Proc. Natl. Acad. Sci.* **90:** 828–832.

Showalter, A.M. 1993. Structure and function of plant cell wall proteins. *Plant Cell* **5:** 9–23.

Showalter, A.M. and J.E. Varner. 1989. Plant hydroxyproline-rich glycoproteins. In *The biochemistry of plants: A comprehensive treatise* (ed. P.K. Stumpf and E.E. Conn), vol. 15, pp. 485–520. Academic Press, New York.

Showalter, A.M., A.D. Butt, and S. Kim. 1992. Molecular details of tomato extensin and glycine-rich protein gene expression. *Plant Mol. Biol.* **19:** 205–215.

Showalter, A.M., J. Zhou, D. Rumeau, S.G. Worst, and J.E. Varner. 1991. Tomato extensin and extensin-like cDNAs: Structure and expression in response to wounding. *Plant Mol. Biol.* **16:** 547–565.

Smith, R.C. and S.C. Fry. 1991. Endotransglycosylation of xyloglucan in plant cell suspension cultures. *Biochem. J.* **279:** 529–535.

Smith, J.J., E.P. Muldoon, and D.T.A. Lamport. 1984. Isolation of extensin precursors by direct elution of intact tomato cell suspension cultures. *Phytochemistry* **23:** 1233–1239.

Staehelin, L.A. and T.H. Giddings. 1982. Membrane-mediated control of cell wall microfibrillar order. In *Developmental order: Its origin and regulation* (ed. S. Subtelny and P.B. Green), pp. 133–147. A.R. Liss, New York.

Stafstrom, J.P. and L.A. Staehelin. 1986. Cross-linking patterns in salt-extractable extensin from carrot cell walls. *Plant Physiol.* **81:** 234–241.

––––––. 1987. A second extensin-like hydroxyproline-rich glycoprotein from carrot cell walls. *Plant Physiol.* **84:** 820–825.

––––––. 1988. Antibody localization of extensin in carrot cell walls. *Planta* **174:** 321–332.

Stevenson, T.T., A.G. Darvill, and P. Albersheim. 1988a. 3-Deoxy-D-*lyxo*-2-heptulosaric acid, a component of the plant cell-wall polysaccharide rhamnogalacturonan II. *Carbohydr. Res.* **179**: 269–288.

———. 1988b. Structural features of the plant cell-wall polysaccharide rhamnogalacturonan-II. *Carbohydr. Res.* **182**: 207–226.

Stiefel, V., L. Pérez-Grau, F. Albericio, E. Giralt, L. Ruiz-Avila, M.D. Ludevid, and P. Puigdomènech. 1988. Molecular cloning of cDNAs encoding a putative cell wall protein from *Zea mays* and immunological identification of related polypeptides. *Plant Mol. Biol.* **11**: 483–493.

Suzuki, H., T. Wagner, and M.L. Tierney. 1993. Differential expression of two soybean (*Glycine max* L.) proline-rich protein genes after wounding. *Plant Physiol.* **101**: 1283–1287.

Swords, K.M.M. and L.A. Staehelin. 1993. Complementary immunolocalization patterns of cell wall hydroxyproline-rich glycoproteins studied with the use of antibodies directed against different carbohydrate epitopes. *Plant Physiol.* **102**: 891–901.

Tagu, D., N. Walker, L. Ruiz-Avila, S. Burgess, J.A. Martìnez-Izquierdo, J.-J. Leguay, P. Netter, and P. Puigdomènech. 1992. Regulation of the maize HRGP gene expression by ethylene and wounding. mRNA accumulation and qualitative expression analysis of the promoter by microprojectile bombardment. *Plant Mol. Biol.* **20**: 529–538.

Taiz, L. 1984. Plant cell expansion: Regulation of cell wall mechanical properties. *Annu. Rev. Plant Physiol.* **35**: 585–657.

Takahashi, R., J. Hayashi, and U. Hiura. 1966. Inheritance and linkage studies in barley. III. Linkage studies of the gene for fragile stem-2 and orientation of linkage map on barley chromosome 5. *Ber. Ohara Inst. Landwirtsch. Forsch.* **13**: 199–212.

Takahashi, M., T. Kinoshita, and K. Takeda. 1968. Character expressions and casual genes of some mutants in rice plant. Genetical studies on rice plant XXXIV. *J. Fac. Agric. Hokkaido University* **55**: 496–512.

Takahashi, M., J. Yamamoto, S. Yasuda, and Y. Itano. 1953. Inheritance and linkage studies in barley. *Ber. Ohara Inst. Landwirtsch. Forsch.* **10**: 29–52.

Terry, M.E. and R.L. Jones. 1981. Soluble cell wall polysaccharides released from pea stems by centrifugation. *Plant Physiol.* **68**: 531–537.

Thomas, J.R., A.G. Darvill, and P. Albersheim. 1989. Rhamnogalacturonan I, a pectic polysaccharide that is a component of monocot cell walls. *Carbohydr. Res.* **185**: 279–305.

Tierney, M.L., J. Wiechert, and D. Pluymers. 1988. Analysis of the expression of extensin and p33-related cell wall proteins in carrot and soybean. *Mol. Gen. Genet.* **211**: 393–399.

Vance, C.P., T.K. Kirk, and R.T. Sherwood. 1980. Lignification as a mechanism of disease resistance. *Annu. Rev. Phytopathol.* **18**: 259–288.

van Holst, G.J. and J.E. Varner. 1984. Reinforced polyproline II conformation in a hydroxyproline-rich cell wall glycoprotein from carrot root. *Plant Physiol.* **74**: 247–251.

Varner, J.E. and G.I. Cassab. 1986. A new protein in petunia. *Nature* **323**: 110.

von Schaewen, A., A. Sturm, J. O'Neill, and M.J. Chrispeels. 1993. Isolation of a mutant *Arabidopsis* plant that lacks N-acetyl glucosaminyl transferase I, and is unable to synthesize Golgi-modified complex N-linked glycans. *Plant Physiol.* **102**: 1109–1118.

von Wettstein-Knowles, P.M. 1993. Waxes, cutin and suberin. In *Lipid metabolism in plants* (ed. T.S. Moore, Jr.), pp. 127–167. CRC Press, Boca Raton, Florida.

Warneck, H. and H.U. Seitz. 1993. Inhibition of gibberellic acid-induced elongation growth of pea epicotyls by xyloglucan oligosaccharides. *J. Exp. Bot.* **44**: 1105–1109.

Wilson, L.G. and J.C. Fry. 1986. Extensin—A major cell wall glycoprotein. *Plant Cell Environ.* **9:** 239–260.

Wu, H.-M., J. Zou, B. May, Q. Gu, and A.Y. Cheung. 1993. A tobacco gene family for flower cell wall proteins with a proline-rich domain and a cysteine-rich domain. *Proc. Natl. Acad. Sci.* **90:** 6829–6833.

Ye, Z.-H. and J.E. Varner. 1991. Tissue-specific expression of cell wall proteins in developing soybean tissues. *Plant Cell* **3:** 23–37.

Ye, Z.-H., Y.-R. Song, A. Marcus, and J.E. Varner. 1991. Comparative localization of three classes of cell wall proteins. *Plant J.* **1:** 175–183.

York, W.S., A.G. Darvill, and P. Albersheim. 1984. Inhibition of 2,4-dichlorophenoxyacetic acid-stimulated elongation of pea stem segments by a xyloglucan oligosaccharide. *Plant Physiol.* **75:** 295–297.

York, W.S., A.G. Darvill, M. McNeil, and P. Albersheim. 1985. 3-Deoxy-D-*manno*-2-octulosonic acid (KDO) is a component of rhamnogalacturonan II, a pectic polysaccharide in the primary cell wall of plants. *Carbohydr. Res.* **138:** 109–126.

Zhou, J., D. Rumeau, and A.M. Showalter. 1992. Isolation and characterization of two wound-regulated tomato extensin genes. *Plant Mol. Biol.* **20:** 5–17.

# 36

# Secondary Metabolism in *Arabidopsis*


**Clint C.S. Chapple**
Department of Biochemistry
Purdue University
West Lafayette, Indiana 47907

**Brenda W. Shirley**
Department of Biology
Virginia Polytechnic Institute and State University
Blacksburg, Virginia 24061

**Mike Zook, Ray Hammerschmidt,**
**and Shauna C. Somerville**
Department of Botany and Plant Pathology
Michigan State University
East Lansing, Michigan 48824


Plants cannot remove themselves from environments where they may be subjected to various insults, such as attack by pathogens, irradiation by UV light, or damage by herbivores. It is thought that to provide protection against the adverse effects of their environment, plants have acquired the ability to accumulate secondary metabolites. Many of these same compounds are also important to humans because of their pharmaceutical, insecticidal, and organoleptic qualities. *Arabidopsis* has been reported to accumulate 36 secondary metabolites belonging to four distinct classes. Flavonoids and hydroxycinnamic acid esters are derived from the general phenylpropanoid pathway (Fig. 1). The other two classes of secondary metabolites known to be produced by *Arabidopsis* are the glucosinolates and the indole phytoalexins.

More than 15,000 plant secondary metabolites have been identified, and this may represent only 5–10% of those that occur in nature (Wink 1988). How this diversity has arisen is unknown, as is the extent to which this diversity is represented within a single species, although some hypotheses have been offered (Williams et al. 1989; Jones and Firn 1991). For example, the pathways devoted to cyanogenic glycoside and glucosinolate biosynthesis appear to be closely related, yet they diverge to produce very different secondary metabolites. Cyanogenic glycosides have been found in more than 200 species of angiosperms (Conn 1980) but have never been found in species such as *Arabidopsis* that accumu-

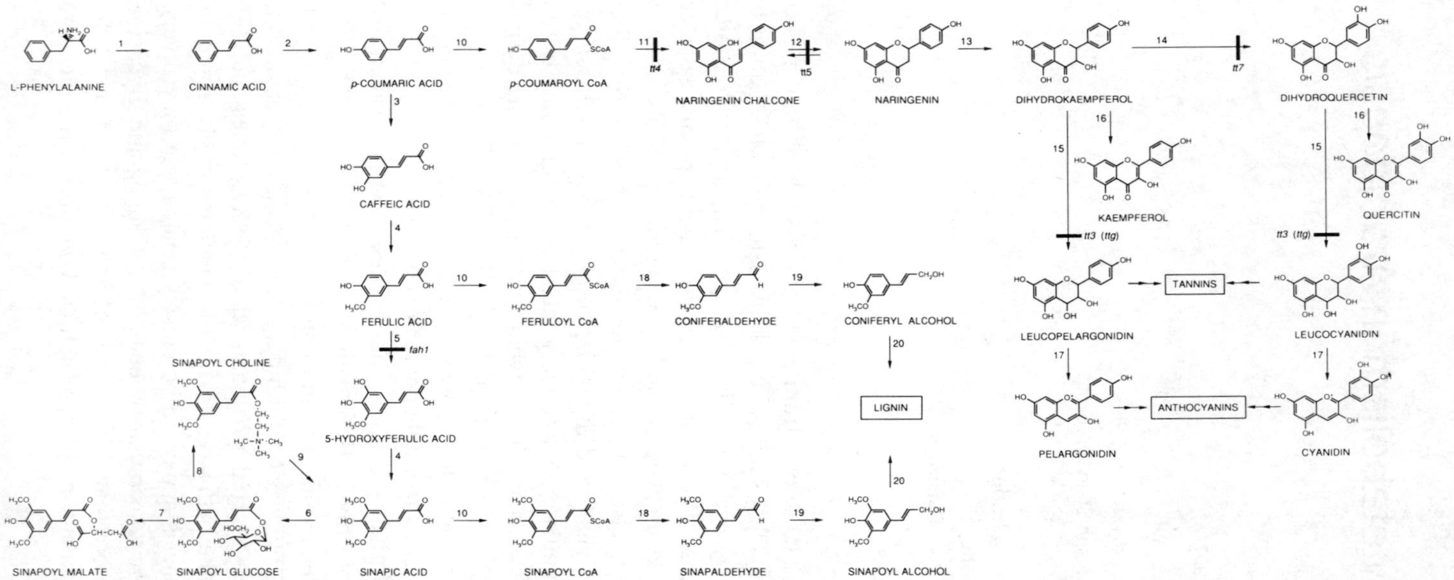

*Figure 1* General phenylpropanoid pathway and branch pathways leading to the synthesis of sinapic acid esters, flavonoids, and lignin in *Arabidopsis*. The positions of metabolic blocks that have been identified in *Arabidopsis* mutants are indicated by heavy lines, and the corresponding mutant locus designation. The steps of the pathway are numbered and are catalyzed by the indicated enzymes: (1) PAL; (2) cinnamate-4-hydroxylase (C4H); (3) *p*-coumarate-3-hydroxylase (C3H); (4) caffeate/5-hydroxyferulate *O*-methyltransferase (OMT); (5) ferulate-5-hydroxylase (F5H); (6) SGT; (7) SMT; (8) SCT; (9) SCE; (10) 4CL; (11) CHS; (12) CHI; (13) flavanone-3-hydroxylase (F3H); (14) flavanone-3′-hydroxylase (F3′H); (15) DFR; (16) flavanol synthase (FS); (17) anthocyanidin synthase (AS); (18) (hydroxy)cinnamoyl CoA reductase (CCR); (19) (hydroxy)cinnamoyl alcohol reductase (CAD); (20) peroxidase (POD), or laccase.

late glucosinolates. What has led to this specialization and mutual exclusivity? Are the genes for cyanogenic glycoside biosynthesis present in *Arabidopsis*, but unexpressed? If so, the biochemical characterization of the enzymes and structural genes of secondary metabolism and the factors that control their expression may reveal how these pathways have evolved, and how certain pathways may have been silenced. By selectively eliminating secondary pathways in *Arabidopsis* by mutation, it may become possible to characterize pathways that are expressed at low levels or to reveal pathways that are normally unexpressed.

In most cases, secondary metabolic pathways can be manipulated because the products of these pathways are dispensable, at least under controlled laboratory conditions. Thus, mutants can readily be obtained that are otherwise phenotypically normal. The fact that many secondary metabolites are either UV-absorbent, colored, or fluorescent makes them ideal endogenous reporters of metabolic pathways and the regulatory elements that control them. In this way, secondary metabolic pathways can serve as convenient models for other essential pathways.

*Arabidopsis* provides a unique opportunity to apply both molecular and classical genetic techniques to the study of the biosynthesis, regulation, and function of secondary metabolites. By applying these techniques, it should be possible to clone all of the genes involved in the major secondary metabolic pathways in *Arabidopsis*. With these tools in hand, it will be possible to dissect the regulation of these pathways in *Arabidopsis* and to explore secondary metabolic pathways and the genetic factors that control their expression in other plant species. An understanding of the control of secondary metabolism may enable the manipulation of pathways to overexpress economically or medicinally valuable secondary metabolites and to improve the resistance of plants to pests, pathogens, and UV light.

## THE FLAVONOID PATHWAY

### Introduction

The branch pathway leading from general phenylpropanoid metabolism to the production of flavonoid compounds is among the best-characterized of all metabolic pathways in plants (for review, see Stafford 1990). Among the reasons often cited for the extensive biochemical, genetic, and molecular analysis of this pathway are that the end products are not essential for plant growth, and at the same time, they provide easily scorable phenotypes such as altered pigmentation of seeds and flowers.

Flavonoids are widely distributed in the plant kingdom: More than

4000 different flavonoids have been identified in vascular plants and in bryophytes (Harborne 1988). Flavonoids are well known as the blue, purple, and red pigments present in flowers, where they function to provide cues for pollinators (Weiss 1991); leaves; fruits; seeds; and many other plant tissues. Flavonoid compounds also function in plants as protective agents against herbivores (Dixon et al. 1983; Hedin and Waage 1986), as UV light-absorbent filters (Stafford 1991; Li et al. 1993), and as signal molecules that induce bacterial nodulation, virulence, and toxin biosynthetic genes (Spencer and Towers 1988; Mo and Gross 1991; Fisher and Long 1992). Flavonoids may also function in regulating the stability and transport of auxins (Galston 1969; Jacobs and Rubery 1988), and recent work has uncovered a requirement for flavonoids in petunia and maize pollen development (Mo et al. 1992; van der Meer et al. 1992). Biological activities for flavonoids have also been identified in animal systems; see, e.g., Fujiki et al. (1986), Lueprasitsakul et al. (1990), De Meyer et al. (1991), and Popp and Schimmer (1991).

The genetics and molecular biology of flavonoid biosynthesis have been studied most extensively in maize, *Antirrhinum majus* (snapdragon), and petunia (Stafford 1990). The pathway provided the basis for Barbara McClintock's work on transposable elements and chromosome structure in maize and for the subsequent development of methods for gene tagging in maize (Fedoroff et al. 1984) and *Antirrhinum* (Coen et al. 1990). Consistent with the variety of functions associated with plant flavonoids, expression of genes encoding flavonoid biosynthetic enzymes is regulated in a tissue-specific and temporal manner (Koes et al. 1990; Schmid et al. 1990; Weiss et al. 1992), and in response to light (Kreuzaler et al. 1983; Chappell and Hahlbrock 1984; Feinbaum and Ausubel 1988; Schmelzer et al. 1988; Taylor and Briggs 1990; Fritze et al. 1991), gibberellic acid (Weiss et al. 1992), and sucrose (Weiss and Halevy 1989; Tsukaya et al. 1991).

Elevated expression of flavonoid biosynthetic genes can be induced by wounding, UV light, and fungal elicitors (Hahlbrock and Scheel 1989; Dooner and Robbins 1991). The chalcone synthase (CHS) gene encodes the first enzyme of flavonoid biosynthesis. Its transcriptional regulation has been extensively characterized in parsley and bean in conjunction with studies on UV induction of the genes encoding phenylalanine ammonia lyase (PAL) and *p*-coumarate Coenzyme A ligase (4CL) (Fig. 1) (Kreuzaler et al. 1983; Reimold et al. 1983; Chappell and Hahlbrock 1984; Ryder et al. 1984; Schmelzer et al. 1988; Lamb et al. 1989; Schulze-Lefert et al. 1989; Loake et al. 1991). From these studies, *cis*- and *trans*-acting regulatory elements required for the induction of these genes by UV and elicitors were identified (for review, see Hahlbrock and

Scheel 1989; Dixon et al. 1990; Dooner and Robbins 1991). Regulatory genes, including members of the R and B gene families in maize and the *delilah* gene in *Antirrhinum*, have also been isolated. The maize genes have homology with the *myc* and *myb* families of transcriptional activators and specifically interact with each other in vivo (Goff et al. 1992; Goodrich et al. 1992). The maize *viviparous-1* gene, which controls a variety of developmental responses associated with seed development, may represent an additional level of regulation. The gene product appears to encode a transcriptional activator required for expression of the flavonoid regulatory gene, *C1* (McCarty et al. 1989, 1991). A very similar protein is encoded by the *Arabidopsis ABI3* locus (Giraudat et al. 1992), although its role, if any, in regulating expression of the flavonoid pathway has not been established (J. Giraudat, pers. comm.).

### Flavonoid Mutants of *Arabidopsis*

The basic flavonoid skeleton is synthesized from *p*-coumaroyl Coenzyme A (CoA) and 3 molecules of malonyl CoA (Fig. 1). The subsequent reactions involve an array of ring closure, hydroxylation, and reduction reactions that lead to the flavanones (naringenin), 3-hydroxyflavanones (dihydrokaempferol and dihydroquercetin), flavonols (kaempferol and quercetin), flavan-3,4-diols (leucopelargonidin and leucocyanidin), and 3-hydroxyanthocyanidins (pelargonidin and cyanidin) that are found in *Arabidopsis*. An *O*-methyltransferase and several glycosyltransferases are required for the further conversion of 3-hydroxyanthocyanidins to the anthocyanins found in *Arabidopsis*.

A fact that has been crucial to the isolation of *Arabidopsis* flavonoid mutants is that the flavan-3,4-diols are also the precursors for the condensed tannins, the pigments found in the *Arabidopsis* seed coat. Thus, mutations affecting the biosynthesis of flavonoids have collectively been named *transparent testa* (*tt*) because they result in a decreased pigmentation of the seed coat (testa). As a result, seeds produced by *transparent testa* homozygotes are yellow (*tt1*, *tt2*, *tt3*, *tt4*, *tt5*, *tt8*, *ttg*) or pale-brown (*tt6*, *tt7*, *tt9*, *tt10*) instead of the dark-brown color of wild-type seeds. The first *Arabidopsis* flavonoid mutations were identified by Bürger (1971). Koornneef subsequently characterized a family of eleven such loci that had been identified in mutant populations generated using chemical mutagenesis and ionizing radiation (Koornneef 1981, 1990; Koornneef et al. 1982). Eight of the loci were mapped and incorporated into the first linkage map of *Arabidopsis* (Koornneef et al. 1983). Four *transparent testa* mutants (*tt1*, *tt2*, *tt9*, *tt10*) appear to produce normal levels of anthocyanins in vegetative tissues, and therefore may identify tissue-specific regulatory genes, members of a differentially expressed structural gene family, or structural genes required only for the synthesis of the condensed tannins. The other *transparent testa* mutants exhibit a reduction (*tt6*, *tt7*, *tt8*) or absence (*tt3*, *tt4*, *tt5*, *ttg*) of pigmentation in leaves, stems, and flowers, indicating that at least part of the pathway involves single loci that function in multiple tissues (Koornneef 1990). One of the *transparent testa* mutants exhibits a particularly interesting pleiotropic phenotype. In *ttg* (transparent testa-glabrous) homozygotes, flavonoid synthesis is disrupted in both seeds and vegetative tissues, and the plants are defective in trichome development (Bürger 1971; Koornneef 1981). This glabrous phenotype appears to be identical to what has been described for the nonpleiotropic glabra (*gl*) mutants. The *ttg* mutation also shares an absence of mucilage on the surface of the seeds with one of the three *gl* mutations. Seven independent alleles of the *ttg* locus have been isolated, and all exhibit this combination of phenotypes, indicating that these diverse phenotypes are determined by the *ttg* gene product. The simplest explanation is that *ttg* encodes a regulatory factor that functions in different pathways. The only other plant in which mutations affecting both flavonoid synthesis and trichome development have been identified is the crucifer, *Matthiola incana* (Koornneef 1990), suggesting that analysis of the *Arabidopsis ttg* locus may elucidate a novel aspect of flavonoid gene regulation that is unique to the Brassicaceae.

Thin layer chromatography has been used to characterize the flavonoids in *Arabidopsis* and the nature of the defects in a number of the *tt* mutants. Koornneef showed that *tt7* was clearly correlated with a defect in flavonoid 3′-hydroxylase activity (Koornneef et al. 1982), whereas *tt3* and *ttg* mutations appeared to affect dihydroflavonol reductase activity (Koornneef 1990). The mutant *tt4*, which contained neither flavonols nor anthocyanins, was blocked at a very early step in the pathway (M. Koornneef, pers. comm.). Whereas cyanidin is the major anthocyanidin in all vegetative tissues of *Arabidopsis*, seeds and flowers preferentially accumulate the flavonols, quercetin and kaempferol, respectively (M. Koornneef, pers. comm.; G. Storz and F.M. Ausubel, unpubl.). In addition, tannins formed from polymerization of cyanidin have been found in *Arabidopsis* seeds, but pelargonidin-containing tannins have not (T. Carron, pers. comm.). One explanation for these findings is the substrate specificity exhibited by dihydroflavonol reductase (DFR) enzymes. The sequence of a domain predicted to control the substrate specificity of DFR is similar in *Arabidopsis* and *Antirrhinum*, both of which accumulate cyanidin derivatives (Shirley et al. 1992). This same domain is significantly different in petunia, where DFR has a strong specificity for the conversion of dihydromyricetin into the precursor of pelargonidin.

**Molecular Biology of *Arabidopsis* Flavonoid Biosynthesis**

CHS is encoded by a single-copy gene in *Arabidopsis* (Feinbaum and Ausubel 1988), similar to the situation in maize and *Antirrhinum* but in contrast to petunia, which contains 8–10 CHS genes (Koes et al. 1989), and bean, which has at least 7 copies (Ryder et al. 1987). Restriction fragment length polymorphism (RFLP) mapping suggested that CHS corresponded to the *tt4* locus. This was subsequently supported by sequence analysis of two independent ethyl methane sulfonate (EMS)-induced alleles, which were found to contain different point mutations in the CHS transcribed region (B.W. Shirley et al., unpubl.).

Chalcone flavanone isomerase (CHI) and DFR also appear to be encoded by single-copy genes in *Arabidopsis* (Shirley et al. 1992), even though multiple copies of CHI and DFR have been found in petunia (van Tunen et al. 1988; Beld et al. 1989). RFLP mapping suggested that these genes corresponded to previously identified *tt* mutants: *tt5* for CHI and *tt3* for DFR. Genomic Southern blots, RFLP analysis, and sequence analysis of radiation-induced *tt3* and *tt5* alleles uncovered substantial chromosomal rearrangements involving these genes, as well as other regions several centimorgans away on the same chromosome (30 cM on chromosome 3 for CHI and 2.7 cM on chromosome 5 for DFR). These results also provided information about the structure of ionizing radiation-induced mutations and the prospects for their use in gene cloning methods such as chromosome walking and genomic subtraction.

The expression of the CHS, CHI, and DFR genes has been examined in the eleven *tt* mutants (B.W. Shirley et al., unpubl.). Lack of expression of CHS, CHI, or DFR does not affect the mRNA levels for the other structural genes, including PAL, despite the altered synthesis of precursor molecules and accumulation of products (Shirley et al. 1992; B.W. Shirley et al., unpubl.). This is the case even in the DFR deletion mutation, *tt3* (M218), which provides a null background for the conversion of flavonols into anthocyanidins (Fig. 1). This result contrasts with findings of effects on CHS promoter activity in response to phenylpropanoid intermediates in cultured bean cells (Loake et al. 1991). It will be of interest to determine whether these or other *transparent testa* mutations affect other steps in the pathway at the level of transcription or enzyme activity.

**Regulation of *Arabidopsis* Flavonoid Biosynthesis**

Analyses of CHS, CHI, and DFR mRNA levels in *Arabidopsis tt* mutations have identified two novel regulatory interactions (B.W. Shirley et al., unpubl.). In *tt8* and *ttg*, CHS and CHI mRNA levels are similar to those in wild-type plants, but expression of the DFR gene is reduced or absent, analogous to what has been observed for the R mutants in maize

(Taylor and Briggs 1990) and *delilah* in *Antirrhinum* (Goodrich et al. 1992). Lloyd et al. (1992) recently introduced R sequences into *Arabidopsis* plants, which resulted in the high-level accumulation of flavonoid pigments in various tissues, as well as overproduction of trichomes on leaves and stems. Taken together, these results have led to the speculation (B.W. Shirley et al., unpubl.) that *ttg* identifies a member of the Myc family of regulators that controls flavonoid biosynthesis and trichome development by interacting with different Myb-like factors, which might even include the product of the *gl1* gene, a Myb homolog involved in *Arabidopsis* trichome development (Oppenheimer et al. 1991). The cloning of the *TTG* gene (M. Walker and J. Gray, pers. comm.) should provide additional insights into the mechanism of flavonoid gene regulation in *Arabidopsis*.

As in other plant species, control of flavonoid biosynthesis in response to light occurs in *Arabidopsis*, at least in part, at the level of transcription of the structural genes. In nuclear run-on assays, transcription of CHS sequences was induced in response to high-intensity light treatment (Feinbaum and Ausubel 1988). Analysis of mRNA levels in *Arabidopsis* plants transformed with CHS promoter-*GUS* fusions (CHSp-*GUS*) was used to show that this promoter was activated in response to blue light, probably via a specific blue-light receptor (Feinbaum and Ausubel 1988; Feinbaum et al. 1991). Analysis of promoter deletion constructs in seedlings indicated that separate elements of the CHS promoter mediate the response to blue and high-intensity white light. Blue light also appears to be the major component of the light spectrum responsible for expression of CHS, CHI, and DFR mRNAs during seedling development (Feinbaum et al. 1991; Kubasek et al. 1992). Expression of these genes is further enhanced when seedlings are exposed to UVB light. Mutations at the *Arabidopsis hy* loci, which interfere with inhibition of hypocotyl elongation in white light, do not affect expression of CHS, CHI, and DFR genes (G. Storz et al., unpubl.); however, two other mutations that interfere with blue-light-dependent inhibition of hypocotyl elongation exhibit altered levels of flavonoid biosynthesis (Liscum and Hangarter 1991). Studies with *Arabidopsis det* and *cop* mutants, which uncouple seedling development from the light response, have identified a putative negative regulatory locus required for the inactivation of flavonoid genes in darkness and in specific tissues (Chory and Peto 1990; Chory et al. 1991; Wei and Deng 1992). Thus, although activation of expression of the flavonoid genes appears to involve a photoreceptor other than phytochrome, the primary photoreceptor for the *rbcS* and *CAB* genes, all of the genes appear to be controlled by a common negative regulator.

*Arabidopsis* PAL, CHS, CHI, and DFR clones have been used to examine coordinate expression during seedling development (Kubasek et al. 1992). All four genes exhibited a peak in mRNA accumulation 3 days after the start of germination, which coincides with localized accumulation of flavonoid pigments in stems and cotyledons. These experiments also uncovered a sequential induction of the PAL, CHS, CHI, and DFR genes when dark-grown seedlings were treated with blue or UVB light.

## Future Prospects

*Arabidopsis* provides new opportunities for extending the analysis of flavonoid biosynthesis in higher plants. In this model plant, the simplicity of the flavonoid structural gene families and the availability of allelic series of mutations in flavonoid biosynthetic loci can be integrated with methods such as chromosome walking and genomic subtraction, which make it possible to isolate novel genes that have been identified solely by a mutant phenotype (Arondel et al. 1992; Giraudat et al. 1992; Sun et al. 1992). These methods are already being applied to several *transparent testa* loci, including *ttg* (M. Walker and J. Gray, pers. comm.), *tt1*, and *tt2* (B. Shirley, unpubl.). Efforts are also being made to identify additional mutations that affect the flavonoid pathway in general or that specifically target the regulation of flavonoid biosynthesis. For example, a genetic system for identifying mutations that affect transcriptional regulation of the flavonoid pathway has been developed. In these experiments, seeds from transgenic *Arabidopsis* plants carrying CHSp-*GUS* fusions are mutagenized, and the resulting $M_2$ seedlings are screened for aberrant expression of *GUS* activity in response to blue light (Feinbaum et al. 1991) or during seedling development (Kubasek et al. 1992). This genetic approach could also elucidate other aspects of regulation such as the stress response in plants, since high-intensity light, UV light, heat shock (Shirley and Goodman 1993), and nutrient stress all induce flavonoid gene expression, perhaps through common regulatory pathways. Novel approaches for analyzing flavonoid synthesis are also being used to screen for plants containing T-DNA insertions that disrupt flavonoid biosynthesis (J. Sheahan and K. Davis, pers. comm.). These include analysis of high-performance liquid chromatography (HPLC) profiles of soluble aromatic secondary metabolites in crude extracts (Graham 1991) and a novel staining technique using diphenyl boric acid 2-amino-ethyl ester that provides differential staining of *tt4*, *tt5*, *tt6*, and *tt7* mutations (Sheahan and Rechnitz 1992, 1993). Novel mutations have also been uncovered in mutant searches for other characters, including embryonic lethals that exhibit dark red cotyledons (Errampalli et al.

1991) and *blu*, *det*, and *cop* mutations that affect both photomorphogenesis and flavonoid biosynthesis (Chory and Peto 1990; Chory et al. 1991; Liscum and Hangarter 1991; Wei and Deng 1992). Recent results showing that the *Arabidopsis* CHS and CHI mutants, *tt4* and *tt5*, are more sensitive than wild-type plants to UVB radiation suggest that screens for UV-sensitive mutants may also uncover mutations in flavonoid biosynthetic loci. These experiments may reveal the importance of flavonoids and other phenylpropanoid compounds involved in UVB protection. A very different approach using the two-hybrid system, in which fusion proteins are used to test for transcriptional activation of reporter genes in yeast, may provide yet another method for isolating novel flavonoid structural genes from *Arabidopsis* (B. Shirley, unpubl.). The potential for isolating mutations in a diversity of structural and regulatory genes, cloning the corresponding genes, and performing reconstruction studies in transgenic plants suggests that *Arabidopsis* may greatly facilitate further analyses of the regulation and function of flavonoids in higher plants.

## PHENYLPROPANOID ESTERS

### Introduction

Many plants are known to accumulate simple phenylpropanoid derivatives in addition to the flavonoids and anthocyanins. Members of the Brassicaceae specialize in the accumulation of hydroxycinnamic acid esters of glucose, malic acid, and choline. These are dominated by sinapic acid esters (Linscheid et al. 1980), although similar esters of *p*-coumarate, caffeate, ferulate (Brandl et al. 1984), and even isoferulate (Gmelin and Kjær 1970) have also been reported. Sinapoyl choline, also known as sinapine, is widely found throughout the Brassicaceae (Bouchereau et al. 1991). Although we have an extensive understanding of the molecular aspects of flavonoid biosynthesis and regulation, our understanding of sinapate ester metabolism is restricted to the enzymology of their biosynthesis.

*Arabidopsis* shows a typical profile of sinapic acid esters. These compounds were first reported by Wintersohl et al. (1979) and were recently conclusively identified as sinapoyl malate, sinapoyl glucose, and sinapoyl choline (Chapple et al. 1992). The biosynthesis of these compounds has been extensively studied in *Raphanus sativus* (radish). The sinapic acid esters are of interest for a number of reasons. First, their accumulation is developmentally regulated. Sinapoyl malate is accumulated only in leaf tissue, whereas in seeds, sinapoyl choline is the dominant ester. In germinating seedlings, sinapoyl glucose accumulates to the highest level (Strack 1982). Second, sinapoyl malate is accumulated primarily in the leaf epidermis (Strack et al. 1985). Given the high UV extinction coefficient of the sinapate moiety, it is reasonable to suspect that sinapate esters may play a role in UV resistance in *Arabidopsis*. Finally, sinapate esters may play a role in plant/pathogen interactions. As signaling molecules, the role of molecules with the syringyl substitution pattern of sinapic acid has been well established in *vir* gene induction in *Agrobacterium* (Stachel et al. 1985; Spencer and Towers 1988) and as haustorium inducers in plant parasitism (Lynn and Chang 1990). As potential precursors of lignin or other polyphenolics, sinapate esters may participate in the *Arabidopsis* plant defense response in ways that have been suggested for phenolics in other species (Nicholson and Hammerschmidt 1992).


## Structure and Biosynthesis

Sinapic acid is synthesized as the final product of the general phenylpropanoid pathway (Fig. 1). The biosynthetic steps specifically devoted to sinapic acid ester biosynthesis have been elucidated from studies in *Raphanus*, and an identical pattern of sinapate ester interconversions occurs in *Arabidopsis* (T. Vogt and B. Ellis, unpubl.).

In the first reaction committed to sinapate ester biosynthesis, sinapic acid is esterified to glucose to form 1-*O*-sinapoyl glucose by a cytoplasmic sinapic acid:UDP-glucose sinapoyl transferase (SGT) (Strack 1980). The resulting 1-*O*-ester is an "energy-rich" compound, with a high free energy of hydrolysis (Mock and Strack 1993); thus, it can act as a donor of the sinapate moiety to various acceptors. In leaf tissue of *Raphanus* and *Arabidopsis*, the acceptor molecule is malic acid. The conversion of sinapoyl glucose to sinapoyl malate is catalyzed by sinapoyl glucose:malate sinapoyl transferase (SMT) (Gräwe et al. 1992), an enzyme localized in the vacuole (Sharma and Strack 1985). In developing seeds, the final product of sinapate ester metabolism is sinapoyl choline. This compound is also synthesized via sinapoyl glucose, with choline rather than malate acting as the final sinapic acid acceptor. This reaction is catalyzed by sinapoyl glucose:choline sinapoyltransferase (SCT) (Strack et al. 1983). In seeds of *Arabidopsis*, this reaction does not proceed to completion, and a substantial pool of sinapoyl glucose accumulates (Chapple et al. 1992).

In the first 3 days following germination, the activity of a specific esterase, sinapoyl choline esterase (SCE), increases and brings about the hydrolysis of sinapoyl choline to sinapate and choline (Strack et al. 1980). In *Raphanus*, this choline supports the synthesis of the phospholipids during early seedling growth (Strack 1981). After a lag period of 1–2 days, the activity of SGT increases, re-esterifying the released sinapate to form sinapoyl glucose. At this time, there is a transient increase in the sinapoyl glucose pool in the young seedling. Four to five days after germination, there is a rise in SMT activity and a subsequent conversion of the accumulated sinapoyl glucose to sinapoyl malate, thus establishing the normal secondary metabolite pool observed in mature plants.

Sinapoyl glucose synthesis has also been reported in *Daucus carota* cell cultures (Halaweish and Dougall 1990), where the major anthocyanin that is accumulated is cyanadin 3-(sinapoylxylosylglucosyl-galactoside) (Harborne et al. 1983). The anthocyanin acylation reaction utilizes sinapoyl glucose as the acyl donor (Glässgen and Seitz 1992), and this acylation is critical for the uptake of the anthocyanins into the vacuole (Hopp and Seitz 1987). The anthocyanins of *Sinapis alba*, a member of the Brassicaceae, are also acylated with sinapic acid (Takeda et al. 1988). Similarly, hydroxycinnamic acid glucose esters are involved in betacyanin (Bokern and Strack 1988) and chlorogenic acid (Villegas and Kojima 1986) biosynthesis, where they function as activated acyl donors in these secondary metabolic pathways.

### Phenylpropanoid Ester Mutants of *Arabidopsis*

Sinapic acid esters are readily identified by their characteristic UV-induced fluorescence after thin-layer chromatography of methanolic *Arabidopsis* leaf or seed extracts. This method has been used to isolate two classes of *Arabidopsis* mutants defective in sinapate ester biosynthesis.

*Arabidopsis* mutants defective at one of these loci, *fah1*, fail to accumulate sinapoyl esters of any kind in either leaf or seed tissue (Chapple et al. 1992) and are thought to be defective in the activity of ferulate-5-hydroxylase, a cytochrome P-450-dependent monooxygenase (Grand 1984). Sequence flanking the T-DNA insertion in a tagged *fah1* mutant has high homology with other plant cytochrome P-450-dependent monooxygenases, suggesting that the insertion is within the structural gene for ferulate-5-hydroxylase (C. Chapple and C. Somerville, unpubl.). Although this mutant was originally referred to as *sin1*, it is now known as *fah1* (*ferulic acid hydroxylase*) to avoid conflict with another mutant previously referred to as *sin1* (Robinson-Beers et al. 1992).

The defect in the *fah1* mutant results in a number of phenotypes. As previously mentioned, sinapoyl malate is accumulated primarily in epidermal vacuoles (Strack et al. 1985). This distribution can be visualized in wild-type *Arabidopsis* when the plants are examined under long-wave UV light. The adaxial (upper) leaf surfaces of wild-type plants appear pale blue/green due to the sinapoyl malate fluorescence in their epidermis, and their abaxial leaf surfaces exhibit a dark red fluorescence due to the chlorophyll in the subtending mesophyll. This difference between upper and lower leaf surfaces is consistent with a role for sinapoyl malate in UV protection in *Arabidopsis*. In the *fah1* mutant, both upper and lower leaf surfaces exhibit red chlorophyll fluorescence (Chapple et al. 1992). Thus, there may be a significantly increased penetration of UVB into the mesophyll of *fah1* mutant leaves, and initial results indicate that the mutant plants exhibit a UVB-sensitive phenotype (R. Last and C. Chapple, unpubl.).

Analysis of another mutant of *Arabidopsis* supports the role of sinapoyl malate in UVB resistance. The *tt5* mutant of *Arabidopsis* is blocked in CHI (Shirley et al. 1992) and therefore does not accumulate flavonoids or anthocyanins. The *tt5* mutant is the most UV-sensitive of all of the *tt* mutants (Li et al. 1993), despite the fact that the *tt4* mutant blocked in the CHS structural gene (Chang et al. 1988) has a similarly attenuated flavonoid accumulation. The UV-sensitive phenotype may be due to the fact that the *tt5* mutant also has significantly reduced levels of sinapoyl malate in its leaves (Li et al. 1993). These data suggest that sinapate esters play an important role in UVB resistance in *Arabidopsis*, despite the fact that flavonoids are most often credited as being the UV-absorptive components involved in UV resistance in plants. It is not clear how a defect in flavonoid biosynthesis leads to lowered levels of sinapoyl malate. If naringenin chalcone accumulates in the *tt5* mutant, it may exert regulatory effects on the general phenylpropanoid pathway similar to those previously reported for cinnamic acid (Mavandad et al. 1990; Loake et al. 1991) and *p*-coumaric acid (Loake et al. 1991, 1992). Interestingly, the *fah1* mutant also has lower levels of flavonoids than wild type (Li et al. 1993), indicating that there may be a number of interactions enforcing coordinate regulation of sinapate ester and flavonoid biosynthesis.

A second mutant with altered sinapoyl ester metabolism accumulates sinapoyl glucose instead of sinapoyl malate in leaf tissue (C. Chapple and C. Somerville, unpubl.). This mutant is less well characterized than *fah1*, but it is likely that this mutant is defective in SMT (Fig. 1), the enzyme responsible for the final step in sinapoyl malate biosynthesis.

**Future Prospects**

It is very likely that sinapoyl malate is important in UVB resistance, but many other questions can be asked about the role of sinapate esters in

*Arabidopsis*. Is there an impact of the lack of sinapoyl esters on plant pathogen interactions, or on feeding by herbivorous insects? Are sinapate esters mobilized as part of the plant defense response against pathogens, and if so, what is the impact of their absence? Do these esters serve as precursors for cell wall cross-linking polymers?

It is noteworthy that the *fah1* mutant does not make feruloyl malate in its leaf tissue. *Raphanus* accumulates a variety of hydroxycinnamoyl malates, including feruloyl malate (Nielsen et al. 1984). The enzymes of sinapate ester biosynthesis in *Arabidopsis* (SGT, SMT, SCT) show significant activity toward ferulate when assayed in vitro and are present at wild-type levels in leaf and seed tissue of the *fah1* mutant (T. Vogt et al., unpubl.). The fact that the mutants do not accumulate feruloyl malate in their leaves may indicate that channeling of intermediates prevents ferulic acid from having access to the SGT active site, and thus feruloyl glucose and feruloyl malate are not accumulated. If this is the case, it would be important to learn how this pathway is structured and to learn how the cell metabolizes ferulic acid that cannot be converted to soluble esters.

The enzymes of sinapate ester biosynthesis are interesting because they catalyze a developmentally regulated pattern of secondary metabolite synthesis and turnover, which may reflect the roles of each of the metabolites within this family of compounds. The examination of these reactions and metabolites in *Arabidopsis* provides an opportunity to test hypotheses concerning the roles of these metabolites and the importance of their regulation. The availability of the cloned *FAH1* gene will greatly facilitate the analysis of this pathway at the molecular genetic level in relation to UVB stress, development, lignification, and its coordinate control with the flavonoid biosynthetic pathway. Finally, because the accumulation of these sinapate esters is developmentally regulated and tissue specific, these metabolites may be useful as endogenous reporters for more global developmental programs.

## GLUCOSINOLATES

### Introduction

Glucosinolates, also known as mustard oil glycosides, are secondary metabolites that are characteristic of all members of the Brassicaceae thus far examined, including *Arabidopsis* (Fig. 2). These compounds are sulfated thioglucosides whose degradation products are responsible for the characteristic flavor of *Brassica* and allied genera such as cabbage, condiment mustards, and horseradish. Many comprehensive reviews of glucosinolate biosynthesis and biological impact (Underhill 1980; Larsen

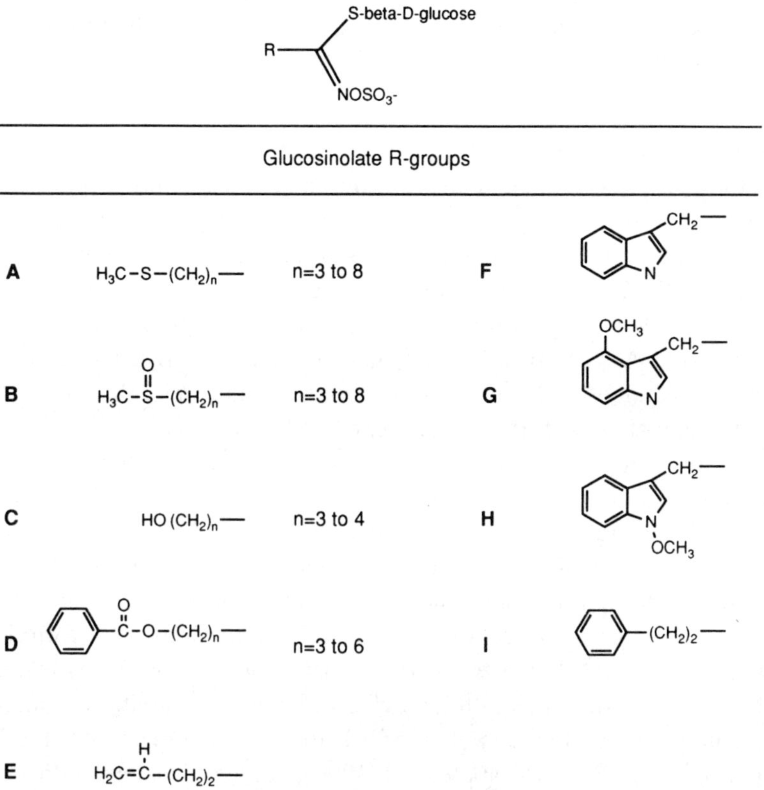

*Figure 2*  Core glucosinolate structure, and the glucosinolates found in *Arabidopsis*. The core glucosinolate structure is shown at top, with an "R" substituent indicating the variable moiety of the glucosinolates. Those "R" groups present in *Arabidopsis* are indicated as *A–I*. (*A*) Methylthioalkyl glucosinolate family; (*B*) methylsulfinylalkyl glucosinolate family; (*C*) ω-hydroxyalkyl glucosinolate family; (*D*) benzoyloxyalkyl glucosinolate family; (*E*) but-3-enyl glucosinolate; (*F*) indolyl-3-methyl glucosinolate; (*G*) 4-methoxyindolyl-3-methyl glucosinolate; (*H*) 1-methoxyindolyl-3-methyl glucosinolate; (*I*) 2-phenethyl glucosinolate.

1981) and a number of extensive surveys of their distribution have been published (Mithen et al. 1987; Sang and Salisbury 1988; Daxenbichler et al. 1991).

Glucosinolates are named according to their side-chain substituent, followed by the suffix glucosinolate to denote the core structure shown in Figure 2. There is also a trivial nomenclature based on the names of the species from which the glucosinolates were first isolated, giving rise to names such as sinigrin, glucotropaeolin, gluconasturtiin, but these names should be abandoned in favor of the systematic nomenclature.

Well over 75 different glucosinolates have been identified, varying from the structure shown in Figure 2 almost exclusively by the nature of their "R" substituent.

## Structure and Biosynthesis

All glucosinolates are synthesized from amino acids; however, many are derived from nonprotein amino acids that are longer chain homologs of the parent amino acids synthesized by a chain extension pathway (Fig. 3) (Underhill 1980). In this series of reactions, the parent amino acid is transaminated to its corresponding $\alpha$-keto acid, condensed with acetyl Coenzyme A to form a 2-substituted malic acid derivative, dehydrated and rehydrated to form a 3-substituted malate, and then oxidatively decarboxylated to form a chain-extended $\alpha$-keto acid (Fig. 3). These reactions are analogous to the conversion of oxaloacetate to $\alpha$-ketoglutarate in the citric acid cycle and the biosynthesis of leucine from $\alpha$-ketoisovalerate, the $\alpha$-keto acid of valine. Following this series of reactions, the chain-extended $\alpha$-keto acid can be transaminated to the corresponding amino acid and directed into glucosinolate biosynthesis, or it can be subjected to further rounds of chain extension to form longer homologs. In *Arabidopsis*, chain extension of up to 6 cycles leading to the synthesis of 8-methylthiooctyl and 8-methylsufinyloctyl glucosinolate may occur (Hogge et al. 1988). Unfortunately, apart from the initial transamination step (Chapple et al. 1990), none of the steps of this chain-extension pathway has been demonstrated in vitro, and the analogous reactions have not even been thoroughly studied in the context of leucine metabolism in plants (Oaks 1965).

The second stage of glucosinolate biosynthesis is the five-step synthesis of the core glucosinolate structure (Larsen 1981). The first two

*Figure 3* Generalized glucosinolate biosynthetic pathway, including the chain extension pathway for the synthesis of precursor amino acid homologs. The steps of the biosynthetic pathway are numbered and are catalyzed by the indicated enzymes: (1) amino acid:$\alpha$-keto acid aminotransferase; (2) 2-alkyl/aryl malate synthase; (3) 2-alkyl/aryl malate isomerase; (4) 3-alkyl/aryl malate dehydrogenase (*Note*: The amino acid R group is now equivalent to the previous R group with one methylene ($-CH_2-$) unit added); (5) amino acid:$\alpha$-keto acid aminotransferase; (6) amino acid N-hydroxylase; (7) N-hydroxyamino acid decarboxylase/dehydrogenase; (8) thiohydroximate synthase? (no name has previously been coined for this enzyme); (9) UDP-glucose:thiohydroximate glucosyltransferase; (10) 3′-phosphoadenosine-5′-phosphosulfate:desulfoglucosinolate sulfotransferase. The degradation of glucosinolates is catalyzed by (11) thioglucoside glucohydrolase (myrosinase); (12) spontaneous reaction.

steps are the N-hydroxylation of the parent amino acid or its chain-extended homolog, probably by a cytochrome P-450-dependent mono-oxygenase (Kindl 1968), and oxidative decarboxylation of the N-hydroxy amino acid to the corresponding aldoxime. The aldoxime is the substrate for the donation of sulfur from a biochemically undefined donor (Wetter 1964; Matsuo 1968; Wetter and Chisholm 1968) to form a thiohydroximate. The enzymes catalyzing the final two steps in gluco-

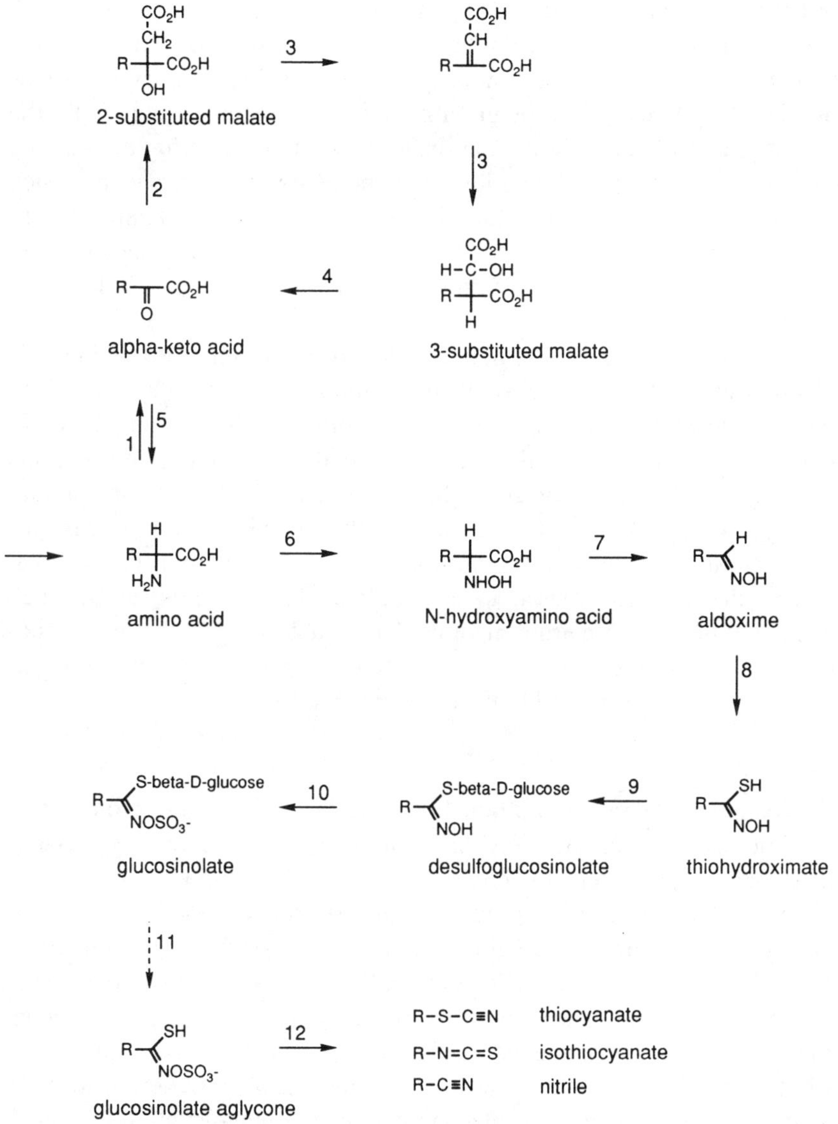

*Figure 3  (See facing page for legend.)*

sinolate biosynthesis, UDP-glucose:thiohydroximate glucosyltransferase and 3′-phosphoadenosine-5′-phosphosulfate:desulfoglucosinolate sulfotransferase catalyze the glucosylation and subsequent sulfation of the thiohydroximate (Matsuo and Underhill 1971; Glendening and Poulton 1990) resulting in the core glucosinolate structure (Fig. 3).

There is often an additional phase to glucosinolate biosynthesis that involves side-chain modification. For example, the methionine-derived glucosinolates found in *Arabidopsis* (Hogge et al. 1988) may terminate in the methylthio group derived from their parent amino acid, a methylsulfinyl or methylsulfonyl group derived through S-oxidation of the parent methylthioalkylglucosinolate (Chisholm 1972), or a terminal double bond synthesized through elimination of methanethiol from the parent glucosinolate (Parry and Naidu 1982) or desulfoglucosinolate (Rossiter et al. 1990). In addition, side chains can be hydroxylated, such as in 2-hydroxybut-3-enylglucosinolate (Rossiter and James 1990). These hydroxylations are important, since it is these compounds that are capable of cyclizing during degradation to form toxic isoxazolidine-2-thiones (van Etten 1969).

Glucosinolates are found in leaves and at even higher levels in seeds of crucifers, but seed glucosinolate accumulation is not determined by the genetic makeup of the embryo (Kondra and Stefansson 1970). In *Arabidopsis*, the *gsm1-1* allele reduces specific leaf glucosinolate levels and has a similar effect on seed glucosinolate levels, but acts maternally (Haughn et al. 1991). Gijzen et al. (1989) studied *Brassica napus* embryos cultured in vitro and found that once removed from the maternal parent, the embryos ceased to accumulate glucosinolates, unless they were supplied in the culture medium. These data suggest that glucosinolates are probably synthesized by the maternal parent, possibly in the silique wall, and transported to the developing seed.

### Degradation

Glucosinolate degradation products, the mustard oils, figure prominently in biological and agronomic considerations of these secondary metabolites because of their toxicity when present at high concentration. Glucosinolate degradation products (Fig. 3) arise through the action of myrosinase, a thioglucoside glucohydrolase that has been found in every glucosinolate-containing plant examined (Björkman 1976). The enzyme hydrolyzes glucosinolates to form glucose and an unstable aglycone. Most aglycones spontaneously rearrange to yield sulfate and a substituted isothiocyanate via a Løssen rearrangement; however, at low pH, glucosinolate aglycones can decompose to give elemental sulfur and a substituted nitrile. Indole glucosinolates degrade via isothiocyanates;

however, these derivatives are highly unstable in aqueous environments (Hanley and Parsley 1990) and in turn decompose to yield indole-3-carbinol and 3,3-diindolylmethane and thiocyanate ion (SCN⁻) (McGregor 1978). Alkyl glucosinolates that are hydroxylated on their "R" group spontaneously cyclize during degradation to form substituted isoxazolidine-2-thiones.

In vivo, the myrosinase-catalyzed degradation of glucosinolates is prevented by compartmentation. Glucosinolates are localized in the vacuole, whereas myrosinase is found within specialized cells (idioblasts) known as myrosin cells (Thangstad et al. 1990). This differential compartmentation has been referred to as "the mustard oil bomb" (Lüthy and Matile 1984) to describe the consequences of cellular disruption leading to the mixing of myrosinase with its substrates and the release of the toxic degradation products. Myrosinase is localized within the idioblasts in protein bodies known as myrosin grains (Höglund et al. 1992). Although these specialized cells appear to contain large quantities of myrosinase, it is not clear whether all cells have a lower and perhaps immunologically undetectable level of myrosinase that may serve to degrade their cellular quota of glucosinolates upon disruption of the tonoplast (Lüthy and Matile 1984), or possibly in healthy tissue under sulfate starvation. Multiple forms of myrosinase have been reported in various plants (Björkman 1976), and in *Arabidopsis*, myrosinase is encoded by a gene family of three members (Xue et al. 1992).

Glucosinolate breakdown products have had significant impact on the agronomic utilization of *Brassica* species (Bell 1984). Thiocyanate ion and isoxazolidine-2-thiones generated from glucosinolate degradation can induce goiter in livestock by inhibiting iodine uptake into the thyroid and the subsequent release of thyroxine (van Etten 1969). Significant improvement in rapeseed germ plasm (Downey et al. 1969) has resulted from the ingression of the "Bronowski" germ plasm, which carries a mutation that reduces the accumulation of alkenyl and hydroxyalkenyl glucosinolates (Josefsson 1971a,b) that lead to the goitrogenic isoxazolidine-2-thiones (goitrins).

Glucosinolate breakdown products may also be important with respect to the biosynthesis of indole acetic acid (IAA) in *Arabidopsis* and other crucifers. A nitrilase has been cloned from *Arabidopsis* that can convert the indolyl-3-methylglucosinolate breakdown product indole-3-acetonitrile to IAA (Bartling et al. 1992). This biosynthetic pathway has also been proposed to be important in the synthesis of IAA in club root disease of *Brassica* (Searle et al. 1982) and may explain the abnormal growth associated with this disease. In addition, indole glucosinolate biosynthetic intermediates may be substrates for IAA synthesis.

Helmlinger et al. (1987) established conditions under which cell-free extracts of *Brassica campestris* would convert indole-3-acetaldoxime to IAA via indole-3-acetaldehyde. The relative significance of any of these routes for IAA synthesis awaits further investigation.

## Roles of Glucosinolates

Because glucosinolate degradation products are toxic, volatile, and produced upon tissue damage, there have been many investigations on the role of glucosinolates in deterring insect attack and in allelopathic interactions. Whereas some insects are completely incapable of surviving on cruciferous plants, many insects are specialist feeders on members of the Brassicaceae. For example, allylglucosinolate is toxic to many insects but has no toxic effects on larvae of the cabbage butterfly, *Pieris rapae* (Blau et al. 1978); however, the mechanism of this resistance is unknown. Although the volatile isothiocyanates function effectively in initial insect attraction (Feeny et al. 1970; Read et al. 1970), Nielsen et al. (1979) reported that another secondary metabolite, kaempferol 3-*O*-xylosylgalactoside, acts synergistically with allylglucosinolate to stimulate feeding of the horseradish flea beetle, *Phyllotreta armoraciae*. Insect feeding in turn induces glucosinolate biosynthesis (Birch et al. 1990), and mechanical wounding induces glucosinolate biosynthesis via a systemically transmissable signal (Bodnaryk 1992). The insect-feeding-induced increase in indole glucosinolates may be mediated by the same mechanism as this wound-induced accumulation and may have evolved to deter further feeding, since Birch et al. (1990) found that high glucosinolate content was correlated with high insect mortality.

Although the suggestion has been made that isothiocyanates act as allelopathic chemicals, the importance of this role of glucosinolates is debatable. Allylisothiocyanate has no allelopathic effect on alfalfa (Choesin and Boerner 1991), but given the structural diversity of glucosinolates, other isothiocyanates may have allelopathic activity toward more sensitive plant species. One example of this is the ability of 2-phenethylisothiocyanate to inhibit the germination of wheat (Bialy et al. 1990). Glucosinolates have also been cited as the reason that members of the Brassicaceae are not mycorrhizal; however, this is speculative (Glenn et al. 1988).

## Glucosinolates in *Arabidopsis* and the Isolation of Glucosinolate Mutants

Glucosinolates can be analyzed either by gas chromatography (GC) of their isothiocyanates (Lockwood and Afsharypuor 1986) or by HPLC

(Helboe et al. 1980). They are often analyzed as desulfoglucosinolates obtained by treatment of the glucosinolate-containing extracts with aryl sulfatase (Minchinton et al. 1982). They are best identified by combining mass spectrometry (MS) in tandem with these techniques (Shaw et al. 1990), since the availability of standard compounds is very limited. The glucosinolates of *Arabidopsis* have been characterized by HPLC-MS (Fig. 2) (Hogge et al. 1988).

Four glucosinolate mutants of *Arabidopsis* have been isolated by screening leaf extracts by HPLC (Haughn et al. 1991). One line, TU1, representing the locus *GSM1*, was characterized in detail and was found to have reduced leaf and seed levels of alkylglucosinolates. Specific reductions were observed for those glucosinolates with methylthioalkyl and methylsulfinylalkyl side chains whose methionine-derived precursor amino acids had been through two, three, or four rounds of chain extension (classes A and B, $n = 4$, 5, or 6 and classes C and D, $n = 4$; Fig. 2). Higher homologs were relatively unaffected, and shorter homologs were accumulated to higher levels. It appears that the *gsm1* mutant is defective in the biosynthesis of specific chain-extended methionine homologs, because when these "missing" amino acids were administered to shoots via the transpiration stream, they were effectively converted to the glucosinolates that were absent in the mutant. Since the higher homologs of the methylthioalkyl and methylsulfinylalkyl glucosinolates are unaffected in the mutant, these data are consistent with the block in the *gsm1* mutant being in a transaminase that converts 2-keto-6-methylthiohexanoic acid, 2-keto-7-methylthioheptanoic acid, and 2-keto-8-methylthiooctanoic acid to their respective amino acids. This proposal would require another aminotransferase capable of acting on the higher homologs of these keto acids. Haughn et al. (1991) also suggested the possibility that the defect in the *gsm1* mutant lies in a gene whose product is required for the release of specific $\alpha$-keto acids from an amino acid chain extension complex, analogous to the function of thioesterases in fatty acid biosynthesis. These two possibilities could be distinguished by a similar feeding experiment to that described above, administering the "missing" $\alpha$-keto acids, rather than their respective amino acids.

Other glucosinolate mutants that were identified include two allelic mutants, TU3 and TU6, deficient in the highest homologs of methylthioalkyl and methylsulfinylalkyl glucosinolates (classes A and B, $n = 8$; Fig. 2). These mutants may also be transaminase or "releasing factor" mutants (Haughn et al. 1991). One line, TU8, is particularly interesting, since it lacks virtually all leaf glucosinolates, including the indole glucosinolates, although its seed glucosinolate profile is normal. In addition, a dwarf phenotype was found to cosegregate with the low-glucosinolate phenotype in this mutant. Line TU7 has decreased levels of all alkyl glucosinolates; however, feeding experiments failed to reveal the nature of the defect in this mutant.

Research by Mithen and colleagues (Mithen and Toroser 1994) has exploited natural variation in the glucosinolate content of *Arabidopsis* ecotypes to identify loci that control glucosinolate composition. One locus, *Gsl-elong-Ar*, appears to be involved in some aspect of methionine elongation, whereas another locus, *Gsl-alk-ar*, governs the side-chain modification that converts methylsulfinyl glucosinolates into prop-2-enyl and but-3-enyl glucosinolate. Finally, from crosses between the Landsberg *erecta* and Columbia ecotypes, a locus required for side-chain hydroxylation was identified called *Gsl-ohp-Ar*. *Gsl-ohp-Ar* maps to the same position as *Gsl-alk-ar* and may only represent an allelic variant (Mithen and Toroser 1994).

## Future Prospects

In summary, although our knowledge of how glucosinolates are synthesized seems relatively complete, much work remains to be done. The chain extension pathway has not been studied in sufficient detail at the enzymatic level, and no gene in the biosynthetic pathway has been cloned. A better understanding of the chain extension pathway will add to our understanding of amino acid biosynthesis in plants or may actually arise from a detailed study of leucine biosynthesis.

The study of *Arabidopsis* glucosinolate mutants will be extremely helpful in elucidating the roles that these secondary metabolites play. For example, it would be extremely interesting to know whether the TU8 mutant, which has very low levels of indole glucosinolates, has altered phytoalexin biosynthesis. The TU8 mutant has also been reported to be somewhat dwarfed. Is this a result of alterations in IAA metabolism that may be linked to glucosinolate biosynthesis? There is also opportunity for the isolation of new mutants. Haughn et al. (1991) screened only 1200 plants, due to the laborious nature of the HPLC screen employed. Many additional loci affecting glucosinolate accumulation may remain to be discovered and characterized.

## PHYTOALEXINS

### Introduction

The currently accepted definition of phytoalexins was formulated by Paxton (1981) who proposed: "Phytoalexins are low molecular weight antimicrobial compounds that are both synthesized by and accumulated in plants after exposure to micro-organisms." This definition does not advance a role for phytoalexins in host/pathogen interactions and represents a compromise of conflicting opinions about the function and significance of phytoalexins (Bailey and Mansfield 1982; Ebel 1986; van Etten et al. 1989).

Given this definition, it can be readily imagined that phytoalexin accumulation is a defense mechanism in plants. For some host/parasite interactions, there does appear to be a strong correlation between the accumulation of phytoalexins and restricted pathogen growth. Nicholson and co-workers have taken advantage of the fact that the sorghum phytoalexins, apigeninidin and luteolinidin, are pigments to monitor the accumulation of these phytoalexins in individual living cells (Nicholson et al. 1987; Snyder and Nicholson 1990). With this method, they were able to demonstrate that the timing and site of deposition of the pigmented phytoalexins were consistent with a significant role for these molecules in disease resistance (Snyder and Nicholson 1990). In addition, the amount of the two deoxyanthocyanidin phytoalexins that accumulated was sufficient to account for the cessation of fungal growth (Snyder et al. 1991). Taken together, these data provide strong circumstantial evidence that phytoalexins play an important role in the resistance of juvenile sorghum to the anthracnose pathogen.

In a complementary set of experiments, van Etten and colleagues have demonstrated that the ability of the pathogen *Nectria haematococca* to detoxify phytoalexins is tightly correlated with pathogenicity and may also be linked to host range (van Etten et al. 1989). These authors demonstrated a correlation between virulence on pea and the ability to demethylate pisatin, the pea phytoalexin, to a less toxic compound (Tegtmeier and van Etten 1982). They then cloned a gene encoding pisatin demethylase and showed that this gene conferred on the maize pathogen *Cochliobolus heterostrophus* a limited ability to colonize and cause symptoms on pea. Thus, for this host/pathogen pair, phytoalexin detoxification appears to be an important determinant of the compatible interaction (Schäfer et al. 1989).

The general significance of phytoalexins in plant defense is still an open question. Due to technical limitations of measuring low-molecular-weight metabolites at highly localized sites of infection, it has been difficult to demonstrate convincingly an important role for phytoalexins in many host/pathogen interactions. In addition, there is the confounding effect of other host responses to inoculation by pathogens. For example, a large and growing number of genes have been described that are induced in plants following inoculation with pathogens or elicitors. Among this number are genes for some of the enzymes of the phenylpropanoid pathway and derivative pathways (Dixon and Lamb 1990), enzymes of the isoprenoid pathway leading to sterols and terpenoids (Zook and Kuć 1991), enzymes associated with oxidative stress such as peroxidases (Lagrimini and Rothstein 1987), putative components of signal transduction mechanisms (e.g., protein kinase regulators, Brandt et al. 1992), and pathogenesis-related proteins (Cutt and Klessig 1992). The induction of some of these genes has been correlated with disease resistance (Kiedrowski et al. 1992; Alexander et al. 1993). Changes in the cell wall such as the deposition of phenolic compounds (Nicholson and Hammerschmidt 1992) or of hydroxyproline-rich glycoproteins (Showalter 1993) and the formation of callose-containing appositions (papillae, Aist 1976) have been widely reported in plants under attack by a variety of pathogens. The hypersensitive necrosis response is also strongly correlated with resistance in some plants (Klement 1982). For those plants in which phytoalexin accumulation precedes the hypersensitive necrosis response, phytoalexins may contribute to cell death (Graham and Graham 1991b; Nicholson and Hammerschmidt 1992). Thus, it is difficult to assess the contribution of phytoalexins to disease resistance.

In *Arabidopsis*, many of the same responses to pathogens, including phytoalexin accumulation, have been documented (Davis and Ausubel 1989; Dong et al. 1991; Kiedrowski et al. 1992; Tsuji et al. 1992; Uknes et al. 1992; Ausubel et al. 1993). Unlike the model phytoalexin systems studied in the past, such as pea, sorghum, and soybean, *Arabidopsis* offers several advantages for critically addressing the relative contribution of phytoalexins to disease resistance through mutational and genetic studies or via genetic engineering of the phytoalexin biosynthetic pathway and its regulatory factors.

## Phytoalexins in Related Brassicaceae

As a group, phytoalexins are classified together only by their biological definition and not by any chemical similarity. However, members of a plant family tend to produce chemically related phytoalexins. The structures of the Brassicaceae phytoalexins and associated stress metabolites are based on an indole ring that is substituted at the 3 position with a sulfur-containing moiety (Fig. 4) (Hammerschmidt et al. 1993).

A limited amount of research has been published describing the relative timing and level of accumulation of phytoalexins in the Brassicaceae in response to inoculation with pathogens. In an early description of the accumulation of phytoalexins following inoculation with the blackspot pathogen *Alternaria brassicae*, Conn et al. (1988) demonstrated that resistant *Brassica* species accumulated higher levels of phytoalexins

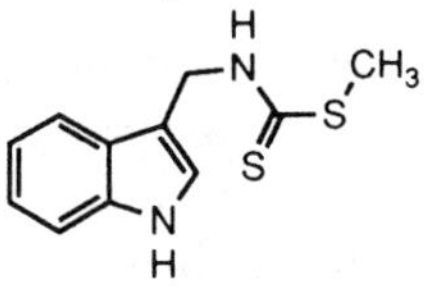

BRASSININ

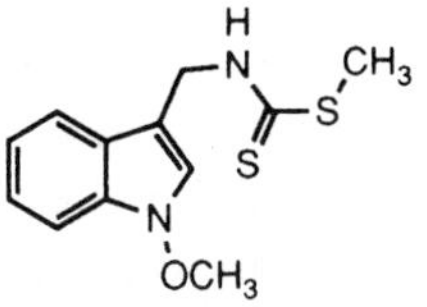

METHOXYBRASSININ

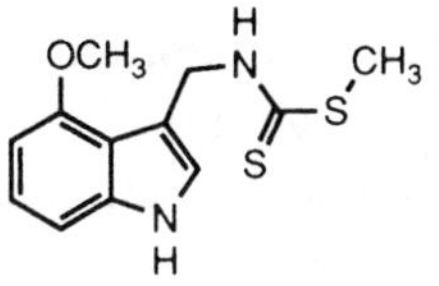

4-METHOXYBRASSININ

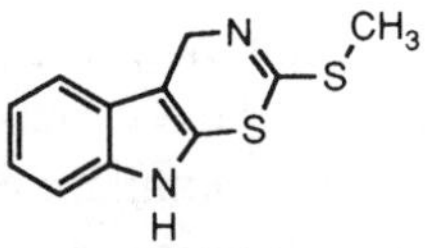

CYCLOBRASSININ

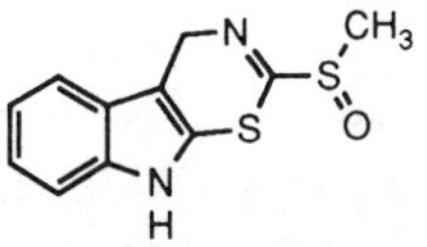

CYCLOBRASSININ SULFOXIDE

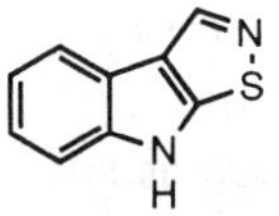

BRASSILEXIN

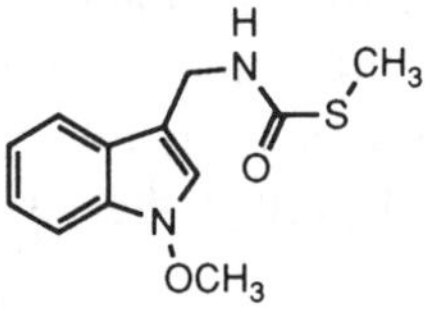

METHOXYBRASSITIN

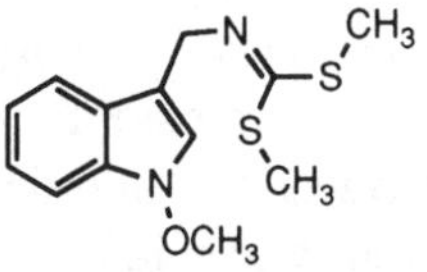

METHOXYBRASSENIN A

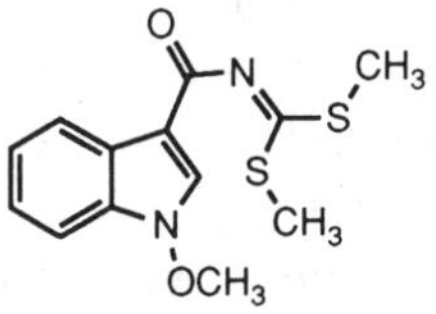

METHOXYBRASSENIN B

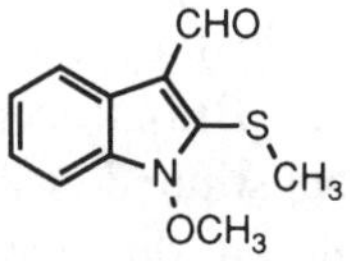

BRASSICANAL A

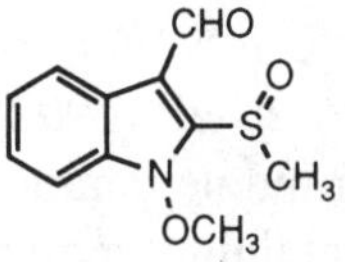

BRASSICANAL C

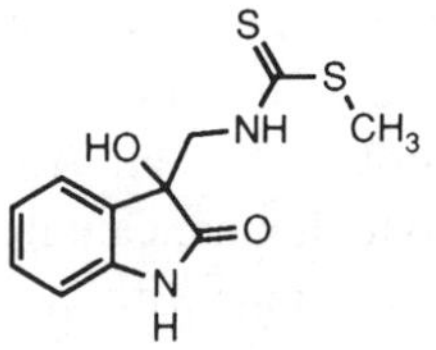

DIOXIBRASSININ

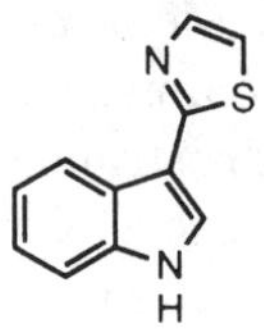

CAMALEXIN

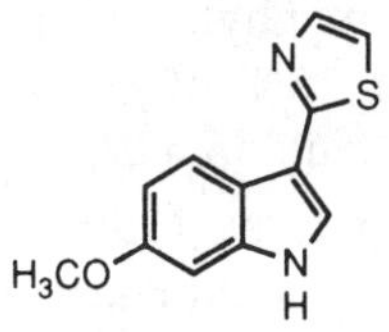

METHOXYCAMALEXIN

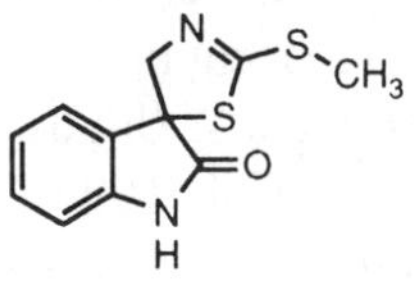

SPIROBRASSININ

*Figure 4* Structures of indole phytoalexins identified to date in members of the Brassicaceae. (The reader is referred to the following articles for additional information: Takasugi et al. 1987, 1988; Devys et al. 1990; Monde et al. 1990, 1991a,b.)

and/or chemically distinct phytoalexins. These authors raised the possibility of improving the resistance of *Brassica* crop species by introducing the ability to synthesize novel phytoalexins from related wild species. In related experiments, Bousquet and colleagues described a correlation between the accumulation of brassilexin and resistance to the black leg pathogen *Leptosphaeria maculans* in *B. juncea* and other B-genome-containing *Brassica* species (Rouxel et al. 1989, 1990). Further work by this group showed that the correlation with black leg resistance did not extend to all *Brassica* species (Rouxel et al. 1991). Highly resistant lines of *B. nigra* and *B. rapa* were identified that did not accumulate high levels of the phytoalexin. This group also demonstrated that heavy metals, such as silver nitrate and copper chloride, induced the accumulation of brassilexin (Rouxel et al. 1989). Thus, the relative importance of phytoalexins as a general defense mechanism in the Brassicaceae is unclear.

There has been one report describing the metabolism of brassinin by a highly virulent isolate of the black leg pathogen (Pedras and Taylor 1991). The authors isolated three metabolites of brassinin from fungal cultures and proposed a pathway for brassinin degradation to the less toxic metabolite indole-3-carboxylic acid.

Conn et al. (1988) describe the accumulation of two novel phytoalexins, camalexin and methoxycamalexin (Fig. 4), in the weedy Brassicaceae species, *Camelina sativa*, following inoculation with *Alternaria brassicae*. The chemical structures of these two phytoalexins were determined and subsequently confirmed by chemical synthesis (Browne et al. 1991; Ayer et al. 1992). Like other Brassicaceae phytoalexins, camalexin is an indole compound that is substituted at the 3 position with a sulfur-containing moiety, a thiazole in this case. Methoxycamalexin is methoxylated at the 6 position, unlike other indole secondary metabolites from Brassicaceae, which commonly exhibit modifications at either the N1 or C4 positions of the indole ring (see Fig. 4). Camalexin is the predominant phytoalexin in *C. sativa*. Methoxycamalexin accumulates to about 15% of camalexin levels (Browne et al. 1991) and exhibits less toxicity to *Cladosporium* sp. than camalexin (Ayer et al. 1992). Jejelowo et al. (1991) suggest that camalexin accumulation plays a significant role in resistance to *A. brassicae* based on the timing, localization, and levels of camalexin that accumulate in leaves following inoculation with this fungal pathogen. For example, in in vitro studies, fungal germination was delayed, germ tube extension was reduced, and the number of abnormal germ tubes was increased by levels of camalexin similar to those found in tissues (e.g., 2–40 µg per ml of water).

## Phytoalexin Accumulation in *Arabidopsis*

In *Arabidopsis*, a toxic, low-molecular-weight metabolite was extracted from leaf tissue with organic solvents and was shown to accumulate following inoculation with pathogens but not in healthy tissue (Slusarenko and Mauch-Mani 1991; Tsuji et al. 1992; Ausubel et al. 1993). By Paxton's definition, this compound can be considered a phytoalexin. Tsuji et al. (1992) developed a purification protocol for recovering sufficient phytoalexin for chemical characterization. By a combination of high-resolution mass-spectral and nuclear magnetic resonance (NMR) analyses, the structure was determined to be 3-thiazol-2'-yl-indole or camalexin (Tsuji et al. 1992).

Due to the structural similarity between camalexin and the common amino acid tryptophan, it was suggested that tryptophan could be the primary metabolic precursor for this compound. Two complementary methods to test this hypothesis were employed: the analysis of camalexin accumulation in tryptophan-deficient mutants and the incorporation of radiolabeled compounds into camalexin in the wild type (Tsuji et al. 1993). In both sets of experiments, anthranilate but not tryptophan served as a biochemical precursor to camalexin. Therefore, it was concluded that the camalexin biosynthetic pathway originates with an intermediate of the tryptophan pathway that lies between anthranilate and indole.

## Camalexin-deficient Mutants of *Arabidopsis*

One of the promises of *Arabidopsis* is the ability to utilize genetic and molecular genetic tools to understand biological problems. The significance of phytoalexin accumulation as a defensive mechanism in host/pathogen interactions can be addressed by generating mutants deficient in phytoalexin accumulation and then testing their resistance to a range of pathogens. J. Glazebrook and F. Ausubel (pers. comm.) have isolated three camalexin-deficient mutants of *Arabidopsis*. The mutants carry mutations in distinct complementation groups, which have been given the names *pad1*, *pad2*, and *pad3* (phytoalexin-deficient). Mutant *pad3* accumulates no camalexin, whereas *pad1* and *pad2* mutants accumulate 30% and 10% of wild-type levels of camalexin, respectively. In preliminary tests for altered disease resistance, Glazebrook measured the ability of the *pad* mutants to support the growth of the bacterial pathogens *Pseudomonas syringae* pv *maculicola* and *P. syringae tomato*. In compatible interactions, *pad1* and *pad2* mutants supported increased levels of growth of virulent strains of the bacteria; these bacteria grew to a similar extent in *pad3* mutants and wild-type *Arabidopsis*. In incompatible interactions, the *pad2, pad3* mutants responded similarly to wild type following inoculation with bacteria carrying the *avrRpt2* gene. Both these results call into question the significance of camalexin accumulation as a defense mechanism.

## Areas for Future Research

### Camalexin Biosynthetic Pathway

Several questions remain unanswered about the biosynthesis of camalexin and other Brassicaceae phytoalexins. Although it has been demonstrated that camalexin biosynthesis originates from the tryptophan biosynthetic pathway, the identity of the camalexin precursor is unknown. Because intermediates of the tryptophan pathway are phosphorylated and in some cases unstable, it will be difficult to directly test their ability to feed into the camalexin biosynthetic pathway in routine feeding studies.

Devys and Barbier (1991) measured significant levels of indole-3-carboxaldehyde (19 mg per kg fresh weight) from the white inner leaves of cabbage (*Brassica oleracea*) and raised the possibility that this compound may serve as a precursor for Brassicaceae phytoalexins (Devys et al. 1988). Browne et al. (1991) state that camalexin could be synthesized via condensation of indole-3-carboxaldehyde with cysteine, cyclization, and decarboxylation. This hypothesis does not address the source of indole-3-carboxaldehyde, but it does provide a useful working model both for feeding studies and for the analysis of biosynthetic lesions in camalexin-deficient mutants.

Browne et al. (1991) propose that the thiazole moiety is derived directly from cysteine. However, the thiazole moiety may be synthesized via the same pathway as the thiazole group found in thiamine (Julliard and Douce 1991). To address the origin of the thiazole moiety, the accumulation of camalexin in the *tz* mutant, a thiamine-deficient mutant that is rescued by exogenously applied thiazole, could be measured to determine whether or not camalexin synthesis is diminished (Li and Rédei 1969).

In prolonged experiments extending up to 5 days postelicitation, camalexin levels in *Arabidopsis* tissues decline (Hammerschmidt et al. 1993). The reduction in phytoalexin levels may reflect catabolism of the phytoalexin by host tissue, as has been suggested for cyclobrassinin in *Brassica napus* tissues (Dahiya and Rimmer 1988). The turnover of camalexin represents one additional point of regulation and, therefore, may influence host/pathogen interactions. This feature of camalexin metabolism has not been addressed.

An additional aspect of phytoalexin metabolism in *Arabidopsis* that has not been fully explored is the range of different phytoalexins that may be produced. Tsuji et al. (1992) mention the presence of a minor, unstable compound that copurified with camalexin until the final purification step. Furthermore, alternate extraction procedures may yield additional phytoalexins. One specific question is whether *Arabidopsis* produces 6-methoxycamalexin, a minor phytoalexin found in *Camelina sativa*.

A final metabolic question is how the synthesis of camalexin affects the accumulation of tryptophan and secondary indole metabolites, such as indole-acetic acid and the indole glucosinolates. Kreps and Town (1992) measured free tryptophan levels of about 5 μg per gram fresh weight, and camalexin accumulates to levels of about 10–15 μg per gram fresh weight (Tsuji et al. 1992, 1993). The first committed step in the tryptophan biosynthetic pathway, the synthesis of anthranilate, is subject to feedback inhibition by the end product tryptophan through the enzyme anthranilate synthase (Bryan 1990; Kreps and Town 1992; Niyogi and Fink 1992). Camalexin levels were reduced about tenfold in the presence of an exogenous source of tryptophan, and it was postulated that this may be due to the feedback inhibition of anthranilate synthase (Tsuji et al. 1993). If significant diversion into camalexin occurs at the expense of other indole-based compounds, then secondary effects on plant metabolism may occur. These postulated metabolic disturbances may influence host/pathogen interactions and contribute to the deleterious nature of the cruciferous phytoalexins. A precedent exists in potato, in which the application of arachidonic acid, an elicitor in this system, leads to both increased activity of sesquiterpene cyclase, an enzyme required for phytoalexin biosynthesis, and decreased activity of squalene synthetase, an enzyme of sterol biosynthesis (Zook and Kuć 1991).

## Regulation of Camalexin Biosynthesis

Although some parallels have been drawn with signal transduction mechanisms described for animal systems, no signal transduction mechanism has been fully characterized in host/pathogen interactions (Sachs et al. 1993). The screen for camalexin-deficient mutants provides one means of identifying components of a signal transduction pathway. If such mutants can be recovered, they will be useful tools for answering a variety of questions about the perception of pathogens by plants. For example, is the signal transduction pathway leading to the induction of camalexin shared with other defense-related responses or with stress-related responses?

A component of this aspect of research is the metabolic regulation of the camalexin biosynthetic pathway. Are all enzymes of the camalexin biosynthetic pathway or only the terminal steps induced by pathogen attack or heavy metal treatment (Graham et al. 1990; Graham and Graham 1991a)? Are any enzymes of the pathway subject to feedback inhibition? Does compartmentation in subcellular compartments and the attendant transport of intermediates from one compartment to the next play a role in the regulation of camalexin biosynthesis? Is camalexin synthesized in plastids like tryptophan (Bryan 1990; Rose et al. 1992)? Since camalexin is a toxic compound, is it sequestered in the vacuole or is it only synthesized in cells destined to die? Is camalexin released from dying cells into the intercellular spaces? Mutations affecting these biochemical processes would affect the level of camalexin that accumulates after elicitation and may also be recovered in the mutant screens described above.

## Role of Camalexin in Defense against Pathogens

In addition to their value in determining the camalexin biosynthetic and signal transduction pathways, *Arabidopsis* mutants deficient in camalexin accumulation will be invaluable for addressing the role of phytoalexins in disease resistance. Among the possible classes unable to accumulate camalexin, only those mutants that are phenotypically normal and do not accumulate any toxic intermediates will be suitable for evaluating disease resistance.

Preliminary experiments with camalexin-deficient mutants suggest that camalexin has no role, or a limited role, in resistance to *P. syringae* pathovars. A large and ever increasing number of fungal, bacterial, nematode, and viral pathogens of *Arabidopsis* are being described. Thus, it is possible to ask whether or not camalexin-deficient mutants show alterations in defense responses to all pathogens or to only a subset of pathogens. Since fungi are more sensitive to camalexin than bacteria in a crude sensitivity assay (Tsuji et al. 1992), one might predict that alterations in camalexin levels will have the most dramatic effect on fungal diseases. From the results of infection studies with a range of pathogens, a broadly based view of the general role of phytoalexins in disease resistance in *Arabidopsis* can be developed.

A second question that could be addressed with the *Arabidopsis* phytoalexin-deficient mutants is, What is the contribution of phytoalexin accumulation to disease resistance relative to other host defensive responses, such as the hypersensitive necrosis response? If an array of defense responses are activated in a coordinate fashion, then reduced phytoalexin accumulation may not have a dramatic effect on the outcome and incompatible interaction. An additional factor to be considered is whether any of the other host defensive responses are enhanced in the camalexin-deficient mutants. Such a compensatory stimulation of alternate defense responses may diminish the impact of restricted phytoalexin synthesis in the mutants.

The analysis of mutants altered in camalexin accumulation promises answers to many of the long-standing questions in the field of phytoalexins. To complement these studies, it would be helpful to know how virulent pathogens avoid this defense mechanism. Do any pathogens exhibit reduced sensitivity to camalexin and, if so, what is the basis for the reduced sensitivity? For example, do any virulent pathogens metabolize camalexin to a less toxic compound? Do any virulent pathogens fail to induce camalexin accumulation? A related question is, What is the target site for camalexin? Browne et al. (1991) have noted the structural similarity between camalexin and thiabenzidole, a putative anti-mitotic agent. Knowledge of the mechanism of toxicity of camalexin may aid in the design of more selective pesticides.

## Future Prospects

The chemical similarity of the Brassicaceae phytoalexins suggests that once the camalexin biosynthetic pathway in *Arabidopsis* has been determined, it can serve as the basis for determining the biosynthetic pathways of other Brassicaceae phytoalexins. If this is indeed true, it offers the possibility that we will be able to "mix and match" biosynthetic enzymes from various species via genetic engineering and thereby expand the chemical arsenal available in *Brassica* crop species. A second potential outcome of such research is the synthesis of novel compounds, which may prove useful in host disease resistance or as compounds of medicinal or other commercial value.

## ACKNOWLEDGMENTS

We thank all our colleagues who kindly provided us with their unpublished results for inclusion in this chapter. This work was supported in part by the Michigan Agriculture Experiment Station (#1648), the U.S. Department of Agriculture (#8080), the U.S. Department of Energy (DE-FG02-90ER20021), and a grant from Hoechst AG. This is journal paper number 13970 from the Purdue University Agricultural Experiment Station.

## REFERENCES

Aist, J.R. 1976. Papillae and related-wound plugs of plant cells. *Annu. Rev. Phytopathol.* **14:** 145–163.

Alexander, D., C. Glascock, J. Pear, J. Stinson, P. Ahl-Goy, M. Gut-Rella, E. Ward, R.M. Goodman, and J. Ryals. 1993. Systemic acquired resistance in tobacco: Use of transgenic expression to study the functions of pathogenesis-related proteins. In *Advances in molecular genetics of plant-microbe interactions* (ed. E.W. Nester and D.P.S. Verma), vol. 2, pp. 527–533. Kluwer Academic, Dordrecht, The Netherlands.

Arondel, V., B. Lemieux, I. Hwang, S. Gibson, H.M. Goodman, and C.R. Somerville. 1992. Map-based cloning of a gene controlling omega-3 fatty acid desaturation in *Arabidopsis. Science* **258:** 1353–1355.

Ausubel, F.M., J. Glazebrook, J. Greenberg, M. Mindrinos, and G.-L. Yu. 1993. Analysis of the *Arabidopsis* defense response to *Pseudomonas* pathogens. In *Advances in molecular genetics of plant-microbe interactions* (ed. E.W. Nester and D.P.S. Verma), vol. 2, pp. 393–403. Kluwer Academic, Dordrecht, The Netherlands.

Ayer, W.A., P.A. Craw, Y.-T. Ma, and S. Miao. 1992. Synthesis of camalexin and related phytoalexins. *Tetrahedron* **48:** 2919–2924.

Bailey, J.A. and J.W. Mansfield. 1982. *Phytoalexins*. Blackie, Glasgow, United Kingdom.

Bartling, D., M. Seedorf, A. Mithöfer, and E.W. Weiler. 1992. Cloning and expression of an *Arabidopsis* nitrilase which can convert indole-3-acetonitrile to the plant hormone, indole-3-acetic acid. *Eur. J. Biochem.* **205:** 417–424.

Beld, M., C. Martin, H. Huits, A.R. Stuitje, and A.G.M. Gerats. 1989. Flavonoid synthesis in Petunia: Partial characterization of dihydroflavonol 4-reductase genes. *Plant Mol. Biol.* **13:** 491–502.

Bell, J.M. 1984. Nutrients and toxicants in rapeseed meal: A review. *J. Anim. Sci.* **58:** 996–1010.

Bialy, Z., W. Oleszek, J. Lewis, and G.R. Fenwick. 1990. Allelopathic potential of glucosinolates (mustard oil glycosides) and their degradation products against wheat. *Plant Soil* **129:** 277–281.

Birch, A.N.E., D.W. Griffits, and W.H.M. Smith. 1990. Changes in forage and oilseed rape (*Brassica napus*) root glucosinolates in response to attack by turnip root fly (*Delia floralis*). *J. Sci. Food Agric.* **51:** 309–320.

Björkman, R. 1976. Properties and function of plant myrosinases. In *The biology and chemistry of the cruciferae* (ed. J.G. Vaughn et al.), pp. 191–205. Academic Press, New York.

Blau, P.A., P. Feeny, L. Contardo, and D.S. Robson. 1978. Allylglucosinolate and herbivorous caterpillars: A contrast in toxicity and tolerance. *Science* **200:** 1296–1298.

Bodnaryk, R.P. 1992. Effects of wounding on glucosinolates in the cotyledons of oilseed rape and mustard. *Phytochemistry* **31:** 2671–2677.

Bokern, M. and D. Strack. 1988. Synthesis of hydroxycinnamic acid esters of betacyanins via 1-*O*-acylglucosides of hydroxycinnamic acids by protein preparations from cell suspension cultures of *Chenopodium rubrun* and petals of *Lampranthus sociorum. Planta* **174:** 101–105.

Bouchereau, A., J. Hamelin, I. Lamour, M. Renard, and F. Larher. 1991. Distribution of sinapine and related compounds in seeds of *Brassica* and allied genera. *Phytochemistry* **30:** 1873–1881.

Brandl, W., K. Herrmann, and L. Grotjahn. 1984. Hydroxycinnamoyl esters of malic acid in small radish (*Raphanus sativus* L. var. *sativus*). *Z. Naturforsch.* **39c:** 515–520.

Brandt, J., H. Thordal-Christensen, K. Vad, P. L. Gergersen, and D.B. Collinge. 1992. A pathogen-induced gene of barley encodes a protein showing high similarity to a protein kinase regulator. *Plant J.* **2:** 815–820.

Browne, L.M., K.L. Conn, W.A. Ayer, and J.P. Tewari. 1991. The camalexins: New phytoalexins produced in the leaves of *Camelina sativa* (Cruciferae). *Tetrahedron* **47:** 3909–3914.

Bryan, J.K. 1990. Advances in the biochemistry of amino acid biosynthesis. In *The biochemistry of plants* (ed. B.J. Miflin and P.J. Lea), vol. 16, pp. 161–195. Academic Press, New York.

Bürger, D. 1971. Die morphologischen mutanten des Göttinger *Arabidopsis*-sortiments, einschlieblich der mutanten mit abweichender samenfarbe. *Arabidopsis Inf. Serv.* **8:** 36–42.

Chang, C., J.L. Bowman, A.W. DeJohn, E.S. Lander, and E.M. Meyerowitz. 1988. Restriction fragment length polymorphism linkage map for *Arabidopsis thaliana*. *Proc. Natl. Acad. Sci.* **85:** 6856–6860.

Chappell, J. and K. Hahlbrock. 1984. Transcription of plant defence genes in response to UV light or fungal elicitor. *Nature* **311:** 76–78.

Chapple, C.C.S., J.R. Glover, and B.E. Ellis. 1990. Purification and characterization of methionine: Glyoxylate aminotransferase from *Brassica carinata* and *Brassica napus*. *Plant Physiol.* **94:** 1887–1896.

Chapple, C.C.S., T. Vogt, B.E. Ellis, and C.R. Somerville. 1992. An *Arabidopsis* mutant defective in the general phenylpropanoid pathway. *Plant Cell* **4:** 1413–1424.

Chisholm, M.D. 1972. Biosynthesis of 3-methylthiopropylglucosinolate and 3-methyl-sulfinylpropylglucosinolate in wallflower *Cheiranthus kewensis*. *Phytochemistry* **11:** 197–202.

Choesin, D.N. and R.E.J. Boerner. 1991. Allyl isothiocyanate release and the allelopathic potential of *Brassica napus* (Brassicaceae). *Am. J. Bot.* **78:** 1083–1090.

Chory, J. and C.A. Peto. 1990. Mutations in the *DET1* gene affect cell-type-specific expression of light-regulated genes and chloroplast development in *Arabidopsis*. *Proc. Natl. Acad. Sci.* **87:** 8776–8780.

Chory, J., P. Nagpal, and C.A. Peto. 1991. Phenotypic and genetic analysis of *det2*, a new mutant that affects light-regulated seedling development in *Arabidopsis*. *Plant Cell* **3:** 445–459.

Coen, E.S., J.M. Romero, S. Doyle, R. Elliott, G. Murphy, and R. Carpenter. 1990. *floricaula*: A homeotic gene required for flower development in *Antirrhinum majus*. *Cell* **63:** 1311–1322.

Conn, E.E. 1980. Cyanogenic compounds. *Annu. Rev. Plant Physiol.* **31:** 433–451.

Conn, K.L., J.P. Tewari, and J.S. Dahiya. 1988. Resistance to *Alternaria brassicae* and phytoalexin-elicitation in rapeseed and other crucifers. *Plant Sci.* **56:** 21–25.

Cutt, J.R. and D.F. Klessig. 1992. Pathogenesis-related proteins. In *Genes involved in plant defense* (ed. T. Boller and F. Meins), pp. 209–243. Springer-Verlag, New York.

Dahiya, J.S. and S.R. Rimmer. 1988. Phytoalexin accumulation in tissues of *Brassica napus* inoculated with *Leptosphaeria maculans*. *Phytochemistry* **27:** 3105–3107.

Davis, K.R. and F.M. Ausubel. 1989. Characterization of elicitor-induced defense responses in suspension-cultured cells of *Arabidopsis*. *Mol. Plant-Microbe Interact.* **2:** 363–368.

Daxenbichler, M.E., G.F. Spencer, D.G. Carlson, G.B. Rose, A.M. Brinker, and R.G. Powell. 1991. Glucosinolate composition of seeds from 297 species of wild plants. *Phytochemistry* **30:** 2623–2638.

De Meyer, N., A. Haemers, L. Mishra, H.-K. Pandey, L.A.C. Pieters, D.A. Vanden

Berghe, and A.J. Vlietinck. 1991. 4′-Hydroxy-3-methoxyflavones with potent antipicornavirus activity. *J. Med. Chem.* **34:** 736–746.

Devys, M. and M. Barbier. 1991. Indole-3-carboxaldehyde in the cabbage *Brassica oleracea*: A systematic determination. *Phytochemistry* **30:** 389–391.

Devys, M., M. Barbier, A. Kollman, T. Rouxel, and J.-F. Bousquet. 1990. Cyclobrassinen sulphoxide, a sulfur-containing phytoalexin from *Brassica juncea*. *Phytochemistry* **29:** 1087–1088.

Devys, M., M. Barbier, I. Loiselet, T. Rouxel, A. Sarniguet, A. Kollman, and J.-F. Bousquet. 1988. Brassilexin, a novel sulphur-containing phytoalexin from *Brassica juncea* L. (Cruciferae). *Tetrahedron Lett.* **29:** 6447–6448.

Dixon, R.A. and C.J. Lamb. 1990. Molecular communication in interactions between plants and microbial pathogens. *Annu. Rev. Plant Physiol. Plant Mol. Biol.* **41:** 339–367.

Dixon, R.A., P.M. Dey, and C.J. Lamb. 1983. Phytoalexins: Enzymology and molecular biology. *Adv. Enzymol. Relat. Areas Mol. Biol.* **55:** 1–135.

Dixon, R.A., M.J. Harrison, S.M. Jenkins, C.J. Lamb, M.A. Lawton, and L. Yu. 1990. *cis*-elements and *trans*-acting factors for regulation of the plant defense gene chalcone synthase. In *Plant gene transfer* (ed. C.J. Lamb and R.N. Beachy), pp. 101–109. A.R. Liss, New York.

Dong, X., M. Mindrinos, K.R. Davis, and F.M. Ausubel. 1991. Induction of *Arabidopsis* defense genes by virulent and avirulent *Pseudomonas syringae* strains and by a cloned avirulence gene. *Plant Cell* **3:** 61–72.

Dooner, H.K. and T.P. Robbins. 1991. Genetic and developmental control of anthocyanin biosynthesis. *Annu. Rev. Genet.* **25:** 173–199.

Downey, R.K., B.M. Craig, and D.G. Youngs. 1969. Breeding rapeseed for oil and meal quality. *J. Am. Oil Chem. Soc.* **46:** 121–123.

Ebel, J. 1986. Phytoalexin synthesis: The biochemical analysis of the induction process. *Annu. Rev. Phytopathol.* **24:** 235–264.

Errampalli, D., D. Patton, L. Castle, L. Michelson, K. Hansen, J. Schnall, K. Feldmann, and D. Meinke. 1991. Embryonic lethals and T-DNA insertional mutagenesis in *Arabidopsis*. *Plant Cell* **3:** 149–157.

Fedoroff, N.V., D.B. Furtek, and O.E. Nelson. 1984. Cloning of the *bronze* locus in maize by a simple and generalizable procedure using the transposable controlling element Activator (Ac). *Proc. Natl. Acad. Sci.* **81:** 3825–3829.

Feeny, P., K.L. Paauwe, and N.J. Demong. 1970. Flea beetles and mustard oils: Host plant specificity of *Phyllotreta cruciferae* and *P. striolata* adults (Coleoptera: Chrysomelidae). *Ann. Entomol. Soc. Am.* **63:** 832–841.

Feinbaum, R.L. and F.M. Ausubel. 1988. Transcriptional regulation of the *Arabidopsis thaliana* chalcone synthase gene. *Mol. Cell. Biol.* **8:** 1985–1992.

Feinbaum, R.L., G. Storz, and F.M. Ausubel. 1991. High intensity and blue light regulated expression of chimeric chalcone synthase genes in transgenic *Arabidopsis thaliana* plants. *Mol. Gen. Genet.* **226:** 449–456.

Fisher, R.F. and S.R. Long. 1992. Rhizobium-plant signal exchange. *Nature* **357:** 655–660.

Fritze, K., D. Staiger, I. Czaja, R. Walden, J. Schell, and D. Wing. 1991. Developmental and UV light regulation of the snapdragon chalcone synthase promoter. *Plant Cell* **3:** 893–905.

Fujiki, H., T. Horiuchi, K. Yamashita, H. Hakii, M. Suganuma, H. Nishino, A. Iwashima, Y. Hirata, and T. Sugimura. 1986. Inhibition of tumor promotion by flavonoids. In *Plant flavonoids in biology and medicine: Biochemical, pharmacological and structure-activity relationships* (ed. V. Cody et al.), vol. 213, pp. 429–440. A.R. Liss, New York.

Galston, A.W. 1969. Flavonoids and photomorphogenesis in peas. In *Perspectives in phytochemistry* (ed. J.B. Harborne and T. Swain), pp. 193–204. Academic Press, New York.

Gijzen, M., I. McGregor, and G. Séguin-Swartz. 1989. Glucosinolate uptake by developing rapeseed embryos. *Plant Physiol.* **89:** 260–263.

Giraudat, J., B.M. Hauge, C. Valon, J. Smalle, F. Parcy, and H.M. Goodman. 1992. Isolation of the *Arabidopsis ABI3* gene by positional cloning. *Plant Cell* **4:** 1251–1261.

Glässgen, W.E. and H.U. Seitz. 1992. Acylation of anthocyanins with hydroxycinnamic acids via 1-*O*-acylglucosides by protein preparations from cell cultures of *Daucus carota* L. *Planta* **186:** 582–585.

Glendening, T.M. and J.E. Poulton. 1990. Partial purification and characterization of a 3′-phosphoadenosine 5′ phosphosulfate: Desulfoglucosinolate sulfotransferase from cress (*Lepidium sativum*). *Plant Physiol.* **94:** 811–818.

Glenn, M.G., F.S. Chew, and P.H. Williams. 1988. Influence of glucosinolate content of *Brassica* (Cruciferae) roots on growth of vesicular-arbuscular mycorrhizal fungi. *New Phytol.* **110:** 217–225.

Gmelin, R. and A. Kjær. 1970. Isoferulic acid choline ester: A new plant constituent. *Phytochemistry* **9:** 667–669.

Goff, S.A., K.C. Cone, and V.I. Chandler. 1992. Functional analysis of the transcriptional activator encoded by the maize B gene: Evidence for a direct functional interaction between two classes of regulatory proteins. *Genes Dev.* **6:** 864–875.

Goodrich, J., R. Carpenter, and E.S. Coen. 1992. A common gene regulates pigmentation pattern in diverse plant species. *Cell* **68:** 955–964.

Graham, T.L. 1991. A rapid, high resolution high performance liquid chromatography profiling procedure for plant and microbial aromatic secondary metabolites. *Plant Physiol.* **95:** 584–593.

Graham, T.L. and M.Y. Graham. 1991a. Glyceollin elicitors induce major but distinctly different shifts in isoflavonoid metabolism in proximal and distal soybean cell populations. *Mol. Plant-Microbe Interact.* **4:** 60–68.

———. 1991b. Cellular coordination of molecular responses in plant defense. *Mol. Plant-Microbe Interact.* **4:** 415–422.

Graham, T.L., J.E. Kim, and M.Y. Graham. 1990. Role of constitutive isoflavone conjugates in the accumulation of glyceollin in soybean infected with *Phytophthora megasperma*. *Mol. Plant-Microbe Interact.* **3:** 157–166.

Grand, C. 1984. Ferulic acid 5-hydroxylase: A new cytochrome P-450-dependent enzyme from higher plant microsomes involved in lignin synthesis. *FEBS Lett.* **169:** 7–11.

Gräwe, W., P. Bachhuber, H.-P. Mock, and D. Strack. 1992. Purification and characterization of sinapoylglucose: Malate sinapoyltransferase from *Raphanus sativus* L. *Planta* **187:** 236–241.

Hahlbrock, K. and D. Scheel. 1989. Physiology and molecular biology of phenylpropanoid metabolism. *Annu. Rev. Plant Physiol. Plant Mol. Biol.* **40:** 347–369.

Halaweish, F.T. and D.K. Dougall. 1990. Sinapoyl glucose synthesis by extracts of wild carrot cell cultures. *Plant Sci.* **71:** 179–184.

Hammerschmidt, R., J. Tsuji, M. Zook, and S. Somerville. 1993. A phytoalexin from *Arabidopsis thaliana* and its relationship to other phytoalexins of Crucifers. In Arabidopsis *as a model system for studying plant-pathogen interactions* (ed. K.R. Davis and R. Hammerschmidt), pp. 73–84. American Phytopathological Society Press, St. Paul, Minnesota.

Hanley, A.B. and K.R. Parsley. 1990. Identification of 1-methoxyindolyl-3-methylisothiocyanate as an indole glucosinolate breakdown product. *Phytochemistry* **29**: 769–771.

Harborne, J.B. 1988. Flavonoids in the environment: Structure-activity relationships. *Prog. Clin. Biol. Res.* **280**: 17–27.

Harborne, J.B., A. M. Mayer, and N. Bar-Nun. 1983. Identification of the major anthocyanin of carrot cells in tissue culture as cyanidin 3-sinapoylxylosylglucosyl-galactoside. *Z. Naturforsch* **38c**: 1055–1056.

Haughn, G.W., L. Davin, M. Giblin, and E.W. Underhill. 1991. Biochemical genetics of plant secondary metabolites in *Arabidopsis thaliana*. The glucosinolates. *Plant Physiol.* **97**: 217–226.

Hedin, P.A. and S.K. Waage. 1986. Roles of flavonoids in plant resistance to insects. In *Plant flavonoids in biology and medicine: Biochemical, pharmacological and structure-activity relationships* (ed. V. Cody et al.), vol. 213, pp. 87–100. A.R. Liss, New York.

Helboe, P., O. Olsen, and H. Sorensen. 1980. Separation of glucosinolates by high-performance liquid chromatography. *J. Chromatogr.* **197**: 199–205.

Helmlinger, J., T. Rausch, and W. Hilgenberg. 1987. A soluble protein factor from Chinese cabbage converts indole-3-acetaldoxime to IAA. *Phytochemistry* **26**: 615–618.

Hogge, L.R., D.W. Reed, E.W. Underhill, and G.W. Haughn. 1988. HPLC separation of glucosinolates from leaves and seeds of *Arabidopsis thaliana* and their identification using thermospray liquid chromatography/mass spectrometry. *J. Chromatogr. Sci.* **26**: 551–556.

Höglund, A.S., M. Lenman, and L. Rask. 1992. Myrosinase is localized to the interior of myrosin grains and is not associated to the surrounding tonoplast membrane. *Plant Sci.* **85**: 165–170.

Hopp, W. and H.U. Seitz. 1987. The uptake of acylated anthocyanin into isolated vacuoles from a cell suspension culture of *Daucus carota*. *Planta* **170**: 74–85.

Jacobs, M. and P.H. Rubery. 1988. Naturally occurring auxin transport regulators. *Science* **241**: 346–349.

Jejelowo, O.A., K.L. Conn, and J.T. Tewari. 1991. Relationship between conidial concentration, germling growth, and phytoalexin production by *Camelina sativa* leaves inoculated with *Alternaria brassicae*. *Mycol. Res.* **95**: 928–934.

Jones, C.G. and R.D. Firn. 1991. On the evolution of plant secondary chemical diversity. *Philos. Trans. Roy. Soc. Lond. B* **333**: 273–280.

Josefsson, E. 1971a. Studies of the biochemical background to differences in glucosinolate content in *Brassica napus* L. I. Glucosinolate content in relation to general chemical composition. *Physiol. Plant.* **24**: 150–159.

———. 1971b. Studies of the biochemical background to differences in glucosinolate content in *Brassica napus* L. II. Administration of some sulphur-35 and carbon-14 compounds and localization of metabolic blocks. *Physiol. Plant.* **24**: 161–175.

Julliard, J.-H., and R. Douce. 1991. Biosynthesis of the thiazole moiety of thiamin (vitamin B1) in higher plant chloroplasts. *Proc. Natl. Acad. Sci.* **88**: 2042–2045.

Kiedrowski, S., P. Kawalleck, K. Halbrock, I.E. Somssich, and J.L. Dangl. 1992. Rapid activation of a novel plant defense gene is strictly dependent on the *Arabidopsis RPM1* disease resistance locus. *EMBO J.* **11**: 4677–4684.

Kindl, H. 1968. Oxydasen und Oxygenasen in höheren Pflanzen. I. Über das Vorkommen von Indolyl-(3)-acetaldehydoxim und seine Bildung aus L-Tryptophan. *Hoppe-Seyler's Z. Physiol. Chem.* **349**: 519–520.

Klement, Z. 1982. Hypersensitivity. In *Phytopathogenic prokaryotes*, vol. 2, pp.

149–177. Academic Press, New York.

Koes, R.E., C.E. Spelt, and J.N.M. Mol. 1989. The chalcone synthase multigene family of *Petunia hybrida* (V30): Differential, light-regulated expression during flower development and UV light induction. *Plant Mol. Biol.* **12:** 213–225.

Koes, R.E., R. van Blokland, R. Quattrocchio, A.J. van Tunen, and J.N.M. Mol. 1990. Chalcone synthase promoters in petunia are active in pigmented and unpigmented cell types. *Plant Cell* **2:** 379–392.

Kondra, Z.P. and B.R. Stefansson. 1970. Inheritance of the major glucosinolates of rapeseed (*Brassica napus*) meal. *Can. J. Plant Sci.* **50:** 643–647.

Koornneef, M. 1981. The complex syndrome of *ttg* mutants. *Arabidopsis Inf. Serv.* **18:** 45–51.

———. 1990. Mutations affecting the testa color in *Arabidopsis*. *Arabidopsis Inf. Serv.* **28:** 1–4.

Koornneef, M., W. Luiten, P. de Vlaming, and A.W. Schram. 1982. A gene controlling flavonoid-3′-hydroxylation in *Arabidopsis*. *Arabidopsis Inf. Serv.* **19:** 113–115.

Koornneef, M., J. van Eden, C.J. Hanhart, P. Stam, F.J. Braaksma, and W.J. Feenstra. 1983. Linkage map of *Arabidopsis thaliana*. *J Hered.* **74:** 265–272.

Kreps, J.A. and C.D. Town. 1992. Isolation and characterization of a mutant of *Arabidopsis thaliana* resistant to α-methyltryptophan. *Plant Physiol.* **99:** 269–275.

Kreuzaler, F., H. Ragg, E. Fautz, D.N. Kuhn, and K. Hahlbrock. 1983. UV-induction of chalcone synthase mRNA in cell suspension cultures of *Petroselinum hortense*. *Proc. Natl. Acad. Sci.* **80:** 2591–2593.

Kubasek, W.L., B.W. Shirley, A. McKillop, H.M. Goodman, W. Briggs, and F.M. Ausubel. 1992. Regulation of flavonoid biosynthetic genes in germinating *Arabidopsis* seedlings. *Plant Cell* **4:** 1229–1236.

Lagrimini, M.L. and S. Rothstein. 1987. Tissue specificity of tobacco peroxidase isozymes and their induction by wounding and tobacco mosaic virus infection. *Plant Physiol.* **84:** 438–442.

Lamb, C.J., M.A. Lawton, M. Dron, and R.A. Dixon. 1989. Signals and transduction mechanisms for activation of plant defenses against microbial attack. *Cell* **56:** 215–224.

Larsen, P.O. 1981. Glucosinolates. In *The biochemistry of plants* (ed. E.E. Conn), vol. 7, pp. 501–525. Academic Press, New York.

Li, J., T.-M. Ou-Lee, R. Raba, R. Amundson, and R.L. Last. 1993. *Arabidopsis* flavonoid mutants are hypersensitive to UV-B irradiation. *Plant Cell* **5:** 171–179.

Li, S.L. and G.P. Rédei. 1969. Gene locus specificity of the glucose effect in the thiamine pathway of the angiosperm, *Arabidopsis*. *Plant Physiol.* **44:** 225–229.

Linscheid, M., D. Wendisch, and D. Strack. 1980. The structures of sinapic acid esters and their metabolism in cotyledons of *Raphanus sativus*. *Z. Naturforsch.* **35c:** 907–914.

Liscum, E. and R.P. Hangarter. 1991. *Arabidopsis* mutants lacking blue light-dependent inhibition of hypocotyl elongation. *Plant Cell* **3:** 685–694.

Lloyd, A.M., V. Walbot, and R.W. Davis. 1992. *Arabidopsis* and *Nicotiana* anthocyanin production activated by maize regulators R and C1. *Science* **258:** 1773–1775.

Loake, G.J., O. Faktor, C.J. Lamb, and R.A. Dixon. 1992. Combination of H-box [CCTACC(N)$_7$CT] and G-box (CACGTG) *cis* elements is necessary for feed-forward stimulation of a chalcone synthase promoter by the phenylpropanoid-pathway intermediate *p*-coumaric acid. *Proc. Natl. Acad. Sci.* **89:** 9230–9234.

Loake, G.J., A.D. Choudary, M.J. Harrison, M. Mavandad, C.J. Lamb, and R.A. Dixon. 1991. Phenylpropanoid pathway intermediates regulate transient expression of a chalcone synthase gene promoter. *Plant Cell* **3:** 829–840.

Lockwood, G.B. and S. Afsharypuor. 1986. Comparative study of the volatile aglucones of glucosinolates from in vivo and in vitro grown *Descurainia sophia* and *Alyssum minimum* using gas chromatography-mass spectrometry. *J. Chromatogr.* **356:** 438–440.

Lueprasitsakul, W., S. Alex, S.L. Fang, S. Pino, K. Irmscher, J. Khrle, and L.E. Braverman. 1990. Flavonoid administration immediately displaces thyroxine (T4) from serum transthyretin, increases serum free T4, and decreases serum thyrotropin in the rat. *Endocrinology* **126:** 2890–2895.

Lüthy, B. and P. Matile. 1984. The mustard oil bomb: Rectified analysis of the subcellular organization of the myrosinase system. *Biochem. Physiol. Pflanz.* **179:** 5–12.

Lynn, D.G. and M. Chang. 1990. Phenolic signals in cohabitation: Implications for plant development. *Annu. Rev. Plant Physiol. Plant Mol. Biol.* **41:** 497–526.

Matsuo, M. 1968. Biosynthesis of sinigrin. V. On the origin of thioglucoside moiety of sinigrin. *Chem. Pharm. Bull.* **16:** 1128–1129.

Matsuo, M. and E.W. Underhill. 1971. Purification and properties of a UDP glucose: Thiohydroximate glucosyltransferase from higher plants. *Phytochemistry* **10:** 2279–2286.

Mavandad, M., R. Edwards, X. Liang, C.J. Lamb, and R.A. Dixon. 1990. Effects of *trans*-cinnamic acid on expression of the bean phenylalanine ammonia-lyase gene family. *Plant Physiol.* **94:** 671–680.

McCarty, D.R., C.B. Carson, P.S. Stinard, and D.S. Robertson. 1989. Molecular analysis of viviparous-1: An abscisic acid-insensitive mutant of maize. *Plant Cell* **1:** 523–532.

McCarty, D.R., T. Hattori, C.B. Carson, V. Vasil, M. Lazar, and I.K. Vasil. 1991. The viviparous-1 developmental gene of maize encodes a novel transcriptional activator. *Cell* **66:** 895–905.

McGregor, D.I. 1978. Thiocyanate ion, a hydrolysis product of glucosinolates from rape and mustard seed. *Can. J. Plant Sci.* **58:** 795–800.

Minchinton, I., J. Sang, D. Burke, and R.J.W. Truscott. 1982. Separation of desulphoglucosinolates by reversed-phase high-performance liquid chromatography. *J. Chromatogr.* **247:** 141–148.

Mithen, R. and D. Toroser. 1994. Biochemical genetics of aliphatic glucosinolates in *Brassica* and *Arabidopsis*. In *Amino acids and their derivatives in higher plants* (ed. R. Wallgrove). Cambridge University Press, Cambridge, United Kingdom.

Mithen, R.F., B.G. Lewis, R.K. Heaney, and G.R. Fenwick. 1987. Glucosinolates of wild and cultivated *Brassica* species. *Phytochemistry* **26:** 1969–1973.

Mo, Y.-Y. and D.C. Gross. 1991. Plant signal molecules activate the *syrB* gene, which is required for syringomycin production by *Pseudomonas syringae* pv. syringae. *J. Bacteriol.* **173:** 5784–5792.

Mo, Y., C. Nagel, and L.P. Taylor. 1992. Biochemical complementation of chalcone synthase mutants defines a role for flavonols in functional pollen. *Proc. Natl. Acad. Sci.* **89:** 7213–7217.

Mock, H.-P. and D. Strack. 1993. Energetics of the uridine 5′-diphosphoglucose: hydroxycinnamic acid acyl-glucosyltransferase reaction. *Phytochemistry* **32:** 575–579.

Monde, K., K. Sasaki, A. Shirata, and M. Takasugi. 1990. 4- Methoxybrassinin, a sulfur containing phytoalexin from *Brassica oleracea*. *Phytochemistry* **29:** 1499–1500.

———. 1991a. Brassicanal C and two dioxindoles from cabbage. *Phytochemistry* **30:** 2915–2917.

———. 1991b. Methoxybrassinens A and B, sulphur containing stress metabolites from *Brassica oleracea* var. capitata. *Phytochemistry* **30:** 3921–3922.

Nicholson, R.L. and R. Hammerschmidt. 1992. Phenolic compounds and their role in disease resistance. *Annu. Rev. Phytopathol.* **30:** 369–389.

Nicholson, R.L., S.S. Kollipara, J.R. Vincent, P.C. Lyons, and G. Cadena-Gomez. 1987.

Phytoalexin synthesis by the sorghum mesocotyl in response to infection by pathogenic and nonpathogenic fungi. *Proc. Natl. Acad. Sci.* **84:** 5520–5524.

Nielsen, J.K., L.M. Larsen, and H. Sørensen. 1979. Host plant selection of the horseradish flea beetle *Phyllotreta armoracieae* (Coleoptera: Chrysomelidae): Identification of two flavonol glycosides stimulating feeding in combination with glucosinolates. *Entomol. Exp. Appl.* **26:** 40–48.

Nielsen, J.K., O. Olsen, L.H. Pedersen, and H. Sørensen. 1984. 2-*O*-(*p*-coumaroyl)-L-malate, 2-*O*-caffeoyl-L-malate and 2-*O*-feruloyl-L-malate in *Raphanus sativus. Phytochemistry* **23:** 1741–1743.

Niyogi, K.K. and G.R. Fink. 1992. Two anthranilate synthase genes in *Arabidopsis*: Defense-related regulation of the tryptophan pathway. *Plant Cell* **4:** 721–733.

Oaks, A. 1965. The synthesis of leucine in maize embryos. *Biochim. Biophys. Acta* **111:** 79–89.

Oppenheimer, D.G., P.L. Herman, S. Sivakumaran, J. Esch, and M.D. Marks. 1991. A *myb* gene required for leaf trichome differentiation in *Arabidopsis* is expressed in stipules. *Cell* **67:** 483–493.

Parry, R.J. and M.V. Naidu. 1982. Biosynthesis of sulfur compounds. Elucidation of the stereochemistry of the conversion of [3-(methylthio)propyl]glucosinolate into allyl-glucosinolate (sinigrin). *J. Am. Chem. Soc.* **104:** 3217–3219.

Paxton, J.D. 1981. Phytoalexins—A working redefinition. *Phytopathol. Z.* **101:** 106–109.

Pedras, M.S.C. and J.L. Taylor. 1991. Metabolic transformation of the phytoalexin brassinin by the "Blackleg" fungus. *J. Org. Chem.* **56:** 2619–2621.

Popp, R. and O. Schimmer. 1991. Induction of sister-chromatid exchanges (SCE), polyploidy, and micronuclei by plant flavonoids in human lymphocyte cultures. A comparative study of 19 flavonoids. *Mutat. Res.* **246:** 205–213.

Read, D.P., P.P. Feeny, and R.B. Root. 1970. Habitat selection by the aphid parasite *Diaeretiella rapae* (Hymenoptera: Braconidae) and hyperparasite *Charips brassicae* (Hymenoptera: Cynipidae). *Can. Entomol.* **102:** 1567–1578.

Reimold, U., M. Kröger, F. Kreuzaler, and K. Hahlbrock. 1983. Coding and 3′ noncoding nucleotide sequence of chalcone synthase mRNA and assignment of amino acid sequence of the enzyme. *EMBO J.* **2:** 1801–1805.

Robinson-Beers, K., R.E. Pruitt, and C.S. Gasser. 1992. Ovule development in wild-type *Arabidopsis* and two female-sterile mutants. *Plant Cell* **4:** 1237–1249.

Rose, A.B., A.L. Casselman, and R.L. Last. 1992. A phosphoribosylanthranilate transferase gene is defective in blue fluorescent *Arabidopsis thaliana* tryptophan mutants. *Plant Physiol.* **100:** 582–592.

Rossiter, J.T. and D.C. James. 1990. Biosynthesis of (*R*)-2-hydroxybut-3-enylglucosinolate (progoitrin) from [3,4-$^3$H]but-3-enylglucosinolate in *Brassica napus. J. Chem. Soc. Perkin Trans.* **1:** 1909–1913.

Rossiter, J.T., D.C. James, and N. Atkins. 1990. Biosynthesis of 2-hydroxy-3-butenyl-glucosinolate and 3-butenylglucosinolate in *Brassica napus. Phytochemistry* **29:** 2509–2512.

Rouxel, T., A. Kollman, L. Boulidard, and R. Mithen. 1991. Abiotic elicitation of indole phytoalexins and resistance to *Leptosphaeria maculans* within Brassicaceae. *Planta* **184:** 271–278.

Rouxel, T., M. Renard, A. Kollman, and J.-F. Bousquet. 1990. Brassilexin accumulation and resistance to *Leptosphaeria maculans* in *Brassica* spp. and progeny of an inter-specific cross *B. junceae* X *B. napus. Euphytica* **46:** 175–181.

Rouxel, T., A. Sarniguet, A. Kollman, and J.-F. Bousquet. 1989. Accumulation of a phytoalexin in *Brassica* spp. in relation to a hypersensitive reaction to *Leptosphaeria maculans. Physiol. Mol. Plant Pathol.* **34:** 507–514.

Ryder, T.B., C.L. Cramer, J.N. Bell, M.P. Robbins, R.A. Dixon, and C.J. Lamb. 1984. Elicitor rapidly induces chalcone synthase mRNA in *Phaseolus vulgaris* cells at the onset of the phytoalexin defense response. *Proc. Natl. Acad. Sci.* **81:** 5724–5728.

Ryder, T.B., S.A. Hedrick, J.N. Bell, X. Liang, S.D. Clouse, and C.J. Lamb. 1987. Organization and differential activation of a gene family encoding the plant defense enzyme chalcone synthase in *Phaseolus vulgaris. Mol. Gen. Genet.* **210:** 219–233.

Sachs, W.R., P. Ferreira, K. Halbrock, T. Jabs, T. Nurnberger, A. Renelt, and D. Scheel. 1993. Elicitor recognition and intracellular signal transduction in plant defense. In *Advances in molecular genetics of plant-microbe interactions* (ed. E.W. Nester and D.P.S. Verma), vol. 2, pp. 485–495. Kluwer Academic, Dordrecht, The Netherlands.

Sang, J.P. and P.A. Salisbury. 1988. Glucosinolate profiles of international rapeseed lines (*Brassica napus* and *Brassica campestris*). *J. Sci. Food Agric.* **45:** 255–261.

Schäfer, W., D. Straney, L. Ciuffetti, H. Van Etten, and O.C. Yoder. 1989. One enzyme makes a fungal pathogen, but not a saprophyte, virulent on a new host plant. *Science* **246:** 247–249.

Schmelzer, E., W. Jahnen, and K. Hahlbrock. 1988. In situ localization of light-induced chalcone synthase mRNA, chalcone synthase, and flavonoid end products in epidermal cells of parsley leaves. *Proc. Natl. Acad. Sci.* **85:** 2989–2993.

Schmid, J., P.W. Doerner, S.D. Clouse, R.A. Dixon, and C.J. Lamb. 1990. Developmental and environmental regulation of a bean chalcone synthase promoter in transgenic tobacco. *Plant Cell* **2:** 619–631.

Schulze-Lefert, P., J.L. Dangl, M. Becker-Andr, K. Hahlbrock, and W. Schulz. 1989. Inducible in vivo DNA footprints define sequences necessary for UV light activation of the parsley chalcone synthase gene. *EMBO J.* **8:** 651–656.

Searle, L.M., K. Chamberlain, T. Rausch, and D.N. Butcher. 1982. The conversion of 3-indolylmethylglucosinolate to 3-indolylacetonitrile by myrosinases, and its relevance to the clubroot disease of the Cruciferae. *J. Exp. Bot.* **33:** 935–942.

Sharma, V. and D. Strack. 1985. Vacuolar localization of 1-sinapoylglucose: L-malate sinapoyltransferase in protoplasts from cotyledons of *Raphanus sativus. Planta* **163:** 563–568.

Shaw, G.J., D. Andrzejewski, J.A.G. Roach, and J.A. Sphon. 1990. Characterization of silylated desulfoglucosinolate mixtures using high-performance capillary gas chromatography (GC)-negative-ion chemical ionization mass spectrometry (NICIMS) and GC-NICIMS/MS. *J. Agric. Food Chem.* **38:** 616–619.

Sheahan, J.J. and G.A. Rechnitz. 1992. Flavonoid-specific staining of *Arabidopsis thaliana. BioTechniques* **13:** 880–883.

———. 1993. Differential visualization of transparent testa mutants in *Arabidopsis thaliana. Anal. Chem.* **65:** 961–963.

Shirley, B.W. and H.M. Goodman. 1993. An *Arabidopsis* gene homologous to mammalian and insect genes encoding the largest proteasome subunit. *Mol. Gen. Genet.* **241:** 586–594.

Shirley, B.W., S. Hanley, and H.M. Goodman. 1992. Effects of ionizing radiation on a plant genome: Analysis of two *Arabidopsis transparent testa* mutations. *Plant Cell* **4:** 333–347.

Showalter, A.M. 1993. Structure and function of plant cell wall proteins. *Plant Cell* **5:** 9–23.

Slusarenko, A.J. and B. Mauch-Mani. 1991. Downy mildew of *Arabidopsis thaliana* caused by *Peronospora parasitica*: A model system for the investigation of the molecular biology of host-pathogen interactions. In *Advances in molecular genetics of plant-microbe interactions* (ed. H. Hennecke and D.P.S. Verma), vol. 1, pp. 280–283. Kluwer Academic, Dordrecht, The Netherlands.

Snyder, B.A. and R.L. Nicholson. 1990. Synthesis of phytoalexins in sorghum as a site-specific response to fungal ingress. *Science* **248:** 1637–1638.

Snyder, B.A., B. Leite, J. Hipskind, L.G. Butler, and R.L. Nicholson. 1991. Accumulation of sorghum phytoalexins induced by *Colletotrichum graminicola* at the infection site. *Physiol. Mol. Plant Pathol.* **39:** 463–470.

Spencer, P.A., and G.H.N. Towers. 1988. Specificity of signal compounds detected by *Agrobacterium tumefaciens*. *Phytochemistry* **27:** 2781–2785.

Stachel, S.E., E. Messens, M. Van Montagu, and P. Zambryski. 1985. Identification of the signal molecules produced by wounded plant cells that activate T-DNA transfer in *Agrobacterium tumefaciens*. *Nature* **318:** 624–629.

Stafford, H.A. 1990. *Flavonoid metabolism*. CRC Press, Boca Raton, Florida.

———. 1991. Flavonoid evolution: An enzymic approach. *Plant Physiol.* **96:** 680–685.

Strack, D. 1980. Enzymatic synthesis of 1-sinapoylglucose from free sinapic acid and UDP-glucose by a cell-free system from *Raphanus sativus* seedlings. *Z. Naturforsch.* **35c:** 204–208.

———. 1981. Sinapine as a supply of choline for the biosynthesis of phosphatidylcholine in *Raphanus sativus* seedlings. *Z. Naturforsch.* **36c:** 215–221.

———. 1982. Development of 1-*O*-sinapoyl-β-D-glucose: L-malate sinapoyltransferase activity in cotyledons of red radish (*Raphanus sativus* L. var. *sativus*). *Planta* **155:** 31–36.

Strack, D., W. Knogge, and B. Dahlbender. 1983. Enzymatic synthesis of sinapine from 1-*O*-sinapoyl-β-D-glucose and choline by a cell-free system from developing seeds of red radish (*Raphanus sativus* L. var. *sativus*). *Z. Naturforsch.* **38c:** 21–27.

Strack, D., G. Nurmann, and G. Sachs. 1980. Sinapine esterase. II. Specificity and change of sinapine esterase activity during germination of *Raphanus sativus*. *Z. Naturforsch.* **35c:** 963–966.

Strack, D., M. Pieroth, H. Scharf, and V. Sharma. 1985. Tissue distribution of phenylpropanoid metabolism in cotyledons of *Raphanus sativus* L. *Planta* **164:** 507–511.

Sun, T.-P., H.M. Goodman, and F.M. Ausubel. 1992. Cloning the *Arabidopsis thaliana* *GA1* locus by genomic subtraction. *Plant Cell* **4:** 119–128.

Takasugi, M., K. Monde, N. Katsui, and A. Shirata. 1987. Spirobrassinin, a novel sulphur-containing phytoalexin from the daikon *Raphanus sativus* L. var. hortensis (Cruciferae). *Chem. Lett.* 1631–1632.

———. 1988. Novel sulfur-containing phytoalexins from the chinese cabbage *Brassica campestris* L. spp. pekinensis (Cruciferae). *Bull. Chem. Soc. Jpn.* **61:** 285–289.

Takeda, K., D. Fischer, and H. Grisebach. 1988. Anthocyanin composition of *Sinapis alba*, light induction of enzymes and biosynthesis. *Phytochemistry* **27:** 1351–1353.

Taylor, L.P. and W.R. Briggs. 1990. Genetic regulation and photocontrol of anthocyanin accumulation in maize seedlings. *Plant Cell* **2:** 115–127.

Tegtmeier, K.J. and H.D. van Etten. 1982. The role of pisatin tolerance and degradation in the virulence of *Nectria haematococca* on peas: A genetic analysis. *Phytopathology* **72:** 608–612.

Thangstad, O.P., T.-H. Iversen, G. Slupphaug, and A. Bones. 1990. Immunocytochemical localization of myrosinase in *Brassica napus* L. *Planta* **180:** 245–248.

Tsuji, J., M. Zook, R. Hammerschmidt, S. Somerville, and R. Last. 1993. Evidence that tryptophan is not a direct biosynthetic intermediate of camalexin in *Arabidopsis thaliana*. *Physiol. Mol. Plant Pathol.* **43:** 221–229.

Tsuji, J., E. Jackson, D. Gage, R. Hammerschmidt, and S.C. Somerville. 1992.

Phytoalexin accumulation in *Arabidopsis thaliana* during the hypersensitive reaction to *Pseudomonas syringae* pv. *syringae*. *Plant Physiol.* **98:** 1304–1309.

Tsukaya, H., T. Ohshima, S. Naito, M. Chino, and Y. Komeda. 1991. Sugar-dependent expression of the CHS-A gene for chalcone synthase from petunia in transgenic *Arabidopsis*. *Plant Physiol.* **97:** 1414–1421.

Uknes, S., B. Mauch-Mani, M. Moyer, S. Potter, S. Williams, S. Dincher, D. Chandler, A. Slusarenko, E. Ward, and J. Ryals. 1992. Acquired resistance in *Arabidopsis*. *Plant Cell* **4:** 645–656.

Underhill, E.W. 1980. Glucosinolates. *Encycl. Plant Physiol. New Ser.* **8:** 493–511.

van der Meer, I.M., M.E. Stam, A.J. van Tunen, J.N.M. Mol, and A.R. Stuitje. 1992. Antisense inhibition of flavonoid biosynthesis in petunia anthers results in male sterility. *Plant Cell* **4:** 253–262.

van Etten, C.H. 1969. Goitrogens. In *Toxic constituents of plant food stuffs* (ed. I.E. Liener), pp. 103–142. Academic Press, New York.

van Etten, H.D., D.E. Matthews, and P.S. Matthews. 1989. Phytoalexin detoxification: Importance for pathogenicity and practical implications. *Annu. Rev. Phytopathol.* **27:** 142–164.

van Tunen, A.J., R.E. Koes, C.E. Spelt, A.R. van der Krol, A.R. Stuitje, and J.N.M. Mol. 1988. Cloning of the two chalcone flavanone isomerase genes from *Petunia hybrida*: Coordinate, light-regulated and differential expression of flavonoid genes. *EMBO J.* **7:** 1257–1263.

Villegas, R.J.A. and M. Kojima. 1986. Purification and characterization of hydroxycinnamoyl D-glucose: Quinate hydroxycinnamoyl transferase in the root of sweet potato, *Ipomoea batatas* Lam. *J. Biol. Chem.* **261:** 8729–8733.

Wei, N. and X.W. Deng. 1992. *COP9:* A new genetic locus involved in light-regulated development and gene expression in *Arabidopsis*. *Plant Cell* **4:** 1507–1518.

Weiss, D. and A.H. Halevy. 1989. Stamens and gibberellin in the regulation of corolla pigmentation and growth in *Petunia hybrida*. *Planta* **179:** 89–96.

Weiss, D., R. van Blokland, J.M. Kooter, J.N.M. Mol, and A.J. van Tunen. 1992. Gibberellic acid regulates chalcone synthase gene transcription in the corolla of *Petunia hybrida*. *Plant Physiol.* **98:** 191–197.

Weiss, M.R. 1991. Floral colour changes as cues for pollinators. *Nature* **354:** 227–229.

Wetter, L.R. 1964. Biosynthesis of mustard oil glucosides. II. The administration of sulphur-35 compounds to horse-radish leaves. *Phytochemistry* **3:** 57–64.

Wetter, L.R. and M.D. Chisholm. 1968. Sources of sulfur in the thioglucosides of various higher plants. *Can. J. Biochem.* **46:** 931–935.

Williams, D.H., M.J. Stone, P.R. Hauck, and S.K. Rahman. 1989. Why are secondary metabolites (natural products) biosynthesized? *J. Nat. Prod.* **52:** 1189–1208.

Wink, M. 1988. Plant breeding: Importance of plant secondary metabolites for protection against pathogens and herbivores. *Theor. Appl. Genet.* **75:** 225–233.

Wintersohl, U., J. Krause, and K. Napp-Zinn. 1979. Phenylpropane derivatives in *Arabidopsis thaliana* (L.) Heynh. *Arabidopsis Inf. Serv.* **16:** 76–77.

Xue, J., M. Lenman, A. Falk, and L. Rask. 1992. The glucosinolate-degrading enzyme myrosinase in Brassicaceae is encoded by a gene family. *Plant Mol. Biol.* **18:** 387–398.

Zook, M. and J.A. Kuć. 1991. Induction of sesquiterpene cyclase and suppression of squalene synthetase activity in elicitor-treated or fungal-infected potato tuber tissue. *Physiol. Mol. Plant Pathol.* **39:** 377–390.

# 37

# Epicuticular Wax and *eceriferum* Mutants

**Bertrand Lemieux**
Department of Biology
York University, North-York
Ontario M3J-1P3, Canada

**Maarten Koornneef**
Department of Genetics
Wageningen Agricultural University
6703 HA Wageningen, The Netherlands

**Kenneth A. Feldmann**
Department of Plant Sciences
University of Arizona
Tucson, Arizona 85721

Like most plants, the aboveground surface of *Arabidopsis thaliana* is covered by a layer of lipids such as fatty acids, fatty aldehydes, primary alcohols, alkanes, secondary alcohols, ketones, and esters. Figure 1 illustrates the chemical structures of some of the principal epicuticular wax components found on the surface of *Arabidopsis*. The majority of these wax components can be separated in a single chromatographic separation by capillary gas chromatography with a low polarity solid phase (Yang et al. 1992). The use of mass spectrometry coupled to a gas chromatograph has allowed the identification of the individual compounds that make up the epicuticular wax layer of plants (Walton 1990). As shown in Figure 2, *Arabidopsis eceriferum* (*cer*) mutants are characterized by a bright green color when compared to wild-type plants because the reduced amount of wax deposition on the stem alters the reflection of light such that mutants can be isolated by a simple visual inspection.

Epicuticular wax components are derived from very long chain fatty acids. The existence of a fatty acid elongation activity within the endoplasmic reticulum has been demonstrated by the partial purification of a 500-kD enzyme complex that elongates stearyl-CoA (18:0) to eicosanoyl-CoA (20:0) (Bessoule et al. 1989). Partial purifications of fatty acid reductase activities have shown that a fatty acid reductase and

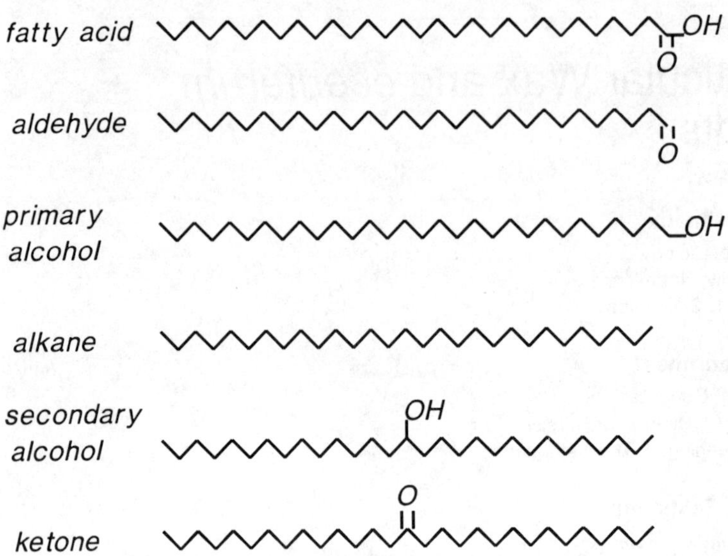



*Figure 1*   Chemical structures of common epicuticular wax compounds. The fatty acid is triacontanoic acid (C30), the aldehyde is triacontanal, the primary alcohol is 1-triacontanol, the alkane is *n*-nonacosane (C29), the secondary alcohol is 15-nonacosanol, and the ketone is 15-nonacosanone.

a fatty aldehyde reductase activity can be separated by protein fractionation (Kolattukudy 1971). Experiments with particulate cell wall fractions of pea have demonstrated that a fatty aldehyde decarbonylation activity is presumably responsible for the biosynthesis of alkanes (Cheesbrough and Kolattukudy 1984). Four principal pathways of wax biosynthesis have emerged from these and other studies (von Wettstein-Knowles 1993); two of these appear to take place in *Arabidopsis*: (1) an elongation reduction pathway that leads to the formation of even chain length primary alcohols and (2) an elongation decarbonylation pathway that supplies odd chain length alkanes which can undergo oxidation to secondary alcohols and ketones.

In *Arabidopsis*, 22 different loci have been implicated in the regulation of epicuticular wax biosynthesis to date (Dellaert et al. 1979; Koornneef et al. 1989; McNevin et al. 1993). This is substantially fewer than the 84 loci that regulate wax accumulation in barley (Sogaard and von Wettstein-Knowles 1987; von Wettstein-Knowles 1993). Mutants at some of these *Arabidopsis* loci (*cer1*, *cer3*, *cer6*, *cer8*, and *cer10*) show reduced fertility under conditions of low humidity (Koornneef et al. 1989). Recently, Preuss et al. (1993) have shown that this reduced fertility phenotype is caused by the erosion of the pollen tryphine layer owing to the absence of pollen surface waxes. Microscopic observations of the

*Figure 2* Visual glossiness of *eceriferum* mutants. A 13x magnification of the stem of a wild-type plant (*top*) compared to the stem of a *cer2* mutant plant (*bottom*).

structure of wax blooms reveal a wide array of structures, which are sometimes found on the surface of the same plant cell (Jeffree et al. 1975). Recrystallization experiments with extracted epicuticular wax components and with fractionated wax compounds have shown that the morphology of the wax blooms is, to a large extent, determined by the chemical composition of the epicuticular wax (Jeffree and Sandford 1982). Our analyses of the morphological (Koornneef et al. 1989) and phytochemical (Hannoufa et al. 1993) variations of the epicuticular wax of the *Arabidopsis* mutants have revealed that these mutants have less wax than the wild-type plants, and some have dramatically different epicuticular wax chemical compositions when compared to that of the wild-type.

The hydrophobic nature of epicuticular wax and the distribution of wax blooms on the plant surface have been implicated in a variety of physiological phenomena involved in water relations. For example, the acquisition of water in conifers may be aided by the epicuticular wax found on the central portion of the needle, because wax blooms are absent on the edges of the pine needles where water is absorbed (Leyton and Juniper 1963). The filamentous wax structures at the entrance of stomata may also help keep water out of the intracellular air spaces by partially blocking the stomatal entrance (Jeffree et al. 1971). In the case of glossy mutants of *Brassica oleracea*, the reduced levels of epicuticular lipids lead to a higher rate of transpiration with stomata closed when compared to glaucous plants (Denna 1970). Commercial grapes also provide a good system to study transpiration with regard to surface wax composition, as grapes have no stomata. The surface of grapes is covered by a layer of soft wax consisting of long-chain alcohols, aldehydes, esters, free acids, and hydrocarbons, which, when removed by organic solvents, leads to a faster drying of grapes due to increased water loss (Martin and Juniper 1970). As a high proportion of alkanes in the epicuticular wax results in a higher water drop contact angle than on plant surfaces whose wax is rich in hydrophilic substances like fatty alcohols and free fatty acids (Holloway 1969), the water-repellent properties of surface lipids are dependent on the chemical composition of this epicuticular wax layer.

We have recently found that *n*-nonacosane is a component of pollen lipids (Preuss et al. 1993). Previous studies on maize pollen lipid composition also suggest that pollen surfaces contain some of the same compounds as epicuticular wax (Bianchi et al. 1990). Our observations of pollen grain morphology suggest that the tryphine layer of the pollen grain is significantly reduced in some *cer* mutants with reduced levels of fertility (Preuss et al. 1993). The pollen waxes of these mutants appear to be lacking *n*-nonacosane, 14-nonacosanol, 15-nonacosanol, and 15-nonacosanone. Aniline blue staining of stigmatic papillae after pollination with grains from *cer6* mutants suggests that callose production is induced owing to the absence of unknown pollen surface components (Preuss et al. 1993). Although pollen waxes probably play a structural role in maintaining the integrity of the tryphine layer after dehiscence, there is a possibility that compounds derived from wax hydrocarbons might play a more active role in the signaling between pollen and stigma.

In this chapter, we compare the alterations in the morphology of the wax blooms with the alterations in the chemical composition of the epicuticular wax in the *cer* mutants. On the basis of the known enzyme activities involved in wax biosynthesis, we have assigned tentative biochemical functions to the gene products of some of the *cer* loci of *Arabidopsis*. Knowledge obtained from the analysis of the *cer* mutants may help elucidate the physiological roles of the epicuticular wax layer as well as the mechanisms through which plant waxes are deposited onto the plant surface. Finally, we describe some of the possible biotechnological applications that may result from work with *Arabidopsis cer* mutants.

## CHARACTERIZATION OF WILD-TYPE *ARABIDOPSIS*: DEVELOPMENTAL AND ENVIRONMENTAL INFLUENCES

Table 1 lists the chemical composition of the epicuticular wax layer of *Arabidopsis thaliana* ecotype Landsberg at three different developmental stages and under two different growth conditions (i.e., growth at 15°C and 25°C under continuous light). The most notable developmental variation in the chemical composition of the wax is the increase in the level of alkanes, secondary alcohols, and ketones on stems and siliques compared to leaf surfaces. The quantitative differences in the total amount of epicuticular wax on the plant surface correlate with the more glaucous appearance of the stems compared to leaves. Indeed, *cer* mutants are readily scored by the visual inspection of the stems and siliques but are impossible to identify by the appearance of their leaves. Koornneef et al. (1989) observed that tube-like wax structures become more abundant on stems as the plant ages and that some of these structures apparently fuse to form bigger structures. As similar tube-like structures are created by the recrystallization of symmetrical secondary alcohols on filters (Jeffree 1986), we believe that these developmental variations in the morphology of wax blooms are related to the increase in the level of secondary alcohols in the wax on different portions of the plant (see Table 1). In the case of maize, a developmentally regulated shutdown in epicuticular wax biosynthesis accompanies the accumulation of silicates within the cells (Poethig 1990), a phenomenon seen in several grasses. The chemical compositions of the epicuticular wax layers at equivalent developmental stages are approximately equivalent over a limited range of environmental growth conditions (Table 1). Variations in growth temperature (15°C, 21°C, and 35°C) and humidity (40%, 70%, and 98%) do not significantly alter the ratio of the different classes of epicuticular wax compounds on the surface of *B. oleracea* (Baker 1974) but do influence the amount of wax produced.

## *ARABIDOPSIS ECERIFERUM* MUTANTS

### Chemical Phenotypes of the *eceriferum* Mutants

Table 2 presents the overall chemical composition (in weight % of wild-type levels) of plants grown at 25°C in continuous illumination for 20 of the *eceriferum* mutants isolated by Koornneef et al. (1989). Most of the classes of compounds are composed of a series of chain length homologs (Hannoufa et al. 1993). On the basis of our studies of the chemical composition of the epicuticular wax of *Arabidopsis cer* mutants, we were able to assign probable defects to only 5 of the 21 loci; i.e., *cer1*, *cer2*, *cer4* (Hannoufa et al. 1993), *cer6*, and *cer8* (B. Lemieux, unpubl.). In

*Table 1* Epicuticular wax composition of *A. thaliana* race Landsberg

| Compound | 15°C leaf 3 weeks | 15°C leaf 6 weeks | 15°C stem 7 weeks | 25°C leaf 2 weeks | 25°C silique 6 weeks | 25°C stem 6 weeks |
|---|---|---|---|---|---|---|
| Tetradecanoic | n.d. | 3.2 ± 0.1 | 0.4 ± 0.1 | 0.62 ± 0.05 | tr | 1.5 ± 0.1 |
| Pentadecanoic | n.d. | 1.2 ± 0.1 | tr | 1.9 ± 0.2 | tr | tr |
| Hexadecenoic | 2.6 ± 0.1 | 2.9 ± 0.1 | 9.8 ± 0.4 | 2.7 ± 0.2 | 26 ± 1 | 1.3 ± 0.1 |
| Hexadecanoic | 1.3 ± 0.1 | 1.5 ± 0.1 | 3.8 ± 0.3 | tr | 7.5 ± 0.3 | 4.7 ± 0.2 |
| Octadecenoic | tr | tr | 3.5 ± 0.3 | tr | 2.7 ± 0.1 | 2.0 ± 0.2 |
| Octadecanoic | tr | tr | 3.5 ± 0.2 | tr | 9.3 ± 0.4 | 2.0 ± 0.2 |
| Eicosanoic | tr | tr | 1.0 ± 0.1 | tr | 3.0 ± 0.2 | tr |
| Docosanoic | tr | tr | 4.8 ± 0.2 | 1.0 ± 0.1 | 7.3 ± 0.3 | 1.2 ± 0.1 |
| *n*-Heptacosane | n.d. | n.d. | 5.3 ± 0.2 | n.d. | 2.5 ± 0.1 | 3.0 ± 0.1 |
| 1-Tetracosanol | 3.2 ± 0.1 | 3.0 ± 0.1 | 4.9 ± 0.2 | 2.0 ± 0.2 | 5.1 ± 0.2 | 3.3 ± 0.1 |
| Tetracosanoic | n.d. | tr | tr | n.d. | tr | 5.7 ± 0.1 |
| *n*-Nonacosane | 1.9 ± 0.1 | 4.5 ± 0.2 | 309 ± 9 | 4.9 ± 0.3 | 668 ± 21 | 371 ± 13 |
| Hexacosanal | n.d. | tr | 1.0 ± 0.1 | tr | 5.2 ± 0.3 | 3.0 ± 0.1 |
| 1-Hexacosanol | 7.7 ± 0.3 | 7.0 ± 0.2 | 36.2 ± 1.2 | 7.2 ± 0.3 | 41 ± 3 | 31.0 ± 0.8 |
| Octacosanal | tr | tr | 0.9 ± 0.05 | tr | 6.5 ± 0.3 | 3.4 ± 0.1 |
| Hexacosanoic | tr | 0.8 ± 0.05 | 1.9 ± 0.1 | tr | 3.8 ± 0.2 | 17.5 ± 0.8 |
| 14- and 15-Nonacosanol | 2.3 ± 0.1 | tr | 41 ± 2 | 6.7 ± 0.3 | 156 ± 8 | 81 ± 1 |
| 15-Nonacosanone | 1.6 ± 0.1 | 5.6 ± 0.2 | 152 ± 7 | 3.9 ± 0.2 | 408 ± 15 | 281 ± 11 |
| 1-Octacosanol | 3.5 ± 0.1 | 3.0 ± 0.1 | 58.0 ± 3 | 2.1 ± 0.1 | 172 ± 6 | 121 ± 5 |
| Octacosanoic | 1.1 ± 0.05 | tr | 10.0 ± 0.6 | 4.7 ± 0.2 | 16.0 ± 1 | 6.2 ± 0.5 |
| Triacontanal | 6.0 ± 0.3 | tr | 22.4 ± 2.2 | 5.8 ± 0.3 | 128.6 ± 6 | 97 ± 3 |
| 1-Triacontanol | 3.4 ± 0.2 | 5.0 ± 0.2 | 25.8 ± 2.3 | 4.7 ± 0.3 | 172 ± 9 | 74 ± 3 |
| β-Amyrin | tr | tr | tr | 3.5 ± 0.2 | 5.1 ± 0.6 | 0.50 ± 0.05 |
| Triacontanoic | 2.7 ± 0.1 | 3.0 ± 0.1 | 2.2 ± 0.2 | 3.6 ± 0.2 | tr | 21.2 ± 0.8 |

Values are in μg of wax compound per gram fresh weight plant tissue and are the mean ± S.D. (*n* = 4). tr indicates less than 0.2 μg/g and n.d. indicates no detectable peak.

*Table 2* Relative changes in the amounts of aliphatic compounds comprising the epicuticular waxes of 20 *eceriferum (cer)* mutants compared to wild-type levels

| Lines | Fatty acids | Primary alcohols | Aldehydes | Alkanes | Secondary alcohols | Ketones |
|---|---|---|---|---|---|---|
| *cer1* | 75 | 30 | 78 | 2.5 | 3.1 | 5.3 |
| *cer2* | 260 | 136 | 0.4 | 6.2 | 10.5 | 0.8 |
| *cer3* | 27.6 | 50 | 1.5 | 7.0 | 7.0 | 13.1 |
| *cer4* | 70.4 | 4.5 | 127 | 112 | 177 | 107 |
| *cer5* | 30.5 | 26 | 5.0 | 10.4 | 24.9 | 25.3 |
| *cer6* | 118 | 60.6 | 2.4 | 7.6 | 4.9 | 9.0 |
| *cer7* | 37.8 | 56.9 | 62.9 | 37.9 | 35.3 | 53.7 |
| *cer8* | 434 | 63.5 | 4.3 | 31.0 | 8.1 | 8.2 |
| *cer9* | 96.1 | 46.5 | 8.1 | 3.6 | 10.6 | 17.5 |
| *cer10* | 15.6 | 31.3 | 72.5 | 43.8 | 49.6 | 47.7 |
| *cer11* | 58.3 | 74 | 61 | 79.7 | 87.7 | 69 |
| *cer12* | 25.3 | 38.4 | 46 | 45.2 | 113 | 72.2 |
| *cer13* | 57 | 72.6 | 55.1 | 85 | 135 | 85 |
| *cer14* | 29 | 42.7 | 29.3 | 36 | 70.4 | 44 |
| *cer15* | 210 | 14 | 46.4 | 40.4 | 76 | 45 |
| *cer16* | 73.5 | 35.7 | 43.7 | 44.1 | 228 | 106 |
| *cer17* | 102 | 32 | 56 | 51.9 | 93 | 69.8 |
| *cer18* | 41.5 | 45.5 | 31.9 | 41.4 | 160 | 89 |
| *cer19* | 129 | 54.3 | 58.3 | 21.5 | 41.1 | 34.3 |
| *cer20* | 91.3 | 54.6 | 49.7 | 81.5 | 41.1 | 35.9 |

Amounts expressed as percent of wild-type levels. Data from Hannoufa et al. (1993) and B. Lemieux (unpubl.).

Figure 3, we summarize the suggested function of some of the CER gene products in a proposed wax biosynthesis pathway for *Arabidopsis*.

The accumulation of long chain length fatty acids, aldehydes, and primary alcohols on the surface of the *cer1* mutant suggests that this gene regulates the decarbonylation of fatty aldehydes to alkanes (Fig. 3). Fatty acid decarbonylase activities have previously been detected in particulate cell wall preparations of *Pisum sativum* (Cheesbrough and Kolattukudy 1984) and in the endoplasmic reticulum of the alga *Botryococcus braunii* (Dennis and Kolattukudy 1992). Our more detailed analysis of the epicuticular wax of this mutant shows that triacontanal is by far the most abundant compound on the surface of the *cer1* mutant. Moreover, although *n*-nonacosane is a major component of the epicuticular wax of wild-type plants, this compound is virtually absent on the surface of the *cer1* mutant plants. A similar chemical phenotype has previously been seen in the case of the maize plants with a mutant allele of the *glossy5* locus (Bianchi et al. 1978).

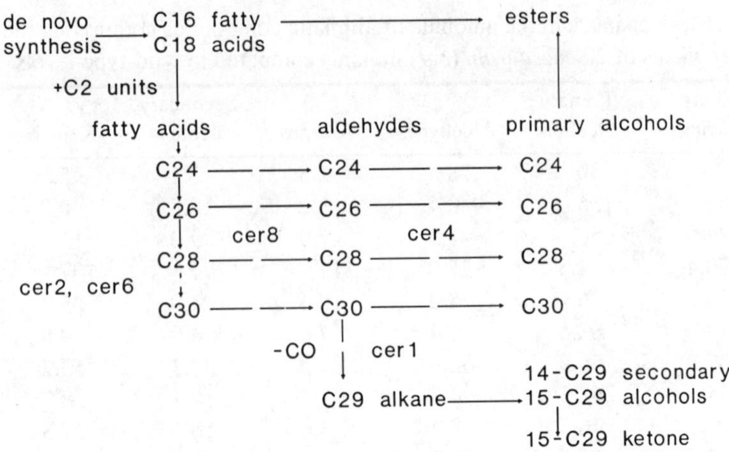


*Figure 3* Proposed metabolic pathway of wax biosynthesis in *Arabidopsis*. Fatty acids are synthesized in the chloroplast and exported to the endoplasmic reticulum, where they are sequentially elongated by the addition of two carbon units up to C24 and C30 chain lengths. At this point, the fatty acids are reduced to fatty aldehydes, which can be decarbonylated to alkanes or reduced to fatty alcohols. The chemical composition of the wax suggests that the decarbonylation (or the availability of substrate to this step) is restricted to 30 carbon chain length aldehydes, whereas the fatty aldehyde reduction system appears to have a more relaxed chain length specificity. Alkanes can be subsequently oxidized to secondary alcohols and ketones. The putative biochemical nature of some of the metabolic blocks in a limited number of *Arabidopsis cer* mutants (*cer1*, *cer2*, *cer4*, *cer6*, and *cer8*) is indicated by the breaks in the pathways.

The chemical composition of the epicuticular wax layer of the *cer4* mutant is characterized by a pronounced reduction in the level of primary alcohols (Table 2). Thus, it appears that this locus regulates the reduction of fatty aldehydes to primary alcohols (Fig. 3). Another locus (*cer8*) appears to regulate the reduction of fatty acids to fatty aldehydes (Fig. 3). Indeed, the surface of this mutant has 434% more free fatty acid than that of the wild-type plant (Table 2). However, the level of primary fatty alcohols does not appear to be as strongly affected in this line (63.5% of wild-type levels). The maize *glossy11* locus may encode a similar gene, as the mutant allele of this locus leads to the absence of long-chain aldehydes (Avato et al. 1985). Previous efforts to purify fatty acid and aldehyde reductase activities suggested that these might be catalyzed by different enzymes (Kolattukudy 1971). More recently, the purification of a jojoba seed fatty acid reductase to homogeneity has led to the isolation of clones of this gene (Anderson et al. 1992). Interestingly, the jojoba fatty acid reductase enzyme was shown to release a fatty alcohol product when expressed in *Escherichia coli* (Anderson et al. 1992).

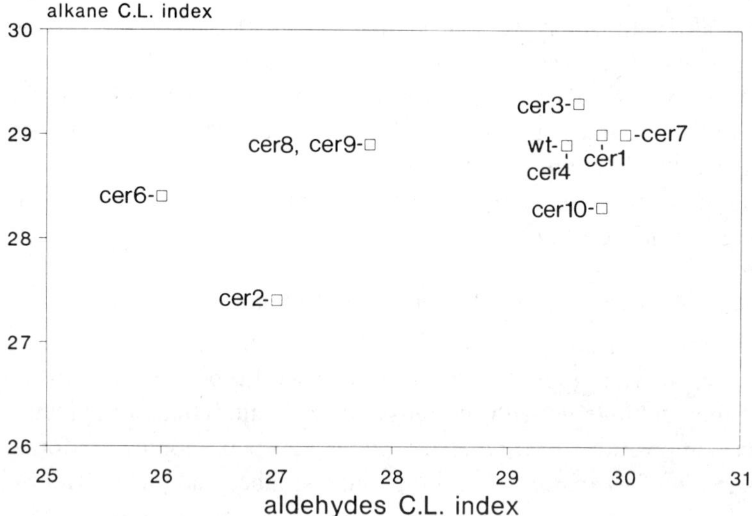


*Figure 4* Chain length indices of *Arabidopsis eceriferum* mutants. The aldehyde chain length index is calculated and plotted relative to the alkane chain length index according to Holloway et al. (1977): $1/100\ \Sigma\ (\% \text{ of } Cn \times n)$, where % of $Cn$ is the composition % of a compound of chain length $n$, and $n$ is the chain length.

In Figure 4, we have plotted the alkane chain length index versus the aldehyde chain length index, according to Holloway et al. (1977), of eight of the *Arabidopsis cer* mutants. The low alkane and aldehyde chain length indices of the *cer2* and *cer6* mutants clearly show that these mutants are defective in fatty acid elongation. Although the *cer2* and *cer6* loci have elevated levels of fatty acids, primary alcohols, and aldehydes, our detailed analysis of their epicuticular wax compositions (Hannoufa et al. 1993) suggests that they regulate the elongation of oc-tacosanoic acid to triacontanoic acid (Fig. 3). The existence of several genes for the regulation of fatty acid elongation in epidermal tissue would be consistent with our previous genetic studies on the elongation of seed fatty acids in *Arabidopsis* (Lemieux et al. 1990), as well as with previous biochemical studies on the partial purification of an epidermal fatty acid elongation complex (Bessoule et al. 1989).

The biochemical function of the majority of the *cer* loci cannot be readily inferred from the alterations in the chemical compositions of their epicuticular waxes (Hannoufa et al. 1993). The chemical phenotypes of these other mutants may be caused by leaky alleles of the corresponding genes. Leaky alleles of some *cer* loci have been isolated before (Koorn-neef et al. 1989). It has previously been suggested that mutations in some of the wax biosynthesis genes of maize which regulate the early steps in fatty acid elongation may lead to similar chemical phenotypes (Avato 1987). Alternatively, these loci may regulate the export of all epicuticular wax compounds to the plant surface rather than enzymes for the synthesis of particular wax compounds. We believe that double mutant analyses and [14]C-labeling studies of the biosynthesis of wax compounds in the mutants will help resolve the biochemical functions of most of the remaining *Arabidopsis cer* mutants.

## Morphology of Wax Blooms on the *eceriferum* Mutants

Figure 5 presents a composite of six scanning electron micrographs of the epicuticular wax blooms on the surface of wild-type *Arabidopsis* and some of the *cer* mutants with the most dramatic alterations in epicuticular wax composition. Two types of epicuticular wax blooms are found on the surface of *Arabidopsis* wild-type plants: tubes and plates (Koornneef et al. 1989). Similar tube-like structures are observed on the surface of *Brassica* spp.; these are believed to be caused by mixtures of symmetrical secondary alcohols (15-nonacosanol) and ketones (15-nonacosanone) (Holloway et al. 1976). Primary alcohols tend to crystallize into plates, whereas aldehydes lead to the production of filamentous wax types (Jeffree et al. 1975; Jeffree and Sandford 1982). Koornneef et al. (1989) observed that plate-like structures (primary alcohols) are present on the younger parts of the plant, whereas the tube-like structures (secondary alcohols and ketones) are more abundant on the older portions of the plant. On the surface of the silique, the tube-like structures are clearly more abundant than on the surface of stems (Fig. 5A, B). The chemical composition of the epicuticular wax on the surface of leaves suggests that leaf waxes have a higher proportion of primary alcohols and fatty acids than alkanes when compared to stem waxes (Table 1). Our previous chemical analyses of the epicuticular wax of *cer* mutants isolated from T-DNA tagged lines also suggested that primary alcohols are formed at the early stages of development of the *cer1* and *cer4* mutants (McNevin et al. 1993). A few filamentous structures are found on the surface of the *cer1-2* mutant, a leaky mutant allele of the aldehyde-rich *cer1* mutant (Fig. 5C). No wax structures could be detected on the surface of the *cer2* mutant (Fig. 5D), and leaky mutant alleles of this locus were never isolated. The surface of the *cer4* mutant is characterized by large sharp-edged plates (Fig. 5E). The high level of alkanes on the surface of this mutant may be responsible for this morphology, as alkanes are known to lead to plates when recrystallized on filters (Jeffree 1986). The fatty-acid-rich surface of the *cer8* mutant (434% of wild-type levels of acids), on the other hand, is characterized by jagged-edged, plate-like structures (Fig. 5F).

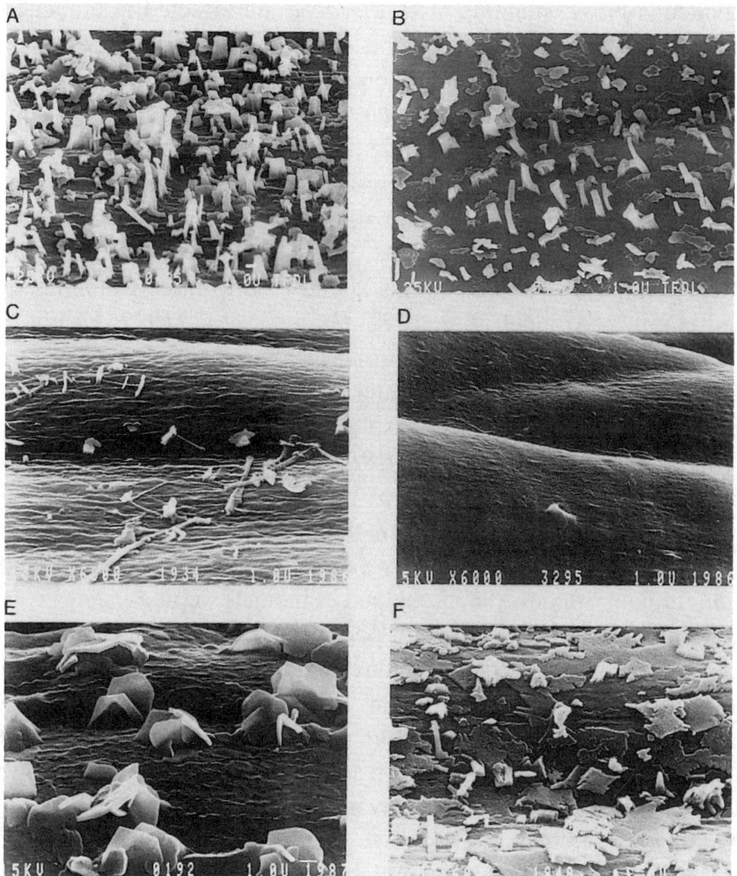

*Figure 5*  Scanning electron microscopy of the plant surface. (*A*) Wild-type stem, (*B*) wild-type silique, (*C*) *cer1-2* stem, (*D*) *cer2* stem, (*E*) *cer4* stem, and (*F*) *cer8* stem.

## CONCLUSIONS AND PERSPECTIVES

We have shown that the glossy bright green appearance of the stems of *Arabidopsis eceriferum* mutants is caused both by reductions in the density of epicuticular wax blooms and by defects in the biosynthesis or deposition of epicuticular wax compounds on the plant surface. The existence of sub-glaucous *glossy* mutant alleles in *B. oleracea* (Baker 1974) suggests that mutant alleles of some of the genes involved in the production of the epicuticular wax layer in *Arabidopsis* may not give a visible phenotype. One of the notable differences between *Arabidopsis* and *Brassica* spp., however, is the apparent absence of branched chain components in the epicuticular wax of *Arabidopsis* (McNevin et al. 1993). As the sub-glaucous *glossy* mutants of *B. oleracea* are affected in branched compounds, it is possible that there is no *Arabidopsis* equivalent to these sub-glaucous mutants. A systematic screen for *Arabidopsis eceriferum* mutants that have an altered epicuticular wax chemical composition but do not display a visual phenotype should help answer this question.

As several *cer* loci have already been mapped (Koornneef 1987), we believe map-based cloning will eventually be used to isolate clones of several *Arabidopsis CER* genes. Indeed, the application of chromosome walking technology to *Arabidopsis* has already led to three examples of map-based cloning (Arondel et al. 1992; Giraudat et al. 1992; Chang et al. 1993; Leyser et al. 1993), as well as to the physical mapping of one third of the *Arabidopsis* genome (Hwang et al. 1991). In the short term, however, we believe that the recent isolation of six T-DNA-tagged *cer* mutants (*BRL1, BRL5, BRL9, BRL11, BRL12,* and *BRL13*) will have a far greater impact on our understanding of the molecular mechanisms that regulate epicuticular wax biosynthesis in plants (McNevin et al. 1993).

Although $^{14}$C-labeling studies of wax compounds in wild-type and *eceriferum* mutant plants have helped elucidate epicuticular wax biosynthetic pathways (von Wettstein-Knowles 1993), the mechanism(s) through which epicuticular wax compounds are deposited onto the plant surface is still the subject of speculation. In 1871, de Bary proposed that epicuticular waxes are extruded through pores on the plant surface (de Bary 1871), and Wiesner argued that wax compounds were transported to the plant surface in solvents which evaporate when they reach the surface, thus allowing the wax to crystallize on the surface (Wiesner 1871). More recently, von Wettstein-Knowles (1982) suggested that wax biosynthesis complexes composed of fatty acid elongases and reductases are associated with the plasma membrane or the plant cell wall. As pores have previously been implicated in the export of wax compounds out of the plant cell (Hall and Donaldson 1962), it is possible that these wax biosynthesis complexes are associated with the pores. Thus, the subunit composition of a given complex would dictate the chemical composition of the wax extruded through the pore. On the other hand, since the morphology of epicuticular wax blooms on the plant surface can be recreated by recrystallizing wax extracts (Jeffree et al. 1975; Jeffree and Sandford 1982), it is possible that wax compounds are transported to the cell wall by vesicles of lipoprotein or glycoproteins before crystallizing on the plant surface. The results of subcellular protein targeting and immunocytochemical localization experiments with lipid transfer proteins suggest that these proteins may be involved in some kind of extracellular lipid transport (Madrid 1991; Thoma et al. 1993). In the absence of localization data on known wax biosynthesis enzymes, however, it is difficult to say which model of wax export best represents the real mechanism(s) involved. Once *CER* genes have been isolated, it will be possible to raise antibodies against their gene products and to immunocytochemically localize wax biosynthesis enzymes. Results obtained from these experiments may help elucidate the mechanisms through which wax compounds are excreted from the interior of the plant cell by indicating if wax biosynthesis enzymes are associated into complexes and if these complexes are localized close to the cell wall or to the endoplasmic reticulum lumen. Another potentially interesting approach may be to express several wax biosynthesis enzyme-encoding genes in yeast in order to reconstruct the wax biosynthesis enzyme complex(es) found in plant cells. As strains of *Candida lipolytica* that excrete fatty acids have been reported (Miyakawa et al. 1984), attempts could be made to create wax-excreting yeast cells by introducing wax biosynthesis enzymes into this yeast.

Some specific epicuticular wax compounds are also believed to play important roles in plant/insect interactions; i.e, as insect anti-feedants (Eigenbrode et al. 1991) or feeding stimulants (Mori 1982), or as settling behavior stimulants (Thompson 1963) or deterrents (Phelan and Miller 1982). Moreover, the ingestion of plant epicuticular wax compounds by herbivorous insects has been shown to influence the insect cuticular lipid composition, thus camouflaging these animals from potential predators and parasitoids (Espelie et al. 1991). Plant surface waxes may also have a direct effect by influencing the release of plant volatile compounds (Spence and Tucknott 1982) that are known to attract predators and parasitoids (Turlings et al. 1990) as well as herbivores (Mitchell 1988). The microscopic wax structures found on plant surfaces may also play a key role in obstructing the movement of insects on the plant surface. One of the best examples of this is found on the inner surface of the pitcher of the carnivorous plant *Nepenthes*. In these plants, wax is deposited around the digestive pitcher in the form of overlapping crystals 1 μm in diameter which are easily dislodged (Juniper and Burras 1962). Another good example of the importance of the morphology of wax blooms in insect adhesion to plant surfaces can be found in the differential ability of beetles to move on the surface of glaucous versus glossy cultivars of *B. oleracea* (Stork 1980; Bodnaryk 1992).

We believe it may be possible to genetically engineer plants for increased resistance to some insect pests which exhibit a preference for a particular epicuticular wax chemical composition on the host plant surface. This might be done by introducing combinations of anti-sense and positive-strand constructs of different *CER* genes under the control of epidermis-specific promoters. It will also be interesting to test the role of pollen waxes in maintaining the integrity of the pollen grain's tryphine layer. This might be done by complementing the reduced fertility phenotype of a *cer* mutant with reduced fertility by introducing the corresponding *CER* gene driven by a tapetal specific promoter. The resulting plant would still have the *eceriferum* phenotype and should have normal fertility under arid growth conditions. Moreover, the introduction of anti-sense *CER* genes under the control of a tapetum-specific promoter should allow us to explore the role of pollen waxes in fertilization in other dry stigma plants like *B. napus*.

## ACKNOWLEDGMENTS

This work was supported by grants OGP-0046649 and STR-0133924 from the Natural Sciences and Engineering Research Council of Canada and by grants-in-aid of research from the University Research Incentive Fund and Pioneer Hi-Bred Production (Georgetown, Ontario) and Pioneer Hi-Bred International (Johnston, Iowa) to B.L. This work was also supported by National Science Foundation grant DIR-9108442, a DOE/NSF/USDA interdisciplinary training grant, and a gift from Monsanto to K.A.F.

## REFERENCES

Anderson, L., C. Fan, C. Lardizabal, T. Sakamoto, M. Pollard, and J. Metz. 1992. Characterization of enzymes involved in biosynthesis of long chain liquid waxes of jojoba (*Simmondsia chinensis*). *Plant Physiol.* **99s:** 77.

Arondel, V., B. Lemieux, I. Hwang, S. Gibson, H.M. Goodman, and C.R. Somerville. 1992. Map-based cloning of a gene controlling omega-3 fatty acid desaturation in *Arabidopsis*. *Science* **258:** 1353–1355.

Avato, P. 1987. Chemical genetics of epicuticular wax formation in maize. *Plant Physiol. Biochemistry* **25:** 179–190.

Avato, P., G. Bianchi, and F. Salamini. 1985. Absence of long chain aldehydes in the wax of the *glossy 11* mutant of maize. *Phytochemistry* **24:** 1995–1997.

Baker, E.A. 1974. The influence of environment on leaf wax development in *Brassica oleracea* var. *gemmifera*. *New Phytol.* **73:** 955–966.

de Bary, A. 1871. Ueber die Wachsuberzuge der Epidermis. *Bot. Zeitung.* **29:** 145–154.

Bessoule, J.J., R. Lessire, and C. Cassagne. 1989. Partial purification of the acyl-CoA elongase of *Allium porrum* leaves. *Arch. Biochem. Biophys.* **268:** 475–484.

Bianchi, G., P. Avato, and F. Salamini. 1978. Glossy mutants of maize. VIII. Accumulation of fatty aldehydes in surface waxes of *gl5* maize seedlings. *Biochem. Genet.* **16:** 1015–1021.

Bianchi, G., C. Murelli, and E. Ottaviano. 1990. Maize pollen lipids. *Phytochemistry* **29:** 739–744.

Bodnaryk, R.P. 1992. Leaf epicuticular wax, an antixenotic factor in Brassicaceae that affects the rate and pattern of feeding of flea beetles, *Phyllotreta cruciferae* (Goeze).

*Can. J. Plant Sci.* **72:** 1295–1303.

Chang, C., S.F. Kwok, A.B. Bleecker, and E.M. Meyerowitz. 1993. *Arabidopsis* ethylene-response gene *ETR-1*: Similarity of product to two-component regulators. *Science* **262:** 539–544.

Cheesbrough, T.M. and P.E. Kolattukudy. 1984. Alkane biosynthesis by decarbonylation of aldehydes catalyzed by a particulate preparation from *Pisum sativum*. *Proc. Natl. Acad. Sci.* **81:** 6613–6617.

Dellaert, L.M.W., J.Y.P. van Es, and M. Koornneef. 1979. *Eceriferum* mutants in *Arabidopsis thaliana* (L.) Heynh. II. Phenotypic and genetic analysis. *Arabidopsis Inf. Serv.* **16:** 10–26.

Denna, D.W. 1970. Transpiration and the waxy bloom in *Brassica oleracea* (L). *Aust. J. Biol. Sci.* **23:** 27–31.

Dennis, M. and P.E. Kolattukudy. 1992. A cobalt-porphyrin enzyme converts a fatty aldehyde to a hydrocarbon and CO. *Proc. Natl. Acad. Sci.* **89:** 5306–5310.

Eigenbrode, S.D., K.E. Espelie, and A.M. Shelton. 1991. Behavior of neonate diamondback moth larvae [*Plutella xylostella* (L.)] on leaves and on extracted leaf waxes of resistant and susceptible cabbages. *J. Chem. Ecol.* **17:** 1691–1704.

Espelie, K.E., E.A. Bernays, and J.J. Brown. 1991. Plant and insect cuticular lipids serve as behavioral cues for insects. *Arch. Insect Biochem. Physiol.* **17:** 223–233.

Giraudat, J., B.M. Hauge, C. Valon, J. Smalle, F. Parcy, and H.M. Goodman. 1992. Isolation of the *Arabidopsis ABI3* gene by positional cloning. *Plant Cell* **4:** 1251–1261.

Hall, D.M. and L.A. Donaldson. 1962. Secretion from pores of the surface wax on plant leaves. *Nature* **197:** 1196.

Hannoufa, A., J.P. McNevin, and B. Lemieux. 1993. Epicuticular waxes of *eceriferum* mutants of *Arabidopsis thaliana*. *Phytochemistry* **33:** 851–855.

Holloway, P.J. 1969. Chemistry of leaf wax in relation to wetting. *J. Sci. Food Agric.* **20:** 124–128.

Holloway, P.J., C.E. Jeffree, and E.A. Baker. 1976. Structural determination of secondary alcohols from plant epicuticular waxes. *Phytochemistry* **15:** 1768–1770.

Holloway, P.J., G.A. Brown, E.A. Baker, and M.J.K. Macey. 1977. Chemical composition and ultrastructure of the epicuticular wax in three lines of *Brassica napus* (L). *Chem. Phys. Lipids* **19:** 14–127.

Hwang, I., T. Kohchi, B. Hauge, H.M. Goodman, R. Schmidt, G. Cnops, C. Dean, S. Gibson, K. Iba, B. Lemieux, V. Arondel, L. Danhoff, and C.R. Somerville. 1991. Identification and map position of YAC clones comprising one-third of the *Arabidopsis* genome. *Plant J.* **1:** 367–374.

Jeffree, C.E. 1986. The cuticle, epicuticular waxes and trichomes of plants, with reference to their structure, functions and evolution. In *Insects and the plant surface* (ed. R. Southwood and B.E. Juniper), pp. 23–64, Edward Arnold, London.

Jeffree, C.E. and A.P. Sandford. 1982. Crystalline structure of plant epicuticular waxes demonstrated by cryostage scanning electron microscopy. *New Phytol.* **91:** 549–559.

Jeffree, C.E., E.A. Baker, and P.J. Holloway. 1975. Ultrastructure and recrystallization of plant epicuticular waxes. *New Phytol.* **75:** 539–549.

Jeffree, C.E., R.P.C. Johnson, and P.G. Jarvis. 1971. Epicuticular wax in the stomatal antechambers of Sitka spruce, and its effects on the diffusion of water vapour and carbon dioxide. *Planta* **98:** 1–10.

Juniper, B.E. and J.K. Burras. 1962. How pitcher plants trap insects. *New Sci.* **73:** 75–77.

Kolattukudy, P.E. 1971. Enzymatic synthesis of fatty alcohols in *Brassica oleracea*. *Arch. Biochem. Biophys.* **142:** 701–709.

Koornneef, M. 1987. Linkage map of *Arabidopsis thaliana*. In *Genetic maps 1987* (ed.

S.J. O'Brien), pp. 742–745. Cold Spring Harbor Laboratory, Cold Spring Harbor, New York.

Koornneef, M., C.J. Hanhart, and F. Thiel. 1989. A genetic and phenotypic description of *eceriferum* (*cer*) mutants of *Arabidopsis thaliana*. *J. Hered.* **80:** 118–122.

Lemieux, B., M. Miquel, C.R. Somerville, and J.A. Browse. 1990. Mutants of *Arabidopsis* with alterations in seed lipid fatty acid composition. *Theor. Appl. Genet.* **80:** 234–240.

Leyser, H.M., C.A. Lincoln, C. Timpte, D. Lammer, and M. Estelle. 1993. *Arabidopsis* auxin-resistance gene *AXR1* encodes a protein related to ubiquitin-activating enzyme E1. *Nature* **364:** 161–164.

Leyton, L. and B.E. Juniper. 1963. Cuticle structure and water relations of pine needles. *Nature* **198:** 770–771.

Madrid, S.M. 1991. The barley lipid transfer protein is targeted into the lumen of the endoplasmic reticulum. *Plant Physiol. Biochem.* **29:** 695–703.

Martin, J.T. and B.E. Juniper. 1970. *The cuticle of plants*. Edward Arnold, London.

McNevin, J.P., W. Woodward, A. Hannoufa, K.A. Feldmann, and B. Lemieux. 1993. Isolation and characterization of *eceriferum* (*cer*) mutants induced by T-DNA insertions in *Arabidopsis thaliana*. *Genome* **36:** 610–618.

Mitchell, B.K. 1988. Adult leaf beetles as models for exploring the chemical basis of host-plant recognition. *J. Insect Physiol.* **34:** 213–225.

Mori, M. 1982. *n*-Hexacosanol and *n*-octacosanol: Feeding stimulants for larvae of the silkworm, *Bombyx mori*. *J. Insect Physiol.* **28:** 969–973.

Miyakawa, T., H. Nakajima, K. Hamada, E. Tsuchiya, T. Kamiryo, and S. Kukui. 1984. Isolation and characterization of a mutant of *Candida lipolytica* which excretes long-chain fatty acids. *Agric. Biol. Chem.* **48:** 499–503.

Phelan, P.L. and J.R. Miller. 1982. Post-landing behavior of alate *Myzus persicae* as altered by (E)-beta-farnesene and three carboxylic acids. *Entomol. Exp. Appl.* **32:** 46–53.

Poethig, R.S. 1990. Phase change and the regulation of shoot morphogenesis in plants. *Science* **250:** 923–930.

Preuss, D., B. Lemieux, G. Yen, and R.W. Davis. 1993. A conditional sterile mutation eliminates surface components from *Arabidopsis* pollen and disrupts cell signaling during fertilization. *Genes Dev.* **7:** 974–985.

Sogaard, B. and P. von Wettstein-Knowles. 1987. Barley genes and chromosomes. *Carlsberg Res. Commun.* **52:** 123–196.

Spence, R.-M.M. and O.G. Tucknott. 1982. Volatiles from the epicuticular wax of watercress [*Roroppa Nasturtium aquaticum*]. *Phytochemistry* **22:** 2521–2523.

Stork, N.E. 1980. Role of wax blooms in preventing attachment to *Brassicas* by the mustard beetle, *Phaedon cochleariae*. *Entomol. Exp. Appl.* **28:** 100–107.

Thoma, S., Y. Kaneko, and C.R. Somerville. 1993. A non-specific lipid transfer protein from *Arabidopsis* is a cell wall protein. *Plant J.* **3:** 427–436.

Thompson, K.F. 1963. Resistance to the cabbage aphid (*Brevicoryne brassicae*) in *Brassica* plants. *Nature* **198:** 209.

Turlings T.C.J., J.H. Tumlinson, and W.J. Lewis. 1990. Exploitation of herbivore induced plant odors by host seeking parasitic wasps. *Science* **250:** 1251–1253.

von Wettstein-Knowles, P.M. 1982. Elongases and epicuticular wax biosynthesis. *Physiol. Veg.* **20:** 797–809.

———. 1993. Waxes, cutin, and suberin. In *Lipid metabolism in plants* (ed. T.S. Moore, Jr.), pp. 127–166. CRC Press, Boca Raton, Florida.

Walton, T.J. 1990. Waxes, cutin, and suberin. In *Methods in plant biochemistry* (ed. J.L.

Harwood and J.R. Bowyer), vol. 4, pp. 105–158, Academic Press, New York.

Wiesner, J. 1871. Beobachtungen uber die Wachsuberzuge der Epidermis. *Bot. Zeitung.* **29**: 769–774.

Yang, G., B.R. Wiseman, and K.E. Espelie. 1992. Cuticular lipids from silks of seven corn genotypes and their effect on development of corn earworm larvae [*Helicoverpa zea* (Boddie)]. *J. Agric. Food Chem.* **40**: 1058–1061.

# 38

# The Plant Cytoskeleton

**Richard B. Meagher**
Department of Genetics
University of Georgia
Athens, Georgia 30602

**Richard E. Williamson**
Plant Cell Biology Group & Plant Science Centre
Research School of Biological Sciences
Australian National University
Canberra ACT 2601, Australia

The cytoskeleton is a system of fibrous polymers (F-actin, microtubules, and intermediate filaments) that provide spatial organization to many crucial subcellular processes. The plant polypeptides related to those forming intermediate filaments in animal cells are not discussed here, since their in vivo aggregation state and biological functions are not yet known (Shaw et al. 1991; McNulty and Saunders 1992). In contrast, it is known that F-actin and microtubules participate in mitosis, cytokinesis, cellulose microfibril alignment, and organelle movements. Although the cytoskeleton is an entirely cytoplasmic system, many cytoskeletal components lie near the plasma membrane, where they are well placed to respond to chemical and physical stimuli transmitted across the plasma membrane (Luna 1991). Importantly, in plants, they influence wall placement at cytokinesis and wall architecture during cell expansion. The subcellular organization of the cytoskeleton during cell division and cell expansion and the characteristic planes of cell division seen in various organs are reflected in cell and organ shape. Thus, the effects on cell-wall structure amplify and propagate the cytoskeleton's influence on development.

The subunits of the cytoskeleton's major fibrous polymers (actins and tubulins) are quite highly conserved among all eukaryotes, even though the roles they perform vary enormously through the eukaryotic kingdoms and even within single cells during ontogeny. Such versatility probably requires the participation of many less highly conserved proteins that interact with these primary cytoskeletal polymers, such as microtubule-associated proteins (MAPs) and actin-binding proteins (ABPs), together with motor proteins specific for both actin and microtubules. In addition,

*Arabidopsis*
© 1994 Cold Spring Harbor Laboratory Press  0-87969-428-9/94 $5 + .00

many regulatory systems control these interactions. Information regarding such accessory proteins and regulatory systems in plants is very limited indeed, and knowledge of the plant cytoskeleton in general is much less detailed than is knowledge of the animal and protist cytoskeleton. Therefore, the future scope for molecular and genetic approaches in plant systems is particularly large.

In this chapter, we provide a brief overview of the cytoskeleton of higher plants. In addition, we describe cell biological and molecular biological studies that have used *Arabidopsis*. Several detailed reviews of the plant cytoskeleton exist and should be consulted for further information (Baskin and Cande 1990; Emons et al. 1991; Lloyd 1991; Meagher 1991; Staiger and Lloyd 1991; Williamson 1991, 1993).

## COMPONENTS OF THE CYTOSKELETON

Hundreds of proteins are involved in the dynamic cytoskeletal structures in eukaryotic cells, and many are well conserved in amino acid sequence among all four eukaryotic kingdoms. Among these, F-actin and microtubules are the major fibrous polymers. Because of their abundance, more is known about the actin and tubulin proteins that comprise these structures than about any of the other proteins in the cytoskeleton.

### The F-actin System

The 43-kD G-actin polypeptide polymerizes into F-actin (Fig. 1) at physiological salt concentrations (G and F denote globular and filamentous, respectively). The G-actin monomers have a flattened globular structure measuring about 5.5 nm x 5.5 nm x 3.5 nm (Kabsch et al. 1990; Bremer and Aebi 1992). They are helically arranged into two parallel strands to make one actin filament (Fig. 1). The average filament diameter is about 7 nm. The subunits on opposite strands are staggered about half of the subunit length and oriented with their flat surfaces in parallel planes. Unlike double helices made from globular subunits, the actin filament appears narrowest where the flattened monomers in different strands are both viewed end-on, adjacent to each other, and widest where the monomers are viewed one on top of the other and fanned outward (Holmes et al. 1990). It takes about 13 monomers on each strand to make one turn of the filament helix (36 nm). The filaments of F-actin are less distinctive than microtubules in electron micrographs, partially due to their smaller size. They are definitively recognized as actin by their ability to bind actin antibodies, the fungal toxin phalloidin, or fragments of the actin motor protein, myosin. The distinctive arrowhead appearance

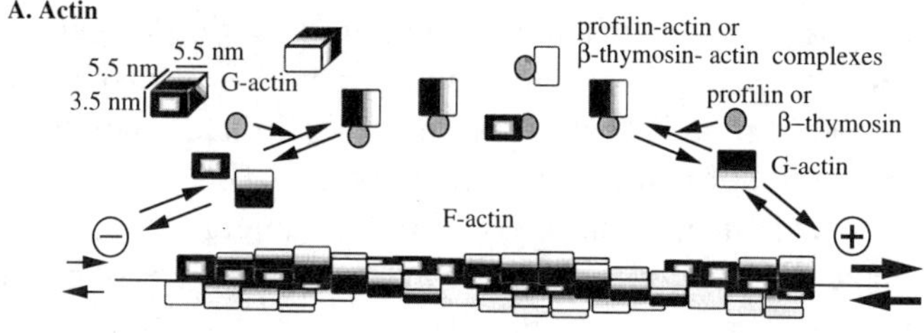


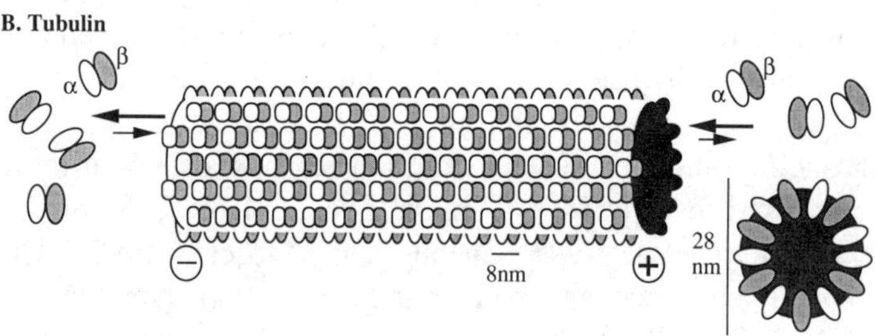


*Figure 1*  Polymerization and depolymerization of actin and tubulin polymers. (*A*) The 43-kD G-actin monomer is normally bound to monomer-binding proteins like profilin or β-thymosin. It is transferred rapidly on and off the plus (+) end of filamentous F-actin but only slowly from the minus (–) end. Each G-actin monomer is about 5.5 nm x 5.5 nm x 3.5 nm and is polymerized into parallel strands which are wound helically around each other with about 13.5 pairs of residues per turn. The filaments have an average diameter of 7 nm. (*B*) The approximately 50-kD α-tubulin and β-tubulin monomers form heterodimers and then assemble into a hollow microtubule with a diameter of about 28 nm. Assembly is most rapid on the plus (+) end and depolymerization is most rapid from the minus (–) end. Although heterodimers are added directly to the microtubule, each microtubule is composed of 13 parallel protofilaments. Each protofilament is composed of a string of α/β-tubulin heterodimers.

of filaments coated with myosin reveals the underlying polarity of each filament that also expresses itself in different rates of assembly at the two filament ends. Assembly and disassembly are rapid at the barbed or plus filament end and slow at the minus or pointed end. A group of fungal toxins termed cytochalasins cap barbed filament ends (i.e., block addition or removal or monomers), although it is not clear if this explains all their in vivo effects (Cooper 1987).

The assembly and disassembly of F-actin and its organization into bundles and networks are controlled in animal cells by a battery of ABPs. ABPs have been categorized functionally into monomer-binding, filament-capping, -severing, -side-binding, and -cross-linking proteins (Vandekerckhove 1990; Hartwig and Kwiatkowski 1991). Plants have been shown to possess the monomer-binding protein profilin (Valenta et al. 1991; Kim et al. 1993) and an immunoreactive homolog of β-spectrin (Michaud et al. 1991), a protein that cross-links F-actin and sometimes microtubules in the cortex of many animal cells. Some ABPs incorporate multiple functions and many, such as profilin, interact with the phosphoinositide and $Ca^{++}$ regulatory pathways to control cytoskeletal structure.

Myosins are ATPases that interact with actin and generate movement or tension according to the way in which the proteins are assembled and anchored. They are a diverse family of proteins in which the head regions, which interact with actin and have ATPase activity, show significant amino acid homology, whereas the tail regions are much more variable and are used to categorize them as myosins I or II (Korn and Hammer 1988; Pollard et al. 1991; Cheney and Mooseker 1992). The tails of myosins II dimerize and can aggregate to higher-molecular-weight filaments. Tails of the myosin I type do not self-associate, vary in length, and show diverse properties such as membrane association, calmodulin-binding, or actin-binding. Unconventional myosins that do not fit these categories have also been recognized. They dimerize only through short regions of their tails and do not aggregate to high-molecular-weight filaments. Recently, more complex classifications based on sequences of the head region have been proposed (Cheney et al. 1993; Goodson and Spudich 1993).

Myosins have characteristic ATPase activities: $K^{+}$EDTA > $Ca^{++}$ > $Mg^{++}$ at high ionic strengths and a $Mg^{++}$-ATPase that is actin-activated at low ionic strengths. By monitoring these ATPase activities, myosins have been purified from the alga *Nitella* (Kato and Tonomura 1977) and from the flowering plants *Lycopersicon* (Vahey et al. 1982) and *Heracleum* (Turkina et al. 1987) and partially purified from *Egaria*, *Luffa*, and *Pisum* (Ohsuka and Inoue 1979; Ma and Yan 1988; Ma and Yen 1989). *Lilium* myosin was partially purified by monitoring its catalysis of ATP-dependent movements of actin (Kohno et al. 1992; Yokota et al. 1993). The $M_r$ of the heavy chains of these plant myosins vary widely (205 kD in *Nitella*, 180 kD in *Egaria*; 170 kD in *Lilium*; 165 kD in *Pisum*, and 100 kD in *Lycopersicon*), as do values determined by immunoblotting using anti-myosin antibodies; ≥ 200 kD in *Allium* (Parke et al. 1986) and in the algae *Chara* (Grolig et al. 1988) and *Ernodesmis* (La Claire 1991); 160–170 kD in *Phaseolus, Triticum, Pisum* (Lin et al. 1989), and *Nicotiana* (Tang et al. 1989); and ≤ 110 kD in the algae *Nitella* (Grolig et al. 1988) and *Ernodesmis* (La Claire 1991). Non-plant myosins show a similar range of $M_r$ values from the low-$M_r$ myosins I to the high-$M_r$ myosins II (Korn and Hammer 1988; Pollard et al. 1991;

Cheney and Mooseker 1992). It is therefore tempting to believe that the range of $M_r$ values for plant myosins indicate that a similar range of structural variation exists. The plant values have to be regarded cautiously, however, whether based on biochemistry (where partial purification or proteolysis may be problems) or on immunoblotting (where identification may depend on a single epitope that cannot rigorously identify a whole protein as a functional myosin). Given the general difficulties encountered during protein purification from plants, gene cloning probably will supply much of the data from which to predict the molecular properties of plant myosins. Recently, a large number of myosin genes have been cloned from *Arabidopsis* and are being characterized (J. Schiefelbein, pers. comm.). A gene encoding an unconventional myosin from *Arabidopsis* has been sequenced (Kinkema and Schiefelbein 1992), as have two myosin-head-related cDNAs from *Arabidopsis* and *Anemia* (Moepps et al. 1993).

## The Microtubule System

Microtubules are rigid tubular strands 24 nm in diameter (Fig. 1). The tube is constructed from heterodimers of α- and β-tubulin. Each tubulin monomer is approximately 50 kD. The 8-nm heterodimers wrap around the surface of the microtubule in a helical array and are stacked one on top of the other along the length of the microtubule in protofilaments, with 13 protofilaments aligned in parallel around the circumference of the tube. Although these protofilaments are an integral part of the microtubule, protofilaments do not exist in isolation from the microtubule. The microtubule itself is built directly from either or both the plus and minus ends from the pool of heterodimers. It can be built rapidly from the plus end but only slowly from the minus end. Microtubules interact with the rest of the cell cytoarchitecture through a large number of MAPs.

Few MAPs are documented from higher plants. Microtubule bundling has been ascribed to 77-kD (Cyr and Palevitz 1989) and 65-kD (Chang-Jie and Sonobe 1993) polypeptides. A MAP fraction from cultured maize cells that promotes tubulin polymerization and microtubule bundling includes an 83-kD polypeptide that reacts with antibodies to neuronal τ factor (Vartard et al. 1991; Schellenbaum et al. 1993). A 34-kD polypeptide has been purified from mung bean that binds to a carboxy-terminal sequence of β-tubulin (P.P. Jablonsky et al., in prep.). This region of the tubulin protein is implicated in controlling polymerization and binding the majority of MAPs (see below). Several families of motor proteins (notably dyneins and kinesins) exist in animal cells that can move microtubules relative to each other and other structures relative to microtubules

(Vallee and Shpetner 1990; Endow 1991). An immunoreactive homolog of kinesin has been reported in plants (Tiezzi et al. 1992). Putative cDNA clones for dynamin1 and kinesin have been identified by partial sequencing of randomly isolated *Arabidopsis* cDNA clones (DBEST sequences #34C6T7 and #34H8T7, respectively).

## CYTOSKELETAL ASSEMBLY AND ORGANIZATION

The assembly of F-actin and microtubules in cells must be spatially and temporally regulated to generate the arrays that direct successive phases of the cell cycle and cell differentiation. The polymerization of actin and tubulin in vitro provides the background against which the behavior of F-actin and microtubules in cells must be seen. Detailed polymerization properties are known only from experiments with mammalian and protist proteins, but sequence conservation probably gives plant subunits similar properties. The polymerization of both G-actin and tubulin shows many common features (Carlier 1989, 1991b). Assembly of both polymers begins with a nucleation reaction in which several monomers associate by an energetically unfavorable reaction. The polymer then grows by adding monomer at its two ends. Both monomers bind nucleotide triphosphates (XTP), GTP in the case of tubulin and ATP in the case of actin. Nucleotide hydrolysis lags behind polymerization so that subunits still bound to XTP cap the ends of growing polymers while internal subunits are bound to XDP. The length of the XTP cap increases during rapid extension and so is bigger at the fast-growing plus end of each polymer than at the slow-growing minus end. XTP subunits dissociate less readily from the polymer end than do XDP subunits, so that fast-growing polymers with large XTP caps are more stable than slow-growing polymers. Chance loss of the XTP cap initiates rapid shortening by the ready loss of exposed XDP subunits. Because of such dynamic instability, a steady-state population of polymers comprises mainly slowly growing polymers with a smaller number of rapidly shortening polymers, rather than an equilibrium population with polymers of characteristic and unchanging length. Polymer extension requires that free monomer concentrations exceed a critical concentration, which is greater for the minus end than for the plus end. If the monomer concentration permits addition only at the plus end and causes net loss at the minus end, subunits in both polymers move from the minus to the plus end. This property, known as "treadmilling," results in net movement of the polymer in the plus direction.

These properties of dynamic instability and treadmilling are seen in vivo to varying degrees in different animal and protist cells where many proteins that modify all phases of polymerization and depolymerization exist. Looking first at microtubules, accessory proteins may promote nucleation and, in animal and many lower plant cells, are confined to one or more well-defined microtubule centers (Kimble and Kuriyama 1992). These centers also stabilize the slow-growing minus ends so that microtubule growth and shrinkage occur only at the distal plus end. Actin assembly rarely relies on a small number of centers: G-actin-binding proteins reduce monomer concentration, soluble and membrane-associated proteins promote actin nucleation, capping proteins prevent loss or addition at filament ends, and severing proteins cut existing filaments to expose uncapped regions (Cooper 1991). All classes of interaction can be regulated by factors such as $Ca^{++}$ and inositol phosphates. It is argued that dynamic instability affords the cell considerable flexibility (Kirschner and Mitchison 1986), because the most-favored arrangement of polymers is selectively stabilized, whereas unsuitable arrangements are rapidly lost. Such flexibility has significant costs in XTP hydrolysis, however, and in many cells actin- and microtubule-binding proteins stabilize the polymers so that dynamic instability is absent.

A large gap separates the detailed mechanisms studied in animal and protist systems from what is known of cytoskeletal assembly and disassembly in plant cells. Figure 2 summarizes the changes that occur in actin and microtubule organization in meristematic cells, and Figure 3 summarizes the situation in vacuolated cells during interphase and when stimulated to divide either during normal ontogeny or after wounding. The diagrams emphasize that cytoskeletal elements assemble and disassemble at precisely determined times and sites, and that microtubules and F-actin coexist in some but not all arrays.

Efforts to understand how microtubule arrays change in higher plant cells have focused on cytological detection of where microtubule assembly begins when arrays change during the cell cycle and when microtubules reassemble after experimental depolymerization. Unlike lower plant and animal cells, there are no obvious and centralized microtubule organizing centers in higher plant cells. Several assembly sites have been proposed: cell corners where interphase microtubules reassemble after cytokinesis (Gunning et al. 1978); the perinuclear region when interphase (Wick 1985) and preprophase band microtubules (Flanders et al. 1990) assemble; the cortical region when the interphase array recovers from experimentally imposed disassembly (Falconer and Seagull 1987). Unfortunately, interpreting such observations is difficult (Williamson 1991), and none of the evidence is compelling. When purified tubulin is derivatized with a fluorescent tag and microinjected into living plant cells, it first appears in microtubules that are part of the interphase or preprophase arrays rather than arriving in those arrays after assembling

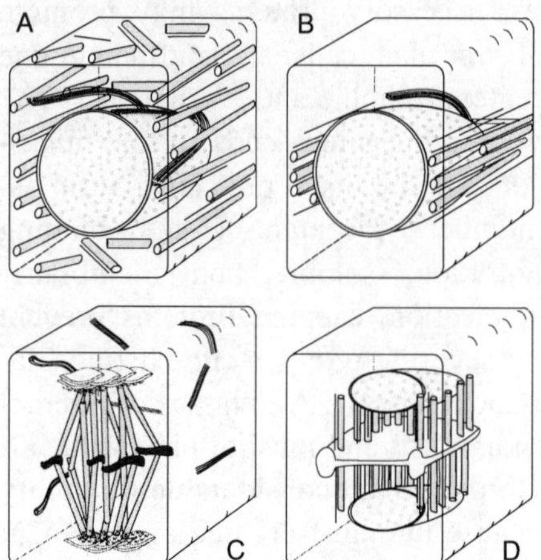

*Figure 2*   The arrangement of F-actin and microtubules in meristematic plant cells. Diagrammatic view of F-actin and microtubules in meristematic cells bisected at different stages of the cell cycle. (*A*) Interphase cell; microtubules and single actin filaments lie in the cortex, and larger bundles of F-actin occur deeper in the cytoplasm. (*B*) Preprophase cell; microtubules and F-actin are concentrated in the preprophase band with some actin bundles remaining in the deeper cytoplasm. (*C*) Mitotic cell; spindle microtubules are loosely focused on the poles where endoplasmic reticulum and granular material are concentrated. Actin bundles in the cytoplasm are reduced in number and length. (*D*) Cytokinetic cell; microtubules and F-actin are concentrated in the phragmoplast around the growing cell plate.

elsewhere in the cell (e.g., the perinuclear region) (Cleary et al. 1992).

The short life span of the microtubule arrays that form at mitosis and cytokinesis sets an upper limit of minutes to the longevity of individual microtubules, but measurements using fluorescence-photobleaching show that spindle microtubules in animal cells turn over in seconds (Salmon et al. 1984). The interphase array of plant cells remains for several hours, however, particularly in cells that have entered their elongation phase. What is the life span of its component microtubules? Photobleaching experiments (J.M. Hush, pers. comm.) indicate half-lives measured in tens of seconds. Whether these and other plant microtubules conform in detail to the dynamic instability model remains unclear, and further information regarding microtubule dynamics in diverse plants is urgently needed.

Actin assembly has rarely been studied in plant cells. In the alga *Chara*, considered to be a close protist ancestor of higher plants, severed actin bundles extend only when the plus ends of filaments are exposed (Williamson et al. 1986). Very low cytochalasin concentrations block

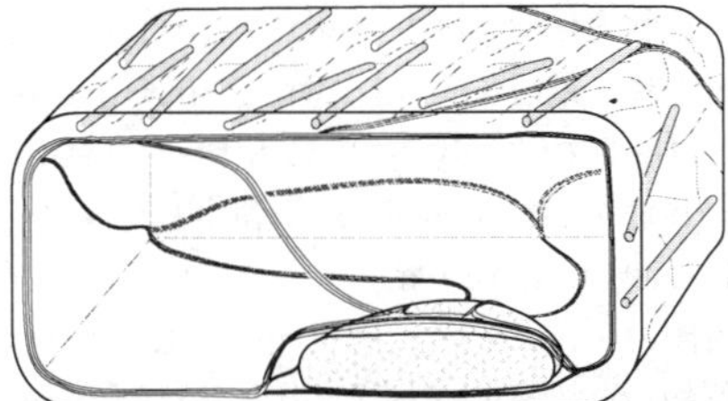

*Figure 3*   The arrangement of F-actin and microtubules in large, vacuolated plant cells. Diagrammatic view of the cytoskeleton in a cell bisected after enlarging and vacuolating. Bundles of actin filaments traverse the cytoplasm and enmesh the nucleus. Smaller bundles or single actin filaments are present in the cortical cytoplasm together with aligned cortical microtubules.


that extension, consistent with cytochalasin capping the plus ends of filaments. Failure to extend exposed minus ends requires either a G-actin concentration that prevents net assembly or disassembly at minus ends or capping of the minus ends. No studies using higher plants address where in the cell and from which ends actin filaments or filament bundles grow. Cytochalasins reversibly fragment actin bundles in higher plant cells (Parthasarathy 1985), but how the system reassembles when cytochalasin is washed out remains unknown.


## CYTOSKELETAL FUNCTION

### Division Plane Determination

Division plane determination is of major developmental significance in plant cells which, unlike animal cells, do not migrate and are more limited in shape changes by their cell walls. The division plane commits a cell to a limited number of positions in the three-dimensional structure of a developing organ. In vegetative cells of flowering plants, the preprophase band of microtubules precisely predicts where the cell plate will fuse with the parental cell wall to end cytokinesis (Fig. 2B,D), even in cases such as the stomatal mother cells in grasses where the new wall is curved (Wick 1991). The microtubule band assembles in cells that have replicated their DNA and disassembles before the spindle forms (Gunning and Sammut 1990). Assembly and disassembly of the band may both involve phosphorylation of unknown proteins (Mineyuki et al. 1991; Katsuta and Shibaoka 1992), but it is not known whether intact, interphase microtubules migrate to form the band or whether interphase microtubules disassemble and new microtubules assemble at the site of the band. The microtubule band often, but perhaps not always, lies within a wider band of F-actin (Wick 1991). The microtubule band forms but does not narrow properly when actin is disrupted with cytochalasin (Eleftheriou and Palevitz 1992), whereas the actin band forms only if microtubules are intact (Katsuta et al. 1990; McCurdy and Gunning 1990; Mineyuki and Palevitz 1990). The division site (and preprophase band site) may already be approximately determined before microtubules assemble, since nuclei migrate to the future division site in vacuolated cells of *Nautilocalyx* before the band of preprophase microtubules forms (Venverloo and Libbenga 1987).

At the end of cytokinesis, the behavior of the cell plate suggests that it is guided to the site previously occupied by the preprophase band, which subsequently facilitates flattening of the small undulations in the new wall (Mineyuki and Gunning 1990). The division site must remain marked after microtubules have disassembled. Strands of actin (Lloyd and Traas 1988) or the absence of actin (Cleary et al. 1992) have both been invoked as mechanisms guiding growth of the new wall toward the site formerly occupied by the preprophase band.

### Mitosis and Cytokinesis

Microtubules are the most prominent components of the spindle, the structure that partitions chromosomes. There is little evidence that actin participates in spindle function, even though filaments can often be found within the spindle, as reviewed by Baskin and Cande (1990). Microtubule assembly is probably nucleated at the spindle poles, which are broad and ill-defined in higher plants, lacking the obvious microtubule centers of animal and lower plant cells (Brown and Lemmon 1990). Observations of cells containing fluorescently tagged tubulin (Mitchison and Sawin 1990) and mitotic mutants deficient in cellular components such as microtubule motor proteins (Endow 1993) seem likely to resolve many long-standing controversies regarding how chromosomes move.

In cytokinesis, the cell plate grows centrifugally as Golgi vesicles containing cell wall materials fuse. Actin (Clayton and Lloyd 1985) and myosin (Parke et al. 1986) localize to the phragmoplast in addition to microtubules. Therefore, vesicles potentially could move along either F-actin or microtubules to the cell-plate site. Unfortunately, vesicle movements are not directly observable in living cells, and inhibitor experiments do not clearly show whether one or both cytoskeletal systems participate (Baskin and Cande 1990).

### Wall Architecture and Cell-shape Determination

Cell-shape determination involves the primary cell wall yielding under the stresses imposed on it by cell turgor. After determination of the division plane, cell expansion and cell wall deposition are the most important factors directing plant cell and organ development. Asymmetrically expanding cells (e.g., cells in roots and internodes) have walls that are mechanically highly anisotropic (i.e., they stretch much more readily in some directions than others). Such cells (including those of *Arabidopsis*) respond to microtubule depolymerization by pronounced radial swelling (Green 1962; Giddings and Staehelin 1991). In contrast, cells such as pollen tubes and root hairs in which all extension is localized to the tip (see tip growth below) show a different form of organization and do not usually swell after such treatments. The swelling is not a direct mechanical weakening; the cortical cytoskeleton itself is probably mechanically insignificant compared to the wall. The effect of microtubule depolymerization on cell shape probably reflects a role for microtubules in aligning cellulose microfibrils, one of the major determinants of mechanical anisotropy in walls (Richmond et al. 1980). Newly synthesized microfibrils coalign with cortical microtubules. Depolymerizing microtubules either causes loss of microfibril alignment (Richmond 1983; Giddings and Staehelin 1991) or maintains current alignments while preventing new ones from starting (Robinson and Quader 1981). This view gives microtubules a major role in cell-shape determination and in morphogenetic events, such as lateral-root initiation, where microfibril alignment must change to allow a new axis of elongation to develop (Green 1980; Lyndon 1990). Although many, perhaps all, cortical microtubules have an associated filament, it is not yet clear that this is actin (Lancelle and Hepler 1991). Cytochalasin does not cause pronounced radial swelling of elongating cells, and any role for actin remains unknown.

The mechanisms by which microtubules align (Williamson 1991) and by which they in turn align microfibrils (Giddings and Staehelin 1991) remain unknown. Furthermore, the view that microtubules align microfibrils has its critics, who observe that microtubule depolymerization does not always realign cellulose (Emons et al. 1992) and that the liquid crystalline properties of wall polymers allow self-organization without direction from a cytoplasmic template (Reis et al. 1991).

The deposition of aligned cellulose in secondary walls continues in the absence of microtubules in some cell types, such as root hairs (Emons et al. 1992). In other cell types where microfibril deposition is localized and strongly aligned (e.g., xylem vessel elements), the submembranous microtubules parallel extracellular microfibril deposition, and microtubule depolymerization prevents ordered cellulose deposition (Seagull and Falconer 1991). F-actin is also thought to participate in this process, but by complementing tubulin localization and not necessarily by paralleling microtubules.

## Organelle Movements

Almost all categories of plant organelles move. Movements can involve all organelle types (cytoplasmic streaming) or be specific for a single type (Williamson 1993). Streaming probably distributes within the cell organelles and macromolecules whose diffusion is restricted by their high $M_r$. Movements of specific organelles position nuclei in relation to cytokinesis and chloroplasts in relation to light. They also ensure transmission of organelle genomes and their deletion from certain cell lines.

The favored model for the mechanism of plant organelle movements is that organelles associate with myosins which enable them to move over bundles of actin filaments (Williamson 1993). Three main pieces of evidence support this model: Organelles isolated from characean algae and pollen tubes move along bundles of characean actin filaments when placed in contact with them (Shimmen and Tazawa 1982; Kohno and Shimmen 1988); myosins localize to motile organelles in these two (and probably other) cell types (Grolig and Wagner 1988; Tang et al. 1989); and characean organelles bind tightly to actin bundles when cells are depleted of ATP, a response most readily interpreted as a rigor response involving organelle myosin binding to actin (Williamson 1975). ATP catalyzes movement and allows easy release from the bundles.

Other mechanisms may exist (Williamson 1993). For example, transvacuolar strands link nuclei to the cell cortex and generate tension, since the surface of protoplasts is indented over these strands (Hahne and Hoffman 1984) and nuclei recoil when strands are severed (Goodbody et al. 1991). Therefore, nuclei could be pulled into position prior to mitosis. Microtubule depolymerization abolishes, and actin disruption slows, premitotic movements in *Nautilocalyx* (Venverloo and Libbenga 1987).

Organelle movements are inhibited when the concentration of free $Ca^{++}$ in the cytoplasm rises from $10^{-7}$ to $10^{-6}$ M (Williamson 1993). In characean algae, the response occurs in less than 1 second and probably involves phosphorylation of a myosin light chain by a $Ca^{++}$-dependent protein kinase (McCurdy and Harmon 1992). The movement of pollen tube organelles over characean actin is also $Ca^{++}$-sensitive, again probably via myosin (Kohno and Shimmen 1987), but pollen actin itself also may be regulated, since the arrangement of F-actin is disrupted in pollen tubes when the $Ca^{++}$ concentration rises (Kohno and Shimmen 1987). Profilin provides a potential route for inositol phosphates to control cytoskeletal function (Goldschmidt-Clermont et al. 1990).

## INITIAL CELL BIOLOGY ON THE *ARABIDOPSIS* CYTOSKELETON

Microtubule organization has been studied in *Arabidopsis* roots and during megasporogenesis and early zygote development. Seedling roots show microtubule arrays corresponding to all stages of the cell cycle that are similar to those familiar from many species (Baskin et al. 1992b). One difference is that many microtubules run between the nucleus and the cell surface in interphase meristematic cells. In other species observed by different immunocytochemical techniques, such microtubules were reported only during reestablishment of the interphase array following cytokinesis. Cortical microtubules are transverse through the elongation zone but become oblique in mature cells (Baskin et al. 1992b; Baskin and Williamson 1992). High-pressure freezing/freeze substitution shows that many of the cortical microtubules are linked by cross-bridges (Kiss et al. 1990). However, this technique did not preserve F-actin, although actin was successfully immunolabeled in sectioned root cells (Baskin et al. 1992a). A full study of the location of *Arabidopsis* actin is awaited (Baskin and Williamson 1992).

Complete microtubule depolymerization causes extensive swelling of the *Arabidopsis* root, consistent with a requirement for microtubules to achieve highly anisotropic growth (Baskin et al. 1992a). Although microtubules are necessary for highly anisotropic growth, they may not closely regulate the degree to which growth is anisotropic (Baskin and Williamson 1992; T.I. Baskin et al., in prep.). At one extreme, cortical microtubules undergo considerable disassembly and disorientation at low oryzalin concentrations with very limited effects on growth anisotropy. At the other extreme, elevated ethylene concentrations and taxol-enhanced microtubule assembly cause considerable swelling even though cortical microtubules remain transversely oriented. The small *Arabidopsis* root facilitates growth measurements, and the availability of large numbers of radial swelling mutants (Baskin et al. 1992a) will further explain growth anisotropy and the role played by cortical microtubules (R.E. Williamson et al., unpubl.; P. Benfey, pers. comm.).

During megasporogenesis, microtubules extend through the cytoplasm and may position organelles rather than determine cell shape. As in many other meiotic cells, no preprophase band forms (Webb and Gunning 1990). After fertilization, the zygote quickly reestablishes microtubule patterns typical of vegetative cells: Cortical microtubules align transverse to the zygote's direction of extension, and the preprophase band is reinstated at the first division (Webb and Gunning 1991).

## STRUCTURE AND CONSERVATION OF ACTIN AND TUBULIN POLYPEPTIDES

### Actin

The actin polypeptide sequences from all eukaryotes are extremely well conserved over their entire length of approximately 376 amino acids, and the actins in different eukaryotic kingdoms usually differ by only 11–17% in amino acid sequence. The deletion or insertion of a single amino acid is quite rare. When the amino acid sequence variability among all kingdoms is considered, the observed changes are distributed randomly over the entire sequence. It is uncommon to find a stretch of 10 amino acids with three substitutions in any pairwise comparison of distant actins. The one exception is the variability in sequence and length of the first several amino-terminal amino acids among distantly related actins (Meagher and McLean 1990; Meagher 1991).

The homogeneity of this conservation over the entire sequence of the actin polypeptide suggests that divergence of most of the actin sequence is constrained by important protein-protein and/or protein-ligand interactions. The binding of numerous factors to particular residues in the actin sequence and the description of the atomic structure of actin (Kabsch et al. 1990) partially confirm this view (Fig. 4), in that there may be few long stretches of sequence without such interactions (Bremer et al. 1991; Carlier 1991a). These interactions can be placed into at least three categories. First, the four subdomains of the atomic structure are held together by salt bridges, hydrogen bonds, and interaction with phosphate groups on the bound nucleotide and its associated divalent cation. Just the residues in the ATP and $Ca^{++}$ pocket at the center of the monomer were mapped to amino acids ranging from positions 11 to 335 (Fig. 4). Second, the interactions of one actin monomer with other actin monomers in the same strand (polymerization of filament) and with actin in the adjacent strand of the helix (adjacent strand contacts) present in F-actin (Bremer et al. 1991; Wertman and Drubin 1992) map to a different subset of sequences. Third, and most complex, are the interactions of G- and F-actin with the numerous ABPs. Many ABPs bind residues from three distant regions of the sequence, which fold together to form subdomain I of the monomer, with the strongest contacts often associated with the amino-terminal sequence (Vandekerckhove and Vancompernolle 1992). For example, the interaction with myosin occurs at four positions in the sequence which are all part of subdomain I (Fig. 4, myosin binding). The interactions of widely dispersed actin sequences with ATP, $Ca^{++}$, with the same and adjacent strands, and with ABPs would be expected to constrain actin evolution. However, mutagenesis studies on yeast actin have not revealed the reason for the extreme level of sequence conservation over the entire length of the actin polypeptide (Wertman et al. 1992). When adjacent charged residues are substituted with alanines, there is usually a lethal or temperature-sensitive phenotype in yeast. However, several pairs of residues, which are conserved among actins in all kingdoms, can be altered with no apparent phenotype. These

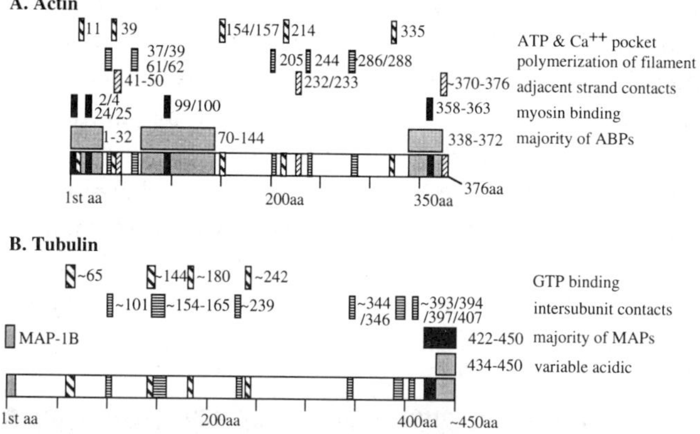


*Figure 4*  Proposed functional domains within the tubulin and actin polypeptides. (*A*) The actin polypeptide is typically 376 amino acids long; amino acid regions involved in forming a few important functional domains are indicated. For example, the ATP & Ca++ pocket is a conformational domain, constructed from distant regions of polypeptide sequence. Residues or regions known to be involved in polymerization of F-actin filaments, adjacent strand contacts in the filament, myosin binding, and binding site for the majority of actin binding proteins (ABPs) are indicated. (*B*) The α- and β-tubulin polypeptides are ~450 amino acids long. Amino acid residues or regions identified as involved in GTP binding, intersubunit contacts in the heterodimer, or contacts with adjacent protofilaments, binding the majority of microtubule associated proteins (MAPs), and the variable acidic domain are indicated. The construction of numerous conformational domains from many remote amino acid residues probably accounts for the exceptional conservation of actin and tubulin sequences.

results are enigmatic in light of the extreme sequence conservation observed in more than 100 actin sequences from diverse species.

## Tubulin

Tubulin polypeptides are approximately 450 amino acids long (Fig. 4). The α- and β-tubulins represent two ancient classes of sequences that diverged from a common ancestral sequence prior to the divergence of the four eukaryotic kingdoms. The equally divergent γ-tubulins are thought to be involved in nucleating microtubules, but very little is known about the γ-tubulins in plants. The α-tubulins from any eukaryote are far more like the α-tubulins from any distant organism (e.g., ~66% identity) than they are like any β-tubulins, even those in the same organism (e.g., ~36% identity and 60% similarity in most α-/β-comparisons). For example, human α-tubulin (Hsa) groups with plant α-tubulins, and human β-tubulin (Hsa) groups with *Arabidopsis* β-tubulins in a comparison of both tubulin classes (Fig. 5). A high percentage of isofunctional amino acid substitution exists between the two classes. Although the α- and β-tubulins are approximately the same overall length, any pairwise comparison contains a few insertion-deletion differences. Sequence-structure analysis of all the tubulins suggests that they all may fold into essentially the same structure (Burns 1991). The MAP-binding domain (e.g., β-tubulin residues 422–434 in Fig. 4) is highly conserved, but the sequence similarity among the α- and β-tubulins breaks down on the carboxy-terminal side of this domain for approximately the last 20 amino acids.

There also is some extreme variation in sequence and length among the carboxy-terminal 10 amino acids within a class. In vertebrates the unique carboxy-terminal-encoding region found in any one β-tubulin isovariant turns out to be conserved among distant animal species, and

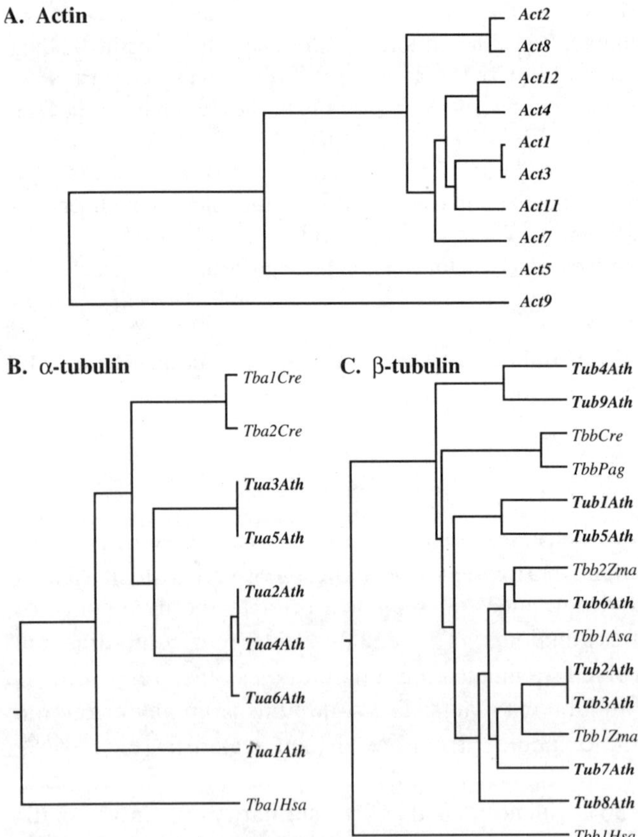


*Figure 5 (See facing page for legend.)*

this sequence was used to define ancient classes and isotypes of β-tubulin (Cleveland 1987). Since multiple isotypes appear to be expressed concurrently in the same cells, the necessity for these different isovariants is not immediately obvious. Other internal tubulin residues near positions 35, 56, and 124 are often conserved in association with particular carboxy-terminal sequences representing "hot spots" for sequence variation in the various isovariants and may represent significant alterations in protein conformation. These and other data suggest that these variable regions might be responsible for the conformational changes involved in the dynamic instability of microtubules (Burns and Surridge 1990; Burns 1991).

Within the α- and β-tubulin families, therefore, polypeptides are nearly as conserved as the actins, and their sequences are constrained by the same three kinds of physical contacts. Although the three-dimensional structure for tubulin is not yet known at the atomic level, these interactions and their constraints on sequence and structure of tubulin have been well documented (Burns 1991). A few are illustrated in Figure 4. For example, several small blocks of residues from amino acids 65 to 242 form the GTP-binding site (Luduena et al. 1992). Similarly, groups of residues

*Figure 5* Phylogenetic relationships among the *A. thaliana* actin and α- and β-tubulin amino acid sequences. Gene trees are based on amino acid sequence comparisons. (*A*) The ten known *Arabidopsis* actins (*ATH1, 2, 3, 4, 5, 7, 8, 9, 11,* and *12*) are compared relative to each other. The distant *ACT9* is a pseudogene. The two most divergent expressed sequences, *ACT2* and *ACT5*, differ in 58 of the 376 amino acids compared (15%). (*B*) The six known *A. thaliana* α-tubulins (*Tba1Ath-Tba6Ath*) are shown in relation to two α-tubulins from the green alga, *Chlamydomonas reinhardtii* (*Tba1Cre* and *Tba2Cre*), and one human α-tubulin for comparison (*Tba1Hsa*). The lengths of the branches in this tree are approximately proportional to the number of amino acid changes between sequences. There are 32 amino acid differences between the most distant Ath α-tubulin sequences, *Tua1Ath* and *Tua3Ath* in the first 444 amino acids that can be aligned (7%). (*C*) The nine known *A. thaliana* β-tubulins (*Tub1Ath-Tub9Ath*) are shown in relation to the monocots *Zea mays* (*Tbb1Zma* and *Tbb2Zma*) and *Avena sativa* (*Tbb1Asa*), the green algal sequences *TbbCre* and*TbbPag* (*Polytomella agilis*), and human (*Tbb1Hsa*) amino acid sequences. There are 38 amino acid differences between the most distant Ath α-tubulin sequences, *Tub4Ath* and *Tub6Ath*, in the first 445 amino acids that can be compared (8.5%). (*B,C*) A single sequence tree including both α- and β-tubulins based on amino acid differences was constructed; the branches joining the two groups of tubulins are not shown. The α- and β-tubulin gene family members typically differ in ~263 amino acids out of the 444 amino acids that can be aligned (59%). A large percentage of these differences are isofunctional amino acid interchanges.

ranging from positions 101 to 407 are involved in intersubunit contacts (Burns 1991). The majority of the MAPs bind the acidic and variable carboxy-terminal 20–30 amino acids. In contrast, MAP-1B binds in the amino-terminal region, and other regions are also involved (e.g., τ protein and MAP 2 have repeated tubulin-binding sequences mapping to diverse locations in the sequence; Chapin and Bulinski 1992; Luduena et al. 1992). Some of these regions are also implicated in controlling micro-tubule assembly and stability (intersubunit contacts). As for actin, the divergence of most tubulin sequences may be limited by the large num-ber of complex interactions with nucleotides, with subunits in the same and adjacent protofilaments, and with numerous binding proteins.

## DIVERSITY AND EVOLUTION WITHIN PLANT
## CYTOSKELETAL GENE FAMILIES

Genes encoding the actins and α- and β-tubulins have been cloned from many organisms including *Arabidopsis* (see below). Since these two proteins are well conserved among all eukaryotes, many of their physicochemical properties can be predicted by comparing the sequence of the plant proteins with those characterized in other species. The plant actins and tubulins are both encoded by relatively large gene families. This has been carried to an extreme in petunia, where data suggest there are hundreds of functional actin genes with multiple copies of the mem-bers from several subfamilies located at discrete sites on the chromosome (Baird and Meagher 1987; M. McLean et al. 1990). Initial data suggest that plant profilins are also encoded by a complex gene family (Staiger et al. 1993; S. Huang et al., in prep.).

Controversies about the need for multiple actin and tubulin genes in animals have continued for decades. Although there is no doubt that the various forms are differentially expressed, the questions have centered around whether or not one tubulin protein or gene could substitute for any other (Cleveland 1987; Rubenstein 1990; Murphy 1991; Herman 1993). Do the various α- and β-tubulins and isoforms of actins have truly different functions? Are these different genes and proteins needed? Do the various cytoplasmic and muscle actin genes encode functionally dif-ferent proteins, or are the genes just units for differential expression and regulation? It is clear that some α- and β-tubulins do have different func-tions (Hoyle and Raff 1990). Retinal pericytes express both cytoplasmic and muscle actins simultaneously, and the two classes of actin sort inde-pendently to different sites in the cell (De Nofrio et al. 1989). This sug-gests that class-specific positional information is stored in the actin se-quences. On the basis of both sequence changes in charged residues and isoelectric behavior of the plant actin isoforms, there is far greater diversity among the isovariants found in plants than in those found in animals (B.G. McLean et al. 1990b; Meagher and McLean 1990). Thus, this dilemma on gene and protein redundancy takes on greater importance for the plant gene families.


### Actin Gene Family

Plant actin genes have highly conserved intron/exon structures, which are summarized in Figure 6A. The coding region is interrupted by three small introns, which follow Lys-20, split Gly-152, and follow Gln-355 (Shah et al. 1983; Hightower and Meagher 1985). These introns have been observed in actin genomic clones from a dozen highly divergent plant species and are easily identified as interruptions in the highly conserved protein-coding region of actin. It appears that most plant actin genes contain a large leader exon 9–11 nucleotides upstream of the ATG initiation codon (McKnight 1983; McElroy et al. 1990a; Pearson and Meagher 1990) that has no well-defined sequence characteristics. The only known exceptions to this intron-exon structure are a potato actin pseudogene that appears to be a nearly complete reverse transcript of an actin RNA (Drouin and Dover 1987) and one highly expressed *Arabidopsis* actin gene, *ACT2*, that lacks one internal intron (Y.-Q. An et al., in prep.).

Due to the presence of the leader intron and the intron at Gly-152 in both plant and vertebrate actin genes, it was argued that these introns must be of very ancient origin (Gilbert et al. 1986; Dorit et al. 1990). However, this would require that numerous independent phyla which have branched off from the common ancestors of plants and vertebrates each independently lost these introns. Therefore, a more parsimonious argument has been made to account for the presence of introns in the same locations in the plant and animal kingdoms. It is based on the idea that protosplice sites act to attract intron insertion and that some of the introns in actin and tubulin are newly derived within conserved sequence domains (Dibb and Newman 1989). This hypothesis requires that only two intron insertion events occurred at each site, one in vertebrates and one in plants.

The rate at which sequence changes accumulate as two actin gene-coding sequences diverge is approximately constant (Hightower and Meagher 1986; Meagher 1991). Two actin proteins diverging from a common ancestral sequence usually acquire less than 1% in amino acid sequence difference every 100 millions years (MY) ± 50 MY. For exam-

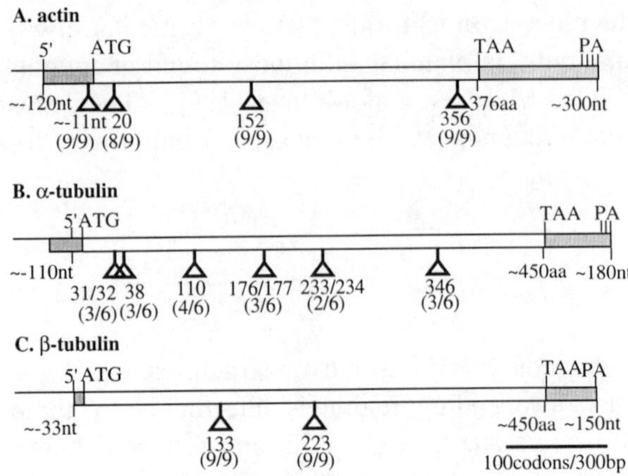



*Figure 6*  Overall physical structure of the actin and tubulin genes in *Arabidopsis*. The location and conservation of intron positions among *Arabidopsis* actin, α-tubulin, and β-tubulin genes are indicated. (*A*) Among the nine functional *Arabidopsis* actin genes only one gene, *ACT2*, lacks an intron at position 20. The plant actin intron positions are extremely well conserved among distant plants (not shown) with only a few exceptions. Most of the introns are small (~90 bp) with the exception of the intron in the mRNA leader, which is always large (200–500 bp in *Arabidopsis* and much larger in the other species where it has been described). (*B*) The α-tubulin intron positions are only partially conserved among the six *Arabidopsis* sequences or among more distant plant species (not shown). One *Arabidopsis* α-tubulin has been described with an exceptionally long 5' untranslated leader region. (*C*) The β-tubulin intron positions are well conserved among the nine *Arabidopsis* sequences characterized and among distant plant species (not shown).

ple, the distance between actins in the different eukaryotic kingdoms is about 10–17%, and they are thought to be about 1–1.5 billion years apart. The distance between actin in a deuterostome like a sea urchin and a protostome like *Drosophila*, which have not had a common ancestor for 680 MY, is typically 5–6%.

Plant actins fit these rates fairly well, differing by about 11–20% in amino acid sequence from animal, fungal, and most protist actins. Animal versus fungal and animal versus protist comparisons typically show 15% divergence (see above). Thus, it is likely that the plant actins and actins in these other kingdoms have evolved at similar rates (~1% per 50–100 MY) from the common ancestral actin genes (Hightower and Meagher 1986; McElroy et al. 1990a; Meagher 1991; Fletcher et al. 1994; J.M. McDowell et al., in prep.).

A partial characterization of the soybean, maize, petunia, rice, and

*Arabidopsis* gene families reveals that there may be 6–10 ancient plant actin subclasses which differ by more than 5% in amino acid sequence. This is about equal to the level of divergence between cytoplasmic and muscle vertebrate actins. Considering the slow rate of actin evolution, this suggests that plants contain an ancient gene family with a complex origin (Drouin and Dover 1990; Meagher 1991). This diversity appears to far predate the evolution of angiosperms, which originated about 140 MY before the present, perhaps dating back to the origin of vascular plants and the diversification of plant-cell types and development (Meagher 1994). These observations raise the possibility that the ancient diversity in the plant actin gene family has been preserved due to a requirement for diverse actin proteins or for diverse regulation of these genes in different cell types or stages of development.

If the amplification and diversification of the higher plant actin-gene family occurred early in vascular plant evolution, and plant actins evolve at the slow rate suggested above, then some of the ancient lineages of actins diverged prior to the divergence of the major groups of higher plants (i.e., before the split between monocots and dicots or angiosperms and gymnosperms from common ancestors). Thus, one would expect to find actins in one plant species that are more closely related to actins in distant vascular plant than they are to some members of the same plant gene family (Hightower and Meagher 1986; Meagher 1991), much like the relationship between cytoplasmic and muscle actins in distant vertebrates. These comparisons of distantly related sequences have to be made at the amino acid sequence or nonsynonymous nucleotide sequence level, because changes in the synonymous positions of codons would obscure such ancient relationships (i.e., synonymous positions that comprise about 25% of the coding sequence are rapidly randomized by mutation, even among relatively closely related genes). Pairs of actins from distant plant species with these close sequence relationships in nonsynonymous positions and amino acid sequences have been found. For example, rice *Rac1* is far more similar to *Arabidopsis Act1* and tobacco *Tac25* than *Rac1* is to the other known rice actins (Thangavelu et al. 1993). Soybean actin *Sac3* is more similar to pine actin *Pac1* than it is to several other soybean actins (McElroy et al. 1990b; Meagher 1991). Tobacco actin *Tac25* (Thangavelu et al. 1993) is closely related to *Arabidopsis* genes *ACT1*, *ACT3*, *ACT4*, and *ACT12*, and all five show pollen-specific expression (Y.-Q. An et al.; S. Huang et al.; both in prep.). Considering the slow rate of actin sequence change and the large distance between these plant species, it is clear that we are dealing with ancient lineages. However, the large size of the actin gene families and the relatively small number of actins characterized in most species limits the ability to define these ancient gene lineages further at this time. To identify these ancient gene classes, it is essential that a few complete actin gene families be characterized from distant plant species.

A gene tree based on the amino acid differences encoded by the 10 diverse *A. thaliana* actin genes (J.M. McDowell et al., in prep.) reveals this complexity (Fig. 5A). The amount of sequence divergence in this one gene family is nearly equal to the diversity observed among all known plant actin genes, as is expected from the ancient origin of the plant actin genes (Meagher 1991). The gene family members encode six distinct ancient classes of proteins. For example, *ACT1*, *ACT2*, *ACT4*, *ACT5*, *ACT7*, and *ACT11* are all more than 6% divergent from each other (Y.-Q. An et al.; S. Huang et al.; J.M. McDowell et al.; all in prep.). Considering the slow rate of nonsynonymous nucleotide substitution resulting in amino acid change among actins (1% per 100 MY ± 50), the deep branching in the *Arabidopsis* actin tree is impressive. The tree also reveals three subfamilies of *Arabidopsis* actin genes: *ACT1* and *ACT3* (Y.-Q. An et al., in prep.); *ACT4* and *ACT12* (S. Huang et al., in prep.); and *ACT2* and *ACT8* (Y.-Q. An et al., in prep.). The two genes in each of these pairs encode very similar proteins. However, within each of these pairs, there is almost no detectable sequence homology in 3′ flanking sequences and introns. Since the mutation rate in plants and other eukaryotes is about 1% per 1–2 MY (Meagher et al. 1989; Wolfe et al. 1989), this divergence of noncoding sequence suggests that these actin gene pairs have been separate for tens of millions of years.

If the hypothesis is correct that this ancient diversity is preserved because different actin proteins or regulating systems are needed, then this tree should also relate to the evolution of diversity of functions and/or regulatory patterns for the different genes. In other words, the topography of this tree undoubtedly reveals a great deal about the temporal and spatial relationships of expression for the various actin genes and perhaps something about their functions. For example, one should expect members of each of the various actin subfamilies to share common aspects of function and/or regulation, as is in fact the case (see below).

## Tubulin Gene Family

Tubulins are encoded by multigene families in all multicellular eukaryotes that have been examined. In vertebrates there are approximately six α- and six β-tubulin genes. *Arabidopsis* contains at least six α-tubulin (Kopczak et al. 1992) and nine β-tubulin genes (Snustad et al. 1992). Although it is statistically possible that there are more α- and β-tubulin genes, they were not isolated in these rather extensive searches. A third class of tubulin, named γ-tubulin, which is associated with the centrosome in animals and with microtubule organizing centers in fungi, has been reported recently in maize, tobacco, and *Arabidopsis* (Stern et al. 1991; Liu et al. 1993). Its expression is controlled in a cell-cycle-specific manner in plant cells.

The plant α- and β-tubulin genes have different intron/exon structures, as can be seen by examining the *Arabidopsis* gene structures summarized in Figure 6, B and C, respectively. The β-tubulin genes examined so far have two introns interrupting Gly-133 and Thr-223. Whereas most of the α-tubulin genes have four introns, their positions are only partially conserved among the α-tubulin gene family members. None of the introns is conserved among all six members, and none is coincident with the positions in the β-tubulin sequences, reflecting the ancient origin of these genes. Tubulin genes also have been part of the debate over the ancient or recent origin of introns (see references in section above on actin).

The plant α- and β-tubulins are 60–70% homologous in amino acid sequence with the animal, fungal, and protist tubulins of the same type as mentioned above. There are some unique α-tubulin and unique β-tubulin sequences in plants which suggest that each plant gene family is derived from a single common α-tubulin and common β-tubulin gene in a primitive plant or algal ancestor (Fosket and Morejohn 1992). As observed in animals, there are a few insertion/deletion differences between these two classes. For example, in *Arabidopsis*, the α-tubulins contain two single amino acid insertions near the amino-terminal end and a large insertion of 8 amino acids at position 359 relative to β-tubulin sequences. Similarly, the β-tubulins contain an extra 9–10 amino acids on the carboxy-terminal end, making the two polypeptides about the same overall length (Fosket and Morejohn 1992).

Figure 5, B and C, presents sequence trees for a few of the available plant and algal α- and β-tubulins and is based on amino acid sequence divergence. It suggests that the various members of a tubulin family may belong to ancient classes. For example, the *Arabidopsis* β-tubulin gene *TUB6* is closer in overall amino acid sequence to maize gene *Tbb2* and the oat gene *Tbb1* than it is to other *Arabidopsis* gene family members (Fig. 5C). Similarly, *Arabidopsis TUB2* and *TUB3* are closer to maize *Tbb1* than to other *Arabidopsis* β-tubulins. In each case, a significant amount of the sequence similarity between species is found in the "variable" carboxy-terminal region as seen for animal tubulins. Thus, ancient subclasses of tubulin may exist in plants, although this will be easier to confirm when larger numbers of higher plant sequences are available.

## EXPRESSION OF RNA AND PROTEIN

### Distinguishing Expression of the Various Gene Family Members

Studies on the expression of *Arabidopsis* actins and tubulins are just beginning. Because these proteins are encoded by large gene families, it is necessary to use gene-specific methods to examine patterns of expression of the individual gene family members. Many tubulin and actin genes have been assayed at the organ level on Northern blots, but very few have been assayed at the tissue-specific level by in situ RNA localization or by immunolocalization of protein. The most powerful gene-specific approach used so far has been to examine tissue-specific transcription using a 5′ "promoter" fragment fused to a reporter gene encoding an easily assayable protein (e.g., β-glucuronidase) in transgenic plants.

Gene-specific probes have been used to measure the steady-state expression of an individual tubulin or actin RNA and to distinguish it from the expression of other family members. In general, the 3′ untranslated portions of the transcripts for different tubulins or actins are gene-specific. The great evolutionary distance among these plant gene family members sometimes provides sufficient diversity to allow the actin and tubulin coding regions to function as unique probes if used at high stringency (Hightower and Meagher 1985; Baird and Meagher 1987; M. McLean et al. 1990; Fosket and Morejohn 1992). However, it is difficult to control and verify gene-specific hybridization conditions during in situ localizations of RNA in tissue sections, and, for these experiments, more defined probes are necessary. Gene-specific synthetic oligonucleotides also can be used as hybridization probes for Northerns, for in situ hybridization, or with even greater specificity as primers to amplify specific transcripts using reverse transcriptase followed by the polymerase chain reaction (RT-PCR).

Examining the expression of specific protein isovariants with antisera has met with limited success, due to the high level of conservation among the different actins and tubulins (B.G. McLean et al. 1990a,b). Actin peptides from the most diverse genes are 80–90% homologous in amino acid sequence, rarely having more than two or three amino acid substitutions in any 10-residue sequence. In contrast, the diverse carboxy-terminal 20 amino acids found among the *Arabidopsis* tubulin family members may provide unique peptide immunogens and, hence, the tubulin isovariant-specific antisera needed to approach this problem.

### Actin

Initial research on the actins of plants shows that the various members of an actin gene family are often differentially expressed. On the basis of predicted protein sequence, the six soybean actin genes characterized (*Sac1, 2, 3, 4, 6, 7*) fell into three ancient groups (κ-, λ-, and μ-actins), which probably represent three pairs of alloalleles in the soybean genome (Hightower and Meagher 1986). They showed more than 100-fold variation in the levels of RNA expressed in various organs (Hightower and Meagher 1985), with *Sac6* and *Sac7* expressed at high levels (~0.5% of mRNA) and *Sac2* and *Sac4* (λ-actin class) at the lowest levels. The μ-actins, *Sac3* and *Sac7*, are differentially expressed in response to treatment with auxin or light. Peptide-specific antisera were used to examine expression of subsets of isovariants in soybean tissues (B.G. McLean et al. 1990a,b). In roots, the expression of *Sac2* and/or *Sac4* appears to be localized in root protoderm. In contrast, the κ-actins, *Sac1* and/or *Sac6*, are expressed in leaf and root tissues, with the exception of root cap.

Rice contains a large actin-gene family (McElroy et al. 1990b), and two of the four rice actin genes characterized so far show a developmentally regulated pattern of expression. High levels of *Rac2* and *Rac3* RNA are expressed in seedlings for the first few days after germination, but this drops fivefold by 13 days. In contrast, *Rac7* is expressed at fairly constant levels during the developmental time course. Efficient monocot promoters for expressing genes in transgenic rice and maize have been difficult to find. The rice *Act1* (*Rac1*) gene is transcribed from a strong constitutive promoter (McElroy et al. 1990c; Zhang et al. 1991; Wang et al. 1992); several transformation vectors designed to express foreign proteins in monocots use the rice *Act1* 5′ region (McElroy et al. 1991).

Nine of the ten *Arabidopsis* actin genes are expressed, and one is a pseudogene (J.M. McDowell et al., in prep.). The 5′ regions from the nine expressed genes were fused to the β-glucuronidase (*GUS*) reporter gene and have been examined in transgenic *A. thaliana* plants. On the basis of initial data for *GUS* expression in several independent transgenic plants, each of the six actin gene subfamilies has a unique and complex pattern of transcriptional expression. Furthermore, the topography in the gene tree is a reasonable predictor of the degree of similarity in patterns of expression among the subfamilies. *ACT2* and *ACT8* are constitutive in vegetative tissues, being expressed at high levels in nearly all organs, including siliques, leaves, shoots, roots, and all flower organs with the exception of hypocotyl, seed coat, pollen sacs, and carpels. They are expressed in all tissues, including vascular, epidermal, and cortical tissues (Y.-Q. An et al., in prep.). *ACT7* is expressed in a developmental pattern, being constitutive at very high levels in seedlings including cotyledon leaves, hypocotyl, roots, and seed coat (J.M. McDowell et al., in prep.). *ACT7* shows a more restricted pattern in mature plants, being expressed primarily in early organ development (J.M. McDowell et al., in prep.).

*ACT11* is expressed in developing leaves and vascular tissue and is constitutive in all floral organs in young infloresences but is restricted to anthers and carpels in older infloresences (S. Huang et al., in prep.). The *ACT1* and *ACT3* subfamily is expressed at high levels in pollen and all organ root primordia, and at low levels in young vascular tissue (Y.-Q. An et al., in prep.). The *ACT4* and *ACT12* subfamily is expressed at high levels in pollen and low levels in vascular tissue. *ACT12* is also expressed in root meristems (S. Huang et al., in prep.). Expression of the most distant of the functional genes, *ACT5*, was detected with RT-PCR in roots. Analysis of the expression of *ACT5 5′GUS* fusion is incomplete, but initial data suggest a very restricted expression pattern.

## Tubulin

Initial data on the differential expression of tubulins in maize, soybean, and carrot suggested a complex pattern of differential expression in different organs and tissues and stimulated speculation that the different plant genes must be required for different functions in plants. The expression of six $\beta$-tubulin protein isovariants is observed on two-dimensional gels in carrot. Between two and five of these are expressed in any one of the various organs and tissues that were analyzed (Hussey et al. 1988). Some of these are unique to, or preferentially expressed in, particular organs or stages in development. $\beta_6$ is specific to seedlings, $\beta_5$ is found only in vegetative tissues of the whole plant, and $\beta_1$ and $\beta_3$ are absent from pollen. $\beta_2$ and $\beta_4$ appear constitutive, with $\beta_4$ being the predominant form. In maize, $\beta_1$ RNA is expressed in seedlings, root tips, and shoots and in tissue-culture cells, but not in immature tassel or mature leaf (Hussey et al. 1990). In contrast, $\beta_2$-tubulin RNA is expressed in tissue-culture cells but not in seedlings, immature tassels, or mature leaves. By comparison with an in-vitro-synthesized $\beta_2$ protein as a standard, it appears that the $\beta_2$ isovariant is expressed at high levels in mature pollen. The soybean $\beta_1$-tubulin RNA is preferentially expressed at high levels in the hypocotyl of etiolated seedlings and shut off by exposure to light (Han et al. 1991). Two closely related maize $\alpha$-tubulin genes express their RNA primarily in young root tissues (Montoliu et al. 1989). All of these data on the complex differential expression of tubulins suggest that the various isovariants are required for diverse functions in plants.

In *Arabidopsis*, the expression of one $\alpha$-tubulin gene, *TUA1*, and one $\beta$-tubulin gene, *TUB1*, has been examined in detail using promoter-*GUS* fusions. They appear to be specific for pollen and roots, respectively (see below). A few genes are preferentially expressed in just a limited number of organs based on Northern analysis of RNA levels. For example, *TUA5* and *TUB9* RNAs are expressed at much lower levels in roots than in leaves or flowers, whereas *TUB3*, *TUB4*, and *TUB7* are more highly expressed in flowers than in other organs. The remaining genes are expressed in most organs examined (i.e., roots, leaves, floral organ complex).

*TUA1* encodes the most divergent α-tubulin isovariant (Fig. 5B) (Kopczak et al. 1992), and steady-state levels of its transcripts accumulate to the highest levels during anthesis in flowers. Steady-state RNA levels accumulate to the highest levels in stamen and pollen, but not in perianth or pistil (Carpenter et al. 1990, 1992). It is not expressed in the leaves or roots of mature plants (Ludwig et al. 1988). A 5′ fragment from *TUA1* containing 533 bp upstream from the transcription start site and 56 bp of the transcribed leader fused to the β-glucuronidase reporter coding sequence were examined in transgenic plants. Expression is detected in young anthers with uninucleate microspores, in mature pollen grains (pollen with a vegetative nucleus and two polar nuclei), and in receptacle, with the highest levels in mature pollen. In plants with large numbers of gene copies (5–10), activity is occasionally detected at lower levels in other parts of the plant. This is presumed to be improper expression of the transgene.

Two identical copies of the *TUB1* genes are separated by approximately 1 kb of DNA in the *Arabidopsis* genome. The *TUB1* sequence is quite distinct from the other β-tubulin genes (i.e., *TUB2Ath-9Ath*; Fig. 5C). *TUB1* RNA is more highly expressed in roots than in other organs, but some RNA is detected in leaves and flowers (Snustad et al. 1992). Among the *Arabidopsis* β-tubulins, only the *TUB1* genes show a unique pattern of steady-state RNA expression. The maize $\beta_1$ gene, which shows higher levels of expression in seedlings, roots, and shoots (Hussey et al. 1990), does not group with *Arabidopsis TUB1* (Fig. 5C) in the gene tree. If *TUB1* is part of a unique ancient lineage with a conserved pattern of expression, its counterpart in other species has not yet been isolated.


## FUTURE RESEARCH

The last decade of research on the plant cytoskeleton has focused on localizing these cytoskeletal systems to particular cell types and on characterizing the numerous actin and tubulin genes. The analysis of large plant gene families will continue to be a problem as other cytoskeletal proteins like the profilins and myosins are examined. The small genome size and more compact gene structures in *Arabidopsis* will undoubtedly simplify this effort. It is anticipated that several distinct complements of cytoskeletal proteins will show a common mode of regulation in particular organs or cell types and will clarify the phylogenetic relationships among all the different gene family members in diverse species (Meagher 1994).

Present work on the molecular cytology of the plant cytoskeleton is focused on characterizing the differential expression of the various genes and proteins and on identifying the unique structures of microtubules and actin filaments in different organs and cell types. The future of the plant cytoskeletal field is dependent on connecting the patterns of gene expression and the diverse cytoskeletal structures observed with particular molecular functions. The physical biochemistry of plant cytoskeletal components is still limited by the inability to purify large quantities of plant proteins that are functional in vitro. In yeast and protists, the analysis of the phenotypes of actin mutants has been extremely helpful in dissecting functions and functional domains (Noegel and Schleicher 1991; Brown 1993). *Arabidopsis* is an ideal higher plant to make and screen cytoskeletal mutations and, perhaps, to express altered cytoskeletal proteins. These approaches could provide the badly needed connection between the molecular genetics, cell biology, and physical biochemistry of the plant cytoskeleton. They also may help to answer the increasingly complex questions about the necessity for such diverse cytoskeletal genes and proteins in plants.

## ACKNOWLEDGMENTS

John McDowell, Shurong Huang, Yong-Quiang An, and Libby McKinney kindly provided unpublished data on *Arabidopsis* actin gene expression and gene family structure. Marcus Fechheimer and John McDowell generously assisted with criticism of the manuscript. A portion of this work was funded by a grant from the Molecular Cytology Study Section within the National Institutes of Health.

## REFERENCES

Baird, W.V. and R.B. Meagher. 1987. A complex gene superfamily encodes actin in petunia. *EMBO J.* **6:** 3223–3231.

Baskin, T.I. and W.Z. Cande. 1990. The structure and function of the mitotic spindle in flowering plants. *Annu. Rev. Plant Physiol. Plant Mol. Biol.* **41:** 277–315.

Baskin, T.I. and R.E. Williamson. 1992. Ethylene, microtubules and root morphology in wild-type and mutant *Arabidopsis* seedlings. *Curr. Top. Plant Biochem. Physiol.* **11:** 118–130.

Baskin, T.I., A.S. Betzner, R. Hoggart, A. Cork, and R.E. Williamson. 1992a. Root morphology mutants in *Arabidopsis thaliana. Aust. J. Plant. Physiol.* **19:** 427–437.

Baskin, T.I., C.H. Busby, L.C. Fowke, M. Sammut, and F. Gubler. 1992b. Improvements in immunostaining samples embedded in methacrylate: Localization of microtubules and other antigens throughout developing organs in plants of diverse taxa. *Planta* **187:** 405–413.

Bremer, A. and U. Aebi. 1992. The structure of the F-actin filament and the actin molecule. *Curr. Opin. Cell Biol.* **4:** 20–26.

Bremer, A., R.C. Millonig, R. Sutterlin, A. Engel, T.D. Pollard, and U. Aebi. 1991. The structural basis for the intrinsic disorder of the actin filament: The "lateral slipping" model. *J. Cell Biol.* **115:** 689–703.

Brown, R.C. and B.E. Lemmon. 1990. Monoplastidic cell division in lower land plants. *Am. J. Bot.* **77:** 559–571.

Brown, S.S. 1993. Phenotypes of cytoskeletal mutants. *Curr. Opin. Cell Biol.* **5:** 129–134.

Burns, R.G. 1991. α-, β- and γ- Tubulins: Sequence comparisons and structural constraints. *Cell Motil. Cytoskeleton* **20:** 181–189.

Burns, R.G. and C. Surridge. 1990. Analysis of β-tubulin sequences reveals highly conserved coordinated amino acid substitutions: Evidence that these "hot spots" are directly involved in the conformational change required for dynamic instability. *FEBS Lett.* **271:** 1–8.

Carlier, M.-F. 1989. Role of nucleotide hydrolysis in the dynamics of actin filaments and microtubules. *Int. Rev. Cytol.* **115:** 139–170.

———. 1991a. Nucleotide hydrolysis in cytoskeletal assembly. *Curr. Opin. Cell Biol.* **3:** 12–17.

———. 1991b. Actin: Protein structure and filament dynamics. *J. Biol. Chem.* **266:** 1–4.

Carpenter, J.L., D.P. Snustad, and C.D. Silflow. 1990. Analysis of tubulin gene expression patterns in *Arabidopsis thaliana*. In *Abstracts from the 4th International Conference on* Arabidopsis *Research*, Vienna (ed. D. Schweizer et al.), p. 54. University of Vienna, Austria.

Carpenter, J.L., S. Ploense, D.P. Snustad, and C.D. Silflow. 1992. Preferential expression of an α-tubulin gene of *Arabidopsis* in pollen. *Plant Cell* **4:** 557–571.

Chang-Jie, J. and S. Sonobe. 1993. Identification and preliminary characterization of a 65 kDa higher-plant microtubule-associated protein. *J. Cell Sci.* **105:** 891–901.

Chapin, S.J. and J.C. Bulinski. 1992. Microtubule-stabilization by assembly-promoting microtubule-associated proteins: A repeat performance. *Cell Motil. Cytoskeleton* **23:** 236–243.

Cheney, R.E. and M.S. Mooseker. 1992. Unconventional myosins. *Curr. Opin. Cell Biol.* **4:** 27–35.

Cheney, R.E., M.A. Riley, and M.S. Mooseker. 1993. Phylogenetic analysis of the myosin superfamily. *Cell Motil. Cytoskeleton* **24:** 215–223.

Clayton, L. and C.W. Lloyd. 1985. Actin organization during the cell cycle in meristematic plant cells. *Exp. Cell Res.* **156:** 231–238.

Cleary, A.L., R.C. Brown, and B.E. Lemmon. 1992. Microtubule arrays during mitosis in monoplastidic root tip cells of *Isoetes*. *Protoplasma* **167:** 123–133.

Cleveland, D.W. 1987. The multitubulin hypothesis revisited: What have we learned? *J. Cell Biol.* **104:** 381–383.

Cooper, J.A. 1987. Effects of cytochalasin and phalloidin on actin. *J. Cell Biol.* **105:** 1473–1478.

———. 1991. The role of actin polymerization in cell motility. *Annu. Rev. Physiol.* **53:** 585–605.

Cyr, R.J. and B.A. Palevitz. 1989. Microtubule-binding proteins from carrot. I. Initial characterization and microtubule bundling. *Planta* **177**: 245–260.

De Nofrio, D., T.C. Hoock, and I.M. Herman. 1989. Functional sorting of actin isoforms in microvascular pericytes. *J. Cell Biol.* **109**: 191–202.

Dibb, N.J. and A.J. Newman. 1989. Evidence that introns arose at proto-splice sites. *EMBO J.* **8**: 2015–2021.

Dorit, R.L., L. Schoenbach, and W. Gilbert. 1990. How big is the universe of exons? *Science* **250**: 1377–1382.

Drouin, G. and G.A. Dover. 1987. A plant processed pseudogene. *Nature* **328**: 557–558.

———. 1990. Independent gene evolution in the potato actin gene family demonstrated by phylogenetic procedures for resolving gene conversions and the phylogeny of angiosperm actin genes. *J. Mol. Evol.* **31**: 132–150.

Eleftheriou, E.P. and B.A. Palevitz. 1992. The effect of cytochalasin D on preprophase band organization in root tip cells of *Allium. J. Cell Sci.* **103**: 989–998.

Emons, A.M.C., J. Derksen, and M.M.A. Sasson. 1992. Do microtubules orient plant cell wall microfibrils. *Physiol. Plant* **84**: 486–493.

Emons, A.M.C., E. Pierson, and J. Derksen. 1991. Cytoskeleton and intracellular movements in plant cells. *Biotechnol. Curr. Prog.* **1**: 311–335.

Endow, S.A. 1991. The emerging kinesin family of microtubule motor proteins. *Trends Biochem. Sci.* **16**: 221–225.

———. 1993. Chromosome distribution, molecular motors and the claret protein. *Trends Genet.* **9**: 52–55.

Falconer, M.M. and R.W. Seagull. 1987. Amiprophos-methyl (APM): A rapid, reversible anti-microtubule agent for plant cell cultures. *Protoplasma* **136**: 118–124.

Flanders, D.J., D.J. Rawlins, P.J. Shaw, and C.W. Lloyd. 1990. Nucleus-associated microtubules help determine the division plane of plant epidermal cells: Avoidance of four-way junctions and the role of cell geometry. *J. Cell Biol.* **110**: 1111–1122.

Fletcher, L.D., J.M. McDowell, R.R. Tidwell, R.R. Meagher, and C.C. Dykstra. 1994. Structure, expression and phylogenetic analysis of the gene encoding actin I in *Pneumocystis carinii. Genetics* **137**: 1–8.

Fosket, D.E. and L.C. Morejohn. 1992. Structural and functional organization of tubulin. *Annu. Rev. Plant Physiol. Plant Mol. Biol.* **43**: 201–240.

Giddings, T.H. and L.A. Staehelin. 1991. Microtubule-mediated control of microfibril deposition: A re-examination of the hypothesis. In *The cytoskeletal basis of plant growth and form* (ed. C.W. Lloyd), pp. 85–99. Academic Press, London.

Gilbert, W., M. Marchionni, and G. McKnight. 1986. On the antiquity of introns. *Cell* **46**: 151–154.

Goldschmidt-Clermont, P.J., L.M. Machesky, J.J. Baldassare, and T.D. Pollard. 1990. The actin-binding protein profilin binds to PIP2 and inhibits its hydrolysis by phospholipase C. *Science* **247**: 1575–1578.

Goodbody, K.C., C.J. Venverloo, and C.W. Lloyd. 1991. Laser microsurgery demonstrates that cytoplasmic strands anchoring the nucleus across the vacuole of premitotic plant cells are under tension. Implications for division plane alignment. *Development* **113**: 931–939.

Goodson, H.V. and J.A. Spudich. 1993. Molecular evolution of the myosin family: Relationships derived from comparisons of amino acid sequences. *Proc. Natl. Acad. Sci.* **90**: 659–663.

Green, P.B. 1962. Mechanism for plant cellular morphogenesis. *Science* **138**: 1404–1405.

———. 1980. Organogenesis—A biophysical view. *Annu. Rev. Plant Physiol.* **31**: 51–82.

Grolig, F. and G. Wagner. 1988. Light-dependent chloroplast reorientation in *Mougeotia* and *Mesotaenium*: Biased by pigment regulated plasmalemma anchorage sites to actin filaments? *Bot. Acta* **101:** 2–6.

Grolig, F., R.E. Williamson, J. Parke, C. Miller, and B.H. Anderton. 1988. Myosin and Ca²⁺-sensitive streaming in the alga *Chara*: Two polypeptides reacting with a monoclonal anti-myosin and their localization in the streaming endoplasm. *Eur. J. Cell Biol.* **47:** 22–31.

Gunning, B.E.S. and M. Sammut. 1990. Rearrangements of microtubules involved in establishing cell division planes start immediately after DNA synthesis and are completed just before mitosis. *Plant Cell* **2:** 1273–1282.

Gunning, B.E.S., A.R. Hardham, and J.E. Hughes. 1978. Evidence for initiation of microtubules in discrete regions of the cell cortex in *Azolla* root-tip cells, and an hypothesis on the development of cortical arrays of microtubules. *Planta* **143:** 161–179.

Hahne, G. and F. Hoffman. 1984. The effect of laser microsurgery on cytoplasmic strands and cytoplasmic streaming in isolated plant protoplasts. *Eur. J. Cell Biol.* **33:** 175–179.

Han, I.-S., I. Jongewaard, and D.E. Fosket. 1991. Limited expression of a diverged β-tubulin gene during soybean (*Glycine max* [L.] Merr.) development. *Plant Mol. Biol.* **16:** 225–234.

Hartwig, J.H. and D.J. Kwiatkowski. 1991. Actin-binding proteins. *Curr. Opin. Cell Biol.* **3:** 87–97.

Herman, I.M. 1993. Actin isoforms. *Curr. Opin. Cell Biol.* **5:** 48–55.

Hightower, R.C. and R.B. Meagher. 1985. Divergence and differential expression of soybean actin genes. *EMBO J.* **4:** 1–8.

———. 1986. The molecular evolution of actin. *Genetics* **114:** 315–332.

Holmes, K.C., D. Popp, W. Gebhard, and W. Kabsch. 1990. Atomic model of the actin filament. *Nature* **347:** 44–49.

Hoyle, H.D. and E.C. Raff. 1990. Two *Drosophila* beta tubulin isoforms are not functionally equivalent. *J. Cell Biol.* **111:** 1009–1026.

Hussey, P.J., C.W. Lloyd, and K. Gull. 1988. Differential and developmental expression of β-tubulins in a higher plant. *J. Biol. Chem.* **263:** 5474–5479.

Hussey, P.J., N. Haas, J. Hunsperger, J. Larkin, P. Snustad, and C.D. Silflow. 1990. The β-tubulin gene family in *Zea mays*: Two differentially expressed β-tubulin genes. *Plant Mol. Biol.* **15:** 957–972.

Kabsch, W., H.G. Mannherz, D. Suck, E.F. Pai, and K.C. Holmes. 1990. Atomic structure of the actin: DNase I complex. *Nature* **347:** 37–44.

Kato, T. and Y. Tonomura. 1977. Identification of myosin in *Nitella flexilis. J. Biochem.* **82:** 777–782.

Katsuta, J. and H. Shibaoka. 1992. Inhibition by kinase inhibitors of the development and the disappearance of the preprophase band of microtubules in tobacco BY-2 cells. *J. Cell Sci.* **103:** 397–405.

Katsuta, J., Y. Hashiguchi, and H. Shibaoka. 1990. The role of the cytoskeleton in positioning of the nucleus in premitotic tobacco BY-2 cells. *J. Cell Sci.* **95:** 413–422.

Kim, S.-R., Y. Kim, and G. An. 1993. Molecular cloning a characterization of anther-preferential cDNA encoding a putative actin-depolymerizing factor. *Plant Mol. Biol.* **21:** 39–45.

Kimble, M. and R. Kuriyama. 1992. Functional components of microtubule-organizing centers. *Int. Rev. Cytol.* **136:** 1–50.

Kinkema, M.D. and J.W. Schiefelbein. 1992. Isolation of a complementary DNA encoding an unconventional myosin from *Arabidopsis thaliana. Mol. Biol. Cell.* **3:** 46a.

Kirschner, M. and T. Mitchison. 1986. Beyond self-assembly: From microtubules to morphogenesis. *Cell* **45:** 329–342.

Kiss, J.Z., T.H. Giddings, L.A. Staehelin, and F.D. Sack. 1990. Comparison of the ultrastructure of conventionally fixed and high pressure frozen/freeze substituted root tips of *Nicotiana* and *Arabidopsis*. *Protoplasma* **157**: 64–74.

Kohno, T. and T. Shimmen. 1987. $Ca^{2+}$-induced fragmentation of actin filaments in pollen tubes. *Protoplasma* **141**: 177–179.

———. 1988. Accelerated sliding of pollen tube organelles along *Characeae* actin bundles regulated by $Ca^{2+}$. *J. Cell Biol.* **106**: 1539–1543.

Kohno, T., R. Ishikawa, T. Nagata, K. Kohama, and T. Shimmen. 1992. Partial purification of myosin from lily pollen tubes by monitoring with in vitro motility assay. *Protoplasma* **170**: 77–85.

Kopczak, S.D., N.A. Haas, P.J. Hussey, C.D. Silflow, and D.P. Snustad. 1992. The small genome of *Arabidopsis* contains at least six expressed α-tubulin genes. *Plant Cell* **4**: 539–547.

Korn, E.D. and J.A. Hammer. 1988. Myosins of non-muscle cells. *Annu. Rev. Biophys. Biophys. Chem.* **17**: 23–45.

La Claire, J.W.I. 1991. Immunolocalization of myosin in intact and wounded cells of the green alga *Ernodesmis verticillata* (Kützing) Børgesen. *Planta* **184**: 209–217.

Lancelle, S.A. and P.K. Hepler. 1991. Association of actin with cortical microtubules revealed by immunogold electron microscopy in *Nicotiana* pollen tubes. *Protoplasma* **165**: 167–172.

Lin, Q., F. Grolig, P.P. Jablonsky, and R.E. Williamson. 1989. Myosin heavy chains: Detection by immunoblotting in higher plants and localization by immunofluorescence in the alga *Chara*. *Cell Biol. Int. Rep.* **13**: 107–117.

Liu, B., J. Marc, H.C. Joshi, and B.A. Palevitz. 1993. A γ-tubulin related protein associated with the microtubule arrays of higher plants in a cell cycle dependent manner. *J. Cell Sci.* **104**: 1217–1228.

Lloyd, C.W., ed. 1991. *The cytoskeletal basis of plant growth and form*. Academic Press, London.

Lloyd, C.W. and J.A. Traas. 1988. The role of F-actin in determining the division plane of carrot suspension cells. *Development* **102**: 211–221.

Luduena, R.F., A. Banerjee, and I.A. Khan. 1992. Tubulin structure and biochemistry. *Curr. Opin. Cell Biol.* **4**: 53–57.

Ludwig, S.R., D.G. Oppenheimer, C.D. Silflow, and D.P. Snustad. 1988. The α1-tubulin gene of *Arabidopsis thaliana*: Primary structure and preferential expression in flowers. *Plant Mol. Biol.* **10**: 311–321.

Luna, E.J. 1991. Molecular links between the cytoskeleton and membranes. *Curr. Opin. Cell Biol.* **3**: 120–126.

Lyndon, R.F. 1990. *Plant development: The cellular basis*. Unwin Hyman, London.

Ma, Y.-Z. and L.-F. Yan. 1988. The presence of contractile proteins in pollens and their role in cytoplasmic streaming. *Acta Bot. Sin.* **30**: 285–291.

Ma, Y. and L. Yen. 1989. Actin and myosin in pea tendrils. *Plant Physiol.* **89**: 586–589.

McCurdy, D.W. and B.E.S. Gunning. 1990. Reorganization of cortical actin microfilaments and microtubules at preprophase and mitosis in wheat root-tip cells: A double label immunofluorescence study. *Cell Motil. Cytoskeleton* **15**: 76–87.

McCurdy, D.W. and A.C. Harmon. 1992. Phosphorylation of a putative myosin light chain in *Chara* by calcium-dependent protein kinase. *Protoplasma* **171**: 85–88.

McElroy, D., M. Rothenberg, and R. Wu. 1990a. Structural characterization of a rice actin gene. *Plant Mol. Biol.* **14**: 163–171.

McElroy, D., A.D. Blowers, B. Jenes, and R. Wu. 1991. Construction of rice actin *Act1*-based expression vectors for use in monocot transformation. *Mol. Gen. Genet.* **230**:

150–160.

McElroy, D., M. Rothenberg, K.S. Reece, and R. Wu. 1990b. Characterization of the rice (*Oryza sativa*) actin gene family. *Plant Mol. Biol.* **15:** 257–268.

McElroy, D., W. Zhang, J. Cao, and R. Wu. 1990c. Isolation of an efficient actin promoter for use in rice transformation. *Plant Cell* **2:** 163–171.

McKnight, T.D. 1983. Sequences up stream from the first translated exon of plant actin genes suggest there may be an intervening sequence in the 5′ untranslated region. In "DNA sequences at the 5′ ends of plant genes." Ph.D. thesis, University of Georgia, Athens, pp. 77–93.

McLean, B.G., S. Eubanks, and R.B. Meagher. 1990a. Tissue specific expression of divergent actins in soybean root. *Plant Cell* **2:** 335–344.

McLean, B.G., S. Huang, E.C. McKinney, and R.B. Meagher. 1990b. Plants contain highly divergent actin isovariants. *Cell Motil.* **17:** 276–290.

McLean, M., A.G.M. Gerats, W.V. Baird, and R.B. Meagher. 1990. Six actin gene sub-families map to five chromosomes of *Petunia hybrida*. *J. Hered.* **81:** 341–346.

McNulty, A.K. and M.J. Saunders. 1992. Purification and immunological detection of pea nuclear intermediate filaments: Evidence for plant nuclear lamins. *J. Cell Sci.* **103:** 407–414.

Meagher, R.B. 1991. Divergence and differential expression of actin gene families in higher plants. *Int. Rev. Cytol.* **125:** 139–163.

———. 1994. The impact of historical contingency on gene phylogeny: Plant actin diversity. *Evol. Biol.* **28:** (in press).

Meagher, R.B. and B.G. McLean. 1990. Diversity of plant actins. *Cell Motil.* **16:** 164–166.

Meagher, R.B., S. Berry-Lowe, and K. Rice. 1989. Molecular evolution of the small sub-unit of ribulose bisphosphate carboxylase: Nucleotide substitution and gene conversion. *Genetics* **123:** 845–863.

Michaud, D., G. Guillet, P.A. Rogers, and P.M. Charest. 1991. Identification of a 220 kDa membrane-associated plant spectrin related to human β-spectrin. *FEBS Lett.* **294:** 77–80.

Mineyuki, Y. and B.E.S. Gunning. 1990. A role for preprophase bands of microtubules in maturation of new cell walls, and a general proposal on the function of preprophase band sites in cell division in higher plants. *J. Cell Sci.* **97:** 527–537.

Mineyuki, Y. and B.A. Palevitz. 1990. Relationship between preprophase band organization, F-actin and the division site in *Allium*. Fluorescence and morphometric studies on cytochalasin-treated cells. *J. Cell Sci.* **97:** 283–295.

Mineyuki, Y., M. Yamashita, and Y. Nagahama. 1991. p34[cdc2] homologue in the preprophase band. *Protoplasma* **162:** 182–186.

Mitchison, T.J. and K.E. Sawin. 1990. Tubulin flux in the mitotic spindle: Where does it come from, where is it going? *Cell Motil. Cytoskeleton* **16:** 93–98.

Moepps, B., S. Conrad, and H. Schraudolf. 1993. PCR-dependent amplification and se-quence characterization of partial cDNAs encoding mycosin-like proteins in *Anemia phyllitidis* (L.) Sw. and *Arabidopsis thaliana* (L.) Heynh. *Plant Mol. Biol.* **21:** 1077–1083.

Montoliu, L., J. Rigau, and P. Puigdomenech. 1989. A tandem of α-tubulin genes prefer-entially expressed in radicular tissues from *Zea mays*. *Plant Mol. Biol.* **14:** 1–15.

Murphy, D.B. 1991. Functions of tubulin isoforms. *Curr. Opin. Cell Biol.* **3:** 43–51.

Noegel, A.A. and M. Schleicher. 1991. Phenotypes of cells with cytoskeletal mutations. *Curr. Opin. Cell Biol.* **3:** 18–26.

Ohsuka, K. and A. Inoue. 1979. Identification of myosin in a flowering plant *Egeria densa. J. Biochem.* **85:** 375–378.

Parke, J., C. Miller, and B.H. Anderson. 1986. Higher plant myosin heavy-chain identified using a monoclonal antibody. *Eur. J. Cell Biol.* **41:** 9–13.

Parthasarathy, M.V. 1985. F-actin architecture in coleoptile epidermal cells. *Eur. J. Cell Biol.* **39:** 1–12.

Pearson, L. and R.B. Meagher. 1990. Diverse soybean actin transcripts contain a large intron in the 5′ untranslated leader: Structural similarity to vertebrate muscle actin genes. *Plant Mol. Biol.* **14:** 513–526.

Pollard, T.D., S.K. Doberstein, and H.G. Zot. 1991. Myosin-I. *Annu. Rev. Physiol.* **53:** 653–681.

Reis, D., B. Vian, H. Chanzy, and J.-C. Roland. 1991. Liquid crystal-type assembly of native cellulose-glucuronoxylans extracted from plant cell walls. *Biol. Cell* **73:** 173–178.

Richmond, P.A. 1983. Patterns of cellulose microfibril deposition and rearrangement in *Nitella*: In vivo analysis by a birefringence index. *J. Appl. Polymer Sci.* **37:** 107–122.

Richmond, P.A., J.-P. Métraux, and L. Taiz. 1980. Cell expansion patterns and directionality of wall mechanical properties in *Nitella. Plant Physiol.* **65:** 211–217.

Robinson, D.G. and H. Quader. 1981. Structure, synthesis, and orientation of microfibrils. IX. A freeze-fracture investigation of the *Oocystis* plasma membrane after inhibitor treatments. *Eur. J. Cell Biol.* **25:** 278–288.

Rubenstein, P.A. 1990. The functional importance of multiple actin isoforms. *BioEssays* **12:** 309–315.

Salmon, E.D., R.J. Leslie, W.M. Saxton, M.L. Karaw, and J.R. McIntosh. 1984. Spindle microtubule dynamics in sea urchin embryos: Analysis using a fluorescein-labelled tubulin and measurements using fluorescence redistribution after laser photobleaching. *J. Cell Biol.* **99:** 2165–2174.

Schellenbaum, P., M. Vartard, C. Peter, A. Fellows, and A.-M. Lambert. 1993. Coassembly properties of higher plant microtubule-associated proteins with purified brain and plant tubulins. *Plant J.* **3:** 253–260.

Seagull, R.W. and M.M. Falconer. 1991. In vitro xylogenesis. In *The cytoskeletal basis of plant growth and form* (ed. C.W. Lloyd), pp. 183–194. Academic Press, London.

Shah, D.M., R.C. Hightower, and R.B. Meagher. 1983. Genes encoding actin in higher plants are highly conserved but the coding sequences are not. *J. Mol. Appl. Genet.* **2:** 111–126.

Shaw, P.J., D.J. Fairbairn, and C.W. Lloyd. 1991. Cytoplasmic and nuclear intermediate filament antigens in higher plants. In *The cytoskeletal basis of plant growth and form* (ed. C.W. Lloyd), pp. 69–81. Academic Press, London.

Shimmen, T. and M. Tazawa. 1982. Reconstitution of cytoplasmic streaming in *Characeae. Protoplasma* **113:** 127–131.

Snustad, D.P., N.A. Haas, S.D. Kopczak, and C.D. Silflow. 1992. The small genome of *Arabidopsis thaliana* contains at least nine expressed β-tubulin genes. *Plant Cell* **4:** 549–556.

Staiger, C.J. and C.W. Lloyd. 1991. The plant cytoskeleton. *Curr. Opin. Cell Biol.* **3:** 33–42.

Staiger, C.J., K.C. Goodbody, P.J. Hussey, R. Valenta, B.K. Drøbak, and C.W. Lloyd. 1993. The profilin multigene family of maize: Differential expression of three isoforms. *Plant J.* **4:** 631–641.

Stern, T., L. Evans, and M. Kirshner. 1991. γ-Tubulin is a highly conserved component of the centrosome. *Cell* **65:** 825–836.

Tang, X., P.K. Hepler, and S.P. Scordilis. 1989. Immunochemical and immunocytochemical identification of a myosin heavy chain polypeptide in *Nicotiana* pollen tubes. *J. Cell Sci.* **92:** 569–574.

Thangavelu, M., D. Belostotsky, M.W. Bevan, R.B. Flavell, H.J. Rogers, and D.M. Lonsdale. 1993. Partial characterization of the *Nicotiana tabacum* actin gene family: Evidence for pollen specific expression of one of the gene family members. *Mol. Gen. Genet.* **240:** 290–295.

Tiezzi, A., A. Moscatelli, G. Cai, A. Bartelesi, and M. Cresti. 1992. An immunoreactive homolog of mammalian kinesin in *Nicotiana tabacum* pollen tubes. *Cell Motil. Cytoskeleton* **21:** 132–137.

Turkina, M.V., A.L. Kulikova, O.I. Sokolov, T.A. Bogatyrev, and A.L. Kursanov. 1987. Actin and myosin filaments from the conducting tissues of *Heracleum sosnowskyi*. *Plant Physiol. Biochem.* **25:** 689–696.

Vahey, M., M. Titus, R. Trautwein, and S. Scordilis. 1982. Tomato actin and myosin: Contractile proteins from a higher land plant. *Cell Motil.* **2:** 131–147.

Valenta, R., M. Duchene, K. Pettenburger, C. Sillaber, P. Valent, P. Bettelheim, M. Breitenbach, H. Rumpold, D. Kraft, and O. Scheiner. 1991. Identification of profilin as a novel pollen allergen; IgE autoreactivity in sensitized individuals. *Science* **253:** 557–560.

Vallee, R.D. and H.S. Shpetner. 1990. Motor proteins of cytoplasmic microtubules. *Annu. Rev. Biochem.* **59:** 909–932.

Vandekerckhove, J. 1990. Actin-binding proteins. *Curr. Opin. Cell Biol.* **2:** 41–50.

Vandekerckhove, J. and K. Vancompernolle. 1992. Structural relationships of actin-binding proteins. *Curr. Opin. Cell Biol.* **4:** 36–42.

Vartard, M., P. Schellenbaum, A. Fellows, and A.-M. Lambert. 1991. Characterization of maize microtubule-associated proteins one of which is immunologically related to tau. *Biochemistry* **30:** 9334–9340.

Venverloo, C.J. and K.R. Libbenga. 1987. Regulation of the plane of cell division in vacuolated cells. I. The function of nuclear positioning and phragmosome formation. *J. Plant Physiol.* **131:** 267–284.

Wang, Y., W. Zhang, J. Cao, D. McElroy, and R. Wu. 1992. Characterization of *cis*-acting elements regulating transcription from the promoter of a constitutively active rice actin gene. *Mol. Cell. Biol.* **12:** 3399–3406.

Webb, M.C. and B.E.S. Gunning. 1990. Embryo sac development in *Arabidopsis thaliana*. I. Megasporogenesis including the microtubular cytoskeleton. **184:** 244–256.

———. 1991. The microtubule cytoskeleton during the development of the zygote, proembryo and free-nuclear endosperm in *Arabidopsis*. *Planta* **184:** 244–256.

Wertman, K.F. and D.G. Drubin. 1992. Actin constitution: Guaranteeing the right to assemble. *Science* **258:** 759–760.

Wertman, K.F., D.G. Drubin, and D. Botstein. 1992. Systematic mutational analysis of the yeast *ACT1* gene. *Genetics* **132:** 337–350.

Wick, S.M. 1985. The higher plant mitotic apparatus: Redistribution of microtubules, calmodulin and microtubule initiation material during its establishment. *Cytobios* **43:** 285–294.

———. 1991. The preprophase band. In *The cytoskeletal basis of plant growth and form* (ed. C.W. Lloyd), pp. 231–244. Academic Press, London.

Williamson, R.E. 1975. Cytoplasmic streaming in *Chara*: A cell model activated by ATP and inhibited by cytochalasin B. *J. Cell Sci.* **17:** 655–668.

———. 1991. Orientation of cortical microtubules in interphase plant cells. *Int. Rev. Cytol.* **129:** 135–206.

———. 1993. Organelle movements. *Annu. Rev. Plant Physiol. Plant Mol. Biol.* **44:**

181–202.

Williamson, R.E., U.A. Hurley, and J.L. Perkin. 1986. Regeneration of actin bundles in *Chara*: Polarized growth and orientation by endoplasmic flow. *Eur. J. Cell Biol.* **34:** 221–228.

Wolfe, K.H., P.M. Sharp, and W.-H. Li. 1989. Rates of synonymous substitution in plant nuclear genes. *J. Mol. Evol.* **29:** 208–211.

Yokota, E., T. Kohno, and T. Shimmen. 1993. Mechanism of cytoplasmic streaming isolation and characterization of motor protein myosin. *Plant Cell Physiol.* **34:** s3.

Zhang, W., D. McElroy, and R. Wu. 1991. Analysis of rice *Act1* 5′ region activity in transgenic rice plants. *Plant Cell* **3:** 1155–1165.

# 39

# Calcium, Protons, and Potassium as Inorganic Second Messengers in the Cytoplasm of Plant Cells

**Michael R. Sussman and Natalie D. DeWitt**
Cell and Molecular Biology Program
and Department of Horticulture
University of Wisconsin
Madison, Wisconsin 53706

**Jeffrey F. Harper**
Department of Cell Biology, MB8
Scripps Research Institute
La Jolla, California 92037

The cyclic nucleotides and phosphoinositides are second messengers that play recognized roles in mediating the effects of hormones in animals. Changes in the concentrations of these messengers are mediated by changes in the activities of enzymes that catalyze their synthesis or breakdown (e.g., adenyl cyclase, cAMP phosphodiesterase, phosphatidylinositol kinase, phospholipase C). In higher plants, despite several decades of physiological and biochemical research, the role of these potential signal-transducing messengers remains unclear.

It is also recognized that in all eukaryotes, the cytoplasmic concentration of calcium is maintained at a very low resting level (~100 nM), which fluctuates in response to regulatory or developmental cues. Recent results demonstrate that calcium plays an important role as a second messenger in higher plants. Unlike cAMP and the phosphoinositides, however, calcium is nonmetabolizable and, thus, changes in its concentration are mediated exclusively by alterations in transport across membranes, into or out of the cytoplasm. In this chapter, we review the state of our knowledge on calcium transport in higher plants. We also review the state of our knowledge on proton and potassium transport, since the cytoplasmic concentrations of these monovalent cations may also play key growth-regulating roles. Although it is not yet clear whether cytoplasmic proton and potassium ions can be considered true plant second messengers, we propose here a model in which changes in their concentra-

*Arabidopsis*
© 1994 Cold Spring Harbor Laboratory Press 0-87969-428-9/94 $5 + .00

tions cause changes in cell growth, possibly mediated by an amplification scheme involving the proton motive force and turgor pressure.

Central to this model is the recognition that all plants possess one fundamental difference from their animal eukaryotic counterparts. In animals, the plasma membrane *sodium* pump ($Na^+,K^+$-ATPase) transports sodium out of the cell to generate a sodium motive force that drives the inward flux of solutes via sodium cotransport proteins. In contrast, plants utilize a primary plasma membrane *proton* pump ($H^+$-ATPase) that transports a proton outside the cell to generate a proton motive force (PMF) (for recent reviews, see Maathuis and Sanders 1992; Sussman 1994). This PMF, in turn, drives the inward flux of sugars, amino acids, and other solutes via proton cotransport proteins (Bush 1990, 1993; Li and Bush 1991).

Another unique characteristic of plants is the cell wall. Because the cellulosic wall restrains any increase in volume caused by increases in the concentration of cytoplasmic osmotica, it pushes on the plasma membrane with a physical force called turgor pressure. Anyone who has left celery or lettuce in the refrigerator too long understands turgor. Wilty vegetables are simply tissues in which the plasma membrane is no longer as highly pressed onto the outer wall. Animal cells, which lack a rigid cell wall, simply increase in volume with increasing concentrations of cytoplasmic solutes. In plants, the cells cannot increase in volume without more long-term changes in the wall structure, and hence, this pressure of the membrane on the wall causes an increased rigidity. In a sense, this turgor pressure is the plant's skeleton, since it gives the plant an erect habit. The turgor pressure can be a formidable force, as exemplified by the ability of bamboo shoots to grow through concrete. Although the woody bamboo shoot tip provides the piercing point, the actual pressure that drives the bamboo through concrete is turgor pressure in the shoot cells. In young plant tissues, the turgor pressure alone maintains the plant's erect shape, but as the plant gets older, the wall becomes lignified and more permanently rigid, resulting in a plant structure that maintains its shape even when turgor is lost (e.g., dried corn stalks or tree bark).

Potassium is the main osmoticum in living plant cells, with typical cytoplasmic concentrations in the 100–200 mM range. Thus, any change in the cytoplasmic potassium concentration that results in swelling or shrinking of the protoplast also results in changes in turgor pressure at the cell surface. There are some cells and organs in which turgor changes play obvious biological roles as, for example, stomatal guard cells and rapidly moving parts of touch-sensitive plants such as *Mimosa pudica* and the Venus flytrap. In light of recent studies on pressure-dependent changes in plasma membrane ion channels (cf. Cosgrove and Hedrich 1991), a role for turgor pressure as a unique means of transmitting information across the cytoplasm of normal "immobile" cells should not be overlooked.

Electrophysiological measurements of mechanosensitive ion channels in a wide variety of species (plant and animal) indicate that plasma membrane proteins can sense and respond to changes in pressure at the cell surface. A recent convergence of electrophysiological (Gustin et al. 1988) and molecular biological (Braam and Davis 1990; Ling et al. 1991) results indicate that calcium channels in the plasma membrane mediate, at least in part, the effects of such pressure. These cellular responses include mechanosensitive changes in growth rate, which could play a key role in the smooth orchestration of cell shape and growth in complex, organized plant tissues.

In this chapter, we examine what is known about how plant cells control the cytoplasmic concentration of calcium, protons, and potassium ($pCa_{cyto}$, $pH_{cyto}$, and $pK_{cyto}$). We also touch on the state of our knowledge about whether these parameters are involved in plant signal transduction pathways. Due to the large number of reports on $pCa_{cyto}$ as a second messenger in animal cells, it is easy to imagine similar molecular mechanisms by which this could occur in plants. However, as described below, there may be some fairly major differences between plants and animals, e.g., in the compartments in which calcium is stored and in the protein kinases involved in the transduction of these changes.

In contrast to $pCa_{cyto}$, there is little definitive information pertaining to a role for $pH_{cyto}$ and $pK_{cyto}$ in animal signal transduction. In plants and animals, changes in $pH_{cyto}$ may alter cell metabolism simply by altering the catalytic rate of enzymes with discrete pH optima. In addition, in plants, changes in $pH_{cyto}$ could alter the activity of the proton pump and proton symport or antiport systems, resulting in changes in the membrane potential that could trigger the opening or closure of rapid voltage-gated channels. In animals, where the primary ion pump is a $Na^+$ pump, and most porters are $Na^+$ coupled, such "indirect" effects of $pH_{cyto}$ might not occur.

Along the same vein, changes in $pK_{cyto}$ could cause changes in turgor pressure which, via mechanosensitive ion channels, would translate into changes of cellular growth rates and/or differentiation. In the leaves of plants, there are specialized pairs of cells in the outer epidermal layer called guard cells. Between these two cells is an open space called a stoma, through which most of the gases enter and leave the plant. For a plant, the two most important gases are water and carbon dioxide. At noonday sun, the plant must compromise its desire to maintain large stomata for maximal carbon dioxide uptake and fixation into sugars via photosynthesis, with a desire to not lose water. Thus, the regulation of stomatal aperture is one of the most fundamental and unique aspects of plant growth and development. The stomatal aperture is controlled exclusively by the degree of turgor and volume of the guard cells and this, in turn, is controlled completely by the fluxes of potassium across the guard cell plasma membrane. Guard cells are thus an obvious example in which K fluxes actually do work; i.e., they cause volume changes. However, other instances may occur in which K fluxes result in turgor changes without a volume change. A pressure- or voltage-induced change in ion channels (especially one that transports calcium) could then act as an amplifier to cause larger changes in many cellular activities. "Action potentials" similar, but not identical, to those found in animal nerve cells are frequently observed in plant and fungal cells (Sussman 1992b). At present, these rapid fluctuations in membrane potential have no clear function in the plant kingdom, but a role in signal transduction within an individual cell, or even in transmitting information between distant cells within the plant, warrants serious examination (Wildon et al. 1992).

From the above discussion, it should be evident that the membrane potential, $pCa_{cyto}$, $pH_{cyto}$, $pK_{cyto}$, and turgor pressure are all interconnected, and each may carry important information for the cell's sensory response system. It is unlikely that changes in one of these components will not affect the others. Thus, even though it is technically quite difficult, one must be careful to consider *each* of these components in any attempt to define the molecular pathway for sensory transduction within a particular plant cell.

### TRANSPORTER DEFINITIONS

When the plasma membrane proton pump ($H^+$-ATPase) ejects a proton out of the cytoplasm into the extracellular space, it makes two products: a proton chemical gradient (abbreviated as $\Delta pH$) and an electric potential (abbreviated as $\Psi$ or EMF, for electromotive force). Together, the $\Delta pH$ and $\Psi$ constitute the PMF. The proton pump is the primary active pump of plant and fungal cells. By this, we mean that the PMF is used by all other secondary transporters, including the proton-coupled symporters and antiporters (e.g., the $H^+$/sucrose transporter) and ion channels (e.g., the inward rectifying $K^+$ channel). In animal cells, the plasma membrane primary active pump is a sodium pump (a $Na^+,K^+$-ATPase) that creates a sodium motive force composed of a sodium chemical gradient and an electric potential. This sodium motive force in turn drives the sodium-coupled cotransporters (e.g., the $Na^+$/glucose transporter) as well as the

Na$^+$,K$^+$, and Ca$^+$ channels that create the propagating electrical signals that underlie the mammalian nervous system. Despite large physiological differences in transport processes (e.g., sodium versus proton-based transporters), there is a surprising similarity in the amino acid sequence of membrane cation transporters between *Arabidopsis thaliana* and mammals, as described in greater detail below.

There are four basic types of known transporters: (1) pumps, (2) carriers, (3) channels, and (4) ABC-type hybrid transporters. These four types differ in electrophysiological properties, such as the type of energy utilized and speed of transport. Pumps are the slowest (~100–500 ions transported per protein per second), and they perform energy transduction; i.e., they convert one type of energy into another. The plant plasma membrane proton pump (H$^+$-ATPase) converts the chemical energy of ATP into electrical energy. Bacteriorhodopsin, the primary active pump of some bacterial cells, converts light energy into electrical energy.

Carriers are the second slowest (~$10^3$ to $10^5$ ions transported per protein per second) and always simultaneously transport at least two solutes. Thus, symport and antiport cotransporters are carriers, which convert one type of solute gradient into another type of solute gradient. Examples of plant carriers include proton symporters that transport sucrose (Sauer et al. 1990; Riesmeier et al. 1993), amino acids (Frommer et al. 1993; Hsu et al. 1993), and nitrate (Tsay et al. 1993). A carrier can be electrogenic or electrophoretic. In other words, if there is an unequal number of charges passing in the two directions, the carrier can both *create* an electric potential (i.e., be electrogenic) and *respond* (i.e., be electrophoretic) to an electric potential. This is an important point. All eukaryotic cells uniformly expend energy to maintain a "resting" electric potential that is negative inside the cell. In plants and fungi, this membrane potential is especially large and can exceed 0.2–0.3 volts (roughly 1/5 the voltage of commercial batteries that power your flashlights!). To understand the power that this gradient or PMF contains, consider that the Nernst equation states that each 60 millivolts is equivalent to a 10-fold chemical gradient of a univalent charged molecule. Thus, a −0.24 volt potential across the root hair plasma membrane can be used to create a 10,000-fold gradient of K$^+$. For example, with a 20 μM potassium concentration in the extracellular space, using *ONLY* the Ψ at the root hair plasma membrane, a K$^+$ channel could drive the cytoplasmic potassium concentration to an equilibrium concentration of 20 x 10,000 = 200,000 μM, or 0.2 M concentration. In a like manner, a H$^+$-coupled sucrose carrier that carries one proton into the cytoplasm, along with each sucrose molecule, will drive a 20 μM external sucrose concentration into a 0.2 M cytoplasmic sucrose concentration.

What about the negatively charged solutes that need to be accumulated in the cytoplasm? For example, if chloride is the predominant anion transported along with the high concentrations of $K^+$ (>0.2 M) required for osmoregulation in the plant cytoplasm, how is it accumulated to such high concentrations with a high internal negative potential? One solution, which has been documented in *Chara* and other plants, is a proton-coupled chloride carrier that takes in two protons along with every chloride ion (i.e., a $2H^+/Cl^-$ symporter) (Sanders 1990).

The third type of transporters, channels, are also the fastest and are capable of moving $>10^5$ to $10^6$ solutes per protein per second. Because they are so fast, their abundance in the membrane may be very low, a situation that may necessitate the use of a genetic approach for their molecular characterization. In fact, as described below, a chromosome walk to clone and sequence the genetic loci that underlie "behavioral" mutants of *Drosophila* provided the first molecular structure for plasma membrane potassium channels. In a surprising result, plant potassium channels homologous to the *Drosophila* channels were cloned for the first time by "rescuing" yeast potassium transport mutants with *A. thaliana* cDNAs (for references, see Sussman 1992b). Another important and surprising plant channel whose molecular structure was determined this year was a protein that appears to catalyze the transport of water itself, rather than any charged or neutral solute (cf. Maurel et al. 1993). Although we are unaccustomed to thinking that water needs the assistance of a protein to move through the membrane, the data are fairly convincing that this family of channel proteins performs that function. Water certainly can move across membranes without the assistance of a specific protein, but it is reasonable to think that there may be instances where more rapid movement is needed. For example, touch-sensitive plants that have organs which move very fast due to turgor changes in "motor" cells may need a water channel protein to catalyze the millisecond fast changes in water content of specific cells (for discussion of motor cells, see Simons 1992). It is also interesting to speculate that if water channel proteins are not evenly distributed in the plasma membrane or tonoplast, then specific regions of the cytoplasm may become "diluted" faster than others. This could have profound implications for signal transduction events that rely on the inhomogeneity of plant cytoplasm.

One of my (M. Sussman) wisest mentors in graduate school once said that if you asked a biochemist to determine how a car worked, he/she would first homogenize that car in phosphate buffer and purify and characterize each of its parts. Although the homogenization and purification approach has obviously worked for elucidating the structure and function of cytoplasmic soluble proteins central to metabolism and cellular regulation, rare membrane proteins do not lend themselves as readily to this treatment. The last frontier of cell biology lies in characterizing proteins that make up the cell's insoluble machinery, such as the cytoskeleton, membranes, and centrosomes. The bidirectional genetic approach, i.e., *classical genetics*, from mutant to clone via chromosome walk or gene tagging and *reverse genetics*, from clone to mutant via gene disruptions and antisense expression, provides an excellent strategy to elucidate the structure and function of these machines. In addition, recent advances in patch-clamp technology, combined with a genetic approach using *A. thaliana*, promise great advances in the coming years.

The fourth and newest category of transporters has its origins in such a genetic approach. This fourth family of transporters comprises proteins that appear to be hybrid between a pump and a channel. This group includes three well-known membrane proteins: (1) the mammalian P-glycoprotein that is responsible for secreting a variety of drugs and whose gene, when mutated, underlies the phenomenon of multidrug resistance; (2) the yeast *ste6* gene, whose protein product mediates the export of the peptide **a** mating factor; and (3) the cystic fibrosis gene, whose product is a transmembrane chloride transporter in humans. All of these proteins contain 6–10 membrane-spanning domains, as well as a nucleotide-binding domain which is highly conserved in amino acid sequence. This group is also called the ABC-type membrane transporters, for *ATP-b*inding *c*assette. Electrophysiological data indicate that members of this group act as channels with ATP-dependent transport (cf. Wine and Silverstein 1992). The hybrid nature of this activity has led to the proposal that this single type of polypeptide can shift its catalytic activity from a pump-type to a channel-type, depending on physiological conditions (e.g., osmolarity) (Gill et al. 1992). In yeast, this type of protein has been found in the plasma membrane, as well as in the vacuolar membrane. Recently, the sequence of an *A. thaliana* genomic DNA clone encoding the first plant ABC-type transporter was reported by Dudler and Hertig (1992). One may speculate that such clones encode proteins responsible for xenobiotic secretion and/or herbicide resistance, functions that may be unique to plants and that are of potential agronomic importance.

### THE PLASMA MEMBRANE PROTON CYCLE

As illustrated by the above discussion, there is a constant cycling of protons going in both directions across the plasma membrane. It is the challenge of molecular biologists and electrophysiologists to reveal the molecular properties of these proton transporters. Once the molecular structure of these proteins is known, the next question becomes how they are regulated to satisfy the nutritional and mechanical needs for each unique cell, tissue, and organism. Transport processes at the plant plasma membrane are dominated by the proton pump ($H^+$-ATPase), since it alone seems to be the protein responsible for transducing the biochemical fuel (ATP) into a form useful for transport.

The plasma membrane proton pump ($H^+$-ATPase) contains a single catalytic polypeptide of $M_r$ = 100,000 daltons, with 8–10 transmembrane domains (cf. reviews by Serrano 1989; Sussman and Harper 1989; Briskin 1990). This ATPase is known as a P-type ATPase because it forms a phosphorylated intermediate at an aspartyl residue located within a highly conserved domain of the nucleotide binding/hydrolysis site. The P-type ATPases are distinct from three other known proton pumps: (1) the $F_0F_1$ $H^+$-ATPase located in the bacterial, mitochondrial, and chloroplast membranes, (2) the V-type $H^+$-ATPase in plant vacuoles and animal lysosomes, and (3) a newly described proton-translocating pyrophosphatase found in the vacuole of plants. The P-type proton pump is found only in the plasma membrane of fungi, plants, and some mammalian parasites (for the most recent sequence alignment, see Wach et al. 1992). Because of the power of yeast genetics and the high abundance of this enzyme in the fungal plasma membrane, a great deal is known concerning structure/function relationships of the plasma membrane proton pump in yeast. The reader is referred to Gaber (1992) for a recent review and interpretation of these yeast mutational studies.

In higher plants, the first structural studies with the plasma membrane proton pump ($H^+$-ATPase) were reported by Schaller and Sussman (1988), who used automated Edman degradations to determine the amino acid sequence for randomly chosen tryptic peptides from the $H^+$-ATPase polypeptide purified from oat root plasma membranes. These results clearly established that the plant enzyme shared sequence identity with the P-type family of ATPases, which includes the $Na^+,K^+$-ATPase, the $H^+,K^+$-ATPase, and the $Ca^{++}$-ATPase of mammalian plasma membranes; the $K^+$-ATPase of bacterial cells; the $H^+$-ATPase of fungal plasma membrane; and the $Ca^{++}$-ATPase of animal sarcoplasmic (endoplasmic) reticulum. The interpretation based on protein sequence data was confirmed by cloning and sequencing full-length cDNA clones for the $M_r$ = 100,000 plant ATPase polypeptide. The cloning studies were all performed with *A. thaliana*, to set the stage for future efforts in isolating mutants. Harper et al. (1989) reported the complete sequence for one gene isoform called *AHA-1* (for *Arabidopsis H^+-ATPase*) as well as the identification of a second incomplete cDNA corresponding to a similar but distinct gene, *AHA-2*, and in the same year, Pardo and Serrano (1989) reported the complete sequence for a third (*AHA-3*). A follow-up report by Harper et al. (1990) provided the complete sequence for *AHA-2*, as well as Northern blots using gene-specific probes (from 5′ untranslated regions) to measure the mRNA levels for each of the three isoforms in roots and shoots. These studies indicated that *AHA-3* was found in both roots and shoots, whereas *AHA-1* and *AHA-2* were predominantly expressed in roots. The study also established that expression of *AHA-1* is probably environmentally controlled, since variable levels of *AHA-1* mRNA were detected in shoots from plants grown under different conditions of humidity, temperature, and light. Organ-specific differences in the expression of specific gene isoforms encoding the plasma membrane $H^+$-ATPase have also been reported in tobacco (Perez et al. 1992), and changes in the level of mRNA corresponding to a plasma membrane $H^+$-ATPase gene have been reported following alterations in water supply to germinating soybeans (Surowy and Boyer 1991).

Northern blots thus demonstrated that at least one of the three *AHA* isoform genes shows organ-specific expression. To determine whether expression is cell-specific, i.e., localized to particular cells in tissues within the organ, it is necessary to perform in situ hybridization with gene-specific mRNA probes. Since we already had indications that the *AHA* gene family was larger than just these three, we could not be confident that our probes were gene-specific.

An alternative approach to this problem of measuring expression of specific genes within a large gene family is to use a reporter gene such as the bacterial gene encoding an enzyme, β-glucuronidase (*GUS*), as a replacement for the *AHA* coding sequences. Thus, we constructed a plasmid in which a gene encoding *GUS* was fused to 4 kb of potential upstream regulatory sequences of the *AHA-3* gene isoform. The *GUS* was fused in-frame to the eighth codon (Ile) of the AHA-3 polypeptide. Using an anchored polymerase chain reaction (PCR) method (Harper et al. 1990), the transcription start site was mapped to 199 bp 5′ upstream of the *AHA-3* initiating ATG. This region includes a 144-bp intron and a TATA box at −27 bp.

When *Arabidopsis* and *Nicotiana* plants were transformed with *Agrobacterium* containing the *AHA-3* promoter-*GUS* construct, all of the transgenic plants showed a similar pattern of expression (DeWitt et al. 1991). A histochemical microscopic examination showed that in vegetative tissues, *GUS* was only expressed in specific vascular cells, the phloem cells. In contrast to two other promoters that show increased expression in vascular tissue (cauliflower mosaic virus and *Arabidopsis* ubiquitin promoters), a biochemical analysis of extracted *GUS* activity demonstrated that the *AHA-3* promoter gives essentially *no* activity in nonvascular vegetative tissue. In vegetative structures, the phloem plays a fundamental role in moving sucrose from sites of synthesis (source) to sites of use (sinks). Phloem loading and unloading depend on a sucrose/$H^+$ cotransporter that is driven by the PMF. Thus, high activity of this *AHA-3* $H^+$-ATPase promoter in phloem implicates this gene isoform in generating the PMF used to drive long-range transport of sugar and other nutrients. In immunocytochemical measurements of overall plasma membrane $H^+$-ATPase content, the phloem is one of the richest tissues in plants (Villalba et al. 1991). Further studies will be needed to establish whether *AHA-3* alone provides this high expression or whether additional *AHA* isoforms are involved.

As part of an exhaustive study to identify *all* members of the *AHA* gene family, PCR with *A. thaliana* DNA was performed utilizing degenerate DNA oligonucleotides corresponding to the two most conserved domains of the eukaryotic P-type ATPase family of enzymes (Sussman 1992a). By this strategy, a total of 12 distinct genes encoding P-type ATPases were identified in the *Arabidopsis* genome. In addition to the three *AHA*s already published, 7 new genes that encode polypeptides approximately 80–90% identical to *AHA-1* were identified (*AHA-4* through *AHA-10*) (Harper et al. 1994). Two additional genomic clones predict P-type ATPases of much lower identity to *AHA-1*, and these were called *AXA* to underscore the sequence differences between these two clones and the remaining ten. The AXA clones may represent *Arabidopsis* P-type ATPases that translocate ions other than $H^+$ (e.g., $Ca^{++}$, see below) and/or are located in endomembranes rather than the plasma membrane.

An *AHA-2* promoter driving *GUS* expression in transgenic plants containing *GUS* fusion constructs displays *GUS* activity mainly in epidermis and cortical cell layers of the root (J. Harper, pers. comm.). Nutrient uptake by the root epidermis is an obvious important specialized transport function and is presumably being driven by the *AHA-2* proton pump gene isoform. Thus, our preliminary results with *Arabidopsis* indicate that cells with specialized plasma membrane functions seem to utilize a specialized isoform. A prediction of this model is that guard cells should contain pump polypeptides produced by one of the remaining *AHA* genes. Multiple genes encoding P-type ATPases from tobacco (Perez et al. 1992) and tomato (Ewing et al. 1990) have been reported, but cell-specific expression has not yet been examined in these plant species.

Estimates derived from electrophysiological measurements of current carried by the proton pump in actively transporting cells such as mustard root hairs (Felle 1982) indicate that as much as 1/4 to 1/3 of the cellular ATP is consumed by this one enzyme. Measurements of growth rates in yeast proton pump mutants expressing a range of ATPase activities clearly demonstrate that the growth rate is closely tied to catalytic activity of this one enzyme, and mutants expressing less than 10–20% of the wild-type $H^+$-ATPase activity are lethal. It is thus not surprising that the $H^+$-ATPase is regulated posttranslationally.

External glucose concentrations exert a strong control on the level of plasma membrane $H^+$-ATPase activity in yeast, and this occurs without a change in $H^+$-ATPase content of the membrane. When glucose-starved fungal cells are placed in glucose, there is a large rapid depolarization presumably caused by the inward rush of protons that drives the $H^+$/glucose symporter. Within 10–20 seconds of glucose addition, however, the electric potential recovers its normal resting highly negative potential. The time course of recovery of the potential correlates with a rapid activation of the proton pump, as measured in crude cell homogenates prepared at 5-second intervals after glucose addition.

Several lines of evidence point to the hydrophilic carboxy-terminal 10-kD domain as an important "inhibitory" domain, whose function is modulated posttranslationally after glucose addition. In this model, kinase-mediated phosphorylation, or changes in associated lipids, interfere with this domain's inhibitory effects on the pump's catalytic activity. Thus, when the carboxy-terminal 11 amino acids were genetically deleted, the yeast proton pump was constitutively activated and non-responsive to glucose stimulation in situ (Serrano and Portillo 1990). Single base-pair substitutions that alter a conserved carboxy-terminal threonine reduce, but do not completely eliminate, the response to glucose (Portillo et al. 1991; Serrano et al. 1992). Together with direct studies of yeast $H^+$-ATPase phosphorylation (Kolarov et al. 1988; Chang and Slayman 1991), these results support a model in which the yeast proton pump is posttranslationally regulated by protein kinases and phosphatases, but additional regulation by changes in associated lipids or other unrecognized factors has not been ruled out.

In higher plants, there are two instances in which a rapid change in catalytic activity of the plasma membrane proton pump has been observed. Fusicoccin (FC) is a remarkable alkaloid produced by a fungus that causes rapid increases in the elongation rate of all plant cells. In many ways, the effects of FC resemble those of a plant hormone, indoleacetic acid (IAA), but unlike the hormone, FC acts practically with no lag period (IAA requires several minutes for a first noticeable effect) but is not sustained for long periods of time (IAA continues to act for many hours). A current-voltage analysis performed with microelectrodes inserted into the cytoplasm demonstrates that within 10 seconds after root hair cells are treated with FC, the pump is activated (Felle 1982). De

Michelis et al. (1989) observed an in vitro stimulation of the ATPase after treating membrane vesicles with FC, but this has been difficult to reproduce in other laboratories (Schulz et al. 1990). In situ, FC induces a robust and rapid increase in proton secretion. Simultaneously, the membrane potential hyperpolarizes and there is an increased elongation rate. It is important to recognize that there is no direct evidence that all of the physiological effects of FC result from activation of the pump. Although FC seems to rapidly activate the pump, a simultaneous effect on ion channels and other plasma membrane proteins has not been ruled out.

A second instance of rapid pump activation occurs in guard cells. Patch-clamp electrophysiological studies in the whole-cell configuration demonstrate that the proton pump undergoes a dramatic activation following light treatment. In fact, this activation precedes, and presumably is the cause of, the opening of a voltage-gated potassium channel that causes guard cells to swell and stomata to open. In contrast, abscisic acid (ABA)-induced closure of guard cells seems to occur via a transduction pathway independent of the pump. In this case, current models indicate that ABA treatment opens a calcium channel, which causes an anion channel to open, leading to a depolarization of the electric potential and a concomitant activation of a potassium efflux channel preceding guard cell shrinkage and stomatal closure.

What is the molecular mechanism of pump activation in higher plants? Limited proteolysis of the carboxyl terminus of the plant $M_r =$ 100,000 ATPase polypeptide induces a significant activation of catalytic activity in vitro (Palmgren et al. 1990, 1991). Recent success in expression of the AHA polypeptides in yeast provides a genetic strategy to analyze "regulatory" amino acid residues in the carboxy-terminal domain of the plant plasma membrane $H^+$-ATPase (Nakamoto et al. 1991; Villalba et al. 1992; Palmgren and Christensen 1993, 1994). Although Schaller and Sussman (1988) demonstrated that radioactive phosphate fed to whole cells is incorporated into an $M_r =$ 100,000 polypeptide coincident with the ATPase protein on one-dimensional SDS-PAGE of purified oat plasma membranes, changes in the phosphorylation status following in vivo treatment with FC or light have not yet been investigated. These experiments are technically difficult for two reasons. First, the $H^+$-ATPase in plant extracts is at least 10-fold less abundant than it is in yeast and will need to be purified by immunoprecipitation from plants fed large amounts of radioactivity. Second, in yeast there is only one essential yeast isoform (which makes up at least 10–20% of the total plasma membrane protein), whereas in plants, there is a large family of isoforms being expressed in different cells and tissues. We have recently found that one *AHA* isoform (*AHA-10*) has dramatic sequence differences concentrated in the carboxyl terminus (J. Harper, pers. comm.), and it is interesting to speculate that this carboxy-terminal difference may reflect differences in regulation.

More extensive in vitro studies have been performed to study the protein kinase(s) that phosphorylates the plant plasma membrane $H^+$-ATPase. Schaller and Sussman (1987) reported results of studies in which isolated oat root plasma membrane vesicles were treated with $[\gamma\text{-}^{32}P]ATP$, and phosphorylation of the $H^+$-ATPase and other polypeptides was measured by autoradiography after one-dimensional SDS-PAGE. This in vitro study demonstrated that the oat root proton pump is phosphorylated in a calcium-dependent manner, at both serine and threonine residues. In situ evidence for calcium-dependent regulation of the proton pump was provided by Lew (1989), who observed a rapid hyperpolarization attributed to proton pump activation when calcium was microinjected into intact hyphal cells of the fungus, *Neurospora crassa*.

Although phosphorylation of the plant proton pump is easy to demonstrate in vitro and in situ, there is not yet concrete evidence that such phosphorylation actually regulates catalytic activity. These higher plant studies are again hampered by low abundance of the pump and multiplicity of isoforms, as well as a large degree of denaturation that occurs whenever the $H^+$-ATPase is detergent-solubilized and purified away from other proteins. In purified plasma membranes, the pump is fairly stable, but its catalytic activity is very sensitive to its lipid environment. Many detergents and lipids increase the $H^+$-ATPase activity in vitro in vesicles. Although some of this activation can be accounted for by a breakdown in the permeability to substrates (ATP) in sealed right-side-out plasma membrane vesicles, there is a component of the activation that may be caused by a bona fide "regulatory" activation. Separating the regulatory activation component from the sealed vesicle problem is not trivial, and thus, results of biochemical experiments with membrane vesicles where chemical agents are used to activate the ATPase should be interpreted with caution. Finally, when $[\gamma\text{-}^{32}P]ATP$ is added to plasma membrane vesicles in vitro, more than 99% of the radioactivity is incorporated into lipids, rather than protein. Thus, it is difficult to conclude that protein phosphorylation alone is responsible for observed changes in catalytic activity. Evidence has been reported for a direct role of lipids such as the phosphoinositides (PIP and $PIP_2$) (Memon and Boss 1990) and lysolecithin (Palmgren et al. 1988; Pedchenko et al. 1990; Andre and Scherer 1991) in regulating the plant plasma membrane proton pump.

On an optimistic note, the recent discovery that a functioning plant ATPase protein can be expressed from the AHA clones in large amounts in yeast provides a means to rigorously test whether serine/threonine phosphorylation of the proton pump is a viable regulatory mechanism (Palmgren and Christensen 1994). In preparation for these experiments using a yeast expression system to circumvent problems inherent in plant biochemistry, we have focused most of our efforts on characterizing and cloning genes encoding a plant calcium-activated protein kinase that phosphorylates the ATPase. In a collaborative study with Dr. Alice Harmon (University of Florida), Drs. Jeff Harper and Eric Schaller in the laboratory of Dr. Sussman were able to demonstrate that the *calcium dependent protein kinase* (abbreviated hereafter as CDPK) associated with the plant plasma membrane is a new type of calcium-regulated protein kinase not previously observed in yeast or animals (Harper et al. 1991; Roberts and Harmon 1992; Schaller et al. 1992). Previously, protein kinase C (PKC) and calmodulin-dependent protein kinase (CaMK) were the only kinases known to be regulated by calcium. The plant CDPK is unique because it contains, in a single polypeptide, an amino-terminal $M_r = 30,000$ kinase catalytic domain fused to a $M_r = 20,000$ regulatory carboxy-terminal domain that resembles calmodulin and contains four calcium-binding sites. Thus, in contrast to the two animal calcium-activated protein kinases, the plant CDPK only requires a binary complex (kinase·$Ca^{++}$) for activity, although lipid is required for maximal activity in some circumstances. We have now established that the *Arabidopsis* CDPK is encoded by a gene family containing several distinct genes. Properties of CDPK are described in more detail in the calcium section below.

Is the cytoplasmic pH a second messenger in plant cells, or in any other eukaryote, for that matter? The answer to this question is still up in the air. Although the cytoplasmic pH certainly varies with differing environmental conditions or developmental states, there is no direct proof that such changes are a causal agent for the subsequent biological response. In plants and fungi, the plasma membrane proton pump responds to changes in cytoplasmic pH in a manner analogous to the way in which normal enzymes respond to changes in substrate concentration, but in addition, there may be more complicated modes of response. In oat root plasma membrane, Schaller and Sussman (1987) found that the ATPase is especially resistant to phosphorylation by CDPK above pH 7. In contrast, other polypeptides were still readily phosphorylated in a calcium-dependent fashion above pH 7. This may indicate that the conformation of the pump which predominates above pH 7 is incapable of acting as a substrate for calcium-dependent phosphorylation and hints at the possibility that $pH_{cyto}$ and $pCa_{cyto}$ may be *interacting* variables, particularly in their effects on the proton pump. A more recent study in which $pH_{cyto}$ and $pCa_{cyto}$ were measured simultaneously in intact plant tissues supports this concept (Gehring et al. 1990). When auxin was added, the cytoplasm became acidified and the calcium level increased, both of which would result in increased phosphorylation state of the $H^+$-ATPase and presumably, a more active proton pump. In contrast, when ABA was added, although the cytoplasmic calcium level also increased, the cytoplasm became more alkaline. The in vitro results of Schaller and Sussman (1987) predict that the $H^+$-ATPase would be resistant to increased phosphorylation under these conditions and, hence, little activation of the pump would ensue. Gehring et al. (1990) suggest that the opposite effect of auxin and ABA on cytoplasmic pH accounts for their antagonistic effects on cell elongation, even though both agents cause an increased level of calcium in the cytoplasm. Few investigators are technically capable of performing simultaneous measurements of cytoplasmic pH and pCa. If the effects of a change in one depend on the context of changes in the other, it may prove impossible to relate such changes to biological phenomena until more facile measuring techniques are available.

There is genetic evidence in favor of a role for cytoplasmic pH in the signal transduction pathway in non-plant systems. When a gene encoding the yeast plasma membrane proton pump is expressed in animal fibroblast cell lines, the cytoplasm is constitutively alkalinized compared to wild-type cells. The transgenic cells then become cancer-like; i.e., they lose their contact inhibition and form foci which, when injected into mice, form tumors (Perona and Serrano 1988; Perona et al. 1990). Transgenic cell lines containing a mutant proton pump gene that expresses catalytically inactive enzyme do not show the oncogenic phenotype. These results indicate that the yeast proton pump behaves as an oncogene.

An association between cytoplasmic alkalinization and the onset of cell division is further suggested by the observation of cytoplasmic alkalinization at the "start" point in the yeast cell cycle (Anand and Prasad 1989). One would thus expect that if cytoplasmic pH were a regulator of cell division and/or growth, the yeast mutants with decreased pump activity would show interesting phenotypes, perhaps with defects in cell cycle progression. Although there are no reports of a cell cycle defect in proton pump mutants, a "multi-budded" phenotype was observed, suggestive of a loss in coordination between the processes of cell division and cell elongation (Cid et al. 1987; McCusker et al. 1987).

In conclusion, there is no universal acceptance or direct proof of the notion that changes in cytoplasmic pH play a significant role in biological processes in eukaryotes. There are two major problems in this field. First is the difficulty in accepting the idea that such a pleiotropic regulator of enzymes (pH) could be a useful second messenger to regulate specific cellular processes. A second problem is the difficulty in being certain that effects of $pH_{cyto}$ changes are physiologically significant, rather than a "pharmacological" manifestation of an indirect response to changes in pH. In any case, as Gaber (1992) points out, transgenic animal experiments with plant/fungal proton pumps and proton-coupled transporters provide an excellent means to clarify the regulatory role, if any, that changes in cytoplasmic pH play in animals. Likewise, experiments with transgenic plants containing genetically increased or decreased levels of activity of the plasma membrane $H^+$-ATPase could provide a new strategy to more directly test the long-standing "acid-growth" hypothesis that changes in proton secretion regulate plant cell growth (cf. Rayle and Cleland 1992).

### CYTOPLASMIC CALCIUM

In contrast to the confusing situation with cyclic nucleotides, the evidence for a regulatory role of cytoplasmic calcium in plant growth and development is substantive. In particular, there is evidence now for a direct role of cytoplasmic calcium in early steps of the transduction pathways for photomorphogenesis and thigmomorphogenesis, as well as the response to cold shock and microbial pathogens (Knight et al. 1991, 1992; Shacklock et al. 1992; Bowler et al. 1994). During the past 3 years, one of the most fruitful areas of research on the role of cytoplasmic calcium in plants has been an electrophysiological analysis of the guard cells that control rapid opening and closure of stomata. Light, auxin, and fusicoccin all induce opening of stomata, whereas ABA causes closure. The opening and closure responses are both rapid (within seconds and minutes) and mediated by a sequential series of changes in ion transporters that can be observed "in real time" by patch-clamp studies with protoplasts derived from the guard cells. Because of the rapidity of these responses, it is likely that the observed changes in catalytic activity of plasma membrane proteins are caused by allosteric effectors, protein-protein interactions, or posttranslational modifications, possibly mediated by protein kinases and phosphatases.

The opening of stomata is caused by an increase in turgor pressure of guard cells mediated by an increase in potassium uptake through an inward-rectifying potassium channel. The channel is opened by hyperpolarization caused by light-induced activation of the proton pump (Assmann et al. 1985; Serrano and Zeiger 1989). There is no convincing evidence that cytoplasmic calcium is involved in this response. In contrast, there is substantive evidence that an increase in cytoplasmic calcium is causally associated with ABA-induced closure of stomata (Keller et al. 1989; Schroeder and Hagiwara 1989, 1990; Blatt et al. 1990; Gilroy et al. 1990, 1991; Hedrich et al. 1990; McAinsh et al. 1990; Marten et al. 1991, 1992; Cao et al. 1992; Schroeder and Keller 1992). In this case, the sequence of events seems to be as follows: (1) ABA opens a plasma membrane calcium channel, leading to increased cytoplasmic calcium; (2) the cytoplasmic calcium rise causes an anion channel to open, which leads to depolarization of the membrane potential; (3) the depolarization activates an outward rectifying potassium channel which allows potassium efflux and loss of turgor.

In animals, there are two types of protein kinases that are activated in response to the increased cytoplasmic calcium concentration: PKC and CaMK. In plants, there is no clear biochemical evidence for a cal-modulin-dependent protein kinase. A predominant calcium-sensing protein kinase in plants seems to be CDPK, the third type of protein kinase described above. A CDPK-like enzyme is also present in *Paramecium* (Gundersen and Nelson 1987) and *Plasmodium* (Zhao et al. 1993), and thus, it is possible that a CDPK-like enzyme is present in all phylogenetic kingdoms. Perhaps PKC and CaMK are so abundant in animals that CDPK has been overlooked. Likewise, there may be a bona fide PKC or CaMK homolog in plants, but the greater abundance of CDPK renders their discovery difficult.

In animals, PKC is activated by lipids such as diacylglycerol (DAG) and phosphatidylserine (PS) and, hence, plays a role in mediating the action of such membrane-bound hydrophobic messengers. There have been numerous reports of lipid-stimulated protein kinase activity in plant cell extracts. However, since the plant CDPK is also activated by lipids, it is unclear how many of these reports represent a bona fide plant PKC homolog. CDPK solubilized from oat root plasma membrane (Schaller et al. 1992) or produced by expression in *Escherichia coli* via a cDNA clone (Harper et al. 1993; Binder et al. 1994) is stimulated severalfold by lysolecithin and phosphoinositides (including phosphatidyl inositol [PI] phosphatidyl inositol phosphate [PIP], and phosphatidyl inositol biphosphate [$PIP_2$]), but not by other lipids such as DAG, PS, or phosphatidylcholine. In this respect, the activation of CDPK by lysolecithin and phosphoinositides could be significant, since these lipids are known activators of the proton pump when applied to purified plant plasma membrane vesicles (see earlier discussion). In addition, these lipids are synthesized by enzymes (phospholipase $A_2$ and PI kinases) that are emerging as significant players in animal signal transduction (cf. Bazenet et al. 1990a,b; Brockman and Anderson 1991; Jenkins et al. 1991; Brockman et al. 1992; Nishizuka 1992). However, since there is no direct evidence that activation of CDPK by lipids is physiologically significant, further speculation is unwarranted at this time.

A key difference between known animal PKC isoforms and the plant CDPK is the lack of effect of DAG on the plant enzyme. Thus, it is conceivable that reports of an effect of DAG on plasma membrane function (Lee and Assmann 1991) and cell cycle progression (Larsen et al. 1989; Larsen and Wolniak 1990) in plants may represent responses mediated by a true plant PKC homolog that has yet to be identified.

Are other components of the animal phosphoinositide pathway present in plants? In vitro measurements with purified plasma membrane vesicles from oat roots demonstrate the presence of phospholipase C activity in plants (Tate et al. 1989). However, there is still no clear demonstration for the coupling of phospholipase C activation to occupation of a plant plasma membrane receptor protein, as occurs in animals. It is interesting to note that the plant phospholipase C is activated by low micromolar concentrations of calcium. Thus, the plant plasma membrane phospholipase C may be important for amplifying a transient channel-induced increase in cytoplasmic calcium just under the plasma membrane into a global wave of increased calcium throughout the plant cytoplasm. Several laboratories have identified plant inositol triphosphate ($IP_3$)-activated calcium channels that release calcium from internal stores, although these channels seem to be located on the tonoplast rather than endoplasmic reticulum (Schroeder and Thuleau 1991). Thus, apart from the lack of a clear demonstration of plant PKC, other components of the animal phosphoinositide signaling pathway seem to be present in higher plants.

Although the cell fractionation data demonstrating a CDPK-like enzyme attached to the plasma membrane are fairly clear, it is conceivable that CDPK association with the membrane is caused by its attachment to membrane-associated cytoskeletal elements. This interpretation is consistent with immunocytochemical results in which CDPK was visualized on actin bundles in cells of soybean (Putnam-Evans et al. 1989) and, more recently, in *Chara* (McCurdy and Harmon 1992). The inhibition of cytoplasmic streaming caused by an increase in cytoplasmic calcium is well documented in *Chara*, and CDPK localization on the subcortical actin bundles of the streaming endoplasm provides a ready explanation for how the calcium signal is transduced to the streaming motors (i.e., cytoskeletal actinomyosin). Tan and Boss (1992) recently reported that PI and $PIP_2$ kinases were associated with actin filaments isolated from biochemically purified cytoskeleton from carrot cells. Similar fractionation studies using Western blots with CDPK antibodies should help to clarify the subcellular location(s) for CDPK. Finally, our preliminary cloning studies with *Arabidopsis* cDNA indicate that the CDPK gene family may be large. Thus, it is conceivable that various CDPK isoforms are present in different cell fractions and respond to different types of lipid signals, as has been reported for the PKC gene family in animals (Nishizuka 1992).

After cytoplasmic calcium rises in animal cells, two transporters act to reduce the concentration to its "resting" level. One is a plasma membrane-bound P-type $Ca^{++}$-ATPase and the second is another P-type $Ca^{++}$-ATPase located in the endoplasmic reticulum. The two ATPases differ in several characteristic properties. The plasma membrane enzyme is (1) slightly larger ($M_r$ = 140,000 vs. 105,000), (2) activated by calmodulin (endoplasmic reticulum enzyme is not), and (3) shows a stricter nucleotide specificity (the endoplasmic reticulum enzyme uses GTP and ITP as well as ATP, whereas the plasma membrane enzyme does not).

$Ca^{++}$-stimulated ATPase activity has been identified attached to plant membranes in vitro. However, studies with membrane vesicles isolated from various plants have produced conflicting conclusions concerning whether the $Ca^{++}$-ATPases are located in the plasma membrane, tonoplast, or endoplasmic reticulum. Recent studies indicate that plants may be distinctly different from animals, since the most abundant $Ca^{++}$-ATPase associated with plant microsomes is calmodulin-activated like the animal plasma membrane enzyme, but according to sucrose density gradient centrifugation, it seems to be associated with an internal membrane rather than the plasma membrane (Askerlund and Evans 1994). Given the heterogeneous cell population of most plant tissues, and because of the possibility that unique domains of the plasma membrane give it an unusual buoyant density with centrifugation experiments, it will be necessary to definitively identify the membranous location for these transporters using immunocytochemical techniques. Adding to this confusion is the possibility that plants are also utilizing calcium/proton antiporters as calcium pumps at both the plasma membrane and vacuole (Maathuis and Sanders 1992). As suggested by Ghislain et al. (1990), it may be that the bulk movement of $Ca^{++}$ is handled by antiporters and channels, since these have a low affinity but high capacity for $Ca^{++}$. The $Ca^{++}$-ATPases, because of their high affinity and lower capacity for $Ca^{++}$, may act to fine-tune the $Ca^{++}$ concentration.

On the basis of sequence similarity to the animal $Ca^{++}$-transporting ATPases, several genes encoding P-type $Ca^{++}$-ATPases have been isolated from the genomes of yeast (Rudolph et al. 1989; Ghislain et al. 1990) and higher plants (Perez-Prat et al. 1992; Wimmers et al. 1992). The plant clone has approximately 50% identity with the mammalian endoplasmic reticulum $Ca^{++}$-ATPase, and less than 35% identity with any of the putative three fungal Ca$^{++}$-ATPase genes. Since the animal endoplasmic reticulum Ca$^{++}$-ATPase is not activated by calmodulin, it is not clear whether the plant clone encodes the calmodulin-activated Ca$^{++}$-ATPase already observed in endoplasmic reticulum purified from plant extracts.

Northern blots indicate that the mRNA level for one plant Ca$^{++}$-ATPase gene is increased after exposure of the cells to 50 mM NaCl (Perez-Prat et al. 1992; Wimmers et al. 1992). This would seem to suggest that salt stress causes a sustained higher cytoplasmic calcium level, resulting in a need for a long-term increase in levels of the endomembranous Ca$^{++}$-ATPase. However, a study of genetic disruptants in yeast suggests that a fungal ATPase gene that was thought to encode a Ca$^{++}$-ATPase based solely on sequence identity to purified animal Ca$^{++}$-ATPases may actually be a sodium pump (Na$^{+}$-ATPase) (Hario et al. 1991). This report serves to underscore the importance for direct biochemical or electrophysiological demonstrations that a putative Ca$^{++}$-ATPase DNA clone actually produces a protein that displays calcium-pumping activity. Alternatively, protein sequence from the purified plant protein can be used to help correlate function with structure of the cloned genes.

Calcium entry into the cytoplasm across the plasma membrane of plant cells is mediated by one or more calcium channels that have been characterized electrophysiologically using patch-clamp procedures with protoplasts. Plant plasma membrane calcium channels have been reported to be gated by voltage, by ABA, and by pressure (stretching) (Schroeder and Thuleau 1991). Pressure-sensitive calcium channels have also been observed in yeast and other microbes and may be a universal means of osmoregulation in protoplasts surrounded by a cell wall (Gustin et al. 1988; Martinac et al. 1990; Sukharev et al. 1994). Despite their ubiquitous presence in patch-clamp recordings, no DNA clone encoding a candidate calcium channel has yet been reported for plants, fungi, or any other microbe.

Recent studies in the laboratory of T. Trewavas at the University of Edinburgh have opened an exciting new strategy for investigating fluctuations in cytoplasmic calcium levels in situ in higher plants (Knight et al. 1991, 1992). In these experiments, a gene encoding the bioluminescent protein, aequorin, was expressed in transgenic tobacco plants. After the uptake of a hydrophobic luminophore (coelenterazine, $M_r$ = 423), visible light was emitted in transgenic cells whenever cytoplasmic calcium levels increased. Most interesting, the application of plant hormones or other growth substances to the intact plant did not cause an overall increase in light emission, but the application of cold shock, touch, and fungal elicitors did.

This elegant demonstration of a rise in calcium level after mechanostimulation supports the earlier suggestion by Braam and Davis (1990), on the basis of touch-induced expression of *A. thaliana* calmodulin-like genes, that changes in intracellular calcium mediate the effect of touch on plants. Ling et al. (1991) recently confirmed the observation of increased calmodulin expression following touch stimulation.

A more recent study by Trewavas's group suggests that there are multiple mechanisms for increased cytoplasmic calcium. In particular, they observed that after repetitive stimulations by touch, the cells adapted and showed little change in cytoplasmic calcium (Knight et al. 1992). After a refractory period of several minutes, the cells recovered and touch elicited a typical cytoplasmic calcium increase. When the cells adapted to touch, the ability of cold shock to elicit a calcium rise was unaffected, possibly indicating that the two environmental signals were transduced by separate mechanisms. The cold shock response was inhibited by lanthanum and gadolinium, compounds that selectively inhibit plasma membrane calcium channels in animal cells, whereas the touch response was inhibited by ruthenium red, an inhibitor of animal mitochondrial calcium channels. This inhibitor specificity, and the earlier observation of cold-shock-induced action potentials in higher plants (Minorsky and Spanswick 1989), led Knight et al. (1992) to suggest that the cold shock response was mediated by a plasma membrane calcium channel, whereas the touch response was mediated by calcium released from internal organelles. These investigators envisioned a system of microfilaments or microtubules attached to the organelles as the touch-sensing response internal element. This view is contrary to the notion that the plasma membrane itself senses touch via mechanosensitive calcium channels. Considering the complexity of different cell types in the transgenic aequorin plant, as well as the lack of firm data on the pharmacological mechanism of action of the putative calcium channel blockers in plants, at this point it is difficult to be certain which cellular compartments are responsible for calcium release into the cytoplasm following touch and cold shock.

In animal cells, the concept of localized domains with increased calcium levels that are not in equilibrium with the bulk cytoplasm is well accepted. In fact, a recent study by Allbritton et al. (1992) indicates that calcium is so immobile in the cytoplasm, its effective range for diffusion may be considerably less than 5 $\mu$m. Thus, since many plant and animal cells are many times larger than 5 $\mu$m, how is the increase in cytoplasmic calcium propagated through the cytoplasm? The answer to this question lies with inositol triphosphate (IP$_3$), a compound produced from PIP$_2$ by phospholipase C at the plasma membrane surface, and which can activate calcium channels in endomembranes, thus releasing calcium from internal compartments. Since phospholipase C is calcium-activated, a small transient increase in calcium can be amplified by positive feedback between calcium and IP$_3$ (Harootunian et al. 1991). In contrast to calcium's restricted diffusibility, IP$_3$ is highly mobile and displays an effective range of 20 μm or more. Thus, a wave of increased calcium propagating through the cytoplasm may actually be a wave of IP$_3$ causing the release of calcium stored in internal compartments within each cell. This wave can even move across separate cells within a tissue, as demonstrated in a recent report with animal cells which documents the movement of IP$_3$ through gap junctions (Boitano et al. 1992).

The lack of mobility of calcium may be an essential aspect of its role as a second messenger in higher plants. In 1982, Williamson and Ashley first documented the fact that a transient rise in cytoplasmic calcium levels causes an inhibition of cytoplasmic streaming in the giant alga, *Chara*. This explains why cytoplasmic streaming characteristically ceases after initiation of the action potential generated by electrical or mechanical stimulation. Together with Trewavas's demonstration of a mechano-induced increase in cytoplasmic calcium in plants, these results may provide a molecular explanation for striking observations made in simple experiments by Lintilhac and Vesecky (1984). These investigators applied compressive forces in one direction to dividing cells in callus cultures. Normally, these tissues are disorganized masses of randomly dividing cells, but under the compressive forces, conspicuous areas of organized parallel cell divisions appeared just under the point of application of pressure. Trewavas's results indicate that a membrane transporter regulating cytoplasmic calcium is the "intracellular mechanical sensor" that ultimately directs the plane of the new cell partition.

This interpretation has far-reaching implications for understanding the molecular basis of plant development. As suggested by Knight et al. (1992), most cells within the plant are constantly under restraining forces defined by the spatial orientation of contiguous cells. A simple demonstration of this internal pressure is provided by the rapid splitting apart that occurs when a turgid stem section is pierced by a razor blade or needle. Thus, the observed increase in cytoplasmic calcium due to mechanical perturbation of the intact plant by wind or beasts may be minuscule compared to the calcium fluxes induced by the internal pressure imposed during organized divisions within the plant meristem.

It is a fascinating possibility that the constraints imposed by turgor pressure in cells of the meristem may regulate the geometry of cell division and differentiation in a complex, organized tissue. Such biophysical models are not new (cf. Green 1969) and, although attractive for their simplicity, have been very difficult to test rigorously using physiological and biochemical techniques. This is probably one of the most important areas of cell biology in which the application of a genetic approach could be useful. In yeast, mutants have been obtained in which the site of bud formation is altered, and the mutant gene loci encode proteins involved in organizing and linking the actin cytoskeleton to the cell surface (Chenevert et al. 1992). It is a fitting challenge for the bold new generation of molecular biologists to come up with the right mutant screen/selection strategy that will confirm or refute the notion that changes in the concentration of a simple inorganic molecule like calcium are essential in the morphogenesis of a complex multicellular eukaryote like *A. thaliana.*

## POTASSIUM TRANSPORT AND TURGOR PRESSURE

In bacterial, animal, and fungal cells, there are a myriad of proteins involved in the uptake of potassium. One of the earliest to be characterized at the molecular level was the $K^+$-ATPase of *E. coli*, encoded by the *kdp* operon (Polarek et al. 1988). Expression of this gene is induced by potassium starvation. However, an in-depth reporter gene study of the *kdp* promoter clearly demonstrated that it is the *loss of turgor* induced by the lack of intracellular K, and not the actual K concentration, which activates transcription (Walderhaug et al. 1992). The $K^+$-ATPase encoded by *kdp* is a P-type ATPase with a high affinity for K that operates as a scavenger system for taking up K from external solutions with very low K concentrations. In animal cells, another P-type ATPase, the plasma membrane $Na^+,K^+$-ATPase, is likewise the major route for the uptake of K.

In plants and fungi, until recently, there was no information available on the molecular structure of plasma membrane proteins involved in K uptake. When the primary cation translocating ATPase of plant and fungal plasma membranes was first biochemically characterized a decade ago, investigators considered this to be a likely uptake system for K since ATP hydrolysis was stimulated in vitro by this monovalent cation. With all of the other P-type ATPases known, ion transport is obligately required for ATP hydrolysis; i.e., the enzyme does not turn over unless the substrate ion being transported is present. Despite these early observations of K-stimulated ATPase activity in plant plasma membrane vesicles, it was striking that unlike all of the other pumps known, the plant/fungal enzyme showed significant basal level of ATPase activity in the absence of K, and was only stimulated 2- to 3-fold by this cation. This observation, along with other observations on the effect of ion gradients on ATPase activity, led most investigators to conclude that the plant/fungal plasma membrane ATPase is a pure and simple proton pump ($H^+$-ATPase) and that potassium transport is indirectly coupled to the ATPase activity by the membrane potential (cf. Cao et al. 1992) and possibly also by the pH gradient.

So what then is the molecular identity of the plant/fungal plasma membrane potassium transporter(s)? Success in answering this question was made possible by a genetic analysis of potassium utilization in yeast performed in the laboratory of R. Gaber at Northwestern University. These studies applied basically the same strategy used in the early 1980s with *E. coli*; i.e., to isolate mutants that require more potassium for growth. The first yeast mutant, called *trk1* (Gaber et al. 1988), does not grow at the 0.2 mM K concentration that fully supports the growth of wild-type yeast. When *trk1* cells were mutagenized and screened for new mutations which caused cells to require even higher $K^+$ concentrations, a second locus called *trk2* was identified. Transformation studies allowed the cloning of the mutant genes, and their sequence demonstrated that *trk1* and *trk2* both encode hydrophobic proteins of similar amino acid sequence and contain 8–10 transmembrane domains (Ko and Gaber 1991). Most interestingly, these proteins lacked homology with any known ion channels or carriers or with any protein in the protein sequence database. Thus, *trk1/trk2* could be $K^+/H^+$ symporters or $K^+$ channels, but in either case, they would have to be of a type not previously identified in other organisms.

To determine whether a gene homologous to *trk* exists in plants, Gaber transformed yeast *trk* mutant cells with an *A. thaliana* cDNA library designed for expression in yeast. From this experiment, one cDNA clone was discovered that complements the *trk* defect. A surprising result, however, was that the DNA sequence of this *A. thaliana* clone predicted a protein completely unlike *trk*, but very close in sequence to a family of genes encoding $K^+$ channels in the plasma membrane of other eukaryotes such as *Drosophila* and mammals. In fact, this class of $K^+$ channels was first identified via a chromosome walk to the mutant loci underlying behavioral mutants of *Drosophila melanogaster* (Jan and Jan 1992).

The observation that an *A. thaliana* cDNA clone could rescue *trk*-yeast mutants and that it has sequence identity to known K channels does not necessarily prove that it is an ion channel responsible for K uptake in plants. In fact, all known members of this K channel family in advanced eukaryotes behave as outward rectifiers; i.e., they mediate K efflux during the propagating action potential in nerve cells. The hypothesis that this clone encodes a bona fide potassium uptake system in *A. thaliana* is supported by electrophysiological patch-clamp studies with bean guard cell protoplasts and with *Xenopus* cells expressing the protein encoded by the cDNA. Both measurements reveal the catalytic properties of a bona fide potassium channel, with identical transport properties observed in the intact plant (Schroeder and Fang 1991; Schachtman et al. 1992).

As with the plasma membrane proton pump ($H^+$-ATPase), the *A. thaliana* potassium channel seems to be encoded by a multigene family. Simultaneous with Gaber's discovery of the *A. thaliana* K channel clone, a French group performed similar studies and identified a second *A. thaliana* K channel gene (Sentenac et al. 1992). Most interesting, the second gene predicted a protein with an additional potential ankryn-binding domain not observed in Gaber's clone. In nerve cells, this domain seems to be involved in the polarized location of potassium channels. Many plant cells are highly polarized, i.e., show differences in function and shape at either end. Previous studies with a special ion measuring electrode known as the vibrating probe have identified asymmetric ion gradients across individual plant cells (cf. Speksnijder et al. 1989 and references therein). It is thus tempting to speculate that an ankryn-binding domain in one *A. thaliana* K channel clone may be one way in which K channels acquire a polar distribution to help generate current loops involved in establishing or maintaining polar growth of plant cells.

Higher plants contain many different cell types with different transport functions and needs. Hence, it is likely that other potassium transporters with amino acid sequences different from known K channels could exist. It is interesting to note that an exhaustive mutational analysis with the double mutant *trk1⁻ trk2⁻* indicates that there exists yet an additional gene that allows growth of yeast at still higher K concentrations (Ko and Gaber 1991). So many genes, and so few graduate students!

## CONCLUSIONS AND PERSPECTIVES

Experiments to identify the molecular structure and function of plasma membrane transporters in plants and fungi are now bearing fruit, although this field is in its infancy. Historically, this research has been much more refractory to a molecular analysis, especially in comparison to the studies of soluble, cytoplasmic enzymes involved in basic cellular metabolism. With the advent of sophisticated electrophysiological techniques, as well as facile DNA cloning and sequencing technology, ion transporters are no longer as difficult to characterize either functionally or structurally. With soluble, cytoplasmic enzymes that make and break covalent bonds, it is possible to follow the flow of carbon or nitrogen through anabolic and catabolic pathways. We suspect that as the molecular identity of more plant transporters becomes known, new inhibitors and other physiological reagents will become available to help trace the precise movement of ions into and out of each cell. Perhaps this analysis will reveal ubiquitous "pathways" or "cycles" of ions which will have to be memorized by graduate students preparing for preliminary exams in the 21st century.

Three of the most recent discoveries on the molecular structure of important signal-transducing proteins in higher plants indicate that yeast and other model systems may be inadequate for predicting the structure of important elements in the plant transduction pathway. Thus, the predominant calcium-activated protein kinase in green plants is a fusion protein containing a calmodulin-like domain, not previously seen in yeast or any other eukaryote (Harper et al. 1991). *COP1*, an *A. thaliana* mutant with abnormal light-regulated development, encodes a protein that represents a fusion between a $G_\beta$ protein and a transcription factor (Deng et al. 1992), a fused protein motif that was discovered first in *Arabidopsis*. Finally, there is an *A. thaliana* gene family encoding proteins closely resembling potassium channel genes seen in animals, but not at all in yeast (Sussman 1992b). Genetics is one of the few fields of biology that requires no prior understanding of the underlying mechanisms in order to correlate protein structure with biological function. With the advent of facile *A. thaliana* gene cloning, via insertional mutagenesis or chromosome walks, it is hoped and likely that more such "surprises" will be revealed.

## ACKNOWLEDGMENTS

The authors gratefully acknowledge a critical reading of the manuscript by R. Amasino. Financial support from the Department of Energy, the U.S. Department of Agriculture, and the National Science Foundation for research in the laboratory of the authors is gratefully acknowledged.

## REFERENCES

Albritton, N.L., T. Meyer, and L. Stryer. 1992. Range of messenger action of calcium ion and inositol 1,4,5-triphosphate. *Science* **258:** 1812–1815.

Anand, S. and R. Prasad. 1989. Rise in intracellular pH is concurrent with "start" progression of *Saccharomyces cerevisiae*. *J. Gen. Microbiol.* **135:** 2173–2179.

Andre, B. and G.F.E. Scherer. 1991. Stimulation by auxin of phospholipase A in membrane vesicles from an auxin-sensitive tissue is mediated by an auxin receptor. *Planta* **185:** 209–214.

Askerlund, P. and D.E. Evans. 1994. Reconstitution and characterization of a calmodulin-stimulated $Ca^{2+}$-pumping ATPase purified from *Brassica oleracea* L. *Plant Physiol.*

**100:** 1670–1681.

Assmann, S.M., L. Simoncini, and J.I. Schroeder. 1985. Blue light activates electrogenic ion pumping in guard cell protoplasts of *Vicia faba. Nature* **318:** 285–287.

Bazenet, C.E., A.R. Ruano, J.L. Brockman, and R.A. Anderson. 1990a. The human erythrocyte contains two forms of phosphatidylinositol-4-phosphate 5-kinase which are differentially active toward membranes. *J. Biol. Chem.* **265:** 18012–18022.

Bazenet, C.E., J.L. Brockman, D. Lewis, C. Chan, and R.A. Anderson. 1990b. Erythroid membrane-bound protein kinase binds to a membrane component and is regulated by phosphatidylinositol 4,5-bisphosphate. *J. Biol. Chem.* **265:** 7369–7376.

Binder, B.M., J.F. Harper, and M.R. Sussman. 1994. Characterization of an *Arabidopsis* calmodulin-like domain protein kinase purified from *Escherichia coli* using an affinity sandwich technique. *Biochemistry* **33:** 2033–2041.

Blatt, M.R., G. Thiel, and D.R. Trentham. 1990. Reversible inactivation of $K^+$ channels of *Vicia* stomatal guard cells following the photolysis of caged inositol-1,4,5-triphosphate. *Nature* **346:** 766–769.

Boitano, S., E.R. Dirksen, and M.J. Sanderson. 1992. Intercellular propagation of calcium wares mediated by inositol triphosphate. *Science* **258:** 292–295.

Bowler, C., G. Neuhaus, H. Yamagata, and N.H. Chua. 1994. Cyclic GMP and calcium mediate phytochrome transduction. *Cell* **77:** 73–81.

Braam, J. and R.W. Davis. 1990. Rain-, wind-, and touch-induced expression of calmodulin and calmodulin-related genes in *Arabidopsis. Cell* **60:** 357–364.

Briskin, D.P. 1990. The plasma membrane $H^+$-ATPase of higher plant cells: Biochemistry and transport function. *Biochim. Biophys. Acta* **1019:** 95–109.

Brockman, J.L. and R.A. Anderson. 1991. Casein kinase I is regulated by phosphatidylinositol 4,5-bisphosphate in native membranes. *J. Biol. Chem.* **266:** 2508–2512.

Brockman, J.L., S.D. Gross, M.R. Sussman, and R.A. Anderson. 1992. Cell cycle-dependent localization of casein kinase I to mitotic spindles. *Proc. Natl. Acad. Sci.* **89:** 9454–9458.

Bush, D.R. 1990. Electrogenicity, pH-dependence, and stoichiometry of the proton-sucrose symport. *Plant Physiol.* **93:** 1590–1596.

———. 1993. Proton-coupled sugar and amino acid transporters in plants. *Annu. Rev. Plant Physiol. Plant Mol. Biol.* **44:** 513–542.

Cao, Y., M. Anderova, N.M. Crawford, and J.I. Schroeder. 1992. Expression of an outward-rectifying potassium channel from maize mRNA. *Plant Cell* **4:** 961–969.

Chang, A. and C.W. Slayman. 1991. Maturation of the yeast plasma membrane [$H^+$]ATPase involves phosphorylation during intracellular transport. *J. Cell Biol.* **115:** 289–295.

Chenevert, J., K. Corrado, A. Bender, J. Pringle, and I. Herskowitz. 1992. A yeast gene (*BEM1*) necessary for cell polarization whose product contains two SH3 domains. *Nature* **356:** 77–79.

Cid, A., R. Perona, and R. Serrano. 1987. Replacement of the promoter of the yeast plasma membrane ATPase gene by a galactose-dependent promoter and its physiological consequences. *Curr. Genet.* **12:** 105–110.

Cosgrave, D.J. and R. Hedrich. 1991. Stretch-activated chloride, potassium, and calcium channels coexisting in plasma membranes of guard cells of *Vicia faba* L. *Planta* **186:** 143–153.

Deng, X.-W., M. Matsui, N. Wei, D. Wagner, A.M. Chu, K.A. Feldmann, and P. Quail. 1992. *COP1*, an *Arabidopsis* regulatory gene, encodes a protein with both a zinc-binding motif and a $G-_{\text{beta}}$ homologous domain. *Cell* **71:** 791–801.

DeMichelis, M.I., M.C. Pugliarello, and F. Rosi-Caldogno. 1989. Fusicoccin binding to its plasma membrane receptor and activation of the plasma membrane H⁺-ATPase. *Plant Physiol.* **90:** 133–139.

DeWitt, N.D., J.F. Harper, and M.R. Sussman. 1991. Evidence for a plasma membrane proton pump in phloem cells of higher plants. *Plant J.* **1:** 121–128.

Dudler, R. and C. Hertig. 1992. Structure of an *mdr*-like gene from *Arabidopsis thaliana*: Evolutionary implications. *J. Biol. Chem.* **267:** 5882–5888.

Ewing, N.N., L.E. Wimmers, D.J. Meyer, R.T. Chetelat, and A.B. Bennett. 1990. Molecular cloning of tomato plasma membrane H⁺-ATPase. *Plant Physiol.* **94:** 1874–1881.

Felle, H. 1982. Effects of fusicoccin upon membrane potential, resistance and current-voltage characteristics in root hairs of *Sinapis alba*. *Plant Sci. Lett.* **25:** 219–225.

Frommer, W.B., S. Hummel, and J.W. Riesmeier. 1993. Expression cloning in yeast of a cDNA encoding a broad specificity amino acid permease from *Arabidopsis thaliana*. *Proc. Natl. Acad. Sci.* **90:** 5944–5948.

Gaber, R. 1992. Molecular genetics of yeast ion transport. *Int. Rev. Cytol.* **137A:** 299–353.

Gaber, R.F., C.A. Styles, and G.R. Fink. 1988. *Trk1* encodes a plasma membrane protein required for high affinity potassium transport in *Saccharomyces cerevisiae*. *Mol. Cell. Biol.* **8:** 2848–2859.

Gehring, C.A., H.R. Irving, and R.W. Parish. 1990. Effects of auxin and abscisic acid on cytosolic calcium and pH in plant cells. *Proc. Natl. Acad. Sci.* **87:** 9645–9649.

Ghislain, M., A. Goffeau, D. Halachmi, and Y. Eilam. 1990. Calcium homeostasis and transport are affected by disruption of *cta 3*, a novel gene encoding Ca²⁺-ATPase in *Schizosaccharomyces pombe*. *J. Biol. Chem.* **265:** 18400–18407.

Gill, D.R., S.C. Hyde, C.F. Higgins, M.A. Valverde, G.M. Mintenig, and F.V. Sepulveda. 1992. Separation of drug transport and chloride channel functions of the human multidrug resistance P-glycoprotein. *Cell* **71:** 23–32.

Gilroy, S., N.D. Read, and A.J. Trewavas. 1990. Elevation of cytoplasmic calcium by caged calcium or caged inositol triphosphate initiates stomatal closure. *Nature* **346:** 769–771.

Gilroy, S., M.D. Fricker, N.D. Read, and A.J. Trewavas. 1991. Role of calcium in signal transduction of *Commelina* guard cells. *Plant Cell* **3:** 333–344.

Green, P.B. 1969. Cell morphogenesis. *Annu. Rev. Plant Physiol.* **20:** 365–394.

Gundersen, R.E. and D.L. Nelson. 1987. A novel Ca²⁺-dependent protein kinase from *Paramecium tetraurelia*. *J. Biol. Chem.* **262:** 4602–4609.

Gustin, M.C., X.-L. Zhou, B. Martinac, and C. Kung. 1988. A mechanosensitive ion channel in the yeast plasma membrane. *Science* **242:** 762–765.

Hario, R., B. Garciadeblas, and A. Rodriguez-Navarro. 1991. A novel P-type ATPase from yeast involved in sodium transport. *FEBS Lett.* **291:** 189–191.

Harootunian, A.T., J.P.Y. Kao, S. Paranjape, and R. Tsien. 1991. Generation of calcium oscillations in fibroblasts by positive feedback between calcium and IP₃. *Science* **251:** 75–78.

Harper, J.F., B.M. Binder, and M.R. Sussman. 1993. A calcium-dependent protein kinase from *Arabidopsis thaliana* expressed in *Escherichia coli*. *Biochemistry* **32:** 3282–3290.

Harper, J. F., L. Manney, and M.R. Sussman. 1994. Evidence for at least ten plasma membrane H⁺-ATPase genes in *Arabidopsis* and sequence of *AHA10* which is expressed primarily in developing seeds. *Mol. Gen. Genet.* (in press).

Harper, J.F., T.K. Surowy, and M.R. Sussman. 1989. Molecular cloning and sequence of a cDNA encoding the plasma membrane proton pump (H⁺-ATPase) of *Arabidopsis thaliana. Proc. Natl. Acad. Sci.* **86:** 1234–1238.

Harper, J.F., L. Manney, N.D. DeWitt, M.H. Yoo, and M.R. Sussman. 1990. The *Arabidopsis thaliana* plasma membrane $H^+$-ATPase multigene family. *J. Biol. Chem.* **265:** 13601–13608.

Harper, J.F., M.R. Sussman, G.E. Schaller, C. Putnam-Evans, H. Charbonneau, and A.C. Harmon. 1991. A calcium-dependent protein kinase with a regulatory domain similar to calmodulin. *Science* **252:** 951–954.

Hedrich, R., H. Busch, and K. Raschke. 1990. $Ca^{2+}$ and nucleotide dependent regulation of voltage dependent anion channels in the plasma membrane of guard cells. *The EMBO J.* **9:** 3889–3892.

Hsu, L. -C., T.-J. Chiou, L. Chen, and D.R. Bush. 1993. Cloning a plant amino acid transporter by functional complementation of a yeast amino acid transport mutant. *Proc. Natl. Acad. Sci.* **90:** 7441–7445.

Jan, L.Y. and Y.N. Jan. 1992. Tracing the roots of ion channels. *Cell* **69:** 715–718.

Jenkins, G.H., G. Subrahmanyam, and R.A. Anderson. 1991. Purification and reconstitution of phosphatidyl 4-kinase from human erythrocytes. *Biochim. Biophys. Acta* **1080:** 11–18.

Keller, B.V., R. Hedrich, and K. Raschke. 1989. Voltage dependent anion channels in the plasma membrane of guard cells. *Nature* **341:** 450–453.

Knight, M.R., S.M. Smith, and A.J. Trewavas. 1992. Wind-induced plant motion immediately increases cytosolic calcium. *Proc. Natl. Acad. Sci.* **89:** 4967–4971.

Knight, M.R., A.K. Campbell, S.M. Smith, and A.J. Trewavas. 1991. Transgenic plant aequorin reports the effects of touch and cold-shock and elicitors on cytoplasmic calcium. *Nature* **352:** 524–526.

Ko, C.H. and R.F. Gaber. 1991. *Trk1* and *Trk2* encode structurally related potassium ion transporters in *Saccharomyces cerevisiae. Mol. Cell Biol.* **11:** 4266–4273.

Kolarov, J., J. Kulpa, M. Baijot, and A. Goffeau. 1988. Characterization of a protein serine kinase from yeast plasma membrane. *J. Biol. Chem.* **263:** 10613–10619.

Larsen, P.M. and S. Wolniak. 1990. 1,2-dioctanoyl glycerol accelerates or retards mitotic progression in *Trodescantia* stamen hair cells as a function of the time of its addition. *Cell Motil. Cytoskeleton* **16:** 190–203.

Larsen, P.M., T.L. Chen, and S.M. Wolniak. 1989. Quin 2-induced metaphase arrest in stamen hair cells can be reversed by 1,2-dioctanoyl glycerol but not by 1,2-dioctanoyl glycerol. *Eur. J. Cell Biol.* **48:** 212–219.

Lee, Y. and S.M. Assmann. 1991. Diacylglycerols induce both ion pumping in patch-clamped guard-cell protoplasts and opening of intact stomata. *Proc. Natl. Acad. Sci.* **88:** 2127–2131.

Lew, R.R. 1989. Calcium activates an electrogenic proton pump in *Neurospora* plasma membrane. *Plant Physiol.* **91:** 213–216.

Li, Z. and D.R. Bush. 1991. pH-dependent amino acid transport into plasma membrane vesicles isolated from sugar beet (*Beta vulgaris* L.) leaves. *Plant Physiol.* **96:** 1338–1344.

Ling, V., I. Perera, and R.E. Zielinski. 1991. Primary structures of *Arabidopsis* calmodulin isoforms deduced from the sequence of cDNA clones. *Plant Physiol.* **96:** 1196–1202.

Lintilhac, P.M. and T.B. Vesecky. 1984. Stress-induced alignment of division plane in plant tissues grown *in vitro. Nature* **307:** 363–364.

Maathuis, F.J.M. and D. Sanders. 1992. Plant membrane transport. *Curr. Opin. Cell Biol.* **4:** 661–669.

Marten, I., G. Lohse, and R. Hedrich. 1991. Plant growth hormones control voltage-dependent activity of anion channels in plasma membrane of guard cells. *Nature* **353:** 758–762.

Marten, I., C. Zeilinger, C. Redhead, D.W. Landry, Q. Al-Awqati, and R. Hedrich. 1992. Identification and modulation of a voltage-dependent anion channel in the plasma membrane of guard cells by high-affinity ligands. *EMBO J.* **11:** 3569–3575.

Martinac, B., J. Adler, and C. Kung. 1990. Mechanosensitive ion channels of *E. coli* activated by amphipaths. *Nature* **348:** 261–263.

Maurel, C., J. Reizer, J.I. Schroeder, and M.J. Chrispeels.1993. The vacuolar membrane protein gamma-TIP creates water specific channels in *Xenopus* oocytes. *EMBO J.* **12:** 2241–2247.

McAinsh, M.R., C. Brownlee, and A.M. Hetherington. 1990. Abscisic acid-induced elevation of guard cell cytosolic $Ca^{2+}$ precedes stomatal closure. *Nature* **343:** 186–188.

McCurdy, D.W. and A.C. Harmon. 1992. Calcium-dependent protein kinase in the green algae, *Chara*. *Planta* **188:** 54–61.

McCusker, J.H., D.S. Perlin, and J.E. Haber. 1987. Pleiotropic plasma membrane ATPase mutations of *Saccharomyces cerevisiae*. *Mol. Cell. Biol.* **7:** 4082–4088.

Memon A. and W.F. Boss. 1990. Rapid light-induced changes in phosphoinositide kinases and ATPase activity in sunflower hypocotyls. *J. Biol. Chem.* **265:** 14817–14821.

Minorsky, P.V. and R.M. Spanswick. 1989. Electrophysiological evidence for a role of calcium in temperature sensing by roots of cucumber seedlings. *Plant Cell Environ.* **12:** 137–143.

Nakamoto, R.K., R. Rao, and C.W. Slayman. 1991. Expression of the yeast plasma membrane [$H^+$]ATPase in secretory vesicles. *J. Biol. Chem.* **266:** 7940–7949.

Nishizuka, Y. 1992. Intracellular signaling by hydrolysis of phospholipids and activation of protein kinase C. *Science* **258:** 607–614.

Palmgren, M.G. and G. Christensen. 1993. Complementation *in situ* of the yeast plasma membrane $H^+$-ATPase gene *pma1* by an $H^+$-ATPase gene from a heterologous species. *FEBS Lett.* **317:** 216–222.

———. 1994. Functional comparison between plant plasma membrane $H^+$-ATPase isoforms expressed in yeast. *J. Biol. Chem.* **269:** 3027–3033.

Palmgren, M.G., C. Larsson, and M. Sommarin. 1990. Proteolytic activation of the plant plasma membrane $H^+$-ATPase by removal of a terminal segment. *J. Biol. Chem.* **265:** 13423–13426.

Palmgren, M.G., M. Sommarin, R. Serrano, and C. Larsson. 1991. Identification of an autoinhibitory domain in the C-terminal region of the plant plasma membrane $H^+$-ATPase. *J. Biol. Chem.* **266:** 20470–20475.

Palmgren, M.G., M. Sommarin, P. Ulvskov, and P.L. Jorgensen. 1988. Modulation of plasma membrane $H^+$-ATPase from oat roots by lysophosphatidylcholine, free fatty acids and phospholipase Az. *Physiol. Plant.* **74:** 11–19.

Pardo, J.M. and R. Serrano. 1989. Structure of a plasma membrane $H^+$-ATPase gene from the plant *Arabidopsis thaliana*. *J. Biol. Chem.* **264:** 8557–8562.

Pedchenko, U.K., G.F. Nasirova, and T.A. Nalladina. 1990. Lysophosphatidylcholine specifically stimulates plasma membrane $H^+$-ATPase from corn roots. *FEBS Lett.* **275:** 205–208.

Perez, C., B. Michelet, V. Ferrant, P. Bogaerts, and M. Boutry. 1992. Differential expression within a three-gene subfamily encoding a plasma membrane proton ATPase in *Nicotiana plumbaginofolia*. *J. Biol. Chem.* **267:** 1204–1211.

Perez-Prat, E., M.L. Narasimhan, M.L. Binzel, M.A. Botella, Z. Chen, V. Valpuesta, R.A. Bressan, and P.M. Hasegawa. 1992. Induction of a putative $Ca^{2+}$-ATPase mRNA in NaCl-adapted cells. *Plant Physiol.* **100:** 1471–1478.

Perona, R. and R. Serrano. 1988. Increased pH and tumorigenicity of fibroblasts expressing a yeast proton pump. *Nature* **334:** 438–440.

Perona, R., F. Portillo, F. Giraldez, and R. Serrano. 1990. Transformation and pH homeostasis of fibroblasts expression yeast H$^+$-ATPase containing site-directed mutations. *Mol. Cell Biol.* **10:** 4110–4115.

Polarek, J.W., M.O. Walderhaug, and W. Epstein. 1988. Genetics of Kdp, the K$^+$-transport ATPase of *Escherichia coli. Methods Enzymol.* **157:** 655–667.

Portillo, F., P. Eraso, and R. Serrano. 1991. Analysis of the regulatory domain of yeast plasma membrane H$^+$-ATPase by directed mutagenesis and intragenic suppression. *FEBS Lett.* **287:** 71–74.

Putnam-Evans, C., A.C. Harmon, B.A. Palevitz, M. Fechheimer, and M. Cormier. 1989. Calcium-dependent protein kinase is localized with F-actin in plant cells. *Cell Motil. Cytoskeleton* **12:** 12–22.

Rayle, D.L. and R.E. Cleland. 1992. The acid growth theory of auxin-induced cell elongation is alive and well. *Plant Physiol.* **99:** 1271–1274.

Riesmeier, J.W., L. Willmitzer, and W.B. Frommer. 1993. Isolation and characterization of a sucrose carrier cDNA from spinach by functional expression in yeast. *EMBO J.* **11:** 4705–4713.

Roberts, D.M. and A.C. Harmon. 1992. Calcium-modulated proteins: Targets of intracellular calcium signals in higher plants. *Annu. Rev. Plant. Physiol. Plant Mol. Biol.* **43:** 375–414.

Rudolph, H.K., A. Antebi, G.A. Fink, C.M. Buckley, T.E. Dorman, J. LeVitre, L.S. Davidow, J. Mao, and D.T. Moir. 1989. The yeast secretory pathway is perturbed by mutations in PMR1, a member of a Ca$^{2+}$-ATPase family. *Cell* **58:** 133–145.

Sanders, D. 1990. Kinetic modeling of plant and fungal membrane transport systems. *Annu. Rev. Plant Physiol. Plant Mol. Biol.* **41:** 77–107.

Sauer, N., K. Friedlander, and U. Graml-Wicke. 1990. Primary structure, genomic organization and heterologous expression of a glucose transporter from *Arabidopsis thaliana. EMBO J.* **9:** 3045–3050.

Schachtman, D.P., J.I. Schroeder, W.J. Lucas, J.A. Anderson, and R.F. Gaber. 1992. Expression of an inward-rectifying potassium channel. *Science* **258:** 1654–1658.

Schaller, G.E. and M.R. Sussman. 1987. Kinase-mediated phosphorylation of the oat plasma membrane H$^+$-ATPase. *UCLA Symp. Mol. Cell. Biol. New Ser.* **63:** 419–429.

———. 1988. Isolation and sequence of tryptic peptides from the oat plasma membrane H$^+$-ATPase. *Plant Physiol.* **86:** 512–516.

Schaller, G.E., A.C. Harmon, and M.R. Sussman. 1992. Characterization of a calcium- and lipid-dependent protein kinase associated with the plasma membrane of oat. *Biochemistry* **31:** 1721–1727.

Schroeder, J.I. and H.H. Fang. 1991. Inward-rectifying potassium channels in guard cells provide a mechanism for low affinity potassium uptake. *Proc. Natl. Acad. Sci.* **88:** 11583–11587.

Schroeder, J.I. and S. Hagiwara. 1989. Cytosolic calcium regulates ion channels in the plasma membrane of *Vicia faba* guard cells. *Nature* **338:** 427–430.

———. 1990. Repetitive increases in cytosolic Ca$^{2+}$ of guard cells by abscisic acid activation of nonselective Ca$^{2+}$ permeable channels. *Proc. Natl. Acad. Sci.* **87:** 9305–9309.

Schroeder, J.I. and B.U. Keller. 1992. Two types of anion channel currents in guard cells with distinct voltage regulation. *Proc. Natl. Acad. Sci.* **89:** 5025–5029.

Schroeder, J.I. and P. Thuleau. 1991. Ca$^{2+}$ channels in higher plant cells. *Plant Cell* **3:**

555–559.

Schulz, S., E. Oelgemoeller, and E.W. Weiler. 1990. Fusicoccin action in cell-suspension cultures of *Corydalis sempervirens. Planta* **183:** 83–91.

Sentenac, H., N. Bonneaud, M. Minet, F. Lacroute, J.-M. Salmon, F. Gaymard, and C. Grignon. 1992. Cloning and expression in yeast of a plant potassium ion transport system. *Science* **256:** 663–666.

Serrano, E.E. and E. Zeiger. 1989. Sensory transduction and electrical signaling in guard cells. *Plant Physiol.* **91:** 795–799.

Serrano, R. 1989. Structure and function of plasma membrane ATPase. *Annu. Rev. Plant Physiol.* **40:** 61–94.

Serrano, R. and F. Portillo. 1990. Catalytic and regulatory sites of yeast plasma membrane H$^+$-ATPase studied by directed mutagenesis. *Biochim. Biophys. Acta* **1018:** 195–199.

Serrano, R., F. Portillo, B.C. Monk, and M.G. Palmgren. 1992. The regulatory domain of fungal and plant plasma membrane H$^+$-ATPase. *Acta Physiol. Scand.* **146:** 131–136.

Shacklock, P.S., N.D. Read, and A.J. Trewavas. 1992. Cytosolic free calcium mediates red light-induced photomorphogenesis. *Nature* **358:** 753–755.

Simons, P. 1992. *The action plant.* Blackwell, Oxford, United Kingdom.

Speksnijder, J.E., A.L. Miller, M.H. Weisenseel, T.-H. Chen, and L.F. Jaffee. 1989. Calcium buffer injections block fucoid egg development by facilitating calcium diffusion. *Proc. Natl. Acad. Sci.* **86:** 6607–6611.

Sukharev, S.I., P. Blount, B. Martinac, F.R. Blattner, and C. Kung. 1994. A large-conductance mechanosensitive channel in *E. coli* encoded by *mscL* alone. *Nature* **368:** 265–268.

Surowy, T.K. and J.S. Boyer. 1991. Low water potentials affect expression of genes encoding vegetative storage proteins and plasma membrane proton ATPase in soybean. *Plant Mol. Biol.* **16:** 251–262.

Sussman, M.R. 1992a. A plethora of plant plasmalemma proton pumps. In *Transport and receptor proteins of plant membranes* (ed. D.T. Cooke and D.T. Clarkson), pp. 5–11. Plenum Press, New York.

———. 1992b. Shaking *Arabidopsis thaliana. Science* **256:** 619.

———. 1994. Molecular analysis of proteins in the plant plasma membrane. *Annu. Rev. Plant Physiol. Plant Mol. Biol.* **45:** 211–234.

Sussman, M.R. and J.F. Harper. 1989. Molecular biology of the plasma membrane of higher plants. *Plant Cell* **1:** 953–960.

Tan, Z. and W.F. Boss. 1992. Association of PI kinase, P1P kinase and diacylglycerol kinase with the cytoskeleton and F-actin fractions of carrot cells grown in suspension cultures. *Plant Physiol.* **100:** 2116–2120..

Tate, B.F., G.E. Schaller, M.R. Sussman, and R.C. Crain. 1989. Characterization of a polyphosphoinositide phospholipase C from the plasma membrane of *Avena sativa. Plant Physiol.* **91:** 1275–1279.

Tsay, Y. F., J.I. Schroeder, K.A. Feldmann, and N.M. Crawford. 1993. The herbicide sensitivity gene *CHL1* of *Arabidopsis* encodes a nitrate-inducible nitrate transporter. *Cell* **72:** 705–713.

Villalba, J.M., M. Lutzelschwab, and R. Serrano. 1991. Immunocytolocalization of plasma membrane H$^+$-ATPase in maize coleoptiles and enclosed leaves. *Planta* **185:** 458–461.

Villalba, J.M., M.G. Palmgren, G.E. Berberian, C. Ferguson, and R. Serrano. 1992. Functional expression of plant plasma membrane H$^+$-ATPase in yeast endoplasmic reticulum. *J. Biol. Chem.* **267:** 12341–12349.

Wach, A., A. Schlesser, and A. Goffeau. 1992. An alignment of 17 deduced protein sequences from plant, fungi, and ciliate H⁺-ATPase genes. *J. Bioenerg. Biomembr.* **24:** 309–317.

Walderhaug, M.O., J.W. Polarek, P. Voelkner, J.M. Daniel, J.E. Hessell, K. Altendorf, and E. Epstein. 1992. KdpD and kdpE, proteins that control expression of the kdpABC operon, are members of the two-component sensor-effector class of regulators. *J. Bacteriol.* **174:** 2152–2159.

Wildon, D.C., J.F. Thain, P.E.H. Minchin, I.R. Gubb, A.J. Reilly, Y.D. Skipper, H.M. Doherty, P.J. O'Donnell, and D.J. Bowles. 1992. Electrical signalling and systemic proteinase inhibitor induction in the wounded plant. *Nature* **360:** 62–65.

Williamson, R.E. and C.C. Ashley. 1982. Free $Ca^{2+}$ and cytoplasmic streaming in the algae, *Cara*. *Nature* **296:** 647–651.

Wimmers, L.E., N.N. Ewing, and A.B. Bennett. 1992. Higher plant $Ca^{2+}$-ATPase: Primary structure and regulation of mRNA abundance by salt. *Proc. Natl. Acad. Sci.* **89:** 9205–9209.

Wine, J.J. and S.C. Silverstein. 1992. ATP and chloride conductance. *Nature* **360:** 18.

Zhao, Y., B. Kappes, and R.M. Franklin. 1993. Gene structure and expression of an unusual protein kinase from *Plasmodium falciparum* homologous at its carboxyl terminus with the EF hand calcium-binding proteins. *J. Biol. Chem.* **268:** 4347–4354.

# 40

# Metabolic and Genetic Control of Nitrate, Phosphate, and Iron Assimilation in Plants

**Nigel M. Crawford**

Department of Biology and Center for Molecular Genetics
University of California at San Diego
La Jolla, California 92093-0116

Seventeen elements are known to be essential for plant growth and development. These elements are C, H, O, N, K, P, Mg, Ca, S, Fe, B, Mn, Cu, Zn, Mo, Cl, and Ni (Salisbury and Ross 1992). Some of these elements (e.g., Mo) are required at concentrations less than one part per million, but if they are absent, the plant cannot complete its life cycle. Other nutrients (the macronutrients) are needed at high levels and reach concentrations of 0.1–3% of dry weight (~3 mM–100 mM) in plant tissues (for examples, see Tables 1 and 2). These nutrients are applied as fertilizers in ever-increasing quantities as demand for more food has accelerated during this century. In the United States, \$10 billion is now spent to produce 20 million tons of fertilizer every year (Glass 1989). Worldwide, more than 140 million tons of nitrogen, phosphorus, and potassium fertilizer are consumed annually (Conway and Pretty 1988). Unfortunately, much of this fertilizer leaches into ground and surface water, causing serious contamination (Conway and Pretty 1988). The study of plant mineral nutrition has become a vital component of efforts to increase the world's food production as well as to decrease the environmental costs of our conventional agricultural practices.

Of the nutrients that plants must obtain from the soil, N, P, and Fe are often the most limited. Nitrogen is available primarily as nitrate or ammonium ions in the soil solution, although organic forms as well as nitrogen from the atmosphere can also be important sources. Phosphorus is available as phosphate, and iron is present as ferrous or ferric ions. Assimilation of these nutrients requires ion transport across membranes, oxidation/reduction reactions, group transfer reactions, and an intricate and complex regulatory system that integrates these processes into the overall physiology of the plant. In this chapter, I discuss the assimilation of nitrate, phosphate, and iron. For the assimilation of other ions, one can

*Arabidopsis*
© 1994 Cold Spring Harbor Laboratory Press 0-87969-428-9/94 \$5 + .00

*Table 1*  Elemental analysis of maize and *Arabidopsis*

| Element | Maize leaf (% of dry weight) | *Arabidopsis* (% of dry weight) |
|---|---|---|
| Nitrogen | 3.2 | 2.8 |
| Potassium | 2.1 | 4.7 |
| Calcium | 0.52 | 1.6 |
| Phosphorus | 0.20 | 0.9 |
| Magnesium | 0.32 | 0.8 |
| Sulfur | 0.17 | n.d. |
| Iron | 0.012 | 0.0093 |

Data for maize leaf from 1982 data of P. Soltanpour et al., Colorado State University Soil Testing Laboratory (unpubl. [cited in Salisbury and Ross 1992]). Data for *Arabidopsis* plants grown on fertilized soil from Y. Poirier and C. Somerville (unpubl.).

refer to M. Sussman (this volume) and other excellent reviews and text books (Marschner 1986; Glass 1989; Taiz and Zeiger 1991; Salisbury and Ross 1992).

Besides serving as essential nutrients, nitrate, phosphate, and iron act as environmental signals that can dramatically alter the physiology and development of a plant (for review, see Marschner 1986; Glass 1989). For example, plants respond to changes in the extracellular concentration of these ions by altering the growth rate and branching of roots (Drew 1975) and by activating the expression of specific genes (e.g., transporters and reductases) (Bienfait 1988b; Goldstein et al. 1989; Crawford and Campbell 1990; Duff et al. 1991; Redinbaugh and Campbell 1991). Some plants initiate symbiotic relationships with soil fungi and bacteria to enhance nutrient acquisition (primarily nitrogen and phosphorus) if the nutrient concentration becomes too low. Plants must be able to respond to nutrient deficiencies by accelerating the mobilization or acquisition of

*Table 2*  Ion concentration in pea roots

| Ion | External concentration (mM) | Internal concentration (mM) |
|---|---|---|
| $K^+$ | 1.0 | 75 |
| $NO_3^-$ | 2.0 | 28 |
| $H_2PO_4^-$ | 1.0 | 21 |
| $SO_4^-$ | 0.25 | 19 |
| $Na^+$ | 1.0 | 8 |
| $Cl^-$ | 1.0 | 7 |
| $Mg^{++}$ | 0.25 | 3 |
| $Ca^{++}$ | 1.0 | 2 |

External concentrations refer to the ionic concentrations in the nutrient media provided to the plant roots. (Adapted from Higinbotham et al. 1967.)

the limited ion from the environment. The response is specific to the limited nutrient, as each ion presents very different challenges. Nitrate diffuses freely in the soil solution (it is soluble and not retained by the negatively charged soil particles) and can be easily lost by leaching or by bacterial denitrification. Phosphate and iron, on the other hand, can be present in the soil at adequate concentrations but not be soluble. Iron precipitates out of the soil solution with hydroxyl ions, especially in alkaline soils, and phosphate precipitates as Mn, Fe, or Al salts in acidic soils or as Ca and Mg salts in alkaline soils. Plants employ a diversity of strategies to cope with these mineral deficiencies.

Genetic and physiological studies have provided much of what we know about nutrient assimilation in plants and have elucidated the biochemical pathways that are involved. Cellular and molecular approaches are now being employed to characterize structural and regulatory genes, signal transduction pathways, and cellular responses necessary for ion mobilization, transport, and metabolism. *Arabidopsis* and several other key plants, including barley, maize, tobacco, and tomato, have been invaluable for these studies, which are described below.

**NITRATE**

**Overview**

Under certain environmental conditions (e.g., well-aerated soils), nitrate can be the predominant or even sole source of nitrogen. Plants have evolved mechanisms to utilize this nitrate to satisfy their demand for nitrogen. The assimilation of nitrate, however, presents several difficulties for plants. The concentration of nitrate in the soil solution can vary from 50 $\mu$M to 100 mM, and plants must accommodate these variable and often fluctuating conditions. The uptake of nitrate is energetically unfavorable and often requires transport up very steep electrochemical gradients. Significant quantities of reduced carbon must be expended for nitrate assimilation to provide carbon skeletons, energy, and protons, the latter needed to neutralize the alkaline products produced during nitrate reduction. Finally, two intermediates in the nitrate assimilation pathway (nitrite and ammonium ions) cannot be allowed to accumulate because of their toxic effects. To overcome these obstacles, plants employ a simple pathway that is controlled by an intricate and complicated regulatory network which is sensitive to several key environmental and internal signals.

The nitrate assimilation pathway begins with the uptake of nitrate from the soil solution by the root. Root epidermal and cortical cells are the first to actively transport nitrate into their cytosols. Nitrate then crosses the endodermis of the root and is released into the xylem. After long-distance transport up the xylem, nitrate is again actively transported into the cells of the leaf. Once in the cell, nitrate can be stored in the vacuole or reduced to nitrite by the enzyme nitrate reductase (NR). Nitrite is transported into the chloroplast, where it is reduced to ammonia by nitrite reductase (NiR). Ammonia is then incorporated into glutamine primarily by the action of glutamine synthetase. The overall pathway can be summarized as:

$$NO_3^- \xrightarrow{\text{Up}} NO_3^- \xrightarrow{\text{NR}} NO_2^- \xrightarrow{\text{NiR}} NH_4^+ \xrightarrow[\text{Glu}]{\text{GS}} Gln$$

A cell's eye view of the pathway is shown in Figure 1 for a leaf cell. For many plants, most of the nitrate (80–90%) taken up by the roots is transported to the shoot, where it is reduced. However, nitrate is also reduced in the root, producing amino acids (primarily glutamine and asparagine) that are loaded into the xylem for transport to the shoot.

All the redox enzymes and their structural genes in this pathway have been isolated and characterized (Caboche and Rouze 1990; Campbell and Kinghorn 1990; Kleinhofs and Warner 1990; Solomonson and Barber 1990; Crawford and Arst 1993). Many mutations have been identified in the NR gene. Many of these mutants were isolated using chlorate, the chlorine analog of nitrate. Chlorate is taken up into the cell and then reduced by NR to toxic chlorite (Åberg 1947). Mutants impaired in chlorate (nitrate) uptake or reduction do not produce chlorite and are thus resistant to chlorate treatment. In practice, almost all chlorate-resistant mutants are defective in nitrate reduction, due to mutations in NR genes or in genes required for the synthesis of MoCo, an NR cofactor (Caboche and Rouze 1990; Kleinhofs and Warner 1990; Crawford 1992; Crawford and Arst 1993). As yet, no regulatory mutants in the nitrate assimilation pathway of higher plants have been found (one has been described for *Chlamydomonas*; Fernandez and Cardenas 1989; Fernandez et al. 1989); however, both transcriptional and posttranscriptional regulation of the nitrate reductase gene in response to nitrate, light, sugars, $CO_2$, nitrogen metabolites, plastidic signals, phytohormones, and circadian rhythms have been documented (Caboche and Rouze 1990; Crawford and Campbell 1990; Kleinhofs and Warner 1990; Cheng et al. 1991, 1992; Crawford et al. 1992; Vincentz et al. 1993). Only one nitrate uptake mutant has been described in higher plants, and it is the *chl1* mutant of *Arabidopsis* (Doddema et al. 1978; Scholten and Feenstra 1986; Larsson and Ingemarsson 1989).

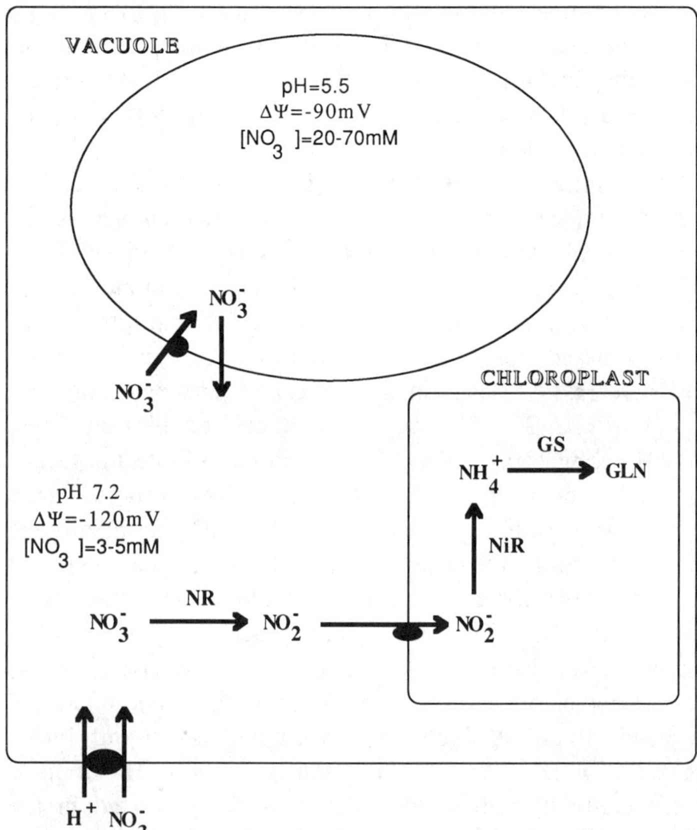


*Figure 1* Schematic diagram of a leaf cell showing steps of the nitrate assimilation pathway. Outer line represents plasma membrane. Filled circles represent membrane transport proteins. (NR) Nitrate reductase; (NiR) nitrite reductase; (GS) glutamine synthetase; (GLN) glutamine. This figure was adapted from several sources (Taiz and Zeiger 1991; Zhen et al. 1991).

## Nitrate Uptake

The first step in the assimilation of nitrate is the transport of nitrate into the cell. Much is known about nitrate uptake in plants at the physiological and cellular level, and very recently two nitrate transporter genes were described (Tsay et al. 1993; Quesada et al. 1994). Nitrate uptake is an active, multiphasic process that is regulated in response to changes in the nitrate concentration of the external medium. The initial uptake of nitrate into plants occurs across the plasma membrane of the epidermal and cortical cells of the root where nitrate enters the symplastic compartment (Larsson and Ingemarsson 1989). However, nitrate transport also occurs across the tonoplast membrane (Granstedt and Huffaker 1982; Blumwald and Poole 1985; Miller and Smith 1992) and the plasma membrane of other cells, including those of the vascular system and leaf (Jackson et al. 1986). This discussion is limited to transport of nitrate across the plasma membrane of root cells.

Nitrate uptake requires that nitrate be transported against a large electrochemical gradient (see Fig. 1). The electrical potential across the plasma membrane of plants is usually between −100 mV and −200 mV (more negative inside). For a typical value of −110 mV and an external nitrate concentration of 2 mM, one would expect an internal nitrate concentration of 28 μM based on the Nernst equation. However, using microelectrodes and other assays, one finds cytoplasmic concentrations between 1 mM and 10 mM, 35–350 times the predicted equilibrium value (Zhen et al. 1991; King et al. 1992). To concentrate nitrate up such a steep gradient, it has been proposed that nitrate is cotransported with protons into the cell. A proton gradient is created by plasma membrane $H^+$-ATPases, which acidify the external media to pH between 5.0 and 6.0. The energy stored in the proton gradient could be used directly to take up nitrate via a proton/nitrate cotransport process as described below. Proton symport has also been reported recently for sulfate uptake in *Brassica napus* (Hawkesford et al. 1993). Other mechanisms for nitrate uptake have been proposed, including hydroxide/nitrate and bicarbonate/nitrate exchange (Hodges 1972; Jackson et al. 1986). Bicarbonate exchange is especially intriguing. Organic acids are produced in the shoot to neutralize the hydroxide ions released by nitrate reduction. These organic acids (especially malate) are transported down the phloem to the root. It has been proposed that these organic acids are decarboxylated, producing bicarbonate, which can exchange for nitrate outside the cell (Touraine et al. 1994).

Current evidence strongly supports a proton/nitrate symport as the mechanism for nitrate uptake. Chlorate is often used as a nitrate analog to measure uptake, as radioactive chlorate can be readily synthesized from $^{36}Cl^-$, which is stable. Radioactive nitrogen is difficult to obtain and is very unstable ($t_{1/2}$ = 10 min). It has been shown that the transport of chlorate can be driven into plasma membrane vesicles by acidifying the external solution even if the vesicles are preloaded with chlorate (Ruiz-Cristin and Briskin 1991). Thus, external acidification leads to active transport of chlorate into the vesicles. Further evidence has come from monitoring the electrical potential across the plasma membrane of root cells with a microelectrode. When roots are exposed to nitrate, a rapid and transient depolarization of the membrane potential occurs (Ullrich and Novacky 1981; McClure et al. 1990a,b; Glass et al. 1992).

Depolarization indicates a positive current into the cell. The nitrate-induced depolarization is dependent on a pH gradient across the membrane (more acidic outside) (Ullrich and Novacky 1981; McClure et al. 1990a,b; Ruiz-Cristin and Briskin 1991). To explain these results, it has been proposed that nitrate is cotransported with two protons for every one nitrate molecule (McClure et al. 1990a,b; Glass et al. 1992), so that when nitrate is taken up by the cell, one detects a net positive current into the cell and a depolarization of the membrane potential. Thus, nitrate uptake is both active and electrogenic.

Analysis of nitrate uptake kinetics reveals a multiphasic process, which is typical of nutrient uptake systems in plants and other organisms (Doddema and Telkamp 1979; Goyal and Huffaker 1986; MacKown and McClure 1988; Siddiqi et al. 1989; Hole et al. 1990; Siddiqi et al. 1990; Glass et al. 1992). For barley, wheat, maize, and *Arabidopsis*, biphasic kinetics have been observed (Neyra and Hageman 1975; Doddema and Telkamp 1979; Goyal and Huffaker 1986; Siddiqi et al. 1990). At low concentrations of nitrate, the rate of nitrate uptake displays Michaelis-Menten kinetics, reaching a plateau (saturating) at 0.1 mM–0.5 mM. As the concentration of nitrate increases, the uptake rate again increases, following either a linear progression (Siddiqi et al. 1990; Aslam et al. 1992) or a hyperbolic curve to saturation (Neyra and Hageman 1975; Doddema and Telkamp 1979). These results suggest that nitrate uptake involves (1) a high-affinity, saturable system ($K_m$ between 10 μM and 300 μM) and (2) a low-affinity system with linear or saturable kinetics above 0.5 mM for these plants. Figure 2 shows nitrate uptake kinetics measured for *Arabidopsis* wild-type and *chl1* mutant plants. Both high- and low-affinity systems show rapid depolarization of the membrane in barley roots in response to nitrate, suggesting that both utilize proton/nitrate symport (Glass et al. 1992). Figure 3 shows an example of depolarization across the plasma membrane of a barley root cell. Recently, another high-affinity system was described in nitrate-induced barley roots with a $K_m$ of 7 μM (Aslam et al. 1992); this result suggests that there may be three transport systems. How these transport systems relate to individual transport proteins is as yet unknown. Each transport system may correspond to a distinct transporter or may arise from a single transporter with changing kinetic properties. Characterization of individual transporter proteins and genes will be needed to formulate an accurate molecular model for nitrate uptake.

Another interesting feature of nitrate uptake is that it is subject to both positive and negative feedback inhibition (Larsson and Ingemarsson 1989; Redinbaugh and Campbell 1991; Touraine et al. 1994). Plants can take up nitrate even if they have never been exposed to nitrate. If a plant

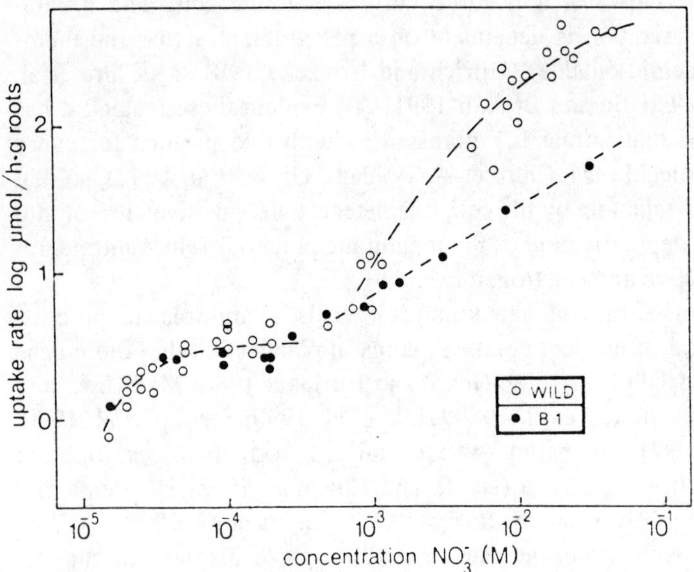



*Figure 2* Nitrate uptake rate as a function of nitrate concentration for wild type (open circles) and *chl1* mutant (B1, closed circles) *Arabidopsis* plants. The part of the curves below 1 mM represents the high-affinity system and above 1 mM the low-affinity system. Graph was copied from Doddema and Telkamp (1979).

is exposed to nitrate, then nitrate uptake capacity is increased after a lag of several hours (Heimer and Filner 1971; Jackson et al. 1973; Doddema and Otten 1979; MacKown and McClure 1988; Siddiqi et al. 1989; Hole et al. 1990). This response is specific; ammonia, an alternative nitrogen source, does not enhance the nitrate uptake system. The nitrate-induced enhancement of nitrate uptake involves synthesis of new proteins, as RNA and protein synthesis inhibitors block the enhancement (Jackson et al. 1973, 1986; Hole et al. 1990), and several new plasma membrane proteins are synthesized within several hours after nitrate treatment of maize roots (McClure et al. 1987; Dhugga et al. 1988). In addition, new high-affinity nitrate transport systems distinguishable by their kinetic properties have been detected in nitrate-induced barley and sugar beet plants (Mack and Tischner 1990; Aslam et al. 1992; Glass et al. 1992). These responses are indicative of positive regulation and resemble classic substrate-induced transport systems.

The regulation of nitrate uptake is not restricted to simple induction by nitrate leading to enhanced influx. Plants also increase the rate of nitrate influx in response to a decrease in external nitrate concentration. When plants are starved for nitrogen after they have been growing on

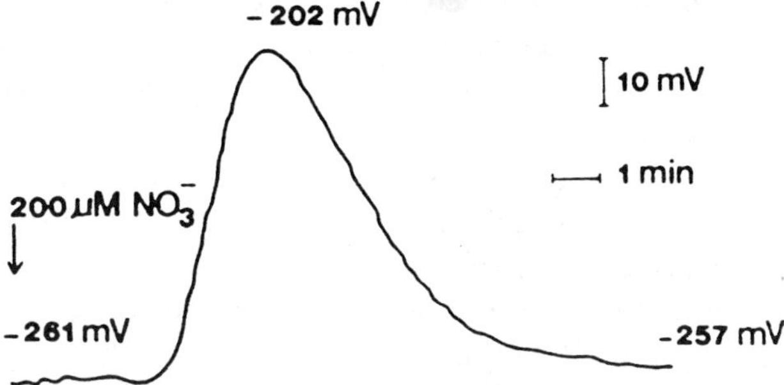


*Figure 3* Measurement of electrical potential across the plasma membrane of a barley epidermal or cortical cell in response to exposure to nitrate. Plants were pretreated with 100 μM nitrate for 18 hr before cell was impaled with electrode. Upward deflection represents depolarization (more positive inside the cell). After the initial depolarization, repolarization occurs, presumably due to the stimulation of the plasma membrane proton pump. Graph was copied from Siddiqi et al. (1990).

nitrate, the rate of nitrate uptake increases severalfold within 24 hours (Jackson et al. 1986; Teyker et al. 1988; Larsson and Ingemarsson 1989). The converse is also true: If plants are exposed to nitrate so that the level of nitrate increases inside the cell (after 6–15 hr), the rate of uptake decreases (Larsson and Ingemarsson 1989; Siddiqi et al. 1989). Nitrate uptake also decreases when the demand for nitrogen decreases in plants (e.g., when their growth is slowed). It has been proposed that transport of amino acids from the shoot to the root increases with decreasing demand for nitrogen and that this leads to a repression of nitrate uptake (Touraine et al. 1994). Thus, plants adapt to the environmental and internal conditions by regulating their uptake systems to compensate for changes in external nitrate concentrations and in nitrogen demand. What is especially interesting about this system is how versatile it can be. For example, optimal growth of perennial ryegrass occurs at 1.4 mM nitrate, but if the nitrate concentration is lowered 100-fold, the dry weight of the ryegrass decreases only 10% (Clement et al. 1978). Nitrate concentrations as low as 5 μM can sustain growth without deficiency symptoms in plants such as wheat seedlings, as long as the nitrate concentration is maintained (Edwards and Asher 1974). These results suggest that nitrate uptake can be very efficient in scavenging nitrate and can maintain some homeostasis.

To understand how nitrate uptake is regulated at the molecular level and what constitutes a nitrate uptake system, it is essential to identify nitrate transporter proteins and genes. In fungi (*Aspergillus*), a gene (*crnA*) has been identified that is involved in nitrate uptake (Brownlee and Arst 1983). *crnA* mutants are resistant to chlorate and have a 4-fold reduction in nitrate uptake. The *CRNA* gene encodes a 52-kD membrane protein with 10 putative membrane-spanning segments (Unkles et al. 1991). This gene is thought to encode a nitrate transporter, although no direct evidence on the activity of the protein has been published. In higher plants, uptake-deficient lines of barley (Wallsgrove et al. 1989) and a chlorate-resistant mutant of *Arabidopsis* (*chl1*) defective in chlorate and nitrate uptake (Doddema et al. 1978; Doddema and Otten 1979; Doddema and Telkamp 1979; Scholten and Feenstra 1986) have been described. The *Arabidopsis* mutant *chl1* appears to be defective in the low-affinity nitrate uptake system (see Fig. 2) (Doddema and Telkamp 1979). Surprisingly, the *chl1* mutant presently provides the only genetic handle with which to isolate a nitrate transporter gene in plants. The *CHL1* gene was isolated recently from a T-DNA-tagged *chl1* mutant. The gene encodes a 65-kD membrane protein that has 12 putative membrane-spanning segments (Tsay et al. 1993). The *CHL1* protein when expressed in *Xenopus* oocytes behaves as an electrogenic nitrate transporter. Nitrate treatment of *CHL1*-expressing oocytes leads to a depolarization of the membrane potential and an inward current that depend on acidification of the external bath solution. The *CHL1* gene is predominantly expressed in roots of *Arabidopsis* and is induced within minutes by nitrate, with *CHL1* mRNA levels reaching a maximum at 2 hours, then declining thereafter. Thus, the *CHL1* gene has all the hallmarks of a plant nitrate transporter. This protein must also transport chlorate, as *chl1* mutants are very resistant to chlorate. Uptake of other ions (chloride and potassium but not sulfate) is also impaired in *chl1* mutants (Scholten and Feenstra 1986). As yet, it is not clear whether the *CHL1* protein transports other ions as well as nitrate, or whether reduced chloride and potassium uptake in the *chl1* mutant is due indirectly to the *chl1* mutation. It is hoped that other nitrate transporter genes will now be identified so that a complete molecular characterization of nitrate transport in plants can be obtained.

**Nitrate Reduction**

Once inside the cell, nitrate is reduced to nitrite by NR. This reaction is thought to be the first committed and rate-limiting step in the nitrate assimilation pathway within the cell (Layzell 1990). Many studies have focused on elucidating the structure, function, and regulation of NR and have led to some important insights. NR is located in the cytosol in both roots and shoots. Several different forms of the enzyme exist that are distinguished by their kinetic properties, and many plants have more than one NR structural gene. The enzyme is tightly regulated at the transcriptional, posttranscriptional, and enzyme levels. Given these findings, it is surprising that the level of NR is in such excess of what is needed for normal vegetative growth of plants.

The overall outlines of the structure of NR are now known. Based on sequences from at least 11 species of plants, 2 species of algae, and 2 species of fungi, and on extensive biochemical studies, it has been shown that nitrate reductase is a flavometallo-oxidoreductase with two or four identical subunits, each composed of three functional domains (Crawford et al. 1988; Campbell and Kinghorn 1990; Crawford and Campbell 1990; Solomonson and Barber 1990; Hoff et al. 1992). Each domain corresponds to a redox center that binds one of two prosthetic groups (FAD and Heme) or a molybdenum cofactor (MoCo) composed of molybdopterin and molybdate. Electrons are transferred from the reductant (NADH or NADPH) to the flavin domain, then shuttled via the heme to the MoCo domain where the nitrate is reduced (Campbell and Smarelli 1986; Campbell and Kinghorn 1990). It is interesting that homologous domains have been commandeered by other enzymes, and the entire group comprises the cytochrome $b_5$ superfamily of proteins (Guiard and Lederer 1979; Le and Lederer 1983; Calza et al. 1987; Crawford and Davis 1988).

The number of genes that encode NR is low but varies from plant to plant (Caboche and Rouze 1990). *Nicotiana plumbaginifolia* has only one gene. Barley, *Nicotiana tabacum*, and *Arabidopsis* have two. Other plants, such as soybean, are likely to have three or more. It is unclear why plants have more than one gene. In barley, the *nar1* gene encodes a NADH-specific NR that is expressed primarily in shoots (Warner and Kleinhofs 1981). This gene encodes most of the NR activity (90%) in barley. Inactivation of the *nar1* gene has little effect on the growth of barley (Warner et al. 1981). Search for a second NR gene in barley led to the discovery of *nar7*, which encodes a NADH/NADPH bispecific enzyme that is expressed only in roots of wild-type plants (Kleinhofs and Warner 1990). In *nar1* mutants, the *nar7* gene appears to be activated in the shoots, as the bispecific enzyme can be detected in leaves of *nar1* mutants.

*Arabidopsis* also has two NR genes, *NIA1* and *NIA2* (Cheng et al. 1988; Crawford et al. 1988; Wilkinson and Crawford 1991, 1993). Genomic and cDNA clones for these genes were originally obtained using anti-NR antibodies from barley (Cheng et al. 1986) and NR cDNA clones from squash (Crawford et al. 1986). The original genetic characterization of chlorate-resistant mutants of *Arabidopsis* had also indicated that there were two NR genes corresponding to the *CHL2* and *CHL3* genes (Braaksma and Feenstra 1982). It was subsequently shown that *CHL3* is the *NIA2* NR structural gene (Chang et al. 1988; Cheng et al. 1988; Wilkinson and Crawford 1991). The *CHL2* gene, however, is not *NIA1* but was found to be involved in synthesis of the MoCo (LaBrie et al. 1992). The *NIA1* NR gene encodes only 10–15% of the NR activity in *Arabidopsis* (Wilkinson and Crawford 1991, 1993; LaBrie et al. 1992) and maps nowhere near any known chlorate-resistance locus (Cheng et al. 1988). Mutations in the *NIA1* gene were ultimately found by mutagenizing *chl3* mutants (with a deletion of the *NIA2* gene) and selecting for plants that are more resistant to chlorate than the *chl3* parent (Wilkinson and Crawford 1993). *nia1 nia2* double mutants were identified and found to have less than 1% of the wild-type level of NR activity and to grow very poorly on nitrate as the sole source of nitrogen; thus, these genes are likely to represent all of the NR structural genes of *Arabidopsis*. Subtle differences in the regulation of these genes have been noted (Cheng et al. 1991), but overall they appear to be expressed in both roots and shoots and to encode NADH-specific enzymes.

One of the interesting findings from the study of NR genes in *Arabidopsis* is that the *NIA2* gene, which encodes 90% of the NR activity in *Arabidopsis*, is dispensable for growth even when limiting levels of nitrate were the only source of nitrogen (Wilkinson and Crawford 1991; Crawford et al. 1992). This result suggests that the level of NR enzyme is in 10-fold excess of what is needed for normal vegetative growth. Similar results have been obtained for barley and *Nicotiana* (Warner et al. 1981; Vaucheret et al. 1990). It is still a mystery why such a highly regulated enzyme that catalyzes such a critical step in the pathway is present in such excess.

The regulation of nitrate reduction in plants is complex, responding to many environmental and internal signals and involving several levels of control. The key signal is nitrate. NR is one of the few substrate-inducible enzymes in plants. Nitrate treatment activates the expression of the NR structural genes within minutes, leading to a 5- to 100-fold increase in NR mRNA levels (Melzer et al. 1989; Friemann et al. 1992). NR mRNA and protein levels are also regulated in response to light, plastidic signals, and circadian rhythms, even though the protein is not chloroplast-targeted (Crawford et al. 1992). NR mRNA levels vary 5- to 10-fold in a diurnal cycle, peaking at the end of the dark period or beginning of the light period (Deng et al. 1990; Cheng et al. 1991). Phytochrome plays a role in inducing NR gene expression in etiolated plants (Rajasekhar et al. 1988). Reduced carbon (the identity of the signaling molecule is not known) also activates NR gene expression even in the dark (Aslam and Huffaker 1984; Crawford et al. 1992). This carbon response has been shown to effect the promoter activity of an NR gene from *Arabidopsis* (Cheng et al. 1992). Posttranscriptional regulation of NR has also been documented (Caboche and Rouze 1990). The most dramatic effect is a 5- to 10-fold drop in NR activity within several minutes in response to dark treatment or to a drop in $CO_2$ levels (Kaiser and Brendle-Behnisch 1991; J.L. Huber et al. 1992). The drop in NR activity is manifest only if magnesium is present in the assay buffer and correlates with phosphorylation of the NR protein on serine residues (J.L. Huber et al. 1992). Restoring the light or the $CO_2$ levels leads to a rapid increase in NR activity and a decrease in phosphorylation. It has been proposed that phosphorylation coordinates carbon metabolism with nitrogen assimilation, as critical enzymes in carbon metabolism are also under phosphorylation control (S.C. Huber et al. 1992).

**Nitrite Reduction**

Once nitrite is produced, it is transported into the chloroplast, where it is reduced by nitrite reductase (for review, see Siegel and Wilderson 1989; Wray 1989; Campbell and Kinghorn 1990; Crawford and Campbell 1990). Nitrite uptake into chloroplasts has been reported to be carrier-mediated (exhibiting saturation kinetics) and to require light (Brunswick and Cresswell 1988). Nitrite reduction to ammonia requires six electrons from reduced ferredoxin. NiR is a monomeric protein of molecular mass between 60 kD and 70 kD that contains a transit sequence used in targeting the enzyme to the chloroplast after synthesis in the cytosol (Back et al. 1988). The transit sequence (~30 amino acids long in the case of spinach) is cleaved off to produce the mature protein (Gupta and Beevers 1987; Back et al. 1988). The enzyme has two redox centers made up of a tetranuclear iron-sulfur center ($Fe_4S_4$) and a siroheme-Fe. These prosthetic groups are thought to reside in the carboxy-terminal half of the protein, where one finds homology with the only other $Fe_4S_4$-siroheme reductase known, bacterial NADPH sulfite reductase. The amino terminus of the enzyme is thought to bind ferredoxin, based on sequence similarities to ferredoxin-NADPH reductase, which is also thought to bind ferredoxin in its amino terminus (Campbell and Kinghorn 1990).

NiR plays a vital role in the assimilation of nitrate. Most of the energy (75%) consumed in the reduction of nitrate to ammonium occurs at this step. In addition, nitrite cannot be allowed to accumulate, as it is toxic. Recently, a transgenic tobacco plant was engineered to inactivate NiR genes by antisense expression (Vaucheret et al. 1992). This plant had no detectable NiR mRNA or activity but did have five times the level of nitrite (487 nmole/gm fresh weight in the transgenic plant). This plant grew normally on ammonia as the sole source of nitrogen but displayed severe symptoms when exposed to nitrate (drastically reduced development and chlorotic leaves). In wild-type plants, the level of NiR activity is approximately ten times higher than the level of NR activity, so that nitrite will not accumulate in the plant cell (Layzell 1990).

The regulation of NiR is very similar to the regulation of nitrate reductase. In fact, it has been proposed that both genes are controlled by a common regulatory system (Faure et al. 1991). NiR gene expression is activated by nitrate, nitrite, and light and is controlled by circadian rhythms (Gupta and Beevers 1987; Lahners et al. 1988; Privalle et al. 1989; Schuster and Mohr 1990). Analysis of the spinach NiR promoter in transgenic tobacco plants has shown that the promoter can confer nitrate inducibility to a marker gene ($\beta$-glucuronidase) (Back et al. 1991). Thus, nitrate regulation includes activation of the NiR promoter.

## PHOSPHATE

Phosphate is another essential macronutrient of plants that makes up about 0.3–0.5% of the dry weight. Phosphate is an integral component of membranes and nucleic acids and plays a vital role in energy transactions and regulatory systems of the plant cell. About 75% of cellular phosphate is found as free inorganic phosphate (~20 mM) in *Arabidopsis* given ample phosphate fertilizer, and the remainder is divided evenly among sugar phosphates, lipids, and nucleic acids (Poirier et al. 1991). Kinases and key allosteric enzymes use phosphate to regulate primary metabolic pathways in plant cells from carbon fixation to starch synthesis. To acquire phosphate from the soil, plants encounter very different problems than with nitrate (Marschner 1986). Phosphate binds tightly to aluminum and iron oxides that make up the soil particles, especially at lower pH. At higher pH, the phosphate precipitates out of solution with calcium and magnesium ions. The concentration of soluble phosphate in the soil is often no greater than 10 μM (Marschner 1986). In tropical and subtropical regions, phosphate is normally the most limiting nutrient, especially for leguminous crops (Ae et al. 1990). To access phosphate from the soil solution, plants have active transport systems for accumulating phosphate. In response to phosphate deprivation, they can undergo physiological and developmental responses to scavenge limited phosphate from the environment. In addition, mutualistic relationships can be established between roots of certain plant species and mycorrhizal fungi in the soil to increase the acquisition of phosphate.

To take up phosphate from the soil solution, plants must be able to accommodate enormous variability in phosphate concentration. Without fertilization, phosphate levels are usually 1 $\mu$M–10 $\mu$M (Marschner 1986). With added fertilizer, the levels can be 1 mM or greater. To deal with such variable concentrations, plants employ a multiphasic transport process that is active and carrier-mediated. If absorption rate is measured over a 25,000-fold range of phosphate concentration (from 3 $\mu$M to 75 mM), one can detect five distinct saturable kinetic classes (see Fig. 4) (Nandi et al. 1987). Each class may represent a unique and distinct transporter or a single transporter that changes its kinetic properties in response to the external phosphate concentration. The highest-affinity class has a $K_m$ of 1 $\mu$M–5 $\mu$M (Nandi et al. 1987; Schmidt et al. 1992). This system allows plants to deplete phosphate from the soil down to 1 $\mu$M and in solution culture down to 0.01 $\mu$M (for review, see Marschner 1986). These $C_{min}$ values vary from species to species. $C_{min}$ for tomato is 0.12 $\mu$M, for soybean 0.04 $\mu$M, and for ryegrass 0.01 $\mu$M (for review, see Marschner 1986). For 25 different races of *Arabidopsis*, $C_{min}$ varied from 0.03 $\mu$M to 0.12 $\mu$M (Krannitz et al. 1991).

Having multiple kinetic classes helps plants adjust to different phosphate concentrations and concentrate phosphate in the cell. One would expect the intracellular concentration of phosphate to be 14 $\mu$M, given an external concentration of 1 mM and a membrane electropotential difference of –110 mV; however, one measures in pea roots a tissue concentration of 25 mM, 1810-fold higher than predicted by the Nernst equation (Higinbotham et al. 1967). Some of this difference may be due to sequestering of the phosphate (e.g., in the vacuole), but it is clear that energy must be provided to concentrate phosphate to such a level in the cell. The required energy comes from the proton gradient provided by the $H^+$-ATPase of the plasma membrane (Schmidt et al. 1992). The transport of phosphate into the cell results in depolarization of the plasma membrane potential (becomes more positive inside) (Ullrich-Eberius et al. 1981, 1984) and rapid alkalinization of the medium (e.g., from pH 3.5 to pH 5.8 for cultured *Catharanthus roseus* cells) (Sakano 1990). Concomitantly, the cytoplasmic pH drops 0.2 units (Ullrich and Novacky 1990; Sakano et al. 1992). Quantification of these results reveals that four protons are transported per phosphate ion. Thus, phosphate uptake is accomplished by proton cotransport.

If a plant begins to experience phosphate deprivation, several physiological and developmental responses occur in an attempt to adapt (Clarkson and Scattergood 1982; Goldstein et al. 1989). Root growth is enhanced relative to shoot growth, and phosphate uptake increases. Because phosphate uptake is subject to negative feedback inhibition, reducing the phosphate levels in the medium or within the plant derepresses phosphate uptake, leading to increase in $V_{max}$ up to 5-fold (Lefebvre and

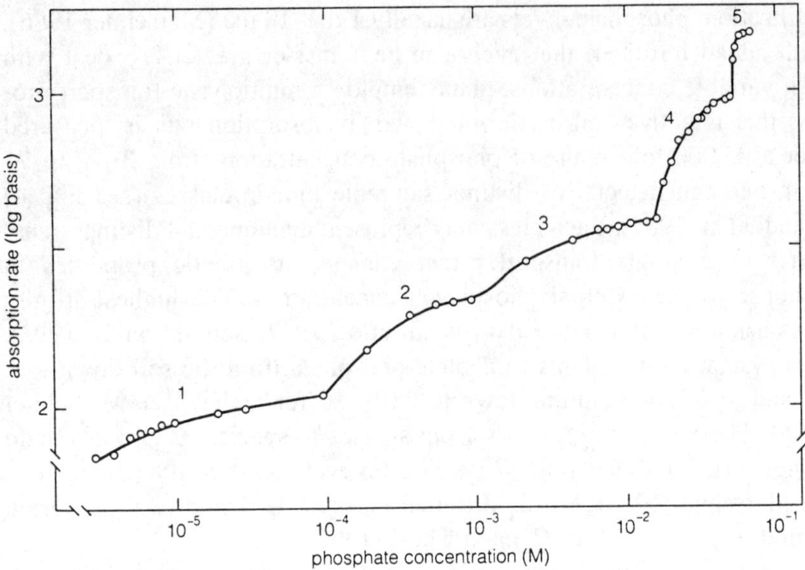

*Figure 4* Graph of phosphate uptake rate as a function of phosphate concentration in the media surrounding maize root sections. Five separate kinetic classes are observed (numbered 1–5) over a 25,000 concentration range. Graph was copied from Salisbury and Ross (1992); original data was from Nandi et al. (1987).

Glass 1982; Cogliatti and Clarkson 1983; Goldstein et al. 1989; Jungk et al. 1990; Goldstein 1991). In addition, plant cells secrete new molecules into the rhizosphere to mobilize phosphate from the soil. Such molecules include citric, malic, and piscidic acids, which release phosphorus from soil by chelating iron and aluminum (Marschner 1986; Ae et al. 1990), and acid phosphatase, which frees phosphate groups from organic molecules (Goldstein et al. 1988; Lefebvre et al. 1990; Duff et al. 1991). Plants can also mobilize stored phosphate reserves in the vacuole and transport it via the phloem to young and growing tissue if needed.

Some plants are naturally more efficient at adapting to low levels of phosphate. Different varieties within a species tolerate low levels of phosphate better than others in that their growth is not impeded (Whiteaker et al. 1976), or they compete more effectively for the available phosphate (Caldwell et al. 1985). This greater tolerance could be due to production of large root systems or to greater mobilization or transport of phosphate in response to starvation. These differences in phosphate utilization can also lead to arsenate tolerance. Arsenate can leach from mine sites and poison plants. Arsenate competes with phosphate for uptake. Arsenate-tolerant *Holcus* plants have been shown to be deficient in high-affinity phosphate transport (Meharg and MacNair 1990, 1992).

Many plants can grow and thrive in natural conditions where phosphate concentrations are very low, due to their association with mycorrhizal fungi (Smith and Gianinazzi-Pearson 1988). This interaction is especially important for forest trees and crops grown in the tropics. Certain soil fungi infect the roots of a plant, forming a mutually beneficial relationship. One of the primary benefits for the plant is that the surface area of the root is greatly expanded by the fungal hyphae so that nutrient uptake, especially of phosphate, is increased several times per unit length of root (Marschner 1986). In practice, this interaction allows some plants to grow without any added phosphate fertilizer. For example, alfalfa plants grown in the absence of mycorrhizae and phosphate fertilizer (in a phosphorus-deficient soil) had a shoot dry weight of 32 mg/plant in one study (Lambert et al. 1980). In the presence of mycorrhizae, the weight was 2177 mg/plant. It took 80 mg P/kg phosphate treatment in the absence of the fungi to support similar growth of the alfalfa plants. Plants appear to be able to influence the extent of fungal infection in their roots depending on the phosphate content in their tissues. When phosphate concentrations become low, roots secrete more reducing sugars and amino acids (3- to 5-fold increase), and the infection rate by mycorrhizal fungi increases (18-fold) (Graham et al. 1981). Foliar application of phosphate to increase the phosphate levels in the plant depresses the fungal infection rate. The importance of mycorrhizal fungi cannot be overstated, as they have been found associated with plant roots in fossils from the Triassic period and are thought to have played a critical role in the colonization of land by early plants (Pirozynski and Hawksworth 1988).

Once taken up by the cell, phosphate is not reduced but is incorporated into ATP or is stored primarily in the vacuole. If a plant experiences phosphate deprivation, phosphate reserves can be mobilized to sink tissues, especially young leaves (Clarkson and Scattergood 1982). Recently, a screen was developed to identify *Arabidopsis* mutants that are deficient in phosphate accumulation in leaves (Poirier et al. 1991). In this screen, 2200 $M_2$ plants were screened for reduced levels of inorganic phosphate in leaf disks, and a mutant (*pho1*) that accumulated only 5% of the wild-type level of inorganic phosphate in the aerial portions of the plant was identified. The growth of this mutant was retarded, producing about 5% of the fresh weight of wild-type plants, but the uptake of phosphate into the roots was not affected. It was proposed that this mutant is deficient in the loading of phosphate into the xylem, and it is hoped that it may help in identifying transporters involved in inter-organ ion transport in plants.

## IRON

Iron is an essential nutrient for plants and, in many circumstances, can limit the plant growth (Marschner 1986). Iron deficiency is a worldwide problem, especially in regions where the soil is alkaline or calcareous. Iron deficiency arises from the fact that even though 2–6% of the soil is iron by weight, very little of the iron is in a form available to plants. Most iron is present as ferric ($Fe^{+++}$) ion bound in very stable reddish-brown oxides (rust) ($K_{sp}$ of $10^{-39}$). The concentration of iron in the soil solution is usually $10^{-10}$ M or lower, yet most crop plants require concentrations above $10^{-7}$ M to obtain the 1–2 kg iron per hectare they need annually for optimal growth (Marschner 1986). Iron is required by many enzymes involved in oxidation and reduction reactions. These enzymes contain either heme groups or iron-sulfur centers and include cytochromes, catalase, nitrogenase, ferredoxin, NR, and NiR. The iron in these proteins is coordinated to the thiol group of cysteine or to inorganic sulfur.

Much of what is known about the uptake and assimilation of iron has come from studies of plant responses to iron deprivation (Yi et al. 1994). Iron deficiency leads to chlorosis of young leaves. Plants can store iron (primarily in stroma) as a ferric oxide/phosphate complex with phytoferritin, a protein which is made up of 24 identical subunits (total mass of 480,000 D) and forms a hollow sphere around a crystalline iron core (Bienfait and Van der Mark 1983; Marschner 1986). In contrast to phosphate, iron cannot be mobilized from mature to young leaves via the phloem. Thus, when a plant can no longer access iron from the soil, the young leaves become chlorotic due to inhibition of chloroplast development and chlorophyll synthesis, even though the mature leaves may have sufficient iron.

Two types of strategies have been identified in plants to adapt to iron deprivation (for review, see Marschner 1986; Bienfait 1988a; Brown and Jolly 1989). Strategy I is found primarily in dicotyledonous species but also in some monocots. When these plants are deprived of iron, they acidify the external solution by pumping protons and secreting organic acids (phenolics) such as caffeic acid. The more acidic solution helps solubilize the iron in the soil, and caffeic acid chelates the soluble iron. Ferric forms of the iron are reduced to the ferrous form. The caffeic acid can perform this reaction, but most of the reduction is thought to occur at the surface of the root cells. A NADH reductase in the plasma membrane of root cells is induced upon iron deprivation, and an increase in membrane electron transfer is observed (Buckhout et al. 1989). This increased reduction of $Fe^{+++}$ to $Fe^{++}$ helps keep the iron soluble and facilitates the transport of iron across the plasma membrane. These physiological responses have been associated with specific developmental changes as well. In tomato, for example, the roots of plants deprived of iron for 8 days reduced chelated iron seven times faster than roots from control plants (Buckhout et al. 1989). This accelerated activity was localized to root hairs that developed on secondary roots that formed during the iron stress period. In another example, certain species of plants develop a new cell type called rhizodermal transfer cells in the epidermis of the root (Kramer et al. 1980). These cells have a distinct cell wall, a high metabolic rate, and are thought to be the sites of rapid proton efflux and phenolic compound secretion in response to iron deprivation. If the iron is restored to sufficient levels, the cells degenerate within 2 days. In tomato, only iron-efficient cultivars develop transfer cells whereas iron-inefficient varieties do not (Landsberg 1984).

*Arabidopsis* has been shown to respond to iron deprivation (Tingey et al. 1982). When grown on vermiculite with a nutrient solution containing a limiting amount of iron (0–10 µM), *Arabidopsis* plants became stunted and chlorotic. After several weeks, however, the plants recovered and became green and healthy again. These recovered plants had 2–3 times the amount of iron in their tissues compared to plants grown on optimal levels of iron (~100 µM), as well as increased levels of magnesium and aluminum. The interpretation of these data is that the deficient plants responded by mobilizing (mining) the metals from the vermiculite, which included iron, magnesium, and aluminum. In support of this conclusion, iron-deprived *Arabidopsis* have been shown to induce ferric reductase and to increase acidification of the rhizosphere (Yi et al. 1994).

Strategy II for the acquisition of iron is found only in graminaceous species, including cereals. These plants respond to iron deprivation by secreting phytosiderophores, which serve to chelate iron (Marschner 1986; Neilands and Leong 1986). The most-studied phytosiderophores are nonproteinogenic amino acids related to nicotianamine, such as mugineic acid and avenic acid. Iron complexed to these molecules will be taken up 100–1000 times faster than iron chelated with EDTA or microbial siderophores, suggesting that there is a specific uptake system for these molecules (Romheld and Marschner 1986).

To begin elucidating these iron-deprivation responses at the genetic level, several mutants have been identified in dicots, including *Arabidopsis*. A tomato mutant (*fer*) has been isolated that develops severe chlorosis on normal soils and lacks the typical iron-deprivation responses of the wild-type parent (formation of extra root hairs and rhizodermal transfer cells and enhancement of excretion of protons and ferric reductase activity) (Brown et al. 1971; Wann and Hills 1973; Bienfait 1988b). The responsible mutation is recessive and monogenic. In peas, a mutant (E107) that accumulates high levels of iron in its leaves has been described (Grusak et al. 1990; Kneen et al. 1990; Welch and LaRue 1990). The phenotypic defect is localized in the roots and results in a 4- to 7-fold increase in extracellular ferric reductase activity and a 3-fold increase in iron influx, regardless of the iron status. *Arabidopsis* mutants have been found by screening for plants that have root ferric reductase activity in the presence of ample iron (Yi et al. 1994). Two potential mutants have been found in a pool of 1500 $M_2$ seeds. In addition, potential mutants that display leaf freckling, a symptom of excess iron accumulation, have been identified. These mutants should provide clues about the regulation of iron-deprivation responses in plants.

## CONCLUSION

Exciting progress has been made, and will continue to be made, in the field of plant mineral nutrition. The fundamental pathways and physiological responses to key environmental signals have been elucidated for the assimilation of iron, phosphate, and nitrogen. Present and future work will provide new insights into the molecular structure and regulation of the ion transporters and signal transduction pathways that are essential for nutrient acquisition and plant responses to nutrient deprivation. Ultimately, it should be possible to genetically engineer plants to modify these processes for the benefit of people and the environment.

## ACKNOWLEDGMENTS

I thank the members of my research group, Jack Wilkinson, Sam LaBrie, Yongwei Cao, Mary Frank, and Yi-Fang Tsay, as well as my colleagues, Julian Schroeder, Ken Feldmann, Wilbur Campbell, Bob Schmidt, and Marty Yanofsky. I also thank Mary Lou Guerinot for providing her unpublished manuscript. Research in my laboratory is supported by grants from the Powell Foundation, National Institutes of Health (GM-40672), and the National Science Foundation (MCB-9219374).

## REFERENCES

Åberg, B. 1947. On the mechanism of the toxic action of chlorates and some related substances upon young wheat plants. *Ann. R. Agric. Coll. Swed.* **15:** 37–107.

Ae, N., J. Arihara, K. Okada, T. Yoshihara, and C. Johansen. 1990. Phosphorus uptake by pigeon pea and its role in cropping systems of the Indian subcontinent. *Science* **248:** 477–480.

Aslam, M. and R.C. Huffaker. 1984. Dependency of nitrate reduction on soluble carbohydrates in primary leaves of barley under aerobic conditions. *Plant Physiol.* **75:** 623–628.

Aslam, M., R.L. Travis, and R.C. Huffaker. 1992. Comparative kinetics and reciprocal inhibition of nitrate and nitrite uptake in roots of uninduced and induced barley seedlings. *Plant Physiol.* **99:** 1124–1133.

Back, E., W. Burkhart, M. Moyer, L. Privalle, and S. Rothstein. 1988. Isolation of cDNA clones coding for spinach nitrite reductase: Complete sequence and nitrate induction. *Mol. Gen. Genet.* **212:** 20–26.

Back, E., W. Dunne, A. Schneiderbauer, A. Framond, R. Rastogi, and S. Rothstein. 1991. Isolation of the spinach nitrite reductase gene promoter which confers nitrate inducibility on GUS gene expression in transgenic tobacco. *Plant Mol. Biol.* **17:** 9–18.

Bienfait, H.F. 1988a. Mechansims in Fe-efficiency reactions of higher plants. *J. Plant Nutr.* **11:** 605–632.

———. 1988b. Proteins under the control of the gene for Fe efficiency in tomato. *Plant Physiol.* **88:** 785–787.

Bienfait, H.F. and F. Van der Mark. 1983. Phytoferritin and its role in iron metabolism. In *Metals and micronutrients: Uptake and utilization by plants* (ed. D.A. Robb and W.J. Pierpoint), pp. 58–76. Academic Press, San Diego, California.

Blumwald, E. and R.J. Poole. 1985. Nitrate storage and retrieval in *Beta vulgaris*: Effects of nitrate and chloride on proton gradients in tonoplast vesicles. *Proc. Natl. Acad. Sci.* **82:** 3683–3687.

Braaksma, F.J. and W.J. Feenstra. 1982. Isolation and characterization of nitrate reductase-deficient mutants of *Arabidopsis thaliana. Theor. Appl. Genet.* **64:** 83–90.

Brown, J.C. and V.D. Jolly. 1989. Plant metabolic responses to iron-deficiency stress. *BioScience* **39:** 546–551.

Brown, J.C., R.L. Chaney, and J.E. Ambler. 1971. A new tomato mutant inefficient in the transport of iron. *Physiol. Plant.* **25:** 48–53.

Brownlee, A.G. and H.N. Arst, Jr. 1983. Nitrate uptake in *Aspergillus nidulans* and involvement of the third gene of the nitrate assimilation gene cluster. *J. Bacteriol.* **155:** 1138–1146.

Brunswick, P. and C.F. Cresswell. 1988. Nitrite uptake into intact pea chloroplasts. I. Kinetics and relationship with nitrite assimilation. *Plant Physiol.* **86:** 378–383.

Buckhout, T.J., P.F. Bell, D.G. Luster, and R.L. Chaney. 1989. Iron stress induced redox activity in tomato is localized on the plasma membrane. *Plant Physiol.* **90:** 151–156.

Caboche, M. and P. Rouze. 1990. Nitrate reductase: A target for molecular and cellular studies in higher plants. *Trends Genet.* **6:** 187–192.

Caldwell, M.M., D.M. Eissenstat, J.H. Richards, and M.F. Allen. 1985. Competition for phosphorus: Differential uptake from dual isotope-labeled soil interspaces between shrub and grass. *Science* **229:** 384–387.

Calza, R., E. Huttner, M. Vincentz, P. Touze, F. Galangau, H. Vaucheret, I. Cherel, C. Meyer, J. Kronenberger, and M. Caboche. 1987. Cloning of DNA fragments complementary to tobacco nitrate reductase mRNA and encoding epitopes common to the nitrate reductases from higher plants. *Mol. Gen. Genet.* **209:** 552–562.

Campbell, W.H. and J.R. Kinghorn. 1990. Functional domains of assimilatory nitrate reductases and nitrite reductases. *Trends Biochem. Sci.* **15:** 315–319.

Campbell, W.H. and J. Smarelli. 1986. Nitrate reductase: Biochemistry and regulation. In *Biochemical basis of plant breeding* (ed. C.A. Neyra), pp. 1–39. CRC Press, Boca Raton, Florida.

Chang, C., J.L. Bowman, A.W. DeJohn, E.S. Lander, and E.M. Meyerowitz. 1988. Restriction fragment length polymorphism linkage map for *Arabidopsis thaliana. Proc.*

*Natl. Acad. Sci.* **85**: 6856–6860.

Cheng, C.-L., G.N. Acedo, M. Cristinsin, and M.A. Conkling. 1992. Sucrose mimics the light induction of *Arabidopsis* nitrate reductase gene transcription. *Proc. Natl. Acad. Sci.* **89**: 1861–1864.

Cheng, C.-L., J. Dewdney, A. Kleinhofs, and H.M. Goodman. 1986. Cloning and nitrate induction of nitrate reductase mRNA. *Proc. Natl. Acad. Sci.* **83**: 6825–6828.

Cheng, C.-L., G.N. Acedo, J. Dewdney, H.M. Goodman, and M.A. Cankling. 1991. Differential expression of the two *Arabidopsis* nitrate reductase genes. *Plant Physiol.* **96**: 275–279.

Cheng, C.-L., J. Dewdney, H. Nam, B.G.W. Den Boer, and H.M. Goodman. 1988. A new locus (*NIA1*) in *Arabidopsis thaliana* encoding nitrate reductase. *EMBO J.* **7**: 3309–3314.

Clarkson, D.D. and C.B. Scattergood. 1982. Growth and phosphate transport in barley and tomato plants during development of and recovery from phosphate stress. *J. Exp. Bot.* **33**: 865–875.

Clement, C.R., M.J. Hopper, L.H. Jones, and E.L. Leafe. 1978. The uptake of nitrate by *Lolium perenne* from flowing nutrient solution. *J. Exp. Bot.* **29**: 1173–1178.

Cogliatti, D.H. and D.T. Clarkson. 1983. Physiological changes in and phosphate uptake by potato plants during development of and recovery from phosphate deficiency. *Physiol. Plant.* **58**: 287–294.

Conway, G.R. and J.N. Pretty. 1988. Fertilizer risks in the developing countries. *Nature* **334**: 207–208.

Crawford, N.M. 1992. Study of chlorate resistant mutants of *Arabidopsis*: Insights into nitrate assimilation and ion metabolism of plants. In *Genetic engineering, principles and methods* (ed. J.K. Setlow), pp. 89–98. Plenum Press, New York.

Crawford, N.M. and H.N.J. Arst. 1993. The molecular genetics of nitrate assimilation in fungi and plants. *Annu. Rev. Genet.* **27**: 115–146.

Crawford, N.M. and W.H. Campbell. 1990. Fertile fields. *Plant Cell* **2**: 829–835.

Crawford, N.M. and R.W. Davis. 1988. Plant nitrate reductase is a tripartite electron transfer protein that belongs to the cytochrome b5 superfamily of enzymes. In *Current topics in plant biochemistry and physiology* (ed. D.G. Blevins and W.H. Campbell), pp. 16–26. University of Missouri, Columbia.

Crawford, N.M., W.H. Campbell, and R.W. Davis. 1986. Nitrate reductase from squash: cDNA cloning and nitrate regulation. *Proc. Natl. Acad. Sci.* **83**: 8073–8076.

Crawford, N.M., J.Q. Wilkinson, and S.T. LaBrie. 1992. Metabolic control of nitrate reduction in *Arabidopsis thaliana*. *Aust. J. Plant Physiol.* **19**: 377–385.

Crawford, N.M., M. Smith, D. Bellissimo, and R.W. Davis. 1988. Sequence and nitrate regulation of the *Arabidopsis thaliana* mRNA encoding nitrate reductase, a metalloflavoprotein with three functional domains. *Proc. Natl. Acad. Sci.* **85**: 5006–5010.

Deng, M.-D., T. Moureaux, M.-T. Leydecker, and M. Caboche. 1990. Nitrate-reductase expression is under the control of a circadian rhythm and is light inducible in *Nicotiana tabacum* leaves. *Planta* **180**: 281–285.

Dhugga, K.S., J.G. Waines, and R.T. Leonard. 1988. Correlated induction of nitrate uptake and membrane polypeptides in corn roots. *Plant Physiol.* **87**: 120–125.

Doddema, H. and H. Otten. 1979. Uptake of nitrate by mutants of *Arabidopsis thaliana*, disturbed in uptake or reduction of nitrate. III. Regulation. *Physiol. Plant.* **45**: 339–346.

Doddema, H. and G.P. Telkamp. 1979. Uptake of nitrate by mutants of *Arabidopsis thaliana*, disturbed in uptake or reduction of nitrate. II. Kinetics. *Physiol. Plant.* **45**: 332–338.

Doddema, H., J. Hofstra, and W. Feenstra. 1978. Uptake of nitrate by mutants of

*Arabidopsis thaliana*, disturbed in uptake or reduction of nitrate. I. Effect of nitrogen source during growth on uptake of nitrate and chlorate. *Physiol. Plant.* **43:** 343–350.

Drew, M.C. 1975. Comparison of the effects of a localized supply of phosphate, nitrate, ammonium and potassium on the growth of the seminal root system, and the shoot, in barley. *New Phytol.* **75:** 479–490.

Duff, S.M.G., W.C. Plaxton, and D.D. Lefebvre. 1991. Phosphate starvation response in plant cells: *De novo* synthesis and degradation of acid phosphatases. *Proc. Natl. Acad. Sci.* **88:** 9538–9542.

Edwards, D.G. and C.J. Asher. 1974. The significance of solution flow rate in flowing culture experiments. *Plant Soil* **41:** 161–166.

Faure, J.D., M. Vincentz, J. Kronenberger, and M. Caboche. 1991. Co-regulated expression of nitrate and nitrite reductase. *Plant J.* **1:** 107–114.

Fernandez, E. and J. Cardenas. 1989. Genetics and regulatory aspects of nitrate assimilation in algae. In *Molecular and genetic aspects of nitrate assimilation* (ed. J.L. Wray and J.R. Kinghorn), pp. 101–124. Oxford Science Publications, Oxford, United Kingdom.

Fernandez, E., R. Schnell, L.P. Ranum, S.C. Hussey, C.D. Silflow, and P.A. Lefebvre. 1989. Isolation and characterization of the nitrate reductase structural gene of *Chlamydomonas reinhardtii. Proc. Natl. Acad. Sci.* **86:** 6449–6453.

Friemann, A., M. Lange, W. Hachtel, and K. Brinkmann. 1992. Induction of nitrate assimilatory enzymes in the tree *Betula pendula. Plant Physiol.* **99:** 837–842.

Glass, A.D.M. 1989. *Plant nutrition: An introduction to current concepts.* Jones and Bartlett, Boston, Massachusetts.

Glass, A.D., J.E. Shaff, and L.V. Kochian. 1992. Studies of the uptake of nitrate in barley. IV. Electrophysiology. *Plant Physiol.* **99:** 456–463.

Goldstein, A.H. 1991. Plant cells selected for resistance to phosphate starvation show enhanced P use efficiency. *Theor. Appl. Genet.* **82:** 191–194.

Goldstein, A.H., D.A. Baertlein, and A. Danon. 1989. Phosphate starvation stress as a system for molecular analysis. *Plant Mol. Biol. Rep.* **7:** 7–16.

Goldstein, A.H., D.A. Baertlein, and R.G. McDaniel. 1988. Phosphate starvation inducible metabolism in *Lycopersicon esculentum.* I. Excretion of acid phosphatase by tomato plants and suspension cultured cells. *Plant Physiol.* **87:** 711–715.

Goyal, S.S. and R.C. Huffaker. 1986. The uptake of $NO_3^-$, $NO_2^-$, and $NH_4^+$ by intact wheat (*Triticum aestivum*) seedlings. I. Induction and kinetics of transport systems. *Plant Physiol.* **82:** 1051–1056.

Graham, J.H., R.T. Leonard, and J.A. Menge. 1981. Membrane-mediated decrease in root exudation responsible for phosphorus inhibiton of vesicular-arbuscular mycorrhiza formation. *Plant Physiol.* **68:** 548–552.

Granstedt, R.C. and R.C. Huffaker. 1982. Identification of the leaf vacuole as a major nitrate storage pool. *Plant Physiol.* **70:** 410–413.

Grusak, M.A., R.M. Welch, and L.V. Kochian. 1990. Physiological characterization of a single gene mutant of *Pisum sativum* exhibiting excess iron accumulation. 1. Root iron reduction and iron uptake. *Plant Physiol.* **93:** 976–981.

Guiard, B. and G. Lederer. 1979. The cytochrome $b_5$ fold: Structure of a novel protein superfamily. *J. Mol. Biol.* **135:** 639–650.

Gupta, S.C. and L. Beevers. 1987. Regulation of nitrite reductase, cell-free translation and processing. *Plant Physiol.* **83:** 750–754.

Hawkesford, M., J.-C. Davidian, and C. Grignon. 1993. Sulphate/proton cotransport in plasma-membrane vesicles isolated from roots of *Brassica napus* L.: Increased transport in membranes isolated from sulphur-starved plants. *Planta* **190:** 297–304.

Heimer, Y.M. and P. Filner. 1971. Regulation of the nitrate assimilation pathway in cultured tobacco cells. III. The nitrate uptake system. *Biochim. Biophys. Acta* **230:** 362–372.

Higinbotham, N., B. Etherton, and R.J. Foster. 1967. Mineral ion contents and cell transmembrane electropotentials of pea and oat seedling tissue. *Plant Physiol.* **42:** 37–46.

Hodges, T.K. 1972. Ion adsorption by plant roots. *Adv. Agron.* **25:** 163–207.

Hoff, T., B.M. Stummann, and K.W. Henningsen. 1992. Structure, function and regulation of nitrate reductase in higher plants. *Physiol. Plant.* **84:** 616–624.

Hole, D., A.M. Emran, Y. Fares, and M.C. Drew. 1990. Induction of nitrate transport in maize roots, and kinetics of influx, measured with nitrogen-13. *Plant Physiol.* **93:** 642–647.

Huber, J.L., S.C. Huber, W.H. Campbell, and M.G. Redinbaugh. 1992. Reversible light/dark modulation of spinach leaf nitrate reductase activity involves protein phosphorylation. *Arch. Biochem. Biophys.* **296:** 58–65.

Huber, S.C., J.L. Huber, W.H. Campbell, and M.G. Redinbaugh. 1992. Comparative studies of the light modulation of nitrate reductase and sucrose-phosphate synthase activities in spinach leaves. *Plant Physiol.* **100:** 706–712.

Jackson, W.A., D. Flesher, and R.H. Hageman. 1973. Nitrate uptake by dark-grown corn seedlings. *Plant Physiol.* **51:** 120–127.

Jackson, W.A., W.A. Pan, R.H. Moll, and E.J. Kamprath. 1986. Uptake, translocation and reduction of nitrate. In *Biochemical basis of plant breeding* (ed. C.A. Neyra), pp. 73–108. CRC Press, Boca Raton, Florida.

Jungk, A., C.J. Asher, D.G. Edwards, and D. Meyer. 1990. Influence of phosphate status on phosphate uptake kinetics of maize and soybean. *Plant Soil* **124:** 175–182.

Kaiser, W.M. and E. Brendle-Behnisch. 1991. Rapid modulation of spinach leaf nitrate reductase activity by photosynthesis. I. Modulation in vivo by $CO_2$ availability. *Plant Physiol.* **96:** 363–367.

King, B.J., M.Y. Siddiqi, and A.D. Glass. 1992. Studies of the uptake of nitrate in barley. V. Estimation of root cytoplasmic nitrate concentration using nitrate reductase activity-implications for nitrate influx. *Plant Physiol.* **99:** 1582–1589.

Kleinhofs, A. and R.L. Warner. 1990. Advances in nitrate assimilation. In *The biochemistry of plants* (ed. B.J. Miflin and P.J. Lea), pp. 89–120. Academic Press, San Diego, California.

Kneen, B.E., T.A. LaRue, R.M. Welch, and N.F. Weeden. 1990. Pleiotropic effects of *brz*. A mutation in *Pisum sativum* cv "Sparkle" conditioning decreased nodulation, increased iron uptake and leaf necrosis. *Plant Physiol.* **93:** 717–722.

Kramer, D., V. Romheld, E. Landsberg, and H. Marschner. 1980. Induction of transfer cell formation by iron deficiency in the root epidermis of *Helianthus annuus*. *Planta* **147:** 335–339.

Krannitz, P.G., L.W. Aarssen, and D.D. Lefebvre. 1991. Relationships between physiological and morphological attributes related to phosphate uptake in 25 genotypes of *Arabidopsis thaliana*. *Plant Soil* **133:** 169–175.

LaBrie, S.T., J.Q. Wilkinson, Y.-F. Tsay, K.A. Feldmann, and N.M. Crawford. 1992. Identification of two tungstate-sensitive molybdenum cofactor mutants, *chl2* and *chl7*, of *Arabidopsis thaliana*. *Mol. Gen. Genet.* **233:** 169–176.

Lahners, K., V. Kramer, E. Back, L. Privalle, and S. Rothstein. 1988. Molecular cloning of complementary DNA encoding maize nitrite reductase. *Plant Physiol.* **88:** 741–746.

Lambert, D.H., H. Cole, and D.E. Baker. 1980. Variation in the response of alfalfa clones and cultivars to mycorrhizae and phosphorus. *Crop Sci.* **20:** 615–618.

Landsberg, E.C. 1984. Regulation of iron stress response by whole plant activity. *J. Plant*

*Nutr.* **7:** 609–621.

Larsson, C.-M. and B. Ingemarsson. 1989. Molecular aspects of nitrate uptake in higher plants. In *Molecular and genetic aspects of nitrate assimilation* (ed. J.L. Wray and J.R. Kinghorn), pp. 3–14. Oxford Science Publications, Oxford, United Kingdom.

Layzell, D.B. 1990. $N_2$ fixation, nitrate reduction and ammonium assimilation. In *Plant physiology, biochemistry and molecular biology* (ed. D.T. Dennis and D.H. Turpin), pp. 389–406. Longman Scientific & Technical, New York.

Le, K.H.D. and F. Lederer. 1983. On the presence of a heme-binding domain homologous to cytochrome b5 in *Neurospora crassa* assimilatory nitrate reductase. *EMBO J.* **2:** 1909–1914.

Lefebvre, D.D. and A.D.M. Glass. 1982. Regulation of phosphate influx in barley roots; effects of phosphate deprivation and reduction of influx with provision of orthophosphate. *Physiol. Plant.* **54:** 199–206.

Lefebvre, D.D., S.M. Duff, C. Fife, C. Julien-Inalsingh, and W.C. Plaxton. 1990. Response to phosphate deprivation in *Brassica nigra* suspension cells-enhancement of intracellular, cell surface and secreted phosphatase activities compared to increases in Pi. *Plant Physiol.* **93:** 505–511.

Mack, G. and R. Tischner. 1990. The effect of endogenous and externally supplied nitrate on nitrate uptake and reduction in sugarbeet seedlings. *Planta* **182:** 169–173.

MacKown, C.T. and P.R. McClure. 1988. Development of accelerated net nitrate uptake, effects of nitrate concentration and exposure time. *Plant Physiol.* **87:** 162–166.

Marschner, H. 1986. *Mineral nutrition of higher plants.* Academic Press, San Diego, California.

McClure, P.R., L.V. Kochian, R.M. Spanswick, and J.E. Shaff. 1990a. Evidence for cotransport of nitrate and protons in maize roots. I. Effects of nitrate on the membrane potential. *Plant Physiol.* **93:** 281–289.

———. 1990b. Evidence for cotransport of nitrate and protons in maize roots. II. Measurement of $NO_3^-$ and $H^+$ fluxes with ion-selective microelectrodes. *Plant Physiol.* **93:** 290–294.

McClure, P.R., T.E. Omholt, G.M. Pace, and P.Y. Bouthyette. 1987. Nitrate-induced changes in protein synthesis and translation of RNA in maize roots. *Plant Physiol.* **84:** 52–57.

Meharg, A.A. and M.R. MacNair. 1990. An altered phosphate uptake system in arsenate tolerant *Holcus lanatus. New Phytol.* **116:** 29–35.

———. 1992. Suppression of the high affinity phosphate uptake system: A mechanism of arsenate tolerance in *Holcus lanatus. J. Exp. Bot.* **43:** 519–524.

Melzer, J.M., A. Kleinhofs, and R.L. Warner. 1989. Nitrate reductase regulation: Effects of nitrate and light on nitrate reductase mRNA accumulation. *Mol. Gen. Genet.* **217:** 341–346.

Miller, A.J. and S.J. Smith. 1992. The mechanism of nitrate transport across the tonoplast of barley root cells. *Planta* **187:** 554–557.

Nandi, S.K., R.C. Pant, and P. Nissen. 1987. Multiphasic uptake of phosphate by corn roots. *Plant Cell Environ.* **10:** 463–474.

Neilands, J.B. and S.A. Leong. 1986. Siderophores in relation to plant growth and disease. *Annu. Rev. Plant Physiol.* **37:** 187–208.

Neyra, C.A. and R.H. Hageman. 1975. Nitrate uptake and induction of nitrate reductase in excised corn roots. *Plant Physiol.* **56:** 692–695.

Pirozynski, K.A. and D.L. Hawksworth, eds. 1988. *Coevolution of fungi with plants and animals.* Academic Press, San Diego, California.

Poirier, Y., S. Thoma, C. Somerville, and J. Schiefelbein. 1991. A mutant of *Arabidopsis* deficient in xylem loading of phosphate. *Plant Physiol.* **97:** 1087–1093.

Privalle, L.S., K.N. Lahners, M.A. Mullins, and S. Rothstein. 1989. Nitrate effects nitrate reductase activity and nitrite reductase mRNA levels in maize suspension cultures. *Plant Physiol.* **90:** 962–967.

Quesada, A., A. Galvan, P.A. Lefebvre, and E. Fernandez. 1994. Identification of nitrate transporter genes in *Chlamydomonas reinhardtii*. *Plant J.* **5:** 407–419.

Rajasekhar, V.K., G. Gowri, and W.H. Campbell. 1988. Phytochrome-mediated light regulation of nitrate reductase expression in squash cotyledons. *Plant Physiol.* **88:** 242–244.

Redinbaugh, M.G. and W.H. Campbell. 1991. Higher plant responses to environmental nitrate. *Physiol. Plant.* **82:** 640–650.

Romheld, V. and H. Marschner. 1986. Evidence for a specific uptake system for iron phytosiderophores in roots of grasses. *Plant Physiol.* **80:** 175–180.

Ruiz-Cristin, J. and D.P. Briskin. 1991. Characterization of a $H^+/NO_3^-$ symport associated with plasma membrane vesicles of maize roots using $^{36}ClO_3^-$ as a radiotracer analog. *Arch. Biochem. Biophys.* **285:** 74–82.

Sakano, K. 1990. Proton/phosphate stoichiometry in uptake of inorganic phosphate by cultured cells of *Catharanthus roseus*. *Plant Physiol.* **93:** 479–483.

Sakano, K., Y. Yazake, and T. Mimura. 1992. Cytoplasmic acidification induced by inorganic phosphate uptake in suspension cultured *Catharanthus roseus* cells. *Plant Physiol.* **99:** 672–680.

Salisbury, F.B. and C.W. Ross. 1992. *Plant physiology*. Wadsworth, Belmont, California.

Schmidt, M.-E., S. Heim, C. Wylegalla, C. Helmbrecht, and K.G. Wagner. 1992. Characterization of phosphate uptake by suspension cultured *Catharanthus roseus* cells. *J. Plant Physiol.* **140:** 179–184.

Scholten, H.J. and W.J. Feenstra. 1986. Uptake of chlorate and other ions in seedlings of the nitrate uptake mutant B1 of *Arabidopsis thaliana*. *Physiol. Plant.* **66:** 265–269.

Schuster, C. and H. Mohr. 1990. Appearance of nitrite reductase mRNA in mustard seedling cotyledons is regulated by phytochrome. *Planta* **181:** 327–334.

Siddiqi, M.Y., A.D.M. Glass, T.J. Ruth, and T.W. Rufty. 1990. Studies of the uptake of nitrate in barley. *Plant Physiol.* **93:** 1426–1432.

Siddiqi, M.Y., A.D.M. Glass, T.J. Ruth, and M. Fernando. 1989. Studies on the regulation of nitrate influx by barley seedlings using $^{13}NO_3^{-1}$. *Plant Physiol.* **90:** 806–813.

Siegel, L.M. and J.O. Wilderson. 1989. Structure and function of spinach ferredoxin-nitrite reductase. In *Molecular and genetic aspects of nitrate assimilation* (ed. J.L. Wray and J.R. Kinghorn), pp. 263–283. Oxford Science Publications, Oxford, United Kingdom.

Smith, S.E. and V. Gianinazzi-Pearson. 1988. Physiological interactions between symbionts in vesicular-arbuscular mycorrhizal plants. *Annu. Rev. Plant Physiol. Plant Mol. Biol.* **39:** 221–244.

Solomonson, L.P. and M.J. Barber. 1990. Assimilatory nitrate reductase: Functional properties and regulation. *Annu. Rev. Plant Physiol. Plant Mol. Biol.* **41:** 225–253.

Taiz, L. and E. Zeiger. 1991. *Plant physiology*. Benjamin/Cummings, Menlo Park, California.

Teyker, R.H., W. Jackson, R.J. Volk, and R.H. Moll. 1988. Exogenous $^{15}NO^-_3$ influx and endogenous $^{14}NO^-_3$ efflux by two maize (*Zea mays* L.) inbreds during nitrogen deprivation. *Plant Physiol.* **86:** 778–781.

Tingey, D.T., S. Raba, K.D. Rodecap, and J.J. Wagner. 1982. Vermiculite, a source of metals for *Arabidopsis thaliana*. *J. Am. Soc. Hortic. Sci.* **107:** 465–468.

Touraine, B., D.T. Clarkson, and B. Muller. 1994. Regulation of ntirate uptake at the whole plant level. In *A whole plant perspective of carbon-nitrogen interactions* (ed. J. Roy and E. Garnier), pp. 11–30. SPB Academic Publishing, The Hague, The Netherlands.

Tsay, Y.-F., J.I. Schroeder, K.A. Feldmann, and N.M. Crawford. 1993. A herbicide sensitivity gene *CHL1* of *Arabidopsis* encodes a nitrate-inducible nitrate transporter. *Cell* **72:** 705–713.

Ullrich, C.I. and A.J. Novacky. 1990. Extra- and intracellular pH and membrane potential changes induced by $K^+$, $Cl^-$, $H_2PO_4^-$, $NO_3^-$ uptake and fusicoccin in root hairs of *Limnobium storoniferum*. *Plant Physiol.* **94:** 1561–1567.

Ullrich, W.R. and A. Novacky. 1981. Nitrate-dependent membrane potential changes and their induction in *Lemna gibba* G1. *Plant Science Letters* **22:** 221–217.

Ullrich-Eberius, C., C. Novacky, and A.J. van Bell. 1984. Phosphate uptake in *Lemna gibba* G1: Energetics and kinetics. *Planta* **161:** 43–52.

Ullrich-Eberius, C., C. Novacky, E. Fischer, and U. Luttge. 1981. Relationship between energy-dependent phosphate uptake and the electrical membrane potential in *Lemna gibba* G1. *Plant Physiol.* **67:** 797–801.

Unkles, S.E., K.L. Hawker, C. Grieve, E.I. Campbell, P. Montague, and J.R. Kinghorn. 1991. *crnA* encodes a nitrate transporter in *Aspergillus nidulans*. *Proc. Natl. Acad. Sci.* **88:** 204–208.

Vaucheret, H., M. Chabaud, J. Kronenberger, and M. Caboche. 1990. Functional complementation of tobacco and *Nicotiana plumbaginofolia* nitrate reductase deficient mutants by transformation with the wild-type alleles of the tobacco structural genes. *Mol. Gen. Genet.* **220:** 468–474.

Vaucheret, H., J. Kronenberger, A. Lepingle, F. Vilaine, J.-P. Boutin, and M. Caboche. 1992. Inhibition of tobacco nitrite reductase activity by expression of antisense RNA. *Plant J.* **2:** 559–569.

Vincentz, M., T. Moureaux, M.-T. Leydecker, H. Vaucheret, and M. Caboche. 1993. Regulation of nitrate and nitrite reductase expression in *Nicotiana plumbaginifolia* leaves by nitrogen and carbon metabolites. *Plant J.* **3:** 315–324.

Wallsgrove, R.M., H. Hasegawa, A.C. Kendall, and J.C. Turner. 1989. The genetics of nitrate uptake in higher plants. In *Molecular and genetic aspects of nitrate assimilation* (ed. J.L. Wray and J.R. Kinghorn), pp. 15–26. Oxford Science Publications, Oxford, United Kingdom.

Wann, E.V. and W.A. Hills. 1973. The genetics of boron and iron transport in the tomato. *J. Hered.* **64:** 370–371.

Warner, R.L. and A. Kleinhofs. 1981. Nitrate utilization by nitrate reductase-deficient barley mutants. *Plant Physiol.* **67:** 740–743.

Warner, R.L., C.J. Lin, and A. Kleinhofs. 1981. Nitrate reductase-deficient mutants in barley. *Nature* **269:** 406–407.

Welch, R.M. and T.A. LaRue. 1990. Physiological characteristics of Fe accumulation in the "bronze" mutant of *Pisum sativum*, cv "Sparkle" E107. *Plant Physiol.* **93:** 723–729.

Whiteaker, G., G.C. Gerloff, W.H. Gabelman, and D. Lindgren. 1976. Intraspecific differences in growth of beans at stress levels of phosphorus. *J. Am. Soc. Hortic. Sci.* **101:** 472–475.

Wilkinson, J. and N. Crawford. 1991. Identification of the *Arabidopsis CHL3* gene as the nitrate reductase structural gene *NIA2*. *Plant Cell* **3:** 461–471.

———. 1993. Identification and characterization of a chlorate resistant mutant of *Arabidopsis* with mutations in both *NIA1* and *NIA2* nitrate reductase structural genes. *Mol. Gen. Genet.* **239:** 289–297.

Wray, J.L. 1989. Molecular and genetic aspects of nitrite reduction in higher plants. In

*Molecular and genetic aspects of nitrate assimilation* (ed. J.L. Wray and J.R. Kinghorn), pp. 244–262. Oxford Science Publications, Oxford, United Kingdom.

Yi, Y., J.A. Saleeba, and M.L. Guerinot. 1994. Iron uptake in *Arabidopsis thaliana*. In *The biochemistry of metal micronutrients in the rhizosphere* (ed. J. Manthey et al.), pp. 295–307. Lewis Publishers, Boca Raton, Florida.

Zhen, R.G., H.-W. Koyro, R.A. Leigh, A.D. Tomos, and A.J. Miller. 1991. Compartmental nitrate concentration in barley root cells measure with nitrate-selective microelectrodes and by single-cell sap sampling. *Planta* **185:** 356–361.

# APPENDICES
## Genetic Resources

# APPENDIX A
# The Internet and Electronic
# *Arabidopsis* Information Resources

**J. Michael Cherry**
Department of Genetics
Stanford University
Stanford, California 94305-5120

## THE INTERNET AND INTERNET SOFTWARE

The ability of public resources to collect, maintain, and electronically distribute information on *Arabidopsis* is expanding with each passing year. There are two major reasons for this change. First, the worldwide Internet computer network has become a ubiquitous feature of the academic and industrial biological research environment. Second, several software development efforts have devised simple but effective software that enables users to search for and retrieve information. Much of this new software has been designed specifically to allow academic information to be distributed freely and easily. This chapter presents a very brief introduction to the Internet and some useful software, and a summary of the specific electronic resources available to the *Arabidopsis* community.

The Internet computer network is a worldwide collection of networks using a common mechanism to allow computers to communicate. Each of the participating networks is initially built and subsequently maintained by the institution it serves. A university network is a small version of the Internet. For example, a group of computers are linked together on a floor of a building. This group or network is then linked to other floors, thus creating an inter-network or internet. The building's network is further linked to the main university network where many other building and dormitory networks are also connected. The worldwide Internet is simply the interconnection of national and regional networks that contain academic and commercial networks. Just as a university network allows communication between the different buildings on campus, the Internet allows communication between computers located around the planet. One important aspect of the Internet is that the computers interact with each other directly. No intermediate computer is involved. To reach a computer on the Internet it is sufficient to know the computer's name, what is called the host name. The user or the computer does not need to know an explicit path between the computers. The special devices that connect networks handle the routing of the communications between the networks. With all the computers interconnected in this manner, it is just as easy for the user to communicate with a computer on the other side of the country as with a computer in another part of the university. However, the speed of the interaction depends on the number of networks crossed and the rate that information is transmitted to each of these networks. To learn more details about the Internet, seek out one of the many books now available at most libraries and bookstores that carry computer-related information. Books are available on how to obtain an Internet connection for your home (Estrada 1993), lists of resources available on the Internet (Engst 1993), and general nontechnical starter manuals on what networking is all about (Krol 1992; Dern 1993; Engst 1993).

It is becoming increasingly easier to obtain access to the Internet. Many universities and companies provide network access through normal telephone lines with modems and special software or digital telephone lines such as ISDN. Several of the commercial electronic computer information services, sometimes called bulletin board systems, also provide access to the Internet. Even cable television networks are beginning to provide the Internet to their subscribers in many regions.

Much of the software that has been designed for use with the Internet is based on the client-server model of software design. A server is provided by specifically designed software on a computer that listens to the network for requests from its clients. The client software is installed on the user's computer. The client-server model allows efficient use of computing resources by localizing the computer-intensive tasks on the server. Generally, the server is provided on a fast computer that can quickly perform searches or store large collections of files that are needed by the clients. The client receives requested information from the server and presents the information to the user.

The Internet resources that are available for scientists are evolving rapidly; thus any list of resources will quickly become out of date. Presented below are brief descriptions of Internet software used to provide academic information. It is my hope that a general understanding of what these services can provide will allow the *Arabidopsis* community to find and use new services even after this chapter is out of date.

Five types of general Internet software are currently used to provide scientific information on the Internet. FTP (File Transfer Protocol) allows retrieval of files through the network. Telnet permits interactive user sessions through the Internet. WAIS (Wide Area Information Service) provides a mechanism to query and retrieve information from remote databases. Gopher presents a menu of options that can contain a wide variety of types of information and services. WWW (World Wide Web), the newest of the resources, provides hypertext, graphics, and access to remote databases. All of this software requires a computer that is connected to the Internet: A simple modem connection is not sufficient. Special networking software called SLIP or PPP can be used to connect a home computer to the Internet. Even if an Internet connection is not available, all of the above software is available in a form that functions with a terminal emulator program and a modem, thus allowing a user to explore the Internet using an institutional account from a home computer.

**FTP and Telnet**

FTP and Telnet are the oldest examples of this type of software and are available on almost every computer connected to the Internet. FTP and Telnet have long been standard components of software packages that provide computers the ability to communicate with an inter-networked computer. FTP allows a user to retrieve files from a distant computer and save them on the local computer's disk drive. Therefore, FTP is the networking analog to file transfer programs such as Kermit or XModem that are commonly used to transfer files over a modem connection. Many hundreds of computers on the Internet currently act as Anonymous FTP servers. This means that these computers allow anyone access to a special directory of files on their disk drives. This service is called Anonymous FTP because the user name employed is simply "anonymous." Telnet allows a user to connect with another computer on the Internet as a full interactive user. To successfully enter a computer with Telnet, a valid account name and password must be known. Many people are familiar with Telnet as the basic method available to access institutional computer facilities. Telnet is also used by some database resources when a special interactive software program is needed to access their service.

**WAIS**

The software programs described below are relatively new additions to the Internet. They were created to simplify public access to academic information. This type of software is sometimes referred to as Internet discovery software because it allows users to search and discover information they desire. The WAIS (Wide Area Information Server) software developed by Thinking Machines, Inc. was created to provide an easy method to search databases of documents such as textbooks, newspapers, or technical journals. The software creates an index containing the location of every word in the original material. The user can search for any word in the original document, rather than just keywords that someone else thought were sufficient to describe the document. WAIS-indexed collections allow the user to directly explore all parts of a document or database of interest. Searching a WAIS index is as simple as entering the word or words that describe the information being sought. To restrict a search, the modifiers "and" and "not" can be added to the query. The WAIS software assumes every word of a question is separated by the modifier "or." To expand a search, the wild card character asterisk ('*') can be added, although unfortunately, only at the end of a word. For example, to locate information that contains a word which begins with "homeo" and is associated with the flower, the question to an appropriate WAIS index could be "homeo* and flower". WAIS clients are available for many types of computers through Anonymous FTP from the host name think.com. WAIS indices are very common today because of the speed at which they provide an answer. However, the use of WAIS clients has largely given way to the Gopher and WWW software mentioned below.

### Gopher

The Internet Gopher software, created at the University of Minnesota, allows a variety of information to be presented in menus, including, but not limited to, plain text files, graphic image files, and links to documents and folders on other Gopher servers. Gopher also allows links to other types of resources. The WAIS software mentioned above is one of the most powerful of these. A WAIS index is recognized by a question mark ('?') icon on the Gopher menu. In text-only Gopher clients, the name of the WAIS index typically has "<?>" added to the name. Gopher also provides access to FTP servers. A menu of the files provided by an FTP server is created, thus simplifying the task of finding and retrieving a file. In general, Gopher provides an inexpensive method for the distribution of large amounts of information to anyone connected to the Internet. The Gopher software that can be installed on your computer is called the "client." The client interacts through the Internet with any of several hundred different Gopher "servers." Gopher clients also work with other software on the local computer to present non-text information, such as graphic images and sound recordings. Gopher clients are available for most computers. Ask your computer administrator for information or use Anonymous FTP to the host name boombox.micro.umn.edu to retrieve the specific client needed.

## WWW

The World Wide Web (WWW), initially developed at CERN (European Laboratory for Particle Physics, Geneva, Switzerland), is the latest addition to the family of Internet discovery software. WWW provides hypertext links to be created by a document's author. Hypertext is the idea that a portion of a document, a word or collection of words, even a figure or part of a figure, can be associated with another document on the same computer or on a completely different computer. For example, if this paragraph was being read through a WWW client, a link could be provided for every occurrence of the abbreviation "WWW" to a document containing a description of WWW. By selecting the "WWW" abbreviation, the link would be followed to the description of the WWW software. The result of following this link could be the presentation of a short text definition, a portion of a long paper, or a figure illustrating the topic. The WWW software can also work with the Gopher, WAIS, and FTP servers. Thus, a WWW client is sufficient to access Gopher, WAIS, and FTP servers. There are several WWW clients available today; one of the most popular, called Mosaic, has been created by the National Center for Supercomputing Applications (NCSA) at the University of Illinois. The NCSA Mosaic software is available in versions for UNIX X-windows, Apple Macintosh, IBM OS/2, and Microsoft Windows via Anonymous FTP from ftp.ncsa.uiuc.edu.

As this chapter is being written (early 1994), there are a growing number of Gopher and WWW servers available for biological information. However, only recently have servers been created that use the full power of the WWW and Mosaic software. An excellent WWW server on protein sequences was created by Dan Jacobson of the Genome Database group (GDB) at The Johns Hopkins University. The Hopkins BioInformatics WWW server (see access information below) provides several protein databases with links to other servers housing such information as three-dimensional protein structure files, pictures of the 3D structure, and Medline citations, as well as links among the various protein databases. This is truly an important use of the WWW technology for molecular biological information and has set a new standard. Future enhancements of the WWW and Mosaic software will allow electronic books and genomic information to become commonplace on the Internet. Hypertext will allow parts of figures and tables to be linked with descriptive text, associated information, or the data used to create the illustration.

Resources on the Internet change quickly as new software is created and existing software is enhanced to provide new functionality. Sometimes the new software is so good that the resource administrators decide to abandon the older software for the new. This creates the problem that any written account of the resources becomes rapidly out of date. With that disclaimer, several of the resources that are currently of value to *Arabidopsis* researchers are described below.

**BioSci Electronic Conferences**

New information resources on *Arabidopsis* and molecular biology will become available in the future. The best source of current information is the BioSci electronic conference on the *Arabidopsis* Genome. To subscribe to the BioSci conference via electronic mail, send a message to biosci-server@net.bio.net and simply include the words "subscribe arabidopsis" as the body of the message. You will then begin to receive all the messages submitted to the *Arabidopsis* Genome conference. If you would like more information on BioSci send the word "help" to the biosci-server address. A subscription to the BioSci conference through electronic mail will cause the BioSci messages to be intermixed with all your other electronic mail. The *Arabidopsis* BioSci conference has a fairly low volume of messages, but some people dislike any extra messages. BioSci conferences are also available through the communications system known as Usenet. Usenet takes several forms and is not something that an individual sets up. It is typically provided on a departmental or institutional level. Consult your local computer administrator for more information on the possibility of using Usenet. Usenet has many advantages over electronic mail. The messages are not copied to the user's electronic mailbox; instead, the messages only exist once for the whole organization. Thus, it provides a more efficient method of distributing information. A user can easily save or print a message if desired. The other major advantage is that special software has been developed to interact with the Usenet environment. This software, called a news reader, allows the user to subscribe to many different Usenet conferences or news groups and to keep up with the messages in an effortless manner. A more detailed description of Usenet news and the other software mentioned above is beyond the scope of this chapter. For more information, consult your location computer administrator or one of the many useful nontechnical books about the Internet and its software (for a partial list of available books, see Krol 1992; Dern 1993; Engst 1993; Estrada 1993).

*ARABIDOPSIS* INFORMATION

Electronic *Arabidopsis* information resources are concentrated at two locations, the AAtDB Project and AIMS database. Both efforts were created around an *Arabidopsis* database project. The AAtDB (An *Arabidop-*

*sis thaliana* Data Base) Project at Massachusetts General Hospital is supported by the U.S. Department of Agriculture Plant Genome Research Program through the National Agricultural Library. The AIMS (*Arabidopsis* Information Management System) database at Michigan State University is part of the ABRC (*Arabidopsis* Biological Resource Center) at Ohio State University and is funded by the National Science Foundation.

### The AAtDB Project

The AAtDB database (Cherry et al. 1992) was created using specialized software developed for the *Caenorhabditis elegans* genome effort (Durbin and Thierry-Mieg 1991). This specialized interactive software, called ACEDB, allows the user to browse through information by simply pointing and clicking with the computer mouse. A powerful query facility is also provided. The ACEDB software is available for UNIX and Macintosh computers. The AAtDB form of this software can be installed on a local computer or accessed directly from the AAtDB Project's computer in Boston if the X-windows environment is available. X-windows is typically available on UNIX workstations, providing the presentation of graphics and text plus allowing the use of the keyboard and mouse. X-windows emulators are available for Macintosh, MS Windows, and DOS computers. For more information on obtaining a copy or direct access to AAtDB, contact the AAtDB curator listed below. The AAtDB database contains a wide variety of information about *Arabidopsis* including:

- Physical map of cosmid and lambda clones
- Genetic maps of classical and molecular markers
- Primary information for $F_2$ and RI population recombination data as well as data from two-point crosses obtained from the literature or by direct submission
- Seed stocks and clone resources available from the ABRC at Ohio State or the NASC (Nottingham *Arabidopsis* Stock Centre) in England
- *Arabidopsis* DNA sequences from GenBank and EMBL data banks as well as the EST (expressed sequence tags, i.e., cDNA) sequences from the NCBI's (National Center for Biotechnology Information) dbEST database
- Protein sequence similarities between the known *Arabidopsis* DNA sequences and all known protein sequences
- Phenotype descriptions
- Bibliographic citations

- Scanned images of RFLP autoradiograms, mutant plants, and restriction enzyme digests of clones used as RFLP probes
- A list of *Arabidopsis* researchers with their postal and electronic addresses, as well as research interests

The AAtDB Project provides electronic access to all of the above AAtDB database information and more through FTP and Gopher. Some of the additional *Arabidopsis* information that is provided through the AAtDB Gopher server, the AAtDB Research Companion, and has been separately indexed with the WAIS software includes:

- Archive of the BioSci *Arabidopsis* Genome conference
- The Compleat Guide, a collection of protocols on how to work with *Arabidopsis* compiled by Caroline Dean and David Flanders
- Nottingham *Arabidopsis* Stock Centre Catalog
- Electronic version of the *Arabidopsis* Information Service, edited by A.R. Kranz, which published reports on *Arabidopsis* from 1964 until 1989; figures and tables are available as image files
- Reports from the NCBI on similarities of the *Arabidopsis* EST sequences and all other known DNA sequences

The AAtDB FTP and Gopher servers are available from the Internet host name weeds.mgh.harvard.edu. If a Gopher client is not available on your computer, you can use a public access Gopher service. The Department of Molecular Biology at Massachusetts General Hospital provides such a service. You can simply use Telnet to the host name weeds.mgh.harvard.edu with the user name "gopher." No password is required. This supplies access to the AAtDB Research Companion, allowing movement throughout the Internet and exploration of the wealth of publicly available information through Gopher.

A new WWW server is available containing the *Arabidopsis* Information Service. This experimental server is being developed by the AAtDB Project and provides access to 25 years of the *Arabidopsis* Information Service in a hypertext format with figures and tables included as scanned images. The WWW specific access code, which is called a Universal Resource Locator (URL), for this service is as follows: "http://weeds.mgh.harvard.edu/index.html".

For more information on AAtDB or the AAtDB Research Companion, contact John Morris via e-mail at curator@frodo.mgh.harvard.edu, FAX at 617-726-6893, or postal mail at Department of Molecular Biology, Massachusetts General Hospital, Boston, Massachusetts 02114.

**The AIMS Database**

The AIMS database, located at Michigan State University, is the information management resource of the *Arabidopsis* Biological Resource Center at Ohio State. AIMS provides information on all seed and DNA stocks maintained by the ABRC. In addition, on-line ordering of any stock is also provided. The AIMS database includes:

- Genetic map of *Arabidopsis*
- Literature references
- Cosmid and lambda RFLP clones
- YAC libraries from several sources
- Variety of seed stocks
- Bibliographic information on *Arabidopsis* researchers
- Scanned images of RFLP autoradiograms, mutant plants, and restriction enzyme digests of clones used as RFLP probes

Specialized software available on the AIMS central computer provides access to the AIMS relational database. This software can be used to search the variety of information contained within the database. Once a clone or seed stock of interest is found, an option can be selected to create an order for that item from the ABRC. The primary method available to access AIMS is through Telnet. AIMS is reached by using Telnet to the host name aims.msu.edu with the user name guest1; no password is required. The AIMS database can then be viewed in two forms by selecting options on the initial display. The first is for a Telnet terminal session where only the keyboard is used. The second form requires access to the X-windows environment on your computer. With either form of AIMS, an account is required. To obtain an account send your name, address, phone/FAX telephone number, and e-mail address in an electronic mail message to aims-manager@aims.msu.edu. A tutorial describing the use of the AIMS system is available on request from the AIMS manager. The tutorial is also available through Gopher from the Internet host name gopher.cps.msu.edu. Once connected, move into the folder called "Department of Computer Science," and then the folder called "AIMS." Besides the tutorial, the AIMS Gopher provides text files of the ABRC catalog as well as images of plants and gels mentioned above. This information is also available through anonymous FTP from the host name aims.msu.edu.

For more information on AIMS or ABRC, contact Randy Scholl through electronic mail at arabidopsis+@osu.edu, FAX at 614-292-0603, or postal mail at *Arabidopsis* Biological Resource Center, Ohio State University, 1735 Neil Avenue, Columbus, Ohio 43210.

**GENERAL MOLECULAR BIOLOGICAL INFORMATION**

The number of information resources for DNA, protein, and genome information is becoming quite large. Many of the resources are included in a variety of the major Gopher or WWW servers. Therefore, by simply installing a Gopher or WWW client, a wealth of molecular biology information and software will be at your fingertips. A few significant Gopher and WWW servers are listed below that provide information on plant genomic or general molecular biological information.

**Gopher Servers**

The resource's title and host name of interesting Gopher servers are included below. For UNIX and VMS Gopher clients start the software with the command "gopher *hostname*," where *hostname* is replaced by one of the host names shown below. For the Macintosh TurboGopher software, double click on the TurboGopher icon to start the software, then select "Another Gopher..." from the File menu and enter one of the host names. These servers also provide links to other Gopher services of interest.

*Chlamydomonas* Genetics Center, Duke University
atlas.acpub.duke.edu

EMBnet BioInformation Resource (Finland)
gopher.csc.fi

GRIN, National Genetic Resources Program, USDA-ARS
gopher.ars-grin.gov

Indiana University Biology Software and Database Archive
ftp.bio.indiana.edu

Maize Genome Database, MaizeDB, University of Missouri
teosinte.agron.missouri.edu

*Saccharomyces* Genome Information
genome-gopher.stanford.edu

The *Triticeae* genome, GrainGenes, Cornell University
greengenes.cit.cornell.edu

USDA National Agricultural Library Plant Genome
probe.nalusda.gov

**WWW Servers**

Resource title and Universal Resource Locator (URL) are listed below for some of the notable WWW servers providing information on genetics or molecular biology. For the Mosaic software, select Open or Open URL and enter the URL shown below. All of these servers provide links to many other WWW servers.

Agricultural Genome, National Agricultural Library
http://probe.nalusda.gov:8000/index.html

The Australian National Botanical Gardens
http://155.187.10.12:80/anbg.html

The ExPASy Molecular Biology Server, Geneva, Switzerland
http://expasy.hcuge.ch/

Harvard Biological Laboratories - Genome Research
http://golgi.harvard.edu/homepage.genome

The Johns Hopkins University BioInformatics
http://www.gdb.org/hopkins.html

*Saccharomyces* Genome Information
genome-www.stanford.edu

National Center for Biotechnology Information
http://www.ncbi.nlm.nih.gov/

**REFERENCES**

Cherry, J.M., S.W. Cartinhour, and H.M. Goodman. 1992. AAtDB, An *Arabidopsis thaliana* database. *Plant Mol. Biol. Rep.* **10:** 308–309, 409–410.

Dern, D.P. 1993. *The internet guide for new users.* McGraw-Hill, New York.

Durbin, R. and J. Thierry-Mieg. 1991. *A C.* elegans *database, documentation code and data* (available through the Internet from Anonymous FTP servers at lirmm.lirmm.fr, cele.mrc-lmb.cam.ac.uk and ncbi.nlm.nih.gov.).

Engst, A.C. 1993. *Internet starter kit for Macintosh.* Hayden Books, Carmel, Indiana.

Estrada, S. 1993. *Connecting to the Internet.* O'Reilly and Associates, Sebastopol, California.

Krol, E. 1992. *The whole internet user's guide and catalog.* O'Reilly and Associates, Sebastopol, California.

# APPENDIX B
# Genetic Variations of
## *Arabidopsis thaliana*

**Elliot M. Meyerowitz**
Division of Biology
California Institute of Technology
Pasadena, California 91125

**Hong Ma**
Cold Spring Harbor Laboratory
Cold Spring Harbor, New York 11724

This Appendix is an updated version of the Green Book, a list of *Arabidopsis thaliana* genetic variants published by Meyerowitz and Pruitt in 1984. Since that time, many old mutations have been characterized in detail; new mutations have been isolated and analyzed, thus, new genes have been defined. Furthermore, a number of the genetically defined genes have been cloned. The purpose of this Appendix is to provide a list of published *Arabidopsis* mutations and brief summaries of relevant information on them. When the genes have been cloned, the molecular results are also included.

The genetic nomenclature used here is based on the recommendations of the *Arabidopsis* Nomenclature Committee at the 1987 Third International *Arabidopsis* Meeting, East Lansing, Michigan. The following is a brief summary: (1) genotypes are italicized or underlined; (2) the wild-type genotype is capitalized; (3) mutant genotypes are in lower case; (4) gene symbols consist of three letters with a number; for example, *CLV2* for *CLAVATA 2*; existing gene symbols with two letters and/or without a number continue to be used to avoid confusion; (5) alleles are denoted with a number, separated from the gene symbol by a dash, regardless of the presence or absence of a number in the gene symbol; for example, *ag-1* for allele 1 of *agamous*, and *ch1-1* for allele 1 of *chlorina* 1; (6) phenotypes are designated by the gene symbol, which is not italicized but has the first letter capitalized.

Unless otherwise noted, the mutations listed here were induced by mutagenic treatment of seeds, they are nuclear and recessive to wild type, and their locations are from the Koornneef and Meinke genetic map in the chapter by Koornneef, this volume. Arabidopsis Information Ser-

*Arabidopsis*
© 1994 Cold Spring Harbor Laboratory Press 0-87969-428-9/94 $5 + .00

vice, a publication that was the original reference for many of the earlier mutations discovered in *Arabidopsis*, is abbreviated *AIS*. This newsletter spanned the years 1964 to 1990, in 27 volumes. Volume 24, 1987, is a comprehensive list and description of the characteristics and origins of the many available wild-type strains (ecotypes) of *Arabidopsis*. We do not repeat this information here.

In the preface to the 1984 version of this list, the authors stated "we have undoubtedly omitted important information, and have probably made mistakes in description and attribution of mutants. We welcome any correction or new information, and will strive to include them in future editions." We want to emphasize that this is true of this edition as well, and we look forward to comments and corrections from our colleagues.

a: see *alb1*

A11: see *tt5*

A26: see *aba-1*

A73: see *aba-4*

A482: see *ga1-8*

AII: see *abi1-1*

**aba: abscisic acid deficient**

LOCATION: 5-97.5

ORIGIN: induced with EMS in Landsberg *erecta gl1 ga1*

ALLELES: *aba-1* (line A26); *aba-3* (line G4); *aba-4* (line A73)

REFERENCES: Koornneef et al. (1980) *AIS 17:* 99-102; Koornneef et al. (1982) *Theor. Appl. Genet. 61:* 385-393; Karssen et al. (1983) *Planta 157:* 158-165; Duckham et al. (1989) *J. Exp. Botany 40:* 901-905; Koornneef et al. (1989) *Plant Physiol. 90:* 463-469; Duckham et al. (1991) *Plant, Cell and Environment 14:* 601-606; Rock and Zeevaart (1991) *Proc. Natl. Acad. Sci. USA 88:* 7496-7499; Rock et al. (1992) *Plant Physiol. 100:* 1796-1801; Finkelstein (1993) *MGG 238:* 401-408

PHENOTYPE: second-site revertant of *ga1;* little or no endogenous ABA; seeds germinate in mature silique on plant if plants are in a humid atmosphere; plants are wilted and strongly reduced in vigor; fresh seeds are not dormant and therefore germinate at high frequency (unlike wild type); *aba-1* and *aba-3* plants deficient in epoxidation of zeaxanthin, a presumed precursor of ABA, but *aba-1* plants are normal in their ability to convert ABA-aldehyde to ABA, indicating that the *aba-1* defect is before this last step in ABA biosynthesis; reduced fluorescence yields and reduced thylakoid stacking

OTHER INFORMATION: the *gl1* mutation was removed in backcrosses to wild type; in combination with a nongerminating *ga1* mutant allele gives seeds that germinate but grow to dwarfed plants that are more dwarfed than plants homozygous for germinating *ga1* alleles

aberrant testa shape: see *ats*

**abi1: abscisic acid insensitive 1**

LOCATION: 4-55.0

ORIGIN: induced with EMS in Landsberg *erecta*

ALLELES: *abi1-1* (line AII); *abi1-2* (line CVI)

REFERENCES: Koornneef et al. (1984) *Physiol. Plant. 61:* 377-383; Koornneef and Hanhart (1984) *AIS 21:* 5-9; Koornneef et al. (1989) *Plant Physiol. 90:* 463-469; Finkelstein and Somerville (1990) *Plant Physiol. 93:* 1172-1179; Finkelstein (1993) *MGG 238:* 401-408; Ooms et al. (1993) *Plant Physiol. 102:* 1185-1191; Finkelstein (1994) *Plant Physiol. 105:* 1203-1208; Leung et al. (1994) *Science 264:* 1448-1452; Meyer et al. (1994) *Science 264:* 1452-1455

PHENOTYPE: dominant over wild type; germinates in the presence of 10 μM ABA (unlike wild type); fresh seeds germinate at high frequency (also unlike wild type); plants wilt in conditions of mild water stress

GENE: isolated positionally by Leung et al. (1994) and Meyer et al. (1994);

encodes a protein of 434 amino acids, with a putative $Ca^{++}$-binding site near the N-terminus and similarity (35% identity) to protein phosphatases (2C); the *abi1-1* mutation is a $G_{970}$ to A transition ($Gly_{180}$ to Asp)

### *abi2: ab*scisic acid *i*nsensitive 2

**LOCATION:** chromosome 5 (D. Meinke, pers. comm.)

**ORIGIN:** induced with EMS in Landsberg *erecta*

**ALLELE:** *abi2-1* (line EII)

**REFERENCES:** Koornneef et al. (1984) *Physiol. Plant. 61:* 377-383; Finkelstein and Somerville (1990) *Plant Physiol. 93:* 1172-1179; Finkelstein (1993) *MGG 238:* 401-408; Ooms et al. (1993) *Plant Physiol. 102:* 1185-1191; Finkelstein (1994) *Plant Physiol. 105:* 1203-1208

**PHENOTYPE:** germinates in 10 µM ABA; fresh seeds germinate at high frequency; young plants are wilted

**OTHER INFORMATION:** wilting enhanced by 10 µM ABA spray

### *abi3: ab*scisic acid *i*nsensitive 3

**LOCATION:** 3-37.5

**ORIGIN:** induced with EMS in Landsberg *erecta* (*abi3-1,2*) or in Columbia (*abi3-3*); *abi3-4* induced with EMS in *abi3-1* (Landsberg *erecta*); *abi3-5* induced with DEB in Landsberg *erecta*

**ALLELES:** *abi3-1* (line CIV) (Koornneef et al. 1984); *abi3-2* (line DV) (Koornneef et al. 1984); *abi3-3* (Nambara et al. 1992); *abi3-4* (Giraudat et al. 1992); *abi3-5* (Ooms et al. 1993)

**REFERENCES:** Koornneef et al. (1984) *Physiol. Plant. 61:* 377-383; Koornneef and Hanhart (1984) *AIS 21:* 5-9; Koornneef et al. (1989) *Plant Physiol. 90:* 463-469; Finkelstein and Somerville (1990) *Plant Physiol. 93:* 1172-1179; Nambara et al. (1992) *Plant J. 2:* 435-441; Giraudat et al. (1992) *Plant Cell 4:* 1251-1261; Finkelstein (1993) *MGG 238:* 401-408; Ooms et al. (1993) *Plant Physiol. 102:* 1185-1191

**PHENOTYPE:** *abi3-1* and *abi3-2:* germinate in the presence of 10 µM ABA *abi3-3:* germinates in the presence of uniconazol, a GA biosynthetic inhibitor; defective in seed development and storage protein accumulation; seeds germinate well when green and immature; seeds of the extreme alleles (*abi3-3,4,5*) are still green 20 days after fertilization and do not germinate after they are fully desiccated

**GENE:** cloned by Giraudat et al. (1992); predicted protein (720 amino acids, 79.5 kD) shows similarity with the VP1 protein from maize (McCarty et al. 1991, *Cell 66:* 895-905) and is a potential transcription factor with P/Q/N- and S/T-rich regions; the most severe allele, *abi3-4,* is a C to T change, resulting in a TAA stop codon truncating the protein by 40%

**OTHER INFORMATION:** not wilted, unlike *abi1* and *abi2*

### *abi4: ab*scisic acid *i*nsensitive 4

**LOCATION:** chromosome 2 (13.0 ± 4.0 south of *py*) (Finkelstein 1994)

**ORIGIN:** induced by γ-ray in Columbia

**ALLELE:** one

**REFERENCE:** Finkelstein (1994) *Plant J. 5:* 765-771

**PHENOTYPE:** germinates in the presence of 5 μM ABA; dormancy similar to the Columbia wild type; accumulation of mRNA of the late-embryogenesis abundant *AtEM6* is reduced; double mutant with *abi1*, but not with *abi2* or *abi3*, is less sensitive to ABA than single mutants

### *abi5: abscisic acid insensitive 5*

**LOCATION:** chromosome 2 (between *hy1* and *py*) (Finkelstein 1994)

**ORIGIN:** isolated from T-DNA insertional populations in Wassilewskija (Ws)

**ALLELES:** *abi5-1; abi5-2*

**REFERENCE:** Finkelstein (1994) *Plant J. 5:* 765-771

**PHENOTYPE:** germinates in the presence of 3 μM ABA; dormancy similar to the Ws wild type; accumulation of mRNA of the late-embryogenesis abundant *AtEM6* is greatly reduced; double mutant with *abi1* or *abi2*, but not with *abi3*, is less sensitive to ABA than single mutants

abscisic acid deficient: see *aba*

abscisic acid insensitive: see *abi*

acaulis: see *acl*

accelerated cell death: see *acd1*

### *acd1: accelerated cell death 1*

**LOCATION:** 3-68.2

**ORIGIN:** induced with EMS in Columbia

**ALLELES:** *acd1-12; acd1-20*

**REFERENCE:** Greenberg and Ausubel (1993) *Plant J. 4:* 327-341

**PHENOTYPE:** accelerated cell death when exposed to any of a number of stimuli, including virulent and avirulent *Pseudomonas syringae* pv *maculicola* or pv. *tomato* pathogens, and ethylene; susceptible to opportunistic pathogens; sensitive to mechanical stress and develops necrotic lesions as the plant ages

### *acd2: accelerated cell death 2*

**LOCATION:** 4-67.6

**REFERENCE:** mentioned in Greenberg and Ausubel (1993) *Plant J. 4:* 327-341 as in preparation

acetohydroxy acid synthase: AHAS, see *csr1*

acetolactate synthase: ALS, see *csr1*

### *acl1: acaulis 1*

**LOCATION:** 4-55.1

**ORIGIN:** *acl1-1* induced by X-ray in Erhlangen; *acl1-2* and *acl1-3* induced with EMS in Columbia

**ALLELES:** *acl1-1* (G. Rédei); *acl1-2,3* (Tsukaya et al. 1993)

**REFERENCE:** Tsukaya et al. (1993) *Development 118:* 751-764

**PHENOTYPE:** flower stalks (stems) absent or much reduced in length; number of flowers reduced (*acl1-1* and *acl1-2* produce only two to three flowers on primary inflorescence); defective in cell elongation; wild-type phenotype at 28ºC

### *acl2: acaulis 2*

**LOCATION:** not mapped

ORIGIN: induced with EMS in Columbia
ALLELE: *acl2-1*
REFERENCE: Tsukaya et al. (1993) *Development 118:* 751-764
PHENOTYPE: similar to *acl1*; not characterized in detail

### *act1*: glycerol-3-phosphate *acyl*transferase deficient 1

LOCATION: 1-35.2
ORIGIN: induced with EMS in Columbia
ALLELES: *act1-1* (line JB25); *act1-2* (line LK8); and those in lines JB3 and JB28
REFERENCES: Kunst et al. (1988) *Proc. Natl. Acad. Sci. USA 85:* 4143-4147; Kunst et al. (1989) *Plant Physiol. 90:* 846-853; Nishida et al. (1993) *Plant Mol. Biol. 21:* 267-277
PHENOTYPE: lack glycerol-3-phosphate acyltransferase activity; deficient in C16:3 lipids; altered chloroplast structure and function
GENE: cloned by Nishida et al. (1993); contains 12 introns, one of which is in the 3' untranslated region; the predicted protein has 459 amino acids; the first 65 amino acids are characteristic of those in a plastid transit peptide; protein expressed in *E. coli* has the expected enzyme activity

adenine phosphoribosyl transferase deficient: see *apt*

### *adg1*: ADPglucose pyrophosphorylase deficient 1

LOCATION: not mapped
ORIGIN: induced with EMS in Columbia
ALLELES: *adg1-1* (line TL25: TL255 was derived from TL25 by backcrossing); *adg1-2* (line TL24)
REFERENCES: Lin et al. (1988) *Plant Physiol. 86:* 1131-1135; Caspar et al. (1989) *Proc. Natl. Acad. Sci. USA 86:* 5830-5833; Murgia et al. (1993) *Am. J. Bot. 80:* 824-838; Caspar, this volume
PHENOTYPE: *adg1-1* has no detectable ADPglucose pyrophosphorylase activity and no detectable starch in leaves; iodine staining fails to detect any starch in root, petiole, floral stem and flower; growth rate normal in continuous light but greatly reduced in 12 hrs-day/12 hrs-night photoperiod
OTHER INFORMATION: both of the ADPglucose pyrophosphorylase subunits are absent from the *adg1-1* mutant and the heterozygote has normal specific activity for the enzyme, suggesting that *adg1-1* affects a regulatory protein

### *adg2*: ADPglucose pyrophosphorylase deficient 2

LOCATION: not mapped
ORIGIN: induced with EMS in Columbia
ALLELES: *adg2-1* (line TL46: TL46BC1 was derived from TL46 by backcrossing); *adg2-2* (line TL3)
REFERENCES: Lin et al. (1988) *Plant Physiol. 88:* 1175-1181; Caspar et al. (1989) *Proc. Natl. Acad. Sci. USA 86:* 5830-5833; Caspar, this volume
PHENOTYPE: *adg2-1* has about 5% of the normal level of ADPglucose pyrophosphorylase activity and about 40% of the normal level of starch in leaves; growth rate normal in continuous light and in 12 hrs-day/12 hrs-night photoperiod

**OTHER INFORMATION:** one (54 kD) of the two ADPglucose pyrophosphorylase subunits is absent from the *adg2-1* mutant and the heterozygote has 50–60% of the normal level of enzyme activity, suggesting that *ADG2* is the structural gene for the 54-kD subunit

## *adh*: *a*lcohol *deh*ydrogenase

**LOCATION:** 1-113.6

**ORIGIN:** induced with EMS (Jacobs et al. 1988) in Bensheim (Be-0) ecotype (11 alleles) or in Tossa de mar (Ts-1) ecotype (21 alleles)

**ALLELES:** lines R001 through R004 and R006

**REFERENCES:** Dolferus and Jacobs (1984) *Biochem. Genet. 22:* 817-837; Chang and Meyerowitz (1986) *Proc. Natl. Acad. Sci. USA 83:* 1408-1412; Jacobs et al. (1988) *Biochem. Genet. 26:* 105-122; Dolferus et al. (1990) *MGG 224:* 297-302

**PHENOTYPE:** resistant to allyl alcohol; line R002 has no detectable ADH mRNA or protein; line R006 has close to normal amounts of mRNA and protein but no activity

**GENE:** cloned by Chang and Meyerowitz (1986) from Landsberg *erecta* and by Dolferus et al. (1990) from Bensheim; both of these wild-type alleles encode the same electrophoretic variant (ADH-A, Dolferus and Jacobs 1984); the mutant alleles in lines R002 and R006 were isolated by Dolferus et al. (1990); R002 has a single C to T change ($Gln_{106}$ to Stop), leading to a truncated protein that is apparently unstable; R006 has a single G to A change, causing a change from a highly conserved $Cys_{105}$ to Tyr; the Cys is known to be required for binding Zn in active enzyme

**OTHER INFORMATION:** three electrophoretic variants were observed among different ecotypes: S, slow; F, fast; A, superfast (Dolferus and Jacobs 1984)

ADPglucose pyrophosphorylase deficient: see *adg1* and *adg2*

## *ag*: *ag*amous

**LOCATION:** 4-42.1

**ORIGIN:** *ag-1* and *ag-3* induced with EMS in Landsberg *erecta*; *ag-2* induced by T-DNA insertion in Wassilewskija (Ws)

**ALLELES:** *ag-1* (Koornneef 1980); *ag-2* (Feldmann et al. 1989; Yanofsky et al. 1990); *ag-3* (Bowman et al. 1991a)

**REFERENCES:** Koornneef (1980) *AIS 17:* 11-18; Bowman et al. (1989) *Plant Cell 1:* 37-52; Feldmann et al. (1989) *Science 243:* 1351-1354; Meyerowitz et al. (1989) *Development 106:* 209-217; Yanofsky et al. (1990) *Nature 346:* 35-39; Bowman et al. (1991a) *Development 112:* 1-20; Bowman et al. (1991b) *Plant Cell 3:* 749-758; Drews et al. (1991) *Cell 65:* 991-1002; Mandel et al. (1992) *Cell 71:* 133-143; Mizukami and Ma (1992) *Cell 71:* 119-131; Bowman et al. (1993) *Development 119:* 721-743; Clark et al. (1993) *Development 119:* 397-418; Weigel and Meyerowitz (1993) *Science 261:* 1723-1726

**PHENOTYPE:** sterile; petals develop where stamens normally do; another flower (four sepals and ten petals) replaces pistil and this repeats a few times so that the total number of organs is more than 70

**GENE:** cloned by Yanofsky et al. (1990); encodes a MADS-box protein, putative transcription factor; member of the MADS-box gene family (Ma et al. 1991, *Genes Dev. 5:* 484-495); *ag-1* is a G to A change at the acceptor site of intron 4; *ag-2* is a T-DNA insertion in intron 2; *AG* is most likely the homolog of the *Antirrhinum majus* floral gene *PLENA* (Carpenter and Coen 1990, *Genes Dev. 4:* 1483-1493; Bradley et al. 1993, *Cell 72:* 85-95); ectopic expression of *AG* leads to *ap2*-like floral phenotypes

**OTHER INFORMATION:** may be allelic with multipetala, see Conrad (1971) *Biol. Zentralbl. 90:* 137-144; probably allelic with the earliest described *Arabidopsis* mutant, see Braun (1873) *Gesell. Naturforsch. Freunde Z. Berlin,* 75.

agamous: see *ag*

***agr1: agr*avitropic 1, allelic with *wav6***

**LOCATION:** not mapped

**ORIGIN:** induced with EMS in Landsberg *erecta*

**ALLELES:** *agr1-1* (line 932), *agr1-2* (line 857), *agr1-3* (line 858); *wav6-52*

**REFERENCES:** Bell and Maher (1990) *MGG 220:* 289-293; Okada and Shimura (1990) *Science 250:* 274-276

**PHENOTYPE:** agravitropic; non-auxin resistant

*wav6-52:* lack of root tip rotation and wavy growth of root on angled agar surface; gravitropism normal

AHAS: acetohydroxy acid synthase, see *csr1*

***ain1:* 1-aminocyclopropane-1-carboxylate *in*sensitive 1**

**LOCATION:** 1-69.7

**ORIGIN:** induced with EMS in the Columbia C24 ecotype

**ALLELES:** *ain1-1* through *ain1-6*

**REFERENCE:** Van Der Straeten et al. (1993) *Plant Physiol. 102:* 401-408

**PHENOTYPE:** isolated as insensitive to 500 μM 1-aminocyclopropane-1-carboxylate (ACC); insensitive to ethylene; *ain1* plants have larger than normal rosettes and slightly delayed bolting; *ain1-1* plants have reduced gravitropism

AL63: see *fad2-4*

***alb1: alb*ina 1**

**LOCATION:** 1-11.7

**ORIGIN:** induced with 10 mM EMS and 0.2% colchicine for 24 hrs in Landsberg *erecta* (*alb1-1*) or induced in race En-2 (Enkheim) (*alb1-v*)

**ALLELES:** *alb1-1* (= *a* or *alb-1*) (van der Veen et al. 1973); *alb1-v* (= *alb-1*^v, V157, or chlorina V157) (Röbbelen 1972)

**REFERENCES:** Röbbelen (1972) *AIS 9:* 21-25; van der Veen et al. (1973) *AIS 10:* 11-12; Kranz and Scheidemann (1978) *AIS 15:* 31-34

**PHENOTYPE:** *alb1-1:* white embryo, white seedling, lethal

*alb1-v:* uniform yellow-green color, viable

**OTHER INFORMATION:** *alb1-1* was recovered in tetraploid; +/*alb1-1*/*alb1-1*/*alb1-1* tetraploid light green, +/+/*alb1-1*/*alb1-1* normal

***alb2: alb*ina 2**

LOCATION: 5-14.5

ORIGIN: induced with 0.05 mM *N*-nitroso-*N*-methylurea in S96 (derived from a cross between Li-2 and Dijon)

SYNONYM: M 4-6-18

REFERENCES: Fischerova (1975) *Biol. Plantarum 17:* 182-188; Relichova (1976) *AIS 13:* 25-28

PHENOTYPE: white embryo and seedling, lethal

### *alb3: alb*ina 3

LOCATION: chromosome 2, 9.5 cM below *er*

ORIGIN: induced by *Ds* insertion in Landsberg *erecta*

ALLELE: *alb3-1*

REFERENCE: Long et al. (1993) *90:* 10370-10374

PHENOTYPE: albino

albina: see *alb*

### *ali: ali*en

LOCATION: not mapped

ORIGIN: induced with EMS in Landsberg *erecta*

ALLELE: one

REFERENCE: Hülskamp et al. (1994) *Cell 76:* 555-566

PHENOTYPE: trichomes distorted

alien: see *ali*

ALS: acetolactate synthase, see *csr1*

altered meristem program: see *amp1*

1-aminocyclopropane-1-carboxylate insensitive: see *ain1*

### *amp1: altered meristem program 1*

LOCATION: chromosome 3, linked to *cer7* (Chaudhury et al. 1993)

ORIGIN: induced with EMS in Columbia

ALLELES: *amp1-1; amp1-2*

REFERENCE: Chaudhury et al. (1993) *Plant J. 4:* 907-916

PHENOTYPE: up to 20% of seedlings are tricots or tetracots; dark-grown seedlings have light-grown morphology; leaf initiated and flowering is accelerated; 6x normal amount of cytokinin; bushy plant and more rosette leaves

### *amt1:* α-*methyltryptophan resistant 1*

LOCATION: 5-11.3

ORIGIN: induced with EMS in Columbia

REFERENCE: Kreps and Town (1992) *Plant Physiol. 99:* 269-275

PHENOTYPE: dominant over wild type; selected on the basis of resistance to α-methyltryptophan; tryptophan levels 5x normal; anthranilate synthase about 2x normal

### *an: an*gustifolia

LOCATION: 1-0.4

ORIGIN: original allele (Rédei 1962) induced by 12 kr X-ray to presoaked Landsberg seeds; others (Koornneef et al. 1982; Hülskamp et al. 1994) induced with EMS in Landsberg *erecta*

REFERENCES: Rédei (1962) *Z. Vererbungsl. 93:* 164-170; Lee-Chen and Steinitz-Sears (1967) *Can. J. Genet. Cytol. 9:* 381-384; Koornneef et al. (1982) *Mut. Res. 93:* 109-123; Hülskamp et al. (1994) *Cell 76:* 555-566

PHENOTYPE: narrow leaves, slightly twisted seed pods; new alleles have trichomes with only two branches

OTHER INFORMATION: autonomous in genetic mosaics

angustifolia: see *an*

## *ap1:* apetala 1

LOCATION: 1-99.3

ORIGIN: induced with EMS in Landsberg *erecta,* except for *ap1-7* induced with EMS in Columbia and *ap1-8* induced with EMS or by X-ray in Estland or Limburg

ALLELES: see Bowman et al. (1993) for allele assignment: *ap1-1* (McKelvie 1963; Koornneef et al. 1982; Irish and Sussex 1990); *ap1-2,3,4,5,7* (lines S1504, S1350, S1735, S1732, S2723, respectively) (isolated by J. Alvarez and D. Smyth; see Mandel et al. 1992; Bowman et al. 1993); *ap1-6* (line La77313/2) (isolated by Paddy Maher; see Bowman et al. 1993); *ap1-8* (axillaris) (McKelvie 1962); *ap1-9* (line LC500) (L. Comai, see Bowman et al. 1993); *ap1-10* through *ap1-14* (Schultz and Haughn 1993); possible additional alleles were induced by McKelvie (1963); other alleles induced by Koornneef et al. (1982)

REFERENCES: McKelvie (1962) *Radiat. Bot. 1:* 233-241; McKelvie (1963) *Radiat. Bot. 3:* 105-123; Koornneef et al. (1982) *Mut. Res. 93:* 109-123; Irish and Sussex (1990) *Plant Cell 2:* 741-753; Bowman (1992) *Flowering Newsletter 14:* 7-19; Weigel et al. (1992) *Cell 69:* 843-859; Mandel et al. (1992) *Nature 360:* 273-277; Bowman et al. (1993) *Development 119:* 721-743; Clark et al. (1993) *Development 119:* 397-418; Shannon and Meeks-Wagner (1993) *Plant Cell 5:* 639-655; Schultz and Haughn (1993) *Development 119:* 745-765; Weigel and Meyerowitz (1993) *Science 261:* 1723-1726; Gustafson-Brown et al. (1994) *Cell 76:* 131-143

PHENOTYPE: *ap1-1:* homeotic conversion of sepals to bracts and formation of secondary flowers in the axil of the transformed sepals; petals are absent; tertiary flowers are found within the secondary flower

*ap1-2:* mosaic floral organs and single axillary flower are present in the first and second whorl

*ap1-3:* second whorl organs are petal-stamen mosaic or filamentous structures

more detailed descriptions of these and other *ap1* mutants can be found in Bowman (1992)

GENE: *AP1* gene was cloned by Mandel et al. (1992); *AP1* is a member of the MADS-box gene family; multiple mutant alleles were sequenced; *AP1* expression is negatively regulated by *TFL1* and *AG* (Gustafson-Brown et al. 1994); *AP1* is most likely the homolog of the *Antirrhinum majus* floral gene *SQUAMOSA* (Huijser et al. 1992, *EMBO J. 11:* 1239-1249)

## *ap2:* apetala 2, allelic with *flo2, flo3,* and *flo4*

**LOCATION:** 4-67.5

**ORIGIN:** induced with EMS in Landsberg *erecta* (*ap2-1,2,3,4,8,9*) or Columbia (*ap2-5,6,7*); *ap2-10* induced by T-DNA insertion in Columbia C24 (*gl1/gl1*) background

**ALLELES:** *ap2-1* (Koornneef et al. 1980); *ap2-2* (Bowman et al. 1989); *ap2-3*, *ap2-4* (lines Fl-40, Fl-48, respectively) (Komaki et al. 1988); *ap2-5, ap2-6, ap2-7* (*flo2,3,4*, respectively) (Kunst et al. 1989); *ap2-8, ap2-9* (Bowman et al. 1991a); *ap2-10* (line T10, previously called *flower1*) (Boerjan et al. 1992; Jofuku et al. 1994)

**REFERENCES:** Koornneef et al. (1980) *AIS 17:* 11-18; Haughn and Somerville (1988) *Dev. Genet. 9:* 73-89; Komaki et al. (1988) *Development 104:* 195-203; Bowman et al (1989) *Plant Cell 1:* 37-52; Kunst et al. (1989) *Plant Cell 1:* 1195-1208; Meyerowitz et al. (1989) *Development 106:* 209-217; Bowman et al. (1991a) *Development 112:* 1-20; Bowman et al. (1991b) *Plant Cell 3:* 749-758; Drews et al. (1991) *Cell 65:* 991-1002; Boerjan et al. (1992) *Int. J. Dev. Biol. 36:* 59-66; Mandel et al. (1992) *Cell 71:* 133-143; Mizukami and Ma (1992) *Cell 71:* 119-131; Bowman et al. (1993) *Development 119:* 721-743; Clark et al. (1993) *Development 119:* 397-418; Shannon and Meeks-Wagner (1993) *Plant Cell 5:* 639-655; Schultz and Haughn (1993) *Development 119:* 745-765; Jofuku et al. (1994) *Plant Cell 6:* 1211-1225; Modrusan et al. (1994) *Plant Cell 6:* 333-349

**PHENOTYPE:** sepals converted to leaves (*ap2-1*), carpels (*ap2-2, ap2-3, ap2-4, ap2-5, ap2-6, ap2-7, ap2-8*), or missing (*ap2-2, ap2-6, ap2-7, ap2-8*); petals converted to stamens or staminoid petals (*ap2-1, ap2-5*) or missing (*ap2-2, ap2-3, ap2-4, ap2-7, ap2-8*); stamens normal (*ap2-1*) or reduced in number (*ap2-2, ap2-3, ap2-4, ap2-6, ap2-7*); carpels normal or sometimes unfused (*ap2-2, ap2-3, ap2-7*); mutants show more severe phenotypes in later flowers than in earlier ones; mutant phenotypes are more pronounced at higher temperature; ovule development abnormal, form carpel-like structures (*ap2-6,7*)

**GENE:** cloned by Jofuku et al. (1994); expressed in all floral organs, the inflorescence meristem, and vegetative organs; encodes a protein of 432 amino acids (48 kD) with a serine-rich acidic region and a putative nuclear localization signal; the AP2 protein has two copies of a 68-amino acid repeat (AP2 repeat) that contains a region capable of forming an amphipathic $\alpha$-helix; three mutations have been sequenced: *ap2-1* has a $Gly_{251}$ to Ser change; *ap2-5* has two alterations, $Gly_{159}$ to Glu and $Gln_{420}$ to Glu; *ap2-10* is an insertion into the exon coding for a part of the second copy of the AP2 repeats

**OTHER INFORMATION:** the expression of *AG* was observed to expand in *ap2* mutants, suggesting a negative effect of AP2 on *AG* expression; ectopic expression of *AG* produces an *ap2* phenotype

## *ap3:* apetala 3

**LOCATION:** 3-81.4

**ORIGIN:** induced with EMS in Landsberg *erecta*

**ALLELES:** *ap3-1* (Koornneef, see Bowman et al. 1989); *ap3-3,4,5* (Jack et al.

1992)

REFERENCES: Bowman et al. (1989) *Plant Cell 1:* 37-52; Meyerowitz et al. (1989) *Development 106:* 209-217; Bowman et al. (1991) *Development 112:* 1-20; Coen and Meyerowitz (1991) *Nature 353:* 31-37; Schultz et al. (1991) *Plant Cell 3:* 1221-1237; Bowman et al. (1992) *Development 114:* 599-615; Jack et al. (1992) *Cell 68:* 683-697; Clark et al. (1993) *Development 119:* 397-418; Weigel and Meyerowitz (1993) *Science 261:* 1723-1726

PHENOTYPE: conversion of petals to sepals and stamens to carpels; some alleles cause reduction of number of second whorl organs

GENE: cloned by Jack et al. (1992); homolog of the *Antirrhinum majus DEF A* gene (Sommer et al. 1990, *EMBO J. 9:* 605-613); member of the MADS-box family; several alleles sequenced by Jack et al. (1992)

OTHER INFORMATION: phenotype similar to that of *pi* mutants; AP3 and PI proteins probably function as a heterodimer to regulate gene expression; *AP3* expression regulated by the products of the *AP1, LFY,* and *SUP* genes

apetala: see *ap*

**apt: adenine *p*hosphoribosyl *t*ransferase deficient**

LOCATION: not mapped

ORIGIN: induced with EMS in Columbia

ALLELES: *apt-1* (line BM1); *apt-2* (line BM2); *apt-3* (line BM3)

REFERENCES: Moffatt and Somerville (1988) *Plant Physiol. 86:* 1150-1154; Regan and Moffatt (1990) *Plant Cell 2:* 877-889

PHENOTYPE: selected on the basis of resistance to 0.1 mM 2,6-diaminopurine; *apt-1, 2,* and *3* homozygotes have 10–15%, 2%, and 1%, respectively, of the normal level of adenine phosphoribosyl transferase activity; *apt-3* homozygote has an appreciably slower than normal growth rate; *apt-1* plants have 2% of normal fertility (partial male sterility), and *apt-2* and *apt-3* plants are completely male sterile; the male sterility is due to defects in microsporogenesis after the tetrad stage in later stages of microspore formation; empty microspores are seen in mutant flowers

AR119: see *trp1*

**ara1: sensitive to exogenous L-*arabinose***

LOCATION: 4-36.6

ORIGIN: induced with EMS in Columbia

REFERENCE: Dolezal and Cobbett (1991) *Plant Physiol. 96:* 1255-1260

PHENOTYPE: growth inhibited by L-arabinose and has reduced levels of arabinose kinase activity

**arc1: *a*ccumulation and *r*eplication of *c*hloroplast 1**

LOCATION: not mapped

ORIGIN: induced with EMS in Landsberg *erecta*

REFERENCE: Pyke and Leech (1992) *Plant Physiol. 99:* 1005-1008

PHENOTYPE: increased number of chloroplasts in unit mesophyll cell area; reduced chloroplast size

**arc2: *a*ccumulation and *r*eplication of *c*hloroplast 2**

LOCATION: not mapped

**ORIGIN:** induced with EMS in Landsberg *erecta*
**REFERENCE:** Pyke and Leech (1992) *Plant Physiol. 99:* 1005-1008
**PHENOTYPE:** decreased number of chloroplasts in unit mesophyll cell area; increased chloroplast size

*arc3:* **accumulation and replication of chloroplast 3**
**LOCATION:** not mapped
**ORIGIN:** induced with EMS in Landsberg *erecta*
**REFERENCE:** Pyke and Leech (1992) *Plant Physiol. 99:* 1005-1008
**PHENOTYPE:** reduced number of chloroplasts in unit mesophyll cell area — fewer than *arc2*; increased chloroplast size — larger than *arc2* ones

*arc5:* **accumulation and replication of chloroplast 5**
**LOCATION:** not mapped
**ORIGIN:** induced with EMS in Landsberg *erecta*
**REFERENCE:** Pyke and Leech (1994) *Plant Physiol. 104:* 201-207
**PHENOTYPE:** reduced number of chloroplasts in mesophyll cells (average 13 chloroplasts per cell, normal is about 120 per cell); number of chloroplasts is constant as mesophyll cells increase in size; increased chloroplast size — larger than *arc2* ones

*as1:* **asymmetric leaves 1**
**LOCATION:** 2-59.9
**REFERENCE:** Rédei and Hirono (1964) *AIS 1:* 9-10
**PHENOTYPE:** asymmetric and lobed leaves
**OTHER INFORMATION:** allelic with *magnifica* of Reinholz (1947, *Fiat Rep. 1006:* 1-70; see Barabas and Rédei 1971, *AIS 8:* 7-8)

asymmetric leaves: see *as*

*ats:* **aberrant *testa* shape**
**LOCATION:** 5-56.5
**ORIGIN:** induced with EMS in Landsberg *erecta ttg/ttg*
**REFERENCE:** Léon-Kloosterziel et al. (1994) *Plant Cell 6:* 385-392
**PHENOTYPE:** reduced number of testa layers, resulting in heart-shaped seeds; slight reduction of seed dormancy; slight reduction of female fertility and silique length

*aux1:* **auxin-herbicide resistant 1, allelic with *wav5***
**LOCATION:** 2-61.9
**ORIGIN:** induced with EMS in Landsberg *erecta* (*aux1-1,2,* Maher and Martindale 1980; *wav5-33,* Okada and Shimura 1990) or Columbia (*aux1-7,12,15,19,* Pickett et al. 1990; others, Wilson et al. 1990)
**ALLELES:** *aux1-1* (line P83, *aux-1*); *aux1-2* (line P391, *aux-2*); *aux1-7,12,15,19,* and others isolated as resistant to IAA; *wav5-33*
**REFERENCES:** Maher (1977) *AIS 14:* 18-21; Mirza and Maher (1980) *AIS 17:* 103-107; Maher and Martindale (1980) *Biochem. Genet. 18:* 1041-1053; Martindale and Maher (1981) *AIS 18:* 75-84; Mirza and Maher (1981) *AIS 18:* 85-99; Mirza et al. (1984) *Physiol. Plant. 60:* 516-522; Olsen et al. (1984) *Physiol. Plant. 60:* 523-531; Mirza and Maher (1987) *Plant Growth Regul. 5:* 41-49; Okada and Shimura (1990) *Science 250:* 274-276; Pickett et

al. (1990) *Plant Physiol. 94:* 1462-1466; Wilson et al. (1990) *MGG 222:* 377-383

**PHENOTYPE:** *aux1-1:* agravitropic seedling roots, moderate resistance to the auxin 2,4-dichlorophenoxyacetic acid (2,4-D) (50% root inhibition at 14-fold higher 2,4-D level than wild type)

*aux1-2:* slight resistance to 2,4-D (50% root inhibition at 3-fold higher 2,4-D level than wild type) but roots are gravitropic

*aux1-7,12,15,19:* mutants isolated as resistant to ethylene precursor 1-aminocyclopropane-1-carboxylic acid (ACC); roots agravitropic and resistant to ethylene and auxins (2,4-D and IAA); hypocotyl and leaf sensitive to ethylene

*wav5-33:* root tip rotates continuously in left-hand direction so that the roots form clockwise circles on agar surface

aux-1: see *aux1-1*

aux-2: see *aux1-2*

auxin-herbicide resistant: see *aux*

auxin resistant dwarf: see *dwf*

ax: see *ap1*

axillaris: see *ap1*

**axr1: auxin resistant 1**

**LOCATION:** 1-12.0

**ORIGIN:** *axr1-1* through *axr1-22* induced with EMS; *axr1-23* induced by γ-ray in Columbia

**ALLELES:** *axr1-1* through *axr1-23* (*axr1-1* through *axr1-12,* Estelle and Somerville 1987; *axr1-13* through *axr1-23,* Lincoln et al. 1990)

**REFERENCES:** Estelle and Somerville (1987) *MGG 206:* 200-206; Lincoln et al. (1990) *Plant Cell 2:* 1071-1080; Lincoln and Estelle (1991) *J. Iowa Acad. Sci. 98:* 68-71; Leyser et al. (1993) *Nature 364:* 161-164

**PHENOTYPE:** selected by resistance to the auxin 2,4-dichlorophenoxyacetic acid (2,4-D); rosettes smaller than wild type; rosette leaves crinkled and petioles shorter than normal; roots thinner and less branched than normal; mature plants are shorter and much bushier, with a large number of primary inflorescences, suggesting a reduction in apical dominance

**GENE:** cloned by Leyser et al. (1993), encodes a protein of 539 amino acids; protein shows similarity (22–27%) to the N-terminal region of ubiquitin-activating enzyme (E1) from wheat, yeast, and human but AXR1 protein lacks a cysteine residue essential for E1 activity

**axr2: auxin resistant 2**

**LOCATION:** 3-33.4

**ORIGIN:** induced with EMS in Columbia

**ALLELE:** *axr2-1*

**REFERENCES:** Wilson et al. (1990) *MGG 222:* 377-383; Timpte et al. (1992) *Planta 188:* 271-278

**PHENOTYPE:** selected as resistant to IAA; dwarf; dark green; fertile; roots have no root hair and abnormal gravitropism; reduced elongation of cells in the hypocotyl and floral stem

B-1: see *chl1*

B2-1: see *chl2*

B25: see *chl4*

B25R1, B25R2, B25R3, B25R4, B25R5, B25R6, B25R7: see *su*

B29, B31-2, B33, B35: see *chl3*

B31-1: see *chl5*

**B36: chlorate resistant**

ORIGIN: induced with EMS in Landsberg *erecta*

REFERENCES: Braaksma and Feenstra (1975) *AIS 12:* 16-17; Braaksma and Feenstra (1982) *Theor. Appl. Genet. 64:* 83-90

PHENOTYPE: chlorate resistant; normal growth on nitrate; nitrate reductase induction by nitrate is reduced 4- to 5-fold in wild type to 1.5- to 2.5-fold; nitrite reductase induction by nitrate is nearly normal

OTHER INFORMATION: not allelic with *chl2,3,4,5,6,* or line B40

**B40: chlorate resistant**

ORIGIN: induced with EMS in Landsberg *erecta*

REFERENCES: Braaksma and Feenstra (1975) *AIS 12:* 16-17; Braaksma and Feenstra (1982) *Theor. Appl. Genet. 64:* 83-90

PHENOTYPE: chlorate resistant; normal growth on nitrate; nitrate reductase and nitrite reductase induction by nitrate is relatively normal

OTHER INFORMATION: not allelic with *chl2,3,4,5,6,* or line B36

B73: see *chl6*

**bel1: bel1 1**

LOCATION: 5-50.1

ORIGIN: induced with EMS in Landsberg *erecta* (*bel1-1,6,8*); segregated from T-DNA lines in Wassilewskija (Ws) (*bel1-2,3*); induced with ethylnitrosourea in Landsberg *erecta* (*bel1-4,5,7*)

ALLELES: *bel1-1* (Robinson-Beers et al. 1992); *bel1-2,3* (Modrusan et al. 1994); *bel1-4,5,6,7,8* (Ray et al. 1994)

REFERENCES: Robinson-Beers et al. (1992) *Plant Cell 4:* 1237-1249; Modrusan et al. (1994) *Plant Cell 6:* 333-349; Ray et al. (1994) *Proc. Natl. Acad. Sci. USA 91:* 5761-5765

PHENOTYPE: female sterile; ovules malformed, only a single integument-like structure formed in place of integuments; this structure develops into a collar of tissue; sometimes carpel-like structures develop in place of ovules

**bio1: biotin auxotroph 1**

LOCATION: 5-73.6

ORIGIN: induced with EMS in Columbia

ALLELES: line 122G-E

REFERENCES: Marsden and Meinke (1985) *Am. J. Bot. 72:* 1801-1812; Meinke (1985) *Theor. Appl. Genet. 69:* 543-552; Franzmann et al. (1989) *MGG 77:* 609-616; Schneider et al. (1989) *Dev. Biol. 131:* 161-167; Patton et al. (1991) *MGG 227:* 337-347; Shellhammer and Meinke (1991) *Plant Physiol. 93:* 1162-1167

PHENOTYPE: embryo development arrests at globular to cotyledon stages without exogenous biotin; seed creamy white to pale green and embryo creamy white to very pale yellow-green; normal plants develop when supplemented with biotin

biotin auxotroph: see *bio1*

BL1: see *fad3*

### *blu1: blue light uninhibited 1*

LOCATION: unknown

ORIGIN: induced with EMS in Columbia (homozygous for *gl1-1*)

REFERENCES: Liscum and Hangarter (1991) *Plant Cell 3:* 685-694; Liscum et al. (1992) *Plant Physiol. 100:* 267-271; Young et al. (1992) *Planta 188:* 106-114; Liscum and Hangarter (1993) *Planta 191:* 214-221

PHENOTYPE: seedling hypocotyl elongation is not inhibited by blue light but is inhibited by far-red, red, and UV-A light

OTHER INFORMATION: not allelic with any of the *hy1* through *hy6* mutants

### *blu2: blue light uninhibited 2*

LOCATION: unknown

ORIGIN: induced with EMS in Columbia (homozygous for *gl1-1*)

REFERENCES: Liscum and Hangarter (1991) *Plant Cell 3:* 685-694; Liscum et al. (1992) *Plant Physiol. 100:* 267-271; Liscum and Hangarter (1993) *Planta 191:* 214-221

PHENOTYPE: seedling hypocotyl elongation is not inhibited by blue light but is inhibited by far-red light

OTHER INFORMATION: not allelic with any of the *hy1* through *hy6* mutants

### *blu3: blue light uninhibited 3*

LOCATION: unknown

ORIGIN: induced with EMS in Columbia (homozygous for *gl1-1*)

ALLELES: *blu3-1; blu3-2*

REFERENCES: Liscum and Hangarter (1991) *Plant Cell 3:* 685-694; Liscum et al. (1992) *Plant Physiol. 100:* 267-271; Liscum and Hangarter (1993) *Planta 191:* 214-221

PHENOTYPE: seedling hypocotyl elongation is not inhibited by blue light but is inhibited by far-red light

OTHER INFORMATION: not allelic with any of the *hy1* through *hy6* mutants

BM1, BM2, BM3: see *apt1*

Bo27: see *ga1-7*

Bo64: see *hy3*

### *bp: brevipedicellus*

LOCATION: 4-15.0

REFERENCE: Koornneef et al. (1983) *J. Hered. 74:* 265-272

PHENOTYPE: very short pedicels, siliques pointed downward

brevipedicellus: see *bp*

BRL1: see *cer3*

BRL2: see *cer4*

BRL3: see *cer1*

BRL4: see *cer6*
BRL5: see *cer2*
BRL6: see *cer3*
BRL7: see *cer2*
BRL8: see *cer10*
BRL9: see *cer2*
BRL11: see *cer1*
BRL12: see *cer4*
BRL13: see *cer21*
C5, C6: see *cgl*
CIV: see *abi3-1*
CVI: see *abi1-2*

***ca: ca*espitosa**

LOCATION: not mapped

ORIGIN: induced by X-ray or chemical mutagenesis of Estland or Limburg (Li 5)

REFERENCE: McKelvie (1962) *Radiat. Bot. 1:* 233-241

PHENOTYPE: bushy habit, with many short stems

OTHER INFORMATION: Napp-Zinn and Bonzi (1970, *AIS 7:* 8-9) found no effect of gibberellin on the habit or stem length of *ca* plants, though Bose (1971, *AIS 8:* 19-20) finds somewhat longer (still less than wild type) stems to result from treatment with higher levels of gibberellins than those used by Napp-Zinn and Bonzi

Ca2: see *hcs2*

***cad1: cad*mium sensitive 1**

LOCATION: 5-57.2

ORIGIN: induced with EMS in Columbia

ALLELES: *cad1-1* (line CC5); *cad1-2* (line CC10)

REFERENCE: Howden and Cobbett (1992) *Plant Physiol. 99:* 100-107

PHENOTYPE: root growth inhibited by 90 ml of $CdSO_4$; more sensitive to $HgSO_4$; slightly more sensitive to $CuSO_4$ and $ZnSO_4$

caespitosa: see *ca*

***cal: cal*uliflower**

LOCATION: 1-46.0

ORIGIN: spontaneous (?) in Wassilewskija (Ws) (*cal-1/cal-1*)

ALLELE: *cal-1*

REFERENCES: Bowman (1992) *Flowering Newsletter 14:* 7-19; Bowman et al. (1993) *Development 119:* 721-743

PHENOTYPE: no visible phenotype when alone; enhances *ap1-1* mutant phenotype to transform floral meristem to inflorescence meristem; inflorescence of *ap1 cal* double mutant similar to that of cauliflower or broccoli

cauliflower: see *cal*
CC5, CC10: see *cad1*
CC126: see *hcr1-2*

*cdo:* **chardonnay**

LOCATION: not mapped
ORIGIN: induced with EMS in Landsberg *erecta*
ALLELES: three
REFERENCE: Hülskamp et al. (1994) *Cell 76:* 555-566
PHENOTYPE: trichomes glassy

*cer1:* **eceriferum 1**

LOCATION: 1-1.4
ORIGIN: line F4, induced with EMS in Landsberg *erecta*; line BRL3, segregated from a T-DNA line of Wassilewskija (Ws); line BRL11, cosegregate with a T-DNA insertion in Ws
ALLELES: line F4 (Dellaert et al. 1979); lines BRL3, BRL11 (McNevin et al. 1993)
REFERENCES: Dellaert et al. (1979) *AIS 16:* 10-26; Koornneef et al. (1989) *J. Hered. 80:* 118-122; McNevin et al. (1993) *Génome 36:* 610-618
PHENOTYPE: bright green siliques and stem due to abnormal wax layer — thick organized wax layer and no plate-like structures or rodlet-shaped crystals in epidermal wax; semi-sterile; reduced plant height

*cer2:* **eceriferum 2**

LOCATION: 4-51.9
ORIGIN: line BRL7, segregated from a T-DNA line in Wassilewskija (Ws); lines BRL5 and BRL9, each cosegregate with an independent T-DNA insertion in Ws
ALLELES: line G5, *virescens-2, vc-2* (Rédei 1965); line G8 (Dellaert et al. 1979; see Koornneef et al. 1980, *AIS 17:* 11-18); lines BRL5, BRL7, BRL9 (McNevin et al. 1993)
REFERENCES: Rédei (1965) in Arabidopsis *Research: Proc. Int. Symp. Göttingen Univ.* (ed. G. Röbbelen), pp. 207-210; Dellaert et al. (1979) *AIS 16:* 10-26; Koornneef et al. (1989) *J. Hered. 80:* 118-122; Hannoufa et al. (1993) *Phytochem. 33:* 851-855; McNevin et al. (1993) *Génome 36:* 610-618; Preuss et al. (1993) *Genes Dev. 7:* 974-985
PHENOTYPE: as *cer1* but fully fertile under humid conditions; completely male sterile under dry conditions; pollen fail to hydrate on stigma

*cer3:* **eceriferum 3**

LOCATION: 5-82.0
ORIGIN: line F5, induced by EMS in Landsberg *erecta*; line BRL1, cosegregate with a T-DNA insertion in Wassilewskija (Ws); line BRL6, segregated from a T-DNA line in Ws
ALLELES: line F5 (Dellaert et al. 1979); lines BRL1, BRL6 (McNevin et al. 1993)
REFERENCES: Dellaert et al. (1979) *AIS 16:* 10-26; Koornneef et al. (1989) *J. Hered. 80:* 118-122; Hannoufa et al. (1993) *Phytochem. 33:* 851-855; McNevin et al. (1993) *Génome 36:* 610-618
PHENOTYPE: like *cer1* but plate-like structures are present, although reduced in size and/or frequency in epidermal wax

## *cer4:* **eceriferum 4**

LOCATION: 4-60.9

ORIGIN: line G7, induced with EMS in Landsberg *erecta*; line BRL12, cosegregate with a T-DNA insertion in Wassilewskija (Ws); line BRL2, segregated from a T-DNA line in Ws

ALLELES: line G7 (Dellaert et al. 1979); lines BRL2, BRL12 (McNevin et al. 1993)

REFERENCES: Dellaert et al. (1979) *AIS 16:* 10-26; Koornneef et al. (1989) *J. Hered. 80:* 118-122; Hannoufa et al. (1993) *Phytochem. 33:* 851-855; McNevin et al. (1993) *Génome 36:* 610-618

PHENOTYPE: like *cer2* but resembles wild type in lacking an organized wax layer on stems and siliques, with plate-like structures rare, and rods absent in epidermal wax

## *cer5:* **eceriferum 5**

LOCATION: 1-70.1

ORIGIN: induced by 4.7 kr fast neutrons in Landsberg *erecta* (seeds submerged in tap water at 0.1 kr/min, 22°C)

ALLELES: line D3

REFERENCES: Dellaert et al. (1979) *AIS 16:* 10-26; Koornneef et al. (1989) *J. Hered. 80:* 118-122

PHENOTYPE: like *cer2* but resembles wild type in lacking an organized wax layer on siliques and stems, with plate-like structures reduced, and rods rare in epidermal wax

## *cer6:* **eceriferum 6, allelic with** *pop1*

LOCATION: 1-95.6

ORIGIN: *cer6-1* induced by 6.7 kr fast neutrons in Landsberg *erecta* (seeds submerged in 1.2% DTT at 0.1 kr/min, 22°C); *cer6-2* (*pop1*) induced with EMS in Landsberg *erecta;* line BRL4, segregated from a T-DNA line in Wassilewskija (Ws)

ALLELES: *cer6-1* (line G2) (Dellaert et al. 1979); *cer6-2* (= *pop1*) (Preuss et al. 1993); line BRL4 (McNevin et al. 1993)

REFERENCES: Dellaert et al. (1979) *AIS 16:* 10-26; Koornneef et al. (1989) *J. Hered. 80:* 118-122; Hannoufa et al. (1993) *Phytochem. 33:* 851-855; McNevin et al. (1993) *Génome 36:* 610-618; Preuss et al. (1993) *Genes Dev. 7:* 974-985

PHENOTYPE: like *cer1*: thick organized wax layer on stems and siliques, plate-like structures rare and rods absent in epidermal wax; pollen lacking tryphine (extracellular coat); sterile under dry conditions due to failure to hydrate; partially fertile under humid conditions; *cer6-2* pollen can be rescued by wild-type pollen present on the stigma

OTHER INFORMATION: *cer6-2* (*pop1*) was isolated as a male sterile mutant

## *cer7:* **eceriferum 7**

LOCATION: 3-87.8

ORIGIN: induced by 9.3 kr fast neutrons in Landsberg *erecta* (seeds submerged in 1.2% DTT at 0.1 kr/min, 22°C)

ALLELES: line G3

REFERENCES: Dellaert et al. (1979) *AIS 16:* 10-26; Koornneef et al. (1989) *J. Hered. 80:* 118-122

PHENOTYPE: like *cer2* but plates and rods of epidermal wax rare

### *cer8:* eceriferum 8

LOCATION: 2-71.9

ORIGIN: induced with EMS (four alleles), by γ-ray (one allele), or by fast neutrons (one allele) in Landsberg *erecta*

ALLELES: six

REFERENCE: Koornneef et al. (1989) *J. Hered. 80:* 118-122

PHENOTYPE: like *cer1* but slightly less bright green

### *cer9:* eceriferum 9

LOCATION: 4-76.0

ORIGIN: induced with EMS in Landsberg *erecta*

ALLELES: three

REFERENCE: Koornneef et al. (1989) *J. Hered. 80:* 118-122

PHENOTYPE: like *cer1* but fertile

### *cer10:* eceriferum 10

LOCATION: not mapped

ORIGIN: Koornneef alleles induced with EMS or by γ-ray (one allele each) in Landsberg *erecta*; line BRL8, segregated from a T-DNA line in Wassilewskija (Ws)

ALLELES: two (Koornneef et al. 1989); line BRL8 (McNevin et al. 1993)

REFERENCES: Koornneef et al. (1989) *J. Hered. 80:* 118-122; McNevin et al. (1993) *Génome 36:* 610-618

PHENOTYPE: like *cer1*

### *cer11:* eceriferum 11

LOCATION: not mapped

ORIGIN: induced with EMS or by fast neutrons (one allele each) in Landsberg *erecta*

ALLELES: two

REFERENCE: Koornneef et al. (1989) *J. Hered. 80:* 118-122

PHENOTYPE: like *cer8* but fertile; slender

### *cer12:* eceriferum 12

LOCATION: not mapped

ORIGIN: induced by γ-ray in Landsberg *erecta*

ALLELE: one

REFERENCE: Koornneef et al. (1989) *J. Hered. 80:* 118-122

PHENOTYPE: like *cer8* but fertile

### *cer13:* eceriferum 13

LOCATION: not mapped

ORIGIN: induced with EMS (four alleles) or by fast neutrons (one allele) in Landsberg *erecta*

ALLELES: five

REFERENCE: Koornneef et al. (1989) *J. Hered. 80:* 118-122

PHENOTYPE: like *cer12*

## cer14: eceriferum 14

LOCATION: not mapped

ORIGIN: induced by fast neutrons in Landsberg *erecta*

ALLELE: one

REFERENCE: Koornneef et al. (1989) *J. Hered. 80:* 118-122

PHENOTYPE: like *cer12*

## cer15: eceriferum 15

LOCATION: not mapped

ORIGIN: induced with EMS (two alleles) or by fast neutrons (one allele) in Landsberg *erecta*

ALLELES: three

REFERENCES: Koornneef et al. (1989) *J. Hered. 80:* 118-122; Hannoufa et al. (1993) *Phytochem. 33:* 851-855

PHENOTYPE: like *cer12*

## cer16: eceriferum 16

LOCATION: not mapped

ORIGIN: induced by fast neutrons in Landsberg *erecta*

ALLELE: one

REFERENCES: Koornneef et al. (1989) *J. Hered. 80:* 118-122; Hannoufa et al. (1993) *Phytochem. 33:* 851-855

PHENOTYPE: like *cer12*

## cer17: eceriferum 17

LOCATION: not mapped

ORIGIN: induced with EMS (two alleles) or by fast neutrons (two alleles) in Landsberg *erecta*

ALLELES: four

REFERENCES: Koornneef et al. (1989) *J. Hered. 80:* 118-122; Hannoufa et al. (1993) *Phytochem. 33:* 851-855

PHENOTYPE: like *cer12*

## cer18: eceriferum 18

LOCATION: not mapped

ORIGIN: induced by fast neutrons in Landsberg *erecta*

ALLELE: one

REFERENCES: Koornneef et al. (1989) *J. Hered. 80:* 118-122; Hannoufa et al. (1993) *Phytochem. 33:* 851-855

PHENOTYPE: like *cer12*

## cer19: eceriferum 19

LOCATION: not mapped

ORIGIN: induced with EMS in Landsberg *erecta*

ALLELE: one

REFERENCES: Koornneef et al. (1989) *J. Hered. 80:* 118-122; Hannoufa et al. (1993) *Phytochem. 33:* 851-855

PHENOTYPE: like *cer12*

## cer20: eceriferum 20

LOCATION: not mapped

ORIGIN: induced by fast neutrons in Landsberg *erecta*

ALLELE: one

REFERENCES: Koornneef et al. (1989) *J. Hered. 80:* 118-122; Hannoufa et al. (1993) *Phytochem. 33:* 851-855

PHENOTYPE: like *cer12*

### *cer21: ecer*iferum 21

LOCATION: not mapped

ORIGIN: line BRL13, cosegregate with a T-DNA insertion in Wassilewskija (Ws)

ALLELES: line BRL13

REFERENCE: McNevin et al. (1993) *Génome 36:* 610-618

PHENOTYPE: bright green stems

cesium insensitive: see *csi-32* and *csi-52*

### *cgl: c*omplex *gl*ycan minus

LOCATION: not mapped

ORIGIN: induced with EMS in Columbia

ALLELES: two (lines C5, C6)

REFERENCE: von Schaewen et al. (1993) *Plant Physiol. 102:* 1109-1118

PHENOTYPE: lack complex glycans; accumulate $Man_5$ $GlcNac_2$ glycans; deficient in *N*-acetyl glucosamine transferase I; no visible phenotype

*ch-2:* see *ch1*

### *ch1: ch*lorina 1

LOCATION: 1-58.4

ALLELES: *ch1-1; ch1-12* (= *ch2, ch-2, ch-12*)

REFERENCES: Rédei and Hirono (1964) *AIS 1:* 9-10; Hirono and Rédei (1963) *Nature 197:* 1324-1325; Kranz (1970) *AIS 7:* 28-30; Murray and Kohorn (1991) *Plant Mol Biol. 16:* 71-79

PHENOTYPE: *ch1-1:* green-yellow

*ch1-12:* light green

OTHER INFORMATION: chlorophyll B is absent from *ch1-1/ch1-1* plants; chlorophyll B content is reduced in *ch1-12/ch1-12* plants; LHCIIP mRNAs are present at normal levels and can be translated in vitro

*ch2:* see *ch1*

### *ch5: ch*lorina 5

LOCATION: 5-42.5

REFERENCE: Koornneef et al. (1983) *J. Hered. 74:* 265-272

PHENOTYPE: green-yellow

### *ch6: ch*lorina 6

LOCATION: 3-84.0

ORIGIN: induced with 0.1 mM *N*-nitroso-*N*-methylurea in line S96 (derived from a cross between Li-2 and Dijon)

ALLELE: *ch6-1* (line M33)

REFERENCES: Fischerova (1975) *Biol. Plantarum 17:* 182-188; Relichova (1976) *AIS 13:* 25-28

PHENOTYPE: pale green embryos, green-yellow plants

OTHER INFORMATION: some alleles and/or growing conditions may render *ch6/ch6* inviable

*ch-12:* see *ch1*

## *ch42: chlorina 42*

LOCATION: 4-43.6

ORIGIN: original allele induced by X-ray treatment of unknown background; a T-DNA insertional allele (*cs*, line N6H-14) was isolated from transgenic plants in Columbia (Koncz et al. 1990)

REFERENCES: Fischerova (1975) *Biol. Plantarum 17:* 182-188; Relichova (1976) *AIS 13:* 25-28; Koncz et al. (1990) *EMBO J. 9:* 1337-1346

PHENOTYPE: pale green embryos, pale yellow-green seedlings; lethal

GENE: cloned by Koncz et al. (1990); predicted protein is a novel chloroplast protein

## *cha: chablis*

LOCATION: not mapped

ORIGIN: induced with EMS in Landsberg *erecta*

ALLELES: three

REFERENCE: Hülskamp et al. (1994) *Cell 76:* 555-566

PHENOTYPE: trichomes glassy

chablis: see *cha*

chalcone synthase: see *tt4*

chardonnay: see *cdo*

## *chl1: chlorate resistance 1*

LOCATION: 1-14.3

ORIGIN: *chl1-1* induced with EMS in Landsberg *erecta*; *chl1-2* induced by T-DNA insertion in Wassilewskija (Ws); *chl1-3* induced with EMS in a Landsberg *erecta* line carrying a *chl3* allele; *chl1-4,5* induced by γ-ray in Columbia; *chl1-6* induced by insertion of an *Arabidopsis thaliana* transposon *Tag1* (Tsay et al. 1993b)

ALLELES: *chl1-1* (and line B-1) (Doddema and Telkamp 1979); *chl1-2,3,4,5* (Tsay et al. 1993a); *chl-6* (Tsay et al. 1993b); other mutants exist (e.g., Braaksma and Feenstra 1975, *AIS 12:* 16-17)

REFERENCES: Oostindier-Braaksma and Feenstra (1973) *Mut. Res. 19:* 175-185; Doddema et al. (1978) *Physiol. Plant. 43:* 343-350; Doddema and Telkamp (1979) *Physiol. Plant. 45:* 332-338; Crawford (1992) *Genet. Engineering* (ed. J.K. Setlow) *14:* 89-98; Tsay et al. (1993a) *Cell 72:* 705-713; Tsay et al. (1993b) *Science 260:* 342-344; Crawford, this volume

PHENOTYPE: chlorate resistant due to decreased chlorate uptake

GENE: cloned (Tsay et al. 1993a) from a T-DNA tagged line; codes for 65-kD membrane protein with 12 transmembrane segments; protein acts as an electrogenic nitrate transporter; the *CHL1* mRNA is found predominantly in roots

## *chl2: chlorate resistance 2*

LOCATION: 2-32.3

ORIGIN: induced with 1 mM nitrosoguanidine in Landsberg *erecta* (4–4.5 hrs, 25°C, seeds imbibed 7 days at 4°C before treatment)

ALLELES: line B2-1

REFERENCES: Oostindier-Braaksma and Feenstra (1973) *Mut. Res. 19:* 175-185; Braaksma and Feenstra (1982) *Theor. Appl. Genet. 64:* 83-90; LaBrie et al. (1992) *MGG 233:* 169-176; Crawford (1992) *Genet. Engineering* (ed. J.K. Setlow) *14:* 89-98

PHENOTYPE: chlorate resistant due to decreased nitrate reductase activity; reduced levels of the nitrate reductase cofactor molybdenum-pterin; sensitive to 0.1 mM tungstate

OTHER INFORMATION: 20–30% of wild-type nitrate reductase activity, poor growth on nitrate medium

### *chl3: chl*orate resistance 3, allelic with *nia2*

LOCATION: 1-56.4

ORIGIN: induced with EMS in Landsberg *erecta* (*chl3-1,2,3,4*) or Columbia (*chl3-5,6,7*)

ALLELES: *chl3-1,2,3,4* (lines B29, B31-2, B33, B35, respectively) (Braaksma and Feenstra 1975)

*chl3-5,6,7* (lines G-5, G-6, G9-2, respectively) (Wilkinson and Crawford 1991)

REFERENCES: Braaksma and Feenstra (1975) *AIS 12:* 16-17; Braaksma and Feenstra (1982) *Theor. Appl. Genet. 64:* 83-90; Crawford et al. (1988) *Proc. Natl. Acad. Sci. USA 85:* 5006-5010; Cheng et al. (1988) *EMBO J. 7:* 3309-3314; Wilkinson and Crawford (1991) *Plant Cell 3:* 461-471; Crawford (1992) *Genet. Engineering* (ed. J.K. Setlow) *14:* 89-98

PHENOTYPE: chlorate resistant due to reduced nitrate reductase activity

GENE: encodes nitrate reductase (NIA2); has been cloned by Wilkinson and Crawford (1991); immunoblot detected in *chl3-1,3,4* plants normal levels of nitrate reductase protein; the *chl3-5,6,7* plants all carry deletion at the *CHL3* locus

OTHER INFORMATION: about 20% of wild-type nitrate reductase activity, grows reasonably well on nitrate medium

### *chl4: chl*orate resistant 4, same as *rgn*

LOCATION: 1-0.5

ORIGIN: induced in Landsberg *erecta*

ALLELES: line B25; also known as *rgn*

REFERENCES: Oostindier-Braaksma and Feenstra (1974) *AIS 11:* 8; Braaksma and Feenstra (1975) *AIS 12:* 16-17; Brandsma and Doddema (1982) *AIS 19:* 19-24; Braaksma and Feenstra (1982) *Physiol. Plant. 54:* 351-360; Braaksma and Feenstra (1982) *Theor. Appl. Genet. 64:* 83-90; Braaksma and Feenstra (1982) *Theor. Appl. Genet. 61:* 263-271; Crawford (1992) *Genet. Engineering* (ed. J.K. Setlow) *14:* 89-98

PHENOTYPE: selected for chlorate resistance, little or no nitrate reductase activity, dies on nitrate medium

OTHER INFORMATION: induces nitrite reductase normally when exposed to nitrate; cytochrome-*c* reductase activity is induced to a degree by nitrate (as in wild type) but wild-type cytochrome-*c* reductase runs at 8S on sucrose gradients and *rgn* cytochrome-*c* reductase runs at 4S; many second-site suppressors of *chl4* (*rgn*) exist — see *su*

### *chl5: chlorate resistant 5*

LOCATION: chromosome 2

ORIGIN: induced in Landsberg *erecta*

ALLELES: line B31-1

REFERENCES: Braaksma and Feenstra (1975) *AIS 12:* 16-17; Braaksma and Feenstra (1982) *Theor. Appl. Genet. 64:* 83-90; Crawford (1992) *Genet. Engineering* (ed. J.K. Setlow) *14:* 89-98; LaBrie et al. (1992) *MGG 233:* 169-176

PHENOTYPE: chlorate resistant

### *chl6: chlorate resistant 6, same as cnx*

LOCATION: 5-21.8

ORIGIN: induced with EMS in Landsberg *erecta*

ALLELES: line B73 (= *cnx*)

REFERENCES: Braaksma et al. (1979) *AIS 16:* 68-70; Braaksma and Feenstra (1982) *Theor. Appl. Genet. 64:* 83-90; Jacobsen and Feenstra (1984) *Z. Pflanzenphysiol. 113:* 183-188; Crawford (1992) *Genet. Engineering* (ed. J.K. Setlow) *14:* 89-98; LaBrie et al. (1992) *MGG 233:* 169-176

PHENOTYPE: chlorate resistant, about 10% of normal nitrate reductase activity, similar to *chl2;* defective in molybdenum-pterin cofactor; grows poorly on nitrate medium, enhanced nitrate uptake, 16% of normal xanthine dehydrogenase activity at 3 days and none at 8 days, partially reverted by high molybdenum concentration

OTHER INFORMATION: probably molybdenum cofactor gene; reverts at high frequency (Braaksma and Feenstra 1982, *AIS 19:* 111-112)

### *chl7: chlorate resistant 7*

LOCATION: not mapped

ORIGIN: induced by T-DNA insertion in Wassilewskija (Ws)

REFERENCE: LaBrie et al. (1992) *MGG 233:* 169-176

PHENOTYPE: similar to *chl2*; defective in molybdenum-pterin cofactor

chlorate resistance: see *chl*

chlorina: see *ch* and *chp*

chlorina V157: see *alb1*

chloroplast mutator: see *chm*

chlorsulfuron resistant: see *csr1*

### *chm: chloroplast mutator*

LOCATION: 3-32.2

ORIGIN: induced with EMS in Columbia (with *gl1* for *chm-1* or *GL1* for *chm-2*)

ALLELES: *chm-1; chm-2*

REFERENCES: Rédei (1971) *AIS 8:* 28; Rédei (1973) *Mut. Res. 18:* 149-162; Martínez-Zapater et al. (1992) *Plant Cell 4:* 889-899; Mourad and White

(1992) *Theor. Appl. Genet. 84:* 906-914
PHENOTYPE: *chm-1:* plant variegated green, yellow, and white; *chm-2:* like *chm-1* but sharper borders on sectors
OTHER INFORMATION: originally proposed to be a nuclear mutator gene affecting the chloroplast genome, but now shown to be a nuclear mutator of the mitochondrial genome that causes specific mitochondrial DNA rearrangements (Martínez-Zapater et al. 1992); apparently homoplastidic lines (having either rough leaf or solid pale green phenotype) have been generated by stabilizing the mutations caused by the *chm-1* defect after segregating away the *chm-1* mutation through genetic crosses (Mourad and White 1992), although the genetic basis for the alterations in these lines is not known

### *chp7:* **chlorina 7**
LOCATION: 5-7.8
ORIGIN: present in a *co* mutant line (probably Columbia) from Rédei
REFERENCE: Koornneef et al. (1991) *MGG 229:* 57-66
PHENOTYPE: pale green

### *chs1:* **chilling sensitive 1**
LOCATION: 1-32.9
ORIGIN: induced with EMS (0.3% v/v in water for 16 hrs at room temperature) in Columbia
ALLELE: *chs1-1* (line PM11)
REFERENCE: Hugly et al. (1990) *Plant Physiol. 93:* 1053-1062
PHENOTYPE: becomes chlorotic 3 days after shifting from 22°C to 13°C; no growth at temperatures below 18°C; altered steryl-ester metabolism

### *ckr1:* **cytokinin resistant 1**
LOCATION: chromosome 5 (Su and Howell 1992), linked to marker 211 (5-91.7) of Chang et al. (1988, *Proc. Natl. Acad. Sci. USA 85:* 6856-6860)
ORIGIN: induced with EMS in Columbia
ALLELES: *ckr1-7,8,12,50,109*
REFERENCE: Su and Howell (1992) *Plant Physiol. 99:* 1569-1574
PHENOTYPE: primary root elongation insensitive to 2.5 μM of benzyladenine; in the absence of cytokinin, cotyledon larger and slightly more yellow than normal and rosette leaves slightly cupped and more yellow under constant lighting

clavata: see *clv*

### *clv1:* **clavata 1, allelic with** *flo5* **and** *fas3*
LOCATION: 1-109.8
ORIGIN: induced with EMS in Landsberg *erecta* (*clv1-1* and *clv1-3*) or Columbia (*clv1-2*), induced by X-ray in Estland or Limburg (*clv1-4*) or in Columbia pollen (*clv1-6,7*), or spontaneous (?) in Landsberg *erecta* (*clv1-5*)
ALLELES: *clv1-1* (McKelvie, see Koornneef et al. 1983); *clv1-2* (= *flo5*) (Haughn and Somerville 1988); *clv1-3* (= *fas3*) (Leyser and Furner 1992); *clv1-4* (McKelvie 1962); *clv1-5,6,7* (Clark et al. 1993)
REFERENCES: McKelvie (1962) *Radiat. Bot. 1:* 233-241; Krickhan and Napp-Zinn (1975) *AIS 12:* 9; Koornneef et al. (1983) *J. Hered. 74:* 265-272;

Haughn and Somerville (1988) *Devel. Genet. 9:* 73-89; Leyser and Furner (1992) *Development 116:* 397-403; Clark et al. (1993) *Development 119:* 397-418; Wilson and Lord (1993) *Am. J. Bot. 80:* 1419-1426

PHENOTYPE: club-like pods; more than two carpels in an ovary; large shoot apical meristems; plants are sometimes fasciated; severe alleles (*clv1-1* and *clv1-4*) show increased number of floral organs in all whorls

### clv2: clavata 2

LOCATION: 1-88.8

REFERENCE: Koornneef et al. (1983) *J. Hered. 74:* 265-272

PHENOTYPE: club-shaped pods, phenotype less pronounced than in *clv1*

cnx: cofactor of NR and XDH, see *chl6*

### co: constans, allelic with fg

LOCATION: 5-13.1

ORIGIN: induced by 8–12 kr X-ray to presoaked Columbia seeds (*co-1*) (Rédei 1962) or with EMS in Landsberg *erecta* (*co-2,3,4,5,6,7,* and three others) (Koornneef et al. 1991)

REFERENCES: Rédei (1962) *Genetics 47:* 443-460; Koornneef et al. (1983) *J. Hered. 74:* 265-272; Martínez-Zapater and Somerville (1990) *Plant Physiol. 92:* 770-776; Koornneef et al. (1991) *MGG 229:* 57-66; Putterill et al. (1993) *MGG 239:* 145-157

PHENOTYPE: like *fb* but heterozygotes are intermediate; relatively insensitive to vernalization or incandescent light treatment and short days

### cob: cobra

LOCATION: not mapped

ORIGIN: induced with EMS in Columbia

REFERENCES: Benfey et al. (1993) *Development 119:* 57-70; Hauser and Benfey (1994) *Proc. NATO-ASI, Plant Molecular Biology Series: Molecular Genetic Analysis of Plant Development and Metabolism* (eds. G. Coruzzi and P. Puigdoménech), pp. 31-40, Springer-Verlag, Berlin

PHENOTYPE: conditional abnormal cell expansion in the root, more dramatic on high (4.5%) sucrose, but no expansion at 14°C; expansion most severe in epidermis (15x), less in stele (4x), and least in cortex and endodermis (2.5x)

OTHER INFORMATION: double mutant with *lit* has stunted shoots and abnormal amounts of anthocyanin; sterile

cobra: see *cob*

cofactor of NR and XDH: see *chl6*

### coi1: coronatine insensitive 1

LOCATION: not mapped

ORIGIN: induced with EMS in Columbia

REFERENCE: Feys et al. (1994) *Plant Cell 6:* 751-759

PHENOTYPE: isolated as seedlings insensitivity to coronatine, a toxin produced by *Pseudomonas syringae* pathogens; male sterile and pollen development abnormal; 2D gel revealed that two normally abundant proteins (~61 and ~31 kD) are absent among the soluble proteins of the mutant flowers; in addition, a ~55-kD protein was detected in the mutant extract but not in wild-type extract; insensitive to methyl jasmonate; resistant to the pathogen *Pseudomonas syringae* pv. *atropurpurea*

compacta: see *cp*

complex glycan minus: see *cgl*

constans: see *co*

constitutively photomorphogenic: see *cop*

constitutive triple response: see *ctr1*

**cop1: constitutively photomorphogenic 1, allelic with emb168 and fus1**

LOCATION: 2-55.8

ORIGIN: *cop1-1* through *cop1-4* induced with EMS in Columbia; *cop1-5* induced by T-DNA insertion in Wassilewskija (Ws); *fus1* alleles induced with EMS in Landsberg *erecta* or Di-G (Miséra et al. 1994)

ALLELES: *cop1-1,2,3,4,5; emb168; fus1,* ten in Landsberg *erecta* and three in Di-G (33 lost)

REFERENCES: Deng et al. (1991) *Genes Dev. 5:* 1172-1182; Deng and Quail (1992) *Plant J. 2:* 83-95; Deng et al. (1992) *Cell 71:* 791-801; Ang and Deng (1994) *Plant Cell 6:* 613-628; Castle and Meinke (1994) *Plant Cell 6:* 25-41; Miséra et al. (1994) *MGG 244:* 242-252

PHENOTYPE: dark-grown seedlings resemble light-grown wild-type seedlings; activation of light-inducible promoters in dark; *cop1-4* has less severe phenotypes; not allelic with *det1* or *det2*

GENE: *COP1* cloned by Deng et al. (1992) and it encodes a protein of 658 amino acids containing a zinc-binding motif and a domain with WD-40 repeats (similar to G protein β subunit and other proteins)

OTHER INFORMATION: genetic studies suggest specific interaction with *HY5,* possibly at the protein level

**cop2: constitutively photomorphogenic 2**

LOCATION: not mapped

ORIGIN: induced with EMS in Columbia

REFERENCE: Hou et al. (1993) *Plant Cell 5:* 329-339

PHENOTYPE: dark-grown seedlings have normal elongated hypocotyl of etiolated seedlings but have open and enlarged cotyledons; *cop2* partially suppress *hy1* long hypocotyl phenotype in light; no effect on plastid differentiation or on light-regulated gene expression

**cop3: constitutively photomorphogenic 3**

LOCATION: not mapped

ORIGIN: induced with EMS in Columbia

REFERENCE: Hou et al. (1993) *Plant Cell 5:* 329-339

PHENOTYPE: similar to *cop2:* dark-grown seedlings have normal elongated hypocotyl of etiolated seedlings but have open and enlarged cotyledons; *cop3* partially suppress *hy1* long hypocotyl phenotype in light

**cop4: constitutively photomorphogenic 4**

LOCATION: not mapped

ORIGIN: induced with EMS in Columbia

REFERENCE: Hou et al. (1993) *Plant Cell 5:* 329-339

PHENOTYPE: dark-grown seedlings have normal elongated hypocotyl of etio-lated seedlings but have open and enlarged cotyledons; *cop4* partially sup-press *hy1* and *hy3* long hypocotyl phenotype in light; no effect on plastid dif-ferentiation; higher expression of the light-induced *rbcS*, *cab*, and *fedA* gene expression in dark; defective in gravitropism

*cop8:* constitutively photomorphogenic 8, allelic with *fus8*, see *fus8*

**cop9: constitutively photomorphogenic 9, allelic with *fus7* and *emb143***

LOCATION: 4-28.6

ORIGIN: *cop9-1* induced by T-DNA insertion in Wassilewskija (Ws); *fus7* in-duced with EMS in Landsberg *erecta* (four alleles) and Di-G (one allele, seven others lost)

ALLELES: *cop9-1;* many *fus7* alleles (Miséra et al. 1994)

REFERENCES: Wei and Deng (1992) *Plant Cell 4:* 1507-1518; Castle and Meinke (1994) *Plant Cell 6:* 25-41; Meinke, this volume; Miséra et al. (1994) *MGG 244:* 242-252; Wei et al. (1994) *Cell 78:* 117-124

PHENOTYPE: dark-grown seedlings have the morphology of wild-type light-grown seedlings: short hypocotyl, open and expanded cotyledons; high level expression of light-induced genes in dark; adult lethal, severely dwarfed and unable to reach maturity or flower; these phenotypes are similar to those of *cop1* mutants

GENE: cloned by Wei et al. (1994) using the T-DNA tagged *cop9-1* allele; en-codes a hydrophilic protein of 197 amino acids (22.5 kD); the T-DNA inser-tion occurred between amino acids 182 and 183; *COP9* expression is not regulated by light; COP9 protein seems to be part of a large complex that fails to form in *cop8* (*fus8*) and *cop11* (*fus6*) mutants

OTHER INFORMATION: allelic with *fus7* and *emb143* (Meinke, this volume; Miséra et al. 1994)

*cop10:* constitutively photomorphogenic 10, allelic with *fus9*, see *fus9*

*cop11:* constitutively photomorphogenic 11, allelic with *fus6*, see *fus6*

coronatine insensitive: see *coi*

**cp1: compacta 1**

LOCATION: 4-28.2

ORIGIN: induced with EMS in Landsberg *erecta*

REFERENCES: Koornneef et al. (1980) *AIS 17:* 11-18; Koornneef et al. (1983) *J. Hered. 74:* 265-272

PHENOTYPE: compact semi-dwarf

**cp2: compacta 2**

LOCATION: 2-33.6

REFERENCE: Koornneef et al. (1983) *J. Hered. 74:* 265-272

PHENOTYPE: compact semi-dwarf

**cp3: compacta 3**

LOCATION: 4-77.8

ORIGIN: induced with EMS in Landsberg *erecta*

REFERENCES: Koornneef et al. (1980) *AIS 17:* 11-18; Koornneef et al. (1983) *J. Hered. 74:* 265-272

    PHENOTYPE: compact semi-dwarf with round, dark green leaves

*crk: crooked*

    LOCATION: not mapped

    ORIGIN: induced with EMS in Landsberg *erecta*

    ALLELES: two

    REFERENCE: Hülskamp et al. (1994) *Cell 76:* 555-566

    PHENOTYPE: trichomes distorted

*crl: curly*

    LOCATION: not mapped

    ORIGIN: induced by T-DNA insertion in Wassilewskija (Ws)

    REFERENCE: Medford et al. (1992) *Plant Cell 4:* 631-643

    PHENOTYPE: cotyledons and leaves expand abnormally

crooked: see *crk*

cs: see *ch42*

CS30: see *gluS*

CS42: see *stm*

CS51: see *sat*

CS64: see *stm*

CS103: see *gluS*

CS108: see *sat*

CS113: see *gluS*

CS115: see *stm*

CS116: see *glyD*

CS117: see *sat*

CS119: see *pcoA*

CS156: see *dct*

CS208: see *hcr1-1*

CS2071: see *rca*

*csi32: cesium insensitive 32*

    LOCATION: not mapped

    ORIGIN: induced with EMS in Columbia

    REFERENCE: Sheahan et al. (1993) *Plant J. 3:* 647-656

    PHENOTYPE: seedlings insensitive to 2 M cesium; green cotyledons and chlorotic true leaves; reduced uptake of $^{32}$P and $^{86}$Rb

*csi52: cesium insensitive 52*

    LOCATION: 4-10.5

    ORIGIN: induced with EMS in Columbia

    REFERENCE: Sheahan et al. (1993) *Plant J. 3:* 647-656

    PHENOTYPE: seedlings insensitive to 2 M cesium; reduced uptake of $^{32}$P and $^{86}$Rb

*csr1: chlorsulfuron resistant 1, allelic with imr*

    LOCATION: 3-70.1

    ORIGIN: induced with EMS in Columbia

    ALLELES: *csr1-1* (*csr1* = allele in line GH50); two additional lines, GH51 and GH52 (not independent of *csr1-1*), also showed resistance to chlorsulfuron;

*csr1-2* (*imr*, line GH90) (Haughn and Somerville 1990); *csr1-3*, *csr1-4* (*csr1-3* was originally designated *csr1-2* in Mourad and King 1992 but was changed to *csr1-3* in Mourad et al. 1993); line VAL-2 (Wu et al. 1994)

**REFERENCES:** Haughn and Somerville (1986) *MGG 204:* 430-434; Haughn and Somerville (1987) *AIS 25:* 43-48; Mazur et al. (1987) *Plant Physiol. 85:* 1110-1117; Haughn et al. (1988) *MGG 211:* 266-271; Haughn and Somerville (1990) *Plant Physiol. 92:* 1081-1085; Sathasivan et al. (1990) *Nuc. Acid Res. 18:* 2188; Sathasivan et al. (1991) *Plant Physiol. 97:* 1044-1050; Hattori et al. (1992) *MGG 232:* 167-173; Li et al. (1992) *Plant Physiol. 100:* 662-668; Mourad and King (1992) *Planta 188:* 491-497; Mourad et al. (1993) *J. Hered. 84:* 91-96; Wu et al. (1994) *Planta 192:* 249-255

**PHENOTYPE:** *csr1-1:* dominant over wild type; selected on the basis of resistance to chlorsulfuron, which inhibits acetolactate synthase (ALS); 50% inhibition of ALS activity from mutant requires 1000x higher concentration of chlorsulfuron than the wild-type enzyme; also resistant to triazoloprimidine but not resistant to imidazolinone or pyrimidyl-oxy-benzoate
*csr1-2:* dominant over wild type; selected on the basis of resistance to imidazolinone; the ALS enzyme activity is resistant to imidazolinone and pyrimidyl-oxy-benzoate; not resistant to chlorsulfuron
*csr1-3:* isolated as resistant to triazoloprimidine; also resistant to chlorsulfuron; not resistant to imidazolinone or to pyrimidyl-oxy-benzoate
*csr1-4:* an intragenic recombinant between *csr1-1* and *csr1-2* and is resistant to both chlorsulfuron (and triazoloprimidine) and imidazolinone (and pyrimidyl-oxy-benzoate)
line VAL-2, most likely an allele of *CSR1:* 4-fold more resistant to valine than wild type; ALS (acetolactate synthase) is less sensitive to feedback control by valine, leucine, isoleucine, or valine plus leucine

**GENE:** the *ALS* gene (the enzyme is also called acetohydroxy acid synthase, AHAS) was isolated by Mazur et al. (1987) and used as a probe to isolate the *ALS* gene from the *csr1-1* mutant (Haughn et al. 1988); the *ALS* gene from the *csr1-1* mutant differs from wild type by a single base change (Pro to Ser substitution in the product) and confers chlorsulfuron resistance to tobacco transgenic plants and transformed rice protoplasts; the *csr1-2* allele was cloned by Sathasivan et al. (1990) and Hattori et al. (1992) and a single base change (causing a Ser to Asn change) was found in the coding region; a chimeric *ALS* gene carrying both *csr1-1* and *csr1-2* alleles confers to tobacco transgenic plants resistance to both sulfonylurea and imidazolinone herbicides

### *ctr1*, constitutive *tr*iple response 1

**LOCATION:** 5-2.1
**ORIGIN:** *ctr1-1*, *ctr1-2*, *ctr1-3*, and *ctr1-4* induced with diepoxybutane, by X-ray, with EMS, and with EMS, respectively, in Columbia; *ctr1-5* induced by T-DNA insertion in Wassilewskija (Ws)
**ALLELES:** *ctr1-1* through *ctr1-5*
**REFERENCE:** Kieber et al. (1993) *Cell 72:* 427-441

PHENOTYPE: in absence of exogenous ethylene, dark-grown seedlings show triple response: shortening and radial swelling of hypocotyl, inhibition of root elongation, and exaggeration of the curvature of the apical hook; no increased production of endogenous ethylene; rosette leaves and adult plant much smaller than wild type; 1–2-week delay in flowering and early flowers are infertile; the mutant phenotypes are similar to wild type grown in constant exogenous ethylene; ethylene-induced genes are constitutively expressed in the mutants

GENE: cloned by Kieber et al. (1993) and encodes a member of the Raf family of serine/threonine protein kinases; the gene contains 15 exons and 14 introns; the diepoxybutane-induced *ctr1-1* mutation is a T to A change (Asp→Glu), the X-ray-induced *ctr1-2* mutation is a small deletion of 17 nucleotides, the EMS-induced *ctr1-3* and *ctr1-4* mutations are C to T (Arg→Stop) and G to A (Glu→Lys) changes

curly: see *crl*

D1: see *trp4-1*

D3: see *cer5*

d69: see *ga1-10*

d352: see *ga1-6*

d501: see *tt7*

DV: see *abi3-2*

dark overexpression of cab: see *doc*

*dct: d*icarboxylate *t*ransport

LOCATION: nuclear, not mapped

ORIGIN: EMS mutagenesis (12 hrs, 0.2% EMS) of Columbia seeds

ALLELES: line CS156

REFERENCES: Somerville and Ogren (1983) *Proc. Natl. Acad. Sci. USA 80:* 1290-1294; Somerville and Somerville (1985) *Plant Sci. Lett. 37:* 217-220

PHENOTYPE: defective in chloroplast dicarboxylate transport; deficient in a chloroplast envelope protein

OTHER INFORMATION: selected by inability to survive at normal atmospheric levels of $O_2$ and $CO_2$ but has normal growth at 1% $CO_2$, which represses photorespiration

*ddm1: d*ecrease in *D*NA *m*ethylation 1

LOCATION: not mapped

ORIGIN: induced with EMS in Columbia

ALLELES: *ddm1-1; ddm1-2*

REFERENCE: Vongs et al. (1993) *Science 260:* 1926-1928

PHENOTYPE: isolated by screening for hypomethylation of repetitive DNA by Southern analysis with a 180-bp centromere repeat; also have hypomethylation for 18S, 5.8S, and 25S ribosomal genes, as well as 5S rRNA gene and two telomere-associated sequences; normal methylation for two random single-copy sequences; reduced levels (26–28% of normal) of 5 methylcytosine

OTHER INFORMATION: the heterozygotes show a slow increase in methylation levels through successive backcrosses with wild type, indicating that de novo methylation is low

decrease in DNA methylation: see *ddm1*

de-etiolated: see *det*

deficient in xylem loading of phosphate: see *pho1*

deformed roots and leaves: see *drl1*

***det1*: de-etiolated 1, allelic with *fus2***

LOCATION: 4-20.0

ORIGIN: *det1-1,2,3,4* induced with EMS in Columbia; *det1-5* from a T-DNA insertional line in Wassilewskija (Ws); *det1-6,7,8,9* induced with EMS in Landsberg *erecta* or Di-G (*fus2*)

ALLELES: *det1-1,2* (Chory et al. 1989, 1991); *det1-3,4,5* (Pepper et al. 1994); *det1-6,7,8,9* (*fus2*) (Miséra et al. 1994; Pepper et al. 1994)

REFERENCES: Chory et al. (1989) *Cell 58:* 991-999; Chory and Peto (1990) *Proc. Natl. Acad. Sci. USA 87:* 8776-8780; Chory et al. (1991) *Plant Cell 3:* 445-459; Deng et al. (1991) *Genes Dev. 5:* 1172-1182; Chory (1992) *Development 115:* 337-354; Chory (1993) *TIG 9:* 167-172; Sun et al. (1993) *Plant Cell 5:* 109-121; Castle and Meinke (1994) *Plant Cell 6:* 25-41; Chory et al. (1994) *Plant Physiol. 104:* 339-347; Li et al. (1994) *Genes Dev. 8:* 339-349; Miséra et al. (1994) *MGG 244:* 242-252; Pepper et al. (1994) *Cell 78:* 109-116

PHENOTYPE: 7-day-old dark-grown seedlings have short and thick hypocotyls and pigmented and expanded cotyledons; when grown in light, smaller and paler green than wild type; delayed flowering and reduced apical dominance; smaller flowers and reduced male fertility; light-regulated genes are constitutively expressed in dark; extracts lack the activity of a DNA-binding phosphoprotein, CA-1, which interacts with the promoter of the *cab140* gene; embryos homozygous for severe alleles are purple (*fusca*), and germinate into purple seedlings, which die before flowering

GENE: cloned positionally by Pepper et al. (1994); encodes a nuclear protein of 541 amino acids (62.2 kD) with nuclear localization signals; expressed in both light- and dark-grown seedlings; both *det1-1* and *det1-3* alleles have a G→A change at the fifth base of intron 1, and the *det1-1* mutant has a large *DET1* mRNA; *det1-4* has a $Gly_{58}$→Arg missense mutation and is the weakest of all of the alleles; *det1-5* has a 9-bp deletion that removes $Pro_{519}$ $Leu_{520}$ $Ala_{521}$ and also has a weak phenotype; *det1-6* is a G→A transition that affects the 3' acceptor site of intron 1 and no *DET1* mRNA is detected; *det1-7,8,9* have C→T changes that result in stop codons at positions 141, 14, and 106, respectively

***det2*: de-etiolated 2**

LOCATION: 2-59.5

ORIGIN: *det2-1,2,3,4,5* induced with EMS in Columbia

REFERENCES: Chory et al. (1989) *Cell 58:* 991-999; Chory et al. (1991) *Plant Cell 3:* 445-459; Deng et al. (1991) *Genes Dev. 5:* 1172-1182; Chory (1992) *Development 115:* 337-354; Chory (1993) *TIG 9:* 167-172; Chory et al.

(1994) *Plant Physiol. 104:* 339-347

**PHENOTYPE:** 7-day-old dark-grown seedlings have short and thick hypocotyls and pigmented and expanded cotyledons; when grown in light, smaller and darker green than wild type; delayed flowering and reduced apical dominance; smaller flowers and reduced male fertility; light-regulated genes are constitutively expressed in dark

### *det3: de-etiolated 3*

**LOCATION:** 1-57.3

**ORIGIN:** induced with diepoxybutane in Columbia

**ALLELE:** one

**REFERENCES:** Poch et al. (1993) *Plant J. 4:* 671-682; Li et al. (1994) *Genes Dev. 8:* 339-349

**PHENOTYPE:** seedlings have light-grown morphology when grown in dark; develop true leaves in dark; etioplasts (not chloroplasts) of dark-grown seedling leaves rich in starch granules, suggesting that they are metabolically active; light-dependent genes are not expressed in *det3* as in wild type; when grown in light, *det3* plants are short and have reduced apical dominance

DH47, DH48: see *ixr1*

dicarboxylate transport: see *dct*

### *dip: disrupted*

**LOCATION:** not mapped

**ORIGIN:** induced by T-DNA insertion in Wassilewskija (Ws)

**REFERENCE:** Medford et al. (1992) *Plant Cell 4:* 631-643

**PHENOTYPE:** apical meristem misshapen; leaf shape abnormal

### *dis1: distorted trichomes 1*

**LOCATION:** 1-18.7

**ORIGIN:** a new allele was induced with EMS in Landsberg *erecta* (Hülskamp et al. 1994)

**REFERENCES:** Feenstra (1978) *AIS 15:* 35-38; Hülskamp et al. (1994) *Cell 76:* 555-566

**PHENOTYPE:** trichomes on leaves and stems are short, bent, more or less club-shaped, and less often branched than wild type

**OTHER INFORMATION:** non-autonomous in genetic mosaics

### *dis2: distorted trichomes 2*

**LOCATION:** 1-44.6

**ORIGIN:** a new allele was induced with EMS in Landsberg *erecta* (Hülskamp et al. 1994)

**REFERENCES:** Feenstra (1978) *AIS 15:* 35-38; Hülskamp et al. (1994) *Cell 76:* 555-566

**PHENOTYPE:** like *dis1*

**OTHER INFORMATION:** autonomous in genetic mosaics

disrupted: see *dip*

distorted trichomes: see *dis*

### *doc1: dark overexpression of CAB 1*

**LOCATION:** chromosome 3—between *GapC* and *myb4* (Li et al. 1994)

ORIGIN: induced with EMS in a transgenic line (pOCA107-2, carrying *CAB3-Hph* and *CAB3-uidA* fusions) of Columbia background

ALLELE: one

REFERENCE: Li et al. (1994) *Genes Dev. 8:* 339-349

PHENOTYPE: resistant to hygromycin (thus has *CAB3*-driven expression) in the dark yet still has long hypocotyl, indicating normal dark morphogenesis; *CAB3-uidA* expression more than 6-fold higher than normal

OTHER INFORMATION: other mutants were isolated but not assigned to *doc1*, *doc2*, or *doc3* loci (Li et al. 1994), including a dominant mutant *doc0-6*, which was mapped to chromosome 2

### *doc2:* **dark overexpression of CAB 2**

LOCATION: chromosome 5 — between *tt3* (*DFR*) and *ASA1* (Li et al. 1994)

ORIGIN: induced with EMS (*doc2-1*) and by γ-ray (*doc2-2*) in a transgenic line (pOCA107-2, carrying *CAB3-Hph* and *CAB3-uidA* fusions) of Columbia background

ALLELES: *doc2-1; doc2-2*

REFERENCE: Li et al. (1994) *Genes Dev. 8:* 339-349

PHENOTYPE: resistant to hygromycin (thus has *CAB3*-driven expression) in the dark yet still has long hypocotyl, indicating normal dark morphogenesis; *CAB3-uidA* expression to 7-fold higher than normal

### *doc3:* **dark overexpression of CAB 3**

LOCATION: chromosome 1 — between *GapB* and *ADH* (Li et al. 1994)

ORIGIN: induced with EMS in a transgenic line (pOCA107-2, carrying *CAB3-Hph* and *CAB3-uidA* fusions) of Columbia background

ALLELE: one

REFERENCE: Li et al. (1994) *Genes Dev. 8:* 339-349

PHENOTYPE: resistant to hygromycin (thus has *CAB3*-driven expression) in the dark yet still has long hypocotyl, indicating normal dark morphogenesis; *CAB3-uidA* expression more than 3-fold higher than normal

### *drl1:* **deformed roots and leaves 1**

LOCATION: chromosome 1 (Bancroft et al. 1993)

ORIGIN: induced by *Ds* insertion in Landsberg *erecta*

REFERENCE: Bancroft et al. (1993) *Plant Cell 5:* 631-638

PHENOTYPE: stunted roots and few root hairs; lanceolate leaves; shoot apex extremely large and disorganized; no floral inflorescence; show somatic reversion and revertant sectors are fertile, producing normal progeny

### *dw1:* **dwarf 1**

LOCATION: 1-22.8

REFERENCE: Koornneef et al. (1983) *J. Hered. 74:* 265-272

PHENOTYPE: very short dwarf, heterozygote intermediate (semi-dominant)

### *dw2:* **dwarf 2, T-DNA insertional**

LOCATION: not mapped

ORIGIN: induced by T-DNA insertion in Wassilewskija (Ws)

REFERENCE: Feldmann et al. (1989) *Science 243:* 1351-1354

PHENOTYPE: dwarf, height about 1/5 (8 cm) of normal; short hypocotyl;

square cotyledons; short and dark leaves; large number of inflorescences (more than 40) form by time of senescence, indicating reduction of apical dominance; seed set reduced and seeds small

dwarf: see *dw1*

***dwf:*** auxin resistant ***dwarf***

LOCATION: 3-25.7

ORIGIN: induced in Landsberg *erecta*

ALLELES: line P445/10

REFERENCES: Mirza et al. (1984) *Physiol. Plant. 60:* 516-522; Olsen et al. (1984) *Physiol. Plant. 60:* 523-531; Mirza and Maher (1985) *AIS 22:* 23-33; Mirza and Maher (1985) *AIS 22:* 35-41; Mirza and Maher (1987) *Plant Growth Regul. 5:* 41-49; Wilson et al. (1990) *MGG 222:* 377-383

PHENOTYPE: homozygous lethal, heterozygous plants have a dwarf rosette and inflorescence, single unbranched and agravitropic roots, and show 50% root inhibition due to exogenous application of the auxin 2,4-dichloro-phenoxyacetic acid (2,4-D) at a 2,4-D level 2000-fold higher than wild type

E17: see *trp4-2*

EII: see *abi2-1*

eceriferum: see *cer*

*ein1:* ethylene insensitive 1, allelic with *etr,* see *etr*

***ein2:* ethylene *insensitive* 2**

LOCATION: 5-1.3

ORIGIN: induced with EMS in Columbia

ALLELES: *ein2-1,2,3,4,5*

REFERENCES: Guzman and Ecker (1990) *Plant Cell 2:* 513-523; Lawton et al. (1994) *Plant Cell 6:* 581-588.

PHENOTYPE: seedlings insensitive to ethylene; larger than normal rosette; delayed bolting; normal systemic acquired resistance

***ein3:* ethylene *insensitive* 3**

LOCATION: not mapped

ORIGIN: induced with EMS in Columbia

REFERENCES: Kieber et al. (1993) *Cell 72:* 427-441; Rothenberg and Ecker (1993) *Sem. Dev. Biol.: Plant Dev. Genet. 4:* 3-13

PHENOTYPE: seedlings insensitive to ethylene

***ela1:* enhanced *l*inolenate *accumulation***

LOCATION: not known

ORIGIN: induced with EMS in Columbia

ALLELE: *ela1-1* (line JB11)

REFERENCES: Lemieux et al. (1990) *Theor. Appl. Genet. 80:* 234-240; James and Dooner (1991) *Theor. Appl. Genet. 82:* 409-412

PHENOTYPE: increased level of 18:3 fatty acid and decreased levels of 18:1 and 18:2 fatty acids

***emb1:* emb*ryo defective 1**

LOCATION: 3-5.0

ORIGIN: induced with EMS in Columbia

ALLELES: line 53D-4A

REFERENCES: Meinke (1985) *Theor. Appl. Genet. 69:* 543-552; Franzmann et al. (1989) *Theor. Appl. Genet. 77:* 609-616

PHENOTYPE: embryo development arrested at early preglobular stage; seed creamy white

### *emb2: emb*ryo **defective 2**

LOCATION: 5-27.4

ORIGIN: induced with EMS in Columbia

ALLELES: line 124D

REFERENCES: Meinke and Sussex (1979) *Dev. Biol. 72:* 62-72; Meinke (1982) *Theor. Appl. Genet. 63:* 381-386; Franzmann et al. (1989) *Theor. Appl. Genet. 77:* 609-616

PHENOTYPE: embryo development arrested at preglobular stage; seed creamy white

### *emb5: emb*ryo **defective 5**

LOCATION: not mapped

ORIGIN: induced with EMS in Columbia

ALLELES: line 127AX-A

REFERENCES: Meinke (1985) *Theor. Appl. Genet. 69:* 543-552; Franzmann et al. (1989) *Theor. Appl. Genet. 77:* 609-616

PHENOTYPE: embryo development arrested at preglobular stage; embryo and seed creamy white

### *emb6: emb*ryo **defective 6**

LOCATION: not mapped

ORIGIN: induced with EMS in Columbia

ALLELES: line 113K-1B

REFERENCES: Meinke (1985) *Theor. Appl. Genet. 69:* 543-552; Franzmann et al. (1989) *Theor. Appl. Genet. 77:* 609-616

PHENOTYPE: embryo development arrested at preglobular stage; embryo and seed creamy white

### *emb7: emb*ryo **defective 7**

LOCATION: 1-11.2

ORIGIN: induced with EMS in Columbia

ALLELES: line 112E-2A

REFERENCES: Meinke (1985) *Theor. Appl. Genet. 69:* 543-552; Franzmann et al. (1989) *Theor. Appl. Genet. 77:* 609-616

PHENOTYPE: embryo development arrested at preglobular stage; embryo and seed creamy white

### *emb8: emb*ryo **defective 8**

LOCATION: not mapped

ORIGIN: induced with EMS in Columbia

ALLELES: line 113J-4A

REFERENCES: Meinke (1985) *Theor. Appl. Genet. 69:* 543-552; Franzmann et al. (1989) *Theor. Appl. Genet. 77:* 609-616

PHENOTYPE: embryo development arrested at preglobular stage; embryo and

seed creamy white
### *emb9: emb*ryo defective 9
LOCATION: 5-59.2
ORIGIN: induced with EMS in Columbia
ALLELES: line 79A
REFERENCES: Meinke and Sussex (1979) *Dev. Biol. 72:* 62-72; Meinke (1982) *Theor. Appl. Genet. 63:* 381-386; Franzmann et al. (1989) *Theor. Appl. Genet. 77:* 609-616; Patton et al. (1991) *MGG 227:* 337-347
PHENOTYPE: embryo development arrested at early globular stage; seed and embryo creamy white; no growth in culture
### *emb10: emb*ryo defective 10
LOCATION: 1-10.9
ORIGIN: induced with EMS.in Columbia
ALLELES: line 112E-1B
REFERENCES: Meinke (1985) *Theor. Appl. Genet. 69:* 543-552; Franzmann et al. (1989) *Theor. Appl. Genet. 77:* 609-616
PHENOTYPE: embryo development arrested at early globular stage; embryo creamy white and seed white to very pale yellow-green
### *emb11: emb*ryo defective 11
LOCATION: not mapped
ORIGIN: induced with EMS in Columbia
ALLELES: line 111H-2B1
REFERENCES: Meinke (1985) *Theor. Appl. Genet. 69:* 543-552; Franzmann et al. (1989) *Theor. Appl. Genet. 77:* 609-616
PHENOTYPE: embryo development arrested at early globular stage; embryo creamy white and seed white to very pale yellow-green
### *emb12: emb*ryo defective 12
LOCATION: not mapped
ORIGIN: induced with EMS in Columbia
ALLELES: line 130BA-1
REFERENCES: Meinke (1985) *Theor. Appl. Genet. 69:* 543-552; Franzmann et al. (1989) *Theor. Appl. Genet. 77:* 609-616
PHENOTYPE: embryo development arrested at early globular stage; embryo creamy white and seed white to very pale yellow-green
### *emb13: emb*ryo defective 13
LOCATION: not mapped
ORIGIN: induced with EMS in Columbia
ALLELES: line 111B-5E
REFERENCES: Meinke (1985) *Theor. Appl. Genet. 69:* 543-552; Franzmann et al. (1989) *Theor. Appl. Genet. 77:* 609-616
PHENOTYPE: embryo development arrested at early globular stage; embryo and seed white to very pale yellow-green
*emb14:* embryo defective 14, allelic with *emb177,* see *emb177*
### *emb15: emb*ryo defective 15
LOCATION: 5-85.4

ORIGIN: induced with EMS in Columbia

ALLELES: line 57B-4C

REFERENCES: Meinke (1985) *Theor. Appl. Genet. 69:* 543-552; Franzmann et al. (1989) *Theor. Appl. Genet. 77:* 609-616; Patton et al. (1991) *MGG 227:* 337-347

PHENOTYPE: embryo development arrested at globular stage; seed and embryo creamy white; limited callus formation in culture

*emb16: emb*ryo defective 16

LOCATION: 5-89.7

ORIGIN: induced with EMS in Columbia

ALLELES: line 123B

REFERENCES: Meinke and Sussex (1979) *Dev. Biol. 72:* 62-72; Meinke (1982) *Theor. Appl. Genet. 63:* 381-386; Franzmann et al. (1989) *Theor. Appl. Genet. 77:* 609-616; Patton et al. (1991) *MGG 227:* 337-347

PHENOTYPE: embryo development arrested at globular stage; seed and embryo creamy white; limited callus formation in culture

*emb17: emb*ryo defective 17

LOCATION: not mapped

ORIGIN: induced with EMS in Columbia

ALLELES: line 109A-1B

REFERENCES: Meinke (1985) *Theor. Appl. Genet. 69:* 543-552; Franzmann et al. (1989) *Theor. Appl. Genet. 77:* 609-616

PHENOTYPE: embryo development arrested at globular stage; embryo and seed creamy white; limited callus formation in culture

*emb18: emb*ryo defective 18

LOCATION: 2-60.5

ORIGIN: induced with EMS in Columbia

ALLELES: line 109F-5D

REFERENCES: Meinke (1985) *Theor. Appl. Genet. 69:* 543-552; Franzmann et al. (1989) *Theor. Appl. Genet. 77:* 609-616; Patton et al. (1991) *MGG 227:* 337-347; Yeung and Meinke (1993) *Plant Cell 5:* 1371-1381

PHENOTYPE: embryo development arrested at globular to heart stages; seed creamy white to very pale yellow-green; embryo creamy white; limited callus formation in culture

*emb19: emb*ryo defective 19

LOCATION: not mapped

ORIGIN: induced with EMS in Columbia

ALLELES: line 112G-1A

REFERENCES: Meinke (1985) *Theor. Appl. Genet. 69:* 543-552; Franzmann et al. (1989) *Theor. Appl. Genet. 77:* 609-616; Yeung and Meinke (1993) *Plant Cell 5:* 1371-1381

PHENOTYPE: embryo development arrested at globular to heart stages; embryo and seed creamy white to very pale yellow-green

*emb20: emb*ryo defective 20

LOCATION: 4-54.4

ORIGIN: induced with EMS in Columbia

ALLELES: *emb20-1* (line 117N-1B); *emb20-2* (line 87A); *emb20-3* (formerly *emb26* = line 63A-1A)

REFERENCES: Meinke and Sussex (1979) *Dev. Biol. 72:* 62-72; Meinke (1982) *Theor. Appl. Genet. 63:* 381-386; Meinke (1985) *Theor. Appl. Genet. 69:* 543-552; Franzmann et al. (1989) *Theor. Appl. Genet. 77:* 609-616; Patton et al. (1991) *MGG 227:* 337-347

PHENOTYPE: *emb20-1* and *emb20-2:* embryo development arrested at globular to heart stages; seed and embryo creamy white; form limited callus and distorted shoot in culture

*emb20-3:* embryo development arrested at globular to cotyledon stages; seed and embryo creamy white to very pale yellow-green; plants with pale flowers develop in culture

### *emb21: emb*ryo defective 21

LOCATION: not mapped

ORIGIN: induced with EMS in Columbia

ALLELES: line 129AX-2A

REFERENCES: Meinke (1985) *Theor. Appl. Genet. 69:* 543-552; Franzmann et al. (1989) *Theor. Appl. Genet. 77:* 609-616

PHENOTYPE: embryo development arrested at globular to heart stages; embryo and seed creamy white

### *emb22: emb*ryo defective 22

LOCATION: 1-62.1

ORIGIN: induced with EMS in Columbia

ALLELES: line 115D-4A

REFERENCES: Meinke (1985) *Theor. Appl. Genet. 69:* 543-552; Franzmann et al. (1989) *Theor. Appl. Genet. 77:* 609-616; Patton et al. (1991) *MGG 227:* 337-347

PHENOTYPE: embryo development arrested as green blimp; seed and embryo pale green to green; abnormal green shoots form in culture

### *emb23: emb*ryo defective 23

LOCATION: not mapped

ORIGIN: induced with EMS in Columbia

ALLELES: line 126E-B

REFERENCES: Meinke (1985) *Theor. Appl. Genet. 69:* 543-552; Franzmann et al. (1989) *Theor. Appl. Genet. 77:* 609-616

PHENOTYPE: embryo development arrested at globular to linear stages; embryo creamy white to green and seed creamy white to pale green

### *emb24: emb*ryo defective 24

LOCATION: 5-1.1

ORIGIN: induced with EMS in Columbia

ALLELES: line 109F-1C

REFERENCES: Meinke (1985) *Theor. Appl. Genet. 69:* 543-552; Franzmann et al. (1989) *Theor. Appl. Genet. 77:* 609-616; Patton et al. (1991) *MGG 227:* 337-347

**PHENOTYPE:** embryo development arrested at globular to cotyledon stages; seed very pale yellow-green to pale green and embryo creamy white to pale green; normal plants develop in culture

***emb25: emb*ryo defective 25**

LOCATION: 1-99.6

ORIGIN: induced with EMS in Columbia

ALLELES: line 115J-4A

REFERENCES: Meinke (1985) *Theor. Appl. Genet. 69:* 543-552; Franzmann et al. (1989) *Theor. Appl. Genet. 77:* 609-616; Patton et al. (1991) *MGG 227:* 337-347

**PHENOTYPE:** embryo development arrested at globular to cotyledon stages; seed and embryo creamy white; distorted pale plantlets form in culture

*emb26:* embryo defective 26, allelic with *emb20,* see *emb20*

***emb27: emb*ryo defective 27**

LOCATION: not mapped

ORIGIN: induced with EMS in Columbia

ALLELES: line 111B-5B

REFERENCES: Meinke (1985) *Theor. Appl. Genet. 69:* 543-552; Franzmann et al. (1989) *Theor. Appl. Genet. 77:* 609-616

**PHENOTYPE:** embryo development arrested at globular to cotyledon stages; embryo and seed creamy white; very pale yellow-green shoots with abnormal leaves and flowers form in culture

***emb28: emb*ryo defective 28**

LOCATION: 4-58.2

ORIGIN: induced with EMS in Columbia

ALLELES: line 115H-1A

REFERENCES: Meinke (1985) *Theor. Appl. Genet. 69:* 543-552; Franzmann et al. (1989) *Theor. Appl. Genet. 77:* 609-616; Patton et al. (1991) *MGG 227:* 337-347

**PHENOTYPE:** embryo development arrested at globular to cotyledon stages; seed and embryo creamy white; pale green plantlets form in culture

***emb29: emb*ryo defective 29, allelic with *emb95***

LOCATION: 3-24.0

ORIGIN: induced with EMS in Columbia; a second allele (*emb95*) was found to be tightly linked to a T-DNA insertion (T-1793) in Wassilewskija (Ws)

ALLELES: line 115C-1C

REFERENCES: Meinke (1985) *Theor. Appl. Genet. 69:* 543-552; Franzmann et al. (1989) *Theor. Appl. Genet. 77:* 609-616; Errampalli et al. (1991) *Plant Cell 3:* 149-157; Patton et al. (1991) *MGG 227:* 337-347

**PHENOTYPE:** embryo development arrested at globular to cotyledon stages; seed and embryo creamy white; distorted albino plantlets form in culture

*emb30:* embryo defective 30, allelic with *gn,* see *gn*

***emb31: emb*ryo defective 31**

LOCATION: not mapped

ORIGIN: induced with EMS in Columbia

ALLELES: line 130BA-2

REFERENCES: Meinke (1985) *Theor. Appl. Genet. 69:* 543-552; Franzmann et al. (1989) *Theor. Appl. Genet. 77:* 609-616; Patton and Meinke (1990) *Am. J. Bot. 77:* 653-661

PHENOTYPE: embryo has normal hypocotyl and abnormal cotyledons with variable size and shape; embryo pale green; accumulation of protein bodies delayed in cotyledons; embryos germinate in culture, as well as occasionally in soil, to develop into plants that are normal except for the slow growth rate and high frequency of aborted seeds, suggesting that the allele may be leaky

*emb33:* embryo defective 33, allelic to *emb177,* see *emb177*

***emb39: emb*ryo defective 39**

LOCATION: 2-72.8

ORIGIN: induced with EMS in Columbia

REFERENCE: Patton et al. (1991) *MGG 227:* 337-347

PHENOTYPE: embryo development arrested as distorted cotyledon; seed creamy white and embryo creamy white to very pale yellow-green; distorted green plantlets form in culture

*emb60:* embryo defective 60, allelic with *emb76,* see *emb76*

***emb76: emb*ryo defective 76, allelic with *emb60***

LOCATION: 1-0.0

ORIGIN: *emb60* induced with EMS in Columbia; *emb76* induced by T-DNA insertion (T-76) in Wassilewskija (Ws)

REFERENCES: Errampalli et al. (1991) *Plant Cell 3:* 149-157; Patton et al. (1991) *MGG 227:* 337-347; Castle et al. (1993) *MGG 241:* 504-514; Yeung and Meinke (1993) *Plant Cell 5:* 1371-1381

PHENOTYPE: embryo development arrested at globular stage; seed and embryo creamy white; limited callus formation in culture

***emb77: emb*ryo defective 77**

LOCATION: 4-66.8

ORIGIN: segregated from a T-DNA insertional line (T-77) in Wassilewskija (Ws)

REFERENCE: Errampalli et al. (1991) *Plant Cell 3:* 149-157

PHENOTYPE: embryo development arrested at late preglobular stage; seed white

*emb78:* embryo defective 78; allelic with *fus6,* see *fus6*

***emb79: emb*ryo defective 79**

LOCATION: not mapped

ORIGIN: segregated from a T-DNA insertional line (T-79) in Wassilewskija (Ws)

REFERENCE: Errampalli et al. (1991) *Plant Cell 3:* 149-157

PHENOTYPE: embryo development arrested at late preglobular stage; seed white

***emb80: emb*ryo defective 80**

LOCATION: not mapped

ORIGIN: segregated from a T-DNA insertional line (T-98) in Wassilewskija

(Ws)

**REFERENCE:** Errampalli et al. (1991) *Plant Cell 3:* 149-157

**PHENOTYPE:** embryo development arrested at late preglobular stage; seed pale yellow-green

### *emb81: emb*ryo defective 81

**LOCATION:** 4-38.5

**ORIGIN:** segregated from a T-DNA insertional line (T-668) in Wassilewskija (Ws)

**REFERENCE:** Errampalli et al. (1991) *Plant Cell 3:* 149-157

**PHENOTYPE:** embryo development arrested at early heart stage; seed white

### *emb82: emb*ryo defective 82

**LOCATION:** not mapped

**ORIGIN:** tightly linked to a T-DNA insertion (T-739) in Wassilewskija (Ws)

**REFERENCE:** Errampalli et al. (1991) *Plant Cell 3:* 149-157

**PHENOTYPE:** embryo albino; seed white

### *emb83: emb*ryo defective 83

**LOCATION:** not mapped

**ORIGIN:** tightly linked to a T-DNA insertion (T-869) in Wassilewskija (Ws)

**REFERENCE:** Errampalli et al. (1991) *Plant Cell 3:* 149-157

**PHENOTYPE:** embryo arrested at globular to heart stages; seed white to pale yellow-green

### *emb84: emb*ryo defective 84

**LOCATION:** not mapped

**ORIGIN:** tightly linked to a T-DNA insertion (T-508) in Wassilewskija (Ws)

**REFERENCES:** Errampalli et al. (1991) *Plant Cell 3:* 149-157; Castle et al. (1993) *MGG 241:* 504-514; Yeung and Meinke (1993) *Plant Cell 5:* 1371-1381

**PHENOTYPE:** embryo arrested at globular stage, with an abnormal suspensor; seed pale yellow-green

### *emb86: emb*ryo defective 86

**LOCATION:** 5-24

**ORIGIN:** tightly linked to a T-DNA insertion (T-731) in Wassilewskija (Ws)

**REFERENCES:** Errampalli et al. (1991) *Plant Cell 3:* 149-157; Castle et al. (1993) *MGG 241:* 504-514

**PHENOTYPE:** embryo arrested at globular to heart stages; seed white

### *emb87: emb*ryo defective 87

**LOCATION:** 5-84.7

**ORIGIN:** tightly linked to a T-DNA insertion (T-811) in Wassilewskija (Ws)

**REFERENCES:** Errampalli et al. (1991) *Plant Cell 3:* 149-157; Castle et al. (1993) *MGG 241:* 504-514

**PHENOTYPE:** embryo arrested at globular stage; seed white

### *emb88: emb*ryo defective 88

**LOCATION:** not mapped

**ORIGIN:** tightly linked to a T-DNA insertion (T-1002) in Wassilewskija (Ws)

**REFERENCES:** Errampalli et al. (1991) *Plant Cell 3:* 149-157; Castle et al.

(1993) *MGG 241:* 504-514; Yeung and Meinke (1993) *Plant Cell 5:* 1371-1381

PHENOTYPE: embryo arrested at globular stage with an abnormal suspensor; seed white

*emb89: emb*ryo defective 89

LOCATION: not mapped

ORIGIN: segregated from a T-DNA insertional line (T-543) in Wassilewskija (Ws)

REFERENCE: Errampalli et al. (1991) *Plant Cell 3:* 149-157

PHENOTYPE: embryo development arrested at globular to mature stages; seed white

*emb90: emb*ryo defective 90

LOCATION: not mapped

ORIGIN: segregated from a T-DNA insertional line (T-749) in Wassilewskija (Ws)

REFERENCE: Errampalli et al. (1991) *Plant Cell 3:* 149-157

PHENOTYPE: embryo development arrested at globular to heart stages; seed white

*emb91: emb*ryo defective 91

LOCATION: not mapped

ORIGIN: segregated from a T-DNA insertional line (T-685) in Wassilewskija (Ws)

REFERENCE: Errampalli et al. (1991) *Plant Cell 3:* 149-157

PHENOTYPE: embryo development arrested at early globular stage; seed white

*emb93: emb*ryo defective 93

LOCATION: 2-7.3

ORIGIN: possibly tagged with a T-DNA insertion in line T-562 in Wassilewskija (Ws), line has additional inserts

REFERENCES: Errampalli et al. (1991) *Plant Cell 3:* 149-157; Castle et al. (1993) *MGG 241:* 504-514

PHENOTYPE: embryo arrested at globular to mature stages; seed white

*emb94: emb*ryo defective 94

LOCATION: not mapped

ORIGIN: segregated from a T-DNA insertional line (T-1790) in Wassilewskija (Ws)

REFERENCE: Errampalli et al. (1991) *Plant Cell 3:* 149-157

PHENOTYPE: embryo development arrested at globular to heart stages; seed white

*emb95:* embryo defective 95, allelic with *emb29,* see *emb29*

*emb104: emb*ryo defective 104

LOCATION: 4-63.3

ORIGIN: induced by T-DNA insertion in Wassilewskija (Ws)

REFERENCE: Castle et al. (1993) *MGG 241:* 504-514

PHENOTYPE: embryo with elongated and distorted cotyledons; seed and embryo white to pale yellow-green

*emb107:* **embryo defective 107**
  LOCATION: 4-61.5
  ORIGIN: induced by T-DNA insertion in Wassilewskija (Ws)
  REFERENCE: Castle et al. (1993) *MGG 241:* 504-514
  PHENOTYPE: embryo arrested at preglobular stage; seed and embryo white
*emb111:* **embryo defective 111**
  LOCATION: 2-64.2
  ORIGIN: induced by T-DNA insertion in Wassilewskija (Ws)
  REFERENCE: Castle et al. (1993) *MGG 241:* 504-514
  PHENOTYPE: elongated globular embryo; embryo white and seed pale yellow-green
*emb115:* **embryo defective 115**
  LOCATION: chromosome 4 (Castle et al. 1993)
  ORIGIN: induced by T-DNA insertion in Wassilewskija (Ws)
  REFERENCE: Castle et al. (1993) *MGG 241:* 504-514
  PHENOTYPE: globular embryo; embryo white and seed pale yellow-green
*emb116:* **embryo defective 116**
  LOCATION: 5-5.7
  ORIGIN: induced by T-DNA insertion in Wassilewskija (Ws)
  REFERENCE: Castle et al. (1993) *MGG 241:* 504-514
  PHENOTYPE: embryo arrested at late preglobular stage; seed and embryo white
*emb123:* **embryo defective 123**
  LOCATION: on a translocated chromosome with portions of chromosome 1 and 2; linked to *clv2* (7.6±1.6 cM) and *as* (1.9±0.8 cM) (Castle et al. 1993)
  ORIGIN: induced by T-DNA insertion in Wassilewskija (Ws)
  REFERENCE: Castle et al. (1993) *MGG 241:* 504-514
  PHENOTYPE: embryo arrested at late preglobular stage; seed and embryo white
*emb129:* **embryo defective 129**
  LOCATION: on a translocated chromosome with portions of chromosome 3 and 5; linked to *gl1* (2.6±1.0 cM) and *yi* (1.6±0.7 cM) (Castle et al. 1993)
  ORIGIN: induced by T-DNA insertion in Wassilewskija (Ws)
  REFERENCE: Castle et al. (1993) *MGG 241:* 504-514
  PHENOTYPE: embryo arrested at late preglobular stage; seed and embryo white
*emb130:* **embryo defective 130**
  LOCATION: 4-46.1
  ORIGIN: induced by T-DNA insertion in Wassilewskija (Ws)
  REFERENCE: Castle et al. (1993) *MGG 241:* 504-514
  PHENOTYPE: embryo arrested at early preglobular stage; seed and embryo white
*emb134:* embryo defective 134, allelic with *fus8,* see *fus8*
*emb140:* **embryo defective 140**
  LOCATION: 4-51.9
  ORIGIN: induced by T-DNA insertion in Wassilewskija (Ws)
  REFERENCE: Castle et al. (1993) *MGG 241:* 504-514
  PHENOTYPE: embryo arrested at early globular stage; seed pale yellow-green

and embryo white to pale yellow-green

***emb142: emb*ryo defective 142**

LOCATION: 1-22.0

ORIGIN: induced by T-DNA insertion in Wassilewskija (Ws)

REFERENCE: Castle et al. (1993) *MGG 241:* 504-514

PHENOTYPE: embryo arrested at early globular stage; seed white to pale yellow-green and embryo white

***emb143:*** embryo defective 143, allelic with *cop9,* see *cop9*

***emb144:*** embryo defective 144, allelic with *fus9,* see *fus9*

***emb145: emb*ryo defective 145**

LOCATION: 3-0.0

ORIGIN: induced by T-DNA insertion in Wassilewskija (Ws)

REFERENCE: Castle et al. (1993) *MGG 241:* 504-514

PHENOTYPE: embryo arrested at globular stage; irregular protoderm; seed pale yellow-green and embryo white to pale yellow-green

***emb146: emb*ryo defective 146**

LOCATION: chromosome 4 (Castle et al. 1993)

ORIGIN: induced by T-DNA insertion in Wassilewskija (Ws)

REFERENCE: Castle et al. (1993) *MGG 241:* 504-514

PHENOTYPE: embryo arrested at preglobular stage; seed and embryo white

***emb151: emb*ryo defective 151**

LOCATION: on a translocated chromosome with portions of chromosome 3 and 5; linked to *hy2* (0.9±0.5 cM) and *yi* (6.2±1.6 cM) (Castle et al. 1993)

ORIGIN: induced by T-DNA insertion in Wassilewskija (Ws)

REFERENCE: Castle et al. (1993) *MGG 241:* 504-514

PHENOTYPE: embryo arrested at globular to heart stages; seed pale yellow-green and embryo white to pale yellow-green

***emb167: emb*ryo defective 167**

LOCATION: 3-63.2

ORIGIN: induced by T-DNA insertion in Wassilewskija (Ws)

REFERENCE: Castle et al. (1993) *MGG 241:* 504-514

PHENOTYPE: embryo arrested at preglobular stage; seed and embryo white

***emb168:*** embryo defective 168, allelic with *cop1,* see *cop1*

***emb175: emb*ryo defective 175**

LOCATION: 5-6.4

ORIGIN: induced by T-DNA insertion in Wassilewskija (Ws)

REFERENCE: Castle et al. (1993) *MGG 241:* 504-514

PHENOTYPE: embryo arrested at globular stage; seed and embryo white

***emb177: emb*ryo defective 177, allelic with *emb14* and *emb33***

LOCATION: 1-115.8

ORIGIN: the original *emb14* and *emb33* induced with EMS in Columbia; *emb177* induced by T-DNA insertion in Wassilewskija (Ws)

ALLELES: *emb14* (line 95A-2B); *emb33; emb177*

REFERENCES: Meinke (1985) *Theor. Appl. Genet. 69:* 543-552; Franzmann et al. (1989) *Theor. Appl. Genet. 77:* 609-616; Patton et al. (1991) *MGG 227:*

337-347; Castle et al. (1993) *MGG 241:* 504-514

**PHENOTYPE:** *emb14* and *emb33:* embryo development arrested at globular stage; seed very pale yellow-green and embryo creamy white to very pale yellow-green

*emb33:* limited callus formation in culture

*emb177:* embryo arrested at globular to heart stages; huge suspensor; seed pale yellow-green and embryo white to pale yellow-green

*emb179: emb***ryo defective 179**

**LOCATION:** 1-98.7

**ORIGIN:** induced by T-DNA insertion in Wassilewskija (Ws)

**REFERENCE:** Castle et al. (1993) *MGG 241:* 504-514

**PHENOTYPE:** globular or abnormal liner embryo; seed pale yellow-green and embryo pale yellow-green to pale green

*emb201: emb***ryo defective 201**

**LOCATION:** 3-23.3

**ORIGIN:** induced by T-DNA insertion in Wassilewskija (Ws)

**REFERENCE:** Castle et al. (1993) *MGG 241:* 504-514

**PHENOTYPE:** embryo arrested at linear stage; reduced cotyledons; seed pale yellow-green and embryo white to pale yellow-green

*emb210: emb***ryo defective 210**

**LOCATION:** chromosome 4 (Castle et al. 1993)

**ORIGIN:** induced by T-DNA insertion in Wassilewskija (Ws)

**REFERENCE:** Castle et al. (1993) *MGG 241:* 504-514

**PHENOTYPE:** embryo arrested at globular to linear stages; monocot; seed pale yellow-green and embryo white to pale green

*emb212:* embryo defective 212, allelic with *lec1,* see *lec1*

*emb215: emb***ryo defective 215**

**LOCATION:** not mapped

**ORIGIN:** induced by T-DNA insertion in Wassilewskija (Ws)

**REFERENCE:** Castle et al. (1993) *MGG 241:* 504-514

**PHENOTYPE:** embryo arrested at globular stage; seed and embryo white

*emb224 emb***ryo defective 224**

**LOCATION:** 3-29.7

**ORIGIN:** induced by T-DNA insertion in Wassilewskija (Ws)

**REFERENCE:** Castle et al. (1993) *MGG 241:* 504-514

**PHENOTYPE:** embryo arrested at large heart stage; seed and embryo white

*emb229: emb***ryo defective 229**

**LOCATION:** not mapped

**ORIGIN:** induced by T-DNA insertion in Wassilewskija (Ws)

**REFERENCE:** Castle et al. (1993) *MGG 241:* 504-514

**PHENOTYPE:** embryo arrested at globular stage; seed and embryo white

*emb233: emb***ryo defective 233**

**LOCATION:** not mapped

**ORIGIN:** induced by T-DNA insertion in Wassilewskija (Ws)

**REFERENCE:** Castle et al. (1993) *MGG 241:* 504-514

PHENOTYPE: embryo arrested at globular stage; seed and embryo white

*emb236: emb***ryo defective 236**

LOCATION: 4-32.6

ORIGIN: induced by T-DNA insertion in Wassilewskija (Ws)

REFERENCE: Castle et al. (1993) *MGG 241:* 504-514

PHENOTYPE: embryo arrested at globular stage; seed and embryo pale yellow-green

*emb244: emb***ryo defective 244**

LOCATION: not mapped

ORIGIN: induced by T-DNA insertion in Wassilewskija (Ws)

REFERENCE: Castle et al. (1993) *MGG 241:* 504-514

PHENOTYPE: embryo arrested at globular stage; abnormal suspensor; seed and embryo white

*emb251: emb***ryo defective 251**

LOCATION: not mapped

ORIGIN: induced by T-DNA insertion in Wassilewskija (Ws)

REFERENCE: Castle et al. (1993) *MGG 241:* 504-514

PHENOTYPE: embryo arrested at late preglobular stage; seed and embryo white

*emb262: emb***ryo defective 262**

LOCATION: 5-17.6

ORIGIN: induced by T-DNA insertion in Wassilewskija (Ws)

REFERENCE: Castle et al. (1993) *MGG 241:* 504-514

PHENOTYPE: embryo arrested at linear to curled cotyledon stage; seed pale green and embryo white to pale yellow-green

*emb270: emb***ryo defective 270**

LOCATION: on a translocated chromosome with portions of chromosome 1 and 4; linked to *ch1* (12.2±1.6 cM) and *cer2* (14.3±1.7 cM) (Castle et al. 1993)

ORIGIN: induced by T-DNA insertion in Wassilewskija (Ws)

REFERENCE: Castle et al. (1993) *MGG 241:* 504-514

PHENOTYPE: embryo arrested at globular stage and distorted; tricot; seed and embryo white to pale green

*emb286: emb***ryo defective 286**

LOCATION: not mapped

ORIGIN: induced by T-DNA insertion in Wassilewskija (Ws)

REFERENCE: Castle et al. (1993) *MGG 241:* 504-514

PHENOTYPE: embryo arrested at early globular stage; seed and embryo white

*emb287: emb***ryo defective 287**

LOCATION: not mapped

ORIGIN: induced by T-DNA insertion in Wassilewskija (Ws)

REFERENCE: Castle et al. (1993) *MGG 241:* 504-514

PHENOTYPE: embryo arrested at late preglobular stage; seed and embryo white

embryo defective: see *bio1* and *emb*

embryonic flower: see *emf*

*emf*: **embryonic *flower***

LOCATION: not mapped

**ORIGIN**: induced with EMS or by γ-ray in Columbia

**ALLELES**: four independent mutants were isolated; phenotypes described for one EMS mutant

**REFERENCE**: Sung et al. (1992) *Science 258:* 1645-1647

**PHENOTYPE**: inflorescence produced after cotyledons without formation of vegetative shoot; cotyledons lack petioles; shoot apical meristem in mutant embryo differs from wild-type one, showing floral inflorescence meristem characteristics; a few leaves form that are morphologically similar to cauline leaves; early flowering not affected by photoperiod; reproductive shoots, but not rosette shoots, were induced from mutant calli; although mutant flowers have pistils, they usually lack petals and often are without anthers; the anthers that do appear lack pollen and the ovules fail to develop into viable seeds; normal roots

EMS142: see *hy3*

### *er: erecta*

**LOCATION**: 2-43.5

**ORIGIN**: induced in presoaked Landsberg seeds by 12 kr X-ray (*er*) or with 40 mM EMS for 8 hrs (*er*-cit, isolated by R.E. Pruitt at Caltech)

**REFERENCES**: Rédei (1962) *Z. Vererbungsl. 93:* 164-170; Rédei and Hirono (1964) *AIS 1:* 9-10; Lang et al. (1994) *Genetics 137:* 1101-1110

**PHENOTYPE**: compact rosette, short petioles, blunt fruits

**OTHER INFORMATION**: the *er* marker homozygous in the Landsberg background is used by a number of labs as their standard wild-type strain; *er* may be the same as *mod1*

erecta: see *er*

ethylene insensitive: see *ain1, ein,* and *etr*

ethylene overproducer: see *eto*

ethylene resistant: see *ain1, aux1, ein,* and *etr*

### *eto1: ethylene overproducer 1*

**LOCATION**: 5-82.9

**ORIGIN**: induced with EMS in Columbia

**ALLELES**: *eto1-1; eto1-2;* additional alleles reported in Kieber et al. (1993) but not described

**REFERENCES**: Guzman and Ecker (1990) *Plant Cell 2:* 513-523; Kieber et al. (1993) *Cell 72:* 427-441.

**PHENOTYPE**: recessive to wild type; constitutive seedling triple response to ethylene; overproduction of ethylene; in air, rosette smaller than wild type

### *eto2: ethylene overproducer 2*

**LOCATION**: chromosome 5, near *yi* (2.2 ± 0.8 cM) (Kieber et al. 1993)

**ORIGIN**: induced with diepoxybutane in Columbia

**REFERENCE**: Kieber et al. (1993) *Cell 72:* 427-441

**PHENOTYPE**: dominant over wild type; in the absence of exogenous ethylene, dark-grown seedlings show triple response: shortening and radial swelling of hypocotyl, inhibition of root elongation, and exaggeration of the curvature of the apical hook; increased production of endogenous ethylene in seedlings but not in adult; the constitutive triple response eliminated by inhibitors of ethylene production and binding

**eto3: *et*hylene *o*verproducer 3**

LOCATION: not known

ORIGIN: induced with diepoxybutane in Columbia

REFERENCE: Kieber et al. (1993) *Cell 72:* 427-441

PHENOTYPE: similar to *eto2*

OTHER INFORMATION: maps to a different chromosome than *eto2* (J. Ecker, pers. comm.)

**etr1: *et*hylene resistant, allelic with *ein1***

LOCATION: 1-92.7

ORIGIN: induced with EMS in Columbia

ALLELES: *etr1-1* (= *etr*) (Bleecker et al. 1988); *etr1-2* (Chang et al. 1993); *etr1-3* (formerly *ein1*) (Guzman and Ecker 1990); *etr1-4* (G. Roman and J. Ecker, unpubl., see Chang et al. 1993)

REFERENCES: Bleecker et al. (1988) *Science 241:* 1086-1089; Guzman and Ecker (1990) *Plant Cell 2:* 513-523; Chang et al. (1993) *Science 262:* 539-544; Lawton et al. (1994) *Plant Cell 6:* 581-588

PHENOTYPE: dominant over wild type; root and hypocotyl elongation insensitive to ethylene; leaf chlorophyll content not reduced in ethylene; ethylene also fails to enhance peroxidase activity in leaf and stem and fails to increase the frequency of seed germination in dark; reduced ethylene binding activity; larger than normal rosette; delayed bolting; normal systemic acquired resistance

GENE: cloned by Chang et al. (1993) using chromosome walking; encodes a protein of 738 amino acids (82.5 kD); ETR1 is a protein with similarity to bacterial two-component regulators: the region from residue 326 to 562 of ETR1 shows similarity with the histidine kinase domains of the sensor components and the region from 610 to 729 shows similarity with the response regulator domains; four mutant alleles were sequenced: *etr1-1*, $Cys_{65}$ (TGT) to Tyr (TAT); *etr1-2*, $Ala_{102}$ (GCG) to Thr (ACG); *etr1-3*, $Ala_{31}$ (GCG) to Val (GTG); and *etr1-4*, $Ile_{62}$ (ATC) to Phe (TTC); the positions of the dominant ethylene resistant mutations are all within the first 110 amino acid residues, away from the conserved domains, suggesting that the N-terminal domain may be involved in ethylene recognition

F4: see *cer1*

F5: see *cer3*

F17: see *tt8*

F31: see *ttg*

F107: see *tt8*

F114, F142: see *tt4*

fackel: see *fk*

**fab1: *f*atty *a*cid *b*iosynthesis 1**

LOCATION: not known

ORIGIN: induced with EMS in Columbia

ALLELES: line 1A9

REFERENCES: James and Dooner (1990) *Theor. Appl. Genet. 80:* 241-245; James and Dooner (1991) *Theor. Appl. Genet. 82:* 409-412

PHENOTYPE: increased levels of 16:0, 16:1, and 20:0 fatty acids; decreased levels of 18:1 and 18:2 fatty acids

### *fab2: fatty acid biosynthesis 2*

LOCATION: not known

ORIGIN: induced with EMS in Columbia

ALLELES: line 2A11

REFERENCES: James and Dooner (1990) *Theor. Appl. Genet. 80:* 241-245; James and Dooner (1991) *Theor. Appl. Genet. 82:* 409-412

PHENOTYPE: increased levels of 18:0 and 20:0 fatty acids; decreased levels of 18:1 fatty acid

### *fad2: fatty acid desaturation 2, eukaryotic*

LOCATION: 3-20.1

ORIGIN: *fad2-1, fad2-2,* and *fad2-3* induced with EMS in Columbia, *fad2-4* in Landsberg *erecta; fad2-5* induced by a T-DNA insertion in Wassilewskija (Ws)

ALLELES: *fad2-1* (line JB9), *fad2-2* (line 4A5), *fad2-3* (line JB12), *fad2-4* (line AL63), *fad2-5* (Okuley et al. 1994)

REFERENCES: James and Dooner (1990) *Theor. Appl. Genet. 80:* 241-245; Lemieux et al. (1990) *Theor. Appl. Genet. 80:* 234-240; Miquel and Browse (1992) *J. Biol. Chem. 267:* 1052-1059; Okuley et al. (1994) *Plant Cell 6:* 147-158

PHENOTYPE: high levels of 18:1 and reduced levels of 18:2 and 18:3 fatty acids; characterized in detail for *fad2-1*: heterozygote has slightly higher levels of 18:1 fatty acid; fatty acid composition different from those of *fad6* mutants; probably deficient in a lipid desaturase outside chloroplast

GENE: isolated by Okuley et al. (1994) using a T-DNA insertional allele (*fad2-5*); a *FAD2* cDNA expressed under the control of the 35S promoter complements the *fad2-1* allele; the encoded protein shares similarity with other plant desaturases and contains a histidine-rich region that could be involved in iron binding

### *fad3: fatty acid desaturation 3, eukaryotic*

LOCATION: 2-53.7

ORIGIN: line BL1, segregant from a multiply marked line in Landsberg *erecta;* lines G30 and 1E5, induced with EMS in Columbia

REFERENCES: James and Dooner (1990) *Theor. Appl. Genet. 80:* 241-245; Lemieux et al. (1990) *Theor. Appl. Genet. 80:* 234-240; James and Dooner (1991) *Theor. Appl. Genet. 82:* 409-412; Arondel et al. (1992) *Science 258:* 1353-1355

PHENOTYPE: deficient in C18:2 desaturation

GENE: cloned by Arondel et al. (1992) using chromosome walking; codes for ω-3 desaturase

### *fad4: fatty acid desaturation 4, prokaryotic; same as fadA*

LOCATION: 4-69.0

ORIGIN: induced with EMS in Columbia

ALLELE: *fad4-1* (line JB60)

REFERENCES: Browse et al. (1985) *Science 227:* 763-765; McCourt et al. (1985) *Plant Physiol. 78:* 853-855

PHENOTYPE: lack the chloroplast-specific lipid $\Delta 3$-trans-hexadecenoate; has increased levels of palmitic acid; growth normal

*fad5: fatty acid desaturation 5, prokaryotic; same as fadB*

LOCATION: 3-15.5

ORIGIN: induced with EMS in Columbia

ALLELE: *fad5-1* (line JB67)

REFERENCES: Kunst et al. (1989) *Plant Physiol. 90:* 943-947; Kunst et al. (1989) *Plant Physiol. 91:* 401-408

PHENOTYPE: accumulates high levels of palmitic acid and is deficient in unsaturated 16-carbon fatty acids; altered lipid compositions; deficient in a chloroplast $\omega$-9 fatty acid desaturase; enhanced thermal tolerance and higher growth rate than wild type at 28°C

*fad6: fatty acid desaturation 6, prokaryotic; same as fadC*

LOCATION: 4-53.1

ORIGIN: induced with EMS in Columbia

ALLELE: *fad6-1* (line LK3)

REFERENCES: Browse et al. (1989) *Plant Physiol. 90:* 522-529; Hugly et al. (1989) *Plant Physiol. 90:* 1134-1142; Falcone, D., S. Gibson, B. Lemieux, C.R. Somerville *Plant Physiol.,* submitted

PHENOTYPE: deficient in chloroplast $\omega$-6 fatty acid desaturase; accumulate high levels of 16:1 and 18:1 lipids and low levels of polyunsaturated lipids; abnormal chloroplast ultrastructure and thylakoid membrane protein and chlorophyll content; enhanced thermal tolerance of photosynthesis

GENE: cloned by Falcone et al. (submitted)

*fad7: fatty acid desaturation 7, prokaryotic; same as fadD*

LOCATION: 3-8.4

ORIGIN: induced with EMS in Columbia

ALLELE: *fad7-1* (line JB1)

REFERENCES: Browse et al. (1986) *Plant Physiol. 81:* 859-864; McCourt et al. (1987) *Plant Physiol. 84:* 353-360; Iba et al. (1993) *J. Biol. Chem. 268:* 24099-24105

PHENOTYPE: deficient in C18:3 and C16:3 leaf lipids; increased levels of C18:2 and C16:2 lipids; the mutant defect is temperature-sensitive; both chloroplast and non-chloroplast lipids are affected; alteration of chloroplast ultrastructure (more than 40% reduction of cross-sectional area in mutant plastids) and increased number of chloroplasts; normal levels of photosynthesis

GENE: cloned by Iba et al. (1993); encodes a chloroplast $\omega$-3 fatty acid desaturase

*fad8: fatty acid desaturation 8, prokaryotic*

LOCATION: not mapped
ORIGIN: induced with EMS in Columbia
ALLELE: one
REFERENCES: Gibson, S., V. Arondel, K. Iba, C.R. Somerville, *Plant Physiol.,* submitted; McConn, M., S. Hugly, J. Browse, C.R. Somerville, *Plant Physiol.,* submitted
PHENOTYPE: deficient in a chloroplast ω-3 desaturase, which is a temperature-inducible isozyme of FAD7
GENE: cloned by Gibson et al. (submitted)

*fadA:* fatty acid desaturation A, prokaryotic, see *fad4*

*fadB:* fatty acid desaturation B, prokaryotic, see *fad5*

*fadC:* fatty acid desaturation C, prokaryotic, see *fad6*

*fadD:* fatty acid desaturation D, prokaryotic, see *fad7*

**fae1: fatty acid elongation 1**

LOCATION: not known
ORIGIN: line JB20, induced with EMS in Columbia
ALLELES: *fae1-1* (line JB20); line 9A1
REFERENCES: James and Dooner (1990) *Theor. Appl. Genet. 80:* 241-245; Lemieux et al. (1990) *Theor. Appl. Genet. 80:* 234-240; James and Dooner (1991) *Theor. Appl. Genet. 82:* 409-412
PHENOTYPE: deficient in C20:1 fatty acid (0.2% compared with 17% in wild type)

**fah1: ferulic acid hydroxylase (previously designated as *sin1*)**

LOCATION: not known
ORIGIN: induced with EMS in Columbia
ALLELES: *fah1-1* through *fah1-5* (previously designated as *sin1-1* through *sin1-5*)
REFERENCE: Chapple et al. (1992) *Plant Cell 4:* 1413-1424
PHENOTYPE: reduced sinapoyl malate, probably due to loss of ferulate-5-hydroxylase activity
GENE: cloned by C. Chapple and C.R. Somerville (pers. comm.)

far-red elongated: see *phyA*

far-red elongated hypocotyl: see *phyA*

**fas1: fasciata 1**

LOCATION: 1-88.1
ORIGIN: induced by X-ray in Enkheim
REFERENCES: Reinholz (1966) *AIS 3:* 19-20; Leyser and Furner (1992) *Development 116:* 397-403
PHENOTYPE: fasciated stems; disrupted phyllotaxy; enlarged shoot apical meristem; reduced numbers of petals and stamens and extra sepals

**fas2: fasciata 2**

LOCATION: 5-56.2
ORIGIN: induced with EMS in Landsberg *erecta*
REFERENCE: Leyser and Furner (1992) *Development 116:* 397-403
PHENOTYPE: fasciated stems; disrupted phyllotaxy; enlarged shoot apical

meristem; reduced numbers of petals and stamens and extra sepals; appearance of some fused flowers

***fas3: fasciata* 3, allelic with *clv1*, see *clv1-3***

*fasciata:* see *fas*

*fass:* see *fs*

fatty acid desaturation: see *fad*

*fb:* late flowering, allelic with *gi*, see *gi*

***fca:* late flowering**

LOCATION: 4-32.4

ORIGIN: induced with EMS (*fca-1,5,6*, and two others), by fast neutrons (*fca-2*), or by X-ray (*fca-3,4*) in Landsberg *erecta*

REFERENCES: H.A.S. Hussein (1968) Ph.D. Thesis, Agriculture University, Wageningen; Martínez-Zapater and Somerville (1990) *Plant Physiol. 92:* 770-776; Koornneef et al. (1991) *MGG 229:* 57-66

PHENOTYPE: late flowering, like *co*; vernalization and, to a lesser extent, incandescent light reduce delay in flowering

***fd:* late flowering**

LOCATION: 4-63.5

ORIGIN: induced with EMS in Landsberg *erecta*

ALLELE: *fd-1*

REFERENCES: Koornneef (1990) *Genetic Maps: Locus Maps of Complex Genomes,* 5th Edition (ed. S.J. O'Brien), Book 6, pp. 6.94-6.97, Cold Spring Harbor Laboratory Press, Cold Spring Harbor, NY; Koornneef et al. (1991) *MGG 229:* 57-66

PHENOTYPE: like *co;* late flowering not altered by vernalization; flowering delayed further by short-day photoperiod

***fdh: fiddlehead***

LOCATION: not known

ORIGIN: induced with EMS in Landsberg *erecta*

REFERENCE: Lolle et al. (1992) *Dev. Biol. 152:* 383-392

PHENOTYPE: fusion occurs between any of leaves and floral organs that come in contact

***fe:* late flowering**

LOCATION: 1-121.9

ORIGIN: induced with EMS in Landsberg *erecta*

ALLELE: *fe-1*

REFERENCES: Koornneef et al. (1983) *J. Hered. 74:* 265-272; Martínez-Zapater and Somerville (1990) *Plant Physiol. 92:* 770-776; Koornneef et al. (1991) *MGG 229:* 57-66

PHENOTYPE: like *co;* vernalization and incandescent light reduce delay in flowering

female gametophyte factor: see *gf*

female sterile: see *bel1* and *sin1*

ferulic acid hydroxylase: see *fah1*

***fey: forever young***

LOCATION: not mapped
ORIGIN: induced by T-DNA insertion in Wassilewskija (Ws)
REFERENCE: Medford et al. (1992) *Plant Cell 4:* 631-643
PHENOTYPE: abnormal development of the shoot apex; normal cotyledons, hypocotyl, and root; phenotype less severe at 18°C than at 25°C

*fg:* late flowering, allelic with *co,* see *co*

### *fha:* **late flowering**

LOCATION: 1-11.5
ORIGIN: induced with EMS (*fha-1,2*) or by X-ray (*fha-3*) in Landsberg *erecta*
ALLELES: *fha-1,2,3*
REFERENCE: Koornneef et al. (1991) *MGG 229:* 57-66
PHENOTYPE: delayed flowering; both long-day and vernalization slightly reduce delay of flowering

### *fhy1:* *f*ar-red elongated *h*ypocotyl 1

LOCATION: not mapped
ORIGIN: induced by γ-ray in Columbia *gl* (*gl1?*)
REFERENCES: Whitelam et al. (1993) *Plant Cell 5:* 757-768; Johnson et al. (1994) *Plant Physiol. 105:* 141-149
PHENOTYPE: elongated hypocotyl in far-red light—about half as long as in dark; normal in red, blue, or white light; normal levels of phytochrome A in dark-grown seedlings by both spectrophotometric and immunological analyses; abnormal day length perception

*fhy2:* far-red elongated hypocotyl 2, alleles of *PHYA,* see *phyA*

### *fhy3:* *f*ar-red elongated *h*ypocotyl 3

LOCATION: not mapped
ORIGIN: induced with EMS in Columbia *gl* (*gl1?*)
REFERENCE: Whitelam et al. (1993) *Plant Cell 5:* 757-768
PHENOTYPE: elongated hypocotyl in far-red light—about 80% as long as in dark; normal in red, blue, or white light; normal levels of phytochrome A in dark-grown seedlings by both spectrophotometric and immunological analyses

fiddlehead: see *fdh*

### *fk:* *f*ackel

LOCATION: 3-76.6
ORIGIN: five alleles induced with EMS (0.3%, 8 hrs) in Landsberg *erecta* seeds; one allele induced by 20,000 R X-ray in Landsberg *erecta* pollen
REFERENCES: Mayer et al. (1991) *Nature 353:* 402-407; U. Mayer (1993) Ph.D. Thesis, Universität München
PHENOTYPE: seedling pattern mutant; deletion of hypocotyl; cotyledons directly attached to root; cotyledons are abnormally shaped and often more than two are in a seedling

Fl-40: see *ap2-3*

Fl-48: see *ap2-4*

### **Fl-54**

LOCATION: not mapped

ORIGIN: induced with EMS in Landsberg *erecta*

REFERENCE: Komaki et al. (1988) *Development 104:* 195-203

PHENOTYPE: flowers abnormal; shape, number, and position of floral organs affected; variable phenotypes between different flowers; abnormal inflorescence structure

### Fl-82

LOCATION: not mapped

ORIGIN: induced with EMS in Landsberg *erecta*

REFERENCES: Komaki et al. (1988) *Development 104:* 195-203; Okada et al. (1989) *Cell, Diff. and Dev. 28:* 27-37

PHENOTYPE: large pistil usually consisting of three or more carpels; carpels sometimes unfused; anthers sometimes attached to the side of unfused carpel; some flowers have two pistils; female semi-sterile

OTHER INFORMATION: not allelic with *clv1*

### Fl-89

LOCATION: chromosome 4 (Okada et al. 1989)

ORIGIN: induced with EMS in Landsberg *erecta*

REFERENCES: Komaki et al. (1988) *Development 104:* 195-203; Okada et al. (1989) *Cell, Diff. and Dev. 28:* 27-37

PHENOTYPE: pistil has two clusters of stigmatic papillae and two green horn-shaped projections; septum abnormal

### Fl-165

LOCATION: not mapped

ORIGIN: induced with EMS Landsberg *erecta* (?)

REFERENCE: Okada et al. (1989) *Cell, Diff. and Dev. 28:* 27-37

PHENOTYPE: pistils have thin ovaries and long style; septum not fused, medium vascular bundles are often divided into two separate bundles; nearly female sterile

### *fla:* late flowering

LOCATION: 4-0.0

ORIGIN: dominant alleles of this locus confer lateness in late ecotypes Sf-2 and Le-0 (Lee et al. 1993)

REFERENCES: Napp-Zinn (1957) *Z. Indukt. Abstammungs and Vererbungsl. 88:* 253-285; Lee et al. (1993) *MGG 237:* 171-176

PHENOTYPE: delayed flowering, vernalization reduces delay of flowering

OTHER INFORMATION: *FLA* is probably allelic with *Fri* (*frigida*) described by Napp-Zinn (1957)

### *flo2, flo3, flo4:* floral mutant 2, 3, 4, allelic with *ap2*, see *ap2*

### *flo5:* floral mutant 5, allelic with *clv1*, see *clv1*

### *flo8:* floral mutant 8

LOCATION: not mapped

REFERENCE: Haughn and Somerville (1988) *Dev. Genet. 9:* 73-89

PHENOTYPE: similar to *pin*

### *flo9:* floral mutant 9

LOCATION: not mapped

**REFERENCE:** Haughn and Somerville (1988) *Dev. Genet. 9:* 73-89

**PHENOTYPE:** third whorl floral organs are petal-like

*flo10:* floral mutant 10, allelic with *sup,* see *sup*

forever young: see *fey*

***fpa:* late flowering**

**LOCATION:** 2-83.0

**ORIGIN:** induced with EMS in Landsberg *erecta*

**ALLELES:** *fpa-1; fpa-2*

**REFERENCE:** Koornneef et al. (1991) *MGG 229:* 57-66

**PHENOTYPE:** delayed flowering; both long-day and vernalization reduce delay of flowering

*fre1:* far-red elongated 1, alleles of *PHYA,* see *phyA*

***fs:* fass**

**LOCATION:** chromosome 5, near *ttg* (D. Meinke, pers. comm.; G. Jürgens, pers. comm.)

**ORIGIN:** 12 alleles induced with EMS (0.3%, 8 hrs) in Landsberg *erecta* seeds

**REFERENCES:** Mayer et al. (1991) *Nature 353:* 402-407; R.A. Torres Ruis and G. Jürgens (pers. comm.)

**PHENOTYPE:** seedling pattern mutant; seedlings are stout and short, compressed in the apical-basal axis; all structures of a seedling are present but they have abnormal shapes; cells are abnormally shaped and arranged from an early stage

***ft:* late flowering**

**LOCATION:** 1-88.2

**ORIGIN:** induced with EMS in Landsberg *erecta*

**ALLELES:** *ft-1,2,3* (Koornneef et al. 1991)

**REFERENCES:** Koornneef et al. (1983) *J. Hered. 74:* 265-272; Martínez-Zapater and Somerville (1990) *Plant Physiol. 92:* 770-776; Koornneef et al. (1991) *MGG 229:* 57-66

**PHENOTYPE:** delayed flowering; long-day photoperiod, but not vernalization, reduces delay of flowering

***fuf: fully fasciated***

**LOCATION:** not mapped

**ORIGIN:** induced with EMS in Columbia

**REFERENCE:** Medford et al. (1992) *Plant Cell 4:* 631-643

**PHENOTYPE:** increased number of rosette leaves; flattened (fasciated) floral stem; larger than normal apical meristem

fully fasciated: see *fuf*

*fus1:* fusca 1, allelic with *cop1,* see *cop1*

*fus2:* fusca 2, allelic with *det1,* see *det1*

***fus3: fusca 3***

**LOCATION:** 3-40.6

**ORIGIN:** *fus3-1,2* induced in Di-G; *fus3-3* induced with EMS in Columbia

**ALLELES:** *fus3-1,2* (20 others lost) (Miséra et al. 1994); *fus3-3* (Keith et al. 1994)

**REFERENCES:** Keith et al. (1994) *Plant Cell 6:* 589-600; Meinke et al. (1994) *Plant Cell 6:* 1049-1064; Miséra et al. (1994) *MGG 244:* 242-252

**PHENOTYPE:** accumulation of anthocyanin in developing embryo; dry seed shrunken and does not germinate; conditional-lethal

*fus3-3:* seeds desiccation sensitive; seeds germinate and leaf primordia appear before desiccation; cotyledons have trichomes and have both cotyledon and leaf characteristics

**OTHER INFORMATION:** linked to, but not allelic with, *abi3* (S. Miséra, pers. comm.)

## *fus4: fus*ca 4

**LOCATION:** not mapped

**ORIGIN:** induced in Di-G

**ALLELE:** one (25 others lost)

**REFERENCE:** Miséra et al. (1994) *MGG 244:* 242-252

**PHENOTYPE:** accumulation of anthocyanin in developing embryo; dry seed almost wild type; late-seedling-lethal

## *fus5: fus*ca 5

**LOCATION:** not mapped

**ORIGIN:** induced with EMS in Landsberg *erecta* (three alleles) and Di-G (two alleles) (11 others lost)

**ALLELES:** five

**REFERENCE:** Miséra et al. (1994) *MGG 244:* 242-252

**PHENOTYPE:** accumulation of anthocyanin in developing embryo; dry seed (two-colored) and seedling; seedling-lethal; dark-grown seedlings show light-grown morphology

## *fus6: fus*ca 6, allelic with *cop11* and *emb78*

**LOCATION:** 3-87.2

**ORIGIN:** *fus6-1* and *fus6-2* induced by T-DNA insertions in Wassilewskija (Ws); others induced in Di-G (one allele, six lost) or with EMS in Landsberg *erecta* (four alleles)

**ALLELES:** *fus6-1, fus6-2* (Patton et al. 1991); five other alleles isolated by Miséra et al. (1994)

**REFERENCES:** Errampalli et al. (1991) *Plant Cell 3:* 149-157; Patton et al. (1991) *MGG 227:* 337-347; Castle and Meinke (1994) *Plant Cell 6:* 25-41; Miséra et al. (1994) *MGG 244:* 242-252; Wei et al. (1994) *Plant Cell 6:* 629-643

**PHENOTYPE:** embryo development arrested as mature cotyledon; seed and embryo green with dark red regions and have fusca phenotype; limited growth in culture; seedlings assume light-grown morphology in the dark (*fus6-2 = cop11-1*), accumulate COP1 protein in dark

**GENE:** isolated by Castle and Meinke (1994) using the T-DNA insertional alleles; encodes a novel protein

*fus7:* fusca 7, allelic with *cop9,* see *cop9*

## *fus8: fus*ca 8, allelic with *cop8* and *emb134*

**LOCATION:** 5-63.5

ORIGIN: *fus8-1* segregated from a T-DNA insertional line in Wassilewskija (Ws) (Castle and Meinke 1994); others induced with EMS in Landsberg *erecta* or Di-G

ALLELES: *fus8-1* (*emb134*) = *cop8-1*; others: nine in Landsberg *erecta* and one in Di-G (eight others lost)

REFERENCES: Castle and Meinke (1994) *Plant Cell 6:* 25-41; Miséra et al. (1994) *MGG 244:* 242-252; Wei et al. (1994) *Plant Cell 6:* 629-634

PHENOTYPE: like *fus5:* seedlings assume light-grown morphology in the dark (*fus6-2* = *cop11-1*), accumulate COP1 protein in dark

### *fus9: fus*ca 9, allelic with *cop9* and *emb144*

LOCATION: 3-16.4

ORIGIN: *fus9-1* segregated from a T-DNA insertional line in Wassilewskija (Ws); others induced with EMS in Landsberg *erecta* or Di-G

ALLELES: *fus9-1* (*emb144*) = *cop9-1*; others: seven in Landsberg *erecta* and one in Di-G (six others lost)

REFERENCES: Castle and Meinke (1994) *Plant Cell 6:* 25-41; Miséra et al. (1994) *MGG 244:* 242-252; Wei et al. (1994) *Plant Cell 6:* 629-634

PHENOTYPE: like *fus5:* seedlings assume light-grown morphology in the dark (*fus6-2* = *cop11-1*), accumulate COP1 protein in dark

### *fus10: fus*ca 10

LOCATION: not mapped

ORIGIN: induced with EMS in Di-G

ALLELE: one (six others lost)

REFERENCE: Miséra et al. (1994) *MGG 244:* 242-252

PHENOTYPE: like *fus3*

### *fus11: fus*ca 11

LOCATION: not mapped

ORIGIN: induced with EMS in Landsberg *erecta* or Di-G

ALLELES: one in Landsberg *erecta* and two in Di-G (one lost)

REFERENCE: Miséra et al. (1994) *MGG 244:* 242-252

PHENOTYPE: like *fus5*

### *fus12: fus*ca 12

LOCATION: chromosome 2, about halfway between *sti* and *as*

ORIGIN: induced with EMS in Landsberg *erecta*

ALLELES: three

REFERENCE: Miséra et al. (1994) *MGG 244:* 242-252

PHENOTYPE: like *fus5*

OTHER INFORMATION: all alleles induced in Di-G lost

fusca: see *fus, cop1, cop9, det1*

### *fve:* late flowering

LOCATION: 2-27.5

ORIGIN: induced with EMS in Landsberg *erecta*

ALLELES: *fve-1; fve-2*

REFERENCES: Koornneef et al. (1991) *MGG 229:* 57-66

PHENOTYPE: delayed flowering; both long-day and vernalization reduce delay of flowering

***fwa*: late flowering**

LOCATION: 4-54.7

ORIGIN: induced with EMS (*fwa-1*) or by fast neutrons (*fwa-2*) in Landsberg *erecta*

ALLELES: *fwa-1; fwa-2*

REFERENCES: Koornneef et al. (1991) *MGG 229:* 57-66

PHENOTYPE: delayed flowering; long-day, but not vernalization, reduces delay of flowering

***fy*: late flowering**

LOCATION: 5-12.1

ORIGIN: induced with EMS in Landsberg *erecta*

ALLELE: *fy-1*

REFERENCES: Koornneef et al. (1983) *J. Hered. 74:* 265-272; Martínez-Zapater and Somerville (1990) *Plant Physiol. 92:* 770-776; Koornneef et al. (1991) *MGG 229:* 57-66

PHENOTYPE: like *fb:* vernalization and incandescent light reduce delay in flowering

G2: see *cer6*

G3: see *cer7*

G4: see *aba-3*

G5: see *cer2*

G7: see *cer4*

G8: see *cer2*

G30: see *fad3*

***ga1*: gibberellin requiring 1**

LOCATION: 4-4.8

ORIGIN: *ga1-1,6,7,8,9,10* induced with EMS in Landsberg *erecta; ga1-2,3,4* induced by fast neutrons in Landsberg *erecta*

ALLELES: *ga1-1* (line NG5); *ga1-2* (line 6.59); *ga1-3* (line 31.89); *ga1-4* (line 29.9); *ga1-5* (line 29.423); *ga1-6* (line d352); *ga1-7* (line Bo27); *ga1-8* (line A428); *ga1-9* (line NG4); *ga1-10* (line d69)

REFERENCES: Koornneef (1978) *AIS 15:* 17-20; Koornneef (1979) *AIS 16:* 41-47; Koornneef and van der Veen (1980) *Theor. Appl. Genet. 58:* 257-263; Barendse and Koornneef (1982) *AIS 19:* 25-28; Koornneef et al. (1982) *Mut. Res. 93:* 109-123; Koornneef et al. (1983) *Genet. Res. (Camb.) 41:* 57-68; Barendse et al. (1986) *Physiol. Plant. 67:* 315-319; Zeevaart and Talon (1991) *Current Plant Sci. & Biotech. in Agricul. 13:* 34-42; Sun et al. (1992) *Plant Cell 4:* 119-128; Wilson et al. (1992) *Plant Physiol. 100:* 403-408

PHENOTYPE: dwarfed unless treated with gibberellins; most alleles do not germinate unless treated with gibberellin; delayed flowering under short-day lighting; *ga1-5,6* are leaky

GENE: *GA1* has been cloned by Sun et al. (1992); *ga1-2* is an insertion or inversion within the last known intron; *ga1-3* is a deletion of about 5 kb; *ga1-6, ga1-8* (in penultimate exon) and *ga1-7* (in the last exon) are point mutations (Sun et al. 1992)

**OTHER INFORMATION:** recombination between alleles has been shown (Koornneef 1979; Koornneef et al. 1983); deficient in *ent*-kaurene synthase in some organs but not others (Zeevaart and Talon 1991), which, along with the fact that a large deletion mutant (*ga1-3*) still produces some GA, suggests that *GA1* may encode a regulatory protein or that there is a duplicate gene

## *ga2:* gibberellin requiring 2

**LOCATION:** 1-119.0

**ORIGIN:** induced in Landsberg *erecta*

**REFERENCES:** Koornneef and van der Veen (1980) *Theor. Appl. Genet. 58:* 257-263; Barendse and Koornneef (1982) *AIS 19:* 25-28; Koornneef et al. (1982) *Mut. Res. 93:* 109-123; Zeevaart and Talon (1991) *Current Plant Sci. & Biotech. in Agricul. 13:* 34-42

**PHENOTYPE:** dwarfed unless treated with gibberellins; some alleles require gibberellin to germinate

**OTHER INFORMATION:** deficient in B activity of *ent*-kaurene synthase (Zeevaart and Tolan 1991)

## *ga3:* gibberellin requiring 3

**LOCATION:** 5-32.5

**ORIGIN:** induced in Landsberg *erecta*

**ALLELES:** *ga3-1; ga3-2*

**REFERENCES:** Koornneef and van der Veen (1980) *Theor. Appl. Genet. 58:* 257-263; Barendse and Koornneef (1982) *AIS 19:* 25-28; Koornneef et al. (1982) *Mut. Res. 93:* 109-123; Zeevaart and Talon (1991) *Current Plant Sci. & Biotech. in Agricul. 13:* 34-42

**PHENOTYPE:** dwarfed unless treated with gibberellins; some alleles (e.g., *ga3-1*) require gibberellin to germinate; *ga3-2* is leaky

**OTHER INFORMATION:** deficient in oxidation of *ent*-kaurene to *ent*-kaurenoic acid (Zeevaart and Talon 1991)

## *ga4:* gibberellin requiring 4

**LOCATION:** 1-22.3

**ORIGIN:** induced in Landsberg *erecta*

**REFERENCES:** Koornneef and van der Veen (1980) *Theor. Appl. Genet. 58:* 257-263; Barendse and Koornneef (1982) *AIS 19:* 25-28; Koornneef et al. (1982) *Mut. Res. 93:* 109-123; Talon et al. (1990) *Proc. Natl. Acad. Sci. USA 87:* 7983-7987; Zeevaart and Talon (1991) *Current Plant Sci. & Biotech. in Agricul. 13:* 34-42

**PHENOTYPE:** dwarfed unless treated with gibberellin, gibberellin not required for germination

**OTHER INFORMATION:** many alleles, all called *ga4;* contains changed composition of endogenous gibberellins; deficient in 3β-hydroxylation of GAs (Talon et al. 1990)

## *ga5:* gibberellin requiring 5

**LOCATION:** 4-52.4

**ORIGIN:** induced in Landsberg *erecta*

REFERENCES: Koornneef and van der Veen (1980) *Theor. Appl. Genet. 58:* 257-263; Barendse and Koornneef (1982) *AIS 19:* 25-28; Koornneef et al. (1982) *Mut. Res. 93:* 109-123; Talon et al. (1990) *Proc. Natl. Acad. Sci. USA 87:* 7983-7987; Zeevaart and Talon (1991) *Current Plant Sci. & Biotech. in Agricul. 13:* 34-42

PHENOTYPE: dwarfed unless treated with gibberellin, germinates normally without GA

OTHER INFORMATION: contains endogenous gibberellins and deficient in the elimination of C-20 at the aldehyde level and possibly deficient in C-20 hydroxylation (Talon et al. 1990)

**gai: gibberellin insensitive, possibly allelic with gai-d**

LOCATION: 1-22.4

ORIGIN: *gai* induced by X-ray in Landsberg *erecta; gai-d1,d2,d3,d4* induced by γ-ray in Landsberg *erecta gai tt1*

REFERENCES: Koornneef et al. (1985) *Physiol. Plant. 65:* 33-39; Wilson et al. (1992) *Plant Physiol. 100:* 403-408; Peng and Harberd (1993) *Plant Cell 5:* 351-360

PHENOTYPE: never observed to flower in short-day photoperiod; semi-dominant dwarf; late flowering and dwarfism cannot be relieved by exogenous GA; normal flowering in continuous lighting

OTHER INFORMATION: secondary mutations (*gai-d1,d2,d3,d4,* suppressors of dwarf phenotype) were isolated (Peng and Harberd 1993) that are genetically inseparable from *gai,* suggesting that they are in the *GAI* gene; *gai-3d* is unable to be transmitted through pollen and may be homozygous-lethal, suggesting that it is a large deletion; that several linked mutations, including a possible deletion, reverted the *gai* mutant phenotype suggests that *gai* is a gain-of-function mutation and that the *gai-d* mutations eliminate *GAI* gene function

*gai-d1* through *gai-d4:* derivative of *gai,* see *gai*

genomes uncoupled: see *gun*

germination under saline conditions: see RS17, RS19, RS20

**gf: female gametophyte factor**

LOCATION: chromosome 2 (Koornneef 1990, *Genetic Maps: Locus Maps of Complex Genomes,* 5th Edition (ed. S.J. O'Brien), Book 6, pp. 6.94-6.97, Cold Spring Harbor Laboratory Press, Cold Spring Harbor, NY)

ORIGIN: induced by X-ray treatment of Columbia (12 kr to 24-hr presoaked seeds)

REFERENCES: Rédei (1965) *Genetics 51:* 857-872; Rédei (1968) *AIS 5:* 25

PHENOTYPE: not transmitted through the female gamete, fair transmission through the pollen, reduced recombination in vicinity of *gf* in female

GH50, GH51, GH52: see *csr1*

GH90, GH91, GH92: see *imr* and *csr1*

**gi: gigantea, allelic with fb**

LOCATION: 1-33.3

ORIGIN: induced by X-ray treatment (8–12 kr to 24-hr presoaked seeds) of

Landsberg *erecta* (*gi-1, gi-2*); by X-ray treatment of Landsberg *erecta* (*gi-5*); with EMS in Landsberg *erecta* (*gi-3,4,6*)

ALLELES: *gi-1* and *gi-2* (gi$^1$ and gi$^2$, respectively) (Rédei 1962); *gi-3,4,5,6* and four other alleles (also known as *fb* alleles) (Koornneef et al. 1991)

REFERENCES: Rédei (1962) *Genetics 47:* 443-460; Rédei (1964) *AIS 1:* 9-10; Kranz and Scheidemann (1978) *AIS 15:* 31-34; Koornneef et al. (1983) *J. Hered. 74:* 265-272; Martínez-Zapater and Somerville (1990) *Plant Physiol. 92:* 770-776; Koornneef et al. (1991) *MGG 229:* 57-66; Araki and Komeda (1993) *Plant J. 3:* 231-239

PHENOTYPE: delayed flowering (*gi-1*, 2–3x; *gi-2*, 4–6x), produces large rosettes; relatively insensitive to vernalization or incandescent light treatment

OTHER INFORMATION: *gi-1* dominant to *gi-2*

gibberellin requiring: see *ga*

gigantea: see *gi*

## *gk: gurke*

LOCATION: chromosome 1 (R.A. Torres Ruis and T. Fischer, pers. comm.)

ORIGIN: eight strong and several weak alleles induced with EMS in Landsberg *erecta;* one allele induced with 20,000 R X-ray in Landsberg *erecta* pollen

REFERENCES: Mayer et al. (1991) *Nature 353:* 402-407; R.A. Torres Ruis and G. Jürgens, pers. comm.

PHENOTYPE: seedling pattern mutant; deletion of cotyledons and shoot apical meristem

## *gl1: glabra 1*

LOCATION: 3-46.2

SYNONYM: *gl;* glabrous

ALLELES: numerous (Koornneef 1981; Koornneef et al. 1982); *gl1-43* is the T-DNA tagged allele; the Columbia C24 ecotype is also *gl1* (M. Koornneef, pers. comm.); four new alleles induced with EMS in Landsberg *erecta* (Hülskamp et al. 1994)

REFERENCES: Lee-Chen and Steinitz-Sears (1967) *Can. J. Genet. Cytol. 9:* 381-384; Koornneef (1981) *AIS 18:* 45-51; Koornneef et al. (1982) *Mut. Res. 93:* 109-123; Marks and Feldmann (1989) *Plant Cell 1:* 1043-1050; Herman and Marks (1989) *Plant Cell 1:* 1051-1055; Oppenheimer et al. (1991) *Cell 67:* 483-493; Larkin et al. (1993) *Plant Cell 5:* 1739-1748; Hülskamp et al. (1994) *Cell 76:* 555-566; Larkin et al. (1994a) *Results and Problems in Cell Differentiation: Plant Promoters and Transcription Factors* (ed. L. Nover), vol. 20, pp. 259-275, Springer-Verlag, Berlin; Larkin et al. (1994b) *Plant Cell 6:* 1065-1076; Marks (1994) *Curr. Biology 4:* 621-623

PHENOTYPE: trichomes absent on leaf surface and stems, some rudimentary trichomes in leaf margins

GENE: cloned by T-DNA tagging (Marks and Feldmann 1989; Herman and Marks 1989; Oppenheimer et al. 1991); gene product contains a *myb* homologous domain; analysis of GUS fusions (Larkin et al. 1993) indicates that there is a regulatory element(s) downstream of the coding region; expression of a 35S-*GL1* fusion (Larkin et al. 1994b) results in fewer than normal number of trichomes in both wild-type and *gl1-1* backgrounds; expression of both 35S-*GL1* and 35S-*R* (from maize) fusions results in increased number of trichomes on leaves, and ectopic trichomes on cotyledons, but not roots; 35S-*R* does not suppress *gl1* mutant phenotype

OTHER INFORMATION: autonomous in genetic mosaics

### *gl2: glabra* 2

LOCATION: 1-118.1

ORIGIN: induced with EMS in Landsberg *erecta*

ALLELES: original allele from Koornneef et al. (1982); three new alleles from Hülskamp et al. (1994)

REFERENCES: Koornneef et al. (1982) *Mut. Res. 93:* 109-123; Hülskamp et al. (1994) *Cell 76:* 555-566

PHENOTYPE: trichomes rudimentary on first two leaves and reduced in number on higher leaves; (new alleles) no local outgrowth of trichome cell; enhanced by *gl3,* suppressed by *Try*

OTHER INFORMATION: autonomous in genetic mosaics

### *gl3: glabra* 3

LOCATION: 5-53.4

ORIGIN: induced with EMS in Landsberg *erecta*

ALLELES: original allele from Koornneef et al. (1982); two new alleles from Hülskamp et al. (1994)

REFERENCES: Koornneef et al. (1982) *Mut. Res. 93:* 109-123; Hülskamp et al. (1994) *Cell 76:* 555-566

PHENOTYPE: trichomes unbranched and reduced in number; trichome cell half the normal size

glabra: see *gl*

glabrous: see *gl1*

### *gld: glycine decarboxylase,* same as *glyD*

LOCATION: chromosome 2 (~40 cM from *er;* C.R. Somerville, pers. comm.)

ORIGIN: EMS mutagenesis of seeds of Columbia strain

ALLELES: line CS116

REFERENCE: Somerville and Ogren (1982a) *Biochem. J. 202:* 373-380

PHENOTYPE: lacks mitochondrial glycine decarboxylase activity

OTHER INFORMATION: five alleles are reported (Somerville and Ogren 1982b, *Methods in Chloroplast Molecular Biology,* eds. Edelman et al., pp. 129-138, Elsevier Biomedical Press, New York); selected by ability to grow normally in 1% $CO_2$ (i.e., with photorespiration blocked) but phenotype of chlorosis and cessation of growth in normal atmospheric conditions

### *gld2: glycine* decarboxylation 2

LOCATION: chromosome 5 (21±5 cM from *tt3;* C.R. Somerville, pers. comm.)

ORIGIN: induced with EMS in Columbia

ALLELE: one

REFERENCE: Artus et al. (1994) *Plant Cell Physiol.,* in press

PHENOTYPE: deficient in photorespiratory glycine decarboxylation

**gls: glutamate synthase, same as gluS**
LOCATION: not mapped
ORIGIN: induced with EMS in Columbia
ALLELES: lines CS113, CS103, CS30 carry different alleles
REFERENCES: Somerville and Ogren (1980) *Nature 286:* 257-259; Morris et al. (1989) *Plant Physiol. 89:* 498-500
PHENOTYPE: isolated by inability to grow in normal atmospheric conditions but normal growth in 1% $CO_2$ (i.e., with photorespiration suppressed); deficient in leaf chloroplast glutamate synthase activity

glucosinolate metabolism: see *gsm1*

gluS: see *gls*

glutamate synthase: see *gls*

glycine decarboxylase: see *gld, gly2*

glyD: see *gld*

**gm: male gametophyte factor**
LOCATION: 1- between *ch1* and *le*
REFERENCE: Rédei and Hirono (1964) *AIS 1:* 9-10
PHENOTYPE: no transmission in male and poor transmission in female

GM11b: see *omr1*

**gn: gnom, allelic with emb30**
LOCATION: 1-21.0
ORIGIN: 24 *gn* alleles induced with EMS (8 hrs) in Landsberg *erecta; emb30-1* (line 112A-2A) and *emb30-2* induced with EMS in Columbia; *emb30-3* and *emb30-4* induced by T-DNA insertion in Wassilewskija (Ws)
ALLELES: *gn* alleles (Mayer et al. 1991, 1993); *emb30-1, emb30-2* (Meinke 1985; Patton et al. 1991); *emb30-3, emb30-4* (Shevell et al. 1994)
REFERENCES: Meinke (1985) *Theor. Appl. Genet. 69:* 543-552; Franzmann et al. (1989) *Theor. Appl. Genet. 77:* 609-616; Mayer et al. (1991) *Nature 353:* 402-407; Patton et al. (1991) *MGG 227:* 337-347; Mayer et al. (1993) *Development 117:* 149-162; Shevell et al. (1994) *Cell 77:* 1051-1062
PHENOTYPE: *gn* alleles: seedling pattern mutant; deletion of root and severe reduction or elimination of cotyledons; seedlings are cone- or ball-shaped; zygote divides nearly symmetrically, variable division plane of enlarged apical daughter cell; callus, but not root, formation on root-inducing medium
*emb30* alleles: embryo development arrested as fused cotyledon; seed pale green to green and embryo green; rootless green plantlets form in culture; cell shapes and divisions are abnormal
GENE: cloned by Shevell et al. (1994) using T-DNA tag from the *emb30-3* line; encodes a protein with 1451 amino acids (163 kD) and having sequence similarity with the yeast Sec7 protein involved in secretion; expressed in all organs tested including most vegetative and reproductive structures; *emb30-3* has a mutation that causes a $Glu_{658} \rightarrow Lys$ change; *emb30-3* and *emb30-4* have insertions at amino acid residues 68 and 119, respectively

gnarled: see *grl*

gnom: see *gn*

### *grl: gnarled*
   LOCATION: 2-59.8
   ORIGIN: induced with EMS in Landsberg *erecta*
   ALLELES: four
   REFERENCE: Hülskamp et al. (1994) *Cell 76:* 555-566
   PHENOTYPE: trichome distorted

### *gsm1: glucosinolate metabolism*
   LOCATION: not mapped
   ORIGIN: induced with EMS in Columbia
   ALLELE: *gsm1-1*
   REFERENCE: Haughn et al. (1991) *Plant Physiol. 97:* 217-226
   PHENOTYPE: isolated by analyzing the leaf glucosinolates content using HPLC; *gsm1-1* shows greatly reduced levels of several glucosinolates with aliphatic side chains, particularly 4-methylsulphinylbutyl glucosinolate; no visible growth defects
   OTHER INFORMATION: additional mutants were reported in Haughn et al. (1991) but not described.

### *gun1: genomes uncoupled 1*
   LOCATION: chromosome 1 (J. Chory, pers. comm.)
   ORIGIN: induced with EMS in a transgenic line (pOCA107-2) of Columbia containing fusions of the *CAB3* promoter to the *uidA* (GUS) and *hph* (hygromycin resistance) reporter genes
   ALLELES: *gun1-1,2,3,4*
   REFERENCE: Susek et al. (1993) *Cell 74:* 787-799
   PHENOTYPE: isolated by screening for hygromycin resistance and GUS ex pression in the presence of Norflurazon (a herbicide that inhibits the carotenoid biosynthetic enzyme phytoene desaturase, thus blocking chloroplast development); elevated levels (10x wild type) of *CAB* and *RBCS* mRNAs in photobleached plants on Norflurazon but 10x lower than wild-type green plants; *CAB* gene regulation by light and circadian control is normal; reduced inhibition (from 39x in wild type to 3x) of CAB expression by chloramphenicol; the ability to de-etiolate following growth in dark is lost faster than in wild type; de-etiolate at a slower rate than wild type
   OTHER INFORMATION: additional mutants have been isolated from the same screen: *gun0-4, gun0-6, gun0-10, gun0-34,* and *gun0-38; gun0-10* and *gun0-34* not allelic with *gun1* or each other; allelism tests for others with *gun1,* or for any with *gun2,* have not been completed

### *gun2: genomes uncoupled 2*
   LOCATION: not mapped
   ORIGIN: induced with EMS in a transgenic line (pOCA107-2) of Columbia containing fusions of the *CAB3* promoter to the *uidA* (GUS) and *hph* (hygromycin resistance) reporter genes
   ALLELE: *gun2-1*
   REFERENCE: Susek et al. (1993) *Cell 74:* 787-799
   PHENOTYPE: isolated by screening for hygromycin resistance and GUS ex-

pression in the presence of Norflurazon (a herbicide that inhibits the carotenoid biosynthetic enzyme phytoene desaturase, thus blocking chloroplast development)

**OTHER INFORMATION**: additional mutants have been isolated from the same screen: *gun0-4, gun0-6, gun0-10, gun0-34,* and *gun0-38*; allelism tests for these with *gun2* have not been completed

## *gun3*: genomes *un*coupled 3

**LOCATION**: not mapped

**ORIGIN**: induced with EMS in a transgenic line (pOCA107-2) of Columbia containing fusions of the *CAB3* promoter to the *uidA* (GUS) and *hph* (hygromycin resistance) reporter genes

**ALLELE**: *gun3-1*

**REFERENCE**: Susek et al. (1993) *Cell 74:* 787-799

**PHENOTYPE**: isolated by screening for hygromycin resistance and GUS expression in the presence of Norflurazon (a herbicide which inhibits the carotenoid biosynthetic enzyme phytoene desaturase, thus blocking chloroplast development)

**OTHER INFORMATION**: additional mutants have been isolated from the same screen: *gun0-4, gun0-6, gun0-10, gun0-34,* and *gun0-38*; allelism tests for these with *gun3* have not been completed

gurke: see *gk*

## *hcr1*: high $CO_2$ requiring 1

**LOCATION**: not mapped

**ORIGIN**: induced with EMS in Columbia

**ALLELES**: *hcr1-1* (line CS208); *hcr1-2* (line CC126)

**REFERENCE**: Artus and Somerville (1988) *Plant Physiol. 87:* 83-88

**PHENOTYPE**: growth normal in 1% $CO_2$ but become chlorotic in air; lighter in color in 2% $CO_2$ but otherwise healthy; chlorophyll and carotenoids in plants grown in 2% $CO_2$ are 80% of wild-type levels and reduced by 60% and 40%, respectively, after 5 hrs in air

## *hcs1*: high $CO_2$ sensitive 1

**LOCATION**: not mapped

**ORIGIN**: induced with EMS in Columbia

**ALLELES**: line NA73

**REFERENCE**: Artus (1990) *Plant, Cell and Environment 13:* 575-580

**PHENOTYPE**: become chlorotic when grown in atmosphere enriched with $CO_2$ (20,000 $cm^3$ $CO_2$ $m^{-3}$); slightly lighter in color and grows slower in air

## *hcs2*: high $CO_2$ sensitive 2

**LOCATION**: not mapped

**ORIGIN**: induced with EMS in Columbia

**ALLELES**: line Ca2

**REFERENCE**: Artus (1990) *Plant, Cell and Environment 13:* 575-580

**PHENOTYPE**: become chlorotic when grown in atmosphere enriched with $CO_2$ (20,000 $cm^3$ $CO_2$ $m^{-3}$); grows slightly slower in air

high $CO_2$ requiring: see *hcr*

high $CO_2$ sensitive: see *hcs*
high hypocotyl: see *hy*
**hls1: *hookless* 1**
LOCATION: chromosome 4 (D. Meinke, pers. comm.)
ORIGIN: induced with EMS in Columbia
ALLELE: *hls1-1*
REFERENCE: Guzman and Ecker (1990) *Plant Cell 2:* 513-523
PHENOTYPE: recessive to wild type; seedlings lack hooks in the absence of exogenous ethylene; hypocotyl and root elongation inhibited by exogenous ethylene; in air, bolt and senesce much earlier than wild type
**hls2: *hookless* 2**
LOCATION: chromosome 4 ( D. Meinke, pers. comm.)
ORIGIN: induced with EMS in Columbia
ALLELE: *hls2-1*
REFERENCE: Guzman and Ecker (1990) *Plant Cell 2:* 513-523
PHENOTYPE: recessive to wild type; seedlings lack hooks in the absence of exogenous ethylene; complements *hls1-1*
hookless: see *hls*
**hy1: long *hypocotyl* 1**
LOCATION: 2-43.8
ORIGIN: first allele, Rédei (1964); others induced in Landsberg *erecta* with EMS or by fast neutrons
ALLELES: *hy1-1* (*hy,* high hypocotyl); seven others from Koornneef et al. (1980)
REFERENCES: Rédei and Hirono (1964) *AIS 1:* 9-10; Rédei (1965) *Genetics 51:* 857-872; Koornneef et al. (1980) *Z. Pflanzenphysiol. 100:* 147-160; Spruit et al. (1980) *AIS 17:* 137-141; Parks et al. (1989) *Plant Mol. Biol. 12:* 425-437; Goto et al. (1991) *Physiol. Plant. 83:* 209-215; Parks and Quail (1991) *Plant Cell 3:* 1177-1186; Chory (1992) *Development 115:* 337-354; Chory (1993) *TIG 9:* 167-172; Liscum and Hangarter (1993) *Planta 191:* 214-221
PHENOTYPE: plants are yellow-green; in white light hypocotyls are long, in UV and blue light hypocotyls behave as wild type (do not elongate much), and in red and far-red light elongation is not inhibited as it is in wild type; defective in phytochrome chromophore biosynthesis
OTHER INFORMATION: line d412 is green, not yellow-green; all alleles except line d412 are reported to have less than 10% of normal phytochrome level
**hy2: long *hypocotyl* 2**
LOCATION: 3-11.5
ORIGIN: induced with EMS in Landsberg *erecta*
ALLELES: four alleles (Koornneef et al. 1980)
REFERENCES: Koornneef et al. (1980) *Z. Pflanzenphysiol. 100:* 147-160; Spruit et al. (1980) *AIS 17:* 137-141; Parks et al. (1989) *Plant Mol. Biol. 12:* 425-437; Goto et al. (1991) *Physiol. Plant. 83:* 209-215; Parks and Quail (1991) *Plant Cell 3:* 1177-1186; Chory (1992) *Development 115:* 337-354;

Chory (1993) *TIG 9:* 167-172; Liscum and Hangarter (1993) *Planta 191:* 214-221; Liscum and Hangarter (1993) *Plant Phsiol. 103:* 15-19; S. Miséra, A. Müller, U. Heidedeck, and G. Jürgens, pers. comm.

**PHENOTYPE:** like *hy1* in color but more green; action spectrum like *hy1*; defective in phytochrome chromophore biosynthesis

**OTHER INFORMATION:** less than 10% of wild-type phytochrome level reported in mutants; suppressed by *fusca* mutations

*hy3:* long hypocotyl 3, same as *phyB,* see *phyB*

## *hy4:* long *h*ypocotyl 4

**LOCATION:** 4-9.5

**ORIGIN:** induced with EMS (*hy4-1*) or by fast neutrons (*hy4-2.23N*) in Landsberg *erecta*; induced by T-DNA insertion in Wassilewskija (Ws) (*hy4-2*) or segregated from a T-DNA line in Ws (*hy4-3*); others induced with EMS, by fast neutrons, or by X-ray in Landsberg *erecta*

**ALLELES:** *hy4-1,2,3,4* (Ahmad and Cashmore 1993); *hy4-2.23N* and other alleles (Koornneef et al. 1980)

**REFERENCES:** Koornneef et al. (1980) *Z. Pflanzenphysiol. 100:* 147-160; Spruit et al. (1980) *AIS 17:* 137-141; Goto et al. (1991) *Physiol. Plant. 83:* 209-215; Chory (1992) *Development 115:* 337-354; Ahmad and Cashmore (1993) *Nature, 366:* 162-166; Chory (1993) *TIG 9:* 167-172; Liscum and Hangarter (1993) *Planta 191:* 214-221

**PHENOTYPE:** slightly elongated hypocotyl, elongation inhibited (as in wild type) by UV, green, red, and far-red light, but elongates abnormally in blue light

**GENE:** cloned (Ahmad and Cashmore 1993) using a T-DNA insertional allele in Ws; *HY4* has 3 introns; encodes a protein (681 amino acids, 75.8 kD) with similarity to blue-light-stimulated photolyases; *hy4-1* has a $Gly_{340}$ to Glu substitution, and *hy4-4* has a $Gly_{337}$ to Asp substitution; *hy4-3* has a 5-bp deletion (nucleotides 1636-1640) resulting in a frameshift; *hy4-2.23N* has a deletion starting within the third intron (which is one nucleotide 5′ of the termination codon) and removing 3′ untranslated region

**OTHER INFORMATION:** elongation phenotype semi-dominant

## *hy5:* long *h*ypocotyl 5

**LOCATION:** 5-7.5

**ORIGIN:** induced with EMS (10 mM, 24 hrs at 24ºC) in Landsberg *erecta*

**ALLELES:** two (Koornneef et al. 1980)

**REFERENCES:** Koornneef et al. (1980) *Z. Pflanzenphysiol. 100:* 147-160; Spruit et al. (1980) *AIS 17:* 137-141; Goto et al. (1991) *Physiol. Plant. 83:* 209-215; Chory (1992) *Development 115:* 337-354; Chory (1993) *TIG 9:* 167-172; Liscum and Hangarter (1993) *Planta 191:* 214-221; Ang and Deng (1994) *Plant Cell 6:* 613-628

**PHENOTYPE:** like *hy4* but elongation not inhibited by far-red light and somewhat inhibited by blue light

**OTHER INFORMATION:** genetic studies suggest specific interaction with *COP1,* possibly at the protein level

*hy6:* **long *h*ypocotyl 6**

LOCATION: 2-43.9

ORIGIN: induced with EMS in Columbia

REFERENCES: Chory et al. (1989) *Plant Cell 1:* 867-880; Chory (1992) *Development 115:* 337-354; Young et al. (1992) *Planta 188:* 106-114; Chory (1993) *TIG 9:* 167-172; Liscum and Hangarter (1993) *Planta 191:* 214-221

PHENOTYPE: long hypocotyl; more severe than other *hy* mutants; adult plant paler than normal; smaller leaves and increased apical dominance; contain phytochrome apoprotein but deficient in phytochrome activity; hypocotyl elongation sensitive to blue and UV-A light inhibition

*hy8:* long hypocotyl 8, alleles of *PHYA,* see *phyA*

*im:* ***im*mutans**

LOCATION: 4-48.3

ORIGIN: induced by X-ray (10 kr) in presoaked Columbia seeds (*im-1*); induced with alkylating agent (EMS, ethylene imine, or ethylene oxide) in En-2 (Enkheim) wild type (*im-52, im-56, im-245*); induced with EMS in Columbia (*im-*"spotty")

ALLELES: *im-1; im-52* (= V52, 63/3/1); *im-56; im-245*

REFERENCES: Rédei (1964) *AIS 1:* 11; Rédei (1965) *AIS 2:* 26; Rédei (1967) *Genetics 56:* 431-443; Rédei (1967) *J. Hered. 58:* 229-235; Röbbelen (1968) *Planta 80:* 237-254; Chung and Rédei (1974) *Biochem. Genet. 11:* 441-453; Kranz (1977) *AIS 14:* 31-39; Kranz (1979) *AIS 16:* 47-56; Wetzel et al. (1994) *Plant J. 6:* 161-175

PHENOTYPE: *im-1:* plant variegated green and white

*im-52:* like *im-1,* more white than *im-1*

*im-56:* like *im-1,* more green than *im-1*

*im-245:* like *im-1,* more green than *im-56*

*im-*"spotty": under continuous light, more green than white sectors in older rosette leaves and more white than green sectors in later leaves; white sectors accumulate phytoene, a precursor of carotinoids, suggesting that the mutant defect is in the phytoene desaturase; however, molecular analyses indicate that the mutation does not affect the structural gene for phytoene desaturase

OTHER INFORMATION: 6-azauracil ($1.6 \times 10^{-5}$ M) suppresses the phenotype of *im-1,* as does low light intensity; phenotype dependent on light quality

imidazolinone resistance: see *csr1-2*

immutans: see *im*

*imr: im*idazolinone *r*esistant, see *csr1-2*

*ixr1:* ***iso*xaben *r*esistant 1**

LOCATION: chromosome 5, tightly linked (less than 3 cM) to *lu* (at 5-17.6)

ORIGIN: induced with EMS in Columbia

ALLELES: *ixr1-1* (*ixrA1,* line DH48) and *ixr1-2* (*ixrA2,* line DH47)

REFERENCE: Heim et al. (1989) *Plant Physiol. 90:* 146-150

PHENOTYPE: selected on the basis of resistance to isoxaben; homozygous cells resistant in tissue culture and heterozygous cells partially resistant

*ixrA1:* see *ixr1-1*

*ixrA2:* see *ixr1-2*
JB1: see *fad7-1*
JB3: see *act1*
JB9: see *fad2-1*
JB11: see *ela1*
JB12: see *fad2-3*
JB25: see *act1-1*
JB28: see *act1*
JB60: see *fad4-1*
JB67: see *fad5-1*

## JK218: reduced phototropism
ORIGIN: induced with EMS in Estland
REFERENCES: Khurana et al. (1989) *Planta 178:* 400-406; Konkevic et al. (1992) *Photochem. Photobiol. 55:* 789-792; Liscum et al. (1992) *Plant Physiol. 100:* 267-271
PHENOTYPE: no phototropic response after 2 hrs of unilateral irradiation

## JK224: reduced phototropism
ORIGIN: induced with EMS in Estland
REFERENCES: Khurana et al. (1989) *Planta 178:* 400-406; Konkevic et al. (1992) *Photochem. Photobiol. 55:* 789-792; Liscum et al. (1992) *Plant Physiol. 100:* 267-271; Reymond et al. (1992) *Proc. Natl. Acad. Sci. USA 89:* 47-18-4721
PHENOTYPE: threshold and saturation of the first phototropic response is shifted to higher fluences; normal second positive curvature; reduced phosphorylation of crude membrane fraction

## *kak: kaktus*
LOCATION: not mapped
ORIGIN: induced with EMS in Landsberg *erecta*
ALLELES: four
REFERENCE: Hülskamp et al. (1994) *Cell 76:* 555-566
PHENOTYPE: trichome enlarged; supernumerary branches

kaktus: see *kak*

## *keu: keule*
LOCATION: 1-13.9
ORIGIN: induced with EMS (0.3%, 8 hrs) in Landsberg *erecta*
ALLELES: nine
REFERENCES: Mayer et al. (1991) *Nature 353:* 402-407; U. Mayer (1993) Ph.D. Thesis, Universität München
PHENOTYPE: seedling pattern mutant; defective seedling epidermis; rough surface; seedling shape variable, strongly reduced cotyledons

keule: see *keu*

## *klk: klunker*
LOCATION: chromosome 5, near *ttg* (G. Jürgens, pers. comm.)
ORIGIN: induced with EMS in Landsberg *erecta*
ALLELES: six

REFERENCE: Hülskamp et al. (1994) *Cell 76:* 555-566
PHENOTYPE: trichome distorted

## *kn: knolle*

LOCATION: chromosome 1, between *an* and *dis1* (G. Jürgens, pers. comm.)
ORIGIN: one allele induced with EMS (0.3%, 8 hrs) in Landsberg *erecta*; another allele induced with 20,000 R X-ray in Landsberg *erecta* pollen
ALLELES: two
REFERENCES: Mayer et al. (1991) *Nature 353:* 402-407; U. Mayer (1993) Ph.D. Thesis, Universität München
PHENOTYPE: seedling pattern mutant; seedlings are round or tuber-shaped; surface appears rough due to lack of well-formed epidermis; abnormally large cells surround lumps of vascular tissue

## *knf: knopf*

LOCATION: not mapped
ORIGIN: induced with EMS in Landsberg *erecta*
ALLELES: six
REFERENCE: Mayer et al. (1991) *Nature 353:* 402-407
PHENOTYPE: seedling pattern mutant; seedlings are small, round, and pale; epidermal cells are columnar and densely packed; vascular tissue replaced by ground tissue (normally found between epidermis and vascular tissue

knolle: see *kn*

knopf: see *knf*

late flowering: see *co* (*fg*), *fca, fd, fe, fha, fla, fpa, ft, fve, fwa, fy, gi* (*fb*), and *ld*

LC500: see *ap1-9*

## *ld: luminidependans*

LOCATION: chromosome 4, between *fla* and the Meyerowitz RFLP marker 506 (Lee et al. 1994)
ORIGIN: *ld-1* induced by X-ray mutagenesis of Columbia (8–12 kr to presoaked seeds); *ld-2* segregated from a T-DNA insertional line in Wassilewskija (Ws); *ld-3* induced by a T-DNA insertion in Ws
ALLELES: *ld-1* (Rédei 1962); *ld-2, ld-3* (Lee et al. 1994)
REFERENCES: Rédei (1962) *Genetics 47:* 443-460; Lee et al. (1994) *Plant Cell 6:* 75-83
PHENOTYPE: late flowering in short-day (natural winter) conditions
GENE: isolated by Lee et al. (1994) using a T-DNA insertional allele; encodes a novel protein with a glutamine-rich region and nuclear localization signals
OTHER INFORMATION: in continuous illumination flower primordia appear only a few days later than in wild type

## *le: lepida,* **allelic with** *pa*

LOCATION: 1-66.4
ORIGIN: induced by X-ray in Limburg (Li 5) or Estland
ALLELES: two isolated by McKelvie (1963); *pa* (Kranz and Scheidemann 1978)
REFERENCES: McKelvie (1962) *Radiat. Bot. 1:* 233-243; McKelvie (1963) *Radiat. Bot. 3:* 105-123; Kranz and Scheidemann (1978) *AIS 15:* 31-34

PHENOTYPE: dwarf, with round leaves and small pods; dark green

OTHER INFORMATION: Napp-Zinn and Bonzi (1970, *AIS 7:* 8-9) found that gibberellin slightly increases stalk length in *le* but not to normal height

leafy: see *lfy*

leafy cotyledon: see *lec*

### *lec1:* **leafy cotyledon 1, allelic with** *emb212*

LOCATION: 1-29.3

ORIGIN: *lec1-1* segregated from a T-DNA insertional line in Wassilewskija (Ws), not linked to the T-DNA; *lec1-2* isolated from a population of T-DNA insertional line in Ws but linkage to a T-DNA remains to be determined (aberrant segregation of the T-DNA-encoded Kan$^R$ marker)

ALLELES: *lec1-1* (Meinke 1992); *lec1-2* (Meinke et al. 1994)

REFERENCES: Meinke (1992) *Science 258:* 1647-1650; Meinke et al. (1994) *Plant Cell 6:* 1049-1064

PHENOTYPE: heterozygote produces 25% defective embryos; mutant embryos are normal through the heart stage but have distorted shape starting at the torpedo stage; hypocotyl has highly vacuolated cells and fails to elongate at the cotyledon stage; embryos remain green late in development and are occasionally (5%) viviparous; dried seeds fail to germinate; mutant embryos can be germinated in culture if removed from the seeds before desiccation and seedling cotyledons have trichomes; the trichomes on mutant cotyledons fail to form in *gl1* or *ttg* background; the mutant seedlings produced in culture can develop into normal plants in soil with the exception that they produce 100% defective seeds; mutant cotyledons have a vascular system intermediate between normal cotyledons and leaves

### *lec2:* **leafy cotyledon 2**

LOCATION: chromosome 1 (Meinke et al. 1994)

ORIGIN: isolated from a population of T-DNA insertional line in Wassilewskija (Ws)

ALLELE: *lec2-1*

REFERENCE: Meinke et al. (1994) *Plant Cell 6:* 1049-1064

PHENOTYPE: cotyledons have leaf characteristics, including trichomes, stomata, and pattern of the vascular tissues; other aspects of seed maturation normal, including activation of late embryo developmental program, desiccation tolerance, and lack of vivipary; lower portion of the cotyledons contains more protein bodies and less starch than the tip portion

lepida: see *le*

lesions stimulating disease resistance response: see *lsd*

### *lfy:* **leafy**

LOCATION: 5-82.3

ORIGIN: *lfy-1, lfy-9, lfy-10,* and *lfy-12* through *lfy-14* induced with EMS in Columbia (Haughn and Somerville 1988; Huala and Sussex 1992; Weigel et al. 1992); *lfy-3* through *lfy-6, lfy-8,* and *lfy-11* induced with EMS in Landsberg *erecta* (Weigel et al. 1992); *lfy-7* found in a T-DNA insertion line in Wassilewskija (Ws) (Feldmann 1991, *Plant J. 1:* 71-82); *lfy-15* found in a

T-DNA insertion line in Columbia C24 (Weigel et al. 1992); see table of alleles in Weigel et al. (1992)

**REFERENCES:** Haughn and Somerville (1988) *Dev. Genet. 9:* 73-89; Schultz and Haughn (1991) *Plant Cell 3:* 771-781; Huala and Sussex (1992) *Plant Cell 4:* 901-913; Weigel et al. (1992) *Cell 69:* 843-859; Bowman et al. (1993) *Development 119:* 721-743; Clark et al. (1993) *Development 119:* 397-418; Shannon and Meeks-Wagner (1993) *Plant Cell 5:* 639-655; Schultz and Haughn (1993) *Development 119:* 745-765; Weigel and Meyerowitz (1993) *Science 261:* 1723-1726

**PHENOTYPE:** altered inflorescence structure characterized by partial conversion of flower/floral meristem to inflorescence/inflorescence meristem

**GENE:** cloned by Weigel et al. (1992); homolog of *Antirrhinum majus FLO* gene (Coen et al. 1990, *Cell 63:* 1311-1322); multiple alleles have been sequenced; mRNA is expressed in very early floral primordia, as well as in later stages of floral development

lion's tail: see *lit*

***lit: lion's tail***

**LOCATION:** not mapped

**ORIGIN:** induced with EMS in Columbia

**REFERENCES:** Benfey et al. (1993) *Development 119:* 57-70; Hauser and Benfey (1994) *Proc. NATO-ASI, Plant Molecular Biology Series: Molecular Genetic Analysis of Plant Development and Metabolism* (eds. G. Coruzzi and P. Puigdoménech), pp. 31-40, Springer-Verlag, Berlin

**PHENOTYPE:** conditional abnormal root expansion, more severe on 4.5% sucrose, no expansion at 14°C; expansion most severe in stele (9x), less in epidermis (5x) and cortex (4x), and least in endodermis (2x); very slightly stunted growth of shoot

**OTHER INFORMATION:** double mutant with *cob* has stunted shoots and abnormal amount of anthocyanin; sterile

LK3: see *fad6-1*

LK8: see *act1-2*

L-*O*-methylthreonine resistant: see *omr1*

long hypocotyl: see *hy*

***lsd1: lesions stimulating disease resistance response 1***

**LOCATION:** not mapped

**ORIGIN:** segregated from a T-DNA line (T4855) in Wassilewskija (Ws)

**REFERENCE:** Dietrich et al. (1994) *Cell 77:* 565-577

**PHENOTYPE:** spontaneous necrotic lesions formed under long-day, but not short-day, conditions; lesions appear on leaves, stem, and flowers but not on cotyledons; seedlings die under long-day conditions due to extensive lesion formation; constitutive expression of PR-1, which is associated with disease resistance response; reduced susceptibility to pathogens; lesions can be induced on prelesion leaves by biotic or chemical agents that normally induce disease response

***lsd2: lesions stimulating disease resistance response 2***

**LOCATION:** not mapped
**ORIGIN:** induced with EMS in Columbia
**REFERENCE:** Dietrich et al. (1994) *Cell 77:* 565-577
**PHENOTYPE:** spontaneous lesions form on leaves under short-day, but not long-day, conditions; lesions first appear on mature leaves of 2–3-week-old plants as pairs of symmetric chlorotic regions on each side of leaf margins; these lesions enlarge toward the midvein and develop into bands of necrosis across the leaf; constitutive expression of PR-1, which is associated with disease resistance response; reduced susceptibility to pathogens
**OTHER INFORMATION:** dominant

### *lsd3: l*esions *s*timulating *d*isease resistance response 3

**LOCATION:** not mapped
**ORIGIN:** segregated from a T-DNA line (T656) in Wassilewskija (Ws)
**REFERENCE:** Dietrich et al. (1994) *Cell 77:* 565-577
**PHENOTYPE:** lesions form under long-day, but not short-day, conditions; both rosette and cauline leaves form lesions 7–10 days after shift to long-day; the lesions begin as gray loci, and occasionally form V-shaped area, with distinct margin; constitutive expression of PR-1, which is associated with disease resistance response; reduced susceptibility to pathogens

### *lsd4: l*esions *s*timulating *d*isease resistance response 4

**LOCATION:** not mapped
**ORIGIN:** segregated from a T-DNA line (T6142) in Wassilewskija (Ws)
**REFERENCE:** Dietrich et al. (1994) *Cell 77:* 565-577
**PHENOTYPE:** lesions appear on very young plants under both long- and short-day conditions; plants have diffuse areas of chlorosis, without a clear margin; plants slightly smaller than normal; constitutive expression of PR-1, which is associated with disease resistance response; reduced susceptibility to pathogens
**OTHER INFORMATION:** dominant

### *lsd5: l*esions *s*timulating *d*isease resistance response 5

**LOCATION:** not mapped
**ORIGIN:** segregated from a T-DNA line in Wassilewskija (Ws)
**REFERENCE:** Dietrich et al. (1994) *Cell 77:* 565-577
**PHENOTYPE:** lesion phenotype more severe under low light intensity, and there is some seedling lethality under short-day conditions; lesions form on all leaves, starting on the primary leaves and often at the tip; plants are stunted; constitutive expression of PR-1, which is associated with disease resistance response; reduced susceptibility to pathogens

luminidependans: see *ld*

### *lu: l*utescens

**LOCATION:** 5-10.4
**SYNONYM:** lutescent
**REFERENCE:** Lee-Chen and Steinitz-Sears (1967) *Can. J. Genet. Cytol. 9:* 381-384
**PHENOTYPE:** center of rosette yellow-green, serrated leaf margin

lutescens: see *lu*

lutescent: see *lu*

M 4-6-18: see *alb2*

M33: see *ch6*

M218: see *tt3*

male gametophyte factor: see *gm*

male sterile: see *ms*

methionine overaccumulation: see *mto1*

*mic: mickey*

    LOCATION: not mapped

    ORIGIN: induced with EMS in Landsberg *erecta*

    ALLELES: eight

    REFERENCE: Mayer et al. (1991) *Nature 353:* 402-407

    PHENOTYPE: seedling pattern mutant; cotyledons are thick, disc-shaped, and disproportionately large compared to the short hypocotyl and root; cotyledons have abnormal texture; hypocotyl contains bloated epidermal cells and fuzzy vascular strands

mickey: see *mic*

*min: miniature*

    LOCATION: 5-89.6

    REFERENCE: Koornneef et al. (1983) *J. Hered. 74:* 265-272

    PHENOTYPE: small plant with greyish-green leaves

miniature: see *min*

*mod1: modifier 1*

    LOCATION: inseparable from *er* (2-43.5) (Lang et al. 1994)

    ORIGIN: uncovered in Landsberg *erecta* ecotype, normal in Columbia ecotype

    REFERENCE: Lang et al. (1994) *Genetics 137:* 1101-1110

    PHENOTYPE: affect the phenotype of *sin1* homozygotes, see *sin1*

    OTHER INFORMATION: tightly linked to, and maybe is, *er*

monopteros: see *mp*

*mp: monopteros*

    LOCATION: 1-21.6

    ORIGIN: induced with EMS in Landsberg *erecta*

    ALLELES: ten strong, three weak

    REFERENCES: Mayer et al. (1991) *Nature 353:* 402-407; Berleth and Jürgens (1993) *Development 118:* 575-587

    PHENOTYPE: seedling pattern mutant; abnormal development of lower tier of embryo proper and hypophysis; deletion of hypocotyl and root; seedling has only cotyledons and apical meristem; some seedlings have fused cotyledons or a single cotyledon and lack apical meristem; root formation on root-inducing medium

*ms1: male sterile 1*

    LOCATION: 5-22.5

    REFERENCES: Van der Veen and Wirtz (1968) *Euphytica 17:* 371-377; Koornneef et al. (1983) *J. Hered. 74:* 265-272; Dawson et al. (1993) *Can. J. Bot.*

*71:* 629-638

**PHENOTYPE:** male sterile, pollen decays and anthers fail to open; pollen wall fails to develop after microspore release from tetrads

### *ms2: male sterile 2*

**LOCATION:** 3-27.3

**ORIGIN:** induced by insertion of maize *Inhibitor/defective Suppressor-Mutator* in Landsberg *erecta*

**ALLELES:** insertional allele and several derivative alleles

**REFERENCE:** Aarts et al. (1993) *Nature 363:* 715-717

**PHENOTYPE:** male sterile; no pollen grains; produce a small number of self-fertilized seeds from late flowers; sectors of usually single (fertile) flower size are produced in the presence of transposase; tapetal layer seems to be defective

**GENE:** cloned by Aarts et al. (1993); encodes a predicted protein of 616 amino acids with a short region of similarity with a wheat mitochondrial open reading frame

### *ms31: male sterile 31, same as msK*

**LOCATION:** chromosome 1 (Mulligan et al. 1994)

**ORIGIN:** induced with EMS in Landsberg *erecta*

**ALLELE:** one

**REFERENCES:** Dawson et al. (1993) *Can. J. Bot. 71:* 629-638; Mulligan et al. (1994) *Flowering Newsletter 17:* 12-20

**PHENOTYPE:** male sterile; abnormal pollen mother cell and tetrads; abnormal dissolution of callose

**OTHER INFORMATION:** F2 segregate for normal to sterile plants 7.9:1

### *ms32: male sterile 32, same as msW*

**LOCATION:** chromosome 1 (Mulligan et al. 1994)

**ORIGIN:** induced with EMS in Landsberg *erecta*

**ALLELE:** one

**REFERENCES:** Dawson et al. (1993) *Can. J. Bot. 71:* 629-638; Mulligan et al. (1994) *Flowering Newsletter 17:* 12-20

**PHENOTYPE:** male sterile; granular cytoplasm in pollen mother cells, callose reduced or absent; abnormalities in meiosis; irregularly shaped meiotic products that degenerate

**OTHER INFORMATION:** F2 segregate for normal to sterile plants 5.2:1

### *ms33: male sterile 33, same as msZ*

**LOCATION:** chromosome 2 (Mulligan et al. 1994)

**ORIGIN:** induced with EMS in Landsberg *erecta*

**ALLELE:** one

**REFERENCES:** Dawson et al. (1993) *Can. J. Bot. 71:* 629-638; Mulligan et al. (1994) *Flowering Newsletter 17:* 12-20

**PHENOTYPE:** male sterile; elongation of anther filament blocked; pollen maturation halted at a late stage; infertile pollen grains produced and eventually released

### *ms34: male sterile 34*

LOCATION: chromosome 2 (Mulligan et al. 1994)
ORIGIN: induced by γ-ray in Landsberg *erecta*
ALLELE: one
REFERENCE: Mulligan et al. (1994) *Flowering Newsletter 17:* 12-20
PHENOTYPE: male sterile; sterile pollen grains of different sizes; well-developed pollen walls; abnormal tetrads with thin callose

***ms35: male sterile 35, same as msH***
LOCATION: chromosome 3 (Mulligan et al. 1994)
ORIGIN: induced by X-ray in Landsberg *erecta*
ALLELE: one
REFERENCES: Dawson et al. (1993) *Can. J. Bot. 71:* 629-638; Mulligan et al. (1994) *Flowering Newsletter 17:* 12-20
PHENOTYPE: male sterile; final stage of anther dehiscence blocked; fertile pollen produced but not released

***ms36: male sterile 36***
LOCATION: chromosome 3 (Mulligan et al. 1994)
ORIGIN: induced by γ-ray in Landsberg *erecta*
ALLELE: one
REFERENCE: Mulligan et al. (1994) *Flowering Newsletter 17:* 12-20
PHENOTYPE: male sterile; similar to *ms1;* pollen wall development fails after microspores released from the tetrad

***ms37: male sterile 37, same as msY***
LOCATION: chromosome 5 (Mulligan et al. 1994)
ORIGIN: induced with EMS in Landsberg *erecta*
ALLELE: one
REFERENCES: Dawson et al. (1993) *Can. J. Bot. 71:* 629-638; Mulligan et al. (1994) *Flowering Newsletter 17:* 12-20
PHENOTYPE: male sterile; abnormally granular pollen mother cells, abnormalities in meiosis; callose reduced or absent; any apparently normal microspores degenerate

***ms38: male sterile 38***
LOCATION: chromosome 5 (Mulligan et al. 1994)
ORIGIN: induced by γ-ray in Landsberg *erecta*
ALLELE: one
REFERENCE: Mulligan et al. (1994) *Flowering Newsletter 17:* 12-20
PHENOTYPE: male sterile; stamen elongation delayed; pollen decays at a late stage

*msH:* male sterile H, see *ms35*
*msK:* male sterile K, see *ms31*
*msW:* male sterile W, see *ms32*
*msY:* male sterile Y, see *ms37*
*msZ:* male sterile Z, see *ms33*
***mto1: methionine overaccumulation 1***
LOCATION: chromosome 3 (Inaba et al. 1994)
ORIGIN: induced with EMS in Columbia *gl1/gl1*

**ALLELE:** one

**REFERENCE:** Inaba et al. (1994) *Plant Physiol. 104:* 881-887

**PHENOTYPE:** isolated as seedling resistant to ethionine, a toxic analog of methionine; semi-dominant over wild type; accumulate about 40-fold the normal levels of methionine; some reduction of fresh weight

multipetala: see *ag*

## *mur1*

**LOCATION:** not mapped

**ORIGIN:** induced with EMS in Columbia

**ALLELES:** *mur1-1; mur1-2*

**REFERENCE:** Reiter et al. (1993) *Science 261:* 1032-1035

**PHENOTYPE:** shoot cell wall has greatly reduced levels of L-fucose; plant dwarf and the dwarfism cannot be corrected by phytohormones; cell wall more fragile than wild type

N6H-14: see *ch42*

NA73: see *hcs1*

NG4: see *ga1-9*

NG5: see *ga1-1*

## *nia1:* **nitrate reductase 1**

**LOCATION:** 1-115.6

**ORIGIN:** induced with EMS in *chl3-5* (Columbia) mutant seeds

**ALLELE:** *nia1-1* (Wilkinson and Crawford 1993)

**REFERENCES:** Cheng et al. (1988) *EMBO J. 7:* 3309-3314; Wilkinson and Crawford (1993) *MGG 239:* 289-297

**PHENOTYPE:** increased chlorate sensitivity of *chl3* mutant line (Columbia); *nia1-1 chl3-5* double mutant has only 0.5% of wild-type level of nitrate reductase activity; very poor growth on nitrate as the sole nitrogen source

**GENE:** cloned by Cheng et al. (1988), appears by homology to encode nitrate reductase (as does *CHL3*); *nia1-1* mutation analyzed by Wilkinson and Crawford (1993) and it is a point mutation resulting in the substitution of an alanine with a threonine in a conserved molybdenum binding domain

*nia2:* see *chl3*

nitrate reductase: see *nia1* and *chl3*

## *omr1:* L-*O*-**methylthreonine** *resistant* **1**

**LOCATION:** 3-17.2

**ORIGIN:** induced with EMS in Columbia

**ALLELES:** *omr1-1* (line GM11b); four other mutants were isolated

**REFERENCES:** Mourad and King (1993) *Plant Physiol. 102:* 38S, poster #196; G. Mourad, pers. comm.

**PHENOTYPE:** dominant; resistant to the isoleucine analog L-*O*-methylthreonine (at 0.5 mM, which completely inhibits wild-type growth), which when incorporated into proteins in place of isoleucine causes cell death; threonine dehydratase, the target enzyme, is 20- to 50-fold less sensitive to feedback control by isoleucine than wild type, thus the level of isoleucine in cells is much higher than normal and the elevated level of isoleucine out-competes

L-*O*-methylthreonine for incorporation into proteins

P83: see *aux1*

P391: see *aux1*

P445/10: see *dwf*

*pa:* allelic with *le,* see *le*

*pac: pale cress*

> LOCATION: chromosome 2 (2 cM from *cer8* and 21.6 cM from *er*) (Reiter et al. 1994)
>
> ORIGIN: induced by T-DNA insertion in Wassilewskija (Ws)
>
> ALLELE: one
>
> REFERENCE: Reiter et al. (1994) *Plant Cell 6:* 1253-1264
>
> PHENOTYPE: seedling has pale green cotyledons and leaves; less than 3% of normal levels of chlorophylls *a* and *b* and carotenoids; chloroplasts contain much reduced thylakoid membranes; leaf development aberrant
>
> GENE: cloned by Reiter et al. (1994); encodes three alternatively spliced mRNAs; the largest ORF contained in these mRNAs predicts a very acidic protein; *PAC* expression is enhanced at least 10-fold by light

pale cress: see *pac*

*pcoA:* **phosphoglycolate phosphatase**

> LOCATION: nuclear, not mapped
>
> ORIGIN: induced with EMS in Columbia
>
> SYNONYM: CS119
>
> REFERENCE: Somerville and Ogren (1979) *Nature 280:* 833-836
>
> PHENOTYPE: deficient in phosphoglycolate phosphatase activity (a chloroplast enzyme); grows in 1% $CO_2$ (with photorespiration blocked) but not in 0.03% $CO_2$ (photorespiration occurs); plants are chlorotic in 0.03% $CO_2$

PD114, PD378: see *tom1*

*pfl:* **pointed first leaf**

> LOCATION: chromosome 1 (~19 cM below *an*) (Van Lijsebettens et al. 1994)
>
> ORIGIN: induced by a T-DNA insertion
>
> REFERENCE: Van Lijsebettens et al. (1994) *EMBO J. 13:* 3378-3388
>
> PHENOTYPE: seedlings have retarded growth when grown at 20°C; at 13°C mutant first leaves are very narrow and chlorotic and late leaves remain pointed but less chlorotic; mutant plants are fertile at 13°C but more than normal number of seeds abort
>
> GENE: isolated by Van Lijsebettens et al. (1994), encodes an S18 ribosomal protein (RPS18A); two other related genes (*RPS18B* and *RPS18C*) exist in *Arabidopsis;* GUS fusion analysis indicates that PFL is expressed mainly in meristems, organ primordia, vascular tissues, anther and pollen grains, ovules, and embryos

*pgm:* **phosphoglucomutase**

> LOCATION: 5-63.4
>
> ORIGIN: induced with EMS in Columbia
>
> ALLELES: *pgm-1* (= *pgmP,* line TC7: TC75 was derived from TC7 by backcrossing to wild type); *pgm-2* (line TC9)

**REFERENCES:** Caspar et al. (1985) *Plant Physiol. 79:* 11-17; Sivak and Rowell (1988) *Plant Physiol. Biochem. 26:* 493-501; Caspar and Pickard (1989) *Planta 177:* 185-197; Caspar et al. (1989) *Proc. Natl. Acad. Sci. USA 86:* 5830-5833; Moore (1989) *Annals Bot. 64:* 271-277; Sack and Kiss (1989) *Am. J. Bot. 76:* 454-464; Caspar, this volume

**PHENOTYPE:** isolated on the basis of lack of starch in leaf; deficient in chloroplast phosphoglucomutase activity

**OTHER INFORMATION:** another mutant, TC135, was isolated as starchless mutant but it complements *pgm-1* or *pgm-2* (Caspar et al. 1985)

*pgmP:* see *pgm*

**pho1: deficient in xylem loading of *pho*sphate**

**LOCATION:** 3-27.2

**ORIGIN:** induced with EMS in Columbia

**REFERENCE:** Poirier et al. (1991) *Plant Physiol. 97:* 1087-1093

**PHENOTYPE:** mutant plants accumulate 5% of normal inorganic phosphate in aerial portions; root phosphate uptake normal; phosphate transfer to shoot reduced to 3–10% of normal in media with 200 µM phosphate or less; thought to be defective in xylem loading of phosphate

phosphoglucomutase: see *pgm*

phosphoglycolate phosphatase: see *pcoA*

**phototropic defect: ZR8 and ZR19**

**LOCATION:** not mapped

**ORIGIN:** induced with EMS in Estland

**ALLELES:** lines ZR8 and ZR19 probably carry alleles of the same gene

**REFERENCE:** Khurana et al. (1989) *Plant Physiol. 91:* 685-689

**PHENOTYPE:** decreased phototropism to 450 nm light; decreased gravitropism

**phyA: phytochrome A, allelic with *fhy2, fre1*, and *hy8***

**LOCATION:** 1-15.7

**ORIGIN:** *phyA-1,2 (fhy2-1,2)* induced by γ-ray in Columbia *gl (gl1?)*; *phyA-101,102,103 (hy8-1,2,3)* induced with EMS in ecotype RLD; *phyA-201,202 (fre1-1, fre1-2)* induced with EMS in Landsberg *erecta; phyA-203* through *208* induced in Landsberg *erecta; phyA-209,210,211* induced in Columbia transgenic line pOCA107-2 (Susek et al. 1993, *Cell 74:* 787-799) by γ-ray

**ALLELES:** *phyA-1,2* (Whitelam et al. 1993); *phyA-101,102,103* (Parks and Quail 1993); *phyA-201,202* (Nagatani et al. 1993); *phyA-203* through *phyA-211* (Reed et al. 1994)

**REFERENCES:** Sharrock and Quail (1989) *Genes Dev. 3:* 1745-1757; Dehesh et al. (1993) *Plant Cell 5:* 1081-1088; Liscum and Hangarter (1993) *Plant Physiol. 103:* 15-19; Nagatani et al. (1993) *Plant Physiol. 102:* 269-277; Parks and Quail (1993) *Plant Cell 5:* 39-48; Whitelam et al. (1993) *Plant Cell 5:* 757-768; Boylan et al. (1994) *Plant Cell 6:* 449-460; Johnson et al. (1994) *Plant Physiol. 105:* 141-149; Quail et al. (1994) *Plant Cell 6:* 468-471; Reed et al. (1994) *Plant Physiol. 104:* 1139-1149

**PHENOTYPE:** *phyA-1,2:* elongated hypocotyl in far-red light, as long as in dark; normal in red, blue, or white light; lack phytochrome A in dark-grown seedlings by both spectrophotometric and immunological analyses; semi-dominant; abnormal day length perception

*phyA-101,102,103:* long hypocotyl under continuous far-red light; partially elongated hypocotyl and open and expanded cotyledons under continuous red light, like the wild type but unlike *hy1* mutants; phytochrome A protein is undetectable in *phyA-101,102* mutants, whereas phytochrome B and C proteins are normal; *phyA-103* has normal levels and normal light-induced turnover of phytochrome A

*phyA-201,202:* hypocotyl elongation in far-red light, as long as in dark; the elongation is altered by the addition of biliverdin, a precursor of the phytochrome chromophore; the ability to de-etiolate after transfer from dark to light is lost faster than in wild type—this loss can be delayed by adding sucrose to medium; no immunologically detectable PHYA protein

GENE: isolated by Sharrock and Quail (1989), encodes a 124-kD hydrophilic protein; mutant sequences: *phyA-1* has rearrangement at the *PHYA* locus, no detectable *PHYA* mRNA in *phyA-1,* and much reduced level in *phyA-2;* *phyA-101* is a C→T substitution resulting in a stop codon at residue 25; *phyA-102* is a G→A substitution resulting in a defective acceptor site of intron 4; *phyA-103* is a G→A substitution resulting in a change of the invariant $Gly_{727}$→Glu; *phyA-201* is a C→T substitution resulting in a stop codon at residue 980; *phyA-205* is a G→A substitution resulting in a change of $Met_{631}$→Val

## *phyB: phytochrome B*

LOCATION: 2-30.9

ORIGIN: *phyB-1* through *phyB-8* induced with EMS in Landsberg *erecta; phyB-9* induced with EMS in Columbia; *phyB-10* is a T-DNA insertion mutant in Wassilewskija (Ws)

ALLELES: *phyB-1* (line Bo64); *phyB-2* (line V197); *phyB-3* (line d504); *phyB-4* (line 4-117); *phyB-5* (line 8-36); *phyB-6* (line 548); phyB-7 (line 1053); *phyB-8* (line M4084) (Koornneef et al. 1980); *phyB-9* (line EMS142) and *phyB-10* (line 464-19) (Reed et al. 1993)

REFERENCES: Koornneef et al. (1980) *Z. Pflanzenphysiol. 100:* 147-160; Spruit et al. (1980) *AIS 17:* 137-141; Sharrock and Quail (1989) *Genes Dev. 3:* 1745-1757; Goto et al. (1991) *Physiol. Plant. 83:* 209-215; Nagatani et al. (1991) *Plant Cell Physiol. 32:* 1119-1122; Somers et al. (1991) *Plant Cell 3:* 1263-1274; Chory (1992) *Development 115:* 337-354; Chory (1993) *TIG 9:* 167-172; Reed et al. (1993) *Plant Cell 5:* 147-157; Robson et al. (1993) *Plant Physiol. 102:* 1179-1184; Liscum and Hangarter (1993) *Planta 191:* 214-221; Liscum and Hangarter (1993) *Plant Physiol. 103:* 15-19; Quail et al. (1994) *Plant Cell 6:* 468-471; Reed et al. (1994) *Plant Physiol. 104:* 1139-1149; S. Miséra, A. Müller, U. Heidedeck, and G. Jürgens, pers. comm.

PHENOTYPE: light green to green in color, elongates abnormally in red but not in far-red light; deficient in *HPYB* mRNA and phytochrome B protein

GENE: wild-type gene (*PHYB*) cloned by Sharrock and Quail (1989) from

Columbia and by Reed et al. (1993) from Landsberg *erecta;* mutant alleles cloned and sequenced by Reed et al. (1993); alleles *phyB-1,5,9* are nonsense mutations, CAG→TAG (Gln$_{448}$→Stop), TGG→TGA (Trp$_{552}$→Stop), and TGG→TGA (Trp$_{397}$→Stop), respectively; *phyB-4* is a missense mutation: CAT→TAC (His$_{283}$→Tyr); the T-DNA insertional allele *phyB-10* does indeed have an insertion in the *PHYB* gene

**OTHER INFORMATION:** elongation phenotype semi-dominant; suppressed by *fusca* mutations

### *pi: pi*stillata

**LOCATION:** 5-21.3

**ORIGIN:** induced with EMS in Landsberg *erecta*

**ALLELES:** *pi-1,2,3* (see Bowman et al. 1991)

**REFERENCES:** Koornneef et al. (1983) *J. Hered. 74:* 265-272; Bowman et al (1989) *Plant Cell 1:* 37-52; Hill and Lord (1989) *Can. J. Bot. 67:* 2922-2936; Meyerowitz et al. (1989) *Development 106:* 209-217; Bowman et al. (1991) *Development 112:* 1-20; Coen and Meyerowitz (1991) *Nature 353:* 31-37; Schultz et al. (1991) *Plant Cell 3:* 1221-1237; Bowman et al. (1992) *Development 114:* 599-615; Jack et al. (1992) *Cell:* 683-697; Clark et al. (1993) *Development 119:* 397-418; Weigel and Meyerowitz (1993) *Science 261:* 1723-1726; Goto and Meyerowitz (1994) *Genes Dev. 8:* 1548-1560

**PHENOTYPE:** anthers and petals absent or altered in some alleles, converted to carpels and sepals, respectively, in others

**GENE:** cloned by Goto and Meyerowitz (1994); member of the MADS-box gene family; most likely homolog (58% identity) of the *Antirrhinum majus* floral gene *GLOBOSA* (Tröbner et al. 1992, *EMBO J. 11:* 4693-4704) and is closely related in sequence to MADS-box genes from petunia (*FBP1* and *pMADS2*) and tobacco (*NTGLO*); encodes a protein with 208 amino acid residues; PI and AP3 proteins co-immunoprecipitate; expressed in flowers, initially in progenitor cells for whorls 2–4 but later restricted to whorls 2 and 3, which are affected by *pi* mutations

### *pin1: pin*form 1

**LOCATION:** not mapped

**ORIGIN:** induced with EMS in Enkheim (*pin1-1*) or Landsberg *erecta* (*pin1-2*)

**ALLELES:** *pin1-1* (Goto et al. 1987); *pin1-2* (Okada et al. 1991)

**REFERENCES:** Goto et al. (1987) *AIS 23:* 66-71; Haughn and Somerville (1988) *Dev. Genet. 9:* 73-89; Goto et al. (1991) *Jpn. J. Genet. 66:* 551-567; Okada et al. (1991) *Plant Cell 3:* 677-684

**PHENOTYPE:** abnormal inflorescence and flower structures; floral stems usually have no flower, although occasionally a flower is present at the top of the stem; it has wide petals, often more than four, no stamens, and one or more pistil-like structures lacking ovules; phenotype similar to that of wild type treated with an auxin polar transport inhibitor, 9-hydroxyfluorene-9-carboxylic acid or *N*-(1-naphthyl)phthalamic acid; *pin1-1* and *pin1-2* plants have 14% and 7%, respectively, of the wild-type level of auxin polar transport activity

pinform: see *pin1*

pistillata: see *pi*

PM11: see *chs1*

pollen separation: see *qrt*

**pom1: pom pom 1**

    LOCATION: not mapped

    ORIGIN: identified as a T-DNA line in Wassilewskija (Ws)

    REFERENCE: Hauser and Benfey (1994) *Proc. NATO-ASI, Plant Molecular Biology Series: Molecular Genetic Analysis of Plant Development and Metabolism* (eds. G. Coruzzi and P. Puigdoménech), pp. 31-40, Springer-Verlag, Berlin

    PHENOTYPE: conditional (similar to *cob* and *lit*) expansion of root: most severe in epidermis (14x), less in cortex (5.5x), and least in endodermis (2x) and stele (3.5x)

*pop1:* pollen-pistil interaction 1, allelic with *cer6;* see *cer6*

**py: *py*rimidine requiring**

    LOCATION: 2-49.1

    ORIGIN: induced by 50 kr X-ray to dry seeds or either 10–12 kr X-ray or 16 mM EMS to soaked seeds; different alleles induced in different genetic backgrounds

    ALLELES: several from Feenstra (1964, 1965) and many from Li and Rédei (1968)

    REFERENCES: Rédei (1960) *Genetics 45:* 1007; Feenstra (1964) *Genetica 35:* 259-269; Rédei (1965) *Am. J. Bot. 52:* 834-841; Feenstra (1965) *AIS 2:* 24; Li and Rédei (1968) *Biochem. Genet. 3:* 163-170

    PHENOTYPE: leaves, but not cotyledons, white, lethal; restored to normal by thiamine or 2,5-dimethyl-4-aminopyrimidine

    OTHER INFORMATION: alleles induced by Li and Rédei (1968) include both cold- and heat-sensitive conditional mutations

pyrimidine requiring: see *py*

**qrt1: quartet 1**

    LOCATION: chromosome 5 (84 ± 1.5) (Preuss et al. 1994)

    ORIGIN: induced with EMS in Landsberg *erecta*

    REFERENCE: Preuss et al. (1994) *Science 264:* 1458-1460

    PHENOTYPE: outer walls of the four meiotic products of the pollen mother cell are fused and the pollen grains fail to separate, forming tetrads; pollen viable and fertile, and fertilization with a single tetrad usually yields four seeds

**qrt2: quartet 2**

    LOCATION: chromosome 3 (3 ± 2) (Preuss et al. 1994)

    ORIGIN: induced with EMS in Landsberg *erecta*

    REFERENCE: Preuss et al. (1994) *Science 264:* 1458-1460

    PHENOTYPE: similar to *qrt1*

quartet: see *qrt*

R001 through R004 and R006: see *adh*

**r2: rosula 2**

LOCATION: 1-40 (McKelvie, see Kranz and Scheidemann 1978)

ORIGIN: X-ray mutagenesis of Estland or Limburg (Li 5)

ALLELES: two isolated by McKelvie (1963)

REFERENCES: McKelvie (1962) *Radiat. Bot. 1:* 233-241; McKelvie (1963) *Radiat. Bot. 3:* 105-123; Kranz and Scheidemann (1978) *AIS 15:* 31-34

PHENOTYPE: rosette small, compact, and glossy; leaves sessile; flowers small; subfertile; height 2 inches

OTHER INFORMATION: alleles mapped 37 map units left of *le* by McKelvie (Kranz and Scheidemann 1978), so location conceivably around 1-40

## *rca:* **regulation of *carboxylase* activation**

LOCATION: nuclear, not mapped

ORIGIN: induced with EMS in Columbia

ALLELES: line CS2071

REFERENCES: Somerville et al. (1982) *Plant Physiol. 70:* 381- 387; Werneke et al. (1988) *Proc. Natl. Acad. Sci. USA 85:* 787-791

PHENOTYPE: causes ribulose-1,5-bisphosphate carboxylase to be non-activatable by light in vivo

OTHER INFORMATION: RuBP carboxylase purified from the mutant appears normal, mutant affects a factor necessary for light-activation of the enzyme in vivo; mutant must be grown in high (i.e., 1%) $CO_2$, presumably because the non-activated form of RuBP carboxylase has too low a Km for $CO_2$ to support photosynthesis in atmospheres with normal $CO_2$ levels

GENE: cloned by Werneke et al. (1988), encodes RUBP carboxylase activase

## *re:* *re*ticulata

LOCATION: 2-58.6

REFERENCES: Rédei (1964) *AIS 1:* 9-10; Rédei (1965) *Am. J. Bot. 52:* 834-841

PHENOTYPE: veins darker than interveinal tissues

## **reduced response to fusicoccin**

LOCATION: not mapped

ORIGIN: induced with EMS in Landsberg *erecta*

ALLELE: one described (5-2)

REFERENCE: Gomarasca et al. (1993) *Plant Physiol. 103:* 165-170

PHENOTYPE: isolated as green seedlings on media containing both Paraquat (Pq), a toxic cation, and fusicoccin (FC), a fungal toxin, which promote solute uptake; the concentration of Pq used was such that it was not toxic to wild type when used alone but caused seedling bleaching when combined with FC; mutant seedlings bleached normally if 5x concentration of Pq was used; although the sensitivity of the mutant seedlings to hygromycin is similar to that of the wild type in the absence of FC, the presence of FC increases the sensitivity of the wild type but not that of the 5-2 mutant; the effect of FC on stomatal opening and $H^+$ extrusion is greatly reduced; normal morphology

OTHER INFORMATION: another mutant was isolated (Gomarasca et al. 1993) but not described

regulating nitrate reductase: see *chl4* (*rgn*)

regulation of carboxylase activity: see *rca*

reticulata: see *re*

retsina: see *rts*

*rgn:* regulating nitrate reductase, see *chl4*

***rhd1: root hair development 1***

    LOCATION: chromosome 1 (J. Schiefelbein, pers. comm.)

    ORIGIN: induced with EMS in Columbia

    ALLELE: *rhd1-1*

    REFERENCE: Schiefelbein and Somerville (1990) *Plant Cell 2:* 235-243

    PHENOTYPE: root hair length and distribution similar to wild type; root hair base enlarged

***rhd2: root hair development 2***

    LOCATION: chromosome-5 (J. Schiefelbein, pers. comm.)

    ORIGIN: induced with EMS in Columbia

    ALLELES: *rhd2-1,2,3*

    REFERENCE: Schiefelbein and Somerville (1990) *Plant Cell 2:* 235-243

    PHENOTYPE: short root hairs

***rhd3: root hair development 3***

    LOCATION: chromosome 3 (J. Schiefelbein, pers. comm.)

    ORIGIN: induced with EMS in Columbia

    ALLELES: *rhd3-1,2*

    REFERENCE: Schiefelbein and Somerville (1990) *Plant Cell 2:* 235-243

    PHENOTYPE: root hair shorter than wild type and wavy, occasionally branched

***rhd4: root hair development 4***

    LOCATION: chromosome 3 (J. Schiefelbein, pers. comm.)

    ORIGIN: induced with EMS in Columbia

    ALLELES: *rhd4-1,2*

    REFERENCE: Schiefelbein and Somerville (1990) *Plant Cell 2:* 235-243

    PHENOTYPE: root hair shorter than wild type, diameter varies

root hair development: see *rhd*

root morphology: see *cob, lit, pom pom, sab, shr*

rosula 2: see *r2*

**RP07: resistant to proline analog**

    LOCATION: unknown

    ORIGIN: induced with EMS in Bensheim

    REFERENCE: Verbruggen and Jacobs (1987) *AIS 23:* 15-23

    PHENOTYPE: selected for resistance to proline analog trans-4-hydryx-L-proline; soluble proline content is increased 3-fold

**RP14: resistant to proline analog**

    LOCATION: unknown

    ORIGIN: induced with EMS in Bensheim

    REFERENCE: Verbruggen and Jacobs (1987) *AIS 23:* 15-23

    PHENOTYPE: selected for resistance to proline analog azetidine-2-carboxylic acid

## RP18: resistant to proline analog

LOCATION: unknown

ORIGIN: induced with EMS in Bensheim

REFERENCE: Verbruggen and Jacobs (1987) *AIS 23:* 15-23

PHENOTYPE: selected for resistance to proline analog azetidine-2-carboxylic acid

## *rpm1:* resistance to *Pseudomonas* syringae pv. *maculicola* 1, allelic with *rps3*

LOCATION: 3-8.2

ORIGIN: naturally occurring difference between Col-0, Oy-0, and Nd-0

REFERENCES: Debener et al. (1991) *Plant J. 1:* 289-302; Dangl et al. (1992) *Plant Cell 4:* 1359-1369

PHENOTYPE: Col-0 and Oy-0 are resistant to *Pseudomonas syringae* pv. *maculicola* strains carrying an *avr* (avirulence) gene from the strain m2 (LMG5071); Nd-0 is not resistant to *Pseudomonas* strains carrying this *avr* gene; the resistance is characterized by limited bacterial growth in plants and the hypersensitive response, which involves localized cell death and tissue necrosis at the infection site

OTHER INFORMATION: the resistance in Col-0 and Oy-0 map to the same region of chromosome 3; therefore, these may be alleles

## *rps2:* resistance to *Pseudomonas syringae* 2

LOCATION: 4-54.8

ORIGIN: *rps2-201C, rps2-202C, rps2-203C,* and *rps2-301C* induced with diepoxybutane in Columbia

ALLELES: *rps2-201C, rps2-202C, rps2-203C, rps2-301C,* the last two may not be independent (Kunkel et al. 1993); *rps2-101C, rps2-102C, rps2-101N* (Mindrinos et al. 1994a,b)

REFERENCES: Kunkel et al. (1993) *Plant Cell 5:* 865-875; Yu et al. (1993) *Mol. Plant-Microbe Interact. 6:* 434-443; Bent et al. (1994) *Science 265:* 1856-1860; Mindrinos et al. (1994a) *Cell 78:* 1089-1099; Mindrinos et al. (1994b) *Adv. Mol. Genet. Plant-Microbe Interact. 3:* in press

PHENOTYPE: lost resistance to *Pseudomonas syringae* carrying the avirulence gene *avrRpt2*

GENE: isolated by Bent et al. (1994) and Mindrinos et al. (1994a); encodes a protein of 909 amino acids (104.6 kD) with a leucine zipper near the N-terminus, a potential nucleotide binding region in the N-terminal half, a putative membrane anchoring peptide near the middle, and 14 imperfect leucine-rich repeats in the C-terminal half; the latter two suggest that it may be a receptor; *rps2-101C* is a transition of $G_{704}$ to A ($Trp_{235}$→Stop); *rps2-102C* is a transition of $G_{1427}$ to A ($Arg_{476}$→Lys); *rps2-201C* is a transversion of $A_{2002}$ to C ($Thr_{668}$→Pro); *rps2-101N* has a 10-bp insertion after nucleotide 1744

OTHER INFORMATION: *RPS2* hybridizes to multiple bands under low stringency; the RPS2 protein shares extensive similarity with the product of the tobacco *N* gene (Whitham et al. 1994, *Cell 78:* 1101-1115), which confers resistance to tobacco mosaic virus

**rps3: resistance to Pseudomonas syringae 3, allelic with rpm1**

LOCATION: chromosome 3

ORIGIN: resistance naturally occurring in Columbia-0 (Co-0); sensitivity found in the Bla-2 ecotype; mutants were isolated from Co-0 after mutagenesis by fast neutrons (*rps3-1*), with diepoxybutane (*rps3-2*), with ethylnitrosourea (*rps3-3*), or with EMS (*rps3-3*)

ALLELES: many (e.g., Ler, Ws-0, Oy-0) ecotypes are resistant, some (Nd-0, Co-1, Bs-1, etc.) are sensitive (Innes et al. 1993); *rps3-1,2,3,4* (Bisgrove et al. 1994)

REFERENCES: Innes et al. (1993) *Plant J. 4:* 813-820; Bisgrove et al. (1994) *Plant Cell 6:* 927-933

PHENOTYPE: Co-0 is resistant to *Pseudomonas syringae* strain carrying the avirulence gene *avrB* from the soybean pathogen *Pseudomonas syringae* pv. *glycinea*

*rps3-1,2,3,4* mutants are sensitive to both *Pseudomonas syringae* with *avrB* and *Pseudomonas syringae* with *avrRpm1*

OTHER INFORMATION: closely linked to the *RPM1* locus; mutant analysis indicates that *rps3* and *rpm1* are allelic

**RS17**

LOCATION: not known

ORIGIN: induced with EMS in Columbia

REFERENCE: Saleki et al. (1993) *Plant Physiol. 101:* 839-845

PHENOTYPE: germinate on media with up to 225 mM NaCl

**RS19**

LOCATION: not known

ORIGIN: induced with EMS in Columbia

REFERENCE: Saleki et al. (1993) *Plant Physiol. 101:* 839-845

PHENOTYPE: germinate on media with 125-170 mM NaCl

OTHER INFORMATION: not simple Mendelian inheritance of a single gene

**RS20**

LOCATION: not known

ORIGIN: induced with EMS in Columbia

REFERENCE: Saleki et al. (1993) *Plant Physiol. 101:* 839-845

PHENOTYPE: germinate on media with up to 225 mM NaCl

**rts: retsina**

LOCATION: not mapped

ORIGIN: induced with EMS in Landsberg *erecta*

ALLELES: four

REFERENCE: Hülskamp et al. (1994) *Cell 76:* 555-566

PHENOTYPE: trichome glassy, fragile

**rxc1: reaction to Xanthomonas campestris pv. campestris**

LOCATION: chromosome 2—between markers 220 and 323 (R. Buell and S. Somerville, pers. comm.)

ORIGIN: *rxc1-1*, naturally occurring in Columbia; *rxc1-2*, naturally occurring in Pr-0

**ALLELES:** *rxc1-1,2*

**REFERENCE:** Tsuji et al. (1991) *Physiol. Mol. Plant Pathol. 38:* 57-65

**PHENOTYPE:** *rxc1-1:* no disease development following inoculation with the bacterial pathogen *Xanthomonas campestris* pv. *campestris,* race 2D520

*rxc1-2:* black rot disease development following inoculation with the same *Xanthomonas campestris* pv. *campestris* race

**OTHER INFORMATION:** *rxc1-1* is dominant over *rxc1-2* in crosses between Columbia and Pr-0

## *sab: sabre*

**LOCATION:** not mapped

**ORIGIN:** segregated from a T-DNA line in Wassilewskija (Ws)

**REFERENCES:** Benfey et al. (1993) *Development 119:* 57-70; Hauser and Benfey (1994) *Proc. NATO-ASI, Plant Molecular Biology Series: Molecular Genetic Analysis of Plant Development and Metabolism* (eds. G. Coruzzi and P. Puigdoménech), pp. 31-40, Springer-Verlag, Berlin

**PHENOTYPE:** abnormal cell expansion of the root: epidermis (5x), cortex (8x), endodermis (2x), and stele (2.5x); shoot small; semi-sterile

sabre: see *sab*

salt-tolerant germination: see RS17, RS19, RS20

## *sat:* **serine-glyoxylate *amino*transferase**

**LOCATION:** nuclear, not mapped

**ORIGIN:** EMS mutagenesis of Columbia

**ALLELES:** lines CS51, CS108, CS117

**REFERENCE:** Somerville and Ogren (1980) *Proc. Natl. Acad. Sci. USA 77:* 2684-2687

**PHENOTYPE:** serine-glyoxylate aminotransferase activity absent (peroxisomal enzyme)

**OTHER INFORMATION:** lines CS51, CS108, and CS117 carry alleles that are probably independent; mutants were selected as plants that grew under non-photorespiratory conditions (1% $CO_2$, 21% $O_2$) but not under photorespiratory conditions (0.032% $CO_2$, 21% $O_2$)

schizoid: see *shz*

## *se: serrate*

**LOCATION:** chromosome 2, near *er* (Rédei and Hirono 1964)

**REFERENCE:** Rédei and Hirono (1964) *AIS 1:* 9-10

**PHENOTYPE:** serrated leaves

**OTHER INFORMATION:** that *se* stands for serrate is a guess, McKelvie (1962, *Radiat. Bot. 1:* 233-241) had *se* mutants that he called serrata

serine-glyoxylate aminotransferase: see *sat*

serine transhydroxymethylase: see *stm*

serrate: see *se*

## *sex1:* **starch *excess* 1, same as *sop1***

**LOCATION:** 1-12.8

**ORIGIN:** induced with EMS in Columbia

**ALLELES:** *sex1-1* (= *sop1-1,* line TC26: TC265 was derived from TC26 by backcrossing to wild type; *sex1-2* (= *sop1-2,* line TL52)

REFERENCES: Caspar et al. (1989) *Proc. Natl. Acad. Sci. USA 86:* 5830-5833; Caspar et al. (1991) *Plant Physiol. 95:* 1181-1188; Caspar, this volume

PHENOTYPE: excess accumulation of starch in dark, apparently due to a defect in starch degradation; growth much slower than normal in 12-hr-day photoperiod but normal in constant light

### *sex2: starch excess 2*

LOCATION: not mapped

ORIGIN: induced with EMS in Columbia

ALLELE: *sex2-1* (line TL50)

REFERENCES: Caspar et al. (1991) *Plant Physiol. 95:* 1181-1188; Caspar, this volume

PHENOTYPE: excess accumulation of starch in dark, apparently due to a defect in starch degradation

### *sex3: starch excess 3*

LOCATION: not mapped

ORIGIN: induced with EMS in Columbia

ALLELE: *sex3-1* (line TL54)

REFERENCES: Caspar et al. (1991) *Plant Physiol. 95:* 1181-1188; Caspar, this volume

PHENOTYPE: excess accumulation of starch in dark, apparently due to a defect in starch degradation

shoot redifferentiation: see *srd*

short integuments: see *sin1*

short root: see *shr*

### *shr: short root*

LOCATION: not mapped

ORIGIN: cosegregate with a T-DNA insertion in Wassilewskija (Ws)

REFERENCE: Benfey et al. (1993) *Development 119:* 57-70

PHENOTYPE: root much shorter than wild type; mature plants grown on plates have large number of secondary roots from the junction of hypocotyl and root; much reduced in the number of cells in root meristem and elongation zone in the longest roots but normal in newly emerging roots; lack endodermis and reduced stele and this defect is present in the embryo

### *shz: schizoid*

LOCATION: not mapped

ORIGIN: induced by T-DNA insertion in Wassilewskija (Ws)

REFERENCE: Medford et al. (1992) *Plant Cell 4:* 631-643

PHENOTYPE: multiple vegetative shoot apices

### *sin1: short integuments 1*

LOCATION: not mapped

ORIGIN: induced with EMS in Landsberg *erecta*

ALLELE: *sin1-1*

REFERENCES: Robinson-Beers et al. (1992) *Plant Cell 4:* 1237-1249; Lang et al. (1994) *Genetics 137:* 1101-1110

**PHENOTYPE:** female sterile; morphological development of ovules arrested; integuments fail to cover nucellus; it appears that meiosis does not occur

**OTHER INFORMATION:** phenotype enhanced by a recessive allele, *mod1*, which is tightly linked to *er* (see *mod1*); *sin1 mod1* (Landsberg *erecta*) plants have short outer and inner integuments, whereas *sin1 MOD1* (Columbia) have uniform ovule morphology before anthesis but variable abnormal mature ovules/seeds; about 3/4 (Type I) of *sin1 MOD1* ovules have short outer integuments and greatly expanded cylindrical inner integuments; about 1/4 (Type II) of the mature ovules have near normal outer integuments and massively expanded inner integuments; Type II ovules occasionally have gametophyte development; very few (Type III) of the ovules are morphologically nearly normal, but they are larger than wild-type ones, and they have arrested embryo development; surfaces of Type II and III ovules resemble immature seed coat

sinapoyl ester accumulation: previously *sin1*, see *fah1* (ferulic acid hydroxylase)

*sop1:* see *sex1*

**spi: spirrig**

    **LOCATION:** not mapped

    **ORIGIN:** induced with EMS in Landsberg *erecta*

    **ALLELES:** eight

    **REFERENCE:** Hülskamp et al. (1994) *Cell 76:* 555-566

    **PHENOTYPE:** trichome distorted

spindly: see *spy*

**spy: spindly**

    **LOCATION:** not mapped

    **ORIGIN:** *spy-1,2,3* induced with EMS in Columbia; *spy-4* induced by T-DNA insertion (from the ABRC collection)

    **ALLELES:** *spy-1,2,3,4*

    **REFERENCE:** Jacobsen and Olszewski (1993) *Plant Cell 5:* 887-896

    **PHENOTYPE:** resemble wild-type plants that have been repeatedly treated with GA, longer than normal hypocotyl, light green, increased stem elongation, partially male sterile

    **OTHER INFORMATION:** suppress GA deficiency phenotypes, which are caused by either the *ga1-2* mutation or the GA biosynthesis inhibitor paclobutrazol

**srd1: shoot redifferentiation 1**

    **LOCATION:** not mapped

    **ORIGIN:** induced with EMS in Landsberg *erecta*

    **ALLELE:** one (line L1045)

    **REFERENCE:** Yasutani et al. (1994) *Plant Physiol. 105:* 815-822

    **PHENOTYPE:** isolated as individual with reduced frequency of shoot redifferentiation from root segments at 27°C than at 22°C; initial callus formation and chloroplast biogenesis similar to wild type (not temperature-sensitive) in the redifferentiation culture media

**srd2: shoot redifferentiation 2**

    **LOCATION:** not mapped

ORIGIN: induced with EMS in Landsberg *erecta*

ALLELE: one (line L131)

REFERENCE: Yasutani et al. (1994) *Plant Physiol. 105:* 815-822

PHENOTYPE: isolated as individual with reduced frequency of shoot redifferentiation from root segments at 27°C than at 22°C; initial callus formation and chloroplast biogenesis similar to wild type (not temperature-sensitive) in the redifferentiation culture media; seedling growth is temperature-sensitive

*srd3: shoot redifferentiation 3*

LOCATION: not mapped

ORIGIN: induced with EMS in Landsberg *erecta*

ALLELE: one (line L1919)

REFERENCE: Yasutani et al. (1994) *Plant Physiol. 105:* 815-822

PHENOTYPE: isolated as individual with reduced frequency of shoot redifferentiation from root segments at 27°C than at 22°C; initial callus formation and chloroplast biogenesis similar to wild type (not temperature-sensitive) in the redifferentiation culture media; seedling growth is temperature-sensitive

*sta: stachel*

LOCATION: not mapped

ORIGIN: induced with EMS in Landsberg *erecta*

ALLELE: one

REFERENCE: Hülskamp et al. (1994) *Cell 76:* 555-566

PHENOTYPE: trichome with only two branches, primary branch missing

stachel: see *sta*

starch free or starchless: see *adg1*, *adg2*, *pgm*, and *stf1*

starch excess: see *sex*

starch overproducer (*sop*): see *sex1*

*stf1: starch free 1*

LOCATION: not mapped

ORIGIN: induced with EMS in Columbia

ALLELE: *stf1-1*

REFERENCES: Caspar et al. (1989) *Proc. Natl. Acad. Sci. USA 86:* 5830-5833; Caspar, this volume

PHENOTYPE: lack starch; elevated β-amylase activity in 12-hr-day photoperiod

*sti: stichel*

LOCATION: 2-4.3

ORIGIN: induced with EMS (five alleles) or by X-ray (one allele) in Landsberg *erecta*

ALLELES: six

REFERENCE: Hülskamp et al. (1994) *Cell 76:* 555-566

PHENOTYPE: trichomes long, unbranched

stichel: see *sti*

*stm: serine transhydroxymethylase*

LOCATION: nuclear, not mapped

ORIGIN: induced with EMS in Columbia

ALLELES: three

REFERENCE: Somerville and Ogren (1981) *Plant Physiol. 67:* 666-671

PHENOTYPE: absence of mitochondrial serine transhydroxymethylase activity

OTHER INFORMATION: lines CS42, CS64, and CS115 carry alleles of the *stm* locus; mutants were selected by their inability to grow in photorespiratory conditions (atmospheric $CO_2$ and $O_2$) but ability to grow normally in non-photorespiratory conditions (1% $CO_2$, normal $O_2$)

### *stm-1:* **shoo*t* meristemless 1**

LOCATION: chromosome 1 (~19.5 cM from *chl*) (Barton and Poethig 1993)

ORIGIN: induced with EMS in Landsberg *erecta*

ALLELE: one

REFERENCE: Barton and Poethig (1993) *Development 119:* 823-831

PHENOTYPE: seedling and embryo lack a shoot apical meristem; mutant tissue fails to regenerate shoots in tissue culture

### *su1:* **suppressor of *chl4* (*rgn*) 1**

LOCATION: 5-34.9

ORIGIN: induced with EMS in Landsberg *erecta chl4* line B25

ALLELES: *su1-1* (line B25R1); *su1-2* (line B25R2); *su1-3* (line B25R3); *su1-4* (line B25R4); *su1-5* (line B25R5); *su1-6* (line B25R6); *su1-7* (line B25R7)

REFERENCES: Braaksma and Feenstra (1979) *AIS 16:* 66-67; Braaksma and Feenstra (1982) *Theor. Appl. Genet. 61:* 263-271

PHENOTYPE: *su1-1,5,6:* dominant suppressor of *chl4* phenotype of inability to grow on nitrate medium and recessive suppressor of *chl4* phenotype of chlorate resistance

*su1-2,7:* dominant partial suppressor of *chl4* phenotype of inability to grow on nitrate medium and dominant suppressor of *chl4* phenotype of chlorate resistance

*su1-3:* recessive suppressor of *chl4* phenotypes of chlorate resistance and growth on nitrate medium

*su1-4:* like *su3* except *chl4/chl4; su1-4/su1-4* plants are small and have narrow leaves when grown on ammonium nitrate medium

OTHER INFORMATION: *su1-2* and *su1-7* are considered allelic to *su1-1* on the basis of similar phenotype and tight linkage with each other and with *su1-5*

### *su:* see *sul*

### *sul:* **su*l*furata**

LOCATION: 2-69.6

ALLELE: *sul-1* (= *su*)

REFERENCES: Rédei (1962) *Z. Vererbungsl. 93:* 164-170; Rédei and Hirono (1964) *AIS 1:* 9-10; Rédei (1965) *Genetics 51:* 857-872

PHENOTYPE: bright yellow-green

### sulfurata: see *sul*

### *sup:* **su*p*erman, allelic with *flo10***

LOCATION: 3-34.6

ORIGIN: induced with EMS in Landsberg *erecta* (*sup-1*) and Columbia (*sup-2,3,4*)

ALLELES: *sup-1* (= *flo10-2*); *sup-2* (= *flo10-1*); *sup-3* (= *flo10-3*); *sup-4*

**REFERENCES:** Schultz et al. (1991) *Plant Cell 3:* 1221-1237; Bowman et al. (1992) *Development 114:* 599-615

**PHENOTYPE:** additional whorl(s) of stamens are formed at the expense of carpels; *AG* expression is not affected by *sup* mutations but *AP3* expression expands toward center of the flower; double mutants with *ap3* or *pi* show the same phenotypes as *ap3* or *pi* mutants

superman: see *sup*

suppressor of *rgn* (*chl4*): see *su1*

**T5121**

**LOCATION:** not mapped

**ORIGIN:** segregated from a T-DNA line (T5121) in Wassilewskija (Ws)

**REFERENCE:** Dietrich et al. (1994) *Cell 77:* 565-577

**PHENOTYPE:** spontaneous lesions of mottled chlorosis on all true leaves; unlike *lsd* mutants, T5121 plants do not express PR-1 mRNA

TC7, TC9, TC75: see *pgm*

TC26, TC265: see *sex1-1*

TC135: starchless mutant, see *pgm*

***tfl1*: terminal *flower* 1**

**LOCATION:** 5-0.0

**ORIGIN:** induced with EMS in Columbia (*tfl1-1, tfl8-14*) or Landsberg *erecta* (*tfl1-2,3,4,5,6,7*)

**ALLELES:** *tfl1-1* (Shannon and Meeks-Wagner 1991); *tfl1-2* through *tfl1-9* (Alvarez et al. 1992); *tfl1-10* (Shannon and Meeks-Wagner 1993); *tfl1-11* through *tfl1-14* (Schultz and Haughn 1993)

**REFERENCES:** Shannon and Meeks-Wagner (1991) *Plant Cell 3:* 877-892; Alvarez et al. (1992) *Plant J. 2:* 103-106; Bowman et al. (1993) *Development 119:* 721-743; Shannon and Meeks-Wagner (1993) *Plant Cell 5:* 639-655; Schultz and Haughn (1993) *Development 119:* 745-765

**PHENOTYPE:** early flowering; have a determinant inflorescence with terminal flowers; terminal flowers usually consist of two or three, sometimes incomplete, flowers; mutants with strong phenotype have on average only 2–5 normal flowers; *tfl1-3* and *tfl1-4* have more flowers (about 50 and 20, respectively) than others

***th1*: *th*iamine requiring 1**

**LOCATION:** 1-33.3

**ORIGIN:** induced either by 50 kr X-ray to dry seeds or with 16 mM EMS; different alleles induced in different genetic backgrounds

**ALLELE:** *th1* (= *thi1-1*) (Komeda et al. 1988)

**REFERENCES:** Li and Rédei (1968) *Biochem. Genet. 3:* 163-170; Komeda et al. (1988) *Plant Physiol. 88:* 248-250

**PHENOTYPE:** leaves, but not cotyledons, yellow-white, lethal, restored to wild type by thiamine; deficient in thiamine phosphate pyrophosphorylase

***th2*: *th*iamine requiring 2**

**LOCATION:** 5-46.6

**ORIGIN:** original allele is 1018/6 (Langridge 1955); other alleles induced by

X-ray or with EMS (Li and Rédei 1968)

**ALLELES:** line 1018/6

**REFERENCES:** Langridge (1955) *Nature 176:* 260-261; Langridge (1958) *Austr. J. Biochem. Sci. 11:* 58-63; Rédei (1965) *AIS 2:* 25; Li and Rédei (1968) *Biochem. Genet. 3:* 163-170; Rédei (1975) *Ann. Rev. Genet. 9:* 111-127; Koornneef and Hanhart (1981) *AIS 18:* 52-58

**PHENOTYPE:** variegated, restored to wild type by thiamine

**OTHER INFORMATION:** alleles 1018/6 and *th-21ts* are temperature-sensitive (Li and Rédei 1968)

**th3:** *th*iamine requiring 3

**LOCATION:** 4-34.1

**ORIGIN:** induced with EMS in Landsberg *erecta* (?)

**ALLELES:** line V345

**REFERENCE:** Koornneef and Hanhart (1981) *AIS 18:* 52-58

**PHENOTYPE:** green to pale green leaves with irregular chlorotic and occasionally necrotic spots, viable, restored to wild type by thiamine

**OTHER INFORMATION:** restored to wild type by $10^{-4}$ M thiamine-HCl, partially restored by thiazole (necrotic spots do not disappear); 4-amino-5-aminomethyl-pyrimidine-di HCl has a deleterious effect on growth at $10^{-4}$ M

*thi1-1:* see *th1*

thiamine requiring: see *th*

thiazole requiring: see *tz*

TL3, TL46, TL46BC1: see *adg2*

TL24, TL25, TL255: see *adg1*

TL50: see *sex2*

TL52: see *sex1-2*

TL54: see *sex3*

**tom1:** *to*bacco-mosaic-virus *m*ultiplication 1

**LOCATION:** not mapped

**ORIGIN:** induced with EMS in Columbia

**ALLELES:** *tom1-1* (line PD114); *tom1-2* (line PD378)

**REFERENCE:** Ishikawa et al. (1991) *MGG 230:* 33-38

**PHENOTYPE:** the level of tobacco mosaic virus coat protein decreased by 4-fold in infected rosette leaves; levels of coat protein of other viruses (turnip crinkle or turnip yellow mosaic viruses) are normal

transparent testa: see *tt*

transparent testa glabra: see *ttg*

trichome development: see *ali, an, cdo, cha, crk, dis1, dis2, gl1, gl2, gl3, grl, kak, klk, rts, spi, sta, sti, try, ttg, wrm, zwi*

**trp1:** *tr*yptophan requiring 1

**LOCATION:** 5-25.9 (Hauge et al. 1993, *Plant J. 3:* 745-754)

**ORIGIN:** induced with EMS in Columbia

**ALLELES:** *trp1-1* (Last and Fink 1988); *trp1-100* (Rose et al. 1992)

**REFERENCES:** Last and Fink (1988) *Science 240:* 305-310; Berlyn et al. (1989) *Proc. Natl. Acad. Sci. USA 86:* 4604-4608; Last et al. (1991) *Plant*

*Cell 3:* 345-358; Rose et al. (1992) *Plant Physiol. 100:* 582-592

PHENOTYPE: *trp1-1:* resistant to 5-methylanthranilic acid + tryptophan; tryptophan requiring; can grow on either tryptophan- or indole-supplemented media; plants have amethyst fluorescence under UV light; fail to develop true leaves or roots when germinated without TRP; grow more slowly than wild type in soil with TRP; have small and crinkled leaves; more bushy than wild type; most flowers are female sterile and have few pollen grains; deficient in anthranilate phosphoribosyl (PR)-transferase activity

*trp1-100:* resistant to 5-methylanthranilic acid, blue fluorescent under UV light but less than *trp1-1* plants; grow normally on medium without TRP; fertile

GENE: cloned by Rose et al. (1992); encodes phosphoribosylanthranilate transferase (PAT), designated as AR119 on RFLP maps

## *trp2:* **tryptophan requiring 2**

LOCATION: 5-81.0 (Hauge et al. 1993, *Plant J. 3:* 745-754)

ORIGIN: induced with EMS in Columbia

ALLELE: *trp2-1*

REFERENCES: Last and Fink (1988) *Science 240:* 305-310; Berlyn et al. (1989) *Proc. Natl. Acad. Sci. USA 86:* 4604-4608; Last et al. (1991) *Plant Cell 3:* 345-358

PHENOTYPE: resistant to 5-methylanthranilic acid + tryptophan; tryptophan requiring; fail to develop true leaves or roots when germinated without TRP; grow more slowly than wild type in soil with TRP; more bushy than wild type; larger and less bushy than *trp1-1*; fertile; seeds yellow-green (early) to light brown (late); deficient in tryptophan synthase β activity (15% of wild type level); not requiring tryptophan under low light

GENE: *TRP2* is linked to and complemented by *TSB1* (tryptophan synthase β1), which was cloned by Berlyn et al. (1989), suggesting the two are the same; there is a homologous gene (*TSB2*) which is 85% similar to *TSB1* at the nucleotide level (Last et al. 1991); *TSB2* maps to 4-64.9

## *trp4:* **tryptophan requiring 4**

LOCATION: chromosome 1, between RFLP markers GAPB (glyceraldehyde-3-phosphate dehydrogenase, about 15 cM) and CHL1 (about 22 cM) (Niyoki et al. 1993)

ORIGIN: induced with EMS in Columbia *trp1-100/trp1-100, gl1-1/gl1-1, pgm/pgm*

ALLELES: *trp4-1* (line D1) and *trp4-2* (line E17); the lines are still *trp1-100/trp1-100*

REFERENCE: Niyoki et al. (1993) *Plant Cell 5:* 1011-1027

PHENOTYPE: *trp4 trp1-100:* isolated as suppressors of the blue fluorescence phenotype of the *trp1-100* mutant; weakly blue fluorescent; can grow on media supplemented with anthranilate, indole, or tryptophan; fail to grow on media supplemented with chorismate, PABA, phenylalanine, tyrosine, or phenylalanine + tyrosine; the growth phenotypes suggest a defect in anthranilate synthase (AS); auxotrophic phenotype partially suppressed under low-light growth conditions; although small, the double mutants are fertile and produce viable seeds; double mutant has a female-specific transmission defect

*trp4-1:* reduced AS activity; not auxotrophic; grows normally in soil

**GENE:** *TRP4* encodes the β subunit of AS; the gene (*ASB1*) was isolated by complementing an *E. coli* mutant; the *Arabidopsis* ASB1 protein has 276 amino acids, shows 34–45% indentity to other microbial counterparts, and has a predicted animo-terminal chloroplast transit peptide; *trp4-1* is a G→A substitution resulting in a change of an invariant $Gly_{154}$ to a Glu; *trp4-2* is a G→A substitution resulting in a defective acceptor site of the second last intron; *ASB* mRNA is induced by virulent and, to a lesser degree, avirulent strains of *Pseudomonas syringae*

**OTHER INFORMATION:** another suppressor (A2) of *trp1-100* blue fluorescence was isolated (Niyoki et al. 1993) and it is not auxotropic; two other genes (*ASB2* and *ASB3*) coding for AS β subunit were isolated; *ASB2* maps to chromosome 5 near *lfy*

**try: *triptychon***

**LOCATION:** chromosome 3, about halfway between *gl1* and *tt5*

**ORIGIN:** induced with EMS in Landsberg *erecta*

**ALLELES:** two

**REFERENCE:** Hülskamp et al. (1994) *Cell 76:* 555-566

**PHENOTYPE:** nests of trichomes (recessive); trichomes with more branches (semi-dominant)

*TSB1*: see *trp2*

*TSB2*: see *trp2*

***tsl: tousled***

**LOCATION:** 5-20.0

**ORIGIN:** induced by T-DNA insertion in Wassilewskija (Ws) (*tsl-1*) or with diepoxybutane in Columbia (*tsl-2,3*)

**ALLELES:** *tsl-1,2,3*

**REFERENCE:** Roe et al. (1993) *Cell 75:* 939-950

**PHENOTYPE:** delayed flowering by about one week; slightly increased number of rosette leaves; margins of leaves show deep serrations; more secondary inflorescences; *tsl-1* plants have cauline leaves tightly curled around the secondary inflorescence; abnormal flowers, with fewer organs: 2–4 sepals, 0–4 petals, 0–5 stamens, and 1–2 carpels; organ identity is not altered; floral organs are misshaped, some have reduced size or are filamentous; gynoecium is always unfused at the upper portion; defective organ primordia initiation

**GENE:** cloned by Roe et al. (1993), encodes a protein (78 kD) with an N-terminal domain exhibiting coiled-coil structure features and a C-terminal domain similar to protein kinases; highly expressed in developing floral meristems; *tsl-1* was due to T-DNA insertion into intron 14 of a total of 15 introns; *tsl-2* is a $T_{1369}$ to A substitution resulting in a missense mutation of $Val_{355}$ to Asp; *tsl-3* is a $A_{912}$ to T substitution resulting in a nonsense mutation from Lys$_{203}$

***tt1*: *transparent testa* 1**

LOCATION: 1-55.2

ORIGIN: induced by X-ray in Landsberg *erecta*

ALLELE: *tt1-1*

REFERENCES: Koornneef (1981) *AIS 18:* 45-51; Koornneef et al. (1983) *J. Hered. 74:* 265-272; Koornneef (1990) *AIS 27:* 1-4

PHENOTYPE: yellow seeds due to absence of brown pigment in seed coat

***tt2*: *transparent testa* 2**

LOCATION: 5-44.3

ORIGIN: induced by X-ray (*tt2-1*) in Landsberg *erecta*

ALLELE: *tt2-1*

REFERENCES: Koornneef (1981) *AIS 18:* 45-51; Koornneef (1990) *AIS 27:* 1-4

PHENOTYPE: like *tt1*, yellow seeds due to absence of brown pigment in testa

***tt3*: *transparent testa* 3**

LOCATION: 5-57.4

ORIGIN: the Koornneef alleles induced by X-ray (*tt3-1*) or with EMS (two alleles) in Landsberg *erecta*

ALLELES: *tt3-1* (line M218) and two more (Koornneef 1990); one (line V220) (Bürger 1971)

REFERENCES: Bürger (1971) *AIS 8:* 36-42; Koornneef (1981) *AIS 18:* 45-51; Koornneef (1990) *AIS 27:* 1-4; Shirley et al.(1992) *Plant Cell 4:* 333-347

PHENOTYPE: yellow seeds due to transparent testa, absence of anthocyanins in leaves and stems

GENE: *TT3* encodes dihydroflavonol 4-reductase, has been cloned by Shirley et al. (1992); the M218 allele has a 2.8-cM inversion with a 7.4-kb deletion at the *TT3* end

OTHER INFORMATION: absence of anthocyanins is conspicuous in older plants when chlorophyll starts breaking down: in wild type, old plants have a purple reddish appearance, mutant plants are yellowish

***tt4*: *transparent testa* 4**

LOCATION: 5-12.9

ORIGIN: induced with EMS (*tt4-1* and three other alleles) or by X-ray (one allele) in Landsberg *erecta;* origin for Bürger alleles not available

ALLELES: *tt4-1* and four others (Koornneef 1990); two (lines F114, F142) (Bürger 1971)

REFERENCES: Bürger (1971) *AIS 8:* 36-42; Koornneef (1981) *AIS 18:* 45-51; Feinbaum and Ausubel (1988) *Mol. Cell. Biol. 8:* 1985-1992; Koornneef (1990) *AIS 27:* 1-4; Li et al. (1993) *Plant Cell 5:* 171-179

PHENOTYPE: like *tt3*, yellow seeds, no anthocyanin in stems and leaves; more sensitive than wild type to high UV-B irradiance (13 kJ m$^{-2}$ day$^{-1}$); near normal growth under lower UV-B irradiance (8 kJ m$^{-2}$ day$^{-1}$)

GENE: *TT4* codes for chalcone synthase (see Hauge et al. 1993, *Plant J. 3:* 745-754)

***tt5*: *transparent testa* 5**

LOCATION: 3-80.6

ORIGIN: induced by fast neutrons in Landsberg *erecta*

ALLELE: *tt5-1* (line A11)

REFERENCES: Dellaert et al. (1979) *AIS 16:* 10-26; Koornneef (1981) *AIS 18:* 45-51; Koornneef et al. (1983) *J. Hered. 74:* 265-272; Koornneef et al. (1989) *J. Hered. 80:* 118-122; Koornneef (1990) *AIS 27:* 1-4; Shirley et al., (1992) *Plant Cell 4:* 333-347; Li et al. (1993) *Plant Cell 5:* 171-179

PHENOTYPE: like *tt4,* yellow seeds and no anthocyanin in stems and leaves, but brighter green; growth sensitive to low irradiance of UV-B (50% less growth under 2.3 kJ m$^{-2}$ day$^{-1}$ and 70% decrease under 4.5 kJ m$^{-2}$ day$^{-1}$)

GENE: *TT5* encodes chalcone flavanone isomerase, has been cloned by Shirley et al. (1992); the allele in line 40.443 has an inversion of the *TT5* gene with a 272-bp insertion at one end

***tt6:* *t*ransparent *t*esta 6**

LOCATION: 3-96.4

ORIGIN: induced with EMS in Landsberg *erecta*

ALLELE: *tt6-1*

REFERENCES: Koornneef et al. (1983) *J. Hered. 74:* 265-272; Koornneef (1990) *AIS 27:* 1-4

PHENOTYPE: brownish yellow seed and reduced anthocyanin content in leaves

***tt7:* *t*ransparent *t*esta 7**

LOCATION: 5-6.0

ORIGIN: induced with EMS in Landsberg *erecta*

ALLELES: line d501 and another allele

REFERENCES: Koornneef et al. (1982) *AIS 19:* 113-115; Koornneef (1990) *AIS 27:* 1-4

PHENOTYPE: pale brown to ochre-yellow seeds, reduced anthocyanin in leaves (as determined by visual observation only), blocked in 3′ hydroxylation of flavonols and anthocyanidins, thus having pelargonidin instead of cyanidin and kaempferol instead of a mixture of both kaempferol and quercitin

***tt8:* *t*ransparent *t*esta 8**

LOCATION: not mapped

ORIGIN: induced in Enkheim by Röbbelen (see Bürger 1971)

ALLELES: lines F17 and F107 (Bürger 1971)

REFERENCES: Bürger (1971) *AIS 8:* 36-42; Koornneef (1990) *AIS 27:* 1-4

PHENOTYPE: pale brown to ochre seeds, reduced anthocyanins

***tt9:* *t*ransparent *t*esta 9**

LOCATION: not mapped

ORIGIN: induced with EMS in Landsberg *erecta*

REFERENCE: Koornneef (1990) *AIS 27:* 1-4

PHENOTYPE: pale brown to ochre seeds, normal anthocyanins

***tt10:* *t*ransparent *t*esta 10**

LOCATION: not mapped

ORIGIN: induced with EMS in Landsberg *erecta*

REFERENCE: Koornneef (1990) *AIS 27:* 1-4

PHENOTYPE: pale brown to ochre seeds, normal anthocyanins

**ttg: transparent testa glabra**

LOCATION: 5-26.3

ORIGIN: one allele (line F31) induced in Enkheim or Antwerp (Bürger 1971); several alleles induced with EMS and by fast neutrons in Landsberg *erecta* (Koornneef et al. 1982)

ALLELES: line F31; seven alleles from Koornneef (1981, 1982); two new EMS alleles in Landsberg *erecta* from Hülskamp et al. (1994); additional one in AIS collection

REFERENCES: Bürger (1971) *AIS 8:* 36-42; Koornneef (1981) *AIS 18:* 45-51; Koornneef et al. (1982) *Mut. Res. 93:* 109-123; Koornneef (1990) *AIS 27:* 1-4; Lloyd et al. (1992) *Science 258:* 1773-1775; Hülskamp et al. (1994) *Cell 76:* 555-566; Larkin et al. (1994a) *Results and Problems in Cell Differentiation: Plant Promoters and Transcription Factors* (ed. L. Nover), vol. 20, pp. 259-275, Springer-Verlag, Berlin; Larkin et al. (1994b) *Plant Cell 6:* 1065-1076; Marks (1994) *Curr. Biology 4:* 621-623

PHENOTYPE: transparent seed coat gives seeds yellow color, absence of the ruthenium red-staining seed mucilage, abnormal appearance of cells of dry seed coat in scanning electron microscope, no trichomes on leaf surfaces or stem base, no anthocyanins in leaves and stems

GENE: the mutant phenotypes are corrected by the introduction of the maize *R* cDNA, suggesting that *TTG* encodes a homolog of *R* or an activator of such a homolog; studies with 35S-*GL1* and 35S-*R* fusions (Larkin et al. 1994b) suggest that *GL1* and *TTG* are not components of a linear pathway but may interact with each other to regulate trichome development; in addition, *TTG* may be involved in suppressing trichome development in neighboring cells

OTHER INFORMATION: presumably involved in trichome cell specification

**tz: thiazole requiring**

LOCATION: 5-76.7

ORIGIN: Rédei's (1962, 1965; Li and Rédei 1969) alleles from X-ray, EMS, or 5-bromouracil treatment of seeds

ALLELES: *tz-1*, *tz-2*, and *tz-3* (Rédei 1962, 1965; Li and Rédei 1969) and others (Feenstra 1964, 1965)

REFERENCES: Rédei (1962) *Genetics 47:* 979; Feenstra (1964) *Genetica 35:* 259-269; Feenstra (1965) *AIS 2:* 24; Rédei (1965) *Am. J. Bot. 52:* 834-841; Li and Rédei (1969) *Biochem. Genet. 3:* 163-170

PHENOTYPE: leaves except cotyledons white; 4-methyl-5-hydroxyethyl-thiazole and thiamine restore to normal; viability as low as 10% in presence of thiazole (R.E. Pruitt, pers. comm.) but 100% on thiamine

ultraviolet hypersensitive: see *uvh*

ultraviolet repair defective: see *uvr1*

**uvh1: ultraviolet hypersensitive 1**

LOCATION: chromosome 3

ORIGIN: induced with EMS in Columbia

ALLELE: one

REFERENCES: Harlow et al. (1993) *Frontiers of Photobiol., Proc. 11th Internatl. Cong. on Photobiol.* (eds. Shima et al.), pp. 319-321, Elsevier, Amsterdam; Harlow et al. (1994) *Plant Cell 6:* 227-235

PHENOTYPE: 8-fold more sensitive to UV-B and UV-C than wild type; 5-fold more sensitive to γ irradiation

OTHER INFORMATION: not defective in UV-filtering capacity; not defective in photoreactivation

### *uvh2: ultraviolet hypersensitive 2*

LOCATION: not mapped

ORIGIN: induced with EMS in Columbia

ALLELE: one

REFERENCE: Harlow et al. (1993) *Frontiers of Photobiol., Proc. 11th Internatl. Cong. on Photobiol.* (eds. Shima et al.), pp. 319-321, Elsevier, Amsterdam; G. Harlow (manuscript in preparation)

PHENOTYPE: more sensitive to UV-B and UV-C than wild type

### *uvh3: ultraviolet hypersensitive 3*

LOCATION: chromosome 3

ORIGIN: induced with EMS in Columbia

ALLELE: one

REFERENCE: Harlow et al. (1993) *Frontiers of Photobiol., Proc. 11th Internatl. Cong. on Photobiol.* (eds. Shima et al.), pp. 319-321, Elsevier, Amsterdam; G. Harlow (manuscript in preparation)

PHENOTYPE: more sensitive to UV-B and UV-C than wild type

### *uvh4: ultraviolet hypersensitive 4*

LOCATION: not mapped

ORIGIN: induced with EMS in Columbia

ALLELE: one

REFERENCE: Harlow et al. (1993) *Frontiers of Photobiol., Proc. 11th Internatl. Cong. on Photobiol.* (eds. Shima et al.), pp. 319-321, Elsevier, Amsterdam; G. Harlow (manuscript in preparation)

PHENOTYPE: more sensitive to UV-B and UV-C than wild type

### *uvh5: ultraviolet hypersensitive 5*

LOCATION: not mapped

ORIGIN: induced with EMS in Columbia

ALLELE: one

REFERENCE: Harlow et al. (1993) *Frontiers of Photobiol., Proc. 11th Internatl. Cong. on Photobiol.* (eds. Shima et al.), pp. 319-321, Elsevier, Amsterdam; G. Harlow (manuscript in preparation)

PHENOTYPE: more sensitive to UV-B and UV-C than wild type

### *uvh6: ultraviolet hypersensitive 6*

LOCATION: chromosome 3

ORIGIN: induced with EMS in Columbia

ALLELE: one

REFERENCE: Harlow et al. (1993) *Frontiers of Photobiol., Proc. 11th Internatl. Cong. on Photobiol.* (eds. Shima et al.), pp. 319-321, Elsevier, Amsterdam; G. Harlow (manuscript in preparation)

PHENOTYPE: more sensitive to UV-B and UV-C than wild type; hypersensitive to heat stress; plants yellow-green

### *uvr1: ultraviolet repair defective*

LOCATION: 3-52.5

ORIGIN: induced with EMS in Landsberg *erecta tt5*

ALLELE: one

REFERENCES: Britt et al. (1993) *Science 261:* 1571-1574; Chen et al. (1994) *Plant Cell 6:* 1311-1317

PHENOTYPE: screened on the basis of sensitivity of immature root elongation to UV irradiation; under white light with some UV the illuminated surface of stem blackens and leaves gradually turn brown and wither, with the leaf edge curling upward and inward; normal when grown under white light with the UV filtered out; defective in the dark repair of UV-induced pyrimidine-pyrimidinone (6-4) dimer

V52: see *im*

V157: see *alb1*

V220: see *tt3*

V345: see *th3*

VAL-2: see *csr1*

valine resistant: see *csr1*

### *var1: variegated 1*

LOCATION: 5-65.1

ORIGIN: segregated in the progeny of a plant regenerated from tissue culture (Columbia)

ALLELE: *var1-1*

REFERENCE: Martínez-Zapater (1993) *J. Hered. 84:* 138-140

PHENOTYPE: normal when grown at temperatures below 20°C; chlorotic sectors seen exclusively on leaves of plants grown at 25°C; when plants were shifted from low to high temperature only the leaves differentiated at the high temperatures were chlorotic; the cells of the chlorotic sectors remain chlorotic even after the temperature was reduced to 20°C

### *var2: variegated 2*

LOCATION: 2-71.2

ORIGIN: induced with EMS in Columbia

ALLELES: *var2-1; var2-2*

REFERENCE: Martínez-Zapater (1993) *J. Hered. 84:* 138-140

PHENOTYPE: green sectors of different shades on a pale or albino background in all organs that are normally green (leaves, stems, sepals, and siliques) except cotyledons; under dim light about 10% of seedlings show complete bleaching initially in their first leaves; under high light over 40% of the seedlings show bleaching in the first two leaves; green sectors appear late in these albino or pale leaves; variegated phenotype is irreversible; have normal carotinoid composition

OTHER INFORMATION: reciprocal crosses with wild type indicate the *var2* mutations are nuclear recessive

variegated: see *var*

*vc-2:* see *cer2*

virescens-2: see *vr2* and *cer2*

*vr-2:* see *vr2*

**vr2: *virescens* 2**

   LOCATION: 2-64 (Rédei 1964, 1965)

   ALLELE: *vr2-1* (= *vr-2*)

   REFERENCES: Rédei (1964) *AIS 1:* 9-10; Rédei (1965) *Am. J. Bot. 52:* 834-841

   PHENOTYPE: virescent; pale green

   OTHER INFORMATION: map location given by Rédei as 2 mu right of *py,* thus 2-64; not the same as *vc-2:* virescens-2 which is a synonym of *cer2* and which maps on chromosome 4 (see *cer2* and Kranz and Scheidemann 1978, *AIS 15:* 31-34)

**wav1: *wavy* growth 1**

   LOCATION: not mapped

   ORIGIN: induced with EMS in Landsberg *erecta*

   ALLELE: *wav1-1*

   REFERENCE: Okada and Shimura (1990) *Science 250:* 274-276

   PHENOTYPE: lack of root tip rotation and wavy growth of root on angled agar surface; gravitropism normal

**wav2: *wavy* growth 2**

   LOCATION: not mapped

   ORIGIN: induced with EMS in Landsberg *erecta*

   ALLELE: *wav2-1*

   REFERENCE: Okada and Shimura (1990) *Science 250:* 274-276

   PHENOTYPE: root wavy growth on angled agar surface has a shorter pitch than normal, possibly due to higher rate of root tip rotation

**wav3: *wavy* growth 3**

   LOCATION: not mapped

   ORIGIN: induced with EMS in Landsberg *erecta*

   ALLELE: *wav3-1*

   REFERENCE: Okada and Shimura (1990) *Science 250:* 274-276

   PHENOTYPE: similar to *wav2-1*

**wav4: *wavy* growth 4**

   LOCATION: not mapped

   ORIGIN: induced with EMS in Landsberg *erecta*

   ALLELE: *wav4-1*

   REFERENCE: Okada and Shimura (1990) *Science 250:* 274-276

   PHENOTYPE: waving pattern of root on angled agar surface is rectangular

*wav5:* wavy growth 5, allelic with *aux1,* see *aux1*

*wav6:* wavy growth 6, allelic with *agr1,* see *agr1*

wavy growth (of roots): see *wav, aux1,* and *agr1*

**wrm: *wurm***

LOCATION: not mapped
ORIGIN: induced with EMS in Landsberg *erecta*
ALLELES: three
REFERENCE: Hülskamp et al. (1994) *Cell 76:* 555-566
PHENOTYPE: trichome distorted
wurm: see *wrm*

### *xav: xantha-variegation*

LOCATION: nuclear, not mapped
ORIGIN: induced with ethylene imine in Enkheim
ALLELE: *xav* (= Xav)
REFERENCE: Mednik and Usmanov (1982) *AIS 19:* 87-92
PHENOTYPE: homozygous lethal, with seedlings light yellow and dying at cotyledon stage; heterozygotes have the first pair of real leaves yellow-green, which is also the color of rosette leaves until bud formation, with the center of the rosette lighter than the outer part
OTHER INFORMATION: penetrance almost 100%, expressivity better in brighter illumination

Xav: see *xav*
Xantha-variegation: see *xav*
xanthaviridis: see *xv*

### *xv: xanthaviridis*

LOCATION: 1-30 (Rédei, see Kranz and Scheidemann 1978)
REFERENCES: Lee-Chen and Steinitz Sears (1967) *Can. J. Genet. Cytol. 9:* 381-384; Kranz and Scheidemann (1978) *AIS 15:* 31-34
PHENOTYPE: yellow-green cotyledons, leaf margins yellow
OTHER INFORMATION: 28 map units left of *chl* (Rédei, see Kranz and Scheidemann 1978), so around 1-30
yellow inflorescence: see *yi*

### *yi: yellow inflorescence*

LOCATION: 5-85.1
ORIGIN: induced with EMS in Landsberg *erecta*
REFERENCE: Koornneef et al. (1983) *J. Hered. 74:* 265-272
PHENOTYPE: yellowish flower buds and yellow-grayish sharper leaves

### *zll: zwille*

LOCATION: chromosome 5
ORIGIN: induced with EMS in Landsberg *erecta*
ALLELES: seven
REFERENCES: Jürgens et al. (1994) *Proc. NATO-ASI, Plant Molecular Biology Series: Molecular Genetic Analysis of Plant Development and Metabolism* (eds. G. Coruzzi and P. Puigdoménech), pp. 95-104, Springer-Verlag, Berlin; T. Laux and G. Jürgens, pers. comm.
PHENOTYPE: seedlings lack functional shoot meristem, can form secondary shoots

### *zwi: zwichel*

LOCATION: not mapped

ORIGIN: induced with EMS in Landsberg *erecta* (six alleles) or Columbia (one allele)

ALLELES: two strong and five weak

REFERENCE: Hülskamp et al. (1994) *Cell 76:* 555-566

PHENOTYPE: trichome asymmetric ("monospike," strong) or with two branches (weak)

zwichel: see *zwi*

zwille see *zll*

ZR8: see phototropic defect

ZR19: see phototropic defect

1A9: see *fab1*

1E5: see *fad3*

109A-1B: see *emb17*

109F-1C: see *emb24*

109F-5D: see *emb18*

111B-5B: see *emb27*

111B-5E: see *emb13*

111H-2B1: see *emb11*

112A-2A: see *emb30-1* and *gnom*

112E-1B: see *emb10*

112E-2A: see *emb7*

112G-1A: see *emb19*

113J-4A: see *emb8*

113K-1B: see *emb6*

115C-1C: see *emb29*

115D-4A: see *emb22*

115H-1A: see *emb28*

115J-4A: see *emb25*

117N-1B: see *emb20-1*

122G-E: see *bio1*

123B: see *emb16*

124D: see *emb2*

126E-B: see *emb23*

127AX-A: see *emb5*

129AX2-A: see *emb21*

130BA-1: see *emb12*

130BA-2: see *emb31*

2A11: see *fab2*

4A5: see *fad2-2*

53D-4A: see *emb1*

57B-4C: see *emb15*

63A-1A: see *emb20-3*

79A: see *emb9*

87A: see *emb20-2*

95A-2B: see *emb177*

1018/6: see *th2*
1053: see *hy3*
29.423: see *ga1-5*
29.9: see *ga1-4*
31.86: see *ga1-3*
40.443: see *tt5*
4-117: see *hy3*
464-19: see *hy3*
5-2: see reduced response to fusicoccin
548: see *hy3*
6.59: see *ga1-2*
8-36: see *hy3*
857: see *agr1-2*
858: see *agr1-3*
932: see *agr1-1*


## TRISOMICS

Trisomics have been made and analyzed three times, by three groups. There are thus three sets of trisomics, three sets of references, and three nomenclatures. The origins and references are:

### Columbia trisomics

ORIGIN: progeny of 4N x 2N plants of Columbia strain, presumably by selfing of 3N F1

REFERENCES: Steinitz-Sears (1963) *Genetics 48:* 483-490; Lee-Chen and Bürger (1967) *AIS 4:* 4-5; Lee-Chen and Steinitz-Sears (1967) *Can. J. Genet. Cytol. 9:* 381-384; Lee-Chen and Sears (1969) *AIS 6:* 22; Sears and Lee-Chen (1970) *Can. J. Genet. Cytol. 12:* 217-223

### Göttingen trisomics

ORIGIN: cross of 3N x 2N plants of En-2 background

REFERENCES: Röbbelen (1966) *AIS 3:* 16-17; Lee-Chen and Bürger (1967) *AIS 4:* 4-5

### Wageningen trisomics

ORIGIN: cross of 3N x 2N plants of Landsberg *erecta* background

REFERENCES: Koornneef and van der Veen (1978) *AIS 15:* 38-43; Koornneef and den Besten (1979) *AIS 16:* 35-40; Koornneef and van der Beck (1979) *AIS 16:* 103-106; Koornneef and van der Veen (1983) *Genetica 61:* 41-46

The trisomics are listed under the terminology of the Wageningen trisomics, with the synonyms from the other sets given. The identity of the Göttingen trisomic II was given incorrectly in some of the earlier papers, see Sears and Lee-Chen (1970) and Koornneef and van der Veen (1983).

The lines with an ABRC stock number (with the ecotype given in the parentheses) are available from the *Arabidopsis* Biological Resource Center at the Ohio State University at Columbus, Ohio. Their descriptions have been confirmed and kindly provided by Dr. R. Scholl.

## Tr1

**TRISOMIC FOR:** chromosome 1

**SYNONYMOUS WITH:** F (fragilis) of Columbia trisomics

**PHENOTYPE:** small rosette which becomes necrotic, fragile plant, nearly sterile

**GENES SHOWN TO BE TRIPLOID:** *ch1, dis2*

## Tr1A

**ABRC STOCK #:** CS2358 (Landsberg *erecta*)

**TRISOMIC FOR:** chromosome arm 1 L

**PHENOTYPE:** small rosettes; dark green leaves that are slightly lozenge shaped, semi-dwarf, petals irregular

**GENES SHOWN TO BE TRIPLOID:** *dis2, an, dis1, cer1, th1, alb1*

## Tr1B

**ABRC STOCK #:** CS2359 (Landsberg *erecta*)

**TRISOMIC FOR:** chromosome arm 1 R

**PHENOTYPE:** small rosette leaves which become necrotic, slender plant

**GENES SHOWN TO BE TRIPLOID:** *ch1, ap1, clv1*

## Tr2

**ABRC STOCK #:** CS3234 (Columbia) (Rédei Stock #: T22)

**TRISOMIC FOR:** chromosome 2

**SYNONYMOUS WITH:** R (round leaf) of Columbia trisomics, III of Göttingen trisomics

**PHENOTYPE:** compact rosette; rosette leaves round, dark green, with wide petiole, often slightly bent, and with sharp shoulders; late flowering

**GENES SHOWN TO BE TRIPLOID:** *py, hy1*

## Tr3

**ABRC STOCK #:** CS3227 (Columbia) (Rédei Stock #: T20)

**TRISOMIC FOR:** chromosome 3

**SYNONYMOUS WITH:** Y (light yellow) of Columbia trisomics, I of Göttingen trisomics

**PHENOTYPE:** yellow-green and elongated rosette leaves without shoulders; irregular petals, highly reduced fertility

**GENES SHOWN TO BE TRIPLOID:** *gl1, ch6, hy2*

## Tr3A

**TRISOMIC FOR:** chromosome arm 3 R

**PHENOTYPE:** like Tr3 but less extreme and more fertile

**GENES SHOWN TO BE TRIPLOID:** *ch6*

## Tr4

**ABRC STOCK #:** CS3229 (Enkheim) (Rédei Stock #: Conc.)

**TRISOMIC FOR:** chromosome 4

**SYNONYMOUS WITH:** C (concave) of Columbia trisomics, II of Göttingen trisomics

**PHENOTYPE:** small flat rosette; concave and blunt leaves; thickened flower buds giving the top of the inflorescence a flattened appearance; late flowering

**GENES SHOWN TO BE TRIPLOID:** *cer2, ga1, ch42, ga5, cp1, fca*

## Tr5

**ABRC STOCK #:** CS3174 (Columbia) (Rédei Stock #: T31)

**TRISOMIC FOR:** chromosome 5

**SYNONYMOUS WITH:** N (narrow) of Columbia trisomics, IV of Göttingen trisomics

**PHENOTYPE:** narrow leaves, plant semi-dwarf, greatly reduced fertility

**GENES SHOWN TO BE TRIPLOID:** *lu, tz, tt3*

## Tr5A

**ABRC STOCK #:** CS3228 (Columbia) (Rédei Stock #: T24)

**TRISOMIC FOR:** chromosome arm 5 R

**SYNONYMOUS WITH:** Nc (narrow, curved leaf) of Columbia trisomics

**PHENOTYPE:** like Tr5 but less extreme, leaf margins more serrated, fertility much better than Tr5

**GENES SHOWN TO BE TRIPLOID:** *tz, tt3*


## DEFICIENCIES

**ORIGIN:** induced by γ-ray in Landsberg *erecta* pollen

**REFERENCE:** Vizir et al. (1994) *Genetics 137:* 1111-1119

**DESCRIPTION:** F1 plants from crosses between visibly marked lines and irradiated Landsberg *erecta* pollen were examined for embryo lethals and/or gaps (indicating gametophyte lethality) in siliques and for the uncovering of one of the visible markers; many of the F1 carrying lethal mutations but normal for the markers had segregation ratios from 0:1 to 2:1 for normal to mutant for the visible markers in F2 and F3 progeny, suggesting loss of the irradiated chromosome at different frequencies and/or varying linkage between the deletion and the marker; the deletions were estimated to be 160 kb or less on average


## ACKNOWLEDGMENTS

We thank Mark Aarts, Ian Bancroft, Phil Benfey, Anthony Bleecker, Caren Chang, Anthony Cashmore, Tim Caspar, Joanne Chory, George Coupland, Nigel Crawford, Greg Harlow, Roger Innes, Steve Jacobsen, Gerd Jürgens, Maarten Koornneef, Barbara Kunkel, David Meinke, George Mourad, Bernard Mulligan, Chris Somerville, Shauna Somerville, Hirokazu Tsukaya, and Antje von Schaewen for communicating results prior to publication; Gerd Jürgens, Maarten Koornneef, David Meinke, Randy Scholl, and Chris Somerville for valuable comments and suggestions; and Yi Hu for assistance in preparation of this manuscript.

# Index